ANALYTICAL MECHANICS

Analytical Mechanics

A Comprehensive Treatise
on the Dynamics of Constrained Systems;
for Engineers, Physicists, and Mathematicians

John G. Papastavridis, Ph.D.

OXFORD
UNIVERSITY PRESS

2002

OXFORD
UNIVERSITY PRESS

Oxford New York
Athens Auckland Bangkok Bogotá Buenos Aires Cape Town
Chennai Dar es Salaam Dehli Florence Hong Kong Istanbul Karachi
Kolkata Kuala Lumpur Madrid Melbourne Mexico City Mumbai
Nairobi Paris São Paulo Singapore Taipei Tokyo Toronto Warsaw

and associated companies in
Berlin Ibadan

Published by Oxford University Press, Inc.
198 Madison Avenue, New York, New York 10016
www.oup.com

Library of Congress Cataloging-in-Publication Data
Papastavridis, J. G. (John G.)
Analytical mechanics: a comprehensive treatise on the
dynamics of constrained systems: for engineers, physicists,
and mathematicians/John G. Papastavridis.
p. cm.
Includes bibliographical references and indexes.
ISBN 0-19-512697-1
1. Mechanics, Analytic. I. Title.
QA805 .P355 2001
531—dc21 00-027884

Photograph on page v: The author's father is shown here in the traditional Greek *foustanella*, ca. 1930.

9 8 7 6 5 4 3 2 1

Printed in the United States of America
on acid-free paper

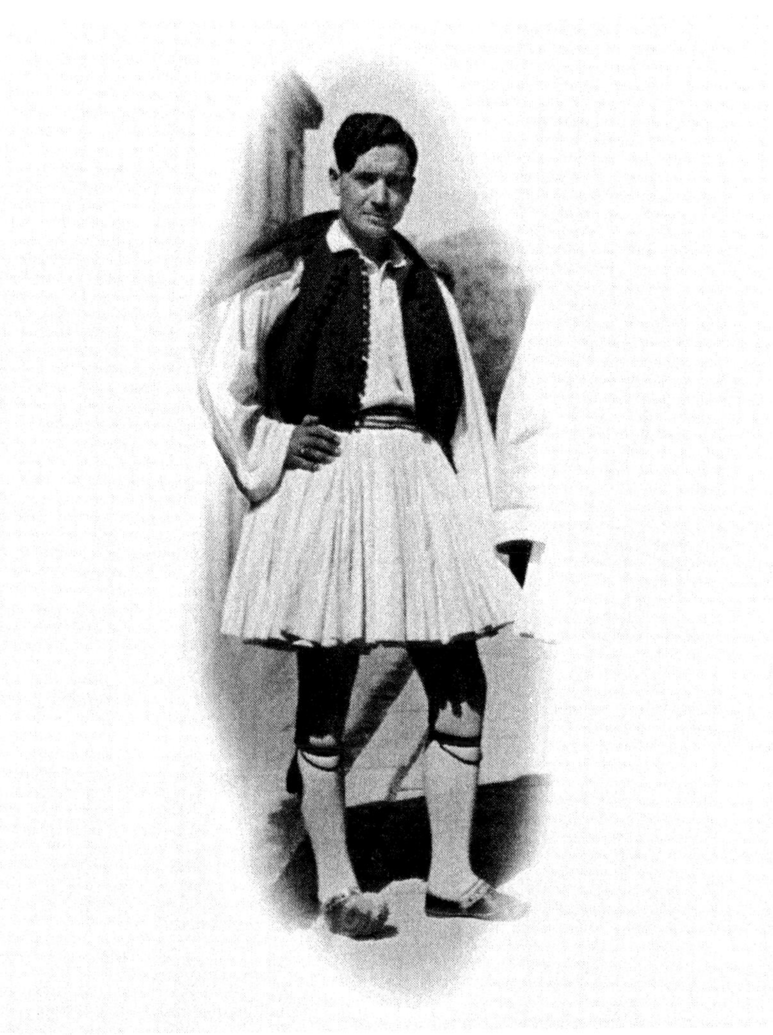

To the living and loving memory of my father,

GEORGE S. PAPASTAVRIDIS
(ΓΕΩΡΓΙΟΥ Σ. ΠΑΠΑΣΤΑΥΡΙΔΗ)

A lawyer and fearless maverick, who, throughout
his life, fought with fortitude, conviction, and class
to better his world;
and to whom I owe a critical part of my *Weltanschauung*.

PREFACE

Many of the scientific treatises today are formulated in a half-mystical language, as though to impress the reader with the uncomfortable feeling that he is in the permanent presence of a superman. The present book is conceived in a humble spirit and is written for humble people.

(Lanczos, 1970, pp. vii–viii)

GENERAL DESCRIPTION

This book is a classical and detailed introduction to advanced analytical mechanics (AM), with special emphasis on its basic principles and equations of motion, as they apply to the most general constrained mechanical systems with a finite number of degrees of freedom (this term is explained in Chapter 2). For the reasons detailed below, and in spite of the age of the subject, I think that *no other single volume exists, in English and in print, that is comparable to the one at hand in breadth and depth of the material covered—and, in this nontrivial sense, this ca. 1400-page and 174-figure long work is unique.*

The book is addressed to graduate students, professors, and researchers, in the areas of applied mechanics, engineering science, and mechanical, aerospace, structural, (even) electrical engineering, as well as physics and applied mathematics. Advanced undergraduates are also very welcome to browse, and thus get initiated into higher dynamics. The sole technical prerequisite here, a relatively modest one, is a solid working knowledge of "elementary/intermediate" (i.e., undergraduate) dynamics; roughly, equivalent to the (bulk of the) material covered in, say, Spiegel's *Theoretical Mechanics*, part of the well-known Schaum's outline series. Also, familiarity with the simplest aspects of Lagrange's equations, that is, how to take the partial and total derivatives of scalar energetic functions, would be helpful; although, strictly, it is not necessary. [See also "Suggestions to the Reader" (Introduction, §3).]

CONTENTS

Specifically, the book covers in what I consider to be a most logical and pedagogical sequence, the following topics:

Introduction: *Introduction* to analytical mechanics, brief summary of the history of theoretical mechanics; suggestions to the reader; and list of symbols/notations, abbreviations, and basic formulae.

Chapter 1: *Background*: Algebra of vectors and Cartesian tensors, and basic concepts and equations of Newton–Euler (or momentum) mechanics of particles and rigid bodies; that is, a highly selective compendium of undergraduate dynamics, and (some of) its mathematics, from a mature viewpoint.

Chapter 2: *Kinematics* of constrained systems (i.e., Lagrangean kinematics); including the general theory of up to linear velocity (i.e., Pfaffian) *constraints*, in both holonomic (or true) and nonholonomic (or quasi) *coordinates*; and a uniquely readable account of the fundamental *theorem of Frobenius*, for testing the nonholonomicity of such constraints.

Chapter 3: *Kinetics* of constrained systems (i.e., Lagrangean kinetics); including the fundamental principles of AM; that is, those of *d'Alembert–Lagrange* and of *relaxation of the constraints*, the *central equation* of Heun–Hamel; equations of motion with or without reactions, with or without multipliers, in true or quasi system variables; an introduction to *servoconstraints* (theories of Appell–Beghin, et al.); and rigid-body applications. This is the key chapter of the entire book, as far as engineering readers are concerned.

Chapter 4: *Impulsive motion*, under ideal constraints; including the associated extremum theorems of Carnot, Kelvin, Bertrand, Robin, et al.

Chapter 5: *Nonlinear nonholonomic constraints*; that is, kinematics and kinetics under nonlinear, and generally nonholonomic, *velocity* constraints.

Chapter 6: *Differential variational principles*, of Jourdain, Gauss, Hertz, et al., and their derivative higher-order equations of motion of Nielsen, Tsenov, et al.

Chapter 7: *Time-integral theorems and variational principles*, of Lagrange, Hamilton, Jacobi, O. Hölder, Voss, Suslov, Voronets, Hamel, et al.; for linear and nonlinear velocity constraints in true and quasi variables, with or without multipliers; plus energy and virial theorems.

Chapter 8: Introduction to *Hamiltonian/Canonical methods*; that is, equations of Hamilton and Routh–Helmholtz, cyclic systems, steady motion and its stability, variation of constants, canonical transformations and Poisson's brackets, Hamilton–Jacobi integration theory, integral invariants, Noether's theorem, and action–angle variables and their applications to adiabatic invariants and perturbation theory.

Chapters 2–8 each contain a large number of completely solved examples, and problems with their answers (and, occasionally, hints), to illustrate and extend the previous theories; short ones are integrated within each chapter section, and longer, more synthetic, ones are collected at each chapter's end; and also, critical comments/references for further study. The exposition ends with a relatively extensive, cumulative, and alphabetical list of References and Suggested Reading, including everything from standard textbooks all the way to epoch-making memoirs of the last (more than) two hundred years. This list complements those found in such well-known references as Neimark and Fufaev (1967/1972) and Roberson and Schwertassek (1988).

Parts of the text have, unavoidably, state-of-the-art flavor. However, as far as fundamental ideas go, very little, if anything, of the topics covered is truly new—today, no one can claim much originality in classical mechanics! The newness here, a nontrivial one, I think, consists in restoring, clarifying, putting together, and presenting, in what I hope is a readable form, material most of which has appeared over the past one hundred fifty, or so, years; frequently in little known, and/or hard to find and decipher, sources. (In view of the thousands of books, lecture notes, articles, and so on, used in the writing of this work, failure to acknowledge an author's particular contribution is not intentional, merely an oversight.) But, given the astonishing unfamiliarity, confusion, and intellectual provincialism so prevalent in many theoretical and applied mechanics circles today, even in the fundamental concepts and principles of analytical dynamics (like virtual displacements/work and principle of d'Alembert–Lagrange, which is, by far, the most misunderstood "principle" of physics!), I felt very strongly that this noble, beautiful, and powerful body of knowledge, that diamond of our cultural heritage, should be accurately preserved and represented, so as to benefit present and future workers in dynamics.

No single volume can even pretend to cover satisfactorily all aspects of this vast and fascinating subject; in particular, both its theoretical and applied aspects, let alone the currently popular computational ones. Since this is not an encyclopedia of theoretical and applied dynamics, an inescapable and necessary selection has operated, and so, the following important topics are *not* covered: applications of differential forms/exterior calculus (of Cartan, Gallissot, et al.) and symplectic geometry to Lagrangean and Hamiltonian mechanics; group theoretic applications; stability of motion (nonlinear theory) and theory of orbits; and computational/numerical techniques. For all these, there already exists an enormous and competent literature (see "Suggestions to the Reader"). However, with the help of this treatise, the conscientious reader will be able to move quickly and confidently into any particular theoretical and/or computational area of modern dynamics. In this sense, *the work at hand constitutes an optimal investment of the reader's precious energies.*

RAISON D'ÊTRE, AND SOME PHILOSOPHY

The customary words of explanation, or apology, for writing "another" book on advanced dynamics are now in order. The main theme of this work, like a Wagnerian leitmotiv, is deductive order, formal structure, and physical ideas, as they pertain to that particular energetic form of mechanics of constrained systems founded by Lagrange and known as analytical (= deductive) mechanics; to be differentiated from the also analytical but momentum, or "elementary," form of system mechanics, founded by Euler. *It is a book for people who place theory (theories), ideas, knowledge, and understanding above all else—and do not apologize for it.* Here, AM is studied not as the "maid" of some other more important discipline, but as a subject worth knowing in its own right; that is, as a "king or queen." As such, it will attract those with a qualitative and theoretical bent of mind; while it may not be as agreeable to those with purely computational and/or intellectually local predilections. [In the words of the late Professor R. M. Rosenberg (University of California, Berkeley): "The field of dynamics is plowed by two classes of people: those who enjoy the inherent beauty, symmetry and consistency of this discipline, and those who are satisfied with having a machine that manufactures equations of motion of complex mechanisms" (private communication, 1986).] Generally, science is more than a collection of particular problems and special techniques, even involved ones—it is much more than mere information. However, practical people should be reminded that theory and application are mutually *complementary* rather than adversarial; in fact, contemporary important practical problems and the availability of powerful computational capabilities have made the thorough understanding of the fundamental principles of mechanics more necessary today than before. Applications and computers have, among other things, helped resurrect, restore, and sharpen old academic curiosities (for engineers anyway), such as the *differential* variational principles of Jourdain and Gauss (which have found applications in such "unrelated" areas as multibody dynamics, nonlinear oscillations, even the elasto-plastic buckling of shells); and Hamilton's canonical equations in quasi variables (which have found applications in robotic manipulators).

A more concrete reason for writing this book is that, outside of the truly monumental British treatise of Pars (1965) and the English translations of the beautiful (former) Soviet monographs of Neimark and Fufaev (1967/1972) and Gantmacher (1966/1970), *there is no comprehensive exposition of advanced engineering-oriented dynamics in print, in the entire English language literature!* True, the famous treatise of Whittaker (1904/1917/1927/1937), for many years out of print, has recently been reprinted (1988). However, even Whittaker, although undeniably a classic and in many respects the single most influential dynamics volume of the twentieth century (primarily, to celestial and quantum mechanics), nevertheless leaves a lot to be desired in matters of logic, fundamental principles, and their earthly applications; for example, there is no clear and general formulation of the principle of d'Alembert–Lagrange and its use, in connection with Hamel's method of quasi variables, to uncouple the equations of motion and obtain constraint reactions; also, Whittaker would be totally unacceptable with the better of today's educational philosophies. Such drawbacks have plagued most British texts of that era; for example, the otherwise excellent works of Thomson/Tait, Routh, Lamb, Ramsey, Smart, and many of their U.S.-made descendants. [In a way, Whittaker et al. have been pretty lucky in that most of the great continental European works on advanced dynamics—for example, those of Boltzmann, Heun, Maggi, Appell, Marcolongo, Suslov, Nordheim et al. (vol. 5 of *Handbuch der Physik*, 1927), Winkelmann (vol. 1 of *Handbuch der Physikalischen und Technischen Mechanik*, 1929), Prange (vol. 4 of *Encyclopädie der Mathematischen Wissenschaften*, 1935), Rose, Hamel, Pérès, Lur'e, et al. were never translated into English.] Next, the comprehensive three-volume work of MacMillan (late 1920s to early 1930s) and the encyclopedic treatise of Webster (early 1900s), probably the two best U.S.-made mechanics texts, are, unfortunately, out of print. The very lively and deservedly popular monograph of Lanczos (1949–1970) does not go far enough in areas of engineering importance; for example, nonholonomic variables and constraints; and, also, lacks in examples and problems. Only the excellent encyclopedic article of Synge (1960) comes close to our objectives; but, that, too, has Lanczos' drawbacks for engineering students and classroom use.

The existing contemporary expositions on advanced dynamics, in English and in print, fall roughly into the following three groups:

Formalistic/Abstract, of the by-and-for-mathematicians variety, and, as such, of next to zero relevance and/or usefulness to most nonmathematicians. Considering the high mental effort and time that must be expended toward their mastery vis-à-vis their meager results in understanding mechanics better and/or solving new and nontrivial problems, these works represent a pretty poor investment of ever scarce intellectual resources; that is, they are not worth their "money." The effort should be commensurate to the returns. And, contrary to the impression given by authors of this group, even in the most exact sciences, books are written by and for concrete people; not by superlogical, detached, and cold machines. As Winner puts it: "The accepted form of 'objectivity' in scientific and technical reports (one can also include books and articles in social science) requires that the prose read as if there were no person in the room when the writing took place" (1986, p. 71). Also, I categorically reject soothing apologies of the type "oh, well, that is a book for mathematicians"; that is, the book has little or no consideration for ordinary folk. The distinguished physicist F. J. Dyson confirms our suspicions that "the marriage of mathematics and physics [about which we have been told so many nice things since our high school days] has ended in divorce" (quoted in M. Kline's *Mathematics, The Loss of Certainty*, Oxford University Press, 1980, pp. 302–303).

Applied, which either emphasize the numerical/computational aspects of mechanics, but, perhaps unavoidably, are soft and/or sketchy on its fundamental principles; or are so theoretically/conceptually impoverished and unmotivated that the reader is soon led to a narrow and dead-end view of mechanics. [Notable and refreshing *exceptions* to this style are the recent compact but rich-in-ideas works by Bremer et al. [1988(a), (b), 1992] in dynamics/control/flexible multibody systems.]

Mainstream or traditionalist; for example, those by (alphabetically): Arya, Baruh, Calkin, Crandall et al., Corben et al., Desloge, Goldstein, Greenwood, Kilmister et al., Konopinski, Kuypers, Lanczos, Marion, McCauley, Meirovitch, Park, Rosenberg, Woodhouse. The problem with this group, however, is that its representatives either do not go far and deep enough (somehow, the more advanced topics seem to be monopolized by the expositions of the first group); or they could use some improvements in the quality and/or quantity of their engineeringly relevant examples and problems.

The book at hand belongs squarely and unabashedly to this last group, and aims to remedy its above-mentioned shortcomings by bridging the space between it and some of the earlier-mentioned classics, such as (chronologically): Heun (1906, 1914), Prange (1933–1935), Hamel (1927, 1949), Pérès (1953), Lur'e (1961/1968), Gantmacher (1966/1970), Neimark and Fufaev (1967/1972), Dobronravov (1970, 1976), and Novoselov (1966, 1967, 1979). Hence, my earlier claim that this treatise is unique in the entire contemporary literature; and my strong belief that it does meet real and long overdue needs of students and teachers of advanced (engineering) dynamics of the international community. I have sought to combine the best of the old and new—no age discrimination here—and I hope that this work will help counter the very real and disturbing trend, brought about by the proponents of the first two groups, toward a dynamical tower of Babel.

ON NOTATION

To make the exposition accessible to as many willing and able readers as possible, and following the admirable and ever applicable example of Lanczos (1949–1970), I have chosen, wherever possible, an *informal approach*; and I have, thus, deliberately avoided all set-theoretic and functional-analytic formalisms, all unnecessary rigor ("epsilonics") and similar ahistorical/unmotivating/intuition-deadening tools and methods. For the same reasons, I have also avoided the currently popular direct/dyadic (coordinate-free) and matrix notations (except in a very small number of truly useful situations); and I have, instead, chosen good old-fashioned elementary/geometrical (undergraduate) form, for vectors, and/or indicial Cartesian tensorial notation for vectors, tensors, etc.

The ad nauseam advertised "advantages" of the coordinate-free ("direct") notation and matrices are vastly exaggerated and misguiding. To begin with, it is no accident that the solution of all concrete physical problems is intimately connected with a specific and convenient (or natural, or canonical) system of coordinates. Indicial tensorial notation seems to kill two birds with one stone: it combines both coordinate invariance (generality) and coordinate specificity; that is, one knows exactly what to do in a given set of coordinates/axes; see, for example, Korenev

(1979), Maißer (1988) for robotics applications. However, the systematic use of general tensors in dynamics has been kept out of this book. [That is carried out in my monograph, *Tensor Calculus and Analytical Dynamics* (CRC Press, 1999).] The only thing tensorial used here amounts to nothing more than the earlier-mentioned indicial Cartesian tensor notation; and for reasons that will become clear later, *not even the well-known summation convention is employed.* Indicial tensorial notation turns out to be the best tool in "unknown and rugged terrain"; and frequently it is the only available notation, for example, in dealing with nonvectorial/tensorial "geometrical objects," such as the Christoffel symbols and the Ricci/Boltzmann/Hamel coefficients. Once the fundamental theory is thoroughly understood, and the numerical implementation of a (frequently large-scale) concrete problem is sought, then one can profitably use matrices, and so on. Heavy use of matrices, with their non-commutativity "straitjacket," at an early stage [e.g., Haug, 1992(a)] is likely to restrict creativity and replace physical understanding with the local mechanical manipulation of symbols.

FURTHER PHILOSOPHY: On Computerization, Applications, and Ultimate Goals of Research

I do not think that the author of a book on analytical mechanics (AM) should be constantly defending it as simply a means to some other allegedly higher ends [e.g., a prerequisite to quantum mechanics, as Goldstein (1980) does], or in terms of its current "real life" applications in space or earth (e.g., artificial satellites, rocketry, robotics, etc.; i.e., in terms of dollars to be made); although, clearly, such connections do exist and can be helpful. What should worry us is that these days, under what B. Schwartz calls "economic imperialism," or what R. Bellah calls "market totalitarianism" (i.e., the penetration of purely monetary values into every aspect of social life; or, to regard all aspects of human relations as matters of economic self-interest, and model them after the market) every activity is fast becoming a means for something else, preferably quantifiable and monetary. In the process, daily work, craftsmanship, and the pleasure derived from the practice of that activity, have all been degraded. Unless we restore some *internal, or intrinsic*, goals and rewards to our subject and disseminate them to our young students, pretty soon such an activity will be no different from clerical or assembly-line work; that is, just a paycheck. As stated earlier, we view AM as a course worth pursuing in its own terms. We study it because it is *worth learning*, and because it is a grand and glorious part of our intellectual/cultural heritage—those who do not care about the past cannot possibly care about the present, let alone the future.

On a more practical level, a few years from now such applied areas as multibody dynamics, a subject with which so many dynamicists are preoccupied today, will be exhausted—some say that that has already happened. What are the practical mechanicians going to do then? Most of their expositions (second of the earlier groups) are too narrow and do not prepare the reader for the long haul. But there is a more fundamental reason for adopting "my" particular approach to mechanics: I strongly believe that *every generation has to rediscover (better, reinvent) AM, and most other areas of knowledge for that matter, anew and on its own terms;* that is, replow the soil and not just be handed down from their predecessors, discontinuously, prepackaged and predigested "information" in a diskette (the electronic equivalent of ashes in an urn). To squeeze the "entire" mechanics into a huge master computer program, which (according to common but nevertheless vulgar advertisements) "does everything," and makes it available to the reader ("user") in the form of data inputs, is not only dangerous for the present (e.g., accidents, screw-ups, which are especially consequential in today's large-scale systems — recall the omnipresent Murphy's laws), but also, being a degradation and dehumanization of knowledge, it guarantees the intellectual death of our society. If the job makes the person (mentally, psychologically, and physically), then how are we going to answer the question "What are people for?"

Typical of such contemporary one-dimensional, or "digital," approaches to dynamics are sweeping statements like: "pre-computer analytical methods for deriving the system equations must be replaced by systematic computer oriented formalisms, which can be translated conveniently into efficient computer codes for * *generating* the system equations based on simple user data describing the system model, * *solving* those complex equations yielding results ready for design evaluation" and "Emphasis is on computer based derivation of the system equations thus freeing the user from the time consuming and error-prone task of developing equations of motion for various problems again and again." [From advertisement of Roberson and Schwertassek (1988) in *Ingenieur-Archiv*, **59**, p.A.3, 1989.] Here, the advertisers hide the well-known fact of how much error prone is the formulating and running of any complicated program; and how the combination of this with the absence of any physically simple and meaningful checks for finding errors — something of a certainty for the structureless/formless mechanics of Newton–Euler, on which so much of multibody dynamics rests — is a recipe for chaos (⇒ arbitrariness)! Our reading of this ad is that the whole process will, eventually, "free" the user from thinking at all — first, we replace the human functions and then we replace humans altogether [first industrial revolution: *mechanization* of muscles, second (current) industrial revolution: *automation* of both muscles and brains]; and anyone who dares to criticize, or inquire about *choices* (i.e., politics), is summarily and arrogantly dismissed as a technophobe or, worse, a neo-Luddite!

As the mathematicians Davis and Hersh put it accurately:

> By turning attention away from underlying physical mechanisms and towards the possibility of once-for-all algorithmization, it encourages the feeling that the purpose of computation is to spare mankind of the necessity of thinking deeply.... Excessive computerization would lead to a life of formal actions devoid of meaning, for the computer lives by precise languages, precise recipes, abstract and general programs wherein the underlying significance of what is done becomes secondary. [Inimitably captured in M. McLuhan's well-known dictum: The medium *is* the message.] It fosters a spirit-sapping formalism. The computer is often described as a neutral but willing slave. The danger is not that the computer is a robot but that humans will become robotized as they adapt to its abstractions and rigidities (1986, pp. 293, 16).

And, in a similar vein, H. R. Post adds: "You understand a subject when you have grasped its structure, not when you are merely informed of specific numerical results" (quoted in Truesdell, 1984, p. 601).

The issue is not whether the complete computer codification of (some version of) dynamics can be achieved or not; it clearly *can*, somehow. The issue is the desirability of it; that is, the *could* versus the *should*, its scale compared with the other approaches, and the *temporal order* of such a presentation to the student ("user"). The only safe way for using such heavily computerized schemes is for the student to already possess a very thorough grounding in the fundamentals of mechanics — like vaccination against a virus! There is no painless and short way to bypass several centuries of hard thinking by a handful of great fellow humans—no royal road to mechanics! Otherwise, we are headed for more confusion, degradation, errors, and accidents, and eventual *disengagement* from our subject. [For iconoclastic, devastating, and sobering critiques of the contemporary mindless and rabid computeritis, see, for example, Truesdell (1984, pp. 594–631), Davis and Hersh (1986), and Mander (1985).]

As for the applications of mechanics, there is nothing wrong with them; as long as they do not hurt or exploit people and nature—alas, several such contemporary applications do just that. Those preoccupied with them rarely, if ever, ask the natural question: What are the (most likely) applications of the applications; namely, their social/environmental consequences? In this light, common statements like "the computer is only a tool" are utterly naive and meaningless. I should also add that the current relentless emphasis, even in the academia, on applied research with quick tangible results—that is, dollars at the expense of every other nonmonetary aspect—is a *relatively recent* phenomenon imposed on us from *outside*; it is neither intrinsic nor accidental to science, but instead is an intensely socio-economic activity—technology is neither autonomous nor neutral! [And as Truesdell concurs, with depressing accuracy (1987, p. 91): "It is not philosophers of science who will enforce one kind of research or another. No, it will be the national funding agencies, the sources of manna, nectar, and ambrosia for the corrupted scientists. The directors of funds are birds of a feather, chattering mainly to each other and at any one moment singing more or less the same cacophonous tune. There may come a time when even the scholarly foundations will give preference to those who claim to promote national 'defense' by research on the basic principles governing some new, as yet totally secret—that is, known only to the directors of war in the U.S. and Russia—allegedly secret idea for a broader and more effective death by torture in a world full of humanitarians and their -isms."]

If applications, even worthwhile ones, are but one motive for studying mechanics, and science in general, then what else is? Here are some plausible (existential?) reasons offered by Einstein, which I have found particularly inspiring, since my high school years:

> Man tries to make for himself in the fashion that suits him best a simplified and intelligible picture of the world; he then tries to some extent to substitute this cosmos of his for the world of experience, and thus to overcome it. This is what the painter, the poet, the speculative philosopher, and the natural philosopher do, each in his own fashion. *Each makes this cosmos and its construction the pivot of his emotional life, in order to find in this way the peace and security which he cannot find in the narrow whirlpool of personal experience* (emphasis added; from "Principles of Research," an address delivered in 1918, on the occasion of M. Planck's sixtieth birthday).

From a broader perspective, I am convinced that the quality of our lives depends not so much on specific gadgets/artifacts, no matter how technically advanced they may be (e.g., from artificial hearts to space stations), but on our collective abilities to formulate simple, clear, and unifying ideas that will allow us to understand (and then change gently and gracefully—sustainably) our increasingly complicated, unstable and fragile societies; *and, in the process, understand ourselves*. The resulting psychological and intellectual peace of mind from such a liberal arts (= liberating) approach cannot be overstated. It is this kind of activity and attitude that gives human life meaning—we do not do science just to make money, merely to exchange and consume. This book is intended as a small but tangible contribution to this lofty goal.

SOME PERSONAL HISTORY

My interest in AM began during my undergraduate studies (mid-to-late 1960s) upon reading in Hamel (1949, pp. 233–236, 367) about the *differences* between the calculus of variations (mathematics) and Hamilton's variational principle (mechanics) for nonholonomic systems. The need for a deeper understanding of the underlying kinematical concepts led me, about twenty years later, to the study of the original epoch-making memoirs of such mechanics masters as Appell, Boltzmann, Heun, and Hamel. Then, in the spring of 1986, in related studies on variational calculus, I had the good fortune to stumble upon the virtually unknown but excellent papers of Schaefer (1951) and Stückler (1955), which, along with my earlier acquaintance with tensors, showed me the way toward the correct understanding of everything virtual: virtual displacements and virtual work/Lagrange's principle; that is, I arrived at AM via the calculus of variations, just like Lagrange in the 1760s! Finally, the emphasis on the fundamental distinction between *particle* and *system* quantities I owe to the writings of Heun, the founder of theoretical engineering dynamics (early 20th century), and especially to those of his students: Winkelmann and the great Hamel.

In closing, I would like to recommend the reading of the preface(s) of Lanczos (1949–1970); the present work has been conceived and driven by a similar overall philosophy.

May this book make many and loyal friends!

EVERLIVINGFIRE@netscape.net

Atlanta, Georgia — *J. G. P.*
Autumn 2001

ACKNOWLEDGMENTS

(Where the author recognizes, with gratitude and pleasure, the social dimension of his activity)

Every book on analytical mechanics is better off the closer it comes to the simplicity, clarity, and thoroughness of Georg Hamel's classic *Theoretische Mechanik*, arguably the best (broadest and deepest) single work on mechanics; and, secondarily, of Anatolii I. Lur'e's outstanding *Analiticheskaya Mekhanika.* I hope that this treatise follows closely and loyally the tradition created by these great masters. My indebtedness to their monumental works is hereby permanently registered.

Next, I express my deep appreciation and thanks to

Ms. Katharine L. Calhoun, of the Georgia Tech library, for her most courteous and capable help in locating and obtaining for me, over the past several years (1986 to present), hundreds of rare and critically needed references, from all over the country. Katharine is an oasis of humanity and graciousness in an otherwise arid and grim campus.

Drs. Feng Xiang Mei, professor at the Beijing Institute of Technology, China, and Zhen Wu, professor at the University of Tsiao Tong, China; and Drs. Sergei A. Zegzhda and Mikhail P. Yushkov, professors at the Mathematics and Mechanics faculty of St. Petersburg University, Russia [birthplace of the *first* treatise on theoretical/analytical mechanics (Euler's, *Mechanica Sive Motus Scientia...*, 2 vols, 1736)] for making available to me copies of their excellent textbooks and papers on advanced dynamics, which are virtually unavailable in the West (see the list of References and Suggested Reading).

Dr. John L. Junkins, chaired (and distinguished) professor of aeronautical engineering at Texas A&M University, for early encouragement on a previous version (outline) of the manuscript (1986)—what John aptly dubbed "the zeroth approximation."

Dr. Donald T. Greenwood, professor of aerospace engineering at the University of Michigan (Ann Arbor) and a true Nestor among American dynamicists, also author of internationally popular and instructive graduate texts on dynamics, for his detailed and mature (and very time- and energy-consuming) comments on the entire technical part of the manuscript; and for sharing with me his (soon to appear in book form) notes on *Special Advanced Topics in Dynamics.*

Dr. Wolfram Stadler, professor and scholar of mechanics at San Francisco State University and author of uniquely encyclopaedic work on robotics/mechatronics, for qualitative and constructive criticism on Chapter 1 (Background). Wolf remains a staunch and creative individualist, in an age of ruthless academic collectivization.

A very special expression of indebtedness to my friend Dr. Hartmut Bremer (O. Univ.-Prof. Dr.-Ing. habil.), professor of mechatronics at the University of Linz, Austria (formerly, professor of mechanics at the Technical University of Munich, Germany) and author of two dense and comprehensive textbooks on dynamics/control/multibody systems, for extensive, thoughtful and critical discussions/comments on several topics of theoretical and applied dynamics, including their historical, cultural, and educational aspects; and for supplying me with rare and precious references on the subject. To Hartmut's persistent efforts (1989–1993), I owe a very rare photograph of Karl Heun (to appear in a special gallery of photos of mechanics masters, which will be included in my forthcoming *Elementary Mechanics*).

In addition, I consider myself very fortunate to have benefited from the "mechanical" knowledge and wisdom of my friend Dr. Leon Y. Bahar, professor of mechanical engineering and mechanics at Drexel University and another American Nestor. Leon, a veritable engineering science scholar, craftsman, and above all dedicated teacher (i.e., representative of an academic species that is somewhere between endangered and extinct, as professionalism goes up and scholarship goes down), has over the past several years selflessly provided me with critical and enlightening quanta of knowledge and insight (lecture notes, papers, and extensive letters), and much valued mentorship.

These are not the best of times for writing "another" book on advanced theoretical mechanics — to put it mildly. However, and this provides a certain consolation, even such all-time titans of mathematics and mechanics as Euler, Lagrange, and Gauss had considerable difficulties in publishing, respectively, their *Theoria Motus* ... (1765), *Méchanique Analitique* (1788), and *Disquisitiones Arithmeticae* (1801). [According to Truesdell (1984, p. 352), all that Euler received for his masterpiece on rigid-body dynamics (1765) was ... twelve free copies of it!] Aspiring academic writers in this area are forewarned that the contemporary "research" university is not particularly supportive to such scholarly activities; these latter, obviously "interfere" with the more lucrative business (to the university bureaucrats, but not necessarily to students and society at large) of contracts and grants from big business and big government. Such an inimical "academic" environment makes it much more natural than usual that I reserve the strongest expression of gratitude, by far, to my family, here in Atlanta, Georgia: my wife Kim Ann and daughter Julia Constantina; and in my native Athens, Greece: my mother Konstantina and brother Stavros (in my nonobjective but fair view, one of the brightest mathematicians of contemporary Greece, and an island of moral and intellectual nobility, in an archipelago of petty greed and irrationality) — for their continual and critical moral and material support throughout the several long and solitary years of writing of the book.

Last, this volume (as well as my other two mechanics books) could not have been written without (i) the institution of academic *tenure* (much maligned and curtailed recently by reactionary ideologues, demagogues, and ignoramuses) and (ii) the (alas, fast disappearing) policy of most university libraries, of *open*, *direct*, and *free access* to books and journals. Regrettably, and in spite of high-tech millennarian promises, the next generation of scholarly authors will not be as lucky as I have been, in both these areas!

Contents

Words of Wisdom and Beauty

On Rigor

It is not so much important to be rigorous as to be right.

—A. N. KOLMOGOROV

On Theory

There is nothing more practical than a good theory.

—L. BOLTZMANN

We have no access to a *theory-independent* world — that is, a world unconditioned by our point of view .... The world we see is ... *theory-laden*: it already bears the ineliminable mark of our involvement in it .... Knowledge is always a representation of reality from within a particular perspective .... We cannot assume ... "the view from nowhere."

—T. W. CLARK

I really do not at all like the now fashionable "positivistic" tendency of clinging to what is observable ... I think ... that theory cannot be fabricated out of the results of observation, but that it can only be invented.

—A. EINSTEIN

On Method

In the sciences the subject is not only set by the method; at the same time it is set into the method and remains subordinate to the method .... *In the method lies all the power of knowledge. The subject belongs to the method.* (emphasis added)

—M. HEIDEGGER

The core of the practice of science — the thread that keeps it going as a coherent and developing activity — lies in the actions of those whose goals are internal to the practice. And *these internal goals are all noneconomic.* (emphasis added.)

—B. SCHWARTZ

On Beauty

My own students, few they have been, I have tried to teach how to ask questions humbly and to see ways to some taste in a vulgar, obscene epoch. Taste is acquired by those who can face questions, especially insoluble questions.

—C. A. TRUESDELL

It is by the steady elimination of everything which is ugly — thoughts and words no less than tangible objects — and by the substitution of things of true and lasting beauty that the whole progress of humanity proceeds.

—A. PAVLOVA

ANALYTICAL MECHANICS

Introduction

ΚΟΣΜΟΝ ΤΟΝΔΕ, ΤΟΝ ΑΥΤΟΝ ΑΠΑΝΤΩΝ, ΟΥΤΕ ΤΙΣ ΘΕΩΝ ΟΥΤΕ ΑΝΘΡΩΠΩΝ ΕΠΟΙΗΣΕΝ, ΑΛΛ'ΗΝ ΑΕΙ ΚΑΙ ΕΣΤΙΝ ΚΑΙ ΕΣΤΑΙ ΠΥΡ ΑΕΙΖΩΟΝ, ΑΠΤΟΜΕΝΟΝ ΜΕΤΡΑ ΚΑΙ ΑΠΟΣΒΕΝΝΥΜΕΝΟΝ ΜΕΤΡΑ.

[ΗΡΑΚΛΕΙΤΟΣ (Herakleitos, Greek philosopher; Ephesos, Ionia, late 6th century B.C.)]

[Translation: "This world [order], which is the same for all [beings], no one of gods or humans have created; but it was ever, is now, and ever shall be an ever-living Fire, that starts and goes out according to certain rules [laws]."
This magnificent statement marks the beginning of science—one of the countless, fundamental, and original gifts of Greece to the world. (See, e.g., Burnet, 1930, p. 134; also Frankfort et al., 1946, chap. 8.)]

Dynamics or Mechanics is the science of motion . . . . The problem of dynamics according to Kirchhoff, is to describe all motions occurring in nature in an unambiguous and the simplest manner. In addition it is our object to classify them and to arrange them on the basis of the simplest possible laws.
The success which has attended the efforts of physicists, mathematicians, and astronomers in achieving this object from the time of Galileo and Newton through that of Lagrange and Laplace to that of Helmholtz and Kelvin, constitutes one of the greatest triumphs of the human intellect.

(Webster, 1912, p. 3, emphasis added)

Die Mechanik ist die Wissenschaft von der Bewegung; als ihre Aufgabe bezeichnen wir: die in der Natur vor sich gehenden Bewegungen *vollständig* und auf die *einfachste Weise* zu beschreiben.
(Translation: Mechanics is the science of motion; we define as its task the *complete* description and in the *simplest possible manner* of such motions as occur in nature.)

(Kirchhoff, 1876, p. 1, author's emphasis)

Die Mechanik ist ein Teil der Physik.
(Translation: Mechanics is a part of physics.)

(Föppl, 1898, vol. 1, p. 1)

1 INTRODUCTION TO ANALYTICAL MECHANICS

What Is Analytical Mechanics?

Classical mechanics (CM)—that is, the exact science of nonrelativistic and non-quantum *motion* (effects) and *forces* (causes)—was founded in the 17th century (Galileo, 1638; Newton, 1687), and was brought to fruition and generality during the next century, almost single-handedly, by Euler (1752: principle of linear momentum; 1775: principle of *angular* momentum). [D'Alembert too had formulated separate laws of linear and angular momentum (1743, 1758), but his approach came nowhere near that of Euler in generality and power.] That was the *first complete dynamical theory* in history. We shall call it, conveniently (even though not quite accurately), the *Newton–Euler* method of mechanics (NEM).

The *second* such theory was also initiated in the (late) 17th century, this time by Huygens and Jakob Bernoulli; it was further developed during the 18th century by Johann Bernoulli (Jakob's brother) and d'Alembert (early 1740s), and was finally brought to relative mathematical and physical completion by the other great mathematician of that century, Lagrange (1760: principle of "least" action; 1764: principle of d'Alembert in Lagrange's form, or *Lagrange's principle*; 1780: *central equation* and *Lagrange's equations*; 1788: *Méchanique Analitique*; 1811–1812: *transitivity equations*). This second approach, what we shall call the method of *d'Alembert–Lagrange*, or, simply and more accurately, the method of *Lagrange*, forms the basis of what has come to be known as *analytical mechanics* (AM); or, equivalently, *Lagrangean mechanics* (LM). Although both these methods are, roughly, theoretically equivalent, since there is only one classical mechanics, the second approach proved much more influential and fertile to the subsequent development not only of mechanics, but also of practically all areas of physics: from generalized coordinates and configuration space to Riemannian geometry and tensors, and from there to general relativity; and similarly for quantum mechanics.

Analytical mechanics proved particularly significant and useful to engineers, although it took another century after Lagrange for this to be fully realized (see §2). The reason for this is that AM was specifically designed by its inventors to handle constrained (earthly) systems—*the concept of constraint is central to AM*. Not that NEM could not handle such systems, but AM proved incomparably more expedient both for formulating their simplest (or minimal) equations of motion, and also for offering numerous theoretical and practical insights and tools for their solution (e.g., theorems of conservation and invariance, variational "principles" and associated *direct* methods of approximation, etc.—detailed in chaps. 3–8).

In NEM, the basic principles (or axioms) are those of *linear* and *angular momentum*, and, secondarily, that of action–reaction, for the internal (or mutual) forces (see chap. 1); that is,

NEM is a mechanics of systems based on *momentum* principles.

In LM, on the other hand, the primary axiom is the *kinetic principle of virtual work* for the constraint reactions [=Lagrange's principle (LP)] and, secondarily, the *principle of relaxation* (or liberation, or freeing) of the constraints (see chap. 3); that is,

LM is a mechanics of systems based on *energetic* principles.

With the help of his LP, Lagrange and many others later (see §2) formulated the most general equations of motion of systems subject to general positional and/or motional constraints. [The former are called *holonomic*, while the latter, if they cannot be brought (integrated) to positional form are called *nonholonomic* (see chap. 2).]

Last, from the viewpoint of applications, AM constitutes the *theoretical foundation of advanced engineering dynamics*; which, in turn, is very useful to the following: *structural* dynamics (e.g., bridges, airport runways); *machine* dynamics (e.g., internal combustion engines); *vehicle* dynamics (e.g., automobiles, locomotives); *rotor* dynamics (e.g., turbines); *robot* dynamics (e.g., robotic manipulators); *aero-/astro-*dynamics (e.g., airplanes, artificial satellites); *control theory/system* dynamics (e.g., electromechanical systems, valves); *celestial* dynamics (e.g., astronomy), and so on.

Comments on the Methodology of AM

1. From the otherwise physically complete (particle) mechanics of Newton *two* things were missing: *rotation* and *constraints* (and, of course, *deformation*, but we do not deal with continua here). The first was taken care of by Euler, Mozzi, Cauchy, Chasles, Rodrigues et al. (1750s to the mid 19th century), and the second by Lagrange (1760s–1780s) and later many others (1870–1910). Of course, *special* cases of both problems had been examined earlier: for example, Newton discussed motions on specified curves and the associated forces, and, as Heun points out, with the help of his third law of action/reaction, he could have built a constrained particle mechanics, had he pursued that possibility; d'Alembert worked with particles "constrained in rigid body connections"; and even Huygens had such pendula involving several constrained particles. Much later (early 1810s), Lagrange brought rotation under his energetic plan (genesis of *nonholonomic*, or *quasi*-coordinates; special *transitivity* equations—see bridge between Euler and Lagrange below).

2. *Analytical versus synthetical, Euler versus Lagrange.* To begin with, CM holds quite satisfactorily for *sizes*, or lengths, from 10^{-10} m (atom) to 10^{20} m (galaxy), and for *speeds* up to $c/10$ (c = speed of light $\approx$ 300,000 km/s). Outside of these broad ranges, CM is replaced by *relativity* (high speeds) and *quanta* (small sizes) (see, e.g., French, 1971, p. 8). Now, depending on the method adopted, CM can be classified as follows:

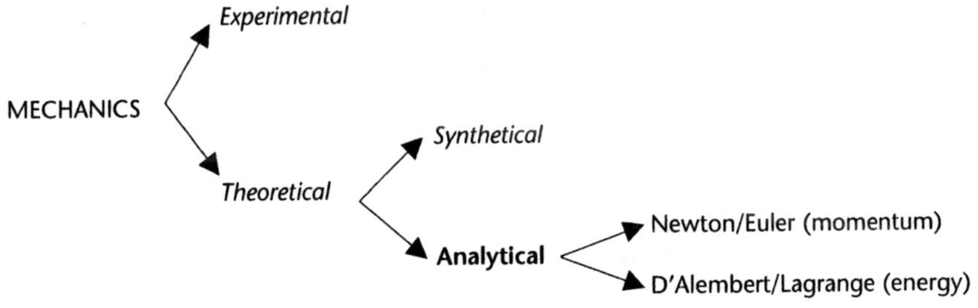

This classification, a logically possible one out of many (see below; e.g., Hamel, 1917), stresses the following:

2(a). Contrary to popular declarations, and Lagrange himself is partly to blame for this, AM does not mean mechanics via mathematical *analysis*; that is, it does not mean an ageometrical and figureless mechanics. [Even such 20th century mechanics authorities as Whittaker state that "The name Analytical Dynamics is given to that branch of knowledge in which the motions of material bodies, . . . , are discussed by the aid of mathematical analysis" (1937, p. 1).] Instead, and in the sense used in philosophy/logic, AM means a *deductive* mechanics: everything flowing from a few selected initial postulates/principles/axioms by logical (mathematical) reasoning; that is, from the general to the particular—as contrasted with *inductive*, or *synthetic*, mechanics; that is, from the particular to the general. As such, AM is by no means ageometrical (and, similarly, synthetic mechanics does not necessarily mean geometrical and nonmathematical mechanics). Also, in the past (mainly 19th century) the terms *theoretical*, *rational*, and *analytical* have frequently been used synonymously.

2(b). In such a classification, the mechanics of Euler also deserves to be called analytical! The reason that we in this book, and most everybody else, do not have more to do with *historical* tradition and usage rather than with strict logic: today AM has come to mean, specifically,

Lagrangean method = Principle of Lagrange

(= Principle of d'Alembert + Johann Bernoulli's principle of virtual work)

+ Principle of relaxation of the constraints (Hamel's *Befreiungsprinzip*)

After more than 200 years, AM is defined by its *practice*—that is, by its methods, tools, and range of problems dealt with by its practitioners—and because, contrary to the mechanics of Newton–Euler, it is capable of extending to other areas of physics: for example, statistical mechanics, electrodynamics. As the distinguished applied mathematician Gantmacher puts it

> [A]nalytical mechanics is characterized both by a specific system of presentation and also by a definite range of problems investigated. The characteristic feature . . . is that general principles (differential or integral) serve as the foundation, then the basic differential equations of motion are derived from these principles analytically. The basic content of analytical mechanics consists in describing the general principles of mechanics, deriving from them the fundamental differential equations of motion, and investigating the equations obtained and methods of integrating them (1970, p. 7).

2(c). Frequently, one is left with the impression that Eulerian mechanics is vectorial, whereas Lagrangean mechanics is scalar. This, however, is only superficially true: LM can be quite geometrical and vectorial, but in generalized *nonphysical/non-Euclidean* (*Riemannian and beyond*) *spaces* [see, e.g., Papastavridis (1999), Synge (1926–1927, 1936), and references therein].

2(d). Euler and Lagrange should be viewed as mutually complementary, not as adversarial—as some historians of mechanics do. And although it is undeniably true that, of the two, Euler was the greater "geometer" in both quantity and quality, yet it was the *method of Lagrange* that shaped and drove the subsequent epoch-making developments of theoretical physics and a fair part of applied mathematics (i.e., differential geometry/tensors → relativity; Hamiltonian mechanics/phase space → statistical mechanics, quantum mechanics). Lagrange himself, shortly before his death (in 1813), succeeded in building the bridge between his method and that of Euler (rigid-body equations) by obtaining a special case of "transitivity equations" [so named by Heun (early 20th century) because they allow the *transition* from Lagrangean to Eulerian], which appeared in the second volume of the second edition of his *Mécanique Analytique* (1815). And that is why the great mechanician Hamel, in 1903–1904, dubbed his own famous equations the "Lagrange–Euler equations"; and in his magnum opus *Theoretische Mechanik* (1949) he founded the entire mechanics on Lagrangean principles. [With the exception of Neimark and Fufaev (1972), the transitivity equations are completely and conspicuously absent from the entire English and French literature!]

3. *Newtonian particles versus Eulerian continua.* There is a certain viewpoint, particularly popular among celestial dynamicists/astronomers, (particle) physicists, and some applied mathematicians, according to which classical mechanics is the study of the motions of systems of particles under mutually attractive/repulsive forces, whose intensities depend only on the distances among these particles (molecules, etc.); and that, eventually, all physical phenomena are to be explained by such a "mechanistic" model. This Newtonian mindset dominated 19th century mechanics and physics almost completely, and obscured the fact that such a "central force + particle(s)" mechanics [launched, mainly, by P. S. de Laplace in his monumental five-volume *Traité de Mécanique Celeste* (1799–1825)] is but one possibility, even within the nonrelativistic and nonquantum confines of the 19th century. Under other, physically more realistic, possibilities the total interparticle force, generally, consists of a *reaction* to the geometrical and/or kinematical constraints imposed, and an *impressed*, or *physical*, part that can depend explicitly on time, position(s), and velocity(-ies) of some or all of the system particles. However, the introduction of such forces to mechanics creates effects that *cannot* be accounted by mechanics alone, such as thermal and/or electromagnetic phenomena; whereas, the consequences of Newtonian forces stay within mechanics.

The "mechanistic theory of matter"—namely, to explain all nonmechanical phenomena via simple models of internal nonvisible (concealed) motions of the system's molecules (second half of 19th century, proposed by physicists like W. Thomson, J. Thomson, Helmholtz, Hertz et al.)—was only partially successful, and eventually evolved to *statistical* mechanics and physics (Boltzmann, Gibbs) and *quantum* mechanics [Planck, Einstein, Bohr, Born, Heisenberg, Schrödinger, Dirac et al.; see also Stäckel (1905, p. 453 ff.)].

Finally, as Hamel (1917) points out, it should be remembered that AM is not restricted to particles: even though Lagrange himself starts with particles, that fact is totally unimportant to his method; he could have just as well spoken of "volume elements."

4. *Theory versus experiment.* The logical consequences of the principles of AM should *not* contradict experience. This, however, does not mean that these consequences (theorems, corollaries, etc.) should be derived from experiments; the latter cannot supply missing mathematics, or be used to prove and/or verify something, but they can be used to disprove a hypothesis. As H. R. Post puts it:

> [There are] three items of religious worship *inside* present-day science, the third of which is experiment. [I]n the main the role of experiment constitutes a harmless *myth* in the philosophy of scientists. The myth considers experiment to be a generator of theories. In fact the role of experiment ... is solely to decide between two or more existing theories ... Experiment does not generate theories but rather is suggested by them. [As quoted in Truesdell (1987, p. 83). And, in a similar vein, Einstein declares: "Experiment never responds with a 'yes' to theory. At best, it says 'maybe' and, most frequently, simply 'no.' When it agrees with theory, this means 'maybe' and, if it does not, the verdict is 'no.' "]

But if the axioms of mechanics do not flow simply ("mechanically") and uniquely out of experiments, then where do they come from? Paraphrasing Hamel, Einstein et al., we may say that *these axioms are erected from the facts of experience (the object) by the human mind (the subject) as an equal and imaginative partner, from a little observation, a lot of thought and eventually a rather sudden (qualitative) understanding and insight into nature.* In other words, humans are not passive at all in the formation of scientific theories, but because of the enormous difficulty involved, the creation of a successful set of axioms is the rare act of genius [e.g. (chronologically): Euclid, Archimedes, Newton, Euler, Lagrange, Maxwell, Gibbs, Boltzmann, Planck, Einstein, Heisenberg, Schrödinger].

In CM, although open and nontrivial problems still remain, yet they are to be solved by the adoped principles; namely, we do not risk much in stating that this science is essentially closed, and that is why here the analytical/deductive method is possible. Otherwise, we would have to adopt a synthetic/inductive approach and change it slowly, depending on the new empirical facts.

5. In addition to the Lagrangean (and Hamiltonian) analytical formulation of mechanics—namely, the classical tradition of Whittaker, Hamel, Lur'e, Pars, Gantmacher et al. followed here, and depending on the emphasis laid on their most significant aspects, the following *complementary* formulations of CM also exist:

Variational (e.g., Lanczos, Rund).

Vako-nomic (= **V**ariational **A**xiomatic **K**ind; e.g., Arnold, Kozlov).

Algebraic (= infinitesimal and finite canonical transformations, Lie algebras and groups, symmetries and conservation theorems; e.g., McCauley, Mittelstaedt, Saletan and Cromer, Sudarshan and Mukunda).

Nonlinear dynamics (= regular and stochastic/chaotic motion; e.g., Gabor, Guggenheimer and Holmes, Lichtenberg and Lieberman, McCauley).

Geometrical (= symplectic geometry, canonical structure; e.g., Arnold, Abraham and Marsden, MacLane).

Statistical and thermodynamical (= Liouville's theorem, equilibrium and nonequilibrium statistical mechanics, irreversible processes, entropy, etc.; e.g., Gibbs, Katz, Fürth, Sommerfeld, Tolman).

Many-body and celestial mechanics (= orbits and their stability, many-body problem; e.g., Charlier, Hagihara, Happel, Siegel and Moser, Szebehely, Wintner).

All these, and other, formulations testify once more to the vitality and importance of CM for the entire natural science, even today.

6. For *engineering* purposes, the following (nonunique) partitioning of mechanics seems useful:

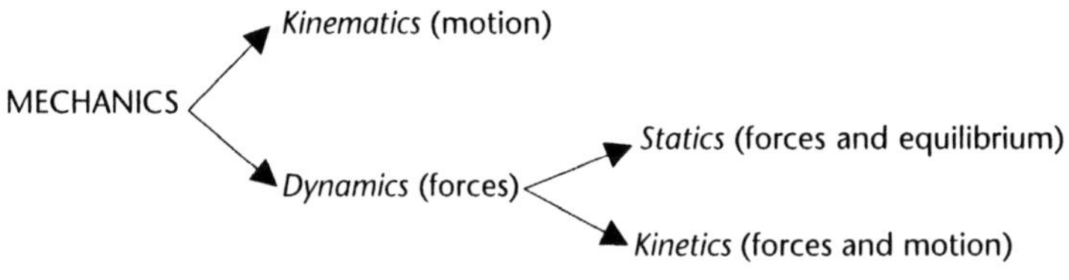

[We consider this preferable to the following partitioning, customary in the U.S. undergraduate engineering education:

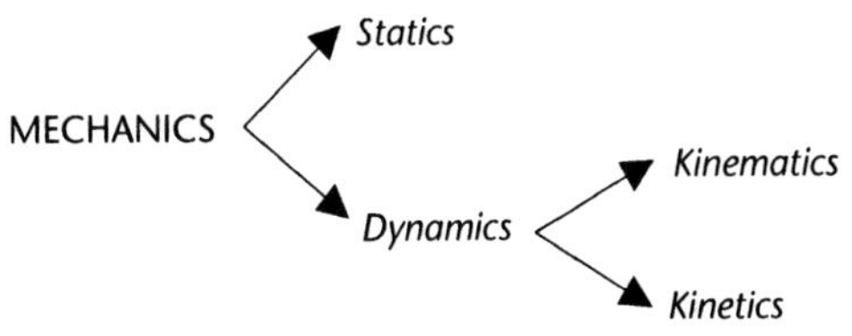

In addition, we will be using the following, not so common, terms:

Stereomechanics: mechanics of rigid bodies (and, accordingly, stereostatics, stereokinetics, etc.—mainly, after Maggi, late 1800s to early 1900s);

Kinetostatics: study of internal and external reactions in rigid bodies in motion (after Heun, early 1900s).

7. Finally, the *problem of AM* consists in the following:

(a). Formulation of the smallest, or minimal, number of equations of motion without (external and/or internal) constraint reactions; namely, the so-called *kinetic* equations; and also the ability to retrieve these reactions if needed; namely, the so-called *kinetostatic* equations. And then,

(b). The ability to solve these equations for the motion and unknown forces, respectively, either analytically (exactly or approximately) or numerically (computationally or symbolically). This is aided by the possible existence of first integrals; for example, energy, momentum, and conservation/invariance principles; more on these in chapter 3.

2 HISTORY OF THEORETICAL MECHANICS: A BIRD'S-EYE VIEW

> For us believing physicists the distinction between past, present, and future is only an illusion, even if a stubborn one.
>
> (A. Einstein, *Aphorisms*)

> The past is intelligible to us only in the light of the present; and we can fully understand the present only in the light of the past . . . . Past, present, and future are linked together in the endless chain of history.
>
> (E. H. Carr, *What is History?*, 1961)

> But it is from the Greeks, and not from any other ancient society, that we derive our interest in history and our belief that events in the past have relevance for the present.
>
> (M. Lefkowitz, 1996, p. 6)

> For in a real sense, history isn't the past—it's a posture in the present toward the future.
>
> (L. Weschler, American author/journalist, 1986)

> Rootless men and women take no more interest in the future than they take in the past.
>
> (C. Lasch, *The Minimal Self*, 1984)

> The devaluation of history is a prerequisite for the free exercise of pure power.
>
> (J. Rifkin, *Time Wars*, 1988)

The complete history of analytical mechanics, including 20th century contributions, has not been written yet—in English, anyway—and lack of space prevents us from doing so here. However, we hope that the following brief, selective, subjective, and unavoidably incomplete (but essentially correct and fair) summary, and references, will inspire others to pursue such a worthwhile and long overdue task more fully.

Most Important Milestones in the Evolution of Theoretical Engineering Dynamics (from the Viewpoint of Analytical Mechanics)

Unconstrained System Mechanics (*Momentum* mechanics of Newton–Euler)

1638:	Special particle motions (Galileo)
1687:	*Physical* foundations of mechanics (Newton—no equations of motion for systems of *more than two unconstrained*, or *more than one constrained particle* appear in his *Principia*)
1730s:	Mechanics of a *particle* (Euler)
1740s:	Mechanics of a *system of particles* (Euler, late 1740s: *Newtonian* equations of motion!)
1750s:	General principle of *linear momentum* (Euler); kinetics of *rigid bodies* (Euler)
1770s:	General principle of *angular momentum* (Euler); kinematics and geometry of rigid body motion (Euler)

Constrained System Mechanics (*Energetic* mechanics of Lagrange)

1743:	Principle of *d'Alembert* (Jakob Bernoulli → Jean Le Rond d'Alembert)
1760:	Principle of *least action* (Maupertuis → Euler → Lagrange)
1764:	*Principle of Lagrange* [Principle of virtual work (Johann Bernoulli, 1717, published 1725) + Principle of d'Alembert]
1780:	*Equations of Lagrange*
1788:	*Méchanique Analitique* (1st ed.; note old spelling)
1811:	Special *transitivity equations* (Lagrangean derivation of rigid-body Eulerian equations)
1811–1815:	*Mécanique Analytique* (2nd ed.; 3rd ed.: 1853–1855; 4th ed.: 1888–1889; English translation, from 2nd ed.: 1997!)
1829:	Gauss' Principle of *least constraint* (or *least deviation*, or *extreme compulsion*)
1830s:	*Canonical* formulation of mechanics (Hamilton)
1840s:	*Transformation/integration* theory of dynamics (Jacobi)
1860s:	*Gyroscopic* systems [Thomson (Lord Kelvin), Tait]
1870s:	*Cyclic coordinates, steady motion*, and its stability (Routh)
1873:	Earliest reactionless Lagrange-like equations for nonholonomic systems (Ferrers)

1879:	Gauss' Principle for *inequality* constraints; Gibbs–Appell equations for unconstrained systems, but in general nonholonomic velocities (quasi velocities; Gibbs)
1870s–1910s:	Dynamics of *nonholonomic systems*, under linear (or Pfaffian) velocity constraints, possibly nonholonomic (Routh, Appell, Chaplygin, Voronets, Maggi, Heun, Hamel et al.—see below)
1903–1904:	Definitive and general study of nonholonomic systems (Pfaffian constraints) in nonholonomic variables; *Lagrange–Euler equations* (Hamel)
1910s–1930s:	Dynamics of nonholonomic systems, under *nonlinear* velocity constraints (Appell, Delassus, Chetaev, Johnsen, Hamel); Study of nonholonomic systems via *general tensor calculus* (Schouten, Synge, Vranceanu, Wundheiler, Horák, Vagner et al.)
1970s–present:	Applications of the above to *multi-body* dynamics, computational dynamics [(alphabetically): Bremer, Haug, Huston, Maiβer, Roberson and Schwertassek, Schiehlen, Wittenburg et al.]

Let us elaborate a little on the dynamics of nonholonomic systems. The original Lagrangean equations (1780) were limited to holonomically constrained systems. At that time, and for several decades afterwards, *velocity* constraints (holonomic or not) were only a theoretical possibility; though one that could be easily handled by the Lagrangean *method* (i.e., principles of Lagrange and of the relaxation of the constraints (detailed in chap. 3)). But it was not until about a century later that such constraints were studied systematically. However, that necessitated a thorough re-examination of the entire edifice of Lagrangean mechanics: roughly between 1870 and 1910, what may be accurately called *the second golden age of analytical mechanics*, a host of first-rate mathematicians (Ferrers, Lindelöf, Hadamard, *Appell*, Volterra, Poincaré, Klein, Jourdain, Stäckel, Maurer), physicists (Gibbs, C. Neumann, Korteweg, Boltzmann), mechanicians (*Routh*, *Maggi*, Chaplygin, Voronets, Suslov, *Heun*, *Hamel*), and engineering scientists (Vierkandt, Beghin) developed the modern AM of constrained systems, including nonholonomic ones; and, also, the unified theory of *differential* variational principles of Lagrange, Jourdain, Gauss, Hertz et al. Up until then (ca. 1900), AM was used almost exclusively, by mathematicians and physicists, to study *unconstrained* systems: for example, celestial mechanics. The Promethean transition from heavens down to earth (i.e., constraints) was led by the great German mechanician Heun (1859–1929), who can be fairly considered as the *founder of modern engineering dynamics*; and, also, by his more famous student Hamel (1877–1954), arguably the greatest *mechaniker* of the 20th century. For instance, to these two we owe the correct formulation and interpretation of the d'Alembert–Lagrange principle (i.e., LP), and its successful application (along with additional geometrical and kinematical concepts, already in embryonic or special forms in Lagrange's works) to systems under general holonomic and/or linear *velocity* (or Pfaffian, possibly nonholonomic) constraints. Therein lie the roots of all correct treatments of the subject. [Heun also made important contributions to applied mathematics. For example, the well-known Runge–Kutta method in ordinary differential equations should be called method of *Runge–Kutta–Heun*; see, for example, Renteln (1995).]

Between the two world wars, on the basis of the so-accumulated powerful insights into the mathematical structure of LM (especially from the differential variational principles), its methods were extended to *nonlinear* nonholonomic constraints; first by Appell (1911–1925) and his student Delassus (1910s) [also by Prange and Müller (1923)] and then by Chetaev (1920s), Johnsen (1936–1941), and Hamel (1938). During the post World War II era, the entire field was summarized by Hamel himself in his magnum opus *Theoretische Mechanik* (1949); and then elaborated upon by a new generation of Soviet mechanicians [(alphabetically) Dobronravov, Fradlin, Fufaev, Lobas, Lur'e, Novoselov, Neimark, Poliahov, Rumyantsev (or Rumiantsev), et al.], whose efforts culminated in the unique and classic monograph *Dynamics of Nonholonomic Systems* by Neimark and Fufaev (1967, transl. 1972). Both of these works are most highly recommended to all serious dynamicists.

[(a) On the history of the nonholonomic equations of motion, see also chapter 3, appendix I. (b) Nonlinear (possibly nonholonomic) constraints are an area that, probably, constitutes the last theoretical frontier of LM and is of potential interest to nonlinear control theory. Also, the differential variational principles have rendered important services in the numerical treatment of problems of multibody dynamics, and promise to do more in the future.]

Guide Through the Literature on the History of Mechanics

1. *General (Mechanics and Physics):*

D'Abro (1939, 2nd ed.): Qualitative and quantitative tracing of the evolution of ideas from antiquity to modern quantum mechanics; excellent. **Hoppe** [1926(a), (b)]: Concise history of physics, with some quantitative detail; good place to start. **Hund** (1972): Panoramic, competent and compact history of physics, from antiquity to modern quantum mechanics, cosmology, and so on; one of the best places to start. **Simonyi** (1990): Comprehensive and sufficiently quantitative history of physics from antiquity to modern; beautifully and richly illustrated; most highly recommended.

2. *Mechanics—General:*

Dugas (1955): Comprehensive and quantitative history of classical and modern mechanics, from a French physicist's viewpoint; quite useful. **Dühring** (1887): Comprehensive treatment of the history of mechanics from antiquity to the middle of the 19th century; difficult to read due to its complete absence of figures and almost complete absence of mathematics; for specialists/scholars. **Haas** (1914): Detailed and pedagogical treatment of the principles of classical mechanics, from antiquity to the early 19th century; very warmly recommended, especially for undergraduates in science/engineering. **Mach** (1883–1933): Leisurely and mostly qualitative history of the principles of classical mechanics, from antiquity to the end of the 19th century; interesting and influential, but in some respects incomplete and misleading. **Szabó** (1979): Selective history of entire mechanics, with lots of beautiful photographs and diagrams; combines features of Mach, Dugas, and Truesdell. **Tiolina** (1979) and **Vesselovskii** (1974): General histories of mechanics, with detailed accounts of Russian contributions; very highly recommended for both their contents and references.

3. *Mechanics—Specialized:*

Cayley (1858, 1863): Excellent and authoritative summaries of theoretical developments until the mid 19th century; by a very famous mathematician. **Hankins** (1970): Detailed account of the life and work of d'Alembert; highly recommended to mechanics historians/scholars. **Hankins** (1985): Physics during the 18th century (of enlightenment). **Kochina** (1985): Life and works of S. Kovalevskaya. **Oravas and**

McLean (1966): Detailed account of the development of energetic/variational principles, mainly of *elastostatics*. **Polak** (1959, 1960): Detailed and lively history of differential and integral variational principles of mechanics and classical/modern physics, from antiquity to the 20th century; most highly recommended. **Stäckel** (1905): Excellent quantitative history of particle and rigid-body dynamics (elementary to intermediate), from antiquity to the early 20th century; a must for mature dynamicists; complements Voss's article. **Truesdell** (1968, 1984, 1987): Authoritative and lively detailing of the life and contributions of Euler; but invariably unfair/misleading to Lagrange and to anything remotely connected to particles and physics; for mature mechanicians/physicists. **Voss** (1901): Detailed and quantitative history of the principles of theoretical mechanics; with extensive lists of original references, from antiquity to ca. 1900; very highly recommended to mechanics and physics specialists. **Wheeler** (1952): Life and works of J. W. Gibbs. **Wintner** (1941, pp. 410–443): Notes and references on the history of analytical mechanics, with special emphasis on the mathematical aspects of *celestial* mechanics—the book, in general, is not recommended to anyone but specialists in theoretical astronomy. **Ziegler** (1985): Detailed and quantitative history of geometrical approach to rigid-body mechanics; primarily for kinematicians, not dynamicists.

4. *Histories of Mathematics:*

Bell (1937): Lively and enjoyable; concentrates on the lives and times of famous mathematicians. **Bochner** (1966): Informative, unconventional. **Klein** (1926(b), 1927): Detailed and authoritative. **Kline** (1972): Arguably, the best of its kind in English; encyclopedic, reliable, insightful, witty; a scholarly masterpiece. **Kramer** (1970): Like a more elementary version of Kline's book; interesting account of the evolution of *determinism* in physics (pp. 204–245). **Struik** (1987): Compact, dependable; includes socioeconomic explanations of mathematical inventions. Also Dictionary of Scientific Biography (ed. Gillispie, 1970s).

3 SUGGESTIONS TO THE READER

A cumulative and alphabetical bibliography is located at the end of the book (see References and Suggested Reading, pp. 1323–1370). The following grouping of textbooks and treatises aims to better orient the reader relative to (some of) the best available international mechanics/dynamics literature, and thus obtain maximum benefit from this work. References in **bold**, below, happen to be our personal favorites, and have influenced us the most in the writing of this book.

1. *For Background (Elementary to Intermediate Level):*

Butenin et al. (1985), Coe (1938), Crandall et al. (1968), Easthope (1964), **Fox** (1967), **Hamel** (1912, 1st ed.; 1927), **Loitsianskii** and **Lur'e** (1983), Milne (1948), **Nielsen** (1935), Osgood (1937), Papastavridis (EM, in preparation), **Parkus** (1966), Rosenberg (1977), Smith (1982), **Sommerfeld** (1964), Spiegel (1967), **Stäckel** (1905), **Suslov** (1946), **Synge** and **Griffith** (1959), Wells (1967).

2. *For Concurrent Reading (Intermediate to Advanced Level):*

Boltzmann (1902, 1904), **Butenin** (1971), **Dobronravov** (1970, 1976), **Gantmacher** (1970), Gray (1918), **Greenwood** (1977, 2000), **Hamel** (1949), Heil and Kitzka (1984), **Heun** (1906), **Lamb** (1943), **Lanczos** (1970), **Lur'e** (1968), MacMillan (1927, 1936), **Mei** (1985, 1987), **Neimark** and **Fufaev** (1972), **Nordheim** (1927), Pars (1965),

Päsler (1968), Pérès (1953), **Poliahov** et al. (1985), **Prange** (1935), **Rose** (1938), **Synge** (1960), Winkelman (1929, 1930).

3. *For Further Reading:*

Theoretical Physics, Nonlinear Dynamics, and so on:

Arnold (1989), Arnold et al. (1988), **Bakay** and **Stepanovskii** (1981), Birkhoff (1927), **Born** (1927), Corben and Stehle (1960), **Dittrich** and **Reuter** (1994), **Dobronravov** (1976), **Fues** (1927), **Hagihara** (1970), **Lichtenberg** and **Lieberman** (1983/1992), **McCauley** (1997), Mittelstaedt (1970), **Nordheim** and **Fues** (1927), Pars (1965), **Prange** (1935), Santilli (1978, 1980), **Straumann** (1987), **Synge** (1960), **Tabor** (1989), van Vleck (1926), Vujanovic and Jones (1989), Whittaker (1937).

Special Topics (Analytical):

Altmann (1986), Arhangelskii (1977), Chertkov (1960), Korenev (1967, 1979), **Koshlyakov** (1985), Leimanis (1965), **Lobas** (1986), **Lur'e** (1968), **Merkin** (1974, 1987), **Neimark** and **Fufaev** (1972), Novoselov (1969), **Timerding** (1908).

Applied (Multibody Dynamics/Computational/Numerical, etc.):

Battin (1987), **Bremer** [1988(a)], **Bremer** and **Pfeiffer** (1992), Haug (1992), Hughes (1986), **Huston** (1990), **Junkins** and **Turner** (1986), **Magnus** (1971), McCarthy (1990), Roberson and Schwertassek (1988), Schiehlen (1986), Shabana (1989), **Wittenburg** (1977).

4 ABBREVIATIONS, SYMBOLS, NOTATIONS, FORMULAE

These are the customary meanings; but, of course, some, hopefully easily understood, exceptions are possible. The reader is urged always to keep common sense handy!

Numbering of Equations, Examples, and Problems

Chapters are divided into sections; for example, §3.4 means chapter 3, section 4. Equations are numbered consecutively within each section. For example, reference to eq. (3.4.2) means equation (2) of chapter 3, section 4. Related equations are indicated, further, by letters; for example, eq. (3.4.2a) follows eq. (3.4.2) and somehow complements or explains it.

In chapters 2–8, examples and problems are placed anywhere within a section, and are numbered consecutively within it; for example, ex. 5.7.2 means the second example of chapter 5, section 7 and prob. 5.7.3 means the third problem of the same section. Within examples/problems, equations are numbered consecutively alphabetically; for example, reference to (ex. 5.7.3: b) means equation (b) of the third example of chapter 5, section 7. Related equations in examples/problems are followed by numbers; for example, (ex. 5.7.2: k2) is related to or explains (ex. 5.7.2: k).

Abbreviations

AD	Analytical dynamics
AM	Analytical mechanics
CM	Classical mechanics
GP	Gauss' principle (§6.4, §6.6)
H	Holonomic (coordinate/constraint/system)
HP	Hamilton's principle (ch. 7)
HZP	Hertz's principle (§6.7)
JP	Jourdain's principle (§6.3)
LP	Lagrange's principle (or D'Alembert's principle in Lagrange's form, §3.2)
NP	Nonholonomic (coordinate/constraint/system)
VD	Virtual displacement (§2.5)
PVW	Principle of virtual work (§3.2)

Chapter 1: Background

Scalars in *italics*: for example, a, A, ω, Ω
Vectors in ***boldface italics***: for example, $\boldsymbol{a}$, $\boldsymbol{A}$, $\boldsymbol{\omega}$, $\boldsymbol{\Omega}$
Matrices, tensors (usually but not always) in **boldface roman/plain text**: for example, $\mathbf{a}$, $\mathbf{A}$, $\boldsymbol{\omega}$, $\boldsymbol{\Omega}$
General symbols

N	Number of particles of a system ($P = 1, \ldots, N$)
h	Number of holonomic constraints ($H = 1, \ldots, h$)
n	Number of Lagrangean (or global) coordinates ($= 3N - h$)
m	Number of Pfaffian (holonomic and/or nonholonomic) constraints
f	Number of (local or global) degrees of freedom ($\equiv n - m$)
$k, l, p, r, \ldots$	General (system) indices ($= 1, \ldots, n$)
$I, I', I'', \ldots$	*Independent* variable indices ($= m + 1, \ldots, n$)
$D, D', D'', \ldots$	*Dependent* variable indices ($= 1, \ldots, m < n$)
$A \Rightarrow B$	A implies, or leads to, B ($A \Leftrightarrow B$, for both "directions")
$\sum$	*Discrete* summation; usually, over a pair of indices (one $\sum$ for each such pair)
$S(\ldots)$	*Summation over all the material points (particles) of a system, for a fixed time*; a three-dimensional material Stieltjes integral, equivalent to Lagrange's famous integration sign $\boldsymbol{S}\ldots$
$(\ldots)^{\cdot} \equiv d(\ldots)/dt$	Total/inertial time derivative
$(\ldots)'$	The $(\ldots)$ have been subjected to some kind of transformation
$(\ldots)_o \equiv (\ldots)$	Evaluated at some special value: for example, *initial* or *equilibrium*; or with some *constraints enforced* in it
$(\ldots)^* \equiv (\ldots)$	Expressed as function of the variables t, q, ω (quasi velocities)
$(\ldots)^{\mathrm{T}} \equiv$	Transpose of matrix $(\ldots)$
$(\ldots)^{-1} \equiv$	Inverse of matrix $(\ldots)$
δ_{kl}	Kronecker delta
ε_{krs}	Permutation symbol ($\rightarrow$ tensor, in rectangular Cartesian coordinates)

$\boldsymbol{r}/\boldsymbol{v}/\boldsymbol{a}$	(Instantaneous, and usually *inertial*) *position/velocity/acceleration* of a particle P
$\boldsymbol{r}_{A/B}/\boldsymbol{v}_{A/B}/\boldsymbol{\alpha}_{A/B}$	Position/velocity/acceleration vector of particle A relative to particle B
dm	Mass of a particle P (Continuum approach)
m_P	Mass of a particle P (Discrete approach)
$d\boldsymbol{f}$	Total force on particle P $(= d\boldsymbol{f}_e + d\boldsymbol{f}_i$; Newton–Euler) $(= d\boldsymbol{F} + d\boldsymbol{R}$; D'Alembert–Lagrange)
$d\boldsymbol{f}_e$	Total *external* force on particle P
$d\boldsymbol{f}_i$	Total *internal*, or mutual, force on particle P
$d\boldsymbol{F}$	Total *impressed*, or physical, force on particle P
$d\boldsymbol{R}$	Total *constraint reaction*, or geometrical/kinematical force on particle P
$O - XYZ\ (O - \boldsymbol{IJK})$ or $O - x_{k'}\ (O - \boldsymbol{u}_{k'})$	Space-fixed; namely, *inertial*, axes (basis) at O
$\blacklozenge - xyz\ (\blacklozenge - \boldsymbol{ijk})$ or $\blacklozenge - x_k\ (\blacklozenge - \boldsymbol{u}_k)$	Body-fixed; namely, *moving*, axes (basis) at body point $\blacklozenge$
$\blacklozenge - XYZ\ (\blacklozenge - \boldsymbol{IJK})$ or $\blacklozenge - x_{k'}\ (\blacklozenge - \boldsymbol{u}_{k'})$	Comoving, *translating but nonrotating* (or intermediate, or accessory) axes at body point $\blacklozenge$
$d(\ldots)/dt$	Rate of change of vector or tensor $(\ldots)$ relative to *fixed* (inertial) axes
$d'(\ldots)/dt$	Rate of change of vector or tensor $(\ldots)$ relative to *moving* (noninertial) axes
$\left.\begin{aligned} d\boldsymbol{a}/dt &= d'\boldsymbol{a}/dt + \boldsymbol{\Omega} \times \boldsymbol{a} \\ d\boldsymbol{T}/dt &= d'\boldsymbol{T}/dt + \boldsymbol{\Omega} \times \boldsymbol{T} - \boldsymbol{T} \times \boldsymbol{\Omega} \end{aligned}\right\}$	$\boldsymbol{a}$ arbitrary *vector*, $\boldsymbol{T}$ arbitrary (second-order) *tensor*, $\boldsymbol{\Omega}$ *angular velocity vector* of moving axes relative to fixed ones
$\mathbf{A} = (A_{k'k})$	Matrix of direction cosines between moving (e.g., body-fixed) axes and space-fixed ones; a (proper) *orthogonal* matrix (*passive* interpretation): $A_{k'k} = A_{kk'} \equiv \text{cosine}\ [(\text{fixed})_{k'},\ (\text{moving})_k]$ $\equiv \text{cosine}\ [x_{k'}, x_k]$ $\equiv \text{cosine}\ [x_k, x_{k'}]$
$\mathbf{A}^{-1} = \mathbf{A}^{\mathrm{T}}$	Determinant of $\mathbf{A} \equiv \text{Det}\ \mathbf{A} = +1$
$\mathbf{R}$	Rotation tensor (*active* interpretation of $\mathbf{A}$)
$\mathbf{R}(\boldsymbol{n}, \chi)$	Rotation tensor about a point O and axis through it specified by the unit vector $\boldsymbol{n}$, by an angle χ
$\mathbf{R}' \equiv \mathbf{R} - \mathbf{1}$	Rotator tensor
χ	Angle of finite rotation about a point O and axis specified by the unit vector $\boldsymbol{n}$
$\boldsymbol{\gamma} \equiv \tan(\chi/2)\boldsymbol{n}$	Gibbs vector of finite rotation
$\boldsymbol{\gamma} = (\gamma_{1,2,3}) \equiv (\gamma_{X,Y,Z})$	Rodrigues parameters relative to $O - XYZ$
ϕ, θ, ψ	Eulerian angles (sequence $3 \to 1 \to 3$): precession $(\phi) \to$ nutation $(\theta) \to$ proper (or eigen-) spin (ψ)
$\boldsymbol{\omega}/\boldsymbol{\omega} \equiv \mathbf{A}^{\mathrm{T}} \cdot (\mathrm{d}\mathbf{A}/\mathrm{dt})$	Angular velocity vector/tensor (moving axes components)
$\boldsymbol{\alpha} \equiv d\boldsymbol{\omega}/dt/\boldsymbol{\alpha} \equiv \mathrm{d}\boldsymbol{\omega}/\mathrm{dt}$	Angular acceleration vector/tensor (moving axes components)
$\boldsymbol{\omega}'/\boldsymbol{\omega}' \equiv (\mathrm{d}\mathbf{A}/\mathrm{dt}) \cdot \mathbf{A}^{\mathrm{T}}$	Angular velocity vector/tensor (space-fixed axes components)

$\boldsymbol{\alpha}'/\boldsymbol{\alpha}' \equiv d\boldsymbol{\omega}'/dt$	Angular acceleration vector/tensor (space-fixed axes components)
$\boldsymbol{\varepsilon} \equiv \boldsymbol{\alpha} + \boldsymbol{\omega} \cdot \boldsymbol{\omega}$	Sometimes referred to as tensor of angular acceleration
G	Center of mass of a rigid body
C	Contact point between two bodies
◆	Generic/arbitrary *body* point
•	Generic/arbitrary *space* point
$\boldsymbol{H}_{\bullet,\text{absolute}} \equiv \boldsymbol{H}_{\bullet} \equiv \mathcal{S}(\boldsymbol{r} - \boldsymbol{r}_{\bullet}) \times dm\,\boldsymbol{v}$	*Absolute* angular momentum about •
$\boldsymbol{H}_{\bullet,\text{relative}} \equiv \boldsymbol{h}_{\bullet} \equiv \mathcal{S}(\boldsymbol{r} - \boldsymbol{r}_{\bullet}) \times dm\,(\boldsymbol{v} - \boldsymbol{v}_{\bullet}) \equiv \mathcal{S}\boldsymbol{r}_{/\bullet} \times dm\,\boldsymbol{v}_{/\bullet}$	*Relative* angular momentum about •
$\boldsymbol{M}\ldots$	Moment of a force (or couple) about ...
$\mathbf{I}\ldots$	Moment of inertia tensor about ...
$T \equiv (1/2)\mathcal{S}\boldsymbol{v} \cdot \boldsymbol{v}\,dm$	(Usually inertial) kinetic energy of a system

Chapter 2: Kinematics

$q \equiv (q_1, \ldots, q_n)$	Holonomic, or global, or *Lagrangean*, or *system*, coordinates; otherwise known as *generalized coordinates*
$\boldsymbol{r} = \boldsymbol{r}(t, q)$	Fundamental *Lagrangean representation* of position of typical system particle P
$\boldsymbol{r}(t, q + \delta q) - \boldsymbol{r}(t, q) \approx \delta\boldsymbol{r} \equiv \sum (\partial\boldsymbol{r}/\partial q_k)\delta q_k$	(First-order) virtual displacement of P
$\boldsymbol{e}_k \equiv \partial\boldsymbol{r}/\partial q_k$, $\boldsymbol{e}_0 \equiv \boldsymbol{e}_{n+1} \equiv \partial\boldsymbol{r}/\partial t$	Fundamental holonomic particle and system vectors (Heun's *begleitvektoren*)
$v \equiv (dq_1/dt \equiv \dot{q}_1 \equiv v_1, \ldots, dq_n/dt \equiv \dot{q}_n \equiv v_n)$	Holonomic, or global, or *Lagrangean*, or *system*, velocities; otherwise known as generalized velocities
$\boldsymbol{v} = \sum \boldsymbol{e}_k\dot{q}_k + \boldsymbol{e}_0$	Particle velocity expressed in holonomic variables
$\boldsymbol{a} = \sum \boldsymbol{e}_k\ddot{q}_k + \text{No other } \ddot{q}_k \text{ terms}$	Particle acceleration expressed in holonomic variables
$\partial\boldsymbol{r}/\partial q_k = \partial\boldsymbol{v}/\partial\dot{q}_k = \partial\boldsymbol{a}/\partial\ddot{q}_k = \cdots = \boldsymbol{e}_k$	Basic kinematical identity (holonomic variables)
$\omega_D \equiv \sum a_{Dk}\dot{q}_k + a_D = 0$	Pfaffian constraints in velocity form (a_{Dk}, a_D: constraint coefficients, functions of t and q; ω: quasi velocities)
$d\theta_D \equiv \sum a_{Dk}dq_k + a_D dt = 0$	Pfaffian constraints in *kinematically admissible*, or *possible*, form (θ: quasi coordinates)
$\delta\theta_D \equiv \sum a_{Dk}\delta q_k = 0$	Pfaffian constraints in *virtual* form

General, kinematically admissible, variations of (...):

$$d(\ldots) \equiv \sum (\partial\ldots/\partial q_k)\,dq_k + (\partial\ldots/\partial t)\,dt = \sum (\partial\ldots/\partial\theta_k)\,d\theta_k + (\partial\ldots/\partial\theta_{n+1})\,dt$$

Virtual variation of (...):

$$\delta(\ldots) \equiv \sum(\partial\ldots/\partial q_k)\delta q_k = \sum (\partial\ldots/\partial\theta_k)\delta\theta_k$$

Quasi chain rules

$$\partial\ldots/\partial\theta_k \equiv \sum(\partial\ldots/\partial q_l)(\partial\dot{q}_l/\partial\omega_k) = \sum A_{lk}(\partial\ldots/\partial q_l)$$
$$\partial\ldots/\partial q_l \equiv \sum(\partial\ldots/\partial\theta_k)(\partial\omega_k/\partial\dot{q}_l) = \sum a_{kl}(\partial\ldots/\partial\theta_k)$$
$$\partial\ldots/\partial\theta_{n+1} \equiv \sum(\partial\ldots/\partial q_l)(\partial\dot{q}_l/\partial\omega_{n+1}) + \partial\ldots/\partial t$$
$$= \sum A_l(\partial\ldots/\partial q_l) + \partial\ldots/\partial t \equiv \partial\ldots/\partial(t) + \partial\ldots/\partial t$$
$$\partial\ldots/\partial t = \sum a_k(\partial\ldots/\partial\theta_k) + \partial\ldots/\partial\theta_{n+1} = -\sum A_k(\partial\ldots/\partial q_k) + \partial\ldots/\partial\theta_{n+1}$$

GENERAL (LOCAL) QUASI-VELOCITY TRANSFORMATIONS

Velocity form

$$\omega_D \equiv \sum a_{Dk}\dot{q}_k + a_D = 0, \qquad \omega_I \equiv \sum a_{Ik}\dot{q}_k + a_I \neq 0, \qquad \omega_{n+1} \equiv \dot{q}_{n+1} = \dot{t} = 1$$

Kinematically admissible (or possible) form

$$d\theta_D \equiv \sum a_{Dk}\,dq_k + a_D\,dt = 0, \qquad d\theta_I \equiv \sum a_{Ik}\,dq_k + a_I\,dt \neq 0,$$
$$d\theta_{n+1} \equiv dq_{n+1} = dt \neq 0$$

Virtual form

$$\delta\theta_D \equiv \sum a_{Dk}\delta q_k = 0, \qquad \delta\theta_I \equiv \sum a_{Ik}\delta q_k \neq 0, \qquad \delta\theta_{n+1} \equiv \delta q_{n+1} = \delta t = 0$$

HOLONOMIC VELOCITIES EXPRESSED IN TERMS OF QUASI VELOCITIES [(a_{kl}) and (A_{kl}) are *inverse* matrices]

Velocity form

$$\dot{q}_k = v_k = \sum A_{kl}\omega_l + A_k = \sum A_{kI}\omega_I + A_k \neq 0$$

Kinematically admissible (or possible) form

$$dq_k = \sum A_{kl}\,d\theta_l + A_k\,dt = \sum A_{kI}\,d\theta_I + A_k\,dt \neq 0 \qquad (\text{under } d\theta_D = 0)$$

Virtual form

$$\delta q_k = \sum A_{kl}\delta\theta_l = \sum A_{kI}\,\delta\theta_I \neq 0 \qquad (\text{under } \delta\theta_D = 0)$$

PARTICLE KINEMATICS IN TERMS OF QUASI VARIABLES (θ, ω, etc.)

Virtual displacement

$$\delta\mathbf{r} \equiv \sum \mathbf{e}_k\,\delta q_k = \sum \boldsymbol{\varepsilon}_k\,\delta\theta_k = \sum \boldsymbol{\varepsilon}_I\,\delta\theta_I$$

Velocity

$$\boldsymbol{v} = \sum \omega_I \boldsymbol{\varepsilon}_I + \boldsymbol{\varepsilon}_{n+1} \equiv \sum \omega_I \boldsymbol{\varepsilon}_I + \boldsymbol{\varepsilon}_0$$

Acceleration

$$\boldsymbol{a} = \sum \dot{\omega}_k \boldsymbol{\varepsilon}_k + \text{No other } \dot{\omega} \text{ terms}$$
$$= \sum \dot{\omega}_I \boldsymbol{\varepsilon}_I + \text{No other } \dot{\omega} \text{ terms}$$

Basic kinematical identity [where $f = f(t, q, \dot{q}) = f^*(t, q, \omega) = f^*$]

$$\partial \boldsymbol{r}/\partial \theta_k = \partial \boldsymbol{v}^*/\partial \omega_k = \partial \boldsymbol{a}^*/\partial \dot{\omega}_k = \cdots = \boldsymbol{\varepsilon}_k$$

Transformation relations between the holonomic and nonholonomic bases $\boldsymbol{e}_{\ldots}$, $\boldsymbol{\varepsilon}_{\ldots}$

$$\boldsymbol{\varepsilon}_k = \sum (\partial \dot{q}_l/\partial \omega_k)\boldsymbol{e}_l = \sum A_{lk}\boldsymbol{e}_l$$
$$\boldsymbol{\varepsilon}_{n+1} \equiv \sum \boldsymbol{\varepsilon}_0 = \sum A_l \boldsymbol{e}_l + \boldsymbol{e}_{n+1} = -\sum a_k \boldsymbol{\varepsilon}_k + \boldsymbol{e}_{n+1}$$
$$\boldsymbol{e}_l = \sum (\partial \omega_k/\partial \dot{q}_l)\boldsymbol{\varepsilon}_k = \sum a_{kl}\boldsymbol{\varepsilon}_k$$
$$\boldsymbol{e}_{n+1} \equiv \boldsymbol{e}_0 \equiv \boldsymbol{e}_t = \sum a_k \boldsymbol{\varepsilon}_k + \boldsymbol{\varepsilon}_{n+1} = -\sum A_l \boldsymbol{e}_l + \boldsymbol{\varepsilon}_{n+1}$$

FROM PARTICLE TO SYSTEM VECTORS
(i.e., vectors characterizing, or expressing, system variables)

$\mathcal{S}$ (*particle* vector) $\cdot\, \boldsymbol{e}_k$ = (*system* vector)$_k$ (*holonomic* components)

$\mathcal{S}$ (*particle* vector) $\cdot\, \boldsymbol{\varepsilon}_k$ = (*system* vector)$_k$ (*nonholonomic* components)

SPECIAL FORMS OF PFAFFIAN CONSTRAINTS

Chaplygin

$$\omega_D \equiv \dot{q}_D - \sum b_{DI}\dot{q}_I = 0; \qquad b_{DI}\text{: functions of } q_I \equiv (q_{m+1}, \ldots, q_n)$$
$$\omega_I \equiv \dot{q}_I \neq 0$$

Voronets

$$\omega_D \equiv \dot{q}_D - \sum b_{DI}\dot{q}_I - b_D = 0; \qquad b_{DI}, b_D\text{: functions of } t \text{ and } all\ q\text{s}$$
$$\omega_I \equiv \dot{q}_I \neq 0$$
$$\left[\Rightarrow \dot{q}_D = \sum b_{DI}\omega_I + b_D, \qquad \dot{q}_I \equiv \omega_I\right]$$

Corresponding particle virtual displacement

$$\delta \boldsymbol{r} \equiv \sum \boldsymbol{e}_k \delta q_k = \sum \boldsymbol{\beta}_I \delta q_I$$

Corresponding particle velocity

$$\boldsymbol{v} = \sum \dot{q}_l \boldsymbol{\beta}_l + \boldsymbol{\beta}_{n+1} \equiv \boldsymbol{v}_o$$

HAMEL COEFFICIENTS

$$\begin{aligned}
\gamma^k{}_{rs} = -\gamma^k{}_{sr} &= \sum\sum (\partial a_{kb}/\partial q_c - \partial a_{kc}/\partial q_b) A_{br} A_{cs} \\
&= \sum\sum a_{kb}[A_{cr}(\partial A_{bs}/\partial q_c) - A_{cs}(\partial A_{br}/\partial q_c)] \\
&= \sum\sum (A_{br}A_{cs} - A_{cr}A_{bs})(\partial a_{kb}/\partial q_c)
\end{aligned}$$

$$\begin{aligned}
\gamma^k{}_{r,n+1} = -\gamma^k{}_{n+1,r} \equiv \gamma^k{}_r &= \sum\sum (\partial a_{kb}/\partial q_c - \partial a_{kc}/\partial q_b) A_{br} A_c \\
&+ \sum (\partial a_{kb}/\partial t - \partial a_k/\partial q_b) A_{br}
\end{aligned}$$

$$\gamma^{n+1}{}_{kl} = 0, \qquad \gamma^{n+1}{}_{k,n+1} = -\gamma^{n+1}{}_{n+1,k} = 0, \qquad \gamma^{n+1}{}_{n+1,n+1} = 0$$

TRANSITIVITY EQUATIONS

$$(\delta\theta_k)^{\cdot} - \delta\omega_k = \sum a_{kl}[(\delta q_l)^{\cdot} - \delta(\dot{q}_l)] + \sum\sum \gamma^k{}_{bs}\omega_s\delta\theta_b + \sum \gamma^k{}_b\delta\theta_b$$

$$(\delta q_k)^{\cdot} - \delta(\dot{q}_k) = \sum A_{kl}\Big\{[(\delta\theta_l)^{\cdot} - \delta\omega_l] - \sum\sum \gamma^l{}_{bs}\omega_s\delta\theta_b - \sum \gamma^l{}_b\delta\theta_b\Big\}$$

$$(\delta\theta_{n+1})^{\cdot} - \delta\omega_{n+1} \equiv (\delta q_{n+1})^{\cdot} - \delta(\dot{q}_{n+1}) \equiv (\delta t)^{\cdot} - \delta(\dot{t}) = 0$$

or, equivalently,

$$d(\delta\theta_k) - \delta(d\theta_k) = \sum a_{kl}[d(\delta q_l) - \delta(dq_l)] + \sum\sum \gamma^k{}_{bs}\, d\theta_s\, \delta\theta_b + \sum \gamma^k{}_b\, dt\, \delta\theta_b$$

$$d(\delta q_k) - \delta(dq_k) = \sum A_{kl}\Big\{[d(\delta\theta_l) - \delta(d\theta_l)] - \sum\sum \gamma^l{}_{bs}\, d\theta_s\, \delta\theta_b - \sum \gamma^l{}_b\, dt\, \delta\theta_b\Big\}$$

$$d(\delta\theta_{n+1}) - \delta(d\theta_{n+1}) = d(\delta t) - \delta(dt) = d(0) - \delta(dt) = 0 - 0 = 0;$$

or, assuming (Hamel viewpoint)

$$(\delta q_k)^{\cdot} = \delta(\dot{q}_k) \qquad \text{or} \qquad d(\delta q_k) = \delta(dq_k)$$

$$(\delta\theta_k)^{\cdot} - \delta\omega_k = \sum\sum \gamma^k{}_{bs}\, \omega_s\, \delta\theta_b + \sum \gamma^k{}_b\, \delta\theta_b \equiv \sum h^k{}_b\, \delta\theta_b$$

$$\begin{aligned}
d(\delta\theta_k) - \delta(d\theta_k) &= \sum\sum \gamma^k{}_{bs}\, d\theta_s\, \delta\theta_b + \sum \gamma^k{}_b\, dt\, \delta\theta_b \\
&= \sum\sum{}' \gamma^k{}_{bs}(d\theta_s\, \delta\theta_b - \delta\theta_s\, d\theta_b) + \sum \gamma^k{}_b\, dt\, \delta\theta_b
\end{aligned}$$

(where $\sum\sum{}'$ means that the summation extends over b and s only *once*; say, $s < b$)

Generally [with o, $\bullet = 1, \ldots, n$; $\delta\theta_{n+1} \equiv \delta t = 0$]

$$d(\delta\theta_*) - \delta(d\theta_*) = \sum\sum \gamma^*{}_{\bullet o}\, d\theta_o\, \delta\theta_\bullet + \sum \gamma^*{}_\bullet\, dt\, \delta\theta_\bullet$$

FROBENIUS' THEOREM

(Necessary and sufficient conditions for holonomicity = complete integrability of a system of m Pfaffian constraints in the $n+1$ variables $q_1, \ldots, q_n; q_{n+1}$)

$$\gamma^D{}_{II'} = 0, \qquad \gamma^D{}_{I,n+1} \equiv \gamma^D{}_I = 0 \qquad (D = 1, \ldots, m; I, I' = m+1, \ldots, n)$$

Chapter 3: Kinetics

BASIC QUANTITIES

$T \equiv (1/2) S\, \boldsymbol{v} \cdot \boldsymbol{v}\, dm$ (Usually inertial) Kinetic Energy of system

$S \equiv (1/2) S\, \boldsymbol{a} \cdot \boldsymbol{a}\, dm$ (Usually inertial) Gibbs–Appell function of system, or simply Appellian

NOTATION

$$f(t, q, \dot{q}) = f[t, q, \dot{q}_D(t, q, \omega)] \equiv f^*(t, q, \omega) \equiv f^* \quad \text{[arbitrary function]};$$

for example,

$$T(t, q, \dot{q}) = T[t, q, \dot{q}_D(t, q, \omega)] \equiv T^*(t, q, \omega) \equiv T^*$$
$$\Rightarrow T^*(t, q, \omega_D = 0, \omega_I) \equiv T^*{}_o(t, q, \omega_I) \equiv T^*{}_o$$
$$T(t, q, \dot{q}) = T[t, q, \dot{q}_D(t, q, \dot{q}_I), \dot{q}_I] \equiv T_o(t, q, \dot{q}_I) \equiv T_o \qquad \text{(and similarly for } S\text{)}$$

LAGRANGE'S PRINCIPLE

$$\delta' W_R \geq 0 \quad \Rightarrow \quad \delta I \geq \delta' W$$

(for *unilateral* constraints; for *bilateral* constraints, $\geq$ is replaced by $=$)

Particle (or raw) forms

$$\delta' W_R \equiv S\, d\boldsymbol{R} \cdot \delta \boldsymbol{r}, \qquad \delta' W \equiv S\, d\boldsymbol{F} \cdot \delta \boldsymbol{r}, \qquad \delta I \equiv S\, dm\, \boldsymbol{a} \cdot \delta \boldsymbol{r}$$

Holonomic variable forms

$$\delta' W_R = \sum R_k\, \delta q_k, \qquad R_k \equiv S\, d\boldsymbol{R} \cdot \boldsymbol{e}_k,$$
$$\delta' W = \sum Q_k\, \delta q_k, \qquad Q_k \equiv S\, d\boldsymbol{F} \cdot \boldsymbol{e}_k,$$
$$\delta I = \sum [(\partial T/\partial \dot{q}_k)^{\cdot} - \partial T/\partial q_k]\, \delta q_k = \sum (\partial S/\partial \ddot{q}_k)\, \delta q_k \equiv \sum E_k\, \delta q_k$$

Nonholonomic variable forms

$$\delta' W_R = \sum \Lambda_k\, \delta\theta_k = \sum \Lambda_l\, \delta\theta_l, \qquad \Lambda_k \equiv \mathsf{S}\, d\boldsymbol{R} \cdot \boldsymbol{\varepsilon}_k,$$

$$\delta' W = \sum \Theta_k\, \delta\theta_k, \qquad \Theta_k \equiv \mathsf{S}\, d\boldsymbol{F} \cdot \boldsymbol{\varepsilon}_k,$$

$$\delta I = \sum [(\partial T^*/\partial\omega_k)^{\cdot} - \partial T^*/\partial\theta_k - \Gamma_k]\, \delta\theta_k = \sum (\partial S^*/\partial\dot{\omega}_k)\, \delta\theta_k \equiv \sum I_k\, \delta\theta_k$$

INERTIAL "FORCES" IN HOLONOMIC VARIABLES

$$E_k \equiv \mathsf{S}\, dm\, \boldsymbol{a} \cdot \boldsymbol{e}_k$$

$$= (\partial T/\partial \dot{q}_k)^{\cdot} - \partial T/\partial q_k \qquad \text{(Lagrangean form)}$$

$$= \partial S/\partial \ddot{q}_k \qquad \text{(Appellian form)}$$

$$= \partial \dot{T}/\partial \dot{q}_k - 2(\partial T/\partial q_k) \equiv N_k(T) \equiv N_k \qquad \text{(Nielsen form; see chap. 5)}$$

INERTIAL "FORCES" IN NONHOLONOMIC VARIABLES

$$I_k \equiv \mathsf{S}\, dm\, \boldsymbol{a} \cdot \boldsymbol{\varepsilon}_k$$

$$= (\partial T^*/\partial\omega_k)^{\cdot} - \partial T^*/\partial\theta_k - \Gamma_k \equiv E_k{}^*(T^*) - \Gamma_k \qquad \text{(Volterra–Hamel form)}$$

$$= \partial S^*/\partial\dot{\omega}_k \qquad \text{(Gibbs–Appell form)}$$

$$= \sum (\partial \dot{q}_l/\partial\omega_k) E_l = \sum A_{lk} E_l \qquad \text{(Maggi form)}$$

Nonholonomic deviation

$$\Gamma_k \equiv \mathsf{S}\, dm\, \boldsymbol{v}^* \cdot [(\partial \boldsymbol{v}^*/\partial\omega_k)^{\cdot} - \partial \boldsymbol{v}^*/\partial\theta_k]$$

$$\equiv \mathsf{S}\, dm\, \boldsymbol{v}^* \cdot E_k{}^*(\boldsymbol{v}^*) \equiv \mathsf{S}\, dm\, \boldsymbol{v}^* \cdot \boldsymbol{\gamma}_k \qquad \text{(particle/raw form)}$$

$$= -\sum\sum \gamma^l_{ks}(\partial T^*/\partial\omega_l)\omega_s - \sum \gamma^l_k(\partial T^*/\partial\omega_l)$$

$$= -\sum h^l_k(\partial T^*/\partial\omega_l) \qquad \left[h^l_k \equiv \sum \gamma^l_{ks}\, \omega_s + \gamma^l_k\right]$$

TRANSFORMATION EQUATIONS

$$R_l = \sum a_{kl}\Lambda_k \;\Leftrightarrow\; \Lambda_k = \sum A_{lk} R_l$$

$$Q_l = \sum a_{kl}\Theta_k \;\Leftrightarrow\; \Theta_k = \sum A_{lk} Q_l$$

$$E_l = \sum a_{kl} I_k \;\Leftrightarrow\; I_k = \sum A_{lk} E_l$$

THE CENTRAL EQUATION

(Lagrange–Heun–Hamel Zentralgleichung)

First Form

$$\delta T + \delta' W + \delta D = (\delta P)^{\cdot}$$

Second Form

$$\sum \dot{P}_k\, \delta\theta_k + \sum P_k[(\delta\theta_k)^{\cdot} - \delta\omega_k] - \sum (\partial T^*/\partial\theta_k)\, \delta\theta_k = \sum \Theta_k\, \delta\theta_k$$

where

$$\begin{aligned} \delta T = \mathcal{S}\, dm\, \boldsymbol{v} \cdot \delta \boldsymbol{v} &= \sum [(\partial T/\partial q_k)\, \delta q_k + (\partial T/\partial \dot{q}_k)\, \delta(\dot{q}_k)] \\ &= \sum [(\partial T^*/\partial q_k)\, \delta q_k + (\partial T^*/\partial \omega_k)\, \delta\omega_k] \\ &\equiv \sum [(\partial T^*/\partial \theta_k)\, \delta\theta_k + (\partial T^*/\partial \omega_k)\, \delta\omega_k] = \delta T^* \end{aligned}$$

$$\delta' W \equiv \mathcal{S}\, d\boldsymbol{F} \cdot \delta \boldsymbol{r} = \sum Q_k\, \delta q_k = \sum \Theta_k\, \delta\theta_k$$

$$\begin{aligned} \delta D &\equiv \mathcal{S}\, dm\, \boldsymbol{v} \cdot [(\delta \boldsymbol{r})^{\cdot} - \delta \boldsymbol{v}] \\ &= \sum (\partial T^*/\partial \omega_k)[(\delta\theta_k)^{\cdot} - \delta\omega_k] - \sum h^l_k (\partial T^*/\partial \omega_l)\, \delta\theta_k \end{aligned}$$

$$\delta P = \mathcal{S}\, dm\, \boldsymbol{v} \cdot \delta \boldsymbol{r} = \sum p_k\, \delta q_k = \sum P_k\, \delta\theta_k$$

$p_k \equiv \mathcal{S}\, dm\, \boldsymbol{v} \cdot \boldsymbol{e}_k = \partial T/\partial v_k$ (*holonomic* momentum)

$P_k \equiv \mathcal{S}\, dm\, \boldsymbol{v}^* \cdot \boldsymbol{\varepsilon}_k = \partial T^*/\partial \omega_k$ (*nonholonomic* momentum)

$p_l = \sum a_{kl} P_k \;\Leftrightarrow\; P_k = \sum A_{lk} p_l$ (*transformation formulae*)

EQUATIONS OF MOTION

COUPLED

Routh–Voss (adjoining of constraints via multipliers)

$E_k = Q_k + R_k$ (multipliers; holonomic variables)

UNCOUPLED

Maggi (projections)

Kinetostatic: $\sum A_{Dk} E_D = \sum A_{Dk} Q_D + \Lambda_D$ (multipliers; holonomic variables)

Kinetic: $\sum A_{Ik} E_I = \sum A_{Ik} Q_I$ (no multipliers; holonomic variables)

Hamel (embedding of constraints via quasi variables)

Kinetostatic: $E_D{}^*(T^*) - \Gamma_D = \Theta_D + \Lambda_D$ (multipliers; nonholonomic variables)

Kinetic: $E_I{}^*(T^*) - \Gamma_I = \Theta_I$ (no multipliers; nonholonomic variables)

SPECIAL FORMS (constraints of form $\dot{q}_D = \sum b_{DI}\dot{q}_I + b_D$; b_{DI}, b_D functions of t, q)

Maggi → Hadamard

$E_D = Q_D + \lambda_D$ (kinetostatic) $\qquad E_I = Q_I - \sum b_{DI}\lambda_D$

$\Rightarrow E_I + \sum b_{DI} E_D = Q_I + \sum b_{DI} Q_D \equiv Q_{I,o} \equiv Q_{Io}$ (kinetic)

Hamel $\rightarrow$ Voronets

$$\left[T_o = T_o(t, q, \dot{q}_I), \qquad \dot{q}_D = \sum b_{DI}\dot{q}_I + b_D\right]$$

$$(\partial T_o/\partial \dot{q}_I)^{\cdot} - \partial T_o/\partial q_I - \sum b_{DI}(\partial T_o/\partial q_D)$$
$$- \sum\sum w^D{}_{II'}(\partial T/\partial \dot{q}_D)_o \dot{q}_{I'} - \sum w^D{}_I(\partial T/\partial \dot{q}_D)_o = Q_I + \sum b_{DI}Q_D,$$

$$w^D{}_{II'} \equiv \left[\partial b_{DI}/\partial q_{I'} + \sum b_{D'I}(\partial b_{DI'}/\partial q_{D'})\right] - \left[\partial b_{DI}/\partial q_{I'} + \sum b_{D'I'}(\partial b_{DI}/\partial q_{D'})\right]$$

$$w^D{}_I \equiv w^D{}_{I,n+1} \equiv \left[\partial b_D/\partial q_I + \sum b_{D'I}(\partial b_D/\partial q_{D'})\right] - \left[\partial b_{DI}/\partial t + \sum b_{D'}(\partial b_{DI}/\partial q_{D'})\right]$$

Voronets $\rightarrow$ Chaplygin

$$\left[T_o = T_o(q_I, \dot{q}_I), \qquad \dot{q}_D = \sum b_{DI}(q_{m+1}, \ldots, q_n)\dot{q}_I; \qquad \text{i.e., } b_D = 0\right]$$

$$(\partial T_o/\partial \dot{q}_I)^{\cdot} - \partial T_o/\partial q_I$$
$$- \sum\sum t^D{}_{II'}(\partial T/\partial \dot{q}_D)_o \dot{q}_{I'} = Q_I + \sum b_{DI}Q_D$$

$$t^D{}_{II'} \equiv \partial b_{DI}/\partial q_{I'} - \partial b_{DI'}/\partial q_I$$

POWER (OR ENERGY RATE) THEOREMS

Holonomic variables

$$dh/dt = -\partial L/\partial t + \sum Q_{k,\text{nonpotential}}\, \dot{q}_k - \sum \lambda_D a_D,$$
$$h \equiv \sum (\partial L/\partial \dot{q}_k)\, \dot{q}_k - L = T_2 + (V_0 - T_0), \qquad L \equiv T - V;$$
$$dE/dt = -\partial L/\partial t + d(T_1 + 2T_0)/dt + \sum Q_{k,\text{nonpotential}}\, \dot{q}_k - \sum \lambda_D a_D,$$
$$E \equiv T + V_0, \qquad L = T - V = T - (V_0 + V_1), \qquad h \equiv E - (T_1 + 2T_0)$$

Nonholonomic variables

$$dh^*/dt = -\partial L^*/\partial\theta_{n+1} + \sum \Theta_{I,\text{nonpotential}}\, \omega_I - R,$$
$$h^* \equiv \sum (\partial L^*/\partial\omega_I)\omega_I - L^* = T^*_2 + (V_0 - T^*_0)$$
$$\partial L^*/\partial\theta_{n+1} \equiv \partial L^*/\partial t + \sum A_k(\partial L^*/\partial q_k)$$
$$R \equiv \sum\sum \gamma^r{}_I(\partial L^*/\partial\omega_r)\omega_I \qquad (\textit{Rheonomic}\ \text{nonholonomic power})$$

EXPLICIT FORMS OF THE EQUATIONS OF MOTION

Lagrangean equations: with

$$T = T_2 + T_1 + T_0; \qquad 2T_2 \equiv \sum\sum M_{kr}\dot{q}_r\dot{q}_k, \qquad T_1 \equiv \sum M_r\dot{q}_r, \qquad 2T_0 \equiv M_0,$$
$$M_{kl} = M_{lk}, \qquad M_{k,n+1} = M_{n+1,k} \equiv M_{k0} = M_{0k} \equiv M_k,$$
$$M_{n+1,n+1} \equiv M_{00} \equiv M_0\text{: } \textit{Inertia coefficients},$$

$2\Gamma_{k,rs} \equiv 2\Gamma_{k,sr} \equiv \partial M_{kr}/\partial q_s + \partial M_{ks}/\partial q_r - \partial M_{rs}/\partial q_k$: *1st kind Christoffels*,

$G_k \equiv \sum g_{kr}\,\dot{q}_r \equiv \sum (\partial M_r/\partial q_k - \partial M_k/\partial q_r)\dot{q}_r$: *Gyroscopic "force"*,

$Q_k = Q_{k,\text{nonpotential}} + (\partial V/\partial \dot{q}_k)^{\cdot} - \partial V/\partial q_k$,

$V = \sum V_k(t,q)\dot{q}_k + V_0(t,q) \equiv V_1(t,q,\dot{q}) + V_0(t,q)$: *Generalized* potential,

the Lagrangean-type equations, say $E_k = Q_k$, assume the form

$$E_k(T_2) + E_k(T_1) + E_k(T_0) = Q_k,$$
$$E_k(T_2) = \sum M_{kr}\ddot{q}_r + \sum\sum \Gamma_{k,rs}\dot{q}_r\dot{q}_s + \sum (\partial M_{kr}/\partial t)\dot{q}_r,$$
$$E_k(T_1) = \partial M_k/\partial t - G_k,$$
$$E_k(T_0) = -(1/2)(\partial M_0/\partial q_k).$$

Hamel equations (stationary case, no constraints), with

$$2T^* = 2T^*{}_2 = \sum\sum M^*{}_{kr}\,\omega_r\,\omega_k,$$
$$2\Gamma^*{}_{k,rs} \equiv 2\Gamma^*{}_{k,sr} \equiv \partial M^*{}_{kr}/\partial\theta_s + \partial M^*{}_{ks}/\partial\theta_r - \partial M^*{}_{rs}/\partial\theta_k,$$
$$\Lambda_{k,lp} \equiv \Gamma^*{}_{k,lp} + \sum \gamma^r{}_{kl} M^*{}_{rp} \qquad \text{("nonholonomic Christoffels")}$$

Hamel-type equations, say $I_k = \Theta_k$, assume the form

$$\sum M^*{}_{kl}\dot{\omega}_l + \sum\sum \Lambda_{k,lp}\,\omega_l\,\omega_p = \Theta_k.$$

APPELLIAN FUNCTION

Holonomic variables

$$2S = \sum M_{kr}\ddot{q}_r\ddot{q}_k + 2\sum\sum\sum \Gamma_{k,lp}\ddot{q}_k\,\dot{q}_l\,\dot{q}_p$$
$$+ 4\sum\sum \Gamma_{k,l,n+1}\ddot{q}_k\,\dot{q}_l + 2\sum \Gamma_{k,n+1,n+1}\ddot{q}_k$$

Nonholonomic variables (stationary case)

$$2S^* = \sum\sum M^*{}_{kr}\,\dot{\omega}_k\,\dot{\omega}_r + 2\sum\sum\sum \Lambda_{k,lp}\,\dot{\omega}_k\,\omega_l\,\omega_p$$

LAGRANGEAN TREATMENT OF THE RIGID BODY

Kinetic energy

$$T = T_{\text{translation}} + T_{\text{rotation}} + T_{\text{coupling}}$$
$$2T_{\text{translation}} = m\,v_\blacklozenge{}^2 \qquad (\blacklozenge\text{: arbitrary } \textit{body} \text{ point; } m\text{: mass of body})$$
$$2T_{\text{rotation}} = \boldsymbol{\omega}\cdot \mathop{S} dm(\boldsymbol{r}_{/\blacklozenge}\times \boldsymbol{v}_{/\blacklozenge})$$
$$= \boldsymbol{\omega}\cdot \mathop{S} dm\,[\boldsymbol{r}_{/\blacklozenge}\times(\boldsymbol{\omega}\times\boldsymbol{r}_{/\blacklozenge})] \equiv \boldsymbol{\omega}\cdot\boldsymbol{h}_\blacklozenge = \boldsymbol{\omega}\cdot\boldsymbol{I}_\blacklozenge\cdot\boldsymbol{\omega}$$
$$T_{\text{coupling}} = \boldsymbol{\omega}\cdot \mathop{S} dm\,(\boldsymbol{r}_{/\blacklozenge}\times\boldsymbol{v}_\blacklozenge) = m\,\boldsymbol{v}_\blacklozenge\cdot(\boldsymbol{\omega}\times\boldsymbol{r}_{G/\blacklozenge}) = m\,\boldsymbol{v}_\blacklozenge\cdot\boldsymbol{v}_{G/\blacklozenge}$$

Momentum vectors

$$\delta P = \mathcal{S}\, dm\, \boldsymbol{v} \cdot \delta \boldsymbol{r} = \boldsymbol{p} \cdot \delta \boldsymbol{r}_{\blacklozenge} + \boldsymbol{H}_{\blacklozenge} \cdot \delta \boldsymbol{\chi} \qquad (\delta\boldsymbol{\chi}\text{: elementary rotation vector})$$

$\boldsymbol{p} \equiv \mathcal{S}\, dm\, \boldsymbol{v} = m\, \boldsymbol{v}_G$: *linear momentum* of body

$\boldsymbol{H}_{\blacklozenge} \equiv \mathcal{S}\, \boldsymbol{r}_{/\blacklozenge} \times (dm\, \boldsymbol{v}) = \boldsymbol{h}_{\blacklozenge} + \boldsymbol{r}_{G/\blacklozenge} \times (m\, \boldsymbol{v}_{\blacklozenge})$: *absolute angular momentum* of body about $\blacklozenge$

$$\boldsymbol{H}_O \equiv \mathcal{S}\, \boldsymbol{r} \times (dm\, \boldsymbol{v}) = \boldsymbol{H}_{\blacklozenge} + \boldsymbol{r}_{\blacklozenge/O} \times \boldsymbol{p} \qquad (O\text{: fixed point})$$

Kinetic energy in terms of the momentum vectors

$$2T = \boldsymbol{p} \cdot \boldsymbol{v}_{\blacklozenge} + \boldsymbol{H}_{\blacklozenge} \cdot \boldsymbol{\omega}, \qquad \boldsymbol{p} = \partial T / \partial \boldsymbol{v}_{\blacklozenge}, \qquad \boldsymbol{H}_{\blacklozenge} = \partial T / \partial \boldsymbol{\omega}$$

Kinematico-inertial (KI) acceleration vectors

$$\delta I = \mathcal{S}\, dm\, \boldsymbol{a} \cdot \delta \boldsymbol{r} = \boldsymbol{I} \cdot \delta \boldsymbol{r}_{\blacklozenge} + \boldsymbol{A}_{\blacklozenge} \cdot \delta \boldsymbol{\chi}$$

$\boldsymbol{I} \equiv \mathcal{S}\, dm\, \boldsymbol{a} = m\, \boldsymbol{a}_G$: *linear KI acceleration* of body

$\boldsymbol{A}_{\blacklozenge} \equiv \mathcal{S}\, \boldsymbol{r}_{/\blacklozenge} \times (dm\, \boldsymbol{a})$: *angular KI acceleration* of body about $\blacklozenge$

Eulerian principles in Lagrangean form

Linear momentum ($\boldsymbol{\Omega}$: angular velocity of moving axes)

$$\boldsymbol{I} = d\boldsymbol{p}/dt = d'\boldsymbol{p}/dt + \boldsymbol{\Omega} \times \boldsymbol{p} = d'/dt\,(\partial T / \partial \boldsymbol{v}_{\blacklozenge}) + \boldsymbol{\Omega} \times (\partial T / \partial \boldsymbol{v}_{\blacklozenge})$$

Angular momentum

$$\begin{aligned} \boldsymbol{A}_{\blacklozenge} &= d\boldsymbol{H}_{\blacklozenge}/dt + \boldsymbol{v}_{\blacklozenge} \times \boldsymbol{p} \\ &= (d'\boldsymbol{H}_{\blacklozenge}/dt + \boldsymbol{\Omega} \times \boldsymbol{H}_{\blacklozenge}) + \boldsymbol{v}_{\blacklozenge} \times \boldsymbol{p} \\ &= d'/dt\,(\partial T / \partial \boldsymbol{\omega}) + \boldsymbol{\Omega} \times (\partial T / \partial \boldsymbol{\omega}) + \boldsymbol{v}_{\blacklozenge} \times (\partial T / \partial \boldsymbol{v}_{\blacklozenge}) \end{aligned}$$

(also $\boldsymbol{A}_{\bullet} = d\boldsymbol{H}_{\bullet}/dt + \boldsymbol{v}_{\bullet} \times \boldsymbol{p}$; $\bullet$: any point)

APPELLIAN FUNCTION (to within acceleration-proportional terms)

$$\begin{aligned} 2S &= m\, {\boldsymbol{a}_{\blacklozenge}}^2 + 2m\, \boldsymbol{r}_{G/\blacklozenge} \cdot (\boldsymbol{a}_{\blacklozenge} \times \boldsymbol{\alpha}) + 2m\, (\boldsymbol{\omega} \times \boldsymbol{r}_{G/\blacklozenge}) \cdot (\boldsymbol{a}_{\blacklozenge} \times \boldsymbol{\omega}) \\ &\quad + \boldsymbol{\alpha} \cdot \boldsymbol{I}_{\blacklozenge} \cdot \boldsymbol{\alpha} + 2(\boldsymbol{\alpha} \times \boldsymbol{\omega}) \cdot \boldsymbol{I}_{\blacklozenge} \cdot \boldsymbol{\omega} \\ &= m\, a_G^2 + \boldsymbol{\alpha} \cdot \boldsymbol{I}_G \cdot \boldsymbol{\alpha} + 2(\boldsymbol{\alpha} \times \boldsymbol{\omega}) \cdot \boldsymbol{I}_G \cdot \boldsymbol{\omega} \end{aligned}$$

(Appellian counterpart of König's theorem)

RELATIVE MOTION (I: *inertial* origin; O: *moving* origin)

Positions

$$\boldsymbol{r}_I = \boldsymbol{r}_O(t) + \boldsymbol{r}(q_1, \ldots, q_n) \qquad (\text{motion of } O \text{ known, } q\text{: } \textit{noninertial} \text{ coordinates})$$

Velocities

$$\boldsymbol{v} = \boldsymbol{v}_O + \boldsymbol{v}_{\text{relative}} + \boldsymbol{\Omega} \times \boldsymbol{r}, \qquad \boldsymbol{v}_{\text{relative}} \equiv d'\boldsymbol{r}/dt = \sum (\partial \boldsymbol{r}/\partial q_k)\, \dot{q}_k$$

Virtual displacements

$$\delta \boldsymbol{r}_I = \delta \boldsymbol{r}_O + \delta \boldsymbol{r} = \delta \boldsymbol{r}_O + \delta' \boldsymbol{r} + \delta \boldsymbol{X} \times \boldsymbol{r} \qquad (\boldsymbol{\Omega} \equiv d\boldsymbol{X}/dt\text{: frame angular velocity})$$

Kinetic energy

$$T = T_{\text{transport}} + T_{\text{relative}} + T_{\text{coupling}}$$

$$2T_{\text{transport}} = m\, v_O{}^2 + 2m\, \boldsymbol{v}_O \cdot (\boldsymbol{\Omega} \times \boldsymbol{r}_G) + \boldsymbol{\Omega} \cdot \boldsymbol{I}_O \cdot \boldsymbol{\Omega} \equiv 2T_0 \qquad [\sim \dot{q}^0]$$

$$2T_{\text{relative}} = S\, dm\, \boldsymbol{v}_{\text{rel've}} \cdot \boldsymbol{v}_{\text{rel've}} \equiv 2T_2 \qquad [\sim \dot{q}^2]$$

$$T_{\text{coupling}} = \boldsymbol{p}_{\text{rel've}} \cdot \boldsymbol{v}_O + \boldsymbol{H}_{O,\text{rel've}} \cdot \boldsymbol{\Omega} \equiv T_1 \qquad [\sim \dot{q}^1]$$

$$\boldsymbol{p}_{\text{rel've}} \equiv S\, dm\, \boldsymbol{v}_{\text{rel've}} = m(d'\boldsymbol{r}_G/dt) = \sum \Big(S\, dm (\partial \boldsymbol{r}/\partial q_k) \Big)\, \dot{q}_k$$

(*noninertial* linear momentum)

$$\boldsymbol{H}_{O,\text{rel've}} \equiv S\, \boldsymbol{r} \times (d'\boldsymbol{r}/dt) = \sum \Big(S\, dm\, \boldsymbol{r} \times (\partial \boldsymbol{r}/\partial q_k) \Big) \dot{q}_k$$

(*noninertial* absolute angular momentum)

LAGRANGEAN TREATMENT OF RELATIVE MOTION
[equations of *carried* body; say, $E_k(T) = Q_k$]

$$E_k(T_2) = Q_k + Q_{k,\text{transport}\,\textbf{transl'n}}$$
$$+ Q_{k,\text{transport}\,\textbf{rotat'n}} + Q_{k,\text{transport rotat'n}\,\textbf{centrifugal}} + Q_{k,\textbf{Coriolis}},$$

$$Q_{k,\text{transport}\,\textbf{transl'n}} \equiv -\partial V_{\text{translation}}/\partial q_k, \qquad V_{\text{translation}} \equiv m\, \boldsymbol{a}_O \cdot \boldsymbol{r}_G;$$

$$Q_{k,\text{transport}\,\textbf{rotat'n}} \equiv -(d\boldsymbol{\Omega}/dt) \cdot (\partial \boldsymbol{H}_{O,\text{rel've}}/\partial q_k)$$
$$= -(d\boldsymbol{\Omega}/dt) \cdot \Big(S\, dm\, \boldsymbol{r} \times (\partial \boldsymbol{r}/\partial q_k) \Big);$$

$Q_{k,\text{transport rotat'n}\,\textbf{centrifugal}}$

$$2V_{\text{centrifugal}} \equiv -S\, dm\, (\boldsymbol{\Omega} \times \boldsymbol{r})^2 = -\boldsymbol{\Omega} \cdot \boldsymbol{I}_O \cdot \boldsymbol{\Omega};$$

$$Q_{k,\textbf{Coriolis}} \equiv -2 S\, \boldsymbol{\Omega} \times (dm\, \boldsymbol{v}_{\text{rel've}}) \cdot (\partial \boldsymbol{r}/\partial q_k) = \sum g_{kl} \dot{q}_l,$$

$$g_{kl} \equiv \boldsymbol{g}_{kl} \cdot \boldsymbol{\Omega}, \qquad \boldsymbol{g}_{kl} \equiv 2 S\, dm\, (\partial \boldsymbol{r}/\partial q_k) \times (\partial \boldsymbol{r}/\partial q_l)$$

Chapter 4: Impulsive Motion

Fundamental impulsive variational equation (*impulsive principle of Lagrange—LIP*):

$$\widehat{\delta I} = \widehat{\delta' W},$$

where

$$\widehat{\delta I} \equiv \widehat{S\, dm\, \boldsymbol{a} \cdot \delta \boldsymbol{r}} = S\, \Delta(dm\, \boldsymbol{v}) \cdot \delta \boldsymbol{r}:$$

(first-order) *virtual work of impulsive momenta*, and

$$\widehat{\delta' W} \equiv \widehat{S\, d\boldsymbol{F} \cdot \delta \boldsymbol{r}} = S\, \widehat{d\boldsymbol{F}} \cdot \delta \boldsymbol{r}:$$

(first-order) *virtual work of impulsive impressed "forces."*

$$\widehat{\delta' W_R} = S\, \widehat{d\boldsymbol{R}} \cdot \delta \boldsymbol{r} = \sum \left(S\, \widehat{d\boldsymbol{R}} \cdot \boldsymbol{e}_k \right) \delta q_k \equiv \sum \hat{R}_k\, \delta q_k,$$

$$\widehat{\delta' W} = S\, \widehat{d\boldsymbol{F}} \cdot \delta \boldsymbol{r} = \sum \left(S\, \widehat{d\boldsymbol{F}} \cdot \boldsymbol{e}_k \right) \delta q_k \equiv \sum \hat{Q}_k\, \delta q_k,$$

$$\widehat{\delta I} = S\, \Delta(dm\, \boldsymbol{v}) \cdot \delta \boldsymbol{r} = \sum \left(S\, dm\, \Delta \boldsymbol{v} \cdot \boldsymbol{e}_k \right) \delta q_k$$
$$= \sum \Delta \left(S\, dm\, \boldsymbol{v} \cdot \boldsymbol{e}_k \right) \delta q_k \equiv \sum \Delta p_k\, \delta q_k,$$

and

$$p_k \equiv S\, (dm\, \boldsymbol{v} \cdot \boldsymbol{e}_k) \equiv \partial T / \partial \dot{q}_k$$
$$\Rightarrow \Delta p_k = \Delta \left(S\, dm\, \boldsymbol{v} \cdot \boldsymbol{e}_k \right) = S\, [\Delta(dm\, \boldsymbol{v}) \cdot \boldsymbol{e}_k]:$$

[holonomic (k)th component] impulsive system *momentum change*,

$$\hat{Q}_k \equiv \widehat{S\, d\boldsymbol{F} \cdot \boldsymbol{e}_k} = S\, \widehat{d\boldsymbol{F}} \cdot \boldsymbol{e}_k:$$

[holonomic (k)th component] impulsive system *impressed* force; or, simply, impressed system *impulse*,

$$\hat{R}_k \equiv \widehat{S\, d\boldsymbol{R} \cdot \boldsymbol{e}_k} = S\, \widehat{d\boldsymbol{R}} \cdot \boldsymbol{e}_k:$$

[holonomic (k)th component] impulsive system *constraint reaction* force, we finally obtain LIP in *holonomic system variables*:

$$\sum \hat{R}_k\, \delta q_k = 0, \qquad \sum \Delta(\partial T / \partial \dot{q}_k)\, \delta q_k = \sum \hat{Q}_k\, \delta q_k,$$

and similarly in *quasi variables*.

Energetic theorem

$$\Delta T \equiv T^+ - T^- = W_{-/+},$$

where

$$2T^+ \equiv S\, dm\, \boldsymbol{v}^+ \cdot \boldsymbol{v}^+, \qquad 2T^- \equiv S\, dm\, \boldsymbol{v}^- \cdot \boldsymbol{v}^-,$$

and

$$W_{-/+} \equiv \mathsf{S}\, d\widehat{F} \cdot (v^+ + v^-)/2$$

In words: The sudden change of the kinetic energy of a moving system, due to arbitrary impressed impulses, equals the sum of the dot products of these impulses with the mean (average) velocities of their material points of application, immediately before and after their action.

APPELLIAN CLASSIFICATION OF IMPULSIVE CONSTRAINTS, AND CORRESPONDING EQUATIONS OF IMPULSIVE MOTION

At a given initial instant t', new constraints are suddenly introduced into the system and/or some old constraints are removed, or suppressed. As a result, mutual percussions are generated, which, in the very short time interval $\tau \equiv t'' - t'$ over which they are supposed to act and during which the shock lasts, produce finite velocity changes, but, according to our "first" approximation, produce negligible position changes; that is, for $\tau \to 0$: $\Delta q = 0$, $\Delta(dq/dt) \neq 0$. The constraints existing at the shock moment are either *persistent* or *nonpersistent*. By persistent we mean constraints that, existing at the shock "moment," exist also after it, so that the actual postimpact displacements are compatible with them; whereas by nonpersistent we mean constraints that, existing at the shock moment, do not exist after it, so that the actual postimpact displacements are incompatible with them.

The constraints that exist at the shock instant can be classified into the following four distinct kinds or types:

1. Constraints that exist *before, during, and after the shock*; that is, the latter neither introduces new constraints, nor does it change the old ones; the system, however, is acted on by impulsive forces. An example of such a constraint is the striking of a physical pendulum with a nonsticking (or nonplastic) hammer at one of its points, and the resulting communication to it of a specified impressed impulsive force.
2. Constraints that exist *during and after the shock, but not before it*; that is, the latter introduces suddenly new constraints on the system. Examples: (a) A rigid bar that falls freely, until the two inextensible slack strings that connect its endpoints to a fixed ceiling become taut (during) and do not break (after). (b) The inelastic central collision of two solid spheres ("coefficient of restitution" $\equiv e = 0$—see below). (c) In a ballistic pendulum, the pendulum is constrained to rotate about a fixed axis, which is a constraint that exists before, during, and after the percussion of the pendulum with a projectile (i.e., first-type constraint). The projectile, however, originally independent of the pendulum, strikes it and becomes embedded into it, which is a case of a new constraint whose sudden realization produces the shock, and which exists during and after the shock but not before it (i.e., second-type constraint).
3. Constraints that exist *before and during the shock, but not after it*. For example, let us imagine a system that consists of two particles connected by a light and inextensible bar, or thread, thrown up into the air. Then, let us assume that one of these particles is suddenly seized (persistent constraint introduced abruptly; i.e., second type), and, at the same time, the bar breaks (constraint that exists before the shock but does not exist after it; i.e., third type).
4. Constraints that exist *only during the shock, but neither before nor after it*. For example, when two solids collide, since their bounding surfaces come into contact, a constraint is abruptly introduced into this two-body system. If these bodies are

elastic ($e = 1$—see coefficient of restitution, below), they separate after the collision, which is a case of a constraint that exists during the percussion but neither before nor after it (i.e., fourth type); while if they are *plastic* ($e = 0$), they do not separate (projectile and pendulum, above; i.e., second type). If $0 < e < 1$, the bodies separate; that is, we have a fourth kind constraint.

Clearly, the first two types contain the persistent constraints, while the last two contain the nonpersistent ones. Schematically, we have the classification shown in table 1.

Table 1 Appellian Classification of Impulsive Constraints

	Preshock (before)	Shock (during)	Postshock (after)
1 (persistent)	▬	▬	▬
2 (persistent)		▬	▬
3 (nonpersistent)	▬	▬	
4 (nonpersistent)		▬	

In impulsive problems: the excess of the number of unknowns (postimpact velocities and constraint reactions) over that of the available equations [those obtained from Lagrange's impulsive principle; plus preimpact velocities, impressed impulsive forces, constraints, and, sometimes, knowledge of the postimpact state (second type; e.g., $e = 0$)]—namely, the degree of its indeterminancy—equals the number of its constraints, which, having existed before or during the shock, cease to do so at the end of it; that is:

$$\textit{Degree of indeterminacy} = \textit{Number of nonpersistent constraints};$$

that is, the persistent types 1 and 2 are determinate, while the nonpersistent ones 3 and 4 are indeterminate.

COEFFICIENT OF RESTITUTION (e)

$$e = -\frac{(\boldsymbol{v}_{2/1} \cdot \boldsymbol{n})^+}{(\boldsymbol{v}_{2/1} \cdot \boldsymbol{n})^-} \equiv -\frac{{v_{2/1,n}}^+}{{v_{2/1,n}}^-} = -\frac{\text{Relative velocity of } \textit{separation}}{\text{Relative velocity of } \textit{approach}}$$

where 1 and 2 are the two points of bodies A and B that come into contact during the collision, and $\boldsymbol{n}$ is the unit vector along the common normal to their bounding surfaces there, say from A to B. This coefficient ranges from 0 (*plastic* impact, no separation) to 1 (*elastic* impact, no energy loss); that is, $0 \leq e \leq 1$.

ANALYTICAL EXPRESSION OF THE APPELLIAN CLASSIFICATION; PERSISTENCY VERSUS DETERMINACY

1. In terms of *elementary* dynamics: Consider a system that consists of N solids, in contact with each other at K points, out of which C are of the nonpersistent type, and/or with a number of foreign solid obstacles that are either fixed or have *known* motions. Assuming *frictionless* collisions, we shall have a total of $6N + K$ *unknowns* ($6N$ postshock velocities, plus K percussions at the smooth contacts, along the common normals), and $6N + K - C$ *equations* ($6N$ impulsive momentum equations,

plus $K - C$ persistent-type constraints); and therefore *the degree of indeterminacy equals the number of nonpersistent contacts C* (i.e., the kind that disappear after the shock).

Hence: (a) a *free* (i.e., *unconstrained*) solid subjected to *given* percussions or (b) a system subjected only to *persistent* constraints are impulsively determinate.

2. From the Lagrangean viewpoint: (a) A number of constraints, imposed on a system originally defined by n Lagrangean coordinates, can *always* be put in the *equilibrium* form:

$$q_1 = 0, \quad q_2 = 0, \ldots, \quad q_m = 0 \qquad (m\text{: number of such constraints} < n).$$

(b) *Within our impulsive approximations*, even Pfaffian constraints (including nonholonomic ones) can be brought to the holonomic form; that is, in impulsive motion, *all* constraints behave as holonomic; and to solve them, either we use impulsive multipliers, or we avoid them by choosing the above equilibrium coordinates; or we use quasi variables.

Assuming, henceforth, such a choice of Lagrangean coordinates for all our impulsive constraints (and, for convenience, re-denoting these new equilibrium coordinates by $q_1, \ldots, q_m; \ldots, q_n$), we can quantify the four Appellian types of impulsive constraints as follows:

- *First-type constraints* (existing *before*, *during*, and *after* the shock). As a result of these constraints, let the system configurations depend on n, hitherto independent, Lagrangean parameters: $q \equiv (q_1, \ldots, q_n)$. During the shock interval (t', t''), the corresponding velocities $\dot{q} \equiv (\dot{q}_1, \ldots, \dot{q}_n)$ pass suddenly from the known values $(\dot{q})^-$, at t', to other values $(\dot{q})^+$, while the q's remain practically unchanged; that is, here we have

$$(q_k)_{\text{before}} = 0, \qquad (q_k)_{\text{during}} = 0, \qquad (q_k)_{\text{after}} = 0,$$
$$\Delta\dot{q}_k \equiv (\dot{q}_k)^+ - (\dot{q}_k)^- \neq 0 \qquad [(\dot{q}_k)^+\text{: unknown, } (\dot{q}_k)^-\text{: known}].$$

- *Second-type constraints* (additional constraints existing *during* and *after* the shock, but not before it). Here, with $q_{D''} \equiv (q_1, \ldots, q_{m''})$, where $m'' < n$, we have

$$(q_{D''})_{\text{before}} \neq 0, \qquad (q_{D''})_{\text{during}} = 0, \qquad (q_{D''})_{\text{after}} = 0;$$
$$(\dot{q}_{D''})^- \neq 0, \qquad (\dot{q}_{D''})^+ = 0 \Rightarrow \Delta(\dot{q}_{D''}) = -(\dot{q}_{D''})^- \neq 0.$$

- *Third-type constraints* (additional constraints existing *before and during* the shock, but not after it). Here, with $q_{D'''} \equiv (q_{m''+1}, \ldots, q_{m'''})$, where $m''' < n$, we have

$$(q_{D'''})_{\text{before}} = 0, \qquad (q_{D'''})_{\text{during}} = 0, \qquad (q_{D'''})_{\text{after}} \neq 0,$$
$$(\dot{q}_{D'''})^- = 0, \qquad (\dot{q}_{D'''})^+ \neq 0 \Rightarrow \Delta(\dot{q}_{D'''}) = (\dot{q}_{D'''})^+ \neq 0.$$

- *Fourth-type constraints* (additional constraints existing only *during* the shock, but neither before nor after it). Here, with $q_{D''''} \equiv (q_{m'''+1}, \ldots, q_{m''''})$, where $m'''' < n$, we have

$$(q_{D''''})_{\text{before}} \neq 0, \qquad (q_{D''''})_{\text{during}} = 0, \qquad (q_{D''''})_{\text{after}} \neq 0;$$
$$(\dot{q}_{D''''})^- \neq 0, \qquad (\dot{q}_{D''''})^+ \neq 0 \Rightarrow \Delta(\dot{q}_{D''''}) = (\dot{q}_{D''''})^+ - (\dot{q}_{D''''})^- \neq 0.$$

Hence, if no fourth-type constraints exist, $m''' = m''''$; and if no third-type constraints exist, $m'' = m'''$; etc.

Next, arguing as in the case of continuous motion (chap. 3), during the shock interval, we may view the constraints of the second, third, and fourth types as absent, provided that, in the spirit of the *impulsive principle of relaxation* (LIP), we add to the system the corresponding constraint reactions. All relevant equations of motion are contained in the LIP:

$$\sum \Delta(\partial T/\partial \dot{q}_k)\,\delta q_k = \sum \hat{Q}_k\,\delta q_k \qquad (k = 1, \ldots, n).$$

If the *virtual displacements* $\delta q \equiv (\delta q_1, \ldots, \delta q_n)$ are arbitrary, the right side of the above equation contains the impulsive virtual works of the reactions stemming from the second, third, and fourth type constraints, and operating during the shock interval (t', t''). Therefore, to eliminate these "forces," and thus produce $n - m''''$ reactionless, or kinetic, impulsive equations, we choose δq's that are compatible with *all constraints holding at the shock moment*; that is, we take

$$\delta q_1, \ldots, \delta q_{m''}; \qquad \delta q_{m''+1}, \ldots, \delta q_{m'''}; \qquad \delta q_{m'''+1}, \ldots, \delta q_{m''''} = 0;$$
$$\delta q_{m''''+1}, \ldots, \delta q_n \neq 0.$$

Corresponding *two* (uncoupled) sets of equations:

Impulsive *kinetostatic*: $\Delta(\partial T/\partial \dot{q}_D) = \hat{Q}_D + \hat{\lambda}_D \quad (D = 1, \ldots, m'''')$, (a)

Impulsive *kinetic*: $\Delta(\partial T/\partial \dot{q}_I) = \hat{Q}_I \quad (I = m'''' + 1, \ldots, n)$. (b)

Further, since the velocity jumps $\Delta\dot{q}$ are produced only by the very large impulsive constraint reactions, operating during the very small interval $t'' - t'$, within our approximations, the $\hat{Q}_I$ [since they derive only from ordinary (i.e., finite, nonimpulsive) forces, like gravity] vanish: $\hat{Q}_I = 0$; and so eq. (b) reduces to *Appell's rule*:

$$\Delta(\partial T/\partial \dot{q}_I) = 0 \;\Rightarrow\; (\partial T/\partial \dot{q}_I)^+ = (\partial T/\partial \dot{q}_I)^-.$$

In words: The partial derivatives of the kinetic energy relative to the velocities of those system coordinates q's that are not forced to vanish at the shock instant (i.e., $q_{\text{during}} \neq 0$) have the same values before and after the impact; or, these $n - m''''$ unconstrained momenta, $p_I \equiv \partial T/\partial \dot{q}_I$, are conserved.

To make the problem determinate, in the presence of nonpersistent-type constraints, we must make particular *constitutive* (*i.e.*, *physical*) *hypotheses*: for example, elasticity assumptions about the postshock state.

EXTREMUM THEOREMS OF IMPULSIVE MOTION

All based on the following master equation (impulsive Lagrange's principle):

$$S\, dm\,(\boldsymbol{v}^+ - \boldsymbol{v}^-)\cdot\delta\boldsymbol{r} = S\, d\widehat{\boldsymbol{F}}\cdot\delta\boldsymbol{r}$$

Carnot (first part—collisions)

$$\delta\boldsymbol{r} \sim \boldsymbol{v}^+, \qquad d\widehat{\boldsymbol{F}} = \boldsymbol{0} \;\rightarrow\; T^+ - T^- < 0$$

Carnot (second part—explosions)

$$\delta \boldsymbol{r} \sim \boldsymbol{v}^-, \qquad \widehat{d\boldsymbol{F}} = \boldsymbol{0} \;\rightarrow\; T^+ - T^- > 0$$

Kelvin (prescribed velocities)

$$\delta \boldsymbol{r} \sim \boldsymbol{v}^+, \qquad \delta \boldsymbol{r} \sim \boldsymbol{v}^+ + \delta_K \boldsymbol{v} = \boldsymbol{v}, \qquad \boldsymbol{v}^- = \boldsymbol{0} \;\rightarrow\; T(\boldsymbol{v}) - T(\boldsymbol{v}^+) > 0, \qquad \delta_K T^+ = 0$$

Bertrand–Delaunay (prescribed impulses)

$$\delta \boldsymbol{r} \sim \boldsymbol{v}^+, \qquad \delta \boldsymbol{r} \sim \boldsymbol{v}^+ + \delta_{B/D} \boldsymbol{v} = \boldsymbol{v} \quad \rightarrow \quad T(\boldsymbol{v}) - T(\boldsymbol{v}^+) < 0, \qquad \delta_{B/D} T^+ = 0$$

[Taylor: $T_{\text{Kelvin}}(\boldsymbol{v}) - T(\boldsymbol{v}^+) > T(\boldsymbol{v}^+) - T(\boldsymbol{v})_{\text{Bertrand–Delaunay}}$]

Robin (prescribed impulses and constraints)

$$\delta \boldsymbol{r} \sim \boldsymbol{v}^+, \qquad \delta \boldsymbol{r} \sim \boldsymbol{v}^+ + \delta_R \boldsymbol{v} = \boldsymbol{v}$$

$$\rightarrow P \equiv \mathcal{S}\,(dm/2)(\boldsymbol{v} - \boldsymbol{v}^-)^2 - \mathcal{S}\,\widehat{d\boldsymbol{F}} \cdot (\boldsymbol{v} - \boldsymbol{v}^-)\text{: stationary and minimum}$$

Gauss (impulsive compulsion)

$$\hat{Z} \equiv \mathcal{S}\,(dm/2)(\boldsymbol{v} - \boldsymbol{v}^- - \widehat{d\boldsymbol{F}}/dm)^2 \;=\; P + \mathcal{S}\,(\widehat{d\boldsymbol{F}})^2/2dm\text{: stationary and minimum}$$

Chapter 5: Nonlinear Nonholonomic Constraints

CONSTRAINTS

$$f_D(t, q, \dot{q}) = 0$$

QUASI VARIABLES

Velocity form

$$\omega_D \equiv f_D(t, q, \dot{q}) = 0, \qquad \omega_I \equiv f_I(t, q, \dot{q}) \neq 0, \qquad \omega_{n=1} \equiv \dot{q}_{n+1} = \dot{t} = 1$$

Virtual form (by Maurer–Appell–Chetaev–Johnsen–Hamel)

$$\begin{aligned}
\delta\theta_D &= \sum (\partial f_D/\partial \dot{q}_k)\,\delta q_k = \sum (\partial \omega_D/\partial \dot{q}_k)\,\delta q_k = 0,\\
\delta\theta_I &= \sum (\partial f_I/\partial \dot{q}_k)\,\delta q_k = \sum (\partial \omega_I/\partial \dot{q}_k)\,\delta q_k \neq 0,\\
\delta q_k &= \sum (\partial \dot{q}_k/\partial \omega_l)\,\delta\theta_l = \sum (\partial \dot{q}_k/\partial \omega_I)\,\delta\theta_I
\end{aligned}$$

Compatibility

$$\begin{aligned}
&\sum (\partial f_k/\partial \dot{q}_b)(\partial \dot{q}_b/\partial \omega_l) \equiv \sum (\partial \omega_k/\partial \dot{q}_b)(\partial \dot{q}_b/\partial \omega_l) = \partial \omega_k/\partial \omega_l = \delta_{kl},\\
&\sum (\partial F_k/\partial \omega_b)(\partial \omega_b/\partial \dot{q}_l) \equiv \sum (\partial \dot{q}_k/\partial \omega_b)(\partial \omega_b/\partial \dot{q}_l) = \partial \dot{q}_k/\partial \dot{q}_l = \delta_{kl}
\end{aligned}$$

PARTICLE KINEMATICS

Virtual displacements

$$\delta \boldsymbol{r} = \sum (\partial \boldsymbol{r}^*/\partial\theta_l)\,\delta\theta_l \equiv \sum \boldsymbol{\varepsilon}_l\,\delta\theta_l \equiv \delta\boldsymbol{r}^*,$$

where

$$\boldsymbol{\varepsilon}_l = \sum (\partial \boldsymbol{r}/\partial q_k)(\partial \dot{q}_k/\partial\omega_l) \equiv \sum (\partial \dot{q}_k/\partial\omega_l)\boldsymbol{e}_k,$$
$$\boldsymbol{e}_k = \sum (\partial \boldsymbol{r}^*/\partial\theta_l)(\partial\omega_l/\partial \dot{q}_k) \equiv \sum (\partial\omega_l/\partial \dot{q}_k)\boldsymbol{\varepsilon}_l;$$

that is,

$$\partial(\ldots)/\partial\theta_l \equiv \sum [\partial(\ldots)/\partial q_k](\partial \dot{q}_k/\partial\omega_l),$$
$$\partial(\ldots)/\partial q_k \equiv \sum [\partial(\ldots)/\partial\theta_l](\partial\omega_l/\partial \dot{q}_k)$$

[nonlinear symbolic (nonvectorial/tensorial) quasi chain rules]

Velocities

$$\boldsymbol{v} = \sum \dot{q}_k(t, q, \omega)\boldsymbol{e}_k + \boldsymbol{e}_0 \qquad [t \equiv q_{n+1}]$$
$$= \sum \omega_k(t, q, \dot{q})\boldsymbol{\varepsilon}_k + \boldsymbol{\varepsilon}_0 \equiv \boldsymbol{v}^*(t, q, \omega) \equiv \boldsymbol{v}^*,$$

where

$$\boldsymbol{\varepsilon}_0 \equiv \partial\boldsymbol{r}/\partial\theta_{n+1} \equiv \sum (\partial\boldsymbol{r}/\partial q_\alpha)(\partial\dot{q}_\alpha/\partial\omega_{n+1}) \qquad [\alpha = 1, \ldots, n+1]$$
$$= \sum (\partial \dot{q}_k/\partial\omega_{n+1})\boldsymbol{e}_k + \boldsymbol{e}_0$$
$$= \sum (\dot{q}_k\boldsymbol{e}_k - \omega_k\boldsymbol{\varepsilon}_k) + \boldsymbol{e}_0$$
$$= \boldsymbol{e}_0 + \sum \Big(\dot{q}_k - \sum (\partial\dot{q}_k/\partial\omega_l)\omega_l\Big)\boldsymbol{e}_k,$$

and, inversely,

$$\boldsymbol{e}_0 \equiv \partial\boldsymbol{r}/\partial t \equiv \sum (\partial\boldsymbol{r}/\partial\theta_\alpha)(\partial\omega_\alpha/\partial\dot{q}_{n+1}) \qquad [\alpha = 1, \ldots, n+1]$$
$$= \boldsymbol{\varepsilon}_0 + \sum \Big(\omega_k - \sum (\partial\omega_k/\partial\dot{q}_l)\dot{q}_l\Big)\boldsymbol{\varepsilon}_k.$$

For any function $f^* = f^*(t, q, \omega)$,

$$\partial f^*/\partial\theta_{n+1} \equiv \sum (\partial f^*/\partial q_k)\Big(\dot{q}_k - \sum (\partial\dot{q}_k/\partial\omega_l)\omega_l\Big) + \partial f^*/\partial t;$$

which in the Pfaffian case reduces to the earlier

$$\partial f^*/\partial\theta_{n+1} = \sum (\partial f^*/\partial q_k)A_k + \partial f^*/\partial t \equiv \partial f^*/\partial(t) + \partial f^*/\partial t.$$

In particular, for $f^* = q_b$ we find

$$\partial q_b/\partial\theta_s = \partial\dot{q}_b/\partial\omega_s,$$
$$\partial q_b/\partial\theta_{n+1} = \partial\dot{q}_b/\partial\omega_{n+1} = \dot{q}_b - \sum (\partial\dot{q}_b/\partial\omega_l)\omega_l$$
$$[= \dot{q}_b - \sum A_{bl}\omega_l = A_b, \text{ in the Pfaffian case}];$$

and, inversely,

$$\partial\theta_k/\partial t \equiv \partial\omega_k/\partial\dot{q}_{n+1} = \omega_k - \sum (\partial\omega_k/\partial\dot{q}_l)\dot{q}_l.$$

Accelerations

$$\boldsymbol{a} \equiv d\boldsymbol{v}/dt = \sum (\partial\boldsymbol{v}/\partial\dot{q}_k)\ddot{q}_k + \textit{No other } \ddot{q}/\dot{\omega} \text{ terms}$$
$$\equiv \sum (\partial\boldsymbol{v}^*/\partial\omega_l)\dot{\omega}_l + \cdots$$
$$= \sum \boldsymbol{\varepsilon}_l\dot{\omega}_l + \cdots = \sum (\partial\boldsymbol{a}^*/\partial\dot{\omega}_l)\dot{\omega}_l + \cdots$$
$$\equiv \boldsymbol{a}^*(t, q, \omega, \dot{\omega}) \equiv \boldsymbol{a}^*,$$

where

$$\partial\boldsymbol{v}^*/\partial\omega_l = \sum (\partial\boldsymbol{v}/\partial\dot{q}_k)(\partial\dot{q}_k/\partial\omega_l) \qquad \text{or} \qquad \boldsymbol{\varepsilon}_l = \sum \boldsymbol{e}_k(\partial\dot{q}_k/\partial\omega_l)$$

(which is a vectorial transformation equation, and not some quasi chain rule).

BASIC KINEMATIC IDENTITIES

Holonomic variables

$$\partial\boldsymbol{r}/\partial q_k = \partial\boldsymbol{v}/\partial\dot{q}_k = \partial\boldsymbol{a}/\partial\ddot{q}_k = \cdots = \boldsymbol{e}_k$$

Nonholonomic variables

$$\partial\boldsymbol{r}^*/\partial\theta_k = \partial\boldsymbol{v}^*/\partial\omega_k = \partial\boldsymbol{a}^*/\partial\dot{\omega}_k = \cdots = \boldsymbol{\varepsilon}_k$$

System forms

$$\partial q_k/\partial\theta_l \equiv \partial\dot{q}_k/\partial\omega_l = \partial\ddot{q}_l/\partial\dot{\omega}_l = \cdots$$
$$\partial\theta_l/\partial q_k \equiv \partial\omega_l/\partial\dot{q}_k = \partial\dot{\omega}_l/\partial\ddot{q}_k = \cdots$$

NONINTEGRABILITY RELATIONS

Nonholonomic deviation (vector)

$$\boldsymbol{\gamma}_k \equiv E_k^*(\boldsymbol{v}^*) = \sum E_k{}^*(\dot{q}_l)\boldsymbol{e}_l \equiv \sum V^l{}_k\boldsymbol{e}_l = -\sum H^b{}_k\boldsymbol{\varepsilon}_b$$

where

Nonlinear Voronets–Chaplygin coefficients

$$V^l{}_k \equiv (\partial\dot{q}_l/\partial\omega_k)^{\cdot} - \partial\dot{q}_l/\partial\theta_k \equiv E_k{}^*(\dot{q}_l).$$

Nonlinear Hamel coefficients

$$H^k_{\ b} \equiv \sum (\partial\dot{q}_l/\partial\omega_b)[(\partial\omega_k/\partial\dot{q}_l)^{\cdot} - \partial\omega_k/\partial q_l] \equiv \sum (\partial\dot{q}_l/\partial\omega_b)E_l(\omega_k)$$

$$\left[\Rightarrow h^k_{\ b} \equiv \sum \gamma^k_{\ b\alpha}\omega_\alpha = \sum \gamma^k_{\ bs}\omega_s + \gamma^k_{\ b,n+1} \text{ (in the } Pfaffian \text{ case)}\right]$$

$$H^b_{\ k} \equiv -\sum (\partial\omega_b/\partial\dot{q}_l)V^l_{\ k} \Leftrightarrow V^l_{\ k} = -\sum (\partial\dot{q}_l/\partial\omega_b)H^b_{\ k}$$

$$E_l(\omega_k) = -\sum\sum (\partial\omega_b/\partial\dot{q}_l)(\partial\omega_k/\partial\dot{q}_s)E_b{}^*(\dot{q}_s)$$

$$E_b{}^*(\dot{q}_s) = -\sum\sum (\partial\dot{q}_l/\partial\omega_b)(\partial\dot{q}_s/\partial\omega_k)E_l(\omega_k)$$

For a general function $f^* = f^*(t, q, \omega)$, the following *noncommutativity* relations hold:

$$\partial/\partial\theta_l(\partial f^*/\partial\theta_k) - \partial/\partial\theta_k(\partial f^*/\partial\theta_l) = \sum\sum\sum [(\partial^2\dot{q}_b/\partial q_s\,\partial\omega_k)(\partial\dot{q}_s/\partial\omega_l) - (\partial^2\dot{q}_b/\partial q_s\,\partial\omega_l)(\partial\dot{q}_s/\partial\omega_k)](\partial\omega_p/\partial\dot{q}_b)(\partial f^*/\partial\theta_p)$$

THE NONLINEAR TRANSITIVITY EQUATIONS

$$\begin{aligned}(\delta\theta_k)^{\cdot} - \delta\omega_k &= \sum (\partial\omega_k/\partial\dot{q}_l)[(\delta q_l)^{\cdot} - \delta(\dot{q}_l)] + \sum E_l(\omega_k)\,\delta q_l \\ &= \sum (\partial\omega_k/\partial\dot{q}_l)[(\delta q_l)^{\cdot} - \delta(\dot{q}_l)] + \sum H^k_{\ b}\,\delta\theta_b \\ &= \sum (\partial\omega_k/\partial\dot{q}_l)[(\delta q_l)^{\cdot} - \delta(\dot{q}_l)] - \sum\sum V^l_{\ b}(\partial\omega_k/\partial\dot{q}_l)\,\delta\theta_b \\ (\delta q_l)^{\cdot} - \delta(\dot{q}_l) &= \sum (\partial\dot{q}_l/\partial\omega_k)[(\delta\theta_k)^{\cdot} - \delta\omega_k] + \sum V^l_{\ k}\,\delta\theta_k \\ &= \sum (\partial\dot{q}_l/\partial\omega_k)[(\delta\theta_k)^{\cdot} - \delta\omega_k] - \sum\sum (\partial\dot{q}_l/\partial\omega_b)H^b_{\ k}\,\delta\theta_k\end{aligned}$$

SPECIAL CHOICE OF QUASI VELOCITIES

$$\omega_D \equiv f_D(t, q, \dot{q}) = 0, \qquad \omega_I = f_I(t, q, \dot{q}) = \dot{q}_I \neq 0;$$

$$\Rightarrow \dot{q}_D = \dot{q}_D(t, q, \dot{q}_I) \equiv \phi_D(t, q, \dot{q}_I)$$

System virtual displacements

$$\begin{aligned}\delta q_k\colon \delta q_D &= \sum (\partial\phi_D/\partial\dot{q}_I)\,\delta q_I, \\ \delta q_I &= \sum (\partial\dot{q}_I/\partial\dot{q}_{I'})\,\delta q_{I'} = \sum (\delta_{II'})\,\delta q_{I'} = \delta q_I\end{aligned}$$

Particle virtual displacements

$$\delta\boldsymbol{r} = \sum \boldsymbol{e}_k\,\delta q_k = \sum \boldsymbol{B}_I\,\delta q_I,$$

where

$$\boldsymbol{B}_I \equiv \partial\boldsymbol{r}/\partial(q_I) \equiv \partial\boldsymbol{r}/\partial q_I + \sum (\partial\boldsymbol{r}/\partial q_D)(\partial\phi_D/\partial\dot{q}_I) \equiv \boldsymbol{e}_I + \sum (\partial\phi_D/\partial\dot{q}_I)\boldsymbol{e}_D;$$

and, in general,

$$\partial \boldsymbol{B}_I/\partial q_{I'} \neq \partial \boldsymbol{B}_{I'}/\partial q_I \qquad \text{(i.e., the } \boldsymbol{B}_I \text{ are } \textit{nongradient} \text{ vectors)}$$

Particle velocities and accelerations

$$\boldsymbol{v} \to \boldsymbol{v}_o = \sum \boldsymbol{B}_I \dot{q}_I + \text{No other } \dot{q} \text{ terms},$$
$$\boldsymbol{a} \to \boldsymbol{a}_o = \sum \boldsymbol{B}_I \ddot{q}_I + \text{No other } \ddot{q} \text{ terms};$$
$$\Rightarrow \partial \boldsymbol{r}/\partial(q_I) \equiv \partial \boldsymbol{v}_o/\partial \dot{q}_I \equiv \partial \boldsymbol{a}_o/\partial \ddot{q}_I \equiv \cdots \equiv \boldsymbol{B}_I$$

Special transitivity relations

$$\delta q_D = \sum (\partial\phi_D/\partial\dot{q}_I)\,\delta q_I \qquad \text{and} \qquad \dot{q}_D = \dot{q}_D(t, q, \dot{q}_I) \equiv \phi_D(t, q, \dot{q}_I)$$
$$(\delta q_D)^{\cdot} - \delta\dot{q}_D = \sum (\partial\phi_D/\partial\dot{q}_I)[(\delta q_I)^{\cdot} - \delta(\dot{q}_I)]$$
$$+ \sum [(\partial\phi_D/\partial\dot{q}_I)^{\cdot} - \partial\phi_D/\partial(q_I)]\,\delta q_I,$$

where

$$\partial\phi_D/\partial(q_I) \equiv \partial\phi_D/\partial q_I + \sum (\partial\phi_D/\partial q_{D'})(\partial\phi_{D'}/\partial\dot{q}_I),$$
$$\partial(\ldots)/\partial(q_I) \equiv \partial(\ldots)/\partial q_I + \sum [\partial(\ldots)/\partial q_D](\partial\dot{\phi}_D/\partial\dot{q}_I)$$

Nonlinear Suslov transitivity relations

$$(\delta q_k)^{\cdot} - \delta(\dot{q}_k): \qquad (\delta q_D)^{\cdot} - \delta\dot{q}_D = \sum W^D{}_I\,\delta q_I \qquad (\neq 0)$$
$$(\delta q_I)^{\cdot} - \delta\dot{q}_I = 0 \qquad [= 0;\ \text{i.e., } W^{I'}{}_I = 0],$$

where

$$W^D{}_I \equiv E_I(\phi_D) - \sum (\partial\phi_D/\partial q_{D'})(\partial\phi_{D'}/\partial\dot{q}_I)$$
$$\equiv (\partial\phi_D/\partial\dot{q}_I)^{\cdot} - \partial\phi_D/\partial(q_I) \equiv E_{(I)}(\phi_D)$$

[special nonlinear Voronets coefficients]

Nonlinear Chaplygin system

$$\dot{q}_D = \dot{q}_D(q_I, \dot{q}_I) \equiv \phi_D(q_I, \dot{q}_I)$$
$$W^D{}_I \to T^D{}_I \equiv (\partial\phi_D/\partial\dot{q}_I)^{\cdot} - \partial\phi_D/\partial q_I \equiv E_I(\phi_D)$$

[special nonlinear Chaplygin coefficients]

KINETIC PRINCIPLES ($P_k \equiv \partial T^*/\partial\omega_k$)

Central equation

$$\sum (dP_k/dt)\,\delta\theta_k - \sum (\partial T^*/\partial\theta_k)\,\delta\theta_k + \sum P_k[(\delta\theta_k)^{\cdot} - \delta\omega_k] = \sum \Theta_k\,\delta\theta_k$$

Lagrange's principle in NNH variables

$$\sum \left(dP_k/dt - \partial T^*/\partial\theta_k + \sum H^l{}_k P_l - \Theta_k \right) \delta\theta_k = 0$$

EQUATIONS OF MOTION ($D = 1, \ldots, m$; $I = m+1, \ldots, n$)

Coupled

$$E_k(T) = Q_k + \sum \lambda_D (\partial f_D/\partial \dot{q}_k) \qquad (\textit{Routh–Voss} \text{ form})$$

Uncoupled

$$I_D = \Theta_D + \Lambda_D \quad \text{(Kinetostatic)} \qquad I_I = \Theta_I \quad \text{(Kinetic)}$$

where

$$\begin{aligned} I_k &\equiv \mathsf{S}\, dm\, \boldsymbol{a}^* \cdot \boldsymbol{\varepsilon}_k && (\textit{Raw} \text{ form}) \\ &= \sum (\partial \dot{q}_l/\partial \omega_k) E_l && (\textit{Maggi} \text{ form}) \\ &= \partial S^*/\partial \dot{\omega}_k && (\textit{Appell} \text{ form}) \\ &= (\partial T^*/\partial \omega_k)^{\cdot} - \partial T^*/\partial \theta_k - \Gamma_k && (\textit{Johnsen–Hamel} \text{ form}) \end{aligned}$$

and

$$\Gamma_k \equiv \mathsf{S}\, dm\, \boldsymbol{v}^* \cdot E_k{}^*(\boldsymbol{v}^*) = \sum V^l{}_k p_l = -\sum H^l{}_k P_l$$

[nonholonomic correction term]

Transformation equations between holonomic and nonholonomic components

$$\begin{aligned} I_k &= \sum (\partial \dot{q}_l/\partial \omega_k) E_l \Leftrightarrow E_l = \sum (\partial \omega_k/\partial \dot{q}_l) I_k, \\ \Theta_k &= \sum (\partial \dot{q}_l/\partial \omega_k) Q_l \Leftrightarrow Q_l = \sum (\partial \omega_k/\partial \dot{q}_l) \Theta_k, \\ E_{k'}{}^* - \Gamma_{k'} &= \sum (\partial \omega_k/\partial \omega_{k'})(E_k{}^* - \Gamma_k) \end{aligned}$$

Transformation equations of $E_k{}^*(T^*)$ and Γ_k between the quasi velocities $\omega \leftrightarrow \omega'$

$$\begin{aligned} E_{k'}{}^*(T^{*\prime}) &= \sum (\partial \omega_k/\partial \omega_{k'}) E_k{}^*(T^*) + \sum (\partial T^*/\partial \omega_k) E_{k'}{}^*(\omega_k), \\ \Gamma_{k'} &= \sum (\partial T^*/\partial \omega_k) E_{k'}{}^*(\omega_k) + \sum (\partial \omega_k/\partial \omega_{k'}) \Gamma_k \end{aligned}$$

Johnsen–Hamel forms in extenso

$$\begin{aligned} I_k &= dP_k/dt - \partial T^*/\partial \theta_k + \sum H^l{}_k P_l \\ &= dP_k/dt - \partial T^*/\partial \theta_k - \sum \sum (\partial \omega_l/\partial \dot{q}_b) V^b{}_k P_l \\ &= dP_k/dt - \partial T^*/\partial \theta_k - \sum V^b{}_k p_b{}^* \end{aligned}$$

$$\left[\sum (\partial \omega_l/\partial \dot{q}_k) P_l = p_k = p_k(t, q, \dot{q}) = p_k{}^*(t, q, \omega) \equiv (\partial T/\partial \dot{q}_k)^* \right]$$

SPECIAL FORMS OF THE EQUATIONS OF MOTION FOR THE CHOICE

$$\omega_D \equiv f_D(t, q, \dot{q}) = \dot{q}_D - \phi_D(t, q, \dot{q}_I) = 0, \qquad \omega_I \equiv f_I(t, q, \dot{q}) = \dot{q}_I \neq 0$$

and its inverse

$$\dot{q}_D = \omega_D + \phi_D(t, q, \dot{q}_I) = \omega_D + \phi_D(t, q, \omega_I), \qquad \dot{q}_I = \omega_I,$$

and with the notation

$$E_k \equiv E_k(T) \equiv (\partial T/\partial \dot{q}_k)^{\cdot} - \partial T/\partial q_k = \partial S/\partial \ddot{q}_k$$

Maggi equations $\Rightarrow$ nonlinear Hadamard equations

Kinetostatic: $\quad E_D = Q_D + \lambda_D$

Kinetic: $\quad E_I + \sum (\partial \phi_D/\partial \dot{q}_I) E_D = Q_I + \sum (\partial \phi_D/\partial \dot{q}_I) Q_D$

or

$$\partial S_o/\partial \ddot{q}_I = Q_I + \sum (\partial \phi_D/\partial \dot{q}_I) Q_D \qquad (\equiv Q_{I,o} \equiv Q_{Io}),$$

where

$$S = S(t, q, \dot{q}, \ddot{q}) = \cdots = S_o(t, q, \dot{q}_I, \ddot{q}_I) = S_o, \textit{ constrained } \text{Appellian } S_o$$

Hamel equations $\Rightarrow$ nonlinear Voronets equations

$$E_I(T_o) - \sum (\partial \phi_D/\partial \dot{q}_I)(\partial T_o/\partial q_D) - \Gamma_{Io} \equiv E_{(I)}(T_o) - \Gamma_{Io} = Q_{Io},$$
$$H^D{}_I \to -E_{(I)}(\phi_D) = -W^D{}_I,$$
$$\Gamma_I \to \Gamma_{I,o} \equiv \Gamma_{Io} = \sum W^D{}_I (\partial T/\partial \dot{q}_D)_o \equiv \sum W^D{}_I p_{Do}$$

Voronets equations $\Rightarrow$ Chaplygin equations

$$\Big[\dot{q}_D = \dot{q}_D(q_I, \dot{q}_I) \equiv \phi_D(q_I, \dot{q}_I) \quad \text{and} \quad T_o = T_o(q_I, \dot{q}_I)$$
$$\Rightarrow W^D{}_I \equiv E_{(I)}(\phi_D) \to E_I(\phi_D) \equiv (\partial \phi_D/\partial \dot{q}_I)^{\cdot} - \partial \phi_D/\partial q_I \equiv T^D{}_I$$
$$\Rightarrow \Gamma_{Io} \to \sum T^D{}_I (\partial T/\partial \dot{q}_D)_o \Big]$$
$$(\partial T_o/\partial \dot{q}_I)^{\cdot} - \partial T_o/\partial q_I - \sum T^D{}_I (\partial T/\partial \dot{q}_D)_o = Q_{Io}$$

Transformation of the nonlinear Hamel and Voronets coefficients $V^l{}_k, H^l{}_k$ under

$$\omega_{b'} = \omega_{b'}(t, q, \dot{q}) \;\Leftrightarrow\; \dot{q}_l = \dot{q}_l(t, q, \omega'):$$
$$\Gamma_k = \sum V^l{}_k p_l = -\sum H^l{}_k P_l,$$
$$V^l{}_{k'} = \sum (\partial \omega_k/\partial \omega_{k'}) V^l{}_k + \sum [(\partial \omega_k/\partial \omega_{k'})^{\cdot} - \partial \omega_k/\partial \theta_{k'}](\partial \dot{q}_l/\partial \omega_k),$$

$$H'^{l'}_{k'} = \sum\sum (\partial\omega_k/\partial\omega_{k'})(\partial\omega_{l'}/\partial\omega_l)H^l_k - \sum[(\partial\omega_k/\partial\omega_{k'})^{\cdot} - \partial\omega_k/\partial\theta_{k'}](\partial\omega_{l'}/\partial\omega_k),$$

$$H'^{l'}_{k'} = -\sum(\partial\omega_{l'}/\partial\dot{q}_l)V^l_{k'} \quad \Leftrightarrow \quad V^l_{k'} = -\sum(\partial\dot{q}_l/\partial\omega_{l'})H'^{l'}_{k'}$$

Chapter 6: Differential Variational Principles

PRINCIPLE OF LAGRANGE

$$S\,(dm\,\boldsymbol{a} - d\boldsymbol{F})\cdot\delta\boldsymbol{r} = 0, \qquad \text{with} \qquad \delta t = 0$$

PRINCIPLE OF JOURDAIN

$$S\,(dm\,\boldsymbol{a} - d\boldsymbol{F})\cdot\delta\boldsymbol{v} = 0, \qquad \text{with} \qquad \delta t = 0 \qquad \textit{and} \qquad \delta\boldsymbol{r} = \boldsymbol{0}$$

PRINCIPLE OF GAUSS

$$S\,(dm\,\boldsymbol{a} - d\boldsymbol{F})\cdot\delta\boldsymbol{a} = 0, \qquad \text{with} \qquad \delta t = 0, \qquad \delta\boldsymbol{r} = \boldsymbol{0}, \qquad \textit{and} \qquad \delta\boldsymbol{v} = \boldsymbol{0}$$

PRINCIPLE OF MANGERON–DELEANU

$$S\,(dm\,\boldsymbol{a} - d\boldsymbol{F})\cdot\delta\overset{(s)}{\boldsymbol{r}} = 0, \qquad (s = 1, 2, \ldots)$$

with

$$\delta t = 0, \qquad \textit{and} \qquad \delta\boldsymbol{r} = \boldsymbol{0}, \qquad \delta(\dot{\boldsymbol{r}}) = \boldsymbol{0}, \qquad \delta(\ddot{\boldsymbol{r}}) = \boldsymbol{0}, \ \ldots,$$

$$\delta\overset{(s-1)}{(r)} = \boldsymbol{0} \qquad (s - 1 \geq 0)$$

NIELSEN IDENTITY

$$N_k(T) \equiv \partial\dot{T}/\partial\dot{q}_k - 2(\partial T/\partial q_k) = (\partial T/\partial\dot{q}_k)^{\cdot} - \partial T/\partial q_k \equiv E_k(T)$$

TSENOV IDENTITIES

Second kind

$$E_k(T) = C_k^{(2)}(T) \equiv (1/2)[\partial\ddot{T}/\partial\ddot{q}_k - 3(\partial T/\partial q_k)]$$

Third kind

$$E_k(T) \equiv C_k^{(3)}(T) \equiv (1/3)[\partial\dddot{T}/\partial\dddot{q}_k - 4(\partial T/\partial q_k)]$$

MANGERON–DELEANU IDENTITIES

$$E_k(T) = C_k^{(s)}(T) \equiv (1/s)\left[\partial \overset{(s)}{T}/\partial \overset{(s)}{q}_k - (s+1)(\partial T/\partial q_k)\right]$$

$$[C_k^{(1)}(T) = N_k(T)]$$

VARIOUS KINEMATICO-INERTIAL IDENTITIES

$$\partial \overset{(s-1)}{T}/\partial \overset{(s)}{q}_k = \partial T/\partial \dot{q}_k \qquad \left[= S\, dm\, \dot{\boldsymbol{r}}\cdot(\partial \dot{\boldsymbol{r}}/\partial \dot{q}_k)\right],$$

$$\partial \overset{(s-1)}{T}/\partial \overset{(s-1)}{q}_k = s(\partial T/\partial q_k) \qquad \left[= s\, S\, dm\, \dot{\boldsymbol{r}}\cdot(\partial \dot{\boldsymbol{r}}/\partial q_k)\right],$$

$$\overset{(s)}{T} = S\, dm\, \dot{\boldsymbol{r}}\cdot \overset{(s+1)}{\boldsymbol{r}} + s\, S\, dm\, \ddot{\boldsymbol{r}}\cdot \overset{(s)}{\boldsymbol{r}} + \text{no } \overset{(s+1)}{\boldsymbol{r}} \text{ terms},$$

$$d/dt\left[\partial \overset{(s-1)}{T_o}/\partial \overset{(s)}{q}_I\right] - \partial T/\partial q_I$$

$$= d/dt(\partial T/\partial \dot{q}_I) - \partial T/\partial q_I + \sum d/dt\left[(\partial T/\partial \dot{q}_D)(\partial \overset{(s)}{q}_D/\partial \overset{(s)}{q}_I)\right]$$

$$\partial \overset{(s-1)}{T_o}/\partial \overset{(s)}{q}_I = \partial \overset{(s-1)}{T}/\partial \overset{(s)}{q}_I + \sum\left(\partial \overset{(s-1)}{T}/\partial \overset{(s)}{q}_D\right)\left(\partial \overset{(s)}{q}_D/\partial \overset{(s)}{q}_I\right)$$

VIRTUAL DISPLACEMENTS NEEDED TO PRODUCE THE CORRECT EQUATIONS OF MOTION

Constraints	*Lagrange*	*Jourdain*	*Gauss*
$f(t,q)=0$: $\partial f/\partial q$	$\delta f = (\partial f/\partial q)\,\delta q$	$\delta' f = 0$ $\delta' \dot{f} = (\partial f/\partial q)\,\delta\dot{q}$	$\delta'' f = 0,$ $\delta'' \dot{f} = 0$ $\delta'' \ddot{f} = (\partial f/\partial q)\,\delta\ddot{q}$
$f(t,q,\dot{q})=0$: $\partial f/\partial \dot{q}$	—	$\delta' f = (\partial f/\partial \dot{q})\,\delta\dot{q}$	$\delta'' f = 0$ $\delta'' \dot{f} = (\partial f/\partial \dot{q})\,\delta\ddot{q}$
$f(t,q,\dot{q},\ddot{q})=0$: $\partial f/\partial \ddot{q}$	—	—	$\delta'' f = (\partial f/\partial \ddot{q})\,\delta\ddot{q}$

CORRECT EQUATIONS OF MOTION

[Notation: $M_k \equiv E_k(T) - Q_k \equiv N_k(T) - Q_k \equiv \partial S/\partial \ddot{q}_k - Q_k$.
Principle: $\sum M_k\,\delta *_k = 0$, $\delta *_k = \delta q_k, \delta \dot{q}_k, \delta \ddot{q}_k, \ldots$]

Constraints	*Virtual Displacements*	*Equations of Motion*
$f_D(t,q)=0$	$\delta f_D = \sum(\partial f_D/\partial q_k)\,\delta q_k$	$M_k = \sum \lambda_D(\partial f_D/\partial q_k)$
$f_D(t,q,\dot{q})=0$	$\delta' f_D = \sum(\partial f_D/\partial \dot{q}_k)\,\delta \dot{q}_k$	$M_k = \sum \lambda_D(\partial f_D/\partial \dot{q}_k)$
$f_D(t,q,\dot{q},\ddot{q})=0$	$\delta'' f_D = \sum(\partial f_D/\partial \ddot{q}_k)\,\delta \ddot{q}_k$	$M_k = \sum \lambda_D(\partial f_D/\partial \ddot{q}_k)$

SPECIAL FORM OF CONSTRAINTS

$$\dot{q}_D = \phi_D(t,q,\dot{q}_I) \qquad (D = 1,\ldots,m; I = m+1,\ldots,n).$$

For an arbitrary differentiable function

$$f = f(t,q,\dot{q}) = f[t,q,\phi_D(t,q,\dot{q}_I),\dot{q}_I] = f_o(t,q,\dot{q}_I) \equiv f_o,$$

the following identity holds:

$$N_I(f_o) = E_I(f_o) + \sum (\partial f_o/\partial q_D)(\partial \phi_D/\partial \dot{q}_I)$$
$$\Rightarrow N_I(T_o) = E_I(T_o) + \sum (\partial T_o/\partial q_D)(\partial \phi_D/\partial \dot{q}_I),$$
$$N_I(\dot{q}_D) = E_I(\dot{q}_D) + \sum (\partial \dot{q}_D/\partial q_{D'})(\partial \phi_{D'}/\partial \dot{q}_I).$$

NIELSEN FORM OF SPECIAL NONLINEAR VORONETS EQUATIONS

$$N_I(T_o) - \sum (\partial T/\partial \dot{q}_D)_o N_I(\dot{q}_D) - 2\sum (\partial T/\partial \dot{q}_D)_o(\partial \phi_D/\partial \dot{q}_I) = Q_{Io}$$

NIELSEN FORM OF SPECIAL NONLINEAR CHAPLYGIN EQUATIONS

$$\partial \dot{T}_o/\partial \dot{q}_I - 2(\partial T_o/\partial q_I) - \sum (\partial T/\partial \dot{q}_D)_o[\partial \ddot{q}_D/\partial \dot{q}_I - 2(\partial \dot{q}_D/\partial q_I)] = Q_{Io}$$

Special Pfaffian $\rightarrow$ Voronets form

$$\dot{q}_D = \sum b_{DI}(t, q)\dot{q}_I + b_D(t, q)$$

Then the above Voronets equations assume the special Nielsen form:

$$\partial \dot{T}_o/\partial \dot{q}_I - 2(\partial T_o/\partial q_I) - \sum (\partial T/\partial \dot{q}_D)_o\left\{\sum [b^D{}_{II'} - 2(\partial b_{DI'}/\partial q_I)]\dot{q}_{I'} + [b^D{}_I - 2(\partial b_D/\partial q_I)]\right\} - 2\sum (\partial T/\partial q_D)_o b_{DI} = Q_{Io},$$

where

$$b^D{}_{II'} \equiv \sum [(\partial b_{DI}/\partial q_{D'})b_{D'I'} + (\partial b_{DI'}/\partial q_{D'})b_{D'I}] + (\partial b_{DI}/\partial q_{I'} + \partial b_{DI'}/\partial q_I),$$
$$b^D{}_I \equiv b^D{}_{I,n+1} \equiv \sum [(\partial b_D/\partial q_{D'})b_{D'I} + (\partial b_{DI}/\partial q_{D'})b_{D'}] + (\partial b_{DI}/\partial t + \partial b_D/\partial q_I).$$

Special Voronets $\rightarrow$ Chaplygin form

$$\dot{q}_D = \sum b_{DI}(q_I)\dot{q}_I, \quad \text{and} \quad \partial T/\partial q_D = 0$$

Then the above Chaplygin equations assume the special Nielsen form:

$$\partial \dot{T}_o/\partial \dot{q}_I - 2(\partial T_o/\partial q_I) - \sum\sum (\partial T/\partial \dot{q}_D)_o(\partial b_{DI}/\partial q_{I'} - \partial b_{DI'}/\partial q_I)\dot{q}_{I'} = Q_{Io}$$

$$[b_D = 0, \quad b^D{}_I = 0, \quad b^D{}_{II'} = \partial b_{DI}/\partial q_{I'} + \partial b_{DI'}/\partial q_I]$$

NIELSEN FORMS OF HIGHER-ORDER EQUATIONS

Let

$$N_k{}^{(s)}(\ldots) \equiv \partial(\overset{(s)}{\ldots})/\partial \overset{(s)}{q}_k - 2\left[\partial(\overset{(s-1)}{\ldots})/\partial \overset{(s-1)}{q_k}\right],$$

$$E_k{}^{(s)}(\ldots) \equiv d/dt\left[\partial(\overset{(s-1)}{\ldots})/\partial \overset{(s)}{q}_k\right] - \left[\partial(\overset{(s-1)}{\ldots})/\partial \overset{(s-1)}{q_k}\right].$$

Then, for any sufficiently differentiable function $f = f(t, q, \dot{q})$, and any $k = 1, 2, \ldots, n$; $s = 1, 2, 3, \ldots$,

$$N_k{}^{(s)}(f) = E_k{}^{(s)}(f).$$

Let

$$N_k{}^{*(s)}(\ldots) \equiv \partial(\overset{(s)}{\ldots})/\partial \overset{(s)}{\theta}_k - 2\left[\partial(\overset{(s-1)}{\ldots})/\partial \overset{(s-1)}{\theta_k}\right],$$

$$E_k{}^{*(s)}(\ldots) \equiv d/dt\left[\partial(\overset{(s-1)}{\ldots})/\partial \overset{(s)}{\theta}_k\right] - \left[\partial(\overset{(s-1)}{\ldots})/\partial \overset{(s-1)}{\theta_k}\right],$$

where

$$\partial(\overset{(s-1)}{\ldots})/\partial \overset{(s-1)}{\theta_k} \equiv \sum \left[\partial(\overset{(s-1)}{\ldots})/\partial \overset{(s-1)}{q_l}\right]\left[\partial \overset{(s)}{q}_l/\partial \overset{(s)}{\theta}_k\right]$$

[(*s*)th-order quasi chain rule].

Then, for any sufficiently differentiable function $f^* = f^*(t, q, \omega)$, and any $k = 1, 2, \ldots, n$; $s = 1, 2, 3, \ldots$,

$$N_k{}^{*(s)}(f^*) = E_k{}^{*(s)}(f^*),$$

where

$$f(t, q, \dot{q}) \Rightarrow \dot{f} \Rightarrow \ldots \overset{(s-1)}{f} \Rightarrow \overset{(s)}{f},$$

$$\overset{(s-1)}{f^*} = \overset{(s-1)}{f}\left[t, q, \dot{q} \equiv \overset{(1)}{q}, \ldots, \overset{(s-1)}{q}; \overset{(s)}{q}\left(t, q, \overset{(1)}{q}, \ldots, \overset{(s-1)}{q}, \overset{(s)}{\theta}\right)\right]$$

$$= \overset{(s-1)}{f}\left(t, q, \overset{(1)}{q}, \ldots, \overset{(s-1)}{q}, \overset{(s)}{\theta}\right),$$

$$\overset{(s)}{f^*} = \overset{(s)}{f}\left[t, q, \overset{(1)}{q}, \ldots, \overset{(s-1)}{q}; \overset{(s)}{q}\left(t, q, \overset{(1)}{q}, \ldots, \overset{(s-1)}{q}, \overset{(s)}{\theta}\right), \overset{(s+1)}{q}\left(t, q, \overset{(1)}{q}, \ldots, \overset{(s-1)}{q}, \overset{(s)}{\theta}, \overset{(s+1)}{\theta}\right)\right]$$

$$= \overset{(s)}{f}\left(t, q, \overset{(1)}{q}, \ldots, \overset{(s-1)}{q}, \overset{(s)}{\theta}, \overset{(s+1)}{\theta}\right).$$

Hamel-type equations ($s = 1, 2, 3, \ldots$)

$$(s)\,d/dt\left[\partial\overset{(s-1)}{T}{}^*/\partial\overset{(s)}{\theta}_I\right] - \partial\overset{(s-1)}{T}{}^*/\partial\overset{(s-1)}{\theta}_I$$

$$- \sum\left[s\left(\partial\overset{(s)}{q}_k/\partial\overset{(s)}{\theta}_I\right)^{\bullet} - \partial\overset{(s)}{q}_k/\partial\overset{(s-1)}{\theta}_I\right]\left(\partial\overset{(s-1)}{T}/\partial\overset{(s)}{q}_k\right)^*$$

$$= \sum\left(\partial\overset{(s)}{q}_k/\partial\overset{(s)}{\theta}_I\right)Q_k \equiv \Theta_I$$

Nielsen-type equations

$$(s)\left(\partial\overset{(s)}{T}{}^*/\partial\overset{(s)}{\theta}_I\right) - (s+1)\left(\partial\overset{(s-1)}{T}{}^*/\partial\overset{(s-1)}{\theta}_I\right)$$

$$- \sum\left[s\left(\partial\overset{(s+1)}{q}_k/\partial\overset{(s)}{\theta}_I\right) - (s+1)\left(\partial\overset{(s)}{q}_k/\partial\overset{(s-1)}{\theta}_I\right)\right]\left(\partial\overset{(s-1)}{T}/\partial\overset{(s)}{q}_k\right)^* = \Theta_I$$

For $s = 1$, the above yield, respectively,

$$d/dt(\partial T^*/\partial\dot{\theta}_I) - \partial T^*/\partial\theta_I$$

$$- \sum\left[(\partial\dot{q}_k/\partial\dot{\theta}_I)^{\cdot} - \partial\dot{q}_k/\partial\theta_I\right](\partial T/\partial\dot{q}_k)^* = \Theta_I,$$

$$\partial\dot{T}^*/\partial\dot{\theta}_I - 2(\partial T^*/\partial\theta_I)$$

$$- \sum\left[\partial\ddot{q}_k/\partial\dot{\theta}_I - 2(\partial\dot{q}_k/\partial\theta_I)\right](\partial T/\partial\dot{q}_k)^* = \Theta_I;$$

and, for $s = 2$,

$$2(\partial\dot{T}^*/\partial\ddot{\theta}_I)^{\cdot} - \partial\dot{T}^*/\partial\dot{\theta}_I$$

$$- \sum\left[2(\partial\ddot{q}_k/\partial\ddot{\theta}_I)^{\cdot} - \partial\ddot{q}_k/\partial\dot{\theta}_I\right](\partial\dot{T}/\partial\ddot{q}_k)^* = \Theta_I,$$

$$2(\partial\ddot{T}^*/\partial\ddot{\theta}_I) - 3(\partial\dot{T}^*/\partial\dot{\theta}_I)$$

$$- \sum\left[2(\partial\dddot{q}_k/\partial\ddot{\theta}_I) - 3(\partial\ddot{q}_k/\partial\dot{\theta}_I)\right](\partial\dot{T}/\partial\ddot{q}_k)^* = \Theta_I.$$

GAUSS' PRINCIPLE

Compulsion

$$Z \equiv (1/2)\mathbf{S}\,dm\,[\boldsymbol{a} - (d\boldsymbol{F}/dm)]^2 \equiv (1/2)\mathbf{S}\,(1/dm)(dm\,\boldsymbol{a} - d\boldsymbol{F})^2$$

$$\left[\equiv (1/2)\mathbf{S}\,(d\boldsymbol{R})^2/dm = \mathbf{S}\,(-d\boldsymbol{R})^2/2\,dm = \mathbf{S}\,(\textit{Lost force})^2/2\,dm \geq 0\right]$$

$$= S - \mathbf{S}\,d\boldsymbol{F}\cdot\boldsymbol{a} + \text{terms not containing accelerations,}$$

where

$$S = (1/2)\mathbf{S}\,dm\,\boldsymbol{a}\cdot\boldsymbol{a}\text{: Appellian.}$$

Gauss' principle

$$\delta'' Z = 0,$$

where

$$\delta'' t = 0, \qquad \delta'' \boldsymbol{r} = \boldsymbol{0}, \qquad \delta'' \boldsymbol{v} = \boldsymbol{0}, \qquad \delta''(d\boldsymbol{F}) = \boldsymbol{0}, \qquad \text{but } \delta'' \boldsymbol{a} \neq \boldsymbol{0}$$

$$[d\boldsymbol{F} = d\boldsymbol{F}(t, \boldsymbol{r}, \boldsymbol{v}) \Rightarrow \delta''(d\boldsymbol{F}) = \boldsymbol{0}, \qquad \delta'' Q_k = 0]$$

$$\boldsymbol{a} = \sum \boldsymbol{e}_k \ddot{q}_k + \text{no } \ddot{q}\text{-terms} = \sum \boldsymbol{\varepsilon}_I \dot{\omega}_I + \text{no } \dot{\omega}\text{-terms},$$

$$\Rightarrow \delta'' \boldsymbol{a} = \sum \boldsymbol{e}_k \, \delta \ddot{q}_k = \sum \boldsymbol{\varepsilon}_I \, \delta \dot{\omega}_I,$$

and so, explicitly,

$$\begin{aligned}
\delta'' Z &= (1/2) S\, dm\, 2\,[\boldsymbol{a} - (d\boldsymbol{F}/dm)] \cdot \delta'' \boldsymbol{a} \\
&= S\, (dm\, \boldsymbol{a} - d\boldsymbol{F}) \cdot \delta'' \boldsymbol{a} \\
&= S\, (d\boldsymbol{R}/dm) \cdot \delta''(d\boldsymbol{R}) = S\, (d\boldsymbol{R}/dm) \cdot \delta''(dm\, \boldsymbol{a} - d\boldsymbol{F}) \\
&= S\, (d\boldsymbol{R}/dm) \cdot dm\, \delta'' \boldsymbol{a} = S\, d\boldsymbol{R} \cdot \delta'' \boldsymbol{a} = 0.
\end{aligned}$$

COMPATIBILITY BETWEEN THE PRINCIPLES OF GAUSS AND LAGRANGE

$$\begin{aligned}
\delta q_k &\equiv \sum (\partial \dot{q}_k / \partial \omega_I)\, \delta\theta_I, \\
\delta\theta_I &\equiv \sum (\partial \omega_I / \partial \dot{q}_k)\, \delta q_k \equiv \sum (\partial f_I / \partial \dot{q}_k)\, \delta q_k \neq 0
\end{aligned}$$

Also,

$$\delta f_D \equiv \delta\omega_D = S\, (\partial f_D / \partial \boldsymbol{v}) \cdot \delta \boldsymbol{r} = \sum (\partial f_D / \partial \dot{q}_k)\, \delta q_k = 0,$$

instead of the formal (calculus of variations) definition

$$\delta f_D = \sum (\partial f_D / \partial q_k)\, \delta q_k + \sum (\partial f_D / \partial \dot{q}_k)\, \delta \dot{q}_k = 0.$$

The same conclusion can be reached by requiring compatibility between the principles of Lagrange and Jourdain.

EQUATIONS OF MOTION

$$\delta'' Z + \sum \lambda_D\, \delta'' \dot{f}_D = 0,$$

where

$$\begin{aligned}
\delta'' Z &= \sum [E_k(T) - Q_k]\, \delta \ddot{q}_k && (\textit{Holonomic} \text{ system variables}) \\
&= \sum (\partial S^* / \partial \dot{\omega}_k - \Theta_k)\, \delta(\dot{\omega}_k) && (\textit{Nonholonomic} \text{ system variables})
\end{aligned}$$

$$\begin{aligned}\delta''(\dot{f}_D) &= \delta''\left\{\partial f_D/\partial t + \textstyle\mathcal{S}\,[(\partial f_D/\partial \boldsymbol{r})\cdot \boldsymbol{v} + (\partial f_D/\partial \boldsymbol{v})\cdot \boldsymbol{a}]\right\}\\ &= \textstyle\mathcal{S}\,(\partial f_D/\partial \boldsymbol{v})\cdot \delta \boldsymbol{a} \qquad\qquad (\textit{Particle}\ \text{form})\\ &= \delta''\left\{\partial f_D/\partial t + \textstyle\sum [(\partial f_D/\partial q_k)\dot{q}_k + (\partial f_D/\partial \dot{q}_k)\ddot{q}_k]\right\}\\ &= \textstyle\sum (\partial f_D/\partial \dot{q}_k)\,\delta\ddot{q}_k \qquad\qquad (\textit{Holonomic}\ \text{system variables})\end{aligned}$$

[which, does *not* equal $(\delta'' f_D)^{\cdot} = (\sum (\partial f_D/\partial \ddot{q}_k)\,\delta\ddot{q}_k)^{\cdot} = (\sum (0)\,\delta\ddot{q}_k)^{\cdot} = 0$].

MINIMALITY OF THE COMPULSION

$$\begin{aligned}\Delta'' Z &\equiv Z(\boldsymbol{a} + \delta''\boldsymbol{a}) - Z(\boldsymbol{a})\\ &= (1/2)\textstyle\mathcal{S}\,dm\,[(\boldsymbol{a}+\delta''\boldsymbol{a}) - (d\boldsymbol{F}/dm)]^2 - (1/2)\mathcal{S}\,dm\,[\boldsymbol{a} - (d\boldsymbol{F}/dm)]^2\\ &= \delta'' Z + (1/2)\,\delta''^{\,2} Z \ge 0,\end{aligned}$$

where

$$\begin{aligned}\delta'' Z &= \textstyle\mathcal{S}\,(dm\,\boldsymbol{a} - d\boldsymbol{F})\cdot\delta''\boldsymbol{a} \qquad (=0),\\ \delta''^{\,2} Z &= \textstyle\mathcal{S}\,(dm\,\delta''\boldsymbol{a}\cdot\delta''\boldsymbol{a}) \qquad (\ge 0).\end{aligned}$$

Chapter 7: Time-Integral Theorems and Variational Principles

GENERALIZED HOLONOMIC VIRIAL IDENTITY

$$\int\left\{\textstyle\sum (\partial T/\partial \dot{q}_k)\dot{z}_k + \sum\left(\partial T/\partial q_k + Q_k + \sum \lambda_D a_{Dk}\right) z_k\right\} dt = \left\{\textstyle\sum (\partial T/\partial \dot{q}_k) z_k\right\}_1^2$$

[$z_k = z_k(t)$: arbitrary functions, but as well behaved as needed; and integral extends from t_1 to t_2 (arbitrary time limits).

Specializations

$z_k \to \delta q_k$ [*virtual displacement of* q_k; and assuming $\delta\dot{q}_k = (\delta q_k)^{\cdot}$]:

$$\int (\delta T + \delta' W)\,dt = \left\{\textstyle\sum p_k\,\delta q_k\right\}_1^2,$$

[Hamilton's law of virtually/vertically varying action]

$z_k \to \Delta q_k = \delta q_k + \dot{q}_k\,\Delta t$ (*noncontemporaneous, or skew, or oblique, variation of* q_k):

$$\int \left\{ \sum (\partial T/\partial \dot{q}_k)(\Delta q_k)^{\cdot} + \sum (\partial T/\partial q_k + Q_k)\, \Delta q_k - \sum \lambda_D a_D\, \Delta_D \right\} dt = \left\{ \sum p_k\, \Delta \dot{q}_k \right\}_1^2$$

[Hamilton's law of skew-varying action]

$$(\Delta q_k)^{\cdot} - \Delta(\dot{q}_k) = \dot{q}_k(\Delta t)^{\cdot} \qquad \text{[i.e., } \Delta(\ldots) \text{ and } (\ldots)^{\cdot} \text{ do not commute]}$$

$z_k \to q_k$ (*actual system coordinate*):

$$\int \left\{ \sum (\partial T/\partial \dot{q}_k)\dot{q}_k + \sum \left(\partial T/\partial q_k + Q_k + \sum \lambda_D a_{Dk} \right) q_k \right\} dt = \left\{ \sum p_k q_k \right\}_1^2$$

[Virial theorem (of Clausius, Szily et al.)]

$z_k \to \dot{q}_k$ (*actual system velocity*): power theorem in holonomic variables.

GENERALIZED NONHOLONOMIC VIRIAL IDENTITY

$$\int \left(\sum (\partial T^*/\partial \omega_k)\dot{z}_k + \sum (\partial T^*/\partial \theta_k) z_k - \sum\sum h^b{}_k (\partial T^*/\partial \omega_b) z_k + \sum (\Theta_k + \Lambda_k) z_k \right) dt = \left\{ \sum (\partial T^*/\partial \omega_k) z_k \right\}_1^2$$

Specializations

$z_k \to \delta\theta_k$ (recalling that $\delta\theta_D = 0$, $\delta\theta_{n+1} \equiv \delta t = 0$, while $\delta\theta_I \neq 0$):

$$\int \left(\delta T^* + \sum \Theta_I\, \delta\theta_I \right) dt = \left\{ \sum P_I\, \delta\theta_I \right\}_1^2$$

[Hamilton's law of virtual and nonholonomic action].

$z_k \to \dot{\theta}_k \equiv \omega_k$ (recalling that $\omega_D = 0$): power theorem in nonholonomic variables.

$z_k \to \theta_k$: This case is meaningless because there is no such thing as θ_k.

$z_k \to \dot{\omega}_k$: This case does not seem to lead to any readily useful and identifiable result.

$z_k \to \Delta\theta_k$:

$$\Big[\equiv \delta\theta_b + \dot{\theta}_b\, \Delta t \equiv \delta\theta_b + \omega_b\, \Delta t \Rightarrow (\Delta\theta_b)^{\cdot} - \Delta\omega_b = (\delta\theta_b)^{\cdot} - \delta(\dot{\theta}_b) + \omega_b(\Delta t)^{\cdot} = \sum h^b{}_k\, \delta\theta_k + \omega_b(\Delta t)^{\cdot} \Big]$$

$$\int \Big\{ \sum (\partial T^*/\partial \theta_k)\, \Delta\theta_k + \sum (\partial T^*/\partial \omega_k)\, \Delta\omega_k + \sum (\partial T^*/\partial \omega_k) \left[\omega_k (\Delta t)^{\cdot} - \sum \gamma^k{}_b \omega_b\, \Delta t \right] + \sum (\Theta_k + \Lambda_k)\, \Delta\theta_k \Big\} dt = \left\{ \sum (\partial T^*/\partial \omega_k)\, \Delta\theta_k \right\}_1^2$$

[Hamilton's law of skew-varying action in nonholonomic variables].

NONLINEAR NONHOLONOMIC CONSTRAINTS;
HOLONOMIC VARIABLES

$$\int\Big\{\sum(\partial T/\partial\dot q_k)\dot z_k+\sum\Big[\partial T/\partial q_k+Q_k+\sum\lambda_D(\partial f_D/\partial\dot q_k)\Big]z_k\Big\}\,dt$$
$$=\Big\{\sum(\partial T/\partial\dot q_k)z_k\Big\}_1^2$$

Specializations

$z_k\to q_k$ (Virial theorem):

$$\int\Big\{\sum(\partial T/\partial\dot q_k)\dot q_k+\sum\Big[\partial T/\partial q_k+Q_k+\sum\lambda_D(\partial f_D/\partial\dot q_k)\Big]q_k\Big\}\,dt$$
$$=\Big\{\sum(\partial T/\partial\dot q_k)q_k\Big\}_1^2$$

$z_k\to\dot q_k$ (Nonlinear (nonpotential) generalized power equation):

$$d/dt\Big(\sum(\partial T/\partial\dot q_k)\dot q_k-T\Big)$$
$$=-\partial T/\partial t+\sum Q_k\dot q_k+\sum\sum\lambda_D(\partial f_D/\partial\dot q_k)\dot q_k$$

$z_k\to\delta q_k$ (Hamilton's law of varying action);

$z_k\to\Delta q_k$ (Hamilton's law of skew-varying action):

$$\int\Big\{\sum(\partial T/\partial\dot q_k)(\Delta q_k)^{\cdot}+\sum(\partial T/\partial q_k+Q_k)\,\Delta q_k$$
$$+\Big(\sum\sum\lambda_D(\partial f_D/\partial\dot q_k)\dot q_k\Big)\,\Delta t\Big\}\,dt$$
$$=\Big\{\sum(\partial T/\partial\dot q_k)z_k\Big\}_1^2$$

NONLINEAR NONHOLONOMIC CONSTRAINTS;
NONHOLONOMIC VARIABLES

$$\int(\delta T^*+\delta' W)\,dt=\Big\{\sum P_k\,\delta\theta_k\Big\}_1^2,$$

where

$$\delta T^*=\cdots=\Big(\sum(\partial T^*/\partial\omega_k)\,\delta\theta_k\Big)^{\cdot}-\sum(\partial T^*/\partial\omega_k)^{\cdot}\,\delta\theta_k$$
$$-\sum\sum H^k{}_b(\partial T^*/\partial\omega_k)\,\delta\theta_b+\sum(\partial T^*/\partial\theta_k)\,\delta\theta_k,$$

$$\Rightarrow\int\delta T^*\,dt$$
$$=-\int\sum\Big[(\partial T^*/\partial\omega_k)^{\cdot}-\partial T^*/\partial\theta_k+\sum H^b{}_k(\partial T^*/\partial\omega_b)\Big]\,\delta\theta_k\,dt$$
$$+\Big\{\sum(\partial T^*/\partial\omega_k)\,\delta\theta_k\Big\}_1^2,$$

and

$$(\delta\theta_b)^{\cdot} - \delta\omega_b = \sum E_s(\omega_b)\,\delta q_s = \sum\sum E_s(\omega_b)(\partial\dot{q}_s/\partial\omega_k)\,\delta\theta_k \equiv \sum H^b{}_k\,\delta\theta_k$$

$$= -\sum\sum E_k(\dot{q}_l)(\partial\omega_b/\partial\dot{q}_l)\,\delta\theta_k \equiv -\sum\sum V^l{}_k(\partial\omega_b/\partial\dot{q}_l)\,\delta\theta_k,$$

$$\Gamma_k = -\sum H^b{}_k(\partial T^*/\partial\omega_b) = \sum V^b{}_k(\partial T/\partial\dot{q}_b)^*$$

[assuming again that $(\delta q_k)^{\cdot} = \delta(\dot{q}_k)$].

GENERAL INTEGRAL EQUATIONS

$$\int\left\{\delta T + \sum(\partial T/\partial\dot{q}_k)[(\delta q_k)^{\cdot} - \delta(\dot{q}_k)] + \delta' W\right\} dt$$

$$= \int\left\{\delta T + \delta' W + \sum P_k[(\delta\theta_k)^{\cdot} - \delta\omega_k] + \sum\sum V^k{}_b\, p_k\,\delta\theta_b\right\} dt$$

$$= \int\left\{\delta T + \delta' W + \sum P_k[(\delta\theta_k)^{\cdot} - \delta\omega_k] - \sum\sum H^k{}_b P_k\,\delta\theta_b\right\} dt$$

$$= \left\{\sum(\partial T/\partial\dot{q}_k)\,\delta q_k\right\}_1^2;$$

where

$$(\delta q_k)^{\cdot} - \delta(\dot{q}_k) = \sum(\partial\dot{q}_k/\partial\omega_b)[(\delta\theta_b)^{\cdot} - \delta\omega_b] + \sum V^k{}_b\,\delta\theta_b$$

$$= \sum(\partial\dot{q}_k/\partial\omega_b)[(\delta\theta_b)^{\cdot} - \delta\omega_b] - \sum\sum(\partial\dot{q}_k/\partial\omega_l)H^l{}_b\,\delta\theta_b$$

$$T = T[t, q, \dot{q}(t, q, \omega)] \equiv T^*(t, q, \omega) \equiv T^*.$$

The above yield the "equation of motion forms" [without the assumption $(\delta q_k)^{\cdot} = \delta(\dot{q}_k)$]:

$$-\int\sum\Big[(\partial T^*/\partial\omega_k)^{\cdot} - \partial T^*/\partial\theta_k - \sum(\partial T/\partial\dot{q}_b)^* V^b{}_k - \Theta_k\Big]\,\delta\theta_k\,dt = 0,$$

$$-\int\sum\Big[(\partial T^*/\partial\omega_k)^{\cdot} - \partial T^*/\partial\theta_k + \sum(\partial T^*/\partial\omega_b)H^b{}_k - \Theta_k\Big]\,\delta\theta_k\,dt = 0.$$

HÖLDER–VORONETS–HAMEL VIEWPOINT

$(\delta q_k)^{\cdot} = \delta\dot{q}_k$, whether the δq_k are further constrained or not. Then, with: $\delta' W^* \equiv \sum \Theta_k\,\delta\theta_k$, the above yield

$$\int(\delta T + \delta' W)\,dt = \int(\delta T^* + \delta' W^*)\,dt$$

$$= \left\{\sum(\partial T^*/\partial\omega_k)\,\delta\theta_k\right\}_1^2$$

CONSTRAINED INTEGRAL FORMS
[i.e., in terms of $T^* \to T^*_o = T^*(t, q, \omega_I)$]

Generally:

$$\delta T^* = \delta T^*_o + \sum (\partial T^*/\partial \omega_D)_o \, \delta\omega_D,$$
$$\delta T^*_o = \sum (\partial T^*_o/\partial \theta_I)\, \delta\theta_I + \sum (\partial T^*_o/\partial \omega_I)\, \delta\omega_I$$

Under the Hölder–Voronets–Hamel viewpoint:

$$\delta(\dot{q}_k) = (\delta q_k)^{\cdot}, \qquad \delta\theta_D = 0, \qquad d(\delta\theta_D) = 0 \Rightarrow (\delta\theta_D)^{\cdot} = 0;$$

but $\quad \delta(d\theta_D) \neq 0, \qquad \delta\omega_D = \sum\sum V^k{}_I (\partial\omega_D/\partial\dot{q}_k)\, \delta\theta_I = -\sum H^D{}_I \, \delta\theta_I \neq 0,$

and $\quad \delta\omega_I = (\delta\theta_I)^{\cdot} - \sum H^I{}_{I'}\, \delta\theta_{I'};$

we obtain the constrained integral equation

$$\int \left[\delta T^*_o + \sum\sum (\partial T/\partial \dot{q}_k)_o V^k{}_I \, \delta\theta_I + \delta' W^*_o\right] dt$$
$$= \int \left[\delta T^*_o - \sum\sum (\partial T^*/\partial \omega_D)_o H^D{}_I \, \delta\theta_I + \delta' W^*_o\right] dt$$
$$= \left\{\sum (\partial T/\partial \dot{q}_k)\, \delta q_k\right\}_1^2.$$

Special form of the constraints:

$$\dot{q}_D = \phi_D(t, q, \dot{q}_I)$$
$$\Rightarrow \omega_D \equiv \dot{q}_D - \phi_D(t, q, \dot{q}_I) = 0, \qquad \omega_I \equiv \dot{q}_I \neq 0,$$
$$\dot{q}_D = \omega_D + \phi_D[t, q, \dot{q}_I(t, q, \omega_I)] = \omega_D + \phi_D(t, q, \omega_I),$$
$$\delta\theta_D = \delta q_D - \sum (\partial\phi_D/\partial\dot{q}_I)\, \delta q_I = 0, \qquad \delta\theta_I = \delta q_I \neq 0.$$

Suslov transitivity assumptions and integral equation:

$$(\delta q_D)^{\cdot} \neq \delta(\dot{q}_D), \qquad (\delta q_I)^{\cdot} - \delta(\dot{q}_I) = 0;$$

but

$$\delta(d\theta_D) = 0$$

or

$$\delta\omega_D = \delta(\dot{q}_D - \phi_D) = \delta(\dot{q}_D) - \delta\phi_D = 0 \qquad [\text{and } (\delta\theta_D)^{\cdot} = 0]$$
$$\Rightarrow \delta(\dot{q}_D) = \delta\phi_D \qquad [\text{definition of } \delta(\dot{q}_D)];$$
$$\Rightarrow (\delta q_D)^{\cdot} - \delta(\dot{q}_D) = \left(\sum (\partial\phi_D/\partial\dot{q}_I)\, \delta q_I\right) - \delta\phi_D$$
$$= \cdots = \sum E_{(I)}(\phi_D)\, \delta q_I \equiv \sum W^D{}_I \, \delta q_I \neq 0;$$
$$\Rightarrow \delta T = \delta T_o.$$

Suslov principle:

$$\int \Big\{\delta T_o + \sum (\partial T/\partial \dot{q}_D)_o[(\delta q_D)^{\cdot} - \delta \phi_D] + \delta' W_o\Big\}\,dt$$
$$= \int \Big\{\delta T_o + \sum\sum (\partial T/\partial \dot{q}_D) W^D{}_I\,\delta q_I + \delta' W_o\Big\}\,dt$$
$$= \Big\{\sum (\partial T/\partial \dot{q}_k)\,\delta q_k\Big\}_1^2$$

Hölder–Voronets–Hamel transitivity assumptions:

$$\delta(\dot{q}_k) = (\delta q_k)^{\cdot}, \qquad \delta\theta_D = 0, \qquad d(\delta\theta_D) = 0 \Rightarrow (\delta\theta_D)^{\cdot} = 0;$$

but

$$\delta(d\theta_D) \neq 0$$

or

$$\delta\omega_D = \delta(\dot{q}_D - \phi_D) = \delta(\dot{q}_D) - \delta\phi_D = (\delta q_D)^{\cdot} - \delta\phi_D$$
$$= \sum E_{(I)}(\phi_D)\,\delta q_I \equiv \sum W^D{}_I\,\delta q_I \neq 0 \qquad [\text{definition of } \delta(\dot{q}_D)];$$
$$\Rightarrow \delta T = \delta T_o + \sum\sum (\partial T/\partial \dot{q}_D) W^D{}_I\,\delta q_I.$$

Voronets principle:

$$\int \Big[\delta T_o + \sum\sum (\partial T/\partial \dot{q}_D) W^D{}_I\,\delta q_I + \delta' W_o\Big]\,dt$$
$$= \Big\{\sum (\partial T/\partial \dot{q}_k)\,\delta q_k\Big\}_1^2.$$

In both cases:

$$T \to T_o(t, q, \dot{q}_I) \to \delta T_o \text{ (variation of constrained } T\text{)},$$
$$\partial T/\partial \dot{q}_D \to (\partial T/\partial \dot{q}_D)_o = p_D[t, q, \dot{q}_I, \phi_D(t, q, \dot{q}_I)] \equiv p_{D,o}(t, q, \dot{q}_I) \equiv p_{Do},$$
$$\Big\{\sum (\partial T/\partial \dot{q}_k)\,\delta q_k\Big\}_1^2 = \cdots = \Big\{\sum (\ldots)_I\,\delta q_I\Big\}_1^2,$$
$$\delta' W_o \equiv \sum Q_{Io}\,\delta q_I.$$

NONCONTEMPORANEOUS VARIATIONS AND RELATED THEOREMS

Definition:

$\Delta(\ldots) \equiv \delta(\ldots) + [d(\ldots)/dt]\,\Delta t$: *noncontemporaneous variation operator*

$$\Rightarrow \Delta q_k = \delta q_k + \dot{q}_k\,\Delta t, \qquad \Delta t = \delta t + (dt/dt)\,\Delta t = 0 + (1)\,\Delta t = \Delta t.$$

Basic identities:

$$\Delta \int (\ldots)\, dt = \int \delta(\ldots)\, dt + \{(\ldots)\, \Delta t\}_1^2$$
$$= \int \{\Delta(\ldots) + (\ldots)[d(\Delta t)/dt]\}\, dt$$
$$= \int [\Delta(\ldots)\, dt + (\ldots)\, d(\Delta t)],$$

$$\int \Delta(\ldots)\, dt = \int \{\delta(\ldots) - (\ldots)[d(\Delta t)/dt]\}\, dt + \{(\ldots)\, \Delta t\}_1^2$$
$$= \int [\delta(\ldots)\, dt - (\ldots)\, d(\Delta t)] + \{(\ldots)\, \Delta t\}_1^2;$$

$$\Rightarrow \Delta \int (\ldots)\, dt - \int \Delta(\ldots)\, dt = \int (\ldots)\, d(\Delta t)$$
$$= \int \{(\ldots)[d(\Delta t)/dt]\}\, dt;$$

$$\Delta \int (\ldots)\, dt = \cdots = -\int \sum E_k(\ldots)\, \Delta q_k\, dt$$
$$+ \int [dh(\ldots)/dt + \partial(\ldots)/\partial t]\, \Delta t\, dt$$
$$+ \Big\{\sum (\partial\ldots/\partial \dot{q}_k)\, \Delta q_k - h(\ldots)\, \Delta t\Big\}_1^2$$
$$= -\int \sum E_k(\ldots)\, \delta q_k\, dt + \Big\{\sum (\partial\ldots/\partial \dot{q}_k)\, \Delta q_k - h(\ldots)\, \Delta t\Big\}_1^2$$
$$= -\int \sum E_k(\ldots)\, \delta q_k\, dt + \Big\{\sum (\partial\ldots/\partial \dot{q}_k)\, \delta q_k + (\ldots)\, \Delta t\Big\}_1^2;$$

where

$$h(\ldots) \equiv \sum [\partial(\ldots)/\partial \dot{q}_k]\dot{q}_k - (\ldots)\text{: generalized energy operator.}$$

With

$h \equiv h(L) \equiv \sum p_k \dot{q}_k - L = h(t, q, \dot{q})$: generalized energy,

$A_H \equiv \int (T - V)\, dt \equiv \int L\, dt$: Hamiltonian action (functional),

$A_L \equiv \int 2T\, dt$: Lagrangean action (functional),

$E \equiv T + V$: total energy of the system;

we have the following mechanical integral theorems:

$$\Delta \int T\, dt + \int \delta' W\, dt = \Big\{\sum p_k\, \Delta q_k + \Big(T - \sum p_k \dot{q}_k\Big)\, \Delta t\Big\}_1^2,$$

$$
\begin{aligned}
\Delta A_H + \int \delta' W_{np}\, dt &= \Big\{\sum p_k\, \Delta q_k - h\, \Delta t\Big\}_1^2 \\
&= \Big\{\sum p_k\, \delta q_k + L\, \Delta t\Big\}_1^2, \\
\Delta A_L - \int (\delta E - \delta' W_{np})\, dt &= \Big\{\sum p_k\, \Delta q_k - \Big(\sum p_k \dot{q}_k - 2T\Big)\, \Delta t\Big\}_1^2 \\
&= \Big\{\sum p_k\, \delta q_k + 2T\, \Delta t\Big\}_1^2, \\
\Delta \int E\, dt &= \int \delta E\, dt + \{E\, \Delta t\}_1^2, \\
\Delta \int 2T\, dt &= \int (\delta E - \delta' W_{np})\, dt \\
&\quad + \Big\{\sum p_k\, \Delta q_k + \Big(2T - \sum p_k \dot{q}_k\Big)\, \Delta t\Big\}_1^2, \\
\Delta \int 2V\, dt &= \int (\delta E + \delta' W_{np})\, dt \\
&\quad + \Big\{-\sum p_k\, \Delta q_k + \Big(2V + \sum p_k \dot{q}_k\Big)\, \Delta t\Big\}_1^2, \\
\int [\Delta T + 2T(\Delta t)^{\cdot} &+ \dot{T}\, \Delta t]\, dt + \int \delta' W\, dt \\
&= \int [\Delta T\, dt + 2T\, d(\Delta t) + dT\, \Delta t + \delta' W\, dt] \\
&= \Big\{\sum p_k\, \delta q_k + (2T)\, \Delta t\Big\}_1^2 \\
&= \Big\{\sum p_k\, \Delta q_k - \Big(\sum p_k \dot{q}_k - 2T\Big)\, \Delta t\Big\}_1^2
\end{aligned}
$$

SECOND (VIRTUAL) VARIATION OF A_H

Total (virtual) variation:

$$\delta^T A_H \equiv A_H(q + \delta q) - A_H(q) = \delta A_H + (1/2)\, \delta^2 A_H + \cdots$$

First (virtual) variation:

$$\delta A_H = \int \delta L\, dt = \cdots = -\int E(q)\, \delta q\, dt + \{p\, \delta q\}_1^2$$

Second (virtual) variation (*one* Lagrangean coordinate):

$$
\begin{aligned}
\delta^2 A_H \equiv \delta(\delta A_H) &= \int \delta^2 L\, dt \\
&= \cdots = -\int J(\delta q)\, \delta q\, dt + \{\delta p\, \delta q\}_1^2,
\end{aligned}
$$

where

$$\delta^2 L \equiv \delta(\delta L) = [(\partial/\partial q)\,\delta q + (\partial/\partial \dot{q})\,\delta(\dot{q})]^2 L$$
$$= \cdots = (\partial^2 L/\partial \dot{q}^2)(\delta\dot{q})^2 + 2(\partial^2 L/\partial q\,\partial\dot{q})\,\delta q\,\delta\dot{q} + (\partial^2 L/\partial q^2)(\delta q)^2$$

Jacobi's variational equation:

$$J(\delta q) = \{d/dt[\partial/\partial(\delta\dot{q})] - [\partial/\partial(\delta q)]\}(1/2)\,\delta^2 L$$
$$= (\partial^2 L/\partial\dot{q}^2)\,\delta\ddot{q} + (\partial^2 L/\partial\dot{q}^2)^{\cdot}\,\delta\dot{q} + [(\partial^2 L/\partial q\,\partial\dot{q})^{\cdot} - (\partial^2 L/\partial q^2)]\,\delta q$$
$$= d/dt[(\partial^2 L/\partial\dot{q}^2)\,\delta\dot{q}] - \left[\partial^2 L/\partial q^2 - d/dt(\partial^2 L/\partial q\,\partial\dot{q})\right]\delta q = 0$$

Equivalently:

$$E[L(t, q + \delta q, \dot{q} + \delta\dot{q})] - E[L(t, q, \dot{q})] \approx \delta E(q, \delta q) \qquad \text{(to first-order)}$$
$$= J(\delta q; q) \equiv J(\delta q)$$

Chapter 8: Hamiltonian/Canonical Methods

CONJUGATE (HAMILTONIAN) KINETIC ENERGY

$$T' \equiv \left(\sum p_k \dot{q}_k - T\right)\Big|_{\dot{q}=\dot{q}(t,q,p)} = \sum p_k \dot{q}_k(t, q, p) - T_{(qp)} \equiv T'(t, q, p)$$
$$\Big[= \sum (\partial T/\partial \dot{q}_k)\dot{q}_k - T = (2T_2 + T_1) - (T_2 + T_1 + T_0) = T_2 - T_0;$$
$$\text{i.e., if } T = T_2 \text{ (e.g., stationary constraints), then } T' = T\Big]$$

CANONICAL, OR HAMILTONIAN, CENTRAL EQUATION

$$\sum (dp_k/dt + \partial T'/\partial q_k - Q_k)\,\delta q_k + \sum (dq_k/dt - \partial T'/\partial p_k)\,\delta p_k = 0$$

CANONICAL, OR HAMILTONIAN, EQUATIONS OF MOTION
(for unconstrained variations)

$$dp_k/dt = -(\partial T'/\partial q_k) + Q_k \qquad (= \partial T/\partial q_k + Q_k \Rightarrow \partial T/\partial q_k = -\partial T'/\partial q_k),$$
$$dq_k/dt = \partial T'/\partial p_k$$

If $Q_k = -\partial V(t, q)/\partial q_k$, the above assume the *antisymmetrical* form:

$$dp_k/dt = -\partial H/\partial q_k, \qquad dq_k/dt = \partial H/\partial p_k,$$

where

$$H \equiv T' + V = \left(\sum p_k \dot{q}_k - T + V\right)\Big|_{\dot{q}=\dot{q}(t,q,p)} \equiv \sum p_k \dot{q}_k(t,q,p) - (T_{(qp)} - V)$$

$$= \left(\sum p_k \dot{q}_k - L\right)\Big|_{\dot{q}=\dot{q}(t,q,p)} \equiv \sum p_k \dot{q}_k(t,q,p) - L_{(qp)}$$

$\equiv H(t,q,p)$: *Hamiltonian* of system (function of $2n+1$ arguments).

If both potential *and* nonpotential forces (Q_k) are present, the above are replaced by

$$dp_k/dt = -\partial H/\partial q_k + Q_k, \qquad dq_k/dt = \partial H/\partial p_k;$$

also,

$$\partial H/\partial q_k = -\partial L/\partial q_k \qquad \text{and} \qquad \partial L/\partial t = -\partial H/\partial t.$$

For *stationary* (holonomic) constraints,

$H = T(t,q,p) + V_0(t,q) \equiv E(t,q,p) =$ *total energy*, in Hamiltonian variables.

In all cases, the following kinematico-inertial identities hold:

$$\partial T'/\partial t = -\partial T/\partial t, \qquad \partial T'/\partial q_k = -\partial T/\partial q_k, \qquad \partial T'/\partial p_k = dq_k/dt;$$

$$\partial H/\partial t = -\partial L/\partial t, \qquad \partial H/\partial q_k = -\partial L/\partial q_k, \qquad \partial H/\partial p_k = dq_k/dt.$$

LEGENDRE TRANSFORMATION (LT)

An LT transforms a function $Y(\ldots, y, \ldots)$ into its *conjugate* function $Z(\ldots, z, \ldots)$, where $z = \partial Y/\partial y$, so that $\partial Z/\partial z = y$. Here in dynamics we have the following identifications:

$$Y(\ldots) \to L, \qquad \ldots \to q, t, \qquad y \to \dot{q}, \qquad z = \partial Y/\partial y \to p = \partial L/\partial \dot{q},$$

$$Z(\ldots) \to H, \qquad \partial Z/\partial z = y \to \partial H/\partial p = \dot{q}.$$

POWER THEOREM

$$dH/dt = \partial H/\partial t + \sum Q_k \dot{q}_k$$

If $\partial H/\partial t = 0$ (e.g., stationary constraints) and $Q_k = 0$ (e.g., potential forces), then the Hamiltonian energy of the system is conserved:

$$H = H(q,p) = constant.$$

CANONICAL ROUTH–VOSS EQUATIONS

Under the m Pfaffian constraints

$$\sum a_{Dk}\, \delta q_k = 0,$$

the canonical equations are

$$dp_k/dt = -\partial T'/\partial q_k + Q_k + \sum \lambda_D a_{Dk} = -\partial H/\partial q_k + Q_{k,\text{nonpotential}} + \sum \lambda_D a_{Dk},$$
$$dq_k/dt = \partial T'/\partial p_k \; (= \partial H/\partial p_k).$$

ROUTH'S EQUATIONS

Ignorable (or cyclic) coordinates and momenta

$$(q_1, \ldots, q_M) \equiv (\psi_1, \ldots, \psi_M) \equiv (\psi_i) \equiv \psi, \quad (p_1, \ldots, p_M) \equiv (\Psi_1, \ldots, \Psi_M) \equiv (\Psi_i) \equiv \Psi$$

Positional (or palpable) coordinates and velocities

$$(q_{M+1}, \ldots, q_n) \equiv (q_p) \equiv q \qquad (\dot{q}_{M+1}, \ldots, \dot{q}_n) \equiv (\dot{q}_p) \equiv \dot{q}$$

Kinetic energy

$$\begin{aligned} T &\equiv T(t; \psi_1, \ldots, \psi_M; q_{M+1}, \ldots, q_n; \dot{\psi}_1, \ldots, \dot{\psi}_M; \dot{q}_{M+1}, \ldots, \dot{q}_n) \\ &\equiv T(t, \psi, q; \dot{\psi}, \dot{q}) = T[t, \psi, q; \dot{\psi}(t, \psi, q; \Psi, \dot{q}), \dot{q}] \\ &= T(t, \psi, q; \Psi, \dot{q}) \equiv T_{\psi\Psi} \end{aligned}$$

Modified (Routhian) kinetic energy

$$T'' \equiv \left(T - \sum \Psi_i \dot{\psi}_i\right)\Big|_{\dot{\psi}=\dot{\psi}(t;\psi,q;\Psi,\dot{q})} = T''(t, \psi, q; \Psi, \dot{q})$$

Routhian central equation

$$\sum (dp_k/dt - \partial T''/\partial q_k - Q_k)\, \delta q_k + \sum (p_p - \partial T''/\partial \dot{q}_p)\, \delta \dot{q}_p - \sum (d\psi_i/dt + \partial T''/\partial \Psi_i)\, \delta \Psi_i = 0$$

Routh's equations (for unconstrained variations)

$$\begin{aligned} &dp_k/dt = \partial T''/\partial q_k + Q_k: \quad && d\Psi_i/dt = \partial T''/\partial \psi_i + Q_i \quad && (i = 1, \ldots, M), \\ & && dp_p/dt = \partial T''/\partial q_p + Q_p \quad && (p = M+1, \ldots, n); \\ &d\psi_i/dt = -\partial T''/\partial \Psi_i && && (i = 1, \ldots, M), \\ &p_p = \partial T''/\partial \dot{q}_p && && (p = M+1, \ldots, n) \end{aligned}$$

Hamilton-like Routh's equations

$$d\Psi_i/dt = -\partial(-T'')/\partial \psi_i + Q_i, \qquad d\psi_i/dt = \partial(-T'')/\partial \Psi_i$$

Lagrange-like Routh's equations

$$dp_p/dt = \partial T''/\partial q_p + Q_p, \qquad p_p = \partial T''/\partial \dot{q}_p \qquad (= \partial T/\partial \dot{q}_p)$$
$$\Rightarrow \; (\partial T''/\partial \dot{q}_p)^{\cdot} - \partial T''/\partial q_p = Q_p$$

Additional Routhian kinematico-inertial identities

$$\partial T/\partial q_k = \partial T''/\partial q_k: \qquad \partial T/\partial \psi_i = \partial T''/\partial \psi_i \qquad (i = 1, \ldots, M),$$
$$\partial T/\partial q_p = \partial T''/\partial q_p \qquad (p = M+1, \ldots, n)$$

In sum, we have the following *two groups* of such kinematico-inertial identities:

$$\partial T''/\partial \psi_i = \partial T/\partial \psi_i \quad \text{and} \quad \partial T''/\partial \Psi_i = -d\psi_i/dt;$$
$$\partial T''/\partial q_p = \partial T/\partial q_p \quad \text{and} \quad \partial T''/\partial \dot{q}_p = \partial T/\partial \dot{q}_p \qquad (= p_p).$$

If $p_k \equiv \partial L/\partial \dot{q}_k$, the above are replaced by the following:

Hamilton-like Routh's equations

$$d\Psi_i/dt = \partial R/\partial \psi_i + Q_i, \qquad d\psi_i/dt = -\partial R/\partial \Psi_i$$

and Lagrange-like Routh's equations

$$dp_p/dt = \partial R/\partial q_p + Q_p, \qquad p_p = \partial R/\partial \dot{q}_p \qquad (= \partial L/\partial \dot{q}_p)$$
$$\Rightarrow \; E_p(R) \equiv (\partial R/\partial \dot{q}_p)^{\cdot} - \partial R/\partial q_p = Q_p;$$

where

$$R \equiv \left(L - \sum \Psi_i \dot{\psi}_i\right)\Big|_{\dot{\psi}=\dot{\psi}(t;\psi,q;\Psi,\dot{q})} = R(t;\psi,q;\Psi,\dot{q})$$

$= \textit{Routhian function}$, or *modified Lagrangean*,

$$L = L(t;\psi,q;\Psi,\dot{q}) \equiv T_{\psi\Psi} - V \equiv L_{\psi\Psi}$$

$=$ Lagrangean expressed in Routhian variables;

that is, the Routhian is a Hamiltonian [times (-1)] for the ψ_i, and a Lagrangean for the q_p.

Relation between Routhian and Hamiltonian

$$H \equiv \sum p_k \dot{q}_k - L, \qquad R = \sum p_p \dot{q}_p - H = L - \sum \Psi_i \dot{\psi}_i$$

STRUCTURE OF THE ROUTHIAN

Decomposition of T (scleronomic system):

$$T = T_{\dot{q}\dot{q}} + T_{\dot{q}\dot{\psi}} + T_{\dot{\psi}\dot{\psi}} = T(\psi, q; \dot{\psi}, \dot{q}),$$

where

$$2T_{\dot{q}\dot{q}} \equiv \sum\sum a_{pq}\dot{q}_p\dot{q}_q = \textit{homogeneous quadratic} \text{ in the } \dot{q}\text{'s}$$
$$(a_{pq} = a_{qp}\text{: positive definite}),$$
$$T_{\dot{q}\dot{\psi}} \equiv \sum\sum b_{pi}\dot{q}_p\dot{\psi}_i = \textit{homogeneous bilinear} \text{ in the } \dot{q}\text{'s and } \dot{\psi}\text{'s}$$
$$(\text{in general: } b_{pi} \neq b_{ip}\text{; sign indefinite}),$$

$$2T_{\dot\psi\dot\psi} \equiv \sum\sum c_{ij}\dot\psi_i\dot\psi_j = \textit{homogeneous quadratic} \text{ in the } \dot\psi\text{'s}$$
$$(c_{ij} = c_{ji}\text{: positive definite})$$

$[i, j = 1, \ldots, M;\ p, q = M+1, \ldots, n$; and the coefficients are functions of all n q_k's].

Next,

$$\Psi_i \equiv \partial T/\partial\dot\psi_i = \sum c_{ji}\dot\psi_j + \sum b_{pi}\dot q_p \Rightarrow \sum c_{ji}\dot\psi_j = \Psi_i - \sum b_{pi}\dot q_p,$$
$$d\psi_j/dt = \sum C_{ji}\Big(\Psi_i - \sum b_{pi}\dot q_p\Big)$$

(since $T_{\dot\psi\dot\psi}$ is positive definite $\Rightarrow c_{ij}$ is nonsingular), where

$$C_{ji} = [\text{cofactor of element } c_{ji} \text{ } in \text{ Det}(c_{ji})]/\text{Det}(c_{ji}) = C_{ij}$$

(=known function of the q's and ψ's).

Then

$$T = T_{2,0} + T_{0,2} = T(\psi, q; \Psi, \dot q),$$

where

$$2T_{2,0} \equiv \sum\sum\Big(a_{pq} - \sum\sum C_{ji}b_{pj}b_{qi}\Big)\dot q_p\dot q_q, \qquad 2T_{0,2} \equiv \sum\sum\sum C_{ji}\Psi_j\Psi_i;$$

that is, $T = T(\psi, q; \Psi, \dot q)$ does *not* contain any bilinear terms in the $\dot q$'s and Ψ's; and so

$$T'' \equiv T - \sum \Psi_i\dot\psi_i = T - \sum \Psi_i\Big(\sum C_{ij}\Big(\Psi_j - \sum b_{pj}\dot q_p\Big)\Big)$$
$$= T_{2,0} + T''_{1,1} - T_{0,2} \equiv T''_{2,0} + T''_{1,1} + T''_{0,2}$$
$$= T''(\psi, q; \Psi, \dot q),$$

where

$$2T''_{2,0} \equiv \sum\sum\Big(a_{pq} - \sum\sum C_{ji}b_{pj}b_{qi}\Big)\dot q_p\dot q_q \equiv \sum\sum r_{pq}(q)\dot q_p\dot q_q$$
$$= 2T_{2,0} \text{ (= positive definite in the } \dot q\text{'s)},$$
$$T''_{1,1} \equiv \sum\sum\Big(\sum C_{ji}b_{pi}\Big)\Psi_j\dot q_p \equiv \sum r_p(q, \Psi)\dot q_p$$

[No counterpart in $T = T(\psi, q; \Psi, \dot q)$, i.e., $T_{1,1} = 0$; sign indefinite],

$$2T''_{0,2} \equiv -\sum\sum C_{ji}\Psi_j\Psi_i = 2T''_{0,2}(q, \Psi)$$
$$= -2T_{0,2} \text{ (= negative definite in the } \Psi\text{'s)}.$$

Conversely,

$$\begin{aligned} T &\equiv T'' + \sum \Psi_i \dot{\psi}_i = T'' - \sum \Psi_i (\partial T''/\partial \Psi_i) \\ &= (T''_{2,0} + T''_{1,1} + T''_{0,2}) - (T''_{1,1} + 2T''_{0,2}) \\ &= T''_{2,0} - T''_{0,2} = T(\psi, q; \Psi, \dot{q}). \end{aligned}$$

Hence,

$$\begin{aligned} L &= T - V = (T_{2,0} + T_{0,2}) - V = T_{2,0} - (V - T_{0,2}) \\ &= (T''_{2,0} - T''_{0,2}) - V = T''_{2,0} - (V + T''_{0,2}) = L(\psi, q; \Psi, \dot{q}) \\ &\Rightarrow R = L - \sum \Psi_i \dot{\psi}_i = L + \sum (\partial T''/\partial \Psi_i) \Psi_i \\ &\qquad = (T''_{2,0} - T''_{0,2} - V) + (2T''_{0,2} + T''_{1,1}) \\ &\qquad \equiv R_2 + R_1 + R_0 = R(\psi, q; \Psi, \dot{q}), \end{aligned}$$

where

$$R_2 \equiv T''_{2,0} = T_{2,0}, \qquad R_1 \equiv T''_{1,1}, \qquad R_0 \equiv T''_{0,2} - V = -T_{0,2} - V.$$

Additional results

(i) With

$$\begin{aligned} T &= T_{\dot{q}\dot{q}} + T_{\dot{q}\dot{\psi}} + T_{\dot{\psi}\dot{\psi}} = T(\psi, q; \dot{\psi}, \dot{q}) \\ &\Rightarrow T'' = T - \sum \Psi_i \dot{\psi}_i = T - \sum (\partial T/\partial \dot{\psi}_i)\dot{\psi}_i = T_{\dot{q}\dot{q}} - T_{\dot{\psi}\dot{\psi}} = T''(t, \psi, q; \dot{\psi}, \dot{q}); \end{aligned}$$

(ii) $$d\psi_i/dt = -\partial T''/\partial \Psi_i = \cdots = \partial T_{0,2}/\partial \Psi_i - \partial K_{2,2}/\partial \pi_i,$$

where

$$2T_{0,2} = -2T''_{0,2} \equiv \sum \sum C_{ji} \Psi_j \Psi_i,$$

and

$$2K_{2,2} \equiv \sum \sum C_{ji} \Big(\sum b_{pj}\dot{q}_p\Big)\Big(\sum b_{qi}\dot{q}_q\Big) \equiv \sum \sum C_{ji} \pi_j \pi_i.$$

Matrix form of these results:

$$\dot{\mathbf{q}}^{\mathrm{T}} = (\dot{q}_{M+1}, \ldots, \dot{q}_n), \qquad \dot{\boldsymbol{\psi}}^{\mathrm{T}} = (\dot{\psi}_1, \ldots, \dot{\psi}_M), \qquad \boldsymbol{\Psi}^{\mathrm{T}} = (\Psi_1, \ldots, \Psi_M),$$

$$\mathbf{a} = (a_{pq}) = (a_{qp}) = \mathbf{a}^{\mathrm{T}}, \qquad \mathbf{b} = (b_{ip}) \neq (b_{pi}) = \mathbf{b}^{\mathrm{T}}, \qquad \mathbf{c} = (c_{ij}) = (c_{ji}) = \mathbf{c}^{\mathrm{T}},$$

$$2T = \dot{\mathbf{q}}^{\mathrm{T}}\mathbf{a}\dot{\mathbf{q}} + 2\dot{\boldsymbol{\psi}}^{\mathrm{T}}\mathbf{b}^{\mathrm{T}}\dot{\mathbf{q}} + \dot{\boldsymbol{\psi}}^{\mathrm{T}}\mathbf{c}\,\dot{\boldsymbol{\psi}},$$

$$\partial T/\partial \dot{\boldsymbol{\psi}} = \mathbf{b}^{\mathrm{T}}\dot{\mathbf{q}} + \mathbf{c}\dot{\boldsymbol{\psi}} = \boldsymbol{\Psi}$$

$$\Rightarrow \dot{\boldsymbol{\psi}} = \mathbf{c}^{-1}(\boldsymbol{\Psi} - \mathbf{b}^{\mathrm{T}}\dot{\mathbf{q}}) \equiv \mathbf{C}(\boldsymbol{\Psi} - \mathbf{b}^{\mathrm{T}}\dot{\mathbf{q}}) \Rightarrow \dot{\boldsymbol{\psi}}^{\mathrm{T}} = (\boldsymbol{\Psi}^{\mathrm{T}} - \dot{\mathbf{q}}^{\mathrm{T}}\mathbf{b})\mathbf{C}$$

[since $\mathbf{c}$ is symmetric, so is its inverse $\mathbf{C} \equiv (C_{ji})$: $\mathbf{C} \equiv \mathbf{c}^{-1} = (\mathbf{c}^{-1})^{\mathrm{T}} \equiv \mathbf{C}^{\mathrm{T}}$],

$$T = \cdots = (1/2)\dot{\mathbf{q}}^{\mathrm{T}}(\mathbf{a} - \mathbf{b}\,\mathbf{C}\,\mathbf{b}^{\mathrm{T}})\dot{\mathbf{q}} + (1/2)\mathbf{\Psi}^{\mathrm{T}}\,\mathbf{C}\,\mathbf{\Psi} \equiv T_{2,0} + T_{0,2} = T''_{2,0} - T''_{0,2},$$
$$[\text{since } \mathbf{\Psi}^{\mathrm{T}}\,\mathbf{C}\,\mathbf{b}^{\mathrm{T}}\,\dot{\mathbf{q}} = \dot{\mathbf{q}}^{\mathrm{T}}\,\mathbf{b}\,\mathbf{C}\mathbf{\Psi}]$$

$$\mathbf{\Psi}^{\mathrm{T}}\dot{\boldsymbol{\psi}} = \cdots = \mathbf{\Psi}^{\mathrm{T}}\,\mathbf{C}\,\mathbf{\Psi} - \mathbf{\Psi}^{\mathrm{T}}\,\mathbf{C}\,\mathbf{b}^{\mathrm{T}}\dot{\mathbf{q}} = -2T''_{0,2} - \mathbf{\Psi}^{\mathrm{T}}\,\mathbf{C}\,\mathbf{b}^{\mathrm{T}}\dot{\mathbf{q}},$$

$$R = (T - V) - \mathbf{\Psi}^{\mathrm{T}}\dot{\boldsymbol{\psi}} = \cdots = R_2 + R_1 + R_0$$

$$R_2 \equiv T''_{2,0} = T_{2,0} = (1/2)\,\dot{\mathbf{q}}^{\mathrm{T}}\,(\mathbf{a} - \mathbf{b}\,\mathbf{C}\,\mathbf{b}^{\mathrm{T}})\dot{\mathbf{q}},$$

$$R_1 \equiv T''_{1,1} = \mathbf{\Psi}^{\mathrm{T}}\,\mathbf{C}\,\mathbf{b}^{\mathrm{T}}\,\dot{\mathbf{q}},$$

$$R_0 \equiv T''_{0,2} - V = -(V + T_{0,2}) = -(1/2)\mathbf{\Psi}^{\mathrm{T}}\,\mathbf{C}\,\mathbf{\Psi} - V.$$

If $\mathbf{b} = 0$ (i.e., $\dot{q}$'s and $\dot{\psi}$'s *uncoupled* in the original T), R reduces to

$$R = (1/2)\dot{\mathbf{q}}^{\mathrm{T}}\,\mathbf{a}\,\dot{\mathbf{q}} - (1/2)\mathbf{\Psi}^{\mathrm{T}}\,\mathbf{C}\,\mathbf{\Psi} - V.$$

CYCLIC (OR GYROSTATIC) SYSTEMS

(i) $$(q_1, \ldots, q_M) \equiv (\psi_1, \ldots, \psi_M) \equiv (\psi_i) \equiv \psi$$

do *not* appear explicitly, neither in its kinetic energy nor in its nonvanishing impressed forces; only the corresponding Lagrangean velocities

$$(\dot{q}_1, \ldots, \dot{q}_M) \equiv (\dot{\psi}_1, \ldots, \dot{\psi}_M) \equiv (\dot{\psi}_i) \equiv \dot{\psi}$$

appear there, and, of course, time t and the remaining coordinates and/or velocities

$$(q_{M+1}, \ldots, q_n) \equiv (q_p) \equiv q \qquad \text{and} \qquad (\dot{q}_{M+1}, \ldots, \dot{q}_n) \equiv (\dot{q}_p) \equiv \dot{q};$$

respectively; that is,

$$\partial T/\partial \psi_i = 0$$

but, in general,

$$\partial T/\partial \dot{\psi}_i \neq 0 \Rightarrow T = T(t; q, \dot{\psi}, \dot{q}).$$

(ii) The corresponding impressed forces vanish; that is,

$$Q_i = 0, \qquad \text{but} \qquad Q_p = Q_p(q) \neq 0.$$

If all impressed forces are wholly *potential*, the above requirements are replaced, respectively, by

$$\partial L/\partial \psi_i = 0 \qquad \text{and} \qquad \partial L/\partial \dot{\psi}_i \neq 0 \Rightarrow L = L(t; q, \dot{\psi}, \dot{q}).$$

The coordinates ψ, and corresponding velocities $\dot{\psi}$, are called *cyclic* (Helmholtz), or *absent* (Routh), or *kinosthenic*, or *speed* (J. J. Thomson), or *ignorable* (Whittaker). The remaining coordinates q, and corresponding velocities $\dot{q}$, are called *palpable*, or *positional*. Then the Lagrangean equations corresponding to the cyclic coordinates/variables, become

$$(\partial T/\partial\dot{\psi}_i)^{\cdot} - \partial T/\partial\psi_i = Q_i: \qquad (\partial T/\partial\dot{\psi}_i)^{\cdot} = 0 \Rightarrow \partial T/\partial\dot{\psi}_i \equiv \Psi_i = \text{constant} \equiv C_i;$$

that is, the momenta Ψ_i corresponding to the cylic coordinates ψ_i are *constants of the motion*. [Conversely, however, if $\partial T/\partial\dot{\psi}_i = 0$, then $\partial T/\partial\psi_i = 0$, and, as a result, $T = T(t; q, \dot{q})$; that is, the evolution of the ψ's does not affect that of the q's.] Hence, the Routhian of a cyclic system is a function of t, q, $\dot{q}$ and $\Psi \equiv (\Psi_i)$; that is, with $C \equiv (C_i)$,

$$R \equiv \Big(L - \sum \Psi_i\dot{\psi}_i\Big)\Big|_{\dot{\psi}=\dot{\psi}(t;q;\dot{q},C)}$$

[after solving $\partial T/\partial\dot{\psi}_i \equiv \Psi_i = C_i$ for the $\dot{\psi}$ in terms of $t, q, \dot{q}, C$]

$$= L[t, q, \dot{q}, \dot{\psi}(t; q; C, \dot{q}); C] - \sum \Psi_i\dot{\psi}_i(t; q, \dot{q}; C)$$

$$= R(t; q, \dot{q}; C)$$

$$\Big[\Rightarrow L = \sum C_i\dot{\psi}_i(t; q, \dot{q}; C) + R(t; q, \dot{q}; C)];$$

that is, the system has been reduced to one with only $n - M$ Lagrangean coordinates, new "reduced Lagrangean" R, and, therefore, Lagrange-type Routhian equations for the positional coordinates and the "palpable motion" $q_p(t)$:

$$(\partial R/\partial\dot{q}_p)^{\cdot} - \partial R/\partial q_p = Q_{p,\text{nonpotential impressed positional forces}}.$$

Then,

$$R = \text{known function of time}$$

$$\Rightarrow \partial R/\partial C_i = \text{known function of time} \equiv -f_i(t; C),$$

$$\Rightarrow \psi_i = -\int(\partial R/\partial\Psi_i)\,dt + \text{constant} = \int f_i(t; C)\,dt + \text{constant}$$

$$= \psi_i(t, C) + \text{constant}.$$

EQUATIONS OF KELVIN–TAIT

Let

$$T = T(q, \dot{q}, \dot{\psi}) = \text{homogeneous quadratic in the } \dot{\psi} \text{ and } \dot{q},$$

$$\Rightarrow R = R_2 + R_1 + R_0,$$

where

$$R_2 \equiv T''_{2,0} = (1/2)\sum\sum r_{pq}(q)\dot{q}_p\dot{q}_q (= T_{2,0})$$

$= R_2(q, \dot{q}) =$ homogeneous *quadratic* in the nonignorable velocities $\dot{q}$,

$$R_1 \equiv T''_{1,1} = \sum r_p(q, C)\dot{q}_p$$

$= R_1(q, \dot{q}, C) =$ homogeneous *linear* in the nonignorable velocities $\dot{q}$, [*apparent kinetic* energy $T''_{1,1}$];

and

$$r_p = \sum \rho_{pi} C_i \qquad \left[\rho_{pi} \equiv \sum C_{ij} b_{pj} = \rho_{pi}(q)\right],$$

$$R_0 \equiv T''_{0,2} - V = -(V - T''_{0,2}) \equiv -(1/2) \sum\sum C_{ji} C_j C_i - V \qquad [= -(V + T_{0,2})]$$

$$= R_0(q; C) = \text{homogeneous } \textit{quadratic} \text{ in the constant ignorable momenta } \Psi = C$$
$$[\textit{apparent potential} \text{ energy } T''_{0,2} = -T_{0,2}(< 0)].$$

Hence, the situation is *mathematically identical* to that of *relative motion* (§3.16) Lagrangean equations of palpable motion:

$$(\partial R/\partial \dot{q}_p)^{\cdot} - \partial R/\partial q_p = Q_{p,\text{nonpotential impressed positional forces}}.$$

From the above we obtain the following.

Kelvin–Tait equations (with $p, p' = M + 1, \ldots, n$)

$$E_p(R) \equiv E_p(R_2 + R_1 + R_0) = E_p(R_2) + E_p(R_1) + E_p(R_0) = Q_p,$$

or

$$E_p(R_2) = Q_p - E_p(R_1) - E_p(R_0),$$

or, explicitly,

$$(\partial R_2/\partial \dot{q}_p)^{\cdot} - \partial R_2/\partial q_p = Q_p + \partial R_0/\partial q_p - [(\partial R_1/\partial \dot{q}_p)^{\cdot} - \partial R_1/\partial q_p]$$
$$= Q_p - \partial(V - T''_{0,2})/\partial q_p + \sum (\partial r_{p'}/\partial q_p - \partial r_p/\partial q_{p'}) \dot{q}_{p'}$$
$$= Q_p - \partial(V - T''_{0,2})/\partial q_p + G_p,$$

where

$$G_p = -[(\partial R_1/\partial \dot{q}_p)^{\cdot} - \partial R_1/\partial q_p]$$
$$= \sum (\partial r_{p'}/\partial q_p - \partial r_p/\partial q_{p'}) \dot{q}_{p'} \equiv \sum G_{pp'} \dot{q}_{p'}$$

[*Gyroscopic Routhian "force,"* since $G_{pp'} = -G_{p'p} = G_{pp'}(q; C)$].

These are the equations of motion of a *fictitious* scleronomic system (sometimes referred to as "conjugate" to the original, or reduced, system) with $n - M$ positional coordinates q, and subject, in addition to the impressed forces Q_p (nonpotential) and $-\partial V/\partial q_p$ (potential), to two special constraint forces: a centrifugal-like $\partial T''_{0,2}/\partial q_p$, and a gyroscopic one G_p.

Ignorable motion, once the palpable motion has been determined:

$$q_p(t) \Rightarrow d\psi_i/dt = -\partial R/\partial C_i = -\partial R_1/\partial C_i - \partial R_0/\partial C_i = -\partial T''_{0,2}/\partial C_i - \sum \rho_{pi} \dot{q}_p$$

Gyroscopic uncoupling $G_{pp'} = 0$

$$\Rightarrow \; E_p(R_2) \equiv (\partial R_2/\partial \dot{q}_p)^{\cdot} - \partial R_2/\partial q_p = Q_p + \partial R_0/\partial q_p$$

A system is gyroscopically uncoupled if, and only if, $R_1\,dt \equiv \sum r_p(q; C)\,dq_p$ is an *exact differential*. [A similar uncoupling occurs if all the C_i vanish: $r_p = 0 \Rightarrow R_1 = 0$; *and* $R_0 = -V(q)$.]

A cyclic power theorem

$$dh_R/dt = \sum Q_p\,\dot{q}_p,$$

where

$$\begin{aligned} h_R &\equiv R_2 - R_0 = T''_{2,0} + (V - T''_{0,2}) \\ &= T_{2,0} + (V + T_{0,2}) = T(q, \dot{q}, C) + V(q) \equiv E(q, \dot{q}, C) \\ &= \text{Modified (or cyclic) generalized energy;} \end{aligned}$$

if $\sum Q_p\,\dot{q}_p = 0$:

$$h_R \equiv T''_{2,0} + (V - T''_{0,2}) \equiv T(q, \dot{q}, C) + V(q) = \text{constant}.$$

Alternatively,

$$\begin{aligned} H &\equiv \sum (\partial L/\partial \dot{q}_k)\dot{q}_k - L \qquad (= \text{constant, if } Q_p = 0 \text{ and } \partial L/\partial t = \partial R/\partial t = 0) \\ &= -R + \sum (\partial R/\partial \dot{q}_p)\dot{q}_p \\ &= -(R_2 + R_1 + R_0) + (2R_2 + R_1) \\ &= R_2 - R_0 = H(q, \dot{q}, C) \qquad (= h_R). \end{aligned}$$

For *rheonomic* cyclic systems; that is, $L = L(t, q, \dot{q}, C)$

$$\Rightarrow R = L(t, q, \dot{q}, C) - \sum C_i \dot{\psi}_i(t, q, \dot{q}, C) = R(t, q, \dot{q}, C).$$

STEADY MOTION (OR CYCLIC SYSTEMS)

$$\dot{\psi}_i = \text{constant} \equiv c_i \text{ (in addition to } \Psi_i = \text{constant} \equiv C_i),$$

$$\text{and } q_p = \text{constant} \equiv s_P \; (\Rightarrow \dot{q}_p = 0)$$

$$(\text{with } i = 1, \ldots, M;\; p = M + 1, \ldots, n);$$

that is, *all* velocities are *constant* (and, hence, all accelerations vanish); and, for scleronomic such systems, the Lagrangean has the form $L = L(c_i, s_p)$.

Conditions for steady motion [necessary and sufficient conditions for the steady motion of an originally (scleronomic and holonomic) system; or, equivalently, for the equilibrium of the corresponding reduced q-system]:

$$Q_p + \partial R_0/\partial q_p \equiv Q_p + (\partial T''_{0,2}/\partial q_p - \partial V/\partial q_p) = 0,$$

or, if the forces are *wholly potential*:

$$\partial R_0/\partial q_p = 0, \qquad \text{or} \qquad \partial T''_{0,2}/\partial q_p = \partial V/\partial q_p.$$

Equivalently, since

$$R = R_2(\textit{homogeneous quadratic} \text{ in the } \dot{q}\text{'s})$$
$$+ R_1\ (\textit{homogeneous bilinear} \text{ in the } \Psi\text{'s and } \dot{q}\text{'s})$$
$$+ R_0\ (\textit{homogeneous quadratic} \text{ in the } \Psi\text{'s})$$

and

$$\partial R/\partial q_p = \partial L/\partial q_p,$$

the above equations can be rewritten as

$$(\partial R/\partial q_p)_o = (\partial L/\partial q_p)_o = 0 \qquad [(\ldots)_o \equiv (\ldots)|_{\dot\psi=c,\,q=s}],$$

expressing q's $\equiv s$'s in terms of the arbitrarily chosen Ψ's $\equiv C$'s. The $\dot\psi$'s can then be found from the *second* (Hamiltonian) group of Routh's equations:

$$d\psi_i/dt = -(\partial R/\partial \Psi_i)_o = -(\partial R_o/\partial \Psi_i)_o = -(\partial T''_{0,2}/\partial \Psi_i)_o$$
$$= \sum C_{ij} C_j = \text{constant} \equiv c_i \qquad [\text{with } \dot{q}_p = 0]$$
$$= \text{Function of the } s\text{'s and the (arbitrarily chosen) } C\text{'s},$$
$$\Rightarrow \psi_i(t) = -c_i(t - t_{\text{initial}}) + \psi_{i,\text{initial}}$$
$$= \text{Function of the } s\text{'s and the (now) arbitrarily chosen } c_i\text{'s and } \psi_{\text{initial}}\text{'s};$$

i.e., *in steady motion, the cyclic coordinates vary linearly with time.*

If we initially choose arbitrarily the Ψ's, then the above equations relate them to the q's. If, on the other hand, we choose the $\dot\psi$'s $\equiv c$'s, then, to relate them directly to the q's: first, we take $T''_{0,2}$, and, using $\Psi_i = \sum c_{ji}\dot\psi_j$, change it to a *homogeneous quadratic* function in the $\dot\psi$'s (with i, j, j', j'': $1, \ldots, M$):

$$2T''_{0,2} \equiv 2T''_{\Psi\Psi} \equiv -\sum\sum C_{ji}\Psi_j\Psi_i \qquad \left[\text{recalling that } \sum C_{ji}c_{j'j} = \delta_{ij'}\right]$$
$$= \cdots = -\sum\sum c_{ij}\dot\psi_i\dot\psi_j = 2T''_{\dot\psi\dot\psi} = -2T_{\dot\psi\dot\psi};$$

or, since

$$\partial T''_{\Psi\Psi}/\partial q_p = -(\partial T''_{\dot\psi\dot\psi}/\partial q_p) = \partial T_{\dot\psi\dot\psi}/\partial q_p,$$

we can, finally, replace the steady motion conditions by

$$-(\partial T''_{\dot\psi\dot\psi}/\partial q_p) = \partial V/\partial q_p, \qquad \text{or} \qquad \partial T_{\dot\psi\dot\psi}/\partial q_p = \partial V/\partial q_p,$$

relating the q's to the $\dot\psi$'s; and, using $\Psi_i = \sum c_{ji}\dot\psi_j$, we can relate both to the Ψ's.

VARIATION OF CONSTANTS (OR PARAMETERS)

Theorem of Lagrange–Poisson:

Equations of motion:

$$dp_k/dt = f_k(t, q, p) \qquad \text{and} \qquad dq_k/dt = g_k(t, q, p),$$
$$[f_k = -\partial H/\partial q_k + Q_k \qquad \text{and} \qquad g_k = \partial H/\partial p_k, \text{ for a Hamiltonian system}];$$

general solutions:

$$p_k = p_k(t; c) \qquad \text{and} \qquad q_k = q_k(t; c),$$

where

$$c \equiv (c_1, \ldots, c_{2n}) \equiv (c_\nu; \nu = 1, \ldots, 2n)\text{: constants of integration.}$$

Adjacent trajectory, $II = I + \delta(I)$,

$$\delta p_k = \sum (\partial p_k/\partial c_\nu)\, \delta c_\nu \qquad \text{and} \qquad \delta q_k = \sum (\partial q_k/\partial c_\nu)\, \delta c_\nu.$$

Linear *variational*, or perturbational, equations:

$$(\delta p_k)^{\cdot} = \delta(\dot{p}_k) = \sum [(\partial f_k/\partial p_l)\, \delta p_l + (\partial f_k/\partial q_l)\, \delta q_l],$$
$$(\delta q_k)^{\cdot} = \delta(\dot{q}_k) = \sum [(\partial g_k/\partial p_l)\, \delta p_l + (\partial g_k/\partial q_l)\, \delta q_l].$$

Then, for a Hamiltonian system,

$$d/dt\Big(\sum (\partial_1 p_k\, \delta_2 q_k - \delta_2 p_k\, \delta_1 q_k)\Big) = \sum (\delta_1 Q_k\, \delta_2 q_k - \delta_2 Q_k\, \delta_1 q_k).$$

Theorem of Lagrange–Poisson: In a holonomic and potential (i.e., $Q_k = 0$, or $\partial Q_k/\partial q_l = \partial Q_l/\partial q_k$, for all k, $l = 1, \ldots, n$), but possibly rheonomic, system, the bilinear expression

$$I \equiv \sum (\delta_1 p_k\, \delta_2 q_k - \delta_2 p_k\, \delta_1 q_k)$$

is *time-independent*; that is, it is a constant of the motion.

Lagrange's brackets (LB):

$$I = \sum \sum [c_\mu, c_\nu]\, \delta_1 c_\mu\, \delta_2 c_\nu,$$

where

$$[c_\mu, c_\nu] \equiv \sum [(\partial p_k/\partial c_\mu)(\partial q_k/\partial c_\nu) - (\partial p_k/\partial c_\nu)(\partial q_k/\partial c_\mu)]$$
$$= \textit{Lagrangean bracket} \text{ of } c_\mu, c_\nu.$$

Properties of LB:

$$[c_\mu, c_\mu] = 0; \qquad [c_\mu, c_\nu] = -[c_\nu, c_\mu];$$
$$\partial[c_\mu, c_\nu]/\partial c_\lambda + \partial[c_\nu, c_\lambda]/\partial c_\mu + \partial[c_\lambda, c_\mu]/\partial c_\nu = 0,$$
$$[c_\mu, c_\nu] = \partial/\partial c_\nu\Big(\sum q_k(\partial p_k/\partial c_\mu)\Big) - \partial/\partial c_\mu\Big(\sum q_k(\partial p_k/\partial c_\nu)\Big).$$

PERTURBATION EQUATIONS

Unperturbed problem and its solution

$$dp_k/dt = -\partial H/\partial q_k, \qquad dq_k/dt = \partial H/\partial p_k; \qquad p_k = p_k(t; c), \qquad q_k = q_k(t; c)$$

Slightly perturbed problem

$$dp_k/dt = -\partial H/\partial q_k + X_k, \qquad dq_k/dt = \partial H/\partial p_k,$$

where

$$X_k = X_k(t, q, p) = \text{given function of its arguments}$$
$$\approx X_k^{(1)}(t; c) \text{ [first-order approximation, upon substitution of } \textit{unperturbed} \text{ solution in it]}$$

$2n$ first-order differential equations for the $c_\mu = constant \rightarrow c_\mu(t)$:

$$\sum (\partial p_k/\partial c_\mu)(dc_\mu/dt) = X_k^{(1)}, \qquad \sum (\partial q_k/\partial c_\mu)(dc_\mu/dt) = 0.$$

Lagrangean form of the perturbation equations:

$$\sum [c_\nu, c_\mu](dc_\nu/dt) = \sum X_k^{(1)}(\partial q_k/\partial c_\mu).$$

If the perturbations are *potential*—that is, if $X_k = -\partial\Omega/\partial q_k$—then, since $q_k = q_k(t; c)$, the above specializes to

$$\sum [c_\nu, c_\mu](dc_\nu/dt) = -\partial\Omega/\partial c_\mu.$$

Inverting, we obtain

$$c_\mu = h_\mu(t, q, p) = \text{first integral (constant) of the unperturbed problem,}$$
$$dc_\mu/dt = \sum (\partial h_\mu/\partial p_k)X_k = \sum (\partial c_\mu/\partial p_k)X_k^{(1)}.$$

Poisson's brackets. If the perturbations are *potential*—that is, if

$$X_k = -\partial\Omega/\partial q_k = -\sum (\partial\Omega/\partial c_\nu)(\partial c_\nu/\partial q_k),$$

then

$$dc_\mu/dt = -\sum (\partial\Omega/\partial c_\nu)(c_\mu, c_\nu),$$

where

$$(c_\mu, c_\nu) \equiv \sum [(\partial c_\mu/\partial p_k)(\partial c_\nu/\partial q_k) - (\partial c_\mu/\partial q_k)(\partial c_\nu/\partial p_k)]$$
$$= \textit{Poisson bracket} \text{ of } c_\mu, c_\nu.$$

Compatibility with LB:

$$\sum [c_\nu, c_\mu](c_\nu, c_\lambda) = \delta_{\mu\lambda}.$$

First-order corrections. Setting in $c_\mu = c_{\mu o} + c_{\mu 1}$, where $c_{\mu o}$ = unperturbed values and $c_{\mu 1}$ = corresponding first-order corrections, we have

$$dc_{\mu 1}/dt = -\sum (\partial\Omega_o/\partial c_{\nu o})(c_{\mu o}, c_{\nu o}) \qquad [\text{where } \Omega_o \equiv \Omega(c_o)].$$

Lagrange's result. Let

$$q_k = q_{k0} + q_{k1}t + q_{k2}t^2 + \cdots, \qquad p_k = p_{k0} + p_{k1}t + p_{k2}t^2 + \cdots.$$

Then, with

$$c_k = q_{k0} \qquad \text{and} \qquad c_{n+l} = p_{l0} \qquad (k, l = 1, \ldots, n),$$

the perturbation equations assume the canonical form:

$$dc_k/dt = \partial\Omega/\partial c_{n+k}, \qquad dc_{n+k}/dt = -\partial\Omega/\partial c_k \qquad (k = 1, \ldots, n).$$

CANONICAL TRANSFORMATIONS

Transformations

$$q = q(t, q', p') \leftrightarrow q' = q'(t, q, p); \qquad p = p(t, q', p') \leftrightarrow p' = p'(t, q, p),$$

[with nonvanishing Jacobian $|\partial(q', p')/\partial(q, p)|$] that leave Hamilton's equations form invariant.

Requirements:

$$\begin{aligned} L\,dt &= L'\,dt + dF \\ &\Rightarrow \sum p_k\,dq_k - H\,dt = \sum p_{k'}\,dq_{k'} - H'\,dt + dF, \\ &\Rightarrow \sum p_k\,dq_k - \sum p_{k'}\,dq_{k'} = (H - H')\,dt + dF, \end{aligned}$$

where F is the *generating function* of the transformation (an arbitrary differentiable function of the coordinates, momenta, and time); and H' satisfies the Hamiltonian equations in the new variables.

Alternatively,

$$\sum p_k\,dq_k - H\,dt = df(t, q, p) \qquad \text{and} \qquad \sum p_{k'}\,dq_{k'} - H'\,dt = df'(t, q', p'),$$

$$\begin{aligned} &\Rightarrow \sum p_k\,dq_k - \sum p_{k'}\,dq_{k'} - (H - H')\,dt \\ &= df(t, q, p) - df'(t, q', p') \equiv dF. \end{aligned}$$

Virtual form of a canonical transformation:

$$\sum p_k\,\delta q_k - \sum p_{k'}\,\delta q_{k'} = \delta F.$$

Forms of F and their relations with the corresponding conjugate variables:

$$
\begin{array}{llll}
F = F_1(t,q,q'): & p_k = \partial F_1/\partial q_k, & p_{k'} = -\partial F_1/\partial q_{k'}; & H' = H + \partial F_1/\partial t; \\
F = F_2(t,q,p'): & p_k = \partial F_2/\partial q_k, & q_{k'} = \partial F_2/\partial p_{k'}; & H' = H + \partial F_2/\partial t; \\
F = F_3(t,p,q'): & q_k = -\partial F_3/\partial p_k, & p_{k'} = -\partial F_3/\partial q_{k'}; & H' = H + \partial F_3/\partial t; \\
F = F_4(t,p,p'): & q_k = -\partial F_4/\partial p_k, & q_{k'} = \partial F_4/\partial p_{k'}; & H' = H + \partial F_4/\partial t;
\end{array}
$$

$$F_2 = F_1 + \sum p_{k'} q_{k'},$$

$$F_3 = F_1 - \sum p_k q_k,$$

$$F_4 = F_1 + \sum p_{k'} q_{k'} - \sum p_k q_k = F_2 - \sum p_k q_k = F_3 + \sum p_{k'} q_{k'}.$$

POISSON'S BRACKETS (PB) AND CANONICITY CONDITIONS

The PB of f, g (where f, g, h are arbitrary differentiable dynamical quantities) is

$$(f,g) \equiv \sum [(\partial f/\partial p_k)(\partial g/\partial q_k) - (\partial f/\partial q_k)(\partial g/\partial p_k)] \equiv \sum \partial(f,g)/\partial(p_k, q_k).$$

Then

$$df/dt = \partial f/\partial t + (H,f) + \sum (\partial f/\partial p_k) Q_k;$$

and so for f to be an *integral of the motion*, we must have

$$\partial f/\partial t + \sum (\partial f/\partial p_k) Q_k + (H,f) = 0 \;\Rightarrow\; (H,f) = 0, \quad \text{if } f = f(q,p) \text{ and } Q_k = 0,$$

that is, *its PB with the Hamiltonian of its variables must be zero.*

[Remarks on notation: A number of authors define PBs as the *opposite* of ours; that is, as

$$(f,g) \equiv \sum [(\partial f/\partial q_k)(\partial g/\partial p_k) - (\partial f/\partial p_k)(\partial g/\partial q_k)].$$

Therefore, a certain caution should be exercised when comparing references. Also, others denote our Lagrangean brackets, [...], by {...}; and our Poisson brackets, (...), by [...].]

Properties/theorems of PBs

$(f,g) = -(g,f) = (-g,f) \;\Rightarrow f,f) = 0$	(anti-symmetry)
$(f,c) = 0$	(c = a constant)
$(f_1 + f_2, g) = (f_1, g) = (f_2, g)$	(distributivity)
$(f_1 f_2, g) = f_1(f_2, g) + f_2(f_1, g)$	
$\Rightarrow (cf, g) = c(f,g)$	(c = a constant)
$\Rightarrow$ If $f = \sum c_k f_k$, then $(f,g) = \sum c_k (f_k, g)$	(c_k = constants)

$\partial/\partial t(f,g) = (\partial f/\partial t, g) + (f, \partial g/\partial t)$ ("Leibniz rule")

[Actually, $\partial/\partial x(f,g) = (\partial f/\partial x, g) + (f, \partial g/\partial x)$; $x =$ any variable]

$(f, q_k) = \partial f/\partial p_k,$

$(f, p_k) = -\partial f/\partial q_k,$

$(q_k, q_l) = 0,$

$(p_k, p_l) = 0,$

$(p_k, q_l) = \delta_{kl}$ (= *Kronecker delta*).

[The last three types of brackets are called *fundamental*, or *basic*, PB]

$(f,(g,h)) + (g,(h,f)) + (h,(f,g)) = 0,$

$((f,g),h) + ((g,h),f) + ((h,f),g) = 0$ (Poisson–Jacobi identity)

Theorem of Poisson–Jacobi: If f and g are any two integrals of the motion, so is their PB; that is, if $f = c_1$ and $g = c_2$, then $(f,g) = c_3$ ($c_{1,2,3}$ = constants).

Theorem: The *PBs are invariant under CT*; that is, $(f,g)_{q,p} = (f,g)_{q',p'} = \cdots$; where f and g keep their *value*, but not necessarily their *form*, in the various canonical coordinates involved.

Canonicity conditions via PB

$$[p_{l'}, p_{k'}] = 0, \qquad [q_{l'}, q_{k'}] = 0, \qquad [p_{k'}, q_{l'}] = \delta_{kl},$$
$$(p_{l'}, p_{k'}) = 0, \qquad (q_{l'}, q_{k'}) = 0, \qquad (p_{l'}, q_{k'}) = \delta_{lk},$$

since *both* Poisson and Lagrange brackets are canonically invariant.

Theorem of Jacobi

(i) The integration of the canonical equations

$$dq_k/dt = \partial H/\partial p_k, \qquad dp_k/dt = -\partial H/\partial q_k,$$

is reduced to the integration of the Hamilton–Jacobi equation ($H - J$):

$$H(t, q, \partial A/\partial q) + \partial A/\partial t = 0,$$

$A = A(t, q, p')$: generating function (Hamiltonian action).

(ii) If we have a complete solution of $H - J$; that is, a solution of the form

$$A = A(t; q_1, \ldots, q_n; \beta_1, \ldots, \beta_n) \equiv A(t; q, \beta),$$

where $\beta \equiv (\beta_1, \ldots, \beta_n) = n$ essential arbitrary constants, and $|\partial^2 A/\partial q\, \partial\beta| \neq 0$ (non-vanishing Jacobian), then the solution of the algebraic system:

$\partial A/\partial \beta_k = \alpha_k$

[Finite equations of motion, α: new arbitrary constants $\Rightarrow q_k = q_k(t, \alpha, \beta)$],

$\partial A/\partial q_k = p_k$

$[\Rightarrow p_k = p_k(t, \alpha, \beta)$: canonically conjugate (finite) equations of motion],

constitutes a complete solution of the canonical equations. Schematically, these are as follows.

Hamilton: *Differential* equations of motion:

$$dq/dt = \partial H/\partial p, \qquad dp/dt = -\partial H/\partial q$$

(If these equations can be integrated, an action function can be obtained)

Hamilton–Jacobi:

$$H(t, q, \partial A/\partial q) + \partial A/\partial t = 0 \Rightarrow A = A(t, q, \beta)$$

Jacobi: *Finite* equations of motion:

$$\partial A/\partial \beta = \alpha \rightarrow q = q(t, \alpha, \beta);$$
$$\partial A/\partial q = p \rightarrow p = p(t, \alpha, \beta)$$

(If an action function can be obtained, then Hamilton's equations can be integrated.)

1

Background

Basic Concepts and Equations of Particle and Rigid-Body Mechanics

> Therefore it would seem right that any systematic treatment of classical dynamics should start with axioms carefully laid down, on which the whole structure would rest as a house rests on its foundations. The analogy to a house is, however, a false one. *Theories are created in mid-air, so to speak, and develop both upward and downward.* Neither process is ever completed. Upward, the ramifications can extend indefinitely, downward, the axiomatic base must be rebuilt continually as our views change as to what constitutes logical precision. Indeed, there is little promise of finality here, as we seem to be moving towards the idea that logic is a man-made thing, a game played according to rules to some extent arbitrary.
>
> (Synge, 1960, p. 5, emphasis added)

In this chapter we summarize, without detailed proofs and/or elaborate discussions, in a *handbook* (*not textbook*) fashion, like a first-aid kit, but in a hopefully accurate and serviceable form, the basic concepts, definitions, axioms, and theorems of "elementary" (or momentum/Newton–Euler, or general) *theoretical mechanics*. This compact, highly selective, perhaps nonhomogeneous, and unavoidably incomplete account should help to establish a common background with readers, and thus enhance their understanding of the rest of this relatively self-contained book.

For complementary reading, we recommend (alphabetically):

Fox (1967): one of the best, and most economically written, U.S. texts on general elementary–intermediate mechanics.

Hamel (1909), (1912, 1st ed.; 1922, 2nd ed.): arguably the best text on elementary mechanics written to date), (1927), (1949).

Hund (1972): concise, insightful.

Langner (1996–1997): dense, clear; "best buy."

Loitsianskii and Lur'e (1982, 1983): excellent.

Marcolongo (1905, 1911/1912): rigorous, comprehensive.

Milne (1948): interesting vectorial treatment of rigid dynamics.

Parkus (1966): an educational classic.

Synge and Griffith (1959): clear, reliable.

Synge (1960): comprehensive, encyclopedic, mature.

Winkelmann (1929, 1930): concise, comprehensive.

Papastavridis: Elementary Mechanics (EM for short), in preparation: encyclopedia/handbook of undergraduate mechanics; includes elements of *continuum* mechanics.

Additional references, at particular sections, and so on, will also be given, as deemed beneficial.

1.1 VECTOR AND (CARTESIAN) TENSOR ALGEBRA

Vectors: Basic Concepts/Definitions and Algebra

Geometrically, vectors are straight line segments that, in the most general case, have the following *five* characteristics: (i) *length*, (ii) *direction*, (iii) *sense*, (iv) *line of action* (or *carrier*), and (v) *origin* (or *point of application*) on carrier; (iv) and (v) can be replaced with *spatial origin*. Also, vectors obey the well-known *parallelogram law* of addition ($\Rightarrow$ commutativity); that is, not all line segments with characteristics (i)–(v) are vectors (e.g., finite rotations, §1.10). Next, if only characteristics (i)–(iii) matter, but (iv) and (v) do not, the vector is called *free*; if characteristics (i)–(iv) matter, but (v) does not, the vector is called *line bound* or *sliding*; and if all five characteristics matter, the vector is called *point bound*. As a rule, the *vectors* of *continuum mechanics* and the *system vectors* of *analytical mechanics* (chap. 2 ff.) are *point bound*; while those of rigid-body mechanics are *line bound*.

Notation for vectors: $\boldsymbol{a}, \boldsymbol{b}, \ldots$ (bold italic).

Length, or magnitude, or modulus, or intensity, or norm, of $\boldsymbol{a}$: $|\boldsymbol{a}| \equiv a \geq 0$. If $a = 0$, the vector is called *null*; if $a = 1$, the vector is called *unit* (or *normalized*).

The physical space of classical mechanics is a three-dimensional Euclidean *point* space, denoted by E_3 or E; while the associated (also Euclidean) *vector* space is denoted by $\boldsymbol{E}_3$ or $\boldsymbol{E}$.

An *orthonormal basis* (i.e., one whose vectors are unit *and* mutually orthogonal—see below)

$$\{\boldsymbol{u}_1, \boldsymbol{u}_2, \boldsymbol{u}_3\} \equiv \{\boldsymbol{u}_{1,2,3}\} \equiv \{\boldsymbol{u}_k; k = 1, 2, 3\} \equiv \{\boldsymbol{u}_k\}$$
$$\equiv \{\boldsymbol{u}_x, \boldsymbol{u}_y, \boldsymbol{u}_z\} \equiv \{\boldsymbol{u}_{x,y,z}\} \equiv \{\boldsymbol{i}, \boldsymbol{j}, \boldsymbol{k}\}, \tag{1.1.1}$$

together with an "origin," O, make up a (local) rectangular Cartesian *frame*: $\{O, \boldsymbol{u}_k\}$. If the origin is not important, we simply write $\{\boldsymbol{u}_k\}$.

[Since E is *flat* (noncurved), a single such frame, and associated rectilinear and mutually rectangular axes of coordinates O–$123 \equiv O$–xyz, can be extended to cover, or represent, the entire space: *local* frame $\rightarrow$ *global* frame. For details, see, for example, Papastavridis (1999, pp. 84–91, 211–218), or Lur'e (1968, p. 807).]

In such a basis, a vector $\boldsymbol{a}$ can be represented by its rectangular Cartesian *components*

$$\{a_1, a_2, a_3\} \equiv \{a_{1,2,3}\} \equiv \{a_k; k = 1, 2, 3\} \equiv \{a_k\} \equiv \{a_x, a_y, a_z\} \equiv \{a_{x,y,z}\}, \tag{1.1.2a}$$

or

$$\boldsymbol{a} = a_1 \boldsymbol{u}_1 + a_2 \boldsymbol{u}_2 + a_3 \boldsymbol{u}_3 = a_x \boldsymbol{u}_x + a_y \boldsymbol{u}_y + a_z \boldsymbol{u}_z = \sum a_k \boldsymbol{u}_k. \tag{1.1.2b}$$

In terms of the famous Einsteinian *summation convention* [= lone, or *free*, subscripts range over the integers 1, 2, 3, or x, y, z, while *summation* is implied over repeated (i.e., pairs) of *subscripts*], we can simply write $\boldsymbol{a} = a_k \boldsymbol{u}_k$. In this book, however, and for reasons that will gradually become clear (chap. 2), we shall *NOT* use this convention!

Dotting (1.1.2b) with $\boldsymbol{u}_k$, and noting the *six orthonormality* (metric!) conditions or constraints:

$$\boldsymbol{u}_k \cdot \boldsymbol{u}_l\text{: } \textit{scalar}\text{, or } \textit{dot}\text{, or } \textit{inner}\text{, product of } \boldsymbol{u}_k, \boldsymbol{u}_l = \delta_{kl} = \delta_{lk} \text{ (Kronecker delta)}$$
$$= 1 \quad \text{if } k = l, \qquad = 0 \quad \text{if } k \neq l \qquad (k, l = 1, 2, 3, \text{ or } x, y, z), \tag{1.1.3}$$

in extenso:

$$\boldsymbol{i} \cdot \boldsymbol{j} = \boldsymbol{j} \cdot \boldsymbol{j} = \boldsymbol{k} \cdot \boldsymbol{k} = 1 \text{ (normality)}, \qquad \boldsymbol{i} \cdot \boldsymbol{j} = \boldsymbol{i} \cdot \boldsymbol{k} = \boldsymbol{j} \cdot \boldsymbol{k} = 0 \text{ (orthogonality)}. \tag{1.1.3a, b}$$

we obtain the following expression for the $\boldsymbol{a}$-components:

$$a_k = \boldsymbol{a} \cdot \boldsymbol{u}_k. \tag{1.1.2c}$$

In such a basis, the dot product of two vectors $\boldsymbol{a}$ and $\boldsymbol{b}$ is expressed as

$$\boldsymbol{a} \cdot \boldsymbol{b} = \boldsymbol{b} \cdot \boldsymbol{a} = \left(\sum a_k \boldsymbol{u}_k\right) \cdot \left(\sum b_l \boldsymbol{u}_l\right) = \cdots = \sum a_k b_k. \tag{1.1.4}$$

For $\boldsymbol{a} = \boldsymbol{b}$, the above yields the *length*, or *norm*, or *magnitude*, of $\boldsymbol{a}$:

$$N(\boldsymbol{a}) \equiv a = |\boldsymbol{a}| = (\boldsymbol{a} \cdot \boldsymbol{a})^{1/2} = \left(\sum a_k a_k\right)^{1/2} \geq 0 \qquad \text{(this book)}. \tag{1.1.5}$$

The basis $\{\boldsymbol{u}_{1,2,3}\}$ is called O$_{\text{rtho}}$N$_{\text{ormal}}$D$_{\text{extral (i.e., right-handed)}}$ ≡ OND, if, in addition to (1.1.3), it satisfies

$$\boldsymbol{u}_k \cdot (\boldsymbol{u}_r \times \boldsymbol{u}_s) \equiv (\boldsymbol{u}_k, \boldsymbol{u}_r, \boldsymbol{u}_s) \equiv \varepsilon_{krs} \text{ (permutation symbol, or alternator, of Levi–Civita)}$$
$$= +1/-1/0 \quad \text{according as } k, r, s \text{ are an } \textit{even/odd/no} \text{ permutation of } 1, 2, 3$$
$$[\text{i.e., } \varepsilon_{123} = \varepsilon_{231} = \varepsilon_{312} = +1, \quad \varepsilon_{132} = \varepsilon_{213} = \varepsilon_{321} = -1,$$
$$\varepsilon_{112} = \varepsilon_{122} = \varepsilon_{313} = \cdots = 0 \text{ (two or more indices equal)}], \tag{1.1.6}$$

or, equivalently, if

$$\boldsymbol{u}_r \times \boldsymbol{u}_s = \sum \varepsilon_{rsk} \boldsymbol{u}_k = \sum \varepsilon_{krs} \boldsymbol{u}_k \Leftrightarrow \boldsymbol{u}_k = (1/2) \sum \sum \varepsilon_{krs} (\boldsymbol{u}_r \times \boldsymbol{u}_s); \tag{1.1.6a}$$

that is, $(\boldsymbol{u}_r \times \boldsymbol{u}_s)_k = \varepsilon_{rsk}$; otherwise $\{\boldsymbol{u}_{1,2,3}\}$ is *left-handed*, or *sinister*, in which case $(\boldsymbol{u}_k, \boldsymbol{u}_r, \boldsymbol{u}_s) \equiv -\varepsilon_{krs}$. Henceforth, only OND bases will be used.

- The symbols of Kronecker and Levi–Civita are connected by the following "**ed** identity":

$$\sum \varepsilon_{krs} \varepsilon_{lms} = \sum \varepsilon_{skr} \varepsilon_{slm} = \delta_{kl} \delta_{rm} - \delta_{km} \delta_{rl}; \tag{1.1.6b}$$

which, for $r = m$ (and then summation over repeated subscripts), produces

$$\sum\sum \varepsilon_{krs}\varepsilon_{lrs} = 2\delta_{kl}; \tag{1.1.6c}$$

and this, for $k = l$, etc., yields

$$\sum\sum\sum \varepsilon_{krs}\varepsilon_{krs} = 2\left(\sum \delta_{kk}\right) = 2(3) = 6. \tag{1.1.6d}$$

- The dextrality of the orthonormal basis $(\boldsymbol{i}, \boldsymbol{j}, \boldsymbol{k})$ (i.e., $\boldsymbol{i} \times \boldsymbol{i} = \boldsymbol{j} \times \boldsymbol{j} = \boldsymbol{k} \times \boldsymbol{k} = \boldsymbol{0}$), is expressed by

$$\boldsymbol{i} \times \boldsymbol{j} = -(\boldsymbol{j} \times \boldsymbol{i}) = \boldsymbol{k}, \ \ \boldsymbol{j} \times \boldsymbol{k} = -(\boldsymbol{k} \times \boldsymbol{j}) = \boldsymbol{i}, \ \ \boldsymbol{k} \times \boldsymbol{i} = -(\boldsymbol{i} \times \boldsymbol{k}) = \boldsymbol{j}. \tag{1.1.6e}$$

With the help of the above, we express the *vector*, or *cross*, or *outer*, product of $\boldsymbol{a}$ and $\boldsymbol{b}$ as

$$\boldsymbol{a} \times \boldsymbol{b} = -(\boldsymbol{b} \times \boldsymbol{a}) = \sum\sum\sum \varepsilon_{klr} a_k b_l \boldsymbol{u}_r, \tag{1.1.7a}$$

that is,

$$(\boldsymbol{a} \times \boldsymbol{b})_r = \sum\sum \varepsilon_{klr} a_k b_l = \sum\sum \varepsilon_{rkl} a_k b_l. \tag{1.1.7b}$$

It can be shown that

$$|\boldsymbol{u}_1 \times \boldsymbol{u}_2|^2 = |\boldsymbol{u}_2 \times \boldsymbol{u}_3|^2 = |\boldsymbol{u}_3 \times \boldsymbol{u}_1|^2 = (\boldsymbol{u}_1, \boldsymbol{u}_2, \boldsymbol{u}_3)^2 = +1, \tag{1.1.8a}$$

where

$$\begin{aligned}(\boldsymbol{a}, \boldsymbol{b}, \boldsymbol{c}) &\equiv \boldsymbol{a} \cdot (\boldsymbol{b} \times \boldsymbol{c}) = \boldsymbol{b} \cdot (\boldsymbol{c} \times \boldsymbol{a}) = \boldsymbol{c} \cdot (\boldsymbol{a} \times \boldsymbol{b}) \\ &= (\boldsymbol{a} \times \boldsymbol{b}) \cdot \boldsymbol{c} = (\boldsymbol{b} \times \boldsymbol{c}) \cdot \boldsymbol{a} = (\boldsymbol{c} \times \boldsymbol{a}) \cdot \boldsymbol{b} \\ &= \sum\sum\sum \varepsilon_{krs} a_k b_r c_s:\end{aligned} \tag{1.1.8b}$$

[+, if $(\boldsymbol{a}, \boldsymbol{b}, \boldsymbol{c})$ is *right*; −, if $(\boldsymbol{a}, \boldsymbol{b}, \boldsymbol{c})$ is *left*; 0, if $(\boldsymbol{a}, \boldsymbol{b}, \boldsymbol{c})$ are *coplanar* or *zero*]: *scalar* triple product of $\boldsymbol{a}$, $\boldsymbol{b}$, $\boldsymbol{c}$ = *signed volume of parallelepiped* having $\boldsymbol{a}$, $\boldsymbol{b}$, $\boldsymbol{c}$ as sides; also

$$\begin{aligned}&[\boldsymbol{a}, \boldsymbol{b}, \boldsymbol{c}] \equiv \boldsymbol{a} \times (\boldsymbol{b} \times \boldsymbol{c}) = (\boldsymbol{a} \cdot \boldsymbol{c})\boldsymbol{b} - (\boldsymbol{a} \cdot \boldsymbol{b})\boldsymbol{c} \\ &\qquad [\neq (\boldsymbol{a} \times \boldsymbol{b}) \times \boldsymbol{c} = -\boldsymbol{c} \times (\boldsymbol{a} \times \boldsymbol{b}) = (\boldsymbol{a} \cdot \boldsymbol{c})\boldsymbol{b} - (\boldsymbol{b} \cdot \boldsymbol{c})\boldsymbol{a}]:\end{aligned} \tag{1.1.8c}$$

vector triple product of $\boldsymbol{a}$, $\boldsymbol{b}$, $\boldsymbol{c}$.

The *dyadic*, or *direct*, or *open*, or *tensor* product of two vectors $\boldsymbol{a}$ and $\boldsymbol{b}$,

$$\boldsymbol{a}\boldsymbol{b} \equiv \boldsymbol{a} \otimes \boldsymbol{b} \qquad (\neq \boldsymbol{b} \otimes \boldsymbol{a}, \text{ in general}), \tag{1.1.9a}$$

is defined as (the *tensor* — see below):

$$\boldsymbol{a}\boldsymbol{b} \equiv \boldsymbol{a} \otimes \boldsymbol{b} = \left(\sum a_k \boldsymbol{u}_k\right) \otimes \left(\sum b_l \boldsymbol{u}_l\right) = \sum\sum a_k b_l (\boldsymbol{u}_k \otimes \boldsymbol{u}_l). \tag{1.1.9b}$$

- This product can also be defined as the tensor that assigns to each vector $\boldsymbol{x}$ the vector $\boldsymbol{a}(\boldsymbol{b} \cdot \boldsymbol{x})$:

$$(\boldsymbol{a} \otimes \boldsymbol{b}) \cdot \boldsymbol{x} = \boldsymbol{a}\,(\boldsymbol{b} \cdot \boldsymbol{x}) = (\boldsymbol{b} \cdot \boldsymbol{x})\boldsymbol{a}, \tag{1.1.9c}$$

and also

$$\boldsymbol{x} \cdot (\boldsymbol{a} \otimes \boldsymbol{b}) = (\boldsymbol{x} \cdot \boldsymbol{a})\boldsymbol{b} = \boldsymbol{b}\,(\boldsymbol{x} \cdot \boldsymbol{a}). \tag{1.1.9d}$$

In components, these read, respectively,

$$(\boldsymbol{a} \otimes \boldsymbol{b}) \cdot \boldsymbol{x} = \sum\sum (a_k b_l x_l)\boldsymbol{u}_k, \qquad \boldsymbol{x} \cdot (\boldsymbol{a} \otimes \boldsymbol{b}) = \sum\sum (x_l a_l b_k)\boldsymbol{u}_k. \tag{1.1.9e}$$

- It can be shown that

$$[\boldsymbol{a}, \boldsymbol{b}, \boldsymbol{c}] = [(\boldsymbol{b} \otimes \boldsymbol{c}) - (\boldsymbol{c} \otimes \boldsymbol{b})] \cdot \boldsymbol{a}. \tag{1.1.8d}$$

Tensors: Basic Concepts/Definitions and Algebra

[For a detailed classical treatment of general tensors, see, for example, Papastavridis (1999).]

A second-order (or rank) *tensor* (or *dyadic*, from the Greek ΔYO = *two*) or, here, simply tensor $\boldsymbol{T}$ (bold, in italics *or* roman) is defined as a linear transformation from V to V; or as a linear mapping assigning to each vector $\boldsymbol{a}$ another vector $\boldsymbol{b}$:

$$\boldsymbol{b} = \boldsymbol{T} \cdot \boldsymbol{a}, \tag{1.1.10a}$$

or in components

$$\sum b_k \boldsymbol{u}_k = \sum\sum T_{kl} a_l \boldsymbol{u}_k \;\Rightarrow\; b_k = \sum T_{kl} a_l, \tag{1.1.10b}$$

or as

$$\boldsymbol{b} = \boldsymbol{a} \cdot \boldsymbol{T} = \sum\sum a_k T_{kl} \boldsymbol{u}_l \;\Rightarrow\; b_l = \sum T_{kl} a_k,$$

where

$$T_{kl} \equiv \boldsymbol{u}_k \cdot (\boldsymbol{T} \cdot \boldsymbol{u}_l) = (\boldsymbol{T} \cdot \boldsymbol{u}_l) \cdot \boldsymbol{u}_k = \boldsymbol{T} \cdot (\boldsymbol{u}_k \otimes \boldsymbol{u}_l), \tag{1.1.10c}$$

are the Cartesian components of $\boldsymbol{T}$ (see tensor *products*, below). Alternatively, a *vector/tensor//(n)th order tensor* associates a *scalar/vector//(n − 1)th order tensor* with each spatial direction $\boldsymbol{u}_d = (u_{(d)k}$: direction cosines of unit vector $\boldsymbol{u}_d)$, via a *linear* and *homogeneous* expression in the $u_{(d)k}$; that is, for a (second-order) tensor:

$$\boldsymbol{T} \rightarrow \boldsymbol{v}_d = \boldsymbol{T} \cdot \boldsymbol{u}_d \quad \text{(direct notation)}, \qquad v_{(d)k} = \sum T_{kl} u_{(d)l} \quad \text{(component notation)}.$$

Thus (and in addition to the well-known 3 × 3 matrix form), $\boldsymbol{T}$ has the following representations:

$$\begin{aligned} \boldsymbol{T} &= \sum\sum T_{kl} \boldsymbol{u}_k \otimes \boldsymbol{u}_l && \text{(\textit{Dyadic} or \textit{nonion} representation)} \\ &= \sum \boldsymbol{u}_k \otimes \boldsymbol{t}_k, && \text{where } \boldsymbol{t}_k \equiv \sum T_{kl} \boldsymbol{u}_l, && (1.1.10\text{d}) \\ &= \sum \boldsymbol{t}'_l \otimes \boldsymbol{u}_l, && \text{where } \boldsymbol{t}'_l \equiv \sum T_{kl} \boldsymbol{u}_k. && (1.1.10\text{e}) \end{aligned}$$

The nine tensors $\{\boldsymbol{u}_k \otimes \boldsymbol{u}_l\}$ span the set of all (second-order) tensors; they form an orthonormal "tensor basis" there. If $T_{12} = T_{21} (= -T_{21})$, etc., then $\boldsymbol{T}$ is called *symmetric* (*anti-*, or *skew*-symmetric). Generally [see definition of *transpose*, $(\ldots)^{\mathrm{T}}$, below]:

Symmetric tensor: $\boldsymbol{T} = \boldsymbol{T}^{\mathrm{T}}, \quad T_{kl} = T_{lk};$ (1.1.11a)

Antisymmetric tensor: $\boldsymbol{T} = -\boldsymbol{T}^{\mathrm{T}}, \quad T_{kl} = -T_{lk} \quad (\Rightarrow T_{kk} = 0,$ no sum!) (1.1.11b)

Algebra of Tensors: Basic Operations

- *Sum/difference* of tensors $\boldsymbol{T}$ and $\boldsymbol{S}$:

$$\boldsymbol{T} \pm \boldsymbol{S} = \sum\sum (T_{kl} \pm S_{kl}) \boldsymbol{u}_k \otimes \boldsymbol{u}_l. \tag{1.1.12a}$$

- *Product* of $\boldsymbol{T}$ with a scalar (number) λ, $\lambda\boldsymbol{T}$:

$$\lambda\boldsymbol{T} = \sum\sum (\lambda T_{kl}) \boldsymbol{u}_k \otimes \boldsymbol{u}_l. \tag{1.1.12b}$$

- *Tensor product* of $\boldsymbol{T}$ and $\boldsymbol{S}$, $\boldsymbol{T}\cdot\boldsymbol{S}$, is defined by

$$\boldsymbol{T}\cdot\boldsymbol{S} = \sum\sum\sum T_{kr}S_{rl}\boldsymbol{u}_k \otimes \boldsymbol{u}_l \qquad (\neq \boldsymbol{S}\cdot\boldsymbol{T}, \text{ in general});$$

that is,

$$(\boldsymbol{T}\cdot\boldsymbol{S})_{kl} = \sum T_{kr}S_{rl} \neq (\boldsymbol{S}\cdot\boldsymbol{T})_{kl} = \sum S_{kr}T_{rl}. \tag{1.1.12c}$$

- *Inner*, or *dot*, *scalar* product of $\boldsymbol{T}$ and $\boldsymbol{S}$, $\boldsymbol{T} : \boldsymbol{S}$, is defined by (see *trace* below)

$$\begin{aligned}\boldsymbol{T}:\boldsymbol{S} &\equiv \sum\sum T_{kl}S_{kl} = Tr(\boldsymbol{T}\cdot\boldsymbol{S}^{\mathrm{T}}) \\ &= \sum\sum S_{kl}T_{kl} = Tr(\boldsymbol{S}\cdot\boldsymbol{T}^{\mathrm{T}}) \equiv \boldsymbol{S}:\boldsymbol{T},\end{aligned} \tag{1.1.12d}$$

where Tr means "trace of." If $\boldsymbol{T} = \boldsymbol{S}$,

$$T \equiv |\boldsymbol{T}| = (\boldsymbol{T}:\boldsymbol{T})^{1/2}: \textit{magnitude} \text{ of } \boldsymbol{T} (> 0, \text{ unless } \boldsymbol{T} = \boldsymbol{0}). \tag{1.1.12e}$$

If either of $\boldsymbol{T}$, $\boldsymbol{S}$ is *symmetric* (as is almost always the case in mechanics), then,

$$\begin{aligned}\boldsymbol{T}:\boldsymbol{S} &\equiv \sum\sum T_{kl}S_{kl} = \sum\sum T_{kl}S_{lk} \\ &\Big[= Tr(\boldsymbol{T}\cdot\boldsymbol{S}) \equiv \boldsymbol{T}\cdot\cdot\,\boldsymbol{S} \\ &= \sum\sum S_{lk}T_{kl} = Tr(\boldsymbol{S}\cdot\boldsymbol{T}) \equiv \boldsymbol{S}\cdot\cdot\,\boldsymbol{T}\Big].\end{aligned} \tag{1.1.12f}$$

In sum, we have defined the following *three* tensorial products:

$$(\boldsymbol{T}\cdot\boldsymbol{S})_{kl} = \sum T_{kr}S_{rl} \quad \text{(Tensor)},$$

$$\boldsymbol{T}:\boldsymbol{S} \equiv \sum\sum T_{kl}S_{kl} \quad \text{(Scalar)}, \qquad \boldsymbol{T}\cdot\cdot\,\boldsymbol{S} \equiv \sum\sum T_{kl}S_{lk} \quad \text{(Scalar)}. \tag{1.1.12g}$$

The reader should be warned that these notations are by no means uniform, and so caution should be exercised in comparing various references.

- *Transpose* of $\boldsymbol{T}$, $\boldsymbol{T}^{\mathrm{T}}$, is defined uniquely by

$$(\boldsymbol{T}\cdot\boldsymbol{a})\cdot\boldsymbol{b} = \boldsymbol{a}\cdot(\boldsymbol{T}^{\mathrm{T}}\cdot\boldsymbol{b}), \text{ for } all\ \boldsymbol{a}, \boldsymbol{b}. \tag{1.1.12h}$$

- *Trace* of $\boldsymbol{T}$ is defined by

$$Trace\ of\ \boldsymbol{T} \equiv Tr(\boldsymbol{T}) \equiv T_{11} + T_{22} + T_{33} \equiv \sum T_{kk}. \tag{1.1.12i}$$

- *Determinant* of $\boldsymbol{T}$ is defined by

$$Determinant\ of\ \boldsymbol{T} \equiv Det(\boldsymbol{T}) = Det(T_{kl}) \equiv |T_{kl}|. \tag{1.1.12j}$$

It can be shown that:

(i) $$Tr(\boldsymbol{T}) = Tr(\boldsymbol{T}^{\mathrm{T}}), \qquad Det(\boldsymbol{T}) = Det(\boldsymbol{T}^{\mathrm{T}}), \tag{1.1.12k}$$

(ii) For any two vectors $\boldsymbol{a}$ and $\boldsymbol{b}$:

$$(\boldsymbol{a}\otimes\boldsymbol{b})^{\mathrm{T}} = \boldsymbol{b}\otimes\boldsymbol{a}, \qquad Tr(\boldsymbol{a}\otimes\boldsymbol{b}) = \boldsymbol{a}\cdot\boldsymbol{b} = \sum a_k b_k, \qquad Det(\boldsymbol{a}\otimes\boldsymbol{b}) = 0, \tag{1.1.12l}$$

(iii) For any two tensors $\boldsymbol{T}$ and $\boldsymbol{S}$:

$$(\boldsymbol{T}\cdot\boldsymbol{S})^{\mathrm{T}} = \boldsymbol{S}^{\mathrm{T}}\cdot\boldsymbol{T}^{\mathrm{T}}, \qquad Tr(\boldsymbol{T}\cdot\boldsymbol{S}) = Tr(\boldsymbol{S}\cdot\boldsymbol{T}) = \boldsymbol{T}\cdot\cdot\boldsymbol{S}, \tag{1.1.12m}$$

$$Det(\boldsymbol{T}\cdot\boldsymbol{S}) = Det(\boldsymbol{T})\,Det(\boldsymbol{S}); \tag{1.1.12n}$$

also (in *three* dimensions):

$$Det(t\boldsymbol{T}) = t^3\,Det(\boldsymbol{T}), \text{ for any real number } t. \tag{1.1.12o}$$

- *Inverse* of $\boldsymbol{T}$, $\boldsymbol{T}^{-1}$, is defined uniquely by:

$$\boldsymbol{T}\cdot\boldsymbol{T}^{-1} = \boldsymbol{T}^{-1}\cdot\boldsymbol{T} = \boldsymbol{1}\ (unit\ \text{tensor}), \qquad [Det(\boldsymbol{T}) \neq 0]. \tag{1.1.12p}$$

From the above, we can easily deduce that

(i) $$Det(\boldsymbol{T}^{-1}) = (Det\ \boldsymbol{T})^{-1} \tag{1.1.12q}$$

(ii) $$(\boldsymbol{T}\cdot\boldsymbol{S})^{-1} = \boldsymbol{S}^{-1}\cdot\boldsymbol{T}^{-1} \qquad (\boldsymbol{T}, \boldsymbol{S}\text{: invertible}) \tag{1.1.12r}$$

(iii) $$d/dx(Det\ \boldsymbol{T}) = (Det\ \boldsymbol{T})Tr[(d\boldsymbol{T}/dx)\cdot\boldsymbol{T}^{-1}], \tag{1.1.12s}$$

where $\boldsymbol{T} = \boldsymbol{T}(x) = invertible$, $x = real\ parameter$, and $d\boldsymbol{T}/dx \equiv (dT_{kl}/dx)$.

- A tensor can be built from two vectors; but, in general, it cannot be decomposed into two vectors.
- Every tensor can be decomposed uniquely into a sum of a *symmetric* part (T'_{kl}) and an *antisymmetric* part (T''_{kl}):

$$T_{kl} = T'_{kl} + T''_{kl},$$

$$2T'_{kl} \equiv T_{kl} + T_{lk} = 2T'_{lk}, \qquad 2T''_{kl} \equiv T_{kl} - T_{lk} = -2T''_{lk}; \tag{1.1.13a}$$

that is,

$$\boldsymbol{T} = \boldsymbol{T}' + \boldsymbol{T}'', \qquad \boldsymbol{T}' = (\boldsymbol{T}')^{\mathrm{T}}, \qquad \boldsymbol{T}'' = -(\boldsymbol{T}'')^{\mathrm{T}}. \tag{1.1.13b}$$

- For any tensor $\boldsymbol{T}$ and any three vectors $\boldsymbol{a}$, $\boldsymbol{b}$, $\boldsymbol{c}$, the following identities hold:

(i) $\boldsymbol{a}\cdot(\boldsymbol{T}\cdot\boldsymbol{b}) = \boldsymbol{T}:(\boldsymbol{a}\otimes\boldsymbol{b}),\ \sum a_k\left(\sum T_{kl}b_l\right) = \sum\sum T_{kl}(a_k b_l)$ (in components). (1.1.14a)

(ii) Since

$$\boldsymbol{T}\cdot\boldsymbol{a} = \sum\sum (T_{kl}a_l)\boldsymbol{u}_k, \qquad \boldsymbol{a}\cdot\boldsymbol{T} = \sum\sum (a_l T_{lk})\boldsymbol{u}_k,$$

we will have $\boldsymbol{T}\cdot\boldsymbol{a} = \boldsymbol{a}\cdot\boldsymbol{T}$, only if $\boldsymbol{T}$ is *symmetric*; from which we also conclude that

$$(\boldsymbol{u}_k\otimes\boldsymbol{u}_l) : (\boldsymbol{u}_r\otimes\boldsymbol{u}_s) = \delta_{kr}\,\delta_{ls}. \tag{1.1.14b}$$

(iii) $$(\boldsymbol{a}\times\boldsymbol{T})\cdot\boldsymbol{b} = \boldsymbol{a}\times(\boldsymbol{T}\cdot\boldsymbol{b}), \qquad (\boldsymbol{T}\times\boldsymbol{a})\cdot\boldsymbol{b} = \boldsymbol{T}\cdot(\boldsymbol{a}\times\boldsymbol{b}), \tag{1.1.14c}$$

where

$$\boldsymbol{T}\times\boldsymbol{a} = \sum\sum\sum\sum (T_{kr}a_s\varepsilon_{rsl})\boldsymbol{u}_k\otimes\boldsymbol{u}_l;$$

that is,

$$(\boldsymbol{T}\times\boldsymbol{a})_{kl} = \sum\sum \varepsilon_{lrs}T_{kr}a_s, \tag{1.1.14d}$$

and

$$\boldsymbol{a}\times\boldsymbol{T} = \sum\sum\sum\sum (T_{sl}a_r\varepsilon_{rsk})\boldsymbol{u}_k\otimes\boldsymbol{u}_l, \quad (\boldsymbol{a}\times\boldsymbol{T})_{kl} = \sum\sum \varepsilon_{krs}a_r T_{sl}. \tag{1.1.14e}$$

(iv) $$(\boldsymbol{T}\cdot\boldsymbol{a}, \boldsymbol{T}\cdot\boldsymbol{b}, \boldsymbol{T}\cdot\boldsymbol{c}) = (Det\ \boldsymbol{T})(\boldsymbol{a}, \boldsymbol{b}, \boldsymbol{c}). \tag{1.1.14f}$$

(v) $$\boldsymbol{T}^{\mathrm{T}}\cdot(\boldsymbol{T}\cdot\boldsymbol{a}\times\boldsymbol{T}\cdot\boldsymbol{b}) = (Det\ \boldsymbol{T})(\boldsymbol{a}\times\boldsymbol{b}). \tag{1.1.14g}$$

Special Tensors

Zero tensor $\boldsymbol{O}$:

$$\boldsymbol{O}\cdot\boldsymbol{a} = \boldsymbol{0}, \qquad \text{for every vector } \boldsymbol{a}. \tag{1.1.15a}$$

Unit, or *identity*, tensor $\mathbf{1}$:

$$\mathbf{1}\cdot\boldsymbol{a} = \boldsymbol{0}, \qquad \text{for every vector } \boldsymbol{a}, \tag{1.1.15b}$$

$$\mathbf{1} = \sum\sum \delta_{kl}\boldsymbol{u}_k\otimes\boldsymbol{u}_l = \boldsymbol{u}_1\otimes\boldsymbol{u}_1 + \boldsymbol{u}_2\otimes\boldsymbol{u}_2 + \boldsymbol{u}_3\otimes\boldsymbol{u}_3 \qquad \text{(Dyadic form)}$$

$$= (\delta_{kl}) = diagonal(1, 1, 1) \qquad \text{(Matrix form)},$$

$$\Rightarrow Det\ \mathbf{1} = +1. \tag{1.1.15c}$$

Diagonal tensor $\boldsymbol{D}$:

$$\boldsymbol{D} = D_{11}\boldsymbol{u}_1 \otimes \boldsymbol{u}_1 + D_{22}\boldsymbol{u}_2 \otimes \boldsymbol{u}_2 + D_{33}\boldsymbol{u}_3 \otimes \boldsymbol{u}_3 \qquad \text{(Dyadic form)}$$
$$= \textit{diagonal}\ (D_{11}, D_{22}, D_{33}) \qquad \text{(Matrix form).} \qquad (1.1.15d)$$

If $D_{11} = D_{22}$, $\boldsymbol{D}$ reduces to

$$\boldsymbol{D} = D_{11}\mathbf{1} + (D_{33} - D_{11})\boldsymbol{u}_3 \otimes \boldsymbol{u}_3, \qquad (1.1.15e)$$

a result that is useful in the representation of *moments of inertia of bodies of revolution*.

Alternator tensor $\boldsymbol{\varepsilon}$:

$$\boldsymbol{\varepsilon} = \sum\sum\sum \varepsilon_{klm}\, \boldsymbol{u}_k \otimes \boldsymbol{u}_l \otimes \boldsymbol{u}_m. \qquad (1.1.15f)$$

It can be shown that

(i) $\quad \textit{Det}\ \boldsymbol{T} \equiv |T_{kl}| = \sum\sum\sum\sum\sum\sum (1/6)\varepsilon_{klm}\varepsilon_{pqr}T_{kp}T_{lq}T_{mr}.$ (1.1.15g)

(ii) If $\boldsymbol{S}$ is *symmetric*, then $\boldsymbol{T}:\boldsymbol{S} = \boldsymbol{T}^{\mathrm{T}}:\boldsymbol{S} = (1/2)(\boldsymbol{T} + \boldsymbol{T}^{\mathrm{T}}):\boldsymbol{S}$, (1.1.15h)

If $\boldsymbol{S}$ is *antisymmetric*, then $\boldsymbol{T}:\boldsymbol{S} = -(\boldsymbol{T}^{\mathrm{T}}:\boldsymbol{S}) = (1/2)(\boldsymbol{T} - \boldsymbol{T}^{\mathrm{T}}):\boldsymbol{S}$, (1.1.15i)

If $\boldsymbol{S}$ is *symmetric* and $\boldsymbol{T}$ is *antisymmetric*, then $\boldsymbol{T}:\boldsymbol{S} = 0$. (1.1.15j)

(iii) If $\boldsymbol{T}:\boldsymbol{S} = 0$ for every tensor $\boldsymbol{S}$, then $\boldsymbol{T} = \mathbf{0}$, (1.1.15k)

If $\boldsymbol{T}:\boldsymbol{S} = 0$ for every *symmetric* tensor $\boldsymbol{S}$, then $\boldsymbol{T} =$ *antisymmetric*, (1.1.15l)

If $\boldsymbol{T}:\boldsymbol{S} = 0$ for every *antisymmetric* tensor $\boldsymbol{S}$, then $\boldsymbol{T} =$ *symmetric*. (1.1.15m)

Axial Vectors

There exists a one-to-one correspondence between antisymmetric tensors and vectors: given a (any) antisymmetric tensor $\boldsymbol{W}$—that is, $\boldsymbol{W} = -\boldsymbol{W}^{\mathrm{T}}$—there exists a unique vector $\boldsymbol{w}$, its *axial* (or *dual*) *vector* or *axis*, such that for every vector $\boldsymbol{a}$:

$$\boldsymbol{W} \cdot \boldsymbol{a} = \boldsymbol{w} \times \boldsymbol{a}; \qquad (1.1.16a)$$

that is, (recalling the earlier definitions of products, etc.)

$$\boldsymbol{W} \cdot (\ldots) = (\boldsymbol{w} \times \mathbf{1}) \cdot (\ldots) = \boldsymbol{w} \times (\ldots) \qquad [\Rightarrow \boldsymbol{a} \cdot (\boldsymbol{W} \cdot \boldsymbol{a}) = 0]. \qquad (1.1.16b)$$

And, conversely, given a vector $\boldsymbol{w}$, there exists a unique antisymmetric tensor $\boldsymbol{W}$, such that (1.1.16a,b) hold. In components, the above read:

$$w_k = -(1/2)\sum\sum \varepsilon_{klm} W_{lm} = (1/2)\sum\sum \varepsilon_{lkm} W_{lm}, \qquad (1.1.16c)$$

$$W_{lm} = -\sum \varepsilon_{lmk} w_k = \sum \varepsilon_{lkm} w_k; \qquad (1.1.16d)$$

or, in matrix form:

$$\boldsymbol{W} = (W_{lm}) = \left(\begin{array}{c|c|c} 0 & W_{12} = -w_3 & W_{13} = w_2 \\ \hline W_{21} = w_3 & 0 & W_{23} = -w_1 \\ \hline W_{31} = -w_2 & W_{32} = w_1 & 0 \end{array}\right). \qquad (1.1.16e)$$

[Sometimes (especially in general indicial tensorial treatments) w_k is defined as the *negative of the above*; that is,

$$w_k = (1/2)\sum\sum \varepsilon_{klm} W_{lm} \Leftrightarrow W_{lm} = \sum \varepsilon_{lmk} w_k, \tag{1.1.16f}$$

or

$$\boldsymbol{W}\cdot\boldsymbol{a} = -\boldsymbol{w}\times\boldsymbol{a} = \boldsymbol{a}\times\boldsymbol{w}, \tag{1.1.16g}$$

and so, here too, the reader should be careful when comparing references.]

It can be shown that:

(i) The axial vector of a general *nonsymmetric* tensor equals the axial vector of its antisymmetric part; that is, the axial vector of its symmetric part (and, generally, of any symmetric tensor) vanishes; and, conversely, the vanishing of that vector shows that that tensor is symmetric.

(ii) The axial vector of $\boldsymbol{T}$, $\boldsymbol{t}$, can be expressed as

$$\begin{aligned}-2\boldsymbol{t} &= (T_{23} - T_{32})\boldsymbol{u}_1 + (T_{31} - T_{13})\boldsymbol{u}_2 + (T_{12} - T_{21})\boldsymbol{u}_3\\ &= \boldsymbol{u}_1\times\boldsymbol{t}_1 + \boldsymbol{u}_2\times\boldsymbol{t}_2 + \boldsymbol{u}_3\times\boldsymbol{t}_3.\end{aligned} \tag{1.1.16h}$$

(iii) The axial vector of

$$\boldsymbol{W} = \sum\sum W_{kl}\,\boldsymbol{u}_k\otimes\boldsymbol{u}_l = \sum\sum (1/2)W_{kl}(\boldsymbol{u}_k\otimes\boldsymbol{u}_l - \boldsymbol{u}_l\otimes\boldsymbol{u}_k),$$

$\boldsymbol{w}$ has the following dyadic representation (note k, l order):

$$\boldsymbol{w} = \sum\sum\sum[-(1/2)\varepsilon_{rkl}W_{kl}]\boldsymbol{u}_r = \cdots = \sum\sum(1/2)W_{kl}(\boldsymbol{u}_l\times\boldsymbol{u}_k). \tag{1.1.16i}$$

(iv) Let $\boldsymbol{w} = w\boldsymbol{u}_1$. Then,

$$\boldsymbol{W} = w(\boldsymbol{u}_3\otimes\boldsymbol{u}_2 - \boldsymbol{u}_2\otimes\boldsymbol{u}_3), \qquad w = \boldsymbol{u}_3\cdot(\boldsymbol{W}\cdot\boldsymbol{u}_2); \tag{1.1.16j}$$

and cyclically for $\boldsymbol{w} = w\boldsymbol{u}_2$, $\boldsymbol{w} = w\boldsymbol{u}_3$

(v) The antisymmetric part of the tensor $\boldsymbol{a}\otimes\boldsymbol{b}$ equals (in matrix form)

$$\begin{pmatrix} 0 & -w_3 & w_2 \\ w_3 & 0 & -w_1 \\ -w_2 & w_1 & 0 \end{pmatrix}, \tag{1.1.16k}$$

where

$$\boldsymbol{w} = (1/2)\,\boldsymbol{b}\times\boldsymbol{a} \qquad \text{(note order)}. \tag{1.1.16l}$$

(vi) The tensor $\boldsymbol{a}\otimes\boldsymbol{b} - \boldsymbol{b}\otimes\boldsymbol{a}$, where $\boldsymbol{a}$, $\boldsymbol{b}$ are arbitrary vectors, is antisymmetric; and its axial vector is $\boldsymbol{b}\times\boldsymbol{a}$ (note order).

(vii) Let $\boldsymbol{w}_1$, $\boldsymbol{w}_2$ be the axial vectors of the antisymmetric tensors $\boldsymbol{W}_1$, $\boldsymbol{W}_2$, respectively. Then,

$$\boldsymbol{W}_1\cdot\boldsymbol{W}_2 = \boldsymbol{w}_2\otimes\boldsymbol{w}_1 - (\boldsymbol{w}_1\cdot\boldsymbol{w}_2)\mathbf{1}, \qquad Tr(\boldsymbol{W}_1\cdot\boldsymbol{W}_2) = -2(\boldsymbol{w}_1\cdot\boldsymbol{w}_2). \tag{1.1.16m}$$

$$\Rightarrow \boldsymbol{W}\cdot\boldsymbol{W} = \boldsymbol{w}\otimes\boldsymbol{w} - (\boldsymbol{w}\cdot\boldsymbol{w})\mathbf{1}, \text{ or } \boldsymbol{W}^2 = \boldsymbol{w}\otimes\boldsymbol{w} - w^2\mathbf{1}. \tag{1.1.16n}$$

Spectral Theory of Tensors

DEFINITION

A scalar λ is a *principal*, or *characteristic*, or *proper* value, or *eigenvalue*, of $\boldsymbol{T}$ if there exists a unit vector $\boldsymbol{n} = (n_1, n_2, n_3)$ such that

$$\boldsymbol{T}\cdot\boldsymbol{n} = \lambda\boldsymbol{n}, \qquad \text{or in components } \sum T_{kl}n_l = \lambda n_k. \tag{1.1.17a}$$

Then $\boldsymbol{n}$ is called a *principal*, or *characteristic*, or *proper*, or *eigen-direction* of $\boldsymbol{T}$ corresponding to that value of λ.

DEFINITION

The *principal*, or *characteristic*, or *proper*, or *eigen-space* of $\boldsymbol{T}$ corresponding to λ is the subspace of V consisting of *all* vectors $\boldsymbol{a}$ satisfying (1.1.17a): $\boldsymbol{T}\cdot\boldsymbol{a} = \lambda\boldsymbol{a}$; that is, *the subspace of all the eigenvectors of* $\boldsymbol{T}$.

If $\boldsymbol{T}$ is *positive definite*—that is, if $\boldsymbol{a}\cdot(\boldsymbol{T}\cdot\boldsymbol{a}) > 0$ for all $\boldsymbol{a} \neq \boldsymbol{0}$—then its eigenvalues are *strictly positive*.

THEOREM OF SPECTRAL DECOMPOSITION (of $\boldsymbol{T}$)

If $\boldsymbol{T} = \boldsymbol{T}^{\mathrm{T}}$ (i.e., *symmetric*) then there exists an *orthonormal* basis $\{\boldsymbol{n}_1, \boldsymbol{n}_2, \boldsymbol{n}_3\}$ for V and three real, but not necessarily distinct, eigenvalues λ_1, λ_2, λ_3 of $\boldsymbol{T}$ such that

$$\boldsymbol{T}\cdot\boldsymbol{n}_k = \lambda_k\boldsymbol{n}_k \qquad (k = 1, 2, 3;\ \text{no sum}), \tag{1.1.17b}$$

and

$$\begin{aligned}
\boldsymbol{T} = \boldsymbol{T}\cdot\boldsymbol{1} = \boldsymbol{T}\cdot\left(\sum \boldsymbol{n}_k \otimes \boldsymbol{n}_k\right) &= \sum (\boldsymbol{T}\cdot\boldsymbol{n}_k)\otimes\boldsymbol{n}_k \\
&= \sum \lambda_k(\boldsymbol{n}_k\otimes\boldsymbol{n}_k) \qquad (\textit{Dyadic}\ \text{representation}) \\
&= \textit{diagonal}(\lambda_1, \lambda_2, \lambda_3) \qquad (\textit{Matrix}\ \text{representation}); \qquad (1.1.17c)
\end{aligned}$$

$$\Rightarrow \boldsymbol{n}_k\cdot(\boldsymbol{T}\cdot\boldsymbol{n}_l) = \boldsymbol{T}\cdot\boldsymbol{n}_k\cdot\boldsymbol{n}_l = \lambda_k\,\delta_{kl}\ (= \lambda_k \text{ or } 0, \text{ according as } k = l \text{ or } k \neq l);$$

that is, with

$$\boldsymbol{n}_k = (n_{(k)l}\text{: components of } \boldsymbol{n}_k),$$

$$T_{kl} = \lambda_1 n_{(1)k}n_{(1)l} + \lambda_2 n_{(2)k}n_{(2)l} + \lambda_3 n_{(3)k}n_{(3)l}. \tag{1.1.17d}$$

Conversely, if $\boldsymbol{T} = \sum \lambda_k(\boldsymbol{n}_k\otimes\boldsymbol{n}_k)$, with $\{\boldsymbol{n}_k\}$ = orthonormal, then $\boldsymbol{T}\cdot\boldsymbol{n}_k = \lambda_k\boldsymbol{n}_k$ (no sum).

Depending on the relative sizes of the three eigenvalues, we distinguish the following *three* cases:

(i) If λ_1, λ_2, λ_3 = *distinct*, then the eigendirections of $\boldsymbol{T}$ are the three mutually orthogonal lines, through the origin, spanned by $\boldsymbol{n}_1$, $\boldsymbol{n}_2$, $\boldsymbol{n}_3$.

(ii) If $\lambda_1 \neq \lambda_2 = \lambda_3$ (i.e., *two* distinct eigenvalues), then the spectral decomposition (1.1.17c) reduces to the following (with $|\boldsymbol{n}_1| = 1$):

$$\boldsymbol{T} = \lambda_1(\boldsymbol{n}_1\otimes\boldsymbol{n}_1) + \lambda_2(\boldsymbol{1} - \boldsymbol{n}_1\otimes\boldsymbol{n}_1). \tag{1.1.17e}$$

Conversely, if (1.1.17e) holds with $\lambda_1 \neq \lambda_2 = \lambda_3$, then λ_1 and λ_2 are the sole distinct eigenvalues of $\boldsymbol{T}$; which, in this case, has the *two* distinct eigenspaces: (a) the *line* spanned by $\boldsymbol{n}_1$, and (b) the *plane* perpendicular to $\boldsymbol{n}_1$.

(iii) If $\lambda_1 = \lambda_2 = \lambda_3 = \lambda$, in which case

$$\boldsymbol{T} = \lambda \mathbf{1} = \lambda(\boldsymbol{n}_1 \otimes \boldsymbol{n}_1 + \boldsymbol{n}_2 \otimes \boldsymbol{n}_2 + \boldsymbol{n}_3 \otimes \boldsymbol{n}_3) \qquad (\textit{Dyadic}\ \text{representation})$$
$$= \textit{diagonal}(\lambda, \lambda, \lambda) \qquad (\textit{Matrix}\ \text{representation}), \qquad (1.1.17\text{f})$$

then the eigenspace of $\boldsymbol{T}$ is the entire space V. Conversely, if V is the eigenspace of $\boldsymbol{T}$, then $\boldsymbol{T}$ has the form (1.1.17f). [For extensions of the theorem to *polynomial functions of* $\boldsymbol{T}$ see books on linear algebra; also Bradbury (1968, pp. 113–116).] The requirement of nontrivial solutions for $\boldsymbol{n}$, in (1.1.17a), leads, in well-known ways, to the *characteristic* (polynomial) equation for $\boldsymbol{T}$:

$$-Det(\boldsymbol{T} - \lambda\mathbf{1}) = Det(\lambda\mathbf{1} - \boldsymbol{T}) \equiv D(\lambda) \equiv \lambda^3 - I_1\lambda^2 + I_2\lambda - I_3 = 0, \qquad (1.1.18\text{a})$$

where the coefficients, or *principal invariants* of $\boldsymbol{T}$ (i.e., quantities independent of the choice of the basis used for the representation of $\boldsymbol{T}$), are given by

$$I_1(\boldsymbol{T}) \equiv I_1 \equiv Tr(\boldsymbol{T}) = \sum T_{kk} = \lambda_1 + \lambda_2 + \lambda_3,$$
$$I_2(\boldsymbol{T}) \equiv I_2 \equiv (1/2)[(Tr\,\boldsymbol{T})^2 - Tr(\boldsymbol{T}^2)]$$
$$= (1/2)\Big[\Big(\sum T_{kk}\Big)\Big(\sum T_{ll}\Big) - \Big(\sum\sum T_{kl}T_{lk}\Big)\Big] = \lambda_1\lambda_2 + \lambda_1\lambda_3 + \lambda_2\lambda_3,$$
$$I_3(\boldsymbol{T}) \equiv I_3 \equiv Det\,\boldsymbol{T} = |T_{kl}| = \sum\sum\sum \varepsilon_{klm} T_{k1} T_{l2} T_{m3} = \lambda_1\lambda_2\lambda_3$$
$$= (1/6)\Big[(Tr\,\boldsymbol{T})^3 - 3(Tr\,\boldsymbol{T})(Tr\,\boldsymbol{T}^2) + 2\,Tr(\boldsymbol{T}^3)\Big]; \qquad (1.1.18\text{b})$$

also

$$I_1^{\,2} - 2I_2 = \lambda_1^{\,2} + \lambda_2^{\,2} + \lambda_3^{\,2} = Tr(\boldsymbol{T}^2). \qquad (1.1.18\text{c})$$

[(a) It is shown in linear algebra/matrix theory that:

- In general, that is, $\boldsymbol{T} =$ *nonsymmetric*, eq. (1.1.18a) has either *three real roots*; or *one real* and *two complex* (*conjugate*) *roots*.
- Every tensor $\boldsymbol{T}$ satisfies its own characteristic equation; that is, eq. (1.1.18a) with λ replaced by $\boldsymbol{T}$: $\boldsymbol{T}^3 - I_1\boldsymbol{T}^2 + I_2\boldsymbol{T} - I_3\mathbf{1} = \mathbf{0}$ (*Cayley–Hamilton* theorem). And, more generally, if $f(\lambda) =$ *real* polynomial in an eigenvalue λ of $\boldsymbol{T}$, then $f(\lambda)$ is an eigenvalue of $f(\boldsymbol{T})$; and, an eigenvector of $\boldsymbol{T}$ corresponding to λ is also an eigenvector of $f(\boldsymbol{T})$ corresponding to $f(\lambda)$.

(b) The above show that $Tr\,\boldsymbol{T}$, $Tr(\boldsymbol{T}^2)$, $Tr(\boldsymbol{T}^3)$ may also be considered as principal invariants of $\boldsymbol{T}$.]

Further, it can be shown, that:

(i) If $\boldsymbol{N}_{1,2,3}$ are the antisymmetric tensors whose axial vectors are, respectively, the three orthonormal eigenvectors of (the *symmetric* tensor) $\boldsymbol{T}$: $\boldsymbol{n}_{1,2,3}$, then $\boldsymbol{T}$ has, in addition to (1.1.17c), the following spectral decomposition:

$$\boldsymbol{T} = \lambda_1(\boldsymbol{N}_1 \cdot \boldsymbol{N}_1) + \lambda_2(\boldsymbol{N}_2 \cdot \boldsymbol{N}_2) + \lambda_3(\boldsymbol{N}_3 \cdot \boldsymbol{N}_3) + Tr(\boldsymbol{T})\mathbf{1}; \qquad (1.1.18\text{d})$$

and, therefore, for an arbitrary vector $\boldsymbol{a}$,

$$\boldsymbol{T}\cdot\boldsymbol{a} = \lambda_1(\boldsymbol{N}_1\cdot\boldsymbol{N}_1)\cdot\boldsymbol{a} + \lambda_2(\boldsymbol{N}_2\cdot\boldsymbol{N}_2)\cdot\boldsymbol{a} + \lambda_3(\boldsymbol{N}_3\cdot\boldsymbol{N}_3)\cdot\boldsymbol{a} + Tr(\boldsymbol{T})\boldsymbol{a}; \quad (1.1.18\text{e}1)$$

also,

$$Tr(\boldsymbol{N}_1\cdot\boldsymbol{N}_1) = Tr(\boldsymbol{N}_2\cdot\boldsymbol{N}_2) = Tr(\boldsymbol{N}_3\cdot\boldsymbol{N}_3) = -2. \quad (1.1.18\text{e}2)$$

(ii) If $\boldsymbol{a}$ = *axial vector* of $\boldsymbol{A}$, then

$$\boldsymbol{T}\cdot\boldsymbol{a} = \textit{axial vector}\text{ of } [-(\boldsymbol{T}\cdot\boldsymbol{A} + \boldsymbol{A}\cdot\boldsymbol{T}) + Tr(\boldsymbol{T})\boldsymbol{A}]. \quad (1.1.18\text{f})$$

(iii) The principal invariants of

$$\boldsymbol{T} = \sum\sum T_{kl}\boldsymbol{u}_k\otimes\boldsymbol{u}_l = \sum \boldsymbol{u}_k\otimes\boldsymbol{t}_k, \qquad \text{where} \quad \boldsymbol{t}_k \equiv \sum T_{kl}\boldsymbol{u}_l, \quad (1.1.18\text{g})$$

can be expressed as

$$I_1 = \boldsymbol{u}_1\cdot\boldsymbol{t}_1 + \boldsymbol{u}_2\cdot\boldsymbol{t}_2 + \boldsymbol{u}_3\cdot\boldsymbol{t}_3, \quad (1.1.18\text{h})$$

$$I_2 = \boldsymbol{u}_1\cdot(\boldsymbol{t}_2\times\boldsymbol{t}_3) + \boldsymbol{u}_2\cdot(\boldsymbol{t}_3\times\boldsymbol{t}_1) + \boldsymbol{u}_3\cdot(\boldsymbol{t}_1\times\boldsymbol{t}_2), \quad (1.1.18\text{i})$$

$$I_3 = \boldsymbol{t}_1\cdot(\boldsymbol{t}_2\times\boldsymbol{t}_3). \quad (1.1.18\text{j})$$

(iv) The principal invariants of an *antisymmetric* tensor $\boldsymbol{W}$ are

$$I_1 = Tr\,\boldsymbol{W} = 0, \quad (1.1.18\text{k})$$

$$I_2 = W_{23}{}^2 + W_{31}{}^2 + W_{12}{}^2$$
$$= (-w_1)^2 + (-w_2)^2 + (-w_3)^2 = w_1{}^2 + w_2{}^2 + w_3{}^2, \quad (1.1.18\text{l})$$

$$I_3 = Det\,\boldsymbol{W} = 0 \qquad \left[|\boldsymbol{w}|^2 = w^2 = (\textit{axial vector}\text{ of } \boldsymbol{W})^2\right]; \quad (1.1.18\text{m})$$

from which, and from (1.1.18c), we can deduce that $\boldsymbol{W}$ has a *single* real eigenvector $\lambda = 0$.

(v) If $\boldsymbol{T}$ is a *symmetric* and *positive definite* tensor with (⇒ positive) eigenvalues, then

$$Det\,\boldsymbol{T} > 0 \quad (\text{i.e., } \boldsymbol{T} \text{ is invertible}), \qquad \boldsymbol{T}^{-1} = \sum \lambda_k{}^{-1}(\boldsymbol{u}_k\otimes\boldsymbol{u}_k). \quad (1.1.18\text{n})$$

(vi) If $\boldsymbol{T}$ is an invertible tensor, and the characteristic equation of $\boldsymbol{T}^{-1}$ is

$$Det(\boldsymbol{T}^{-1} - \mu\boldsymbol{1}) = 0 \;\Rightarrow\; \mu^3 - I'_1\mu^2 + I'_2\mu - I'_3 = 0, \quad (1.1.18\text{o})$$

then

$$\mu = 1/\lambda; \text{ i.e., the eigenvalues of } \boldsymbol{T}^{-1} \text{ are the inverse of those of } \boldsymbol{T}, \quad (1.1.18\text{p})$$

$$I'_1 = I_2/I_3, \qquad I'_2 = I_1/I_3, \qquad I'_3 = 1/I_3, \quad (1.1.18\text{q})$$

$$\boldsymbol{T}^{-1} = (\boldsymbol{T}^2 - I_1\boldsymbol{T} + I_2\boldsymbol{1})/I_3. \quad (1.1.18\text{r})$$

Orthogonal Transformations

A tensor $\boldsymbol{T}$ is called *orthogonal* (or *length-preserving*) if it satisfies

$$\boldsymbol{T}\cdot\boldsymbol{T}^{\mathrm{T}} = \boldsymbol{T}^{\mathrm{T}}\cdot\boldsymbol{T} = \boldsymbol{1} \;\Rightarrow\; \boldsymbol{T}^{-1} = \boldsymbol{T}^{\mathrm{T}}; \quad (1.1.19\text{a})$$

or, in components,

$$\sum T_{kl}(\boldsymbol{T}^{\mathrm{T}})_{lr} = \sum T_{kl}T_{rl} = \delta_{kr}, \tag{1.1.19b}$$

$$\sum (\boldsymbol{T}^{\mathrm{T}})_{kl}T_{lr} = \sum T_{lk}T_{lr} = \delta_{kr}; \tag{1.1.19c}$$

from which, since $Det\,\boldsymbol{T} = Det\,\boldsymbol{T}^{\mathrm{T}}$ (always), and $Det(\boldsymbol{T}\cdot\boldsymbol{T}^{\mathrm{T}}) = (Det\,\boldsymbol{T})(Det\,\boldsymbol{T}^{\mathrm{T}})$ and $Det\,\mathbf{1} = 1$, it follows that

$$(Det\,\boldsymbol{T})^2 = 1 \;\Rightarrow\; Det\,\boldsymbol{T} = \pm 1. \tag{1.1.19d}$$

THEOREM

The set of all orthogonal tensors forms the (full) *orthogonal group*; and the set of all orthogonal tensors with $Det\,\boldsymbol{T} = +1$ forms the *proper* orthogonal (sub) group.

THEOREM (transformation of bases and preservation of their dextrality)

If $\boldsymbol{A} = (A_{k'k} = A_{kk'})$ is a proper orthogonal tensor, or a *rotation*, and the basis $\{\boldsymbol{u}_k;\ k = 1, 2, 3\}$ is ortho–normal–dextral (OND), the new basis $\{\boldsymbol{u}_{k'};\ k' = 1, 2, 3\}$ defined by

$$\boldsymbol{u}_{k'} = \sum A_{k'k}\boldsymbol{u}_k \;\Leftrightarrow\; \boldsymbol{u}_k = \sum A_{kk'}\boldsymbol{u}_{k'} \tag{1.1.19e}$$

is also OND. Conversely, if both $\{\boldsymbol{u}_k\}$ and $\{\boldsymbol{u}_{k'}\}$ are OND, then there exists a unique proper orthogonal tensor such that (1.1.19e) holds. It is not hard to see that

$$A_{k'k} = \cos(\boldsymbol{u}_{k'}, \boldsymbol{u}_k) = \cos(\boldsymbol{u}_k, \boldsymbol{u}_{k'}) = A_{kk'}; \tag{1.1.19f}$$

and in this *commutativity* of the indices lies one of the advantages of the non-accented/accented index notation: one does not have to worry about their *order*. [However, in a *matrix* representation of $\boldsymbol{A}$, in general, $(A_{k'k}) \neq (A_{kk'})$.] Also, in view of the earlier orthonormality conditions (or constraints):

$$\boldsymbol{u}_{k'}\cdot\boldsymbol{u}_{l'} = \delta_{k'l'} \qquad \text{and} \qquad \boldsymbol{u}_k\cdot\boldsymbol{u}_l = \delta_{kl}, \tag{1.1.19g}$$

[which, due to (1.1.19e) are none other than (1.1.19a): $\boldsymbol{A}\cdot\boldsymbol{A}^{\mathrm{T}} = \boldsymbol{A}^{\mathrm{T}}\cdot\boldsymbol{A} = \mathbf{1}$] only *three* of the *nine* elements (direction cosines) of $\boldsymbol{A}$ are independent.

- For a *vector* $\boldsymbol{a}$, we have the following component representations in $\{\boldsymbol{u}_k\}$, $\{\boldsymbol{u}_{k'}\}$:

$$\boldsymbol{a} = \sum a_k\boldsymbol{u}_k = \sum a_{k'}\boldsymbol{u}_{k'}; \tag{1.1.19h}$$

and from this, using the *basis transformation equations* (1.1.19e), we readily obtain the corresponding *component transformation equations*:

$$a_{k'} = \sum A_{k'k}a_k = \sum A_{kk'}a_k \;\Leftrightarrow\; a_k = \sum A_{kk'}a_{k'} = \sum A_{k'k}a_{k'}. \tag{1.1.19i}$$

- *Polar versus axial vectors*: In general tensor algebra, the word axial (vector, tensor) is frequently used in the following broader sense:

(a) Vectors that transform as (1.1.19i) under any/all orthogonal transformations $\{\boldsymbol{u}_k\} \Leftrightarrow \{\boldsymbol{u}_{k'}\}$ *proper or not*, are called *polar* (or *genuine*); whereas,

(b) Vectors that, under such transformations, transform as

$$a_{k'} = (Det\, A)^{-1} \sum A_{k'k} a_k = (Det\, A) \sum A_{k'k} a_k \Leftrightarrow$$
$$a_k = (Det\, A^{-1})^{-1} \sum A_{kk'} a_{k'} = (Det\, A^{\mathrm{T}})^{-1} \sum A_{kk'} a_{k'} = (Det\, A) \sum A_{kk'} a_{k'},$$

are called *axial* (or *pseudo-*) vectors. Hence, under a change from a right-hand system to a left-hand system (a reflection), in which case $Det\, A = Det(A_{k'k}) = -1$, the components of the axial vectors are unaffected; while those of polar vectors are multiplied by -1. Since only proper orthogonal transformations are used in this book, this difference disappears—*all our vectors will be polar, in that sense.* This polar/axial distinction is of importance in other areas of physics; for example, relativity, electrodynamics (see, e.g., Bergmann, 1942, p. 56; Malvern, 1969, pp. 25–29).

• Every orthogonal tensor is either a *rotation*, $\boldsymbol{A} \to \boldsymbol{R}$, or the product of a rotation with $-\mathbf{1}$; that is, $\boldsymbol{R}$ or $-\mathbf{1} \cdot \boldsymbol{R}$ ($\mathbf{1}$: 3×3 unit tensor).

• The eigenvectors of $\boldsymbol{R}$—that is, the set of vectors satisfying $\boldsymbol{R} \cdot \boldsymbol{x} = \boldsymbol{x}$ ($\boldsymbol{R} \neq \mathbf{1}$)—build a *one-dimensional* subspace of V called the *axis* (*of rotation*) of $\boldsymbol{R}$.

• Under $\{\boldsymbol{u}_k\} \Leftrightarrow \{\boldsymbol{u}_{k'}\}$ transformations, the components of a tensor $\boldsymbol{T} = (T_{kl}) = (T_{k'l'})$ transform as follows:

$$T_{k'l'} = \sum\sum A_{k'k} A_{l'l} T_{kl} = \sum\sum A_{kk'} A_{ll'} T_{kl}, \tag{1.1.19j}$$
$$T_{kl} = \sum\sum A_{kk'} A_{ll'} T_{k'l'} = \sum\sum A_{k'k} A_{l'l} T_{k'l'}; \tag{1.1.19k}$$

or, in *matrix* form (also shown, frequently, in bold but *roman*),

$$(1.1.19j):\quad (T_{k'l'}) = (A_{k'k})(T_{kl})(A_{l'l}) \quad \text{or} \quad \boldsymbol{T}' = \boldsymbol{A} \cdot \boldsymbol{T} \cdot \boldsymbol{A}^{\mathrm{T}}, \tag{1.1.19l}$$
$$(1.1.19k):\quad (T_{kl}) = (A_{kk'})(T_{k'l'})(A_{ll'}) \quad \text{or} \quad \boldsymbol{T} = \boldsymbol{A}^{\mathrm{T}} \cdot \boldsymbol{T}' \cdot \boldsymbol{A}. \tag{1.1.19m}$$

[(a) $\boldsymbol{T}'$ should not be confused with the symmetrical part of $\boldsymbol{T}$, (1.1.13a,b). The precise meaning should be clear from the context.

(b) We do not see much advantage of (1.1.19l,m) over (1.1.19j,k), especially as a working tool in new and nontrivial situations. However, (1.1.19l,m) could be useful once the general theory has been thoroughly understood and is about to be applied to a concrete/numerical problem.]

It can be shown that:

(i) If $\boldsymbol{W}$ is antisymmetric, then

(a) $\mathbf{1} + \boldsymbol{W}$ is *nonsingular*; that is, $Det(\mathbf{1} + \boldsymbol{W}) \neq 0$; and

(b) $(\mathbf{1} - \boldsymbol{W}) \cdot (\mathbf{1} + \boldsymbol{W})^{-1}$ is *orthogonal*
(a result useful in rigid-body rotations). (1.1.19n)

(ii) If O–$\boldsymbol{u}_{123}$ and O–$\boldsymbol{u}_{1'2'3'}$ originally coincide, then the rotation tensor of a counterclockwise (positive) rotation of O–$\boldsymbol{u}_{123}$ through an angle ϕ about $\boldsymbol{u}_3 = \boldsymbol{u}_{3'}$ has the matrix form (with $c\phi \equiv \cos\phi, s\phi \equiv \sin\phi$):

$$\boldsymbol{A} \to \boldsymbol{R} = \begin{pmatrix} c\phi & -s\phi & 0 \\ s\phi & c\phi & 0 \\ 0 & 0 & 1 \end{pmatrix}. \tag{1.1.19o}$$

Moving Axes Theorems for Vectors and Tensors

Let us consider the following representation of a vector $\boldsymbol{a}$ and a tensor $\boldsymbol{T}$, measured relative to inertial, or *fixed*, OND axes $\{\boldsymbol{u}_{k'}\}$, but expressed in terms of their components along (also OND) *moving* axes $\{\boldsymbol{u}_k\}$ rotating with *angular velocity* $\boldsymbol{\omega}$ relative to $\{\boldsymbol{u}_{k'}\}$:

$$\boldsymbol{a} = \sum a_k \boldsymbol{u}_k, \qquad \boldsymbol{T} = \sum\sum T_{kl} \boldsymbol{u}_k \otimes \boldsymbol{u}_l. \tag{1.1.20a}$$

Let us calculate their inertial rates of change [i.e., relative to the fixed axes, $d\boldsymbol{a}/dt$, $d\boldsymbol{T}/dt$ ($t = t'$: time)], but in terms of their moving axes representations (1.1.20a) and *their* rates of change.

(i) By $d(\ldots)/dt$-differentiating the first of (1.1.20a) and invoking the fundamental kinematical result (most likely known from undergraduate dynamics)—a result which, along with the concept of angular velocity, is detailed in §1.7:

$$d\boldsymbol{u}_k/dt = \boldsymbol{\omega} \times \boldsymbol{u}_k, \tag{1.1.20b}$$

we obtain

$$d\boldsymbol{a}/dt = \sum \big[(da_k/dt)\boldsymbol{u}_k + a_k(\boldsymbol{\omega} \times \boldsymbol{u}_k)\big] = d'\boldsymbol{a}/dt + \boldsymbol{\omega} \times \boldsymbol{a}, \tag{1.1.20c}$$

where

$$d'\boldsymbol{a}/dt \equiv \sum (da_k/dt)\boldsymbol{u}_k\text{: rate of change of } \boldsymbol{a} \text{ relative to the moving axes.} \tag{1.1.20d}$$

(ii) Repeating this process for the second of (1.1.20a) we obtain

$$\begin{aligned} d\boldsymbol{T}/dt &= \sum\sum \{(dT_{kl}/dt)\boldsymbol{u}_k \otimes \boldsymbol{u}_l + T_{kl}[(\boldsymbol{\omega} \times \boldsymbol{u}_k) \otimes \boldsymbol{u}_l + \boldsymbol{u}_k \otimes (\boldsymbol{\omega} \times \boldsymbol{u}_l)]\} \\ &= d'\boldsymbol{T}/dt + \boldsymbol{\omega} \times \boldsymbol{T} - \boldsymbol{T} \times \boldsymbol{\omega}, \end{aligned} \tag{1.1.20e}$$

where

$$d'\boldsymbol{T}/dt \equiv \sum (dT_{kl}/dt)\boldsymbol{u}_k \otimes \boldsymbol{u}_l\text{: rate of change of } \boldsymbol{T} \text{ relative to the moving axes (or } \textit{Jaumann}\text{, or } \textit{corotational}\text{, derivative of } \boldsymbol{T}\text{).} \tag{1.1.20f}$$

Recalling the earlier results on the algebra of vectors/tensors and axial vectors [eqs (1.1.12), (1.1.14), (1.1.16)] we can rewrite (1.1.20c,e) in $\boldsymbol{u}_k$-components as follows:

(i)
$$\begin{aligned} (d\boldsymbol{a}/dt)_k &= da_k/dt + (\boldsymbol{\omega} \times \boldsymbol{a})_k \qquad (\neq da_k/dt) \\ &= da_k/dt + \sum\sum \varepsilon_{krs}\,\omega_r a_s = da_k/dt + \sum \omega_{ks} a_s, \end{aligned} \tag{1.1.20g}$$

(ii)
$$\begin{aligned} (d\boldsymbol{T}/dt)_{kl} &= (d'\boldsymbol{T}/dt)_{kl} + (\boldsymbol{\omega} \times \boldsymbol{T})_{kl} - (\boldsymbol{T} \times \boldsymbol{\omega})_{kl} \\ &= dT_{kl}/dt + \sum\sum \varepsilon_{krs}\,\omega_r T_{sl} - \sum\sum \varepsilon_{lrs}\,\omega_s T_{kr} \\ &= dT_{kl}/dt + \sum \omega_{ks} T_{sl} + \sum \omega_{lr} T_{kr} \end{aligned}$$

[after some index renaming in the last (third) group of terms, and noting that $\omega_{ls} = -\omega_{sl}$]

$$= dT_{kl}/dt + \sum \omega_{ks} T_{sl} - \sum T_{ks}\,\omega_{sl}, \tag{1.1.20h}$$

where

$\boldsymbol{\omega} = \sum\sum \omega_{kl}\boldsymbol{u}_k \otimes \boldsymbol{u}_l$: *moving axes* representation of *angular velocity tensor* (of these axes relative to the fixed ones);
i.e., antisymmetric tensor whose axial vector is ω: $\boldsymbol{\omega}\cdot\boldsymbol{a} = \omega \times \boldsymbol{a}$,

in components:

$$\omega_k = -(1/2)\sum\sum \varepsilon_{krs}\,\omega_{rs} \;\Leftrightarrow\; \omega_{rs} = -\sum \varepsilon_{rsk}\,\omega_k. \tag{1.1.20i}$$

Thus, in dyadic/matrix notation (see table 1.1), eq. (1.1.20e) reads

$$d\boldsymbol{T}/dt = d'\boldsymbol{T}/dt + \boldsymbol{\omega}\cdot\boldsymbol{T} - \boldsymbol{T}\cdot\boldsymbol{\omega} \tag{1.1.20j}$$

$$\left[= d'\boldsymbol{T}/dt + \boldsymbol{\omega}\cdot\boldsymbol{T} + (\boldsymbol{\omega}\cdot\boldsymbol{T})^{\mathrm{T}}, \quad \text{if } \boldsymbol{T} = \boldsymbol{T}^{\mathrm{T}}\right]. \tag{1.1.20k}$$

REMARKS

(i) Overdots, like $(\ldots)^{\cdot}$, are unambiguous only when applied to well-defined *components* of vectors/tensors; that is, $\dot{a}_k$, $\dot{a}_{k'}$, $\dot{T}_{kl}$, $\dot{T}_{k'l'}, \ldots$; *not* when applied to their *direct* or *dyadic*, and/or *matrix* representations; that is, does $\dot{\boldsymbol{a}}$ mean $d\boldsymbol{a}/dt$ or $d'\boldsymbol{a}/dt$? This is a common source of confusion in rigid-body dynamics.

(ii) We hope that this has convinced the reader of the superiority of the *indicial* notation over the (currently popular but nevertheless cumbersome and after-the-factish) *dyadic/matrix* notations.

Coordinate Transformations versus Frame of Reference Transformations

See also §1.2, §1.5. Let $\boldsymbol{a}'$ and $\boldsymbol{a}$ be the values of a vector as measured, respectively, in the *fixed* (inertial) and *moving* (noninertial) frames. Then [recalling (1.1.19e–i)], we have

$$\text{Inertial:} \quad \boldsymbol{a}' = \sum a'_k \boldsymbol{u}_k = \sum a'_{k'} \boldsymbol{u}_{k'}; \tag{1.1.20l}$$

$$\Rightarrow a'_{k'} = \sum A_{k'k} a'_k \;\Leftrightarrow\; a'_k = \sum A_{kk'} a'_{k'} \quad (\text{definition of } a'_{k'}, a'_k) \tag{1.1.20m}$$

$$\text{Noninertial:} \quad \boldsymbol{a} = \sum a_k \boldsymbol{u}_k = \sum a_{k'} \boldsymbol{u}_{k'}; \tag{1.1.20n}$$

$$\Rightarrow a_{k'} = \sum A_{k'k} a_k \;\Leftrightarrow\; a_k = \sum A_{kk'} a_{k'} \quad (\text{definition of } a_{k'}, a_k). \tag{1.1.20o}$$

Table 1.1 Common Tensor Notations

Direct/Dyadic	Matrix	Indicial/Component
$\boldsymbol{a}\cdot\boldsymbol{b} = \boldsymbol{b}\cdot\boldsymbol{a}$ (Dot product)	$\mathbf{a}^{\mathrm{T}}\cdot\mathbf{b} = \mathbf{b}^{\mathrm{T}}\cdot\mathbf{a}$	$\sum a_k b_k$
$\boldsymbol{T} = \boldsymbol{a}\otimes\boldsymbol{b}$ (Outer product)	$\mathbf{T} = \mathbf{a}\cdot\mathbf{b}^{\mathrm{T}}$	$T_{kl} = a_k b_l$
$\boldsymbol{b} = \boldsymbol{T}\cdot\boldsymbol{a}$	$\mathbf{b} = \mathbf{T}\cdot\mathbf{a}$	$b_k = \sum T_{kl} a_l$
$\boldsymbol{b} = \boldsymbol{a}\cdot\boldsymbol{T}$	$\mathbf{b}^{\mathrm{T}} = \mathbf{a}^{\mathrm{T}}\cdot\mathbf{T}$ or $\mathbf{b} = \mathbf{T}^{\mathrm{T}}\cdot\mathbf{a}$	$b_k = \sum a_l T_{lk}$
$\boldsymbol{a}\cdot\boldsymbol{T}\cdot\boldsymbol{b}$ (Bilinear form)	$\mathbf{a}^{\mathrm{T}}\cdot\mathbf{T}\cdot\mathbf{b}$	$\sum\sum T_{kl} a_k b_l$
$\boldsymbol{T}\cdot\boldsymbol{S}$ (Tensor product)	$\mathbf{T}\cdot\mathbf{S}$	$\sum T_{kr} S_{rl}$
$\boldsymbol{T}\cdot\boldsymbol{S}^{\mathrm{T}}$ (Tensor product)	$\mathbf{T}\cdot\mathbf{S}^{\mathrm{T}}$	$\sum T_{kr} S_{lr}$
$\boldsymbol{T}:\boldsymbol{S} = \boldsymbol{S}:\boldsymbol{T}$ (Dot product)	$\mathrm{Tr}(\mathbf{T}\cdot\mathbf{S}^{\mathrm{T}}) = \mathrm{Tr}(\mathbf{S}\cdot\mathbf{T}^{\mathrm{T}})$	$\sum\sum T_{kl} S_{kl}$
$\boldsymbol{T}\cdot\cdot\boldsymbol{S} = \boldsymbol{S}\cdot\cdot\boldsymbol{T}$ (Dot product)	$\mathrm{Tr}(\mathbf{T}\cdot\mathbf{S}) = \mathrm{Tr}(\mathbf{S}\cdot\mathbf{T})$	$\sum\sum T_{kl} S_{lk}$

Note: In *matrix* notation, the *product dot* is, frequently, omitted.

However, to relate the *noninertial* components $a_{k'}$, a_k to the *inertial* components $a'_{k'}$, a'_k, say, to be able to write something like

$$a_k = a'_k \Leftrightarrow a_{k'} = a'_{k'}, \tag{1.1.20p}$$

we need *additional* assumptions (postulates) or derivations—eqs. (1.1.20p) express frame of reference (*physical*) transformations; that is, they do *not* follow from eqs. (1.1.20m,o), which are simply coordinate system (*geometrical/projection*) transformations; (1.1.20p) have to be either postulated or derived from these postulates! Mathematically, a frame of reference transformation is equivalent to an *explicitly time-dependent* transformation between coordinate systems representing the two frames: $x_{k'} = x_{k'}(x_k, t) \Leftrightarrow x_k = x_k(x_{k'}, t)$, while an ordinary coordinate transformation is explicitly time-independent: $x_{k'} = x_{k'}(x_k) \Leftrightarrow x_k = x_k(x_{k'})$.

For example, let us consider an inertial frame represented by the (fixed) axes O–$x_{k'}$ and a noninertial one represented by the (moving) axes O–x_k, related by the homogeneous transformation (common origin!)

$$x_{k'} = \sum A_{k'k} x_k \Leftrightarrow x_k = \sum A_{kk'} x_{k'}, \tag{1.1.20q}$$

where

$$A_{k'k} = A_{kk'} = A_{k'k}(t).$$

Clearly, from geometry [i.e., (1.1.20p)-type postulates]:

$$x'_{k'} = x_{k'}, \qquad x'_k = x_k. \tag{1.1.20r}$$

By $(\ldots)^{\cdot}$-differentiating the first of (1.1.20q), and since $dx'_{k'}/dt = dx_{k'}/dt \equiv v'_{k'}$: inertial velocity of particle (with inertial coordinates $x_{k'}$) resolved along inertial axes, $dx'_k/dt = dx_k/dt \equiv v_k$: noninertial velocity of same particle (with noninertial coordinates x_k) resolved along noninertial axes, we get

$$v'_{k'} = \sum A_{k'k} v_k + \sum (dA_{k'k}/dt) x_k = v_{k'} + \sum (dA_{k'k}/dt) x_k, \tag{1.1.20s}$$

[invoking (1.1.20o)], where $dA_{k'k}/dt = \sum\sum \omega_{k'l'} A_{l'k} = \sum\sum A_{k'l}\omega_{lk}$ (see §1.7); that is, $v'_{k'} \neq v_{k'}$, even if the x_k and $x_{k'}$ are, instantaneously, aligned (i.e., $A_{k'k} = \delta_{k'k}$—see §1.7); and, similarly, from the second of (1.1.20q), $v'_k \neq v_k$, where $v'_k = \sum A_{kk'} v'_{k'}$. As eq. (1.1.20s) shows, $v'_{k'}$ depends on both the relative *orientation* between x_k and $x_{k'}$ (term $\sum A_{k'k} v_k = v_{k'}$: noninertial particle velocity, but resolved along inertial axes—a *geometrical* effect) as well as on their relative *motion* [term $\sum (dA_{k'k}/dt) x_k$—a *kinematical* effect]. There is more on moving axes theorems/applications in §1.7. Vectors transforming between *frames* as (1.1.20p) are called *objective*—namely, frame-independent; otherwise they are called *nonobjective*. Similarly for tensors: if $T'_{k'l'} = T_{k'l'}$, or $T'_{kl} = T_{kl}$, where $T'_{k'l'} = \sum\sum A_{k'k} A_{l'l} T'_{kl}$ and $T_{k'l'} = \sum\sum A_{k'k} A_{l'l} T_{kl}$, that tensor is called objective.

These concepts are important in continuum mechanics: the *constitutive* (*physical*) *equations*—namely, those relating stresses with strains/deformations and their time rates of change—*must be objective*. They also constitute the fundamental, or guiding, philosophical principle of the "Theory of Relativity" [A. Einstein, 1905 (special theory); 1916 (general theory)]. Classical mechanics does not admit of a fully physically invariant formulation (although its geometrically invariant formulation is easy via tensor calculus), and the reason is that it *is based on Euclidean geometry and on a sharp separation between space and (absolute, or Newtonian) time.* Hence, to obtain such a physically invariant mechanics, one had to change these concepts—and this was the great achievement of relativity: The latter replaced classical space and time with a more general non-Euclidean "space-time," a fusion of both space *and* time (and gravity). In this new "space," physical invariance is again expressed as geometrical invariance, via a "physical tensor calculus." (See, e.g., Bergmann, 1942.)

Table 1.1 summarizes, for the readers' convenience, common vector and tensor operations in all three notations. [We are reminded that in *matrix* notation, vectors are displayed as 3×1 *column* matrices, so that, in order to save space, we write $\boldsymbol{a} \to \boldsymbol{a}^{\mathrm{T}} = (a_1, a_2, a_3)^{\mathrm{T}}$.]

Differential Operators (Field Theory)

The most important differential operators of scalar (f)/vector $(\boldsymbol{a})$/tensor $(\boldsymbol{T})$ field theory, needed not so much in analytical mechanics as in continuum mechanics/physics, are

$$(\partial/\partial \boldsymbol{r})f \equiv \boldsymbol{grad}\, f \equiv \partial f/\partial \boldsymbol{r} = \sum (\partial f/\partial x_k)\boldsymbol{u}_k; \tag{1.1.21a}$$

$$(\partial/\partial \boldsymbol{r}) \otimes \boldsymbol{a} \equiv \boldsymbol{grad}\ \boldsymbol{a} \equiv \partial \boldsymbol{a}/\partial \boldsymbol{r} = \sum\sum (\partial a_l/\partial x_k)(\boldsymbol{u}_k \otimes \boldsymbol{u}_l), \tag{1.1.21b}$$

$$(\partial/\partial \boldsymbol{r}) \cdot \boldsymbol{a} = Tr(\boldsymbol{grad}\ \boldsymbol{a}) \equiv div\, \boldsymbol{a} \equiv \sum (\partial a_k/\partial x_k), \tag{1.1.21c}$$

$$(\partial/\partial \boldsymbol{r}) \times \boldsymbol{a} \equiv \boldsymbol{curl}\ \boldsymbol{a} \equiv \sum\sum\sum \varepsilon_{krs}(\partial a_s/\partial x_r)\boldsymbol{u}_k; \tag{1.1.21d}$$

$$(\partial/\partial \boldsymbol{r}) \otimes \boldsymbol{T} \equiv \boldsymbol{grad}\ \boldsymbol{T} = \sum\sum\sum (\partial T_{rs}/\partial x_k)[\boldsymbol{u}_k \otimes (\boldsymbol{u}_r \otimes \boldsymbol{u}_s)], \tag{1.1.21e}$$

$$(\partial/\partial \boldsymbol{r}) \cdot \boldsymbol{T} = Tr(\boldsymbol{grad}\ \boldsymbol{T}) \equiv div\, \boldsymbol{T} \equiv \sum\sum (\partial T_{ks}/\partial x_k)\boldsymbol{u}_s; \tag{1.1.21f}$$

where $\boldsymbol{r} = (x, y, z)$: position vector, from some origin O, on which f, $\boldsymbol{a}$, $\boldsymbol{T}$ depend; and (a_k), (T_{kl}) are the respective components of $\boldsymbol{a}$, $\boldsymbol{T}$ relative to an OND basis $\{O, \boldsymbol{u}_k\}$.

1.2 SPACE-TIME AXIOMS; PARTICLE KINEMATICS

Space, Time, Events

Classical mechanics (CM), the only kind of mechanics studied here, and that of which analytical mechanics is the most illustrious exponent, studies the motions of material *bodies*, or systems, under the action of mechanical *loads* (forces, moments). Hence, *bodies*, *forces*, and *motions* are its fundamental ingredients. Before examining them, however, we must postulate a certain *space-time*, or stage, where these phenomena occur, so that we may describe them via numbers assigned to elements of *length/area/volume/time interval*. In CM: (i) space is assumed to be *three-dimensional* and *Euclidean* (E_3); that is, in good experimental agreement with the Pythagorean theorem, *both locally and globally*; and (ii) there is a definite method for assigning numbers to *time intervals*, which is based on the existence of *perfect clocks*; that is, on completely *periodic* physical systems (i.e., such that a certain of their configurations is repeated indefinitely; e.g., an oscillating pendulum in vacuo, or our Earth in its daily rotation about its axis). Further, we assume that space and time are *homogeneous* (i.e., no preferred positions), and that space is also *isotropic* (i.e., no preferred directions). A physical phenomenon that is more or less sharply localized spatially and temporally (i.e., one that is occurring in the immediate neighborhood of a space point at a definite time: e.g., the arrival of a train at a certain station at a certain time) is called an *event*. Geometrically, events can be viewed as points in *space-time*, or *event space*; that is, in a four-dimensional mathematical space formed jointly by three-dimensional space *and* time. There, the four coordinates of an event, three for space and one for time, are measured by *observers* using geometrically invariant, or rigid, yardsticks (space) and the earlier postulated perfect clocks (time). [Fuller understanding of this measurement process requires elaboration of the concepts of *immediate* (*spatial*) *neighborhood* and (*temporal*) *simultaneity*. This is done in relativistic physics. Here, we take them with their intuitive meaning.]

Frame of Reference

A frame of reference is a rigid material framework, or rigid body, relative to which spatial and temporal measurements of events are made, by a team of (equivalent) observers, distributed over that body (at rest relative to it), equipped with rigid yardsticks and mutually synchronized perfect clocks. Clearly, some, if not all, of these measurements will depend on the state of motion of the frame (relative to some other frame!); that is, this "coordinate-ization of events" is, generally, nonunique. The relation between the measurements of the same event(s), as registered in two such frames, *in relative motion to each other*, is called a *frame of reference transformation*; and the latter is expressed, mathematically, by an *explicitly time-dependent* coordinate transformation — one coordinate system rigidly embedded to each frame and "representing" it.

Inertial Frame of Reference

This is a frame determined by the center of mass ("origin") of our Sun and the so-called *fixed* stars (directions of axes of frame). This *primary*, or *astronomical*, frame is Newton's *absolute space*; and, like a cosmic substratum, is assumed to exist (in Newton's words) "in its own nature, and without reference to anything external, remains always similar and immovable." Similarly, Newton assumes the existence of *absolute time*, which is measured by standard clocks, and flows uniformly and *independently of any physical phenomena or processes* — something that, today, is considered physically absurd: "[I]t is contrary to the mode of thinking in science to conceive of a thing (the space-time continuum) which acts itself, but which cannot be acted upon" (Einstein, 1956, pp. 55–56). In spite of its logically/epistemologically crude and no longer tenable foundations, CM is astonishingly accurate in several areas. For example, the planet Mercury in its motion around our Sun sweeps out a total angle of *150,000°/century*; which is only *43″* more than the Newtonian prediction! In this sense, of Machean *Denkökonomie* (≈ Principle of economy, in the formation of concepts), CM is an extremely economical intellectual and practical investment.

As the mathematical structure of the Newton–Euler laws of motion shows (§1.4,5), any other frame moving with (vectorially) constant velocity, relative to the primary frame, is also inertial; so we have a family, or group, of *secondary* inertial frames. In inertial frames, the laws of motion have their simplest form [the familiar "force equals mass times acceleration (relative to that frame)"].

Particle Kinematics

The instantaneous *position*, or *place*, of a particle P relative to an *origin*, or *reference point*, O, fixed in a, say, inertial frame F in E_3, is given by its *position* vector $\boldsymbol{r} = (x, y, z)$; where x, y, z are at least twice (piecewise) continuously differentiable functions of time t. Clearly, $\boldsymbol{r}$ depends on O while x, y, z depend on the kind of coordinates used in F. (Also, we are reminded that in kinematics, the frame does not really matter; any frame is as good as any other.) At time t, a collection of particles, or body B, occupies in E_3 a certain shape, or *configuration*, described by the single-valued and invertible mapping

$$\boldsymbol{r} = \boldsymbol{f}(P, t)\text{: Place of } P, \text{ in } F, \text{ at time } t; \qquad (1.2.1a)$$

from which, inverting, we obtain

$$P = \boldsymbol{f}^{-1}(\boldsymbol{r}, t). \qquad (1.2.1b)$$

(As mathematicians put it, in their characteristically counterintuitive jargon: *a configuration is a smooth homeomorphism of B onto a region of* E_3*!*)

A *motion* of B is a change of its configuration with time; that is, it is the locus of $\boldsymbol{r}$ of each and every P of B, for all time in a certain interval. Formally, this is a one-parameter family $\boldsymbol{f}$ of configurations with time as the (real) parameter.

Often, especially in continuum mechanics, the motion of P is described as

$$\boldsymbol{r} = \boldsymbol{f}(\boldsymbol{r}_o, t) \equiv \boldsymbol{r}(\boldsymbol{r}_o, t), \qquad (1.2.2)$$

where $\boldsymbol{r}_o$ is the position of P at some "initial or reference" time; that is, *a reference configuration used as the name of P* (see also §2.2 ff.). The above representation [in addition to being single-valued, continuous, and twice (piecewise) continuously differentiable in t] must also be single-valued and invertible in $\boldsymbol{r}_o$; that is, one-to-one in both directions.

The *velocity* and *acceleration* of P, relative to a frame F, are defined, respectively, by (assuming rectangular Cartesian coordinates)

$$\boldsymbol{v} \equiv d\boldsymbol{r}/dt = (dx/dt, dy/dt, dz/dt),$$
$$\boldsymbol{a} \equiv d\boldsymbol{v}/dt = d^2\boldsymbol{r}/dt^2 = (d^2x/dt^2, d^2y/dt^2, d^2z/dt^2). \qquad (1.2.3)$$

Clearly, $\boldsymbol{v}$ and $\boldsymbol{a}$ depend on the frame, but not on its chosen fixed origin O. The representation of the velocity and acceleration of P, relative to F, moving on a general space, or skew, curve C, along its *natural*, or *intrinsic*, ortho–normal–dextral and comoving triad $\{\boldsymbol{u}_t, \boldsymbol{u}_n, \boldsymbol{u}_b\} \equiv \{\boldsymbol{t}, \boldsymbol{n}, \boldsymbol{b}\}$ (see fig. 1.1 for definitions, etc.) is

$$\boldsymbol{v} = d\boldsymbol{r}/dt = (d\boldsymbol{r}/ds)(ds/dt) \equiv (ds/dt)\boldsymbol{t} \equiv v_t\boldsymbol{t} \qquad (= v_t\boldsymbol{t} + 0\boldsymbol{n} + 0\boldsymbol{b}), \qquad (1.2.3a)$$

$$\boldsymbol{a} = d\boldsymbol{v}/dt = (d^2s/dt^2)\boldsymbol{t} + [(ds/dt)^2/\rho]\boldsymbol{n} \equiv (dv_t/dt)\boldsymbol{t} + (v_t{}^2/\rho)\boldsymbol{n}$$
$$= (dv_t/dt)\boldsymbol{t} + (v^2/\rho)\boldsymbol{n} \qquad (= a_t\boldsymbol{t} + a_n\boldsymbol{n} + 0\boldsymbol{b}, \text{ see below}). \qquad (1.2.3b)$$

- The *speed* of P is defined as the magnitude of its velocity:

$$\text{Speed} \equiv v \equiv |\boldsymbol{v}| = |v_t| = |ds/dt| = +\left[(dx/dt)^2 + (dz/dt)^2 + (dz/dt)^2\right]^{1/2}$$
$$\equiv d\sigma/dt, \qquad \text{where: } d\sigma \equiv |ds| \to v_t = \pm v. \qquad (1.2.3c)$$

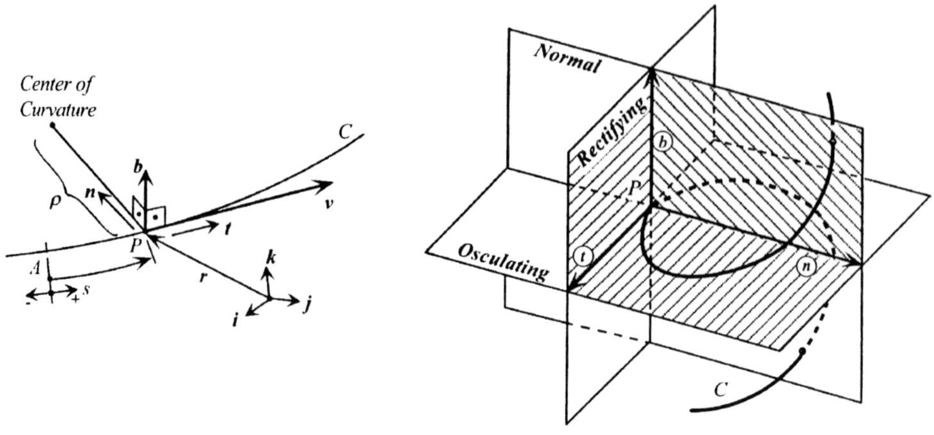

Figure 1.1 Natural, or intrinsic, triad representation in particle kinematics. s: Arc length along C, measured (positive or negative) from some origin A on C; ρ: radius of curvature of C at $P(0 \leq \rho \leq \infty)$; orthonormal and dextral (OND) triad: $\{\boldsymbol{u}_t, \boldsymbol{u}_n, \boldsymbol{u}_b\} \equiv \{\boldsymbol{t}, \boldsymbol{n}, \boldsymbol{b}\}$; tangent: $\boldsymbol{u}_t \equiv \boldsymbol{t} \equiv d\boldsymbol{r}/ds$ (always pointing toward increasing arc lengths); (first) normal: $\boldsymbol{u}_n \equiv \boldsymbol{n} \equiv \rho(d\boldsymbol{t}/ds)$ (always in sense of concavity, toward center of curvature); (second) normal, or binormal: $\boldsymbol{u}_b \equiv \boldsymbol{b} \equiv \boldsymbol{t} \times \boldsymbol{n}$; osculating plane: plane spanned by $\boldsymbol{t}$ and $\boldsymbol{n}$ (locus of tip of acceleration vector); rectifying plane: plane spanned by $\boldsymbol{t}$ and $\boldsymbol{b}$; normal plane: plane spanned by $\boldsymbol{n}$ and $\boldsymbol{b}$.

Hence, in general, since $v^2 = v_t v_t$,

$$a \equiv |\boldsymbol{a}| \equiv |d\boldsymbol{v}/dt| = [(dv/dt)^2 + (v^4/\rho^2)]^{1/2} \neq |dv_t/dt| \equiv |d^2s/dt^2| = dv/dt. \quad (1.2.3\text{d})$$

REMARK

We notice [(1.2.3c)] that the speed (v) does not always equal the tangential component of the velocity (v_t). The latter may be positive, negative, or zero; whereas the former is always positive or zero—that is, nonnegative: $\boldsymbol{t} \equiv d\boldsymbol{r}/ds$ has the same direction and sense whether $ds > 0$ or $ds < 0$, but

$$\text{(a)} \quad ds > 0 \Rightarrow v_t = \boldsymbol{v} \cdot \boldsymbol{t} > 0, \qquad \text{(b)} \quad ds < 0 \Rightarrow v_t = \boldsymbol{v} \cdot \boldsymbol{t} < 0.$$

The absence of recognition of the difference between these two quantities is a common (but, fortunately, nonfatal) error encountered in the literature. On the other hand, the positive *length* of the arc $P(t_1)P(t_2) = \sigma_2 - \sigma_1 \equiv \sigma_{1\to 2}$—namely, what one reads in a car's *odometer*—is defined by the *nonnegative* integral:

$$\text{Arc length} \equiv \sigma_{1\to 2} = \int_1^2 d\sigma = \int_1^2 |ds/dt|\, dt = \int_1^2 |ds| = \int_1^2 v\, dt = \int_1^2 |v_t|\, dt$$

$$= \int_1^2 |d\boldsymbol{r}/dt|\, dt = \int_1^2 \left[(dx/d\tau)^2 + (dy/d\tau)^2 + (dz/d\tau)^2\right]^{1/2} d\tau, \qquad (1.2.3\text{e})$$

where τ is an *arbitrary* curve parameter, so that $x = x(\tau)$, $y = y(\tau)$, $z = z(\tau)$.

It can be shown that [with the additional notation $(\ldots)' \equiv d(\ldots)/d\tau$]:

(i) $\boldsymbol{t} = \boldsymbol{r}'/s' = \boldsymbol{r}'/(\boldsymbol{r}'\cdot\boldsymbol{r}')^{1/2} = (dx/ds)\boldsymbol{i} + (dy/ds)\boldsymbol{j} + (dz/ds)\boldsymbol{k},$ (1.2.4a)

(ii) $$\begin{aligned}\boldsymbol{n} &= \rho(d\boldsymbol{t}/ds) = \rho(d^2\boldsymbol{r}/ds^2) = \rho[(d^2x/ds^2)\boldsymbol{i} + (d^2y/ds^2)\boldsymbol{j} + (d^2z/ds^2)\boldsymbol{k}] \\ &= \rho(\boldsymbol{t}'/s') = \rho\boldsymbol{t}'/(\boldsymbol{r}'\cdot\boldsymbol{r}')^{1/2} \\ &= [\rho/(\boldsymbol{r}'\cdot\boldsymbol{r}')^{3/2}][(\boldsymbol{r}'\cdot\boldsymbol{r}')^{1/2}\boldsymbol{r}'' - (\boldsymbol{r}'\cdot\boldsymbol{r}')^{-1/2}(\boldsymbol{r}'\cdot\boldsymbol{r}'')\boldsymbol{r}'] \\ &= [\rho/(\boldsymbol{r}'\cdot\boldsymbol{r}')^2][(\boldsymbol{r}'\cdot\boldsymbol{r}')\boldsymbol{r}'' - (\boldsymbol{r}'\cdot\boldsymbol{r}'')\boldsymbol{r}'];\end{aligned} \tag{1.2.4b}$$

(iii) $\kappa \equiv 1/\rho = |d^2\boldsymbol{r}/ds^2| = [(x'')^2 + (y'')^2 + (z'')^2]$:
(*first*) *curvature* of C, (1.2.4c)

(iv) $$\begin{aligned}\kappa^2 &= 1/\rho^2 = (\boldsymbol{r}'\times\boldsymbol{r}'')^2/(\boldsymbol{r}'\cdot\boldsymbol{r}')^3 = [(\boldsymbol{r}'\cdot\boldsymbol{r}')(\boldsymbol{r}''\cdot\boldsymbol{r}'') - (\boldsymbol{r}'\cdot\boldsymbol{r}'')^2]\big/(\boldsymbol{r}'\cdot\boldsymbol{r}')^3 \\ &[= (d^2x/ds^2)^2 + (d^2y/ds^2)^2 + (d^2z/ds^2)^2, \qquad \text{if } \tau = s];\end{aligned} \tag{1.2.4d}$$

(v) $$\begin{aligned}\boldsymbol{r}' &= s'\boldsymbol{t}, \\ \boldsymbol{r}'' &= s''\boldsymbol{t} + s'\boldsymbol{t}' = s''\boldsymbol{t} + s'[(d\boldsymbol{t}/ds)(ds/d\tau)] \qquad [= s''\boldsymbol{t} + (s')^2(d\boldsymbol{t}/ds)] \\ &= s''\boldsymbol{t} + \kappa(s')^2\boldsymbol{n} = s''\boldsymbol{t} + [(s')^2/\rho]\boldsymbol{n} \\ &[= (dv_t/dt)\boldsymbol{t} + (v^2/\rho)\boldsymbol{n}, \qquad v_tv_t = vv; \qquad \text{if } \tau = t];\end{aligned} \tag{1.2.4e}$$

(vi) $\boldsymbol{t} = \boldsymbol{v}/(ds/dt) \equiv \boldsymbol{v}/v_t,$ (1.2.4f)

$\boldsymbol{n} = \rho[v^2\boldsymbol{a} - (\boldsymbol{v}\cdot\boldsymbol{a})\boldsymbol{v}]/v^4,$ (1.2.4g)

$$\begin{aligned}\boldsymbol{b} &= \boldsymbol{t}\times\boldsymbol{n} = \rho[(d\boldsymbol{r}/ds)\times(d^2\boldsymbol{r}/ds^2)] = \rho(\boldsymbol{r}'\times\boldsymbol{r}'')/(\boldsymbol{r}'\cdot\boldsymbol{r}')^{3/2} \\ &= \rho(\boldsymbol{v}\times\boldsymbol{a})/v_t^{\,3} = \rho(\boldsymbol{v}\times\boldsymbol{a})/v_tv^2;\end{aligned} \tag{1.2.4h}$$

(vii) $\kappa^2 = 1/\rho^2 = (\boldsymbol{v}\times\boldsymbol{a})^2/v^6 = [v^2a^2 - (\boldsymbol{v}\cdot\boldsymbol{a})^2]/v^6;$ (1.2.4i)

(viii) $a_t \equiv \boldsymbol{a}\cdot\boldsymbol{t} = [v_x(dv_x/dt) + v_y(dv_y/dt) + v_z(dv_z/dt)]\big/v_t;$ (1.2.4j)

(ix) $$\begin{aligned}a_n = |\boldsymbol{a}\times\boldsymbol{t}| &= \{[v_x(dv_y/dt) - v_y(dv_x/dt)]^2 + [v_y(dv_z/dt) - v_z(dv_y/dt)]^2 \\ &+ [v_z(dv_x/dt) - v_x(dv_z/dt)]^2\}^{1/2}\big/(v_x^{\,2} + v_y^{\,2} + v_z^{\,2})^{1/2}.\end{aligned} \tag{1.2.4k}$$

• In plane *polar* coordinates, the *position/velocity/acceleration* of a particle P are (where $\boldsymbol{u}_r$, $\boldsymbol{u}_\phi$: unit vectors *along* OP and *perpendicular* to it, in the sense of increasing r, ϕ respectively):

$$d\boldsymbol{u}_r/dt = (d\phi/dt)\boldsymbol{u}_\phi \quad \text{and} \quad d\boldsymbol{u}_\phi/dt = -(d\phi/dt)\boldsymbol{u}_r;$$
$$\text{or} \quad d\boldsymbol{u}_r = d\phi\,\boldsymbol{u}_\phi \quad \text{and} \quad d\boldsymbol{u}_\phi = -d\phi\,\boldsymbol{u}_r,$$

$\boldsymbol{r} = r\,\boldsymbol{u}_r \qquad [= (r)\boldsymbol{u}_r + (0)\boldsymbol{u}_\phi],$ (1.2.5a)

$\boldsymbol{v} = (dr/dt)\boldsymbol{u}_r + r(d\phi/dt)\boldsymbol{u}_\phi \equiv v_r\boldsymbol{u}_r + v_\phi\boldsymbol{u}_\phi,$ (1.2.5b)

$$\begin{aligned}\boldsymbol{a} &= [d^2r/dt^2 - r(d\phi/dt)^2]\boldsymbol{u}_r + \{r^{-1}\,d/dt[r^2(d\phi/dt)]\}\boldsymbol{u}_\phi \\ &= [d^2r/dt^2 - r(d\phi/dt)^2]\boldsymbol{u}_r + [2(dr/dt)(d\phi/dt) + r(d^2\phi/dt^2)]\boldsymbol{u}_\phi \\ &\equiv a_{(r)}\boldsymbol{u}_r + a_{(\phi)}\boldsymbol{u}_\phi.\end{aligned} \tag{1.2.5c}$$

The vectors $(d\phi/dt)\boldsymbol{k}$ and $(d^2\phi/dt^2)\boldsymbol{k}$ are, respectively, the *angular velocity* and *angular acceleration* of the radius $\boldsymbol{OP} = \boldsymbol{r}$ relative to $O\text{–}xy$. It can be shown that

$$\text{(i)}\quad a_t \equiv \boldsymbol{a}\cdot\boldsymbol{t} = \pm(v_r a_r + v_\phi a_\phi)\big/(v_r{}^2 + v_\phi{}^2)^{1/2} \qquad [+ \text{ if } v_t > 0,\ - \text{ if } v_t < 0]. \tag{1.2.5d}$$

(ii) The rectangular Cartesian components of the velocity and acceleration are, respectively,

$$dx/dt = (dr/dt)\cos\phi - [r(d\phi/dt)]\sin\phi, \quad dy/dt = (dr/dt)\sin\phi + [r(d\phi/dt)]\cos\phi; \tag{1.2.5e}$$

$$d^2x/dt^2 = [d^2r/dt^2 - r(d\phi/dt)^2]\cos\phi - [2(dr/dt)(d\phi/dt) + r(d^2\phi/dt^2)]\sin\phi,$$

$$d^2y/dt^2 = [d^2r/dt^2 - r(d\phi/dt)^2]\sin\phi + [2(dr/dt)(d\phi/dt) + r(d^2\phi/dt^2)]\sin\phi, \tag{1.2.5f}$$

and, inversely,

$$dr/dt = (xv_x + yv_y)/(x^2 + y^2)^{1/2}, \qquad d\phi/dt = (xv_y - yv_x)/(x^2 + y^2), \text{ etc.} \tag{1.2.5g}$$

[A more precise notation of vector components along various bases of orthogonal curvilinear (i.e., nonrectangular Cartesian) coordinates is introduced below.]

• In general (i.e., not necessarily plane) motion, the *areal velocity* dA/dt of a particle equals

$$dA/dt = (1/2)|\boldsymbol{r}\times\boldsymbol{v}| = (1/2)|\textit{angular momentum of particle about origin, per unit mass}|. \tag{1.2.6a}$$

It can be shown that (assuming $\boldsymbol{r} \neq \boldsymbol{0}$)

$$d^2A/dt^2 = (\boldsymbol{r}\times\boldsymbol{v})\cdot(\boldsymbol{r}\times\boldsymbol{a})/2|\boldsymbol{r}\times\boldsymbol{v}|. \tag{1.2.6b}$$

Velocity and Acceleration in Orthogonal Curvilinear Coordinates

(A certain familiarity with the latter is assumed—otherwise, this topic can be omitted at this point.) In such coordinates, say $q \equiv (q_1, q_2, q_3) \equiv (q_{1,2,3})$ [see fig. 1.2(a)] the position vector $\boldsymbol{r}$, of a particle P, is expressed as:

$$\boldsymbol{r} = x(q)\boldsymbol{i} + y(q)\boldsymbol{j} + z(q)\boldsymbol{k} \equiv \boldsymbol{r}(q), \tag{1.2.7a}$$

and so the corresponding *unit* tangent vectors along the coordinate lines $q_{1,2,3}$, $\boldsymbol{u}_{1,2,3}$, are

$$\boldsymbol{u}_1 = (1/h_1)(\partial\boldsymbol{r}/\partial q_1) \equiv \boldsymbol{e}_1/h_1,$$

$$\boldsymbol{u}_2 = (1/h_2)(\partial\boldsymbol{r}/\partial q_2) \equiv \boldsymbol{e}_2/h_2,$$

$$\boldsymbol{u}_3 = (1/h_3)(\partial\boldsymbol{r}/\partial q_3) \equiv \boldsymbol{e}_3/h_3,$$

where

$$\boldsymbol{u}_k\cdot\boldsymbol{u}_l = \delta_{kl} \qquad (k, l = 1, 2, 3); \tag{1.2.7b}$$

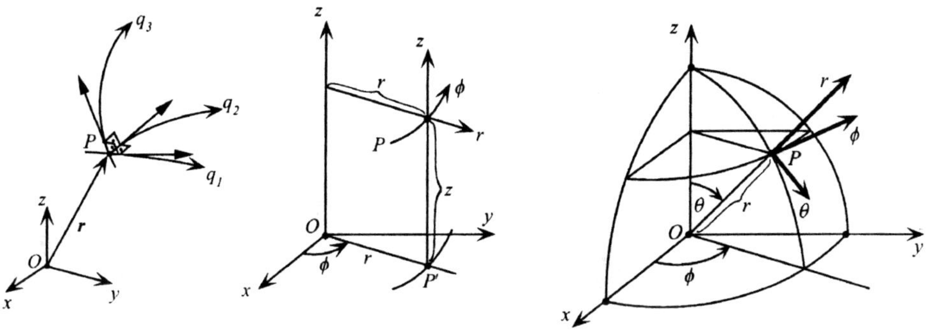

Figure 1.2 (a) General orthogonal curvilinear coordinates;
(b) cylindrical (polar) coordinates: $x = r\cos\phi$, $y = r\sin\phi$, $z = z$, $r = |OP'|$; $h_{1,2,3} \equiv h_{r,\phi,z} = 1, r, 1$;
(c) spherical coordinates: $x = (r\sin\theta)\cos\phi$, $y = (r\cos\theta)\sin\phi$, $z = r\cos\theta$, $r = |OP|$;
$h_{1,2,3} \equiv h_{r,\theta,\phi} = 1, r, r\sin\theta$.

and since

$$\partial\boldsymbol{r}/\partial q_1 = (\partial x/\partial q_1)\boldsymbol{i} + (\partial y/\partial q_1)\boldsymbol{j} + (\partial z/\partial q_1)\boldsymbol{k}, \qquad \partial\boldsymbol{r}/\partial q_2 = \cdots, \qquad \partial\boldsymbol{r}/\partial q_3 = \cdots, \tag{1.2.7c}$$

the (normalizing) *Lamé coefficients* $h_{1,2,3}$ are given by

$$h_1 \equiv |\partial\boldsymbol{r}/\partial q_1| = [(\partial x/\partial q_1)^2 + (\partial y/\partial q_1)^2 + (\partial z/\partial q_1)^2]^{1/2}, \quad h_2 = \cdots, h_3 = \cdots. \tag{1.2.7d}$$

We notice that

$$\cos(\boldsymbol{u}_k, x) = \boldsymbol{u}_k \cdot \boldsymbol{i} = \left[(1/h_k)(\partial\boldsymbol{r}/\partial q_k)\right] \cdot \boldsymbol{i} = (1/h_k)(\partial x/\partial q_k), \text{ etc.},$$

or, generally, with $x \equiv x_1$, $y \equiv x_2$, $z \equiv x_3$, and $\boldsymbol{i} \equiv \boldsymbol{i}_1$, $\boldsymbol{j} \equiv \boldsymbol{i}_2$, $\boldsymbol{k} \equiv \boldsymbol{i}_3$,

$$\cos(\boldsymbol{u}_k, x_l) = \boldsymbol{u}_k \cdot \boldsymbol{i}_l = (1/h_k)(\partial x_l/\partial q_k). \tag{1.2.7e}$$

As a result of the above, and since $\partial\boldsymbol{r}/\partial q_k \equiv \boldsymbol{e}_k = h_k\boldsymbol{u}_k$, the *arc length element* ds, *velocity* $\boldsymbol{v}$, and *speed* $|\boldsymbol{v}|$ of P are given, respectively, by

(i) $$ds = |d\boldsymbol{r}| = \left|\sum (\partial\boldsymbol{r}/\partial q_k)\, dq_k\right| = ({h_1}^2\, {dq_1}^2 + {h_2}^2\, {dq_2}^2 + {h_3}^2\, {dq_3}^2)^{1/2}, \tag{1.2.7f}$$

(ii) $$\boldsymbol{v} \equiv d\boldsymbol{r}/dt = \sum (\partial\boldsymbol{r}/\partial q_k)(dq_k/dt) \equiv \sum v_k\boldsymbol{e}_k = \sum v_k(h_k\boldsymbol{u}_k) \equiv \sum v_{(k)}\boldsymbol{u}_k, \tag{1.2.7g}$$

(iii) $$|\boldsymbol{v}| \equiv v = ({h_1}^2{v_1}^2 + {h_2}^2{v_2}^2 + {h_3}^2{v_3}^2)^{1/2}, \tag{1.2.7h}$$

where

$dq_k/dt \equiv v_k$: "contravariant" or *generalized* component of $\boldsymbol{v}$ along q_k, (1.2.7i)

$v_{(k)} \equiv h_k(dq_k/dt) \equiv h_k v_k$: corresponding *physical* component (with units of *length/time*), (1.2.7j)

Next, we define the *generalized* and *physical* components of the particle acceleration $\boldsymbol{a}$ as

$$a_k \equiv \boldsymbol{a}\cdot\boldsymbol{e}_k = (d\boldsymbol{v}/dt)\cdot(\partial\boldsymbol{r}/\partial q_k), \qquad a_{(k)} \equiv \boldsymbol{a}\cdot\boldsymbol{u}_k = \boldsymbol{a}\cdot(\boldsymbol{e}_k/h_k) = a_k/h_k. \quad (1.2.7\text{k})$$

REMARKS

(i) For an arbitrary vector $\boldsymbol{b}$, in general orthogonal curvilinear coordinates, we have the following representations:

$$\boldsymbol{b} = \sum b^k\boldsymbol{e}_k = \sum b_k\boldsymbol{e}^k = \sum b_k(\boldsymbol{e}_k/h_k{}^2) = \sum (b_k/h_k)(\boldsymbol{e}_k/h_k) \equiv \sum b_{(k)}\boldsymbol{u}_k$$

where

$$\boldsymbol{e}_k\cdot\boldsymbol{e}_l \equiv g_{kl} = 0, \text{ if } k \neq l; \quad = h_k{}^2 \text{ if } k = l; \quad \boldsymbol{e}^k\cdot\boldsymbol{e}_l = \delta^k{}_l = \delta_{kl}, \quad \boldsymbol{e}^k\cdot\boldsymbol{e}^l = g^{kl} = g^{lk}$$

$$\Rightarrow g^{kk} = 1/h_k{}^2, \quad g_{kk} = 1/g^{kk} = h_k{}^2, \quad Det(g_{kl}) = h_1{}^2h_2{}^2h_3{}^2; \quad \boldsymbol{e}^k = \boldsymbol{e}_k/h_k{}^2;$$

$$b_k \equiv \boldsymbol{b}\cdot\boldsymbol{e}_k = \boldsymbol{b}\cdot(h_k\boldsymbol{u}_k) = h_k(\boldsymbol{b}\cdot\boldsymbol{u}_k) \equiv h_kb_{(k)}; \quad b^k \equiv \boldsymbol{b}\cdot\boldsymbol{e}^k = \boldsymbol{b}\cdot(\boldsymbol{u}_k/h_k) = b_{(k)}/h_k;$$

that is,

$$b_{(k)} = b^kh_k = b_k/h_k\text{: physical components of } \boldsymbol{b},\ b^k = b_k/h_k{}^2; \boldsymbol{u}_k = \boldsymbol{e}_k/h_k = h_k\boldsymbol{e}^k.$$

(ii) Strictly speaking, q_k should have been written as q^k; and, consequently, v_k as v^k!

(iii) In *rectangular Cartesian coordinates/axes* (this book), clearly, $h_k = 1 \Rightarrow b_{(k)} = b_k = b^k$.

(iv) For the extension of the above to *general* curvilinear coordinates, see books on tensor calculus; for example, Papastavridis (1999, chap. 2, especially §2.10).

From the first of (1.2.7k) we obtain successively (what are, in essence, the famous *Lagrangean kinematico-inertial transformations*, to be generalized and detailed in chaps. 2 and 3):

$$a_k \equiv \boldsymbol{a}\cdot\boldsymbol{e}_k \equiv (d\boldsymbol{v}/dt)\cdot(\partial\boldsymbol{r}/\partial q_k) = d/dt\,[\boldsymbol{v}\cdot(\partial\boldsymbol{r}/\partial q_k)] - \boldsymbol{v}\cdot d/dt(\partial\boldsymbol{r}/\partial q_k)$$

{and, using the basic kinematical *identities*:

(a) $\partial\boldsymbol{r}/\partial q_k = \partial\boldsymbol{v}/\partial v_k$ [from (1.2.7g)]

(b) $d/dt(\partial\boldsymbol{r}/\partial q_k)$

$$= \sum \partial/\partial q_l(\partial\boldsymbol{r}/\partial q_k)(dq_l/dt) + \partial/\partial t(\partial\boldsymbol{r}/\partial q_k)$$

$$= \partial/\partial q_k\Big(\sum (\partial\boldsymbol{r}/\partial q_l)(dq_l/dt) + \partial\boldsymbol{r}/\partial t\Big)$$

$$= \partial\boldsymbol{v}/\partial q_k;$$

i.e., $d/dt(\partial\boldsymbol{v}/\partial v_k) - \partial\boldsymbol{v}/\partial q_k = 0$}

$$= d/dt\big[\boldsymbol{v}\cdot(\partial\boldsymbol{v}/\partial v_k)\big] - \boldsymbol{v}\cdot(\partial\boldsymbol{v}/\partial q_k)$$

$$= d/dt[\partial/\partial v_k(v^2/2)] - \partial/\partial q_k(v^2/2)$$

$$(\text{since } \boldsymbol{v}\cdot\boldsymbol{v} = v^2); \quad (1.2.7\text{l})$$

and, invoking the second of (1.2.7k), we get, finally, the *Lagrangean form*:

$$a_{(k)} = a_k/h_k = (1/h_k)\left[d/dt(\partial T/\partial v_k) - \partial T/\partial q_k\right], \qquad (1.2.7\text{m})$$

where

$$\begin{aligned} T \equiv v^2/2 &= (1/2)[h_1{}^2(dq_1/dt)^2 + h_2{}^2(dq_2/dt)^2 + h_3{}^2(dq_3/dt)^2]^{1/2} \\ &\equiv (1/2)(h_1{}^2 v_1{}^2 + h_2{}^2 v_2{}^2 + h_3{}^2 v_3{}^2)^{1/2}: \end{aligned}$$

kinetic energy of a particle of unit mass (i.e., $m = 1$). (1.2.7n)

Application

(i) *Cylindrical* (*polar*) coordinates [fig. 1.2(b)]. Here, $x = r\cos\phi$, $y = r\sin\phi$, $z = z$, and, therefore,

$$ds^2 = ds_r{}^2 + ds_\phi{}^2 + ds_z{}^2 = dr^2 + r^2\,d\phi^2 + dz^2, \qquad (1.2.8\text{a})$$

from which we immediately read off the following Lamé coefficients:

$$h_1 \to h_r = 1, \qquad h_2 \to h_\phi = r, \qquad h_3 \to h_z = 1. \qquad (1.2.8\text{b})$$

Hence, the "unit kinetic energy" equals

$$2T = (ds/dt)^2 = v^2 = [(dr/dt)^2 + r^2(d\phi/dt)^2 + (dz/dt)^2] \equiv v_r{}^2 + r^2 v_\phi{}^2 + v_z{}^2, \qquad (1.2.8\text{c})$$

and so, by (1.2.7l), the (physical) components of the acceleration are

$$a_{(1)} \to a_{(r)} = d/dt(\partial T/\partial v_r) - \partial T/\partial r = d^2r/dt^2 - r(d\phi/dt)^2, \qquad (1.2.8\text{d})$$

$$\begin{aligned} a_{(2)} \to a_{(\phi)} &= (1/r)\left[d/dt(\partial T/\partial v_\phi) - \partial T/\partial\phi\right] \\ &= (1/r)\left\{d/dt[r^2(d\phi/dt)]\right\} = r(d^2\phi/dt^2) + 2(dr/dt)(d\phi/dt), \qquad (1.2.8\text{e}) \end{aligned}$$

$$a_{(3)} \to a_{(z)} = d/dt(\partial T/\partial v_z) - \partial T/\partial z = d^2z/dt^2. \qquad (1.2.8\text{f})$$

(ii) *Spherical* coordinates. Here, $x = (r\sin\theta)\cos\phi$, $y = (r\cos\theta)\sin\phi$, $z = r\cos\theta$ [fig. 1.2(c)]. Using similar steps, we can show that

$$a_{(1)} \to a_{(r)} = d/dt(\partial T/\partial v_r) - \partial T/\partial r = d^2r/dt^2 - r(d\theta/dt)^2 - r(d\phi/dt)^2\sin^2\theta; \qquad (1.2.8\text{g})$$

$$\begin{aligned} a_{(2)} \to a_{(\theta)} &= (1/r)[\partial T/\partial v_\theta - \partial T/\partial\theta] \\ &= (1/r)\left\{d/dt[r^2(d\theta/dt)] - r^2(d\phi/dt)^2\sin\theta\cos\theta\right\}; \qquad (1.2.8\text{h}) \end{aligned}$$

$$\begin{aligned} a_{(3)} \to a_{(\phi)} &= (1/r\sin\theta)\left[d/dt(\partial T/\partial v_\phi) - \partial T/\partial\phi\right] \\ &= (1/r\sin\theta)\left\{d/dt[r^2(d\phi/dt)\sin^2\theta]\right\}; \qquad (1.2.8\text{i}) \end{aligned}$$

$$v_x = dx/dt = (dr/dt)\sin\theta\cos\phi + r(d\theta/dt)\cos\theta\cos\phi - r(d\phi/dt)\sin\theta\sin\phi, \qquad (1.2.8\text{j})$$

$$v_y = dy/dt = (dr/dt)\sin\theta\sin\phi + r(d\theta/dt)\cos\theta\sin\phi + r(d\phi/dt)\sin\theta\cos\phi, \tag{1.2.8k}$$

$$v_z = dz/dt = (dr/dt)\cos\theta - r(d\theta/dt)\sin\theta; \tag{1.2.8l}$$

$$\begin{aligned} a_x = d^2x/dt^2 = {} & [d^2r/dt^2 - r(d\theta/dt)^2 - r(d\phi/dt)^2]\sin\theta\cos\phi \\ & + [r(d^2\theta/dt^2) + 2(dr/dt)(d\theta/dt)]\cos\theta\cos\phi \\ & - [r(d^2\phi/dt^2) + 2(dr/dt)(d\phi/dt)]\sin\theta\sin\phi \\ & - 2r(d\phi/dt)(d\theta/dt)\cos\theta\sin\phi; \end{aligned} \tag{1.2.8m}$$

$$\begin{aligned} a_y = d^2y/dt^2 = {} & [d^2r/dt^2 - r(d\theta/dt)^2 - r(d\phi/dt)^2]\sin\theta\sin\phi \\ & + [r(d^2\theta/dt^2) + 2(dr/dt)(d\theta/dt)]\cos\theta\sin\phi \\ & + [r(d^2\phi/dt^2) + 2(dr/dt)(d\phi/dt)]\sin\theta\cos\phi \\ & + 2r(d\phi/dt)(d\theta/dt)\cos\theta\cos\phi, \end{aligned} \tag{1.2.8n}$$

$$\begin{aligned} a_z = d^2z/dt^2 = {} & [d^2r/dt^2 - r(d\theta/dt)^2]\cos\theta \\ & - [r(d^2\theta/dt^2) + 2(dr/dt)(d\theta/dt)]\sin\theta; \end{aligned} \tag{1.2.8o}$$

and, inversely,

$$dr/dt = [x(dx/dt) + y(dy/dt) + z(dz/dt)]\big/(x^2 + y^2 + z^2)^{1/2}, \tag{1.2.8p}$$

$$d\theta/dt = \{[x(dx/dt) + y(dy/dt)]z - (x^2 + y^2)(dz/dt)\}\big/(x^2 + y^2)^{1/2}(x^2 + y^2 + z^2), \tag{1.2.8q}$$

$$d\phi/dt = [x(dy/dt) - y(dx/dt)]\big/(x^2 + y^2); \tag{1.2.8r}$$

and

$$d^2r/dt^2 = \cdots, \qquad d^2\theta/dt^2 = \cdots, \qquad d^2\phi/dt^2 = \cdots,$$

in complete agreement with (1.2.5).

REMARK

From now on, *parentheses around subscripts* (employed to denote physical components) will, normally, be omitted; that is, unless absolutely necessary, we shall simply write a_r, a_θ, a_ϕ for $a_{(r)}$, $a_{(\theta)}$, $a_{(\phi)}$, respectively, etc.

1.3 BODIES AND THEIR MASSES

Body or System

A body or system is an ordinary three-dimensional material object whose points fill a spatial region completely; or a *continuous* connected three-dimensional set of *material points*, or *mass points*, or *particles*, such that any part of it, no matter how small, possesses the same physical properties as the entire object. The interactions of bodies, under the action of forces/fields, produces the various physical phenomena.

Bodies are usually classified as *solids*, *fluids*, and *gases*.

- The *rigid body* is a special solid whose *deformation* (or *strain*), relative to its other motions, can be neglected; and whose geometric form/shape and spatial material distribution are invariable.
- The *particle* is a special rigid body whose *rotation*, relative to its other motions, can be neglected; it is small relative to its distance from other bodies, and its motion as a whole is virtually unaffected by its *internal* motion. It is a special *localized continuum* of infinite material density (see below).

The complete characterization of a particle requires specification of its *spatial position* and of the values of its associated *parameters* (e.g., mass, electric charge). The former varies with time but the latter, since they describe the internal constitution of our particle, do not; if they did, we would have a more complex system.

Whether one and the same body or system will be modeled as deformable continuum, or rigid, or particle, etc., depends on the problem at hand. Below, we show such a problem to model correspondence for the system Earth:

Problem	*Mathematical Model*
Orbit around the Sun	Particle
Tides and/or lunar eclipses	Rigid sphere
Precession of the equinoxes	Rigid ellipsoid
Earthquakes	Elastic sphere
etc.	

Mass

To each body, B, that instantaneously occupies continuously a spatial region of volume V, we assign, or order, a real, positive and *time-independent* number expressing the *quantity of matter in B*, its mass m; a primitive concept with dimensions independent of the (also primitives) length and time. Symbolically, we have

$$B \rightarrow m(B) \equiv m = S_B\, dm = \int_V (dm/dV)\, dV \equiv \int_V \rho\, dV > 0, \tag{1.3.1}$$

where (continuity hypothesis)

$$\rho \equiv [\lim(\Delta m/\Delta V)]_{\Delta V \to 0} \equiv dm/dV: \quad \text{mass } \textit{density}, \text{ or } \textit{specific mass}, \text{ of } B$$
$$\text{(a piecewise continuous function of } t \text{ and } \mathbf{r}) \tag{1.3.2}$$

and $m = constant$, for a given body (*conservation of mass*).

The above imply that the mass is *additive*: the mass of a body, or system, equals the sum of the masses of its parts; with some intuitively obvious notation:

$$m(B) = m(B_1 + B_2) = m(B_1) + m(B_2) = m_1 + m_2. \tag{1.3.3}$$

REMARKS

(i) For so-called "variable mass problems" (clearly, a misleading term); for example, rockets, chemical reactions, see Fox (1967, pp. 321–326) and, particularly, Novoselov (1969).

(ii) To describe several bodies, including possible gaps, via (1.3.1) and (1.3.2), we may have to assume that in some regions $\rho = 0$.

(iii) Mathematically, mass additivity can be expressed as follows: Consider an arbitrary subset of the body B, b. If we can associate with b a nonnegative real number $m(b)$, with physical dimensions independent of those of time and length, and such that

$$m(b_1 \cup b_2) = m(b_1) + m(b_2) \qquad [\cup \equiv \textit{union} \text{ of two sets}]$$

for all pairs b_1 and b_2 of disjoint subsets of b; and

$$m(b) \to 0,$$

as the volume occupied by b goes to zero; then we call B a *material body* with *mass function m*. The *additive set function* $m(b)$ is the mass of b; or the *mass content* of the corresponding set of points occupied by b. The above properties of $m(\ldots)$ imply the existence of a scalar field ρ = *mass density* of B, defined over the configuration of B, such that (1.3.1) holds.

Impenetrability Axiom (and One-to-One Event Occurrence)

Not more than one particle may occupy any position in space, at any given time. More generally (continuum form), if, during its motion, the material system initially occupies the spatial region V_o, and later the region V, then the relation between V_o and V is mutually one-to-one, and piecewise continuously differentiable (for the associated field functions). Discontinuities (e.g., rupture, impact) and accompanying loss of uniqueness can occur only across certain (two-dimensional) boundary surfaces.

Remarks on Particles, Bodies, Mathematical Modeling, and so on

(i) A finite, or *extended*, body B or system S can be treated exactly, or approximately, as a particle in the following three cases:

(a) If B undergoes pure *translation*; that is, all its points describe congruent paths with (vectorially) equal velocities and accelerations. In this case, *any point* of B can play the role of that particle.
(b) If the description of the kinetic properties of B requires only the investigation of the motion of its center of mass (§1.4).
(c) If B is such that its dimensions are so small (or its distances from other bodies, its environment, are so large) that its size can be neglected; and its motion can be represented satisfactorily by the motion of either its mass center or any other internal point of it. Such bodies we call *small*.

- In cases (b) (always) and (c) (usually) that particle is the mass center.
- Cases (a, b) are exact, while (c) is only approximate.
- In case (a), that particle describes the motion of B completely, in (b) only partially (the motion *about* the mass center is neglected), and in (c) with an *error* depending on the neglected dimensions of B.

From such a continuum viewpoint, a particle is viewed not as the building block of matter, but as *a rigid and rotationless body*! As Hamel (1909, p. 351) aptly summarizes: "What one understands, in practice, by particle mechanics

(Punktmechanik) is none other than the theorem of the center of mass (Schwerpunktsatz)."

(ii) Both models of a body—that is, the one based on the *atomistic* hypothesis (body as a finite, discrete, set of material points, or particles; namely, small hard balls with no rotational characteristics) and the other based on the *continuity* hypothesis (body as a family of measurable sets, with associated *additive set functions* representing the mass of that set)—have advantages and disadvantages; and both are useful for various purposes. The sometimes (in some engineering circles) fierce debate for/against one or the other viewpoint, we consider counterproductive and petty hair-splitting; and so we will use both models as needed. Such dualisms are no strangers to physics (e.g., particles/corpuscules vs. waves/fields in atomic phenomena) and constitute a creative, dialectical, stress in it.

Thus, we will view the rigid body (§1.8 ff.) either as a (finite or infinite) set of particles whose mutual distances are constrained to remain invariable (i.e., fixed in time); or, more conveniently, as *a rigid continuum*, and accept the Newton–Euler law of motion for its differential mass elements as for a particle (§1.4, §1.6). In the discrete model, the building block is the single "sizeless," but possibly quite "massive," particle of mass $m_k > 0$ $(k = 1, 2, \ldots)$; while, in the continuum model, it is the differential element with mass $dm = \rho\, dV > 0$. In sum, we shall adopt the logically unorthodox, but quite fertile and successful, dialectical compromise: *particle language and continuum notation*; and eventually (chap. 3 ff.) we will end up with *ordinary differential equations*.

[In general, it is extremely difficult, if not impossible, to go by a limiting process from a statement about particles to one about continua; whereas, conversely, continuum statements formulated in terms of Stieltjes' integrals, like our earlier $\mathcal{S}(\ldots)$:

$$\mathcal{S}(\ldots)\, dm: \quad \sum (\ldots)_k m_k \ \text{(discrete)}, \quad \text{or} \quad \int_B (\ldots)\, dm \ \text{(continuum)},$$

lead to the same statements for discrete systems without much difficulty, almost automatically. See, for example, Kilmister and Reeve (1966, pp. 129–131).]

1.4 FORCE; LAW OF NEWTON–EULER

> [I]n the concept of force lies the chief difficulty in the whole of mechanics.
>
> (Hamel, 1952; as quoted in Truesdell, 1984, p. 527)

> Jeder weiβ aus der Erfahrung, was Schwerkraft ist; jede gerichtete Physikalische Gröβe, die sich mit der Schwerkraft in Gleichgewicht befinden kann, ist eine Kraft!
> [Approximate translation: Everyone knows from experience what gravity is; *every directed physical quantity that can be in equilibrium with gravity is a force!* (emphasis added).]
>
> (How Hamel used to begin his mechanics lectures; quoted in Szabó, 1954, p. 26)

> The fundamental law of mechanics [i.e. mass × acceleration = force] is a blank form which acquires a concrete content only when the conception of force occurring in it is filled in by physics.
>
> (Weyl, 1922, pp. 66–67)

Local Form of Newton–Euler Law

To each and every material particle P of elementary mass dm and inertial acceleration $\boldsymbol{a}$, of a body B or system S, we associate a *total* elementary force vector $d\boldsymbol{f}$ acting on it, such that

$$dm\,\boldsymbol{a} = d\boldsymbol{f}, \tag{1.4.1}$$

where $d\boldsymbol{f}$ itself is the resultant of other "partial" elementary forces of various origins (to be examined later); that is,

$$d\boldsymbol{f} = \sum d\boldsymbol{f}_k \qquad (k = 1, 2, \ldots). \tag{1.4.2}$$

Equation (1.4.1) is not simply a definition of one vector ($d\boldsymbol{f}$) in terms of another ($dm\,\boldsymbol{a}$), but is an equality of two physically very different vectors: one, *the effect* or *kinetic reaction* ($dm\,\boldsymbol{a}$), *depending only on the properties of the particle P itself*; and another, *the cause* ($d\boldsymbol{f}$), *depending on the interaction between P and the rest of the universe*—that is, on the action of the *external* world on the moving system, and the *mutual*, or *internal*, actions of the body parts on each other. Paraphrasing Hamel (1927, p. 3) slightly, we may state: The forces are determined by their "causes"; that is, by variables that represent the *geometrical*, *kinematical*, and *physical* state of the matter surrounding P (local causes) and away from it (global causes). This dependence is single-valued and, in general, continuous and differentiable; and, in addition, these forces are *objective*—that is, independent of the frame of reference (see also Hamel, 1949, pp. 509–512). In practice, this leads to *constitutive equations* for the forces (stresses) that, when combined with the *field*, or *ponderomotive*, equations (1.4.1) lead to relations of the form:

$$\boldsymbol{a} = \boldsymbol{a}(t, \boldsymbol{r}, \boldsymbol{v};\ \text{physical constants}); \tag{1.4.3}$$

where $\boldsymbol{a}$ may also depend on the $\boldsymbol{r}$'s and $\boldsymbol{v}$'s of *other system* (and even *external*) particles, but *not* on accelerations or other higher (than the first) $d/dt(\ldots)$-derivatives. Such an $\boldsymbol{a}$-dependence would introduce an *additional* constitutive, or constraint, equation of the form: $dm\,\boldsymbol{a} = d\boldsymbol{f}\ (\ldots, \boldsymbol{a}, \ldots)$. However, and this does not contradict (1.4.1), such equations can occur as *part of the solution process*; namely, through elimination of variables from the complete set of equations of the problem; that is, *elimination of forces related to the accelerations of other parts of the body*, so that the acceleration of point P depends on, among other things, the accelerations of points Q, R, ... On this delicate and sometimes confusing point, see Hamel (1949, p. 49). In view of such difficulties in defining the force, a number of authors (mostly continuum mechanicians) consider it as a *primitive concept*—along with *space*, *time*, and *mass*.

Force Classification

[This also includes moments; and, in analytical mechanics, both forces and moments are replaced by *system*, or *generalized*, *forces* (§3.4).]

The most important such classifications are as follows:

Newton–Euler (or momentum) mechanics:

Internal: originating wholly from *within* the system; in pairs. They depend on the spatial limits of the system.

External: originating, even partially, from *outside* the system. Only such forces appear in the corresponding equations of equilibrium/motion.

Lagrangean (or energetic) mechanics:

Impressed: depending, even partially, on physical (material) coefficients (chap. 3).
Constraint reactions: depending exclusively on the constraints; geometrical and/or kinematical forces (chap. 3).

Continuum mechanics:

Surface, or *contact*: continuously distributed over material *surfaces* (and/or *lines* and *points*).
Volume, or *body*: continuously distributed over material *volumes*.

Usually, a given force is a combination of the above, and more. For example:

Gravity: external, impressed, body;

Stresses in rigid bodies: internal, reactions, surface;

Stresses in elastic bodies: internal, impressed, surface;

Dry rolling friction: internal or external, reaction, surface;

Dry sliding friction: internal or external, impressed, surface.

Other, more specialized force classifications are the following: *potential/nonpotential*, *conservative/nonconservative*, *gyroscopic/nongyroscopic*, *circulatory/noncirculatory*, *autonomous/nonautonomous*, etc. They will be introduced later, if and when needed. Occasionally, forces are classified with the help of the momentum principles as follows:

Linear or translatory loads: forces;

Angular or rotatory loads: moments of forces and moments of couples;

but such terminology is not uniform. For example, the authoritative Truesdell and Toupin (1960, p. 531) states that, in the general case, *the (total) torque consists of two parts: the moment of the force(s) and the couple*; also, virtually alone among mechanics works, it refuses to use the term *internal* forces, opting instead for the term *mutual* (loc. cit., pp. 533–535).

On Centers of Gravity and Mass, and Centroid

The *center of gravity* (*CG*) of a material system in a *parallel* gravitational field is a point defined uniquely by

$$\mathbf{r}_{CG} = S\, \mathbf{r}\, dG \Big/ S\, dG, \qquad (1.4.4)$$

where dG = elementary gravity force = $g\,dm = \rho g\,dV \equiv \gamma\,dV$; g = acceleration of gravity; ρ = density of matter; γ = specific weight; dm = element of mass; dV = element of volume; and CG is *independent of the orientation of the system*, and through it passes the resultant gravity force, or weight, of the system, and: $S(\ldots)$: *material summation*, for a fixed time, and valid for discrete and/or continuous distributions (Stieltjes' integral). This helpful notation, originated informally by Lagrange, is used a lot in the main body of this work.

The *center of mass*, or *inertial center*, (CM) of a material distribution is defined uniquely by

$$\boldsymbol{r}_{CM} \equiv \boldsymbol{r}_G = \int \boldsymbol{r}\, dm \Big/ \int dm. \tag{1.4.5}$$

The *centroid* (or *geometrical center*, or *geometrical* center of gravity) (C) of a figure is defined uniquely by

$$\boldsymbol{r}_C = \int \boldsymbol{r}\, dV \Big/ \int dV. \tag{1.4.6}$$

- If g = constant, the gravitational field is *uniform*. Then, $\boldsymbol{g} = g\boldsymbol{u} = constant$, $\boldsymbol{u}$ = vertical unit vector (positive downward).
- If $\rho = constant$, the body (matter) is *homogeneous*.

In a uniform field:

$$\boldsymbol{r}_{CG} = \boldsymbol{r}_{CM} \equiv \boldsymbol{r}_G; \tag{1.4.7a}$$

For a homogeneous body:

$$\boldsymbol{r}_{CM} \equiv \boldsymbol{r}_G = \boldsymbol{r}_C; \tag{1.4.7b}$$

For a homogeneous body in a uniform field:

$$\boldsymbol{r}_{CG} = \boldsymbol{r}_{CM} = \boldsymbol{r}_C. \tag{1.4.7c}$$

REMARK

In *nonuniform* fields, eq. (1.4.7a) is no longer true: the parts of the body closer to the attracting earth experience stronger gravity forces than those farther from it; and, therefore, upon rotation of the body, *the point of application of the resultant of such forces changes relative to the body*; that is, *the center of gravity is no longer definable as a unique body-fixed point, independent of the orientation of the body relative to the field.* The center of mass and centroid, however, are still defined uniquely by (1.4.5) and (1.4.6), respectively. Such complications may arise in problems of astronautics/spacecraft dynamics; there, we replace the constant $\boldsymbol{g}$ with a *central–symmetric* gravitational *field.*

1.5 SPACE-TIME AND THE PRINCIPLE OF GALILEAN RELATIVITY

Galilean Transformations (GT)

These are frame of reference transformations that leave the Newton–Euler law (1.4.1) *form invariant*. The most general such transformations have the following form (fig. 1.3):

$$\boldsymbol{r}' = \boldsymbol{A} \cdot \boldsymbol{r} + \boldsymbol{b}t + \boldsymbol{c} \qquad (\textit{Direct/matrix}\ \text{notation}) \tag{1.5.1a}$$

or

$$x_{k'} = \sum A_{k'k} x_k + b_{k'} t + c_{k'} \qquad (\textit{Component}\ \text{notation}), \tag{1.5.1b}$$

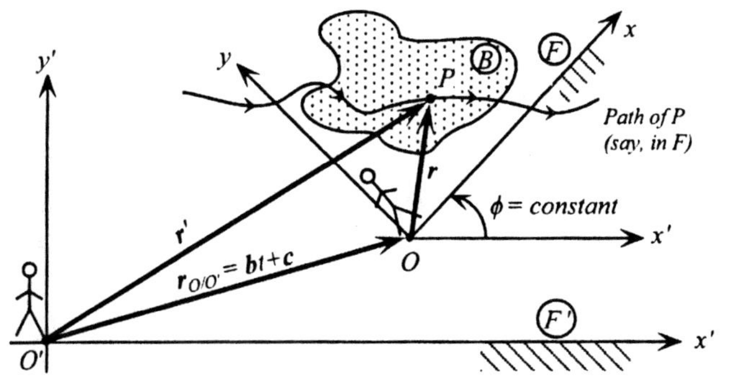

Figure 1.3 On the geometry of Galilean transformations.

where $A = (A_{k'k})$ is a *proper orthogonal tensor with constant components*—that is, $A^{-1} = A^{\mathrm{T}}$; *Det* $A = +1$; and $b = (b_{k'})$ and $c = (c_{k'})$ are *constant vectors*—that is, F and F' are in *nonrotating uniform motion* (*uniform translation*) relative to each other, with velocity b; and

$$t' = \alpha t + \beta, \tag{1.5.1c}$$

where t is measured in F and t' in F', and α, β are constant scalars; α depends on the *units of time*, while β depends on its *origin* in the two systems of time measurement. Hence, if these units are taken to be the same, and these origins are made to coincide, then $\alpha = 1$ and $\beta = 0$; in which case (henceforth assumed in this book),

$$t' = t; \tag{1.5.1d}$$

that is, *in classical (Newtonian) mechanics there is, essentially, only one time scale.*

From the transformation equations (1.5.1a–d) we immediately obtain the following:

$$d^2r'/dt^2 = A \cdot (d^2r/dt^2) \qquad \text{or} \qquad a' = A \cdot a, \tag{1.5.2a}$$

or, explicitly, with some easily understood notation,

$$d^2x'/dt^2 = \cos(x', x)(d^2x/dt^2) + \cos(x', y)(d^2y/dt^2) + \cos(x', z)(d^2z/dt^2), \text{ etc.;} \tag{1.5.2b}$$

that is, the accelerations of a particle P as measured in F and F' differ only by an ordinary (time-independent) *geometrical* transformation due to the, possibly, different orientation of their axes; and, therefore, they are *physically* equal: that is, unaffected by the relative motion of F and F'. Hence, we may take, with no loss in physical generality, the corresponding axes of F and F' to be *ever parallel*, in which case $A = \mathbf{1}$ (unit tensor), in which case (1.5.1a) simplifies to

$$r' = r + bt + c \;\Rightarrow\; a' = a. \tag{1.5.2c}$$

Since $dm|_F = dm|_{F'} \equiv dm$, and assuming that from $dm\, a = df(t, r, v)$ and (1.5.2c) it follows that

$$dm\, a' = df\,(t, r' - bt - c, dr'/dt - b) \equiv df\,'(t, r', dr'/dt \equiv v') \equiv df\,'; \tag{1.5.3}$$

that is, df is also invariant under *GT*, and, therefore, as far as the law of motion (1.4.1) is concerned, *there is no one (absolute) frame in which it holds, but, in fact, once one such "inertial" frame is established, there is a whole family of them dynamically equivalent to it.* More precisely, there is a (continuous linear) *group* that depends on *ten* (10) parameters: three for $\boldsymbol{A}$ [out of its nine components (direction cosines), due to the six orthonormality constraints, only three are independent], three for $\boldsymbol{b}$, three for $\boldsymbol{c}$, and one for β [equations (1.5.1c, d), $\alpha = 1$, with no loss in generality]. This Galilean, or Newtonian, *principle of relativity* can be summed up as follows: *an inertial frame—that is, one in which $dm(d^2\boldsymbol{r}/dt^2) = d\boldsymbol{f}$ holds—is determined only to within a Galilean transformation* (1.5.1a–d).

REMARKS

(i) The linear transformation (1.5.1c) can also be obtained by requiring that if

$$\boldsymbol{a} = d^2\boldsymbol{r}/dt^2 = \boldsymbol{0}, \tag{1.5.4a}$$

then also

$$d^2\boldsymbol{r}/d(t')^2 = \boldsymbol{0}, \tag{1.5.4b}$$

for arbitrary values of $\boldsymbol{r}$ and $d\boldsymbol{r}/dt$. Indeed, using chain rule, we find: $d\boldsymbol{r}/dt' = (d\boldsymbol{r}/dt)/(dt'/dt)$

$$\Rightarrow\ d^2\boldsymbol{r}/d(t')^2 = [(dt'/dt)(d^2\boldsymbol{r}/dt^2) - (d\boldsymbol{r}/dt)(d^2t'/dt^2)]\big/(dt'/dt)^3, \tag{1.5.4c}$$

and so, due to (1.5.4a), the requirement (1.5.4b) translates to

$$(d\boldsymbol{r}/dt)(d^2t'/dt^2) = \boldsymbol{0}, \qquad \text{for arbitrary } d\boldsymbol{r}/dt; \tag{1.5.4d}$$

that is,

$$d^2t'/dt^2 = 0 \ \Rightarrow\ t' = \alpha t + \beta, \qquad \alpha, \beta\text{: integration constants; Q.E.D.} \tag{1.5.4e}$$

(ii) The logical circularity involved in the classical mechanics definition of inertial frames (i.e., "if $dm\,\boldsymbol{a} = d\boldsymbol{f}$ holds, the frame is inertial" and "if the frame is inertial frame then $dm\,\boldsymbol{a} = d\boldsymbol{f}$ holds") can be resolved only by *relativistic* physics. Here, we are content to postulate the existence of frames in which $dm\,\boldsymbol{a} = d\boldsymbol{f}$ holds exactly (or, equivalently, of frames in which *forceless motions are also unaccelerated motions*; i.e., the position vectors are linear functions of time, and vice versa); and to call such frames inertial. For detailed discussions of this important topic, see any good text on the physical foundations of relativity; e.g., Bergmann, 1942; Nevanlina, 1968.

1.6 THE FUNDAMENTAL PRINCIPLES (OR BALANCE LAWS) OF GENERAL SYSTEM MECHANICS

> An Axiom is a proposition, the truth of which must be admitted as soon as the terms in which it is expressed are clearly understood ... physical axioms are axiomatic to those only who have sufficient knowledge of the action of physical causes to enable them to see their truth.
>
> (Thomson and Tait, 1912, part 1, section 243, p. 240)

Conservation of Mass (Euler, Early 1760s)

$$dm(B)/dt \equiv dm/dt = d/dt\Big(\mathcal{S}\, dm\Big) = d/dt\Big(\int \rho\, dV\Big) = \int d/dt(\rho\, dV) = 0. \tag{1.6.1a}$$

(Henceforth, we shall, usually, omit the subscripts V, ∂V, etc., in the various integrals.)

In the absence of discontinuities, the above leads to the local (differential) form:

$$d/dt(\rho\, dV) = 0 \;\Rightarrow\; \rho\, dV = constant = \rho_o\, dV_o$$

[*Material*, or *Lagrangean*, or *referential*, *equation of continuity*] (1.6.1b)

where $\rho_o(dV_o)$ = density (element of volume) in some initial or *reference* configuration.

Principle of Linear Momentum (Euler, Early 1750s)

$$d/dt\Big(\mathcal{S}\, \boldsymbol{v}\, dm\Big) = \mathcal{S}\, d\boldsymbol{f} \qquad \text{or} \qquad d\boldsymbol{p}/dt = \boldsymbol{f}, \tag{1.6.2a}$$

where

$$\boldsymbol{p}(B) \equiv \boldsymbol{p} \equiv \mathcal{S}\, \boldsymbol{v}\, dm = \int \rho \boldsymbol{v}\, dV: \quad \textit{Linear momentum of } B, \tag{1.6.2b}$$

a system vector that depends on the frame, but not on the (fixed) origin in it; equivalent to Newton's "quantitas motus"; and $\mathcal{S}\, d\boldsymbol{f} \equiv \boldsymbol{f}$. From the above, and invoking mass conservation [§1.3:(1.3.1)ff.), (1.6.1a, b)] and the definition of mass center (§1.4), we obtain

$$\boldsymbol{p} = m\boldsymbol{v}_G \;\Rightarrow\; m\boldsymbol{a}_G = \boldsymbol{f}, \tag{1.6.2c}$$

where $\boldsymbol{r}_G/\boldsymbol{v}_G/\boldsymbol{a}_G$ are, respectively, the position/velocity/acceleration vectors of the center of mass of B, G. Equation (1.6.2c) shows that *the motion of the center of mass G, of a body (or any material system, rigid or not), B, is identical to that of a fictitious particle of mass m located at G and acted upon by the body resultant on B,* $\boldsymbol{f}$; that is, by the vector sum of all (→ external) forces transported parallel to themselves to G. Thus, the motion of G is taken care of by this simple principle → theorem. But the remaining problem of the *motion of B about G* (and, generally, of the motion of other body points) is far more difficult, and, unlike the motion of G, does depend on the specific material constitution of B (e.g., rigid, elastic), as well as on its motion (i.e., 1-, 2-, 3-dimensional); and, therefore, that problem necessitates additional considerations, such as the following.

Principle of Angular Momentum (Euler, mid-1770s)

$$d/dt\Big(\mathcal{S}\,(\boldsymbol{r} \times \boldsymbol{v}\, dm)\Big) = \mathcal{S}\,(\boldsymbol{r} \times d\boldsymbol{f}) \qquad \text{or} \qquad d\boldsymbol{H}_O/dt = \boldsymbol{M}_O, \tag{1.6.3a}$$

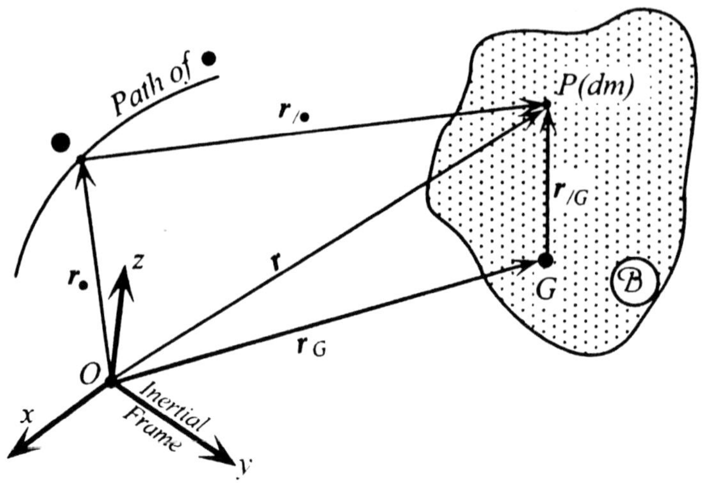

Figure 1.4 On the meaning of absolute and relative angular momentum.

where

$$H_O \equiv S(r \times v\, dm): \textit{absolute angular momentum}, \text{ about the \textit{fixed} point } O, \tag{1.6.3b}$$

and

$$M_O \equiv S(r \times df): \textit{total moment} \text{ about } O \text{ (fig. 1.4)}. \tag{1.6.3c}$$

Other angular momenta, and their interrelations, are detailed in "Additional Forms of the Angular Momentum," below.

External and Internal Loads

In the Newton–Euler approach to system mechanics, whether discrete or continuous, we classify body and/or surface forces and moments as *internal* or *mutual* (i.e., those due *exclusively* to internal causes) and *external* [i.e., those whose cause(s) lie, even *partially*, outside of the body or system]. Stresses are caused by one or more of the following: (i) *deformations* (solids); (ii) *flows* (gases, liquids); (iii) *geometrical/kinematical constraints* [e.g., incompressibility, inextensibility (= incompressibility in one or two dimensions)].

Analytical mechanics necessitates a different force/moment classification (chap. 3).

Principle of Action–Reaction

(i) *Discrete version.* Let us consider a system of N particles $\{P_k;\ k = 1, \ldots, N\}$. Each particle P_k is acted upon by a total *external* (to that system) force $f_{k,\text{ext}}$ and a total *internal* force $f_{k,\text{int}}$ due to the other $N - 1$ particles:

$$f_{k,\text{int}} = \sum f_{kl}, \qquad \text{with } l \neq k;\ \text{i.e., } f_{kk} \text{ is, as yet, undefined(!)} \tag{1.6.4a}$$

Now, by *Newton's third law of motion* (*action–reaction*) we shall understand the *constitutive* (i.e., *physical*) postulate;

(a) $\boldsymbol{f}_{kl} = -\boldsymbol{f}_{lk}$ and $\boldsymbol{f}_{kk} = \mathbf{0}$ (i.e., the particle cannot act on itself!) (1.6.4b)

and

(b) $(\boldsymbol{r}_k - \boldsymbol{r}_l) \times \boldsymbol{f}_{kl} = \mathbf{0}$ (i.e., the internal forces are *central* and *opposite*; or *oppositely directed pair by pair* and *collinear*). (1.6.4c)

[The *second* of (1.6.4b) is *not* included in the original Newtonian formulation. We follow Hamel (1949, p. 51).]

In the discrete/particle model, so popular among physicists and such an anathema among certain mechanicians, this postulate, plus the principle of linear momentum, lead to the *theorem* of angular momentum for the *external* loads only. However, the *converse* is not necessarily true; that is, *the angular momentum equation for a finite body* $d\boldsymbol{H}_O/dt = \boldsymbol{M}_{O,\text{external}}$ *does not necessarily lead to* (1.6.4b, c); other combinations of the internal forces may lead to the same effect (e.g., a sum of terms may vanish in a number of different ways). The converse *may* hold if we assume the validity of the angular momentum equation for *any* part of the system, or for any size subsystem.

(ii) *Continuum version.* For every pair of particles P_1 and P_2, with respective positions $\boldsymbol{r}_1$ and $\boldsymbol{r}_2$, the mutual forces and moments satisfy the following *constitutive* postulate:

$$d\boldsymbol{f}(\boldsymbol{r}_1, \boldsymbol{r}_2) = -d\boldsymbol{f}(\boldsymbol{r}_2, \boldsymbol{r}_1) \quad \text{and} \quad d\boldsymbol{M}(\boldsymbol{r}_1, \boldsymbol{r}_2) = -d\boldsymbol{M}(\boldsymbol{r}_2, \boldsymbol{r}_1). \tag{1.6.4d}$$

Without (1.6.4d), or something equivalent that supplies knowledge of the internal loads, the problem of mechanics would, in general, be *indeterminate* (i.e., the adopted model would produce more unknowns than the number of scalar equations furnished by its laws).

Additional Forms of the Angular Momentum

Although the results derived below hold for any body or system, they become useful only for *rigid* ones. We define the following *two* kinds of (inertial) angular momentum (fig. 1.4):

$$\boldsymbol{H}_{\bullet,\text{absolute}} \equiv \boldsymbol{H}_\bullet \equiv \mathcal{S}\,(\boldsymbol{r} - \boldsymbol{r}_\bullet) \times dm\,\boldsymbol{v} \equiv \mathcal{S}\,\boldsymbol{r}_{/\bullet} \times dm\,\boldsymbol{v}: \qquad [\boldsymbol{v} \equiv d\boldsymbol{r}/dt]$$

Absolute angular momentum of body B, about the arbitrarily moving point •

[because it involves the absolute (inertial) velocity $\boldsymbol{v} \equiv d\boldsymbol{r}/dt$], (1.6.5a)

and

$$\boldsymbol{H}_{\bullet,\text{relative}} \equiv \boldsymbol{h}_\bullet \equiv \mathcal{S}\,(\boldsymbol{r} - \boldsymbol{r}_\bullet) \times dm\,(\boldsymbol{v} - \boldsymbol{v}_\bullet) \equiv \mathcal{S}\,\boldsymbol{r}_{/\bullet} \times dm\,\boldsymbol{v}_{/\bullet}:$$

Relative angular momentum of body B, about the arbitrarily moving point •

[because it involves the relative (inertial) velocity $\boldsymbol{v} - \boldsymbol{v}_\bullet \equiv \boldsymbol{v}_{/\bullet}$]. (1.6.5b)

REMARKS

(i) Although these kinematico-inertial definitions hold for any frame of reference (with $\boldsymbol{r}$, $\boldsymbol{r}_\bullet$, $\boldsymbol{v}$, $\boldsymbol{v}_\bullet$ denoting the positions and velocities relative to points fixed or moving with respect to that frame—see §1.7), they will normally be understood to refer to a specific *inertial* frame, unless explicitly stated to the contrary.

(ii) Some authors define absolute angular momentum as in our (1.6.5a), but only for *fixed* points (i.e., $\boldsymbol{v}_\bullet = \boldsymbol{0}$); in which case, clearly, (1.6.5a) and (1.6.5b) coincide. Unfortunately, here too, there is no uniformity of terminology and or notation in the literature; but, as will be seen in kinetics, some angular momenta are more useful than others. The connection between the above two angular momenta is given by the following basic theorem.

THEOREM

The angular momenta $\boldsymbol{H}_\bullet$ and $\boldsymbol{h}_\bullet$, defined by equations (1.6.5a, b), are related by

$$\boldsymbol{H}_\bullet - \boldsymbol{h}_\bullet = m(\boldsymbol{r}_G - \boldsymbol{r}_\bullet) \times \boldsymbol{v}_\bullet \equiv m\boldsymbol{r}_{G/\bullet} \times \boldsymbol{v}_\bullet. \tag{1.6.5c}$$

PROOF

Subtracting (1.6.5b) from (1.6.5a) side by side, and then utilizing the properties of the center of mass of B, G, we obtain

$$\begin{aligned}\boldsymbol{H}_\bullet - \boldsymbol{h}_\bullet &= S\, \boldsymbol{r}_{/\bullet} \times (\boldsymbol{v} - \boldsymbol{v}_{/\bullet})\, dm = S\, (\boldsymbol{r}_{/\bullet} \times \boldsymbol{v}_\bullet)\, dm \\ &= S\, \boldsymbol{r} \times (dm\, \boldsymbol{v}_\bullet) - S\, (\boldsymbol{r}_\bullet \times \boldsymbol{v}_\bullet)\, dm = (m\boldsymbol{r}_G) \times \boldsymbol{v}_\bullet - \boldsymbol{r}_\bullet \times (m\boldsymbol{v}_\bullet), \quad \text{Q.E.D.}\end{aligned} \tag{1.6.5d}$$

Equations (1.6.5c, d) show immediately that, in the following three cases, the difference between absolute and relative angular momentum disappears:

(i) $\boldsymbol{r}_{G/\bullet} = \boldsymbol{0}$, i.e., $\bullet = G$: $\quad \boldsymbol{H}_G = \boldsymbol{h}_G = S\, \boldsymbol{r}_{/G} \times (dm\, \boldsymbol{v}_{/G})$, (1.6.5e)

(ii) $\boldsymbol{v}_\bullet = \boldsymbol{0}$, i.e., $\bullet$ = fixed origin, say O: $\quad \boldsymbol{H}_O = \boldsymbol{h}_O = S\, \boldsymbol{r} \times (dm\, \boldsymbol{v})$, (1.6.5f)

(iii) $\boldsymbol{r}_{G/\bullet}$ parallel to $\boldsymbol{v}_\bullet$. (1.6.5g)

The first and second cases, (1.6.5e, f), are, by far, the most important; (1.6.5g) may be hard to check *before* solving the (kinetic) problem.

Next, let us relate $\boldsymbol{H}_\bullet$ and $\boldsymbol{h}_\bullet$ with $\boldsymbol{H}_O$ (which appears in the basic Eulerian form of the angular momentum principle). We have, successively,

$$\begin{aligned}\boldsymbol{H}_O &= S\, \boldsymbol{r} \times (dm\, \boldsymbol{v}) \qquad \text{(introducing positions/velocities } \textit{relative} \text{ to } \bullet) \\ &= S\, [(\boldsymbol{r}_\bullet + \boldsymbol{r}_{/\bullet}) \times dm\, (\boldsymbol{v}_\bullet + \boldsymbol{v}_{/\bullet})] \\ &= \cdots = \boldsymbol{h}_\bullet + m(\boldsymbol{r}_\bullet \times \boldsymbol{v}_G) + m(\boldsymbol{r}_{G/\bullet} \times \boldsymbol{v}_\bullet), \qquad (1.6.5\text{h}) \\ &= \boldsymbol{H}_\bullet + m(\boldsymbol{r}_\bullet \times \boldsymbol{v}_G) \qquad [\text{thanks to } (1.6.5\text{c})]. \qquad (1.6.5\text{i})\end{aligned}$$

The above leads easily to the following corollaries:

(i) If $\bullet = \textit{fixed} \Rightarrow \boldsymbol{v}_\bullet = \boldsymbol{0}$, then

$$\boldsymbol{H}_O = \boldsymbol{H}_\bullet + \boldsymbol{r}_\bullet \times (m\boldsymbol{v}_G) = \boldsymbol{h}_\bullet + m(\boldsymbol{r}_\bullet \times \boldsymbol{v}_G) \qquad [\boldsymbol{r}_\bullet \equiv \boldsymbol{r}_{\bullet/O},\ \boldsymbol{H}_\bullet = \boldsymbol{h}_\bullet]; \tag{1.6.5j}$$

a slight generalization over (1.6.5f).

(ii) If $\bullet = G$, then

$$\boldsymbol{H}_O = \boldsymbol{H}_G + \boldsymbol{r}_G \times (m\boldsymbol{v}_G) = \boldsymbol{h}_G + m(\boldsymbol{r}_G \times \boldsymbol{v}_G)$$
$$[\boldsymbol{r}_G \equiv \boldsymbol{r}_{G/O},\ \boldsymbol{v}_G \equiv d\boldsymbol{r}_G/dt; \quad \boldsymbol{H}_G = \boldsymbol{h}_G]. \tag{1.6.5k}$$

By comparing (1.6.5h,i) with (1.6.5k), it can be seen that

$$\boldsymbol{H}_\bullet = \boldsymbol{H}_G + \boldsymbol{r}_{G/\bullet} \times (m\boldsymbol{v}_G), \qquad \boldsymbol{h}_\bullet = \boldsymbol{H}_G + \boldsymbol{r}_{G/\bullet} \times (m\boldsymbol{v}_{G/\bullet}). \tag{1.6.5l}$$

(Interpret these "transfer" equations geometrically. What happens if • is *fixed*; say, an origin O?) Finally, by applying the transfer equations (1.6.5h, i) between O and the arbitrarily moving points 1 and 2, and then comparing, we can obtain the relation between the absolute, relative, and absolute–relative angular momenta of a body: $\boldsymbol{H}_1 \leftrightarrow \boldsymbol{H}_2$, $\boldsymbol{H}_1 \leftrightarrow \boldsymbol{h}_2$, $\boldsymbol{h}_1 \leftrightarrow \boldsymbol{h}_2$.

Additional Forms of the Principle of Angular Momentum

With the help of the preceding kinematico-inertial identities/results, and the purely geometrical theorem of transfer of moments (hopefully well known from elementary statics)

$$\begin{aligned} \boldsymbol{M}_\bullet &= \boldsymbol{M}_G + \boldsymbol{r}_{G/\bullet} \times \boldsymbol{f} && [\text{where the force resultant } \boldsymbol{f} \text{ goes through } G] \\ &= \boldsymbol{M}_G + \boldsymbol{r}_{G/\bullet} \times (m\boldsymbol{a}_G) && [\text{by the principle of } \textit{linear} \text{ momentum}], \end{aligned} \tag{1.6.6a}$$

the Eulerian principle of angular momentum

$$\mathsf{S}\, \boldsymbol{r} \times (dm\, \boldsymbol{a}) = d/dt\Big(\mathsf{S}\, \boldsymbol{r} \times (dm\, \boldsymbol{v})\Big) = \mathsf{S}\, \boldsymbol{r} \times d\boldsymbol{f};$$

that is,

$$d\boldsymbol{H}_O/dt = \boldsymbol{M}_O \tag{1.6.6b}$$

[$\Rightarrow \boldsymbol{M}_{O,\text{external}}$, by action–reaction (plus, in the continuum version, of *Boltzmann's axiom* $\Rightarrow$ *symmetry* of the stress tensor)],

assumes the following forms:

Center of Mass Form

By (1.6.5k):

$$d\boldsymbol{H}_O/dt = d/dt[\boldsymbol{H}_G + \boldsymbol{r}_G \times (m\boldsymbol{v}_G)] = d\boldsymbol{H}_G/dt + m(\boldsymbol{r}_G \times \boldsymbol{a}_G), \tag{1.6.6c}$$

and by (1.6.6a), for $\bullet \to O$:

$$\boldsymbol{M}_O = \boldsymbol{M}_G + \boldsymbol{r}_G \times (m\boldsymbol{a}_G); \tag{1.6.6d}$$

and comparing these expressions with (1.6.6b), we obtain the fundamental form

$$\boldsymbol{M}_G = d\boldsymbol{H}_G/dt \ (= d\boldsymbol{h}_G/dt). \tag{1.6.6e}$$

Absolute Form

Using the above, and (1.6.5l), we obtain, successively,

$$\begin{aligned} \boldsymbol{M}_\bullet &= \boldsymbol{M}_G + \boldsymbol{r}_{G/\bullet} \times (m\boldsymbol{a}_G) = d\boldsymbol{H}_G/dt + \boldsymbol{r}_{G/\bullet} \times (m\boldsymbol{a}_G) \\ &= d/dt\,[\boldsymbol{H}_\bullet - \boldsymbol{r}_{G/\bullet} \times (m\boldsymbol{v}_G)] + \boldsymbol{r}_{G/\bullet} \times (m\boldsymbol{a}_G) \\ &= d\boldsymbol{H}_\bullet/dt - \boldsymbol{v}_{G/\bullet} \times (m\boldsymbol{v}_G) - \boldsymbol{r}_{G/\bullet} \times (m\boldsymbol{a}_G) + \boldsymbol{r}_{G/\bullet} \times (m\boldsymbol{a}_G); \end{aligned}$$

that is, finally,

$$\boldsymbol{M}_\bullet = d\boldsymbol{H}_\bullet/dt - \boldsymbol{v}_{G/\bullet} \times (m\boldsymbol{v}_G) \qquad (\text{using } \boldsymbol{v}_{G/\bullet} \equiv \boldsymbol{v}_G - \boldsymbol{v}_\bullet)$$
$$= d\boldsymbol{H}_\bullet/dt + \boldsymbol{v}_\bullet \times (m\boldsymbol{v}_G) = d\boldsymbol{H}_\bullet/dt + \boldsymbol{v}_\bullet \times (m\boldsymbol{v}_{G/\bullet}). \qquad (1.6.6f)$$

Relative Form

Similarly, using the above, and (1.6.5l), we obtain, successively,

$$\boldsymbol{M}_\bullet = \boldsymbol{M}_G + \boldsymbol{r}_{G/\bullet} \times (m\boldsymbol{a}_G) = d\boldsymbol{H}_G/dt + \boldsymbol{r}_{G/\bullet} \times (m\boldsymbol{a}_G)$$
$$= d/dt\left(\boldsymbol{h}_\bullet - \boldsymbol{r}_{G/\bullet} \times (m\boldsymbol{v}_{G/\bullet})\right) + \boldsymbol{r}_{G/\bullet} \times (m\boldsymbol{a}_G)$$
$$= d\boldsymbol{h}_\bullet/dt - \boldsymbol{v}_{G/\bullet} \times (m\boldsymbol{v}_{G/\bullet}) - \boldsymbol{r}_{G/\bullet} \times (m\boldsymbol{a}_{G/\bullet}) + \boldsymbol{r}_{G/\bullet} \times (m\boldsymbol{a}_G);$$

that is, finally,

$$\boldsymbol{M}_\bullet = d\boldsymbol{h}_\bullet/dt + \boldsymbol{r}_{G/\bullet} \times (m\boldsymbol{a}_\bullet). \qquad (1.6.6g)$$

In particular, if $\bullet$ is *fixed*, then (1.6.6f,g) reduce at once to

$$\boldsymbol{M}_\bullet = d\boldsymbol{H}_\bullet/dt \quad (= d\boldsymbol{h}_\bullet/dt); \qquad (1.6.6h)$$

which, since it holds for *any* fixed point, is a slight generalization of (1.6.6b). These forms show clearly the importance of *fixed points* and of the *center of mass*, above all other points, in rotational dynamics, especially rigid-body dynamics. All these forms of the principle of angular momentum, and many more flowing from them, can be quite confusing, they are almost impossible to remember, and may be error-prone in concrete applications. They are stated here only for comparison purposes with the existing literature. From them, the most useful in both theoretical and practical situations, are, by far, (1.6.6b,e), and, secondarily, (1.6.6a) with (1.6.6e). We summarize them here:

$$\boldsymbol{M}_O = d\boldsymbol{H}_O/dt \quad \left\{\equiv d/dt\left(\textstyle\int \boldsymbol{r} \times (dm\boldsymbol{v})\right)\right\}, \qquad O\text{: fixed origin}; \qquad (1.6.6i)$$

$$\boldsymbol{M}_G = d\boldsymbol{H}_G/dt \quad \left\{\equiv d/dt\left(\textstyle\int \boldsymbol{r}_{/G} \times (dm\boldsymbol{v}_{/G})\right)\right\}, \qquad G\text{: center of mass}; \qquad (1.6.6j)$$

$$\boldsymbol{M}_\bullet = d\boldsymbol{H}_G/dt + \boldsymbol{r}_{G/\bullet} \times (m\boldsymbol{a}_G), \qquad \bullet\text{: arbitrarily moving spatial point}, \qquad (1.6.6k)$$

or, compactly,

$$\text{Kinetic vectors ("torsor") at } G\text{: } (m\boldsymbol{a}_G, d\boldsymbol{H}_G/dt)$$
$$\sim \text{Kinetic torsor at } \bullet\text{: } \left(m\boldsymbol{a}_G, d\boldsymbol{H}_G/dt + \boldsymbol{r}_{G/\bullet} \times (m\boldsymbol{a}_G)\right);$$

and we are reminded that their left sides, by action–reaction (plus Boltzmann's axiom, i.e., *symmetry* of stress tensor), include only *external* moments and couples.

By comparing the absolute and relative forms of the principle of angular momentum, eqs. (1.6.6f, g) [or by $d/dt(\ldots)$, eq. (1.6.5c)], we can show that

$$d\boldsymbol{H}_\bullet/dt - d\boldsymbol{h}_\bullet/dt = \boldsymbol{r}_{G/\bullet} \times (m\boldsymbol{a}_\bullet) + \boldsymbol{v}_{G/\bullet} \times (m\boldsymbol{v}_\bullet)$$
$$= \boldsymbol{r}_{G/\bullet} \times (m\boldsymbol{a}_\bullet) + \boldsymbol{v}_G \times (m\boldsymbol{v}_\bullet) = \boldsymbol{r}_{G/\bullet} \times (m\boldsymbol{a}_\bullet) + \boldsymbol{v}_G \times (m\boldsymbol{v}_{\bullet/G}). \qquad (1.6.6l)$$

Finally, crossing the local law of motion $dm\,\boldsymbol{a} = d\boldsymbol{f}$ with $\boldsymbol{r}_{/\bullet} \equiv \boldsymbol{r} - \boldsymbol{r}_{\bullet}$, and then integrating over the body, etc., we obtain the following additional form of the principle of angular momentum:

$$\boldsymbol{M}_{\bullet} = d\boldsymbol{H}_O/dt - \boldsymbol{r}_{\bullet} \times (m\boldsymbol{a}_G) \quad (= \boldsymbol{M}_O - \boldsymbol{r}_{\bullet} \times \boldsymbol{f}, \text{ with } \boldsymbol{f} \text{ applied at } \bullet). \qquad (1.6.6\text{m})$$

1.7 ACCELERATED (NONINERTIAL) FRAMES OF REFERENCE (OR RELATIVE MOTION, OR MOVING AXES); ANGULAR VELOCITY AND ACCELERATION

The theory of moving axes, a subject indispensable to rigid-body dynamics and other key areas of mechanics (including the transition to relativity), is based on the following fundamental kinematical theorem.

Theorem (of Moving Axes)

Let us consider two frames of reference in arbitrary relative motion, each represented by an ortho–normal–dextral (OND) basis and associated coordinate axes, rigidly attached to the frame; say, for concreteness but no loss in generality, one *fixed* or *inertial F*:

$$(O_F - \boldsymbol{I}, \boldsymbol{J}, \boldsymbol{K}/X, Y, Z) \equiv (O_F - \boldsymbol{u}_X, \boldsymbol{u}_Y, \boldsymbol{u}_Z/X, Y, Z) \equiv (O_F - \boldsymbol{u}_{X,Y,Z}/X, Y, Z)$$
$$\equiv (O_F - \boldsymbol{u}_{k'}/x_{k'};\ k' = 1, 2, 3/X, Y, Z), \qquad (1.7.1\text{a})$$

and one *moving* or *noninertial M*:

$$(O_M - \boldsymbol{i}, \boldsymbol{j}, \boldsymbol{k}/x, y, z) \equiv (O_M - \boldsymbol{u}_x, \boldsymbol{u}_y, \boldsymbol{u}_z/x, y, z) \equiv (O_M - \boldsymbol{u}_{x,y,z}/x, y, z)$$
$$\equiv (O_M - \boldsymbol{u}_k/x_k;\ k = 1, 2, 3/x, y, z), \qquad (1.7.1\text{b})$$

and an arbitrary (say free) vector $\boldsymbol{p}$ [fig. 1.5(a)]. Then its rate of change in F and M, $d\boldsymbol{p}/dt$ and $d'\boldsymbol{p}/dt$, respectively, are related by

$$d\boldsymbol{p}/dt = d'\boldsymbol{p}/dt + \boldsymbol{\omega} \times \boldsymbol{p}, \qquad (1.7.2\text{a})$$

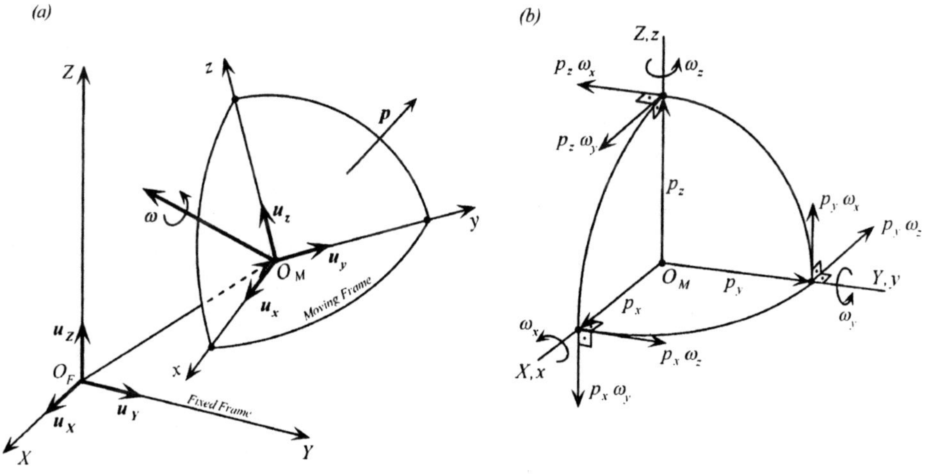

Figure 1.5 (a) Geometry of moving frames; (b) geometrical proof of (1.7.3c).

where (recalling the moving axes theory, §1.1)

$$p = p_X u_X + p_Y u_Y + p_Z u_Z \equiv \sum p_{k'} u_{k'} = p_x u_x + p_y u_y + p_z u_z \equiv \sum p_k u_k,$$

[*assumed* instantaneous representation of p in F and M]; (1.7.2b)

$$dp/dt \equiv (dp_X/dt)u_X + (dp_Y/dt)u_Y + (dp_Z/dt)u_Z = \sum (dp_{k'}/dt)u_{k'}:$$

Absolute rate of change of p (or *time flux*); i.e., relative to F; (1.7.2c)

$$d'p/dt \equiv (dp_x/dt)u_x + (dp_y/dt)u_y + (dp_z/dt)u_z = \sum (dp_k/dt)u_k:$$

Relative rate of change of p; i.e., relative to M; (1.7.2d)

$$\omega = \omega_X u_X + \omega_Y u_Y + \omega_Z u_Z \equiv \sum \omega_{k'} u_{k'} = \omega_x u_x + \omega_y u_y + \omega_z u_z \equiv \sum \omega_k u_k$$

$$\equiv [(du_y/dt)\cdot u_z]u_x + [(du_z/dt)\cdot u_x]u_y + [(du_x/dt)\cdot u_y]u_z:$$

Angular velocity (vector) of $M_{\text{moving frame}}$ *relative to* $F_{\text{fixed frame}}$;

i.e., *of* $(O_M - u_{k'})$ *relative to* $(O_F - u_k)$; (1.7.2e)

$\omega \times p =$ *Transport rate of change of* p *relative to* F. (1.7.2f)

NOTATIONAL CLARIFICATION

In this book, *primed derivatives*, $d'(\ldots)/dt$, are always associated with *moving* frame(s); while, for simplicity, primed *subscripts* signify *fixed* axes/components.

To express this theorem in components, which is the best way to understand it, the simplest way is to *choose the axes* O_F–XYZ and O_M–xyz so that, instantaneously, *either they coincide or are parallel.* Then, since in such a case,

$$(dp/dt)_X \equiv (dp/dt)\cdot u_X \equiv dp_X/dt = (dp/dt)\cdot u_x \equiv (dp/dt)_x, \quad \text{etc., cyclically,} \tag{1.7.3a}$$

$$(d'p/dt)_x \equiv (d'p/dt)\cdot u_x \equiv dp_x/dt = (d'p/dt)\cdot u_X \equiv (d'p/dt)_X, \quad \text{etc., cyclically,} \tag{1.7.3b}$$

the theorem assumes the component form:

$$\begin{aligned} dp_X/dt &= dp_x/dt + \omega_y p_z - \omega_z p_y, \\ dp_Y/dt &= dp_y/dt + \omega_z p_x - \omega_x p_z, \\ dp_Z/dt &= dp_z/dt + \omega_x p_y - \omega_y p_x; \end{aligned} \tag{1.7.3c}$$

and gives inertial rates of change, but expressed in terms of noninertial (relative) and transport rates. The above show clearly that

$$(dp/dt)_k \neq dp_k/dt \qquad (k = x, y, z); \tag{1.7.3d}$$

even though, instantaneously,

$$p_X = p_x, \quad \text{etc., cyclically,} \tag{1.7.3e}$$

unless $\omega \times p = 0$ ($\Rightarrow \omega = 0$, or $p = 0$, or ω parallel to p).

A geometrical interpretation of (1.7.3c) is shown in fig. 1.5(b): the moving axes O_M–xyz momentarily coincide with the axes O_M–XYZ; the latter are always translating relative to O_F–XYZ—that is, they are "rotationally equivalent" to them.

PROOF OF EQUATION (1.7.2a)

By $d/dt(\ldots)$-differentiating (1.7.2b), we obtain

$$d\boldsymbol{p}/dt = (dp_x/dt)\boldsymbol{u}_x + p_x(d\boldsymbol{u}_x/dt) + \cdots = d'\boldsymbol{p}/dt + \sum p_k(d\boldsymbol{u}_k/dt). \qquad (1.7.4a)$$

To transform the key second term in the above, we begin by $d/dt(\ldots)$-differentiating the six *geometrical orthonormality* ($\Rightarrow$ *rigidity*) *constraints* of these basis vectors $\boldsymbol{u}_k \cdot \boldsymbol{u}_l = \delta_{kl}$ $(k, l = x, y, z)$, thus translating them into the following six *kinematical constraints*:

$$(d\boldsymbol{u}_k/dt) \cdot \boldsymbol{u}_l + \boldsymbol{u}_k \cdot (d\boldsymbol{u}_l/dt) = 0; \qquad (1.7.4b)$$

that is, from the nine components of $\{d\boldsymbol{u}_k/dt\}$ only $9 - 6 = 3$ are *independent.*

Let us find them. By (1.7.4b) for k, $l = x$, $d\boldsymbol{u}_x/dt$ is perpendicular to $\boldsymbol{u}_x$; that is, it must lie in the plane of $\boldsymbol{u}_y$, $\boldsymbol{u}_z$. Therefore, we can write

$$d\boldsymbol{u}_x/dt = l_1\boldsymbol{u}_y + l_2\boldsymbol{u}_z; \qquad (1.7.4c)$$

and, cyclically,

$$d\boldsymbol{u}_y/dt = l_3\boldsymbol{u}_z + l_4\boldsymbol{u}_x, \qquad d\boldsymbol{u}_z/dt = l_5\boldsymbol{u}_x + l_6\boldsymbol{u}_y; \qquad (1.7.4d)$$

where $l_{1,\ldots,6}$ are scalar functions of time. Substituting these representations back into (1.7.4b) for $k = x$, $l = y$, and taking into account the geometrical constraints, we obtain

$$(d\boldsymbol{u}_x/dt) \cdot \boldsymbol{u}_y + \boldsymbol{u}_x \cdot (d\boldsymbol{u}_y/dt) = 0 \;\Rightarrow\; l_1 + l_4 = 0; \qquad (1.7.4e)$$

and, cyclically,

$$(d\boldsymbol{u}_y/dt) \cdot \boldsymbol{u}_z + \boldsymbol{u}_y \cdot (d\boldsymbol{u}_z/dt) = 0 \;\Rightarrow\; l_3 + l_6 = 0, \qquad (1.7.4f)$$

$$(d\boldsymbol{u}_z/dt) \cdot \boldsymbol{u}_x + \boldsymbol{u}_z \cdot (d\boldsymbol{u}_x/dt) = 0 \;\Rightarrow\; l_5 + l_2 = 0. \qquad (1.7.4g)$$

Hence, (1.7.4c,d) can be rewritten in terms of the following *three* independent (unconstrained) l's, or in terms of the three equivalent parameters ω_x, ω_y, ω_z:

$$l_1 = -l_4 \equiv \omega_z, \qquad l_3 = -l_6 \equiv \omega_x, \qquad l_5 = -l_2 \equiv \omega_y, \qquad (1.7.4h)$$

as

$$\begin{aligned} d\boldsymbol{u}_x/dt &= \omega_z\boldsymbol{u}_y - \omega_y\boldsymbol{u}_z = \boldsymbol{\omega} \times \boldsymbol{u}_x, \\ d\boldsymbol{u}_y/dt &= \omega_x\boldsymbol{u}_z - \omega_z\boldsymbol{u}_x = \boldsymbol{\omega} \times \boldsymbol{u}_y, \qquad (1.7.4i) \\ d\boldsymbol{u}_z/dt &= \omega_y\boldsymbol{u}_x - \omega_x\boldsymbol{u}_y = \boldsymbol{\omega} \times \boldsymbol{u}_z; \end{aligned}$$

where

$$\boldsymbol{\omega} = \omega_x \boldsymbol{u}_x + \omega_y \boldsymbol{u}_y + \omega_z \boldsymbol{u}_z$$
$$= \boldsymbol{u}_x[(d\boldsymbol{u}_y/dt)\cdot\boldsymbol{u}_z] + \boldsymbol{u}_y[(d\boldsymbol{u}_z/dt)\cdot\boldsymbol{u}_x] + \boldsymbol{u}_z[(d\boldsymbol{u}_x/dt)\cdot\boldsymbol{u}_y]$$
[a form that shows the *cyclicity* of the subscripts x, y, z]
$$= -\boldsymbol{u}_x[(d\boldsymbol{u}_z/dt)\cdot\boldsymbol{u}_y] - \boldsymbol{u}_y[(d\boldsymbol{u}_x/dt)\cdot\boldsymbol{u}_z] - \boldsymbol{u}_z[(d\boldsymbol{u}_y/dt)\cdot\boldsymbol{u}_x]. \qquad (1.7.4j)$$

Finally, substituting these results into (1.7.4a), we obtain (1.7.2a):

$$d\boldsymbol{p}/dt = d'\boldsymbol{p}/dt + \sum p_k(\boldsymbol{\omega}\times\boldsymbol{u}_k) = d'\boldsymbol{p}/dt + \boldsymbol{\omega}\times\left(\sum p_k\boldsymbol{u}_k\right)$$
$$= d'\boldsymbol{p}/dt + \boldsymbol{\omega}\times\boldsymbol{p}. \qquad (1.7.4k)$$

REMARKS

(i) Frequently, and with some good reason, the notation $\delta\boldsymbol{p}/\delta t$ is employed for our $d'\boldsymbol{p}/dt$. Here, however, we chose the latter because in analytical mechanics $\delta(\ldots)$ is reserved for *virtual* changes, under which $\delta t = 0$ (chap. 2 ff.). Other popular notations for the relative rate of change are $\partial\boldsymbol{p}/\partial t$ or $\partial^*\boldsymbol{p}/\partial t$ (British authors; but some German authors use $\partial\boldsymbol{p}/\partial t$ for our $\boldsymbol{\omega}\times\boldsymbol{p}$), $(d\boldsymbol{p}/dt)_M$ or $(d\boldsymbol{p}/dt)_{\mathrm{rel}}$ or $d^*\boldsymbol{p}/dt$; or with a tilde over d (Soviet/Russian authors) $\tilde{d}$. Also recall remarks made regarding eq. (1.1.20i) about the *overdot* notation.

(ii) The *vector* equation (1.7.2a) can be expressed in component form (i.e., it can be projected) along any axes, fixed or moving, by eqs. (1.7.3c), if O_M–xyz and O_M–XYZ momentarily coincide; and, if they do not, by

$$(d\boldsymbol{p}/dt)_x = \cos(x,X)(dp_X/dt) + \cos(x,Y)(dp_Y/dt) + \cos(x,Z)(dp_Z/dt) \quad (\neq dp_x/dt)$$
$$= \cos(x,X)(dp_1/dt + \omega_2 p_3 - \omega_3 p_2) + \cdots, \qquad (1.7.5a)$$

where the new axes O_M–123 coincide momentarily with O_M–XYZ, but, in general, have an angular velocity $\boldsymbol{\omega}' = (\omega_1, \omega_2, \omega_3)$ relative to them.

(iii) The above show that as long as no rates of change are involved, the components of a vector along the various axes (fixed or moving) are related by ordinary *coordinate* transformations, with possibly *time-dependent coefficients*—that is, like the *first* line of (1.7.5a), or (1.7.5b), below; all such axes are mechanically (though not mathematically) equivalent. But when rates of change between such moving axes (→ frames) are compared, then, in general, *a component of a vector derivative* $(d\boldsymbol{p}/dt)_x$ *does not equal the derivative of that component* dp_x/dt [(1.7.3d, e)]; these quantities are related by a *frame of reference* transformation—that is, like the *second* line of (1.7.5a). Mathematically, this is equivalent to an *explicitly time-dependent* coordinate transformation: $x = x(X, Y, Z; t), \ldots \Leftrightarrow X = X(x, y, z; t), \ldots$ (recall discussion following eq. (1.1.20k)). In such cases, to obtain equations like (1.7.3c), we begin with O_M–XYZ and O_M–xyz in *arbitrary relative orientations*, then we $d/dt(\ldots)$-differentiate the component transformations, like

$$p_x = \cos(x,X)p_X + \cos(x,Y)p_Y + \cos(x,Z)p_Z, \quad \text{etc., cyclically,} \qquad (1.7.5b)$$

(*not* like $p_x = p_X$) and *then* we make O_M–XYZ and O_M–xyz coincide.

(iv) In kinematics, all frames are theoretically equivalent; and thus during the 17th century *both* Galileo and the Catholic church were ... kinematically correct! This is expressed by the following *geometrical*, or *Euclidean*, and *kinematical principle of relativity*: any system of rectangular Cartesian coordinates can be replaced by any other such system that moves in an arbitrary fashion relative to the first; or, alternatively, *the form of geometrical relationships must be invariant under the proper orthogonal group of rotations*—and this, in effect, constitutes a definition of Euclidean geometry—that is, any two such sets of coordinates $x_{k'}$ and x_k are related by

$$x_{k'} = \sum A_{k'k}(t)x_k + A_{k'}(t), \tag{1.7.6a}$$

where

$$\sum A_{k'k}(t)A_{l'k}(t) = \delta_{k'l'}, \qquad \sum A_{k'k}(t)A_{k'l}(t) = \delta_{kl},$$

and

$$\mathrm{Det}(A_{k'k}(t)) = +1, \tag{1.7.6b}$$

and $A_{k'k}(t)$, $A_{k'}(t)$ are continuous functions of time, with first and second time derivatives. Such transformations include all frames/motions produced from the moving frame M by a *continuous* rigid-body movement (*translations* and *rotations*, but not mirror reflections).

(v) If the moving triad $\boldsymbol{u}_{x,y,z}$ is *non-OND*, then its inertial angular velocity is, instead of (1.7.4j),

$$\boldsymbol{\omega} = \{\boldsymbol{u}_x[(d\boldsymbol{u}_y/dt)\cdot\boldsymbol{u}_z] + \boldsymbol{u}_y[(d\boldsymbol{u}_z/dt)\cdot\boldsymbol{u}_x] + \boldsymbol{u}_z[(d\boldsymbol{u}_x/dt)\cdot\boldsymbol{u}_y]\}/[\boldsymbol{u}_x\cdot(\boldsymbol{u}_y\times\boldsymbol{u}_z)]. \tag{1.7.6c}$$

[See, for example, Truesdell and Toupin (1960, p. 437). In case such angular velocity vector definitions seem unmotivated, another more natural one, based on the *linearization of the finite rotation equation*, is detailed in §1.10.]

Corollaries of the Moving Axes Theorem

Applying (1.7.2a) for $\boldsymbol{\omega}$, we get

$$d\boldsymbol{\omega}/dt = d'\boldsymbol{\omega}/dt + \boldsymbol{\omega}\times\boldsymbol{\omega} = d'\boldsymbol{\omega}/dt \equiv \boldsymbol{\alpha}:$$

Angular acceleration of moving axes relative to fixed axes. (1.7.7a)

This result shows the special position of $\boldsymbol{\omega}$ in moving axes theory.

From eq. (1.7.2a) and its derivation, we easily obtain the following general *operator* form:

$$d(\ldots)/dt = d'(\ldots)/dt + \boldsymbol{\omega}\times(\ldots), \qquad (\ldots)\text{: any } \textit{vector}. \tag{1.7.7b}$$

Applying (1.7.7b) to (1.7.2a), and invoking (1.7.7a), we obtain the following expression for the *second* absolute rate of $\boldsymbol{p}$, $d/dt(d\boldsymbol{p}/dt) \equiv d^2\boldsymbol{p}/dt^2$:

$$\begin{aligned} d^2\boldsymbol{p}/dt^2 &= d(\ldots)/dt(d'\boldsymbol{p}/dt + \boldsymbol{\omega}\times\boldsymbol{p}) \\ &= [d'(\ldots)/dt + \boldsymbol{\omega}\times(\ldots)](d'\boldsymbol{p}/dt) + (d\boldsymbol{\omega}/dt)\times\boldsymbol{p} + \boldsymbol{\omega}\times(d\boldsymbol{p}/dt) \\ &= \cdots = d'^2\boldsymbol{p}/dt^2 + [\boldsymbol{\alpha}\times\boldsymbol{p} + \boldsymbol{\omega}\times(\boldsymbol{\omega}\times\boldsymbol{p})] + 2\boldsymbol{\omega}\times(d'\boldsymbol{p}/dt), \end{aligned} \tag{1.7.7c}$$

where

$$d'^2\boldsymbol{p}/dt^2 = (d^2p_x/dt^2)\boldsymbol{u}_x + (d^2p_y/dt^2)\boldsymbol{u}_y + (d^2p_z/dt^2)\boldsymbol{u}_z. \qquad (1.7.7\text{d})$$

In general, if $\boldsymbol{a} \to \boldsymbol{b} = d\boldsymbol{a}/dt \to \boldsymbol{c} = d\boldsymbol{b}/dt \to \cdots$, then we shall have for their components:

$$b_X = b_x = da_x/dt + \omega_y a_z - \omega_z a_y, \qquad (1.7.7\text{e})$$

$$\begin{aligned} c_X = c_x &= db_x/dt + \omega_y b_z - \omega_z b_y \\ &= d/dt(da_x/dt + \omega_y a_z - \omega_z a_y) + \omega_y(da_z/dt + \omega_x a_y - \omega_y a_x) \\ &\qquad - \omega_z(da_y/dt + \omega_z a_x - \omega_x a_z), \quad \text{etc., cyclically.} \end{aligned} \qquad (1.7.7\text{f})$$

For example, application of (1.7.7c, d) to the moving basis vectors $\boldsymbol{u}_{x,y,z}$ yields

$$\begin{aligned} d^2\boldsymbol{u}_x/dt^2 &= d'^2\boldsymbol{u}_x/dt^2 + [\boldsymbol{\alpha} \times \boldsymbol{u}_x + \boldsymbol{\omega} \times (\boldsymbol{\omega} \times \boldsymbol{u}_x)] + 2\boldsymbol{\omega} \times (d'\boldsymbol{u}_x/dt) \\ &= \boldsymbol{0} + [\boldsymbol{\alpha} \times \boldsymbol{u}_x + \boldsymbol{\omega} \times (\boldsymbol{\omega} \times \boldsymbol{u}_x)] + \boldsymbol{0} \\ &= \boldsymbol{\alpha} \times \boldsymbol{u}_x + \boldsymbol{\omega} \times (\boldsymbol{\omega} \times \boldsymbol{u}_x), \quad \text{etc., cyclically.} \end{aligned} \qquad (1.7.7\text{g})$$

Since (1.7.2a) is a purely kinematical result, the roles of the frames F and M can be interchanged. Indeed, from it, we immediately obtain

$$d'\boldsymbol{p}/dt = d\boldsymbol{p}/dt + (-\boldsymbol{\omega}) \times \boldsymbol{p}, \qquad (1.7.7\text{h})$$

where $-\boldsymbol{\omega}$ is the angular velocity of F relative to M.

In particular, *if* $\boldsymbol{p}$ *remains constant* (i.e., *fixed*) *relative to* F, (1.7.2a) and (1.7.7h) yield

$$d'\boldsymbol{p}/dt = (-\boldsymbol{\omega}) \times \boldsymbol{p}; \qquad (1.7.7\text{i})$$

that is, an observer, stationed in M, sees the tip of $\boldsymbol{p}$ rotate relative to that frame with an angular velocity $-\boldsymbol{\omega}$. Application of (1.7.7i) to the fixed basis $\boldsymbol{u}_{X,Y,Z}$ gives

$$\begin{aligned} d'\boldsymbol{u}_X/dt &= (-\boldsymbol{\omega}) \times \boldsymbol{u}_X = -(\omega_X, \omega_Y, \omega_Z) \times (1, 0, 0) \\ &= \cdots = (0)\boldsymbol{u}_X + (-\omega_Z)\boldsymbol{u}_Y + (\omega_Y)\boldsymbol{u}_Z, \\ d'\boldsymbol{u}_Y/dt &= (-\boldsymbol{\omega}) \times \boldsymbol{u}_Y = \cdots = (\omega_Z)\boldsymbol{u}_X + (0)\boldsymbol{u}_Y + (-\omega_X)\boldsymbol{u}_Z, \\ d'\boldsymbol{u}_Z/dt &= (-\boldsymbol{\omega}) \times \boldsymbol{u}_Z = \cdots = (-\omega_Y)\boldsymbol{u}_X + (\omega_X)\boldsymbol{u}_Y + (0)\boldsymbol{u}_Z; \end{aligned} \qquad (1.7.7\text{j})$$

and, therefore,

$$\begin{aligned} &(d'\boldsymbol{u}_X/dt) \cdot \boldsymbol{u}_Y = -\omega_Z \ (= -\omega_z, \text{ for coinciding axes}) \\ &(d'\boldsymbol{u}_X/dt) \cdot \boldsymbol{u}_Z = +\omega_Y \ (= +\omega_y, \text{ for coinciding axes}), \quad \text{etc., cyclically.} \end{aligned} \qquad (1.7.7\text{k})$$

Alternative Definition of Angular Velocity

(i) Below, we show that

$$\boldsymbol{\omega} = \sum (1/2)\left[\boldsymbol{u}_k \times (d\boldsymbol{u}_k/dt)\right] \qquad (\text{where } k = 1, 2, 3 \to x, y, z), \qquad (1.7.8\text{a})$$

which can be viewed as an alternative to (1.7.2e, 6c) definition of angular velocity.

Indeed, using the fundamental equations (1.7.4i), we obtain, successively,

$$\sum [\boldsymbol{u}_k \times (d\boldsymbol{u}_k/dt)] = \sum [\boldsymbol{u}_k \times (\boldsymbol{\omega} \times \boldsymbol{u}_k)] = \sum [(\boldsymbol{u}_k \cdot \boldsymbol{u}_k)\boldsymbol{\omega} - (\boldsymbol{u}_k \cdot \boldsymbol{\omega})\boldsymbol{u}_k)]$$
$$= \boldsymbol{\omega}\Big(\sum (\boldsymbol{u}_k \cdot \boldsymbol{u}_k)\Big) - \sum (\omega_k \boldsymbol{u}_k) = \boldsymbol{\omega}(3) - \boldsymbol{\omega} = 2\boldsymbol{\omega}, \quad \text{Q.E.D.} \tag{1.7.8b}$$

From the above, and using the results of §1.1, we can show that the (inertial) *angular velocity tensor* of the moving frame $\boldsymbol{\omega}$ [i.e., the antisymmetric tensor whose axial vector is the (inertial) angular velocity of that frame $\boldsymbol{\omega}$] can be expressed as

$$\boldsymbol{\omega} = (1/2)\sum \big[(d\boldsymbol{u}_k/dt) \otimes \boldsymbol{u}_k - \boldsymbol{u}_k \otimes (d\boldsymbol{u}_k/dt)\big]. \tag{1.7.8c}$$

(ii) Next, if the (orthonormal) basis vectors $\boldsymbol{u}_k$ are functions of the curvilinear coordinates $q = (q_1, q_2, q_3)$—that is, $\boldsymbol{u}_k = \boldsymbol{u}_k(q)$—then, applying (1.7.8a), we find, successively (with all Latin subscripts running from 1 to 3; i.e., x, y, z),

$$\boldsymbol{\omega} = \sum (1/2)\Big\{\boldsymbol{u}_k \times \Big(\sum (\partial \boldsymbol{u}_k/\partial q_l)(dq_l/dt)\Big)\Big\} = \cdots = \sum \boldsymbol{c}_k(dq_l/dt), \tag{1.7.9a}$$

where

$$\boldsymbol{c}_l \equiv \sum (1/2)\big[\boldsymbol{u}_k \times (\partial \boldsymbol{u}_k/\partial q_l)\big] \qquad (\text{“Eulerian basis” for } \boldsymbol{\omega}); \tag{1.7.9b}$$

that is, the dq_l/dt are the (contravariant) components of $\boldsymbol{\omega}$ in the (covariant) basis $\boldsymbol{c}_l$.

By formally comparing (1.7.8a) and the earlier equations (1.7.4i), (1.7.2e, 4j), with (1.7.9a, b) [i.e., $\boldsymbol{\omega} \to \boldsymbol{c}_l$ and $d\boldsymbol{u}_k/dt \to \partial \boldsymbol{u}_k/\partial q_l$], it is easy to conclude that

$$\partial \boldsymbol{u}_k/\partial q_l = \boldsymbol{c}_l \times \boldsymbol{u}_k, \tag{1.7.9c}$$

$$\boldsymbol{c}_l = \boldsymbol{u}_1[(\partial \boldsymbol{u}_2/\partial q_l) \cdot \boldsymbol{u}_3] + \boldsymbol{u}_2[(\partial \boldsymbol{u}_3/\partial q_l) \cdot \boldsymbol{u}_1] + \boldsymbol{u}_3[(\partial \boldsymbol{u}_1/\partial q_l) \cdot \boldsymbol{u}_2]. \tag{1.7.9d}$$

We leave it to the reader to extend the above to the “rheonomic” case: $\boldsymbol{u}_k = \boldsymbol{u}_k(q, t)$.

EXAMPLES

1. The absolute (i.e., inertial) components of the angular acceleration of a rigid body rotating with angular velocity $\boldsymbol{\omega}_B$ are (with the hitherto used notations)

$$d\omega_{B,X}/dt = d\omega_{B,x}/dt + \omega_y\, \omega_{B,z} - \omega_z\, \omega_{B,y}, \quad \text{etc., cyclically.} \tag{1.7.10a}$$

What happens if $\boldsymbol{\omega}_B = \boldsymbol{\omega}$?

2. The conditions for a straight line with direction cosines (relative to moving axes) l_x, l_y, l_z to have a *fixed inertial direction* are

$$dl_x/dt + \omega_y l_z - \omega_z l_y = 0, \quad \text{etc., cyclically.} \tag{1.7.10b}$$

How many of these three conditions are independent? Hint: $l_x{}^2 + l_y{}^2 + l_z{}^2 = 1$.

3. The moving axis theorem (1.7.2a), applied to the generic vector $\boldsymbol{p}$ expressed in *plane polar coordinates*:

$$\boldsymbol{p} = p_r \boldsymbol{u}_r + p_\phi \boldsymbol{u}_\phi, \tag{1.7.10c}$$

yields

$$d\boldsymbol{p}/dt = [dp_r/dt - p_\phi(d\phi/dt)]\boldsymbol{u}_r + [dp_\phi/dt + p_r(d\phi/dt)]\boldsymbol{u}_\phi. \tag{1.7.10d}$$

Apply (1.7.10d) for $\boldsymbol{p}$ = *position* vector of a particle $\boldsymbol{r}$, and *velocity* vector of a particle $\boldsymbol{v}$.

Hint: The angular velocity of the moving polar ortho–normal–dextral triad $\boldsymbol{u}_{r,\phi,z=Z}$, relative to the inertial one $\boldsymbol{u}_{X,Y,Z}$, is

$$\boldsymbol{\omega} = (d\phi/dt)\boldsymbol{u}_z = (d\phi/dt)\boldsymbol{u}_Z. \tag{1.7.10e}$$

Particle Kinematics in Moving Frames

Velocities

Application of the fundamental formula (1.7.2a) to the motion of a particle P, of inertial position vector $\mathfrak{R} = \boldsymbol{r}_O + \boldsymbol{r}$ (fig. 1.6) (i.e., for $\boldsymbol{p} \to \boldsymbol{r}$), yields

$$\begin{aligned} \boldsymbol{v} \equiv d\mathfrak{R}/dt &= d(\boldsymbol{r}_O + \boldsymbol{r})/dt = d\boldsymbol{r}_O/dt + d\boldsymbol{r}/dt \\ &= d\boldsymbol{r}_O/dt + (d'\boldsymbol{r}/dt + \boldsymbol{\omega} \times \boldsymbol{r}), \end{aligned} \tag{1.7.11a}$$

(since, in general, $\boldsymbol{r}$ is known only along the moving axes) or, rearranging,

$$\boldsymbol{v} = (d\boldsymbol{r}_O/dt + \boldsymbol{\omega} \times \boldsymbol{r}) + d'\boldsymbol{r}/dt, \tag{1.7.11b}$$

or

$$\boldsymbol{v}_{\text{abs}} = \boldsymbol{v}_{\text{trans}} + \boldsymbol{v}_{\text{rel}}, \tag{1.7.11c}$$

where

$$\boldsymbol{v}_{\text{abs}} \equiv \boldsymbol{v} \equiv d\mathfrak{R}/dt = (dX/dt)\boldsymbol{u}_X + (dY/dt)\boldsymbol{u}_Y + (dZ/dt)\boldsymbol{u}_Z:$$

$$\textit{Absolute velocity of } P, \tag{1.7.11d}$$

$$\boldsymbol{v}_{\text{rel}} \equiv d'\boldsymbol{r}/dt \equiv (dx/dt)\boldsymbol{u}_x + (dy/dt)\boldsymbol{u}_y + (dz/dt)\boldsymbol{u}_z:$$

$$\textit{Relative velocity of } P, \tag{1.7.11e}$$

$$\boldsymbol{v}_{\text{trans}} \equiv d\boldsymbol{r}_O/dt + \boldsymbol{\omega} \times \boldsymbol{r} = d\boldsymbol{r}_O/dt + [x(d\boldsymbol{u}_x/dt) + y(d\boldsymbol{u}_y/dt) + z(d\boldsymbol{u}_z/dt)]:$$

$$\textit{Transport velocity of } P. \tag{1.7.11f}$$

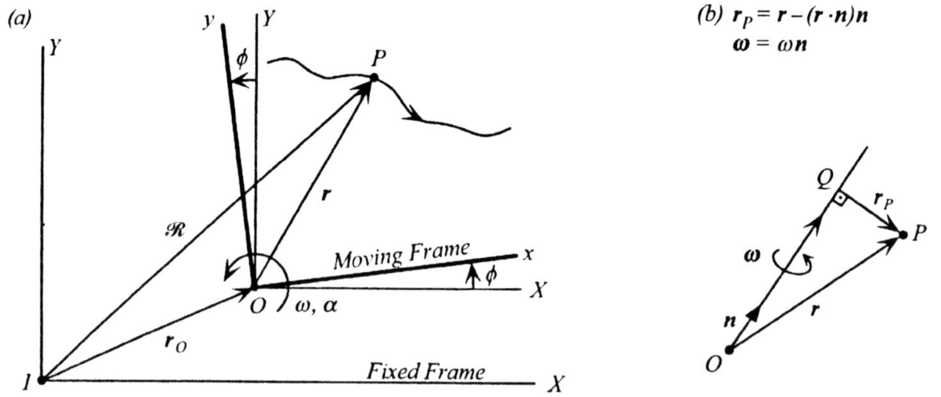

Figure 1.6 (a) Relative kinematics of particle P in two dimensions; (b) geometry of centripetal acceleration.

Clearly, if P is rigidly attached to M_{moving} frame (e.g., if it is one of the particles of the rigid body M), then $\boldsymbol{v}_{\text{rel}} = \mathbf{0}$ and $\boldsymbol{v} = \boldsymbol{v}_{\text{trans}}$; that is, generally, $\boldsymbol{v}_{\text{trans}}$ is the velocity of a particle rigidly attached to M and instantaneously coinciding with P.

Accelerations

Application of (1.7.2a) to (1.7.11a–f) yields

$$\boldsymbol{a}_{\text{abs}} = \boldsymbol{a}_{\text{rel}} + \boldsymbol{a}_{\text{trans}} + \boldsymbol{a}_{\text{cor}}, \tag{1.7.12a}$$

where

$$\boldsymbol{a}_{\text{abs}} \equiv \boldsymbol{a} \equiv d^2\Re/dt^2 = (d^2X/dt^2)\boldsymbol{u}_X + (d^2Y/dt^2)\boldsymbol{u}_Y + (d^2Z/dt^2)\boldsymbol{u}_Z:$$

Absolute acceleration of P, (1.7.12b)

$$\boldsymbol{a}_{\text{rel}} \equiv d'\boldsymbol{v}_{\text{rel}}/dt = d'^2\boldsymbol{r}/dt^2 \equiv (d^2x/dt^2)\boldsymbol{u}_x + (d^2y/dt^2)\boldsymbol{u}_y + (d^2z/dt^2)\boldsymbol{u}_z:$$

Relative acceleration of P, (1.7.12c)

$$\boldsymbol{a}_{\text{trans}} \equiv d^2\boldsymbol{r}_O/dt^2 + \boldsymbol{\alpha}\times\boldsymbol{r} + \boldsymbol{\omega}\times(\boldsymbol{\omega}\times\boldsymbol{r})$$
$$= d^2\boldsymbol{r}_O/dt^2 + [x(d^2\boldsymbol{u}_x/dt^2) + y(d^2\boldsymbol{u}_y/dt^2) + z(d^2\boldsymbol{u}_z/dt^2)]:$$

Transport (or drag) acceleration of P

[= Inertial acceleration of a particle fixed relative to M, and momentarily coinciding with P; its *first* term,

$$d^2\boldsymbol{r}_O/dt^2 = d\boldsymbol{v}_O/dt = d'\boldsymbol{v}_O/dt + \boldsymbol{\omega}\times\boldsymbol{v}_O,$$

is due to the inertial acceleration of the origin of M; its *second*, $\boldsymbol{\alpha}\times\boldsymbol{r}$, to the inertial angular acceleration of M; and its *last* term,

$$\boldsymbol{\omega}\times(\boldsymbol{\omega}\times\boldsymbol{r}) \equiv (\boldsymbol{\omega}\cdot\boldsymbol{r})\boldsymbol{\omega} - \omega^2\boldsymbol{r} \equiv -\omega^2\boldsymbol{r}_p,$$

where $\boldsymbol{r}_p$ = vector of perpendicular distance from $\boldsymbol{\omega}$ – axis (through O) to P, (fig. 1.6(b)), is called *centripetal* acceleration of P],

(1.7.12d)

$$\boldsymbol{a}_{\text{cor}} \equiv 2\boldsymbol{\omega}\times\boldsymbol{v}_{\text{rel}} \equiv 2\boldsymbol{\omega}\times(d'\boldsymbol{r}/dt) = 2\big[(dx/dt)(d\boldsymbol{u}_x/dt) + (dy/dt)(d\boldsymbol{u}_y/dt)$$
$$+ (dz/dt)(d\boldsymbol{u}_z/dt)\big]:$$

Coriolis (or complementary) acceleration of P

[= Acceleration due to the coupling between the relative motion of the particle P, $\boldsymbol{v}_{\text{rel}}$, and the absolute rotation (transport motion) of the frame $M, \boldsymbol{\omega}$; it vanishes if $\boldsymbol{v}_{\text{rel}} = \mathbf{0}$, or if $\boldsymbol{\omega}$ is parallel to $\boldsymbol{v}_{\text{rel}}$]. (1.7.12e)

If $\boldsymbol{\omega} = \mathbf{0}$ and $\boldsymbol{\alpha} = \mathbf{0}$—that is, if M *translates* relative to F—these equations reduce to

$$\boldsymbol{v} = \boldsymbol{v}_{\text{rel}} + \boldsymbol{v}_O = d'\boldsymbol{r}/dt + d\boldsymbol{r}_O/dt = d\boldsymbol{r}/dt + d\boldsymbol{r}_O/dt, \tag{1.7.12f}$$

$$\boldsymbol{a} = \boldsymbol{a}_{\text{rel}} + \boldsymbol{a}_O = d'^2\boldsymbol{r}/dt^2 + d^2\boldsymbol{r}_O/dt^2 = d^2\boldsymbol{r}/dt^2 + d^2\boldsymbol{r}_O/dt^2. \tag{1.7.12g}$$

Component Forms

To appreciate eqs. (1.7.11) and (1.7.12) better, and prepare the reader for the key concept of *nonholonomic coordinates*, and so on (§2.9 ff.), we present them below in terms of their components. In the general case of nonaligned axes we can project them on an arbitrary, fixed, or moving axis; that is, each of their terms can be resolved along any set of axes.

(i) The position relation $\mathfrak{R} = \boldsymbol{r}_O + \boldsymbol{r}$, with $\boldsymbol{r}_O = (X_O, Y_O, Z_O)$, reads

$$X = X_O + \cos(X,x)x + \cos(X,y)y + \cos(X,z)z, \quad \text{etc., cyclically.} \tag{1.7.13a}$$

(ii) The velocity equations (1.7.11a ff.) assume the following forms, along the *fixed* axes:

$$\begin{aligned} dX/dt &= dX_O/dt + \cos(X,x)(dx/dt + \omega_y z - \omega_z y) \\ &\qquad + \cos(X,y)(dy/dt + \omega_z x - \omega_x z) + \cos(X,z)(dz/dt + \omega_x y - \omega_y x) \\ &= dX_O/dt + d/dt(X - X_O), \quad \text{etc., cyclically;} \end{aligned} \tag{1.7.13b}$$

and, along the *moving* axes:

$$\boldsymbol{v}\cdot\boldsymbol{u}_x \equiv v_x = v_{O,x} + dx/dt + \omega_y z - \omega_z y, \quad \text{etc., cyclically,} \tag{1.7.13c}$$

where

$v_{O,x} \equiv \boldsymbol{v}_O \cdot \boldsymbol{u}_x = \cos(x,X)(dX_O/dt) + \cos(x,Y)(dY_O/dt) + \cos(x,Z)(dZ_O/dt)$:

component of inertial velocity of moving origin O, along the moving axis Ox [in general, *not equal to the* $d/dt(\ldots)$*-derivative of a coordinate*, like dX_O/dt or dx/dt, and hence a *quasi velocity* (§2.9 ff.)], etc., cyclically.

(1.7.13d)

(iii) The acceleration equations (1.7.12a ff.) read, along the *fixed* axes:

$$\begin{aligned} d^2X/dt^2 &= d^2X_O/dt^2 + \cos(X,x)[d/dt(dx/dt + \omega_y z - \omega_z y) \\ &\qquad + \omega_y(dz/dt + y\omega_x - x\omega_y) - \omega_z(dy/dt + x\omega_z - z\omega_x)] + \cdots \\ &= d^2X_O/dt^2 + \cos(X,x)\{(d^2x/dt^2) + [z(d\omega_y/dt) - y(d\omega_z/dt)] \\ &\qquad + \omega_y(\omega_x y - \omega_y x) - \omega_z(\omega_z x - \omega_x z) \\ &\qquad + 2[\omega_y(dz/dt) - \omega_z(dy/dt)]\} + \cdots \\ &= d^2X_O/dt^2 + d^2/dt^2(X - X_O), \quad \text{etc., cyclically;} \end{aligned} \tag{1.7.13e}$$

$$= (d^2X/dt^2)_{\text{rel}} + (d^2X/dt^2)_{\text{trans}} + (d^2X/dt^2)_{\text{cor}}, \tag{1.7.13f}$$

where

$$\begin{aligned} (d^2X/dt^2)_{\text{rel}} &= \cos(X,x)(d^2x/dt^2) + \cos(X,y)(d^2y/dt^2) + \cos(X,z)(d^2z/dt^2), \\ (d^2X/dt^2)_{\text{trans}} &= \cos(X,x)\{d^2X_O/dt^2 + [z(d\omega_y/dt) - y(d\omega_z/dt)] \\ &\qquad + \omega_y(\omega_x y - \omega_y x) - \omega_z(\omega_z x - \omega_x z)\} + \cdots, \\ (d^2X/dt^2)_{\text{cor}} &= \cos(X,x)\{2[\omega_y(dz/dt) - \omega_z(dy/dt)]\} + \cdots, \quad \text{etc., cyclically;} \end{aligned} \tag{1.7.13g}$$

and, along the *moving* axes:

$$\boldsymbol{a}\cdot\boldsymbol{u}_x \equiv a_x = a_{O,x} + [d/dt(dx/dt + \omega_y z - \omega_z y) + \omega_y(dz/dt + y\,\omega_x - x\,\omega_y) - \omega_z(dy/dt + x\,\omega_z - z\,\omega_x)]$$
$$= a_{x,\text{rel}} + a_{x,\text{trans}} + a_{x,\text{cor}}, \qquad (1.7.13\text{h})$$

where

$$a_{x,\text{rel}} = d^2x/dt^2,$$
$$a_{x,\text{trans}} = a_{O,x} + [z(d\omega_y/dt) - y(d\omega_z/dt)] + \omega_y(\omega_x\, y - \omega_y\, x) - \omega_z(\omega_z\, x - \omega_x\, z),$$
$$a_{x,\text{cor}} = 2[\omega_y(dz/dt) - \omega_z(dy/dt)], \text{ and} \qquad (1.7.13\text{i})$$
$$a_{O,x} \equiv \boldsymbol{a}_O\cdot\boldsymbol{u}_x = \cos(x,X)(d^2X_O/dt^2 + \cos(x,Y)(d^2Y_O/dt^2) + \cos(x,Z)(d^2Z_O/dt^2),$$
(in general, a *quasi acceleration*), etc., cyclically. (1.7.13j)

EXAMPLES

1. It is not hard to show that the conditions for a particle, with coordinates x, y, z, relative to moving axes, to be stationary relative to absolute space are

$$u + dx/dt + z\omega_y - y\omega_z = 0, \quad \text{etc., cyclically,} \qquad (1.7.14)$$

where (u, v, w) = inertial components of velocity of origin of moving frame.

2. *Plane Rotation Case.* Let us find the components of velocity and acceleration of a particle P in motion on a plane described by the two sets of *momentarily coincident* rectangular Cartesian axes, a fixed O–XY and a second O–xy rotating relative to the first so that always $OZ = Oz$, with angular velocity $\boldsymbol{\omega} = (0, 0, \omega_z = \omega_Z \equiv \omega)$. Here, momentarily,

$$X = x, \qquad Y = y. \qquad (1.7.15\text{a, b})$$

Application of the moving axes theorem (1.7.2a), or (1.7.3c), (1.7.7e), with $\omega_{x,y} = 0$ and $\omega_z = \omega$, yields the *velocity* components:

$$dX/dt = dx/dt - y\,\omega, \qquad dY/dt = dy/dt + x\,\omega; \qquad (1.7.15\text{c, d})$$

and application of that theorem, or (1.7.3c), (1.7.7f), to the above gives the *acceleration* components:

$$d^2X/dt^2 = d/dt(dx/dt - y\,\omega) - (dy/dt + x\,\omega)\omega$$
$$= d^2x/dt^2 - y(d\omega/dt) - x\,\omega^2 - 2(dy/dt)\omega$$
$$(= \textit{relative} + \textit{transport} + \textit{Coriolis}), \qquad (1.7.15\text{e})$$
$$d^2Y/dt^2 = d/dt(dy/dt + x\,\omega) + (dx/dt - y\,\omega)\omega$$
$$= d^2y/dt^2 + x(d\omega/dt) - y\,\omega^2 + 2(dx/dt)\omega$$
$$(= \textit{relative} + \textit{transport} + \textit{Coriolis}); \qquad (1.7.15\text{f})$$

and similarly for higher $d/dt(\ldots)$-derivatives.

[Alternatively, we may start from the geometrical $O\text{–}XY/O\text{–}xy$ relationship for a *generic angle of orientation* $\phi = \phi(t)$:

$$X = (\cos\phi)x + (-\sin\phi)y, \qquad Y = (\sin\phi)x + (\cos\phi)y, \tag{1.7.15g}$$

$d/dt(\ldots)$-differentiate it, and *then* set $\phi = 0$ $(d\phi/dt \equiv \omega \neq 0)$, thus obtaining (1.7.15c, d); then $d/dt(\ldots)$-differentiate once more, for general ϕ, and *then* set $\phi = 0$ $(\omega \neq 0,\ d\omega/dt \equiv \alpha \neq 0)$, thus obtaining (1.7.15e, f). The details of this straightforward calculation are left to the reader. In this way we do not have to remember any kinematical theorems—differential calculus does it for us!]

3. *Velocity and Acceleration in Plane Polar Coordinates via the Moving Axes Theorem* [continued from (1.7.10c–e)]. Here, with the usual notations,

$$\boldsymbol{r} = r\boldsymbol{u}_r \quad \text{and} \quad \boldsymbol{\omega} = (d\phi/dt)\boldsymbol{u}_z = (d\phi/dt)\boldsymbol{u}_Z, \tag{1.7.16a}$$

and, therefore, by direct $d/dt(\ldots)$-differentiation and then use of (1.7.4i)—that is, *treating the corresponding OND basis/axes through P,* $P - \boldsymbol{u}_r\boldsymbol{u}_\phi/r,\ \phi$, *as the moving frame*—we obtain

$$\begin{aligned}
\text{(i)}\quad \boldsymbol{v} &= d\boldsymbol{r}/dt = (dr/dt)\boldsymbol{u}_r + r(d\boldsymbol{u}_r/dt) = (dr/dt)\boldsymbol{u}_r + r(\boldsymbol{\omega}\times\boldsymbol{u}_r)\\
&= (dr/dt)\boldsymbol{u}_r + r(d\phi/dt)(\boldsymbol{u}_z\times\boldsymbol{u}_r) = (dr/dt)\boldsymbol{u}_r + r(d\phi/dt)\boldsymbol{u}_\phi \equiv v_r\,\boldsymbol{u}_r + rv_\phi\,\boldsymbol{u}_\phi;
\end{aligned} \tag{1.7.16b}$$

$$\begin{aligned}
\text{(ii)}\quad \boldsymbol{a} &= d\boldsymbol{v}/dt = (dv_r/dt)\boldsymbol{u}_r + v_r(d\boldsymbol{u}_r/dt) + [d(rv_\phi)/dt)]\boldsymbol{u}_\phi + (rv_\phi)(d\boldsymbol{u}_\phi/dt)\\
&= (dv_r/dt)\boldsymbol{u}_r + v_r(\boldsymbol{\omega}\times\boldsymbol{u}_r) + [d(rv_\phi)/dt)]\boldsymbol{u}_\phi + (rv_\phi)(\boldsymbol{\omega}\times\boldsymbol{u}_\phi)\\
&= (dv_r/dt)\boldsymbol{u}_r + v_r[(d\phi/dt)\boldsymbol{u}_\phi] + [d(rv_\phi)/dt)]\boldsymbol{u}_\phi + (rv_\phi)[(-d\phi/dt)\boldsymbol{u}_r]\\
&= [dv_r/dt - (d\phi/dt)(rv_\phi)]\boldsymbol{u}_r + [v_r(d\phi/dt) + d(rv_\phi)/dt]\boldsymbol{u}_\phi\\
&= [d^2r/dt^2 - r(d\phi/dt)^2]\boldsymbol{u}_r + \{(dr/dt)(d\phi/dt) + d/dt[r(d\phi/dt)]\}\boldsymbol{u}_\phi\\
&= [d^2r/dt^2 - r(d\phi/dt)^2]\boldsymbol{u}_r + [2(dr/dt)(d\phi/dt) + r(d^2\phi/dt^2)]\boldsymbol{u}_\phi\\
&\equiv a_{(r)}\boldsymbol{u}_r + a_{(\phi)}\boldsymbol{u}_\phi.
\end{aligned} \tag{1.7.16c}$$

4. *Velocity and Acceleration in Spherical Coordinates via the Moving Axes Theorem.* Proceeding as in the preceding example, and since here $\boldsymbol{r} = r\boldsymbol{u}_r$ (not the r of the polar cylindrical case) and $\boldsymbol{\omega} = (d\phi/dt)\boldsymbol{u}_Z + (d\theta/dt)\boldsymbol{u}_\phi$, $\boldsymbol{u}_Z = -\sin\theta\boldsymbol{u}_\theta + \cos\theta\boldsymbol{u}_r$, we can show that the velocity and acceleration are given, respectively, by

$$\boldsymbol{v} = (dr/dt)\boldsymbol{u}_r + [r(d\theta/dt)]\boldsymbol{u}_\theta + [r(d\phi/dt)\sin\theta]\boldsymbol{u}_\phi \equiv v_r\,\boldsymbol{u}_r + rv_\theta\,\boldsymbol{u}_\theta + v_\phi\,\boldsymbol{u}_\phi, \tag{1.7.17a}$$

$$\begin{aligned}
\boldsymbol{a} &= [d^2r/dt^2 - r(d\theta/dt)^2 - r(d\phi/dt)^2\sin^2\theta]\boldsymbol{u}_r\\
&\quad + [2(dr/dt)(d\theta/dt) + r(d^2\theta/dt^2) - r(d\phi/dt)^2\sin\theta\cos\theta]\boldsymbol{u}_\theta\\
&\quad + [2(dr/dt)(d\phi/dt)\sin\theta + r(d^2\phi/dt^2)\sin\theta + 2r(d\phi/dt)(d\theta/dt)\cos\theta]\boldsymbol{u}_\phi\\
&\equiv a_{(r)}\boldsymbol{u}_r + a_{(\theta)}\boldsymbol{u}_\theta + a_{(\phi)}\boldsymbol{u}_\phi.
\end{aligned} \tag{1.7.17b}$$

The above are, naturally, in agreement with (1.2.8a ff.)

5. *Inertial Angular Velocity of the Natural, or Intrinsic, OND Triad* O_M–$\boldsymbol{u}_t\boldsymbol{u}_n\boldsymbol{u}_b \equiv$ O_M–$\boldsymbol{tnb}$; *Frenet–Serret Equations* (fig. 1.7). We have already seen (§1.2) that

$$d\boldsymbol{t}/ds = \boldsymbol{n}/\rho \Rightarrow d\boldsymbol{t}/dt = (d\boldsymbol{t}/ds)(ds/dt) = [(ds/dt)/\rho]\boldsymbol{n} \equiv (v_t/\rho)\boldsymbol{n}, \qquad (1.7.18a)$$

also

$$\boldsymbol{b} = \boldsymbol{t} \times \boldsymbol{n}. \qquad (1.7.18b)$$

Next, $d/dt(\ldots)$-differentiating $\boldsymbol{b} \cdot \boldsymbol{t} = 0$, we obtain

$$0 = (d\boldsymbol{b}/dt) \cdot \boldsymbol{t} + \boldsymbol{b} \cdot (d\boldsymbol{t}/dt) = (d\boldsymbol{b}/dt) \cdot \boldsymbol{t} + \boldsymbol{b} \cdot [(v_t/\rho)\boldsymbol{n}] = (d\boldsymbol{b}/dt) \cdot \boldsymbol{t}; \qquad (1.7.18c)$$

and, similarly, $d/dt(\ldots)$-differentiating $\boldsymbol{b} \cdot \boldsymbol{b} = 1$ we readily conclude that

$$(d\boldsymbol{b}/dt) \cdot \boldsymbol{b} = 0. \qquad (1.7.18d)$$

Equations (1.7.18c, d) show that $d\boldsymbol{b}/dt$ must be perpendicular to both $\boldsymbol{t}$ and $\boldsymbol{b}$. Hence, we can set

$$d\boldsymbol{b}/ds = -(1/\tau)\boldsymbol{n} \Rightarrow d\boldsymbol{b}/dt = (d\boldsymbol{b}/ds)(ds/dt) = -(v_t/\tau)\boldsymbol{n}, \qquad (1.7.18e)$$

where τ = *radius of torsion* (or *second curvature*) of the curve C, traced by the moving origin $O_M \equiv O$, at O; positive (negative) whenever the tip of $d\boldsymbol{b}/dt$ turns around $\boldsymbol{t}$ positively (negatively); that is, like a right- (left-)hand screw; or, according as $d\boldsymbol{b}/dt$ has the opposite (same) direction as $\boldsymbol{n}$. [Some authors use τ for our $1/\tau$, and $1/\rho_\tau$ for our τ; and ρ_κ for our ρ.]

Now, the angular velocity of O–$\boldsymbol{tnb}$, relative to some background fixed triad O_F–$\boldsymbol{u}_X\boldsymbol{u}_Y\boldsymbol{u}_Z$, is found by application of the basic formulae (1.7.4j), with the identification O_M–$\boldsymbol{u}_x\boldsymbol{u}_y\boldsymbol{u}_z = O$–$\boldsymbol{tnb}$, and eqs. (1.7.18a–e). Thus, we find

$$\textit{Tangent:}\ \ \omega_t \to \omega_x = \boldsymbol{u}_z \cdot (d\boldsymbol{u}_y/dt) = -\boldsymbol{u}_y \cdot (d\boldsymbol{u}_z/dt) = -\boldsymbol{n} \cdot (d\boldsymbol{b}/dt) = v_t/\tau; \qquad (1.7.18f)$$

$$\textit{Normal:}\ \ \omega_n \to \omega_y = \boldsymbol{u}_x \cdot (d\boldsymbol{u}_z/dt) = -\boldsymbol{u}_z \cdot (d\boldsymbol{u}_x/dt) = -\boldsymbol{b} \cdot (d\boldsymbol{t}/dt) = 0; \qquad (1.7.18g)$$

$$\textit{Binormal:}\ \omega_b \to \omega_z = \boldsymbol{u}_y \cdot (d\boldsymbol{u}_x/dt) = -\boldsymbol{u}_x \cdot (d\boldsymbol{u}_y/dt) = \boldsymbol{n} \cdot (d\boldsymbol{t}/dt) = v_t/\rho. \qquad (1.7.18h)$$

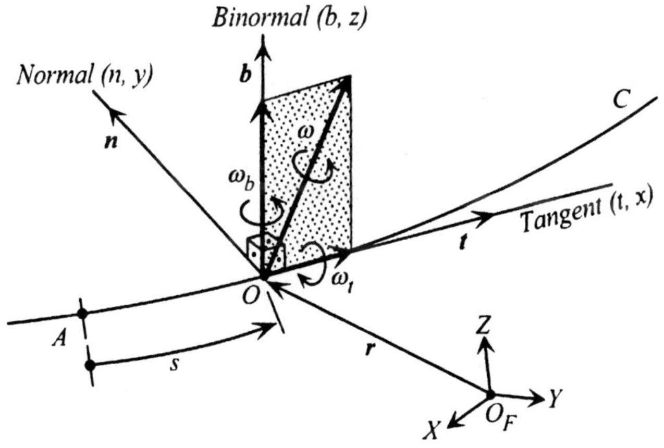

Figure 1.7 On the geometry and kinematics of the Frenet–Serret triad O–$\boldsymbol{tnb}$.

In sum, the triad O–$\boldsymbol{tnb}$ rotates with inertial angular velocity:

$$\boldsymbol{\omega} = (v_t/\tau)\boldsymbol{t} + (0)\boldsymbol{n} + (v_t/\rho)\boldsymbol{b} = v_t(\boldsymbol{t}/\tau + \boldsymbol{b}/\rho). \tag{1.7.18i}$$

In general, $\boldsymbol{\omega} \neq d\boldsymbol{\chi}/dt$, where $\boldsymbol{\chi}$ is some vector expressing angular displacement/rotation; that is, $\boldsymbol{\chi}$ is a *quasi vector* (chap. 2). Further, from (1.7.18a–e) we also conclude that

$$\begin{aligned} d\boldsymbol{n}/ds &= d/ds(\boldsymbol{b} \times \boldsymbol{t}) = (d\boldsymbol{b}/ds) \times \boldsymbol{t} + \boldsymbol{b} \times (d\boldsymbol{t}/ds) \\ &= -(1/\tau)(\boldsymbol{n} \times \boldsymbol{t}) + (1/\rho)(\boldsymbol{b} \times \boldsymbol{n}) \\ &= -(1/\tau)(-\boldsymbol{b}) + (1/\rho)(-\boldsymbol{t}) = (-1/\rho)\boldsymbol{t} + (1/\tau)\boldsymbol{b}. \end{aligned} \tag{1.7.18j}$$

Equations (1.7.18a, e, j) are the famous *Frenet–Serret formulae* for a *space* (or *skew*, or *twisted*) curve.

They can also be written in the following memorable "antisymmetric form":

$$d\boldsymbol{t}/dt = (0)\boldsymbol{t} + (v_t/\rho)\boldsymbol{n} + (0)\boldsymbol{b}, \tag{1.7.18k}$$

$$d\boldsymbol{n}/dt = (-v_t/\rho)\boldsymbol{t} + (0)\boldsymbol{n} + (v_t/\tau)\boldsymbol{b}, \tag{1.7.18l}$$

$$d\boldsymbol{b}/dt = (0)\boldsymbol{t} + (-v_t/\tau)\boldsymbol{n} + (0)\boldsymbol{b}. \tag{1.7.18m}$$

The above allow us to calculate the torsion, $1/\tau$. From (1.7.18j, l), with $(\ldots)' \equiv d(\ldots)/ds$, we get

$$\boldsymbol{b}/\tau = \boldsymbol{t}/\rho + (\rho\boldsymbol{t}')' = \boldsymbol{t}/\rho + \rho'\boldsymbol{t}' + \rho\boldsymbol{t}'' = \boldsymbol{t}/\rho + \rho'(\boldsymbol{n}/\rho) + \rho\boldsymbol{r}''', \tag{1.7.18n}$$

and so, dotting this equation with $\boldsymbol{b}$, we find

$$1/\tau = \rho(\boldsymbol{b} \cdot \boldsymbol{r}''') = \rho[(\boldsymbol{t} \times \boldsymbol{n}) \cdot \boldsymbol{r}'''] = \rho^2[(\boldsymbol{r}' \times \boldsymbol{r}'') \cdot \boldsymbol{r}'''], \tag{1.7.18o}$$

or, since [recalling (1.2.4c)]

$$1/\rho^2 = \boldsymbol{r}'' \cdot \boldsymbol{r}'' = |\boldsymbol{r}''|^2, \tag{1.7.18p}$$

finally,

$$\textit{Torsion} \equiv 1/\tau = [(\boldsymbol{r}' \times \boldsymbol{r}'') \cdot \boldsymbol{r}''']/|\boldsymbol{r}''|^2. \tag{1.7.18q}$$

With the help of the above, we can easily show that

(i) The Frenet–Serret equations can be put in the following kinematical form:

$$d\boldsymbol{t}/dt = \boldsymbol{\omega} \times \boldsymbol{t}, \qquad d\boldsymbol{n}/dt = \boldsymbol{\omega} \times \boldsymbol{n}, \qquad d\boldsymbol{b}/dt = \boldsymbol{\omega} \times \boldsymbol{b}$$

$$(\boldsymbol{\omega}\text{: kinematical } \textit{Darboux} \text{ vector, (1.7.18i)}). \tag{1.7.19a}$$

(ii) If $\boldsymbol{t}, \boldsymbol{n}, \boldsymbol{b}$ can be expressed, in terms of their direction cosines along a fixed *OND* triad, as

$$\boldsymbol{t} = (t_1, t_2, t_3), \qquad \boldsymbol{n} = (n_1, n_2, n_3), \qquad \boldsymbol{b} = (b_1, b_2, b_3), \tag{1.7.19b}$$

then

$$dt_1/ds = n_1/\rho, \qquad dn_1/ds = b_1/\tau - t_1/\rho, \qquad db_1/ds = -n_1/\tau; \tag{1.7.19c}$$

and similarly for the other components.

(iii) The (inertial) *angular acceleration of the Frenet–Serret triad*, $\boldsymbol{\alpha} \equiv d\boldsymbol{\omega}/dt$, is given by

$$\boldsymbol{\alpha} = [(dv_t/dt)/\tau - (v_t^2/\tau^2)(d\tau/ds)]\boldsymbol{t} + [(dv_t/dt)/\rho - (v_t^2/\rho^2)(d\rho/ds)]\boldsymbol{b}. \quad (1.7.19d)$$

(iv) The (inertial) ***jerk*** vector of a particle, $\boldsymbol{j} \equiv d\boldsymbol{a}/dt$ (or hyperacceleration, or velocity of the acceleration) is expressed along the Frenet–Serret triad as

$$\begin{aligned}\boldsymbol{j} &= [d^2 v_t/dt^2 - (v_t^3/\rho^2)]\boldsymbol{t} + \{v_t^2[d/dt(1/\rho)] + 3v_t(dv_t/dt)/\rho)]\}\boldsymbol{n} + (v_t^3/\rho\tau)\boldsymbol{b}\\ &= [d^2 v_t/dt^2 - (v_t^3/\rho^2)]\boldsymbol{t} + [d/dt(v_t^3/\rho)/v_t]\boldsymbol{n} + (v_t^3/\rho\tau)]\boldsymbol{b}\\ &= [d^2 v_t/dt^2 - (v_t^3/\rho^2)]\boldsymbol{t} + [(3v_t^2/\rho)(dv_t/ds) - (v_t^3/\rho^2)(d\rho/ds)]\boldsymbol{n} + (v_t^3/\rho\tau)]\boldsymbol{b},\end{aligned} \quad (1.7.19e)$$

where

$$d(\ldots)/dt = [d(\ldots)/ds](ds/dt) = v_t[d(\ldots)/ds];$$

that is, contrary to the acceleration, $\boldsymbol{a} = (dv_t/dt)\boldsymbol{t} + (v_t^2/\rho)\boldsymbol{n}$, the jerk vector has a $\boldsymbol{b}$-component, and also involves τ.

(v) The following kinematic formulae hold for the curvature and torsion:

$$1/\rho = [(\boldsymbol{v} \times \boldsymbol{a})^2]^{1/2}/v_t^3, \qquad 1/\tau = [\boldsymbol{j} \cdot (\boldsymbol{v} \times \boldsymbol{a})]/(\boldsymbol{v} \times \boldsymbol{a})^2. \quad (1.7.19f)$$

HISTORICAL

The theory of accelerations of *any order* (along general curvilinear coordinates) is due to the Russian mathematician/mechanician Somov (1860s), who also gave recurrence formulae, from the $(n-1)$th order to the (n)th order; and to the French mathematician Bouquet (1879). The *second* order shown above is due to the French mechanician Resal (1862), although the earliest such investigations seem to be due to a certain Transon (1845) (see, e.g., Schönflies and Grübler, 1902: 1901–1908). The jerk vector is called "accéleration du second ordre" (Resal), or "Beschleunigung $\boldsymbol{a}^{(2)}$" (Schönflies/Grübler), where the ordinary acceleration (of the first order) is denoted by $\boldsymbol{a}^{(1)} \equiv \boldsymbol{a}$. Clearly, such derivations are enormously aided with the use of vectors. These results allowed Möbius (1846, 1848) to give a geometrical interpretation to Taylor's expansion (with some standard notations):

$$\begin{aligned}\Delta\boldsymbol{r} \equiv \boldsymbol{r}(t) - \boldsymbol{r}(0) &= \boldsymbol{v}t + \boldsymbol{a}^{(1)}(t^2/2) + \boldsymbol{a}^{(2)}(t^3/1.2.3) + \cdots\\ &= \textit{chord} \text{ of particle trajectory between the times 0 and } t.\end{aligned}$$

Particle Kinetics in Moving Frames

Substituting the inertial acceleration $\boldsymbol{a}$ of a particle P of mass m, in terms of its moving axes representation, into its Newton–Euler equation of motion

$$m\,\boldsymbol{a} = \boldsymbol{f} \quad (= \textit{total noninertial, or real, or objective, force on } P), \quad (1.7.20a)$$

and, rearranging slightly, we obtain its *fundamental equation of relative motion* (fig. 1.8):

$$m\boldsymbol{a}_{\text{rel}} = \boldsymbol{f} + \boldsymbol{f}_{\text{trans}} + \boldsymbol{f}_{\text{cor}}; \quad (1.7.20b)$$

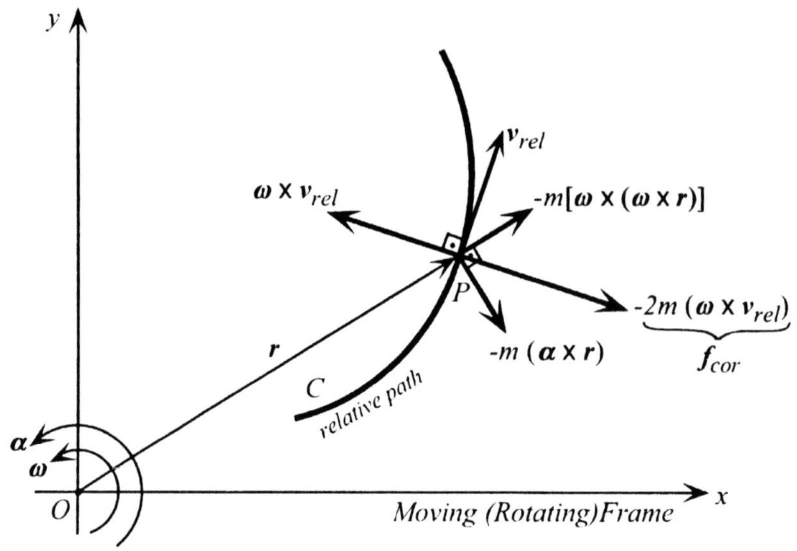

Figure 1.8 Geometry and forces in two-dimensional relative motion.

in words:

mass $\times$ *relative acceleration* $(\boldsymbol{a}_{\text{rel}})$ = *total real* $(\boldsymbol{f})$ *plus inertial* $(\boldsymbol{f}_{\text{trans}} + \boldsymbol{f}_{\text{cor}})$ *force*, (1.7.20c)

where

$\boldsymbol{a}_{\text{rel}} \equiv d'\boldsymbol{v}_{\text{rel}}/dt \equiv d'^2\boldsymbol{r}/dt^2$: *apparent or relative acceleration of P*, (1.7.20d)

$\boldsymbol{f}_{\text{trans}} \equiv -m\boldsymbol{a}_{\text{trans}} \equiv -m[d^2\boldsymbol{r}_O/dt^2 + \boldsymbol{\alpha} \times \boldsymbol{r} + \boldsymbol{\omega} \times (\boldsymbol{\omega} \times \boldsymbol{r})]$:

total inertial force of transport on P

$= -m(d^2\boldsymbol{r}_O/dt^2)$ [due to the inertial acceleration of the origin of the moving

$-m(\boldsymbol{\alpha} \times \boldsymbol{r})$ [due to the inertial angular acceleration of M]

$-m[\boldsymbol{\omega} \times (\boldsymbol{\omega} \times \boldsymbol{r})] = -m[(\boldsymbol{\omega} \cdot \boldsymbol{r})\boldsymbol{\omega} - \omega^2 \boldsymbol{r}] = \cdots = m\omega^2 \boldsymbol{r}_p$

[*centrifugal* force on P, due to the inertial angular velocity of M; always perpendicular to the axis of $\boldsymbol{\omega}$, in the plane of P and that axis, and directed away from it], (1.7.20e)

$\boldsymbol{f}_{\text{cor}} \equiv -m\boldsymbol{a}_{\text{cor}} \equiv -2m(\boldsymbol{\omega} \times \boldsymbol{v}_{\text{rel}}) = -2m[\boldsymbol{\omega} \times (d'\boldsymbol{r}/dt)]$:

inertial force of Coriolis (*or composite centrifugal force*) *on P* [due to the interaction of the relative motion of P $(\boldsymbol{v}_{\text{rel}} = d'\boldsymbol{v}/dt)$ with the absolute rotation of the moving frame $(\boldsymbol{\omega})$; normal to both $\boldsymbol{v}_{\text{rel}}$, $\boldsymbol{\omega}$, and such that $\boldsymbol{v}_{\text{rel}}$, $\boldsymbol{\omega}$, and $\boldsymbol{f}_{\text{cor}} = 2m(\boldsymbol{v}_{\text{rel}} \times \boldsymbol{\omega})$, in that order, form a right-hand system]. (1.7.20f)

REMARKS

(i) In classical mechanics, only $\boldsymbol{f}$ is a real (i.e., frame independent, or absolute) force; $\boldsymbol{f}_{\text{trans}}$ and $\boldsymbol{f}_{\text{cor}}$ are relative (i.e., frame dependent). At most, $\boldsymbol{f}$ can depend explicitly on relative positions (displacements), relative velocities, and time; but not on relative accelerations (as an independent constitutive equation). [In relativity *all* forces are relative, and hence can be eliminated by proper frame choice. On the classical objectivity requirements for $\boldsymbol{f}$, see, for example, Pars (1965, pp. 11–12), Rosenberg (1977, pp. 12–16).] In addition, in general, *the relative forces are not additive*; for example, the total force acting on a particle P due to two or more attracting masses, each exerting separately on it the absolute forces $\boldsymbol{f}_1$ and $\boldsymbol{f}_2$, equals $(\boldsymbol{f}_1 + \boldsymbol{f}_2) + (\boldsymbol{f}_{\text{trans}} + \boldsymbol{f}_{\text{cor}})$; *not* $(\boldsymbol{f}_1 + \boldsymbol{f}_{\text{trans}} + \boldsymbol{f}_{\text{cor}}) + (\boldsymbol{f}_2 + \boldsymbol{f}_{\text{trans}} + \boldsymbol{f}_{\text{cor}})$. As for the Coriolis "force" $\boldsymbol{f}_{\text{cor}} = -2m(\boldsymbol{\omega} \times \boldsymbol{v}_{\text{rel}})$, even for the same problem (i.e., same m and $\boldsymbol{f}$) that term obviously does depend on the particular noninertial frame used. This, however, does not mean that its effects on people, property, and so on, are any less physically/technically real than those of the real force $\boldsymbol{f}$. [In fact, the study of such similarities between these forces led to the general theory of relativity (mid-1910s).]

For the *comoving* (noninertial) observer, both $\boldsymbol{f}_{\text{trans}}$ and $\boldsymbol{f}_{\text{cor}}$ are very real! Some of the most spectacular Coriolis effects occur in the atmospheric sciences (meteorology, etc.); that is, in phenomena involving the coupling between the motion of large liquid and/or gas masses and the Earth's rotation about its axis. A prime such example is *Baer's law of river displacements*: The inertia force on the northbound flowing water, along a meridian, presses against the right (left) bank in the northern (southern) hemisphere. The effects of this pressure are a stronger erosion of the right embankment; and a slightly but measurably higher water level at the right shore of the river. [In view of these realities, statements like the following cannot be taken seriously: "From the foregoing it is clear that the Coriolis-acceleration term arises from the description adopted, namely, via moving observers, and hence, contrary to popular belief *it bears no physical significance*" [Angeles, 1988, p. 74 (the italics are that author's)].]

Finally, since $\boldsymbol{f}_{\text{cor}}$ is perpendicular to $\boldsymbol{v}_{\text{rel}}$, its "relative power" $\boldsymbol{f}_{\text{cor}} \cdot \boldsymbol{v}_{\text{rel}}$ vanishes (more on such "gyroscopic forces" in §3.9).

(ii) In the case of a finite body, $\boldsymbol{v}_{\text{rel}}$ ($\boldsymbol{a}_{\text{rel}}$) in (1.7.20b) refers to the relative velocity (acceleration) of its *center of mass* G; and $\boldsymbol{r}$ is the position of G relative to the origin of the moving frame.

Power Theorem in Relative Motion

This constitutes the vector/particle form of theorems treated in detail in §3.9. Let us consider a system S in motion relative to the *noninertial* axes O–xyz. To find its power equation in relative variables, we start with the equation of relative motion of a generic particle P of S, of mass dm [recall (1.7.20b ff.)]

$$dm\, \boldsymbol{a}_{\text{rel}} = d\boldsymbol{f} + d\boldsymbol{f}_{\text{trans}} + d\boldsymbol{f}_{\text{cor}}, \tag{1.7.21a}$$

where

$$\boldsymbol{a}_{\text{rel}} \equiv d'\boldsymbol{v}_{\text{rel}}/dt \equiv d'^2\boldsymbol{r}/dt^2, \tag{1.7.21b}$$

$$d\boldsymbol{f} = d\boldsymbol{F} + d\boldsymbol{R} \ (\textit{impressed} + \textit{constraint reaction} - \text{see §3.2}) \tag{1.7.21c}$$

$$d\boldsymbol{f}_{\text{trans}} \equiv -dm\, \boldsymbol{a}_{\text{trans}} \equiv -dm\, [d^2\boldsymbol{r}_O/dt^2 + \boldsymbol{\alpha} \times \boldsymbol{r} + \boldsymbol{\omega} \times (\boldsymbol{\omega} \times \boldsymbol{r})], \tag{1.7.21d}$$

$$d\boldsymbol{f}_{\text{cor}} \equiv -dm\, \boldsymbol{a}_{\text{cor}} \equiv -2\, dm\, (\boldsymbol{\omega} \times \boldsymbol{v}_{\text{rel}}) = -2\, dm\, [\boldsymbol{\omega} \times (d'\boldsymbol{r}/dt)]. \tag{1.7.21e}$$

Now, the system power equation corresponding to the particle equation (1.7.21a) is

$$S\, dm\, \boldsymbol{a}_{\rm rel} \cdot \boldsymbol{v}_{\rm rel} = S\, d\boldsymbol{f} \cdot \boldsymbol{v}_{\rm rel} + S\, d\boldsymbol{f}_{\rm rel} \cdot \boldsymbol{v}_{\rm rel} + S\, d\boldsymbol{f}_{\rm cor} \cdot \boldsymbol{v}_{\rm rel}. \tag{1.7.21f}$$

Let us transform each of its terms:

(i)
$$S\, dm\, \boldsymbol{a}_{\rm rel} \cdot \boldsymbol{v}_{\rm rel} = S\, dm\, \boldsymbol{v}_{\rm rel} \cdot (d'\boldsymbol{v}_{\rm rel}/dt)$$
$$= d'/dt\left(S\,(1/2)\, dm\, \boldsymbol{v}_{\rm rel} \cdot \boldsymbol{v}_{\rm rel}\right) \equiv d'T_{\rm rel}/dt, \tag{1.7.21g}$$

or, since

$$\boldsymbol{v}_{\rm rel} \cdot (d\boldsymbol{v}_{\rm rel}/dt) = \boldsymbol{v}_{\rm rel} \cdot (d'\boldsymbol{v}_{\rm rel}/dt + \boldsymbol{\omega} \times \boldsymbol{v}_{\rm rel}) = \boldsymbol{v}_{\rm rel} \cdot (d'\boldsymbol{v}_{\rm rel}/dt),$$

finally,

$$S\, dm\, \boldsymbol{a}_{\rm rel} \cdot \boldsymbol{v}_{\rm rel} = d'T_{\rm rel}/dt = dT_{\rm rel}/dt. \tag{1.7.21h}$$

(ii) We define

$$S\, d\boldsymbol{f} \cdot \boldsymbol{v}_{\rm rel} = S\, d\boldsymbol{f} \cdot (d'\boldsymbol{r}/dt) \equiv d'W/dt; \tag{1.7.21i}$$

where, in general, no W exists (i.e., W is a quasi variable—more on this in §2.9). If

$$S\, d\boldsymbol{R} \cdot \boldsymbol{v}_{\rm rel} = 0, \qquad \text{then} \qquad d'W/dt = S\, d\boldsymbol{F} \cdot \boldsymbol{v}_{\rm rel}.$$

(iii) Clearly,

$$S\, d\boldsymbol{f}_{\rm cor} \cdot \boldsymbol{v}_{\rm rel} = S\left[-2\, dm\, (\boldsymbol{\omega} \times \boldsymbol{v}_{\rm rel})\right] \cdot \boldsymbol{v}_{\rm rel} = 0. \tag{1.7.21j}$$

(iv)
$$S\, d\boldsymbol{f}_{\rm rel} \cdot \boldsymbol{v}_{\rm rel} = -S\, dm\left[\boldsymbol{a}_O + \boldsymbol{\alpha} \times \boldsymbol{r} + \boldsymbol{\omega} \times (\boldsymbol{\omega} \times \boldsymbol{r})\right] \cdot \boldsymbol{v}_{\rm rel}. \tag{1.7.21k}$$

(a)
$$-S\, dm\, \boldsymbol{a}_O \cdot \boldsymbol{v}_{\rm rel} = \boldsymbol{a}_O \cdot \left(S\, dm\, \boldsymbol{v}_{\rm rel}\right) = -m\, \boldsymbol{v}_{G,\rm rel} \cdot \boldsymbol{a}_O \qquad (\boldsymbol{v}_{G,\rm rel} \equiv d'\boldsymbol{r}_G/dt). \tag{1.7.21l}$$

(b)
$$-S\, dm\, [\boldsymbol{v}_{\rm rel} \cdot (\boldsymbol{\alpha} \times \boldsymbol{r})] = -\boldsymbol{\alpha} \cdot \left(S\, dm\, (\boldsymbol{r} \times \boldsymbol{v}_{\rm rel})\right) \equiv -\boldsymbol{\alpha} \cdot \boldsymbol{H}_{O,\rm rel}. \tag{1.7.21m}$$

(c) We have, successively,

$$\boldsymbol{v}_{\rm rel} \cdot [\boldsymbol{\omega} \times (\boldsymbol{\omega} \times \boldsymbol{r})] = (\boldsymbol{\omega} \times \boldsymbol{r}) \cdot (\boldsymbol{v}_{\rm rel} \times \boldsymbol{\omega}) = -(\boldsymbol{\omega} \times \boldsymbol{r}) \cdot [\boldsymbol{\omega} \times (d'\boldsymbol{r}/dt)]$$
$$= -d'/dt\,[(\boldsymbol{\omega} \times \boldsymbol{r})^2/2] = -d'/dt\left[|\boldsymbol{\omega} \times \boldsymbol{r}|^2/2\right] = -d/dt\left[|\boldsymbol{\omega} \times \boldsymbol{r}|^2/2\right], \tag{1.7.21n}$$

and, therefore,

$$-S\, dm\, \boldsymbol{v}_{\rm rel} \cdot [\boldsymbol{\omega} \times (\boldsymbol{\omega} \times \boldsymbol{r})] = d'/dt\left(S\, dm\left[|\boldsymbol{\omega} \times \boldsymbol{r}|^2/2\right]\right)$$
$$= d/dt\left(S\, dm\left[|\boldsymbol{\omega} \times \boldsymbol{r}|^2/2\right]\right). \tag{1.7.21o}$$

In view of (1.7.21g–o), eq. (1.7.21f) takes the following definitive form:

$$dT_{\mathrm{rel}}/dt = d'W/dt - (m\boldsymbol{v}_{G,\mathrm{rel}}) \cdot \boldsymbol{a}_O - \boldsymbol{H}_{O,\mathrm{rel}} \cdot \boldsymbol{\alpha} + d/dt\Big(S\, dm\, \big[|\boldsymbol{\omega} \times \boldsymbol{r}|^2/2\big]\Big). \quad (1.7.21\mathrm{p})$$

Specializations

If O–xyz spins about a *fixed* axis through O, Ol, then $\boldsymbol{a}_O = \boldsymbol{0}$ and eq. (1.7.21p) reduces to

$$dT_{\mathrm{rel}}/dt = d'W/dt - \boldsymbol{H}_{O,\mathrm{rel}} \cdot \boldsymbol{\alpha} + d/dt(I\omega^2/2), \quad (1.7.21\mathrm{q})$$

where

$$I \equiv S\, dm\, r^2 = \textit{moment of inertia of } S \textit{ about } Ol. \quad (1.7.21\mathrm{r})$$

If, further, $\omega = constant$, then (1.7.21q) simplifies to

$$dT_{\mathrm{rel}}/dt = d'W/dt + (dI/dt)\omega^2/2. \quad (1.7.21\mathrm{s})$$

Finally, if $d'W/dt = -dV_O(\boldsymbol{r})/dt$, where $V_O = V_O(\boldsymbol{r}) =$ *potential of impressed forces*, then (1.7.21s) yields the *conservation* theorem:

$$d/dt\big[T_{\mathrm{rel}} + (V_O - I\omega^2/2)\big] = 0 \;\Rightarrow\; T_{\mathrm{rel}} + (V_O - I\omega^2/2) = constant. \quad (1.7.21\mathrm{t})$$

The above is a special case of the *Jacobi–Painlevé integral* (§3.9). As with the equations of motion, the "Newton–Euler" power equation (1.7.21p) may be physically clearer than its Lagrangean counterparts, but the latter have the same *form* in both inertial and noninertial frames, and hence are easier to remember and apply. For further details and insights, see Hamel (1912, pp. 440–443).

The Angular Velocity Tensor

Moving Axes Components

Let us consider two *OND* frames/axes with common origin $O_F \equiv O_M \equiv O$ (no loss in generality here), in arbitrary relative motion (rotation): one *fixed* O–$\boldsymbol{u}_X\boldsymbol{u}_Y\boldsymbol{u}_Z$/–$XYZ$ and another *moving* O–$\boldsymbol{u}_x\boldsymbol{u}_y\boldsymbol{u}_z$/–$xyz$; or, compactly (in view of the heavy indicial notation that follows), O–$\boldsymbol{u}_{k'}$/–$x_{k'}$ and O–$\boldsymbol{u}_k$/–x_k, respectively.

Now, $d/dt(\ldots)$-differentiating their transformation relations,

$$\boldsymbol{u}_k = \sum A_{kk'}\boldsymbol{u}_{k'}, \qquad A_{kk'} \equiv \boldsymbol{u}_k \cdot \boldsymbol{u}_{k'} = \cos(x_k, x_{k'}) = \cos(x_{k'}, x_k) \equiv A_{k'k}, \quad (1.7.22\mathrm{a})$$

and then employing their inverses, we find (since $d\boldsymbol{u}_{k'}/dt = \boldsymbol{0}$):

$$d\boldsymbol{u}_k/dt = \sum (dA_{kk'}/dt)\boldsymbol{u}_{k'} = \sum (dA_{kk'}/dt)\Big(\sum A_{k'l}\boldsymbol{u}_l\Big) \equiv \sum \omega_{lk}\boldsymbol{u}_l, \quad (1.7.22\mathrm{b})$$

where

$$\begin{aligned} \omega_{lk} &\equiv \sum A_{k'l}(dA_{kk'}/dt) = \sum A_{lk'}(dA_{kk'}/dt) = \sum (dA_{kk'}/dt)A_{lk'} = \cdots \\ &= \sum \big\{\cos(x_l, x_{k'})\, d/dt[\cos(x_k, x_{k'})]\big\} \\ &= \boldsymbol{u}_l \cdot (d\boldsymbol{u}_k/dt) = (d\boldsymbol{u}_k/dt) \cdot \boldsymbol{u}_l \qquad [= (l)\textit{th component of } d\boldsymbol{u}_k/dt]: \end{aligned}$$

Tensor of angular velocity of moving axes relative to the fixed axes; but resolved along the *moving axes*. (1.7.22c)

[As already pointed out (§1.1), this *commutativity of subscripts* in $A_{..}$ constitutes one of the big advantages of the accented indices over other notations, such as A_{kl}, A'_{kl}.] Below we show that this tensor is *antisymmetric*: $\omega_{lk} = -\omega_{kl}$. Indeed, $d/dt(\ldots)$-differentiating the orthonormality conditions (constraints!),

$$\boldsymbol{u}_k \cdot \boldsymbol{u}_l = \left(\sum A_{kk'}\boldsymbol{u}_{k'}\right) \cdot \left(\sum A_{ll'}\boldsymbol{u}_{l'}\right) = \cdots = \sum A_{kk'}A_{lk'} = \delta_{kl}, \qquad (1.7.22\text{d})$$

and then invoking the definition (1.7.22c) we obtain

$$0 = \sum (dA_{kk'}/dt)A_{lk'} + \sum A_{kk'}(dA_{lk'}/dt) \qquad [= \boldsymbol{u}_l \cdot (d\boldsymbol{u}_k/dt) + \boldsymbol{u}_k \cdot (d\boldsymbol{u}_l/dt)]$$
$$= \omega_{lk} + \omega_{kl} \;\Rightarrow\; \omega_{lk} = -\omega_{kl}, \quad \text{Q.E.D.}; \qquad (1.7.22\text{e})$$

that is, due to the six constraints (1.7.22d), only three of the nine components of ω_{kl} are independent. Hence, we can replace this tensor by its axial vector (1.1.16a ff.)

$$\omega_k = -\sum\sum (1/2)\varepsilon_{krs}\,\omega_{rs} = -\sum\sum\sum (1/2)\varepsilon_{krs}[A_{rp'}(dA_{sp'}/dt)], \qquad (1.7.22\text{f})$$

and, inversely,

$$\omega_{rs} = -\sum \varepsilon_{krs}\,\omega_k = -\sum \varepsilon_{rsk}\,\omega_k. \qquad (1.7.22\text{g})$$

In extenso, and recalling the properties of ε_{krs} (§1.1), eqs. (1.7.22f) yield

$$\omega_1 \equiv \omega_x = -(1/2)(\varepsilon_{123}\omega_{23} + \varepsilon_{132}\omega_{32}) = -\omega_{23} = \omega_{32}$$
$$= -\sum A_{2k'}(dA_{3k'}/dt) \qquad [= -\boldsymbol{u}_2 \cdot (d\boldsymbol{u}_3/dt) \equiv -\boldsymbol{u}_y \cdot (d\boldsymbol{u}_z/dt)]$$
$$= \sum A_{3k'}(dA_{2k'}/dt) \qquad [= \boldsymbol{u}_3 \cdot (d\boldsymbol{u}_2/dt) \equiv \boldsymbol{u}_z \cdot (d\boldsymbol{u}_y/dt)]$$
$$\{\text{with } 1, 2, 3 \to x, y, z;\ 1', 2', 3' \to X, Y, Z:$$
$$= -[A_{Xy}(dA_{Xz}/dt) + A_{Yy}(dA_{Yz}/dt) + A_{Zy}(dA_{Zz}/dt)]$$
$$= A_{Xz}(dA_{Xy}/dt) + A_{Yz}(dA_{Yy}/dt) + A_{Zz}(dA_{Zy}/dt)\}; \qquad (1.7.23\text{a})$$

$$\omega_2 \equiv \omega_y = -(1/2)(\varepsilon_{231}\omega_{31} + \varepsilon_{213}\omega_{13}) = -\omega_{31} = \omega_{13}$$
$$= -\sum A_{3k'}(dA_{1k'}/dt) \qquad [= -\boldsymbol{u}_3 \cdot (d\boldsymbol{u}_1/dt) \equiv -\boldsymbol{u}_z \cdot (d\boldsymbol{u}_x/dt)]$$
$$= \sum A_{1k'}(dA_{3k'}/dt) \qquad [= \boldsymbol{u}_1 \cdot (d\boldsymbol{u}_3/dt) \equiv \boldsymbol{u}_x \cdot (d\boldsymbol{u}_z/dt)]$$
$$\{= -[A_{Xz}(dA_{Xx}/dt) + A_{Yz}(dY_{Yx}/dt) + A_{Zz}(dA_{Zx}/dt)]$$
$$= A_{Xx}(dA_{Xz}/dt) + A_{Yx}(dA_{Yz}/dt) + A_{Zx}(dA_{Zz}/dt)\}; \qquad (1.7.23\text{b})$$

$$\omega_3 \equiv \omega_z = -(1/2)(\varepsilon_{312}\omega_{12} + \varepsilon_{321}\omega_{21}) = -\omega_{12} = \omega_{21}$$
$$= -\sum A_{1k'}(dA_{2k'}/dt) \quad [= -\boldsymbol{u}_1 \cdot (d\boldsymbol{u}_2/dt) \equiv -\boldsymbol{u}_x \cdot (d\boldsymbol{u}_y/dt)]$$
$$= \sum A_{2k'}(dA_{1k'}/dt) \qquad [= \boldsymbol{u}_2 \cdot (d\boldsymbol{u}_1/dt) \equiv \boldsymbol{u}_y \cdot (d\boldsymbol{u}_x/dt)]$$
$$\{= -[A_{Xx}(dA_{Xy}/dt) + A_{Yx}(dA_{Yy}/dt) + A_{Zx}(dA_{Zy}/dt)]$$
$$= A_{Xy}(dA_{Xx}/dt) + A_{Yy}(dA_{Yx}/dt) + A_{Zy}(dA_{Zx}/dt)\}; \qquad (1.7.23\text{c})$$

which are in complete agreement with (1.7.4j), and justify the name *angular velocity tensor* for (1.7.22c). In terms of *matrices*, the above assume the following memorable form:

$$(\omega_{kl}) = \begin{pmatrix} 0 & -\omega_3 & \omega_2 \\ \omega_3 & 0 & -\omega_1 \\ -\omega_2 & \omega_1 & 0 \end{pmatrix}$$

$$= \left(\begin{array}{c|c|c} 0 & \sum A_{1k'}(dA_{2k'}/dt) & \sum A_{1k'}(dA_{3k'}/dt) \\ \hline \sum A_{2k'}(dA_{1k'}/dt) & 0 & \sum A_{2k'}(dA_{3k'}/dt) \\ \hline \sum A_{3k'}(dA_{1k'}/dt) & \sum A_{3k'}(dA_{2k'}/dt) & 0 \end{array}\right). \qquad (1.7.23d)$$

REMARKS

(i) The formulae (1.7.22f ff.) can be combined into the following useful form:

$$\omega_k = (d\boldsymbol{u}_r/dt) \cdot \boldsymbol{u}_s, \qquad (1.7.24)$$

where

$$k, r, s = cyclic\ (even)\ permutation\ of\ 1, 2, 3\ (\equiv x, y, z).$$

(ii) The final expressions (1.7.23d) would have resulted if we had employed the following common angular velocity tensor definitions:

$$\omega_{kl} \equiv (d\boldsymbol{u}_k/dt) \cdot \boldsymbol{u}_l = -(d\boldsymbol{u}_l/dt) \cdot \boldsymbol{u}_k = \sum (dA_{kk'}/dt)A_{lk'} = -\sum (dA_{lk'}/dt)A_{kk'}, \qquad (1.7.25a)$$

but in connection with the also common *axial vector definition*:

$$\omega_k = \sum\sum (1/2)\varepsilon_{krs}\,\omega_{rs}. \qquad (1.7.25b)$$

Then, we would have

$$\omega_1 = (1/2)(\varepsilon_{123}\omega_{23} + \varepsilon_{132}\omega_{32}) = \omega_{23} = -\omega_{32}$$
$$= \sum (dA_{2k'}/dt)A_{3k'} = -\sum (dA_{3k'}/dt)A_{2k'}, \text{ etc.} \qquad (1.7.25c)$$

Fixed Axes Components

Let us express the above inertial angular velocity tensor in terms of their components along the fixed axes $O\text{–}\boldsymbol{u}_X\boldsymbol{u}_Y\boldsymbol{u}_Z/\text{–}XYZ \equiv O\text{–}\boldsymbol{u}_{k'}/\text{–}x_{k'}$. Dotting the representations of the position vector of a typical particle P,

$$\boldsymbol{r} = \sum x_k\boldsymbol{u}_k = \sum x_{k'}\boldsymbol{u}_{k'}, \qquad (1.7.26a)$$

with $\boldsymbol{u}_l$ and $\boldsymbol{u}_{l'}$, respectively, and taking into account the orthonormality constraints of their basis vectors:

$$\boldsymbol{u}_k \cdot \boldsymbol{u}_l = \left(\sum A_{kk'}\boldsymbol{u}_{k'}\right) \cdot \left(\sum A_{ll'}\boldsymbol{u}_{l'}\right) = \sum A_{kk'}A_{lk'} = \delta_{kl}, \qquad (1.7.26b)$$

$$\boldsymbol{u}_{k'} \cdot \boldsymbol{u}_{l'} = \left(\sum A_{k'k}\boldsymbol{u}_k\right) \cdot \left(\sum A_{l'l}\boldsymbol{u}_l\right) = \sum A_{k'k}A_{l'k} = \delta_{k'l'}, \qquad (1.7.26c)$$

we easily obtain the component transformation equation

$$x_{k'} = \sum A_{k'k} x_k \Leftrightarrow x_k = \sum A_{kk'} x_{k'}. \tag{1.7.26d}$$

If the two sets of axes do *not* have a common origin, but [recalling fig. 1.6(a)]

$$\mathfrak{R} = \boldsymbol{r}_O + \boldsymbol{r}, \tag{1.7.26e}$$

where

$$\mathfrak{R} = \sum x_{k'} \boldsymbol{u}_{k'}, \tag{1.7.26f}$$

$$\boldsymbol{r}_O \equiv \boldsymbol{r}_{\text{moving origin/fixed origin}} = \sum b_k \boldsymbol{u}_k = \sum b_{k'} \boldsymbol{u}_{k'}$$

$$\Rightarrow \quad b_{k'} = \sum A_{k'k} b_k \quad \Leftrightarrow \quad b_k = \sum A_{kk'} b_{k'}, \tag{1.7.26g}$$

$$\boldsymbol{r} = \sum x_k \boldsymbol{u}_k, \tag{1.7.26h}$$

then (1.7.26d) are replaced by

$$x_{k'} = \sum A_{k'k} x_k + b_{k'} \equiv \sum A_{k'k}(x_k + b_k)$$
$$\Leftrightarrow x_k = \sum A_{kk'}(x_{k'} - b_{k'}) = \sum A_{kk'} x_{k'} - b_k. \tag{1.7.26i}$$

Now, let us consider P to be *rigidly* attached to the moving axes. Then $d/dt(\ldots)$-differentiating the $x_{k'}$, while recalling that in this case $x_k = constant \Rightarrow dx_k/dt = 0$, we obtain, successively,

$$dx_{k'}/dt = \sum (dA_{k'k}/dt) x_k = \sum (dA_{k'k}/dt)\left(\sum A_{kl'} x_{l'}\right) \equiv \sum \omega_{k'l'} x_{l'},$$

[which is none other than the familiar $\boldsymbol{v} = \boldsymbol{\omega} \times \boldsymbol{r}$, resolved along the *fixed* axes] (1.7.26j)

where

$$\omega_{k'l'} \equiv \sum (dA_{k'k}/dt) A_{kl'} \equiv \sum (dA_{k'k}/dt) A_{l'k}$$
$$= \sum \{\cos(x_{l'}, x_k)\ d/dt[\cos(x_{k'}, x_k)]\}:$$

Tensor of angular velocity of moving axes relative to the fixed axes;
but resolved along the *fixed axes* [Note order of *accented* indices, and compare with order of unaccented indices in expression (1.7.22c, 25a).](1.7.26k)

The components $\omega_{k'l'}$, just like the ω_{kl}, are *antisymmetric*. Indeed, $d/dt(\ldots)$-differentiating (1.7.26c), we obtain

$$0 = \sum (dA_{k'k}/dt) A_{l'k} + \sum A_{k'k}(dA_{l'k}/dt) = \omega_{l'k'} + \omega_{k'l'}$$
$$\Rightarrow \omega_{l'k'} = -\omega_{k'l'}, \quad \text{Q.E.D.} \tag{1.7.26l}$$

$$\omega_{k'} = -\sum\sum (1/2)\varepsilon_{k'r's'} \omega_{r's'} = -\sum\sum\sum (1/2)\varepsilon_{k'r's'} \left[A_{r'r}(dA_{s'r}/dt)\right], \tag{1.7.26m}$$

and, inversely,

$$\omega_{r's'} = -\sum \varepsilon_{k'r's'}\,\omega_{k'} = -\sum \varepsilon_{r's'k'}\,\omega_{k'}; \tag{1.7.26n}$$

or, in extenso,

$$\begin{aligned}
\omega_{1'} \equiv \omega_{x'} \equiv \omega_X &= -(1/2)(\varepsilon_{1'2'3'}\,\omega_{2'3'} + \varepsilon_{1'3'2'}\,\omega_{3'2'}) = -\omega_{2'3'} = \omega_{3'2'} \\
&\{\text{with } 1,2,3 \to x,y,z;\ 1',2',3' \to X,Y,Z: \\
&= -[A_{Zx}(dA_{Yx}/dt) + A_{Zy}(dA_{Yy}/dt) + A_{Zz}(dA_{Yz}/dt)] \\
&= A_{Yx}(dA_{Zx}/dt) + A_{Yy}(dA_{Zy}/dt) + A_{Yz}(dA_{Zz}/dt)\};
\end{aligned} \tag{1.7.27a}$$

$$\begin{aligned}
\omega_{2'} \equiv \omega_{y'} \equiv \omega_Y &= -(1/2)(\varepsilon_{2'3'1'}\,\omega_{3'1'} + \varepsilon_{2'1'3'}\,\omega_{1'3'}) = -\omega_{3'1'} = \omega_{1'3'} \\
&= -\sum A_{1'k}(dA_{3'k}/dt) = \sum A_{3'k}(dA_{1'k}/dt) \\
\{ &= -[A_{Xx}(dA_{Zx}/dt) + A_{Xy}(dA_{Zy}/dt) + A_{Xz}(dA_{Zz}/dt)] \\
&= A_{Zx}(dA_{Xx}/dt) + A_{Zy}(dA_{Xy}/dt) + A_{Zz}(dA_{Xz}/dt)\};
\end{aligned} \tag{1.7.27b}$$

$$\begin{aligned}
\omega_{3'} \equiv \omega_{z'} \equiv \omega_Z &= -(1/2)(\varepsilon_{3'1'2'}\,\omega_{1'2'} + \varepsilon_{3'2'1'}\,\omega_{2'1'}) = -\omega_{1'2'} = \omega_{2'1'} \\
&= -\sum A_{2'k}(dA_{1'k}/dt) = \sum A_{1'k}(dA_{2'k}/dt) \\
\{ &= -[A_{Yx}(dA_{Xx}/dt) + A_{Yy}(dA_{Xy}/dt) + A_{Yz}(dA_{Xz}/dt)] \\
&= A_{Xx}(dA_{Yx}/dt) + A_{Xy}(dA_{Yy}/dt) + A_{Xz}(dA_{Yz}/dt)\};
\end{aligned} \tag{1.7.27c}$$

or, finally, in the following memorable matrix form:

$$(\omega_{k'l'}) = \begin{pmatrix} 0 & -\omega_{3'} & \omega_{2'} \\ \omega_{3'} & 0 & -\omega_{1'} \\ -\omega_{2'} & \omega_{1'} & 0 \end{pmatrix}$$

$$= \left(\begin{array}{c|c|c} 0 & \sum (dA_{1'k}/dt)A_{2'k} & \sum (dA_{1'k}/dt)A_{3'k} \\ \hline \sum (dA_{2'k}/dt)A_{1'k} & 0 & \sum (dA_{2'k}/dt)A_{3'k} \\ \hline \sum (dA_{3'k}/dt)A_{1'k} & \sum (dA_{3'k}/dt)A_{2'k} & 0 \end{array}\right); \tag{1.7.27d}$$

or

$$\omega_{k'l'} = \sum\sum A_{k'k}A_{l'l}\omega_{kl} \Leftrightarrow \omega_{kl} = \sum\sum A_{kk'}A_{ll'}\,\omega_{k'l'}. \tag{1.7.27e}$$

A Special Case

If the axes x_k and $x_{k'}$ *coincide momentarily*—that is, if instantaneously $A_{k'k} = \delta_{k'k}$ (Kronecker delta), then eqs. (1.7.23) and (1.7.27) yield

$$\omega_x = dA_{Zy}/dt = -dA_{Yz}/dt, \qquad \omega_y = dA_{Xz}/dt = -dA_{Zx}/dt,$$
$$\omega_z = dA_{Yx}/dt = -dA_{Xy}/dt; \tag{1.7.28a}$$

$$\omega_X = dA_{Zy}/dt = -dA_{Yz}/dt, \qquad \omega_Y = dA_{Xz}/dt = -dA_{Zx}/dt,$$
$$\omega_Z = dA_{Yx}/dt = -dA_{Xy}/dt. \tag{1.7.28b}$$

Rates of Change of Direction Cosines

Let us calculate $dA_{k'k}/dt$ in terms of ω_{kl}, $\omega_{k'l'}$.

(i) *Fixed axes representation*: Multiplying both sides of (1.7.22c) with $A_{l'l}$ and summing over l, we obtain

$$\sum \omega_{lk} A_{l'l} = \sum (dA_{k'k}/dt)\left(\sum A_{k'l} A_{l'l}\right) = \sum (dA_{k'k}/dt)(\delta_{k'l'}) = dA_{l'k}/dt. \tag{1.7.29a}$$

(ii) *Moving axes representation*: Multiplying both sides of (1.7.26k) with $A_{l's}$ and summing over l', we obtain

$$\sum \omega_{k'l'} A_{l's} = \sum (dA_{k'k}/dt)\left(\sum A_{kl'} A_{l's}\right) = \sum (dA_{k'k}/dt)(\delta_{ks}) = dA_{k's}/dt; \tag{1.7.29b}$$

$$dA_{k'k}/dt = \sum A_{k'l}\omega_{lk} = \sum \omega_{k'l'} A_{l'k}; \tag{1.7.29c}$$

$$dA_{k'k}/dt = A_{k'1}\omega_{1k} + A_{k'2}\omega_{2k} + A_{k'3}\omega_{3k}$$

$$\Rightarrow dA_{k'1}/dt = A_{k'2}\omega_3 - A_{k'3}\omega_2; \qquad \text{i.e., } dA_{k'x}/dt = A_{k'y}\omega_z - A_{k'z}\omega_y,$$

$$dA_{k'2}/dt = A_{k'3}\omega_1 - A_{k'1}\omega_3; \qquad \text{i.e., } dA_{k'y}/dt = A_{k'z}\omega_x - A_{k'x}\omega_z,$$

$$dA_{k'3}/dt = A_{k'1}\omega_2 - A_{k'2}\omega_1; \qquad \text{i.e., } dA_{k'z}/dt = A_{k'x}\omega_y - A_{k'y}\omega_x$$

$$(k' = X, Y, Z); \tag{1.7.29d}$$

$$dA_{k'k}/dt = A_{1'k}\omega_{k'1'} + A_{2'k}\omega_{k'2'} + A_{3'k}\omega_{k'3'}$$

$$\Rightarrow dA_{1'k}/dt = A_{3'k}\omega_{2'} - A_{2'k}\omega_{3'}; \qquad \text{i.e., } dA_{Xk}/dt = A_{Zk}\omega_Y - A_{Yk}\omega_Z,$$

$$dA_{2'k}/dt = A_{1'k}\omega_{3'} - A_{3'k}\omega_{1'}; \qquad \text{i.e., } dA_{Yk}/dt = A_{Xk}\omega_Z - A_{Zk}\omega_X,$$

$$dA_{3'k}/dt = A_{2'k}\omega_{1'} - A_{1'k}\omega_{2'}; \qquad \text{i.e., } dA_{Zk}/dt = A_{Yk}\omega_X - A_{Xk}\omega_Y$$

$$(k = x, y, z). \tag{1.7.29e}$$

Additional Useful Results

(i) By $d/dt(\ldots)$-differentiating the fixed basis vectors:

$$\mathbf{0} = d\boldsymbol{u}_{k'}/dt = \sum \left[(dA_{k'k}/dt)\boldsymbol{u}_k + A_{k'k}(d\boldsymbol{u}_k/dt)\right] = \cdots, \tag{1.7.30a}$$

it can be shown that

$$d\boldsymbol{u}_k/dt = \sum \omega_{k'k}\boldsymbol{u}_{k'}, \tag{1.7.30b}$$

where

$$\omega_{k'k} \equiv \sum A_{l'k}\,\omega_{k'l'} = \sum (\partial x_{l'}/\partial x_k)\,\omega_{k'l'} = \cdots = dA_{k'k}/dt \qquad (\textit{mixed}\ \text{"tensor"}). \tag{1.7.30c}$$

[Similarly, we can define the following *mixed angular velocity "tensor"*:

$$\omega_{kk'} \equiv \sum \omega_{kl}A_{lk'} = \sum (\partial x_l/\partial x_{k'})\,\omega_{kl} = \sum A_{lk'}\left(\sum\sum A_{kp'}A_{lq'}\omega_{p'q'}\right)$$

$$= \cdots = \sum A_{kl'}\omega_{l'k'}.]$$

(ii) By $d/dt(\ldots)$-differentiating $x'_{k'} = x_{k'} = \sum A_{k'k} x_k$, and noticing that $v'_k = \sum A_{kk'} v'_{k'}$, it can be shown that the *inertial* velocity of a particle permanently fixed in the moving frame (i.e., $dx_k/dt \equiv v_k = 0 \Rightarrow v_{k'} = 0$) equals:

$$v'_k = \sum \omega_{kl} x_l \qquad \text{(along the } moving \text{ axes)}, \tag{1.7.30d}$$

$$dx_{k'}/dt \equiv v'_{k'} = \sum \omega_{k'l'} x_{l'} \qquad \text{(along the } fixed \text{ axes)}. \tag{1.7.30e}$$

(iii) Let us define the following *matrices*:

$\boldsymbol{\omega} = (\omega_{kl})$: matrix of angular velocity *tensor*, along the *moving* axes, (1.7.30f)

$\boldsymbol{\omega}' = (\omega_{k'l'})$: matrix of angular velocity *tensor*, along the *fixed* axes, (1.7.30g)

$\mathbf{A} = (A_{k'k})$: matrix of direction cosines between moving and fixed axes. (1.7.30h)

It can be shown that the earlier relations among them (i.e., among their elements) can be put in the following matrix forms [recalling that $\mathbf{A}^{-1} = \mathbf{A}^{\mathrm{T}}$ and $(\ldots)^{\mathrm{T}} \equiv$ *Transpose* of $(\ldots)$]:

$$\boldsymbol{\omega} = \mathbf{A}^{\mathrm{T}} \cdot (\mathrm{d}\mathbf{A}/\mathrm{dt}) = -(\mathrm{d}\mathbf{A}/\mathrm{dt})^{\mathrm{T}} \cdot \mathbf{A} \;\Leftrightarrow\; \mathrm{d}\mathbf{A}/\mathrm{dt} = \mathbf{A} \cdot \boldsymbol{\omega}, \tag{1.7.30i}$$

$$\boldsymbol{\omega}' = (\mathrm{d}\mathbf{A}/\mathrm{dt}) \cdot \mathbf{A}^{\mathrm{T}} = -\mathbf{A} \cdot (\mathrm{d}\mathbf{A}/\mathrm{dt})^{\mathrm{T}} \;\Leftrightarrow\; \mathrm{d}\mathbf{A}/\mathrm{dt} = \boldsymbol{\omega}' \cdot \mathbf{A}, \tag{1.7.30j}$$

$$\boldsymbol{\omega}' = \mathbf{A} \cdot \boldsymbol{\omega} \cdot \mathbf{A}^{\mathrm{T}} \;\Leftrightarrow\; \boldsymbol{\omega} = \mathbf{A}^{\mathrm{T}} \cdot \boldsymbol{\omega}' \cdot \mathbf{A}. \tag{1.7.30k}$$

[(a) Equation (1.7.30j) expresses the following important general theorem: *for an arbitrary (differentiable) orthogonal matrix (or tensor)* $\mathbf{A} = \mathbf{A}(\mathrm{t})$,

$$\mathrm{d}\mathbf{A}/\mathrm{dt} = (\text{matrix of } second\text{-}order\ antisymmetric\ tensor) \cdot \mathbf{A}; \tag{1.7.30l}$$

and similarly for equation (1.7.30i).
(b) Notice that for matrix operations, the *order* of the indices of the elements (direction cosines) $A_{k'k}$ *does* matter; that is, in this algebra, accented (nonaccented) indices are always first (second) and denote rows (columns).]

Angular Velocity Vector in General Orthogonal Curvilinear Coordinates

[This section may be omitted in a first reading. For background, see (1.2.7a ff.).]

In such coordinates, say $q \equiv (q_1, q_2, q_3) \equiv (q_{1,2,3})$, the inertial position vector of a particle $\boldsymbol{r}$ becomes

$$\boldsymbol{r} = X(q)\boldsymbol{u}_X + Y(q)\boldsymbol{u}_Y + Z(q)\boldsymbol{u}_Z \equiv \sum x_{k'}(q)\boldsymbol{u}_{k'}, \tag{1.7.31a}$$

and so the corresponding moving *OND* basis along $q_{1,2,3}$ (i.e., the earlier x_k) is

$$\boldsymbol{u}_k = (\partial\boldsymbol{r}/\partial q_k)\big/|\partial\boldsymbol{r}/\partial q_k| \equiv (1/h_k)(\partial\boldsymbol{r}/\partial q_k) \qquad (k = x, y, z), \tag{1.7.31b}$$

with

$$\boldsymbol{u}_k \cdot \boldsymbol{u}_l = \delta_{kl} \qquad (k, l = x, y, z). \tag{1.7.31c}$$

Next, $d/dt(\dots)$-differentiating (1.7.31b), we obtain, successively,

$$d/dt(\partial \boldsymbol{r}/\partial q_r) = d/dt(h_r \boldsymbol{u}_r) = (dh_r/dt)\boldsymbol{u}_r + h_r(d\boldsymbol{u}_r/dt)$$
$$= (dh_r/dt)\boldsymbol{u}_r + h_r(\boldsymbol{\omega} \times \boldsymbol{u}_r) \qquad \text{[by (1.7.4i)]}, \qquad (1.7.31\text{d})$$

and dotting this with $\partial \boldsymbol{r}/\partial q_s = h_s \boldsymbol{u}_s$ ($\equiv \boldsymbol{e}_s$, where $r \neq s$), in order to isolate ω_k, we get

$$\big[d/dt(\partial \boldsymbol{r}/\partial q_r)\big] \cdot (\partial \boldsymbol{r}/\partial q_s) = (dh_r/dt)h_s(\boldsymbol{u}_r \cdot \boldsymbol{u}_s) + h_r h_s[(\boldsymbol{\omega} \times \boldsymbol{u}_r) \cdot \boldsymbol{u}_s]$$
$$= 0 + h_r h_s[(\boldsymbol{\omega} \cdot (\boldsymbol{u}_r \times \boldsymbol{u}_s)] = h_r h_s(\boldsymbol{\omega} \cdot \boldsymbol{u}_k) \equiv h_r h_s \omega_k$$

[definition of ω_k's; where k, r, s = even (cyclic) permutation of 1, 2, 3 $\equiv x$, y, z],

that is, finally,

$$\omega_k = (1/h_r h_s)\big[d/dt(\partial \boldsymbol{r}/\partial q_r) \cdot (\partial \boldsymbol{r}/\partial q_s)\big]$$
$$\big\{ = (d\boldsymbol{u}_r/dt) \cdot \boldsymbol{u}_s = d/dt\big[(1/h_r)(\partial \boldsymbol{r}/\partial q_r)\big] \cdot \big[(1/h_s)(\partial \boldsymbol{r}/\partial q_s)\big]\big\}. \qquad (1.7.31\text{e})$$

Additional forms for these components exist in the literature; for example, with the help of the differential-geometric identities:

$$\partial \boldsymbol{u}_r/\partial q_s = (1/h_r)(\partial h_s/\partial q_r)\boldsymbol{u}_s \qquad (r \neq s), \qquad (1.7.31\text{f})$$
$$\partial \boldsymbol{u}_r/\partial q_r = -(1/h_s)(\partial h_r/\partial q_s)\boldsymbol{u}_s - (1/h_k)(\partial h_r/\partial q_k)\boldsymbol{u}_k \qquad (r \neq s \neq k \neq r), \qquad (1.7.31\text{g})$$

and applying the second line of (1.7.31e), we can easily show that

$$\omega_1 = (1/h_2)(\partial h_3/\partial q_2)(dq_3/dt) - (1/h_3)(\partial h_2/\partial q_3)(dq_2/dt), \qquad (1.7.31\text{h})$$
$$\omega_2 = (1/h_3)(\partial h_1/\partial q_3)(dq_1/dt) - (1/h_1)(\partial h_3/\partial q_1)(dq_3/dt), \qquad (1.7.31\text{i})$$
$$\omega_3 = (1/h_1)(\partial h_2/\partial q_1)(dq_2/dt) - (1/h_2)(\partial h_1/\partial q_2)(dq_1/dt). \qquad (1.7.31\text{j})$$

[See Richardson (1992), also Ames and Murnaghan (1929, pp. 26–34, 94–98), for an alternative derivation based on the direction cosines between the moving and fixed axes:

$$A_{k'k} = A_{kk'} \equiv \boldsymbol{u}_{k'} \cdot \boldsymbol{u}_k = (\partial \boldsymbol{r}/\partial x_{k'}) \cdot \big[(1/h_k)(\partial \boldsymbol{r}/\partial q_k)\big]$$
$$= (1/h_k)\Big\{(\partial \boldsymbol{r}/\partial x_{k'}) \cdot \Big(\sum (\partial \boldsymbol{r}/\partial x_{l'})(\partial x_{l'}/\partial q_k)\Big)\Big\}$$
$$= (1/h_k)\Big(\sum (\boldsymbol{u}_{k'} \cdot \boldsymbol{u}_{l'})(\partial x_{l'}/\partial q_k)\Big) \qquad (\text{since } \boldsymbol{u}_{k'} \cdot \boldsymbol{u}_{l'} = \delta_{k'l'})$$
$$= (1/h_k)(\partial x_{k'}/\partial q_k), \qquad (1.7.31\text{k})$$

and their $d/dt(\dots)$-derivatives.]

1.8 THE RIGID BODY: INTRODUCTION

The following material relies heavily on the preceding theory of moving axes (§1.7). The reason for this is that *every set of such axes can be thought of as a moving rigid body*; and, conversely, *every rigid body in motion carries along with it one or more sets of axes rigidly attached to it, or embedded in it.* To describe the translatory and

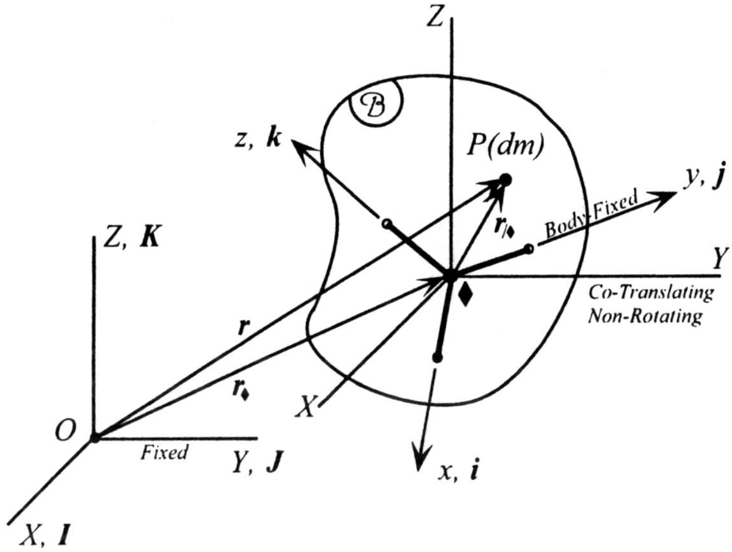

Figure 1.9 Axes used to describe rigid-body motion.
O–XYZ/IJK: fixed axes/basis; ◆–xyz/ijk: moving (body-fixed) axes/basis;
◆–XYZ/IJK: moving, translating but nonrotating axes/basis.

angular motion of a rigid body B, we consider (at least) *two* sets of rectangular Cartesian axes and associated bases:

(i) a *fixed*: that is, inertial, O–$XYZ/\boldsymbol{IJK}$ or compactly O–$x_{k'}/\boldsymbol{u}_{k'}$; and
(ii) a *moving*: that is, noninertial, and *body-fixed* set ◆–$xyz/\boldsymbol{ijk}$ or compactly ◆–$x_k/\boldsymbol{u}_k$, at the arbitrary body point ◆ (fig. 1.9).

In the language of constraints (chap. 2), a *free* rigid body in space is a mechanical system with *six* degrees of *global* freedom; that is, six independent possibilities of *finite* spatial mobility: (i) three for the location of its body point ◆, say its O–XYZ coordinates

$$X_\blacklozenge = f_1(t), \qquad Y_\blacklozenge = f_2(t), \qquad Z_\blacklozenge = f_3(t); \tag{1.8.1a}$$

and (ii) three for its *orientation*—that is, of ◆–xyz relative to either O–XYZ or ◆–XYZ; where the latter are a translating frame at ◆ *ever parallel* to O–XYZ—that is, one that is nonrotating but translating and hence is, generally, *noninertial*. Such "rotational freedoms" can be described via the nine direction cosines of ◆–xyz relative to ◆–XYZ (of which, as explained in §1.7, only three are independent); or via their three *attitude* angles: for example, their Eulerian or Cardanian angles

$$\phi = f_4(t), \qquad \theta = f_5(t), \qquad \psi = f_6(t); \tag{1.8.1b}$$

or via a directed line segment called *rotation "vector"* [or via *four* parameter formalisms (plus one constraint among them); for example, Hamiltonian quaternions, Euler–Rodrigues parameters, or complex numbers; detailed in kinematics treatises]. With the help of the six *positional system parameters*, or *system coordinates*, $f_{1,\ldots,6}(t)$, the location/motion of any other body point P can be determined:

$$\boldsymbol{r} = \boldsymbol{r}(P, t) = \boldsymbol{r}(P; f_1, \ldots, f_6) = \boldsymbol{r}_\blacklozenge(f_1, f_2, f_3) + \boldsymbol{r}_{/\blacklozenge}(P; f_4, f_5, f_6), \tag{1.8.2a}$$

or, in components,

$$X = X_\blacklozenge + \cos(X,x)x_{/\blacklozenge} + \cos(X,y)y_{/\blacklozenge} + \cos(X,z)z_{/\blacklozenge}, \quad \text{etc., cyclically,} \quad (1.8.2b)$$

where

$$\mathbf{r}_{/\blacklozenge} = (x_{/\blacklozenge}, y_{/\blacklozenge}, z_{/\blacklozenge}):$$

constant rectangular Cartesian coordinates of P relative to $\blacklozenge$*–xyz*, (1.8.2c)

or, in compact (self-explanatory) indicial notation,

$$x_{k'} = x_{\blacklozenge,k'} + \sum A_{k'k} x_k. \quad (1.8.2d)$$

In addition to $\blacklozenge$–xyz and $\blacklozenge$–XYZ, we occasionally use other *intermediate* axes (or *accessory* axes, in Routh's terminology) that, like $\blacklozenge$–XYZ, are neither space- nor body-fixed, but have their own special translatory and/or rotatory motion.

1.9 THE RIGID BODY: GEOMETRY OF MOTION AND KINEMATICS (SUMMARY OF BASIC THEOREMS)

Sections §1.9–1.13 cover material that is due to Euler, Mozzi, Cauchy, Chasles, Poinsot, Rodrigues, Cayley et al. (late 18th to mid-19th century). For detailed discussions, proofs, insights, and so on, see for example (alphabetically): Alt (1927), Altmann (1986), Beyer (1929, 1963), Bottema and Roth (1979), Coe (1938), Garnier (1951, 1956, 1960), Hunt (1978), McCarthy (1990), Schönflies and Grübler (1902), Timerding (1902, 1908).

The *position*, or *configuration*, of a rigid body B is known when the positions of any three *noncollinear* of its points are known; hence, *six* independent parameters are needed to specify it [e.g., $3 \times 3 = 9$ rectangular Cartesian coordinates of these points, minus the three independent constraints of distance invariance (i.e., rigidity) among them; or six coordinates for two of its points defining an axis of rotation, minus one invariance constraint between them, plus the angle of rotation of a body-fixed plane with a space-fixed plane, both through that axis]. If the body is further constrained, this number is less than six. It follows that the most general change of position, or *displacement*, of B is determined by the displacements of any three noncollinear of its points; that is, given their initial and final positions and the initial (final) position of a fourth, fifth, and so on, we can find their final (initial) positions with no additional data.

Special Rigid-Body Displacements

(i) *Plane, or planar, displacement*: One in which the paths of all body points are plane curves on planes parallel to each other and to a fixed plane f [fig. 1.10(a)]: the body fiber $P'PP''$ remains perpendicular to f, and the distance $P\blacklozenge$ remains constant, so that we need to study only *the motion of a typical body section, or rigid lamina, b imagined superimposed on f and sliding on it*.

THEOREM

Every displacement of a rigid lamina in its plane is equivalent to a rotation about some plane point I [fig. 1.10(b)].

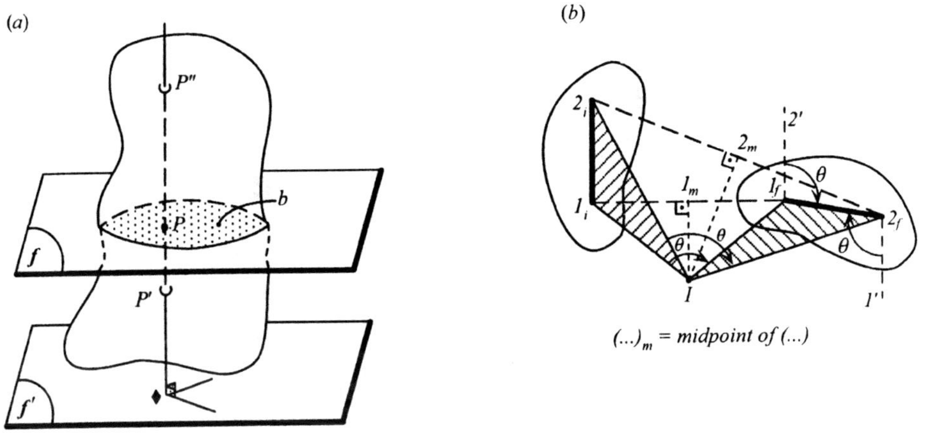

Figure 1.10 (a) Plane displacement of a rigid body. (b) The plane displacement of a rigid lamina on its plane is equivalent to a rotation about I; if $I \to \infty$, that displacement degenerates to a translation.

(ii) *Translational displacement*: One in which all body points have vectorially equal velocities. Translations can be either *rectilinear* or *curvilinear*, and can be represented by a *free* vector (three components).

(iii) *Rotational displacement*: One in which at least *two* points remain *fixed.* These points define the *axis of rotation*; and either they are actual body points, or belong to its appropriate *fictitious rigid extensions.* Rotations are, by far, the more complex and interesting part of rigid-body displacements/motions.

The rotation is specified by its axis (i.e., its *line of action*) and by its *angle of rotation*; and since a line is specified by, say, its two points of intersection with two coordinate planes—that is, four coordinates—and an angle is specified by one coordinate, *the complete characterization of rotation requires* $4 + 1 = 5$ *positional parameters.*

THEOREM

Every translation can be decomposed into rotations.

COROLLARY

All rigid displacements can be reduced to rotations. The above special displacements (plane, translations, rotations) are all examples of *constrained* motions; that is, they result from special geometrical [or finite, or holonomic (chap. 2)] *restrictions on the global mobility of the body*; as contrasted with *local* restrictions of its mobility [by nonholonomic constraints (chap. 2)].

EULER'S THEOREM (1775–1776)

Any displacement of a rigid body, one point of which is fixed but is otherwise free to move, can be achieved by a single rotation, of 180° or less, about some axis through that point; that is, any displacement of such a system is equivalent to a rotation. Or: any rigid displacement of a spherical surface into itself leaves two (diametrically opposed) points of that surface fixed; and hence, in such a displacement, an infinite number of points, lying on the axis of rotation defined by the preceding two points, remain fixed. (Under certain conditions this theorem extends to *deformable* bodies: one body-fixed line remains invariant.)

To understand this fundamental theorem, let us consider a body-fixed *unit sphere* S_B with center the fixed point ◆, representing the body, and let us follow its motion as it slides over another unit sphere S_S concentric to S_B but space-fixed and representing fixed space. (This is the spatial equivalent of the earlier plane motion problem where a representative rigid lamina slides over another fixed lamina.) Now, since this is a *three* degree-of-freedom system, its position can be specified by the coordinates of *two* of its points on S_B, P, and Q [fig. 1.11(a)]: $2 \times 2 = 4$ coordinates [of which, since the distance between P and Q (= length of arc of great circle joining P and Q) remains invariable, only *three* can be varied independently]. Hence, to study two positions of the body—that is, a displacement of it—it suffices to study two positions of an arbitrary pair of surface points of it: an initial PQ and a final $P'Q'$ [fig. 1.11(b)]. Then we join P and P', and Q and Q' by great arcs and draw the two symmetry planes of the arcs PP' and QQ'; that is, the two great circle planes that halve these two arcs. Their intersection, ◆C (which, contrary to the plane

$C \to C$, $P \to P'$, $Q \to Q'$

$\overset{\triangle}{CPQ}$, $\overset{\triangle}{CP'Q'}$: Congruent Triangles

$[CP = CP', CQ = CQ', PQ = P'Q']$

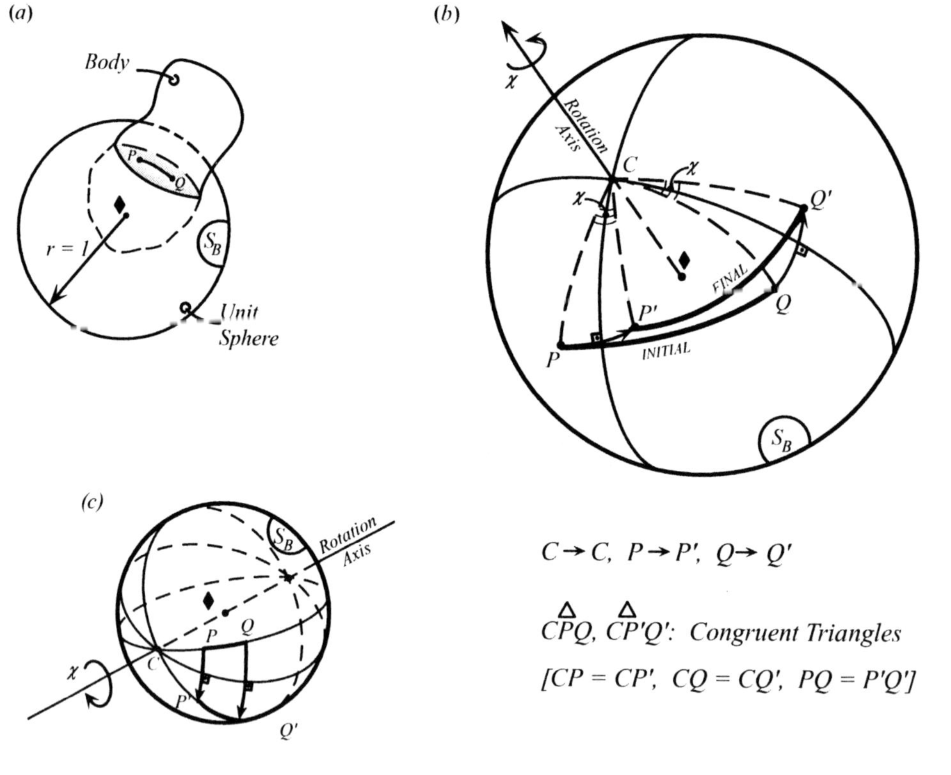

Figure 1.11 (a, b) The motion of a rigid body about a fixed point ◆ can be found by studying the motion of a pair of its points on the unit sphere with center ◆: from PQ to $P'Q'$; (c) special case of (b) where the planes of symmetry of the arcs PP' and QQ' coincide.

motion case, always lies a *finite* distance away), defines the axis of rotation; and their angle, χ, defines the angle of rotation (around ◆C) that brings the spherical triangle CPQ into coincidence with its congruent triangle $CP'Q'$; and hence *arc* (PP') into coincidence with *arc* (QQ'); and ◆PQ into coincidence with ◆$P'Q'$, and similarly for any other point of S_B. In the special case where these two symmetry planes coincide [fig. 1.11(c)], the rotation axis is the intersection of the planes defined by ◆PQ and ◆$P'Q'$.

FUNDAMENTAL THEOREM OF GEOMETRY OF RIGID-BODY MOTION

Any rigid-body displacement can be reduced to a succession of translations and rotations. Specifically, any such displacement can be produced by the translation of an arbitrary "base point," or "pole," of the body, from its initial to its final position, followed by a rotation about an axis through the final position of the chosen pole—and this is the most general rigid-body displacement. The translatory part varies with the pole, but the rotatory part (i.e., the axis direction and angle of rotation) is independent of it (fig. 1.12).

COROLLARY FOR PLANE MOTION

Any rigid planar displacement can be produced by a single rotation about a certain axis perpendicular to the plane of the motion; in the translation case, that axis recedes to infinity [fig. 1.10(b)].

THEOREMS OF CHASLES (1830) AND POINSOT (1830s, 1850s)

Any rigid-body displacement can be reduced, by a certain choice of pole, to a screw displacement; that is, to a rotation about an axis and a translation along that axis. In special cases, either of these two displacements may be missing.

In a screw displacement: (a) The axis of rotation is called *central axis*, and (for given initial and final body positions) it is unique, except when the displacement is a pure translation; (b) The ratio of the translation (l) to the rotation angle (χ), which

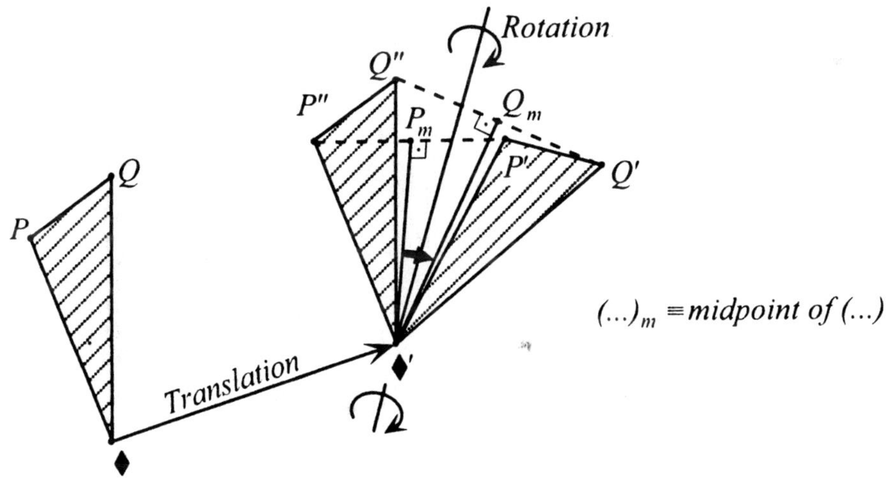

Figure 1.12 The most general displacement of the rigid body ◆PQ can be effected by a translation of the pole ◆, from ◆PQ to ◆$'P''Q''$; followed by a rotation about an axis through ◆', from ◆$'P''Q''$ to ◆$'P'Q'$.

equals the *advance (p) per revolution (2π)*, is called *pitch* of the screw: $p/2\pi = l/\chi \Rightarrow p = 2\pi(l/\chi)$; and (c) The translation and rotation commute.

EXTENSION TO DEFORMABLE BODIES
(Chasles' Theorem + Deformation = Cauchy's Theorem)

The total displacement of a generic point of a continuous medium, say a small deformable sphere (fig. 1.13), is the result of a *translation*, a rigid *rotation* [of the local *principal axes* (or directions) of *strain*], and *stretches* along these axes; that is, the sphere becomes a general ellipsoid. Hence, rigid-body kinematics is of interest to continuum mechanics too; the latter, however, will not be pursued any further here.

Rigid-Body Kinematics

Thus far, no restrictions have been placed on the size of the displacements; the above theorems hold whether the translations and rotations are *finite* or *infinitesimal*. The finite case is detailed quantitatively in the following sections.

Next, let us examine the important case of sequence of rigid infinitesimal displacements in time, namely, *rigid motion*. In particular, let us return to the motion about a fixed point (Euler's theorem) and consider the case where the initial and final positions of the arcs PQ (at time t) and $P'Q'$ (at time $t' = t + \Delta t$) *are very close to each other*. Now, as $\Delta t \to 0$ the earlier (great circle) planes that halve the arcs PP' and QQ' coincide with the normal planes to the *directions* of motion of P and Q, respectively, at time t; and their intersection yields the *instantaneous axis of rotation*. Then the velocity of the generic body point P equals

$$\boldsymbol{v}_P \equiv \boldsymbol{v} = [\lim(\boldsymbol{PP'}/\Delta t)]_{\Delta t \to 0} = \boldsymbol{\omega} \times \boldsymbol{r}_{P/\blacklozenge} \equiv \boldsymbol{\omega} \times \boldsymbol{r}, \tag{1.9.1}$$

since $v_P \equiv v \equiv |\boldsymbol{v}|$ equals the magnitude of the *angular velocity* of that rotation, $|\boldsymbol{\omega}| \equiv |d\chi/dt|$, times the perpendicular distance of P from the rotation axis. [Euler (1750s), Poisson (1831); of course, in components.] Hence, the instantaneous rotation of the body B about the fixed point $\blacklozenge$ is described by the single vector $\boldsymbol{\omega} = \omega\boldsymbol{n}$ ($\boldsymbol{n}$ = *unit vector along axis of rotation*), which combines all three characteristics of rotation: *axis*, *magnitude*, and *sense*. As the motion proceeds, and since only the point $\blacklozenge$ is fixed, the axis of rotation (carrier of $\boldsymbol{\omega}$) traces, or generates, two general and generally *open conical surfaces with common center* $\blacklozenge$: one fixed on the *body*, the

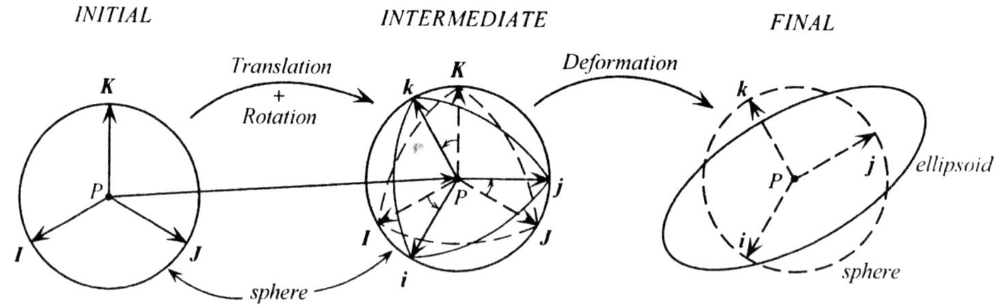

Figure 1.13 General displacement of a small deformable sphere:
Translation → Rotation → Strain (Sphere → Ellipsoid).

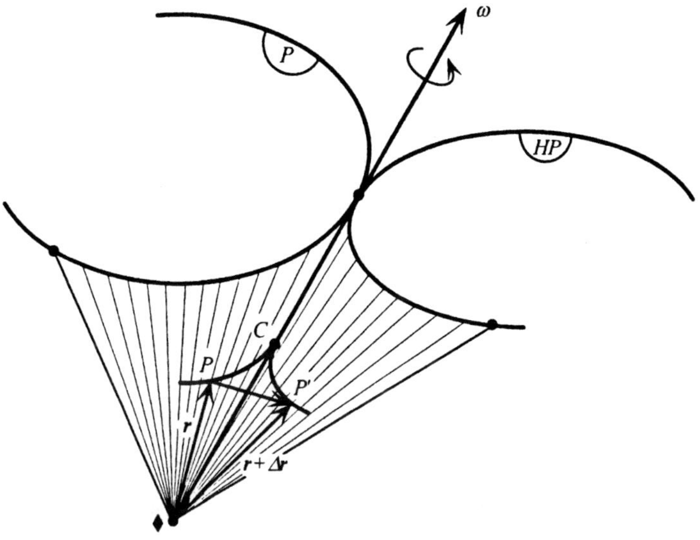

Figure 1.14 Rolling of body cone (P_{Polhode}) on space cone ($H_{er}P_{\mathrm{Polhode}}$).

polhode cone; and one fixed in space, the *herpolhode* cone (fig. 1.14). Hence, the following theorem:

THEOREM

Every finite motion of a rigid body, having one of its points ◆ fixed, can be described by the pure (or slippingless) rolling of the polhode cone on the herpolhode cone; and, at every moment, their common generator (through ◆) gives the direction of the instantaneous axis of rotation/angular velocity. If ◆ recedes to infinity, these two cones reduce to cylinders and their normal sections become, respectively, the body and space centrodes.

Velocity Field (Mozzi, 1763)

Since, for the first-order geometrical changes involved here ("infinitesimal displacements") superposition holds, we conclude that the velocity of a generic body point P in general motion, $\boldsymbol{v}_P \equiv \boldsymbol{v}$, is given by the following fundamental formula of rigid body kinematics:

$$\boldsymbol{v} = \boldsymbol{v}_\blacklozenge + \boldsymbol{\omega} \times (\boldsymbol{r} - \boldsymbol{r}_\blacklozenge) \equiv \boldsymbol{v}_\blacklozenge + \boldsymbol{\omega} \times \boldsymbol{r}_{/\blacklozenge} \equiv \boldsymbol{v}_\blacklozenge + \boldsymbol{v}_{/\blacklozenge}$$

$$[\boldsymbol{v}_{/\blacklozenge} = \text{velocity of } P \text{ relative to } \blacklozenge \text{ (both measured in the same frame)}] \qquad (1.9.2)$$

where ◆ is any body point (pole) (fig. 1.15); or, in terms of components (fig. 1.9) as follows:

Space-Fixed Axes

$$dX/dt \equiv dX_\blacklozenge/dt + \omega_Y(Z - Z_\blacklozenge) - \omega_Z(Y - Y_\blacklozenge), \quad \text{etc., cyclically,} \qquad (1.9.2a)$$

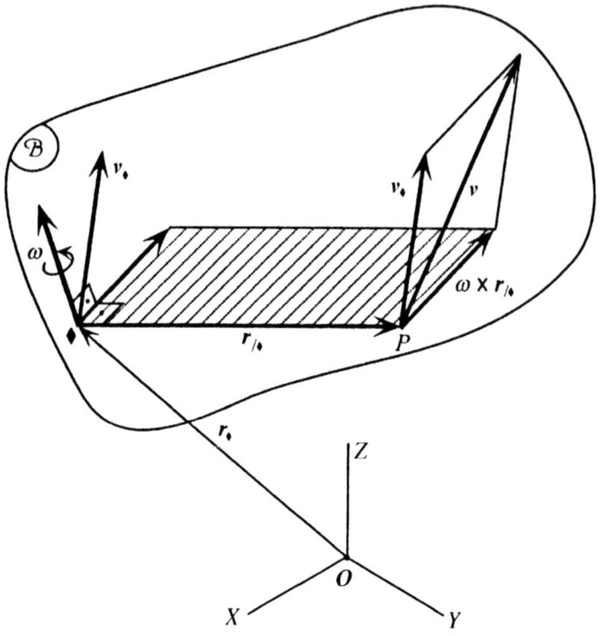

Figure 1.15 Geometrical interpretation of eq. (1.9.2).

or, equivalently,

$$v_X \equiv v_{\blacklozenge,X} + \omega_Y Z_{/\blacklozenge} - \omega_Z Y_{/\blacklozenge}, \quad \text{etc., cyclically.} \tag{1.9.2b}$$

Body-Fixed Axes

$$v_x \equiv v_{\blacklozenge,x} + \omega_y z_{/\blacklozenge} - \omega_z y_{/\blacklozenge}, \quad \text{etc., cyclically;} \tag{1.9.2c}$$

where

$$v_{\blacklozenge,x} = \cos(x, X) v_{\blacklozenge,X} + \cos(x, Y) v_{\blacklozenge,Y} + \cos(x, Z) v_{\blacklozenge,Z}, \quad \text{etc., cyclically;} \tag{1.9.2d1}$$

and, inversely,

$$v_{\blacklozenge,X} = \cos(X, x) v_{\blacklozenge,x} + \cos(X, y) v_{\blacklozenge,y} + \cos(X, z) v_{\blacklozenge,z}, \quad \text{etc., cyclically.} \tag{1.9.2d2}$$

The *six* functions of time $v_{\blacklozenge;x,y,z}, \omega_{x,y,z}$ (or $v_{\blacklozenge;X,Y,Z}, \omega_{X,Y,Z}$) characterize the rigid-body motion completely. The line-bound vectors $\boldsymbol{\omega}$ and $\boldsymbol{v}_\blacklozenge$ constitute the *torsor of motion*, or *velocity torsor*, at $\blacklozenge$, from which the rigid-body *velocity field* can be determined uniquely. [Just as, in elementary statics, the resultant force $\boldsymbol{f}$ (or $\boldsymbol{R}$) and moment $\boldsymbol{M}_\blacklozenge$ of a system of forces constitute the *force system torsor* at $\blacklozenge$ (see "Formal Analogies ..." section that follows.] In the case of motion about a fixed point $\blacklozenge$, that torsor reduces there to $(\boldsymbol{\omega}, \boldsymbol{0})$.

Now, from the displacement viewpoint, the *velocity transfer* equation (1.9.2) states that:

(i) The state of motion of the body consists of an *elementary translation* ($d\boldsymbol{r}_\blacklozenge \equiv \boldsymbol{v}_\blacklozenge\, dt$) of a *base point* (or *pole*) $\blacklozenge$, and an *elementary rotation* ($d\boldsymbol{\chi} \equiv \boldsymbol{\omega}\, dt$) about that point. Therefore, applying the earlier theorem of Chasles, we deduce that:

(ii) Any *infinitesimal rigid (nontranslatory) displacement* can be reduced uniquely to an *infinitesimal screw*; that is, an infinitesimal translation plus an infinitesimal rotation about a (central) axis parallel to the translation. (The *location* of that axis and the *pitch of the screw* are given in the "Formal Analogies ..." section below.) As the motion proceeds, that axis traces two (ruled) surfaces with it as common generator: one *fixed in space* (Γ_S) and another *fixed in the body* (Γ_B)—which constitute the "no fixed point" generalization of the herpolhode and polhode, respectively. Hence, the following theorem:

(iii) The general finite motion of a rigid body can be produced by the *rolling and sliding* of Γ_B over Γ_S. (In plane motion, sliding is absent.) Next, we prove that

(iv) The angular velocity vector $\boldsymbol{\omega}$ is *independent of the choice of the pole*. Applying the fundamental formula (1.9.2) for the two arbitrary and distinct poles $\blacklozenge$ and $\blacklozenge'$, we have

$$\boldsymbol{v} = \boldsymbol{v}_\blacklozenge + \boldsymbol{\omega} \times (\boldsymbol{r} - \boldsymbol{r}_\blacklozenge) \equiv \boldsymbol{v}_\blacklozenge + \boldsymbol{\omega} \times \boldsymbol{r}_{/\blacklozenge}$$

$$= \boldsymbol{v}_{\blacklozenge'} + \boldsymbol{\omega}' \times (\boldsymbol{r} - \boldsymbol{r}_{\blacklozenge'}) \equiv \boldsymbol{v}_{\blacklozenge'} + \boldsymbol{\omega}' \times \boldsymbol{r}_{/\blacklozenge'}, \tag{1.9.2e}$$

where initially, we assume that $\boldsymbol{\omega}$ and $\boldsymbol{\omega}'$ are different and go through $\blacklozenge$ and $\blacklozenge'$, respectively. We shall show that

$$\boldsymbol{\omega} = \boldsymbol{\omega}'. \tag{1.9.2f}$$

Indeed, since

$$\boldsymbol{r}_{/\blacklozenge} = \boldsymbol{r}_{/\blacklozenge'} + \boldsymbol{r}_{\blacklozenge'/\blacklozenge} \quad \text{and} \quad \boldsymbol{v}_{\blacklozenge'} = \boldsymbol{v}_\blacklozenge + \boldsymbol{\omega} \times \boldsymbol{r}_{\blacklozenge'/\blacklozenge}, \tag{1.9.2g}$$

equating the right sides of (1.9.2e) we obtain

$$\boldsymbol{\omega} \times \boldsymbol{r}_{/\blacklozenge} = \boldsymbol{\omega} \times \boldsymbol{r}_{\blacklozenge'/\blacklozenge} + \boldsymbol{\omega}' \times \boldsymbol{r}_{/\blacklozenge'} \;\Rightarrow\; \boldsymbol{\omega} \times \boldsymbol{r}_{/\blacklozenge'} = \boldsymbol{\omega}' \times \boldsymbol{r}_{/\blacklozenge'}, \tag{1.9.2h}$$

from which, since $\boldsymbol{r}_{/\blacklozenge'}$ is arbitrary, (1.9.2f) follows.

[Since $\boldsymbol{\omega}$ is a *body* quantity (a *system* vector), it carries no body point subscripts (like $\boldsymbol{v}_{\ldots}$), just like a force resultant. The only "insignia" it may carry are those needed to specify a particular body and/or frame of reference. Perhaps this supposed "base point invariance" of it may have given rise to the false notion that "$\boldsymbol{\omega}$ [of a body-fixed basis relative to a space-fixed basis] is a *free* vector, not bound to any point or line in space" (Likins, 1973, p. 105, near page bottom); emphasis added. A correct interpretation of (1.9.2e,f), however, shows that $\boldsymbol{\omega}$ is a *line-bound*, or *sliding*, vector, not a *free* one (just like the force on a rigid body); hence, $\boldsymbol{\omega}$ in eq. (1.9.2), is understood to be going through point $\blacklozenge$.]

A USEFUL RESULT

Let $\boldsymbol{r}_1$ and $\boldsymbol{r}_2$ be the position vectors of two arbitrary points of a rigid body. Then, its angular velocity $\boldsymbol{\omega}$ equals

$$\boldsymbol{\omega} = (\boldsymbol{v}_1 \times \boldsymbol{v}_2)/(\boldsymbol{v}_1 \cdot \boldsymbol{v}_2), \qquad \text{where} \quad \boldsymbol{v}_{\ldots} \equiv d\boldsymbol{r}_{\ldots}/dt. \tag{1.9.2i}$$

Formal Analogies Between Forces/Moments and Linear/Angular Velocities

Comparing (1.9.2), rewritten as $\boldsymbol{v}_2 = \boldsymbol{v}_1 + \boldsymbol{r}_{1/2} \times \boldsymbol{\omega}$ (1, 2: two arbitrary body points) with the well-known moment transfer theorem of elementary statics (with some, hopefully, self-explanatory notation): $\boldsymbol{M}_2 = \boldsymbol{M}_1 + \boldsymbol{r}_{1/2} \times \boldsymbol{f}$, we may say that the velocity $\boldsymbol{v}_2$ is the moment of the *motion, or velocity torsor* $(\boldsymbol{\omega}, \boldsymbol{v}_1)$ about point 2; that is, $\boldsymbol{\omega}$ is the *kinematic counterpart of the force resultant* (*f* or ***R***), and hence is a *line-bound*, or *sliding vector;* while $\boldsymbol{v}\ldots$ is the *counterpart of the point–dependent moment of the torsor* ***M***. Hence, recalling the (presumably, well-known) theorems of elementary statics, we can safely state the following:

- An elementary rotation $d\boldsymbol{\chi} \equiv \boldsymbol{\omega}\, dt$ about an axis can always be replaced with an elementary rotation of equal angle about another arbitrary but parallel axis, plus an elementary translation $d\boldsymbol{r} = \boldsymbol{v}\, dt$, where $\boldsymbol{v} = \boldsymbol{\omega} \times \boldsymbol{r}$ is perpendicular to (the plane of) both axes of rotation, and $\boldsymbol{r}$ is the vector from an arbitrary point of the original axis to an arbitrary point of the second axis; that is, an elementary rotation here is equivalent to an equal rotation plus an elementary perpendicular translation there.
- Several elementary rotations about a number of arbitrary axes can be replaced by a resultant motion as follows: (a) We choose a reference point O, and transport all these elementary rotations parallel to themselves to O, and then add them geometrically there. Then, (b) We combine the corresponding translational velocities, created by the parallel transport of the rotations in (a) (according to the preceding statement), to a single translational velocity at O. For example, two equal and opposite elementary rotations about *parallel* axes can be replaced by a single elementary translation perpendicular to (the plane of) both axes. These formal analogies between forces/moments and linear/angular velocities (also, linear/angular momenta), which are quite useful from the viewpoint of economy of *thought* (elimination of unnecessary duplication of proofs), are summarized in table 1.2.

Table 1.2 Formal Analogies Among Vectors/Forces/Rigid-Body Velocities

Vector Systems	*Forces/Moments* (On Rigid Bodies)	*Rigid-Body Velocities* (Instantaneous Geometry)
Single vector $\boldsymbol{a}$	Single force $\boldsymbol{f}$ (along line of action)	Angular velocity $\boldsymbol{\omega}$ (about axis of rotation)
Moment of $\boldsymbol{a}$ about point O	Moment of $\boldsymbol{f}$ about O	Linear velocity of body point O$\boldsymbol{v}_O$
Vector couple $(\boldsymbol{a}_1, \boldsymbol{a}_2 = -\boldsymbol{a}_1)$ $\Rightarrow$ Constant moment	Force couple $(\boldsymbol{f}_1, \boldsymbol{f}_2 = -\boldsymbol{f}_1)$ $\Rightarrow$ Constant moment; or couple	Rotational pair $(\boldsymbol{\omega}_1, \boldsymbol{\omega}_2 = -\boldsymbol{\omega}_1)$ $\Rightarrow$ Constant translational velocity
Vector resultant $\boldsymbol{R}$	Force resultant $\boldsymbol{R}$	Rotation resultant $\boldsymbol{\omega}$
Vector torsor $(\boldsymbol{R}, \boldsymbol{M}_O)$	Force torsor $(\boldsymbol{R}, \boldsymbol{M}_O)$	Motion torsor $(\boldsymbol{\omega}, \boldsymbol{v}_O)$
	Spatial Variation (or Transfer) Theorem: $O \to O'(\boldsymbol{R}, \boldsymbol{\omega}$ at $O)$	
$\boldsymbol{M}_{O'} = \boldsymbol{M}_O + \boldsymbol{r}_{O/O'} \times \boldsymbol{R}$ Invariants: $\boldsymbol{R}\cdot\boldsymbol{R}$, $\boldsymbol{R}\cdot\boldsymbol{M}\ldots$	$\boldsymbol{M}_{O'} = \boldsymbol{M}_O + \boldsymbol{r}_{O/O'} \times \boldsymbol{R}$ Invariants: $\boldsymbol{R}\cdot\boldsymbol{R}$, $\boldsymbol{R}\cdot\boldsymbol{M}\ldots$	$\boldsymbol{v}_{O'} = \boldsymbol{v}_O + \boldsymbol{r}_{O/O'} \times \boldsymbol{\omega}$ Invariants: $\boldsymbol{\omega}\cdot\boldsymbol{\omega}$, $\boldsymbol{\omega}\cdot\boldsymbol{v}\ldots$
	Simplest Representation of Torsor	
Vector wrench (or screw) $(\boldsymbol{R}, \boldsymbol{M}_c)$	Force wrench $(\boldsymbol{R}, \boldsymbol{M}_c)$	Motion screw $(\boldsymbol{\omega}, \boldsymbol{v}_c)$
	Central Axis of Wrench/Screw	
$\boldsymbol{r} = \lambda\boldsymbol{R} + (\boldsymbol{R} \times \boldsymbol{M}_O)/R^2$ $[\lambda \equiv (\boldsymbol{r}\cdot\boldsymbol{R})/R^2]$	$\boldsymbol{r} = \lambda\boldsymbol{R} + (\boldsymbol{R} \times \boldsymbol{M}_O)/R^2$	$\boldsymbol{r} = \mu\boldsymbol{\omega} + (\boldsymbol{\omega} \times \boldsymbol{v}_O)/\omega^2$ $[\mu \equiv (\boldsymbol{r}\cdot\boldsymbol{\omega})/\omega^2]$
Pitch $\equiv p = M_c/R = \boldsymbol{R}\cdot\boldsymbol{M}_O/R^2$	$p = M_c/R = \boldsymbol{R}\cdot\boldsymbol{M}_O/R^2$	$p = v_c/\omega = \boldsymbol{\omega}\cdot\boldsymbol{v}_O/\omega^2$
• $p = 0$:		
Vector resultant $\boldsymbol{R}$	Pure force (resultant) $\boldsymbol{R}$	Pure rotation $\boldsymbol{\omega}$
• $p = \infty$:		
Couple	Pure couple	Pure translation*

* See also Hunt (1974).

Acceleration Field

By $d/dt(\ldots)$-differentiating (1.9.2), we readily obtain the *acceleration field* of a rigid body in *general motion*:

$$\begin{aligned} \boldsymbol{a} &= \boldsymbol{a}_\blacklozenge + \boldsymbol{\alpha} \times \boldsymbol{r}_{/\blacklozenge} + \boldsymbol{\omega} \times (\boldsymbol{\omega} \times \boldsymbol{r}_{/\blacklozenge}) = \boldsymbol{a}_\blacklozenge + \boldsymbol{\alpha} \times \boldsymbol{r}_{/\blacklozenge} + [(\boldsymbol{\omega} \cdot \boldsymbol{r}_{/\blacklozenge})\boldsymbol{\omega} - \omega^2 \boldsymbol{r}_{/\blacklozenge}] \\ &= \boldsymbol{a}_\blacklozenge + (\boldsymbol{a}_{/\blacklozenge})_{\text{tangent}} + (\boldsymbol{a}_{/\blacklozenge})_{\text{normal}} \qquad (\equiv \boldsymbol{a}_\blacklozenge + \boldsymbol{a}_{/\blacklozenge}); \end{aligned} \tag{1.9.3}$$

or in terms of components (figure 1.9):

Space-Fixed Axes

$$\begin{aligned} a_X \equiv a_{\blacklozenge,X} &+ (\alpha_Y Z_{/\blacklozenge} - \alpha_Z Y_{/\blacklozenge}) \\ &+ [\omega_X(\omega_X X_{/\blacklozenge} + \omega_Y Y_{/\blacklozenge} + \omega_Z Z_{/\blacklozenge}) - \omega^2 X_{/\blacklozenge}], \quad \text{etc., cyclically.} \end{aligned} \tag{1.9.3a}$$

Body-Fixed Axes

$$\begin{aligned} a_x \equiv a_{\blacklozenge,x} &+ (\alpha_y z_{/\blacklozenge} - \alpha_z y_{/\blacklozenge}) \\ &+ [\omega_x(\omega_x x_{/\blacklozenge} + \omega_y y_{/\blacklozenge} + \omega_z z_{/\blacklozenge}) - \omega^2 x_{/\blacklozenge}], \quad \text{etc., cyclically;} \end{aligned} \tag{1.9.3b}$$

where

$$\begin{aligned} a_{\blacklozenge,x} &= \cos(x, X) a_{\blacklozenge,X} + \cos(x, Y) a_{\blacklozenge,Y} + \cos(x, Z) a_{\blacklozenge,Z} \\ &\equiv \cos(x, X)(d^2 X_\blacklozenge/dt^2) + \cos(x, Y)(d^2 Y_\blacklozenge/dt^2) + \cos(x, Z)(d^2 Z_\blacklozenge/dt^2), \\ &\qquad \text{etc., cyclically;} \end{aligned} \tag{1.9.3c}$$

and, inversely,

$$a_{\blacklozenge,X} = \cos(X, x) a_{\blacklozenge,x} + \cos(X, y) a_{\blacklozenge,y} + \cos(X, z) a_{\blacklozenge,z}, \quad \text{etc., cyclically;} \tag{1.9.3d}$$

and

$$\alpha_X = (d\boldsymbol{\omega}/dt)_X = d\omega_X/dt, \quad \text{etc., cyclically,} \tag{1.9.3e}$$

$$\begin{aligned} \alpha_x &= (d\boldsymbol{\omega}/dt) \cdot \boldsymbol{i} = d(\boldsymbol{\omega} \cdot \boldsymbol{i})/dt - \boldsymbol{\omega} \cdot (d\boldsymbol{i}/dt) \\ &= d\omega_x/dt - \boldsymbol{\omega} \cdot (\boldsymbol{\omega} \times \boldsymbol{i}) = d\omega_x/dt, \quad \text{etc., cyclically.} \end{aligned} \tag{1.9.3f}$$

Plane Motion

The distances of all body points from a fixed, say inertial, plane f' remain constant; and so the body B moves parallel to f' (fig. 1.10a). [For extensive discussions of this pedagogically and technically important topic, see, for example, Pars (1953, pp. 336–356), Loitsianskii and Lur'e (1982, pp. 227–261).] A rigid body in plane (but otherwise free) motion is a system with *three global, or finite, degrees of freedom.* As such, we choose (fig. 1.16): (a) The two positional coordinates of an arbitrary body point (pole) ◆ (that is, of a point belonging to the cross section of B with a generic

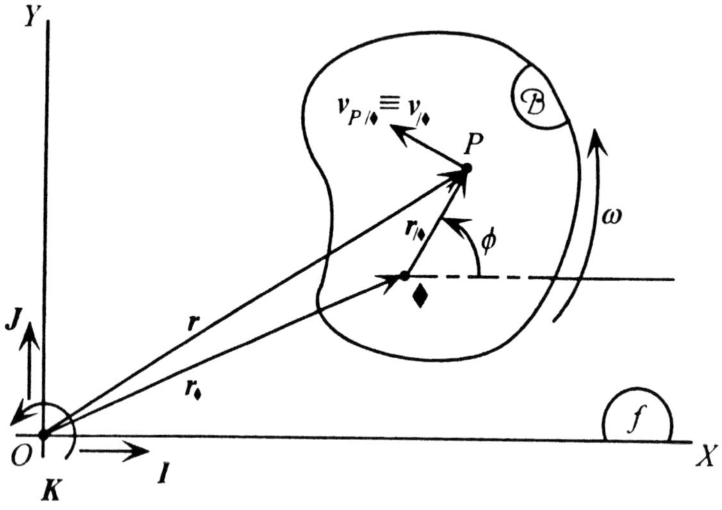

Figure 1.16 Plane motion of a rigid body *B*.

plane f ever parallel to f') relative to arbitrary but f-fixed rectangular Cartesian coordinates O–XY, $(X_\blacklozenge, Y_\blacklozenge)$; and (b) The angle between an arbitrary f-fixed line, say the axis OX, and an arbitrary B-fixed line, say $\blacklozenge P$, where P is a generic body point.

(i) The velocity field (i.e., the instantaneous spatial distribution of velocity)

Here,

$$\boldsymbol{\omega} = \omega_z \boldsymbol{k} = \omega_Z \boldsymbol{K} \equiv \omega \boldsymbol{K} = (d\phi/dt)\boldsymbol{K} \qquad \text{(i.e., } \boldsymbol{\omega} \text{ is perpendicular to } \boldsymbol{v}\text{)}, \qquad (1.9.4a)$$

and so the general velocity formula (1.9.2) becomes

$$\boldsymbol{v}_P \equiv d\boldsymbol{r}_{P/O} \equiv d\boldsymbol{r}/dt$$
$$\equiv \boldsymbol{v} = \boldsymbol{v}_\blacklozenge + \boldsymbol{v}_{P/\blacklozenge} \equiv \boldsymbol{v}_\blacklozenge + \boldsymbol{v}_{/\blacklozenge} = \boldsymbol{v}_\blacklozenge + \boldsymbol{\omega} \times \boldsymbol{r}_{P/\blacklozenge} \equiv \boldsymbol{v}_\blacklozenge + \boldsymbol{\omega} \times \boldsymbol{r}_{/\blacklozenge}, \qquad (1.9.4b)$$

or, in components [along space-fixed (inertial) axes]

$$(dX/dt, dY/dt, 0) = (dX_\blacklozenge/dt, dY_\blacklozenge/dt, 0) + (0, 0, \omega) \times (X_{/\blacklozenge}, Y_{/\blacklozenge}, 0),$$
$$\Rightarrow dX/dt = dX_\blacklozenge/dt - \omega Y_{/\blacklozenge}, \qquad dY/dt = dY_\blacklozenge/dt + \omega X_{/\blacklozenge}. \qquad (1.9.4c)$$

The above show that, in plane motion, there exists—in every configuration—a point, either belonging to the body or to its fictitious rigid extension, called *instantaneous center of zero velocity*, or *velocity pole* (*IC*, or *I*, for short), whose velocity, at least momentarily, vanishes; that is, locally, at least, *the motion can be viewed as an elementary rotation about that point* (local version of fig. 1.10b). Indeed, setting in (1.9.4b,c)

$$\boldsymbol{v} \rightarrow \boldsymbol{v}_I = \boldsymbol{0}, \qquad \text{i.e., choosing } P = I, \qquad (1.9.4d)$$

we obtain its inertial instantaneous coordinates relative to our originally chosen pole $\blacklozenge$:

$$X_{I/\blacklozenge} \equiv X_I - X_\blacklozenge = -(dY_\blacklozenge/dt)\big/\omega, \qquad Y_{I/\blacklozenge} \equiv Y_I - Y_\blacklozenge = +(dX_\blacklozenge/dt)\big/\omega. \quad (1.9.4e)$$

From these equations we conclude that, as long as $\omega \neq 0, I$ is located at a finite distance from the body and is unique; if $\omega = 0$, then I recedes to infinity, and the motion becomes a *translation*; and if we choose I as our pole—that is, $\blacklozenge = I$—then (1.9.4b, c) yield

$$dX/dt = -\omega Y_{/I}, \quad dY/dt = \omega X_{/I}, \quad \text{or } v = \omega\, r_{/I} \quad [v^2 = (dX/dt)^2 + (dY/dt)^2]. \tag{1.9.4f}$$

[In the case of translation, eq. (1.9.4f) can be written *qualitatively/symbolically* as

finite velocity = (*zero angular velocity*) × (*infinite radius of rotation*)].

As the body moves, I traces *two* curves: one fixed on the *body* (*space centrode*) and one fixed in the *plane* (*space centrode*); so that the general plane motion can be described as *the slippingless rolling of the body centrode on the space centrode, with angular velocity* ω.

(ii) The acceleration field

Here,

$$\boldsymbol{\alpha} \equiv d\boldsymbol{\omega}/dt = (d\omega/dt)\boldsymbol{k} \equiv \alpha\,\boldsymbol{k} = \alpha\,\boldsymbol{K}, \tag{1.9.4g}$$

and $\boldsymbol{\omega} \cdot \boldsymbol{r}_{/\blacklozenge} = 0$, and so the general acceleration formula (1.9.3) becomes

$$\begin{aligned} \boldsymbol{a}_P \equiv \boldsymbol{a} \equiv \boldsymbol{a}_\blacklozenge + \boldsymbol{a}_{/\blacklozenge} &= \boldsymbol{a}_\blacklozenge + \boldsymbol{\alpha} \times \boldsymbol{r}_{/\blacklozenge} + \boldsymbol{\omega} \times (\boldsymbol{\omega} \times \boldsymbol{r}_{/\blacklozenge}) \\ &= \boldsymbol{a}_\blacklozenge + \boldsymbol{\alpha} \times \boldsymbol{r}_{/\blacklozenge} - \omega^2 \boldsymbol{r}_{/\blacklozenge}, \end{aligned} \tag{1.9.4h}$$

or, in components [along space-fixed (inertial) axes],

$$\begin{aligned} (d^2X/dt^2, d^2Y/dt^2, 0) &= (d^2X_\blacklozenge/dt^2,\, d^2Y_\blacklozenge/dt^2,\, 0) \\ &\quad + (0, 0, \alpha) \times (X_{/\blacklozenge},\, Y_{/\blacklozenge}, 0) - \omega^2 (X_{/\blacklozenge},\, Y_{/\blacklozenge}, 0), \\ \Rightarrow d^2X/dt^2 = d^2X_\blacklozenge/dt^2 - \alpha Y_{/\blacklozenge} - \omega^2 X_{/\blacklozenge}, &\quad d^2Y/dt^2 = d^2Y_\blacklozenge/dt^2 + \alpha X_{/\blacklozenge} - \omega^2 Y_{/\blacklozenge}. \end{aligned} \tag{1.9.4i}$$

Along *body-fixed* axis $\blacklozenge$–xy, eq. (1.9.4h) yields the components (with some easily understood notation):

$$a_x = (\boldsymbol{a}_\blacklozenge)_x - \alpha y_{/\blacklozenge} - \omega^2 x_{/\blacklozenge}, \qquad a_y = (\boldsymbol{a}_\blacklozenge)_y + \alpha x_{/\blacklozenge} - \omega^2 y_{/\blacklozenge}; \tag{1.9.4j}$$

where

$$(\boldsymbol{a}_\blacklozenge)_x \equiv \boldsymbol{a}_\blacklozenge \cdot \boldsymbol{i} = \cos(x, X)(d^2X_\blacklozenge/dt^2) + \cos(x, Y)(d^2Y_\blacklozenge/dt^2), \text{ etc.;}$$

and similarly for the velocity field (1.9.4b), if needed.

Here, too, there exists an *instantaneous center of zero acceleration*, or *acceleration pole*, I', whose coordinates are found by setting in (1.9.4i) $d^2X/dt^2 = 0$, $d^2Y/dt^2 = 0$ and then solving for $X_{/\blacklozenge}$, $Y_{/\blacklozenge}(P \to I')$:

$$X_{I'/\blacklozenge} \equiv X_{I'} - X_\blacklozenge = \left[\omega^2(d^2X_\blacklozenge/dt^2) - \alpha(d^2Y_\blacklozenge/dt^2)\right]\big/(\alpha^2 + \omega^4),$$
$$Y_{I'/\blacklozenge} \equiv Y_{I'} - Y_\blacklozenge = \left[\omega^2(d^2Y_\blacklozenge/dt^2) + \alpha(d^2X_\blacklozenge/dt^2)\right]\big/(\alpha^2 + \omega^4). \tag{1.9.4k}$$

These equations show that as long as $\alpha^2 + \omega^4 \neq 0$ (i.e., not both ω and α vanish), the acceleration pole I' exists and is unique. If $\omega, \alpha = 0$ (i.e., if the body translates), then I' (as well as I) recedes to infinity. Finally, with the choice $\blacklozenge = I'$ eqs. (1.9.4h,i) specialize to

$$\boldsymbol{a} \equiv \boldsymbol{a}_\blacklozenge + \boldsymbol{a}_{/\blacklozenge} = \boldsymbol{\alpha} \times \boldsymbol{r}_{/I'} + \boldsymbol{\omega} \times (\boldsymbol{\omega} \times \boldsymbol{r}_{/I'}) = \boldsymbol{\alpha} \times \boldsymbol{r}_{/I'} - \omega^2 \boldsymbol{r}_{/I'}, \tag{1.9.4l}$$

or, in components

$$d^2X/dt^2 = -\alpha Y_{/I'} - \omega^2 X_{/I'}, \qquad d^2Y/dt^2 = +\alpha X_{/I'} - \omega^2 Y_{/I'}. \tag{1.9.4m}$$

For the geometrical properties of I', the reader is referred to texts on kinematics.

Additional Useful Results

(i) Crossing $\mathbf{0} = \boldsymbol{v}_\blacklozenge + \boldsymbol{\omega} \times (\boldsymbol{r}_I - \boldsymbol{r}_\blacklozenge)$ with $\boldsymbol{\omega}$, expanding, and so on, it can be shown that the position of the instantaneous velocity center is given by

$$\boldsymbol{r}_{I/\blacklozenge} \equiv \boldsymbol{r}_I - \boldsymbol{r}_\blacklozenge = (\boldsymbol{\omega} \times \boldsymbol{v}_\blacklozenge)/\omega^2; \tag{1.9.4n}$$

and similarly for the location of the acceleration pole I'.

(ii) The location of the instantaneous center of zero velocity I, and zero acceleration I', in *body-fixed* coordinates $\blacklozenge$–xy, are given, respectively, by (fig. 1.17)

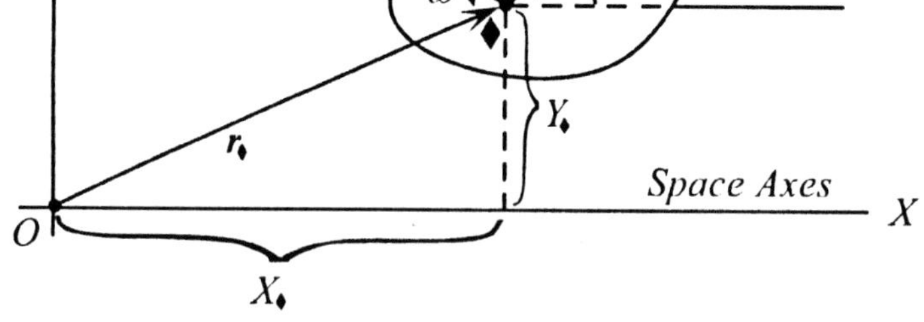

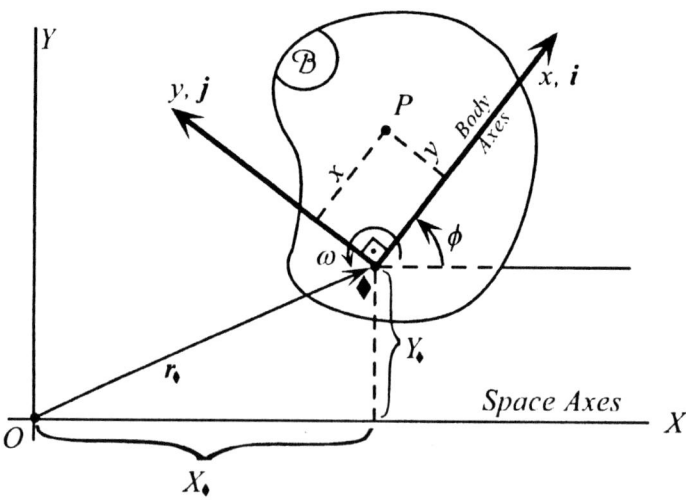

Figure 1.17 Body-fixed axes in plane motion.

$$x_I = (1/\omega)\big[(dX_\bullet/dt)\sin\phi - (dY_\bullet/dt)\cos\phi\big] = -(\boldsymbol{v}_\bullet)_y/\omega, \tag{1.9.4o}$$

$$y_I = (1/\omega)\big[(dX_\bullet/dt)\cos\phi + (dY_\bullet/dt)\sin\phi\big] = (\boldsymbol{v}_\bullet)_x/\omega, \tag{1.9.4p}$$

$$x_{I'} = [\omega^2(\boldsymbol{a}_\bullet)_x - \alpha(\boldsymbol{a}_\bullet)_y]/(\omega^4 + \alpha^2), \qquad y_{I'} = [\alpha(\boldsymbol{a}_\bullet)_x + \omega^2(\boldsymbol{a}_\bullet)_y]/(\omega^4 + \alpha^2), \tag{1.9.4q}$$

where

$$(\boldsymbol{v}_\bullet)_x \equiv \boldsymbol{v}_\bullet \cdot \boldsymbol{i} = \cos(x, X)(dX_\bullet/dt) + \cos(x, Y)(dY_\bullet/dt), \quad \text{etc.}$$

Contact of Two Rigid Bodies; Slipping, Rolling, Pivoting

Let us consider a system of rigid bodies forced to remain in mutual *contact* at points, or along curves or surfaces of their boundaries. For simplicity and concreteness, we restrict the discussion to two rigid bodies, B' (fixed) and B (moving), in contact at a space point C; that is, a certain point P of the bounding surface of B, S, is in contact with a point P' of the bounding surface of B', S'; that is, then, $C = P = P'$ (fig. 1.18).

Now: (i) If C is fixed on both bodies, we call such a "bilateral constraint" (i.e., one expressible by equalities) a *hinge*, and we say that the bodies are *pivoting* about it.

(ii) If, on the other hand, C is not fixed on one (both) of the bodies, we say that it is *wandering* on it (them). In this case, we call the relative velocity of P and P', which are instantaneously at C, the *slip* velocity there:

$$\boldsymbol{v}_{P/P'} \equiv \boldsymbol{v}_P - \boldsymbol{v}_{P'} \equiv \boldsymbol{v}_s. \tag{1.9.5a}$$

If we view the motion of C relative to B', C/B', as the resultant of C/B and B/B', then, since the velocities of the latter are tangent to the surfaces S and S', respectively, at C we conclude that $\boldsymbol{v}_s$ lies on their common tangent plane there, p. Analytically,

$$\boldsymbol{v}_s = \boldsymbol{v}_{s,T} + \boldsymbol{v}_{s,N} = \boldsymbol{v}_{s,T}, \tag{1.9.5b}$$

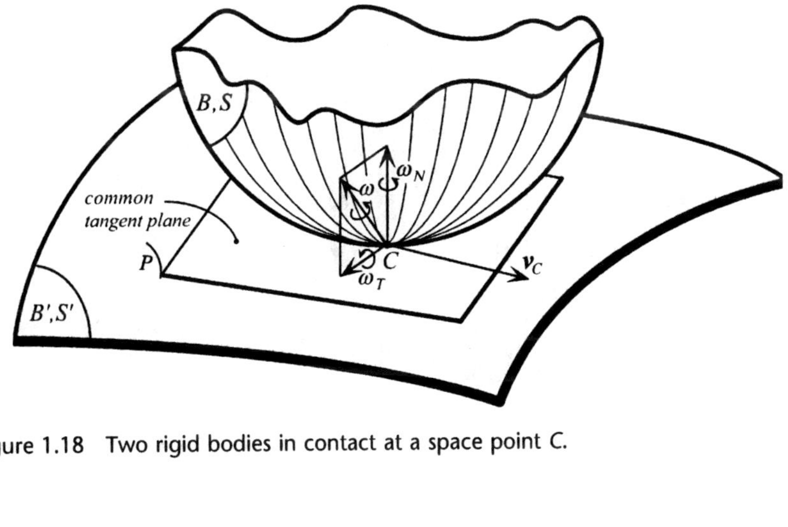

Figure 1.18 Two rigid bodies in contact at a space point C.

where

$\boldsymbol{v}_{s,T}$ = *component of* $\boldsymbol{v}_s$ *along* p, $\boldsymbol{v}_{s,N}$ = *component of* $\boldsymbol{v}_s$ *normal to* p

(= $\mathbf{0}$; i.e., contact is preserved; the two bodies cannot penetrate each other);

(1.9.5c)

and if, at that instant, B and B' *separate*, then $\boldsymbol{v}_{s,N}$ lies on the side of B'.

Next, if the angular velocity of B relative to B', at C, is $\boldsymbol{\omega}$ with components *along* and *normal* to p: $\boldsymbol{\omega}_T$, $\boldsymbol{\omega}_N$, respectively; that is,

$$\boldsymbol{\omega} = \boldsymbol{\omega}_T + \boldsymbol{\omega}_N, \qquad (1.9.5d)$$

then we can say that the most general infinitesimal motion of B relative to B', B/B', is a superposition of the following special motions:

a *pure slipping*: $\boldsymbol{v}_s \neq \mathbf{0}$, $\boldsymbol{\omega}_T = \mathbf{0}$, $\boldsymbol{\omega}_N = \mathbf{0}$; (1.9.5e)

a *pure rolling*: $\boldsymbol{v}_s = \mathbf{0}$, $\boldsymbol{\omega}_T \neq \mathbf{0}$, $\boldsymbol{\omega}_N = \mathbf{0}$; (1.9.5f)

a *pure pivoting*: $\boldsymbol{v}_s = \mathbf{0}$, $\boldsymbol{\omega}_T = \mathbf{0}$, $\boldsymbol{\omega}_N \neq \mathbf{0}$. (1.9.5g)

If $\boldsymbol{v}_s = \mathbf{0}$ and $\boldsymbol{\omega} \neq \mathbf{0}$, the motion B/B' is an instantaneous rotation called *rolling and pivoting*; which results in two (scalar) equations of constraint. In this case, the point C has *identical* velocities relative to both B and B'; and hence its trajectories, or loci, on their bounding surfaces, denoted by γ and γ', are continuously tangent, and are *traced at the same pace*; that is, if, starting from C, we grade them in, say centimeters, then the points that will come into contact during the subsequent motion will have the same arc-coordinates numerically. Such a B/B' rolling is expressed by saying that P and P', both at C at the moment under consideration, have *equal velocities* relative to a (third) arbitrary body, or frame or reference; and the velocities of B about B' are the same as if B had only a rotation $\boldsymbol{\omega}$ about an axis through the "instantaneous hinge" C. If the locus of $\boldsymbol{\omega}$ on B is the ruled surface Σ, and on B' the also ruled surface Σ', then the slippingless motion B/B' can be obtained by rolling Σ on Σ' [The earlier curve $\gamma(\gamma')$ is the intersection of Σ with $S(\Sigma'$ with $S')$].

If B and B' are in contact at *two* points, say C and C', and if $\boldsymbol{v}_s = \boldsymbol{v}_{s'} = \mathbf{0}$, then the motion B/B' is an *instantaneous rotation* about the line CC'; that is, $\boldsymbol{\omega}$ is along it. And if B, B' contact each other at several points $C, C', C'', \ldots$, then slipping cannot vanish at all of them *unless they all lie on a straight line*. If, in addition, $\boldsymbol{\omega}_N = \mathbf{0}$ (or $\boldsymbol{\omega}_T = \mathbf{0}$), we have pure rolling (or pure pivoting). In sum, slippingless rolling along a curve can happen only if that curve is a straight line carrying $\boldsymbol{\omega}$ (like a long hinge).

Some Analytical Remarks on Rolling

(i) The *contact* among rigid bodies is expressed analytically by one or more equations of the form

$$f(t; q_1, q_2, \ldots, q_n) = 0, \qquad (1.9.6a)$$

where $q \equiv (q_1, \ldots, q_n)$ are geometrical *parameters* that determine the position, or configuration, of the bodies of the system; hence, their alternative name: *system coordinates*. Equation (1.9.6a) is called a *holonomic* constraint.

(ii) If, in addition to contact, there is also slippingless *rolling*, and possibly *pivoting*, then equating the velocities of the two (or more pairs of) material points in contact, we obtain constraints of the form

$$a_1 dq_1 + a_2 dq_2 + \cdots + a_n dq_n + a_{n+1} dt = 0, \tag{1.9.6b}$$

or, (roughly) equivalently,

$$a_1(dq_1/dt) + a_2(dq_2/dt) + \cdots + a_n(dq_n/dt) + a_{n+1} = 0, \tag{1.9.6c}$$

where $a_k = a_k(t, q)$ $(k = 1, \ldots, n)$. If (1.9.6b,c) is not *integrable* [i.e., if it cannot be replaced, through mathematical manipulations, by a finite (1.9.6a)-like equation], it is called *nonholonomic*. In mechanical terms, *holonomic constraints* restrict the mobility of a system in the *large* (i.e., *globally*); whereas *nonholonomic constraints* restrict its mobility in the *small* (i.e., *locally*). The systematic study of both these types of constraints (chap. 2) and their fusion with the general principles and equations of motion (chap. 3 ff.) is the object of Lagrangean analytical mechanics.

1.10 THE RIGID BODY: GEOMETRY OF ROTATIONAL MOTION; FINITE ROTATION

> The peculiarities of the algebra of finite rotations are just the peculiarities of matrix multiplication.
>
> (Crandall et al., 1968, p. 58)

Recommended for concurrent reading with this section are (alphabetically): Bahar (1987), Coe (1938, pp. 157 ff.), Hamel (1949, pp. 103–117), Shuster (1993), Timerding (1908).

The Fundamental Equation of Finite Rotation

Since, by the fundamental theorem of the preceding section, the rotatory part of a general displacement of a rigid body is independent of the base point (pole), let us examine first, with no loss in generality, the *finite rotation of a rigid body B about the (body- and space-) fixed point O*; and later we will add to it the translatory displacement of O. Specifically, let us examine the finite rotation of B about an axis through O, with positive direction (unit) vector $\boldsymbol{n}$, by an angle χ that is counted positive in accordance with the right-hand (screw) rule (fig. 1.19).

As a result of such an angular displacement, a generic body point P moves from an *initial* position P_i to a *final* position P_f; or, symbolically,

$$(\boldsymbol{r}_i, \boldsymbol{p}_i) \rightarrow (\boldsymbol{r}_f, \boldsymbol{p}_f), \tag{1.10.1a}$$

where $\boldsymbol{p}$ is the projection, or component, of the actual position vector of P, $\boldsymbol{r}$, on the plane through it normal to the axis of rotation; that is, to $\boldsymbol{n}$. Our objective here is to express $\boldsymbol{r}_f$ in terms of $\boldsymbol{r}_i$, $\boldsymbol{n}$, and χ. To this end, we decompose the displacement $\Delta\boldsymbol{r} \equiv \boldsymbol{r}_f - \boldsymbol{r}_i = \boldsymbol{p}_f - \boldsymbol{p}_i \equiv \Delta\boldsymbol{p}$, which lies on the plane of the triangle AP_iP_f, into two components: one *along* $\boldsymbol{p}_i$, $\boldsymbol{P_iB} = \Delta\boldsymbol{r}_1$, and one *perpendicular* to it, $\boldsymbol{BP_f} = \Delta\boldsymbol{r}_2$:

$$\Delta\boldsymbol{r} = \Delta\boldsymbol{r}_1 + \Delta\boldsymbol{r}_2. \tag{1.10.1b}$$

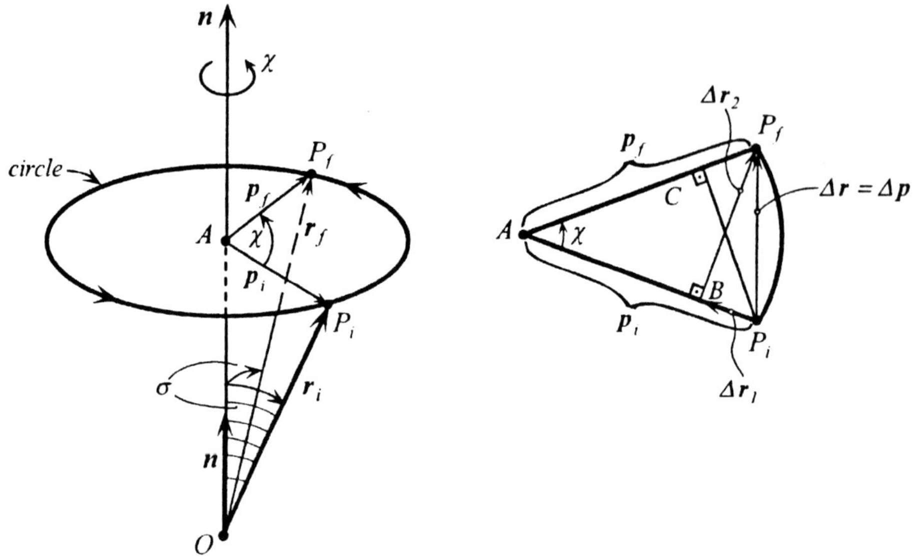

Figure 1.19 Finite rigid rotation about a fixed point O (axis $\boldsymbol{n}$, angle χ).

Now, from fig. 1.19 and some simple geometry, we find, successively,

(i) $\Delta\boldsymbol{r}_1 = -(\boldsymbol{AP}_i - \boldsymbol{AB}) = -(\boldsymbol{p}_i - \boldsymbol{p}_i \cos\chi) = -\boldsymbol{p}_i(1 - \cos\chi) = -2\boldsymbol{p}_i \sin^2(\chi/2)$; or, since $\Delta\boldsymbol{r}_1$ is perpendicular to both $\boldsymbol{n} \times \boldsymbol{r}_i$ and $\boldsymbol{n}$, and

$$\boldsymbol{n} \times (\boldsymbol{n} \times \boldsymbol{r}_i) = (\boldsymbol{n} \cdot \boldsymbol{r}_i)\boldsymbol{n} - (\boldsymbol{n} \cdot \boldsymbol{n})\boldsymbol{r}_i = \boldsymbol{OA} - \boldsymbol{r}_i = \boldsymbol{P}_i\boldsymbol{A} = -\boldsymbol{p}_i,$$

finally,

$$\Delta\boldsymbol{r}_1 = \boldsymbol{n} \times (\boldsymbol{n} \times \boldsymbol{r}_i) 2\sin^2(\chi/2). \tag{1.10.1c}$$

(ii) The component $\Delta\boldsymbol{r}_2$ is perpendicular to the plane OAP_i, and lies along $\boldsymbol{n} \times \boldsymbol{r}_i$; and since the length of the latter equals

$$|\boldsymbol{n} \times \boldsymbol{r}_i| = |\boldsymbol{n}||\boldsymbol{r}_i| \sin\sigma = |\boldsymbol{r}_i| \sin\sigma = |\boldsymbol{p}_i|,$$

and

$$|\boldsymbol{p}_i| \sin\chi = |\boldsymbol{CP}_i| = |\boldsymbol{BP}_f| \equiv |\Delta\boldsymbol{r}_2| \qquad \text{(the triangle } AP_iP_f \text{ being isosceles!)},$$

finally

$$\Delta\boldsymbol{r}_2 = (\boldsymbol{n} \times \boldsymbol{r}_i) \sin\chi. \tag{1.10.1d}$$

Substituting the expressions (1.10.1c, d) into (1.10.1b), we obtain the following *fundamental equation of finite rotation*:

$$\Delta\boldsymbol{r} \equiv \boldsymbol{r}_f - \boldsymbol{r}_i = (\boldsymbol{n} \times \boldsymbol{r}_i) \sin\chi + \boldsymbol{n} \times (\boldsymbol{n} \times \boldsymbol{r}_i) 2\sin^2(\chi/2). \tag{1.10.1e}$$

All subsequent results on this topic are based on it.

Alternative Forms of the Fundamental Equation

(i) With the help of the so-called "*Gibbs* vector of finite rotation"

$$\boldsymbol{\gamma} \equiv \tan(\chi/2)\boldsymbol{n} \equiv (\gamma_1, \gamma_2, \gamma_3) \equiv (\gamma_X, \gamma_Y, \gamma_Z) = \textit{Rodrigues parameters}, \tag{1.10.2a}$$

relative to some background axes, say O–XYZ [Rodrigues (*1840*) – Gibbs (*late 1800s*) 'vector'] and, since by simple trigonometry,

$$\sin\chi = 2\sin(\chi/2)\cos(\chi/2) = 2\tan(\chi/2)/[1+\tan^2(\chi/2)]$$

$$= 2\gamma/(1+\gamma^2), \qquad \text{where } \gamma = |\boldsymbol{\gamma}| = |\tan(\chi/2)|, \tag{1.10.2b}$$

$$\sin^2(\chi/2) = \tan^2(\chi/2)/[1+\tan^2(\chi/2)] = (1-\cos\chi)/2 = \gamma^2/(1+\gamma^2), \tag{1.10.2c}$$

we can easily rewrite (1.10.1e) as

$$\Delta\boldsymbol{r} = [2/(1+\gamma^2)][\boldsymbol{\gamma}\times\boldsymbol{r}_i + \boldsymbol{\gamma}\times(\boldsymbol{\gamma}\times\boldsymbol{r}_i)]; \tag{1.10.2d}$$

and from this, since $\boldsymbol{\gamma}\times(\boldsymbol{\gamma}\times\boldsymbol{r}_i) = -\gamma^2\boldsymbol{r}_i + (\boldsymbol{\gamma}\cdot\boldsymbol{r}_i)\boldsymbol{\gamma}$, we obtain the additional form

$$\boldsymbol{r}_f = [2/(1+\gamma^2)][\boldsymbol{\gamma}\times\boldsymbol{r}_i + (\boldsymbol{\gamma}\cdot\boldsymbol{r}_i)\boldsymbol{\gamma}] + [(1-\gamma^2)/(1+\gamma^2)]\boldsymbol{r}_i; \tag{1.10.2e}$$

which, clearly, has a singularity at $\gamma = \pm i$.

Further, in terms of the normal projection of $\boldsymbol{r}_i$ to the rotation axis $\boldsymbol{n}$, $\boldsymbol{r}_{i,n}$, defined by

$$\boldsymbol{r}_{i,n} \equiv \boldsymbol{r}_i - (\boldsymbol{\gamma}\cdot\boldsymbol{r}_i)\boldsymbol{\gamma}/\gamma^2 = \boldsymbol{r}_i - [(\boldsymbol{\gamma}\otimes\boldsymbol{\gamma})\cdot\boldsymbol{r}_i]/\gamma^2, \tag{1.10.2f}$$

we can rewrite (1.10.2e) successively as

$$\begin{aligned}\boldsymbol{r}_f &= \boldsymbol{r}_i + [2/(1+\gamma^2)](\boldsymbol{\gamma}\times\boldsymbol{r}_{i,n} - \gamma^2\boldsymbol{r}_{i,n})\\ &= \boldsymbol{r}_i + [2/(1+\gamma^2)][\boldsymbol{\gamma}\times\boldsymbol{r}_i - \gamma^2\boldsymbol{r}_i + (\boldsymbol{\gamma}\otimes\boldsymbol{\gamma})\cdot\boldsymbol{r}_i]\\ &= \boldsymbol{r}_i + [2/(1+\gamma^2)][\boldsymbol{\gamma}\times\boldsymbol{r}_i + \boldsymbol{\gamma}\times(\boldsymbol{\gamma}\times\boldsymbol{r}_i)]\\ &= \boldsymbol{r}_i + [2\boldsymbol{\gamma}/(1+\gamma^2)]\times(\boldsymbol{r}_i + \boldsymbol{\gamma}\times\boldsymbol{r}_i);\end{aligned} \tag{1.10.2g}$$

that is, express $\boldsymbol{r}_f$ in terms of $\boldsymbol{r}_i$ and the single vector $\boldsymbol{\gamma}$.

{It is not hard to show that the components, or projections, of a vector $\boldsymbol{a}$ *along* ($\boldsymbol{a}_{\text{along}} \equiv \boldsymbol{a}_l$) and *perpendicular to* ($\boldsymbol{a}_{\text{perpendicular/normal}} \equiv \boldsymbol{a}_n$) another vector $\boldsymbol{b}$ (of common origin) are

$$\boldsymbol{a}_l = (\boldsymbol{a}\cdot\boldsymbol{b})\boldsymbol{b}/b^2, \qquad \boldsymbol{a}_n = \boldsymbol{a} - \boldsymbol{a}_l = \boldsymbol{a} - (\boldsymbol{a}\cdot\boldsymbol{b})\boldsymbol{b}/b^2 = [\boldsymbol{b}\times(\boldsymbol{a}\times\boldsymbol{b})]/b^2\}.$$

Inversion of Eqs. (1.10.2e,g)

Since a rotation $-\boldsymbol{\gamma}$ should bring $\boldsymbol{r}_f$ back to $\boldsymbol{r}_i$, if in (1.10.2g) we swap the roles of $\boldsymbol{r}_i$ and $\boldsymbol{r}_f$ and replace $\boldsymbol{\gamma}$ with $-\boldsymbol{\gamma}$, we obtain the initial position in terms of the final one and its rotation:

$$\boldsymbol{r}_i = \boldsymbol{r}_f - [2\boldsymbol{\gamma}/(1+\gamma^2)]\times(\boldsymbol{r}_f - \boldsymbol{\gamma}\times\boldsymbol{r}_f); \tag{1.10.3}$$

and thus avoid complicated vector-algebraic inversions.

Rodrigues' Formula (1840)

Adding $\boldsymbol{r}_i$ to both sides of (1.10.2e), we obtain

$$\boldsymbol{r}_i + \boldsymbol{r}_f = [2/(1+\gamma^2)][\boldsymbol{r}_i + \boldsymbol{\gamma}\times\boldsymbol{r}_i + (\boldsymbol{\gamma}\cdot\boldsymbol{r}_i)\boldsymbol{\gamma}], \tag{1.10.4a}$$

and crossing both sides of the above with γ, and then using simple vector identities and (1.10.2g) [or, adding (1.10.2g) and (1.10.3) and setting the coefficient of $2\gamma/(1+\gamma^2)$ equal to zero, since it cannot be nonzero and parallel to γ], we arrive at the *formula of Rodrigues*:

$$\boldsymbol{r}_f - \boldsymbol{r}_i = \boldsymbol{\gamma} \times (\boldsymbol{r}_i + \boldsymbol{r}_f) \equiv 2\boldsymbol{\gamma} \times \boldsymbol{r}_m = 2\boldsymbol{n} \times \boldsymbol{r}_m \tan(\chi/2), \tag{1.10.4b}$$

where

$$2\boldsymbol{r}_m \equiv \boldsymbol{r}_i + \boldsymbol{r}_f = 2\ (\textit{position vector of midpoint of } P_i P_f); \tag{1.10.4c}$$

or, rearranging,

$$\boldsymbol{r}_f + \boldsymbol{r}_f \times \boldsymbol{\gamma} = \boldsymbol{r}_i + \boldsymbol{\gamma} \times \boldsymbol{r}_i. \tag{1.10.4d}$$

Finally, dotting both sides of this equation with γ (or $\boldsymbol{n}$), we obtain

$$\boldsymbol{\gamma} \cdot \boldsymbol{r}_f = \boldsymbol{\gamma} \cdot \boldsymbol{r}_i, \tag{1.10.4e}$$

as expected.

(ii) With the help of the *finite rotation vector*

$$\boldsymbol{\chi} \equiv \chi \boldsymbol{n}, \tag{1.10.5a}$$

which is, obviously, related to the earlier Gibbs vector γ by

$$\boldsymbol{\gamma} = \tan(\chi/2)(\boldsymbol{\chi}/\chi), \tag{1.10.5b}$$

and since

$$1 + \gamma^2 = 1/\cos^2(\chi/2), \qquad 1 - \gamma^2 = \cos\chi/\cos^2(\chi/2), \tag{1.10.5c}$$

the preceding rotation equations yield

$$\boldsymbol{r}_f = 2\cos^2(\chi/2)\left[\tan(\chi/2)(\boldsymbol{\chi} \times \boldsymbol{r}_i)(1/\chi) + \tan^2(\chi/2)(\boldsymbol{\chi} \cdot \boldsymbol{r}_i)(\boldsymbol{\chi}/\chi^2)\right] + \cos\chi\, \boldsymbol{r}_i, \tag{1.10.5d}$$

or finally,

$$\boldsymbol{r}_f = \boldsymbol{r}_i \cos\chi + (\boldsymbol{\chi} \times \boldsymbol{r}_i)(\sin\chi/\chi) + (\boldsymbol{\chi} \cdot \boldsymbol{r}_i)\boldsymbol{\chi}[(1 - \cos\chi)/\chi^2], \tag{1.10.5e}$$

a form that is symmetrical and (integral) transcendental function of $\boldsymbol{\chi} \cdot \boldsymbol{\chi} = \chi^2$.

The above can also be rewritten as

$$\begin{aligned} \boldsymbol{r}_f - \boldsymbol{r}_i &= (\sin\chi)(\boldsymbol{n} \times \boldsymbol{r}_i) + (1 - \cos\chi)[\boldsymbol{n} \times (\boldsymbol{n} \times \boldsymbol{r}_i)] \\ &= (\sin\chi)(\boldsymbol{n} \times \boldsymbol{r}_i) + (1 - \cos\chi)[(\boldsymbol{n} \cdot \boldsymbol{r}_i)\boldsymbol{n} - (\boldsymbol{n}^2)\boldsymbol{r}_i], \end{aligned} \tag{1.10.5f}$$

or, slightly rearranged (since $\boldsymbol{n}^2 = 1$),

$$\begin{aligned} \boldsymbol{r}_f &= \boldsymbol{r}_i \cos\chi + (\boldsymbol{n} \times \boldsymbol{r}_i)\sin\chi + (\boldsymbol{n} \cdot \boldsymbol{r}_i)\boldsymbol{n}(1 - \cos\chi) \\ &= \boldsymbol{r}_i + \sin\chi(\boldsymbol{n} \times \boldsymbol{r}_i) + (\cos\chi - 1)[\boldsymbol{r}_i - (\boldsymbol{r}_i \cdot \boldsymbol{n})\boldsymbol{n}] \\ &[= \boldsymbol{r}_i + \sin\chi(\boldsymbol{n} \times \boldsymbol{r}_i) + (\cos\chi - 1)\ (\textit{component of } \boldsymbol{r}_i \textit{ perpendicular to } \boldsymbol{n})]. \end{aligned} \tag{1.10.5g}$$

REMARK

The preceding rotation equations give the final position vector $\boldsymbol{r}_f$ in terms of the initial position vector $\boldsymbol{r}_i$ and the various rotation vectors $\boldsymbol{\gamma}$, $\boldsymbol{\chi}$, $\boldsymbol{n}$ (and χ). It is shown later in this section that, despite appearances, $\boldsymbol{\gamma}$ is *not* a vector in all respects, but simply a *directed line segment*; that is, it has some *but not all* of the vector characteristics (§ 1.1). This is a crucial point in the theory of finite rotations.

Additional Useful Results

(i) In the preceding rotation formulae:

(a) For $\chi = \pm 2\pi n$ $(n = 1, 2, 3, \ldots)$ they yield

$$\boldsymbol{r}_f = \boldsymbol{r}_i, \tag{1.10.6a}$$

that is, the body point returns to its initial position, as it should; and

(b) If $\boldsymbol{r}_i \cdot \boldsymbol{n} = 0$, and $\chi = \pi/2$, then

$$\boldsymbol{r}_f = \boldsymbol{n} \times \boldsymbol{r}_i, \tag{1.10.6b}$$

that is, $\boldsymbol{n}$, $\boldsymbol{r}_i$, $\boldsymbol{r}_f$ form an orthogonal and dextral triad at O.

(ii) By swapping the roles of $\boldsymbol{r}_f$ and $\boldsymbol{r}_i$ and replacing χ with $-\chi$ in (1.10.5g) (i.e., inverting it), we get

$$\boldsymbol{r}_i = \boldsymbol{r}_f \cos\chi - (\boldsymbol{n} \times \boldsymbol{r}_f)\sin\chi + (\boldsymbol{n} \cdot \boldsymbol{r}_f)\boldsymbol{n}(1 - \cos\chi). \tag{1.10.6c}$$

(iii) For *small* χ, eqs. (1.10.5d, e) linearize to the earlier "Euler–Mozzi" formula:

$$\boldsymbol{r}_f = \boldsymbol{r}_i + \boldsymbol{\chi} \times \boldsymbol{r}_i \;\Rightarrow\; \Delta\boldsymbol{r} \equiv \boldsymbol{r}_f - \boldsymbol{r}_i = \boldsymbol{\chi} \times \boldsymbol{r}_i. \tag{1.10.6d}$$

Finite Rotation of a Line

By using the rotation formulae, one can show that the final position of a *body-fixed* straight fiber joining two arbitrary such points P_1 and P_2, or 1 and 2 (fig. 1.20), is given by

$$\begin{aligned}(\boldsymbol{r}_{2/1})_f &\equiv \boldsymbol{r}_{2,f} - \boldsymbol{r}_{1,f} \\ &= \cdots = (\sin\chi)\boldsymbol{n} \times (\boldsymbol{r}_{2/1})_i + (\cos\chi)(\boldsymbol{r}_{2/1})_i + (1 - \cos\chi)[\boldsymbol{n} \cdot (\boldsymbol{r}_{2/1})_i]\boldsymbol{n},\end{aligned} \tag{1.10.7a}$$

where

$$\textit{Initial position} \equiv (\boldsymbol{r}_{2/1})_i \to \textit{Final position} \equiv (\boldsymbol{r}_{2/1})_f, \tag{1.10.7b}$$

and

$$\boldsymbol{r}_{2/1} \equiv \boldsymbol{r}_2 - \boldsymbol{r}_1, \qquad \text{for both } i \text{ and } f. \tag{1.10.7c}$$

Finite Rotation of an Orthonormal Basis

By employing the finite rotation equations, let us find the relations between the two ortho–normal–dextral (OND) bases of common origin, O–$\boldsymbol{u}_{k'}$ (space-fixed) and

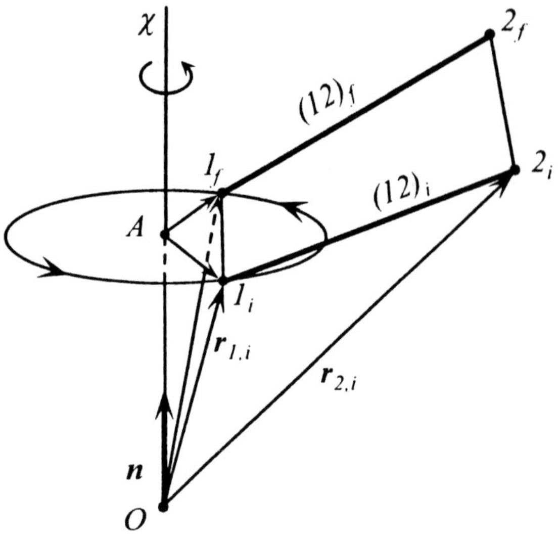

Figure 1.20 Finite rotation of straight segment *12*, from $(12)_i$ to $(12)_f$.

O–$\boldsymbol{u}_k$ (body-fixed), if the latter results from the former by a rotation χ about an axis $\boldsymbol{n}$; that is, symbolically,

$$\boldsymbol{u}_{k'} \xrightarrow{(\boldsymbol{n},\chi)} \boldsymbol{u}_k. \tag{1.10.8a}$$

Applying the earlier rotation equations to this transformation, with $\boldsymbol{r}_i = \boldsymbol{u}_{k'}$ and $\boldsymbol{r}_f = \boldsymbol{u}_k$, we obtain the following equivalent expressions:

(i)
$$\boldsymbol{u}_k = \boldsymbol{u}_{k'} + (\sin\chi)(\boldsymbol{n}\times\boldsymbol{u}_{k'}) + (\cos\chi - 1)\boldsymbol{u}_{k',n}, \tag{1.10.8b}$$

where

$$\begin{aligned} \boldsymbol{u}_{k',n} &\equiv \boldsymbol{u}_{k'} - (\boldsymbol{u}_{k'}\cdot\boldsymbol{n})\boldsymbol{n} = \boldsymbol{u}_{k'} - (\boldsymbol{n}\otimes\boldsymbol{n})\cdot\boldsymbol{u}_{k'} = (\mathbf{1} - \boldsymbol{n}\otimes\boldsymbol{n})\cdot\boldsymbol{u}_{k'} \\ &\equiv \mathbf{P}\cdot\boldsymbol{u}_{k'} = \textit{Component of } \boldsymbol{u}_{k'} \textit{ normal to } \boldsymbol{n} \\ &[\mathbf{P} = \textit{projection operator},\ \mathbf{1} = \textit{unit tensor}\ (\S1.1)]. \end{aligned} \tag{1.10.8c}$$

(ii)
$$\begin{aligned} \boldsymbol{u}_k &= (\cos\chi)\boldsymbol{u}_{k'} + (\sin\chi)(\boldsymbol{n}\times\boldsymbol{u}_{k'}) + (1-\cos\chi)(\boldsymbol{n}\cdot\boldsymbol{u}_{k'})\boldsymbol{n} \\ &= [(\cos\chi)\mathbf{1} + (\sin\chi)(\boldsymbol{n}\times\mathbf{1}) + (1-\cos\chi)(\boldsymbol{n}\otimes\boldsymbol{n})]\cdot\boldsymbol{u}_{k'} \\ &\equiv (\textit{rotation tensor})\cdot\boldsymbol{u}_{k'} \qquad \text{[Examined in detail below]} \\ &= \boldsymbol{u}_{k'} + (\chi\boldsymbol{n})\times\boldsymbol{u}_{k'} + [\textit{Terms of order } \chi^2 \equiv O(\chi^2)] \\ &\approx \boldsymbol{u}_{k'} + (\chi\boldsymbol{n})\times\boldsymbol{u}_{k'} \\ &= \boldsymbol{u}_{k'} + \boldsymbol{\chi}\times\boldsymbol{u}_{k'} \qquad \text{[Euler–Mozzi formula for small rotations].} \end{aligned} \tag{1.10.8d}$$

(iii)
$$\begin{aligned} \boldsymbol{u}_k &= \boldsymbol{u}_{k'} + [2/(1+\gamma^2)][\boldsymbol{\gamma} \times \boldsymbol{u}_{k'} - \gamma^2 \boldsymbol{u}_{k'} + (\boldsymbol{\gamma} \otimes \boldsymbol{\gamma}) \cdot \boldsymbol{u}_{k'}] \\ &= \boldsymbol{u}_{k'} + [2/(1+\gamma^2)][\boldsymbol{\gamma} \times \boldsymbol{u}_{k'} + \boldsymbol{\gamma} \times (\boldsymbol{\gamma} \times \boldsymbol{u}_{k'})] \\ &= \boldsymbol{u}_{k'} + [2\boldsymbol{\gamma}/(1+\gamma^2)] \times (\boldsymbol{u}_{k'} + \boldsymbol{\gamma} \times \boldsymbol{u}_{k'}). \end{aligned} \tag{1.10.8e}$$

To express the initial basis vectors $\boldsymbol{u}_{k'}$ in terms of the final ones $\boldsymbol{u}_k$, we simply replace in any of the above, say (1.10.8e), $\boldsymbol{\gamma}$ with $-\boldsymbol{\gamma}$. The result is

$$\boldsymbol{u}_{k'} = \boldsymbol{u}_k - [2\boldsymbol{\gamma}/(1+\gamma^2)] \times (\boldsymbol{u}_{k'} - \boldsymbol{\gamma} \times \boldsymbol{u}_{k'}). \tag{1.10.8f}$$

From the above, we can easily deduce that

$$\boldsymbol{\gamma} \cdot \boldsymbol{u}_k = \boldsymbol{\gamma} \cdot \boldsymbol{u}_{k'}, \tag{1.10.8g}$$

as expected; or setting

$$\boldsymbol{\gamma} = \sum \gamma_k \boldsymbol{u}_k = \sum \gamma_{k'} \boldsymbol{u}_{k'}, \tag{1.10.8h}$$

in the component form

$$\gamma_k = \gamma_{k'}. \tag{1.10.8i}$$

The Tensor of Finite Rotation

Let us express the earlier rotation equations in direct/matrix and component forms. Along the rectangular Cartesian axes $O\text{–}XYZ \equiv O\text{–}X_k$, common to all vectors and tensors involved here, and with the component notations ($k = X, Y, Z$):

$$\begin{aligned} &\boldsymbol{r}_i \equiv (X_k), \qquad \boldsymbol{r}_f \equiv (Y_k), \\ &\boldsymbol{\gamma} = (\gamma_k\text{: } \textit{Rodrigues parameters}) \ \Rightarrow\ \gamma^2 = \sum {\gamma_k}^2 = (\gamma_X)^2 + (\gamma_Y)^2 + (\gamma_Z)^2, \\ &\boldsymbol{n} = (n_k\text{: } \textit{direction cosines of unit vector defining the axis of rotation}), \end{aligned} \tag{1.10.9a}$$

our rotation equations become

$$\boldsymbol{r}_f = \mathbf{R} \cdot \boldsymbol{r}_i, \qquad Y_k = \sum R_{kl} X_l = \sum [r_{kl}/(1+\gamma^2)] X_l, \tag{1.10.9b}$$

where, recalling (1.10.2e ff.) and the simple tensor algebra of §1.1, the (nonsymmetrical but proper orthogonal) *tensor of finite rotation*,

$$\mathbf{R} \equiv \mathbf{R}(\boldsymbol{n}, \chi) \equiv (R_{kl}) \equiv \big(r_{kl}/(1+\gamma^2)\big), \tag{1.10.9c}$$

has the following equivalent representations.

(i) *Direct/matrix form* (with $\mathbf{N}$: antisymmetric tensor of vector $\boldsymbol{n}$):

$$\mathbf{R} = \mathbf{1}\cos\chi + \mathbf{N}\sin\chi + \boldsymbol{n}\otimes\boldsymbol{n}(1-\cos\chi)$$

$$= \begin{pmatrix} 1 & 0 & 0 \\ 0 & 1 & 0 \\ 0 & 0 & 1 \end{pmatrix}\cos\chi + \begin{pmatrix} 0 & -n_Z & n_Y \\ n_Z & 0 & -n_X \\ -n_Y & n_X & 0 \end{pmatrix}\sin\chi$$

$$+ \begin{pmatrix} {n_X}^2 & n_X n_Y & n_X n_Z \\ n_Y n_X & {n_Y}^2 & n_Y n_Z \\ n_Z n_X & n_Z n_Y & {n_Z}^2 \end{pmatrix}(1-\cos\chi)$$

$$= \begin{pmatrix} c\chi + {n_X}^2(1-c\chi) & -n_Z s\chi + n_X n_Y(1-c\chi) & n_Y s\chi + n_X n_Z(1-c\chi) \\ n_Z s\chi + n_X n_Y(1-c\chi) & c\chi + {n_Y}^2(1-c\chi) & -n_X s\chi + n_Y n_Z(1-c\chi) \\ -n_Y s\chi + n_X n_Z(1-c\chi) & n_X s\chi + n_Y n_Z(1-c\chi) & c\chi + {n_Z}^2(1-c\chi) \end{pmatrix}$$

$$= \mathbf{R}(n_X, n_Y, n_Z; \chi), \qquad \text{under } {n_X}^2 + {n_Y}^2 + {n_Z}^2 = 1, \tag{1.10.10a}$$

where, as usual, $c(\ldots) \equiv \cos(\ldots)$, $s(\ldots) \equiv \sin(\ldots)$.

(ii) *Indicial (Cartesian tensor) form:*

$$R_{kl} \equiv R_{kl}(n_r, \chi) = (\delta_{kl})\cos\chi + (n_{kl})\sin\chi + n_k n_l(1-\cos\chi)$$

$$= (\delta_{kl})\cos\chi + \left(\sum \varepsilon_{krl} n_r\right)\sin\chi + n_k n_l(1-\cos\chi). \tag{1.10.10b}$$

Occasionally, the rotation formula is written as

$$\Delta\boldsymbol{r} = \mathbf{R}'\cdot\boldsymbol{r}, \qquad \text{where } \Delta\boldsymbol{r} \equiv \boldsymbol{r}_f - \boldsymbol{r}_i, \qquad \boldsymbol{r} \equiv \boldsymbol{r}_i, \tag{1.10.10c}$$

and

$$\mathbf{R}' \equiv (R'_{kl}) \equiv \mathbf{R} - \mathbf{1} = (R_{kl} - \delta_{kl}); \qquad \textit{rotator} \text{ tensor},$$

$$R'_{kl} \equiv R_{kl} - \delta_{kl} = \cdots = \left(\sum \varepsilon_{krl} n_r\right)\sin\chi + (n_k n_l - \delta_{kl})(1-\cos\chi). \tag{1.10.10d}$$

We notice that the representation (1.10.10d) coincides with the decomposition of R'_{kl} into its *antisymmetric* part:

$$\sum(\varepsilon_{krl} n_r)\sin\chi = n_{kl}\sin\chi,$$

and *symmetric* part:

$$(n_k n_l - \delta_{kl})(1-\cos\chi);$$

of which, the former is of the *first order* in χ, while the latter is of the *second order*; a result that explains the antisymmetry of the *angular velocity tensor* [(1.7.22e)].

(iii) *In terms of the Rodrigues parameters* (a form, most likely, due to G. Darboux):

$$(r_{kl}) = \begin{pmatrix} 1+\gamma_X{}^2-(\gamma_Y{}^2+\gamma_Z{}^2) & 2(\gamma_X\gamma_Y-\gamma_Z) & 2(\gamma_X\gamma_Z+\gamma_Y) \\ 2(\gamma_X\gamma_Y+\gamma_Z) & 1+\gamma_Y{}^2-(\gamma_Z{}^2+\gamma_X{}^2) & 2(\gamma_Y\gamma_Z-\gamma_X) \\ 2(\gamma_X\gamma_Z-\gamma_Y) & 2(\gamma_Y\gamma_Z+\gamma_X) & 1+\gamma_Z{}^2-(\gamma_X{}^2+\gamma_Y{}^2) \end{pmatrix} \tag{1.10.10e}$$

The properties of $\mathbf{R}$ can be summarized as follows:

(i) $$\lim \mathbf{R}(\boldsymbol{n},\chi)\Big|_{\chi\to 0} = \mathbf{R}(\boldsymbol{n},0) = \mathbf{1}, \qquad \text{for all } \boldsymbol{n}; \tag{1.10.11a}$$

that is, $\mathbf{R}(\boldsymbol{n},\chi)$ is a continuous function of χ.

(ii) $$\mathbf{R}(\boldsymbol{n},\chi)\cdot\boldsymbol{n} = \boldsymbol{n}; \qquad \boldsymbol{n} = \textit{axis of rotation}. \tag{1.10.11b}$$

(iii) $$\mathbf{R}(\boldsymbol{n},\chi_1)\cdot\mathbf{R}(\boldsymbol{n},\chi_2) = \mathbf{R}(\boldsymbol{n},\chi_1+\chi_2). \tag{1.10.11c}$$

(iv) $$\mathbf{R}(\boldsymbol{n},\chi)\cdot\mathbf{R}^{\mathrm{T}}(\boldsymbol{n},\chi) = \mathbf{1}, \tag{1.10.11d}$$

$$\mathbf{R}^{\mathrm{T}}(\boldsymbol{n},\chi) = \mathbf{R}^{-1}(\boldsymbol{n},\chi) = \mathbf{R}(\boldsymbol{n},-\chi). \tag{1.10.11e}$$

Also, since the elements of $\mathbf{R}$, R_{kl}, depend continuously and differentiably on three independent parameters—for example, Euler's angles (§1.12)—we can say that the

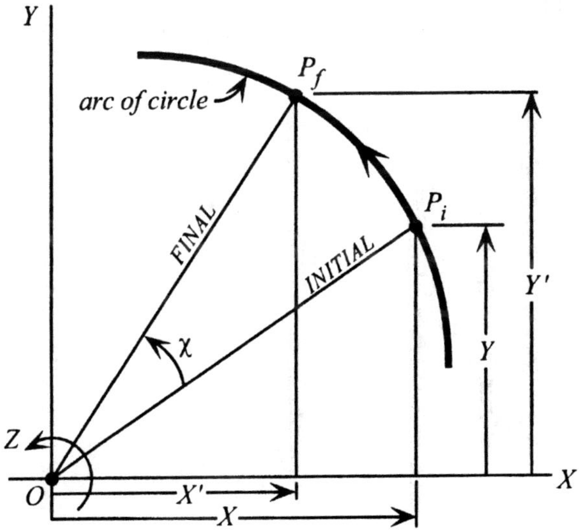

Figure 1.21 Plane rotation about Oz, through an angle χ.

rotation group is a *continuous* one; or a *Lie group*; see, for example, Argyris and Poterasu (1993).

Plane Rotation

This is a special rotation in which

$$\gamma = (\gamma_X = 0, \gamma_Y = 0, \gamma_Z = \tan(\chi/2) \equiv \lambda) = \tan(\chi/2)\boldsymbol{n} \;\Rightarrow\; \boldsymbol{n} = \boldsymbol{K}. \qquad (1.10.12a)$$

Then, with $X_k \equiv X,\ Y$ and $Y_k \equiv X',\ Y'$ (fig. 1.21), the rotational equations, (1.10.2g), and so on, specialize to

$$X' = [(1-\gamma^2)/(1+\gamma^2)]X - [2\gamma/(1+\gamma^2)]Y = \cdots = (\cos\chi)X + (-\sin\chi)Y, \qquad (1.10.12b)$$

$$Y' = [2\gamma/(1+\gamma^2)]X + [(1-\gamma^2)/(1+\gamma^2)]Y = \cdots = (\sin\chi)X + (\cos\chi)Y, \qquad (1.10.12c)$$

$$Z' = Z. \qquad (1.10.12d)$$

Additional Useful Results

(i) *Alternative expressions of the rotation tensor:*

(a) Indicial notation:

$$\begin{aligned} R_{kl} &= \delta_{kl} + \left(\sum \varepsilon_{krl} n_r\right)\sin\chi + (n_k n_l - \delta_{kl})(1-\cos\chi) \\ &= \delta_{kl} + n_{kl}\sin\chi + \sum n_{ks} n_{sl}(1-\cos\chi). \end{aligned} \qquad (1.10.13a)$$

(b) Direct/matrix form [$\mathbf{N} = (n_{kl})$ antisymmetric tensor of vector $\boldsymbol{n} = (n_k)$]:

$$\begin{aligned} \mathbf{R} &= \mathbf{1} + \mathbf{N}\sin\chi + 2\mathbf{N}\cdot\mathbf{N}\sin^2(\chi/2) & (1.10.13b) \\ &= \mathbf{1} + (\sin\chi)\mathbf{N} + [2\sin^2(\chi/2)]\mathbf{N}^2 & (1.10.13c) \\ &= \mathbf{1} + (\sin\chi)\mathbf{N} + (1-\cos\chi)\mathbf{N}^2 & (1.10.13d) \\ &= \mathbf{1} + 2\mathbf{N}\sin(\chi/2)[\mathbf{1}\cos(\chi/2) + \mathbf{N}\sin(\chi/2)] & (1.10.13e) \end{aligned}$$

[Notice that $1 - \cos\chi = 2\sin^2(\chi/2)$ and $\sum n_{ks}n_{sl} = n_k n_l - \delta_{kl}$, or, in direct notation, $\mathbf{N}\cdot\mathbf{N} = \boldsymbol{n}\otimes\boldsymbol{n} - \mathbf{1}$. See also Bahar (1970)].

(ii) By swapping the roles of $\boldsymbol{r}_i$ and $\boldsymbol{r}_f$ and setting $\chi \to -\chi$, in the preceding rotation formulae, one can show that

$$\boldsymbol{r}_i = \mathbf{R}^{-1}\cdot\boldsymbol{r}_f, \qquad (1.10.14a)$$

where

$$\mathbf{R}^{-1} = \mathbf{1} - \mathbf{N}\sin\chi + 2\mathbf{N}\cdot\mathbf{N}\sin^2(\chi/2) = \mathbf{R}^{\mathrm{T}} = \mathbf{R}(\boldsymbol{n}, -\chi); \qquad (1.10.14b)$$

that is, the rotation tensor is indeed *orthogonal*.

(iii) Let $\mathbf{\Gamma} \equiv \mathbf{N}\tan(\chi/2)$: antisymmetric tensor of the Gibbs vector γ. By applying the Cayley–Hamilton theorem to $\mathbf{\Gamma}$ [i.e., *every tensor satisfies its own characteristic equation* (§1.1)],

$$\Delta(\lambda) \equiv |\mathbf{\Gamma} - \lambda\mathbf{1}| = 0 \Rightarrow \Delta(\mathbf{\Gamma}) = -\mathbf{\Gamma}^3 - [\tan^2(\chi/2)]\mathbf{\Gamma} = \mathbf{0}, \tag{1.10.15a}$$

(since $Tr\ \mathbf{\Gamma} = 0$ and $Det\ \mathbf{\Gamma} = 0$), one can show that

$$\mathbf{R} = \mathbf{1} + 2\cos^2(\chi/2)(\mathbf{\Gamma} + \mathbf{\Gamma}^2), \qquad \mathbf{R} = (\mathbf{1} - \mathbf{\Gamma})^{-1} \cdot (\mathbf{1} + \mathbf{\Gamma}). \tag{1.10.15b}$$

Next, expanding (1.10.15b) symbolically in powers of $\mathbf{\Gamma}$, we obtain the representation

$$\mathbf{R} = (\mathbf{1} + \mathbf{\Gamma} + \cdots) \cdot (\mathbf{1} + \mathbf{\Gamma}) = \mathbf{1} + 2\mathbf{\Gamma}, \qquad \text{to first } \mathbf{\Gamma}\text{-order}; \tag{1.10.15c}$$

$$\Rightarrow \mathbf{R}' \equiv \mathbf{R} - \mathbf{1} = 2\mathbf{\Gamma}, \qquad \text{to first } \mathbf{\Gamma}\text{-order}. \tag{1.10.15d}$$

[Equations (1.10.15c, d) shed some light into the meaning of γ and $\mathbf{\Gamma}$; and prepare us for the treatment of *angular velocity* later in this section.] Similar results can be obtained in terms of $\mathbf{N}$.

The Mathematical Problem of Finite Rotation

Usually, this takes one of the following two forms: (i) given χ and $\boldsymbol{n}$, find $\mathbf{R}$; or (ii) given $\mathbf{R}$, find χ and $\boldsymbol{n}$. Now, from the preceding indicial forms, we easily obtain (with $k = X, Y, Z$):

$$\text{(i)}\ Tr\ \mathbf{R} \equiv \sum R_{kk} = \cos\chi\Big(\sum \delta_{kk}\Big) + \sin\chi\Big(\sum\sum \varepsilon_{krk} n_r\Big) + (1 - \cos\chi)\Big(\sum\sum n_k n_k\Big)$$
$$= \cos\chi(3) + \sin\chi(0) + (1 - \cos\chi)(1) = 2\cos\chi + 1. \tag{1.10.16a}$$

$$\text{(ii)}\ \sum\sum \varepsilon_{skl} R_{kl} = \cos\chi\Big(\sum\sum \varepsilon_{skl}\delta_{kl}\Big) + \sin\chi\Big(\sum\sum\sum \varepsilon_{skl}\varepsilon_{krl} n_r\Big) + (1 - \cos\chi)\Big(\sum\sum \varepsilon_{skl} n_k n_l\Big)$$
$$= \cos\chi(0) + \sin\chi\Big(\sum (-2\delta_{rs}) n_r\Big) + (1 - \cos\chi)(\boldsymbol{n} \times \boldsymbol{n})_s$$
$$= -2(\sin\chi) n_s \qquad \text{[Thanks to the } \varepsilon\text{-identities (1.1.6b ff.)]}. \tag{1.10.16b}$$

In sum,

$$I_1 \equiv Tr\ \mathbf{R} \equiv \sum R_{kk} = 1 + 2\cos\chi = \textit{First invariant of } \mathbf{R}, \tag{1.10.16c}$$

$$-\sum\sum \varepsilon_{skl} R_{kl} = 2R_s = 2(\textit{Axial vector of } \mathbf{R}) = 2(\sin\chi) n_s \Rightarrow R_k = (\sin\chi) n_k, \tag{1.10.16d}$$

or, explicitly,

$$R_1 = (-1/2)(\varepsilon_{123}R_{23} + \varepsilon_{132}R_{32}) = (R_{32} - R_{23})/2 = (\sin\chi)n_1, \qquad (1.10.16e)$$

$$R_2 = (-1/2)(\varepsilon_{231}R_{31} + \varepsilon_{213}R_{13}) = (R_{13} - R_{31})/2 = (\sin\chi)n_2, \qquad (1.10.16f)$$

$$R_3 = (-1/2)(\varepsilon_{312}R_{12} + \varepsilon_{321}R_{21}) = (R_{21} - R_{12})/2 = (\sin\chi)n_3. \qquad (1.10.16g)$$

Now, the *first* problem of rotation is, clearly, answered by the earlier rotation formulae (1.10.10 ff.); while the *second* is answered by solving the system of the four equations (1.10.16c, e–g) for the four unknowns χ; $n_{1,2,3}$. Indeed,

(i) From (1.10.16c), we obtain

$$\cos\chi = (I_1 - 1)/2 \equiv (\mathrm{Tr}\,\mathbf{R} - \mathbf{1})/2. \qquad (1.10.17a)$$

(a) From (1.10.16e–g), if $\sin\chi \neq 0$,

$$n_1 = (R_{32} - R_{23})/2\sin\chi, \qquad n_2 = (R_{13} - R_{31})/2\sin\chi, \qquad n_3 = (R_{21} - R_{12})/2\sin\chi, \qquad (1.10.17b)$$

or, vectorially,

$$\boldsymbol{n} = (1/n')\big[(R_{32} - R_{23})\boldsymbol{I} + (R_{13} - R_{31})\boldsymbol{J} + (R_{21} - R_{12})\boldsymbol{K}\big],$$

where

$$n' \equiv 2\sin\chi = \cdots = \big[(1 + \mathrm{Tr}\,\mathbf{R})\cdot(3 + \mathrm{Tr}\,\mathbf{R})\big]^{1/2}: \quad \textit{normalizing factor}. \qquad (1.10.17c)$$

(b) If $\sin\chi = 0$, then $\chi = 0$ or $\pm\pi$ (or some integral multiple thereof);
(b.1) If $\chi = 0$, then, as (1.10.11a) shows, $\mathbf{R} = (R_{kl}) = (\delta_{kl}) \equiv \mathbf{1}$; that is, $\boldsymbol{n}$ becomes undetermined: *no rotation occurs*; while
(b.2) If $\chi = \pm\pi \Rightarrow \cos\chi = -1$, then, as (1.10.10 ff.) show,

$$\mathbf{R} = (R_{kl}) = (2n_k n_l - \delta_{kl}) \qquad \text{(a \textit{symmetric} tensor)}$$

$$= \begin{pmatrix} 2n_1{}^2 - 1 & 2n_1 n_2 & 2n_1 n_3 \\ 2n_2 n_1 & 2n_2{}^2 - 1 & 2n_2 n_3 \\ 2n_3 n_1 & 2n_3 n_2 & 2n_3{}^2 - 1 \end{pmatrix}, \qquad (1.10.17d)$$

or, explicitly,

$$R_{11} = 2n_1{}^2 - 1 \Rightarrow n_1 = \pm[(1 + R_{11})/2]^{1/2}, \qquad (1.10.17e)$$

$$R_{22} = 2n_2{}^2 - 1 \Rightarrow n_2 = \pm[(1 + R_{22})/2]^{1/2}, \qquad (1.10.17f)$$

$$R_{33} = 2n_3{}^2 - 1 \Rightarrow n_3 = \pm[(1 + R_{33})/2]^{1/2}, \qquad (1.10.17g)$$

and the ultimate signs of $n_{1,2,3}$ are chosen so that (1.10.17e–g) are consistent with the rest of (1.10.17d):

$$n_1 n_2 = R_{12}/2 = R_{21}/2, \qquad n_1 n_3 = R_{13}/2 = R_{31}/2, \qquad n_2 n_3 = R_{23}/2 = R_{32}/2.$$

The angle χ can also be obtained from the off-diagonal elements of $\mathbf{R}$ as follows: multiplying (1.10.17b) with n_1, n_2, n_3, respectively, adding together, and invoking the normalization constraint $n_1^2 + n_2^2 + n_3^2 = 1$, we find

$$\sin\chi = (1/2)\left[n_1(R_{32} - R_{23}) + n_2(R_{13} - R_{31}) + n_3(R_{21} - R_{12})\right]. \qquad (1.10.17\text{h})$$

Rotation as an Eigenvalue Problem

(This subsection relies heavily on the spectral theory of § 1.1.) In view of the rotation formula

$$\boldsymbol{r}_f = \mathbf{R}\cdot\boldsymbol{r}_i, \qquad (1.10.18\text{a})$$

the earlier fundamental Eulerian theorem (§1.9: The most general displacement of a rigid body about a fixed point can be effected by a rotation about an axis through that point $\Rightarrow$ that axis is carried onto itself: $\mathbf{R}\cdot\boldsymbol{n} = \boldsymbol{n}$) translates to the following algebraic statement: *The real proper orthogonal tensor of rotation* $\mathbf{R}$ *has always the eigenvalue* $+1$; that is, at least one of the eigenvalues of the eigenvalue problem

$$(\boldsymbol{r}_f =)\ \mathbf{R}\cdot\boldsymbol{r}_i = \lambda\boldsymbol{r}_i, \qquad (1.10.18\text{b})$$

equals $+1$; or, *every rotation has an invariant vector*, which is Euler's theorem.

Let us examine these eigenvalues more systematically. The latter are the three roots of

$$|\mathbf{R} - \lambda\mathbf{1}| = 0, \qquad \lambda_{1,2,3}, \qquad (1.10.18\text{c})$$

and it is shown in linear algebra that:

(a) *They all have unit magnitude* [Since $\boldsymbol{r}_f\cdot\boldsymbol{r}_f = (\mathbf{R}\cdot\boldsymbol{r}_i)\cdot(\mathbf{R}\cdot\boldsymbol{r}_i) = (\boldsymbol{r}_i\cdot\mathbf{R}^{\mathrm{T}})\cdot(\mathbf{R}\cdot\boldsymbol{r}_i) = \boldsymbol{r}_i\cdot\boldsymbol{r}_i = \boldsymbol{r}_i^2$, the eigenvalue equation (1.10.18b) becomes

$$\boldsymbol{r}_f\cdot\boldsymbol{r}_f = \boldsymbol{r}_i\cdot\boldsymbol{r}_i = \lambda^2\boldsymbol{r}_i\cdot\boldsymbol{r}_i \Rightarrow \lambda^2 = 1 \qquad (\text{for } \boldsymbol{r}_i \neq \mathbf{0})];$$

(b) *At least one of them is real* [From the corresponding characteristic equation:

$$\Delta(\lambda) \equiv |\mathbf{R} - \lambda\mathbf{1}| = (-1)^3\lambda^3 + \cdots + (Det\ \mathbf{R})\lambda^0 = 0,$$

we readily see that

$$\lim \Delta(\lambda)\Big|_{\lambda\to-\infty} = +\infty, \qquad \text{and} \qquad \lim \Delta(\lambda)\Big|_{\lambda\to+\infty} = -\infty.$$

Hence, $\Delta(\lambda)$ crosses the λ axis *at least once*; that is, $\Delta(\lambda) = 0$ has *at least one real root*; and, by (i), that root is either $+1$ or -1.]

(c) *Complex eigenvalues occur in pairs of complex conjugate numbers* [since the coefficients of $\Delta(\lambda) = 0$ are real];

(d) $I_3(\mathbf{R}) \equiv I_3 = Det\ \mathbf{R} \equiv |R_{kl}| \equiv R = \lambda_1\lambda_2\lambda_3 = +1$. [Initially, that is before the rotation, $\boldsymbol{r}_f = \mathbf{R}\cdot\boldsymbol{r}_i = \boldsymbol{r}_i \Rightarrow \mathbf{R} = \mathbf{1} \Rightarrow Det\ \mathbf{1} = +1$, and since thereafter $\mathbf{R}$ evolves continuously from $\mathbf{1}$, it must be a *proper orthogonal tensor*; that is, $|\mathbf{R}| \equiv Det\ \mathbf{R} = +1 = \Delta(0)$. This expresses the "obvious" kinematical fact that, as long as we remain inside our Euclidean three-dimensional space, *a right-handed coordinate system cannot change to a left-handed one by a continuous rigid-body motion of its axes*; such "polarity" changes, called *inversions* or *reflections*, require continuous transformations in a higher dimensional space; for example, right-handed *two*-dimensional axes can be changed to left-handed *two*-dimensional axes by a continuous rotation inside the surrounding *three*-dimensional space.]

Combining these results, we conclude that either: (i) All three eigenvalues of **R** are real and equal to +1; which is the trivial case of the *identity transformation*; or, and this is the case of main interest (Euler's theorem), (ii) Only one of these eigenvalues is real and equals +1 [$\Rightarrow \Delta(1) \equiv |\mathbf{R} - \mathbf{1}| = 0$]; while the other two are the complex conjugate numbers: $\cos \chi \pm i \sin \chi \equiv \exp(\pm i\chi)$. As a result of the above:

(a) The direction cosines of the axis of rotation $\mathbf{n} = (n_X, n_Y, n_Z)$ can be obtained by setting in eq. (1.10.18b) $\lambda = 1$, $r_i = \mathbf{n}$:

$$(\mathbf{R} - \lambda \mathbf{1}) \cdot r_i = 0 \Rightarrow \mathbf{R} \cdot \mathbf{n} = \mathbf{n}, \quad (1.10.19a)$$

and then solving for $n_{X,Y,Z}$ under the constraint $n_X^2 + n_Y^2 + n_Z^2 = 1$; and

(b) The invariants of **R** can be summarized as follows:

$$I_1(\mathbf{R}) = Tr \, \mathbf{R} \equiv R_{11} + R_{22} + R_{33}$$
$$= \lambda_1 + \lambda_2 + \lambda_3 = 1 + \exp(+i\chi) + \exp(-i\chi) = 1 + 2\cos\chi; \quad (1.10.19b)$$
$$I_2(\mathbf{R}) = [(Tr \, \mathbf{R})^2 - Tr \, (\mathbf{R}^2)]/2 = (Det \, \mathbf{R})(Tr \, \mathbf{R}^{-1})$$
$$= (+1)(Tr \, \mathbf{R}^T) = (+1)(Tr \, \mathbf{R}) = I_1(\mathbf{R})$$
$$[= \lambda_1 \lambda_2 + \lambda_1 \lambda_3 + \lambda_2 \lambda_3$$
$$= (1)[\exp(i\chi)] + (1)[\exp(-i\chi)] + \exp(i\chi)\exp(-i\chi) = 2\cos\chi + 1]; \quad (1.10.19c)$$
$$I_3(\mathbf{R}) = Det \, \mathbf{R} = \lambda_1 \lambda_2 \lambda_3 = +1; \quad (1.10.19d)$$

that is, **R** has only *two* independent invariants.

Composition of Finite Rotations

Here we show that finite rotations are noncommutative; specifically, that two or more successive finite rotations of a rigid body with a fixed point O (or, generally, about axes *intersecting* at the real or fictitious rigid extension of the body) can be reproduced by a single rotation about an axis through O; but that resultant or equivalent single rotation *does* depend on the *order of the component or constituent rotations*.

Quantitatively, let the rotation vector γ_1 carry the generic body point position vector from r_1 to r_2; and, similarly, let γ_2 carry r_2 to r_3. We are seeking to express the vector of the *resultant* rotation $\gamma_{1,2}$ (i.e., of the one carrying r_1 to r_3) in terms of its "components" γ_1 and γ_2. Schematically,

$$r_1 \xrightarrow{\gamma_1} r_2 \xrightarrow{\gamma_2} r_3$$
$$\gamma_{1,2} \quad (1.10.20a)$$

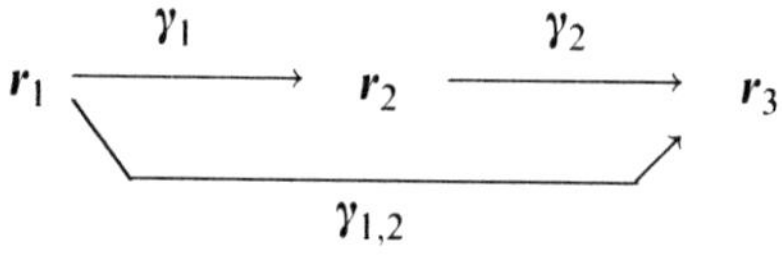

By Rodrigues' formula (1.10.4b), applied to $\boldsymbol{r}_1 \to \boldsymbol{r}_2$ and $\boldsymbol{r}_2 \to \boldsymbol{r}_3$, we obtain

$$\boldsymbol{r}_2 - \boldsymbol{r}_1 = \boldsymbol{\gamma}_1 \times (\boldsymbol{r}_2 + \boldsymbol{r}_1), \qquad \boldsymbol{r}_3 - \boldsymbol{r}_2 = \boldsymbol{\gamma}_2 \times (\boldsymbol{r}_3 + \boldsymbol{r}_2), \tag{1.10.20b}$$

respectively. Now, on these two basic equations we perform the following operations:

(i) We dot the first of the above with $\boldsymbol{\gamma}_1$ and the second with $\boldsymbol{\gamma}_2$:

$$\boldsymbol{\gamma}_1 \cdot (\boldsymbol{r}_2 - \boldsymbol{r}_1) = \boldsymbol{\gamma}_1 \cdot [\boldsymbol{\gamma}_1 \times (\boldsymbol{r}_2 + \boldsymbol{r}_1)] = 0 \Rightarrow \boldsymbol{\gamma}_1 \cdot \boldsymbol{r}_2 = \boldsymbol{\gamma}_1 \cdot \boldsymbol{r}_1, \tag{1.10.20c}$$

$$\boldsymbol{\gamma}_2 \cdot (\boldsymbol{r}_3 - \boldsymbol{r}_2) = \boldsymbol{\gamma}_2 \cdot [\boldsymbol{\gamma}_2 \times (\boldsymbol{r}_3 + \boldsymbol{r}_2)] = 0 \Rightarrow \boldsymbol{\gamma}_2 \cdot \boldsymbol{r}_3 = \boldsymbol{\gamma}_2 \cdot \boldsymbol{r}_2. \tag{1.10.20d}$$

(ii) We cross the first of (1.10.20b) with $\boldsymbol{\gamma}_2$ and the second with $\boldsymbol{\gamma}_1$ and subtract side by side:

$$\begin{aligned}
\boldsymbol{\gamma}_2 \times (\boldsymbol{r}_2 - \boldsymbol{r}_1) - \boldsymbol{\gamma}_1 \times (\boldsymbol{r}_3 - \boldsymbol{r}_2) &= (\boldsymbol{\gamma}_1 + \boldsymbol{\gamma}_2) \times \boldsymbol{r}_2 - \boldsymbol{\gamma}_2 \times \boldsymbol{r}_1 - \boldsymbol{\gamma}_1 \times \boldsymbol{r}_3 \\
&= \boldsymbol{\gamma}_2 \times [\boldsymbol{\gamma}_1 \times (\boldsymbol{r}_2 + \boldsymbol{r}_1)] - \boldsymbol{\gamma}_1 \times [\boldsymbol{\gamma}_2 \times (\boldsymbol{r}_3 + \boldsymbol{r}_2)] \\
&= \{\boldsymbol{\gamma}_1[\boldsymbol{\gamma}_2 \cdot (\boldsymbol{r}_2 + \boldsymbol{r}_1)] - (\boldsymbol{\gamma}_1 \cdot \boldsymbol{\gamma}_2)(\boldsymbol{r}_2 + \boldsymbol{r}_1)\} \\
&\qquad - \{\boldsymbol{\gamma}_2[\boldsymbol{\gamma}_1 \cdot (\boldsymbol{r}_3 + \boldsymbol{r}_2)] - (\boldsymbol{\gamma}_1 \cdot \boldsymbol{\gamma}_2)(\boldsymbol{r}_3 + \boldsymbol{r}_2)\}
\end{aligned}$$

[expanding, and then rearranging while taking into account (1.10.20c, d)]

$$\begin{aligned}
&= [(\boldsymbol{\gamma}_2 \cdot \boldsymbol{r}_2 + \boldsymbol{\gamma}_2 \cdot \boldsymbol{r}_1)\boldsymbol{\gamma}_1 - (\boldsymbol{\gamma}_1 \cdot \boldsymbol{\gamma}_2)\boldsymbol{r}_2 - (\boldsymbol{\gamma}_1 \cdot \boldsymbol{\gamma}_2)\boldsymbol{r}_1] \\
&\qquad - [(\boldsymbol{\gamma}_1 \cdot \boldsymbol{r}_3 + \boldsymbol{\gamma}_1 \cdot \boldsymbol{r}_2)\boldsymbol{\gamma}_2 - (\boldsymbol{\gamma}_1 \cdot \boldsymbol{\gamma}_2)\boldsymbol{r}_3 - (\boldsymbol{\gamma}_1 \cdot \boldsymbol{\gamma}_2)\boldsymbol{r}_2] \\
&= [(\boldsymbol{\gamma}_2 \cdot \boldsymbol{r}_3 + \boldsymbol{\gamma}_2 \cdot \boldsymbol{r}_1)\boldsymbol{\gamma}_1 - (\boldsymbol{\gamma}_1 \cdot \boldsymbol{\gamma}_2)\boldsymbol{r}_2 - (\boldsymbol{\gamma}_1 \cdot \boldsymbol{\gamma}_2)\boldsymbol{r}_1] \\
&\qquad - [(\boldsymbol{\gamma}_1 \cdot \boldsymbol{r}_3 + \boldsymbol{\gamma}_1 \cdot \boldsymbol{r}_1)\boldsymbol{\gamma}_2 - (\boldsymbol{\gamma}_1 \cdot \boldsymbol{\gamma}_2)\boldsymbol{r}_3 - (\boldsymbol{\gamma}_1 \cdot \boldsymbol{\gamma}_2)\boldsymbol{r}_2] \\
&= [(\boldsymbol{\gamma}_2 \cdot \boldsymbol{r}_3 + \boldsymbol{\gamma}_2 \cdot \boldsymbol{r}_1)\boldsymbol{\gamma}_1] - [(\boldsymbol{\gamma}_1 \cdot \boldsymbol{r}_3 + \boldsymbol{\gamma}_1 \cdot \boldsymbol{r}_1)\boldsymbol{\gamma}_2] - (\boldsymbol{\gamma}_1 \cdot \boldsymbol{\gamma}_2)(\boldsymbol{r}_1 - \boldsymbol{r}_3) \\
&= [(\boldsymbol{\gamma}_2 \cdot \boldsymbol{r}_1)\boldsymbol{\gamma}_1 - (\boldsymbol{\gamma}_1 \cdot \boldsymbol{r}_1)\boldsymbol{\gamma}_2] + [(\boldsymbol{\gamma}_2 \cdot \boldsymbol{r}_3)\boldsymbol{\gamma}_1 - (\boldsymbol{\gamma}_1 \cdot \boldsymbol{r}_3)\boldsymbol{\gamma}_2] \\
&\qquad - (\boldsymbol{\gamma}_1 \cdot \boldsymbol{\gamma}_2)(\boldsymbol{r}_1 - \boldsymbol{r}_3) \\
&= (\boldsymbol{\gamma}_2 \times \boldsymbol{\gamma}_1) \times \boldsymbol{r}_1 + (\boldsymbol{\gamma}_2 \times \boldsymbol{\gamma}_1) \times \boldsymbol{r}_3 - (\boldsymbol{\gamma}_1 \cdot \boldsymbol{\gamma}_2)(\boldsymbol{r}_1 - \boldsymbol{r}_3) \\
&= (\boldsymbol{\gamma}_2 \times \boldsymbol{\gamma}_1) \times (\boldsymbol{r}_1 + \boldsymbol{r}_3) + (\boldsymbol{\gamma}_1 \cdot \boldsymbol{\gamma}_2)(\boldsymbol{r}_3 - \boldsymbol{r}_1),
\end{aligned} \tag{1.10.20e}$$

or, equating the *right side of the first line* with the *last* line of (1.10.20e) and rearranging,

$$\begin{aligned}
(\boldsymbol{\gamma}_1 + \boldsymbol{\gamma}_2) \times \boldsymbol{r}_2 &= \boldsymbol{\gamma}_2 \times \boldsymbol{r}_1 + \boldsymbol{\gamma}_1 \times \boldsymbol{r}_3 \\
&\qquad + (\boldsymbol{\gamma}_2 \times \boldsymbol{\gamma}_1) \times (\boldsymbol{r}_1 + \boldsymbol{r}_3) + (\boldsymbol{\gamma}_1 \cdot \boldsymbol{\gamma}_2)(\boldsymbol{r}_3 - \boldsymbol{r}_1).
\end{aligned} \tag{1.10.20f}$$

(iii) We add (1.10.20b) side by side and rearrange to obtain

$$\begin{aligned}
\boldsymbol{r}_3 - \boldsymbol{r}_1 &= \boldsymbol{\gamma}_1 \times (\boldsymbol{r}_2 + \boldsymbol{r}_1) + \boldsymbol{\gamma}_2 \times (\boldsymbol{r}_3 + \boldsymbol{r}_2) = \boldsymbol{\gamma}_1 \times \boldsymbol{r}_2 + \boldsymbol{\gamma}_1 \times \boldsymbol{r}_1 + \boldsymbol{\gamma}_2 \times \boldsymbol{r}_3 + \boldsymbol{\gamma}_2 \times \boldsymbol{r}_2 \\
&\Rightarrow (\boldsymbol{\gamma}_1 + \boldsymbol{\gamma}_2) \times \boldsymbol{r}_2 = \boldsymbol{r}_3 - \boldsymbol{r}_1 - \boldsymbol{\gamma}_1 \times \boldsymbol{r}_1 - \boldsymbol{\gamma}_2 \times \boldsymbol{r}_3.
\end{aligned} \tag{1.10.20g}$$

(iv) Finally, equating the two expressions for $(\boldsymbol{\gamma}_1 + \boldsymbol{\gamma}_2) \times \boldsymbol{r}_2$, right sides of (1.10.20f) and (1.10.20g), and rearranging, we obtain the Rodrigues-like formula [i.e., à la (1.10.4b)]

$$\boldsymbol{r}_3 - \boldsymbol{r}_1 = \boldsymbol{\gamma}_{1,2} \cdot (\boldsymbol{r}_3 + \boldsymbol{r}_1), \tag{1.10.20h}$$

where

$$\gamma_{1,2} \equiv \gamma_{1\to 2} \equiv [\gamma_1 + \gamma_2 + \gamma_2 \times \gamma_1]/(1 - \gamma_1 \cdot \gamma_2)$$

$$= \textit{Resultant single rotation "vector," that brings } \mathbf{r}_1 \textit{ to } \mathbf{r}_3. \quad (1.10.20\text{i})$$

This is the sought fundamental formula for the *composition of finite rigid rotations*. [For additional derivations of (1.10.20h, i) see, for example, Hamel (1949, pp. 107–117; via complex number representations and quaternions), Lur'e (1968, pp. 101–104; via spherical trigonometry); also, Ames and Murnaghan (1929, pp. 82–85). The above vectorial proof seems to be due to Coe (1938, p. 170); see also Fox (1967, p. 8); and, for a simpler proof, Chester (1979, pp. 246–248).]

In terms of the corresponding rotation tensors, we would have (with some ad hoc notations),

$$\mathbf{r}_i \to \mathbf{r}_{f'}: \qquad \mathbf{r}_{f'} = \mathbf{R}_1 \cdot \mathbf{r}_i, \qquad (1.10.21\text{a})$$

$$\mathbf{r}_{f'} \to \mathbf{r}_f: \qquad \mathbf{r}_f = \mathbf{R}_2 \cdot \mathbf{r}_{f'} = \mathbf{R}_2 \cdot (\mathbf{R}_1 \cdot \mathbf{r}_i) \equiv \mathbf{R}_{1,2} \cdot \mathbf{r}_i, \qquad (1.10.21\text{b})$$

where

$$\mathbf{R}_{1,2} \equiv \mathbf{R}_2 \cdot \mathbf{R}_1 \qquad (\neq \mathbf{R}_1 \cdot \mathbf{R}_2 \equiv \mathbf{R}_{2,1}): \textit{ resultant rotation } \text{tensor}. \qquad (1.10.21\text{c})$$

REMARKS ON $\gamma_{1,2}$

(i) Equation (1.10.20i) readily shows that the γ's are *not* genuine vectors; as the presence of $\gamma_2 \times \gamma_1$ there makes clear [or the noncommutativity in (1.10.21c)], in general, *finite rotations are noncommutative*. Indeed, had we applied γ_2 first, and γ_1 second, the resultant would have been [swap the order of γ_1 and γ_2 in (1.10.20i)]

$$(\gamma_2 + \gamma_1 + \gamma_1 \times \gamma_2)/(1 - \gamma_2 \cdot \gamma_1) \equiv \gamma_{2,1} \equiv \gamma_{2\to 1} \neq \gamma_{1,2} \equiv \gamma_{1\to 2}. \qquad (1.10.22\text{a})$$

For rotations to commute, like genuine vectors, the term $\gamma_2 \times \gamma_1$ must vanish, either exactly or approximately. The former happens for rotations about the *same* axis; and the latter for *infinitesimal* (i.e., linear) rotations: there, $\gamma_2 \times \gamma_1 =$ *second*-order quantity $\approx \mathbf{0}$.

(ii) If $\gamma_1 \cdot \gamma_2 = 1$, the composition formula (1.10.20i), obviously, fails. Then, the corresponding "resultant angle" $\chi_{1,2}$ is an integral multiple of π.

(iii) From (1.10.20i) it is not hard to show that

$$1/(1 + \gamma_{1,2}{}^2)^{1/2} = (1 - \gamma_2 \cdot \gamma_1)/[(1 + \gamma_1{}^2)^{1/2}(1 + \gamma_2{}^2)^{1/2}], \qquad (1.10.22\text{b})$$

and combining this, again, with (1.10.20i) we readily obtain

$$\gamma_{1,2}/(1 + \gamma_{1,2}{}^2)^{1/2} = [\gamma_1 + \gamma_2 + \gamma_2 \times \gamma_1]/[(1 + \gamma_1{}^2)^{1/2}(1 + \gamma_2{}^2)^{1/2}] \qquad (1.10.22\text{c})$$

[which is the formula for the vector part of a product of two (unit) quaternions; see Papastavridis (Elementary Mechanics, in preparation)].

Finite rotations may not be commutative, but they are *associative*: the sequence of rotations, expressed in terms of their γ vectors—for example, $\gamma_1 \to \gamma_2 \to \gamma_3$—can be achieved either by combining the resultant of $\gamma_1 \to \gamma_2$ with γ_3, or by combining γ_1 with the resultant of $\gamma_2 \to \gamma_3$. In view of this, the sequence $-\gamma_1 \to \gamma_1 \to \gamma_2$ is equivalent to the rotation γ_2, and also to the sequence $-\gamma_1 \rightarrow \gamma_{1,2}$. Therefore, if in the fundamental "addition" formula (1.10.20i) we make the following replacements:

$$\gamma_1 \rightarrow -\gamma_1, \qquad \gamma_2 \rightarrow \gamma_{1,2}, \qquad \gamma_{1,2} \rightarrow \gamma_2, \tag{1.10.23a}$$

we obtain the "subtraction" formula:

$$\gamma_2 = [-\gamma_1 + \gamma_{1,2} + \gamma_{1,2} \times (-\gamma_1)]/[1 - (-\gamma_1)\cdot\gamma_{1,2}], \tag{1.10.23b}$$

or, finally,

$$\gamma_2 = [\gamma_{1,2} - \gamma_1 + \gamma_1 \times \gamma_{1,2}]/(1 + \gamma_1 \cdot \gamma_{1,2}), \tag{1.10.23c}$$

which allows us to find the *second* rotation "vector" from a knowledge of the *first* and the *compounded* rotation "vectors." Similarly, to find γ_1 from γ_2 and $\gamma_{1,2}$, we consider the rotation sequence $\gamma_{1,2} \rightarrow -\gamma_2$, which, clearly, is equivalent to the rotation γ_1. Hence, with the following replacements:

$$\gamma_1 \rightarrow \gamma_{1,2}, \qquad \gamma_2 \rightarrow -\gamma_2, \qquad \gamma_{1,2} \rightarrow \gamma_1, \tag{1.10.23d}$$

in (1.10.20i) we obtain the "subtraction" formula:

$$\gamma_1 = (\gamma_{1,2} - \gamma_2 + \gamma_{1,2} \times \gamma_2)/(1 + \gamma_2 \cdot \gamma_{1,2}). \tag{1.10.23e}$$

With such simple (and obviously *nonunique*) geometrical arguments, we can avoid solving (1.10.20i) for γ_1, γ_2. (These results prove useful in relating γ to the angular velocity $\boldsymbol{\omega}$.)

Infinitesimal (Linearized) Rotations Commute

First, let us apply the infinitesimal rotation $\boldsymbol{\chi}_1$ to $\boldsymbol{r}_i$ [recalling (1.10.6d)]:

$$\boldsymbol{r}_i \rightarrow \boldsymbol{r}_1{}' = \boldsymbol{r}_i + d\boldsymbol{r}_i = \boldsymbol{r}_i + \boldsymbol{\chi}_1 \times \boldsymbol{r}_i. \tag{1.10.24a}$$

Next, let us apply $\boldsymbol{\chi}_2$ to $\boldsymbol{r}_1{}'$:

$$\begin{aligned} \boldsymbol{r}_1{}' \rightarrow \boldsymbol{r}_f{}' &= \boldsymbol{r}_1{}' + d\boldsymbol{r}_1{}' = \boldsymbol{r}_1{}' + \boldsymbol{\chi}_2 \times \boldsymbol{r}_1{}' \\ &= (\boldsymbol{r}_i + \boldsymbol{\chi}_1 \times \boldsymbol{r}_i) + \boldsymbol{\chi}_2 \times (\boldsymbol{r}_i + \boldsymbol{\chi}_1 \times \boldsymbol{r}_i) \\ &= \boldsymbol{r}_i + (\boldsymbol{\chi}_1 + \boldsymbol{\chi}_2) \times \boldsymbol{r}_i + \boldsymbol{\chi}_2 \times (\boldsymbol{\chi}_1 \times \boldsymbol{r}_i). \end{aligned} \tag{1.10.24b}$$

Reversing the order of the process—that is, applying $\boldsymbol{\chi}_2$ first to $\boldsymbol{r}_i$, and then $\boldsymbol{\chi}_1$ to the result—we obtain

$$\begin{aligned} \boldsymbol{r}_f{}'' = \boldsymbol{r}_1{}'' + d\boldsymbol{r}_1{}'' &= \boldsymbol{r}_1{}'' + \boldsymbol{\chi}_1 \times \boldsymbol{r}_1{}'' \\ &= (\boldsymbol{r}_i + \boldsymbol{\chi}_2 \times \boldsymbol{r}_i) + \boldsymbol{\chi}_1 \times (\boldsymbol{r}_i + \boldsymbol{\chi}_2 \times \boldsymbol{r}_i) \\ &= \boldsymbol{r}_i + (\boldsymbol{\chi}_2 + \boldsymbol{\chi}_1) \times \boldsymbol{r}_i + \boldsymbol{\chi}_1 \times (\boldsymbol{\chi}_2 \times \boldsymbol{r}_i); \end{aligned} \tag{1.10.24c}$$

and, therefore, subtracting (1.10.24c) from (1.10.24b) side by side, we obtain

$$\boldsymbol{r}_f{}' - \boldsymbol{r}_f{}'' = \boldsymbol{\chi}_2 \times (\boldsymbol{\chi}_1 \times \boldsymbol{r}_i) - \boldsymbol{\chi}_1 \times (\boldsymbol{\chi}_2 \times \boldsymbol{r}_i) = \textit{second-order} \text{ vector in } \boldsymbol{\chi}_1, \boldsymbol{\chi}_2; \tag{1.10.24d}$$

that is, to the *first order* in $\boldsymbol{\chi}_1$, $\boldsymbol{\chi}_2$:

$$\boldsymbol{r}_f{}' = \boldsymbol{r}_f{}'', \quad \text{Q.E.D.} \tag{1.10.24e}$$

Similarly, for an arbitrary number of infinitesimal rotations $\boldsymbol{\chi}_1, \boldsymbol{\chi}_2, \ldots$, to the first order:

$$\boldsymbol{r}_f = \boldsymbol{r}_i + (\boldsymbol{\chi}_1 + \boldsymbol{\chi}_2 + \cdots) \times \boldsymbol{r}_i. \tag{1.10.24f}$$

Angular Velocity

(i) Angular Velocity from Finite Rotation

Expanding the rotation tensor (1.10.10e) [with (1.10.9b)] in powers of $\gamma_{X,Y,Z}$, and since (with customary calculus notations)

$$\boldsymbol{\gamma} \equiv \tan(\chi/2)\boldsymbol{n} = (\chi/2)\boldsymbol{n} + \boldsymbol{O}(\chi^3) = \boldsymbol{\chi}/2 + \boldsymbol{O}(\chi^3), \tag{1.10.25a}$$

we find

$$\mathbf{R} = \begin{pmatrix} 1 & -2\gamma_Z & 2\gamma_Y \\ 2\gamma_Z & 1 & -2\gamma_X \\ -2\gamma_Y & 2\gamma_X & 1 \end{pmatrix} + \boldsymbol{O}(\gamma^2)$$

[*Linear* rotation tensor $\equiv \mathbf{R}_{\text{o}}$]

$$= \begin{pmatrix} 1 & 0 & 0 \\ 0 & 1 & 0 \\ 0 & 0 & 1 \end{pmatrix} + \begin{pmatrix} 0 & -2\gamma_Z & 2\gamma_Y \\ 2\gamma_Z & 0 & -2\gamma_X \\ -2\gamma_Y & 2\gamma_X & 0 \end{pmatrix} + \boldsymbol{O}(\gamma^2)$$

[*Identity* tensor] [*Linear* rotator tensor $\equiv \mathbf{R}_{\text{o}}{}'$ (recall (1.10.10d, 15d))]

(1.10.25b)

$$= \mathbf{1} + \begin{pmatrix} 0 & -n_Z & n_Y \\ n_Z & 0 & -n_X \\ -n_Y & n_X & 0 \end{pmatrix} \chi + \boldsymbol{O}(\chi^2), \tag{1.10.25c}$$

$$= \mathbf{1} + \begin{pmatrix} 0 & -\chi_Z & \chi_Y \\ \chi_Z & 0 & -\chi_X \\ -\chi_Y & \chi_X & 0 \end{pmatrix} + \boldsymbol{O}(\chi^2); \tag{1.10.25d}$$

and, with the notations

$$\boldsymbol{r}_i = (X, Y, Z) \equiv \boldsymbol{r},$$

$$\boldsymbol{r}_f = \boldsymbol{r}_i + \Delta\boldsymbol{r}_i = (X + \Delta X, Y + \Delta Y, Z + \Delta Z) \equiv \boldsymbol{r} + \Delta\boldsymbol{r}, \tag{1.10.25e}$$

we obtain, to the first order in the rotation angle,

$$\boldsymbol{r} + \Delta\boldsymbol{r} = \mathbf{R}_{\text{o}} \cdot \boldsymbol{r} = (\mathbf{1} + \mathbf{R}_{\text{o}}{}') \cdot \boldsymbol{r} \;\Rightarrow\; \Delta\boldsymbol{r} = \mathbf{R}_{\text{o}}{}' \cdot \boldsymbol{r}, \tag{1.10.25f1}$$

or, in extenso,

$$\begin{pmatrix} \Delta X \\ \Delta Y \\ \Delta Z \end{pmatrix} = \begin{pmatrix} 0 & -\chi_Z & \chi_Y \\ \chi_Z & 0 & -\chi_X \\ -\chi_Y & \chi_X & 0 \end{pmatrix} \begin{pmatrix} X \\ Y \\ Z \end{pmatrix}. \qquad (1.10.25\text{f}2)$$

This basic kinematical result states that *any orthogonal tensor that differs infinitesimally from the identity tensor, that is, to within linear terms, differs from it by an antisymmetric tensor.*

Finally, dividing (1.10.25f1, 2) by Δt, during which $\Delta \boldsymbol{r}$ occurs, assuming continuity and with the following notations:

$$\lim(\Delta X/\Delta t)\big|_{\Delta t\to 0} = dX/dt \equiv v_X, \text{ etc.,} \qquad \lim(\chi_X/\Delta t)\big|_{\Delta t\to 0} = d\chi_X/dt \equiv \omega_X, \text{ etc.,}$$

we obtain the earlier found (1.9.1) fundamental kinematical equation of Poisson:

$$\begin{pmatrix} v_X \\ v_Y \\ v_Z \end{pmatrix} = \begin{pmatrix} 0 & -\omega_Z & \omega_Y \\ \omega_Z & 0 & -\omega_X \\ -\omega_Y & \omega_X & 0 \end{pmatrix} \begin{pmatrix} X \\ Y \\ Z \end{pmatrix}; \qquad (1.10.25\text{g})$$

or, in direct notation,

$$\boldsymbol{v} \equiv d\boldsymbol{r}/dt = \boldsymbol{\omega} \cdot \boldsymbol{r} = \boldsymbol{\omega} \times \boldsymbol{r}, \qquad (1.10.25\text{h})$$

where

$$\boldsymbol{\omega} \equiv \lim(\mathbf{R}_o{}'/\Delta t)\big|_{\Delta t\to 0}: \qquad \textit{angular velocity tensor}, \qquad (1.10.25\text{i})$$

$$\boldsymbol{\omega} \equiv \lim(2\boldsymbol{\gamma}/\Delta t)\big|_{\Delta t\to 0} = \lim(\boldsymbol{\chi}/\Delta t)\big|_{\Delta t\to 0}: \quad \textit{angular velocity vector}$$
$$(\textit{axial vector of } \boldsymbol{\omega} - \textit{a genuine vector!}); \qquad (1.10.25\text{j})$$

that is, the velocities of the points of a rigid body moving with one point fixed are, at any instant, the same as they would be if the body were rotating in the positive sense about a fixed axis through the fixed point, in the direction and sense of $\boldsymbol{\omega}$ and with an *angular speed* equal to $|\boldsymbol{\omega}|$; and, since both $\boldsymbol{r}$ and $\boldsymbol{v}$ are genuine vectors, so is $\boldsymbol{\omega}$ (a fact that is re-established below). From all existing definitions of the angular velocity, this seems to be the most natural; but, in return, it requires knowledge of finite rotation.

(ii) $\boldsymbol{\omega}$ is a Genuine Vector

Using the Rodrigues equation (1.10.4b):

$$\boldsymbol{r}_f - \boldsymbol{r}_i = \boldsymbol{\gamma} \times (\boldsymbol{r}_i + \boldsymbol{r}_f), \qquad (1.10.26\text{a})$$

let us prove directly that the angular velocity $\boldsymbol{\omega}$, defined as

$$\boldsymbol{\omega} \equiv \lim(2\boldsymbol{\gamma}/\Delta t)\big|_{\Delta t\to 0}, \qquad \text{where } \boldsymbol{\gamma} \equiv \tan(\chi/2)\boldsymbol{n}, \qquad (1.10.26\text{b})$$

is a genuine vector, even though $\boldsymbol{\gamma}$ is not.

PROOF

With this in mind, we introduce the following renaming:

$$\boldsymbol{r}_i = \boldsymbol{r}, \qquad \boldsymbol{r}_f = \boldsymbol{r}_i + \Delta\boldsymbol{r} = \boldsymbol{r} + \Delta\boldsymbol{r}. \qquad (1.10.26\text{c})$$

Then, eq. (1.10.26a) yields

$$\Delta \boldsymbol{r} = \boldsymbol{\gamma} \times [(\boldsymbol{r} \times \Delta \boldsymbol{r}) + \boldsymbol{r}] = \boldsymbol{\gamma} \times (2\boldsymbol{r} + \Delta \boldsymbol{r}) = (2\boldsymbol{\gamma}) \times (\boldsymbol{r} + \Delta \boldsymbol{r}/2). \tag{1.10.26d}$$

Dividing both sides of the above by Δt, and then letting $\Delta t \to 0$ (while assuming existence of a unique limit as $\Delta \boldsymbol{r} \to \boldsymbol{0}$), we obtain

$$\boldsymbol{v} \equiv \lim(\Delta \boldsymbol{r}/\Delta t)\big|_{\Delta t \to 0} = \lim[(2\boldsymbol{\gamma}/\Delta t) \times \boldsymbol{r}]\big|_{\Delta t \to 0} + \lim[2\boldsymbol{\gamma} \times (\Delta \boldsymbol{r}/2)]\big|_{\Delta t \to 0}$$
$$= \boldsymbol{\omega} \times \boldsymbol{r} + \boldsymbol{0} = \boldsymbol{\omega} \times \boldsymbol{r}. \tag{1.10.26e}$$

Alternatively, if we use definition (1.10.26b) [and the Taylor series expansion of $\tan(\chi/2)$ at $\chi/2 = 0$],

$$\boldsymbol{\omega} \equiv \lim(2\boldsymbol{\gamma}/\Delta t)\big|_{\Delta t \to 0} = \lim[2\tan(\chi/2)\boldsymbol{n}/\Delta t]\big|_{\Delta t \to 0}$$
$$= \lim(\chi\, \boldsymbol{n}/\Delta t)\big|_{\Delta t \to 0} \equiv (d\chi/dt)\boldsymbol{n} \equiv \omega\, \boldsymbol{n}, \tag{1.10.26f}$$

[$\boldsymbol{n}$: instantaneous axis of rotation], we finally arrive at $\boldsymbol{v} = \boldsymbol{\omega} \times \boldsymbol{r}$.

To complete the proof, let us next show that the line segments $\boldsymbol{\omega}$ indeed commute. Dividing the composition of rotations equation (1.10.20i)

$$\boldsymbol{\gamma}_3 \equiv \boldsymbol{\gamma}_{1,2} = (\boldsymbol{\gamma}_1 + \boldsymbol{\gamma}_2 + \boldsymbol{\gamma}_2 \times \boldsymbol{\gamma}_1)\big/(1 - \boldsymbol{\gamma}_1 \cdot \boldsymbol{\gamma}_2) \tag{1.10.26g}$$

by $\Delta t/2$, we get

$$2\boldsymbol{\gamma}_3/\Delta t = \big[2\boldsymbol{\gamma}_1/\Delta t + 2\boldsymbol{\gamma}_2/\Delta t$$
$$+ (\Delta t/2)(2\boldsymbol{\gamma}_2/\Delta t) \times (2\boldsymbol{\gamma}_1/\Delta t)\big]\big/\big[1 - (\Delta t/2)^2(2\boldsymbol{\gamma}_2/\Delta t)\cdot(2\boldsymbol{\gamma}_1/\Delta t)\big];$$

and then letting $\Delta t \to 0$, while recalling the earlier $\boldsymbol{\omega}$-definition (1.10.26b, f), we find

$$\boldsymbol{\omega}_3 \equiv \boldsymbol{\omega}_{1,2} = \boldsymbol{\omega}_1 + \boldsymbol{\omega}_2 = \boldsymbol{\omega}_2 + \boldsymbol{\omega}_1 \equiv \boldsymbol{\omega}_{2,1}, \tag{1.10.26h}$$

that is, simultaneous $\boldsymbol{\omega}$'s obey the parallelogram law for their addition and decomposition, Q.E.D.

(iii) $\boldsymbol{\omega} \leftrightarrow \boldsymbol{\gamma}$ Differential Equation

Let us consider a rigid body B with the fixed point O. Its instantaneous angular velocity $\boldsymbol{\omega}$ is related to its Gibbs "vector" $\boldsymbol{\gamma}$, which carries a typical B-particle

$$\text{from} \quad \boldsymbol{r}_i \equiv \boldsymbol{r}(t) \quad \text{to} \quad \boldsymbol{r}_f \equiv \boldsymbol{r}(t + \Delta t), \tag{1.10.27a}$$

by a differential equation. The latter is obtained as follows: in the composition of rotations equation (1.10.20i) and in order to create the difference $\Delta \boldsymbol{\gamma}$ there, we choose the rotation sequence

$$\boldsymbol{\gamma}_1 = -\boldsymbol{\gamma} \ \to\ \boldsymbol{\gamma}_2 = \boldsymbol{\gamma} + \Delta \boldsymbol{\gamma}, \tag{1.10.27b}$$

which, clearly, is equivalent to the single rotation $\gamma_{1,2} = \Delta\gamma$, and occurs in time Δt. With these identifications in (1.10.20i), the earlier angular velocity definition yields

$$\begin{aligned}\boldsymbol{\omega} &= \lim(2\Delta\gamma/\Delta t)\big|_{\Delta t\to 0} = 2\{\lim(\Delta\gamma/\Delta t)\big|_{\Delta t\to 0}\}\\ &= 2\lim\big[(1/\Delta t)\{[(-\gamma) + (\gamma + \Delta\gamma) + (\gamma \times \Delta\gamma)]\big/[1 - (-\gamma)\cdot(\gamma + \Delta\gamma)]\}\big]\big|_{\Delta t\to 0}\\ &= 2\lim\{\big[(\Delta\gamma/\Delta t) + \gamma \times (\Delta\gamma/\Delta t)\big]\big/\big[1 + \gamma\cdot\gamma + \gamma\cdot\Delta\gamma\big]\}\big|_{\Delta t\to 0},\end{aligned}$$

or, finally,

$$\boldsymbol{\omega} = [2/(1 + \gamma^2)][d\gamma/dt + \gamma \times (d\gamma/dt)]. \qquad (1.10.27c)$$

This remarkable formula, due to A. Cayley (*Cambridge and Dublin J.*, vol. 1, 1846), shows that, in general, $\boldsymbol{\omega}$ and $d\gamma/dt$ are not parallel!

REMARK

Equation (1.10.27c) also results if we apply to the formula for the subtraction of rotations (1.10.23c), the sequence

$$\gamma_1 = \gamma - \Delta\,\gamma \;\rightarrow\; \gamma_2 = \Delta\gamma', \qquad (1.10.27d)$$

which is equivalent to $\gamma_{1,2} = \gamma$. Thus, we obtain

$$\begin{aligned}\Delta\gamma' &= [\gamma - (\gamma - \Delta\gamma) + (\gamma - \Delta\gamma) \times \gamma]\big/[1 + (\gamma - \Delta\gamma)\cdot\gamma]\\ &= (\Delta\gamma + \gamma \times \Delta\gamma)\big/(1 + \gamma^2 - \gamma\cdot\Delta\gamma),\end{aligned} \qquad (1.10.27e)$$

then divide by Δt and take the limit as $\Delta t \rightarrow 0$ to obtain

$$\begin{aligned}\boldsymbol{\omega} &= 2\lim(\Delta\gamma'/\Delta t)\big|_{\Delta t\to 0}\\ &= \{[\lim(\Delta\gamma/\Delta t) + \gamma \times \lim(\Delta\gamma/\Delta t)]\big/(1 + \gamma^2 - \gamma\cdot\Delta\gamma)\}\big|_{\Delta t\to 0}\\ &= [2/(1 + \gamma^2)][d\gamma/dt + \gamma \times (d\gamma/dt)],\end{aligned} \qquad (1.10.27f)$$

as before. The reader may verify that the sequence $\gamma_1 = \gamma \rightarrow \gamma_2 = \Delta\gamma'$, which is equivalent to $\gamma_{1,2} = \gamma + \Delta\gamma$, also leads to the same formula.

(iv) Inversion of the Preceding Formula $\boldsymbol{\omega} = \boldsymbol{\omega}(\gamma, d\gamma/dt)$

First Derivation. Dotting both sides of that equation, (1.10.27c), by γ yields

$$\gamma\cdot\boldsymbol{\omega} = [2/(1 + \gamma^2)][\gamma\cdot(d\gamma/dt)]; \qquad (1.10.28a)$$

while crossing it with γ gives

$$\begin{aligned}\gamma \times \boldsymbol{\omega} &= [2/(1 + \gamma^2)]\{\gamma \times (d\gamma/dt) + \gamma \times [\gamma \times (d\gamma/dt)]\}\\ &= [2/(1 + \gamma^2)]\{\gamma \times (d\gamma/dt) + [\gamma\cdot(d\gamma/dt)]\gamma - \gamma^2(d\gamma/dt)\}.\end{aligned} \qquad (1.10.28b)$$

Eliminating $\gamma \cdot (d\gamma/dt)$ between (1.10.28a, b) produces

$$\gamma \times \omega = [2/(1+\gamma^2)]\gamma \times (d\gamma/dt)$$
$$- [2\gamma^2/(1+\gamma^2)](d\gamma/dt) + (\gamma \cdot \omega)\gamma$$

[expressing the *first* right-side term of the above via (1.10.27c)]

$$= \{\omega - [2/(1+\gamma^2)](d\gamma/dt)\} - [2\gamma^2/(1+\gamma^2)](d\gamma/dt) + (\gamma \cdot \omega)\gamma$$
$$= \omega - 2(d\gamma/dt) + (\gamma \cdot \omega)\gamma, \qquad (1.10.28c)$$

or, rearranging, finally gives

$$2(d\gamma/dt) = \omega + (\gamma \cdot \omega)\gamma + \omega \times \gamma; \qquad (1.10.28d)$$

which, for a given $\omega(t)$, is a vector *first-order nonlinear* (*second-degree*) differential equation for $\gamma(t)$ (and can be further reduced to a "Ricatti-type equation").

Equations (1.10.27d), and (1.10.27c) clearly demonstrate the one-to-one relation between ω and $d\gamma/dt$: *if one of them vanishes, so does the other.*

Second Derivation. Applying the earlier rotation sequence

$$\gamma_1 = \gamma \to \gamma_2 = \Delta\gamma', \qquad (1.10.28e)$$

which is equivalent to $\gamma_{1,2} = \gamma + \Delta\gamma$, both occurring in time Δt, to the composition formula (1.10.20i) we obtain

$$\gamma + \Delta\gamma = (\gamma + \Delta\gamma' + \Delta\gamma' \times \gamma)/(1 - \Delta\gamma' \cdot \gamma), \qquad (1.10.28f)$$

from which, subtracting γ, we get

$$\Delta\gamma = [\Delta\gamma' + (\gamma \cdot \Delta\gamma')\gamma + \Delta\gamma' \times \gamma]/(1 - \gamma \cdot \Delta\gamma'), \qquad (1.10.28g)$$

and from this, dividing by Δt and taking the limit as $\Delta t \to 0$, while recalling that [eq. (1.10.27f)] $\omega = 2\lim(\Delta\gamma'/\Delta t)|_{\Delta t \to 0}$, we re-obtain (1.10.28d).

For still alternative derivations of the $\omega \leftrightarrow \gamma$ equations, via the compatibility of the Eulerian kinematic relation $\boldsymbol{v} \equiv d\boldsymbol{r}/dt = \omega \times \boldsymbol{r}$ with the $d/dt(\ldots)$-derivative of the finite rotation equation $\boldsymbol{r}_f = \boldsymbol{r}_f(\gamma; \boldsymbol{r}_i)$ [eqs. (1.10.2–4)], see, for example, Coe (1938, chap. 5), Ferrarese (1980, pp. 122–137), Hamel (1949, pp. 106–107; pp. 391–393).

(v) Additional Useful Results

(a) Starting with

$$\gamma = \boldsymbol{n}\tan(\chi/2)$$
$$\Rightarrow d\gamma/dt = (d\boldsymbol{n}/dt)\tan(\chi/2) + \boldsymbol{n}[(d\chi/dt)/2]\sec^2(\chi/2), \quad \text{etc.},$$

and then using the $\omega \leftrightarrow \gamma$ equation, we can show that

$$\omega = (d\chi/dt)\boldsymbol{n} + (\sin\chi)(d\boldsymbol{n}/dt) + (1 - \cos\chi)\boldsymbol{n} \times (d\boldsymbol{n}/dt). \qquad (1.10.29a)$$

(What happens if $\boldsymbol{n} = constant$?)

(b) Again, starting with

$$\boldsymbol{\gamma} = \boldsymbol{n}\tan(\chi/2) = \tan(\chi/2)(\boldsymbol{\chi}/\chi) = [\tan(\chi/2)/\chi]\boldsymbol{\chi} \Rightarrow d\boldsymbol{\gamma}/dt = \cdots, \text{etc.},$$

and then using the $\boldsymbol{\omega} \leftrightarrow \boldsymbol{\gamma}$ equation, we can show that

$$\begin{aligned}\boldsymbol{\omega} &= (\sin\chi/\chi)(d\boldsymbol{\chi}/dt) - [(1-\cos\chi)/\chi^2][\boldsymbol{\chi}\times(d\boldsymbol{\chi}/dt)] \\ &\qquad + [(1/\chi) - (\sin\chi/\chi^2)](d\chi/dt)\boldsymbol{\chi} \\ &= d\boldsymbol{\chi}/dt - [(1-\cos\chi)/\chi^2][\boldsymbol{\chi}\times(d\boldsymbol{\chi}/dt)] \\ &\qquad + [(\chi - \sin\chi)/\chi^3)]\{\boldsymbol{\chi}\times[\boldsymbol{\chi}\times(d\boldsymbol{\chi}/dt)]\}. \qquad (1.10.29\text{b})\end{aligned}$$

(c) By inverting (1.10.29b), we can show that

$$d\boldsymbol{\chi}/dt = \boldsymbol{\omega} + (\boldsymbol{\chi}\times\boldsymbol{\omega})/2 + (1/\chi^2)[1 - (\chi/2)\cot(\chi/2)][\boldsymbol{\chi}\times(\boldsymbol{\chi}\times\boldsymbol{\omega})]. \qquad (1.10.29\text{c})$$

General Rigid-Body Displacement (i.e., no point fixed)

We have already seen (§1.9) that the most general rigid-body displacement can be effected by the translation of an arbitrary *base point* or *pole* of it, from its initial to its final position, followed by a rotation about an axis through the final position of that point (see figs 1.12 and 1.22). Here, we show that *the translational part of the above total displacement does depend on the base point, but the rotational part — that is, the rotation tensor — does not.*

Referring to fig. 1.22, let

$$\mathbf{11'} \equiv \boldsymbol{r}_{1'/1}, \quad \boldsymbol{PP''} \equiv \boldsymbol{r}_{f/i}, \quad \mathbf{1}\boldsymbol{P} \equiv \boldsymbol{r}_{/1}, \quad \mathbf{1'}\boldsymbol{P'} \equiv \boldsymbol{r}_{/1'}, \quad \mathbf{1''}\boldsymbol{P''} \equiv \mathbf{1'}\boldsymbol{P''} \equiv \boldsymbol{r}_{/1''},$$

$$\mathbf{R}_1 \equiv \text{rotation tensor bringing } \mathbf{1'}\boldsymbol{P'} \text{ to } \mathbf{1'}\boldsymbol{P''}; \text{ i.e., } \boldsymbol{r}_{/1''} = \mathbf{R}_1\cdot\boldsymbol{r}_{/1'}. \qquad (1.10.30\text{a})$$

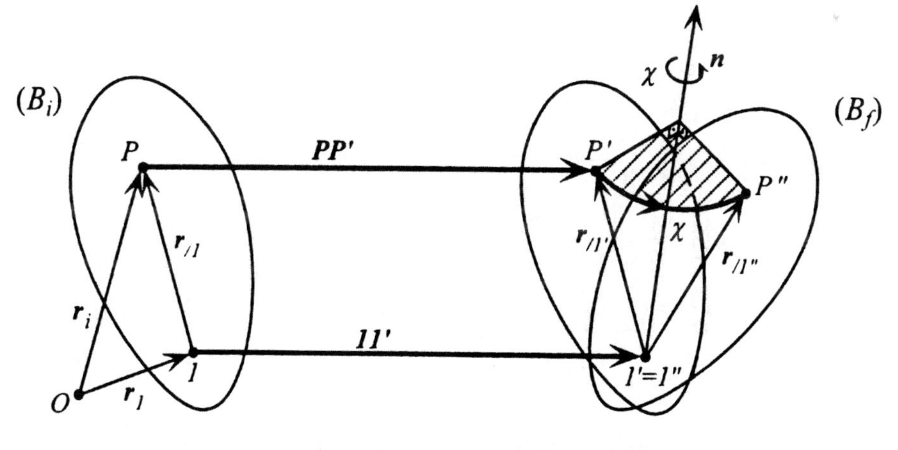

Figure 1.22 Most general rigid-body displacement; the rotation tensor is independent of the base point (or pole).

$\boldsymbol{r}_{/1} \to \boldsymbol{r}_{/1'} \to \boldsymbol{r}_{/1''} = \mathbf{R}_1\cdot\boldsymbol{r}_{/1'}$;

$\boldsymbol{r}_i = \boldsymbol{r}_1 + \boldsymbol{r}_{/1} \to \boldsymbol{r}_{i'} = \boldsymbol{r}_{1'} + \boldsymbol{r}_{/1'} \to \boldsymbol{r}_f = \boldsymbol{r}_{1''} + \boldsymbol{r}_{/1''} = \boldsymbol{r}_{1''} + \mathbf{R}_1\cdot\boldsymbol{r}_{/1'}$ $(\boldsymbol{r}_{1'} = \boldsymbol{r}_{1''})$.

Then, successively,

$$PP'' = P1 + 11' + 1'P'' = -r_{/1} + r_{1'/1} + \mathbf{R}_1 \cdot r_{/1'}, \tag{1.10.30b}$$

or, since $1P = 1'P'$ (i.e., $r_{/1} = r_{/1'}$),

$$r_{f/i} = r_{1'/1} + (\mathbf{R}_1 - \mathbf{1}) \cdot r_{/1}. \tag{1.10.30c}$$

Had we chosen another base point, say 2, then reasoning as above we would have found (with some easily understood notations)

$$r_{f/i} = r_{2'/2} + (\mathbf{R}_2 - \mathbf{1}) \cdot r_{/2}. \tag{1.10.30d}$$

But also, applying (1.10.30c) for $P = 2$, we have (since $r_{2'} = r_{2''}$)

$$r_{2'/2} = r_{1'/1} + (\mathbf{R}_1 - \mathbf{1}) \cdot r_{2/1}. \tag{1.10.30e}$$

Therefore, substituting (1.10.30e) in (1.10.30d) and equating its right side to that of (1.10.30c), we obtain

$$r_{1'/1} + (\mathbf{R}_1 - \mathbf{1}) \cdot r_{2/1} + (\mathbf{R}_2 - \mathbf{1}) \cdot r_{/2} = r_{1'/1} + (\mathbf{R}_1 - \mathbf{1}) \cdot r_{/1},$$

from which, rearranging, we get

$$(\mathbf{R}_1 - \mathbf{1}) \cdot (r_{/1} - r_{2/1}) \equiv (\mathbf{R}_1 - \mathbf{1}) \cdot r_{/2} = (\mathbf{R}_2 - \mathbf{1}) \cdot r_{/2}, \tag{1.10.30f}$$

and since this must hold for all body point pairs P and 2 (i.e., it must be an identity in them), we finally conclude that

$$\mathbf{R}_1 = \mathbf{R}_2 = \cdots \equiv \mathbf{R}. \tag{1.10.30g}$$

In words: the rotation tensor is independent of the chosen base point; it is a *position-independent tensor*. This fundamental theorem simplifies rigid-body geometry enormously and brings out the intrinsic character of rotation. (In *kinetics*, however, as the reader probably knows, such a decoupling between translation and rotation is far more selective.)

1.11 THE RIGID BODY: ACTIVE AND PASSIVE INTERPRETATIONS OF A PROPER ORTHOGONAL TENSOR; SUCCESSIVE FINITE ROTATIONS

A 3×3 proper orthogonal tensor may be interpreted in the following consistent ways:

(i) As the matrix of the direction cosines orienting two orthonormal and dextral (OND) triads, or bases, and associated axes; say, a *body-fixed*, or *moving*, triad $\mathbf{t}$:

$$\mathbf{t} \equiv (u_k) \equiv \begin{pmatrix} u_1 \\ u_2 \\ u_3 \end{pmatrix} \equiv \begin{pmatrix} u_x \\ u_y \\ u_z \end{pmatrix} \equiv \begin{pmatrix} i \\ j \\ k \end{pmatrix}, \tag{1.11.1a}$$

relative to a *space-fixed* triad $\mathbf{T}$:

$$\mathbf{T} \equiv (\boldsymbol{u}_{k'}) \equiv \begin{pmatrix} \boldsymbol{u}_{1'} \\ \boldsymbol{u}_{2'} \\ \boldsymbol{u}_{3'} \end{pmatrix} \equiv \begin{pmatrix} \boldsymbol{u}_X \\ \boldsymbol{u}_Y \\ \boldsymbol{u}_Z \end{pmatrix} \equiv \begin{pmatrix} \boldsymbol{I} \\ \boldsymbol{J} \\ \boldsymbol{K} \end{pmatrix}. \tag{1.11.1b}$$

(ii) Then, since

$$\boldsymbol{I} = (\boldsymbol{I}\cdot\boldsymbol{i})\boldsymbol{i} + (\boldsymbol{I}\cdot\boldsymbol{j})\boldsymbol{j} + (\boldsymbol{I}\cdot\boldsymbol{k})\boldsymbol{k} \equiv A_{Xx}\boldsymbol{i} + A_{Xy}\boldsymbol{j} + A_{Xz}\boldsymbol{k}, \quad \text{etc., cyclically,}$$

$$\boldsymbol{i} = (\boldsymbol{i}\cdot\boldsymbol{I})\boldsymbol{I} + (\boldsymbol{i}\cdot\boldsymbol{J})\boldsymbol{J} + (\boldsymbol{i}\cdot\boldsymbol{K})\boldsymbol{K} \equiv A_{Xx}\boldsymbol{I} + A_{Yx}\boldsymbol{J} + A_{Zx}\boldsymbol{K}, \quad \text{etc., cyclically,}$$

the two triads are related by

$$\mathbf{T} = \mathbf{A}\cdot\mathbf{t} \Leftrightarrow \mathbf{t} = \mathbf{A}^{-1}\cdot\mathbf{T} = \mathbf{A}^{\mathrm{T}}\cdot\mathbf{T}, \tag{1.11.1c}$$

where

$$\mathbf{A} \equiv \begin{pmatrix} \boldsymbol{I}\cdot\boldsymbol{i} & \boldsymbol{I}\cdot\boldsymbol{j} & \boldsymbol{I}\cdot\boldsymbol{k} \\ \boldsymbol{J}\cdot\boldsymbol{i} & \boldsymbol{J}\cdot\boldsymbol{j} & \boldsymbol{J}\cdot\boldsymbol{k} \\ \boldsymbol{K}\cdot\boldsymbol{i} & \boldsymbol{K}\cdot\boldsymbol{j} & \boldsymbol{K}\cdot\boldsymbol{k} \end{pmatrix} \equiv \begin{pmatrix} A_{Xx} & A_{Xy} & A_{Xz} \\ A_{Yx} & A_{Yy} & A_{Yz} \\ A_{Zx} & A_{Zy} & A_{Zz} \end{pmatrix} \tag{1.11.1d}$$

$$= (A_{k'k}), \quad A_{k'k} \equiv \cos(x_{k'}, x_k) = \boldsymbol{u}_{k'}\cdot\boldsymbol{u}_k \; [= \cos(x_k, x_{k'}) = A_{kk'}]. \tag{1.11.1e}$$

The rotation of an OND triad, equation (1.11.1c), $\mathbf{T} \to \mathbf{t}$, constitutes the *second* interpretation of a proper orthogonal tensor.

(iii) The *third* such interpretation is that of a *coordinate* transformation from the **T**-axes: $O\text{–}x_{k'} \equiv O\text{–}XYZ$ to the **t**-axes: $O\text{–}x_k \equiv O\text{–}xyz$ (of common origin, with no loss in generality). In this interpretation, known as *passive* or *alias* (meaning otherwise known as), the point P is fixed in **T**-space and the **t**-axes rotate. Then [fig. 1.23(a)],

$$\boldsymbol{OP} \equiv \boldsymbol{r} = \sum x_{k'}\boldsymbol{u}_{k'} = X\boldsymbol{I} + Y\boldsymbol{J} + Z\boldsymbol{K} = \sum x_k\boldsymbol{u}_k = x\boldsymbol{i} + y\boldsymbol{j} + z\boldsymbol{k}, \tag{1.11.2a}$$

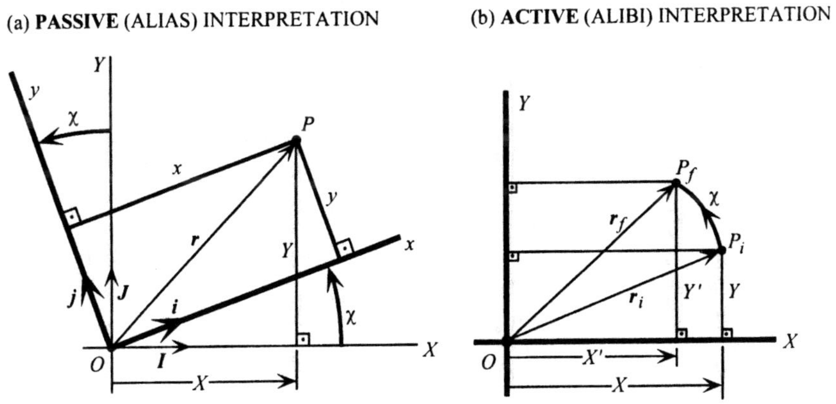

Figure 1.23 (a) Passive and (b) active interpretation of a proper orthogonal tensor (two dimensions).

and so we easily find

$$x_{k'} \equiv \boldsymbol{r} \cdot \boldsymbol{u}_{k'} = \cdots = \sum A_{k'k} x_k,$$

$$x_k \equiv \boldsymbol{r} \cdot \boldsymbol{u}_k = \cdots = \sum A_{kk'} x_{k'} \qquad \left(= \sum A_{k'k} x_{k'}\right), \tag{1.11.2b}$$

or explicitly, in matrix form,

$$\begin{pmatrix} X \\ Y \\ Z \end{pmatrix} = \begin{pmatrix} A_{Xx} & A_{Xy} & A_{Xz} \\ A_{Yx} & A_{Yy} & A_{Yz} \\ A_{Zx} & A_{Zy} & A_{Zz} \end{pmatrix} \begin{pmatrix} x \\ y \\ z \end{pmatrix}$$

$$\boldsymbol{r}' = \mathbf{A} \cdot \boldsymbol{r}$$

Old axes *New* axes, (1.11.2c)

$$\begin{pmatrix} x \\ y \\ z \end{pmatrix} = \begin{pmatrix} A_{Xx} & A_{Yx} & A_{Zx} \\ A_{Xy} & A_{Yy} & A_{Zy} \\ A_{Xz} & A_{Yz} & A_{Zz} \end{pmatrix} \begin{pmatrix} X \\ Y \\ Z \end{pmatrix}$$

$$\boldsymbol{r} = \mathbf{A}^{\mathrm{T}} \cdot \boldsymbol{r}'$$

New axes *Old* axes. (1.11.2d)

For example, in two dimensions [fig. 1.23(a)], the above yield

$$\begin{pmatrix} X \\ Y \end{pmatrix} = \begin{pmatrix} \cos\chi & -\sin\chi \\ \sin\chi & \cos\chi \end{pmatrix} \begin{pmatrix} x \\ y \end{pmatrix} \qquad \begin{pmatrix} x \\ y \end{pmatrix} = \begin{pmatrix} \cos\chi & \sin\chi \\ -\sin\chi & \cos\chi \end{pmatrix} \begin{pmatrix} X \\ Y \end{pmatrix}$$

$$\boldsymbol{r}' = \mathbf{A} \cdot \boldsymbol{r}, \qquad \boldsymbol{r} = \mathbf{A}^{\mathrm{T}} \cdot \boldsymbol{r}' \tag{1.11.2e}$$

(iv) Under the *fourth* interpretation, known as *active* or *alibi* (meaning elsewhere), the axes remain fixed in space, say $\mathbf{I}' = \mathbf{t}$, and the point P rotates about O, from an *initial* position $\boldsymbol{r}_i = X\boldsymbol{I} + Y\boldsymbol{J} + Z\boldsymbol{K}$ to a *final* one $\boldsymbol{r}_f = X'\boldsymbol{I} + Y'\boldsymbol{J} + Z'\boldsymbol{K}$. Then, following §1.10, and with $\mathbf{A} \to \mathbf{R}$ (rotation tensor),

$$\begin{pmatrix} X' \\ Y' \\ Z' \end{pmatrix} = \mathbf{R} \begin{pmatrix} X \\ Y \\ Z \end{pmatrix} = \mathbf{A} \begin{pmatrix} X \\ Y \\ Z \end{pmatrix}$$

$$\boldsymbol{r}_f = \mathbf{R} \cdot \boldsymbol{r}_i$$

Final position *Initial* position, (1.11.2f)

$$\begin{pmatrix} X \\ Y \\ Z \end{pmatrix} = \mathbf{R}^{\mathrm{T}} \begin{pmatrix} X' \\ Y' \\ Z' \end{pmatrix} = \mathbf{A}^{\mathrm{T}} \begin{pmatrix} X' \\ Y' \\ Z' \end{pmatrix}$$

$$\boldsymbol{r}_i = \mathbf{R}^{\mathrm{T}} \cdot \boldsymbol{r}_f$$

Initial position *Final* position. (1.11.2g)

Equations (1.11.2f, g) hold about any *common* axes; and, clearly, the components of **R** depend on the *particular axes* used. For example, in two dimensions [fig. 1.23(b)], the above yield

$$\begin{pmatrix} X' \\ Y' \end{pmatrix} = \begin{pmatrix} \cos\chi & -\sin\chi \\ \sin\chi & \cos\chi \end{pmatrix} \begin{pmatrix} X \\ Y \end{pmatrix} \qquad \begin{pmatrix} X \\ Y \end{pmatrix} = \begin{pmatrix} \cos\chi & \sin\chi \\ -\sin\chi & \cos\chi \end{pmatrix} \begin{pmatrix} X' \\ Y' \end{pmatrix}$$

$$\boldsymbol{r}_f = \mathbf{R}\cdot\boldsymbol{r}_i, \qquad \boldsymbol{r}_i = \mathbf{R}^{\mathrm{T}}\cdot\boldsymbol{r}_f; \tag{1.11.2h}$$

and for the new triad (actually a *dyad*) $\boldsymbol{i}, \boldsymbol{j}$ in terms of the old triad $\boldsymbol{I}, \boldsymbol{J}$ [along the same (old) axes], they readily yield

$$\begin{pmatrix} \cos\chi \\ \sin\chi \end{pmatrix} = \begin{pmatrix} \cos\chi & -\sin\chi \\ \sin\chi & \cos\chi \end{pmatrix} \begin{pmatrix} 1 \\ 0 \end{pmatrix} \qquad \begin{pmatrix} -\sin\chi \\ \cos\chi \end{pmatrix} = \begin{pmatrix} \cos\chi & -\sin\chi \\ \sin\chi & \cos\chi \end{pmatrix} \begin{pmatrix} 0 \\ 1 \end{pmatrix}$$

$$\boldsymbol{i} = \mathbf{R}\cdot\boldsymbol{I}, \qquad \boldsymbol{j} = \mathbf{R}\cdot\boldsymbol{J}. \tag{1.11.2i}$$

The passive and active interpretations are based on the fact that: The rigid body rotation relative to space-fixed axes (*active* interpretation), and the axes rotation relative to a fixed body (*passive* interpretation) are mutually reciprocal motions. Hence [fig. 1.24(a, b)]: The coordinates of a rotated body-fixed vector along the old axes (*final* position, *active* interpretation), equal the coordinates of the unrotated rigid body along the inversely rotated axes (*new* axes, *passive* interpretation).

It follows that if the *body is fixed* relative to the new axes and $\boldsymbol{r}' = X\boldsymbol{I} + Y\boldsymbol{J}$, $\boldsymbol{r} = x\boldsymbol{i} + y\boldsymbol{j}$, then the rotation equations—for example, (1.10.2e)—yields (with $\boldsymbol{r}_i \to \boldsymbol{r}_{\text{new (body-fixed) axes}} \equiv \boldsymbol{r}$ and $\boldsymbol{r}_f \to \boldsymbol{r}'_{\text{old axes}} \equiv \boldsymbol{r}'$)

$$\boldsymbol{r}' = [2/(1+\gamma^2)][\boldsymbol{\gamma}\times\boldsymbol{r} + (\boldsymbol{\gamma}\cdot\boldsymbol{r}_i)\boldsymbol{\gamma}] + [(1-\gamma^2)/(1+\gamma^2)]\boldsymbol{r}. \tag{1.11.3}$$

A correct understanding of the above four interpretations—in particular, the interchange of **A** with $\mathbf{A}^{\mathrm{T}} = \mathbf{A}(-\chi)$ [and **R** with $\mathbf{R}^{\mathrm{T}} = \mathbf{R}(-\chi)$] in single, and, especially, successive rotations (see below)—is crucial to spatial rigid-body kinematics. Lack of it, as Synge (1960, p. 16) accurately puts it "can be a source of such petty confusion."

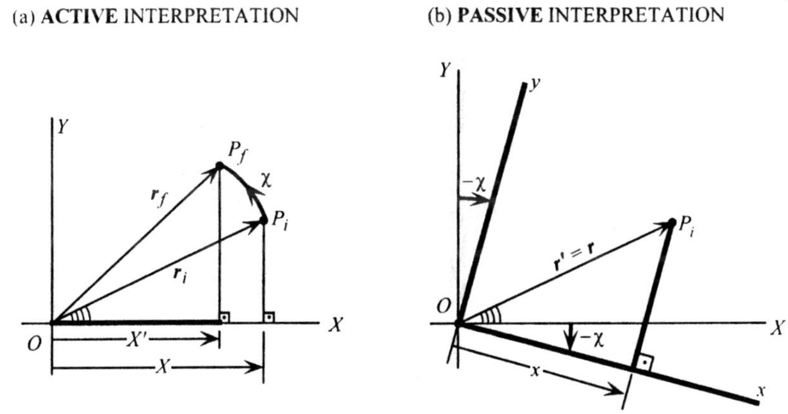

Figure 1.24 The *final* coordinates under χ [*active* interpretation (a)] equal the *new* coordinates under $-\chi$ [*passive* interpretation (b)], and vice versa. ($X'_{\chi\text{-rotated vector, old axes}} = x_{\text{unrotated vector, }-\chi\text{-rotated axes}}$, etc.)

Below, we summarize these four interpretations of an orthogonal tensor **A**:

(i) $\mathbf{A} = (A_{k'k}) = (\boldsymbol{u}_{k'} \cdot \boldsymbol{u}_k)$: *Direction cosine* matrix;

(ii)
$$\begin{pmatrix} \boldsymbol{I} \\ \boldsymbol{J} \\ \boldsymbol{K} \end{pmatrix} = \mathbf{A} \begin{pmatrix} \boldsymbol{i} \\ \boldsymbol{j} \\ \boldsymbol{k} \end{pmatrix}$$

or $\mathbf{T} = \mathbf{A} \cdot \mathbf{t}$: *Triad* rotation;

(iii)
$$\begin{pmatrix} X \\ Y \\ Z \end{pmatrix} = \mathbf{A} \begin{pmatrix} x \\ y \\ z \end{pmatrix}$$

or $\boldsymbol{r}_{old\,\text{axes}} \equiv \boldsymbol{r}' = \mathbf{A} \cdot \boldsymbol{r}_{new\,\text{axes}} \equiv \mathbf{A} \cdot \boldsymbol{r}$: *Passive* interpretation

(*Vector* fixed; *axes* rotated);

(iv)
$$\begin{pmatrix} X' \\ Y' \\ Z' \end{pmatrix} = \mathbf{R} \begin{pmatrix} X \\ Y \\ Z \end{pmatrix}$$

or $\boldsymbol{r}_f = \mathbf{R} \cdot \boldsymbol{r}_i$: *Active* interpretation $\mathbf{A} = \mathbf{R}$

(*Vector* rotated; *axes* fixed, and common).

REMARKS

(i) In the *passive* interpretation, we denote the components of **A** as $A_{k'k}$; whereas, in the *active* one, we denote them, in an arbitrary but common set of axes, as R_{kl} (or $R_{k'l'}$). This is an extra advantage of the accented indicial notation, especially in cases where both interpretations are needed.

(ii) The passive interpretation also holds for the components of *any* other vector; for example, angular velocity.

Successive Rotations

Let us consider a sequence of rotations compounded according to the following scheme:

$$\mathbf{T} \underset{\mathbf{A}_1}{\rightarrow} \mathbf{T}_1 \underset{\mathbf{A}_2}{\rightarrow} \mathbf{T}_2 \underset{\mathbf{A}_3}{\rightarrow} \cdots \underset{\mathbf{A}_{n-1}}{\rightarrow} \mathbf{T}_{n-1} \underset{\mathbf{A}_n}{\rightarrow} \mathbf{T}_n \equiv \mathbf{t} \tag{1.11.4a}$$

Then we shall have the following *composition formulae*, for the various interpretations.

(i) Triad Rotation

$$\mathbf{T} = (\mathbf{A}_1 \cdot \mathbf{A}_2 \cdot \cdots \cdot \mathbf{A}_n) \cdot \mathbf{t} \quad \Leftrightarrow \quad \mathbf{t} = (\mathbf{A}_n{}^{\mathrm{T}} \cdot \mathbf{A}_{n-1}{}^{\mathrm{T}} \cdot \cdots \cdot \mathbf{A}_1{}^{\mathrm{T}}) \cdot \mathbf{T}; \tag{1.11.4b}$$

or, in extenso,

$$\begin{pmatrix} \boldsymbol{I} \\ \boldsymbol{J} \\ \boldsymbol{K} \end{pmatrix} = (\mathbf{A}_1 \cdot \mathbf{A}_2 \cdot \cdots \cdot \mathbf{A}_n) \begin{pmatrix} \boldsymbol{i} \\ \boldsymbol{j} \\ \boldsymbol{k} \end{pmatrix}$$

Initial triad (natural order of component matrices) *Final* triad, (1.11.4c)

$$\begin{pmatrix} \boldsymbol{i} \\ \boldsymbol{j} \\ \boldsymbol{k} \end{pmatrix} = (\mathbf{A}_n{}^{\mathrm{T}} \cdot \mathbf{A}_{n-1}{}^{\mathrm{T}} \cdot \cdots \cdot \mathbf{A}_1{}^{\mathrm{T}}) \begin{pmatrix} \boldsymbol{I} \\ \boldsymbol{J} \\ \boldsymbol{K} \end{pmatrix}$$

Final triad *Initial* triad. (1.11.4d)

(ii) Passive Interpretation

Here, with some easily understood ad hoc notations, we will have

$$\boldsymbol{r}_{\text{old axes}} \equiv \boldsymbol{r}' = \mathbf{A}_1 \cdot \boldsymbol{r}_1 = \mathbf{A}_1 \cdot (\mathbf{A}_2 \cdot \boldsymbol{r}_2) = \cdots = (\mathbf{A}_1 \cdot \mathbf{A}_2 \cdot \cdots \cdot \mathbf{A}_n) \cdot \boldsymbol{r}, \qquad (1.11.4e)$$

$$\boldsymbol{r}_{\text{new axes}} \equiv \boldsymbol{r} = (\mathbf{A}_1 \cdot \mathbf{A}_2 \cdot \cdots \cdot \mathbf{A}_n)^{\mathrm{T}} \cdot \boldsymbol{r}' = (\mathbf{A}_n{}^{\mathrm{T}} \cdot \mathbf{A}_{n-1}{}^{\mathrm{T}} \cdot \cdots \cdot \mathbf{A}_1{}^{\mathrm{T}}) \cdot \boldsymbol{r}'; \qquad (1.11.4f)$$

or, in extenso,

$$\begin{pmatrix} X \\ Y \\ Z \end{pmatrix} = (\mathbf{A}_1 \cdot \mathbf{A}_2 \cdot \cdots \cdot \mathbf{A}_n) \begin{pmatrix} x \\ y \\ z \end{pmatrix}$$

Old axes *New* axes, (1.11.4g)

$$\begin{pmatrix} x \\ y \\ z \end{pmatrix} = (\mathbf{A}_n{}^{\mathrm{T}} \cdot \mathbf{A}_{n-1}{}^{\mathrm{T}} \cdot \cdots \cdot \mathbf{A}_1{}^{\mathrm{T}}) \begin{pmatrix} X \\ Y \\ Z \end{pmatrix}$$

New axes *Old* axes. (1.11.4h)

(iii) Active Interpretation

Here, choosing common axes corresponding to **T**; that is,

$$\boldsymbol{r}_i = X\boldsymbol{I} + Y\boldsymbol{J} + Z\boldsymbol{K} \;\rightarrow\; \boldsymbol{r}_f = X'\boldsymbol{I} + Y'\boldsymbol{J} + Z'\boldsymbol{K} \quad (= X\boldsymbol{i} + Y\boldsymbol{j} + Z\boldsymbol{k}), \qquad (1.11.4i)$$

we obtain, successively,

$$\boldsymbol{r}_i = \mathbf{A}_1{}^{\mathrm{T}} \cdot \boldsymbol{r}_{f,1} = \mathbf{A}^{\mathrm{T}}{}_1 \cdot (\mathbf{A}^{\mathrm{T}}{}_2 \cdot \boldsymbol{r}_{f,2}) = \cdots = (\mathbf{A}_1{}^{\mathrm{T}} \cdot \mathbf{A}_2{}^{\mathrm{T}} \cdot \cdots \cdot \mathbf{A}_n{}^{\mathrm{T}}) \cdot \boldsymbol{r}_f$$
$$\equiv (\mathbf{R}_1{}^{\mathrm{T}} \cdot \mathbf{R}_2{}^{\mathrm{T}} \cdot \cdots \cdot \mathbf{R}_n{}^{\mathrm{T}}) \cdot \boldsymbol{r}_f, \qquad (1.11.4j)$$

$$\Rightarrow \boldsymbol{r}_f = (\mathbf{A}_n \cdot \mathbf{A}_{n-1} \cdot \cdots \cdot \mathbf{A}_1) \cdot \boldsymbol{r}_i \equiv (\mathbf{R}_n \cdot \mathbf{R}_{n-1} \cdot \cdots \cdot \mathbf{R}_1) \cdot \boldsymbol{r}_i; \qquad (1.11.4k)$$

or, in extenso,

$$\begin{pmatrix} X' \\ Y' \\ Z' \end{pmatrix} = (\mathbf{R}_n \cdot \mathbf{R}_{n-1} \cdot \cdots \cdot \mathbf{R}_1) \begin{pmatrix} X \\ Y \\ Z \end{pmatrix}$$

Final position *Initial* position, (1.11.4l)

$$\begin{pmatrix} X \\ Y \\ Z \end{pmatrix} = (\mathbf{R}_1{}^{\mathrm{T}} \cdot \mathbf{R}_2{}^{\mathrm{T}} \cdot \cdots \cdot \mathbf{R}_n{}^{\mathrm{T}}) \begin{pmatrix} X' \\ Y' \\ Z' \end{pmatrix}$$

Initial position *Final* position. (1.11.4m)

Body-Fixed versus Space-Fixed Axes

The moving triad **t** and associated axes (O–xyz) may be considered as a rigid body going through a sequence of rotations, either about these *body-fixed* axes themselves, or about the *space-fixed* axes O–XYZ with which it originally coincided. Either of these two types of sequences may be used (although the tensors/matrices of rotations about body-fixed axes have simpler structure than those about space-fixed axes), and their outcomes are related by the following remarkable theorem: The sequence of rotations about Ox, Oy, Oz has the same effect as the sequence of rotations of equal amounts about OX, OY, OZ, but carried out in the *reverse order*. Symbolically,

$$(\mathbf{R}_1\mathbf{R}_2)_{\textit{body}\text{-fixed axes}} = (\mathbf{R}_2\mathbf{R}_1)_{\textit{space}\text{-fixed axes}}.$$

This nontrivial result will be proved in §1.12.

Thus, for a sequence about space-fixed axes, eq. (1.11.4h) (which expresses the passive interpretation for a body-fixed sequence) should be replaced by

$$\begin{pmatrix} x \\ y \\ z \end{pmatrix} = (\mathbf{S}_1{}^{\mathrm{T}} \cdot \mathbf{S}_2{}^{\mathrm{T}} \cdot \cdots \cdot \mathbf{S}_n{}^{\mathrm{T}}) \begin{pmatrix} X \\ Y \\ Z \end{pmatrix}$$
$$= (\mathbf{S}_n \cdot \mathbf{S}_{n-1} \cdot \cdots \cdot \mathbf{S}_1)^{\mathrm{T}} \begin{pmatrix} X \\ Y \\ Z \end{pmatrix}$$

New axes *Old* axes, (1.11.4n)

where the $\mathbf{S}_k$ are the space-fixed axes counterparts (of equal angle of rotation) of the $\mathbf{R}_k$; and similarly for the other compounded rotation equations.

REMARKS

(i) In algebraic terms, we say that such successive rotations form the Special Orthogonal (Unit Determinant)—Three Dimensional *group* of Real Matrices [≡ SO(3, R)], and are representable by *three independent* parameters; for example, Eulerian angles (§1.12).

[By group, we mean, briefly, that (a) an identity rotation exists (i.e., one that leaves the body unchanged); (b) the product of two successive rotations is also a rotation; (c) every rotation has an inverse; and (d) these rotations are associative. See books on algebra/group theory.]

(ii) Some authors call rotation tensor/matrix the *transpose* of this book's, while others, in addition, fail to mention the distinction between active and passive interpretations. Hence, a certain caution is needed when comparing various references. Our choice was based on the fact that when the rotation tensor of the *active* interpretation is expanded à la Taylor around the identity tensor, and so on (1.10.25a ff.), it leads to an angular velocity compatible with the definition of the axial vector (ω) of an antisymmetric tensor (1.1.16a ff.) $\boldsymbol{\omega}$: $\boldsymbol{\omega}\cdot\boldsymbol{r} = \boldsymbol{\omega}\times\boldsymbol{r}$; otherwise we would have $\boldsymbol{\omega}\cdot\boldsymbol{r} = -\boldsymbol{\omega}\times\boldsymbol{r}$.

Tensorial Derivation of the Finite Rotation Tensor **R**

Let us consider the following two rectangular Cartesian sets of axes, $O\text{–}x_{k'}$ ($\equiv O\text{–}XYZ$, *fixed*) and $O\text{–}x_k$ ($\equiv O\text{–}xyz$, *moving*), related by the proper orthogonal transformation:

$$x_{k'} = \sum A_{k'k}x_k \Leftrightarrow x_k = \sum A_{kk'}x_{k'}, \qquad A_{k'k} = A_{kk'} = \cos(x_{k'}, x_k). \quad (1.11.5a)$$

The corresponding components of the rotation tensor, $R_{k'l'}$ and R_{kl}, respectively, will be related by the well-known transformation rule for second-order tensors (1.1.19j ff.):

$$R_{k'l'} = \sum\sum A_{k'k}A_{l'l}R_{kl} \quad \Leftrightarrow \quad R_{kl} = \sum\sum A_{k'k}A_{l'l}R_{k'l'}, \qquad (1.11.5b)$$

or, in matrix form,

$$\mathbf{R}' = \mathbf{A}\cdot\mathbf{R}\cdot\mathbf{A}^{\mathrm{T}} \quad \Leftrightarrow \quad \mathbf{R} = \mathbf{A}^{\mathrm{T}}\cdot\mathbf{R}'\cdot\mathbf{A}, \qquad (1.11.5c)$$

where

$$\mathbf{R}' = (R_{k'l'}), \qquad \mathbf{R} = (R_{kl}), \qquad \mathbf{A} = (A_{k'k}). \qquad (1.11.5d)$$

Here, choosing axes $O\text{–}x_k$ in which R_{kl} have the *simplest form possible*, and then applying (1.11.5b, c), we will obtain the rotation tensor components in the general axes $O\text{–}x_{k'}$, $R_{k'l'}$; that is, eq. (1.10.10a). To this end, we select Ox_k so that $x_1 \equiv x$ is along the positive sense of the rotation axis $\boldsymbol{n}$, while $x_2 \equiv y$, $x_3 \equiv z$ are on the plane through O perpendicular to $\boldsymbol{n}$ (fig. 1.25). For such special axes, the finite rotation is a

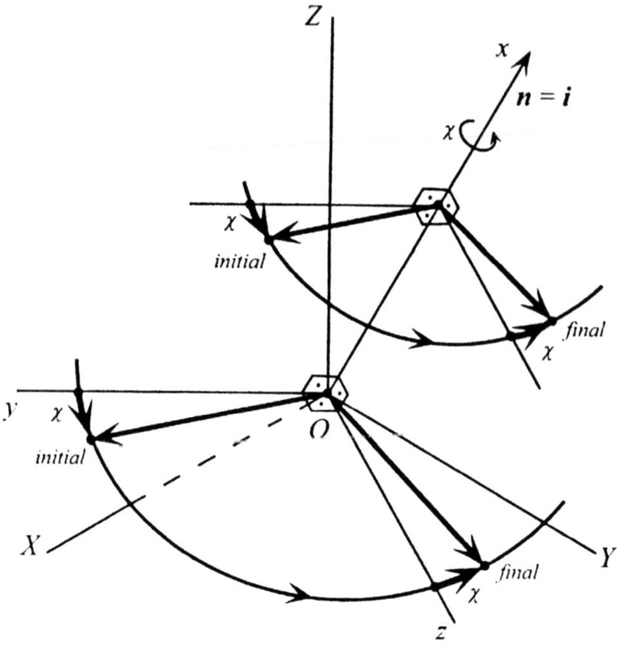

Figure 1.25 Tensor transformation of rotation tensor, between the general fixed axes O–XYZ and the special moving axes O–xyz; Ox axis of rotation.

plane rotation of (say, right-hand rule) angle χ about Ox, and, hence, there the rotation tensor has the following simple planar form:

$$\mathbf{R} = \begin{pmatrix} 1 & 0 & 0 \\ 0 & \cos\chi & -\sin\chi \\ 0 & \sin\chi & \cos\chi \end{pmatrix}. \tag{1.11.5e}$$

Now, to apply (1.11.5c) we need **A**. The latter, since

$$\boldsymbol{i} = A_{xX}\boldsymbol{I} + A_{xY}\boldsymbol{J} + A_{xZ}\boldsymbol{K} = \boldsymbol{n} \equiv n_X\boldsymbol{I} + n_Y\boldsymbol{J} + n_Z\boldsymbol{K}, \tag{1.11.5f}$$

becomes

$$\mathbf{A} = \begin{pmatrix} n_X & A_{Xy} & A_{Xz} \\ n_Y & A_{Yy} & A_{Yz} \\ n_Z & A_{Zy} & A_{Zz} \end{pmatrix}; \tag{1.11.5g}$$

and so, with the abbreviations $\cos(\ldots) \equiv c(\ldots)$, $\sin(\ldots) \equiv s(\ldots)$, (1.11.5c) specializes to

$$\mathbf{R}' = \begin{pmatrix} n_X & A_{Xy} & A_{Xz} \\ n_Y & A_{Yy} & A_{Yz} \\ n_Z & A_{Zy} & A_{Zz} \end{pmatrix} \begin{pmatrix} 1 & 0 & 0 \\ 0 & c\chi & -s\chi \\ 0 & s\chi & c\chi \end{pmatrix} \begin{pmatrix} n_X & n_Y & n_Z \\ A_{Xy} & A_{Yy} & A_{Zy} \\ A_{Xz} & A_{Yz} & A_{Zz} \end{pmatrix},$$

or, carrying out the matrix multiplications, and recalling that $R_{1'1'} \equiv R_{XX}$, $R_{1'2'} \equiv R_{XY}$, and so on,

$$
\begin{aligned}
R_{XX} &= n_X{}^2 + (A_{Xy}{}^2 + A_{Xz}{}^2)\cos\chi,\\
R_{XY} &= n_X n_Y + (A_{Xy}A_{Yy} + A_{Xz}A_{Yz})\cos\chi - (A_{Xy}A_{Yz} - A_{Yy}A_{Xz})\sin\chi,\\
R_{XZ} &= n_X n_Z + (A_{Xy}A_{Zy} + A_{Xz}A_{Zz})\cos\chi + (A_{Zy}A_{Xz} - A_{Xy}A_{Zz})\sin\chi;\\
R_{YX} &= n_Y n_X + (A_{Yy}A_{Xy} + A_{Yz}A_{Xz})\cos\chi + (A_{Xy}A_{Zz} - A_{Yy}A_{Xz})\sin\chi,\\
R_{YY} &= n_Y{}^2 + (A_{Yy}{}^2 + A_{Yz}{}^2)\cos\chi,\\
R_{YZ} &= n_Y n_Z + (A_{Yy}A_{Zy} + A_{Yz}A_{Zz})\cos\chi - (A_{Yy}A_{Zz} - A_{Zy}A_{Yz})\sin\chi;\\
R_{ZX} &= n_Z n_X + (A_{Zy}A_{Xy} + A_{Zz}A_{Xz})\cos\chi - (A_{Zy}A_{Xz} - A_{Xy}A_{Zz})\sin\chi,\\
R_{ZY} &= n_Z n_Y + (A_{Zy}A_{Yy} + A_{Zz}A_{Yz})\cos\chi + (A_{Yy}A_{Zz} - A_{Zy}A_{Yz})\sin\chi,\\
R_{ZZ} &= n_Z{}^2 + (A_{Zy}{}^2 + A_{Zz}{}^2)\cos\chi. \qquad (1.11.5h)
\end{aligned}
$$

However, the nine $A_{k'k}$ are constrained by the *six* orthonormality conditions:

$$
\begin{aligned}
\boldsymbol{I}\cdot\boldsymbol{J} &= n_X n_Y + A_{Xy}A_{Yy} + A_{Xz}A_{Yz} = 0,\\
\boldsymbol{J}\cdot\boldsymbol{K} &= n_Y n_Z + A_{Yy}A_{Zy} + A_{Yz}A_{Zz} = 0,\\
\boldsymbol{K}\cdot\boldsymbol{I} &= n_Z n_X + A_{Zy}A_{Xy} + A_{Zz}A_{Xz} = 0;\\
\boldsymbol{I}\cdot\boldsymbol{I} &= n_X{}^2 + A_{Xy}{}^2 + A_{Xz}{}^2 = 1,\\
\boldsymbol{J}\cdot\boldsymbol{J} &= n_Y{}^2 + A_{Yy}{}^2 + A_{Yz}{}^2 = 1,\\
\boldsymbol{K}\cdot\boldsymbol{K} &= n_Z{}^2 + A_{Zy}{}^2 + A_{Zz}{}^2 = 1; \qquad (1.11.5i)
\end{aligned}
$$

and also $\boldsymbol{n} = \boldsymbol{u}_y \times \boldsymbol{u}_z$, or, in components,

$$
n_X = A_{Yy}A_{Zz} - A_{Zy}A_{Yz}, \qquad n_Y = A_{Zy}A_{Xz} - A_{Xy}A_{Zz}, \qquad n_Z = A_{Xy}A_{Yz} - A_{Yy}A_{Xz}. \qquad (1.11.5j)
$$

As a result of the above, it is not hard to verify that the $R_{k'l'}$, (1.11.5h), reduce to

$$
\begin{aligned}
R_{XX} &= n_X{}^2 + (1 - n_X{}^2)\cos\chi,\\
R_{XY} &= n_X n_Y + (-n_X n_Y)\cos\chi + (n_Z)\sin\chi,\\
R_{XZ} &= n_X n_Z + (-n_X n_Z)\cos\chi + (n_Y)\sin\chi;\\
R_{YX} &= n_Y n_X + (-n_X n_Y)\cos\chi + (n_Z)\sin\chi,\\
R_{YY} &= n_Y{}^2 + (1 - n_Y{}^2)\cos\chi,\\
R_{YZ} &= n_Y n_Z + (-n_Y n_Z)\cos\chi + (-n_X)\sin\chi;\\
R_{ZX} &= n_Z n_X + (-n_Z n_X)\cos\chi + (-n_Y)\sin\chi,\\
R_{ZY} &= n_Z n_Y + (-n_Z n_Y)\cos\chi + (n_X)\sin\chi,\\
R_{ZZ} &= n_Z{}^2 + (1 - n_Z{}^2)\cos\chi; \qquad (1.11.5k)
\end{aligned}
$$

and when put to matrix form is none other than eq. (1.10.10a). We notice that the components $R_{k'l'}$ are *independent of the orientation of the O–xyz axes*, as expected.

Angular Velocity via the Passive Interpretation

Let us consider a generic body point P fixed in the moving frame **t**: O–$\boldsymbol{ijk}/O$–xyz, and hence representable by

$$\boldsymbol{r}' = X\boldsymbol{I} + Y\boldsymbol{J} + Z\boldsymbol{K} \qquad \text{(space-fixed frame } \mathbf{T}\text{: } O\text{–}\boldsymbol{IJK}/O\text{–}XYZ\text{)}, \tag{1.11.6a}$$

$$\boldsymbol{r} = x\boldsymbol{i} + y\boldsymbol{j} + z\boldsymbol{k} \qquad \text{(body-fixed frame; i.e. } x, y, z = \text{constant)}; \tag{1.11.6b}$$

or, in matrix form,

$$\boldsymbol{r}'^{\mathrm{T}} = (X, Y, Z), \qquad \boldsymbol{r}^{\mathrm{T}} = (x, y, z). \tag{1.11.6c}$$

According to the passive interpretation (1.10.2c) (with **A** replaced by the rotation tensor matrix **R**),

$$\boldsymbol{r}' = \mathbf{R} \cdot \boldsymbol{r}, \tag{1.11.6d}$$

and, therefore, the inertial velocity of P, resolved along the fixed axes O–XYZ equals

$$\begin{aligned} \boldsymbol{v}' \equiv d\boldsymbol{r}'/dt &= (\mathrm{d}\mathbf{R}/\mathrm{dt}) \cdot \boldsymbol{r} + \mathbf{R} \cdot (d\boldsymbol{r}/dt) = (\mathrm{d}\mathbf{R}/\mathrm{dt}) \cdot \boldsymbol{r} + \mathbf{R} \cdot \mathbf{0} \\ &= (\mathrm{d}\mathbf{R}/\mathrm{dt}) \cdot (\mathbf{R}^{\mathrm{T}} \cdot \boldsymbol{r}') = \boldsymbol{\omega}' \cdot \boldsymbol{r}' \equiv \omega' \times \boldsymbol{r}', \end{aligned} \tag{1.11.6e}$$

where [recalling (1.7.30f ff.), with $\mathbf{A} \to \mathbf{R}$]

$\boldsymbol{\omega}' \equiv (\mathrm{d}\mathbf{R}/\mathrm{dt}) \cdot \mathbf{R}^{\mathrm{T}}$ = *angular velocity tensor* of body frame **t** relative to the fixed frame **T**, but resolved along the fixed axis O–XYZ, (1.11.6f)

ω' = *axial vector* of $\boldsymbol{\omega}'$; angular veclocity *vector* of **t** relative to **T**. (1.11.6g)

The components of the angular velocity along the moving axes can then be found easily from the vector transformation (passive interpretation):

$\boldsymbol{v}$ = *inertial* velocity of P, but resolved along the moving axes (*not* to be confused with the velocity of P relative to **t**, which is zero: $d\boldsymbol{r}/dt = \mathbf{0}$)

$$= \mathbf{R}^{\mathrm{T}} \cdot \boldsymbol{v}' = \mathbf{R}^{\mathrm{T}} \cdot [(\mathrm{d}\mathbf{R}/\mathrm{dt}) \cdot \boldsymbol{r}] = \boldsymbol{\omega} \cdot \boldsymbol{r} \equiv \omega \times \boldsymbol{r}, \tag{1.11.6h}$$

where

$\boldsymbol{\omega} \equiv \mathbf{R}^{\mathrm{T}} \cdot (\mathrm{d}\mathbf{R}/\mathrm{dt})$ = *angular velocity tensor* of body frame **t** relative to the fixed frame **T**, but resolved along the moving axes O–xyz

$\{ = [\mathbf{R}^{\mathrm{T}} \cdot (\mathrm{d}\mathbf{R}/\mathrm{dt})] \cdot (\mathbf{R}^{\mathrm{T}} \cdot \mathbf{R}) = \mathbf{R}^{\mathrm{T}} \cdot [(\mathrm{d}\mathbf{R}/\mathrm{dt}) \cdot \mathbf{R}^{\mathrm{T}}] \cdot \mathbf{R}$

$= \mathbf{R}^{\mathrm{T}} \cdot \boldsymbol{\omega}' \cdot \mathbf{R}$; a second-order tensor transformation, as it should be$\}$, (1.11.6i)

ω = *axial vector of* $\boldsymbol{\omega}$; angular velocity of **t** relative to **T** $[= \mathbf{R}^{\mathrm{T}} \cdot \omega']$. (1.11.6j)

REMARK

If $\mathbf{R} = \mathbf{R}(q_1, q_2, q_3) \equiv \mathbf{R}(q_\alpha)$, where the q_α are system *rotational* parameters (e.g., the three Eulerian angles, §1.12), then $\boldsymbol{\omega}'$ and ω' can be expressed, respectively as follows:

$$\text{Tensor: } \boldsymbol{\omega}' = \sum \boldsymbol{\omega}'_\alpha (dq_\alpha/dt), \qquad \text{Vector: } \omega' = \sum \omega'_\alpha (dq_\alpha/dt), \qquad (1.11.6k)$$

where

$$\boldsymbol{\omega}'_\alpha \equiv (\partial \mathbf{R}/\partial q_\alpha) \cdot \mathbf{R}^{\mathrm{T}} \qquad \text{and} \qquad \boldsymbol{\omega}'_\alpha \cdot \boldsymbol{x} = \omega'_\alpha \times \boldsymbol{x}, \qquad (1.11.6l)$$

for an arbitrary vector $\boldsymbol{x}$; that is, $\boldsymbol{\omega}'$ can be expressed in terms of the local basis $\{\boldsymbol{\omega}'_\alpha;\ \alpha = 1, 2, 3\}$; and similarly for $\boldsymbol{\omega}$ and ω.

Additional Useful Results

1. Consider the following two successive (component) rotations: First, from the "fixed" frame 0 to the moving frame 1, $\mathbf{R}_{1/0} \equiv \mathbf{R}_1$, and, next, from 1 to the also moving frame 2, $\mathbf{R}_{2/1} \equiv \mathbf{R}_2$. Then, by (1.11.4a ff.), the resultant rotation from 0 to 2 will be $\mathbf{R} = \mathbf{R}_1 \cdot \mathbf{R}_2$. Now, let:

$\boldsymbol{\omega}_{1/0,0} \equiv (\mathrm{d}\mathbf{R}_1/\mathrm{dt}) \cdot \mathbf{R}_1^{\mathrm{T}}$: angular velocity tensor of frame 1 relative to frame 0, *along 0-axes*;
$\boldsymbol{\omega}_{1/0,1} \equiv \mathbf{R}_1^{\mathrm{T}} \cdot (\mathrm{d}\mathbf{R}_1/\mathrm{dt})$: angular velocity tensor of frame 1 relative to frame 0, *along 1-axes*;
$\boldsymbol{\omega}_{2/1,1} \equiv (\mathrm{d}\mathbf{R}_2/\mathrm{dt}) \cdot \mathbf{R}_2^{\mathrm{T}}$: angular velocity tensor of frame 2 relative to frame 1, *along 1-axes*;
$\boldsymbol{\omega}_{2/1,2} \equiv \mathbf{R}_2^{\mathrm{T}} \cdot (\mathrm{d}\mathbf{R}_2/\mathrm{dt})$: angular velocity tensor of frame 2 relative to frame 1, *along 2-axes*;
$\boldsymbol{\omega}_{2/0,0} \equiv (\mathrm{d}\mathbf{R}/\mathrm{dt}) \cdot \mathbf{R}^{\mathrm{T}}$: angular velocity of frame 2 relative to frame 0, *along 0-axes*;
$\boldsymbol{\omega}_{2/0,2} \equiv \mathbf{R}^{\mathrm{T}} \cdot (\mathrm{d}\mathbf{R}/\mathrm{dt})$: angular velocity tensor of frame 2 relative to frame 0, *along 2-axes*; (1.11.7a)

(this or some similar intricate notation is a must in matrix territory!) and therefore

$$\boldsymbol{\omega}_{1/0,0} = \mathbf{R}_1 \cdot \boldsymbol{\omega}_{1/0,1} \cdot \mathbf{R}_1^{\mathrm{T}} \Leftrightarrow \boldsymbol{\omega}_{1/0,1} = \mathbf{R}_1^{\mathrm{T}} \cdot \boldsymbol{\omega}_{1/0,0} \cdot \mathbf{R}_1,$$

$$\boldsymbol{\omega}_{2/1,0} = \mathbf{R}_1 \cdot \boldsymbol{\omega}_{2/1,1} \cdot \mathbf{R}_1^{\mathrm{T}} \Leftrightarrow \boldsymbol{\omega}_{2/1,1} = \mathbf{R}_1 \cdot \boldsymbol{\omega}_{2/1,0} \cdot \mathbf{R}_1^{\mathrm{T}},$$

$$\boldsymbol{\omega}_{2/1,1} = \mathbf{R}_2 \cdot \boldsymbol{\omega}_{2/1,2} \cdot \mathbf{R}_2^{\mathrm{T}} \Leftrightarrow \boldsymbol{\omega}_{2/1,2} = \mathbf{R}_2^{\mathrm{T}} \cdot \boldsymbol{\omega}_{2/1,1} \cdot \mathbf{R}_2,$$

$$\boldsymbol{\omega}_{2/0,1} = \mathbf{R}_1^{\mathrm{T}} \cdot \boldsymbol{\omega}_{2/0,0} \cdot \mathbf{R}_1 \;=\; \mathbf{R}_2 \cdot \boldsymbol{\omega}_{2/0,2} \cdot \mathbf{R}_2^{\mathrm{T}}:$$

angular velocity of frame 2 relative to frame 0, *but expressed along 1-axes;* etc.; i.e., *the multiplications* $\mathbf{R}_1 \cdot (\ldots) \cdot \mathbf{R}_1^{\mathrm{T}}$ *convert components from 1-frame axes to 0-frame axes; while* $\mathbf{R}_1^{\mathrm{T}}(\ldots) \cdot \mathbf{R}_1$ *convert components from 0-frame axes to 1-frame axes;* and analogously for $\mathbf{R}_2 \cdot (\ldots) \cdot \mathbf{R}_2^{\mathrm{T}}$, $\mathbf{R}_2^{\mathrm{T}} \cdot (\ldots) \cdot \mathbf{R}_2$.
Then, and since $\mathbf{R}$, $\mathbf{R}_1$, $\mathbf{R}_2$ are orthogonal tensors,

(a) $\boldsymbol{\omega}_{2/0,0} = (\mathrm{d}\mathbf{R}/\mathrm{dt}) \cdot \mathbf{R}^{\mathrm{T}} = [\mathrm{d}/\mathrm{dt}(\mathbf{R}_1 \cdot \mathbf{R}_2)] \cdot (\mathbf{R}_1 \cdot \mathbf{R}_2)^{\mathrm{T}}$
$= \cdots = (\mathrm{d}\mathbf{R}_1/\mathrm{dt}) \cdot \mathbf{R}_1^{\mathrm{T}} + \mathbf{R}_1 \cdot [(\mathrm{d}\mathbf{R}_2/\mathrm{dt}) \cdot \mathbf{R}_2^{\mathrm{T}}] \cdot \mathbf{R}_1^{\mathrm{T}}$
$= \boldsymbol{\omega}_{1/0,0} + \mathbf{R}_1 \cdot \boldsymbol{\omega}_{2/1,1} \cdot \mathbf{R}_1^{\mathrm{T}} \equiv \boldsymbol{\omega}_{1/0,0} + \boldsymbol{\omega}_{2/1,0}$

(theorem of additivity of angular velocities, along *0-axes*); (1.11.7b)

$$\omega_{2/0,1} = \mathbf{R}_1{}^{\mathrm{T}} \cdot \omega_{2/0,0} \cdot \mathbf{R}_1 = \mathbf{R}_1{}^{\mathrm{T}} \cdot \omega_{1/0,0} \cdot \mathbf{R}_1 + \omega_{2/1,1} \equiv \omega_{1/0,1} + \omega_{2/1,1}$$

(theorem of additivity of angular velocities, along *1-axes*); (1.11.7c)

$$\omega_{2/0,2} = \mathbf{R}^{\mathrm{T}} \cdot (\mathrm{d}\mathbf{R}/\mathrm{dt}) \quad [= \mathbf{R}_2{}^{\mathrm{T}} \cdot \omega_{2/0,1} \cdot \mathbf{R}_2 = \mathbf{R}^{\mathrm{T}} \cdot \omega_{2/0,0} \cdot \mathbf{R}]$$

$$\equiv (\mathbf{R}_1 \cdot \mathbf{R}_2)^{\mathrm{T}} \cdot [\mathrm{d}/\mathrm{dt}(\mathbf{R}_1 \cdot \mathbf{R}_2)]$$

$$= \cdots = \mathbf{R}_2{}^{\mathrm{T}} \cdot [\mathbf{R}_1{}^{\mathrm{T}} \cdot (\mathrm{d}\mathbf{R}_1/\mathrm{dt})] \cdot \mathbf{R}_2 + \mathbf{R}_2{}^{\mathrm{T}} \cdot (\mathrm{d}\mathbf{R}_2/\mathrm{dt})$$

$$= \mathbf{R}_2{}^{\mathrm{T}} \cdot \omega_{1/0,1} \cdot \mathbf{R}_2 + \omega_{2/1,2} \equiv \omega_{1/0,2} + \omega_{2/1,2}$$

(theorem of additivity of angular velocities, along *2-axes*). (1.11.7d)

(b) Next, d(...)/dt-differentiating the above, say (1.11.7b), it is not hard to show that:

$$\mathrm{d}\omega_{2/0,0}/\mathrm{dt} = \mathrm{d}\omega_{1/0,0}/\mathrm{dt} + \mathrm{d}/\mathrm{dt}(\mathbf{R}_1 \cdot \omega_{2/1,1} \cdot \mathbf{R}_1{}^{\mathrm{T}})$$

$$= \mathrm{d}\omega_{1/0,0}/\mathrm{dt} + \mathbf{R}_1 \cdot (\mathrm{d}\omega_{2/1,1}/\mathrm{dt}) \cdot \mathbf{R}_1{}^{\mathrm{T}} + \mathbf{R}_1 \cdot (\omega_{1/0,1} \cdot \omega_{2/1,1} - \omega_{2/1,1} \cdot \omega_{1/0,1}) \cdot \mathbf{R}_1{}^{\mathrm{T}}$$

(theorem of *non-additivity* of angular accelerations, along *0-axes*); (1.11.7e)

and similarly for $\mathrm{d}\omega_{2/0,1}/\mathrm{dt}$, $\mathrm{d}\omega_{2/0,2}/\mathrm{dt}$. The last (third) term of (1.11.7e) shows that *if the elements of the matrices* $\omega_{1/0,0}$, $\omega_{2/1,1}$ *are constant, then, in general, the elements of* $\omega_{2/0,0}$ *will also be constant if* $\omega_{1/0,1}$ *and* $\omega_{2/1,1}$ *commute*, a well-known result from vectorial (undergraduate) kinematics. The extension of the above to three or more successive rotations is obvious.

[As Professor D. T. Greenwood has aptly remarked: "Equations (1.11.7b–e) illustrate how the use of matrix notation can make the simple seem obscure."]

2. *Matrix forms of relative motion of a particle, in two frames with common origin.* By *d/dt*(...)-differentiating the passive interpretation (1.11.2c),

$$\underset{\textit{Fixed axes}}{\begin{pmatrix} X \\ Y \\ Z \end{pmatrix}} = \mathbf{A} \cdot \underset{\textit{Moving axes},}{\begin{pmatrix} x \\ y \\ z \end{pmatrix}} \tag{1.11.8a}$$

we can show that

(i)
$$\mathrm{d}/\mathrm{dt}\begin{pmatrix} X \\ Y \\ Z \end{pmatrix} = \mathbf{A} \cdot \left[\mathrm{d}/\mathrm{dt}\begin{pmatrix} x \\ y \\ z \end{pmatrix} + \omega \cdot \begin{pmatrix} x \\ y \\ z \end{pmatrix}\right]$$

$$\{= \mathbf{A} \cdot [\textit{relative}\ \text{velocity} + \textit{transport}\ \text{velocity}]\}. \tag{1.11.8b}$$

(ii)
$$\mathrm{d}^2/\mathrm{dt}^2\begin{pmatrix}X\\Y\\Z\end{pmatrix} = \mathbf{A}\cdot\left[\mathrm{d}^2/\mathrm{dt}^2\begin{pmatrix}x\\y\\z\end{pmatrix} + (\mathrm{d}\boldsymbol{\omega}/\mathrm{dt})\cdot\begin{pmatrix}x\\y\\z\end{pmatrix} + \boldsymbol{\omega}^2\cdot\begin{pmatrix}x\\y\\z\end{pmatrix} + 2\boldsymbol{\omega}\cdot\mathrm{d}/\mathrm{dt}\begin{pmatrix}x\\y\\z\end{pmatrix}\right]$$

$\{ = \mathbf{A}\cdot[$*relative* acceleration $(d'^2\boldsymbol{r}/dt^2)$ + *transport* acceleration $(\boldsymbol{\alpha}\times\boldsymbol{r} + \boldsymbol{\omega}\times(\boldsymbol{\omega}\times\boldsymbol{r}))$ + *Coriolis* acceleration $(2\boldsymbol{\omega}\times(d'\boldsymbol{r}/dt))]\}$. (1.11.8c)

Equation (1.11.8b) holds for any vector.

(iii) If the position of the origin of the moving axes, relative to that of the fixed ones, is $\boldsymbol{r}_o = (X_o, Y_o, Z_o)^{\mathrm{T}}$, so that [instead of (1.11.8a)]

$$\begin{pmatrix}X\\Y\\Z\end{pmatrix} = \begin{pmatrix}X_o\\Y_o\\Z_o\end{pmatrix} + \mathbf{A}\cdot\begin{pmatrix}x\\y\\z\end{pmatrix}, \qquad (1.11.8d)$$

then we simply add $\mathrm{d}/\mathrm{dt}(X_o, Y_o, Z_o)^{\mathrm{T}}$ to the right side of (1.11.8b) and $\mathrm{d}^2/\mathrm{dt}^2(X_o, Y_o, Z_o)^{\mathrm{T}}$ to the right side of (1.11.8c).

3. *Tensor of Angular Acceleration*, and so on.

(i) By d(...)/dt-differentiating (1.7.30i, j): $\mathrm{d}\mathbf{A}/\mathrm{dt} = \mathbf{A}\cdot\boldsymbol{\omega} = \boldsymbol{\omega}'\cdot\mathbf{A}$, we can show that

$$\mathrm{d}^2\mathbf{A}/\mathrm{dt}^2 = \mathbf{A}\cdot\boldsymbol{\varepsilon} \Rightarrow \boldsymbol{\varepsilon} = \mathbf{A}^{\mathrm{T}}\cdot(\mathrm{d}^2\mathbf{A}/\mathrm{dt}^2), \qquad (1.11.9a)$$

where

$$\boldsymbol{\varepsilon} \equiv \boldsymbol{\alpha} + \boldsymbol{\omega}\cdot\boldsymbol{\omega} \equiv \boldsymbol{\alpha} + \boldsymbol{\omega}^2, \qquad (1.11.9b)$$

$\boldsymbol{\alpha} \equiv \mathrm{d}\boldsymbol{\omega}/\mathrm{dt}$: (Matrix of components, along the *moving* axes, of the) *tensor of angular acceleration* of the moving axes relative to the fixed ones (1.11.9c)

$$\{ = \mathrm{d}/\mathrm{dt}[\mathbf{A}^{\mathrm{T}}\cdot(\mathrm{d}\mathbf{A}/\mathrm{dt})] = (\mathrm{d}\mathbf{A}^{\mathrm{T}}/\mathrm{dt})\cdot(\mathrm{d}\mathbf{A}/\mathrm{dt}) + \mathbf{A}^{\mathrm{T}}\cdot(\mathrm{d}^2\mathbf{A}/\mathrm{dt}^2) = -\boldsymbol{\omega}\cdot\boldsymbol{\omega} + \boldsymbol{\varepsilon}\}. \qquad (1.11.9d)$$

[In fact, both $\boldsymbol{\alpha}$ and $\boldsymbol{\varepsilon}$ appear in (1.11.8c). Also, some authors call $\boldsymbol{\varepsilon}$ the angular acceleration tensor, but we think that that term should apply to $\mathrm{d}\boldsymbol{\omega}/\mathrm{dt}$; that is, definition (1.11.9c).]

(ii) Both $\boldsymbol{\varepsilon}$ and $\boldsymbol{\alpha}$ are (second-order) tensors; that is,

$$\boldsymbol{\varepsilon}'(= \boldsymbol{\alpha}' + \boldsymbol{\omega}'\cdot\boldsymbol{\omega}') = \mathbf{A}\cdot\boldsymbol{\varepsilon}\cdot\mathbf{A}^{\mathrm{T}} \Leftrightarrow \boldsymbol{\varepsilon} = \mathbf{A}^{\mathrm{T}}\cdot\boldsymbol{\varepsilon}'\cdot\mathbf{A}, \qquad (1.11.9e)$$

$$\boldsymbol{\alpha}'(= \mathrm{d}\boldsymbol{\omega}'/\mathrm{dt}) = \mathbf{A}\cdot\boldsymbol{\alpha}\cdot\mathbf{A}^{\mathrm{T}} \Leftrightarrow \boldsymbol{\alpha} = \mathbf{A}^{\mathrm{T}}\cdot\boldsymbol{\alpha}'\cdot\mathbf{A}; \qquad (1.11.9f)$$

where, as before, an *accent* (prime) denotes matrix of components along the *fixed* axes.

(iii) The *fixed* axes counterpart of (1.11.9a) is:

$$\mathrm{d}^2\mathbf{A}/\mathrm{d}t^2 = \boldsymbol{\varepsilon}'\cdot\mathbf{A} \ \Rightarrow\ \boldsymbol{\varepsilon}' = (\mathrm{d}^2\mathbf{A}/\mathrm{d}t^2)\cdot\mathbf{A}^{\mathrm{T}}. \tag{1.11.9g}$$

(iv) It can be verified, independently of (1.11.9a–d) and (1.11.9e–g), that

$$\begin{aligned}\omega^2 &\equiv \boldsymbol{\omega}\cdot\boldsymbol{\omega} = -\boldsymbol{\omega}^{\mathrm{T}}\cdot\boldsymbol{\omega} = -\boldsymbol{\omega}\cdot\boldsymbol{\omega}^{\mathrm{T}}\\ &= \cdots = -(\mathrm{d}\mathbf{A}/\mathrm{d}t)^{\mathrm{T}}\cdot(\mathrm{d}\mathbf{A}/\mathrm{d}t) = -(\mathrm{d}\mathbf{A}^{\mathrm{T}}/\mathrm{d}t)\cdot(\mathrm{d}\mathbf{A}/\mathrm{d}t),\end{aligned} \tag{1.11.9h}$$

$$\begin{aligned}(\omega')^2 &\equiv \boldsymbol{\omega}'\cdot\boldsymbol{\omega}' = -(\boldsymbol{\omega}')^{\mathrm{T}}\cdot\boldsymbol{\omega}' = -\boldsymbol{\omega}'\cdot(\boldsymbol{\omega}')^{\mathrm{T}}\\ &= \cdots = -(\mathrm{d}\mathbf{A}/\mathrm{d}t)\cdot(\mathrm{d}\mathbf{A}/\mathrm{d}t)^{\mathrm{T}} = -(\mathrm{d}\mathbf{A}/\mathrm{d}t)\cdot(\mathrm{d}\mathbf{A}^{\mathrm{T}}/\mathrm{d}t).\end{aligned} \tag{1.11.9i}$$

(v) Since $\mathrm{d}\boldsymbol{\omega}/\mathrm{d}t$ is *antisymmetric*, and $\boldsymbol{\omega}\cdot\boldsymbol{\omega}$ is *symmetric* (explain this), show that the axial vectors of (the *nonsymmetric*) $\boldsymbol{\varepsilon}$ and $\boldsymbol{\alpha}$ coincide, and are both equal to none other than the *vector of angular acceleration* $\boldsymbol{\alpha}$; thus justifying calling $\boldsymbol{\alpha}$ the tensor of angular acceleration.

Finally, if the moving axes are fixed relative to a body B, then $\boldsymbol{\omega}/\boldsymbol{\omega}'$ and $\boldsymbol{\alpha}/\boldsymbol{\alpha}'$ are respectively, the tensors of angular velocity and acceleration of that body relative to the space-fixed axes; and if the earlier particle is frozen (fixed) relative to B (i.e., $dx/dt = 0$, $d^2x/dt^2 = 0$, etc.), then (1.11.8b, c) give, respectively, the matrix forms of the well-known formulae for the distribution of velocities and acceleration of the various points of B (from body-axes components to space-axes components). [For an indicial treatment of these tensors, and recursive formulae for their higher rates, see Truesdell and Toupin (1960, pp. 439–440).]

1.12 THE RIGID BODY: EULERIAN ANGLES

We recommend for concurrent reading with this section: Junkins and Turner (1986, chap. 2), Morton (1984).

As explained already (§1.7, §1.11), the *nine* elements of the proper orthogonal tensor $\mathbf{A}$ (or $\mathbf{R}$), in all its four interpretations, depend on only *three* independent parameters. A particularly popular such parametrization is afforded by the three (generalized) Eulerian angles. These latter appear naturally as we describe the general orientation of an ortho–normal–dextral (OND) *body-fixed* triad, or local frame $\mathbf{t} = \{\boldsymbol{u}_k\} \equiv (\boldsymbol{i}, \boldsymbol{j}, \boldsymbol{k})$ relative to an OND *space-fixed* frame $\mathbf{T} = \{\boldsymbol{u}_{k'}\} \equiv (\boldsymbol{I}, \boldsymbol{J}, \boldsymbol{K})$, with which it originally coincides, via the following sequence of *three, possibly hypothetical, simple planar rotations* (i.e., in each of them, the two triads have one axis in common, or parallel, and so the corresponding "partial rotation tensor" depends on a *single* angle):

(i) Rotation about the (i)th body axis through an angle $\chi_{(i)} \equiv \chi_1 \equiv \phi$; followed by a
(ii) Rotation about the (j)th body axis ($j \neq i$) through an angle $\chi_{(j)} \equiv \chi_2 \equiv \theta$; followed by a
(iii) Rotation about the (k)th body axis ($k \neq j$) through an angle $\chi_{(k)} \equiv \chi_3 \equiv \psi$.

The angles $\chi_1 = \phi$ (about the original $\boldsymbol{u}_i = \boldsymbol{u}_{i'}$), $\chi_2 = \theta$ (about the ϕ-rotated $\boldsymbol{u}_j \to \boldsymbol{u}_{j'}$), and $\chi_3 = \psi$ (about the θ-rotated $\boldsymbol{u}_k \to \boldsymbol{u}_{k''}$) are known as the $i \to j \to k$ *Eulerian angles*.

Of the *twelve* possible such angle triplets, *six* form a group for which $i \neq j \neq k = i$ (*two*-axes group):

$$1 \to 2 \to 1, \quad 1 \to 3 \to 1, \quad 2 \to 1 \to 2, \quad 2 \to 3 \to 2, \quad 3 \to 1 \to 3, \quad 3 \to 2 \to 3;$$

and *six* form a group for which $i \neq j \neq k \neq i$ (*three*-axes group):

$$1 \to 2 \to 3, \quad 1 \to 3 \to 2, \quad 2 \to 1 \to 3, \quad 2 \to 3 \to 1, \quad 3 \to 1 \to 2, \quad 3 \to 2 \to 1.$$

[Similar results, but with more complicated rotation tensors, would hold for rotations about the *space-fixed* axes $\{\boldsymbol{u}_{k'}: \boldsymbol{I}, \boldsymbol{J}, \boldsymbol{K}\}$. If the partial rotations were about arbitrary (body- or space-fixed) axes, then, due to the infinity of their possible directions, we would have an infinity of angle triplets. It is the restriction that these rotations are about the *body-fixed* axes $\{\boldsymbol{u}_k\}$ that brings them down to twelve.]

Eulerian Angles

The sequence $3 \to 1 \to 3$, shown and described in fig. 1.26 [with the customary abbreviations: $\cos(\ldots) \equiv c(\ldots)$, $\sin(\ldots) \equiv s(\ldots)$] is considered to be the classical Eulerian angle description, originated and frequently used in astronomy and physics, [although, as Likins points out "In his original work in 1760, Euler used a combination of right-handed and left-handed rotations, a convention unacceptable today" (1973, p. 97)].

Using the *passive* interpretation and fig. 1.26, we readily find that the corresponding coordinates of the compounded transformation resulting from the above sequence of partial rotations about the *nonmutually orthogonal* axes OZ, Ox', Oz'' [i.e., the (originally assumed coinciding) space-fixed O–XYZ and body-fixed O–xyz] are related by

$$\begin{pmatrix} X \\ Y \\ Z \end{pmatrix} = \begin{pmatrix} c\phi & -s\phi & 0 \\ s\phi & c\phi & 0 \\ 0 & 0 & 1 \end{pmatrix} \begin{pmatrix} x' \\ y' \\ z' \end{pmatrix}$$

$$= \begin{pmatrix} c\phi & -s\phi & 0 \\ s\phi & c\phi & 0 \\ 0 & 0 & 1 \end{pmatrix} \begin{pmatrix} 1 & 0 & 0 \\ 0 & c\theta & -s\theta \\ 0 & s\theta & c\theta \end{pmatrix} \begin{pmatrix} x'' \\ y'' \\ z'' \end{pmatrix}$$

$$= \begin{pmatrix} c\phi & -s\phi & 0 \\ s\phi & c\phi & 0 \\ 0 & 0 & 1 \end{pmatrix} \begin{pmatrix} 1 & 0 & 0 \\ 0 & c\theta & -s\theta \\ 0 & s\theta & c\theta \end{pmatrix} \begin{pmatrix} c\psi & -s\psi & 0 \\ s\psi & c\psi & 0 \\ 0 & 0 & 1 \end{pmatrix} \begin{pmatrix} x \\ y \\ z \end{pmatrix}$$

$$\mathbf{R}(\boldsymbol{K}, \phi) \cdot \mathbf{R}(\boldsymbol{i}', \theta) \cdot \mathbf{R}(\boldsymbol{k}'', \psi) \equiv \mathbf{R}_\phi \cdot \mathbf{R}_\theta \cdot \mathbf{R}_\psi \equiv \mathbf{R}, \qquad (1.12.1\text{a})$$

$$= \begin{pmatrix} c\phi\, c\psi - s\phi\, c\theta\, s\psi & -c\phi\, s\psi - s\phi\, c\theta\, c\psi & s\phi\, s\theta \\ s\phi\, c\psi + c\phi\, c\theta\, s\psi & -s\phi\, s\psi + c\phi\, c\theta\, c\psi & -c\phi\, s\theta \\ s\theta\, s\psi & s\theta\, c\psi & c\theta \end{pmatrix} \begin{pmatrix} x \\ y \\ z \end{pmatrix}$$

$$\mathbf{R} \text{ or } \mathbf{A} = (A_{k'k}) \quad (= \mathbf{0}, \quad \text{if } \phi, \theta, \psi = 0). \tag{1.12.1b}$$

Classical Eulerian Sequence: $(\boldsymbol{I}, \boldsymbol{J}, \boldsymbol{K})$: $3(\phi) \rightarrow 1(\theta) \rightarrow 3(\psi)$: $(\boldsymbol{i}, \boldsymbol{j}, \boldsymbol{k})$ $0 \leq \phi$ (*precession*, or *azimuth*, angle) $< 2\pi$, $0 \leq \theta$ [*nutation* (i.e., nodding), or pole, angle] $\leq \pi$, $0 \leq \psi$ [*proper*, or *intrinsic*, rotation angle; or (*eigen-*) *spin*] $< 2\pi$.

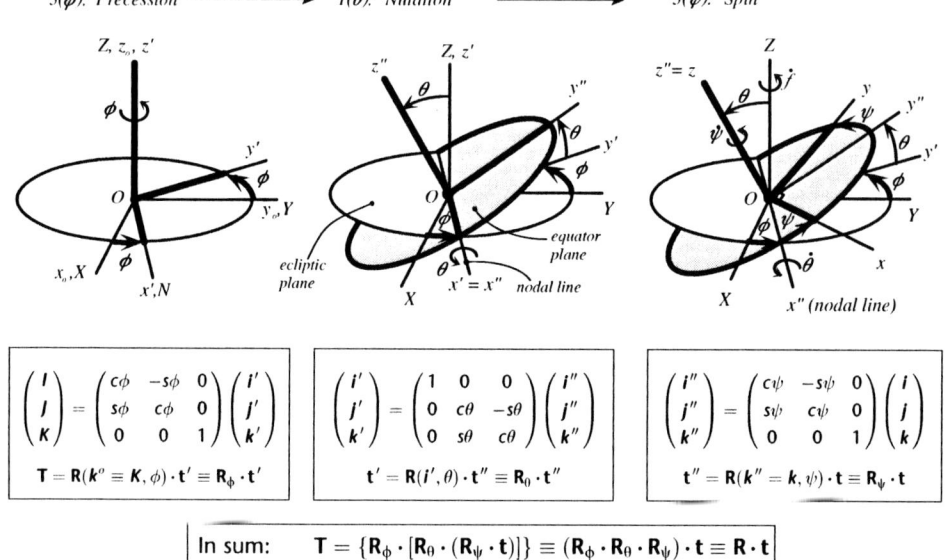

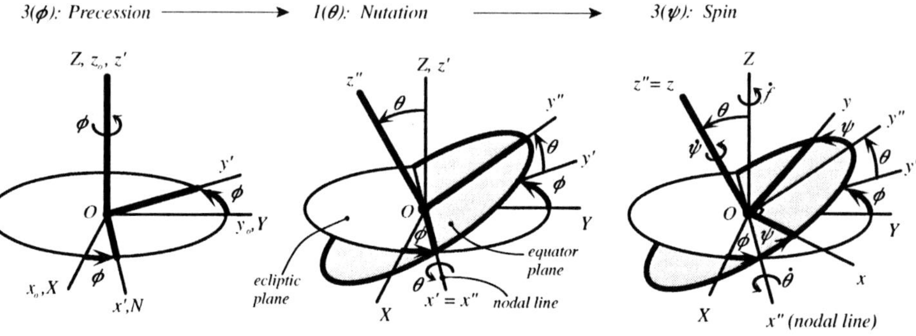

$$\begin{pmatrix} I \\ J \\ K \end{pmatrix} = \begin{pmatrix} c\phi & -s\phi & 0 \\ s\phi & c\phi & 0 \\ 0 & 0 & 1 \end{pmatrix} \begin{pmatrix} i' \\ j' \\ k' \end{pmatrix}$$
$$\mathbf{T} = \mathbf{R}(\boldsymbol{k}^o \equiv \boldsymbol{K}, \phi) \cdot \mathbf{t}' \equiv \mathbf{R}_\phi \cdot \mathbf{t}'$$

$$\begin{pmatrix} i' \\ j' \\ k' \end{pmatrix} = \begin{pmatrix} 1 & 0 & 0 \\ 0 & c\theta & -s\theta \\ 0 & s\theta & c\theta \end{pmatrix} \begin{pmatrix} i'' \\ j'' \\ k'' \end{pmatrix}$$
$$\mathbf{t}' = \mathbf{R}(\boldsymbol{i}', \theta) \cdot \mathbf{t}'' \equiv \mathbf{R}_\theta \cdot \mathbf{t}''$$

$$\begin{pmatrix} i'' \\ j'' \\ k'' \end{pmatrix} = \begin{pmatrix} c\psi & -s\psi & 0 \\ s\psi & c\psi & 0 \\ 0 & 0 & 1 \end{pmatrix} \begin{pmatrix} i \\ j \\ k \end{pmatrix}$$
$$\mathbf{t}'' = \mathbf{R}(\boldsymbol{k}'' = \boldsymbol{k}, \psi) \cdot \mathbf{t} \equiv \mathbf{R}_\psi \cdot \mathbf{t}$$

In sum: $\quad \mathbf{T} = \{\mathbf{R}_\phi \cdot [\mathbf{R}_\theta \cdot (\mathbf{R}_\psi \cdot \mathbf{t})]\} \equiv (\mathbf{R}_\phi \cdot \mathbf{R}_\theta \cdot \mathbf{R}_\psi) \cdot \mathbf{t} \equiv \mathbf{R} \cdot \mathbf{t}$

Figure 1.26 Partial, or elementary, rotations of classical Eulerian sequence: $\phi \rightarrow \theta \rightarrow \psi$ (originally: O—$xyz = O$—$x_o y_o z_o \equiv O$—XYZ).

REMARKS

(i) Equation (1.12.1b) readily shows that if the direction cosines $A_{k'k}$ are known, the three Eulerian angles can be calculated from

$$\phi = \tan^{-1}(-A_{1'3}/A_{2'3}), \qquad \theta = \cos^{-1}(A_{3'3}), \qquad \psi = \tan^{-1}(A_{3'1}/A_{3'2}). \tag{1.12.1c}$$

(ii) If the origin of the body-fixed axes ♦ is moving relative to the space-fixed frame O–XYZ, then in the above we simply replace X with $X - X_\blacklozenge$ and so on, cyclically. Then, x, y, z [or $x_{/\blacklozenge}, y_{/\blacklozenge}, z_{/\blacklozenge}$ (§1.8)] are the particle coordinates relative to ♦–xyz. In this case, eq. (1.12.1b) shows clearly that a *free* (i.e., *unconstrained*) rigid body has *six (global) degrees of freedom*:

$q_{1,2,3} = X_\blacklozenge, Y_\blacklozenge, Z_\blacklozenge$: inertial coordinates of base point (pole) $\blacklozenge$,

$q_{4,5,6} = \phi, \theta, \psi$: Eulerian angles of body-fixed $\blacklozenge$–xyz relative to $\blacklozenge$–XYZ;

and the constant x, y, z is the "name" of a generic body particle [more on this in chap. 2].

Inverting (1.12.1b)—while noting that, since all three component matrices $\mathbf{R}_{\phi,\theta,\psi}$ are orthogonal, the inverse of each equals its transpose (or using the passive interpretation equations in §1.11)—we readily obtain

$$\begin{pmatrix} x \\ y \\ z \end{pmatrix} = \mathbf{R}^{\mathrm{T}} \cdot \begin{pmatrix} X \\ Y \\ Z \end{pmatrix}, \tag{1.12.2}$$

where

$$\mathbf{R}^{\mathrm{T}} = (\mathbf{R}_\phi \cdot \mathbf{R}_\theta \cdot \mathbf{R}_\psi)^{\mathrm{T}} = \mathbf{R}_\psi{}^{\mathrm{T}} \cdot \mathbf{R}_\theta{}^{\mathrm{T}} \cdot \mathbf{R}_\phi{}^{\mathrm{T}} = \mathbf{R}_{-\psi} \cdot \mathbf{R}_{-\theta} \cdot \mathbf{R}_{-\phi}$$

$$= \begin{pmatrix} c\psi & s\psi & 0 \\ -s\psi & c\psi & 0 \\ 0 & 0 & 1 \end{pmatrix} \begin{pmatrix} 1 & 0 & 0 \\ 0 & c\theta & s\theta \\ 0 & -s\theta & c\theta \end{pmatrix} \begin{pmatrix} c\phi & s\phi & 0 \\ -s\phi & c\phi & 0 \\ 0 & 0 & 1 \end{pmatrix}. \tag{1.12.2a}$$

By adopting the *active* interpretation, we can show that (along arbitrary but common axes)

(a) $$\boldsymbol{r}_f = \mathbf{R}(\boldsymbol{k}'', \psi) \cdot \mathbf{R}(\boldsymbol{i}', \theta) \cdot \mathbf{R}(\boldsymbol{k}^o = \boldsymbol{K}, \phi) \cdot \boldsymbol{r}_i = (\mathbf{R}_\psi \cdot \mathbf{R}_\theta \cdot \mathbf{R}_\phi) \cdot \boldsymbol{r}_i, \tag{1.12.3a}$$

(b) $$\boldsymbol{r}_i = (\mathbf{R}_\psi \cdot \mathbf{R}_\theta \cdot \mathbf{R}_\phi)^{\mathrm{T}} \cdot \boldsymbol{r}_f = \mathbf{R}_\phi{}^{\mathrm{T}} \cdot \mathbf{R}_\theta{}^{\mathrm{T}} \cdot \mathbf{R}_\psi{}^{\mathrm{T}} \cdot \boldsymbol{r}_f = (\mathbf{R}_{-\phi} \cdot \mathbf{R}_{-\theta} \cdot \mathbf{R}_{-\psi}) \cdot \boldsymbol{r}_f$$
$$= \left[\mathbf{R}(\boldsymbol{K}, -\phi) \cdot \mathbf{R}(\boldsymbol{i}', -\theta) \cdot \mathbf{R}(\boldsymbol{k}'', -\psi)\right] \cdot \boldsymbol{r}_i; \tag{1.12.3b}$$

while, by adopting the *rotation of a triad* interpretation, we can show that

(a) $$\mathbf{T} = (\mathbf{R}_\phi \cdot \mathbf{R}_\theta \cdot \mathbf{R}_\psi) \cdot \mathbf{t}, \tag{1.12.4a}$$

(b) $$\mathbf{t} = (\mathbf{R}_{-\psi} \cdot \mathbf{R}_{-\theta} \cdot \mathbf{R}_{-\phi}) \cdot \mathbf{T}; \tag{1.12.4b}$$

where $\mathbf{T} = (\boldsymbol{I}, \boldsymbol{J}, \boldsymbol{K})^{\mathrm{T}}$, $\mathbf{t} = (\boldsymbol{i}, \boldsymbol{j}, \boldsymbol{k})^{\mathrm{T}}$.
Next, we prove the following remarkable theorem.

THEOREM (on Compounded Rotations about Body-fixed versus Space-Fixed Axes)

$$\mathbf{R}_\psi \cdot \mathbf{R}_\theta \cdot \mathbf{R}_\phi \equiv \mathbf{R}(\boldsymbol{k}'', \psi) \cdot \mathbf{R}(\boldsymbol{i}', \theta) \cdot \mathbf{R}(\boldsymbol{k}^o = \boldsymbol{K}, \phi)$$
$$= \mathbf{R}(\boldsymbol{K}, \phi) \cdot \mathbf{R}(\boldsymbol{I}, \theta) \cdot \mathbf{R}(\boldsymbol{K}, \psi). \tag{1.12.5a}$$

In words: *the resultant rotation tensor of the classical Eulerian sequence about the body-fixed axes: $\phi(\boldsymbol{k} \equiv \boldsymbol{k}^o = \boldsymbol{K}) \to \theta(\boldsymbol{i}') \to \psi(\boldsymbol{k}'')$, equals the resultant rotation of the reverse-order sequence about the corresponding space-fixed axes: $\psi(\boldsymbol{K}) \to \theta(\boldsymbol{I}) \to \phi(\boldsymbol{K})$.*

(i) To this end, we first prove the following auxiliary theorem.

Shift of the Axis Theorem

Let us consider two concurrent axes of rotation described by the unit vectors $\boldsymbol{n}$ and $\boldsymbol{n}'$, and related by a rotation through an angle μ about a third (also concurrent) axis described by the unit vector $\boldsymbol{m}$; that is,

$$\boldsymbol{n}' = \mathbf{R}(\boldsymbol{m}, \mu) \cdot \boldsymbol{n} = \boldsymbol{n} \cdot \mathbf{R}^{\mathrm{T}}(\boldsymbol{m}, \mu). \tag{1.12.5b}$$

Then, the corresponding rotation tensors about $\boldsymbol{n}$ and $\boldsymbol{n}'$, but with a common angle χ, are related by the tensor-like (or, generally, "similarity") transformation:

$$\mathbf{R}(\boldsymbol{n}', \chi) = \mathbf{R}(\boldsymbol{m}, \mu) \cdot \mathbf{R}(\boldsymbol{n}, \chi) \cdot \mathbf{R}^{\mathrm{T}}(\boldsymbol{m}, \mu). \tag{1.12.5c}$$

PROOF

Applying the rotation formula (1.10.10a) for $\boldsymbol{n} \rightarrow \boldsymbol{n}'$ and χ, we obtain, successively,

$$\mathbf{R}(\boldsymbol{n}', \chi) = \mathbf{R}[\mathbf{R}(\boldsymbol{m}, \mu) \cdot \boldsymbol{n}, \chi]$$

$$= (\cos\chi)\mathbf{1} + (\sin\chi)[\mathbf{R}(\boldsymbol{m}, \mu) \cdot \boldsymbol{n}] \times \mathbf{1}$$
$$+ (1 - \cos\chi)[\mathbf{R}(\boldsymbol{m}, \mu) \cdot \boldsymbol{n}] \otimes [\mathbf{R}(\boldsymbol{m}, \mu) \cdot \boldsymbol{n}]$$

[using the fact that, for any vector, $\boldsymbol{v}$: $(\mathbf{R} \cdot \boldsymbol{v}) \times \mathbf{1} = \mathbf{R} \cdot (\boldsymbol{v} \times \mathbf{1}) \cdot \mathbf{R}^{\mathrm{T}}$ — see proof below]

$$= (\cos\chi)\mathbf{1} + (\sin\chi)[\mathbf{R}(\boldsymbol{m}, \mu) \cdot (\boldsymbol{n} \times \mathbf{1}) \cdot \mathbf{R}^{\mathrm{T}}(\boldsymbol{m}, \mu)]$$
$$+ (1 - \cos\chi)[\mathbf{R}(\boldsymbol{m}, \mu) \cdot (\boldsymbol{n} \otimes \boldsymbol{n}) \cdot \mathbf{R}^{\mathrm{T}}(\boldsymbol{m}, \mu)]$$

[recalling that $\mathbf{R}(\boldsymbol{m}, \mu) \cdot \mathbf{R}^{\mathrm{T}}(\boldsymbol{m}, \mu) = \mathbf{1}$]

$$= \mathbf{R}(\boldsymbol{m}, \mu) \cdot [(\cos\chi)\mathbf{1} + (\sin\chi)(\boldsymbol{n} \times \mathbf{1}) + (1 - \cos\chi)(\boldsymbol{n} \otimes \boldsymbol{n})] \cdot \mathbf{R}^{\mathrm{T}}(\boldsymbol{m}, \mu)$$

$$= \mathbf{R}(\boldsymbol{m}, \mu) \cdot [(\cos\chi)\mathbf{1} + (\sin\chi)\mathbf{N} + (1 - \cos\chi)(\boldsymbol{n} \otimes \boldsymbol{n})] \cdot \mathbf{R}^{\mathrm{T}}(\boldsymbol{m}, \mu)$$

[recalling again (1.10.10a)]

$$= \mathbf{R}(\boldsymbol{m}, \mu) \cdot \mathbf{R}(\boldsymbol{n}, \chi) \cdot \mathbf{R}^{\mathrm{T}}(\boldsymbol{m}, \mu), \quad \text{Q.E.D.} \tag{1.12.5d}$$

[PROOF that $(\mathbf{R} \cdot \boldsymbol{v}) \times \mathbf{1} = \mathbf{R} \cdot (\boldsymbol{v} \times \mathbf{1}) \cdot \mathbf{R}^{\mathrm{T}}$

According to the *passive interpretation*, $\boldsymbol{v}$ and its corresponding antisymmetric tensor $\boldsymbol{V} = \boldsymbol{v} \times \mathbf{1}$ transform as follows:

$$\mathbf{R} \cdot \boldsymbol{v} = \text{components of } \boldsymbol{v} \text{ along the } \textit{old} \text{ axes} \equiv \boldsymbol{v}',$$

$$\mathbf{R} \cdot \boldsymbol{V} \cdot \mathbf{R}^{\mathrm{T}} = \text{components of } \boldsymbol{V} \text{ along the } \textit{old} \text{ axes} \equiv \boldsymbol{V}'.$$

Therefore,

$$(\mathbf{R} \cdot \boldsymbol{v}) \times \mathbf{1} = \boldsymbol{v}' \times \mathbf{1} = \boldsymbol{V}' = \mathbf{R} \cdot \boldsymbol{V} \cdot \mathbf{R}^{\mathrm{T}} = \mathbf{R} \cdot (\boldsymbol{v} \times \mathbf{1}) \cdot \mathbf{R}^{\mathrm{T}}, \quad \text{Q.E.D.]}$$

This theorem allows one to relate the rotation tensors about the initial ($\boldsymbol{n}$) and final (i.e., rotated) ($\boldsymbol{n}'$) positions of a body-fixed axis.

(ii) Now, back to the proof of (1.12.5a). Applying the preceding shift of axis theorem (1.12.5b, c), we get

(a) $$\mathbf{R}(\boldsymbol{k}'', \psi) = \mathbf{R}(\boldsymbol{i}', \theta) \cdot \mathbf{R}(\boldsymbol{k}', \psi) \cdot \mathbf{R}^{\mathrm{T}}(\boldsymbol{i}', \phi), \tag{1.12.5e}$$

where

$$\boldsymbol{k}'' = \mathbf{R}(\boldsymbol{i}', \theta) \cdot \boldsymbol{k}'. \tag{1.12.5f}$$

(b)

$$\mathbf{R}(\boldsymbol{i}', \theta) = \mathbf{R}(\boldsymbol{K}, \phi) \cdot \mathbf{R}(\boldsymbol{I}, \theta) \cdot \mathbf{R}^{\mathrm{T}}(\boldsymbol{K}, \phi), \tag{1.12.5g}$$

$$\Rightarrow \mathbf{R}^{\mathrm{T}}(\boldsymbol{i}', \phi) = \mathbf{R}(\boldsymbol{K}, \phi) \cdot \mathbf{R}^{\mathrm{T}}(\boldsymbol{I}, \theta) \cdot \mathbf{R}^{\mathrm{T}}(\boldsymbol{K}, \phi), \tag{1.12.5h}$$

where

$$\boldsymbol{i}' = \mathbf{R}(\boldsymbol{K}, \phi) \cdot \boldsymbol{I}; \tag{1.12.5i}$$

(c)

$$\mathbf{R}(\boldsymbol{k}', \psi) = \mathbf{R}(\boldsymbol{K}, \phi) \cdot \mathbf{R}(\boldsymbol{K}, \psi) \cdot \mathbf{R}^{\mathrm{T}}(\boldsymbol{K}, \phi), \tag{1.12.5j}$$

where

$$\boldsymbol{k}' = \mathbf{R}(\boldsymbol{K}, \phi) \cdot \boldsymbol{K}. \tag{1.12.5k}$$

Substituting (1.12.5g, h, j) into the right side of (1.12.5e), while recalling that all these **R**'s are orthogonal tensors, yields

$$\begin{aligned} \mathbf{R}(\boldsymbol{k}'', \psi) &= [\mathbf{R}(\boldsymbol{K}, \phi) \cdot \mathbf{R}(\boldsymbol{I}, \theta) \cdot \mathbf{R}^{\mathrm{T}}(\boldsymbol{K}, \phi)] \\ &\quad \cdot [\mathbf{R}(\boldsymbol{K}, \phi) \cdot \mathbf{R}(\boldsymbol{K}, \psi) \cdot \mathbf{R}^{\mathrm{T}}(\boldsymbol{K}, \phi)] \\ &\quad \cdot [\mathbf{R}(\boldsymbol{K}, \phi) \cdot \mathbf{R}^{\mathrm{T}}(\boldsymbol{I}, \theta) \cdot \mathbf{R}^{\mathrm{T}}(\boldsymbol{K}, \phi)] \\ &= \mathbf{R}(\boldsymbol{K}, \phi) \cdot \mathbf{R}(\boldsymbol{I}, \theta) \cdot \mathbf{R}(\boldsymbol{K}, \psi) \cdot \mathbf{R}^{\mathrm{T}}(\boldsymbol{I}, \theta) \cdot \mathbf{R}^{\mathrm{T}}(\boldsymbol{K}, \phi). \end{aligned} \tag{1.12.5l}$$

In view of (1.12.5g) and (1.12.5l), the left side of (1.12.5a) transforms successively to

$$\begin{aligned} &\mathbf{R}(\boldsymbol{k}'', \psi) \cdot \mathbf{R}(\boldsymbol{i}', \theta) \cdot \mathbf{R}(\boldsymbol{K}, \phi) \\ &\quad = \left[\mathbf{R}(\boldsymbol{K}, \phi) \cdot \mathbf{R}(\boldsymbol{I}, \theta) \cdot \mathbf{R}(\boldsymbol{K}, \psi) \cdot \mathbf{R}^{\mathrm{T}}(\boldsymbol{I}, \theta) \cdot \mathbf{R}^{\mathrm{T}}(\boldsymbol{K}, \phi)\right] \\ &\quad \quad \cdot \left[\mathbf{R}(\boldsymbol{K}, \phi) \cdot \mathbf{R}(\boldsymbol{I}, \theta) \cdot \mathbf{R}^{\mathrm{T}}(\boldsymbol{K}, \phi)\right] \cdot \mathbf{R}(\boldsymbol{K}, \phi) \\ &\quad = \mathbf{R}(\boldsymbol{K}, \phi) \cdot \mathbf{R}(\boldsymbol{I}, \theta) \cdot \mathbf{R}(\boldsymbol{K}, \psi), \quad \text{Q.E.D.} \end{aligned} \tag{1.12.5m}$$

Generally, consider a body-fixed frame $O\text{–}xyz$ originally coinciding with the space-fixed frame $O\text{–}XYZ$. Then the sequence of rotations about Ox (*first*, χ_1) $\rightarrow$ Oy (*second*, χ_2) $\rightarrow$ Oz (*third*, χ_3) has the same final orientational effect as the sequence about OZ (*first*, χ_3) $\rightarrow$ OY (*second*, χ_2) $\rightarrow$ OX (*third*, χ_1). [See also Pars, 1965, pp. 103–105.]

Angular Velocity via Eulerian Angle Rates

Let us calculate the vector of angular velocity of the body frame $O\text{–}xyz$ relative to the space frame $O\text{–}XYZ$, in terms of the Eulerian angles ϕ, θ, ψ and their rates $\omega_\phi \equiv d\phi/dt$, $\omega_\theta \equiv d\theta/dt$, $\omega_\psi \equiv d\psi/dt$; both along the body- and the space-fixed axes. We present several approaches.

(i) Geometrical Derivation

By inspection of fig. 1.26 we easily find that

$$\boldsymbol{\omega} = \omega_\phi \boldsymbol{K} + \omega_\theta \boldsymbol{i}' + \omega_\psi \boldsymbol{k}''. \tag{1.12.6a}$$

But, again by inspection, along the *space* basis,

$$\begin{aligned} \boldsymbol{K} &= (0)\boldsymbol{I} + (0)\boldsymbol{J} + (1)\boldsymbol{K}, \\ \boldsymbol{i}' &= (\cos\phi)\boldsymbol{I} + (\sin\phi)\boldsymbol{J} + (0)\boldsymbol{K}, \\ \boldsymbol{k}'' &= (-\sin\theta)\boldsymbol{j}' + (\cos\theta)\boldsymbol{k}' \\ &= (-\sin\theta)[(-\sin\phi)\boldsymbol{I} + (\cos\phi)\boldsymbol{J}] + (\cos\theta)\boldsymbol{K} \\ &= (\sin\theta\sin\phi)\boldsymbol{I} + (-\sin\theta\cos\phi)\boldsymbol{J} + (\cos\theta)\boldsymbol{K}; \end{aligned} \tag{1.12.6b}$$

and along the *body* basis,

$$\boldsymbol{K} = \boldsymbol{j}''(\sin\theta) + \boldsymbol{k}''(\cos\theta) = (\boldsymbol{i}\sin\psi + \boldsymbol{j}\cos\psi)\sin\theta + \boldsymbol{k}\cos\theta,$$
$$\boldsymbol{i}' = \boldsymbol{i}\cos\psi - \boldsymbol{j}\sin\psi, \qquad \boldsymbol{k}'' = \boldsymbol{k}. \tag{1.12.6c}$$

Inserting (1.12.6b, c) in (1.12.6a) and rearranging, we obtain the representations

$$\boldsymbol{\omega} = \omega_X \boldsymbol{I} + \omega_Y \boldsymbol{J} + \omega_Z \boldsymbol{K} = \omega_x \boldsymbol{i} + \omega_y \boldsymbol{j} + \omega_z \boldsymbol{k}, \tag{1.12.7a}$$

where, in matrix form

$$\begin{pmatrix} \omega_X \\ \omega_Y \\ \omega_Z \end{pmatrix} = \begin{pmatrix} 0 & c\phi & s\phi\, s\theta \\ 0 & s\phi & -c\phi\, s\theta \\ 1 & 0 & c\theta \end{pmatrix} \begin{pmatrix} \omega_\phi \\ \omega_\theta \\ \omega_\psi \end{pmatrix} \tag{1.12.7b}$$

Space axes $\qquad \mathbf{E}_{\mathbf{s}(\text{pace})}(\phi, \theta)$ [no ψ-dependence],

$$\begin{pmatrix} \omega_x \\ \omega_y \\ \omega_z \end{pmatrix} = \begin{pmatrix} s\theta\, s\psi & c\psi & 0 \\ s\theta\, c\psi & -s\psi & 0 \\ c\theta & 0 & 1 \end{pmatrix} \begin{pmatrix} \omega_\phi \\ \omega_\theta \\ \omega_\psi \end{pmatrix} \tag{1.12.7c}$$

Body axes $\qquad \mathbf{E}_{\mathbf{b}(\text{ody})}(\theta, \psi)$ [no ϕ-dependence],

Inverting (1.12.7b, c) (noting that, since the axes of $\omega_{\phi,\theta,\psi}$ are *non-orthogonal*, the transformation matrices $\mathbf{E_s}$, $\mathbf{E_b}$ are *nonorthogonal* also; that is, *their inverses do not equal their transposes*), we obtain, respectively,

$$\begin{pmatrix} \omega_\phi \\ \omega_\theta \\ \omega_\psi \end{pmatrix} = (1/\sin\theta) \begin{pmatrix} -s\phi\, c\theta & c\phi\, c\theta & s\theta \\ c\phi\, s\theta & s\phi\, s\theta & 0 \\ s\phi & -c\phi & 0 \end{pmatrix} \begin{pmatrix} \omega_X \\ \omega_Y \\ \omega_Z \end{pmatrix} \tag{1.12.7d}$$

$\mathbf{E_s}^{-1}(\phi, \theta)$,

$$= (1/\sin\theta) \begin{pmatrix} s\psi & c\psi & 0 \\ s\theta\, c\psi & -s\theta\, s\psi & 0 \\ -c\theta\, s\psi & -c\theta\, c\psi & s\theta \end{pmatrix} \begin{pmatrix} \omega_x \\ \omega_y \\ \omega_z \end{pmatrix} \tag{1.12.7e}$$

$\mathbf{E_b}^{-1}(\theta, \psi)$;

from which we can also calculate the $\omega_{X,Y,Z} \Leftrightarrow \omega_{x,y,z}$ (orthogonal!) transformation matrices.

REMARKS

(a) The transformations (1.12.7b–e) readily reveal a serious drawback of the $3 \to 1 \to 3$ Eulerian angle description, for $\theta = 0$ (or $\pm\pi$); that is, when Oz coincides with OZ (or $-OZ$), in which case the nodal line ON disappears, $\sin\theta = 0$, and, so, assuming ϕ, $\psi \neq 0$, eqs. (1.12.7b, c) yield, respectively,

$$\omega_X = (c\phi)\omega_\theta, \qquad \omega_Y = (s\phi)\omega_\theta, \qquad \omega_Z = \omega_\phi + \omega_\psi \Rightarrow \omega_X{}^2 + \omega_Y{}^2 = \omega_\theta{}^2; \quad (1.12.7\text{f})$$

$$\omega_x = (-c\psi)\omega_\theta, \qquad \omega_y = (-s\psi)\omega_\theta, \qquad \omega_z = \omega_\phi + \omega_\psi \Rightarrow \omega_x{}^2 + \omega_y{}^2 = \omega_\theta{}^2; \quad (1.12.7\text{g})$$

which means that knowing $\omega_{X,Y,Z/x,y,z}(t)$ [say, after solving the kinetic Eulerian equations (§1.17)], we can determine ω_θ *uniquely*, but not ω_ϕ and ω_ψ!

Actually, *all* twelve generalized Eulerian angle descriptions mentioned earlier, $\chi_1 \to \chi_2 \to \chi_3$, exhibit such *singularities* for some value(s) of their *second* rotation angle χ_2; in which case, *the planes of the other two angles become indistinguishable*! From the numerical viewpoint, this means that in the close neighborhood of these values of χ_2, it becomes difficult to integrate for the rates $d\chi_k/dt$ $(k = 1, 2, 3)$. This is the main reason that, in rotational (or "attitude") rigid-body dynamics, (singularity free) *four-parameter formalisms* are sought, and the reason that the classical Eulerian sequence $3 \to 1 \to 3$ has been of much use in astronomy (where x, y, z have origin at the center of the Earth, and point to three distant stars) and physics; whereas other Eulerian sequences, such as $1 \to 2 \to 3$ or $3 \to 2 \to 1$ [associated with the names of Cardan (1501–1576) (continental European literature), Tait (1869), Bryan (1911) (British literature); and examined below] are more preferable in engineering rigid-body dynamics; for example, airplanes, ships, railroads, satellites, and so on. [Similarly, the position $(\phi, \theta, \psi) = (0, 0, 0)$ represents a singular "gimbal lock": the motions ω_ϕ and ω_ψ are indistinguishable since each is about the vertical axis Z; only $\omega_\phi + \omega_\psi$ is known. The ω_θ motion is about the X-axis, and so it is impossible to represent rotations about the Y-axis; it is "locked out"; that is (0, 0, 0) introduces artificially a constraint, $\omega_Y = 0$, $\omega_y = 0$ that *mechanically* is not there (then, $\omega_X = \omega_\theta$, $\omega_Y = 0$, $\omega_Z = \omega_\phi + \omega_\psi$; $\omega_x = \omega_\theta$, $\omega_z = \omega_\phi + \omega_\psi$). These remarks are due to Professor D. T. Greenwood, private communication.]

(b) Equations (1.12.7b, c) also show that the components $\omega_{X,Y,Z/x,y,z}$ are *quasi* or *nonholonomic velocities*; that is, although they are linear and homogeneous combinations of the Eulerian angle rates $\omega_\phi \equiv d\phi/dt$, $\omega_\theta \equiv d\theta/dt$, $\omega_\psi \equiv d\psi/dt$, they do not equal the rates of other angles. Indeed, if, for example, $\omega_X = d\chi_X/dt$, where $\chi_X = \chi_X(\phi, \theta, \psi)$, then we should have

$$\begin{aligned} d\chi_X/dt &= (\partial\chi_X/\partial\phi)(d\phi/dt) + (\partial\chi_X/\partial\theta)(d\theta/dt) + (\partial\chi_X/\partial\psi)(d\psi/dt) \\ &= (0)(d\phi/dt) + (c\phi)(d\theta/dt) + (s\phi\, s\theta)(d\psi/dt) \qquad [\text{by } (1.12.7\text{b})], \end{aligned} \quad (1.12.7\text{h})$$

that is,

$$\partial\chi_X/\partial\phi = 0, \qquad \partial\chi_X/\partial\theta = c\theta, \qquad \partial\chi_X/\partial\psi = s\phi\, s\theta. \quad (1.12.7\text{i})$$

But, from (1.12.7i), it follows that, *in general*,

$$\partial/\partial\theta(\partial\chi_X/\partial\phi) = 0 \neq \partial/\partial\phi(\partial\chi_X/\partial\theta) = -s\phi. \quad (1.12.7\text{j})$$

Hence, no such χ_X exists; and similarly for the other ω's. (An introduction to quasi coordinates is given in §1.14; and a detailed treatment is given in chap. 2.)

(ii) Passive Interpretation Derivation

(a) Body-fixed axes representation. Since $\boldsymbol{\omega}$ is a vector, we can express it as the sum of its three *Eulerian angular velocities*:

$$\boldsymbol{\omega} = \boldsymbol{\omega}_\phi + \boldsymbol{\omega}_\theta + \boldsymbol{\omega}_\psi, \tag{1.12.8a}$$

where

$$\boldsymbol{\omega}_\phi = (d\phi/dt)\boldsymbol{K}, \qquad \boldsymbol{\omega}_\theta = (d\theta/dt)\boldsymbol{i}', \qquad \boldsymbol{\omega}_\psi = (d\psi/dt)\boldsymbol{k}''. \tag{1.12.8b}$$

Then, using the passive interpretation, (1.11.4h, 7a ff.), we can express (1.12.8a, b) along the (new) *body* axes basis $(\boldsymbol{i}, \boldsymbol{j}, \boldsymbol{k})$. Since the Eulerian basis $(\boldsymbol{K}, \boldsymbol{i}', \boldsymbol{k}'')$ is *non-orthogonal*, we carry out this transformation, not for the entire $\boldsymbol{\omega}$, but for each of its above components $\boldsymbol{\omega}_\phi, \boldsymbol{\omega}_\theta, \boldsymbol{\omega}_\psi$, and then, adding the results, we obtain

$$\boldsymbol{\omega}_{\phi,\,body\text{ components}} = \mathbf{R}_\psi{}^{\mathrm{T}} \cdot \mathbf{R}_\theta{}^{\mathrm{T}} \cdot \begin{pmatrix} 0 \\ 0 \\ \omega_\phi \end{pmatrix}_{(\boldsymbol{I}\,\boldsymbol{J}\,\boldsymbol{K})} = \mathbf{R}_{-\psi} \cdot \mathbf{R}_{-\theta} \cdot \begin{pmatrix} 0 \\ 0 \\ \omega_\phi \end{pmatrix}$$

$$= \begin{pmatrix} c\psi & s\psi & 0 \\ -s\psi & c\psi & 0 \\ 0 & 0 & 1 \end{pmatrix} \begin{pmatrix} 1 & 0 & 0 \\ 0 & c\theta & s\theta \\ 0 & -s\theta & c\theta \end{pmatrix} \begin{pmatrix} 0 \\ 0 \\ \omega_\phi \end{pmatrix} = \begin{pmatrix} (s\theta\, s\psi)\omega_\phi \\ (s\theta\, c\psi)\omega_\phi \\ (c\theta)\omega_\phi \end{pmatrix}, \tag{1.12.8c}$$

$$\boldsymbol{\omega}_{\theta,\,body\text{ components}} = \mathbf{R}_\psi{}^{\mathrm{T}} \cdot \begin{pmatrix} \omega_\theta \\ 0 \\ 0 \end{pmatrix}_{(\boldsymbol{i}'\,\boldsymbol{j}'\,\boldsymbol{k}')} = \mathbf{R}_{-\psi} \cdot \begin{pmatrix} \omega_\theta \\ 0 \\ 0 \end{pmatrix}$$

$$= \begin{pmatrix} c\psi & s\psi & 0 \\ -s\psi & c\psi & 0 \\ 0 & 0 & 1 \end{pmatrix} \begin{pmatrix} \omega_\theta \\ 0 \\ 0 \end{pmatrix} = \begin{pmatrix} (c\psi)\omega_\theta \\ (-s\psi)\omega_\theta \\ (0)\omega_\theta \end{pmatrix}, \tag{1.12.8d}$$

$$\boldsymbol{\omega}_{\psi,\,body\text{ components}} = \begin{pmatrix} 0 \\ 0 \\ \omega_\psi \end{pmatrix}_{(\boldsymbol{i}\,\boldsymbol{j}\,\boldsymbol{k})}. \tag{1.12.8e}$$

Adding (1.12.8c–e), we obtain the body axes components, equations (1.12.7c), as expected.

(b) Space-fixed axes representation. Proceeding similarly, we find

$$\omega = \begin{pmatrix} 0 \\ 0 \\ \omega_\phi \end{pmatrix} + \underset{\mathbf{R}_\phi}{\begin{pmatrix} c\phi & -s\phi & 0 \\ s\phi & c\phi & 0 \\ 0 & 0 & 1 \end{pmatrix}} \begin{pmatrix} \omega_\theta \\ 0 \\ 0 \end{pmatrix}$$

$$+ \underset{\mathbf{R}_\phi}{\begin{pmatrix} c\phi & -s\phi & 0 \\ s\phi & c\phi & 0 \\ 0 & 0 & 1 \end{pmatrix}} \underset{\mathbf{R}_\theta}{\begin{pmatrix} 1 & 0 & 0 \\ 0 & c\theta & -s\theta \\ 0 & s\theta & c\theta \end{pmatrix}} \begin{pmatrix} 0 \\ 0 \\ \omega_\psi \end{pmatrix}$$

$$= \begin{pmatrix} 0 \\ 0 \\ 1 \end{pmatrix} \omega_\phi + \begin{pmatrix} c\phi \\ s\phi \\ 0 \end{pmatrix} \omega_\theta + \begin{pmatrix} s\phi\, s\theta \\ -c\phi\, s\theta \\ c\theta \end{pmatrix} \omega_\psi, \tag{1.12.8f}$$

which is none other than (1.12.7b).

Let the reader verify that the *space*-axes representation (1.12.8f) can also be rewritten as

$$\omega = \mathbf{R}_\phi \cdot \begin{pmatrix} 0 \\ 0 \\ \omega_\phi \end{pmatrix} + \mathbf{R}_\phi \cdot \mathbf{R}_\theta \cdot \begin{pmatrix} \omega_\theta \\ 0 \\ 0 \end{pmatrix} + \mathbf{R}_\phi \cdot \mathbf{R}_\theta \cdot \mathbf{R}_\psi \cdot \begin{pmatrix} 0 \\ 0 \\ \omega_\psi \end{pmatrix}, \tag{1.12.8g}$$

while the *body*-axes representation (1.12.8c–e) can be rewritten as

$$\omega = \mathbf{R}_{-\psi} \cdot \mathbf{R}_{-\theta} \cdot \mathbf{R}_{-\phi} \cdot \begin{pmatrix} 0 \\ 0 \\ \omega_\phi \end{pmatrix} + \mathbf{R}_{-\psi} \cdot \mathbf{R}_{-\theta} \cdot \begin{pmatrix} \omega_\theta \\ 0 \\ 0 \end{pmatrix} + \mathbf{R}_{-\psi} \cdot \begin{pmatrix} 0 \\ 0 \\ \omega_\psi \end{pmatrix}. \tag{1.12.8h}$$

(iii) Tensor (Matrix) Derivation

We have already seen [(1.7.27e) and (1.7.30i–k)] that the space-axes components of the angular velocity tensor (vector) $\boldsymbol{\omega}'(\omega')$ are related to its body-axes components $\boldsymbol{\omega}(\omega)$ by the tensor (vector) transformation

$$\boldsymbol{\omega}' = \mathbf{R} \cdot \boldsymbol{\omega} \cdot \mathbf{R}^{\mathrm{T}} \Leftrightarrow \boldsymbol{\omega} = \mathbf{R}^{\mathrm{T}} \cdot \boldsymbol{\omega}' \cdot \mathbf{R}$$
$$(\omega' = \mathbf{R} \cdot \omega \Leftrightarrow \omega = \mathbf{R}^{\mathrm{T}} \cdot \omega'), \tag{1.12.9a}$$

where $\mathbf{R}$, or $\mathbf{A}$, is the matrix of the direction cosines between these axes; and also that

$$\boldsymbol{\omega}' = (\mathrm{d}\mathbf{R}/\mathrm{dt}) \cdot \mathbf{R}^{\mathrm{T}} = -\mathbf{R} \cdot (\mathrm{d}\mathbf{R}/\mathrm{dt})^{\mathrm{T}} \qquad [\text{due to } \mathrm{d}/\mathrm{dt}(\mathbf{R} \cdot \mathbf{R}^{\mathrm{T}}) = \mathrm{d}\mathbf{1}/\mathrm{dt} = \mathbf{0}]$$
$$\boldsymbol{\omega} = \mathbf{R}^{\mathrm{T}} \cdot (\mathrm{d}\mathbf{R}/\mathrm{dt}) = -(\mathrm{d}\mathbf{R}/\mathrm{dt})^{\mathrm{T}} \cdot \mathbf{R},$$
$$\mathrm{d}\mathbf{R}'/\mathrm{dt} = \boldsymbol{\omega}' \cdot \mathbf{R} \quad [= (\mathbf{R} \cdot \boldsymbol{\omega} \cdot \mathbf{R}^{\mathrm{T}}) \cdot \mathbf{R}] = \mathbf{R} \cdot \boldsymbol{\omega}. \tag{1.12.9b}$$

(a) Space-fixed axes representation. As we have seen, in the case of the classical Eulerian sequence $\phi \to \theta \to \psi$: $\mathbf{R} \equiv \mathbf{R}_\phi \cdot \mathbf{R}_\theta \cdot \mathbf{R}_\psi$, and therefore (1.12.9b) yields, successively,

$$\begin{aligned}
\boldsymbol{\omega}' &= (d\mathbf{R}/dt) \cdot \mathbf{R}^T = d/dt(\mathbf{R}_\phi \cdot \mathbf{R}_\theta \cdot \mathbf{R}_\psi) \cdot (\mathbf{R}_\phi \cdot \mathbf{R}_\theta \cdot \mathbf{R}_\psi)^T \\
&= [(d\mathbf{R}_\phi/dt) \cdot \mathbf{R}_\theta \cdot \mathbf{R}_\psi + \mathbf{R}_\phi \cdot (d\mathbf{R}_\theta/dt) \cdot \mathbf{R}_\psi + \mathbf{R}_\phi \cdot \mathbf{R}_\theta \cdot (d\mathbf{R}_\psi/dt)] \cdot (\mathbf{R}_\psi{}^T \cdot \mathbf{R}_\theta{}^T \cdot \mathbf{R}_\phi{}^T) \\
&= (d\mathbf{R}_\phi/dt) \cdot [\mathbf{R}_\theta \cdot (\mathbf{R}_\psi \cdot \mathbf{R}_\psi{}^T) \cdot \mathbf{R}_\theta{}^T] \cdot \mathbf{R}_\phi{}^T \\
&\qquad + \mathbf{R}_\phi \cdot [(d\mathbf{R}_\theta/dt) \cdot (\mathbf{R}_\psi \cdot \mathbf{R}_\psi{}^T) \cdot \mathbf{R}_\theta{}^T] \cdot \mathbf{R}_\phi{}^T \\
&\qquad + \mathbf{R}_\phi \cdot \{\mathbf{R}_\theta \cdot [(d\mathbf{R}_\psi/dt) \cdot \mathbf{R}_\psi{}^T] \cdot \mathbf{R}_\theta{}^T\} \cdot \mathbf{R}_\phi{}^T \\
&= (d\mathbf{R}_\phi/dt) \cdot \mathbf{R}_\phi{}^T + \mathbf{R}_\phi \cdot [(d\mathbf{R}_\theta/dt) \cdot \mathbf{R}_\theta{}^T] \cdot \mathbf{R}_\phi{}^T + \mathbf{R}_\phi \cdot \mathbf{R}_\theta \cdot [(d\mathbf{R}_\psi/dt) \cdot \mathbf{R}_\psi{}^T] \cdot (\mathbf{R}_\phi \cdot \mathbf{R}_\theta)^T
\end{aligned}$$

[recalling the definition of tensor transformation (1.12.9a), and (1.12.9b)], (1.12.9c)

$$= \boldsymbol{\omega}'_\phi + \mathbf{R}_\phi \cdot \boldsymbol{\omega}'_\theta \cdot \mathbf{R}_\phi{}^T + \mathbf{R}_\phi \cdot \mathbf{R}_\theta \cdot \boldsymbol{\omega}'_\psi \cdot (\mathbf{R}_\phi \cdot \mathbf{R}_\theta)^T$$

[$\boldsymbol{\omega}'_{\phi,\theta,\psi}$: "partial" rotation tensors, along the space-fixed axes], (1.12.9d)

from which, after some long but straightforward algebra, we obtain [recalling (1.12.1a ff.)]

$$\begin{aligned}
&\omega_{1'1'} \equiv \omega_{XX} = 0, \\
&\omega_{1'2'} = -\omega_{2'1'} \equiv \omega_{XY} = -\omega_{YX} = -\omega_Z = -\big[d\phi/dt + (c\theta)(d\psi/dt)\big], \\
&\omega_{1'3'} = -\omega_{3'1'} \equiv \omega_{XZ} = -\omega_{ZX} = \omega_Y = (s\phi)(d\theta/dt) - (c\phi\, s\theta)(d\psi/dt), \\
&\omega_{2'2'} \equiv \omega_{YY} = 0, \\
&\omega_{2'3'} = -\omega_{3'2'} \equiv \omega_{YZ} = -\omega_{ZY} = -\omega_X = -\big[(c\phi)(d\theta/dt) + (s\phi\, s\theta)(d\psi/dt)\big], \\
&\omega_{3'3'} \equiv \omega_{ZZ} = 0, \qquad (1.12.9e)
\end{aligned}$$

which coincide with (1.12.7b), as expected.

(b) Body-fixed axes representation. Proceeding analogously, we obtain

$$\begin{aligned}
\boldsymbol{\omega} &= \mathbf{R}^T \cdot (d\mathbf{R}/dt) = (\mathbf{R}_\phi \cdot \mathbf{R}_\theta \cdot \mathbf{R}_\psi)^T \cdot [d/dt(\mathbf{R}_\phi \cdot \mathbf{R}_\theta \cdot \mathbf{R}_\psi)] \\
&= \cdots = \mathbf{R}_\psi{}^T \cdot \mathbf{R}_\theta{}^T \cdot [\mathbf{R}_\phi{}^T \cdot (d\mathbf{R}_\phi/dt)] \cdot \mathbf{R}_\theta \cdot \mathbf{R}_\psi \\
&\qquad + \mathbf{R}_\psi{}^T \cdot [\mathbf{R}_\theta{}^T \cdot (d\mathbf{R}_\theta/dt)] \cdot \mathbf{R}_\psi + \mathbf{R}_\psi{}^T \cdot (d\mathbf{R}_\psi/dt) \\
&\equiv \mathbf{R}_\psi{}^T \cdot \mathbf{R}_\theta{}^T \cdot \boldsymbol{\omega}_\phi \cdot \mathbf{R}_\theta \cdot \mathbf{R}_\phi + \mathbf{R}_\psi{}^T \cdot \boldsymbol{\omega}_\theta \cdot \mathbf{R}_\psi + \boldsymbol{\omega}_\psi
\end{aligned}$$

[$\boldsymbol{\omega}_{\phi,\theta,\psi}$: "partial" rotation tensors, along the body-fixed axes]. (1.12.9f)

We leave it to the reader to verify that the above coincides with (1.12.7c). Alternatively, one can use the transformation equations (1.12.9a) to calculate $\boldsymbol{\omega}/\omega$ from $\boldsymbol{\omega}'/\omega'$. (See also Hamel, 1949, pp. 735–739.)

Cardanian Angles

This is the Eulerian rotation sequence $3 \to 2 \to 1$ (fig. 1.27). The angles $\chi_1 = \gamma(3) \to \chi_2 = \beta(2) \to \chi_3 = \alpha(1)$ are commonly (but not uniformly) referred to as *Cardanian*

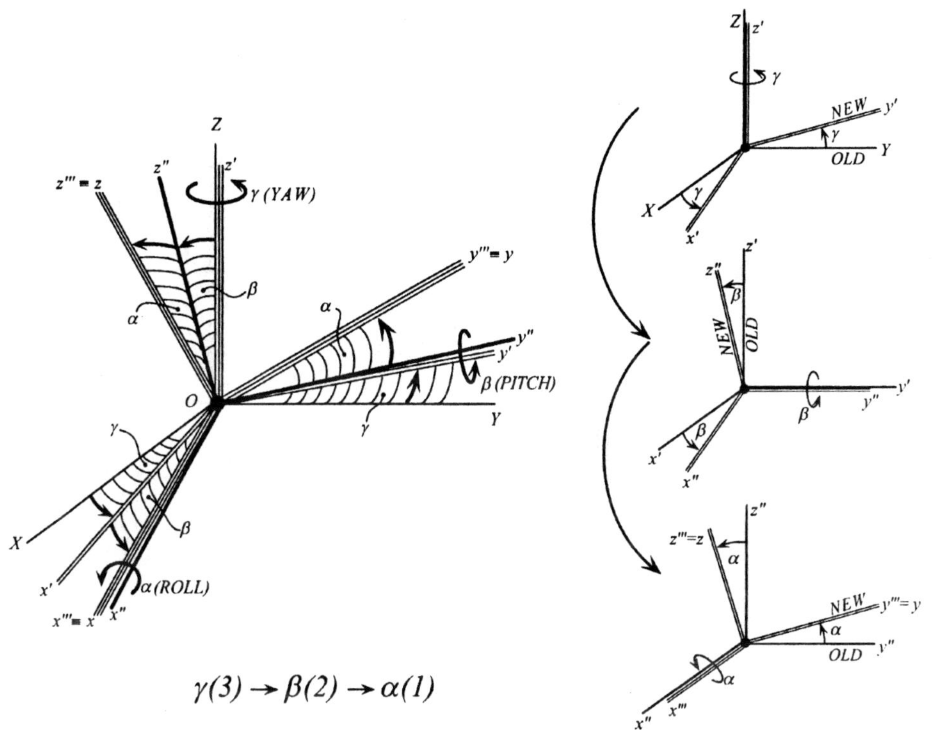

Figure 1.27 Cardanian angles: $\chi_1 = \gamma(3) \rightarrow \chi_2 = \beta(2) \rightarrow \chi_3 = \alpha(1)$.
(i) Rotation (OZ, $\chi_1 = \gamma$): O–XYZ (space axes) = O–$x_o y_o z_o$ (initial body axes) → O–$x'y'z'$.
(ii) Rotation (Oy', $\chi_2 = \beta$): O–$x'y'z'$ → O–$x''y''z''$.
(iii) Rotation (Ox'', $\chi_3 = \alpha$): O–$x''y''z''$ → O–xyz (final body axes).

angles. In vehicle and aeronautical dynamics, where such an attitude representation is popular, they are called *yaw* (γ), *pitch* (β), and *roll* (α).

Following the passive interpretation, we readily obtain

$$\begin{pmatrix} X \\ Y \\ Z \end{pmatrix} = \mathbf{R} \cdot \begin{pmatrix} x \\ y \\ z \end{pmatrix} = \mathbf{R}_\gamma \cdot \left\{ \mathbf{R}_\beta \cdot \left[\mathbf{R}_\alpha \cdot \begin{pmatrix} x \\ y \\ z \end{pmatrix} \right] \right\}$$

$$= \underset{\mathbf{R}_\gamma}{\begin{pmatrix} c\gamma & -s\gamma & 0 \\ s\gamma & c\gamma & 0 \\ 0 & 0 & 1 \end{pmatrix}} \cdot \underset{\mathbf{R}_\beta}{\begin{pmatrix} c\beta & 0 & s\beta \\ 0 & 1 & 0 \\ -s\beta & 0 & c\beta \end{pmatrix}} \cdot \underset{\mathbf{R}_\alpha,}{\begin{pmatrix} 1 & 0 & 0 \\ 0 & c\alpha & -s\alpha \\ 0 & s\alpha & c\alpha \end{pmatrix}} \begin{pmatrix} x \\ y \\ z \end{pmatrix} \tag{1.12.10a}$$

$$= \begin{pmatrix} c\beta\, c\gamma & s\alpha\, s\beta\, c\gamma - c\alpha\, s\gamma & c\alpha\, s\beta\, c\gamma + s\alpha\, s\gamma \\ c\beta\, s\gamma & s\alpha\, s\beta\, s\gamma + c\alpha\, c\gamma & c\alpha\, s\beta\, s\gamma - s\alpha\, c\gamma \\ -s\beta & s\alpha\, c\beta & c\alpha\, c\beta \end{pmatrix} \begin{pmatrix} x \\ y \\ z \end{pmatrix}, \tag{1.12.10b}$$

and, inversely, since $\mathbf{R}_{\alpha,\beta,\gamma}$ are proper orthogonal,

$$\begin{pmatrix} x \\ y \\ z \end{pmatrix} = \mathbf{R}^{\mathrm{T}} \cdot \begin{pmatrix} X \\ Y \\ Z \end{pmatrix} = (\mathbf{R}_\alpha{}^{\mathrm{T}} \cdot \mathbf{R}_\beta{}^{\mathrm{T}} \cdot \mathbf{R}_\gamma{}^{\mathrm{T}}) \cdot \begin{pmatrix} X \\ Y \\ Z \end{pmatrix}$$

$$= (\mathbf{R}_{-\alpha} \cdot \mathbf{R}_{-\beta} \cdot \mathbf{R}_{-\gamma}) \cdot \begin{pmatrix} X \\ Y \\ Z \end{pmatrix}. \qquad (1.12.10c)$$

Angular Velocity Tensors

Using the basic relations (1.12.9a, b), we can show, after some long and careful but straightforward algebra, that (with $\omega_\gamma \equiv d\gamma/dt$, $\omega_\beta \equiv d\beta/dt$, $\omega_\alpha \equiv d\alpha/dt$)

$$\begin{pmatrix} \omega_X \\ \omega_Y \\ \omega_Z \end{pmatrix} = \begin{pmatrix} 0 & -s\gamma & c\gamma\, c\beta \\ 0 & c\gamma & s\gamma\, c\beta \\ 1 & 0 & -s\beta \end{pmatrix} \begin{pmatrix} \omega_\gamma \\ \omega_\beta \\ \omega_\alpha \end{pmatrix} \qquad (1.12.10d)$$

Space axes $\qquad \mathbf{E}_{\mathbf{s}(\mathrm{pace})}(\gamma, \beta) \qquad$ [no α-dependence],

$$\begin{pmatrix} \omega_x \\ \omega_y \\ \omega_z \end{pmatrix} = \begin{pmatrix} -s\beta & 0 & 1 \\ c\beta\, s\alpha & c\alpha & 0 \\ c\alpha\, c\beta & -s\alpha & 0 \end{pmatrix} \begin{pmatrix} \omega_\gamma \\ \omega_\beta \\ \omega_\alpha \end{pmatrix} \qquad (1.12.10e)$$

Body axes $\qquad \mathbf{E}_{\mathbf{b}(\mathrm{ody})}(\beta, \alpha) \qquad$ [no γ-dependence].

Inverting (1.12.10d, e) (noting that, since the axes of $\omega_{\alpha,\beta,\gamma}$ are *non-orthogonal*, the transformation matrices $\mathbf{E}_\mathbf{s}(\gamma,\beta)$, $\mathbf{E}_\mathbf{b}(\beta,\alpha)$ are *nonorthogonal* also; that is, *their inverses do not equal their transposes*), we obtain respectively,

$$\begin{pmatrix} \omega_\gamma \\ \omega_\beta \\ \omega_\alpha \end{pmatrix} = (1/\cos\beta) \begin{pmatrix} s\beta\, c\gamma & s\beta\, s\gamma & c\beta \\ -s\gamma\, c\beta & c\gamma\, c\beta & 0 \\ c\gamma & s\gamma & 0 \end{pmatrix} \begin{pmatrix} \omega_X \\ \omega_Y \\ \omega_Z \end{pmatrix} \qquad (1.12.10f)$$

$$\mathbf{E}_\mathbf{s}{}^{-1}(\beta, \gamma),$$

$$= (1/\cos\beta) \begin{pmatrix} 0 & s\alpha & c\alpha \\ 0 & c\alpha\, c\beta & -s\alpha\, c\beta \\ c\beta & s\beta\, s\alpha & s\beta\, c\alpha \end{pmatrix} \begin{pmatrix} \omega_x \\ \omega_y \\ \omega_z \end{pmatrix} \qquad (1.12.10g)$$

$$\mathbf{E}_\mathbf{b}{}^{-1}(\alpha, \beta);$$

from which it is clear that the Cardanian sequence $3(\gamma) \to 2(\beta) \to 1(\alpha)$ has a singularity at $\beta = \pm(\pi/2)$. There, (1.12.10d, e) become, respectively (for $\beta = \pi/2$),

$$\omega_X = (-s\gamma)\omega_\beta, \quad \omega_Y = (c\gamma)\omega_\beta, \quad \omega_Z = \omega_\gamma - \omega_\alpha \Rightarrow \omega_X{}^2 + \omega_Y{}^2 = \omega_\beta{}^2, \quad (1.12.10h)$$

$$\omega_x = -\omega_\gamma + \omega_\alpha, \quad \omega_y = (c\alpha)\omega_\beta, \quad \omega_z = (-s\alpha)\omega_\beta \Rightarrow \omega_x{}^2 + \omega_y{}^2 = \omega_\beta{}^2; \quad (1.12.10i)$$

that is, a unique determination of ω_γ and ω_α from ω_Z, or ω_x, is impossible.

Finally, using (1.12.10f, g), we can obtain the $\omega_{X,Y,Z} \leftrightarrow \omega_{x,y,z}$ transformation. For a complete listing of the transformations between $\omega_{x,y,z} \equiv \omega_{1,2,3}$ (body-fixed axes) and the Eulerian rates $d\chi_{1,2,3}/dt \equiv \nu_{1,2,3}$ (and corresponding singularities), for all body-/space-axis Eulerian rotation sequences, see the next section.

1.13 THE RIGID BODY: TRANSFORMATION MATRICES (DIRECTION COSINES) BETWEEN SPACE-FIXED AND BODY-FIXED TRIADS; AND ANGULAR VELOCITY COMPONENTS ALONG BODY-FIXED AXES, FOR ALL SEQUENCES OF EULERIAN ANGLES

Summary of Theory, Notations

$$\mathbf{T} = (\boldsymbol{u}_{k'})^{\mathrm{T}} \equiv (\boldsymbol{u}_{1'}, \boldsymbol{u}_{2'}, \boldsymbol{u}_{3'})^{\mathrm{T}} \equiv (\boldsymbol{I}, \boldsymbol{J}, \boldsymbol{K})^{\mathrm{T}}: \ \textit{Space-fixed (fixed) triad.}$$

$$\mathbf{t} = (\boldsymbol{u}_{k})^{\mathrm{T}} \equiv (\boldsymbol{u}_{1}, \boldsymbol{u}_{2}, \boldsymbol{u}_{3})^{\mathrm{T}} \equiv (\boldsymbol{i}, \boldsymbol{j}, \boldsymbol{k})^{\mathrm{T}}: \ \textit{Body-fixed (moving) triad.}$$

All triads are assumed *ortho–normal–dextral* (OND), and such that, initially, $\mathbf{T} = \mathbf{t}$. Eulerian angles (see §1.12): χ_1, χ_2, χ_3 (the earlier ϕ, θ, ψ; or α, β, γ).

1. Basic Triad Transformation Formula

$$\mathbf{T} = \mathbf{R} \cdot \mathbf{t} \ \Leftrightarrow \ \mathbf{t} = \mathbf{R}^{\mathrm{T}} \cdot \mathbf{T},$$

where

$$\mathbf{R} = (R_{k'k}) \equiv (\boldsymbol{u}_{k'} \cdot \boldsymbol{u}_k) \qquad [\text{or } (A_{k'k})]: \ \textit{Tensor/Matrix of rotation}$$
$$= \mathbf{R}(\boldsymbol{u}_i, \chi_1) \cdot \mathbf{R}(\boldsymbol{u}_j, \chi_2) \cdot \mathbf{R}(\boldsymbol{u}_k, \chi_3) \equiv [i(\chi_1), j(\chi_2), k(\chi_3)]$$

[Rotation sequence $\chi_1 \to \chi_2 \to \chi_3$ about the *body*-fixed axes $\boldsymbol{u}_i \to \boldsymbol{u}_j \to \boldsymbol{u}_k$]

$$= \mathbf{R}(\boldsymbol{u}_{k'}, \chi_3) \cdot \mathbf{R}(\boldsymbol{u}_{j'}, \chi_2) \cdot \mathbf{R}(\boldsymbol{u}_{i'}, \chi_1) \equiv [k'(\chi_3), j'(\chi_2), i'(\chi_1)]$$

[Rotation sequence $\chi_3 \to \chi_2 \to \chi_1$ about the *space*-fixed axes $\boldsymbol{u}_{k'} \to \boldsymbol{u}_{j'} \to \boldsymbol{u}_{i'}$]

$$[i, j, k = 1, 2, 3; \ i', j', k' = 1', 2', 3'];$$

and, by the basic theorem on compounded rotations (§1.12), the *inverse* rotation

$$\mathbf{R}^{-1} = \mathbf{R}^{\mathrm{T}} = \mathbf{R}(\boldsymbol{u}_k, -\chi_3) \cdot \mathbf{R}(\boldsymbol{u}_j, -\chi_2) \cdot \mathbf{R}(\boldsymbol{u}_i, \chi_1)$$
$$= \mathbf{R}(\boldsymbol{u}_{i'}, -\chi_1) \cdot \mathbf{R}(\boldsymbol{u}_{j'}, -\chi_2) \cdot \mathbf{R}(\boldsymbol{u}_{k'}, -\chi_3)$$

returns the body-triad $\mathbf{t}$ to its original position, i.e. realigns it with the space-triad $\mathbf{T}$.

How to obtain *space-axis* rotations; i.e., $[k'(\chi_1), j'(\chi_2), i'(\chi_3)]$, from a knowledge of *body-axis* rotations with the *same rotation sequence*: $\chi_1 \to \chi_2 \to \chi_3$; i.e., from $[i(\chi_1), j(\chi_2), k(\chi_3)]$, and vice versa. An example should suffice; by the above theorem, we will have

$$[2(\chi_1), 3(\chi_2), 1(\chi_3)] = [1'(\chi_3), 3'(\chi_2), 2'(\chi_1)]$$

and, therefore, swapping in the latter χ_3 with χ_1 (and vice versa), we obtain $[1'(\chi_1), 3'(\chi_2), 2'(\chi_3)]$, which appears in the listing below. Similarly, we have

$$[2'(\chi_1), 3'(\chi_2), 1'(\chi_3)] = [1(\chi_3), 3(\chi_2), 2(\chi_1)]$$

and swapping in there χ_3 with χ_1 (and vice versa) we obtain $[1(\chi_1), 3(\chi_2), 2(\chi_3)]$. Abbreviations: $s_i(\ldots) \equiv \sin(\chi_i)$, $c_i(\ldots) \equiv \cos(\chi_i)$.

2. Angular Velocity Components

Body-fixed (moving) axes components:

$$\boldsymbol{\omega} = \mathbf{R}^{\mathrm{T}} \cdot (\mathrm{d}\mathbf{R}/\mathrm{dt}) = -(\mathrm{d}\mathbf{R}/\mathrm{dt})^{\mathrm{T}} \cdot \mathbf{R}, \qquad [\text{due to } \mathrm{d}/\mathrm{dt}(\mathbf{R} \cdot \mathbf{R}^{\mathrm{T}}) = \mathrm{d}\mathbf{1}/\mathrm{dt} = \mathbf{0}]$$

Space-fixed (fixed) axes components:

$$\boldsymbol{\omega}' = (\mathrm{d}\mathbf{R}/\mathrm{dt}) \cdot \mathbf{R}^{\mathrm{T}} = -\mathbf{R} \cdot (\mathrm{d}\mathbf{R}/\mathrm{dt})^{\mathrm{T}};$$

with mutual transformations:

$$\boldsymbol{\omega}' = \mathbf{R} \cdot \boldsymbol{\omega} \cdot \mathbf{R}^{\mathrm{T}} \Leftrightarrow \boldsymbol{\omega} = \mathbf{R}^{\mathrm{T}} \cdot \boldsymbol{\omega}' \cdot \mathbf{R}, \qquad \omega' = \mathbf{R} \cdot \omega \Leftrightarrow \omega = \mathbf{R}^{\mathrm{T}} \cdot \omega'$$

where $\omega = \textit{axial}$ vector of $\boldsymbol{\omega}$, $\omega' = \textit{axial}$ vector of $\boldsymbol{\omega}'$

[i.e. $\boldsymbol{\omega} \cdot (\textit{vector}) = \omega \times (\textit{vector})$, etc.];

Rotation tensor derivative:

$$\mathrm{d}\mathbf{R}/\mathrm{dt} = \boldsymbol{\omega}' \cdot \mathbf{R} \quad [= (\mathbf{R} \cdot \boldsymbol{\omega} \cdot \mathbf{R}^{\mathrm{T}}) \cdot \mathbf{R}] \quad = \mathbf{R} \cdot \boldsymbol{\omega}.$$

Listing of Transformation Matrices; and Angular Velocity Components

(Body-fixed vs. Eulerian rates; and corresponding singularities. Notation: $d\chi_{1,2,3}/dt \equiv v_{1,2,3}$)

1(a) $[1(\chi_1), 2(\chi_2), 3(\chi_3)] = [3'(\chi_3), 2'(\chi_2), 1'(\chi_1)]$ [Singularity at $\chi_2 = \pm(\pi/2)$]:

$$\begin{pmatrix} c_2c_3 & -c_2s_3 & s_2 \\ s_1s_2c_3 + s_3c_1 & -s_1s_2s_3 + c_1c_3 & -s_1c_2 \\ -c_1s_2c_3 + s_1s_3 & c_1s_2s_3 + s_1c_3 & c_1c_2 \end{pmatrix};$$

$$\omega_1 = (c_2c_3)v_1 + (s_3)v_2 + (0)v_3 \qquad v_1 = (c_2)^{-1}[(c_3)\omega_1 + (-s_3)\omega_2 + (0)\omega_3]$$

$$\omega_2 = (-c_2s_3)v_1 + (c_3)v_2 + (0)v_3 \qquad v_2 = (c_2)^{-1}[(c_2s_3)\omega_1 + (c_2c_3)\omega_2 + (0)\omega_3]$$

$$\omega_3 = (s_2)v_1 + (0)v_2 + (1)v_3 \qquad v_3 = (c_2)^{-1}[(-s_2c_3)\omega_1 + (s_2s_3)\omega_2 + (c_2)\omega_3].$$

1(b) $[1'(\chi_1), 2'(\chi_2), 3'(\chi_3)] = [3(\chi_3), 2(\chi_2), 1(\chi_1)]$ [Singularity at $\chi_2 = \pm(\pi/2)$]:

$$\begin{pmatrix} c_2c_3 & s_1s_2c_3 - c_1s_3 & c_1s_2c_3 + s_1s_3 \\ c_2s_3 & s_1s_2s_3 + c_1c_3 & c_1s_2s_3 - s_1c_3 \\ -s_2 & s_1c_2 & c_1c_2 \end{pmatrix};$$

$$\omega_1 = (1)\nu_1 + (0)\nu_2 + (-s_2)\nu_3 \quad \Big| \quad \nu_1 = (c_2)^{-1}[(c_2)\omega_1 + (s_1s_2)\omega_2 + (c_1s_2)\omega_3]$$
$$\omega_2 = (0)\nu_1 + (c_1)\nu_2 + (s_1c_2)\nu_3 \quad \Big| \quad \nu_2 = (c_2)^{-1}[(0)\omega_1 + (c_1c_2)\omega_2 + (-s_1c_2)\omega_3]$$
$$\omega_3 = (0)\nu_1 + (-s_1)\nu_2 + (c_1c_2)\nu_3 \quad \Big| \quad \nu_3 = (c_2)^{-1}[(0)\omega_1 + (s_1)\omega_2 + (c_1)\omega_3].$$

2(a) $[2(\chi_1), 3(\chi_2), 1(\chi_3)] = [1'(\chi_3), 3'(\chi_2), 2'(\chi_1)]$ [Singularity at $\chi_2 = \pm(\pi/2)$]:

$$\begin{pmatrix} c_1c_2 & -c_1s_2c_3 + s_1s_3 & c_1s_2s_3 + s_1c_3 \\ s_2 & c_2c_3 & -c_2s_3 \\ -s_1c_2 & s_1s_2c_3 + c_1s_3 & -s_1s_2s_3 + c_1c_3 \end{pmatrix};$$

$$\omega_1 = (s_2)\nu_1 + (0)\nu_2 + (1)\nu_3 \quad \Big| \quad \nu_1 = (c_2)^{-1}[(0)\omega_1 + (c_3)\omega_2 + (-s_3)\omega_3]$$
$$\omega_2 = (c_2c_3)\nu_1 + (s_3)\nu_2 + (0)\nu_3 \quad \Big| \quad \nu_2 = (c_2)^{-1}[(0)\omega_1 + (c_2s_3)\omega_2 + (c_2c_3)\omega_3]$$
$$\omega_3 = (-c_2s_3)\nu_1 + (c_3)\nu_2 + (0)\nu_3 \quad \Big| \quad \nu_3 = (c_2)^{-1}[(c_2)\omega_1 + (-s_2c_3)\omega_2 + (s_2s_3)\omega_3].$$

2(b) $[2'(\chi_1), 3'(\chi_2), 1'(\chi_3)] = [1(\chi_3), 3(\chi_2), 2(\chi_1)]$ [Singularity at $\chi_2 = \pm(\pi/2)$]:

$$\begin{pmatrix} c_1c_2 & -s_2 & s_1c_2 \\ c_1s_2c_3 + s_1s_3 & c_2c_3 & s_1s_2c_3 - c_1s_3 \\ c_1s_2s_3 - s_1c_3 & c_2s_3 & s_1s_2s_3 + c_1c_3 \end{pmatrix};$$

$$\omega_1 = (0)\nu_1 + (-s_1)\nu_2 + (c_1c_2)\nu_3 \quad \Big| \quad \nu_1 = (c_2)^{-1}[(c_1s_2)\omega_1 + (c_2)\omega_2 + (s_1s_2)\omega_3]$$
$$\omega_2 = (1)\nu_1 + (0)\nu_2 + (-s_2)\nu_3 \quad \Big| \quad \nu_2 = (c_2)^{-1}[(-s_1c_2)\omega_1 + (0)\omega_2 + (c_1c_2)\omega_3]$$
$$\omega_3 = (0)\nu_1 + (c_1)\nu_2 + (s_1c_2)\nu_3 \quad \Big| \quad \nu_3 = (c_2)^{-1}[(c_1)\omega_1 + (0)\omega_2 + (s_1)\omega_3].$$

3(a) $[3(\chi_1), 1(\chi_2), 2(\chi_3)] = [2'(\chi_3), 1'(\chi_2), 3'(\chi_1)]$ [Singularity at $\chi_2 = \pm(\pi/2)$]:

$$\begin{pmatrix} -s_1s_2s_3 + c_1c_3 & -s_1c_2 & s_1s_2c_3 + c_1s_3 \\ c_1s_2s_3 + s_1c_3 & c_1c_2 & -c_1s_2c_3 + s_1s_3 \\ -c_2s_3 & s_2 & c_2c_3 \end{pmatrix};$$

$$\omega_1 = (-c_2s_3)\nu_1 + (c_3)\nu_2 + (0)\nu_3 \quad \Big| \quad \nu_1 = (c_2)^{-1}[(-s_3)\omega_1 + (0)\omega_2 + (c_3)\omega_3]$$
$$\omega_2 = (s_2)\nu_1 + (0)\nu_2 + (1)\nu_3 \quad \Big| \quad \nu_2 = (c_2)^{-1}[(c_2c_3)\omega_1 + (0)\omega_2 + (c_2s_3)\omega_3]$$
$$\omega_3 = (c_2c_3)\nu_1 + (s_3)\nu_2 + (0)\nu_3 \quad \Big| \quad \nu_3 = (c_2)^{-1}[(s_2s_3)\omega_1 + (c_2)\omega_2 + (-s_2c_3)\omega_3].$$

3(b) $[3'(\chi_1), 1'(\chi_2), 2'(\chi_3)] = [2(\chi_3), 1(\chi_2), 3(\chi_1)]$ [Singularity at $\chi_2 = \pm(\pi/2)$]:

$$\begin{pmatrix} s_1s_2s_3 + c_1c_3 & c_1s_2s_3 - s_1c_3 & c_2s_3 \\ s_1c_2 & c_1c_2 & -s_2 \\ s_1s_2c_3 - c_1s_3 & c_1s_3c_3 + s_1s_3 & c_2c_3 \end{pmatrix};$$

$$\begin{array}{l|l} \omega_1 = (0)v_1 + (c_1)v_2 + (s_1c_2)v_3 & v_1 = (c_2)^{-1}[(s_1s_2)\omega_1 + (c_1s_2)\omega_2 + (c_2)\omega_3] \\ \omega_2 = (0)v_1 + (-s_1)v_2 + (c_1c_2)v_3 & v_2 = (c_2)^{-1}[(c_1c_2)\omega_1 + (-s_1c_2)\omega_2 + (0)\omega_3] \\ \omega_3 = (1)v_1 + (0)v_2 + (-s_2)v_3 & v_3 = (c_2)^{-1}[(s_1)\omega_1 + (c_1)\omega_2 + (0)\omega_3]. \end{array}$$

4(a) $[1(\chi_1), 3(\chi_2), 2(\chi_3)] = [2'(\chi_3), 3'(\chi_2), 1'(\chi_1)]$ [Singularity at $\chi_2 = \pm(\pi/2)$]:

$$\begin{pmatrix} c_2c_3 & -s_2 & c_2s_3 \\ c_1s_2c_3 + s_1s_3 & c_1c_2 & c_1s_2s_3 - s_1c_3 \\ s_1s_2c_3 - c_1s_3 & s_1c_2 & s_1s_2s_3 + c_1c_3 \end{pmatrix};$$

$$\begin{array}{l|l} \omega_1 = (c_2c_3)v_1 + (-s_3)v_2 + (0)v_3 & v_1 = (c_2)^{-1}[(c_3)\omega_1 + (0)\omega_2 + (s_3)\omega_3] \\ \omega_2 = (-s_2)v_1 + (0)v_2 + (1)v_3 & v_2 = (c_2)^{-1}[(-c_2s_3)\omega_1 + (0)\omega_2 + (c_2c_3)\omega_3] \\ \omega_3 = (c_2s_3)v_1 + (c_3)v_2 + (0)v_3 & v_3 = (c_2)^{-1}[(s_2c_3)\omega_1 + (c_2)\omega_2 + (s_2s_3)\omega_3]. \end{array}$$

4(b) $[1'(\chi_1), 3'(\chi_2), 2'(\chi_3)] = [2(\chi_3), 3(\chi_2), 1(\chi_1)]$ [Singularity at $\chi_2 = \pm(\pi/2)$]:

$$\begin{pmatrix} c_2c_3 & -c_1s_2c_3 + s_1s_3 & s_1s_2c_3 + c_1s_3 \\ s_2 & c_1c_2 & -s_1c_2 \\ -c_2s_3 & c_1s_2s_3 + s_1c_3 & s_1s_2s_3 + c_1c_3 \end{pmatrix};$$

$$\begin{array}{l|l} \omega_1 = (1)v_1 + (0)v_2 + (s_2)v_3 & v_1 = (c_2)^{-1}[(c_2)\omega_1 + (-c_1s_2)\omega_2 + (s_1s_2)\omega_3] \\ \omega_2 = (0)v_1 + (s_1)v_2 + (c_1c_2)v_3 & v_2 = (c_2)^{-1}[(0)\omega_1 + (s_1c_2)\omega_2 + (c_1c_2)\omega_3] \\ \omega_3 = (0)v_1 + (c_1)v_2 + (-s_1c_2)v_3 & v_3 = (c_2)^{-1}[(0)\omega_1 + (c_1)\omega_2 + (-s_1)\omega_3]. \end{array}$$

5(a) $[2(\chi_1), 1(\chi_2), 3(\chi_3)] = [3'(\chi_3), 1'(\chi_2), 2'(\chi_1)]$ [Singularity at $\chi_2 = \pm(\pi/2)$]:

$$\begin{pmatrix} s_1s_2s_3 + c_1c_3 & s_1s_2c_3 - c_1s_3 & s_1c_2 \\ c_2s_3 & c_2c_3 & -s_2 \\ c_1s_2s_3 - s_1c_3 & c_1s_2c_3 + s_1s_3 & c_1c_2 \end{pmatrix};$$

$$\begin{array}{l|l} \omega_1 = (c_2s_3)v_1 + (c_3)v_2 + (0)v_3 & v_1 = (c_2)^{-1}[(s_3)\omega_1 + (c_3)\omega_2 + (0)\omega_3] \\ \omega_2 = (c_2c_3)v_1 + (-s_3)v_2 + (0)v_3 & v_2 = (c_2)^{-1}[(c_2c_3)\omega_1 + (-c_2s_3)\omega_2 + (0)\omega_3] \\ \omega_3 = (-s_2)v_1 + (0)v_2 + (1)v_3 & v_3 = (c_2)^{-1}[(s_2s_3)\omega_1 + (s_2c_3)\omega_2 + (c_2)\omega_3]. \end{array}$$

5(b) $[2'(\chi_1), 1'(\chi_2), 3'(\chi_3)] = [3(\chi_3), 1(\chi_2), 2(\chi_1)]$ [Singularity at $\chi_2 = \pm(\pi/2)$]:

$$\begin{pmatrix} -s_1s_2s_3 + c_1c_3 & -c_2s_3 & c_1s_2s_3 + s_1c_3 \\ s_1s_2c_3 + c_1s_3 & c_2c_3 & -c_1s_2c_3 + s_1s_3 \\ -s_1c_2 & s_2 & c_1c_2 \end{pmatrix};$$

$$\begin{array}{l|l} \omega_1 = (0)v_1 + (c_1)v_2 + (-s_1c_2)v_3 & v_1 = (c_2)^{-1}[(s_1s_2)\omega_1 + (c_2)\omega_2 + (-c_1s_2)\omega_3] \\ \omega_2 = (1)v_1 + (0)v_2 + (s_2)v_3 & v_2 = (c_2)^{-1}[(c_1c_2)\omega_1 + (0)\omega_2 + (s_1c_2)\omega_3] \\ \omega_3 = (0)v_1 + (s_1)v_2 + (c_1c_2)v_3 & v_3 = (c_2)^{-1}[(-s_1)\omega_1 + (0)\omega_2 + (c_1)\omega_3]. \end{array}$$

6(a) $[3(\chi_1), 2(\chi_2), 1(\chi_3)] = [1'(\chi_3), 2'(\chi_2), 3'(\chi_1)]$ [Singularity at $\chi_2 = \pm(\pi/2)$]:

$$\begin{pmatrix} c_1c_2 & c_1s_2s_3 - s_1c_3 & c_1s_2c_3 + s_1s_3 \\ s_1c_2 & s_1s_2s_3 + c_1c_3 & s_1s_2c_3 - c_1s_3 \\ -s_2 & c_2s_3 & c_2c_3 \end{pmatrix};$$

$$\begin{array}{l|l} \omega_1 = (-s_2)v_1 + (0)v_2 + (1)v_3 & v_1 = (c_2)^{-1}[(0)\omega_1 + (s_3)\omega_2 + (c_3)\omega_3] \\ \omega_2 = (c_2s_3)v_1 + (c_3)v_2 + (0)v_3 & v_2 = (c_2)^{-1}[(0)\omega_1 + (c_2c_3)\omega_2 + (-c_2s_3)\omega_3] \\ \omega_3 = (c_2c_3)v_1 + (-s_3)v_2 + (0)v_3 & v_3 = (c_2)^{-1}[(c_2)\omega_1 + (s_2s_3)\omega_2 + (s_2c_3)\omega_3]. \end{array}$$

6(b) $[3'(\chi_1), 2'(\chi_2), 1'(\chi_3)] = [1(\chi_3), 2(\chi_2), 3(\chi_1)]$ [Singularity at $\chi_2 = \pm(\pi/2)$]:

$$\begin{pmatrix} c_1c_2 & -s_1c_2 & s_2 \\ c_1s_2s_3 + s_1c_3 & -s_1s_2s_3 + c_1c_3 & -c_2s_3 \\ -c_1s_2c_3 + s_1s_3 & s_1s_2c_3 + c_1s_3 & c_2c_3 \end{pmatrix};$$

$$\begin{array}{l|l} \omega_1 = (0)v_1 + (s_1)v_2 + (c_1c_2)v_3 & v_1 = (c_2)^{-1}[(-c_1s_2)\omega_1 + (s_1s_2)\omega_2 + (c_2)\omega_3] \\ \omega_2 = (0)v_1 + (c_1)v_2 + (-s_1c_2)v_3 & v_2 = (c_2)^{-1}[(s_1c_2)\omega_1 + (c_1c_2)\omega_2 + (0)\omega_3] \\ \omega_3 = (1)v_1 + (0)v_2 + (s_2)v_3 & v_3 = (c_2)^{-1}[(c_1)\omega_1 + (-s_1)\omega_2 + (0)\omega_3]. \end{array}$$

7(a) $[1(\chi_1), 2(\chi_2), 1(\chi_3)] = [1'(\chi_3), 2'(\chi_2), 1'(\chi_1)]$ [Singularities at $\chi_2 = 0, \pm\pi$]:

$$\begin{pmatrix} c_2 & s_2s_3 & s_2c_3 \\ s_1s_2 & -s_1c_2s_3 + c_1c_3 & -s_1c_2c_3 - c_1s_3 \\ -c_1s_2 & c_1c_2s_3 + s_1c_3 & c_1c_2c_3 - s_1s_3 \end{pmatrix};$$

$$\begin{array}{l|l} \omega_1 = (c_2)v_1 + (0)v_2 + (1)v_3 & v_1 = (s_2)^{-1}[(0)\omega_1 + (s_3)\omega_2 + (c_3)\omega_3] \\ \omega_2 = (s_2s_3)v_1 + (c_3)v_2 + (0)v_3 & v_2 = (s_2)^{-1}[(0)\omega_1 + (s_2c_3)\omega_2 + (-s_2s_3)\omega_3] \\ \omega_3 = (s_2c_3)v_1 + (-s_3)v_2 + (0)v_3 & v_3 = (s_2)^{-1}[(s_2)\omega_1 + (-c_2s_3)\omega_2 + (-c_2c_3)\omega_3]. \end{array}$$

7(b) $[1'(\chi_1), 2'(\chi_2), 1'(\chi_3)] = [1(\chi_3), 2(\chi_2), 1(\chi_1)]$ [Singularities at $\chi_2 = 0, \pm\pi$]:

$$\begin{pmatrix} c_2 & s_1 s_2 & c_1 s_2 \\ s_2 s_3 & -s_1 c_2 s_3 + c_1 c_3 & -c_1 c_2 s_3 - s_1 c_3 \\ -s_2 c_3 & s_1 c_2 c_3 + c_1 s_3 & c_1 c_2 c_3 - s_1 s_3 \end{pmatrix};$$

$$\begin{aligned} \omega_1 &= (1)\nu_1 + (0)\nu_2 + (c_2)\nu_3 & \nu_1 &= (s_2)^{-1}[(s_2)\omega_1 + (-s_1 c_2)\omega_2 + (c_1 c_2)\omega_3] \\ \omega_2 &= (0)\nu_1 + (c_1)\nu_2 + (s_1 s_2)\nu_3 & \nu_2 &= (s_2)^{-1}[(0)\omega_1 + (c_1 s_2)\omega_2 + (-s_1 s_2)\omega_3] \\ \omega_3 &= (0)\nu_1 + (-s_1)\nu_2 + (c_1 s_2)\nu_3 & \nu_3 &= (s_2)^{-1}[(0)\omega_1 + (s_1)\omega_2 + (c_1)\omega_3]. \end{aligned}$$

8(a) $[1(\chi_1), 3(\chi_2), 1(\chi_3)] = [1'(\chi_3), 3'(\chi_2), 1'(\chi_1)]$ [Singularities at $\chi_2 = 0, \pm\pi$]:

$$\begin{pmatrix} c_2 & -s_2 c_3 & s_2 s_3 \\ c_1 s_2 & c_1 c_2 c_3 - s_1 s_3 & -c_1 c_2 s_3 - s_1 c_3 \\ s_1 s_2 & s_1 c_2 c_3 + c_1 s_3 & -s_1 c_2 s_3 - c_1 c_3 \end{pmatrix};$$

$$\begin{aligned} \omega_1 &= (c_2)\nu_1 + (0)\nu_2 + (1)\nu_3 & \nu_1 &= (s_2)^{-1}[(0)\omega_1 + (-c_3)\omega_2 + (s_3)\omega_3] \\ \omega_2 &= (-s_2 c_3)\nu_1 + (s_3)\nu_2 + (0)\nu_3 & \nu_2 &= (s_2)^{-1}[(0)\omega_1 + (s_2 s_3)\omega_2 + (s_2 c_3)\omega_3] \\ \omega_3 &= (s_2 s_3)\nu_1 + (c_3)\nu_2 + (0)\nu_3 & \nu_3 &= (s_2)^{-1}[(s_2)\omega_1 + (c_2 c_3)\omega_2 + (-c_2 s_3)\omega_3]. \end{aligned}$$

8(b) $[1'(\chi_1), 3'(\chi_2), 1'(\chi_3)] = [1(\chi_3), 3(\chi_2), 1(\chi_1)]$ [Singularities at $\chi_2 = 0, \pm\pi$]:

$$\begin{pmatrix} c_2 & -c_1 s_2 & s_1 s_2 \\ s_2 c_3 & c_1 c_2 c_3 - s_1 s_3 & -s_1 c_2 c_3 - c_1 s_3 \\ s_2 s_3 & c_1 c_2 s_3 + s_1 c_3 & -s_1 c_2 s_3 + c_1 c_3 \end{pmatrix};$$

$$\begin{aligned} \omega_1 &= (1)\nu_1 + (0)\nu_2 + (c_2)\nu_3 & \nu_1 &= (s_2)^{-1}[(s_2)\omega_1 + (c_1 c_2)\omega_2 + (-s_1 c_2)\omega_3] \\ \omega_2 &= (0)\nu_1 + (s_1)\nu_2 + (-c_1 s_2)\nu_3 & \nu_2 &= (s_2)^{-1}[(0)\omega_1 + (s_1 s_2)\omega_2 + (c_1 s_2)\omega_3] \\ \omega_3 &= (0)\nu_1 + (c_1)\nu_2 + (s_1 s_2)\nu_3 & \nu_3 &= (s_2)^{-1}[(0)\omega_1 + (-c_1)\omega_2 + (s_1)\omega_3]. \end{aligned}$$

9(a) $[2(\chi_1), 1(\chi_2), 2(\chi_3)] = [2'(\chi_3), 1'(\chi_2), 2'(\chi_1)]$ [Singularities at $\chi_2 = 0, \pm\pi$]:

$$\begin{pmatrix} -s_1 c_2 s_3 + c_1 c_3 & s_1 s_2 & s_1 c_2 c_3 + c_1 s_3 \\ s_2 s_3 & c_2 & -s_2 c_3 \\ -c_1 c_2 s_3 - s_1 c_3 & c_1 s_2 & c_1 c_2 c_3 - s_1 s_3 \end{pmatrix};$$

$$\begin{aligned} \omega_1 &= (s_2 s_3)\nu_1 + (c_3)\nu_2 + (0)\nu_3 & \nu_1 &= (s_2)^{-1}[(s_3)\omega_1 + (0)\omega_2 + (-c_3)\omega_3] \\ \omega_2 &= (c_2)\nu_1 + (0)\nu_2 + (1)\nu_3 & \nu_2 &= (s_2)^{-1}[(s_2 c_3)\omega_1 + (0)\omega_2 + (s_2 s_3)\omega_3] \\ \omega_3 &= (-s_2 c_3)\nu_1 + (s_3)\nu_2 + (0)\nu_3 & \nu_3 &= (s_2)^{-1}[(-c_2 s_3)\omega_1 + (s_2)\omega_2 + (c_2 s_3)\omega_3]. \end{aligned}$$

9(b) $[2'(\chi_1), 1'(\chi_2), 2'(\chi_3)] = [2(\chi_3), 1(\chi_2), 2(\chi_1)]$ [Singularities at $\chi_2 = 0, \pm\pi$]:

$$\begin{pmatrix} -s_1c_2s_3 + c_1c_3 & s_2s_3 & c_1c_2s_3 + s_1c_3 \\ s_1s_2 & c_2 & -c_1s_2 \\ -s_1c_2c_3 - c_1s_3 & s_2c_3 & c_1c_2c_3 - s_1s_3 \end{pmatrix};$$

$\omega_1 = (0)v_1 + (c_1)v_2 + (s_1s_2)v_3$	$v_1 = (s_2)^{-1}[(-s_1c_2)\omega_1 + (s_2)\omega_2 + (c_1c_2)\omega_3]$
$\omega_2 = (1)v_1 + (0)v_2 + (c_2)v_3$	$v_2 = (s_2)^{-1}[(c_1s_2)\omega_1 + (0)\omega_2 + (s_1s_2)\omega_3]$
$\omega_3 = (0)v_1 + (s_1)v_2 + (-c_1s_2)v_3$	$v_3 = (s_2)^{-1}[(s_1)\omega_1 + (0)\omega_2 + (-c_1)\omega_3]$.

10(a) $[2(\chi_1), 3(\chi_2), 2(\chi_3)] = [2'(\chi_3), 3'(\chi_2), 2'(\chi_1)]$ [Singularities at $\chi_2 = 0, \pm\pi$]:

$$\begin{pmatrix} c_1c_2c_3 - s_1s_3 & -c_1s_2 & c_1c_2s_3 + s_1c_3 \\ s_2c_3 & c_2 & s_2s_3 \\ -s_1c_2c_3 - c_1s_3 & s_1s_2 & -s_1c_2s_3 + c_1c_3 \end{pmatrix};$$

$\omega_1 = (s_2c_3)v_1 + (-s_3)v_2 + (0)v_3$	$v_1 = (s_2)^{-1}[(c_3)\omega_1 + (0)\omega_2 + (s_3)\omega_3]$
$\omega_2 = (c_2)v_1 + (0)v_2 + (1)v_3$	$v_2 = (s_2)^{-1}[(-s_2s_3)\omega_1 + (0)\omega_2 + (s_2c_3)\omega_3]$
$\omega_3 = (s_2s_3)v_1 + (c_3)v_2 + (0)v_3$	$v_3 = (s_2)^{-1}[(-c_2c_3)\omega_1 + (s_2)\omega_2 + (-c_2s_3)\omega_3]$.

10(b) $[2'(\chi_1), 3'(\chi_2), 2'(\chi_3)] = [2(\chi_3), 3(\chi_2), 2(\chi_1)]$ [Singularities at $\chi_2 = 0, \pm\pi$]:

$$\begin{pmatrix} c_1c_2c_3 - s_1s_3 & -s_2c_3 & s_1c_2c_3 + c_1s_3 \\ c_1s_2 & c_2 & s_1s_2 \\ -c_1c_2s_3 - s_1c_3 & s_2s_3 & -s_1c_2s_3 + c_1c_3 \end{pmatrix};$$

$\omega_1 = (0)v_1 + (-s_1)v_2 + (c_1s_2)v_3$	$v_1 = (s_2)^{-1}[(-c_1c_2)\omega_1 + (s_2)\omega_2 + (-s_1c_2)\omega_3]$
$\omega_2 = (1)v_1 + (0)v_2 + (c_2)v_3$	$v_2 = (s_2)^{-1}[(-s_1s_2)\omega_1 + (0)\omega_2 + (c_1s_2)\omega_3]$
$\omega_3 = (0)v_1 + (c_1)v_2 + (s_1s_2)v_3$	$v_3 = (s_2)^{-1}[(c_1)\omega_1 + (0)\omega_2 + (s_1)\omega_3]$.

11(a) $[3(\chi_1), 1(\chi_2), 3(\chi_3)] = [3'(\chi_3), 1'(\chi_2), 3'(\chi_1)]$ [Singularities at $\chi_2 = 0, \pm\pi$]:

$$\begin{pmatrix} -s_1c_2s_3 + c_1c_3 & -s_1c_2c_3 - c_1s_3 & s_1s_2 \\ c_1c_2s_3 + s_1c_3 & c_1c_2c_3 - s_1s_3 & -c_1s_2 \\ s_2s_3 & s_2c_3 & c_2 \end{pmatrix};$$

$\omega_1 = (s_2s_3)v_1 + (c_3)v_2 + (0)v_3$	$v_1 = (s_2)^{-1}[(s_3)\omega_1 + (c_3)\omega_2 + (0)\omega_3]$
$\omega_2 = (s_2c_3)v_1 + (-s_3)v_2 + (0)v_3$	$v_2 = (s_2)^{-1}[(s_2c_3)\omega_1 + (-s_2s_3)\omega_2 + (0)\omega_3]$
$\omega_3 = (c_2)v_1 + (0)v_2 + (1)v_3$	$v_3 = (s_2)^{-1}[(-c_2s_3)\omega_1 + (-c_2c_3)\omega_2 + (s_2)\omega_3]$.

11(b) $[3'(\chi_1), 1'(\chi_2), 3'(\chi_3)] = [3(\chi_3), 1(\chi_2), 3(\chi_1)]$ [Singularities at $\chi_2 = 0, \pm\pi$]:

$$\begin{pmatrix} -s_1c_2s_3 + c_1c_3 & -c_1c_2s_3 - s_1c_3 & s_2s_3 \\ s_1c_2c_3 + c_1s_3 & c_1c_2c_3 - s_1s_3 & -s_2c_3 \\ s_1s_2 & c_1s_2 & c_2 \end{pmatrix};$$

$$\begin{array}{l|l} \omega_1 = (0)v_1 + (c_1)v_2 + (s_1s_2)v_3 & v_1 = (s_2)^{-1}[(-s_1c_2)\omega_1 + (-c_1c_2)\omega_2 + (s_2)\omega_3] \\ \omega_2 = (0)v_1 + (-s_1)v_2 + (c_1s_2)v_3 & v_2 = (s_2)^{-1}[(c_1s_2)\omega_1 + (-s_1s_2)\omega_2 + (0)\omega_3] \\ \omega_3 = (1)v_1 + (0)v_2 + (c_2)v_3 & v_3 = (s_2)^{-1}[(s_1)\omega_1 + (c_1)\omega_2 + (0)\omega_3]. \end{array}$$

12(a) $[3(\chi_1), 2(\chi_2), 3(\chi_3)] = [3'(\chi_3), 2'(\chi_2), 3'(\chi_1)]$ [Singularities at $\chi_2 = 0, \pm\pi$]:

$$\begin{pmatrix} c_1c_2c_3 - s_1s_3 & -c_1c_2s_3 - s_1c_3 & c_1s_2 \\ s_1c_2c_3 + c_1s_3 & -s_1c_2s_3 + c_1c_3 & s_1s_2 \\ -s_2c_3 & s_2s_3 & c_2 \end{pmatrix};$$

$$\begin{array}{l|l} \omega_1 = (-s_2c_3)v_1 + (s_3)v_2 + (0)v_3 & v_1 = (s_2)^{-1}[(-c_3)\omega_1 + (s_3)\omega_2 + (0)\omega_3] \\ \omega_2 = (s_2s_3)v_1 + (c_3)v_2 + (0)v_3 & v_2 = (s_2)^{-1}[(s_2s_3)\omega_1 + (s_2c_3)\omega_2 + (0)\omega_3] \\ \omega_3 = (c_2)v_1 + (0)v_2 + (1)v_3 & v_3 = (s_2)^{-1}[(c_2c_3)\omega_1 + (-c_2s_3)\omega_2 + (s_2)\omega_3]. \end{array}$$

12(b) $[3'(\chi_1), 2'(\chi_2), 3'(\chi_3)] = [3(\chi_3), 2(\chi_2), 3(\chi_1)]$ [Singularities at $\chi_2 = 0, \pm\pi$]:

$$\begin{pmatrix} c_1c_2c_3 - s_1s_3 & -s_1c_2c_3 - c_1s_3 & s_2c_3 \\ c_1c_2s_3 + s_1c_3 & -s_1c_2s_3 + c_1c_3 & s_2s_3 \\ -c_1s_2 & s_1s_2 & c_2 \end{pmatrix};$$

$$\begin{array}{l|l} \omega_1 = (0)v_1 + (s_1)v_2 + (-c_1s_2)v_3 & v_1 = (s_2)^{-1}[(c_1c_2)\omega_1 + (-s_1c_2)\omega_2 + (s_2)\omega_3] \\ \omega_2 = (0)v_1 + (c_1)v_2 + (s_1s_2)v_3 & v_2 = (s_2)^{-1}[(s_1s_2)\omega_1 + (c_1s_2)\omega_2 + (0)\omega_3] \\ \omega_3 = (1)v_1 + (0)v_2 + (c_2)v_3 & v_3 = (s_2)^{-1}[(-c_1)\omega_1 + (s_1)\omega_2 + (0)\omega_3]. \end{array}$$

1.14 THE RIGID BODY: AN INTRODUCTION TO QUASI COORDINATES

As an introduction to quasi coordinates, and quasi variables in general (a topic to be detailed in chap. 2), we show in this section that the angular velocity, although a vector, does not result by simple $d/dt(\ldots)$-differentiation of an angular displacement; *its components along space-/body-fixed axes, say* ω_k, *do not equal the total time derivatives of angles or any other genuine (global) rotational coordinates/parameters, say* θ_k; *that is,* $\omega_k \neq d\theta_k/dt$. This is another complication of rotational mechanics, one that is intimately connected with the noncommutativity of finite rotations; and it necessitates the hitherto search for connections of the ω_k's with genuine angular coordinates and their rates, like the Eulerian angles ϕ, θ, ψ.

Let us consider, for concreteness, the *body-axes* components of the angular velocity. From

$$\boldsymbol{\omega} = \mathbf{A}^{\mathrm{T}} \cdot (\mathrm{d}\mathbf{A}/\mathrm{dt}) = -(\mathrm{d}\mathbf{A}^{\mathrm{T}}/\mathrm{dt}) \cdot \mathbf{A}, \tag{1.14.1a}$$

(1.7.30f ff.) we have

$$\omega_x = A_{Xz}(dA_{Xy}/dt) + A_{Yz}(dA_{Yy}/dt) + A_{Zz}(dA_{Zy}/dt), \quad \text{etc., cyclically,} \tag{1.14.1b}$$

or, multiplying through by dt and setting $\omega_x dt \equiv d\theta_x$ (just a suggestive shorthand),

$$d\theta_x = A_{Xz} dA_{Xy} + A_{Yz} dA_{Yy} + A_{Zz} dA_{Zy}, \quad \text{etc., cyclically.} \tag{1.14.1c}$$

We shall show that

$$\delta(d\theta_x) - d(\delta\theta_x) \neq 0, \quad \text{etc., cyclically,} \tag{1.14.2a}$$

where, for our purposes, $\delta(\ldots)$ can be viewed as just a differential of $(\ldots)$, along a different direction from $d(\ldots)$; that is, with $d(\ldots) \equiv d_1(\ldots)$ and $\delta(\ldots) \equiv d_2(\ldots)$,

$$\delta(d\theta_x) \equiv d_2(d_1\theta_x), \qquad d(\delta\theta_x) \equiv d_1(d_2\theta_x); \tag{1.14.2b}$$

and

$$\delta\theta_x = A_{Xz}\delta A_{Xy} + A_{Yz}\delta A_{Yy} + A_{Zz}\delta A_{Zy}, \quad \text{etc., cyclically.} \tag{1.14.2c}$$

Now, $\delta(\ldots)$-differentiating $d\theta_x$ and $d(\ldots)$-differentiating $\delta\theta_x$, and then subtracting side by side, while noticing that

$$\delta(dA_{k'k}) = d(\delta A_{k'k}) \qquad (k' = X, Y, Z;\ k = x, y, z), \tag{1.14.3a}$$

we get

$$\delta(d\theta_x) - d(\delta\theta_x) = \delta A_{Xz} dA_{Xy} - dA_{Xz}\delta A_{Xy} + \delta A_{Yz} dA_{Yy} - dA_{Yz}\delta A_{Yy} + \delta A_{Zz} dA_{Zy} - dA_{Zz}\,\delta A_{Zy}. \tag{1.14.3b}$$

Next, in order to express $\delta A_{k'k}$, $dA_{k'k}$ in terms of $\delta\theta_k$ and $d\theta_k$, we multiply the components of $\mathrm{d}\mathbf{A}/\mathrm{dt} = \mathbf{A} \cdot \boldsymbol{\omega}$ (1.7.30i) with dt, thus obtaining

$$dA_{Xz} = A_{Xx} d\theta_y - A_{Xy} d\theta_x \;\Rightarrow\; \delta A_{Xz} = A_{Xx}\delta\theta_y - A_{Xy}\delta\theta_x, \tag{1.14.4a}$$

$$dA_{Yz} = A_{Yx} d\theta_y - A_{Yy} d\theta_x \;\Rightarrow\; \delta A_{Yz} = A_{Yx}\delta\theta_y - A_{Yy}\delta\theta_x, \tag{1.14.4b}$$

$$dA_{Zz} = A_{Zx} d\theta_y - A_{Zy} d\theta_x \;\Rightarrow\; \delta A_{Zz} = A_{Zx}\delta\theta_y - A_{Zy}\delta\theta_x, \tag{1.14.4c}$$

$$dA_{Xy} = A_{Xz} d\theta_x - A_{Xx} d\theta_z \;\Rightarrow\; \delta A_{Xy} = A_{Xz}\delta\theta_x - A_{Xx}\delta\theta_z, \tag{1.14.4d}$$

$$dA_{Yy} = A_{Yz} d\theta_x - A_{Yx} d\theta_z \;\Rightarrow\; \delta A_{Yy} = A_{Yz}\delta\theta_x - A_{Yx}\delta\theta_z, \tag{1.14.4e}$$

$$dA_{Zy} = A_{Zz} d\theta_x - A_{Zx} d\theta_z \;\Rightarrow\; \delta A_{Zy} = A_{Zz}\delta\theta_x - A_{Zx}\delta\theta_z. \tag{1.14.4f}$$

Substituting (1.14.4a–f) into the right side of (1.14.3b), and invoking the orthogonality of $\mathbf{A} = (A_{k'k})$ [e.g., (1.7.6a, b; 1.7.22d)], we find, after some straightforward algebra, the noncommutativity equation

$$\delta(d\theta_x) - d(\delta\theta_x) = d\theta_y\delta\theta_z - d\theta_z\delta\theta_y. \tag{1.14.5a}$$

Working in complete analogy with the above, we obtain

$$\delta(d\theta_y) - d(\delta\theta_y) = d\theta_z\delta\theta_x - d\theta_x\delta\theta_z, \tag{1.14.5b}$$

$$\delta(d\theta_z) - d(\delta\theta_z) = d\theta_x\delta\theta_y - d\theta_y\delta\theta_x. \tag{1.14.5c}$$

These remarkable *transitivity* equations [because they allow for a smooth *transition* from Lagrangean mechanics (chap. 2 ff.) to Eulerian mechanics (§1.15 ff.)] show clearly that the $\theta_{x,\,y,\,z}$ are not ordinary (or genuine, or holonomic, or global; or as Lagrange puts it "variables finies") coordinates, like the Eulerian angles ϕ, θ, ψ; that is why they are called *pseudo-* or *quasi coordinates*. Their general theory, along with a simpler derivation of the above, are detailed in chap. 2.

Similarly, we can show that in terms of *space-axes* components, the transitivity equations are

$$\delta(d\theta_X) - d(\delta\theta_X) = d\theta_Z\delta\theta_Y - d\theta_Y\delta\theta_Z, \tag{1.14.6a}$$

$$\delta(d\theta_Y) - d(\delta\theta_Y) = d\theta_X\delta\theta_Z - d\theta_Z\delta\theta_X, \tag{1.14.6b}$$

$$\delta(d\theta_Z) - d(\delta\theta_Z) = d\theta_Y\delta\theta_X - d\theta_X\delta\theta_Y; \tag{1.14.6c}$$

or, in compact vector form,

$$\delta(d\boldsymbol{\theta}) - d(\delta\boldsymbol{\theta}) = \delta\boldsymbol{\theta} \times d\boldsymbol{\theta}, \tag{1.14.6d}$$

where $d\boldsymbol{\theta} \equiv d\theta_X\boldsymbol{I} + d\theta_Y\boldsymbol{J} + d\theta_Z\boldsymbol{K} \Rightarrow \delta\boldsymbol{\theta} \equiv \delta\theta_X\boldsymbol{I} + \delta\theta_Y\boldsymbol{J} + \delta\theta_Z\boldsymbol{K}$; that is, $\boldsymbol{\theta}$ is a *quasi vector*. Here (recall 1.7.30j), $\boldsymbol{\omega}' = (d\mathbf{A}/dt) \cdot \mathbf{A}^T$ and $d\mathbf{A}/dt = \boldsymbol{\omega}' \cdot \mathbf{A}$.

HISTORICAL

Equations (1.14.5a–c), along with the systematic use of direction cosines to rigid-body dynamics, are due to Lagrange. They appeared posthumously in the 2nd edition of the 2nd volume of his *Mécanique Analytique* (1815/1816). See also (alphabetically): Funk (1962, pp. 334–335), Kirchhoff (1876, sixth lecture, §2), Mathieu (1878, pp. 138–139).

1.15 THE RIGID BODY: TENSOR OF INERTIA, KINETIC ENERGY

Introduction, Basic Definitions

To get motivated, let us begin by calculating the (inertial) kinetic energy T of a rigid body B rotating about a fixed point O; the extension to the case of general motion follows easily. If $\boldsymbol{\omega}$ is the inertial angular velocity of B, then, since the inertial velocity of a genetic body particle P, of inertial position $\boldsymbol{r}$, is $\boldsymbol{\omega} \times \boldsymbol{r} = \boldsymbol{v}$, we have, successively,

$$\begin{aligned} 2T &\equiv \int dm\, \boldsymbol{v}\cdot\boldsymbol{v} = \int dm(\boldsymbol{\omega}\times\boldsymbol{r})\cdot(\boldsymbol{\omega}\times\boldsymbol{r}) \\ &= \int dm\big[(\boldsymbol{\omega}\cdot\boldsymbol{\omega})(\boldsymbol{r}\cdot\boldsymbol{r}) - (\boldsymbol{\omega}\cdot\boldsymbol{r})(\boldsymbol{\omega}\cdot\boldsymbol{r})\big] \qquad \text{(by simple vector algebra)} \\ &= \int dm[\omega^2 r^2 - (\boldsymbol{\omega}\cdot\boldsymbol{r})^2], \end{aligned} \tag{1.15.1a}$$

or, in terms of components along arbitrary (i.e., not necessarily body-fixed) rectangular Cartesian axes $O\text{–}xyz \equiv O\text{–}x_k$, in which $\boldsymbol{r} = (x, y, z) \equiv (x_k)$, $\boldsymbol{\omega} = (\omega_x, \omega_y, \omega_z) \equiv (\omega_k)$ $(k = 1, 2, 3)$,

$$2T = S\, dm\Big[\Big(\sum \omega_k{}^2\Big)\Big(\sum x_k{}^2\Big) - \Big(\sum \omega_k x_k\Big)\Big(\sum \omega_l x_l\Big)\Big]$$

$$= S\, dm\Big[\Big(\sum\sum \delta_{kl}\,\omega_k\,\omega_l\Big)\Big(\sum x_r{}^2\Big) - \Big(\sum\sum \omega_k\,\omega_l\,x_k x_l\Big)\Big]$$

$$= \sum\sum I_{kl}\,\omega_k\,\omega_l \qquad (\textit{Indicial}\ \text{notation}) \qquad (1.15.1b)$$

$$= \boldsymbol{\omega}\cdot\mathbf{I}\cdot\boldsymbol{\omega} \qquad (\textit{Direct}\ \text{notation}) \qquad (1.15.1c)$$

$$= \boldsymbol{\omega}^{\mathrm{T}}\cdot\mathbf{I}\cdot\boldsymbol{\omega} \qquad (\textit{Matrix}\ \text{notation};\ \boldsymbol{\omega}\text{: } \textit{column}\ \text{vector}), \qquad (1.15.1d)$$

where

$$\mathbf{I}_O \equiv \mathbf{I} = (I_{O,kl}) \equiv (I_{kl}), \qquad (1.15.2a)$$

$$I_{kl} \equiv S\, dm(r^2\delta_{kl} - x_k x_l):$$

Components of *tensor of inertia* of B, $\mathbf{I}$, at O, along $O\text{–}x_k$, (1.15.2b)

$$r^2 \equiv \sum x_k{}^2 = \sum x_k x_k; \qquad (1.15.2c)$$

or, equivalently,

$$I_{kl} = J_o\,\delta_{kl} - J_{kl}, \qquad (1.15.2d)$$

where

$$\mathbf{J}_O \equiv \mathbf{J} = (J_{O,kl}) \equiv (J_{kl}), \qquad (1.15.2e)$$

$$J_{kl} \equiv S\, x_k x_l\, dm$$

$\equiv$ Components of *Binet's tensor* of B, $\mathbf{J}$, at O, along $O\text{–}x_k$, (1.15.2f)

$$J_o \equiv J_{11} + J_{22} + J_{33} = S\, r^2 dm = \mathrm{Tr}\,\mathbf{J}. \qquad (1.15.2g)$$

In direct notation, the above read

$$\mathbf{I} = S\,[(\boldsymbol{r}\cdot\boldsymbol{r})\mathbf{1} - \boldsymbol{r}\otimes\boldsymbol{r}]\,dm, \qquad \mathbf{J} = S\,(\boldsymbol{r}\otimes\boldsymbol{r})\,dm. \qquad (1.15.2h)$$

That $\mathbf{I}$ is a (second-order) tensor follows from the fact that, under rotations of the axes, T is a scalar invariant while $\boldsymbol{\omega}$ is a vector (what, in effect, constitutes a simple application of the tensorial "quotient rule"). This means that the components of $\mathbf{I}$ along $O\text{–}x_k, I_{kl}$, and along $O\text{–}x_{k'}$, $I_{k'l'}$, where $x_{k'} = \sum A_{k'k}x_k$ (proper orthogonal transformation), are related by

$$I_{k'l'} = \sum\sum A_{k'k}A_{l'l}I_{kl} = \sum\sum A_{k'k}I_{kl}A_{ll'}$$

$$= (\mathbf{A}\cdot\mathbf{I}\cdot\mathbf{A}^{\mathrm{T}})_{k'l'} \qquad [\text{recalling eqs. (1.1.19j ff.)}]$$

$$\Big[\Leftrightarrow I_{kl} = \sum\sum A_{kk'}A_{ll'}I_{k'l'} \qquad (\text{Since, indicially, } A_{k'k} = A_{kk'})\Big]. \qquad (1.15.2i)$$

Properties of the Inertia Tensor

Clearly (and like most mechanics tensors), $\mathbf{I}$ is *symmetric*: $I_{kl} = I_{lk}$; that is, at most *six*, of its nine components, are independent. In extenso, (1.15.2a, b) read

$$\mathbf{I} = \begin{pmatrix} I_{xx} & I_{xy} & I_{xz} \\ I_{yx} & I_{yy} & I_{yz} \\ I_{zx} & I_{zy} & I_{zz} \end{pmatrix}$$

$$= \begin{pmatrix} S\, dm\,(y^2 + z^2) & -S\, dm\; x\, y & -S\, dm\; x\, z \\ -S\, dm\; y\, x & S\, dm\,(z^2 + x^2) & -S\, dm\; y\, z \\ -S\, dm\; z\, x & -S\, dm\; z\, y & S\, dm\,(x^2 + y^2) \end{pmatrix}. \qquad (1.15.3)$$

The *diagonal* elements of $\mathbf{I}$, I_{xx}, I_{yy}, I_{zz}, are called *moments of inertia* of B about O–xyz; and they are *nonnegative*; that is, $I_{xx,\, yy,\, zz} \geq 0$. The *off-diagonal* elements of $\mathbf{I}$, $I_{xy} = I_{yx}$, $I_{xz} = I_{zx}$, $I_{yz} = I_{zy}$, are called *products of inertia* of B about O–xyz, and they are *sign-indefinite*; that is, they may be >0, <0, or $=0$.

In view of the above, T can be rewritten as

$$2T = I_{xx}\,\omega_x{}^2 + I_{yy}\,\omega_y{}^2 + I_{zz}\,\omega_z{}^2 + 2I_{xy}\,\omega_x\,\omega_y + 2I_{xz}\,\omega_x\,\omega_z + 2I_{yz}\,\omega_y\,\omega_z. \qquad (1.15.4)$$

Now, evidently, the choice of the axes O–xyz is *nonunique*. Not only can they be *non–body-fixed* (in which case, the I_{kl} are, in general, time-dependent); but even if they are taken as body-fixed, (1.15.3, 4) are still fairly complicated. Hence, to simplify matters as much as possible, and since the kinetic energy is so central to analytical mechanics, we, in general, strive to choose *principal axes* at O: O–$xyz \rightarrow O$–123; usually, but not always, body-fixed. Since $\mathbf{I}$ is symmetric, such (mutually orthogonal) axes exist always; and along them $\mathbf{I}$ becomes

$$\mathbf{I} = \begin{pmatrix} I_1 & 0 & 0 \\ 0 & I_2 & 0 \\ 0 & 0 & I_3 \end{pmatrix} = \textit{Principal axes representation of inertia tensor}, \qquad (1.15.5)$$

where the *principal moments of inertia*, at O, $I_{1,2,3}$ are the eigenvalues of

$$\sum I_{kl}\,\omega_l = \lambda\,\omega_k, \qquad (1.15.6a)$$

that is, they are the roots of its characteristic equation:

$$D(\lambda) \equiv -\mathrm{Det}(I_{kl} - \lambda\delta_{kl}) = 0; \qquad \lambda_{1,2,3} \equiv I_{1,2,3}. \qquad (1.15.6b)$$

Using basic theorems of the spectral theory of tensors [(1.1.17a ff.)] we can show the following:

(i) At each point of a rigid body B there exists *at least one set of principal axes*.
(ii) Since, by (1.15.1b–d), the inertia tensor is not only *symmetric*, but also *positive definite* [i.e., $\sum\sum I_{kl} a_k a_l > 0$, for all vectors $\boldsymbol{a} = (a_k) \neq \mathbf{0}$], all three roots of (1.15.6b) are not only *real* but also *strictly positive*.

Further

- If $\lambda_1 \neq \lambda_2 \neq \lambda_3 \neq \lambda_1$ (all three eigenvalues distinct), then *O–123* is unique.
- If $\lambda_1 \neq \lambda_2 = \lambda_3$ (two distinct eigenvalues), there exists a *single* infinity of such sets of principal axes; *O1* and every line perpendicular to it are principal axes; that is, the direction of either *O2* or *O3*, in the plane perpendicular to *O1*, can be chosen arbitrarily (e.g., a homogeneous right circular cylinder, with O on its axis of symmetry).
- If $\lambda_1 = \lambda_2 = \lambda_3$ (only one distinct eigenvalue), there exists a *double* infinity: any three mutually perpendicular axes can be chosen arbitrarily as *O–123* (e.g., O being the center of a homogeneous sphere). Along principal axes, T, (1.15.4), with $\boldsymbol{\omega} = (\omega_1, \omega_2, \omega_3)$, reduces to

$$2T = I_1\omega_1{}^2 + I_2\omega_2{}^2 + I_3\omega_3{}^2. \tag{1.15.6c}$$

The Generalized Parallel Axis Theorem (Huygens–Steiner)

This explains how $\mathbf{I}$ changes from point to point, among parallel sets of axes.

THEOREM

Let O–xyz and G–xyz be two sets of mutually parallel axes, and let the coordinates of the center of mass of B, G, relative to O, be

$$\boldsymbol{OG} \equiv \boldsymbol{r}_G \equiv (x_G, y_G, z_G) \equiv (G_1, G_2, G_3) \equiv (-a, -b, -c), \tag{1.15.7a}$$

that is, $a, b, c =$coordinates of O relative to G–xyz. Then, the components of the inertia tensor of B at O, $I_{O,kl}$, and at G, $I_{G,kl}$, are related by

Direct notation:
$$\mathbf{I}_O = \mathbf{I}_G + m(r_G{}^2\mathbf{1} - \boldsymbol{r}_G \otimes \boldsymbol{r}_G), \tag{1.15.7b}$$

Indicial notation:
$$I_{O,kl} = I_{G,kl} + m\Big[\Big(\sum G_r G_r\Big)\delta_{kl} - G_k G_l\Big]; \tag{1.15.7c}$$

or, in extenso,

$$\mathbf{I}_O = \mathbf{I}_G + \begin{pmatrix} m(b^2+c^2) & -mab & -mca \\ -mab & m(c^2+a^2) & -mbc \\ -mca & -mbc & m(a^2+b^2) \end{pmatrix}. \tag{1.15.7d}$$

PROOF

We have, successively,

$$\begin{aligned}
\text{(i)}\quad I_{O,xx} &= \int dm[(y-b)^2 + (z-c)^2] \\
&= \int dm(y^2+z^2) - 2b\Big(\int dm\, y\Big) - 2b\Big(\int dm\, z\Big) + \int dm(b^2+c^2) \\
&= I_{G,xx} + 0 + 0 + m(b^2+c^2); \quad \text{etc., cyclically, for } I_{O,yy}, I_{O,zz}.
\end{aligned} \tag{1.15.7e}$$

$$\begin{aligned}
\text{(ii)}\quad I_{O,yz} &= -\int dm[(y-b)(z-c)] \\
&= -\int dm\, yz + c\Big(\int dm\, y\Big) + b\Big(\int dm\, z\Big) - \int dm\, bc \\
&= I_{G,yz} + 0 + 0 - mbc; \quad \text{etc., cyclically, for } I_{O,xz}, I_{O,xy}.
\end{aligned} \tag{1.15.7f}$$

More generally, it can be shown that between *any two* points A, B (with some ad hoc but, hopefully, self-explanatory notation),

$$\mathbf{I}_B = \mathbf{I}_A + m({r_{A/B}}^2\mathbf{1} - \boldsymbol{r}_{A/B}\otimes\boldsymbol{r}_{A/B}) + 2m\big[(\boldsymbol{r}_{A/B}\cdot\boldsymbol{r}_{G/A})\mathbf{1} - (1/2)(\boldsymbol{r}_{A/B}\otimes\boldsymbol{r}_{G/A} + \boldsymbol{r}_{G/A}\otimes\boldsymbol{r}_{A/B})\big]; \quad (1.15.7g)$$

which, when $A \to G$ and $\boldsymbol{r}_{G/A} \to \mathbf{0}$, reduces to (1.15.7b) (see below). (See, e.g., Lur'e, 1968, p. 143; also Crandall et al., 1968, pp. 180–182, Magnus, 1974, pp. 200–201.)

It should be noted that the transfer formulae (1.15.7b, g) are based on definitions of moments of inertia about *points*, like (1.15.2h), not axes, and therefore hold for any set of axes through these points; that is, they are independent of the axes orientation at A, B. If, however, these axes are parallel, certain simplifications occur; indeed, (1.15.7g) then yields the component form,

$$I_{B,kl} = I_{A,kl} + m\big[({x_{A/B,1}}^2 + {x_{A/B,2}}^2 + {x_{A/B,3}}^2)\delta_{kl} - x_{A/B,k}x_{A/B,l}\big] + 2m\big[(x_{A/B,1}\,x_{G/A,1} + x_{A/B,2}\,x_{G/A,2} + x_{A/B,3}\,x_{G/A,3})\delta_{kl} - (1/2)(x_{A/B,k}\,x_{G/A,l} + x_{G/A,k}\,x_{A/B,l})\big], \quad (1.15.7h)$$

where $\boldsymbol{r}_{A/B} \equiv (x_{A/B,1}, x_{A/B,2}, x_{A/B,3})$ = coordinates of A relative to B, along axes $B\text{–}xyz \equiv B\text{–}x_k$, and $\boldsymbol{r}_{G/A} \equiv (x_{G/A,1}, x_{G/A,2}, x_{G/A,3})$ = coordinates of G relative to A, along axes $A\text{–}xyz \equiv A\text{–}x_k$ (parallel to $B\text{–}x_k$); or, in extenso, with $x_{A/B,1} \equiv x_{A/B}$, $x_{A/B,2} \equiv y_{A/B}$, $x_{A/B,3} \equiv z_{A/B}$, $x_{G/A,1} \equiv x_{G/A}$, etc.,

$$I_{B,xx} = I_{A,xx} + m({y_{A/B}}^2 + {z_{A/B}}^2) + 2m(y_{A/B}\,y_{G/A} + z_{A/B}z_{G/A}), \quad \text{etc., cyclically,} \quad (1.15.7i)$$

$$I_{B,xy} = I_{A,xy} - m(y_{A/B}x_{G/A} + x_{A/B}\,y_{G/A}) - m(x_{A/B}\,y_{A/B}), \quad \text{etc., cyclically,} \quad (1.15.7j)$$

If $A \to G$, then $\boldsymbol{r}_{G/A} \to \mathbf{0}$, $\boldsymbol{r}_{A/B} \to \boldsymbol{r}_{G/B}$, and the above reduces to

$$I_{B,kl} = I_{G,kl} + m\big[({x_{G/B,1}}^2 + {x_{G/B,2}}^2 + {x_{G/B,3}}^2)\delta_{kl} - x_{G/B,k}x_{G/B,l}\big], \quad (1.15.7k)$$

from which, in extenso,

$$I_{B,xx} = I_{G,xx} + m({y_{G/B}}^2 + {z_{G/B}}^2) = I_{G,xx} + m[(-b)^2 + (-c)^2], \quad \text{etc., cyclically,} \quad (1.15.7l)$$

$$I_{B,xy} = I_{G,xy} - m(x_{G/B}\,y_{G/B}) = I_{G,xy} - m[(-a)(-b)], \quad \text{etc., cyclically; i.e., (1.15.7b–f).} \quad (1.15.7m)$$

Ellipsoid of Inertia

Let us consider a rectangular Cartesian coordinate system/basis $O\text{–}xyz/\boldsymbol{ijk}$, and an axis u through O defined by the unit vector $\boldsymbol{u} = (u_x, u_y, u_z)$. Then, as the transformation equations against rotations (1.15.2i) show, the moment of inertia of a body B about u, $I_{uu} \equiv I$, will be (with $k' = l' = u; k, l = x, y, z$; $A_{k'k} = u_k = u_{1,2,3}$, $A_{k'l} = u_l = u_{1,2,3}$, etc.)

$$I = I_{xx}{u_x}^2 + I_{yy}{u_y}^2 + I_{zz}{u_z}^2 + 2I_{xy}u_xu_y + 2I_{yz}u_yu_z + 2I_{xz}u_xu_z. \quad (1.15.8a)$$

[For *nontensorial* derivations of (1.15.8a), see, for example, Lamb (1929, pp. 66–67), Spiegel (1967, pp. 263–264)]. We notice that if $\boldsymbol{\omega} = \omega \boldsymbol{u} = (\omega u_x, \omega u_y, \omega u_z) = (\omega_x, \omega_y, \omega_z)$, then the kinetic energy of B, moving about the fixed point O, eq. (1.15.4), becomes

$$2T = I\omega^2. \tag{1.15.8b}$$

Now, by defining the radius vector

$$\boldsymbol{r} \equiv \boldsymbol{u}/I^{1/2} = x\boldsymbol{i} + y\boldsymbol{j} + z\boldsymbol{k} \qquad [\text{i.e., } |\boldsymbol{r}| \equiv r = (1/I)^{1/2}], \tag{1.15.8c}$$

we can rewrite (1.15.8a) as

$$I_{xx}x^2 + I_{yy}y^2 + I_{zz}z^2 + 2I_{xy}xy + 2I_{yz}yz + 2I_{zx}zx = 1. \tag{1.15.8d}$$

Since I is positive [except when all the mass lies on u; then one of the principal moments of inertia, roots of (1.15.6b), is zero and the other two are equal and positive], every radius through O meets the quadric surface represented by (1.15.8d), in O–xyz, in real points located a distance $r = (1/I)^{1/2}$ from O, and therefore (1.15.8d) is an ellipsoid; appropriately called *ellipsoid of inertia* or *momental ellipsoid.* [A term most likely introduced by Cauchy (1827), who also carried out similar investigations in the theory of stress in continuous media ("stress quadric").]

If the axes are rotated so as to coincide with the principal axes of the ellipsoid—that is, O–$xyz \to O$–123—then (1.15.8d) simplifies to

$$I_1 r_1{}^2 + I_2 r_2{}^2 + I_3 r_3{}^2 = 1,$$

or

$$[r_1/(1/I_1)^{1/2}]^2 + [r_2/(1/I_2)^{1/2}]^2 + [r_3/(1/I_3)^{1/2}]^2 = 1, \tag{1.15.8e}$$

where $r_{1,2,3}$ are the "principal" coordinates of $\boldsymbol{r}$, and $(1/I_{1,2,3})^{1/2}$ are the semidiameters of the ellipsoid. [Some authors (mostly British) define the radius of the momental ellipsoid along $\boldsymbol{u}$ (i.e., our r) as

$$r = m\varepsilon^4/I^{1/2} \sim I^{1/2}, \tag{1.15.8c.1}$$

where m = mass of body, and ε = any linear magnitude (taken in the fourth power for purely dimensional purposes), so that the ellipsoid equations (1.15.8d) and (1.15.8e) are replaced, respectively, by

$$I_{xx}x^2 + I_{yy}y^2 + I_{zz}z^2 + 2I_{xy}xy + 2I_{yz}yz + 2I_{zx}zx = m\varepsilon^4, \tag{1.15.8d.1}$$

$$I_1 r_1{}^2 + I_2 r_2{}^2 + I_3 r_3{}^2 = m\varepsilon^4. \tag{1.15.8e.1}$$

Also, for a discussion of the closely related concept of the *ellipsoid of gyration* (introduced by MacCullagh, 1844), see, for example, Easthope (1964, p. 134 ff.), Lamb (1929, p. 68 ff.)] However, it should be remarked that *not every ellipsoid* can represent an inertia ellipsoid; in view of the "triangle inequalities" (see below), certain restrictions apply on the relative magnitudes of the semidiameters, and hence the possible forms of the momental ellipsoid.

Now, and these constitute a geometrical sequel to the discussion of the roots of the characteristic equation (1.15.6b):

- If $I_1 = I_2 = I_3$, the momental ellipsoid is a sphere. All axes through O are principal, and all moments of inertia are mutually equal. Such a body is called *kinetically symmetrical about O*.
- If, say, $I_2 = I_3$, the ellipsoid is one of revolution about Ox—all perpendicular diameters to Ox are principal axes. Such a body is called *kinetically symmetric about that axis*; or simply *uniaxial* (Routh).

The above show that, in general, the ellipsoid of inertia, at a point, is nonunique. The ellipsoid of inertia of a body at its mass center G, commonly referred to as its *central ellipsoid* (Poinsot), is of particular importance: As the parallel axis theorem shows, if the moment of inertia about an axis through G is known, I_G, then the moment of inertia about any other axis parallel to it is obtained by adding to I_G the nonnegative quantity md^2, where d is the distance between the two axes.

Finally, the momental ellipsoid interpretation, plus the above parallel axis theorem, allow us to conclude the following *extremum* (i.e., *maximum/minimum*) *properties* of the principal axes:

- The principal axes of inertia, at a point O, are those with the larger or smaller moment of inertia than those about any other line through that point, I. Quantitatively, if

$$I_1 \equiv I_{max} \geq I_2 \geq I_3 \equiv I_{min} \tag{1.15.8f}$$

$[\Rightarrow (1/I_1)^{1/2} \leq (1/I_2)^{1/2} \leq (1/I_3)^{1/2}]$, something that can always be achieved by appropriate numbering of the principal axes, then

$$I_{max} \geq I \geq I_{min}. \tag{1.15.8g}$$

- The smallest *central* principal moment of inertia of a body, say $I_{G,3} \equiv I_{G,min}$, is smaller than or equal to *any* other possible moment of inertia of the body (i.e., moment of inertia about any other space point and direction there); that is, $I_{G,min} \geq I_{...,...}$.

Additional Useful Results

(i) It can be shown that

$$I_1 \leq I_2 + I_3, \qquad I_2 \leq I_3 + I_1, \qquad I_3 \leq I_1 + I_2; \tag{1.15.8h}$$

that is, no principal moment of inertia can exceed the *sum of the other two*. Equations (1.15.8h) are referred to as the *triangle inequalities* (since similar relations hold for the sides of a plane triangle). Actually, this theorem holds for the moments of inertia about *any* mutually orthogonal axes (McKinley, 1981).

(ii) Let $\rho_{1,2,3}$ be the semidiameters (semiaxes) of the ellipsoid of inertia; that is, $\rho_{1,2,3} \equiv (I_{1,2,3})^{-1/2}$. Then, the third and second of (1.15.8h) lead, respectively, to the following *lower and upper bounds* for ρ_3, if $\rho_{1,2}$ are given,

$$(\rho_2{}^{-2} + \rho_1{}^{-2})^{-1/2} \leq \rho_3 \leq |\rho_2{}^{-2} - \rho_1{}^{-2}|^{-1/2}; \tag{1.15.8i}$$

and, cyclically, for $\rho_{1,2}$; that is, *arbitrary inertia tensors*, upon *diagonalization*, may yield (mathematically correct but) physically impossible principal moments of inertia!

As a result of the above, if two axes, say ρ_1 and ρ_2, are approximately equal, the corresponding inertia ellipsoid can be quite *prolate* (longer in the third direction, cigar shaped), but not too *oblate* (shorter in the third direction; flattened at the poles, like the Earth).

(iii) The quantity $\mathrm{Tr}\,\mathbf{I} \equiv I_{xx} + I_{yy} + I_{zz}$ (first invariant of $\mathbf{I}$) depends on the origin of the coordinates, but *not on their orientation.*

$$\text{(iv)}\quad -I_{xx}/2 \le I_{yz} \le I_{xx}/2, \qquad -I_{yy}/2 \le I_{zx} \le I_{yy}/2, \qquad -I_{zz}/2 \le I_{xy} \le I_{zz}/2. \tag{1.15.8j}$$

(v) Consider the following three sets of axes: (a) O–XYZ: arbitrary "background" (say, inertial) axes; (b) G–$XYZ \equiv G$–xyz: translating but nonrotating axes, at center of mass G; and (c) G–123: *principal* axes at G. By combining the transformation formulae for I_{kl}, between parallel axes of differing origins (like O–XYZ and G–xyz) and arbitrary oriented axis of common origin (like G–xyz and G–123), we can show that

$$I_{XX} = m(Y_G{}^2 + Z_G{}^2) + A_{X1}{}^2 I_1 + A_{X2}{}^2 I_2 + A_{X3}{}^2 I_3, \tag{1.15.8k}$$

$$I_{YY} = m(Z_G{}^2 + X_G{}^2) + A_{Y1}{}^2 I_1 + A_{Y2}{}^2 I_2 + A_{Y3}{}^2 I_3, \tag{1.15.8l}$$

$$I_{ZZ} = m(X_G{}^2 + Y_G{}^2) + A_{Z1}{}^2 I_1 + A_{Z2}{}^2 I_2 + A_{Z3}{}^2 I_3; \tag{1.15.8m}$$

$$I_{XY} = -mX_G Y_G + A_{X1}A_{Y1}(I_3 - I_1) + A_{X2}A_{Y2}(I_3 - I_2), \tag{1.15.8n}$$

$$I_{YZ} = -mY_G Z_G + A_{Y1}A_{Z1}(I_3 - I_1) + A_{Y2}A_{Z2}(I_3 - I_2), \tag{1.15.8o}$$

$$I_{ZX} = -mZ_G X_G + A_{Z1}A_{X1}(I_3 - I_1) + A_{Z2}A_{X2}(I_3 - I_2); \tag{1.15.8p}$$

where $A_{Y1} \equiv \cos(OX, G1) = \cos(Gx, G1)$, etc., and X_G, Y_G, Z_G are the coordinates of G relative to O–XYZ. The usefulness of (1.15.8k–p) lies in the fact that they yield the moments/products of inertia about arbitrary axes, once the principal moments of inertia at the center of mass are known.

(a) If a body has a *plane of symmetry*, then (α) its center of mass and (β) two of its principal axes of inertia there lie on that plane; while the third principal axis is perpendicular to it.
(b) If a body has an *axis of symmetry*, then (α) its center of a mass lies there, and (β) that axis is one of its principal axes of inertia; while the other two are perpendicular to it.
(c) If two perpendicular axes, through a body point, are axes of symmetry, then they are principal axes there. (But principal axes are not necessarily axes of symmetry!)
(d) If the products of inertia vanish, for three mutually perpendicular axes at a point, these axes are principal axes there. [For a general discussion of the relations between principal axes and symmetry (via the concept of *covering operation*), see, for example, Synge and Griffith (1959, p. 288 ff.).]
(e) A principal axis at the center of mass of a body is a principal axis at all points of that axis.
(f) If an axis is principal at any two of its points, then it passes through the center of mass of the body, and is a principal axis at all its points.

(vii) *Centrifugal forces*: whence the products of inertia originate. Let us consider an arbitrary rigid body rotating about a *fixed* axis OZ with constant angular velocity ω. Then, since the centripetal acceleration of a generic particle of it P, of mass dm, equals $v^2/r = \omega^2 r$, where $r =$ distance of P from OZ, the associated centrifugal force

df_c has magnitude $df_c = dm(\omega^2 r)$, and hence components along a, say, *body-fixed* set of axes $O\text{–}xyz(OZ = Oz)$ will be

$$df_{c,x} = df_c(x/r) = dm\, x\omega^2, \qquad df_{c,y} = df_c(y/r) = dm\, y\omega^2, \qquad df_{c,z} = 0; \quad (1.15.9a)$$

where x, y, z are the coordinates of P. Therefore, the components of the *moment* of df_c along these axes are

$$\begin{aligned} dM_{c,x} &= y\, df_{c,z} - z\, df_{c,y} = -dm\, yz\omega^2, \\ dM_{c,y} &= z\, df_{c,x} - x\, df_{c,z} = +dm\, xz\omega^2, \\ dM_{c,z} &= x\, df_{c,y} - y\, df_{c,x} = 0. \end{aligned} \qquad (1.15.9b)$$

From the above, it follows that these centrifugal forces, when summed over the entire body and reduced to the origin O, yield a *resultant centrifugal force* $\boldsymbol{f}_c$:

$$\begin{aligned} f_{c,x} &\equiv \textstyle\int df_{c,x} = \omega^2 \int dm\, x \equiv \omega^2 m\, x_G, \\ f_{c,y} &\equiv \textstyle\int df_{c,y} = \omega^2 \int dm\, y \equiv \omega^2 m\, y_G, \\ f_{c,z} &\equiv \textstyle\int df_{c,z} = 0, \end{aligned} \qquad (1.15.9c)$$

where x_G, y_G are the coordinates of the mass center of B, G; and a *resultant centrifugal moment* $\boldsymbol{M}_c$:

$$\begin{aligned} M_{c,x} &\equiv \textstyle\int dM_{c,x} = -\omega^2 \int dm\, yz \equiv +\omega^2 I_{yz}, \\ M_{c,y} &\equiv \textstyle\int dM_{c,y} = \omega^2 \int dm\, zx \equiv -\omega^2 I_{xz}, \\ M_{c,z} &\equiv \textstyle\int dM_{c,z} = 0. \end{aligned} \qquad (1.15.9d)$$

Equations (1.15.9c, d) show clearly that if G lies on the $Z = z$ axis, then $\boldsymbol{f}_c$ vanishes, but $\boldsymbol{M}_c$ does not. For the moment to vanish, we must have $I_{yz} = 0$ *and* $I_{xz} = 0$; that is, Oz must be a *principal axis*. In sum: *The centrifugal forces on a spinning body tend to change the orientation of its instantaneous axis of rotation, unless the latter goes through the center of mass of the body and is a principal axis there.* Such kinetic considerations led to the formulation of the concept of principal axes of inertia, at a point of a rigid body [Euler, Segner (1750s)]; and to the alternative term *deviation moments*, for the products of inertia. We shall return to this important topic in §1.17.

1.16 THE RIGID BODY: LINEAR AND ANGULAR MOMENTUM

(i) The *inertial*, or *absolute*, linear momentum of a rigid body B (or system S), relative to an inertial frame F, represented by the axes $I\text{–}XYZ$ (fig. 1.28), is defined as

$$\boldsymbol{p} \equiv \textstyle\int dm\, \boldsymbol{v} = m\boldsymbol{v}_G \qquad (G\text{: } \textit{center of mass} \text{ of } B). \qquad (1.16.1a)$$

Substituting in the above [recalling (1.7.11a ff.)]

$$\boldsymbol{v}_G = \boldsymbol{v}_O + \boldsymbol{v}_{G/O} = \boldsymbol{v}_O + \boldsymbol{v}_{G,\text{rel}} + \boldsymbol{\Omega} \times \boldsymbol{r}_G \qquad (\boldsymbol{v}_{G,\text{rel}} \equiv d'\boldsymbol{r}/dt) \qquad (1.16.1b)$$

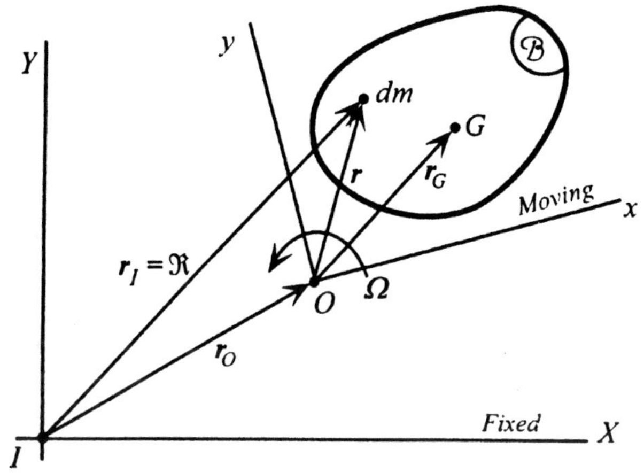

Figure 1.28 Rigid body (*B*), or system (*S*), in general motion relative to the noninertial frame *O–xyz*; **Ω**: inertial angular velocity of *O–xyz* (two-dimensional case).

readily yields

$$\boldsymbol{p} \equiv \boldsymbol{p}_{\text{trans}} + \boldsymbol{p}_{\text{rel}}, \qquad (1.16.1c)$$

$$\boldsymbol{p}_{\text{trans}} \equiv m(\boldsymbol{v}_O + \boldsymbol{\Omega} \times \boldsymbol{r}_G): \text{ Linear momentum of } \textit{transport}, \qquad (1.16.1d)$$

$$\boldsymbol{p}_{\text{rel}} \equiv m\boldsymbol{v}_{G,\text{rel}} \equiv m(d'\boldsymbol{r}/dt): \text{ Linear momentum of } \textit{relative motion}. \qquad (1.16.1e)$$

If B is rigidly attached to the moving frame M, represented by the axes O–xyz (fig. 1.28), then, clearly, $\boldsymbol{p}_{\text{rel}} = \boldsymbol{0}$.

(ii) The *inertial and absolute* angular momentum of B, relative to the inertial origin I, $\boldsymbol{H}_{I,\text{abs}} \equiv \boldsymbol{H}_I$, is defined as

$$\begin{aligned} \boldsymbol{H}_I &\equiv \mathbf{S}\, [\boldsymbol{r}_I \times dm\,(d\boldsymbol{r}_I/dt)] \equiv \mathbf{S}\, [\mathfrak{R} \times dm\,(d\mathfrak{R}/dt)] \\ &\quad [\text{substituting } \mathfrak{R} = \boldsymbol{r}_{O/I} + \boldsymbol{r} \equiv \boldsymbol{r}_O + \boldsymbol{r} \Rightarrow d\mathfrak{R}/dt = \boldsymbol{v}_O + \boldsymbol{v}_{\text{rel}} + \boldsymbol{\Omega} \times \boldsymbol{r}, \boldsymbol{v}_{\text{rel}} \equiv d'\boldsymbol{r}/dt] \\ &= m\boldsymbol{r}_O \times (\boldsymbol{v}_O + \boldsymbol{\Omega} \times \boldsymbol{r}_G) + m\boldsymbol{r}_G \times \boldsymbol{v}_O \\ &\qquad + \mathbf{S}\, dm\,[\boldsymbol{r} \times (\boldsymbol{\Omega} \times \boldsymbol{r})] + \mathbf{S}\, dm\,(\boldsymbol{r} \times \boldsymbol{v}_{\text{rel}}), \end{aligned} \qquad (1.16.2a)$$

or, since

$$\begin{aligned} \mathbf{S}\, dm\,[\boldsymbol{r} \times (\boldsymbol{\Omega} \times \boldsymbol{r})] &= \mathbf{S}\, dm\,[r^2\boldsymbol{\Omega} - (\boldsymbol{r} \otimes \boldsymbol{r}) \cdot \boldsymbol{\Omega}] \\ &= \mathbf{S}\, dm\,[r^2\mathbf{1} - (\boldsymbol{r} \otimes \boldsymbol{r})] \cdot \boldsymbol{\Omega} \equiv \mathbf{I}_O \cdot \boldsymbol{\Omega}, \end{aligned} \qquad (1.16.2b)$$

and calling

$$\boldsymbol{H}_{O,\text{rel}} \equiv \mathbf{S}\, \boldsymbol{r} \times (dm\, \boldsymbol{v}_{\text{rel}}): \textit{Noninertial and absolute angular momentum of } B, \textit{ about } O, \qquad (1.16.2c)$$

we finally obtain the following general kinematico-inertial result:

$$\boldsymbol{H}_I = \boldsymbol{H}_O + \boldsymbol{r}_O \times \boldsymbol{p} + m\boldsymbol{r}_G \times \boldsymbol{v}_O, \qquad \boldsymbol{H}_O \equiv \mathbf{I}_O \cdot \boldsymbol{\Omega} + \boldsymbol{H}_{O,\text{rel}}. \tag{1.16.2d}$$

Special Cases

(i) If the body B is rigidly attached to the moving frame M, then

$$\boldsymbol{v}_{\text{rel}} = \mathbf{0}, \quad \boldsymbol{p}_{\text{rel}} = \mathbf{0}, \quad \boldsymbol{H}_{O,\text{rel}} = \mathbf{0}, \quad \boldsymbol{\Omega} = \boldsymbol{\omega} = \text{inertial angular velocity of } B, \tag{1.16.3a}$$

and, therefore,

$$\boldsymbol{p} = m(\boldsymbol{v}_O + \boldsymbol{\omega} \times \boldsymbol{r}_G) = m\boldsymbol{v}_G, \qquad \boldsymbol{H}_I = \mathbf{I}_O \cdot \boldsymbol{\omega} + \boldsymbol{r}_O \times \boldsymbol{p} + m\boldsymbol{r}_G \times \boldsymbol{v}_O. \tag{1.16.3b}$$

(ii) If, *further*, $O = G$, then $\boldsymbol{r}_G = \mathbf{0}$, $\boldsymbol{r}_O = \boldsymbol{r}_G$, and, therefore,

$$\boldsymbol{p} = m\boldsymbol{v}_G, \qquad \boldsymbol{H}_I = \mathbf{I}_O \cdot \boldsymbol{\omega} + \boldsymbol{r}_O \times \boldsymbol{p}. \tag{1.16.3c}$$

(iii) If $I = O$ (i.e., rigid-body motion with one point, O, fixed), then $\boldsymbol{r}_O = \mathbf{0}$, $\boldsymbol{v}_O = \mathbf{0}$, and, therefore,

$$\boldsymbol{p} = m(\boldsymbol{\omega} \times \boldsymbol{r}_G) = m\boldsymbol{v}_G, \qquad \boldsymbol{H}_I = \boldsymbol{H}_O = \mathbf{I}_O \cdot \boldsymbol{\omega} \qquad \left[\equiv \int \boldsymbol{r} \times dm\, \boldsymbol{v}\right]. \tag{1.16.3d1, 2}$$

It should be pointed out that the above hold for any set of axes, including *non–body-fixed* ones, at the fixed point O; but along such axes the components of $\mathbf{I}_O$ will, in general, not be constant. Equation (1.16.3d2) would then yield, in components [omitting the subscript O and with $\boldsymbol{r} = (x_k)$],

$$H_k = \sum I_{kl}\,\omega_l, \tag{1.16.4a}$$

$$I_{kl} = \int dm\big[(\boldsymbol{r}\cdot\boldsymbol{r})\mathbf{1} - (\boldsymbol{r}\otimes\boldsymbol{r})\big]_{kl} = \int dm\Big[\Big(\sum x_r x_r\Big)\,\delta_{kl} - x_k x_l\Big]. \tag{1.16.4b}$$

If the axes at O are *body-fixed*, then the I_{kl} are constant; and, further, if they are *principal*, then

$$H_k = I_k\,\omega_k. \tag{1.16.4c}$$

Linear Momentum of a Rotating Body

To dispel possible notions that the linear momentum is associated only with translation, let us calculate the linear momentum of a rigid body rotating about a *fixed point* ◆. We have, successively, with the usual notations,

$$\begin{aligned}\boldsymbol{p} = m\boldsymbol{v}_G &= m(\boldsymbol{\omega} \times \boldsymbol{r}_{G/\blacklozenge}) \equiv m(\boldsymbol{\omega} \times \boldsymbol{r}_G)\\ &= m(\omega_x, \omega_y, \omega_z) \times (x_G, y_G, z_G) \qquad \text{[components along any ◆-axes]}\\ &= m(\omega_y z_G - \omega_z y_G, \quad \omega_z x_G - \omega_x z_G, \quad \omega_x y_G - \omega_y x_G).\end{aligned} \tag{1.16.5a}$$

In particular, if the body rotates about a *fixed axis through* ◆, say ◆z [recalling (1.15.9a ff.)], then $\boldsymbol{\omega} = \omega_z \boldsymbol{k} \equiv (d\phi/dt)\boldsymbol{k}$, and so (1.16.5a) reduces to

$$p_x = -(my_G)\omega_z \equiv -m_y\,\omega_z, \qquad p_y = (mx_G)\omega_z \equiv m_x\,\omega_z, \qquad p_z = 0. \tag{1.16.5b}$$

These expressions appear in the problem of rotation of a rigid body about a fixed axis, treated via body-fixed axes ◆–xyz [see, e.g., Butenin et al. (1985, pp. 266–278) and Papastavridis (EM, in preparation)].

It is not hard to show that, in this case, the (inertial and absolute) angular momentum of the body

$$\boldsymbol{H}_\blacklozenge \equiv \mathcal{S}\, \boldsymbol{r} \times dm\,\boldsymbol{v} = (H_x, H_y, H_z), \tag{1.16.5c}$$

reduces to

$$H_x = I_{xz}\,\omega_z, \qquad H_y = I_{yz}\,\omega_z, \qquad H_z = I_{zz}\,\omega_z \equiv I_z\,\omega_z. \tag{1.16.5d}$$

1.17 THE RIGID BODY: KINETIC ENERGY AND KINETICS OF TRANSLATION AND ROTATION (EULERIAN "GYRO EQUATIONS")

We recommend, for this section, the concurrent reading of a good text on rigid-body dynamics; for example (alphabetically): Grammel (1950), Gray (1918), Hughes (1986), Leimanis (1965), Magnus (1971, 1974), Mavraganis (1987), Stäckel (1905, pp. 556–563).

(i) The inertial kinetic energy of a rigid body B in general motion, T, is defined as the sum of the (inertial) kinetic energies of its particles:

$$2T \equiv \mathcal{S}\, dm\,\boldsymbol{v} \cdot \boldsymbol{v}, \tag{1.17.1}$$

or, since

$$\boldsymbol{v} = \boldsymbol{v}_\blacklozenge + \boldsymbol{\omega} \times \boldsymbol{r}_{/\blacklozenge} \qquad (\blacklozenge\text{: arbitrary } \textit{body-fixed} \text{ point}), \tag{1.17.2}$$

$$2T = \mathcal{S}\, dm(\boldsymbol{v}_\blacklozenge + \boldsymbol{\omega} \times \boldsymbol{r}_{/\blacklozenge}) \cdot (\boldsymbol{v}_\blacklozenge + \boldsymbol{\omega} \times \boldsymbol{r}_{/\blacklozenge}) = 2(T_{\text{transl'n}} + T_{\text{rot'n}} + T_{\text{cpl'g}}), \tag{1.17.3}$$

where

$$2T_{\text{transl'n}} \equiv m\boldsymbol{v}_\blacklozenge \cdot \boldsymbol{v}_\blacklozenge = m{v_\blacklozenge}^2:$$

(twice of) *Translatory* (or sliding) kinetic energy of B, (1.17.3a)

$$2T_{\text{rot'n}} \equiv \mathcal{S}\, dm\,(\boldsymbol{\omega} \times \boldsymbol{r}_{/\blacklozenge}) \cdot (\boldsymbol{\omega} \times \boldsymbol{r}_{/\blacklozenge}) = \boldsymbol{\omega} \cdot \mathbf{I}_\blacklozenge \cdot \boldsymbol{\omega}: \qquad \text{(recalling §1.15)}$$

(twice of) *Rotatory* kinetic energy of B, (1.17.3b)

$$T_{\text{cpl'g}} \equiv m(\boldsymbol{\omega} \times \boldsymbol{r}_{G/\blacklozenge}) \cdot \boldsymbol{v}_\blacklozenge = m\boldsymbol{v}_{G/\blacklozenge} \cdot \boldsymbol{v}_\blacklozenge = (m\boldsymbol{r}_{G/\blacklozenge})^{\cdot} \cdot \boldsymbol{v}_\blacklozenge \equiv (d\boldsymbol{m}_{G/\blacklozenge}/dt) \cdot \boldsymbol{v}_\blacklozenge,$$

or

$$T_{\text{cpl'g}} \equiv m\boldsymbol{r}_{G/\blacklozenge} \cdot (\boldsymbol{v}_\blacklozenge \times \boldsymbol{\omega}) \equiv \boldsymbol{m}_{G/\blacklozenge} \cdot (\boldsymbol{v}_\blacklozenge \times \boldsymbol{\omega}): \tag{1.17.3c}$$

Kinetic energy of *coupling* between $\boldsymbol{v}_\blacklozenge$ and $\boldsymbol{\omega}$ (where $\boldsymbol{m}_{G/\blacklozenge} \equiv m\boldsymbol{r}_{G/\blacklozenge}$) [$= 0$; if $G = \blacklozenge$, or if $\boldsymbol{v}_\blacklozenge$ and $\boldsymbol{\omega}$ are parallel (screw motion), or if $\boldsymbol{v}_\blacklozenge$ is normal to $\boldsymbol{v}_G/\blacklozenge$; in which case, T decouples into translatory and rotatory kinetic energy]. (1.17.3d)

• These expressions hold for any axes, either body-fixed or moving in an arbitrary manner, or even inertial. But if they are *non-body-fixed*, the components of $\boldsymbol{r}_{G/\blacklozenge}$ and $\mathbf{I}_\blacklozenge$ will, in general, not be constant.

• We also notice that, in there, the mass m appears as a *scalar* (m: $T_{\text{transl'n}}$), as a *vector* of a first-order moment ($\boldsymbol{m}_{G/\blacklozenge}$: $T_{\text{cpl'g}}$), and as a second-order *tensor* ($\mathbf{I}_\blacklozenge$: $T_{\text{rot'n}}$).

• From (1.17.3b), we obtain, successively,

$$\begin{aligned} \boldsymbol{grad}_\omega T_{\text{rot'n}} &\equiv \partial T_{\text{rot'n}}/\partial \boldsymbol{\omega} \\ &= \mathcal{S}\, dm\, \boldsymbol{v}_{/\blacklozenge} \cdot (\partial \boldsymbol{v}_{/\blacklozenge}/\partial \boldsymbol{\omega}) = \mathcal{S}\, \boldsymbol{r}_{/\blacklozenge} \times (\boldsymbol{\omega} \times \boldsymbol{r}_{/\blacklozenge})\, dm = \mathcal{S}\, \boldsymbol{r}_{/\blacklozenge} \times (dm\, \boldsymbol{v}_{/\blacklozenge}) \\ &\equiv \boldsymbol{H}_{\blacklozenge,\text{relative}} \equiv \boldsymbol{h}_\blacklozenge \qquad \text{[recalling (1.6.5b)];} \end{aligned} \tag{1.17.4}$$

that is, the angular momentum is normal to the surface $T_{\text{rot'n}} = constant$, in the space of the ω's.

• If $\boldsymbol{v}_\blacklozenge = \mathbf{0}$—for example, gyro spinning about a fixed point—(1.17.3b) yield

$$2T \;\Rightarrow\; 2T_{\text{rot'n}} = I_{xx}\,\omega_x{}^2 + \cdots + 2I_{xy}\,\omega_x\,\omega_y + \cdots = \boldsymbol{H}_\blacklozenge \cdot \boldsymbol{\omega} = \boldsymbol{h}_\blacklozenge \cdot \boldsymbol{\omega} \geq 0; \tag{1.17.4a}$$

that is, since T is positive definite, the *angle between* $\boldsymbol{H}_\blacklozenge = \boldsymbol{h}_\blacklozenge$ *and* $\boldsymbol{\omega}$ is *never obtuse*:

$$0^\circ \leq \text{angle}\,(\boldsymbol{H}_\blacklozenge, \boldsymbol{\omega}) < 90^\circ. \tag{1.17.4b}$$

• Similarly, we can express T in terms of relative velocities; that is, with

$$\boldsymbol{v} = \boldsymbol{v}_\blacklozenge + \boldsymbol{\Omega} \times \boldsymbol{r}_{/\blacklozenge} + \boldsymbol{v}_{/\blacklozenge,\text{relative}}, \qquad \boldsymbol{v}_{/\blacklozenge,\text{relative}} \equiv d'\boldsymbol{r}_{/\blacklozenge}/dt, \tag{1.17.5}$$

where $\boldsymbol{\Omega}$ is the inertial angular velocity of the moving axes.

Another Useful T-Representation

We have, successively,

$$\begin{aligned} 2T &= \mathcal{S}\, dm\, \boldsymbol{v} \cdot \boldsymbol{v} = \mathcal{S}\, \boldsymbol{v} \cdot (dm\, \boldsymbol{v}) = \mathcal{S}\, (\boldsymbol{v}_\blacklozenge + \boldsymbol{\omega} \times \boldsymbol{r}_{/\blacklozenge}) \cdot (dm\, \boldsymbol{v}) \\ &= \boldsymbol{v}_\blacklozenge \cdot \left(\mathcal{S}\, dm\, \boldsymbol{v}\right) + \mathcal{S}\, d\boldsymbol{p} \cdot (\boldsymbol{\omega} \times \boldsymbol{r}_{/\blacklozenge}) \qquad (\text{since } dm\, \boldsymbol{v} \equiv d\boldsymbol{p}) \\ &= \boldsymbol{v}_\blacklozenge \cdot \boldsymbol{p} + \boldsymbol{\omega} \cdot \left(\mathcal{S}\, \boldsymbol{r}_{/\blacklozenge} \times d\boldsymbol{p}\right) = \boldsymbol{v}_\blacklozenge \cdot \boldsymbol{p} + \boldsymbol{\omega} \cdot \boldsymbol{H}_{\blacklozenge,\text{absolute}}. \end{aligned} \tag{1.17.6}$$

Kinetic Energy of a Thin Plate of Mass m
in Plane Motion on its Own Plane (fig. 1.17)

Using *plate-fixed* axes $\blacklozenge$–xy, we find [with $\cos(\ldots) \equiv c(\ldots)$, $\sin(\ldots) \equiv s(\ldots)$]

$$\boldsymbol{r}_{G/\blacklozenge} = x_G\,\boldsymbol{i} + y_G\,\boldsymbol{j} \equiv x\,\boldsymbol{i} + y\,\boldsymbol{j}, \qquad (x, y\text{: } constant) \tag{1.17.7a}$$

$$\begin{aligned}\boldsymbol{v}_\blacklozenge &= (dX_\blacklozenge/dt)\boldsymbol{I} + (dY_\blacklozenge/dt)\boldsymbol{J} \equiv (dX/dt)\boldsymbol{I} + (dY/dt)\boldsymbol{J}\\ &= (dX/dt)(c\phi\,\boldsymbol{i} - s\phi\,\boldsymbol{j}) + (dY/dt)(s\phi\,\boldsymbol{i} + c\phi\,\boldsymbol{j})\\ &= \big[(dX/dt)c\phi + (dY/dt)s\phi\big]\boldsymbol{i} + \big[-(dX/dt)s\phi + (dY/dt)c\phi\big]\boldsymbol{j}\\ &\equiv v_x\,\boldsymbol{i} + v_y\,\boldsymbol{j}, \qquad \boldsymbol{\omega} = (d\phi/dt)\boldsymbol{K} = (d\phi/dt)\boldsymbol{k};\end{aligned} \tag{1.17.7b, c}$$

and so, successively,

$$2T_{\text{transl'n}} \equiv m\,\boldsymbol{v}_\blacklozenge \cdot \boldsymbol{v}_\blacklozenge = m\big[(dX/dt)^2 + (dY/dt)^2\big]; \tag{1.17.7d}$$

$$\begin{aligned}T_{\text{cpl'g}} &\equiv m\,\boldsymbol{r}_{G/\blacklozenge} \cdot (\boldsymbol{v}_\blacklozenge \times \boldsymbol{\omega}) = m(x, y, 0) \cdot [(v_x, v_y, 0) \times (0, 0, d\phi/dt)]\\ &= m(d\phi/dt)(v_y\,x - v_x\,y)\\ &= m(d\phi/dt)\big\{[(dY/dt)x - (dX/dt)y]c\phi - [(dX/dt)x + (dY/dt)y]\,s\phi\big\};\end{aligned} \tag{1.17.7e}$$

$$2T_{\text{rot'n}} \equiv \boldsymbol{\omega} \cdot \mathbf{I}_\blacklozenge \cdot \boldsymbol{\omega} = I_{\blacklozenge,zz}\,\omega_z{}^2 \equiv I(d\phi/dt)^2; \tag{1.17.7f}$$

that is,

$$\begin{aligned}2T &= 2T(dX/dt, dY/dt, d\phi/dt)\\ &= m[(dX/dt)^2 + (dY/dt)^2] + I(d\phi/dt)^2\\ &\quad + 2m(d\phi/dt)\big\{[(dY/dt)x - (dX/dt)y]c\phi - [(dX/dt)x + (dY/dt)y]s\phi\big\}.\end{aligned} \tag{1.17.7g}$$

An Application

It is shown in chap. 3 that for this *three* degrees of freedom (*DOF*) (unconstrained) system, defined by the positional coordinates $q_1 = X$, $q_2 = Y$, $q_3 = \phi$, the Lagrangean equations of motion $d/dt\big[\partial T/\partial(dq_k/dt)\big] - (\partial T/\partial q_k) = Q_k$ [$=$ *system* (*impressed*) *force* corresponding to q_k]; or, explicitly, *angular equation* (with $M =$ total external moment about $\blacklozenge$):

$$\begin{aligned}I(d^2\phi/dt^2) + m\big\{&[(d^2Y/dt^2)x - (d^2X/dt^2)y]c\phi\\ &- [(d^2X/dt^2)x + (d^2Y/dt^2)y]s\phi\big\} = M;\end{aligned} \tag{1.17.7h}$$

which is none other than the (not-so-common form of the) angular momentum equation:

$$I_\blacklozenge\alpha_z + (\boldsymbol{r}_{G/\blacklozenge} \times m\,\boldsymbol{a}_\blacklozenge)_z = I(d^2\phi/dt^2) + m\big[x(\boldsymbol{a}_\blacklozenge)_y - y(\boldsymbol{a}_\blacklozenge)_x\big] = M, \tag{1.17.7i}$$

where (fig. 1.17),

$$\begin{aligned} \boldsymbol{r}_{G/\blacklozenge} &= x_G \boldsymbol{i} + y_G \boldsymbol{j} \equiv x \boldsymbol{i} + y \boldsymbol{j} = \cdots \\ &= (x\,c\phi - y\,s\phi)\boldsymbol{I} + (x\,s\phi + y\,c\phi)\boldsymbol{J} \equiv X\boldsymbol{I} + Y\boldsymbol{J}, \end{aligned} \tag{1.17.7j}$$

$$\begin{aligned} \boldsymbol{a}_\blacklozenge &= (dX_\blacklozenge/dt^2)\boldsymbol{I} + (d^2Y_\blacklozenge/dt^2)\boldsymbol{J} \equiv (d^2X/dt^2)\boldsymbol{I} + (d^2Y/dt^2)\boldsymbol{J} \\ &= (d^2X/dt^2)(c\phi \boldsymbol{i} - s\phi \boldsymbol{j}) + (d^2Y/dt^2)(s\phi\,\boldsymbol{i} + c\phi \boldsymbol{j}) \\ &= [(d^2X/dt^2)c\phi + (d^2Y/dt^2)s\phi]\,\boldsymbol{i} + [-(d^2X/dt^2)s\phi + (d^2Y/dt^2)c\phi]\,\boldsymbol{j} \\ &\equiv (\boldsymbol{a}_\blacklozenge)_x \boldsymbol{i} + (\boldsymbol{a}_\blacklozenge)_y \boldsymbol{j} \equiv a_x \boldsymbol{i} + a_y \boldsymbol{j}; \end{aligned} \tag{1.17.7k}$$

x, y-equations (with $f_{x,y}$ = components of total external force about x, y-axes, respectively):

$$m[d^2X/dt^2 - (d^2\phi/dt^2)(x\,s\phi + y\,c\phi) - (d\phi/dt)^2(x\,c\phi - y\,s\phi)] = f_x, \tag{1.17.7l}$$

$$m[d^2Y/dt^2 + (d^2\phi/dt^2)(x\,c\phi - y\,s\phi) - (d\phi/dt)^2(x\,s\phi + y\,c\phi)] = f_y, \tag{1.17.7m}$$

which are none other than

$$m(\boldsymbol{a}_\blacklozenge)_x - m[(d^2\phi/dt^2)Y + (d\phi/dt)^2 X] = f_x, \tag{1.17.7n}$$

$$m(\boldsymbol{a}_\blacklozenge)_y + m[(d^2\phi/dt^2)X - (d\phi/dt)^2 Y] = f_y. \tag{1.17.7o}$$

For additional related plane motion problems, see, for example, Wells (1967, pp. 150–152).

"British Theorem"

It can be shown that the (inertial) kinetic energy of a thin homogeneous bar AB of mass m equals

$$T = (m/6)(\boldsymbol{v}_A \cdot \boldsymbol{v}_A + \boldsymbol{v}_B \cdot \boldsymbol{v}_B + \boldsymbol{v}_A \cdot \boldsymbol{v}_B) = (m/6)({v_A}^2 + {v_B}^2 + \boldsymbol{v}_A \cdot \boldsymbol{v}_B). \tag{1.17.8}$$

(This useful result appears almost exclusively in British texts on dynamics; hence, the name; see, for example, Chorlton, 1983, pp. 165–166.)

Principle of Linear Momentum; Motion of Mass Center

Since

$$\boldsymbol{v}_G = \boldsymbol{v}_\blacklozenge + \boldsymbol{\omega} \times \boldsymbol{r}_{G/\blacklozenge} \qquad (\blacklozenge\text{: } \textit{body-fixed} \text{ point}), \tag{1.17.9a}$$

the principle of linear momentum (§1.6)

$$m(d\boldsymbol{v}_G/dt) = \boldsymbol{f} \qquad \text{(total external force, acting at } G\text{)}, \tag{1.17.9b}$$

along *body-fixed* axes (i.e., $\boldsymbol{\omega} = \boldsymbol{\Omega}$) yields, successively,

$$m\, d/dt(\boldsymbol{v}_\blacklozenge + \boldsymbol{\omega} \times \boldsymbol{r}_{G/\blacklozenge})$$
$$= m(d'\boldsymbol{v}_\blacklozenge/dt + \boldsymbol{\omega} \times \boldsymbol{v}_\blacklozenge) + m\big[(d\boldsymbol{\omega}/dt) \times \boldsymbol{r}_{G/\blacklozenge} + \boldsymbol{\omega} \times (d\boldsymbol{r}_{G/\blacklozenge}/dt)\big]$$
$$[\text{with } d\boldsymbol{\omega}/dt \equiv \boldsymbol{\alpha}, \qquad d\boldsymbol{r}_{G/\blacklozenge}/dt \equiv \boldsymbol{v}_{G/\blacklozenge} = \boldsymbol{\omega} \times \boldsymbol{r}_{G/\blacklozenge}]$$
$$= m(d'\boldsymbol{v}_\blacklozenge/dt) + m(\boldsymbol{\omega} \times \boldsymbol{v}_\blacklozenge) + \boldsymbol{\alpha} \times (m\boldsymbol{r}_{G/\blacklozenge}) + \boldsymbol{\omega} \times [\boldsymbol{\omega} \times (m\boldsymbol{r}_{G/\blacklozenge})] = \boldsymbol{f}, \tag{1.17.9c}$$

or, in terms of the *center of mass vector of the mass moment* $\boldsymbol{m}_{G/\blacklozenge} \equiv m\boldsymbol{r}_{G/\blacklozenge}$ [as in (1.17.3c, d)],

$$m(d'\boldsymbol{v}_\blacklozenge/dt) + m(\boldsymbol{\omega} \times \boldsymbol{v}_\blacklozenge) + \boldsymbol{\alpha} \times \boldsymbol{m}_{G/\blacklozenge} + \boldsymbol{\omega} \times (\boldsymbol{\omega} \times \boldsymbol{m}_{G/\blacklozenge}) = \boldsymbol{f}. \tag{1.17.9d}$$

Along *body-fixed* axes $\blacklozenge$–xyz, and with $\boldsymbol{m}_{G/\blacklozenge} = (m_{x,y,z})$, $\boldsymbol{v}_\blacklozenge = (v_{x,y,z})$ there, the x-component of (1.17.9d) is

$$m[dv_x/dt + \omega_y v_z - \omega_z v_y] + [m_z(d\omega_y/dt) - m_y(d\omega_z/dt)]$$
$$+ [\omega_y(m_y\omega_x - m_x\omega_y) - \omega_z(m_x\omega_z - m_z\omega_x)] = f_x,$$
$$\text{etc., cyclically.} \tag{1.17.9e}$$

Special Cases

(i) If $\blacklozenge = G$, then $\boldsymbol{m}_{G/\blacklozenge} = \boldsymbol{0}$ and, clearly, (1.17.9d) reduces to

$$m(d'\boldsymbol{v}_G/dt + \boldsymbol{\omega} \times \boldsymbol{v}_G) = \boldsymbol{f}. \tag{1.17.9f}$$

(ii) Along *non–body-fixed* axes at G, rotating with inertial angular velocity $\boldsymbol{\Omega}$, (1.17.9b) yields

$$m(d'\boldsymbol{v}_G/dt + \boldsymbol{\Omega} \times \boldsymbol{v}_G) = \boldsymbol{f}\,; \tag{1.17.9g}$$

or, in components, with $\boldsymbol{v}_G = (v_{G;\,x,y,z})$,

$$m(d\boldsymbol{v}_G/dt)_x = m(dv_{G,x}/dt + \Omega_y\, v_{G,z} - \Omega_z v_{G,y}) = f_x, \quad \text{etc., cyclically}, \tag{1.17.9h}$$

where $(d\boldsymbol{v}_G/dt)_x$ = component of $\boldsymbol{a}_G$ along an inertial axis that instantaneously coincides with the moving axis Gx, and so on. In general, the $v_{G;\,x,y,z}$ are quasi velocities.

Principle of Angular Momentum; Motion (Rotation) about the Mass Center

Along *body-fixed* axes $\blacklozenge$–xyz, the principle of angular momentum [§1.6, with $\bullet$ (arbitrary spatial point) $\rightarrow$ $\blacklozenge$ (arbitrary body point), and $\boldsymbol{H}_{\blacklozenge,\text{relative}} \equiv \boldsymbol{h}_\blacklozenge$],

$$d\boldsymbol{h}_\blacklozenge/dt + \boldsymbol{r}_{G/\blacklozenge} \times [m(d\boldsymbol{v}_\blacklozenge/dt)] = \boldsymbol{M}_\blacklozenge, \tag{1.17.10a}$$

becomes

$$d'\boldsymbol{h}_\blacklozenge/dt + \boldsymbol{\omega} \times \boldsymbol{h}_\blacklozenge + \boldsymbol{m}_{G/\blacklozenge} \times (d'\boldsymbol{v}_\blacklozenge/dt) + \boldsymbol{m}_{G/\blacklozenge} \times (\boldsymbol{\omega} \times \boldsymbol{v}_\blacklozenge) = \boldsymbol{M}_\blacklozenge; \tag{1.17.10b}$$

or, in components, with $\boldsymbol{h}_\blacklozenge = (h_{x,y,z})$, $\boldsymbol{m}_{G/\blacklozenge} \equiv m\boldsymbol{r}_{G/\blacklozenge} = (m_{x,y,z})$, $\boldsymbol{v}_\blacklozenge = (v_{x,y,z})$, and so on,

$$dh_x/dt + \omega_y h_z - \omega_z h_y + m_y[dv_z/dt + \omega_x v_y - \omega_y v_x] \\ - m_z[dv_y/dt + \omega_z v_x - \omega_x v_z] = M_x, \quad \text{etc., cyclically.} \quad (1.17.10c)$$

[The forms (1.17.9d, e) and (1.17.10b–c) seem to be due to Heun (1906, 1914); see also Winkelmann and Grammel (1927) for a concise treatment via von Mises' (not very popular) "motor calculus."]

Special Cases

(i) If $\blacklozenge = G$, then $\boldsymbol{m}_{G/\blacklozenge} = \boldsymbol{0}$, and (1.17.10b) reduces to

$$d'\boldsymbol{h}_G/dt + \boldsymbol{\omega} \times \boldsymbol{h}_G = \boldsymbol{M}_G. \quad (1.17.10d)$$

(ii) Along *non–body-fixed* axes at G, rotating with inertial angular velocity $\boldsymbol{\Omega}$, (1.17.10a) yields

$$d'\boldsymbol{h}_G/dt + \boldsymbol{\Omega} \times \boldsymbol{h}_G = \boldsymbol{M}_G; \quad (1.17.10e)$$

or, in components,

$$(d\boldsymbol{h}_G/dt)_x = dh_{G,x}/dt + \Omega_y h_{G,z} - \Omega_z h_{G,y} = M_{G,x}, \quad \text{etc., cyclically.} \quad (1.17.10f)$$

(iii) If the axes are *body-fixed*, then $\boldsymbol{\Omega} = \boldsymbol{\omega}$; and if they are also *principal axes*, then, since (omitting the subscript G throughout) $\boldsymbol{h} = \mathbf{I} \cdot \boldsymbol{\omega} : h_k = I_k \omega_k$, $k = 1, 2, 3$, eqs. (1.17.10f) assume the famous *Eulerian form* (1758, publ. 1765):

$$\begin{aligned} I_1(d\omega_1/dt) - (I_2 - I_3)\omega_2\omega_3 &= M_1, \\ I_2(d\omega_2/dt) - (I_3 - I_1)\omega_3\omega_1 &= M_2, \\ I_3(d\omega_3/dt) - (I_1 - I_2)\omega_1\omega_2 &= M_3; \end{aligned} \quad (1.17.11a)$$

or, alternatively,

$$d\omega_1/dt - [(I_2 - I_3)/I_1]\omega_2\omega_3 = M_1/I_1, \quad \text{etc., cyclically.} \quad (1.17.11b)$$

From the above we readily conclude that:

- A force-free rigid body in space can rotate permanently [i.e., $d\boldsymbol{\omega}/dt = \boldsymbol{0} \Rightarrow \boldsymbol{\omega} = (\omega_1, 0, 0)$, or $(0, \omega_2, 0)$, or $(0, 0, \omega_3) =$ *constant*] only about a central principal axis of inertia. Or, if a free rigid body under no external forces begins to rotate about one of its central principal axes, it will continue to rotate uniformly about that axis; and, if a rigid body *with a fixed point*, and zero torque about that point, begins to rotate about a principal axis through that point, it will continue to do so uniformly about that axis.
- The principle of angular momentum takes the "elementary" form $M = d/dt\,(I\omega)$ only for principal axes of inertia, or if the body rotates about a (body- and space-) fixed axis. That is why a central principal axis was called a *permanent* axis (Ampère, 1823).

REMARKS

(i) Equations (1.17.10a ff.) also hold for any *fixed point O*.

(ii) The principles of linear and angular momentum are summarized as follows: the vector system of mechanical loads, or inputs [$\boldsymbol{f}$ at G, $\boldsymbol{M}_G$ (*moments of forces and couples*)] is equivalent to the kinematico-inertial vector system of the responses, or outputs, ($m\boldsymbol{a}_G$ at G, $d\boldsymbol{h}_G/dt$); and this equivalence, holding about any other space point •, can be expressed via the (hopefully familiar from elementary statics) purely geometrical transfer theorem:

$$\boldsymbol{M}_\bullet = \boldsymbol{M}_G + \boldsymbol{r}_{G/\bullet} \times \boldsymbol{f}_{\text{at}\,G} = \boldsymbol{M}_G + \boldsymbol{r}_{G/\bullet} \times m\boldsymbol{a}_G \equiv d\boldsymbol{h}_G/dt + \boldsymbol{m}_{G/\bullet} \times \boldsymbol{a}_G. \qquad (1.17.12)$$

(iii) In general, the direct application of the *vectorial* forms of the principle of angular momentum, either about the mass center G, or a fixed point O, and then taking components of all quantities involved about common axes in which the inertia tensor components remain constant, is much preferable to trying to match a (any) particular problem to the various scalar components forms of the principle.

(iv) The *relative magnitudes* of the principal moments of inertia of a rigid body at, say its mass center G, $I_{G;1,2,3} \equiv I_{1,2,3}$ (i.e., its *mass distribution* there) provide an important means of classifying such systems. Thus, we have the following classification (§1.15: subsection "Ellipsoid of Inertia"):

- If $I_1 = I_2 = I_3 \equiv I$, we have a *spherical top*, or a *kinetically symmetrical* body. Then,

$$\boldsymbol{H}_G = \boldsymbol{h}_G = \mathbf{I}_G \cdot \boldsymbol{\omega} = (I\mathbf{1}) \cdot \boldsymbol{\omega} = I\boldsymbol{\omega}.$$

- If $I_1 = I_2 \neq I_3$, the body (or "top") is *symmetric*; if $I_1 > I_3$, it is *elongated*, and if $I_1 < I_3$, it is *flattened*.
- If $I_1 \neq I_2 \neq I_3 \neq I_1$, the body is *unsymmetric*.

For further details and insights on these fascinating equations, see Cayley (1863, pp. 230–231), Dugas (1955, pp. 276–278), Stäckel (1905, pp. 581–589).

Energy Rate, or Power, Theorem for a Rigid Body

By $d/dt(\ldots)$-differentiating the kinetic energy definition $2T = S\, dm\, \boldsymbol{v} \cdot \boldsymbol{v}$, and then utilizing in there the rigid-body kinetic equation $\boldsymbol{v} = \boldsymbol{v}_\blacklozenge + \boldsymbol{\omega} \times \boldsymbol{r}_{/\blacklozenge}$, we obtain, successively,

$$\begin{aligned} dT/dt &= S\, dm\, \boldsymbol{v} \cdot (d\boldsymbol{v}/dt) = S\, dm\, \boldsymbol{v} \cdot \boldsymbol{a} = S\, dm\, (\boldsymbol{v}_\blacklozenge + \boldsymbol{\omega} \times \boldsymbol{r}_{/\blacklozenge}) \cdot \boldsymbol{a} \\ &= S\, dm\, \boldsymbol{v}_\blacklozenge \cdot \boldsymbol{a} + S\, dm (\boldsymbol{\omega} \times \boldsymbol{r}_{/\blacklozenge}) \cdot \boldsymbol{a} = \boldsymbol{v}_\blacklozenge \cdot \Big(S\, dm\, \boldsymbol{a} \Big) + \boldsymbol{\omega} \cdot \Big(S\, dm\, \boldsymbol{r}_{/\blacklozenge} \times \boldsymbol{a} \Big) \\ &= \boldsymbol{v}_\blacklozenge \cdot (m\, \boldsymbol{a}_G) + \boldsymbol{\omega} \cdot \Big\{ S\, dm\, [d/dt(\boldsymbol{r}_{/\blacklozenge} \times \boldsymbol{v}) + \boldsymbol{v}_\blacklozenge \times \boldsymbol{v}] \Big\} \\ &= \boldsymbol{v}_\blacklozenge \cdot (d\boldsymbol{p}/dt) + \boldsymbol{\omega} \cdot (d\boldsymbol{H}_\blacklozenge/dt + \boldsymbol{v}_\blacklozenge \times \boldsymbol{p}), \end{aligned} \qquad (1.17.13a)$$

where (recalling the definitions in §1.6),

$\boldsymbol{p} \equiv S\, dm\, \boldsymbol{v} = m\, \boldsymbol{v}_G$: Linear momentum of body, (1.17.13b)

$\boldsymbol{H}_\blacklozenge \equiv S\, dm\, (\boldsymbol{r}_{/\blacklozenge} \times \boldsymbol{v})$: Absolute (and inertial) angular momentum of body, about the body-fixed point ♦. (1.17.13c)

Invoking the principles of linear and angular momentum (§1.6), we can rewrite (1.17.13a) as

$$dT/dt = \boldsymbol{v}_\blacklozenge \cdot \boldsymbol{f} + \boldsymbol{\omega} \cdot \boldsymbol{M}_\blacklozenge. \tag{1.17.13d}$$

On the other hand, the *power* of all forces, $dW/dt = \int d\boldsymbol{f} \cdot \boldsymbol{v}$, transforms, successively, to

$$dW/dt = \int d\boldsymbol{f} \cdot (\boldsymbol{v}_\blacklozenge + \boldsymbol{\omega} \times \boldsymbol{r}_{/\blacklozenge}) = \boldsymbol{v}_\blacklozenge \cdot \left(\int d\boldsymbol{f}\right) + \boldsymbol{\omega} \cdot \left(\int \boldsymbol{r}_{/\blacklozenge} \times d\boldsymbol{f}\right)$$
$$= \boldsymbol{v}_\blacklozenge \cdot \boldsymbol{f} + \boldsymbol{\omega} \cdot \boldsymbol{M}_\blacklozenge; \tag{1.17.13e}$$

that is,

$$dT/dt = dW/dt, \tag{1.17.13f}$$

which is the well-known power theorem, proved here for a rigid system.

Special Case

If $\boldsymbol{v}_\blacklozenge = \boldsymbol{0}$ (i.e., rotation about a *fixed point*), (1.17.13d–f) reduce to

$$dT/dt = dW/dt = \boldsymbol{\omega} \cdot \boldsymbol{M}_\blacklozenge. \tag{1.17.13g}$$

If, in addition, $\boldsymbol{M}_\blacklozenge = \boldsymbol{0}$ (*torque-free* motion), then $dW/dt = 0$ and $T = constant$ (energy integral), and $\boldsymbol{M}_\blacklozenge = d\boldsymbol{H}_\blacklozenge/dt = d\boldsymbol{h}_\blacklozenge/dt = \boldsymbol{0} \Rightarrow \boldsymbol{H}_\blacklozenge = \boldsymbol{h}_\blacklozenge = constant$ (angular momentum integral). These *two* integrals of the torque-free and fixed-point motion form the basis of an interesting geometrical interpretation of rigid-body motion, due to Poinsot (1850s). For details see, for example (alphabetically): MacMillan (1936, pp. 204–216), Webster (1912, pp. 252–270), Winkelmann and Grammel (1927, pp. 392–398).

Additional Useful Results

(i) By multiplying the Eulerian (rotational) equations with $\omega_{x,y,z}$, respectively, and then adding them, we obtain the following power equation:

$$d/dt[(A\omega_x{}^2 + B\omega_y{}^2 + C\omega_z{}^2)/2] = M_x\omega_x + M_y\omega_y + M_z\omega_z,$$

i.e., d/dt (*Rotational kinetic energy*) = *Power of external moments*. (1.17.14)

(ii) *Plane motion*: Principle of angular momentum for a rigid body B, about its instantaneous center of rotation I. We have already seen (1.9.4d ff.) that the inertial coordinates of the *instantaneous center* (*of zero velocity*) I, relative to the center of mass G, are

$$\boldsymbol{r}_{I/G} = (X_{I/G}, Y_{I/G}, 0) = (-dY_G/dt\big/\omega,\ +dX_G/dt\big/\omega,\ 0). \tag{1.17.15a}$$

Therefore, application of the principle of angular momentum about I:

$$M_I = I_G(d\omega/dt) + (\boldsymbol{r}_{G/I} \times m\boldsymbol{a}_G)_Z \tag{1.17.15b}$$

yields, successively (with $I_G \equiv mk^2$),

$$\begin{aligned} M_I &= I_G(d\omega/dt) + (m/\omega)\big[(dY_G/dt, -dX_G/dt, 0) \times (d^2X_G/dt^2, d^2Y_G/dt^2, 0)\big] \\ &= I_G(d\omega/dt) + (m/\omega)\big[(dY_G/dt)(d^2Y_G/dt^2) - (-dX_G/dt)(d^2X_G/dt^2)\big] \\ &= I_G(d\omega/dt) + (m/\omega)\big[(dX_G/dt)(d^2X_G/dt^2) + (dY_G/dt)(d^2Y_G/dt^2)\big] \\ &= (m/\omega)\big[k^2\omega(d\omega/dt) + (dX_G/dt)(d^2X_G/dt^2) + (dY_G/dt)(d^2Y_G/dt^2)\big] \\ &= (m/\omega)\big(d/dt\{(1/2)[k^2\omega^2 + (dX_G/dt)^2 + (dY_G/dt)^2]\}\big) \\ &\qquad \big[\text{noting that } (dX_G/dt)^2 + (dY_G/dt)^2 = v_G{}^2 = r^2\omega^2, \qquad r = |\boldsymbol{r}_{G/I}|\big] \\ &= (1/2\omega)\{d/dt[m(k^2 + r^2)\omega^2]\}, \end{aligned}$$

or, finally, with $I_I \equiv m(k^2 + r^2) \equiv mK^2$: *moment of inertia of B about I* (by the parallel axis theorem),

$$\begin{aligned} M_I = (1/2\omega)[d/dt\,(I_I\omega^2)] &= I_I(d\omega/dt) + (1/2)\omega(dI_I/dt) \\ &= I_I(d\omega/dt) + mr(dr/dt)\omega. \end{aligned} \tag{1.17.15c}$$

Special Cases

(a) If B is turning about a *fixed* axis, or if I is at a constant distance from G, then $dr/dt = 0$ and (1.17.15c) reduces to

$$M_I = I_I(d\omega/dt). \tag{1.17.15d}$$

(b) If the axis of rotation is mobile, but *the body starts from rest*, then, since initially $\omega = 0$ and $dr/dt = 0$, the *initial* value of its angular acceleration is given by (1.17.15d):

$$d\omega/dt = M_I/I_I. \tag{1.17.15e}$$

(c) If the body undergoes *small* angular oscillations about a position of equilibrium, then the term $dI_I/dt = 2mr(dr/dt)$ is of the order of the rate dr/dt, and therefore $(dI_I/dt)\omega$ is of the order of the *square* of a small velocity and so, to the first order (linear angular oscillations), it can be neglected; thus reducing (1.17.15c) to (1.17.15d), with I_I given by its equilibrium value.

In sum, eq. (1.17.15d) holds if the instantaneous axis of rotation is either *fixed*, or remains at a *constant distance* from the *center of mass*; or if the problem is one of *initial motion*, or of a *small oscillation*. In all other cases of moments about I, we must use (1.17.15c). For further details and applications, see, for example (alphabetically): Besant (1914, pp. 310–314), Loney (1909, pp. 287, 346–347), Pars (1953, pp. 403–404), Ramsey (1933, part I, pp. 241–242), Routh (1905(a), pp. 103–104, 171–172). Somehow this topic is treated only in older British treatises!

Rigid-Body Mechanics in Matrix Form

[We are reminded (§ 1.1) that vectors are shown in bold italics, and (matrix forms of) tensors in bold roman (plain text); for example, $\boldsymbol{a}$, $\boldsymbol{A}$ (vectors); $\mathbf{a}$, $\mathbf{A}$ (tensors). This material (notation) is presented here not because we think that it adds anything significant to our *conceptual* understanding of mechanics, but because it happens to be fashionable among some contemporary applied dynamicists.]

By recalling the tensor results of §1.1, and the earlier definitions and notations, (1.15.2a ff.),

$$\mathbf{I} \equiv S\, [(\boldsymbol{r}\cdot\boldsymbol{r})\mathbf{1} - \boldsymbol{r}\otimes\boldsymbol{r}]\, dm = -S\,(\mathbf{r}\cdot\mathbf{r})\, dm = (1/2)(\mathrm{Tr}\,\mathbf{I})\mathbf{1} - \mathbf{J} \qquad (1.17.16\text{a}1)$$

$$[\Rightarrow \mathrm{Tr}\,\mathbf{I} = 2\,\mathrm{Tr}\,\mathbf{J}],$$

$$\mathbf{J} \equiv S\,(\boldsymbol{r}\otimes\boldsymbol{r})\, dm, \qquad (1.17.16\text{a}2)$$

[$\boldsymbol{r}$ = axial vector of tensor $\mathbf{r}$ and $d(\ldots)/dt$ is inertial rate of change], we can verify the following matrix forms of the earlier (§1.15–1.17) basic equations of rigid-body mechanics [while assuming that, in a given equation, all moments of inertia and moments of forces are taken either about the body's center of mass, or about a body-and-space-fixed point (if one exists), and along body-fixed axes; and *suppressing all such point-dependence* for notational simplicity, except in eqs. (1.17.16b1–3) for obvious reasons]:

(i) $\mathbf{I}_O = \mathbf{I}_G - m\,\mathbf{r}_G\cdot\mathbf{r}_G = \mathbf{I}_G + m\big[(\boldsymbol{r}_G\cdot\boldsymbol{r}_G)\mathbf{1} - \boldsymbol{r}_G\otimes\boldsymbol{r}_G\big]$

[$\boldsymbol{r}_G \equiv \boldsymbol{r}_{G/O}$, etc., parallel axis theorem in terms of $\mathbf{I}$: (1.15.7b)],(1.17.16b1)

$$\Rightarrow \mathrm{Tr}\,\mathbf{I}_O = \mathrm{Tr}\,\mathbf{I}_G + 2m\,\boldsymbol{r}_G\cdot\boldsymbol{r}_G, \qquad (1.17.16\text{b}2)$$

$$\mathbf{J}_O = \mathbf{J}_G + m\,\boldsymbol{r}_G\otimes\boldsymbol{r}_G = (\mathrm{Tr}\,\mathbf{I}_G/2)\mathbf{1} - \mathbf{I}_G + m\,\boldsymbol{r}_G\otimes\boldsymbol{r}_G$$

[Parallel axis theorem in terms of $\mathbf{J}$]; (1.17.16b3)

(ii) $$d\mathbf{I}/dt = \boldsymbol{\omega}\cdot\mathbf{I} + \mathbf{I}\cdot\boldsymbol{\omega}^{\mathrm{T}} = \boldsymbol{\omega}\cdot\mathbf{I} - \mathbf{I}\cdot\boldsymbol{\omega}$$

[recalling results of 1.1.20a ff.; ω = *axial vector of tensor* $\boldsymbol{\omega}$]; (1.17.16c)

(iii) $$d\mathbf{I}/dt = -(d\mathbf{J}/dt) \qquad [= -(\boldsymbol{\omega}\cdot\mathbf{J} - \mathbf{J}\cdot\boldsymbol{\omega})] \qquad (1.17.16\text{d})$$

(iv) $$\boldsymbol{H} = \mathbf{I}\cdot\omega = -\mathbf{J}\cdot\omega + (\mathrm{Tr}\,\mathbf{J})\omega$$

(v) $$\mathbf{H} = \mathbf{I}\cdot\boldsymbol{\omega}^{\mathrm{T}} - \boldsymbol{\omega}\cdot\mathbf{I} + (\mathrm{Tr}\,\mathbf{I})\cdot\boldsymbol{\omega} = (\boldsymbol{\omega}\cdot\mathbf{I})^{\mathrm{T}} - \boldsymbol{\omega}\cdot\mathbf{I} + (\mathrm{Tr}\,\mathbf{I})\cdot\boldsymbol{\omega} \qquad (1.17.16\text{e}1)$$

$$= \mathbf{J}\cdot\boldsymbol{\omega} + \boldsymbol{\omega}\cdot\mathbf{J} = \mathbf{J}\cdot\boldsymbol{\omega} - (\mathbf{J}\cdot\boldsymbol{\omega})^{\mathrm{T}} = \mathbf{J}\cdot\boldsymbol{\omega} - \boldsymbol{\omega}^{\mathrm{T}}\cdot\mathbf{J}; \qquad (1.17.16\text{e}2)$$

[$\boldsymbol{H}$ = *axial vector of* $\mathbf{H}$ (*angular momentum tensor*)]

(vi) $$\boldsymbol{M} = d/dt\,(\mathbf{I}\cdot\omega) = (d\mathbf{I}/dt)\cdot\omega + \mathbf{I}\cdot(d\omega/dt) \quad [\text{then invoking (1.17.16c)}]$$

$$= \mathbf{I}\cdot(d\omega/dt) + \boldsymbol{\omega}\cdot(\mathbf{I}\cdot\omega) = \mathbf{I}\cdot(d\omega/dt) + \omega\times(\mathbf{I}\cdot\omega) \qquad (1.17.16\text{f}1)$$

$$= -\big[\mathbf{J}\cdot(d\omega/dt) + \boldsymbol{\omega}\cdot(\mathbf{J}\cdot\omega)\big] + (\mathrm{Tr}\,\mathbf{J})(d\omega/dt); \qquad (1.17.16\text{f}2)$$

(vii) $$\mathbf{M} = (\boldsymbol{\varepsilon}\cdot\mathbf{I})^{\mathrm{T}} - \boldsymbol{\varepsilon}\cdot\mathbf{I} + (\mathrm{Tr}\,\mathbf{I})\cdot(d\boldsymbol{\omega}/dt) \qquad (1.17.16\text{g}1)$$

$$= \boldsymbol{\varepsilon}\cdot\mathbf{J} - (\boldsymbol{\varepsilon}\cdot\mathbf{J})^{\mathrm{T}} \qquad (1.17.16\text{g}2)$$

[$\boldsymbol{M}$ = *axial vector of* $\mathbf{M}$ (*moment, or torque, tensor*);

recalling (1.11.9a ff.): $\boldsymbol{\varepsilon} \equiv d\boldsymbol{\omega}/dt + \boldsymbol{\omega}\cdot\boldsymbol{\omega} \equiv \boldsymbol{\alpha} + \boldsymbol{\omega}\cdot\boldsymbol{\omega}$].

Additional forms of the above are, of course, possible.

A Comprehensive Example: The Rolling Disk

Let us discuss the motion of a thin homogeneous disk D (or coin, or hoop) of mass m and radius r, on a fixed, horizontal, and rough plane P (fig. 1.29).

Kinematics

Relative to the *intermediate* axes/basis $G\text{–}xyz/\boldsymbol{ijk}$ (defined so that $\boldsymbol{k}$ is perpendicular to D, at its center of mass G; $\boldsymbol{i}$ is continuously horizontal and parallel to the tangent to D, at its contact point C; and $\boldsymbol{j}$ goes through G, along the steepest diameter of D, and is such that $\boldsymbol{ijk}$ form an ortho–normal–dextral triad), whose inertial angular velocity $\boldsymbol{\Omega}$ is

$$\boldsymbol{\Omega} = \Omega_x\,\boldsymbol{i} + \Omega_y\,\boldsymbol{j} + \Omega_z\,\boldsymbol{k} = (\omega_\theta)\boldsymbol{i} + (\omega_\phi \sin\theta)\boldsymbol{j} + (\omega_\phi \cos\theta)\boldsymbol{k}, \tag{1.17.17a}$$

[where $\omega_\phi \equiv d\phi/dt$, $\omega_\theta \equiv d\theta/dt$, $\omega_\psi \equiv d\psi/dt$] the inertial angular velocity of D, $\boldsymbol{\omega}$, equals

$$\begin{aligned}\boldsymbol{\omega} = \omega_x\,\boldsymbol{i} + \omega_y\,\boldsymbol{j} + \omega_z\,\boldsymbol{k} &= (\omega_\theta)\boldsymbol{i} + (\omega_\phi \sin\theta)\boldsymbol{j} + (\omega_\phi \cos\theta + \omega_\psi)\boldsymbol{k} \\ &= \boldsymbol{\Omega} + \omega_\psi\,\boldsymbol{k}.\end{aligned} \tag{1.17.17b}$$

In view of the above, the rolling constraint $\boldsymbol{v}_C = \boldsymbol{0}$, becomes

$$\begin{aligned}\boldsymbol{v}_C = \boldsymbol{v}_G + \boldsymbol{\omega} \times \boldsymbol{r}_{C/G} &= v_x\,\boldsymbol{i} + v_y\,\boldsymbol{j} + v_z\boldsymbol{k} + (\omega_x, \omega_y, \omega_z) \times (0, -r, 0) \\ &= (v_x + r\omega_z)\boldsymbol{i} + (v_y)\boldsymbol{j} + (v_z - \omega_x r)\boldsymbol{k} = \boldsymbol{0},\end{aligned} \tag{1.17.17c}$$

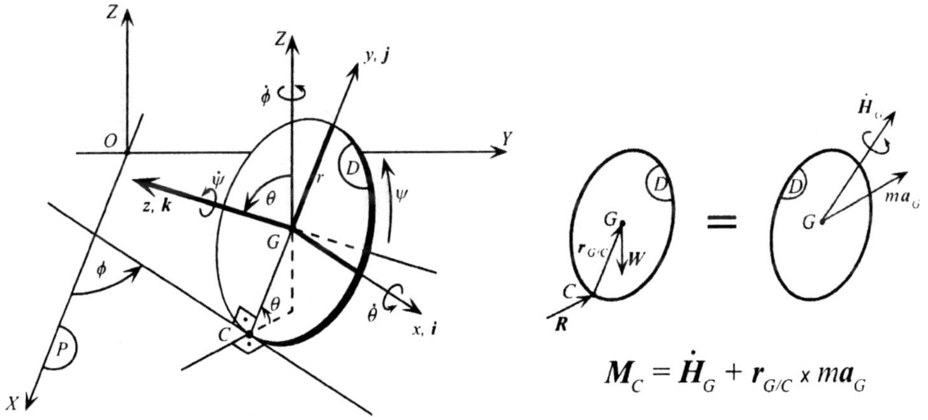

Figure 1.29 Rolling of thin disk/coin D on a fixed, rough, and horizontal plane P. $O\text{–}XYZ$: *space-fixed* (inertial) axes; $G\text{–}xyz$: *intermediate* axes (of angular velocity $\boldsymbol{\Omega}$) A, $B = A$, C: *principal* moments of inertia at G. For our disk: $A = mr^2/4$, $C = mr^2/2$.

from which it follows that

$$v_x + r\omega_z = 0 \Rightarrow v_x = -r\omega_z = -r(\omega_\phi \cos\theta + \omega_\psi), \quad (1.17.17d)$$

$$v_y = 0, \quad (1.17.17e)$$

$$v_z - \omega_x r = 0 \Rightarrow v_z = r\omega_x = r\omega_\theta. \quad (1.17.17f)$$

These equations connect the velocity of G with the angular velocity and the rates of the Eulerian angles.

Kinetics

To eliminate the rolling contact reaction $\boldsymbol{R}$, we apply the principle of angular momentum about C; that is, we take moments of all forces and couples [including inertial ones at G; i.e., $-m(d\boldsymbol{v}_G/dt)$ and $-d\boldsymbol{h}_G/dt$] about G (recalling 1.6.6a ff.) to give

$$\boldsymbol{M}_C = d\boldsymbol{h}_G/dt + \boldsymbol{r}_{G/C} \times [m(d\boldsymbol{v}_G/dt)]. \quad (1.17.18a)$$

But, with W = weight of disk, and $\sin(\ldots) \equiv s(\ldots)$, $\cos(\ldots) \equiv c(\ldots)$, we have

(i) $\boldsymbol{M}_C = \boldsymbol{r}_{G/C} \times \boldsymbol{W} = (0, r, 0) \times (0, -Ws\theta, -Wc\theta) = (-rWc\theta)\boldsymbol{i}$; (1.17.18b)

(ii) $d\boldsymbol{v}_G/dt = d'\boldsymbol{v}_G/dt + \boldsymbol{\Omega} \times \boldsymbol{v}_G$ [with the ad hoc notation $dv_{x,y,z}/dt \equiv a_{x,y,z}$]

$$\begin{aligned}
&= a_x\boldsymbol{i} + a_y\boldsymbol{j} + a_z\boldsymbol{k} + (\Omega_x, \Omega_y, \Omega_z) \times (v_x, v_y, v_z) \\
&= (a_x + \Omega_y v_z - \Omega_z v_y)\boldsymbol{i} \\
&\qquad + (a_y + \Omega_z v_x - \Omega_x v_z)\boldsymbol{j} + (a_z + \Omega_x v_y - \Omega_y v_x)\boldsymbol{k} \\
&= (a_x + v_z\omega_\phi s\theta - v_y\omega_\phi c\theta)\boldsymbol{i} + (a_y + v_x\omega_\phi c\theta - v_z\omega_\theta)\boldsymbol{j} \\
&\qquad + (a_z + v_y\omega_\theta - v_x\omega_\phi s\theta)\boldsymbol{k}; \quad (1.17.18c)
\end{aligned}$$

(iii) $d\boldsymbol{h}_G/dt = d'\boldsymbol{h}_G/dt + \boldsymbol{\Omega} \times \boldsymbol{h}_G$ [with the ad hoc notation $d\omega_{x,y,z}/dt \equiv \alpha_{x,y,z}$]

$$\begin{aligned}
&= \big[(A\alpha_x)\boldsymbol{i} + (A\alpha_y)\boldsymbol{j} + (C\alpha_z)\boldsymbol{k}\big] + (\Omega_x, \Omega_y, \Omega_z) \times (A\omega_x, B\omega_y, C\omega_z) \\
&= (A\alpha_x + C\Omega_y\omega_z - A\Omega_z\omega_y)\boldsymbol{i} + (A\alpha_y + A\Omega_z\omega_x - C\Omega_x\omega_z)\boldsymbol{j} \\
&\qquad + (C\alpha_z + A\Omega_x\omega_y - A\Omega_y\omega_x)\boldsymbol{k} \\
&= (A\alpha_x + C\omega_z\omega_\phi s\theta - A\omega_y\omega_\phi c\theta)\boldsymbol{i} + (A\alpha_y + A\omega_x\omega_\phi c\theta - C\omega_z\omega_\theta)\boldsymbol{j} \\
&\qquad + (C\alpha_z + A\omega_y\omega_\theta - A\omega_x\omega_\phi s\theta)\boldsymbol{k}, \quad (1.17.18d)
\end{aligned}$$

and so (1.17.18a) yields the three component equations of angular motion:

$$mr(a_z + v_y\omega_\theta - v_x\omega_\phi s\theta) + (A\alpha_x + C\omega_z\omega_\phi s\theta - A\omega_y\omega_\phi c\theta) = -Wr\,c\theta, \quad (1.17.18e)$$

$$A\alpha_y + A\omega_x\omega_\phi c\theta - C\omega_z\omega_\theta = 0, \quad (1.17.18f)$$

$$-mr(a_x + v_z\omega_\phi s\theta - v_y\omega_\phi c\theta) + (C\alpha_z + A\omega_y\omega_\theta - A\omega_x\omega_\phi s\theta) = 0. \quad (1.17.18g)$$

The nine equations (1.17.18e, f, g) + (1.17.17d, e, f) + (1.17.17b, in components) constitute a determinate system for the nine functions (of time): ϕ, θ, ψ; $\omega_{x,y,z}$ (quasi velocities); $v_{x,y,z}$ (quasi velocities). We may reduce it further to two steps:

(i) Using (1.17.17d, e, f) in (1.17.18e, f, g) (i.e., eliminating $v_{x,y,z}$), we obtain

$$mr(r\alpha_x + r\omega_z\,\omega_\phi\, s\theta) + A\alpha_x + C\omega_z\,\omega_\phi\, s\theta - A\omega_y\,\omega_\phi\, c\theta = -Wr\,c\theta, \tag{1.17.19a}$$

$$A\alpha_y + A\omega_x\,\omega_\phi\, c\theta - C\omega_z\,\omega_\theta = 0, \tag{1.17.19b}$$

$$-mr(-r\alpha_z + r\omega_x\,\omega_\phi\, s\theta) + C\alpha_z + A\omega_y\,\omega_\theta - A\omega_x\,\omega_\phi\, s\theta = 0. \tag{1.17.19c}$$

(ii) Using (1.17.17b) in (1.17.19a, b, c) (i.e., eliminating $\omega_{x,y}$), we get three equations of *rotational motion* in terms of θ, the *rates of* ϕ, θ, and the *total spin* $\omega_z = \omega_\psi + \omega_\phi c\theta$:

$$(A + mr^2)(d^2\theta/dt^2) + (C + mr^2)\omega_z(d\phi/dt)s\theta - A(d\phi/dt)^2 c\theta s\theta = -Wr\,c\theta, \tag{1.17.20a}$$

$$A\,d/dt\,[(d\phi/dt)s\theta] + A(d\phi/dt)(d\theta/dt)c\theta - C\omega_z(d\theta/dt) = 0, \tag{1.17.20b}$$

$$(C + mr^2)(d\omega_z/dt) - mr^2(d\phi/dt)(d\theta/dt)s\theta = 0; \tag{1.17.20c}$$

or, since $A = B = mr^2/4 = (1/2)(mr^2/2) = C/2$,

$$\theta\text{:}\quad 5r(d^2\theta/dt^2) + 6r\omega_z(d\phi/dt)\sin\theta - r(d\phi/dt)^2\sin\theta\cos\theta + 4g\cos\theta = 0, \tag{1.17.21a}$$

$$\phi\text{:}\quad 2\omega_z(d\theta/dt) - 2(d\phi/dt)(d\theta/dt)\cos\theta - (d^2\phi/dt^2)\sin\theta = 0, \tag{1.17.21b}$$

$$\omega_z\text{:}\quad 3(d\omega_z/dt) - 2(d\phi/dt)(d\theta/dt)\sin\theta = 0. \tag{1.17.21c}$$

These three nonlinear coupled equations contain an enormous variety of disk motions. For simple particular solutions of them, see, for example, MacMillan (1936, pp. 276–281); also Fox (1967, pp. 263–267). Once $\phi(t)$, $\theta(t)$, $\psi(t)$ have been found, the rolling contact reaction $\boldsymbol{R} = (R_{x,y,z})$ can be easily obtained from the principle of linear momentum:

$$m\boldsymbol{a}_G = m(d'\boldsymbol{v}_G/dt + \boldsymbol{\Omega}\times\boldsymbol{v}_G) = \boldsymbol{W} + \boldsymbol{R} \;\Rightarrow\; \boldsymbol{R} = \cdots = \boldsymbol{R}(t). \tag{1.17.22}$$

The details are left to the reader.

1.18 THE RIGID BODY: CONTACT FORCES, FRICTION

Recommended for concurrent reading with this section are (alphabetically): Beghin (1967, pp. 139–145), Kilmister and Reeve (1966, pp. 81–84, 141–143, 164–177), Pérès (1953, pp. 62–66).

Introduction and Constitutive Equations

The forces between two rigid bodies, B and B_1, at a mutual contact point C (actually, a small area around C that is practically independent of the macroscopic shape of the bodies and increases with pressure), say from B to B_1, reduce, in general, to a resultant *force* $\boldsymbol{R}$ and a *couple* $\boldsymbol{C}$; frequently, $\boldsymbol{C}$ can be neglected. Decomposing $\boldsymbol{R}$

and $\boldsymbol{C}$ along the common *normal* to the bounding surfaces of B, B_1, say from B towards B_1, and along the common tangent plane, at C, we obtain

$$\begin{aligned} \boldsymbol{R} &= \boldsymbol{R}_N + \boldsymbol{R}_T \\ &= \textit{Normal reaction} \text{ (opposing mutual penetration)} \\ &\quad + \textit{Tangential reaction} \text{ (opposing relative slipping)}, \end{aligned} \tag{1.18.1}$$

$$\begin{aligned} \boldsymbol{C} &= \boldsymbol{C}_N + \boldsymbol{C}_T \\ &= \textit{Pivoting couple} \text{ (opposing mutual pivoting)} \\ &\quad + \textit{Rolling couple} \text{ (opposing relative rolling)}. \end{aligned} \tag{1.18.2}$$

These components satisfy the following "laws" (better, *constitutive* equations) of *dry* friction; that is, for a solid rubbing against solid, without lubrication:

(i) As soon as an existing contact ceases, $\boldsymbol{R} = \boldsymbol{0}$.

(ii) Whenever there is *slipping*—that is, relative motion of B and B_1 ($\boldsymbol{v}_C \neq \boldsymbol{0}$)—$\boldsymbol{R}_N$ points toward B_1; and $\boldsymbol{R}_T$ and $\boldsymbol{v}_C$ are *collinear* and in *opposite* directions:

$$\boldsymbol{R}_T \times \boldsymbol{v}_C = \boldsymbol{0}, \qquad \boldsymbol{R}_T \cdot \boldsymbol{v}_C < 0, \tag{1.18.3}$$

and

$$R_T = f(R_N, v_C), \tag{1.18.4}$$

or, approximately (for small relative velocities),

$$|R_T/R_N| = \mu\text{: } \textit{coefficient} \text{ of friction between } B \text{ and } B_1\text{; a } \textit{nonnegative} \text{ constant.} \tag{1.18.5}$$

Frequently, we use the following notation:

$$R_T \equiv F \quad \text{and} \quad R_N \equiv N. \tag{1.18.6}$$

Then, with $|F| = \mu|N|$, the above read

$$\boldsymbol{R} = \boldsymbol{R}_N + \boldsymbol{R}_T; \qquad \boldsymbol{R}_N = N\boldsymbol{n}, \qquad \boldsymbol{R}_T = |F|\boldsymbol{t} = -\mu|N|\boldsymbol{t}, \tag{1.18.7}$$

where

$$\boldsymbol{n} = \text{common unit normal vector, from } B \text{ towards } B_1, \tag{1.18.7a}$$

$$\boldsymbol{t} = \text{unit tangent vector, in direction of } \textit{slipping} \text{ velocity.} \tag{1.18.7b}$$

(iii) When $\boldsymbol{v}_C = \boldsymbol{0}$ (no slipping—relative rest), $\boldsymbol{R}_N$ points toward B_1, while $\boldsymbol{R}_T$ can have any arbitrary direction and value on the common tangent plane, as long as

$$|R_T/R_N| \equiv |F/N| \leq \mu; \qquad \text{or, vectorially,} \qquad |\boldsymbol{R} \times \boldsymbol{n}| \leq \mu|\boldsymbol{R} \cdot \boldsymbol{n}|; \tag{1.18.8}$$

with the equality sign holding for *impending* tangential motion. Actually, the μ in (1.18.8) is called coefficient of *static* friction, μ_S; and the μ in (1.18.5) is called coefficient of *kinetic* friction, μ_K; and, generally,

$$\mu_S \geq \mu_K. \tag{1.18.9}$$

Here, unless specified otherwise, μ will mean μ_K.

The friction coefficient μ is, in general, not a constant but a function of: (a) the *nature of the contacting surfaces*; (b) the *conditions of contact* (e.g., dry vs. lubricated surfaces); (c) the *normal* forces (pressure) between the surfaces; and (d) the velocity of slipping. Further, in the dry friction case (solid/solid, no lubricant), μ *increases with pressure*, and *decreases with* v_C; and this dependence is particularly pronounced for *small values of* v_C, so that, if $\mu = \mu(v_C)$, then $\mu < \mu_o$, where $\mu_o \equiv \mu(0)$. In most such applications, we assume that μ is, approximately, a *positive constant* (*rough* surface). Then the relation $\mu = \tan\phi$ defines the "angle of friction." If $\mu \approx 0$ (*smooth* surfaces), then $\boldsymbol{R} \approx \boldsymbol{R}_N \equiv N$, $\boldsymbol{R}_T \approx 0$. If, on the other end, $\mu \to \infty$ (*perfect* roughness), then $\boldsymbol{v}_C = \boldsymbol{0}$ throughout the motion, and $\boldsymbol{R}$ can have *any* direction, as long as $\boldsymbol{R}_N \equiv \boldsymbol{N}$ points toward B_1.

(iv) The contact couple $\boldsymbol{C}$ is included in the cases of *small* μ and/or *slippingless* motion as follows:

(a) If at a given instant and immediately afterwards $\boldsymbol{\omega}_N = \boldsymbol{0}$ (i.e., no instantaneous pivoting), then

$$|C_N| \le |C_{N,\max}|, \qquad C_{N,\max} \equiv f_p R_N = \text{limiting } \textit{pivoting} \text{ moment}, \tag{1.18.10}$$

$$f_p \equiv \textit{pivoting} \text{ friction/resistance coefficient.} \tag{1.18.10a}$$

(b) If at a given instant $\boldsymbol{\omega}_N \neq \boldsymbol{0}$, or if it stops being zero at that instant, then

$$|C_N| = |C_{N,\max}|; \tag{1.18.11}$$

and $\boldsymbol{C}_N$ and $\boldsymbol{\omega}_N$ have opposite senses.

(c) If $\boldsymbol{\omega}_T = \boldsymbol{\omega}_{\text{rolling}} \equiv \boldsymbol{\omega}_R = \boldsymbol{0}$, then

$$|C_T| \le |C_{T,\max}|, \qquad C_{T,\max} \equiv f_r R_N = \text{limiting } \textit{rolling} \text{ moment}, \tag{1.18.12}$$

$$f_r \equiv \textit{rolling} \text{ friction/resistance coefficient.} \tag{1.18.12a}$$

(d) If at a given instant $\boldsymbol{\omega}_T \neq \boldsymbol{0}$, or if it stops being zero at that instant, then

$$|C_T| = |C_{T,\max}|; \tag{1.18.13}$$

and $\boldsymbol{C}_T$ and $\boldsymbol{\omega}_T$ have opposite senses.

The coefficients f_p and f_r have dimensions of *length* (whereas μ is dimensionless!), and their values are to be determined experimentally. Theoretically, f_p can be related to μ, if $\boldsymbol{C}_N$ is viewed as resulting from the slipping friction over a small area around the contact point C—something requiring use of the theory of elasticity (no such relationship can be established for f_r). It turns out that f_p is, generally, five to ten times smaller than f_r; in general, *pivoting is produced faster than rolling.*

In closing this brief summary, we point out that the above "friction laws" supply only *indirect* criteria for relative rest or motion (rolling and slipping); that is, if, for example, we assume rest and the resulting equations are consistent with it, it means that rest is possible, not that it will happen. And if we end up with an inconsistency, it means that the particular assumption(s) that led to it is (are) false. Thus, to show that two contacting bodies roll (slip) on each other, all we can do is show that the assumptions of their slipping (rolling) lead to a contradiction. [For detailed examples illustrating these points, see, for example (alphabetically): Hamel (1949, pp. 543–549, 629–639), Kilmister and Reeve (1966, pp. 165–177); also Pöschl (1927, pp. 484–497).]

Work of Contact Forces

Under a *kinematically possible* infinitesimal displacement of B_1 relative to B (assumed fixed) that preserves their mutual contact at C, the total elementary (first-order) work of the contact actions (of B on B_1) is:

$$d'W = \boldsymbol{R} \cdot d\boldsymbol{r}_C + \boldsymbol{C} \cdot d\boldsymbol{\chi}, \tag{1.18.14}$$

where

$d\boldsymbol{r}_C$ = elementary *translatory* displacement of the B_1-fixed point, at contact, relative to B

$$(\equiv \boldsymbol{v}_C\, dt, \text{ in an } \textit{actual} \text{ such displacement}), \tag{1.18.14a}$$

$d\boldsymbol{\chi}$ = elementary *rotatory* displacement of B_1 relative to B

$$(\equiv \boldsymbol{\omega}\, dt, \text{ in an } \textit{actual} \text{ such displacement}), \tag{1.18.14b}$$

• Since $d\boldsymbol{r}_C$ preserves the B/B_1 contact, it lies on their *common tangent plane* at C. Then: (α) if $\boldsymbol{R}_T \approx \boldsymbol{0}$ (i.e., negligible slipping friction), or (β) if $d\boldsymbol{r}_C = \boldsymbol{0}$ (i.e., no slipping) and $d\boldsymbol{\chi} = \boldsymbol{0}$ (i.e., no rotating), then:

$$d'W = 0. \tag{1.18.15}$$

• If $d\boldsymbol{r}_C$ violates contact, but remains compatible with the unilateral constraints, it makes an *acute* angle with the normal toward B_1. In this case, if $\boldsymbol{R}_T \approx \boldsymbol{0} \Rightarrow \boldsymbol{R} \approx \boldsymbol{R}_N$, and therefore

$$d'W > 0; \tag{1.18.16}$$

while for elementary displacements *incompatible* with the constraints,

$$d'W < 0. \tag{1.18.17}$$

• In a real, or actual, displacement $d'W$ becomes

$$d'W = (\boldsymbol{R} \cdot \boldsymbol{v}_C + \boldsymbol{C} \cdot \boldsymbol{\omega})\, dt.$$

From the earlier constitutive laws, we see that, as long as $\boldsymbol{v}_C$, ω_N, ω_T *do not vanish*, the pairs

$$(\boldsymbol{R}_T, \boldsymbol{v}_C), \qquad (\boldsymbol{C}_N, \boldsymbol{\omega}_N), \qquad (\boldsymbol{C}_T, \boldsymbol{\omega}_T),$$

are *collinear* and *oppositely directed*. Hence, *frictions do negative work*; that is, in general,

$$d'W \leq 0. \tag{1.18.18}$$

If, as commonly assumed, $\boldsymbol{C} \approx \boldsymbol{0}$, then

$$
\begin{aligned}
d'W &= (\boldsymbol{R}_T \cdot \boldsymbol{v}_C)\, dt \equiv (\boldsymbol{F} \cdot \boldsymbol{v}_C)\, dt \\
&= 0; \quad \text{if } \boldsymbol{F} = \boldsymbol{0} \qquad \text{(frictionless, or } \textit{smooth}\text{, contact)} \\
&= 0; \quad \text{if } \boldsymbol{v}_C = \boldsymbol{0} \qquad \text{(slippingless, or } \textit{rough}\text{, contact).}
\end{aligned}
\tag{1.18.19}
$$

It should be stressed that, in all these considerations, the relevant *velocities* are those of *material particles*, and not those of geometrical points of application of the loads.

2

Kinematics of Constrained Systems

(i.e., Lagrangean Kinematics)

I cannot too strongly urge that a kinematical result is a result valid forever, no matter how time and fashion may change the "laws" of physics.

(Truesdell, 1954, p. 2)

It is my belief that students have difficulty with mechanics because of an inadequate knowledge of kinematics.

(Fox, 1967, p. xi)

2.1 INTRODUCTION

As complementary reading for this chapter, we recommend the following (alphabetically):

General: Hamel (1904(a), (b)), Heun (1906, 1914), Lur'e (1968), Neimark and Fufaev (1972), Novoselov (1979), Papastavridis (1999), Prange (1935).

Special problems, extensions: Carvallo (1900, 1901), Lobas (1986), Lur'e (1968), Stückler (1955), Synge (1960).

Research journals (see the references at the end of this book): *Acta Mechanica Sinica* (Chinese), *Applied Mathematics and Mechanics* (Chinese), *Archive of Applied Mechanics* (former *Ingenieur Archiv*; German), *Journal of Applied Mechanics* (ASME; American), *Applied Mechanics* (Soviet → Ukrainian), *Journal of Guidance, Control, and Dynamics* (AIAA; American), *PMM* (Soviet → Russian), *ZAMM* (German), *ZAMP* (Swiss); also the various journals on kinematics, mechanisms, machine theory, design, robotics, etc.

In this chapter we begin the study of analytical mechanics proper with a detailed treatment of Lagrangean kinematics, i.e., the theory of *position* and linear *velocity* constraints (or Pfaffian constraints) in mechanical systems with a finite number of *degrees of freedom*; that is, a finite number of movable parts; as opposed to continuous systems that have a *countably infinite* set of such freedoms. All relevant fundamental concepts/definitions/equations—such as velocity, acceleration, constraint, *holonomicity* versus *nonholonomicity*, constraint stationarity (or *scleronomicity*) versus nonstationarity (or *rheonomicity*)—are detailed in both particle and system variables, along with elaborate discussions of *quasi coordinates* and the associated *transitivity equations* and *Hamel coefficients*; and as well as a direct and readable (and very rare) treatment of *Frobenius*' fundamental necessary and sufficient conditions for the holonomicity, or lack thereof, of a system of Pfaffian constraints.

The examples and problems, some at the ends of the paragraphs and some (the more comprehensive ones) at the end of the chapter, are an indispensable part of the material; several secondary theoretical points and results are presented there.

This chapter, and the next one on Kinetics, constitute the fundamental essence and core of Lagrangean analytical mechanics.

2.2 INTRODUCTION TO CONSTRAINTS AND THEIR CLASSIFICATIONS

Positions, Configurations, Motions

Let us consider a general finite mechanical system S consisting of N (=positive integer), *free*, or *unconstrained*, material particles. The *position* $\boldsymbol{r}$ of a generic S-particle, P, at the generic time instant, t, relative to an "origin" fixed in a, say inertial, frame of reference, F, is defined by the vector function

$$\boldsymbol{r} = \boldsymbol{f}(P, t) \equiv \boldsymbol{r}(P, t). \tag{2.2.1}$$

The collection of all these particle vectors, at a *current* instant t, make up a current *system position*, or current *configuration* of $S, C(t)$, and its evolution in time constitutes a *motion* of S. The latter, clearly, depends on the frame of reference. Thus, the complete description of a motion of S, if the latter is modeled as a collection of N particles, requires (at most) knowledge of $3N$ functions of time; for example, the $3N$ rectangular Cartesian components = coordinates of the N $\boldsymbol{r}$'s:

$$(x_1, y_1, z_1; \ldots; x_N, y_N, z_N) \equiv (x, y, z) \equiv (\xi_1, \ldots, \xi_{3N}) \equiv \boldsymbol{\xi}. \tag{2.2.1a}$$

These numbers can be viewed as the rectangular Cartesian coordinates of the $3N$-dimensional position vector of a *single* fictitious, or *figurative*, particle representing S, in a $3N$-dimensional Euclidean space, E_{3N}, henceforth called the system's *unconstrained configuration space*; and, therefore, a motion of S can be visualized as the path traced by the tip of that *system position vector* in E_{3N}. Equation (2.2.1) can be replaced by

$$\boldsymbol{r} = \boldsymbol{f}(\boldsymbol{r}_o, t; t_o) \equiv \boldsymbol{r}(\boldsymbol{r}_o, t; t_o), \tag{2.2.2}$$

where (fig. 2.1): $\boldsymbol{r}_o$ = "reference position" of P at the "reference time" $t = t_o$, is used to distinguish, or label, the various S-particles; and the totality of $\boldsymbol{r}_o$'s constitutes the *reference* configuration of S at t_o, $C(t_o)$. For a fixed $\boldsymbol{r}_o$ and variable t (i.e., a motion

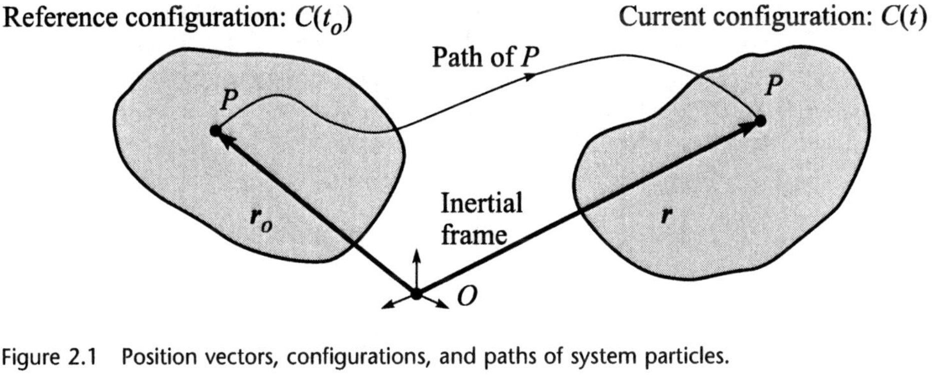

Figure 2.1 Position vectors, configurations, and paths of system particles.

of P), eqs. (2.2.1, 2) give the *path* of a particle P that was initially at $\boldsymbol{r}_o$. (The same equations for fixed t and variable $\boldsymbol{r}_o$ would give us the *transformation* of the spatial region initially occupied by the system, to its current position at time t.)

The one-to-one correspondence between $\boldsymbol{r}$ (and t) and $\boldsymbol{r}_o$ (and t_o), of the same particle P—that is, the physical fact that "initially distinct particles must remain distinct throughout the motion"—requires that (2.2.2) has an *inverse*:

$$\boldsymbol{r}_o = \boldsymbol{f}^{-1}(\boldsymbol{r}, t; t_o) \equiv \boldsymbol{g}(\boldsymbol{r}, t_o; t): \text{ reference configuration at (variable) time } t_o. \tag{2.2.2a}$$

Switching the roles of $(\boldsymbol{r}, t)$ and $(\boldsymbol{r}_o, t_o)$, we can view (2.2.2a) as expressing the "current" position $\boldsymbol{r}_o$ in terms of the "reference" position and time $(\boldsymbol{r}, t)$ and "current" time t_o. From now on, for simplicity, we shall drop, in the above, the explicit $(\boldsymbol{r}_o, t_o)$ and/or P-dependence [also, replace the rigorous notation $\boldsymbol{f}(\ldots)$ with $\boldsymbol{r}(\ldots)$, as done frequently in engineering mathematics, except whenever extra clarity is needed], and write (2.2.1) simply as

$$\boldsymbol{r} = \boldsymbol{r}(t). \tag{2.2.3}$$

REMARKS

(i) For (2.2.2) and (2.2.2a) to be mutually consistent, we must have

$$\text{(2.2.2) for } t = t_o \Rightarrow \text{(2.2.2a)}: \quad \boldsymbol{r} = \boldsymbol{f}(\boldsymbol{r}_o, t; t_o) \Rightarrow \boldsymbol{r}_o = \boldsymbol{f}(\boldsymbol{r}_o, t_o; t_o) = \boldsymbol{f}^{-1}(\boldsymbol{r}, t; t_o);$$

$$\text{(2.2.2a) for } t_o = t \Rightarrow \text{(2.2.2)}: \quad \boldsymbol{r}_o = \boldsymbol{f}^{-1}(\boldsymbol{r}, t; t_o) \Rightarrow \boldsymbol{r} = \boldsymbol{f}^{-1}(\boldsymbol{r}, t; t) = \boldsymbol{f}(\boldsymbol{r}_o, t; t_o); \tag{2.2.2b}$$

also

$$\boldsymbol{r} = \boldsymbol{f}\big[\boldsymbol{f}(\boldsymbol{r}_o, t_1; t_o), t; t_1\big] = \boldsymbol{f}(\boldsymbol{r}_o, t; t_o), \tag{2.2.2c}$$

where t_1 is another reference time, say $t_o \leq t_1 \leq t$.

(ii) In continuum mechanics, $(\boldsymbol{r}_o, t)$ and $(\boldsymbol{r}, t)$ are called, respectively, *material* (or *Lagrangean*) and *spatial* (or *Eulerian*) variables; with the former preferred in *solid* mechanics (e.g., nonlinear elasticity), and the latter dominating *fluid* mechanics (e.g., hydrodynamics). (See, e.g., Truesdell and Toupin, 1960, and Truesdell and Noll, 1965.)

(iii) For systems with a finite number of particles, the dependence on the latter is, frequently, expressed by the discrete subscript notation (i.e., $\boldsymbol{r}_o \to$ positive integer denoting the "name" of the particle):

$$\boldsymbol{r}_P = \boldsymbol{r}_P(t) = \{x_P(t), y_P(t), z_P(t)\} \qquad (P = 1, \ldots, N). \tag{2.2.4}$$

The simpler continuum mechanics notation, eqs. (2.2.1, 3), dispenses with all unnecessary *particle* indices, and allows one to concentrate on the *system indices* (as we begin to show later), which is the essence of the method of analytical mechanics. It also allows for a more general exposition; for example, a unified treatment of systems containing both rigid (discrete) and flexible (continuous) parts.

Constraints

If the N vectors $\boldsymbol{r}$, and/or corresponding (inertial) velocities $\boldsymbol{v} \equiv d\boldsymbol{r}/dt$, are *functionally unrelated* and *uninfluenced* from each other (*internally*) or from their environment (*externally*), apart from continuity and consistency requirements, like (2.2.2b,c)—

something we will normally assume—that is, if, and prior to any *kinetic considerations*, the $\boldsymbol{r}$'s and $\boldsymbol{v}$'s are free to vary *arbitrarily and independently* from each other, then S is called (internally and/or externally) *free or unconstrained*; if not, S is called (internally and/or externally) *constrained.* In the latter case, certain configurations and/or (velocities $\Rightarrow$) motions are unattainable, or inadmissible; or, alternatively, if we know the positions and velocities of some of the particles of the system, we can deduce those of the rest, without recourse to kinetics. [Outside of areas like astronomy/celestial mechanics, ballistics, etc., almost all other Earthly systems of relevance, and a lot of non-Earthly ones, are constrained—hence, the importance of analytical mechanics, especially to engineers.]

Such restrictions, or *constraints*, on the positions and/or velocities of S are expressed analytically by one or more ($< 3N$) scalar functional relations of the form

$$f(t, \boldsymbol{r}_1, \ldots, \boldsymbol{r}_N; \boldsymbol{v}_1, \ldots, \boldsymbol{v}_N) = 0, \qquad \text{or, compactly,} \qquad f(t, \boldsymbol{r}, \boldsymbol{v}) = 0. \tag{2.2.5}$$

These equalities are assumed to be: (i) *continuous* and as many times *differentiable in their arguments* as needed (usually, continuity of the zeroth, first-, and second-order partial derivatives will suffice), in some region of the $(x, y, z; dx/dt, dy/dt, dz/dt; t)$; (ii) *mutually consistent* (i.e., kinematically possible, or admissible); (iii) *independent* [i.e., not connected by additional functional relations like $F(f_1, f_2, \ldots) = 0$]; and (iv) *valid for any forces acting on S, any motions of it, and any temporal boundary/initial conditions on these motions* (see also *semiholonomic* systems below).

Following ordinary differential equation terminology, we call (2.2.5) a first-order (nonlinear) constraint, or nonlinear velocity constraint. With few exceptions [as in chaps. 5 and 6, where generally nonlinear constraints of the form $f(\boldsymbol{r}, \boldsymbol{v}, \boldsymbol{a}, t) = 0$ ($\boldsymbol{a}$: accelerations) are discussed], the *velocity constraint* (2.2.5) is the most general constraint examined here.

[Other, perhaps more suggestive terms, for constraints are *conditions* (Victorian English: *equations of condition*; German: *bedingungen*), and *connections* or *couplings* (French: *liaisons*; German: *bindungen*; Greek: *σύνδεσμοι*; Russian: *svyaz'*).]

Special Cases of Equation (2.2.5)

(i) Constraints like

$$\phi(t, \boldsymbol{r}) = 0, \qquad \text{or [recalling (2.2.1a)],} \qquad \phi(t, \xi) = 0, \tag{2.2.6}$$

are called *finite*, or *geometrical*, or *positional*, or *configurational*, or *holonomic*. [From the Greek: *hólos* = complete, whole, integral; that is, finite, nondifferential; and *nómos* = law, rule, (here) condition, constraint. After Hertz (early 1890s); also C. Neumann (mid-1880s).]

(ii) Again, with the exception of chapters 5, 6, and 7, all velocity constraints treated here have the practically important *linear velocity*, or *Pfaffian*, form

$$f \equiv \mathsf{S}\,(\boldsymbol{B} \cdot \boldsymbol{v}) + B = 0, \tag{2.2.7}$$

where $\boldsymbol{B} = \boldsymbol{B}(t, \boldsymbol{r})$, $B = B(t, \boldsymbol{r})$ are known functions of the $\boldsymbol{r}$'s and t, and *Lagrange's symbol* $\mathsf{S}(\ldots)$ signifies summation over all the material particles of S, at a given instant, like a Stieltjes' integral (so it can handle uniformly both continuous and discrete situations). Those uncomfortable with it may replace it with the more familiar Leibnizian $\int(\ldots)$.

Multiplying (2.2.7) by dt, which does not interact with $\mathcal{S}(\ldots)$, we obtain the *kinematically possible*, or *kinematically admissible*, form of the Pfaffian constraint,

$$f\,dt \equiv \mathcal{S}(\boldsymbol{B}\cdot d\boldsymbol{r}) + B\,dt = 0. \tag{2.2.7a}$$

Degrees of Freedom

A system of N particles subject to h (independent) positional constraints:

$$\phi_H(t, \boldsymbol{r}) = 0 \qquad (H = 1, \ldots, h), \tag{2.2.8}$$

and m (independent) Pfaffian constraints:

$$f_D \equiv \mathcal{S}(\boldsymbol{B}_D\cdot \boldsymbol{v}) + B_D = 0 \qquad (D = 1, \ldots, m), \tag{2.2.9}$$

that is, a total of $h + m$ constraints, is said to have a total of $3N - (h + m)(>0)$ *degrees of freedom* (*DOF*). This is a fundamental concept whose significance to both kinematics and kinetics (of constrained systems) will emerge gradually in what follows.

[Quick preview: DOF = Number of independent components of system vector of virtual displacement (§ 2.3–7)
= Number of kinetic (i.e., reactionless) equations of motion of system (chap. 3).]

Holonomicity versus Nonholonomicity

A positional constraint like (2.2.6), since it holds identically during all system motions, can *always* be brought to the velocity form (2.2.7) by $d(\ldots)/dt$-differentiation:

$$d\phi/dt = \mathcal{S}(\partial\phi/\partial\boldsymbol{r})\cdot\boldsymbol{v} + \partial\phi/\partial t = 0; \tag{2.2.10}$$

that is, $\boldsymbol{B} \to \partial\phi/\partial\boldsymbol{r} \equiv \boldsymbol{grad}\,\phi$ (normal to the E_{3N}-surface $\phi = 0$) and $B \to \partial\phi/\partial t$. However, the converse is *not* always true: the velocity constraint (2.2.7) may or may not be (able to be) brought to the positional form (2.2.6); that is, by integration and with no additional knowledge of the motion of the system; namely, without recourse to kinetics. If (2.2.7) *can* be brought to the form (2.2.6), then it is called *completely integrable*, or *holonomic* (H); if it *cannot*, it is called nonintegrable, or *nonholonomic* (NH); or, sometimes, anholonomic. This holonomic/nonholonomic distinction of velocity constraints is fundamental to analytical mechanics; it is by far the most important of all other constraint classifications. [The term anholonomic, more consistent than the term nonholonomic seems to be due to Schouten (1954).]

Schematically, we have

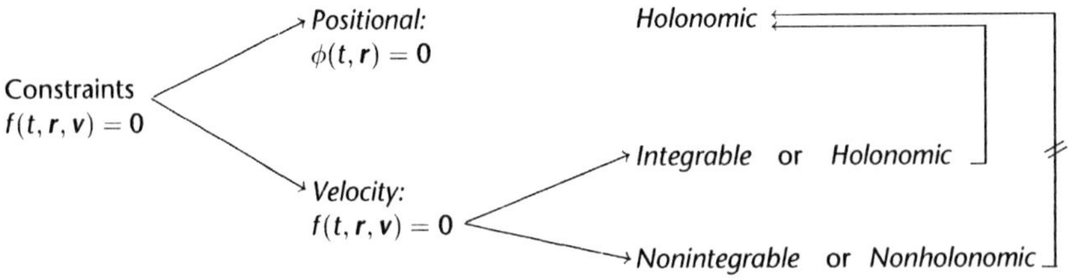

Hence, a H velocity constraint, like (2.2.10), is actually a positional constraint disguised in kinematical form. Before embarking into a detailed study of H/NH constraints, we will mention some additional, secondary but useful, constraint classifications.

Scleronomicity versus Rheonomicity

Velocity constraints of the form

$$f(\boldsymbol{r}, \boldsymbol{v}) = 0 \;\Rightarrow\; \partial f/\partial t = 0 \tag{2.2.7b}$$

are called *stationary*; otherwise (i.e., if $\partial f/\partial t \neq 0$), they are called *nonstationary*. If *all* the constraints of a system are stationary, the system is called *scleronomic*; if not, the system is called *rheonomic*. [From the Greek: *sclerós* = hard, rigid, invariable; *rhéo* = to flow; and the earlier *nómos* = law, rule, decree, (here) condition; that is, *scleronomic* = invariable constraint, *rheonomic* = variable/fluid constraint. After Boltzmann (1897–1904).] For positional constraints and Pfaffian constraints, stationarity means, respectively,

$$\phi(\boldsymbol{r}) = 0 \qquad \text{and} \qquad S\, \boldsymbol{B}(\boldsymbol{r}) \cdot \boldsymbol{v} = 0. \tag{2.2.11}$$

Catastaticity versus Acatastaticity

Pfaffian constraints of the form

$$S\, \boldsymbol{B}(t, \boldsymbol{r}) \cdot \boldsymbol{v} + B(t, \boldsymbol{r}) = 0, \tag{2.2.11a}$$

are called *acatastatic*; while those of the form

$$S\, \boldsymbol{B}(t, \boldsymbol{r}) \cdot \boldsymbol{v} = 0 \qquad [\text{i.e., } B(t, \boldsymbol{r}) = 0] \tag{2.2.11b}$$

are called *catastatic*. It is this classification [due to Pars (1965, pp. 16, 24) and, obviously, having meaning only for Pfaffian constraints], and not the earlier one of scleronomicity versus rheonomicity, that is important in the kinetics of systems under such constraints.

REMARKS

(i) The reason for calling the second of (2.2.11) scleronomic, instead of

$$S\, \boldsymbol{B}(\boldsymbol{r}) \cdot \boldsymbol{v} + B(\boldsymbol{r}) = 0, \tag{2.2.11c}$$

that is, for requiring that scleronomic constraints linear in the velocities be also *homogeneous* in them (i.e., have $B = 0 \Rightarrow$ catastaticity), is so that it matches the kinematic *form* generated by $d/dt(\ldots)$-differentiating the scleronomic positional constraint (first of 2.2.11):

$$\phi(\boldsymbol{r}) = 0 \;\Rightarrow\; d\phi/dt = S\, (\partial\phi/\partial\boldsymbol{r}) \cdot \boldsymbol{v} = 0. \tag{2.2.11d}$$

Geometrical interpretation of this requirement: Otherwise, the corresponding constraint surface, in "velocity space," would be a plane with distance from the origin proportional to B. That term, representing the (negative of the) velocity of the constraint plane normal to itself, is clearly a rheonomic effect. (Remark due to Prof. D. T. Greenwood, private communication.)

(ii) Clearly, every scleronomic Pfaffian constraint is catastatic ($B = 0$); but catastatic Pfaffian constraints may be scleronomic [$\boldsymbol{B} = \boldsymbol{B}(\boldsymbol{r})$, second of (2.2.11)] or rheonomic [$\boldsymbol{B} = \boldsymbol{B}(t, \boldsymbol{r})$, (2.2.11b)].

Bilateral versus Unilateral Constraints

Equality constraints of the form (2.2.5) are called *bilateral*, or *two-sided*, or *equality*, or *reversible*, or *unchecked* (after Langhaar, 1962, p. 16); while constraints of the form

$$f(t, \boldsymbol{r}, \boldsymbol{v}) \geq 0 \qquad \text{or} \qquad f(t, \boldsymbol{r}, \boldsymbol{v}) \leq 0 \tag{2.2.11e}$$

are called *unilateral*, or *one-sided*, or *inequality*, or *irreversible*. Physically, bilateral constraints occur when the bodies in contact *cannot* separate from each other: for example, a rigid sphere moving between two parallel fixed planes, in continuous contact with both. In the unilateral case, the bodies in contact *can* separate: for example, a sphere in contact with only one plane, or a system of two particles connected by an inextensible string—their distance cannot exceed the string's length. Following Gantmacher (1970, p. 12), we can state that the general motion of a unilaterally constrained motion may be divided into segments, such that: (i) in certain segments the constraint is "taut" [(2.2.11e) with the = sign; e.g., particle on a light, inextensible, and taut string], and motion occurs as if the constraint were *bilateral*; and (ii) in other segments, the constraint is not taut, it is "loose," and motion occurs as if the constraint were *absent*. Concisely, a *unilateral constraint* is either *replaced by a bilateral one*, or is *eliminated altogether*. Hence, in what follows, we shall limit ourselves to bilateral constraints.

REMARKS

(i) A small number of authors call all constraints of the form (2.2.6) holonomic, as well as those reducible to that form; and call all others nonholonomic. According to such a definition, bilateral constraints like (2.2.11e) would be nonholonomic! The reader should be aware of such historically unorthodox practices.

(ii) The equations $\phi(\boldsymbol{r}, t) = 0$ and $d\phi/dt \equiv S\,(\partial\phi/\partial\boldsymbol{r}) \cdot \boldsymbol{v} + \partial\phi/\partial t = 0$ restrict a system's positions and velocities; equation $d\phi/dt = 0$ is the compatibility of velocities with $\phi = 0$. Similarly, the equation

$$d^2\phi/dt^2 = S\,[d/dt(\partial\phi/\partial\boldsymbol{r}) \cdot \boldsymbol{v} + (\partial\phi/\partial\boldsymbol{r}) \cdot \boldsymbol{a}] + d/dt(\partial\phi/\partial t) = 0$$

is the compatibility of *accelerations* with $\phi = 0$, $d\phi/dt = 0$; and likewise for higher such derivatives.

(iii) In the case of *unilateral* constraints, if at a certain time t: $f > 0$, then, as explained earlier, that constraint plays no role in the system's motion. But if $f = 0$, then, as a Taylor expansion around t shows, motion that satisfies either of these two relations may occur; in the former case $df/dt = 0$, and in the latter $df/dt \geq 0$. Thus, the simultaneous conditions $f = 0$ and $df/dt < 0$ allow us to detect a *possible incompatibility* between velocities and $f \geq 0$. Usually, such conditions occur in impact problems (chap. 4; also Kilmister and Reeve, 1966, pp. 67–68).

(iv) *Geometrical/physical remarks:* In a system S consisting of several rigid bodies, and its *environment* (i.e., other bodies/foreign obstacles, massless coupling elements:

e.g., springs, cables) the following conditions apply:

(a) Every condition expressing the direct *contact* of two rigid bodies of S, or the contact of one of its bodies with a foreign obstacle (environment) that is either *fixed* or has *known motion* (i.e., its position coordinates are known functions of time only), results in a holonomic equation of the form (2.2.6); and the corresponding contact forces are the reactions of that constraint.

(b) If, further, at those contact points, *friction is high enough* to guarantee us (in advance of kinetic considerations) *slippinglessness*, then the positions and velocities there satisfy (2.2.7)-like Pfaffian equations (usually, but not always, nonholonomic). These conditions express the vanishing of a component of (relative) slipping velocity in a certain direction; and, therefore, there are as many as the number of independent such nonslipping directions.

(c) If, in addition, friction there is *very high*, so that not only slipping but also *pivoting* vanishes, then we have additional (usually nonholonomic) (2.2.7)-like equations; that is, linear velocity constraints arise quite naturally and frequently in daily life. [Nonslipping and nonpivoting are maintained by constraint forces (and couples), just like contact. All these constraint forces are examples of *passive* reactions; for more general, *active*, constraint reactions, see, for example, § 3.17.]

(v) Holonomic and/or nonholonomic constraints due exclusively to the mutual interaction of the system bodies are called *internal* (or *mutual*); while those arising, even partially, from the interaction of the system with its environment are called *external*. The associated constraint reactions are called, respectively, *internal* (or *mutual*) and *external*.

(vi) Finally, we repeat that such holonomic and/or nonholonomic constraints express restrictions among positions and velocities *independently* of the equations of motion and associated (temporal) initial/boundary conditions, and *before* the complete solution of the problem is carried out. Solving the problem means finding $\boldsymbol{r} = \boldsymbol{r}(t)$: known function of time; then $\boldsymbol{v} = d\boldsymbol{r}/dt = \boldsymbol{v}(t)$: known function of time; and these $\boldsymbol{r}$'s and $\boldsymbol{v}$'s automatically satisfy the constraints. Under such a viewpoint, integrals of the system, like those of linear/angular momentum and energy, assuming they exist, *do not qualify* as constraint equations.

The (bilateral) constraints, discussed above, are summarized as follows:

General first-order constraints

$f(\boldsymbol{r}) = 0$:	Holonomic (integrable) and scleronomic (stationary)
$f(t, \boldsymbol{r}) = 0$:	Holonomic (integrable) and rheonomic (nonstationary)
$f(\boldsymbol{r}, \boldsymbol{v}) = 0$:	Nonholonomic (if nonintegrable) and scleronomic (stationary)
$f(t, \boldsymbol{r}, \boldsymbol{v}) = 0$:	Nonholonomic (if nonintegrable) and rheonomic (nonstationary)

Pfaffian velocity constraints

$\mathcal{S}\, \boldsymbol{B}(t, \boldsymbol{r}) \cdot \boldsymbol{v} + B(t, \boldsymbol{r}) = 0$:	Rheonomic and acatastatic
$\mathcal{S}\, \boldsymbol{B}(\boldsymbol{r}) \cdot \boldsymbol{v} + B(\boldsymbol{r}) = 0$:	Rheonomic and acatastatic
$\mathcal{S}\, \boldsymbol{B}(t, \boldsymbol{r}) \cdot \boldsymbol{v} = 0$:	Rheonomic and catastatic
$\mathcal{S}\, \boldsymbol{B}(\boldsymbol{r}) \cdot \boldsymbol{v} = 0$:	Scleronomic and catastatic

(There is no such thing as scleronomic and acatastatic Pfaffian constraint.)

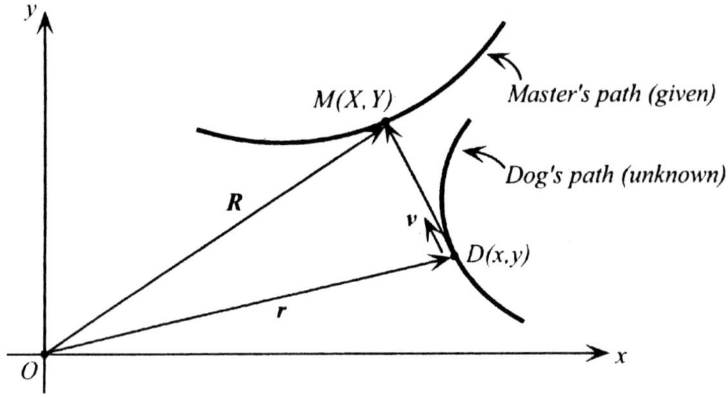

Figure 2.2 Plane pursuit problem: a dog (D) moving continuously toward its master (M).

Example 2.2.1 *Plane Pursuit Problem—Catastatic but Rheonomic (or Nonstationary) Pfaffian constraint.* The master (M) of a dog (D) walks along a *given* plane curve: $\boldsymbol{R} = \boldsymbol{R}(t) = \{X = X(t), Y = Y(t)\}$. Let us find the differential equation of the path of D: $\boldsymbol{r} = \boldsymbol{r}(t) = \{x = x(t), y = y(t)\}$, if D moves, with instantaneous velocity $\boldsymbol{v}$, to meet M, so that at every instant its velocity is directed toward M (fig. 2.2).

We must have:

$$\boldsymbol{v} = \text{parallel to } \boldsymbol{R} - \boldsymbol{r} = v\big[(\boldsymbol{R} - \boldsymbol{r})/|\boldsymbol{R} - \boldsymbol{r}|\big] \equiv v\boldsymbol{e},$$

or, in components,

$$dx/dt = v\big[(X - x)/|\boldsymbol{R} - \boldsymbol{r}|\big], \qquad dy/dt = v\big[(Y - y)/|\boldsymbol{R} - \boldsymbol{r}|\big]; \tag{a}$$

or, eliminating v between them,

$$[Y(t) - y](dx/dt) - [X(t) - x](dy/dt) = 0. \tag{b}$$

It is not hard to show that this pursuit problem in *space* leads to the following constraints (with some obvious notation):

$$[Y(t) - y](dx/dt) - [X(t) - x](dy/dt) = 0, \tag{c}$$

$$[Z(t) - z](dx/dt) - [X(t) - x](dz/dt) = 0, \tag{d}$$

$$[Z(t) - z](dy/dt) - [Y(t) - y](dz/dt) = 0. \tag{e}$$

See also Hamel (1949, pp. 770–773).

Example 2.2.2 *Acatastatic Constraints.* Let us consider the rolling of a sphere S of radius r and center G on the rough inner surface of a vertical circular cylinder A of radius $R(> r)$. Let us introduce the following convenient *intermediate* axes/basis G–123/G–$\boldsymbol{ijk}$ (fig. 2.3): Let ϕ be the *azimuth*, or *precession*-like, angle of the plane G–13, and z = vertical coordinate of G (positive upward from some fixed

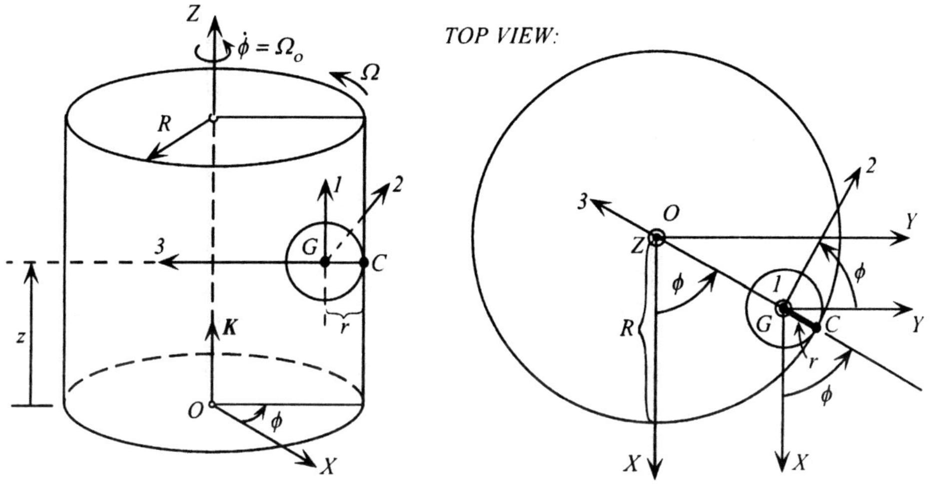

Figure 2.3 Rolling of a sphere on a vertical circular cylinder. *G1*: vertically upward; *G3*: horizontally intersects the (vertical) cylinder axis; *G2*: horizontal, so that *G–123* is orthogonal–normalized–dextral (OND).

plane, perpendicular to the cylinder axis). Then,

$$\boldsymbol{v}_G = \text{inertial velocity of } G = (v_1 = dz/dt, v_2 = (R-r)(d\phi/dt) \equiv (R-r)\omega_\phi, v_3 = 0); \tag{a}$$

or, alternatively, if $\boldsymbol{OG} = z\boldsymbol{K} + (R-r)(-\boldsymbol{k})$, then (with $d\phi/dt \equiv \omega_\phi$)

$$\boldsymbol{v}_G = d(\boldsymbol{OG})/dt = (dz/dt)\boldsymbol{K} + (R-r)(-d\boldsymbol{k}/dt) = (dz/dt)\boldsymbol{i} + (R-r)(\omega_\phi\,\boldsymbol{j}). \tag{b}$$

If $\boldsymbol{\omega}$ = inertial angular velocity of sphere = $(\omega_1, \omega_2, \omega_3)$, then the inertial velocity of the contact point C, $\boldsymbol{v}_C$, is

$$\boldsymbol{v}_C = \boldsymbol{v}_G + \boldsymbol{\omega} \times \boldsymbol{r}_{C/G} = (v_1, v_2, v_3) + (\omega_1, \omega_2, \omega_3) \times (0, 0, -r)$$
$$= (v_1 - \omega_2 r, \quad v_2 + \omega_1 r, \quad v_3). \tag{c}$$

Therefore: (i) If the cylinder is *stationary* (i.e., fixed), the rolling constraint is $\boldsymbol{v}_C = \boldsymbol{0}$, or, in components,

$$v_1 - \omega_2 r = 0 \Rightarrow \omega_2 = (dz/dt)\big/r, \quad v_2 + \omega_1 r = 0 \Rightarrow \omega_1 = [1 - (R/r)]\omega_\phi, \quad v_3 = 0. \tag{d}$$

(ii) If the cylinder is made to rotate about its axis with an (inertial) angular velocity $\boldsymbol{\Omega} = \boldsymbol{\Omega}(t)$ = *given function of time*, the rolling constraint is

$$\boldsymbol{v}_C = \boldsymbol{\Omega} \times \boldsymbol{r}_{C/O} = (\Omega\boldsymbol{K}) \times (-R\boldsymbol{k}) = (\Omega\boldsymbol{i}) \times (-R\boldsymbol{k}) = (\Omega R)\boldsymbol{j} \equiv (0, \Omega R, 0),$$

or, in components [invoking (c)],

$$v_1 - \omega_2 r = 0 \Rightarrow \omega_2 = (dz/dt)\big/r \equiv v_z/r,$$
$$v_2 + \omega_1 r = \Omega(t)R \Rightarrow \omega_1 = \omega_\phi + (R/r)\omega_r, \qquad v_3 = 0, \tag{e}$$

where $\omega_r \equiv \Omega - \omega_\phi$ = relative angular velocity of cylinder about meridian plane G–13. The first of the constraints (e) is nonstationary and acatastatic, even if $\Omega = constant$. [As explained in §2.5 ff., the *virtual* form of that constraint is $\delta p_2 + \delta\theta_1 r = 0$, where $dp_2 \equiv v_2\, dt$ and $d\theta_1 \equiv \omega_1\, dt$; and this coincides with the virtual form of the catastatic second of the constraints (d). In general, p_2 and θ_1 are "quasi coordinates"—see §2.9 ff.]

First and second of the constraints (*d*) in terms of the Eulerian angles of the sphere Φ, Θ, Ψ, relative to the "semiinertial" (translating but nonrotating) axes G–XYZ

We have, successively (recalling §1.12,13),

$$v_1 = dz/dt \equiv v_Z,$$

$$\begin{aligned}\omega_2 &- \cos(2, X)\omega_X + \cos(2, Y)\omega_Y + \cos(2, Z)\omega_Z \\ &= (-\sin\phi)\omega_X + (\cos\phi)\omega_Y + (0)\omega_Z \\ &= (-\sin\phi)[\cos\Phi(d\Theta/dt) + \sin\Phi\sin\Theta(d\Psi/dt)] \\ &\quad + (\cos\phi)[\sin\Phi(d\Theta/dt) - \cos\Phi\sin\Theta(d\Psi/dt)] \\ &= \cdots = \sin(\Phi - \phi)(d\Theta/dt) - \cos(\Phi - \phi)\sin\Theta(d\Psi/dt),\end{aligned} \tag{f}$$

that is, the familiar ω_Y component but with ϕ replaced by $\Phi - \phi$;

$$v_2 = (R - r)(d\phi/dt),$$

$$\begin{aligned}\omega_1 &= \cos(1, X)\omega_X + \cos(1, Y)\omega_Y + \cos(1, Z)\omega_Z \\ &= (0)\omega_X + (0)\omega_Y + (1)\omega_Z = d\Phi/dt + \cos\Theta(d\Psi/dt).\end{aligned} \tag{g}$$

Therefore, the first and second constraints (d) transform to

$$v_1 - \omega_2 r = dz/dt - r\big[\sin(\Phi - \phi)(d\Theta/dt) - \cos(\Phi - \phi)\sin\Theta(d\Psi/dt)\big] = 0, \tag{h}$$

$$v_2 + \omega_1 r = (R - r)(d\phi/dt) + r\big[d\Phi/dt + \cos\Theta(d\Psi/dt)\big] = 0; \tag{i}$$

and similarly for the first two of (e).

Example 2.2.3 *Acatastatic Constraints.* Let us consider the rolling of a sphere S of radius r and center G on a rough surface of revolution with a vertical axis. Let us introduce the convenient frame/axes/basis G–123/G–**ijk** shown in fig. 2.4. Further, let ϕ be the *azimuth*, or *precession*-like, angle of the meridian plane (and of plane G–23); and θ be the *nutation*-like angle between the positive surface axis and the common (outward) normal. Then, with $d\phi/dt \equiv \omega_\phi$, $d\theta/dt \equiv \omega_\theta$, we will have

$$\begin{aligned}\boldsymbol{\Omega}_o &= \text{inertial angular velocity of } G\text{–}123 \equiv (\Omega_1, \Omega_2, \Omega_3) \\ &= (-\omega_\phi\sin\theta,\ \omega_\theta,\ \omega_\phi\cos\theta),\end{aligned} \tag{a}$$

$$\boldsymbol{v}_G = \text{inertial velocity of } G \equiv (v_1, v_2, v_3) = (\rho\omega_\theta,\ R\omega_\phi = \rho\sin\theta\,\omega_\phi,\ 0), \tag{b}$$

where ρ = radius of curvature of meridian curve of parallel surface at G.

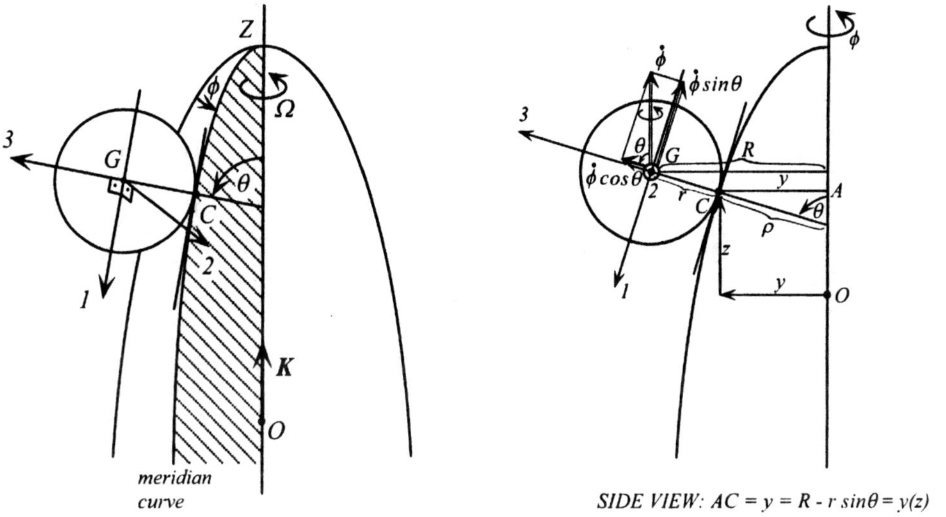

Figure 2.4 Rolling of a sphere on a vertical surface of revolution. *G3*: along common normal, outward; *G1*: parallel to tangent to meridian curve, at contact point *C*; *G2*: parallel to tangent to circular section through *C* (or, so that *G–123* is OND).

If $\boldsymbol{\omega}$ = inertial angular velocity of sphere = $(\omega_1, \omega_2, \omega_3)$, then the inertial velocity of the contact point C, $\boldsymbol{v}_C$, equals

$$\boldsymbol{v}_C = \boldsymbol{v}_G + \boldsymbol{\omega} \times \boldsymbol{r}_{C/G}$$
$$= (v_1, v_2, v_3) + (\omega_1, \omega_2, \omega_3) \times (0, 0, -r) = (v_1 - \omega_2 r,\ v_2 + \omega_1 r,\ v_3). \qquad \text{(c)}$$

Therefore: (i) If the surface is *stationary*, the rolling constraint is $\boldsymbol{v}_C = \boldsymbol{0}$, or, in components,

$$v_1 - \omega_2 r = 0, \qquad v_2 + \omega_1 r = 0, \qquad v_3 = 0. \qquad \text{(d)}$$

(ii) If the surface is compelled to rotate about its axis with (inertial) angular velocity $\boldsymbol{\Omega} = \boldsymbol{\Omega}(t)$ = *given function of time*, the rolling constraint is

$$\boldsymbol{v}_C = \boldsymbol{\Omega} \times \boldsymbol{r}_{C/O} = \big(0, \Omega(R - r\sin\theta), 0\big), \qquad \text{(e)}$$

or, in components,

$$v_1 - \omega_2 r = 0 \ \Rightarrow\ \omega_2 = v_1/r = \rho\,\omega_\theta/r, \qquad \text{(f)}$$
$$v_2 + \omega_1 r = \Omega(R - r\sin\theta) \ \Rightarrow\ \omega_1 = (R/r)\omega_r - \Omega\sin\theta, \qquad v_3 = 0, \qquad \text{(g)}$$

where $\omega_r \equiv \Omega - \omega_\phi$ = relative angular velocity of surface about meridian plane G–13. The first constraint (g) is nonstationary and acatastatic, even if $\Omega = constant$. [As explained in §2.5 ff., the *virtual* form of that constraint is $\delta p_2 + \delta\theta_1 r = 0$, where $dp_2 \equiv v_2\,dt$ and $d\theta_1 \equiv \omega_1\,dt$; and it coincides with the virtual form of the catastatic second constraint (d). In general, p_2 and θ_1 are "quasi coordinates"— see §2.9 ff.]

SPECIALIZATIONS

(i) If the surface of revolution is another *sphere* with radius $\rho_o \equiv \rho - r = constant$, since then $v_1 = (\rho_o + r)\omega_\theta$, $v_2 = [(\rho_o + r)\sin\theta]\omega_\phi$, the constraints (f) and the second of (g) reduce, respectively, to

$$\omega_2 = [(\rho_o + r)/r]\omega_\theta = [(\rho_o/r) + 1]\omega_\theta, \tag{h}$$

$$\omega_1 = [(\rho_o/r) + 1]\sin\theta\,\omega_r - \Omega\sin\theta. \tag{i}$$

(ii) If the surface of revolution is another *sphere* with radius $\rho_o \equiv \rho - r$, that is free (i.e., *unconstrained*) to rotate about its fixed center with inertial angular velocity $\boldsymbol{\omega}' \equiv (\omega'_1, \omega'_2, \omega'_3)$, then, reasoning as earlier, we obtain the catastatic constraint equations

$$v_1 - \omega_2 r = \rho_o\omega'_2, \qquad v_2 + \omega_1 r = -\rho_o\omega'_1, \qquad v_3 = 0. \tag{j}$$

However, if the ω'_1, ω'_2, ω'_3 are *prescribed functions of time*, then the first and second of (j) become *nonstationary* (and *acatastatic*).

For additional such rolling examples, including the corresponding Newton–Euler (kinetic) equations, and so on, see the older British textbooks: for example, Atkin (1959, pp. 253–259), Besant (1914, pp. 353–359), Lamb (1929, pp. 162–170), Milne (1948, chaps. 15, 17).

Example 2.2.4 *Problem of Ishlinsky (or Ishlinskii).* Let us consider the rolling of a circular rough cylinder of radius R on top of two other identical circular and rough cylinders, each of radius r, rolling on a rough, fixed, and horizontal plane (fig. 2.5).

Let O–xyz and O–$x'y'z'$ be inertial axes, such that O–xy and O–$x'y'$ are both on that plane, while their axes Ox and Ox' are parallel to the lower cylinder generators

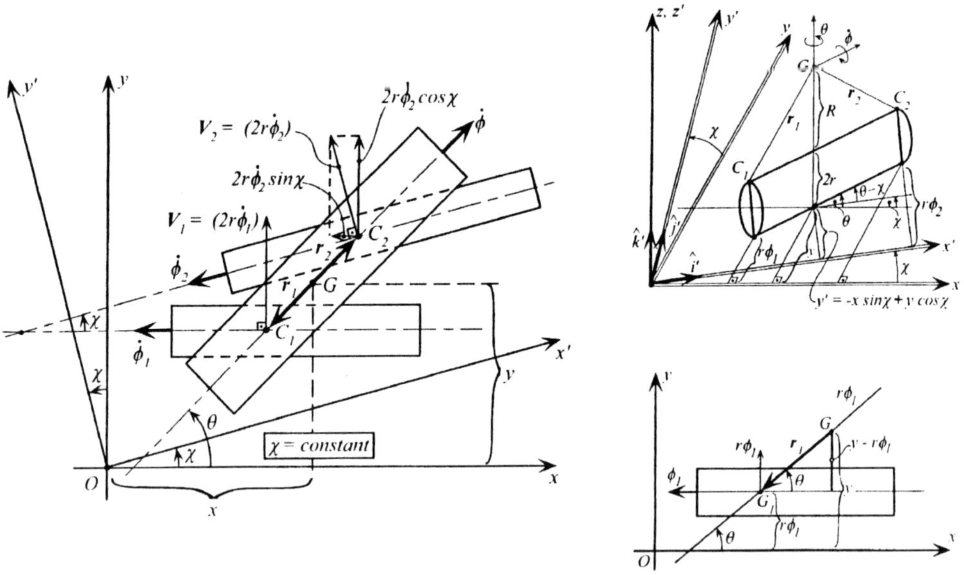

Figure 2.5 Rolling of a cylinder on top of two other rolling cylinders. Transformation equations: $x = x'\cos\chi - y'\sin\chi$, $y = x'\sin\chi + y'\cos\chi$.

and make, with each other, a *constant* angle χ. To describe the (global) system motion, let us choose the following *six* position coordinates: (i) (x, y) = inertial coordinates of *mass center* of upper cylinder G (as for its third, vertical, coordinate we have $z = 2r + R$); (ii) θ = angle between $+Ox$ and upper cylinder generator; (iii) ψ, ψ_1, ψ_2 = spin angles of the upper and two lower cylinders, respectively. Finally, let $\boldsymbol{r}_1$ and $\boldsymbol{r}_2$ be the *position vectors* of the contact points of the lower cylinders with the upper one, relative to G, and $\boldsymbol{v}_1$, $\boldsymbol{v}_2$ be the corresponding (inertial) velocities.

The rolling constraints are

$$\boldsymbol{v}_G + \boldsymbol{\omega} \times \boldsymbol{r}_1 = \boldsymbol{v}_1 \qquad \text{and} \qquad \boldsymbol{v}_G + \boldsymbol{\omega} \times \boldsymbol{r}_2 = \boldsymbol{v}_2. \tag{a}$$

Let us express them in terms of components along $O\text{–}xyz$. We have

$$\boldsymbol{v}_G = (dx/dt, dy/dt, 0) \equiv (v_x, v_y, 0)$$

$\boldsymbol{\omega}$: inertial angular velocity of upper cylinder

$$\begin{aligned}
&= \big((d\phi/dt)\cos\theta, (d\phi/dt)\sin\theta, d\theta/dt\big) \equiv (\omega_\phi\cos\theta, \omega_\phi\sin\theta, \omega_\theta),\\
\boldsymbol{r}_1 &= \big(-(y - r\phi_1)\cot\theta, -(y - r\phi_1), -R\big),\\
\boldsymbol{r}_2 &= [r\phi_2 - (y\cos\chi - x\sin\chi)]\cot(\theta - \chi)\boldsymbol{i}' + [r\phi_2 - (y\cos\chi - x\sin\chi)]\boldsymbol{j}' - R\boldsymbol{k}'\\
&= \big((r\phi_2 + x\sin\chi - y\cos\chi)\cos\theta/\sin(\theta - \chi),\\
&\qquad (r\phi_2 + x\sin\chi - y\cos\chi)\sin\theta/\sin(\theta - \chi), -R\big),\\
\boldsymbol{v}_1 &= (0, 2r\omega_1, 0) \qquad [\text{where } \omega_{1,2} \equiv d\phi_{1,2}/dt],\\
\boldsymbol{v}_2 &= (-2r\omega_2\sin\chi, 2r\omega_2\cos\chi, 0).
\end{aligned} \tag{b}$$

Substituting the above into (a), we obtain the following *four* constraint components:

$$\begin{aligned}
&v_x - R\omega_\phi\sin\theta - \omega_\theta(r\phi_1 - y) = 0,\\
&v_y + R\omega_\phi\cos\theta + \omega_\theta(r\phi_1 - y)\cot\theta - 2r\omega_1 = 0;\\
&v_x\sin(\theta - \chi) - R\omega_\phi\sin\theta\sin(\theta - \chi) - \omega_\theta(r\phi_2 + x\sin\chi - y\cos\chi)\sin\theta\\
&\qquad\qquad + 2r\omega_2\sin\chi\sin(\theta - \chi) = 0,\\
&v_y\sin(\theta - \chi) + R\omega_\phi\cos\theta\sin(\theta - \chi) + \omega_\theta(r\phi_2 + x\sin\chi - y\cos\chi)\cos\theta\\
&\qquad\qquad - 2r\omega_2\cos\chi\sin(\theta - \chi) = 0.
\end{aligned} \tag{c}$$

For further details, see, for example, Mei (1985, pp. 33–35), Neimark and Fufaev (1972, pp. 99–101). It can be shown (§2.11, 12) that these constraints are nonholonomic. Therefore, the system has $n = 6$ *global DOF*, and $n - m = 6 - 4 = 2$ *local DOF* (concepts explained in §2.3 ff.).

Example 2.2.5 *When is Rolling Holonomic?* So as to dispell the possible notion that all problems of (slippingless) rolling among rigid bodies lead to nonholonomic constraints, let us summarize below the cases of rolling that lead to holonomic constraints. It has been shown by Beghin (1967, pp. 436–438) that these are the following *two* kinds:

(i) The paths of the contact point(s) of the rolling bodies are known *ahead of time*; that is, before any dynamical consideration of the system involved and as function of its original position, on these bodies. Consider two such bodies whose bounding surfaces, S_1 and S_2, are described by the curvilinear surface (Gaussian) coordinates (u_1, v_1) and (u_2, v_2), respectively, in contact at a point C. Their relative positions, say of S_1 relative to S_2, are determined by the values of these coordinates at C and the angle ϕ formed by the tangents to the lines $u_1 = constant$ and $u_2 = constant$ (or v_1, $v_2 = constant$) there. Knowledge of the paths of C on both S_1 and S_2 translates to knowledge of the four holonomic functional relations:

$$u_1 = u_1(v_1), \qquad u_2 = u_2(v_2), \qquad \phi = \phi(u_1, u_2), \qquad s_1(u_1) = s_2(u_2) \pm c; \qquad \text{(a)}$$

where s_1 and s_2 are the arc lengths (or curvilinear abscissas) of the contact point paths S_1 and S_2, and c is an integration constant. It follows that, out of the *five* surface positional parameters, u_1, v_1, u_2, v_2, ϕ, only *one* is independent; the other four can be expressed in terms of that one by finite (holonomic) relations.

(ii) The bounding surfaces S_1 and S_2 are applicable on each other; they touch at homologous points and their homologous curves (trajectories of the contact point C on them) join together there. This is expressed by the condition of contact, and by

$$u_1 = u_2, \qquad v_1 = v_2, \qquad \phi = 0, \qquad \text{(b)}$$

at C; that is, again, a total of four holonomic equations. This condition is guaranteed to hold continuously if it holds initially and, afterwards, the pivoting vanishes. Such conditions are met in the following examples:

(a) Rolling of two plane curves (or normal cross sections of cylindrical surfaces S_1 and S_2) on each other, and expressed by $s_1 = s_2 \pm c$.
(b) Rolling of a body on a fixed surface, which it touches on only two points. For example, the rolling of a sphere on a system made up of a fixed circular cylinder and a fixed plane perpendicular to it [fig. 2.6(a)]. (If the cylinder rotates about its axis in a known fashion, the trajectories of the contact points on both plane and cylinder are known, but they are unknown on the sphere and, hence, such rolling is non-holonomic.)
(c) Rolling of two equal bodies of revolution whose axes are constrained to meet and, initially, are in contact along homologous parallels, or meridians [fig. 2.6(b)]. The pivoting of such applicable surfaces vanishes.

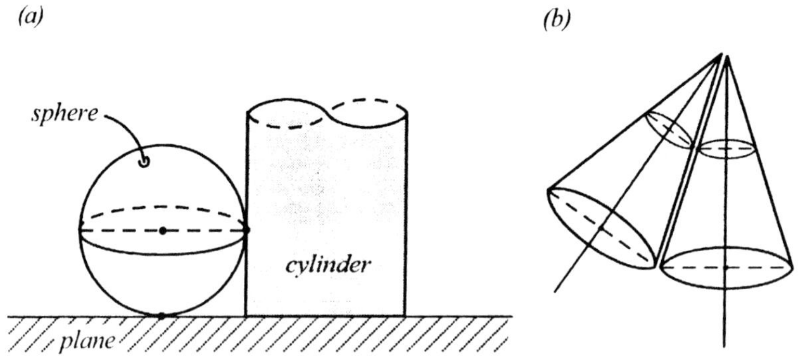

Figure 2.6 Examples of holonomic rolling: (a) rolling of a sphere on a fixed circular cylinder and a fixed plane perpendicular to it; (b) rolling of a cone on another equal fixed cone.

2.3 QUANTITATIVE INTRODUCTION TO NONHOLONOMICITY

Let us examine the differences between holonomic and nonholonomic constraints, in some mathematical detail, for the simplest possible case: a single particle, with (inertial) rectangular Cartesian coordinates x, y, z, moving in space under the Pfaffian equation

$$a\,dx + b\,dy + c\,dz = 0, \tag{2.3.1}$$

where a, b, c = continuously differentiable functions of x, y, z.

[The Pfaffian *expression* $a\,dx + b\,dy + c\,dz$ is a special *differential form* of the *first* degree. The *total* or Pfaffian *differential equation* (2.3.1) is a specialization of the *Monge* form:

$0 = f(x, y, z; dx, dy, dz)$ = *stationary* and *homogeneous* in the velocity components $(dx/dt, dy/dt, dz/dt)$, and hence (since t is absent) only *path restricting*.

The Monge form is, in turn, a specialization of the general first-order partial differential equation:

$$F(t; x, y, z; dx/dt, dy/dt, dz/dt) = 0.]$$

Now, the constraint (2.3.1) may be nonholonomic or it may be holonomic in differential (or velocity) form; specifically, if (2.3.1) can become, through multiplication with an appropriate *integrating factor*, $\mu = \mu(x, y, z)$, an *exact*, or *perfect*, or *total* differential $d\phi = d\phi(x, y, z)$ of a scalar function $\phi = \phi(x, y, z)$:

$$\mu(a\,dx + b\,dy + c\,dz) = d\phi, \tag{2.3.1a}$$

from which, by integration, we may obtain the (rigid and stationary) surface:

$$\phi(x, y, z) = constant, \qquad \text{or} \qquad z = z(x, y), \tag{2.3.1b}$$

then (2.3.1) is holonomic; if not, it is nonholonomic.

[Since, as is well known, the *two*-variable Pfaffian $a(x, y)\,dx + b(x, y)\,dy$ has *always* an integrating factor (in fact, an infinity of them), eq. (2.3.1) is the simplest possibly nonholonomic constraint. More on this below.]

In particular, if $\mu = 1$ (i.e., $d\phi = a\,dx + b\,dy + c\,dz$), the integrable Pfaffian $d\phi$ is *exact*. Then,

$$a = \partial\phi/\partial x, \qquad b = \partial\phi/\partial y, \qquad c = \partial\phi/\partial z, \tag{2.3.2}$$

and so the *necessary and sufficient* conditions for (2.3.1) to be exact are that the first partial derivatives of a, b, c, exist and satisfy (by equating the second mixed ϕ-derivatives):

$$\partial a/\partial y = \partial b/\partial x, \qquad \partial a/\partial z = \partial c/\partial x, \qquad \partial b/\partial z = \partial c/\partial y. \tag{2.3.3}$$

Equations (2.3.3) are *sufficient* for (2.3.1) to be completely integrable = holonomic; but they are *not necessary: every exact Pfaffian equation is integrable, but every integrable Pfaffian equation need not be exact*; in general, a $\mu \neq 1$ may exist, even though not all of (2.3.3) hold. In mechanics, we are interested in the holonomicity ($\equiv$ *complete* or *unconditional*) *integrability*, or absence thereof, of the constraints.

Let us now make a brief detour to the *general case*: the system of m Pfaffian constraints in the $n(>m)$ variables $x = (x_1, \ldots, x_n)$,

$$d'\chi_D \equiv \sum a_{Dk}(x)\, dx_k = 0 \qquad [D = 1, \ldots, m(<n)], \tag{2.3.4}$$

where $rank(a_{Dk}) = m$ (i.e., these equations are linearly *independent* in a certain x-region), is called *completely* (or *unconditionally*) *integrable*, or *complete*, or *holonomic*, if either (i) it is *immediately integrable*, or *exact*; that is, if the m $d'\chi_D$'s are the exact, or total, or perfect, differentials of m functions $\phi_D = \phi_D(x)$:

$$\sum a_{Dk}(x)\, dx_k = d\phi_D(x); \tag{2.3.4a}$$

or (ii) each $d'\chi_D$, although not immediately integrable, nevertheless admits a (nonzero) *integrating factor* $\Phi_D(x)$; that is, if the $2m$ (not all zero) functions $\{\Phi_D(x), \phi_D(x);\ D = 1, \ldots, m(<n)\}$ and (2.3.4) satisfy

$$\begin{aligned} \Phi_1\, d'\chi_1 &= \Phi_1(a_{11}\, dx_1 + \cdots + a_{1n}\, dx_n) = d\phi_1(x), \\ &\cdots\cdots\cdots\cdots\cdots\cdots\cdots\cdots\cdots\cdots \\ \Phi_m\, d'\chi_m &= \Phi_m(a_{m1}\, dx_1 + \cdots + a_{mn}\, dx_n) = d\phi_m(x); \end{aligned} \tag{2.3.4b}$$

or, compactly, $\Phi_D\, d'\chi_D = \Phi_D(\sum a_{Dk}\, dx_k) = d\phi_D(x)$, where the $\{d\phi_D\}$ are (linearly) independent. Summing (2.3.4b), over D, we also obtain its following consequence:

$$\sum \Phi_D\, d'\chi_D = \sum \Phi_D\Big(\sum a_{Dk}\, dx_k\Big) = \sum d\phi_k \equiv d\phi = \sum (\partial\phi/\partial x_k)\, dx_k$$
$$\Rightarrow \sum \Phi_D a_{Dk} = \partial\phi/\partial x_k.$$

Clearly, in both cases, (2.3.4a, b), the constraints (2.3.4) are equivalent to the *holonomic* equations

$$\phi_1(x) = C_1, \ldots, \phi_m(x) = C_m, \tag{2.3.4c}$$

where the m constants $\{C_D;\ D = 1, \ldots, m\}$ are fixed throughout the motion of the system. (Elaboration of this leads to the concept of *semiholonomic* constraints, treated later in this section.) If the constraints (2.3.4) are nonintegrable, neither immediately nor with integrating factors, they are called *nonholonomic*; and the mechanical system whose motion obeys, in addition to the kinetic equations, such nonholonomic constraints, either *internally* (constitution of its bodies) or *externally* (interaction with its environment, obstacles, etc.), is called a nonholonomic system.

An alternative definition of complete integrability of the system (2.3.4), equivalent to (2.3.4b), is the existence of m *independent, that is, distinct, linear, combinations of the m $d'\chi_D$* that are exact differentials of the m independent functions $f_D(x)$:

$$\mu_{11}\, d'\chi_1 + \cdots + \mu_{1m}\, d'\chi_m = df_1, \ldots, \quad \mu_{m1}\, d'\chi_1 + \cdots + \mu_{mm}\, d'\chi_m = df_m, \tag{2.3.4d}$$

where $\mu_{DD'} = \mu_{DD'}(x)$, or, compactly,

$$\sum \mu_{DD'}\, d'\chi_{D'} = df_D \Leftrightarrow d'\chi_D = \sum M_{DD'}\, df_{D'} \qquad (D, D' = 1, \ldots, m), \tag{2.3.4e}$$

[where $(M_{DD'})$ is the inverse matrix of $(\mu_{DD'})$, and both $(m \times m)$ matrices are nonsingular] and, hence, yield the m *independent* integrals (hypersurfaces):

$f_1 = c_1, \ldots, f_m = c_m$; that is, the system of eqs. (2.3.4) is completely integrable if there exists an *m-parameter* $(n-m)$-dimensional manifold that solves them. [Frobenius (1877) has shown that if $m = n$, or $n - 1$, then the system (2.3.4) is *always* completely integrable—more on this later.]

Finally, calling the determinant of the coefficients $\mu_{DD'}$ the *multiplicator* of (2.3.4) [i.e., $|\mu_{DD'}| \equiv \mu(\neq 0)$], and generalizing from the single constraint case (2.3.1), we can state that every multiplicator has always the form $\mu\, F(f_1, \ldots, f_D)$, where $F(\ldots)$ is an arbitrary differentiable function of the f's; that is, there exists an *infinity* of multiplicators.

From the above, it immediately follows that in the case of a *single* Pfaffian equation in the n variables $x = (x_1, \ldots, x_n)$ (i.e., for $m = 1$), complete integrability, in a certain x-domain, means that there exists, locally at least, a *one-parameter family of* $(n-1)$*-dimensional manifolds* $f(x) \equiv \phi(x) - constant = 0$, which solves that equation.

[We remark that the solutions of $d'\chi \equiv \sum a_k(x)\, dx_k = 0$ are *always one-dimensional* manifolds, or *curves*: $x_k = x_k(u)$, where $u =$ curve parameter. And, generally, if the x are functions of the $m(< n)$ new variables $(u_1, \ldots, u_m)$, then $x_k = x_k(u_1, \ldots, u_m)$ is called an *m-dimensional* solution manifold of $d'\chi = 0$, if, upon substitution into it, identical satisfaction results.]

Problem 2.3.1 Verify that the *sufficient* (but *non-necessary*!) conditions for the complete integrability of the system of m Pfaffian equations [essentially the discrete version of (2.2.9) for a system of N particles],

$$f_D\, dt \equiv \sum (a_{Dk}\, dx_k + b_{Dk}\, dy_k + c_{Dk}\, dz_k) + e_D\, dt = 0, \tag{a}$$

where $D = 1, \ldots, m(< 3N)$, $k = 1, \ldots, N$, and $(a, b, c, e) =$ continuously differentiable functions of (x, y, z, t), are that

$$\partial a_{Dk}/\partial x_l = \partial a_{Dl}/\partial x_k, \qquad \partial a_{Dk}/\partial y_l = \partial b_{Dl}/\partial x_k, \qquad \partial a_{Dk}/\partial z_l = \partial c_{Dl}/\partial x_k,$$
$$\partial a_{Dk}/\partial t = \partial e_D/\partial x_k; \tag{b}$$

$$\partial b_{Dk}/\partial y_l = \partial b_{Dl}/\partial y_k, \qquad \partial b_{Dk}/\partial z_l = \partial c_{Dl}/\partial y_k, \qquad \partial b_{Dk}/\partial t = \partial e_D/\partial y_k; \tag{c}$$

$$\partial c_{Dk}/\partial z_l = \partial c_{Dl}/\partial z_k, \qquad \partial c_{Dk}/\partial t = \partial e_D/\partial z_k; \tag{d}$$

for *all* $k, l = 1, \ldots, N$, for a fixed D. [In fact, the (obvious) choice: $a_{Dk} = \partial\phi_D/\partial x_k$, $b_{Dk} = \partial\phi_D/\partial y_k$, $c_{Dk} = \partial\phi_D/\partial z_k$, $e_D = \partial\phi_D/\partial t$; $\phi_D = \phi_D(t; x, y, z)$ satisfies (b–d).] Then, (a) simply states that $d\phi_D = 0$; and the latter integrates immediately to the holonomic constraints: $\phi_D = \phi_D(t; x, y, z) = (constant)_D$.

Introduction to Necessary and Sufficient Conditions for Holonomicity

Let us, for the time being, postpone the discussion of the general case and return to the *single* Pfaffian equation in *three* variables, eq. (2.3.1), and find the necessary and sufficient conditions for its holonomicity. Assuming that this is indeed the case, then from (2.3.1) and the second of (2.3.1b) we readily see that

$$dz = (\partial z/\partial x)\, dx + (\partial z/\partial y)\, dy = (-a/c)\, dx + (-b/c)\, dy \tag{2.3.5}$$

must hold for all dx, dy, dz. Therefore, equating the coefficients of dx and dy of both sides, we obtain [assuming $c \neq 0$, and that $z(x, y)$ is substituted for z in a, b, c]

$$\partial z/\partial x = -(a/c) \qquad \text{and} \qquad \partial z/\partial y = -(b/c), \tag{2.3.5a}$$

and since $\partial/\partial y(\partial z/\partial x) = \partial/\partial x(\partial z/\partial y)$, we obtain $\partial/\partial y(a/c) = \partial/\partial x(b/c)$, or, explicitly,

$$c[\partial a/\partial y + (\partial a/\partial z)(\partial z/\partial y)] - a[\partial c/\partial y + (\partial c/\partial z)(\partial z/\partial y)]$$
$$= c[\partial b/\partial x + (\partial b/\partial z)(\partial z/\partial x)] - b[\partial c/\partial x + (\partial c/\partial z)(\partial z/\partial x)],$$

and inserting in it the $\partial z/\partial x$- and $\partial z/\partial y$-values from (2.3.5a), and simplifying, we finally find

$$I \equiv a(\partial b/\partial z - \partial c/\partial y) + b(\partial c/\partial x - \partial a/\partial z) + c(\partial a/\partial y - \partial b/\partial x) = 0. \tag{2.3.6}$$

Equation (2.3.6), being a direct consequence of the earlier mixed partial derivative equality, is the *necessary and sufficient* condition for (2.3.1) to be holonomic. If $I = 0$ *identically* (i.e., for arbitrary x, y, z), then (2.3.1) is holonomic; if $I \neq 0$ *identically*, then (2.3.1) is nonholonomic.

REMARKS

(i) The form I is *symmetric* in (x, y, z) and (a, b, c); that is, it remains unchanged under simultaneous cyclic changes of (x, y, z) *and* (a, b, c).

(ii) *Alternative derivation of equation (2.3.6):* The mixed partial derivatives rule applied to (2.3.1a) readily yields

$$\partial(\mu b)/\partial x = \partial(\mu a)/\partial y, \qquad \partial(\mu c)/\partial x = \partial(\mu a)/\partial z, \qquad \partial(\mu c)/\partial y = \partial(\mu b)/\partial z.$$

Multiplying the above equalities with c, b, a, respectively, and adding them together, we obtain (2.3.6); so, clearly, the latter is *necessary and sufficient* for the existence of an integrating factor (for further details, see, e.g., Forsyth, 1885 and 1954, pp. 247 ff.).

(iii) *A special case:* If $a = a(x, y)$, $b = b(x, y)$, and $c = 0$, then, clearly, $I = 0$; which proves the earlier claim that the two-variable Pfaffian equation $a(x, y)\,dx + b(x, y)\,dy = 0$ is *always holonomic*; that is, for nonholonomicity, we need at least three variables.

(iv) *A special form:* If (2.3.1) has the equivalent form

$$dz = (-a/c)\,dx + (-b/c)\,dy \equiv A(x, y, z)\,dx + B(x, y, z)\,dy$$
$$= A[x, y, z(x, y)]\,dx + B[x, y, z(x, y)]\,dy$$
$$\equiv A^*(x, y)\,dx + B^*(x, y)\,dy, \tag{2.3.7}$$

(or, similarly, $dx = \cdots, dy = \cdots$; depending on analytical convenience and/or avoidance of singularities), then the mixed partial derivative rule

$$\partial A^*(x, y)/\partial y = \partial B^*(x, y)/\partial x, \tag{2.3.7a}$$

due to the chain rule (one should be extra careful here):

$$\partial A^*/\partial y = \partial A/\partial y + (\partial A/\partial z)(\partial z/\partial y) = \partial A/\partial y + (\partial A/\partial z)B, \tag{2.3.7b}$$

$$\partial B^*/\partial x = \partial B/\partial x + (\partial B/\partial z)(\partial z/\partial x) = \partial B/\partial x + (\partial B/\partial z)A, \tag{2.3.7c}$$

finally yields

$$\partial A/\partial y + (\partial A/\partial z)B = \partial B/\partial x + (\partial B/\partial z)A; \qquad (2.3.7d)$$

whose *identical* satisfaction in x, y, z, is the necessary and sufficient condition for the complete integrability, or holonomicity, of (2.3.7).

It is not hard to verify that (i) replacing, in (2.3.7d), A with $-a/c$ and B with $-b/c$, we recover (2.3.6); and, conversely, (ii) since (2.3.7) can be written in the (2.3.1)-like form: $A\,dx + B\,dy + (-1)\,dz = 0$, replacing, in (2.3.6), a, b, c, with A, B, -1, respectively, we recover (2.3.7d). If, in (2.3.7), $\partial A/\partial z = 0$ and $\partial B/\partial z = 0$, then (2.3.7d) reduces to $\partial A/\partial y = \partial B/\partial x$. Finally, the sole analytical requirement here is the *continuity* of all partial derivatives appearing in these conditions (but not those of the nonappearing ones, such as $\partial A/\partial x$ and $\partial B/\partial y$).

Example 2.3.1 Let us test, for complete integrability, the following constraints:

$$\text{(i)}\ dz = (z)\,dx + (z^2 + a^2)\,dy; \qquad \text{(ii)}\ dz = z(dx + x\,dy).$$

(i) Here, $A = z$ and $B = z^2 + a^2$, and therefore (2.3.7d) yields

$$(1)(z^2 + a^2) = (2z)z \Rightarrow z^2 = a^2,$$

that is, no identical satisfaction; or, our constraint is not completely integrable—it is nonholonomic. Then, the original equation becomes

$$dz = z\,dx + 2z^2\,dy;$$

and so (a) if $a = 0$, then $z = 0$ is a constraint integral; but (b) if $a \neq 0$, then there is no integral. For complete integrability, we should have an infinity of integrals depending on an arbitrary integration constant.

(ii) Here, the test (2.3.7d) gives $xz = z + xz \Rightarrow z = 0$; that is, no identical satisfaction, and therefore no holonomicity. As the original equation shows, this is the sole integral.

Problem 2.3.2 Show that the constraint of the plane pursuit problem (ex. 2.2.1):

$$[Y(t) - y](dx/dt) - [X(t) - x](dy/dt) = 0, \qquad \text{(a)}$$

or, equivalently,

$$[Y(t) - y]\,dx - [X(t) - x]\,dy + (0)\,dt = 0, \qquad \text{(b)}$$

is holonomic if and only if

$$[X(t) - x]/[Y(t) - y] = (dX/dt)/(dY/dt) \qquad [= (dx/dt)/(dy/dt)]. \qquad \text{(c)}$$

Problem 2.3.3 Show that under a general one-to-one (nonsingular) coordinate transformation $(x, y, z) \Leftrightarrow (u, v, w)$: $x = x(u, v, w)$, $y = \cdots, z = \cdots$,

$$I = [\partial(u, v, w)/\partial(x, y, z)] \cdot I'; \qquad \text{(a)}$$

where (with subscripts denoting partial derivatives)

$$d\chi \equiv a\,dx + b\,dy + c\,dz = p\,du + q\,dv + r\,dw, \tag{b}$$

$$I = I(x, y, z) \equiv a(b_z - c_y) + b(c_x - a_z) + c(a_y - b_x), \tag{c}$$

$$I' = I'(u, v, w) \equiv p(q_w - r_v) + q(r_u - p_w) + r(p_v - q_u), \tag{d}$$

and $\partial(u, v, w)/\partial(x, y, z) =$ *Jacobian* of the transformation ($\neq 0$); that is, I and I' vanish simultaneously; or, the holonomicity of $d\chi = 0$, or absence thereof, is *coordinate invariant*, and hence an *intrinsic* property of the constraint (a proof of this fundamental fact, for a general Pfaffian system, will be given later).

[Incidentally, the transformation law (a) also shows that scalars like I are *not necessarily invariants* ($I \neq I'$, in general); in fact, in the more precise language of tensor calculus, they are called *relative scalars of weight* +1, or *scalar densities*; see, e.g., Papastavridis (1999, pp. 46–49).]

Geometrical Interpretation of the Pfaffian Equation (2.3.1)

The latter, rewritten with the help of the vectors $d\boldsymbol{r} = (dx, dy, dz)$ and $\boldsymbol{h} = (a, b, c)$ as

$$\boldsymbol{h} \cdot d\boldsymbol{r} = 0, \tag{2.3.8}$$

means that, at each specified point $Q(x, y, z)$, $d\boldsymbol{r}$ must lie on a local plane *perpendicular* to the "constraint coefficient vector" $\boldsymbol{h}$ there; or, that the particle P can move only along those curves, emanating from Q, whose tangent is perpendicular to $\boldsymbol{h}$. Such curves are called *kinematically admissible*, or *kinematically possible*. If (2.3.1, 8) is holonomic, then all motions lie on the integral surface (2.3.1b); that is, (2.3.6) is the necessary and sufficient condition for the existence of an orthogonal surface through Q, for the field $\boldsymbol{h} = (a, b, c)$ [actually, *a family of surfaces* $\phi = \phi(x, y, z) =$ *constant*, *everywhere normal to* $\boldsymbol{h}$—see below]. We also notice that, with the help of $\boldsymbol{h}$, the condition (2.3.6) takes the memorable (invariant) form:

$$I \equiv \boldsymbol{h} \cdot \boldsymbol{curl}\ \boldsymbol{h} = 0; \qquad \text{or, symbolically,} \qquad \begin{vmatrix} a & b & c \\ \partial/\partial x & \partial/\partial y & \partial/\partial z \\ a & b & c \end{vmatrix} = 0; \tag{2.3.8a}$$

that is, at every field point, $\boldsymbol{h}$ is parallel to the plane of its rotation, or perpendicular to that rotation and tangent to the surface $\phi =$ *constant* there [W. Thomson (Lord Kelvin) called such fields *doubly lamellar*]; while (2.3.7d), with $\boldsymbol{h} \to \boldsymbol{H} \equiv (A, B, -1)$, becomes

$$\boldsymbol{H} \cdot \boldsymbol{curl}\ \boldsymbol{H} = \partial A/\partial y + B(\partial A/\partial z) - \partial B/\partial x - A(\partial B/\partial z) = (1/c^2)\boldsymbol{h} \cdot \boldsymbol{curl}\ \boldsymbol{h} = 0. \tag{2.3.8b}$$

Vectorial Derivation of Equation (2.3.8a)

We recall from vector analysis that a (continuously differentiable) vector is called *irrotational*, or *singly lamellar*, if (a) its line integral around every closed circuit vanishes, or, equivalently, (b) if its ***curl*** (rotation) vanishes, or (c) if it equals the gradient of a scalar.

Now: (i) If $\boldsymbol{h} = (a, b, c)$ is irrotational, then there is a $\phi = \phi(x, y, z)$ such that $\boldsymbol{h} = \boldsymbol{grad}\ \phi$, and, therefore, $\boldsymbol{h} \cdot \boldsymbol{dr} = \boldsymbol{grad}\ \phi \cdot \boldsymbol{dr} = d\phi =$ *exact differential.*

(ii) If $\boldsymbol{h} \neq$ irrotational, still an *integrating factor (IF)* $\mu = \mu(x, y, z)$ may exist so that $\mu\, \boldsymbol{h} = \boldsymbol{grad}\ \phi$. Then, as before, $\mu\, \boldsymbol{h} \cdot \boldsymbol{dr} = \boldsymbol{grad}\ \phi \cdot \boldsymbol{dr} = d\phi =$ *exact differential.*

(iii) Conversely, if $\mu = IF$, then $\mu\, \boldsymbol{h} = \boldsymbol{grad}\ \phi =$ irrotational; and "curling" both sides of this latter, we obtain: $\boldsymbol{0} = \boldsymbol{curl}(\boldsymbol{grad}\ \phi) = \mu\, \boldsymbol{curl}\ \boldsymbol{h} + \boldsymbol{grad}\ \mu \times \boldsymbol{h}$, and dotting this with $\boldsymbol{h}$: $0 = \mu(\boldsymbol{h} \cdot \boldsymbol{curl}\ \boldsymbol{h})$, from which, since $\mu \neq 0$, we finally get (2.3.8a). In this case, since $\boldsymbol{h}$ and $\boldsymbol{grad}\ \phi$ are parallel: $\boldsymbol{h} = (1/\mu)\ \boldsymbol{grad}\ \phi \equiv \nu(\boldsymbol{grad}\ \phi)$, and, therefore, $\boldsymbol{curl}\ \boldsymbol{h} = \boldsymbol{curl}(\nu\ \boldsymbol{grad}\ \phi) = \boldsymbol{grad}\ \nu \times \boldsymbol{grad}\ \phi$, so that

$$\boldsymbol{h} \cdot \boldsymbol{curl}\ \boldsymbol{h} = \nu\ \boldsymbol{grad}\ \phi \cdot (\boldsymbol{grad}\ \nu \times \boldsymbol{grad}\ \phi) = 0; \tag{2.3.8c}$$

that is, the *doubly lamellar field* $\boldsymbol{h}$ is *perpendicular* to its *rotation* $\boldsymbol{curl}\ \boldsymbol{h}$. [This condition is *necessary* for the existence of an *IF*. For its *sufficiency*, see, for example, Brand (1947, pp. 200, 230–231), Sneddon (1957, pp. 21–23); also Coe (1938, pp. 477–478), for an *integral* vector calculus treatment.] These derivations are based on a general vector field theorem according to which an *arbitrary* vector field can be written as the sum of a *simple* and a *complex* (or *doubly*) *lamellar* field: $\boldsymbol{h} = \boldsymbol{grad}\, f + \nu\ \boldsymbol{grad}\ \phi$.

Finally, if the Pfaffian constraint is, nonholonomic, then (2.3.1, 7) yield one-dimensional "nonholonomic manifolds"; that is, space curves orthogonal to the field $\boldsymbol{h}$ (or $\boldsymbol{H}$), and constituting a one-parameter family on an arbitrary surface.

Accessibility

The restrictions on the motion of the particle P in the two cases $I = 0$ (holonomic) and $I \neq 0$ (nonholonomic) are of entirely different nature. If $I = 0$, then P is obliged to *move on the surface* $\phi = \phi(x, y, z) = 0$. If, on the other hand, $I \neq 0$, then the constraint (2.3.1) does *not* restrict the (x, y, z), but does restrict the *direction* (*velocity*) of the curves through a given point (x, y, z). The cumulative effect of these local restrictions in the direction of motion (velocity) is that the transition between two arbitrary points is *not arbitrary*; P can move (or be guided through) from an *arbitrary initial* (analytically possible) position, to any other *arbitrary final* (analytically possible) position, while at every point of its path satisfying (2.3.1, 8); that is, the particle can move from "anywhere" to "anywhere," not via any route we want, but along restricted paths. As Langhaar puts it, the particle is "constrained to follow routes that coincide with a *certain dense network of paths*" (1962, pp. 5–6); like kinematically possible tracks guiding the system.

In sum: (i) Holonomic constraints *do* reduce the dimension of the space of accessible configurations, but do not restrict motion and paths in there; in Hertz's words: "all conceivable continuous motions [between two arbitrary accessible positions] are also possible motions."

(ii) Nonholonomic constraints do *not* affect the dimension of the space of accessible configurations, but do restrict the motions locally (and, cumulatively, also globally) in there; *not all conceivable continuous motions (between two arbitrary accessible positions) are possible motions* (Hertz, 1894, p. 78 ff.).

These geometrical interpretations and associated concepts are extended to general systems in §2.7.

Degrees of Freedom

The above affect the earlier DOF definition: they force us to distinguish between *DOF in the large* (measure of global accessibility, or global mobility) and *DOF in the small* (measure of local/infinitesimal mobility). We define the former, $DOF(L)$, as the number of independent global positional (or holonomic) "parameters," or Lagrangean coordinates $\equiv n (= 3$ in our examples, so far); and the latter, $DOF(S) \equiv f$, as n minus the number of additional (possibly nonholonomic) independent Pfaffian constraints: $f = n - m (> 0)$. In the absence of the latter, $DOF(L) = DOF(S)$: $f = n$. This fine distinction between DOFs rarely appears in the literature, where, as a rule, DOF means DOF in the *small*. {For enlightening *exceptions*, see, for example, Sommerfeld (1964, pp. 48–51); also Roberson and Schwertassek (1988, p. 96), who call these DOFs, respectively, *positional*(L) and *motional*(S); and the pioneering Korteweg (1899, p. 134), who states that "Die anzahl der Freiheitsgrade sei bei ihr eine andere (kleinere) für unendlich kleine wie für endliche Verrückungen" [Translation: The number of degrees of freedom is different (smaller) for infinitesimal displacements than for finite displacements.]}

As explained later in this chapter (§2.5 ff.), $DOF(S) \equiv f$ equals the *number of independent virtual displacements of the system*; and this, in turn (chap. 3), equals the smallest, or minimal, number of kinetic (i.e., reactionless) equations of motion of it. In view of this, from now on by DOF we shall understand DOF *in the small*; that is, $DOF \equiv DOF(S) \equiv n - m \equiv f$, unless explicitly specified otherwise. The concept of DOF in the large is more important in pure kinematics (mechanisms).

Finally, in the general constraint case, all these results hold intact, but for the figurative system "particle" in a higher dimensional space—more on this later.

Semiholonomic Constraints

We stated earlier that if $I = 0$, the Pfaffian constraint (2.3.1) is holonomic; that is, it can be brought to the form

$$d\phi/dt = 0 \rightarrow \phi = constant = c. \tag{2.3.9}$$

Such situations necessitate an additional, albeit minor, classification of holonomic constraints into *proper* holonomic, or simply holonomic, and *improper* holonomic, or *semiholonomic* ones. In both cases, the constraints are finite (i.e., holonomic), but, in the proper case, the constraint constants have a priori fixed values, independent of the system's position/motion; whereas, in the semiholonomic case, those constants depend on the arbitrarily specified values of the system coordinates at some "initial" instant; that is, semiholonomic constraints are completely integrable velocity (Pfaffian) constraints $\Rightarrow$ (generally) initial condition-depending holonomic constraints. In the proper holonomic case, the initial values of the coordinates must be determined *in conjunction* with the given constraints and their constants; that is, *they* must be compatible with the latter. However, semiholonomic constraints, being essentially holonomic, can be used to reduce the number of independent global/Lagrangean coordinates; and, thus, differ profoundly from the nonholonomic ones. Clearly, the proper/semiholonomic distinction applies to *rheonomic* holonomic constraints, like $\phi(x, y, z, t) = c$. For further details, see Pérès (1953, pp. 60–62, 218–219), who seems to have introduced the concept of semiholonomicity; also Moreau (1971, pp. 228–232).

Critical Comments on Nonholonomic Constraints

The concept of nonholonomicity (in mechanics) has been around since the 1880s, and has been thoroughly studied and expounded by some of the greatest mathematicians, physicists, and mechanicians, for example (approximately chronologically): Voss, Hertz, Hadamard, Appell, Chaplygin, Voronets, Maggi, Boltzmann, Hamel, Heun, Delassus, Carathéodory, Schouten, Struik, Goursat, Cartan, Synge, Vranceanu, Vagner, Dobronravov, Lur'e, Neimark, Fufaev, et al. Direct definitions of nonholonomicity and analytical tests have been available, on a large and readable scale, at least since the 1920s. And yet, on this topic, there exists widespread misunderstanding and confusion; especially in the engineering literature. For example, some authors state that constraints that can be represented by equations like $\phi(\mathbf{r}, t) = 0$, or $\phi(x, y, z, t) = 0$, are called holonomic, and that *all* others are called nonholonomic; for example, Goldstein (1980, p. 12 ff.), Kane (1968, p. 14); Kane and Levinson (1985, p. 43), Likins (1973, pp. 184, 295), Matzner and Shepley (1991, pp. 23–24). Under such an indirect, vague, negative definition, *inequality* constraints like $\phi \geq 0$, or (perhaps?!) holonomic ones, but in velocity form, like

$$d\phi/dt = S\,(\partial\phi/\partial\mathbf{r}) \cdot \mathbf{v} + \partial\phi/\partial t = 0, \qquad (2.2.10)$$

would be called nonholonomic! Or, we read blatantly contradictory and erroneous statements like "With nonholonomic systems the generalized coordinates are not independent of each other, and it is not possible to reduce them further by means of equations of constraint of the form $f(q_1, \ldots, q_n, t) = 0$. *Hence it is no longer true that the* q_j*'s are independent*" (Goldstein (1980, p. 45), emphasis added). Others call nonholonomic all velocity constraints that *cannot* be written in the above form $\phi = 0$, which is correct; but they fail to supply the reader with analytical (or geometrical, or even numerical) tools on how to test this; for example, Roberson and Schwertassek (1988, p. 96), Shabana (1989, pp. 123, 128). The more careful of this last group talk clearly about integrability, exactness, and so on, but restrict themselves to only *one* velocity constraint; for example, Haug (1992, pp. 87–89). Still others mix nonholonomic *coordinates* (*quasi coordinates*, etc.) with nonholonomic *constraints*, and *exactness* of Pfaffian *forms* with (complete) *integrability* of a system of Pfaffian *equations*, without ever supplying clear and general definitions, let alone analytical tests. And this results in defective definitions of the concept of DOF; for example, Angeles (1988, pp. 80, 103). Even the (otherwise monumental) treatise of Pars (1965, pp. 16–19, 22–24, 35–37, 64–72, 196) is limited to an introduction to the subject, albeit a careful and precise one. Finally, there is the recent crop of texts on "modern" dynamics, where the problem of nonholonomicity is "solved" by ignoring it altogether; for example, Rasband (1983). Only Neimark and Fufaev (1967/1972) discuss the nonholonomicity issue clearly, competently, and in sufficient generality and completeness to be useful. We hope that our treatment complements and extends their beautiful work.

Extensions/Generalizations of the Integrability Conditions (May be omitted in a first reading)

(i) Single Pfaffian Equation in the n Variables $x = (x_1, \ldots, x_n)$:

$$d'\chi \equiv \sum a_k\, dx_k = 0, \qquad a_k = a_k(x). \qquad (2.3.10)$$

It can be shown that the necessary and sufficient condition for the complete integrability = holonomicity of (2.3.10) is the identical satisfaction of the following "symmetric" equations:

$$I_{klp} \equiv a_k(\partial a_l/\partial x_p - \partial a_p/\partial x_l) + a_l(\partial a_p/\partial x_k - \partial a_k/\partial x_p) \\ + a_p(\partial a_k/\partial x_l - \partial a_l/\partial x_k) = 0, \qquad (2.3.10a)$$

for *all* combinations of the indices k, l, $p = 1, \ldots, n$. [For example, one may start with the integrability condition of (2.3.1), (2.3.6) (i.e., $n = 3$) and then use the method of induction; or perform similar steps as in the three-dimensional case; see, for example, Forsyth (1885 and 1954, pp. 259–260).] Further, it can be shown (e.g., again, by induction) that out of a total of $n(n-1)(n-2)/6$ equations (2.3.10a), equal to the number of triangles that can be formed with n given points as corners, *only* $n_I \equiv (n-1)(n-2)/2$ *are independent*. For $n = 3$, that number is indeed 1: eqs. (2.3.6) or (2.3.8a). Also, if $a_k \neq 0$, it suffices to apply (2.3.10) only for l and p different from k. Finally, with appropriate extension of the ***curl*** of a vector to n-dimensional spaces, (2.3.10) can be cast into a (2.3.8a)-like form (see, e.g., Papastavridis, 1999, chaps. 3, 6).

Problem 2.3.4 (i) Specialize (2.3.10a) to the *acatastatic* constraint ($n = 4$):

$$a(t, x, y, z)\, dx + b(t, x, y, z)\, dy + c(t, x, y, z)\, dy + e(t, x, y, z)\, dt = 0. \qquad \text{(a)}$$

(ii) Show that (a) is *holonomic* if, and only if, the *symbolic* matrix

$$\begin{pmatrix} a & b & c & e \\ \partial/\partial x & \partial/\partial y & \partial/\partial z & \partial/\partial t \\ a & b & c & e \end{pmatrix}, \qquad \text{(b)}$$

has rank 2 (actually, less than 3); that is, all possible four of its 3×3 symbolic subdeterminants, each to be developed along its first row, vanish.

(iii) Further, show that if all such 2×2 subdeterminants of (b) vanish, then (a) is *exact*.

(iv) Specialize the preceding result to the catastatic case $e \equiv 0$; verify that, then, we obtain (2.3.6).

Problem 2.3.5 For the Pfaffian equation (2.3.10), define the $(n+1) \times n$ matrix

$$\mathbf{P} = \begin{pmatrix} a_1 & \cdots & a_n \\ \hline a_{11} & \cdots & a_{1n} \\ \cdots & \cdots & \cdots \\ a_{n1} & \cdots & a_{nn} \end{pmatrix}, \qquad \text{(a)}$$

where $a_{kl} \equiv \partial a_k/\partial x_l - \partial a_l/\partial x_k \ (= -a_{lk})$; $k, l = 1, \ldots, n$. Clearly, $a_{11} = \cdots = a_{nn} = 0$. Now, it is shown in differential equations/differential geometry that for the *holonomicity* of (2.3.10), it is necessary and sufficient that the rank of $\mathbf{P}$ *equal* 1 *or* 2.

Show that (i) *rank* $\mathbf{P} = 1$ (i.e., *all* its 2×2 subdeterminants vanish) leads to the *exactness* conditions

$$a_{kl} = 0; \qquad \text{(b)}$$

(ii) *rank* $\mathbf{P} = 2$ (i.e., *all* its 3×3 subdeterminants vanish) leads to the earlier *complete integrability* conditions (2.3.10a)

$$a_k a_{lp} + a_l a_{pk} + a_p a_{kl} = 0. \tag{c}$$

Problem 2.3.6 Show that for $n = 3$, equations (b, c) of the preceding problem become, respectively,

$$a_{kl} = 0 \qquad (k, l = 1, 2, 3), \tag{a}$$

and

$$a_1 a_{23} + a_2 a_{31} + a_3 a_{12} = 0 \qquad \text{[i.e., (2.3.6)]}. \tag{b}$$

Problem 2.3.7 Consider the Pfaffian equation (2.3.10). Subject its variables x to the invertible coordinate transformation (with nonvanishing Jacobian) $x \to x'$; in extenso:

$$x_k = x_k(x_{k'}) \Leftrightarrow x_{k'} = x_{k'}(x_k) \qquad (k, k' = 1, \ldots, n). \tag{a}$$

Show that the requirement that, under that transformation, the Pfaffian form $d'\chi$ remain (form) *invariant*; that is,

$$d'\chi \to (d'\chi)' \equiv \sum a_{k'}\, dx_{k'} = d'\chi \quad (= 0), \qquad a_{k'} = a_{k'}(x'), \tag{b}$$

leads to the following (covariant vector) transformations for the form coefficients:

$$a_{k'} = \sum (\partial x_k / \partial x_{k'}) a_k \Leftrightarrow a_k = \sum (\partial x_{k'} / \partial x_k) a_{k'}. \tag{c}$$

Problem 2.3.8 Continuing from the previous problem, define the antisymmetric quantities

$$a_{kl} = \partial a_k / \partial x_l - \partial a_l / \partial x_k \qquad (= -a_{lk}), \tag{a}$$

$$a_{k'l'} = \partial a_{k'} / \partial x_{l'} - \partial a_{l'} / \partial x_{k'} \qquad (= -a_{l'k'}), \qquad (k', l' = 1, \ldots, n). \tag{b}$$

Show that under the earlier invariance requirement $d'\chi \to (d'\chi)' = d'\chi$, the above quantities transform as (second-order covariant tensors):

$$a_{k'l'} = \sum\sum (\partial x_k / \partial x_{k'})(\partial x_l / \partial x_{l'}) a_{kl} \Leftrightarrow a_{kl} = \sum\sum (\partial x_{k'} / \partial x_k)(\partial x_{l'} / \partial x_l) a_{k'l'}. \tag{c}$$

Problem 2.3.9 Continuing from the preceding problems, assume that the x (and, therefore, also the x') depend on *two* parameters u_1 and u_2:

$$x_k = x_k(u_1, u_2) \qquad \text{and} \qquad x'_k = x'_k(u_1, u_2). \tag{a}$$

Introducing the simpler notation $d'\chi \equiv d\chi$ and $(d'\chi)' \equiv d\chi'$, show that

$$d_2(d_1\chi) - d_1(d_2\chi) = d_2(d_1\chi') - d_1(d_2\chi'), \tag{b}$$

where

$$d_1\chi = \sum a_k\, d_1 x_k = \sum a_k[(\partial x_k/\partial u_1)\, du_1],$$
$$d_2\chi = \sum a_k\, d_2 x_k = \sum a_k[(\partial x_k/\partial u_2)\, du_2],$$

are equivalent to

$$\sum\sum (\partial x_k/\partial u_1)(\partial x_l/\partial u_2)a_{kl} = \sum\sum (\partial x_{k'}/\partial u_1)(\partial x_{l'}/\partial u_2)a_{k'l'}. \qquad \text{(c)}$$

(ii) If the Pfaffian Constraint (2.3.10) has the Equivalent, (2.3.5, 7)-like, Special Form:

$$dz = \sum b_k(x, z)\, dx_k \qquad (k = 1, \ldots, n), \qquad \text{(2.3.10b)}$$

then, proceeding as in the three-dimensional case, or specializing (2.3.10a), we can show that the necessary and sufficient integrability conditions are the $n(n-1)/2$ independent identities [replacing n with $n+1$ in the earlier n_I, following (2.3.10a)]:

$$\partial b_k/\partial x_l + (\partial b_k/\partial z)b_l = \partial b_l/\partial x_k + (\partial b_l/\partial z)b_k \qquad (k, l = 1, \ldots, n). \qquad \text{(2.3.10c)}$$

Here, too, only the *existence* and *continuity* of the partial derivatives involved is needed.

(iii) General Case of $m(< n)$ Independent Pfaffian Equations in n Variables

[In the slightly special *total differential equation* form, with $x \equiv (x_D, x_I)$]:

$$dx_D = \sum b_{DI}(x)\, dx_I \quad \text{or} \quad \partial x_D/\partial x_I = b_{DI}(x) \quad \text{(general form)}, \qquad \text{(2.3.11)}$$

where (*here and throughout this book*)

$D = 1, \ldots, m$ (for ***Dependent***) and $I = m + 1, \ldots, n$ (for ***Independent***),

b_{DI} = given (continuously differentiable) functions of the m $x_D = (x_1, \ldots, x_m)$, and the $(n - m)$ $x_I = (x_{m+1}, \ldots, x_n)$. (2.3.11a)

The system (2.3.11) is called holonomic or completely integrable [i.e., functions $x_D(x_I)$ can be found whose total differentials are given by (2.3.11)], if, for any set of initial values $x_{I,o}$, $x_{D,o}$, for which the b_{DI} are analytic, there exists one, and only one, set of D functions $x_D(x_I)$ satisfying (2.3.11) and taking on the initial values $x_{D,o}$ at $x_{I,o}$. It is shown in the theory of partial (total) differential equations—see references below—that:

For the system (2.3.11) to be *holonomic*, it is necessary and sufficient that the following conditions hold:

$$\partial b_{DI}/\partial x_{I'} + \sum b_{D'I'}(\partial b_{DI}/\partial x_{D'}) = \partial b_{DI'}/\partial x_I + \sum b_{D'I}(\partial b_{DI'}/\partial x_{D'})$$
$$[D, D' = 1, \ldots, m; \quad I, I' = m + 1, \ldots, n], \qquad \text{(2.3.11b)}$$

identically in the x_D, x_I's [i.e., not just for some particular motion(s)] and for all combinations of the above values of their indices; if they hold for some, but not all, m values of D, then the system (2.3.11) is called "partially integrable."

Now, and this is very important, as the *second (sum)* term, on each side of (2.3.11b), shows, the *integrability* of the Dth constraint equation of (2.3.11) depends, through the coupling with $b_{D'I'}$ and $b_{D'I}$, on *all* the other constraint equations of that system; that is, each (2.3.11b) tests the integrability of the corresponding constraint equation (i.e., same D) against the *entire* system—in general, holonomicity/nonholonomicity are *system* not *individual* constraint properties.

Geometrically, integrability means that the system (2.3.11) yields a field of *$(n-m)$-dimensional surfaces in the n-dimensional space of the* x's; that is, mechanically, the system has $(n-m)$ *global* positional/Lagrangean coordinates, namely, $DOF(L) = DOF(S) = n - m$.

Further:

- With the notation

$$b_{DI} = b_{DI}(x_D, x_I) = b_{DI}[x_D(x_I), x_I] \equiv \beta_{DI}(x_I) \equiv \beta_{DI}, \qquad (2.3.11c)$$

  and since, by careful application of chain rule to the above,

$$\begin{aligned}\partial\beta_{DI}/\partial x_{I'} &= \partial b_{DI}/\partial x_{I'} + \sum (\partial b_{DI}/\partial x_{D'})(\partial x_{D'}/\partial x_{I'}) \\ &= \partial b_{DI}/\partial x_{I'} + \sum (\partial b_{DI}/\partial x_{D'}) b_{D'I'},\end{aligned}$$

  [if $x_D = x_D(x_I)$, then $dx_D = \sum (\partial x_D/\partial x_I)\, dx_I = \sum b_{DI}(x)\, dx_I$] the holonomicity conditions (2.3.11b) can also be expressed in the following perhaps more intelligible/memorable ("exactness") form:

$$\partial\beta_{DI}/\partial x_{I'} = \partial\beta_{DI'}/\partial x_I \qquad (I' = m+1, \ldots, n); \qquad (2.3.11d)$$

- It is not hard to verify that the system (2.3.11b, d) stands for a total of $m(n-1)(n-2)/2$ identities, out of which, however, only $m(n-m)(n-m-1)/2 \equiv mf(f-1)/2$ are independent [$f \equiv n - m$; as in the general case of the first of (2.12.5)].
- In the special case where $b_{DI} = b_{DI}(x_I)$ [Chaplygin systems (§3.8)], (2.3.11b) reduce to the conditions:

$$\partial b_{DI}/\partial x_{I'} = \partial b_{DI'}/\partial x_I \qquad \text{[compare with (2.3.11d)]}, \qquad (2.3.11e)$$

  which, being *uncoupled*, test each constraint equation (2.3.11) *independently* of the others. Last, we point out that all these holonomicity conditions are special cases of the general theorem of Frobenius, which is discussed in §2.8–2.11.
- Equations (2.3.11b, d) also appear as necessary and sufficient conditions for a Riemannian ("curved") space to be Euclidean ("flat") $\Rightarrow$ vanishing of Riemann–Christoffel "curvature tensor"; and in the related *compatibility* conditions in nonlinear theory of strain—see, for example, Sokolnikoff (1951, pp. 96–100), Truesdell and Toupin (1960, pp. 271–274).
- *Historical*: The fundamental partial differential equations (2.3.11b) are due to the German mathematician H. W. F. Deahna [*J. für die reine und angewandte Mathematik* (*Crelle's Journal*) **20**, 340–349, 1840] and, also, the French mathematician J. C. Bouquet [*Bull. Sci. Math. et Astron.*, **3**(1), 265 ff., 1872]. For extensive and readable discussions, proofs, and so on, see, for example, De la Valée Poussin (1912,

pp. 312–336), Levi-Civita (1926, pp. 13–33), and the earlier Forsyth (1885/1954). Regrettably, most contemporary treatments of Pfaffian system integrability are written in the language of Cartan's "exterior forms," and so are virtually inaccessible to the average nonmathematician.

2.4 SYSTEM POSITIONAL COORDINATES AND SYSTEM FORMS OF THE HOLONOMIC CONSTRAINTS

So far, we have examined constraints in terms of *particle* vectors, and so on. Here, we begin to move into the main task of this chapter: to describe constrained systems in terms of general *system* variables. Let us assume that our originally free, or *unconstrained*, mechanical system S, consisting of N particles with inertial position vectors [recalling (2.2.4)]

$$\boldsymbol{r}_P = \boldsymbol{r}_P(t) = \{x_P(t),\, y_P(t), z_P(t)\} \qquad (P = 1, \ldots, N), \tag{2.4.1}$$

is now subject to $h(< 3N)$ *independent positional/geometrical/holonomic* (internal and/or external) constraints

$$\phi_H(t, \boldsymbol{r}_P) \equiv \phi_H(t, \boldsymbol{r}) \equiv \phi_H(t; x, y, z) = 0 \qquad [H = 1, \ldots, h(< 3N)], \tag{2.4.2}$$

or, in extenso,

$$\begin{aligned} &\phi_1(t; x_1, y_1, z_1, \ldots, x_N, y_N, z_N) = 0, \\ &\ldots\ldots\ldots\ldots\ldots\ldots\ldots\ldots\ldots\ldots\ldots \\ &\phi_h(t; x_1,\, y_1, z_1, \ldots, x_N, y_N, z_N) = 0; \end{aligned} \tag{2.4.2a}$$

where *independent* means that the $\phi_1, \ldots, \phi_h$ are not related by a(ny) functional equation of the form $\Phi(\phi_1, \ldots, \phi_h) = 0$ {In that case we would have, e.g., $\phi_h = F(t; \phi_1, \ldots, \phi_{h-1})$, so that one of the constraints (2.4.2, 2a), i.e., here $\phi_h = 0$, would either be a consequence of the rest of them [if $F(t; 0, \ldots, 0) \equiv 0$, while $\phi_h = 0$], or it would contradict them [if $F(t; 0, \ldots, 0) \neq 0$, while $\phi_h = 0$]}.

At this point, to simplify our discussion and improve our understanding, we rename the particle coordinates (x, y, z) as follows [recalling (2.2.1a)]:

$$x_1 \equiv \xi_1, \quad y_1 \equiv \xi_2, \quad z_1 \equiv \xi_3, \ldots, \quad x_N \equiv \xi_{3N-2}, \quad y_N \equiv \xi_{3N-1}, \quad z_N \equiv \xi_{3N}, \tag{2.4.3}$$

or, compactly,

$$x_P \equiv \xi_{3P-2}, \qquad y_P \equiv \xi_{3P-1}, \qquad z_P \equiv \xi_{3P} \qquad (P = 1, \ldots, N); \tag{2.4.3a}$$

in which case, the constraints (2.4.2a) read simply

$$\phi_H(t; \xi_1, \ldots, \xi_{3N}) \equiv \phi_H(t, \xi_*) = 0 \qquad [H = 1, \ldots, h(< 3N); \quad * = 1, \ldots, 3N]. \tag{2.4.3b}$$

Therefore, using the h constraints (2.4.2a, 3b), we can express h out of the $3N$ coordinates $\xi \equiv (x, y, z)$, say the first h of them ("dependent") in terms of the remaining $n \equiv 3N - h$ ("independent"), and time:

$$\xi_d = \Xi_d(t; \xi_{h+1}, \ldots, \xi_{3N}) \equiv \Xi_d(t; \xi_i) \qquad [d = 1, \ldots, h; \quad i = h + 1, \ldots, 3N]; \tag{2.4.4}$$

and so it is now clear that our system has n (*global*) *DOF*, h down from the previous $3N$ of the unconstrained situation. Further, since for $h = 3N$ (i.e., $n = 0$) the solutions of (2.4.2a) would, in general, be incompatible with the equations of motion and/or initial conditions, while for $h = 0$ (i.e., $n = 3N$) we are back to the original unconstrained system; therefore, we should always assume tacitly that

$$0 < h < 3N \qquad \text{or} \qquad 0 < n < 3N. \tag{2.4.5}$$

Now, to express this n-parameter freedom of our system, we can use either the last n of the ξ's [i.e., the earlier $\xi_i \equiv (\xi_{h+1}, \ldots, \xi_{3N})$], or, more generally, any other set of n independent (or unconstrained, or minimal), and generally *curvilinear*, coordinates, or *holonomic positional parameters*

$$\boldsymbol{q} \equiv [q_1 = q_1(t), \ldots, q_n = q_n(t)] \equiv \{q_k = q_k(t);\ k = 1, \ldots, n\},$$

or, simply,

$$q = (q_1, \ldots, q_n), \tag{2.4.6}$$

related to the ξ_i via invertible transformations of the type

$$\xi_i = \xi_i(t; q) \;\Leftrightarrow\; q_k = q_k(t; \xi_i). \tag{2.4.6a}$$

[The reader has, no doubt, already noticed that sometimes we use ξ_i for the *totality* of the independent ξ's; i.e., $(\xi_{h+1}, \ldots, \xi_{3N})$, and sometimes for *a generic one* of them; and similarly for other variables. We hope the meaning will be clear from the context.] In view of (2.4.6a), eq. (2.4.4) can be rewritten as

$$\xi_d = \Xi_d(t; \xi_i) = \Xi_d\big[t; \xi_i(t; q)\big] = \Xi_d(t; q), \tag{2.4.6b}$$

that is, in toto, $\xi_* = \xi_*(t, q)$, $* = 1, \ldots, 3N$; and so (2.2.4), (2.4.1) can be replaced by

$$x_P = x_P(t, q), \qquad y_P = y_P(t, q), \qquad z_P = z_P(t, q),$$

or

$$\boldsymbol{r}_P = \boldsymbol{r}_P(t, q), \tag{2.4.6c}$$

or, finally, by the definitive continuum notation,

$$\boldsymbol{r} = \boldsymbol{r}(t, q). \tag{2.4.7}$$

Let us pause and re-examine our findings.

(i) The $n \equiv 3N - h$ independent positional parameters $q = q(t)$ are, at every instant t, common to all particles of the system (even though not every particle, necessarily, depends on all of them); that is, the q's are *system* coordinates; but once known as functions of time they allow us, through (2.4.7), to calculate the motion of the *individual* particles of our system S. The q's are also called holonomic (or true, or genuine, or global), independent (or unconstrained, or minimal) coordinates, although they might be constrained later (!); for short, *Lagrangean coordinates*; and the problem of analytical mechanics (AM) is to calculate them as *functions of time*. Most authors call them "generalized coordinates" (and, similarly, "generalized velocities, accelerations, forces, momenta, etc."). This pretensorial/Victorian terminology, introduced (most likely) by Thomson and Tait [1912, pp. 157–60, 286 ff.; also 1867 (1st ed.)], though inoffensive, we think is misguiding, because it directs attention away from the true role of the q's: the key word here is not generalized but *system* (coordinates)! The fact that they are, or can be, general—that is, *curvilinear* (nonrectangular Cartesian, nonrectilinear)—which is the meaning intended by Thomson and Tait, is, of course, very welcome but secondary to AM, whose task is, among others, to express all its concepts, principles, and theorems in terms of system variables. Nevertheless, to avoid breaking with such an entrenched tradition, we shall be using both terms, generalized and system coordinates, and the earlier compact expression, Lagrangean coordinates.

(ii) The ability to represent by (2.4.7) the most general position (and, through it, motion) of every system particle (i.e., in terms of a *finite* number of parameters), before any other kinetic consideration, is absolutely critical ("nonnegotiable") to AM; without it, no further progress toward the derivation of (the smallest possible number of) equations of motion could be made.

(iii) Further, as pointed out by Hamel, as long as the representation (2.4.7) holds, the original assumption of discrete mass-points/particles is not really necessary. We could, just as well, have modeled our system as a rigid *continuum*; for example, a rigid body moving about a fixed point, whether assumed discrete or continuum, needs three q's to describe its most general (angular) motion, such as its three Eulerian angles (§1.12). In sum, as long as (2.4.7) is valid, AM does not care about the molecular structure/constitution of its systems. [However, as $n \to \infty$ (continuum mechanics), the description of motion changes so that the corresponding differential equations of motion experience a "qualitative" change from ordinary to partial.]

(iv) Even though, so far, $\boldsymbol{r}$ has been assumed inertial, nevertheless, the q's do *not* have to be inertial; they may define the system's configuration(s) relative to a *noninertial* body, or frame, of known or unknown motion, and that (on top of the possible curvilinearity of the q's) is an additional advantage of the Lagrangean method. (As shown later, the $\boldsymbol{r}$'s may also be noninertial.) For example, in the double pendulum of fig. 2.7, ϕ_1, ϕ_2, θ_1 are inertial angles, whereas θ_2 is not.

If the constraints are *stationary* ($\to$ *scleronomic system*), then we can choose the q's so that (2.4.7) assumes the stationary form [recalling (2.2.2 ff.)]:

$$\boldsymbol{r}_P = \boldsymbol{r}_P(q) \qquad \text{or} \qquad \boldsymbol{r} = \boldsymbol{r}(\boldsymbol{r}_\alpha, q) \equiv \boldsymbol{r}(q); \tag{2.4.7a}$$

and, therefore, scleronomicity/rheonomicity (=absence/presence of $\partial \boldsymbol{r}/\partial t$) are q-dependent properties, unlike holonomicity/nonholonomicity.

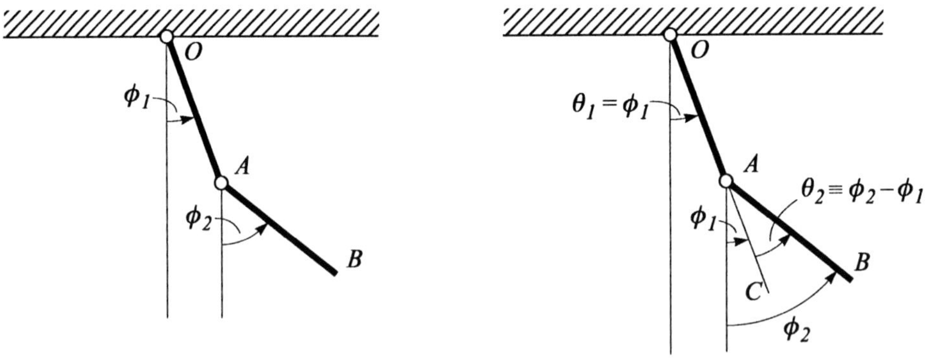

Figure 2.7 Inertial and noninertial descriptions of a double pendulum: *OA*, *AB*. Coordinates: $\phi_{1,2}$: inertial; $\theta_1 = \phi_1$: inertial; $\theta_2 \equiv \phi_2 - \phi_1$: noninertial; *O*, *A*, *C*: collinear.

Analytical Requirements on Equations (2.4.6a–c, 7)

The n q's are arbitrary, that is, nonunique, except that when the representations (2.4.6c, 7) are inserted back into the constraints (2.4.2, 2a) they must satisfy them identically in the q's, which, analytically, means that

$$\phi_H(t; \xi_*) = 0 \;\Rightarrow\; \phi_H[t; \xi_*(t; q)] \equiv 0 \;\Rightarrow\; \sum (\partial\phi_H/\partial\xi_*)(\partial\xi_*/\partial q_k) \equiv \partial\phi_H/\partial q_k \equiv 0,$$

where

$$H = 1, \ldots, h; \qquad * = 1, \ldots, 3N; \qquad k = 1, \ldots, n\,(\equiv 3N - h); \tag{2.4.8}$$

and where, due to the constraint independence and to (2.4.5), the Jacobians of the transformations $\phi_H \Leftrightarrow \xi_*$ and $\xi_* \Leftrightarrow q_k$ must satisfy

$$rank(\partial\phi_H/\partial\xi_*) = h, \qquad rank(\partial\xi_*/\partial q_k) = n \tag{2.4.8a}$$

[and since $|\partial\xi_i/\partial q_k| \neq 0 \Rightarrow rank(\partial\xi_i/\partial q_k) = n$], in the region of definition of the ξ and t. In addition, the functions in the transformations (2.4.6a, b) must be of class C^2 (i.e., have continuous partial derivatives of *zeroth*, *first*, and *second* order, at least, to accommodate accelerations) in the region of definition of the q's and t.

Last, conditions (2.4.8a) imply that the representations (2.4.6a, b) have a (nonunique) inverse:

$$q_k = q_k(t, \xi) \equiv q_k(t, x, y, z) = q_k = q_k(t, \boldsymbol{r}). \tag{2.4.8b}$$

Additional "regularity" requirements are presented in §2.7.

Example 2.4.1 Let us express the above analytical requirements in *particle* variables. Indeed, substituting into (2.2.8) and (2.2.10):

$$\boldsymbol{v} = d\boldsymbol{r}/dt = \sum (\partial\boldsymbol{r}/\partial q_k)(dq_k/dt) + \partial\boldsymbol{r}/\partial t \qquad (k = 1, \ldots, n), \tag{a}$$

we obtain, successively,

$$\begin{aligned} 0 = d\phi_H/dt &= S\,(\partial\phi_H/\partial\boldsymbol{r}) \cdot \Big(\sum (\partial\boldsymbol{r}/\partial q_k)(dq_k/dt) + \partial\boldsymbol{r}/\partial t\Big) + \partial\phi_H/\partial t \\ &= \sum \Big(S\,(\partial\phi_H/\partial\boldsymbol{r}) \cdot (\partial\boldsymbol{r}/\partial q_k)\Big)(dq_k/dt) \\ &\qquad + \Big(S\,(\partial\phi_H/\partial\boldsymbol{r}) \cdot (\partial\boldsymbol{r}/\partial t) + \partial\phi_H/\partial t\Big) \\ &\equiv \sum (\partial\Phi_H/\partial q_k)(dq_k/dt) + \partial\Phi_H/\partial t, \end{aligned} \tag{b}$$

from which, since the *holonomic system velocities* dq_k/dt are independent,

$$\partial\Phi_H/\partial q_k = 0, \tag{c}$$

$$\partial\Phi_H/\partial t = 0. \tag{d}$$

Constraint Addition and Constraint Relaxation

The n q's (just like the h ϕ_H's) are *independent*; that is, we *cannot* couple them by *nontrivial* functions $\Phi(q) = 0$, independent of the problem's initial conditions, and such that upon substitution of the q's from (2.4.8b) into them they vanish identically in the ξ's and t (i.e., $\Phi[t, q(t, \xi)] \equiv \Phi(t, \xi) = 0$ is impossible). Thus, as in differential calculus, when all the q's except (any) one of them remain fixed, we are still left with a "nonempty" continuous numerical range for the nonfixed q's; and these latter correspond to a "nonempty" continuous kinematically admissible range of system configurations (a similar conception of independence will apply to the various q-differentials, dq, $\delta q, \ldots$, to be introduced later). However, upon subsequent imposition of *additional* holonomic constraints to the system, the n q's will no longer be independent, or minimal. To elaborate: in the "beginning," the system of particles is free, or unconstrained ("brand new"); then, its q's are the $3N$ ξ's. Next, it is subjected to a mix of constraints; say, h holonomic ones like (2.4.2, 2a), and m Pfaffian (possibly nonholonomic) ones like (2.2.7, 9). Now, the introduction of the $n = 3N - h$ q's, as explained above, allows us to *absorb*, or *build in*, or *embed*, the h holonomic constraints into our description; the representations (2.4.6c, 7) guarantee automatically the satisfaction of the holonomic constraints, and thus achieve the primary goal of Lagrangean kinematics, which is the expression of the system's configurations, at every constrained stage, by the *smallest*, or *minimal*, number of positional coordinates needed [which, in turn (chap. 3) results in the smallest number of equations of motion. The corresponding embedding of the Pfaffian constraints, which is the next important objective of Lagrangean kinematics (to be presented later, §2.11 ff.), follows a conceptually identical methodology, but requires new "nonholonomic, or quasi, coordinates"]. Specifically, if at a later stage, $h'(< n)$ *additional*, or residual, or non–built-in, independent holonomic constraints, say of the form

$$\Phi_{H'}(t, q) = 0 \qquad (H' = 1, \ldots, h'), \tag{2.4.9}$$

are imposed on our already constrained system, then, repeating the earlier procedure, we express the n q's in terms of $n' \equiv n - h'$ new positional parameters $q' \equiv (q_{k'}; k' = 1, \ldots, n')$:

$$q_k = q_k(t, q_{k'}), \qquad rank(\partial q/\partial q') = n', \tag{2.4.10}$$

so that, now, (2.4.7) may be replaced by

$$\mathbf{r} = \mathbf{r}(t, q) = \mathbf{r}[t, q(t, q')] \equiv \mathbf{r}(t, q'); \tag{2.4.11}$$

the representation (2.4.7) still holds, no matter how many holonomic and nonholonomic constraints are imposed on the system; but then our q's will not be independent: they have become the earlier ξ's.

This process of adding holonomic constraints to an already constrained system, one or more at a time, can be continued until the number of (global) DOF reduces to zero: $3N - (h + h' + h'' + \cdots) \to 0$. Also, no matter what the actual sequence (history) of constraint imposition is, it helps to imagine that they are applied successively, one or more at a time, in any desired order, until we reach the current, or last, state of "constrainedness" of the system. It helps to think of a given constrained system as being somewhere "in the middle of the constraint scale": when we first encounter it, it already has some constraints built into it; say, it was not born yesterday. Then, as part of a problem's requirements, it is being added new constraints that reduce its $DOF(L)$, eventually to zero; and, similarly, proceeding in the opposite direction, we may subtract some of its built-in constraints, thus *relaxing* the system and increasing its $DOF(L)$, eventually to $3N$. [Usually, such a (mental) relaxation of one or more built-in constraints is needed to calculate the reaction forces caused by them (→ principle of "relaxation," §3.7).]

In sum: Any given system may be viewed as having evolved from a former "relaxed" (younger) one by imposition of constraints; and it is capable of becoming a more "rigid" (older) one by imposition of additional constraints.

For example, let us consider a "newborn" free rigid body. The meaning of rigidity is that our system is *internally* constrained; and the meaning of free(dom) is that, when presented to us and unless additionally constrained later, the system is *externally* unconstrained; that is, at this point, its built-in constraints are all internal: hence, $n = 6$. If, from there on, we require it to have, say, one of its points fixed (or move in a prescribed way), then, essentially, we add to it three external (holonomic) constraints; that is, $n' = n - 3 = 6 - 3 = 3$. If, further, we require it to have one more point fixed, then we add two more such constraints; that is, $n'' = n' - 2 = 3 - 2 = 1$. And if, finally, we require that one more of its points (noncollinear with its previous two) be fixed, then we add one more such constraint; that is, $n''' = n'' - 1 = 1 - 1 = 0$. But if, on the other hand, we, mentally or actually, separate the original single free rigid body into two free rigid bodies, then we subtract from it six internal built-in constraints (in Hamel's terminology, we "liberate" the system from those constraints) so that this new relaxed system has $n + 6 = 6 + 6 = 12$ (global) *DOF*.

Equilibrium, or Adapted, Coordinates

Frequently, we choose, in E_{3N}, the following "equilibrium," or "adapted (to the constraints)" curvilinear coordinates:

$$\chi_1 \equiv \phi_1(t; x, y, z) \quad (= 0), \ldots, \qquad \chi_h \equiv \phi_h(t; x, y, z) \quad (= 0);$$
$$\chi_{h+1} \equiv \phi_{h+1}(t; x, y, z) \quad (\neq 0), \ldots, \qquad \chi_{3N} \equiv \phi_{3N}(t; x, y, z) \quad (\neq 0);$$

or, compactly,

$$\chi_d \equiv \phi_d(t; x, y, z) \quad (= 0) \qquad (d = 1, \ldots, h);$$
$$\chi_i \equiv \phi_i(t; x, y, z) \quad (\neq 0) \qquad (i = h + 1, \ldots, 3N), \tag{2.4.12}$$

and $\chi_{3N+1} \equiv \phi_{3N+1} \equiv t$ ($\neq 0$); where $\phi_d \equiv (\phi_1, \ldots, \phi_h)$ are the given constraints, and $\phi_i \equiv (\phi_{h+1}, \ldots, \phi_{3N})$ are n new and arbitrary functions, but such that when (2.4.12) are solved for the $3N + 1$ $(t; x, y, z)$, in terms of $(t; \chi_1, \ldots, \chi_{3N})$, and the results are substituted back into the constraints $\phi_d = 0$, they satisfy them identically in these variables. In terms of the latter, which are indeed a special case of q's, the constraints take the simple *equilibrium* forms:

$$\chi_d \equiv (\chi_1 = 0, \ldots, \chi_h = 0), \tag{2.4.12a}$$

and so (2.4.7), with $q \to \chi_i$, reduces to

$$\mathbf{r} = \mathbf{r}(t, \chi_{h+1}, \ldots, \chi_{3N}) \equiv \mathbf{r}(t, \chi_i). \tag{2.4.12b}$$

Clearly, the earlier choice (2.4.4) corresponds to the following special χ-case (assuming nonvanishing Jacobian of the transformation):

$$\chi_d = \chi_d(t, \xi) \equiv \xi_d - \Xi_d(t, \xi_i) = 0 \qquad (d = 1, \ldots, h),$$
$$\chi_i = \chi_i(t, \xi) \equiv \xi_i \neq 0 \qquad (i = h + 1, \ldots, 3N). \qquad (2.4.12c)$$

In practice, the transition from ξ to q, χ_i is frequently suggested "naturally" by the geometry of the particular problem. However, the general method described above [but in differential forms; i.e., as $d\chi_d \equiv d\phi_d$ $(= 0)$ and $d\chi_i \equiv d\phi_i$ $(\neq 0)$] will allow us, later (§2.11 ff.), to build in Pfaffian (possibly nonholonomic) constraints.

Finally, such equilibrium q, χ_i's extend to the case of the earlier described $n'(> 0)$ *additional* constraints. There we may choose the new equilibrium coordinates:

$$\chi'_{d'} \equiv \Phi_{d'} \quad (= 0) \qquad (d' = 1, \ldots, h'),$$
$$\chi'_{i'} \equiv \Phi_{i'} \quad (\neq 0) \qquad (i' = h' + 1, \ldots, n), \qquad (2.4.12d)$$

so that

$$\boldsymbol{r} = \boldsymbol{r}(t, q) \to \boldsymbol{r}(t, \chi'_{i'}). \qquad (2.4.12e)$$

Excess Coordinates

Sometimes, in a system possessing n minimal Lagrangean coordinates, $q = (q_1, \ldots, q_n)$, we introduce, say for mathematical convenience, e additional *excess*, or *surplus*, Lagrangean coordinates $q_E = (q_{n+1}, \ldots, q_{n+e})$. Since the $n + e$ positional coordinates q and q_E are nonminimal—that is, mutually dependent—they satisfy e constraints of the type

$$F_E(t; q_1, \ldots, q_n; q_{n+1}, \ldots, q_{n+e}) \equiv F_E(t; q, q_E) = 0 \qquad (E = 1, \ldots, e); \qquad (2.4.13)$$

and then we may have

$$\boldsymbol{r} = \boldsymbol{r}(t, q, q_E). \qquad (2.4.13a)$$

If we do not need the q_E's, we can easily get rid of them: solving the e equations (2.4.13) for them, we obtain $q_E = q_E(t; q)$, and substituting these expressions back into (2.4.13a) we recover (2.4.7). For this to be analytically possible we must have (see any book on advanced calculus)

$$|\partial F_E / \partial q_{E'}| \neq 0 \qquad (E = 1, \ldots, e;\ E' = n + 1, \ldots, n + e). \qquad (2.4.13b)$$

Example 2.4.2 Let us consider the planar three-bar mechanism shown in fig. 2.8. The O–xy coordinates of a generic point on bars OA_1 and A_2A_3 can be expressed in terms of the angles ϕ_1 and ϕ_3, respectively; similarly, for a generic point P on A_1A_2, such that $A_1P = l$, we have

$$x = l_1 \cos\phi_1 + l\cos\phi_2, \qquad y = l_1 \sin\phi_1 + l\sin\phi_2. \qquad (a)$$

So, ϕ_1, ϕ_2, ϕ_3 express the configurations of this system; but they are *not* minimal (i.e., independent). Indeed, taking the x, y components of the obvious vector equation

$$\boldsymbol{OA}_1 + \boldsymbol{A}_1\boldsymbol{A}_2 + \boldsymbol{A}_2\boldsymbol{A}_3 + \boldsymbol{A}_3\boldsymbol{O} = \boldsymbol{0},$$

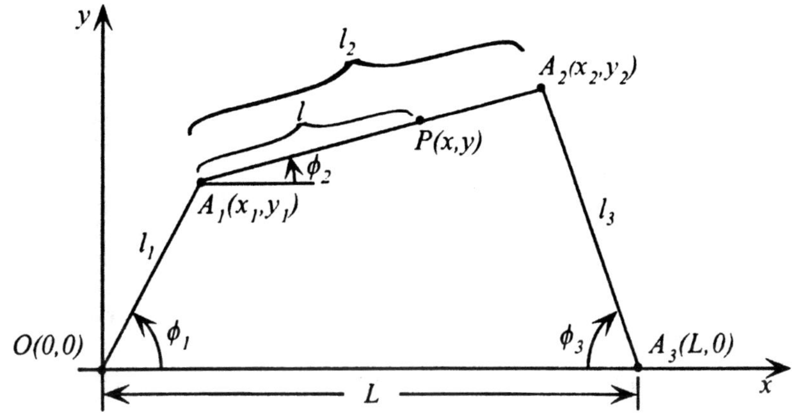

Figure 2.8 Excess coordinates in a planar three-bar mechanism.

we obtain the two constraints:

$$F_1 \equiv l_1 \cos\phi_1 + l_2 \cos\phi_2 + l_3 \cos\phi_3 - L = 0,$$
$$F_2 \equiv l_1 \sin\phi_1 + l_2 \sin\phi_2 - l_3 \sin\phi_3 = 0. \tag{b}$$

Therefore, here, $n = 1$ and $e = 2$; knowledge of any *one* of these three angles determines the mechanism's configuration.

However, it is preferable to work with the representation (a), under (b), because if we tried to use the latter to express x and y in terms of either ϕ_1, or ϕ_2, or ϕ_3 only (wherever the corresponding Jacobian does not vanish), we would end up with fewer but *very* complicated looking equations of motion. It is preferable to have more but simpler equations (of motion and of constraint); that is, requiring minimality of coordinates, and thus embedding all holonomic constraints into the equations of motion, may be highly impractical. [See books on multibody dynamics; or Alishenas (1992). On the other hand, minimal formulations have numerical advantages (computational robustness).]

Another "excess representation" of this mechanism would be to use the *four* O–xy coordinates of A_1 and A_2, (x_1, y_1) and (x_2, y_2), respectively. Clearly, these latter are subject to the *three* constraints (so that, again, $n = 1$ but $e = 3$):

$$(x_1)^2 + (y_1)^2 = (l_1)^2; \qquad (x_2 - x_1)^2 + (y_2 - y_1)^2 = (l_2)^2;$$
$$(L - x_2)^2 + (0 - y_2)^2 = (l_3)^2. \tag{c}$$

Example 2.4.3 Let us consider the planar double pendulum shown in fig. 2.9. The *four* bob coordinates x_1, y_1 and x_2, y_2 are constrained by the *two* equations

$$(x_1)^2 + (y_1)^2 = (l_1)^2, \qquad (x_2 - x_1)^2 + (y_2 - y_1)^2 = (l_2)^2; \tag{a}$$

that is, here $N = 2 \Rightarrow 2N = 4$, and so the number of *holonomic constraints* $\equiv$ $H = 2 \Rightarrow n = 2N - H = 2$. A convenient minimal representation of the pendulum's

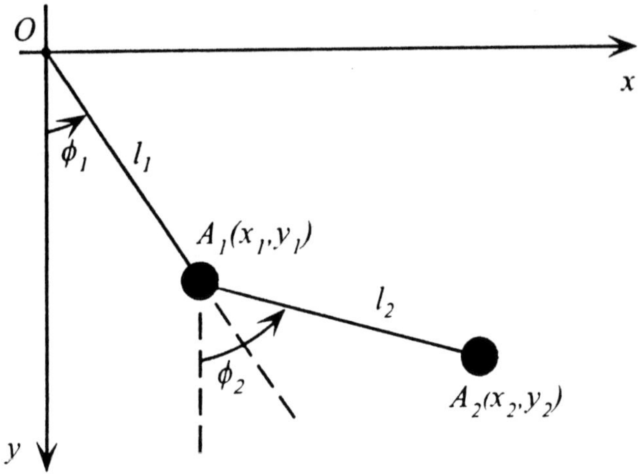

Figure 2.9 Excess coordinates in a planar double pendulum.

configurations is

$$x_1 = l_1 \cos\phi_1, \qquad y_1 = l_1 \sin\phi_1;$$
$$x_2 = l_1 \cos\phi_1 + l_2 \cos\phi_2, \qquad y_2 = l_1 \sin\phi_1 + l_2 \sin\phi_2. \tag{b}$$

2.5 VELOCITY, ACCELERATION, ADMISSIBLE AND VIRTUAL DISPLACEMENTS; IN SYSTEM VARIABLES

Velocity and Acceleration

We begin with the fundamental representation of the inertial position of a typical system particle P in Lagrangean variables (2.4.7):

$$\boldsymbol{r} = \boldsymbol{r}(t, q). \tag{2.5.1}$$

[Again, the inertialness of $\boldsymbol{r}$ is not essential, and is stated here just for concreteness. The methodology developed below applies to inertial and noninertial position vectors alike; and this, along with the possible curvilinearity (nonrectangular Cartesianness) and possible noninertialness of the coordinates, are the two key advantages of Lagrangean kinematics (and, later, kinetics) over that of Newton–Euler. This will become evident in the Lagrangean treatment of relative motion (§3.16).]

From this, it readily follows that the (inertial) *velocity* and *acceleration* of P, in these variables, are, respectively,

$$\boldsymbol{v} \equiv d\boldsymbol{r}/dt = \sum (\partial\boldsymbol{r}/\partial q_k)(dq_k/dt) + \partial\boldsymbol{r}/\partial t \equiv \sum v_k \boldsymbol{e}_k + \boldsymbol{e}_0, \tag{2.5.2}$$

$$\begin{aligned}\boldsymbol{a} \equiv d\boldsymbol{v}/dt &= \sum (\partial\boldsymbol{r}/\partial q_k)(d^2 q_k/dt^2) + \sum\sum (\partial^2\boldsymbol{r}/\partial q_k\,\partial q_l)(dq_k/dt)(dq_l/dt)\\ &\quad + 2\sum (\partial^2\boldsymbol{r}/\partial q_k\,\partial t)(dq_k/dt) + \partial^2\boldsymbol{r}/\partial t^2\\ &\equiv \sum w_k \boldsymbol{e}_k + \sum\sum v_k v_l\, \boldsymbol{e}_{k,l} + 2\sum v_k \boldsymbol{e}_{k,0} + \boldsymbol{e}_{0,0},\end{aligned} \tag{2.5.3}$$

where

$$dq_k/dt \equiv v_k, \qquad v = (v_1, \ldots, v_n) \equiv (v_k; k = 1, \ldots, n), \tag{2.5.2a}$$

$$d^2q_k/dt^2 \equiv dv_k/dt \equiv w_k, \qquad w = (w_1, \ldots, w_n) \equiv (w_k; k = 1, \ldots, n)$$

$$\left[\text{but, in general,} \quad \boldsymbol{a} \neq \sum w_k \boldsymbol{e}_k + w_0 \boldsymbol{e}_0; \qquad \text{see (2.5.4–6) below}\right], \tag{2.5.3a}$$

associated with these q's; and the fundamental (holonomic) basis vectors $\boldsymbol{e}_k$, $\boldsymbol{e}_0$, also associated with the q's, are defined by

$$\boldsymbol{e}_k = \boldsymbol{e}_k(t, q) \equiv \partial\boldsymbol{r}/\partial q_k, \qquad \boldsymbol{e}_0 = \boldsymbol{e}_0(t, q) \equiv \partial\boldsymbol{r}/\partial t \qquad (\text{or, sometimes}, \boldsymbol{e}_{n+1}, \text{ or } \boldsymbol{e}_t); \tag{2.5.4}$$

and the commas signify *partial derivatives* with respect to the q's, t:

$$\boldsymbol{e}_{k,l} \equiv \partial\boldsymbol{e}_k/\partial q_l = \partial\boldsymbol{e}_l/\partial q_k = \boldsymbol{e}_{l,k} \qquad [\text{i.e., } \partial/\partial q_l(\partial\boldsymbol{r}/\partial q_k) = \partial/\partial q_k(\partial\boldsymbol{r}/\partial q_l)], \tag{2.5.4a}$$

$$\boldsymbol{e}_{k,0} \equiv \partial\boldsymbol{e}_k/\partial t = \partial\boldsymbol{e}_0/\partial q_k = \boldsymbol{e}_{0,k} \qquad [\text{i.e., } \partial/\partial t(\partial\boldsymbol{r}/\partial q_k) = \partial/\partial q_k(\partial\boldsymbol{r}/\partial t)]; \tag{2.5.4b}$$

we reserve the notation a_k for the representation $\boldsymbol{a} = \sum a_k \boldsymbol{e}_k + a_0 \boldsymbol{e}_0$.

Also, note that with the help of the formal (nonrelativistic) notations:

$$t \equiv q_0 \equiv q_{n+1} \Rightarrow dt/dt \equiv dq_0/dt \equiv dq_{n+1}/dt \equiv v_0 \equiv v_{n+1} = 1, \tag{2.5.5a}$$

and

$$d^2t/dt^2 \equiv dv_0/dt \equiv dv_{n+1}/dt \equiv w_0 \equiv w_{n+1} = 0, \tag{2.5.5b}$$

we can rewrite (2.5.2, 3), respectively, in the "stationary" forms:

$$\boldsymbol{v} = \sum v_\alpha \boldsymbol{e}_\alpha, \qquad \boldsymbol{a} = \sum w_\alpha \boldsymbol{e}_\alpha + \sum\sum v_\alpha v_\beta \boldsymbol{e}_{\alpha,\beta}, \tag{2.5.6}$$

where, *here and throughout the rest of the book*, Greek subscripts range from 1 to $n + 1$ ($\equiv$"0").

The $v_k \equiv dq_k/dt$ are the *holonomic* (and *contravariant*, in the sense of tensor algebra) *components*, in the q-coordinates, of the system *velocity* or, simply, *Lagrangean velocities* or, briefly, but not quite accurately, "generalized velocities." The key point here is that the velocity and acceleration of each *particle*, $\boldsymbol{v}$ and $\boldsymbol{a}$, respectively, are expressed in terms of *system* velocities $v \equiv dq/dt$ and their rates $w \equiv dv/dt \equiv d^2q/dt^2$, which are *common* to all particles, via the (generally, neither unit nor orthogonal) "mixed" = *particle and system*, basis vectors $\boldsymbol{e}_k$, $\boldsymbol{e}_0$. The latter, since they effect the *transition from particle to system quantities*, are fundamental to Lagrangean mechanics.

HISTORICAL

These vectors, most likely introduced by Somoff (1878, p. 155 ff.), were brought to prominence by Heun (in the early 1900s; e.g., Heun, 1906, p. 67 ff., 78 ff.), and were called by him *Begleitvektoren* ≈ accompanying, or attendant, vectors. Perhaps a better term would be "H(olonomic) mixed basis vectors" (see also Clifford, 1887, p. 81).

From the above, we readily deduce the following basic kinematical *identities*:

(i) $$\partial\boldsymbol{r}/\partial q_k = \partial\boldsymbol{v}/\partial v_k = \partial\boldsymbol{a}/\partial w_k = \cdots \equiv \boldsymbol{e}_k, \tag{2.5.7}$$

that is, [with $(\ldots)^{\cdot} \equiv d(\ldots)/dt$],

$$\partial \boldsymbol{r}/\partial q_k = \partial \dot{\boldsymbol{r}}/\partial \dot{q}_k = \partial \ddot{\boldsymbol{r}}/\partial \ddot{q}_k = \cdots = \boldsymbol{e}_k \qquad \text{("cancellation of the (over)dots");} \tag{2.5.7a}$$

and

(ii) $$d/dt(\partial \boldsymbol{r}/\partial q_k) \equiv d/dt\,(\partial \boldsymbol{v}/\partial v_k) \equiv d\boldsymbol{e}_k/dt = \partial/\partial q_k\,(d\boldsymbol{r}/dt) \equiv \partial \boldsymbol{v}/\partial q_k,$$

or, with the help of the *Euler–Lagrange operator* in holonomic coordinates:

$$E_k(\ldots) \equiv d/dt(\ldots/\partial \dot{q}_k) - \partial\ldots/\partial q_k \equiv d/dt(\ldots/\partial v_k) - \partial\ldots/\partial q_k, \tag{2.5.9}$$

finally,

$$E_k(\boldsymbol{v}) \equiv d/dt(\partial \boldsymbol{v}/\partial v_k) - \partial \boldsymbol{v}/\partial q_k = \boldsymbol{0}. \tag{2.5.10}$$

In fact, for any well-behaved function $f = f(t, q)$, we have

$$\dot{f} \equiv df/dt \equiv \sum (\partial f/\partial q_k)(dq_k/dt) + \partial f/\partial t, \qquad \ddot{f} \equiv d^2 f/dt^2 = \cdots,$$

$$\Rightarrow \partial f/\partial q_k = \partial \dot{f}/\partial \dot{q}_k = \partial \ddot{f}/\partial \ddot{q}_k = \cdots; \tag{2.5.8}$$

and

$$E_k(f) \equiv d/dt\,(\partial f/\partial \dot{q}_k) - \partial f/\partial q_k \equiv d/dt\,(\partial f/\partial v_k) - \partial f/\partial q_k = 0. \tag{2.5.11}$$

The *integrability* conditions (2.5.7, 10) are crucial to Lagrangean kinetics (chap. 3); and, just like (2.5.2, 3), have nothing to do with constraints; that is, they hold the same, even if holonomic and/or nonholonomic constraints are later imposed on the system, as long as the q's are holonomic (genuine) coordinates (i.e., $q \neq$ nonholonomic or quasi coordinates; see §2.6, §2.9).

Admissible and Virtual Displacements

Proceeding as with the velocities, (2.5.2), we define the (first-order and inertial) *kinematically admissible*, or *possible*, and *virtual* displacements of a generic system particle P, respectively, by

$$d\boldsymbol{r} = \sum (\partial \boldsymbol{r}/\partial q_k)\,dq_k + (\partial \boldsymbol{r}/\partial t)\,dt \equiv \sum \boldsymbol{e}_k\,dq_k + \boldsymbol{e}_0\,dt, \tag{2.5.12a}$$

$$\delta \boldsymbol{r} = \sum (\partial \boldsymbol{r}/\partial q_k)\,\delta q_k \equiv \sum \boldsymbol{e}_k\,\delta q_k; \tag{2.5.12b}$$

whether the q-increments, or differentials, dq, δq, and dt are *independent or not* (say, by imposition of additional holonomic and nonholonomic constraints, later).

As the above show:

(i) if $dq_k = (dq_k/dt)\,dt \equiv v_k\,dt$, then $d\boldsymbol{r} = \boldsymbol{v}\,dt$;
(ii) if *all* the dq's and dt (δq's) vanish, then $d\boldsymbol{r} = \boldsymbol{0}$ ($\delta \boldsymbol{r} = \boldsymbol{0}$); and
(iii) $\partial(d\boldsymbol{r})/\partial(dq_k) = \partial(\delta \boldsymbol{r})/\partial(\delta q_k) = \boldsymbol{e}_k$.

These identities (in unorthodox but highly instructive notation) are useful in preparing the reader to understand, later, the nonholonomic coordinates.

REMARKS ON THE VIRTUAL DISPLACEMENT

Let us, now, pause to examine carefully this fundamental concept. First, we notice that $\delta\boldsymbol{r}$ is the *linear* (or *first-order*) and *homogeneous*, in the δq's, part of the "total virtual displacement" $\Delta\boldsymbol{r}$, which is defined by the following Taylor-like $\boldsymbol{r}$-expansion in the first-order increments δq, from a generic system configuration corresponding to the values q, but for *a fixed time t*:

$$\Delta\boldsymbol{r} \equiv \boldsymbol{r}(t, q + \delta q) - \boldsymbol{r}(t, q) \equiv \delta\boldsymbol{r} + (1/2)\,\delta^2\boldsymbol{r} + \cdots. \tag{2.5.13}$$

In other words, $\delta\boldsymbol{r}$ is a *special first position differential*, mathematically equivalent to $d\boldsymbol{r}$ with $t = constant \to dt = 0$ (i.e., completely equivalent to it for *stationary* constraints); hence, the special notation $\delta(\ldots)$:

$$d\boldsymbol{r} \to \delta\boldsymbol{r}, \qquad dq \to \delta q, \qquad \text{and} \qquad dt \to \delta t = 0 \text{ (isochrony, } always\text{)}. \tag{2.5.13a}$$

One could have denoted it as $d^*\boldsymbol{r}$, or $(d\boldsymbol{r})^*$, or $\boldsymbol{z}$, and so on; but since we do not see anything wrong with $\delta(\ldots)$, and to keep with the best traditions of analytical mechanics [originated by Lagrange himself and observed by all mechanics masters, such as Kirchhoff, Routh, Schell, Thomson and Tait, Gibbs, Appell, Volterra, Poincaré, Maggi, Webster, Heun, Hamel, Prange, Whittaker, Chetaev, Lur'e, Synge, Gantmacher, Pars et al.], we shall stick with it. Readers who feel uncomfortable with $\delta(\ldots)$ may devise their own suggestive notation; $d\boldsymbol{r}$ and dq won't do!

The above definitions also show the following:

(i) $\delta\boldsymbol{r}$ is mathematically equivalent to the *difference* between two possible/admissible displacements, say $d_1\boldsymbol{r}$ and $d_2\boldsymbol{r}$, taken along *different directions* but at the *same time* (and *same dt*); that is, skipping summation signs and subscripts, for simplicity,

$$\begin{aligned} d_2\boldsymbol{r} - d_1\boldsymbol{r} &= [(\partial\boldsymbol{r}/\partial q)\, d_2 q + (\partial\boldsymbol{r}/\partial t)\, dt] - [(\partial\boldsymbol{r}/\partial q)\, d_1 q + (\partial\boldsymbol{r}/\partial t)\, dt] \\ &= (\partial\boldsymbol{r}/\partial q)(d_2 q - d_1 q) \sim (\partial\boldsymbol{r}/\partial q)\, \delta q = \delta\boldsymbol{r}. \end{aligned} \tag{2.5.14}$$

(ii) For any well-behaved function $f = f(t)$: $\delta f = (\partial f/\partial t)\, \delta t = 0$; but if $f = f(t, q)$, then $\delta f = (\partial f/\partial q)\, \delta q \neq 0$ [even though, *after* the problem is solved, $q = q(t)$!].

(iii) The virtual displacements of mechanics do not always coincide with those of mathematics (i.e., calculus of variations). For example, even though, in general, $d\boldsymbol{r} \neq \delta\boldsymbol{r}$, for *catastatic* systems [i.e., $d\boldsymbol{r} = \sum \boldsymbol{e}_k(t, q)\, dq_k$, $\delta\boldsymbol{r} = \sum \boldsymbol{e}_k(t, q)\, \delta q_k$] the equality $dq_k = \delta q_k \Rightarrow d\boldsymbol{r} = \delta\boldsymbol{r}$ is kinematically possible [and in (q, t)-space $d\boldsymbol{r}$ and $\delta\boldsymbol{r}$ are "orthogonal" to the t-axis, even though $dt \neq 0$, $\delta t = 0$]; whereas, in variational calculus we are explicitly warned that dq (*parallel* to the t-axis) $\neq \delta q$ (*perpendicular* to it). These differences, rarely mentioned in mechanics and/or mathematics books, are very consequential, especially in *integral* variational principles for nonholonomic systems (chap. 7).

As definitions (2.5.12, 13), and so on, show, the (particle and/or system) virtual displacement is a simple, direct, and, as explained in chapter 3 and elsewhere, *indispensable* concept—without it Lagrangean mechanics would be impossible! Yet, since its inception (in the early 18th century), this concept has been surrounded with mysticism and confusion; and even today it is frequently misunderstood and/or ignorantly maligned. For instance, it has been called "too vague and cumbersome to be of practical use" by D. A. Levinson, in discussion in Borri et al. (1992); "ill-defined, nebulous, and hence objectionable" by T. R. Kane, in rebuttal to Desloge (1986); or, at best, has been given the impression that it has to be defined, or "chosen properly" (Kane and Levinson (1983)), in an ad hoc or a posteriori fashion to fit the facts, that is, to produce the correct equations of motion. For an extensive rebuttal of these false and misleading statements, from the viewpoint of kinetics, see chap. 3, appendix II. Others object to the arbitrariness of the δq's. But it is precisely in this arbitrariness that their strength and effectiveness lies: they do the job (e.g., yielding of the equations of motion) and then, modestly, retreat to the background leaving behind the mixed basis vectors $\boldsymbol{e}_k$. It is these latter [and their nonholonomic counterparts (§2.9)] that appear in the final equations of motion (chap. 3), just as in the derivation of differential ("field") equations in other areas of mathematical physics. For example, in continuum mechanics, for better visualization, we may employ a small spatial element (e.g., a "control volume"), of "infinitesimal" dimensions dx, dy, dz, to derive the local field equations of balance of mass, momentum, energy, and so on; but the ultimate differential equations never contain lone differentials—only combinations of finite limits of ratios among them; that is, *combinations of derivatives*. Moreover, differentials, actual/admissible/virtual, in addition to being easier to visualize than derivatives, are *invariant* under coordinate transformations; whereas derivatives are not. [Such invariance ideas led the Italian mathematicians G. Ricci and T. Levi-Civita to the development of *tensor* calculus (late 19th to early 20th century); see, for example, Papastavridis (1999).] For example, taking for simplicity, a one (global) *DOF* system, under the transformation $q \to q' = q'(t, q)$, we find, successively,

$$\delta \boldsymbol{r} = \boldsymbol{e}\, \delta q = (\partial \boldsymbol{r}/\partial q)\, \delta q = (\partial \boldsymbol{r}/\partial q)[(\partial q/\partial q')\, \delta q'] = [(\partial \boldsymbol{r}/\partial q)(\partial q/\partial q')]\, \delta q' \equiv \boldsymbol{e}'\, \delta q',$$

that is,

$$\boldsymbol{e}' \equiv \partial \boldsymbol{r}/\partial q' = (\partial q/\partial q')\boldsymbol{e} \Leftrightarrow \boldsymbol{e} \equiv \partial \boldsymbol{r}/\partial q = (\partial q'/\partial q)\boldsymbol{e}'. \tag{2.5.15}$$

But there is an additional, deeper, reason for the representation (2.5.12b): the position vectors $\boldsymbol{r}(t, q)$ and (possible) additional constraints, say $\psi_{H'}(t, \boldsymbol{r}) = 0 \to \psi_{H'}(t, q) = 0$, *cannot* be attached in these finite forms to the general kinetic principles of analytical mechanics, which are differential, and lead to the equations of motion (§3.2 ff.). Only virtual forms of $\boldsymbol{r}$ and $\psi_{H'} = 0$—special first differentials of them, linear and homogeneous in the δq's [like (2.5.12b)]—can be attached, or *adjoined*, to the Lagrangean variational equation of motion via the well-known method of *Lagrangean multipliers* (§3.5 ff.), and similarly for nonlinear (non-Pfaffian) velocity constraints, an area that shows clearly that nonvirtual schemes (as well as those based on the calculus of variations) break down (chap. 5)! Hence, the older admonition that the virtual displacements must be "small" or "infinitesimal." For example, to incorporate the *nonlinear* holonomic constraint $\phi(x, y) \equiv x^2 + y^2 = constant$ to the kinetic principles, we must attach to them its first virtual differential, $\delta\phi = 2(x\, \delta x + y\, \delta y) = 0$; which is the linear and homogeneous part of the total constraint change, between the system configurations (x, y) and $(x + \delta x,\ y + \delta y)$:

$$\Delta\phi \equiv \phi(x + \delta x,\ y + \delta y) - \phi(x, y) = [\delta\phi + (1/2)\, \delta^2\phi]_{\text{for small } \delta x, \delta y} \approx \delta\phi = 0.$$

But in the case of the *linear* holonomic constraint $\phi \equiv x + y = constant$, that total constraint change equals

$$\Delta\phi = \delta\phi = \delta x + \delta y = 0, \qquad \text{no matter what the size of } \delta x,\ \delta y;$$

and both equations, $\phi = 0$ and $\delta\phi = 0$, have the same coefficients ($\to$ slopes).

In sum: As long as we take the first virtual differentials of the constraints, the size of the δq's is inconsequential, whether they are one millimeter or ten million miles! It is the holonomic (or "gradient," or "natural") basis vectors $\{\boldsymbol{e}_k;\ k = 1, \ldots, n\}$, that matter.

As Coe puts it: "We often speak of displacements, both virtual and real, as being *arbitrarily small or infinitesimal*. This means that we are concerned only with the limiting directions of these displacements and the limiting values of the ratios of their lengths as they approach zero. Thus any two systems of virtual displacements are for our purposes identical if they have the same limiting directions and length ratios as they approach zero" (1938, p. 377). Coe's seems to be the earliest correct and vectorial exposition of these concepts in English; most likely, following the exposition of Burali-Forti and Boggio (1921, pp. 136 ff.). See also Lamb (1928, p. 113).

The earlier mentioned indispensability of the virtual displacements for kinetics will become clearer in chapter 3. Nevertheless, here is a preview: it is the *virtual* work of the forces maintaining the holonomic and/or nonholonomic constraints (i.e., the *constraint reactions*) that vanishes, and not just any work, admissible or actual; in fact, the latter would supply only *one* equation. This vanishing-of-the-virtual-work-of-constraint-reactions (principle of d'Alembert–Lagrange) is a physical postulate that generates not just one equation of motion (like the actual work/power equation does), but as many as the number of (local) *DOFs*; and, additionally, it allows us to eliminate/calculate these constraint forces. A more specialized virtual displacement → virtual work-based postulate is used to characterize the more general "servo/control" constraints (§3.17).

Example 2.5.1 *Differences Between Kinematically Admissible/Possible and Virtual Displacements.*

(i) Let us assume that we seek to determine the motion of a particle P capable of sliding along an ever straight line l rotating on the plane O–xy about O. The configurations of l and of P relative to that plane are determined, respectively, by ϕ and r, ϕ (fig. 2.10). Since $\boldsymbol{r} = \boldsymbol{r}(r, \phi)$: position of P in O–xy, we will have, in the most general case,

$$d\boldsymbol{r} = (\partial\boldsymbol{r}/\partial r)\, dr + (\partial\boldsymbol{r}/\partial\phi)\, d\phi\text{: kinematically admissible displacement of } P, \text{ in } O\text{–}xy, \tag{a}$$

$$\delta\boldsymbol{r} = (\partial\boldsymbol{r}/\partial r)\, \delta r + (\partial\boldsymbol{r}/\partial\phi)\, \delta\phi\text{: virtual displacement of } P, \text{ in } O\text{–}xy. \tag{b}$$

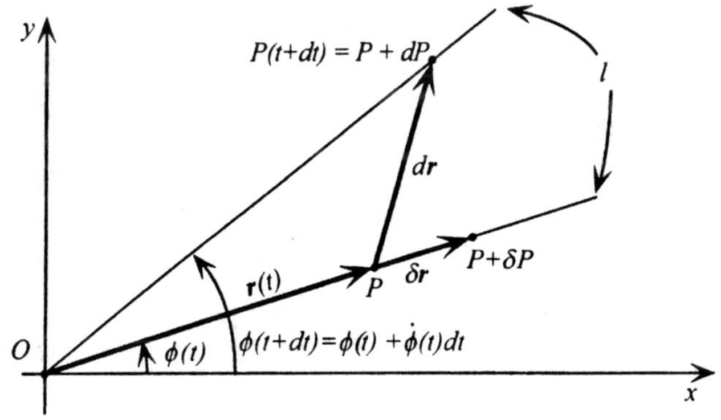

Figure 2.10 On the difference between possible/admissible and virtual displacements (ex. 2.5.1: a, b).

(a) If the rotation of l is influenced by the motion of P relative to it, then ϕ is another unknown Lagrangean coordinate, like r, waiting to be found from the equations of motion of the system P *and* l (i.e., $n = 2$). Then $d\mathbf{r}$ and $\delta\mathbf{r}$ are given by (a, b), respectively, and are mathematically equivalent. (b) If, however, the motion of l is known ahead of time (i.e., if it is constrained to rotate in a specified way, uninfluenced by the, yet unknown, motion of P), then

$$\phi = f(t)\text{: given function of time} \Rightarrow$$
$$d\phi = df(t) = [df(t)/dt]\, dt \equiv \omega(t)\, dt \neq 0, \qquad \text{but} \qquad \delta\phi = \delta f(t) = \omega(t)\, \delta t = 0.$$

As a result, (a, b) yield

$$d\mathbf{r} = (\partial\mathbf{r}/\partial r)\, dr + (\partial\mathbf{r}/\partial\phi)\, d\phi = (\partial\mathbf{r}/\partial r)\, dr + (\partial\mathbf{r}/\partial\phi)\, \omega(t)\, dt = d\mathbf{r}(t, r; dt, dr), \quad \text{(c)}$$
$$\delta\mathbf{r} = (\partial\mathbf{r}/\partial r)\, \delta r = \delta\mathbf{r}(t, r; \delta r). \quad \text{(d)}$$

(ii) Let us consider the motion of a particle P along the inclined side of a wedge W that moves with a *given* horizontal motion: $x = f(t)$ (fig. 2.11). Here, we have

$$\mathbf{MM}_1 = \mathbf{M}_3\mathbf{M}_2 = (\partial\mathbf{r}/\partial x)\, dx = (\partial\mathbf{r}/\partial x)[df(t)/dt]\, dt = (\partial\mathbf{r}/\partial t)\, dt \sim dt; \quad \text{(e)}$$
$$\mathbf{MM}_3 = \mathbf{M}_1\mathbf{M}_2 = \delta\mathbf{r} = (\partial\mathbf{r}/\partial q)\, \delta q \sim \delta q; \quad \text{(f)}$$
$$\mathbf{MM}_2 \sim d\mathbf{r} = (\partial\mathbf{r}/\partial q)\, dq + (\partial\mathbf{r}/\partial t)\, dt; \qquad \text{but } \delta x = 0. \quad \text{(g)}$$

(iii) Let us consider the motion of a particle P on the fixed and rigid surface $\phi(x, y, z) = 0$ or $z = z(x, y)$. Then, $\mathbf{r} = \mathbf{r}(x, y, z) = \mathbf{r}[x, y, z(x, y)] \equiv \mathbf{r}(x, y)$, and the classes of $d\mathbf{r}$ and $\delta\mathbf{r}$ are equivalent. But, on the moving and possibly deforming surface $\phi(t; x, y, z) = 0$ or $z = z(x, y; t)$, $\mathbf{r} = \cdots = \mathbf{r}(t; x, y)$, and so $d\mathbf{r} \neq \delta\mathbf{r}$: $\delta\mathbf{r}$ still lies on the instantaneous tangential plane of the surface at P, whereas $d\mathbf{r}$ does not.

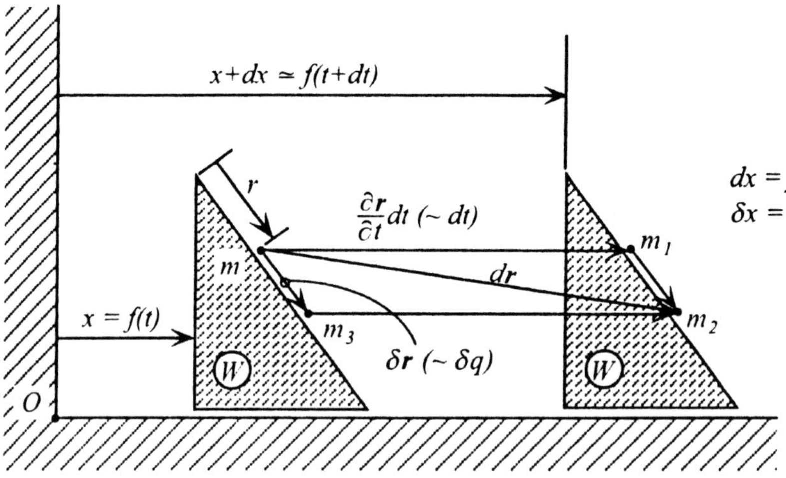

Figure 2.11 On the difference between possible/admissible and virtual displacements (ex. 2.5.1: b).

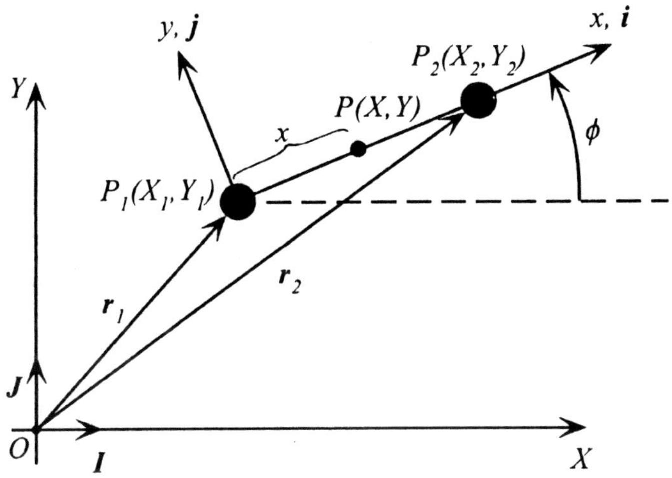

Figure 2.12 Two-particle system connected by a light rod.

Example 2.5.2 *Lagrangean Coordinates and Virtual Displacements.* Let us determine the Lagrangean description $\boldsymbol{r} = \boldsymbol{r}(\boldsymbol{r}_o; t, q)$ and corresponding virtual displacements $\delta\boldsymbol{r} = \cdots$ for the following systems:

(i) Two particles, P_1 and P_2, are connected by a massless rod of length l, in plane motion (fig. 2.12). For an arbitrary rod point $P(X, Y)$, including P_1 and P_2, we have

$$X = X_1 + x\cos\phi = X(x; X_1, \phi), \qquad Y = Y_1 + x\sin\phi = Y(x; Y_1, \phi), \tag{a}$$

or, vectorially,

$$\boldsymbol{r} = \boldsymbol{r}_1 + x\boldsymbol{i} = \boldsymbol{r}(x; X_1, Y_1, \phi), \qquad 0 \le x \le l. \tag{b}$$

Therefore, $\boldsymbol{r}_o = x\boldsymbol{i}$; while the (inertial) positions of P_1 and P_2 are given, respectively, by

$$\boldsymbol{r}_1 = \boldsymbol{r}(0; X_1, Y_1, \phi) = X_1\boldsymbol{I} + Y_1\boldsymbol{J}, \tag{c}$$

$$\begin{aligned}\boldsymbol{r}_2 = \boldsymbol{r}(l; X_1, Y_1, \phi) &= (X_1 + l\cos\phi)\boldsymbol{I} + (Y_1 + l\sin\phi)\boldsymbol{J}\\ &= (X_1\boldsymbol{I} + Y_1\boldsymbol{J}) + l(\cos\phi\boldsymbol{I} + \sin\phi\boldsymbol{J}) = \boldsymbol{r}_1 + l\boldsymbol{i}.\end{aligned} \tag{d}$$

Hence, this is a (holonomic) *three DOF* system: $q_1 = X_1$, $q_2 = Y_1$, $q_3 = \phi$. From (a) we obtain, for the virtual displacements,

$$\delta X = \delta X_1 + x(-\sin\phi)\,\delta\phi, \qquad \delta Y = \delta Y_1 + x(\cos\phi)\,\delta\phi; \tag{e}$$

or, vectorially,

$$\delta\boldsymbol{r} = \delta\boldsymbol{r}_1 + x\,\delta\boldsymbol{i} = \delta\boldsymbol{r}_1 + x[(\delta\phi\,\boldsymbol{k}) \times \boldsymbol{i}] = \delta\boldsymbol{r}_1 + (x\,\delta\phi)\boldsymbol{j}. \tag{f}$$

(ii) A rigid body in plane motion (fig. 2.13). For this *three DOF* system we have

$$\begin{aligned}X &= X_\blacklozenge + x\cos\phi - y\sin\phi = X(x, y; X_\blacklozenge, \phi),\\ Y &= Y_\blacklozenge + x\sin\phi + y\cos\phi = Y(x, y; Y_\blacklozenge, \phi),\end{aligned} \tag{g}$$

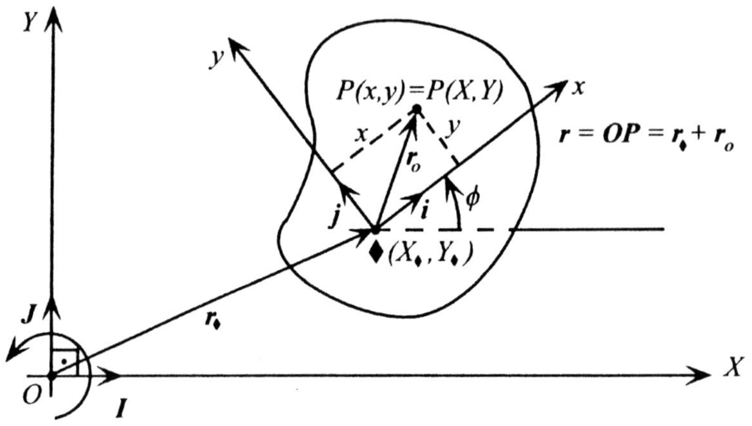

Figure 2.13 Rigid body in plane motion.

(and $Z = Z_\blacklozenge = 0$, say), or, vectorially,

$$\boldsymbol{r} = X\boldsymbol{I} + Y\boldsymbol{J} = \boldsymbol{r}(x, y; X_\blacklozenge, Y_\blacklozenge, \phi); \tag{h}$$

that is, $\boldsymbol{r}_o = x\boldsymbol{i} + y\boldsymbol{j}$ and $q_1 = X_\blacklozenge$, $q_2 = Y_\blacklozenge$, $q_3 = \phi$. Therefore, the virtual displacements are

$$\begin{aligned} \delta X &= \delta X_\blacklozenge - x \sin\phi\, \delta\phi - y\cos\phi\, \delta\phi = \delta X_\blacklozenge - \delta\phi(Y - Y_\blacklozenge), \\ \delta Y &= \delta Y_\blacklozenge + x\cos\phi\, \delta\phi - y \sin\phi\, \delta\phi = \delta Y_\blacklozenge + \delta\phi(X - X_\blacklozenge), \end{aligned} \tag{i}$$

(and $\delta Z = \delta Z_\blacklozenge = 0$), or, vectorially,

$$\delta\boldsymbol{r} = \delta\boldsymbol{r}_\blacklozenge + \delta\boldsymbol{\phi} \times (\boldsymbol{r} - \boldsymbol{r}_\blacklozenge), \qquad \delta\boldsymbol{\phi} = \delta\phi\, \boldsymbol{k}. \tag{j}$$

The extension to a rigid body in general *spatial* motion (with the help of the Eulerian angles, §1.12, and recalling discussion in §1.8) is straightforward.

2.6 SYSTEM FORMS OF LINEAR VELOCITY (PFAFFIAN) CONSTRAINTS

Stationarity/Scleronomicity/Catastaticity for Positional/Geometrical (⇒ Holonomic) Constraints in System Variables

We begin by extending the discussion of §2.2 to general system variables, inertial or not. Positional constraints of the form $\phi(q) = 0$ ($\Rightarrow \partial\phi/\partial t = 0$) are called *stationary*; otherwise—that is, if $\phi(t, q) = 0$ ($\Rightarrow \partial\phi/\partial t \neq 0$)—they are called *nonstationary*; and if all constraints of a system are (or can be reduced to) such stationary (nonstationary) forms, the system is called *scleronomic* (*rheonomic*). Clearly, such a classification is *nonobjective*—that is, it depends on the particular mode and/or frame of reference used for the description of position/configuration: for example, substituting $\boldsymbol{r}(t, q)$ into the stationary constraint $\phi(\boldsymbol{r}) = 0$ turns it to a nonstationary constraint, $\phi[\boldsymbol{r}(t, q)] = \phi(t, q) = 0$ (and this is a reason that certain authors prefer to base this classification only for constraints expressed in system variables); or, a constraint that is stationary when expressed in terms of inertial coordinates (q) may very well become nonstationary when expressed in terms of noninertial coordinates (q'): under the frame of reference (i.e., explicitly time-dependent!) transformation $q \to q'(t,q) \Leftrightarrow q' \to q(t,q')$, the stationary constraint $\phi(q) = 0$ transforms to the nonstationary one $\phi(t,q') = 0$. Hence, a *scleronomic constraint* $\phi(q) = 0$ remains scleronomic under all coordinate (not frame of reference) transformations $q \to q'(q) \Leftrightarrow q' \to q(q')$; that is, its scleronomicity under such transformations is an objective property.

Stationarity/Scleronomicity/Catastaticity for Pfaffian Constraints in System Variables

Next, let us assume that the h holonomic constraints have been embedded into our system by the $n \equiv 3N - h$ Lagrangean coordinates q. To embed the *additional*, $m(< n)$ mutually independent and possibly nonholonomic Pfaffian constraints (2.2.9) into our Lagrangean kinematics and kinetics: first, we express them in system variables. Indeed, substituting $\boldsymbol{v}$ from (2.5.2) into (2.2.9), we obtain the Pfaffian constraints in system (holonomic) variables:

$$f_D \equiv S(\boldsymbol{B}_D \cdot \boldsymbol{v}) + B_D = \cdots = \sum c_{Dk} v_k + c_D = 0 \quad (D = 1, \ldots, m), \tag{2.6.1}$$

where

$$c_{Dk} = c_{Dk}(t,q) \equiv S\, \boldsymbol{B}_D \cdot (\partial \boldsymbol{r}/\partial q_k) \equiv S\, \boldsymbol{B}_D \cdot \boldsymbol{e}_k, \tag{2.6.1a}$$

$$c_D \equiv c_{D,n+1} \equiv c_{D,0} = c_D(t,q) \equiv S\, \boldsymbol{B}_D \cdot (\partial \boldsymbol{r}/\partial t) + B_D \equiv S\, \boldsymbol{B}_D \cdot \boldsymbol{e}_0 + B_D; \tag{2.6.1b}$$

and $rank(c_{Dk}) = m$. Similarly, substituting $d\boldsymbol{r}$ from (2.5.12a) into the differential form of (2.2.9), $f_D\, dt = 0$, we obtain the *kinematically admissible*, or *possible*, form of these constraints in (holonomic) system variables:

$$d'\chi_D \equiv f_D\, dt = \sum c_{Dk}\, dq_k + c_D\, dt = 0; \tag{2.6.2}$$

with $d'\chi_D$: *not necessarily an exact differential*; that is, χ_D may not exist, it may be a "quasi variable" (§2.9) and, in view of what has already been said about virtualness, namely, $dt \to \delta t = 0$, the *virtual* form of these constraints in particle variables is

$$\delta'\chi_D \equiv S\, \boldsymbol{B}_D \cdot \delta \boldsymbol{r} = 0, \tag{2.6.3}$$

and, accordingly, invoking (2.5.12b), in system variables,

$$\delta'\chi_D \equiv \sum c_{Dk}\, \delta q_k = 0. \tag{2.6.4}$$

The above show that, as in the particle variable case, the virtual displacements are mathematically equivalent to the difference between two systems of possible displacements, $d_1 q$ and $d_2 q$, occurring at the same position and for the same time, but in different directions: apply (2.6.2) at (t,q), for $d_1 q \neq d_2 q$, and subtract side by side and a (2.6.4)-like equation results.

And, as in (2.5.12a,b), once the constraints have been brought to these Pfaffian forms, the size of the δq's does not matter; it is the *constraint coefficients* c_{Dk} that do.

Now, if in (2.6.1–2),

(i)
$$\partial c_{Dk}/\partial t = 0 \;\Rightarrow\; c_{Dk} = c_{Dk}(q) \tag{2.6.5a}$$

and

(ii)
$$c_D \equiv c_{D,n+1} \equiv c_{D,0} = 0, \tag{2.6.5b}$$
$$\Rightarrow \text{constraint: } f_D = \sum c_{Dk}(q) v_k = 0,$$

for *all* $D = 1, \ldots, m$ and $k = 1, \ldots, n$, these constraints are called *stationary*; otherwise they are *nonstationary*; and a system with even one nonstationary constraint is called *rheonomic*; otherwise it is *scleronomic*. The inclusion of (2.6.5b) in the stationarity definition is made so that the velocity *form* of stationary *position* constraints coincides with that of the stationary *velocity* constraints:

$$\phi_D(q) = 0 \;\Rightarrow\; d\phi_D/dt = \sum (\partial\phi_D/\partial q_k) v_k = \sum \phi_{Dk}(q) v_k = 0. \tag{2.6.5c}$$

If only $c_D \equiv c_{D,n+1} \equiv c_{D,0} = 0$, for *all* D, but $\partial c_{Dk}/\partial t \neq 0 \;\Rightarrow\; c_{Dk} = c_{Dk}(t, q)$ even for one value of D and k, the Pfaffian constraints are called *catastatic* [≈ calm, orderly (Greek)]; otherwise they are called *acatastatic*. We notice that stationary constraints are catastatic, but catastatic constraints may not be stationary; we may still have $\partial c_{Dk}/\partial t \neq 0$ for some D and k. As mentioned earlier (2.2.11a ff.), it is the *castastatic/acatastatic classification*, having meaning only for Pfaffian constraints, that is the important one for analytical kinetics, not the stationary/nonstationary one.

Finally, as (2.6.1b) shows, the acatastatic coefficients c_D result from the nonstationary part of $\boldsymbol{v}$ (i.e., $\partial \boldsymbol{r}/\partial t$), and the acatastatic part of (2.2.9) (i.e., B_D). From this comes the search for frames of reference/Lagrangean coordinates where the Pfaffian constraint coefficients take their simplest possible form; a problem that, in turn, leads us to the investigation of the following.

Transformation Properties of c_{Dk} and c_D, under a General Frame-of-Reference Transformation

The latter is mathematically equivalent to an *explicitly* time-dependent coordinate transformation: $q \to q' = q'(t, q)$ and $t \to t' = t$. Then (2.6.1–1b) become

$$f_D = \sum c_{Dk}\Big(\sum (\partial q_k/\partial q_{k'}) v_{k'} + \partial q_k/\partial t\Big) + c_D$$
$$= \cdots = \sum c_{Dk'} v_{k'} + c'_D \quad (= 0) \qquad (k, k' = 1, \ldots, n;\; D = 1, \ldots, m); \tag{2.6.6}$$

where

$$c_{Dk'} \equiv \sum (\partial q_k/\partial q_{k'}) c_{Dk} \qquad \text{(covariant vector-like transformation in } k\text{)}, \tag{2.6.6a}$$

$$c'_D \equiv \sum (\partial q_k/\partial t) c_{Dk} + c_D \qquad \text{(covariant vector-like transformation in } t \equiv n+1,$$
$$\text{with } q'_{n+1} \equiv t' = t \;\Rightarrow\; \partial t'/\partial t = 1). \tag{2.6.6b}$$

The above readily show that: (i) if $\partial q_k/\partial t = 0$ [i.e., $q = q(q')$] (= *coordinate transformation*; in the same frame of reference), then $c'_D = c_D$; and (ii) we *can* choose a frame of reference in which $c'_D = 0$; that is, catastaticity/acatastaticity (and stationarity/nonstationarity) are *frame-dependent properties*.

Holonomicity versus Nonholonomicity

The $m(< n)$ constraints (2.6.1) are *independent* if the $m \times n$ *constraint matrix* (c_{Dk}) has maximal rank (i.e., m) at each point in the region of definition of the q's and t.

Now, if these constraints are completely integrable $\equiv$ holonomic [i.e., either they are exact: $c_{Dk} = \partial h_D/\partial q_k$ and $c_D = \partial h_D/\partial t$, where $h_D = h_D(t, q)$ $(= 0)$; or they possess integrating factors, as explained in §2.2], then there exists a set of n "equilibrium," or "adapted (to the constraints)" system coordinates $\chi = (\chi_1, \ldots, \chi_n)$ in which these constraints take the simple *uncoupled* form:

$$\chi_1 \equiv h_1(t, q) = 0, \ldots, \chi_m \equiv h_m(t, q) = 0; \tag{2.6.7a}$$

$$\chi_{m+1} \equiv h_{m+1}(t, q) \neq 0, \ldots, \chi_n \equiv h_n(t, q) \neq 0; \tag{2.6.7b}$$

where, as in §2.4, the $n - m$ functions $h_{m+1}(t, q), \ldots, h_n(t, q)$ are arbitrary, except that when (2.6.7a, b) are solved for the n q's in terms of the $(n - m)$ $\chi_I \equiv (\chi_{m+1}, \ldots, \chi_n)$ and these expressions are inserted back into the m holonomic constraints $h_1(t, q) = 0, \ldots, h_m(t, q) = 0$, they satisfy them identically in the χ_I's and t. The χ_I's are the new positional system coordinates of this $3N - (h + m) = (3N - h) + m = n - m \equiv n'$ (both global and local) *DOF*:

$$q \to q' \equiv (\chi_{m+1}, \ldots, \chi_n) \equiv (q'_1, \ldots, q'_{n'}). \tag{2.6.7c}$$

This process of *adaptation to the constraints via new equilibrium coordinates* can be repeated if additional holonomic constraints are imposed on the system; and with some nontrivial modifications it carries over to the case of additional nonholonomic constraints (§2.11: essentially, by *expressing this adaptation ... idea in the small*; i.e., locally, via "equilibrium quasi coordinates"). The importance of this method to AM lies in its ability to *uncouple* constraints, and thus to simplify significantly the equations of motion (chap. 3).

If, on the other hand, the constraints (2.6.1) are *noncompletely integrable* $\equiv$ *nonholonomic*, then the number of independent Lagrangean coordinates (= number of *global* DOF) remains n, but the system has $n - m \equiv f$ *DOF* (in the *small*, or *local* case); that is, under the additional m nonholonomic constraints [(2.6.1), (2.6.2)], *the n q's remain independent* (unlike the holonomic case!), *but the n $v/dq/\delta q$'s do not*—or, if the differential increments δq are *arbitrary* (if, for example, we let q_k become $q_k + \delta q_k$ while all the other q's *remain constant*), then they will *no longer be virtual*; that is, they will not be compatible with the virtual form of the constraints (2.6.4); and similarly for the v's, dq's. [Of course, if $m = 0$, then the n q's are independent *and* their arbitrary increments δq are *virtual*; that is, both q's and δq's satisfy the existing (initial) h holonomic constraints. For example, in the case of a sphere rolling on, say, a fixed plane: (a) if the plane is smooth (i.e., $m = 0$), both the arbitrary q's and the arbitrary $(q + dq)$'s, are kinematically possible; while (b) if the plane is sufficiently rough so that the sphere rolls on it (i.e., $m \neq 0$, and the additional (rolling) constraints are nonholonomic), only the q's are still arbitrary (independent), the $(q + dq)$'s are not—or, if they are, the sphere does not roll. For details, see exs. 2.13.4, 2.13.5, 2.13.6.]

To find the number of independent δq's under the additional m (holonomic or nonholonomic) constraints (2.6.1, 2, 4) we must now turn to the examination of the following.

Introduction to Virtual Displacements under Pfaffian Constraints (Introduction to Quasi Variables)

In this case, the particle virtual displacement is still represented by (2.5.12b):

$$\delta \boldsymbol{r} = \sum (\partial \boldsymbol{r}/\partial q_k)\, \delta q_k \equiv \sum \boldsymbol{e}_k\, \delta q_k, \tag{2.6.8}$$

but, due to the virtual constraints (2.6.4), out of the n δq's only $n - m$ are independent; that is, if, now, all n δq's vary arbitrarily, the resulting $\delta \boldsymbol{r}$, via (2.6.8), will *not* be virtual—denoting a differential increment of a system coordinate by δq does not necessarily make it virtual; it must also be constraint compatible. For example, solving (2.6.4) for the first m δq's,

$$\delta q_D \equiv (\delta q_1, \ldots, \delta q_m) = \textit{Dependent}\ \delta q\text{'s}, \tag{2.6.9a}$$

in terms of the last $n - m$ of them,

$$\delta q_I \equiv (\delta q_{m+1}, \ldots, \delta q_n) = \textit{Independent}\ \delta q\text{'s}, \tag{2.6.9b}$$

we obtain

$$\delta q_D = \sum b_{DI}\, \delta q_I \qquad (D = 1, \ldots, m;\ I = m+1, \ldots, n), \tag{2.6.9}$$

where $b_{DI} = b_{DI}(q, t) =$ known functions of (generally, all) the q's and t. Substituting (2.6.9) into (2.6.8), we obtain, successively,

$$\delta \boldsymbol{r} = \sum \boldsymbol{e}_k\, \delta q_k = \sum \boldsymbol{e}_D\, \delta q_D + \sum \boldsymbol{e}_I\, \delta q_I = \sum \boldsymbol{e}_D \Big(\sum b_{DI}\, \delta q_I\Big) + \sum \boldsymbol{e}_I\, \delta q_I;$$

finally

$$\text{either}\quad \delta \boldsymbol{r} = \sum \boldsymbol{e}_k\, \delta q_k,\ \text{under} \sum c_{Dk}\, \delta q_k = 0 \qquad (\delta q_k,\ \text{nonarbitrary}), \tag{2.6.10a}$$

$$\text{or}\quad \delta \boldsymbol{r} = \sum \boldsymbol{\beta}_I\, \delta q_I \qquad (\delta q_I,\ \text{arbitrary}), \tag{2.6.10b}$$

where

$$\boldsymbol{\beta}_I \equiv \boldsymbol{e}_I + \sum b_{DI} \boldsymbol{e}_D \equiv \partial \boldsymbol{r}/\partial q_I + \sum b_{DI}\, (\partial \boldsymbol{r}/\partial q_D)\ \text{[see also (2.11.13a ff.)]}; \tag{2.6.10c}$$

that is, the most general particle virtual displacement under (2.6.4) can be expressed as a *linear and homogeneous combination of the "narrower" basis* $\{\boldsymbol{\beta}_I;\ I = m+1, \ldots, n\}$, whose vectors are, *in general* [and unlike the $\boldsymbol{e}_k$'s—recalling (2.5.4a ff.)], *nongradient*, or *nonholonomic*:

$$\boldsymbol{\beta}_I \neq \partial \boldsymbol{r}/\partial q_I \ \Rightarrow\ \partial \boldsymbol{\beta}_I/\partial q_{I'} \neq \partial \boldsymbol{\beta}_{I'}/\partial q_I \qquad (I, I' = m+1, \ldots, n); \tag{2.6.11}$$

as can be verified directly by using (2.6.10c) in (2.6.11).

The number of independent δq's, here $n - m \equiv f$, equals the earlier defined number of local DOFs; and, inversely, we can redefine the number of DOFs in the small, henceforth called simply DOF, as the smallest number of independent parameters $\dot{q}_I \equiv v_I/dq_I/\delta q_I$ needed to determine $\boldsymbol{v}/d\boldsymbol{r}/\delta \boldsymbol{r}$, for all system particles and any admissible, and so on, local motion; that is, the number of DOFs (in the small) = minimum number of independent "local positional", or *motional*, parameters. Just as the number of DOFs in the large, $F = n$ (here), is the minimum number of independent *positional* parameters needed to determine the configurations of all system particles in any admissible, and so on, global motion.

REMARKS

(i) The f δq_I can, in turn, be expressed as linear and homogeneous combinations of another set of f-independent motional parameters, say η_I: $\delta q_I = \sum H_{II'}(t, q)\, \eta_{I'}$ $(I, I' = m+1, \ldots, n)$; in which case (2.6.10b) becomes

$$\delta \boldsymbol{r} = \sum \boldsymbol{\beta}_I\, \delta q_I = \sum \boldsymbol{\beta}_I \left(\sum H_{II'}\eta_{I'}\right) = \sum \left(\sum H_{II'}\boldsymbol{\beta}_I\right)\eta_{I'}$$

$$\equiv \sum \boldsymbol{h}_{I'}\eta_{I'} = \sum \boldsymbol{h}_I\eta_I. \qquad \text{(2.6.10d)}$$

(ii) As already mentioned, the importance of these considerations lies in kinetics (chap. 3), where it is shown that the number of *independent kinetic equations of motion* (=equations not containing forces of constraint) equals the number of *independent δq's*.

Problem 2.6.1 Show that due to the m Pfaffian constraints (2.6.1) (expressed in terms of the notation $dq_k/dt \equiv v_k$):

$$\sum c_{Dk}v_k + c_D = 0 \qquad (D = 1, \ldots, m;\ k = 1, \ldots, n), \qquad \text{(a)}$$

or, equivalently, in the (2.6.9)-like form, in the velocities,

$$v_D = \sum b_{DI}v_I + b_D \qquad (I = m+1, \ldots, n), \qquad \text{(b)}$$

the additional holonomic constraint $\phi(t, q) = 0$ satisfies the following $(n - m) + 1$ conditions:

$$\partial\phi/\partial q_I + \sum b_{DI}(\partial\phi/\partial q_D) = 0 \quad \text{and} \quad \partial\phi/\partial t + \sum b_D(\partial\phi/\partial q_D) = 0; \qquad \text{(c)}$$

which, in terms of the notation $\phi(t, q_D, q_I) = \phi[t, q_D(t, q_I), q_I] \equiv \phi_o(t, q_I) = 0$, read simply

$$\partial\phi_o/\partial q_I = 0 \quad \text{and} \quad \partial\phi_o/\partial t = 0, \qquad \text{(d)}$$

respectively (compare with example 2.4.1.).

Before embarking into the detailed study of nonholonomic constraints and associated "coordinates" (to embed them), and the most general $\boldsymbol{v}/d\boldsymbol{r}/\delta\boldsymbol{r}$-representations in terms of $n - m$ arbitrary motional system parameters, of which the previous $v_I \equiv \dot{q}_I/dq_I/\delta q_I$ are a special case, let us pause to geometrize our analytical findings; and in the process dispel the incorrect impressions, held by many, that analytical mechanics is, somehow, only numbers (analysis), no pictures—an impression initiated, ironically, by Lagrange himself!

2.7 GEOMETRICAL INTERPRETATION OF CONSTRAINTS

Configuration Spaces

As explained in §2.2, *before* the imposition of any constraints, the configurations of a mechanical system S are described by the motion of its representative, or figurative, particle $P(S) \equiv P$ in a (clearly, nonunique) $3N$-dimensional Euclidean, or noncurved/

flat, space, E_{3N}, called *unconstrained*, or *free, configuration space*. [Briefly, *Euclidean*, or noncurved, or flat, means that, in it, the *Pythagorean theorem* ("distance squared = sum of squares of coordinate differences") *holds globally*; that is, between *any* two space points, no matter how far apart they may be; see, for example, Lur'e (1968, p. 807 ff.), Papastavridis (1999, §2.12.3).]

The position vector of P, in terms of its rectangular Cartesian coordinates/components relative to some orthonormal basis of fixed origin O, in there, is [recall (2.4.3 ff.)]

$$\boldsymbol{\xi} = [\xi_1 = \xi_1(t), \ldots, \xi_{3N} = \xi_{3N}(t)]. \tag{2.7.1}$$

However, as detailed in §2.4, upon imposition on S of h holonomic constraints and subsequent introduction of $n \equiv 3N - h$ Lagrangean coordinates $\boldsymbol{q} \equiv [q_1 = q_1(t), \ldots, q_n = q_n(t)]$, or simply $q = (q_1, \ldots, q_n)$, the above assumes the parametric representation

$$\boldsymbol{\xi} = \boldsymbol{\xi}(t, \boldsymbol{q}) = [\xi_1 = \xi_1(t; q), \ldots, \xi_{3N} = \xi_{3N}(t; q)], \tag{2.7.1a}$$

which, in geometrical terms, means that, as a result of these constraints, P can no longer roam throughout E_{3N}, but is forced to remain on its time-dependent *n-dimensional surface* defined by (2.7.1a), called *reduced*, or *constrained configuration space* of the system; actually the portion of that surface corresponding to the mathematically and physically allowable range of its curvilinear coordinates q. In differential–geometric/tensorial terms, that space, described by the *surface coordinates* q, when equipped with a physically motivated metric, becomes, at every instant t, a generally non-Euclidean (or curved, or nonflat) *metric manifold*, $M_n(t) \equiv M_n$, usually a Riemannian one, embedded in E_{3N}; and this explains the importance of Riemannian geometry to theoretical dynamics. [*Riemannian manifold* means one in which the square of the infinitesimal distance ("line element") is quadratic, homogeneous, and (usually) *positive–definite* in the coordinate differentials dq_k. In dynamics, the *manifold metric* is built from the system's *kinetic energy* (§3.9). See, for example, Lur'e (1968, pp. 810 ff.), Papastavridis (1999, §2.12, §5.6 ff.)] Schematically, we have

Configuration spaces
- *Unconstrained: E_{3N} (Euclidean manifold)*
- *Constrained: M_n (Non-Euclidean manifold)*

[N = number of particles, h = number of holonomic constraints, $n \equiv 3N - h$]

Now, as S moves in any continuous, or finite, way in the ordinary physical (three-dimensional and Euclidean) space, or some portion of it, P moves along a continuous M_n-curve, $q = q(t)$. The relevant analytical requirements on such q's (§2.4) are summarized as follows:

(i) The correspondence between the q *n-tuples* and some region of M_n must be *one-to-one and continuous* (additional holonomic constraints would exclude some parts of that region from the possible configurations).

(ii) If Δs = displacement, in M_n, corresponding to the q-increment Δq, we must have $\lim(\Delta s/\Delta q_k) \neq 0$, as $\Delta q_k \to 0$ $(k = 1, \ldots, n)$; or dq_k/ds (= "*direction cosines*" of unit tangent vector to system path in M_n) = finite. The q's are then called *regular*. (See also Langhaar, 1962, p. 16.)

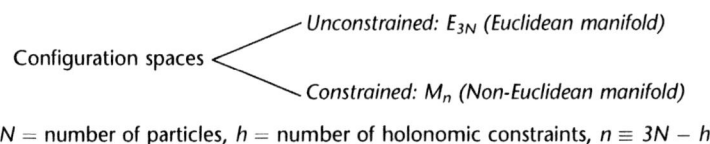

Event Spaces

Instead of the "dynamical" spaces E_{3N} and M_n, we may use their (formal and nonrelativistic) "union" with time $t \equiv q_0 \equiv q_{n+1}$; symbolically,

$$E_{3N+1} \equiv E_{3N} \times T(ime) \qquad \text{and} \qquad M_{n+1} \equiv M_n \times T(ime). \tag{2.7.2}$$

These latter are called (unconstrained and constrained, respectively): *manifolds of configuration and time* (or *of extended configuration*), or "*geometrical*" *space-time manifolds*, or *film spaces*; or, simply, *event spaces*. $M_n(M_{n+1})$ is suitable for the study of scleronomic (rheonomic) systems. (One more such "generalized space," the *phase space* of Lagrangean coordinates and momenta, is examined in chap. 8.)

Constrained Configuration Spaces and their Tangent Planes

The h *stationary* and holonomic constraints define, in E_{3N}, a *stationary* (nonmoving) and *rigid* (nondeforming) n-*dimensional* surface M_n; while h *nonstationary* holonomic constraints define, in E_{3N}, a *nonstationary* (moving) and *nonrigid* (deforming) n-*dimensional* surface $M_n(t)$. However, these same *nonstationary* constraints also define, in E_{3N+1}, a stationary and rigid $(n+1)$-*dimensional* surface M_{n+1}; hence, the relativity of these terms! The equations $t = constant$ define ∞^1 privileged surfaces $M_n(t)$ in M_{n+1}. Thus, the motion of the system can be viewed either as (i) *a stationary curve in the geometrical space* M_{n+1}; or (ii) as *the motion of the representative system point in the deformable, or* "breathing," dynamical space $M_n(t)$. Further, through each M_n-point $q(t)$ there passes a $(n-1)$-ple infinity of kinematically possible system paths, on each of which the "rate of traverse" dq/dt is arbitrary; and through each M_{n+1}-point (q, t) there passes an n-ple infinity of such paths, but these latter, since there is no motion in M_{n+1}, are not traversed. The *kinetic* paths of a system in M_n and M_{n+1} are called its *trajectories/orbits* and *world lines*, respectively. Additional M_n/M_{n+1} differences are given below, in connection with nonholonomic constraints.

Next, and as differential geometry teaches, (i) the set of all $(n+1)$-ples (dq_α) make up the *tangent point space (hyperplane)* to M_{n+1} at P, $T_{n+1}(P)$; while (ii) the vectors $\{\boldsymbol{E}_\alpha \equiv \boldsymbol{E}_\alpha(P);\ \alpha = 1, \ldots, n;\ n+1\}$, defined by $d\boldsymbol{P} \equiv d\boldsymbol{\xi} \equiv d\boldsymbol{q} \equiv \sum \boldsymbol{E}_\alpha\, dq_\alpha$: vector of elementary system displacement determined by $P(q)$ and $P(q + dq)$ (each $\boldsymbol{E}_\alpha$ being tangent to the coordinate line q_α through P) constitute a "natural" basis for the *tangent vector space associated with, or corresponding to*, $T_{n+1}(P)$; and similarly for M_n. For simplicity, we shall denote both these point and vector spaces by $T_{n+1}(P)$, $T_n(P)$.

REMARKS

(i) Without a metric, these tangent spaces are *affine*. After they become equipped with one, they become Euclidean; *properly Euclidean* if the metric is positive definite, and *pseudo-Euclidean* if the metric is indefinite. As mentioned earlier, in mechanics the metric is based on the kinetic energy, and, therefore, it is either positive definite or positive semidefinite.

(ii) It is shown in differential geometry that the condition that $d\boldsymbol{E}_\alpha = \sum(\cdots)_{\alpha\beta}\boldsymbol{E}_\beta$ be an exact differential [i.e., $\partial/\partial q_\gamma(\partial \boldsymbol{E}_\alpha/\partial q_\beta) = \partial/\partial q_\beta(\partial \boldsymbol{E}_\alpha/\partial q_\gamma)$] leads to the requirement that M_{n+1}/M_n be a Riemannian manifold. For details, see, for example, Papastavridis (1999, p. 135).

Pfaffian Constraints

Let us begin with a system subjected to h holonomic constraints (2.4.2), and, therefore, described by the $n \equiv 3N - h$ holonomic coordinates q. Then, a motion of the system in the physical space E_3 corresponds to a certain curve in M_n(*trajectory or orbit*)/M_{n+1}(*world line*) traced by the figurative system particle P; and, conversely, admissible M_n/M_{n+1} curves represent some system motion. Now, let us impose on it the additional m Pfaffian constraints:

Kinematically admissible form: $$d'\chi_D \equiv \sum c_{Dk}\, dq_k + c_D\, dt = 0, \qquad (2.7.3)$$

Virtual form: $$\delta'\chi_D \equiv \sum c_{Dk}\, \delta q_k = 0. \qquad (2.7.4)$$

As a result of the above, we have the following geometrical picture:

(i) At each admissible M_{n+1}-point $P \equiv (q, t)$, the m constraints (2.7.3) define (or order, or map, or form), the $[(n+1) - m] = [(n-m)+1]$-dimensional "element" $T_{(n+1)-m}(P) \equiv T_{(n-m)+1}(P) \equiv T_{l+1}(P)$: *tangent space (plane) of kinematically admissible displacements (motions)*, of the earlier tangent plane $T_{n+1}(P)$, on which the kinematically admissible displacements of the system, dq, and dt lie. Therefore, at every P, only world lines with velocities $v_\alpha \equiv dq_\alpha/dt$ on that plane are possible—the system can only move along directions compatible with (2.7.3).

(ii) At each such point P, the m constraints (2.7.4) define the $(n-m)$-dimensional plane $T_{n-m}(P)$: *tangent space of virtual displacements (motions)*, or *virtual plane*, on which the virtual displacements of the system, δq, lie. Clearly, $T_{n-m}(P)$ is the intersection of $T_{(n-m)+1}(P)$ with the hyperplane $dt \to \delta t = 0$ there; symbolically, $T_{n-m}(P) = T_{(n-m)+1}|_{\delta t=0} \equiv V_{n-m}(P)$ (V for virtual). {And a manifold M_n/M_{n+1} whose *tangential bundle* (i.e., totality of its tangential spaces) is restricted by the m nonholonomic equations (2.7.3) [assuming that (2.7.3), (2.7.4) are nonholonomic] is called *nonholonomic manifold* $M_{n,n-m}/M_{n+1,n-m}$. Some authors call the so-restricted bundle, $T_{(n-m)+1}$ or T_{n-m}, *nonholonomic space embedded in* M_n, or M_{n+1}. See also Maißer (1983–1984), Papastavridis (1999, chap. 6), Prange (1935, pp. 557–560), Schouten (1954, p. 196).}

(iii) The given constraint coefficients (c_{Dk}, c_D) define, at P, an $(m+1)$-dimensional *kinematically admissible constraint plane* (*element*) $C_{m+1}(P)$ perpendicular to $T_{m+1}(P)$ (with orthogonality defined in terms of the kinetic energy-based metric); while the (c_{Dk}) define an m-dimensional *virtual constraint plane* (i.e., of the virtual form of the constraints) $C_m(P)$ perpendicular to $V_{n-m}(P)$. Sometimes, $C_m(P)$ is referred to as the *orthogonal complement* of $V_{n-m}(P)$ relative to $T_n(P)$. The c_{Dk} can be viewed as the covariant (in the sense of tensor calculus) and holonomic components of the m *virtual constraint vectors* $\mathbf{c}_D = (c_{Dk})$, which, by (2.7.4), are *orthogonal to the virtual displacements* δq_k: $\mathbf{c}_D \cdot \delta\mathbf{q} = \sum c_{Dk}\, \delta q_k = 0$. Hence since the $\mathbf{c}_D$ are independent, they constitute a basis (span) for the earlier space $C_m(P)$. These two local planes are frequently called the *null*$[V_{n-m}(P)]$ and *range*$[C_n(P)]$ spaces of the $m \times n$ constraint matrix (c_{Dn}). These geometrical results are shown in fig. 2.14 (see also fig. 3.1).

Let us consolidate our findings:

(i) Under n *initial* holonomic constraints, a system can go from any admissible *initial* M_n/M_{n+1}-point, P_i, to any other *final* such point, P_f, along any chosen (M_n/M_{n+1})-lying path joining P_i and P_f.

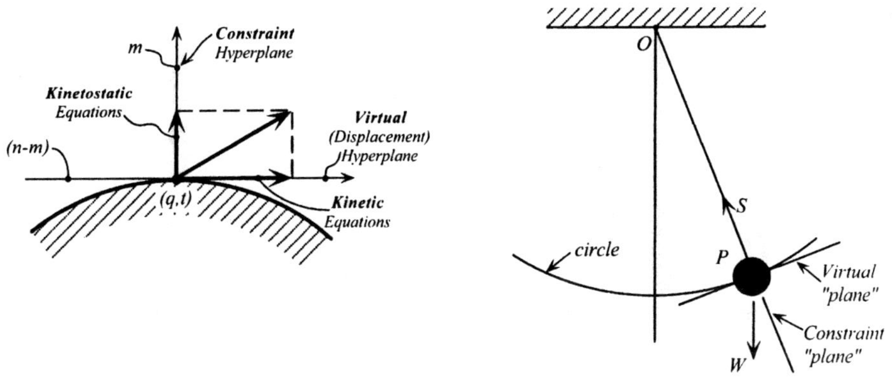

Example of mathematical pendulum

Figure 2.14 Virtual displacement (V_{n-m}) and constraint (C_m) hyperplanes in configuration space (see also fig. 3.1).

(ii) If the additional m Pfaffian constraints (2.7.3, 4) are holonomic, disguised in kinematical form, the local tangent planes become the earlier local tangent planes to reduced, or "smaller," configuration/event manifolds $M_{n-m}/M_{(n+1)-m}$, inside M_n/M_{n+1}. These reduced but finite surfaces contain all possible system motions through a given P_i—the system can go from any admissible *initial* $M_{n-m}/M_{(n+1)-m}$-point, P_i, to any other *final* such point, P_f, along any chosen (M_n/M_{n+1})-*lying path joining* P_i *and* P_f; that is, $DOF(local) = DOF(global) = n - m$.

(iii) On the other hand, if the additional m Pfaffian constraints are nonholonomic, we cannot construct these $M_{n-m}/M_{(n+1)-m}$. The global configuration/event manifolds of the system are still M_n/M_{n+1}, but these constraints have created, in there, a certain *path-dependence*: any (M_n/M_{n+1})-point P_f (in the admissible portions of M_n/M_{n+1}) is, again, accessible from any other (M_n/M_{n+1})-point P_i but only along a certain *kinematical family*, or "*network*," of tracks that is "narrower" than that of case (i); that is, the transition $P_i \to P_f$ is no longer arbitrary because of direction-of-motion constraints, at every point of those paths. Or, under such constraints, all configurations/events are still possible, *but not all velocities* (and, hence, *not all paths*); only certain M_n/M_{n+1}-curves correspond to physically realizable motions—the system is restricted locally, not globally; that is, $n = DOF(global) \neq DOF(local) = n - m$. We continue this geometrical interpretation of constrained systems in §2.11.

Kinetic Preview, Quasi Coordinates

The importance of these considerations, and especially of the concept of virtualness, to contrained system mechanics arises from the fact that most of the constraint forces dealt by AM (the so-called "passive," or contact, ones; i.e., those satisfying the d'Alembert–Lagrange principle, chap. 3) are perpendicular to the virtual displacement plane V_{n-m}, and so lie on the *virtual constraint plane* C_m. And this, as detailed in chapter 3, allows us to bring the system equations of motion into their *simplest* form; that is (i) to their smallest possible, or minimal, number (n), and (ii) to a complete *decoupling* of them into $(n-m)$ purely *kinetic* equations—that is, equations not containing constraint forces—by projecting them onto the *local virtual hyperplane*, and (m) *kinetostatic* equations—equations containing constraint forces—by projecting them onto the *local constraint hyperplane*, which is perpendicular to the virtual hyperplane there. This is the raison d'être of virtualness, and the essence of Lagrangean analytical mechanics. In all cases, under given initial/boundary conditions and forces, the system will follow a unique path (a trajectory, or orbit) determined, or singled out among the problem's kinematically admissible paths, by solving the full set of its kinetic and kinematic equations.

Schematically, our strategic plan is as the following:

Virtual displacements ⇒ Simplest form of eqs. of motion
- → Minimal number of eqs. of motion (n)
- → Uncoupling of eqs. of motion:
  - → *Kinetic* $(n - m)$
  - → *Kinetostatic* (m)

Now, if the m Pfaffian constraints are holonomic, their uncoupling (and that of the corresponding equations of motion) is easily achieved by "adaptation to the constraints," as explained in §2.4 and §2.6; but, if they are nonholonomic this "adaptation" can be achieved only *locally*, via "equilibrium" *nonholonomic coordinates*, or *quasi coordinates*.

We begin the study of these fundamental kinematical concepts by first examining one of their important features: the possible *commutativity/noncommutativity* of the *virtual and possible operations*, $\delta(\ldots)$ *and* $d(\ldots)$, respectively, when applied to this new breed of "coordinates"; that is, we investigate the relation between $d[\delta(\textit{quasi coordinate})]$ and $\delta[d(\textit{quasi coordinate})]$.

2.8 NONCOMMUTATIVITY VERSUS NONHOLONOMICITY; INTRODUCTION TO THE THEOREM OF FROBENIUS

Let us recall the admissible (d) and virtual (δ) forms of the Pfaffian constraints (2.7.3, 4):

$$d\chi_D \equiv \sum c_{Dk}\, dq_k + c_D\, dt = 0 \qquad \text{and} \qquad \delta\chi_D \equiv \sum c_{Dk}\, \delta q_k = 0, \tag{2.8.1}$$

where $D = 1, \ldots, m$; $I = m + 1, \ldots, n$; k (*and all other small Latin indices*) $= 1, \ldots, n$. Now, $\delta(\ldots)$-varying the first of (2.8.1), and $d(\ldots)$-varying the second, and then subtracting them side by side, we find, after some straightforward differentiations and dummy index changes,

$$\begin{aligned} d(\delta\chi_D) - \delta(d\chi_D) = &\sum c_{Dk}[d(\delta q_k) - \delta(dq_k)] \\ &+ \sum \Big(\sum (\partial c_{Dk}/\partial q_l - \partial c_{Dl}/\partial q_k)\, dq_l \\ &+ (\partial c_{Dk}/\partial t - \partial c_D/\partial q_k)\, dt \Big)\, \delta q_k \\ &+ c_D[d(\delta t) - \delta(dt)], \end{aligned} \tag{2.8.1a}$$

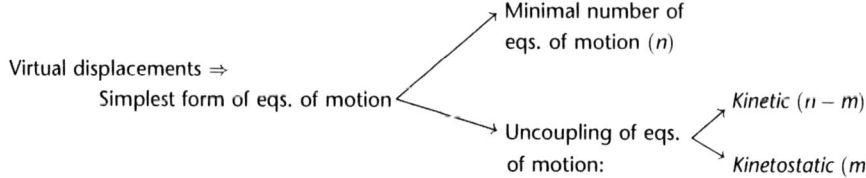

or, since the last term is zero [$\delta t = 0 \Rightarrow d(\delta t) = 0$, and, during $\delta(\ldots)$ time is kept constant $\Rightarrow \delta(dt) = 0$], and with the earlier notations $q_0 \equiv q_{n+1} \equiv t \Rightarrow \delta q_0 = \delta q_{n+1} = \delta t = 0, c_D \equiv c_{D0} \equiv c_{D,n+1}$, and *Greek subscripts* running from 1 to $n+1$ (or from 0 to n):

$$d(\delta\chi_D) - \delta(d\chi_D) = \sum c_{Dk}[d(\delta q_k) - \delta(dq_k)] + \sum\sum (\partial c_{Dk}/\partial q_\alpha - \partial c_{D\alpha}/\partial q_k)\, dq_\alpha\, \delta q_k. \tag{2.8.2}$$

A final simplification occurs with the useful notations $d(\delta\ldots) - \delta(d\ldots) \equiv D(\ldots)$, and

$$C^D{}_{\beta\alpha} \equiv \partial c_{D\beta}/\partial q_\alpha - \partial c_{D\alpha}/\partial q_\beta = -C^D{}_{\alpha\beta}, \tag{2.8.2a}$$

$F_D \equiv \sum\sum C^D{}_{k\alpha}\, dq_\alpha\, \delta q_k$: Frobenius' *bilinear*, or *antisymmetric*, *covariant* of the Pfaffian forms (2.8.1). (2.8.2b)

Thus, (2.8.2) transforms to

$$D\chi_D = \sum c_{Dk}\, Dq_k + F_D. \tag{2.8.2c}$$

Problem 2.8.1 Starting with eqs. (2.5.12a,b):

$$d\mathbf{r} = \sum \mathbf{e}_k\, dq_k + \mathbf{e}_0\, dt, \qquad \delta\mathbf{r} = \sum \mathbf{e}_k\, \delta q_k, \tag{a}$$

and repeating the above process, show that

$$D\mathbf{r} = \sum Dq_k \mathbf{e}_k. \tag{b}$$

From the above basic kinematical identities, we draw the following conclusions:

(i) If $C^D{}_{k\alpha} = 0$, identically in the q's and t, and for all values of D, k, α then, since

$$Dq_k \equiv d(\delta q_k) - \delta(dq_k) = 0, \tag{2.8.3a}$$

the q_k being genuine = *holonomic* coordinates, it follows that

$$D\chi_D \equiv d(\delta\chi_D) - \delta(d\chi_D) = 0; \tag{2.8.3b}$$

that is, the χ_D are also holonomic coordinates, the $d\chi_D/\delta\chi_D$ are exact differentials. In this case (2.8.1) may be replaced by m holonomic constraints; which, in turn, may be embedded into the system via $n' = n - m$ new equilibrium coordinates, as explained in §2.4.

(ii) If $F_D \neq 0$, then $D\chi_D \neq 0$; or, more generally, we *cannot* assume that *both* $d(\delta q_k) = \delta(dq_k)$ *and* $d(\delta\chi_D) = \delta(d\chi_D)$ hold; it is *either the one or the other*. (As detailed in chap. 7, this realization helps one understand the fundamental differences that exist between variational *mathematics* and variational *mechanics*. See also pr. 2.12.5.) If we assume (2.8.3a) for all holonomic coordinates, *constrained or not*, then $D\chi_D \neq 0$; that is, the χ_D are nonholonomic coordinates; and, as Frobenius' theorem shows (see below), the constraints (2.8.1) are nonholonomic.

(iii) If, however, $F_D = 0$, since the $dq/\delta q$ are not independent, it does *not* necessarily follow that $C^D{}_{k\alpha} = 0$. To make further progress—that is, to establish *necessary and sufficient holonomicity/nonholonomicity conditions* in terms of the constraint coefficients, c_{Dk} and c_D, we need nontrivial help from differential equations/differential geometry; and this leads us directly to the following fundamental theorem of Frobenius (1877). First, let us formulate it in simple and general mathematical terms, and then we will tailor it to our kinematical context.

Theorem of Frobenius

The necessary and sufficient condition for the *complete* (or *unrestricted*) *integrability* ≡ *holonomicity* of the Pfaffian system:

$$X_D \equiv \sum X_{DK}\, dx_K = 0 \qquad [D = 1, \ldots, m(<F);\ K, L = 1, \ldots, F], \tag{2.8.4}$$

where $X_{DK} = X_{DK}(x_1, \ldots, x_F) \equiv X_{DK}(x) =$ given and well-behaved functions of their arguments, and $\operatorname{rank}(X_{DK}) = m(<F)$; that is, for it to have m independent integrals $f_D(x) = C_D =$ constants, is the vanishing of the corresponding m bilinear forms:

$$F_D \equiv \sum\sum (\partial X_{DK}/\partial x_L - \partial X_{DL}/\partial x_K) u_K v_L, \tag{2.8.5}$$

identically (in the x's) and simultaneously (for all D's), for any/all solutions $u = (u_1, \ldots, u_F)$, and $v = (v_1, \ldots, v_F)$ of the m constraints $\sum X_{DK}\eta_K = 0$; that is, for any/all $\eta_K \to u_K, v_K$ satisfying

$$\sum X_{DK} u_K = 0 \qquad \text{and} \qquad \sum X_{DK} v_K \equiv \sum X_{DL} v_L = 0. \tag{2.8.6}$$

[Also, recall comments following eqs. (2.3.11e).]

[If the system (2.8.4) is completely integrable, then, since its finite form depends on the integration constants C_D (i.e., ultimately, on the *initial values* of the x's), then it is *semiholonomic* (§2.3).]

Adapted to our kinematical problem—that is, with the identifications $F \to n + 1$, $u_K \to \delta q_k$, $v_L \to dq_\alpha$, $x \to t$, q, and recalling that $q_{n+1} \equiv t$ satisfies the additional holonomic constraint $\delta q_{n+1} \equiv \delta t = 0$—Frobenius theorem states that: If

$$\begin{aligned} F_D &\equiv d(\delta x_D) - \delta(d x_D) = d\Big(\sum c_{Dk}\,\delta q_k\Big) - \delta\Big(\sum c_{D\alpha}\, dq_\alpha\Big) \\ &= \sum\Big(\sum (\partial c_{Dk}/\partial q_\alpha - \partial c_{D\alpha}/\partial q_k)\, dq_\alpha\Big)\,\delta q_k \equiv \sum\sum C^D{}_{k\alpha}\, dq_\alpha\,\delta q_k = 0, \end{aligned} \tag{2.8.7}$$

for arbitrary $dq_\alpha = dq_k$, $dq_{n+1} \equiv dq_0 \equiv dt$ and δq_k, solutions of the constraints:

$$\sum c_{D\alpha}\, dq_\alpha = \sum c_{Dk}\, dq_k + c_D\, dt = 0 \qquad \text{and} \qquad \sum c_{Dk}\,\delta q_k = 0, \tag{2.8.1}$$

then these constraints are holonomic.

The above show that since our dq's and δq's are not independent, the vanishing of the F_D's does *not* necessarily lead to

$$C^D{}_{k\alpha} = 0, \tag{2.8.8}$$

as holonomicity conditions. For this to be the case, eqs. (2.8.8) are, clearly, *sufficient but not necessary*; they would be if the dq's and δq's were independent; namely, unconstrained.

This observation leads to the following implementation of Frobenius' theorem: we express each of the (n) nonindependent dq's and δq's as a linear and homogeneous combination of a new set of $n - m$ *independent* parameters (and dt, for the dq's), insert these representations in $F_D = 0$, and then, in each of the so resulting m bilinear covariants (in these new parameters), set its $n - m$ coefficients equal to zero. We shall see in §2.12 that, in the general case, this approach leads to a direct and usable form of Frobenius' theorem, due to Hamel. But before proceeding in that direction, we need to examine in sufficient detail the necessary tools: *nonholonomic coordinates*, or *quasi coordinates* (§2.9), and the associated *transitivity relations* (§2.10).

REFERENCES ON PFAFFIAN SYSTEMS AND FROBENIUS' THEOREM

(for proofs, and so on, in *decreasing* order of readability for nonmathematicians):

Klein (1926(a), pp. 207–214): introductory, quite insightful.

De la Vallée Poussin (1912, vol. 2, chap. 7): most readable classical exposition.

Guldberg (1927, pp. 573–576) and Pascal (1927, pp. 579–588): outstanding handbook summaries.

Forsyth (1890/1959, especially chaps. 2 and 11): detailed classical treatment.

Lovelock and Rund (1975/1989, chap. 5): excellent balance between classical and modern approaches.

Cartan (1922, chaps. 4–10): the foundation of modern treatments.

Weber [1900(a), (b)]: older encyclopedic treatise (a) and article (b, pp. 317–319).

Heil and Kitzka (1984, pp. 264–295): relatively readable modern summary.

Chetaev (1987/1989, pp. 319–326): happens to be in English (not particularly enlightening).

Frobenius (1877, pp. 267–287; also, in his *Collected Works*, pp. 249–334): the original exposition; not for beginners.

Hartman (1964, chap. 6): quite advanced; for ordinary differential equations specialists.

Outside of Lovelock et al., we are unaware of any contemporary *readable* exposition of these topics in English; i.e., without Cartanian exterior forms, and so on.

Example 2.8.1 *Necessary and Sufficient Condition(s) for the Holonomicity of the Single Pfaffian Constraint (2.3.1) via Frobenius' Theorem:*

$$d\chi \equiv a(x, y, z)\,dx + b(x, y, z)\,dy + c(x, y, z)\,dz \equiv a\,dx + b\,dy + c\,dz = 0, \tag{a}$$

or, since it is *catastatic*,

$$\delta\chi \equiv a\,\delta x + b\,\delta y + c\,\delta z = 0. \tag{b}$$

By d-varying (b), and δ-varying (a), and then subtracting side by side, we find, after some straightforward differentiations:

$$\begin{aligned} d(\delta\chi) - \delta(d\chi) &= a[d(\delta x) - \delta(dx)] + b[d(\delta y) - \delta(dy)] + c[d(\delta z) - \delta(dz)] \\ &\quad + [(da\,\delta x - \delta a\,dx) + (db\,\delta y - \delta b\,dy) + (dc\,\delta z - \delta c\,dz)] \\ &= (\partial a/\partial y - \partial b/\partial x)(dy\,\delta x - \delta y\,dx) + (\partial a/\partial z - \partial c/\partial x)(dz\,\delta x - \delta z\,dx) \\ &\quad + (\partial b/\partial z - \partial c/\partial y)(dz\,\delta y - \delta z\,dy) \end{aligned}$$

[substituting into this, $dz = (-a/c)\,dx + (-b/c)\,dy$ and $\delta z = (-a/c)\,\delta x + (-b/c)\,\delta y$, solutions of the constraints (a, b), respectively; since here $n - m = 3 - 1 = 2 =$ number of *independent differentials* (for each form of the constraint); we could, just as well, substitute $dx = \cdots dy + \cdots dz$ and $\delta x = \cdots \delta y + \cdots \delta z$, or $dy = \cdots$ and $\delta y = \cdots$]

$$\begin{aligned} &= \cdots = (\partial a/\partial y - \partial b/\partial x)(dy\,\delta x - \delta y\,dx) \\ &\qquad + (\partial a/\partial z - \partial c/\partial x)(-b/c)(dy\,\delta x - \delta y\,dx) \\ &\qquad + (\partial b/\partial z - \partial c/\partial y)(-a/c)(dx\,\delta y - \delta x\,dy) \\ &= [(\partial a/\partial y - \partial b/\partial x) + (b/c)(\partial c/\partial x - \partial a/\partial z) \\ &\qquad + (a/c)(\partial b/\partial z - \partial c/\partial y)](dy\,\delta x - \delta y\,dx). \end{aligned}$$

Setting $d(\delta\chi) - \delta(d\chi) = 0$, and since now the bilinear terms $dy\,\delta x$ and $\delta y\,dx$ are independent, we recover the earlier holonomicity condition (2.3.6).

Vectorial Considerations

Equations (a)/(b), in terms of the vector notation

$$\boldsymbol{h} = (a, b, c), \qquad d\boldsymbol{r} = (dx, dy, dz), \qquad \text{and} \qquad \delta\boldsymbol{r} = (\delta x, \delta y, \delta z), \tag{d}$$

state that

$$\boldsymbol{h} \cdot d\boldsymbol{r} = 0 \qquad \text{and} \qquad \boldsymbol{h} \cdot \delta\boldsymbol{r} = 0; \tag{e}$$

that is, $\boldsymbol{h}$ is *perpendicular* to the plane defined by the two (generally independent) directions $d\boldsymbol{r}$ and $\delta\boldsymbol{r}$, through $\boldsymbol{r} = (x, y, z)$. On the other hand, the second of (c) states that

$$d(\delta\chi) - \delta(d\chi) = \boldsymbol{curl}\ \boldsymbol{h} \cdot (d\boldsymbol{r} \times \delta\boldsymbol{r}) = 0, \tag{f}$$

that is, $\boldsymbol{curl}\ \boldsymbol{h}$ is perpendicular to the normal to that plane; and, hence, excluding the trivial case $d\boldsymbol{r} \times \delta\boldsymbol{r} = \boldsymbol{0}$, $\boldsymbol{curl}\ \boldsymbol{h}$ lies on that plane. Accordingly, $\boldsymbol{h}$ and $\boldsymbol{curl}\ \boldsymbol{h}$ are *perpendicular* to each other:

$$\boldsymbol{h} \cdot \boldsymbol{curl}\ \boldsymbol{h} = 0. \qquad \text{i.e., (2.3.8a).} \tag{g}$$

Example 2.8.2 *The Two Independent and Catastatic Pfaffian Constraints:*

$$d\chi \equiv a(x, y, z)\,dx + b(x, y, z)\,dy + c(x, y, z)\,dz \equiv a\,dx + b\,dy + c\,dz = 0, \tag{a}$$

$$dX \equiv A(x, y, z)\,dx + B(x, y, z)\,dy + C(x, y, z)\,dz \equiv A\,dx + B\,dy + C\,dz = 0, \tag{b}$$

when taken *together* (i.e., $n = 3$, $m = 2$) will always make up a *holonomic system*; even if each one of them separately (i.e., $n = 3$, $m = 1$) may be nonholonomic!

Solving (a) and (b) for any two of the dx, dy, dz in terms of the third, say dx and dy in terms of dz, we obtain

$$dx \equiv e(x, y, z)\, dz \qquad \text{and} \qquad dy \equiv f(x, y, z)\, dz; \tag{c}$$

and, similarly, since (a) and (b) are catastatic,

$$\delta x \equiv e(x, y, z)\, \delta z \qquad \text{and} \qquad \delta y \equiv f(x, y, z)\, \delta z. \tag{d}$$

Therefore, we find, successively,

$$\begin{aligned} d(\delta\chi) - \delta(d\chi) &= \\ &= \cdots = (\cdots)(dy\,\delta x - \delta y\, dx) + (\cdots)(dz\,\delta x - \delta z\, dx) + (\cdots)(dz\,\delta y - \delta z\, dy) \\ &= [\text{using (c) and (d)}] = \cdots = (\cdots)(dz\,\delta z - \delta z\, dz) = (\cdots)0 = 0; \end{aligned} \tag{e}$$

and, similarly,

$$d(\delta X) - \delta(dX) = \cdots = (\cdots)(dz\,\delta z - \delta z\, dz) = (\cdots)0 = 0; \quad \text{Q.E.D.} \tag{f}$$

Proceeding in a similar fashion, we can show that: *a system of $n - 1$ (or n) independent Pfaffian equations, in n variables [like* (2.8.1) *with $m = n - 1$ or n] is always holonomic.* This theorem illustrates the interesting kinematical fact that *additional constraints* may turn an originally (individually) nonholonomic constraint into a holonomic one (as part of a system of constraints); see also §2.12.

2.9 QUASI COORDINATES, AND THEIR CALCULUS

Let us, again, consider a holonomic system S described by the hitherto minimal, or independent, n Lagrangean coordinates $q = (q_1, \ldots, q_n)$, and hence having *kinematically admissible/possible* system displacements $(dq, dt) \equiv (dq_1, \ldots, dq_n; dt)$. Now, at a generic admissible point of S's configuration or event space (q, t), we can describe these local displacements via a new set of general differential positional and time parameters $(d\theta, dt) \equiv (d\theta_1, \ldots, d\theta_n; d\theta_{n+1} \equiv d\theta_0)$, defined by the $n + 1$ linear, homogeneous, and invertible transformations:

$$d\theta_k \equiv \sum a_{kl}\, dq_l + a_k\, dt, \quad d\theta_{n+1} \equiv d\theta_0 \equiv dq_{n+1} \equiv dq_0 \equiv dt, \tag{2.9.1}$$

$$rank(a_{kl}) = n \;\Rightarrow\; Det(a_{kl}) \neq 0, \qquad (k, l = 1, \ldots, n), \tag{2.9.1a}$$

where the coefficients a_{kl} and $a_k \equiv a_{k,n+1} \equiv a_{k0}$ are given functions of the q's and t (and as well-behaved as needed; say, continuous and once piecewise continuously differentiable, in some region of interest of their variables). Inverting (2.9.1), we obtain

$$dq_l = \sum A_{lk}\, d\theta_k + A_l\, dt, \qquad dq_{n+1} \equiv d\theta_{n+1} \equiv d\theta_0 \equiv dt, \tag{2.9.2}$$

$$rank(A_{lk}) = n \;\Rightarrow\; Det(A_{lk}) \neq 0, \qquad (k, l = 1, \ldots, n), \tag{2.9.2a}$$

where the "inverted coefficients" A_{lk} and $A_l \equiv A_{l,n+1} \equiv A_{l0}$ become known functions of the q's and t, and are also well-behaved. Clearly, since the transformations (2.9.1) and (2.9.2) are mutually inverse, their coefficients must satisfy certain consistency, or compatibility, conditions; so that, given the a's, one can determine the A's and vice versa. Indeed, substituting dq_l from (2.9.2) into (2.9.1), and $d\theta_k$ from (2.9.1) into (2.9.2), and with δ_{kl} = *Kronecker delta* (= 1 or 0, according as $k = l$, or $k \neq l$), we obtain the inverseness relations:

$$\sum a_{kr}A_{rl} \equiv \sum A_{rl}a_{kr} = \delta_{kl}, \qquad \sum a_{kr}A_r \equiv \sum A_r a_{kr} = -a_k, \tag{2.9.3a}$$

$$\sum A_{lr}a_{rk} \equiv \sum a_{rk}A_{lr} = \delta_{kl}, \qquad \sum A_{lr}a_r \equiv \sum a_r A_{lr} = -A_l. \tag{2.9.3b}$$

Further, with the help of the unifying notations $a_k \equiv a_{k,n+1}$ and $A_l \equiv A_{l,n+1}$, the definitions $a_{n+1,k} \equiv \delta_{n+1,k}$ (= 0) and $A_{n+1,l} \equiv \delta_{n+1,l}$ (= 0), and recalling that Greek subscripts have been agreed to run from 1 to $n+1$, the transformation coefficient matrices in (2.9.1) and (2.9.2) take the $(n+1) \times (n+1)$ "**Spatio-Temporal**" forms:

$$\mathbf{a} = \left(\begin{array}{ccc|c} a_{11} & \cdots & a_{1n} & a_{1,n+1} \\ \cdots & \cdots & \cdots & \cdots \\ \cdots & \cdots & \cdots & \cdots \\ a_{n1} & \cdots & a_{nn} & a_{n,n+1} \\ \hline 0 & \cdots & 0 & 1 \end{array}\right) \equiv \left(\begin{array}{c|c} a_{kl} & a_k \\ \hline 0 & 1 \end{array}\right) \equiv \left(\begin{array}{c|c} \mathbf{a}_\mathrm{S} & \mathbf{a}_\mathrm{T} \\ \hline \mathbf{0} & \mathbf{1} \end{array}\right) \equiv (a_{\beta\gamma}), \tag{2.9.4a}$$

$$\mathbf{A} = \left(\begin{array}{ccc|c} A_{11} & \cdots & A_{1n} & A_{1,n+1} \\ \cdots & \cdots & \cdots & \cdots \\ \cdots & \cdots & \cdots & \cdots \\ A_{n1} & \cdots & A_{nn} & A_{n,n+1} \\ \hline 0 & \cdots & 0 & 1 \end{array}\right) \equiv \left(\begin{array}{c|c} A_{kl} & A_k \\ \hline 0 & 1 \end{array}\right) \equiv \left(\begin{array}{c|c} \mathbf{A}_\mathrm{S} & \mathbf{A}_\mathrm{T} \\ \hline \mathbf{0} & \mathbf{1} \end{array}\right) \equiv (A_{\beta\gamma}), \tag{2.9.4b}$$

Then (2.9.1, 2) assume the simpler (homogeneous) forms:

$$d\theta_\gamma = \sum a_{\gamma\beta}\, dq_\beta \Leftrightarrow dq_\beta = \sum A_{\beta\gamma}\, d\theta_\gamma, \tag{2.9.5}$$

while the consistency relations (2.9.3a) read simply

$$\mathbf{a}\,\mathbf{A} = \mathbf{1} \quad \text{or} \quad \sum a_{\beta\delta}\, A_{\delta\gamma} \equiv \sum A_{\delta\gamma}\, a_{\beta\delta} = \delta_{\beta\gamma}; \tag{2.9.6a}$$

and from this we obtain the "spatio-temporally partitioned" matrix multiplications:

$$\left(\begin{array}{c|c} \mathbf{a}_\mathrm{S} & \mathbf{a}_\mathrm{T} \\ \hline \mathbf{0} & \mathbf{1} \end{array}\right)\left(\begin{array}{c|c} \mathbf{A}_\mathrm{S} & \mathbf{A}_\mathrm{T} \\ \hline \mathbf{0} & \mathbf{1} \end{array}\right) = \left(\begin{array}{c|c} \mathbf{a}_\mathrm{S}\mathbf{A}_\mathrm{S} & \mathbf{a}_\mathrm{S}\mathbf{A}_\mathrm{T} + \mathbf{a}_\mathrm{T} \\ \hline \mathbf{0} & \mathbf{1} \end{array}\right) = \left(\begin{array}{c|c} \mathbf{1} & \mathbf{0} \\ \hline \mathbf{0} & \mathbf{1} \end{array}\right)$$

that is,

$$\mathbf{a}_\mathrm{S}\mathbf{A}_\mathrm{S} = \mathbf{1} \quad \text{and} \quad \mathbf{a}_\mathrm{S}\mathbf{A}_\mathrm{T} + \mathbf{a}_\mathrm{T} = \mathbf{0}; \tag{2.9.7a}$$

and, similarly, the consistency relations (2.9.3b) read

$$(\mathbf{a}\,\mathbf{A})^{\mathrm{T}} = \mathbf{A}^{\mathrm{T}}\mathbf{a}^{\mathrm{T}} = \mathbf{1} \quad \text{or} \quad \sum A_{\gamma\delta}\, a_{\delta\beta} = \sum a_{\delta\beta}\, A_{\gamma\delta} = \delta_{\beta\gamma}, \tag{2.9.6b}$$

from which

$$\left(\begin{array}{c|c} \mathbf{A}_{\mathrm{S}} & \mathbf{0} \\ \hline \mathbf{A}_{\mathrm{T}} & 1 \end{array}\right)\left(\begin{array}{c|c} \mathbf{a}_{\mathrm{S}} & \mathbf{0} \\ \hline \mathbf{a}_{\mathrm{T}} & 1 \end{array}\right) = \left(\begin{array}{c|c} \mathbf{A}_{\mathrm{S}}\mathbf{a}_{\mathrm{S}} & \mathbf{0} \\ \hline \mathbf{A}_{\mathrm{T}}\mathbf{a}_{\mathrm{S}} + \mathbf{a}_{\mathrm{T}} & 1 \end{array}\right) = \left(\begin{array}{c|c} \mathbf{1} & \mathbf{0} \\ \hline \mathbf{0} & 1 \end{array}\right)$$

that is,

$$\mathbf{A}_{\mathrm{S}}\mathbf{a}_{\mathrm{S}} = \mathbf{1} \quad \text{and} \quad \mathbf{A}_{\mathrm{T}}\mathbf{a}_{\mathrm{S}} + \mathbf{a}_{\mathrm{T}} = \mathbf{0}. \tag{2.9.7b}$$

Let us recapitulate the *notations* used here:

(i) Matrices are shown in roman and bold; vectors in italic and bold;
(ii) $(\ldots)^{\mathrm{T}} \equiv$ *transpose* of square matrix $(\ldots)$;
(iii) $\mathbf{a}_{\mathrm{S}}, \mathbf{A}_{\mathrm{S}} = (n \times n)$ *spatial*, or *catastatic*, submatrices of $\mathbf{a}$ and $\mathbf{A}$, respectively; and $\mathbf{a}_{\mathrm{T}}, \mathbf{A}_{\mathrm{T}} = (n \times 1)$ *temporal*, or *acatastatic*, submatrices of $\mathbf{a}$ and $\mathbf{A}$, respectively;
(iv) $\mathbf{1} =$ square *unit*, or *identity*, matrix (of appropriate dimensions);
(v) $\mathbf{0} =$ *zero* matrix (column or row vector of appropriate dimension); and
(vi) Here, *commas in subscripts*—for example, $a_{k,n+1}, A_l \equiv A_{l,n+1}$—are used only to separate the spatial from the temporary of these subscripts, for better visualization; that is, no partial differentiations are implied, unless explicitly specified to that effect.

Thus, for example, for $\beta \to k$ and $\gamma \to l$ eqs. (2.9.6a) yield

$$\sum a_{kr}A_{rl} + a_{k,n+1}A_{n+1,l} = \delta_{kl} \to \sum a_{kr}A_{rl} = \delta_{kl}, \qquad \text{i.e., first of (2.9.3a);}$$

for $\beta \to n+1$ and $\gamma \to l$ they yield

$$\sum a_{n+1,r}\, A_{rl} + a_{n+1,n+1}\, A_{n+1,l} = \delta_{n+1,l}, \qquad \text{i.e., } 0 + 0 = 0;$$

while for $\beta \to k$ and $\gamma \to n+1$ they yield

$$\sum a_{kr}\, A_{r,n+1} + a_{k,n+1}\, A_{n+1,n+1} = \delta_{k,n+1} = 0, \qquad \text{i.e., second of (2.9.3a);}$$

and similarly with (2.9.6b).

Specializations, Remarks

(i) If (a_{kl}) is an *orthogonal* matrix—that is, if

$$a_{kl} = A_{lk} \quad \text{and} \quad \mathrm{Det}(a_{kl}) = \pm 1, \tag{2.9.8a}$$

then the spatial parts of (2.9.3a, 3b) are replaced, respectively, by

$$\sum a_{kr}\, a_{lr} \equiv \sum a_{lr}\, a_{kr} = \delta_{kl} \quad \text{and} \quad \sum a_{rl}\, a_{rk} \equiv \sum a_{rk}\, a_{rl} = \delta_{kl}; \tag{2.9.8b}$$

and, similarly, for the full $(n+1) \times (n+1)$ $\mathbf{a}$ and $\mathbf{A}$ matrices.

(ii) As shown in chap. 3, and foreshadowed below, it is the *spatial/catastatic submatrices* $\mathbf{a}_{\mathrm{S}}$ and $\mathbf{A}_{\mathrm{S}}$ that enter the equations of motion; not the *temporary/acatastatic submatrices* $\mathbf{a}_{\mathrm{T}}$ and $\mathbf{A}_{\mathrm{T}}$. The latter, however, enter the rate of energy, or power, equations (§3.9). In what follows, we shall have the opportunity to use all these, mutually equivalent and complementary notations; primarily the *indicial* and secondarily the matrix ones. All have relative advantages/drawbacks, depending on the task at hand.

Velocities and Virtual Displacements

Just as we defined new general kinematically admissible/possible system displacements via (2.9.1, 2), etc., we next define the following:

(i) The corresponding general system *velocities* ($d\theta \to \omega\, dt$);

$$\omega_k \equiv \sum a_{kl}(dq_l/dt) + a_k \equiv \sum a_{kl}\dot{q}_l + a_k \equiv \sum a_{kl}v_l + a_k,$$

$$\omega_{n+1} \equiv \omega_0 \equiv dq_{n+1}/dt \equiv dt/dt = 1 \qquad (\textit{isochrony}), \tag{2.9.9}$$

or, compactly,

$$\omega_\beta \equiv \sum a_{\beta\gamma}(dq_\gamma/dt) \equiv \sum a_{\beta\gamma}\, v_\gamma; \tag{2.9.9a}$$

and, inversely,

$$dq_l/dt \equiv \dot{q}_l \equiv v_l = \sum A_{lk}\,\omega_k + A_l, \qquad dq_{n+1}/dt \equiv \omega_{n+1} \equiv dt/dt = 1, \tag{2.9.10}$$

or, compactly,

$$dq_\gamma/dt \equiv \dot{q}_\gamma \equiv v_\gamma = \sum A_{\gamma\beta}\,\omega_\beta; \tag{2.9.10a}$$

and

(ii) The corresponding general system *virtual displacements* ($d\theta \to \delta\theta$, $d\theta_{n+1} \to \delta\theta_{n+1} = \delta t = 0$):

$$\delta\theta_k \equiv \sum a_{kl}\,\delta q_l, \qquad \delta\theta_{n+1} \equiv \delta q_{n+1} \equiv \delta t = 0; \tag{2.9.11}$$

and, inversely,

$$\delta q_l = \sum A_{lk}\,\delta\theta_k, \qquad \delta q_{n+1} \equiv \delta q_0 \equiv \delta\theta_{n+1} \equiv \delta\theta_0 \equiv \delta t = 0. \tag{2.9.12}$$

If the $d\theta$ and dt describe an *actual* motion, then $d\theta_k = \omega_k\, dt$. But it would be incorrect to set $\delta\theta_k = \omega_k\, \delta t$, because of the ever present (better, ever assumed) *virtual time constraint* $\delta t = 0$; whereas, in general, $\delta\theta_k \neq 0$!

Next, let us examine the integrability of these Pfaffian *forms* (not constraints!) (2.9.1, 11), of our hitherto n *DOF* system.

Bilinear Covariants, Integrability, Quasi Coordinates

Indeed, proceeding as in §2.8, and assuming that $d(\delta q_k) = \delta(dq_k)$, constraints or not, we find that the *Frobenius bilinear covariants* of (2.9.1, 11), $d(\delta\theta_k) - \delta(d\theta_k)$, equal

$$\begin{aligned} d(\delta\theta_k) - \delta(d\theta_k) &= \sum\sum (\partial a_{kl}/\partial q_s - \partial a_{ks}/\partial q_l)\, dq_s \delta q_l \\ &\quad + \sum (\partial a_{kl}/\partial t - \partial a_k/\partial q_l)\, dt\, \delta q_l \\ &\equiv \sum d\psi_{kl}\, \delta q_l. \end{aligned} \tag{2.9.13}$$

Now, with the help of these expressions, and since the dq's and δq's are (as yet) unconstrained (i.e., $m = 0$), like the q's, we can enunciate the following "obvious" theorems, in increasing order of specificity:

- The necessary and sufficient conditions for the *particular* Pfaffian *form* (not constraint!)

$$d\theta_k \equiv \sum a_{kl}\, dq_l + a_k\, dt, \qquad \text{or in virtual form} \qquad \delta\theta_k \equiv \sum a_{kl}\, \delta q_l, \quad (2.9.14)$$

  to be an *exact* differential—that is, for the hitherto shorthand symbols $d\theta_k$ and $\delta\theta_k$ to be the genuine (first and total) differentials of a bona fide function $\theta_k = \theta_k(q, t)$ ($\rightarrow$ *holonomic coordinate*)—is that its bilinear covariant (2.9.13), vanish.

- The necessary and sufficient condition for *a* Pfaffian form (2.9.14) to be the exact differential of θ_k [since its n δq's in (2.9.13) are arbitrary] is that its *associated n Pfaffian forms*

$$d\psi_{kl} \equiv \sum (\partial a_{kl}/\partial q_s - \partial a_{ks}/\partial q_l)\, dq_s + (\partial a_{kl}/\partial t - \partial a_k/\partial q_l)\, dt, \qquad (2.9.15)$$

  all vanish; that is, $d\psi_{kl} = 0$ for all l $(= 1, \ldots, n)$.

- The necessary and sufficient condition for *a* Pfaffian form (2.9.14) to be the exact differential of θ_k [since the n dq's and dt in (2.9.15) are arbitrary] is that the following $n(n+1)/2$ integrability (or exactness) conditions hold:

$$\partial a_{kl}/\partial q_s - \partial a_{ks}/\partial q_l = 0 \qquad \textit{and} \qquad \partial a_{kl}/\partial t - \partial a_k/\partial q_l = 0, \qquad (2.9.16)$$

  identically in the q's and t, and for *all* values of l, s $(= 1, \ldots, n)$. [For additional insights and details, see, for example, Hagihara (1970, pp. 42–46), Whittaker (1937, p. 296 ff.).]

Hence, if (2.9.16) hold for *all* $k = 1, \ldots, n$, the n θ's are just another minimal set of Lagrangean coordinates, like the q's: $\theta_k = \theta_k(q_1, \ldots, q_n; t)$; and $\omega_k \equiv d\theta_k/dt$ are the corresponding holonomic Lagrangean (generalized) velocities. But if, and this is the case of interest to AM,

$$\partial a_{kl}/\partial q_s - \partial a_{ks}/\partial q_l \neq 0 \qquad \text{or} \qquad \partial a_{kl}/\partial t - \partial a_k/\partial q_l \neq 0, \qquad (2.9.17)$$

even for one l, s, then ω_k is *not* a total time derivative, and $d\theta_k$ is *not* a genuine differential of a holonomic coordinate θ_k; *only* the $d\theta_k/\delta\theta_k/\omega_k$ are defined through (2.9.1, 9, 11). Such undefined quantities, θ_k, are called *pseudo- or quasi coordinates* [a term, most likely, due to Whittaker (1904)], or *nonholonomic coordinates*; and the ω_k, depicted by some authors by *symbolic* $(\ldots)^{\cdot}$-derivatives, like

$$\omega_k \equiv d'\theta_k/dt \equiv \overset{*}{\theta}_k \equiv \overset{o}{\theta}_k, \quad \text{etc.}, \qquad (2.9.18)$$

instead of $d\theta_k/dt$, are called quasi velocities. From now on we shall assume, with no loss in generality, that all (2.9.17) hold, and therefore all θ_k are quasi coordinates. {We notice that, the *isochrony* choice $d\theta_{n+1} \equiv dq_{n+1} \equiv dt$, resulting in [recalling (2.9.4a, b)]

$$a_{n+1,k} \equiv \delta_{n+1,k} = 0, \qquad a_{n+1,n+1} = \delta_{n+1,n+1} = 1, \qquad (2.9.19a)$$

and

$$A_{n+1,k} \equiv \delta_{n+1,k} = 0, \qquad A_{n+1,n+1} = \delta_{n+1,n+1} = 1, \qquad (2.9.19b)$$

guarantees that θ_{n+1} remains holonomic.}

REMARKS

(i) Let us consider, for simplicity, the catastatic version of (2.9.9),

$$\omega_k = \sum a_{kl}(t, q) v_l = \omega_k(t, q, \dot{q} \equiv v). \tag{2.9.20}$$

If the q's are *known/specified functions of time t*, then integrating (2.9.20) between an initial instant t_o and a current one t we obtain the line integral

$$\theta_k(t) - \theta_k(t_o) = \int_{t_o}^{t} \omega_k[\tau, q(\tau), v(\tau)]\, d\tau = \int_{t_o}^{t} \left(\sum a_{kl}[\tau, q(\tau)] v_l(\tau)\right) d\tau; \tag{2.9.20a}$$

similar to the work integral of general mechanics and thermodynamics. Since this is the integral of an inexact differential, as calculus/vector field theory teach, $\theta_k(t)$ depends on both t (current configuration) and the particular *path of integration/history* followed from t_o to t; it is *point- and path-dependent*. If it was a genuine global coordinate, it would be *point-dependent, but path-independent*. $\theta_k(t; t_o)$ is a *functional* of the particular curves/motion $\{q(\tau), t_o \le \tau \le t\}$!

(ii) As will be explained in §2.11, the satisfaction of (2.9.16) guarantees that θ_k, as defined by (2.9.14), is a holonomic coordinate; and that property will hold even if, at a later stage, the $dq_k/\delta q_k/v_k$ become *holonomically* and/or *nonholonomically constrained*. One the other hand, if θ_k is originally [i.e., as defined by (2.9.14)] *nonholonomic*, then upon imposition on the latter's right side of a sufficient number of additional *holonomic* and/or *nonholonomic* constraints, later, it will become holonomic; but that would be a different Pfaffian form.

In sum: once a holonomic coordinate, *always* a holonomic coordinate; but once a nonholonomic coordinate, *not always* a nonholonomic coordinate.

(iii) The local transformations $\boldsymbol{a}_\eta \equiv \sum A_{\beta\eta} \boldsymbol{E}_\beta \Leftrightarrow \boldsymbol{E}_\beta = \sum a_{\eta\beta} \boldsymbol{a}_\eta$, where [recalling discussion in (§2.7)] each $\boldsymbol{E}_\beta$ is tangent to the coordinate line dq_β at (q, t) and all together they constitute a holonomic basis for the local tangent space T_{n+1}, and the coefficients satisfy the earlier (2.9.3a, 3b), define a new but, generally nonholonomic basis there: that is, $\sum \boldsymbol{a}_\eta\, d\theta_\eta$: *nonexact differential* $\Rightarrow \partial \boldsymbol{u}_\eta/\partial\theta_\beta \neq \partial \boldsymbol{a}_\beta/\partial\theta_\eta$ [where the nonholonomic gradients, $\partial/\partial\theta_\beta$, are defined in (2.9.27 ff.)]. And, in view of

$$\sum \dot{q}_\beta \boldsymbol{E}_\beta \equiv \sum v_\beta \boldsymbol{E}_\beta = \sum v_\beta \left(\sum a_{\eta\beta}\, \boldsymbol{a}_\eta\right) = \sum \left(\sum a_{\eta\beta} v_\beta\right) \boldsymbol{a}_\eta = \sum \omega_\eta\, \boldsymbol{a}_\eta$$
$$= \sum \omega_\beta\, a_\beta,$$

the ω_β are simply the *nonholonomic components* of the system velocity vector, while the v_β are its *holonomic components*. [The system basis $\{\boldsymbol{a}_\eta\}$ plays a key role in the geometrical interpretation of Pfaffian constraints (§2.11.19a ff.)]

(iv) The precise term for the θ_k's is "nonholonomic (local) system coordinates," and for the ω_k's "nonholonomic system velocity parameters," or "(*contravariant*) *nonholonomic components of the system velocity*" (Schouten, 1954/1989, pp. 194–197). We shall call them collectively *quasi variables*; and their symbolic calculus, if proper precautions are taken, is quite useful. As Synge puts it: "In the theory of quasi-coordinates in dynamics, however, it pays to live dangerously and to use the notation $d\theta_k$ [in our notation]. Otherwise we shall be depriving ourselves of a very neat formal expression of the equations of motion" (1936, p. 29). On the symbolic calculus of quasi variables, see also Johnsen (1939).

Example 2.9.1 The most common example of quasi velocities in mechanics is the components of the (inertial) angular velocity of a rigid body moving, with no loss in generality here, about a fixed point O, resolved along either space-fixed (inertial) axes O–XYZ, $\omega_X, \omega_Y, \omega_Z$; or body-fixed (moving) axes O–$xyz, \omega_x, \omega_y, \omega_z$. If $\phi \to \theta \to \psi$ are the three Eulerian angles $3 \to 1 \to 3$, then for body-axes, and with the convenient notations $s(\ldots) \equiv \sin(\ldots)$ and $c(\ldots) \equiv \cos(\ldots)$, and $d\phi/dt \equiv \omega_\phi$, $d\theta/dt \equiv \omega_\theta$, $d\psi/dt \equiv \omega_\psi$, we have (§1.12)

$$\omega_x = (s\psi\, s\theta)\omega_\phi + (c\psi)\omega_\theta + (0)\omega_\psi, \tag{a}$$

$$\omega_y = (c\psi\, s\theta)\omega_\phi + (-s\psi)\omega_\theta + (0)\omega_\psi, \tag{b}$$

$$\omega_z = (c\theta)\omega_\phi + (0)\omega_\theta + (1)\omega_\psi; \tag{c}$$

that is, with $k = x \to 1$, $y \to 2$, $z \to 3$; and $l = \phi \to 1$, $\theta \to 2$, $\psi \to 3$, the nonvanishing elements of (a_{kl}) are

$$a_{11} = s\psi\, s\theta, \quad a_{12} = c\psi; \qquad a_{21} = c\psi\, s\theta, \quad a_{22} = -s\psi; \qquad a_{31} = c\theta, \quad a_{33} = 1. \tag{d}$$

Clearly, not all (2.9.16) hold identically here. For example,

$$\partial a_{12}/\partial q_3 \neq \partial a_{13}/\partial q_2: \quad \partial(c\psi)/\partial\psi \neq \partial(0)/\partial\theta: \quad -s\psi \neq 0; \tag{e}$$

except in the *special* (*nonidentical!*) case: $\psi = 0,\ 2\pi$. If we set $\omega_x = d\theta_x/dt$, then

$$\theta_x(t) \equiv \int_{t_o}^{t} \omega_x\big[\theta(\tau), \psi(\tau); \omega_\phi(\tau), \omega_\theta(\tau)\big]\, d\tau + \theta_x\,(\textit{initial}): \ \textit{path dependent}; \tag{f}$$

that is, θ_x is an (angular) quasi coordinate, and ω_x an (angular) quasi velocity; and similarly for $\theta_y, \theta_z; \omega_y, \omega_z$; that is, they are quasi variables (if the ϕ, θ, ψ *are unconstrained*). However, if we impose *additional* constraints, for example, $\phi = \textit{constant}$, $\theta = \textit{constant}$ (fixed-axis rotation), then (a–c) reduce to

$$\omega_x = 0, \quad \omega_y = 0, \quad \omega_z = d\psi/dt \ \Rightarrow\ \omega_x, \omega_y, \omega_z: \textit{holonomic velocities}; \tag{g}$$

$$\theta_z(t) - \theta_z\,(t_o\text{: } \textit{initial}) = \int_{t_o}^{t} [d\psi(\tau)/d\tau]\, d\tau = \psi(t) - \psi\,(t_o\text{: } \textit{initial}): \textit{path independent}. \tag{h}$$

Problem 2.9.1 Let the reader verify that the corresponding *space-fixed* components $\theta_X, \theta_Y, \theta_Z$ and $\omega_X, \omega_Y, \omega_Z$ (such that $\omega_X \equiv d\theta_X/dt$, etc.) are also, respectively, quasi coordinates and quasi velocities; and that under additional constraints they too may become holonomic variables.

Particle Kinematics in Quasi Variables

Due to the $\theta \leftrightarrow q$ transformation relations (2.9.1, 2, 9, 10, 11, 12), the (inertial) velocity, acceleration, kinematically admissible/possible displacement, and virtual displacement, of a typical system particle, obtained in §2.5 in holonomic variables, assume the following quasi-variable representations, respectively:

(i) Velocity:

$$\boldsymbol{v} = \sum \boldsymbol{e}_k \Big(\sum A_{kl}\,\omega_l + A_k\Big) + \boldsymbol{e}_0 = \cdots = \sum \boldsymbol{\varepsilon}_k \omega_k + \boldsymbol{\varepsilon}_{n+1} \equiv \sum \boldsymbol{\varepsilon}_k \omega_k + \boldsymbol{\varepsilon}_0; \quad (2.9.21)$$

(ii) Acceleration:

$$\boldsymbol{a} = \cdots = \sum \boldsymbol{\varepsilon}_k (d\omega_k/dt) + \text{terms not containing } (d\omega/dt)\text{'s};$$
$$\equiv \sum \boldsymbol{\varepsilon}_k \dot{\omega}_k + \text{terms not containing } \dot{\omega}\text{'s}, \quad (2.9.22)$$

(iii) Kinematically possible/admissible displacement:

$$d\boldsymbol{r} = \sum \boldsymbol{e}_k \Big(\sum A_{kl}\, d\theta_l + A_k\, dt\Big) + \boldsymbol{e}_0\, dt = \cdots = \sum \boldsymbol{\varepsilon}_k\, d\theta_k + \boldsymbol{\varepsilon}_0\, dt; \quad (2.9.23)$$

(iv) Virtual displacement:

$$\delta\boldsymbol{r} = \sum \boldsymbol{e}_k \Big(\sum A_{kl}\, \delta\theta_l\Big) = \cdots = \sum \boldsymbol{\varepsilon}_k\, \delta\theta_k; \quad (2.9.24)$$

where the fundamental, *generally nongradient*, $n+1$ *particle and system* vectors $\boldsymbol{\varepsilon}_k$ and $\boldsymbol{\varepsilon}_{n+1} \equiv \boldsymbol{\varepsilon}_0$, corresponding to the θ's, nonholonomic counterparts of the gradient vectors $\boldsymbol{e}_k$ and $\boldsymbol{e}_{n+1} \equiv \boldsymbol{e}_0$, which correspond to the q's [recalling (2.5.4–4b)], and *defined* naturally by (2.9.21–24), obey the following basic (covariant vector-like) transformation equations:

$$\boldsymbol{\varepsilon}_k \equiv \sum (\partial v_l/\partial\omega_k)\boldsymbol{e}_l = \sum A_{lk}\boldsymbol{e}_l, \quad (2.9.25a)$$

$$\boldsymbol{e}_k \equiv \sum (\partial\omega_l/\partial v_k)\boldsymbol{\varepsilon}_l = \sum a_{lk}\boldsymbol{\varepsilon}_l \qquad \text{[comparing with (2.9.11, 12)];} \quad (2.9.25b)$$

$$\boldsymbol{\varepsilon}_0 \equiv \sum A_k \boldsymbol{e}_k + \boldsymbol{e}_0 = -\sum a_k \boldsymbol{\varepsilon}_k + \boldsymbol{e}_0, \quad (2.9.26a)$$

$$\boldsymbol{e}_0 \equiv \sum a_k \boldsymbol{\varepsilon}_k + \boldsymbol{\varepsilon}_0 = -\sum A_k \boldsymbol{e}_k + \boldsymbol{\varepsilon}_0 \qquad \text{[recalling (2.9.3a, 3b)].} \quad (2.9.26b)$$

Clearly, if the $\boldsymbol{e}$ vectors are linearly independent (and $|a_{kl}|, |A_{kl}| \neq 0$), so are the $\boldsymbol{\varepsilon}$ vectors; even if the q's and/or $dq/dt \equiv v$'s get constrained later. And, as with the $\delta\boldsymbol{r}$-representation (2.5.12b), so with (2.9.24): *the size of the $\delta\theta$'s is unimportant; it is the $\boldsymbol{\varepsilon}$'s that matter, because they are the ones entering the equations of motion* (chap. 3)!

Quasi Chain Rule, Symbolic Notations

The above, especially (2.9.24), suggest the adoption of the following very useful symbolic *quasi-chain rule* for quasi variables:

$$\partial\boldsymbol{r}/\partial\theta_k \equiv \sum (\partial\boldsymbol{r}/\partial q_l)(\partial v_l/\partial\omega_k) \equiv \sum (\partial\boldsymbol{r}/\partial q_l)\big[\partial(dq_l)/\partial(d\theta_k)\big]$$
$$= \sum (\partial\boldsymbol{r}/\partial q_l)\big[\partial(\delta q_l)/\partial(\delta\theta_k)\big],$$

or, simply,

$$\partial \mathbf{r}/\partial \theta_k \equiv \sum A_{lk}(\partial \mathbf{r}/\partial q_l): \quad \text{i.e., (2.9.25a);} \tag{2.9.27}$$

and, inversely,

$$\partial \mathbf{r}/\partial q_k = \sum (\partial \mathbf{r}/\partial \theta_l)(\partial \omega_l/\partial v_k) = \sum a_{lk}(\partial \mathbf{r}/\partial \theta_l): \quad \text{i.e., (2.9.25b).} \tag{2.9.28}$$

Similarly, for a general well-behaved function $f = f(q, t)$, and recalling (2.9.12), we obtain, successively, (i) for its *virtual variation* δf:

$$\delta f = \sum (\partial f/\partial q_k)\, \delta q_k = \sum (\partial f/\partial q_k)\Big(\sum (\partial v_k/\partial \omega_l)\, \delta\theta_l\Big) \equiv \sum (\partial f/\partial \theta_l)\, \delta\theta_l; \tag{2.9.29}$$

that is,

$$\partial f/\partial \theta_l \equiv \sum (\partial f/\partial q_k)(\partial v_k/\partial \omega_l) = \sum A_{kl}(\partial f/\partial q_k), \tag{2.9.30a}$$

and, inversely,

$$\partial f/\partial q_k = \sum (\partial f/\partial \theta_l)(\partial \omega_l/\partial v_k) = \sum a_{lk}(\partial f/\partial \theta_l); \tag{2.9.30b}$$

and (ii) for its *total differential* df [recalling (2.9.2)]:

$$\begin{aligned} df &= \sum (\partial f/\partial q_\beta)\, dq_\beta = \sum (\partial f/\partial q_k)\, dq_k + (\partial f/\partial t)\, dt \\ &= \sum (\partial f/\partial q_k)\Big(\sum A_{kl}\, d\theta_l + A_k\, dt\Big) + (\partial f/\partial t)\, dt \\ &= \sum \Big(\sum A_{kl}(\partial f/\partial q_k)\Big)\, d\theta_l + \Big(\sum A_k(\partial f/\partial q_k) + \partial f/\partial t\Big)\, dt \\ &\equiv \sum (\partial f/\partial \theta_l)\, d\theta_l + (\partial f/\partial \theta_0)\, dt, \end{aligned} \tag{2.9.31}$$

where we have introduced the additional symbolic notation [recalling that $\dot{\theta}_0 \equiv \dot{\theta}_{n+1} \equiv \omega_{n+1} = 1$]:

$$\begin{aligned} \partial \ldots/\partial \theta_{n+1} &\equiv \sum (\partial \ldots/\partial q_\beta)(\partial v_\beta/\partial \omega_{n+1}) \\ &= \sum (\partial \ldots/\partial q_k)(\partial v_k/\partial \omega_{n+1}) + (\partial \ldots/\partial t)(\partial v_{n+1}/\partial \omega_{n+1}) \\ &= \sum A_k(\partial \ldots/\partial q_k) + \partial \ldots/\partial t; \end{aligned} \tag{2.9.32}$$

instead of the formal extension of (2.9.30a) for $\theta_l \to \theta_{n+1}$. This latter we shall denote by $\partial \ldots/\partial(t)$:

$$\partial \ldots/\partial(t) \equiv \sum (\partial \ldots/\partial q_k)(\partial v_k/\partial \omega_{n+1}) = \sum A_k(\partial \ldots/\partial q_k); \tag{2.9.32a}$$

so that (2.9.32) assumes the final symbolic form

$$\partial \ldots/\partial \theta_{n+1} \equiv \partial \ldots/\partial \theta_0 \equiv \partial \ldots/\partial(t) + \partial \ldots/\partial t. \tag{2.9.32b}$$

Inversely, we have

$$\partial\ldots/\partial t = \sum (\partial\omega_\alpha/\partial v_{n+1})(\partial\ldots/\partial\theta_\alpha) = \partial\ldots/\partial\theta_{n+1} + \sum a_k(\partial\ldots/\partial\theta_k), \tag{2.9.32c}$$

and, comparing this with (2.9.30a, b), we readily conclude that

$$\partial\ldots/\partial(t) = \sum A_k(\partial\ldots/\partial q_k) = -\sum a_k(\partial\ldots/\partial\theta_k). \tag{2.9.32d}$$

Such (by no means uniform) symbolic notations are useful in *energy rate/power* theorems in nonholonomic variables (§3.9).

Some Fundamental Kinematical Identities

From the above (2.9.21 ff.), we readily obtain the following fundamental kinematical identities, nonholonomic counterparts of (2.5.7–10), and like them, holding independently of any subsequent holonomic and/or nonholonomic constraints.

(i)
$$\partial\boldsymbol{r}/\partial\theta_k = \partial\dot{\boldsymbol{r}}/\partial\dot{\theta}_k = \partial\ddot{\boldsymbol{r}}/\partial\ddot{\theta}_k \equiv \partial\ddot{\boldsymbol{r}}/\partial\dot{\omega}_k = \cdots \equiv \boldsymbol{\varepsilon}_k,$$

or

$$\partial\boldsymbol{r}/\partial\theta_k = \partial\boldsymbol{v}/\partial\omega_k = \partial\boldsymbol{a}/\partial\dot{\omega}_k = \cdots \equiv \boldsymbol{\varepsilon}_k; \tag{2.9.33}$$

(ii)
$$\partial q_k/\partial\theta_l \equiv \partial\dot{q}_k/\partial\dot{\theta}_l = \partial\ddot{q}_k/\partial\ddot{\theta}_l \equiv \partial\ddot{q}_k/\partial\dot{\omega}_l = \cdots \equiv A_{kl},$$

or

$$\partial q_k/\partial\theta_l = \partial v_k/\partial\omega_l = \partial w_k/\partial\dot{\omega}_l = \cdots \equiv A_{kl}; \qquad (\text{where } dv_k/dt \equiv w_k) \tag{2.9.34}$$

(iii)
$$\partial\theta_k/\partial q_l = \partial\omega_k/\partial v_l = \partial\dot{\omega}_k/\partial w_l = \cdots \equiv a_{kl}, \tag{2.9.35}$$

with formal extensions for $\theta_{n+1} \equiv q_{n+1} \equiv t$. The $d\omega_k/dt \equiv d^2\theta_k/dt^2$ are called (not quite correctly) *quasi accelerations*; while the $\theta/\omega/\dot{\omega}/\ldots$ are referred to, collectively, as (system) *quasi variables*.

(iv) We have, successively,

$$\begin{aligned} d(\partial\boldsymbol{r}/\partial\theta_k)/dt &= d(\partial\boldsymbol{v}/\partial\omega_k)/dt = d\boldsymbol{\varepsilon}_k/dt \\ &= d\Big(\sum A_{lk}\boldsymbol{e}_l\Big)\Big/dt = \sum \big[(dA_{lk}/dt)\boldsymbol{e}_l + A_{lk}(d\boldsymbol{e}_l/dt)\big] \\ &= \sum (dA_{lk}/dt)\boldsymbol{e}_l + \sum A_{lk}(\partial\boldsymbol{v}/\partial q_l) \quad [\text{recalling (2.5.7, 10)}]. \end{aligned} \tag{2.9.36}$$

But by partial ∂q_l-differentiation of $\boldsymbol{v}(q, v, t) = \boldsymbol{v}[q, v(q,\omega,t), t] \equiv \boldsymbol{v}^*(q,\omega,t)$, we find

$$\partial\boldsymbol{v}^*/\partial q_l = \partial\boldsymbol{v}/\partial q_l + \sum (\partial\boldsymbol{v}/\partial v_r)(\partial v_r/\partial q_l) = \partial\boldsymbol{v}/\partial q_l + \sum (\partial v_r/\partial q_l)\boldsymbol{e}_r,$$

and so

$$\sum A_{lk}(\partial \boldsymbol{v}/\partial q_l) = \sum A_{lk}(\partial \boldsymbol{v}^*/\partial q_l) - \sum\sum A_{lk}(\partial v_r/\partial q_l)\boldsymbol{e}_r$$
$$= \partial \boldsymbol{v}^*/\partial \theta_k - \sum\sum A_{lk}(\partial v_r/\partial q_l)\boldsymbol{e}_r.$$

Therefore, returning to (2.9.36), we see that it yields

$$d\boldsymbol{\varepsilon}_k/dt - \partial \boldsymbol{v}^*/\partial \theta_k = \sum \left[dA_{lk}/dt - \sum A_{rk}(\partial v_l/\partial q_r)\right]\boldsymbol{e}_l \neq \boldsymbol{0}, \qquad (2.9.36a)$$

that is, *unlike the H coordinate case* (2.5.10),

$$E_k{}^*(\boldsymbol{v}^*) \equiv (\partial \boldsymbol{v}^*/\partial \dot{\theta}_k)^{\cdot} - \partial \boldsymbol{v}^*/\partial \theta_k \equiv d/dt(\partial \boldsymbol{v}^*/\partial \omega_k) - \partial \boldsymbol{v}^*/\partial \theta_k \neq \boldsymbol{0}. \qquad (2.9.37)$$

This *nonintegrability* relation is a first proof that, in general, the $\boldsymbol{\varepsilon}_k$ *basis vectors are nongradient*, or *nonholonomic*. More comprehensible and useful forms of $E_k{}^*(\boldsymbol{v}^*)$ are presented in the next section.

[Some authors call the $\boldsymbol{\varepsilon}_k$ vectors "partial velocities." However, in view of (2.9.33), they could just as well have been called *partial positions*, or *partial accelerations*, or even *partial jerks* (recall that $d\boldsymbol{a}/dt \equiv \boldsymbol{j} =$ *jerk vector*, and therefore $\partial \boldsymbol{j}/\partial \ddot{\omega}_k = \boldsymbol{\varepsilon}_k$), etc. Perhaps a better term would be *nonholonomic mixed basis vectors* (i.e., nonholonomic counterpart of Heun's *Begleitvektoren*).]

A Useful Nonholonomic-Variable Notation

Frequently, for extra clarity, we will be using the following "(...)*-notation":

$$f = f(t, q, dq/dt \equiv v) = f[t, q, v(t, q, \omega)] \equiv f^*(t, q, \omega) \equiv f^*. \qquad (2.9.38)$$

With its help:

(i) Equations (2.9.21), (2.9.22), (2.9.33) become, respectively,

$$\boldsymbol{v}(t, q, v) = \sum \boldsymbol{e}_k(t, q)v_k + \boldsymbol{e}_0(t, q) = \sum \boldsymbol{\varepsilon}_k(t, q)\omega_k + \boldsymbol{\varepsilon}_0(t, q) \equiv \boldsymbol{v}^*(t, q, \omega); \qquad (2.9.39)$$

$$\boldsymbol{a}(t, q, v, w) = \sum \boldsymbol{e}_k(t, q)w_k + \text{no other } \ddot{q} \equiv w\text{-terms}$$
$$= \sum \boldsymbol{\varepsilon}_k(t, q)\dot{\omega}_k + \text{no other } \dot{\omega}\text{-terms} \equiv \boldsymbol{a}^*(t, q, \omega, \dot{\omega}); \qquad (2.9.40)$$

$$\partial \boldsymbol{r}/\partial \theta_k = \partial \boldsymbol{v}^*/\partial \omega_k = \partial \boldsymbol{a}^*/\partial \dot{\omega}_k = \cdots \equiv \boldsymbol{\varepsilon}_k; \qquad (2.9.41)$$

(ii) The quasi-chain rule (2.9.30a) and its inverse (2.9.30b) *generalize*, respectively, to

$$\partial f^*/\partial \theta_l \equiv \sum (\partial f^*/\partial q_k)(\partial v_k/\partial \omega_l) = \sum A_{kl}(\partial f^*/\partial q_k), \qquad (2.9.42a)$$

and

$$\partial f^*/\partial q_k = \sum (\partial f^*/\partial \theta_l)(\partial \omega_l/\partial v_k) = \sum a_{lk}(\partial f^*/\partial \theta_l); \qquad (2.9.42b)$$

also, we easily obtain the related chain rules [recall derivation of (2.9.37), and (2.9.42a)]

$$\partial f^*/\partial q_k = \partial f/\partial q_k + \sum (\partial f/\partial v_l)(\partial v_l/\partial q_k), \tag{2.9.43a}$$

$$\Rightarrow \partial f^*/\partial \theta_l = \sum A_{kl}(\partial f^*/\partial q_k) = \sum A_{kl}\Big[\partial f/\partial q_k + \sum (\partial f/\partial v_r)(\partial v_r/\partial q_k)\Big]. \tag{2.9.43b}$$

(iii) The following *genuine* (i.e., *ordinary calculus*) *chain rule*, and its inverse, hold:

$$\partial f^*/\partial \omega_l \equiv \sum (\partial f/\partial v_k)(\partial v_k/\partial \omega_l) = \sum A_{kl}(\partial f/\partial v_k), \tag{2.9.44a}$$

$$\partial f/\partial v_k = \sum (\partial f^*/\partial \omega_l)(\partial \omega_l/\partial v_k) = \sum a_{lk}(\partial f^*/\partial \omega_l). \tag{2.9.44b}$$

We notice the *difference* between (2.9.42a, b) and (2.9.44a, b); the former are *non-vectorial* transformations, just symbolic definitions; while (for those familiar with tensors) the latter are genuine *covariant vector* transformations.

(iv) Finally, invoking (2.9.11, 12, 42a, b), it is not hard to see that

$$\sum (\partial f^*/\partial \theta_k)\, \delta\theta_k = \sum (\partial f^*/\partial q_k)\, \delta q_k. \tag{2.9.45}$$

Some Closing Comments on Quasi Coordinates

The theory of nonholonomic coordinates and constraints is, by now, a well established and well understood part of differential geometry/tensor calculus and mechanics, with many fertile applications in those areas. Its long and successful history has been created by several famous mathematicians, such as (chronologically): Gibbs, Volterra, Poincaré, Heun, Hamel, Synge, Schouten, Struik, Vranceanu, Vagner, Kron, Kondo, Dobronravov et al. And yet, we encounter contemporary statements of appalling ignorance and confusion, like the following from an *advanced* "Tract in Natural Philosophy" devoted to rigid kinematics: "It appears that the reason why many a book on classical dynamics follows Kirchhoff's approach is a lack of understanding of the kinematics of rigid bodies. Thus, one finds extensive discussions on ill-defined — or, sometimes, totally undefined — esoteric quantities such as *quasi-coordinates* and *virtual displacements*," (Angeles, 1988, p. 2, the italics are that author's).

2.10 TRANSITIVITY, OR TRANSPOSITIONAL, RELATIONS; HAMEL COEFFICIENTS

So far, our system remains a holonomic (H) one, with $n \equiv 3N - h$ *DOF*. Now, to be able to either (i) embed to it *additional Pfaffian* (*possibly nonholonomic*) *constraints* in their "simplest possible form" or, even if no such additional constraints are imposed, (ii) express the equations of the problem in quasi variables, or (iii) do both, we need to represent the right sides of the Frobenius bilinear covariants of the Pfaffian forms of its quasi variables, $(\ldots)\, dq\, \delta q$ [recall (2.9.13)], in terms of the latter's differentials, $(\ldots)\, d\theta\, \delta\theta$. [By simplest possible form we mean *uncoupled* from each other; and, as detailed in chap. 3, this leads to the simplest possible form of the equations of motion.] To this end, we insert expressions (2.9.2 and 12) into the right side of (2.9.13), and group the terms appropriately. The result is the following *generalized transitivity, or transpositional, equations* (Hamel's *Übergangs-*, or *Transitivitäts-gleichungen*):

$$\begin{aligned} d(\delta\theta_k) - \delta(d\theta_k) &= \sum a_{kl}[d(\delta q_l) - \delta(dq_l)] + \sum\sum \gamma^k{}_{\alpha\beta}\, d\theta_\beta\, \delta\theta_\alpha \\ &= \sum a_{kl}[d(\delta q_l) - \delta(dq_l)] + \sum\sum \gamma^k{}_{r\beta}\, d\theta_\beta\, \delta\theta_r \quad [\text{since } \delta\theta_{n+1} \equiv \delta t = 0] \\ &= \sum a_{kl}[d(\delta q_l) - \delta(dq_l)] + \sum\sum \gamma^k{}_{rs}\, d\theta_s\, \delta\theta_r + \sum \gamma^k{}_r\, dt\, \delta\theta_r, \end{aligned} \tag{2.10.1}$$

(again, we recall that all Latin (Greek) indices run from 1 to n (1 to $n+1$)) where the so-defined γ's, known as *Hamel (three-index) coefficients*, are explicitly given (and sometimes also defined) by

$$\begin{aligned} \gamma^k{}_{rs} &= \sum\sum (\partial a_{k\beta}/\partial q_\varepsilon - \partial a_{k\varepsilon}/\partial q_\beta) A_{\beta r} A_{\varepsilon s} \\ &= \sum\sum (\partial a_{kb}/\partial q_c - \partial a_{kc}/\partial q_b) A_{br} A_{cs} \\ &\quad + \sum (\partial a_{kb}/\partial t - \partial a_{k,n+1}/\partial q_b) A_{br} A_{n+1,s} \\ &\quad + \sum (\partial a_{k,n+1}/\partial q_c - \partial a_{kc}/\partial t) A_{n+1,r} A_{cs} \\ &\quad + (\partial a_{k,n+1}/\partial t - \partial a_{k,n+1}/\partial t) A_{n+1,r} A_{n+1,s}, \end{aligned} \tag{2.10.1a}$$

or, due to $A_{n+1,r} = \delta_{n+1,r} = 0$ which leads to the *vanishing of the last three groups/ sums of terms*, finally,

$$\gamma^k{}_{rs} = \sum\sum (\partial a_{kb}/\partial q_c - \partial a_{kc}/\partial q_b) A_{br} A_{cs}; \tag{2.10.2}$$

and

$$\gamma^k{}_{r,n+1} = -\gamma^k{}_{n+1,r} \equiv \gamma^k{}_r \equiv \sum\sum (\partial a_{k\beta}/\partial q_\varepsilon - \partial a_{k\varepsilon}/\partial q_\beta) A_{\beta r} A_{\varepsilon,n+1}, \tag{2.10.3}$$

or, with $a_{k,n+1} \equiv a_k$, $A_{k,n+1} \equiv A_k$, and since $A_{n+1,n+1} \equiv \delta_{n+1,n+1} = 1$, finally,

$$\gamma^k{}_r \equiv \sum\sum (\partial a_{kb}/\partial q_c - \partial a_{kc}/\partial q_b) A_{br} A_c + \sum (\partial a_{kb}/\partial t - \partial a_k/\partial q_b) A_{br}. \tag{2.10.4}$$

[The γ's are a significant generalization of coefficients introduced by Ricci (mid-1890s), Volterra (1898), Boltzmann (1902) et al.; and, hence, they are also referred as "Ricci/Boltzmann/Hamel (rotation) coefficients." See, for example, Papastavridis (1999, chaps. 3, 6).]

It is not hard to show [with the help of (2.9.3a, b)] that (2.10.1) inverts to

$$\begin{aligned} d(\delta q_k) - \delta(dq_k) = \sum A_{kl}\Big\{&[d(\delta\theta_l) - \delta(d\theta_l)] \\ &- \sum\sum \gamma^l{}_{rs}\, d\theta_s\, \delta\theta_r - \sum \gamma^l{}_r\, dt\, \delta\theta_r\Big\}. \end{aligned} \tag{2.10.5}$$

For an actual motion, dividing both sides of (2.10.1) and (2.10.5) with dt [which does not interact with $\delta(\ldots)$], we obtain, respectively, the (system) *velocity transitivity equation* and its inverse:

$$(\delta\theta_k)^{\cdot} - \delta\omega_k = \sum a_{kl}[(\delta q_l)^{\cdot} - \delta v_l] + \sum\sum \gamma^k{}_{rs}\,\omega_s\,\delta\theta_r + \sum \gamma^k{}_r\,\delta\theta_r, \tag{2.10.6}$$

$$(\delta q_k)^{\cdot} - \delta v_k = \sum A_{kl}\Big\{[(\delta\theta_l)^{\cdot} - \delta\omega_l] - \sum\sum \gamma^l{}_{rs}\,\omega_s\,\delta\theta_r - \sum \gamma^l{}_r\,\delta\theta_r\Big\}. \tag{2.10.7}$$

Properties of the Hamel Coefficients

(i) Clearly, these coefficients depend, through the transformation coefficients $a_{\beta\varepsilon}$ and $A_{\beta\varepsilon}$, on the particular $v \leftrightarrow \omega$ choice; that is, *they do not depend on any particular system motion.*

(ii) The $\gamma^k{}_r$ contain the contributions of (a) the *acatastatic* terms a_k and A_k, and of (b) the *explicit time-dependence of the homogeneous coefficients* of the $v \Leftrightarrow \omega$ transformation. Hence, for *scleronomic* such transformations (i.e., $a_k = 0 \Rightarrow A_k = 0$, and $\partial a_{kl}/\partial t = 0 \Rightarrow \partial A_{kl}/\partial t = 0$) they vanish; but for *catastatic* ones, in general, they do not. In fact then, as (2.10.4) shows, they reduce to

$$\gamma^k{}_r = \sum (\partial a_{kb}/\partial t) A_{br} \qquad \text{(for catastatic Pfaffian transformations).} \tag{2.10.4a}$$

(iii) The matrix $\boldsymbol{\gamma}^k = (\gamma^k{}_{rs})$ is, obviously, *antisymmetric*; that is,

$$\gamma^k{}_{rs} = -\gamma^k{}_{sr} \Rightarrow \gamma^k{}_{rr}:\ \textit{diagonal elements} = 0 \qquad (k, r, s = 1, \ldots, n;\ \text{also } n+1). \tag{2.10.8}$$

To stress this antisymmetry in r and s, we chose to raise k; that is, we wrote $\gamma^k{}_{rs}$ instead of γ_{rks}, or γ_{krs}, or γ_{rsk}, and so on. [Nothing tensorial is implied here, although this happens to be the tensorially correct index positioning; see, for example, Papastavridis (1999, chaps. 3, 6).] Hence, each matrix $\boldsymbol{\gamma}^k$ can have at most $n(n-1)/2$ nonzero (nondiagonal) elements.

(iv) From the above, we readily conclude that

$$\gamma^{n+1}{}_{\varepsilon\beta} = 0 \Rightarrow \gamma^{n+1}{}_{kl} = 0, \qquad \gamma^{n+1}{}_{k,n+1} = -\gamma^{n+1}{}_{n+1,k} = 0, \qquad \gamma^{n+1}{}_{n+1,n+1} = 0$$

$$[k, l = 1, \ldots, n;\ \varepsilon, \beta = 1, \ldots, n; n+1], \tag{2.10.9}$$

and from this (recalling that $a_{n+1,k} = \delta_{n+1,k} = 0$), that

$$(\delta\theta_{n+1})^{\cdot} - \delta\omega_{n+1} \equiv d/dt(\delta q_{n+1}) - \delta(dq_{n+1}/dt) \equiv d/dt(\delta t) - \delta(dt/dt)$$

$$= \sum\sum \gamma^{n+1}{}_{rs}\,\omega_s\,\delta\theta_r + \sum \gamma^{n+1}{}_{r,n+1}\,\delta\theta_r = 0 + 0 = 0, \tag{2.10.10}$$

which, essentially, states that

$$d(\delta\theta_{n+1}) - \delta(d\theta_{n+1}) = d(\delta t) - \delta(dt) = d(0) - \delta(dt) = 0 - 0 = 0, \tag{2.10.10a}$$

as it should, and also shows that (2.10.1) and (2.10.2) also hold for $k = n+1$.

(v) In concrete problems, the analytical calculation of the nonvanishing γ's is best done, as Hamel et al. have pointed out, not by applying (2.10.1a–4), which are admittedly laborious and error prone, but by reading them off as *coefficients of the bilinear covariant* (2.10.1, 6), in terms of the general subindices: $o, \blacklozenge = 1, \ldots, n; n+1$:

$$d(\delta\theta_*) - \delta(d\theta_*) = \cdots + (\gamma^*{}_{o\blacklozenge})\, d\theta_\blacklozenge\, \delta\theta_o + \cdots. \tag{2.10.11}$$

Also, this task is independent of any particular assumptions about $d(\delta q) - \delta(dq)$; and, hence, assuming that for *all* holonomic coordinates $d(\delta q_k) = \delta(dq_k)$, or equivalently $(\delta q_k)^{\cdot} = \delta(\dot{q}_k) \equiv \delta v_k$ (Hamel viewpoint—see also pr. 2.12.5), *even if they (or their differentials) become constrained later*, we may safely and conveniently calculate all the nonvanishing γ's from the simplified, and henceforth definitive, transitivity equation:

$$d(\delta\theta_k) - \delta(d\theta_k) = \sum\sum \gamma^k{}_{rs}\, d\theta_s\, \delta\theta_r + \sum \gamma^k{}_r\, dt\, \delta\theta_r. \tag{2.10.12}$$

Finally, dividing the above with dt, and so on, we obtain its velocity form:

$$(\delta\theta_k)^{\cdot} - \delta\omega_k = \sum\sum \gamma^k{}_{rs}\omega_s\, \delta\theta_r + \sum \gamma^k{}_r\, \delta\theta_r; \tag{2.10.13}$$

a representation useful in Hamilton's time integral "principle" in quasi variables (chap. 7). Unfortunately, the transitivity equations, and their relations with the γ's, are nowhere to be found in the English language literature (with the exception of Neimark and Fufaev, 1967 and 1972, p. 126. ff.); although the definition of the γ's via (2.10.1a, 2) appears in a number of places. This unnatural situation produces an incomplete understanding of these basic quantities.

REMARK (A PREVIEW)

As will become clear in chapter 3, the expression for the system kinetic energy (and the Appellian "acceleration energy") are simpler in terms of quasi variables, such as the ω's and $d\omega/dt$'s, than in terms of holonomic variables like the v's and dv/dt's. And this leads to formally simpler equations of motion in the former variables than in the latter; for example, the well-known Eulerian rotational rigid-body equations (§1.17) are simpler in terms of such quasi variables than, say, in terms of Eulerian angles and their $(\ldots)^{\cdot}$-derivatives. But there is a catch: to obtain such simpler-looking Lagrange-type equations of motion—that is, equations based on the kinetic energy and its various gradients—we must calculate the corresponding γ's; something that, even with utilization of (2.10.11–13) and other practice-based short cuts, requires some labor and skill. On the positive side, however, the γ's supply an important "amount" of understanding into the kinematical structure of the particular problem; and Appellian-type equations in quasi variables may not contain the γ's, but they have other calculational difficulties. In sum, there is no painless way to obtain simple-looking equations of motion in quasi variables.

Problem 2.10.1 Verify that the transitivity equations, say (2.10.12), can be rewritten as

$$d(\delta\theta_k) - \delta(d\theta_k) = \sum\sum{}' \gamma^k{}_{rs}(d\theta_s\, \delta\theta_r - \delta\theta_s\, d\theta_r) + \sum \gamma^k{}_r\, dt\, \delta\theta_r, \tag{a}$$

where $\sum\sum{}'$ means that the summation extends over r and s only *once*; say, for $s<r$. [We point out the following interesting geometrical interpretation of (a): each of its

double summation terms is proportional to a 2×2 determinant, which, in turn, equals the area of the infinitesimal parallelogram with sides two vectors on the local "$\theta_s\theta_r$-plane," at its origin (q, t) in configuration/event space, of respective rectangular Cartesian components $(d\theta_s, d\theta_r)$ and $(\delta\theta_s, \delta\theta_r)$ there; with the factor of proportionality being $\gamma^k{}_{rs}$. That parallelogram is the projection of the generalized parallelogram with sides $d\theta \equiv (d\theta_1, \ldots, d\theta_n)$ and $\delta\theta \equiv (\delta\theta_1, \ldots, \delta\theta_n)$, at (q, t), on the "$\theta_s\theta_r$-plane" (see, e.g., Boltzmann, 1904, pp. 104–107; Webster, 1912, pp. 84–87, 381–383; also Papastavridis, 1999, §3.14).]

Other Expressions for the γ's

By ∂q-differentiating (2.9.3a) and then rearranging so as to go from the $(\partial a/\partial q)$'s to the $(\partial A/\partial q)$'s, we obtain

$$\sum (\partial a_{kb}/\partial q_c) A_{br} = -\sum a_{kb}(\partial A_{br}/\partial q_c), \tag{2.10.14a}$$

$$\sum (\partial a_{kc}/\partial q_b) A_{cs} = -\sum a_{kc}(\partial A_{cs}/\partial q_b); \tag{2.10.14b}$$

then, substituting the above into (2.10.2), and renaming some dummy indices, we obtain the equivalent γ-expression:

$$\gamma^k{}_{rs} = \sum\sum a_{kb}\left[A_{cr}(\partial A_{bs}/\partial q_c) - A_{cs}(\partial A_{br}/\partial q_c)\right]. \tag{2.10.15}$$

For $s \to n+1$, the above yields an alternative to the (2.10.3), (2.10.4) expression for $\gamma^k{}_{r,n+1} \equiv \gamma^k{}_r$.

Problem 2.10.2 Show that yet another γ-expression is

$$\gamma^k{}_{rs} = \sum\sum (A_{br}A_{cs} - A_{cr}A_{bs})(\partial a_{kb}/\partial q_c), \tag{a}$$

and similarly for $\gamma^k{}_{r,n+1} \equiv \gamma^k{}_r$ (see also Stückler, 1955; Lobas, 1986, pp. 34–36).

Some Transformation Properties of the γ's

(i) With the help of the following useful notation:

$$a^k{}_{bc} \equiv \partial a_{kb}/\partial q_c - \partial a_{kc}/\partial q_b = -a^k{}_{cb}, \tag{2.10.16a}$$

$$a^k{}_{b,n+1} \equiv a^k{}_b \equiv \partial a_{kb}/\partial t - \partial a_k/\partial q_b \tag{2.10.16b}$$

[recalling (2.9.16); also similar notation in (2.8.2a)], the γ-definitions (2.10.2)–(2.10.4) are rewritten, respectively, as

$$\gamma^k{}_{rs} = \sum\sum a^k{}_{bc}A_{br}A_{cs}, \qquad \gamma^k{}_r = \sum\sum a^k{}_{bc}A_{br}A_c + \sum a^k{}_b A_{br}. \tag{2.10.17a, b}$$

With the help of the inverseness conditions (2.9.3a, 3b) and a number of dummy index changes, it is not too hard to show that (2.10.17a,b) invert, respectively, to

$$a^k{}_{bc} = \sum\sum \gamma^k{}_{rs}a_{rb}a_{sc}, \qquad a^k{}_b = \sum\sum \gamma^k{}_{rs}a_{rb}a_s + \sum \gamma^k{}_r a_{rb}. \tag{2.10.18a, b}$$

The above transformation equations show that if the $a^k{}_{bc}$ and $a^k{}_b$ vanish [recall conditions (2.9.16)], so do the $\gamma^k{}_{rs}$ and $\gamma^k{}_r$; and vice versa; that is, the vanishing of $\gamma^k\ldots$ constitutes the necessary and sufficient condition for $d\theta_k/\delta\theta_k$ to be an *exact differential*, and hence, for θ_k to be a *holonomic coordinate*. If the $dq/\delta q/v$ are *unconstrained*, as is the case so far (i.e., $m = 0$), this new set of exactness conditions in terms of the γ's does not offer any advantages over (2.9.16); the $a^k{}_{bc}$ and $a^k{}_b$ are easier to calculate than $\gamma^k{}_{rs}$ and $\gamma^k{}_r$. As shown in the next section, the real value of the γ's, in questions of holonomicity, appears whenever the $dq/\delta q/v$ are *constrained* ($m \neq 0$).

REMARK

For those familiar with tensors, the transformation equations (2.10.17a, b) show that the $\gamma^k{}_{\ldots}$ and $a^k{}_{\ldots}$ transform as *covariant tensors* in their two subscripts; that is, both are components of the *same geometrical entity*: the a's, its holonomic components in the local "coordinates" $dq/\delta q$, and the γ's, its nonholonomic components in the local "coordinates" $d\theta/\delta\theta$, at (q, t). In precise tensor notation, using, for example, *accented* (*unaccented*) indices for *nonholonomic* (*holonomic*) *components*, summation convention over pairs of diagonal indices of the same kind (i.e., both holonomic, or both nonholonomic), and with the notational changes: $A_{br} \to A^b{}_{r'} \to A^r{}_{r'}$, $A_{cs} \to A^c{}_{s'} \to A^s{}_{s'}$, and $a^k{}_{bc} \to a^{k'}{}_{bc} \to a^{k'}{}_{rs} \to \gamma^{k'}{}_{rs}$ (= holonomic components), $\gamma^k{}_{rs} \to \gamma^{k'}{}_{r's'}$ (= nonholonomic components), the transformation equations (2.10.17a) read

$$\gamma^{k'}{}_{r's'} = A^r{}_{r'} A^s{}_{s'} \gamma^{k'}{}_{rs}; \tag{2.10.17c}$$

and similarly for (2.10.17b)–(2.10.18b). Such elaborate notation is a must in advanced differential-geometric investigations of nonholonomic systems. Fortunately, it will not be needed here.

(ii) The invariant definition of the γ's via the transitivity equations (2.10.1) and (2.10.12) readily shows that, contrary to what one might conclude by casually inspecting their derivative definition via (2.10.2–4), these *nontensorial* coefficients, known in tensor calculus as *geometrical objects of nonholonomicity* (or *anholonomicity*), are independent of the original holonomic coordinates q, and thus express geometric properties of the local/differential basis $d\theta/\delta\theta/\omega$. In particular, it follows that *if the γ's do (not) vanish, when based on some (q, t) frame of reference, they will (not) vanish in any other frame (q', t), obtainable from the original frame by an admissible transformation.*

(iii) However, under a *local* transformation $d\theta_k \Leftrightarrow d\theta_{k'}$, that is, at the same (q, t)-point, the γ's, *do* change, in the earlier mentioned nontensorial fashion.

[(a) For further details on tensorial nonholonomic dynamics see, for example, Dobronravov (1948, 1970, 1976), Kil'chevskii (1972, 1977), Maißer (1981, 1982, 1983–1984, 1991(b), 1997), Papastavridis (1999), Schouten (1954), Synge (1936), Vranceanu (1936); and references cited there. (b) For *transitivity* equation–based proofs of these statements, see, for (ii): ex. 2.12.2, and for (iii): ex. 2.10.1; and for a *derivative* definition–based proof, see, for example, Golab (1974, pp. 140–141).]

Noncommutativity of Mixed Partial Quasi Derivatives

Below we show that the second mixed partial symbolic quasi derivatives of an arbitrary well-behaved function $f = f(q, t, \ldots)$, in general, do *not* commute:

$$\partial/\partial\theta_k(\partial f/\partial\theta_l) \neq \partial/\partial\theta_l(\partial f/\partial\theta_k). \tag{2.10.19}$$

Invoking the basic quasi-derivative definition (2.9.30a, b), we obtain, successively,

$$\begin{aligned}\partial^2 f/\partial\theta_k\,\partial\theta_l &\equiv \partial/\partial\theta_k(\partial f/\partial\theta_l) \equiv \sum A_{rk}\Big\{\partial/\partial q_r\Big(\sum A_{sl}(\partial f/\partial q_s)\Big)\Big\}\\ &= \sum\sum\big[A_{rk}A_{sl}(\partial^2 f/\partial q_r\,\partial q_s) + A_{rk}(\partial A_{sl}/\partial q_r)(\partial f/\partial q_s)\big]\\ &= \sum\sum A_{rk}A_{sl}(\partial^2 f/\partial q_r\,\partial q_s) + \sum\Big(\sum\sum a_{bs}A_{rk}(\partial A_{sl}/\partial q_r)\Big)(\partial f/\partial\theta_b),\end{aligned}$$

and, analogously (with $k \to l$ and $l \to k$ in the above),

$$\begin{aligned}\partial^2 f/\partial\theta_l\,\partial\theta_k &\equiv \partial/\partial\theta_l(\partial f/\partial\theta_k) = \cdots\\ &= \sum\sum A_{rl}A_{sk}(\partial^2 f/\partial q_r\,\partial q_s) + \sum\Big(\sum\sum a_{bs}A_{rl}(\partial A_{sk}/\partial q_r)\Big)(\partial f/\partial\theta_b),\end{aligned}$$

and therefore subtracting these two side by side, and recalling the γ-definition (2.10.15), we obtain the following alternative *transitivity/noncommutativity* relation:

$$\begin{aligned}\partial^2 f/\partial\theta_k\,\partial\theta_l - \partial^2 f/\partial\theta_l\,\partial\theta_k &\equiv \partial/\partial\theta_k(\partial f/\partial\theta_l) - \partial/\partial\theta_l(\partial f/\partial\theta_k)\\ &= \sum\Big\{\sum\sum a_{bs}\big[A_{rk}(\partial A_{sl}/\partial q_r) - A_{rl}(\partial A_{sk}/\partial q_r)\big]\Big\}(\partial f/\partial\theta_b)\\ &= \sum \gamma^b{}_{kl}(\partial f/\partial\theta_b); \qquad (2.10.20)\end{aligned}$$

which expresses noncommutativity in terms of $(\partial\ldots/\partial\theta)$-derivatives, rather than $(d\ldots/\delta\ldots)$-differentials, as (2.10.1) and (2.10.12) do.

REMARK

In the theory of continuous (or Lie) groups, it is customary to write $X_k f$ for our $\partial f/\partial\theta_k$, (2.9.30a); that is,

$$\partial\ldots/\partial\theta_k \equiv X_k\cdots \equiv \sum(\partial\ldots/\partial q_l)(\partial v_l/\partial\omega_k) = \sum A_{lk}(\partial\ldots/\partial q_l). \qquad (2.10.21)$$

The differential operators X_k are called the *generators* of that group. In this notation, equation (2.10.20) is rewritten as

$$[X_k, X_l]f = \sum \gamma^b{}_{kl}(X_b f), \qquad (2.10.22)$$

where $[X_k, X_l] \equiv X_k X_l - X_l X_k \equiv \sum \gamma^b{}_{kl}(X_b)$: *commutator of group*. For further details, see texts on Lie groups, and so on; also Hamel (1904(a), (b)), Hagihara (1970), McCauley (1997).

Problem 2.10.3 Extend (2.10.20) to the case where one or both of θ_k, θ_l are the (θ_{n+1})th "coordinate", that is, $\theta \to t$.

Problem 2.10.4 The choice $f \to q_r$ in (2.10.20), and then use of (2.9.34), yields the symbolic identity

$$\partial^2 q_r/\partial\theta_k\,\partial\theta_l - \partial^2 q_r/\partial\theta_l\,\partial\theta_k = \sum \gamma^b{}_{kl}(\partial q_r/\partial\theta_b) = \sum A_{rb}\gamma^b{}_{kl}. \qquad \text{(a)}$$

Solving (a) for the γ's, derive the following alternative symbolic expression/definition for γ:

$$\gamma^b{}_{kl} = \sum a_{br}(\partial^2 q_r/\partial\theta_k\,\partial\theta_l - \partial^2 q_r/\partial\theta_l\,\partial\theta_k). \tag{b}$$

HINT

Multiply (a) with a_{sr} and sum over r, and so on.

Nonintegrability Conditions for a Nonholonomic Basis

Since (2.10.20) holds for an arbitrary f, let us apply it for $f \to \boldsymbol{r} = \boldsymbol{r}(t, q)$. In this case, $\partial f/\partial\theta_b \to \partial\boldsymbol{r}/\partial\theta_b \equiv \boldsymbol{\varepsilon}_b$, and thus we obtain the basic *nonintegrability* conditions for the nonholonomic basis $\{\boldsymbol{\varepsilon}_k; k = 1, \ldots, n\}$:

$$\partial\boldsymbol{\varepsilon}_l/\partial\theta_k - \partial\boldsymbol{\varepsilon}_k/\partial\theta_l = \sum \gamma^b{}_{kl}\,\boldsymbol{\varepsilon}_b, \tag{2.10.23}$$

or, compactly,

$$[\boldsymbol{\varepsilon}_k, \boldsymbol{\varepsilon}_l] \equiv \sum \gamma^b{}_{kl}\,\boldsymbol{\varepsilon}_b \equiv \textit{commutator of basis } \{\boldsymbol{\varepsilon}_k\}. \tag{2.10.23a}$$

In differential geometry, such bases are called *nonholonomic*, or *noncoordinate*, or *nongradient*; that is, they are *not* parts of a global coordinate system; like the $\{\boldsymbol{e}_k \equiv \partial\boldsymbol{r}/\partial q_k\}$ for which, clearly [recalling (2.5.4a)],

$$\partial\boldsymbol{e}_l/\partial q_k - \partial\boldsymbol{e}_k/\partial q_l \equiv [\boldsymbol{e}_k, \boldsymbol{e}_l] \equiv \boldsymbol{0}. \tag{2.10.23b}$$

In sum: the vanishing of the γ's is the necessary and sufficient condition for the corresponding basis to be *holonomic*; or *gradient*, or *coordinate*.

We leave it to the reader to show that (2.10.23) also hold for $k, l = n + 1$; that is, $\theta \to t$.

A Fundamental Kinematical Identity

Here, with the help of (2.10.23), we will complete the derivation of the basic identity (2.9.37). Indeed, since $\boldsymbol{\varepsilon}_k = \partial\boldsymbol{v}/\partial\omega_k \equiv \partial\boldsymbol{v}^*/\partial\omega_k = \boldsymbol{\varepsilon}_k(t, q)$, and [recalling (2.9.21)] $\boldsymbol{v} = \boldsymbol{v}^*(t, q, \omega) = \sum \boldsymbol{\varepsilon}_k\omega_k + \boldsymbol{\varepsilon}_{n+1} \equiv \sum \boldsymbol{\varepsilon}_k\omega_k + \boldsymbol{\varepsilon}_0$, we obtain, successively,

(i)
$$\begin{aligned}
d/dt(\partial\boldsymbol{v}/\partial\omega_k) &\equiv d/dt(\partial\boldsymbol{v}^*/\partial\omega_k) \equiv d\boldsymbol{\varepsilon}_k/dt \\
&= \sum (\partial\boldsymbol{\varepsilon}_k/\partial q_l)v_l + \partial\boldsymbol{\varepsilon}_k/\partial t \\
&\qquad \text{[recalling the inverse quasi chain rule (2.9.30b)]} \\
&= \sum \left(\sum a_{rl}(\partial\boldsymbol{\varepsilon}_k/\partial\theta_r)\right)v_l + \partial\boldsymbol{\varepsilon}_k/\partial t \qquad \text{[recalling (2.9.9)]} \\
&= \sum (\partial\boldsymbol{\varepsilon}_k/\partial\theta_r)(\omega_r - a_r) + \partial\boldsymbol{\varepsilon}_k/\partial t \\
&= \sum (\partial\boldsymbol{\varepsilon}_k/\partial\theta_r)\omega_r - \sum (\partial\boldsymbol{\varepsilon}_k/\partial\theta_r)a_r + \partial\boldsymbol{\varepsilon}_k/\partial t.
\end{aligned} \tag{2.10.24a}$$

(ii)
$$\partial\boldsymbol{v}/\partial\theta_k \equiv \partial\boldsymbol{v}^*/\partial\theta_k = \sum (\partial\boldsymbol{\varepsilon}_r/\partial\theta_k)\omega_r + \partial\boldsymbol{\varepsilon}_0/\partial\theta_k. \tag{2.10.24b}$$

Therefore, subtracting the above side by side, and recalling (2.9.32a, d), we obtain

$$d\boldsymbol{\varepsilon}_k/dt - \partial\boldsymbol{v}/\partial\theta_k = \sum(\partial\boldsymbol{\varepsilon}_k/\partial\theta_r - \partial\boldsymbol{\varepsilon}_r/\partial\theta_k)\omega_r + (\partial\boldsymbol{\varepsilon}_k/\partial t - \partial\boldsymbol{\varepsilon}_0/\partial\theta_k) - \sum(\partial\boldsymbol{\varepsilon}_k/\partial\theta_r)a_r$$

$$= \sum(\partial\boldsymbol{\varepsilon}_k/\partial\theta_r - \partial\boldsymbol{\varepsilon}_r/\partial\theta_k)\omega_r$$

$$+ \left(\partial\boldsymbol{\varepsilon}_k/\partial\theta_0 - \sum A_s(\partial\boldsymbol{\varepsilon}_k/\partial q_s)\right)$$

$$- \left(\partial\boldsymbol{\varepsilon}_0/\partial\theta_k - \sum(\partial\boldsymbol{\varepsilon}_k/\partial\theta_r)a_r\right)$$

$$= \sum(\partial\boldsymbol{\varepsilon}_k/\partial\theta_\beta - \partial\boldsymbol{\varepsilon}_\beta/\partial\theta_k)\omega_\beta$$

$$- \sum A_s\left(\sum a_{rs}(\partial\boldsymbol{\varepsilon}_k/\partial\theta_r)\right) - \sum a_r(\partial\boldsymbol{\varepsilon}_k/\partial\theta_r)$$

{for the first sum we use (2.10.23), with $l \to k$, $k \to \beta$, $b \to r$ [recalling (2.10.9)]; and by the second of (2.9.3a) the last two sums add up to zero}

$$= \sum\left(\sum \gamma^r_{\beta k}\boldsymbol{\varepsilon}_r\right)\omega_\beta; \tag{2.10.24c}$$

and so, finally,

$E_k{}^*(\boldsymbol{v}) \equiv E_k{}^*(\boldsymbol{v}^*)$: Hamel vector of *nonholonomic deviation* of a particle

$$\equiv d/dt(\partial\boldsymbol{v}/\partial\omega_k) - \partial\boldsymbol{v}/\partial\theta_k \equiv d/dt(\partial\boldsymbol{v}^*/\partial\omega_k) - \partial\boldsymbol{v}^*/\partial\theta_k \equiv d\boldsymbol{\varepsilon}_k/dt - \partial\boldsymbol{v}/\partial\theta_k$$

$$= \sum\sum \gamma^r_{\beta k}\boldsymbol{\varepsilon}_r\,\omega_\beta = \sum\sum \gamma^r_{lk}\,\omega_l\boldsymbol{\varepsilon}_r + \sum \gamma^r_{n+1,k}\,\omega_{n+1}\boldsymbol{\varepsilon}_r \quad \text{[swapping } k \text{ and } l\text{]}$$

$$= -\sum\sum \gamma^r_{kl}\,\omega_l\boldsymbol{\varepsilon}_r - \sum \gamma^r_k\boldsymbol{\varepsilon}_r = -\sum\left(\sum \gamma^r_{kl}\,\omega_l + \gamma^r_k\right)\boldsymbol{\varepsilon}_r$$

$$\equiv -\sum h^r_k\,\boldsymbol{\varepsilon}_r; \tag{2.10.25}$$

where

$$h^r_k \equiv \sum \gamma^r_{kl}\,\omega_l + \gamma^r_k = \sum \gamma^r_{k\beta}\,\omega_\beta: \textit{ Two-index Hamel symbols.} \tag{2.10.25a}$$

This fundamental kinematical identity, in its various equivalent forms, like the transitivity equations (2.10.1, etc.), shows clearly the difference between holonomic and nonholonomic *coordinates* (not constraints): for the former, $E_k(\boldsymbol{v}) = \boldsymbol{0}$; while for the latter, $E_k{}^*(\boldsymbol{v}) \equiv E_k{}^*(\boldsymbol{v}^*) \neq \boldsymbol{0}$. It is indispensable in the derivation of equations of motion in quasi variables (§3.3).

Problem 2.10.5 *Transitivity Relations for System Velocities.*

(i) Show that for the general nonstationary transformation (with $\dot{q}_l \equiv v_l$)

$$\omega_k \equiv \sum a_{kl}v_l + a_k \Leftrightarrow v_l = \sum A_{lk}\omega_k + A_l, \tag{a}$$

the following transitivity identities hold:

$$E_l(\omega_k) \equiv d/dt(\partial\omega_k/\partial v_l) - \partial\omega_k/\partial q_l = \sum\left(\sum \gamma^k_{r\beta}\,\omega_\beta\right)a_{rl}$$

$$= \sum\left(\sum \gamma^k_{rs}\,\omega_s + \gamma^k_r\right)a_{rl} \equiv \sum h^k_r\,a_{rl}. \tag{b}$$

(ii) Then show that, in the *stationary* case, (b) specializes to

$$E_l(\omega_k) \equiv d/dt(\partial\omega_k/\partial v_l) - \partial\omega_k/\partial q_l = \sum\sum \gamma^k{}_{rs}\,\omega_s a_{rl}; \tag{c}$$

that is, the first line of (b) with $\beta \to s$.

(iii) Show that, as a result of the above, the transitivity equations (2.10.13), become

$$\begin{aligned}(\delta\theta_k)^{\cdot} - \delta\omega_k &= \sum\sum \gamma^k{}_{rs}\,\omega_s\,\delta\theta_r + \sum \gamma^k{}_r\,\delta\theta_r = \sum h^k{}_r\,\delta\theta_r \\ &= \sum\sum E_l(\omega_k)A_{lr}\,\delta\theta_r = \sum\sum (\partial v_l/\partial\omega_r)E_l(\omega_k)\delta\theta_r,\end{aligned} \tag{d}$$

where the $\sum(\partial v_l/\partial\omega_r)E_l(\omega_k)$ can be viewed as the *nonlinear* generalization of the $h^k{}_r$ (§5.2).

Problem 2.10.6 By direct d/δ-differentiations of $\delta\boldsymbol{r} = \sum \boldsymbol{\varepsilon}_k\,\delta\theta_k$ and $d\boldsymbol{r} = \sum \boldsymbol{\varepsilon}_k\,d\theta_k$, respectively (assume stationary systems, for algebraic simplicity but no loss in generality), and then use of

$$d\boldsymbol{\varepsilon}_k = d\Big(\sum A_{lk}\,\boldsymbol{e}_l\Big) = \sum (dA_{lk}\,\boldsymbol{e}_l + A_{lk}\,d\boldsymbol{e}_l), \tag{a}$$

and

$$d\boldsymbol{e}_l = \sum (\partial\boldsymbol{e}_l/\partial q_r)\,dq_r = \sum\sum (\partial\boldsymbol{e}_l/\partial q_r)A_{rs}\,d\theta_s,$$

$$dA_{lk} = \sum (\partial A_{lk}/\partial q_r)\,dq_r = \sum\sum (\partial A_{lk}/\partial q_r)A_{rs}\,d\theta_s, \tag{b}$$

and similarly for $\delta\boldsymbol{\varepsilon}_k = \delta(\sum A_{lk}\boldsymbol{e}_l) = \ldots$, and then recalling the γ-definitions, obtain the following basic *particle/vectorial transitivity equation*:

$$d(\delta\boldsymbol{r}) - \delta(d\boldsymbol{r}) = \sum \Big\{[d(\delta\theta_k) - \delta(d\theta_k)] + \sum\sum \gamma^k{}_{rs}\,d\theta_r\,\delta\theta_s\Big\}\boldsymbol{\varepsilon}_k; \tag{c}$$

or, dividing by dt, its equivalent velocity form

$$(\delta\boldsymbol{r})^{\cdot} - \delta\boldsymbol{v} = \sum \Big\{[(\delta\theta_k)^{\cdot} - \delta\omega_k] + \sum\sum \gamma^k{}_{rs}\,\omega_r\,\delta\theta_s\Big\}\,\boldsymbol{\varepsilon}_k. \tag{d}$$

Replacing in the above r with $\beta = 1,\ldots,n+1$, extends it to the *nonstationary/rheonomic* case.

[Note that (c) and (d) are independent of any $d(\delta q) - \delta(dq)$ assumptions. Therefore, since

$$d(\delta\boldsymbol{r}) - \delta(d\boldsymbol{r}) = \sum [d(\delta q_l) - \delta(dq_l)]\boldsymbol{e}_l = \sum\sum [d(\delta q_l) - \delta(dq_l)]a_{kl}\boldsymbol{\varepsilon}_k, \tag{e}$$

if we assume $d(\delta q_l) - \delta(dq_l) = 0$ (Hamel viewpoint), then $d(\delta\boldsymbol{r}) - \delta(d\boldsymbol{r}) = \boldsymbol{0}$, and this leads us back to the transitivity equations (2.10.12) and (2.10.13).]

Example 2.10.1 *Local Transformation Properties of the Hamel Coefficients.* Let us find how the γ's transform under the admissible (and, for simplicity, but with no loss in generality) stationary quasi-variable transformation $\theta \to \theta'$:

$$d\theta_{k'} = \sum a_{k'k}\,d\theta_k \Leftrightarrow d\theta_k = \sum A_{kk'}\,d\theta_{k'}, \tag{a}$$

where $a_{k'k} = a_{k'k}(q)$, $A_{kk'} = A_{kk'}(q)$, and all Latin indices run from 1 to n. We find, successively,

$$
\begin{aligned}
d(\delta\theta_{k'}) - \delta(d\theta_{k'}) &= d\left(\sum a_{k'k}\,\delta\theta_k\right) - \delta\left(\sum a_{k'k}\,d\theta_k\right) \\
&= \sum \left[da_{k'k}\,\delta\theta_k + a_{k'k}\,d(\delta\theta_k) - \delta a_{k'k}\,d\theta_k - a_{k'k}\,\delta(d\theta_k)\right] \\
&= \sum \left\{\left(\sum (\partial a_{k'k}/\partial q_p)\,dq_p\right)\delta\theta_k + a_{k'k}\,d(\delta\theta_k)\right. \\
&\qquad \left. - \left(\sum (\partial a_{k'k}/\partial q_p)\,\delta q_p\right) d\theta_k - a_{k'k}\,\delta(d\theta_k)\right\} \\
&\qquad\qquad \left[\text{recalling that } dq_p = \sum A_{pr}\,d\theta_r\,,\ \text{etc.}\right] \\
&= \sum a_{k'k}[d(\delta\theta_k) - \delta(d\theta_k)] \\
&\quad + \sum\sum\sum \left[(\partial a_{k'k}/\partial q_p)A_{pr}\,d\theta_r\,\delta\theta_k - (\partial a_{k'k}/\partial q_p)A_{pr}\,d\theta_k\,\delta\theta_r\right] \\
&= \sum a_{k'k}\left(\sum\sum \gamma^k{}_{bc}\,d\theta_c\,\delta\theta_b\right) \\
&\quad + \sum\sum\sum \left[(\partial a_{k'k}/\partial q_p)A_{pr} - (\partial a_{k'r}/\partial q_p)A_{pk}\right] d\theta_r\,\delta\theta_k \\
&= \sum\sum\sum a_{k'k}\gamma^k{}_{bc}\left(\sum A_{cc'}\,d\theta_{c'}\right)\left(\sum A_{bb'}\,\delta\theta_{b'}\right) \\
&\quad + \sum\sum\sum \left[(\partial a_{k'k}/\partial q_p)A_{pr} - (\partial a_{k'r}/\partial q_p)A_{pk}\right]\left(\sum A_{rr'}\,d\theta_{r'}\right)\left(\sum a_{kl'}\ \delta\right. \\
&= \sum\sum\sum\sum\sum (a_{k'k}A_{rr'}A_{ll'}\gamma^k{}_{lr})\,d\theta_{r'}\,\delta\theta_{l'} \\
&\quad + \sum\sum\sum\sum (\partial a_{k'k}/\partial\theta_r - \partial a_{k'r}/\partial\theta_k)A_{rr'}A_{kl'}\,d\theta_{r'}\,\delta\theta_{l'}; \qquad \text{(b)}
\end{aligned}
$$

and since, by definition,

$$
d(\delta\theta_{k'}) - \delta(d\theta_{k'}) = \sum\sum \gamma^{k'}{}_{l'r'}\,d\theta_{r'}\,\delta\theta_{l'}, \qquad \text{(c)}
$$

we conclude that

$$
\gamma^{k'}{}_{l'r'} = \sum\sum\sum a_{k'k}A_{ll'}A_{rr'}\gamma^k{}_{lr} + \sum\sum (\partial a_{k'k}/\partial\theta_r - \partial a_{k'r}/\partial\theta_k)A_{kl'}A_{rr'}. \qquad \text{(d)}
$$

In tensor calculus language, the transformation equation (d) shows that *the* $\gamma^k{}_{lr}$ *do not constitute a tensor*: if $\gamma^k{}_{lr} = 0$ (i.e., if the θ_k are holonomic coordinates), it does not necessarily follow that $\gamma^{k'}{}_{l'r'} = 0$; and that is why these quantities are called, instead, components of a *geometrical object*. However, if the *second* group of terms (double sum) in (d), which looks (symbolically) like a Hamel coefficient between the $d\theta_k$ and $d\theta_{k'}$, vanishes, the $\gamma^k{}_{lr}$ transform tensorially. In such a case, we call the $d\theta_k$ and $d\theta_{k'}$ *relatively holonomic*; that happens, for example, if the coefficients $a_{k'k}$ are constant.

For futher details, and the relation of the γ's to the *Christoffell symbols* (§3.10) and the *Ricci rotation coefficients*, both of which are also geometrical objects, see, for example (alphabetically): Papastavridis (1999), Schouten (1954), Synge (1936), Vranceanu (1936); also, for an alternative derivation of (d), see Golab (1974, pp. 141–142), Lynn (1963, pp. 201–203).

We have developed all the necessary analytical tools of Lagrangean kinematics. In the following sections, we will show how to apply them to the handling of *additional* Pfaffian (possibly nonholonomic) constraints.

For quick comparison, when working with other references, we present below the following, admittedly incomplete, but hopefully helpful, list of common γ-notations in the literature:

(i) *Our notation* (also in Papastavridis, 1999): $\gamma_{a\ c}^{\ b} \equiv \gamma^{b}_{\ ac}$ (sometimes, for extra clarity, a *subscript dot* is added between a and c, directly below b).

(ii) *Authors whose notation coincides with ours*: Dobronravov (1948, 1970, 1976), Golomb and Marx (1961), Gutowski (1971), Kil'chevskii (1972, 1977), Koiller (1992): $\gamma_{a\ c}^{\ b}$.

(iii) *Authors whose notation differs from ours*: Butenin (1971), Fischer and Stephan (1972), Neimark and Fufaev (1967/1972), Whittaker (1937; but his a_{kl} is our a_{lk}): γ_{abc}; Corben and Stehle (1960): γ_{acb}; Nordheim (1927): γ_{cba}; Rose (1938): γ_{bac}; Päsler (1968): $-\gamma_{bac}$; Djukic (1976), Funk (1962), Lur'e (1961/1968), Mei (1985), Prange (1935): $\gamma_{c\ a}^{\ b}$; Kilmister (1964, 1967): $\gamma^{a\ c}_{\ b}$; Maißer (1981): $A_{c\ a}^{\ b}$; Desloge (1982): α_{abc}; Stückler (1955): β_{abc}; Heun (1906): β_{acb}; Winkelmann and Grammel (1927): β_{cab}; Morgenstern and Szabó (1961): $\beta_{b,ac}$; Hamel (1904(a), (b)): $\beta_{a,c,b}$; Hamel (1949): $\beta_b^{\ a,c}$; Schaefer (1951): $\beta_{c\ a}^{\ b}$; Vranceanu (1936): $w_{a\ c}^{\ b}$; Wang (1979): $K_{A\ C}^{\ B}$; Schouten (1954): $2\Omega_{c\ a}^{\ b}$; Levi-Civita and Amaldi (1927): $\eta_{b|ca}$.

2.11 PFAFFIAN (VELOCITY) CONSTRAINTS VIA QUASI VARIABLES, AND THEIR GEOMETRICAL INTERPRETATION

Let us, now, assume that our hitherto holonomic $n\,(\equiv 3N - h)$-DOF system is subjected to the additional m independent Pfaffian constraints [recalling (2.7.3 and 2.7.4)]:

Kinematically admissible/possible form:

$$\sum c_{Dk}\, dq_k + c_D\, dt = 0, \tag{2.11.1a}$$

Virtual form:

$$\sum c_{Dk}\, \delta q_k = 0, \tag{2.11.1b}$$

Velocity form (with $dq_k/dt \equiv v_k$):

$$\sum c_{Dk} v_k + c_D = 0; \tag{2.11.1c}$$

where $D = 1, \ldots, m\,(<n), k = 1, \ldots, n$; and the constraint independence is expressed by the algebraic requirement $rank(c_{Dk}) = m$. Since additional holonomic constraints (in any form) can always be embedded, or built in, with a new set of fewer q's, we can, with no loss of generality, assume that all constraints (2.11.1) are nonholonomic.

Now, and in what constitutes a direct and natural extension of the method of holonomic equilibrium coordinates (§2.4) to the embedding Pfaffian constraints, we introduce the following *equilibrium quasi variables* (Hamel's choice):

Kinematically admissible/possible form:

$$d\theta_D \equiv \sum a_{Dk}\, dq_k + a_D\, dt \quad (=0), \tag{2.11.2a}$$

$$d\theta_I \equiv \sum a_{Ik}\, dq_k + a_I\, dt \quad (\neq 0), \tag{2.11.2b}$$

$$d\theta_{n+1} \equiv d\theta_0 \equiv dq_{n+1} \equiv dq_0 \equiv dt \quad (\neq 0); \tag{2.11.2c}$$

Virtual form:

$$\delta\theta_D \equiv \sum a_{Dk}\, \delta q_k \quad (=0), \tag{2.11.2d}$$

$$\delta\theta_I \equiv \sum a_{Ik}\, \delta q_k \quad (\neq 0), \tag{2.11.2e}$$

$$\delta\theta_{n+1} \equiv \delta q_{n+1} \equiv \delta t \quad (=0); \tag{2.11.2f}$$

Velocity form:

$$\omega_D \equiv \sum a_{Dk} v_k + a_D \quad (=0), \tag{2.11.2g}$$

$$\omega_I \equiv \sum a_{Ik} v_k + a_I \quad (\neq 0), \tag{2.11.2h}$$

$$\omega_{n+1} \equiv \omega_0 \equiv v_{n+1} \equiv v_0 \equiv dt/dt = 1 \quad (\neq 0); \tag{2.11.2i}$$

where (here and throughout the rest of the book): $D = 1, \ldots, m\ (< n) =$ *Dependent*, $I = m+1, \ldots, n =$ *Independent* [additional *dependent* (*independent*) indices will be denoted by $D', D'', \ldots (I', I'', \ldots)$]; and the coefficients a_{kl}, a_k are chosen as follows:

(i) $a_{Dk} \equiv c_{Dk}$ and $a_D \equiv c_D$ [i.e., $\theta_D \equiv \chi_D$, recall (2.6.2–4; 2.8.1)], (2.11.3)

(ii) The a_{Ik} and a_I are *arbitrary*, except that when eqs. (2.11.2) are solved (inverted) for the $dq/\delta q/v$ in terms of the independent $d\theta/\delta\theta/\omega$, respectively; that is,

Kinematically admissible/possible form:

$$dq_k \equiv \sum A_{kI}\, d\theta_I + A_I\, dt \quad (\neq 0), \tag{2.11.4a}$$

$$dq_{n+1} \equiv dq_0 \equiv d\theta_{n+1} \equiv d\theta_0 \equiv dt \quad (\neq 0); \tag{2.11.4b}$$

Virtual form:

$$\delta q_k \equiv \sum A_{kI}\, \delta\theta_I \quad (\neq 0), \tag{2.11.4c}$$

$$\delta q_{n+1} \equiv \delta q_0 \equiv \delta\theta_{n+1} \equiv \delta\theta_0 \equiv \delta t = 0; \tag{2.11.4d}$$

Velocity form:

$$v_k \equiv \sum A_{kI}\omega_I + A_I \quad (\neq 0), \tag{2.11.4e}$$

$$v_{n+1} \equiv v_0 \equiv \omega_{n+1} \equiv \omega_0 \equiv dt/dt = 1 \quad (\neq 0); \tag{2.11.4f}$$

and then these results are substituted back into (2.11.1a–c) and (2.11.3), they *satisfy them identically*. Other choices of θ's and a's are, of course, possible (see special forms/choices, below), but Hamel's choice (2.11.2) is the simplest and most natural, because then our Pfaffian constraints assume the simple and *uncoupled* form:

Kinematically admissible/possible form:

$$d\theta_D = 0, \tag{2.11.5a}$$

Virtual form:

$$\delta\theta_D = 0, \tag{2.11.5b}$$

Velocity form:

$$\omega_D = 0; \tag{2.11.5c}$$

and, as a result (already described in § 2.7 and detailed in ch. 3), the equations of motion decouple into $n - m$ *kinetic* equations (no constraint forces) and m *kinetostatic* equations (constraint forces).

Constrained Particle Kinematics

In view of the constraints (2.11.5), the particle kinematical quantities (2.9.23–26) reduce to the following:

Kinematically admissible/possible displacement:

$$d\boldsymbol{r} = \sum \boldsymbol{\varepsilon}_I \, d\theta_I + \boldsymbol{\varepsilon}_{n+1} \, dt \equiv \sum \boldsymbol{\varepsilon}_I \, d\theta_I + \boldsymbol{\varepsilon}_0 \, dt; \tag{2.11.6a}$$

Virtual displacement:

$$\delta\boldsymbol{r} = \sum \boldsymbol{\varepsilon}_I \, \delta\theta_I ; \tag{2.11.6b}$$

Velocity:

$$\boldsymbol{v} = \sum \boldsymbol{\varepsilon}_I \, \omega_I + \boldsymbol{\varepsilon}_{n+1} \equiv \sum \boldsymbol{\varepsilon}_I \, \omega_I + \boldsymbol{\varepsilon}_0 ; \tag{2.11.6c}$$

Acceleration:

$$\boldsymbol{a} = \sum \boldsymbol{\varepsilon}_I \, \dot{\omega}_I + \text{terms not containing } \dot{\omega}. \tag{2.11.6d}$$

Special Forms/Choices of Quasi Variables

1. Once we have chosen the equilibrium quasi variables $d\theta/\delta\theta/\omega$, we can move to any other such set $d\theta'/\delta\theta'/\omega'$, defined via linear (invertible) transformations of the following type:

$$d\theta_{k'} \equiv \sum a_{k'k} \, d\theta_k + a_{k'} \, dt = \sum a_{k'I} \, d\theta_I + a_{k'} \, dt \quad (\neq 0), \tag{2.11.7a}$$

$$d\theta_{(n+1)'} \equiv d\theta_{n+1} \equiv dq_{n+1} \equiv dt \quad (\neq 0); \tag{2.11.7b}$$

and, inversely $[(a_{k'k}), (A_{kk'})$: nonsingular matrices],

$$d\theta_k \equiv \sum A_{kk'} \, d\theta_{k'} + A_k \, dt \quad \rightarrow \quad d\theta_D = 0 \quad \text{and} \quad d\theta_I \neq 0; \tag{2.11.7c}$$

and similarly for $\delta\theta_{k'}$, $\omega_{k'}$.

2. If the Pfaffian nonholonomic *constraints* are given in the quasi-variable forms:

$$\sum a_{D'k}\, d\theta_k + a_{D'}\, dt = 0, \quad \text{or} \quad \sum a_{D'k}\, \delta\theta_k = 0, \quad \text{or} \quad \sum a_{D'k}\, \omega_k + a_{D'} = 0, \tag{2.11.8a}$$

then, proceeding à la Hamel again, we may introduce new quasi variables by

$$d\theta_{D'} \equiv \sum a_{D'k}\, d\theta_k + a_{D'}\, dt = 0, \qquad d\theta_{I'} \equiv \sum a_{I'k}\, d\theta_k + a_{I'}\, dt \neq 0; \tag{2.11.8b}$$

or

$$\delta\theta_{D'} \equiv \sum a_{D'k}\, \delta\theta_k = 0, \qquad \delta\theta_{I'} \equiv \sum a_{I'k}\, \delta\theta_k \neq 0; \tag{2.11.8c}$$

or

$$\omega_{D'} \equiv \sum a_{D'k}\omega_k + a_{D'} = 0, \qquad \omega_{I'} \equiv \sum a_{I'k}\, \omega_k + a_{I'} \neq 0; \tag{2.11.8d}$$

where, again, the coefficients $a_{I'k}$, $a_{I'}$ are arbitrary; but when (2.11.8b–d) are solved for the $d\theta/\delta\theta/\omega$ in terms of the $d\theta'/\delta\theta'/\omega'$, and the results are substituted back into (2.11.8a), they satisfy them identically (see also their specialization in item 4, below).

3. Frequently, the Pfaffian constraints (2.11.1) are given, or can be easily brought to, the special form [recalling (2.6.9–11), and, using the notation $dq_k/dt \equiv v_k$]:

$$dq_D = \sum b_{DI}\, dq_I + b_D\, dt, \quad \text{or} \quad \delta q_D = \sum b_{DI}\, \delta q_I, \quad \text{or} \quad v_D = \sum b_{DI} v_I + b_D, \tag{2.11.9}$$

where the coefficients b_{DI}, b_D are known functions of q and t; that is, the *first* m (or *dependent*) $dq_D/\delta q_D/v_D$ are expressed in terms of the *last* $n - m$ (*independent*) $dq_I/\delta q_I/v_I$. [In terms of the elements of the *original* $m \times n$ constraint matrix $(c_{Dk}) \equiv (a_{Dk})$, we, clearly, have $(b_{DI}) = -(a_{DD'})^{-1}(a_{DI})$, and so on. See also pr. 2.11.2.]

Now, the transformations (2.11.9) can be viewed as the following *special choice* of $d\theta/\delta\theta/\omega$:

$$d\theta_D = dq_D - \sum b_{DI}\, dq_I - b_D\, dt = 0, \quad d\theta_I \equiv dq_I \neq 0, \quad d\theta_{n+1} \equiv dq_{n+1} \equiv dt \neq 0; \tag{2.11.10a}$$

$$\delta\theta_D \equiv \delta q_D - \sum b_{DI}\, \delta q_I = 0, \qquad \delta\theta_I \equiv \delta q_I \neq 0, \qquad \delta\theta_{n+1} \equiv \delta q_{n+1} \equiv \delta t = 0; \tag{2.11.10b}$$

$$\omega_D \equiv v_D - \sum b_{DI} v_I - b_D = 0, \quad \omega_I \equiv v_I \neq 0, \quad \omega_{n+1} \equiv v_{n+1} \equiv dt/dt = 1 \neq 0. \tag{2.11.10c}$$

The above invert easily to

$$dq_D = d\theta_D + \sum b_{DI}\, d\theta_I + b_D\, dt = \sum b_{DI}\, d\theta_I + b_D\, dt,$$

$$dq_I = d\theta_I, \qquad dq_{n+1} \equiv d\theta_{n+1} \equiv dt; \tag{2.11.11a}$$

$$\delta q_D = \delta\theta_D + \sum b_{DI}\,\delta q_I = \sum b_{DI}\,\delta q_I,$$

$$\delta q_I = \delta\theta_I, \qquad \delta q_{n+1} \equiv \delta\theta_{n+1} \equiv \delta t = 0; \tag{2.11.11b}$$

$$v_D = \omega_D + \sum b_{DI}\omega_I + b_D = \sum b_{DI}\omega_I + b_D,$$

$$v_I = \omega_I, \qquad v_{n+1} \equiv v_0 = \omega_{n+1} \equiv \omega_0 = dt/dt = 1. \tag{2.11.11c}$$

Comparing (2.11.10, 11) with (2.11.2, 4) we readily conclude that, in this case, the (mutually inverse) transformation matrices **a** and **A** [recalling (2.9.4a ff.)] have the following special forms:

$$\mathbf{a} = \left(\begin{array}{cc|c} 1 & -\mathbf{b} & -\mathbf{b}_{n+1} \\ 0 & 1 & 0 \\ \hline 0 & 0 & 1 \end{array}\right) \qquad \mathbf{A} = \left(\begin{array}{cc|c} 1 & \mathbf{b} & \mathbf{b}_{n+1} \\ 0 & 1 & 0 \\ \hline 0 & 0 & 1 \end{array}\right), \tag{2.11.12}$$

that is,

$$\mathbf{a}_S = \left(\begin{array}{c|c} 1 & -\mathbf{b} \\ \hline 0 & 1 \end{array}\right) \qquad \mathbf{a}_T = \left(\begin{array}{c} -\mathbf{b}_{n+1} \\ \hline 0 \end{array}\right), \tag{2.11.12a}$$

$$\mathbf{A}_S = \left(\begin{array}{c|c} 1 & \mathbf{b} \\ \hline 0 & 1 \end{array}\right) \qquad \mathbf{A}_T = \left(\begin{array}{c} \mathbf{b}_{n+1} \\ \hline 0 \end{array}\right); \tag{2.11.12b}$$

where $\mathbf{b} = (b_{DI})$, $\mathbf{b}_{n+1} = (b_{D,n+1} \equiv b_D)$; and, of course, satisfy the consistency relations (2.9.3a, b). For a slight generalization of the choice (2.11.10c), see pr. 2.11.2.

Particle Kinematics

In this case, the particle kinematical quantities [recalling (2.5.2 ff.) and (2.11.6a ff.), and that $\boldsymbol{e}_{n+1} \equiv \boldsymbol{e}_0 \equiv \partial \boldsymbol{r}/\partial t$] specialize to

$$\begin{aligned} d\boldsymbol{r} &= \sum \boldsymbol{e}_k\,dq_k + \boldsymbol{e}_{n+1}\,dt = \sum \boldsymbol{e}_D\,dq_D + \sum \boldsymbol{e}_I\,dq_I + \boldsymbol{e}_{n+1}\,dt \\ &= \sum \boldsymbol{e}_D\Big(\sum b_{DI}\,dq_I + b_D\,dt\Big) + \sum \boldsymbol{e}_I\,dq_I + \boldsymbol{e}_{n+1}\,dt \\ &\equiv \sum \boldsymbol{\beta}_I\,dq_I + \boldsymbol{\beta}_{n+1}\,dt \equiv \sum \boldsymbol{\beta}_I\,dq_I + \boldsymbol{\beta}_0\,dt, \end{aligned} \tag{2.11.13a}$$

$$\delta\boldsymbol{r} = \cdots = \sum \boldsymbol{\beta}_I\,\delta q_I, \tag{2.11.13b}$$

$$\boldsymbol{v} = \sum \boldsymbol{\beta}_I v_I + \boldsymbol{\beta}_{n+1} = \boldsymbol{v}(t, q, v_I) \equiv \boldsymbol{v}_o, \tag{2.11.13c}$$

$$\boldsymbol{a} = \sum \boldsymbol{\beta}_I \dot{v}_I + \text{terms not containing } \dot{v}_I = \boldsymbol{a}(t, q, v_I, \dot{v}_I) \equiv \boldsymbol{a}_o, \tag{2.11.13d}$$

where ($\boldsymbol{\varepsilon}_I \to \boldsymbol{\beta}_I$):

$$\boldsymbol{\beta}_I \equiv \boldsymbol{e}_I + \sum b_{DI}\boldsymbol{e}_D, \qquad \boldsymbol{\beta}_{n+1} \equiv \boldsymbol{\beta}_0 \equiv \boldsymbol{e}_{n+1} + \sum b_D\boldsymbol{e}_D \equiv \boldsymbol{e}_0 + \sum b_D\boldsymbol{e}_D. \tag{2.11.13e}$$

REMARK

It should be pointed out that under the quasi-variable choice (2.11.9), and, according to an unorthodox yet internally consistent interpretation [advanced, mainly, by Ukrainian/Soviet/Russian authors, like Suslov, Voronets, Rumiantsev; and at odds with the earlier statement (§2.9) that the q's are always holonomic coordinates], the q_I, and hence also the q_D, are no longer genuine ≡ holonomic coordinates, but have instead become quasi-, or nonholonomic coordinates; even though one could not tell that very well from their notation. To avoid errors in this slippery terrain, some authors have introduced the particular notation (q) (Johnsen, 1939); we shall use it occasionally, for extra clarity. Thus, specializing (2.9.27), while recalling the first of (2.11.12b), we can write

$$\begin{aligned}\partial\boldsymbol{r}/\partial(q_I) &\equiv \sum(\partial\boldsymbol{r}/\partial q_k)(\partial v_k/\partial v_I)\\ &= \sum(\partial\boldsymbol{r}/\partial q_D)(\partial v_D/\partial v_I) + \sum(\partial\boldsymbol{r}/\partial q_{I'})(\partial v_{I'}/\partial v_I)\\ &= \partial\boldsymbol{r}/\partial q_I + \sum b_{DI}(\partial\boldsymbol{r}/\partial q_D) = \sum A_{DI}\boldsymbol{e}_D + \sum A_{I'I}\boldsymbol{e}_{I'}\\ &= \sum b_{DI}\boldsymbol{e}_D + \sum \delta_{I'I}\boldsymbol{e}_{I'} = \sum b_{DI}\boldsymbol{e}_D + \boldsymbol{e}_I \equiv \boldsymbol{\beta}_I;\end{aligned} \tag{2.11.14a}$$

and analogously for $\boldsymbol{\beta}_{n+1} \equiv \boldsymbol{\beta}_0$. Similarly, with the helpful notation [(2.11.13c)]: $\boldsymbol{v} = \boldsymbol{v}(t, q, v) = \cdots = \boldsymbol{v}_o(t, q, v_I) \equiv \boldsymbol{v}_o$, chain rule, and recalling (2.11.9), we obtain

$$\partial\boldsymbol{v}_o/\partial v_I \equiv \partial\boldsymbol{v}/\partial v_I + \sum(\partial\boldsymbol{v}/\partial v_D)(\partial v_D/\partial v_I) = \boldsymbol{e}_I + \sum \boldsymbol{e}_D b_{DI} = \boldsymbol{\beta}_I; \tag{2.11.14b}$$

that is, the fundamental identities (2.9.33) specialize to

$$\partial\boldsymbol{r}/\partial(q_I) = \partial\boldsymbol{v}_o/\partial v_I = \partial\boldsymbol{a}_o/\partial\dot{v}_I = \cdots = \boldsymbol{\beta}_I = \boldsymbol{\beta}_I(t, q) \tag{2.11.14c}$$

[*not* to be confused with the analogous holonomic identities (2.5.7, 7a)].

Equation (2.11.14a) gives rise to the *special symbolic quasi chain rule* (see also chap. 5):

$$\begin{aligned}\partial\ldots/\partial(q_I) &\equiv \sum(\partial\ldots/\partial q_k)(\partial v_k/\partial v_I)\\ &= \sum(\partial\ldots/\partial q_D)(\partial v_D/\partial v_I) + \sum(\partial\ldots/\partial q_{I'})(\partial v_{I'}/\partial v_I)\\ &= \partial\ldots/\partial q_I + \sum b_{DI}(\partial\ldots/\partial q_D);\end{aligned} \tag{2.11.15a}$$

which, when applied to v_D, yields

$$\begin{aligned}\partial v_D/\partial(q_I) &\equiv \sum(\partial v_D/\partial q_{D'})(\partial v_{D'}/\partial v_I) + \sum(\partial v_D/\partial q_{I'})(\partial v_{I'}/\partial v_I)\\ &= \partial v_D/\partial q_I + \sum b_{D'I}(\partial v_D/\partial q_{D'}).\end{aligned} \tag{2.11.15b}$$

Generally, applying chain rule to

$$f = f(t, q, v) = f[t, q, v_D(t, q, v_I), v_I] \equiv f_o(t, q, v_I) = f_o, \qquad (2.11.15c)$$

we obtain the useful formulae

$$\partial f_o/\partial v_I = \partial f/\partial v_I + \sum (\partial f/\partial v_D)(\partial v_D/\partial v_I) = \partial f/\partial v_I + \sum b_{DI}(\partial f/\partial v_D); \qquad (2.11.15d)$$

and

$$\partial f_o/\partial q_I = \partial f/\partial q_I + \sum (\partial f/\partial v_D)(\partial v_D/\partial q_I); \qquad (2.11.15e)$$

while (2.11.15a,b) are seen as specializations of

$$\partial f_o/\partial(q_I) \equiv \partial f_o/\partial q_I + \sum (\partial f_o/\partial q_D)(\partial v_D/\partial v_I)$$
$$\equiv \partial f_o/\partial q_I + \sum b_{DI}(\partial f_o/\partial q_D) \qquad \text{[\textit{notation}, not chain rule!]}. \qquad (2.11.15f)$$

Problem 2.11.1 With the help of the above symbolic identities [recall (2.11.12 ff.)] show that:

(i) $\partial q_k/\partial\theta_I \equiv \partial v_k/\partial\omega_I \to \partial q_k/\partial(q_I)$:

$$\partial q_D/\partial(q_{D'}) = A_{DD'} = \delta_{DD'}, \quad \partial q_D/\partial(q_I) = A_{DI} = b_{DI}, \quad \partial q_I/\partial(q_D) = A_{ID} = 0,$$
$$\partial q_I/\partial(q_{I'}) = A_{II'} = \delta_{II'}. \qquad (a)$$

(ii) $\partial\theta_k/\partial q_I \equiv \partial\omega_k/\partial v_I \to \partial(q_k)/\partial q_I$:

$$\partial(q_D)/\partial q_{D'} = a_{DD'} = \delta_{DD'}, \quad \partial(q_D)/\partial q_I = a_{DI} = -b_{DI}, \quad \partial(q_I)/\partial q_D = a_{ID} = 0,$$
$$\partial(q_I)/\partial q_{I'} = a_{II'} = \delta_{II'}. \qquad (b)$$

[Notice that $\partial q_D/\partial(q_I) = b_{DI} \neq \partial(q_D)/\partial q_I = -b_{DI}$.]

(iii) $\partial\ldots/\partial\theta_{n+1} \to \partial\ldots/\partial(q_{n+1})$ [recall (2.9.32 ff.), and since $A_{k,n+1} \equiv A_k$]

$$= \sum A_k(\partial\ldots/\partial q_k) + \partial\ldots/\partial t$$
$$= \sum A_D(\partial\ldots/\partial q_D) + \sum A_I(\partial\ldots/\partial q_I) + \partial\ldots/\partial t$$
$$= \sum b_D(\partial\ldots/\partial q_D) + 0 + \partial\ldots/\partial t \equiv \partial\ldots/\partial(t) + \partial\ldots/\partial t; \qquad (c)$$

which for $\boldsymbol{r}$ yields the earlier (2.11.13e).

(iv) $\partial\boldsymbol{r}/\partial(q_D) \equiv \sum (\partial\boldsymbol{r}/\partial q_k)(\partial v_k/\partial v_D) = \sum \boldsymbol{e}_k A_{kD} = \cdots = \partial\boldsymbol{r}/\partial q_D$,

$$\text{i.e., } \boldsymbol{\beta}_D = \boldsymbol{e}_D. \qquad (d)$$

(v) $\partial\boldsymbol{\beta}_I/(q_{I'}) \neq \partial\boldsymbol{\beta}_{I'}/\partial(q_I)$; (e)

which is a specialization of (2.10.23), and shows clearly that the basis $\{\boldsymbol{\beta}_I\}$ is non-gradient.

---

4. Occasionally, the constraints appear in the (2.11.9)-like form, but in the quasi variables $d\theta/\delta\theta/\omega$ [special case of (2.11.8a)]:

$$d\theta_D = \sum B_{DI}\, d\theta_I + B_D\, dt; \quad \text{or} \quad \delta\theta_D = \sum B_{DI}\, \delta\theta_I; \quad \text{or} \quad \omega_D = \sum B_{DI}\omega_I + B_D, \tag{2.11.16}$$

where the coefficients B_{DI}, B_D are known functions of q and t.

To uncouple them, proceeding as before, we introduce the following *new* equilibrium quasi variables $d\theta'/\delta\theta'/\omega'$ (to avoid accented indices, we accent the quasi variables themselves):

$$d\theta'_D \equiv d\theta_D - \sum B_{DI}\, d\theta_I - B_D\, dt = 0,\ d\theta'_I \equiv d\theta_I \neq 0,\ d\theta'_{n+1} \equiv d\theta_{n+1} \equiv dt \neq 0; \tag{2.11.17a}$$

$$\delta\theta'_D \equiv \delta\theta_D - \sum B_{DI}\, \delta\theta_I = 0, \quad \delta\theta'_I \equiv \delta\theta_I \neq 0, \quad \delta\theta'_{n+1} \equiv \delta\theta_{n+1} \equiv \delta t = 0; \tag{2.11.17b}$$

$$\omega'_D \equiv \omega_D - \sum B_{DI}\omega_I - B_D = 0, \quad \omega'_I \equiv \omega_I \neq 0, \quad \omega'_{n+1} \equiv \omega_{n+1} \equiv dt/dt = 1 \neq 0; \tag{2.11.17c}$$

which invert easily to

$$d\theta_D = d\theta'_D + \sum B_{DI}\, d\theta'_I + B_D\, dt = \sum B_{DI}\, d\theta'_I + B_D\, dt,$$
$$d\theta_I = d\theta'_I, \qquad d\theta_{n+1} \equiv d\theta'_{n+1} \equiv dt; \tag{2.11.18a}$$

$$\delta\theta_D = \delta\theta'_D + \sum B_{DI}\, \delta\theta'_I = \sum B_{DI}\, \delta\theta'_I,$$
$$\delta\theta_I = \delta\theta'_I, \qquad \delta\theta_{n+1} \equiv \delta\theta'_{n+1} \equiv \delta t = 0; \tag{2.11.18b}$$

$$\omega_D = \omega'_D + \sum B_{DI}\, \omega'_I + B_D = \sum B_{DI}\, \omega'_I + B_D,$$
$$\omega_I = \omega'_I, \qquad \omega_{n+1} = \omega'_{n+1} = dt/dt = 1. \tag{2.11.18c}$$

Clearly, (2.11.16)–(2.11.18) bear the same formal relation to (2.11.8a) that (2.11.9)–(2.11.11) bear to (2.11.2)–(2.11.4).

In sum, the possibilities are endless and, in practice, they are dictated by the specific features and needs of the problem at hand. The essential point in all these descriptions is that, ultimately, they express the n $dq/\delta q/v$ in terms of $n - m$ *independent* parameters $d\theta_I/\delta\theta_I/\omega_I$; and if the nonholonomic constraints are in coupled form, either among the $dq/\delta q/v$ or among another set of n quasi variables $d\theta/\delta\theta/\omega$ then, following Hamel, we introduce new *equilibrium* quasi variables $d\theta'/\delta\theta'/\omega'$ such that $d\theta'_D/\delta\theta'_D/\omega'_D = 0$ and $d\theta'_I/\delta\theta'_I/\omega'_I \neq 0$. And, as already stated, this *uncoupling* of the Pfaffian constraints is the main advantage of the method.

Problem 2.11.2 Consider the homogeneous Pfaffian constraints,

$$\omega_D = \sum a_{Dk} v_k = \sum a_{DD'} v_{D'} + \sum a_{DI'} v_{I'} \quad (= 0), \tag{a}$$

$$\omega_I = v_I = \sum \delta_{ID'} v_{D'} + \sum \delta_{II'} v_{I'} \quad \left(= \sum a_{Ik} v_k \neq 0\right), \tag{b}$$

where D, $D' = 1, \ldots, m; I$, $I' = m+1, \ldots, n$; that is, (with some easily understood notations)

$$\mathbf{a} \Rightarrow \mathbf{a}_S = (a_{kl}) \equiv \begin{pmatrix} (a_{DD'}) & (a_{DI'}) \\ (a_{ID'}) & (a_{II'}) \end{pmatrix} = \begin{pmatrix} (a_{DD'}) & (a_{DI'}) \\ (0_{ID'}) & (\delta_{II'}) \end{pmatrix}. \tag{c}$$

(i) Verify that its inverse (assuming that $\mathbf{a}$ is nonsingular) equals

$$\mathbf{A} \Rightarrow \mathbf{A}_S = (A_{kl}) \equiv \begin{pmatrix} (A_{DD'}) & (A_{DI'}) \\ (A_{ID'}) & (A_{II'}) \end{pmatrix} = \begin{pmatrix} (a_{DD'})^{-1} & -(a_{DD'})^{-1}(a_{D'I'}) \\ (0_{ID'}) & (\delta_{II'}) \end{pmatrix}. \tag{d}$$

(ii) Extend the above to the nonhomogeneous case; that is, $\omega_D = \sum a_{Dk} v_k + a_D \ (=0)$, $\omega_I = v_I \ (\neq 0)$.

(iii) Verify that the earlier particular choice (2.11.9 ff.) is a specialization of the above.

Geometrical Interpretation of Constraints

(May be omitted in a first reading.) We begin by partitioning the mutually inverse $n \times n$ matrices of the virtual transformation between $\delta q \leftrightarrow \delta\theta$, $\mathbf{a}_S = (a_{kl})$ and $\mathbf{A}_S = (A_{kl})$, into their *dependent* and *independent* parts:

$$\mathbf{a}_S = \begin{pmatrix} \mathbf{a}_D \\ \mathbf{a}_I \end{pmatrix} = \begin{pmatrix} a_{Dk} \\ a_{Ik} \end{pmatrix}, \tag{2.11.19a}$$

$$\mathbf{A}_S = (\mathbf{A}_D \mid \mathbf{A}_I) = (A_{kD} \mid A_{kI}). \tag{2.11.19b}$$

Clearly,

$$\mathbf{a}_S\,\mathbf{A}_S = \begin{pmatrix} \mathbf{a}_D\mathbf{A}_D & \mathbf{a}_D\mathbf{A}_I \\ \mathbf{a}_I\mathbf{A}_D & \mathbf{a}_I\mathbf{A}_I \end{pmatrix} = \begin{pmatrix} \mathbf{1} & \mathbf{0} \\ \mathbf{0} & \mathbf{1} \end{pmatrix}. \tag{2.11.19c}$$

Next, we partition these submatrices in terms of their *dependent* and *independent* (column) vectors as follows [with $(\ldots)^{\mathrm{T}} \equiv$ transpose of $(\ldots)$, and using strict matrix notation for vectors and their dot products, instead of the customary vector notation used before and after this subsection]:

$$\mathbf{a}_D = \begin{pmatrix} \boldsymbol{a}_1{}^{\mathrm{T}} \\ \cdots \\ \boldsymbol{a}_m{}^{\mathrm{T}} \end{pmatrix}, \qquad \mathbf{a}_D{}^{\mathrm{T}} = (\boldsymbol{a}_1, \ldots, \boldsymbol{a}_m), \qquad \boldsymbol{a}_D{}^{\mathrm{T}} = (a_{D1}, \ldots a_{Dn}), \tag{2.11.20a}$$

$$\mathbf{a}_I = \begin{pmatrix} \boldsymbol{a}_{m+1}{}^{\mathrm{T}} \\ \cdots \\ \boldsymbol{a}_n{}^{\mathrm{T}} \end{pmatrix}, \qquad \mathbf{a}_I{}^{\mathrm{T}} = (\boldsymbol{a}_{m+1}, \ldots, \boldsymbol{a}_n), \qquad \boldsymbol{a}_I{}^{\mathrm{T}} = (a_{I1}, \ldots a_{In}), \tag{2.11.20b}$$

$$\mathbf{A}_\mathrm{D} = (\boldsymbol{A}_1, \ldots, \boldsymbol{A}_m), \qquad \mathbf{A}_\mathrm{D}{}^\mathrm{T} = \begin{pmatrix} \boldsymbol{A}_1{}^\mathrm{T} \\ \cdots \\ \boldsymbol{A}_m{}^\mathrm{T} \end{pmatrix}, \qquad \boldsymbol{A}_D{}^\mathrm{T} = (A_{1D}, \ldots, A_{nD}), \quad (2.11.20\mathrm{c})$$

$$\mathbf{A}_\mathrm{I} = (\boldsymbol{A}_{m+1}, \ldots, \boldsymbol{A}_n), \qquad \mathbf{A}_\mathrm{I}{}^\mathrm{T} = \begin{pmatrix} \boldsymbol{A}_{m+1}{}^\mathrm{T} \\ \cdots \\ \boldsymbol{A}_n{}^\mathrm{T} \end{pmatrix}, \qquad \boldsymbol{A}_I{}^\mathrm{T} = (A_{1I}, \ldots, A_{nI}), \quad (2.11.20\mathrm{d})$$

Also, since $(\mathbf{a}_\mathrm{D} \cdot \mathbf{a}_\mathrm{D}{}^\mathrm{T})^{-1} \cdot (\mathbf{a}_\mathrm{D} \cdot \mathbf{a}_\mathrm{D}{}^\mathrm{T}) = \mathbf{1}$ and $\mathbf{A}_\mathrm{D}{}^\mathrm{T} \cdot \mathbf{a}_\mathrm{D}{}^\mathrm{T} = \mathbf{1}$, it follows that

$$\mathbf{A}_\mathrm{D}{}^\mathrm{T} = (\mathbf{a}_\mathrm{D} \cdot \mathbf{a}_\mathrm{D}{}^\mathrm{T})^{-1} \cdot \mathbf{a}_\mathrm{D}. \quad (2.11.20\mathrm{e})$$

Now, in linear algebra terms, the virtual form of the constraint equations

$$\sum a_{Dk}\, \delta q_k = 0 \qquad [rank(a_{Dk}) = m\,(<n)], \quad (2.11.21\mathrm{a})$$

[we note, in passing, that $rank(a_{Dk})_{m\times n} = rank(a_{Dk}|a_D)_{[m\times(n+1)]}$] or, in the above-introduced matrix notation,

$$\mathbf{a}_\mathrm{D} \cdot \delta\boldsymbol{q} = \mathbf{0}\,(\textit{one}\text{ matrix eq.}), \qquad \boldsymbol{a}_D{}^T \cdot \delta\boldsymbol{q} = 0\,[m \text{ vector (dot product) eqs.}], \quad (2.11.21\mathrm{b})$$

state that *every virtual displacement* (*column*) *vector* $\delta\boldsymbol{q}^T = (\delta q_1, \ldots, \delta q_n)$, at the point (q, t), lies on the *local* $(n-m)$*-dimensional tangent/null/virtual plane of the* (*virtual form of the*) *constraint matrix* $\mathbf{a}_\mathrm{D} = (a_{Dk})$, $T_{n-m}(P) \equiv V_{n-m}(P) \equiv V_{n-m}$ (§2.7, suppressing the point dependence); or, equivalently, that $\delta\boldsymbol{q}$ is always *orthogonal* to the *local* m*-dimensional range space/constraint plane of* $\mathbf{a}_\mathrm{D}{}^\mathrm{T}$, $C_m(P) \equiv C_m$ (which is *orthogonally complementary to* V_{n-m}).

Next, in view of our quasi-variable choice, that is, $\delta\boldsymbol{q} = \mathbf{A}_\mathrm{I} \cdot \delta\boldsymbol{\theta}_I$, where $\delta\boldsymbol{\theta}_I{}^T = (\delta\theta_{m+1}, \ldots, \delta\theta_n)$, the $(n-m)$ vectors $(\boldsymbol{A}_{m+1}, \ldots, \boldsymbol{A}_n) \equiv \{\boldsymbol{A}_I\}$ constitute a basis for V_{n-m}; while the m constraint vectors $(\boldsymbol{a}_1, \ldots, \boldsymbol{a}_m) \equiv \{\boldsymbol{a}_D\}$ constitute a basis for C_m. Or, all (δq_k) satisfying (2.11.21a, b), at (q, t), form a local vector space V_{n-m}, which is orthogonal to the local vector space C_m built (spanned) by the m constraint vectors $\boldsymbol{a}_D{}^T = (a_{D1}, \ldots, a_{Dn})$.

More precisely, expressing (2.9.3a, b) in the above matrix/vector notation, we have

(i) $$\sum a_{Ik} A_{kI'} = \boldsymbol{a}_I{}^T \cdot \boldsymbol{A}_{I'} = \delta_{II'} \qquad (I, I' = m+1, \ldots, n), \quad (2.11.22\mathrm{a})$$

$$\text{or} \quad \mathbf{a}_\mathrm{I} \cdot \mathbf{A}_\mathrm{I} = \mathbf{1}, \quad \text{or} \quad \mathbf{A}_\mathrm{I}{}^\mathrm{T} \cdot \mathbf{a}_\mathrm{I}{}^\mathrm{T} = \mathbf{1}; \quad (2.11.22\mathrm{b})$$

that is, the columns of $\mathbf{a}_\mathrm{I}{}^\mathrm{T}$ and $\mathbf{A}_\mathrm{I}$, or the rows of $\mathbf{a}_\mathrm{I}$ and $\mathbf{A}_\mathrm{I}{}^\mathrm{T}$, namely, the vectors $\{\boldsymbol{a}_I\}$ and $\{\boldsymbol{A}_I\}$, are mutually *dual*, or *reciprocal*, bases of V_{n-m}; and

(ii) $$\sum a_{Dk} A_{kD'} = \boldsymbol{a}_D{}^T \cdot \boldsymbol{A}_{D'} = \delta_{DD'} \qquad (D, D' = 1, \ldots, m), \quad (2.11.22\mathrm{c})$$

$$\text{or} \quad \mathbf{a}_\mathrm{D} \cdot \mathbf{A}_\mathrm{D} = \mathbf{1}, \quad \text{or} \quad \mathbf{A}_\mathrm{D}{}^\mathrm{T} \cdot \mathbf{a}_\mathrm{D}{}^\mathrm{T} = \mathbf{1}; \quad (2.11.22\mathrm{d})$$

that is, the columns of $\mathbf{a_D}^{\mathrm{T}}$ and $\mathbf{A_D}$, or the rows of $\mathbf{a_D}$ and $\mathbf{A_D}^{\mathrm{T}}$, namely, the vectors $\{\boldsymbol{a}_D\}$ and $\{\boldsymbol{A}_D\}$, are mutually *dual* bases of C_m. Clearly, if the $\{\boldsymbol{a}_D\}$ are *orthonormal*, so are the $\{\boldsymbol{A}_D\}$, and the two bases coincide; and similarly for the bases $\{\boldsymbol{a}_I\}$, $\{\boldsymbol{A}_I\}$.

Likewise, from (2.9.3a, b) we obtain

$$\text{(iii)} \quad \sum a_{Dk} A_{kI} = \boldsymbol{a}_D{}^T \cdot \boldsymbol{A}_I = \delta_{DI} = 0 \qquad (D = 1, \ldots, m;\ I = m+1, \ldots, n), \tag{2.11.22e}$$

$$\text{or} \qquad \mathbf{a_D} \cdot \mathbf{A_I} = \mathbf{0}, \qquad \text{or} \qquad \mathbf{A_I}^{\mathrm{T}} \cdot \mathbf{a_D}^{\mathrm{T}} = \mathbf{0}; \tag{2.11.22f}$$

that is, the vectors $\{\boldsymbol{a}_D\}$ and $\{\boldsymbol{A}_I\}$ are mutually orthogonal.

$$\text{(iv)} \quad \sum a_{Ik} A_{kD} = \boldsymbol{a}_I{}^T \cdot \boldsymbol{A}_D = \delta_{ID} = 0 \qquad (I = m+1, \ldots, n;\ D = 1, \ldots, m), \tag{2.11.22g}$$

$$\text{or} \qquad \mathbf{a_I} \cdot \mathbf{A_D} = \mathbf{0}, \qquad \text{or} \qquad \mathbf{A_D}^{\mathrm{T}} \cdot \mathbf{a_I}^{\mathrm{T}} = \mathbf{0}. \tag{2.11.22h}$$

that is, the vectors $\{\boldsymbol{a}_I\}$ and $\{\boldsymbol{A}_D\}$ are mutually orthogonal. Equations (2.11.22f) and (2.11.22h) state, in linear algebra terms, that the "virtual displacement matrix" $\mathbf{A_I}$ is the *orthogonal complement* of the "constraint matrix" $\mathbf{a_D}$.

[Hence, the projections of an arbitrary system vector $\boldsymbol{M} = (M_1, \ldots, M_n)$ on the local mutually orthogonal (complementary) subspaces V_{n-m} and C_m, are, respectively,

Null/Virtual space projection $P\ldots(\cdots)$:

$$\sum A_{kI} M_k = (\mathbf{A_I}^{\mathrm{T}} \cdot \boldsymbol{M})_I = \boldsymbol{A}_I{}^T \cdot \boldsymbol{M} \equiv M_I \equiv P_{N(\mathrm{Null})}(\boldsymbol{M}) \equiv P_{V(\mathrm{Virtual})}(\boldsymbol{M}); \tag{2.11.23a}$$

Range/constraint space projection $P\ldots(\cdots)$:

$$\sum A_{kD} M_k = (\mathbf{A_D}^{\mathrm{T}} \cdot \boldsymbol{M})_D = \boldsymbol{A}_D{}^T \cdot \boldsymbol{M} \equiv M_D \equiv P_{R(\mathrm{Range})}(\boldsymbol{M}) \equiv P_{C(\mathrm{Constraint})}(\boldsymbol{M}).] \tag{2.11.23b}$$

The above hold, locally at least, for any velocity constraints, be they holonomic or nonholonomic. However: (a) If the constraints are nonholonomic, the corresponding null and range spaces are only local; at each admissible point of the system's constrained configuration (or event) space; but (b) If they are holonomic, then these spaces become global; that is, the hitherto n-dimensional configuration space is replaced by a new "smaller" such space described by $n - m$ Lagrangean coordinates, as detailed in §2.4 and §2.7.

Tensorial Hors d'Oeuvre

These *projection* ideas, originated by G. A. Maggi (1890s) and elaborated, via tensors, by J. L. Synge, G. Vranceanu, V. V. Vagner, G. Prange, G. Ferrarese, P. Maißer, N. N. Poliahov et al. (1920s–1980s), are very useful in interpreting the general problem of AM [i.e., of *decoupling* its equations of constrained motion into those *containing* the forces resulting from these constraints and those *not containing* these forces], in terms of simple geometrical pictures of the motion of a single "particle" in a generalized system space. They have become quite popular among multibody dynamicists, in recent decades; but, predominantly as exercises in linear algebra/matrix manipulations, that is, without the geometrical understanding and insight resulting from the full use of general tensors.

To show the advantages of the powerful tensorial indicial notation, over the *noncommutative* straightjacket of matrices, we summarize below some of the above results. With the help of the *summation convention* [over pairs of indices, one up and one down, from 1 to n; and where, here, capital indices (accented and/or unaccented), signify nonholonomic components], we have the following:

(a) Equations (2.11.21a, b), and their inverses:

$$\delta\theta^D \equiv a^D{}_k\, \delta q^k = 0, \qquad \delta q^k = A^k{}_I\, \delta\theta^I, \tag{2.11.24a}$$

(b) Equations (2.11.22a, b):

$$a^I{}_k A^k{}_{I'} = \delta^I{}_{I'}, \tag{2.11.24b}$$

(c) Equations (2.11.22c, d):

$$a^D{}_k A^k{}_{D'} = \delta^D{}_{D'}, \tag{2.11.24c}$$

(d) Equations (2.11.23a):

$$P_V(\boldsymbol{M}) \equiv A^k{}_I M_k = M_I, \tag{2.11.24d}$$

(e) Equations (2.11.23b):

$$P_C(\boldsymbol{M}) \equiv A^k{}_D M_k = M_D. \tag{2.11.24e}$$

The summation convention explains why, in order to project the (covariant) M_k, above, we dot them with $A^k{}_I$ and $A^k{}_D$, instead of $a^I{}_k$, $a^D{}_k$, respectively. [Briefly, the $\boldsymbol{a}^I(\boldsymbol{A}_I)$ build a nonholonomic contravariant (covariant) basis in V_{n-m}, while the $\boldsymbol{a}^D(\boldsymbol{A}_D)$ build a nonholonomic contravariant (covariant) basis in C_m.]

Last, a higher level of tensorial formalism may be achieved, if, as described briefly in (2.10.17c), we use accented (unaccented) indices to denote nonholonomic (holonomic) components; for example, successively: $a_{kl} \to a^k{}_l \to A^{k'}{}_k$, $A_{kl} \to A^k{}_l \to A^k{}_{k'}$; so that $\mathbf{a} \cdot \mathbf{A} = \mathbf{1}$ reads $A^{k'}{}_k A^k{}_{l'} = \delta^{k'}{}_{l'}$, and similarly for the other equations. For further details on tensorial nonholonomic dynamics, see, for example, Papastavridis (1999) and references cited therein.

2.12 CONSTRAINED TRANSITIVITY EQUATIONS, AND HAMEL'S FORM OF FROBENIUS' THEOREM

Constrained Transitivity Equations

Let us begin by examining the transitivity relations (2.10.1) under the Pfaffian constraints (2.11, 2a ff.), $d\theta_D = 0$, $\delta\theta_D = 0$, and their implications for the latter's holonomicity/nonholonomicity. Indeed, assuming $d(\delta q_k) = \delta(dq_k)$ for all $k = 1, \ldots, n$, whether the $dq/\delta q$ are constrained or not (what is known as the *Hamel viewpoint*, see pr. 2.12.5), the general transitivity equations (2.10.1) reduce to

$$d(\delta\theta_D) - \delta(d\theta_D) = \sum\sum \gamma^D{}_{II'}\, d\theta_{I'}\, \delta\theta_I + \sum \gamma^D{}_I\, dt\, \delta\theta_I, \qquad (2.12.1a)$$

$$d(\delta\theta_I) - \delta(d\theta_I) = \sum\sum \gamma^I{}_{I'I''}\, d\theta_{I''}\, \delta\theta_{I'} + \sum \gamma^I{}_{I'}\, dt\, \delta\theta_{I'}. \qquad (2.12.1b)$$

From the above we conclude that, even though $\omega_D(t) = 0$ (or a constant), or $d\theta_D(t) = 0$, or $\delta\theta_D(t) = 0$, from which it follows that $(\delta\theta_D)^{\cdot} = 0$ or $d(\delta\theta_D) = 0$, yet, in general, $d(\delta\theta_D) - \delta(d\theta_D) \neq 0 \Rightarrow -\delta(d\theta_D) \neq 0$! Specifically, as (2.12.1a) shows,

$$-\delta(d\theta_D) = \sum\sum \gamma^D{}_{II'}\, d\theta_{I'}\, \delta\theta_I + \sum \gamma^D{}_I\, dt\, \delta\theta_I \neq 0 \qquad \text{(in general)}; \qquad (2.12.1c)$$

that is, we cannot assume that both $d(\delta q_k) = \delta(dq_k)$ and $d(\delta\theta_D) = \delta(d\theta_D)\ (= 0)$. This is a delicate point that has important consequences in time-integral variational principles for nonholonomic systems (see Hamel, 1949, pp. 476–477; and this book, chapter 7; also pr. 2.12.5).

The Frobenius Theorem Revisited (and Made Easier to Implement)

We have already stated (§2.8) that the necessary and sufficient condition for the holonomicity of the system of m Pfaffian constraints

$$d\theta_D \equiv \sum a_{Dk}\, dq_k + a_D\, dt = 0, \qquad \delta\theta_D \equiv \sum a_{Dk}\, \delta q_k = 0 \qquad (D = 1, \ldots, m), \qquad (2.12.2)$$

that is, for the existence of m linear combinations of the $d\theta_D = 0$, or $\delta\theta_D = 0$, that equal m independent exact differential equations $df_1 = 0, \ldots, df_m = 0 \Rightarrow f_1 = constant, \ldots, f_m = constant$, is the identical vanishing of their Frobenius bilinear covariants [recall (2.9.13)]

$$d(\delta\theta_D) - \delta(d\theta_D)$$
$$= \sum\sum (\partial a_{Dk}/\partial q_l - \partial a_{Dl}/\partial q_k)\, dq_l\, \delta q_k + \sum (\partial a_{Dk}/\partial t - \partial a_D/\partial q_k)\, dt\, \delta q_k, \qquad (2.12.3)$$

for all dq_k, dt, δq_k solutions of (2.12.2). From this fundamental theorem we draw the following conclusions:

(i) If the dq_k, dt, δq_k are *unconstrained*, that is, if $m = 0$, then the identical satisfaction of the conditions

$$a^D{}_{kl} \equiv \partial a_{Dk}/\partial q_l - \partial a_{Dl}/\partial q_k = 0, \qquad a^D{}_k \equiv \partial a_{Dk}/\partial t - \partial a_D/\partial q_k = 0, \qquad (2.12.3a)$$

for all $k, l = 1, \ldots, n$, is both *necessary* and *sufficient* for the holonomicity of θ_D (§2.9).

(ii) But, if the dq_k, dt, and δq_k are *constrained* by (2.12.2), then the vanishing of $d(\delta\theta_D) - \delta(d\theta_D)$ does not necessarily lead to (2.12.3a).

To obtain necessary conditions for the holonomicity of the system (2.12.2), we must express the dq_k, dt, δq_k, on the right side of (2.12.3), as linear and homogeneous combinations of $n - m$ *independent* parameters (Maggi's idea); that is, we must take the constraints (2.12.2) themselves into account. Indeed, substituting into (2.12.3) the general solutions of (2.12.2) [recalling (2.11.4)]:

$$dq_k \equiv \sum A_{kI}\, d\theta_I + A_I\, dt, \qquad \delta q_k \equiv \sum A_{kI}\, \delta\theta_I, \tag{2.12.4}$$

we obtain (2.12.1a). From this, it follows that (and this is the crux of this argument), since the $2(n-m)$ differentials $d\theta_I$, $\delta\theta_I$ are *independent/unconstrained*, the conditions

$$\gamma^D{}_{II'} = 0 \qquad \text{and} \qquad \gamma^D{}_{I,n+1} \equiv \gamma^D{}_I = 0, \tag{2.12.5}$$

for all $D = 1, \ldots, m$; $I, I' = m+1, \ldots, n$, are both *sufficient* and *necessary* for the *holonomicity of the Pfaffian system* (2.12.2).

- Since [recalling the γ-definition (2.10.2 ff.)] (2.12.5) can be rewritten as

$$\gamma^D{}_{II'} = \sum\sum (\partial a_{Db}/\partial q_c - \partial a_{Dc}/\partial q_b) A_{bI} A_{cI'} \equiv \sum\sum a^D{}_{bc} A_{bI} A_{cI'} = 0, \tag{2.12.5a}$$

$$\gamma^D{}_I \equiv \sum\sum (\partial a_{Db}/\partial q_c - \partial a_{Dc}/\partial q_b) A_{bI} A_c + \sum (\partial a_{Db}/\partial t - \partial a_D/\partial q_b) A_{bI}$$

$$\equiv \sum\sum a^D{}_{bc} A_{bI} A_c + \sum a^D{}_b A_{bI} = 0, \tag{2.12.5b}$$

we readily recognize that the (identical) vanishing of (all) the $\gamma^D{}_{..}$'s does not necessarily lead to the vanishing of (all) the $a^D{}_{bc}$, $a^D{}_b$, while the vanishing of all the latter leads to the vanishing of all the $\gamma^D{}_{..}$'s; that is, (2.12.3a) lead to (2.12.5, 5a, b) but not the other way around. Hence, (2.12.3a) are sufficient for holonomicity but not necessary, whereas (2.12.5, 5a, b) are both necessary and sufficient.

- Since, as (2.12.5a, b) make clear, each $\gamma^D{}_{..}$ ($\gamma^D{}_{.}$) depends, in general, on *all* the coefficients a_{Dk}, A_{kI} ($a_{Dk}, A_{kI}; a_D, A_k$), the holonomicity/nonholonomicity of a(ny) particular constraint, of the given system (2.12.2), depends on all the others; that is, on the entire system of constraints. In other words: eqs. (2.12.5) check the holonomicity, or absence thereof, of each equation $d\theta_D, \delta\theta_D = 0$ against the entire system; that is, there is no such thing as testing an individual Pfaffian constraint, of a given system of such constraints, for holonomicity; doing that would be testing the new system consisting of that Pfaffian equation alone (i.e., $m = 1$) for holonomicity. In short, holonomicity/nonholonomicity is a *system* property.

As Neimark and Fufaev put it "the existence of a *single* nonintegrable constraint (in a system of constraints) does *not* necessarily mean a system is nonholonomic, since this constraint may prove to be integrable by virtue of the *remaining* constraint equations" (1972, p. 6, italics added). However [and recalling (2.10.16a–18b)], we can see that the identical vanishing of all coefficients $\gamma^k{}_{rs}$ and $\gamma^k{}_{r,n+1} \equiv \gamma^k{}_r$ (for *all* $r, s = 1, \ldots, n$) in

$$d(\delta\theta_k) - \delta(d\theta_k) = \cdots + (\gamma^k{}_{..})\, d\theta\, \delta\theta + (\gamma^k{}_{..})\, dt\, \delta\theta, \tag{2.12.5c}$$

independently of the constraints $d\theta_D, \delta\theta_D = 0$ (or, as if no constraints had been applied; and which is equivalent to $a^k{}_{rs} = 0$, $a^k{}_r = 0$, identically), is the necessary and sufficient condition for that particular θ_k to be a genuine/Lagrangean *coordinate*; that is, $a_{kr} = \partial\theta_k/\partial q_r$, $a_k = \partial\theta_k/\partial t$.

Let us recapitulate/summarize our findings:

(i) Pfaffian *forms* (not equations), like

$$d\theta_k \equiv \sum a_{kl}(q)\, dq_l \qquad (k = 1, \ldots, n';\ l = 1, \ldots, n;\ n \text{ and } n' \text{ unrelated}) \tag{2.12.6}$$

(for algebraic simplicity, but no loss in generality, we consider the stationary case), are either *exact* differentials, or *inexact* differentials.

If their dq's are *unconstrained*, then the necessary and sufficient conditions for $d\theta_k$ to be exact, and hence for θ_k to be a *holonomic coordinate*, are

$$a^k{}_{rs} = 0 \qquad (r, s = 1, \ldots, n). \tag{2.12.7}$$

In this case, each of the n' forms $d\theta_k$ is tested for *exactness* independently of the others; k, in (2.12.7), is a free index, uncoupled to both r and s. If $n = n'$, then, as already stated, conditions (2.12.7) can be replaced by

$$\gamma^k{}_{rs} = 0 \qquad (r, s = 1, \ldots, n); \tag{2.12.8}$$

but since calculating the γ's requires inverting (2.12.6) for the n dq's in terms of the n $d\theta$'s, eqs. (2.12.8) offer no advantage over eqs. (2.12.7).

If eqs. (2.12.7) hold, then $d\theta_k$ remains exact *no matter how many additional constraints may be imposed on its dq's later.* For, then, we have

$$d(\delta\theta_k) - \delta(d\theta_k) = \sum\sum (\partial a_{kr}/\partial q_s - \partial a_{ks}/\partial q_r)\, dq_s\, \delta q_r = 0; \tag{2.12.9}$$

that is, if θ_k is a holonomic coordinate, it *remains holonomic* if additional constraints be imposed among its dq, δq's, later. This is the meaning of Hamel's rule: $d(\delta q_k) = \delta(dq_k)$, for all q's, constrained or not.

If eqs. (2.12.7) do not hold, $d\theta_k$ is inexact; but it can be made exact by additional constraints among its dq, δq's; that is, if θ_k is a quasi coordinate, it may *become a holonomic coordinate* by imposition of additional appropriate dq, δq constraints. For example, let us consider the Pfaffian form (not constraint)

$$d\chi \equiv a(x, y, z)\, dx + b(x, y, z)\, dy + c(x, y, z)\, dz.$$

Under the additional constraints $y = constant \Rightarrow dy = 0$ and $z = constant \Rightarrow dz = 0$, it becomes $d\chi \equiv a(x, y, z)\, dx \equiv f(x)\, dx = exact\ differential$, even if, originally, χ was a quasi coordinate.

(ii) Pfaffian *systems* of *constraints*

$$d\theta_D \equiv \sum a_{Dk}\, dq_k = 0, \qquad \delta\theta_D \equiv \sum a_{Dk}\, \delta q_k = 0, \tag{2.12.10a}$$

are either *holonomic* or they are nonholonomic. The necessary and sufficient conditions for holonomicity are

$$d(\delta\theta_D) - \delta(d\theta_D) = \sum\sum \gamma^D{}_{II'}\, d\theta_{I'}\, \delta\theta_I = 0, \tag{2.12.10b}$$

or, since the $d\theta_{I'}$, $\delta\theta_I$ are independent,

$$\gamma^D{}_{II'} = 0 \qquad (D = 1, \ldots, m;\ I, I' = m + 1, \ldots, n); \tag{2.12.10c}$$

that is, *the "dependent" γ's relative to their "independent" indices (subscripts) should vanish; or, the components of the dependent (constrained) Hamel coefficients along the independent (unconstrained) directions vanish.*

{This, more easily implementable, form of Frobenius' theorem seems to be due to Hamel (1904(a), 1935); also Cartan (1922, p. 105), Synge [1936, p. 19, eq. (4.16)], and Vranceanu [1929, p. 17, eq. (9′); 1936, p. 13].}

In closing this section, we repeat that Frobenius' theorem is about the integrability of *systems* of Pfaffian *equations*, like (2.12.2), not about the exactness of individual Pfaffian *forms*, like (2.12.6).

Example 2.12.1 *Special Case of the Hamel Coefficients, via Frobenius' Theorem.* Let us calculate the Hamel coefficients corresponding to the special constraint form

$$dq_D = \sum b_{DI}\, dq_I, \qquad \delta q_D = \sum b_{DI}\, \delta q_I, \tag{a}$$

where $b_{DI} = b_{DI}(q)$, and formulate the necessary/sufficient conditions for their holonomicity. We begin by viewing (a) as the following special Hamel choice [stationary version of (2.11.10a–12b)]:

$$d\theta_D = dq_D - \sum b_{DI}\, dq_I = 0, \qquad d\theta_I \equiv dq_I \neq 0, \qquad d\theta_{n+1} \equiv dq_{n+1} \equiv dt \neq 0; \tag{b}$$

$$\delta\theta_D \equiv \delta q_D - \sum b_{DI}\, \delta q_I = 0, \qquad \delta\theta_I \equiv \delta q_I \neq 0, \qquad \delta\theta_{n+1} \equiv \delta q_{n+1} \equiv \delta t = 0; \tag{c}$$

$$dq_D = d\theta_D + \sum b_{DI}\, d\theta_I = \sum b_{DI}\, d\theta_I, \quad dq_I = d\theta_I, \quad dq_{n+1} \equiv d\theta_{n+1} \equiv dt; \tag{d}$$

$$\delta q_D = \delta\theta_D + \sum b_{DI}\, \delta q_I = \sum b_{DI}\, \delta q_I, \quad \delta q_I = \delta\theta_I, \quad \delta q_{n+1} \equiv \delta\theta_{n+1} \equiv \delta t = 0; \tag{e}$$

also, since here $dq_I = d\theta_I$, we can rewrite the system (a) as

$$dq_k = \sum B_{kI}\, dq_I \equiv \sum A_{kI}\, d\theta_I,$$

where

$$(B_{kI}) = \begin{pmatrix} b_{1,m+1} & \cdots & b_{1n} \\ \cdots & \cdots & \cdots \\ b_{m,m+1} & \cdots & b_{mn} \\ \hline 1 & \cdots & 0 \\ \cdots & \cdots & \cdots \\ 0 & \cdots & 1 \end{pmatrix}. \tag{f}$$

Since $\theta_I = q_I$, we shall have $\gamma^I{}_{\alpha\beta} = 0$; while (2.12.9), with $k \to D$ and (f), becomes, successively,

$$\begin{aligned} d(\delta\theta_D) - \delta(d\theta_D) &= \sum\sum a^D{}_{rs}\, dq_s\, \delta q_r = \sum\sum a^D{}_{rs}\Big(\sum B_{sI}\, dq_I\Big)\Big(\sum B_{rI'}\, \delta q_{I'}\Big) \\ &= \cdots = \sum\sum\sum\sum (a^D{}_{rs}\, b_{sI}\, b_{rI'})\, dq_I\, \delta q_{I'} \\ &= \cdots = \sum\sum \gamma^D{}_{I'I}\, dq_I\, \delta q_{I'} \\ &\Big[= \sum\sum \gamma^D{}_{II'}\, dq_{I'}\, \delta q_I = \sum\sum \gamma^D{}_{II'}\, d\theta_{I'}\, \delta\theta_I\Big], \end{aligned} \tag{g}$$

where (expanding the sums in r and s, with $D, D', D'' = 1, \ldots, m$; $I, I' = m+1, \ldots, n$)

$$\gamma^D{}_{I'I} \equiv \sum\sum a^D{}_{D'D''}\, b_{D''I}\, b_{D'I'} + \sum a^D{}_{D'I}\, b_{D'I'} + \sum a^D{}_{I'D'}\, b_{D'I} + a^D{}_{I'I}; \tag{h}$$

or, since $a^D{}_{D'D''} \equiv a_{DD',D''} - a_{DD'',D'}$, where commas denote partial differentiations relative to the indicated q's and [by (2.11.12–12b)] $a_{DD'} \to \delta_{DD'}$, $a_{DI} \to -b_{DI}$, $a_{ID} \to 0$, $a_{II'} \to \delta_{II'}$:

$$\gamma^D{}_{I'I} = \sum\sum (0) b_{D''I} b_{D'I'} + \sum \left[0 - (-\partial b_{DI}/\partial q_{D'})\right] b_{D'I'} \\ + \sum \left[(-\partial b_{DI'}/\partial q_{D'}) - 0\right] b_{D'I} + \left[(-\partial b_{DI'}/\partial q_I) - (-\partial b_{DI}/\partial q_{I'})\right],$$

or finally,

$$\gamma^D{}_{I'I} = \left[\partial b_{DI}/\partial q_{I'} + \sum (\partial b_{DI}/\partial q_{D'}) b_{D'I'}\right] - \left[\partial b_{DI'}/\partial q_I + \sum (\partial b_{DI'}/\partial q_{D'}) b_{D'I}\right]$$
$$\equiv -w^D{}_{I'I} = w^D{}_{II'} = \textit{Voronets (or Woronetz) coefficients}; \tag{i}$$

clearly, a specialization of $\gamma^D{}_{I'I}$. Thus, (g) becomes

$$d(\delta\theta_D) - \delta(d\theta_D) = \sum\sum \gamma^D{}_{I'I}\, dq_I\, \delta q_{I'} = \sum\sum w^D{}_{II'}\, dq_I\, \delta q_{I'}, \tag{j}$$

and, since the dq_I and δq_I are independent, by Frobenius' theorem, the necessary and sufficient conditions for the holonomicity of the system (a) are

$$w^D{}_{II'} = 0, \tag{k}$$

which are none other than the earlier Deahna–Bouquet conditions (2.3.11b ff.).

REMARKS

(i) With the help of the symbolic notation (2.11.15a), we can rewrite (k) in the more memorable form,

$$\gamma^D{}_{I'I} = w^D{}_{II'} = \partial b_{DI}/\partial(q_{I'}) - \partial b_{DI'}/\partial(q_I). \tag{l}$$

(ii) In the special "Chaplygin (or Tchapligine) case" (§3.8), where $b_{DI} = b_{DI}(q_{m+1}, \ldots, q_n) \equiv b_{DI}(q_D)$, the above reduce to

$$\gamma^D{}_{I'I} = \partial b_{DI}/\partial q_{I'} - \partial b_{DI'}/\partial q_I \equiv t^D{}_{II'}. \tag{m}$$

Problem 2.12.1 Continuing from the previous example, show that eqs. (i) for the Voronets coefficients also result by direct application of the definition (2.10.2)

$$\gamma^D{}_{II'} = \sum\sum (\partial a_{Dk}/\partial q_r - \partial a_{Dr}/\partial q_k) A_{kI} A_{rI'} \tag{a}$$

to the constraints (ex. 2.12.1:a) in the equilibrium forms (ex. 2.12.1:b–e).

HINT

Here [recalling again (2.11.12–12b)]: $a_{DD'} = \delta_{DD'}$, $a_{DI} = -b_{DI}$, $a_{ID} = 0$, $a_{II'} = \delta_{II'}$; $A_{DD'} = \delta_{DD'}$, $A_{DI} = b_{DI}$, $A_{ID} = 0$, $A_{II'} = \delta_{II'}$.

Problem 2.12.2 Continuing from the above, show that in the general *nonstationary* case

$$dq_D = \sum b_{DI}\, dq_I + b_D\, dt, \qquad dq_I = \sum \delta_{II'}\, dq_{I'} = dq_I, \tag{a}$$

$$\delta q_D = \sum b_{DI}\, \delta q_I, \;\; \delta q_I = \sum \delta_{II'}\, \delta q_{I'} = \delta q_I; \;\;\; b_{DI} = b_{DI}(t, q), \;\; b_D = b_D(t, q), \tag{b}$$

the $\gamma^D{}_{II'}$ *remain unchanged*, but we have, the additional *nonstationary Voronets* coefficients:

$$\gamma^D{}_{I,n+1} \equiv \gamma^D{}_I = -w^D{}_{I,n+1} \equiv -w^D{}_I$$
$$= \left[\partial b_D/\partial q_I + \sum(\partial b_D/\partial q_{D'})b_{D'I}\right] - \left[\partial b_{DI}/\partial t + \sum(\partial b_{DI}/\partial q_{D'})b_{D'}\right]$$
$$= \partial b_D/\partial(q_I) - \partial b_{DI}/\partial(q_{n+1}) \tag{c}$$

[recalling the symbolic (2.9.32 ff.), (2.11.15): $A_k \to b_D$, and $\sum(\partial\ldots/\partial q_D)b_D = \partial\ldots/\partial(t)$].

REMARK

In concrete problems, use of the above definitions to calculate the w-coefficients is *not* recommended. Instead, the safest way to do this is to read them off directly as coefficients of the following bilinear difference/covariant:

$$d(\delta\theta_D) - \delta(d\theta_D) = \cdots + \gamma^D{}_{II'}\, d\theta_{I'}\,\delta\theta_I + \cdots + \gamma^D{}_I\, dt\,\delta\theta_I + \cdots$$
$$= \cdots - w^D{}_{II'}\, dq_{I'}\,\delta q_I + \cdots - w^D{}_I\, dt\,\delta q_I + \cdots. \tag{d}$$

Problem 2.12.3 Continuing from the preceding problem, verify that:

(i) in the *catastatic Voronets* case, the $w^D{}_{II'}$ remain unchanged, while $w^D{}_I = \partial b_{DI}/\partial t$; and

(ii) in the *stationary Voronets* case, the $w^D{}_{II'}$ remain unchanged, while $w^D{}_I = 0$.

Problem 2.12.4 Continuing from the above problems, verify that

(i) $$\gamma^I{}_{\beta\varepsilon} = 0 \qquad (I = m+1,\ldots,n;\ \beta,\varepsilon = 1,\ldots,n;\ n+1); \tag{a}$$

(ii) $$\gamma^D{}_{D'\varepsilon} = 0 \qquad (D, D' = 1,\ldots,m;\ \varepsilon = 1,\ldots,n;\ n+1); \tag{b}$$

(recall that $\theta_I = q_I$ is a holonomic coordinate).

Problem 2.12.5 Continuing from the above example and problems, consider again the nonstationary constraints in the special form (2.11.10a ff.):

$$dq_D = \sum b_{DI}\, dq_I + b_D\, dt, \qquad \delta q_D = \sum b_{DI}\,\delta q_I, \qquad v_D = \sum b_{DI} v_I + b_D, \tag{a}$$

where $b_{DI} = b_{DI}(t,q)$, $b_D = b_D(t,q)$, and, as usual, $v_k \equiv dq_k/dt$.

Show by direct d/δ-differentiations of the above, and assuming that $d(\delta q_I) - \delta(dq_I) = 0$, that

$$d(\delta q_D) - \delta(dq_D) = \sum\sum w^D{}_{II'}\, dq_{I'}\,\delta q_I + \sum w^D{}_I\, dt\,\delta q_I, \tag{b}$$

or, dividing both sides by dt,

$$(\delta q_D)^{\cdot} - \delta(\dot{q}_D) \equiv (\delta q_D)^{\cdot} - \delta v_D = \sum\left(\sum w^D{}_{II'}\, v_{I'} + w^D{}_I\right)\delta q_I \equiv \sum v^D{}_I\,\delta q_I; \tag{c}$$

that is, in general, and *contrary to the hitherto adopted Hamel viewpoint* (§2.12), $d(\delta q_D) \neq \delta(dq_D)$, as if the q_D are no longer holonomic coordinates!

REMARKS

The alternative (and, as shown below, internally consistent) viewpoint exhibited by (c) [originally advanced by Suslov (1901–1902), (1946, pp. 596–600), and continued by Levi-Civita (and Amaldi), Neimark and Fufaev, Rumiantsev, and others], is based on the following assumptions:

(i) If the n differentials/velocities $dq/\delta q/v$ are unconstrained, then we assume that the Hamel viewpoint holds for all of them; that is, $d(\delta q_k) = \delta(dq_k)$ $(k = 1, \ldots, n)$.

(ii) But, if these differentials/velocities are subject to m (a)-like constraints, then we assume that the Hamel viewpoint holds only for the *independent* of them, say the last $n - m$, but not for the *dependent* of them, that is for the remaining (first) m:

$$\text{Suslov viewpoint:} \qquad d(\delta q_I) - \delta(dq_I) = 0, \quad \text{but} \quad d(\delta q_D) - \delta(dq_D) \neq 0. \tag{d}$$

Let us examine this quantitatively, from the earlier generalized transitivity equations (2.10.1, 5):

$$d(\delta\theta_k) - \delta(d\theta_k) = \sum a_{kl}[d(\delta q_l) - \delta(dq_l)] + \sum\sum \gamma^k{}_{rs}\, d\theta_s\, \delta\theta_r + \sum \gamma^k{}_r\, dt\, \delta\theta_r, \tag{e}$$

$$d(\delta q_k) - \delta(dq_k) = \sum A_{kl}\Big\{[d(\delta\theta_l) - \delta(d\theta_l)] - \sum\sum \gamma^l{}_{rs}\, d\theta_s\, \delta\theta_r - \sum \gamma^l{}_r\, dt\, \delta\theta_r\Big\}. \tag{f}$$

(a) *Hamel viewpoint*: $d(\delta q_k) = \delta(dq_k)$, always. Then, since $d\theta_D$, $\delta\theta_D = 0$, (e) yields

$$d(\delta\theta_D) - \delta(d\theta_D) = \sum\sum \gamma^D{}_{II'}\, d\theta_{I'}\, \delta\theta_I + \sum \gamma^D{}_I\, dt\, \delta\theta_I \quad \text{[by (pr. 2.12.4: b)]} \tag{g}$$

$$= -\sum\sum w^D{}_{II'}\, dq_{I'}\, \delta q_I - \sum w^D{}_I\, dt\, \delta q_I \quad \text{[by (ex. 2.12.1: g, j)]}; \tag{h}$$

$$d(\delta\theta_I) - \delta(d\theta_I) = \sum\sum \gamma^I{}_{I'I''}\, d\theta_{I''}\, \delta\theta_{I'} + \sum \gamma^I{}_{I'}\, dt\, \delta\theta_{I'} = 0 \quad \text{[by (pr. 2.12.4: a)]}. \tag{i}$$

(b) *Suslov viewpoint* [for the Voronets-type constraints (a)]. Since here, $A_{DD'} = \delta_{DD'}$, $A_{ID} = 0$, $A_{II'} = \delta_{II'}$, eq. (f) yields, successively,

(1) $0 = d(\delta q_I) - \delta(dq_I)$

$$= \sum A_{II'}\Big\{[d(\delta\theta_{I'}) - \delta(d\theta_{I'})] - \sum\sum \gamma^I{}_{I'I''}\, d\theta_{I''}\, \delta\theta_{I'} - \sum \gamma^I{}_{I'}\, dt\, \delta\theta_{I'}\Big\}$$

$$= [d(\delta\theta_I) - \delta(d\theta_I)] - \sum\sum \gamma^I{}_{I'I''}\, d\theta_{I''}\, \delta\theta_{I'} - \sum \gamma^I{}_{I'}\, dt\, \delta\theta_{I'}$$

$$= d(\delta\theta_I) - \delta(d\theta_I) \quad \text{[by (pr. 2.12.4: a)]}; \tag{j}$$

(2) $d(\delta q_D) - \delta(dq_D)$

$$= \sum A_{DD'}\Big\{[d(\delta\theta_{D'}) - \delta(d\theta_{D'})] - \sum\sum \gamma^{D'}{}_{II'}\, d\theta_{I'}\, \delta\theta_I - \sum \gamma^{D'}{}_I\, dt\, \delta\theta_I\Big\}$$

$$= [d(\delta\theta_D) - \delta(d\theta_D)] - \sum\sum \gamma^D{}_{II'}\, d\theta_{I'}\, \delta\theta_I - \sum \gamma^D{}_I\, dt\, \delta\theta_I$$

$$= [d(\delta\theta_D) - \delta(d\theta_D)] + \sum\sum w^D{}_{II'}\, d\theta_{I'}\, \delta\theta_I + \sum w^D{}_I\, dt\, \delta\theta_I$$

[by (ex. 2.12.1: i), (pr. 2.12.2: c)]

$$= \sum\sum w^D{}_{II'}\, dq_{I'}\, \delta q_I + \sum w^D{}_I\, dt\, \delta q_I \quad \text{[by (b)]},$$

and comparing the last two expressions of $d(\delta q_D) - \delta(dq_D)$, we immediately conclude that

$$d(\delta\theta_D) - \delta(d\theta_D) = 0. \tag{k}$$

Hence: In the Suslov viewpoint we must assume that $d(\delta\theta_k) = \delta(d\theta_k)$ $(k = 1, \ldots, n)$.

Both viewpoints are internally consistent; but, if applied improperly, they may give rise to contradictions/paradoxes. Hamel's viewpoint, however, has the advantage of being in agreement with variational calculus (more on this in §7.8).

Problem 2.12.6 Consider the special stationary $d\theta \Leftrightarrow dq$ transformation:

$$d\theta_D \equiv \sum a_{DD'}\, dq_{D'}\ (= 0) \quad \text{and} \quad d\theta_I \equiv dq_I\ (\neq 0;\ \text{the } \theta_I \text{ are } \textit{holonomic} \text{ coordinates}), \tag{a}$$

where $a_{DD'} = a_{DD'}(q_1, \ldots, q_m) \equiv a_{DD'}(q_D)$. Show that, in this case, the Hamel coefficients are

(i) $$\gamma^D{}_{D'D''} = \sum\sum (\partial a_{Dd'}/\partial q_{d''} - \partial a_{Dd''}/\partial q_{d'}) A_{d'D'} A_{d''D''}, \tag{b}$$

where $D, D', D'', d', d'' = 1, \ldots, m$ and $dq_{D'} = \sum A_{D'D}\, d\theta_D$; and

(ii) $$\gamma^k{}_{rs} = 0, \text{ for any one of } k, r, s \text{ greater than } m. \tag{c}$$

Example 2.12.2 *Transformation of the Hamel Coefficients under Frame of Reference Transformations.* Let us again consider, for algebraic simplicity but no loss in generality, the stationary Pfaffian constraint system:

$$d\theta_D \equiv \sum a_{Dk}\, dq_k = 0, \qquad \delta\theta_D \equiv \sum a_{Dk}\, \delta q_k = 0. \tag{a}$$

Further, let us assume that (a) is nonholonomic; that is, $\gamma^D{}_{II'} \neq 0$. Now we ask the question: Is it possible, by a frame of reference transformation $q \to q'(t, q)$, to make the constraints (a) holonomic? In other words, is it possible to find new Lagrangean coordinates $q_{k'} = q_{k'}(t, q_k)$, in which the corresponding $(dq_{k'} \Leftrightarrow d\theta_k)$ Hamel coefficients $\gamma(q')^D{}_{II'} \equiv \gamma'^D{}_{II'}$ vanish? Below we show that the answer to this is *no*; that is, if a system of constraints is nonholonomic in one frame of reference, it remains nonholonomic in all other frames of reference obtainable from the original via admissible frame of reference transformations.

Indeed, we find, successively [with $\gamma(q)^D{}_{II'} \equiv \gamma^D{}_{II'}$],

$$\begin{aligned}
0 \neq d(\delta\theta_D) - \delta(d\theta_D) &= \sum\sum \gamma^D{}_{I'I}\, d\theta_I\, \delta\theta_{I'} = \sum\sum a^D{}_{rs}\, dq_s\, \delta q_r \\
&= \sum\sum a^D{}_{rs} \Big(\sum (\partial q_s/\partial q_{s'})\, dq_{s'} + (\partial q_s/\partial t)\, dt\Big) \Big(\sum (\partial q_r/\partial q_{r'})\, \delta q_{r'}\Big) \\
&= \sum\sum\sum\sum [(\partial q_s/\partial q_{s'})(\partial q_r/\partial q_{r'}) a^D{}_{rs}]\, dq_{s'}\, \delta q_{r'} \\
&\quad + \sum\sum\sum [(\partial q_s/\partial t)(\partial q_r/\partial q_{r'}) a^D{}_{rs}]\, dt\, \delta q_{r'} \\
&\equiv \sum\sum a^D{}_{r's'}\, dq_{s'}\, \delta q_{r'} + \sum a^D{}_{r'}\, dt\, \delta q_{r'}
\end{aligned}$$

[where $a^D{}_{r's'} \equiv \partial a_{Dr'}/\partial q_{s'} - \partial a_{Ds'}/\partial q_{r'}$, etc.]

$$= \sum\sum a^D{}_{r's'} \Big(\sum A_{s'I}\, d\theta_I + A_{s'}\, dt\Big)\Big(\sum A_{r'I'}\, \delta\theta_{I'}\Big) + \sum a^D{}_{r'} \Big(\sum A_{r'I'}\, \delta\theta_{I'}\Big)\, dt$$

$$= \sum\sum\sum\sum (a^D{}_{r's'} A_{s'I} A_{r'I'})\, d\theta_I\, \delta\theta_{I'} + \sum\Big(\sum\sum a^D{}_{r's'} A_{r'I'} A_{s'} + \sum a^D{}_{r'} A_{r'I'}\Big)\, dt\, \delta\theta_{I'}$$

$$= \sum\sum \gamma'^D{}_{I'I}\, d\theta_I\, \delta\theta_{I'} + \sum \gamma'^D{}_{I'}\, dt\, \delta\theta_{I'}, \qquad \text{(b)}$$

from which, comparing with the first line of this equation, we readily conclude that

$$\gamma(q)^D{}_{I'I} = \gamma(q')^D{}_{I'I} \qquad \text{and} \qquad \gamma(q')^D{}_{I'} = 0; \qquad \text{(c)}$$

that is, *the Hamel coefficients remain invariant under frame of reference transformations*; or, these coefficients depend on the nonholonomic "coordinates" θ_k but they are independent of the particular holonomic coordinates frame used for their derivation.

Incidentally, this derivation also demonstrates that the γ-definition (2.10.1) is both practically and theoretically superior to the more common (2.10.2–4).

REMARKS

(i) We are reminded that the transformation properties of the γ's under *local* transformations: $d\theta_k \Leftrightarrow d\theta_{k'}$, at (q, t), have already been given in ex. 2.10.1.

(ii) The reader can easily verify that if, instead of (a), we had chosen a general *nonstationary* $d\theta \Leftrightarrow dq$ transformation, we would have found $\gamma(q')^D{}_{I'} = \gamma(q)^D{}_{I'}$, instead of the second of (c). Also, then,

$$d\theta_r = \sum a_{rs}\, dq_s + a_r\, dt = \sum a_{rs}\Big(\sum (\partial q_s/\partial q_{s'})\, dq_{s'} + (\partial q_s/\partial t)\, dt\Big) + a_r\, dt$$

$$\equiv \sum a_{rr'}\, dq_{r'} + a'_r\, dt,$$

from which we can readily deduce the transformation relations among the coefficients $a(q)$, $a(q')$ [recall (2.6.6 ff.)].

Problem 2.12.7 (see Forsyth, 1890, p. 54.) Verify that a system of n independent Pfaffian constraints in the n (or even $n+1$) variables; that is,

$$d\theta_k \equiv \sum a_{kl}\, dq_l = 0 \qquad (k, l = 1, \ldots, n), \qquad \text{(a)}$$

is always holonomic.

Problem 2.12.8 *Alternative Formulation of Frobenius' Theorem.* It has been shown, by Frobenius and others (see, e.g., Pascal, 1927, p. 584), that the Pfaffian system:

$$d\theta_D \equiv \sum a_{Dk}(q)\, dq_k = 0 \qquad (D = 1, \ldots, m;\ k = 1, \ldots, n), \qquad \text{(a)}$$

is holonomic if, and only if, each of its m $(n+m)\times(n+m)$ antisymmetric "Frobenius matrices":

$$\mathbf{F}_\mathrm{D} \equiv \left(\begin{array}{ccccc|ccc}
0 & a^D{}_{12} & a^D{}_{13} & \cdots & a^D{}_{1n} & a_{11} & \cdots & a_{m1} \\
a^D{}_{21} & 0 & a^D{}_{23} & \cdots & a^D{}_{2n} & a_{12} & \cdots & a_{m2} \\
\cdots & \cdots & \cdots & \cdots & \cdots & \cdots & \cdots & \cdots \\
a^D{}_{n1} & a^D{}_{n2} & \cdots & \cdots & 0 & a_{1n} & \cdots & a_{mn} \\
\hline
a_{11} & a_{12} & \cdots & \cdots & a_{1n} & 0 & \cdots & 0 \\
\cdots & \cdots & \cdots & \cdots & \cdots & \cdots & \cdots & \cdots \\
a_{m1} & a_{m2} & \cdots & \cdots & a_{mn} & 0 & \cdots & 0
\end{array}\right)$$

where $a^D{}_{kl} \equiv \partial a_{Dk}/\partial q_l - \partial a_{Dl}/\partial q_k \equiv a_{Dk,l} - a_{D,lk} = -a^D{}_{lk}$ (e.g., $a^D{}_{12} = -a^D{}_{21}$, $a^D{}_{11} = -a^D{}_{11} \Rightarrow a^D{}_{11} = 0$, etc.), has rank $2m$.

Apply this theorem for various simple cases: for example, $m = 0$ (i.e., dq_k unconstrained), $m = 1$ (one constraint), and $m = 2$ (two constraints).

Example 2.12.3 *Geometrical Interpretations of the Frobenius Conditions* (May be omitted in a first reading.) In terms of the earlier (2.11.20a ff.) *m constraint vectors* $\boldsymbol{a}_D = (a_{D1}, \ldots, a_{Dn})$ and *n − m virtual vectors* $\boldsymbol{A}_I = (A_{I1}, \ldots, A_{In})$ (in ordinary vector, *nonmatrix* notation), Frobenius' conditions first of (2.12.5) assume the following forms:

(i) *First interpretation:* From the $\boldsymbol{a}_D$ we build the antisymmetric tensor:

$$(a^D{}_{kl}) = (-a^D{}_{lk}) \equiv (\partial a_{Dk}/\partial q_l - \partial a_{Dl}/\partial q_k). \tag{a}$$

These can be viewed as the holonomic (covariant) components of the "*rotation* or *curl(ing)* of $\boldsymbol{a}_D$": $a^D{}_{kl} = -(\boldsymbol{curl}\,\boldsymbol{a}_D)_{kl}$. Also, we recall that $A_{kl} \equiv \partial v_k/\partial \omega_l$. As a result of the above, (2.12.5):

$$\gamma^D{}_{II'} = \sum\sum (\partial a_{Dk}/\partial q_l - \partial a_{Dl}/\partial q_k) A_{kI} A_{lI'} \equiv \sum\sum a^D{}_{kl} A_{kI} A_{lI'} = 0, \tag{b}$$

assumes the (covariant) tensor transformation form, in k, l:

$$\begin{aligned}\gamma^D{}_{II'} &= \sum\sum (\partial v_k/\partial \omega_I)(\partial v_l/\partial \omega_{I'}) a^D{}_{kl} \\ &= \boldsymbol{A}_{I'} \cdot \boldsymbol{curl}\,\boldsymbol{a}_D \cdot \boldsymbol{A}_I = \sum\sum (A_{I'})^l (\boldsymbol{curl}\,\boldsymbol{a}_D)_{lk} (A_I)^k = 0;\end{aligned} \tag{c}$$

that is, *the (covariant) components of the* ***curl*** *of the dependent/constraint vectors* $\boldsymbol{a}_D$ *along the independent nonholonomic directions* $\boldsymbol{A}_I$ *should vanish.*

(ii) *Second interpretation:* The Frobenius conditions (first of 2.12.5), rewritten with the help of the alternative expression (2.10.15) and the quasi chain rule

(2.9.30a) as

$$\gamma^D{}_{II'} = \sum\sum \{a_{Db}[A_{cI}(\partial A_{bI'}/\partial q_c) - A_{cI'}(\partial A_{bI}/\partial q_c)]\}$$
$$\equiv \sum a_{Db}(\partial A_{bI'}/\partial\theta_I - \partial A_{bI}/\partial\theta_{I'}) = 0, \tag{d}$$

state that the constraint vectors $\boldsymbol{a}_D$ should be *perpendicular* to the $(n-m)\,(n-m-1)/2$ vectors:

$$\boldsymbol{A}_{II'} = \left(\sum A_{cI}(\partial A_{bI'}/\partial q_c) - \sum A_{cI'}(\partial A_{bI}/\partial q_c)\right) \equiv (\partial A_{bI'}/\partial\theta_I - \partial A_{bI}/\partial\theta_{I'}) = -\boldsymbol{A}_{I'I},$$

that is,

$$\gamma^D{}_{II'} = \boldsymbol{a}_D \cdot \boldsymbol{A}_{II'} = 0. \tag{e}$$

Similarly for the nonstationary/rheonomic Frobenius conditions (second of 2.12.5): $\gamma^D{}_I = 0$.

For further details, including the precise positioning of indices, as practiced in general tensor analysis (and not observed in the above discussion!), see, for example, Papastavridis (1999, §6.9).

2.13 GENERAL EXAMPLES AND PROBLEMS

Example 2.13.1 *Introduction to the Simplest Nonholonomic Problem: Knife, Sled, Scissors, and so on.* Let us consider the motion of a knife S, whose rigid blade remains perpendicular to the fixed plane $O\text{–}xy$, and in contact with it at the point $C(x, y)$, and whose mass center G lies a distance b ($\neq 0$) from C along the blade (fig. 2.15). The instantaneous angular orientation of S is given by its blade's angle with the $+Ox$ axis ϕ.

Let us choose as Lagrangean coordinates: $q_1 = x$, $q_2 = y$, $q_3 = \phi$. If $\boldsymbol{v} = (dx/dt, dy/dt, dz/dt = 0) \equiv (v_x, v_y, 0) =$ (inertial) velocity of C, and $\boldsymbol{u} \equiv (\cos\phi, \sin\phi, 0)$: unit vector along the blade, then the velocity constraint is

$$\boldsymbol{v} \times \boldsymbol{u} = \boldsymbol{0} \Rightarrow (\sin\phi)v_x + (-\cos\phi)v_y = 0, \qquad \text{or} \qquad dy/dx = \tan\phi. \tag{a}$$

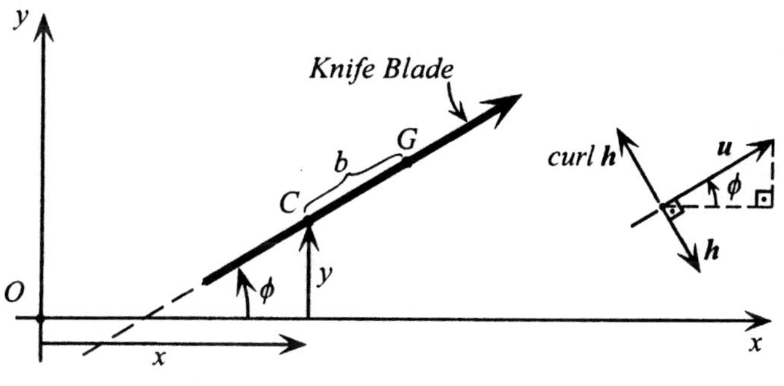

Figure 2.15 Knife in motion on fixed plane.

Since this is a *stationary* (and, of course, catastatic) constraint, we will also have, for its kinematically admissible/possible and virtual forms, respectively,

$$(\sin\phi)\,dx + (-\cos\phi)\,dy = 0 \qquad \text{and} \qquad (\sin\phi)\,\delta x + (-\cos\phi)\,\delta y = 0. \tag{b}$$

Other physical problems leading to such a constraint are:

(i) A racing boat with thin, deep, and wide keel, sailing on a still sea. Since the water resistance to the boat's longitudinal motion is *much larger* than the resistance to its transverse motion, the direction of the boat's instantaneous velocity must be always parallel to its keel's instantaneous heading;
(ii) a lamina moving on its plane, with a short and very stiff razor blade (or some similar rigid and very thin object: e.g., a small knife) embedded on its underside. Again, the lamina can move only along the instantaneous direction of its guiding blade;
(iii) a sled;
(iv) a pair of scissors cutting through a piece of paper;
(v) a pizza cutter, etc.

Application of the holonomicity criterion (2.3.6) or (2.3.8a) to (a), (b), with $\boldsymbol{h} = (\sin\phi, -\cos\phi, 0)$ and $d\boldsymbol{r} = (dx, dy, d\phi)$ [as if x, y, ϕ were rectangular Cartesian right-handed coordinates] yields

$$\begin{aligned} I \equiv \boldsymbol{h}\cdot \boldsymbol{curl\,h} &= \boldsymbol{h}\cdot\big[(\partial/\partial x, \partial/\partial y, \partial/\partial\phi) \times (\sin\phi, -\cos\phi, 0)\big] \\ &= (\sin\phi, -\cos\phi, 0)\cdot(-\sin\phi, \cos\phi, 0) = -1 \neq 0; \end{aligned} \tag{c}$$

that is, the constraint (a), (b) is *nonholonomic*. This means that, although the general (global) configuration of S is specified completely by the *three* independent coordinates x, y, ϕ, *not all three of them* can be given, simultaneously, small arbitrary variations; that is, although there is no functional restriction of the type $f(x, y, \phi) = 0$, there is one of the type $g(dx, dy, d\phi; x, y, \phi) = 0$, namely the Pfaffian constraint (a), (b). Put geometrically: the blade has three *global* freedoms (x, y, ϕ), but only *two local* freedoms (any *two* of $dx, dy, d\phi$). Since $n = 3$ and $m = 1$, this is the *simplest* nonholonomic problem; and, accordingly, it has been studied extensively (by Bahar, Carathéodory, Chaplygin, et al.).

The independence of x, y, ϕ can be demonstrated as follows: we keep any two of them constant, and then show that varying the third results in a nontrivial (or nonempty) range of kinematically admissible positions:

(i) keep x and y fixed and vary ϕ continuously; the constraint (a), (b) is not violated [fig. 2.16(a)];
(ii) keep y and ϕ fixed. Varying x we can achieve other admissible configurations with different x's but the same y and ϕ;
but to go from one of them to another we have to vary all three coordinates [fig. 2.16(b)];
(iii) similarly when x and ϕ are fixed and y varies [fig. 2.16(c)].
The precise *kinetic* path followed in each case, among the kinematically possible/admissible ones, depends on the system's equations of (constrained) motion and on its initial conditions.

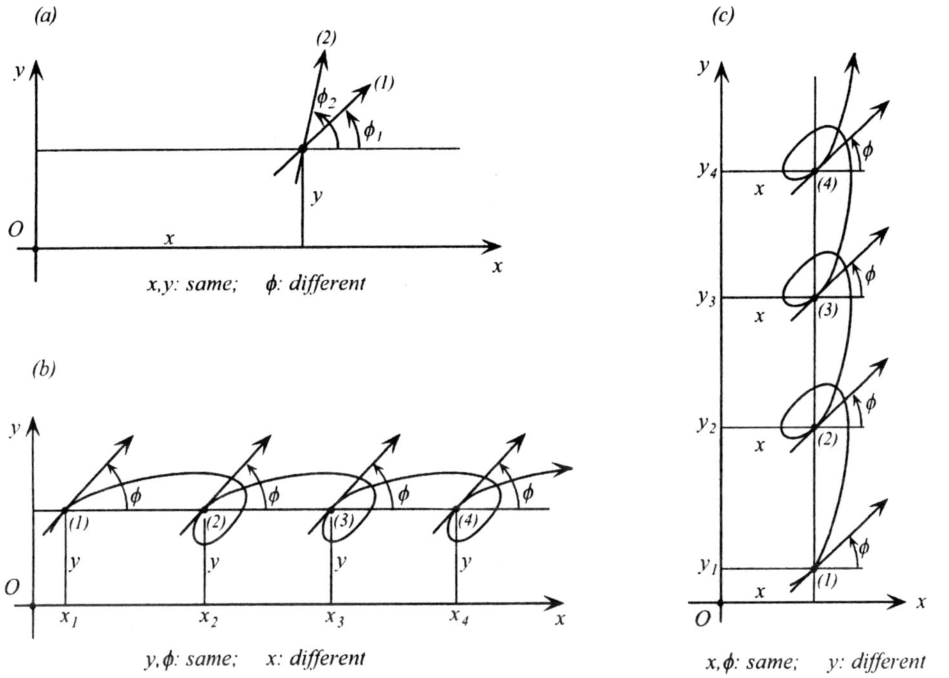

Figure 2.16 Global motions of a knife showing the independence of its three positional coordinates. We can always, through a suitable *finite* motion, bring the knife to a position and orientation as close as we want to any specified *original* position and orientation; that is, the relation among x, y, ϕ is nonunique.

An Ad Hoc Proof of the Impossibility of Obtaining a Relation $f(x, y, \phi) = 0$

Let us assume that such a constraint exists. Then d-varying it, and with subscripts for partial derivatives, yields

$$df = f_x\, dx + f_y\, dy + f_\phi\, d\phi = 0, \tag{d}$$

or, taking into account the constraint in the form: $dy = (\tan\phi)\, dx$,

$$df = (f_x + f_y \tan\phi)\, dx + (f_\phi)\, d\phi = 0, \tag{e}$$

where now dx and $d\phi$ are *independent*. Equation (e) leads immediately to

$$f_\phi = 0 \Rightarrow f = f(x, y) \qquad \text{and} \qquad f_x + f_y \tan\phi = 0. \tag{f}$$

By $(\partial/\partial\phi)$-differentiating the second of (f), while observing the first of (f), we obtain

$$f_y(1/\cos^2\phi) = 0, \tag{g}$$

from which, since in general $1/\cos^2\phi \neq 0$, it follows that $f_y = 0 \Rightarrow f = f(x)$. But then the second of (f) leads to $f_x = 0 \Rightarrow f = constant$ (*independent of* x, y, ϕ), and as such it cannot enforce the constraint $f(x, y, \phi) = 0$. Hence, no such f exists (with or without integrating factors).

However, if the knife was constrained to move along a *prescribed* path, on the O–xy plane, the system would be holonomic! In that case, we would have in *advance* the path's equations, say in the parametric form:

$$x = x(s) \quad \text{and} \quad y = y(s) \quad (s = \text{arc length}), \tag{h}$$

from which ϕ could be uniquely determined for every s [i.e., $\phi = \phi(s)$], via

$$dy/dx = dy/ds \Big/ dx/ds \equiv y'(s)/x'(s) = \tan\phi(s). \tag{i}$$

This is somewhat analogous to the basic variables of Lagrangean mechanics q_k, $dq_k/dt \equiv v_k$, which, *before* the problem is solved, are considered as *independent*, and then, *after* the problem is completely solved, become dependent through time.

Example 2.13.2 *The Knife Problem: Hamel Coefficients.* Continuing from the preceding example: in view of the constraint (a), (b) there, and following Hamel's methodology ("equilibrium quasi velocities," §2.11), let us introduce the following three quasi velocities:

$$\begin{aligned}
\omega_1 &\equiv (-\sin\phi)v_x + (\cos\phi)v_y + (0)v_\phi \quad (= 0),\\
\omega_2 &\equiv (\cos\phi)v_x + (\sin\phi)v_y + (0)v_\phi = v \quad (\neq 0),\\
\omega_3 &\equiv (0)v_x + (0)v_y + (1)v_\phi = v_\phi \quad (\neq 0),
\end{aligned} \tag{a}$$

where v = velocity component of the knife's contact point C; and hence $v_x = v\cos\phi$, $v_y = v\sin\phi$, and the constraint is simply $\omega_1 = 0$. Clearly, since

$$\partial(\cos\phi)/\partial\phi \neq \partial(0)/\partial x \quad \text{and} \quad \partial(\sin\phi)/\partial\phi \neq \partial(0)/\partial y, \tag{b}$$

$\omega_2 = v$ is a *quasi velocity*; that is, $v \neq$ total time derivative of a genuine position coordinate, or of any function of x, y, ϕ. Inverting (a), we obtain

$$\begin{aligned}
v_x &= (-\sin\phi)\omega_1 + (\cos\phi)\omega_2 + (0)\omega_3,\\
v_y &= (\cos\phi)\omega_1 + (\sin\phi)\omega_2 + (0)\omega_3,\\
v_\phi &= (0)\omega_1 + (0)\omega_2 + (1)\omega_3.
\end{aligned} \tag{c}$$

If **a** and **A** are the matrices of the transformations (a) and (c), respectively, then we easily verify that $\mathbf{a} = \mathbf{A}$, and $\operatorname{Det}\mathbf{a} = \operatorname{Det}\mathbf{A} = -\sin^2\phi - \cos^2\phi = -1$ (i.e., nonsingular transformations). Further, we notice that (a), (c) hold with $\omega_{1,2,3}$ and v_x, v_y, v_ϕ replaced, respectively, with $d\theta_{1,2,3} = \omega_{1,2,3}\,dt$ and $(dx, dy, d\phi) = (v_x, v_y, v_\phi)\,dt$; and, since they are *stationary*, also for $\delta\theta_{1,2,3}$ and $\delta x, \delta y, \delta\phi$.

Next, by direct d/δ-differentiations of $\delta\theta_1$, $d\theta_1$, and then subtraction, we find, successively,

$$\begin{aligned}
d(\delta\theta_1) - \delta(d\theta_1) &= d[(-\sin\phi)\,\delta x + (\cos\phi)\,\delta y + (0)\,\delta\phi]\\
&\quad - \delta[(-\sin\phi)\,dx + (\cos\phi)\,dy + (0)\,d\phi]\\
&= (-\sin\phi)(d\delta x - \delta dx) + (\cos\phi)(d\delta y - \delta dy)\\
&\quad - \cos\phi\, d\phi\,\delta x - \sin\phi\, d\phi\,\delta y + \cos\phi\, dx\,\delta\phi + \sin\phi\, dy\,\delta\phi\\
&= 0 + 0 - \cos\phi(1)\,d\theta_3[(-\sin\phi)\,\delta\theta_1 + (\cos\phi)\,\delta\theta_2 + (0)\,\delta\theta_3] - \cdots
\end{aligned}$$

[i.e., expressing dx, δx, dy, δy, $d\phi$, $\delta\phi$ from (c), with $\omega_{1,2,3}$ replaced with $d\theta_{1,2,3}$, $\delta\theta_{1,2,3}$] and so we, finally, obtain the differential transitivity equation:

$$d(\delta\theta_1) - \delta(d\theta_1) = d\theta_2\,\delta\theta_3 - d\theta_3\,\delta\theta_2, \tag{d}$$

[i.e., $d(\delta\theta_1) - \delta(d\theta_1) \neq 0$, even though $\delta\theta_1 = 0$ and $d\theta_1 = 0$]; and, also, dividing this by dt, which does not couple with $\delta(\ldots)$, we obtain its (equivalent) velocity transitivity equation:

$$(\delta\theta_1)^{\cdot} - \delta\omega_1 = (0)\,\delta\theta_1 + (-1)\omega_3\,\delta\theta_2 + (1)\omega_2\,\delta\theta_3 \quad (\neq 0). \tag{e}$$

Similarly, after some straightforward differentiations, we find

$$(\delta\theta_2)^{\cdot} - \delta\omega_2 = (1)\omega_3\,\delta\theta_1 + (0)\,\delta\theta_2 + (-1)\omega_1\,\delta\theta_3 \quad (= 0), \tag{f}$$

$$(\delta\theta_3)^{\cdot} - \delta\omega_3 = (0)\,\delta\theta_1 + (0)\,\delta\theta_2 + (0)\,\delta\theta_3 \quad (= 0). \tag{g}$$

From (e, f, g) we readily read off the *nonvanishing* Hamel's coefficients:

$$\gamma^D{}_{II'}\ (D = 1;\ I, I' = 2, 3): \qquad \gamma^1{}_{23} = -\gamma^1{}_{32} = -1; \tag{h}$$

$$\gamma^I{}_{kl}\ (I = 2;\ k, l = 1, 3): \qquad \gamma^2{}_{13} = -\gamma^2{}_{31} = 1. \tag{i}$$

REMARKS

(i) Since not all $\gamma^D{}_{II'} \to \gamma^1{}_{kl}$ $(k, l = 2, 3)$ vanish, we conclude, by Frobenius' theorem (§2.12), that our constraint, in any one of the following three forms:

Velocity:

$$\omega_1 \equiv (-\sin\phi)v_x + (\cos\phi)v_y + (0)v_\phi \quad (= 0), \tag{j}$$

Kinematically admissible:

$$d\theta_1 \equiv (-\sin\phi)\,dx + (\cos\phi)\,dy + (0)\,d\phi \quad (= 0), \tag{k}$$

Virtual:

$$\delta\theta_1 \equiv (-\sin\phi)\,\delta x + (\cos\phi)\,\delta y + (0)\,\delta\phi \quad (= 0), \tag{l}$$

is nonholonomic.

(ii) The fact that *upon imposition of the constraints* $\delta\theta_1 = 0$, $\omega_1 = 0$, the transitivity equation (f) yields $(\delta\theta_2)^{\cdot} - \delta\omega_2 = 0$ does *not* mean that

$$d\theta_2 \equiv (\cos\phi)\,dx + (\sin\phi)\,dy + (0)\,d\phi = v\,dt \quad (\neq 0), \tag{m}$$

or

$$\delta\theta_2 \equiv (\cos\phi)\,\delta x + (\sin\phi)\,\delta y + (0)\,\delta\phi \quad (\neq 0), \tag{n}$$

are exact; it does not mean that θ_2 is a genuine (Lagrangean) coordinate. For exactness, we should have $\gamma^2{}_{kl} = 0\,(k, l = 1, 2, 3) \Rightarrow (\delta\theta_2)^{\cdot} - \delta\omega_2 = 0$, *independently of the constraints* $\omega_1/d\theta_1/\delta\theta_1 = 0$. [We recall (§2.12) that Frobenius' theorem tests the holonomicity, or absence thereof, of a *system* of Pfaffian *equations of constraint*; whereas the *exactness*, or inexactness, of a particular Pfaffian *form*, like $d\theta_2$ and $d\theta_3 (\neq 0)$ is a property of that form; that is, it is ascertained by examination of that form alone, independently of other constraint equations. In sum: *constraint holonomicity is a system (coupled) property; while coordinate holonomicity is an individual (uncoupled) property*.]

(iii) Since $\omega_3 = d\phi/dt \equiv v_\phi$ is a genuine velocity, $\gamma^3{}_{kl} = 0$ $(k, l = 1, 2, 3)$; as expected.

Hamel Viewpoint versus Suslov Viewpoint

So far, we have assumed Hamel's viewpoint; that is,

$$d(\delta x) = \delta(dx), \qquad d(\delta y) = \delta(dy), \qquad d(\delta\phi) = \delta(d\phi); \tag{o}$$

and $d(\delta\theta_1) \neq \delta(d\theta_1)$, in spite of the constraint $\delta\theta_1 = 0$ and $d\theta_1 = 0$ [and that even if $d(\delta\theta_1) = 0$, still $-\delta(d\theta_1) \neq 0$!].

Let us now examine the Suslov viewpoint: with the analytically convenient choice, $q_D = y$ and $q_I = x, \phi$, we can rewrite the constraint as

$$d\theta_1 \equiv dy - (\tan\phi)\, dx = 0 \qquad \text{and} \qquad \delta\theta_1 \equiv \delta y - (\tan\phi)\, \delta x = 0 \qquad [\text{instead of (a)}],$$

or

$$dy = (\tan\phi)\, dx + (0)\, d\phi \qquad \text{and} \qquad \delta y = (\tan\phi)\, \delta x + (0)\, \delta\phi; \tag{p}$$

and, therefore, the corresponding transitivity equations become [instead of (d)–(g)]

$$\begin{aligned} \text{Dependent:}\quad d(\delta y) - \delta(dy) &= d(\delta x \tan\phi) - \delta(dx \tan\phi) = \cdots \\ &= [d(\delta x) - \delta(dx)]\tan\phi + (1/\cos^2\phi)(d\phi\, \delta x - dx\, \delta\phi) \\ &= (1/\cos^2\phi)(d\phi\, \delta x - dx\, \delta\phi) \neq 0, \end{aligned} \tag{q}$$

$$\text{Independent:}\quad d(\delta x) - \delta(dx) = 0, \qquad d(\delta\phi) - \delta(d\phi) = 0; \tag{r}$$

from which we readily read off the sole nonvanishing *Voronets* symbol:

$$w^y{}_{x\phi} = -w^y{}_{\phi x} = 1/\cos^2\phi, \tag{s}$$

Under Hamel's viewpoint, using the same variables, from $\delta y = (\tan\phi)\, \delta x$ (i.e., $\delta\theta_1 = 0$) it follows that $d(\delta y) = d(\delta x)\tan\phi + (1/\cos^2\phi)\, d\phi\, \delta x$ [i.e., $d(\delta\theta_1) = 0$]; but from $dy = (\tan\phi)\, dx$ (i.e., $d\theta_1 = 0$) it does *not* follow that $\delta(dy) = \delta(dx)\tan\phi - (1/\cos^2\phi)\, \delta\phi\, dx$ [i.e., $\delta(d\theta_1) \neq 0$].

Problem 2.13.1 Consider a knife (or sled, or scissors, etc.) moving on a *uniformly* rotating turntable T (fig. 2.17). In *T-fixed* (moving) coordinates O–$xy\phi$, its constraint is

$$(\sin\phi)v_x + (-\cos\phi)v_y = 0 \qquad [v_x \equiv dx/dt,\ v_y \equiv dy/dt]. \tag{a}$$

Show that in *inertial* (fixed) coordinates O–$XY\Phi$, where

$$\begin{aligned} X &= (\cos\theta)x + (-\sin\theta)y + (0)\phi, \\ Y &= (\sin\theta)x + (\cos\theta)y + (0)\phi, \\ \Phi &= (0)x + (0)y + (1)\phi + \theta, \end{aligned} \tag{b}$$

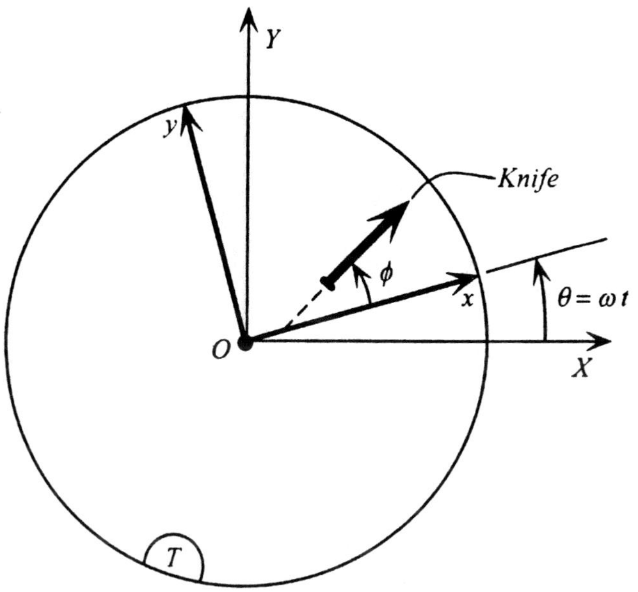

Figure 2.17 Knife moving on a uniformly rotating turntable.

and $\theta = \omega t, \omega$: constant angular velocity of O–xy relative to O–XY [i.e., say, $X = X(x, y, t)$, etc.], the constraint takes the (acatastatic) form (with $v_X \equiv dX/dt$, $v_Y \equiv dY/dt$),

$$(\sin \Phi)v_X + (-\cos \Phi)v_Y + \omega\big[(\cos \Phi)X + (\sin \Phi)Y\big] = 0. \qquad \text{(c)}$$

Example 2.13.3 *Rolling Disk—Vertical Case.* Let us consider a circular thin disk D, of center G and radius r, rolling while remaining vertical on a fixed, rough, and horizontal plane P (fig. 2.18). (The general *nonvertical* case is presented later in ex. 2.13.7.) This system has *four* Lagrangean coordinates (or global *DOF*): the $(x, y, z = r)$ coordinates of G, and the Eulerian angles ϕ (precession) and ψ (spin). The constraints $z = r$ (contact) and $\theta = \pi/2$ are, clearly, holonomic (H). The velocity constraint is $\boldsymbol{v}_C = \boldsymbol{0}$ (where C is the contact point); or, since along the fixed axes O–XYZ [with the notation $dx/dt \equiv v_x$, $dy/dt \equiv v_y$; $d\phi/dt \equiv \omega_\phi$, $d\psi/dt \equiv \omega_\psi$]:

$$\boldsymbol{v}_G = (v_x, v_y, 0), \qquad \boldsymbol{\omega} = (-\omega_\psi \sin\phi,\ \omega_\psi \cos\phi,\ \omega_\phi), \qquad \text{and} \qquad \boldsymbol{r}_{C/G} = (0, 0, -r),$$

$$\Rightarrow \boldsymbol{v}_C = \boldsymbol{v}_G + \boldsymbol{\omega} \times \boldsymbol{r}_{C/G} = \cdots = (v_x - r\omega_\psi \cos\phi,\ v_y - r\omega_\psi \sin\phi,\ 0) = \boldsymbol{0}, \qquad \text{(a)}$$

or, in components, in the following equivalent forms:

Velocity : $\quad v_x = r\omega_\psi \cos\phi \quad$ and $\quad v_y = r\omega_\psi \sin\phi,$ (b)

Kinematically admissible: $\quad dx = (r\cos\phi)\, d\psi \quad$ and $\quad dy = (r\sin\phi)\, d\psi,$ (c)

Virtual: $\quad \delta x = (r\cos\phi)\, \delta\psi \quad$ and $\quad \delta y = (r\sin\phi)\, \delta\psi.$ (d)

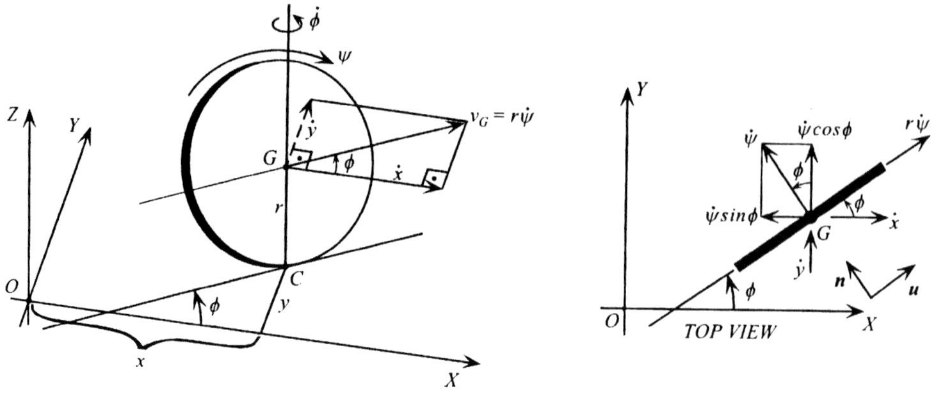

$$\dot x = v_G \cos\phi = r\dot\psi \cos\phi,\ \dot y = v_G \sin\phi = r\dot\psi \sin\phi;\quad \mathbf{u} = (\cos\phi, \sin\phi),\ \mathbf{n} = (-\sin\phi, \cos\phi)$$

Figure 2.18 Rolling of vertical disk on a fixed plane.

As shown below, these constraints are nonholonomic (NH). Hence, the disk is a *scleronomic* NH system with $f \equiv n - m = 4 - 2 = 2$ *DOF in the small.*

It is not hard to see that imposition, on (b–d), of the *additional* H constraint $d\phi = 0 \Rightarrow \phi = constant$, say $\phi = 0$, would reduce them to the well-known H case of *plane rolling*: $dx = r\,d\psi \Rightarrow x = r\psi + constant$, and $dy = 0 \Rightarrow y = constant$. {Also, the problem would become H if the disk was forced to roll along a *prescribed* O–XY path. For, then, the rolling condition would be [with s: arc-length along (c)] $ds = r\,d\psi \Rightarrow s = r\psi + constant$, and (c) would yield the parametric equations $x = x(s)$ and $y = y(s)$; that is, for each s there would correspond a unique x, y, ψ, and ϕ [from (b–d)], and that would make the disk a *1 (global) DOF H* system.}

Ad Hoc Proof of the Nonholonomicity of the Constraints (b–d)

Let us assume that we could find a finite relation $f(x, y, \phi, \psi) = 0$, compatible with (b–d). Then (with subscripts denoting partial derivatives), we would have

$$df = f_x\,dx + f_y\,dy + f_\phi\,d\phi + f_\psi\,d\psi = 0. \tag{e}$$

Substituting dx and dy from (c) into (e)—that is, *embedding* the constraints into it—yields

$$(r f_x \cos\phi + r f_y \sin\phi + f_\psi)\,d\psi + (f_\phi)\,d\phi = 0, \tag{f}$$

which, since now $d\psi$ and $d\phi$ are independent, gives

$$f_\phi = 0 \Rightarrow f = f(x, y, \psi) \qquad \text{and} \qquad r f_x \cos\phi + r f_y \sin\phi + f_\psi = 0. \tag{g}$$

Next, $(\partial/\partial\phi)$-differentiating the second of (g) once, while taking into account the first of (g), yields

$$-r f_x \sin\phi + r f_y \cos\phi = 0, \tag{h}$$

and repeating this procedure on (h), while again observing the first of (g), produces

$$-r f_x \cos\phi - r f_y \sin\phi = 0. \tag{i}$$

[Further $(\partial/\partial\phi)$-differentiations would not produce anything new.] The system (h), (i) has the unique solution,

$$f_x = 0 \quad \text{and} \quad f_y = 0, \tag{j}$$

due to which the second of (g) reduces to $f_\psi = 0$. It is clear that the above result in $f = constant$, and such a functional relation, obviously, *cannot* produce the constraints (b–d)—no $f(x, y, \phi, \psi)$ exists. Geometrically, this nonholonomicity has the following consequences: Starting from a certain *initial* configuration, we can roll the disk along two different paths to two *final* configurations with the same contact point—namely, same final (x, y), but *rotated relative to each other*; that is, with different final (ϕ, ψ). If the constraints were H, then ϕ and ψ would be functions of (x, y) and the two final positions of the disk would coincide completely.

Proof that the Constraints (b–d) are NH via Frobenius' Theorem

Let us rewrite the two constraints (c, d) in the equilibrium forms:

Kinematically admissible :

$$d\theta_1 \equiv dx - (r\cos\phi)\, d\psi = 0, \qquad d\theta_2 \equiv dy - (r\sin\phi)\, d\psi = 0, \tag{k}$$

Virtual :

$$\delta\theta_1 \equiv \delta x - (r\cos\phi)\, \delta\psi = 0, \qquad \delta\theta_2 \equiv \delta y - (r\sin\phi)\, \delta\psi = 0. \tag{l}$$

It follows that the corresponding bilinear covariants (2.8.2 ff.) are

$$d(\delta\theta_1) - \delta(d\theta_1) = \cdots = (r\sin\phi)(d\phi\,\delta\psi - d\psi\,\delta\phi), \tag{m}$$

$$d(\delta\theta_2) - \delta(d\theta_2) = \cdots = (-r\cos\phi)(d\phi\,\delta\psi - d\psi\,\delta\phi), \tag{n}$$

and, clearly, these vanish for arbitrary values of the *independent* differentials $d\phi, \delta\phi, d\psi, \delta\psi$, if $\sin\phi = 0$ and $\cos\phi = 0$. But then the constraints (c) reduce to $dx = 0 \Rightarrow x = constant$ and $dy = 0 \Rightarrow y = constant$, which is, in general, impossible. Hence, the constraints are NH [one can arrive at the same conclusion with the help of the γ's (§2.12), but that is more laborious].

Problem 2.13.2 Continuing from the previous problem (vertically rolling disk), show that its velocity constraints can be expressed in the equivalent form:

$$\boldsymbol{v}_G \cdot \boldsymbol{u} = v_x \cos\phi + v_y \sin\phi = r\,\omega_\psi, \tag{a}$$

$$\boldsymbol{v}_G \cdot \boldsymbol{n} = -v_x \sin\phi + v_y \cos\phi = 0, \tag{b}$$

where $\boldsymbol{u}$ and $\boldsymbol{n}$ are unit vectors on the disk plane (parallel to O–XY) and perpendicular to it, respectively (fig. 2.18). [Notice that (b) coincides, formally, with the *knife* problem constraint.]

Example 2.13.4 *Rolling Sphere—Introduction.* Let us consider a sphere of center G and radius r, rolling without slipping on a fixed, rough and, say, horizontal plane P (fig. 2.19). The complete specification of a generic sphere configuration requires *five* independent (minimal) Lagrangean coordinates. As such, we could take the (inertial)

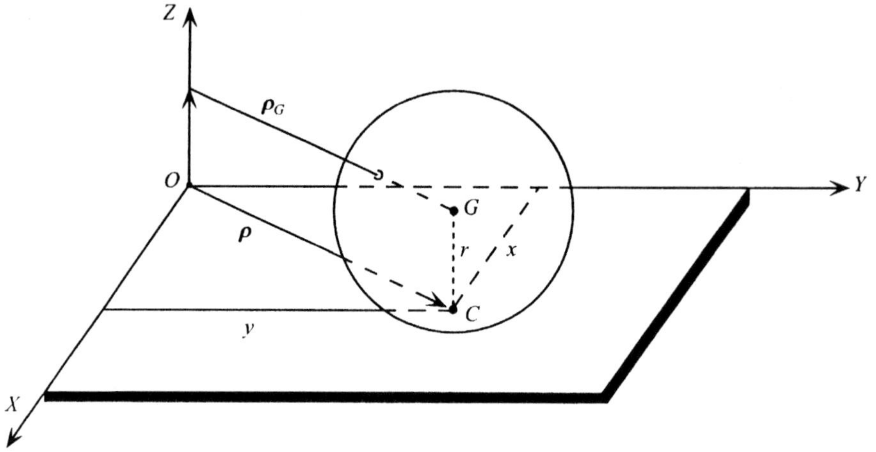

Figure 2.19 Rolling of a sphere on a fixed horizontal plane.

coordinates of C (X, Y), and the three Eulerian angles (ϕ, θ, ψ) of body-fixed axes G–xyz relative to translating (nonrotating) axes G–XYZ. The *contact* constraint is expressed by the holonomic (H) equation, $Z \equiv$ vertical coordinate of $G = r$. The *rolling* constraint is found by equating the (inertial) velocity of C with that of its adjacent plane point, which here is zero, with $\boldsymbol{\omega} \equiv$ inertial angular velocity of sphere. Using components along O–XYZ axes throughout, we find

$$\boldsymbol{v}_C = \boldsymbol{v}_G + \boldsymbol{\omega} \times \boldsymbol{r}_{C/G} = d\boldsymbol{\rho}_G/dt + \boldsymbol{\omega} \times (-r\boldsymbol{K})$$
$$= (v_X, v_Y, 0) + (\omega_X, \omega_Y, \omega_Z) \times (0, 0, -r) = \cdots = (v_X - r\omega_Y, v_Y + r\omega_X, 0) = \mathbf{0}.$$

Hence, the rolling conditions are

$$v_X - r\omega_Y = 0, \qquad v_Y + r\omega_X = 0; \tag{a}$$

or, expressing the *space-fixed* components $\omega_{X,Y}$ in terms of their Eulerian angle rates (§1.12),

$$v_X - r(\sin\phi\,\omega_\theta - \sin\theta\cos\phi\,\omega_\psi) = 0, \qquad v_Y + r(\cos\phi\,\omega_\theta + \sin\theta\sin\phi\,\omega_\psi) = 0; \tag{b}$$

or, further, in *kinematically admissible* form,

$$dX - r(\sin\phi\,d\theta - \sin\theta\cos\phi\,d\psi) = 0, \qquad dY + r(\cos\phi\,d\theta + \sin\theta\sin\phi\,d\psi) = 0; \tag{c}$$

or, finally, since these constraints are catastatic, in *virtual* form,

$$\delta X - r(\sin\phi\,\delta\theta - \sin\theta\cos\phi\,\delta\psi) = 0, \qquad \delta Y + r(\cos\phi\,\delta\theta + \sin\theta\sin\phi\,\delta\psi) = 0. \tag{d}$$

[Absence of *pivoting* would have meant the following additional constraint:

$$(\boldsymbol{\omega})_{\text{normal to sphere at } C} = \omega_Z = \omega_\phi + \cos\theta\,\omega_\psi = 0,$$

or

$$d\phi + \cos\theta\,d\psi = 0 \Rightarrow d\phi/d\psi = -\cos\theta \equiv h(\theta)]. \tag{e}$$

As shown later, the constraints (a–d) are nonholonomic (NH). [We already notice that (c), for example, do *not* involve $d\phi$, and yet the constraints feature $\sin\phi$ and $\cos\phi$.] Mathematically, this means that it is impossible to obtain them by differentiating two finite constraint equations of the form $F(X, Y, \phi, \theta, \psi) = 0$ and $E(X, Y, \phi, \theta, \psi) = 0$; that is, the coordinates X, Y, ϕ, θ, ψ are independent. But their differentials $dX, dY, d\phi, d\theta, d\psi$, in view of (a–d), are not independent; that is, in general, only *three* of them can be varied simultaneously and arbitrarily. We say that *the sphere has five DOF in the large, but only three DOF in the small:* $f \equiv n - m = 5 - 2 = 3$. (Had we added pivoting, we would have $f = 2$.)

Kinematically, the above mean that the sphere may roll from an initial configuration, along two different routes, to two final configurations, which have both the same contact point and center location (i.e., same X, Y), but different angular orientations relative to each other (i.e., different ϕ, θ, ψ). If the constraints (a–d) were holonomic—for example, if the plane was *smooth*—it would be possible to vary all X, Y, ϕ, θ, ψ independently and arbitrarily without violating the (then) constraints; namely, the sphere's rigidity and the constancy of distance between G and C. Further, the sphere can roll from any *initial* configuration, with the sphere point C_i in contact with the plane point P_i, to any other *final* configuration, with the sphere point C_f in contact with the plane point P_f. To see this property, known as *accessibility* (§ 2.3), we draw on the plane a curve (γ) joining C_i and P_f, and another curve on the sphere (δ), of equal length to (γ), joining C_i and C_f. Now, a pivoting of the sphere can make the two arcs (γ) and (δ) tangent, at $C_i = P_i$. Then, we bring C_f to P_f by rolling (δ) on (γ). A final pivoting of the sphere brings it to its final configuration (see also Rutherford, 1960, pp. 161–162).

A Special Case

Assume, next, that the sphere rolls *without* pivoting, and also moves so that $\theta = constant \equiv \theta_o$. Let us find the path of G. With $\theta = constant \Rightarrow d\theta = 0$, the rolling constraints (c) reduce to

$$dX + r(\sin\theta_o)\cos\phi\, d\psi = 0, \qquad dY + r(\sin\theta_o)\sin\phi\, d\psi = 0; \tag{f}$$

and the no-pivoting constraint (e) to

$$d\phi/d\psi = -\cos\theta_o = constant. \tag{g}$$

This leaves only $n - m = 5 - 4 = 1$ *DOF* in the small. Taking ϕ as the independent coordinate and eliminating $d\psi$ between (f), with the help of (g), yields

$$dX = r(\tan\theta_o)\cos\phi\, d\phi, \qquad dY = r(\tan\theta_o)\sin\phi\, d\phi, \tag{h}$$

which integrates readily to the curve (with X_o and Y_o as integration constants):

$$X - X_0 = r(\tan\theta_o)\sin\phi, \qquad Y - Y_0 = -r(\tan\theta_o)\cos\phi; \tag{i}$$

that is, G describes, on the plane $Z = r$, a circle of radius $r\tan\theta_o$.

[These considerations also show how imposition of a sufficient number of additional holonomic and/or nonholonomic constraints turns an originally nonholonomic system into a holonomic one.]

Example 2.13.5 *Rolling Sphere on a Spinning Table—Introduction.* Let us extend the previous example to the case where the plane P is not fixed, but rotates about a fixed axis OZ perpendicular to it with, say, constant (inertial) angular velocity $\boldsymbol{\Omega}$. In this case, the rolling condition expresses the fact that the contact points of the sphere and the plane, C, have *equal inertial velocities*:

$$(\boldsymbol{v}_C)_{\text{sphere}} = (\boldsymbol{v}_C)_{\text{plane}}: \quad \boldsymbol{v}_G + \boldsymbol{\omega} \times \boldsymbol{r}_{C/G} = \boldsymbol{\Omega} \times \boldsymbol{r}_{C/O} \quad (= \boldsymbol{\Omega} \times \boldsymbol{\rho}); \tag{a}$$

or, in terms of their *components along inertial (background) axes O–XYZ/O–$\boldsymbol{IJK}$*:

$$(v_X, v_Y, 0) + (\omega_X, \omega_Y, \omega_Z) \times (0, 0, -r) = (0, 0, \Omega) \times (X, Y, 0), \tag{b}$$

from which we easily obtain the two rolling conditions:

$$v_X - r\omega_Y = -\Omega Y, \qquad v_Y + r\omega_X = \Omega X. \tag{c}$$

Next, expressing ω_X, ω_Y in terms of their Eulerian angles (between translating/nonrotating axes G–XYZ and sphere-fixed axes G–xyz) and their time rates, as in the preceding example, we transform (c) to

$$v_X - r(\sin\phi\,\omega_\theta - \sin\theta\cos\phi\,\omega_\psi) + \Omega Y = 0,$$
$$v_Y + r(\cos\phi\,\omega_\theta + \sin\theta\sin\phi\,\omega_\psi) - \Omega X = 0. \tag{d}$$

The Ω-proportional terms in (d) are the *acatastatic* parts of these constraints, and arise out of our use of *inertial* coordinates to describe the kinematics in a *noninertial* frame; had we used *plane-fixed* (noninertial) coordinates, the constraints would have been catastatic in them. It is not hard to see that the *kinematically admissible/possible* and *virtual* forms of these constraints are, respectively (note differences between them resulting from constraint $\delta t = 0$),

$$dX - r(\sin\phi\,d\theta - \sin\theta\cos\phi\,d\psi) + (\Omega Y)\,dt = 0,$$
$$dY + r(\cos\phi\,d\theta + \sin\theta\sin\phi\,d\psi) - (\Omega X)\,dt = 0; \tag{e}$$
$$\delta X - r(\sin\phi\,\delta\theta - \sin\theta\cos\phi\,\delta\psi) = 0,$$
$$\delta Y + r(\cos\phi\,\delta\theta + \sin\theta\sin\phi\,\delta\psi) = 0. \tag{f}$$

Example 2.13.6 *Rolling Sphere on Spinning Table—the Transitivity Equations.* Continuing from the preceding example, let us show that its rolling constraints (c–f); as well as those of its previous, *stationary* table case) are nonholonomic; that is, the system has $n = 5$ *DOF in the large*, and $f \equiv n - m = 5 - 2 = 3$ *DOF in the small*.

In view of the structure of these constraints, we choose the following equilibrium quasi velocities (with the usual notations: $dX/dt \equiv v_X, \ldots, d\phi/dt \equiv \omega_\phi, \ldots$):

Dependent:

$$\omega_1 \equiv v_X - r\,\omega_Y + \Omega Y = v_X - r(\sin\phi\,\omega_\theta - \cos\phi\sin\theta\,\omega_\psi) + \Omega Y = v_X - r\,\omega_4 + \Omega Y\ (= 0), \tag{a}$$

$$\omega_2 \equiv v_Y + r\,\omega_X - \Omega X = v_Y + r(\cos\phi\,\omega_\theta + \sin\phi\sin\theta\,\omega_\psi) - \Omega X = v_Y + r\,\omega_3 - \Omega X\ (= 0), \tag{b}$$

Independent:

$$\omega_3 \equiv \omega_X = (\cos\phi)\omega_\theta + (\sin\phi\sin\theta)\omega_\psi \quad (\neq 0), \tag{c}$$

$$\omega_4 \equiv \omega_Y = (\sin\phi)\omega_\theta + (-\cos\phi\sin\theta)\omega_\psi \quad (\neq 0), \tag{d}$$

$$\omega_5 \equiv \omega_Z = (1)\omega_\phi + (\cos\theta)\omega_\psi \quad (\neq 0), \tag{e}$$

$$\omega_6 \equiv dt/dt = 1 \quad (\textit{isochrony}). \tag{f}$$

Recalling results from §1.12, we readily see that these partially decoupled equations invert to

$$v_1 \equiv v_X = \omega_1 + r\,\omega_4 - \Omega Y \qquad \text{(without enforcement of constraints } \omega_{1,2} = 0), \tag{g1}$$

$$v_2 \equiv v_Y = \omega_2 - r\,\omega_3 + \Omega X \qquad \text{(without enforcement of constraints } \omega_{1,2} = 0), \tag{g2}$$

$$v_3 \equiv \omega_\phi = (-\cot\theta\sin\phi)\omega_3 + (\cot\theta\cos\phi)\omega_4 + \omega_5, \tag{g3}$$

$$v_4 \equiv \omega_\theta = (\cos\phi)\omega_3 + (\sin\phi)\omega_4, \tag{g4}$$

$$v_5 \equiv \omega_\psi = (\sin\phi/\sin\theta)\omega_3 + (-\cos\phi/\sin\theta)\omega_4, \tag{g5}$$

$$v_6 \equiv dt/dt = \omega_6 = 1. \tag{g6}$$

The *virtual* forms of (a–g6) are as follows [note absence of acatastatic terms in (h1, 2)]:

Dependent:

$$\delta\theta_1 \equiv \delta X - r\delta\theta_Y = \delta X + (-r\sin\phi)\,\delta\theta + (r\cos\phi\sin\theta)\,\delta\psi = \delta X - r\,\delta\theta_4 \quad (= 0), \tag{h1}$$

$$\delta\theta_2 \equiv \delta Y + r\delta\theta_X = \delta Y + (r\cos\phi)\,\delta\theta + (r\sin\phi\sin\theta)\,\delta\psi = \delta Y + r\,\delta\theta_3 \quad (= 0), \tag{h2}$$

Independent:

$$\delta\theta_3 \equiv \delta\theta_X = (\cos\phi)\,\delta\theta + (\sin\phi\sin\theta)\,\delta\psi \quad (\neq 0), \tag{h3}$$

$$\delta\theta_4 \equiv \delta\theta_Y = (\sin\phi)\,\delta\theta + (-\cos\phi\sin\theta)\,\delta\psi \quad (\neq 0), \tag{h4}$$

$$\delta\theta_5 \equiv \delta\theta_Z = (1)\,\delta\phi + (\cos\theta)\,\delta\psi \quad (\neq 0), \tag{h5}$$

$$\delta\theta_6 \equiv \delta q_6 \equiv \delta t = 0 \quad (\textit{isochrony}); \tag{h6}$$

$$\delta q_1 \equiv \delta X = \delta\theta_1 + r\,\delta\theta_4, \tag{i1}$$

$$\delta q_2 \equiv \delta Y = \delta\theta_2 - r\,\delta\theta_3, \tag{i2}$$

$$\delta q_3 \equiv \delta\phi = (-\cot\theta\sin\phi)\,\delta\theta_3 + (\cot\theta\cos\phi)\,\delta\theta_4 + \delta\theta_5, \tag{i3}$$

$$\delta q_4 \equiv \delta\theta = (\cos\phi)\,\delta\theta_3 + (\sin\phi)\,\delta\theta_4, \tag{i4}$$

$$\delta q_5 \equiv \delta\psi = (\sin\phi/\sin\theta)\,\delta\theta_3 + (-\cos\phi/\sin\theta)\,\delta\theta_4, \tag{i5}$$

$$\delta q_6 \equiv \delta t = \delta\theta_6 = 0. \tag{i6}$$

Now we are ready to calculate Hamel's coefficients from the transitivity equations (§2.10):

$$(\delta\theta_k)^{\cdot} - \delta\omega_k = \sum\sum \gamma^k{}_{r\beta}\,\omega_\beta\,\delta\theta_r = \sum\sum \gamma^k{}_{rs}\,\omega_s\,\delta\theta_r + \sum \gamma^k{}_r\,\delta\theta_r, \tag{j}$$

where $k, r, s = 1, \ldots, 5$; $\beta = 1, \ldots, 6$; $\gamma^k{}_r \equiv \gamma^k{}_{r,n+1} = \gamma^k{}_{r6}$.

By direct differentiations, use of the above, and the indicated shortcuts [and noting that, even if $\Omega = \Omega(t) =$ given function of time, still $\delta\Omega = 0$], we obtain, successively,

$$\begin{aligned}(\delta\theta_1)^{\cdot} - \delta\omega_1 &= (\delta X - r\,\delta\theta_Y)^{\cdot} - \delta(v_X - r\,\omega_Y + \Omega Y)\\ &= [(\delta X)^{\cdot} - \delta v_X] - r[(\delta\theta_Y)^{\cdot} - \delta\omega_Y] - \Omega\,\delta Y\\ &= 0 - r[(\delta\theta_4)^{\cdot} - \delta\omega_4] - \Omega\,\delta Y\end{aligned}$$

[invoking the rotational transitivity equations (§1.14 and ex. 2.13.9), and (i2)]

$$\begin{aligned}&= -r(\omega_Z\,\delta\theta_X - \omega_X\,\delta\theta_Z) - \Omega(\delta\theta_2 - r\delta\theta_3)\\ &= -r(\omega_5\delta\theta_3 - \omega_3\delta\theta_5) - \Omega(\delta\theta_2 - r\delta\theta_3),\end{aligned}$$

or, finally,

$$(\delta\theta_1)^{\cdot} - \delta\omega_1 = (-r)\omega_5\,\delta\theta_3 + (r)\omega_3\,\delta\theta_5 + (-\Omega)\,\delta\theta_2 + (r\Omega)\,\delta\theta_3; \tag{k1}$$

$$\begin{aligned}(\delta\theta_2)^{\cdot} - \delta\omega_2 &= (\delta Y + r\,\delta\theta_X)^{\cdot} - \delta(v_Y + r\,\omega_X - \Omega X)\\ &= [(\delta Y)^{\cdot} - \delta v_Y] + r[(\delta\theta_X)^{\cdot} - \delta\omega_X] + \Omega\,\delta X\\ &= 0 + r[(\delta\theta_3)^{\cdot} - \delta\omega_3] + \Omega\,\delta X\end{aligned}$$

[invoking again the rotational transitivity equations and (i1)]

$$\begin{aligned}&= r(\omega_Y\delta\theta_Z - \omega_Z\delta\theta_Y) + \Omega(\delta\theta_1 + r\,\delta\theta_4)\\ &= r(\omega_4\delta\theta_5 - \omega_5\delta\theta_4) + \Omega(\delta\theta_1 + r\,\delta\theta_4),\end{aligned}$$

or, finally,

$$(\delta\theta_2)^{\cdot} - \delta\omega_2 = (-r)\omega_5\delta\theta_4 + (r)\omega_4\delta\theta_5 + (\Omega)\,\delta\theta_1 + (r\Omega)\,\delta\theta_4; \tag{k2}$$

and, again, the rotational transitivity equations (with $X \to 3$, $Y \to 4$, $Z \to 5$) give

$$(\delta\theta_3)^{\cdot} - \delta\omega_3 = \omega_4\,\delta\theta_5 - \omega_5\,\delta\theta_4, \tag{k3}$$

$$(\delta\theta_4)^{\cdot} - \delta\omega_4 = \omega_5\,\delta\theta_3 - \omega_3\,\delta\theta_5, \tag{k4}$$

$$(\delta\theta_5)^{\cdot} - \delta\omega_5 = \omega_3\,\delta\theta_4 - \omega_4\,\delta\theta_3. \tag{k5}$$

Comparing (j) with (k1–5) we readily find that the nonvanishing γ's are

$$\gamma^1{}_{35} = -\gamma^1{}_{53} = -r, \quad \gamma^1{}_{26} = -\gamma^1{}_{62} \equiv \gamma^1{}_2 = -\Omega, \quad \gamma^1{}_{36} = -\gamma^1{}_{63} \equiv \gamma^1{}_3 = r\,\Omega; \tag{11}$$

$$\gamma^2{}_{45} = -\gamma^2{}_{54} = -r, \quad \gamma^2{}_{16} = -\gamma^2{}_{61} \equiv \gamma^2{}_1 = \Omega, \quad \gamma^2{}_{46} = -\gamma^2{}_{64} \equiv \gamma^2{}_4 = r\,\Omega; \tag{12}$$

$$\gamma^3{}_{45} = -\gamma^3{}_{54} = \gamma^4{}_{53} = -\gamma^4{}_{35} = \gamma^5{}_{34} = -\gamma^5{}_{43} = -1 \quad [= -1\,(\textit{permutation symbol})]. \tag{13}$$

Here, $D(ependent) = 1, 2$ and $I, I'(ndependent)) = 3, 4, 5$. Therefore,

$$\gamma^D{}_{II'} : \gamma^1{}_{35} = -r \neq 0 \quad \text{and} \quad \gamma^2{}_{45} = -r \neq 0; \tag{m}$$

and so, according to Frobenius' theorem (§2.12), the system of Pfaffian constraints $\omega_1 = 0$ and $\omega_2 = 0$ is *nonholonomic*, in both the catastatic (rolling on fixed plane) and acatastatic (rolling on rotating plane) cases; that is, for any given $\Omega = \Omega(t)$.

Example 2.13.7 *Rolling Disk on Fixed Plane.* Let us consider a thin circular disk (or coin, or ring, or hoop), of radius r and center G, rolling on a fixed horizontal and rough plane (fig. 2.20). A generic configuration of the disk is determined by the following *six* Lagrangean coordinates:

X, Y, Z: inertial coordinates of G;

ϕ, θ, ψ: Eulerian angles of *body-fixed* axes G–xyz relative to the cotranslating but nonrotating axes G–XYZ (similar to the rolling sphere case).

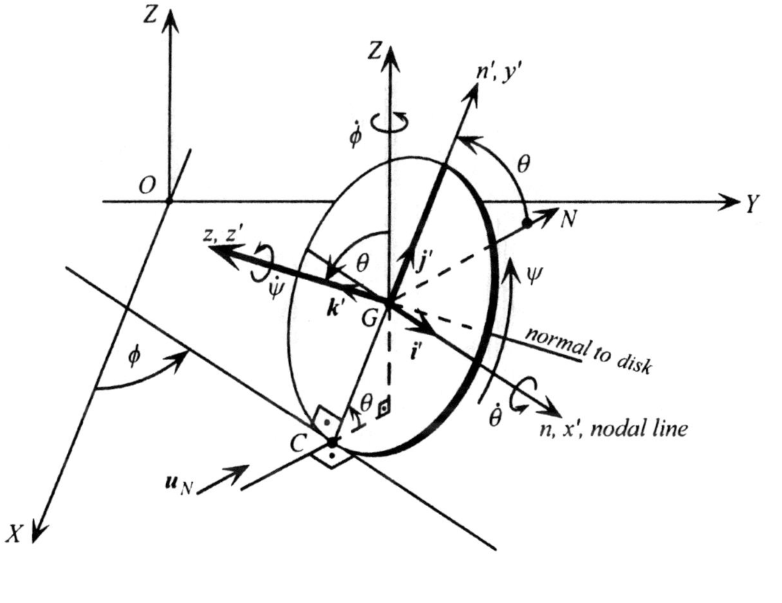

$i' = \cos\phi\, I + \sin\phi\, J,\ \ j' = \cos\theta\, u_N + \sin\theta\, K,\ \ k' = -\sin\theta\, u_N + \cos\theta\, K;$
$u_N = -\sin\phi\, I + \cos\phi\, J,\ \ u_n \equiv i'$

Figure 2.20 Geometry and kinematics of circular disk rolling on fixed rough plane. Axes: G–$nNZ \equiv G$–$x'NZ$: semifixed; G–$nn'z \equiv G$–$x'y'z'$: semimobile; G–xyz: body axes (not shown, but easily pictured); G–XYZ: space axes.

[In view of the complicated geometry, we avoid all ad hoc, possibly shorter, treatments, in favor of a fairly general and uniform approach. An alternative description is shown later.]

The vertical coordinate of G, clearly, satisfies the holonomic constraint

$$Z = r\sin\theta, \tag{a}$$

and this brings the number of independent Lagrangean coordinates down to *five*: X, Y, ϕ, θ, ψ; that is, $n = 5$.

The rolling constraint becomes, successively,

$$\begin{aligned}
\mathbf{0} = \boldsymbol{v}_C &= \boldsymbol{v}_G + \boldsymbol{\omega}\times\boldsymbol{r}_{C/G} \qquad (\boldsymbol{\omega}\text{: inertial angular velocity of disk})\\
&= \boldsymbol{v}_G - r\boldsymbol{\omega}\times\boldsymbol{j}'\\
&= \boldsymbol{v}_G - r(\omega_{x'}\boldsymbol{i}' + \omega_{y'}\boldsymbol{j}' + \omega_{z'}\boldsymbol{k}')\times\boldsymbol{j}' \qquad (\text{semimobile } \boldsymbol{\omega}\text{-decomposition})\\
&= \boldsymbol{v}_G - r\omega_{x'}(\boldsymbol{i}'\times\boldsymbol{j}') - r\omega_{z'}(\boldsymbol{k}'\times\boldsymbol{j}')\\
&= \boldsymbol{v}_G - r\omega_{x'}(\boldsymbol{k}') - r\omega_{z'}(-\boldsymbol{i}'),
\end{aligned} \tag{b}$$

from which we obtain the constraint components along the two "natural" (semi-fixed) directions $n(\boldsymbol{i}')$ and $N(\boldsymbol{u}_N)$:

(i) $$0 = \boldsymbol{v}_C\cdot\boldsymbol{i}' = \boldsymbol{v}_G\cdot\boldsymbol{i}' + r\,\omega_{z'} \qquad \text{or} \qquad v_{G,n} + r\,\omega_{z'} = 0; \tag{c1}$$

(ii) $$0 = \boldsymbol{v}_C\cdot\boldsymbol{u}_N = \boldsymbol{v}_G\cdot\boldsymbol{u}_N - r\,\omega_{x'}(\boldsymbol{k}'\cdot\boldsymbol{u}_N) + r\,\omega_{z'}(\boldsymbol{i}'\cdot\boldsymbol{u}_N) = \boldsymbol{v}_G\cdot\boldsymbol{u}_N - r\,\omega_{x'}(\boldsymbol{k}'\cdot\boldsymbol{u}_N),$$

or

$$v_{G,N} - r\,\omega_{x'}\cos(\pi/2+\theta) = v_{G,N} + r\,\omega_{x'}\sin\theta = 0. \tag{c2}$$

The *third* semifixed direction component gives the earlier constraint (a):

$$0 = \boldsymbol{v}_C\cdot\boldsymbol{K} = \boldsymbol{v}_G\cdot\boldsymbol{K} - r\,\omega_{x'}(\boldsymbol{k}'\cdot\boldsymbol{K}) + r\,\omega_{z'}(\boldsymbol{i}'\cdot\boldsymbol{K}) = \boldsymbol{v}_G\cdot\boldsymbol{K} - r\,\omega_{x'}(\boldsymbol{k}'\cdot\boldsymbol{K}),$$

or, since $\omega_{x'} = \omega_\theta$,

$$0 = v_{G,Z} - r\,\omega_\theta\cos\theta \;\Rightarrow\; dZ - r\cos\theta\, d\theta = 0 \;\Rightarrow\; Z - r\sin\theta = \textit{constant} \to 0.$$

Equations (c1, 2) contain nonholonomic velocities. Let us express them in terms of holonomic velocities exclusively. It is not hard to see that

(i) $$\omega_{x'} = \omega_\theta, \qquad \omega_{y'} = (\sin\theta)\omega_\phi, \qquad \omega_{z'} = (\cos\theta)\omega_\phi + \omega_\psi; \tag{d1}$$

(ii) $$\begin{aligned} v_{G,n} &= \boldsymbol{v}_G\cdot\boldsymbol{i}' \equiv \boldsymbol{v}_G\cdot\boldsymbol{u}_n\\ &= (v_X\boldsymbol{I} + v_Y\boldsymbol{J} + v_Z\boldsymbol{K})\cdot(\cos\phi\,\boldsymbol{I} + \sin\phi\,\boldsymbol{J}) = (\cos\phi)v_X + (\sin\phi)v_Y; \end{aligned} \tag{d2}$$

(iii) $$\begin{aligned} v_{G,N} &= \boldsymbol{v}_G\cdot\boldsymbol{u}_N\\ &= (v_X\boldsymbol{I} + v_Y\boldsymbol{J} + v_Z\boldsymbol{K})\cdot(-\sin\phi\,\boldsymbol{I} + \cos\phi\,\boldsymbol{J}) = (-\sin\phi)v_X + (\cos\phi)v_Y. \end{aligned} \tag{d3}$$

With the help of (d1–3), the constraints (c1, 2) take, respectively, the holonomic velocities form:

(i) $$\boldsymbol{v}_C\cdot\boldsymbol{i}' \equiv v_{C,n} = (\cos\phi)v_X + (\sin\phi)v_Y + r(\omega_\psi + \cos\theta\,\omega_\phi) = 0, \tag{e1}$$

(ii) $$\boldsymbol{v}_C\cdot\boldsymbol{u}_N \equiv v_{C,N} = (-\sin\phi)v_X + (\cos\phi)v_Y + r\sin\theta\,\omega_\theta = 0. \tag{e2}$$

In view of (e1, 2), we introduce the following equilibrium quasi velocities:

Dependent:

$$\omega_1 \equiv v_{C,n} = v_{G,n} + r\,\omega_{z'} = (\cos\phi)v_X + (\sin\phi)v_Y + (r\cos\theta)\omega_\phi + (r)\omega_\psi \quad (=0), \quad \text{(f1)}$$

$$\omega_2 \equiv v_{C,N} = v_{G,N} + r\sin\theta\,\omega_{x'} = (-\sin\phi)v_X + (\cos\phi)v_Y + (r\sin\theta)\omega_\theta \quad (=0); \quad \text{(f2)}$$

Independent (semimobile components of ω):

$$\omega_3 \equiv \omega_n \equiv \omega_{x'} = \omega_\theta \quad (\neq 0), \quad \text{(f3)}$$

$$\omega_4 \equiv \omega_{n'} \equiv \omega_{y'} = (\sin\theta)\omega_\phi \quad (\neq 0), \quad \text{(f4)}$$

$$\omega_5 \equiv \omega_z \equiv \omega_{z'} = (\cos\theta)\omega_\phi + \omega_\psi \quad (\neq 0). \quad \text{(f5)}$$

These catastatic, and partially uncoupled, equations invert easily to

$$v_1 \equiv v_X = (\cos\phi)\omega_1 + (-\sin\phi)\omega_2 + (r\sin\theta\sin\phi)\omega_3 + (-r\cos\phi)\omega_5, \quad \text{(g1)}$$

$$v_2 \equiv v_Y = (\sin\phi)\omega_1 + (\cos\phi)\omega_2 + (-r\sin\theta\cos\phi)\omega_3 + (-r\sin\phi)\omega_5, \quad \text{(g2)}$$

$$v_3 \equiv \omega_\phi = (1/\sin\theta)\omega_4, \quad \text{(g3)}$$

$$v_4 \equiv \omega_\theta = \omega_3, \quad \text{(g4)}$$

$$v_5 \equiv \omega_\psi = \omega_5 - (\cot\theta)\omega_4. \quad \text{(g5)}$$

Below, we show that the constraints $\omega_1 = 0$ and $\omega_2 = 0$ are nonholonomic; that is, $n = 5$ *global DOF*, $m = 2 \rightarrow f \equiv n - m = 3$ *local DOF*.

Indeed, by direct d- and δ-operations on (f1–g5), and their virtual forms (which can be obtained from the above velocity forms in, by now, obvious ways), and combination of simple shortcuts with some straightforward algebra, we find, successively,

$$\begin{aligned}
(\delta\theta_1)^{\cdot} - \delta\omega_1 &= \left[(\delta p_{G,n})^{\cdot} - \delta v_{G,n}\right] + r[(\delta\theta_5)^{\cdot} - \delta\omega_5] \qquad [\text{where } dp_{G,n} \equiv v_{G,n}\,dt] \\
&= \cdots = \{[(\cos\phi)\,\delta X + (\sin\phi)\,\delta Y]^{\cdot} - \delta[(\cos\phi)v_X + (\sin\phi)v_Y]\} \\
&\qquad + r(\omega_4\,\delta\theta_3 - \omega_3\,\delta\theta_4) \\
&= \cdots = \{\omega_\phi[(-\sin\phi)\,\delta X + (\cos\phi)\,\delta Y] - \delta\phi[(-\sin\phi)v_X + (\cos\phi)v_Y]\} \\
&\qquad + r(\omega_4\,\delta\theta_3 - \omega_3\,\delta\theta_4) \\
&= (\omega_\phi\,\delta p_{G,N} - v_{G,N}\,\delta\phi) + r(\omega_4\,\delta\theta_3 - \omega_3\,\delta\theta_4) \qquad [\text{where } dp_{G,N} \equiv v_{G,N}\,dt] \\
&= [(\omega_4/\sin\theta)\,\delta p_{G,N} - v_{G,N}(\delta\theta_4/\sin\theta)] + r(\omega_4\,\delta\theta_3 - \omega_3\,\delta\theta_4) \\
&= (\omega_4/\sin\theta)(\delta p_{G,N} + r\sin\theta\,\delta\theta_3) - (\delta\theta_4/\sin\theta)(v_{G,N} + r\sin\theta\,\omega_3) \\
&= (1/\sin\theta)(\omega_4\,\delta p_{C,N} - v_{C,N}\,\delta\theta_4) \qquad [\text{where } dp_{C,N} \equiv v_{C,N}\,dt] \\
&= (1/\sin\theta)(\omega_4\,\delta\theta_2 - \omega_2\,\delta\theta_4) = 0
\end{aligned}$$

(after enforcing the constraints $\delta\theta_2, \omega_2 = 0$), (h1)

$$(\delta\theta_2)^\cdot - \delta\omega_2 = [(\delta p_{G,N})^\cdot - \delta v_{G,N}] + [(r\sin\theta\,\delta\theta_3)^\cdot - \delta(r\sin\theta\,\omega_3)]$$

[since $\omega_3 \sin\theta = \omega_\theta \sin\theta$ is integrable, the *second* bracket term vanishes]

$$= [(-\sin\phi)\,\delta X + (\cos\phi)\,\delta Y]^\cdot - \delta[(-\sin\phi)v_X + (\cos\phi)v_Y]$$

$$= -\omega_\phi(\cos\phi\,\delta X + \sin\phi\,\delta Y) + \delta\phi[(\cos\phi)v_X + (\sin\phi)v_Y]$$

$$= -\omega_\phi\,\delta p_{G,n} + \delta\phi\,v_{G,n}$$

$$= -(\omega_4/\sin\theta)(\delta\theta_1 - r\,\delta\theta_5) + (\delta\theta_4/\sin\theta)(\omega_1 - r\omega_5)$$

$$= -(1/\sin\theta)(\omega_4\,\delta\theta_1 - \omega_1\,\delta\theta_4) + (r/\sin\theta)(\omega_4\,\delta\theta_5 - \omega_5\,\delta\theta_4)$$

$$= (r/\sin\theta)(\omega_4\,\delta\theta_5 - \omega_5\,\delta\theta_4) \neq 0$$

$$\text{(even after enforcing the constraints } \delta\theta_1, \omega_1 = 0); \qquad \text{(h2)}$$

$$(\delta\theta_3)^\cdot - \delta\omega_3 = 0 \text{ (independently of constraints)} \Rightarrow \theta_3 = \textit{holonomic coordinate}, \qquad \text{(h3)}$$

$$(\delta\theta_4)^\cdot - \delta\omega_4 = (\sin\theta\,\delta\theta)^\cdot - \delta(\sin\theta\,\omega_\phi) = (\cos\theta)(\omega_\theta\,\delta\phi - \omega_\phi\,\delta\theta)$$

$$= (\cot\theta)(\omega_3\,\delta\theta_4 - \omega_4\,\delta\theta_3), \qquad \text{(h4)}$$

$$(\delta\theta_5)^\cdot - \delta\omega_5 = (\delta\psi + \cos\theta\,\delta\phi)^\cdot - \delta(\omega_\psi + \cos\theta\,\omega_\phi) = (\sin\theta)(\omega_\phi\,\delta\theta - \omega_\theta\,\delta\phi)$$

$$= \omega_4\,\delta\theta_3 - \omega_3\,\delta\theta_4; \qquad \text{(h5)}$$

and since $Z = r\sin\theta$, with $\omega_6 \equiv v_Z - r\cos\theta\,\omega_\theta \Rightarrow \delta\theta_6 = \delta Z - r\cos\theta\,\delta\theta$, we get

$$(\delta\theta_6)^\cdot - \delta\omega_6 = (\delta Z - r\cos\theta\,\delta\theta)^\cdot - \delta(v_Z - r\cos\theta\,\omega_\theta)$$

$$= (\delta Z)^\cdot - r(\cos\theta)^\cdot\delta\theta - r\cos\theta(\delta\theta)^\cdot - \delta v_Z + r(-\sin\theta)\,\delta\theta\,\omega_\theta + r\cos\theta\,\delta\omega_\theta = 0,$$

(independently of the other constraints) as expected.

From the above, we immediately read off the nonvanishing γ's:

$$\gamma^1{}_{24} = -\gamma^1{}_{42} = 1/\sin\theta;$$

$$\gamma^2{}_{41} = -\gamma^2{}_{14} = 1/\sin\theta, \qquad \gamma^2{}_{54} = -\gamma^2{}_{45} = r/\sin\theta;$$

$$\gamma^4{}_{43} = -\gamma^4{}_{34} = \cot\theta;$$

$$\gamma^5{}_{34} = -\gamma^5{}_{43} = 1. \qquad \text{(i)}$$

Here, $D(\textit{ependent}) = 1, 2$ and $I, I'(\textit{ndependent}) = 3, 4, 5$. Therefore,

$$\gamma^D{}_{II'}: \quad \gamma^2{}_{54} = r/\sin\theta \neq 0; \qquad \text{(j)}$$

and so, according to Frobenius' theorem (§2.12), the system of Pfaffian constraints $\omega_1 = 0$ and $\omega_2 = 0$ is *nonholonomic*. We also notice that to calculate all nonvanishing γ's, we must refrain from enforcing the constraints ω_1, $\delta\theta_1 = 0$ and ω_2, $\delta\theta_2 = 0$, in the earlier bilinear covariants.

Rolling Constraints via Components along Space Axes

With reference to fig. 2.20, we have, successively,

$$\boldsymbol{r}_{C/G} = -(r\cos\theta)\boldsymbol{u}_N - (r\sin\theta)\boldsymbol{K} = (-r\cos\theta)(-\sin\phi\,\boldsymbol{I} + \cos\phi\,\boldsymbol{J}) + (-r\sin\theta)\boldsymbol{K}$$

$$= (r\cos\theta\sin\phi)\boldsymbol{I} + (-r\cos\theta\cos\phi)\boldsymbol{J} + (-r\sin\theta)\boldsymbol{K}, \qquad \text{(k1)}$$

$$\boldsymbol{\omega} = \omega_X \boldsymbol{I} + \omega_Y \boldsymbol{J} + \omega_Z \boldsymbol{K} \qquad \text{[recalling formulae in §1.12]}$$
$$= [(\cos\phi)\omega_\theta + (\sin\phi\sin\theta)\omega_\psi]\boldsymbol{I} + [(\sin\phi)\omega_\theta + (-\cos\phi\sin\theta)\omega_\psi]\boldsymbol{J}$$
$$+ [\omega_\phi + (\cos\theta)\omega_\psi]\boldsymbol{K}, \qquad \text{(k2)}$$

and, of course,

$$\boldsymbol{v}_G = v_X \boldsymbol{I} + v_Y \boldsymbol{J} + v_Z \boldsymbol{K}. \qquad \text{(k3)}$$

Substituting these fixed-axes representations into the constraint (b): $\boldsymbol{0} = \boldsymbol{v}_C = \boldsymbol{v}_G + \boldsymbol{\omega} \times \boldsymbol{r}_{C/G}$, and setting its components along $\boldsymbol{I}, \boldsymbol{J}, \boldsymbol{K}$, equal to zero, we obtain the scalar conditions:

$$v_X + r(\cos\phi\cos\theta\,\omega_Z - \sin\theta\,\omega_Y)$$
$$= v_X + r[(\cos\phi\cos\theta)\omega_\phi - (\sin\phi\sin\theta)\omega_\theta + (\cos\phi)\omega_\psi] = 0,$$

$$v_Y + r(\sin\theta\,\omega_X + \sin\phi\cos\theta\,\omega_Z)$$
$$= v_Y + r[(\sin\phi\cos\theta)\omega_\phi + (\cos\phi\sin\theta)\omega_\theta + (\sin\phi)\omega_\psi] = 0,$$

$$v_Z - r\cos\theta(\cos\phi\,\omega_X + \sin\phi\,\omega_Y)$$
$$= v_Z - r\cos\theta\,\omega_\theta = 0 \;\Rightarrow\; Z = r\sin\theta \quad \text{(i.e., holonomic).} \qquad \text{(k4, 5, 6)}$$

We leave it to the reader to verify that (k4, 5) are equivalent to the earlier (e1–f2); and, also, that they can be brought to the (perhaps simpler) form,

$$[(X/r) + \sin\phi\cos\theta]^{\cdot} + (\cos\phi)\omega_\psi = 0, \qquad [(Y/r) + \cos\phi\cos\theta]^{\cdot} + (\sin\phi)\omega_\psi = 0. \qquad \text{(k7)}$$

Constraints and Transitivity Equations in Terms of the (Inertial) Coordinates of the Contact Point of the Disk (X_C, Y_C)

[This is a popular choice among mechanics authors (e.g., Hamel, 1949, pp. 470 ff., 478–479; Rosenberg, 1977, pp. 265 ff.) but our choice—that is, in terms of the coordinates of the disk center, G—shows more clearly the connection with the Eulerian angles.]

Taking the fixed-axes components of the obvious relation $\boldsymbol{r}_G = \boldsymbol{r}_C + \boldsymbol{r}_{G/C}$, and then $d/dt(\ldots)$-differentiating them, we obtain (consulting again fig. 2.20, and with $v_{C,X} \equiv dX_C/dt$, $v_{C,Y} \equiv dY_C/dt$)

(i) $X = X_C - (r\cos\theta)\sin\phi \Rightarrow v_X = v_{C,X} - (r\cos\phi\cos\theta)\omega_\phi + (r\sin\phi\sin\theta)\omega_\theta$, (l1)

(ii) $Y = Y_C + (r\cos\theta)\cos\phi \Rightarrow v_Y = v_{C,Y} - (r\sin\phi\cos\theta)\omega_\phi - (r\cos\phi\sin\theta)\omega_\theta$; (l2)

and substituting these v_X, v_Y expressions into (k4, 5), respectively, we eventually obtain the simpler forms

$$v_{C,X} + (r\cos\phi)\omega_\psi = 0 \qquad \text{and} \qquad v_{C,Y} + (r\sin\phi)\omega_\psi = 0. \qquad \text{(l3)}$$

(The above can, also, be obtained by ad hoc knife problem–type considerations.)

In view of (13), we introduce the following *new* equilibrium quasi velocities:

Dependent:

$$\omega_1 \equiv v_{C,X} + (r\cos\phi)\omega_\psi \quad (= 0), \tag{m1}$$

$$\omega_2 \equiv v_{C,Y} + (r\sin\phi)\omega_\psi \quad (= 0); \tag{m2}$$

Independent:

$$\omega_3 \equiv \omega_\phi(\neq 0) \;\Rightarrow\; \gamma^3_{\cdot\cdot} = 0, \tag{m3}$$

$$\omega_4 \equiv \omega_\theta(\neq 0) \;\Rightarrow\; \gamma^4_{\cdot\cdot} = 0, \tag{m4}$$

$$\omega_5 \equiv (\cos\phi)v_{C,X} + (\sin\phi)v_{C,Y} \quad [= -r\,\omega_\psi, \text{ by (13)}]; \tag{m5}$$

or, instead, the equivalent but simpler, knife-type, quasi velocities:

Dependent:

$$\Omega_1 \equiv (-\sin\phi)v_{C,X} + (\cos\phi)v_{C,Y} \quad [= 0, \text{ by (13)}], \tag{n1}$$

$$\Omega_2 \equiv r\,\omega_\psi + (\cos\phi)v_{C,X} + (\sin\phi)v_{C,Y} = r\,\omega_\psi + \omega_5\,(= 0); \tag{n2}$$

Independent:

$$\Omega_3 \equiv \omega_3 \equiv \omega_\phi \quad (\neq 0), \tag{n3}$$

$$\Omega_4 \equiv \omega_4 \equiv \omega_\theta \quad (\neq 0), \tag{n4}$$

$$\Omega_5 \equiv \omega_5 \equiv (\cos\phi)v_{C,X} + (\sin\phi)v_{C,Y} \quad [= -r\,\omega_\psi + \Omega_2 = -r\,\omega_\psi]. \tag{n5}$$

Inverting the above yields

$$v_1 \equiv v_{C,X} = (-\sin\phi)\Omega_1 + (\cos\phi)\Omega_5, \tag{o1}$$

$$v_2 \equiv v_{C,Y} = (\cos\phi)\Omega_1 + (\sin\phi)\Omega_5, \tag{o2}$$

$$v_3 \equiv \omega_\phi = \Omega_3, \tag{o3}$$

$$v_4 \equiv \omega_\theta = \Omega_4, \tag{o4}$$

$$v_5 \equiv \omega_\psi = (1/r)(\Omega_2 - \Omega_5). \tag{o5}$$

By direct d/δ-differentiations of (n1–5), use of (o1–5), and the obvious notation $d\Theta_k \equiv \Omega_k dt$; $k = 1, \ldots, 5$, we obtain the corresponding transitivity equations as follows:

$$\begin{aligned}
(\delta\Theta_1)^{\cdot} - \delta\Omega_1 &= [(-\sin\phi)\,\delta X_C + (\cos\phi)\,\delta Y_C]^{\cdot} - \delta[(-\sin\phi)v_{C,X} + (\cos\phi)v_{C,Y}] \\
&= \cdots = \cos\phi(v_{C,X}\,\delta\phi - \omega_\phi\,\delta X_C) + \sin\phi(v_{C,Y}\,\delta\phi - \omega_\phi\,\delta Y_C) \\
&= \cos\phi\big[(-\sin\phi\,\Omega_1 + \cos\phi\,\Omega_5)\,\delta\Theta_3 - \Omega_3(-\sin\phi\,\delta\Theta_1 + \cos\phi\,\delta\Theta_5)\big] \\
&\quad + \sin\phi\big[(\cos\phi\,\Omega_1 + \sin\phi\,\Omega_5)\,\delta\Theta_3 - \Omega_3(\cos\phi\,\delta\Theta_1 + \sin\phi\,\delta\Theta_5)\big] \\
&= \Omega_5\,\delta\Theta_3 - \Omega_3\,\delta\Theta_5,
\end{aligned} \tag{p1}$$

$$
\begin{aligned}
(\delta\Theta_2)^{\cdot} - \delta\Omega_2 &= [r\,\delta\psi + (\cos\phi)\,\delta X_C + (\sin\phi)\,\delta Y_C]^{\cdot} - \delta[r\,\omega_\psi + (\cos\phi)v_{C,X} + (\sin\phi)v_{C,Y}] \\
&= \cdots = \sin\phi(v_{C,X}\,\delta\phi - \omega_\phi\,\delta X_C) + \cos\phi(\omega_\phi\,\delta Y_C - v_{C,Y}\,\delta\phi) \\
&= \sin\phi\big[(-\sin\phi\,\Omega_1 + \cos\phi\,\Omega_5)\,\delta\Theta_3 - \Omega_3(-\sin\phi\,\delta\Theta_1 + \cos\phi\,\delta\Theta_5)\big] \\
&\quad + \cos\phi\big[\Omega_3(\cos\phi\,\delta\Theta_1 + \sin\phi\,\delta\Theta_5) - (\cos\phi\,\Omega_1 + \sin\phi\,\Omega_5)\,\delta\Theta_3\big] \\
&= \Omega_3\,\delta\Theta_1 - \Omega_1\,\delta\Theta_3 \\
&\qquad (= 0, \text{ upon imposition of the constraints } \delta\Theta_1, \Omega_1 = 0), \qquad \text{(p2)}
\end{aligned}
$$

$$(\delta\Theta_3)^{\cdot} - \delta\Omega_3 = (\delta\phi)^{\cdot} - \delta\omega_\phi = 0 \qquad (\Theta_3 = \textit{holonomic}), \qquad \text{(p3)}$$

$$(\delta\Theta_4)^{\cdot} - \delta\Omega_4 = (\delta\theta)^{\cdot} - \delta\omega_\theta = 0 \qquad (\Theta_4 = \textit{holonomic}), \qquad \text{(p4)}$$

$$
\begin{aligned}
(\delta\Theta_5)^{\cdot} - \delta\Omega_5 &= [(\cos\phi)\,\delta X_C + (\sin\phi)\,\delta Y_C]^{\cdot} - \delta[(\cos\phi)v_{C,X} + (\sin\phi)v_{C,Y}] \\
&= (-r\,\delta\psi)^{\cdot} - \delta(-r\,\omega_\psi) + (\delta\Theta_2)^{\cdot} - \delta\Omega_2 = 0 + \Omega_3\,\delta\Theta_1 - \Omega_1\delta\Theta_3 \\
&= \Omega_3\,\delta\Theta_1 - \Omega_1\,\delta\Theta_3 \\
&\qquad (= 0, \text{ upon imposition of the constraints } \delta\Theta_1, \Omega_1 = 0); \qquad \text{(p5)}
\end{aligned}
$$

that is, just like the knife problem (ex. 2.13.2), all the γ's are either ± 1 or 0.
Finally, here, $D(ependent) = 1, 2$ and $I, I'(ndependent) = 3, 4, 5$. Therefore,

$$\gamma^{D}{}_{II'}: \quad \gamma^{1}{}_{35} = -\gamma^{1}{}_{53} = 1 \neq 0; \qquad \text{(j)}$$

and so, by Frobenius' theorem (§2.12), the constraint system $\Omega_1 = 0$ and $\Omega_2 = 0$ is nonholonomic.

Problem 2.13.3 *Rolling Disk in Accelerating Plane.* Continuing from the preceding example, show that if the plane *translates* (i.e., no rotation), relative to inertial space, with a *given* velocity $(v_X(t), v_Y(t), v_Z(t))$, and the new inertial axes $O{-}XYZ$ are chosen so that *OZ is always perpendicular to the translating plane*, and X, Y, Z are the new *inertial* coordinates of the center of the disk G, then the rolling constraints take the rheonomic form

$$(\cos\phi)[V_X - v_X(t)] + (\sin\phi)[V_Y - v_Y(t)] + (r\cos\theta)\omega_\phi + (r)\omega_\psi = 0, \qquad \text{(a)}$$

$$(-\sin\phi)[V_X - v_X(t)] + (\cos\phi)[V_Y - v_Y(t)] + (r\sin\theta)\omega_\theta = 0, \qquad \text{(b)}$$

where $v_X \equiv dX/dt$, $v_Y \equiv dY/dt$; $\omega_\phi \equiv d\phi/dt$, and so on; that is, they are the same as in the fixed plane case, but with $\boldsymbol{v}_G$ replaced with $\boldsymbol{v}_G - \boldsymbol{v}_C$ (where $\boldsymbol{v}_C$ is the inertial velocity of contact point of disk with plane).

Example 2.13.8 *Pair of Rolling Wheels on an Axle.* Let us discuss the kinematics of a pair of two thin identical wheels, each of radius r, connected by a light axle and able to turn freely about its ends (fig. 2.21), rolling on a fixed, horizontal, and rough plane. For its description, we choose the following (*six* $\rightarrow$) *five* Lagrangean coordinates:

- $(X, Y, Z = r)$: inertial coordinates of midpoint of axle, G;
- ϕ: angle between the $O{-}XY$ projection of the axle (say, from G'' toward G') and $+OX$;
- ψ', ψ'': spin angles of the two wheels.

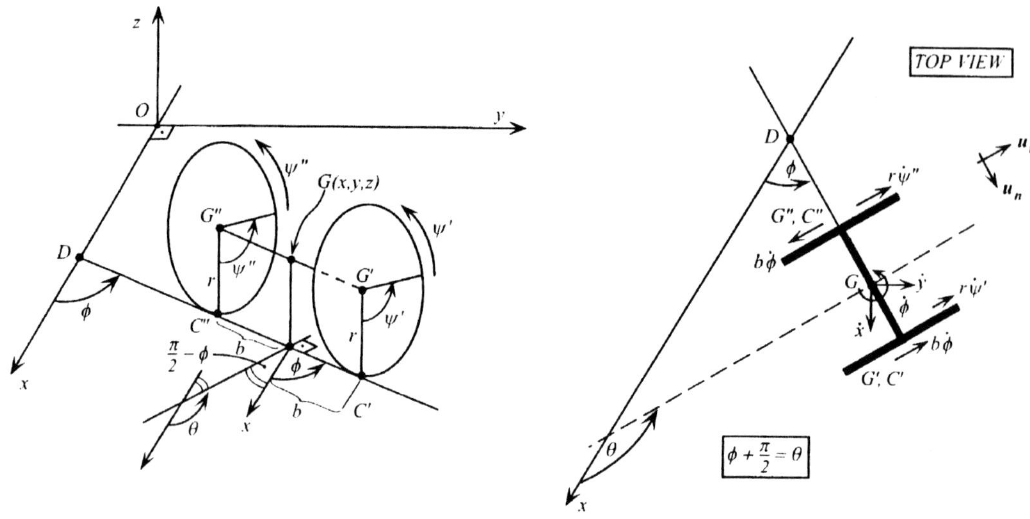

Figure 2.21 Rolling of two wheels on an axle, on fixed plane.

Here, the constraints are $\boldsymbol{v}_{C'} = \mathbf{0}$ and $\boldsymbol{v}_{C''} = \mathbf{0}$, where C' and C'' are the contact points of the two wheels. However, due to the constancy of $G''G'$ (and $C''C' = 2b$) and the continuous perpendicularity of the wheels to the axle, these conditions translate to *three* independent component equations, not four; say, the vanishing of $\boldsymbol{v}_{C'}$ and $\boldsymbol{v}_{C''}$ along and perpendicularly to the axle (the "natural" directions of the problem). Let us express this analytically: since

$$\boldsymbol{v}_{C'} = \boldsymbol{v}_{G'} + \boldsymbol{\omega}_{w'} \times \boldsymbol{r}_{C'/G'} = (\boldsymbol{v}_G + \boldsymbol{\omega}_A \times \boldsymbol{r}_{G'/G}) + \boldsymbol{\omega}_{w'} \times \boldsymbol{r}_{C'/G'}$$

[$\boldsymbol{\omega}_{w'}$ and $\boldsymbol{\omega}_A$: inertial angular velocities of first wheel and axle, respectively]

$$= (v_X, v_Y, 0) + (0, 0, \omega_\phi) \times (b\cos\phi, b\sin\phi, 0) + (\omega_{\psi'}\cos\phi, \omega_{\psi'}\sin\phi, \omega_\phi) \times (0, 0, -r)$$

$$= (v_X - b\,\omega_\phi \sin\phi - r\,\omega_{\psi'}\sin\phi,\; v_Y + b\,\omega_\phi\cos\phi + r\,\omega_{\psi'}\cos\phi,\; 0); \qquad \text{(a)}$$

and similarly, for the second wheel [whose inertial angular velocity is $\boldsymbol{\omega}_{w''} = (\omega_{\psi''}\cos\phi,\, \omega_{\psi''}\sin\phi,\, \omega_\phi)$],

$$\boldsymbol{v}_{C''} = (v_X + b\,\omega_\phi \sin\phi - r\,\omega_{\psi''}\sin\phi,\; v_Y - b\,\omega_\phi\cos\phi + r\,\omega_{\psi''}\cos\phi,\; 0); \qquad \text{(b)}$$

the constraints are (with $\boldsymbol{u}\ldots$ for *unit* vector):

$$0 = v_{C',n} \equiv \boldsymbol{v}_{C'} \cdot \boldsymbol{u}_n = \boldsymbol{v}_{C'} \cdot (\cos\phi, \sin\phi, 0) = v_{C'',n} \equiv \boldsymbol{v}_{C''} \cdot \boldsymbol{u}_n = \boldsymbol{v}_{C''} \cdot (\cos\phi, \sin\phi, 0),$$

or

$$v_{C',n} = v_{C'',n} = v_X\cos\phi + v_Y\sin\phi = 0; \qquad \text{(c1)}$$

and

$$v_{C',t} \equiv \boldsymbol{v}_{C'} \cdot \boldsymbol{u}_t = \boldsymbol{v}_{C'} \cdot (-\sin\phi, \cos\phi, 0) = -v_X\sin\phi + v_Y\cos\phi + b\,\omega_\phi + r\,\omega_{\psi'} = 0, \qquad \text{(c2)}$$

$$v_{C'',t} \equiv \boldsymbol{v}_{C''} \cdot \boldsymbol{u}_t = \boldsymbol{v}_{C''} \cdot (-\sin\phi, \cos\phi, 0) = -v_X\sin\phi + v_Y\cos\phi - b\,\omega_\phi + r\,\omega_{\psi''} = 0. \qquad \text{(c3)}$$

[The above can also be obtained by simple geometrical considerations based on fig. 2.21.]

By inspection, we see that (c2, 3) yield the *integrable* combination

$$2b\,\omega_\phi + r(\omega_{\psi'} - \omega_{\psi''}) = 0 \;\Rightarrow\; 2b\,\phi = c - r(\psi' - \psi''),$$
$$[c = \textit{integration constant}, \text{ depending on the initial values of } \phi, \psi', \psi'']. \tag{d}$$

Hence, we may take X, Y, ψ', ψ'', as the minimal Lagrangean coordinates of our system, subject to the two knife-like nonholonomic (to be shown below) constraints

$$v_{C',n} = v_{C'',n} = v_X \cos\phi + v_Y \sin\phi = 0; \tag{e1}$$

$$v_{C',t} = -v_X \sin\phi + v_Y \cos\phi + (r/2)(\omega_{\psi'} + \omega_{\psi''}) = 0; \tag{e2}$$

that is, $n = 4, m = 2 \;\Rightarrow\; f \equiv n - m = 4 - 2 = 2$ *DOF in the small*, and *4 DOF in the large*.

In view of (e1, 2), we introduce the following equilibrium quasi velocities:

$$\omega_1 \equiv (\cos\phi)v_X + (\sin\phi)v_Y \quad (= 0); \tag{f1}$$

$$\omega_2 \equiv (-\sin\phi)v_X + (\cos\phi)v_Y \quad (\neq 0); \tag{f2}$$

$$\omega_3 \equiv \omega_\phi \quad (\neq 0); \tag{f3}$$

$$\omega_4 \equiv 2(-v_X \sin\phi + v_Y \cos\phi) + r(\omega_{\psi'} + \omega_{\psi''}) \quad (= 0); \tag{f4}$$

$$\omega_5 \equiv 2b\,\omega_\phi + r(\omega_{\psi'} - \omega_{\psi''}) \quad (= 0;\ \omega_5 = \textit{holonomic}\text{ velocity}); \tag{f5}$$

which invert easily to

$$v_X = (\cos\phi)\omega_1 + (-\sin\phi)\omega_2, \tag{f6}$$

$$v_Y = (\sin\phi)\omega_1 + (\cos\phi)\omega_2, \tag{f7}$$

$$\omega_\phi = (0)\omega_1 + (0)\omega_2 + (1)\omega_3, \tag{f8}$$

$$\omega_{\psi'} = (1/2r)(-2\omega_2 - 2r\,\omega_3 + \omega_4 + \omega_5), \tag{f9}$$

$$\omega_{\psi''} = (1/2r)(-2\omega_2 + 2r\,\omega_3 + \omega_4 - \omega_5). \tag{f10}$$

Comparing the above with the quasi velocities of the knife problem (ex. 2.13.2), to be denoted in this example by $\omega^K{}_{\dots}$, we readily see that we have the following correspondences:

$$\omega_1 \to \omega^K{}_2, \qquad \omega_2 \to \omega^K{}_1, \qquad \omega_3 \to \omega^K{}_3. \tag{f11}$$

Hence, and recalling the transitivity equations of that example, we find

$$(\delta\theta_1)^\cdot - \delta\omega_1 = (\delta\theta^K{}_2)^\cdot - \delta\omega^K{}_2 = \omega^K{}_3\,\delta\theta^K{}_1 - \omega^K{}_1\,\delta\theta^K{}_3 = \omega_3\,\delta\theta_2 - \omega_2\,\delta\theta_3 \quad (\neq 0), \tag{g1}$$

$$(\delta\theta_2)^\cdot - \delta\omega_2 = (\delta\theta^K{}_1)^\cdot - \delta\omega^K{}_1 = \omega^K{}_2\,\delta\theta^K{}_3 - \omega^K{}_3\,\delta\theta^K{}_2 = \omega_1\,\delta\theta_3 - \omega_3\,\delta\theta_1 \quad (\neq 0),$$
$$(= 0, \text{ after enforcement of the constraints } \delta\theta_1, \omega_1 = 0), \tag{g2}$$

$$(\delta\theta_3)^\cdot - \delta\omega_3 = 0 \qquad (\theta_3 = \phi = \textit{holonomic}\text{ coordinate}), \tag{g3}$$

$$(\delta\theta_4)^{\cdot} - \delta\omega_4 = 2[(\delta\theta_2)^{\cdot} - \delta\omega_2] + r[(\delta\psi' + \delta\psi'')^{\cdot} - \delta(\omega_{\psi'} + \omega_{\psi''})]$$
$$= 2(\omega_1 \delta\theta_3 - \omega_3\, \delta\theta_1) + 0 \quad (= 0, \text{ after enforcing } \delta\theta_1, \omega_1 = 0), \tag{g4}$$
$$(\delta\theta_5)^{\cdot} - \delta\omega_5 = 2b[(\delta\phi)^{\cdot} - \delta\omega_\phi] + r[(\delta\psi' - \delta\omega'')^{\cdot} - \delta(\omega_{\psi'} - \omega_{\psi''})]$$
$$[= 0 \Rightarrow \theta_5 = 2b\phi + r(\psi' - \psi'') = \textit{holonomic}\text{ coordinate}]. \tag{g5}$$

The above immediately show that the nonvanishing γ's equal ± 1, as in the knife problem; and since here $D = 1, 4$; $I, I' = 2, 3$ and

$$\gamma^D{}_{II'}: \quad \gamma^1{}_{23} = -\gamma^1{}_{32} = 1 \neq 0, \tag{h}$$

the system of Pfaffian constraints $\omega_1 = 0$ and $\omega_4 = 0$ is *nonholonomic*.

For additional wheeled vehicle applications, see also Lobas (1986), Lur'e (1968, pp. 27–31), Mei (1985, pp. 35–36, 168–175, 437–439), Stückler (1955—excellent treatment).

Example 2.13.9 *Transitivity Equations for a Rigid Body in General (Uncontrained) Motion.* As explained in §1.8 ff., to describe the general spatial motion of a rigid body B we employ, among others, the following *two* sets of rectangular Cartesian axes [and associated *orthogonal–normalized–dextral (OND)* bases]: (i) a *body-fixed* set ◆–xyz/◆–$\mathbf{ijk}$ (noninertial), where ◆ is a generic body point (pole); and (ii) a *space-fixed* one O–XYZ/O–$\mathbf{IJK}$ (inertial), where O is a generic fixed origin. Frequently (recalling §1.17, "A Comprehensive Example: The Rolling Coin"), we also use other "intermediate" axes/bases that are neither space- nor body-fixed: ◆–$x'y'z'$/◆–$\mathbf{i'j'k'}$; for example, axes ◆–XYZ translating, or comoving, with B but nonrotating (i.e., ever parallel to O–XYZ).

Let us examine the transitivity equations associated with the translation of B with pole ◆, and its rotation about ◆ (earliest systematic treatment in Kirchhoff, 1883, pp. 56–59).

(i) *Rotation.* As shown in §1.12, the transformation relations among the spatial and body components of the inertial angular velocity of B, $\boldsymbol{\omega}$, and the Eulerian angles (and their rates) between ◆–xyz and O–XYZ (or ◆–XYZ) are [with $s\ldots \equiv \sin\ldots$, $c\ldots \equiv \cos\ldots$] as follows:

Body axes components (assuming $\sin\theta \neq 0$):

$$\omega_x = (s\theta\, s\psi)\omega_\phi + (c\psi)\omega_\theta, \quad \omega_y = (s\theta\, c\psi)\omega_\phi + (-s\psi)\omega_\theta, \quad \omega_z = (c\theta)\omega_\phi + \omega_\psi; \tag{a1}$$

$$\Rightarrow \omega_\phi = (s\psi/s\theta)\omega_x + (c\psi/s\theta)\omega_y, \qquad \omega_\theta = (c\psi)\omega_x + (-s\psi)\omega_y,$$
$$\omega_\psi = (-\cot\theta\, s\phi)\omega_x + (-\cot\theta\, c\psi)\omega_y + \omega_z. \tag{a2}$$

Space axes components (assuming $\sin\theta \neq 0$):

$$\omega_X = (c\phi)\omega_\theta + (s\phi\, s\theta)\omega_\psi, \quad \omega_Y = (s\phi)\omega_\theta + (-c\phi\, s\theta)\omega_\psi, \quad \omega_Z = \omega_\phi + (c\theta)\omega_\psi; \tag{b1}$$

$$\Rightarrow \omega_\phi = (-\cot\theta\, s\phi)\omega_X + (\cot\theta\, c\phi)\omega_Y + \omega_Z, \qquad \omega_\theta = (c\phi)\omega_X + (s\phi)\omega_Y,$$

$$\omega_\psi = (s\phi/s\theta)\omega_X + (-c\phi/s\theta)\omega_Y. \tag{b2}$$

Since these transformations are *stationary*, they also hold with the $\omega_{x,y,z;X,Y,Z}$ replaced by the $d\theta_{x,y,z;X,Y,Z}$ or $\delta\theta_{x,y,z;X,Y,Z}$ and the $\omega_{\phi,\theta,\psi}$ replaced by $d\phi, \ldots,$ or $\delta\phi, \ldots,$ respectively.

Transitivity Equations

(a) *Body axes:* Differentiating/varying the first of (a1), while invoking the "$d\delta = \delta d$" rule for ϕ, θ, ψ, we obtain, successively,

$$\begin{aligned}(\delta\theta_x)^{\cdot} - \delta\omega_x &= [(s\theta\, s\psi)\,\delta\phi + (c\psi)\,\delta\theta]^{\cdot} - \delta\big[(s\theta\, s\psi)\omega_\phi + (c\psi)\omega_\theta\big]\\ &= c\theta\, s\psi(\omega_\theta\,\delta\phi - \omega_\phi\,\delta\theta) + s\theta\, c\psi(\omega_\psi\,\delta\phi - \omega_\phi\,\delta\psi) + s\psi(\omega_\theta\,\delta\psi - \omega_\psi\,\delta\theta),\end{aligned}$$

and substituting $\omega_\phi, \ldots/\delta\phi, \ldots$ in terms of $\omega_x, \ldots/\delta\theta_x, \ldots$, from (a2), we eventually find

$$(\delta\theta_x)^{\cdot} - \delta\omega_x = \omega_z\,\delta\theta_y - \omega_y\,\delta\theta_z, \tag{c1}$$

and similarly, for the other two,

$$(\delta\theta_y)^{\cdot} - \delta\omega_y = \omega_x\,\delta\theta_z - \omega_z\,\delta\theta_x, \tag{c2}$$

$$(\delta\theta_z)^{\cdot} - \delta\omega_z = \omega_y\,\delta\theta_x - \omega_x\,\delta\theta_y. \tag{c3}$$

Hence, the nonvanishing γ's are

$$\gamma^x{}_{yz} = -\gamma^x{}_{zy} = 1, \qquad \gamma^y{}_{zx} = -\gamma^y{}_{xz} = 1, \qquad \gamma^z{}_{xy} = -\gamma^z{}_{yx} = 1; \tag{d1}$$

or, compactly [with $k, r, s \to x, y, z : 1, 2, 3$],

$$\begin{aligned}\gamma^k{}_{rs} &= \varepsilon_{krs} \equiv (k - r)(r - s)(s - k)/2\\ &= \pm 1, \text{according as } k, r, s/x, y, z \text{ are an } \textit{even} \text{ or } \textit{odd} \text{ permutation of } 1, 2, 3;\\ &\quad\text{and} = 0 \text{ in all other cases: } \textit{Levi-Civita permutation symbol } (1.1.6 \text{ ff.}).\end{aligned} \tag{d2}$$

(b) *Space axes:* Applying similar steps to (b1, 2), we eventually obtain

$$(\delta\theta_X)^{\cdot} - \delta\omega_X = \omega_Y\,\delta\theta_Z - \omega_Z\,\delta\theta_Y, \tag{e1}$$

and similarly, for the other two,

$$(\delta\theta_Y)^{\cdot} - \delta\omega_Y = \omega_Z\,\delta\theta_X - \omega_X\,\delta\theta_Z, \tag{e2}$$

$$(\delta\theta_Z)^{\cdot} - \delta\omega_Z = \omega_X\,\delta\theta_Y - \omega_Y\,\delta\theta_X. \tag{e3}$$

Hence, the nonvanishing γ's are

$$\gamma^X{}_{YZ} = -\gamma^X{}_{ZY} = -1, \qquad \gamma^Y{}_{ZX} = -\gamma^Y{}_{XZ} = -1, \qquad \gamma^Z{}_{XY} = -\gamma^Z{}_{YX} = -1; \tag{f1}$$

or, compactly (with $k, r, s \to X, Y, Z$: 1, 2, 3),

$$\gamma^K{}_{RS} = -\varepsilon_{KRS} = \varepsilon_{RKS}. \tag{f2}$$

The above show clearly that the *orthogonal* components of $\boldsymbol{\omega}$, $\omega_{x,y,z}$ and $\omega_{X,Y,Z}$ are nonholonomic; while the *nonorthogonal* components $\omega_{\phi,\theta,\psi}$ are holonomic (and, again, this has nothing to do with constraints, but is a mathematical consequence of the *noncommutativity* of rigid rotations).

REMARK

These transitivity relations and γ-values, (c1–f2), are independent of the particular $\omega_{x,y,z;X,Y,Z} \Leftrightarrow \omega_{\phi,\theta,\psi}$ relationships (a1–b2); they express, in component form, invariant noncommutativity properties between the differentials of the vectors of infinitesimal rotation and angular velocity. A direct vectorial proof of these properties is presented in the next example.

(c) *Intermediate axes:* Such sets are the following axes of ex. 2.13.7: (i) $G\text{–}nn'z \equiv G\text{–}x'y'z'$, with OND basis $G\text{–}\boldsymbol{i}'\boldsymbol{j}'\boldsymbol{k}' \equiv G\text{–}\boldsymbol{u}_n\boldsymbol{u}_{n'}\boldsymbol{k}$; and (ii) $G\text{–}nNZ$, with OND basis $G\text{–}\boldsymbol{u}_n\boldsymbol{u}_N\boldsymbol{K}$ ($\boldsymbol{u}_n \equiv \boldsymbol{i}' =$ unit vector along $+$*nodal* line); and they are called by some authors *semimobile* (SM) and *semifixed* (SF), respectively.

Below, we collect some kinematical data pertinent to them. Since their inertial angular velocities are (consult fig. 2.20)

$$\begin{aligned}\boldsymbol{\omega}_{SM} \equiv \boldsymbol{\omega}' &= \omega_\phi\,\boldsymbol{K} + \omega_\theta\,\boldsymbol{i}' = \omega_\theta\,\boldsymbol{i}' + \omega_\phi(\sin\theta\,\boldsymbol{j}' + \cos\theta\,\boldsymbol{k}') \\ &= (\omega_\theta)\boldsymbol{i}' + (\omega_\phi\sin\theta)\boldsymbol{j}' + (\omega_\phi\cos\theta)\boldsymbol{k}' \\ &= \omega_\phi\,\boldsymbol{K} + \omega_\theta\,\boldsymbol{u}_n = \boldsymbol{\omega}_{SF} + \omega_\theta\,\boldsymbol{u}_n \equiv \boldsymbol{\omega}'' + \omega_\theta\,\boldsymbol{u}_n,\end{aligned} \tag{g1}$$

$$\boldsymbol{\omega}_{SF} \equiv \boldsymbol{\omega}'' = \omega_\phi\,\boldsymbol{K}, \tag{g2}$$

we will have the following relations for the rates of change of their bases:

$$d\boldsymbol{u}_n/dt = \boldsymbol{\omega}' \times \boldsymbol{u}_n = \omega_\phi\,\boldsymbol{u}_N = \omega_\phi(\cos\theta\,\boldsymbol{u}_{n'} - \sin\theta\,\boldsymbol{k}'); \tag{g3}$$

$$d\boldsymbol{u}_{n'}/dt = \boldsymbol{\omega}' \times \boldsymbol{u}_{n'} = (-\omega_\phi\cos\theta)\boldsymbol{u}_n + \omega_\theta\,\boldsymbol{k}'; \tag{g4}$$

$$d\boldsymbol{k}'/dt = \boldsymbol{\omega}' \times \boldsymbol{k}' = (\omega_\phi\sin\theta)\boldsymbol{u}_n - \omega_\theta\,\boldsymbol{u}_{n'}; \tag{g5}$$

$$d\boldsymbol{u}_n/dt = \boldsymbol{\omega}'' \times \boldsymbol{u}_n = \omega_\phi\,\boldsymbol{u}_N; \tag{g6}$$

$$d\boldsymbol{u}_N/dt = \boldsymbol{\omega}'' \times \boldsymbol{u}_N = -\omega_\phi\,\boldsymbol{u}_n; \tag{g7}$$

$$d\boldsymbol{K}/dt = \boldsymbol{\omega}'' \times \boldsymbol{K} = \boldsymbol{0}. \tag{g8}$$

Finally, the body angular velocity along the SM axes, thanks to the second line of (g1), equals

$$\begin{aligned}\boldsymbol{\omega} = \boldsymbol{\omega}' + \omega_\psi\,\boldsymbol{k}' &= (\omega_\theta)\boldsymbol{i}' + (\omega_\phi\sin\theta)\boldsymbol{j}' + (\omega_\psi + \omega_\phi\cos\theta)\boldsymbol{k}' \\ &\equiv \omega_{x'}\,\boldsymbol{i}' + \omega_{y'}\,\boldsymbol{j}' + \omega_{z'}\,\boldsymbol{k}';\end{aligned} \tag{g9}$$

and since this is a scleronomic system, (g9) holds with $\omega_{x',y',z'}$, replaced with $d\theta_{x',y',z'}$ and $\delta\theta_{x',y',z'}$; and $\omega_{\phi,\theta,\psi}$ replaced with $d\phi, d\theta, d\psi$ and $\delta\phi, \delta\theta, \delta\psi$, respectively.

From the above, by straightforward differentiations, we obtain the *rotational transitivity equations* in terms of *semimobile components*:

$$(\delta\theta_{x'})^{\cdot} - \delta\omega_{x'} = 0 \qquad (\theta_{x'} \equiv \theta = \textit{holonomic}\text{ coordinate} \Rightarrow \gamma^{x'}{}_{..} = 0), \tag{h1}$$

$$(\delta\theta_{y'})^{\cdot} - \delta\omega_{y'} = \cot\theta(\omega_{x'}\,\delta\theta_{y'} - \omega_{y'}\,\delta\theta_{x'}), \tag{h2}$$

$$(\delta\theta_{z'})^{\cdot} - \delta\omega_{z'} = (\omega_{y'}\,\delta\theta_{x'} - \omega_{x'}\,\delta\theta_{y'}); \tag{h3}$$

and hence the nonvanishing γ's are (assuming $\cot\theta = \textit{finite}$)

$$\gamma^{y'}{}_{y'x'} = -\gamma^{y'}{}_{x'y'} = \cot\theta, \qquad \gamma^{z'}{}_{x'y'} = -\gamma^{z'}{}_{y'x'} = 1. \tag{h4}$$

Note that (h1–4) are none other than (h3–5) and (i) of ex. 2.13.7.

(ii) *Translation of pole (or basepoint)* ◆. Let us assume that ◆ has inertial position:

$$\boldsymbol{OP} = \boldsymbol{\rho} = \rho_X\,\boldsymbol{I} + \rho_Y\,\boldsymbol{J} + \rho_Z\,\boldsymbol{K}, \tag{h5}$$

and, therefore, inertial velocity:

$$\begin{aligned}
\boldsymbol{v} &\equiv d\boldsymbol{\rho}/dt\\
&= (d\rho_X/dt)\boldsymbol{I} + (d\rho_Y/dt)\boldsymbol{J} + (d\rho_Z/dt)\boldsymbol{K} \equiv v_X\,\boldsymbol{I} + v_Y\,\boldsymbol{J} + v_Z\,\boldsymbol{K}\\
&\qquad [\text{along } \textit{space}\text{-axes: } v_X \equiv d\rho_X/dt = \boldsymbol{v}\cdot\boldsymbol{I}, \text{etc.};\\
&\qquad\qquad \rho_{X,Y,Z}(v_{X,Y,Z})\text{: } \textit{holonomic} \text{ coordinates (velocities) of } \blacklozenge.]\\
&= v_x\,\boldsymbol{i} + v_y\,\boldsymbol{j} + v_z\,\boldsymbol{k} \equiv (dp_x/dt)\boldsymbol{i} + (dp_y/dt)\boldsymbol{j} + (dp_z/dt)\boldsymbol{k}\\
&\qquad [\text{along } \textit{body}\text{-axes: } v_x \equiv dp_x/dt \equiv \boldsymbol{v}\cdot\boldsymbol{i}, \text{ etc.};\\
&\qquad\qquad p_{x,y,z}(v_{x,y,z})\text{: } \textit{nonholonomic} \text{ coordinates (velocities) of } \blacklozenge.]
\end{aligned} \tag{h6}$$

Clearly, the above velocity components are related by the following vector transformations:

$$\begin{aligned}
v_x &= \cos(x, X)v_X + \cos(x, Y)v_Y + \cos(x, Z)v_Z, \quad \text{etc.}\\
v_X &= \cos(X, x)v_x + \cos(X, y)v_y + \cos(X, z)v_z, \quad \text{etc.,}
\end{aligned} \tag{h7}$$

and, since this is a scleronomic system, their differentials are related by

$$\begin{aligned}
dp_x &\equiv v_x\,dt = (\boldsymbol{v}\cdot\boldsymbol{i})\,dt \equiv d\boldsymbol{\rho}\cdot\boldsymbol{i} = \cos(x, X)\,d\rho_X + \cos(x, Y)\,d\rho_Y + \cos(x, Z)\,d\rho_Z, \quad \text{etc.}\\
\delta p_x &\equiv \delta\boldsymbol{\rho}\cdot\boldsymbol{i} = \cos(x, X)\,\delta\rho_X + \cos(x, Y)\,\delta\rho_Y + \cos(x, Z)\,\delta\rho_Z, \quad \text{etc.}
\end{aligned} \tag{h8}$$

Next, since

$$d\boldsymbol{i} = d\boldsymbol{\chi}\times\boldsymbol{i}, \qquad \delta\boldsymbol{i} = \delta\boldsymbol{\chi}\times\boldsymbol{i}, \quad \text{etc.,} \tag{i1}$$

where $d\boldsymbol{\chi} \equiv \boldsymbol{\omega}\,dt$ =*infinitesimal (inertial) kinematically admissible/possible rotation vector*, and $\delta\boldsymbol{\chi} = \delta\chi_x\,\boldsymbol{i} + \delta\chi_y\,\boldsymbol{j} + \delta\chi_z\,\boldsymbol{k} = \delta\phi\,\boldsymbol{K} + \delta\theta\,\boldsymbol{u}_n + \delta\psi\,\boldsymbol{k}$ = *(inertial) virtual rotation vector*, we find by direct calculation [with $(\ldots)^{\cdot}$ = *inertial rate of change*, for vectors]

$$\begin{aligned}
(\delta p_x)^{\cdot} &= (\delta\boldsymbol{\rho}\cdot\boldsymbol{i})^{\cdot} = (\delta\boldsymbol{\rho})^{\cdot}\cdot\boldsymbol{i} + \delta\boldsymbol{\rho}\cdot(\boldsymbol{i})^{\cdot} = (\delta\boldsymbol{\rho})^{\cdot}\cdot\boldsymbol{i} + \delta\boldsymbol{\rho}\cdot(\boldsymbol{\omega}\times\boldsymbol{i}),\\
\delta(dp_x/dt) &\equiv \delta v_x = \delta(\boldsymbol{v}\cdot\boldsymbol{i}) = \delta\boldsymbol{v}\cdot\boldsymbol{i} + \boldsymbol{v}\cdot\delta\boldsymbol{i} = \delta\boldsymbol{v}\cdot\boldsymbol{i} + \boldsymbol{v}\cdot(\delta\boldsymbol{\chi}\times\boldsymbol{i}),
\end{aligned} \tag{i2}$$

and subtracting the above side by side, while noting that $(\delta\boldsymbol{\rho})^{\cdot} - \delta\boldsymbol{v} = \delta(d\boldsymbol{\rho}/dt) - \delta\boldsymbol{v} = \delta\boldsymbol{v} - \delta\boldsymbol{v} = \boldsymbol{0}$, we obtain the x-component of the *pole velocity transitivity equation*:

$$\begin{aligned}
(\delta p_x)^{\cdot} - \delta v_x &= \delta\boldsymbol{\rho}\cdot(\boldsymbol{\omega}\times\boldsymbol{i}) - \boldsymbol{v}\cdot(\delta\boldsymbol{\chi}\times\boldsymbol{i}) = (\delta\boldsymbol{\rho}\times\boldsymbol{\omega})\cdot\boldsymbol{i} - (\boldsymbol{v}\times\delta\boldsymbol{\chi})\cdot\boldsymbol{i}\\
&= (\delta\boldsymbol{\rho}\times\boldsymbol{\omega} - \boldsymbol{v}\times\delta\boldsymbol{\chi})\cdot\boldsymbol{i};
\end{aligned} \tag{i3}$$

and similarly for its y and z components. Hence, our pole transitivity equation can be written in the following *vector* form [with $d'(\ldots)$ and $\delta'(\ldots)$ denoting differentials of vectors, and so on, relative to the moving, here *body*-fixed, axes]:

$$d'(\delta\boldsymbol{\rho})/dt - \delta'(d\boldsymbol{\rho}/dt) = \delta\boldsymbol{\rho} \times \boldsymbol{\omega} - \boldsymbol{v} \times \delta\boldsymbol{\chi}, \tag{i4a}$$

or

$$\delta'\boldsymbol{v} = d'(\delta\boldsymbol{\rho})/dt + \boldsymbol{\omega} \times \delta\boldsymbol{\rho} + \boldsymbol{v} \times \delta\boldsymbol{\chi}. \tag{i4b}$$

To obtain the counterpart of (i4) in terms of *fixed*-axes differentials, we may apply to it the well-known operator identities $d(\ldots) = d'(\ldots) + d\boldsymbol{\chi} \times (\ldots)$ and $\delta(\ldots) = \delta'(\ldots) + \delta\boldsymbol{\chi} \times (\ldots)$, where $(\ldots)$ = arbitrary *vector*. The details are left to the reader.

In component form, along ◆–xyz, (i4a) reads

$$(\delta p_x)^{\cdot} - \delta v_x = (\omega_z\, \delta p_y - \omega_y\, \delta p_z) - (v_y\, \delta\theta_z - v_z\, \delta\theta_y), \tag{i5}$$

$$(\delta p_y)^{\cdot} - \delta v_y = (\omega_x\, \delta p_z - \omega_z\, \delta p_x) - (v_z\, \delta\theta_x - v_x\, \delta\theta_z), \tag{i6}$$

$$(\delta p_z)^{\cdot} - \delta v_z = (\omega_y\, \delta p_x - \omega_x\, \delta p_y) - (v_x\, \delta\theta_y - v_y\, \delta\theta_x); \tag{i7}$$

and, therefore, the nonvanishing γ's are [with accented (unaccented) indices for the components $\delta p_{x,y,z}(\delta\theta_{x,y,z})$]

$$\gamma^{x'}{}_{y'z} = -\gamma^{x'}{}_{zy'} = 1 \quad \text{and} \quad \gamma^{x'}{}_{yz'} = -\gamma^{x'}{}_{z'y} = 1, \tag{i8}$$

$$\gamma^{y'}{}_{z'x} = -\gamma^{y'}{}_{xz'} = 1 \quad \text{and} \quad \gamma^{y'}{}_{zx'} = -\gamma^{y'}{}_{x'z} = 1, \tag{i9}$$

$$\gamma^{z'}{}_{x'y} = -\gamma^{z'}{}_{yx'} = 1 \quad \text{and} \quad \gamma^{z'}{}_{xy'} = -\gamma^{z'}{}_{y'z} = 1. \tag{i10}$$

Semifixed axes ◆–$\boldsymbol{u}_n\boldsymbol{u}_N\boldsymbol{K}$. Here, we have

$$v_n \equiv (p_n)^{\cdot} = \boldsymbol{v} \cdot \boldsymbol{u}_n \equiv \boldsymbol{v} \cdot \boldsymbol{i}' = (\rho_X)^{\cdot} \cos\phi + (\rho_Y)^{\cdot} \sin\phi, \tag{j1}$$

$$v_N \equiv (p_N)^{\cdot} = \boldsymbol{v} \cdot \boldsymbol{u}_N = -(\rho_X)^{\cdot} \sin\phi + (\rho_Y)^{\cdot} \cos\phi, \tag{j2}$$

$$v_Z \equiv (p_Z)^{\cdot} = \boldsymbol{v} \cdot \boldsymbol{K} = (\rho_Z)^{\cdot} \;\Rightarrow\; \gamma^{Z}{}_{..} = 0 \quad (\text{i.e., } v_Z = \textit{holonomic} \text{ velocity}). \tag{j3}$$

We leave it to the reader to show that (recalling the earlier *semimobile* axes kinematics)

$$\begin{aligned}(\delta p_n)^{\cdot} - \delta v_n &= -[\omega_\phi\, \delta\rho_X - (\rho_X)^{\cdot}\, \delta\phi] \sin\phi + [\omega_\phi\, \delta\rho_Y - (\rho_Y)^{\cdot}\, \delta\phi] \cos\phi \\ &= \omega_\phi\, \delta p_N - v_N\, \delta\phi = (1/\sin\phi)(\omega_y\, \delta p_N - v_N\, \delta\theta_y),\end{aligned} \tag{j4}$$

$$\begin{aligned}(\delta p_N)^{\cdot} - \delta v_N &= -[\omega_\phi\, \delta\rho_X - (\rho_X)^{\cdot}\, \delta\phi] \cos\phi + [(\rho_Y)^{\cdot}\, \delta\phi - \omega_\phi\, \delta\rho_Y] \sin\phi \\ &= -\omega_\phi\, \delta p_n + (p_n)^{\cdot}\, \delta\phi = (1/\sin\phi)[(p_n)^{\cdot}\, \delta\theta_y - \omega_y\, \delta p_n];\end{aligned} \tag{j5}$$

and hence that the nonvanishing γ's are (with some, easily understood, ad hoc notation; and assuming that $\sin\theta \neq 0$)

$$\gamma^{n}{}_{Ny} = -\gamma^{n}{}_{yN} = 1/\sin\theta, \qquad \gamma^{N}{}_{yn} = -\gamma^{N}{}_{ny} = 1/\sin\theta. \tag{j6}$$

[Recalling ex. 2.13.7 (rolling disk problem), eqs. (d2, 3), (h1, 2), (i), etc.]

For related discussions of the rigid-body transitivity equations, see also Bremer (1988(b)) and Moiseyev and Rumyantsev (1968, pp. 7–8).

Example 2.13.10 *Cardanian Suspension of a Gyroscope.* Let us consider a gyroscope suspended à le Cardan (fig. 2.22). The rotation sequence

$$q_1 \equiv \phi(precession) \rightarrow q_2 \equiv \theta(nutation) \rightarrow q_3 \equiv \psi(spin)$$

(i.e., $3 \rightarrow 2 \rightarrow 1$, in the Eulerian angle sense of §1.12) brings the original axes $G\text{–}XYZ$, through the intermediate position $G\text{–}x'y'z'$ (outer gimbal), to the also intermediate position $G\text{–}xyz$ (inner gimbal).

Now: (i) The inertial angular velocity of the *outer* gimbal $\boldsymbol{\omega}_O$, *along outer gimbal–fixed axes*, is

$$\omega_{O,x'} = 0, \qquad \omega_{O,y'} = 0, \qquad \omega_{O,z'} = \omega_\phi; \tag{a}$$

(ii) the inertial angular velocity of the *inner* gimbal $\boldsymbol{\omega}_I$, *along inner gimbal–fixed axes*, is

$$\omega_{I,x} = -\omega_\phi \sin\theta, \qquad \omega_{I,y} = \omega_\theta, \qquad \omega_{I,z} = \omega_\phi \cos\theta; \tag{b}$$

and (iii) the inertial angular velocity of the *gyroscope* $\boldsymbol{\omega}$, *along inner gimbal–fixed axes*, is

$$\omega_x = \omega_\psi - \omega_\phi \sin\theta, \qquad \omega_y = \omega_\theta, \qquad \omega_z = \omega_\phi \cos\theta. \tag{c}$$

Let us find the transitivity equations corresponding to these quasi velocities. Equations (c) can be rewritten as

$$\omega_1 \equiv \omega_x \equiv (-\sin\theta)\omega_\phi + (0)\omega_\theta + (1)\omega_\psi \quad (\neq 0), \tag{d}$$

$$\omega_2 \equiv \omega_y \equiv (0)\omega_\phi + (1)\omega_\theta + (0)\omega_\psi \quad (\neq 0), \tag{e}$$

$$\omega_3 \equiv \omega_z \equiv (\cos\theta)\omega_\phi + (0)\omega_\theta + (0)\omega_\psi \quad (\neq 0); \tag{f}$$

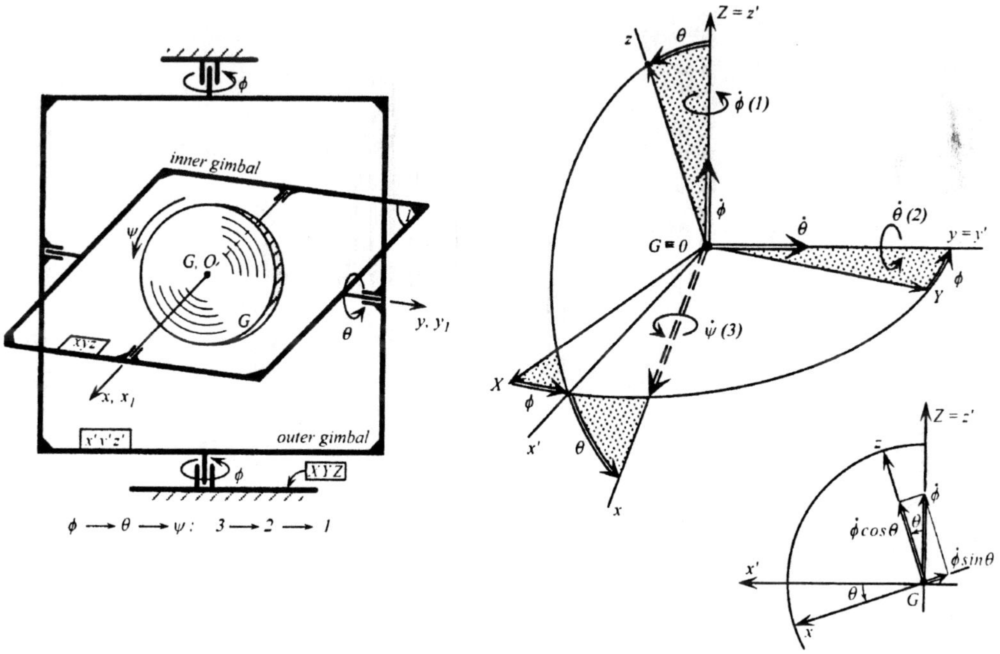

Figure 2.22 Kinematics of Cardanian suspension of a gyroscope.

and their inverses are readily found to be

$$v_1 \equiv \omega_\phi = (0)\omega_x + (0)\omega_y + (1/\cos\theta)\omega_z, \tag{g}$$

$$v_2 \equiv \omega_\theta = (0)\omega_x + (1)\omega_y + (0)\omega_z, \tag{h}$$

$$v_3 \equiv \omega_\psi = (1)\omega_x + (0)\omega_y + (\sin\theta/\cos\theta)\omega_z. \tag{i}$$

From these stationary relations, and assuming $d(\delta q_k) = \delta(dq_k)$ $(k = x, y, z)$, we obtain, successively,

(i) $d(\delta\theta_x) - \delta(d\theta_x) = d[(-\sin\theta)\,\delta\phi + \delta\psi] - \delta[(-\sin\theta)\,d\phi + d\psi]$

$$= \cdots = (\cos\theta)(d\phi\,\delta\theta - d\theta\,\delta\phi) = \cdots = d\theta_z\,\delta\theta_y - d\theta_y\,\delta\theta_z; \tag{j}$$

(ii) $d(\delta\theta_y) - \delta(d\theta_y) = 0$ $\quad(\Rightarrow \theta_y =$ *holonomic* coordinate); (k)

(iii) $d(\delta\theta_z) - \delta(d\theta_z) = d[(\cos\theta)\,\delta\phi] - \delta[(\cos\theta)\,d\phi]$

$$= \cdots = (\sin\theta)(d\phi\,\delta\theta - d\theta\,\delta\phi) = \cdots = (\tan\theta)(d\theta_z\,\delta\theta_y - d\theta_y\,\delta\theta_z); \tag{l}$$

and so the nonvanishing γ's are (assuming $\theta \neq \pm\pi/2$)

$$\gamma^x{}_{yz} = -\gamma^x{}_{zy} \equiv \gamma^1{}_{23} = -\gamma^1{}_{32} = +1, \tag{m}$$

$$\gamma^z{}_{yz} = -\gamma^z{}_{zy} \equiv \gamma^3{}_{23} = -\gamma^3{}_{32} = \tan\theta. \tag{n}$$

Example 2.13.11 *An Elementary ad hoc Vectorial Derivation of the Rotational Rigid-Body Transitivity Equations.* Let us consider, with no loss of generality, a free rigid body B rotating with (inertial) angular velocity $\boldsymbol{\omega}$ about a fixed point O. Then, as is well known (§1.9 ff.), and since this is an internally scleronomic system, the (inertial) velocity/kinematically admissible displacements/virtual displacements of a typical B-particle of (inertial) position vector $\boldsymbol{r}$, are, respectively,

$$\boldsymbol{v} = \boldsymbol{\omega} \times \boldsymbol{r} \Rightarrow d\boldsymbol{r} = d\boldsymbol{\chi} \times \boldsymbol{r}, \qquad \delta\boldsymbol{r} = \delta\boldsymbol{\chi} \times \boldsymbol{r}, \tag{a}$$

where $d\boldsymbol{\chi} \equiv \boldsymbol{\omega}\,dt$, and $d(\ldots)/\delta(\ldots)$ are *kinematically admissible/virtual* (*inertial*) *variation operators*. Now, $d(\ldots)$-varying the last of (a), $\delta(\ldots)$-varying the second, and then subtracting the results side by side, while invoking (a) and the rule $d(\delta\boldsymbol{r}) - \delta(d\boldsymbol{r}) = \boldsymbol{0}$, we obtain, successively,

$$\begin{aligned}\boldsymbol{0} = d(\delta\boldsymbol{r}) - \delta(d\boldsymbol{r}) &= [d(\delta\boldsymbol{\chi}) \times \boldsymbol{r} + \delta\boldsymbol{\chi} \times d\boldsymbol{r}] - [\delta(d\boldsymbol{\chi}) \times \boldsymbol{r} + d\boldsymbol{\chi} \times \delta\boldsymbol{r}]\\ &= [d(\delta\boldsymbol{\chi}) \times \boldsymbol{r} + \delta\boldsymbol{\chi} \times (d\boldsymbol{\chi} \times \boldsymbol{r})] - [\delta(d\boldsymbol{\chi}) \times \boldsymbol{r} + d\boldsymbol{\chi} \times (\delta\boldsymbol{\chi} \times \boldsymbol{r})]\\ &= [d(\delta\boldsymbol{\chi}) - \delta(d\boldsymbol{\chi})] \times \boldsymbol{r} + [\delta\boldsymbol{\chi} \times (d\boldsymbol{\chi} \times \boldsymbol{r}) - d\boldsymbol{\chi} \times (\delta\boldsymbol{\chi} \times \boldsymbol{r})]\end{aligned}$$

[and applying to the second bracket (last two triple cross-products) the cyclic vector identity, holding for any three vectors $\boldsymbol{a}, \boldsymbol{b}, \boldsymbol{c}$: $\boldsymbol{a} \times (\boldsymbol{b} \times \boldsymbol{c}) + \boldsymbol{b} \times (\boldsymbol{c} \times \boldsymbol{a}) + \boldsymbol{c} \times (\boldsymbol{a} \times \boldsymbol{b}) = \boldsymbol{0}$, with the identifications: $\boldsymbol{a} \to \delta\boldsymbol{\chi}$, $\boldsymbol{b} \to d\boldsymbol{\chi}$, $\boldsymbol{c} \to \boldsymbol{r}$]

$$= [d(\delta\boldsymbol{\chi}) - \delta(d\boldsymbol{\chi})] \times \boldsymbol{r} + (\delta\boldsymbol{\chi} \times d\boldsymbol{\chi}) \times \boldsymbol{r},$$

from which, since $\boldsymbol{r}$ is arbitrary, we finally get the fundamental and general *inertial* rotational transitivity equation:

$$d(\delta\boldsymbol{\chi}) - \delta(d\boldsymbol{\chi}) = d\boldsymbol{\chi} \times \delta\boldsymbol{\chi}. \tag{b}$$

Dividing the above with dt, which does no interact with these differentials (and noting that, by Newtonian relativity, $dt = d't$), we also obtain the equivalent transitivity equation in terms of the angular velocities:

$$d(\delta\boldsymbol{\chi})/dt - \delta\boldsymbol{\omega} = \boldsymbol{\omega} \times \delta\boldsymbol{\chi}. \tag{c}$$

Next, let us find the counterparts of (b, c) in terms of differentials/variations relative to *body-fixed* (moving) axes; to be denoted by $d'(\ldots)/\delta'(\ldots)$. Applying the well-known kinematical operator identities [(§1.7 ff.)]

$$d(\ldots) = d'(\ldots) + d\boldsymbol{\chi} \times (\ldots), \qquad \delta(\ldots) = \delta'(\ldots) + \delta\boldsymbol{\chi} \times (\ldots), \tag{d}$$

(which immediately yield $d'\boldsymbol{r} = \boldsymbol{0}$ and $\delta'\boldsymbol{r} = \boldsymbol{0}$, as expected) to $\delta\boldsymbol{\chi}$ and $d\boldsymbol{\chi}$, respectively, and then substracting side by side, we find, successively,

$$\begin{aligned} d(\delta\boldsymbol{\chi}) - \delta(d\boldsymbol{\chi}) &= d'(\delta\boldsymbol{\chi}) - \delta'(d\boldsymbol{\chi}) + (d\boldsymbol{\chi} \times \delta\boldsymbol{\chi} - \delta\boldsymbol{\chi} \times d\boldsymbol{\chi}) \\ &= d'(\delta\boldsymbol{\chi}) - \delta'(d\boldsymbol{\chi}) + 2(d\boldsymbol{\chi} \times \delta\boldsymbol{\chi}), \end{aligned} \tag{e}$$

or, invoking (b) for its left side and rearranging slightly, we get, finally,

$$d'(\delta\boldsymbol{\chi}) - \delta'(d\boldsymbol{\chi}) = \delta\boldsymbol{\chi} \times d\boldsymbol{\chi}; \tag{f}$$

and dividing by dt, we also obtain its velocity equivalent,

$$d'(\delta\boldsymbol{\chi})/dt - \delta'\boldsymbol{\omega} = \delta\boldsymbol{\chi} \times \boldsymbol{\omega}. \tag{g}$$

The kinematical identities (f, g) are the noninertial counterparts of (b, c).

The *difference* between (b, c) and (f, g) often goes unnoticed in the literature. To understand it better, let us write them down in component form, along *space-fixed* axes ◆–XYZ and *body-fixed* axes ◆–xyz. Only the first equations are shown (i.e., X, x); the rest follow cyclically:

Space-fixed (inertial) axes:

$$d(\delta\theta_X) - \delta(d\theta_X) = d\theta_Y\,\delta\theta_Z - d\theta_Z\,\delta\theta_Y, \qquad \text{or} \qquad (\delta\theta_X)^{\cdot} - \delta\omega_X = \omega_Y\,\delta\theta_Z - \omega_Z\,\delta\theta_Y; \tag{h1}$$

Body-fixed (noninertial) axes:

$$d(\delta\theta_x) - \delta(d\theta_x) = d\theta_z\,\delta\theta_y - d\theta_y\,\delta\theta_z, \qquad \text{or} \qquad (\delta\theta_x)^{\cdot} - \delta\omega_x = \omega_z\,\delta\theta_y - \omega_y\,\delta\theta_z; \tag{h2}$$

which, naturally, coincide with equations (c1–f2) of ex. 2.13.9, and §1.14.

[When dealing with derivatives/differentials of *components*, we may safely use the same notation $(\ldots)^{\cdot}/d(\ldots)/\delta(\ldots)$ for both space and body such changes; here, the intended meaning is conveyed unambiguously].

Additional Special Results

(i) Applying the *second* of (d) for $\boldsymbol{\omega}$, and then equating the resulting $\delta\boldsymbol{\omega}$-expression with that obtained from (c), we get $d(\delta\boldsymbol{\chi})/dt - \boldsymbol{\omega} \times \delta\boldsymbol{\chi} = \delta'\boldsymbol{\omega} + \delta\boldsymbol{\chi} \times \boldsymbol{\omega}$, or, simplifying, $d(\delta\boldsymbol{\chi})/dt = \delta'\boldsymbol{\omega}$; or, equivalently (multiplying with dt),

$$d(\delta\boldsymbol{\chi}) = \delta'(d\boldsymbol{\chi}). \tag{i}$$

(ii) Starting from (c), and then invoking the first of (d), we obtain, successively,

$$\delta\boldsymbol{\omega} = d(\delta\boldsymbol{\chi})/dt - \boldsymbol{\omega} \times \delta\boldsymbol{\chi} = \left[d'(\delta\boldsymbol{\chi})/dt + \boldsymbol{\omega} \times \delta\boldsymbol{\chi}\right] - \boldsymbol{\omega} \times \delta\boldsymbol{\chi}$$
$$= d'(\delta\boldsymbol{\chi})/dt \qquad [= \delta'\boldsymbol{\omega} + \delta\boldsymbol{\chi} \times \boldsymbol{\omega}, \quad \text{by (g)}];$$

that is,

$$\delta\boldsymbol{\omega} = d'(\delta\boldsymbol{\chi})/dt, \qquad \text{or, equivalently,} \qquad \delta(d\boldsymbol{\chi}) = d'(\delta\boldsymbol{\chi}); \tag{j}$$

which is "symmetrical" to (i).

(iii) Applying the *first* of (d) for $\boldsymbol{\omega}$ yields

$$d\boldsymbol{\omega} = d'\boldsymbol{\omega}, \qquad \text{or, equivalently,} \qquad d(d\boldsymbol{\chi}) = d'(d\boldsymbol{\chi}); \tag{k}$$

but the *second* of (d) shows that

$$\delta\boldsymbol{\omega} \neq \delta'\boldsymbol{\omega}, \qquad \text{or, equivalently;} \qquad \delta(d\boldsymbol{\chi}) \neq \delta'(d\boldsymbol{\chi}). \tag{l}$$

Problem 2.13.4 *Rigid-body Transitivity Equations.* Using the results of the preceding example and its notations, show that, for a rigid body rotating about a fixed point,

$$d'(\delta\mathbf{r}) - \delta'(d\mathbf{r}) = (\delta\boldsymbol{\chi} \times d\boldsymbol{\chi}) \times \mathbf{r} \neq \mathbf{0}, \tag{a}$$

or, equivalently (dividing by $dt = d't$),

$$d'(\delta\mathbf{r})/dt - \delta'\mathbf{v} = (\delta\boldsymbol{\chi} \times \boldsymbol{\omega}) \times \mathbf{r} \neq \mathbf{0}; \tag{b}$$

even though $d(\delta\mathbf{r}) - \delta(d\mathbf{r}) = \mathbf{0}$; that is, *the rule* $d(\delta\ldots) = \delta(d\ldots)$ *is not frame-invariant.*

Example 2.13.12 *A Special Rigid-Body Transitivity Equation—Holonomic Coordinates.* Continuing from the above examples, we show below that, for a rigid body rotating about a fixed point, the following transitivity/nonintegrability identity holds:

$$E_k(\boldsymbol{\omega}) \equiv d/dt(\partial\boldsymbol{\omega}/\partial v_k) - \partial\boldsymbol{\omega}/\partial q_k = \boldsymbol{\omega} \times (\partial\boldsymbol{\omega}/\partial v_k). \tag{a}$$

For such a system (with $k = 1, 2, 3$; and q_k = *angular Lagrangean coordinates*; e.g., Eulerian angles ϕ, θ, ψ) we will have

$$\boldsymbol{\omega} = \boldsymbol{\omega}(q_k, dq_k/dt \equiv v_k) \equiv \boldsymbol{\omega}(q, v)$$
$$= \textit{linear} \text{ and (for our system, also) } \textit{homogeneous} \text{ function of the } v_k\text{'s}$$
$$= \sum (\partial\boldsymbol{\omega}/\partial v_k) v_k \text{ (by Euler's homogeneous function theorem)} \equiv \sum \mathbf{c}_k v_k, \tag{b}$$

[definition of the $\mathbf{c}_k$'s; also, recalling (1.7.9a, b)] from which it follows that

$$d\boldsymbol{\chi} \equiv \boldsymbol{\omega}\, dt = \sum \mathbf{c}_k\, dq_k,$$

and since this is a scleronomic system

$$\delta\boldsymbol{\chi} = \sum \boldsymbol{c}_k \, \delta q_k; \tag{c}$$

and so the basis (quasi) vectors $\boldsymbol{c}_k \equiv \partial\boldsymbol{\omega}/\partial v_k$ (independent of the v_k's) can also be defined *symbolically* by

$$\boldsymbol{c}_k \equiv \partial\boldsymbol{\chi}/\partial q_k \equiv \partial(d\boldsymbol{\chi})/\partial(dq_k) \equiv \partial(\delta\boldsymbol{\chi})/\partial(\delta q_k). \tag{d}$$

Now, let us substitute the above representations into the earlier (inertial) transitivity equation

$$d(\delta\boldsymbol{\chi})/dt - \delta\boldsymbol{\omega} = \boldsymbol{\omega} \times \delta\boldsymbol{\chi}. \tag{e}$$

We find, successively,

(i) *Left* side [we assume that $(\delta q)^{\cdot} = \delta(dq/dt) \equiv \delta v$]:

$$d(\delta\boldsymbol{\chi})/dt - \delta\boldsymbol{\omega} = d/dt\Big(\sum (\partial\boldsymbol{\omega}/\partial v_k)\, \delta q_k\Big) - \sum \big[(\partial\boldsymbol{\omega}/\partial q_k)\delta q_k + (\partial\boldsymbol{\omega}/\partial v_k)\, \delta v_k\big]$$

$$= \cdots = \sum [d/dt(\partial\boldsymbol{\omega}/\partial v_k) - \partial\boldsymbol{\omega}/\partial q_k]\delta q_k = \sum E_k(\boldsymbol{\omega})\, \delta q_k. \tag{f}$$

(ii) *Right* side:

$$\boldsymbol{\omega} \times \delta\boldsymbol{\chi} = \boldsymbol{\omega} \times \Big(\sum (\partial\boldsymbol{\omega}/\partial v_k)\, \delta q_k\Big) = \sum [\boldsymbol{\omega} \times (\partial\boldsymbol{\omega}/\partial v_k)]\delta q_k; \tag{g}$$

and therefore (since the δq_k are independent—but even if they were constrained that would only affect the equations of motion) equating the right sides of (f) and (g), the identity (a) follows.

In terms of the earlier $\boldsymbol{c}_k$ vectors, (a) reads

$$d\boldsymbol{c}_k/dt = \boldsymbol{\omega} \times \boldsymbol{c}_k + \partial\boldsymbol{\omega}/\partial q_k = \sum (\boldsymbol{c}_l \times \boldsymbol{c}_k + \partial\boldsymbol{c}_l/\partial q_k)v_l. \tag{h}$$

Finally, applying the first of (d) of ex. 2.13.11 to $\partial\boldsymbol{\omega}/\partial v_k$, and inserting the result into (a, h) produces the following interesting result:

$$E_k{}'(\boldsymbol{\omega}) \equiv d'/dt(\partial\boldsymbol{\omega}/\partial v_k) - \partial\boldsymbol{\omega}/\partial q_k = \boldsymbol{0} \qquad \text{or} \qquad d'\boldsymbol{c}_k/dt = \partial\boldsymbol{\omega}/\partial q_k. \tag{i}$$

Problem 2.13.5 Using the well-known kinematical result

$$d\boldsymbol{u}_k/dt = \boldsymbol{\omega} \times \boldsymbol{u}_k, \tag{a}$$

where $\{\boldsymbol{u}_k = \boldsymbol{u}_k(q)\}$ is, say, a *body-fixed* basis rotating with inertial angular velocity $\boldsymbol{\omega}$ (like the earlier $\boldsymbol{i}, \boldsymbol{j}, \boldsymbol{k}$), with the $\boldsymbol{\omega}$-representation (b) of the preceding example:

$$\boldsymbol{\omega} = \boldsymbol{\omega}(q_k, v_k) \equiv \boldsymbol{\omega}(q, v) = \sum \boldsymbol{c}_k v_k, \tag{b}$$

show that

$$\partial\boldsymbol{u}_k/\partial q_l = \boldsymbol{c}_l \times \boldsymbol{u}_k \qquad \text{[note subscript order]}, \tag{c}$$

i.e., (1.7.9c). Clearly, such a result holds for *any* vector $\boldsymbol{b} = \boldsymbol{b}(q)$ rotating with angular velocity $\boldsymbol{\omega}$: $\partial\boldsymbol{b}/\partial q_l = \boldsymbol{c}_l \times \boldsymbol{b} \equiv (\partial\boldsymbol{\omega}/\partial v_k) \times \boldsymbol{b}$. Also: (i) $d/dt(\partial\boldsymbol{b}/\partial q_l) = \partial/\partial q_l(d\boldsymbol{b}/dt)$, and (ii) $\partial\boldsymbol{b}/\partial v_l = \boldsymbol{0}$.

Problem 2.13.6 By direct substitution of the representations

$$d\boldsymbol{\chi} \equiv \boldsymbol{\omega}\, dt = \sum \boldsymbol{c}_k\, dq_k \qquad \text{and} \qquad \delta\boldsymbol{\chi} = \sum \boldsymbol{c}_k\, \delta q_k \tag{a}$$

into the earlier inertial rotational transitivity equation [ex. 2.13.11: eq. (b)].

$$d(\delta\boldsymbol{\chi}) - \delta(d\boldsymbol{\chi}) = d\boldsymbol{\chi} \times \delta\boldsymbol{\chi}, \tag{b}$$

and some simple differentiations, show that

$$\partial \boldsymbol{c}_k/\partial q_l - \partial \boldsymbol{c}_l/\partial q_k = \boldsymbol{c}_l \times \boldsymbol{c}_k. \tag{c}$$

This *nonintegrability* relation shows clearly that the basis $\{\boldsymbol{c}_k\}$ is nonholonomic (nongradient); whereas if $\boldsymbol{c}_k = \partial\boldsymbol{\chi}/\partial q_k$, then $d(\delta\boldsymbol{\chi}) = \delta(d\boldsymbol{\chi}) \Rightarrow \boldsymbol{\chi} =$ genuine angular coordinate. Simplify (c) if the $\{\boldsymbol{c}_k\}$ are an *orthogonal–unit–dextral* basis (see also Brunk, 1981).

Example 2.13.13 *A Special Rigid-Body Transitivity Equation—Nonholonomic Coordinates.* Continuing from ex. 2.13.11, let us substitute the (fully nonholonomic) representations

$$\boldsymbol{\omega} = \sum (\partial\boldsymbol{\omega}/\partial\omega_k)\omega_k \equiv \sum \boldsymbol{\varepsilon}_k\, \omega_k = \boldsymbol{\omega}(q, \omega), \qquad d\boldsymbol{\chi} = \sum \boldsymbol{\varepsilon}_k\, d\theta_k, \qquad \delta\boldsymbol{\chi} = \sum \boldsymbol{\varepsilon}_k\, \delta\theta_k, \tag{a}$$

where, as usual, $\theta_k =$ *quasi coordinates*, $\omega_k \equiv d\theta_k/dt =$ *quasi velocities*, and

$$\boldsymbol{\varepsilon}_k \equiv \partial\boldsymbol{\omega}/\partial\omega_k \equiv \partial(d\boldsymbol{\chi})/\partial(d\theta_k) \equiv \partial(\delta\boldsymbol{\chi})/\partial(\delta\theta_k) \equiv \partial\boldsymbol{\chi}/\partial\theta_k : \textit{nonholonomic}\ \text{basis}, \tag{b}$$

into the fundamental inertial rotational transitivity equation

$$d(\delta\boldsymbol{\chi})/dt - \delta\boldsymbol{\omega} = \boldsymbol{\omega} \times \delta\boldsymbol{\chi}. \tag{c}$$

We find, successively,

(i) *Left* side:

$$d(\delta\boldsymbol{\chi})/dt - \delta\boldsymbol{\omega} = \sum \big[(\partial\boldsymbol{\omega}/\partial\omega_k)^{\cdot}\, \delta\theta_k + (\partial\boldsymbol{\omega}/\partial\omega_k)(\delta\theta_k)^{\cdot}\big]$$
$$\qquad - \sum \big[(\partial\boldsymbol{\omega}/\partial q_k)\, \delta q_k + (\partial\boldsymbol{\omega}/\partial\omega_k)\, \delta\omega_k\big]$$

[and setting $\delta q_k = \sum A_{kl}\, \delta\theta_l \equiv \sum (\partial v_k/\partial\omega_l)\, \delta\theta_l$ (definition of the A_{kl})]

$$= \sum \Big[(\partial\boldsymbol{\omega}/\partial\omega_k)^{\cdot} - \sum A_{lk}(\partial\boldsymbol{\omega}/\partial q_l)\Big]\delta\theta_k + \sum (\partial\boldsymbol{\omega}/\partial\omega_l)[(\delta\theta_l)^{\cdot} - \delta\omega_l]$$

[recalling the $\partial \ldots/\partial\theta_k$ definition (2.9.30a); and setting (as in pr. 2.10.5)

$(\delta\theta_l)^{\cdot} - \delta\omega_l = \sum h^l{}_k\, \delta\theta_k$ (definition of the $h^l{}_k$)]

$$\equiv \sum \big[(\partial\boldsymbol{\omega}/\partial\omega_k)^{\cdot} - \partial\boldsymbol{\omega}/\partial\theta_k\big]\delta\theta_k + \sum\sum (\partial\boldsymbol{\omega}/\partial\omega_l) h^l{}_k\, \delta\theta_k$$
$$\equiv \sum \Big[E_k{}^*(\boldsymbol{\omega}) + \sum h^l{}_k(\partial\boldsymbol{\omega}/\partial\omega_l)\Big]\delta\theta_k. \tag{d}$$

(ii) *Right* side:

$$\boldsymbol{\omega} \times \delta\boldsymbol{\chi} = \boldsymbol{\omega} \times \left(\sum (\partial\boldsymbol{\omega}/\partial\omega_k)\, \delta\theta_k\right) = \sum \left[\boldsymbol{\omega} \times (\partial\boldsymbol{\omega}/\partial\omega_k)\right] \delta\theta_k; \tag{e}$$

and, therefore, equating the right sides of (d) and (e), we obtain the identity

$$d/dt(\partial\boldsymbol{\omega}/\partial\omega_k) - \partial\boldsymbol{\omega}/\partial\theta_k + \sum h'_k(\partial\boldsymbol{\omega}/\partial\omega_l) = \boldsymbol{\omega} \times (\partial\boldsymbol{\omega}/\partial\omega_k), \tag{f}$$

or, in terms of the quasi vectors $\boldsymbol{\varepsilon}_k = \boldsymbol{\varepsilon}_k(q)$,

$$d\boldsymbol{\varepsilon}_k/dt - \partial\boldsymbol{\omega}/\partial\theta_k = \boldsymbol{\omega} \times (\partial\boldsymbol{\omega}/\partial\omega_k) - \sum h^l_k \boldsymbol{\varepsilon}_l. \tag{g}$$

Finally, since $d\boldsymbol{\varepsilon}_k/dt = d'\boldsymbol{\varepsilon}_k/dt + \boldsymbol{\omega} \times \boldsymbol{\varepsilon}_k$, (g) takes the *body-axes* form:

$$d'\boldsymbol{\varepsilon}_k/dt - \partial\boldsymbol{\omega}/\partial\theta_k \equiv d'/dt(\partial\boldsymbol{\omega}/\partial\dot{\theta}_k) - \partial\boldsymbol{\omega}/\partial\theta_k \equiv E_k{}^{*\prime}(\boldsymbol{\omega}) = -\sum h^l_k \boldsymbol{\varepsilon}_l, \tag{h}$$

which is a special case of the transitivity equation (2.10.25).

[Here too, we point out the differences between the *notation*:

$$\partial\boldsymbol{\omega}(q, \omega)/\partial\theta_l \equiv \sum \left[\partial\boldsymbol{\omega}(q, \omega)/\partial q_k\right](\partial v_k/\partial\omega_l), \tag{i}$$

and the vector *transformation* (by chain rule):

$$\partial\boldsymbol{\omega}(q, \omega)/\partial\omega_l = \sum \left[\partial\boldsymbol{\omega}(q, v)/\partial v_k\right](\partial v_k/\partial\omega_l) \qquad \text{or} \qquad \boldsymbol{\varepsilon}_l = \sum A_{kl}\, \boldsymbol{e}_k.] \tag{j}$$

See also Papastavridis, 1992.

Problem 2.13.7 By direct substitution of the representations

$$d\boldsymbol{\chi} \equiv \boldsymbol{\omega}\, dt = \sum \boldsymbol{\varepsilon}_k\, d\theta_k \qquad \text{and} \qquad \delta\boldsymbol{\chi} = \sum \boldsymbol{\varepsilon}_k\, \delta\theta_k \tag{a}$$

into the earlier inertial rotational transitivity equation [ex. 2.13.11, eq. (b)]

$$d(\delta\boldsymbol{\chi}) - \delta(d\boldsymbol{\chi}) = d\boldsymbol{\chi} \times \delta\boldsymbol{\chi}, \tag{b}$$

and some simple differentiations, show that

$$\partial\boldsymbol{\varepsilon}_k/\partial\theta_l - \partial\boldsymbol{\varepsilon}_l/\partial\theta_k + \sum \eta^b{}_{kl}\, \boldsymbol{\varepsilon}_b = \boldsymbol{\varepsilon}_l \times \boldsymbol{\varepsilon}_k, \tag{c}$$

where these special Hamel coefficients $\eta^b{}_{kl}$ are defined by $d(\delta\theta_b) - \delta(d\theta_b) = \sum\sum \eta^b{}_{kl}\, d\theta_l\, \delta\theta_k$.

Example 2.13.14 *Angular Acceleration.* Let us consider *intermediate* axes $\blacklozenge\!-\!\boldsymbol{u}_k$ rotating with inertial angular velocity $\boldsymbol{\Omega} = \sum \Omega_k\, \boldsymbol{u}_k$. If the inertial angular velocity of a rigid body, resolved along these axes, is $\boldsymbol{\omega} = \sum \omega_k\, \boldsymbol{u}_k$ then its inertial angular *acceleration* equals

$$\boldsymbol{\alpha} \equiv d\boldsymbol{\omega}/dt = d'\boldsymbol{\omega}/dt + \boldsymbol{\Omega} \times \boldsymbol{\omega} = d'\boldsymbol{\omega}/dt - \boldsymbol{\omega}_o \times \boldsymbol{\omega}, \tag{a}$$

where $d'\boldsymbol{\omega}/dt = \sum (d\omega_k/dt)\boldsymbol{u}_k$, and $\boldsymbol{\omega}_o \equiv \boldsymbol{\omega} - \boldsymbol{\Omega} =$ angular velocity of body relative to the *intermediate axes*.

Applying this result to the earlier case of *semimobile* axes $\blacklozenge\text{-}\boldsymbol{i}'\boldsymbol{j}'\boldsymbol{k}' \equiv \blacklozenge\text{-}\boldsymbol{u}_n\boldsymbol{u}_{n'}\boldsymbol{k}$ (ex. 2.13.9) where

$$\boldsymbol{\omega} = (\omega_\theta)\boldsymbol{u}_n + (\omega_\phi \sin\theta)\boldsymbol{u}_{n'} + (\omega_\psi + \omega_\phi \cos\theta)\boldsymbol{k} = \boldsymbol{\Omega} + \omega_\psi \boldsymbol{k} = \boldsymbol{\Omega} + \boldsymbol{\omega}_o, \tag{b}$$

[with the customary notations: $\omega_\phi \equiv d\phi/dt$, $\omega_\theta \equiv d\theta/dt$, $\omega_\psi \equiv d\psi/dt$]

that is, $\boldsymbol{\omega}_o = \omega_\psi \boldsymbol{k}$, we find, after some straightforward calculations,

$$\boldsymbol{\alpha} \equiv \alpha_n \boldsymbol{u}_n + \alpha_{n'}\boldsymbol{u}_{n'} + \alpha_k \boldsymbol{k}, \tag{c}$$

where

$$\begin{aligned} \alpha_n &\equiv d\omega_\theta/dt + \omega_\phi\,\omega_\psi \sin\theta, \\ \alpha_{n'} &\equiv (d\omega_\phi/dt)\sin\theta + \omega_\phi\,\omega_\theta \cos\theta - \omega_\theta\,\omega_\psi, \\ \alpha_k &\equiv (d\omega_\phi/dt)\cos\theta + d\omega_\psi/dt - \omega_\phi\,\omega_\theta \sin\theta. \end{aligned} \tag{d}$$

Let the reader repeat the above for the *semifixed* axes $\blacklozenge\text{-}\boldsymbol{u}_n\boldsymbol{u}_N\boldsymbol{K}$, where

$$\begin{aligned} \boldsymbol{\omega} &= (\omega_\phi \boldsymbol{K} + \omega_\theta \boldsymbol{u}_n) + \omega_\psi \boldsymbol{k} = (\omega_\phi \boldsymbol{K} + \omega_\theta \boldsymbol{u}_n) + \omega_\psi(-\sin\theta\,\boldsymbol{u}_N + \cos\theta \boldsymbol{K}) \\ &= (\omega_\theta)\boldsymbol{u}_n + (-\omega_\psi \sin\theta)\boldsymbol{u}_N + (\omega_\phi + \omega_\psi \cos\theta)\boldsymbol{K} \\ &\equiv \omega_\phi \boldsymbol{K} + \boldsymbol{\omega}_o \equiv \boldsymbol{\Omega} + \boldsymbol{\omega}_o. \end{aligned} \tag{e}$$

[For *matrix* forms of rigid-body accelerations, see (1.11.9a ff.); also Lur'e (1968, pp. 68–72).]

3

Kinetics of Constrained Systems

(i.e., Lagrangean Kinetics)

Where we may appear to have rashly and needlessly interfered with methods and systems of proof in the present day generally accepted, we take the position of Restorers, and not of Innovators.

(Thomson and Tait, 1867–1912, Preface, p. vi)

[A] work of which the *unity of method* is one of the most striking characteristics.... That which most distinguishes the plan of this treatise from the usual type is the *direct application of the general principle* to each particular case.

(Gibbs, 1879, 3rd footnote, emphasis added; the work/treatise Gibbs refers to is Lagrange's *Mécanique Analytique*, and the principle is Lagrange's Principle)

[T]he author . . . again and again . . . experienced the extraordinary elation of mind which accompanies a preoccupation with the basic principles and methods of analytical mechanics.

(Lanczos, 1970, p. vii)

3.1 INTRODUCTION

This is the key chapter of the entire book; and since it is based on chapter 2, it should be read *after* the latter. We begin with a detailed coverage of the *two* fundamental principles, or pillars, of Lagrangean analytical mechanics:

(i) The principle of Lagrange (and its *velocity* form known as the central equation); and
(ii) The principle of relaxation of the constraints.

From these two, with the help of *virtual displacements*, and so on (§2.5 ff.), we, subsequently, obtain all possible *kinetic energy–based* (*Lagrangean*) and *acceleration energy–based* (*Appellian*) *equations of motion* of holonomic and/or Pfaffian (possibly nonholonomic) systems; in holonomic and/or nonholonomic variables, with/without constraint reactions; such as the equations of Routh–Voss, Maggi, Hamel, and Appell, to name the most important.

Next, applying standard mathematical transformations to these equations, we obtain the *theorem of work–energy* in its various forms; that is, in holonomic and/or nonholonomic variables, with/without constraint reactions, and so on. This concludes the first, general, part of the chapter (§3.1–12). The second and third parts apply the previous Lagrangean and Appellian methods/principles/equations, respectively, to the *rigid body* (§3.13–15) and to *noninertial frames of reference* (*or moving axes*) (§3.16). The chapter ends with (i) a concise discussion of the *servo-*, or *control*, constraints of Beghin–Appell (§3.17); and (ii) two Appendices on the historical evolution of (some of) the above principles/equations of motion, and their relations to virtual displacements and the confusion-laden principle of d'Alembert–Lagrange.

As with the previous chapters, a large number of completely solved examples and problems with their answers and/or helpful hints, many of them kinetic continuations of corresponding kinematical examples and problems of chapter 2, have been appropriately placed throughout this chapter.

For complementary reading, we recommend (alphabetically): Butenin (1971), Dobronravov (1970, 1976), Gantmacher (1966/1970), Hamel (1912/1922(b), 1949), Kil'chevskii (1977), Lur'e (1961/1968), Mei (1985, 1987(a)), Mei and Liu (1987), Neimark and Fufaev (1967/1972), Nordheim (1927), Pars (1965), Poliahov et al. (1985), Prange (1935), Synge (1960). As with chapter 2, we are unaware of any other single exposition, in English, comparable to this one in the range of topics covered. Only Hamel (1949), Mei et al. (1991) and Neimark and Fufaev (1967/1972) cover major portions of the material treated here.

3.2 THE PRINCIPLE OF LAGRANGE (LP)

We begin with a finite mechanical system S consisting of particles $\{P\}$; each of mass dm, inertial acceleration $\boldsymbol{a} \equiv d\boldsymbol{v}/dt \equiv d^2\boldsymbol{r}/dt^2$, and each obeying the Newton–Euler equation of motion (§1.4):

$$dm\, \boldsymbol{a} = d\boldsymbol{f}, \tag{3.2.1}$$

where $d\boldsymbol{f}$ = *total force acting on P*. As explained in chapter 2, the *continuum* notation for particle quantities, employed here, simplifies matters, since it allows us to reserve all indices (to be introduced below) for *system* quantities.

The Force Classification

Now, and here we start parting company with the Newton–Euler mechanics, we decompose $d\boldsymbol{f}$ into *two* parts: (i) a total *physical*, or *impressed, force* $d\boldsymbol{F}$, and (ii) a total *constraint force*, or *constraint reaction*, $d\boldsymbol{R}$:

$$d\boldsymbol{f} = d\boldsymbol{F} + d\boldsymbol{R}. \tag{3.2.2}$$

Let us elaborate on these fundamental concepts:

(i) By constraint reactions, on our particle P, we shall understand (external and/or internal) forces, due solely to the (external and/or internal) geometrical and/or kinematical constitution of the system S; that is, forces caused exclusively by the prescribed (external and/or internal) constraints of S, and whose raison d'être is the preservation of these constraints. As a result, such forces are (a) *passive* (i.e., they appear only when absolutely needed; see below), and (b) expressible only through these constraints (since, by their definition, they contain neither physical constants nor material functions/coefficients). Therefore, these reactions become fully known only *after* the motion of S (under possible additional, nonconstraint forces and initial conditions) has been found. Examples of constraint reactions are: inextensible cable tensions, internal forces in a rigid body, normal forces among contacting (rolling/sliding/pivoting/nonpivoting) rigid bodies, and rolling (or static) friction.

(Generally, constraints and their reactions are classified, on the basis of the precise physical manner by which they are maintained, as *passive*, or as *active*. Except §3.17, where the latter are elaborated, this chapter deals only with *passive constraints/reactions*.)

(ii) By physical or impressed forces, on our particle P, we shall understand *all other (external and/or internal, nonconstraint) forces* acting on it, which means that [since the *total force on P* is determined through *variables describing the geometrical/kinematical and physical state of the rest of the matter surrounding that particle* (recalling §1.4)] the impressed forces depend, *at least partially*, on physical, or material, constants, unrelated to the constraints, and which can be determined only experimentally. Examples of such constants are: mass, gravitational constant, elastic moduli, viscous and/or dry friction coefficients, readings of the scale of a barometer or manometer; and examples of physical/impressed forces are gravity (weight), elastic (spring) forces, viscous damping forces, steam pressure, slipping (or sliding, or kinetic) friction [see remark (iii) below].

In other words, the impressed forces are forces expressed by material, or *constitutive*, equations, that contain those physical constants, and are assumed to be valid for any motion of the system; physical means *physically (functionally) given*—it does not mean that the values of these forces are necessarily known ahead of time.

In sum: *Impressed forces are given by constitutive equations, while reactions are not; but, in general, both these forces require, for their complete determination, knowledge of the subsequent motion of the system* (which, in turn, requires solution of an initial-value problem; namely, that of its equations of motion plus initial conditions).

Impressed forces are also, variously, referred to as *(directly) applied, active, acting, assigned, given, known* (where the last two terms have the meaning described above—see also remarks (iii) and (iv) below). In addition, the great physicist Planck (1928, pp. 101–103) calls our impressed forces "*treibende*" (driving, or propelling), while the highly instructive Langner (1997–1998, p. 49) proposes the rare but conceptually useful terms "*urgente*" (urging) for the impressed forces, and "*cogente*" (cogent, convincing) for the constraint forces. We follow Hamel (1949, pp. 65, 82, 517, 551), who calls impressed forces "*physikalisch gegebene*" (physically given) or "*eingeprägte*"; also Sommerfeld (1964, pp. 53–54), who calls them "forces of physical origin."

REMARKS

(i) From the viewpoint of continuum mechanics, practically all forces are physical (i.e., impressed); for example, an inextensible cable tension can be viewed as the limit of the tension of an elastic cable, or rubber band, whose modulus is getting higher and higher ($\rightarrow \infty$); and a rigid body can be viewed as a very stiff, practically strainless, deformable body. But there is also the exactly opposite viewpoint: kinetic and statistical theories of matter explain macroscopic phenomena, such as friction, viscosity, rust, by the motion of large numbers of smooth molecules, atoms, and so on. Their 19th century forerunners (Kelvin, Helmholtz, et al.) even tried to reduce the internal potential energy of bodies to the kinetic energy of a number of spinning "molecular gyrostats" strategically located inside them—see, for example, Gray (1918, chap. 8). And there is, of course, general relativity, which, continuing traditions of forceless mechanics, initiated by Hertz et al., set out to geometrize gravity completely; that is, *replace tactile mechanics by a visual mechanics*, albeit in a four-dimensional "space." For the modest purposes of macroscopic earthly mechanics, the impressed/constraint force division is both logically consistent and practically useful (economical), and so we uphold it throughout this book.

(ii) The decomposition (3.2.2), what Hamel (1949, p. 218) calls "d'Alembertsche Ansatz" ($\sim$ initial proposition), is the hallmark of analytical mechanics. Expressing system accelerations as partial/total derivatives of kinetic energies with respect to system coordinates, velocities, and time (§3.3) is a welcome but *secondary* characteristic of Lagrangean analytical mechanics; the primary one is the decomposition (3.2.2) and its consequences with regard to the equations of motion. By contrast, the Newton–Euler mechanics decomposes $d\boldsymbol{f}$ into (a) a total *external* force $d\boldsymbol{f}_e$ (= force originating, even partially, from *outside* of our system S), and (b) a total *internal*, or *mutual*, force $d\boldsymbol{f}_i$ (= force due *exclusively* to the rest of S, on its generic particle P):

$$d\boldsymbol{f} = d\boldsymbol{f}_e + d\boldsymbol{f}_i. \tag{3.2.3}$$

The connection between (3.2.2) and (3.2.3) is easily seen by decomposing $d\boldsymbol{F}(d\boldsymbol{R})$ into an external part $d\boldsymbol{F}_e(d\boldsymbol{R}_e)$ and an internal part $d\boldsymbol{F}_i(d\boldsymbol{R}_i)$, and then rearranging à la (3.2.3); that is, successively,

$$\begin{aligned} d\boldsymbol{f} &= d\boldsymbol{F} + d\boldsymbol{R} = (d\boldsymbol{F}_e + d\boldsymbol{F}_i) + (d\boldsymbol{R}_e + d\boldsymbol{R}_i) \\ &= (d\boldsymbol{F}_e + d\boldsymbol{R}_e) + (d\boldsymbol{F}_i + d\boldsymbol{R}_i) \equiv d\boldsymbol{f}_e + d\boldsymbol{f}_i, \end{aligned} \tag{3.2.4}$$

where

$$d\boldsymbol{f}_e \equiv d\boldsymbol{F}_e + d\boldsymbol{R}_e \qquad \text{and} \qquad d\boldsymbol{f}_i \equiv d\boldsymbol{F}_i + d\boldsymbol{R}_i. \tag{3.2.4a}$$

The decompositions (3.2.2) and (3.2.3), although physically different, may, for some special systems, coincide. For example, in a *free* (i.e., *externally unconstrained*) *rigid body* all *external forces* are *impressed* (i.e., *external reactions* = 0), and all *internal forces* are *reactions* (i.e., *internal impressed forces* = 0). The coincidence of external forces with impressed forces and of internal forces with reactions in this popular and well-known system is, probably, responsible for the frequent confusion and error accompanying d'Alembert's principle (detailed below), even in contemporary dynamics expositions.

(iii) *Rolling* friction should be counted as a *constraint reaction* because it is expressed by a geometrical/kinematical condition, not by a constitutive equation; while *slipping friction* should be counted as an *impressed force* because, according to the well-known Coulomb–Morin friction "law," it depends both on the contact condition (through the normal force, which is in both cases a constraint reaction) *and* on the physical properties of the contacting surfaces (through the kinetic friction coefficient). (That slipping friction is governed by a physical *inequality* does not affect our force classification.) The above apply to the (possible) rolling/slipping and pivoting/non-pivoting *couples*.

The difference between rolling and slipping friction, from the viewpoint of analytical mechanics (principle of virtual work, etc.), has been a source of considerable confusion and error, even among the better authors on the subject.

(iv) The force decomposition (3.2.2) is completely analogous to that occurring in continuum mechanics. For instance, in an incompressible (i.e., internally constrained) elastic solid, the total stress (force) consists of a "hydrostatic pressure" or "reaction stress" term (constraint reaction), plus an "elastic stress" term (impressed force) expressed by a constitutive equation/function of the elastic moduli (material constants) and the strains (motion → deformation), and it is assumed to be valid for any motion of that system. In general, the values of the stresses, both "incompressible/pressure" and "elastic" parts, for specific initial and boundary conditions, are found after solving that particular "initial- and boundary-value problem"; namely, the equations of motion of the solid plus its initial and boundary conditions.

HISTORICAL

The fundamental decomposition (3.2.2) seems to have been first given by Delaunay (1856); see, for example (alphabetically): Rumyantsev (1990, p. 268), Stäckel (1905, p. 450, footnote 11a); also Hamel (1912, pp. 81–82, 301–302, 457–458, 469–470), Heun [1902 (a, d)], Pars (1953, pp. 447–448), Webster (1912, pp. 41–42, 63–65).

Example 3.2.1 *Let us Find the Most Important Internal/External and Impressed/Constraint Forces in a Diesel-Powered Electric Locomotive, Rolling on Rails.* These are as follows:

(i) *Gravity* and *air resistance* (*drag*) are both *external* (their cause lies outside the system locomotive), and *impressed* (both depend partially on the physical constants: g = *acceleration of gravity* and ρ = *air density*, respectively).

(ii) *Pressure of burnt diesel fuel* is *internal* (it originates within the engine's cylinders) and *impressed* (depends on the gas temperature, density, etc.).

(iii) Forces on *connecting rods* and other moving parts of the engine:

(a) If these bodies are considered *rigid*, the forces are *internal reactions*;
(b) If they are considered *flexible*, say *elastic*, these forces are *internal* but *impressed* (and to calculate them we must know their elastic moduli).

(iv) Forces between *axles and their wheel bearings* are *internal* (for obvious reasons) and *impressed* (due to the relative motion among them—no constraints).

(v) Friction forces between *wheels and rail* are *external* (caused, partially, by an external body, the rail) and *reactions* (due to the slippingless rolling of wheels), and this holds for both their tangential (friction) and normal components; however, in the case of slipping (skidding), the friction changes to an *external impressed* force (it depends, partially, on the wheel–rail friction coefficient).

Example 3.2.2 *Let us Identify and Classify the Key Forces on a Person Walking up a Rough Hilly Road.* The external forces needed to overcome the (also external) forces of gravity and air resistance are those generated by the road friction. The latter are reactions, since there is no relative motion (i.e., constraint) between the walker's shoes and the road surface.

Arguments of the Forces

In classical (Newtonian) mechanics, the force $d\boldsymbol{f}$ on a particle P, of a system S, can depend, at most, on its position, velocity, and time; and on those of other particles of S, or even outside of S; and also, on material functions/coefficients. But, as an *independent constitutive equation* (i.e., not by some artificial control law), $d\boldsymbol{f}$ *cannot* depend on the acceleration $\boldsymbol{a}$ of P (and/or its higher time derivatives). This, however, does not preclude the occurrence of such a dependence by *elimination*: in the course of solving the equations of motion, and so on, of a problem, it is possible to relate functionally *a* force with *an* acceleration; but that is a *mathematical* coupling, not an independent physical one.

[Pars (1965, pp. 11–12; also 24–25) has shown that if df depended on $\boldsymbol{a}$, then the *initial state* of P, that is, its initial position and velocity, would not determine its future uniquely; see also Rosenberg (1977, pp. 10–17); and Hamel (1949, p. 49). But in other areas of classical physics, for instance electrodynamics (e.g., radiation damping), such a non-Newtonian explicit $\boldsymbol{a}$-dependence does not create inconsistencies.]

Lagrange's Principle

Dotting each of (3.2.1) and (3.2.2) with the corresponding particle's inertial virtual displacement $\delta\boldsymbol{r}$ (§2.5 ff.) and then summing the resulting equations over all system particles, for a fixed generic time, we obtain

$$S\, dm\, \boldsymbol{a}\cdot\delta\boldsymbol{r} = S\, d\boldsymbol{F}\cdot\delta\boldsymbol{r} + S\, d\boldsymbol{R}\cdot\delta\boldsymbol{r}, \tag{3.2.5}$$

or, rearranging,

$$S\,(dm\,\boldsymbol{a} - d\boldsymbol{F})\cdot\delta\boldsymbol{r} + S\,(-d\boldsymbol{R})\cdot\delta\boldsymbol{r} = 0; \tag{3.2.6}$$

where [recall (§2.2.7 ff.)] the material sum $S(\ldots)$ is to be understood as a Stieltjes' integral extending over all the *continuously and/or discretely* distributed system particles and their geometric/kinematic/inertial/kinetic variables.

Equations (3.2.5, 6) do not contain anything physically new; that is, they result from (3.2.1, 2) by purely mathematical transformations. To make further progress towards the derivation of reactionless equations of motion, one of the key objectives of analytical mechanics, we now *postulate* that (for bilateral, or equality, or reversible, constraints)

$$-\delta' W_R \equiv S\,(-d\boldsymbol{R})\cdot\delta\boldsymbol{r} \equiv -S\, d\boldsymbol{R}\cdot\delta\boldsymbol{r} = 0; \tag{3.2.7}$$

in words: at each instant, the *(first-order) total virtual work* of the system of (external and internal) "lost" (or forlorn, or accessory) forces $\{-d\boldsymbol{R}\}$, $-\delta' W_R$, *vanishes.* Then, equations (3.2.5, 6) immediately reduce to the new and nontrivial *principle of d'Alembert in Lagrange's form*, or, simply and more accurately, *principle of Lagrange* (*LP*) for such constraints:

$$S\, dm\,\boldsymbol{a}\cdot\delta\boldsymbol{r} = S\, d\boldsymbol{F}\cdot\delta\boldsymbol{r} \qquad \text{or} \qquad S\,(dm\,\boldsymbol{a} - d\boldsymbol{F})\cdot\delta\boldsymbol{r} = 0; \tag{3.2.8}$$

what Lagrange calls "la formule générale de la Dynamique pour le mouvement d'un système quelconque de corps."

This fundamental differential variational equation states that *during the motion of a constrained system whose reactions, at each instant, satisfy the physical postulate (3.2.7), the total (first-order) virtual work of (the negative of) its "inertial forces"* $-\{-dm\,\boldsymbol{a}\} = \{dm\,\boldsymbol{a}\}$,

$$\delta I \equiv S\, dm\,\boldsymbol{a}\cdot\delta\boldsymbol{r}, \tag{3.2.9}$$

equals the *similar virtual work of its (external and internal) impressed forces* $\{d\boldsymbol{F}\}$,

$$\delta' W \equiv S\, d\boldsymbol{F}\cdot\delta\boldsymbol{r}; \tag{3.2.10}$$

that is,

$$\delta' W_R = 0 \Rightarrow \delta I = \delta' W. \qquad (3.2.11)$$

The entire Lagrangean kinetics is based on LP, equations (3.2.7–11). Let us, therefore, examine them closely.

• Another, equivalent, formulation of the above is the following: during the motion, the totality of the lost forces $\{-d\boldsymbol{R} = d\boldsymbol{F} - dm\,\boldsymbol{a}\}$ are, at each instant, in *equilibrium*; not in the elementary sense of zero force and moment, but in that of the virtual work equation (3.2.7) (see also chap. 3, appendix 2).

• Here, we must stress that the above equations, and associated virtual work conception of equilibrium, are the *contemporary* formulation and interpretation of d'Alembert's principle; and they are due, primarily, to Heun and Hamel (early 20th century). As such, they bear practically zero resemblance to the original *workless* exposition of d'Alembert (1743). The latter postulated what, again in contemporary terms, amounts to equilibrium of the $\{-d\boldsymbol{R}\}$ in the elementary (i.e., Newton–Euler) sense of *zero resultant force and moment*:

$$\begin{aligned} \mathcal{S}(-d\boldsymbol{R}) = \boldsymbol{0} &\Rightarrow \mathcal{S}(dm\,\boldsymbol{a} - d\boldsymbol{F}) = \boldsymbol{0}, \\ \mathcal{S}\boldsymbol{r} \times (-d\boldsymbol{R}) = \boldsymbol{0} &\Rightarrow \mathcal{S}\boldsymbol{r} \times (dm\,\boldsymbol{a} - d\boldsymbol{F}) = \boldsymbol{0}. \end{aligned} \qquad (3.2.12)$$

It is not hard to see that for a *rigid* body (what d'Alembert dealt with) (3.2.7) specializes to (3.2.12). Indeed, substituting into (3.2.7) the most general rigid virtual displacement, $\delta\boldsymbol{r} = \delta\boldsymbol{r}_\blacklozenge + \delta\boldsymbol{\chi} \times (\boldsymbol{r} - \boldsymbol{r}_\blacklozenge)$ [where $\blacklozenge$ = *generic body point* and $\delta\boldsymbol{\chi}$ = *virtual rigid body rotation* (recalling §1.10 ff.)] and simple vector algebra, we obtain, successively,

$$\begin{aligned} -\delta' W_R &= \mathcal{S}(-d\boldsymbol{R}) \cdot [\delta\boldsymbol{r}_\blacklozenge + \delta\boldsymbol{\chi} \times (\boldsymbol{r} - \boldsymbol{r}_\blacklozenge)] \\ &= \Big[\mathcal{S}(-d\boldsymbol{R})\Big] \cdot \delta\boldsymbol{r}_\blacklozenge + \Big[\mathcal{S}(\boldsymbol{r} - \boldsymbol{r}_\blacklozenge) \times (-d\boldsymbol{R})\Big] \cdot \delta\boldsymbol{\chi} \\ &\equiv (-\boldsymbol{R}) \cdot \delta\boldsymbol{r}_\blacklozenge + \boldsymbol{M}_\blacklozenge(-\boldsymbol{R}) \cdot \delta\boldsymbol{\chi} = 0, \end{aligned} \qquad (3.2.13)$$

from which, since $\delta\boldsymbol{r}_\blacklozenge$ and $\delta\boldsymbol{\chi}$ are arbitrary, (3.2.12) follows [and if $\mathcal{S}(-d\boldsymbol{R}) = \boldsymbol{0}$, then $\boldsymbol{M}_\blacklozenge(-\boldsymbol{R}) = \boldsymbol{M}_{\text{origin}}(-\boldsymbol{R})$]. If, further, the rigid body is *free*, that is, unconstrained, then, as explained earlier, *all its external* (*internal*) *forces are impressed* (*reactions*) (i.e., $\{d\boldsymbol{f}_e\} = \{d\boldsymbol{F}\}$ and $\{d\boldsymbol{f}_i\} = \{d\boldsymbol{R}\}$), and the above lead to the Eulerian principles of linear and angular momentum (recall §1.8.18):

$$\mathcal{S}d\boldsymbol{f}_e = \mathcal{S}dm\,\boldsymbol{a} \quad \text{and} \quad \mathcal{S}(\boldsymbol{r} - \boldsymbol{r}_\blacklozenge) \times d\boldsymbol{f}_e = \mathcal{S}(\boldsymbol{r} - \boldsymbol{r}_\blacklozenge) \times dm\,\boldsymbol{a}. \qquad (3.2.14)$$

It follows that, in studying the statics of free rigid bodies via virtual work, we only need include their external = impressed forces; and that is why here the methods of Newton–Euler and d'Alembert–Lagrange coincide and supply conditions that are both necessary and sufficient for equilibrium (see also Hamel, 1949, pp. 80–83).

This preoccupation of d'Alembert, and many others since him, with the special case of (systems of) rigid bodies and elementary *vector* equilibrium (3.2.12), has diverted attention from the far more general *scalar* virtual work equilibrium (3.2.7), which constitutes the essence of LP.

• In LP it is the *sum* $\delta' W_R \equiv \mathcal{S}d\boldsymbol{R} \cdot \delta\boldsymbol{r}$ that vanishes, and not necessarily each of its terms $d\boldsymbol{R} \cdot \delta\boldsymbol{r}$ separately; although this latter may happen in special cases.

For example, as explained above, in a free rigid body (3.2.7) reduces to $\delta' W_R \to (\delta' W)_{\text{internal forces}} = 0$, although individually $d\boldsymbol{f}_i \cdot \delta \boldsymbol{r}$ may not vanish.

• While the $dm\,\boldsymbol{a}$ are present wherever a mass is accelerated, the $d\boldsymbol{F}$ may act only at a few system particles.

• In general, $\delta' W_R$ and $\delta' W$ are *not* the exact (or perfect, or total) virtual differentials of some system "work/force functions" W_R and W, respectively; that is, in general, they are *quasi variables*, and that is the purpose of the accented delta δ' (recall §2.9 ff.). The same holds for δI, but here, for convenience, we will make an exception and leave it unaccented.

• For *unilateral* (or *inequality*, or *irreversible*) constraints, LP is enlarged from (3.2.7–11) to

$$S\, d\boldsymbol{R} \cdot \delta \boldsymbol{r} = S\,(dm\,\boldsymbol{a} - d\boldsymbol{F}) \cdot \delta \boldsymbol{r} \geq 0 \;\Rightarrow\; S\, dm\,\boldsymbol{a} \cdot \delta \boldsymbol{r} \geq S\, d\boldsymbol{F} \cdot \delta \boldsymbol{r}, \qquad (3.2.15)$$

or

$$\delta' W_R \geq 0 \;\Rightarrow\; \delta I \geq \delta' W. \qquad (3.2.15\text{a})$$

For example, in the case of a block resting under its own weight on a fixed horizontal table, the sole *impressed* force on the block, gravity, cannot perform *positive* virtual work; while the normal table *reaction* cannot perform *negative* virtual work: $\delta' W_R = -\delta' W \geq 0$.

Lagrange's Principle as a Constitutive Postulate

It must be stressed that LP, eqs. (3.2.7–11), is what is known in continuum mechanics as a *constitutive postulate* for the *nonphysical* part of the $d\boldsymbol{f}$'s, namely, the constraint reactions $\{d\boldsymbol{R}\}$; like Hooke's law in elasticity, or the Navier–Stokes law in fluid mechanics; hence, applying LP to a free (i.e., unconstrained) particle is like, say, applying the theory of elasticity to a rigid body! As such, LP is *not* a law of nature, like the Newton–Euler equation (3.2.1) (and its Cauchy form, in continuum mechanics), but subservient to them; if (3.2.1) can be likened to a constitution article, LP is a secondary law (say, a state law). Just as in continuum mechanics, where not all parts of the stress need be elastic, here in analytical mechanics too, *not all constraint reactions need satisfy (3.2.7)* (see §3.17). Those reactions that do, which is most of this book, we shall call *ideal* (or *perfect*, or *passive*, or *frictionless*).

In view of these facts, the frequently occurring expression "workless, or nonworking, constraints" must be replaced by the more precise one, *virtually workless constraints*. Indeed, under the most general kinematically admissible/possible particle displacement (§2.5)

$$d\boldsymbol{r} = \sum \boldsymbol{e}_k\, dq_k + \boldsymbol{e}_0\, dt, \qquad \text{where} \qquad \boldsymbol{e}_k \equiv \partial \boldsymbol{r}/\partial q_k, \quad \boldsymbol{e}_0 \equiv \partial \boldsymbol{r}/\partial t \; (\equiv \boldsymbol{e}_{n+1}), \qquad (3.2.16)$$

the corresponding (first-order, or elementary) work of the constraint reactions is

$$d' W_R \equiv S\, d\boldsymbol{R} \cdot d\boldsymbol{r} = \cdots = (d' W_R)_1 + (d' W_R)_2, \qquad (3.2.16\text{a})$$

where

$$(d'W_R)_1 \equiv \sum R_k\, dq_k, \qquad R_k \equiv \int dR\cdot e_k, \tag{3.2.16b}$$

$$(d'W_R)_2 \equiv R_0\, dt, \qquad R_0 \equiv \int dR\cdot e_0\ (\equiv R_{n+1}); \tag{3.2.16c}$$

while, under an equally general virtual displacement $\delta r = \sum e_k\, \delta q_k$, the corresponding work is

$$\delta' W_R \equiv \int dR\cdot \delta r = \cdots = \sum R_k\, \delta q_k = 0; \tag{3.2.16d}$$

and therefore, since $(d'W_R)_1$ and $\delta' W_R$ are mathematically equivalent ($dq \sim \delta q$),

$$(d'W_R)_1 = 0 \ \Rightarrow\ d'W_R = (d'W_R)_2 = \left[\int dR\cdot(\partial r/\partial t)\right] dt \neq 0. \tag{3.2.16e}$$

[In view of (3.2.16 ff.), it is, probably, better to think of *first-order* virtual work as *projection of the forces* in certain directions; and to forget all those traditional (and confusion-prone) definitions of it like "work of forces for a constraint compatible infinitesimal movement of the system."]

In sum: in general, the constraint reactions *are* working; even when virtually nonworking. *Actually, that is why the whole concept of virtualness was invented in analytical mechanics.* For example, let us consider a particle P constrained to remain on a rigid surface S, which undergoes a *given* motion. Then, the virtual work of the normal reaction exerted by S on P is zero, while the corresponding $d'W_R$ is not; but, if S is *stationary*, then both $\delta' W_R$ and $d'W_R$ vanish. From the viewpoint of continuum mechanics, the need for LP, or something equivalent, for the constraint reactions is relatively obvious.

Below, we present a simple such mathematical argument from the viewpoint of discrete mechanics. In an N-particle system with equations of motion [discrete counterparts of (3.2.1)],

$$m_P\, a_P = F_P + R_P \qquad (P = 1, \ldots, N), \tag{3.2.17a}$$

and assuming that the impressed F_P's are completely *known* functions of t, r, v (something that may not always be the case: e.g., sliding friction), we have $3N + 3N = 6N$ unknown scalar functions: (i) the $3N$ position vector components/coordinates $\{x_P(t), y_P(t), z_P(t)$: rectangular Cartesian components of $r_P\} \rightarrow \{d^2x_P/dt^2 = a_{P,x},\ d^2y_P/dt^2 = a_{P,y},\ d^2z_P/dt^2 = a_{P,z}$: rectangular Cartesian components of $a_P = d^2r_P/dt^2\}$, plus (ii) the $3N$ reaction force components $\{R_{P,x}, R_{P,y}, R_{P,z}\}$. Against these unknowns, we have available: (i) the $3N$ scalar equations of motion (3.2.17a), and (ii) a total of $h + m$ scalar equations of constraint (recall §2.2 ff.):

h *geometric*: $\phi_H(t, r_P) = 0 \qquad (H = 1, \ldots, h;\ P = 1, \ldots, N),$ (3.2.17b)

m *velocity* (possibly nonholonomic): $f_D(t, r_P, v_P) = 0 \quad (D = 1, \ldots, m;\ P = 1, \ldots, N);$ (3.2.17c)

that is, a total of $3N + h + m$ (differential) equations. Therefore, to make our problem determinate, we need $6N - (3N + h + m) = (3N - h) - m \equiv n - m \equiv f$ ($\equiv$ # *DOF in the small*) additional scalar equations. And here is where LP comes in: as shown later in this chapter, the *single energetic but variational equation*

$\delta' W_R = 0 \Rightarrow \delta I = \delta' W$ *produces precisely these f needed independent scalar equations* (unlike the single *actual, nonvariational*, work/energy theorem, which always produces only *one* such equation!); and the latter, along with initial/boundary conditions make the above constrained dynamical problem determinate, or closed.

This simple argument, *number of equations = number of unknowns* [probably originated by Lur'e (1968, pp. 245–248) and Gantmacher (1970, pp. 16–23)], shows clearly the *impossibility of building a general constrained system mechanics without additional physical postulates*, like LP, or something equivalent (it would be like trying to build a theory of elasticity without Hooke's law, or something similar relating stress to strain!), and thus lays to rest frequent but nevertheless erroneous claims that "analytical mechanics is nothing but a mathematically sophisticated rearrangement of Newton's laws."

In sum, analytical mechanics is both mathematically and physically different from the momentum mechanics of Newton–Euler. Schematically:

Lagrangean analytical mechanics = Newton–Euler laws

+ d'Alembert's physical postulate.

As Lanczos puts it: "Those scientists who claim that analytical mechanics is nothing but a mathematically different formulation of the laws of Newton must assume that [LP] is deducible from the Newtonian laws of motion. The author is unable to see how this can be done. Certainly the third law of motion, "action equals reaction," is not wide enough to replace [LP]" (1970, p. 77).

The above also show clearly that trying to prove LP is meaningless; although, in the past, several scientists have tried to do that (like trying to prove Hooke's law in elasticity!). These considerations also indicate that if we choose to decompose the total force $d\boldsymbol{f}$ according to some other physical characteristic, then we must equip that mechanics with appropriate constitutive postulates for (some of) the forces involved, so as to make the corresponding dynamical problem determinate. Thus, in the Newton–Euler mechanics, where, as we have already seen, $d\boldsymbol{f}$ is decomposed into *external* and *internal* parts, the system equations of motion—that is, the principles of linear and angular momentum—thanks to the additional *constitutive* postulate of action–reaction, contain only the *external* forces (and couples); without that postulate, the equations of motion would involve *all* the forces, and the corresponding problem would be, in general, indeterminate. And in the case of matter–electromagnetic field interactions (e.g., electroelasticity, magneto-fluid-mechanics), we must, similarly, either know all forces involved, or supplement the equations of motion (of Newton–Euler and Maxwell) with special electromechanical constitutive equations, so that we end up again with a determinate system of equations.

More on Lagrange's Principle as a Constitutive Postulate

Here is what the noted mechanics historian E. Jouguet says about the *physical* nature of Lagrange's Principle (freely translated):

> In sum, therefore, Huygens and Jacob Bernoulli implicitly admit that the *forces developed by the constraints in the case of motion, are, like the forces developed by the constraints in the case of equilibrium*, forces that do no work in the virtual displacements

> compatible with the constraints. There is here a new *physical* postulate. It could be quite possible that *the property of not doing work be true for the constraint forces during equilibrium and not for the constraint forces during motion;* the reaction of a fixed surface on a point could be normal if the point was in equilibrium, and inclined if the point was moving; the reaction of a surface on a point could be normal if the surface was fixed and oblique if it was moving or deformable. This new postulate expresses, to use the language of Mr. P. Duhem [a French master (1861–1916), particularly famous for his contributions to continuum thermodynamics/energetics (in the tradition of Gibbs), and the history/axiomatics of theoretical mechanics], that the constraints, that have already been supposed [statically] frictionless, are also without *viscosity*. (1908, pp. 195–196),

and

> The dynamics of systems with constraints rests therefore on the property of forces generated, during the motion, by the constraints, of not doing work in the virtual displacements compatible with the given constraints. This is an *experimental* property, and at the same time an experimental property distinct from those that we have found for the forces developed by the constraints in the case of equilibrium, because it introduces the condition *that the constraints are without viscosity*. (1908, p. 202, emphasis added).

When Are the Methods of Newton–Euler (NE) and d'Alembert–Lagrange (AL) Equivalent?

Since there is only one mechanics, this is a natural question, but not an easy one. To begin with, since NE divides forces into external and internal ("apples"), while AL divides them into impressed and reactions ("oranges"), we should not be surprised if, for general mechanical systems and forces, no such equivalence exists, or should be expected, at all stages of the formulation and solution of a problem.

Equivalence at the highest level of the fundamental principles *may* exist only for special systems and problems: that is, those for which (i) the *internal forces* (NE) coincide with those of *constraint* (AL), and (ii) the *external forces* (NE) coincide with the *impressed ones* (AL). The only such system that we are aware of, satisfying both (i) and (ii), is the earlier-examined *free rigid body*; and there we saw that LP leads to the NE principles of linear and angular momentum. For other systems where the internal forces may be (wholly or partly) impressed, for example, an elastic body, the NE principles do not follow from LP; the latter, as an *independent axiom*, says nothing about *impressed forces*. However, for a given system and forces, both methods of NE and AL do the job pledged by all classical descriptions of motion, which is, given (i) the external (NE) and impressed (AL) forces, along with (ii) the system's state at an "initial" instant (i.e., initial configuration and velocities = initial conditions), and (iii) appropriate constitutive postulates for its internal forces (NE) and constraint reactions (AL), respectively (and possibly other additional geometrical/kinematical/physical facts intrinsic to that problem), then both NE and AL are theoretically equally capable in predicting the subsequent motion of the system and its remaining unknown forces (although both approaches may not be equivalent laborwise, or from the important Machian viewpoint of conceptual economy). On these fundamental issues, see also the masterful treatment of Hamel (1909; 1927, pp. 8–10, 14–18, 23–27, 38–39; 1949, chap. 4 and pp. 513–524).

The above can be summarized in the following:

(i) *Force decomposition:*

- Newton–Euler (NE)
  - *External*
    - External *impressed*
    - External *reactions*
  - *Internal*
    - Internal *impressed*
    - Internal *reactions*
- D'Alembert–Lagrange (AL)
  - *Impressed*
    - Impressed *external*
    - Impressed *internal*
  - *Reactions*
    - *External* reactions
    - *Internal* reactions

(ii) *Whence the need for d'Alembert's principle:*

Unknown forces

NE: Internal $\{d\boldsymbol{f}_i\}$ → Discrete: *action–reaction* $\quad S\, d\boldsymbol{f}_i = \mathbf{0}, \quad S\, \boldsymbol{r} \times d\boldsymbol{f}_i = \mathbf{0}$

[→ Continuum: *Boltzmann's axiom*; i.e., symmetry of stress tensor]

AL: Reactions $\{d\boldsymbol{R}\}$ → *Lagrange's principle* $\quad S\, d\boldsymbol{R} \cdot \delta\boldsymbol{r} = 0$

(iii) *Consequences:*

NE: *Linear momentum*: $\quad S\, d\boldsymbol{f}_i + S\, d\boldsymbol{f}_e = S\, dm\, \boldsymbol{a}$

$$\Rightarrow S\, d\boldsymbol{f}_e = S\, dm\, \boldsymbol{a} \ \Rightarrow \ \boldsymbol{f}_e = m\,\boldsymbol{a}_G \quad (G = \text{mass center})$$

Angular momentum: $\quad S\, [\boldsymbol{r} \times (d\boldsymbol{f}_i + d\boldsymbol{f}_e)] = S\, (\boldsymbol{r} \times dm\, \boldsymbol{a})$

$$\Rightarrow S\, \boldsymbol{r} \times d\boldsymbol{f}_e = d/dt\Big[S\, (\boldsymbol{r} \times dm\, \boldsymbol{v})\Big]$$

AL: *Lagrange's principle*: $\quad [-\delta' W_R \equiv S\, d\boldsymbol{R} \cdot \delta\boldsymbol{r} = 0] + [dm\, \boldsymbol{a} = d\boldsymbol{F} + d\boldsymbol{R}]$

$$\Rightarrow S\, d\boldsymbol{F} \cdot \delta\boldsymbol{r} = S\, dm\, \boldsymbol{a} \cdot \delta\boldsymbol{r}$$

(iv) *Unknown force retrieval:*

NE: Principles of *rigidification* and *cut*
AL: Principle of *constraint relaxation* (Befreiungsprinzip, see below and §3.7)

(v) *Coincidence of NE with d'AL: free rigid body*

Free: External forces = Impressed forces; i.e., $\{d\boldsymbol{f}_e\} = \{d\boldsymbol{F}\} \qquad (\{d\boldsymbol{R}_e\} = \mathbf{0})$
Rigid: Internal forces = Constraint reactions; i.e., $\{d\boldsymbol{f}_i\} = \{d\boldsymbol{R}\} \qquad (\{d\boldsymbol{F}_i\} = \mathbf{0})$

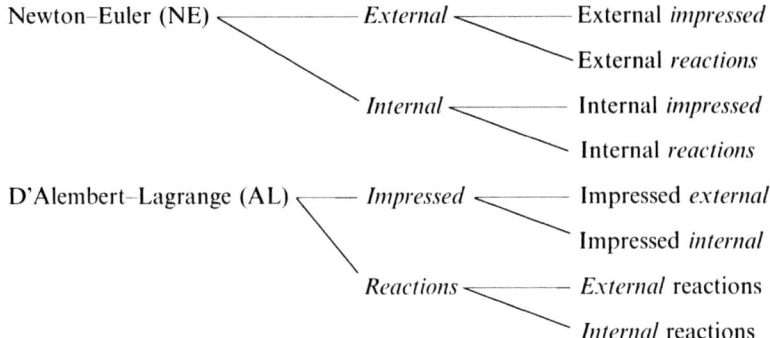

[Briefly (a) *Rigidification principle*: If a system is in equilibrium under impressed and constraint forces, it will remain in equilibrium if additional constraints are imposed on it so as to render it partly or wholly rigid; that is, deformable bodies in equilibrium can be treated just like rigid ones—both satisfy the same (necessary) conditions; (b) *Cut principle*: We can replace the action of two contiguous parts of the body by corresponding force systems (⇒ free body diagrams). Both principles are due to Euler. For details, see books on statics; also Papastavridis (EM, in prep.).]

Example 3.2.3 *Plane Mathematical Pendulum: Comparison Between Principles of Moment (Original d'Alembert) and Virtual Work (Lagrange).* Let us consider the motion of a mathematical pendulum, of length l and mass m, about a fixed point O on a vertical plane.

(i) According to the *original formulation of the principle* (first by *Jakob Bernoulli* and then by *d'Alembert*), the string reaction $\boldsymbol{S}$ on the oscillating particle P must be in equilibrium; that is, its moment about O must vanish:

$$\boldsymbol{M}_O \equiv \boldsymbol{r} \times \boldsymbol{S} = \boldsymbol{0} \;\Rightarrow\; \boldsymbol{S} \text{ must be parallel to the string } OP \; (\boldsymbol{r} \equiv \boldsymbol{OP}). \tag{a}$$

As a result, the second part of the principle—that is, *impressed forces minus inertia forces must be in equilibrium*, yields (with $\boldsymbol{W}$ = *weight of* P)

$$\begin{aligned} \boldsymbol{r} \times \boldsymbol{W} = \boldsymbol{r} \times (m\,\boldsymbol{a}) \;&\Rightarrow\; -(W)(l \sin\phi) = \{m[l(d^2\phi/dt^2)]\}(l) \\ &\Rightarrow\; d^2\phi/dt^2 + (g/l)\sin\phi = 0. \end{aligned} \tag{b}$$

(ii) According to *Lagrange's formulation of the principle*, the virtual work of $\boldsymbol{S}$ must vanish:

$$\delta' W_R = \boldsymbol{S} \cdot \delta\boldsymbol{r} = 0 \;\Rightarrow\; \boldsymbol{S} \text{ must be perpendicular to the virtual displacement of } P, \tag{c}$$

and since the latter is along the instantaneous tangent to P's circular path about O, we conclude that $\boldsymbol{S}$ must be parallel to OP, as before.

Hence, the second part of the principle—that is, *virtual work of impressed forces minus that of inertia forces must vanish*, yields

$$\begin{aligned} \boldsymbol{W} \cdot \delta\boldsymbol{r} = (m\,\boldsymbol{a}) \cdot \delta\boldsymbol{r} \;&\Rightarrow\; -(W \sin\phi)(l\,\delta\phi) = \{m[l(d^2\phi/dt^2)]\}(l\,\delta\phi) \\ &\Rightarrow\; d^2\phi/dt^2 + (g/l)\sin\phi = 0; \end{aligned} \tag{d}$$

that is, the moment condition (b) and the virtual work condition (d) differ only by an inessential factor $\delta\phi$, and thus they produce the same reactionless equation of motion.

In view of the extreme similarity, almost identity, of these two approaches in this and other rigid-body problems, we can see how, over the 19th and 20th centuries, various scientists came to confuse the zero moment method of James (Jakob) Bernoulli-d'Alembert {i.e., $S\,\boldsymbol{r} \times (d\boldsymbol{F} - dm\,\boldsymbol{a}) = \boldsymbol{0}$, in our notation} with the zero virtual work method of Lagrange {i.e., $S\,\delta\boldsymbol{r} \cdot (d\boldsymbol{F} - dm\,\boldsymbol{a}) = 0$}, and to view the former as equivalent to the latter. (Also, the fact that the string tension $\boldsymbol{S}$ is not zero—that is, that the constraint reaction is in equilibrium, not in the elementary sense of *zero moment and force*, but in the sense of *zero virtual work*, demonstrates clearly one of the drawbacks of the original d'Alembertian formulation of the principle.)

Example 3.2.4 *Motion of an Unconstrained System Relative to its Mass Center G, via Lagrange's Principle* (Adapted from Williamson and Tarleton, 1900, pp. 242–293). Substituting $\boldsymbol{r} = \boldsymbol{r}_G + \boldsymbol{r}_{/G} \Rightarrow \boldsymbol{a} = \boldsymbol{a}_G + \boldsymbol{a}_{/G}$ into LP, (3.2.8), and regrouping, we obtain

$$0 = \delta\boldsymbol{r}_G \cdot \left[\mathbf{S}\,(dm\,\boldsymbol{a}_G - d\boldsymbol{F})\right] + \delta\boldsymbol{r}_G \cdot \left(\mathbf{S}\,dm\,\boldsymbol{a}_{/G}\right) + \boldsymbol{a}_G \cdot \left(\mathbf{S}\,dm\,\delta\boldsymbol{r}_{/G}\right) + \mathbf{S}\,(dm\,\boldsymbol{a}_{/G} - d\boldsymbol{F}) \cdot \delta\boldsymbol{r}_{/G}, \tag{a}$$

from which, since $\mathbf{S}\,dm\,\boldsymbol{r}_{/G} = \boldsymbol{0} \Rightarrow \mathbf{S}\,dm\,\delta\boldsymbol{r}_{/G} = \boldsymbol{0}$ and $\mathbf{S}\,dm\,\boldsymbol{a}_{/G} = \boldsymbol{0}$, and the $\delta\boldsymbol{r}_G$, $\delta\boldsymbol{r}_{/G}$ are unrelated, we obtain

(i) $$\delta\boldsymbol{r}_G \cdot \left[\mathbf{S}\,(dm\,\boldsymbol{a}_G - d\boldsymbol{F})\right] = 0 \;\Rightarrow\; \mathbf{S}\,dm\,\boldsymbol{a}_G = \mathbf{S}\,d\boldsymbol{F},$$

that is,

$$m\,\boldsymbol{a}_G = \boldsymbol{F} \qquad (\textit{Principle of linear momentum}), \tag{b}$$

if $\delta\boldsymbol{r}_G$ is unconstrained, and

(ii) $$\mathbf{S}\,(dm\,\boldsymbol{a}_{/G} - d\boldsymbol{F}) \cdot \delta\boldsymbol{r}_{/G} = 0, \qquad \text{under the constraint } \mathbf{S}\,dm\,\delta\boldsymbol{r}_{/G} = \boldsymbol{0}. \tag{c}$$

Combining, or adjoining, the second of (c) into the first of (c) with the vectorial Lagrangean multiplier $\boldsymbol{\lambda} = \boldsymbol{\lambda}(t)$ (see §3.5), we readily get

$$dm\,\boldsymbol{a}_{/G} = d\boldsymbol{F} + \boldsymbol{\lambda}\,dm, \tag{d}$$

and, summing this over the system, we obtain

$$\mathbf{S}\,dm\,\boldsymbol{a}_{/G} = \mathbf{S}\,d\boldsymbol{F} + \boldsymbol{\lambda}\left(\mathbf{S}\,dm\right) \;\Rightarrow\; \boldsymbol{0} = \boldsymbol{F} + \boldsymbol{\lambda}\,m \;\Rightarrow\; \boldsymbol{\lambda} = -\boldsymbol{F}/m, \tag{e}$$

so that, finally, (d) becomes

$$dm\,\boldsymbol{a}_{/G} = d\boldsymbol{F} - dm(\boldsymbol{F}/m) \qquad (= d\boldsymbol{F} - dm\,\boldsymbol{a}_G, \text{ as expected}). \tag{f}$$

Example 3.2.5 *Sufficiency of the Statical Principle of Virtual Work (PVW) for the Equilibrium of Ideally Constrained Systems Deduced from LP.* In analytical statics (i.e., LP with $\boldsymbol{a} = \boldsymbol{0}$), the PVW states that *in a bilaterally constrained and originally motionless system (in an inertial frame), the vanishing of $\delta' W$ is a necessary and sufficient condition for it to remain in equilibrium in that frame.* In concrete applications, what we really employ is the *sufficiency* of the principle; that is, if $\delta' W = 0$, then the originally motionless system remains in equilibrium.

Here, we will start with LP as the basic axiom, set $\delta' W = 0$, and then derive sufficient conditions to maintain equilibrium; that is, go *from kinetics to statics.* Most authors proceed inversely—that is, go from statics to kinetics—and that makes the detection of the importance of the various constraints more difficult.

(i) *Necessary conditions*: If the system is in (inertial) equilibrium, then $\boldsymbol{a} = \boldsymbol{0}$, and therefore

$$\delta' W \equiv \mathbf{S}\,d\boldsymbol{F} \cdot \delta\boldsymbol{r} = 0 \qquad (\Rightarrow \delta' W_R \equiv \mathbf{S}\,d\boldsymbol{R} \cdot \delta\boldsymbol{r} = 0), \tag{a}$$

for $t_i \le t \le t_f$, where $t_i(t_f) = \textit{initial}\ (\textit{final})$ time and $t_f - t_i \equiv \tau$.

(ii) *Sufficiency conditions*: If $\delta' W = 0$, for $t_i \le t \le t_f$, then LP gives

$$\delta I \equiv S\, dm\, \boldsymbol{a} \cdot \delta \boldsymbol{r} = 0, \qquad \text{for} \quad t_i \le t \le t_f. \tag{b}$$

Let us investigate the consequences of (a, b) for equilibrium. Substituting into (b) the particle displacement $d\boldsymbol{r} - \boldsymbol{e}_0\, dt = (\boldsymbol{v} - \boldsymbol{e}_0)\, dt \equiv [\boldsymbol{v} - (\partial \boldsymbol{r}/\partial t)]dt$, which is *mathematically equivalent to its virtual displacement*, and cancelling $dt(\neq 0)$, we obtain

$$S\, dm\, \boldsymbol{a} \cdot \boldsymbol{v} = S\, dm\, \boldsymbol{a} \cdot \boldsymbol{e}_0, \tag{c}$$

and since $2T \equiv S\, dm\, \boldsymbol{v} \cdot \boldsymbol{v} \Rightarrow dT/dt = S\, dm\, \boldsymbol{a} \cdot \boldsymbol{v}$, we are readily led to the following *rheonomic-type power equation*:

$$dT/dt = S\, dm\, \boldsymbol{a} \cdot \boldsymbol{e}_0 = S\, d\boldsymbol{F} \cdot \boldsymbol{e}_0 + S\, d\boldsymbol{R} \cdot \boldsymbol{e}_0. \tag{d}$$

Integrating the above between t_i and $t(\le t_f)$, and setting $T_i \equiv T(t_i)$, $T \equiv T(t)$ yields

$$\Delta T \equiv T - T_i = \int_{t_i}^{t} \left(S\, dm\, \boldsymbol{a} \cdot \boldsymbol{e}_0 \right) dt; \tag{e}$$

which also follows from $\int_{t_i}^{t} \delta' W\, dt = 0$. Equation (e) leads to the following conclusions:

(a) If $\boldsymbol{e}_0 = \boldsymbol{0}$, then $\Delta T = 0$, and since $\boldsymbol{v}_i \equiv \boldsymbol{v}(t_i) = \boldsymbol{0} \Rightarrow T_i = 0$, it follows that $T = 0$ for some time $t - t_i (\le t_f - t_i)$; and from this, since $T =$ *positive definite* in the $\boldsymbol{v} \cdot \boldsymbol{v} = v^2$, we conclude that then *all* the $\boldsymbol{v}$'s vanish for $t - t_i\, (\le t_f - t_i)$; that is, the system remains in equilibrium in that time interval.

Conversely, *if* $\Delta T = 0$ *for any* $t > t_i$, *then* (e) *leads, for arbitrary systems, to* $\boldsymbol{e}_0 = \boldsymbol{0}$. In this case, (c) gives $\boldsymbol{v} = \boldsymbol{0}$; that is, equilibrium [while (a) yields $S\, d\boldsymbol{F} \cdot \boldsymbol{v} = S\, d\boldsymbol{F} \cdot \boldsymbol{e}_0 = 0$]. The consequences of this in the presence of additional Pfaffian constraints are discussed below.

(b) If $\boldsymbol{e}_0 \neq \boldsymbol{0}$, then, in general, $\Delta T \neq 0$; that is, the system moves away from its original equilibrium configuration, even though $\delta' W = 0$, for $t_i \le t \le t_f$, and $\boldsymbol{v}_i = \boldsymbol{0}$. Weaker special assumptions for equilibrium result for the following conditions:

(c) If $\boldsymbol{e}_0 \neq \boldsymbol{0}$, but $\int_{t_i}^{t} (S\, dm\, \boldsymbol{a} \cdot \boldsymbol{e}_0)\, dt = 0$; or

(d) If $\boldsymbol{e}_0 \neq \boldsymbol{0}$ but $\boldsymbol{a} \cdot \boldsymbol{e}_0 = 0$, for $t_i \le t \le t_f$.

Comparison with Gantmacher

Gantmacher's formulation of the PVW is as follows: "For some position (compatible with constraints) of a system to be an equilibrium position, it is necessary and sufficient that in this position the sum of the works of effective forces [our impressed forces] on any virtual displacements of the system be zero" and "If the constraints are nonstationary, then the term 'compatible with constraints' signifies that they are satisfied for any t if in them we put [our notation] $\boldsymbol{r} = \boldsymbol{r}_i$ and $\boldsymbol{v} = \boldsymbol{0}$" and "It is then assumed that [our] equation (a) holds for any value of t if in the expression for $d\boldsymbol{F}$ we put all $\boldsymbol{r} = \boldsymbol{r}_i$ and all $\boldsymbol{v} = \boldsymbol{0}$" (1970, p. 25). Let us relate this formulation with ours. By (2.5.2) $\boldsymbol{v} = \sum \boldsymbol{e}_k v_k + \boldsymbol{e}_0$. Hence, if $\boldsymbol{v} = \boldsymbol{0}$:

(i) If the v_k's are *unconstrained*, and since $\boldsymbol{e}_k \neq \boldsymbol{0}$, then $v_k = 0 \Rightarrow q_k =$ *constant* and $\boldsymbol{e}_0 = \boldsymbol{0}$—that is, the constraints are *stationary*—then the system *will* remain in equilibrium.

(ii) If, on the other hand, the v_k's are *constrained*, then, invoking the convenient representations (2.11.9, 13c, e), we have

$$\boldsymbol{v} \to \boldsymbol{v}_0 = \sum \boldsymbol{\beta}_I v_I + \boldsymbol{\beta}_0 = \mathbf{0} \Rightarrow v_i = 0 \;\Rightarrow\; q_I = constant, \quad \text{and} \quad \boldsymbol{\beta}_0 = \mathbf{0};$$

$$\boldsymbol{\beta}_0 = \boldsymbol{e}_0 + \sum b_D \boldsymbol{e}_D = \mathbf{0} \;\Rightarrow\; \boldsymbol{e}_0 = \mathbf{0}, \quad \text{and} \quad b_D = 0;$$

$$v_D = \sum b_{DI} v_I + b_D \;\Rightarrow\; v_D = 0 \;\Rightarrow\; q_D = constant \qquad (\Rightarrow q_k = constant).$$

In the light of the above, the PVW can be reformulated as follows: An originally motionless system remains in equilibrium if and only if (i) $\delta' W = 0$ and (ii) its *holonomic constraints* are *stationary* ($\boldsymbol{e}_0 \equiv \partial \boldsymbol{r}/\partial t = \mathbf{0}$) and its *Pfaffian constraints* are *catastatic* ($a_D = 0$ or $b_D = 0$). (The latter, however, may be nonstationary; and this explains Gantmacher's statement: "Note that in this case the virtual displacements ... may also be different for different t.")

REMARKS

(i) That "compatibility with constraints (during equilibrium)" leads to the above conclusions about them can be seen more clearly as follows. Let our system be subject to h holonomic constraints and m Pfaffian (holonomic and/or nonholonomic constraints):

$$\phi_H(t, \boldsymbol{r}) = 0, \qquad f_D \equiv \mathcal{S}\, \boldsymbol{B}_D(t, \boldsymbol{r}) \cdot \boldsymbol{v} + B_D(t, \boldsymbol{r}) = 0 \qquad (H = 1, \ldots, h;\; D = 1, \ldots, m). \tag{f}$$

By $d/dt(\ldots)$-differentiating the above, to make them explicit in both velocities and accelerations, we readily obtain [recalling *dot-product-of-tensor-definition* [(see 1.1.12d ff.)], in the first sum in (g2) below]:

$$\text{(a)}\quad d\phi_H/dt = \mathcal{S}\, (\partial \phi_H/\partial \boldsymbol{r}) \cdot \boldsymbol{v} + \partial \phi_H/\partial t = 0, \tag{g1}$$

$$\text{(b)}\quad d^2\phi_H/dt^2 = \mathcal{S}\, [(\partial^2 \phi_H/\partial \boldsymbol{r}\, \partial \boldsymbol{r}) : (\boldsymbol{v} \otimes \boldsymbol{v}) + (\partial \phi_H/\partial \boldsymbol{r}) \cdot \boldsymbol{a} + 2(\partial^2 \phi_H/\partial t\, \partial \boldsymbol{r}) \cdot \boldsymbol{v}] + \partial^2 \phi_H/\partial t^2 = 0, \tag{g2}$$

$$\text{(c)}\quad df_D/dt \equiv \mathcal{S}\, \{[(\partial \boldsymbol{B}_D/\partial \boldsymbol{r}) \cdot \boldsymbol{v} + (\partial \boldsymbol{B}_D/\partial t)] \cdot \boldsymbol{v} + \boldsymbol{B}_D \cdot \boldsymbol{a}\} + \mathcal{S}\, (\partial B_D/\partial \boldsymbol{r}) \cdot \boldsymbol{v} + \partial B_D/\partial t = 0. \tag{g3}$$

Now, since compatibility requires that, for $t_i \le t \le t_f$, eqs. (f–g3) should hold with $\boldsymbol{v} = \mathbf{0}$ *and* $\boldsymbol{a} = \mathbf{0}$ in them (just like the equations of motion), we readily obtain from the above the following conditions on these constraints:

$$\phi_H = 0, \tag{h1}$$

$$d\phi_H/dt = 0 \Rightarrow \partial \phi_H/\partial t = 0, \tag{h2}$$

$$d^2\phi_H/dt^2 = 0 \Rightarrow \partial^2 \phi_H/\partial t^2 = 0; \tag{h3}$$

$$f_D = 0 \Rightarrow B_D = 0, \tag{h4}$$

$$df_D/dt = 0 \Rightarrow \partial B_D/\partial t = 0; \tag{h5}$$

that is, for $t_i \le t \le t_f$, the holonomic constraints must be *stationary*, and the Pfaffian ones must be *catastatic*, as found earlier.

(ii) If we assume Earth to be *inertial*, then an *Earth-bound* system is scleronomic. But if we assume it to have a *given* motion, then our system is *rheonomic*. In both cases, the contact (nongravitational) forces from the Earth to that system are *external reactions*. If, finally, the Earth interacts with our system, then the two *taken together* constitute a *scleronomic* system whose internal forces are impressed (see also Nordheim, 1927, pp. 47–49).

Example 3.2.6 *Nonideal Constraints.* Let us consider a particle P of mass m, moving under an impressed force $\boldsymbol{F}$ and subject to the velocity constraint

$$f(t, \boldsymbol{r}, \boldsymbol{v}) = 0. \tag{a}$$

If the reaction created by (a) is $\boldsymbol{R}$, then the equation of motion of P is

$$m\,\boldsymbol{a} = \boldsymbol{F} + \boldsymbol{R}. \tag{b}$$

To relate the constraint equation to the reaction, so as to incorporate (a) into (b), we $d/dt(\ldots)$-differentiate the former:

$$f = 0 \Rightarrow df/dt = \partial f/\partial t + (\partial f/\partial \boldsymbol{r})\cdot \boldsymbol{v} + (\partial f/\partial \boldsymbol{v})\cdot \boldsymbol{a} = 0, \tag{c}$$

and, therefore,

$$m\,\boldsymbol{a}\cdot(\partial f/\partial \boldsymbol{v}) = -m[\partial f/\partial t + (\partial f/\partial \boldsymbol{r})\cdot \boldsymbol{v}]; \tag{d}$$

but, also, from (b),

$$m\,\boldsymbol{a}\cdot(\partial f/\partial \boldsymbol{v}) = \boldsymbol{F}\cdot(\partial f/\partial \boldsymbol{v}) + \boldsymbol{R}\cdot(\partial f/\partial \boldsymbol{v}). \tag{e}$$

Equating the right sides of (d, e), thus eliminating the acceleration, and rearranging, we obtain

$$\boldsymbol{R}\cdot(\partial f/\partial \boldsymbol{v}) = -[m(\partial f/\partial t) + m(\partial f/\partial \boldsymbol{r})\cdot \boldsymbol{v} + \boldsymbol{F}\cdot(\partial f/\partial \boldsymbol{v})]. \tag{f}$$

Now, the most general solution of (f), for $\boldsymbol{R}$, is

$$\boldsymbol{R} = -(\partial f/\partial \boldsymbol{v})\big[m(\partial f/\partial t) + m(\partial f/\partial \boldsymbol{r})\cdot \boldsymbol{v} + \boldsymbol{F}\cdot(\partial f/\partial \boldsymbol{v})\big]\big/(\partial f/\partial \boldsymbol{v})^2 + \boldsymbol{T}, \tag{g}$$

where $\boldsymbol{T}$ = *arbitrary vector orthogonal to* $\partial f/\partial \boldsymbol{v}$. The above shows that, generally, the constraint reaction consists of *two* parts: (i) one *parallel* to $\partial f/\partial \boldsymbol{v}$:

$$\boldsymbol{N} = -(\partial f/\partial \boldsymbol{v})\big[m(\partial f/\partial t) + m(\partial f/\partial \boldsymbol{r})\cdot \boldsymbol{v} + \boldsymbol{F}\cdot(\partial f/\partial \boldsymbol{v})\big]\big/(\partial f/\partial \boldsymbol{v})^2 \equiv \lambda(\partial f/\partial \boldsymbol{v}) \tag{h}$$

[where λ = *Lagrangean multiplier*—see Lagrange's equations of the *first* kind, (§3.5)]; and (ii) one *normal* to it, $\boldsymbol{T}$.

If $\boldsymbol{T} = \boldsymbol{0}$, the constraint (a) is called *ideal*; and in that case, clearly, the equation of motion of the particle (b), under (a), becomes

$$m\,\boldsymbol{a} = \boldsymbol{F} - \big[\boldsymbol{F}\cdot(\partial f/\partial \boldsymbol{v}) + m(\partial f/\partial \boldsymbol{r})\cdot \boldsymbol{v} + m(\partial f/\partial t)\big]\,\big[(\partial f/\partial \boldsymbol{v})/(\partial f/\partial \boldsymbol{v})^2\big]. \tag{i}$$

To make the problem determinate, we, usually, introduce a *constitutive* equation between $\boldsymbol{N}$ and $\boldsymbol{T}$. For example, in the common case of dry (solid/solid) *sliding* friction, we postulate the following relation between their magnitudes:

$$T = \mu N = \mu|\lambda(\partial f/\partial \boldsymbol{v})|, \qquad \mu = \textit{coefficient of kinetic friction}. \tag{j}$$

Then, and with (g, h), eq. (b) becomes

$$m\boldsymbol{a} = \boldsymbol{F} + \lambda(\partial f/\partial \boldsymbol{v}) - \mu|\lambda(\partial f/\partial \boldsymbol{v})|\boldsymbol{u}, \tag{k}$$

where $\boldsymbol{u} = \boldsymbol{v}/|\boldsymbol{v}|$.

For further details and applications of (k) see Poliahov et al. (1985, pp. 152–170).

Problem 3.2.1 Continuing from the preceding example, show that if the constraint (a) has the *holonomic* form

$$\phi(t, \boldsymbol{r}) = 0, \tag{a}$$

then (h) and (i) reduce, respectively, to

$$\boldsymbol{N} = -(\partial\phi/\partial\boldsymbol{r})\left[m(\partial\dot{\phi}/\partial t) + m(\partial\dot{\phi}/\partial\boldsymbol{r})\cdot\boldsymbol{v} + \boldsymbol{F}\cdot(\partial\phi/\partial\boldsymbol{r})\right]/(\partial\phi/\partial\boldsymbol{r})^2 \equiv \lambda(\partial\phi/\partial\boldsymbol{r}) \tag{b}$$

and

$$m\boldsymbol{a} = \boldsymbol{F} - \left[\boldsymbol{F}\cdot(\partial\phi/\partial\boldsymbol{r}) + m(\partial\dot{\phi}/\partial\boldsymbol{r})\cdot\boldsymbol{v} + m(\partial\dot{\phi}/\partial t)\right]\left[(\partial\phi/\partial\boldsymbol{r})/(\partial\phi/\partial\boldsymbol{r})^2\right]. \tag{c}$$

(See also Lagrange's equations of the *first* kind, in §3.5.)

Introduction to the Principle of Relaxation of the Constraints (PRC)

Before we embark into a detailed quantitative discussion of Lagrange's Principle (LP) and its derivative equations of motion, let us discuss briefly the *second* pillar of analytical mechanics, the *principle of relaxation of the constraints* (*Befreiungsprinzip*; Hamel, 1917). LP allows us to get rid of the constraint forces and, eventually, obtain *reactionless* equations of motion; and, historically, this has been considered (and *is*) one of the advantages of the method, especially in physics. However, in many engineering problems we *do* need to calculate these reactions, and thus the question arises: How do we achieve this with such a reaction-eliminating Lagrangean formalism?

Here is where PRC comes in: to retrieve a(ny) particular, external and/or internal, "lost" reaction we, hypothetically, free, or relax, the system of its particular, external and/or internal, geometrical and/or motional, constraint(s) causing that reaction; that is, we, mentally, allow the formerly rigid, or unyielding, constraint(s) to deform, or become flexible, relaxed, so that the former reaction becomes an *impressed force* that depends on the *deformation of the violated constraint via some constitutive equation*. Then we calculate its virtual work, add it to $\delta' W$, and apply LP: $(\delta I = \delta' W)_{relaxed\ system}$; and so on and so forth, for as many reactions as needed (one, or more, or all, at a time). Last, since in our model the constraints are rigid, we enforce them in the *final* stage of the differential equations of motion. The mathematical expression of PRC is the very well-known and widely applied *method of "undetermined," or Lagrangean, multipliers* (§3.5).

REMARKS

(i) Another, *mixed*, method is, first, to use LP to calculate the *reactionless* equations (and from them the *motion*), and to then use the method of Newton–Euler (NE) to calculate the external and/or internal reactions. This may be practically expedient, but it is not logically/conceptually satisfactory; it makes Lagrangean mechanics look incomplete.

(ii) The counterpart of PRC in the NE method is the following: if, for example, we want to calculate an *internal* force—that is, one that, due to the action–reaction postulate, drops out of the force/moment side in the NE principles of linear/angular momentum—then, applying Euler's *cut principle*, we choose an appropriate *new free-body diagram* so that the former internal force(s)/moment(s) becomes *external*, and then apply to these *new subsystems*, the NE principles.

3.3 VIRTUAL WORK OF INERTIAL FORCES (δI), AND RELATED KINEMATICO-INERTIAL IDENTITIES

Here we transform (3.2.9), $\delta I \equiv S\, dm\, \boldsymbol{a} \cdot \delta \boldsymbol{r}$, from particle variables to *system* variables; both holonomic and nonholonomic. (Actually, δI is the *negative* of the virtual work of the "inertial forces" $\{-dm\, \boldsymbol{a}\}$. We hope that this slight deviation from traditional terminology will not cause any problems.) Understandably, this relies critically on the kinematical results of chapter 2 and, therefore knowledge of that material is *absolutely necessary*. To obtain the most general system equations of motion from LP, we must use the most general expressions for $\boldsymbol{a}$ and $\delta \boldsymbol{r}$. We recall (§2.5 ff.) that these are (with $k, l = 1, \ldots, n$)

$$\begin{aligned}\delta \boldsymbol{r} &= \sum \boldsymbol{e}_k \delta q_k = \textit{holonomic}\ \text{variable representation}\\ &= \sum \boldsymbol{\varepsilon}_l\, \delta\theta_l = \textit{nonholonomic}\ \text{variable representation}\ (\equiv \delta \boldsymbol{r}^*)\\ \Big(&= \sum \boldsymbol{\varepsilon}_I\, \delta\theta_I,\ \text{under the constraints}\ \delta\theta_D = 0;\ D + 1, \ldots, m;\ I = m+1, \ldots, n\Big).\end{aligned} \tag{3.3.1}$$

where the fundamental *mixed basis vectors* $\{\boldsymbol{e}_k\}$ and $\{\boldsymbol{\varepsilon}_l\}$ are related by

$$\boldsymbol{e}_k \equiv \partial \boldsymbol{r}/\partial q_k = \sum a_{lk} \boldsymbol{\varepsilon}_l \ \Leftrightarrow\ \boldsymbol{\varepsilon}_l \equiv \partial \boldsymbol{r}/\partial \theta_l = \sum A_{kl} \boldsymbol{e}_k. \tag{3.3.1a}$$

1. Holonomic System Variables

Substituting the first of (3.3.1) into δI we obtain, successively,

$$\delta I \equiv S\, dm\, \boldsymbol{a} \cdot \delta \boldsymbol{r} = S\, dm\, \boldsymbol{a} \cdot \Big(\sum \boldsymbol{e}_k\, \delta q_k\Big) = \cdots = \sum E_k\, \delta q_k, \tag{3.3.2}$$

where $E_k \equiv S\, dm\, \boldsymbol{a} \cdot \boldsymbol{e}_k$: *holonomic* (k)th component of *system inertial "force"*

$$\begin{aligned}\Big[&\equiv S\, dm\, \boldsymbol{a} \cdot (\partial \boldsymbol{r}/\partial q_k) = S\, dm\, \boldsymbol{a} \cdot (\partial \boldsymbol{v}/\partial \dot{q}_k) = S\, dm\, \boldsymbol{a} \cdot (\partial \boldsymbol{a}/\partial \ddot{q}_k)\\ &\equiv S\, dm\, \boldsymbol{a} \cdot (\partial \boldsymbol{r}/\partial q_k) = S\, dm\, \boldsymbol{a} \cdot (\partial \boldsymbol{v}/\partial v_k) = S\, dm\, \boldsymbol{a} \cdot (\partial \boldsymbol{a}/\partial w_k)\Big].\end{aligned} \tag{3.3.3}$$

Now, E_k transforms, successively, as follows:

$$E_k \equiv S\, dm\, \boldsymbol{a} \cdot \boldsymbol{e}_k = S\, dm(d\boldsymbol{v}/dt) \cdot (\partial \boldsymbol{v}/\partial v_k)$$
$$= d/dt\Big[S\, dm\, \boldsymbol{v} \cdot (\partial \boldsymbol{v}/\partial v_k)\Big] - S\, [dm\, \boldsymbol{v} \cdot (d/dt)(\partial \boldsymbol{v}/\partial v_k)]$$
$$[\text{recalling identity (2.5.10)}:\ E_k(\boldsymbol{v}) \equiv d/dt(\partial \boldsymbol{v}/\partial v_k) - \partial \boldsymbol{v}/\partial q_k = \boldsymbol{0}]$$
$$= d/dt\Big[S\, dm\, \boldsymbol{v} \cdot (\partial \boldsymbol{v}/\partial v_k)\Big] - S\, dm\, \boldsymbol{v} \cdot (\partial \boldsymbol{v}/\partial q_k), \tag{3.3.4}$$

or, finally, with the help of the (inertial) *kinetic energy*

$$T \equiv S\,(1/2)(dm\, \boldsymbol{v} \cdot \boldsymbol{v}) = T(t, q, \dot{q}) \equiv T(t, q, v) \qquad [\text{since } \boldsymbol{v} = \boldsymbol{v}(t, q, v)], \tag{3.3.5}$$

we obtain

$$E_k = d/dt(\partial T/\partial v_k) - \partial T/\partial q_k \equiv d/dt(\partial T/\partial \dot{q}_k) - \partial T/\partial q_k \equiv E_k(T), \tag{3.3.6}$$

where

$$E_k(\ldots) = d/dt(\partial \ldots/\partial v_k) - \partial \ldots/\partial q_k \equiv d/dt(\partial \ldots/\partial \dot{q}_k) - \partial \ldots/\partial q_k:$$
$$(\text{holonomic Euler–Lagrange operator})_k. \tag{3.3.6a}$$

Equation (3.3.6) is a kinematico-inertial *identity*; that is, it holds always, independently of any possible additional constraints, as long as the q's are *holonomic coordinates*. Its cardinal importance to Lagrangean mechanics lies in the fact that it expresses system accelerations in terms of the *partial and total derivatives of a scalar energetic function of the system coordinates and velocities*, $T(t, q, v)$, *as if* the q's and $\dot{q}$'s $\equiv v$'s (and t) were *independent variables*. That is why we have reserved the *special notation* $E_k(T) \equiv E_k$ when that operator is applied to the *kinetic energy*; even though $E_k(\ldots)$ can be applied to any function of the q's, v's, and t. Also, (3.3.1–6a) clearly show the *indispensability of virtual displacements* (i.e., the $\boldsymbol{e}_k$ vectors) to Lagrangean mechanics/equations of motion [i.e., the particular T-based expression for the system inertia/acceleration given by (3.3.6)], whether the constraint reactions are ideal or not.

In sum: no $\boldsymbol{e}_k$'s, no Lagrangean equations, that is, for an arbitrary particle/system vector $\boldsymbol{z}_k \neq \boldsymbol{e}_k$,

$$S\, dm\, \boldsymbol{a} \cdot \boldsymbol{z}_k \neq (d/dt)(\partial T/\partial v_k) - \partial T/\partial q_k. \tag{3.3.6b}$$

This should put to rest once and for all false claims that "one can build Lagrangean mechanics without virtual displacements." The $\delta(\ldots)$ is not the issue; the $\boldsymbol{e}_k$ ($\rightarrow$ *projections*) are!

Let us collect the key kinematico-inertial identities involved here:

(a) $S\, dm\, \boldsymbol{v} \cdot \boldsymbol{e}_k = S\, dm\, \boldsymbol{v} \cdot (\partial \boldsymbol{v}/\partial v_k) = \partial T/\partial v_k \equiv \partial T/\partial \dot{q}_k \equiv p_k(t, q, v) = p_k$:

Holonomic (k)th component of system *momentum*; (3.3.7a)

(b) $S\, dm\, \boldsymbol{v} \cdot (d\boldsymbol{e}_k/dt) = S\, dm\, \boldsymbol{v} \cdot (\partial \boldsymbol{v}/\partial q_k) = \partial T/\partial q_k \equiv r_k(t, q, v) = r_k$:

Holonomic (k)th component of "associated, or momental, inertial force"; (3.3.7b)

(c) $$E_k \equiv dp_k/dt - r_k. \tag{3.3.7c}$$

[$p_k \equiv \partial T/\partial \dot{q}_k$ is the only kind of momentum that there is in analytical (Lagrangean and Hamiltonian) mechanics; and, as shown later, it comprises both the linear and angular momentum of the Newton–Euler mechanics.]

2. Nonholonomic System Variables

Substituting the second of (3.3.1) into δI, we obtain, successively with $\boldsymbol{a} = \boldsymbol{a}^* =$ *particle acceleration in nonholonomic variables* (and similarly for other quantities):

$$\delta I \equiv \mathsf{S}\, dm\, \boldsymbol{a} \cdot \delta \boldsymbol{r} = \mathsf{S}\, dm\, \boldsymbol{a}^* \cdot \left(\sum \boldsymbol{\varepsilon}_k\, \delta\theta_k\right) = \cdots = \sum I_k\, \delta\theta_k, \tag{3.3.8}$$

where $I_k \equiv \mathsf{S}\, dm\, \boldsymbol{a}^* \cdot \boldsymbol{\varepsilon}_k =$ *nonholonomic* (k)th component of *system inertial force*

$$\left[\equiv \mathsf{S}\, dm\, \boldsymbol{a}^* \cdot (\partial \boldsymbol{v}^*/\partial \omega_k) = \mathsf{S}\, dm\, \boldsymbol{a}^* \cdot (\partial \boldsymbol{a}^*/\partial \dot{\omega}_k),\ \text{recalling (2.9.35, 43)}\right]$$

$$= I_k(t, q, \omega, \dot{\omega}) \qquad [\text{since } \boldsymbol{a}^* = \boldsymbol{a}^*(t, q, \omega, \dot{\omega}) \text{ and } \boldsymbol{\varepsilon}_k = \boldsymbol{\varepsilon}_k(t, q)]. \tag{3.3.9}$$

From the invariance of δI: $\sum E_k\, \delta q_k = \sum I_k\, \delta\theta_k$, and [recalling (2.9.11, 12)] $\delta q_k = \sum A_{kl}\, \delta\theta_l \Leftrightarrow \delta\theta_l = \sum a_{lk} \delta q_k$, we readily obtain the basic (covariant vector-like) transformation equations:

$$I_k = \sum A_{lk} E_l \quad \Leftrightarrow \quad E_k = \sum a_{lk} I_l. \tag{3.3.10}$$

The above expresses the nonholonomic inertial components in *holonomic* variables. To express them in terms of *nonholonomic* variables, we transform (3.3.9), successively, as follows:

$$\begin{aligned}
I_k &\equiv \mathsf{S}\, dm\, \boldsymbol{a}^* \cdot \boldsymbol{\varepsilon}_k = \mathsf{S}\, dm (d\boldsymbol{v}^*/dt) \cdot (\partial \boldsymbol{v}^*/\partial \omega_k) \\
&= d/dt\left(\mathsf{S}\, dm\, \boldsymbol{v}^* \cdot (\partial \boldsymbol{v}^*/\partial \omega_k)\right) - \mathsf{S}\, [dm\, \boldsymbol{v}^* \cdot d/dt(\partial \boldsymbol{v}^*/\partial \omega_k)] \\
&\qquad [\text{adding and subtracting } \mathsf{S}\, dm\, \boldsymbol{v}^* \cdot (\partial \boldsymbol{v}^*/\partial \theta_k), \text{ and regrouping}] \\
&= d/dt\left(\mathsf{S}\, dm\, \boldsymbol{v}^* \cdot (\partial \boldsymbol{v}^*/\partial \omega_k)\right) - \mathsf{S}\, dm\, \boldsymbol{v}^* \cdot (\partial \boldsymbol{v}^*/\partial \theta_k) \\
&\qquad - \mathsf{S}\, dm\, \boldsymbol{v}^* \cdot [(d/dt)(\partial \boldsymbol{v}^*/\partial \omega_k) - \partial \boldsymbol{v}^*/\partial \theta_k];
\end{aligned} \tag{3.3.11a}$$

or, invoking the nonintegrability identity (2.10.24, 25) [Greek subscripts run from 1 to $n + 1$ (time)],

$$\begin{aligned}
E_k{}^*(\boldsymbol{v}^*) &\equiv d/dt(\partial \boldsymbol{v}^*/\partial \omega_k) - \partial \boldsymbol{v}^*/\partial \theta_k \equiv d\boldsymbol{\varepsilon}_k/dt - \partial \boldsymbol{v}^*/\partial \theta_k \\
&= -\sum\sum \gamma^r{}_{kl}\, \omega_l \boldsymbol{\varepsilon}_r - \sum \gamma^r{}_k\, \boldsymbol{\varepsilon}_r \qquad [\text{since } \omega_{n+1} \equiv \omega_0 \equiv dt/dt = 1] \\
&= -\sum\sum \gamma^r{}_{k\alpha}\, \omega_\alpha \boldsymbol{\varepsilon}_r = -\sum\sum \gamma^r{}_{k\alpha}\, \omega_\alpha (\partial \boldsymbol{v}^*/\partial \omega_r),
\end{aligned} \tag{3.3.11b}$$

introducing the (inertial) *kinetic energy in quasi variables*

$$T \equiv \mathsf{S}\, 1/2(dm\, \boldsymbol{v}^* \cdot \boldsymbol{v}^*) = T(t, q, \omega) \equiv T^* \qquad [\text{since } \boldsymbol{v}^* = \boldsymbol{v}^*(t, q, \omega)] \tag{3.3.11c}$$

and recalling the *symbolic quasi chain rule* (2.9.32a, 44a)

$$\partial T^*/\partial \theta_k \equiv \sum (\partial T^*/\partial q_l)(\partial v_l/\partial \omega_k) = \sum A_{lk} (\partial T^*/\partial q_l), \tag{3.3.11d}$$

and the (*nonholonomic Euler–Lagrange operator*)$_k$

$$E_k{}^*(\ldots) \equiv d/dt(\partial\ldots/\partial\omega_k) - \partial\ldots/\partial\theta_k, \tag{3.3.11e}$$

we finally obtain the nonholonomic (system) variable counterpart of E_k:

$$\begin{aligned} I_k &= d/dt(\partial T^*/\partial\omega_k) - \partial T^*/\partial\theta_k + \sum\sum \gamma^r{}_{kl}(\partial T^*/\partial\omega_r)\omega_l + \sum \gamma^r{}_k(\partial T^*/\partial\omega_r) \\ &= d/dt(\partial T^*/\partial\omega_k) - \partial T^*/\partial\theta_k + \sum\sum \gamma^r{}_{k\alpha}(\partial T^*/\partial\omega_r)\omega_\alpha \\ &\equiv E_k{}^*(T^*) - \Gamma_k \equiv E_k{}^* - \Gamma_k \qquad \text{[note difference from (3.3.6)]}, \end{aligned} \tag{3.3.12}$$

where [recalling (2.10.25a)]

$$\begin{aligned} -\Gamma_k &\equiv -S\, dm\, \boldsymbol{v}^* \cdot E_k{}^*(\boldsymbol{v}^*) = \sum\sum \gamma^r{}_{k\alpha}(\partial T^*/\partial\omega_r)\omega_\alpha \equiv \sum h^r{}_k(\partial T^*/\partial\omega_r) \\ &= -(\text{System } \textit{nonholonomic deviation}, \text{ or correction, term})_k. \end{aligned} \tag{3.3.12a}$$

We summarize the key kinematico-inertial identities below:

(a) $$S\, dm\, \boldsymbol{v}^* \cdot \boldsymbol{\varepsilon}_k = S\, dm\, \boldsymbol{v}^* \cdot (\partial\boldsymbol{v}^*/\partial\omega_k) = \partial T^*/\partial\omega_k \equiv P_k(t, q, \omega) = P_k:$$
Nonholonomic (k)th component of system *momentum*, (3.3.13a)

(b) $$\begin{aligned} \gamma_k &\equiv E_k{}^*(\boldsymbol{v}^*) \equiv d\boldsymbol{\varepsilon}_k/dt - \partial\boldsymbol{v}^*/\partial\theta_k = \cdots = \sum\sum \gamma^r{}_{\alpha k}\,\omega_\alpha \boldsymbol{\varepsilon}_r \\ &= -(\text{Particle } \textit{nonholonomic deviation}, \text{ or correction, term})_k, \end{aligned} \tag{3.3.13b}$$

$$\begin{aligned} S\, dm\, \boldsymbol{v}^* \cdot (d\boldsymbol{\varepsilon}_k/dt) &\equiv S\, dm\, \boldsymbol{v}^* \cdot (\partial\boldsymbol{v}^*/\partial\theta_k) + S\, dm\, \boldsymbol{v}^* \cdot \gamma_k \\ &= \partial T^*/\partial\theta_k + \sum\sum \gamma^r{}_{\alpha k}(\partial T^*/\partial\omega_r)\omega_\alpha = \partial T^*/\partial\theta_k - \sum\sum \gamma^r{}_{k\alpha}(\partial T^*/\partial\omega_r)\omega_\alpha \\ &= \partial T^*/\partial\theta_k + \Gamma_k \qquad \text{[note difference from (3.3.7b)]}, \end{aligned} \tag{3.3.13c}$$

$$-\Gamma_k = \sum\sum \gamma^r{}_{k\alpha}(\partial T^*/\partial\omega_r)\omega_\alpha = -\Gamma_{k,n} - \Gamma_{k,0}, \tag{3.3.13d}$$

where

$$-\Gamma_{k,n} \equiv \sum\sum \gamma^r{}_{kl}(\partial T^*/\partial\omega_r)\omega_l, \tag{3.3.13e}$$

$$-\Gamma_{k,0} \equiv -\Gamma_{k,n+1} \equiv \sum \gamma^r{}_k(\partial T^*/\partial\omega_r):$$
"nonholonomic rheonomic force". (3.3.13f)

With the help of the above, I_k, (3.3.12), can be rewritten in the *momentum* form:

$$I_k = dP_k/dt - \partial T^*/\partial\theta_k + \sum\sum \gamma^r{}_{k\alpha} P_r\,\omega_\alpha. \tag{3.3.14}$$

[Originally due to Hamel [1904(a),(b)], but for *stationary/scleronomic* transformations; that is, with α replaced by, say, $l = 1, \ldots, n$.]

(c) $$\delta I \equiv S\, dm\, \boldsymbol{a} \cdot \delta\boldsymbol{r} = \sum E_k\,\delta q_k = \sum I_k\,\delta\theta_k, \tag{3.3.15}$$

where

$$E_k \equiv S\, dm\, \boldsymbol{a} \cdot \boldsymbol{e}_k = d/dt(\partial T/\partial v_k) - \partial T/\partial q_k \equiv E_k(T) = \sum a_{lk} I_l, \qquad (3.3.15a)$$

$$I_k \equiv S\, dm\, \boldsymbol{a}^* \cdot \boldsymbol{\varepsilon}_k = d/dt(\partial T^*/\partial \omega_k) - \partial T^*/\partial \theta_k + \sum\sum \gamma^r_{k\alpha}(\partial T^*/\partial \omega_r)\omega_\alpha$$
$$\equiv E_k{}^*(T^*) - \Gamma_k \equiv E_k{}^* - \Gamma_k = \sum A_{lk} E_l; \qquad (3.3.15b)$$

that is, it is $E_k \equiv E_k(T)$ and I_k that transform like covariant vectors; the $E_k{}^* \equiv E_k{}^*(T^*)$ do *not* (or, the terms $E_k{}^*$ and Γ_k, considered separately, do not transform as covariant vectors; but taken *together*, as $E_k{}^* - \Gamma_k \equiv I_k$, *they do!*).

3. Acceleration, or Appellian, Forms

The above expressions for the inertia vector E_k (or I_k) are based on the kinetic energy T (*or* T^*), because for their derivation we used the *velocity* identities $\boldsymbol{e}_k = \partial \boldsymbol{v}/\partial v_k$ (or $\boldsymbol{\varepsilon}_k = \partial \boldsymbol{v}^*/\partial \omega_k$). Let us now find expressions for these vectors using the *acceleration* identities $\boldsymbol{e}_k = \partial \boldsymbol{a}/\partial \ddot{q}_k \equiv \partial \boldsymbol{a}/\partial w_k$ (or $\boldsymbol{\varepsilon}_k = \partial \boldsymbol{a}^*/\partial \dot{\omega}_k$). The results will turn out to be based on a scalar function that depends on the accelerations in a similar way that T (*or* T^*) depend on the velocities. [The choice $\boldsymbol{e}_k = \partial \boldsymbol{r}/\partial q_k$ does not seem to lead to any useful expression for E_k; while the choice $\boldsymbol{\varepsilon}_k = \partial \boldsymbol{r}^*/\partial \theta_k \equiv \sum A_{lk}\boldsymbol{e}_l \equiv \sum A_{lk}(\partial \boldsymbol{v}/\partial v_l)$ will be examined later.]

(i) Holonomic variables

We have, successively,

$$E_k \equiv S\, dm\, \boldsymbol{a} \cdot \boldsymbol{e}_k = S\, dm\, \boldsymbol{a} \cdot (\partial \boldsymbol{a}/\partial \ddot{q}_k) = \partial S/\partial \ddot{q}_k \equiv \partial S/\partial w_k, \qquad (3.3.16a)$$

where

$$S \equiv S\,(1/2)(dm\, \boldsymbol{a} \cdot \boldsymbol{a}) = S\,(1/2)(dm\, a^2) = S(t, q, \dot{q}, \ddot{q}) \equiv S(t, q, v, w):$$

"Gibbs–Appell function," or simply *Appellian*, in holonomic variables [or "acceleration energy" (Saint-Germain, 1901)]. (3.3.16b)

(ii) Nonholonomic variables

Similarly, we obtain

$$I_k \equiv S\, dm\, \boldsymbol{a}^* \cdot \boldsymbol{\varepsilon}_k = S\, dm\, \boldsymbol{a}^* \cdot (\partial \boldsymbol{a}^*/\partial \dot{\omega}_k) = \partial S^*/\partial \dot{\omega}_k, \qquad (3.3.17a)$$

where

$$S^* \equiv S\,(1/2)(dm\, \boldsymbol{a}^* \cdot \boldsymbol{a}^*) = S\,(1/2)[dm(a^*)^2] = S^*(t, q, \omega, \dot{\omega}):$$

Appellian, in nonholonomic variables. (3.3.17b)

To relate the above, we apply chain rule to $S = S^*$. We obtain, successively,

$$\partial S/\partial \ddot{q}_k \equiv \partial S/\partial w_k = \sum (\partial S^*/\partial \dot{\omega}_l)(\partial \dot{\omega}_l/\partial \ddot{q}_k) = \sum a_{lk}(\partial S^*/\partial \dot{\omega}_l), \qquad (3.3.17c)$$

and, inversely,

$$\partial S^*/\partial \dot{\omega}_k = \sum (\partial S/\partial \ddot{q}_l)(\partial \ddot{q}_l/\partial \dot{w}_k) = \sum A_{lk}(\partial S/\partial \ddot{q}_l) \equiv \sum A_{lk}(\partial S/\partial w_l); \tag{3.3.17d}$$

which are none other than the transformation equations (3.3.10).

In sum, we have the following theoretically equivalent expressions for E_k and I_k:

(i)
$$\begin{aligned} E_k &\equiv S\, dm\, \boldsymbol{a}\cdot \boldsymbol{e}_k \Big(= \sum a_{lk} I_l\Big) \\ &= d/dt(\partial T/\partial v_k) - \partial T/\partial q_k && \text{[Lagrange (1780)]} \\ &= \sum a_{lk}[E_l^*(T^*) - \Gamma_l] \\ &= \sum a_{lk}(\partial S^*/\partial \dot{\omega}_l) = \partial S/\partial \ddot{q}_k \equiv \partial S/\partial w_k && \text{[Appell (1899)];} \end{aligned} \tag{3.3.18a}$$

(ii)
$$\begin{aligned} I_k &\equiv S\, dm\, \boldsymbol{a}^*\cdot \boldsymbol{\varepsilon}_k \quad \Big(= \sum A_{lk} E_l\Big) \\ &= d/dt(\partial T^*/\partial \omega_k) - \partial T^*/\partial \theta_k + \sum\sum \gamma^r{}_{k\alpha}(\partial T^*/\partial \omega_r)\omega_\alpha \\ &\equiv E_k^*(T^*) - \Gamma_k && \text{[Volterra (1898), Hamel (1903/1904)]} \\ &= \sum A_{lk}\big[d/dt(\partial T/\partial v_l) - \partial T/\partial q_l\big] && \text{[Maggi (1896, 1901, 1903)]} \\ &= \sum A_{lk}(\partial S/\partial \ddot{q}_l) = \partial S^*/\partial \dot{\omega}_k && \text{[Gibbs (1879)].} \end{aligned} \tag{3.3.18b}$$

REMARKS

(i) We can define $(n+1)$th, or (0)th, "temporal" holonomic and nonholonomic components of the system inertia vector by (with $dq_{n+1}/dt \equiv dq_0/dt \equiv dt/dt \equiv v_{n+1} \equiv v_0 = 1$)

$$\begin{aligned} E_{n+1} &\equiv E_0 \equiv S\, dm\, \boldsymbol{a}\cdot \boldsymbol{e}_{n+1} = S\, dm\, \boldsymbol{a}\cdot \boldsymbol{e}_0 = S\, dm\, \boldsymbol{a}\cdot(\partial \boldsymbol{r}/\partial t) \\ &= \cdots = d/dt(\partial T/\partial v_0) - \partial T/\partial q_0 = d/dt(\partial T/\partial \dot{t}) - \partial T/\partial t, \end{aligned} \tag{3.3.19a}$$

$$I_{n+1} \equiv I_0 \equiv S\, dm\, \boldsymbol{a}^*\cdot \boldsymbol{\varepsilon}_{n+1} = S\, dm\, \boldsymbol{a}^*\cdot \boldsymbol{\varepsilon}_0 = S\, dm\, \boldsymbol{a}^*\cdot(\partial \boldsymbol{r}^*/\partial \theta_0) = \cdots. \tag{3.3.19b}$$

However, such *nonvirtual* components will not be needed in the equations of motion; they could play a role in the formulation of "partial work/energy rate" equations (§3.9).

(ii) Here, as throughout this book [e.g. (2.9.38ff.), ch. 5], superstars $(\ldots)^*$ denote functions of $t, q, \omega, \dot{\omega}, \ldots$:

$$f(t, q, \dot{q}, \ddot{q}, \ldots) = f[t, q, \dot{q}(t, q, \omega), \ddot{q}(t, q, \omega, \dot{\omega}), \ldots] \equiv f^*(t, q, \omega, \dot{\omega}, \ldots).$$

A Special Case

Let us find E_k and I_k for the following special quasi-velocity choice (recalling 2.11.9 ff.)

$$v_D = \sum b_{DI}(t, q) v_I + b_D(t, q), \qquad v_I = \sum \delta_{II'} v_{I'} = v_I, \tag{3.3.20a}$$

and its inverse

$$\omega_D = \sum b_{DI}(t,q)v_I + b_D(t,q), \qquad \omega_I = v_I. \tag{3.3.20b}$$

Here, clearly [recalling (2.11.12b), and with $\delta_{..}$ = *Kronecker delta*];

$$A_{DD'} = \delta_{DD'}, \qquad A_{DI} = b_{DI}, \qquad A_{ID} = 0, \qquad A_{II'} = \delta_{II'}, \tag{3.3.20c}$$

and so the *Maggi form* $I_k = \sum A_{lk} E_l$ specializes to $I_k = \sum A_{Dk} E_D + \sum A_{Ik} E_I$

$$\Rightarrow I_{D'} = \sum \delta_{DD'} E_D + \sum (0) E_I = E_{D'}, \tag{3.3.20d}$$

$$I_{I'} = \sum b_{DI'} E_D + \sum \delta_{II'} E_I = E_{I'} + \sum b_{DI'} E_D. \tag{3.3.20e}$$

In sum, for the special choice (3.3.20a, b) I_k takes the following form, in terms of *holonomic* Lagrangean (T) and Appellian (S) variables (with $D = 1, \ldots, m$; $I = m+1, \ldots, n$ as usual):

$$I_D = E_D \equiv (\partial T/\partial v_D)^{\cdot} - \partial T/\partial q_D = \partial S/\partial \dot{v}_D \equiv \partial S/\partial w_D; \tag{3.3.20f}$$

$$I_I = E_I + \sum b_{DI} E_D \equiv [(\partial T/\partial v_I)^{\cdot} - \partial T/\partial q_I] + \sum b_{DI}[(\partial T/\partial v_D)^{\cdot} - \partial T/\partial q_D]$$

[Chaplygin (1895, publ. 1897), Hadamard (1895)] (3.3.20g)

$$= \partial S/\partial \ddot{q}_I + \sum b_{DI}(\partial S/\partial \ddot{q}_D) \equiv \partial S/\partial w_I + \sum b_{DI}(\partial S/\partial w_D). \tag{3.3.20h}$$

The specialization of I_k, for (3.3.20a, b), to *nonholonomic* variables [due to Chaplygin (1895/1897), *in addition to* his equations (3.3.20g); and Voronets (1901)] and other related results, are given in §3.8.

We have expressed the (*total, first order*) *virtual work of the* (*negative of the*) *inertial "forces,"* δI, in system variables. The kinematico-inertial identities obtained are central to analytical mechanics, and that is why they were deliberately presented before any discussion of system forces and constraints; because, indeed, they are independent of the latter. These identities also show clearly the importance of the *kinetic energy* (primarily) and the *Appellian* (secondarily) to our subject, and so these quantities are examined in detail later (§3.9.11, 13–16).

Now, let us proceed to express the virtual works of the real forces, namely, $\delta' W$ and $\delta' W_R$, in system variables. This will be considerably easier than the task just completed.

3.4 VIRTUAL WORKS OF FORCES: IMPRESSED ($\delta' W$) AND CONSTRAINT REACTIONS ($\delta' W_R$)

1. Holonomic Variables

Substituting $\delta \boldsymbol{r} = \sum \boldsymbol{e}_k \, \delta q_k$ into the earlier expressions for $\delta' W$ and $\delta' W_R$ (3.2.7, 10), we readily obtain

$$\delta' W \equiv S\, d\boldsymbol{F} \cdot \delta \boldsymbol{r} = S\, d\boldsymbol{F} \cdot \left(\sum \boldsymbol{e}_k \, \delta q_k\right) = \cdots = \sum Q_k \, \delta q_k, \tag{3.4.1a}$$

$$\delta' W_R \equiv S\, d\boldsymbol{R} \cdot \delta \boldsymbol{r} = S\, d\boldsymbol{R} \cdot \left(\sum \boldsymbol{e}_k \, \delta q_k\right) = \cdots = \sum R_k \, \delta q_k, \tag{3.4.1b}$$

where

$$Q_k \equiv \mathsf{S}\, d\boldsymbol{F}\cdot\boldsymbol{e}_k\text{: Holonomic }(k)\text{th component of system } \textit{impressed} \text{ force,} \tag{3.4.1c}$$

$$R_k \equiv \mathsf{S}\, d\boldsymbol{R}\cdot\boldsymbol{e}_k\text{: Holonomic }(k)\text{th component of system } \textit{constraint reaction.} \tag{3.4.1d}$$

2. Nonholonomic Variables

Substituting $\delta\boldsymbol{r} = \sum \boldsymbol{\varepsilon}_k\, \delta\theta_k$ $(\equiv \delta\boldsymbol{r}^*)$ into $\delta' W$ and $\delta' W_R$, we, similarly, obtain

$$\delta' W \equiv \mathsf{S}\, d\boldsymbol{F}\cdot\delta\boldsymbol{r}^* = \mathsf{S}\, d\boldsymbol{F}\cdot\left(\sum \boldsymbol{\varepsilon}_k\, \delta\theta_k\right) = \cdots = \sum \Theta_k\, \delta\theta_k, \tag{3.4.2a}$$

$$\delta' W_R \equiv \mathsf{S}\, d\boldsymbol{R}\cdot\delta\boldsymbol{r}^* = \mathsf{S}\, d\boldsymbol{R}\cdot\left(\sum \boldsymbol{\varepsilon}_k\, \delta\theta_k\right) = \cdots = \sum \Lambda_k\, \delta\theta_k, \tag{3.4.2b}$$

where

$$\Theta_k \equiv \mathsf{S}\, d\boldsymbol{F}\cdot\boldsymbol{\varepsilon}_k\text{: Nonholonomic }(k)\text{th component of system } \textit{impressed} \text{ force,} \tag{3.4.2c}$$

$$\Lambda_k \equiv \mathsf{S}\, d\boldsymbol{R}\cdot\boldsymbol{\varepsilon}_k\text{: Nonholonomic }(k)\text{th component of system } \textit{constraint} \text{ force.} \tag{3.4.2d}$$

Here too, these are ever valid definitions/results, no matter how many constraints may be imposed on the system later.

3. Transformation Relations

From the invariance of the virtual differentials $\delta' W$ and $\delta' W_R$, we obtain the following transformation formulae for the various system forces; that is, from

$$\delta' W = \sum Q_k\, \delta q_k = \sum Q_k\left(\sum A_{kl}\, \delta\theta_l\right) = \sum \Theta_l\, \delta\theta_l = \sum \Theta_l\left(\sum a_{lk}\, \delta q_k\right)$$
$$= \sum Q_k\, \delta q_k, \tag{3.4.3a}$$

we conclude

$$\Theta_l = \sum A_{kl} Q_k \qquad \left[= \sum Q_k(\partial v_k/\partial\omega_l)\right] \tag{3.4.3b}$$

and, inversely,

$$Q_k = \sum a_{lk}\Theta_l \qquad \left[= \sum (\partial\omega_l/\partial v_k)\Theta_l\right]; \tag{3.4.3c}$$

and, similarly, from $\delta' W_R = \cdots$, we conclude

$$\Lambda_l = \sum A_{kl} R_k \qquad \left[= \sum R_k(\partial v_k/\partial\omega_l)\right] \tag{3.4.3d}$$

and, inversely,

$$R_k = \sum a_{lk}\Lambda_l \qquad \left[= \sum (\partial\omega_l/\partial v_k)\Lambda_l\right]. \tag{3.4.3e}$$

[These formulae can also be obtained from the $e_k \Leftrightarrow \varepsilon_k$ transformation equations (2.9.25a, b) as follows:

$$\Theta_l \equiv \mathsf{S}\, d\boldsymbol{F}\cdot\boldsymbol{\varepsilon}_l = \mathsf{S}\, d\boldsymbol{F}\cdot\left(\sum A_{kl}\boldsymbol{\varepsilon}_k\right) = \sum A_{kl}\left(\mathsf{S}\, d\boldsymbol{F}\cdot\boldsymbol{e}_k\right) = \sum A_{kl}Q_k,$$

$$Q_k \equiv \mathsf{S}\, d\boldsymbol{F}\cdot\boldsymbol{e}_k = \mathsf{S}\, d\boldsymbol{F}\cdot\left(\sum a_{lk}\boldsymbol{\varepsilon}_l\right) = \sum a_{lk}\left(\mathsf{S}\, d\boldsymbol{F}\cdot\boldsymbol{\varepsilon}_l\right) = \sum a_{lk}\Theta_l.]$$

Rheonomic, or "temporal," $(n+1)$th *nonvirtual* force components can also be defined by

$$Q_{n+1} \equiv Q_0 \equiv \mathsf{S}\, d\boldsymbol{F}\cdot\boldsymbol{e}_{n+1} \equiv \mathsf{S}\, d\boldsymbol{F}\cdot\boldsymbol{e}_0 \equiv \mathsf{S}\, d\boldsymbol{F}\cdot(\partial\boldsymbol{r}/\partial t) \quad \text{(holonomic impressed)}, \tag{3.4.4a}$$

$$R_{n+1} \equiv R_0 \equiv \mathsf{S}\, d\boldsymbol{R}\cdot\boldsymbol{e}_{n+1} \equiv \mathsf{S}\, d\boldsymbol{R}\cdot\boldsymbol{e}_0 \equiv \mathsf{S}\, d\boldsymbol{R}\cdot(\partial\boldsymbol{r}/\partial t) \quad \text{(holonomic reaction)}; \tag{3.4.4b}$$

$$\Theta_{n+1} \equiv \Theta_0 \equiv \mathsf{S}\, d\boldsymbol{F}\cdot\boldsymbol{\varepsilon}_{n+1} \equiv \mathsf{S}\, d\boldsymbol{F}\cdot\boldsymbol{\varepsilon}_0 \equiv \mathsf{S}\, d\boldsymbol{F}\cdot(\partial\boldsymbol{r}^*/\partial\theta_{n+1}) \quad \text{(nonholonomic impressed)}, \tag{3.4.4c}$$

$$\Lambda_{n+1} \equiv \Lambda_0 \equiv \mathsf{S}\, d\boldsymbol{R}\cdot\boldsymbol{\varepsilon}_{n+1} \equiv \mathsf{S}\, d\boldsymbol{R}\cdot\boldsymbol{\varepsilon}_0 \equiv \mathsf{S}\, d\boldsymbol{R}\cdot(\partial\boldsymbol{r}^*/\partial\theta_{n+1}) \quad \text{(nonholonomic reaction)}; \tag{3.4.4d}$$

and, recalling (2.9.26a, b), we can easily deduce the following transformation equations among these components:

$$\begin{aligned} Q_{n+1} \equiv Q_0 &\equiv \mathsf{S}\, d\boldsymbol{F}\cdot\boldsymbol{e}_{n+1} = \mathsf{S}\, d\boldsymbol{F}\cdot\left(\sum a_{k,n+1}\boldsymbol{\varepsilon}_k + \boldsymbol{\varepsilon}_{n+1}\right) \\ &= \sum a_{k,n+1}\left(\mathsf{S}\, d\boldsymbol{F}\cdot\boldsymbol{\varepsilon}_k\right) + \mathsf{S}\, d\boldsymbol{F}\cdot\boldsymbol{\varepsilon}_{n+1} = \sum a_{k,n+1}\Theta_k + \Theta_{n+1}, \end{aligned} \tag{3.4.4e}$$

$$\begin{aligned} R_{n+1} \equiv R_0 &\equiv \mathsf{S}\, d\boldsymbol{R}\cdot\boldsymbol{e}_{n+1} = \mathsf{S}\, d\boldsymbol{R}\cdot\left(\sum a_{k,n+1}\boldsymbol{\varepsilon}_k + \boldsymbol{\varepsilon}_{n+1}\right) \\ &= \sum a_{k,n+1}\left(\mathsf{S}\, d\boldsymbol{R}\cdot\boldsymbol{\varepsilon}_k\right) + \mathsf{S}\, d\boldsymbol{R}\cdot\boldsymbol{\varepsilon}_{n+1} = \sum a_{k,n+1}\Lambda_k + \Lambda_{n+1}; \end{aligned} \tag{3.4.4f}$$

and, conversely,

$$\begin{aligned} \Theta_0 &\equiv \mathsf{S}\, d\boldsymbol{F}\cdot\boldsymbol{\varepsilon}_0 = \mathsf{S}\, d\boldsymbol{F}\cdot\left(\sum A_k\boldsymbol{e}_k + \boldsymbol{e}_0\right) \\ &= \sum A_k\left(\mathsf{S}\, d\boldsymbol{F}\cdot\boldsymbol{e}_k\right) + \mathsf{S}\, d\boldsymbol{F}\cdot\boldsymbol{e}_0 = \sum A_kQ_k + Q_0, \end{aligned} \tag{3.4.4g}$$

$$\begin{aligned} \Lambda_0 &\equiv \mathsf{S}\, d\boldsymbol{R}\cdot\boldsymbol{\varepsilon}_0 = \mathsf{S}\, d\boldsymbol{R}\cdot\left(\sum A_k\boldsymbol{e}_k + \boldsymbol{e}_0\right) \\ &= \sum A_k\left(\mathsf{S}\, d\boldsymbol{R}\cdot\boldsymbol{e}_k\right) + \mathsf{S}\, d\boldsymbol{R}\cdot\boldsymbol{e}_0 = \sum A_kR_k + R_0. \end{aligned} \tag{3.4.4h}$$

Problem 3.4.1 With the help of the second of each of (2.9.3a, b), prove the additional forms of the above transformation equations:

$$\Theta_{n+1} = -\sum a_{k,n+1}\Theta_k + Q_{n+1}, \quad \text{or, simply,} \quad \Theta_0 = -\sum a_k\Theta_k + Q_0, \tag{a}$$

$$\Lambda_{n+1} = -\sum a_{k,n+1}\Lambda_k + R_{n+1}, \quad \text{or, simply,} \quad \Lambda_0 = -\sum a_k\Lambda_k + R_0. \tag{b}$$

REMARK

A little analytical reflection will show that all these transformations can be condensed in the formulae [with Greek subscripts running from 1 to $n+1$, recall (2.9.6a, b)]:

$$\Theta_\alpha = \sum A_{\beta\alpha} Q_\beta \Leftrightarrow Q_\beta = \sum a_{\alpha\beta} \Theta_\alpha, \tag{3.4.5a}$$

$$\Lambda_\alpha = \sum A_{\beta\alpha} R_\beta \Leftrightarrow R_\beta = \sum a_{\alpha\beta} \Lambda_\alpha. \tag{3.4.5b}$$

A Special Case

For the earlier particular case (3.3.20a ff.: $A_{DD'} = \delta_{DD'}$, $A_{DI} = b_{DI}$, $A_{ID} = 0$, $A_{II'} = \delta_{II'}$), the above transformation equations specialize to

$$\Theta_{D'} = \sum A_{kD'} Q_k = \sum \delta_{DD'} Q_D + \sum (0) Q_I = Q_{D'}, \quad \text{i.e.,} \quad \Theta_D = Q_D, \tag{3.4.6a}$$

$$\Theta_{I'} = \sum A_{kI'} Q_k = \sum b_{DI'} Q_D + \sum \delta_{II'} Q_I = Q_{I'} + \sum b_{DI'} Q_D,$$

$$\text{i.e.,} \quad \Theta_I = Q_I + \sum b_{DI} Q_D \equiv Q_{I,o} \equiv Q_{Io}; \tag{3.4.6b}$$

and, similarly,

$$\Theta_0 = \cdots = \sum b_D Q_D + Q_0 \equiv Q_{0,o}. \tag{3.4.6c}$$

Example 3.4.1 *Virtually Workless Forces.* The following are examples of forces that do zero virtual work:

(i) Forces among the particles of a rigid body; generally, the forces among rigidly connected particles and/or bodies.
(ii) Forces on particles that are either at rest (e.g., a fixed pivot, or hinge, about which a system body may turn, or a joint between two system bodies), or are constrained to move in *prescribed* ways; that is, their (inertial) motion is known in advance as a function of time.
(iii) Forces from completely *smooth* curves and/or surfaces that are either at (inertial) rest or have *prescribed* (inertial) motions.
(iv) Forces from perfectly *rough* curves and/or surfaces, either at rest or having prescribed motions. See also Pars (1965, pp. 24–25), Whittaker (1937, pp. 31–32).

Example 3.4.2 If z is a virtual displacement, then $S\, d\boldsymbol{R} \cdot \boldsymbol{z} = 0$. Let us show the *converse*: If for a kinematically admissible/possible vector z we have $S\, d\boldsymbol{R} \cdot \boldsymbol{z} = 0$, then z is a virtual displacement; that is,

$$\boldsymbol{z} = \sum \boldsymbol{\varepsilon}_I \, \delta\theta_I \qquad (I = m+1, \ldots, n). \tag{a}$$

The proof is by contradiction: Let $\boldsymbol{z} = \delta\boldsymbol{r} + \boldsymbol{y}$ ($\neq \delta\boldsymbol{r}$), where the *relaxed* part $\boldsymbol{y}$ may be, at most,

$$\boldsymbol{y} = \sum \boldsymbol{\varepsilon}_D \, \delta'\theta_D + \boldsymbol{\varepsilon}_0 \, \delta' t \qquad (D = 1, \ldots, m; \quad \delta'\theta_D, \delta' t\text{: components of } \boldsymbol{y}). \tag{b}$$

Substituting (b) into LP we get, successively,

$$0 = S\, d\boldsymbol{R} \cdot \boldsymbol{z} = S\, d\boldsymbol{R} \cdot \delta\boldsymbol{r} + S\, d\boldsymbol{R} \cdot \boldsymbol{y} = \cdots = 0 + \sum \Lambda_D \, \delta'\theta_D + \Lambda_0 \, \delta' t,$$

from which, since the $m+1$ $\delta'\theta_D$ and $\delta' t$ are *independent*, we obtain $\Lambda_D, \Lambda_0 = 0$. But, clearly, due to the constraints $\delta\theta_D = 0$ and $\delta t = 0$, this is impossible. Thus, if we assume that $\mathbf{y} \neq \mathbf{0}$, we are led to a contradiction. Hence, $\mathbf{y} = \mathbf{0}$, and $\mathbf{z}$ is a virtual displacement expressible by (a).

3.5 EQUATIONS OF MOTION VIA LAGRANGE'S PRINCIPLE: GENERAL FORMS

Let us now proceed to the final synthesis; that is, the formulation of equations of motion in general system variables. We begin with the "constraint reaction part" of LP, eq. (3.2.7), and so on:

$$\delta' W_R \equiv S\, d\mathbf{R} \cdot \delta\mathbf{r} = \sum R_k\, \delta q_k = \sum \Lambda_k\, \delta\theta_k = 0. \tag{3.5.1}$$

If the n δq's are *unconstrained* (or *independent*, or *free*), so are the n $\delta\theta$'s. Then, (3.5.1) leads to

$$R_k = 0, \qquad \Lambda_k = 0. \tag{3.5.2}$$

If, however, the n δq's are *constrained* by the m ($< n$) Pfaffian, holonomic and/or nonholonomic, constraints

$$\delta\theta_D \equiv \sum a_{Dk}\, \delta q_k = 0 \qquad (D = 1, \ldots, m), \tag{3.5.3}$$

then, introducing m *Lagrangean* (hitherto) *undetermined multipliers* $-\lambda_D = -\lambda_D(t)$ (the minus sign is only for algebraic convenience—see multiplier rule, below), and invoking (3.5.3), we can replace (3.5.1) with

$$\begin{aligned} \delta' W_R + \sum (-\lambda_D)\, \delta\theta_D &= \sum \Lambda_k\, \delta\theta_k + \sum (-\lambda_D)\, \delta\theta_D \\ &= \sum R_k\, \delta q_k + \sum\sum (-\lambda_D) a_{Dk}\, \delta q_k = 0, \end{aligned} \tag{3.5.4}$$

where, now, the n δq's and $\delta\theta$'s (can be treated as if they) are free. Therefore, (3.5.4) leads immediately to the following:

(i) in *holonomic* variables,

$$\sum \left(R_k - \sum \lambda_D\, a_{Dk} \right) \delta q_k = 0 \;\Rightarrow\; R_k = \sum \lambda_D a_{Dk} \qquad [= R_k(q, t)]; \tag{3.5.5}$$

(ii) in *nonholonomic* variables (with $I = m+1, \ldots, n$),

$$\sum \Lambda_k\, \delta\theta_k - \sum \lambda_D\, \delta\theta_D = \sum (\Lambda_D - \lambda_D)\, \delta\theta_D + \sum (\Lambda_I - 0)\, \delta\theta_I = 0, \tag{3.5.6}$$

and from this to the nonholonomic counterpart of (3.5.5),

$$\Lambda_D = \lambda_D \quad (1 \cdot \delta\theta_D = 0) \qquad \text{and} \qquad \Lambda_I = 0 \quad (0 \cdot \delta\theta_I = 0); \tag{3.5.7}$$

that is, the m Lagrangean multipliers associated with the m "equilibrium" constraints $\omega_D = 0$ or $\delta\theta_D = 0$ are, in effect, the first m nonholonomic (covariant) components of the system reaction vector in configuration space. We also notice that whenever $\delta\theta_k = 0$, $\Lambda_k \neq 0$, and vice versa ($k = 1, \ldots, n$; and even $n+1$), that is,

$\Lambda_D\,\delta\theta_D = (\Lambda_D)(0) = 0$ and $\Lambda_I\,\delta\theta_I = (0)(\delta\theta_I) = 0$, so that

$$\delta' W_R = \sum \Lambda_D\,\delta\theta_D + \sum \Lambda_I\,\delta\theta_I = 0 + 0 = 0, \tag{3.5.8}$$

in accordance with LP.

The advantage of the nonholonomic (3.5.7, 8) over the holonomic (3.5.5) is that, in the former, constraints and reactions *decouple* naturally; whereas in the latter they are coupled; that is, in general,

$$\delta q_k \neq 0, \quad R_k \neq 0 \;\Rightarrow\; R_k\,\delta q_k \neq 0, \qquad \textit{but} \qquad \sum R_k\,\delta q_k = 0. \tag{3.5.9}$$

Finally, substituting the first of (3.5.7) into (3.5.5), we recover the earlier transformation equations (3.4.3e): $R_k = \sum \Lambda_D a_{Dk} = \sum \Lambda_I a_{Ik}$, as expected.

REMARK

In the case of *unilateral* constraints $\delta\theta_D \geq 0$ (if, originally, they have the form $\delta\theta_D \leq 0$, we replace $\delta\theta_D$ with $-\delta\theta_D$), from the "unilateral LP" $\delta' W_R \geq 0$ and (3.5.4) we conclude that $\sum \lambda_D\,\delta\theta_D \geq 0$; and since the $\delta\theta_D$ are positive or zero, the λ_D must also be positive or zero.

In sum: If the unilateral constraints are chosen so that $\delta\theta_* > 0$ is possible/admissible, then the corresponding reaction λ_* is positive or zero (see also §3.7).

The Lagrangean Multiplier Rule, or Adjoining of Constraints

This fundamental mathematical theorem [one of the most useful mathematical results of the 18th century, initiated by Euler, but brought to prominence by Lagrange—see Hoppe (1926(a), p. 62)] states that:

The single (differential) variational equation

$$\delta' M \equiv \sum M_k\,\delta q_k = 0, \tag{3.5.10a}$$

where $M_k = M_k(q, \dot{q}, \ddot{q}, \ldots, t)$ and the n δq's are restricted by the m ($<n$) independent Pfaffian constraints

$$\delta\theta_D \equiv \sum a_{Dk}\,\delta q_k = 0 \qquad [\textit{rank}\ (a_{Dk}) = m], \tag{3.5.10b}$$

is completely equivalent to the new variational equation

$$\delta' M + \sum (-\lambda_D)\,\delta\theta_D = \delta' M + \sum\sum (-\lambda_D a_{Dk})\,\delta q_k = 0,$$

or

$$\sum \Big(M_k - \sum \lambda_D a_{Dk}\Big)\,\delta q_k = 0, \tag{3.5.10c}$$

where the n δq's are (better, can be viewed as) unconstrained; that is, (3.5.10a, b) are equivalent to the n equations

$$M_k = \sum \lambda_D a_{Dk} \tag{3.5.10d}$$

which, along with the m constraints (3.5.10b), in velocity form

$$\omega_D \equiv \sum a_{Dk}\dot{q}_k + a_D = 0, \tag{3.5.10e}$$

make up a system of $n+m$ equations for the $n+m$ unknown functions $q(t)$ and $\lambda(t)$.

INFORMAL PROOF

Let us define the $m\,\lambda_D$'s by the m nonsingular equations

$$M_{D'} = \sum \lambda_D a_{DD'} \qquad (D, D' = 1, \ldots, m), \tag{3.5.10f}$$

that is, eqs. (3.5.10d) with $k \to D'$. For such λ's eq. (3.5.10c) reduces to

$$\sum \Big(M_I - \sum \lambda_D a_{DI}\Big)\,\delta q_I = 0, \tag{3.5.10g}$$

where the $(n-m)\,\delta q_I$'s are now free. From the above, we immediately conclude that

$$M_I = \sum \lambda_D a_{DI}; \tag{3.5.10h}$$

that is, eqs. (3.5.10d) with $k \to I$.

[References on the multiplier rule: Gantmacher (1970, pp. 20–23), Hamel (1949, pp. 85–91), Osgood (1937, pp. 316–318), Rosenberg (1977, pp. 132, 212–214). For a linear algebra based proof, see, for example, Woodhouse (1987, pp. 114–115).]

Example 3.5.1 *Lagrange's Equations of the First Kind.* The multiplier rule applied to

$$\delta' W_R \equiv \int d\boldsymbol{R} \cdot \delta\boldsymbol{r} = 0, \tag{a}$$

where the $\delta\boldsymbol{r}$ are restricted by (i) the h holonomic constraints $(H = 1, \ldots, h)$

$$\phi_H(t, \boldsymbol{r}) = 0 \;\Rightarrow\; \delta\phi_H = \int (\partial\phi_H/\partial\boldsymbol{r}) \cdot \delta\boldsymbol{r} = 0, \tag{b1}$$

and (ii) the m Pfaffian (possibly nonholonomic) constraints $[D = 1, \ldots, m$ $(< n \equiv 3N - h)$ and $\boldsymbol{B}_D = \boldsymbol{B}_D(t, \boldsymbol{r})]$

$$\int \boldsymbol{B}_D \cdot \boldsymbol{v} + B_D = 0 \;\Rightarrow\; \int \boldsymbol{B}_D \cdot \delta\boldsymbol{r} = 0, \tag{b2}$$

leads, with the help of the $h+m$ Lagrangean multipliers $\mu_H = \mu_H(t)$ and $\lambda_D = \lambda_D(t)$, to

$$\int \Big(d\boldsymbol{R} - \sum \mu_H(\partial\phi_H/\partial\boldsymbol{r}) - \sum \lambda_D \boldsymbol{B}_D\Big) \cdot \delta\boldsymbol{r} = 0, \tag{c1}$$

from which, since the $\delta\boldsymbol{r}$ can now be treated as free, we obtain the *constitutive equation for the total constraint reaction* on the typical particle P due to all system constraints:

$$d\boldsymbol{R} = \sum \mu_H(\partial\phi_H/\partial\boldsymbol{r}) + \sum \lambda_D \boldsymbol{B}_D. \tag{c2}$$

Then, the *Newton–Euler/d'Alembert* particle equation $dm\,\boldsymbol{a} = d\boldsymbol{F} + d\boldsymbol{R}$ becomes the famous *Lagrange's equation of the first kind*:

$$dm\,\boldsymbol{a} = d\boldsymbol{F} + \sum \mu_H(\partial\phi_H/\partial\boldsymbol{r}) + \sum \lambda_D \boldsymbol{B}_D. \tag{c3}$$

In the more common *discrete* notation, the constraints (b1, 2) become (with $P = 1, \ldots, N \equiv$ *number of system particles*)

$$\delta\phi_H = \sum (\partial\phi_H/\partial \boldsymbol{r}_P) \cdot \delta\boldsymbol{r}_P = 0 \quad \text{and} \quad \sum \boldsymbol{B}_{DP} \cdot \delta\boldsymbol{r}_P = 0, \tag{c4}$$

respectively, while the equation of constrainted motion (c3) assumes the form

$$m_P \boldsymbol{a}_P = \boldsymbol{F}_P + \boldsymbol{R}_P = \boldsymbol{F}_P + \sum \mu_H(\partial\phi_H/\partial\boldsymbol{r}_P) + \sum \lambda_D \boldsymbol{B}_{DP}. \tag{c5}$$

To understand the relation between the *particle* reactions $d\boldsymbol{R}$, $\boldsymbol{R}_P$ and their *system* counterparts R_k, Λ_k, we insert (c2) into their corresponding definitions (3.4.1d, 2d). We find, successively,

(i)
$$\begin{aligned} R_k &\equiv S\, d\boldsymbol{R} \cdot \boldsymbol{e}_k = S\Big(\sum \mu_H(\partial\phi_H/\partial\boldsymbol{r}) + \sum \lambda_D \boldsymbol{B}_D\Big) \cdot \boldsymbol{e}_k \\ &= \sum \mu_H\Big(S\,(\partial\phi_H/\partial\boldsymbol{r}) \cdot (\partial\boldsymbol{r}/\partial q_k)\Big) + \sum \lambda_D\Big(S\,\boldsymbol{B}_D \cdot \boldsymbol{e}_k\Big) \\ &\equiv \sum \mu_H(\partial\phi_H/\partial q_k) + \sum \lambda_D B_{Dk}, \end{aligned} \tag{d1}$$

and, comparing with the second of (3.5.5), $R_k = \sum \lambda_D a_{Dk}$, we readily conclude that

$$\partial\phi_H(t, q)/\partial q_k \equiv S\,[\partial\phi_H(t, \boldsymbol{r})/\partial\boldsymbol{r}] \cdot [\partial\boldsymbol{r}(t, q)/\partial q_k] = 0, \tag{d2}$$

and

$$B_{Dk} \equiv S\,\boldsymbol{B}_D \cdot \boldsymbol{e}_k = a_{Dk}; \tag{d3}$$

recall ex. 2.4.1 and (2.6.1ff.).

REMARK

The above also show that, as long as the quasi variables are chosen so that

$$0 = S\,\boldsymbol{B}_D \cdot \delta\boldsymbol{r} \equiv \delta\theta_D = \sum a_{Dk}\,\delta q_k \;\Rightarrow\; S\,\boldsymbol{B}_D \cdot \boldsymbol{e}_k = a_{Dk}, \tag{e1}$$

the multipliers λ_D in (c2, 3) coincide with those in the second of (3.5.5). Indeed, $\boldsymbol{e}_k$ – dotting (c3) and then S-summing, we obtain the "Routh–Voss" equations [see (3.5.15) below]:

$$S\,dm\,\boldsymbol{a} \cdot \boldsymbol{e}_k = S\,d\boldsymbol{F} \cdot \boldsymbol{e}_k + \sum \mu_H\Big(S\,(\partial\phi_H/\partial\boldsymbol{r}) \cdot \boldsymbol{e}_k\Big) + \sum \lambda_D\Big(S\,\boldsymbol{B}_D \cdot \boldsymbol{e}_k\Big), \tag{e2}$$

or

$$E_k = Q_k + 0 + \sum \lambda_D a_{Dk}. \tag{e3}$$

(ii)
$$\begin{aligned} \Lambda_k &\equiv S\, d\boldsymbol{R} \cdot \boldsymbol{\varepsilon}_k = S\Big(\sum \mu_H(\partial\phi_H/\partial\boldsymbol{r}) + \sum \lambda_D \boldsymbol{B}_D\Big) \cdot \boldsymbol{\varepsilon}_k \\ &= \sum \mu_H\Big(S\,(\partial\phi_H/\partial\boldsymbol{r}) \cdot (\partial\boldsymbol{r}/\partial\theta_k)\Big) + \sum \lambda_D\Big(S\,\boldsymbol{B}_D \cdot \boldsymbol{\varepsilon}_k\Big) \\ &\equiv \sum \mu_H(\partial\phi_H/\partial\phi_k) + \sum \lambda_D {B'}_{Dk}, \end{aligned} \tag{f1}$$

and, comparing with (3.5.7), $\Lambda_D = \lambda_D$ and $\Lambda_I = 0$, we readily conclude that

$$\partial\phi_H/\partial\theta_k \equiv S\,(\partial\phi_H/\partial\boldsymbol{r})\cdot\boldsymbol{\varepsilon}_k = S\,(\partial\phi_H/\partial\boldsymbol{r})\cdot\Big(\sum A_{lk}\boldsymbol{e}_l\Big)$$
$$= \cdots = \sum A_{lk}(\partial\phi_H/\partial q_l) = 0, \tag{f2}$$

and (with $k = 1,\ldots,n$; $D, D' = 1,\ldots,m$; $I = m+1,\ldots,n$)

$$B'_{Dk} \equiv S\,\boldsymbol{B}_D\cdot\boldsymbol{\varepsilon}_k = S\,\boldsymbol{B}_D\cdot\Big(\sum A_{lk}\boldsymbol{e}_l\Big) = \sum A_{lk}\Big(S\,\boldsymbol{B}_D\cdot\boldsymbol{e}_l\Big) = \sum A_{lk}a_{Dl} = \delta_{kD};$$

that is,

$$B'_{DD'} \equiv S\,\boldsymbol{B}_D\cdot\boldsymbol{\varepsilon}_{D'} = \delta_{DD'}, \qquad B'_{DI} \equiv S\,\boldsymbol{B}_D\cdot\boldsymbol{\varepsilon}_I = \delta_{DI} = 0. \tag{f3}$$

Some of the above can also be obtained from the virtual forms of the constraints. Thus, we find, successively,

(a)

$$0 = \delta\phi_H = S\,(\partial\phi_H/\partial\boldsymbol{r})\cdot\delta\boldsymbol{r} = S\,(\partial\phi_H/\partial\boldsymbol{r})\cdot\Big(\sum \boldsymbol{\varepsilon}_k\,\delta\theta_k\Big) = \cdots = \sum(\partial\phi_H/\partial\theta_k)\,\delta\theta_k$$
$$= \sum(\partial\phi_H/\partial\theta_D)\,\delta\theta_D + \sum(\partial\phi_H/\partial\theta_I)\,\delta\theta_I = 0 + \sum(\partial\phi_H/\partial\theta_I)\,\delta\theta_I \Rightarrow \partial\phi_H/\partial\theta_I = 0. \tag{g1}$$

(b)

$$0 = \delta\theta_D \equiv S\,\boldsymbol{B}_D\cdot\delta\boldsymbol{r} = S\,\boldsymbol{B}_D\cdot\Big(\sum \boldsymbol{\varepsilon}_k\,\delta\theta_k\Big) = \cdots = \sum B'_{Dk}\,\delta\theta_k$$
$$= \sum B'_{DD'}\,\delta\theta_{D'} + \sum B'_{DI}\,\delta\theta_I = 0 + \sum B'_{DI}\,\delta\theta_I \;\Rightarrow\; B'_{DI} = 0. \tag{g2}$$

HISTORICAL

The terms *Lagrange's equations of the first kind* (and *second* kind—see below) seem to have originated in Jacobi's famous *Lectures on Dynamics* (winter 1842/1843, publ. 1866), and have been widely used in the German and Russian literature. They are not too well known among English and French authors (see, e.g., Voss, 1901, p. 81, footnote #220).

Example 3.5.2 *Lagrange's Principle and Multipliers: Particle on a Surface* (Kraft, 1885, vol. 2, pp. 194–195). Let us consider a particle P of mass m moving on a smooth surface $\phi(x, y, z, t) = 0$, where x, y, z are inertial rectangular Cartesian coordinates of P, under a total impressed force with rectangular Cartesian components (X, Y, Z). According to LP, the motion is given by (with the customary notations $dx/dt \equiv v_x$, $d^2x/dt^2 \equiv dv_x/dt \equiv a_x, \ldots$)

$$(m\,a_x - X)\,\delta x + (m\,a_y - Y)\,\delta y + (m\,a_z - Z)\,\delta z = 0, \tag{a}$$

under the (virtual form of the surface) constraint

$$\delta\phi = 0: \quad (\partial\phi/\partial x)\,\delta x + (\partial\phi/\partial y)\,\delta y + (\partial\phi/\partial z)\,\delta z = 0. \tag{b}$$

By Lagrange's multipliers, (a) and (b) combine to the *unconstrained* variational equation,

$$[m\,a_x - X - \lambda(\partial\phi/\partial x)]\,\delta x + [m\,a_y - Y - \lambda(\partial\phi/\partial y)]\,\delta y$$
$$+ [m\,a_z - Z - \lambda(\partial\phi/\partial z)]\,\delta z = 0, \tag{c}$$

and this leads directly to the three Lagrangean (Routh–Voss) equations of the first kind:

$$m\,a_x = X + \lambda(\partial\phi/\partial x), \qquad m\,a_y = Y + \lambda(\partial\phi/\partial y), \qquad m\,a_z = Z + \lambda(\partial\phi/\partial z). \quad \text{(d)}$$

Eliminating the multiplier λ among (d) we obtain the *two* reactionless equations (with subscripts denoting partial derivatives):

$$(m\,a_x - X)/\phi_x = (m\,a_y - Y)/\phi_y = (m\,a_z - Z)/\phi_z \qquad (= \lambda). \quad \text{(e)}$$

Next:

(i) either we solve the system consisting of any two of (e), plus the constraint $\phi = 0$, for the three unknown functions $x(t)$, $y(t)$, $z(t)$, and then calculate $\lambda \to \lambda(t)$ from (d), or (e), if needed;
(ii) or we solve the system consisting of (d) and $\phi = 0$ for the four unknown functions $x(t)$, $y(t)$, $z(t)$, and $\lambda(t)$.

Equations (e) can also be obtained as follows: in view of (b), *only two out of the three virtual displacements are independent*; here $n = 3$ and $m = 1$. Taking δx as the *dependent* virtual displacement, and solving (b) for it in terms of the other two (assuming that $\phi_x \neq 0$), we obtain

$$\delta x = -(\phi_y\,\delta y + \phi_z\,\delta z)/\phi_x, \quad \text{(f)}$$

and substituting this into (a), and regrouping terms, we get the new *unconstrained* variational equation of motion

$$[m\,a_y - Y - (m\,a_x - X)(\phi_y/\phi_x)]\,\delta y \\ + [m\,a_z - Z - (m\,a_x - X)(\phi_z/\phi_x)]\,\delta z = 0. \quad \text{(g)}$$

The above, *since δy and δz are now free*, leads immediately to the two *reactionless* $\equiv$ *kinetic* equations,

$$m\,a_y = Y + (m\,a_x - X)(\phi_y/\phi_x), \qquad m\,a_z = Z + (m\,a_x - X)(\phi_z/\phi_x), \quad \text{(h)}$$

which are none other than the earlier eqs. (e).

REMARKS

(i) Equations (h) can be, fairly, called "Maggi $\to$ Hadamard equations of the first kind"; and the extension of this idea to holonomic system variables and corresponding Pfaffian constraints yields "Hadamard's equations (of the second kind)" (§3.8).

(ii) Equations (b–d = "*adjoining* of constraints") and equations (f–h = "*embedding* of constraints") embody the two available ways of handling constrained stationary problems in differential calculus; although, there, the former is discussed much more frequently than the latter.

Specialization

Let the reader verify that if the surface constraint has the special form $z = f(x, y)$, then:

(i) eqs. (d, e) reduce, respectively, to

Routh–Voss equations:

$$m\,a_x = X + \lambda(\partial f/\partial x), \qquad m\,a_y = Y + \lambda(\partial f/\partial y), \qquad m\,a_z = Z - \lambda, \tag{i}$$

Kinetic Maggi → Hadamard equations (of the first kind):

$$(m\,a_x - X)/f_x = (m\,a_y - Y)/f_y = (m\,a_z - Z)/(-1) \qquad (= \lambda), \tag{j}$$

$$\Rightarrow (m\,a_x - X) + (m\,a_z - Z)f_x = 0, \qquad (m\,a_y - Y) + (m\,a_z - Z)f_y = 0. \tag{k}$$

(ii) Substituting into (k): $a_z \equiv \ddot{z} = \cdots = \ddot{x} f_x + \ddot{y} f_y + (\dot{x})^2 f_{xx} + (\dot{y})^2 f_{yy} + 2\dot{x}\,\dot{y}\,f_{xy}$; that is, using the constraint and its $(\ldots)^{\cdot}$-derivatives to eliminate z and its derivatives from them, we obtain the two kinetic equations in x, y and their derivatives alone:

$$\ddot{x}(1 + f_x{}^2) + \ddot{y} f_x f_y + (\dot{x})^2 f_x f_{xx} + 2\dot{x}\,\dot{y}\,f_x f_{xy} + (\dot{y})^2 f_x f_{yy} = (X + f_x Z)/m, \tag{l}$$

$$\ddot{y}(1 + f_y{}^2) + \ddot{x} f_x f_y + (\dot{y})^2 f_y f_{yy} + 2\dot{x}\,\dot{y}\,f_y f_{xy} + (\dot{x})^2 f_y f_{xx} = (Y + f_y Z)/m, \tag{m}$$

[which are the "Chaplygin–Voronets"-type equations of the problem (see §3.8)].

(iii) Solving the last of (j) for λ, and then using into it the earlier expression $\ddot{z} = \cdots$, we obtain the following (kinetostatic) expression:

$$\begin{aligned} \lambda &= Z - m\ddot{z} \\ &= Z - m\Big[\ddot{x} f_x + \ddot{y} f_y + (\dot{x})^2 f_{xx} + (\dot{y})^2 f_{yy} + \dot{x}\,\dot{y}\,f_{xy}\Big]; \end{aligned} \tag{n}$$

which, once the motion has been found: $x = x(t)$, $y = y(t)$, yields the constraint reaction $\lambda = \lambda(t)$.

(iv) Finally, substituting (n) into the first and second of (i), we recover (k, l), respectively.

Example 3.5.3 Let us apply the results of the preceding example to a particle P of mass m moving under gravity on a smooth vertical plane that spins about a vertical of its straight lines, Oz (positive upward), with *constant* angular velocity ω (Kraft, 1885, vol. 2, pp. 194–195). Choosing inertial axes $O\text{–}xyz$ so that Ox coincides with the original intersection of the spinning plane and the horizontal plane $O\text{–}xy$ through the origin, we have, for the impressed forces,

$$X = 0, \qquad Y = 0, \qquad Z = +mg; \tag{a}$$

and, for the constraint,

$$y/x = \sin(\omega t)/\cos(\omega t) \;\Rightarrow\; \phi(t, x, y, z) = y\cos(\omega t) - x\sin(\omega t) = 0. \tag{b}$$

Therefore, equations (e) of the preceding example yield

$$(m\,\ddot{x} - 0)/[-\sin(\omega t)] = (m\,\ddot{y} - 0)/\cos(\omega t) = (m\,\ddot{z} - mg)/0 \quad (= -\lambda),$$

or, rearranging (to *avoid the singularity* caused by $f_z = 0$),

$$(m\,\ddot{x} - 0)\cos(\omega t) = (m\,\ddot{y} - 0)[-\sin(\omega t)] \;\Rightarrow\; \ddot{x}\cos(\omega t) + \ddot{y}\sin(\omega t) = 0, \tag{c}$$

$$(m\,\ddot{x} - 0)(0) = (m\,\ddot{z} - mg)[-\sin(\omega t)] \;\Rightarrow\; \ddot{z} = g, \tag{d}$$

$$(m\,\ddot{y} - 0)(0) = (m\,\ddot{z} - mg)\cos(\omega t) \;\Rightarrow\; \ddot{z} = g. \tag{e}$$

Let us, now, solve (c–e). Using plane polar coordinates (r, ϕ): $x = r\cos(\omega t) \Rightarrow \dot{x} = \cdots \Rightarrow \ddot{x} = \cdots$ and $y = r\sin(\omega t) \Rightarrow \dot{y} = \cdots \Rightarrow \ddot{y} = \cdots$, we can rewrite (c) in the simpler form

$$\ddot{r} - \omega^2 r = 0. \tag{f}$$

The solution of (d) = (e), with initial conditions $z(0) \equiv z_o$ and $\dot{z}(0) \equiv v_o$, is

$$z = (1/2)gt^2 + v_o t + z_o, \tag{g}$$

while that of (f), with initial conditions $r(0) = r_o$ and $\dot{r}(0) = v_{r,o}$, is

$$2\omega r = (\omega r_o + v_{r,o})e^{\omega t} + (\omega r_o - v_{r,o})e^{-\omega t}. \tag{h}$$

Equations (g, h) locate P on the spinning plane at time t, and, with the help of (b), specify its inertial position at the same time. (See also Walton, 1876, pp. 398–411.)

Example 3.5.4 *Lagrange's Equations of the First Kind; Particle on Two Surfaces.* Let us calculate the reactions on a particle P moving in space under the two constraints (where x, y, z are the inertial rectangular Cartesian coordinates of P)

$$\phi_1 \equiv x^2 + y^2 + z^2 - l^2 = 0 \quad \text{and} \quad \phi_2 \equiv z - y\tan\theta = 0; \tag{a}$$

that is, $n = 3 - 2 = 1$: for example, the bob of spherical pendulum of (constant) length l, forced to remain on the plane $\phi_2 = 0$, that makes an angle θ with the plane $z = 0$. Using commas followed by subscripts to denote partial (coordinate) derivatives, we find, from (a),

$$\delta\phi_1 = \phi_{1,x}\,\delta x + \phi_{1,y}\,\delta y + \phi_{1,z}\,\delta z = 0, \qquad \delta\phi_2 = \phi_{2,x}\,\delta x + \phi_{2,y}\,\delta y + \phi_{2,z}\,\delta z = 0. \tag{b}$$

Solving (b) for the two excess virtual displacements in terms of the third, say δy and δz in terms of δx, we obtain

$$\delta y = -(2x/J)\,\delta x \quad \text{and} \quad \delta z = -(2x\tan\theta/J)\,\delta x, \tag{c}$$

where

$$J \equiv \begin{vmatrix} \phi_{1,y} & \phi_{1,z} \\ \phi_{2,y} & \phi_{2,z} \end{vmatrix} = 2(y + z\tan\theta) \qquad (\neq 0, \text{ assumed}). \tag{d}$$

Substituting δy and δz from (c) into the principle of d'Alembert–Lagrange for the particle reaction—that is,

$$R_x\,\delta x + R_y\,\delta y + R_z\,\delta z = 0, \tag{e}$$

results in

$$\left[R_x - (2x/J)R_y - (2x\tan\theta/J)R_z\right]\delta x \equiv R'_x\,\delta x = 0, \tag{f}$$

from which, since δx is independent, we obtain $R'_x = 0$; that is,

$$R_x/(R_y + \tan\theta R_z) = x/(y + z\tan\theta). \tag{g}$$

Further, the ideal reaction postulate for $\boldsymbol{R}$:

$$\boldsymbol{R} = \lambda_1(\partial\phi_1/\partial\boldsymbol{r}) + \lambda_2(\partial\phi_2/\partial\boldsymbol{r}), \tag{h}$$

with (a) and in components, yields

$$R_x = \lambda_1\phi_{1,x} + \lambda_2\phi_{2,x} = \cdots = 2\lambda_1 x, \tag{i}$$

$$R_y = \lambda_1\phi_{1,y} + \lambda_2\phi_{2,y} = \cdots = 2\lambda_1 y - \lambda_2 \tan\theta, \tag{j}$$

$$R_z = \lambda_1\phi_{1,z} + \lambda_2\phi_{2,z} = \cdots = 2\lambda_1 z + \lambda_2; \tag{k}$$

which, of course, are consistent with (g). Finally, since $J \neq 0$, we can use any two of (i–k) to express $\lambda_{1,2}$ uniquely, in terms of $R_{x,y,z}$; for instance, solve (j, k) for $\lambda_{1,2}$ in terms of $R_{y,z}$. (See also Routh, 1891, p. 35.)

Example 3.5.5 *Lagrange's Equations of the First Kind; and Elimination of Reactions.* Let us consider a system of N particles, moving under the $h + m \equiv M$ (possibly nonholonomic but ideal) constraints

$$f_D(t, \boldsymbol{r}, \boldsymbol{v}) = 0 \qquad (D = 1, \ldots, M < 3N), \tag{a}$$

and, therefore (recalling ex. 3.2.6), having Lagrangean equations of motion of the first kind (we revert to continuum notation for convenience),

$$dm\, \boldsymbol{a} = d\boldsymbol{F} + \sum \lambda_D(\partial f_D/\partial \boldsymbol{v}). \tag{b}$$

Now, to obtain *reactionless* = *kinetic* equations of motion, we will combine (b) with the *acceleration form* of (a). To this end, we $(\ldots)^{\cdot}$-differentiate (a) once, thus obtaining

$$df_D/dt = \partial f_D/\partial t + S\big[(\partial f_D/\partial \boldsymbol{r}) \cdot \boldsymbol{v} + (\partial f_D/\partial \boldsymbol{v}) \cdot \boldsymbol{a}\big] = 0, \tag{c}$$

and from this, rearranging, we get

$$S(\partial f_D/\partial \boldsymbol{v}) \cdot \boldsymbol{a} = -S(\partial f_D/\partial \boldsymbol{r}) \cdot \boldsymbol{v} - \partial f_D/\partial t. \tag{d}$$

Now, to be able to use (d) in (b), so as to eliminate $\boldsymbol{a}$, we dot the latter with $\partial f_D/\partial \boldsymbol{v}$ and sum over the particles (with $D, D' = 1, \ldots, M$):

$$S(\partial f_D/\partial \boldsymbol{v}) \cdot \boldsymbol{a} = S[(d\boldsymbol{F}/dm) \cdot (\partial f_D/\partial \boldsymbol{v})] + \sum \lambda_{D'}\Big(S(\partial f_{D'}/\partial \boldsymbol{v}) \cdot (\partial f_D/\partial \boldsymbol{v})/dm\Big), \tag{e}$$

and, comparing the right sides of the above with (d), we readily conclude that

$$\sum \lambda_{D'}\Big\{S[\partial f_{D'}/\partial(dm\boldsymbol{v})] \cdot (\partial f_D/\partial \boldsymbol{v})\Big\}$$
$$= -S\{d\boldsymbol{F} \cdot [\partial f_D/\partial(dm\boldsymbol{v})]\} - S(\partial f_D/\partial \boldsymbol{r}) \cdot \boldsymbol{v} - \partial f_D/\partial t. \tag{f}$$

Since $rank[\partial f_D/\partial(dm\boldsymbol{v})] = rank(\partial f_D/\partial \boldsymbol{v}) = M$, and therefore

$$Det\Big\{S[\partial f_{D'}/\partial(dm\boldsymbol{v})] \cdot (\partial f_D/\partial \boldsymbol{v})\Big\} \neq 0, \tag{g}$$

the M linear nonhomogeneous equations (f) can supply uniquely (locally, at least) the λ_D's as functions of the $\boldsymbol{r}$'s, $\boldsymbol{v}$'s, and t. Finally, substituting the so-calculated λ_D's back into (b), we obtain N second-order equations for the $\boldsymbol{r} = \boldsymbol{r}(t)$. (See also exs. 3.10.2, 5.3.5, and 5.3.6; and Voss, 1885.)

Problem 3.5.1 Continuing from the preceding example, find the form that (f) takes if the constraints (a) have the *holonomic* form

$$\phi_H(t, \mathbf{r}) = 0. \tag{a}$$

HINT

Calculate $f_D \equiv d\phi_D/dt = 0$, and then show that $\partial f_D/\partial \mathbf{v} = \partial \phi_D/\partial \mathbf{r}$.

Let us now turn to the second, and more important, "reactionless part" of LP, (eqs. 3.2.8, 11), and express it in system variables.

1. Holonomic Variables

In this case LP, $\delta I = \delta' W$, assumes the form

$$\sum E_k\, \delta q_k = \sum Q_k\, \delta q_k,$$

or, explicitly,

$$\sum \{[d/dt(\partial T/\partial \dot{q}_k) - \partial T/\partial q_k] - Q_k\}\, \delta q_k$$
$$\equiv \sum \{[d/dt(\partial T/\partial v_k) - \partial T/\partial q_k] - Q_k\}\, \delta q_k = 0. \tag{3.5.11}$$

This *differential variational equation* is fundamental to Lagrangean analytical mechanics; all conceivable/possible Lagrangean equations of motion are based on it and flow from it.

(a) Now, if the n δq's are independent (i.e., $m = 0 \Rightarrow f = n$ DOF), (3.5.11) leads immediately to *Lagrange's equations of the second kind*: $E_k = Q_k$; or explicitly (recalling the kinematico-inertial results of §3.3 in holonomic variables),

$$E_k \equiv d/dt(\partial T/\partial \dot{q}_k) - \partial T/\partial q_k$$
$$\equiv d/dt(\partial T/\partial v_k) - \partial T/\partial q_k = Q_k \qquad \text{[Lagrange (1780)]}, \tag{3.5.12}$$
$$\equiv \partial S/\partial \ddot{q}_k \equiv \partial S/\partial \dot{v}_k \equiv \partial S/\partial w_k = Q_k \qquad \text{[Appell (1899)]}. \tag{3.5.13}$$

Further, substituting $\delta q_k = \sum A_{kl}\, \delta\theta_l$ $(k, l = 1, \ldots, n)$ into (3.5.11) readily yields

$$\sum A_{kl} E_k = \sum A_{kl} Q_k \qquad (\textit{i.e., } I_l = \Theta_l, \textit{ but in holonomic variables})$$
$$\text{[Maggi (1896, 1901, 1903)]}. \tag{3.5.14}$$

However, in this unconstrained case, neither Appell's equations, (3.5.13), nor Maggi's equations, (3.5.14), offer any particular advantages over those of Lagrange, (3.5.12); their real usefulness/advantages over eqs. (3.5.12) lie in the constrained case (see below). Equations (3.5.12) are rightfully considered among the most important ones of the entire mathematical physics and engineering; we shall call them simply *Lagrange's equations*.

(b) If the n δq's are *constrained* by (3.5.3): $\sum a_{Dk}\, \delta q_k = 0$ $(D = 1, \ldots, m;$ $k = 1, \ldots, n)$, that is, $f \equiv n - m =$ number of DOF, then application of the multiplier rule, between these constraints and (3.5.11), leads immediately to the *Routh–Voss equations*

$$E_k = Q_k + \sum \lambda_D a_{Dk} \ (\equiv Q_k + R_k); \tag{3.5.15}$$

or, explicitly, as in (3.5.12, 13),

$$E_k \equiv d/dt(\partial T/\partial \dot{q}_k) - \partial T/\partial q_k = Q_k + \sum \lambda_D a_{Dk}$$

[Routh (1877), Voss (1885)], (3.5.16)

$$\equiv \partial S/\partial \ddot{q}_k = Q_k + \sum \lambda_D a_{Dk} \text{ (Appellian form of the Routh–Voss eqs.).} \tag{3.5.17}$$

The corresponding Maggi form is presented below.

[Equations (3.5.15) are not to be confused with the other, more famous, equations of Routh of *steady motion*, etc. (§8.3 ff.)]

CAUTION

Some authors (e.g., Haug, 1992, pp. 169–170) state, falsely, that if the n q's are independent, the n δq's are arbitrary, and then (3.5.12, 13) follow from (3.5.11). But as we have seen (§2.3, §2.8, and §2.12), if the additional constraints (3.5.3, 10b) are nonholonomic the q's remain independent, whereas, obviously, the δq's are no longer arbitrary, that is, (3.5.12, 13) do *not* always hold for independent q's.

2. Holonomic → Nonholonomic Variables

In this case LP, $\delta I = \delta' W$, assumes the form

$$\sum I_k \,\delta\theta_k = \sum \Theta_k \,\delta\theta_k. \tag{3.5.18}$$

(a) If the n $\delta\theta$'s are *unconstrained* (i.e., if $m = 0 \Rightarrow f \equiv n - m = n = \#\, DOF$), then (3.5.18) leads to $I_k = \Theta_k$, or, due to the kinematico-inertial identities (3.3.10 ff.) for I_k, to the following *three* general forms:

$$I_k \equiv \sum A_{lk} E_l = \sum A_{lk} Q_l,$$

or, in extenso,

$$\sum [d/dt(\partial T/\partial \dot{q}_l) - \partial T/\partial q_l] A_{lk} = \sum A_{lk} Q_l$$

(*Maggi form:* holonomic variables), (3.5.19a)

$$\equiv \sum A_{lk}(\partial S/\partial \ddot{q}_l) = \sum A_{lk} Q_l$$

(*Appellian form of Maggi form:* holonomic variables), (3.5.19b)

$$\equiv \partial S^*/\partial \dot{\omega}_k = \Theta_k$$

[Gibbs (1879): nonholonomic variables, but no constraints!], (3.5.19c)

$$\equiv d/dt(\partial T^*/\partial \omega_k) - \partial T^*/\partial \theta_k + \sum\sum \gamma^r{}_{k\alpha}(\partial T^*/\partial \omega_r)\omega_\alpha = \Theta_k$$

$$\equiv E_k{}^*(T^*) - \Gamma_k = \Theta_k$$

[Volterra (1898), Hamel (1903–1904)]. (3.5.19d)

Equations (3.5.19a, b) have no advantages over Lagrange's equations (3.5.12); but equations (3.5.19c, d) may be truly useful for unconstrained systems in quasi variables, for example, a rigid body moving about a fixed point [→ Eulerian rotational equations (§1.17)].

(b) If the $\delta\theta$'s are *constrained* by $\delta\theta_D = 0$, but $\delta\theta_I \neq 0$ (i.e., if $f \equiv n - m =$ # *DOF*), then the multiplier rule applied to (3.5.18) yields the following *two* groups of equations:

Kinetostatic (i.e., reaction containing) equations:

$$I_D = \Theta_D + \Lambda_D \qquad [= \Theta_D + \lambda_D \quad (D = 1, \ldots, m)], \tag{3.5.20a}$$

Kinetic (i.e., reactionless) equations:

$$I_I = \Theta_I \qquad (I = m + 1, \ldots, n); \tag{3.5.20b}$$

[and in view of the constraint $1 \cdot \delta\theta_{n+1} = 1 \cdot \delta t = 0$, we also have $I_{n+1} = \Theta_{n+1} + \Lambda_{n+1}$, but that nonvirtual relation is more of an *energy rate*–like equation (as in §3.9)].

Alternative Derivation of Equations (3.5.20a, b)

First, with the help of the Kronecker delta (hopefully, not to be confused with the virtual variation symbol $\delta \ldots$), we rewrite the constraints $\delta\theta_D = 0$ as

$$0 = \delta\theta_D = \sum \delta_{DD'}\, \delta\theta_{D'} = \sum \delta_{DD'}\, \delta\theta_{D'} + \sum \delta_{DI}\, \delta\theta_I = \sum \delta_{Dk}\, \delta\theta_k. \tag{3.5.20c}$$

Then, using the method of Lagrangean multipliers, we combine them with (3.5.18): (1) we multiply each constraint $\delta\theta_D(= 0)$ with $-\lambda_D(\neq 0)$ and sum over D; (2) we multiply each "nonconstraint" $\delta\theta_I(\neq 0)$ with $-\lambda_I(= 0)$ and sum over I; and, (3), we add the so-resulting two zeros to (3.5.18), thus obtaining

$$\sum \left(I_k - \Theta_k - \sum \lambda_D\, \delta_{Dk}\right) \delta\theta_k = 0. \tag{3.5.20d}$$

Since the $\delta\theta_k$ can now be viewed as unconstrained, (3.5.20d) decouples to the two sets of equations:

$$k = D': \qquad I_{D'} - \Theta_{D'} = \sum \lambda_D\, \delta_{DD'} = \lambda_{D'} \qquad (\textit{Kinetic equations}), \tag{3.5.20e}$$

$$k = I: \qquad I_I - \Theta_I = \sum \lambda_D\, \delta_{DI} = \lambda_I = 0 \qquad (\textit{Kinetostatic equations}). \tag{3.5.20f}$$

Here, too, as with the unconstrained case (3.5.19a–d), we have the following *three* general forms for (3.5.20a) and (3.5.20b):

- Kinetic equations (with $I, I' = m + 1, \ldots, n$; $k = 1, \ldots, n$; $\gamma^r{}_I \equiv \gamma^r{}_{I,n+1}$):

$$I_I \equiv \sum A_{kI} E_k = \sum A_{kI} Q_k,$$

or, in extenso,

$$\sum [d/dt(\partial T/\partial \dot{q}_k) - \partial T/\partial q_k] A_{kI} = \sum A_{kI} Q_k$$

[Maggi (1896, 1901, 1903): holonomic variables], (3.5.21a)

$$\equiv \sum A_{kI}(\partial S/\partial \ddot{q}_k) = \sum A_{kI} Q_k$$

(Appellian form of Maggi form: holonomic variables), (3.5.21b)

$$\equiv \partial S^*/\partial \dot{\omega}_I = \Theta_I$$

[Appell (1899–1925): special cases of nonholonomic variables], (3.5.21c)

$$\equiv d/dt(\partial T^*/\partial \omega_I) - \partial T^*/\partial \theta_I + \sum\sum \gamma^k{}_{II'}(\partial T^*/\partial \omega_k)\omega_{I'} + \sum \gamma^k{}_I(\partial T^*/\partial \omega_k) = \Theta_I$$

[Hamel (1903–1904): "Lagrange–Euler equations"]; (3.5.21d)

- Kinetostatic equations (with $D = 1, \ldots, m$; $I = m+1, \ldots, n$; $k = 1, \ldots n$; $\gamma^r{}_D \equiv \gamma^r{}_{D,n+1}$) :

$$I_D \equiv \sum A_{kD} E_k = \sum A_{kD} Q_k + \Lambda_D,$$

or, in extenso,

$$\sum [d/dt(\partial T/\partial \dot{q}_k) - \partial T/\partial q_k] A_{kD} = \sum A_{kD} Q_k + \Lambda_D, \quad (3.5.22a)$$

$$\equiv \sum A_{kD}(\partial S/\partial \ddot{q}_k) = \sum A_{kD} Q_k + \Lambda_D, \quad (3.5.22b)$$

$$\equiv \partial S^*/\partial \dot{\omega}_D = \Theta_D + \Lambda_D$$

[Cotton (1907): special variables] (3.5.22c)

$$\equiv d/dt(\partial T^*/\partial \omega_D) - \partial T^*/\partial \theta_D + \sum\sum \gamma^k{}_{DI}(\partial T^*/\partial \omega_k)\omega_I + \sum \gamma^k{}_D(\partial T^*/\partial \omega_k) = \Theta_D + \Lambda_D$$

[Stückler (1955); special case by Schouten (late 1920s, 1954)]. (3.5.22d)

REMARKS

(i) In the absence of constraints, the above n equations in the ω's, plus the n transformation equations $\dot{q}_k = \sum A_{kI}\omega_I + A_k$, constitute a system of $2n$ *first-order* equations in the $2n$ unknown functions $\omega_k = \omega_k(t)$ and $q_k = q_k(t)$. [Or, after using the $\omega \leftrightarrow \dot{q}$ equations in them, thus expressing the ω_k's in terms of the $\dot{q}_k$'s, they constitute a set of n *second-order* equations for the n unknowns $q_k(t)$.] In the presence of m constraints $\omega_D = 0$, the $n - m$ kinetic equations plus the n transformation equations $\dot{q}_k = \sum A_{kI}\omega_I + A_k$ constitute a system of $2n - m$ *first-order* equations in the $2n - m$ functions $\omega_I = \omega_I(t)$ and $q_k = q_k(t)$. Or, equivalently, substituting $\omega_I = \sum a_{Ik}\dot{q}_k + a_I$ $(\neq 0)$ into the $n - m$ kinetic equations, we obtain a system of $n - m$ *second-order* equations for the $q_k = q_k(t)$; and then, pairing them with the m constraints $\sum a_{Dk}\dot{q}_k + a_D = 0$, we finally obtain a system of $(n - m) + m = n$ *second-order* reactionless equations for the $q_k = q_k(t)$. Further, it can be shown that there exists a nonsingular linear transformation $\dot{q}_k = \sum A_{kI}\omega_I + A_k$, or

$\dot{q}_\alpha = \sum A_{\alpha\beta}\omega_\beta$ (recalling that $dq_{n+1}/dt = \omega_{n+1} = dt/dt = 1$) that brings the *nonnegative* kinetic energy $2T = \sum\sum M_{\alpha\beta}\dot{q}_\alpha\dot{q}_\beta$ to the following *sum of squares* form:

$$2T \to 2T^* = \sum \omega_\alpha{}^2 = \sum \omega_k{}^2 + \omega_{n+1}{}^2; \tag{3.5.23a}$$

in which case, since $P_k \equiv \partial T^*/\partial\omega_k = \omega_k$ and $P_{n+1} \equiv \partial T^*/\partial\omega_{n+1} = \omega_{n+1} = 1$, $\partial T^*/\partial\theta_k \equiv \sum(\partial T^*/\partial q_l)A_{lk} = 0$, the nonholonomic system inertia assumes the *Eulerian form* (recall inertia side of Eulerian rigid-body rotational equations, §1.17):

$$I_k = d\omega_k/dt + \sum\sum \gamma^r{}_{k\alpha}\,\omega_r\,\omega_\alpha; \tag{3.5.23b}$$

and that is why Hamel called *his* equations "Lagrange–Euler equations." (However, by choosing the ω_k's so as to nullify the $\partial T^*/\partial\theta_k$'s, we probably end up complicating the $\gamma^r{}_{k\alpha}$'s.)

(ii) The advantage of nonholonomic *variables* in the Hamel "equilibrium form" $\omega_D = 0$ is that then both constraints and equations of motion *decouple* naturally into $n - m$ *purely kinetic* (i.e., *reactionless*) *equations* ($\delta\theta_I \neq 0$; $\Lambda_I = 0 \Rightarrow I_I = \Theta_I$) *and m reaction-containing, or kinetostatic, equations* ($\delta\theta_D = 0$; $\Lambda_D \neq 0 \Rightarrow I_D = \Theta_D + \Lambda_D$). In holonomic variables, by contrast, both (Pfaffian) constraints and (Routh–Voss and Appell) equations of motion are coupled. Solving the $n - m$ *kinetic* equations (plus constraints, etc.) constitutes the lion's share of the difficulty of the problem. Once this has been achieved, then the reactions Λ_D follow immediately from the (now) *algebraic* equations: $\Lambda_D = \Lambda_D(t) = I_D(t) - \Theta_D(t)$.

(iii) When using Hamel's equations under the constraints $\omega_D = 0$, we must enforce the latter *after* all partial differentiations have been carried out, not before; otherwise, we would not, in general, calculate correctly the key nonholonomic terms ($k, r = 1, \ldots, n$; $\alpha = 1, \ldots, n+1$; $I' = m+1, \ldots, n$)

$$-\Gamma_k = \sum\sum \gamma^r{}_{k\alpha}(\partial T^*/\partial\omega_r)\omega_\alpha = \sum\sum \gamma^r{}_{kI'}(\partial T^*/\partial\omega_r)\omega_{I'} + \sum \gamma^r{}_k(\partial T^*/\partial\omega_r); \tag{3.5.24a}$$

and, unfortunately, this drawback holds for both kinetic and kinetostatic equations. Let us see why. Expanding T^* à la Taylor around $\omega_D = 0$, we obtain

$$T^* = T^*{}_o + \sum(\partial T^*/\partial\omega_D)_o\,\omega_D + \text{quadratic terms in } \omega_D, \tag{3.5.24b}$$

where

$$T^*{}_o = T^*(q, \omega_D = 0, \omega_I, t) = T^*{}_o(q, \omega_I, t); \tag{3.5.24c}$$

and, generally, $(\ldots)_o \equiv (\ldots, \omega_D = 0, \ldots)$ (a useful notation, to be utilized *frequently*, for extra clarity); and, therefore,

$$(\partial T^*/\partial\omega_D)_o \neq \partial T^*{}_o/\partial\omega_D = 0, \tag{3.5.24d}$$

$$(\partial T^*/\partial\omega_I)_o = \partial T^*{}_o/\partial\omega_I \;\Rightarrow\; d/dt[(\partial T^*/\partial\omega_I)_o] = d/dt(\partial T^*{}_o/\partial\omega_I), \tag{3.5.24e}$$

$$(\partial T^*/\partial\theta_k)_o \equiv \sum A_{rk}(\partial T^*/\partial q_r)_o = \sum A_{rk}(\partial T^*{}_o/\partial q_r). \tag{3.5.24f}$$

In view of these results, $-\Gamma_k$, (3.5.24a), transforms to

$$-\Gamma_k = \sum\sum \gamma^D{}_{kI'}(\partial T^*/\partial\omega_D)_o\,\omega_{I'} + \sum\sum \gamma^{I''}{}_{kI'}(\partial T^*{}_o/\partial\omega_{I''})\omega_{I'}$$
$$+\sum \gamma^D{}_k(\partial T^*/\partial\omega_D)_o + \sum \gamma^I{}_k(\partial T^*{}_o/\partial\omega_I); \tag{3.5.24g}$$

an expression that shows clearly that the presence of the *first* (*double*) *and third* (*single*) *sums* generally necessitates the use of T^*, instead of $T^*{}_o$.

However, with the help of the above expression we can obtain conditions that tell us when we can use the constrained kinetic energy $T^*{}_o$ in Hamel's equations right from the start. Let us do this, for simplicity, for the common case of the *kinetic* such equations of a *scleronomic* system. Then,

$$-\Gamma_k \to -\Gamma_I = \sum\sum \gamma^D{}_{II'}(\partial T^*/\partial\omega_D)_o\,\omega_{I'} + \sum\sum \gamma^{I''}{}_{II'}(\partial T^*{}_o/\partial\omega_{I''})\omega_{I'}; \tag{3.5.24h}$$

and (3.5.24d, e) make it clear that the sought conditions will result from the (identical) vanishing of the *first* sum in (3.5.24h); that is,

$$\sum\sum \gamma^D{}_{II'}(\partial T^*/\partial\omega_D)_o\,\omega_{I'} = 0. \tag{3.5.24i}$$

But (as made clear in §3.9),

$$2T^* = \sum\sum M^*{}_{kl}(q)\omega_k\,\omega_l = \sum\sum (\partial^2 T^*/\partial\omega_k\,\partial\omega_l)\omega_k\,\omega_l$$
$$\Rightarrow\ (\partial T^*/\partial\omega_D)_o = \sum (\partial^2 T^*/\partial\omega_D\,\partial\omega_I)\omega_I, \tag{3.5.24j}$$

and so (3.5.24i) reduces to

$$\sum\sum\sum \gamma^D{}_{II'}(\partial^2 T^*/\partial\omega_D\,\partial\omega_{I''})\,\omega_{I''}\,\omega_{I'} = 0; \tag{3.5.24k}$$

and from this we easily conclude that the *necessary and sufficient conditions for the use of* $T^*{}_o$ *in Hamel's equations* are

$$\sum \gamma^D{}_{II'}(\partial^2 T^*/\partial\omega_D\,\partial\omega_{I''}) = 0. \tag{3.5.24l}$$

For example, in the case of a *single* Pfaffian constraint, $\omega_1 = 0$ (i.e., $m = 1$), (3.5.24l) yields

$$\gamma^1{}_{II'}(\partial^2 T^*/\partial\omega_1\,\partial\omega_{I''}) = 0 \qquad (I, I', I'' = 2, \ldots, n), \tag{3.5.24m}$$

which means that either all "nonholonomic inertial coefficients" $\partial^2 T^*/\partial\omega_1\,\partial\omega_{I''} \equiv M^*{}_{1I''}$ vanish [i.e., T^* consists of an ω_1-free part and an $\omega_1{}^2$-proportional part; or $\gamma^1{}_{II'} = 0$, which means that constraint is holonomic (by Frobenius' theorem, §2.12)]. The consequences of (3.5.24l) are detailed in Hamel (1904(a), pp. 22–29); see also Hadamard (1895).

In sum, in using the Hamel equations, and even if we are not interested in constraint reactions, we must begin with the unconstrained kinetic energy $T^* = T^*(q, \omega_D, \omega_I, t)$, carry out all required differentiations, and then enforce the constraints $\omega_D = 0$, at the end; and a constraint $\omega_D = 0$ can be enforced ahead of time in T^*-terms that are *quadratic* in that ω_D; namely, in $(\ldots)\omega_D{}^2$ terms.

This inconvenience is a small price to pay for such powerful and conceptually insightful equations. Similarly, a detailed analysis of (3.5.24g) shows that *it is possible to have* $\Gamma_k = 0$ (i.e., Hamel equations $\to$ Lagrange's equations) *even though not all* γ*'s are zero*. An analogous situation occurs in the Maggi equations, even in the kinetic case—that is,

$$I_I \equiv \sum A_{kI} E_k \equiv \sum [d/dt(\partial T/\partial \dot{q}_k) - \partial T/\partial q_k] A_{kI} = \sum A_{kI} Q_k, \qquad (3.5.24n)$$

since $k = 1, \ldots, n$, we have to calculate $T = T(t, q, \dot{q}_D, \dot{q}_I)$; the "reduced," or constrained, kinetic energy

$$T_o \equiv T(t, q, \dot{q}_D = \sum b_{DI} \dot{q}_I + b_D, \dot{q}_I) \equiv T_o(t, q, \dot{q}_I), \qquad (3.5.24o)$$

obviously will not do. This seems to be a drawback of all T-based (i.e., Lagrangean) equations. No such problems appear for the kinetic *Appellian* equations: there, with the convenient notation

$$S^* = S^*(t, q, \omega_D, \omega_I, \dot{\omega}_D, \dot{\omega}_I) = \textit{original, or unconstrained, or relaxed, Appellian}$$
$$\to S^*(t, q, \omega_D = 0, \omega_I, \dot{\omega}_D = 0, \dot{\omega}_I) \equiv S^*{}_o(t, q, \omega_I, \dot{\omega}_I)$$
$$\equiv S^*{}_o = \textit{constrained Appellian}, \qquad (3.5.25a)$$

and the help of the Taylor expansion (with some obvious calculus notations)

$$S^* = S^*{}_o + \sum \left[(\partial S^*/\partial \omega_D)_o \omega_D + (\partial S^*/\partial \dot{\omega}_D)_o \dot{\omega}_D\right] + \textit{quadratic terms in } \omega_D, \dot{\omega}_D, \qquad (3.5.25b)$$

we get the general results [similar to (3.5.24c, e)]

$$(\partial S^*/\partial \dot{\omega}_I)_o = \partial S^*{}_o/\partial \dot{\omega}_I \qquad \text{and} \qquad (\partial S^*/\partial \dot{\omega}_D)_o \neq \partial S^*{}_o/\partial \dot{\omega}_D = 0. \qquad (3.5.25c)$$

Therefore, if we are not interested in finding constraint reactions, we *can* enforce the constraints $\omega_D = 0$ and $\dot{\omega}_D = 0$ into S^* right from the beginning; that is, start working with $S^*{}_o$, and thus save a considerable amount of labor. This property, due to the first of (3.5.25c), marks a key difference between the equations of Appell and Hamel, and their corresponding special cases.

Special Case

If all constraints on the q's are *holonomic* and have the *equilibrium* form $\theta_D \to q_D = \textit{constant} \equiv q_{Do}$, then

$$(\partial S^*/\partial \dot{\omega}_I)_o \to E_I(T)|_o \to E_I(T_o),$$

where $T_o \equiv T(t, q_D = \textit{constant},\ q_I,\ \dot{q}_D = 0,\ \dot{q}_I) \equiv T_o(t, q_I, \dot{q}_I)$, and, similarly for the impressed forces, $Q_I = Q_I(t, q, \dot{q}) \to Q_I(t, q_I, \dot{q}_I) \equiv Q_{Io}$, and so the kinetic equations become $E_I \equiv E_I(T_o) \equiv \partial S_o/\partial \ddot{q}_I = Q_{Io}$.

(iv) *Comparison between Lagrange's equations of the first and second kind, and their respective constraints.* Those of the *first* kind, eq. (ex. 3.5.1: c3), constitute a set of

$$3N + (h + m) = [(3N - h) + m] + 2h \equiv (n + m) + 2h$$

scalar equations, for the $3N + (h + m)$ unknown functions:

$$\{x_P, y_P, z_P;\ P = 1, \ldots, N\}, \qquad \{\mu_H;\ H = 1, \ldots, h\} \qquad \text{and} \qquad \{\lambda_D;\ D = 1, \ldots, m\}.$$

Once the *positions* (and hence accelerations) and *multipliers* become known functions of time, (ex. 3.5.1: c2) supply the reactions.

Those of the *second* kind, actually the Routh–Voss equations (3.5.15, 16), constitute a set of

$$n + m \equiv (3N - h) + m$$

equations for the $n + m$ unknowns:

$$\{q_k;\ k = 1, \ldots, n\} \qquad \text{and} \qquad \{\lambda_D;\ D = 1, \ldots, m\}.$$

Once the q's and λ's have been found as functions of time, then $\boldsymbol{r}_P = \boldsymbol{r}_P(t, q) \to \boldsymbol{r}_P(t)$ $[\to \boldsymbol{a}_P = \boldsymbol{a}_P(t)]$, and, again, (ex. 3.5.1: c2) supply the reactions. From the latter and the (now) known λ's, we can calculate the μ's.

In sum, in the *second-kind case* we have $2h$ *fewer equations*, which is the result of having absorbed the h holonomic constraints into that description with the $n \equiv 3N - h$ q's [see remark (v) below]. Also, even in the presence of additional holonomic and/or nonholonomic constraints, we still work with the *unconstrained* kinetic energy T.

However, and this is a general comment, the ultimate judgement regarding the relative merits of various types of equations of motion must be shaped by several, frequently intangible/nonquantifiable considerations (in the sense of the famous Machian principle of *Denkökonomie*), in addition to the mere tallying of their number of equations, and so on ("bean counting").

(v) *Purpose for appearance of the multipliers.* That the multipliers μ_H, of the h holonomic constraints $\phi_H(t, \boldsymbol{r}) = 0$, are not present in Lagrange's equations of the second kind (and in the Routh–Voss equations) is no accident: the m λ_D's (and this is a general remark) express the reactions of whatever constraints have not been taken care of by our chosen q's; that is, *they are due to the additional holonomic and/or nonholonomic constraints not yet built in (or embedded, or absorbed) into our particular q's description.* Then, the multipliers appear as coefficients in the virtual work of the reactions of these additional constraints.

(vi) *Apparent indeterminacy of Lagrange's equations.* Let us consider a system with equations of motion

$$E_k \equiv d/dt(\partial T/\partial \dot{q}_k) - \partial T/\partial q_k = Q_k. \tag{3.5.26a}$$

Since, as explained earlier, all possible constraints are already built in into the chosen q-description, the corresponding system constraint reactions R_k have been eliminated from the right side of (3.5.26a); the Q_k are wholly impressed. However, occasionally, the latter depend on constraint reactions: for example, the *sliding* Coulomb–Morin friction F on a particle sliding on a rough surface—according to our definition, an impressed force—is given by

$$-\mu N(\boldsymbol{v}/|\boldsymbol{v}|), \tag{3.5.26b}$$

where N = *normal* force from surface to particle (clearly, a contact constraint *reaction*), μ = *sliding friction coefficient*, and $\boldsymbol{v}$ = particle *velocity relative to the surface*. In such a case, if we embed *all* holonomic constraints into our q's, and hence into our T

and Q_k's, the resulting Lagrangean equations (3.5.26a) will, in general, constitute an *indeterminate* system; that is, the total number of equations, including constitutive ones like (3.5.26b), will be smaller than the number of unknowns involved. Such an indeterminacy [what Kilmister and Reeve (1966, p. 215) call "failure" of Lagrange's equations] can be easily removed by *relaxing* the system's constraints, and thus generating the hitherto missing equations (see also "principle of relaxation" in §3.7). Similar "failures" would appear if one used minimal quasi velocities to embed all nonholonomic constraints (see also Rosenberg, 1977, pp. 152–157).

(vii) We have presented the *four* basic types of equations of motion: *Routh–Voss, Maggi, Hamel,* and *Appell.* They can be classified as follows:

Kinetic energy-based equations of motion

Holonomic variables: Routh-Voss (coupled)

Maggi (uncoupled: kinetic, kinetostatic)

Nonholonomic variables: Hamel (uncoupled: kinetic, kinetostatic)

Acceleration-based equations of motion

Holonomic variables: Appell (coupled)

Nonholonomic variables: Appell (uncoupled: kinetic, kinetostatic)

Additional special cases and/or combinations of the above—for example, equations of Ferrers, Hadamard, Chaplygin, Voronets, et al.—are presented in §3.8.

• From all the equations of constrained motion given earlier, *only those by Hamel* (and their special cases—see §3.8), through their γ-proportional terms (recall Hamel's formulation of Frobenius' theorem, §2.12), can distinguish between genuinely nonholonomic Pfaffian constraints and holonomic ones disguised in Pfaffian/velocity form. All other types, that is the equations of Routh–Voss, Maggi, Appell (and their special cases—see §3.8), hold *unchanged* in form whether their Pfaffian constraints are holonomic or nonholonomic; that is, *those equations cannot detect nonholonomicity, only Hamel's equations can do that*

• On the other hand, only Appell's equations preserve their form in *both holonomic and nonholonomic variables*; and, in the kinetic ones, the nonholonomic constraints can be enforced in the Appellian function right from the start.

(viii) The terms *kinetic* and *kinetostatic*, in the particular sense used here (brought to mainstream dynamics by Heun and his students, in the early 20th century), and observed by some of the best contemporary textbooks on engineering dynamics, for example, Butenin et al. (1985, vol. 2, chap. 16, pp. 330–339), Loitsianskii and Lur'e (1983, vol. 2, chap. 28, pp. 345–384), Ziegler (1965, vol. 2, pp. 146–152), are not well known among English language authors, and so one should be careful in comparing various references.

(ix) Finally, we would like to state that we are not partial to any particular set of equations of motion; all have advantages and disadvantages; all are worth learning!

All such conceivable equations (whose combinations and special cases are practically endless; see also §3.8) flow out of the differential variational principles of analytical mechanics; that is, the principles of *Lagrange* and of *relaxation of the constraints*, in their various forms (see also §3.6 and §3.7). These principles, being *invariant*, constitute the sole physical and mathematical glue that holds all these (coordinate and constraint-dependent) equations of motion together—and they keep reminding us that, in spite of appearances, *there is only one* (*classical*) *mechanics!*

Geometrical Interpretation of the Uncoupling of the Equations of Motion into Kinetic and Kinetostatic

The Routh–Voss equations,

$$E_k \equiv d/dt(\partial T/\partial \dot{q}_k) - \partial T/\partial q_k = Q_k + \sum \lambda_D a_{Dk}, \tag{3.5.27a}$$

represent an equation among (covariant) components of vectors at a point (q) in configuration space, or a point (t, q) in event space. Now, we recall from §2.11, eq. (2.11.19a ff.), that the $n - m$ vectors $\boldsymbol{A_I}^T = (A_{1I}, \ldots, A_{nI})$ span, at that point, the *null*, or *virtual*, hyperplane (or affine space) $N_I \equiv V$, of the *constraint matrix* $\mathbf{A_D} = (a_{Dk})$; while the m vectors $\boldsymbol{A_D}^T = (A_{1D}, \ldots, A_{nD})$ span its orthogonal complement, the *range*, or *constraint*, hyperplane (or affine space) C_m. Therefore, multiplying (3.5.27a) with $A_{kI}(A_{kD})$ and then summing over k, from 1 to n, means projecting that equation onto the local *virtual* (*constraint*) space; and since the constraint reactions $R_k = \sum \lambda_D a_{Dk}$ are perpendicular to the virtual space, they disappear from the kinetic Maggi equations. Indeed, we have, successively,

(i)
$$\sum A_{kI} E_k = \sum A_{kI} Q_k + \sum\sum \lambda_D a_{Dk} A_{kI}$$
$$= \sum A_{kI} Q_k + \sum \lambda_D\, \delta_{DI} = \sum A_{kI} Q_k + 0,$$

that is,

$$\sum A_{kI} E_k = \sum A_{kI} Q_k \quad \text{or} \quad I_I = \Theta_I; \tag{3.5.27b}$$

(ii)
$$\sum A_{kD'} E_k = \sum A_{kD'} Q_k + \sum\sum \lambda_D a_{Dk} A_{kD'}$$
$$= \sum A_{kD'} Q_k + \sum \lambda_D\, \delta_{DD'} = \sum A_{kD'} Q_k + \lambda_{D'},$$

that is,

$$\sum A_{kD} E_k = \sum A_{kD} Q_k + \lambda_D \quad \text{or} \quad I_D = \Theta_D + \lambda_D. \tag{3.5.27c}$$

Tensorial Treatment

(Kinetic complement of comments made at the end of §2.11; may be omitted in a first reading.) In the language of tensors (whose general indicial notation begins to show its true simplicity and power here), the I_I, Θ_I, $\Lambda_I = 0$ $(I_D, \Theta_D, \Lambda_D = \lambda_D)$ are *covariant* components of the corresponding system vectors along the *contravariant* basis $\boldsymbol{A}^I(\boldsymbol{A}^D)$, which is *dual* to the earlier basis $\boldsymbol{A}_I(\boldsymbol{A}_D)$. Dotting the vectorial Routh–Voss equations [fig. 3.1(a)]

$$\boldsymbol{E} = \boldsymbol{Q} + \boldsymbol{R}, \tag{3.5.28a}$$

where (with summation convention) $\boldsymbol{R} = R_k \boldsymbol{E}^k = (\lambda_D\, a^D{}_k)\boldsymbol{E}^k = \lambda_D \boldsymbol{A}^D$ (i.e., $\boldsymbol{R}$ is perpendicular to the virtual local plane) with $\boldsymbol{A}_I = A^k{}_I \boldsymbol{E}_k$—that is, projecting it onto the virtual local plane — yields

$$\boldsymbol{E} \cdot \boldsymbol{A}_I = \boldsymbol{Q} \cdot \boldsymbol{A}_I + \boldsymbol{R} \cdot \boldsymbol{A}_I, \tag{3.5.28b}$$

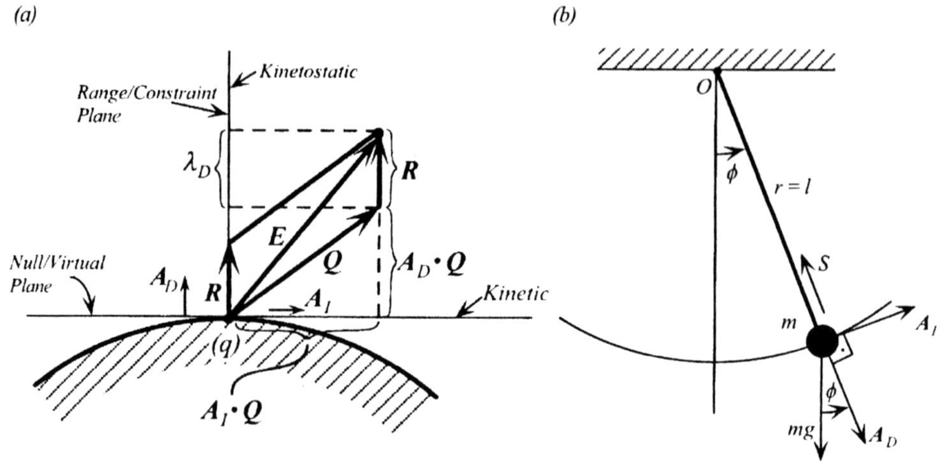

Figure 3.1 (a) Geometrical interpretation of uncoupling of equations of motion ("Method of projections" of Maggi); and (b) its application to the planar mathematical pendulum.

or, since $\boldsymbol{R}\cdot\boldsymbol{A}_I = \lambda_D(\boldsymbol{A}^D\cdot\boldsymbol{A}_I) = \lambda_D\,\delta^D{}_I = 0$, finally $\boldsymbol{E}\cdot\boldsymbol{A}_I = \boldsymbol{Q}\cdot\boldsymbol{A}_I$,

$$\text{i.e.,}\quad A^k{}_I E_k = A^k{}_I Q_k \ \text{(kinetic Maggi)}, \qquad \text{or} \qquad I_I = \Theta_I; \tag{3.5.28c}$$

while dotting them with $\boldsymbol{A}_D = A^k{}_D \boldsymbol{E}_k$—that is, projecting it onto the constraint local plane—yields

$$\boldsymbol{E}\cdot\boldsymbol{A}_D = \boldsymbol{Q}\cdot\boldsymbol{A}_D + \boldsymbol{R}\cdot\boldsymbol{A}_D, \tag{3.5.28d}$$

or, since $\boldsymbol{R}\cdot\boldsymbol{A}_D = \lambda_{D'}(\boldsymbol{A}^{D'}\cdot\boldsymbol{A}_D) = \lambda_{D'}\,\delta^{D'}{}_D = \lambda_D$, finally

$$A^k{}_D E_k = A^k{}_D Q_k + \lambda_D \ \text{(kinetostatic Maggi)}, \qquad \text{or} \qquad I_D = \Theta_D + \lambda_D. \tag{3.5.28e}$$

For the planar mathematical pendulum of length l, mass m, and string tension S [fig. 3.1(b)], $\boldsymbol{A}_I = \partial\boldsymbol{r}/\partial\phi = \textit{along tangent}$, $\boldsymbol{A}_D = \partial\boldsymbol{r}/\partial r = \textit{along normal}$, and so (3.5.28b, d) become

$$\boldsymbol{E}\cdot\boldsymbol{A}_I = \boldsymbol{Q}\cdot\boldsymbol{A}_I: \qquad ml(d^2\phi/dt^2) = -mg\sin\phi \qquad \textit{(kinetic Maggi eq.)}, \tag{3.5.28f}$$

$$\boldsymbol{E}\cdot\boldsymbol{A}_D = \boldsymbol{Q}\cdot\boldsymbol{A}_D + \boldsymbol{R}\cdot\boldsymbol{A}_D: \quad ml(d\phi/dt)^2 = -mg\cos\phi + S \quad \textit{(kinetostatic Maggi eq.)}. \tag{3.5.28g}$$

These geometrical considerations demonstrate the importance of the *method* of projections of Maggi, over and above that of the Maggi *equations*. His method can be applied to any kind of multiplier-containing (mixed) equations.

Example 3.5.6 *Lagrange's Equations* (Williamson and Tarleton, 1900, pp. 437–438). Let us consider a scleronomic system described by the Lagrangean equations

$$d/dt(\partial T/\partial v_k) - \partial T/\partial q_k = Q_k \qquad (k = 1, \ldots, n). \tag{a}$$

Now, the change of the system momentum $p_k \equiv \partial T/\partial v_k$ during an elementary time interval dt is $(dp_k/dt)dt$, and this, according to (a), equals $Q_k\, dt + (\partial T/\partial q_k)\, dt$. Since the system is scleronomic, $\partial T/\partial q_k =$ quadratic homogeneous function of the v_k's (see also §3.9), and therefore if the system is at rest, it vanishes. Hence, the result: *The elementary change of a typical component of the system momentum consists of two parts: one due to the corresponding impressed force, and one due to the (possible) previous motion.*

Problem 3.5.2 *Lagrange's Equations: 1 DOF.* Let us consider the most general holonomic and rheonomic 1 DOF system; that is, $n = 1$ and $m = 0$, with inertial (double) kinetic energy

$$2T = A(t,q)\dot{q}^2 + 2B(t,q)\dot{q} + C(t,q), \quad (A, C \geq 0, \text{ always}) \tag{a}$$

and hence Lagrangean (negative) inertial force

$$\begin{aligned} E_q(T) &\equiv (\partial T/\partial \dot{q})^{\cdot} - \partial T/\partial q \\ &= (1/2)\big[2A\,\ddot{q} + (\partial A/\partial q)\dot{q}^2 + 2(\partial A/\partial t)\dot{q} + 2(\partial B/\partial t) - \partial C/\partial q\big]. \end{aligned} \tag{b}$$

(i) Show that the *new* Lagrangean coordinate x, defined by

$$x \equiv \int [A(t,q)]^{1/2}\, dq = x(t,q) \;\Leftrightarrow\; q = q(t,x), \tag{c}$$

reduces $2T$ to

$$2T = \dot{x}^2 + 2b(x,t)\dot{x} + c(x,t), \tag{d}$$

where

$$b(t,x) \equiv \left\{A^{1/2}(B/A + \partial q/\partial t)\right\}_{\text{evaluated at } q=q(t,x)}, \tag{e}$$

$$c(t,x) \equiv \left\{A(\partial q/\partial t)^2 + 2B(\partial q/\partial t) + C\right\}_{\text{evaluated at } q=q(t,x)}, \tag{f}$$

and generates the following (negative) Lagrangean inertial force:

$$E_x(T) \equiv (\partial T/\partial \dot{x})^{\cdot} - \partial T/\partial x = d^2x/dt^2 + \partial b/\partial t - (1/2)(\partial c/\partial x); \tag{g}$$

that is, *no* (dx/dt)*-proportional* (*i.e., damping/friction*) *terms.* Such coordinate transformations may prove useful in nonlinear oscillation problems.

(ii) Show that in the *scleronomic* case, i.e., when $B, C \equiv 0$ and hence $2T = A(q)\dot{q}^2$, the inertia forces (b) and (g) reduce, respectively, to

$$A(d^2q/dt^2) + (1/2)(dA/dq)(dq/dt)^2 \quad \text{and} \quad d^2x/dt^2. \tag{h}$$

Problem 3.5.3 *Lagrange's Equations: 1 DOF.* Let us consider a 1 DOF system with kinetic and potential energies

$$2T = A(q)(dq/dt)^2 \quad \text{and} \quad V = V(q), \tag{a}$$

respectively, capable of oscillating about its equilibrium position $q = 0$. Show that the *period* of its *small amplitude* (i.e., *linearized*, or *harmonic*) vibration equals [with $(\ldots)' \equiv d(\ldots)/dq$]

$$2\pi[A(0)/V''(0)]^{1/2}. \tag{b}$$

HINT

Here, $A(0) > 0$, $V(0) = 0$, $V'(0) = 0$, $V''(0) > 0$; and, as shown in §3.9 ff.,

$$Q = -\partial V/\partial q = -dV/dq \equiv -V'.$$

Expand T and V à la Taylor about $q = 0$, and keep only up to *quadratic* terms in q and $\dot{q}$, etc.

Problem 3.5.4 *Lagrange's Equation: 1 DOF.* Continuing from the preceding problem, show that if $q = q_o$ is an equilibrium position, instead of $q = 0$, then (b) is replaced by

$$2\pi[A(q_o)/V''(q_o)]^{1/2}. \tag{a}$$

Problem 3.5.5 *Lagrange's Equations: Pendulum of Varying Length.* Show that the planar oscillations of a mathematical pendulum of varying, or variable, length $l = l(t)$ = *given function of time*, on a vertical plane, are governed by the (variable coefficient) equation

$$(l^2\dot{\phi})^{\cdot} + g\,l\sin\phi = 0 \;\Rightarrow\; d^2\phi/dt^2 + 2(\dot{l}/l)(d\phi/dt) + (g/l)\sin\phi = 0, \tag{a}$$

where ϕ = angle of pendulum string with vertical.

For the treatment of special cases, see for example, Lamb (1943, pp. 198–199).

Example 3.5.7 *Lagrange's Equations: Planar Double Pendulum; Work of Impressed Forces.* Let us consider a double mathematical pendulum in vertical plane motion, under gravity [fig. 3.2(a)]. Below we calculate the components of the system impressed force by several methods.

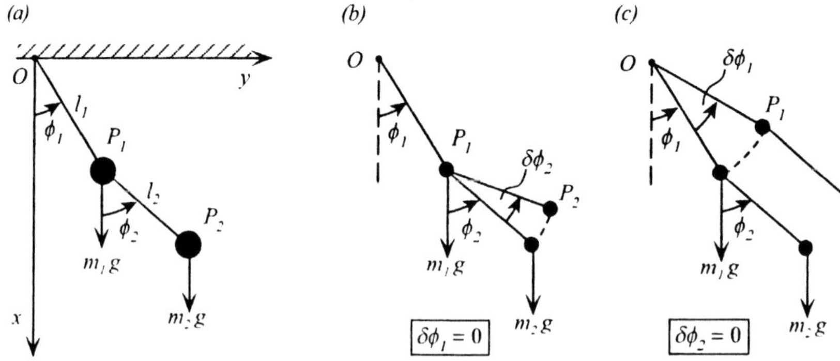

Figure 3.2 (a–c) Double planar mathematical pendulum, under gravity; calculation of impressed system forces.

(i) *From the Q_k-definitions* (§3.4). Here, with $q_1 = \phi_1$, $q_2 = \phi_2$ and some obvious notations, we have

$$\boldsymbol{r}_1 = (l_1 \cos\phi_1, l_1 \sin\phi_1, 0), \qquad \boldsymbol{r}_2 = (l_1 \cos\phi_1 + l_2 \cos\phi_2, l_1 \sin\phi_1 + l_2 \sin\phi_2, 0), \quad \text{(a)}$$
$$\boldsymbol{F}_1 = (m_1 g, 0, 0), \qquad \boldsymbol{F}_2 = (m_2 g, 0, 0), \quad \text{(b)}$$

and, therefore, we obtain

$$Q_1 \equiv S\, d\boldsymbol{F} \cdot (\partial\boldsymbol{r}/\partial q_1) = \boldsymbol{F}_1 \cdot (\partial\boldsymbol{r}_1/\partial q_1) + \boldsymbol{F}_2 \cdot (\partial\boldsymbol{r}_2/\partial q_1)$$
$$= \cdots = -m_1 g l_1 \sin\phi_1 - m_2 g\, l_1 \sin\phi_1 = -(m_1 + m_2) g l_1 \sin\phi_1, \quad \text{(c)}$$
$$Q_2 \equiv S\, d\boldsymbol{F} \cdot (\partial\boldsymbol{r}/\partial q_2) = \boldsymbol{F}_1 \cdot (\partial\boldsymbol{r}_1/\partial q_2) + \boldsymbol{F}_2 \cdot (\partial\boldsymbol{r}_2/\partial q_2)$$
$$= \cdots = -m_2 g\, l_2 \sin\phi_2. \quad \text{(d)}$$

(ii) *Directly from virtual work.* Let us find Q_2; that is, $\delta' W$ for $\delta\phi_1 = 0$ and $\delta\phi_2 \neq 0$: $(\delta' W)_2 \equiv Q_2\, \delta\phi_2$. Referring to fig. 3.2(b), we have

$$(\delta' W)_2 = (m_2 g)\, \delta(l_2 \cos\phi_2) = -m_2 g\, l_2 \sin\phi_2\, \delta\phi_2 \;\Rightarrow\; Q_2 = -m_2 g\, l_2 \sin\phi_2. \quad \text{(e)}$$

Similarly, to find Q_1—that is, $\delta' W$ for $\delta\phi_1 \neq 0$ and $\delta\phi_2 = 0$: $(\delta' W)_1 \equiv Q_1\, \delta\phi_1$, referring to fig. 3.2(c), we find

$$(\delta' W)_1 = (m_1 g)\, \delta(l_1 \cos\phi_1) + (m_2 g)\, \delta(l_1 \cos\phi_1) = (m_1 + m_2) g\, \delta(l_1 \cos\phi_1)$$
$$= -(m_1 + m_2) g\, l_1 \sin\phi_1\, \delta\phi_1 \;\Rightarrow\; Q_1 = -(m_1 + m_2) g\, l_1 \sin\phi_1. \quad \text{(f)}$$

(iii) *From potential energy (see also §3.9).* Here, the total potential energy of gravity ($\rightarrow$ impressed forces), $V = V(\phi_1, \phi_2)$, is

$$V = -(m_1 g)(l_1 \cos\phi_1) - (m_2 g)(l_1 \cos\phi_1 + l_2 \cos\phi_2)$$
$$= -(m_1 + m_2) g\, l_1 \cos\phi_1 - m_2 g\, l_2 \cos\phi_2, \quad \text{(g)}$$

and since $\delta' W = -\delta V$, we obtain

$$Q_1 = -\partial V/\partial\phi_1 = -(m_1 + m_2) g\, l_1 \sin\phi_1, \qquad Q_2 = -\partial V/\partial\phi_2 = -m_2 g\, l_2 \sin\phi_2. \quad \text{(h, i)}$$

REMARK

Had we chosen as system positional coordinates (fig. 3.3)

$$q_1 = \theta_1 \equiv \phi_1 \qquad \text{and} \qquad q_2 = \theta_2 \equiv \phi_2 - \theta_1, \quad \text{(j)}$$

then (g) would assume the form

$$V = V(\theta_1, \theta_2) = -(m_1 + m_2) g\, l_1 \cos\theta_1 - m_2 g\, l_2 \cos(\theta_1 + \theta_2), \quad \text{(k)}$$

and the corresponding Lagrangean forces would be

$$Q_1 = -\partial V/\partial\theta_1 = -(m_1 + m_2) g\, l_1 \sin\theta_1 - m_2 g\, l_2 \sin(\theta_1 + \theta_2), \quad \text{(l)}$$
$$Q_2 = -\partial V/\partial\theta_2 = -m_2 g\, l_2 \sin(\theta_1 + \theta_2). \quad \text{(m)}$$

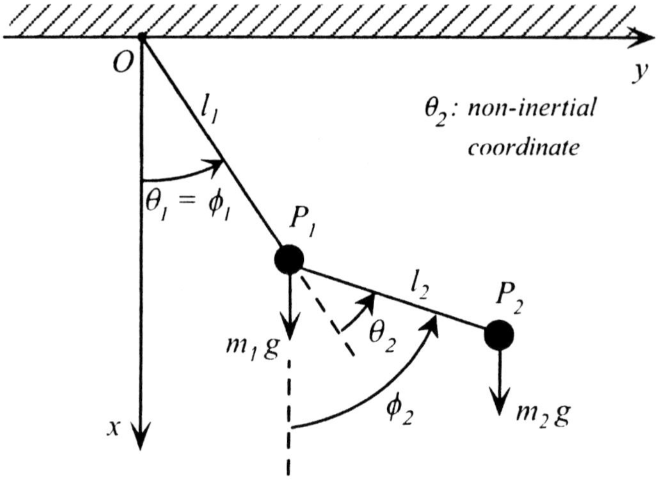

Figure 3.3 Double planar mathematical pendulum under gravity; alternative coordinates.

Example 3.5.8 *Lagrange's Equations: Planar Double Pendulum; Derivation of Equations of Motion.* Continuing from the preceding example (and its figures), let us first calculate the kinetic energy of the pendulum. We find, successively,

$$x_1 = l_1 \cos\phi_1 \Rightarrow \dot{x}_1 = -l_1\dot{\phi}_1 \sin\phi_1, \tag{a}$$

$$y_1 = l_1 \sin\phi_1 \Rightarrow \dot{y}_1 = l_1\dot{\phi}_1 \cos\phi_1, \tag{b}$$

$$x_2 = l_1 \cos\phi_1 + l_2\cos\phi_2 \Rightarrow \dot{x}_2 = -l_1\dot{\phi}_1 \sin\phi_1 - l_2\dot{\phi}_2\sin\phi_2, \tag{c}$$

$$y_2 = l_1 \sin\phi_1 + l_2\sin\phi_2 \Rightarrow \dot{y}_2 = l_1\dot{\phi}_1 \cos\phi_1 + l_2\dot{\phi}_2\cos\phi_2; \tag{d}$$

$$v_1^2 = (\dot{x}_1)^2 + (\dot{y}_1)^2 = \cdots = l_1^2(\dot{\phi}_1)^2, \tag{e}$$

$$v_2^2 = (\dot{x}_2)^2 + (\dot{y}_2)^2 = \cdots = l_1^2(\dot{\phi}_1)^2 + 2l_1l_2\cos(\phi_2 - \phi_1)\dot{\phi}_1\dot{\phi}_2 + l_2^2(\dot{\phi}_2)^2; \tag{f}$$

$$\begin{aligned} 2T &= m_1 v_1^2 + m_2 v_2^2 \\ &= \cdots = (m_1 + m_2)l_1^2(\dot{\phi}_1)^2 + 2m_2l_1l_2\cos(\phi_2 - \phi_1)\dot{\phi}_1\dot{\phi}_2 + m_2l_2^2(\dot{\phi}_2)^2; \end{aligned} \tag{g}$$

and by the preceding example,

$$Q_1 = -(m_1 + m_2)g\,l_1 \sin\phi_1, \qquad Q_2 = -m_2 g\,l_2\sin\phi_2. \tag{h}$$

From the above, we obtain

$$\begin{aligned} \partial T/\partial\dot{\phi}_1 &= (m_1 + m_2)l_1^2\dot{\phi}_1 + m_2l_1l_2\cos(\phi_2 - \phi_1)\dot{\phi}_2, \\ (\partial T/\partial\dot{\phi}_1)^{\cdot} &= (m_1 + m_2)l_1^2\ddot{\phi}_1 \\ &\quad + m_2l_1l_2\cos(\phi_2 - \phi_1)\ddot{\phi}_2 - m_2l_1l_2\sin(\phi_2 - \phi_1)(\dot{\phi}_2 - \dot{\phi}_1)\dot{\phi}_2, \end{aligned} \tag{i}$$

$$\partial T/\partial\phi_1 = m_2l_1l_2\sin(\phi_2 - \phi_1)\dot{\phi}_1\dot{\phi}_2. \tag{j}$$

Therefore, Lagrange's equation for $q_1 = \phi_1$: $(\partial T/\partial\dot{\phi}_1)^{\cdot} - \partial T/\partial\phi_1 = Q_1$, becomes after some simple algebra,

$$(m_1 + m_2)l_1{}^2(d^2\phi_1/dt^2) + m_2 l_1 l_2 \cos(\phi_2 - \phi_1)(d^2\phi_2/dt^2) \\ - m_2 l_1 l_2 \sin(\phi_2 - \phi_1)(d\phi_2/dt)^2 + (m_1 + m_2)g\, l_1 \sin\phi_1 = 0. \qquad \text{(k)}$$

Similarly, we find Lagrange's equation for $q_2 = \phi_2$:

$$m_2 l_2{}^2(d^2\phi_2/dt^2) + m_2 l_1 l_2 \cos(\phi_2 - \phi_1)(d^2\phi_1/dt^2) \\ + m_2 l_1 l_2 \sin(\phi_2 - \phi_1)(d\phi_1/dt)^2 + m_2 g\, l_2 \sin\phi_2 = 0. \qquad \text{(l)}$$

The above constitute a set of two *coupled* nonlinear second-order equations for $\phi_1(t)$ and $\phi_2(t)$.

Constraints

(i) Assume, next, that we impose on our system the constraint

$$f_1 \equiv y_1 = l_1 \sin\phi_1 = 0 \qquad [\Rightarrow \phi_1(t) = 0]; \qquad \text{(m)}$$

that is, we restrict the *upper half* OP_1 to remain vertical, so that the double pendulum reduces to a simple pendulum P_1P_2 oscillating about the fixed point P_1.

Since $\partial f_1/\partial\phi_1 = l_1\cos\phi_1$ and $\partial f_1/\partial\phi_2 = 0$ [$\Rightarrow \delta f_1 = (l_1\cos\phi_1)\,\delta\phi_1 + (0)\,\delta\phi_2$], the equations of motion in this case are (k) and (l), but with the terms $\lambda_1 l_1 \cos\phi_1$ and $\lambda_1 \cdot 0 = 0$ (where λ_1 = multiplier corresponding to the constraint $\delta f_1 = 0$) *added*, respectively, to their right sides; that is, in general, it is *not enough* to simply set in these two equations $\phi_1 = 0$ ($\Rightarrow \dot{\phi}_1 = 0, \ddot{\phi}_1 = 0$)! Indeed, then the equations of the (m)-constrained pendulum motion decouple to the Routh–Voss equations:

$$\phi_1\!: \quad \lambda_1 = m_2 l_2\left[\cos\phi_2(d^2\phi_2/dt^2) - \sin\phi_2(d\phi_2/dt)^2\right] \qquad (\textit{kinetostatic}), \qquad \text{(n)}$$

$$\phi_2\!: \quad d^2\phi_2/dt^2 + (g/l_2)\sin\phi_2 = 0 \qquad (\textit{kinetic}). \qquad \text{(o)}$$

With the initial conditions at, say, $t = 0$: $\phi_2(0) = 0$ and $\dot{\phi}_2(0) = \dot{\phi}_o$, equation (o) readily integrates, in well-known elementary ways, to (the energy equation)

$$(\dot{\phi}_2)^2 = (\dot{\phi}_o)^2 - (2g/l_2)(1 - \cos\phi_2), \qquad \text{(p)}$$

in which case, (n) yields the *constraint reaction* in terms of the angle ϕ_2 and its *initial conditions*

$$\lambda_1 = m_2[(2 - 3\cos\phi_2)g - l_2(\dot{\phi}_o)^2]. \qquad \text{(q)}$$

Finally, since

$$\delta' W_R = [\lambda_1(\partial f_1/\partial\phi_1)]\,\delta\phi_1 \equiv R_1\,\delta\phi_1 \\ = (\lambda_1 l_1 \cos\phi_1)\,\delta\phi_1 = \lambda_1\,\delta(l_1\sin\phi_1) = \lambda_1\,\delta y_1\,(= 0), \qquad \text{(r)}$$

the multiplier represents the (variable) *horizontal force of reaction needed to preserve the constraint (m)*. [Other forms of (m) will result in different, but physically equivalent, forms of the multiplier.]

(ii) Similarly, if ϕ_2 acquires a prescribed motion, say $\phi_2 = f(t) =$ *known function of time*, then, since in that case $\delta\phi_2 = \delta f(t) = 0$ $[= (0)\,\delta\phi_1 + (1)\,\delta\phi_2]$, we must add a term $\lambda_2 \cdot 0 = 0$ to the right side of the ϕ_1-equation, and a term $\lambda_2 \cdot 1$ to the right side of the ϕ_2-equation [where $\lambda_2 =$ multiplier corresponding to the constraint $f_2 \equiv \phi_2 - f(t) = 0 \Rightarrow \delta f_2 = 0$]. The rest of the calculations are left to the reader.

Problem 3.5.6 *Constrained Double Pendulum.* Continuing from the preceding example, assume that we impose on our pendulum the constraint

$$f_2 \equiv y_2 = l_1 \sin\phi_1 + l_2 \sin\phi_2 = 0. \tag{a}$$

(i) Show that in this case, and for the special simplifying choice $l_1 = l_2 \equiv l$ [$\Rightarrow \sin\phi_1 + \sin\phi_2 = 0 \Rightarrow \phi_1 + \phi_2 = \pi/2$, etc.], the equations of motion reduce to

$$(m_1 + m_2)l^2(d^2\phi_1/dt^2) - m_2 l^2\cos(2\phi_1)(d^2\phi_1/dt^2) + m_2 l^2 \sin(2\phi_1)(d\phi_1/dt)^2 \\ + (m_1 + m_2)g\,l\sin\phi_1 = \lambda_1 l\cos\phi_1, \tag{b}$$

$$-m_2 l^2(d^2\phi_1/dt^2) + m_2 l^2\cos(2\phi_1)(d^2\phi_1/dt^2) - m_2 l^2\sin(2\phi_1)(d\phi_1/dt)^2 \\ - m_2 g\,l\sin\phi_1 = \lambda_1 l\cos\phi_1. \tag{c}$$

(ii) From the above, deduce that

$$\lambda_1 = (m_1 l/2)(1/\cos\phi_1)(d^2\phi_1/dt^2) + (m_1 g/2)\tan\phi_1. \tag{d}$$

Interpret the multiplier λ_1 physically.

(iii) Show that the result of eliminating λ_1 between (b) and (c) and then linearizing the ensuing $\phi_1 \equiv \phi$-equation is the single pendulum-like equation:

$$d^2\phi/dt^2 + [(m_1 + 2m_2)/(m_1 + 4m_2)](g/l)\phi = 0. \tag{e}$$

Example 3.5.9 *Small (Linearized) Oscillations of Double Pendulum.* Continuing from the preceding example, let us study the small (linearized) amplitude/velocity/acceleration oscillatory motions of our planar double mathematical pendulum about its equilibrium configuration $\phi_1 = 0$, $\phi_2 = 0$.

There are two ways to proceed. Either (i) we keep up to *quadratic* terms in ϕ_1, ϕ_2 and their derivatives in T and V (or up to *linear* ones in the Q's) so that the corresponding Lagrangean equations end up *linear* in these functions; or (ii) we directly linearize the earlier-found equations of motion (for a more general treatment of linearized motions, see §3.10).

Let us begin with the first way; it is not hard to show that the earlier T, V $(Q_{1,2})$ approximate to the homogeneous *quadratic* (*linear*) forms:

$$2T = (m_1 + m_2){l_1}^2(\dot\phi_1)^2 + 2m_2 l_1 l_2 \dot\phi_1\dot\phi_2 + m_2 {l_2}^2(\dot\phi_2)^2; \tag{a}$$

$$2V = (m_1 + m_2)g\,l_1{\phi_1}^2 + m_2 g\,l_2{\phi_2}^2 + \textit{constant terms}, \tag{b}$$

$$Q_1 = -(m_1 + m_2)g\,l_1\phi_1, \qquad Q_2 = -m_2 g\,l_2\phi_2. \tag{c}$$

Then, with $L \equiv T - V =$ *Lagrangean of the system*, we easily obtain

$$\partial L/\partial\dot{\phi}_1 = (m_1 + m_2){l_1}^2\dot{\phi}_1 + m_2 l_1 l_2 \dot{\phi}_2, \tag{d}$$

$$(\partial L/\partial\dot{\phi}_1)^{\cdot} = (m_1 + m_2){l_1}^2\ddot{\phi}_1 + m_2 l_1 l_2 \ddot{\phi}_2, \tag{e}$$

$$\partial L/\partial\phi_1 = -(m_1 + m_2) g\, l_1 \phi_1 \quad (= Q_1); \tag{f}$$

$$\partial L/\partial\dot{\phi}_2 = m_2 l_1 l_2 \dot{\phi}_1 + m_2 {l_2}^2 \dot{\phi}_2, \tag{g}$$

$$(\partial L/\partial\dot{\phi}_2)^{\cdot} = m_2 l_1 l_2 \ddot{\phi}_1 + m_2 {l_2}^2 \ddot{\phi}_2, \tag{h}$$

$$\partial L/\partial\phi_2 = -m_2 g\, l_2 \phi_2 \quad (= Q_1). \tag{i}$$

Therefore, Lagrange's linearized (but still coupled!) equations are

$$(m_1 + m_2) l_1 (d^2\phi_1/dt^2) + m_2 l_2 (d^2\phi_2/dt^2) + (m_1 + m_2) g\, \phi_1 = 0, \tag{j}$$

$$l_1 (d^2\phi_1/dt^2) + l_2 (d^2\phi_2/dt^2) + g\, \phi_2 = 0. \tag{k}$$

The reader can verify that (j, k) result by direct linearization of (k, l) of the preceding example, respectively.

Solution of System of Equations (j, k)

As the theory of differential equations/linear vibration teaches us, the general solution of this *homogeneous* system is a linear combination, or superposition, of the following *harmonic motions* (*or modes*):

$$\phi_1 = A \sin(\omega t + \varepsilon) \qquad \text{and} \qquad \phi_2 = B \sin(\omega t + \varepsilon), \tag{l}$$

where $A, B =$ *mode amplitudes*, $\omega =$ *mode frequency*, and $\varepsilon =$ *mode phase*. Substituting (l) into (j, k), we are readily led to the algebraic system for the mode amplitudes:

$$[(m_1 + m_2)(g - l_1\omega^2)]A + (-m_2 l_2 \omega^2)B = 0, \tag{m}$$

$$(-l_1\omega^2)A + (g - l_2\omega^2)B = 0. \tag{n}$$

The requirement for *nontrivial* A and B leads, in well-known ways, to the determinantal equation

$$\Delta(\omega^2) \equiv \begin{vmatrix} (m_1 + m_2)(g - l_1\omega^2) & -m_2 l_2 \omega^2 \\ -l_1\omega^2 & g - l_2\omega^2 \end{vmatrix} = 0, \tag{o}$$

which, when expanded, becomes

$$(m_1 l_1 l_2)\omega^4 - [(m_1 + m_2)(l_1 + l_2)g]\omega^2 + (m_1 + m_2)g^2 = 0. \tag{p}$$

To simplify the algebra we, henceforth, assume that $m_1 = m_2 \equiv m$ and $l_1 = l_2 \equiv l$. Then (p) reduces to

$$\omega^4 - 4(g/l)\omega^2 + 2(g/l)^2 = 0, \tag{q}$$

and its positive roots can be easily shown to be

$$\omega_1 = \{[2 - (2)^{1/2}](g/l)\}^{1/2} \qquad (\textit{lower frequency}), \tag{r1}$$

$$\omega_2 = \{[2 + (2)^{1/2}](g/l)\}^{1/2} \qquad (> \omega_1, \textit{ higher frequency}). \tag{r2}$$

For $\omega = \omega_1, \omega_2$, the *amplitude ratios*

$$\mu \equiv B/A = [l_1\omega^2/(g - l_2\omega^2)] = \omega^2/[(g/l) - \omega^2] \quad [= \mu(\omega^2)] \tag{s}$$

[obtained from (n), for $l_1 = l_2$] are found to be

$$\mu_1 = B_1/A_1 = [2 - (2)^{1/2}]/[(2)^{1/2} - 1] = (2)^{1/2}, \tag{s1}$$

$$\mu_2 = B_2/A_2 = -[2 + (2)^{1/2}]/[1 + (2)^{1/2}] = -(2)^{1/2}, \tag{s2}$$

that is, $B_1 = (2)^{1/2}A_1$ and $B_2 = -(2)^{1/2}A_2$, for *any initial conditions*, and therefore the general solution of (j, k) is

$$\phi_1 = \phi_{1,1} + \phi_{1,2}, \qquad \phi_2 = \phi_{2,1} + \phi_{2,2}, \tag{t}$$

where

$$\phi_{1,1} = A_1 \sin(\omega_1 t + \varepsilon_1), \qquad \phi_{2,1} = \mu_1 A_1 \sin(\omega_1 t + \varepsilon_1), \tag{t1}$$

$$\phi_{1,2} = A_2 \sin(\omega_2 t + \varepsilon_2), \qquad \phi_{2,2} = \mu_2 A_2 \sin(\omega_2 t + \varepsilon_2). \tag{t2}$$

The above show that, for each frequency $\omega_k (k = 1, 2)$, the ratio of the corresponding mode amplitudes $\phi_{1,k}$ and $\phi_{2,k}$ is *constant*; that is, *independent of the initial conditions*

$$\phi_{2,1}/\phi_{1,1} = \mu_1 = (2)^{1/2} \quad \text{and} \quad \phi_{2,2}/\phi_{1,2} = \mu_2 = -(2)^{1/2}. \tag{t3}$$

The remaining *four* constants A_1, ε_1, and A_2, ε_2 are determined from the initial conditions.

For example, if at $t = 0$ we choose $\phi_1 = 0$, $\dot{\phi}_1 = 0$, and $\phi_2 = \phi_o$, $\dot{\phi}_2 = 0$, then, since

$$\dot{\phi}_1 = A_1\omega_1 \cos(\omega_1 t + \varepsilon_1) + A_2\omega_2 \cos(\omega_2 t + \varepsilon_2), \tag{u1}$$

$$\dot{\phi}_2 = (2)^{1/2} A_1\omega_1 \cos(\omega_1 t + \varepsilon_1) - (2)^{1/2} A_2\omega_2 \cos(\omega_2 t + \varepsilon_2), \tag{u2}$$

eqs. (t–t2), the above, and the initial conditions lead to the following algebraic system:

$$\phi_1: \quad 0 = A_1 \sin\varepsilon_1 + A_2 \sin\varepsilon_2, \tag{v1}$$

$$\phi_2: \quad \phi_o = (2)^{1/2} A_1 \sin\varepsilon_1 - (2)^{1/2} A_2 \sin\varepsilon_2, \tag{v2}$$

$$\dot{\phi}_1: \quad 0 = A_1\omega_1 \cos\varepsilon_1 + A_2\omega_2 \cos\varepsilon_2, \tag{v3}$$

$$\dot{\phi}_2: \quad 0 = (2)^{1/2} A_1\omega_1 \cos\varepsilon_1 - (2)^{1/2} A_2\omega_2 \cos\varepsilon_2. \tag{v4}$$

From the last two equations, we readily conclude that $\cos\varepsilon_1 = \cos\varepsilon_2 = 0 \Rightarrow \varepsilon_1 = \varepsilon_2 = \pi/2$; and so the first two reduce to $A_1 + A_2 = 0$ and $A_1 - A_2 = [(2)^{1/2}/2]\phi_o$, and from these we easily find $A_1 = [(2)^{1/2}/4]\phi_o$ and $A_2 =$

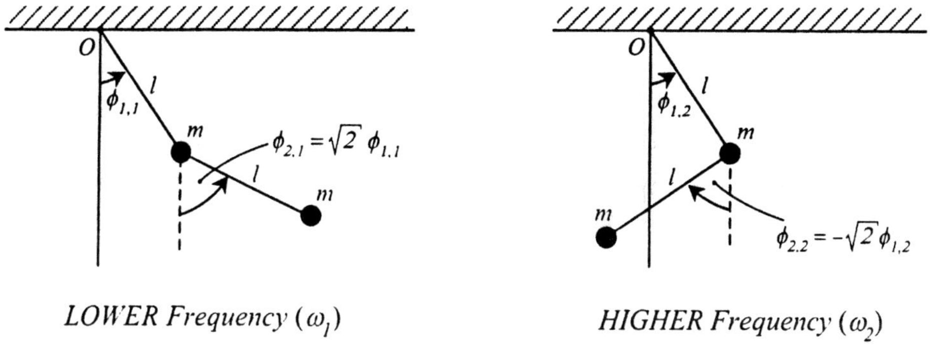

Figure 3.4 Angular modes of planar double pendulum, for its two frequencies: (a) lower frequency, (b) higher frequency. The amplitudes of $\phi_{1,1}$, $\phi_{1,2}$ depend on the initial conditions.

$-[(2)^{1/2}/4]\phi_o$. Hence, the particular solution of our system (j, k), satisfying the earlier chosen initial conditions, is

$$\phi_1 = [\phi_o(2)^{1/2}/4][\cos(\omega_1 t) - \cos(\omega_2 t)], \tag{w1}$$

$$\phi_2 = (\phi_o/2)[\cos(\omega_1 t) + \cos(\omega_2 t)]; \tag{w2}$$

where ω_1, ω_2 are given by (r1, 2).

The relative modal contributions for each frequency are shown in fig. 3.4(a, b).

Problem 3.5.7 *Double Pendulum; Noninertial Coordinates.* Consider the double pendulum of fig. 3.3.

(i) Show that its (Lagrangean) equations of motion in the angles $\theta_1 (\equiv \phi_1)$ and θ_2, under gravity, are

$$\begin{aligned}
&[m_1 l_1{}^2 + m_2(l_1{}^2 + 2l_1 l_2 \cos\theta_2 + l_2{}^2)](d^2\theta_1/dt^2) + m_2 l_2(l_1\cos\theta_2 + l_2)(d^2\theta_2/dt^2)\\
&\quad - (m_2 l_1 l_2 \sin\theta_2)(d\theta_2/dt)^2 - (2m_2 l_1 l_2 \sin\theta_2)(d\theta_1/dt)(d\theta_2/dt)\\
&\quad + (m_1 + m_2)l_1 g\sin\theta_1 + m_2 l_2 g\sin(\theta_1 + \theta_2) = 0,
\end{aligned} \tag{a}$$

$$\begin{aligned}
&(m_2 l_2{}^2)(d^2\theta_2/dt^2) + m_2 l_2(l_1\cos\theta_2 + l_2)(d^2\theta_1/dt^2)\\
&\quad + (m_2 l_1 l_2 \sin\theta_2)(d\theta_1/dt)^2 + m_2 l_2 g\sin(\theta_1 + \theta_2) = 0.
\end{aligned} \tag{b}$$

(ii) Obtain its equations of small motion; that is, linearize (a, b).

(iii) What do (a, b) reduce to for $l_1 = 0$, or $l_2 = 0$, before and after their linearization?

Problem 3.5.8 *Double Physical Pendulum.* A rigid body *I* of mass *M* can rotate freely about a fixed and smooth *vertical* axis. A second rigid body *II* of mass *m* can rotate freely about a second smooth and also vertical axis that is fixed on body *I* (fig. 3.5).

(i) Show that the (double) kinetic energy of this double planar "physical" pendulum is

$$2T = A\dot\phi^2 + 2\Gamma\dot\phi\dot\psi + B\dot\psi^2, \tag{a}$$

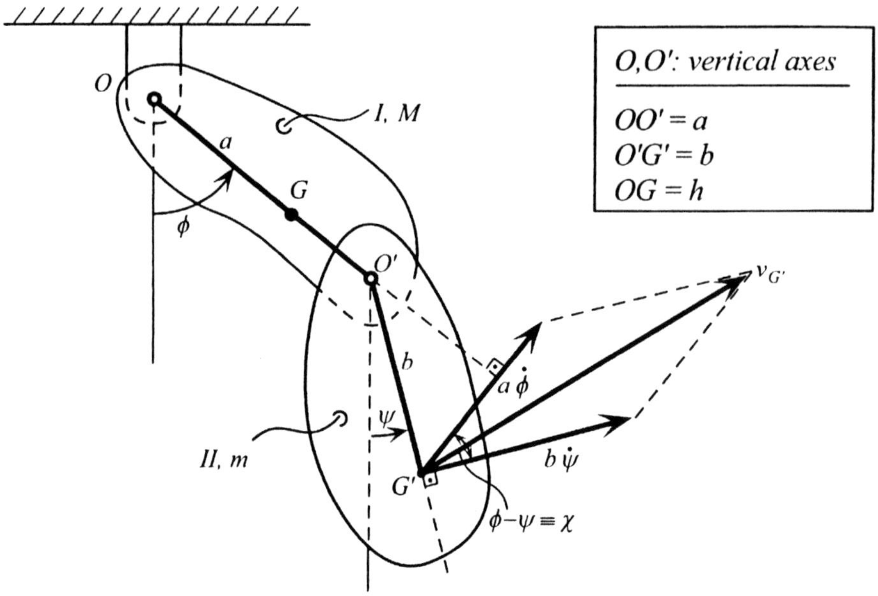

Figure 3.5 Double and planar physical pendulum, moving on horizontal plane.

where $A \equiv MK^2 + ma^2$, $B \equiv m(k^2 + b^2)$, $\Gamma \equiv m\,a\,b\cos(\phi - \psi) = m\,a\,b\cos(\psi - \phi) \equiv m\,a\,b\cos\chi$ (definition of angle χ); $K(k) =$ *radius of gyration of* $I(II)$ *about* O (G').

(ii) Show that, in this (force-free) case,

$$\partial T/\partial\dot{\phi} + \partial T/\partial\dot{\psi} \equiv p_\phi + p_\psi \equiv \text{total angular momentum about } O\text{-axis}$$
$$= constant \equiv c, \tag{b}$$

or

$$(A + \Gamma)(d\phi/dt) + (B + \Gamma)(d\psi/dt) = c; \tag{b1}$$

and

$$2T = E \ (another\ constant). \tag{c}$$

(iii) Show that, with the help of $\chi \equiv \phi - \psi$, eq. (b1) can be further transformed to

$$(A + 2\Gamma + B)(d\psi/dt) = c - (A + \Gamma)(d\chi/dt),$$

or

$$(A + 2\Gamma + B)[A(d\phi/dt) + \Gamma(d\psi/dt)] = (A + \Gamma)c + (AB - \Gamma^2)(d\chi/dt). \tag{d}$$

(iv) With the help of this integral, show that the energy integral (c) can be rewritten as

$$(d\chi/dt)[A(d\phi/dt) + \Gamma(d\psi/dt)] + c(d\psi/dt) = E,$$

or

$$(d\chi/dt)^2(AB - \Gamma^2) + c^2 = (A + 2\Gamma + B)E. \tag{e}$$

(v) Finally, and recalling the Γ-definition, show that (e) transforms to

$$(d\chi/dt)^2 = [(A + B + 2mab\cos\chi)E - c^2]/[AB - (mab)^2\cos^2\chi] \equiv f(\chi), \quad \text{(f)}$$

that is, the problem has been led to a quadrature.

For further discussion of this famous problem, and of its many variations, see, for example (alphabetically): Marcolongo (1912, pp. 213–216), Schell (1880, pp. 549–551), Thomson and Tait (1912, pp. 310, 324–325), Timoshenko and Young (1948, pp. 209–211, 215–216, 249–250, 276–278, 312–314).

Problem 3.5.9 *Double Physical Pendulum: Vertical Axes.* Continuing from the preceding problem (penduli axes through O and O' *vertical*), obtain its Lagrangean equations of motion. What happens to these equations if the center of mass of the entire system $I + II$ is at its *maximum/minimum* distance from O? Assume that O, G, O' are collinear, and $OG = h$.

HINT

Introduce the new angular variables $q_1 = \phi$ and $q_2 = \theta \equiv \psi - \phi\ (= -\chi) =$ inclination of body II relative to I (positive counterclockwise). Then, $T \to T(\theta; \dot{\phi}, \dot{\theta})$, and so on.

Problem 3.5.10 *Double Physical Pendulum: Horizontal Axes.* Consider the preceding double pendulum problem, but now with both axes through O and O' *horizontal*. In addition, assume that the mass center of body I, G, lies in the plane of the axes O and O', and $OG = h$. Show that here T is the same, *in form*, as in the previous vertical axes case, but the *potential* of gravity forces, V, equals (exactly)

$$V = -Mgh\cos\phi - mg(a\cos\phi + b\cos\psi) + \textit{constant}, \quad \text{(a)}$$

and therefore the corresponding Lagrangean impressed forces are

$$Q_\phi = -\partial V/\partial\phi = \cdots \quad \text{and} \quad Q_\psi = -\partial V/\partial\psi = \cdots. \quad \text{(b)}$$

Then write down Lagrange's equations for $q_1 = \phi$ and $q_2 = \psi$.

Problem 3.5.11 *Double Physical Pendulum: Horizontal Axes; Small Oscillations.* Continuing from the preceding problem (O and O' horizontal), show that for *small* oscillations about the vertical equilibrium position ϕ, $\psi = 0$, *linearization* of the exact equations leads to the coupled system

$$(Mh + ma)[L(d^2\phi/dt^2) + g\phi] + mab(d^2\psi/dt^2) = 0,$$
$$a(d^2\phi/dt^2) + L'(d^2\psi/dt^2) + g\psi = 0, \quad \text{(a)}$$

where

$$L \equiv (MK^2 + ma^2)/(Mh + ma) \quad \text{and} \quad L' \equiv (k^2 + b^2)/b. \quad \text{(b)}$$

Interpret L and L' in terms of single pendulum quantities. Then, assume as solutions of (a)

$$\phi = \phi_o \sin(\omega t + \varepsilon) \quad \text{and} \quad \psi = \psi_o \sin(\omega t + \varepsilon), \quad \text{(c)}$$

where ϕ_o, ψ_o = *angular amplitudes*, ε = *initial phase*, and ω = *frequency*, and show that the ω^2 are real, positive, and unequal, and are the roots of

$$(\Omega - L)(\Omega - L') = (ma^2 b)/(Mh + ma), \qquad \text{where } \Omega \equiv g/\omega^2; \tag{d}$$

say $\Omega_1 < \Omega_2$; and thus conclude that

$$\Omega_1 < \min(L, L') \leq \max(L, L') < \Omega_2. \tag{e}$$

Finally, show that

$$\psi_o/\phi_o = a/(\Omega - L') = \cdots, \tag{f}$$

and, therefore, (i) for the *smaller* ω ($\rightarrow$ *larger* $\Omega = \Omega_2$), $\phi \cdot \psi > 0$ (i.e., in the slower mode, the angles have the same sign); while (ii) for the *larger* ω ($\rightarrow$ *smaller* $\Omega = \Omega_1$), $\phi \cdot \psi < 0$ (i.e., in the faster mode, the angles have opposite signs).

[For a discussion of the historically famous case of the *nonringing, or "silent"*, bell of Köln (Cologne), Germany (1876; bell + clapper = double pendulum), based on (a), see, for example, Hamel ([1922(b)] 1912, 1st ed., pp. 514 ff.), Szabó (1977, pp. 89–90), Timoshenko and Young (1948, p. 278).]

Problem 3.5.12 *General Form of Lagrange's Equations for a 2 DOF System.* Consider a *2 DOF* holonomic and scleronomic system; for example, a particle on a fixed surface, or the previous double pendulum, with (double) kinetic energy

$$2T = A(dx/dt)^2 + 2\Gamma(dx/dt)(dy/dt) + B(dy/dt)^2, \tag{a}$$

and such that $\delta' W = X\delta x + Y\delta y$, where $A, B, \Gamma; X, Y$, are functions of x, y.

(i) Show that its Lagrangean equations of motion in $q_1 = x$ and $q_2 = y$ are

$$A(d^2x/dt^2) + \Gamma(d^2y/dt^2) + (1/2)(\partial A/\partial x)(dx/dt)^2 + (\partial A/\partial y)(dx/dt)(dy/dt) + [\partial\Gamma/\partial y - (1/2)(\partial B/\partial x)](dy/dt)^2 = X, \tag{b}$$

$$B(d^2y/dt^2) + \Gamma(d^2x/dt^2) + (1/2)(\partial B/\partial y)(dy/dt)^2 + (\partial B/\partial x)(dx/dt)(dy/dt) + [\partial\Gamma/\partial x - (1/2)(\partial A/\partial y)](dx/dt)^2 = Y, \tag{c}$$

and ponder over the geometrical/kinematical/inertial meaning and origin of each of these terms.

(ii) Show that these equations *linearize* to the (still coupled) system:

$$A_o(d^2x/dt^2) + \Gamma_o(d^2y/dt^2) = (\partial X/\partial x)_o\, x + (\partial X/\partial y)_o\, y, \tag{d}$$

$$B_o(d^2y/dt^2) + \Gamma_o(d^2x/dt^2) = (\partial Y/\partial x)_o\, x + (\partial Y/\partial y)_o\, y, \tag{e}$$

where $(\ldots)_o \equiv (\ldots)$ *evaluated at* $x, y = 0$.

Example 3.5.10 *Lagrange's Equations, 2 DOF: Elastic Pendulum, or Swinging Spring.* Let us derive and discuss the equations of plane motion, under gravity, of a pendulum consisting of a heavy particle (or bob) of mass m suspended by a linearly elastic and massless spring of stiffness k (a positive constant) and unstretched (or natural) length b (fig. 3.6). This is a holonomic and scleronomic

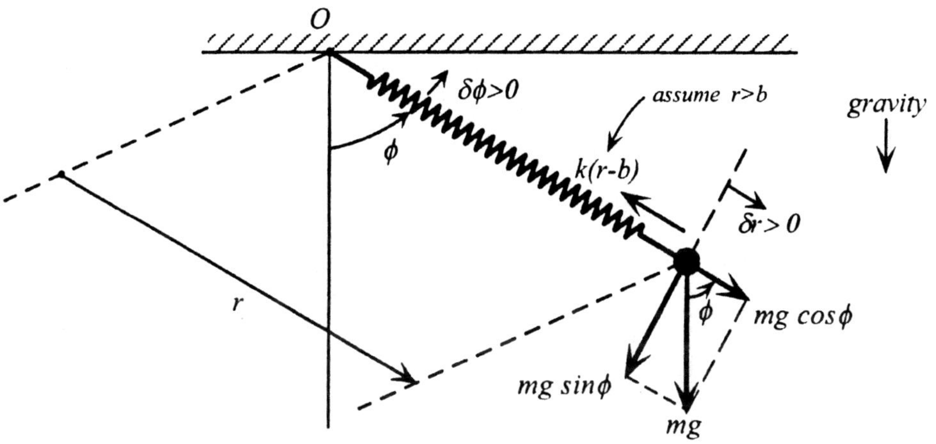

Figure 3.6 Geometry and forces on plane elastic pendulum.

two DOF system; that is, $n = 2$, $m = 0$. With Lagrangean coordinates as the polar coordinates of the bob: $q_1 = r$, $q_1 = \phi$, its (double) kinetic energy is

$$2T = m\,v^2 = m(ds/dt)^2 = m[(\dot{r})^2 + r^2(\dot{\phi})^2], \tag{a}$$

while the virtual work of its impressed forces, gravity and spring force, equals

$$\delta' W = -k(r-b)\,\delta r + (m\,g\cos\phi)\,\delta r - (m\,g\sin\phi)(r\,\delta\phi) \equiv Q_r\,\delta r + Q_\phi\,\delta\phi, \tag{b}$$

that is,

$$Q_r = -k(r-b) + m\,g\cos\phi, \qquad Q_\phi = -m\,g\,r\sin\phi. \tag{c}$$

Alternatively, the potential energy of the system is

$$V = (1/2)k(r-b)^2 - m\,g\,r\cos\phi = V(r,\phi), \tag{d}$$

and so the corresponding Lagrangean forces are $Q_r = -\partial V/\partial r = \cdots$, $Q_\phi = -\partial V/\partial\phi = \cdots$, equations (c). We also notice that for $r > b$: $Q_{r,\text{spring}} \equiv -k(r-b) < 0$, as it should; and analogously for $r < b$. Lagrange's equations, then, are

$$E_r(T) \equiv E_r = Q_r: \qquad (m\dot{r})^{\cdot} - mr(\dot{\phi})^2 = -k(r-b) + m\,g\cos\phi, \tag{e}$$

$$E_\phi(T) \equiv E_\phi = Q_\phi: \qquad (mr^2\dot{\phi})^{\cdot} = -m\,g\,r\sin\phi; \tag{f}$$

or, after some simplifications (since $r \neq 0$),

$$d^2r/dt^2 - r(d\phi/dt)^2 = -(k/m)(r-b) + g\cos\phi, \tag{g}$$

$$r(d^2\phi/dt^2) + 2(dr/dt)(d\phi/dt) = -g\sin\phi. \tag{h}$$

The general solution of this *nonlinear and coupled system* is unknown, and so we will limit ourselves to some simple and physically motivated special solutions of it.

(i) *Equilibrium solution*: Setting all $(\ldots)^{\cdot}$-derivatives in (g, h) equal to zero, we find [with $(\ldots)_o \equiv$ *equilibrium value of* $(\ldots)$]

$$0 = -(k/m)(r_o - b) + g\cos\phi_o, \qquad 0 = -g\sin\phi_o, \tag{i}$$

and, from these algebraic equations, we readily obtain the equilibrium values

$$\phi_o = 0, \qquad r_o - b = mg/k \equiv \rho. \tag{j}$$

Thus, in terms of the new variable

$$x \equiv r - r_o = r - (b + \rho) = r - [b + (mg/k)] = \textit{deviation from vertical equilibrium},$$

and with $\omega_r^2 \equiv k/m$, eqs. (g, h) can be, finally, rewritten as

$$\ddot{x} - (r_o + x)(\dot{\phi})^2 + g(1 - \cos\phi) + \omega_r^2 x = 0, \tag{k}$$

$$(r_o + x)\ddot{\phi} + 2\dot{x}\,\dot{\phi} + g\sin\phi = 0. \tag{l}$$

(ii) *Ordinary (or mathematical) pendulum solution*; that is, $r = \textit{constant} \equiv R$. In this case, (g, h) become [since all forces here are impressed; and, contrary to (ex. 3.5.8: m ff.), no multipliers are involved]:

$$-R(\dot{\phi})^2 = -(k/m)(R - b) + g\cos\phi, \tag{m}$$

$$\ddot{\phi} + \omega_\phi^2 \sin\phi = 0, \qquad \omega_\phi^2 \equiv g/R, \tag{n}$$

and, from these, we get $\phi(t) = \phi_o = 0$ and $R = r_o = b + (mg/k)$; that is, the previous equilibrium case.

(iii) *Linearization of equations (g, h), (k, l)*. We readily obtain the *uncoupled* system:

$$\ddot{r} + \omega_r^2 r = (k/m)b + g$$

$$\Rightarrow \ddot{x} + \omega_r^2 x = 0 \;\Rightarrow\; x = A\sin(\omega_r t) + B\cos(\omega_r t), \tag{o}$$

$$g\sin\phi = 0 \;\Rightarrow\; \phi(t) = 0 \qquad (A, B\text{: integration constants}); \tag{p}$$

that is, a small oscillation of frequency ω_r about the vertical equilibrium $r = r_o$, or $x = 0$.

(iv) *Nearly vertical oscillation*; that is, ϕ small. Then (g, h)/(k, l) reduce to the *coupled* system:

$$\ddot{r} + \omega_r^2 r = (k/m)b + g$$

$$\Rightarrow \ddot{x} + \omega_r^2 x = 0 \;\Rightarrow\; x = A\sin(\omega_r t) + B\cos(\omega_r t), \tag{q = o}$$

$$r\ddot{\phi} + 2\dot{r}\dot{\phi} + g\phi = 0$$

$$\Rightarrow (r_o + x)\ddot{\phi} + 2\dot{x}\dot{\phi} + g\phi = 0 \;\Rightarrow\; (1 + \varepsilon)\ddot{\phi} + 2\dot{\varepsilon}\,\dot{\phi} + \omega_\phi^2\phi = 0, \tag{r}$$

where $\varepsilon = \varepsilon(t) \equiv x/r_o$ (and $\omega_\phi^2 \equiv g/r_o$).

Now, since x = *harmonic in time*, equation (r) is a linear differential equation with harmonically varying coefficients; or, as it is generally called, a *parametrically* excited one [or *rheo-linear* = *rheonomic* + *linear*]. As the theory of these important "Hill/

Floquet/Mathieu" equations shows, the solutions of (r) are stable; that is, $\phi =$ *oscillatory and bounded*, or not, depending on the values of

$$\omega_\phi/\omega_r \equiv \omega, \qquad \text{or} \qquad \omega^2 = (g/r_o)/(k/m) = mg/r_o k = gravity/elasticity. \tag{s}$$

Specifically, it can be shown that:

(i) If $\omega \neq N/2$, or $\omega^2 \neq N^2/4$ $(N = 1, 2, 3, \ldots)$, then both x and ϕ remain small as required by the linearization; but,

(ii) If $\omega \approx N/2$, or $\omega^2 \approx N^2/4$ $(= 1/4, 1, 9/4, \ldots)$, then $\phi \to \infty$, in spite of the absence of external excitation; that is, then, the vertical x-oscillation, acting as *internal forcing*, causes ever larger (nonlinear) angular oscillations; and since the *total energy of the system remains constant*, this phenomenon [commonly known as *parametric, or internal, resonance*] comes at the expense of the x-oscillation; that is, energy flows from the vertical oscillation to the angular one, and (as shown by experiments) *back*. But here, contrary to constant coefficient linearized coupled systems (e.g., the earlier double pendulum), we do need to examine some *nonlinear* version of the problem: either the exact equations (g, h)/(k, l), or some weakly nonlinear system of them, and the linear but parametric equations (q, r).

[The case $N = 1 \Rightarrow \omega_r \approx \omega_\phi$, or $mg \approx kr_o/4$, is the most dangerous one, because, as the "stability chart" of equation (r) shows, that is where *the instability region is at its widest*; and that width is proportional to the amplitude of the "fundamental x-solution," (q = o).]

For further details, see, for example, Nayfeh (1973, pp. 185–189, 214–216, 262–264; and references cited there), Nayfeh and Mook (1979, pp. 369–370, 431–432), Pfeiffer (1989, pp. 209–210); also Dysthe and Gudmestad (1975). For an extensive treatment, see Starzhinskii (1977/1980, pp. 50–55; also pp. 59–75, 79–83, 95–98, 133–135).

Problem 3.5.13 *Elastic Pendulum.* Continuing from the preceding example, let us consider the "fundamental" solution, equations (q = o)

$$\begin{aligned} x &= A\sin(\omega_r t) + B\cos(\omega_r t) = x_o\cos(\omega_r t + \chi), \\ \varepsilon &= \varepsilon(t) \equiv x/r_o = \varepsilon_o\cos(\omega_r t + \chi), \end{aligned} \tag{a}$$

where $A, B, x_o, \chi =$ *integration constants*. Next, and following standard methods of perturbation theory, assume a solution of (r) in the form

$$\phi = \phi_o\cos(\omega_\phi t + \psi) + \Phi_1 \equiv \Phi_o + \Phi_1, \tag{b}$$

where $\Phi_1 =$ *small* relative to Φ_o. Then, insert these ε and ϕ solutions in (r) and, after neglecting all terms containing products of $\varepsilon, \varepsilon_o$ with Φ_1, and its $(\ldots)^\cdot$-derivatives, bring it to the ordinary (i.e., constant coefficient) undamped and forced oscillation form

$$\begin{aligned} \ddot{\Phi}_1 + \omega_\phi{}^2\Phi_1 &= -[(1+\varepsilon)\ddot{\Phi}_o + 2\varepsilon\,\dot{\Phi}_o + \omega_\phi{}^2\Phi_o] \\ &= -[\varepsilon\,\ddot{\Phi}_o + 2\varepsilon\,\dot{\Phi}_o] \qquad\qquad \text{[explain why]} \\ &= \cdots = f(t;\ \omega_r, \omega_f;\ \chi, \psi;\ \varepsilon_o, \phi_o) \end{aligned}$$

(known function; linear superposition of two harmonic excitations of frequencies $\omega_r + \omega_\phi$ and $\omega_r - \omega_\phi$) (c)

Find the particular solution of this equation (nonhomogeneous part), and then establish that:

(i) If $\omega_\phi/\omega_r \neq 1/2$, then both x and ϕ remain small; but
(ii) If $\omega_\phi/\omega_r \approx 1/2$, then Φ_1 (and therefore ϕ) $\to \infty$ [as in the (nonlinear) problem of "small denominators," or "combination tones"—see, e.g., Stoker (1950, pp. 112–114); also §8.16, this volume].

Problem 3.5.14 *Elastic Pendulum.* Continuing from the last example, let us substitute into its exact equations (k, l) [instead of the preceding problem's assumed solution (b)]

$$x = X + \Delta X, \qquad \phi = \Phi + \Delta\Phi, \tag{a}$$

where X and Φ are its following *fundamental* motion/solutions,

$$X = X(t) = x_o \cos(\omega_r t + \chi), \qquad \Phi = \Phi(t) = 0, \tag{b}$$

and $\Delta X(t)$, $\Delta\Phi(t)$ are the *small perturbations* about that state, and keep only up to linear terms in this small (neighboring) motion. Show that, then, we obtain the *uncoupled* linear system:

$$(\Delta X)^{\cdot\cdot} + \omega_r{}^2\,\Delta X = 0 \quad \text{and} \quad (1 + \varepsilon)(\Delta\Phi)^{\cdot\cdot} + 2\dot{\varepsilon}\,(\Delta\Phi)^{\cdot} + \omega_\phi{}^2\,\Delta\Phi = 0, \tag{c}$$

where $\varepsilon \equiv X/r_o$; or, since $|\varepsilon| =$ *much smaller than* 1,

$$(\Delta\Phi)^{\cdot\cdot} + 2\dot{\varepsilon}\,(\Delta\Phi)^{\cdot} + \omega_\phi{}^2(1 - \varepsilon)\Delta\Phi = 0, \tag{d}$$

that is, the neighboring motion $\Delta\Phi(t)$ depends on the fundamental one through $X(t)$, or $\varepsilon(t)$. Solve the first of (c), insert its solution into (d), and then show that, since the coefficients of both $\Delta\Phi$ and $(\Delta\Phi)^{\cdot}$ have the same (parametric) frequency ω_r, the resulting equation (d) can be led to a standard Mathieu equation; that is,

$$\ddot{y} + P(t)y = 0, \tag{e}$$

where $P(t + 2\pi/\omega_r) = P(t)$, $\Delta\phi \sim y\exp(-\varepsilon)$, and $P = \omega_\phi{}^2(1 - \varepsilon) - \ddot{\varepsilon} - \dot{\varepsilon}^2$.

Problem 3.5.15 *Lagrange's Equations: Cylindrical and Spherical Coordinates.* Show that the equations of motion of, say, a free particle P in cylindrical and spherical coordinates, are (fig. 3.7) as follows:

(i) *Cylindrical* ($x = r\cos\phi$, $y = r\sin\phi$, $z = z$; $v_r{}' = \dot{r}$, $v_\phi{}' = r\dot{\phi}$, $v_z{}' = \dot{z}$):

$$\text{Radial:} \quad m[\ddot{r} - r(\dot{\phi})^2] = Q_r \qquad (= F_r;\ m a_r{}' = F_r), \tag{a}$$

$$\text{Transverse:} \quad m(r^2\dot{\phi})^{\cdot} = Q_\phi \qquad (= rF_\phi;\ m a_\phi{}' = F_\phi), \tag{b}$$

$$\text{Vertical:} \quad m\ddot{z} = Q_z \qquad (= F_z;\ m a_z{}' = F_z); \tag{c}$$

where

$$\begin{aligned} \delta' W = (F_r, F_\phi, F_z)\cdot(\delta r, r\,\delta\phi, \delta z) &= F_r\,\delta r + (rF_\phi)\,\delta\phi + F_z\,\delta z \\ &= Q_r\,\delta r + Q_\phi\,\delta\phi + Q_z\,\delta z; \end{aligned} \tag{d}$$

$\boldsymbol{F} = (F_r, F_\phi, F_z) =$ *total force on P*.

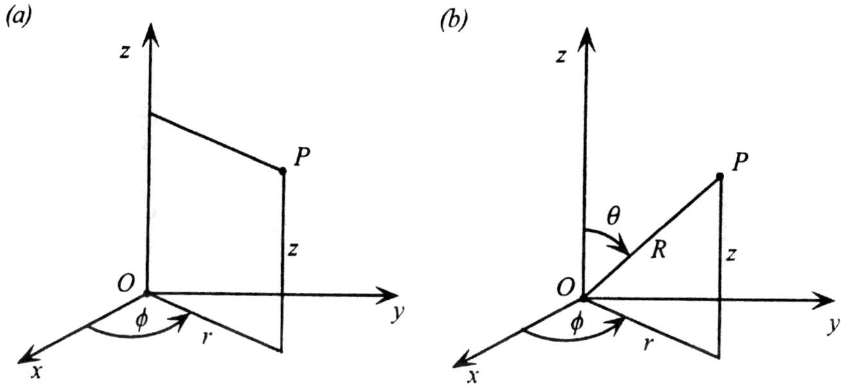

Figure 3.7 Particle in space: (a) cylindrical and (b) spherical coordinates.

(ii) *Spherical* $[x = (R\sin\theta)\cos\phi,\quad y = (R\sin\theta)\sin\phi,\quad z = R\cos\theta;\quad v_R{}' = \dot{R}$, $v_\theta{}' = R\dot{\theta}$, $v_\phi{}' = (R\sin\theta)\dot{\phi}]$:

Radial: $m[\ddot{R} - R(\dot{\theta})^2 - R\sin^2\theta(\dot{\phi})^2] = Q_R \; (= F_R;\; m\,a_R{}' = F_R)$, (e)

Transverse (θ-plane): $m[(R^2\dot{\theta})^{\cdot} - R^2\sin\theta\cos\theta(\dot{\phi})^2] = Q_\theta \; (= R\,F_\theta;\; m\,a_\theta{}' = F_\theta)$, (f)

Normal (to θ-plane): $m(R^2\sin^2\theta\,\dot{\phi})^{\cdot} = Q_\phi \quad (= R\sin\theta\,F_\phi;\; m\,a_\phi{}' = F_\phi)$; (g)

where

$$\begin{aligned}\delta' W &= (F_R,\ F_\theta,\ F_\phi)\cdot(\delta R,\ R\,\delta\theta,\ R\ \sin\theta\ \delta\phi)\\ &= (F_R)\,\delta R + (RF_\theta)\,\delta\theta + (R\sin\theta F_\phi)\,\delta\phi = Q_R\,\delta R + Q_\theta\,\delta\theta + Q_\phi\,\delta\phi;\end{aligned} \tag{h}$$

$\boldsymbol{F} = (F_R,\ F_\theta,\ F_\phi) =$ *total force on P.*

HINT

$2T = m[(\dot{x})^2 + (\dot{y})^2 + (\dot{z})^2];\ \dot{x} = \cdots,\ \dot{y} = \cdots,\ \dot{z} = \cdots$, and so on.

Problem 3.5.16 *Lagrange's Equations: Particle on Sphere, or Spherical Pendulum.* Consider the motion of a heavy particle P of mass m on the inner part of a smooth and stationary spherical surface of radius l, under (constant) gravity (fig. 3.8).

Show that the Routh–Voss equations of this constrained system [constraint: $f_1 \equiv R - l\ (= 0)$; $q_1 = R$, $q_2 = \theta$, $q_3 = \phi$; $n = 3$, $m = 1$], with $+Oz$ taken vertically downwards, are

$$R:\quad (\dot{\theta})^2 + \sin^2\theta\,(\dot{\phi})^2 = -(1/l)[g\cos\theta + (\lambda/m)(\partial f_1/\partial R)_o], \tag{a}$$

$$\theta:\quad \ddot{\theta} - \sin\theta\cos\theta\,(\dot{\phi})^2 = -(g/l)\sin\theta, \tag{b}$$

$$\phi:\quad (\sin^2\theta)\,\dot{\phi} = \textit{constant} \equiv p, \tag{c}$$

[$\Rightarrow (l\sin\theta)^2\dot{\phi} = l^2 p \equiv C$; i.e., the horizontal projection of the motion follows Kepler's second law].

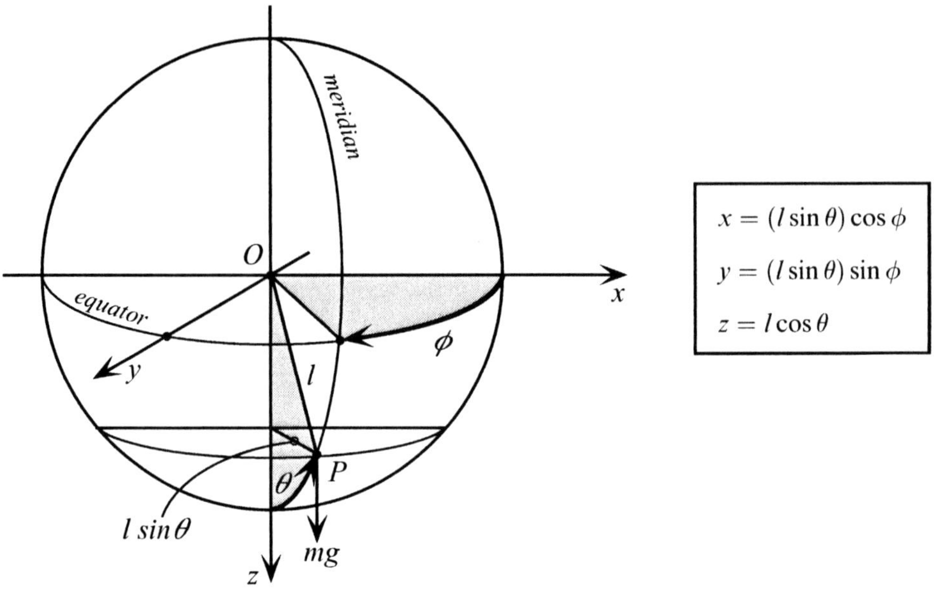

Figure 3.8 Geometry of spherical pendulum.

Solving (b, c) we find the *motion*, and then substituting the results into (a) we obtain the *multiplier* $\lambda = \lambda(t;\ initial\ conditions)$. Show that $\lambda = -S/l$, where $S =$ *sphere reaction on P* (or string tension, in the pendulum case), and $(\ldots)_o \equiv (\ldots)_{R=l}$.

On the analytical treatment of these nonlinear equations (including the stability of special motions), there exists a large literature; see, for example (alphabetically): Hamel (1949, pp. 262–264, 285, 691–692, 710–712), Lamb (1923, pp. 305–307), Landau and Lifshitz (1960, pp. 33–34), MacMillan (1927, pp. 337–344), Müller and Prange (1923, pp. 163–184), Pöschl (1949, pp. 46–49, 141–143), Synge and Griffith (1959, pp. 335–342), Webster (1912, pp. 42–45, 48–55, 124–125); also Corben and Stehle (1960, pp. 105–107), for an application of the perturbation method.

Problem 3.5.17 *Lagrange's Equations: Particle on Sphere, or Spherical Pendulum.* Continuing from the preceding problem:

(i) Show that its integrals of *energy* and *angular momentum* (about the Oz axis) can be written, respectively, as

$$(1/2)ml^2[(\dot{\theta})^2 + \sin^2\theta(\dot{\phi})^2] - m\,g\,l\cos\theta = constant \equiv E, \tag{a}$$

$$(\sin^2\theta)\dot{\phi} = constant \equiv p. \tag{b}$$

(ii) Show that eliminating $\dot{\phi}$ between (a, b) and then $(\ldots)^{\cdot}$-differentiating the resulting equation, and so on, we recover the θ-equation of the last problem.

Problem 3.5.18 *Lagrange's Equations: Particle on Sphere, or Spherical Pendulum; Integration of the θ-Equation.* Continuing from the preceding problem:

(i) Show that its θ- and ϕ-equations combine to the θ-*only* equation:

$$d^2\theta/dt^2 - (C^2/l^4)(\cos\theta/\sin^3\theta) + (g/l)\sin\theta = 0, \tag{a}$$

where $C = (l \sin\theta)^2 \dot{\phi} = constant$. What is the physical meaning of the singularity, in (a), for $\theta = 0$?

(ii) By integrating (a), show that

$$t = \int_{\theta_o}^{\theta} \sin\theta\,[2h \sin^2\theta - (C^2/l^4) + 2(g/l)\cos\theta \sin^2\theta]^{-1/2}\, d\theta. \qquad \text{(b)}$$

HINT

The first integral of the differential equation $\ddot{x} = f(x)$ is

$$(1/2)(dx/dt)^2 + V(x) = constant \equiv h, \qquad \text{where } V(x) = -\int f(x)\, dx; \qquad \text{(c)}$$

and from this we readily obtain

$$t = \int_{x_o = x(0)}^{x} [2h - 2V(x)]^{-1/2}\, dx. \qquad \text{(d)}$$

Here, $x = \theta$, and

$$f(x) \to f(\theta) = (C^2/l^4)(\cos\theta/\sin^3\theta) - (g/l)\sin\theta, \qquad \text{(e)}$$

so that $V(\theta) = -\int f(\theta) d\theta = \cdots$.

(iii) Setting $\cos\theta = z$, and with the abbreviations

$$2h - (C^2/l^4) \equiv \alpha, \qquad -2(g/l) \equiv \beta, \qquad -2h \equiv \gamma, \qquad \text{(f)}$$

reduces (b) to the *elliptic* integral

$$t = -\int_{z_o}^{z} [\alpha - \beta z + \gamma z^2 + \beta z^3]^{-1/2}\, dz, \qquad \text{(g)}$$

which *cannot* be integrated by a combination of elementary functions. For further details, see books on the *asymptotic* integration of ordinary differential equations, elliptic functions, and so on.

Problem 3.5.19 *Lagrange's Equations: Particle on Sphere, or Spherical Pendulum; Steady Motion.* Continuing from the preceding problems, show that the particular solution $\theta = \theta_o$ (i.e., particle describes a horizontal circle), requires that

$$d\phi/dt = C/r^2 = (g/l)^{1/2}(\cos\theta_o)^{-1/2} = constant \equiv \omega_o, \qquad \text{(a)}$$

where $C^2 = (gl^3 \sin^4\theta_o)/\cos\theta_o$; that is, for every θ_o there exists a particular such "steady motion" of constant angular velocity ω_o; and, hence, a period

$$\tau_o \equiv 2\pi/\omega_o = 2\pi(l/g)^{1/2}(\cos\theta_o)^{1/2}$$
$$\approx 2\pi(l/g)^{1/2}, \text{ for small } \theta_o \text{ (as for the \textit{plane} pendulum)}. \qquad \text{(b)}$$

Problem 3.5.20 *Lagrange's Equations: Particle on Sphere, or Spherical Pendulum; Stability of Steady Motion.* Continuing from the preceding problems:

(i) By setting in the exact θ-equation of the pendulum

$$\theta(t) = \theta_o + \Delta\theta(t) \equiv \theta_o + x(t), \qquad d\phi(t)/dt = \omega_o + \Delta[d\phi(t)/dt] \equiv \omega_o + y(t), \qquad \text{(a)}$$

expanding à la Taylor, and keeping only up to first-degree terms in x, y and their $(\ldots)^{\cdot}$-derivatives (i.e., considering small disturbances), and taking into account the equations of the fundamental state θ_o, obtain the *linear perturbation* equations from that state:

$$d^2x/dt^2 + \{[(1 + 2\cos^2\theta_o)/\sin^4\theta_o](C^2/l^4) + (g/l)\cos\theta_o\}x = 0. \tag{b}$$

Then, taking into account the C versus θ_o relation for that state (see preceding problem), show that (b) simplifies to

$$d^2x/dt^2 + k^2x = 0, \qquad \text{where } k^2 \equiv [(g/l)(1 + 3\cos^2\theta_o)]/\sin^4\theta_o, \tag{c}$$

that is, a harmonic oscillation around the constant value θ_o with a period [recall (b) of preceding problem]

$$\tau' = 2\pi/k = \left[2\pi(l/g)^{1/2}(\cos\theta_o)^{1/2}\right]\Big/(1 + 3\cos^2\theta_o)^{1/2}$$
$$\equiv \tau_o/(1 + 3\cos^2\theta_o)^{1/2}. \tag{d}$$

Notice that, since $\theta_o = 0, \pi/2$ are excluded, we will have $\tau_o/2 < \tau' < \tau_o$.

Such motions, where the *linear perturbation equations* around them are *equations with constant coefficients*, we call, after Routh *steady* (for an extensive treatment, see §8.5).

(ii) By carrying out a similar linearization of the exact ϕ-equation, $\dot{\phi} = C/(l\sin\theta)^2$, around θ_o, show that (to the first degree in x)

$$y = -[(2C\cos\theta_o)/(l^2\sin^3\theta_o)]x = -2[(g/l)(\cos\theta_o)/\sin^2\theta_o]^{1/2}x, \tag{e}$$

that is, $\dot{\phi}$ oscillates just like x, but the presence of the minus sign shows that as θ increases $\dot{\phi}$ decreases, and vice versa.

For further details on the integration of the above equations, and the behavior of the perturbed motion, for various values of θ_o, see, for example, Hamel ([1922(b)] 1912, 1st ed., pp. 106–108).

Example 3.5.11 *Constrained Lagrange's Equations → Routh–Voss Equations.* Let us consider the spatial, and initially unconstrained, motion of a particle of mass m in cylindrical coordinates $q_1 = r$, $q_2 = \phi$, $q_3 = z$ (vertical, positive upward), under the action of (constant) gravity (fig. 3.9).

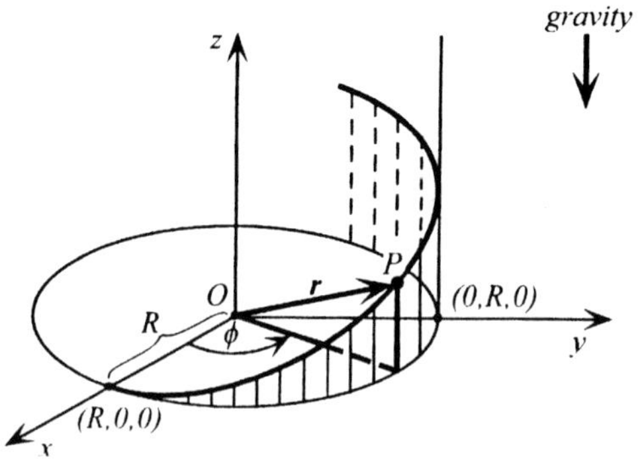

Figure 3.9 Particle on a helical path.

Here, clearly, the (double) kinetic energy and impressed forces are, respectively,

$$2T = m\,v^2 = m(ds/dt)^2 = m[(\dot{r})^2 + r^2(\dot{\phi})^2 + (\dot{z})^2], \tag{a}$$
$$Q_r = 0, \qquad Q_\phi = 0, \qquad Q_z = -mg, \tag{b}$$

and therefore, Lagrange's equations are (recall prob. 3.5.15)

$$E_r \equiv (\partial T/\partial \dot{r})^{\cdot} - \partial T/\partial r = m[\ddot{r} - r(\dot{\phi})^2] = 0, \tag{c}$$
$$E_\phi \equiv (\partial T/\partial \dot{\phi})^{\cdot} - \partial T/\partial \phi = m(r^2\dot{\phi})^{\cdot} = 0, \tag{d}$$
$$E_z \equiv (\partial T/\partial \dot{z})^{\cdot} - \partial T/\partial z = m\,\ddot{z} = -mg. \tag{e}$$

Next, assume that the particle is constrained to move on a smooth circular helix with axis z, radius R, and pitch p. Analytically, this means that now r, ϕ, z are coupled by the *two* constraints (assume that for $\phi = 0$, $z = 0$):

$$f_1 \equiv r - R = 0, \qquad f_2 \equiv z - p\,\phi = 0 \qquad (R, p\text{: positive constants}). \tag{f}$$

In this case, the kinetic energy (a) assumes the constrained form

$$T \to T_o = \cdots = (m/2)(R^2 + p^2)(\dot{\phi})^2 \equiv T_o(\dot{\phi}) = (m/2)[1 + (R/p)^2](\dot{z})^2 \equiv T_o(\dot{z}), \tag{g}$$

and, from $\delta' W = -mg\,\delta z = -mg\,\delta(p\,\phi) = -mgp\,\delta\phi$, we readily conclude that, contrary to (b),

$$Q_{z,o} = -mg, \qquad Q_{\phi,o} = -mgp. \tag{h}$$

Therefore, the *kinetic* = *reactionless* equations are either of the following:

(i)
$$(\partial T_o/\partial \dot{\phi})^{\cdot} - \partial T_o/\partial \phi = Q_{\phi,o}\text{:} \quad m(R^2 + p^2)\ddot{\phi} = -mgp, \tag{i}$$
$$\Rightarrow \phi = -[(gp)/(R^2 + p^2)](t^2/2) + \dot{\phi}(0)t + \phi(0); \tag{j}$$

(ii)
$$(\partial T_o/\partial \dot{z})^{\cdot} - \partial T_o/\partial z = Q_{z,o}\text{:} \quad m[1 + (R/p)^2]\ddot{z} = -mg, \tag{k}$$
$$\Rightarrow z = -[(gp^2)/(R^2 + p^2)](t^2/2) + \dot{z}(0)t + z(0) \quad (= p\,\phi). \tag{l}$$

To calculate the *constraint reactions*, we use the Routh–Voss equations

$$E_k(T) = Q_k + \sum \lambda_D(\partial f_D/\partial q_k) \equiv Q_k + R_k$$
$$(k = 1, 2, 3 \to r, \phi, z;\; D = 1, 2;\; \text{i.e., } n = 3,\; m = 2). \tag{m}$$

These latter here give [using (a), (b) and (f), *not* (g); and *then* enforcing (f)]

$$E_r = \lambda_1(\partial f_1/\partial r) + \lambda_2(\partial f_2/\partial r)\text{:}\;\; m[\ddot{r} - r(\dot{\phi})^2] = \lambda_1(1) + \lambda_2(0) \;\;\text{or}\;\; -m\,R(\dot{\phi})^2 = \lambda_1, \tag{n}$$

$$E_\phi = \lambda_1(\partial f_1/\partial \phi) + \lambda_2(\partial f_2/\partial \phi)\text{:}\;\; m(r^2\dot{\phi})^{\cdot} = \lambda_1(0) + \lambda_2(-p) \;\;\text{or}\;\; m\,R^2\ddot{\phi} = -p\,\lambda_2, \tag{o}$$

$$E_z = Q_z + \lambda_1(\partial f_1/\partial z) + \lambda_2(\partial f_2/\partial z)\text{:}\;\; m\,\ddot{z} = -mg + \lambda_1(0) + \lambda_2(1) \;\;\text{or}\;\; m\,\ddot{z} = -mg + \lambda_2. \tag{p}$$

If (F_r, F_ϕ, F_z) = *vector of constraint reaction on particle, from wire (in polar coordinates)*, then from the reaction virtual work invariance

$$\delta' W = F_r\,\delta r + F_\phi(R\,\delta\phi) + F_z\,\delta z = R_r\,\delta r + R_\phi\,\delta\phi + R_z\,\delta z \quad (=0),$$

we readily obtain

$$R_r = \lambda_1 = F_r, \qquad R_\phi = -p\,\lambda_2 = R\,F_\phi, \qquad R_z = \lambda_2 = F_z, \tag{q}$$

and so (n–p) transform to

$$m\,R(\dot\phi)^2 = -F_r, \qquad m\,R^2\ddot\phi = R\,F_\phi, \qquad m\,\ddot z = -m\,g + F_z, \tag{r}$$

and, from these [since $\lambda_2 = -(R/p)F_\phi = F_z \Rightarrow -R\,F_\phi = p\,F_z$] and the second of (f), we get

$$-m\,R^2\ddot\phi = p(m\,\ddot z + m\,g) \Rightarrow m(R^2 + p^2)\ddot\phi = -m\,g\,p; \quad \text{i.e., eq. (i).} \tag{s}$$

Solving (i = s), we find $\phi(t)$, and then inserting it into (r) we obtain (F_r, F_ϕ, F_z) and λ_1, λ_2 as functions of time and the initial conditions. The reader may wish to discuss the limiting cases:

$$p \to 0 \text{ (i.e., } \textit{helix} \to \textit{circle of radius } R\text{) and } p \to \infty.$$

Problem 3.5.21 *Routh–Voss Equations: Plane Rolling.* Consider two right circular and rough cylinders, C and C', with corresponding masses M and m, radii R and r, and horizontal (mutually parallel) axes O and O', in plane and slippingless rolling on each other. Assume, for simplicity, that C' is stationary (fig. 3.10; initially, P and P' coincide). Here, the constraints are

$$f_1 \equiv \rho - (R + r) \equiv \rho - b = 0 \ (\textit{contact}), \qquad f_2 \equiv b\,\phi - r\,\theta = 0 \ (\textit{rolling}). \tag{a}$$

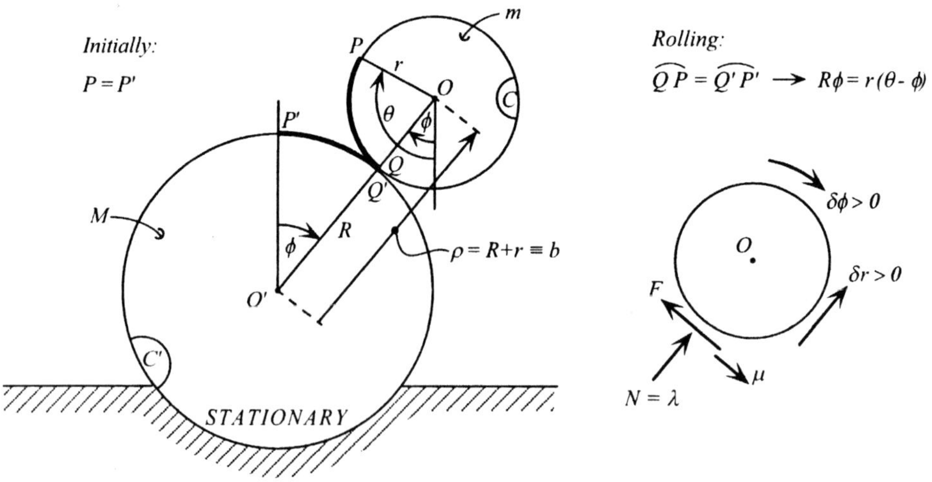

Figure 3.10 Cylinder C in a plane rolling over the fixed cylinder C'. [Initially, $P = P'$; rolling condition: $arc(PQ) = arc(Q'P') \Rightarrow R\phi = r(\theta - \phi) \Rightarrow (R + r)\phi = r\theta$.]

(i) Show that the Routh–Voss equations, in $q_1 = \rho$, $q_2 = \phi$, $q_3 = \theta$, and with $\lambda_1 \equiv \lambda$, $\lambda_2 \equiv \mu$ [i.e., $n = 3$ and $m = 2$, as long as C and C' are in contact], are

$$m\ddot{\rho} = -mg\cos\phi + m\rho(\dot{\phi})^2 + \lambda(1)\big|_{\text{eqs.(a)}} \Rightarrow \lambda = mg\cos\phi - mb(\dot{\phi})^2, \quad \text{(b)}$$

$$(m\rho^2\dot{\phi})^{\cdot} = mg\rho\sin\phi + \mu(b)\big|_{\text{eqs.(a)}} \Rightarrow \mu = mg\ddot{\phi} - mg\sin\phi, \quad \text{(c)}$$

$$(mr^2/2)\ddot{\theta} = \mu(-r)\big|_{\text{eqs.(a)}} \Rightarrow \mu = -(mb/2)\ddot{\phi} = -(mr/2)\ddot{\theta}. \quad \text{(d)}$$

(ii) Eliminating μ between (c, d), obtain the kinetic equation

$$\ddot{\phi} = (2g/3b)\sin\phi. \quad \text{(e)}$$

(iii) Show that (a) $\lambda = N =$ *normal* force from C' to C [> 0, for $g\cos\phi > b(\dot{\phi})^2$, from eq. (b); if not, then $\lambda = 0$]; and (b) $\mu = -F \Rightarrow F = -\mu =$ *tangential* (*frictional*) force from C' to C [$\mu = -(mb/2)\ddot{\phi} = -(mg/3)\sin\phi < 0$, by (e)].

(iv) Finally, find the critical angle at which C loses contact with C'. (See also Fetter and Walecka, 1980, pp. 74–77.)

Example 3.5.12 *Invariance of the Routh–Voss Equations under Frame of Reference Transformations.* Let us consider a system subject to the Pfaffian (holonomic and/or nonholonomic) constraints

$$f_D \equiv \sum a_{Dk}\,\dot{q}_k + a_D = 0 \qquad [D = 1, \ldots, m(< n);\ k = 1, \ldots, n], \quad \text{(a)}$$

and, hence, having the Routh–Voss equations of motion

$$E_k(L) = Q_k + \sum \lambda_D\, a_{Dk} \equiv Q_k + R_k, \quad \text{(b)}$$

where $L \equiv T - V =$ *Lagrangean* of system [$= -(V - T) = -$(*kinetic potential*) of system, in 19th century terminology], and $Q_k =$ *nonpotential impressed forces*.

Now, let us subject its Lagrangean coordinates $q \equiv (q_1, \ldots, q_n)$ to the following general *explicitly time-dependent and nonsingular* (i.e., *uniquely invertible*) *point transformation*:

$$q_k = q_k(t; q_{1'}, \ldots, q_{n'}) \Leftrightarrow q'_{k'} \equiv q_{k'} = q_{k'}(t; q_1, \ldots, q_n), \quad \text{(c)}$$

or, compactly, $q = q(t, q') \Leftrightarrow q' = q'(t, q)$. Let us express eqs. (b) in terms of the $q_{k'}$s. From (c), we readily find

$$dq_s/dt = \sum (\partial q_s/\partial q_{s'})(dq_{s'}/dt) + \partial q_s/\partial t \quad \text{and} \quad \delta q_s = \sum (\partial q_s/\partial q_{s'})\,\delta q_{s'}, \quad \text{(d)}$$

and so the constraints (a) transform to

$$f_D \equiv \sum a_{Dk}\Big(\sum (\partial q_k/\partial q_{k'})(dq_{k'}/dt) + \partial q_k/\partial t\Big) + a_D = \cdots$$
$$= \sum a_{Dk'}(dq_{k'}/dt) + a'_D = 0, \quad \text{(e)}$$

where [recalling (2.6.6–6b)]

$$a_{Dk'} \equiv \sum (\partial q_k/\partial q_{k'})a_{Dk}, \qquad a'_D \equiv \sum (\partial q_s/\partial t)a_{Ds} + a_D. \quad \text{(f)}$$

Next, to the left side of (b). Applying the chain rule to $L(t, q, dq/dt) = L'(t, q', dq'/dt)$: q'-*Lagrangean*, and, with the helpful notations $dq_k/dt \equiv v_k$, $dq_{k'}/dt \equiv v_{k'}$, we obtain

$$\partial L'/\partial v_{s'} = \sum (\partial L/\partial v_s)(\partial v_s/\partial v_{s'}) = \sum (\partial L/\partial v_s)(\partial q_s/\partial q_{s'}), \qquad \text{(g1)}$$

$$\partial L'/\partial q_{s'} = \sum (\partial L/\partial q_s)(\partial q_s/\partial q_{s'}) + \sum (\partial L/\partial v_s)(\partial v_s/\partial q_{s'}). \qquad \text{(g2)}$$

But, from (c, d), and in addition to $\partial v_s/\partial v_{s'} = \partial q_s/\partial q_{s'}$ [utilized in (g1)], we also have

$$\partial v_s/\partial q_{s'} = (\partial/\partial q_{s'})\Big(\sum (\partial q_s/\partial q_{k'})v_{k'} + \partial q_s/\partial t\Big)$$
$$= \sum (\partial^2 q_s/\partial q_{s'}\partial q_{k'})v_{k'} + \partial^2 q_s/\partial q_{s'}\partial t = (\partial q_s/\partial q_{s'})^{\cdot},$$

that is,

$$E_{s'}(v_s) \equiv d/dt(\partial v_s/\partial v_{s'}) - \partial v_s/\partial q_{s'} = 0 \qquad \text{[recalling (2.5.7–10)].} \qquad \text{(h)}$$

Thanks to these identities and (g1, 2) we find, successively,

$$(\partial L'/\partial v_{s'})^{\cdot} - \partial L'/\partial q_{s'} = \Big(\sum (\partial L/\partial v_s)^{\cdot}(\partial q_s/\partial q_{s'}) + \sum (\partial L/\partial v_s)(\partial q_s/\partial q_{s'})^{\cdot}\Big)$$
$$- \Big(\sum (\partial L/\partial q_s)(\partial q_s/\partial q_{s'}) + \sum (\partial L/\partial v_s)(\partial q_s/\partial q_{s'})^{\cdot}\Big)$$
$$= \sum [(\partial L/\partial \dot{q}_s)^{\cdot} - \partial L/\partial q_s](\partial q_s/\partial q_{s'}) \text{ [invoking (b, a) and (f)]}$$
$$= \sum \Big(Q_s + \sum \lambda_D a_{Ds}\Big)(\partial q_s/\partial q_{s'}) \equiv Q_{s'} + \sum \lambda_D a_{Ds'}, \qquad \text{(i)}$$

or, compactly,

$$E_{s'}(L') = Q_{s'} + \sum \lambda_D a_{Ds'} \equiv Q_{s'} + R_{s'}, \qquad \text{(j)}$$

where $Q_{s'} \equiv \sum (\partial q_s/\partial q_{s'})Q_s$. [The latter can also be established from the virtual work invariance: $\delta' W = \sum Q_k\, \delta q_k = \sum Q_{k'}\, \delta q_{k'}$ and the second of eq. (d).]

Notice that (j) amounts to $\lambda'_{D'} \to \lambda_{D'} = \lambda_D$; that is, the multipliers are *invariant*, or *objective*, under the *frame of reference* transformation (c). Equations (j) express the following fundamental theorem.

THEOREM

The Routh–Voss equations transform like a covariant vector under general frame of reference transformations $q \to q'(t, q)$; *that is, these equations are form invariant not only under arbitrary Lagrangean coordinate transformations in a given frame, but also under arbitrary frame of reference transformations* (whereas the Newton–Euler equations are not!).

As already stated on several occasions, this twofold form invariance of the equations of motion constitutes the major advantage of "Lagrange" over "Newton–Euler."

Example 3.5.13 *Uniqueness of the Lagrangean, Introduction to Gyroscopicity, etc.* Let us consider two distinct Lagrangeans, L and L', which, however, produce the same (Lagrangean) equations of motion; that is,

$$E_k(L) = E_k(L') \quad (= 0, \text{ or } Q_k, \quad \text{or} \quad Q_k + R_k). \qquad \text{(a)}$$

We ask the questions: By what amount can L and L' differ at most (so that we work with the simplest of them)? or, How unique is a system Lagrangean? To answer these, we assume that

$$L' - L = f(t, q, dq/dt \equiv v), \tag{b}$$

and then try to find as much as possible about f.

Indeed, since $E_k(\ldots)$ is a *linear* operator, (a) and (b) lead to

$$\begin{aligned} 0 = E_k(L' - L) = E_k(L') - E_k(L) = E_k(f) &\equiv d/dt(\partial f/\partial v_k) - \partial f/\partial q_k \\ = \sum [(\partial/\partial q_s)(\partial f/\partial v_k)]v_s &+ \sum [(\partial/\partial v_s)(\partial f/\partial v_k)](dv_s/dt) \\ &+ (\partial/\partial t)(\partial f/\partial v_k) - \partial f/\partial q_k. \end{aligned} \tag{c}$$

However, since L *and* L' must produce the *same accelerations* (i.e., the same $\sim dv/dt \equiv d^2q/dt^2$ terms), the *corresponding coefficients in (c) must vanish*: $\partial^2 f/\partial v_s \partial v_k = 0$. This leads readily to

$$f = \sum C_s v_s + C, \tag{d}$$

where $C_s = C_s(t, q)$ and $C = C(t, q)$ are arbitrary but sufficiently differentiable functions of the q's and t. Substituting (d) into (c), we obtain

$$E_r(f) = \sum (\partial C_r/\partial q_s - \partial C_s/\partial q_r)(dq_s/dt) + (\partial C_r/\partial t - \partial C/\partial q_r) = 0, \tag{e}$$

and, since this must hold for arbitrary (dq/dt)'s we conclude that

$$\partial C_r/\partial q_s - \partial C_s/\partial q_r = 0 \quad \text{and} \quad \partial C_r/\partial t - \partial C/\partial q_r = 0, \quad \text{for all } r, s = 1, \ldots, n. \tag{f}$$

These *exactness* conditions (recalling §2.3), in turn, imply the existence of a *gauge* function $F(t, q)$, such that

$$C_r = \partial F/\partial q_r \quad \text{and} \quad C = \partial F/\partial t; \tag{g}$$

and so, (d) reduces to

$$f = \cdots = dF(t, q)/dt; \tag{h}$$

that is, L and $L' = L + dF/dt$ will produce the same equations of motion.

In sum: *The Lagrangean function is defined only to within the total time-derivative of an arbitrary function of the Lagrangean coordinates and time.* It follows that *constant terms*, or terms of the form $f(t)$, can be immediately neglected from a Lagrangean with no consequences on the equations of motion.

Generalized Potential, Gyroscopic Forces

A similar argument shows that if some (or all) of the Q_k's are expressible in terms of a *generalized potential* $V = V(t, q, \dot{q} \equiv v)$ as

$$Q_k = d/dt(\partial V/\partial v_k) - \partial V/\partial q_k, \tag{i}$$

then the most general such V must be *linear* in the v's:

$$V = \sum V_s(t, q)\, v_s + V^{(0)}(t, q). \tag{j}$$

Indeed, by (i), $Q_k = \cdots = \sum (\partial^2 V/\partial v_k \partial v_s)(dv_s/dt) + (t, q, v)$-*terms*. But, in classical mechanics, $Q_k = Q_k(t, q, v)$, and therefore $\partial^2 V/\partial v_k \partial v_s = 0$, from which (j) follows. So substituting (j) into (i), we see that such forces take the explicit form

$$Q_k = \cdots = -\partial V^{(0)}/\partial q_k + \sum (\partial V_k/\partial q_l - \partial V_l/\partial q_k)\dot{q}_l + \partial V_k/\partial t. \tag{k}$$

Here, we introduce a new definition.

DEFINITION

The *(non-constraint) forces with vanishing power*—that is, $\sum Q_k \dot{q}_k \equiv \sum Q_k v_k = 0$ —are called *gyroscopic*. Then, since $\sum\sum (\partial V_k/\partial q_l - \partial V_l/\partial q_k) v_l\, v_k = 0$ [due to the antisymmetry of the (...) terms in k, l], it follows that if

$$\sum (\partial V_k/\partial t) v_k = 0 \qquad \text{and} \qquad \sum (\partial V^{(0)}/\partial q_k) v_k = dV^{(0)}/dt - \partial V^{(0)}/\partial t = 0, \tag{l}$$

then the generalized potential forces (i, k) are gyroscopic. (Gyroscopicity is detailed in §3.9 ff.)

Next, let us illustrate the theorem of the uniqueness of the Lagrangean by a few simple examples.

(i) Consider a particle P of mass m free to slide along a massless smooth and rigid rod OA, which rotates, say clockwise, about a horizontal axis through O (i.e., on a vertical plane) with a *given* motion $\phi = \phi(t) =$ *known function of time* (fig. 3.11).

It is not hard to show that *a* Lagrangean for this system is

$$L' = (m/2)[(r_o + r)^2(\dot{\phi})^2 + (\dot{r})^2] - [-mg(r_o + r)\sin\phi], \tag{m}$$

where B is *any* "origin" on OA [$r_o =$ *constant*, $q = r = r(t)$]. By the foregoing theory, the Lagrangean $L = L' - mgr_o \sin\phi(t)$ will result in the same equation of motion for $q_1 = r$ as L':

$$E_r(L) = E_r(L'): \quad \ddot{r} - (\dot{\phi})^2 r = g\sin\phi + r_o(\dot{\phi})^2. \tag{n}$$

That, however, would not be the case if OA was *unconstrained*; then, $q_2 = \phi$.

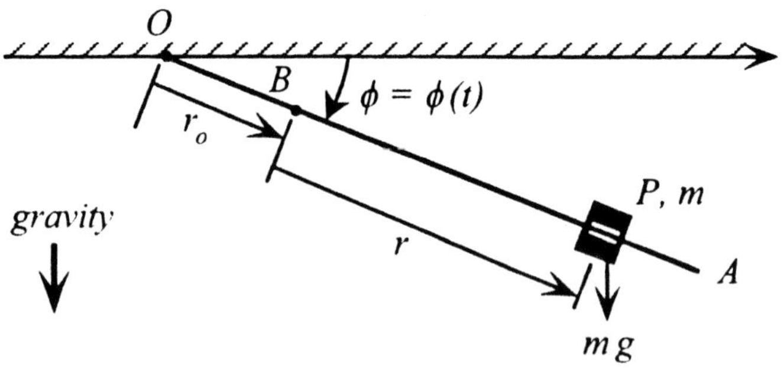

Figure 3.11 Particle P sliding over a rotating rod OA, which rotates in a prescribed way about a horizontal axis through O.

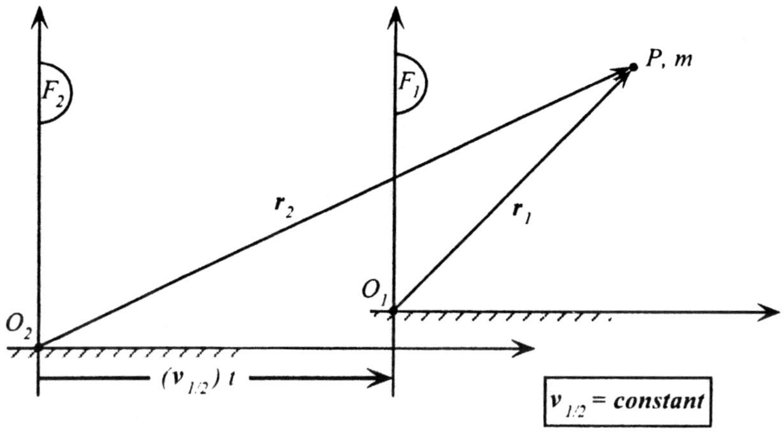

Figure 3.12 Two inertial frames in relative motion, assuming that at time $t = 0$ their origins, O_1 and O_2, coincide.

(ii) Consider the motion of a particle P of mass m in two inertial frames of reference, F_1 and F_2, in relative motion; say, F_1 moving with (vectorially) *constant* velocity $\boldsymbol{v}_{1/2}$ relative to F_2 (fig. 3.12). If we assume, for simplicity, but no loss of generality, that $V = 0$ and/or $Q_k = 0$, then the Lagrangean of P in F_2, which is L_2, equals

$$\begin{aligned} L_2 &= (m/2)\boldsymbol{v}_2 \cdot \boldsymbol{v}_2 = (m/2)(\boldsymbol{v}_1 + \boldsymbol{v}_{1/2}) \cdot (\boldsymbol{v}_1 + \boldsymbol{v}_{1/2}) \\ &= (m/2)\boldsymbol{v}_1 \cdot \boldsymbol{v}_1 + m\,\boldsymbol{v}_1 \cdot \boldsymbol{v}_{1/2} + (m/2)\boldsymbol{v}_{1/2} \cdot \boldsymbol{v}_{1/2} \\ &= L_1 + df/dt \equiv L_1 + F, \end{aligned} \tag{o}$$

where

$$L_1 = (m/2)\boldsymbol{v}_1 \cdot \boldsymbol{v}_1, \qquad f = m\,\boldsymbol{r}_1 \cdot \boldsymbol{v}_{1/2} + (m/2)(\boldsymbol{v}_{1/2} \cdot \boldsymbol{v}_{1/2})t. \tag{o1}$$

Clearly, both Lagrangeans produce the same (i.e., equivalent) equations of motion (*"principle" of Galilean relativity*):

$$(\partial L_2/\partial \boldsymbol{v}_2)^{\cdot} - \partial L_2/\partial \boldsymbol{r}_2 = \boldsymbol{0}: \qquad d(m\boldsymbol{v}_2)/dt = \boldsymbol{0}, \tag{p1}$$

$$(\partial L_1/\partial \boldsymbol{v}_1)^{\cdot} - \partial L_1/\partial \boldsymbol{r}_1 = \boldsymbol{0}: \qquad d(m\boldsymbol{v}_1)/dt = \boldsymbol{0}. \tag{p2}$$

(iii) Let us extend the preceding example to the case of *general translation*; that is,

$$\boldsymbol{v}_{1/2} = \boldsymbol{v}_{1/2}(t). \tag{q}$$

Here, we have, successively,

$$\begin{aligned} L_2 &= (m/2)\boldsymbol{v}_2 \cdot \boldsymbol{v}_2 = (m/2)(\boldsymbol{v}_1 + \boldsymbol{v}_{1/2}) \cdot (\boldsymbol{v}_1 + \boldsymbol{v}_{1/2}) \\ &= (m/2)\boldsymbol{v}_1 \cdot \boldsymbol{v}_1 + m\,\boldsymbol{v}_1 \cdot \boldsymbol{v}_{1/2} + (m/2)\boldsymbol{v}_{1/2} \cdot \boldsymbol{v}_{1/2} \\ &= (m/2)\boldsymbol{v}_1 \cdot \boldsymbol{v}_1 + (m/2)\boldsymbol{v}_{1/2} \cdot \boldsymbol{v}_{1/2} + [(m\,\boldsymbol{r}_1 \cdot \boldsymbol{v}_{1/2})^{\cdot} - m\,\boldsymbol{r}_1 \cdot (d\boldsymbol{v}_{1/2}/dt)] \\ &= L_1 + \textit{given function of time} \text{ (i.e., omittable)} \\ &\qquad + \textit{total derivative of function of position and time} \text{ (i.e., omittable)} \\ &\qquad - m\,\boldsymbol{r}_1 \cdot (d\boldsymbol{v}_{1/2}/dt) \end{aligned}$$

and so, to within "L-important" terms,

$$L_2 = L_1 - m\boldsymbol{r}_1 \cdot (d\boldsymbol{v}_{1/2}/dt) \equiv L_1 - m\boldsymbol{r}_1 \cdot \boldsymbol{a}_{1/2}. \tag{r}$$

[In the earlier, Galilean case, clearly, $\boldsymbol{a}_{1/2} = \mathbf{0}$.] The corresponding Lagrangean equations are

$$(\partial L_2/\partial \boldsymbol{v}_2)^{\cdot} - \partial L_2/\partial \boldsymbol{r}_2 = \mathbf{0}: \qquad d(m\boldsymbol{v}_2)/dt = \mathbf{0} \Rightarrow m\boldsymbol{a}_2 = \mathbf{0}, \tag{s1}$$

$$(\partial L_1/\partial \boldsymbol{v}_1)^{\cdot} - \partial L_1/\partial \boldsymbol{r}_1 = \mathbf{0}: \qquad d(m\boldsymbol{v}_1)/dt - (-m\boldsymbol{a}_{1/2}) = \mathbf{0}$$

$$\Rightarrow m\,\boldsymbol{a}_1 = -m\,\boldsymbol{a}_{1/2} (= \text{"}\textit{transport force}\text{"}); \tag{s2}$$

that is, F_2 is *noninertial*.

Example 3.5.14 *On the Physical Significance of the Lagrangean Multipliers* [May be omitted in a first reading]. Let us consider an n *DOF* system with kinetic and potential energies T and V, respectively, no nonpotential impressed forces, but subject to the holonomic constraints

$$f_D = f_D(q_1, \ldots, q_n) \equiv f_D(q) = 0 \qquad [D = 1, \ldots, m(< n)]. \tag{a}$$

Its Routh–Voss equations of motion (with $k = 1, \ldots, n$)

$$(\partial T/\partial \dot{q}_k)^{\cdot} - \partial T/\partial q_k = -(\partial V/\partial q_k) + \sum \lambda_D(\partial f_D/\partial q_k) \equiv Q_k + R_k, \tag{b}$$

can, clearly, be rewritten in the multiplierless/kinetic form

$$(\partial T/\partial \dot{q}_k)^{\cdot} - \partial T/\partial q_k = -(\partial V_T/\partial q_k), \tag{c}$$

where

$$V_T \equiv V + V_C = V - \sum \lambda_D(t) f_D(q);$$

in words:

$$\text{Total potential} = \text{Ordinary potential} + \text{"Constraint potential"}; \tag{d}$$

that is, the holonomic constraint reactions can be brought to the potential (impressed) force form

$$R_k = -\partial V_C/\partial q_k = -(\partial/\partial q_k)\Big(-\sum \lambda_D f_D\Big) = \sum \lambda_D(\partial f_D/\partial q_k) = R_k(t, q). \tag{e}$$

The apparent contradiction of a *conservative* system [since $\partial V/\partial t = 0$ and $\partial f_D/\partial t = 0$ (more in §3.9)] containing *explicitly time-dependent* forces [i.e., $\partial R_k/\partial t = \sum (d\lambda_D/dt)(\partial f_D/\partial q_K) \neq 0$] is explained by the fact that the constraint potential $V_C = -\sum \lambda_D f_D$ *is known only along the trajectory curve of the figurative system point, in configuration space*; and not throughout the allowable domain of the q's there, like $V(q)$. Below, elaborating the above, we interpret the *constraint reactions* as limiting cases of elastic (potential) forces whose stiffnesses tend to infinity, something which is in agreement with the *principle of relaxation of the constraints* (§3.7); and in the process we obtain an interesting physical interpretation of the Lagrangean multipliers.

Let us consider, for simplicity, but no loss of generality, the case of a single constraint (i.e., $m = 1$)

$$f = f(q_1, \ldots, q_n) \equiv f(q) = 0. \tag{f}$$

Now, since this constraint is maintained by strong forces, *during actual system motions* equation (f) cannot be violated by a large amount. Therefore, the potential of the (reaction turned impressed) forces maintaining (f), $\Pi(f)$, can be written with sufficient accuracy as the following finite Taylor series around $f = 0$ [with $(\ldots)' \equiv d(\ldots)/df$]:

$$\Pi = \Pi(f) = \Pi(0) + \Pi'(0)f + (1/2)\Pi''(0)f^2; \tag{g}$$

and since these forces can be likened, for small f, to very stiff elastic forces, we must also have

$$\Pi'(0) = 0 \qquad \text{and} \qquad \Pi''(0) \equiv 1/\varepsilon > 0, \tag{h}$$

where $\varepsilon =$ *small positive constant*; so that the constraint (f) is maintained by strong forces (theoretically, $\varepsilon \to 0$). Hence, and neglecting the immaterial constant $\Pi(0)$, we have for small f's,

$$\Pi(f) = f^2/2\varepsilon, \tag{i}$$

in which case the corresponding spring-like force equals

$$-\partial\Pi/\partial q_k = -(f/\varepsilon)(\partial f/\partial q_k); \tag{j}$$

and since this must equal the Lagrangean constraint reaction (e), that is,

$$R_k = -\partial V_C/\partial q_k = -(\partial/\partial q_k)(-\lambda f) = \lambda(\partial f/\partial q_k), \tag{k}$$

comparing (j, k), we immediately conclude that

$$\lambda = -f/\varepsilon \equiv \lambda(f;\varepsilon), \tag{l}$$

that is, λ is a measure of the *(time-dependent) violation of the constraint* $f = 0$; and in the theoretical limit of analytical mechanics, $\varepsilon \to 0$ and $f \to 0$,

$$\lambda = -\lim(f/\varepsilon) = \textit{force caused by a linear elastic spring of infinite stiffness}\ (1/\varepsilon). \tag{m}$$

Application of these ideas to the plane motion of a particle of, say, unit mass under the constraint $f(x, y) = 0$ (plane curve, in rectangular Cartesian coordinates x, y), and, for simplicity, no impressed forces, yields the equations

Constrained Lagrangean eqs: $\ddot{x} = \lambda(\partial f/\partial x)$, $\ddot{y} = \lambda(\partial f/\partial y)$, (n)

Unconstrained Newton–Euler eqs: $\ddot{x} = -(1/\varepsilon)(\partial w/\partial x)$, $\ddot{y} = -(1/\varepsilon)(\partial w/\partial y)$, (o)

where (fig. 3.13)

$$\Pi \approx V_C = V_C(f;\varepsilon) = w/\varepsilon = (1/\varepsilon)f^2 = (f/\varepsilon)f = -\lambda f; \tag{p}$$

that is, for small ε, V_C represents a steep potential gully whose bottom coincides with the constraint curve $f = 0$.

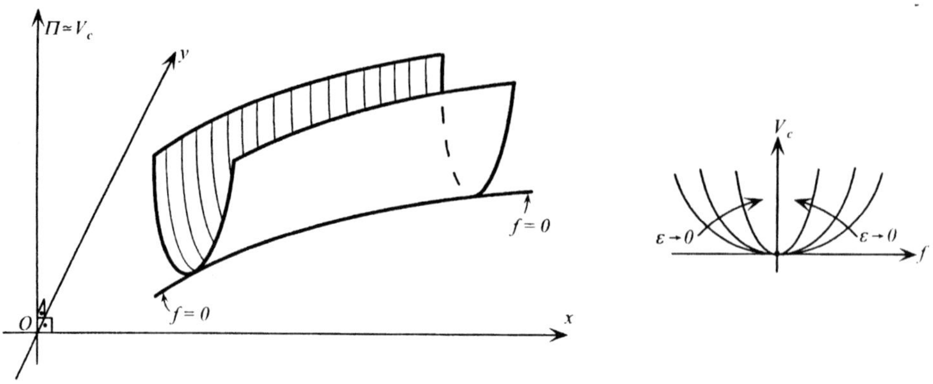

Figure 3.13 Constraint potential as a steep potential gully whose bottom coincides with the constraint curve $f = 0$.

A detailed analysis of whether and how, as $\varepsilon \to 0$ (limit of infinite stiffness), the solutions of (o) tend to those of (n), carried out by Kampen and Lodder, shows that for this to happen

> [T]he applied forces [must] vary smoothly compared with the periods of the internal elastic vibrations in the rods and bodies responsible for the constraints. If that is not satisfied one cannot treat these bodies as rigid, but must include their internal vibrations as additional degrees of freedom in the description of the system. *Lagrange's equations apply to a pendulum that I have set in motion with my hand, but not when I have hit it with a hammer ... or when it is set in motion by an escapement.* (Kampen and Lodder, 1984, pp. 420–421, emphasis added)

For further details and insights, see, for example, Arnold (1974, §17), Kampen and Lodder (1984; and references cited therein); also Gallavotti (1983, p. 168 ff.), Lanczos (1962, chap. 24, pp. 11–12; 1970, pp. 141–145), and Park (1990, pp. 60–61).

Example 3.5.15 *Maggi Equations (Holonomic Constraints).* Let us formulate the kinetic and kinetostatic rotational Maggi equations for a thin homogeneous bar AB [of length $|\boldsymbol{AB}| = 2$, and moment of inertia about a(ny) axis through its mass center G, perpendicular to its length, I], in arbitrary spatial motion, under gravity, in terms of the direction cosines of the bar α, β, γ relative to fixed (inertial) axes O–xyz; or, equivalently, relative to comoving/translating but nonrotating axes G–xyz (fig. 3.14) (Ramsey, 1937, p. 234).

Here, only the *rotational* part of the kinetic energy of the bar is needed. Choosing as Lagrangean rotational coordinates $q_1 = \alpha$, $q_2 = \beta$, $q_2 = \gamma$, we readily obtain from geometry

$$\alpha = \sin\theta\cos\phi, \quad \beta = \sin\theta\sin\phi, \quad \gamma = \cos\theta, \quad \text{and} \quad \alpha^2 + \beta^2 + \gamma^2 = 1 \quad \text{(constraint)}, \tag{a}$$

that is, $n = 3$, $m = 1$. Hence, using König's theorem, the Eulerian angle kinematics (§1.12), and with the usual notations [$(\omega_1, \omega_2, \omega_3) =$ inertial angular velocity of bar, along its principal axes G–123 (G–1: along bar, G–2, 3: perpendicular to it), and

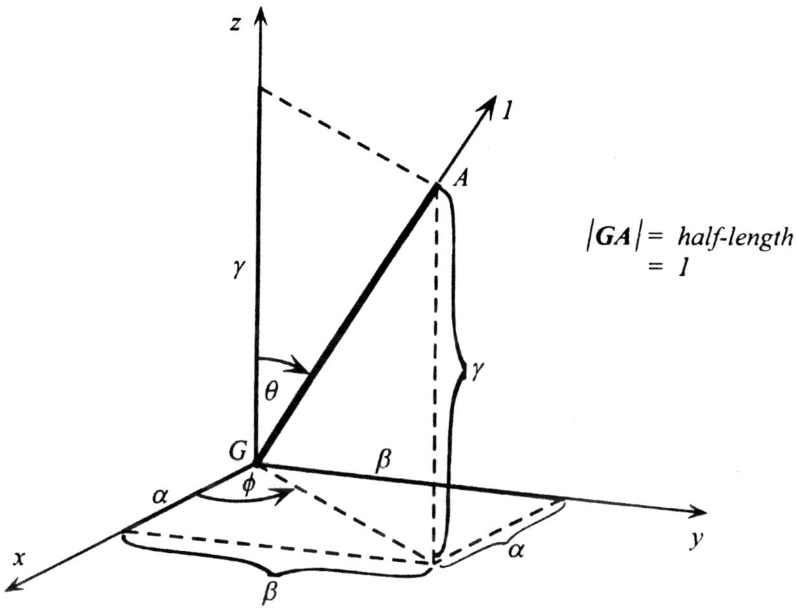

Figure 3.14 Geometry of homogeneous bar in arbitrary spatial motion.

$I_{1,2,3}$ = principal moments of inertia there], we find for the (double) rotational kinetic energy of the bar

$$2T = I_1\,\omega_1{}^2 + I_2\,\omega_2{}^2 + I_3\,\omega_3{}^2$$
$$= (0)(\dot{\psi} + \dot{\phi}\cos\theta)^2 + (I)(\dot{\theta})^2 + (I)(\dot{\phi}\sin\theta)^2 = I\,[(\dot{\theta})^2 + \sin^2\theta(\dot{\phi})^2]. \quad \text{(b)}$$

To implement Maggi's equations, we need to express T in terms of α, β, γ and their $(\ldots)^{\cdot}$-derivatives. Indeed, using (a) we find, successively,

(i) $$\sin^2\theta = 1 - \cos^2\theta = 1 - \gamma^2 = \alpha^2 + \beta^2; \quad \text{(c)}$$

(ii) $$(\dot{\gamma})^2 = (-\sin\theta\,\dot{\theta})^2 = \sin^2\theta(\dot{\theta})^2 = (\alpha^2 + \beta^2)(\dot{\theta})^2; \quad \text{(d)}$$

(iii) $$\begin{aligned}(\alpha\,\dot{\beta} - \dot{\alpha}\,\beta) &= (\sin\theta\cos\phi)[(\cos\theta\sin\phi)\dot{\theta} + \alpha\,\dot{\phi}] \\ &\quad - (\sin\theta\sin\phi)[(\cos\theta\cos\phi)\dot{\theta} - \beta\,\dot{\phi}] \\ &= \cdots = (\alpha^2 + \beta^2)\dot{\phi};\end{aligned} \quad \text{(e)}$$

(iv) $$\chi \equiv (\alpha\,\dot{\beta} - \dot{\alpha}\,\beta)^2 + (\alpha\,\dot{\alpha} + \beta\,\dot{\beta})^2 = \cdots = (\alpha^2 + \beta^2)[(\dot{\alpha})^2 + (\dot{\beta})^2], \quad \text{(f)}$$

but, also [invoking the fourth of eq. (a)]

$$\chi = (\alpha\,\dot{\beta} - \dot{\alpha}\,\beta)^2 + \gamma^2(\dot{\gamma})^2, \quad \text{(g)}$$

and so equating the right sides of (f) and (g), adding $(\dot\gamma)^2$ to both, and rearranging, we get

$$(\alpha\dot\beta - \dot\alpha\beta)^2 + (\dot\gamma)^2 = (\alpha^2+\beta^2)[(\dot\alpha)^2 + (\dot\beta)^2] + (\dot\gamma)^2(1-\gamma^2)$$
$$= \cdots = (\alpha^2+\beta^2)[(\dot\alpha)^2 + (\dot\beta)^2 + (\dot\gamma)^2]. \qquad \text{(h)}$$

As a result of the above [(d) → (c) → (e) → (h)], the kinetic energy (b) becomes, successively,

$$2T = I\Big\{(\dot\gamma)^2/(\alpha^2+\beta^2) + (\alpha^2+\beta^2)[(\alpha\dot\beta - \dot\alpha\beta)^2/(\alpha^2+\beta^2)^2]\Big\}$$
$$= I\Big\{[(\alpha\dot\beta - \dot\alpha\beta)^2 + (\dot\gamma)^2]/(\alpha^2+\beta^2)\Big\}$$
$$= I\Big[(\dot\alpha)^2 + (\dot\beta)^2 + (\dot\gamma)^2\Big]. \qquad \text{(i)}$$

Next, we introduce the new (holonomic) coordinates:

$$f_1 \equiv \alpha^2+\beta^2+\gamma^2-1 \quad (=0), \qquad f_2 \equiv \beta \quad (\neq 0), \qquad f_3 \equiv \gamma \quad (\neq 0), \qquad \text{(j)}$$

which invert easily to (no constraint enforcement yet!)

$$\alpha^2 = 1 + f_1 - (f_2)^2 - (f_3)^2, \qquad \beta = f_2, \qquad \gamma = f_3. \qquad \text{(k)}$$

Now, with the useful notation $M_k \equiv [(\partial T/\partial\dot q_k)^{\cdot} - \partial T/\partial q_k] - Q_k \equiv E_k - Q_k = E_k + \partial V/\partial q_k$, where $V = V(q_k\colon \alpha,\beta,\gamma) =$ *potential energy*, Maggi's equations yield

Kinetostatic:

$$f_1\colon\ (\partial\alpha/\partial f_1)M_\alpha + (\partial\beta/\partial f_1)M_\beta + (\partial\gamma/\partial f_1)M_\gamma = \lambda_1,$$
$$\Rightarrow (1/2\alpha)M_\alpha = \lambda_1 \ \Rightarrow\ (1/\alpha)(I\ddot\alpha + \partial V/\partial\alpha) = 2\lambda_1; \qquad \text{(l)}$$

Kinetic:

$$f_2\colon\ (\partial\alpha/\partial f_2)M_\alpha + (\partial\beta/\partial f_2)M_\beta + (\partial\gamma/\partial f_2)M_\gamma = 0,$$
$$\Rightarrow (-\beta/\alpha)M_\alpha + M_\beta = 0 \Rightarrow (1/\alpha)(I\ddot\alpha + \partial V/\partial\alpha)$$
$$= (1/\beta)(I\ddot\beta + \partial V/\partial\beta); \qquad \text{(m)}$$

$$f_3\colon\ (\partial\alpha/\partial f_3)M_\alpha + (\partial\beta/\partial f_3)M_\beta + (\partial\gamma/\partial f_3)M_\gamma = 0,$$
$$\Rightarrow (-\gamma/\alpha)M_\alpha + M_\gamma = 0 \Rightarrow (1/\alpha)(I\ddot\alpha + \partial V/\partial\alpha)$$
$$= (1/\gamma)(I\ddot\gamma + \partial V/\partial\gamma); \qquad \text{(n)}$$

and from these wc immediately obtain the more symmetric form

$$(1/\alpha)(I\ddot\alpha + \partial V/\partial\alpha) = (1/\beta)(I\ddot\beta + \partial V/\partial\beta) = (1/\gamma)(I\ddot\gamma + \partial V/\partial\gamma) = 2\lambda_1. \qquad \text{(o)}$$

If only the motion is sought, we combine (k, l) with the constraint (a); then, if the reaction $\lambda_1(t)$ is needed, it can be found from (j).

It is not hard to see that the Routh–Voss equations of this problem are

$$M_\alpha = \lambda_1(2\alpha), \qquad M_\beta = \lambda_1(2\beta), \qquad M_\gamma = \lambda_1(2\gamma), \qquad \text{i.e., eqs. (m).} \qquad \text{(p)}$$

Other f-coordinate choices would have led to different, but equivalent, λ's and equations of motion.

Problem 3.5.22 *Maggi Equations.* Formulate both kinetic and kinetostatic Maggi equations for a particle constrained to move on a smooth circular helix (ex. 3.5.11), for the following choice of coordinates:

$$f_1 \equiv r - R\ (= 0), \qquad f_2 \equiv z - p\phi\ (= 0), \qquad f_3 \equiv z\ (\neq 0). \tag{a}$$

HINT

After $(\ldots)^{\cdot}$-differentiating (a), we easily conclude that here the nonvanishing elements of (a_{kl}) are $a_{11} = a_{23} = a_{33} = 1$, $a_{22} = -p$; and, therefore, the nonvanishing elements of its inverse matrix [(i.e., the Maggi equation coefficients) $= (A_{kl})$], are $A_{11} = A_{33} = 1$, $A_{22} = -A_{23} = -1/p$.

Problem 3.5.23 *Maggi Equations.* Repeat the preceding problem, but for the following choice of coordinates:

$$f_1 \equiv r - R\ (= 0), \qquad f_2 \equiv z - p\phi\ (= 0), \qquad f_3 \equiv \phi\ (\neq 0). \tag{a}$$

Problem 3.5.24 *Euler's Equations via Lagrange's Equations.* Consider, for simplicity, but no real loss in generality, the force-free motion of a rigid body B rotating about a fixed point O. With the help of the $\omega_{1,2,3} \Leftrightarrow \dot{\phi}, \dot{\theta}, \dot{\psi}$ relationships (§1.12), where O–123 are principal axes of B at O, show that the Lagrangean equations for ϕ, θ, ψ, lead to Euler's equations (§1.17).

HINT

Since $2T = I_1 {\omega_1}^2 + I_2 {\omega_2}^2 + I_3 {\omega_3}^2 = 2T(\omega_{1,2,3}) = 2T^*$, we find, successively,

$$\begin{aligned}\partial T/\partial \dot{\psi} &= (\partial T^*/\partial \omega_1)(\partial \omega_1/\partial \dot{\psi}) + (\partial T^*/\partial \omega_2)(\partial \omega_2/\partial \dot{\psi}) + (\partial T^*/\partial \omega_3)(\partial \omega_3/\partial \dot{\psi}) \\ &= (I_1\omega_1)(0) + (I_2\omega_2)(0) + (I_3\omega_3)(1) = I_3\omega_3,\end{aligned} \tag{a}$$

$$\begin{aligned}\partial T/\partial \psi &= (\partial T^*/\partial \omega_1)(\partial \omega_1/\partial \psi) + (\partial T^*/\partial \omega_2)(\partial \omega_2/\partial \psi) + (\partial T^*/\partial \omega_3)(\partial \omega_3/\partial \psi) \\ &= (I_1\omega_1)[(s\theta\, c\psi)\dot{\phi} + (-s\psi)\dot{\theta}] + (I_2\omega_2)[(-s\theta\, s\psi)\dot{\phi} + (-c\psi)\dot{\theta}] + (I_3\omega_3)(0) \\ &= (I_1\omega_1)(\omega_2) + (I_2\omega_2)(-\omega_1),\end{aligned} \tag{b}$$

and, therefore,

$$E_\psi(T) \equiv (\partial T/\partial \dot{\psi})^{\cdot} - \partial T/\partial \psi = I_3(d\omega_3/dt) - (I_1 - I_2)\omega_1\omega_2 = 0; \tag{c}$$

and similarly for $E_\theta(T) = 0$ and $E_\phi(T) = 0$. Let the reader ponder about the possible drawbacks of this derivation of *Euler's equations*.

3.6 THE CENTRAL EQUATION (THE *ZENTRALGLEICHUNG* OF HEUN AND HAMEL)

Thus far, the transition from the particle *accelerations* that appear explicitly in Lagrange's principle (LP),

$$\delta I = \delta' W, \qquad \text{or, in extenso,} \qquad S\, dm\, \boldsymbol{a} \cdot \delta \boldsymbol{r} = S\, d\boldsymbol{F} \cdot \delta \boldsymbol{r}, \tag{3.6.1}$$

to system velocities/kinetic energy derivatives, and so on, that appear in the equations of motion deriving from it, is carried out in the components of the (negative of the) system inertial "forces": for example, $E_k \equiv S\, dm\, \boldsymbol{a} \cdot \boldsymbol{e}_k = \cdots = (\partial T/\partial \dot{q}_k)^{\cdot} - \partial T/\partial q_k$, and similarly for $I_k \equiv S\, dm\, \boldsymbol{a} \cdot \boldsymbol{\varepsilon}_k = \cdots$. However, a transition from particle accelerations to particle velocities (i.e., from $\boldsymbol{a}$'s to $\boldsymbol{v}$'s), and from there to system velocities, can also be effected by proceeding directly from the variational equation (3.6.1).

Indeed, in view of the purely *kinematic* identity

$$d(\boldsymbol{v} \cdot \delta\boldsymbol{r})/dt = \boldsymbol{a} \cdot \delta\boldsymbol{r} + \delta(\boldsymbol{v} \cdot \boldsymbol{v}/2) + \boldsymbol{v} \cdot [d(\delta\boldsymbol{r})/dt - \delta(d\boldsymbol{r}/dt)], \tag{3.6.2}$$

δI transforms readily to

$$\begin{aligned} \delta I &\equiv S\, dm\, \boldsymbol{a} \cdot \delta\boldsymbol{r} \\ &= d/dt\Big(S\, dm\, \boldsymbol{v} \cdot \delta\boldsymbol{r}\Big) - S\, dm\, \boldsymbol{v} \cdot \delta\boldsymbol{v} - S\, dm\, \boldsymbol{v} \cdot [d(\delta\boldsymbol{r})/dt - \delta(d\boldsymbol{r}/dt)], \end{aligned}$$

or

$$\delta I \equiv d(\delta P)/dt - \delta T - \delta D, \tag{3.6.3}$$

where

$$\begin{aligned} \delta T \equiv \delta\Big(S\, (1/2) dm\, \boldsymbol{v} \cdot \boldsymbol{v}\Big) &= S\, dm\, \boldsymbol{v} \cdot \delta\boldsymbol{v} = S\, dm\, \boldsymbol{v} \cdot \delta(d\boldsymbol{r}/dt) \\ &= \textit{First virtual variation of (inertial) kinetic energy}, \end{aligned} \tag{3.6.3a}$$

$$\delta P \equiv S\, dm\, \boldsymbol{v} \cdot \delta\boldsymbol{r} = \textit{Total virtual work of (linear) momenta of system particles}, \tag{3.6.3b}$$

$$\begin{aligned} \delta D \equiv S\, dm\, \boldsymbol{v} \cdot [d(\delta\boldsymbol{r})/dt - \delta(d\boldsymbol{r}/dt)] &\equiv S\, dm\, \boldsymbol{v} \cdot [(\delta\boldsymbol{r})^{\cdot} - \delta\boldsymbol{v}] \\ &= \textit{Total "virtual (work of) nonholonomic deviation."} \end{aligned} \tag{3.6.3c}$$

Hence LP, (3.6.1), takes the *velocity* form:

$$\delta T + \delta' W + \delta D = d(\delta P)/dt. \tag{3.6.4}$$

This fundamental *differential* variational equation, on a par with LP, was originally obtained by Lagrange himself (in the course of the derivation of his equations from LP), but its importance was fully appreciated much later by Heun (early 1900s), who dubbed it the *central equation* of AM (*Zentralgleichung*, CE). Specifically, its importance lies in that it replaces a *second*-order invariant, $\delta I \equiv S\, dm\, \boldsymbol{a} \cdot \delta\boldsymbol{r}$, with its *first*-order invariants δT, δP, δD.

We should, also, point out the following:

(i) Multiplying (3.6.4) with dt and then integrating the result over the arbitrary time interval (t_1, t_2), we obtain the following *time integral variational equation:*

$$\int_{t_1}^{t_2} (\delta T + \delta' W + \delta D)\, dt = \Big\{\delta P\Big\}_{t_1}^{t_2} \equiv \Big\{S\, dm\, \boldsymbol{v} \cdot \delta\boldsymbol{r}\Big\}_{t_1}^{t_2}; \tag{3.6.5}$$

commonly known as *Hamilton's principle* (HP—detailed in chap. 7).

(ii) It is not necessary to assume, in CE, that $(\delta \boldsymbol{r})^{\cdot} = \delta \boldsymbol{v}$, or equivalently $d(\delta \boldsymbol{r}) = \delta(d\boldsymbol{r})$. As Hamel has stressed, and as will become clear below, the equations of motion derived from the above are *independent of any such commutation rules.* Thus, assuming in (3.6.4), (3.6.5) $(\delta \boldsymbol{r})^{\cdot} = \delta \boldsymbol{v}$, that is, $\delta D = 0$, reduces them, respectively, to

$$\delta T + \delta' W = d(\delta P)/dt \qquad (\textit{ordinary}\ \text{CE}) \tag{3.6.6}$$

$$\int_{t_1}^{t_2} (\delta T + \delta' W)\, dt = \left\{\delta P\right\}_{t_1}^{t_2} \equiv \left\{ S\, dm\ \boldsymbol{v} \cdot \delta \boldsymbol{r}\right\}_{t_1}^{t_2} \qquad (\textit{ordinary}\ \text{HP}). \tag{3.6.6a}$$

[Heun called (3.6.6) the *central equation*; while Hamel (1949, pp. 233–235) called (3.6.4) the *generalized*, or *general, central equation*. Also, Rosenberg (1977, pp. 167–168), virtually alone in the entire English language literature to handle this matter properly, chose to translate (3.6.4, 6) as the *central principle*.]

Now, the differential variational equation (3.6.4) is expressed in *particle* vectors/variables; and in that form it may be quite useful in, say, rigid-body applications. For the purposes of the general theory, however, we need to express it in general *system* variables. Indeed, proceeding from the invariant definitions (3.6.3a–c) we find, successively,

(a) $$T = T(t, q, \dot{q}) = T^* = T^*(t, q, \omega) \Rightarrow \delta T = \delta T^*$$

$$\delta T = \delta T(t, q, \dot{q}) = \sum [(\partial T/\partial q_k)\, \delta q_k + (\partial T/\partial \dot{q}_k)\, \delta(\dot{q}_k)], \tag{3.6.7a}$$

$$\begin{aligned}\delta T^* = \delta T^*(t, q, \omega) &= \sum [(\partial T^*/\partial \omega_k)\, \delta\omega_k + (\partial T^*/\partial q_k)\, \delta q_k] \\ &= \sum (\partial T^*/\partial \omega_k)\, \delta\omega_k + \sum \left[(\partial T^*/\partial q_k)\left(\sum A_{kl}\, \delta\theta_l\right)\right] \\ &= \cdots \equiv \sum P_k\, \delta\omega_k + \sum (\partial T^*/\partial \theta_k)\, \delta\theta_k;\end{aligned} \tag{3.6.7b}$$

(b) $$\delta' W \equiv S\, d\boldsymbol{F} \cdot \delta \boldsymbol{r} = \sum Q_k\, \delta q_k = \sum \Theta_k\, \delta\theta_k; \tag{3.6.7c}$$

(c) $$\delta P = \cdots = \sum P_k\, \delta\theta_k = \sum p_k\, \delta q_k \left[\text{since}\ S\, dm\ \boldsymbol{v} \cdot \boldsymbol{e}_k \equiv p_k,\ S\, dm\ \boldsymbol{v} \cdot \boldsymbol{\varepsilon}_k \equiv P_k\right] \tag{3.6.7d}$$

$$\Rightarrow p_l = \sum a_{kl} P_k, \quad \text{i.e., } \partial T/\partial \dot{q}_l = \sum (\partial T^*/\partial \omega_k)(\partial \omega_k/\partial \dot{q}_l), \tag{3.6.7e}$$

$$P_k = \sum A_{lk} p_l, \quad \text{i.e., } \partial T^*/\partial \omega_k = \sum (\partial T/\partial \dot{q}_l)(\partial \dot{q}_l/\partial \omega_k); \tag{3.6.7f}$$

$$\Rightarrow (\delta P)^{\cdot} \equiv \left(S\, dm\ \boldsymbol{v} \cdot \delta \boldsymbol{r}\right)^{\cdot}$$

$$= \left(\sum p_k\, \delta q_k\right)^{\cdot} = \sum \dot{p}_k\, \delta q_k + \sum p_k (\delta q_k)^{\cdot}, \tag{3.6.7g}$$

$$= \left(\sum P_k\, \delta\theta_k\right)^{\cdot} = \sum \dot{P}_k\, \delta\theta_k + \sum P_k (\delta\theta_k)^{\cdot}; \tag{3.6.7h}$$

(d) Recalling the general *particle transitivity equation* (prob. 2.10.6; with Greek indices running from 1 to $n + 1$), which holds independently of any $d(\delta q) - \delta(dq)$ rules,

$$\begin{aligned}(\delta \boldsymbol{r})^{\cdot} - \delta \boldsymbol{v} &= \sum [(\delta q_k)^{\cdot} - \delta(\dot{q}_k)] \boldsymbol{e}_k \\ &= \sum [(\delta\theta_k)^{\cdot} - \delta\omega_k] \boldsymbol{\varepsilon}_k - \sum \left(\sum \sum \gamma^k{}_{r\alpha}\, \omega_\alpha\, \delta\theta_r\right) \boldsymbol{\varepsilon}_k,\end{aligned} \tag{3.6.7i}$$

we obtain (invoking the above definitions of p_k, P_k)

$$\delta D \equiv \int dm\ \boldsymbol{v} \cdot [(\delta \boldsymbol{r})^{\cdot} - \delta \boldsymbol{v}]$$
$$= \cdots = \sum p_k[(\delta q_k)^{\cdot} - \delta(\dot{q}_k)] = \sum P_k[(\delta\theta_k)^{\cdot} - \delta\omega_k] - \sum\sum\sum \gamma^k{}_{r\alpha} P_k \omega_\alpha\, \delta\theta_r. \tag{3.6.7j}$$

Next, substituting the above system expressions into the general CE, (3.6.4), rearranged à la LP as $(\delta P)^{\cdot} - \delta T - \delta D = \delta' W$, we obtain its following system forms.

1. Holonomic System Variables

$$\left(\sum p_k\, \delta q_k\right)^{\cdot} - \delta T - \sum p_k[(\delta q_k)^{\cdot} - \delta(\dot{q}_k)] = \sum Q_k\, \delta q_k \quad (\equiv \delta' W), \tag{3.6.8}$$

or (after expanding and collecting terms, and factoring out δq_k)

$$\sum (dp_k/dt - \partial T/\partial q_k)\, \delta q_k = \sum Q_k\, \delta q_k \quad \text{[LP in holonomic variables (3.5.11 ff.)]} \tag{3.6.8a}$$

We notice that (3.6.8a) results always from (3.6.8), *regardless of any assumptions about* $(\delta q_k)^{\cdot} - \delta(\dot{q}_k)$. However, (3.6.8) is more general than (3.6.8a); for example, if we express some of its terms in θ-variables (see below, and ex. 3.8.1), or if we assume that $(\delta q_k)^{\cdot} \neq \delta(\dot{q}_k)$, for some δq, $\dot{q}$'s [see Voronets equations (3.8.14a ff.)].

2. Nonholonomic System Variables

$$\left(\sum P_k\, \delta\theta_k\right)^{\cdot} - \delta T - \left\{\sum P_k[(\delta\theta_k)^{\cdot} - \delta\omega_k] - \sum\sum\sum \gamma^k{}_{r\alpha} P_k\, \omega_\alpha\, \delta\theta_r\right\} = \sum \Theta_k\, \delta\theta_k,$$

or, after expanding and collecting terms, etc.,

$$\sum \dot{P}_k\, \delta\theta_k - \sum (\partial T^*/\partial\theta_k)\, \delta\theta_k + \sum\sum\sum \gamma^k{}_{r\alpha} P_k\, \omega_\alpha\, \delta\theta_r = \sum \Theta_k\, \delta\theta_k, \tag{3.6.9}$$

or (factoring out $\delta\theta_k$, and performing some index changes in the γ-term)

$$\sum \left(dP_k/dt - \partial T^*/\partial\theta_k + \sum\sum \gamma^r{}_{k\alpha} P_r\, \omega_\alpha\right) \delta\theta_k = \sum \Theta_k\, \delta\theta_k, \tag{3.6.9a}$$

which is none other than LP in quasi variables (3.5.18 ff.).

3. Mixed Variable Forms

(i) With the help of (3.6.7a–j) [also, recalling the *general system transitivity equations* (2.10.6, 7)], we can rewrite (3.6.9) as follows:

$$\sum \dot{P}_k\, \delta\theta_k - \sum (\partial T^*/\partial\theta_k)\, \delta\theta_k$$
$$+ \sum P_k\left\{[(\delta\theta_k)^{\cdot} - \delta\omega_k] - \sum a_{kl}[(\delta q_l)^{\cdot} - \delta(\dot{q}_l)]\right\} = \sum \Theta_k\, \delta\theta_k. \tag{3.6.10}$$

From the above we readily see the following.

(ii) If we assume that $(\delta q_k)^{\cdot} - \delta(\dot{q}_k) = 0$ [$\Rightarrow d(\delta \mathbf{r}) - \delta(d\mathbf{r}) = \sum [d(\delta q_k) - \delta(dq_k)]\mathbf{e}_k = \mathbf{0} \Rightarrow \delta D = 0$ (ordinary CE, (3.6.6)) — *Hamel viewpoint*], then (3.6.10) becomes

$$\sum \dot{P}_k\, \delta\theta_k - \sum (\partial T^*/\partial\theta_k)\, \delta\theta_k + \sum P_k[(\delta\theta_k)^{\cdot} - \delta\omega_k] = \sum \Theta_k\, \delta\theta_k; \qquad (3.6.11)$$

which, since in this case $(\delta\theta_k)^{\cdot} - \delta\omega_k = \sum\sum \gamma^k{}_{r\alpha}\, \omega_\alpha\, \delta\theta_r$, is none other than (3.6.9). Hence, even if we had started with (3.6.6), rearranged as $(\delta P)^{\cdot} - \delta T = \delta' W$, with δT, $\delta' W$, $(\delta P)^{\cdot}$ given by (3.6.7b, c, h), respectively, we would *still have arrived at (3.6.11)*. Also, recalling (3.3.12a), and comparing (3.6.11) with (3.6.9), we readily conclude that, then,

$$\sum P_k[(\delta\theta_k)^{\cdot} - \delta\omega_k] = -\sum \Gamma_k\, \delta\theta_k. \qquad (3.6.11a)$$

(iii) However, if we assume that $(\delta\theta_k)^{\cdot} - \delta\omega_k = 0$ (*Suslov viewpoint*), and invoke (3.6.7e), eq. (3.6.10) becomes

$$\sum \dot{P}_k\, \delta\theta_k - \sum (\partial T^*/\partial\theta_k)\, \delta\theta_k - \sum P_k\Big\{\sum a_{kl}[(\delta q_l)^{\cdot} - \delta(\dot{q}_l)]\Big\}$$
$$= \sum \dot{P}_k\, \delta\theta_k - \sum (\partial T^*/\partial\theta_k)\, \delta\theta_k - \sum p_k[(\delta q_k)^{\cdot} - \delta(\dot{q}_k)] = \sum \Theta_k\, \delta\theta_k. \quad (3.6.12)$$

Equations (3.6.4, 6; 8; 9; 10, 11, 12) are mutually equivalent; but, unless properly understood and applied, they may lead to (apparently) contradictory results. [These variational equations are useful in integral variational "principles": chap. 7; also ex. 3.8.1.] Hence, whenever needed, we may safely assume that $d(\delta q_k) = \delta(dq_k)$ for *all* holonomic coordinates, constrained or not. Indeed, in applications to concrete problems, the CE in the form (3.6.11) seems to be the single most useful equation of analytical mechanics. Its implementation requires knowledge of $T \to T^*$, the Θ's, and the $\dot{q} \leftrightarrow \omega$ transformation; the $\gamma^k_{\bullet\blacklozenge}$'s, as already pointed out in § 2.10, are simply read off as the coefficients of

$$(\delta\theta_k)^{\cdot} - \delta\omega_k = \cdots + (\ldots)^k_{\bullet\blacklozenge}\, \omega_\blacklozenge\, \delta\theta_{\bullet\blacklozenge} + \cdots. \qquad (3.6.12a)$$

WHICH ARE THE BEST EQUATIONS OF MOTION?

Over the past couple of decades or so, a debate has been brewing, in applied engineering (multibody) dynamics circles, as to which of all available equations of motion are the best, or "more efficient." We believe that such questions, and attempted answers, are at best counterproductive and myopic; and, at worst, self-serving and wasteful. Fortunately, for dynamics, there is no cure-all set of equations that works best under all circumstances; theoretical and applied, exact and approximate (including computational). Different equations (see also special forms in §3.8) have different uses, advantages and disadvantages, like the various tools in a mechanic's toolbox. Some are more conceptually efficient (a classification almost never heard of by applied dynamicists) and less computationally efficient, and vice versa; some are more fertile and/or beautiful (!) than others. Such a healthy pluralism testifies to the vitality and diversity of the human intellect, keeps our science alive and should be welcome, indeed treasured, by all—*we learn more by solving one problem with several methods than by solving several problems with one method. We must learn them all, especially today!*

It seems to us, however, that one of the best such tools *is not any particular set of equations of motion*, but instead, the *central (variational) equation* in the form (3.6.11) ⇒

$$\sum(\ldots)_D\,\delta\theta_D + \sum(\ldots)_I\,\delta\theta_I = 0. \tag{3.6.12b}$$

The vanishing of its $(\ldots)_D$ terms (appropriately modified, according to the method of Lagrangean multipliers) yields the *m kinetostatic* equations, while the vanishing of its $(\ldots)_I$ terms yields the *n − m kinetic* equations; and these, plus constraints and initial/boundary temporal conditions, constitute a mathematically *determinate* system. [Also, the *differential* variational principles (chap. 6), of which LP, eqs. (3.6.8a) and (3.6.9a), constitute the foundation, seem especially promising, for both finite and impulsive motion, and for both linear (Pfaffian) and nonlinear velocity constraints.] Such a *unifying* approach, acting like a conceptual centripetal force and countering the centrifugal ones of the various equations of motion, should be welcome and psychologically satisfying. After all, *there is only one mechanics*; although the average observer of the contemporary (tower of Babel-like) dynamics literature would not get that impression!

Example 3.6.1 *Holonomic System in Nonholonomic Coordinates: Plane Motion of a Free Particle* (Appell, 1925, pp. 6–7, 17–18; Lur'e, 1968, pp. 401–402). Let us consider a particle P of mass m moving along the plane curve C, on the O–xy plane, under known/given forces (fig. 3.15). The instantaneous position and velocity of this *two DOF* system ($n = 2$, $m = 0$) can be defined in several equivalent ways. Thus, we may choose the following descriptions:

(i) *Holonomic variables*: a convenient such choice are the polar coordinates of P

$$q_1 = r, \quad q_2 = \phi \quad (\text{angle from } Ox) \;\Rightarrow\; \dot{q}_1 = \dot{r}, \quad \dot{q}_2 = \dot{\phi}. \tag{a}$$

(ii) *Nonholonomic variables*: following Appell, we choose as such: (a) $\theta_1 = q_1 = r$; that is, the magnitude of the *position* vector $\mathbf{r} = \mathbf{OP}$, and (b) $\theta_2 = \sigma$ (a quasi coordinate, as shown below), such that $\dot{\theta}_2 \equiv \omega_2 = \dot{\sigma} = r^2\dot{\phi}$, or $d\sigma = r^2 d\phi$ = *twice the area of the elementary sector swept by the radius vector OP between the time instants* $t(\rightarrow r)$

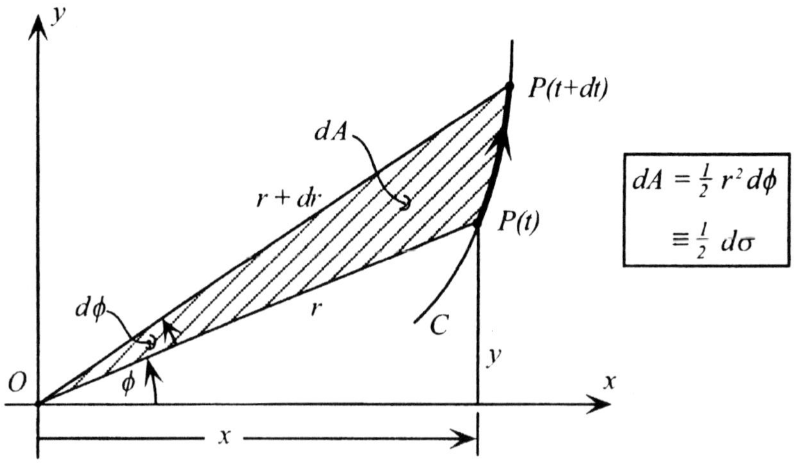

Figure 3.15 Plane motion of a particle under given forces ($2\,dA = r^2\,d\phi \equiv d\sigma$).

and $t + dt\,(\rightarrow r + dr)$; that is, approximately, twice the area of a circular sector of radius r and central angle $d\phi$ — a well-known calculus result. Hence, the following transformation equations, and their inverses:

$$\omega_1 = (1)\dot{q}_1 + (0)\dot{q}_2, \qquad \omega_2(0)\dot{q}_1 + (r^2)\dot{q}_2; \tag{b}$$

$$\dot{q}_1 = (1)\omega_1 + (0)\omega_2, \qquad \dot{q}_2 = (0)\omega_1 + (r^{-2})\omega_2. \tag{c}$$

Transitivity Equations

From (b, c), and since this is a scleronomic system, that is, $\delta q_1 = \delta\theta_1 = \delta r$, $dq_1 = d\theta_1 = dr$; $\delta\theta_2 = \delta\sigma = r^2\,\delta\phi = r^2\,\delta q_2$, $d\theta_2 = d\sigma = r^2\,d\phi = r^2\,dq_2$, and so on, we find

$$(\delta\theta_1)^{\cdot} - \delta\omega_1 = (\delta r)^{\cdot} - \delta(\dot{r}) = 0 \qquad (r = \textit{holonomic coordinate}), \tag{d}$$

$$\begin{aligned}(\delta\theta_2)^{\cdot} - \delta\omega_2 &= (r^2\,\delta\phi)^{\cdot} - \delta(r^2\dot{\phi}) = \cdots = 2r\,\dot{r}\,\delta\phi - 2r\,\dot{\phi}\,\delta r\\ &= 2r\,\omega_1\,\delta\phi - 2r(\omega_2/r^2)\,\delta r = 2r\,\omega_1(r^{-2}\,\delta\theta_2) - (2r^{-1}\omega_2)\,\delta\theta_1\\ &= 2r^{-1}(\omega_1\,\delta\theta_2 - \omega_2\,\delta\theta_1).\end{aligned} \tag{e}$$

Kinetic Energy

From $x = r\cos\phi$, $y = r\sin\phi \Rightarrow \dot{x} = \ldots, \dot{y} = \ldots$, and the above, we find

$$\begin{aligned}2T = m[(\dot{x})^2 + (\dot{y})^2] &= \cdots = m[(\dot{r})^2 + r^2(\dot{\phi})^2]\\ &= \cdots = m[(\omega_1)^2 + r^{-2}(\omega_2)^2] = T^*.\end{aligned} \tag{f}$$

Appellian

From $\ddot{x} = \ldots, \ddot{y} = \ldots$, we find, successively,

$$\begin{aligned}2S = m[(\ddot{x})^2 + (\ddot{y})^2] &= \cdots = m[(a_r)^2 + (a_\phi)^2] \qquad \text{(physical components)}\\ &= m\Big\{[\ddot{r} - r(\dot{\phi})^2]^2 + [r^{-1}(r^2\dot{\phi})^{\cdot}]^2\Big\}\\ &= m\Big\{[\ddot{r} - r^{-3}(\dot{\sigma})^2]^2 + (r^{-1}\ddot{\sigma})^2\Big\},\end{aligned}$$

or, to within "Appell-important" terms $\equiv \cdots$,

$$\begin{aligned}2S \rightarrow 2S^* &= m[(\ddot{r})^2 - 2r^{-3}\ddot{r}(\dot{\sigma})^2 + r^{-2}(\ddot{\sigma})^2] + \cdots\\ &= m[(\dot{\omega}_1)^2 - 2r^{-3}(\omega_2)^2\dot{\omega}_1 + r^{-2}(\dot{\omega}_2)^2] + \cdots.\end{aligned} \tag{g}$$

Virtual Works

Let the physical (polar) components of the total impressed force on P, that is, $\boldsymbol{F}$, be (F_r, F_ϕ). Then,

$$\begin{aligned}\delta' W &= F_r\,\delta r + F_\phi(r\,\delta\phi) \equiv Q_r\,\delta r + Q_\phi\,\delta\phi\\ &= F_r\,\delta\theta_1 + F_\phi(r^{-1}\,\delta\theta_2) = \Theta_1\,\delta\theta_1 + \Theta_2\,\delta\theta_2,\end{aligned} \tag{h}$$

that is,

$$Q_1 \equiv Q_r = F_r, \qquad Q_2 \equiv Q_\phi = rF_\phi \qquad \text{(holonomic components)} \tag{i}$$

$$\Theta_1 = F_r, \qquad \Theta_2 = r^{-1}F_\phi \qquad \text{(nonholonomic components).} \tag{j}$$

Equations of Motion

(i) *Holonomic variables*: The Lagrange and Appell equations are

$$q_1 \equiv r: \quad (\partial T/\partial \dot{r})^{\cdot} - \partial T/\partial r \equiv \partial S/\partial \ddot{r} = Q_r: \quad m[\ddot{r} - r(\dot{\phi})^2] = F_r, \tag{k}$$

$$q_2 \equiv \phi: \quad (\partial T/\partial \dot{\phi})^{\cdot} - \partial T/\partial \phi \equiv \partial S/\partial \ddot{\phi} = Q_\phi: \quad m(r^2\dot{\phi})^{\cdot} = rF_\phi. \tag{l}$$

(ii) *Nonholonomic variables*: Since

$$P_1 \equiv \partial T^*/\partial \omega_1 = m\,\omega_1 \qquad (= m\dot{r}) \;\Rightarrow\; \dot{P}_1 = m\,\dot{\omega}_1 \quad (= m\,\ddot{r}), \tag{m}$$

$$P_2 \equiv \partial T^*/\partial \omega_2 = mr^{-2}\omega_2 \quad (= mr^{-2}\dot{\sigma}) \;\Rightarrow\; \dot{P}_2 = (m\,r^{-2}\omega_2)^{\cdot}\,; \tag{n}$$

and

$$\begin{aligned}\partial T^*/\partial \theta_1 \equiv \partial T^*/\partial r &= A_{11}(\partial T^*/\partial q_1) + A_{21}(\partial T^*/\partial q_2)\\ &= (1)(\partial T^*/\partial r) + (0)(\partial T^*/\partial \phi) = \partial T^*/\partial r\\ &= (m/2)(-2)[r^{-3}(\omega_2)^2] = -m\,r^{-3}(\omega_2)^2,\end{aligned} \tag{o}$$

$$\begin{aligned}\partial T^*/\partial \theta_2 \equiv \partial T^*/\partial \sigma &= A_{12}(\partial T^*/\partial q_1) + A_{22}(\partial T^*/\partial q_2)\\ &= (0)(\partial T^*/\partial r) + (r^2)(\partial T^*/\partial \phi) = 0,\end{aligned} \tag{p}$$

the central equation (3.6.12)

$$\sum \dot{P}_k\,\delta\theta_k - \sum (\partial T^*/\partial \theta_k)\,\delta\theta_k + \sum P_k[(\delta\theta_k)^{\cdot} - \delta\omega_k] = \sum \Theta_k\,\delta\theta_k \tag{q}$$

yields

$$\begin{aligned}&\dot{P}_1\,\delta\theta_1 + \dot{P}_2\,\delta\theta_2 - (\partial T^*/\partial \theta_1)\,\delta\theta_1 - (\partial T^*/\partial \theta_2)\,\delta\theta_2\\ &\qquad + P_1[(\delta\theta_1)^{\cdot} - \delta\omega_1] + P_2[(\delta\theta_2)^{\cdot} - \delta\omega_2] = \Theta_1\,\delta\theta_1 + \Theta_2\,\delta\theta_2,\end{aligned} \tag{r}$$

or collecting $(\ldots)\,\delta\theta_k$ terms,

$$[\dot{P}_1 - \partial T^*/\partial \theta_1 + (-2\omega_2/r)P_2 - \Theta_1]\,\delta\theta_1 + [\dot{P}_2 + (2\omega_1/r)P_2 - \Theta_2]\,\delta\theta_2 = 0, \tag{s}$$

from which, since $\delta\theta_1$ and $\delta\theta_2$ are unconstrained, we obtain the Hamel equations:

$$\theta_1: \qquad \dot{P}_1 - \partial T^*/\partial \theta_1 - (2\omega_2/r)P_2 = \Theta_1, \tag{t1}$$

or

$$m\,\dot{\omega}_1 + m\,r^{-3}(\omega_2)^2 - (2r^{-1}\omega_2)(m\,r^{-2}\omega_2) = \Theta_1,$$

or, finally,

$$m[\ddot{r} - r^{-3}(\dot{\sigma})^2] = F_r, \quad \text{i.e., equation (k);} \tag{t2}$$

and

$$\theta_2: \qquad \dot{P}_2 + (2\omega_1/r)P_2 = \Theta_2, \tag{t3}$$

or

$$(m r^{-2}\omega_2)^{\cdot} + (2r^{-1}\omega_1)(m r^{-2}\omega_2) = \Theta_2,$$

or, finally,

$$m r^{-2}\ddot{\sigma} - r^{-3}(2m\dot{r}\dot{\sigma}) + r^{-3}(2m\dot{r}\dot{\sigma}) = \Theta_2 \Rightarrow m\ddot{\sigma} = rF_\phi, \tag{t4}$$

that is, (l); from which, if $F_\phi = 0$, we obtain Kepler's second "law": $\dot{\sigma} = r^2\dot{\phi} =$ *constant*.

REMARKS

(i) The above Hamel equations coincide with those of Appell, but in nonholonomic variables:

$$\partial S^*/\partial\ddot{r} = \Theta_r \qquad [\partial S^*/\partial\dot{\omega}_1 = \Theta_1], \tag{u1}$$

$$\partial S^*/\partial\ddot{\sigma} = \Theta_\sigma \qquad [\partial S^*/\partial\dot{\omega}_2 = \Theta_2]. \tag{u2}$$

(ii) We also notice that, since

$$[(\partial T^*/\partial\dot{r})^{\cdot} - \partial T^*/\partial r] - \partial S^*/\partial\ddot{r} \equiv \Gamma_r \qquad [= (2\omega_2/r)P_2] \neq 0, \tag{v1}$$

$$[(\partial T^*/\partial\dot{\sigma})^{\cdot} - \partial T^*/\partial\sigma] - \partial S^*/\partial\ddot{\sigma} \equiv \Gamma_\sigma \qquad [= -(2\omega_1/r)P_2] \neq 0, \tag{v2}$$

the Lagrange-type equations in quasi variables

$$E_r(T^*) \equiv (\partial T^*/\partial\dot{r})^{\cdot} - \partial T^*/\partial r = \Theta_r, \qquad E_\sigma(T^*) \equiv (\partial T^*/\partial\dot{\sigma})^{\cdot} - \partial T^*/\partial\sigma = \Theta_\sigma, \tag{v3}$$

would have been incorrect.

3.7 THE PRINCIPLE OF RELAXATION OF THE CONSTRAINTS (THE LAGRANGE–HAMEL *BEFREIUNGSPRINZIP*)

We have already seen (§3.5) how to (i) eliminate the constraint reactions (kinetic equations), and (ii) how to retrieve them, if needed (kinetostatic equations). For the first task, we postulated *Lagrange's principle (LP)*; while for the second, we have utilized the method of Lagrangean multipliers. This latter (recall last part of §3.2) is the mathematical expression of the *principle of relaxation, or freeing, or liberation, of the constraints (PRC)*—the second pillar of analytical mechanics, and its logical counterpart to the Eulerian *cut* principle ("free-body diagram") for calculating internal forces, of the stereomechanical approach.

It was Hamel who, around 1916 (publ. 1917), recognized the cardinal importance of this principle and introduced the term *Befreiungsprinzip* for it. Up until then, it had not been stated explicitly anywhere. Lagrange, in his admirably informal style, had simply said about it, "*Car c'est en quoi consiste l'esprit de la méthode de cette section.* ... Notre méthode donne, comme l'on voit, le moyen de déterminer ces forces et ces résistances; ce qui n'est pas un des moindres avantages de cette méthode". [Lagrange, 1965 (reprint of 4th ed.), vol. 1, p. 73, emphasis added]. Even today, it rarely appears in the English language literature: for example, Lawden (1972, p. 54; who coined our term "method of relaxation of constraints"), Leipholz (1978, 1983; who calls it "relaxation principle of Lagrange"), Serrin (1959, pp. 146–147; refers to it as "Lagrange's Freeing Principle"); while Bahar (1970–1980; describes it, instructively, as "the rubber-band approach"). Such a method was long known and routinely applied in analytical statics; although it was not sufficiently acknowledged, explicitly. That is why its extension to kinetics, a nontrivial matter, was aptly called by Heun (early 1900s) *kinetostatics = the determination of internal and external reactions in moving rigid systems*—an important mechanical engineering problem (see also Stäckel, 1905, pp. 667–670). Let us examine it more closely. Following Hamel (1927, p. 26; 1949, pp. 74, 173, 522):

• In addition to the constrained system, we form (mentally) a relaxed, or freed one, in which a particular, or all, constraints have been eliminated. The equations of motion remain formally the same, except that now the former reactions are *impressed* forces whose virtual work is

$$\sum \lambda_D\, \delta\theta_D = \sum \Big(\sum \lambda_D\, a_{Dk}\Big)\, \delta q_k; \tag{3.7.1}$$

for a particular $\lambda_D = \Lambda_D$, the corresponding "relaxed virtual work" is

$$(\delta' W_R)_D \equiv \lambda_D\, \delta\theta_D = \Lambda_D\, \delta\theta_D, \tag{3.7.1a}$$

as if the constraint $\delta\theta_D = 0$ *did not exist*.

• The former reactions have now become impressed forces that depend on the previously forbidden deformation/motion, i.e., *on those geometrical/kinematical variables that were not allowed to vary in the non-relaxed system* (and, possibly, on other non-mechanical variables — see below).

• A word of caution: When applying this principle to cases where the freed system requires more than mechanics (M) for its description — that is, wherever the additional, relaxed, deformation/motion introduces *additional*, say thermodynamical (T) and/or electrodynamical (E) variables, mutually coupled — we should expect such a dependence to be reflected in the multipliers; that is, symbolically, $\lambda = \lambda(M, T, E)$. As long as we stay within pure mechanics (this book), however, no such problems seem to arise. See for instance Serrin (1959, pp. 146–147), who warns against the blind generalization of the principle from simple mechanics [e.g., the compressible perfect, or ideal, fluid viewed as a freed incompressible perfect fluid, whose stress (pressure), according to the PRC, depends on the formerly forbidden compressibility; i.e., on the density which, before, was not allowed to vary], to more physically complex cases (e.g., a gas where the pressure is a definite thermodynamical variable).

Example 3.7.1 *Applications of PRC.*

(i) In pure (or slippingless) *rolling*, the friction is a *reaction* force. It follows that in *slipping*, the friction depends on the previous constraint; that is, on the *relative velocity* of the two contacting surfaces, and on other material coefficients; and, hence, it has become an impressed force.

(ii) The *tension* in an extensible cable depends on the latter's *stretch*, and other material coefficients.

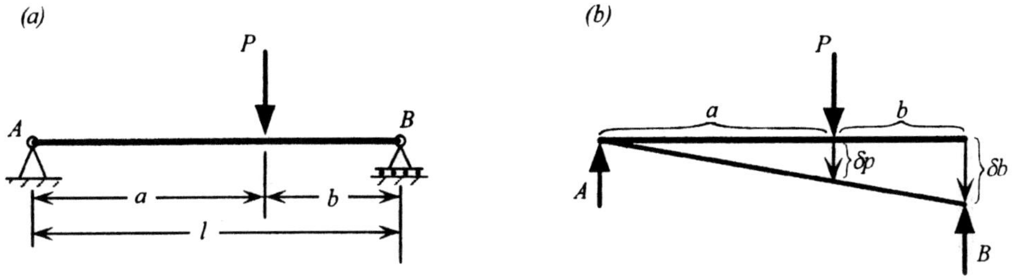

Figure 3.16 Principle of relaxation in statics: reactions on a simply supported beam.

(iii) *Statics*: Let us calculate the reaction B of the simply supported (statically determinate) beam of fig. 3.16(a). We allow, mentally, the formerly unyielding support B to move down (or up) and then calculate the virtual works of all impressed forces on the so relaxed beam, i.e., of the ever impressed P and of the former reaction B, and set it equal to zero [fig. 3.16(b)]:

$$\delta' W = P\,\delta p - B\,\delta b = 0 \;\Rightarrow\; B = (\delta p/\delta b)P = (a/l)P; \tag{a1}$$

(since $\delta p/\delta b =$ *finite*, and independent of the δp, δb magnitudes); and, similarly, $A = (b/l)P$. Because of the linearity of $\delta' W$, we can calculate one or more reactions at a time, or even all of them simultaneously.

(iv) *Dynamics*: Let us calculate the string tension S in the planar mathematical pendulum of mass m and length l (fig. 3.17).

(a) *Relaxed LP version*. Here, we allow the inextensible string to become a rubber band, compute the virtual works of all (old and new) impressed and inertial forces (i.e., apply LP to the relaxed system), and then enforce on the result the old constraint, $r = l$:

$$\begin{aligned}\delta' W|_{r=l} &= (W\cos\phi - S)\,\delta r - (W\sin\phi)(r\,\delta\phi)\big|_{r=l} = (W\cos\phi - S)\,\delta r - (W l\sin\phi)\,\delta\phi\\ &= Q_r\,\delta r + Q_\phi\,\delta\phi \;\Rightarrow\; Q_r = W\cos\phi - S, \qquad Q_\phi = -W l\sin\phi;\end{aligned} \tag{a2}$$

and similarly for $\delta I = E_r\,\delta r + E_\phi\,\delta\phi$, where the relaxed (double) kinetic energy is $2T = m[(\dot r)^2 + r^2(\dot\phi)^2]$, and therefore:

$$E_r \equiv (\partial T/\partial \dot r)^{\cdot} - \partial T/\partial r = (m\dot r)^{\cdot} - m r(\dot\phi)^2\big|_{r=l} = -m l(\dot\phi)^2, \tag{b}$$

$$E_\phi \equiv (\partial T/\partial \dot\phi)^{\cdot} - \partial T/\partial \phi = (m r^2\dot\phi)^{\cdot}\big|_{r=l} = m l^2\ddot\phi. \tag{c}$$

Hence, the relaxed $\rightarrow$ constrained LP, $(\delta I = \delta' W)\big|_{r=l}$, yields

$$m l(\dot\phi)^2 = S - W\cos\phi \quad \text{(kinetostatic)}, \qquad m l^2\ddot\phi = -W l\sin\phi \quad \text{(kinetic)}. \tag{d}$$

First, we solve the kinetic equation, the second of (d) (plus initial conditions), and obtain the motion $\phi = \phi(t)$; and then, substituting the latter into the kinetostatic equation, the first of (d), we get the tension $S = ml(\dot\phi)^2 + W\cos\phi = S(t)$. Had we started with the constrained (double) kinetic energy $ml^2(\dot\phi)^2$, we would not have been able to obtain the kinetostatic equation, just the kinetic one.

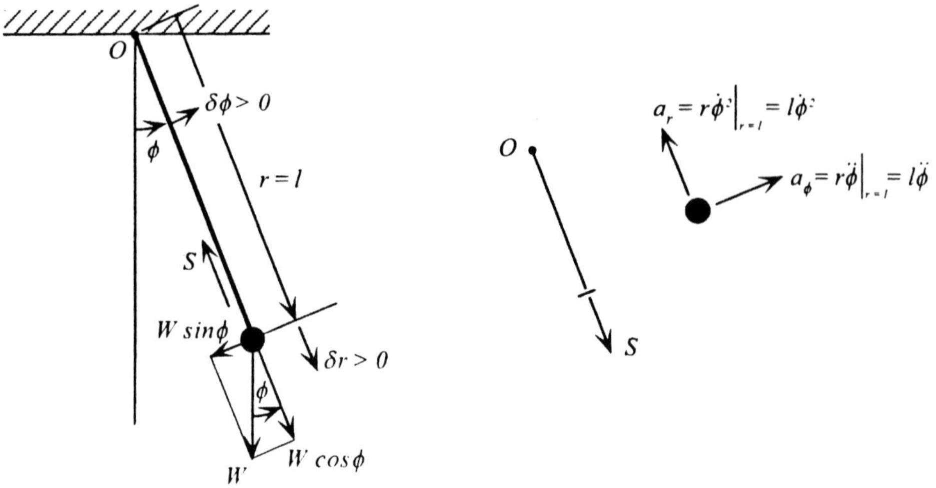

Figure 3.17 Principle of constraint relaxation applied in the calculation of the tension of the inextensible string of a planar pendulum.

(b) *Lagrangean multipliers*. In this version of LP/PRC, we have

$$Q_r = W\cos\phi, \qquad Q_\phi = (Wr\sin\phi)\big|_{r=l} = Wl\sin\phi, \tag{e}$$

since S is a reaction, while $q_1 = r$ and $q_2 = \phi$ are subject to the constraint

$$f \equiv r - l = 0 \Rightarrow \delta f = (\partial f/\partial r)\,\delta r = (1)\,\delta r = 0; \tag{f}$$

that is, $n = 2$, $m = 1$. Hence, the Routh–Voss equations yield

$$E_r = Q_r + \lambda(\partial f/\partial r): \qquad -ml(\dot{\phi})^2 = W\cos\phi + \lambda(1), \tag{g1}$$

$$E_\phi = Q_\phi + \lambda(\partial f/\partial\phi): \qquad ml^2\ddot{\phi} = -Wl\sin\phi; \tag{g2}$$

while from $\delta' W_R = -S\,\delta r = R_r\,\delta r = \lambda(1)\,\delta r$, we conclude that $\lambda = -S$, thus recovering (d). Had we written the constraint as $f \equiv l - r = 0 \Rightarrow (-1)\cdot\delta r = 0$, we would have $R_r = \lambda(\partial f/\partial r) = \lambda(-1) = -\lambda$, and, accordingly, $\delta' W_R = -S\,\delta r = R_r\,\delta r = \lambda(-1)\,\delta r \Rightarrow \lambda = S$, so that the final equations of motion would be unchanged. Also, since $\delta' W_R \geq 0$ for $\delta r \leq 0$ (§3.2.15, 15a), it follows that $S > 0$, as expected.

Example 3.7.2 *Relaxation of Constraints, External Reactions: Pendulum with Horizontally Moving Support* (Butenin, 1971, pp. 73–74). A block A of mass M, capable of translating along the smooth horizontal axis/floor Ox, is smoothly hinged to a massless rod AB of length l. The latter carries at its other end a particle B of mass m [fig. 3.18(a)]. Let us find the floor reaction on A and its motion on the O–xy plane.

Equations of Motion

For the relaxed system [fig. 3.18(b)], we introduce the following equilibrium coordinates:

$$q_1 = y_A \equiv y, \qquad q_2 = x_A \equiv x, \qquad q_3 = \phi, \tag{a}$$

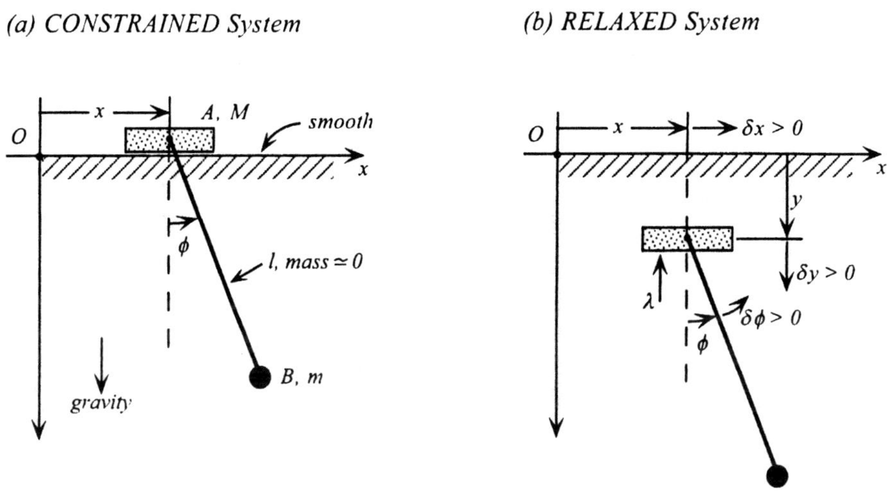

Figure 3.18 Principle of relaxation applied to a two DOF pendulum: (a) constrained, and (b) relaxed.

so that the floor constraint is simply $q_1 = 0$; that is, here $f \equiv n - m = 3 - 1 = 2$ *DOF*. Since the (inertial) coordinates of A and B are, respectively,

$$(x_A, y_A) = (x, y), \qquad (x_B, y_B) = (x + l \sin\phi, y + l\cos\phi), \tag{b}$$

the (double) kinetic energy of the system becomes, successively,

$$\begin{aligned} 2T &= M[(\dot{x}_A)^2 + (\dot{y}_A)^2] + m[(\dot{x}_B)^2 + (\dot{y}_B)^2] \\ &= \cdots = (M+m)(\dot{x})^2 + (M+m)(\dot{y})^2 + (ml^2)(\dot{\phi})^2 \\ &\quad + 2(ml\cos\phi)\dot{x}\dot{\phi} - 2(ml\sin\phi)\dot{y}\dot{\phi}, \end{aligned} \tag{c}$$

and so the *constrained* (double) kinetic energy is

$$2T\big|_{y=0} \equiv 2T_o = (M+m)(\dot{x})^2 + (ml^2)(\dot{\phi})^2 + 2(ml\cos\phi)\dot{x}\dot{\phi}. \tag{d}$$

Next, the relaxed virtual work $\delta' W$ equals, successively,

$$\begin{aligned} \delta' W &= (Mg)\,\delta y + (mg)\,\delta y_B = (Mg)\,\delta y + (mg)\,\delta(y + l\cos\phi) \\ &= [(M+m)g]\,\delta y + (-mgl\sin\phi)\,\delta\phi = Q_x\,\delta x + Q_y\,\delta y + Q_\phi\,\delta\phi, \end{aligned}$$

from which it follows that

$$Q_x = 0, \qquad Q_y = (M+m)g, \qquad Q_\phi = -mgl\sin\phi; \tag{e}$$

that is, Q_y could not have been obtained from the constrained virtual work

$$\delta' W\big|_{y=0} = (\delta' W)_o = (-mgl\sin\phi)\,\delta\phi. \tag{f}$$

Therefore, the Routh–Voss equations [which, due to the chosen special equilibrium relaxed coordinates (a), *decouple*] are

(i) $E_y(T) = Q_y + \lambda_y\big|_o$:

$$[(M + m)\ddot{y} - m\,l\sin\phi\,\ddot{\phi} - m\,l(\dot{\phi})^2\cos\phi]\big|_{y=0} = (M + m)g + \lambda_y, \tag{g}$$

or, finally,

$$\lambda_y \equiv \lambda = -m\,l[\sin\phi\,\ddot{\phi} + \cos\phi(\dot{\phi})^2] - (M + m)g; \tag{h}$$

(ii) $E_x(T) = Q_x\big|_o$: $[(M + m)\dot{x} + m\,l\cos\phi\,\dot{\phi}]^{\cdot}\big|_{y=0} = 0$

$$\Rightarrow (M + m)\dot{x} + (m\,l\,\dot{\phi})\cos\phi = \textit{constant} \tag{i}$$

(i.e., conservation of total linear momentum of system in the x-direction, since the total external force on system vanishes).

(iii) $E_\phi(T) = Q_\phi\big|_o$: $(m\,l^2\dot{\phi} + m\,l\cos\phi\,\dot{x})^{\cdot} + m\,l\sin\phi\,\dot{x}\,\dot{\phi} = -m\,g\,l\sin\phi$,

or, after some simple manipulations,

$$\ddot{\phi} + (\cos\phi/l)\ddot{x} = -(g/l)\sin\phi; \tag{j}$$

and for *small* ϕ's, linearizing in ϕ, we obtain the *forced pendulum-type* equation:

$$\ddot{\phi} + (g/l)\phi = -(1/l)\ddot{x}. \tag{k}$$

SOLUTION

First, we solve the two kinetic (= reactionless) equations (i, j/k), plus *initial* conditions: $\{x_o, \phi_o, \dot{x}_o, \dot{\phi}_o\} \equiv IC$, and thus find $x = x(t, IC)$, $\phi = \phi(t, IC)$; and then we substitute these solutions into the kinetostatic equation (h) and obtain $\lambda = \lambda(t, IC)$. Clearly, thanks to the uncoupling of the equations, all difficulty lies in the first (kinetic) part.

REMARK

Use of the initial conditions, say $\phi(0) = 0$, $\dot{\phi}(0) = \omega_o$, $\dot{x}(0) = v_o$, in the constrained (i.e., actual) energy conservation equation $T_o + V_o = (T_o + V_o)_i$ (*i* for *initial*) where, as found earlier,

$$2T_o = (M + m)(\dot{x})^2 + (ml^2)(\dot{\phi})^2 + 2(ml\cos\phi)\dot{x}\dot{\phi}, \tag{l}$$

$$\Rightarrow 2T_{o,i} = (M + m)(v_o)^2 + (ml^2)(\omega_o)^2 + 2m\,l\,v_o\,\omega_o, \tag{m}$$

and

$$V_o = m\,g\,l(1 - \cos\phi) = -m\,g\,l\cos\phi + \textit{constant}, \tag{n}$$

$$\Rightarrow V_{o,i} = -m\,g\,l + \textit{constant}, \tag{o}$$

yields the additional [to (i)] integral:

$$T_o - m\,g\,l\cos\phi = T_{o,i} - m\,g\,l = (\textit{another})\ \textit{constant}. \tag{p}$$

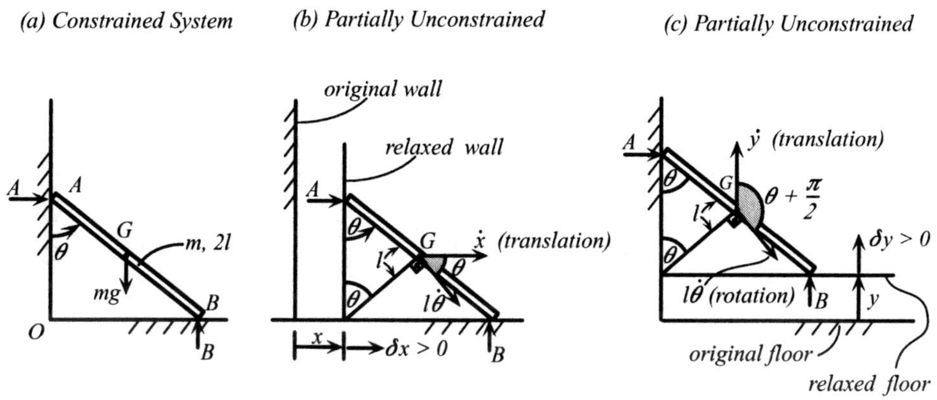

Figure 3.19 Plane motion of ladder: (a) constrained, and (b, c) partially constrained.

Example 3.7.3 *Relaxation of Constraints, External Reactions: Ladder* (Lawden, 1972, pp. 16–17, 54; also Osgood, 1937, pp. 322–325). A homogeneous bar (ladder) of length $2l$ and mass m, with one of its ends, A, constrained to move on a smooth vertical wall and the other, B, on a smooth horizontal floor [fig. 3.19(a)], is released, say with initial conditions (IC) $\theta(0) = \theta_o$, $\dot{\theta}(0) = 0$, and slides on the vertical plane O–xy. Let us find its equations of motion and external reactions: A from the wall and B from the floor, on the ladder. For algebraic simplicity, first, we relax only the wall and find A [fig. 3.19(b)]; then, we relax only the floor and find B [fig. 3.19(c)].

(i) *Wall relaxation*: Assume that the wall can translate horizontally, and let $x = x(t)$ be its generic distance/coordinate from the original wall (or from another fixed wall to its right, or left). With $q_1 = x$, $q_2 = \theta$, the wall constraint is $x = 0$, or $x = constant$; that is, $f \equiv n - m = 2 - 1 = 1$. Next, the (double) relaxed kinetic energy, by König's theorem (G = *center of mass*), is

$$2T = (ml^2/3)(\dot{\theta})^2 + m\,v_G{}^2 \qquad \text{(invoking the law of cosines)}$$
$$= (ml^2/3)(\dot{\theta})^2 + m[(\dot{x})^2 + l^2(\dot{\theta})^2 + 2l\,\dot{x}\,\dot{\theta}\cos\theta], \tag{a}$$

$$\Rightarrow 2T\big|_{x=o} \equiv 2T_o = (4m\,l^2/3)(\dot{\theta})^2 = \textit{constrained (double) kinetic energy}; \tag{b}$$

and the total relaxed virtual work (to the first order in $\delta\theta$, and with A as an impressed force) is

$$\delta' W = A\,\delta x + m\,g\,l[\cos\theta - \cos(\theta + \delta\theta)]$$
$$= A\,\delta x + m\,g\,l[\cos\theta - (\cos\theta - \sin\theta\,\delta\theta + \cdots)]$$
$$= A\,\delta x + (m\,g\,l\sin\theta)\,\delta\theta \;\Rightarrow\; Q_x = A,\; Q_\theta = m\,g\,l\sin\theta; \tag{c}$$

Q_θ can also be found from the constrained system potential $V_o = mgl\cos\theta$:

$$Q_\theta = -\partial V_o/\partial\theta = -\partial(mgl\cos\theta)/\partial\theta = mgl\sin\theta. \tag{d}$$

In view of the above, the equations of motion are

(a) $\quad E_\theta(T)\big|_o = E_\theta(T_o) = Q_\theta\big|_o$:

$$[(4l\ddot{\theta}/3) + \cos\theta\,\ddot{x})]\big|_o = (4l/3)\ddot{\theta} = g\sin\theta, \tag{e}$$

(b) $E_x(T) = Q_x\big|_o$:

$$m[\ddot{x} + (l\cos\theta)\ddot{\theta} - l\sin\theta(\dot{\theta})^2]\big|_o = m\,l[(\cos\theta)\ddot{\theta} - \sin\theta(\dot{\theta})^2] = A. \tag{f}$$

SOLUTION

First, we solve the kinetic equation (e) for $\theta = \theta(t, IC)$, and then insert that value into the kinetostatic equation (f), thus obtaining $A = A(t, IC)$, or $A = A(\theta, IC)$. Indeed, due to the well-known energetic identity $\dot{\theta}\ddot{\theta} = (d/dt)[(\dot{\theta})^2/2]$, (e) integrates to (ladder in contact with wall)

$$2l(\dot{\theta})^2/3 = g(\cos\theta_o - \cos\theta) \;\Rightarrow\; \dot{\theta} = \cdots, \ddot{\theta} = \cdots, \tag{g}$$

and substituting into (f) yields the wall reaction as a function of the angle θ (and the initial condition θ_o as parameter):

$$A = 3mg(3\cos\theta - 2\cos\theta_o)\sin\theta/4; \tag{h}$$

an expression that shows that the ladder loses contact with the wall when $\cos\theta = 2\cos\theta_o/3$; that is, when the end A has descended by 2/3 of its initial height above the floor.

REMARK

Equation (g) also results from energy conservation (i for *initial*):

$$T_o + V_o = (T_o + V_o)_i \;\Rightarrow\; 2m\,l^2(\dot{\theta})^2/3 + m\,g\,l\cos\theta = 0 + m\,g\,l\cos\theta_o. \tag{i}$$

(ii) *Floor relaxation* [fig. 3.19(c)]: By König's theorem (and the theorem of cosines), we find

$$\begin{aligned} 2T &= (ml^2/3)(\dot{\theta})^2 + mv_G{}^2 \\ &= (ml^2/3)(\dot{\theta})^2 + m[(\dot{y})^2 + l^2(\dot{\theta})^2 - 2l\,\dot{y}\,\dot{\theta}\sin\theta] \\ &= (4ml^2/3)(\dot{\theta})^2 + m(\dot{y})^2 - (2m\,l\sin\theta)\dot{y}\,\dot{\theta}, \end{aligned} \tag{j}$$

$$\Rightarrow 2T\big|_{y=o} \equiv 2T_o = (4ml^2/3)(\dot{\theta})^2 = \textit{constrained (double) kinetic energy}; \tag{k}$$

and the total relaxed virtual work (to the first order in $\delta\theta$, and with B as an impressed force) is

$$\begin{aligned} \delta' W &= B\,\delta y - mg\,\delta y - mg\,\delta(l\cos\theta) = (B - mg)\,\delta y + (mgl\sin\theta)\,\delta\theta \\ &\Rightarrow Q_y = B - mg, \qquad Q_\theta = m\,g\,l\sin\theta. \end{aligned} \tag{l}$$

Hence, the equations of motion are

(a) $E_\theta(T)\big|_o = E_\theta(T_o) = Q_\theta\big|_o$:

$$[(ml^2\ddot{\theta}/3) + (ml^2\dot{\theta} - ml\dot{y}\sin\theta)^{\cdot} + ml\dot{\theta}\dot{y}\cos\theta]\big|_o = mgl\sin\theta,$$

$$\Rightarrow (4l/3)\ddot{\theta} = g\sin\theta \qquad \text{(as before)}, \tag{m}$$

(b) $E_y(T) = Q_y|_o \quad [= Q_y|_o + \lambda, \text{ if we set } Q_y = -mg, \lambda = B]$:

$$[m(\dot{y} - l\sin\theta\,\dot{\theta})^{\cdot}]|_o = B - mg$$
$$\Rightarrow -ml[(\sin\theta)\ddot{\theta} + \cos\theta(\dot{\theta})^2] = B - mg, \tag{n}$$

from which [and invoking (g, m)]

$$\begin{aligned} B &= -ml(3g\sin\theta/4\lambda)\sin\theta - ml\cos\theta[(3g\cos\theta_o/2l) - (3g\cos\theta/2l)] + mg \\ &= (mg/4)[-3(1-\cos^2\theta) + 6\cos^2\theta - 6\cos\theta\cos\theta_o] + mg \\ &= (mg/4)[1 - 6\cos\theta\cos\theta_o + 9\cos^2\theta] = \textit{function of } \theta \textit{ and } \theta_o(>0). \end{aligned} \tag{o}$$

For $\cos\theta = 2\cos\theta_o/3$ [loss of contact with the wall, from (h)], the above yields $B = mg/4$.

Example 3.7.4 *Relaxation of Constraints, Internal Reactions: Atwood's Machine.* Two particles, P_1 and P_2, of respective masses m and M, are connected by a light and inextensible string of negligible mass, that passes over a light, smooth, and fixed pulley [fig. 3.20(a)]. Let us find the accelerations of P_1 and P_2 and the (approximately constant) string tension S.

(i) *Original (constrained) system.* This well-known apparatus is subject to the holonomic and stationary constraint

$$x + (h - X) = constant \;\Rightarrow\; x = X \pm constant \;\Rightarrow\; \dot{x} = \dot{X} \quad and \quad \delta x = \delta X; \tag{a}$$

that is, $n = 1$. Choosing as Lagrangean coordinate $q_1 = q = x$, and with the earlier notation $(\ldots)_o \equiv (\ldots)|_{\text{constrained system}}$, we find

$$\begin{aligned} &2T_o = (m + M)(\dot{x})^2, \\ &\delta' W \Rightarrow (\delta' W)_o = (mg)\,\delta x - (Mg)\,\delta x = (m - M)g\,\delta x \;\Rightarrow\; Q_{x,o} = (m - M)g \\ &[= (mg - S)\,\delta x - (Mg - S)\,\delta x, \text{ had we included the constraint reaction}], \end{aligned} \tag{b}$$

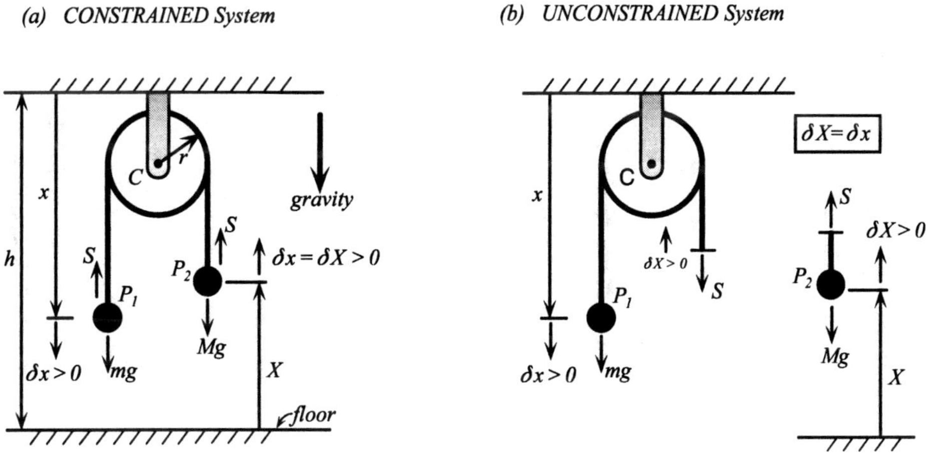

Figure 3.20 Atwood's machine: (a) original (constrained), and (b) relaxed.

and therefore Lagrange's sole (kinetic) equation is

$$(\partial T_o/\partial \dot{x})^{\cdot} - \partial T_o/\partial x = Q_{x,o}: \quad (m + M)\ddot{x} = (m - M)g, \tag{c}$$

from which [plus initial conditions $x(t_o) = x_o$, $\dot{x}(t_o) = v_o$] we obtain the motion $\ddot{x} = \cdots \Rightarrow x = x(t) = \cdots$. Finally, the constraint yields the P_2 motion $X = X(t) = \cdots$.

(ii) *Relaxed (unconstrained) system; calculation of string tension.*

(a) *Rubber-band version.* Since the constraint is the string's *inextensibility*, let us relax it by cutting it into *two separate inextensible substrings* [fig. 3.20(b)], in a manner reminiscent of the "free-body diagram" method ("cut principle"), so that $n = 2$. Choosing as Lagrangean coordinates $q_1 = x$ and $q_2 = X$, we have (now with S considered an impressed force)

$$2T = m(\dot{x})^2 + M(\dot{X})^2,$$

$$\delta' W = (mg - S)\,\delta x + (S - Mg)\,\delta X \;\Rightarrow\; Q_x = mg - S, \quad Q_X = S - Mg, \tag{d}$$

and therefore Lagrange's equations are

$$(\partial T/\partial \dot{x})^{\cdot} - \partial T/\partial x = Q_x: \qquad m\ddot{x} = mg - S, \tag{e}$$

$$(\partial T/\partial \dot{X})^{\cdot} - \partial T/\partial X = Q_X: \qquad m\ddot{X} = S - Mg. \tag{f}$$

These two, plus the constraint (a) (i.e., we return to the original, constrained, system), allow us to find $x(t) = \cdots, X(t) \cdots, S(t) = \cdots$. Eliminating S between (e, f), and then enforcing the constraint (a) $\rightarrow \ddot{x} = \ddot{X}$, produces the earlier equation (c); while utilizing the solution of the latter into either (e) or (f) [or eliminating $\ddot{x} = \ddot{X}$ between (e, f)] yields the sought reaction $S = [2mM/(m + M)]g$.

(b) *Lagrange's multiplier version.* Since the constraint is $f(x, X) \equiv x - X \pm$ *constant* $= 0$, and now, with x and X treated as independent (and S no longer considered an impressed force, but a reaction!), and taking $\delta x,\ \delta X > 0$,

$$\delta' W = (m\,g)\,\delta x - (M\,g)\,\delta X \;\Rightarrow\; Q_x = m\,g, \qquad Q_X = -M\,g, \tag{g}$$

the Routh–Voss equations, with multiplier λ, are

$$(\partial T/\partial \dot{x})^{\cdot} - \partial T/\partial x = Q_x + \lambda(\partial f/\partial x): \qquad m\ddot{x} = mg + \lambda(1), \tag{h}$$

$$(\partial T/\partial \dot{X})^{\cdot} - \partial T/\partial X = Q_X + \lambda(\partial f/\partial X): \qquad M\ddot{X} = -Mg + \lambda(-1). \tag{i}$$

From the above, clearly, $\delta' W_R \equiv R_x\,\delta x + R_X\,\delta X = \lambda(\delta x - \delta X)$ ($= 0$, upon enforcing the constraint $f = 0$; as LP requires). But also, by direct calculation [from fig. 3.20(b)], $\delta' W_R = -S\,\delta x + S\,\delta X = -S(\delta x - \delta X)$. Hence, it follows that $\lambda = -S(< 0)$, and so (h, i) coincide with (e, f), respectively.

REMARKS

(a) Had we written the above constraint as $f(x, X) \equiv X - x \pm$ *constant* $= 0$, the Routh–Voss equations would be

$$(\partial T/\partial \dot{x})^{\cdot} - \partial T/\partial x = Q_x + \lambda(\partial f/\partial x): \qquad m\ddot{x} = mg + \lambda(-1), \tag{j}$$

$$(\partial T/\partial \dot{X})^{\cdot} - \partial T/\partial X = Q_X + \lambda(\partial f/\partial X): \qquad M\ddot{X} = -Mg + \lambda(+1), \tag{k}$$

but since, now,

$$\delta' W_R \equiv R_x\,\delta x + R_X\,\delta X = (-\lambda)\,\delta x + (\lambda)\,\delta X = -\lambda(\delta x - \delta X) = -S(\delta x - \delta X),$$

we conclude that $\lambda = S(>0)$, and so (h, i) coincide with (j, k), respectively.

The lesson of this is that the multiplier adjusts its sign [and/or value, under *mathematically different but physically equivalent* forms of the constraint $f(x, X) = 0$], so that the final equations of motion retain their invariant physical content; and, of course, $\delta' W_R = 0$.

(b) In terms of the equilibrium coordinates $q_1 = x - X \pm constant$ and $q_2 = x (\Rightarrow dq_1 = 0,\ \dot{q}_1 = 0,\ \delta q_1 = 0)$, the constraint is simply $f \equiv q_1 = 0$ (*or a constant*). For such a choice, since now $x = q_2$ and $X = q_2 - q_1 \pm constant$ (no constraint enforcement yet!),

$$2T = m(\dot{x})^2 + M(\dot{X})^2 = \cdots = M(\dot{q}_1)^2 + (m + M)(\dot{q}_2)^2 - 2M\dot{q}_1\dot{q}_2, \tag{l}$$

$$\delta' W = (mg)\,\delta x - (Mg)\,\delta X = \cdots = (Mg)\,\delta q_1 + (m - M)g\,\delta q_2$$

$$\Rightarrow Q_1 = Mg, \qquad Q_2 = (m - M)g, \tag{m}$$

and so, in these coordinates, the Routh–Voss equations *decouple* naturally (upon enforcement of the constraint $q_1 = 0,\ \dot{q}_1 = 0, \ldots$ at the end) to a *kinetostatic* equation:

$$(\partial T/\partial \dot{q}_1)^{\cdot} - \partial T/\partial q_1 = Q_1 + \lambda(\partial f/\partial q_1):$$

$$-M\ddot{q}_2 = Mg + \lambda(+1), \qquad \text{or} \qquad M\ddot{x} = -Mg - \lambda, \tag{n}$$

and a *kinetic* one:

$$(\partial T/\partial \dot{q}_2)^{\cdot} - \partial T/\partial q_2 = Q_2 + \lambda(\partial f/\partial q_2):$$

$$(M + m)\ddot{q}_2 = (m - M)g + \lambda(0), \qquad \text{or} \qquad (M + m)\ddot{x} = (m - M)g, \tag{o}$$

($R_2 = 0$, since q_2 is unconstrained); and since

$$\delta' W_R \equiv R_1\,\delta q_1 + R_2\,\delta q_2 = \lambda\,\delta q_1 = -S(\delta x - \delta X), \tag{p}$$

it follows that now $\lambda = -S$, as observed earlier; that is, (n, o) coincide with (i, c), respectively.

Finally, we notice that $E_2(T_o) = E_2(T)\big|_o = (M + m)\ddot{q}_2 = (M + m)\ddot{x}$; while by (l), $2T\big|_o = 2T_o = (m + M)(\dot{q}_2)^2$, and therefore $E_1(T_o) = 0 \neq E_1(T)\big|_o = -M\ddot{q}_2$.

(iii) *Inclusion of rotary inertia of pulley*. If the pulley is assumed circular and homogeneous with radius r, mass μ, and radius of gyration k about its pin C, then it is not hard to show that the earlier results modify slightly to

$$\ddot{x} = (m - M)g/[(m + M) + \mu(k/r)^2], \quad \text{Left cable} = m(g - \ddot{x}),$$

$$\text{Right cable} = M(g + \ddot{x}). \tag{q}$$

Example 3.7.5 *PRC and the Determinacy versus Indeterminacy of Lagrange's Equations.* Let us consider a homogeneous sphere (or cylinder, or disk) of mass m and radius r, in plane motion (rolling or slipping) down a fixed inclined plane of slope with the horizontal θ (fig. 3.21). In terms of the "natural" Lagrangean coordinates of the problem $q_1 = y$ and $q_2 = x$ (coordinates of center of mass G)

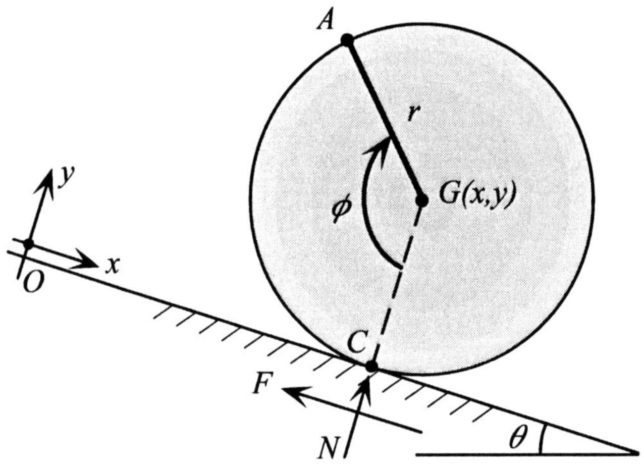

Figure 3.21 Plane motion (rolling/slipping) of a sphere down a fixed incline to illustrate the apparent indeterminacy of the Lagrangean equations. $O(rigin)$ chosen so that $arc(AC) = OC$; that is, during rolling $OC = r\phi$.

and $q_3 = \phi$, the constraints are

$$f_1 \equiv y - r = 0 \Rightarrow \delta f_1 = \delta y = 0 \qquad \text{(contact)}, \tag{a}$$

$$f_2 \equiv x - r\phi = 0 \Rightarrow \delta f_2 = \delta x - r\,\delta\phi = 0 \qquad \text{(if the sphere rolls)}. \tag{b}$$

By König's theorem, the (double) relaxed kinetic energy of the sphere, rolling or slipping, is

$$2T = m[(\dot{x})^2 + (\dot{y})^2] + I(\dot{\phi})^2 \tag{c}$$

(I = moment of inertia of sphere about axis through G, normal to plane of motion $= 2mr^2/5$). Now, and as is well known from the undergraduate dynamics treatment of this problem, we distinguish the following two cases.

(i) *The sphere rolls* [$|F/N| < \mu$ = *coefficient of* (*static*) *friction*].

(a) *Relaxation of constraints; rubber-band approach.* Here, $f \equiv n - m = 3 - 2 = 1$. The virtual work of all forces is (no constraint enforcement!)

$$\begin{aligned} \delta' W &= m g \sin\theta\, \delta x - m g \cos\theta\, \delta y + N\delta y - F(\delta x - r\,\delta\phi) \\ &= Q_x\, \delta x + Q_y\, \delta y + Q_\phi\, \delta\phi \\ \Rightarrow Q_x &= m g \sin\theta - F, \qquad Q_y = N - m g \cos\theta, \qquad Q_\phi = rF. \end{aligned} \tag{d}$$

With eqs. (c, d) Lagrange's equations yield

$$(\partial T/\partial \dot{x})^{\cdot} - \partial T/\partial x = Q_x: \qquad m\ddot{x} = mg\sin\theta - F, \tag{e1}$$

$$(\partial T/\partial \dot{y})^{\cdot} - \partial T/\partial y = Q_y: \qquad m\ddot{y} = N - mg\cos\theta, \tag{e2}$$

$$(\partial T/\partial \dot{\phi})^{\cdot} - \partial T/\partial \phi = Q_\phi: \qquad I\ddot{\phi} = rF, \tag{e3}$$

and along with the constraints (a, b) they constitute a *determinate* system of five equations in the five unknown functions of time: x, y, ϕ, N, F.

(b) *Relaxation of constraints; Lagrangean multiplier approach.* Here, too, $f \equiv n - m = 3 - 2 = 1$. Now the virtual work of *all* forces is [recall eqs. (a, b); the work of the *reactions* F and N appears indirectly as virtual work of the multipliers]

$$\begin{aligned}\delta' W &= mg\sin\theta\,\delta x - mg\cos\theta\,\delta y + \lambda_1\,\delta f_1 + \lambda_2\,\delta f_2\\ &= (mg\sin\theta + \lambda_2)\,\delta x + (-mg\cos\theta + \lambda_1)\,\delta y + (-r\,\lambda_2)\,\delta\phi\\ &= (Q_x + R_x)\,\delta x + (Q_y + R_y)\,\delta y + (Q_\phi + R_\phi)\,\delta\phi \qquad \text{(f1)}\end{aligned}$$

$$\begin{aligned}\Rightarrow Q_x &= mg\sin\theta, & R_x &= \lambda_1(\partial f_1/\partial x) + \lambda_2(\partial f_2/\partial x) = \lambda_1(0) + \lambda_2(1) = \lambda_2,\\ Q_y &= -mg\cos\theta, & R_y &= \lambda_1(\partial f_1/\partial y) + \lambda_2(\partial f_2/\partial y) = \lambda_1(1) + \lambda_2(0) = \lambda_1,\\ Q_\phi &= 0, & R_\phi &= \lambda_1(\partial f_1/\partial\phi) + \lambda_2(\partial f_2/\partial\phi) = \lambda_1(0) + \lambda_2(-r) = -\lambda_2 r.\end{aligned} \qquad \text{(f2)}$$

With eqs. (c, h, i) the Routh–Voss equations yield

$$(\partial T/\partial\dot{x})^{\cdot} - \partial T/\partial x = Q_x + R_x{:} \qquad m\,\ddot{x} = mg\sin\theta + \lambda_2, \qquad \text{(g1)}$$

$$(\partial T/\partial\dot{y})^{\cdot} - \partial T/\partial y = Q_y + R_y{:} \qquad m\,\ddot{y} = -m\,g\cos\theta + \lambda_1, \qquad \text{(g2)}$$

$$(\partial T/\partial\dot{\phi})^{\cdot} - \partial T/\partial\phi = Q_\phi + R_\phi{:} \qquad I\ddot{\phi} = -\lambda_2 r, \qquad \text{(g3)}$$

and along with eqs. (a, b) they constitute a *determinate* system of five equations in the five unknowns: $x, y, \phi, \lambda_1, \lambda_2$. Upon calculating $\delta' W_R \equiv \lambda_1\,\delta f_1 + \lambda_2\,\delta f_2 = \lambda_1\,\delta y + \lambda_2(\delta x - r\,\delta\phi)$ and equating it with $\delta' W_R = (-F)\,\delta x + (N)\,\delta y + (F\,r)\,\delta\phi$, calculated from the relaxed free-body diagram of the sphere, we immediately conclude that $\lambda_1 = N$ and $\lambda_2 = -F$. [Also, a "mixed" relaxation approach; i.e., part rubber band (e.g., including the virtual work of N directly) and part multiplier (e.g., including the virtual work of F via a $\lambda_2\,\delta f_2$ term) would have resulted in completely equivalent results.]

(c) *Embedding of all constraints.* Enforcing both eqs. (a, b) into T and $\delta' W$ and keeping ϕ as the sole Lagrangean coordinate (i.e., $n = 1$, $m = 0$), we obtain

$$2T_o = m(r\dot{\phi})^2 + I(\dot{\phi})^2 = I_C(\dot{\phi})^2 \qquad \text{(h)}$$

($I_C \equiv I + mr^2$ = moment of inertia of sphere about axis through contact point C, normal to plane of motion $= 7mr^2/5$),

$$\delta' W_o = m\,g\sin\theta\,\delta x = Q_{\phi,o}\,\delta\phi \;\Rightarrow\; Q_{\phi,o} = mgr\sin\theta. \qquad \text{(i)}$$

With eqs. (h, i), the sole (kinetic) Lagrangean equation yields

$$I_C\,\ddot{\phi} = m\,g\,r\sin\theta, \qquad \text{or, explicitly,} \qquad \ddot{\phi} - (5g/7r)\sin\theta = 0, \qquad \text{(j)}$$

and with the initial conditions, say $\phi(0) = \phi_o$ and $\dot{\phi}(0) = \omega_o$, readily yields $\phi = \phi(t)$. Then, using the constraints (a, b), we obtain $x = x(t)$ and $y = y(t)$. However, to find the reactions N and F, we either have to apply relaxation [as in parts (i.a) and (i.b) of this example], or go *outside* Lagrangean mechanics and apply the Newton–Euler momentum principles to the sphere. If we *embed only some of the constraints* into

T and $\delta' W$, say eq. (a) but not eq. (b) (i.e., $n = 2$, $m = 1$), then

$$2T = m(\dot{x})^2 + I(\dot{\phi})^2, \tag{k}$$

$$\delta' W = (m g \sin\theta - F)\,\delta x + (F r)\,\delta\phi = Q_x\,\delta x + Q_\phi\,\delta\phi$$

$$\Rightarrow\ Q_x = m g \sin\theta - F, \qquad Q_\phi = rF, \tag{l}$$

and so Lagrange's equations yield

$$(\partial T/\partial \dot{x})^{\cdot} - \partial T/\partial x = Q_x: \qquad m\ddot{x} = m g \sin\theta - F, \tag{m1}$$

$$(\partial T/\partial \dot{\phi})^{\cdot} - \partial T/\partial \phi = Q_\phi: \qquad I\ddot{\phi} = r F, \tag{m2}$$

and along with the constraint (b) constitute a *determinate* system of three equations in the three unknowns: x, ϕ, F. To find y and N we can either use relaxation, as before, or resort to the Newton–Euler momentum principles.

In sum: as long as the sphere rolls, all Lagrangean approaches (zero, partial, or complete embedding of the holonomic constraints) result in determinate systems of equations for their variables.

(ii) *The sphere slips.* In this case, the sole constraint is eq. (a):

$$f_1 \equiv y - r = 0 \ \Rightarrow\ \delta f_1 = \delta y = 0 \qquad \text{(contact)}, \tag{a}$$

while eq. (b) is replaced by the *constitutive* equation

$$|F/N| = \mu, \qquad \text{or} \qquad |F| = \mu|N| \qquad [\mu = \textit{coefficient of (kinetic) friction}]. \tag{n}$$

(a) *Relaxation of constraints; rubber-band approach.* Here, $f \equiv n - m = 3 - 1 = 2$, and so T and $\delta' W$ are given by eqs. (c, d), respectively. Therefore, Lagrange's equations are again eqs. (e–g); and along with eqs. (a), (r) constitute a *determinate* system of five equations in the five unknowns: x, y, ϕ, N, F.

(b) *Relaxation of constraints; Lagrange's multiplier approach.* Here, too, $f \equiv n - m = 3 - 1 = 2$, T is given by eq. (c), while [in the spirit of eq. (f1), the normal *reaction* appears through a *multiplier*]

$$\delta' W = m g \sin\theta\,\delta x - m g \cos\theta\,\delta y - F\,\delta x + (F r)\,\delta\phi + \lambda_1\,\delta f_1$$
$$= (m g \sin\theta - F)\,\delta x + (\lambda_1 - m g \cos\theta)\,\delta y + (F r)\,\delta\phi, \tag{o1}$$

that is,

$$Q_x = m g \sin\theta - F, \qquad R_x = \lambda_1(\partial f_1/\partial x) = 0,$$
$$Q_y = -m g \cos\theta, \qquad R_y = \lambda_1(\partial f_1/\partial y) = \lambda_1,$$
$$Q_\phi = Fr, \qquad R_\phi = \lambda_1(\partial f_1/\partial \phi) = 0. \tag{o2}$$

Hence, the Routh–Voss equations yield

$$(\partial T/\partial \dot{x})^{\cdot} - \partial T/\partial x = Q_x + R_x: \qquad m\ddot{x} = m g \sin\theta - F, \tag{p1}$$

$$(\partial T/\partial \dot{y})^{\cdot} - \partial T/\partial y = Q_y + R_y: \qquad m\ddot{y} = -m g \cos\theta + \lambda_1, \tag{p2}$$

$$(\partial T/\partial \dot{\phi})^{\cdot} - \partial T/\partial \phi = Q_\phi + R_\phi: \qquad I\ddot{\phi} = F r, \tag{p3}$$

and along with eqs. (a, n) constitute a *determinate* system of five equations in the five unknowns: x, y, ϕ, N, F. Here too we can easily show that $\lambda_1 = N$.

(c) *Embedding of all constraints.* Here, $f \equiv n - m = 2 - 1 = 1$. Enforcing eq. (a) into T and $\delta' W$ (= virtual work of all impressed forces), we have

$$2T = m(\dot{x})^2 + I(\dot{\phi})^2, \tag{q}$$

$$\delta' W = (m g \sin\theta - F)\,\delta x + (F\,r)\,\delta\phi = Q_x\,\delta x + Q_\phi\,\delta\phi \tag{r1}$$

$$\Rightarrow Q_x = m g \sin\theta - F, \qquad Q_\phi = r F, \tag{r2}$$

and so Lagrange's equations yield

$$(\partial T/\partial \dot{x})^{\cdot} - \partial T/\partial x = Q_x: \qquad m\ddot{x} = m g \sin\theta - F, \tag{s1}$$

$$(\partial T/\partial \dot{\phi})^{\cdot} - \partial T/\partial \phi = Q_\phi: \qquad I\ddot{\phi} = r F. \tag{s2}$$

But this is an *indeterminate* system, since we have generated *two* equations for our three unknowns: x, ϕ, F; and adding eq. (n) would introduce the extra unknown N.

As the handling of the previous cases has shown, such an apparent failure of the Lagrangean method to produce a "well-posed" problem is, generally, due to (i) our embedding of *all* (here holonomic) constraints into T and $\delta' W$, and (ii) the existence of *impressed forces that depend explicitly on (some or all of) the constraint reactions.* (Similar indeterminacy will result if we embed all *nonholonomic* constraints into T and $\delta' W$ via quasi variables.) To achieve determinacy, either we apply the principle of relaxation of the constraints; or we go outside of Lagrangean mechanics, usually applying the Newton–Euler principles of linear/angular momentum.

REMARK

Similar "indeterminacies" appear often in the Newton–Euler method; for example, by application of the momentum principles to an inappropriate free-body diagram. Here, determinacy is attained through the use of additional judiciously chosen *sub-free-body diagrams*, and subsequent application of the momentum principles to the resulting subbodies; for example, the well-known method of *sections*, of A. Ritter, in the statics of trusses.

Problem 3.7.1 *Determinacy versus Indeterminacy of Lagrange's Equations.* Consider a thin homogeneous bar AB of mass m and length $2l$, in plane motion, sliding on a fixed horizontal rough floor and a fixed vertical rough wall; in both, contacts with the same friction coefficient μ (fig. 3.22).

(i) Show that if we embed *all* (holonomic) constraints into T and $\delta' W$, and keep θ as the sole positional coordinate, then

$$2T_o = (4/3)ml^2(\dot{\theta})^2, \tag{a}$$

$$\delta' W_o = \left[m g l \sin\theta - 2l\sin\theta(\mu A) - 2l\cos\theta(\mu B)\right]\delta\theta, \tag{b}$$

and therefore the (kinetic) Lagrangean equation for θ is

$$4ml(d^2\theta/dt^2) = 3\left[mg\sin\theta - 2\sin\theta(\mu A) - 2\cos\theta(\mu B)\right]. \tag{c}$$

Is this a determinate problem for the equations and unknowns involved?

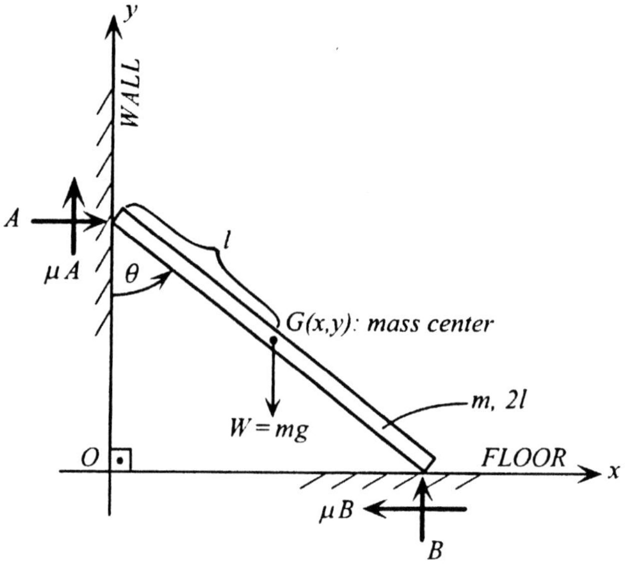

Figure 3.22 Plane sliding of homogeneous bar on rough wall and floor.

(ii) Show that if we apply the principle of relaxation, then (with positional coordinates $x, y =$ *Cartesian coordinates of bar's center of mass*, and θ)

$$2T = m[\dot{x}^2 + \dot{y}^2 + (1/3)l^2\dot{\theta}^2], \tag{d}$$

$$\delta' W = (A - \mu B)\,\delta x + (B + \mu A - mg)\,\delta y$$
$$+ l(B\sin\theta - A\cos\theta - \mu A\sin\theta - \mu B\cos\theta)\,\delta\theta, \tag{e}$$

and therefore Routh–Voss equations yield

$$m\ddot{x} = A - \mu B, \tag{f}$$

$$m\ddot{y} = B + \mu A - mg, \tag{g}$$

$$(ml/3)\ddot{\theta} = B(\sin\theta - \mu\cos\theta) - A(\cos\theta + \mu\sin\theta). \tag{h}$$

Is this system of equations (plus constraints) determinate in its unknowns?

(iii) Show that by eliminating A and B from eq. (h), via eqs. (f, g), and the x, y, θ constraints, we obtain the kinetic θ-equation

$$2l(\mu^2 - 2)(d^2\theta/dt^2) + 6\mu l(d\theta/dt)^2 + 3g[(1 - \mu^2)\sin\theta - 2\mu\cos\theta] = 0. \tag{i}$$

Example 3.7.6 *Motion under Frictionless Unilateral Constraints.* Let us consider a holonomic n *DOF* system under the single unilateral constraint $f(t, q) \geq 0$. Then, two types of motion are possible:

(i) $f > 0$: The system escapes the constraint, and its equations of motion $E_k = Q_k$ $(k = 1, \ldots, n)$ hold until $f = 0$ vanishes.

(ii) $f = 0$: The system obeys the constraint, and its motion is governed by, say its Routh–Voss equations $E_k = Q_k + \lambda(\partial f/\partial q_k)$, to which we must also append $f(t, q) = 0$. As long as $f = 0$, we must also have

$$df/dt = \sum (\partial f/\partial q_k)(dq_k/dt) + \partial f/\partial t = 0 \qquad (\textit{velocity compatibility}), \qquad \text{(a)}$$

$$d^2f/dt^2 = \sum (\partial f/\partial q_k)(d^2q_k/dt^2) + \sum\sum (\partial^2 f/\partial q_l \partial q_k)(dq_l/dt)(dq_k/dt) + 2\sum (\partial^2 f/\partial t\, \partial q_k)(dq_k/dt) + \partial^2 f/\partial t^2 = 0$$
$$(\textit{acceleration compatibility}). \qquad \text{(b)}$$

The $n + 1$ equations [Routh–Voss + eq. (b)] determine, at every instant for which $f = 0$, the n (d^2q/dt^2)'s and t. Further, in such a case, the n δq's may satisfy (a) $\delta f = 0$, or (b) $\delta f > 0$. In the former case, $\delta' W_R \equiv \lambda\, \delta f = 0$; in the latter, $\delta' W_R > 0$ (for every δq compatible with $\delta f > 0) \Rightarrow \lambda > 0$ (i.e., these (normal) reactions have a definite sense; for example, when a body B contacts an obstacle, the reaction from the latter to B must be directed *toward* B).

In sum: an $(f = 0)$-type of motion is physically meaningful as long as λ remains positive.

Finally, let us examine the following three possible cases (at a given instant):

(i) $f = 0$ *and* $df/dt \leq 0$: then we have impact ($\rightarrow$ chap. 4); the velocities at the end of it will be such that $df/dt \geq 0$.

(ii) $f = 0$ *and* $df/dt > 0$: then f *will* take positive values, and we are back to the case $f > 0$.

(iii) $f = 0$ *and* $df/dt = 0$: then we have one of the following two possibilities: (a) *constraint-preserving* motion $(f = 0)$, or (b) *escaping from it* $(f > 0)$. In the first case, we can calculate the (d^2q/dt^2)'s and $\lambda(> 0)$ from the Routh–Voss equations and eq. (b); while in the second, the (d^2q/dt^2)'s are found from $E_k = Q_k$, and then $d^2f/dt^2 \equiv a \geq 0$ (f, after being zero, will become later *positive*). Denoting by $\Delta(d^2q/dt^2)$ the (d^2q/dt^2)-difference in the above two cases, and since the Q's, q's, and (dq/dt)'s are the same for both, we obtain [with kinetic energy: $2T = \sum\sum M_{kl}(dq_k/dt)(dq_l/dt) = \textit{positive definite} \geq 0$]

$$\sum M_{kl}\Delta(d^2q_l/dt^2) = \lambda(\partial f/\partial q_k); \qquad \text{(c)}$$

and similarly from eq. (b)

$$\sum (\partial f/\partial q_k)\Delta(d^2q_k/dt^2) = -a. \qquad \text{(d)}$$

Combining the above we obtain

$$-a\lambda = \sum\sum M_{kl}\, \Delta(dq_k/dt)\Delta(dq_l/dt) = \textit{positive definite} \geq 0,$$

that is, since $\lambda > 0$, it follows that $a \leq 0$. Hence, cases (iii.a, b) are complementary to each other, and thus only *one of them will be physically acceptable*. For example, in a *constraint-preserving* motion $(f = 0)$, as long as $\lambda > 0$, an escape from it is impossible; such an escape will happen if λ, in going from + to −, vanishes.

So, basically, the study of the escape from a unilateral constraint can be made equally well either (i) by looking at the sign of the multiplier λ in the constraint-preserving motion, or (ii) by examining if the escaping assumption leads to $f > 0$.

The general theory of unilateral constraints has been developed by the "French school" of Appell, Delassus, Pérès, Beghin, Bouligand, et al.; see, for example, Pérès (1953, pp. 301–328); also (alphabetically): Glocker (1995), Hamel (1949, pp. 219–220) for a(n implicitly applied) postulate of *continuous transition* from the state where a unilateral constraint holds to the one where that constraint is abandoned (moment of loss of contact, or separation), Zhuravlev et al. (1993).

3.8 EQUATIONS OF MOTION: SPECIAL FORMS

We have established the *four* basic types of equations of motion (§3.5); namely, the equations of

(i) *Lagrange* and *Routh–Voss* (holonomic variables, motion and reactions coupled);
(ii) *Maggi* (holonomic variables, motion and reactions uncoupled);
(iii) *Hamel* (nonholonomic variables, motion and reactions uncoupled); and
(iv) *Appell* (holonomic and/or nonholonomic variables, motion and reactions coupled and/or uncoupled).

Let us now see the various *special* forms that these equations assume, for special choices of the coordinates and forms of the constraints.

3.8.1 Holonomic Constraints

Holonomic Coordinates

If the (additional) constraints are

$$f_D \equiv f_D(t, q) = 0 \qquad \text{(in finite form)} \qquad [D = 1, \ldots, m(< n)], \tag{3.8.1a}$$

or

$$\delta f_D \equiv \sum (\partial f_D/\partial q_k)\, \delta q_k \equiv \sum a_{Dk}\, \delta q_k = 0 \qquad \text{[in virtual (Pfaffian) form]}, \tag{3.8.1b}$$

then
(A) The *Routh–Voss* and *Appell* equations specialize, respectively, to

$$E_k \equiv d/dt(\partial T/\partial \dot{q}_k) - \partial T/\partial q_k = Q_k + \sum \lambda_D(\partial f_D/\partial q_k), \tag{3.8.2a}$$

$$E_k \equiv \partial S/\partial \ddot{q}_k = Q_k + \sum \lambda_D(\partial f_D/\partial q_k); \tag{3.8.2b}$$

or, compactly, with the convenient notation

$$M_k \equiv E_k - Q_k, \tag{3.8.2c}$$

$$M_k = \sum \lambda_D(\partial f_D/\partial q_k). \tag{3.8.2d}$$

(B) Let us find the corresponding *Maggi* equations. To embed (or absorb) equations (3.8.1a) into Lagrange's principle (LP), and following Hamel's choice of quasi variables, we introduce *n new "equilibrium" holonomic coordinates* $e \equiv \{e_k;\ k = 1, \ldots, n\}$ by

$$e_D \equiv f_D(t, q) = 0, \qquad e_I \equiv f_I(t, q) \neq 0, \qquad e_{n+1} \equiv q_{n+1} = t,$$
$$(D = 1, \ldots, m;\ I = m + 1, \ldots, n), \tag{3.8.3a}$$

where the $f_I(t, q)$ are *arbitrary*, but such that when the system (3.8.3a) is solved for the q's in terms of the e's, for each t (assuming this can be done uniquely), and the result is substituted back into the constraints $f_D(t, q) = 0$, it *satisfies them identically*. Clearly, the $\partial e_k/\partial q_l$ $(\partial q_k/\partial e_l)$ are the holonomic counterparts of the Pfaffian coefficients $a_{kl}(A_{kl})$. We also notice that since

$$q_k = q_k(t, e) \Rightarrow \dot{q}_k = \sum (\partial q_k/\partial e_l)\dot{e}_l + \partial q_k/\partial t,$$

$$\ddot{q}_k = \sum (\partial q_k/\partial e_l)\ddot{e}_l + \textit{function of } e,\ \dot{e},\ t, \tag{3.8.3b}$$

we have [as expected, recalling (2.9.35 ff.)]

$$\partial q_k/\partial e_l = \partial \dot{q}_k/\partial \dot{e}_l = \partial \ddot{q}_k/\partial \ddot{e}_l = \cdots \equiv A_{kl}; \tag{3.8.3c}$$

and, similarly, we can show that

$$\partial e_k/\partial q_l = \partial \dot{e}_k/\partial \dot{q}_l = \partial \ddot{e}_k/\partial \ddot{q}_l = \cdots \equiv a_{kl}. \tag{3.8.3d}$$

Inverting (3.8.3a) results in $q_k = q_k(t, e)$ [$\Rightarrow q_k(t, e_I)$, upon enforcing $e_D = 0$], from which we obtain the virtual system displacement representation [holonomic specialization of (2.11.4a ff.)]:

$$\delta q_k = \sum (\partial q_k/\partial e_l)\,\delta e_l \equiv \sum A_{kl}\,\delta e_l$$

$$\left[= \sum (\partial q_k/\partial e_I)\,\delta e_I \equiv \sum A_{kI}\,\delta e_I, \text{ upon enforcing } e_D = 0 \Rightarrow \delta e_D = 0\right]. \tag{3.8.3e}$$

Now, combining LP, in terms of these new coordinates; that is,

$$\begin{aligned}\sum M_k\,\delta q_k &= \sum M_k\Big(\sum (\partial q_k/\partial e_l)\,\delta e_l\Big) = \sum\Big(\sum (\partial q_k/\partial e_l)M_k\Big)\,\delta e_l\\ &= \sum\Big(\sum (\partial q_k/\partial e_D)M_k\Big)\,\delta e_D + \sum\Big(\sum (\partial q_k/\partial e_I)M_k\Big)\,\delta e_I = 0,\end{aligned} \tag{3.8.4a}$$

with the method of Lagrangean multipliers [in effect, rewriting the constraints $\delta e_D = 0$ as $1 \cdot \delta e_D = 0$ (and $1 \cdot \delta e_{n+1} = 1 \cdot \delta q_{n+1} = 1 \cdot \delta t = 0$), and viewing the $\delta e_I \neq 0$ as satisfying the constraints $0 \cdot \delta e_I = 0$], yields

$$\sum\Big(\sum (\partial q_k/\partial e_D)M_k - \lambda_D\Big)\,\delta e_D + \sum\Big(\sum (\partial q_k/\partial e_I)M_k - 0\Big)\,\delta e_I = 0, \tag{3.8.4b}$$

from which, since now the δe_D and δe_I can be viewed as unconstrained, we obtain the following *holonomic* Maggi-type of equations:

$$\sum (\partial q_k/\partial e_D)M_k = \lambda_D \qquad (m\ \textit{kinetostatic}\text{ equations}), \tag{3.8.4c}$$

$$\sum (\partial q_k/\partial e_I)M_k = 0 \qquad (n - m\ \textit{kinetic}\text{ equationsno multipliers}); \tag{3.8.4d}$$

or, in extenso [recalling the notation (3.8.2c)],

$$\sum \big[(\partial T/\partial \dot{q}_k)^{\cdot} - \partial T/\partial q_k\big](\partial q_k/\partial e_D) = \sum (\partial q_k/\partial e_D)Q_k + \lambda_D, \tag{3.8.4e}$$

$$\sum \big[(\partial T/\partial \dot{q}_k)^{\cdot} - \partial T/\partial q_k\big](\partial q_k/\partial e_I) = \sum (\partial q_k/\partial e_I)Q_k. \tag{3.8.4f}$$

The kinetic equations (3.8.4d, f) can also be obtained as follows: substituting $\delta q_k = \sum (\partial q_k/\partial e_I)\,\delta e_I$ (i.e., constraints enforced) into LP we obtain, successively,

$$\sum M_k \delta q_k = \sum M_k \Big(\sum (\partial q_k/\partial e_I)\,\delta e_I\Big) = \sum \Big(\sum (\partial q_k/\partial e_I) M_k\Big)\,\delta e_I = 0,$$

from which, since the δe_I are arbitrary, (3.8.4d, f) result.

REMARKS

(i) In forming the expressions appearing in (3.8.4c–f) we start with the *unconstrained* T and Q_k's (i.e., as functions of both the e_D's and e_I's, etc.), then carry out all relevant differentiations, and then, at the end, set $e_D = 0$ (just as in the nonholonomic variable case); otherwise the $\partial/\partial e_D$-differentiations would be impossible.

Also, using (3.8.3a), we can express these holonomic Maggi equations, exclusively, either in terms of q, $\dot{q}$, t, or in terms of e, $\dot{e}$, t; whichever is more helpful/desirable.

(ii) In view of the kinematico-inertial identity $(\partial T/\partial \dot{q}_k)^{\cdot} - \partial T/\partial q_k \equiv \partial S/\partial \ddot{q}_k$, eqs. (3.8.4c, d) can also be written, respectively, in the Appellian forms

$$\sum (\partial S/\partial \ddot{q}_k)(\partial q_k/\partial e_D) = \sum (\partial q_k/\partial e_D) Q_k + \lambda_D, \tag{3.8.4g}$$

$$\sum (\partial S/\partial \ddot{q}_k)(\partial q_k/\partial e_I) = \sum (\partial q_k/\partial e_I) Q_k. \tag{3.8.4h}$$

Special Case of Coordinates

An interesting special choice of e's, in (3.8.3e), is the following:

$$e_D \equiv f_D(t, q) = 0, \qquad e_I \equiv q_I \neq 0, \qquad e_{n+1} \equiv q_{n+1} = t,$$
$$(D = 1, \ldots, m; \quad I = m+1, \ldots, n); \tag{3.8.5a}$$

that is, the last $n - m$ q's are taken as the new *independent* Lagrangean coordinates. The above invert readily to $q_D = q_D(t, e_I) = q_D(t, q_I)$, $q_I = q_I$, and hence (3.8.4c, d) specialize, respectively, to

$$\sum (\partial q_k/\partial e_D) M_k = \sum (\partial q_{D'}/\partial e_D) M_{D'} + \sum (\partial q_I/\partial e_D) M_I$$
$$= \sum (\delta_{D'D}) M_{D'} + \sum (0) M_I = \lambda_D,$$

or, finally,

$$M_D = \lambda_D \qquad (m \textit{ kinetostatic} \text{ equations}); \tag{3.8.5b}$$

and

$$\sum (\partial q_k/\partial e_I) M_k = \sum (\partial q_D/\partial e_I) M_D + \sum (\partial q_{I'}/\partial e_I) M_{I'}$$
$$= \sum (\partial q_D/\partial q_I) M_D + \sum (\delta_{I'I}) M_{I'} = 0,$$

or, finally,

$$M_I + \sum (\partial q_D/\partial q_I) M_D = 0 \qquad (n - m \textit{ kinetic} \text{ equations}). \tag{3.8.5c}$$

In extenso, eqs. (3.8.5b, c) read, respectively,

$$E_D \equiv (\partial T/\partial \dot{q}_D)^{\cdot} - \partial T/\partial q_D = Q_D + \lambda_D, \tag{3.8.5d}$$

$$\begin{aligned} E_I + \sum (\partial q_D/\partial q_I) E_D &\equiv [(\partial T/\partial \dot{q}_I)^{\cdot} - \partial T/\partial q_I] + \sum [(\partial T/\partial \dot{q}_D)^{\cdot} - \partial T/\partial q_D](\partial q_D/\partial q_I) \\ &= Q_I + \sum (\partial q_D/\partial q_I) Q_D. \end{aligned} \tag{3.8.5e}$$

Appellian forms of (3.8.5b, c) can be immediately written down [$E_D \equiv \partial S/\partial \ddot{q}_D$, $E_I \equiv \partial S/\partial \ddot{q}_I$]. We notice the following

(i) From $q_D = q_D(t, q_I)$, it follows that

$$\dot{q}_D = \sum (\partial q_D/\partial q_I)\dot{q}_I + \partial q_D/\partial t, \qquad \ddot{q}_D = \sum (\partial q_D/\partial q_I)\ddot{q}_I + \textit{function of } t,\ q_I,\ \dot{q}_I, \tag{3.8.6a}$$

and therefore [specialization of (3.8.3c)]

$$\partial q_D/\partial q_I = \partial \dot{q}_D/\partial \dot{q}_I = \partial \ddot{q}_D/\partial \ddot{q}_I = \cdots \equiv b_{DI}(t, q_I). \tag{3.8.6b}$$

Equations (3.8.5b–e) are the holonomic counterparts of *Hadamard*'s equations (see below); and to obtain the latter we simply replace in the above $\partial q_D/\partial q_I$ with $\partial \dot{q}_D/\partial \dot{q}_I \equiv b_{DI}$.

(ii) Equations (3.8.5b, c) can also result if we view the choice (3.8.5a) as the following special case of (3.8.3a):

$$e_D \equiv f_D \equiv q_D - q_D(t, q_I) = 0, \qquad e_I \equiv f_I \equiv q_I \neq 0. \tag{3.8.7a}$$

Because then we have [recalling (2.11.9–12b), and with $D,\ D' = 1, \ldots, m$; $I,\ I' = m+1, \ldots, n$]

$(a_{kl}) \to (\partial e_k/\partial q_l) \equiv (\partial f_k/\partial q_l)$:

$$a_{DD'} = \delta_{DD'}, \qquad a_{DI} = -\partial q_D/\partial q_I, \qquad a_{ID} = 0, \qquad a_{II'} = \delta_{II'}; \tag{3.8.7b}$$

$(A_{kl}) \to (\partial q_k/\partial e_l)$:

$$A_{DD'} = \delta_{DD'}, \qquad A_{DI} = \partial q_D/\partial q_I, \qquad A_{ID} = 0, \qquad A_{II'} = \delta_{II'}; \tag{3.8.7c}$$

and so (3.8.4c, d) specialize to (3.8.5b, c), as shown there. An additional derivation of these last equations, based on a specialization of the Routh–Voss equations, is given later (Hadamard equations).

Special Case of Constraints

If the constraints (3.8.1a), $f_D = 0$, have the special "equilibrium" (or "adapted to the constraints") form $q_D = \textit{constant} \equiv q_{Do}$, then rewriting them as $f_D \equiv q_D - q_{Do} = 0$, we readily find

$$a_{Dk} \equiv \partial f_D/\partial q_k: \qquad a_{DD'} = \partial f_D/\partial q_{D'} = \delta_{DD'}, \qquad a_{DI} = \partial f_D/\partial q_I = 0, \tag{3.8.8a}$$

and so the Routh–Voss equations (3.8.2a, b) readily decouple to

$$E_D = Q_D + \lambda_D \quad (\textit{kinetostatic}), \qquad E_I = Q_I \quad (\textit{kinetic}). \tag{3.8.8b}$$

These are also the corresponding Maggi equations: for, in this case, we have (3.8.8a) and

$$a_{ID} = 0, \quad a_{II'} = \delta_{II'}; \qquad A_{DD'} = \delta_{DD'}, \quad A_{DI} = 0, \quad A_{ID} = 0, \quad A_{II'} = \delta_{II'}; \tag{3.8.8c}$$

and therefore [recalling (3.3.10)]

$$I_D = \sum A_{kD} E_k = \sum A_{D'D} E_{D'} + \sum A_{ID} E_I = \sum \delta_{D'D} E_{D'} + 0 = E_D, \tag{3.8.8d}$$

$$I_I = \sum A_{kI} E_k = \sum A_{DI} E_D + \sum A_{I'I} E_{I'} = 0 + \sum \delta_{I'I} E_{I'} = E_I; \tag{3.8.8e}$$

and similarly for the impressed forces [recalling (3.4.3b ff.)].

Equations (3.8.8b) are sometimes taken as the analytical expression of the principle of relaxation of the constraints; see, for example, Butenin (1971, pp. 70–71), Symon (1971, pp. 370–372).

Further, if we are not interested in calculating the reactions λ_D, then with

$$T = T(t, q, \dot{q}) = T(t, q_D = q_{Do}, q_I, \dot{q}_D = 0, \dot{q}_I) \equiv T_o(t, q_I, \dot{q}_I)$$
$$\equiv T_o\text{: } \textit{constrained} \text{ (or reduced) kinetic energy,}$$

and since (expanding à la Taylor, and with some obvious calculus notations)

$$T = T_o + (\partial T/\partial q_D)_o q_D + (\partial T/\partial \dot{q}_D)_o \dot{q}_D + \cdots, \tag{3.8.9a}$$

from which

$$(\partial T/\partial q_I)_o = \partial T_o/\partial q_I \qquad [(\partial T/\partial q_D)_o \neq \partial T_o/\partial q_D = 0], \tag{3.8.9b}$$

$$(\partial T/\partial \dot{q}_I)_o = \partial T_o/\partial \dot{q}_I \qquad [(\partial T/\partial \dot{q}_D)_o \neq \partial T_o/\partial \dot{q}_D = 0], \tag{3.8.9c}$$

we obtain the following rule: *In the case of holonomic variables and constraints, if we are only interested in the motion (kinetic problem), we may embed/enforce the constraints* $q_D = constant$ *into* $T \to T_o$ *right from the start*; that is, the second of (3.8.8b) can be replaced by $E_I(T_o) \equiv (\partial T_o/\partial \dot{q}_I)^{\cdot} - \partial T_o/\partial q_I = Q_I$; and this saves considerable labor (see also the holonomic Hamel equations below).

(C) Let us find the corresponding *Hamel* equations. In this case (recalling 2.9–2.12), $\theta_k \to e_k$ (*holonomic coordinates*), $\omega_D \equiv \dot{e}_D \equiv \dot{f}_D = 0$, $\omega_I \equiv \dot{e}_I \equiv \dot{f}_I \neq 0$, *the corresponding Hamel coefficients* $\gamma^k_{..}$ *vanish* [$\Rightarrow I_k = E_k{}^*(T^*)$], and so Hamel's equations (3.5.19a ff.) reduce to

$$d/dt(\partial T^*/\partial \dot{e}_D) - \partial T^*/\partial e_D = \sum (\partial q_k/\partial e_D) Q_k + \lambda_D \qquad (\textit{kinetostatic} \text{ eqs.}), \tag{3.8.10a}$$

$$d/dt(\partial T^*/\partial \dot{e}_I) - \partial T^*/\partial e_I = \sum (\partial q_k/\partial e_I) Q_k \qquad (\textit{kinetic} \text{ eqs.}), \tag{3.8.10b}$$

where $T^* = T[t, q(t, e), \dot{q}(t, e, \dot{e})] \equiv T^*(t, e, \dot{e})$; that is, these equations are none other than *Maggi's equations* (3.8.4e, f), respectively, *but expressed in the e-variables.*

We also notice that, here, too, we can replace the $n - m$ kinetic equations (3.8.10b) with

$$d/dt(\partial T^*{}_o/\partial \dot{e}_I) - \partial T^*{}_o/\partial e_I = \sum (\partial q_k/\partial e_I) Q_k \qquad (\textit{kinetic} \text{ eqs.}), \tag{3.8.10c}$$

or, compactly, $E_I(T^*_o) = \Theta_I$, where

$$T^*_o = T^*[t, e_D = 0,\ e_I \neq 0,\ \dot{e}_D = 0,\ \dot{e}_I \neq 0] \equiv T^*_o(t, e_I, \dot{e}_I):$$
constrained kinetic energy T^*.

Thus, for the earlier special choice $q_D = q_D(t, q_I)$, $q_I = q_I$, eqs. (3.8.10c) reduce to

$$E_I(T_o) \equiv d/dt(\partial T_o/\partial \dot{q}_I) - \partial T_o/\partial q_I = Q_{Io}, \tag{3.8.10d}$$

where

$T_o \equiv T[t, q_D(t, q_I), q_I;\ \dot{q}_D(t, q_I, \dot{q}_I), \dot{q}_I] \equiv T_o(t, q_I, \dot{q}_I)$: constrained kinetic energy T,

$Q_{Io} \equiv Q_I + \sum b_{DI} Q_D$: *constrained impressed force*
[a special case of $\Theta_I$recall (3.8.6b)].

Equations (3.8.10d) are none other than the earlier equations (3.8.5e), but expressed in the q_I's only.

Unfortunately, if the constraints $\dot{q}_D = \dot{q}_D(t, q_I, \dot{q}_I)$ are nonholonomic, then eqs. (3.8.10d) do *not* hold; that is, $E_I(T_o) \neq Q_{Io}$ (or even Q_I). It is shown later that, in such a case, if we use the velocity constraints to eliminate the m $\dot{q}_D$'s from the kinetic energy and impressed forces, and then apply these constrained, or reduced, quantities to multiplierless Lagrangean equations, like (3.8.10d), we will get *incorrect* equations of motion; other equations apply there (special cases of Hamel equations: equations of *Chaplygin* and *Voronets*) — equations that, if the velocity constraints are holonomic, reduce to (3.8.10d).

3.8.2 Nonholonomic Constraints

Holonomic Variables

(A) *Equations of Chaplygin–Hadamard.* Let us find the equations of motion corresponding to the additional, possibly nonholonomic, special Pfaffian constraints (2.11.9 ff.)

$$\dot{q}_D = \sum b_{DI} \dot{q}_I + b_D \ \Rightarrow\ \delta q_D = \sum b_{DI}\, \delta q_I. \tag{3.8.11a}$$

In view of the importance of this topic for the entire Lagrangean kinetics, we present *four* derivations.

(i) *Via Lagrange's principle (LP).* With the help of the earlier notation (3.8.2c), $M_k \equiv E_k - Q_k$, LP specializes, successively, to

$$0 = \sum M_k\, \delta q_k = \sum M_D\, \delta q_D + \sum M_I\, \delta q_I = \sum M_D \Big(\sum b_{DI}\, \delta q_I\Big) + \sum M_I\, \delta q_I$$
$$= \sum \Big(M_I + \sum b_{DI} M_D\Big)\, \delta q_I, \tag{3.8.11b}$$

from which, since the $n - m$ δq_I's are independent, we obtain the $n - m$ kinetic equations (Chaplygin, 1895, publ. 1897; Hadamard, 1895)

$$M_I + \sum b_{DI} M_D = 0, \qquad \text{or} \qquad E_I + \sum b_{DI} E_D = Q_I + \sum b_{DI} Q_D \quad (\equiv Q_{Io}), \tag{3.8.11c}$$

or, in extenso, (a) in the (more common) *Lagrangean* form:

$$[(\partial T/\partial \dot{q}_I)^{\cdot} - \partial T/\partial q_I] + \sum [(\partial T/\partial \dot{q}_D)^{\cdot} - \partial T/\partial q_D] b_{DI}$$
$$= Q_I + \sum b_{DI} Q_D \equiv Q_{Io}; \tag{3.8.11d}$$

and, in view of the kinematico-inertial identity $(\partial T/\partial \dot{q}_k)^{\cdot} - \partial T/\partial q_k \equiv \partial S/\partial \ddot{q}_k$, (b) in the *Appellian* form:

$$\partial S/\partial \ddot{q}_I + \sum b_{DI}(\partial S/\partial \ddot{q}_D) = Q_I + \sum b_{DI} Q_D \equiv Q_{Io}. \tag{3.8.11e}$$

(ii) *Via Lagrangean multipliers*. Multiplying each constraint (3.8.11a), $\delta q_D - \sum b_{DI} \delta q_I = 0$, with the multiplier $-\lambda_D$ and adding them to LP, we obtain, successively,

$$0 = \sum M_k\, \delta q_k = \sum M_k\, \delta q_k + \sum (-\lambda_D)\Big(\delta q_D - \sum b_{DI}\, \delta q_I\Big)$$
$$= \cdots = \sum (M_D - \lambda_D)\, \delta q_D + \sum \Big(M_I + \sum \lambda_D b_{DI}\Big)\, \delta q_I, \tag{3.8.11f}$$

from which, since now the δq_D's and δq_I's *can* be treated as independent, we obtain the two groups of equations

$$M_D = \lambda_D \qquad \text{or} \qquad E_D = Q_D + \lambda_D, \tag{3.8.11g}$$

$$M_I = -\sum \lambda_D b_{DI} \qquad \text{or} \qquad E_I = Q_I - \sum b_{DI} \lambda_D; \tag{3.8.11h}$$

and eliminating the m λ_D's among them [solving (3.8.11g) for λ_D and substituting in (3.8.11h), etc.], we recover the Hadamard equations (3.8.11c). Equations (3.8.11g) can be considered as the *kinetostatic complement of the kinetic equations* (3.8.11c).

(iii) *As a specialization of the Routh–Voss equations*. We recall that the special constraint form (3.8.11a) can be viewed as a Pfaffian system with the following coefficients:

$$a_{DD'} = \delta_{DD'}, \qquad a_{DI} = \partial \omega_D/\partial \dot{q}_I = -b_{DI} \qquad (\text{and } a_{ID} = 0,\ a_{II'} = \delta_{II'}). \tag{3.8.11i}$$

In view of these values, the general Routh–Voss equations

$$E_k \equiv (\partial T/\partial \dot{q}_k)^{\cdot} - \partial T/\partial q_k = Q_k + \sum \lambda_D a_{Dk} \tag{3.8.11j}$$

specialize to the two groups:

$$\text{(a)} \quad E_D = Q_D + \sum \lambda_{D'} a_{D'D} = Q_D + \sum \lambda_{D'} \delta_{DD'} = Q_D + \lambda_D,$$
$$\Rightarrow R_D = \sum a_{D'D} \lambda_{D'} = \sum \lambda_{D'} \delta_{DD'} = \lambda_D, \tag{3.8.11k}$$

$$\text{(b)} \quad E_I = Q_I + \sum \lambda_D a_{DI} = Q_I + \sum \lambda_D(-b_{DI}) = Q_I - \sum \lambda_D b_{DI}$$
$$= Q_I - \sum b_{DI}(E_D - Q_D)$$
$$\Rightarrow E_I + \sum b_{DI} E_D = Q_I + \sum b_{DI} Q_D \quad (\equiv Q_{Io}), \tag{3.8.11c}$$
$$\Rightarrow R_I = \sum a_{DI} \lambda_D = -\sum b_{DI} \lambda_D \quad \Big[= -\sum b_{DI}(E_D - Q_D)\Big]. \tag{3.8.11l}$$

Here, too, note that *the multipliers and their interpretation depend on the particular form of the constraints*; that is, if the constraints of a problem are written in two physically equivalent but analytically different forms, the associated multipliers (and, hence, kinetic and kinetostatic equations) will be equivalent but different; although, for theoretical purposes (and because both are components of the same constraint reaction vector, in configuration space), we may designate them both by λ_D. We repeat (§ 3.5), what holds all these descriptions together is the variational equation of Lagrange (LP), plus his method of multipliers (relaxation principle). It is these fundamental *invariant* tools that allow us to interrelate and compare the particular multiplier/constraint representation and equations of motion of the same problem.

(iv) *As a specialization of the Maggi equations.* In this case [recalling (3.8.7c)]

$$A_{DD'} = \delta_{DD'}, \qquad A_{DI} = \partial\dot{q}_D/\partial\dot{q}_I = b_{DI}, \qquad A_{ID} = 0, \qquad A_{II'} = \delta_{II'}. \quad (3.8.11\text{m})$$

Therefore, (a) Maggi's *kinetostatic* equations, $I_D - \Theta_D \equiv \sum A_{kD} M_k = \Lambda_D (= \lambda_D)$, specialize to

$$I_D - \Theta_D = \sum A_{D'D} M_{D'} + \sum A_{ID} M_I = \cdots = M_D + 0 = \lambda_D, \quad \text{i.e., } E_D = Q_D + \lambda_D, \quad (3.8.11\text{n})$$

while (b) Maggi's *kinetic* equations, $I_I - \Theta_I \equiv \sum A_{kI} M_k = 0$, specialize to

$$I_I - \Theta_I = \sum A_{DI} M_D + \sum A_{I'I} M_{I'} = \sum b_{DI} M_D + M_I = 0,$$
$$\text{i.e., } E_I + \sum b_{DI} E_D = Q_I + \sum b_{DI} Q_D.$$

REMARKS

(i) In these equations, T and the Q's (and S) are functions of all n $\dot{q}$'s ($\dot{q}$'s and $\ddot{q}$'s); that is, *no constraints are to be enforced in them yet.* That has to wait until all differentiations have been carried out. The $n - m$ equations (3.8.11o) plus the m constraints (3.8.11a) constitute a system of n equations for the n $q_k(t)$. Had we enforced the constraints in the Appellian

$$S = S(t, q, \dot{q}_D, \dot{q}_I, \ddot{q}_D, \ddot{q}_I) = S[t, q, \dot{q}_D(t, q, \dot{q}_I), \dot{q}_I, \ddot{q}_D(t, q, \dot{q}_I, \ddot{q}_I), \ddot{q}_I]$$
$$\equiv S_o(t, q, \dot{q}_I, \ddot{q}_I) = S_o\text{: } \textit{constrained Appellian}, \quad (3.8.12\text{a})$$

then LP would have given us, not (3.8.11e), but since

$$\delta I = \sum (\partial S/\partial \ddot{q}_k)\, \delta q_k = \sum (\partial S_o/\partial \ddot{q}_I)\, \delta q_I, \quad (3.8.12\text{b})$$

even though $\partial S/\partial \ddot{q}_I \neq \partial S_o/\partial \ddot{q}_I$ and

$$\delta' W = \sum Q_k \delta q_k = \sum \Big(Q_I + \sum b_{DI} Q_D\Big)\, \delta q_I \equiv \sum Q_{Io}\, \delta q_I, \quad (3.8.12\text{c})$$

finally [and this is *Appell's original form of 1899* (scleronomic case), 1900 (rheonomic case)],

$$\partial S_o/\partial \ddot{q}_I = Q_{Io}. \quad (3.8.12\text{d})$$

As stressed earlier, no such simplification (and preservation of form of the equations of motion) holds for $T \to T_o$-based equations.

(ii) Here, too, we first solve the kinetic equations (+ constraints + initial conditions) and obtain the motion $q_k(t)$. Then, the kinetostatic equations immediately yield

$$\lambda_D = E_D(t, q, \dot{q}, \ddot{q}) - Q_D(t, q, \dot{q})$$
$$= \cdots = \textit{known function of time (and initial conditions)}. \tag{3.8.12e}$$

(iii) What is important here is not so much Hadamard's equations themselves, but the *method(s)* for obtaining them. These latter can be applied even for nonholonomic variable constraints: for example,

$$\omega_D = \sum f_{DI}(t, q)\omega_I + f_D(t, q) \Rightarrow \delta\theta_D = \sum f_{DI}(t, q)\, \delta\theta_I, \tag{3.8.12f}$$

(i.e., $\delta\theta_D, \delta\theta_I \neq 0$), and for both Hamel- and Appell-type equations of motion.

(iv) Finally, we point out (what is probably amply clear by now) that if the constraints (3.8.11a) are holonomic ($b_{DI} = \partial q_D/\partial q_I$), then the Hadamard equations reduce to the earlier equations (3.8.5c, e).

Problem 3.8.1 *The Korteweg Equations.* Consider a system subject to the m, possibly nonholonomic, Pfaffian constraints

$$\sum a_{Dk}\, \delta q_k = 0 \qquad [k = 1, \ldots, n;\ D = 1, \ldots, m(< n);\ rank\ (a_{Dk}) = m], \tag{a}$$

and hence, assuming, as usual, *ideal* constraints, that is,

$$\sum R_k\, \delta q_k = 0, \tag{b}$$

having the Routh–Voss equations of motion

$$M_k \equiv E_k - Q_k \equiv (\partial T/\partial \dot{q}_k)^{\cdot} - \partial T/\partial q_k - Q_k = \sum \lambda_D a_{Dk} \quad (= R_k). \tag{c}$$

Show that the above imply that the following $(m+1) \times (m+1)$ determinant

$$\begin{vmatrix} R_1 & \ldots & R_m & \vdots & R_{m+1}\, \delta q_{m+1} + \cdots + R_n\, \delta q_n \\ \ldots & \ldots & \ldots & \vdots & \ldots \\ a_{11} & \ldots & a_{1m} & \vdots & a_{1,m+1}\, \delta q_{m+1} + \cdots + a_{1n}\, \delta q_n \\ \ldots & \ldots & \ldots & \vdots & \ldots \\ a_{m1} & \ldots & a_{mm} & \vdots & a_{m,m+1}\, \delta q_{m+1} + \cdots + a_{mm}\, \delta q_n \end{vmatrix} \tag{d}$$

vanishes, *identically*—that is, for arbitrary $\delta q_I \equiv (\delta q_{m+1}, \ldots, \delta q_n)$; and that this, in turn, thanks to well-known determinant properties, leads to the following $n - m$ determinantal equations:

$$\begin{vmatrix} R_1 & \ldots & R_m & \vdots & R_{m+1} \\ a_{11} & \ldots & a_{1m} & \vdots & a_{1,m+1} \\ \ldots & \ldots & \ldots & \vdots & \ldots \\ a_{m1} & \ldots & a_{mm} & \vdots & a_{m,m+1} \end{vmatrix} = 0, \ldots, \begin{vmatrix} R_1 & \ldots & R_m & \vdots & R_n \\ a_{11} & \ldots & a_{1m} & \vdots & a_{1n} \\ \ldots & \ldots & \ldots & \vdots & \ldots \\ a_{m1} & \ldots & a_{mm} & \vdots & a_{mn} \end{vmatrix} = 0; \tag{e}$$

and, conversely, (e) lead to the vanishing of (d).

REMARKS

(i) In view of (c), eqs. (e) constitute a set of $n - m$ *kinetic* equations, which, along with the m constraints (a) [in velocity form; i.e., $\sum a_{Dk}\dot{q}_k + a_D = 0$], make up a determinate system for the n $q_k(t)$.

(ii) Equations (d, e) seem to be due to Korteweg (1899, pp. 135–136, eqs. (8)]; and also Quanjel (1906, pp. 268–269, eqs. (16)). See also Routh (1891, vol. I, pp. 34–35).

HINT

The m $\delta q_D \equiv (\delta q_1, \ldots, \delta q_m)$, obtained from (a) as functions of the $n - m$ $\delta q_I \equiv (\delta q_{m+1}, \ldots, \delta q_n)$, must satisfy (b) for *arbitrary* δq_I.

Nonholonomic variables

(A) *Equations of Chaplygin (or Tschaplygine)*. Let us find the form that Hamel's equations assume when the, generally nonholonomic, Pfaffian constraints have the special scleronomic/stationary form

$$\dot{q}_D = \sum b_{DI}\dot{q}_I \;\Rightarrow\; \delta q_D = \sum b_{DI}\,\delta q_I, \tag{3.8.13a}$$

where (a) $b_{DI} = b_{DI}(q_{m+1}, \ldots, q_n)$ [a specialization of (3.8.11a)], and (b) $\partial T/\partial q_D = 0 \;\Rightarrow\; T = T(t, q_I, \dot{q}_D, \dot{q}_I)$; that is, *the m* q_D*'s do not appear either in the constraint coefficients or in the original (unconstrained) kinetic energy.*

Such "Chaplygin systems" can be viewed as the following special Hamel case:

$$\omega_D \equiv \dot{q}_D - \sum b_{DI}\dot{q}_I = 0, \qquad \omega_I \equiv \dot{q}_I \neq 0, \tag{3.8.13b}$$

which invert immediately to

$$\dot{q}_D = \omega_D + \sum b_{DI}\omega_I, \qquad \dot{q}_I = \omega_I. \tag{3.8.13c}$$

Equations (3.8.13b, c) readily show that here (a_{kl}) and (A_{kl}) have their earlier special forms:

$$a_{DD'} = \delta_{DD'}, \quad a_{DI} = \partial\omega_D/\partial\dot{q}_I = -b_{DI}, \quad a_{ID} = 0, \quad a_{II'} = \delta_{II'}; \tag{3.8.13d}$$

$$A_{DD'} = \delta_{DD'}, \qquad A_{DI} = \partial\dot{q}_D/\partial\omega_I = b_{DI}, \qquad A_{ID} = 0, \qquad A_{II'} = \delta_{II'}. \tag{3.8.13e}$$

From the above we find, successively, the following specializations for the various Hamel equation terms $(D, D', D'', \ldots = 1, \ldots, m;\; I, I', I'', \ldots = m+1, \ldots, n)$:

(i) $\gamma^I{}_{I'I''} = 0$ (notice that $\theta_I \equiv q_I$; i.e., θ_I is a holonomic coordinate), (3.8.13f)

(ii) $$\gamma^D{}_{II'} = \sum\sum(\partial a_{Dk}/\partial q_r - \partial a_{Dr}/\partial q_k)A_{kI}A_{rI'}$$
$$= \sum\sum[\partial(-b_{DI''})/\partial q_{I'''} - \partial(-b_{DI'''})/\partial q_{I''}]\delta_{I''I}\,\delta_{I'''I'}$$
$$= \sum\sum(\partial b_{DI'''}/\partial q_{I''} - \partial b_{DI''}/\partial q_{I'''})\,\delta_{I''I}\,\delta_{I'''I'}$$
$$= \sum(\partial b_{DI'}/\partial q_{I''} - \partial b_{DI''}/\partial q_{I'})\,\delta_{I''I} \;=\; \partial b_{DI'}/\partial q_I - \partial b_{DI}/\partial q_{I'}$$
$$\equiv -t^D{}_{II'} = t^D{}_{I'I}\text{: \textit{Chaplygin coefficients} (ex. 2.12.1, Remarks).} \tag{3.8.13g}$$

(iii) By chain rule (and recalling that $\partial\dot{q}_k/\partial\omega_r = A_{kr}$),

$$\partial T^*/\partial\omega_D = \sum(\partial T/\partial\dot{q}_k)(\partial\dot{q}_k/\partial\omega_D) = \sum(\partial T/\partial\dot{q}_k)A_{kD}$$
$$= \sum(\partial T/\partial\dot{q}_{D'})A_{D'D} + \sum(\partial T/\partial\dot{q}_I)A_{ID}$$
$$= \sum(\partial T/\partial\dot{q}_{D'})(\delta_{D'D}) + \sum(\partial T/\partial\dot{q}_I)(0)$$
$$= (\partial T/\partial\dot{q}_D)\big|_{\text{enforcing of constraints}} \equiv (\partial T/\partial\dot{q}_D)_o = \textit{function of } t, q_I, \dot{q}_I. \tag{3.8.13h}$$

(iv) Substituting the above into the correction term $-\Gamma_I$, (3.3.12a), we find, successively,

$$-\Gamma_I \equiv \sum\sum \gamma^k{}_{I\alpha}(\partial T^*/\partial\omega_k)\omega_\alpha \to \sum\sum \gamma^D{}_{II'}(\partial T^*/\partial\omega_D)\omega_{I'},$$

or, finally,

$$-\Gamma_I \to -\Gamma_{Io} \equiv \sum\sum t^D{}_{I'I}(\partial T/\partial\dot{q}_D)_o\,\dot{q}_{I'}$$
$$= \sum\sum(\partial b_{DI'}/\partial q_I - \partial b_{DI}/\partial q_{I'})(\partial T/\partial\dot{q}_D)_o\,\dot{q}_{I'}$$
$$= \textit{function of } t, q_I, \dot{q}_I \;(\textit{quadratic in the } \dot{q}_I). \tag{3.8.13i}$$

(v) With

$$T = T(t, q_I, \dot{q}_D, \dot{q}_I) = T[t, q_I, \dot{q}_D(t, q_I, \dot{q}_I), \dot{q}_I]$$
$$= T_o(t, q_I, \dot{q}_I) \equiv T_o = \textit{Chaplygin constrained kinetic energy}, \tag{3.8.13j}$$

we have

$$\partial T^*/\partial\omega_I \to \partial T_o/\partial\dot{q}_I; \tag{3.8.13k}$$

$$\partial T^*/\partial\theta_I \equiv \sum(\partial T^*/\partial q_k)(\partial\dot{q}_k/\partial\omega_I) = \sum(\partial T^*/\partial q_k)A_{kI}$$
$$= \sum(\partial T^*/\partial q_D)A_{DI} + \sum(\partial T^*/\partial q_{I'})A_{I'I}$$
$$= \sum(\partial T^*/\partial q_D)b_{DI} + \sum(\partial T^*/\partial q_{I'})\,\delta_{I'I}$$
$$= \sum(0)b_{DI} + \sum(\partial T^*/\partial q_{I'})\,\delta_{I'I} = \partial T^*/\partial q_I \to \partial T_o/\partial q_I. \tag{3.8.13l}$$

(vi) Recalling (3.8.12c) and (3.8.13b)

$$\delta' W = \sum \Theta_k\,\delta\theta_k = \sum \Theta_I\,\delta\theta_I = \sum Q_{Io}\,\delta q_I. \tag{3.8.13m}$$

(vii) And so,

$$E_I{}^*(T^*) \to E_I(T_o) \equiv (\partial T_o/\partial \dot{q}_I)^{\cdot} - \partial T_o/\partial q_I \quad [\neq E_I(T)]. \tag{3.8.13n}$$

Substituting all these findings into the kinetic Hamel equations (3.5.21d) [or into the central equation (3.6.8 ff.)], we obtain the $n - m$ (kinetic) equations of *Chaplygin*, in the following equivalent forms:

$$\begin{aligned}(\partial T_o/\partial \dot{q}_I)^{\cdot} - \partial T_o/\partial q_I &+ \sum\sum (\partial b_{DI'}/\partial q_I - \partial b_{DI}/\partial q_{I'})(\partial T/\partial \dot{q}_D)_o\, \dot{q}_{I'} \\ &\equiv (\partial T_o/\partial v_I)^{\cdot} - \partial T_o/\partial q_I + \sum\sum t^D{}_{I'I}(\partial T/\partial v_D)_o\, v_{I'} \\ &\equiv (\partial T_o/\partial v_I)^{\cdot} - \partial T_o/\partial q_I - \sum\sum t^D{}_{II'}(\partial T/\partial v_D)_o\, v_{I'} \\ &\equiv E_I(T_o) - \Gamma_{Io} = Q_I + \sum b_{DI} Q_D \equiv Q_{Io}.\end{aligned} \tag{3.8.13o}$$

REMARKS

(i) Chaplygin's equations (1895, publ. 1897) are the earliest (special Hamel-type) equations of motion of nonholonomic systems in terms of T_o (and T) and in special nonholonomic *system* variables. Chaplygin obtained these equations by expressing the T-gradients appearing in his earlier Chaplygin–Hadamard equations (3.8.11c, d) in terms of T_o-gradients [by applying chain rule to (3.8.13j)—an instructive exercise in partial differentiations that the readers are urged to reproduce for themselves]. The importance of (3.8.13o) is primarily theoretical and conceptual, not so much practical: these equations, presented at a time when nonholonomic dynamics was at its infancy, clearly demonstrated that, in general, $E_I(T_o) \neq Q_{Io}$ (or even Q_I); that is, *unless the constraints are holonomic* ($t^D{}_{I'I} = 0$), *or some other special case in which* $\Gamma_{Io} = 0$ *(for some or all* $I = m+1, \ldots, n$*), the ordinary Lagrangean equations do not hold for the constrained, or reduced, system.* [To the best of our knowledge, the first proof of this basic fact, for the special case of a *convex body rolling, under gravity, on a rough plane*, is due to C. Neumann (1885). But Chaplygin's treatment was, simultaneously, more general and easier to follow. See also appendix 3.A1.] Failure to observe this rule led, in the 1890s, to a number of erroneous equations of motion, even by some of the better mathematicians/mechanicians of that epoch (see examples below).

(ii) By their very structure, Chaplygin's equations are only *kinetic*; that is, since $\partial T_o/\partial \dot{q}_D = 0$, they do *not* allow for constraint reaction calculations. However, these $n - m$ equations, when solved, allow us to determine the $n - m$ functions $q_I = q_I(t)$ without recourse to the constraints (3.8.13a); the latter are then used to calculate the m $q_D = q_D(t)$.

(iii) Contrary to Hamel's equations, Chaplygin's equations involve *both* T and T_o, an apparent formal drawback; but, in return, they do not require a(ny) constraint matrix inversion [i.e., $(a_{kl}) \to (A_{kl})$], just the given *nonsquare* matrix (b_{DI}), instead of two.

For ad hoc chain-rule derivations of Chaplygin's equations, see, for example, Butenin (1971, pp. 196–212), Dobronravov (1970, pp. 87–106), Neimark and Fufaev (1972, pp. 106–108, 110–112); also Kil'chevskii (1977, pp. 162–166).

Problem 3.8.2 *Generalized Chaplygin Equations.* Show that in terms of the *general independent quasi velocities* ω_I defined by (in terms of the helpful notation $\dot{q}_k \equiv v_k$)

$$v_I = \sum B_{II'}\omega_{I'}, \qquad B_{II'} = B_{II'}(q_{m+1}, \ldots, q_n) \equiv B_{II'}(q_I), \tag{a}$$

Chaplygin's equations take the [slightly more general than (3.8.13o)] form

$$(\partial T^*{}_o/\partial\omega_I)^{\cdot} - \partial T^*{}_o/\partial\theta_I + \sum\sum (\partial B_{kI'}/\partial\theta_I - \partial B_{kI}/\partial\theta_{I'})(\partial T/\partial\dot{q}_k)_o\, \omega_{I'} = Q^*{}_{Io}, \quad \text{(b)}$$

where:

(i) $$v_k = \sum l_{kI}\omega_I, \qquad l_{kI} = l_{kI}(q_{m+1}, \ldots, q_n) \equiv l_{kI}(q_I),$$
$$v_D = \sum\sum (b_{DI}B_{II'})\omega_{I'} \equiv \sum l_{DI}\omega_I, \qquad v_I = \sum B_{II'}\omega_{I'} \equiv \sum l_{II'}\omega_{I'}; \quad \text{(c)}$$

(ii) $$T = T(t, q_I, v_k) = T\Big(t, q_I, v_k = \sum l_{kI}\omega_I\Big) \equiv T^*{}_o(t, q_I, \omega_I) \equiv T^*{}_o; \quad \text{(d)}$$

(iii) $$\partial T^*{}_o/\partial\theta_I \equiv \sum (\partial T^*{}_o/\partial q_{I'})(\partial v_{I'}/\partial\omega_I) = \sum (\partial T^*{}_o/\partial q_{I'})B_{I'I}, \quad \text{(e)}$$
$$\partial l_{kI'}/\partial\theta_I \equiv \sum (\partial l_{kI'}/\partial q_{I''})(\partial v_{I''}/\partial\omega_I) = \sum (\partial l_{kI'}/\partial q_{I''})B_{I''I}; \quad \text{(f)}$$

(iv) $$\delta' W \equiv \sum Q_k\, \delta q_k = \cdots = \sum Q^*{}_{Io}\, \delta\theta_I \qquad (\text{definition of } Q^*{}_{I'o}), \quad \text{(g)}$$
$$\Rightarrow Q^*{}_{Io} = \sum B_{I'I}\Big(Q_{I'} + \sum b_{DI'}Q_D\Big) \equiv \sum B_{I'I}\, Q_{I'o}. \quad \text{(h)}$$

[See also Neimark and Fufaev (1972, pp. 110–112). In there, on pp. 106–108, eqs. (3.16, 17, 19), it seems that a *tilde* ($\sim$) should be placed on the impressed force Q_I.]

Equations of Voronets (or *Woronetz*). Let us find the form that Hamel's equations assume when the, generally nonholonomic, Pfaffian constraints have the special rheonomic/nonstationary form (again, with $v_k \equiv \dot{q}_k$)

$$v_D = \sum b_{DI}v_I + b_D \;\Rightarrow\; \delta q_D = \sum b_{DI}\, \delta q_I, \quad \text{(3.8.14a)}$$

where $b_{DI} = b_{DI}(t, q_1, \ldots, q_n) \equiv b_{DI}(t, q)$, $b_D = b_D(t, q_1, \ldots, q_n) \equiv b_D(t, q)$; or the Hamel form

$$\omega_D \equiv v_D - \sum b_{DI}v_I - b_D = 0, \qquad \omega_I \equiv v_I \neq 0; \quad \text{(3.8.14b)}$$

with its (easy to obtain) inverse

$$v_D = \omega_D + \sum b_{DI}\omega_I + b_D, \qquad v_I = \omega_I. \quad \text{(3.8.14c)}$$

Clearly, since (3.8.14a–c) are a generalization of the Chaplygin constraints (3.8.13a–c), the associated equations of Voronets, derived below, will constitute a generalization of those of Chaplygin (3.8.13o); but a special case of those of Hamel.

Equations (3.8.14b, c) readily show that, here, the transformation matrices (a_{kl}) and (A_{kl}) have their earlier special forms:

$$a_{DD'} = \delta_{DD'}, \quad a_{DI} = \partial\omega_D/\partial v_I = -b_{DI}, \quad a_{ID} = 0, \quad a_{II'} = \delta_{II'}; \quad \text{(3.8.14d)}$$

$$A_{DD'} = \delta_{DD'}, \quad A_{DI} = \partial v_D/\partial\omega_I = b_{DI}, \quad A_{ID} = 0, \quad A_{II'} = \delta_{II'}. \quad \text{(3.8.14e)}$$

From the above, and the results of ex. 2.12.1 and prob. 2.12.1 ff., we find, successively, the following specializations for the various Hamel equation terms ($D, D', D'', \ldots = 1, \ldots, m$; $I, I', I'', \ldots = m+1, \ldots, n$):

(i) $$\gamma^{I}{}_{I'I''} = 0 \qquad (\theta_I \equiv q_I;\ \text{i.e., } \theta_I \text{ is a holonomic coordinate!}). \tag{3.8.14f}$$

(ii) $$\begin{aligned}\gamma^{D}{}_{II'} &= \sum\sum (\partial a_{Dk}/\partial q_r - \partial a_{Dr}/\partial q_k) A_{kI} A_{rI'} \\ &= \cdots = \left[\partial b_{DI'}/\partial q_I + \sum b_{D'I}(\partial b_{DI'}/\partial q_{D'})\right] \\ &\qquad - \left[\partial b_{DI}/\partial q_{I'} + \sum b_{D'I'}(\partial b_{DI}/\partial q_{D'})\right] \\ &\equiv \partial b_{DI'}/\partial(q_I) - \partial b_{DI}/\partial(q_{I'}) \\ &= t^{D}{}_{I'I} + \sum [b_{D'I}(\partial b_{DI'}/\partial q_{D'}) - b_{D'I'}(\partial b_{DI}/\partial q_{D'})] \\ &\equiv w^{D}{}_{I'I} = -w^{D}{}_{II'} \qquad \text{(recalling ex. 2.12.1)},\end{aligned} \tag{3.8.14g}$$

$$\begin{aligned}\gamma^{D}{}_{I,n+1} \equiv \gamma^{D}{}_{I} &= \cdots = \left[\partial b_D/\partial q_I + \sum b_{D'I}(\partial b_D/\partial q_{D'})\right] \\ &\qquad - \left[\partial b_{DI}/\partial t + \sum b_{D'}(\partial b_{DI}/\partial q_{D'})\right] \\ &\equiv \partial b_D/\partial(q_I) - \partial b_{DI}/\partial(q_{n+1}) \\ &= -t^{D}{}_{I,n+1} + \sum [b_{D'I}(\partial b_D/\partial q_{D'}) - b_{D'}(\partial b_{DI}/\partial q_{D'})] \\ &\equiv -w^{D}{}_{I,n+1} \equiv -w^{D}{}_{I} \qquad \text{(recalling prob. 2.12.2)};\end{aligned} \tag{3.8.14h}$$

or they can be read off from the transitivity equations below (which, incidentally, show clearly that the only surviving γ's are the $\gamma^{D}{}_{II'}$ and $\gamma^{D}{}_{I}$)

$$\begin{aligned}d(\delta\theta_D) - \delta(d\theta_D) &= \sum\sum \gamma^{D}{}_{II'}\, d\theta_{I'}\, \delta\theta_I + \sum \gamma^{D}{}_{I}\, \delta\theta_I \\ &= -\sum\sum w^{D}{}_{II'}\, dq_{I'}\, \delta q_I - \sum w^{D}{}_{I}\, \delta q_I.\end{aligned} \tag{3.8.14i}$$

(iii) With

$$\begin{aligned}T &= T(t, q, v_D, v_I) = T[t, q, v_D(t, q, v_I), v_I] \\ &= T_o(t, q, v_I) \equiv T_o = \textit{Voronets constrained kinetic energy} \\ &\quad \text{[a generalization of Chaplygin's (3.8.13j), and a special case of } T^*(t, q, \omega)],\end{aligned} \tag{3.8.14j}$$

we find, successively,

$$\partial T^*/\partial\omega_I \to \partial T_o/\partial v_I \quad [\neq (\partial T/\partial v_I)_o \equiv p_{Io} = \textit{function of } t, q, v_I]; \tag{3.8.14k}$$

$$\partial T^*/\partial\omega_D = [\text{repeating steps as in (3.8.13h)}] = (\partial T/\partial v_D)_o = \textit{function of } t, q, v_I; \tag{3.8.14l}$$

$$\begin{aligned}\partial T^*/\partial\theta_I &\equiv \sum (\partial T^*/\partial q_k)(\partial v_k/\partial\omega_I) = \sum (\partial T^*/\partial q_k) A_{kI} \\ &= \sum (\partial T^*/\partial q_D) A_{DI} + \sum (\partial T^*/\partial q_{I'}) A_{I'I} \\ &= \sum (\partial T^*/\partial q_D) b_{DI} + \sum (\partial T^*/\partial q_{I'})\, \delta_{I'I} \\ &= \partial T^*/\partial q_I + \sum b_{DI}(\partial T^*/\partial q_D),\end{aligned} \tag{3.8.14m}$$

that is,

$$\partial T^*/\partial\theta_I \to \partial T_o/\partial q_I + \sum b_{DI}(\partial T_o/\partial q_D) \equiv \partial T_o/\partial(q_I) \qquad \text{(symbolic derivative)}. \tag{3.8.14n}$$

(iv) Substituting the above into the correction term $-\Gamma_I$, (3.3.12a), we find, successively,

$$\begin{aligned}-\Gamma_I &\equiv \sum\sum \gamma^b{}_{I\alpha}(\partial T^*/\partial\omega_b)\omega_\alpha \to \sum\sum \gamma^D{}_{II'}(\partial T^*/\partial\omega_D)\omega_{I'} + \sum \gamma^D{}_I(\partial T^*/\partial\omega_D)\\ &= \cdots = -\sum\sum w^D{}_{II'}(\partial T/\partial v_D)_o\, v_{I'} - \sum w^D{}_I(\partial T/\partial v_D)_o \equiv -\Gamma_{Io}.\end{aligned} \tag{3.8.14o}$$

(v) Proceeding as in (3.8.13m),

$$\delta' W = \sum \Theta_k\, \delta\theta_k = \sum \Theta_I\, \delta\theta_I = \sum\left(Q_I + \sum b_{DI}Q_D\right)\delta q_I \equiv \sum Q_{Io}\, \delta q_I. \tag{3.8.14p}$$

Substituting all these findings into the kinetic Hamel equations (3.5.21d) [or into the central equation (3.6.9 or 11), with (3.8.14i); namely, the Hamel approach; *or* into (3.6.8 or 12), but with $d(\delta q_D) - \delta(dq_D) = \sum\sum w^D{}_{II'}\, dq_{I'}\, \delta q_I + \sum w^D{}_I\, \delta q_I$, $d(\delta q_I) - \delta(dq_I) = 0$; namely the Suslov approach (see also ex. 3.8.1, below)], we obtain the $n - m$ (kinetic) equations of *Voronets*, in the following equivalent forms [recalling (3.8.13o)]:

$$\begin{aligned}&(\partial T_o/\partial\dot q_I)^{\cdot} - \partial T_o/\partial q_I - \sum b_{DI}(\partial T_o/\partial q_D)\\ &\qquad - \sum\sum w^D{}_{II'}(\partial T/\partial\dot q_D)_o \dot q_{I'} - \sum w^D{}_I(\partial T/\partial\dot q_D)_o\\ &\quad \equiv (\partial T_o/\partial v_I)^{\cdot} - \partial T_o/\partial(q_I) - \sum\sum w^D{}_{II'}\, p_{Do}\, v_{I'} - \sum w^D{}_I\, p_{Do}, \ [\text{since } v_{n+1} = 1]\\ &\quad \equiv E_I(T_o) - \sum b_{DI}(\partial T_o/\partial q_D) - \Gamma_{Io}\\ &\quad \equiv E_{(I)}(T_o) - \Gamma_{Io} = Q_I + \sum b_{DI}Q_D \equiv Q_{Io}.\end{aligned} \tag{3.8.14q}$$

SPECIALIZATIONS, REMARKS

(i) If the constraints are *catastatic* (i.e., $b_D = 0$), then, as (3.8.14h) readily shows, the $w^D{}_I$ reduce to $\partial b_{DI}/\partial t$; and if they are *stationary*, or scleronomic, they vanish.

(ii) If the Voronets constraints reduce to those of Chaplygin, then (3.8.14q) reduce to (3.8.13o).

(iii) We believe that the above derivation of the Voronets equations (i.e., deductively from those of Hamel) is their clearest presentation in the entire dynamics literature in English; and one of the few anywhere.

(iv) By looking at the constraint forms (3.8.11a), (3.8.14a), we can state that the *Voronets equations bear the same relation to those of Hamel that the Hadamard equations bear relative to those of Maggi* [although it took about 20 years for that to be recognized: Hamel (1924)]. Schematically:

Maggi (1896, 1901, 1903) → Hadamard (1895)

Hamel (1903, 1904) → Voronets (1901).

Problem 3.8.3 *Generalized Voronets' Equations.* Formulate Voronets' equations in the *general quasi velocities* ω_I defined by

$$v_I = \sum B_{II'}\,\omega_{I'} + B_I, \tag{a}$$

where the coefficients $B_{II'}$ and B_I are assumed functions of *all* the q's and t; and the ν_k are constrained, as in (3.8.14a):

$$v_D = \sum b_{DI}\,v_I + b_D. \tag{b}$$

(2C) *Equations in General Nonholonomic Variables; when the nonholonomic constraints have the general form* $[D' = 1, \ldots, m\ (< n), k = 1, \ldots, n]$:

$$\sum a_{D'k}\omega_k + a_{D'} = 0 \;\Rightarrow\; \sum a_{D'k}\,\delta\theta_k = 0, \tag{3.8.15a}$$

where (i) the coefficients $a_{D'k}$ and $a_{D'}$ are functions of all the q's and t [and *rank* $(a_{D'k}) = m$], and (ii) the v's and ω's are related by

$$\omega_k \equiv \sum a_{kl}v_l + a_k \neq 0 \;\Leftrightarrow\; v_l = \sum A_{lk}\omega_k + A_l \neq 0, \tag{3.8.15b}$$

instead of the earlier holonomic variable forms $\omega_D \equiv \sum a_{Dk}\nu_k + a_D = 0$, etc..

Applying the general methods expounded in §3.5, we may proceed in one of the following two ways: either we

(i) *Adjoin* the constraints (3.8.15a) to LP in the ω variables (3.5.18),

$$\sum I_k\,\delta\theta_k = \sum \Theta_k\,\delta\theta_k, \tag{3.8.15c}$$

with m Lagrangean multipliers $\lambda_{D'}$, and thus obtain the n *coupled* Routh–Voss type of equations:

$$I_k = \Theta_k + \sum \lambda_{D'}a_{D'k}, \tag{3.8.15d}$$

where the inertia terms I_k have one of the following basic forms:

$$I_k = \sum (\partial v_l/\partial\omega_k)E_l \equiv \sum A_{lk}[(\partial T/\partial\dot q_l)^{\cdot} - \partial T/\partial q_l] \qquad \text{(Maggi)}, \tag{3.8.15e}$$

$$= (\partial T^*/\partial\omega_k)^{\cdot} - \partial T^*/\partial\theta_k + \sum\sum \gamma^b{}_{k\alpha}(\partial T^*/\partial\omega_b)\omega_\alpha \qquad \text{(Hamel)}, \tag{3.8.15f}$$

$$= \partial S^*/\partial\dot\omega_k \equiv \sum A_{lk}(\partial S/\partial\ddot q_l) \qquad \text{(Appell)}; \tag{3.8.15g}$$

and which, along with the m constraints (3.8.15a) and the n transformation equations (3.8.15b), constitute a system of $n + m + n = 2n + m$ equations for the $2n + m$ unknown functions $q_k(t)$, $\omega_k(t)$, $\lambda_{D'}(t)$; or we

(ii) *Introduce new quasi variables* θ', $\omega' \equiv d\theta'/dt$ by

$$\omega_{D'} \equiv \sum a_{D'k}\,\omega_k + a_{D'} \quad (= 0) \;\Rightarrow\; \delta\theta_{D'} \equiv \sum a_{D'k}\,\delta\theta_k \quad (= 0), \tag{3.8.15h}$$

$$\omega_{I'} \equiv \sum a_{I'k}\,\omega_k + a_{I'} \quad (\neq 0) \;\Rightarrow\; \delta\theta_{I'} \equiv \sum a_{I'k}\,\delta\theta_k \quad (\neq 0), \tag{3.8.15i}$$

and, inversely,

$$\omega_k \equiv \sum A_{kk'}\omega_{k'} + A_k = \sum A_{kI'}\omega_{I'} + A_k \ \Rightarrow\ \delta\theta_k \equiv \sum A_{kI'}\,\delta\theta_{I'} = 0, \quad (3.8.15j)$$

[where, as in §2.11, the $n - m$ $\omega_{I'} \equiv \cdots$ are arbitrary, except that when the system (3.8.15h, i) is solved for the ω in terms of the ω' (and time) and the results are inserted in (3.8.15a), they satisfy them identically], and then, with the help of these Maggi-like representations, apply LP in the ω' variables

$$\sum I_{k'}\,\delta\theta_{k'} = \sum \Theta_{k'}\,\delta\theta_{k'} \ \Rightarrow\ \sum I_{I'}\,\delta\theta_{I'} = \sum \Theta_{I'}\,\delta\theta_{I'}, \quad (3.8.15k)$$

where

$$I_{k'} = \sum (\partial\omega_k/\partial\omega_{k'})I_k \equiv \sum A_{kk'}I_k$$

$$\Leftrightarrow\ I_k = \sum (\partial\omega_{k'}/\partial\omega_k)I_{k'} \equiv \sum a_{k'k}I_{k'}, \quad (3.8.15l)$$

$$\Theta_{k'} = \sum A_{kk'}\Theta_k \Leftrightarrow \Theta_k = \sum a_{k'k}\Theta_{k'}; \quad (3.8.15m)$$

from which, applying the method of Lagrangean multipliers [by now, in well-understood ways; i.e., with the constraints (3.8.15a, h) written as $1 \cdot \delta\theta_{D'} = 0$, and the $\delta\theta_{I'} \neq 0$ viewed as satisfying the constraints $0 \cdot \delta\theta_{I'} = 0$], we readily obtain the following two groups of equations:

$$I_{D'} = \Theta_{D'} + \lambda_{D'} \qquad (n - m \textit{ kinetostatic} \text{ equations}), \quad (3.8.15n)$$

$$I_{I'} = \Theta_{I'} \qquad (m \textit{ kinetic} \text{ equations}); \quad (3.8.15o)$$

where the inertia terms $I_{k'}$ have one of the following basic forms:

$$I_{k'} = \sum (\partial\omega_k/\partial\omega_{k'})I_k \equiv \sum A_{kk'}I_k \qquad \text{(Maggi type)}, \quad (3.8.15p)$$

$$= (\partial T^{*\prime}/\partial\omega_{k'})^{\cdot} - \partial T^{*\prime}/\partial\theta_{k'} + \sum\sum \gamma^{b'}{}_{k'\alpha'}(\partial T^{*\prime}/\partial\omega_{b'})\omega_{\alpha'}$$

$$(\alpha' = m+1, \ldots, n; n+1) \qquad \text{(Hamel-type)}, \quad (3.8.15q)$$

$$= \partial S^{*\prime}/\partial\dot{\omega}_{k'} \equiv \sum A_{kk'}(\partial S^*/\partial\dot{\omega}_k) \qquad \text{(Appell-type)}. \quad (3.8.15r)$$

In these equations:

(i) $T^{*\prime} \equiv T^*(t, q, \omega_k \equiv \sum A_{kk'}\omega_{k'} + A_k) \equiv T^*(t, q, \omega_{D'}, \omega_{I'})$; and the constraints $\omega_{D'} = 0$ are to be enforced *after* all differentiations, not before; otherwise we could not calculate terms like $\gamma^{D'}{}_{k'\alpha'}(\partial T^{*\prime}/\partial\omega_{D'})\omega_{\alpha'}$.

(ii) The $\gamma^{D'}{}_{k'\alpha'}$ can be calculated either from the *transitivity* equations

$$d(\delta\theta_{k'}) - \delta(d\theta_{k'}) = \sum\sum \gamma^{k'}{}_{l'\alpha'}\,d\theta_{\alpha'}\,\delta\theta_{l'} = \sum\sum \gamma^{k'}{}_{l'r'}\,d\theta_{r'}\,\delta\theta_{l'} + \sum \gamma^{k'}{}_{l'}\,\delta\theta_{l'}, \quad (3.8.15s)$$

(which is, usually, the easier way), or from the *transformation* equations (ex. 2.10.1: d)

$$\gamma^{k'}{}_{l'r'} = \sum\sum\sum a_{k'k}A_{ll'}A_{rr'}\gamma^{k}{}_{lr} + \sum\sum (\partial a_{k'k}/\partial\theta_r - \partial a_{k'r}/\partial\theta_k)A_{kl'}A_{rr'}, \quad (3.8.15t)$$

where

$$\partial a_{k'k}/\partial \theta_r \equiv \sum (\partial a_{k'k}/\partial q_b)(\partial \dot{q}_b/\partial \omega_r) = \sum A_{br}(\partial a_{k'k}/\partial q_b); \tag{3.8.15u}$$

and similarly for the $\gamma^{k'}{}_{l'} \equiv \gamma^{k'}{}_{l',(n+1)'}$.

(iii) $S^{*\prime} \equiv S^*[t, q, \omega(t, q, \omega'), \dot{\omega}(t, q, \omega', \dot{\omega}')] \equiv S^{*\prime}(t, q, \omega', \dot{\omega}')$ and
$S^{*\prime}{}_o \equiv S^{*\prime}(t, q, \omega_{D'} = 0, \omega_{I'} \neq 0, \dot{\omega}_{D'} = 0, \dot{\omega}_{I'}) \equiv S^{*\prime}_o(t, q, \omega_{I'}, \dot{\omega}_{I'})$;

and in (3.8.15r) $S^{*\prime}$ can be replaced by $S^{*\prime}{}_o$ for $k' \to I = m+1, \ldots, n$, but not for $k' \to D = 1, \ldots, m$. Let the reader adapt the above to the case where the θ/ω variables are already constrained by, say, the m_1 constraints $\omega_D = 0/\delta\theta_D = 0$ $(D = 1, \ldots, m_1)$, and then are subjected to the *additional* m_2 (3.8.15a)-like constraints

$$\sum a_{D'k}\, \omega_k + a_{D'} = 0 \qquad [D' = 1, \ldots, m_2;\ n - (m_1 + m_2) > 0]. \tag{3.8.15v}$$

Problem 3.8.4 *Hadamard Form of the Hamel Equations.* Let the m Pfaffian constraints have the special Voronets form

$$\omega_D \equiv \sum B_{DI}\omega_I + B_D, \tag{a}$$

with the B_{DI} and B_D assumed known functions of all the q's and t. By viewing (a) as the following special case of the Hamel-type constraints (3.8.15a)

$$\Omega_D \equiv \omega_D - \sum B_{DI}\omega_I - B_D \quad (=0), \qquad \Omega_I \equiv \omega_I \quad (\neq 0), \tag{b, c}$$

with inverse

$$\omega_D \equiv \Omega_D + \sum B_{DI}\Omega_I + B_D \quad \left(= \sum B_{DI}\Omega_I + B_D\right), \tag{d}$$

$$\omega_I \equiv \Omega_I, \tag{e}$$

and applying any one of the above methods used in the derivation of the Hadamard equations, show that, in this case, the equations of motion (in the t, q, ω variables) may take the decoupled Hadamard form as follows:

Kinetostatic: $\quad I_D = \Theta_D + \lambda_D, \qquad$ (f)

Kinetic: $\quad I_I + \sum B_{DI} I_D = \Theta_I + \sum B_{DI}\Theta_D, \qquad$ (g)

where, with our usual notations,

$$I_k = (\partial T^*/\partial \omega_k)^{\cdot} - \partial T^*/\partial \theta_k + \sum\sum \gamma^b{}_{k\alpha}(\partial T^*/\partial \omega_b)\, \omega_\alpha, \tag{h}$$

$$= \partial S^*/\partial \dot{\omega}_k \quad (= \partial S^*{}_o/\partial \dot{\omega}_I, \quad \text{for} \quad I = m+1, \ldots, n). \tag{i}$$

For an application of these equations, see Nikitina (1976).

Problem 3.8.5 *Special Hamel Equations.* Show that if the Pfaffian constraints have the special Hamel form [but slightly more general than Voronets' form (3.8.14a)]

$$\omega_D \equiv \sum a_{Dk} v_k + a_D \quad (=0), \qquad \omega_I \equiv v_I \quad (\neq 0), \tag{a}$$

where the a_{Dk}, a_D are functions of all the q's and t; that is,

$$a_{Dk} = a_{Dk}, \quad a_{D,n+1} = a_D, \quad a_{ID} = \delta_{ID} = 0, \quad a_{II'} = \delta_{II'}, \quad a_{I,n+1} = 0,$$
$$a_{n+1,k} = 0, \quad a_{n+1,n+1} = 1, \tag{b}$$

and, therefore (recalling prob. 2.11.2),

$$(A_{DD'}) = (a_{DD'})^{-1}, \qquad (A_{DI}) = -(a_{DD'})^{-1}(a_{D'I}), \qquad A_{ID} = \delta_{ID} = 0, \qquad A_{II'} = \delta_{II'}, \tag{c}$$

then (i) the Maggi *kinetic* and *kinetostatic* equations specialize, respectively, to

$$M_I + \sum A_{DI} M_D = 0, \qquad \sum A_{D'D} M_{D'} = \lambda_D, \tag{d}$$

where $M_k \equiv E_k - Q_k \equiv [(\partial T/\partial \dot{q}_k)^{\cdot} - \partial T/\partial q_k] - Q_k$ (and for $A_{DI} = b_{DI}$, $A_{D'D} = \delta_{D'D}$, they specialize to the corresponding Hadamard equations), and (ii) the Hamel *kinetic* equations specialize to

$$(\partial T^*/\partial \omega_I)^{\cdot} - \partial T^*/\partial \theta_I + \sum\sum \gamma^D{}_{II'}(\partial T^*/\partial \omega_D)\omega_{I'} + \sum \gamma^D{}_I(\partial T^*/\partial \omega_D) = \Theta_I, \tag{e}$$

where after all differentiations we set $\omega_D = 0$ and $\omega_I \equiv \dot{q}_I$ (Schouten, 1954, pp. 196–197); and similarly for the kinetostatic equations.

HINT

Since $\theta_I \equiv q_I$ (i.e., holonomic coordinates), we will have

$$d(\delta\theta_I) - \delta(d\theta_I) = 0 \Rightarrow \gamma^I{}_{kl}, \gamma^I{}_{k,n+1} \equiv \gamma^I{}_k = 0; \tag{f}$$

and, of course, $\gamma^{n+1}{}_{kl}$, $\gamma^{n+1}{}_k = 0$; and, due to (d, e), the remaining $\gamma^D{}_{II'}$, $\gamma^D{}_I$ simplify further.

Problem 3.8.6 *Special Hamel Equations (continued).* Show that eqs. (e) of the preceding problem can be further simplified to

$$(\partial T^*{}_o/\partial \dot{q}_I)^{\cdot} - \partial T^*{}_o/\partial \theta_I + \sum\sum \gamma^D{}_{II'}(\partial T^*/\partial \omega_D)\dot{q}_{I'} + \sum \gamma^D{}_I(\partial T^*/\partial \omega_D) = \Theta_I, \tag{a}$$

where $T^*{}_o = T^*(t, q, \omega_D = 0, \omega_I = \dot{q}_I) \equiv T^*{}_o(t, q, \dot{q}_I)$; that is, the constraints have been enforced in T^* right from the start (as in the Voronets case) in the *first* and *second* terms, and *after* the differentiations in the *third* and *fourth* terms (sums); or, we replace $\partial T^*/\partial \omega_D$ with its equal:

$$\sum A_{kD}(\partial T/\partial \dot{q}_k) = \sum A_{D'D}(\partial T/\partial \dot{q}_{D'}) + \sum A_{ID}(\partial T/\partial \dot{q}_I) = \sum A_{D'D}(\partial T/\partial \dot{q}_{D'}).$$

REMARKS

(i) The $n - m$ equations (a), plus the m constraints $\sum a_{Dk}\dot{q}_k + a_D = 0$, constitute a determinate system of n equations in the n functions $q_k(t)$.

(ii) Clearly, if the constraints (a) of the preceding problem assume the Voronets form (3.8.14a) (i.e., $A_{DD'} = \delta_{DD'}$, $A_{ID} = \delta_{ID} = 0$, $\gamma^D{}_{II'} \to -w^D{}_{II'}$, etc.) then (a) must reduce to the Voronets equations (3.8.14q). [See also Hamel, 1904(a), pp. 20–21; Prange, 1935, pp. 537–539; Schouten, 1954, pp. 196–197.]

Problem 3.8.7 *Special Hamel Equations (continued).* Write down the *kineto-static* Hamel equations of the preceding problem.

Example 3.8.1 *A Mixed Hamel–Voronets Type of Equations* [May be omitted in a first reading. Adapted from Neimark and Fufaev (1972, pp. 141–143)]. Let us consider a system subject to the Voronets-like Pfaffian constraints (with $\dot{q}_k \equiv v_k$):

$$v_D \equiv \sum b_{DI}\, v_I + b_D \qquad [D = 1, \dots, m\ (< n);\ I = m + 1, \dots, n]. \tag{a}$$

To handle these constraints, we make the following quasi-velocity choice: (i) We express the *first* M $(\leq m)$ of the m v_D's à la Hamel:

$$\omega_d \equiv v_d - \sum b_{dI}\, v_I - b_d \equiv \sum a_{dI}\, v_I + a_I \quad (= 0), \tag{b1}$$

and also

$$\omega_I \equiv \sum a_{II'}\, v_{I'} + a_I \qquad (\neq 0), \tag{b2}$$

where $d = 1, \dots, M$; $I, I' = m + 1, \dots, n$ [and the *new* coefficients $a_{II'}$, a_I are such that upon inversion of (b2) and substitution of the v's in (a), the latter is satisfied identically]; and (ii) express the remaining $m - M$ of the v_D's à la Voronets:

$$v_\delta \equiv \sum b_{\delta I}\, v_I + b_\delta \qquad [\delta = M + 1, \dots, m\ (< n)]. \tag{c}$$

The range of these indices, so important to the understanding of this example, is shown below for quick reference:

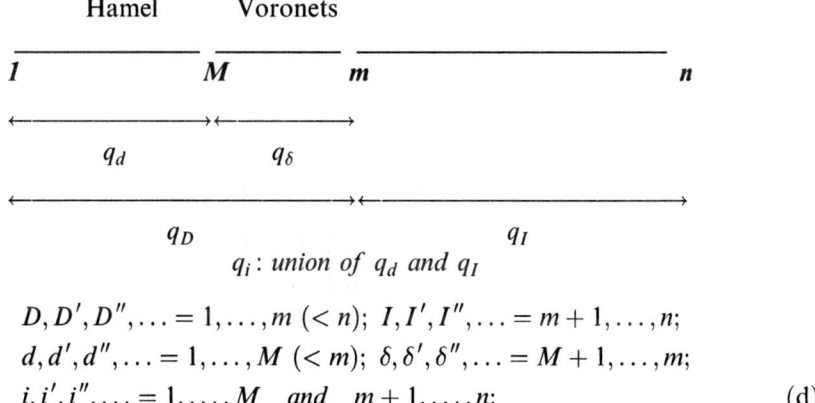

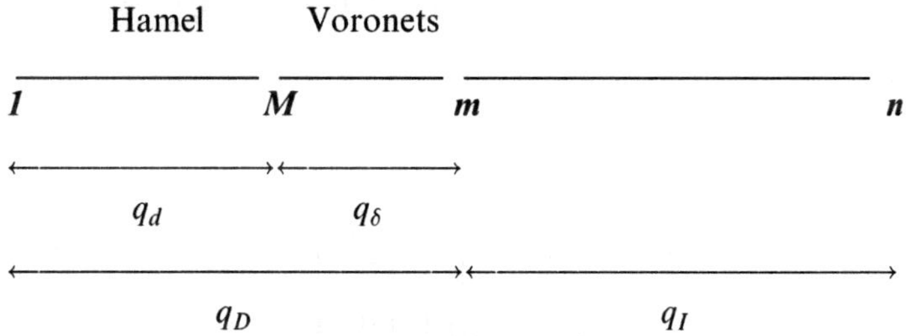

q_i: *union of* q_d *and* q_I

$$\begin{aligned} &D, D', D'', \dots = 1, \dots, m\ (< n);\ I, I', I'', \dots = m + 1, \dots, n; \\ &d, d', d'', \dots = 1, \dots, M\ (< m);\ \delta, \delta', \delta'', \dots = M + 1, \dots, m; \\ &i, i', i'', \dots = 1, \dots, M \quad \textit{and} \quad m + 1, \dots, n; \end{aligned} \tag{d}$$

and, these indices may be used to characterize either a *particular* v or ω of a group, or the *entire* group. Hence, they can be rewritten compactly as $\omega_i \equiv \sum a_{ii'} v_{i'} + a_i$.

Now, let us establish the *transitivity equations*. Recalling the results of prob. 2.12.5, we have the following:

(i) For the Hamel group, eqs. (b):

$$\textit{Hamel viewpoint: } (\delta\theta_i)^{\cdot} - \delta\omega_i = \sum\sum \gamma^i{}_{i'i''}\, \omega_{i''}\, \delta\theta_{i'} + \sum \gamma^i{}_{i'}\, \delta\theta_{i'}, \tag{e}$$

where

$$\sum A_{ii'}a_{i'i''} = \delta_{ii''} \Rightarrow \gamma^{i}_{i'i''} \equiv \sum\sum (\partial a_{ii'''}/\partial q_{i''''} - \partial a_{ii''''}/\partial q_{i'''})A_{i'''i'}A_{i''''i''}, \quad \text{(e1)}$$

and analogously for $\gamma^{i}_{i'} \equiv \gamma^{i}_{i',n+1}$; that is, the $\gamma^{i}_{..}$'s are based on eqs. (b), and $d(\delta q_i) = \delta(dq_i)$.

(ii) For the Voronets group, eq. (c):

Suslov viewpoint: $(\delta q_I)^{\cdot} - \delta(\dot{q}_I) = 0,$

$$(\delta q_\delta)^{\cdot} - \delta(\dot{q}_\delta) = \sum\sum w^{\delta}_{II'}\dot{q}_{I'}\,\delta q_I + \sum w^{\delta}_{I}\,\delta q_I \quad \text{(f)}$$

[i.e., here, too, we may assume $d(\delta q_d) = \delta(dq_d) \Rightarrow d(\delta q_i) = \delta(dq_i)$].

In addition, to implement the central equation (CE), and thus derive the equations of motion:

(i) We invert the virtual form of eqs. (b), thus obtaining [no enforcement of constraints (b1) yet]

$$\delta q_i = \sum A_{ii'}\,\delta\theta_{i'} \quad \left[= \sum A_{iI}\,\delta\theta_I;\ \text{since, by (b1)},\ \delta\theta_d = 0\right], \quad \text{(g)}$$

where $\dot{\theta}_i \equiv \omega_i$.

(ii) From the virtual form of (c), and then use of (g) for $i \to I$, we get

$$\delta q_\delta = \sum b_{\delta I}\,\delta q_I = \sum\left(\sum b_{\delta I}A_{Ii}\right)\delta\theta_i \equiv \sum F_{\delta i}\,\delta\theta_i \quad \left[= \sum F_{\delta I}\,\delta\theta_I\right]. \quad \text{(h)}$$

Equations (g, h) express the n δq's in terms of the $M + (n - m)$ $\delta\theta_i$'s [$\Rightarrow (n - m)$ $\delta\theta_I$'s] introduced by the virtual form of (b).

(iii) Finally, we employ the notation

$$T = T(t, q, \dot{q}_i, \dot{q}_\delta) = \cdots = T^*(t, q, \omega_i) = T^* \quad \text{(no constraint enforcement yet)}, \quad \text{(i)}$$

meaning that T^* is what becomes of T after expressing all its $\dot{q}$'s in terms of the ω_i's [*velocity* forms of (g, h)]; that is, after substituting into it

$$\dot{q}_i = \sum A_{ii'}\omega_{i'} + A_i \quad [\text{obtained after solving (b) for the } \dot{q}_i \equiv v_i],$$

and

$$\dot{q}_\delta = \sum b_{\delta I}\dot{q}_I + b_\delta = \sum b_{\delta I}\left(\sum A_{Ii}\,\omega_i + A_I\right) + b_\delta \equiv \sum F_{\delta i}\,\omega_i + F_\delta. \quad \text{(j)}$$

Next, to obtain the equations of motion, we will utilize the central equation (3.6.8 ff.)

$$\left(\sum p_k\,\delta q_k\right)^{\cdot} - \delta T - \sum p_k[(\delta q_k)^{\cdot} - \delta(\dot{q}_k)] = \sum Q_k\,\delta q_k \quad (\equiv \delta' W). \quad \text{(k)}$$

Indeed, substituting into it the two expressions (f) for $(\delta q)^{\cdot} - \delta(\dot{q})$, we obtain

$$\left(\sum p_k\,\delta q_k\right)^{\cdot} - \delta T - \sum p_\delta[(\delta q_\delta)^{\cdot} - \delta(\dot{q}_\delta)]$$

$$= \left(\sum p_k\,\delta q_k\right)^{\cdot} - \delta T - \sum p_\delta\left(\sum\sum w^{\delta}_{II'}\,v_{I'}\,\delta q_I + \sum w^{\delta}_{I}\,\delta q_I\right) = \delta' W. \quad \text{(l)}$$

Next, due to (i, j) (and this is the important step here)

$$\delta P \equiv S\, dm\, \boldsymbol{v} \cdot \delta \boldsymbol{r} = \sum p_k\, \delta q_k = \sum P_k\, \delta\theta_k = \sum P_i\, \delta\theta_i, \quad P_k \equiv \partial T^*/\partial\omega_k,$$

$$\delta T = \delta T^* = \sum [(\partial T^*/\partial\omega_i)\, \delta\omega_i + (\partial T^*/\partial\theta_i)\, \delta\theta_i], \quad \text{and} \quad \delta' W = \sum \Theta_i\, \delta\theta_i, \qquad \text{(m)}$$

and so, with the help of (g), the variational equation (l) can be rewritten as

$$\Big(\sum P_i\, \delta\theta_i\Big)^{\cdot} - 1 - \delta T^* - \sum\sum\sum\sum w^{\delta}{}_{II'}\, p_\delta\, v_{I'}(A_{Ii}\, \delta\theta_i) - \sum\sum\sum w^{\delta}{}_{I}\, p_\delta(A_{Ii}\, \delta\theta_i)$$
$$= \sum \Theta_i\, \delta\theta_i. \qquad \text{(n)}$$

Finally, transforming the *first two* terms of the above via (m) [or (3.6.7a ff., with $k \to i$)], then applying the transitivity equations (e), and, finally, factoring out the common $\delta\theta_i$, we get

$$\sum (I_i - \Theta_i)\, \delta\theta_i = 0, \qquad \text{(o)}$$

where

$$I_i \equiv (\partial T^*/\partial\omega_i)^{\cdot} - \partial T^*/\partial\theta_i$$
$$+ \Big(\sum\sum \gamma^{i'}{}_{ii''}(\partial T^*/\partial\omega_{i'})\omega_{i''} + \sum \gamma^{i'}{}_{i}(\partial T^*/\partial\omega_{i'})\Big)$$
$$- \Big(\sum\sum\sum w^{\delta}{}_{II'}(\partial T/\partial v_\delta) v_{I'} A_{Ii} + \sum\sum w^{\delta}{}_{I}(\partial T/\partial v_\delta) A_{Ii}\Big), \qquad \text{(p)}$$

and

$$\partial T^*/\partial\theta_i \equiv \sum (\partial T^*/\partial q_k)(\partial v_k/\partial\omega_i)$$
$$= \sum (\partial T^*/\partial q_{i'})(\partial v_{i'}/\partial\omega_i) + \sum (\partial T^*/\partial q_\delta)(\partial v_\delta/\partial\omega_i) \qquad \text{[invoking (j)]}$$
$$= \sum (\partial T^*/\partial q_{i'}) A_{i'i} + \sum\sum (\partial T^*/\partial q_\delta)(b_{\delta I} A_{Ii}). \qquad \text{(q)}$$

From the above we obtain, in by now well-understood ways (i.e., $\delta\theta_d = 0 \Rightarrow$ *multipliers*, Λ_d), the two uncoupled groups of "*mixed*" (or "*intermediate*") *Hamel–Voronets equations*:

$$I_d = \Theta_d + \Lambda_d \ (\textit{kinetostatic} \text{ equations}), \qquad I_I = \Theta_I \ (\textit{kinetic} \text{ equations}). \qquad \text{(r1, 2)}$$

In particular, the *kinetic* equations (r2) are, in extenso,

$$(\partial T^*/\partial\omega_I)^{\cdot} - \partial T^*/\partial\theta_I + \Big(\sum\sum \gamma^{i}{}_{Ii'}(\partial T^*/\partial\omega_i)\omega_{i'} + \sum \gamma^{i}{}_{I}(\partial T^*/\partial\omega_i)\Big)$$
$$- \Big(\sum\sum\sum w^{\delta}{}_{I'I''} A_{I'I}(\partial T/\partial v_\delta) v_{I''} + \sum\sum w^{\delta}{}_{I'} A_{I'I}(\partial T/\partial v_\delta)\Big) = \Theta_I. \qquad \text{(s)}$$

These are, indeed, mixed Hamel–Voronets equations, because:

(i) If $M = m$ (i.e., $\delta = 0$), the Voronets terms (*fourth* group of terms: $-[\ldots]$) disappear and (s) reduce to the kinetic Hamel equations; whereas

(ii) If $M = 0$ (i.e., $d = 0$, $\delta = D$, and $i = I$), and we restrict ourselves to *holonomic* coordinates and the Voronets constraints (a) = (c), then $A_{II'} = \delta_{II'}$ [= Kronecker delta in (g)], all the Hamel symbols $\gamma^{i}{}_{Ii'}$, $\gamma^{i}{}_{I}$ vanish, T^* becomes $T_o(t, q, v_I)$, and (from $\delta' W = \sum \Theta_I\, \delta\theta_I = \sum Q_{Io}\, \delta q_I$) $\Theta_I \to Q_{Io}$; and, as a result, (s) reduce to the familiar Voronets equations (3.8.14q):

$$(\partial T_o/\partial v_I)^{\cdot} - \partial T_o/\partial q_I - \sum (\partial T_o/\partial q_D) b_{DI}$$
$$- \left(\sum\sum w^D{}_{II'}(\partial T/\partial v_D) v_{I'} + \sum w^D{}_I(\partial T/\partial v_D)\right) = \Theta_I; \qquad \text{(t)}$$

and for Chaplygin systems, they reduce to the Chaplygin equations (3.8.13o).

Finally, if $M = 0$ and we choose *nonholonomic* coordinates, (s) lead to the Voronets and Chaplygin equations in quasi variables. The details are left to the reader [recall probs. 3.8.2 and 3.8.3; see also Fradlin (1961)].

Example 3.8.2 *Transformation of the Correction Terms* $-\Gamma_k$ *that Appear in the Hamel Equations.* From the invariance of LP, under local quasi-coordinate differential/velocity transformations

$$\delta\theta_{k'} = \sum (\partial\omega_{k'}/\partial\omega_k)\, \delta\theta_k \equiv \sum a_{k'k}\, \delta\theta_k \;\Leftrightarrow\; \delta\theta_k = \sum (\partial\omega_k/\partial\omega_{k'})\, \delta\theta_{k'} \equiv A_{kk'}\, \delta\theta_{k'},$$

that is,

$$\sum I_k\, \delta\theta_k = \sum \Theta_k\, \delta\theta_k \;\Leftrightarrow\; \sum I_{k'}\, \delta\theta_{k'} = \sum \Theta_{k'}\, \delta\theta_{k'}, \qquad \text{(a1)}$$

we readily conclude that the I_k and Θ_k transform as (covariant) vectors, that is,

$$I_{k'} = \sum A_{kk'} I_k \;\Leftrightarrow\; I_k = \sum a_{k'k} I_{k'}, \quad \text{etc.} \qquad \text{(a2)}$$

However, this does *not* imply that the constituents of $I_k \equiv E_k{}^*(T^*) - \Gamma_k \equiv E_k{}^* - \Gamma_k$, taken individually, transform as such vectors. In fact, as shown below, *neither* $E_k{}^*$ (= nonholonomic Euler–Lagrange part of inertia) nor $-\Gamma_k$ (or Γ_k = nonholonomic deviation/correction) transform vectorially; that is, à la (a2); although their *combination* $E_k{}^* - \Gamma_k$ does!

We begin by examining the transformation properties of the $-\Gamma_k$'s under $\omega_{k'} = \sum a_{k'k}\, \omega_k$, $\omega_k = \sum A_{kk'}\, \omega_{k'}$ (assumed *stationary* for algebraic simplicity, but no loss in generality); that is, relate

$$-\Gamma_{l'} \equiv \sum\sum \gamma^{k'}{}_{l'r'}(\partial T^{*\prime}/\partial\omega_{k'})\omega_{r'} \equiv -\Gamma_{l'}(t, q, \omega'), \qquad \text{(b)}$$

and $-\Gamma_l = -\Gamma_l(t, q, \omega)$. Below, we present two such derivations; one in *system variables* and one in terms of *particle vectors*.

(i) *System variable derivation.* Substituting the γ-transformation equations (3.8.15t)

$$\gamma^{k'}{}_{l'r'} = \sum\sum\sum a_{k'k} A_{ll'} A_{rr'} \gamma^k{}_{lr} + \sum\sum (\partial a_{k'k}/\partial\theta_r - \partial a_{k'r}/\partial\theta_k) A_{kl'} A_{rr'},$$

where

$$\partial a_{k'k}/\partial\theta_r \equiv \sum (\partial a_{k'k}/\partial q_b)(\partial \dot q_b/\partial\omega_r) \equiv \sum A_{br}(\partial a_{k'k}/\partial q_b);$$

into (b), we find, successively,

$$-\Gamma_{l'} \equiv \sum\sum\Big\{\Big(\sum\sum\sum a_{k'k}A_{ll'}A_{rr'}\gamma^{k}{}_{lr} + \cdots\Big)\Big(\sum A_{sk'}(\partial T^*/\partial\omega_s)\Big)\Big(\sum a_{r'b}\,\omega_b\Big)\Big\}$$

$$= \cdots = \sum\sum\sum A_{ll'}\big(\gamma^{s}{}_{lb}(\partial T^*/\partial\omega_s)\omega_b\big)$$

$$+ \sum\sum\sum\sum\sum [A_{ll'}A_{kk'}(\partial a_{k'l}/\partial\theta_r - \partial a_{k'r}/\partial\theta_l)\,\omega_r](\partial T^*/\partial\omega_k)$$

$$= \sum A_{ll'}(-\Gamma_l) + \sum\sum\sum\sum\sum [A_{ll'}A_{kk'}(\partial a_{k'l}/\partial\theta_r - \partial a_{k'r}/\partial\theta_l)\,\omega_r](\partial T^*/\partial\omega_k), \tag{c}$$

which is the sought *nonvectorial* transformation equation.

(ii) *Particle variable derivation.* We begin with the definition of $\Gamma_{k'}$ in terms of the transformed particle variables (3.3.12a):

$$\Gamma_{l'} \equiv S\, dm\, \mathbf{v}^{*\prime}\cdot[(\partial\mathbf{v}^{*\prime}/\partial\omega_{l'})^{\cdot} - \partial\mathbf{v}^{*\prime}/\partial\theta_{l'}]$$

$$= S\, dm\, \mathbf{v}^{*\prime}\cdot(d\boldsymbol{\varepsilon}_{l'}/dt - \partial\mathbf{v}^{*\prime}/\partial\theta_{l'}); \tag{d}$$

where (again, assuming, for algebraic simplicity, stationary constraints and a *stationary* $\omega \Leftrightarrow \omega'$ transformation)

$$\mathbf{v} = \mathbf{v}^* \equiv \sum \omega_l\,\boldsymbol{\varepsilon}_l = \sum(\partial\mathbf{v}^*/\partial\omega_l)\omega_l = \mathbf{v}^{*\prime} \equiv \sum \omega_{l'}\,\boldsymbol{\varepsilon}_{l'} = \sum(\partial\mathbf{v}^{*\prime}/\partial\omega_{l'})\,\omega_{l'}. \tag{e}$$

From the representations (e), we readily deduce the basic $\boldsymbol{\varepsilon} \leftrightarrow \boldsymbol{\varepsilon}'$ transformation equations

$$\boldsymbol{\varepsilon}_{l'} = \sum(\partial\omega_l/\partial\omega_{l'})\boldsymbol{\varepsilon}_l \;\Leftrightarrow\; \boldsymbol{\varepsilon}_l = \sum(\partial\omega_{l'}/\partial\omega_l)\boldsymbol{\varepsilon}_{l'} \qquad \text{[like (a2)]}. \tag{f}$$

From the above, and with an eye toward (d), we obtain, successively,

(a) $$d\boldsymbol{\varepsilon}_{l'}/dt = \sum[(\partial\omega_l/\partial\omega_{l'})^{\cdot}\boldsymbol{\varepsilon}_l + (\partial\omega_l/\partial\omega_{l'})(d\boldsymbol{\varepsilon}_l/dt)]; \tag{g}$$

(b) $$\partial\mathbf{v}^{*\prime}/\partial\theta_{l'} \equiv \sum(\partial\mathbf{v}^{*\prime}/\partial q_l)(\partial\dot{q}_l/\partial\omega_{l'}) \qquad \text{(by definition)}$$

$$= \sum\Big((\partial\mathbf{v}^*/\partial q_l) + \sum(\partial\mathbf{v}^*/\partial\omega_s)(\partial\omega_s/\partial q_l)\Big)(\partial\dot{q}_l/\partial\omega_{l'}) \qquad \text{[by chain rule on (e)]}$$

$$= \sum(\partial\mathbf{v}^*/\partial q_l)(\partial\dot{q}_l/\partial\omega_{l'}) + \sum(\partial\mathbf{v}^*/\partial\omega_s)\Big(\sum(\partial\omega_s/\partial q_l)(\partial\dot{q}_l/\partial\omega_{l'})\Big)$$

$$= \sum\Big(\sum(\partial\mathbf{v}^*/\partial\theta_s)(\partial\omega_s/\partial\dot{q}_l)\Big)(\partial\dot{q}_l/\partial\omega_{l'}) + \sum(\partial\mathbf{v}^*/\partial\omega_s)(\partial\omega_s/\partial\theta_{l'})$$

$$= \sum(\partial\mathbf{v}^*/\partial\theta_l)(\partial\omega_l/\partial\omega_{l'}) + \sum(\partial\omega_l/\partial\theta_{l'})\boldsymbol{\varepsilon}_l. \tag{h}$$

Inserting the expressions (g, h) into (d), and regrouping, we find

$$\Gamma_{l'} = \cdots = S\, dm\, \mathbf{v}^{*\prime}\cdot\Big(\sum[(\partial\omega_l/\partial\omega_{l'})^{\cdot} - \partial\omega_l/\partial\theta_{l'}]\boldsymbol{\varepsilon}_l\Big)$$

$$+ \sum\Big(S\, dm\, \mathbf{v}^{*\prime}\cdot(d\boldsymbol{\varepsilon}_l/dt - \partial\mathbf{v}^*/\partial\theta_l)\Big)(\partial\omega_l/\partial\omega_{l'})$$

$$= \sum\Big\{S\, dm\, \mathbf{v}^{*\prime}\cdot[\partial\mathbf{v}^*/\partial\omega_l - \partial\mathbf{v}^*/\partial\theta_l]\Big\}(\partial\omega_l/\partial\omega_{l'})$$

$$+ \sum[(\partial\omega_l/\partial\omega_{l'})^{\cdot} - \partial\omega_l/\partial\theta_{l'}]\Big(S\, dm\, \mathbf{v}^{*\prime}\cdot\boldsymbol{\varepsilon}_l\Big),$$

or, finally (and recalling that $\boldsymbol{\nu}^{*\prime} = \boldsymbol{\nu}^* = \boldsymbol{\nu}$),

$$\Gamma_{l'} = \sum (\partial\omega_l/\partial\omega_{l'})\Gamma_l + \sum \left[(\partial\omega_l/\partial\omega_{l'})^{\cdot} - \partial\omega_l/\partial\theta_{l'}\right](\partial T^*/\partial\omega_l)', \tag{i}$$

where

$$(\partial T^*/\partial\omega_l)' \equiv (\partial T^*/\partial\omega_l)\big|_{\text{after the differentiations we insert } \omega=\omega(t,q,\omega')} = \textit{function of } t, q, \omega'.$$

We leave it to the reader to show that (i) [which, actually, holds for a general (*nonlinear* and nonstationary) transformation $\omega = \omega(t, q, \omega')$ (see chap. 5)], in our linear and homogeneous case, coincides with (c).

Now, from the transformation law (a2), for the $I_k \equiv E_k{}^* - \Gamma_k$, and (i) for the Γ_l [or following steps entirely similar to those taken in obtaining (g, h), but for $T^* = T^{*\prime}$], it is not hard to see that the nonholonomic Euler–Lagrange terms $E_k{}^* = I_k + \Gamma_k$ must transform as follows:

$$\begin{aligned} E_{k'}{}^* &\equiv (\partial T^{*\prime}/\partial\omega_{k'})^{\cdot} - \partial T^{*\prime}/\partial\theta_{k'} \quad (= I_{k'} + \Gamma_{k'}) \\ &= \sum (\partial\omega_k/\partial\omega_{k'})E_k{}^* + \sum [(\partial\omega_k/\partial\omega_{k'})^{\cdot} - \partial\omega_k/\partial\theta_{k'}](\partial T^*/\partial\omega_k)'; \end{aligned} \tag{j}$$

indeed, subtracting (i) from (j), side by side, yields (a2).

Clearly: (i) if $(\partial\omega_k/\partial\omega_{k'})^{\cdot} - \partial\omega_k/\partial\theta_{k'} = 0$ (in which case, ω and ω' are referred to as "relatively holonomic"), both Γ_k and $E_k{}^*$ transform as vectors; and (ii) if Γ_k, $\Gamma_{k'}, \ldots = 0$ (i.e., holonomic coordinates), then $E_k{}^*$ ($\rightarrow E_k = \textit{holonomic inertia}$) transforms as a vector.

Problem 3.8.8 Show that (assuming a linear and stationary $\dot{q} \leftrightarrow \omega$ relationship)

$$\begin{aligned} -\Gamma_k &= \sum \left[(\partial\dot{q}_l/\partial\omega_k)^{\cdot} - \partial\dot{q}_l/\partial\theta_k\right](\partial T/\partial\dot{q}_l)^* \\ &= \sum\sum \gamma^b{}_{ks}\,\omega_s\Big(\sum (\partial\dot{q}_l/\partial\omega_b)(\partial T/\partial\dot{q}_l)^*\Big) = \sum\sum \gamma^b{}_{ks}(\partial T^*/\partial\omega_b)\,\omega_s; \end{aligned} \tag{a}$$

[for *rheonomic* systems s runs from 1 to $n+1$] and, therefore, also

$$-\Gamma_{k'} = \sum \left[(\partial\dot{q}_l/\partial\omega_{k'})^{\cdot} - \partial\dot{q}_l/\partial\theta_{k'}\right](\partial T/\partial\dot{q}_l)' = \cdots = \sum\sum \gamma^{b'}{}_{k's'}(\partial T^{*\prime}/\partial\omega_{b'})\,\omega_{s'}, \tag{b}$$

where

$$(\partial T/\partial\dot{q}_l)' \equiv (\partial T/\partial\dot{q}_l)\big|_{\text{after the differentiations we insert } \dot{q}=\dot{q}(t,q,\omega')} = \textit{function of } t, q, \omega'.$$

Below, we summarize the most prevalent quasi-velocity/constraint choices in equations of motion:

Hamel (general):	$\omega_D \equiv \sum a_{Dk}\dot{q}_k + a_D = 0,$	ω_I: arbitrary $\neq 0$;
Hamel (special):	$\omega_D \equiv \sum a_{Dk}\dot{q}_k + a_D = 0,$	$\omega_I \equiv \dot{q}_I \neq 0$;
Voronets (special):	$\omega_D \equiv \dot{q}_D - \sum b_{DI}\dot{q}_I - b_D = 0,$	$\omega_I \equiv \dot{q}_I \neq 0$;
Voronets (general):	$\omega_D \equiv \dot{q}_D - \sum b_{DI}\dot{q}_I - b_D = 0,$	ω_I: arbitrary $\neq 0$;
Maggi:	$\dot{q}_k = \sum A_{kI}\omega_I + A_k \neq 0,$	$\omega_D = 0, \quad \omega_I$: arbitrary $\neq 0$.

3.9 KINETIC AND POTENTIAL ENERGIES; ENERGY RATE, OR POWER, THEOREMS

The foregoing theory has shown the importance of kinetic energy, $T \equiv S\,(dm\,\boldsymbol{v}\cdot\boldsymbol{v})/2$, to analytical mechanics and its equations of motion. Let us, therefore, examine in some detail the following topics.

3.9.1 Kinetic Energy in Holonomic (System) Variables

Substituting into it the particle (inertial) velocity representation in holonomic variables (§2.5)

$$\boldsymbol{v} = \sum \boldsymbol{e}_k \dot{q}_k + \boldsymbol{e}_0 \qquad (\boldsymbol{e}_k \equiv \partial\boldsymbol{r}/\partial q_k,\ \ \boldsymbol{e}_0 \equiv \boldsymbol{e}_{n+1} \equiv \partial\boldsymbol{r}/\partial q_{n+1} \equiv \partial\boldsymbol{r}/\partial t), \tag{3.9.1}$$

and grouping terms, we obtain the following basic kinetic energy representation:

$$T = T(t, q, \dot{q}) = T_2 + T_1 + T_0, \tag{3.9.2}$$

where

$$T_2 = T_2(t, q, \dot{q}) \equiv (1/2)\sum\sum M_{kl}\dot{q}_k\dot{q}_l \text{ (quadratic and homogeneous in the } \dot{q}\text{s)},$$

$$M_{kl} = M_{kl}(t, q) \equiv S\,dm\,\boldsymbol{e}_k\cdot\boldsymbol{e}_l \equiv S\,dm\,(\partial\boldsymbol{r}/\partial q_k)\cdot(\partial\boldsymbol{r}/\partial q_l) \quad (= M_{lk}), \tag{3.9.2a}$$

$$T_1 = T_1(t, q, \dot{q}) \equiv \sum M_k\,\dot{q}_k \qquad (\textit{linear} \text{ and homogeneous in the } \dot{q}\text{s}),$$

$$M_k = M_k(t, q) \equiv S\,dm\,\boldsymbol{e}_k\cdot\boldsymbol{e}_0 \equiv S\,dm(\partial\boldsymbol{r}/\partial q_k)\cdot(\partial\boldsymbol{r}/\partial t)$$

$$(= M_{k0} = M_{0k} \equiv M_{k,n+1} = M_{n+1,k}), \tag{3.9.2b}$$

$$T_0 = T_0(t, q) \equiv M_0/2 \quad (\geq 0, \textit{zeroth} \text{ degree in the } \dot{q}\text{'s}),$$

$$M_0 = M_0(t, q) \equiv S\,dm\,\boldsymbol{e}_0\cdot\boldsymbol{e}_0 \equiv S\,dm(\partial\boldsymbol{r}/\partial t)\cdot(\partial\boldsymbol{r}/\partial t)$$

$$(= M_{00} \equiv M_{n+1,n+1} \geq 0), \tag{3.9.2c}$$

that is,

$$\begin{aligned} 2T = {} & M_{11}\dot{q}_1{}^2 + 2M_{12}\,\dot{q}_1\dot{q}_2 + M_{22}\,\dot{q}_2{}^2 + \cdots + M_{nn}\,\dot{q}_n{}^2 \\ & + 2M_1\dot{q}_1 + \cdots + 2M_n\,\dot{q}_n \\ & + M_0; \end{aligned}$$

or, compactly (with $\alpha, \beta = 1, \ldots, n, n+1$; and noting that $\dot{q}_{n+1} = \dot{t} = 1$),

$$2T = \sum\sum M_{\alpha\beta}\,\dot{q}_\alpha\dot{q}_\beta, \qquad M_{\alpha\beta} = M_{\beta\alpha} \equiv S\,dm\,(\partial\boldsymbol{r}/\partial q_\alpha)\cdot(\partial\boldsymbol{r}/\partial q_\beta). \tag{3.9.2d}$$

Clearly, the *holonomic coefficients of inertia* $M_{\alpha\beta} = M_{\alpha\beta}(t, q)$ vary with the system configuration and time, and, of course, the particular q-representation (in some q's, they might even be constant, like the particles' masses).

Some Analytical Considerations

(i) If $\boldsymbol{r} = \boldsymbol{r}(q)$ (i.e., *stationary* holonomic constraints in the q's) $\Rightarrow \boldsymbol{e}_0 \equiv \partial\boldsymbol{r}/\partial t = \boldsymbol{0} \Rightarrow$ *all* the M_k and M_0 vanish, and

$$2T \to 2T_2 = \sum\sum M_{kl}\dot{q}_k\dot{q}_l \quad [= 2(\text{kinetic energy for "frozen constraints"})]; \tag{3.9.2e}$$

and the vanishing of the $\boldsymbol{v}$'s implies that of the $\dot{q}$'s and of T. Also, since, in general,

$$2T_2 = S\, dm(\boldsymbol{v} - \boldsymbol{e}_0)^2 = S\, dm\Big(\sum \boldsymbol{e}_k\dot{q}_k\Big)^2,$$

T_2 represents the kinetic energy of the "virtual velocities" $\boldsymbol{v} - \boldsymbol{e}_0$ (Lur'e, 1968, pp. 10, 135).

(ii) In the general *nonstationary* case ($\partial\boldsymbol{r}/\partial t \neq \boldsymbol{0}$), it can be shown that, as long as the $3N$ coordinates of the system's particles $\boldsymbol{\xi} \equiv \{\xi_*; * = 1, \ldots, 3N\}$ can be expressed in terms of n minimal positional coordinates $q \equiv \{q_k; k = 1, \ldots, n\}$—that is, as long as $rank(\partial\xi_*/\partial q_k) = n$ [recall (2.4.8 ff.)], except possibly at a number of individual singular points where $rank(\partial\xi_*/\partial q_k) < n$ (a situation we shall exclude)—T_2 is *always nondegenerate*, or *nonsingular*; that is, $Det(M_{kl}) \neq 0$; and since $T_2 \geq 0$ it follows that

$$T_2 = \textit{positive definite} \text{ in the } \dot{q}\text{'s} \quad (\text{i.e., } \textit{always nonnegative}, \text{ and } \textit{zero only if all } \dot{q}\text{'s} = 0). \tag{3.9.2f}$$

As is well known, the necessary and sufficient conditions for this are

$$|M_{11}| = M_{11} > 0, \qquad \begin{vmatrix} M_{11} & M_{12} \\ M_{21} & M_{22} \end{vmatrix} > 0, \qquad \begin{vmatrix} M_{11} & \cdots & M_{1n} \\ \vdots & \cdots & \vdots \\ M_{n1} & \cdots & M_{nn} \end{vmatrix} > 0, \tag{3.9.2g}$$

for all q's and t in their domain of definition. The last of the above n inequalities states that the *inertia matrix* $\mathbf{M} = (M_{kl})$ *is nonsingular*; while the one before it states the same for the corresponding matrix of the new system obtained from the given by adding to it the constraint $q_n = \textit{constant}$. [For proofs, see, for example, Gantmacher (1970, pp. 46–47), Lamb (1929, pp. 182–183), Langhaar (1962, pp. 308–313).]

As for the *total* kinetic energy T, it is clear from its definition that it is always *nonnegative*, but it vanishes for a single set of (not necessarily zero) values of the $\dot{q}$'s.

(iii) The terms T_1 and T_0 are called, respectively, *gyroscopic* and *centrifugal* parts of T (§3.16, §8.3 ff.). Clearly, $T_0 \geq 0$; and also $|T_1| < T_2 + T_0$, otherwise we might have $T < 0$.

(iv) If $T_1 = T_0 = 0$, the system is also called *natural*, eq. (3.9.2e).

(v) We notice that T can always be represented as

$$T = T'_2 + T'_0, \tag{3.9.2h}$$

where $2T'_2 \equiv \sum\sum M_{kl}(\dot{q}_k - x_k)(\dot{q}_l - x_l) = \textit{positive definite in the } \dot{q}_k - x_k$, the $x_k = x_k(t)$ have units of Lagrangean velocities; that is, $\dot{q}_k$, and $T'_0 = T'_0(t, q)$. Clearly, if $T'_0 > 0$, then $T > 0$.

Some Useful Identities

Invoking the *homogeneous function theorem* of Euler [according to which, if $f = f(x_1, \ldots, x_n) = \textit{homogeneous of degree } H \textit{ in its variables } x_1, \ldots, x_n$, then

$$\sum (\partial f/\partial x_k)x_k = H \cdot f],$$

we obtain the following useful kinematico-inertial *T-identities*:

(a) $$\sum (\partial T/\partial \dot{q}_k)\dot{q}_k = \sum [\partial(T_2 + T_1 + T_0)/\partial \dot{q}_k]\dot{q}_k$$
$$= (2)T_2 + (1)T_1 + (0)T_0 = 2T_2 + T_1 = T + (T_2 - T_0)$$
$$[= 2T, \text{ if } \partial \boldsymbol{r}/\partial t = \mathbf{0}]; \tag{3.9.3a}$$

(b) $$\sum E_k(T)\dot{q}_k \equiv \sum [(\partial T/\partial \dot{q}_k)^{\cdot} - \partial T/\partial q_k]\dot{q}_k$$
$$= \left(\sum (\partial T/\partial \dot{q}_k)\dot{q}_k\right)^{\cdot} - \sum [(\partial T/\partial \dot{q}_k)\ddot{q}_k + (\partial T/\partial q_k)\dot{q}_k]$$
$$= [T + (T_2 - T_0)]^{\cdot} - (dT/dt - \partial T/\partial t),$$

or, finally, a form that will prove useful in the energy rate theorem below [with $E_k(T) \equiv E_k$],

$$\sum E_k \dot{q}_k = \left(\sum (\partial T/\partial \dot{q}_k)\dot{q}_k - T\right)^{\cdot} + \partial T/\partial t = (T_2 - T_0)^{\cdot} + \partial T/\partial t. \tag{3.9.3b}$$

3.9.2 Kinetic Energy in Nonholonomic (System) Variables

Substituting into T the particle (inertial) velocity representation in nonholonomic variables (2.9.23 ff.)

$$\boldsymbol{v} \to \boldsymbol{v}^* = \sum \boldsymbol{\varepsilon}_k\, \omega_k + \boldsymbol{\varepsilon}_0, \tag{3.9.4}$$

where

$$\boldsymbol{\varepsilon}_k \equiv \partial \boldsymbol{r}/\partial \theta_k,$$
$$\boldsymbol{\varepsilon}_0 \equiv \boldsymbol{\varepsilon}_{n+1} \equiv \partial \boldsymbol{r}/\partial \theta_{n+1} \equiv \sum (\partial \boldsymbol{r}/\partial q_\alpha)(\partial \dot{q}_\alpha/\partial \omega_{n+1})$$
$$= \partial \boldsymbol{r}/\partial t + \sum A_k(\partial \boldsymbol{r}/\partial q_k) \equiv \partial \boldsymbol{r}/\partial t + \partial \boldsymbol{r}/\partial(t) \equiv \boldsymbol{e}_0 + \sum A_k \boldsymbol{e}_k, \tag{3.9.4a}$$

and grouping terms, we obtain the following basic kinetic energy representation:

$$T \to T^* = T^*(t, q, \omega) = T^*{}_2 + T^*{}_1 + T^*{}_0, \tag{3.9.4b}$$

where

$$2T^*{}_2 = 2T^*{}_2(t, q, \omega) \equiv \sum\sum M^*{}_{kl}\, \omega_k\, \omega_l \quad (\textit{quadratic} \text{ and homogeneous in the } \omega\text{'s}),$$
$$M^*{}_{kl} = M^*{}_{kl}(t, q) \equiv S\, dm\, \boldsymbol{\varepsilon}_k \cdot \boldsymbol{\varepsilon}_l \equiv S\, dm(\partial \boldsymbol{r}/\partial \theta_k) \cdot (\partial \boldsymbol{r}/\partial \theta_l) \quad (= M^*{}_{lk}), \tag{3.9.4c}$$
$$T^*{}_1 = T^*{}_1(t, q, \omega) \equiv \sum M^*{}_k\, \omega_k \qquad (\textit{linear} \text{ and homogeneous in the } \omega\text{'s}),$$
$$M^*{}_k = M^*{}_k(t, q) \equiv S\, dm\, \boldsymbol{\varepsilon}_k \cdot \boldsymbol{\varepsilon}_0 \equiv S\, dm(\partial \boldsymbol{r}/\partial \theta_k) \cdot (\partial \boldsymbol{r}/\partial \theta_{n+1})$$
$$(= M^*{}_{k,n+1} = M^*{}_{n+1,k}), \tag{3.9.4d}$$
$$T^*{}_0 = T^*{}_0(t, q) \equiv M^*{}_0/2 \qquad (\textit{zeroth} \text{ degree in the } \omega\text{'s}),$$
$$M^*{}_0 = M^*{}_0(t, q) \equiv S\, dm\, \boldsymbol{\varepsilon}_{n+1} \cdot \boldsymbol{\varepsilon}_{n+1}$$
$$\equiv S\, dm\, \boldsymbol{\varepsilon}_0 \cdot \boldsymbol{\varepsilon}_0 \equiv S\, dm\, (\partial \boldsymbol{r}/\partial \theta_{n+1}) \cdot (\partial \boldsymbol{r}/\partial \theta_{n+1}) \quad (= M^*{}_{n+1,n+1}/2); \tag{3.9.4e}$$

or, compactly (with $\alpha, \beta = 1, \ldots, n, n+1$; and noting that $\dot{\theta}_{n+1} = \dot{t} = 1$),

$$2T^* = \sum\sum M^*_{\alpha\beta}\,\omega_\alpha\,\omega_\beta, \qquad M^*_{\alpha\beta} = M^*_{\beta\alpha} \equiv S\,dm\,(\partial\mathbf{r}/\partial\theta_\alpha)\cdot(\partial\mathbf{r}/\partial\theta_\beta). \tag{3.9.4f}$$

The *nonholonomic coefficients of inertia* $M^*_{\alpha\beta} = M^*_{\alpha\beta}(t, q)$ satisfy similar analytical conditions as the holonomic ones $M_{\alpha\beta}$. Let us relate them; recalling (2.9.25a ff.) we find the following:

(a) $$M^*_{kl} \equiv S\,dm\,\boldsymbol{\varepsilon}_k\cdot\boldsymbol{\varepsilon}_l = S\,dm\Big(\sum A_{rk}\,\mathbf{e}_r\Big)\cdot\Big(\sum A_{sl}\,\mathbf{e}_s\Big)$$
$$= \sum\sum A_{rk}A_{sl}\Big(S\,dm\,\mathbf{e}_r\cdot\mathbf{e}_s\Big); \text{ i.e., } M^*_{kl} = \sum\sum A_{rk}A_{sl}M_{rs}; \tag{3.9.4g}$$

and, inversely,

$$M_{rs} = \cdots = \sum\sum a_{kr}a_{ls}\,M^*_{kl}; \tag{3.9.4h}$$

(b) $$M^*_k \equiv S\,dm\,\boldsymbol{\varepsilon}_k\cdot\boldsymbol{\varepsilon}_0 = S\,dm\Big(\sum A_{rk}\,\mathbf{e}_r\Big)\cdot\Big(\sum A_s\,\mathbf{e}_s + \mathbf{e}_0\Big)$$
$$= \sum\sum A_{rk}A_s\Big(S\,dm\,\mathbf{e}_r\cdot\mathbf{e}_s\Big) + \sum A_{rk}\Big(S\,dm\,\mathbf{e}_r\cdot\mathbf{e}_0\Big), \tag{3.9.4i}$$

that is,

$$M^*_k = \sum\sum A_{rk}\,A_s\,M_{rs} + \sum A_{rk}\,M_r; \tag{3.9.4j}$$

and, inversely,

$$M_r = \cdots = \sum\sum a_{kr}\,a_l\,M^*_{kl} + \sum a_{kr}\,M^*_k; \tag{3.9.4k}$$

(c) $$M^*_0 \equiv S\,dm\,\boldsymbol{\varepsilon}_0\cdot\boldsymbol{\varepsilon}_0 = S\,dm\Big(\sum A_r\mathbf{e}_r + \mathbf{e}_0\Big)\cdot\Big(\sum A_s\mathbf{e}_s + \mathbf{e}_0\Big)$$
$$= \sum\sum A_rA_s\Big(S\,dm\,\mathbf{e}_r\cdot\mathbf{e}_s\Big) + \sum A_r\Big(S\,dm\,\mathbf{e}_r\cdot\mathbf{e}_0\Big)$$
$$+ \sum A_s\Big(S\,dm\,\mathbf{e}_s\cdot\mathbf{e}_0\Big) + S\,dm\,\mathbf{e}_0\cdot\mathbf{e}_0,$$

that is,

$$M^*_0 = \sum\sum A_rA_sM_{rs} + 2\sum A_rM_r + M_0; \tag{3.9.4l}$$

and, inversely,

$$M_0 = \sum\sum a_k\,a_l\,M^*_{kl} + 2\sum a_k\,M^*_k + M^*_0. \tag{3.9.4m}$$

The above show clearly that even if we start with a *homogeneous* (*quadratic*) T, still we may end up with a *nonhomogeneous* (quadratic) T^*, and vice versa. Also, a little reflection (and recollection of the results of §2.9) shows that (3.9.4g–m) can be consolidated into the compact formulae

$$M^*_{\alpha\beta} = \sum\sum A_{\gamma\alpha}\,A_{\delta\beta}\,M_{\gamma\delta} \Leftrightarrow M_{\gamma\delta} = \sum\sum a_{\alpha\gamma}\,a_{\beta\delta}\,M^*_{\alpha\beta}. \tag{3.9.4n}$$

[In tensor language, these are the transformation equations between the holonomic ($M_{\alpha\beta}$) and nonholonomic ($M^*{}_{\alpha\beta}$) *covariant* components of the *metric tensor* in the system *event* space (§2.7), whose arc-length element (squared) is

$$(ds)^2 = 2T(dt)^2 = \sum\sum M_{\alpha\beta}\, dq_\alpha\, dq_\beta = \sum\sum M^*{}_{\alpha\beta}\, d\theta_\alpha\, d\theta_\beta; \qquad (3.9.4o)$$

and similarly in configuration space; see, for example, Lur'e (1968), Papastavridis (1998, 1999), Synge (1936).]

Some Useful Identities

Again, applying Euler's homogeneous function theorem to T^* we easily obtain the following identities:

(a)

$$\sum (\partial T^*/\partial\omega_k)\,\omega_k = \sum [\partial(T^*{}_2 + T^*{}_1 + T^*{}_0)/\partial\omega_k]\,\omega_k$$
$$= (2)T^*{}_2 + (1)T^*{}_1 + (0)T^*{}_0 = 2T^*{}_2 + T^*{}_1 = T^* + (T^*{}_2 - T^*{}_0); \qquad (3.9.5a)$$

(b)

$$\sum E_k{}^*\,\omega_k \equiv \sum [(\partial T^*/\partial\omega_k)^{\cdot} - \partial T^*/\partial\theta_k]\omega_k$$
$$= \left\{\left(\sum (\partial T^*/\partial\omega_k)\omega_k\right)^{\cdot} - \sum (\partial T^*/\partial\omega_k)\dot{\omega}_k\right\} - \sum (\partial T^*/\partial q_l)\left(\sum A_{lk}\omega_k\right)$$

[replacing, in the last term, $\sum A_{lk}\,\omega_k$ with $\dot{q}_l - A_l$]

$$= \left(\sum (\partial T^*/\partial\omega_k)\omega_k - T^*\right)^{\cdot} + \left(\partial T^*/\partial t + \sum A_l(\partial T^*/\partial q_l)\right),$$

or, finally [recalling (2.9.33 ff.) and (3.9.5a)],

$$\sum E_k{}^*\,\omega_k = \left(\sum (\partial T^*/\partial\omega_k)\,\omega_k - T^*\right)^{\cdot} + \partial T^*/\partial\theta_{n+1} = (T^*{}_2 - T^*{}_0)^{\cdot} + \partial T^*/\partial\theta_{n+1},$$

where

$$\partial T^*/\partial\theta_{n+1} \equiv \partial T^*/\partial t + \sum A_l(\partial T^*/\partial q_l); \qquad (3.9.5b)$$

a form that will prove useful in the energy rate theorem below.

3.9.3 Potential Energy

It is frequently possible to express the virtual work of the *impressed* forces, $\delta' W \equiv S\, d\mathbf{F} \cdot \delta\mathbf{r}$, as

$$\delta' W = -\,\delta V + \delta' W_{NP} = \sum [-\partial V/\partial q_k + Q_{k,NP}]\,\delta q_k \equiv \sum Q_k\,\delta q_k$$
$$= \sum [-\partial V^*/\partial\theta_k + \Theta_{k,NP}]\,\delta\theta_k \equiv \sum \Theta_k\,\delta\theta_k, \qquad (3.9.6a)$$

where

$V = V(q, t) = V^*(q, t) =$ *potential* (or *potential energy*), in system variables; (3.9.6b)

$$\partial V^*/\partial\theta_k \equiv \sum (\partial V^*/\partial q_l)(\partial\dot{q}_l/\partial\omega_k) = \sum A_{lk}(\partial V^*/\partial q_l), \qquad (3.9.6c)$$

$$\partial V^*/\partial q_k \equiv \sum (\partial V^*/\partial \theta_l)(\partial \omega_l/\partial \dot{q}_k) = \sum a_{lk}(\partial V^*/\partial \theta_l); \tag{3.9.6d}$$

$$-\partial V/\partial q_k = \textit{holonomic potential} \text{ part of } Q_k, \tag{3.9.6e}$$

$$Q_{k,NP} = \textit{holonomic nonpotential} \text{ part of } Q_k; \tag{3.9.6f}$$

$$-\partial V^*/\partial \theta_k = \textit{nonholonomic potential} \text{ part of } \Theta_k, \tag{3.9.6g}$$

$$\Theta_{k,NP} = \textit{nonholonomic nonpotential} \text{ part of } \Theta_k. \tag{3.9.6h}$$

The connection between δV in particle and system variables is given by

$$\begin{aligned}\delta V &= S(\partial V/\partial \mathbf{r})\cdot\delta\mathbf{r} = S(\partial V/\partial \mathbf{r})\cdot\left(\sum \mathbf{e}_k\,\delta q_k\right)\\ &= \sum\left(S(\partial V/\partial \mathbf{r})\cdot\mathbf{e}_k\right)\delta q_k = \sum(\partial V/\partial q_k)\,\delta q_k,\end{aligned} \tag{3.9.6i}$$

$$\begin{aligned}&= S(\partial V/\partial \mathbf{r})\cdot\left(\sum \boldsymbol{\varepsilon}_k\,\delta\theta_k\right)\\ &= \sum\left(S(\partial V/\partial \mathbf{r})\cdot\boldsymbol{\varepsilon}_k\right)\delta\theta_k = \sum(\partial V^*/\partial \theta_k)\,\delta\theta_k.\end{aligned} \tag{3.9.6j}$$

REMARK ON THE POTENTIAL OF CONSTRAINT REACTIONS

Since, in general,

$$d'W_R \equiv S\,d\mathbf{R}\cdot d\mathbf{r} \neq 0 \qquad \left(\text{whereas } \delta'W_R \equiv S\,d\mathbf{R}\cdot\delta\mathbf{r} = 0\right),$$

it is conceivable that $d'W_R = -dV$, namely, that constraint reactions are, *partly or wholly, potential.* Thus, we may have

$$R_0 \equiv R_{n+1} \equiv S\,d\mathbf{R}\cdot\mathbf{e}_0 = -\partial V/\partial t = -\sum(\partial V/\partial c_k)(dc_k/dt), \tag{3.9.6k}$$

where the c_k $(k = 1, 2, 3, \ldots)$ are certain time-dependent system *parameters*; for example, mass, length, area, and so on. Such "parameteric reactions" appear in areas like *parametric excitation* (*Mathieu–Floquet theory*) and *adiabatic invariance* (§7.9 examples/problems; §8.15).

From now on, generally, we will omit the subscripts NP in the Q's and Θ's. If potential parts exist, they will usually be absorbed into the system's *kinetic potential*, or *Lagrangean function* $L \equiv T - V = L^* \equiv T^* - V^*$; explicitly,

$$L = L(t, q, \dot{q}) \equiv T(t, q, \dot{q}) - V(t, q) \qquad \text{(holonomic variables)}, \tag{3.9.7a}$$

$$L^* = L^*(t, q, \omega) \equiv T^*(t, q, \omega) - V^*(t, q) \qquad \text{(nonholonomic variables)}. \tag{3.9.7b}$$

The only change in the earlier equations of motion is that $T(T^*)$ is replaced, wherever it appears, by $L(L^*)$; then, and unless explicitly stated to the contrary, $Q_k(\Theta_k)$ will stand for the *nonpotential* parts of the corresponding forces.

Generalized Potential

(Recall ex. 3.5.12.) Occasionally, the potential part of Q_k is given by the [slightly more general than (3.9.6e)] expression

$$Q_{k,\text{generalized potential}} \equiv Q_{k,GP} \equiv (\partial V/\partial \dot{q}_k)^{\cdot} - \partial V/\partial q_k \equiv E_k(V), \tag{3.9.8a}$$

where

$$V = V(t, q, \dot{q}) = \textit{generalized (holonomic) potential};$$

or, in extenso,

$$\begin{aligned} Q_{k,GP} &= \sum [(\partial^2 V/\partial q_l\, \partial \dot{q}_k)\dot{q}_l + (\partial^2 V/\partial \dot{q}_l\, \partial \dot{q}_k)\ddot{q}_l] + \partial^2 V/\partial t\, \partial \dot{q}_k - \partial V/\partial q_k \\ &= \sum (\partial^2 V/\partial \dot{q}_l\, \partial \dot{q}_k)\ddot{q}_l + \text{terms not containing accelerations } \ddot{q}. \end{aligned} \tag{3.9.8b}$$

However, since in *classical mechanics* $\partial Q_k/\partial \ddot{q}_l = 0$ (Pars, 1965, pp. 11–12), we conclude from the above that $\partial^2 V/\partial \dot{q}_l\, \partial \dot{q}_k = 0 \Rightarrow V$ can be, at most, *linear* in the $\dot{q}$'s; that is,

$$V = \sum \gamma_k(t, q)\dot{q}_k + V_0(t, q) \equiv V_1(t, q, \dot{q}) + V_0(t, q). \tag{3.9.8c}$$

Substituting (3.9.8c) into (3.9.8a, b), we obtain

$$Q_{k,GP} = d\gamma_k/dt - \partial/\partial q_k\Big(\sum \gamma_l \dot{q}_l + V_0\Big) = \cdots = \sum \gamma_{kl}\, \dot{q}_l + \partial\gamma_k/\partial t - \partial V_0/\partial q_k,$$

where

$$\gamma_{kl} = \gamma_{kl}(t, q) \equiv \partial\gamma_k/\partial q_l - \partial\gamma_l/\partial q_k \quad (= -\,\gamma_{lk}); \tag{3.9.8d}$$

that is, $\gamma_{11} = -\gamma_{11} \;\Rightarrow\; \gamma_{11} = 0$, $\gamma_{12} = -\gamma_{21}$, and so on.

Gyroscopicity

Now we introduce the following important concept.

DEFINITION

Impressed (i.e., nonconstraint) forces Q_k that satisfy the "power" condition

$$\sum Q_k\, \dot{q}_k = 0 \tag{3.9.8e}$$

are called *gyroscopic*. In view of the antisymmetry of the *gyroscopic coefficients* γ_{kl}, it is not hard to see that

$$\sum \Big(\sum \gamma_{kl}\, \dot{q}_l\Big)\dot{q}_k \equiv \sum\sum \gamma_{kl}\, \dot{q}_k \dot{q}_l = \sum\sum [(\gamma_{kl} + \gamma_{lk})/2]\dot{q}_k\, \dot{q}_l = 0. \tag{3.9.8f}$$

Hence, the representation/decomposition (3.9.8d) states that a *generalized potential force* consists, at most, of three parts: a *gyroscopic* $\sum \gamma_{kl}\dot{q}_l$, a *nonstationary* $\partial\gamma_k/\partial t$, and a *purely potential* $-\partial V_0/\partial q_k$. If, further, $\sum (\partial\gamma_k/\partial t)\dot{q}_k = 0$ and $\sum (\partial V_0/\partial q_k)\dot{q}_k = dV_0/dt - \partial V_0/\partial t = 0$, then $Q_{k,GP}$ is (purely) gyroscopic.

REMARKS ON GYROSCOPIC FORCES/TERMS

(i) Typically, but not exclusively, gyroscopic forces appear in problems of *relative motion/moving axes* (§3.16); for example, Coriolis force [see also Gantmacher (1970, pp. 68–69), Goldstein (1980, pp. 21–23), for applications of generalized, or "velocity-dependent," potentials to electrodynamics].

(ii) Let us assume that $dq_k = \dot{q}_k \, dt$ (actual motion). Then, due to (3.9.8e), we have

$$d'W_g \equiv \sum \Big(\sum \gamma_{kl} \, \dot{q}_l\Big) dq_k = \Big(\sum\sum \gamma_{kl} \, \dot{q}_k \, \dot{q}_l\Big) dt = 0; \tag{3.9.9a}$$

but, in general,

$$\delta' W_g \equiv \sum \Big(\sum \gamma_{kl} \, \dot{q}_l\Big) \, \delta q_k = \sum\sum \gamma_{kl} \, \dot{q}_l \, \delta q_k \neq 0; \tag{3.9.9b}$$

that is, the *actual elementary work of gyroscopic forces vanishes*; but their *virtual work, in general, does not* (if it did, such forces would not have the opportunity to appear in Lagrangean equations!).

(iii) *Gyroscopicity of particle variables*: Let us call the force system $\{d\boldsymbol{F}\}$ gyroscopic, if it satisfies $S \, d\boldsymbol{F} \cdot \boldsymbol{v} = 0$. Substituting into this the particle velocity representation (3.9.1), we obtain, successively,

$$\begin{aligned} 0 &= S \, d\boldsymbol{F} \cdot \Big(\sum \boldsymbol{e}_k \, \dot{q}_k + \boldsymbol{e}_0\Big) = \sum \Big(S \, d\boldsymbol{F} \cdot \boldsymbol{e}_k\Big) \dot{q}_k + S \, d\boldsymbol{F} \cdot \boldsymbol{e}_0 \\ &= \sum Q_k \dot{q}_k + Q_0 \;\Rightarrow\; \sum Q_k \dot{q}_k \neq 0; \end{aligned} \tag{3.9.9c}$$

that is, in general, the corresponding *system* forces Q_k are not gyroscopic; and, conversely, even if the Q_k are gyroscopic (i.e., $\sum Q_k \dot{q}_k = 0$), the $d\boldsymbol{F}$ may not be $\big(S \, d\boldsymbol{F} \cdot \boldsymbol{v} \neq 0\big)$. However, if $\boldsymbol{e}_0 \equiv \partial \boldsymbol{r}/\partial t = \boldsymbol{0}$ (stationary holonomic constraints), then the Q_k and $d\boldsymbol{F}$ are gyroscopic simultaneously.

(iv) More generally, let us examine how gyroscopicity is affected by a general *frame of reference* transformation:

$$q \to q' : \; q_k = q_k(t, q_{k'}) \;\Rightarrow\; \dot{q}_k = \sum (\partial q_k / \partial q_{k'}) \dot{q}_{k'} + \partial q_k / \partial t. \tag{3.9.9d}$$

Indeed, substituting (3.9.9d) into (3.9.8e), and recalling that the q'-frame impressed forces $Q_{k'}$ are defined by the frame invariant virtual work relation:

$$\delta' W = \sum Q_k \, \delta q_k = \sum Q_k \Big(\sum (\partial q_k / \partial q_{k'}) \, \delta q_{k'}\Big) \equiv \sum Q_{k'} \, \delta q_{k'}, \tag{3.9.9e}$$

that is,

$$Q_{k'} = \sum (\partial q_k / \partial q_{k'}) Q_k \;\Leftrightarrow\; Q_k = \sum (\partial q_{k'} / \partial q_k) Q_{k'}, \tag{3.9.9f}$$

we obtain

$$\begin{aligned} 0 &= \sum Q_k \, \dot{q}_k = \sum Q_k \Big(\sum (\partial q_k / \partial q_{k'}) \dot{q}_{k'}\Big) + \sum Q_k (\partial q_k / \partial t) \\ &= \sum Q_{k'} \, \dot{q}_{k'} + \sum Q_k (\partial q_k / \partial t) \;\Rightarrow\; \sum Q_{k'} \dot{q}_{k'} \neq 0; \end{aligned} \tag{3.9.9g}$$

that is, in general, the $Q_{k'}$ are nongyroscopic. If, however, $\partial q_k / \partial t = 0$ [in which case (3.9.9d) expresses a *coordinate* (*not a frame of reference*) transformation], the $Q_{k'}$ are also gyroscopic. In sum: *force gyroscopicity is a frame-dependent property.*

Finally, similar results hold for the generalized potential and corresponding forces in *nonholonomic* coordinates; that is,

$$V^* = V^*(t, q, \omega) = \sum \gamma^*{}_k(t, q)\, \omega_k + V^*{}_0(t, q) \equiv V^*{}_1(t, q, \omega) + V^*{}_0(t, q)$$
$$= \textit{generalized (nonholonomic) potential}, \tag{3.9.9h}$$

$$\Rightarrow \Theta_{k,\,\text{generalized potential}} \equiv \Theta_{k,GP} \equiv (\partial V^*/\partial\omega_k)^{\cdot} - \partial V^*/\partial\theta_k$$
$$= \cdots = \sum \gamma^*{}_{kl}\, \omega_l + \partial\gamma^*{}_k/\partial t - \partial V^*{}_0/\partial\theta_k, \tag{3.9.9i}$$

where

$$\gamma^*{}_{kl} = \gamma^*{}_{kl}(t, q) \equiv \partial\gamma^*{}_k/\partial\theta_l - \partial\gamma^*{}_l/\partial\theta_k \quad (= -\gamma^*{}_{lk}). \tag{3.9.9j}$$

Rayleigh's Dissipation Function

Let us consider the linear viscous friction on a particle; that is, the impressed force given by the constitutive equation

$$d\boldsymbol{F} = -f\boldsymbol{v}, \qquad f = \textit{positive constant}. \tag{3.9.10a}$$

Recalling (3.9.1), the corresponding system force $Q_{k,D} \equiv S\, d\boldsymbol{F} \cdot \boldsymbol{e}_k = \cdots$ can be expressed as

$$Q_{k,D} = -\partial F/\partial\dot{q}_k = -\sum f_{kl}\,\dot{q}_l - f_k, \tag{3.9.10b}$$

where the *Rayleigh dissipation function*, or *dissipativity* (Kelvin), F, is defined by

$$F \equiv S\,(f/2)\, \boldsymbol{v} \cdot \boldsymbol{v} = \cdots = F_2 + F_1 + F_0 \quad [= F(t, q, \dot{q})], \tag{3.9.10c}$$
$$F_2 \equiv (1/2) \sum f_{kl}\,\dot{q}_k\,\dot{q}_l \ (\geq 0), \qquad f_{kl} \equiv S f\, \boldsymbol{e}_k \cdot \boldsymbol{e}_l \quad (= f_{lk}), \tag{3.9.10d}$$
$$F_1 \equiv \sum f_k\,\dot{q}_k, \qquad f_k \equiv S f\, \boldsymbol{e}_k \cdot \boldsymbol{e}_0 \quad (= f_{k,n+1} \equiv f_{k,0}), \tag{3.9.10e}$$
$$F_0 \equiv (1/2) S f\, \boldsymbol{e}_0 \cdot \boldsymbol{e}_0 \quad (= f_{n+1,n+1}/2 \equiv f_{0,0}/2); \tag{3.9.10f}$$

and has similar analytical properties with the kinetic energy (in the latter, replace dm with f).

Stationary Case

If $\boldsymbol{e}_0 \equiv \partial\boldsymbol{r}/\partial t = \boldsymbol{0}$ (case dealt by Rayleigh, in 1873), then

$$F = F_2 = (1/2) \sum f_{kl}\,\dot{q}_k\,\dot{q}_l, \tag{3.9.10g}$$

and the power of the corresponding dissipative forces, by Euler's homogeneous function theorem, equals

$$\sum Q_{k,D}\,\dot{q}_k = -\sum (\partial F/\partial\dot{q}_k)\dot{q}_k = -2F_2 \quad (= -2F); \tag{3.9.10h}$$

that is (as detailed below, or may be already known from general mechanics), $2F_2$ *measures the rate of decrease of the system's energy due to such friction.* For more general forms of dissipation functions, see Lur'e (1968, pp. 227–238).

Finally, if we use the nonholonomic particle velocity representation (3.9.4) in F, (3.9.10c), we will obtain the dissipation function in quasi variables: $F \to F^*(t, q, \omega) = \cdots$. The details are left to the reader.

Energy Rate, or Power (or Activity) Theorems

As a rule, such theorems are obtained by, first, multiplying (dotting) the equations of motion by the corresponding velocities and then summing over the entire system (or pairs of indices), for a fixed generic time. The result is a *single scalar equation* whose *inertia side* is (roughly) the rate of change of the kinetic energy of the system, and whose *force side* is the rate of working, or power, of whatever forces appear in the equations of motion.

Contrary to LP, which is a single energetic but *variational* equation (and, as such, can generate as many independent equations of motion as the number of the system's DOF), the energy rate relation is also a *single* energetic but *actual* equation (and, as such, it *cannot*, in general, be used to produce correct equations of motion)—and this is a *fundamental difference between energy and variational theorems/principles of mechanics!* [On this "insidious fallacy," see Pars (1965, pp. 86–87), also ex. 3.9.3.]

Below we derive these theorems (better, theorem in its various forms) in both holonomic and nonholonomic variables.

Holonomic Variables

Multiplying each Routh–Voss equation (3.5.15), say, with free index k, with $\dot{q}_k$ and summing over k, from 1 to n, and then invoking the earlier identity (3.9.3b) and the $m(< n)$ Pfaffian constraints $\sum a_{Dk}\dot{q}_k + a_D = 0$ [from which it follows that

$$\sum \Big(\sum \lambda_D a_{Dk}\Big)\dot{q}_k = \sum \lambda_D\Big(\sum a_{Dk}\dot{q}_k\Big) = -\sum \lambda_D a_D\Big], \tag{3.9.11a}$$

we obtain the *holonomic power equation*

$$\Big(\sum (\partial T/\partial \dot{q}_k)\dot{q}_k - T\Big)^{\cdot} = (T_2 - T_0)^{\cdot} = -\partial T/\partial t + \sum Q_k \dot{q}_k - \sum \lambda_D a_D. \tag{3.9.11b}$$

If some (or all) of the Q_k's are derived partly (or wholly) from a potential function V, then (3.9.3b) is replaced by

$$\sum E_k(L)\dot{q}_k \equiv \sum [(\partial L/\partial \dot{q}_k)^{\cdot} - \partial L/\partial q_k]\dot{q}_k = \partial L/\partial t + dh/dt, \tag{3.9.11c}$$

and, accordingly, (3.9.11b) is replaced by the simpler looking form

$$dh/dt = -\partial L/\partial t + \sum Q_k \dot{q}_k - \sum \lambda_D a_D, \tag{3.9.11d}$$

where

$$h = h(t, q, \dot{q}) \equiv \sum (\partial L/\partial \dot{q}_k)\dot{q}_k - L$$
$$= L_2 - L_0 = T_2 - (T_0 - V_0):$$

Generalized energy of the system in holonomic variables

(= *Hamiltonian function*, when expressed in terms of $t, q,$ and $p \equiv \partial T/\partial \dot{q}$; instead of the $\dot{q}$'s — see chap. 8), (3.9.11e)

and

$$L = L_2 + L_1 + L_0; \qquad L_2 \equiv T_2, \quad L_1 \equiv T_1 - V_1, \quad L_0 \equiv T_0 - V_0, \tag{3.9.11f}$$

$$Q_k = \textit{nonpotential}\ \text{part of virtual work term } (\ldots)\,\delta q_k. \tag{3.9.11g}$$

An additional useful form of the power theorem results if, instead of h, we use the ordinary (or classical) *total energy* of the system E:

$$E \equiv T + V_0 \quad (= T + V, \quad \text{if} \quad V_1 = 0). \tag{3.9.11h}$$

Then, since

$$h = T_2 - (T_0 - V_0) = (T - T_1 - T_0) - (T_0 - V_0) = (T + V_0) - (T_1 + 2T_0)$$

or

$$h = E - (T_1 + 2T_0) \qquad \text{or} \qquad E - h = T_1 + 2T_0, \tag{3.9.11i}$$

the power equations (3.9.11b, d) assume the equivalent *classical energy rate* form (analytical mechanics form of Leibniz's "law of vis viva")

$$dE/dt = -\partial L/\partial t + (T_1 + 2T_0)^{\cdot} + \sum Q_k \dot{q}_k - \sum \lambda_D a_D. \tag{3.9.11j}$$

If, in the above, all forces are *nonpotential*, then E and L must be replaced by T [→ (3.9.11b)].

Problem 3.9.1 Show that in terms of the more general total energy

$$E' \equiv E + V_1 = T + V = T + (V_1 + V_0) \quad (\Rightarrow \dot{E} = \dot{E}' - \dot{V}_1), \tag{a}$$

the power equation (3.9.11j) becomes

$$dE'/dt = -\partial L/\partial t + (T_1 + 2T_0)^{\cdot} + \dot{V}_1 + \sum Q_k \dot{q}_k - \sum \lambda_D a_D. \tag{b}$$

Specializations

(i) If the (initial) holonomic constraints of the system are stationary/scleronomic, then $T_1, T_0 = 0, \partial T/\partial t = 0 \;\Rightarrow\; T = T_2, h = E = T_2 + V_0$, and (3.9.11d, j) reduce to

$$dE/dt = \partial V_0/\partial t + \sum Q_k \dot{q}_k - \sum \lambda_D a_D. \tag{3.9.11k}$$

(ii) If, further, $\partial V/\partial t = 0$ and all additional Pfaffian constraints are *catastatic*—that is, $a_D = 0$ $(D = 1, \ldots, m)$—then the above simplifies to

$$dE/dt = \sum Q_k \dot{q}_k, \tag{3.9.11l}$$

[or $dT/dt = \sum Q_k \dot{q}_k$, if *all* forces are *nonpotential*].

(iii) Finally, if all impressed forces are either *potential* $(Q_k = 0)$, or *gyroscopic*, then (3.9.11l) yields the *theorem of conservation of (classical) energy*:

$$dE/dt = 0 \Rightarrow E = T + V_0 = \textit{constant}. \tag{3.9.11m}$$

Systems that satisfy all the above: namely, (a) *all* their constraints are *stationary* [slightly stronger than just catastatic—in view of (ii), condition (a) is sufficient but nonnecessary], (b) all their forces are either potential or gyroscopic, and (c) their (ordinary or generalized) potential does not depend explicitly on time; are called (classically) *conservative*. Hence, (3.9.11m) expresses the following theorem.

THEOREM

During any actual motion of a conservative system, its (*classical*) *energy*, evaluated at any point of its trajectory (or orbit), remains *constant*.

Equation (3.9.11m) is a *first* integral of the system's equations of motion; that is, it does *not* contain any accelerations; it is called the (classical) *energy integral*. [Equation (3.9.11m) is the reason for the *minus* sign in the potential force definitions (3.9.6a ff.). In the older literature (roughly, until the early 1900s), we frequently encounter the term "force function" for $U \equiv -V_0$: $Q_{k,\text{potential}} \equiv \partial U/\partial q_k$; then (3.9.11m) would read $T = U + \textit{constant}$. Some authors have, *erroneously*, taken this to mean some kind of a scalar function from which we can, by taking its gradients, obtain *all* system forces!]

A slightly more general conservation theorem than (3.9.11m) can be obtained from (3.9.11d) wherever the following conditions apply: (a) $\partial L/\partial t = 0$ (which does *not* necessitate that T_1, T_0 vanish, and/or that $\partial T/\partial t$ or $\partial V/\partial t$ vanish individually), (b) all forces are either potential or gyroscopic, and (c) $a_D = 0$ (catastatic Pfaffian constraints). Then, (3.9.11d) yields the (holonomic) *Jacobi–Painlevé generalized energy integral*:

$$dh/dt = 0 \Rightarrow h = T_2 - (T_0 - V_0) = T_2 + (V_0 - T_0) = \textit{constant}, \tag{3.9.11n}$$

even though in this case, as (3.9.11i) shows, $E \neq \textit{constant}$; if, in addition, $T_0 = 0$, then (3.9.11n) reduces to (3.9.11m). Of course, other combinations of physical circumstances can nullify the right side of (3.9.11d), and thus reproduce (3.9.11n).

Nonholonomic Variables

The power equations (3.9.11b, d, j), just like the equations of motion they came from, have two drawbacks: (i) they contain multipliers (reactions)—that is, they are "mixed power equations," and (ii) they cannot distinguish between nonholonomic Pfaffian constraints and holonomic ones disguised as Pfaffian. Hence the need for nonholonomic power equations. To obtain them, we begin by multiplying the *kinetic*

Hamel equations (3.5.19d, 20b, 21d) with $T^* \to L^* \equiv T^* - V^*$ (to include possible potential forces)

$$I_I \equiv (\partial L^*/\partial\omega_I)^{\cdot} - \partial L^*/\partial\theta_I + \sum\sum \gamma^r{}_{II'}(\partial L^*/\partial\omega_r)\omega_{I'} + \sum \gamma^r{}_I(\partial L^*/\partial\omega_r)$$
$$= \Theta_I \qquad (r = 1, \ldots, n;\ I, I' = m+1, \ldots, n), \tag{3.9.12a}$$

with ω_I and sum over I.

(i) Then notice that by (3.9.5b), with $T^* \to L^*$,

$$\sum E_I{}^*(L^*)\,\omega_I \equiv \sum E^*{}_I\,\omega_I \equiv \sum [(\partial L^*/\partial\omega_I)^{\cdot} - \partial L^*/\partial\theta_I]\omega_I$$
$$= dh^*/dt + \partial L^*/\partial\theta_{n+1}, \tag{3.9.12b}$$

where

$$h^* \equiv \sum (\partial L^*/\partial\omega_I)\,\omega_I - L^* = L^*{}_2 - L^*{}_0 = T^*{}_2 + (V^*{}_0 - T^*{}_0)$$
$$= h^*(t, q, \omega)\text{: \textit{generalized energy of the system, in nonholonomic variables} } (\neq h, \text{ in general}), \tag{3.9.12c}$$

and

$$L^* = L^*(t, q, \omega) \equiv L^*{}_2 + L^*{}_1 + L^*{}_0;$$
$$L^*{}_2 \equiv T^*{}_2, \qquad L^*{}_1 \equiv T^*{}_1 - V^*{}_1, \qquad L^*{}_0 \equiv T^*{}_0 - V^*{}_0, \tag{3.9.12d}$$

$$\Theta_I = \textit{nonpotential} \text{ part of virtual work term } (\ldots)\,\delta\theta_I, \tag{3.9.12e}$$

and

$$\partial L^*/\partial\theta_{n+1} \equiv \partial L^*/\partial t + \sum A_k(\partial L^*/\partial q_k) \tag{3.9.12f}$$

[Note the differences between (3.9.12b) and (3.9.11c); and absence of *linear* $\omega(\dot{q})$ terms in $h^*(h)$, even though $T^*{}_1(T_1)$ appear in $E^*{}_I(E_k)$].

(ii) Since $\gamma^r{}_{II'} = -\gamma^r{}_{I'I}$ (i.e., antisymmetry, or *gyroscopicity*, of Hamel's coefficients),

$$\sum\Big(\sum\sum \gamma^r{}_{II'}(\partial L^*/\partial\omega_r)\omega_{I'}\Big)\omega_I = \sum\Big(\sum\sum \gamma^r{}_{II'}\omega_{I'}\omega_I\Big)(\partial L^*/\partial\omega_r) = 0. \tag{3.9.12g}$$

Collecting these results, we obtain the *nonholonomic* (*multiplierless, or kinetic*) *power equation*

$$\sum I_I\,\omega_I = \sum E_I{}^*(L^*)\,\omega_I + \sum\Big(\sum \gamma^r{}_I(\partial L^*/\partial\omega_r)\Big)\omega_I = \sum \Theta_I\,\omega_I, \tag{3.9.12h}$$

or, finally,

$$dh^*/dt = -\partial L^*/\partial\theta_{n+1} + \sum \Theta_I\,\omega_I - R, \tag{3.9.12i}$$

where

$$R \equiv \sum\sum \gamma^r{}_I(\partial L^*/\partial\omega_r)\,\omega_I\text{: \textit{rheonomic nonholonomic power}.} \tag{3.9.12j}$$

From this important equation, we draw the following special conclusions: by (2.10.4),

$$\gamma^r{}_I = \sum\sum (\partial a_{rk}/\partial q_l - \partial a_{rl}/\partial q_k) A_{kI} A_l + \sum (\partial a_{rk}/\partial t - \partial a_r/\partial q_k) A_{kI}, \quad (3.9.12k)$$

and, therefore, if $a_r \equiv a_{r,n+1} = 0 \Rightarrow A_r \equiv A_{r,n+1} = 0$, then $\partial L^*/\partial \theta_{n+1} = \partial L^*/\partial t$ and

$$R = \cdots = \sum\sum\sum [(\partial a_{rk}/\partial t) A_{kI}](\partial L^*/\partial \omega_r)\, \omega_I = \sum\sum (\partial a_{rk}/\partial t)(\partial L^*/\partial \omega_r)\dot{q}_k; \quad (3.9.12l)$$

if, further, $a_{rk} = a_{rk}(q)$, then $R = 0$, and (3.9.12i) reduces to

$$dh^*/dt = -\partial L^*/\partial t + \sum \Theta_I \omega_I \quad (3.9.12m)$$

(R also vanishes *if all Pfaffian constraints are holonomic*; then $\gamma^r{}_I = 0$); and, if, in addition, $\partial L^*/\partial t = 0$ and $\sum \Theta_I \omega_I = \sum \Theta_k \omega_k = 0$ (all impressed forces are potential; i.e., $\Theta_I = 0$, or gyroscopic), then (3.9.12m) leads immediately to the *nonholonomic Jacobi–Painlevé energy integral*

$$dh^*/dt = 0 \;\Rightarrow\; h^* = T^*{}_2 + (V^*{}_0 - T^*{}_0) = constant. \quad (3.9.12n)$$

[However, other combinations can create the same result—see example of rolling sphere on spinning plane (ex. 3.18.4).]

Finally, as with the Hamel-type equations of motion, the constraints $\omega_D = 0$ should be enforced *after* all pertinent differentiations have been carried out; otherwise we would miss the $\sum\sum \gamma^D{}_I(\partial L^*/\partial \omega_D)\, \omega_I$ terms in R.

REMARKS

(i) As (3.9.12f) shows, the term $\partial L^*/\partial \theta_{n+1}$ derives from the *nonstationarity* of L^* [through $\partial L^*/\partial t$, as in the *holonomic* case (3.9.11d)] and also from the acatastaticity of the Pfaffian constraints [through $\sum A_k(\partial L^*/\partial q_k)$], whether these latter are nonholonomic or not.

(ii) The R term should be expected on analytical and physical grounds: Hamel's equations, through their γ-terms, do distinguish between genuine nonholonomic constraints and holonomic ones disguised in velocity/differential form; and this *unique characteristic* of theirs is carried over to the *corresponding power equation* (3.9.12i).

(iii) The above make clear that the *nonholonomic and kinetic equation* (3.9.12i) can be written down *without knowledge of the solution of the equations of motion* [unlike its holonomic counterpart (3.9.11d, j), which require knowledge of the multipliers].

(iv) Had we used the following definition:

$$d^*L^*/dt \equiv \sum [(\partial L^*/\partial \theta_I)\, \omega_I + (\partial L^*/\partial \omega_I)\, \dot{\omega}_I] + \partial L^*/\partial t, \quad (3.9.12o)$$

instead of the one made here [in view of $L^* = L^*(t, q, \omega)$]:

$$dL^*/dt \equiv \sum [(\partial L^*/\partial q_k)\dot{q}_k + (\partial L^*/\partial \omega_k)\dot{\omega}_k] + \partial L^*/\partial t, \quad (3.9.12p)$$

then

$$dL^*/dt - d^*L^*/dt = \sum (\partial L^*/\partial q_k)\dot{q}_k - \sum (\partial L^*/\partial \theta_I)\, \omega_I$$
$$= \sum (\partial L^*/\partial q_k)\Big(\sum A_{kI}\omega_I + A_k\Big) - \sum (\partial L^*/\partial \theta_I)\, \omega_I$$
$$= \cdots = \sum (\partial L^*/\partial q_k)A_k = \partial L^*/\partial \theta_{n+1} - \partial L^*/\partial t \quad (3.9.12q)$$

[and for a general function $f^*(t, q, \omega)$: $df^*/dt = d^*L^*/dt + \sum (\partial f^*/\partial q_k)A_k$]; and the power equation would be

$$d^*h^*/dt = -\partial L^*/\partial t + \sum \Theta_I\, \omega_I - R, \quad (3.9.12r)$$

where

$$d^*h^*/dt \equiv \Big(\sum (\partial L^*/\partial \omega_I)\, \omega_I\Big)^{\cdot} - d^*L^*/dt. \quad (3.9.12s)$$

(v) Since $\omega_D = 0$, no power equations can result by multiplying Hamel's equations (kinetic and/or kinetostatic) with ω_D.

(vi) A power theorem in terms of the classical *total energy, but in nonholonomic variables*

$$E = E^* \equiv T^* + V^*_0 = \cdots = h^* + (T^*_1 + 2T^*_0) \quad [= T^* + V^*, \quad \text{if} \quad V^*_1 = 0], \quad (3.9.12t)$$

can also be formulated. The details are left to the reader.

(vii) About the possibility of formulating power equations using the remaining two general forms of the equations of motion — namely, those by Appell and Maggi — we note the following:

(a) The equations of Appell *contain accelerations explicitly*, and therefore are pretty inconvenient as a starting point for power equations.

(b) Multiplying each of Maggi's kinetic equations (3.5.19a, 20b, 21a) with ω_I and then summing over I, we obtain, successively,

$$\sum \Big(\sum A_{kI}\, E_k\Big)\omega_I = \sum \Big(\sum A_{kI}\, Q_k\Big)\omega_I, \quad (3.9.12u)$$

or, since $\dot{q}_k = \sum A_{kI}\omega_I + A_k$,

$$\sum E_k(\dot{q}_k - A_k) = \sum Q_k(\dot{q}_k - A_k),$$

or, rearranging,

$$\sum (E_k - Q_k)A_k = \sum (E_k - Q_k)\dot{q}_k. \quad (3.9.12v)$$

It is not hard to show that, since $E_k - Q_k = \sum \lambda_D a_{Dk}$ and [recalling the second of (2.9.3a)] $\sum a_{Dk} A_k = -a_D$, both sides of the above equal $-\sum \lambda_D a_D$, and so no really new power theorem has emerged here.

(viii) The *methodology* of this section can be carried intact to the case of *nonlinear* constraints/coordinates, $\omega = \omega(t, q, \dot{q})$ — see chap. 5.

Example 3.9.1 *Energy Rate Equations in Particle Variables via LP or the Central Equation.* Let us consider a *stationary* system; that is, one whose constraints are

all *scleronomic*. Then, $\delta \boldsymbol{r}$ and $d\boldsymbol{r}$ are mathematically equivalent. Hence:
(i) Substituting $d\boldsymbol{r} = \boldsymbol{v}\, dt$ for $\delta \boldsymbol{r}$ in LP yields

$$\mathcal{S}\, dm\, \boldsymbol{a} \cdot d\boldsymbol{r} = \mathcal{S}\, d\boldsymbol{F} \cdot d\boldsymbol{r} \Rightarrow dT = d'W; \tag{a}$$

(ii) Similarly, with $\delta \boldsymbol{r} \to d\boldsymbol{r} = \boldsymbol{v}\, dt$ and $\delta \boldsymbol{v} \to d\boldsymbol{v}$, the central equation (3.6.6) yields

$$\mathcal{S}\, dm\, \boldsymbol{v} \cdot d\boldsymbol{v} + \mathcal{S}\, d\boldsymbol{F} \cdot d\boldsymbol{r} = d/dt\Big(\mathcal{S}\, dm\, \boldsymbol{v} \cdot d\boldsymbol{r}\Big)$$
$$\Rightarrow dT + d'W = d(2T) \Rightarrow dT = d'W; \tag{b}$$

that is, in both cases we obtain, as a special case, the differential form (in time) of the work–energy theorem.

Problem 3.9.2 Consider a system of N particles, under the ideal constraints

$$\phi_H(\boldsymbol{r}_P, t) = 0 \qquad (H = 1, \ldots, h),$$
$$\sum \boldsymbol{B}_{DP}(\boldsymbol{r}_P, t) \cdot \boldsymbol{v}_P + B_D(\boldsymbol{r}_P, t) = 0 \qquad (D = 1, \ldots, m), \tag{a}$$

and, therefore (recall ex. 3.5.1) having Lagrangean equations of the first kind

$$m_P \boldsymbol{a}_P = \boldsymbol{F}_P + \boldsymbol{R}_P, \qquad \boldsymbol{R}_P = \sum \mu_H (\partial \phi_H / \partial \boldsymbol{r}_P) + \sum \lambda_D \boldsymbol{B}_{DP}, \tag{b}$$

where $\boldsymbol{F}_P$ ($\boldsymbol{R}_P$): *total impressed* (*reaction*) *force on a system particle* $P = 1, \ldots, N$ (= # *particles*), and $3N - (h + m) > 0$. Show that its corresponding *power* equation is

$$dT/dt = \sum \boldsymbol{F}_P \cdot \boldsymbol{v}_P - \sum \mu_H (\partial \phi_H / \partial t) - \sum \lambda_D B_D; \tag{c}$$

and then interpret it physically.

Problem 3.9.3 Continuing from the preceding problem, show that for *stationary* constraints (i.e., scleronomic system) and *potential* impressed forces (i.e., $\boldsymbol{F}_P = -\partial V_0(\boldsymbol{r}_P, t)/\partial \boldsymbol{r}_P$), the power equation (c) reduces to the nonstationary energy rate equation

$$dE/dt = -\partial V_0/\partial t, \qquad E \equiv T + V_0. \tag{a}$$

Example 3.9.2 *Power Equations from Particle Variable Considerations.* Dotting the Newton–Euler equation of particle motion (in continuum form) $dm\, \boldsymbol{a} = d\boldsymbol{F} + d\boldsymbol{R}$ with the inertial particle velocity $\boldsymbol{v}$, and then summing over the system particles, we obtain the "D'Alembert–Lagrange form of the power theorem"

$$\mathcal{S}\, dm\, \boldsymbol{a} \cdot \boldsymbol{v} = \mathcal{S}\, d\boldsymbol{F} \cdot \boldsymbol{v} + \mathcal{S}\, d\boldsymbol{R} \cdot \boldsymbol{v}. \tag{a}$$

[That, in general, $\mathcal{S}\, d\boldsymbol{R} \cdot \boldsymbol{v} \neq 0$ points to another big difference between power theorems and LP]. Next, substituting into (a) the *holonomic* representation (3.9.1),

$\boldsymbol{v} = \sum \boldsymbol{e}_k \dot{q}_k + \boldsymbol{e}_0$, we get

$$\sum \left(S\, dm\, \boldsymbol{a} \cdot \boldsymbol{e}_k \right) \dot{q}_k + S\, dm\, \boldsymbol{a} \cdot \boldsymbol{e}_0 = \sum \left(S\, d\boldsymbol{F} \cdot \boldsymbol{e}_k \right) \dot{q}_k + \sum \left(S\, d\boldsymbol{R} \cdot \boldsymbol{e}_k \right) \dot{q}_k + S\, (d\boldsymbol{F} + d\boldsymbol{R}) \cdot \boldsymbol{e}_0, \quad \text{(b)}$$

and, from this, we immediately obtain the *two* holonomic power equations:

(i)
$$\sum E_k \dot{q}_k = \sum Q_k \dot{q}_k + \sum R_k \dot{q}_k, \quad \text{(c)}$$

and, since $R_k = \sum \lambda_D a_{Dk} \Rightarrow \sum R_k \dot{q}_k = \cdots = -\sum \lambda_D a_D$,

$$\sum E_k \dot{q}_k = \sum Q_k \dot{q}_k - \sum \lambda_D a_D, \quad \text{(d)}$$

that is, eq. (3.9.11c, d); and the "rheonomic power equation"

(ii)
$$S\, dm\, \boldsymbol{a} \cdot \boldsymbol{e}_0 = S\, (d\boldsymbol{F} + d\boldsymbol{R}) \cdot \boldsymbol{e}_0,$$

that is,

$$S\, dm\, \boldsymbol{a} \cdot (\partial \boldsymbol{r} / \partial t) = S\, (d\boldsymbol{F} + d\boldsymbol{R}) \cdot (\partial \boldsymbol{r} / \partial t). \quad \text{(e)}$$

Similarly, inserting in (a) the *nonholonomic* representation (3.9.4), $\boldsymbol{v}^* = \sum \boldsymbol{\varepsilon}_I \omega_I + \boldsymbol{\varepsilon}_0$, we obtain the two power equations

$$\sum I_I \omega_I = \sum \Theta_I \omega_I, \quad \text{(f)}$$

that is, eq. (3.9.12h); and

$$S\, dm\, \boldsymbol{a} \cdot (\partial \boldsymbol{r} / \partial \theta_{n+1}) = S\, (d\boldsymbol{F} + d\boldsymbol{R}) \cdot (\partial \boldsymbol{r} / \partial \theta_{n+1}) \quad \text{or} \quad I_{n+1} = \Theta_{n+1} + \Lambda_{n+1}. \quad \text{(g)}$$

Example 3.9.3 *On the Derivation of Lagrangean Equations of Motion from the Single Power Equation* (Pars' "insidious fallacy"). It is frequently claimed [especially in engineering books on vibration, but also in more theoretical and classy expositions; e.g., Birkhoff (1927, p. 17), Corben and Stehle (1960 and 1994, pp. 78–79)] that the *Lagrangean equations of motion*, say for concreteness, the Routh–Voss equations

$$E_k = Q_k + R_k, \qquad R_k = \sum \lambda_D a_{Dk}, \quad \text{(a)}$$

can be derived, not only from Lagrange's principle (LP), which is variational, but also from a *single power equation*, like (c) of the preceding example:

$$\sum E_k \dot{q}_k = \sum Q_k \dot{q}_k + \sum R_k \dot{q}_k, \quad \text{(b)}$$

and, therefore, one does not need all those strange and annoying concepts like virtual displacements/work, LP, and so on.

Well, such claims are false for the following reasons:

(i) Clearly, the forces in eq. (a) whose power is zero will not appear in eq. (b). How, then, are such forces going to be retrieved in the reverse reasoning from (b) to (a)? Such equations of motion would be "correct to within zero power terms" [just like Lagrangean equations are "correct to within zero *virtual* work terms"; or contain multiplier-proportional terms, like (a)]. The most important such "zero power forces" are the following two: (a) constraint reactions of *catastatic* Pfaffian constraints, and (b) (impressed) *gyroscopic* forces; like the γ-proportional terms of the Hamel-type equations. So the claim that if the Q_k are wholly potential (i.e., $Q_k = -\partial V_0(q,t)/\partial t$), then the equations of motion are $E_k(L) = Q_k + \sum \lambda_D a_{Dk}$, may be correct (for a nongyroscopic system), or it may not (for a gyroscopic system). (See also Ziegler, 1968, pp. 34–35.)

(ii) But there is a more serious objection to a reasoning that "leads" from the *single* equation (b) to the n equations (a), even for catastatic and nongyroscopic systems. We can deduce eq. (a) from LP—that is, $\sum (E_k - Q_k)\,\delta q_k = 0$, under $\sum a_{Dk}\,\delta q_k = 0$ — because of the *arbitrariness* of the δq's. On the other hand, eq. (b), rewritten as $\sum (E_k - Q_k - R_k)\,dq_k = 0$, holds for $dq_k = (\dot q_k)dt =$ *actual motion differentials/velocities*.

(iii) We have seen (§3.6) that LP is equivalent to the central equation

$$\delta T + \delta' W = \left(\sum (\partial T/\partial_k)\,\delta q_k\right)^{\cdot}. \tag{c}$$

On the other hand, as we know from general mechanics, the power equation (b) is equivalent to

$$dT = d'W, \tag{d}$$

where $d'W$ is *actual elementary work, in time, of all forces*; that is, *impressed plus reactions*. Hence, eqs. (c) and (d) are, in general, very different equations; eq. (c) represents much more than eq. (d).

HISTORICAL REMARK

A fair number of (unsuccessful) attempts to derive all the equations of motion from a single energy equation were made in the late 1800s to early 1900s by the so-called school of "Energetics." Specifically, its followers sought to obtain the equations of motion from the energy conservation equation (3.9.11m)

$$E \equiv T + V_0 = \textit{constant}, \qquad V_0 - V_0(r). \tag{e}$$

If the system is *unconstrained*, then $(\ldots)^{\cdot}$-differentiating (e) we obtain

$$S\,[dm\,\boldsymbol{a} + (\partial V_0/\partial \boldsymbol{r})]\cdot \boldsymbol{v} = 0, \tag{f}$$

and further, if this holds for each and every value of $\boldsymbol{v}$, then we are led to the correct equation

$$dm\,\boldsymbol{a} + (\partial V_0/\partial \boldsymbol{r}) = \boldsymbol{0}. \tag{g}$$

If, however, the system if *constrained* then, as Lipschitz remarked, this argument does *not* apply. To circumvent this difficulty, Helm, a leading "energeticist," proposed that, instead of introducing the usual virtual considerations, give the energy "principle" the following form: *the total energy change along any kinematically possible translational and/or rotational direction should vanish.* But, under such an arbitrary variation [assuming $\delta(d\boldsymbol{r}) = d(\delta \boldsymbol{r})$],

$$V_0 \rightarrow V_0 + \delta V_0 = V_0 + S\,(\partial V_0/\partial \boldsymbol{r})\cdot \delta \boldsymbol{r},$$

$$T + \delta T = T + S\,dm\,\boldsymbol{v}\cdot\delta\boldsymbol{v}$$
$$= T + d/dt\left(S\,dm\,\boldsymbol{v}\cdot\delta\boldsymbol{r}\right) - S\,dm\,\boldsymbol{a}\cdot\delta\boldsymbol{v},$$

that is,

$$\delta T = d/dt\left(S\, dm\, \boldsymbol{v}\cdot\delta\boldsymbol{r}\right) - \delta I \quad (\Rightarrow \delta T \neq \delta I).$$

Combining the above with $\delta E \equiv \delta T + \delta V_0 = 0$, we obtain

$$\delta I - \delta V_0 = d/dt\left(S\, dm\, \boldsymbol{v}\cdot\delta\boldsymbol{r}\right),$$

instead of the correct $\delta I + \delta V_0 = 0$. Hence, such a variation of the energy equation does *not* produce the correct equations even for a conservative system; again, the stumbling block is the difference between (c) and (d). If, on the other hand, we had *defined*, a priori,

$$\delta T \equiv S\, dm\, \boldsymbol{a}\cdot\delta\boldsymbol{r} \quad (= \delta I),$$

then $\delta T + \delta V_0 = 0$ would indeed lead to the correct equations of motion, but that would be an *arbitrary formalism* with the sole purpose to show the *equivalence between the energy rate equation and LP*. [At the time, that was a hotly debated issue among some of the best physicists of the day; and it makes us, today, appreciate better the simple and correct formulations of Heun and Hamel.] However, such ideas of *invariance of a certain differential (or integral) variational energetic expression under translations/rotations* proved useful later in supplying classical and nonclassical conservation theorems; for example, integral invariants (§8.12), Noetherian theory (§8.13), and so on; see also Dobronravov (1976, pp. 139–186, 209–249).

Example 3.9.4 Let us consider a homogeneous bar AB of mass m and length l pinned to the vertical shaft S of negligible mass (fig. 3.23). The system is constrained to spin about S with the *constant* angular velocity $\boldsymbol{\Omega}$. Let us discuss its power equation.

We will present two solutions: one with $q_1 = \theta$ as the sole unconstrained Lagrangean coordinate, and one with the *two* Lagrangean coordinates, $q_1 = \phi$ (angle of precession of shaft) and $q_2 = \theta$, but under the holonomic constraint

$$f_1 \equiv \phi - \Omega t \pm constant = 0 \quad \text{(finite form)} \tag{a1}$$

$$\Rightarrow \delta f_1 = \delta\phi = 0 \quad \text{(virtual form, since } \delta t = 0\text{)}, \tag{a2}$$

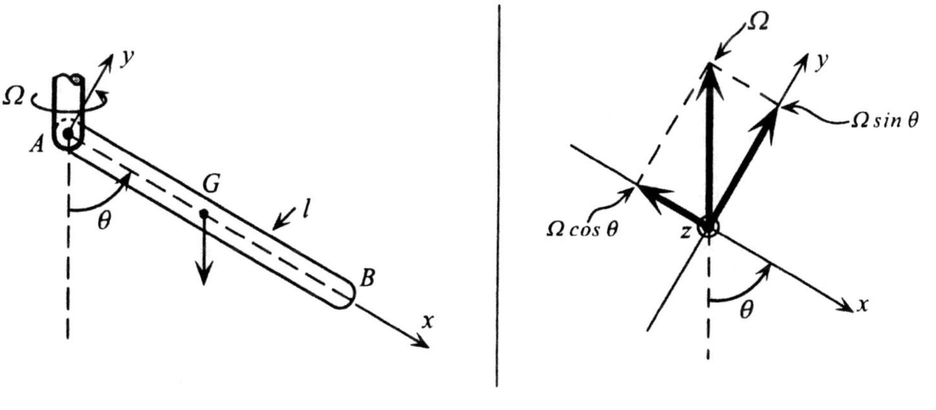

Figure 3.23 Geometry and kinematics of spinning bar *AB*.

or

$$\dot{\phi} + (-\Omega) = 0 \quad \text{(velocity form, acatastatic).} \tag{a3}$$

(i) *First solution (constrained system)*: Here, $\boldsymbol{\omega}$ = *inertial angular velocity of* $AB =$ $(-\Omega\cos\theta,\ \Omega\sin\theta,\ \dot{\theta})$, along the *body-fixed* (*and principal*) axes A–xyz, and $\mathbf{I}$ is the *moment of inertia tensor of* AB *at* A: *diagonal*$(I_x = 0, I_y = ml^2/3, I_z = ml^2/3)$. Hence, by König's theorem, and with $(\ldots)_o$ denoting constrained system quantities (i.e., only one coordinate), we have

$$2T_o = I_x {\omega_x}^2 + I_y {\omega_y}^2 + I_z {\omega_z}^2$$
$$= (ml^2/3)[\Omega^2 \sin^2\theta + (\dot{\theta})^2] = 2T_{o,2} + 2T_{o,1} + 2T_{o,0}, \tag{b1}$$

where

$$2T_{o,2} = (ml^2/3)(\dot{\theta})^2, \qquad 2T_{o,1} = 0, \qquad 2T_{o,0} = (ml^2/3)(\sin^2\theta)\Omega^2, \tag{b2}$$

$$V_o \equiv V = -(mgl/2)\cos\theta \quad (= 0, \text{at horizontal level through } A); \tag{b3}$$

that is, $\partial L_o/\partial t \equiv \partial(T_o - V)/\partial t = 0$; also, $Q_{\theta,\text{nonpotential}} \equiv Q_\theta = 0$, $a_D = 0$ (no constraints $\Rightarrow$ no multipliers). As a result of the above, the power equation (3.9.11d) reduces to the Jacobi–Painlevé integral (3.9.11n):

$$h \to h_o \equiv T_{o,2} + (V - T_{o,0}) = \textit{constant}$$

(evaluated at some initial time instant, and hence function of the initial conditions $\neq E_o \equiv T_o + V$)

or

$$l(\dot{\theta})^2 = 3g\cos\theta + (l\sin^2\theta)\Omega^2 + \textit{constant}; \tag{c}$$

an equation which, for given initial conditions, relates θ and $\dot{\theta}$; but, being a kinetic power equation, cannot supply the reactive couple M enforcing the constraint $\Omega = \textit{constant}$. To find the latter, either we apply the elementary "Newton–Euler" power equation to this constrained system, in which case M appears as an *external* moment; or we apply the generalized power equation to the relaxed system (second solution), in which case M appears either as a Lagrangean multiplier or as an impressed moment. Indeed, we have:

(ii) *Elementary power equation (constrained system)*: here, eq. (3.9.11j)

$$dE/dt = -\partial L/\partial t + (T_1 + 2T_0)^{\cdot} + \sum Q_k \dot{q}_k - \sum \lambda_D a_D, \tag{d1}$$

reduces to

$$dE_o/dt \equiv (T_o + V)^{\cdot} = (2T_{o,0})^{\cdot}, \tag{d2}$$

and therefore the elementary power equation, dE/dt = power of *external* forces and moments, yields

$$(2T_{o,0})^{\cdot} = M\Omega \quad \Rightarrow \quad M = (2T_{o,0})^{\cdot}/\Omega = (2ml^2\Omega/3)\sin\theta\cos\theta\,\dot{\theta}; \tag{d3}$$

that is, $M = \textit{variable}$, even though $\Omega = \textit{constant}$.

Without the benefit of (d1, 2) [or (3.9.11i): $\dot{E}_o - \dot{h}_o = (T_{o,1} + 2T_{o,0})^{\cdot} = (2T_{o,0})^{\cdot}$], the elementary power theorem, $(T_o + V)^{\cdot} = M\Omega$, would have given

$$\left\{(ml^2/6)\left[\Omega^2 \sin^2\theta + (\dot{\theta})^2\right] - (mgl/2)\cos\theta\right\}^{\cdot} = M\Omega, \tag{d4}$$

and this, to eliminate $\ddot{\theta}$ and thus reproduce (d3), would have to be combined with the kinetic θ-equation, or with the $(\ldots)^{\cdot}$ version of (c).

(iii) *Second solution (relaxed system)*: Here, we have

$$2T = (ml^2/3)(\dot{\theta})^2 + (ml^2/3)(\dot{\phi}\sin\theta)^2 = 2T_2 + 2T_1 + 2T_0, \tag{e1}$$

where

$$2T_2 = (ml^2/3)(\sin^2\theta)(\dot{\phi})^2 + (ml^2/3)(\dot{\theta})^2, \qquad 2T_1 = 0, \qquad 2T_0 = 0, \tag{e2}$$

$$V = -(mgl/2)\cos\theta \quad (= 0, \text{ at horizontal level through } A); \tag{e3}$$

that is, $\partial L/\partial t \equiv \partial(T - V)/\partial t = 0$.

Next, and in the sense of the relaxation principle, either we take $Q_{k,\text{nonpotential}} \equiv Q_k = 0$, but [recalling (a1–3), and since now $n = 2$, $m = 1$] *add* the term $-\sum \lambda_D a_D = -\lambda_1 a_1 = -\lambda_1(-\Omega) \equiv \lambda\,\Omega$; or, using the "rubber-band" approach, we take $\lambda_D = 0$, but keep the term $\sum Q_k \dot{q}_k = Q_\phi \dot{\phi}$, where $(\delta' W)_\phi \equiv Q_\phi\,\delta\phi = M\,\delta\phi$, since now M has become an impressed force (moment) and $\delta\phi \neq 0$. Following the first of these two equivalent alternatives, we obtain $\dot{h} = \lambda\,\Omega$, or since [recall (3.9.11i)] $h = E - (T_1 + 2T_0) = E$, we find $(\dot{E})_o = \lambda\,\Omega$, that is, (d3); and similarly for the second approach.

In this problem, the first solution (constrained system) is simpler; but to find the constraint reaction, we had to go outside of Lagrangean mechanics, to the Newton–Euler power equation (3.9.11j). The second solution (relaxed system) could prove more useful in complicated situations, where the application of (d1, 2) might not be so simple.

Problem 3.9.4 Continuing from the preceding example, discuss the power equations if ϕ and θ are connected by the acatastatic Pfaffian constraint $\dot{\phi} + (-c)\theta = 0$, $c = constant$; that is, $\Omega \equiv \dot{\phi} = c\theta$ (variable rate of shaft spinning).

Problem 3.9.5 Consider a particle of mass m moving on a smooth circular tube of radius r (fig. 3.24).

(i) Show that if the tube is *free to rotate* about a vertical diameter, the Lagrangean equations of motion of the particle + tube system, for $q_1 = \phi$ and $q_2 = \theta$, are

$$E_\phi = \left[(I + mr^2 \sin^2\theta)\dot{\phi}\right]^{\cdot} = Q_\phi, \qquad E_\theta = mr^2[\ddot{\theta} - \sin\theta\cos\theta(\dot{\phi})^2] = Q_\theta, \tag{a}$$

where $I =$ *moment of inertia of tube about its vertical diameter*. Calculate and interpret Q_ϕ, Q_θ.

(ii) Show that if the tube is constrained to rotate with *constant* angular velocity, $\dot{\phi} \equiv \Omega =$ *constant*, then the driving moment needed to enforce this constraint, M, equals

$$M = 2mr^2\Omega\sin\theta\cos\theta\,\dot{\theta} = mr^2\Omega\sin(2\theta)\,\dot{\theta}:$$
$$\textit{variable}, \text{ even though } \Omega \text{ is } \textit{constant}. \tag{b}$$

Relate M with Q_ϕ, and interpret the second of (a).

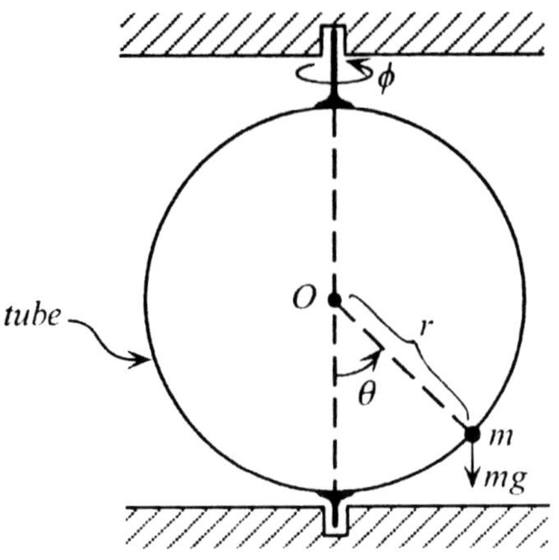

ϕ = *azimuth (precession) angle*
θ = *nutation angle (from lowest tube point)*

Figure 3.24 Particle moving on a smooth circular spinning tube.

HINT

$m\Omega^2(r\sin\theta) = \textit{centrifugal force on particle}$.

(iii) Discuss the power theorem in case (ii), and, with its help, calculate M.

[See also Greenwood (1977, pp. 74–77) for a discussion of the stability of the equilibrium of the particle relative to the tube, in case (ii).]

Problem 3.9.6 (Berezkin, 1968, vol. II, pp. 67–68). Consider a homogeneous bar AB of mass m and length $2l$, whose ends A and B are constrained to slide on the perpendicular and smooth sides of the rigid, plane, and rectangular frame *abcd* [fig. 3.25(a)]. The whole assembly is constrained to rotate about the vertical axis (v) with constant (inertial) angular velocity $\boldsymbol{\omega}$.

(i) Show that the kinetic Lagrangean equation of the (relative) angular motion of the bar is

$$\ddot{\theta} + \omega^2 \sin\theta\cos\theta = -(3g/4l)\cos\theta. \tag{a}$$

(ii) Then show that the corresponding *Jacobi–Painlevé* integral—that is, $(T_2 - T_0) + V_0 = \textit{constant}$ — is

$$(2l/3)\left[(\dot{\theta})^2 - \omega^2\cos^2\theta\right] + g\sin\theta = \textit{constant} \equiv h. \tag{b}$$

(iii) By applying the *principle of relaxation* (§3.7), show that the kinetostatic equation that yields the normal reaction at A, N, is

$$ml\left[\ddot{\theta}\cos\theta - (\dot{\theta})^2\sin\theta\right] = -mg + N. \tag{c}$$

HINT

Introduce the relaxed coordinate y (fig. 3.25); and, at the end, set $y = 0$.

(iv) Finally, show that, substituting into (c): $\ddot{\theta}$ from the kinetic equation (a), and $(\dot{\theta})^2$ from the energy integral (b), we obtain

$$\begin{aligned} N &= mg - ml\left[2\omega^2\sin\theta\cos^2\theta + (3g/4l)(\cos^2\theta - 2\sin^2\theta) + (3h/2m)\sin\theta\right] \\ &= \text{function of } \theta, \omega, \text{ and the initial conditions (through } h). \end{aligned} \tag{d}$$

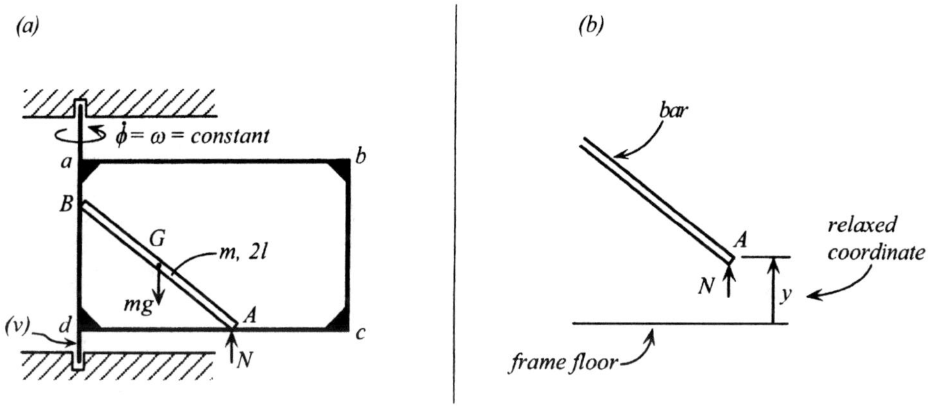

Figure 3.25 (a) Bar *AB* sliding on a uniformly rotating, plane, rigid, and rectangular frame *abcd*; (b) relaxed end *A*, with corresponding coordinate *y* and reaction *N*.

Example 3.9.5 *Introduction to Relative Motion* (for an extensive coverage see §3.16). Let us consider a particle P of mass m constrained to move on, say, the outer surface of a vertical circular and smooth cylinder of radius r (fig. 3.26). We will examine the energy equation of the particle when the cylinder is stationary, and when it spins about OZ with constant angular velocity $\boldsymbol{\omega}$. Then we will examine the case when P moves relative to the cylinder–fixed axes O–xyz.

(i) *Stationary cylinder*. With $q_1 = \Phi$ and $q_2 = Z$ (here, $r = constant \Rightarrow \dot{r} = 0$) and the plane $Z = 0$ for zero potential energy, we have

$$2T = m[r^2(\dot{\Phi})^2 + (\dot{Z})^2], \qquad V = mgZ, \qquad L = T - V, \tag{a}$$

and, therefore,

$$h = (\partial L/\partial\dot{\Phi})\dot{\Phi} + (\partial L/\partial\dot{Z})\dot{Z} - L = \cdots = (m/2)[r^2(\dot{\Phi})^2 + (\dot{Z})^2] + mgZ$$
$$= E \equiv T + V = constant, \text{ since } Q_{k,\,\text{nonpotential}} \equiv Q_k = 0,\ \partial L/\partial t = 0,\ a_D = 0. \tag{b}$$

(ii) *Spinning cylinder*. In terms of the *noninertial* coordinates $r = constant$, ϕ such that $\dot{\phi} = \dot{\Phi} - \omega$, $z = Z$, we readily find

$$2T = m[r^2(\dot{\phi} + \omega)^2 + (\dot{z})^2] \equiv 2T_2 + 2T_1 + 2T_0, \qquad V = m\,g\,z, \qquad L = T - V,$$
$$2T_2 = m[r^2(\dot{\phi})^2 + (\dot{z})^2], \qquad T_1 = m\,r^2\omega\,\dot{\phi}, \qquad 2T_0 = m\,r^2\omega^2, \tag{c}$$

and, therefore,

$$h = (\partial L/\partial\dot{\phi})\dot{\phi} + (\partial L/\partial\dot{z})\dot{z} - L$$
$$= \cdots = (m/2)[r^2(\dot{\phi})^2 + (\dot{z})^2] - (m/2)(r^2\omega^2) + mgz$$
$$= T_2 - T_0 + V = constant, \text{ since } Q_{k,\text{nonpotential}} \equiv Q_k = 0,\ \partial L/\partial t = 0,\ a_D = 0. \tag{d}$$

Clearly, $h \neq E \equiv T + V = (m/2)[r^2(\dot{\phi} + \omega)^2 + (\dot{z})^2] + mgz$.

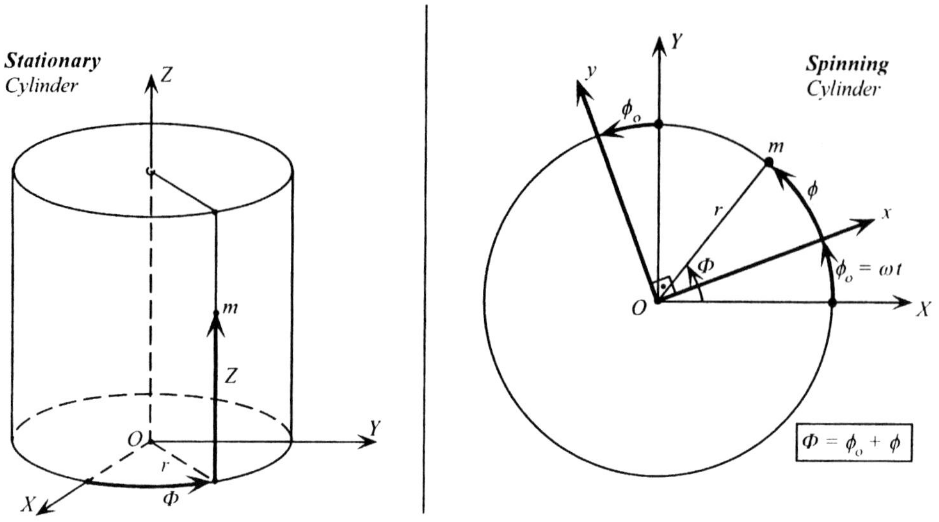

Figure 3.26 Particle moving on the outer surface of (a) a fixed and (b) a moving cylinder.

(iii) *(Introduction to) Relative Motion.* Let us generalize the above for an *arbitrary system moving relative to the uniformly rotating frame Oxyz*; that is, $r = r(t) \neq$ *constant*. Since, in this case,

$$X = r\cos(\phi + \omega t), \qquad Y = r\sin(\phi + \omega t), \qquad Z = z, \tag{e}$$

we obtain, successively,

$$2T = \mathsf{S}\, dm[(\dot{X})^2 + (\dot{Y})^2 + (\dot{Z})^2] = \cdots = T_{(2)} + \omega T_{(1)} + \omega^2 T_{(0)}, \tag{f}$$

where

$$2T_{(2)} \equiv \mathsf{S}\, dm\ [(\dot{r})^2 + r^2(\dot{\phi})^2 + (\dot{z})^2] \qquad (\equiv 2T_2), \tag{g1}$$

$$T_{(1)} \equiv \mathsf{S}\, dm\, r^2\dot{\phi} \qquad (\omega T_{(1)} \equiv T_1), \tag{g2}$$

$$2T_{(0)} \equiv \mathsf{S}\, dm\, r^2 \qquad (\omega^2 T_{(0)} \equiv T_0). \tag{g3}$$

Now, if $r, \phi, z =$ *stationary functions of* n *(noninertial) Lagrangean coordinates* $q \equiv (q_1, \ldots, q_n)$, and the system is further unconstrained, but under a potential $V_0 = V_0(q)$, then (since the Euler–Lagrange operator is linear) the Lagrangean equations of the system in the q's become

$$[(\partial T_{(2)}/\partial \dot{q}_k)^{\cdot} - \partial T_{(2)}/\partial q_k] + \omega[(\partial T_{(1)}/\partial \dot{q}_k)^{\cdot} - \partial T_{(1)}/\partial q_k] = -\partial V_R/\partial q_k, \tag{h}$$

where

$$V_R \equiv V_0 - \omega^2 T_{(0)} = V_0 - T_0 = \textit{relative potential}. \tag{i}$$

For reasons that will become clearer in §3.16, the $\sim \omega$ *(second)* group of terms, in (h), are called *gyroscopic*. If they vanish, the relative motion of the system is the same as if the cylinder was at rest but the system's potential energy was diminished by the "centrifugal potential" $\omega^2 T_{(0)}$. These terms vanish if every $\dot{q}_k$ vanishes; that is, if $q_k = constant$ (*relative equilibrium*). For a general rigid body in relative motion this cannot happen, unless the body *translates parallelly to the $OZ = Oz$ axis*; but it may happen for special systems of particles, or if $T_{(1)}$ is an *exact* differential, say

$$T_{(1)} = df(q, \dot{q})/dt, \qquad \text{where } f = \textit{arbitrary function of its arguments}, \qquad \text{(j)}$$

because then it is not hard to see that

$$(\partial T_{(1)}/\partial \dot{q}_k)^{\cdot} = (\partial \dot{f}/\partial \dot{q}_k)^{\cdot} = (\partial f/\partial q_k)^{\cdot} = \partial/\partial q_k(df/dt) = \partial T_{(1)}/\partial q_k. \qquad \text{(k)}$$

If $k = 1$, this holds always; then $T_{(1)} = F(q_1)\dot{q}_1$, where $F(q_1) =$ *arbitrary function of* q_1*; that is, there are no gyroscopic terms in one DOF systems!* We shall return to this important topic in the examples of §3.16, where it will be shown that (h) has the generalized energy integral

$$h \equiv T_{(2)} + V_R \equiv T_{(2)} + (V - \omega^2 T_{(0)}) = constant. \qquad \text{(l)}$$

Problem 3.9.7 Consider a particle P of mass m constrained to slide inside a smooth and straight tube of negligible mass, and also under the action of a linear spring of constant k and unstretched length r_o (fig. 3.27). The tube spins about a vertical axis OZ with *constant* angular velocity ω.

With $\rho \equiv r - r_o$, show that the system has the following Jacobi–Painlevé integral:

$$\begin{aligned} h &\equiv T_2 - T_0 + V_0 \\ &= (m/2)(\dot{\rho})^2 - (m\omega^2/2)(\rho + r_o)^2 + (k/2)(\rho^2) = constant. \end{aligned} \qquad \text{(a)}$$

Example 3.9.6 *Lagrangean Equations for* q_{n+1}. Let us find the "temporal Lagrangean equation"; that is, (assuming the n δq's are unconstrained)

$$d/dt(\partial T/\partial \dot{q}_0) - \partial T/\partial q_0 = Q_0 + R_0, \qquad \text{(a)}$$

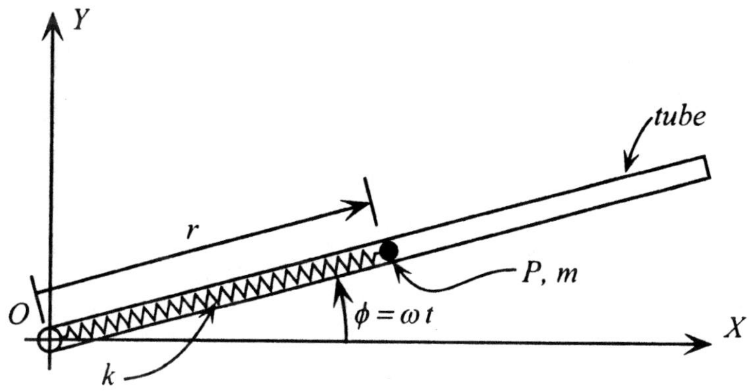

Figure 3.27 Particle on uniformly rotating horizontal tube.

where $q_{n+1} \equiv q_0 \equiv t \Rightarrow \dot{q}_0 = 1$, and [recalling (3.4.4a ff.)]

$$Q_{n+1} \equiv Q_0 \equiv S\, dF \cdot e_0 \equiv S\, dF \cdot (\partial r/\partial t),$$
$$R_{n+1} \equiv R_0 \equiv S\, dR \cdot e_0 \equiv S\, dR \cdot (\partial r/\partial t) \quad (\neq 0, \text{ in general}). \tag{b}$$

Since (recalling that $\alpha, \beta = 1, \ldots, n+1$)

$$\begin{aligned} 2T &= \sum\sum M_{\alpha\beta}\, \dot{q}_\alpha \dot{q}_\beta \\ &= \sum\sum M_{kl}\, \dot{q}_k \dot{q}_l + 2\sum M_{k,n+1}\dot{q}_k\dot{q}_{n+1} + M_{n+1,n+1}\dot{q}_{n+1}\dot{q}_{n+1} \\ &\equiv \sum\sum M_{kl}\, \dot{q}_k \dot{q}_l + 2\sum M_{k0}\dot{q}_k\dot{q}_0 + M_{00}\dot{q}_0\dot{q}_0 \\ &\equiv \sum\sum M_{kl}\, \dot{q}_k \dot{q}_l + 2\sum M_k\dot{q}_k + M_0, \end{aligned} \tag{c}$$

eq. (a) becomes

$$\left(\sum M_k\dot{q}_k + M_0 q_0\right)^{\cdot} - \partial T/\partial q_0 = Q_0 + R_0, \tag{d}$$

or, finally,

$$\left(\sum M_k\dot{q}_k + M_0\right)^{\cdot} - (1/2)\sum\sum(\partial M_{\alpha\beta}/\partial t)\dot{q}_\alpha\dot{q}_\beta = Q_0 + R_0$$
$$\Big[\text{or, if } M_k = 0 \text{ and } M_0 = 0;\ \text{i.e., if } 2T = \sum\sum M_{kl}(t,q)\dot{q}_k\dot{q}_l,$$
$$\text{then } -\partial T/\partial t = Q_0 + R_0\Big]; \tag{e}$$

which is none other than the rheonomic power identity (e) of example 3.9.2, but in system variables.

Extensions to quasi variables (i.e., the $(n+1)$th *Hamel equation*) are easily obtainable. These equations, since they contain the unknown R_0, do not seem to offer any particular advantage and so they will not be pursued any further. [See, e.g., Mattioli (1931–1932), Pastori (1960); also Nadile (1950) for the quasi-variable case, and an alternative derivation of (3.9.12i).] However, the above considerations may prove helpful in understanding better the connection between scleronomic and rheonomic systems. Following Lamb (1910, p. 758), we may consider time as an *additional* $(n+1)$th coordinate of an originally scleronomic system—that is, $q_{n+1} \equiv q_0 \equiv t$; or, start with a *scleronomic* system in the $n+1$ coordinates $(q, q_0) \equiv (q_1, \ldots q_n;\ q_{n+1} \equiv q_0)$ and then let $q_0 = \phi(t) =$ *known function of time*:

$$\begin{aligned} 2T &= \sum\sum M_{\alpha\beta}\, \dot{q}_\alpha\dot{q}_\beta \qquad [\text{where } M_{\alpha\beta} = M_{\alpha\beta}(q, q_0)] \\ &= \sum\sum M_{kl}\, \dot{q}_k\dot{q}_l + 2\sum M_{k0}\, \dot{q}_k\dot{q}_0 + M_{00}\, \dot{q}_0\dot{q}_0 \\ &\equiv \sum\sum M_{kl}\, \dot{q}_k\dot{q}_l + 2\sum M_k\dot{q}_k\dot{\phi} + M_0\dot{\phi}\,\dot{\phi} \\ &\equiv \sum\sum M_{kl}\, \dot{q}_k\dot{q}_l + 2\sum M'_k\, \dot{q}_k + M'_0 \\ &= 2T_2 + 2T_1 + 2T_0 = (\textit{double})\ \textit{kinetic energy of an } n\ \textit{DOF rheonomic system}, \end{aligned} \tag{f}$$

where $M_{kl} = M_{kl}(q, t)$, $M'_k = M'_k(q, t)$, $M'_0 = M'_0(q, t)$.

Finally, since the constraint $\dot{q}_0 = 1 (\Rightarrow \delta q_0 = 0)$ — or, generally, $\Phi[q_0, \phi(t)] = 0$ $(\Rightarrow \delta\Phi = 0)$ — is holonomic it can be enforced in T right from start; that is, before any differentiations (recall remarks in §3.5); and that constraint is maintained by the force $Q_0 + R_0$.

REMARK

The above and ex. 3.9.2 show that, even if we dotted the Newton–Euler particle equation of motion with $\boldsymbol{dr} = \sum \boldsymbol{e}_k \, dq_k + \boldsymbol{e}_0 \, dt$ (instead of with $\delta\boldsymbol{r} = \sum \boldsymbol{e}_k \, \delta q_k$) and then summed over the system particles, (i.e., $S\, dm\, \boldsymbol{a} \cdot \boldsymbol{dr} = S\, d\boldsymbol{F} \cdot \boldsymbol{dr} + S\, d\boldsymbol{R} \cdot \boldsymbol{dr}$), and thus ended up, in system variables, with $\sum (E_k - Q_k - R_k) dq_k + (E_0 - Q_0 - R_0)\, dt = 0$, (i) we would still need an additional postulate for $\sum R_k \, dq_k$ (like the d'Alembert–Lagrange principle), and (ii) the $(\ldots)dt$-terms would not have produced anything new; that is, no additional Lagrangean equation of motion. In sum [and contrary to what some Lagrangean derivations seem to, falsely, imply; e.g., Corben and Stehle (1960, pp. 78–79)]: *there is no way getting around virtual displacements and an independent physical postulate for the constraint reactions!*

Problem 3.9.8 (i) Show that in the Voronets equations case, eqs. (3.8.14a ff.), (a) [recall (3.9.12f)]

$$\partial L^*/\partial\theta_{n+1} = \partial L^*/\partial t + \sum b_D(\partial L^*/\partial q_D), \tag{a}$$

where (by partial differentiation):

$$\partial L^*/\partial q_D = \partial L/\partial q_D + \sum (\partial L/\partial \dot{q}_{D'})(\partial \dot{q}_{D'}/\partial q_D); \tag{b}$$

and (b) the rheonomic nonholonomic power term R, (3.9.12j), reduces to

$$R = \sum \sum w^D{}_I (\partial L/\partial \dot{q}_D)_o \, \dot{q}_I, \tag{c}$$

where [recall (3.8.14h)] $-w^D{}_I = \partial b_D/\partial(q_I) - \partial b_{DI}/\partial(q_{n+1})$.

(ii) Then formulate the "Voronets power equation"; and, finally, specialize the latter to the "Chaplygin power equation."

3.10 LAGRANGE'S EQUATIONS: EXPLICIT FORMS; AND LINEAR VARIATIONAL EQUATIONS (OR METHOD OF SMALL OSCILLATIONS)

Explicit Forms of Lagrange's Equations

We have already seen (§3.9) that the most general expression for T in holonomic system (Lagrangean) variables is

$$2T = \sum \sum M_{kl} \dot{q}_k \dot{q}_l + 2 \sum M_k \dot{q}_k + M_0 \;\equiv\; 2T_2 + 2T_1 + 2T_0. \tag{3.10.1a}$$

Let us find the explicit forms of the corresponding (and, hence, most general) inertia, or Lagrangean acceleration, terms $E_k \equiv E_k(T) \equiv (\partial T/\partial \dot{q}_k)^{\cdot} - \partial T/\partial q_k$. Since $E_k(\ldots)$ is a *linear* operator, we have

$$E_k(T) = E_k(T_2) + E_k(T_1) + E_k(T_0). \tag{3.10.1b}$$

Recalling that $M_{kl} = M_{lk}$, we obtain, successively,

$$\begin{aligned}
\text{(i)}\quad E_k(T_2) &= [(\partial/\partial \dot{q}_k)^{\cdot} - \partial/\partial q_k]\Big[(1/2)\sum\sum M_{kl}\dot{q}_k\dot{q}_l\Big] \\
&= \Big(\sum M_{kr}\,\dot{q}_r\Big)^{\cdot} - (1/2)\sum\sum(\partial M_{rs}/\partial q_k)\dot{q}_r\dot{q}_s \\
&= \sum M_{kr}\,\ddot{q}_r + \sum\sum[\partial M_{kr}/\partial q_s - (1/2)(\partial M_{rs}/\partial q_k)]\dot{q}_r\dot{q}_s \\
&\qquad + \sum(\partial M_{kr}/\partial t)\dot{q}_r. \qquad (3.10.1c)
\end{aligned}$$

But

$$\sum\sum(\partial M_{kr}/\partial q_s)\dot{q}_r\dot{q}_s = (1/2)\sum\sum(\partial M_{kr}/\partial q_s + \partial M_{ks}/\partial q_r)\dot{q}_r\dot{q}_s,$$

and so the middle (double sum) term of (3.10.1c) can be rewritten as

$$\sum\sum(1/2)(\partial M_{kr}/\partial q_s + \partial M_{ks}/\partial q_r - \partial M_{rs}/\partial q_k)\dot{q}_r\dot{q}_s \equiv \sum\sum\Gamma_{k,rs}\,\dot{q}_r\dot{q}_s,$$

where the just introduced quantities ("geometrical objects," in tensorial language)

$$\Gamma_{k,rs} = \Gamma_{k,sr} \equiv (1/2)(\partial M_{kr}/\partial q_s + \partial M_{ks}/\partial q_r - \partial M_{rs}/\partial q_k) \qquad (3.10.1d)$$

are the famous (holonomic) *Christoffel symbols of the first kind*. So, finally, (3.10.1c) becomes

$$E_k(T_2) = \sum M_{kr}\,\ddot{q}_r + \sum\sum\Gamma_{k,rs}\,\dot{q}_r\,\dot{q}_s + \sum(\partial M_{kr}/\partial t)\dot{q}_r. \qquad (3.10.1e)$$

$$\begin{aligned}
\text{(ii)}\quad E_k(T_1) &= \Big[(\partial/\partial \dot{q}_k)\Big(\sum M_r\,\dot{q}_r\Big)\Big]^{\cdot} - (\partial/\partial q_k)\Big(\sum M_r\,\dot{q}_r\Big) \\
&= \partial M_k/\partial t - \sum(\partial M_r/\partial q_k - \partial M_k/\partial q_r)\dot{q}_r \equiv \partial M_k/\partial t - G_k, \qquad (3.10.1f)
\end{aligned}$$

where

$$G_k \equiv \sum g_{kr}\,\dot{q}_r, \qquad g_{kr} = -g_{rk} \equiv \partial M_r/\partial q_k - \partial M_k/\partial q_r; \qquad (3.10.1g)$$

that is, $g_{12} = -g_{21}$, $g_{11} = -g_{11} \Rightarrow g_{11} = 0$, and so on.

$$\text{(iii)}\quad E_k(T_0) = (\partial T_0/\partial \dot{q}_k)^{\cdot} - \partial T_0/\partial q_k = 0 - \partial T_0/\partial q_k = -(1/2)(\partial M_0/\partial q_k). \qquad (3.10.1h)$$

In view of (3.10.1e, f, h), the Lagrangean acceleration (3.10.1b) assumes the definitive form

$$\begin{aligned}
E_k \equiv E_k(T) &= \sum M_{kr}\,\ddot{q}_r + \sum\sum\Gamma_{k,rs}\,\dot{q}_r\dot{q}_s + \sum(\partial M_{kr}/\partial t)\dot{q}_r \\
&\qquad + \partial M_k/\partial t - G_k - (1/2)(\partial M_0/\partial q_k). \qquad (3.10.1i)
\end{aligned}$$

In the theory of *relative motion* (§3.16) where (3.10.1i) primarily applies, it is customary to rearrange and rename the above as follows:

$$E_k = E_{k,R} + E_{k,T} + E_{k,C}, \qquad (3.10.2)$$

where

$$E_{k,R} \equiv E_k(T_2) \equiv (\partial T_2/\partial \dot{q}_k)^{\cdot} - \partial T_2/\partial q_k: \qquad \textit{Relative acceleration}, \tag{3.10.2a}$$

$$E_{k,T} \equiv \partial M_k/\partial t - \partial T_0/\partial q_k: \qquad \textit{Transport acceleration}, \tag{3.10.2b}$$

$$E_{k,C} \equiv \sum (\partial M_k/\partial q_r - \partial M_r/\partial q_k)\dot{q}_r \equiv -\sum g_{kr}\,\dot{q}_r \equiv -G_k:$$
$$\textit{Coriolis (or gyroscopic) acceleration}. \tag{3.10.2c}$$

This *three-part decomposition* of the Lagrangean acceleration E_k, and the recognition of its importance (especially of the Coriolis/gyroscopic part $E_{k,C}$), are due to Thomson and Tait (1867–1912, §319, pp. 318–327).

In view of the above kinematico-inertial identities, the Lagrangean equations of motion in the fairly general case of a holonomic n *DOF* system, with potential $V = V_0(q)$ and under nonpotential forces Q_k (i.e., $E_k = Q_k - \partial V_0/\partial q_k$), assume the explicit form

$$\sum M_{kr}\,\ddot{q}_r + \sum\sum \Gamma_{k,rs}\,\dot{q}_r\dot{q}_s + \sum (\partial M_{kr}/\partial t)\dot{q}_r$$
$$= Q_k + G_k - \partial(V_0 - T_0)/\partial q_k - \partial M_k/\partial t; \tag{3.10.3}$$

and similarly for other T-based equations.

In the common case $\partial T/\partial t = 0$ ("stationary/scleronomic" kinetic energy), (3.10.3) specializes to

$$\sum M_{kr}\ddot{q}_r + \sum\sum \Gamma_{k,rs}\dot{q}_r\,\dot{q}_s = Q_k + G_k - \partial(V_0 - T_0)/\partial q_k; \tag{3.10.3a}$$

while if $\partial \mathbf{r}/\partial t = \mathbf{0}$ (stationary holonomic constraints), then $T \to T_2$ and the above reduces to

$$\sum M_{kr}\ddot{q}_r + \sum\sum \Gamma_{k,rs}\,\dot{q}_r\,\dot{q}_s = Q_k - \partial V_0/\partial q_k. \tag{3.10.3b}$$

Inertial Coupling

In general, *all* the above equations are inertially (or dynamically) coupled; that is, each E_k contains all the system accelerations $\ddot{q}$. To decouple them, we introduce the symmetric inertial quantities $m_{kl} = m_{lk}$, "conjugate" to the $M_{kl} = M_{lk}$, via the definition

$$\sum m_{kl} M_{lr} = \delta_{kr} \quad (= 1, \text{ if } k = r; = 0, \text{ if } k = r). \tag{3.10.4}$$

Multiplying each of (3.10.3) with m_{sk} and adding over k, invoking (3.10.4), and renaming some dummy indices, we obtain the *inertially decoupled* Lagrangean equations

$$\ddot{q}_k + \sum\sum \Gamma^k{}_{rs}\,\dot{q}_r\,\dot{q}_s + \sum\sum m_{ks}(\partial M_{sr}/\partial t)\dot{q}_r$$
$$= \sum m_{ks}[Q_s + G_s - \partial(V_0 - T_0)/\partial q_s - \partial M_s/\partial t], \tag{3.10.5}$$

where we have introduced the following, similar to (3.10.1d), quantities:

$$\Gamma^k{}_{rs} = \Gamma^k{}_{sr} \equiv \sum m_{kl}\Gamma_{l,rs}:$$

(holonomic) Christoffel symbols of the second kind (3.10.5a)

$$\left(\Leftrightarrow \Gamma_{l,rs} = \sum M_{lk}\Gamma^k{}_{rs}\right). \tag{3.10.5b}$$

Velocity-Proportional Terms of (3.10.3, 5)

In general, there are two kinds of such terms: linear ($\sim \dot{q}_r$) and nonlinear ($\sim \dot{q}_r\,\dot{q}_s$). Let us examine the latter first.

(i) *Nonlinear terms* like $\Gamma_{k,rs}\,\dot{q}_r\dot{q}_s$ or $\Gamma^k{}_{rs}\,\dot{q}_r\dot{q}_s$ occur because their coefficients, the Christoffels, are intimately related to the *curvilinearity*; that is, the *nonlinearity*, of the coordinates q. Let us see this: recalling the definitions (3.9.2a) and (3.10.1d), we have, successively,

$$\begin{aligned}
\Gamma_{k,rs} &= (1/2)(\partial M_{kr}/\partial q_s + \partial M_{ks}/\partial q_r - \partial M_{rs}/\partial q_k)\\
&= (1/2)\left\{(\partial/\partial q_s)\left[S\, dm(\partial \mathbf{r}/\partial q_k)\cdot(\partial \mathbf{r}/\partial q_r)\right] + \cdots\right\}\\
&= (1/2)\Big\{S\, dm\big[(\partial^2 \mathbf{r}/\partial q_s\partial q_k)\cdot(\partial \mathbf{r}/\partial q_r) + (\partial \mathbf{r}/\partial q_k)\cdot(\partial^2 \mathbf{r}/\partial q_s\partial q_r)\\
&\quad + (\partial^2 \mathbf{r}/\partial q_r\partial q_k)\cdot(\partial \mathbf{r}/\partial q_s) + (\partial \mathbf{r}/\partial q_k)\cdot(\partial^2 \mathbf{r}/\partial q_r\partial q_s)\\
&\quad - (\partial^2 \mathbf{r}/\partial q_k\partial q_r)\cdot(\partial \mathbf{r}/\partial q_s) - (\partial \mathbf{r}/\partial q_r)\cdot(\partial^2 \mathbf{r}/\partial q_k\partial q_s)\big]\Big\}\\
&= S\, dm(\partial^2 \mathbf{r}/\partial q_r\partial q_s)\cdot(\partial \mathbf{r}/\partial q_k) \equiv S\, dm(\partial \mathbf{e}_r/\partial q_s)\cdot \mathbf{e}_k\\
&= S\, dm(\partial \mathbf{e}_s/\partial q_r)\cdot \mathbf{e}_k = \Gamma_{k,sr} \quad \text{[recalling (2.5.4a, b)]},
\end{aligned} \tag{3.10.6}$$

from which we conclude that if $\partial \mathbf{e}_r/\partial q_s \equiv \partial^2 \mathbf{r}/\partial q_r\partial q_s = \mathbf{0}$ (e.g., rectilinear coordinates), then $\Gamma_{k,rs} = 0 \Rightarrow \Gamma^k{}_{rs} = 0$. Hence, the following theorem.

THEOREM

In general, *Lagrange's equations are nonlinear*; this is part of the price we pay for using *"generalized" (i.e., curvilinear) coordinates.*

(ii) *Linear terms* like:

(a) $(\partial M_{kr}/\partial t)\dot{q}_r$ clearly result from the *nonstationarity of the inertia coefficients;*

(b) $g_{kr}\,\dot{q}_r$ result from the *nonstationarity of the holonomic (built-in) constraints*

$$(\partial \mathbf{r}/\partial t \neq \mathbf{0} \Rightarrow T_1 \neq 0,\ M_k \neq 0);$$

(c) $\gamma_{kr}\,\dot{q}_r$ [recall *generalized (holonomic) potential* (3.9.8a ff.)] result from the part of the potential that is *linear* in the $\dot{q}$'s.

We notice that both forces corresponding to (b) and (c) type of terms—namely, the inertial $G_k = \sum g_{kr}\,\dot{q}_r$ and the potential $Q_{k,GP} - \partial\gamma_k/\partial t = \sum \gamma_{kr}\,\dot{q}_r$ (part of Q_k)—are *gyroscopic*; that is, they have zero power. And this explains the disappearance of T_1 and V_1 from the generalized energy integral $h \equiv T_2 + (V_0 - T_0) = constant.$

Both forces can be combined as follows: with

$$L_1 \equiv T_1 - V_1 = \sum (M_k - \gamma_k)\dot{q}_k \equiv \sum l_k \dot{q}_k, \tag{3.10.7a}$$

[recalling (3.9.8c)] we obtain, successively,

$$-E_k(L_1) = \cdots = -(\partial M_k/\partial t) + G_k + Q_{k,GP} = -(\partial l_k/\partial t) + \sum l_{kr}\dot{q}_r, \tag{3.10.7b}$$

where the *gyroscopic coefficients* of L_1, l_{kr}, are defined by

$$l_{kr} \equiv g_{kr} + \gamma_{kr} \equiv \partial(M_r - \gamma_r)/\partial q_k - \partial(M_k - \gamma_k)/\partial q_r = \partial l_r/\partial q_k - \partial l_k/\partial q_r. \tag{3.10.7c}$$

We also notice that, whenever such terms appear, since $g_{kk} = 0$ and $\gamma_{kk} = 0$, $\dot{q}_k$ *does not appear in the* (k)th *Lagrangean equation*, say $E_k = Q_k$. Instead, for each such term that appears as $g_{rk}\dot{q}_r$ in the (k)th equation $(r \neq k)$, another term, like $g_{kr}\dot{q}_k = -g_{rk}\dot{q}_k$ appears in the (r)th equation $E_r = Q_r$. For example, for $n = 3$, eq. (3.10.2c), $E_{k,C} = -\sum g_{kr}\dot{q}_r$, yields

$$\begin{aligned} E_{1,C} &= (0)\dot{q}_1 + g_{21}\dot{q}_2 + g_{31}\dot{q}_3, \\ E_{2,C} &= -g_{21}\dot{q}_1 + (0)\dot{q}_2 + g_{32}\dot{q}_3, \\ E_{3,C} &= -g_{31}\dot{q}_1 - g_{32}\dot{q}_2 + (0)\dot{q}_3. \end{aligned} \tag{3.10.7d}$$

This property also appears in the theory of *cyclic* systems (§8.4 ff.).

(d) $-f_{kr}\dot{q}_r$ [recall Rayleigh's dissipation function (3.9.10b ff.)] result from linear viscous friction.

A Compact Notation

Since $q_0 \equiv t \Rightarrow \dot{q} \equiv \dot{t} = 1 \Rightarrow \ddot{q}_0 \equiv \ddot{t} = 0$, and with all Greek indices running from 1 to $n+1$, we may rewrite E_k, (3.10.1i), as follows

$$E_k = \sum M_{k\alpha}\ddot{q}_\alpha + \sum\sum \Gamma_{k,\alpha\beta}\dot{q}_\alpha\dot{q}_\beta, \tag{3.10.8a}$$

where

$$\begin{aligned} 2\Gamma_{k,\alpha\beta} = 2\Gamma_{k,\beta\alpha} &\equiv \partial M_{k\alpha}/\partial q_\beta + \partial M_{k\beta}/\partial q_\alpha - \partial M_{\alpha\beta}/\partial q_k \\ &\left[= 2\int dm(\partial \boldsymbol{e}_\alpha/\partial q_\beta)\cdot \boldsymbol{e}_k\right]; \end{aligned} \tag{3.10.8b}$$

a form, most likely, due to T. Levi-Civita (1895), one of the founders of *tensor calculus*. If the index positioning (i.e., sub-/superscripts) appears arbitrary, it is because we have been trying to avoid tensor calculus and its associated simple and helpful conventions. [For an extensive treatment of these topics via this remarkable and beautiful geometrico-analytical tool, see, e.g., Papastavridis (1999) and references cited therein.]

Indeed, expanding the above, we obtain

$$\begin{aligned} E_k &= \sum M_{kr}\ddot{q}_r + \sum\sum \Gamma_{k,rs}\dot{q}_r\dot{q}_s + 2\sum \Gamma_{k,r,n+1}\dot{q}_r + \Gamma_{k,n+1,n+1} \\ &\equiv \sum M_{kr}\ddot{q}_r + \sum\sum \Gamma_{k,rs}\dot{q}_r\dot{q}_s + 2\sum \Gamma_{k,r}\dot{q}_r + \Gamma_k, \end{aligned} \tag{3.10.8c}$$

where

$$\text{(a)}\qquad 2\Gamma_{k,r,n+1} = 2\Gamma_{k,n+1,r} \equiv 2\Gamma_{k,r} \quad \left[= 2\textstyle\int dm(\partial \boldsymbol{e}_r/\partial t)\cdot \boldsymbol{e}_k\right]$$
$$\equiv \partial M_{kr}/\partial q_{n+1} + \partial M_{k,n+1}/\partial q_r - \partial M_{r,n+1}/\partial q_k$$
$$= \partial M_{kr}/\partial t + (\partial M_k/\partial q_r - \partial M_r/\partial q_k) \equiv \partial M_{kr}/\partial t + g_{rk}, \qquad (3.10.8\text{d})$$

so that the corresponding E_k-term becomes

$$2\sum \Gamma_{k,r}\,\dot q_r = \sum (\partial M_{kr}/\partial t)\dot q_r - \sum g_{kr}\,\dot q_r = \sum (\partial M_{kr}/\partial t)\dot q_r - G_k:$$
$$\textit{nonstationary} \text{ and } \textit{gyroscopic/Coriolis} \text{ terms } (\sim \dot q); \qquad (3.10.8\text{e})$$

and

$$\text{(b)}\qquad 2\Gamma_{k,n+1,n+1} \equiv 2\Gamma_k \quad \left[= 2\textstyle\int dm(\partial \boldsymbol{e}_{n+1}/\partial q_{n+1})\cdot \boldsymbol{e}_k = 2\textstyle\int dm(\partial \boldsymbol{e}_0/\partial t)\cdot \boldsymbol{e}_k\right]$$
$$\equiv \partial M_{k,n+1}/\partial q_{n+1} + \partial M_{k,n+1}/\partial q_{n+1} - \partial M_{n+1,n+1}/\partial q_k$$
$$= 2(\partial M_k/\partial t) - \partial M_0/\partial q_k \equiv 2(\partial M_k/\partial t - \partial T_0/\partial q_k):$$
$$\textit{nonstationary} \text{ and } \textit{centrifugal} \text{ terms (no } \dot q\text{'s)}. \qquad (3.10.8\text{f})$$

In this compact notation, the earlier equations (3.10.3) and (3.10.5) can be written, respectively, as

$$\sum M_{kr}\,\ddot q_r + \sum\sum \Gamma_{k,\alpha\beta}\,\dot q_\alpha \dot q_\beta = Q_k - \partial V_0/\partial q_k, \qquad (3.10.8\text{g})$$

and

$$\ddot q_k + \sum\sum \Gamma^k{}_{\alpha\beta}\,\dot q_\alpha \dot q_\beta$$
$$\left[= \ddot q_k + \sum\sum \Gamma^k{}_{rs}\,\dot q_r \dot q_s + 2\sum \Gamma^k{}_{r,n+1}\,\dot q_r + \Gamma^k{}_{n+1,n+1}\right.$$
$$\left.= \ddot q_k + \sum\sum \Gamma^k{}_{rs}\,\dot q_r \dot q_s + 2\sum \Gamma^k{}_r\,\dot q_r + \Gamma^k\right]$$
$$= \sum m_{kr}(Q_r - \partial V_0/\partial q_s), \qquad (3.10.8\text{h})$$

where

$$\Gamma^k{}_{\alpha\beta} \equiv \sum m_{ks}\Gamma_{s,\alpha\beta}$$
$$\Rightarrow \Gamma^k{}_{r,n+1} \equiv \sum m_{ks}\Gamma_{s,r,n+1} \equiv \Gamma^k{}_{r0} \equiv \Gamma^k{}_r,$$
$$\Gamma^k{}_{n+1,n+1} \equiv \sum m_{ks}\Gamma_{s,n+1,n+1} \equiv \Gamma^k{}_{00} \equiv \Gamma^k. \qquad (3.10.8\text{i})$$

We notice that $\Gamma_{n+1,n+1,n+1} \equiv \Gamma_{0,00} \equiv \Gamma_0 = \cdots = (1/2)(\partial M_0/\partial t) = \partial T_0/\partial t$ does not appear in the equations of motion; but it might appear in special forms of power equations.

Explicit Forms of Hamel's Equations

Let us, next, extend the above to *nonholonomic* variables; that is, find the explicit form of the Hamel acceleration (and equations)

$$I_k \equiv (\partial T^*/\partial \omega_k)^{\cdot} - \partial T^*/\partial \theta_k + \sum\sum \gamma^r{}_{k\alpha}(\partial T^*/\partial \omega_r)\omega_\alpha, \qquad (3.10.9\text{a})$$

where

$$2T = \sum\sum M^*{}_{\alpha\beta}\,\omega_\alpha\omega_\beta. \tag{3.10.9b}$$

Nonholonomic Christoffel-Like Symbols

For algebraic simplicity, but no loss in generality, we restrict ourselves to the *stationary* case. By $\partial \ldots / \partial\theta_k$-differentiating the nonholonomic inertia coefficients

$$M^*{}_{kl} = M^*{}_{lk} \equiv S\, dm\, \boldsymbol{\varepsilon}_k \cdot \boldsymbol{\varepsilon}_l \equiv S\, dm(\partial \mathbf{r}/\partial\theta_k)\cdot(\partial \mathbf{r}/\partial\theta_l), \tag{3.10.9c}$$

an operation that we shall also denote here by a subscript comma [i.e., $\partial(\ldots)/\partial\theta_k \equiv (\ldots)_{,k}$], we obtain

$$\partial M^*{}_{kl}/\partial\theta_r = S\, dm(\boldsymbol{\varepsilon}_{k,r}\cdot\boldsymbol{\varepsilon}_l + \boldsymbol{\varepsilon}_k\cdot\boldsymbol{\varepsilon}_{l,r}),$$

and, therefore, the *nonholonomic Christoffel symbol-like* quantities, defined in complete analogy with their holonomic counterparts (3.10.1d) as

$$2\Gamma^*{}_{k,rs} \equiv \partial M^*{}_{ks}/\partial\theta_r + \partial M^*{}_{kr}/\partial\theta_s - \partial M^*{}_{rs}/\partial\theta_k, \tag{3.10.9d}$$

transform, successively, to

$$\begin{aligned}2\Gamma^*{}_{k,rs} &\equiv S\, dm\big[(\boldsymbol{\varepsilon}_k\cdot\boldsymbol{\varepsilon}_{s,r} + \boldsymbol{\varepsilon}_s\cdot\boldsymbol{\varepsilon}_{k,r}) + (\boldsymbol{\varepsilon}_k\cdot\boldsymbol{\varepsilon}_{r,s} + \boldsymbol{\varepsilon}_r\cdot\boldsymbol{\varepsilon}_{k,s}) - (\boldsymbol{\varepsilon}_r\cdot\boldsymbol{\varepsilon}_{s,k} + \boldsymbol{\varepsilon}_s\cdot\boldsymbol{\varepsilon}_{r,k})\big]\\ &= S\, dm\big[\boldsymbol{\varepsilon}_k\cdot(\boldsymbol{\varepsilon}_{s,r} + \boldsymbol{\varepsilon}_{r,s}) + \boldsymbol{\varepsilon}_r\cdot(\boldsymbol{\varepsilon}_{k,s} - \boldsymbol{\varepsilon}_{s,k}) + \boldsymbol{\varepsilon}_s\cdot(\boldsymbol{\varepsilon}_{k,r} - \boldsymbol{\varepsilon}_{r,k})\big];\end{aligned}$$

and recalling the fundamental *noncommutativity/nonintegrability* relations (2.10.23), rewritten here as

$$\begin{aligned}\partial^2\mathbf{r}/\partial\theta_s\partial\theta_k - \partial^2\mathbf{r}/\partial\theta_k\partial\theta_s &\equiv \partial\boldsymbol{\varepsilon}_k/\partial\theta_s - \partial\boldsymbol{\varepsilon}_s/\partial\theta_k \equiv \boldsymbol{\varepsilon}_{k,s} - \boldsymbol{\varepsilon}_{s,k}\\ &= \sum \gamma^l{}_{sk}(\partial\mathbf{r}/\partial\theta_l) \equiv \sum \gamma^l{}_{sk}\,\boldsymbol{\varepsilon}_l,\end{aligned} \tag{3.10.9e}$$

we obtain, finally,

$$\begin{aligned}2\Gamma^*{}_{k,rs} &= S\, dm\Big[\boldsymbol{\varepsilon}_k\cdot(\boldsymbol{\varepsilon}_{s,r} + \boldsymbol{\varepsilon}_{r,s}) + \boldsymbol{\varepsilon}_r\cdot\Big(\sum \gamma^l{}_{sk}\boldsymbol{\varepsilon}_l\Big) + \boldsymbol{\varepsilon}_s\cdot\Big(\sum \gamma^l{}_{rk}\,\boldsymbol{\varepsilon}_l\Big)\Big]\\ &= S\, dm\,\boldsymbol{\varepsilon}_k\cdot(\boldsymbol{\varepsilon}_{s,r} + \boldsymbol{\varepsilon}_{r,s}) + \sum(\gamma^l{}_{sk}M^*{}_{rl} + \gamma^l{}_{rk}M^*{}_{sl}).\end{aligned} \tag{3.10.9f}$$

Nonholonomic Euler–Lagrange Terms

Next, differentiating the stationary version of (3.10.9b)

$$2T = \sum\sum M^*{}_{kl}\,\omega_k\,\omega_l, \tag{3.10.9g}$$

we obtain $\partial T^*/\partial\omega_k = \sum M^*{}_{kl}\omega_l$ and, therefore,

$$(\partial T^*/\partial\omega_k)^{\cdot} = \sum M^*{}_{kl}\dot{\omega}_l + \sum(dM^*{}_{kl}/dt)\,\dot{\omega}_l,$$

or, since

$$dM^*_{kl}/dt = \sum (\partial M^*_{kl}/\partial q_r)\dot{q}_r = \sum (\partial M^*_{kl}/\partial q_r)\left(\sum A_{rs}\omega_s\right)$$
$$= \sum\left(\sum A_{rs}(\partial M^*_{kl}/\partial q_r)\right)\omega_s \equiv \sum (\partial M^*_{kl}/\partial \theta_s)\omega_s,$$

we get,

$$(\partial T^*/\partial \omega_k)^{\cdot} = \sum M^*_{kl}\dot{\omega}_l + \sum\sum (\partial M^*_{kl}/\partial \theta_s)\omega_s\omega_l. \tag{3.10.9h}$$

Introducing the above into the stationary version of (3.10.9a) results in

$$I_k \equiv d/dt(\partial T^*/\partial \omega_k) - \partial T^*/\partial \theta_k + \sum\sum \gamma^r_{kl}(\partial T^*/\partial \omega_r)\,\omega_l,$$
$$= \sum M^*_{kl}(d\omega_l/dt) + \sum\sum\sum \gamma^r_{kl} M^*_{rs}\omega_s\omega_l$$
$$+ \sum\sum [\partial M^*_{kl}/\partial \theta_s - (1/2)(\partial M^*_{sl}/\partial \theta_k)]\,\omega_s\omega_l, \tag{3.10.9i}$$

or, since the third (double sum) term equals

$$(1/2)\sum\sum (\partial M^*_{ks}/\partial \theta_l + \partial M^*_{kl}/\partial \theta_s - \partial M^*_{sl}/\partial \theta_k)\omega_s\omega_l$$
$$\equiv \sum\sum \Gamma^*_{k,ls}\omega_s\omega_l, \tag{3.10.9j}$$

we finally obtain the (3.10.3b)-like form

$$I_k = \sum M^*_{kl}(d\omega_l/dt) + \sum\sum \Lambda_{k,ls}\omega_s\omega_l, \tag{3.10.9k}$$

where

$$\Lambda_{k,ls} \equiv \Gamma^*_{k,ls} + \sum \gamma^r_{kl}M^*_{rs}; \tag{3.10.9l}$$

that is, it is the just introduced quantities $\Lambda_{k,ls}$ that deserve to be called *nonholonomic Christoffel symbols of the first kind*, rather than the formally similar to the holonomic ones $\Gamma^*_{k,ls}$, eqs. (3.10.9d). Finally, it is not hard to see that we can replace in (3.10.9k) the $\Lambda_{k,ls}$ with their *symmetric* parts (1/2) $(\Lambda_{k,ls} + \Lambda_{k,sl})$. Let the reader extend these results to the nonstationary case. [For a detailed tensorial treatment of these topics, see, for example, Papastavridis (1999, chaps. 6, 7).]

Problem 3.10.1 *Explicit Form of Chaplygin's Equations* [recall (3.8.13a ff.)]. Consider a Chaplygin system; that is, one with constraints:

$$\dot{q}_D = \sum b_{DI}\dot{q}_I, \qquad (D, D', D'' \ldots = 1, \ldots, m;\ I, I', I'' \ldots = m+1, \ldots, n), \tag{a}$$

where

$$b_{DI} = b_{DI}(q_1, \ldots, q_m) \equiv b_{DI}(q_I).$$

(i) Show that its (double) kinetic energy becomes

$$2T \equiv \sum\sum M_{kl}\dot{q}_k\dot{q}_l$$
$$= \cdots = \sum\sum M_{II'o}\dot{q}_I\dot{q}_{I'} = \sum\sum m_{II'}\dot{q}_I\dot{q}_{I'} \equiv 2T_o(q_I, \dot{q}_I) = 2T_o, \tag{b}$$

where

$$M_{kl} = M_{kl}(q_I) \qquad (k, l = 1, \ldots, n), \tag{c1}$$

$$M_{II'o} \equiv M_{II'} + 2\sum b_{DI'}M_{DI} + \sum b_{DI}b_{D'I'}M_{DD'} \quad (\neq M_{I'Io}), \tag{c2}$$

$$2m_{II'} = 2m_{I'I} \equiv M_{II'o} + M_{I'Io} \qquad (\textit{functions of the } q_I\text{'s}). \tag{c3}$$

(ii) Then show, by differentiating (b), that Chaplygin's equations [recall (3.8.13o)]

$$(\partial T_o/\partial \dot{q}_I)^\cdot - \partial T_o/\partial q_I - \sum\sum t^D{}_{II'}(\partial T/\partial \dot{q}_D)_o \dot{q}_{I'} = Q_{Io}, \tag{d}$$

where

$$-t^D{}_{II'} \equiv \partial b_{DI'}/\partial q_I - \partial b_{DI}/\partial q_{I'} = \textit{Chaplygin coefficients} \quad (= t^D{}_{I'I}), \tag{d1}$$

$$Q_{Io} \equiv Q_I + \sum b_{DI}Q_D, \tag{d2}$$

assume the explicit, (3.10.9k)-like, form

$$\sum m_{II'}\ddot{q}_{I'} + \sum\sum \lambda_{I,I'I''}\dot{q}_{I'}\dot{q}_{I''} = Q_{Io}, \tag{e}$$

where

$$\lambda_{I,I'I''} \equiv \Gamma'_{I,I'I''} + \sum \left(M_{DI''} + \sum b_{D'I''}M_{DD'}\right)t^D{}_{II'} \quad (\neq \lambda_{I,I''I'}), \tag{e1}$$

$$2\Gamma'_{I,I'I''} \equiv 2\Gamma'_{I,I''I'} = \partial m_{II''}/\partial q_{I'} + \partial m_{II'}/\partial q_{I''} - \partial m_{I'I''}/\partial q_I$$
$$= (\textit{double})\ \textit{first-kind Christoffels, based on the } \{m_{II'}\}. \tag{e2}$$

Note that: (i) in (e), the $\lambda_{I,I'I''}$ may be replaced by their *symmetric* parts in their last two subscripts: $(\lambda_{I,I'I''} + \lambda_{I,I''I'})/2$; and (ii) equations (e) look like the Lagrangean equations of a scleronomic and holonomic system with $n - m$ Lagrangean coordinates q_I, kinetic energy given by (b), first-kind Christoffels = the symmetric parts of the $\lambda_{I,I'I''}$, and under the impressed forces Q_{Io}.

Linear Variational Equations; or Method of Small Oscillations (Routh, Poincaré et al.)

Let us consider, with no real loss of generality, a *scleronomic* and *holonomic* system S with equations of motion

$$(\partial T/\partial \dot{q}_k)^\cdot - \partial T/\partial q_k = Q_k, \qquad \text{where} \ \ 2T = \sum\sum M_{kl}\dot{q}_k\dot{q}_l \qquad (k, l = 1, \ldots, n), \tag{3.10.10}$$

in a completely known state of motion (or equilibrium), henceforth referred to as the *fundamental*, or *undisturbed*, state I; and given by the known particular solution(s) to (3.10.10) $q_k = f_k(t)$. Below, we examine the continuous motions of S in the *neighborhood* of $I, I + \Delta(I) \equiv II$, resulting from *small disturbances* (in some sense) applied to I. Such a study has a twofold usefulness: (i) it informs us about the stability/instability of the original state I; and/or (ii) helps us to understand, approximately, the general motion of S whenever the exact solution of (3.10.10) is beyond our reach. Since this is an approximate method, with no error analysis available, its results should be applied with caution; for example, in the case of *finite* disturbances on I, described by *nonlinear* perturbation equations, it may lead to completely false results.

Linear Perturbation Equations

Let the solution(s) of (3.10.10) for the perturbed state II be $q_k = f_k(t) + x_k(t)$, where $x_k = x_k(t)$ is the small perturbation describing the temporal evolution of $II - I \equiv \Delta(I) \approx \delta(I)$. [Generally, we use $\delta(\ldots)$ for *first*-order changes and $\Delta(\ldots)$ for *total* changes, from I.] Then, with

$$T(II) \equiv T(q, \dot{q}) \equiv T(f + x, \dot{f} + \dot{x}) \equiv T, \qquad T(I) \equiv T(f, \dot{f}) \equiv T_o, \qquad (3.10.10a)$$

and *all* M_{kl}*-derivatives evaluated at* I, we obtain, to the *second* $x, \dot{x}$-order,

$$2T = \sum\sum \Big\{M_{kl} + \Big(\sum (\partial M_{kl}/\partial q_r)x_r + (1/2)\sum\sum (\partial^2 M_{kl}/\partial q_r \partial q_s)x_r x_s\Big)\Big\}(\dot{f}_k + \dot{x}_k)(\dot{f}_l + \dot{x}_l)$$
$$= \cdots = 2(T_o + \Delta T) = 2(T_o + \Delta T_1 + \Delta T_2), \qquad (3.10.10b)$$

where

$$2T_o \equiv \sum\sum M_{kl}\dot{f}_k \dot{f}_l, \qquad (3.10.10c)$$

$$\Delta T_1 \equiv \sum (\alpha_k x_k + \beta_k \dot{x}_k), \qquad (3.10.10d)$$

$$2\alpha_k \equiv \sum\sum (\partial M_{rl}/\partial q_k)\dot{f}_r \dot{f}_l \equiv \sum \varepsilon_{lk}\dot{f}_l, \qquad \beta_k \equiv \sum M_{kl}\dot{f}_l; \qquad (3.10.10e)$$

$$2\Delta T_2 \equiv \sum\sum \mu_{kl}\dot{x}_k \dot{x}_l + 2\sum\sum \varepsilon_{kl}\dot{x}_k x_l + \sum\sum \zeta_{kl}x_k x_l$$
$$(\equiv 2\Delta T_{2,2} + 2\Delta T_{2,1} + 2\Delta T_{2,0}), \qquad (3.10.10f)$$

$$\mu_{kl} \equiv M_{kl} \quad (= \mu_{lk}), \qquad (3.10.10.g)$$

$$\varepsilon_{kl} \equiv \sum (\partial M_{kr}/\partial q_l)\dot{f}_r = \sum (\partial M_{rk}/\partial q_l)\dot{f}_r \quad (\neq \varepsilon_{lk}, \text{ in general}), \qquad (3.10.10h)$$

$$2\zeta_{kl} \equiv \sum\sum (\partial^2 M_{rs}/\partial q_k \partial q_l)\dot{f}_r \dot{f}_s \quad (= 2\zeta_{lk}). \qquad (3.10.10i)$$

Similarly, with

$$Q_k(II) \equiv Q_k(t, q, \dot{q}) = Q_k(t, f + x, \dot{f} + \dot{x}) \equiv Q_k,$$
$$Q_k(I) \equiv Q_k(t, f, \dot{f}) \equiv Q_{k,o}, \qquad (3.10.10j)$$

and *all* Q_k*-derivatives evaluated at* I, we obtain, to the *first* $x, \dot{x}$-order,

$$Q_k = Q_{k,o} + \sum (\eta_{kl}x_l + \theta_{kl}\dot{x}_l), \qquad (3.10.10k)$$

$$\eta_{kl} \equiv \partial Q_k/\partial q_l \quad (\neq \eta_{lk}, \text{ in general}), \qquad \theta_{kl} \equiv \partial Q_k/\partial \dot{q}_l \quad (\neq \theta_{lk}, \text{ in general}). \qquad (3.10.10l)$$

Now, since the f_k and $\dot{f}_k$ (i.e., the fundamental state I), are *known functions of time* (equal to constants or zero in the case of equilibrium), T can be viewed as the *(approximate) kinetic energy of a rheonomic system with hitherto unknown and uncon-*

strained Lagrangean coordinates x_k recalling ex. 3.9.6). Therefore, the *equations of motion of the adjacent state II* are the *Lagrangean equations for the* x_k:

$$(\partial T/\partial \dot{x}_k)^{\cdot} - \partial T/\partial x_k = Q_k. \tag{3.10.11}$$

But, by (3.10.10a–l),

$$\partial T/\partial x_k = \alpha_k + \sum (\zeta_{kl} x_l + \varepsilon_{lk} \dot{x}_l), \tag{3.10.11a}$$

$$\partial T/\partial \dot{x}_k = \beta_k + \sum (\varepsilon_{kl} x_l + \mu_{kl} \dot{x}_l), \tag{3.10.11b}$$

and the undisturbed state I satisfies the equations

$$(\partial T_o/\partial \dot{f}_k)^{\cdot} - \partial T_o/\partial f_k = Q_{k,o},$$

$$\Big(\sum M_{kl} \dot{f}_l\Big)^{\cdot} - (1/2) \sum\sum (\partial M_{rl}/\partial q_k) \dot{f}_r \dot{f}_l \equiv d\beta_k/dt - \alpha_k = Q_{k,o}, \tag{3.10.11c}$$

i.e., (3.10.10) with $x = 0$, $\dot{x} = 0$. Therefore, eqs. (3.10.11), finally, assume the following form of *linear(ized) and homogeneous perturbation equations:*

$$\sum \{\mu_{kl} \ddot{x}_l + [(\varepsilon_{kl} - \varepsilon_{lk}) + \dot{\mu}_{kl}] \dot{x}_l + (\dot{\varepsilon}_{kl} - \zeta_{kl}) x_l\} = \sum (\eta_{kl} x_l + \theta_{kl} \dot{x}_l). \tag{3.10.12}$$

REMARKS

(i) These equations can also be obtained by substituting $q_k = f_k + x_k$, $\dot{q}_k = \dot{f}_k + \dot{x}_k$ in $\partial T/\partial \dot{q}_k$ and $\partial T/\partial q_k$, expanding à la Taylor around I, and keeping only up to *linear* terms in the $x, \dot{x}$:

$$\text{(a)}\quad \partial T/\partial \dot{q}_k = \sum (M_{kl} \dot{f}_l + M_{kl} \dot{x}_l) + \sum\sum [(\partial M_{kl}/\partial q_r) \dot{f}_l] x_r,$$

$$\Rightarrow (\partial T/\partial \dot{q}_k)^{\cdot} = \sum (M_{kl} \ddot{f}_l + \dot{M}_{kl} \dot{f}_l + M_{kl} \ddot{x}_l + \dot{M}_{kl} \dot{x}_l) + \sum\sum \Big\{(\partial M_{kl}/\partial q_r) \dot{f}_l \dot{x}_r + [(\partial M_{kl}/\partial q_r) \dot{f}_l]^{\cdot} x_r\Big\}, \tag{3.10.11d}$$

$$\text{(b)}\quad \partial T/\partial q_k = (1/2) \sum\sum (\partial M_{rl}/\partial q_k) \dot{f}_r \dot{f}_l + \sum\sum (\partial M_{rl}/\partial q_k) \dot{f}_r \dot{x}_l + (1/2) \sum\sum\sum [(\partial^2 M_{rs}/\partial q_l \partial q_k) \dot{f}_r \dot{f}_s] x_l, \tag{3.10.11e}$$

and similarly for Q_k [as in (3.10.10j–l)]; and then inserting these values in (3.10.10) while noting that, since the $f_k(t)$ describe the fundamental state, (3.10.11c) holds:

$$\sum (M_{kl} \ddot{f}_l + \dot{M}_{kl} \dot{f}_l) - (1/2) \sum\sum (\partial M_{rl}/\partial q_k) \dot{f}_r \dot{f}_l = Q_{k,o}. \tag{3.10.11f}$$

Again, the result is eqs. (3.10.12).

(ii) If, during the perturbed motion $\Delta(I)$, an *additional* force X_k, *not provided by the expansion* (3.10.10k) occurs, then such a term should be added to the right side of (3.10.12). Here, we shall assume that $X_k = 0$.

Since $f_r = f_r(t) =$ *known function of time*, so are the coefficients $\mu, \varepsilon, \zeta, \eta, \theta$ in (3.10.12). However, from the mathematical viewpoint, even such a *linear but variable coefficient system* is (or can be) quite complicated. Therefore, to make some headway, from now on we shall restrict ourselves to the special case where all *these coefficients are constant in time.* Then, (3.10.12) reduces to the constant coefficient system:

$$\sum [\mu_{kl}\ddot{x}_l + (\varepsilon_{kl} - \varepsilon_{lk})\dot{x}_l - \zeta_{kl}x_l] = \sum (\eta_{kl}x_l + \theta_{kl}\dot{x}_l); \tag{3.10.13}$$

whose mathematical theory is well known (see below).

Steady Motion

A fundamental state whose *linear perturbational equations* have *constant coefficients*, like (3.10.13), is called a *state of steady motion* (Routh, 1877), or, sometimes (but not quite correctly), *stationary* motion. Common examples of such a state are (i) absolute or relative *equilibrium* (in which case, the f_k are constant or zero); (ii) *cyclic* systems undergoing "isocyclic" motions [i.e., certain of their coordinates (the "nonignorable" ones) and certain of their velocities (the "ignorable" ones) remain constant (§ 8.5)].

We begin our study of steady motion by noting that, in such a state, since the $\mu, \varepsilon, \zeta, \eta, \theta$ are constant, the perturbed motion $x_k(t)$ is *independent* of the particular instant at which the disturbance is applied to that state.

Next, let us examine closely the right (perturbed force) side of (3.10.13). Following Kelvin and Tait, we call the x-proportional terms *positional* forces, and the $\dot{x}$-proportional terms *motional* forces. Each of these terms can be further subdivided into its *symmetric* and *antisymmetric* parts; the latter are defined, respectively, by the following unique decompositions:

$$\eta_{kl} = \eta'_{kl} + \eta''_{kl}, \qquad \theta_{kl} = \theta'_{kl} + \theta''_{kl}, \tag{3.10.14}$$

where the symmetric parts (single accents) are defined by

$$\eta'_{kl} = \eta'_{lk} \equiv (1/2)(\eta_{kl} + \eta_{lk}), \qquad \theta'_{kl} = \theta'_{lk} \equiv (1/2)(\theta_{kl} + \theta_{lk}), \tag{3.10.14a}$$

and the antisymmetric ones (double accents) by

$$\eta''_{kl} = -\eta''_{lk} \equiv (1/2)(\eta_{kl} - \eta_{lk}), \qquad \theta''_{kl} = -\theta''_{lk} \equiv (1/2)(\theta_{kl} - \theta_{lk}); \tag{3.10.14b}$$

that is, $\eta'_{kk} = \eta_{kk}$, $\eta''_{kk} = 0$ and $\theta'_{kk} = \theta_{kk}$, $\theta''_{kk} = 0$. The so-resulting *four* types of forces we classify as follows:

(i)
$$\sum \eta'_{kl}x_l \equiv \eta'_k = \textit{potential positional forces}, \tag{3.10.14c}$$

derivable from the potential:

$$V' = -(1/2)\sum\sum \eta'_{kl}x_k x_l \;\Rightarrow\; -(\partial V'/\partial x_k) = \eta'_k; \tag{3.10.14d}$$

and whose inertial counterparts are the $-\zeta_{kl}x_l$ terms in (3.10.13).

(ii)
$$\sum \eta''_{kl}x_l \equiv \eta''_k = \textit{nonpotential } (\Rightarrow \textit{nonconservative}), \textit{ or circulatory},$$
$$\textit{positional forces} = (1/2)\sum (\partial\eta''_k/\partial x_l - \partial\eta''_l/\partial x_k)x_l, \tag{3.10.14e}$$

where the η''_{kl} are referred to as *vorticity* coefficients. [Such forces are also called *artificial* (Thomson and Tait), since their work over a closed route of configurations is nonzero; and so, upon repetition of that cycle, they can produce unbounded amounts of energy].

(iii) $$\sum \theta'_{kl}\dot{x}_l \equiv \theta'_k = \textit{damping motional forces}, \tag{3.10.14f}$$

derivable from the *Rayleigh dissipation function* (3.9.10a ff.; with $F \to D'$)

$$D' \equiv -(1/2)\sum\sum \theta'_{kl}\dot{x}_k\dot{x}_l \Rightarrow -(\partial D'/\partial \dot{x}_k) = \theta'_k. \tag{3.10.14g}$$

If the (perturbational) power of these forces:

$$\sum \theta'_k\dot{x}_k = \sum\sum \theta'_{kl}\dot{x}_k\dot{x}_l = -2D',$$

is *negative definite* (*in the* $\dot{x}$), then damping is called *complete*; if it is only *negative semidefinite* (i.e., it may vanish for some $\dot{x} \neq 0$), then it is called *pervasive*. (This difference does matter in stability questions.)

(iv) $$\sum \theta''_{kl}\dot{x}_l \equiv \theta''_k = \textit{gyroscopic motional forces}, \tag{3.10.14h}$$

derivable from the gyroscopic *function*

$$\Theta'' = -\sum\sum \theta''_{kl}\,x_k\,\dot{x}_l \;\Rightarrow\; -(\partial\Theta''/\partial x_k) = \theta''_k, \tag{3.10.14i}$$

and whose inertial counterparts are the $(\varepsilon_{kl} - \varepsilon_{lk})\dot{x}_l$ terms in (3.10.13).

To understand these forces and their effects on the disturbance $x_k(t)$ better, let us form the *power equation of the perturbed motion*: multiplying (3.10.13) with $\dot{x}_k$ and summing over k, while noting that the gyroscopic contributions from both sides vanish, we obtain

$$d(\Delta h)/dt = C - 2D', \tag{3.10.15}$$

where [recalling (3.10.10f)]

$$2\Delta h \equiv \sum\sum(\mu_{kl}\dot{x}_k\dot{x}_l - \zeta_{kl}x_kx_l) + 2V'$$
$$\equiv 2[\Delta T_{2,2} + (V' - \Delta T_{2,0})] = 2\ \textit{(generalized) energy of disturbance}, \tag{3.10.15a}$$
$$C \equiv \sum \eta''_k\dot{x}_k = \sum\sum \eta''_{kl}x_l\dot{x}_k = \textit{circulatory power}. \tag{3.10.15b}$$

Hence, if $C = 0$, then $2D'$ represents the *rate of decrease of the perturbational energy* Δh.

For *stability* investigations (see below), it is convenient to bring all terms of (3.10.13) on the same side and group them appropriately as follows (with some renaming, to conform with standard contemporary practices):

$$\sum [M_{kl}\ddot{x}_l + (D_{kl} + G_{kl})\dot{x}_l + (K_{kl} + N_{kl})x_l] = 0, \tag{3.10.16}$$

where

$M_{kl} = M_{lk} \equiv \mu_{kl} =$ *coefficients of inertia/mass*

$[\mathbf{M} = (M_{kl})$: symmetric and positive definite matrix$]$, (3.10.16a)

$D_{kl} = D_{lk} \equiv -\theta'_{kl} =$ *damping coefficients*

$[\mathbf{D} = (D_{kl})$: symmetric matrix; if positive definite: complete damping, if positive semidefinite: pervasive damping$]$, (3.10.16b)

$G_{kl} = -G_{lk} \equiv (\varepsilon_{kl} - \varepsilon_{lk}) - \theta''_{kl} =$ *gyroscopic coefficients*

$[\mathbf{G} = (G_{kl})$: antisymmetric matrix, no general sign properties$]$, (3.10.16c)

$K_{kl} = K_{lk} \equiv -(\zeta_{kl} + \eta'_{kl}) =$ *conservative positional coefficients*

$[\mathbf{K} = (K_{kl})$: symmetric matrix; if positive definite, then static stability$]$, (3.10.16d)

$N_{kl} = -N_{lk} \equiv -\eta''_{kl} =$ *nonconservative positional, or circulatory, coefficients*

$[\mathbf{N} = (N_{kl})$: antisymmetric matrix, no general sign properties$]$. (3.10.16e)

Stability of Steady Motion (see also § 8.6)

Substituting into (3.10.16) $x_k = x_k(t) = X_k \exp(\lambda t)$ ($X_k =$ constant amplitude, depending on the initial conditions, and λ an exponent to be determined), and requiring nontrivial solutions, we are led in well-known ways to the system's *secular* or *characteristic* equation

$$\Delta(\lambda) \equiv \left| M_{kl}\lambda^2 + (D_{kl} + G_{kl})\lambda + (K_{kl} + N_{kl}) \right| = 0, \tag{3.10.17}$$

or, if expanded,

$$\Delta(\lambda) \equiv a_0\lambda^m + a_1\lambda^{m-1} + a_2\lambda^{m-2} + \cdots + a_{m-1}\lambda + a_m = 0, \tag{3.10.18}$$

where $m = 2n$, all coefficients are *real*, and (by Viète's rules, or by induction)

$$a_0 = \left|M_{kl}\right| > 0 \qquad \text{and} \qquad a_m = \left|K_{kl} + N_{kl}\right|. \tag{3.10.18a}$$

Brief Detour/Summary of Relevant Fundamentals of the Theory of Stability of Motion

DEFINITION

A (fundamental) state of motion I is called stable, relative to bounded initial disturbances (i.e., initial condition changes), if the resulting perturbation from it, $\Delta(I)$, also remains bounded for all subsequent time. More precisely, let $y = y(t) \equiv (x, \dot{x})$ and $t_{\text{initial}} \equiv t_i$. Then, I is stable if, for any constant $\varepsilon > 0$, another constant $\delta = \delta(\varepsilon) > 0$ can be found such that, from $|y_i \equiv y(t_i)| < \delta$, it follows that $|y(t)| < \varepsilon$ for all $t > t_i$; that is, I is stable if it is possible to keep y as small as we wish by appropriately restricting its initial value y_i. The intuitive/popular understanding of a stable state of motion (or equilibrium) as one in which "the smaller the initial disturbance, the smaller the subsequent perturbation from it" corresponds, clearly, to the special case where $\delta(\ldots)$ is a *monotonically decreasing* function of ε. (Outside of the absolute value $|\ldots|$, other "norms" $\|\ldots\|$ can be selected.) In many applications, however, such boundedness of $\Delta(I)$ is not enough; there, for stability, the disturbance must also diminish in time, and eventually die away, that is, all so perturbed motions must tend toward I as time increases indefinitely; mathematically: $|y| \to 0$, as $t \to \infty$. This, sharper, type of stability is called *asymptotic stability*, while the earlier one requiring only $\Delta(I)$-boundedness is referred to as *stability in the sense of Lagrange*. If I is stable for any size initial disturbance, then I is called *totally* or *globally* stable; while if it is stable only for "small" initial disturbances, then it is called, simply, stable (e.g., a ship safe for ocean voyages vs. a ship safe only for Mediterranean sea voyages). Clearly, in practice, only the latter type of stability is serviceable. For nonlinear systems in particular, the initial disturbances must be small enough so that the perturbed motions are still controlled by the fundamental motion I. (We should remark that, since, out of nonlinear equations of motion, qualitatively new and unexpected phenomena may emerge, no single definition of stability of motion, that is uniformly physically meaningful and technically useful, is possible or desirable—stability is a human-made *condition*, not an ever valid and exceptionless physical law, like the equations of motion. As the distinguished applied mathematician R. Bellman put it, "stability is a much overburdened word with an unstabilized definition." Below, only the practically important asymptotic stability is examined.)

Usually, the exact equations of a perturbation $\Delta(I)$, from I, consist of a linear part [which here is assumed to (exist and) be the *constant coefficient*, or *autonomous*, system (3.10.16)] and of a nonlinear part. Now, it is shown in the theory of stability [A. M. Lyapounov's "first approximation" (early 1890s); also H. Poincaré's "équations aux variations" (1892)] that for such a system:

1. If the real parts of all roots of its characteristic equation (3.10.17, 18) are negative, then the fundamental state I is asymptotically stable, irrespectively of the nonlinear terms of $\Delta(I)$; that is, our linearized analysis suffices to establish the asymptotic stability of I.
2. If even one of the roots of (3.10.17, 18) has a positive real part, then I is unstable, irrespectively of the nonlinear terms of $\Delta(I)$; again, the linearized analysis suffices.
3. Critical (or neutral, or marginally stable) case: If even one of the roots of (3.10.17, 18) has zero real part while its remaining roots, if any, have negative real parts [i.e., if the linearized perturbations are stable but not asymptotically stable—provided that those zero-real-part roots are *distinct*, so that their contributions to the general solution of (3.10.16) have no t-proportional (*secular*) terms, otherwise I is unstable as in the second case], the stability of I cannot be decided from the first approximation (3.10.16), we must also examine the nonlinear part of the exact perturbation equations; the linearized analysis does not suffice! Physically, the presence of a root with zero real part indicates an exact balance among certain of the system's physical properties/parameters and associated forces. Such systems *may* be *structurally unstable*, that is, they may be such that, if their parameters and forces are subjected to small variations, the nature of their motions changes completely, for example, from oscillatory to nonoscillatory. So, in practical terms (i.e., unavoidable imperfections/irregularities/impurities, etc.), the critical case should be classified as (nonlinearly) unstable!

For these reasons, the behavior of the linearized system (3.10.16) in cases 1 and 2 is called *significant* (i.e., conclusive), while that in case 3 is called *nonsignificant* (i.e. inconclusive). Obviously: (i) If the linear part of $\Delta(I)$ is absent, the above results do not apply; while (ii) if its nonlinear part is absent, then we can safely conclude that the state I is: case 1, asymptotically stable; case 2, unstable; and case 3, stable/not asymptotically stable.]

From the above it follows that, since the imaginary parts of the roots of (3.10.18) do not affect the stability of I, both ordinary and/or asymptotic, it is not necessary to actually solve (3.10.18), just check the sign of the real part of its roots. This is achieved by several (necessary and/or sufficient) criteria of various degrees of generality and ease of application. Below we describe two of the most well-known such criteria: those of *Routh* (1876–1877) and *Hurwitz* (1895) [also *Clifford* (1868) and *Hermite* (1850)].

(i) *Criterion of Routh.* Let us build the following array of *Routh coefficients*:

$$\begin{array}{llll}
a_0 & a_2 & a_4 & \cdots \\
a_1 & a_3 & a_5 & \cdots \\
b_1 \equiv (a_1 a_2 - a_0 a_3)/a_1 & b_2 \equiv (a_1 a_4 - a_0 a_5)/a_1 & b_3 \equiv (a_1 a_6 - a_0 a_7)/a_1 & \cdots \\
c_1 \equiv (b_1 a_3 - a_1 b_2)/b_1 & c_2 \equiv (b_1 a_5 - a_1 b_3)/b_1 & c_3 \equiv (b_1 a_7 - a_1 b_4)/b_1 & \cdots \\
d_1 \equiv (c_1 b_2 - b_1 c_2)/c_1 & d_2 \equiv (c_1 b_3 - b_1 c_3)/c_1 & d_3 \equiv (c_1 b_4 - b_1 c_4)/c_1 & \cdots \\
\cdots\cdots & \cdots\cdots & \cdots\cdots &
\end{array}$$

that is, its first (second) row consists of the *even* (*odd*) coefficients of (3.10.18); also, $a_\bullet = 0$, for $\bullet > m$. Now, all the roots of the characteristic equation have *negative real parts* ($\Rightarrow$ the fundamental state I is *asymptotically stable*) if and only if *all the elements of the first column of the above table are positive*; that is, if and only if

$$a_0 > 0, \qquad a_1 > 0, \qquad b_1 > 0, \qquad c_1 > 0, \qquad d_1 > 0, \ldots; \tag{3.10.18b}$$

or, more generally, *if they have the same sign*—it can be shown that the number of roots with positive real parts ($\Rightarrow$ instability) equals the number of sign changes.

(ii) *Criterion of Hurwitz.* Let us build the following m *Hurwitz determinants*:

$$H_h \equiv \begin{vmatrix} a_1 & a_3 & a_5 & \cdots & a_{2h-1} \\ a_0 & a_2 & a_4 & \cdots & a_{2h-2} \\ 0 & a_1 & a_3 & \cdots & a_{2h-3} \\ 0 & a_0 & a_2 & \cdots & a_{2h-4} \\ \cdots & \cdots & \cdots & \cdots & \cdots \\ 0 & 0 & 0 & \cdots & a_h \end{vmatrix} \qquad (h = 1, 2, \ldots, m), \tag{3.10.18c}$$

that is, we build H_m and its $m - 1$ principal minors, while taking $a_\bullet = 0$ for all $\bullet > m$ or < 0:

$$H_1 = a_1, \qquad H_2 = \begin{vmatrix} a_1 & a_3 \\ a_0 & a_2 \end{vmatrix}, \qquad H_3 = \begin{vmatrix} a_1 & a_3 & a_5 \\ a_0 & a_2 & a_4 \\ 0 & a_1 & a_3 \end{vmatrix}, \ldots. \tag{3.10.18d}$$

Now, assuming that $a_0 > 0$ [if it is not, we multiply (3.10.18) with -1], all the roots of the characteristic equation have *negative real parts* ($\Rightarrow$ the fundamental state I is *asymptotic stability*) if and only if *all m determinants H_h are positive*, that is,

$$a_0 > 0, \qquad \text{and} \qquad H_1 > 0, \ldots, H_{m-1} > 0, \qquad H_m > 0. \tag{3.10.18e}$$

- Since $H_m = a_m H_{m-1}$ (verify this!), these inequalities can be replaced by

$$a_0 > 0, \qquad \text{and} \qquad H_1 > 0, \ldots, H_{m-1} > 0, \qquad a_m > 0; \tag{3.10.18f}$$

that is, there is no need to calculate H_m, just check the signs of the first $m-1$ Hurwitz determinants and a_m (and a_0).

- Further, it can be shown that from (3.10.18e, f) it follows that

$$a_0 > 0, \qquad \text{and} \qquad a_1 > 0, \ldots, a_{m-1} > 0, \qquad a_m > 0, \tag{3.10.18g}$$

and therefore *negativity of even one of the coefficients of (3.10.18) indicates instability.*

REMARKS

(i) For detailed discussions and proofs of these two criteria, see the original works of Routh and Hurwitz; also Bellman and Kalaba (1964: collection of original papers), Chetayev (1961, chap. 4), Di Stefano et al. (1990, chap. 5), Gantmacher (1970, pp. 197–201), Leipholz (1970, §1.3, pp. 21–59), Mansour (1999, pp. R11–R15), McCuskey (1959, pp. 185–187), Synge (1960, pp. 185–188).

(ii) The criteria are theoretically equivalent (as can be verified by, say, the method of induction), but Hurwitz's criterion has the slight advantage over that of Routh of avoiding the calculation of fractions; hence the common term *Routh–Hurwitz criterion.*

(iii) The Routh–Hurwitz criteria are most suitable if all the coefficients of (3.10.18), $a_0, a_1, \ldots, a_{m-1}, a_m$ are *given numbers.* If, however, these coefficients contain parameters, then the implementation of the criteria becomes complicated. For this reason, throughout the 20th century, a number of alternative stability criteria have been formulated, especially criteria that are based directly on the sign properties of the coefficient matrices of (3.10.16); that is, (3.10.16a–e); and thus avoid the calculation of the coefficients of (3.10.18) and associated Routh coefficients (3.10.18b)/Hurwitz determinants (3.10.18c); e.g. criteria of *Liénard–Chipart, Mikhailov* et al.

(iv) On this technically important topic there exists, understandably, a large body of excellent literature; for example, (alphabetically): Bremer (1988, chap. 6), Hiller (1983, chap. 8), Hughes (1986, appendix A, pp. 480–521), Huseyin (1978), Magnus (1970), Merkin (1987), Müller (1977), Müller and Schiehlen (1976/1985), and Pfeiffer (1989).

EXAMPLE

Let us verify the Hurwitz criterion for the simple case of the *linearly damped and undriven oscillator*

$$M\ddot{x} + D\dot{x} + Kx = 0, \tag{3.10.19}$$

where $M = mass\ (> 0)$, $D = damping\ (> 0)$, $K = elasticity\ (> 0)$. It is not hard to see that, here ($m = 2$), the characteristic equation is

$$M\lambda^2 + D\lambda + K = 0, \tag{3.10.19a}$$

and, therefore, the Hurwitz determinants are

$$H_1 = a_1 = D, \qquad H_2 = a_1 a_2 - a_0 a_3 = DK - M0 = DK. \tag{3.10.19b}$$

For asymptotic stability, we must have $H_1 = D > 0$ and $H_2 = DK > 0 \Rightarrow D$, $K > 0$, and hence all solutions of (3.10.19) are asymptotically stable, as is already well known.

Next, let us see some less trivial applications of these criteria.

Example 3.10.1 *Rotating Shaft; Gyroscopic versus Circulatory Forces.* Here, we study a simplified version of the problem of critical speed of rotation of an originally straight shaft (axis OZ), of noncircular cross-section, rotating with constant (inertial) angular velocity ω about OZ, by examining the *equilibrium or small linearized motion* of a particle P of mass m, representing the concentrated mass of a disk (of negligible rotary inertia) mounted on the shaft, *relative to both inertial axes O–XYZ and corotational (shaft-fixed) ones $Oxyz$* ($OZ \equiv Oz$) of angular velocity ω (fig. 3.28).

(i) Let us begin with the *moving axes O–xyz*. The disk/particle P is subjected to the following forces (the $OZ \equiv Oz$ components are omitted if not needed):

(a) *centrifugal* (an *inertial* force):

$$m\,\omega^2(x, y), \tag{a1}$$

(b) *Coriolis* (an *inertial* force):

$$-2m\,\boldsymbol{\omega} \times \boldsymbol{v}_{\text{relative}} = -2m(0, 0, \omega) \times (\dot{x}, \dot{y}, 0) = 2m\,\omega(\dot{y}, -\dot{x}), \tag{a2}$$

(c) *elastic* (assuming the shaft has a *single* flexural rigidity and acts like a linear spring of known constant *stiffness* $k > 0$; a *physical* positional conservative force):

$$-k\,\boldsymbol{r} = -k(x, y), \tag{a3}$$

(d) *external damping* [e.g., aerodynamic forces (drag), bearing forces; a *physical* force]:

$$-(2d_e)m\,\boldsymbol{v}_{\text{relative}} = -2m\,d_e(\dot{x}, \dot{y}) \qquad (d_e\text{: known positive constant}), \tag{a4}$$

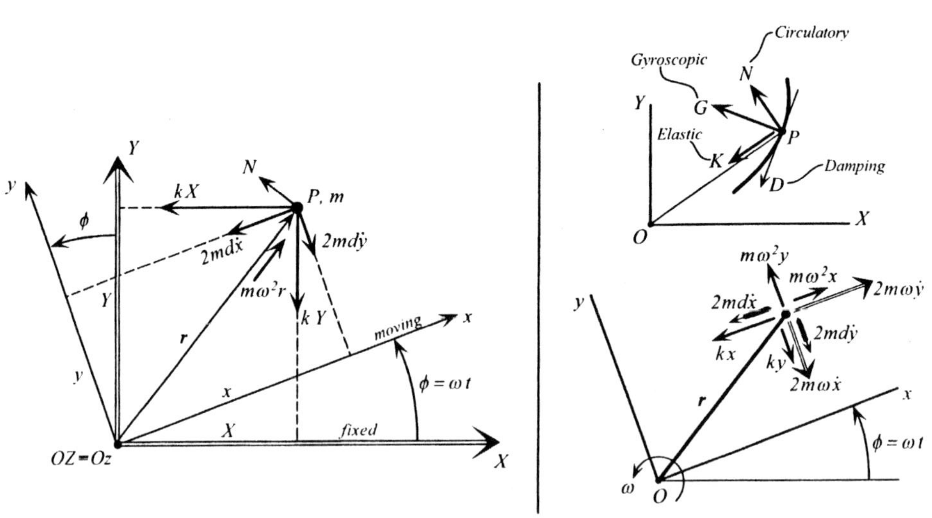

Figure 3.28 Particle model for study of stability of a rotating shaft.

(e) *internal damping* (due to the shaft properties; a *physical* force):

$$-(2d_i)m\,\boldsymbol{v}_{\text{relative}} = -2m\,d_i(\dot{x},\dot{y}) \qquad (d_i\text{: } \textit{known positive constant}). \tag{a5}$$

Applying the principle of linear momentum for relative motion to P (§1.7 ff.), we obtain

$$m\ddot{x} = -kx + m\omega^2 x + 2m\omega\dot{y} - 2m(d_e + d_i)\dot{x}, \tag{b1}$$

$$m\ddot{y} = -ky + m\omega^2 y - 2m\omega\dot{x} - 2m(d_e + d_i)\dot{y}, \tag{b2}$$

or, rearranging (and with $d \equiv d_e + d_i$),

$$\ddot{x} + 2d\dot{x} - 2\omega\dot{y} + (k/m - \omega^2)x = 0, \tag{b3}$$

$$\ddot{y} + 2d\dot{y} + 2\omega\dot{x} + (k/m - \omega^2)y = 0. \tag{b4}$$

These (relative motion) equations contain all types of terms/forces, except *circulatory* ones.

Power Equation

Multiplying (b3) with $\dot{x}$ and (b4) with $\dot{y}$, and adding together, and then transforming à la §3.9, we obtain the *noninertial* power equation

$$dh/dt = -2D_r, \tag{c}$$

where

$$h \equiv T_2 + (V - T_0) = \textit{generalized energy}, \tag{c1}$$

$$2T_2 = m[(\dot{x})^2 + (\dot{y})^2], \qquad 2T_0 = m\omega^2(x^2 + y^2), \qquad 2V = k(x^2 + y^2), \tag{c2}$$

$$2D_r \equiv (2m\,d)[(\dot{x})^2 + (\dot{y})^2] = m\,d[(\dot{x})^2 + (\dot{y})^2]\text{:}$$

$$\textit{relative dissipation (damping) function.} \tag{c3}$$

(ii) Now, let us examine the *fixed axes O–XYZ* description (see also Bahar and Kwatny, 1992). Since the constitutive equations and associated material constants are *objective* = *frame-invariant*, and by a simple moving → fixed axes transformation (§1.7):

$$\begin{aligned} -2m\,d_i\boldsymbol{v}_{\text{relative}} &= -2m\,d_i[\boldsymbol{v}_{\text{absolute}} - (\boldsymbol{\omega} \times \boldsymbol{r})] \\ &= -2m\,d_i[(\dot{X},\dot{Y},0) - (0,0,\omega) \times (X,Y,0)] \\ &= -2m\,d_i(\dot{X} + \omega Y, \dot{Y} - \omega X, 0), \end{aligned} \tag{d}$$

(i.e., $\dot{x} = \dot{X} + \omega Y$ and $\dot{y} = \dot{Y} - \omega X$, if O–XYZ and O–xyz coincide *instantaneously*), the inertial equations of motion of P are

$$m\ddot{X} = -kX - 2m\,d_e\dot{X} - 2m\,d_i(\dot{X} + \omega Y), \tag{e1}$$

$$m\ddot{Y} = -kY - 2m\,d_e\dot{Y} - 2m\,d_i(\dot{Y} - \omega X), \tag{e2}$$

or, rearranging (and with $d \equiv d_e + d_i$, $k/m \equiv \omega_o^2$),

$$\ddot{X} + 2d\dot{X} + \omega_o^2 X + 2d_i\omega Y = 0, \tag{e3}$$

$$\ddot{Y} + 2d\dot{Y} + \omega_o^2 Y - 2d_i\omega X = 0. \tag{e4}$$

Comparing the above with (b3, 4) we see that, here, instead of a gyroscopic force ($\sim \omega$ terms), we have a circulatory one:

$$\boldsymbol{N} \equiv -2m\, d_i\, \omega\, (Y, -X) = 2m\, d_i\, \omega\, (-Y, X), \tag{e5}$$

which, clearly, is perpendicular/transverse to the position vector $\boldsymbol{OP} \equiv \boldsymbol{r}$ (i.e., $\boldsymbol{N} \cdot \boldsymbol{r} = 0$) and rotates, or *circulates*, with it with angular velocity $\boldsymbol{\omega}$; hence its name.

REMARK

Equations (e1, 2) can also be derived in an ad hoc fashion as follows: referring to fig. 3.28, we can write

$$m\ddot{X} = -kX - 2m\, d_e\dot{X} + [-(2m\, d_i\, \dot{x})\cos\phi + (2m\, d_i\, \dot{y})\sin\phi], \tag{e6}$$

$$m\ddot{Y} = -kY - 2m\, d_e\dot{Y} + [-(2m\, d_i\, \dot{x})\sin\phi - (2m\, d_i\, \dot{y})\cos\phi]. \tag{e7}$$

But, as is well known from analytic geometry,

$$x = X\cos\phi + Y\sin\phi, \qquad y = -X\sin\phi + Y\cos\phi,$$

and, therefore (since $\phi = \omega t$),

$$\dot{x} = \dot{X}\cos\phi + \dot{Y}\sin\phi - X\omega\sin\phi + Y\omega\cos\phi, \tag{e8}$$

$$\dot{y} = -\dot{X}\sin\phi + \dot{Y}\cos\phi - X\omega\cos\phi - Y\omega\sin\phi. \tag{e9}$$

Substituting (e8, 9) into (e6, 7), we readily recover (e1, 2).

Power Equation

Multiplying (e3) with $\dot{X}$ and (e4) with $\dot{Y}$, and adding together, and so on, we obtain the *inertial* power equation

$$dE/dt = -2D_a + C, \tag{f}$$

where

$$E \equiv T + V = \textit{total (inertial) energy}, \tag{f1}$$

$$2T = m[(\dot{X})^2 + (\dot{Y})^2], \qquad 2V = k(X^2 + Y^2), \tag{f2}$$

$$2D_a \equiv (2m\, d)[(\dot{X})^2 + (\dot{Y})^2]:$$

$$\textit{absolute dissipation (damping) function}. \tag{f3}$$

$$C \equiv \boldsymbol{N} \cdot \boldsymbol{v} = 2m\, d_i\, \omega(X\dot{Y} - Y\dot{X}): \textit{circulatory power}\ (\boldsymbol{v} \equiv \boldsymbol{v}_{\text{absolute}}).$$

Further, since the quantity

$$X\dot{Y} - Y\dot{X} \equiv 2(dA_Z/dt)$$
$$= 2\ (\textit{areal velocity, OZ-component})\ \textit{swept in inertial space by the radius } OP \textit{ in } dt, \tag{f5}$$

is a *quasi velocity* $[2dA_Z = X\,dY + (-Y)dX \;\Rightarrow\; \partial(-Y)/\partial Y = -1 \neq \partial(X)/\partial X = +1]$, it follows that

$$C = 4m\,d_i\,\omega\,(dA_Z/dt) \neq \textit{total time derivative of a scalar energetic function}; \tag{f6}$$

that is, C is a *path-dependent* quantity, like D_a. Indeed, integrating (f) between two arbitrary instants, from an "initial" t_i to a "final" t_f, we obtain

$$\Delta E \equiv E_f - E_i = \int_{t_i}^{t_f} (-2D_a + C)\,dt = -\int_{t_i}^{t_f} (2D_a)\,dt + (4m\,d_i\,\omega)\Delta A_Z, \tag{g}$$

where

$$\Delta A_Z = \textit{ area swept by } OP \textit{ from } t_i \text{ to } t_f. \tag{g1}$$

[The total inertial power equation (f) holds unchanged even in the presence of *gyroscopic* forces; since these latter are *normal* to P's inertial velocity, we would then have an additional $\sim -\dot{Y}$ term on the right side of (e1) and a $\sim +\dot{X}$ term on that of (e2): two terms whose combined inertial power, clearly, vanishes.]

Stability Investigation

Substituting $X,\ Y \sim \exp(\lambda t)$ in (e3, 4) and requiring nontrivial solutions, we arrive, in well-known ways, at the corresponding characteristic equation

$$a_0\lambda^4 + a_1\lambda^3 + a_2\lambda^2 + a_3\lambda + a_4 = 0, \tag{h}$$

where

$$\begin{aligned}
a_0 &\equiv 1,\\
a_1 &\equiv 4(d_e + d_i) \equiv 4d,\\
a_2 &\equiv 4(d_e + d_i)^2 + 2(k/m) \equiv 4d^2 + 2{\omega_o}^2,\\
a_3 &\equiv 4(d_e + d_i)(k/m) \equiv 4d\,{\omega_o}^2,\\
a_4 &\equiv (k/m)^2 + 4{d_i}^2\omega^2 \equiv {\omega_o}^4 + 4{d_i}^2\omega^2.
\end{aligned} \tag{h1}$$

Hence, the Routh–Hurwitz asymptotic stability conditions (here $m = 4$ — not to be confused with the mass of P)

$$a_0 > 0, \quad a_1 > 0, \quad a_1a_2 - a_0a_3 > 0, \quad (a_1a_2 - a_0a_3)a_3 - {a_1}^2a_4 > 0, \quad a_4 > 0,$$

yield

$$d > 0, \qquad d(4d^2 + {\omega_o}^2), \qquad d^2(d^2{\omega_o}^2 - {d_i}^2\omega^2) > 0, \qquad {\omega_o}^4 + 4{d_i}^2\omega^2 > 0. \tag{h2}$$

Clearly, since d_e, d_i ($\Rightarrow d > 0$) and k ($= m\omega_o^2 > 0$) are positive, the *first, second, and fourth (last)* of conditions (h2) are satisfied; while the *third* of them furnishes the upper ω-bound:

$$\omega^2 < [1 + (d_e/d_i)]^2 (k/m). \tag{h3}$$

This shows that as $d_i \to 0$,

$$\omega_{\text{critical}} \equiv [1 + (d_e/d_i)](k/m)^{1/2} \equiv [1 + (d_e/d_i)]\,\omega_o \to \infty; \tag{h4}$$

and as $d_i \to \infty$, $\omega_{\text{critical}} \to 0$: that is, d_i has a *destabilizing* effect; hence, in rotor design, we should aim at *more* d_e and *less* d_i.

For further details and insights, and discussion of stability using the rotating axes equations (b3, 4), see, for example (alphabetically): Bolotin (1963, chap. 3), Dimentberg (1961, chap. 2), Ziegler (1968, pp. 94–96, 101); and for additional, more realistic, circulatory force examples, see Bremer [1988(a), pp. 144–149]. For further applications of the Routh–Hurwitz criterion to the stability of mechanical systems, see texts on linear and nonlinear vibrations and controls.

Problem 3.10.2 Show that if the fundamental state I of a scleronomic system is one of *equilibrium* in the q_k — that is, if f_k = *constant, or* 0 [recall (3.10.10a ff.)]— the equations of small motion around I are

$$\sum M_{kl}\ddot{x}_l = \sum [(\partial Q_k/\partial q_l)x_l + (\partial Q_k/\partial \dot{q}_l)\dot{x}_l] \tag{a}$$

$$\left[= -\sum (\partial^2 V/\partial q_l \partial q_k)x_l, \quad \text{if } Q_k = -\partial V(q)/\partial q_k \right], \tag{b}$$

where all the $x/\dot{x}/\ddot{x}$-coefficients are evaluated on I and, therefore, are *constant* [something that makes the systems (a, b) always solvable].

Notice that in case (b), or in case (a) with $\partial Q_k/\partial \dot{q}_l = \partial Q_l/\partial \dot{q}_k$, *no gyroscopic terms appear.*

Problem 3.10.3 With the help of the following quadratic and bilinear forms:

$$2T'_2 \equiv \sum\sum M_{kl}\dot{x}_k\dot{x}_l \qquad (\equiv 2\Delta T_{2,2}\text{: “}\textit{contracted}\text{” } \textit{kinetic energy}), \tag{a}$$

$$2D \equiv \sum\sum D_{kl}\dot{x}_k\dot{x}_l \qquad (\equiv -2D'\text{: } \textit{damping function}), \tag{b}$$

$$G \equiv \sum\sum G_{kl}x_k\dot{x}_l \qquad (\equiv -2\Delta T_{2,1} + \Theta''\text{: } \textit{gyroscopic function}), \tag{c}$$

$$2V \equiv \sum\sum K_{kl}x_k x_l \qquad (\equiv 2V' - 2\Delta T_{2,0}\text{: } \textit{potential function}), \tag{d}$$

and

$$N_k \equiv -\sum N_{kl}x_l \qquad (\equiv \eta'_k\text{: } \textit{circulatory force}), \tag{e}$$

show that the linear variational equations of small motion about a fundamental state of steady motion can be rewritten in the Lagrangean (linear vibration) form

$$(\partial T'_2/\partial \dot{x}_k)^{\cdot} + \partial D/\partial \dot{x}_k + \partial (G + V)/\partial x_k = N_k. \tag{f}$$

Problem 3.10.4 Show that if the fundamental state I is one of *equilibrium* — that is, $f_k(t) \equiv 0$, and $\partial Q_k/\partial \dot{q}_l = \partial Q_l/\partial \dot{q}_k$ (e.g., positional forces only) there — then

$$\varepsilon_{kl} = 0, \qquad \zeta_{kl} = 0, \qquad \theta_{kl} = 0 \ (\Rightarrow G_{kl} = 0, \quad K_{kl} = -\eta'_{kl}), \tag{a}$$

and therefore the equations of small motion about such an I reduce to

$$\sum [M_{kl}\ddot{x}_l + D_{kl}\dot{x}_l + (K_{kl} + N_{kl})x_l] = 0. \tag{b}$$

We notice that the *absence of gyroscopic terms* is the key difference between small motion about *absolute* and *relative* equilibrium (and, of course, general motion).

Problem 3.10.5 Consider the *cubic* characteristic equation

$$\lambda^3 + A_2\lambda^2 + A_1\lambda + A_0 = 0, \tag{a}$$

with roots $\lambda_1, \lambda_2, \lambda_3$.

(i) Show that

$$A_2 = -(\lambda_1 + \lambda_2 + \lambda_3), \qquad A_1 = \lambda_1\lambda_2 + \lambda_1\lambda_3 + \lambda_2\lambda_3, \qquad A_0 = -(\lambda_1\lambda_2\lambda_3). \tag{b}$$

(ii) Since one of these roots must be *real* and the other two either *real* or *complex conjugate*, we write

$$\lambda_1 = \rho_1, \qquad \lambda_2 = \rho_2 + i\sigma_2, \qquad \lambda_3 = \rho_2 - i\sigma_2, \tag{c}$$

where $i^2 \equiv -1$, and ρ_1, ρ_2, σ_2 are *real*. Show that

$$A_2 = -(\rho_1 + 2\rho_2), \qquad A_1 = 2\rho_1\rho_2 + {\rho_2}^2 + {\sigma_2}^2, \qquad A_0 = -\rho_1({\rho_2}^2 + {\sigma_2}^2). \tag{d}$$

Problem 3.10.6 Continuing from the preceding problem, show that the (necessary and sufficient) asymptotic stability conditions for a system with the cubic characteristic equation

$$\lambda^3 + A_2\lambda^2 + A_1\lambda + A_0 = 0, \tag{a}$$

are

$$\text{(i) } A_2, A_1, A_0 > 0 \quad \text{(positive coefficients)}, \quad \text{and} \quad \text{(ii) } A_2A_1 > A_0. \tag{b}$$

Problem 3.10.7 Consider the *quartic* characteristic equation

$$\lambda^4 + A_3\lambda^3 + A_2\lambda^2 + A_1\lambda + A_0 = 0, \tag{a}$$

with roots

$$\lambda_1 = \rho_1 + i\sigma_1, \qquad \lambda_2 = \rho_1 - i\sigma_1, \qquad \lambda_3 = \rho_2 + i\sigma_2, \qquad \lambda_4 = \rho_2 - i\sigma_2. \tag{b}$$

Show that

$$\begin{aligned}
A_3 &= -2(\rho_1 + \rho_2),\\
A_2 &= {\rho_1}^2 + {\rho_2}^2 + {\sigma_1}^2 + {\sigma_2}^2 + 4\rho_1\rho_2,\\
A_1 &= -2\rho_1({\rho_2}^2 + {\sigma_2}^2) - 2\rho_2({\rho_1}^2 + {\sigma_1}^2),\\
A_0 &= ({\rho_1}^2 + {\sigma_1}^2)({\rho_2}^2 + {\sigma_2}^2).
\end{aligned} \tag{c}$$

Problem 3.10.8 Consider again the *quartic* characteristic equation

$$\lambda^4 + A_3\lambda^3 + A_2\lambda^2 + A_1\lambda + A_0 = 0. \tag{a}$$

Show that the Routh–Hurwitz criteria applied to (a) produce the following (necessary and sufficient) asymptotic stability conditions:

(i) $A_3, A_2, A_1, A_0 > 0$ (positive coefficients), (b)

(ii) $A_3 A_2 A_1 > A_1{}^2 + A_3{}^2 A_0.$ (c)

Problem 3.10.9 Deduce the Routh–Hurwitz asymptotic stability conditions for the indicated special cases:

(i) $m = 2$, i.e., $a_0\lambda^2 + a_1\lambda + a_2 = 0$:

$$a_0,\ a_1,\ a_2 > 0; \tag{a}$$

(ii) $m = 3$, i.e., $a_0\lambda^3 + a_1\lambda^2 + a_2\lambda + a_3 = 0$:

- $a_0,\ a_1,\ a_2,\ a_3 > 0,$ (b1)
- $a_1 a_2 - a_0 a_3 > 0;$ (b2)

(iii) $m = 4$, i.e., $a_0\lambda^4 + a_1\lambda^3 + a_2\lambda^2 + a_3\lambda + a_4 = 0$:

- $a_0,\ a_1 > 0,$ (c1)
- $A \equiv a_1 a_2 - a_0 a_3 > 0,$ (c2)
- $B \equiv a_3 A - a_1{}^2 a_4 > 0,$ (c3)
- $a_4 > 0.$ (c4)

Due to eqs. (c3, 4), condition (c2) can be replaced by the simpler $a_3 > 0$; then, it follows that $a_2 > 0$.

(iv) $m = 5$, i.e., $a_0\lambda^5 + a_1\lambda^4 + a_2\lambda^3 + a_3\lambda^2 + a_4\lambda + a_5 = 0.$

With the abbreviations $A \equiv a_1 a_2 - a_0 a_3$, $C \equiv a_1 a_4 - a_0 a_5$, and $D \equiv a_3 a_4 - a_2 a_5$, they are

- $a_0,\ a_1,\ A > 0,$ (d1)
- $a_3 A - a_1 C > 0,$ (d2)
- $AD - C^2 > 0,$ (d3)
- $a_5 > 0.$ (d4)

But, since $a_1(AD - C^2) \equiv C(a_3 A - a_1 C) - a_5 A^2$, and due to (d3, 4), condition (d2) can be replaced by the simpler $C > 0$; also, we must have $a_4 > 0$ and $D > 0$.

Notice that, in *all* cases, we must have satisfaction of the *essential conditions*: $a_0 = |M_{kl}| > 0$ and $a_m = |K_{kl} + N_{kl} > 0|$.

Problem 3.10.10 Consider the two-DOF undamped and gyroscopic system with perturbation equations

$$M_1\ddot{x}_1 + G\dot{x}_2 + K_1 x_1 = 0, \qquad M_2\ddot{x}_2 - G\dot{x}_1 + K_2 x_2 = 0, \tag{a}$$

where $M_{1,2}$ = *inertia/mass* (>0), $K_{1,2}$ = *positional noncirculatory coefficients*, and G = *gyroscopicity*.

(i) Show that its characteristic equation is

$$a_0\lambda^4 + a_2\lambda^2 + a_4 = 0, \tag{b}$$

where

$$\begin{aligned} a_0 &\equiv M_1 M_2 \ (>0 - \text{always, on physical grounds}),\\ a_2 &\equiv M_1 K_2 + M_2 K_1 + G^2,\\ a_4 &\equiv K_1 K_2. \end{aligned} \tag{c}$$

(ii) Show that the Routh–Hurwitz criteria applied to this problem ($m = 4$, and $a_1, a_3 = 0$) produce the three asymptotic stability conditions

$$a_0 > 0, \qquad a_2 > 0, \qquad a_4 > 0. \tag{d}$$

(iii) Show that the *second* of (d) can be satisfied for *sufficiently high values of the "spin term"* G^2, no matter what the signs of K_1 and K_2 are. [This is a special case of the famous *gyroscopic stabilization theorem* of Kelvin and Tait. For a more extensive treatment, see §8.6.]

REMARK

The presence of light damping ($\sim \dot{x}$) changes this stability picture considerably. For details and technical applications, for example, see Grammel (1950, vol. 1, pp. 261–262; vol. 2, pp. 230–247).

Problem 3.10.11 Consider a smooth surface S spinning with constant inertial angular velocity $\boldsymbol{\omega}$ about a vertical axis OZ (positive upward), where O is a surface point with horizontal tangential plane to it there. Let the equation of S in *corotating* (*surface-fixed*) coordinates O–xyz, where Ox, Oy are tangent to the lines of principal curvature of S at O and $Oz \equiv OZ$, be, to the *second* order in x, y,

$$2z = x^2/\rho_1 + y^2/\rho_2, \qquad \rho_{1,2} = \textit{principal radii of curvature of } S \text{ at } O. \tag{a}$$

In addition, consider a particle P of mass $m = 1$, moving under gravity on S, in the neighborhood of O.

(i) Show that, *to the second order*, the (double) Lagrangean of P is

$$2L = \Big\{(\dot{x})^2 + (\dot{y})^2 + 2\omega(x\dot{y} - y\dot{x}) + [\omega^2 - (g/\rho_1)]x^2 + [\omega^2 - (g/\rho_2)]y^2\Big\}, \tag{b}$$

and therefore its equations of (relative) motion in the neighborhood of O are

$$\ddot{x} - \gamma\dot{y} + k_1 x = 0, \qquad \ddot{y} + \gamma\dot{x} + k_2 y = 0, \tag{c}$$

where

$$\gamma \equiv 2\omega, \qquad k_1 \equiv (g/\rho_1) - \omega^2, \qquad k_2 \equiv (g/\rho_2) - \omega^2. \tag{d}$$

(ii) The system (c) has the form of eqs. (a) of the preceding problem; with $x_1 = x$, $x_2 = y$, $M_1 = M_2 = 1\,(>0)$, $G = -\gamma$, $K_1 = k_1$, $K_2 = k_2$. Specialize the asymptotic stability conditions (d) established there to this problem.

[See also Whittaker (1937, pp. 207–208); and for the case of small ($\sim \dot{x}, \dot{y}$) friction, see Lamb (1943, pp. 253–254).]

Problem 3.10.12 Using the Routh–Hurwitz criterion, show that in an asymptotically *stable* (linear) system *all the coefficients of the characteristic equation (3.10.18)*, $a_0, a_1, a_2, \ldots, a_{m-1}, a_m$, *have the same sign* ($> 0$)*; that is, none of them vanishes*. (This is a *necessary*, but not sufficient, condition for such stability!)

HINT

Let the roots of that equation be

$$-\varepsilon_1, \ldots, -\varepsilon_* \qquad \text{and} \qquad -\rho_1 \pm i\sigma_1, \ldots, -\rho_\bullet \pm i\sigma_\bullet,$$

where $* + \bullet = m$ (# possible multiple roots being counted individually), and all ε, ρ, σ are real and *positive* (asymptotically stable system). Then, by well-known theorems of the theory of equations,

$$\begin{aligned} \Delta(\lambda) &= a_0\big[(\lambda + \varepsilon_1)\cdots(\lambda + \varepsilon_*)\big]\big[(\lambda^2 + 2\rho_1\lambda + \rho_1{}^2 + \sigma_1{}^2)\cdots(\lambda^2 + 2\rho_\bullet\lambda + \rho_\bullet{}^2 + \sigma_\bullet{}^2)\big] \\ &= a_0(\lambda^m + \cdots) = 0. \end{aligned} \tag{a}$$

Example 3.10.2 *The Jacobi–Synge Equations.* The preceding equations show that the Lagrangean equations of motion, under say, the holonomic constraints

$$\phi_H(t, q) = 0 \Rightarrow \sum (\partial\phi_H/\partial q_k)\dot{q}_k + \partial\phi_H/\partial t = 0 \qquad (H = 1, \ldots, m), \tag{a}$$

have the general form

$$\sum M_{kr}(t, q)\ddot{q}_r = f_k(t, q, \dot{q}) + \sum \lambda_H(\partial\phi_H/\partial q_k), \tag{b}$$

where $f_k(t, q, \dot{q})$ is a known function of its arguments. [If the constraints are given in the general Pfaffian (possibly nonholonomic) form $\sum a_{Hk}\dot{q}_k + a_H = 0$, then we replace the gradients $\partial\phi_H/\partial q_k$ with the constraint coefficients $a_{Hk}(t, q)$.]

It was Jacobi's idea (Jacobi, 1866, p. 55) to $(\ldots)^\cdot$-differentiate the velocity constraints (a) once more, thus bring them into their acceleration form (i.e., $\sim \ddot{q}$ terms), and then combine them, *like additional equations of motion*, with (b). [We are indebted to Dr. F. Pfister for pointing this out to us; see Pfister (1995). This idea was also carried out, independently and slightly differently, by Synge (in 1926) via general tensor calculus, in his pioneering and influential memoir (Synge, 1926–1927, pp. 53–55).] This fusion of constraints, in acceleration form, with the equations of motion in Routh–Voss (multiplier) form, something very popular among applied dynamicistes today, is carried out below in matrix form.

Indeed, first we $(\ldots)^\cdot$-differentiate (a) once more, thus obtaining

$$\sum (\partial\phi_H/\partial q_k)\ddot{q}_k = g_H(t, q, \dot{q}) \qquad (H = 1, \ldots, m), \tag{c}$$

where $g_H(t, q, \dot{q})$ is a known function of its arguments, like $f_k(t, q, \dot{q})$. Next we introduce some simple matrix notation:

$$\mathbf{M} = (M_{kr}): \textit{nonsingular} \text{ and positive definite}, \qquad \mathbf{q}^{\mathrm{T}} = (q_1, \ldots, q_n),$$
$$\mathbf{\Phi_q} = (\partial\phi_H/\partial q_k): \textit{nonsingular}, \qquad \boldsymbol{\lambda}^{\mathrm{T}} = (\lambda_1, \ldots, \lambda_m),$$
$$\mathbf{f}^{\mathrm{T}} = (f_1, \ldots, f_n), \qquad \mathbf{g}^{\mathrm{T}} = (g_1, \ldots, g_n), \quad \text{where} \quad (\ldots)^{\mathrm{T}} \equiv \textit{transpose} \text{ of } (\ldots); \tag{d}$$

so that, with its help, we can rewrite eqs. (b, c) as

$$\mathbf{M}\ddot{\mathbf{q}} = \mathbf{f} + \mathbf{\Phi_q}^{\mathrm{T}}\boldsymbol{\lambda}, \qquad \mathbf{\Phi_q}\ddot{\mathbf{q}} = \mathbf{g}, \tag{e}$$

respectively; and finally, we combine eqs. (e) into the following matrix form:

$$\begin{pmatrix} \mathbf{M} & \mathbf{\Phi_q}^{\mathrm{T}} \\ \mathbf{\Phi_q} & \mathbf{0} \end{pmatrix} \begin{pmatrix} \ddot{\mathbf{q}} \\ -\boldsymbol{\lambda} \end{pmatrix} = \begin{pmatrix} \mathbf{f} \\ \mathbf{g} \end{pmatrix}, \tag{f}$$

where $\mathbf{0}$ is the $m \times n$ zero matrix.

Equations (f) can be justifiably called the *Jacobi form of the Routh–Voss equations*; and, at any instant for which $\mathbf{q}$ and $\dot{\mathbf{q}}$ are known, these constitute a system of $n + m$ algebraic equations that (since $\mathbf{M}$ and $\mathbf{\Phi_q}$ are nonsingular) *can* be solved (numerically) for their linearly appearing $n + m$ unknowns $\ddot{\mathbf{q}}$ and $\boldsymbol{\lambda}$. For further details, see books on computational/multibody dynamics; for example, Nikravesh (1988), Udwadia and Kalaba (1996); while, for a tensorial derivation, see Papastavridis (1998; 1999, pp. 324–325) and Synge (1926–1927).

3.11 APPELL'S EQUATIONS: EXPLICIT FORMS

Holonomic Variables

Let us begin with *holonomic* variables and, for algebraic simplicity, but no loss of generality, *stationary* constraints. Then [recalling (2.5.2 ff.)]

$$\begin{aligned} \boldsymbol{v} &= \sum \boldsymbol{e}_k \dot{q}_k \\ \Rightarrow \boldsymbol{a} \equiv d\boldsymbol{v}/dt &= \sum \boldsymbol{e}_k \ddot{q}_k + \sum (d\boldsymbol{e}_k/dt)\dot{q}_k \\ &= \sum \boldsymbol{e}_k \ddot{q}_k + \sum\sum (\partial \boldsymbol{e}_k/\partial q_l)\dot{q}_l \dot{q}_k, \end{aligned} \tag{3.11.1}$$

and, accordingly (and using subscript commas for partial q-derivatives), the system Appellian becomes

$$\begin{aligned} S \equiv \mathbf{S}(1/2)\, dm\, \boldsymbol{a} \cdot \boldsymbol{a} = (1/2)\mathbf{S}\, dm\Big[\Big(\sum \boldsymbol{e}_k \ddot{q}_k + \sum\sum \boldsymbol{e}_{k,l}\dot{q}_l \dot{q}_k\Big) \\ \cdot \Big(\sum \boldsymbol{e}_r \ddot{q}_r + \sum\sum \boldsymbol{e}_{r,s}\dot{q}_r \dot{q}_s\Big)\Big], \end{aligned} \tag{3.11.2}$$

or, with some dummy index changes, and recalling that $M_{kl} \equiv S\, dm\, \boldsymbol{e}_k \cdot \boldsymbol{e}_l$ and

$$\Gamma_{k,lp} = \Gamma_{k,pl} \equiv S\, dm\, \boldsymbol{e}_k \cdot \boldsymbol{e}_{l,p} = S\, dm\, \boldsymbol{e}_k \cdot \boldsymbol{e}_{p,l}$$
$$= (1/2)(\partial M_{kl}/\partial q_p + \partial M_{kp}/\partial q_l - \partial M_{lp}/\partial q_k) \tag{3.11.3}$$

(§3.9, §3.10), we finally obtain, *to within Appell important terms* (i.e., $\sim \ddot{q}$),

$$S = (1/2) \sum \sum M_{kl} \ddot{q}_k \ddot{q}_l + \sum \sum \sum \Gamma_{k,lp} \ddot{q}_k \dot{q}_l \dot{q}_p. \tag{3.11.4}$$

The above shows how to find the Appellian function for *nonstationary* constraints: (i) since $\ddot{q}_{n+1} = \ddot{t} = \dot{1} = 0$, the *first* group of terms (double sum) remains unchanged; while (ii) in the *second* group of terms (triple sum), k still runs from 1 to n, but l and p must now run from 1 to $n+1$; hence, we replace them, respectively, with the Greek subscripts α and β. The result is

$$S = (1/2) \sum \sum M_{kl} \ddot{q}_k \ddot{q}_l + \sum \sum \sum \Gamma_{k,\alpha\beta} \ddot{q}_k \dot{q}_\alpha \dot{q}_\beta$$
$$= (1/2) \sum \sum M_{kl} \ddot{q}_k \ddot{q}_l + \sum \sum \sum \Gamma_{k,lp} \ddot{q}_k \dot{q}_l \dot{q}_p$$
$$+ 2 \sum \sum \Gamma_{k,l,n+1} \ddot{q}_k \dot{q}_l + \sum \Gamma_{k,n+1,n+1} \ddot{q}_k ; \tag{3.11.5}$$

where $\Gamma_{k,l,n+1} (\Gamma_{k,n+1,n+1})$ is what results from $\Gamma_{k,lp}$ by formally replacing p (p and l) with $n+1$; that is, $q_{n+1} \to t$ [recalling (3.10.8d–f)].

Expressions (3.11.4) and (3.11.5) show clearly how to build S if we know T; that is, if we know its inertial coefficients $M_{\alpha\beta}$: M_{kl}, $M_{k,n+1} \equiv M_k$, $M_{n+1,n+1} \equiv M_0 = 2T_0$; they also reconfirm the kinematico-inertial identity $\partial S/\partial \ddot{q}_k = (\partial T/\partial \dot{q}_k)^{\cdot} - \partial T/\partial q_k$.

Nonholonomic Variables

Next, let us repeat the above, but for quasi variables (i.e., $S \to S^*(q, \omega, \dot{\omega}, t) = S^*$); first, again, for the *stationary* case. Substituting

$$\boldsymbol{a} = \boldsymbol{a}^* \equiv d\boldsymbol{v}^*/dt = \sum \boldsymbol{\varepsilon}_k \dot{\omega}_k + \sum (d\boldsymbol{\varepsilon}_k/dt) \omega_k$$

(and using here subscript commas for partial θ-derivatives)

$$= \sum \boldsymbol{\varepsilon}_k \dot{\omega}_k + \sum \sum \boldsymbol{\varepsilon}_{k,l} \omega_l \omega_k, \tag{3.11.6}$$

into $S^* \equiv S\, (1/2) dm\, \boldsymbol{a}^* \cdot \boldsymbol{a}^*$, we obtain, *to within Appell important terms* (i.e., $\sim \dot{\omega}$)

$$2S^* = \sum \sum \left(S\, dm\, \boldsymbol{\varepsilon}_k \cdot \boldsymbol{\varepsilon}_l \right) \dot{\omega}_k \dot{\omega}_l$$
$$+ \sum \sum \sum \left(S\, dm\, \boldsymbol{\varepsilon}_k \cdot (\boldsymbol{\varepsilon}_{l,p} + \boldsymbol{\varepsilon}_{p,l}) \right) \dot{\omega}_k \omega_l \omega_p,$$

or, since [recalling (3.10.9f)]

$$S\, dm\, \boldsymbol{\varepsilon}_k \cdot (\boldsymbol{\varepsilon}_{l,p} + \boldsymbol{\varepsilon}_{p,l}) = 2\Gamma^*_{k,lp} - \sum (\gamma^r{}_{lk} M^*{}_{pr} + \gamma^r{}_{pk} M^*{}_{lr})$$
$$= 2\Gamma^*_{k,lp} + \sum (\gamma^r{}_{kl} M^*{}_{pr} + \gamma^r{}_{kp} M^*{}_{lr}),$$

finally,

$$S^* = (1/2) \sum \sum M^*{}_{kl} \dot{\omega}_k \dot{\omega}_l + \sum \sum \sum A_{k,lp} \dot{\omega}_k \omega_l \omega_p, \tag{3.11.7}$$

where [recalling (3.10.9l)]

$$\Lambda_{k,lp} \equiv \Gamma^*_{k,lp} + \sum \gamma^r{}_{kl} M^*{}_{pr}. \tag{3.11.7a}$$

from the above we easily see that:

(i) To find S^* we need not just the $M^*{}_{kl}$ (like T^*), but also the $\gamma^r{}_{kl}$ (like I_k); and [recall (3.10.9k)]

(ii) $$\partial S^*/\partial \dot{\omega}_k = \sum M^*{}_{kl}\,\dot{\omega}_l + \sum\sum \Lambda_{k,lp}\,\omega_l\,\omega_p = I_k. \tag{3.11.7b}$$

Let the reader extend the above, eqs. (3.11.6–7a), to the *nonstationary* case.

REMARKS

(i) In general, both kinetic energy and Appellian are more simply expressed in nonholonomic rather than holonomic variables; that is, for the same problem, T^* and S^* are simpler *in form* than T and S, respectively. As a result, for *holonomic systems in holonomic variables*, Appell's equations are trivial; that is, not worth the effort. But for *holonomic systems in nonholonomic variables*, they may offer definite advantages: for example, the Eulerian rigid-body equations (Gibbs, 1879; see example below, and §3.13 ff.); then, the resulting equations of motion are of the *first order* in the ω's.

(ii) For *nonholonomic systems in nonholonomic variables*, the equations of Hamel and Appell, although theoretically equivalent, have the following differences:

(a) In the Hamel case, even if no reactions are sought, we still need the unconstrained (relaxed) kinetic energy T^*; and the coefficients $\gamma^r{}_{II'}$, $\gamma^r{}_{I,n+1}$.
(b) In the Appell case, if no reactions are sought, we may work with the *constrained* Appellian $S^*{}_o$ right from the start, and thus save a considerable amount of labor; otherwise we must calculate the unconstrained Appellian S^*; and, in all cases, the Appellian can be calculated only to within Appell-important (i.e., acceleration-containing) terms.

Also, Appell's equations are *simpler looking* than Hamel's, and *form-invariant in both holonomic and nonholonomic variables*. But calculating the Appellian requires more labor than calculating the kinetic energy. In both cases, as in other areas of science, with constant practice we learn special short cuts, or use ready-made expressions for particular systems.

Example 3.11.1 *Let us Find the Appellian of a Rigid Body Moving about a Fixed Point O.* Using body-fixed principal inertia axes $O{-}xyz \equiv O{-}123$, we find

$$(M^*{}_{kl}) = \textit{diagonal}\ (I_1, I_2, I_3) = \textit{constant components}, \tag{a}$$

and therefore all Γ^*'s vanish. Also, we recall from ex. 2.13.9 that for such axes $\gamma^r{}_{kl} = \varepsilon_{rkl} \equiv \pm 1$, according as r, k, l are an even or odd permutation of 1, 2, 3; and zero in all other cases. Accordingly, the expression (3.11.7), with (3.11.7a), and $\boldsymbol{\omega} = (\omega_1, \omega_2, \omega_3) =$ *inertial angular velocity of body*, specializes to

$$\begin{aligned} S^* = (1/2)\left[I_1(\dot{\omega}_1)^2 + I_2(\dot{\omega}_2)^2 + I_3(\dot{\omega}_3)^2\right] \\ + (I_3 - I_2)\,\dot{\omega}_1\,\omega_2\,\omega_3 + (I_1 - I_3)\,\dot{\omega}_2\,\omega_3\,\omega_1 + (I_2 - I_1)\,\dot{\omega}_3\,\omega_1\,\omega_2, \end{aligned} \tag{b}$$

and from this we immediately obtain the well-known (body-fixed + principal axes) Eulerian expressions for the body inertia (§1.17)

$$\partial S^*/\partial\dot{\omega}_1 = I_1\dot{\omega}_1 + (I_3 - I_2)\,\omega_2\,\omega_3, \quad \text{etc., cyclically.} \tag{c}$$

3.12 EQUATIONS OF MOTION: INTEGRATION AND CONSERVATION THEOREMS

Integrals of the Equations of Motion

Let us consider, without much loss in generality and understanding, a system with equations of motion

$$E_k(L) \equiv (\partial L/\partial\dot{q}_k)^{\cdot} - \partial L/\partial q_k = 0 \qquad (k = 1, \ldots, n), \tag{3.12.1}$$

or, in extenso, since $L = L(q, \dot{q}, t)$ (and with $l = 1, \ldots, n$)

$$\sum (\partial^2 L/\partial\dot{q}_k\,\partial\dot{q}_l)\ddot{q}_l + \sum (\partial^2 L/\partial\dot{q}_k\,\partial q_l)\dot{q}_l + \partial^2 L/\partial\dot{q}_k\,\partial t - \partial L/\partial q_k = 0. \tag{3.12.2}$$

This is a system of n *second*-order equations in the q's, *linear* in the $\ddot{q}$; or, equivalently, a system of *total-order* $2n$ (= sum of orders of highest derivatives of dependent variables). As the theory of differential equations teaches, its general solution (if and when known) will contain (at most) $2n$ *arbitrary constants of integration* $c \equiv (c_1, \ldots, c_{2n})$:

$$q_k = q_k(t, c) = \textit{general solution of } (3.12.1, 2). \tag{3.12.3}$$

Next, and for the purposes of the discussion below, it is expedient to introduce the following transformation of variables $q, \dot{q} \to (x_1, \ldots, x_{2n}) \equiv x$:

$$q_1 = x_1, \ldots, q_n = x_n; \qquad \dot{q}_1 = x_{n+1}, \ldots, \dot{q}_n = x_{2n}. \tag{3.12.4}$$

In terms of them, eqs. (3.12.1, 2), or, equivalently [assuming nonsingular Hessian $|\partial^2 L/\partial\dot{q}_k\,\partial\dot{q}_l|$],

$$\ddot{q}_l = Q_l(q, \dot{q}, t), \tag{3.12.4a}$$

reduce to the $2n$ *first-order* equations

$$\begin{aligned} &\dot{x}_1 = x_{n+1} \equiv X_1(x, t), \ldots, && \dot{x}_n = x_{2n} \equiv X_n(x, t); \\ &\dot{x}_{n+1} = Q_1(x, t) \equiv X_{n+1}(x, t), \ldots, && \dot{x}_{2n} = Q_n(x, t) \equiv X_{2n}(x, t); \end{aligned} \tag{3.12.4b}$$

or, compactly,

$$dx_*/dt = X_*(x, t) \qquad (* = 1, \ldots, 2n). \tag{3.12.4c}$$

[Also, in terms of the Lagrangean *momenta*, conjugate to the q_k,

$$\partial T/\partial\dot{q}_k = \partial L/\partial\dot{q}_k \equiv p_k = p_k(q, \dot{q}, t), \tag{3.12.4d}$$

(assuming $\partial V/\partial\dot{q}_k = 0$), the n *second-order* Lagrangean equations (3.12.1) can be rewritten as the $2n$ *first-order* equations

$$p_k = \partial L/\partial\dot{q}_k \equiv f_k(q, \dot{q}, t)$$

and

$$\dot{p}_k = \partial L/\partial q_k \equiv g_k(q, \dot{q}, t) = G_k(q, p, t), \tag{3.12.4e}$$

where, in the last step, we inverted (3.12.4d) to obtain $\dot{q}_k = \dot{q}_k(q, p, t)$. Such first-order formulations are quite useful in both theoretical and numerical situations (see also §8.2 for the *additional/related*, and very well-known first-order form, known as *Hamiltonian* equations of motion).]

Some Mathematical Background

Consider the $(2n)$th $\equiv (*)$th order differential system, in any of the following equivalent forms:

$$F_*(x, dx/dt, t) = 0 \qquad [\textit{implicit} \text{ form}], \tag{3.12.5a}$$

$$dx_*/dt = X_*(x_\bullet, t) \qquad [\textit{explicit} \text{ form } (*, \bullet = 1, \ldots, 2n)], \tag{3.12.5b}$$

$$dx_1/X_1 = dx_2/X_2 = \cdots = dx_{2n}/X_{2n} = dt, \tag{3.12.5c}$$

where the F_*, X_* are given functions of their arguments. A *solution* of (3.12.5a, b, c), in an open time interval of interest, $\tau \equiv (t_0, t_1)$, is the set of $2n$ (continuously differentiable) functions $x_*(t)$.

THEOREM OF INITIAL CONDITIONS (UNIQUENESS OF SOLUTIONS, LIPSCHITZ CONDITIONS)

If $x(t_{\text{initial}} \equiv t_o) = given$, and if, at every point of τ, all X_* as well as $\partial X_*/\partial x_\bullet$ are continuous, then the system (3.12.5a, b, c) has a *unique* solution in τ. [These conditions are restrictive, so, in practice, we frequently find cases where they do not hold. If the existence conditions hold, but not those of uniqueness, several motions are possible (indeterminate motion). For example, the system (plus initial conditions) $d^2y/dt^2 = 6y^{1/3}$; $y(0) = 0$, $dy(0)/dt = 0$, yields the *three* motions: $y = 0$ (equilibrium), $y = -t^3$, $y = t^3$; i.e., the initial conditions do *not* determine uniquely the ultimate motion. If not even the *existence* conditions hold, worse things may happen.]

In *mechanics* terms, the theorem states that: If the n positions q and n velocities dq/dt are given at an "initial" instant t_o, and if the n forces Q satisfy the above Lipschitz conditions in τ, the subsequent system motion is determined uniquely during that time interval.

The first step in the integration of the system (3.12.5a, b, c) is the search for *first integrals*.

DEFINITION

We call *first integral* of the system (3.12.5a, b, c) every function $f(x, t)$ that, for every one of its solutions $x_* = x_*(t)$, remains constant, for arbitrary t: $f(x, t) = constant \equiv c$, where the constant may change when the particular solution (motion) changes; that is, a first integral is, in general, a function depending on time both explicitly and implicitly, through the x's, that stays constant on account of the equations of motion, independently of initial conditions; but the *value of the constant depends on the initial conditions* (i.e., on the particular solution/motion, and stays the same throughout it).

In mechanics, a first integral will be a function of the n positions q and n velocities dq/dt and time t, that remains constant for every solution of the equations of motion:

$$f(q, \dot{q}, t) = \textit{constant during the motion (depends on initial conditions of that particular motion)}, \tag{3.12.5d}$$

$$\Rightarrow df/dt = \sum (\partial f/\partial x_*)(dx_*/dt) + \partial f/\partial t = \sum (\partial f/\partial x_*)X_* + \partial f/\partial t = 0,$$

identically, for any q, dq/dt, and t satisfying the equations of motion. (3.12.5e)

Hence, a first integral lowers the degree of the differential system by 1; and therein lies their principal usefulness to mechanics. In particular, we may show the following.

THEOREM

Under broad analytical conditions, we can replace one of the equations of motion with a first integral, like (3.12.5d). [Occasionally, for certain values of the initial conditions, such a replacement may introduce foreign solutions.]

Classification of first integrals: (i) *First integrals explicitly independent of time*; that is, $\partial f/\partial t = 0 \Rightarrow f(x) = f(q, dq/dt) = \textit{constant}$. (ii) *First integrals depending explicitly on time*; that is, $f(x, t) = f(q, dq/dt, t) = \textit{constant}$. [In certain areas of physics (e.g., statistical mechanics), other first integral classifications are important.]

Distinct first integrals. If $f = a$ (*constant*) is a first integral, then $F(f) = b$ (*another constant*) is another. More generally, let $f_1 = a_1, f_2 = a_2, \ldots, f_p = a_p$, be p ($\leq 2n$) first integrals. Then, $F(f_1, f_2, \ldots, f_p) = b$ is also a first integral, but not one *distinct* from the f_i $(i = 1, \ldots, p)$. The p integrals $f_1 = a_1$, $f_2 = a_2, \ldots, f_p = a_p$, are called *distinct* if none of them can be expressed as function of the other $p - 1$ in the form

$$f_i = F(f_1, f_2, \ldots, f_{i-1}, f_{i+1}, \ldots, f_p) \tag{3.12.5f}$$

Now, it is shown in the theory of differential equations that *the complete, or general, analytical solution* of the $(2n)$th order system (3.12.4a, c–d, 5a–c) contains $2n$ *independent, or distinct, constants of integration*. One way of expressing such a solution is in the form of $2n$ *first integrals*:

$$f_*(x, t) = c_* = \textit{arbitrary constants (of integration)} \qquad (* = 1, \ldots, 2n). \tag{3.12.5g}$$

A second way is obtained, in principle, by solving (3.12.5g) for the x:

$$x_* = \phi_*(t; c_1, \ldots, c_{2n}) \equiv \phi_*(t; c), \tag{3.12.5h}$$

or, simply,

$$x_* = x_*(t; c_1, \ldots, c_{2n}) \equiv x_*(t; c). \tag{3.12.5i}$$

The constants are usually evaluated by applying the *initial conditions* to (3.12.5g) or (3.12.5h, i). Indeed, applying them to the latter yields $x_{*o} = \phi_*(t_o; c_1, \ldots, c_{2n})$, from which, solving for the c_*, we get

$$c_* = c_*(t_o; x_{1o}, \ldots, x_{2n,o}), \tag{3.12.5j}$$

and, inserting these expressions back into (3.12.5h), we obtain the particular solution

$$x_* = \phi_*(t, t_o; x_{1o}, \ldots, x_{2n,o}) \equiv \phi_*(t, t_o; x_o). \qquad (3.12.5k)$$

Finally, evaluating this for $t = t_o$, or swapping the roles of t, t_o and x, x_o, readily leads to

$$x_{*o} = \phi_*(t_o, t; x_1, \ldots, x_{2n}) \equiv \phi_*(t_o, t; x) \qquad \text{[solution of (3.12.5j) for the } x_o\text{];} \qquad (3.12.5l)$$

that is, the (3.12.5k) are $2n$ first integrals, whose values are the initial values of the variables.

Back to Mechanics

In terms of the variables of Lagrangean mechanics q and $\dot{q}$, the first half of (3.12.5g) translates to the n independent *first integrals* of the equations of motion, or simply *constants of the motion*:

$$f_k(q, \dot{q}, t) = c_k \qquad (k = 1, \ldots, n), \qquad (3.12.6a)$$

which is a system of total order n; whereas the remaining integrals are the integrals of (6a):

$$g_l(q, t; c'_1, \ldots, c'_n) \equiv g_l(q, t; c') = c_l \qquad (l = 1, \ldots, n). \qquad (3.12.6b)$$

The integrals (3.12.6a, b) are equivalent because both contain $2n$ integration constants. The energy integral $E \equiv T(q, \dot{q}, t) + V(q, t) = \textit{constant}$ (§3.9), whenever it exists, is one such first integral. Similarly, the first and second half of (3.12.5h, i) translate, respectively, to

$$q_k = q_k(t; c_1, \ldots, c_{2n}) \equiv q_k(t; c), \qquad \dot{q}_k = \dot{q}_k(t; c_1, \ldots, c_{2n}) \equiv \dot{q}_k(t; c). \qquad (3.12.6c)$$

Each choice of constants c constitutes a different $q_k \leftrightarrow t$ relation and, therefore, a *particular solution/motion.* Inverting (3.12.6c) [assuming that their Jacobian $|\partial(q, \dot{q})/\partial c| \neq 0$], we obtain

$$c_* = f_*(q, \dot{q}, t) \qquad (* = 1, \ldots, 2n). \qquad (3.12.6d)$$

Each c_* is a constant of the motion [and, therefore, along the latter each $f_*(q, \dot{q}, t)$ is conserved] but its value depends on the particular $q_k \leftrightarrow t$; that is, on the particular motion. Further, due to (3.12.6c, d), an integral of motion $F = F(q, \dot{q}, t) = \textit{constant}$ becomes

$$F(q, \dot{q}, t) = F[q(t, c), \dot{q}(t, c), t] \equiv F'(t, c). \qquad (3.12.6e)$$

But $0 = \dot{F} = dF'/dt = \partial F/\partial t$, and therefore F' *does not depend explicitly on time*; that is, finally,

$$F(q, \dot{q}, t) = F'(c) = \textit{function of the independent } c\text{'s}; \qquad (3.12.6f)$$

and every other constant of the motion depends on them.

Usually, we apply (3.12.6d) for some "initial" time t_o, say $t_o = 0$, and thus express the c's for all time in terms of the initial values of the q's and $\dot{q}$'s:

$$q_k(0) = q_{k,o} \qquad \text{and} \qquad \dot{q}_k(0) = \dot{q}_{k,o}; \tag{3.12.7a}$$

or, as we say, in terms of the *initial state* of the system q_o, $\dot{q}_o$. Then, (3.12.6c) can be rewritten, respectively, à la (3.12.5k), as

$$q_k = q_k(t; q_o, \dot{q}_o), \qquad \dot{q}_k = \dot{q}_k(t; q_o, \dot{q}_o); \tag{3.12.7b}$$

a fact that shows that the number of arbitrarily prescribable conditions (data) at our disposal is $2n$.

In sum:

- The determination of the most general motion of an n DOF (holonomic and potential) mechanical system is mathematically equivalent to the determination of the independent integrals, or constants of motion, of its n second-order Lagrangean equations. [However, as Landau and Lifshitz (1960, p. 13) point out, not all such integrals are equally important to mechanics. The most significant ones derive from "the fundamental homogeneity and isotropy of space and time"; and the physical quantities represented by them are said to be *conserved*. Further, such integrals of motion are *additive*; that is, for systems with negligible mutual interaction of their parts (see closed/open systems, below), *their values for the whole system equal the sum of their values for its individual parts*; and therein lies their importance to mechanics; see ex. 3.12.3, below.]
- An initial state of a system—that is, a particular choice of the $2n$ q_o's (positions) and $\dot{q}_o$'s (velocities)—determines a particular motion; and every constant of motion is determined by that initial state, via (3.12.6d). [For a detailed discussion of the geometrical interpretation of the above in generalized spaces, and more, see, for example, Prange (1935, pp. 547–564).]

These fundamental results allow us to express, in principle, the solution of any dynamical problem as the following time power series around $t = 0$:

$$\begin{aligned} q_k(t) &= q_k(0) + \dot{q}_k(0)t + (1/2)\ddot{q}_k(0)t^2 + \cdots \\ &\equiv q_{k,o} + \dot{q}_{k,o}t + (1/2)\ddot{q}_{k,o}t^2 + \cdots. \end{aligned} \tag{3.12.8}$$

To calculate $\ddot{q}_k(0) \equiv \ddot{q}_{k,o}$ we evaluate (3.12.1, 2) at $t = 0$, use the given $q_{k,o}$ and $\dot{q}_{k,o}$, and then solve it for $\ddot{q}_{k,o}$. To calculate $\dddot{q}_k(0) \equiv \dddot{q}_{k,o}$ we $(\ldots)^{\cdot}$-differentiate (3.12.1, 2), evaluate it at $t = 0$, and then solve for $\dddot{q}_{k,o}$ while using $q_{k,o}$, $\dot{q}_{k,o}$, $\ddot{q}_{k,o}$ in it. Continuing this well-known process, we can determine *all* $(\ldots)^{\cdot}$-derivatives of the q_k at $t = 0$ in terms of the $2n$ q_o, $\dot{q}_o$. It can be shown that, for a large number of useful mechanics problems, the conditions of *convergence* of the series (3.12.8) are satisfied, for some time after $t = 0$, and therefore that representation is meaningful. [Also, such series are quite useful in problems of *initial motions*; see, e.g., Whittaker, 1937, pp. 45–46.]

Determinism

The above constitute a quantitative version of the doctrine of classical *determinism* (advanced, especially in connection with problems of *celestial* mechanics, by Laplace et al.). According to this doctrine, if we knew the present state of the universe—that is, the positions and velocities of all its particles, and the forces acting on them, and were able to solve its equations of motion (and exclude internal collisions)—then, we would be able to *predict its entire future* (*and past!*) *uniquely*. However, such *strict causality* = *determinism* is illusory for the following theoretical and practical reasons:

(i) In general, the equations of motion cannot be solved via finite combinations of the known elementary functions (see elementary vs. advanced problems below), and the errors of approximate solutions, due either to truncations of series or to iterations, do not remain small (bounded) for long time intervals.

(ii) The initial state of a system can never be known with infinite accuracy/precision. Therefore, the long-term predictions of *systems that are sensitive to such initial conditions*, due to the cumulative effect of the unavoidable errors in these latter, will be quite erroneous; for example, *chaotic* behavior of nonlinear (deterministic) systems. As McCauley puts it: "*Because* of errors that were made by the computer's roundoff/truncation algorithm, he (E. Lorenz, 1963) discovered what is now called *sensitivity with respect to small changes in initial conditions*: big changes in trajectory patterns occurred at later times owing to shifts in the last digits of the starting conditions" (1993, p. 3, last emphasis added). [The literature on chaotic, or stochastic, motion/dynamics is enormous and growing ... regularly; we recommend Lichtenberg and Lieberman (1992), Tabor (1989).]

(iii) Clearly, our classical (i.e., nonrelativistic, nonquantum, nonprobabilistic, etc.) mathematical model neglects certain factors, or causes, that may prove quite significant in the long (and, sometimes, even short) run; for example, very *high speeds* and *electromagnetic fields* (relativity) and/or very *small spatial regions* (atomic phenomena: Heisenberg's indeterminacy principle, and Born's probabilistic/statistical interpretation of quantum mechanics).

Hence, since, even within classical mechanics, long-term predictions are practically unreliable, and depending on the system at hand and the accuracy sought, we must update our exact or approximate solutions at the end of an(y) appropriate time interval, using data obtained experimentally at that time.

Elementary versus Advanced Problems

On the basis of the principles used for their integration, we divide mechanical problems into *two* kinds: *elementary* and *advanced*; not a clear-cut and/or uniform terminology, by any means.

- *Elementary* are those problems soluble by *quadratures*; namely, those whose solutions are expressible either in terms of known *elementary* functions (solution in "closed form"), or as *indefinite integrals* of such functions.
- *Advanced* are those problems that *cannot* be solved by quadratures.

Unfortunately, as one might have anticipated, most mechanics problems are *not* elementary; and, whenever they are, it is always *because their Lagrangeans possess some special properties*. Below we summarize some of these special properties that are frequently associated with the elementary problems and account for their solvability via quadratures. This ability to predict properties of the solution(s) of a problem by examining its Lagrangean—that is, without acutally solving its equations of motion—is one of the key advantages of the Lagrangean (and Hamiltonian) method over that of Newton–Euler. [For a general discussion of the relation between Lagrangean properties and conservation theorems, see, for example, McCauley (1997), Saletan and Cromer (1971, chap. 3); and for a different definition of the concept of *integrability* of mechanical systems, see also §8.10, §8.14.]

Closed Systems

We begin with an *isolated*, or *closed*, system; that is, a finite system S consisting of a number of rigid bodies and/or particles that interact only with each other; and as such, one *without* time-varying parameters; namely, a system uninfluenced by sources outside itself. If S is also conservative, or reversible, and can be described by a Lagrangean, then the latter must have the general (inertial) form

$$L \equiv T - V = (1/2) \mathcal{S}\, dm\, \boldsymbol{v} \cdot \boldsymbol{v} - V(\boldsymbol{r}), \tag{3.12.9}$$

where $\boldsymbol{r}/\boldsymbol{v}$ = inertial *position*/*velocity* of a generic particle of S. The "force function" $-V(\boldsymbol{r})$ represents the contribution of the mutual *interaction* (forces) to the Lagrangean; if the bodies/particles of S are noninteracting, then $V = 0$.

That V depends only on the $\boldsymbol{r}$'s means that a change in the position of any of the particles of S affects *instantaneously* all its other particles; otherwise, if interactions spread with *finite* velocity, then, due to the law of velocity addition among any two Galilean frames (= inertial frames in relative accelerationless translation), that velocity of propagation would be different in any two such frames. As a result, the equations of motion of these interacting particles in two inertial frames would be *different*; in clear violation of the classical principle of Galilean relativity, which requires form invariance of the equations of motion among inertial frames [recalling §1.4–1.6; see also Landau and Lifshitz (1960, p. 8)].

Now, the equations of motion of a typical particle of S are

$$(\partial L/\partial \boldsymbol{v})^{\cdot} - \partial L/\partial \boldsymbol{r} = \boldsymbol{0}: \quad dm(d\boldsymbol{v}/dt) = -\partial V/\partial \boldsymbol{r}. \tag{3.12.9a}$$

From the above, and since $d\boldsymbol{v}/dt \equiv d^2\boldsymbol{r}/dt^2 \equiv \boldsymbol{a}$, both L and the equations of motion remain unchanged (invariant) under a $t \to -t$ transformation; and, hence, also under $dt \to -dt$. This expresses the *reversibility* of motion of such systems, and the *isotropy* of time (i.e., identical properties in both future and past "directions"), in addition to its *homogeneity*.

In terms of (inertial) Lagrangean coordinates $q = (q_1, \ldots, q_n)$, equations (3.12.9) and (3.12.9a) assume, respectively, the system forms

$$L \equiv T - V = \sum \sum (1/2) M_{kl}(q) \dot{q}_k \dot{q}_l - V(q), \qquad (\partial L/\partial \dot{q}_k)^{\cdot} - \partial L/\partial q_k = 0. \tag{3.12.9b}$$

Open Systems

Next, let us consider a system S_1 that is *not* closed, and interacts with another system S_2 that has a *given* motion; that is, it is unaffected by its interaction with S_1. Then we say that S_1 is *open*, or that it moves in the given *external* field of S_2. If we know the Lagrangean of the combined system $S_1 + S_2 \equiv S$, L, then we can find the Lagrangean of S_1, L_1, by replacing in L the coordinates and velocities of S_2, $q_{(2)}$ and $\dot{q}_{(2)}$ by their given functions of time. In particular, if S is closed, then, since in this case [with $q_{(1)}$ and $\dot{q}_{(1)}$ = *coordinates* and *velocities* of S_1, and corresponding notations for its kinetic and potential energies]

$$L = T_{(1)}\big[q_{(1)}, \dot{q}_{(1)}\big] + T_{(2)}\big[q_{(2)}, \dot{q}_{(2)}\big] - V\big[q_{(1)}, q_{(2)}\big], \tag{3.12.10a}$$

from which (recalling ex. 3.5.13)

$$T_{(2)}\big[q_{(2)}(t), \dot{q}_{(2)}(t)\big] = \textit{given} \text{ function of time, and hence omittable from } L, \tag{3.12.10b}$$

we finally obtain

$$\begin{aligned} L_1 &= T_{(1)}\big[q_{(1)}, \dot{q}_{(1)}\big] - V\big[q_{(1)}, q_{(2)}(t)\big] \\ &\equiv T_{(1)}\big[q_{(1)}, \dot{q}_{(1)}\big] - V\big[q_{(1)}, t\big] \Rightarrow \partial L_1/\partial t = -\partial V/\partial t \neq 0. \end{aligned} \tag{3.12.10c}$$

Integrals of Closed Systems

Here, we prove that *a closed mechanical system with n positional coordinates has* $2n - 1$ *independent integrals*. Indeed, since [as the second of eqs. (3.12.9b) shows] the equations of motion for such a system do *not* contain the time explicitly, the time origin is completely arbitrary and we can take as one of the $2n$ arbitrary constants of integration of the general solution $q_k(t; c_1, \ldots, c_n)$ the *additive* time constant τ. Eliminating $t + \tau$ from the $2n$ functions $q_k = q_k(t + \tau; c_1, \ldots, c_{2n-1})$ and $\dot{q}_k = \dot{q}_k(t + \tau; c_1, \ldots, c_{2n-1})$, we can express the remaining $2n - 1$ constants $c_1, \ldots, c_{2n-1}$ in terms of the $2n$ q's and $\dot{q}$'s:

$$c_l = c_l(q, \dot{q}), \qquad (l = 1, \ldots, n - 1); \tag{3.12.11}$$

that is, our system has $2n - 1$ independent integrals, Q.E.D.

Ignorable Coordinates and Momentum Conservation

We have seen the power theorem and Jacobi–Painlevé generalized energy integral (§3.9). Let us now see some generalized, or Lagrangean, *momentum* integrals. We consider, again, a system with Lagrangean $L = L(t, q, \dot{q})$ and equations of motion

$$(\partial L/\partial \dot{q}_k)^{\cdot} - \partial L/\partial q_k = 0 \qquad (k = 1, \ldots, n). \tag{3.12.12a}$$

If some of the system *coordinates*, say $q_1, \ldots, q_M$ $(M \leq n)$, do not appear explicitly in L (although the corresponding velocities $\dot{q}_1, \ldots, \dot{q}_M$ do) — that is, if

$$L = L(t; q_{M+1}, \ldots, q_n; \dot{q}_1, \ldots, \dot{q}_n), \tag{3.12.12b}$$

then, as (3.12.12a) immediately show, the corresponding (holonomic) Lagrangean *momenta* $p_i \equiv \partial L/\partial \dot{q}_i$ $(i = 1, \ldots, M)$ are conserved:

$$(\partial L/\partial \dot{q}_i)^{\cdot} = 0 \Rightarrow \partial L/\partial \dot{q}_i \equiv p_i = \textit{constant} \equiv c_i; \tag{3.12.12c}$$

where the constants of integration c_i are to be evaluated from the initial conditions. Coordinates like $q_1, \ldots, q_M$ are called *ignorable*, or *cyclic* (after Thomson and Tait, Helmholtz, Routh et al. — see chap. 8); whereas the rest of the q's that do appear in L, and hence satisfy eqs. (3.12.12a) but with $k = M + 1, \ldots, n$, are called *palpable*. The presence (or, rather, absence!) of ignorable coordinates is one of the most common reasons for the solubility of problems by quadratures. We show in chapter 8 how to utilize the M "cyclic integrals" (3.12.12c) to reduce the number of Lagrangean equations (3.12.12a) by the number of ignorable coordinates present in L (method of "ignoration of coordinates" of Routh and Helmholtz).

Example 3.12.1 Consider a system with Lagrangean

$$L = (m/2)\left[(\dot{r})^2 + r^2(\dot{\phi})^2\right] - V(r), \tag{a}$$

[particle of mass m in plane motion ($r, \phi =$ polar coordinates) in potential field $V = V(r)$]. Here, clearly, $\partial L/\partial\phi = 0$; that is, ϕ is ignorable. Therefore, the system possesses the cyclic integral

$$p_\phi \equiv \partial L/\partial\dot{\phi} = mr^2\dot{\phi} = constant, \tag{b}$$

which expresses the conservation of the component of the angular momentum of the particle, about the origin, along an axis perpendicular to the plane of the motion.

Example 3.12.2 *First-Order Form of the Routh–Voss Equations.* Let us find the first-order forms of a system subject to the m Pfaffian constraints

$$\omega_D \equiv \sum a_{Dk}\,\dot{q}_k + a_D = 0 \qquad [k = 1, \ldots, n;\ D = 1, \ldots, m\ (<n)], \tag{a}$$

and, hence, having the Routh–Voss equations of motion

$$E_k(L) = Q_k + \sum \lambda_D a_{Dk} \qquad \left[Q_k = \textit{nonpotential}\ \text{part of } (\ldots)\,\delta q_k \text{ in } \delta' W\right]. \tag{b}$$

With the variable change (3.12.4)

$$q_1 = x_1, \ldots, q_n = x_n; \qquad \dot{q}_1 = x_{n+1} = \dot{x}_1, \ldots, \dot{q}_n = x_{2n} = \dot{x}_n, \tag{c}$$

L becomes $L(t; x_1, \ldots, x_n; x_{n+1}, \ldots, x_{2n})$ and therefore the m first-order equations (a) and n second-order equations (b) in the q's transform, respectively, to the m *finite* equations

$$\sum a_{Dk} x_{n+k} + a_D = 0, \tag{d}$$

and the n *first-order* equations

$$(\partial L/\partial x_{n+k})^{\cdot} - \partial L/\partial x_k = Q_k + \sum \lambda_D a_{Dk}, \tag{e}$$

since $\partial L/\partial x_{n+k} =$ *linear* in the x_{n+k} ($k = 1, \ldots, n$). Equations (d, e) and the second half of (c) constitute a system of $2n + m$ first-order equations for the $2n$ x's and m λ_D's.

Example 3.12.3 *Integrals of a Closed System.* Let us find the integrals of a closed system with (inertial) Lagrangean

$$L = (1/2)\textstyle\int dm\, \boldsymbol{v}\cdot\boldsymbol{v} - V(\boldsymbol{r}) = T(q, \dot{q}) - V(q); \tag{a}$$

or, of a closed system in a *constant* (time-independent) external field.

(i) *Energy integral.* Here, $\partial L/\partial t = 0$, $Q_{k,\text{nonpotential}} = 0$, and $a_D = 0$. Hence, the holonomic power equation (3.9.11d ff.) immediately yields

$$h \equiv \sum (\partial L/\partial \dot{q}_k)\dot{q}_k - L = (2T) - (T - V) = T(q, \dot{q}) + V(q) \equiv E = constant. \quad \text{(b)}$$

The condition $\partial L/\partial t = 0$ is a consequence of the *homogeneity of time* for closed systems. The energy integral (b) is *additive*: the energy of a closed system consisting of several closed subsystems *with negligible mutual interaction* equals the sum of the individual subenergies of these subsystems. This also results from the linearity of $h = E$ in L in (b).

[Whittaker (in 1900) has shown how to use the energy equation (b) to reduce a closed n-*DOF* system into another with $(n-1)$ *DOF* (and a quadrature); see, for example, Whittaker (1937, pp. 64–67); also Butenin (1971, pp. 103–110), MacMillan (1936, pp. 320–322).]

(ii) *Linear momentum integral.* The *homogeneity of space* for such systems leads to the requirement that the first-order change of their Lagrangeans under $\boldsymbol{r} \rightarrow \boldsymbol{r} + \Delta\boldsymbol{r}$, where $\Delta\boldsymbol{r} =$ arbitrary *infinitesimal translation* common to all system particles, and $\boldsymbol{v} \rightarrow \boldsymbol{v}$ (i.e., no velocity change), should vanish.

Since, here, $L = L(\boldsymbol{r}, \boldsymbol{v})$, the condition $\Delta L = 0$, under such changes, translates to

$$\Delta L = S(\partial L/\partial \boldsymbol{r}) \cdot \Delta\boldsymbol{r} = \Delta\boldsymbol{r} \cdot S(\partial L/\partial \boldsymbol{r}) = 0 \;\Rightarrow\; \partial L/\partial \boldsymbol{r} = \boldsymbol{0}. \quad \text{(c)}$$

As a result, Lagrange's equations reduce to

$$S\, d/dt(\partial L/\partial \boldsymbol{v}) = d/dt\Big(S\, \partial L/\partial \boldsymbol{v}\Big) = \boldsymbol{0}, \quad \text{(d)}$$

or

$$\boldsymbol{p} \equiv S\, \partial L/\partial \boldsymbol{v} = S\, dm\, \boldsymbol{v} \equiv S\, d\boldsymbol{p}: \; \textit{system linear momentum} = \textit{constant}. \quad \text{(e)}$$

Clearly, $\boldsymbol{p}$ is additive even for nonnegligible interactions of the constituent subsystems.

Next, the linear momentum $\boldsymbol{p}$/(total) energy E/Lagrangean L of a system in an inertial frame F are related to those in another frame F', translating relative to F with velocity $\boldsymbol{v}_o = d\boldsymbol{r}_o/dt$ (fig. 3.29), $\boldsymbol{p}'/E'/L'$, respectively, as follows: since $\boldsymbol{v} = \boldsymbol{v}_o + \boldsymbol{v}'$ ($\boldsymbol{v}' =$ particle velocities relative to F'), we find, successively,

$$\text{(a)} \quad \boldsymbol{p} = S\, dm\, \boldsymbol{v} = S\, dm(\boldsymbol{v}_o + \boldsymbol{v}') = \boldsymbol{v}_o\Big(S\, dm\Big) + S\, dm\, \boldsymbol{v}' \equiv \boldsymbol{p}' + m\, \boldsymbol{v}_o. \quad \text{(f)}$$

If $\boldsymbol{p}' = \boldsymbol{0}$ (i.e., if the system is *at rest* relative to F'), then (f) yields the *velocity of the system as a whole* relative to F, or *velocity of its mass center* G relative to F:

$$\boldsymbol{v}_o = \boldsymbol{p}/m = S\, dm\, \boldsymbol{v} \Big/ S\, dm$$

$$\Rightarrow S\, dm\, \boldsymbol{r} - \boldsymbol{p}t = m\, \boldsymbol{r}_{Go} \;\Rightarrow\; \boldsymbol{r}_G = \boldsymbol{r}_{Go} + (\boldsymbol{p}/m)t; \quad \text{(g)}$$

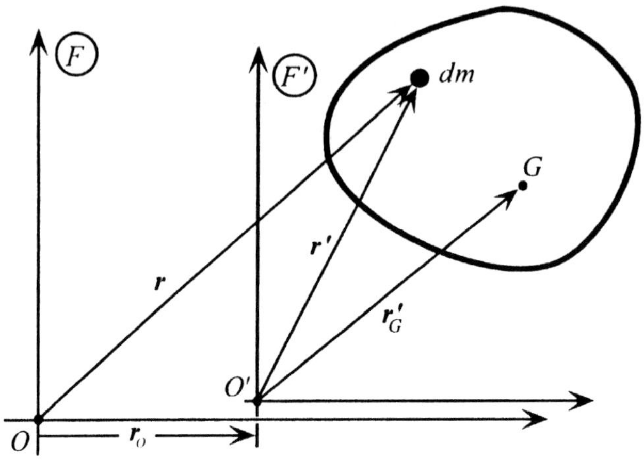

Figure 3.29 Frames F and F' in mutual translation.

that is, the center of mass of a closed system *moves uniformly in a straight line.*

(b)
$$
\begin{aligned}
E &\equiv (1/2)\mathsf{S}\, dm\, \boldsymbol{v}\cdot\boldsymbol{v} + V \\
&= (1/2)\mathsf{S}\, dm\big[(\boldsymbol{v}_o + \boldsymbol{v}')\cdot(\boldsymbol{v}_o + \boldsymbol{v}')\big] + V \\
&= E' + \boldsymbol{p}'\cdot\boldsymbol{v}_o + (1/2)m\, v_o{}^2,
\end{aligned}
\tag{h1}
$$

where

$$E' \equiv (1/2)\mathsf{S}\, dm\, \boldsymbol{v}'\cdot\boldsymbol{v}' + V. \tag{h2}$$

If $\boldsymbol{p}' = \boldsymbol{0}$, as earlier (i.e., if G is at rest in F'), then (h1) and (h2) reduce, respectively, to

$$E = (1/2)m\, v_o{}^2 + E_{\text{internal}}, \tag{h3}$$

and

$$
\begin{aligned}
E' = E_{\text{internal}} &= \textit{internal}\ \text{energy of system} \\
&= \text{kinetic energy of motion of system particles } \textit{relative} \text{ to mass center,} \\
&\qquad \text{plus potential energy of their mutual } \textit{interactions} \\
&= \text{energy of system when at rest as a whole.}
\end{aligned}
\tag{h4}
$$

(c)
$$
\begin{aligned}
L &\equiv (1/2)\mathsf{S}\, dm\, \boldsymbol{v}\cdot\boldsymbol{v} - V \\
&= (1/2)\mathsf{S}\, dm\big[(\boldsymbol{v}_o + \boldsymbol{v}')\cdot(\boldsymbol{v}_o + \boldsymbol{v}')\big] - V \\
&= L' + \boldsymbol{p}'\cdot\boldsymbol{v}_o + (1/2)m\, v_o{}^2,
\end{aligned}
\tag{i1}
$$

where

$$L' \equiv T' - V = (1/2)\mathsf{S}\, dm\, \boldsymbol{v}'\cdot\boldsymbol{v}' - V. \tag{i2}$$

Finally, integrating the above from an "initial" time up to a generic one, t, we obtain the law of transformation of the *Hamiltonian action* of the system (chaps. 7 and 8) between the frames $F(A_H)$ and $F'(A_{H'})$:

$$A_H \equiv \int L\,dt = \cdots = A_{H'} + m\,\mathbf{v}_o \cdot \mathbf{r}_{G'} + (1/2)m\,{v_o}^2 t, \tag{i3}$$

where

$$A_{H'} \equiv \int L'\,dt, \quad \mathbf{r}_{G'} = \textit{position vector of } G \textit{ in } F'. \tag{i4}$$

(iii) *Angular momentum integral.* The *isotropy of space* for such systems leads to the requirement that the first-order change of their Lagrangeans under

$$\mathbf{r} \to \mathbf{r} + \Delta\mathbf{r} = \mathbf{r} + \Delta\boldsymbol{\chi} \times \mathbf{r}, \qquad \mathbf{v} \to \mathbf{v} + \Delta\mathbf{v} = \mathbf{v} + \Delta\boldsymbol{\chi} \times \mathbf{v}, \tag{j1,2}$$

where $\Delta\boldsymbol{\chi}$ = arbitrary *infinitesimal rotation* common to all system particles, should vanish.

Since, by Lagrange's equations $\partial L/\partial\mathbf{r} = d/dt(\partial L/\partial\mathbf{v}) = d/dt(d\mathbf{p})$, the condition $\Delta L = 0$ under (j1, 2), yields, successively,

$$\begin{aligned} \Delta L &= \mathbf{S}\,[(\partial L/\partial\mathbf{r}) \cdot \Delta\mathbf{r} + (\partial L/\partial\mathbf{v}) \cdot \Delta\mathbf{v}] \\ &= \mathbf{S}\,[(d\mathbf{p})^{\cdot} \cdot (\Delta\boldsymbol{\chi} \times \mathbf{r}) + d\mathbf{p} \cdot (\Delta\boldsymbol{\chi} \times \mathbf{v})] \\ &= \Delta\boldsymbol{\chi} \cdot \mathbf{S}\,[\mathbf{r} \times (d\mathbf{p})^{\cdot} + \mathbf{v} \times d\mathbf{p}] \\ &= \Delta\boldsymbol{\chi} \cdot \left(\mathbf{S}\,\mathbf{r} \times d\mathbf{p}\right)^{\cdot} = 0, \end{aligned} \tag{j3}$$

from which, since $\Delta\boldsymbol{\chi}$ is arbitrary, we obtain the principle of *conservation of angular momentum*

$\mathbf{H}_O \equiv \mathbf{S}\,\mathbf{r} \times d\mathbf{p} \equiv \mathbf{S}\,\mathbf{r} \times (dm\,\mathbf{v})$:

(*inertial*) *absolute angular momentum* (*or moment of momentum*) *about an F-fixed point O*

$= \textit{constant}.$ (j4)

Clearly, $\mathbf{H}_O$, like $\mathbf{p}$, is additive even for nonnegligible interactions.

Thus, a closed system has *ten* additive (scalar) integrals:

- Homogeneity of *time*: conservation of *energy* (one),
- Homogeneity of *space*: conservation of *linear* momentum (three),
- Center of mass moves with constant velocity (three), and
- Isotropy of *space*: conservation of *angular* momentum (three).

Finally, let us relate the absolute angular momenta of the system in the two earlier frames $F(\mathbf{H}_O)$ and $F'(\mathbf{H}_{O'})$. With $\mathbf{r}_o$ = position of origin O' of F' relative to origin O of F, and $\mathbf{r}'$ = position of typical system particle relative to O', we have

$$\begin{aligned} \mathbf{H}_O &\equiv \mathbf{S}\,\mathbf{r} \times (dm\,\mathbf{v}) = \mathbf{S}\,[(\mathbf{r}_o + \mathbf{r}') \times dm(\mathbf{v}_o + \mathbf{v}')] \\ &= \mathbf{r}_o \times (m\mathbf{v}_o) + \mathbf{r}_o \times (m\mathbf{v}'_G) + m\mathbf{r}'_G \times \mathbf{v}_o + \mathbf{H}_{O'}, \end{aligned} \tag{j5}$$

where

$$\boldsymbol{H}_{O'} \equiv \mathsf{S}\, \boldsymbol{r}' \times (dm\, \boldsymbol{v}') = absolute\ angular\ momentum\ about\ O', \tag{j6}$$

and

$$\boldsymbol{r}'_G\ (\boldsymbol{v}'_G) = position\ (velocity)\ of\ system\ mass\ center,\ G,\ relative\ to\ O'. \tag{j7}$$

In particular, if the origins of F and F' instantaneously coincide ($\boldsymbol{r}_o = \boldsymbol{0} \Rightarrow \boldsymbol{r}'_G = \boldsymbol{r}_G$), and the system is at rest in F' as a whole ($\boldsymbol{v}_o = \boldsymbol{v}_G \Rightarrow m\boldsymbol{v}_o = m\boldsymbol{v}_G = \boldsymbol{p}$), then (j5, 6) reduce, respectively, to

$$\boldsymbol{H}_O = \boldsymbol{H}_{O,\text{intrinsic}} + \boldsymbol{r}_G \times \boldsymbol{p}, \tag{j8}$$

$$\boldsymbol{H}_{O'} = \boldsymbol{H}_{O,\text{intrinsic}} \equiv \mathsf{S}\, \boldsymbol{r} \times (dm\, \boldsymbol{v}'):$$

$$intrinsic\ angular\ momentum\ of\ system\ in\ F'\ about\ O', \tag{j9}$$

$$\boldsymbol{r}_G \times \boldsymbol{p} = angular\ momentum\ of\ system\ due\ to\ its\ motion\ as\ a\ whole. \tag{j10}$$

For additional special cases see, for example, Landau and Lifshitz (1960, pp. 20–22); also Whittaker (1937, pp. 59–62).

Example 3.12.4 *Separable Systems of Liouville, Stäckel et al.* (see also §8.10). As eqs. (3.12.1, 2) readily show, Lagrange's equations are *coupled* in the q's; that is, in general, the (k)th such equation ($1 \le k \le n$) contains q_k, $\dot{q}_k$, $\ddot{q}_k$, and all the other q's and $\dot{q}$'s. Below we examine some special systems in which *each of their equations of motion contains only one such variable* and its $(\ldots)^{\cdot}$-derivatives. Such *uncoupled*, or *separable*, systems can be solved by quadratures.

(i) Let us consider a system completely describable by

$$2T = v_1(q_1)(\dot{q}_1)^2 + \cdots + v_n(q_n)(\dot{q}_n)^2, \tag{a1}$$

$$V = w_1(q_1) + \cdots + w_n(q_n), \tag{a2}$$

where each $v_k(\ldots)$ (> 0, assumed) and $w_k(\ldots)$ is an arbitrary function of q_k only. Its Lagrangean equations [with $(\ldots)' \equiv d(\ldots)/dq_k$; $k = 1, \ldots, n$]

$$d[v_k(q_k)\dot{q}_k]/dt - (1/2)v_k{}'(q_k)(\dot{q}_k)^2 = -w_k{}'(q_k),$$

or

$$v_k(q_k)\ddot{q}_k + (1/2)v_k{}'(q_k)(\dot{q}_k)^2 = -w_k{}'(q_k), \tag{b}$$

are clearly *uncoupled*. Integrating (b) once, we readily obtain the energy integrals

$$(1/2)v_k(q_k)(\dot{q}_k)^2 + w_k(q_k) = c_k \qquad (c_k\text{: } constants\ of\ integration), \tag{c}$$

and integrating this once more, since q_k and t are separable, we finally obtain the quadrature

$$t = \int \{v_k(q_k)/[2c_k - 2w_k(q_k)]\}^{1/2}\, dq_k + \beta_k \qquad (\beta_k\text{: } new\ integration\ constants). \tag{d}$$

(ii) Let us consider the *Liouville* systems (1849)

$$2T = u\,[v_1(q_1)(\dot{q}_1)^2 + \cdots + v_n(q_n)(\dot{q}_n)^2], \tag{e1}$$

$$V = [w_1(q_1) + \cdots + w_n(q_n)]/u, \tag{e2}$$

where $u \equiv u_1(q_1) + \cdots + u_n(q_n)$ (> 0). Below we show that systems that are, or can be put, in this form can be solved by quadratures, like (d).

Indeed, with the help of the (assumed invertible) transformation of variables $q_k \to x_k$:

$$x_k = \int [v_k(q_k)]^{1/2}\, dq_k \;\Rightarrow\; dx_k = [v_k(q_k)]^{1/2}\, dq_k, \tag{f}$$

we can reduce T (to within u) to a sum of squares in the new velocities:

$$2T = u\,[(\dot{x}_1)^2 + \cdots + (\dot{x}_n)^2], \tag{g1}$$

$$u \equiv u_1[q_1(x_1)] + \cdots + u_n[q_n(x_n)] = u_1(x_1) + \cdots + u_n(x_n); \tag{g2}$$

similarly V, (e2), transforms to

$$V = \{w_1[q_1(x_1)] + \cdots + w_n[q_n(x_n)]\}/u \equiv [w_1(x_1) + \cdots + w_n(x_n)]/u. \tag{g3}$$

Hence, *renaming the* x's *as* q's, we can rewrite (g1, 3) as

$$2T = u\,[(\dot{q}_1)^2 + \cdots + (\dot{q}_n)^2], \tag{h1}$$

$$V = [w_1(q_1) + \cdots + w_n(q_n)]/u. \tag{h2}$$

Now, the typical Lagrangean equation of the above system is

$$(u\,\dot{q}_k)^{\cdot} - (1/2)(\partial u/\partial q_k)[(\dot{q}_1)^2 + \cdots + (\dot{q}_n)^2] = -\partial V/\partial q_k. \tag{i1}$$

To find the corresponding energy equation, we multiply (i1) by $2u\,\dot{q}_k$, and notice that

$$[u^2(\dot{q}_k)^2]^{\cdot} = 2u\,\dot{u}\,(\dot{q}_k)^2 + u^2(2\dot{q}_k\,\ddot{q}_k) = 2u\,\dot{q}_k(\dot{u}\,\dot{q}_k + u\,\ddot{q}_k)$$
$$= 2u\dot{q}_k(u\dot{q}_k)^{\cdot}.$$

The result is

$$[u^2(\dot{q}_k)^2]^{\cdot} - u\dot{q}_k(\partial u/\partial q_k)[(\dot{q}_1)^2 + \cdots + (\dot{q}_n)^2] = -2u\dot{q}_k(\partial V/\partial q_k). \tag{i2}$$

But from the energy integral (of this conservative system), we have

$$T + V \equiv E = h = \textit{constant} \;\Rightarrow\; u[(\dot{q}_1)^2 + \cdots + (\dot{q}_n)^2] = 2(h - V), \tag{j1}$$

and so (i2) can be rewritten, successively, as

$$\begin{aligned}
[u^2(\dot{q}_k)^2]^{\cdot} &= 2(h - V)\dot{q}_k(\partial u/\partial q_k) - 2u\dot{q}_k(\partial V/\partial q_k)\\
&= 2\dot{q}_k\{\partial/\partial q_k[u(h - V)]\}\\
&= 2\dot{q}_k\{\partial/\partial q_k(hu - [w_1(q_1) + \cdots + w_n(q_n)])\}\\
&= 2\dot{q}_k\{d/dq_k[hu_k(q_k) - w_k(q_k)]\}\\
&= 2[hu_k(q_k) - w_k(q_k)]^{\cdot}.
\end{aligned} \tag{j2}$$

Integrating (j2), we immediately obtain

$$(1/2)u^2(\dot{q}_k)^2 = hu_k(q_k) - w_k(q_k) + \gamma_k \quad (\gamma_k\text{: } \textit{integration constants}). \tag{j3}$$

But the n γ_k are *not* independent: summing the n integrals (j3) over all k, and then dividing by u, we obtain

$$(1/2)u\,[(\dot{q}_1)^2 + \cdots + (\dot{q}_n)^2] + [w_1(q_1) + \cdots + w_n(q_n)]/u$$
$$= h + (\gamma_1 + \cdots + \gamma_n)/u,$$

and comparing this with the energy equation (j1), we easily conclude that

$$\gamma_1 + \cdots + \gamma_n = 0. \tag{j4}$$

Hence, the first integration of our system, (j3), has produced n constants, say $\gamma_1, \ldots, \gamma_{n-1}, h$; *not* $n+1$.

Finally, from (j3), we readily obtain the n (separable variable) equations

$$[h\,u_1(q_1) - w_1(q_1) + \gamma_1]^{-1/2}\,dq_1 = \cdots = [h\,u_n(q_n) - w_n(q_n) + \gamma_n]^{-1/2}\,dq_n$$
$$= (2)^{1/2}\,dt/u, \tag{j5}$$

and from these, with the notation $2[h\,u_k(q_k) - w_k(q_k) + \gamma_k] \equiv f_k(q_k)$, we conclude that

(a) $$\sum u_k\Big(dq_k\Big/\big[f_k(q_k)^{1/2}\big]\Big) = \Big(\sum u_k\Big)dt\Big/u = dt,$$

or, integrating,

$$\sum \int \frac{u_k\,dq_k}{\sqrt{f_k(q_k)}} = t + \eta_1 \qquad (\eta_1\text{: } \textit{integration constant}); \tag{k1}$$

and

(b) $$\int \frac{dq_1}{\sqrt{f_1(q_1)}} - \int \frac{dq_l}{\sqrt{f_l(q_l)}} = \eta_l \qquad (l = 2, \ldots, n). \tag{k2}$$

Equation (k1) and the $n-1$ equations (k2) supply the n independent constants of integration $\eta_1, \ldots, \eta_n$, which along with the earlier $n-1$ independent γ's and the energy constant h (i.e., $\gamma_1, \ldots, \gamma_{n-1}, h$), constitute the $2n$ expected independent constants of integration of the system (h1, 2) [we also note that, by (j5), and since $u > 0$, it follows that in (k1, 2) $f_k(q_k)^{1/2}$ has the same sign as dq_k; a fact that becomes important whenever one or more of the q's oscillates between fixed limits (*libration*—see §8.14)]. Lastly, the original variables of (e1, 2) can be recovered from (f) with another integration. For extensive discussions of these systems, including Hamilton–Jacobi methods, see, e.g., Hamel (1949, pp. 302–303, 358–361, 669–688), Lur'e (1968, pp. 538–548), Pars (1965, pp. 291–348); also Whittaker (1937, p. 60, and references therein).

3.13 THE RIGID BODY: LAGRANGEAN–EULERIAN KINEMATICO-INERTIAL IDENTITIES

Kinematical Preliminaries

As we have seen in §1.7 ff. (fig. 3.30), the inertial velocity of a typical body point P equals

$$\boldsymbol{v} \equiv d\boldsymbol{r}/dt \equiv d\boldsymbol{r}_\blacklozenge/dt + d(\boldsymbol{r} - \boldsymbol{r}_\blacklozenge)/dt \equiv d\boldsymbol{r}_\blacklozenge/dt + d\boldsymbol{r}_{/\blacklozenge}/dt$$
$$\equiv \boldsymbol{v}_\blacklozenge + \boldsymbol{v}_{/\blacklozenge} = \boldsymbol{v}_\blacklozenge + \boldsymbol{\omega} \times \boldsymbol{r}_{/\blacklozenge}, \tag{3.13.1}$$

or, in components, with some easily understood notation,

(a) Along the *space-fixed* (inertial) axes/basis $O\text{–}XYZ/\boldsymbol{IJK}$:

$$\boldsymbol{r} = X\boldsymbol{I} + Y\boldsymbol{J} + Z\boldsymbol{K}, \qquad \boldsymbol{r}_{/\blacklozenge} = (X - X_\blacklozenge)\boldsymbol{I} + (Y - Y_\blacklozenge)\boldsymbol{J} + (Z - Z_\blacklozenge)\boldsymbol{K}, \tag{3.13.1a}$$

and, therefore,

$$v_X = v_{\blacklozenge,X} + \omega_Y(Z - Z_\blacklozenge) - \omega_Z(Y - Y_\blacklozenge), \quad \text{etc., cyclically;} \tag{3.13.1b}$$

(b) Along the *body-fixed* (noninertial) axes/basis $\blacklozenge\text{–}xyz/\boldsymbol{ijk}$:

$$\boldsymbol{r}_{/\blacklozenge} = x_{/\blacklozenge}\boldsymbol{i} + y_{/\blacklozenge}\boldsymbol{j} + z_{/\blacklozenge}\boldsymbol{k}, \tag{3.13.1c}$$

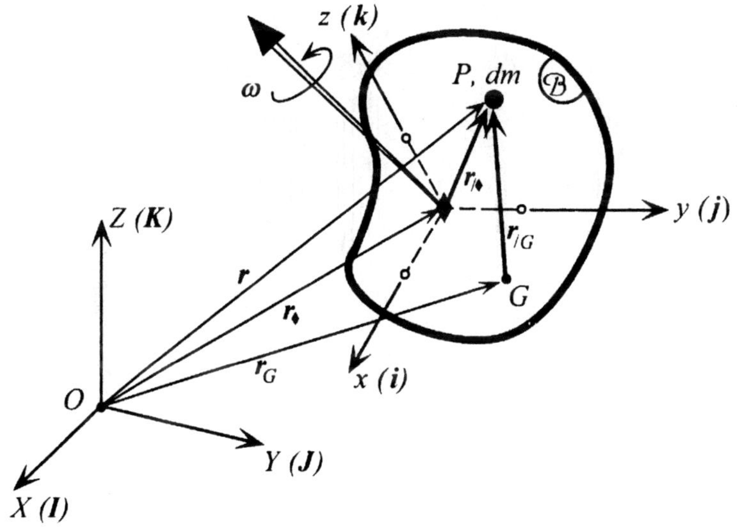

Figure 3.30 Basic notation for general rigid-body motion:
O–XYZ = *space-fixed* (inertial) axes; O–$\boldsymbol{IJK}$ = associated basis;
$\blacklozenge$–xyz = *body-fixed* (noninertial) axes; $\blacklozenge$–$\boldsymbol{ijk}$ = associated basis;
G = center of mass of body B;
$\boldsymbol{r}_{/\blacklozenge}$ = position of representative body particle P relative to arbitrary body point $\blacklozenge$;
$\boldsymbol{r}_{/G}$ = position of representative body particle P relative to G;
$\boldsymbol{\omega}$ = inertial angular velocity of body B.

and therefore,

$$v_x = v_{\blacklozenge,x} + \omega_y\, z_{/\blacklozenge} - \omega_z\, y_{/\blacklozenge}, \quad \text{etc., cyclically.} \tag{3.13.1d}$$

We point out that since $v_{\blacklozenge,X} = \dot{X}$, etc., cyclically, the inertial components $v_{\blacklozenge;X,Y,Z}$ are *holonomic*; whereas, since

$$v_{\blacklozenge,x} = \cos(x, X)\dot{X} + \cos(x, Y)\dot{Y} + \cos(x, Z)\dot{Z}, \quad \text{etc., cyclically,} \tag{3.13.1e}$$

the noninertial components $v_{\blacklozenge;x,y,z}$ are *nonholonomic*, or *quasi velocities*; and so are the $\omega_{x,y,z}$.

Kinetic Energy

Substituting (3.13.1) into $2T \equiv S\, dm\, \boldsymbol{v} \cdot \boldsymbol{v}$, we obtain, successively,

$$\begin{aligned} 2T &\equiv S\, dm(\boldsymbol{v}_\blacklozenge + \boldsymbol{\omega} \times \boldsymbol{r}_{/\blacklozenge}) \cdot (\boldsymbol{v}_\blacklozenge + \boldsymbol{\omega} \times \boldsymbol{r}_{/\blacklozenge}) \\ &= \cdots = 2T_{\text{translation}} + 2T_{\text{rotation}} + 2T_{\text{coupling}}, \end{aligned} \tag{3.13.2}$$

where

$$\begin{aligned} 2T_{\text{translation}} &\equiv 2T_T \equiv S\, dm\, \boldsymbol{v}_\blacklozenge \cdot \boldsymbol{v}_\blacklozenge = m\, \boldsymbol{v}_\blacklozenge \cdot \boldsymbol{v}_\blacklozenge = m\, {v_\blacklozenge}^2 \\ &= m\left({v_{\blacklozenge,X}}^2 + {v_{\blacklozenge,Y}}^2 + {v_{\blacklozenge,Z}}^2\right) = m\left({v_{\blacklozenge,x}}^2 + {v_{\blacklozenge,y}}^2 + {v_{\blacklozenge,z}}^2\right) \\ &= 2(\textit{kinetic energy of translation}), \end{aligned} \tag{3.13.2a}$$

$$\begin{aligned} 2T_{\text{rotation}} &\equiv 2T_R \equiv S\, dm(\boldsymbol{\omega} \times \boldsymbol{r}_{/\blacklozenge}) \cdot (\boldsymbol{\omega} \times \boldsymbol{r}_{/\blacklozenge}) \\ &= \cdots = S\, dm\, \boldsymbol{\omega} \cdot [\boldsymbol{r}_{/\blacklozenge} \times (\boldsymbol{\omega} \times \boldsymbol{r}_{/\blacklozenge})] \equiv \boldsymbol{\omega} \cdot \boldsymbol{h}_\blacklozenge \\ &= 2(\textit{kinetic energy of rotation}), \end{aligned} \tag{3.13.2b}$$

$$\boldsymbol{h}_\blacklozenge \equiv S\, dm[\boldsymbol{r}_{/\blacklozenge} \times (\boldsymbol{\omega} \times \boldsymbol{r}_{/\blacklozenge})]$$

$= (\textit{inertial})\ \textit{relative angular momentum of the body B about}\ \blacklozenge;\ \textit{a system vector},$ (3.13.2c)

$$\begin{aligned} T_{\text{coupling}} &\equiv T_C \equiv S\, dm[\boldsymbol{v}_\blacklozenge \cdot (\boldsymbol{\omega} \times \boldsymbol{r}_{/\blacklozenge})] = \boldsymbol{\omega} \cdot S\, dm(\boldsymbol{r}_{/\blacklozenge} \times \boldsymbol{v}_\blacklozenge) \\ &= \boldsymbol{\omega} \cdot \left[\left(S\, dm\, \boldsymbol{r}_{/\blacklozenge}\right) \times \boldsymbol{v}_\blacklozenge\right] = \boldsymbol{\omega} \cdot (m\, \boldsymbol{r}_{G/\blacklozenge} \times \boldsymbol{v}_\blacklozenge) \\ &= m\, \boldsymbol{v}_\blacklozenge \cdot (\boldsymbol{\omega} \times \boldsymbol{r}_{G/\blacklozenge}) \equiv m\, \boldsymbol{v}_\blacklozenge \cdot \boldsymbol{v}_{G/\blacklozenge} \\ &= \textit{kinetic energy of coupling (of translation of}\ \blacklozenge\ \textit{with rotation of B)}. \end{aligned} \tag{3.13.2d}$$

[Clearly,

$$S\, dm\, \boldsymbol{r}_{/\blacklozenge} = S\, dm(\boldsymbol{r}_{G/\blacklozenge} + \boldsymbol{r}_{/G}) = S\, dm\, \boldsymbol{r}_{G/\blacklozenge} + S\, dm\, \boldsymbol{r}_{/G} = m\, \boldsymbol{r}_{G/\blacklozenge} + \boldsymbol{0}.]$$

Further, since (using simple vector algebra identities; recalling §1.16 ff.)

$$\boldsymbol{h}_\blacklozenge \equiv S\, dm\left[(\boldsymbol{r}_{/\blacklozenge} \cdot \boldsymbol{r}_{/\blacklozenge})\boldsymbol{\omega} - (\boldsymbol{\omega} \cdot \boldsymbol{r}_{/\blacklozenge})\, \boldsymbol{r}_{/\blacklozenge}\right] \equiv \mathbf{I}_\blacklozenge \cdot \boldsymbol{\omega}, \tag{3.13.3}$$

where $\mathbf{I}_\blacklozenge$ = *inertia tensor* of B at $\blacklozenge$; or, in components along, say, $\blacklozenge$–xyz,

$$h_{\blacklozenge,x} = I_{\blacklozenge,xx}\,\omega_x + I_{\blacklozenge,xy}\,\omega_y + I_{\blacklozenge,xz}\,\omega_z, \quad \text{etc., cyclically,} \tag{3.13.3a}$$

the expression (3.13.2b) assumes the form

$$\begin{aligned} 2T_R = \boldsymbol{\omega}\cdot\mathbf{I}_\blacklozenge\cdot\boldsymbol{\omega} &= I_{\blacklozenge,xx}\,\omega_x{}^2 + I_{\blacklozenge,yy}\,\omega_y{}^2 + I_{\blacklozenge,zz}\,\omega_z{}^2 \\ &\quad + 2\big(I_{\blacklozenge,xy}\,\omega_x\,\omega_y + I_{\blacklozenge,xz}\,\omega_x\,\omega_z + I_{\blacklozenge,yz}\,\omega_y\,\omega_z\big). \end{aligned} \tag{3.13.3b}$$

Hence, finally [and with $\omega_x = \omega\, l_x$, etc., where $\boldsymbol{l} = (l_x, l_y, l_z)$ = *unit vector along* $\boldsymbol{\omega}$], $2T$ becomes

$$\begin{aligned} 2T &= m\,v_\blacklozenge{}^2 + I_{\blacklozenge,l}\,\omega^2 + 2\boldsymbol{\omega}\cdot(m\boldsymbol{r}_{G/\blacklozenge}\times\boldsymbol{v}_\blacklozenge) \\ &= m v_\blacklozenge{}^2 + I_{\blacklozenge,l}\,\omega^2 + 2m\boldsymbol{v}_\blacklozenge\cdot(\boldsymbol{\omega}\times\boldsymbol{r}_{G/\blacklozenge}) \\ &\equiv m v_\blacklozenge{}^2 + I_{\blacklozenge,l}\,\omega^2 + 2m\boldsymbol{v}_\blacklozenge\cdot\boldsymbol{v}_{G/\blacklozenge}, \end{aligned} \tag{3.13.3c}$$

(= function of the *six* quasi velocities: $v_{\blacklozenge;x,y,z}$ and $\omega_{x,y,z}$, if *body-fixed* axes are used),

where

$$I_{\blacklozenge,l} \equiv l_x{}^2 I_{\blacklozenge,xx} + \cdots + 2l_x\,l_y\,I_{\blacklozenge,xy} + \cdots$$

= *moment of inertia of B about axis of instantaneous rotation through* $\blacklozenge$. (3.13.3d)

Special Cases of (3.13.3c)

- If we choose our body-fixed axes to be also *principal* axes: $\blacklozenge$–$xyz \to \blacklozenge$–123, then, with some obvious notations,

$$\begin{aligned} I_{\blacklozenge,l} &= l_x{}^2 I_{\blacklozenge,x} + l_y{}^2 I_{\blacklozenge,y} + l_z{}^2 I_{\blacklozenge,z} \\ &\equiv l_1{}^2 I_{\blacklozenge,1} + l_2{}^2 I_{\blacklozenge,2} + l_3{}^2 I_{\blacklozenge,3} \quad (\equiv l_1{}^2 A + l_2{}^2 B + l_3{}^2 C), \end{aligned} \tag{3.13.4a}$$

and so the rotational kinetic energy (3.13.3b) reduces to

$$2T_R = I_{\blacklozenge,1}\,\omega_1{}^2 + I_{\blacklozenge,2}\,\omega_2{}^2 + I_{\blacklozenge,3}\,\omega_3{}^3. \tag{3.13.4b}$$

- If, *further*, $\blacklozenge = G$ then, clearly, $T_C = 0$ and (3.13.3c) assumes the (König) form

$$2T = m v_G{}^2 + I_{G,l}\,\omega^2 = m v_G{}^2 + (I_{G,1}\,\omega_1{}^2 + I_{G,2}\,\omega_2{}^2 + I_{G,3}\,\omega_3{}^2); \tag{3.13.4c}$$

in words: the kinetic energy of a moving rigid body consists of two independent (uncoupled) parts: one depending on the motion of the body's center of mass (G) and another equal to the kinetic energy of motion relative to that center. [This is the kinetic energy analog of the familiar Newton–Euler (momentum) proposition that: (i) the motion of G is indistinguishable from that of a fictitious particle of equal mass placed there and acted on by a force equal to the total external force on the body, through G; and (ii) the motion (rotation) of the body about G is the same as if G were fixed and the body is acted on by the same forces (and/or couples) as in the actual case. These results, clearly, also hold for *impulsive* motion (chap. 4).]

It is also possible to express T in terms of *holonomic* (instead of quasi-) coordinates: specifically, with $\boldsymbol{r}_G = X_G\boldsymbol{I} + Y_G\boldsymbol{J} + Z_G\boldsymbol{K}$ and ϕ, θ, $\psi =$ Eulerian angles of $G\text{-}xyz$ relative to $O\text{-}XYZ$ (or $G\text{-}XYZ$), the (double) kinetic energy transforms to

$$2T = m v_G{}^2 + \boldsymbol{\omega}\cdot \mathsf{S}\, dm[\boldsymbol{r}_{/G}\times(\boldsymbol{\omega}\times\boldsymbol{r}_{/G})]$$
$$= m\Big[(\dot{X}_G)^2 + (\dot{Y}_G)^2 + (\dot{Z}_G)^2\Big] + 2T_R(\phi,\theta,\psi;\dot{\phi},\dot{\theta},\dot{\psi}), \qquad (3.13.4\text{d})$$

[where, since the $\boldsymbol{r}_{/G}$, in the second term of (3.13.4d), depend only on the Eulerian angles, and (recalling results of §1.12): $\boldsymbol{\omega} = \boldsymbol{u}_\phi(\phi,\theta,\psi)\dot{\phi} + \boldsymbol{u}_\theta(\phi,\theta,\psi)\dot{\theta} + \boldsymbol{u}_\psi(\phi,\theta,\psi)\dot{\psi}$; $\boldsymbol{u}_{\phi,\theta,\psi}(\equiv \boldsymbol{K i' k''})$: *nonorthogonal unit vectors*, that term, $2T_R$, becomes a *quadratic homogeneous* function of the Eulerian rates $\dot{\phi}$, $\dot{\theta}$, $\dot{\psi}$, with coefficients functions of the Eulerian angles ϕ, θ, ψ] and, accordingly, the Lagrangean inertial forces (or system accelerations) corresponding to the so-chosen Lagrangean coordinates $q_1 = X_G, \ldots, q_6 = \psi$ are

Motion of G: $\quad E_1 \equiv E_X = m\ddot{X}_G, \qquad E_2 \equiv E_Y = m\ddot{Y}_G, \qquad E_3 \equiv E_Z = m\ddot{Z}_G,$ (3.13.4e)

Motion around G:

$$E_4 \equiv E_\phi = (\partial T_R/\partial\dot{\phi})^{\cdot} - \partial T_R/\partial\phi,$$
$$E_5 \equiv E_\theta = (\partial T_R/\partial\dot{\theta})^{\cdot} - \partial T_R/\partial\theta,$$
$$E_6 \equiv E_\psi = (\partial T_R/\partial\dot{\psi})^{\cdot} - \partial T_R/\partial\psi. \qquad (3.13.4\text{f})$$

However, upon explicit calculation, eqs. (3.13.4f) turn out to be less simple than their quasi-variable counterparts based on eqs. (3.13.2a–4c).

• If B is a *body of revolution* about, say, the $\blacklozenge\text{-}z$ axis, then, since $\blacklozenge\text{-}xyz$ are principal axes and $I_{\blacklozenge,x} = I_{\blacklozenge,y} =$ *perpendicular* (or *transverse*, or *equatorial*) moment of inertia, then

$$2T_R = I_{\blacklozenge,x}\omega_x{}^2 + I_{\blacklozenge,y}\omega_y{}^2 + I_{\blacklozenge,z}\omega_z{}^2 = I_{\blacklozenge,x}(\omega_x{}^2 + \omega_y{}^2) + I_{\blacklozenge,z}\omega_z{}^2$$
$$= I_{\blacklozenge,x}(\omega^2 - \omega_z{}^2) + I_{\blacklozenge,z}\omega_z{}^2 = I_{\blacklozenge,x}\omega^2 + (I_{\blacklozenge,z} - I_{\blacklozenge,x})\,\omega_z{}^2, \qquad (3.13.4\text{g})$$

or, generally, with the helpful notations $I_{\blacklozenge,x} \equiv I_{\blacklozenge,\text{transverse}} \equiv I_{\blacklozenge,T}$, $I_{\blacklozenge,z} \equiv I_{\blacklozenge,\text{axial}} \equiv I_{\blacklozenge,A}$, and $\boldsymbol{k} \equiv \boldsymbol{u}$: *unit* vector along axis of revolution,

$$2T_R = I_{\blacklozenge,T}\,\omega^2 + (I_{\blacklozenge,A} - I_{\blacklozenge,T})(\boldsymbol{\omega}\cdot\boldsymbol{u})^2. \qquad (3.13.4\text{h})$$

The System Momentum Vectors

Since the rigid body is an internally scleronomic system, we have $\delta\boldsymbol{r} = \delta\boldsymbol{r}_\blacklozenge + \delta\boldsymbol{\chi}\times\boldsymbol{r}_{/\blacklozenge}$, where $\boldsymbol{\omega} = d\boldsymbol{\chi}/dt$ ($\boldsymbol{\chi} =$ a *quasi vector*, in general), and, as a result, the total (inertial and first-order) virtual "work" of its linear momenta, δP [recalling (3.6.3b)], specializes to

$$\delta P \equiv \mathsf{S}\, dm\,\boldsymbol{v}\cdot\delta\boldsymbol{r} = \mathsf{S}\, dm\,\boldsymbol{v}\cdot(\delta\boldsymbol{r}_\blacklozenge + \delta\boldsymbol{\chi}\times\boldsymbol{r}_{/\blacklozenge})$$
$$= \cdots = \boldsymbol{p}\cdot\delta\boldsymbol{r}_\blacklozenge + \boldsymbol{H}_\blacklozenge\cdot\delta\boldsymbol{\chi}, \qquad (3.13.5)$$

where

$$p \equiv \int dm\, v = m v_G = (inertial)\ linear\ momentum\ of\ body\ B, \quad (3.13.5a)$$

$$H_\blacklozenge \equiv \int r_{/\blacklozenge} \times (dm\, v) = (inertial)\ absolute\ angular\ momentum\ of\ B\ about\ \blacklozenge. \quad (3.13.5b)$$

These two momenta are the fundamental *system* vectors of Eulerian rigid-body mechanics. They transform, further, as follows:

$$p \equiv \int dm(v_\blacklozenge + \omega \times r_{/\blacklozenge}) = m v_\blacklozenge + \omega \times (m r_{G/\blacklozenge}), \quad (3.13.5c)$$

$$H_\blacklozenge \equiv \int dm\, r_{/\blacklozenge} \times (v_\blacklozenge + \omega \times r_{/\blacklozenge}) = h_\blacklozenge + m(r_{G/\blacklozenge} \times v_\blacklozenge). \quad (3.13.5d)$$

- If $\blacklozenge = G$, then $r_{G/\blacklozenge} = 0$, and (3.13.5c, d) reduce, respectively, to

$$p = m v_G \quad \text{and} \quad H_G = h_G = \int dm\, r_{/\blacklozenge} \times (\omega \times r_{/\blacklozenge}). \quad (3.13.5e)$$

Next, let us relate the above to the (inertial) *absolute angular momentum of B about the origin O*, H_O (recalling §1.6). Since $r = r_\blacklozenge + r_{/\blacklozenge}$, we find, successively,

$$H_O \equiv \int r \times (dm\, v) = \int dm(r_\blacklozenge \times v) + \int dm(r_{/\blacklozenge} \times v)$$

$$= r_\blacklozenge \times \left(\int dm\, v\right) + H_\blacklozenge = r_\blacklozenge \times p + [h_\blacklozenge + m(r_{G/\blacklozenge} \times v_\blacklozenge)] \quad [\text{recalling (3.13.5d)}]$$

$$= \mathbf{I}_\blacklozenge \cdot \omega + m(r_{G/\blacklozenge} \times v_\blacklozenge) + r_\blacklozenge \times p; \quad (3.13.6)$$

and, therefore

- If $\blacklozenge = G$, then

$$H_O = \mathbf{I}_G \cdot \omega + r_G \times p = \mathbf{I}_G \cdot \omega + r_G \times (m v_G); \quad (3.13.6a)$$

- If $\blacklozenge = O$ (i.e., motion about a *fixed* point — rotation), then $r_\blacklozenge = 0$, $v_\blacklozenge = 0$, and the above reduce to

$$p = m v_G = m(\omega \times r_G) \quad \text{and} \quad H_O = h_O = \mathbf{I}_O \cdot \omega. \quad (3.13.6b)$$

More general $H/h/p$ definitions appear in §3.16, in connection with the problem of *relative motion*; that is, when $\blacklozenge$ is *not* a body point but has its own (known or unknown) motion relative to both O-XYZ and the body.

Kinetic Energy via Momentum Vectors

Now we are ready to relate the system quantities $T/p/H_\blacklozenge$. Indeed, since the free (unconstrained) rigid body is a *scleronomic* system, we can replace in (3.13.5) $\delta r/\delta r_\blacklozenge/\delta \chi$ with $dr = v\, dt/dr_\blacklozenge = v_\blacklozenge\, dt/d\chi = \omega\, dt$, respectively, and then divide by dt; thus resulting in

$$2T = p \cdot v_\blacklozenge + H_\blacklozenge \cdot \omega. \quad (3.13.7)$$

Let us examine T more closely. From the earlier representations—that is,

$$2T = m v_\blacklozenge \cdot v_\blacklozenge + \omega \cdot \left\{\int dm[r_{/\blacklozenge} \times (\omega \times r_{/\blacklozenge})]\right\} + \omega \cdot (m r_{G/\blacklozenge} \times v_\blacklozenge),$$

we realize that

$$2T = 2T(\boldsymbol{v}_{\blacklozenge}, \boldsymbol{\omega}) = \textit{function of the velocity variables, or velocity state, of the body,} \tag{3.13.7a}$$

and, therefore, *differentiating T with respect to these variables*, we obtain

$$dT = m\boldsymbol{v}_{\blacklozenge} \cdot d\boldsymbol{v}_{\blacklozenge} + (1/2)\Big\{d\boldsymbol{\omega} \cdot \boldsymbol{h}_{\blacklozenge} + \boldsymbol{\omega} \cdot S\, dm\big[\boldsymbol{r}_{/\blacklozenge} \times (d\boldsymbol{\omega} \times \boldsymbol{r}_{/\blacklozenge})\big]\Big\}$$
$$+ \boldsymbol{\omega} \cdot (m\boldsymbol{r}_{G/\blacklozenge} \times d\boldsymbol{v}_{\blacklozenge}) + d\boldsymbol{\omega} \cdot (m\boldsymbol{r}_{G/\blacklozenge} \times \boldsymbol{v}_{\blacklozenge}); \tag{3.13.7b}$$

or, since $\boldsymbol{\omega} \cdot [\boldsymbol{r}_{/\blacklozenge} \times (d\boldsymbol{\omega} \times \boldsymbol{r}_{/\blacklozenge})] = (r_{/\blacklozenge})^2(\boldsymbol{\omega} \cdot d\boldsymbol{\omega}) - (\boldsymbol{r}_{/\blacklozenge} \cdot d\boldsymbol{\omega})(\boldsymbol{r}_{/\blacklozenge} \cdot \boldsymbol{\omega})$, from which

$$\boldsymbol{\omega} \cdot S\, dm[\boldsymbol{r}_{/\blacklozenge} \times (d\boldsymbol{\omega} \times \boldsymbol{r}_{/\blacklozenge})] = d\boldsymbol{\omega} \cdot S\, dm[(r_{/\blacklozenge})^2\boldsymbol{\omega} - (\boldsymbol{r}_{/\blacklozenge} \cdot \boldsymbol{\omega})\, \boldsymbol{r}_{/\blacklozenge}]$$
$$= d\boldsymbol{\omega} \cdot \boldsymbol{h}_{\blacklozenge}, \tag{3.13.7c}$$

we finally establish that

$$dT = [m\boldsymbol{v}_{\blacklozenge} + m(\boldsymbol{\omega} \times \boldsymbol{r}_{G/\blacklozenge})] \cdot d\boldsymbol{v}_{\blacklozenge} + [\boldsymbol{h}_{\blacklozenge} + m(\boldsymbol{r}_{G/\blacklozenge} \times \boldsymbol{v}_{\blacklozenge})] \cdot d\boldsymbol{\omega}. \tag{3.13.7d}$$

But from (3.13.7a) and the invariant differential definition, we must also have

$$dT = (\partial T/\partial \boldsymbol{v}_{\blacklozenge}) \cdot d\boldsymbol{v}_{\blacklozenge} + (\partial T/\partial \boldsymbol{\omega}) \cdot d\boldsymbol{\omega}. \tag{3.13.7e}$$

The representations (3.13.7d) and (3.13.7e), since the $d\boldsymbol{v}_{\blacklozenge}$ and $d\boldsymbol{\omega}$ are independent, immediately lead to the following basic kinematico-inertial identities:

$$\boldsymbol{p} = \partial T/\partial \boldsymbol{v}_{\blacklozenge} \quad [= m(\boldsymbol{v}_{\blacklozenge} + \boldsymbol{\omega} \times \boldsymbol{r}_{G/\blacklozenge}) = m\boldsymbol{v}_G], \tag{3.13.7f}$$

$$\boldsymbol{H}_{\blacklozenge} = \partial T/\partial \boldsymbol{\omega} \quad [= \boldsymbol{h}_{\blacklozenge} + (m\boldsymbol{r}_{G/\blacklozenge} \times \boldsymbol{v}_{\blacklozenge})]. \tag{3.13.7g}$$

[We notice that $\partial T(\boldsymbol{v}_{\blacklozenge}, \boldsymbol{\omega})/\partial \boldsymbol{v}_{\blacklozenge} = \partial T(\boldsymbol{v}_G, \boldsymbol{\omega})/\partial \boldsymbol{v}_G = \boldsymbol{p}$.] The above translate readily to the following *six* scalar/component equations:

- Along the *body-fixed* axes $\blacklozenge - xyz$:

$$p_x = m(v_{\blacklozenge,x} + \omega_y z_{G/\blacklozenge} - \omega_z y_{G/\blacklozenge}) = \partial T/\partial v_{\blacklozenge,x}, \quad \text{etc., cyclically,} \tag{3.13.7h}$$

$$H_{\blacklozenge,x} = h_{\blacklozenge,x} + m(y_{G/\blacklozenge} v_{\blacklozenge,z} - z_{G/\blacklozenge} v_{\blacklozenge,y}) = \partial T/\partial \omega_x, \quad \text{etc., cyclically,} \tag{3.13.7i}$$

where

$$h_{\blacklozenge,x} = I_{\blacklozenge,xx}\omega_x + I_{\blacklozenge,xy}\omega_y + I_{\blacklozenge,xz}\omega_z = \partial T_R/\partial \omega_x, \quad \text{etc., cyclically,} \tag{3.13.7j}$$

- Along the *space-fixed* axes $\blacklozenge$–XYZ:

$$p_X = m\,[v_{\blacklozenge,X} + \omega_Y(Z_G - Z_{\blacklozenge}) - \omega_Z(Y_G - Y_{\blacklozenge})]$$
$$\equiv m\,(v_{\blacklozenge,X} + \omega_Y Z_{G/\blacklozenge} - \omega_Z Y_{G/\blacklozenge}) = \partial T/\partial v_{\blacklozenge,X}, \quad \text{etc., cyclically,} \tag{3.13.7k}$$

$$H_{\blacklozenge,X} = h_{\blacklozenge,X} + m\,(Y_{G/\blacklozenge} v_{\blacklozenge,Z} - Z_{G/\blacklozenge} v_{\blacklozenge,Y}) = \partial T/\partial \omega_X, \quad \text{etc., cyclically,} \tag{3.13.7l}$$

where the $h_{\blacklozenge;X,Y,Z}$ can be found from the vector transformations

$$h_{\blacklozenge,X} = \cos(X,x)\, h_{\blacklozenge,x} + \cos(X,y)\, h_{\blacklozenge,y} + \cos(X,z)\, h_{\blacklozenge,z}, \quad \text{etc., cyclically.} \tag{3.13.7m}$$

Acceleration Vectors

Let us now calculate the total (inertial and first-order) virtual "work" of the inertial forces of the particles of the body [recall (3.2.9, 3.3.2 ff.)]. We find, successively,

$$\delta I \equiv \mathcal{S}\, dm\, \boldsymbol{a} \cdot \delta \boldsymbol{r} = \mathcal{S}\, dm\, \boldsymbol{a} \cdot (\delta \boldsymbol{r}_\blacklozenge + \delta \boldsymbol{\chi} \times \boldsymbol{r}_{/\blacklozenge})$$
$$= \cdots = \boldsymbol{I} \cdot \delta \boldsymbol{r}_\blacklozenge + \boldsymbol{A}_\blacklozenge \cdot \delta \boldsymbol{\chi}, \tag{3.13.8}$$

where

$$\boldsymbol{I} \equiv \mathcal{S}\, dm\, \boldsymbol{a} = m\, \boldsymbol{a}_G = (inertial)\ linear\ inertia\ of\ body\ B, \tag{3.13.8a}$$

$$\boldsymbol{A}_\blacklozenge \equiv \mathcal{S}\, \boldsymbol{r}_{/\blacklozenge} \times (dm\, \boldsymbol{a}) = (inertial)\ relative\ angular\ inertia\ of\ B\ about\ \blacklozenge. \tag{3.13.8b}$$

Our next task is to relate these two Eulerian system vectors to their *momentum* counterparts, $\boldsymbol{p}$ and $\boldsymbol{H}_\blacklozenge$:

(i) Clearly, $$\boldsymbol{I} = d\boldsymbol{p}/dt; \tag{3.13.8c}$$

(ii) By $(\ldots)^\cdot$-differentiating $\boldsymbol{H}_\blacklozenge$ we obtain, successively,

$$d\boldsymbol{H}_\blacklozenge/dt = d/dt\Big[\mathcal{S}\, \boldsymbol{r}_{/\blacklozenge} \times (dm\, \boldsymbol{v})\Big] = \mathcal{S}\, dm(\boldsymbol{v}_{/\blacklozenge} \times \boldsymbol{v}) + \mathcal{S}\, dm(\boldsymbol{r}_{/\blacklozenge} \times \boldsymbol{a})$$
$$= \mathcal{S}\, dm(\boldsymbol{v} - \boldsymbol{v}_\blacklozenge) \times \boldsymbol{v} + \mathcal{S}\, dm(\boldsymbol{r}_{/\blacklozenge} \times \boldsymbol{a})$$
$$= -\mathcal{S}\, dm(\boldsymbol{v}_\blacklozenge \times \boldsymbol{v}) + \mathcal{S}\, dm(\boldsymbol{r}_{/\blacklozenge} \times \boldsymbol{a}), \tag{3.13.8d}$$

or, finally,

$$\boldsymbol{A}_\blacklozenge = d\boldsymbol{H}_\blacklozenge/dt + \boldsymbol{v}_\blacklozenge \times \boldsymbol{p}. \tag{3.13.8e}$$

For $\blacklozenge = G$ the above specialize, respectively, to

$$\boldsymbol{I} = m\, \boldsymbol{a}_G \quad \text{and} \quad \boldsymbol{A}_G = d\boldsymbol{H}_G/dt = d\boldsymbol{h}_G/dt, \tag{3.13.8f1}$$

where

$$\boldsymbol{H}_G \equiv \mathcal{S}\, \boldsymbol{r}_{/\blacklozenge} \times (dm\, \boldsymbol{v}) = \boldsymbol{h}_G \equiv \mathcal{S}\, \boldsymbol{r}_{/G} \times (dm\, \boldsymbol{v}_{/G}). \tag{3.13.8f2}$$

REMARK

It is not hard to show that (3.13.8e) also holds with $\blacklozenge$ replaced by *any other* (not necessarily body-) point $\bullet$ moving with arbitrary inertial velocity $\boldsymbol{v}_\bullet \equiv \boldsymbol{v}_{\bullet/O}$. Indeed, with

$$\boldsymbol{H}_\bullet \equiv \mathcal{S}\, \boldsymbol{r}_{/\bullet} \times (dm\, \boldsymbol{v}),$$

we readily find

$$d\boldsymbol{H}_\bullet/dt = \mathcal{S}\, \boldsymbol{v}_{/\bullet} \times (dm\, \boldsymbol{v}) + \mathcal{S}\, \boldsymbol{r}_{/\bullet} \times (dm\, \boldsymbol{a})$$
$$= \mathcal{S}\, (\boldsymbol{v} - \boldsymbol{v}_\bullet) \times (dm\, \boldsymbol{v}) + \mathcal{S}\, \boldsymbol{r}_{/\bullet} \times (dm\, \boldsymbol{a})$$
$$= -\mathcal{S}\, \boldsymbol{v}_\bullet \times (dm\, \boldsymbol{v}) + \mathcal{S}\, \boldsymbol{r}_{/\bullet} \times (dm\, \boldsymbol{a}), \tag{3.13.8g}$$

or, finally,

$$\boldsymbol{A}_\bullet = d\boldsymbol{H}_\bullet/dt + \boldsymbol{v}_\bullet \times \boldsymbol{p}, \quad \text{Q.E.D.} \quad \text{(see also §1.6, and appendix 3.A2).} \quad (3.13.8h)$$

Moving Axes

To express the above inertia vectors in terms of the more useful rates relative to (noninertial) axes, either *body-fixed* or *intermediate* (*neither body- nor space-fixed*—chosen so that the inertia tensor components along them remain constant), we simply replace in the right sides of their representations, such as (3.13.8c, e, f), $d(\ldots)/dt$ with $d'(\ldots)/dt + \boldsymbol{\Omega} \times (\ldots)$, where $d'(\ldots)/dt$ = *rate of change of vector ... relative to axes that are rotating with (inertial) angular velocity* $\boldsymbol{\Omega}$ (recalling §1.7 ff.). Thus, expressions (3.13.8c, h) become, respectively,

$$\boldsymbol{I} = d'\boldsymbol{p}/dt + \boldsymbol{\Omega} \times \boldsymbol{p} \quad \text{and} \quad \boldsymbol{A}_\bullet = d'\boldsymbol{H}_\bullet/dt + \boldsymbol{\Omega} \times \boldsymbol{H}_\bullet + \boldsymbol{v}_\bullet \times \boldsymbol{p}. \quad (3.13.9)$$

Special Cases

- If $\bullet = \blacklozenge = G$, then $\boldsymbol{v}_\bullet = \boldsymbol{v}_G$ (or, if $\boldsymbol{v}_\bullet = \boldsymbol{0}$), and so (3.13.9) specialize to

$$\boldsymbol{I} = d'\boldsymbol{p}/dt + \boldsymbol{\Omega} \times \boldsymbol{p} \quad \text{and} \quad \boldsymbol{A}_G = d'\boldsymbol{H}_G/dt + \boldsymbol{\Omega} \times \boldsymbol{H}_G \quad (\text{with } \boldsymbol{H}_G = \boldsymbol{h}_G). \quad (3.13.9a)$$

- If $\boldsymbol{\Omega} = \boldsymbol{\omega}$ and $\bullet = \blacklozenge$ (i.e., for *body-fixed* axes $\blacklozenge$–xyz), then $d'\boldsymbol{r}_{G/\blacklozenge}/dt = \boldsymbol{0}$ and $d'\boldsymbol{\omega}/dt = d\boldsymbol{\omega}/dt \equiv \boldsymbol{\alpha}$ = *inertial angular acceleration of B*, and so (3.13.9), with (3.13.7f, g), reduce to

$$\begin{aligned} \boldsymbol{I} &= d'\boldsymbol{p}/dt + \boldsymbol{\omega} \times \boldsymbol{p} \\ &= m[d'\boldsymbol{v}_\blacklozenge/dt + \boldsymbol{\omega} \times \boldsymbol{v}_\blacklozenge + \boldsymbol{\omega} \times (\boldsymbol{\omega} \times \boldsymbol{r}_{G/\blacklozenge}) + \boldsymbol{\alpha} \times \boldsymbol{r}_{G/\blacklozenge}], \end{aligned} \quad (3.13.9b)$$

$$\begin{aligned} \boldsymbol{A}_\blacklozenge &= d'\boldsymbol{H}_\blacklozenge/dt + \boldsymbol{\omega} \times \boldsymbol{H}_\blacklozenge + \boldsymbol{v}_\blacklozenge \times \boldsymbol{p} \\ &= d'\boldsymbol{h}_\blacklozenge/dt + \boldsymbol{\omega} \times \boldsymbol{h}_\blacklozenge + m\boldsymbol{r}_{G/\blacklozenge} \times [d'\boldsymbol{v}_\blacklozenge/dt + (\boldsymbol{\omega} \times \boldsymbol{v}_\blacklozenge)]. \end{aligned} \quad (3.13.9c)$$

- If $\boldsymbol{\Omega} = \boldsymbol{\omega}$ and $\bullet = \blacklozenge = G$, the last two terms in (3.13.9b) and the last two (of the four) terms of (3.13.9c) vanish, and so these equations reduce, respectively, to

$$\boldsymbol{I} = m(d'\boldsymbol{v}_G/dt + \boldsymbol{\omega} \times \boldsymbol{v}_G), \quad \boldsymbol{A}_G = d'\boldsymbol{h}_G/dt + \boldsymbol{\omega} \times \boldsymbol{h}_G. \quad (3.13.9d)$$

The above show clearly the *decoupling* of the two motions: the *translatory* ($\boldsymbol{v}_G$) from the *rotatory* ($\boldsymbol{\omega}$) in the system inertia vectors: for the rotatory motion, G can be viewed as stationary (Euler, 1749). Indeed, if $\boldsymbol{v}_G = \boldsymbol{0}$, we are left with the *sole* system vector

$$\boldsymbol{A}_G = d'\boldsymbol{h}_G/dt + \boldsymbol{\omega} \times \boldsymbol{h}_G = d'/dt(\partial T_R/\partial \boldsymbol{\omega}) + \boldsymbol{\omega} \times (\partial T_R/\partial \boldsymbol{\omega}). \quad (3.13.9e)$$

Component Representations

A better understanding of the (difficulties involved in the) preceding equations may be achieved if we express them in components. Thus, along *body-fixed* axes $\blacklozenge$–xyz, eqs. (3.13.9b, c) translate to the following system of *six* coupled expressions for the six quasi velocities $v_{\blacklozenge;x,y,z}$ and $\omega_{x,y,z}$:

$$I_x = m\Big\{\dot{v}_{\blacklozenge,x} + (\omega_y v_{\blacklozenge,z} - \omega_z v_{\blacklozenge,y}) + \big[\omega_y(\omega_x y_{G/\blacklozenge} - \omega_y x_{G/\blacklozenge}) - \omega_z(\omega_z x_{G/\blacklozenge} - \omega_x z_{G/\blacklozenge})\big] + (\dot{\omega}_y z_{G/\blacklozenge} - \dot{\omega}_z y_{G/\blacklozenge})\Big\}, \quad \text{etc., cyclically,} \tag{3.13.10a}$$

$$A_{\blacklozenge,x} = I_{\blacklozenge,xx}\dot{\omega}_x + I_{\blacklozenge,xy}\dot{\omega}_y + I_{\blacklozenge,xz}\dot{\omega}_z + \big[-(I_{\blacklozenge,yy} - I_{\blacklozenge,zz})\omega_y\omega_z + I_{\blacklozenge,xy}\omega_x\omega_z + I_{\blacklozenge,xz}\omega_x\omega_y + I_{\blacklozenge,yz}(\omega_y{}^2 + \omega_z{}^2)\big] + m\Big\{(y_{G/\blacklozenge}\,\dot{v}_{\blacklozenge,z} - z_{G/\blacklozenge}\,\dot{v}_{\blacklozenge,y}) + \big[y_{G/\blacklozenge}(\omega_x v_{\blacklozenge,y} - \omega_y v_{\blacklozenge,x}) - z_{G/\blacklozenge}(\omega_y v_{\blacklozenge,z} - \omega_z v_{\blacklozenge,y})\big]\Big\}, \quad \text{etc., cyclically;} \tag{3.13.10b}$$

while (3.13.9d), if the corresponding body-axes $G–xyz$ are also *principal*, $G–123$, take the well-known (decoupled!) Eulerian form as follows:

$$I_x = m[\dot{v}_{G,x} + (\omega_y v_{G,z} - \omega_z v_{G,y})] \equiv m[\dot{v}_{G,1} + (\omega_2 v_{G,3} - \omega_3 v_{G,2})] = I_1, \quad \text{etc., cyclically,} \tag{3.13.10c}$$

$$A_{\blacklozenge,x} = I_{\blacklozenge,x}\,\dot{\omega}_x - (I_{\blacklozenge,y} - I_{\blacklozenge,z})\omega_y\omega_z \equiv I_{\blacklozenge,1}\,\dot{\omega}_1 - (I_{\blacklozenge,2} - I_{\blacklozenge,3})\omega_2\omega_3 = A_{\blacklozenge,1}, \quad \text{etc., cyclically.} \tag{3.13.10d}$$

As (3.13.9c) shows, the expressions (3.13.10d) also hold with G replaced by *any body- and space-fixed point*; if one exists.

Lagrangean Forms

Let us now see the connection of the above with analytical mechanics. In terms of the T-gradients (3.13.7f, g), eqs. (3.13.9 ff.) take the following Lagrangean forms:

(i) $$\boldsymbol{I} = d'/dt(\partial T/\partial \boldsymbol{v}_\blacklozenge) + \boldsymbol{\Omega} \times (\partial T/\partial \boldsymbol{v}_\blacklozenge) \tag{3.13.11a}$$

[recalling that $\partial T(\boldsymbol{v}_\blacklozenge, \boldsymbol{\omega})/\partial \boldsymbol{v}_\blacklozenge = \partial T(\boldsymbol{v}_G, \boldsymbol{\omega})/\partial \boldsymbol{v}_G = \boldsymbol{p}$],

$$\boldsymbol{A}_\bullet = d'/dt(\partial T/\partial \boldsymbol{\omega}) + \boldsymbol{\Omega} \times (\partial T/\partial \boldsymbol{\omega}) + \boldsymbol{v}_\bullet \times (\partial T/\partial \boldsymbol{v}_\blacklozenge), \tag{3.13.11b}$$

and, in components (intermediate axes),

$$I_x = (\partial T/\partial v_{\blacklozenge,x})^{\cdot} + \Omega_y(\partial T/\partial v_{\blacklozenge,z}) - \Omega_z(\partial T/\partial v_{\blacklozenge,y}) \quad \text{etc., cyclically,} \tag{3.13.11c}$$

$$A_{\bullet,x} = (\partial T/\partial \omega_x)^{\cdot} + \Omega_y(\partial T/\partial \omega_z) - \Omega_z(\partial T/\partial \omega_y) + v_{\bullet,y}(\partial T/\partial v_{\blacklozenge,z}) - v_{\bullet,z}(\partial T/\partial v_{\blacklozenge,y}), \quad \text{etc., cyclically;} \tag{3.13.11d}$$

(ii) $$\boldsymbol{I} = d'/dt(\partial T/\partial \boldsymbol{v}_\blacklozenge) + \boldsymbol{\omega} \times (\partial T/\partial \boldsymbol{v}_\blacklozenge), \tag{3.13.11e}$$

$$\boldsymbol{A}_\blacklozenge = d'/dt(\partial T/\partial \boldsymbol{\omega}) + \boldsymbol{\omega} \times (\partial T/\partial \boldsymbol{\omega}) + \boldsymbol{v}_\blacklozenge \times (\partial T/\partial \boldsymbol{v}_\blacklozenge), \tag{3.13.11f}$$

and, in components (body-fixed axes),

$$I_x = (\partial T/\partial v_{\blacklozenge,x})^{\cdot} + \omega_y(\partial T/\partial v_{\blacklozenge,z}) - \omega_z(\partial T/\partial v_{\blacklozenge,y}), \quad \text{etc., cyclically,} \tag{3.13.11g}$$

$$A_{\blacklozenge,x} = (\partial T/\partial \omega_x)^{\cdot} + \omega_y(\partial T/\partial \omega_z) - \omega_z(\partial T/\partial \omega_y) + v_{\blacklozenge,y}(\partial T/\partial v_{\blacklozenge,z}) - v_{\blacklozenge,z}(\partial T/\partial v_{\blacklozenge,y}), \quad \text{etc., cyclically;} \tag{3.13.11h}$$

(iii)

$$\boldsymbol{I} = d'/dt(\partial T/\partial \boldsymbol{v}_G) + \boldsymbol{\Omega} \times (\partial T/\partial \boldsymbol{v}_G), \tag{3.13.11i}$$

$$\boldsymbol{A}_G = d'/dt(\partial T/\partial \boldsymbol{\omega}) + \boldsymbol{\Omega} \times (\partial T/\partial \boldsymbol{\omega}), \tag{3.13.11j}$$

and, in components (intermediate axes),

$$I_x = (\partial T/\partial v_{G,x})^{\cdot} + \Omega_y(\partial T/\partial v_{G,z}) - \Omega_z(\partial T/\partial v_{G,y}), \quad \text{etc., cyclically,} \tag{3.13.11k}$$

$$A_{G,x} = (\partial T/\partial \omega_x)^{\cdot} + \Omega_y(\partial T/\partial \omega_z) - \Omega_z(\partial T/\partial \omega_y), \quad \text{etc., cyclically.} \tag{3.13.11l}$$

For additional forms, see, for example, Heun (1906, pp. 269–271; 1913, pp. 397–401), Suslov (1946, pp. 490–521), Winkelmann and Grammel [1927, pp. 446–449, via the (not very popular) *motor calculus* of R. von Mises (1924)], Winkelmann [1929(b), pp. 14–27]; also Hölder (1939), and §3.16, this volume.

Example 3.13.1 *Kinetic Energy of a Rigid Body.* A thin homogeneous disk D, of mass m and radius r, with fixed center O, rolls without slipping on a fixed rough plane P; its plane thus makes a constant angle (of nutation) θ with P. Let us calculate its kinetic energy if the disk/plane contact point C rotates with a constant angular velocity $\boldsymbol{\omega}_o$ on the circular projection of D on P (fig. 3.31).

Since $\boldsymbol{v}_C = \boldsymbol{0}$ and $\boldsymbol{v}_O = \boldsymbol{0}$, the basic velocity equation $\boldsymbol{v}_C = \boldsymbol{v}_O + \boldsymbol{\omega} \times \boldsymbol{r}_{C/O}$ yields $\boldsymbol{\omega}$: *parallel to the diameter COA*; alternatively, since the velocities of two of its points, C and O, vanish, the disk can only turn about the axis CO. As fig. 3.31(b) shows, $\omega = \omega_o \sin\theta$. [Or, we consider a point B along the disk axis, at distance l from O. During the motion: (i) B traces a circle of radius $l \sin\theta$ (on a plane parallel to P) with angular velocity ω_o, and hence velocity $v_B = \omega_o(l \sin\theta)$; and, simultaneously, (ii) as part of the rotating disk, B turns (instantaneously) about CA with angular velocity

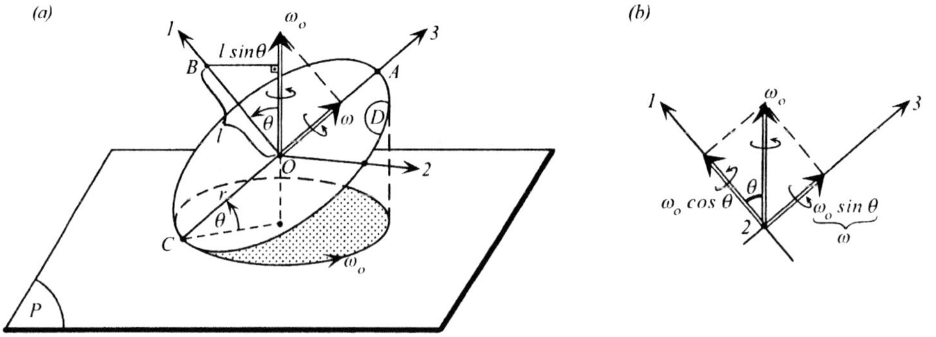

Figure 3.31 (a) Circular disk D rolling at an angle θ (nutation) on a fixed plane P; (b) details of decomposition of ω_o along axes 123.

$\boldsymbol{\omega}$, and therefore has velocity $v_B = \omega l$, Q.E.D.] By König's theorem, and principal central axes O–123 [i.e., $\boldsymbol{\omega} = (0, 0, \omega)$], we readily find

$$\begin{aligned} 2T &= I_{O,1}\omega_1{}^2 + I_{O,2}\omega_2{}^2 + I_{O,3}\omega_3{}^2 \\ &= (mr^2/2)(0) + (mr^2/4)(0) + (mr^2/4)(\omega_o \sin\theta)^2 \\ &= (mr^2 \sin^2\theta/4)\omega_o{}^2 = I_{COA}\,\omega^2. \qquad \text{(a)} \end{aligned}$$

Example 3.13.2 *Kinetic Energy of a Rigid Body.* Let us calculate the kinetic energy of a homogeneous and right circular cone, of radius r, height h, and half angle θ, rolling without slipping on a fixed rough plane P (fig. 3.32), with $\boldsymbol{v}_O = \mathbf{0}$.

Reasoning as in the preceding example—that is, since $\boldsymbol{v}_O = \mathbf{0}$ and $\boldsymbol{v}_B = \mathbf{0}$—we conclude that $\boldsymbol{\omega}$ is parallel to the cone generator OB. If the angular velocity of turning of OB around the perpendicular to the plane is $\boldsymbol{\omega}_o$, then, as fig. 3.32(b) shows,

$$(r\cos\theta)\omega = (h\sin\theta)\omega = (h\cos\theta)\omega_o \;\Rightarrow\; \omega = \omega_o \cot\theta, \qquad \text{(a)}$$

and so along the (intermediate) principal axes O–123,

$$\boldsymbol{\omega} = (-\omega\cos\theta,\ \omega\sin\theta,\ 0) = [-(\omega_o\cot\theta)\cos\theta,\ (\omega_o\cot\theta)\sin\theta, 0]; \qquad \text{(b)}$$

also, from tables,

$$I_{O,1} = (3/10)m\,r^2, \qquad I_{O,2} = I_{O,3} = (3m/5)[h^2 + (r^2/4)].$$

Therefore, König's theorem yields (dropping the subscript O from the I's)

$$\begin{aligned} 2T &= I_1\,\omega_1{}^2 + I_2\,\omega_2{}^2 + I_3\,\omega_3{}^2 \\ &= [(3/10)mr^2](-\omega_o\cos^2\theta/\sin\theta)^2 \\ &\quad + \{(3m/5)[h^2 + (r^2/4)]\}(\omega_o\cos\theta)^2 + \{(3m/5)[h^2 + (r^2/4)]\}(0) \\ &= \{[(3m/20)(r^2 + 6h^2)]\cos^2\theta\}\omega_o{}^2 \\ &= \{[(3m/20)(r^2 + 6h^2)]\sin^2\theta\}\omega^2 \\ &= I_1(-\omega\cos\theta)^2 + I_2(\omega\sin\theta)^2 = (I_1\cos^2\theta + I_2\sin^2\theta)\,\omega^2 = I_{OB}\,\omega^2. \qquad \text{(c)} \end{aligned}$$

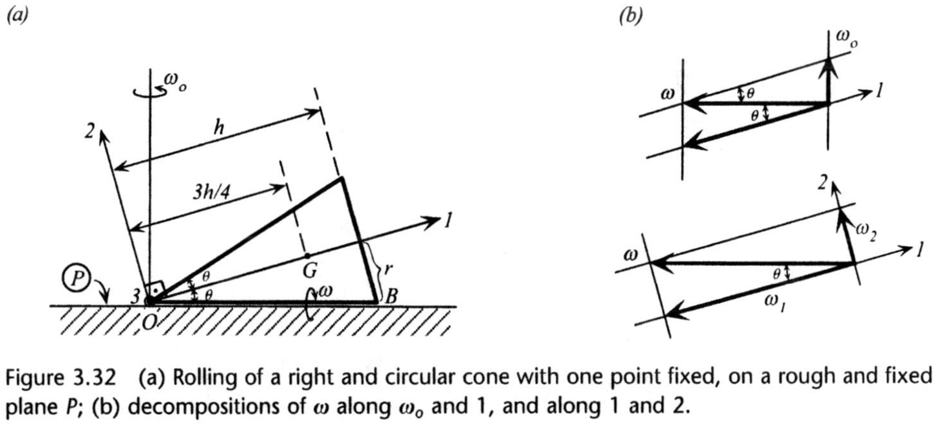

Figure 3.32 (a) Rolling of a right and circular cone with one point fixed, on a rough and fixed plane P; (b) decompositions of $\boldsymbol{\omega}$ along $\boldsymbol{\omega}_o$ and 1, and along 1 and 2.

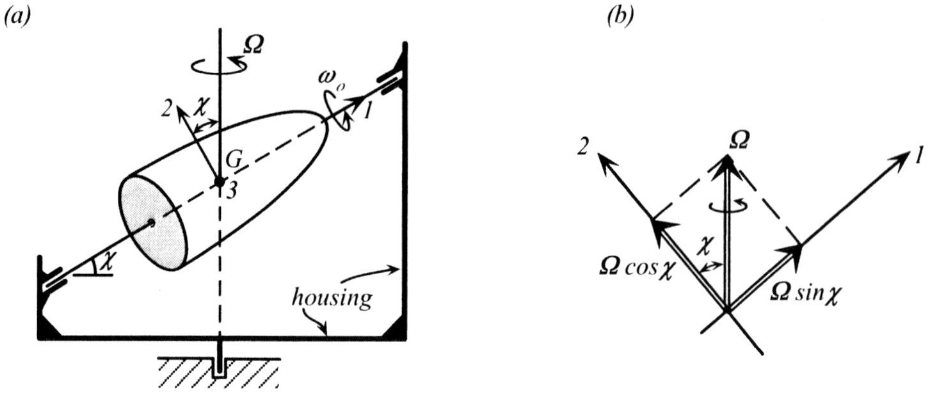

Figure 3.33 (a) Gyrostat spinning inside a housing; (b) details of decomposition of $\boldsymbol{\Omega}$ along 123.

Example 3.13.3 *Kinetic Energy of a Rigid Body.* We consider here a homogeneous rigid body of revolution with central principal moments of inertia $I_1 \equiv A$, $I_2 = I_3 \equiv B$ (fig. 3.33; G = *center of mass*), spinning about its axis of symmetry $G1$ with angular velocity $\boldsymbol{\omega}_o$. This axis is fixed at a constant angle χ in a housing, as shown, and this latter turns about a fixed vertical axis with angular velocity $\boldsymbol{\Omega}$.

Let us calculate the kinetic energy of this body (gyrostat). Here, clearly,

$$\boldsymbol{\omega} = (\omega_1, \omega_2, \omega_3) = (\omega_o + \Omega \sin\chi,\ \Omega\cos\chi,\ 0), \tag{a}$$

and, therefore, by König's theorem ($\boldsymbol{v}_C = \boldsymbol{0}$)

$$2T = I_1\,\omega_1{}^2 + I_2\,\omega_2{}^2 + I_3\,\omega_3{}^2 = A(\omega_o + \Omega\sin\chi)^2 + B(\Omega\cos\chi)^2 + B(0)^2. \tag{b}$$

Example 3.13.4 *Rigid Body: System Force and Kinematico-Inertial Identities.* We consider here a rigid body B in general spatial motion. Let us calculate the components of its Lagrangean (holonomic) forces $\boldsymbol{Q} = \{Q_k;\ k = 1, \ldots, 6\}$ in terms of the corresponding elementary vectorial quantities.

With reference to fig. 3.34, and recalling the results of § 3.4, we find, successively,

$$Q_k \equiv \mathcal{S}\, d\boldsymbol{F} \cdot \boldsymbol{e}_k = \mathcal{S}\, d\boldsymbol{F} \cdot (\partial \boldsymbol{r}/\partial q_k) = \mathcal{S}\, d\boldsymbol{F} \cdot (\partial \boldsymbol{v}/\partial \dot{q}_k)$$
$$= \mathcal{S}\, d\boldsymbol{F} \cdot (\partial/\partial \dot{q}_k)(\boldsymbol{v}_G + \boldsymbol{\omega} \times \boldsymbol{r}_{/G})$$

[but $(\partial/\partial\dot{q}_k)(\boldsymbol{\omega} \times \boldsymbol{r}_{/G}) = (\partial\boldsymbol{\omega}/\partial\dot{q}_k) \times \boldsymbol{r}_{/G} + \boldsymbol{\omega} \times (\partial \boldsymbol{r}_{/G}/\partial\dot{q}_k) = (\partial\boldsymbol{\omega}/\partial\dot{q}_k) \times \boldsymbol{r}_{/G}$ (explain)]

$$= \mathcal{S}\, d\boldsymbol{F} \cdot (\partial \boldsymbol{v}_G/\partial \dot{q}_k) + \mathcal{S}\, d\boldsymbol{F} \cdot [(\partial\boldsymbol{\omega}/\partial\dot{q}_k) \times \boldsymbol{r}_{/G}]$$
$$= (\partial \boldsymbol{v}_G/\partial\dot{q}_k) \cdot \mathcal{S}\, d\boldsymbol{F} + (\partial\boldsymbol{\omega}/\partial\dot{q}_k) \cdot \mathcal{S}\,(\boldsymbol{r}_{/G} \times d\boldsymbol{F})$$
$$\equiv \boldsymbol{F} \cdot (\partial\boldsymbol{v}_G/\partial\dot{q}_k) + \boldsymbol{M}_G \cdot (\partial\boldsymbol{\omega}/\partial\dot{q}_k), \tag{a}$$

where $\boldsymbol{F}(\boldsymbol{M}_G)$ = *resultant force (moment) of all* $d\boldsymbol{F}$, acting at G (about G). Actually, this identity holds for any other chosen *body-fixed* point (pole) ◆.

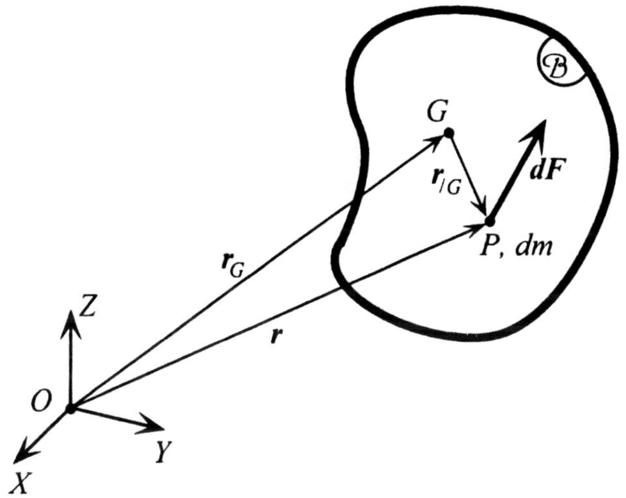

Figure 3.34 Impressed force dF applied to a typical particle P of a rigid body B.

Further, if the body is *unconstrained*, then, by Euler's principles, $\boldsymbol{F} = m\,\boldsymbol{a}_G \equiv d\boldsymbol{p}/dt$ and $\boldsymbol{M}_G = d\boldsymbol{H}_G/dt$, where $\boldsymbol{H}_G = S\,\boldsymbol{r}_{/G} \times (dm\,\boldsymbol{v})$ [(16.5a ff.)], and so we can rewrite (a) in the kinetic form

$$Q_k = (m\,\boldsymbol{a}_G) \cdot (\partial \boldsymbol{v}_G/\partial \dot{q}_k) + (d\boldsymbol{H}_G/dt) \cdot (\partial \boldsymbol{\omega}/\partial \dot{q}_k). \tag{b}$$

[We can also replace in the above $\partial \boldsymbol{v}_G/\partial \dot{q}_k$ with $\partial \boldsymbol{a}_G/\partial \ddot{q}_k$. Then,

$$2[m\,\boldsymbol{a}_G \cdot (\partial \boldsymbol{v}_G/\partial \dot{q}_k)] = \partial(m\,\boldsymbol{a}_G \cdot \boldsymbol{a}_G)/\partial \ddot{q}_k, \text{ etc. (à la Appell)}] \tag{c}$$

Hence, we finally obtain (rather effortlessly!) the following important kinematico-inertial identities:

$$E_k \equiv (\partial T/\partial \dot{q}_k)^{\cdot} - \partial T/\partial q_k = (d\boldsymbol{p}/dt) \cdot (\partial \boldsymbol{v}_G/\partial \dot{q}_k) + (d\boldsymbol{H}_G/dt) \cdot (\partial \boldsymbol{\omega}/\partial \dot{q}_k); \tag{d}$$

which hold even under additional constraints, as long as the q's are *holonomic coordinates.*

Example 3.13.5 *Rigid Body: System Force and Kinematico-Inertial Identities.* Let us prove eq. (d) of the preceding example directly, from general Lagrangean identities. We have, successively,

$$\begin{aligned} E_k &\equiv S\, dm\; \boldsymbol{a} \cdot (\partial \boldsymbol{v}/\partial \dot{q}_k) \\ &= S\, dm\; \boldsymbol{a} \cdot [(\partial \boldsymbol{v}_G/\partial \dot{q}_k) + (\partial \boldsymbol{\omega}/\partial \dot{q}_k) \times \boldsymbol{r}_{/G}] \\ &= \cdots = m\,\boldsymbol{a}_G \cdot (\partial \boldsymbol{v}_G/\partial \dot{q}_k) + \Big[S\,\boldsymbol{r}_{/G} \times (dm\,\boldsymbol{a})\Big] \cdot (\partial \boldsymbol{\omega}/\partial \dot{q}_k) \end{aligned}$$

$$\Big[\text{but } dH_G/dt = \Big[\mathcal{S}\, r_{/G} \times (dm\, v)\Big]^{\cdot} = \mathcal{S}\,[v_{/G} \times (dm\, v) + r_{/G} \times (dm\, a)]$$
$$= \mathcal{S}\,(v - v_G) \times (dm\, v) + \mathcal{S}\, r_{/G} \times (dm\, a)$$
$$= 0 + \mathcal{S}\, r_{/G} \times (dm\, a) \quad \text{(explain)}\Big],$$

and so, finally,

$$E_k = (dp/dt) \cdot (\partial v_G/\partial \dot{q}_k) + (dH_G/dt) \cdot (\partial \omega/\partial \dot{q}_k), \quad \text{Q.E.D.} \tag{a}$$

3.14 THE RIGID BODY: APPELLIAN KINEMATICO-INERTIAL IDENTITIES

Here, we develop *explicit* expressions for the Appellian $S \equiv \mathcal{S}\,(1/2) dm\, a \cdot a$ of a *single* rigid body. (The Appellian of a system of rigid bodies is the sum of the Appellians of its parts; just like the mass and kinetic energy.)

Fixed-Point Rotation

We begin with a rigid body B moving (rotating) about a *fixed point* ◆. Since, then, the inertial acceleration of a typical body particle is

$$a = dv/dt = dv_{/\blacklozenge}/dt = d/dt(\omega \times r_{/\blacklozenge}) = \alpha \times r_{/\blacklozenge} + \omega \times (\omega \times r_{/\blacklozenge}), \tag{3.14.1a}$$

the Appellian of B, *to within acceleration terms* ≡ *Appell-important terms,* becomes

$$S = \mathcal{S}\,(1/2)\, dm[(\alpha \times r_{/\blacklozenge}) \cdot (\alpha \times r_{/\blacklozenge})] + \mathcal{S}\, dm\{(\alpha \times r_{/\blacklozenge}) \cdot [\omega \times (\omega \times r_{/\blacklozenge})]\}. \tag{3.14.1b}$$

Now: (i) The *first* integral in (3.14.1b) equals T_R, eq. (3.13.2b), *but with* ω *replaced with* α. Therefore, reasoning as there, we find

$$\mathcal{S}\,(1/2)\, dm\, [(\alpha \times r_{/\blacklozenge}) \cdot (\alpha \times r_{/\blacklozenge})] = (1/2)\, \alpha \cdot \mathbf{I}_{\blacklozenge} \cdot \alpha. \tag{3.14.1c}$$

(ii) The *second* integral, in view of the transformations [recalling the identities: $a \cdot (b \times c) = b \cdot (c \times a) = c \cdot (a \times b)$ and $a \times (b \times c) = (a \cdot c)b - (a \cdot b)c$, holding for any three vectors a, b, c]:

$$(\alpha \times r_{/\blacklozenge}) \cdot [\omega \times (\omega \times r_{/\blacklozenge})] = (\alpha \times r_{/\blacklozenge}) \cdot [(\omega \cdot r_{/\blacklozenge})\, \omega - \omega^2 r_{/\blacklozenge}]$$
$$= (\omega \times \alpha) \cdot [(\omega \cdot r_{/\blacklozenge})\, r_{/\blacklozenge}]$$
$$[\text{since } (\alpha \times r_{/\blacklozenge}) \cdot r_{/\blacklozenge} = 0 \text{ and } (\alpha \times r_{/\blacklozenge}) \cdot \omega = (\omega \times \alpha) \cdot r_{/\blacklozenge}]$$
$$= (\alpha \times \omega) \cdot [r_{/\blacklozenge} \times (\omega \times r_{/\blacklozenge})]$$
$$[\text{since } (\omega \times \alpha) \cdot \omega = 0], \tag{3.14.1d}$$

reduces to [recalling the definition of $h_{\blacklozenge}$, (3.13.2c)]

$$(\alpha \times \omega) \cdot \mathcal{S}\, dm\, [r_{/\blacklozenge} \times (\omega \times r_{/\blacklozenge})] = (\alpha \times \omega) \cdot h_{\blacklozenge}. \tag{3.14.1e}$$

The above results allow us to rewrite (3.14.1b) in the following equivalent forms:

$$\begin{aligned} S &= (1/2)\,\boldsymbol{\alpha}\cdot\mathbf{I}_\blacklozenge\cdot\boldsymbol{\alpha} + \boldsymbol{\alpha}\cdot(\boldsymbol{\omega}\times\boldsymbol{h}_\blacklozenge) \\ &= (1/2)\,\boldsymbol{\alpha}\cdot\mathbf{I}_\blacklozenge\cdot\boldsymbol{\alpha} + \boldsymbol{\alpha}\cdot[\boldsymbol{\omega}\times(\mathbf{I}_\blacklozenge\cdot\boldsymbol{\omega})] \\ &= (1/2)\,\boldsymbol{\alpha}\cdot\mathbf{I}_\blacklozenge\cdot\boldsymbol{\alpha} + (\boldsymbol{\alpha}\times\boldsymbol{\omega})\cdot(\mathbf{I}_\blacklozenge\cdot\boldsymbol{\omega}); \end{aligned} \qquad (3.14.1\text{f})$$

the second term/sum being a *bilinear* form in the components of $\boldsymbol{\alpha}\times\boldsymbol{\omega}$ and $\boldsymbol{\omega}$, with coefficients the components of the inertia tensor $\mathbf{I}_\blacklozenge$.

Component Forms

It is not hard to see that in terms of the components of $\boldsymbol{\omega}$, $\boldsymbol{\alpha}$, $\mathbf{I}_\blacklozenge$ along *body-fixed* axes, $\blacklozenge$-xyz ($\Rightarrow \alpha_x = \dot{\omega}_x$, etc.), the expression (3.14.1f) assumes the explicit form

$$\begin{aligned} S = (1/2)&(I_{\blacklozenge,xx}\,\alpha_x{}^2 + I_{\blacklozenge,yy}\,\alpha_y{}^2 + I_{\blacklozenge,zz}\,\alpha_z{}^2 \\ &+ 2\,I_{\blacklozenge,xy}\,\alpha_x\alpha_y + 2\,I_{\blacklozenge,xz}\,\alpha_x\alpha_z + 2\,I_{\blacklozenge,yz}\,\alpha_y\alpha_z) \\ &+ [(\alpha_y\,\omega_z - \alpha_z\,\omega_y)(I_{\blacklozenge,xx}\,\omega_x + I_{\blacklozenge,xy}\,\omega_y + I_{\blacklozenge,xz}\,\omega_z) \\ &+ (\alpha_z\,\omega_x - \alpha_x\,\omega_z)(I_{\blacklozenge,yx}\,\omega_x + I_{\blacklozenge,yy}\,\omega_y + I_{\blacklozenge,yz}\,\omega_z) \\ &+ (\alpha_x\,\omega_y - \alpha_y\,\omega_x)(I_{\blacklozenge,zx}\,\omega_x + I_{\blacklozenge,zy}\,\omega_y + I_{\blacklozenge,zz}\,\omega_z)]; \end{aligned} \qquad (3.14.2\text{a})$$

or, if $\blacklozenge$–xyz are also *principal axes* [i.e., $\mathbf{I}_\blacklozenge$ = *diagonal* $(I_{\blacklozenge,x}, I_{\blacklozenge,y}, I_{\blacklozenge,z})$],

$$\begin{aligned} S &= (1/2)(I_{\blacklozenge,x}\,\alpha_x{}^2 + I_{\blacklozenge,y}\,\alpha_y{}^2 + I_{\blacklozenge,z}\,\alpha_z{}^2) \\ &\quad + [(\alpha_y\,\omega_z - \alpha_z\,\omega_y)(I_{\blacklozenge,x}\,\omega_x) + (\alpha_z\,\omega_x - \alpha_x\,\omega_z)(I_{\blacklozenge,y}\,\omega_y) \\ &\quad + (\alpha_x\,\omega_y - \alpha_y\,\omega_x)(I_{\blacklozenge,z}\,\omega_z)] \\ &= (1/2)(I_{\blacklozenge,x}\,\alpha_x{}^2 + I_{\blacklozenge,y}\,\alpha_y{}^2 + I_{\blacklozenge,z}\,\alpha_z{}^2) \\ &\quad - \alpha_x[(I_{\blacklozenge,y} - I_{\blacklozenge,z})\omega_y\,\omega_z] - \alpha_y[(I_{\blacklozenge,z} - I_{\blacklozenge,x})\omega_z\,\omega_x] \\ &\quad - \alpha_z[(I_{\blacklozenge,x} - I_{\blacklozenge,y})\omega_x\,\omega_y]. \end{aligned} \qquad (3.14.2\text{b})$$

From the latter we immediately obtain the well-known Eulerian angular inertia components (Gibbs, 1879)

$$\begin{aligned} A_x &= \partial S/\partial\alpha_x = I_{\blacklozenge,x}\,\alpha_x - (I_{\blacklozenge,y} - I_{\blacklozenge,z})\omega_y\,\omega_z \\ &[= I_{\blacklozenge,x}\,\dot{\omega}_x - (I_{\blacklozenge,y} - I_{\blacklozenge,z})\omega_y\,\omega_z = \partial S/\partial\dot{\omega}_x], \quad \text{etc., cyclically.} \end{aligned} \qquad (3.14.2\text{c})$$

REMARKS

(i) If the axes $\blacklozenge$-xyz are still principal but *non–body-fixed*, rotating with inertial angular velocity $\boldsymbol{\Omega} = (\Omega_x, \Omega_y, \Omega_z)$, then $\boldsymbol{\alpha} = d'\boldsymbol{\omega}/dt + \boldsymbol{\Omega}\times\boldsymbol{\omega}$, or, in components,

$$\alpha_x = \dot{\omega}_x + (\Omega_y\,\omega_z - \Omega_z\,\omega_y), \quad \text{etc., cyclically,}$$

and so (3.14.2b) is replaced by

$$\begin{aligned} S &= (1/2)[I_{\blacklozenge,x}(\dot{\omega}_x)^2 + I_{\blacklozenge,y}(\dot{\omega}_y)^2 + I_{\blacklozenge,z}(\dot{\omega}_z)^2] \\ &\quad - \dot{\omega}_x[(I_{\blacklozenge,y} - I_{\blacklozenge,z})\omega_y\,\omega_z + I_{\blacklozenge,x}(\omega_y\,\Omega_z - \omega_z\,\Omega_y)] - \dot{\omega}_y[\ldots] - \dot{\omega}_z[\ldots]. \end{aligned} \qquad (3.14.2\text{d})$$

(ii) If the axes $\blacklozenge$–xyz are *nonprincipal* and non–body-fixed, then it can be shown (verify it!) that we must add the following terms to the right side of (3.14.2d):

$$-I_{\blacklozenge,yz}\{-\dot{\omega}_y\dot{\omega}_z - \dot{\omega}_x({\omega_y}^2 - {\omega_z}^2) + \dot{\omega}_y[\omega_y(\omega_x - \Omega_x) + \omega_x\Omega_y] - \dot{\omega}_z[\omega_z(\omega_x - \Omega_x) + \omega_x\Omega_z]\} - I_{\blacklozenge,zx}\{\ldots\} - I_{\blacklozenge,xy}\{\ldots\}. \qquad (3.14.2e)$$

For detailed scalar derivations of the above see, for example, Appell [1900(a), (b)].

General Motion

In this case, the inertial acceleration of a typical body particle is

$$\boldsymbol{a} = d\boldsymbol{v}/dt = \boldsymbol{a}_\blacklozenge + \boldsymbol{a}_{/\blacklozenge} = \boldsymbol{a}_\blacklozenge + \boldsymbol{\alpha}\times\boldsymbol{r}_{/\blacklozenge} + \boldsymbol{\omega}\times(\boldsymbol{\omega}\times\boldsymbol{r}_{/\blacklozenge}). \qquad (3.14.3a)$$

Now, to avoid long calculations, we make the following observations:

(i) The difference between the corresponding *velocity* formula and the first *two* terms in (3.14.3a) is that, there, $\boldsymbol{a}_\blacklozenge$ and $\boldsymbol{\alpha}$ are replaced, respectively, by $\boldsymbol{v}_\blacklozenge$ and $\boldsymbol{\omega}$. Therefore, we will obtain the corresponding terms in S if, in the earlier T-expressions (3.13.2 ff.); that is,

$$2T = m\,{v_\blacklozenge}^2 + 2m(\boldsymbol{v}_\blacklozenge\times\boldsymbol{\omega})\cdot\boldsymbol{r}_{G/\blacklozenge} + \boldsymbol{\omega}\cdot\mathbf{I}_\blacklozenge\cdot\boldsymbol{\omega}, \qquad (3.14.3b)$$

we replace $\boldsymbol{v}_\blacklozenge$ and $\boldsymbol{\omega}$ with $\boldsymbol{a}_\blacklozenge$ and $\boldsymbol{\alpha}$, respectively.

(ii) But the product $\boldsymbol{a}\cdot\boldsymbol{a}$ results in *two* additional *Appell-important* terms in S [the square of $\boldsymbol{\omega}\times(\boldsymbol{\omega}\times\boldsymbol{r}_{/\blacklozenge})$ does *not* produce any $(d\boldsymbol{\omega}/dt)$-proportional terms]:

- One from $(\boldsymbol{\alpha}\times\boldsymbol{r}_{/\blacklozenge})\cdot[\boldsymbol{\omega}\times(\boldsymbol{\omega}\times\boldsymbol{r}_{/\blacklozenge})]$, and hence given by (3.14.1d, e) [also (3.14.1f)]; and
- Another that transforms, successively, as follows:

$$S\, dm\boldsymbol{a}_\blacklozenge\cdot[\boldsymbol{\omega}\times(\boldsymbol{\omega}\times\boldsymbol{r}_{/\blacklozenge})] = m\boldsymbol{a}_\blacklozenge\cdot[\boldsymbol{\omega}\times(\boldsymbol{\omega}\times\boldsymbol{r}_{G/\blacklozenge})] = m(\boldsymbol{a}_\blacklozenge\times\boldsymbol{\omega})\cdot(\boldsymbol{\omega}\times\boldsymbol{r}_{G/\blacklozenge}). \qquad (3.14.3c)$$

Collecting all these results, we conclude that in the case of *general motion*, and *to within* $\sim\boldsymbol{\alpha}$ *terms*,

$$2S = m{a_\blacklozenge}^2 + 2m(\boldsymbol{a}_\blacklozenge\times\boldsymbol{\alpha})\cdot\boldsymbol{r}_{G/\blacklozenge} + 2m(\boldsymbol{a}_\blacklozenge\times\boldsymbol{\omega})\cdot(\boldsymbol{\omega}\times\boldsymbol{r}_{G/\blacklozenge}) + \boldsymbol{\alpha}\cdot\mathbf{I}_\blacklozenge\cdot\boldsymbol{\alpha} + 2(\boldsymbol{\alpha}\times\boldsymbol{\omega})\cdot(\mathbf{I}_\blacklozenge\cdot\boldsymbol{\omega}). \qquad (3.14.3d)$$

Specializations

(i) If $\blacklozenge = G$, the *second* and *third* terms in the above vanish, and so (3.14.3d) reduces to the Appellian counterpart of the well-known König's theorem (for T, with $\blacklozenge = G$)

$$2S = m{a_G}^2 + \boldsymbol{\alpha}\cdot\mathbf{I}_G\cdot\boldsymbol{\alpha} + 2(\boldsymbol{\alpha}\times\boldsymbol{\omega})\cdot(\mathbf{I}_G\cdot\boldsymbol{\omega}) = 2\,(\textit{Appellian of translation of } G + \textit{rotation about } G + \textit{coupling of } \boldsymbol{\omega} \textit{ and } \boldsymbol{\alpha}). \qquad (3.14.4a)$$

However, there is no T-counterpart to the last term of (3.14.4a).

(ii) If $\blacklozenge{-}xyz$ ($G{-}xyz$) are *body-fixed* [in which case, $\boldsymbol{\Omega} = \boldsymbol{\omega} \Rightarrow \alpha_{x,y,z} = \dot{\omega}_{x,y,z}$], the expressions (3.14.3d), (3.14.4a) can be simplified further. Since, in this case,

$$\boldsymbol{a}_\blacklozenge \equiv d\boldsymbol{v}_\blacklozenge/dt = d'\boldsymbol{v}_\blacklozenge/dt + \boldsymbol{\omega} \times \boldsymbol{v}_\blacklozenge, \tag{3.14.4b}$$

to within Appell-important terms, $a_\blacklozenge{}^2$ can be replaced by

$$(d'\boldsymbol{v}_\blacklozenge/dt)^2 + 2(d'\boldsymbol{v}_\blacklozenge/dt)\cdot(\boldsymbol{\omega} \times \boldsymbol{v}_\blacklozenge), \tag{3.14.4c}$$

and $\boldsymbol{a}_\blacklozenge \times \boldsymbol{\omega}$ by $(d'\boldsymbol{v}_\blacklozenge/dt) \times \boldsymbol{\omega}$ [where, we recall, $(d'\boldsymbol{v}_\blacklozenge/dt)_{x,y,z} = (\dot{v}_{\blacklozenge;x,y,z})$]; and so, to within $\sim (d'\boldsymbol{v}_\blacklozenge/dt)$ [$(d'\boldsymbol{v}_G/dt)$] and $\sim \boldsymbol{\alpha}$ terms, and after some simple vectorial rearrangement, (3.14.3d) and (3.14.4a) read, respectively,

$$\begin{aligned}2S &= m(d'\boldsymbol{v}_\blacklozenge/dt)^2 + 2m(d'\boldsymbol{v}_\blacklozenge/dt + \boldsymbol{\omega} \times \boldsymbol{v}_\blacklozenge)\cdot(\boldsymbol{\alpha} \times \boldsymbol{r}_{G/\blacklozenge})\\ &\quad + 2m[(d'\boldsymbol{v}_\blacklozenge/dt) \times \boldsymbol{\omega}]\cdot(\boldsymbol{v}_\blacklozenge + \boldsymbol{\omega} \times \boldsymbol{r}_{G/\blacklozenge})\\ &\quad + \boldsymbol{\alpha}\cdot\mathbf{I}_\blacklozenge\cdot\boldsymbol{\alpha} + 2(\boldsymbol{\alpha} \times \boldsymbol{\omega})\cdot(\mathbf{I}_\blacklozenge\cdot\boldsymbol{\omega});\end{aligned} \tag{3.14.4d}$$

and

$$2S = m(d'\boldsymbol{v}_G/dt)^2 + 2m[(d'\boldsymbol{v}_G/dt) \times \boldsymbol{\omega}]\cdot\boldsymbol{v}_G + \boldsymbol{\alpha}\cdot\mathbf{I}_G\cdot\boldsymbol{\alpha} + 2(\boldsymbol{\alpha} \times \boldsymbol{\omega})\cdot(\mathbf{I}_G\cdot\boldsymbol{\omega}). \tag{3.14.4e}$$

Problem 3.14.1 *Appellian Counterpart of the "British Theorem."* Show that, to within "Appell-important terms," the Appellian of a uniform rod AB, of mass m, equals

$$S = (m/6)(\boldsymbol{a}_A{}^2 + \boldsymbol{a}_A\cdot\boldsymbol{a}_B + \boldsymbol{a}_B{}^2), \tag{a}$$

where $\boldsymbol{a}_A$ and $\boldsymbol{a}_B$ are the *accelerations of the endpoints* A and B (see also Bahar, 1994, pp. 1685–1686).

3.15 THE RIGID BODY: VIRTUAL WORK OF FORCES

Introduction, General Results

In the last two sections, we discussed the explicit forms of the virtual work of the *inertia* forces, δI, for a rigid body B, in both Lagrangean and Appellian variables. Here, we present the corresponding forms of the [total (first-order) and inertial] *virtual work of the impressed forces*,

$$\delta' W \equiv \int d\boldsymbol{F}\cdot\delta\boldsymbol{r}, \tag{3.15.1a}$$

[recall (3.2.8 ff.) and §3.4], and thus complete the specialization of *Lagrange's principle*, $\delta I = \delta' W$, to the rigid body.

Using the notations, and so on, of the preceding sections, we obtain, successively,

$$\delta' W = \int d\boldsymbol{F}\cdot(\delta\boldsymbol{r}_\blacklozenge + \delta\boldsymbol{\chi} \times \boldsymbol{r}_{/\blacklozenge}) = \cdots = \boldsymbol{F}\cdot\delta\boldsymbol{r}_\blacklozenge + \boldsymbol{M}_\blacklozenge\cdot\delta\boldsymbol{\chi}, \tag{3.15.1b}$$

where

$$\boldsymbol{F} = \mathbf{S}\, d\boldsymbol{F} = \textit{total impressed force on } B \text{ (acting through } \blacklozenge\text{)},$$

$$\boldsymbol{M}_\blacklozenge = \mathbf{S}\, \boldsymbol{r}_{/\blacklozenge} \times d\boldsymbol{F} = \textit{total impressed moment on } B \textit{ about } \blacklozenge. \tag{3.15.1c}$$

Component Representations

Let us, next, express (3.15.1b) in terms of the components of its vectors along the following useful axes/coordinates:

(i) If the coordinates/components of $\blacklozenge$ relative to the *fixed axes/basis* $O\text{-}XYZ/\boldsymbol{IJK}$ are $X_\blacklozenge$, $Y_\blacklozenge$, $Z_\blacklozenge$, and the Eulerian angles of a *body-fixed* axes/basis $\blacklozenge\text{-}xyz/\boldsymbol{ijk}$ relative to the *cotranslating (nonrotating)* axes/basis $\blacklozenge\text{-}XYZ/\boldsymbol{IJK}$ are ϕ, θ, ψ [recalling §1.12] — that is, if the Lagrangean coordinates of B are $q_{1,2,3} = X_\blacklozenge$, $Y_\blacklozenge$, $Z_\blacklozenge$, and $q_{4,5,6} = \phi$, θ, ψ — then

$$\delta \boldsymbol{r}_\blacklozenge = \delta X_\blacklozenge \boldsymbol{I} + \delta Y_\blacklozenge \boldsymbol{J} + \delta Z_\blacklozenge \boldsymbol{K}, \qquad \delta\boldsymbol{\chi} = \delta\phi\, \boldsymbol{K} + \delta\theta\, \boldsymbol{u}_n + \delta\psi\, \boldsymbol{k} \tag{3.15.1d}$$

[$\boldsymbol{u}_n$: *unit vector along nodal line*] and, substituting them into (3.15.1b), we obtain

$$\delta' W = Q_X\, \delta X_\blacklozenge + Q_Y\, \delta Y_\blacklozenge + Q_Z\, \delta Z_\blacklozenge + M_{\blacklozenge,\phi}\, \delta\phi + M_{\blacklozenge,\theta}\, \delta\theta + M_{\blacklozenge,\psi}\, \delta\psi, \tag{3.15.1e}$$

where

$$
\begin{aligned}
Q_X &\equiv \boldsymbol{F}\cdot\boldsymbol{I} && (\equiv Q_1), \quad \text{etc., cyclically,}\\
M_{\blacklozenge,\phi} &\equiv \boldsymbol{M}_\blacklozenge\cdot\boldsymbol{K} && (\equiv Q_4 \equiv Q_\phi),\\
M_{\blacklozenge,\theta} &\equiv \boldsymbol{M}_\blacklozenge\cdot\boldsymbol{u}_n && (\equiv Q_5 \equiv Q_\theta),\\
M_{\blacklozenge,\psi} &\equiv \boldsymbol{M}_\blacklozenge\cdot\boldsymbol{k} && (\equiv Q_6 \equiv Q_\psi);
\end{aligned} \tag{3.15.1f}
$$

that is, $M_{\blacklozenge;\phi,\theta,\psi}$ are the *components* of $\boldsymbol{M}_\blacklozenge$ along this "natural" unit but *nonorthogonal* axes/basis $\blacklozenge\text{-}Znz/\boldsymbol{K}\boldsymbol{u}_n\boldsymbol{k}$.

(ii) Similarly, using *body-fixed axes/basis*, $\blacklozenge\text{-}xyz/\boldsymbol{ijk}$ and the Eulerian angles ϕ, θ, ψ, we can write

$$\delta' W = Q_x \delta x_\blacklozenge + Q_y \delta y_\blacklozenge + Q_z \delta z_\blacklozenge + M_{\blacklozenge,x}\delta\theta_x + M_{\blacklozenge,y}\delta\theta_y + M_{\blacklozenge,z}\delta\theta_z, \tag{3.15.1g}$$

but, here, both the ($\delta x_\blacklozenge$, $\delta y_\blacklozenge$, $\delta z_\blacklozenge$) and ($\delta\theta_x$, $\delta\theta_y$, $\delta\theta_z$) are *virtual variations of quasi coordinates* [whose $(\ldots)^{\cdot}$-derivatives are the earlier quasi velocities $v_{\blacklozenge;x,y,z}$ and $\omega_{x,y,z}$, respectively; i.e., $dx_\blacklozenge = v_{\blacklozenge,x}dt$, $d\theta_x = \omega_x dt$, etc.].

(iii) Finally, using *cotranslating axes/basis*, $\blacklozenge\text{-}XYZ/\boldsymbol{IJK}$ and the Eulerian angles ϕ, θ, ψ, we have

$$\delta' W = Q_X \delta X_\blacklozenge + Q_Y \delta Y_\blacklozenge + Q_Z \delta Z_\blacklozenge + M_{\blacklozenge,X}\delta\theta_X + M_{\blacklozenge,Y}\delta\theta_Y + M_{\blacklozenge,Z}\delta\theta_Z, \tag{3.15.1h}$$

where the ($X_\blacklozenge$, $Y_\blacklozenge$, $Z_\blacklozenge$) are genuine (holonomic) coordinates, but the (θ_X, θ_Y, θ_Z) are *quasi coordinates* [i.e., $dX_\blacklozenge = (dX_\blacklozenge/dt)dt \equiv v_{\blacklozenge,X}dt$, $d\theta_X \equiv \omega_X dt$, etc.].

Component Transformations

To relate these various $\boldsymbol{M}_\blacklozenge$ components with each other we shall use (i) *basis vector transformations* and, equivalently, (ii) the $\delta' W$ invariance.

(i) Basis Vector Transformations

(a) *Eulerian versus Inertial Components.* With reference to fig. 3.35, we find, successively,

$$M_{\blacklozenge,\phi}(\equiv M_{\blacklozenge,Z}) \equiv \boldsymbol{M}_{\blacklozenge}\cdot\boldsymbol{K} \quad [= (0)M_{\blacklozenge,X} + (0)M_{\blacklozenge,Y} + (1)M_{\blacklozenge,Z}], \tag{3.15.2a}$$

$$\begin{aligned} M_{\blacklozenge,\theta}(\equiv M_{\blacklozenge,n}) &\equiv \boldsymbol{M}_{\blacklozenge}\cdot\boldsymbol{u}_n \\ &= (M_{\blacklozenge,X}\boldsymbol{I} + M_{\blacklozenge,Y}\boldsymbol{J} + M_{\blacklozenge,Z}\boldsymbol{K})\cdot(\cos\phi\,\boldsymbol{I} + \sin\phi\,\boldsymbol{J}) \\ &= (\cos\phi)M_{\blacklozenge,X} + (\sin\phi)M_{\blacklozenge,Y} + (0)M_{\blacklozenge,Z}, \end{aligned} \tag{3.15.2b}$$

$$\begin{aligned} M_{\blacklozenge,\psi}(\equiv M_{\blacklozenge,z}) &\equiv \boldsymbol{M}_{\blacklozenge}\cdot\boldsymbol{k} = \boldsymbol{M}_{\blacklozenge}\cdot(-\sin\theta\,\boldsymbol{u}_N + \cos\theta\,\boldsymbol{K}) \\ &= \boldsymbol{M}_{\blacklozenge}\cdot[-\sin\theta(-\sin\phi\,\boldsymbol{I} + \cos\phi\,\boldsymbol{J}) + \cos\theta\,\boldsymbol{K}] \\ &= (M_{\blacklozenge,X}\boldsymbol{I} + M_{\blacklozenge,Y}\boldsymbol{J} + M_{\blacklozenge,Z}\boldsymbol{K})\cdot(\sin\phi\sin\theta\,\boldsymbol{I} - \cos\phi\sin\theta\,\boldsymbol{J} + \cos\theta\,\boldsymbol{K}) \\ &= (\sin\phi\sin\theta)M_{\blacklozenge,X} + (-\cos\phi\sin\theta)M_{\blacklozenge,Y} + (\cos\theta)M_{\blacklozenge,Z}. \end{aligned} \tag{3.15.2c}$$

Inverting the above, we obtain, after some simple algebra,

$$M_{\blacklozenge,X} = (-\cot\theta\sin\phi)M_{\blacklozenge,\phi} + (\cos\phi)M_{\blacklozenge,\theta} + (\sin\phi/\sin\theta)M_{\blacklozenge,\psi}, \tag{3.15.2d}$$

$$M_{\blacklozenge,Y} = (\cot\theta\cos\phi)M_{\blacklozenge,\phi} + (\sin\phi)M_{\blacklozenge,\theta} + (-\cos\phi/\sin\theta)M_{\blacklozenge,\psi}, \tag{3.15.2e}$$

$$M_{\blacklozenge,Z} = (1)M_{\blacklozenge,\phi} + (0)M_{\blacklozenge,\theta} + (0)M_{\blacklozenge,\psi}. \tag{3.15.2f}$$

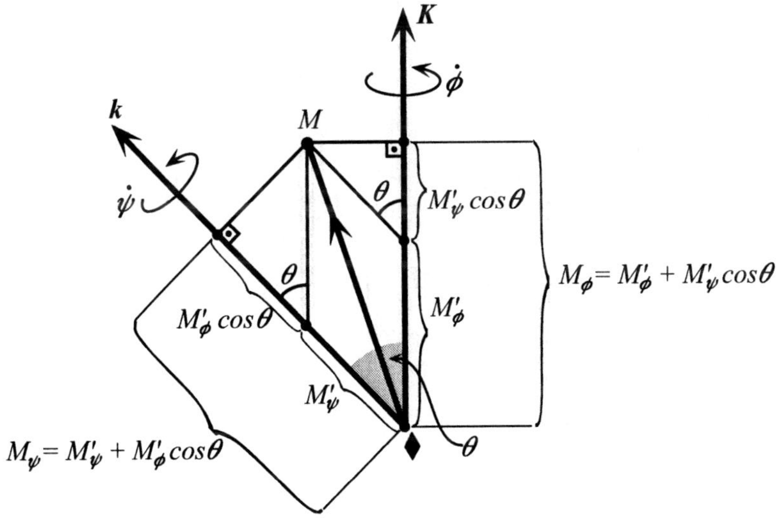

Figure 3.35 Geometrical demonstration of difference between *orthogonal* projections (in nonorthogonal axes) ($M_{\phi,\theta,\psi}$) and *parallel* ($M'_{\phi,\theta,\psi}$) projections (components); $\boldsymbol{M}_{\blacklozenge} \equiv \boldsymbol{M}$.

(b) *Eulerian versus Body-Fixed Components.* Again, with reference to fig. 3.35, we find, successively,

$$\begin{aligned}
M_{\blacklozenge,\phi}(\equiv M_{\blacklozenge,Z}) &\equiv \boldsymbol{M}_{\blacklozenge}\cdot\boldsymbol{K} = \boldsymbol{M}_{\blacklozenge}\cdot(\sin\theta\,\boldsymbol{j}' + \cos\theta\,\boldsymbol{k})\\
&= \boldsymbol{M}_{\blacklozenge}\cdot[\sin\theta(\sin\psi\,\boldsymbol{i} + \cos\psi\,\boldsymbol{j}) + \cos\theta\,\boldsymbol{k}]\\
&= (M_{\blacklozenge,x}\boldsymbol{i} + M_{\blacklozenge,y}\boldsymbol{j} + M_{\blacklozenge,z}\boldsymbol{k})\cdot(\sin\theta\sin\psi\,\boldsymbol{i} + \sin\theta\cos\psi\,\boldsymbol{j} + \cos\theta\,\boldsymbol{k})\\
&= (\sin\theta\sin\psi)M_{\blacklozenge,x} + (\sin\theta\cos\psi)M_{\blacklozenge,y} + (\cos\theta)M_{\blacklozenge,z},
\end{aligned}\tag{3.15.2g}$$

$$\begin{aligned}
M_{\blacklozenge,\theta}(\equiv M_{\blacklozenge,n}) &\equiv \boldsymbol{M}_{\blacklozenge}\cdot\boldsymbol{u}_n = \boldsymbol{M}_{\blacklozenge}\cdot(\cos\psi\,\boldsymbol{i} - \sin\psi\,\boldsymbol{j})\\
&= (\cos\psi)M_{\blacklozenge,x} + (-\sin\psi)M_{\blacklozenge,y} + (0)M_{\blacklozenge,z},
\end{aligned}\tag{3.15.2h}$$

$$M_{\blacklozenge,\psi}(\equiv M_{\blacklozenge,z}) \equiv \boldsymbol{M}_{\blacklozenge}\cdot\boldsymbol{k} = (0)M_{\blacklozenge,x} + (0)M_{\blacklozenge,y} + (1)M_{\blacklozenge,z}. \tag{3.15.2i}$$

Inverting the above, we obtain

$$M_{\blacklozenge,x} = (\sin\psi/\sin\theta)M_{\blacklozenge,\phi} + (\cos\psi)M_{\blacklozenge,\theta} + (-\cot\theta\sin\psi)M_{\blacklozenge,\psi}, \tag{3.15.2j}$$

$$M_{\blacklozenge,y} = (\cos\psi/\sin\theta)M_{\blacklozenge,\phi} + (\sin\psi)M_{\blacklozenge,\theta} + (-\cot\theta\cos\psi)M_{\blacklozenge,\psi}, \tag{3.15.2k}$$

$$M_{\blacklozenge,z} = (0)M_{\blacklozenge,\phi} + (0)M_{\blacklozenge,\theta} + (1)M_{\blacklozenge,\psi}. \tag{3.15.2l}$$

Similarly, we can relate the *Eulerian* axes/basis components with, say, those along the *semimobile* axes/basis $\blacklozenge\text{–}x'y'z'/\boldsymbol{i}'\boldsymbol{j}'\boldsymbol{k}' \equiv \blacklozenge\text{–}nn'z/\boldsymbol{u}_n\,\boldsymbol{j}'\boldsymbol{k}$, or the *semifixed* $\blacklozenge\text{–}x'NZ/\boldsymbol{i}'\boldsymbol{u}_N\boldsymbol{K} \equiv \blacklozenge\text{–}nNZ/\boldsymbol{u}_n\boldsymbol{u}_N\boldsymbol{K}$ ones; and, from (3.15.2a–c) and (3.15.2g–i), we can relate the $M_{\blacklozenge;x,y,z}$ with the $M_{\blacklozenge;X,Y,Z}$. The details are left to the reader.

(ii) $\delta'W$ Invariance

Such derivations are based on the following earlier found kinematic relations (§1.12):

- *Eulerian versus inertial axes*:

$$\begin{aligned}
\delta\theta_X &= (0)\,\delta\phi + (\cos\phi)\,\delta\theta + (\sin\phi\sin\theta)\,\delta\psi,\\
\delta\theta_Y &= (0)\,\delta\phi + (\sin\phi)\,\delta\theta + (-\cos\phi\sin\theta)\,\delta\psi,\\
\delta\theta_Z &= (1)\,\delta\phi + (0)\,\delta\theta + (\cos\theta)\,\delta\psi;
\end{aligned}\tag{3.15.3a}$$

$$\begin{aligned}
\delta\phi &= (-\cot\theta\sin\phi)\,\delta\theta_X + (\cot\theta\cos\phi)\,\delta\theta_Y + (1)\,\delta\theta_Z,\\
\delta\theta &= (\cos\phi)\,\delta\theta_X + (\sin\phi)\,\delta\theta_Y + (0)\,\delta\theta_Z,\\
\delta\psi &= (\sin\phi/\sin\theta)\,\delta\theta_X + (-\cos\phi/\sin\theta)\,\delta\theta_Y + (0)\,\delta\theta_Z.
\end{aligned}\tag{3.15.3b}$$

- *Eulerian versus body-fixed axes*:

$$\begin{aligned}
\delta\theta_x &= (\sin\psi\,\sin\theta)\,\delta\phi + (\cos\psi)\,\delta\theta + (0)\,\delta\psi,\\
\delta\theta_y &= (\cos\psi\,\sin\theta)\,\delta\phi + (-\sin\psi)\,\delta\theta + (0)\,\delta\psi,\\
\delta\theta_z &= (\cos\theta)\,\delta\phi + (0)\,\delta\theta + (1)\,\delta\psi;
\end{aligned}\tag{3.15.3c}$$

$$\begin{aligned}
\delta\phi &= (\sin\psi/\sin\theta)\,\delta\theta_x + (\cos\psi/\sin\theta)\,\delta\theta_y + (0)\,\delta\theta_z,\\
\delta\theta &= (\cos\psi)\,\delta\theta_x + (-\sin\psi)\,\delta\theta_y + (0)\,\delta\theta_z,\\
\delta\psi &= (-\cot\theta\,\sin\phi)\,\delta\theta_x + (-\cot\theta\,\cos\psi)\,\delta\theta_y + (1)\,\delta\theta_z.
\end{aligned}\tag{3.15.3d}$$

From the above, we can also find the relations $\delta\theta_{x,y,z} = (\ldots)\,\delta\theta_{X,Y,Z}$ and its inverse $\delta\theta_{X,Y,Z} = (\ldots)\,\delta\theta_{x,y,z}$.

For obvious reasons, we need consider only the "moment part" of $\delta' W$; that is,

$$\delta' W_M \equiv M_{\blacklozenge,\phi}\,\delta\phi + M_{\blacklozenge,\theta}\,\delta\theta + M_{\blacklozenge,\psi}\,\delta\psi.$$

(a) *Eulerian versus Inertial Components.* With the help of (3.15.3b), we find, successively,

$$\begin{aligned}
\delta' W_M &= M_{\blacklozenge,\phi}[(-\cot\theta\sin\phi)\,\delta\theta_X + (\cot\theta\cos\phi)\,\delta\theta_Y + (1)\,\delta\theta_Z] \\
&\quad + M_{\blacklozenge,\theta}[(\cos\phi)\,\delta\theta_X + (\sin\phi)\,\delta\theta_Y + (0)\,\delta\theta_Z] \\
&\quad + M_{\blacklozenge,\psi}[(\sin\phi/\sin\theta)\,\delta\theta_X + (-\cos\phi/\sin\theta)\,\delta\theta_Y + (0)\,\delta\theta_Z] \\
&= [(-\cot\theta\sin\phi)M_{\blacklozenge,\phi} + (\cos\phi)M_{\blacklozenge,\theta} + (\sin\phi/\sin\theta)M_{\blacklozenge,\psi}]\,\delta\theta_X \\
&\quad + [(\cot\theta\cos\phi)M_{\blacklozenge,\phi} + (\sin\phi)M_{\blacklozenge,\theta} + (-\cos\phi/\sin\theta)M_{\blacklozenge,\psi}]\,\delta\theta_Y \\
&\quad + [(1)M_{\blacklozenge,\phi} + (0)M_{\blacklozenge,\theta} + (0)M_{\blacklozenge,\psi}]\,\delta\theta_Z \\
&= M_{\blacklozenge,X}\,\delta\theta_X + M_{\blacklozenge,Y}\,\delta\theta_Y + M_{\blacklozenge,Z}\,\delta\theta_Z; \qquad \text{that is, eqs. (3.15.2d–f).} \qquad (3.15.4a)
\end{aligned}$$

(b) *Eulerian versus Body-Fixed Components.* With the help of (3.15.3c) we find, successively,

$$\begin{aligned}
\delta' W_M &= M_{\blacklozenge,x}[(0)\,\delta\phi + (\cos\phi)\,\delta\theta + (\sin\phi\sin\theta)\,\delta\psi] \\
&\quad + M_{\blacklozenge,y}[(0)\,\delta\phi + (\sin\phi)\,\delta\theta + (-\cos\phi\sin\theta)\,\delta\psi] \\
&\quad + M_{\blacklozenge,z}[(1)\,\delta\phi + (0)\,\delta\theta + (\cos\phi)\,\delta\psi] \\
&= [(0)M_{\blacklozenge,x} + (0)M_{\blacklozenge,y} + (1)M_{\blacklozenge,z}]\,\delta\phi \\
&\quad + [(\cos\phi)M_{\blacklozenge,x} + (\sin\phi)M_{\blacklozenge,y} + (0)M_{\blacklozenge,z}]\,\delta\theta \\
&\quad + [(\sin\phi\sin\theta)M_{\blacklozenge,x} + (-\cos\phi\sin\theta)M_{\blacklozenge,y} + (\cos\theta)M_{\blacklozenge,z}]\,\delta\psi \\
&= M_{\blacklozenge,\phi}\,\delta\phi + M_{\blacklozenge,\theta}\,\delta\theta + M_{\blacklozenge,\psi}\,\delta\psi; \qquad (3.15.4b)
\end{aligned}$$

that is, eqs. (3.15.2a–c), without inverting eqs. (3.15.2d–f).

Similarly, using the transformations (3.15.3c, d), we can recover the earlier equations (3.15.2g–l).

We hope that the above have demonstrated the simplicity and superiority of the "$\delta' W$ *invariance*" approach. It, clearly, allows us to find the Lagrangean forces in any other "new" system of holonomic/nonholonomic variables—if we know them in an "old" one—plus the differential geometrical equations relating these two systems.

Example 3.15.1 *Eulerian Components versus Projections* [recall (1.2.7a ff.)]. As already known, the Eulerian axes/basis $\blacklozenge$-$Znz/\boldsymbol{K}\boldsymbol{u}_n\boldsymbol{k}$ is *nonorthogonal.* Therefore (and *omitting all subscripts* $\blacklozenge$ for simplicity),

$$\boldsymbol{M} \neq M_\phi\,\boldsymbol{K} + M_\theta\,\boldsymbol{u}_n + M_\psi\,\boldsymbol{k}, \qquad (a)$$

even though, as (3.15.1f) remind us,

$$M_\phi = \boldsymbol{M} \cdot \boldsymbol{K}, \qquad M_\theta = \boldsymbol{M} \cdot \boldsymbol{u}_n, \qquad M_\psi = \boldsymbol{M} \cdot \boldsymbol{k} \tag{b}$$

(components entering virtual work, just like the system momenta $p_{\phi,\theta,\psi}$); that is, in the case of *nonorthogonal axes*, the *(orthogonal) projections of (a vector)* $\boldsymbol{M}$, $M_{\phi,\theta,\psi}$, are *not equal to its components* (i.e., *parallel projections*), *say* $M'_{\phi,\theta,\psi}$ (fig. 3.35).

To find these latter (referred in tensor calculus as *contravariant* components), we set

$$\boldsymbol{M} = M'_\phi \boldsymbol{K} + M'_\theta \boldsymbol{u}_n + M'_\psi \boldsymbol{k}, \tag{c}$$

dot it in succession with $\boldsymbol{K}$, $\boldsymbol{u}_n$, $\boldsymbol{k}$, and then invoke (b) and fig. 3.35.

The results are

$$\begin{aligned} M_\phi &= M'_\phi(\boldsymbol{K} \cdot \boldsymbol{K}) + M'_\theta(\boldsymbol{u}_n \cdot \boldsymbol{K}) + M'_\psi(\boldsymbol{k} \cdot \boldsymbol{K}) \\ &= M'_\phi(1) + M'_\theta(0) + M'_\psi(\cos\theta) = M'_\phi + (\cos\theta)M'_\psi, \end{aligned} \tag{d}$$

$$\begin{aligned} M_\theta &= M'_\phi(\boldsymbol{K} \cdot \boldsymbol{u}_n) + M'_\theta(\boldsymbol{u}_n \cdot \boldsymbol{u}_n) + M'_\psi(\boldsymbol{k} \cdot \boldsymbol{u}_n) \\ &= M'_\phi(0) + M'_\theta(1) + M'_\psi(0) = M'_\theta, \end{aligned} \tag{e}$$

$$\begin{aligned} M_\psi &= M'_\phi(\boldsymbol{K} \cdot \boldsymbol{k}) + M'_\theta(\boldsymbol{u}_n \cdot \boldsymbol{k}) + M'_\psi(\boldsymbol{k} \cdot \boldsymbol{k}) \\ &= M'_\phi(\cos\theta) + M'_\theta(0) + M'_\psi(1) = (\cos\theta)M'_\phi + M'_\psi. \end{aligned} \tag{f}$$

Inverting the above, we easily obtain (see also fig. 3.35)

$$M'_\phi = (1/\sin^2\theta)(M_\phi - \cos\theta M_\psi), \tag{g}$$

$$M'_\theta = M_\theta, \tag{h}$$

$$M'_\psi = (1/\sin^2\theta)(M_\psi - \cos\theta M_\phi). \tag{i}$$

Example 3.15.2 *Equilibrium Conditions; and Accelerationless Rigid-Body Motion.*

(i) *Equilibrium conditions of forces via virtual work.* Let us consider these forces as acting on the various material particles of a rigid body/system, and let us calculate the corresponding (total, first-order, and inertial) virtual work. Reasoning as in (3.15.1b, c), we obtain

$$\begin{aligned} \delta' W_{\text{all forces}} &\equiv \delta' W_f \equiv \mathsf{S}\, d\boldsymbol{f} \cdot \delta\boldsymbol{r} \\ &= \mathsf{S}\, d\boldsymbol{f} \cdot (\delta\boldsymbol{r}_\blacklozenge + \delta\boldsymbol{\chi} \times \boldsymbol{r}_{/\blacklozenge}) = \cdots = \boldsymbol{f} \cdot \delta\boldsymbol{r}_\blacklozenge + \boldsymbol{M}_\blacklozenge \cdot \delta\boldsymbol{\chi}, \end{aligned} \tag{a}$$

where

$$\boldsymbol{f} = \mathsf{S}\, d\boldsymbol{f} = \textit{total force on } B \; (\textit{acting through } \blacklozenge),$$

$$\boldsymbol{M}_\blacklozenge = \mathsf{S}\, \boldsymbol{r}_{/\blacklozenge} \times d\boldsymbol{f} = \textit{total moment on } B \textit{ about } \blacklozenge. \tag{b}$$

Since the virtual displacements $\delta\boldsymbol{r}_\blacklozenge$, $\delta\boldsymbol{\chi}$ are independent/arbitrary, the condition $\delta' W_f = 0$ leads to the well-known force equilibrium equations

$$\boldsymbol{f} = \boldsymbol{0} \quad \text{and} \quad \boldsymbol{M}_\blacklozenge = \boldsymbol{0}. \tag{c}$$

In sum: if $\delta' W_f = 0$, for *every rigid virtual displacement*, the forces are in equilibrium; and, conversely, if the forces are in equilibrium in the sense of (c), then $\delta' W_f = 0$.

Finally, if we invoke the *action–reaction* principle for the internal forces (§ 1.6), eqs. (c) can be replaced by

$$\boldsymbol{f}_{\text{external}} = \mathbf{0} \quad \text{and} \quad \boldsymbol{M}_{\blacklozenge,\text{external}} = \mathbf{0}. \tag{d}$$

[See also Marcolongo, 1911, pp. 266–269; and Heun, 1902(b)].

If $\delta' W \equiv \mathcal{S}\, d\boldsymbol{F} \cdot \delta\boldsymbol{r} = \mathcal{S}\, d\boldsymbol{F} \cdot (\delta\boldsymbol{r}_\blacklozenge + \delta\boldsymbol{\chi} \times \boldsymbol{r}_{/\blacklozenge}) = 0$, then we are led to the following.

(ii) *Accelerationless motion of a rigid body*. The latter is defined as that for which

$$\delta I \equiv \mathcal{S}\, dm\boldsymbol{a} \cdot \delta\boldsymbol{r} = 0, \qquad \text{for } every \quad \delta\boldsymbol{r} = \delta\boldsymbol{r}_\blacklozenge + \delta\boldsymbol{\chi} \times \boldsymbol{r}_{/\blacklozenge}. \tag{e}$$

Then, since [recall (3.13.8 ff.)]

$$\delta I = \mathcal{S}\, dm\boldsymbol{a} \cdot (\delta\boldsymbol{r}_\blacklozenge + \delta\boldsymbol{\chi} \times \boldsymbol{r}_{/\blacklozenge}) = \cdots = \boldsymbol{I} \cdot \delta\boldsymbol{r}_\blacklozenge + \boldsymbol{A}_\blacklozenge \cdot \delta\boldsymbol{\chi}, \tag{f}$$

where

$$\boldsymbol{I} \equiv \mathcal{S}\, dm\, \boldsymbol{a} = m\, \boldsymbol{a}_G = (\textit{inertial})\ \textit{linear inertia of body } B, \tag{g}$$

$$\boldsymbol{A}_\blacklozenge \equiv \mathcal{S}\, \boldsymbol{r}_{/\blacklozenge} \times (dm\, \boldsymbol{a}) = (\textit{inertial})\ \textit{relative angular inertia of } B \textit{ about } \blacklozenge, \tag{h}$$

it follows that

$$\boldsymbol{I} = \mathbf{0} \quad \text{and} \quad \boldsymbol{A}_\blacklozenge = \mathbf{0}; \tag{i}$$

and, therefore, that [choosing in (f) $\delta\boldsymbol{r} \to \boldsymbol{v} = \boldsymbol{v}_\blacklozenge + \boldsymbol{\omega} \times \boldsymbol{r}_{/\blacklozenge}$]

$$\begin{aligned} \delta I \to dT/dt &\equiv \mathcal{S}\, dm\, \boldsymbol{a} \cdot \boldsymbol{v} = \mathcal{S}\, dm\, \boldsymbol{a} \cdot (\boldsymbol{v}_\blacklozenge + \boldsymbol{\omega} \times \boldsymbol{r}_{/\blacklozenge}) \\ &= \cdots = \boldsymbol{I} \cdot \boldsymbol{v}_\blacklozenge + \boldsymbol{A}_\blacklozenge \cdot \boldsymbol{\omega} = 0 \Rightarrow T = \textit{constant}, \end{aligned} \tag{j}$$

that is, *if the body was initially at (inertial) rest, it remains at rest (equilibrium).*

Clearly, the choice of $\blacklozenge$ has no effect on such a motion. In particular, if we select $\blacklozenge = G$, the equations of motion yield

$$m(d\boldsymbol{v}_G/dt) = \mathbf{0} \Rightarrow \boldsymbol{v}_G = \textit{constant} \quad \text{and} \quad d\boldsymbol{h}_G/dt = \mathbf{0} \Rightarrow \boldsymbol{h}_G = \textit{constant}, \tag{k}$$

from which, since $\boldsymbol{v}_G$ and $\boldsymbol{h}_G$ are mutually independent, we conclude that G can be taken as still.

Example 3.15.3 *Analytical Statics: Equilibrium Conditions via Virtual Work.* With the help of the concept of virtual work, and so on, we can summarize statics into the following results/propositions:

THEOREM

Two equivalent force (and/or couple) systems acting on a rigid body produce equal virtual works.

THEOREM

In every *reversible* rigid virtual displacement, the total virtual work of the constraint reactions vanishes (statical principle of Lagrange).

On Irreversible, or Unilateral, Constraints. Consider a particle P and a stationary rigid surface S with equation $f(x, y, z) = 0$. If P moves on S, then its coordinates satisfy the equation $f = 0$. The function f is positive on one side of S and negative on the other. Therefore, if P moves on the positive side of S, and cannot penetrate it or move except on that side, then its constraint is $f \geq 0$. In such cases, we distinguish: (i) *ordinary* positions of P, if $f > 0$, and (ii) *limiting*, or *boundary*, positions, if $f = 0$. Now we can state the general principle of virtual work, as follows.

LEMMA

The total virtual work of the (ideal) constraint reactions of a system in equilibrium (or motion), in every unilateral virtual displacement, is *either positive or zero*:

$$\delta' W_R \equiv \mathsf{S}\, d\boldsymbol{R} \cdot \delta\boldsymbol{r} \geq 0. \tag{a}$$

PRINCIPLE OF VIRTUAL WORK

For equilibrium at a *boundary* configuration of a scleronomic and originally motionless system, it is necessary and sufficient that the total *impressed* virtual work, $\delta' W \equiv \mathsf{S}\, d\boldsymbol{F} \cdot \delta\boldsymbol{r}$, be *zero* for all *reversible* virtual displacements; and *zero* or *negative* for all *nonreversible* virtual displacements (Fourier, 1798).

Problem 3.15.1 *Virtual Work-Like Characterization of Astatic Equilibrium.* Show that the vanishing of the (total, first-order, and inertial) *vector virtual work of all forces*

$$\delta' \boldsymbol{W}_V \equiv \mathsf{S}\, d\boldsymbol{f} \times \delta\boldsymbol{r}, \tag{a}$$

for *every* $\delta\boldsymbol{r} = \delta\boldsymbol{r}_\blacklozenge + \delta\boldsymbol{\chi} \times \boldsymbol{r}_{/\blacklozenge}$, leads to the *astaticity* conditions for these forces (recalling the *tensor product* definition, §1.1)

$$\boldsymbol{f} = \mathsf{S}\, d\boldsymbol{f} = \boldsymbol{0} \quad \text{(vector)}, \qquad \mathsf{S}\, \boldsymbol{r} \otimes d\boldsymbol{f} = \boldsymbol{0} \quad \text{(tensor)}; \tag{b}$$

and, conversely, if (the twelve scalar conditions) (b) hold, then $\delta' \boldsymbol{W}_V = \boldsymbol{0}$.

[This result seems to be due to Heun, 1902(a), p. 69; see also Biezeno, 1927, pp. 253–254.]

Example 3.15.4 *Eulerian Equations of Motion of a Rigid Body B Moving about a Fixed Point* ◆ *via the Central Equation* (recall §3.6):

$$\delta T + \delta' W = d/dt(\delta P). \tag{a}$$

For this special system (with $O = \blacklozenge$; i.e., $\boldsymbol{r}_{/\blacklozenge} = \boldsymbol{r}$, and using *body-fixed* axes at ◆), we have

(i)
$$\delta' W \equiv \mathsf{S}\, d\boldsymbol{F} \cdot \delta\boldsymbol{r} = \mathsf{S}\, d\boldsymbol{F} \cdot (\delta\boldsymbol{\chi} \times \boldsymbol{r}) = \boldsymbol{M} \cdot \delta\boldsymbol{\chi}, \tag{b}$$

$$\boldsymbol{M} \equiv \boldsymbol{M}_\blacklozenge \equiv \mathsf{S}\, \boldsymbol{r} \times d\boldsymbol{F} = \textit{total impressed moment on } B, \textit{ about origin } \blacklozenge; \tag{c}$$

(ii) $$2T = \mathcal{S}\, dm\, \boldsymbol{v}^2 = \mathcal{S}\, dm(\boldsymbol{\omega}\times\boldsymbol{r})^2 = \mathcal{S}\, dm\{[\boldsymbol{r}\times(\boldsymbol{\omega}\times\boldsymbol{r})]\cdot\boldsymbol{\omega}\}$$

$$= \boldsymbol{H}\cdot\boldsymbol{\omega} = \sum H_k\,\omega_k = (\partial T/\partial\boldsymbol{\omega})\cdot\boldsymbol{\omega} = \sum(\partial T/\partial\omega_k)\,\omega_k, \tag{d}$$

where

$$\boldsymbol{H} \equiv \boldsymbol{H}_\blacklozenge \equiv \mathcal{S}\, dm\,[\boldsymbol{r}\times(\boldsymbol{\omega}\times\boldsymbol{r})] = \mathcal{S}\,\boldsymbol{r}\times(dm\,\boldsymbol{v}) = \partial T/\partial\boldsymbol{\omega}$$
$$= (H_x, H_y, H_z) = (\partial T/\partial\omega_x, \partial T/\partial\omega_y, \partial T/\partial\omega_z)$$
$$= (\textit{inertial})\ \textit{absolute angular momentum of } B \textit{ about } \blacklozenge; \tag{e}$$

and $k = 1, 2, 3 \equiv x, y, z$. From the above, and since, here, the *independent kinematical variable* is $\boldsymbol{\omega}$, we obtain

$$\delta T = \mathcal{S}\, dm\boldsymbol{v}\cdot\delta\boldsymbol{v} = \mathcal{S}\, dm(\boldsymbol{\omega}\times\boldsymbol{r})\cdot(\delta\boldsymbol{\omega}\times\boldsymbol{r})$$
$$= \Big\{\mathcal{S}\, dm\,[\boldsymbol{r}\times(\boldsymbol{\omega}\times\boldsymbol{r})]\Big\}\cdot\delta\boldsymbol{\omega}$$
$$= \boldsymbol{H}\cdot\delta\boldsymbol{\omega} = (\partial T/\delta\boldsymbol{\omega})\cdot\delta\boldsymbol{\omega} = \sum H_k\,\delta\omega_k = \sum(\partial T/\partial\omega_k)\,\delta\omega_k$$
$$= \boldsymbol{H}\cdot\delta'\boldsymbol{\omega}\quad [\delta'(\ldots) = \textit{virtual variation of } (\ldots) \textit{ relative to } \blacklozenge\text{–}xyz]. \tag{f}$$

(iii) $$d/dt(\delta P) = d/dt\Big(\mathcal{S}\, dm\boldsymbol{v}\cdot\delta\boldsymbol{r}\Big) = d/dt\Big[\mathcal{S}\, dm\,\boldsymbol{v}\cdot(\delta\boldsymbol{\chi}\times\boldsymbol{r})\Big]$$
$$= d/dt(\boldsymbol{H}\cdot\delta\boldsymbol{\chi}) = d/dt\Big(\sum H_k\,\delta\theta_k\Big) = (d'/dt)(\boldsymbol{H}\cdot\delta\boldsymbol{\chi})$$
$$= (d'\boldsymbol{H}/dt)\cdot\delta\boldsymbol{\chi} + \boldsymbol{H}\cdot[d'(\delta\boldsymbol{\chi})/dt]\qquad [\text{where } d'\boldsymbol{H}/dt = (dH_k/dt)]$$
$$= (d'\boldsymbol{H}/dt)\cdot\delta\boldsymbol{\chi} + \boldsymbol{H}\cdot(\delta'\boldsymbol{\omega} + \delta\boldsymbol{\chi}\times\boldsymbol{\omega})\qquad [\text{recalling ex. 2.3.11: (g)}]$$
$$= (d'\boldsymbol{H}/dt + \boldsymbol{\omega}\times\boldsymbol{H})\cdot\delta\boldsymbol{\chi} + \boldsymbol{H}\cdot\delta'\boldsymbol{\omega}. \tag{g}$$

In view of (b–g), the central equation (a) reduces to

$$(d'\boldsymbol{H}/dt + \boldsymbol{\omega}\times\boldsymbol{H})\cdot\delta\boldsymbol{\chi} = \boldsymbol{M}\cdot\delta\boldsymbol{\chi}; \tag{h}$$

that is, $\delta I = \delta' W$. Now: (a) If $\delta\boldsymbol{X}$ is *unconstrained*, the variational equation (h) leads immediately to the equation of motion

$$d'\boldsymbol{H}/dt + \boldsymbol{\omega}\times\boldsymbol{H} = \boldsymbol{M}; \tag{i}$$

(b) If, on the other hand, $\delta\boldsymbol{\chi}$ is *constrained*, say by the Pfaffian equation $\boldsymbol{B}\cdot\delta\boldsymbol{\chi} = 0$ [where $\boldsymbol{B} = \boldsymbol{B}(t, \boldsymbol{r})$], then the multiplier rule applied to (i) yields the "Routh–Voss-type" equation [with $\lambda = \lambda(t) = \textit{multiplier}$]:

$$d'\boldsymbol{H}/dt + \boldsymbol{\omega}\times\boldsymbol{H} = \boldsymbol{M} + \lambda\boldsymbol{B}. \tag{j}$$

Clearly, the *reaction moment* $\lambda\boldsymbol{B}$ has zero virtual work: $(\lambda\boldsymbol{B})\cdot\delta\boldsymbol{\chi} = 0$.

If the body-fixed axes $\blacklozenge$–xyz are also *principal*, then $H_k = I_k\,\omega_k$ (I_k = *principal moments of inertia at* $\blacklozenge$), and equations (i) readily reduce to the famous Eulerian rotational equations. [The more general forms (i, j) seem to be due to Lagrange. See also Heun (1906, pp. 276–280), and Papastavridis (1992) for alternative derivations and additional insights.]

Problem 3.15.2 Consider a rigid body moving about a fixed point ◆. Relate its Lagrangean and Eulerian momentum and inertia/acceleration vectors.

HINT

With ◆$-xyz$ principal axes at ◆, and Lagrangean coordinates their Eulerian angles relative to fixed axes (ϕ, θ, ψ), we have

$$\delta P \equiv \mathsf{S}\, dm\, \boldsymbol{v} \cdot \delta \boldsymbol{r} = p_\phi\, \delta\phi + p_\theta\, \delta\theta + p_\psi\, \delta\psi \qquad \text{(Lagrangean momenta)},$$

$$\delta I \equiv \mathsf{S}\, dm\, \boldsymbol{a} \cdot \delta \boldsymbol{r} = A_\phi\, \delta\phi + A_\theta\, \delta\theta + A_\psi\, \delta\psi \qquad \text{(Lagrangean inertia/accelerations)},$$

$$2T = 2T^* = I_x \omega_x{}^2 + I_y \omega_y{}^2 + I_z \omega_z{}^2, \tag{a}$$

where

$$A_\phi \equiv I_\phi \equiv (\partial T/\partial \dot{\phi})^{\cdot} - \partial T/\partial \phi = dp_\phi/dt - \partial T/\partial \phi, \quad \text{etc.} \tag{b}$$

But also, using some of the earlier kinematics [eqs. (3.15.3d)],

$$\begin{aligned} \delta P &= p_\phi[(\ldots)\, \delta\theta_x + (\ldots)\, \delta\theta_y + (\ldots)\, \delta\theta_z] + \cdots \\ &\equiv H_x\, \delta\theta_x + \cdots \quad \text{(Eulerian momenta)}, \end{aligned} \tag{c}$$

from which

$$\begin{aligned} H_x &= (\ldots)p_\phi + (\ldots)p_\theta + (\ldots)p_\psi, \qquad H_y = \cdots, \qquad H_z = \cdots; \\ p_\phi &= (\ldots)H_x + (\ldots)H_y + (\ldots)H_z, \qquad p_\theta = \cdots, \; p_\psi = \cdots; \end{aligned} \tag{d}$$

and by chain rule

$$\begin{aligned} \partial T/\partial \phi &= \sum (\partial T^*/\partial \omega_k)(\partial \omega_k/\partial \phi) = \sum (I_k \omega_k)(\partial \omega_k/\partial \phi) = \sum H_k(\partial \omega_k/\partial \phi) \\ &= \cdots, \quad \partial T/\partial \theta = \cdots, \quad \partial T/\partial \psi = \cdots, \text{ etc.} \end{aligned} \tag{e}$$

Hence, using the above, we obtain the sought relations

$$\begin{aligned} A_\phi &= dp_\phi/dt - \partial T/\partial \phi = \cdots = (\ldots)A_x + (\ldots)A_y + (\ldots)A_z, \quad \text{etc.} \\ A_x &= (\ldots)A_\phi + (\ldots)A_\theta + (\ldots)A_\psi = dH_x/dt + \omega_y H_z - \omega_z H_y, \quad \text{etc.} \\ &\qquad \text{(Eulerian inertia/accelerations)}. \end{aligned} \tag{f}$$

3.16 RELATIVE MOTION (OR MOVING AXES/FRAMES) VIA LAGRANGE'S METHOD

In this section, following the rare and masterful treatment of Lur'e (1968, chap. 9, pp. 423–493), we derive the Lagrangean type of equations of motion of a system S relative to a noninertial frame of reference (with associated *moving axes* $O-xyz$), in general known or unknown rigid motion relative to an inertial frame (with associated *fixed axes* $I-XYZ$), or relative to its comoving nonrotating frame (with associated *translating axes* $O-XYZ$)—see fig. 3.36, depicting a convenient two-dimensional such case.

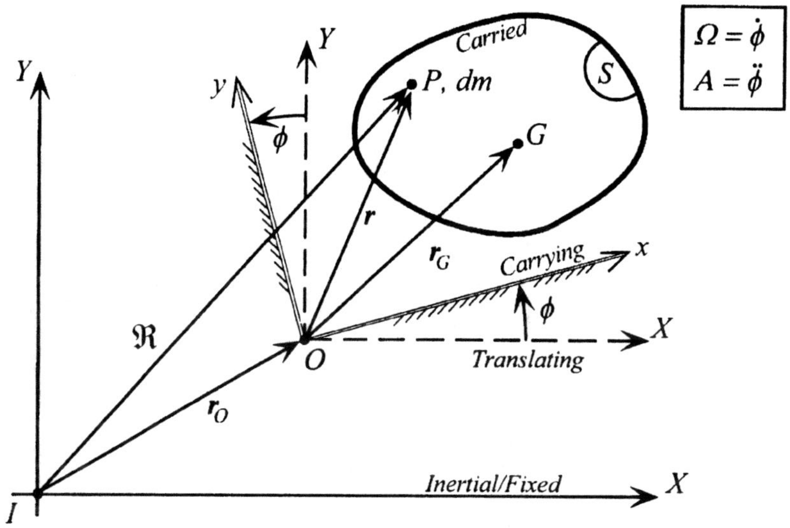

Figure 3.36 Two-dimensional case of system S in general motion relative to the arbitrarily moving axes O–xyz.

Here, we shall use the following terminology and notation:

$O{-}xyz$: *carrying* (or *supporting*, or *transporting*, or *intermediate*, or *housing*) body, or frame; e.g., an airplane. Its origin O is referred to as *moving pole*, or *basis*.
$\boldsymbol{\Omega}$ = (*inertial*) *angular velocity* of $O{-}xyz$; i.e., relative to $I{-}XYZ/O{-}XYZ$;
$d\boldsymbol{\Omega}/dt \equiv \boldsymbol{A}$ = corresponding *angular acceleration*.
$d(\ldots)/dt$ = *inertial rate of change* of $(\ldots)$; i.e., relative to $I{-}XYZ/O{-}XYZ$;
$d'(\ldots)/dt$ = *noninertial rate of change* of $(\ldots)$; i.e., relative to $O{-}xyz$.

As shown in §1.1 and §1.7:

For an arbitrary *vector* $\boldsymbol{b}$,

$$d\boldsymbol{b}/dt = d'\boldsymbol{b}/dt + \boldsymbol{\Omega} \times \boldsymbol{b}; \qquad (3.16.1a)$$

For an arbitrary *second-order tensor* $\mathbf{T}$,

$$d\mathbf{T}/dt = d'\mathbf{T}/dt + \boldsymbol{\Omega} \times \mathbf{T} - \mathbf{T} \times \boldsymbol{\Omega}. \qquad (3.16.1b)$$

Also, as shown there (or, most easily, using Cartesian components), for any two vectors $\boldsymbol{a}$ and $\boldsymbol{b}$,

$$(\boldsymbol{a} \times \mathbf{T}) \cdot \boldsymbol{b} = \boldsymbol{a} \times (\mathbf{T} \cdot \boldsymbol{b}) \qquad \text{and} \qquad (\mathbf{T} \times \boldsymbol{a}) \cdot \boldsymbol{b} = \mathbf{T} \cdot (\boldsymbol{a} \times \boldsymbol{b}). \qquad (3.16.1c)$$

S: *carried* body/system; e.g., a spinning gyroscope inside the (earlier) carrying airplane.

Geometry

We begin with the obvious geometrical relations:

$$\boldsymbol{r}_{/I} = \boldsymbol{r}_{O/I} + \boldsymbol{r}_{/O} \qquad \text{or, simply,} \qquad \mathfrak{R} = \boldsymbol{r}_O + \boldsymbol{r}. \qquad (3.16.2a)$$

Now: (i) Let

$$\boldsymbol{r} = \boldsymbol{r}(q_1, \dots, q_n) \equiv \boldsymbol{r}(q), \tag{3.16.2b}$$

where $n \equiv 3N - h$, N = number of particles of S (*if* we choose to adopt the particle model for it), $h \equiv$ number of holonomic constraints on S. The $q \equiv (q_1, \dots, q_n)$ are *noninertial* Lagrangean coordinates; that is, they specify the configurations of the carried body S relative to the carrying one O–xyz; and, as explained in §2.4, they guarantee the satisfaction of the above holonomic constraints.

(ii) If the motion of O–xyz is *known*, or *prescribed*, and hence unaffected by the motion of S, then $\boldsymbol{r}_O = \boldsymbol{r}_O(t)$; S is *rheonomic*.

(iii) If, on the other hand, the motion of O–xyz is *unknown* [e.g., if the motion of the earlier gyroscope (S) does affect the motion of the airplane (O–xyz)!], then $\boldsymbol{r}_O = \boldsymbol{r}_O$ (*inertial coordinates of pole* O), and hence, not an explicit function of t; S is *scleronomic*.

Kinematics → Kinetic Energy

Since the inertial velocity of a typical S-particle is (recalling §1.7)

$$\begin{aligned} \boldsymbol{v} &\equiv d\boldsymbol{\Re}/dt = d\boldsymbol{r}_O/dt + d\boldsymbol{r}/dt = \boldsymbol{v}_O + (d'\boldsymbol{r}/dt + \boldsymbol{\Omega} \times \boldsymbol{r}) \\ &\equiv \boldsymbol{v}_O + \boldsymbol{v}_{\text{rel}} + \boldsymbol{\Omega} \times \boldsymbol{r}, \end{aligned} \tag{3.16.3a}$$

where

$$\boldsymbol{v}_O = d\boldsymbol{r}_O/dt = \textit{inertial velocity of } \text{pole } O, \tag{3.16.3b}$$

$$\begin{aligned} \boldsymbol{v}_{\text{rel}} &= d'\boldsymbol{r}/dt = \textit{relative velocity } \text{of typical } S\text{-particle} \\ &= \sum (\partial \boldsymbol{r}/\partial q_k)(dq_k/dt) \equiv \sum \boldsymbol{e}_k\, \dot{q}_k \qquad [\text{by } (3.16.2\text{b})], \end{aligned} \tag{3.16.3c}$$

$$(\Rightarrow\ \boldsymbol{\Omega} \times \boldsymbol{r} = \partial \boldsymbol{r}/\partial t),$$

the (inertial) kinetic energy of S, $T \equiv (1/2) \mathsf{S}\, dm\, \boldsymbol{v} \cdot \boldsymbol{v}$, becomes

$$T = T_{\text{trnspt}} + T_{\text{rel}} + T_{\text{cpl}}, \tag{3.16.3d}$$

where

$$\begin{aligned} 2T_{\text{trnspt}} &\equiv m\, {v_O}^2 + \mathsf{S}\, dm (\boldsymbol{\Omega} \times \boldsymbol{r})^2 + 2m\, \boldsymbol{v}_O \cdot (\boldsymbol{\Omega} \times \boldsymbol{r}_G) \\ &= m\, {v_O}^2 + 2m\, \boldsymbol{v}_O \cdot (\boldsymbol{\Omega} \times \boldsymbol{r}_G) + \boldsymbol{\Omega} \cdot \mathbf{I}_O \cdot \boldsymbol{\Omega} \\ &= 2\ (\textit{kinetic energy of transport}), \end{aligned} \tag{3.16.3e}$$

[recall, $T_{\text{rotation}} \equiv T_R$, expressions (3.13.2b, 3b), and see fig. 3.36],

$$\begin{aligned} 2T_{\text{rel}} &\equiv \mathsf{S}\, dm\, {v_{\text{rel}}}^2 = \mathsf{S}\, dm\, (d'\boldsymbol{r}/dt)^2 \\ &= \mathsf{S}\, dm \Big[\sum (\partial \boldsymbol{r}/\partial q_k) \dot{q}_k\Big] \cdot \Big[\sum (\partial \boldsymbol{r}/\partial q_l) \dot{q}_l\Big] \\ &= \sum\sum \Big[\mathsf{S}\, dm\, (\partial \boldsymbol{r}/\partial q_k) \cdot (\partial \boldsymbol{r}/\partial q_l)\Big] \dot{q}_k\, \dot{q}_l \equiv \sum\sum \Big(\mathsf{S}\, dm\, \boldsymbol{e}_k \cdot \boldsymbol{e}_l\Big) \dot{q}_k\, \dot{q}_l \\ &= 2\ (\textit{relative kinetic energy}), \end{aligned} \tag{3.16.3f}$$

$$T_{\text{cpl}} \equiv \boldsymbol{p}_{\text{rel}} \cdot \boldsymbol{v}_O + \boldsymbol{H}_{O,\text{rel}} \cdot \boldsymbol{\Omega}$$
$$= \textit{kinetic energy of coupling (of carrying and carried motions;}$$
$$\textit{or ``Coriolis kinetic energy''}), \tag{3.16.3g}$$

$$\boldsymbol{p}_{\text{rel}} \equiv \mathcal{S}\, dm\, \boldsymbol{v}_{\text{rel}} = \mathcal{S}\, dm(d'\boldsymbol{r}/dt) = m\, \boldsymbol{v}_{G,\text{rel}} \equiv m(d'\boldsymbol{r}_G/dt)$$
$$= \mathcal{S}\, dm\Big[\sum (\partial \boldsymbol{r}/\partial q_k)\dot{q}_k\Big] = \sum \Big[\mathcal{S}\, dm\, (\partial \boldsymbol{r}/\partial q_k)\Big]\dot{q}_k$$
$$= \text{(\textit{noninertial}) \textit{linear momentum of body S}}, \tag{3.16.3h}$$

$$\boldsymbol{H}_{O,\text{rel}} \equiv \mathcal{S}\, \boldsymbol{r} \times (dm\, \boldsymbol{v}_{\text{rel}}) = \mathcal{S}\, \boldsymbol{r} \times [dm(d'\boldsymbol{r}/dt)]$$
$$= \cdots = \sum \Big[\mathcal{S}\, dm\, \boldsymbol{r} \times (\partial \boldsymbol{r}/\partial q_k)\Big]\dot{q}_k$$
$$= \textit{noninertial (absolute) angular momentum of S about O}. \tag{3.16.3i}$$

Before applying the Lagrangean formalism to the above, let us reduce them further to system forms, in the sense of §3.9:

(a) T_{trnspt} is *independent* of the $\dot{q}$'s. It would be the kinetic energy of S if the latter was frozen relative to O–xyz; that is, *if these axes were fixed in S*. In accordance with (3.9.2, 2c), we shall rename it T_0 [$\sim (\dot{q})^0$].
(b) With the notations

$$M_k \equiv \boldsymbol{v}_O \cdot \Big[\mathcal{S}\, dm\, (\partial \boldsymbol{r}/\partial q_k)\Big] + \boldsymbol{\Omega} \cdot \Big[\mathcal{S}\, dm\, \boldsymbol{r} \times (\partial \boldsymbol{r}/\partial q_k)\Big]$$
$$= \boldsymbol{v}_O \cdot \Big(\mathcal{S}\, dm\, \boldsymbol{e}_k\Big) + \boldsymbol{\Omega} \cdot \Big[\mathcal{S}\, (dm\, \boldsymbol{r} \times \boldsymbol{e}_k)\Big], \tag{3.16.4a}$$

$$M_{kl} \equiv \mathcal{S}\, dm\, (\partial \boldsymbol{r}/\partial q_k) \cdot (\partial \boldsymbol{r}/\partial q_l) = \mathcal{S}\, dm\, \boldsymbol{e}_k \cdot \boldsymbol{e}_l, \tag{3.16.4b}$$

and recalling (3.9.2a, b), we can rewrite T_{cpl} and T_{rel} as

$$T_{\text{cpl}} = \sum M_k\, \dot{q}_k \equiv T_1 \quad (\sim \dot{q}),$$
$$2T_{\text{rel}} = \sum\sum M_{kl}\, \dot{q}_k\, \dot{q}_l \equiv 2T_2 \quad [\sim (\dot{q})^2]. \tag{3.16.4c}$$

So, finally, the total kinetic energy, (3.16.3d), assumes the following general Lagrangean notation:

$$T = T_{\text{rel}} + T_{\text{cpl}} + T_{\text{trnspt}} \equiv T_2 + T_1 + T_0. \tag{3.16.4d}$$

Now, as the above expressions show, T depends on the following:

(a) The *six carrying quasi velocities* $\boldsymbol{v}_O = (v_{O,x,y,z})$ and $\boldsymbol{\Omega} = (\Omega_{x,y,z})$, along O–xyz [Along I–XYZ/O–XYZ, we would have $\boldsymbol{r}_O = (X_O, Y_O, Z_O)$, and, therefore, $\boldsymbol{v}_O = (v_{O;X,Y,Z}) \equiv (\dot{X}_O, \dot{Y}_O, \dot{Z}_O) =$ *holonomic* components]; and
(b) The *n carried holonomic velocities* $\dot{q} \equiv (\dot{q}_1, \ldots, \dot{q}_n)$.

Therefore, in the general case, we will obtain *two* groups of equations:

(a) *six* Hamel-type (or Lagrange–Euler) equations for the quasi velocities, *coupled* with
(b) n Lagrange-type equations for the $q/\dot{q}$'s.

The right, or *force*, sides of these equations will contain the corresponding six nonholonomic and n holonomic forces (and/or moments). Indeed, recalling (3.16.2a), we have

$$\delta\mathfrak{R} = \delta\boldsymbol{r}_O + \delta\boldsymbol{r} = \delta\boldsymbol{r}_O + (\delta'\boldsymbol{r} + \delta\boldsymbol{X} \times \boldsymbol{r}) \qquad [\text{where } d\boldsymbol{X}/dt = \boldsymbol{\Omega}]$$
$$= \delta\boldsymbol{r}_O + \delta\boldsymbol{X} \times \boldsymbol{r} + \sum (\partial\boldsymbol{r}/\partial q_k)\,\delta q_k; \qquad (3.16.5a)$$

if $\boldsymbol{r}_O = \boldsymbol{r}_O(t)$ then, clearly, $\delta\boldsymbol{r}_O = \boldsymbol{0}$. Hence, in general, the total (inertial and first-order) impressed virtual work equals

$$\delta'W = \mathsf{S}\, d\boldsymbol{F} \cdot \delta\mathfrak{R} = \cdots = \boldsymbol{F} \cdot \delta\boldsymbol{r}_O + \boldsymbol{M}_O \cdot \delta\boldsymbol{X} + \sum Q_k\,\delta q_k, \qquad (3.16.5b)$$

where

$$\boldsymbol{F} \equiv \mathsf{S}\, d\boldsymbol{F} = \textit{total impressed force (acting at O)}, \qquad (3.16.5c)$$
$$\boldsymbol{M}_O \equiv \mathsf{S}\, \boldsymbol{r} \times d\boldsymbol{F} = \textit{total impressed moment about O}. \qquad (3.16.5d)$$

Let us now find the $6 + n$ equations of motion.

Hamel-Type (or Lagrange–Euler) Carrying Body Equations

(A review of §3.13 is highly recommended.) Since $T_{\text{rel}} = T_2$ is independent of the quasi velocities $v_{O;x,y,z}$ and $\Omega_{x,y,z}$, we can write

$$\partial T/\partial v_{O,k} = \partial T_0/\partial v_{O,k} + \partial T_1/\partial v_{O,k}, \qquad \partial T/\partial \Omega_k = \partial T_0/\partial \Omega_k + \partial T_1/\partial \Omega_k, \qquad (3.16.5e)$$

where $k = x, y, z$. Now, recalling the expressions for T, $\partial T/\partial \boldsymbol{v}_\blacklozenge$, $\partial T/\partial \boldsymbol{\omega}$ from §3.13 (and setting in those formulae $T \to T_O$, $\blacklozenge \to O$, $\boldsymbol{\omega} \to \boldsymbol{\Omega}$, we readily conclude that

$$\partial T_0/\partial v_{O,x} = m(v_{O,x} + \Omega_y z_G - \Omega_z y_G), \quad \text{etc., cyclically,} \qquad (3.16.5f)$$
$$\partial T_0/\partial \Omega_x = m(y_G v_{O,z} - z_G v_{O,y}) + (I_{O,xx}\Omega_x + I_{O,xy}\Omega_y + I_{O,xz}\Omega_z), \quad \text{etc., cyclically,} \qquad (3.16.5g)$$

where $(I_{O,kl};\ k, l = x, y, z)$ are the components of the inertia tensor of the system S about O, along O–xyz; and

$$\partial T_1/\partial v_{O,x} = m(dx_G/dt) = (\boldsymbol{p}_{\text{rel}})_x, \quad \text{etc., cyclically,} \qquad (3.16.5h)$$
$$\partial T_1/\partial \Omega_x = (\boldsymbol{H}_{O,\text{rel}})_x, \quad \text{etc., cyclically.} \qquad (3.16.5i)$$

A moment's reflection will show that the left (inertia/acceleration) sides of the equations for $\boldsymbol{v}_O$ and $\boldsymbol{\Omega}$ are none other than the former expressions (3.13.11a, b) with $\blacklozenge, \bullet \to O$. Hence, for an unconstrained rigid body, we have [since all (symbolic) partial derivatives of T relative to the quasi coordinates vanish]

$$\boldsymbol{I} = d/dt(\partial T/\partial \boldsymbol{v}_O) \equiv d'/dt(\partial T/\partial \boldsymbol{v}_\blacklozenge) + \boldsymbol{\Omega} \times (\partial T/\partial \boldsymbol{v}_\blacklozenge) = \boldsymbol{F}, \qquad (3.16.6a)$$

or, in components,

$$I_x = (\partial T/\partial v_{O,x})^{\cdot} + \Omega_y(\partial T/\partial v_{O,z}) - \Omega_z(\partial T/\partial v_{O,y}) = F_x, \quad \text{etc., cyclically,} \qquad (3.16.6b)$$

and

$$A_O = d/dt(\partial T/\partial \boldsymbol{\Omega}) + \boldsymbol{v}_O \times (\partial T/\partial \boldsymbol{v}_O)$$
$$\equiv d'/dt(\partial T/\partial \boldsymbol{\Omega}) + \boldsymbol{\Omega} \times (\partial T/\partial \boldsymbol{\Omega}) + \boldsymbol{v}_O \times (\partial T/\partial \boldsymbol{v}_O) = \boldsymbol{M}_O, \qquad (3.16.6c)$$

or, in components,

$$A_{O,x} = (\partial T/\partial \Omega_x)^{\cdot} + \Omega_y(\partial T/\partial \Omega_z) - \Omega_z(\partial T/\partial \Omega_y)$$
$$+ v_{O,y}(\partial T/\partial v_{O,z}) - v_{O,z}(\partial T/\partial v_{O,y}) = M_{O,x}, \quad \text{etc., cyclically.} \qquad (3.16.6d)$$

To obtain explicit Euler-type equations from the above, we introduce in (3.16.6a–d) the earlier found expressions for T. In this way:

(i) Equations (3.16.6b) yield

$$m\Big\{\dot{v}_{O,x} + (\dot{\Omega}_y z_G - \dot{\Omega}_z y_G) + (\Omega_y \dot{z}_G - \Omega_z \dot{y}_G) + (\Omega_y v_{O,z} - \Omega_z v_{O,y})$$
$$+ [\Omega_y(\Omega_x y_G - \Omega_y x_G) - \Omega_z(\Omega_z x_G - \Omega_x z_G)]$$
$$+ (\ddot{x}_G + \Omega_y \dot{z}_G - \Omega_z \dot{y}_G)\Big\} = F_x, \quad \text{etc., cyclically;} \qquad (3.16.6e)$$

or, in vector form (appropriately grouped),

$$m[(d'\boldsymbol{v}_O/dt + \boldsymbol{\Omega} \times \boldsymbol{v}_O) + (d\boldsymbol{\Omega}/dt) \times \boldsymbol{r}_G + \boldsymbol{\Omega} \times (\boldsymbol{\Omega} \times \boldsymbol{r}_G) + 2\boldsymbol{\Omega} \times (d'\boldsymbol{r}_G/dt) + d'^2\boldsymbol{r}_G/dt^2]$$
$$\equiv m\big\{[\boldsymbol{a}_O + (d\boldsymbol{\Omega}/dt) \times \boldsymbol{r}_G + \boldsymbol{\Omega} \times (\boldsymbol{\Omega} \times \boldsymbol{r}_G)] + 2\boldsymbol{\Omega} \times \boldsymbol{v}_{G,\mathrm{rel}} + \boldsymbol{a}_{G,\mathrm{rel}}\big\}$$
$$\equiv m(\boldsymbol{a}_{G,\mathrm{transport}} + \boldsymbol{a}_{G,\mathrm{Coriolis}} + \boldsymbol{a}_{G,\mathrm{relative}}) \equiv m\,\boldsymbol{a}_{G,\mathrm{inertial}} \equiv m\,\boldsymbol{a}_G = \boldsymbol{F}; \qquad (3.16.6f)$$

which is the well-known equation of motion of the "inertial center" of the carried system G.

[Also, the above equations and the earlier kinematic relations

$$\boldsymbol{v}_{G,\mathrm{rel}} = d'\boldsymbol{r}_G/dt = \sum (\partial \boldsymbol{r}_G/\partial q_k)\dot{q}_k,$$
$$\boldsymbol{a}_{G,\mathrm{rel}} = d'\boldsymbol{v}_G/dt = \sum (\partial \boldsymbol{r}_G/\partial q_k)\ddot{q}_k + \sum\sum (\partial^2 \boldsymbol{r}_G/\partial q_k \partial q_l)\dot{q}_k\,\dot{q}_l, \qquad (3.16.6g)$$

demonstrate clearly the *coupling* of the holonomic velocities $\dot{q}$ with the nonholonomic ones $v_{O;x,y,z}$ and $\Omega_{x,y,z}$.]

(ii) Equations (3.16.6d) yield

$$m(y_G\,\dot{v}_{O,z} - z_G\,\dot{v}_{O,y}) + (I_{O,xx}\,\dot{\Omega}_x + I_{O,yy}\,\dot{\Omega}_y + I_{O,zz}\,\dot{\Omega}_z$$
$$+ \dot{I}_{O,xx}\,\Omega_x + \dot{I}_{O,xy}\,\Omega_y + \dot{I}_{O,xz}\,\Omega_z)$$
$$+ (\boldsymbol{H}_{O,\mathrm{rel}})_x^{\cdot} + m[\Omega_y(v_{O,y}\,x_G - v_{O,x}\,y_G) - \Omega_z(v_{O,x}\,z_G - v_{O,z}\,x_G)]$$
$$+ \Omega_y(I_{O,zx}\,\Omega_x + I_{O,zy}\,\Omega_y + I_{O,zz}\,\Omega_z) - \Omega_z(I_{O,yx}\,\Omega_x + I_{O,yy}\,\Omega_y + I_{O,yz}\,\Omega_z)$$
$$+ \Omega_y(\boldsymbol{H}_{O,\mathrm{rel}})_z - \Omega_z(\boldsymbol{H}_{O,\mathrm{rel}})_y$$
$$+ m[v_{O,y}(\Omega_x\,y_G - \Omega_y\,x_G) - v_{O,z}(\Omega_z\,x_G - \Omega_x\,z_G)] = M_{O,x}, \quad \text{etc., cyclically;} \qquad (3.16.6h)$$

or, vectorially,

$$\mathbf{I}_O \cdot (d\boldsymbol{\Omega}/dt) + (d'\mathbf{I}_O/dt) \cdot \boldsymbol{\Omega} + \boldsymbol{\Omega} \times (\mathbf{I}_O \cdot \boldsymbol{\Omega}) + (d'\boldsymbol{H}_{O,\mathrm{rel}}/dt + \boldsymbol{\Omega} \times \boldsymbol{H}_{O,\mathrm{rel}}) + m\boldsymbol{r}_G \times \boldsymbol{a}_O = \boldsymbol{M}_O. \qquad (3.16.6\mathrm{i})$$

This equation can be transformed further. Recalling the inertia tensor definition [§1.15, and with $\mathbf{1} = (\delta_{kl}) = 3 \times 3$ *unit* (*Cartesian*) *tensor*, where $k, l = x, y, z$; and $\otimes$ = *tensor product* (§1.1)], $\mathbf{I}_O = S\, dm\, [(\boldsymbol{r} \cdot \boldsymbol{r})\mathbf{1} - \boldsymbol{r} \otimes \boldsymbol{r}]$, we obtain

$$d'\mathbf{I}_O/dt = 2\sum \Big\{ S\, dm \big[\{\boldsymbol{r} \cdot (\partial \boldsymbol{r}/\partial q_k)\}\mathbf{1} - (1/2)\boldsymbol{r} \otimes (\partial \boldsymbol{r}/\partial q_k) - (1/2)(\partial \boldsymbol{r}/\partial q_k) \otimes \boldsymbol{r} \big] \Big\} \dot{q}_k, \qquad (3.16.7\mathrm{a})$$

and, therefore, we find, successively,

$$(d'\mathbf{I}_O/dt) \cdot \boldsymbol{\Omega} + \boldsymbol{\Omega} \times \boldsymbol{H}_{O,\mathrm{rel}}$$

$$= 2\sum \Big\{ S\, dm \big[\{\boldsymbol{r} \cdot (\partial \boldsymbol{r}/\partial q_k)\}\boldsymbol{\Omega} - (1/2)\{(\partial \boldsymbol{r}/\partial q_k) \otimes \boldsymbol{r}\} \cdot \boldsymbol{\Omega} - (1/2)\{\boldsymbol{r} \otimes (\partial \boldsymbol{r}/\partial q_k)\} \cdot \boldsymbol{\Omega} + (1/2)\boldsymbol{\Omega} \times \{\boldsymbol{r} \times (\partial \boldsymbol{r}/\partial q_k)\} \big] \Big\} \dot{q}_k$$

$$= 2\sum \Big\{ S\, dm\, \boldsymbol{r} \times [\boldsymbol{\Omega} \times (\partial \boldsymbol{r}/\partial q_k)] \Big\} \dot{q}_k$$

$$= 2 S\, dm\, \boldsymbol{r} \times (\boldsymbol{\Omega} \times \boldsymbol{v}_{\mathrm{rel}}) = S\, \boldsymbol{r} \times (2dm\, \boldsymbol{\Omega} \times \boldsymbol{v}_{\mathrm{rel}}):$$

$-$ (*Total moment of Coriolis forces, on the carrying body, about O, due to the motion of the carried body relative to it*) $\equiv -\boldsymbol{M}_{O,\mathrm{Coriolis}}$. (3.16.7b)

In view of this result, and using (3.16.6f) to eliminate $\boldsymbol{a}_O$, we can rewrite (3.16.6h) (after some judicious regrouping) in the following Euler-like form:

$$\mathbf{I}_G \cdot (d\boldsymbol{\Omega}/dt) + \boldsymbol{\Omega} \times (\mathbf{I}_G \cdot \boldsymbol{\Omega}) = \boldsymbol{M}_G + \boldsymbol{M}_{G,\mathrm{Coriolis}} - d'\boldsymbol{H}_{G,\mathrm{rel}}/dt, \qquad (3.16.7\mathrm{c})$$

where

$\mathbf{I}_G = \mathbf{I}_O - m\big[(\boldsymbol{r}_G \cdot \boldsymbol{r}_G)\mathbf{1} - \boldsymbol{r}_G \otimes \boldsymbol{r}_G\big]$: *Moment of inertia of S about G* (*direct tensorial form of parallel axis theorem*; *recall §1.15*), (3.16.7d)

$$\boldsymbol{H}_{G,\mathrm{rel}} = \boldsymbol{H}_{O,\mathrm{rel}} - m\boldsymbol{r}_G \times (d'\boldsymbol{r}_G/dt), \qquad (3.16.7\mathrm{e})$$

$$\boldsymbol{M}_{G,\mathrm{Coriolis}} \equiv -[(d'\mathbf{I}_G/dt) \cdot \boldsymbol{\Omega} + \boldsymbol{\Omega} \times \boldsymbol{H}_{G,\mathrm{rel}}] = \boldsymbol{M}_{O,\mathrm{Coriolis}} - \boldsymbol{r}_G \times (2m\, \boldsymbol{\Omega} \times \boldsymbol{v}_{G,\mathrm{rel}})$$ [i.e., same as (3.16.7b), but about G], (3.16.7f)

$\boldsymbol{M}_G = \boldsymbol{M}_O - \boldsymbol{r}_G \times \boldsymbol{F}$ [with $\boldsymbol{F}$ assumed applied at O]. (3.16.7g)

REMARKS, SPECIALIZATIONS

(i) Equations (3.16.7c) result immediately from (3.16.6h) with the choice $O \to G$.

(ii) If S is a rigid body and the $O{-}xyz$ are *body-fixed* axes on it, then the relative rates $d'(\ldots)/dt$ vanish, $\boldsymbol{\Omega} \to \boldsymbol{\omega}$, $d\boldsymbol{\Omega}/dt \to \boldsymbol{\alpha}$, and (3.16.6f, e) reduce, respectively, to

$$m[(d'\boldsymbol{v}_O/dt + \boldsymbol{\omega} \times \boldsymbol{v}_O) + \boldsymbol{\alpha} \times \boldsymbol{r}_G + \boldsymbol{\omega} \times (\boldsymbol{\omega} \times \boldsymbol{r}_G)] = \boldsymbol{F}, \qquad (3.16.8a)$$

$$m\{(\dot{v}_{O,x} + \omega_y\, v_{O,z} - \omega_z\, v_{O,y}) + (\dot{\omega}_y\, z_G - \dot{\omega}_z\, y_G) - [({\omega_y}^2 + {\omega_z}^2)x_G - \omega_x(\omega_y\, y_G + \omega_z\, z_G)]\} = F_x, \quad \text{etc., cyclically;} \qquad (3.16.8b)$$

and, of course, (its left side) coincides with (3.13.10a), with ◆ → O.

If, further, we choose, for algebraic simplicity, (body-fixed) *principal* axes of S at O, eqs. (3.16.6h–7b) reduce to

$$\mathbf{I}_O \cdot \boldsymbol{\alpha} + \boldsymbol{\omega} \times (\mathbf{I}_O \cdot \boldsymbol{\omega}) + m\boldsymbol{r}_G \times \boldsymbol{a}_O = \boldsymbol{M}_O, \qquad (3.16.8c)$$

$$I_{O,xx}\, \dot{\omega}_x + (I_{O,zz} - I_{O,yy})\omega_y\, \omega_z + m[y_G(\dot{v}_{O,z} + \omega_z\, v_{O,y} - \omega_y\, v_{O,z}) - z_G(\dot{v}_{O,y} + \omega_z\, v_{O,x} - \omega_x\, v_{O,z})] = M_{O,x}, \quad \text{etc., cyclically.} \qquad (3.16.8d)$$

If, finally, $O \to G$ then (3.16.8a, c) reduce, respectively, to the well-known

$$m(d'\boldsymbol{v}_G/dt + \boldsymbol{\omega} \times \boldsymbol{v}_G) = \boldsymbol{F} \quad \text{and} \quad \mathbf{I}_G \cdot \boldsymbol{\alpha} + \boldsymbol{\omega} \times (\mathbf{I}_G \cdot \boldsymbol{\omega}) = \boldsymbol{M}_O. \qquad (3.16.8e)$$

All these equations express the Eulerian principles of *linear* and *angular* momentum, for the rigid body S about various points, O, G, and so on.

(iii) We hope that the above lengthy and tedious calculations (especially those in components) have begun to convince the reader that, in this case at least, *the Lagrangean approach (L) is superior to the Eulerian approach (E)*; even though both are, roughly, theoretically equivalent. As Lur'e (1968, pp. 412–413) points out: In E, in order to derive rotational equations we begin with the principle of angular momentum about a fixed point in $I{-}XYZ$ and then transfer both angular momenta and moments of forces to an arbitrary, say, body-fixed point; and in the process we utilize certain kinematico-inertial results. In L, on the other hand, the calculations (of the various partial and total derivatives of the kinetic energy) are almost automatic (... mechanical!); although their final results need "translating" back to the more geometrical Newtonian–Eulerian language. However, as already stressed (Introduction and this chapter), a far more important advantage of L over E, for theoretical work anyway, is that, even in complex problems, *the former (L) preserves the structure/form of the equations of motion, whereas in the latter (E) these equations appear structureless/formless* ("a bunch of terms"), and hence hard to remember, understand, and interpret.

(iv) The preceding equations of motion can, of course, be derived via *Appell's method*. Indeed, from equation (3.14.4d), with ◆ → O and slight rearrangement, we obtain

$$2S = m[(\dot{v}_{O,x})^2 + (\dot{v}_{O,y})^2 + (\dot{v}_{O,z})^2] + 2(d'\boldsymbol{v}_O/dt) \cdot [\boldsymbol{\omega} \times m(\boldsymbol{v}_O + \boldsymbol{\omega} \times \boldsymbol{r}_G)] + 2m\boldsymbol{\alpha} \cdot [\boldsymbol{r}_G \times (d'\boldsymbol{v}_O/dt + \boldsymbol{\omega} \times \boldsymbol{v}_O)] + \boldsymbol{\alpha} \cdot \mathbf{I}_O \cdot \boldsymbol{\alpha} + 2\boldsymbol{\alpha} \cdot [\boldsymbol{\omega} \times (\mathbf{I}_O \cdot \boldsymbol{\omega})]; \qquad (3.16.9a)$$

and from this, using the following simple identities ($\boldsymbol{a}, \boldsymbol{b}$: arbitrary vectors, and $k = x, y, z$):

$$\partial(\boldsymbol{a} \cdot \boldsymbol{b})/\partial a_k = b_k, \qquad \partial(\boldsymbol{a} \cdot \boldsymbol{b})/\partial b_k = a_k, \tag{3.16.9b}$$

and

$$\partial(\boldsymbol{a} \times \boldsymbol{b})/\partial a_x = \boldsymbol{i} \times \boldsymbol{b} = \boldsymbol{k} b_y - \boldsymbol{j} b_z, \quad \text{etc., cyclically}, \tag{3.16.9c}$$

we find

$$\partial S/\partial \dot{v}_{O,k} = m\{\dot{v}_{O,k} + (\boldsymbol{\omega} \times \boldsymbol{v}_O)_k + (\boldsymbol{\alpha} \times \boldsymbol{r}_G)_k + [\boldsymbol{\omega} \times (\boldsymbol{\omega} \times \boldsymbol{r}_G)]_k\} = F_k, \tag{3.16.10a}$$

$$\partial S/\partial \dot{\omega}_k = m[\boldsymbol{r}_G \times (d'\boldsymbol{v}_O/dt + \boldsymbol{\omega} \times \boldsymbol{v}_O)]_k + [\mathbf{I}_O \cdot \boldsymbol{\alpha} + \boldsymbol{\omega} \times (\mathbf{I}_O \cdot \boldsymbol{\omega})]_k = M_{O,k}, \tag{3.16.10b}$$

that is, equations (3.16.8a, c), respectively. This completes the discussion of the quasi-velocity equations of the carrying body. Let us now turn to the q-equations of the carried system S.

Lagrange-Type Carried System Equations

(This is an application of §3.10, and so we recommend a rereading of that section.) First, we notice that, in view of the linearity of the holonomic Euler–Lagrange operator,

$$E_k(T) = E_k(T_0) + E_k(T_1) + E_k(T_2). \tag{3.16.11a}$$

Next, invoking the earlier $T_{0,1,2}$ expressions, (3.16.3d–4c), we find, successively,

(i) $$E_k(T_0) = \cdots = -\partial T_0/\partial q_k = -m(\boldsymbol{v}_O \times \boldsymbol{\Omega}) \cdot (\partial \boldsymbol{r}_G/\partial q_k) - (1/2)\boldsymbol{\Omega} \cdot (\partial \mathbf{I}_O/\partial q_k) \cdot \boldsymbol{\Omega}, \tag{3.16.11b}$$

(ii) $$E_k(T_1) = E_k(\boldsymbol{v}_O \cdot \boldsymbol{p}_{\text{rel}}) + E_k(\boldsymbol{\Omega} \cdot \boldsymbol{H}_{O,\text{rel}}). \tag{3.16.11c}$$

But:

(a) $$E_k(\boldsymbol{v}_O \cdot \boldsymbol{p}_{\text{rel}}) = (d\boldsymbol{v}_O/dt) \cdot (\partial \boldsymbol{p}_{\text{rel}}/\partial \dot{q}_k) + \boldsymbol{v}_O \cdot E_k(\boldsymbol{p}_{\text{rel}}) = m(d\boldsymbol{v}_O/dt) \cdot [\partial/\partial \dot{q}_k(d'\boldsymbol{r}_G/dt)] + m\,\boldsymbol{v}_O \cdot E_k(d'\boldsymbol{r}_G/dt), \tag{3.16.11d}$$

or, in view of the kinematical identities,

$$\partial/\partial \dot{q}_k(d'\boldsymbol{r}_G/dt) = \partial \boldsymbol{r}_G/\partial q_k, \tag{3.16.11e}$$

$$d/dt(\partial \boldsymbol{r}_G/\partial q_k) = \partial/\partial q_k(d\boldsymbol{r}_G/dt) = \partial/\partial q_k(d'\boldsymbol{r}_G/dt) + \partial/\partial q_k(\boldsymbol{\Omega} \times \boldsymbol{r}_G) = \partial/\partial q_k(d'\boldsymbol{r}_G/dt) + \boldsymbol{\Omega} \times (\partial \boldsymbol{r}_G/\partial q_k), \tag{3.16.11f}$$

from which

$$E_k(d'\boldsymbol{r}_G/dt) \equiv [\partial/\partial \dot{q}_k(d'\boldsymbol{r}_G/dt)]^{\cdot} - \partial/\partial q_k(d'\boldsymbol{r}_G/dt) = d/dt(\partial \boldsymbol{r}_G/\partial q_k) - \partial/\partial q_k(d'\boldsymbol{r}_G/dt) = \boldsymbol{\Omega} \times (\partial \boldsymbol{r}_G/\partial q_k), \tag{3.16.11g}$$

so that (3.16.11d) becomes

$$E_k(\boldsymbol{v}_O \cdot \boldsymbol{p}_{\text{rel}}) = m\,\boldsymbol{a}_O \cdot (\partial \boldsymbol{r}_G/\partial q_k) + m(\boldsymbol{v}_O \times \boldsymbol{\Omega}) \cdot (\partial \boldsymbol{r}_G/\partial q_k)$$
$$= m\big[(d'\boldsymbol{v}_O/dt + \boldsymbol{\Omega} \times \boldsymbol{v}_O) + \boldsymbol{v}_O \times \boldsymbol{\Omega}\big] \cdot (\partial \boldsymbol{r}_G/\partial q_k);$$

that is, finally,

$$E_k(\boldsymbol{v}_O \cdot \boldsymbol{p}_{\text{rel}}) = m(d'\boldsymbol{v}_O/dt) \cdot (\partial \boldsymbol{r}_G/\partial q_k). \qquad (3.16.11\text{h})$$

(b) $E_k(\boldsymbol{\Omega} \cdot \boldsymbol{H}_{O,\text{rel}}) = (d\boldsymbol{\Omega}/dt) \cdot (\partial \boldsymbol{H}_{O,\text{rel}}/\partial \dot{q}_k) + \boldsymbol{\Omega} \cdot E_k(\boldsymbol{H}_{O,\text{rel}})$

$$\equiv (d\boldsymbol{\Omega}/dt) \cdot (\partial \boldsymbol{H}_{O,\text{rel}}/\partial \dot{q}_k) + \boldsymbol{\Omega} \cdot [(\partial \boldsymbol{H}_{O,\text{rel}}/\partial \dot{q}_k)^{\cdot} - \partial \boldsymbol{H}_{O,\text{rel}}/\partial q_k]$$
$$= (d\boldsymbol{\Omega}/dt) \cdot (\partial \boldsymbol{H}_{O,\text{rel}}/\partial \dot{q}_k)$$
$$+ \boldsymbol{\Omega} \cdot [d'/dt(\partial \boldsymbol{H}_{O,\text{rel}}/\partial \dot{q}_k) + \boldsymbol{\Omega} \times (\partial \boldsymbol{H}_{O,\text{rel}}/\partial \dot{q}_k) - \partial \boldsymbol{H}_{O,\text{rel}}/\partial q_k];$$

that is, finally,

$$E_k(\boldsymbol{\Omega} \cdot \boldsymbol{H}_{O,\text{rel}}) = (d\boldsymbol{\Omega}/dt) \cdot (\partial \boldsymbol{H}_{O,\text{rel}}/\partial \dot{q}_k) + \boldsymbol{\Omega} \cdot [d'/dt(\partial \boldsymbol{H}_{O,\text{rel}}/\partial \dot{q}_k) - \partial \boldsymbol{H}_{O,\text{rel}}/\partial q_k]$$
$$\equiv (d\boldsymbol{\Omega}/dt) \cdot (\partial \boldsymbol{H}_{O,\text{rel}}/\partial \dot{q}_k) + \boldsymbol{\Omega} \cdot E_{k,\text{rel}}(\boldsymbol{H}_{O,\text{rel}}), \qquad (3.16.11\text{i})$$

where

$E_{k,\text{rel}}(\ldots) \equiv d'/dt(\partial \ldots/\partial \dot{q}_k) - \partial \ldots/\partial q_k$:

Relative Euler–Lagrange operator (for the carrying body). (3.16.11j)

Introducing all these partial results into the Lagrangean equations of motion, say $E_k(T) = Q_k$, or

$$E_k(T_2) = Q_k - E_k(T_1) - E_k(T_0), \qquad (3.16.12\text{a})$$

we obtain the equations of relative motion for the q's:

$$E_k(T_2) = Q_k - m(d'\boldsymbol{v}_O/dt + \boldsymbol{\Omega} \times \boldsymbol{v}_O) \cdot (\partial \boldsymbol{r}_G/\partial q_k) + (1/2)\boldsymbol{\Omega} \cdot (\partial \mathbf{I}_O/\partial q_k) \cdot \boldsymbol{\Omega}$$
$$- (d\boldsymbol{\Omega}/dt) \cdot (\partial \boldsymbol{H}_{O,\text{rel}}/\partial \dot{q}_k) - \boldsymbol{\Omega} \cdot E_{k,\text{rel}}(\boldsymbol{H}_{O,\text{rel}}). \qquad (3.16.12\text{b})$$

However, the inertial, or fictitious (= frame dependent) "forces" on the right side of (3.16.12b) can be further transformed as follows:

(i) $Q_{k,\text{translation}} \equiv Q_{k,T} \equiv -m(d'\boldsymbol{v}_O/dt + \boldsymbol{\Omega} \times \boldsymbol{v}_O) \cdot (\partial \boldsymbol{r}_G/\partial q_k)$

$$= -m\,\boldsymbol{a}_O \cdot (\partial \boldsymbol{r}_G/\partial q_k) = -\partial V_T/\partial q_k:$$

Lagrangean inertial force of translation, (3.16.12c)

where the corresponding potential of the *homogeneous* field of these "forces" is defined by

$$V_T \equiv m\,\boldsymbol{a}_O \cdot \boldsymbol{r}_G = S\, \boldsymbol{a}_O \cdot (dm\,\boldsymbol{r}) = S\, dm(d'\boldsymbol{v}_O/dt + \boldsymbol{\Omega} \times \boldsymbol{v}_O) \cdot \boldsymbol{r}$$
$$= S\, dm\big[(d'\boldsymbol{v}_O/dt) \cdot \boldsymbol{r} + \boldsymbol{r} \cdot (\boldsymbol{\Omega} \times \boldsymbol{v}_O)\big]. \qquad (3.16.12\text{d})$$

(ii)
$$Q_{k,\text{centrifugal}} \equiv Q_{k,CF} \equiv (1/2)[\boldsymbol{\Omega}\cdot(\partial\mathbf{I}_O/\partial q_k)\cdot\boldsymbol{\Omega}]$$
$$= \mathbf{S}\, dm\,(\boldsymbol{\Omega}\times\boldsymbol{r})\cdot[\boldsymbol{\Omega}\times(\partial\boldsymbol{r}/\partial q_k)] = -\,\partial V_{CF}/\partial q_k :$$

Lagrangean centrifugal inertial force, (3.16.12e)

where the corresponding *centrifugal* potential of these "forces" is defined by

$$V_{CF} \equiv -(1/2)\boldsymbol{\Omega}\cdot\mathbf{I}_O\cdot\boldsymbol{\Omega} = -(1/2)\mathbf{S}\, dm\,(\boldsymbol{\Omega}\times\boldsymbol{r})^2. \tag{3.16.12f}$$

(iii) Since

$$\boldsymbol{H}_{O,\text{rel}} \equiv \mathbf{S}\,\boldsymbol{r}\times dm\,(d'\boldsymbol{v}_O/dt) = \sum\Big(\mathbf{S}\, dm\,\boldsymbol{r}\times(\partial\boldsymbol{r}/\partial q_k)\Big)\dot{q}_k, \tag{3.16.12g}$$

we have

$$(d\boldsymbol{\Omega}/dt)\cdot(\partial\boldsymbol{H}_{O,\text{rel}}/\partial\dot{q}_k) = (d\boldsymbol{\Omega}/dt)\cdot\Big(\mathbf{S}\, dm\,\boldsymbol{r}\times(\partial\boldsymbol{r}/\partial q_k)\Big)$$
$$= \mathbf{S}\, dm\,\big((d\boldsymbol{\Omega}/dt)\times\boldsymbol{r}\big)\cdot(\partial\boldsymbol{r}/\partial q_k), \tag{3.16.12h}$$

and, therefore,

$$Q_{k,\text{rotation}} \equiv Q_{k,R} \equiv -(d\boldsymbol{\Omega}/dt)\cdot(\partial\boldsymbol{H}_{O,\text{rel}}/\partial\dot{q}_k)$$
$$= -\mathbf{S}\,[(d\boldsymbol{\Omega}/dt)\times(dm\,\boldsymbol{r})]\cdot(\partial\boldsymbol{r}/\partial q_k) = \mathbf{S}\, dm\,(d\boldsymbol{\Omega}/dt)\cdot[(\partial\boldsymbol{r}/\partial q_k)\times\boldsymbol{r}]$$
$$\equiv \mathbf{S}\, d\boldsymbol{I}_R\cdot\boldsymbol{e}_k\text{: } \textit{Lagrangean rotational inertial force}; \tag{3.16.12i}$$

where

$$d\boldsymbol{I}_{\text{rotation}} \equiv d\boldsymbol{I}_R \equiv -dm\,\big((d\boldsymbol{\Omega}/dt)\times\boldsymbol{r}\big)\text{: } \textit{Rotational inertial force, on a typical particle.} \tag{3.16.12j}$$

(iv) Finally, from (3.16.12g) we obtain

$$\partial\boldsymbol{H}_{O,\text{rel}}/\partial\dot{q}_k = \mathbf{S}\, dm\,\boldsymbol{r}\times(\partial\boldsymbol{r}/\partial q_k),$$

and from this, further,

$$d'/dt(\partial\boldsymbol{H}_{O,\text{rel}}/\partial\dot{q}_k) = \sum\Big\{\mathbf{S}\, dm[(\partial\boldsymbol{r}/\partial q_l)\times(\partial\boldsymbol{r}/\partial q_k) + \boldsymbol{r}\times(\partial^2\boldsymbol{r}/\partial q_l\partial q_k)]\Big\}\dot{q}_l,$$

and

$$\partial\boldsymbol{H}_{O,\text{rel}}/\partial q_k = \sum\Big\{\mathbf{S}\, dm[(\partial\boldsymbol{r}/\partial q_k)\times(\partial\boldsymbol{r}/\partial q_l) + \boldsymbol{r}\times(\partial^2\boldsymbol{r}/\partial q_k\partial q_l)]\Big\}\dot{q}_l.$$

Therefore, recalling (3.16.11j), we find, successively,

$$-\boldsymbol{\Omega}\cdot E_{k,\text{rel}}(\boldsymbol{H}_{O,\text{rel}}) = -2\boldsymbol{\Omega}\cdot\Big(\sum\Big\{\mathbf{S}\, dm[(\partial\boldsymbol{r}/\partial q_l)\times(\partial\boldsymbol{r}/\partial q_k)]\Big\}\Big)\dot{q}_l$$
$$= -2\boldsymbol{\Omega}\cdot\mathbf{S}\, dm[(d'\boldsymbol{r}/dt)\times(\partial\boldsymbol{r}/\partial q_k)] \qquad \text{[recalling (3.16.3c)],}$$
$$= -\mathbf{S}\,[2(\boldsymbol{\Omega}\times dm\,\boldsymbol{v}_{\text{rel}})]\cdot(\partial\boldsymbol{r}/\partial q_k)$$
$$\equiv \sum(\boldsymbol{\Omega}\cdot\boldsymbol{G}_{kl})\,\dot{q}_l \equiv \sum g_{kl}\dot{q}_l, \tag{3.16.12k}$$

where

$$G_{kl} \equiv 2 \mathsf{S}\, dm\, [(\partial \boldsymbol{r}/\partial q_k) \times (\partial \boldsymbol{r}/\partial q_l)] \equiv 2 \mathsf{S}\, dm(\boldsymbol{e}_k \times \boldsymbol{e}_l) \quad (= -\boldsymbol{G}_{lk}), \tag{3.16.12l}$$

$$g_{kl} \equiv \boldsymbol{\Omega} \cdot \boldsymbol{G}_{kl} = 2 \mathsf{S}\, dm[\boldsymbol{\Omega} \cdot (\boldsymbol{e}_k \times \boldsymbol{e}_l)]: \textit{Gyroscopic coefficients} \quad (= -g_{lk}), \tag{3.16.12m}$$

or, finally [recalling (3.10.1f, g)],

$$-\boldsymbol{\Omega} \cdot E_{k,\text{rel}}(\boldsymbol{H}_{O,\text{rel}}) = \sum g_{kl}\, \dot{q}_l$$
$$\equiv Q_{k,\text{gyroscopic}} \equiv Q_{k,\text{Coriolis}} \equiv -E_{k,C} \equiv G_k. \tag{3.16.12n}$$

[Also, recall mathematically similar terms arising out of the coefficients of the $\sim \dot{q}$ terms of the generalized potential, equations (3.9.8a ff.).]

In view of all these partial results, eqs. (3.16.12c–n), and recalling that $T_{\text{relative}} \equiv T_{\text{rel}} \equiv T_2$, we can rewrite (3.16.12b) as

$$E_k(T_{\text{rel}}) = Q_k + Q_{k,R} + G_k - \partial(V_{CF} + V_T)/\partial q_k \equiv Q_k + Q_{k,inertial}. \tag{3.16.13}$$

This completes the discussion of the Lagrange-type carried system equations.

REMARKS

The *left* side of (3.16.13) clearly depends only on quantities describing the configuration and motions of the carried body (bodies) relative to the carrying one; the *four* parts of $Q_{k,inertial}$ are "correction terms," since $O\text{–}xyz$ is noninertial. Hence:

(i) If the motion of $O\text{–}xyz$ is known, or prescribed, *only eqs. (3.16.13) need be considered;* not the earlier Hamel-type carrying equations. Actually, since we have indeed proved that, in the case of relative motion, LP, for the carried system, takes the form

$$\sum \left[E_k(T_{\text{rel}}) - Q_k - Q_{k,inertial}\right] \delta q_k = 0, \tag{3.16.14}$$

any other set of Lagrangean, or Appellian, equations can be employed with the replacements:

$$T \to T_{\text{rel}} \quad \text{and} \quad Q_k \to Q_k + Q_{k,inertial}; \tag{3.16.14a}$$

for example, it is not hard to see that $E_k(T_{\text{rel}}) \equiv \partial S_{\text{rel}}/\partial \ddot{q}_k$, where S_{rel} = part of S that depends solely on $t, q, \dot{q}, \ddot{q}$.

Additional Pfaffian constraints and/or nonholonomic coordinates can be easily handled using the methods described in §3.2–3.8.

(ii) If, on the other hand, the motion of $O\text{–}xyz$ is *not* known, then these equations should be solved together with the earlier ones of the carrying body. Then, we would have a system of $n + 6$ coupled equations of the *second order* in the n q's and of the *first order* in the 6 $v_{O;x,y,z}$ and $\omega_{x,y,z}$. To these we should also add the (linear and homogeneous) relations between the *holonomic* velocity components of $O\text{–}xyz$ relative to $I\text{–}XYZ$, say $\dot{q}_{1,\ldots,6}$ (*not* the q's of S relative to $O\text{–}xyz$), and their *nonholonomic* counterparts $\omega_{1,\ldots,6} \equiv v_{O;x,y,z}/\omega_{x,y,z}$.

(iii) Finally, if we had chosen *inertial (I)* Lagrangean coordinates, say $q_I = q_I(q_{NI}, t)$, where the *noninertial (NI)* coordinates are the earlier q's, then eqs. (3.16.2a) would become

$$\mathfrak{R} = \boldsymbol{r}_O + \boldsymbol{r}(q_{NI}) = \boldsymbol{r}_O + \boldsymbol{r}(t, q_I), \tag{3.16.14b}$$

that is, $\boldsymbol{r}$ would be *nonstationary*! In this case, the relative motion equations (3.16.13) would still hold, but with $T_{\rm rel} \equiv T_2$ replaced by $T_{2,2} + T_{2,1} + T_{2,0}$, where

$$2T_{2,2} \equiv \sum\sum M_{2,kl}\,\dot{q}_k\,\dot{q}_l, \qquad T_{2,1} \equiv \sum M_{2,k}\,\dot{q}_k, \qquad 2T_{2,0} \equiv M_{2,0}, \tag{3.16.14c}$$

$$M_{2,kl} \equiv S\, dm(\partial\boldsymbol{r}/\partial q_k)\cdot(\partial\boldsymbol{r}/\partial q_l), \qquad M_{2,k} \equiv S\, dm(\partial\boldsymbol{r}/\partial q_k)\cdot(\partial\boldsymbol{r}/\partial t),$$
$$M_{2,0} \equiv S\, dm(\partial\boldsymbol{r}/\partial t)\cdot(\partial\boldsymbol{r}/\partial t). \tag{3.16.14d}$$

Example 3.16.1 *Direct Derivation of the Lagrangean Equations of the Carried Body via LP.* Let us assume, for concreteness, that $O\text{–}xyz$ has a *prescribed* motion. Now, the Newton–Euler equation of motion of a typical particle P, of the carried system S, of mass dm and acted upon by a total *impressed* (*reaction*) force $d\boldsymbol{F}(d\boldsymbol{R})$ is (recalling §1.7)

$$dm\,\boldsymbol{a}_{\rm rel} = (d\boldsymbol{F} + d\boldsymbol{R}) + (d\boldsymbol{f}_O + d\boldsymbol{f}_T + d\boldsymbol{f}_C), \tag{a}$$

where

$d\boldsymbol{f} = d\boldsymbol{F} + d\boldsymbol{R}$: *Real (i.e., non–frame-dependent) force,*

$d\boldsymbol{f}_O \equiv -dm\,\boldsymbol{a}_O$: *Transport translational "force"*
(due to the inertial acceleration of the pole O),

$$d\boldsymbol{f}_T \equiv -dm[(d\boldsymbol{\Omega}/dt)\times\boldsymbol{r} + \boldsymbol{\Omega}\times(\boldsymbol{\Omega}\times\boldsymbol{r})]$$
$$= -dm((d\boldsymbol{\Omega}/dt)\times\boldsymbol{r}) - dm[(\boldsymbol{\Omega}\cdot\boldsymbol{r})\boldsymbol{\Omega} - \Omega^2\boldsymbol{r}]:$$

Transport rotational + centrifugal "force" (= purely centrifugal, if $\boldsymbol{\Omega}$ = constant; due to the inertial angular motion of $O\text{–}xyz$),

$d\boldsymbol{f}_C = -2\,dm\,(\boldsymbol{\Omega}\times\boldsymbol{v}_{\rm rel})$:

Coriolis "force" (due to the coupling of the relative motion of S with the inertial angular motion of $O\text{–}xyz$). (b)

[Incidentally, the above show that, in the most general case of motion,

$$dm(d'^2\boldsymbol{r}/dt^2) = d\boldsymbol{f} + (d\boldsymbol{f}_O + d\boldsymbol{f}_T + d\boldsymbol{f}_C) = \textit{function of } t, \boldsymbol{r}, \boldsymbol{v}_{\rm rel} \equiv d'\boldsymbol{r}/dt.]$$

Dotting (a) with $\delta\Re = \delta\boldsymbol{r}_O(t) + \delta\boldsymbol{r} = \delta\boldsymbol{r}$, and then summing over the system particles, yields LP for the carried body:

$$S\, dm\,\boldsymbol{a}_{\rm rel}\cdot\delta\boldsymbol{r} = S\,(d\boldsymbol{F} + d\boldsymbol{R})\cdot\delta\boldsymbol{r} + S\,(d\boldsymbol{f}_O + d\boldsymbol{f}_T + d\boldsymbol{f}_C)\cdot\delta\boldsymbol{r}. \tag{c}$$

Now:

(i) Since $\boldsymbol{a}_{\rm rel} = d'\boldsymbol{v}_{\rm rel}/dt = d'/dt(d'\boldsymbol{r}/dt)$, and during the above virtual variations the axes $O\text{–}xyz$ are held fixed—that is, $\delta\boldsymbol{r} = \delta'\boldsymbol{r} = \sum(\partial\boldsymbol{r}/\partial q_k)\,\delta q_k$—and the q_k are *noninertial* coordinates of S relative to $O\text{–}xyz$, reasoning as in (3.3.3 ff.), we readily conclude that

$$S\, dm\,\boldsymbol{a}_{\rm rel}\cdot\delta\boldsymbol{r} = \sum E_k(T_{\rm rel})\,\delta q_k \equiv \sum\left[(\partial T_{\rm rel}/\partial\dot{q}_k)^{\cdot} - \partial T_{\rm rel}/\partial q_k\right]\delta q_k, \tag{d}$$

where

$$2T_{\text{rel}} = 2T_{\text{rel}}(q, \dot{q}) = S\, dm\, \boldsymbol{v}_{\text{rel}} \cdot \boldsymbol{v}_{\text{rel}}. \tag{e}$$

(ii) The Q_k are defined, as usual, by (recalling §3.4)

$$S\, d\boldsymbol{F} \cdot \delta \boldsymbol{r} = \sum Q_k\, \delta q_k. \tag{f}$$

(iii) The (relative = inertial) virtual work of the constraint reactions vanishes:

$$S\, d\boldsymbol{R} \cdot \delta \boldsymbol{r} = S\, d\boldsymbol{R} \cdot \delta' \boldsymbol{r} = \sum R_k\, \delta q_k = 0. \tag{g}$$

(iv) We define the following fictitious Lagrangean forces:

$$S\, d\boldsymbol{f}_O \cdot \delta \boldsymbol{r} \equiv -S\, dm\, \boldsymbol{a}_O \cdot \delta \boldsymbol{r}$$
$$\equiv \sum Q_{k,translational\ \text{transport}}\, \delta q_k \equiv \sum Q_{k,T}\, \delta q_k, \tag{h}$$
$$S\, d\boldsymbol{f}_T \cdot \delta \boldsymbol{r} \equiv -S\, dm((d\boldsymbol{\Omega}/dt) \times \boldsymbol{r}) \cdot \delta \boldsymbol{r} - \boldsymbol{\Omega} \cdot \left(S\, dm(\boldsymbol{\Omega} \cdot \boldsymbol{r})\, \delta \boldsymbol{r}\right) + \Omega^2 \left(S\, dm\, \boldsymbol{r} \cdot \delta \boldsymbol{r}\right)$$
$$\equiv \sum Q_{k,rotational+centrifugal\ \text{transport}}\, \delta q_k \equiv \sum (Q_{k,R} + Q_{k,CF})\, \delta q_k, \tag{i}$$
$$S\, d\boldsymbol{f}_C \cdot \delta \boldsymbol{r} \equiv -2 S\, dm(\boldsymbol{\Omega} \times \boldsymbol{v}_{\text{rel}}) \cdot \delta \boldsymbol{r}$$
$$\equiv \sum Q_{k,\text{Coriolis/gyroscopic}}\, \delta q_k \equiv \sum Q_{k,C}\, \delta q_k. \tag{j}$$

Now, since the motion of O–xyz is prescribed, $\boldsymbol{a}_O$, $\boldsymbol{\Omega}$, and $d\boldsymbol{\Omega}/dt$ are given functions of time, and, therefore, the "forces" $Q_{k;T,R,CF}$ will be functions of t and the q's, while the $Q_{k,C}$ will be functions of t, q's, and $\dot{q}$'s. If the δq's are independent—that is, if S is *unconstrained relative to* (*the constrained*) O–xyz—then (c), with (d–j), yield the earlier equations (3.16.13):

$$E_k(T_{\text{rel}}) = Q_k + Q_{k,T} + Q_{k,R} + Q_{k,CF} + Q_{k,C}, \tag{k}$$

where $Q_{k,T} = -\partial V_T / \partial q_k$, $Q_{k,CF} = -\partial V_{CF} / \partial q_k$, $Q_{k,C} = G_k$.

If the δq are *constrained*, then we proceed in, by now, well-known ways; that is, either we *adjoin* the constraints to (c ff.) via *multipliers*, or *embed* them via relative *quasi variables*. Let the reader work out the details of this direct approach if the motion of O–xyz is also unknown.

Example 3.16.2 *Cartesian Tensor Derivation of the Lagrangean Equations of the Carried Body.* Let us consider, without much loss in generality, two frames/axes with common origin: an inertial/fixed O–$x_{k'}$, and a noninertial/moving (rotating) O–x_k ($k = 1, 2, 3$; $k' = 1', 2', 3'$). As is well known (§1.1), these two sets of axes are related by the following *orthogonal* transformation:

$$x_{k'} = \sum A_{k'k} x_k \Leftrightarrow x_k = \sum A_{kk'} x_{k'}, \tag{a}$$

where $\{A_{k'k} \equiv \cos(x_{k'}, x_k) = \cos(x_k, x_{k'}) \equiv A_{kk'} = \textit{pure function of time}\}$ is a *proper orthogonal* tensor. The inertial Lagrangean function and corresponding equations of motion of a particle P of unit mass (i.e., $m = 1$, for analytical simplicity) and under ordinary potential forces only (here we are interested in kinematical/frame of reference effects—nonpotential forces can always be added later) are, respectively,

$$L = (1/2)\left(\sum \dot{x}_{k'}\dot{x}_{k'}\right) - V(x_{k'}, t) = L(t, x_{k'}, \dot{x}_{k'}), \tag{b}$$

$$(\partial L/\partial \dot{x}_{k'})^{\cdot} - \partial L/\partial x_{k'} = 0: \quad (\dot{x}_{k'})^{\cdot} + \partial V/\partial x_{k'} = 0. \tag{c}$$

Next, using (a), we will express L in terms of $\{t, x_k, \dot{x}_k, A_{k'k}, \dot{A}_{k'k}\}$ and then, using the frame invariance of the Lagrangean operator/equations *in these new variables*, we will obtain the equations of relative motion of P: $(\partial L/\partial \dot{x}_k)^{\cdot} - \partial L/\partial x_k = 0$. Indeed, $(\ldots)^{\cdot}$-differentiating (a), we obtain

$$\dot{x}_{k'} = \sum (\dot{A}_{k'k}x_k + A_{k'k}\dot{x}_k), \tag{d}$$

and, therefore, successively,

$$\sum \dot{x}_{k'}\dot{x}_{k'} = \sum \left(\sum (\dot{A}_{k'k}x_k + A_{k'k}\dot{x}_k)\right)\left(\sum (\dot{A}_{k'l}x_l + A_{k'l}\dot{x}_l)\right)$$
$$= \sum\sum\sum (A_{k'k}A_{k'l}\dot{x}_k\dot{x}_l + \dot{A}_{k'k}\dot{A}_{k'l}x_k x_l + 2\dot{A}_{k'k}A_{k'l}x_k\dot{x}_l), \tag{e}$$

or, since [(1.7.22a ff.)], $\sum A_{k'k}A_{k'l} = \delta_{kl}$ and $\dot{A}_{k'k} = \sum A_{k'l}\omega_{lk}$, where

$\sum \dot{A}_{k'k}A_{k'l} = \sum \dot{A}_{kk'}A_{lk'} \equiv \omega_{lk} = -\omega_{kl}$:

(inertial) angular velocity tensor of moving axes relative to fixed axes, but resolved along the moving axes, (f)

equation (e) transforms further to

$$\sum \dot{x}_{k'}\dot{x}_{k'} = \sum\sum \delta_{kl}\dot{x}_k\dot{x}_l + \sum\sum\sum \left(\sum A_{k'r}\omega_{rk}\right)\left(\sum A_{k's}\omega_{sl}\right)x_k x_l$$
$$+ 2\sum\sum \omega_{lk}\dot{x}_l x_k, \tag{g}$$

and so, finally, the moving axes Lagrangean becomes

$$L = \sum (1/2)\dot{x}_k\dot{x}_k + \sum\sum\sum (1/2)\,\omega_{sk}\,\omega_{sl}\,x_k\,x_l + \sum\sum \omega_{lk}\,\dot{x}_l\,x_k - V(t, x_k). \tag{h}$$

Now:

- The first term $\sum (1/2)\dot{x}_k\dot{x}_k$, will give rise to the *relative* acceleration;
- The second term, $\sum\sum\sum (1/2)\omega_{sk}\omega_{sl}x_k x_l$: *centrifugal potential*, will give rise to the acceleration of *transport*; and
- The third term, $\sum\sum \omega_{lk}\dot{x}_l x_k$: *Schering potential*, will give rise to the *Coriolis* acceleration.

Indeed, from (h) we obtain

$$\partial L/\partial \dot{x}_k = \dot{x}_k + \sum\sum \omega_{lr}\delta_{lk}x_r = \dot{x}_k + \sum \omega_{kr}x_r, \tag{i}$$

$$\Rightarrow (\partial L/\partial \dot{x}_k)^{\cdot} = \ddot{x}_k + \sum (\dot{\omega}_{kr}x_r + \omega_{kr}\dot{x}_r), \tag{j}$$

$$\partial L/\partial x_k = \sum\sum \omega_{sk}\omega_{sl}x_l + \sum \omega_{rk}\dot{x}_r - \partial V/\partial x_k$$
$$= \sum\sum \omega_{sk}\omega_{sl}x_l - \sum \omega_{kr}\dot{x}_r - \partial V/\partial x_k, \tag{k}$$

and, therefore, the Lagrangean equations of relative motion, $(\partial L/\partial \dot{x}_k)^{\cdot} - \partial L/\partial x_k = 0$, become

$$\ddot{x}_k - \sum\sum \omega_{sk}\,\omega_{sl}\,x_l + \sum \dot{\omega}_{kl} x_l + 2\sum \omega_{kl}\dot{x}_l = -\partial V/\partial x_k. \tag{l}$$

Let the reader show that (l) is none other than [with $\boldsymbol{\omega} = (\omega_k)$, $\boldsymbol{\alpha} = (\dot{\omega}_k)$, $\boldsymbol{r} = (x_k)$, $\boldsymbol{v}_{\text{rel}} = (\dot{x}_k)$]

$$\ddot{x}_k + [\boldsymbol{\omega} \times (\boldsymbol{\omega} \times \boldsymbol{r})]_k + (\boldsymbol{\alpha} \times \boldsymbol{r})_k + (2\boldsymbol{\omega} \times \boldsymbol{v}_{\text{rel}})_k = -\partial V/\partial x_k. \tag{m}$$

HINT

Recall that (1.1.16a ff.) $(\boldsymbol{\omega} \cdot \boldsymbol{r})_k = (\boldsymbol{\omega} \times \boldsymbol{r})_k$: $\sum \omega_{kl} x_l = \sum\sum \varepsilon_{ksl}\,\omega_s\,x_l$; that is, $\boldsymbol{\omega} = (\omega_{kl} = -\omega_{lk} = \sum \varepsilon_{ksl}\,\omega_s)$. See also Morgenstern and Szabó (1961, pp. 7–9).

Problem 3.16.1 Extend the tensorial method of the preceding example to the most general case of relative motion (i.e., no common origin of relatively moving frames):

$$x_{k'} = \sum A_{k'k} x_k + b_{k'} \Leftrightarrow x_k = \sum A_{kk'} x_{k'} + b_k, \tag{a}$$

where $b_{k'} = -\sum A_{k'k} b_k$ = components of position vector of moving origin O relative to fixed origin, along the *fixed* axes; $b_k = -\sum A_{kk'}, b_{k'}$; and $\{A_{k'k}, b_{k'}, b_k\}$ = *pure functions of time*. This will give rise to "forces" due to the inertial motion ("transport") of the origin O.

Example 3.16.3 *Direct Lagrangean Treatment of Gyroscopic Couple.* Let us consider an axisymmetric (carried) body, spinning with angular velocity $\boldsymbol{\omega}_o$ about its axis of symmetry. The latter is fixed relative to the body's "housing" (carrying body). It is shown in gyrodynamics that $\boldsymbol{\omega}_o$ gives rise to an additional moment, or "gyroscopic couple," *on the housing + fixed (nonspinning) gyro* system equal to $(C\boldsymbol{\omega}_o) \times \boldsymbol{\Omega}$, where C = *moment of inertia of carried body about its spinning axis*, and $\boldsymbol{\Omega}$ = *inertial angular velocity of housing*. Let us find the Lagrangean expression of that couple.

Here,

$$\boldsymbol{\Omega} = \sum (\partial\boldsymbol{\Omega}/\partial\dot{q}_k)\dot{q}_k + \partial\boldsymbol{\Omega}/\partial t \Rightarrow \delta\boldsymbol{X} = \sum (\partial\boldsymbol{\Omega}/\partial\dot{q}_k)\,\delta q_k, \tag{a}$$

where $d\boldsymbol{X}/dt \equiv \boldsymbol{\Omega}$. Therefore, the virtual work of the gyro-couple equals, successively,

$$\begin{aligned}
\delta' W_G &\equiv [(C\boldsymbol{\omega}_o) \times \boldsymbol{\Omega}] \cdot \delta\boldsymbol{X} \\
&= \Big\{(C\boldsymbol{\omega}_o) \times \Big(\sum (\partial\boldsymbol{\Omega}/\partial\dot{q}_l)\dot{q}_l + \partial\boldsymbol{\Omega}/\partial t\Big)\Big\} \cdot \Big(\sum (\partial\boldsymbol{\Omega}/\partial\dot{q}_k)\,\delta q_k\Big) \\
&\equiv \sum\Big(\sum g_{kl}\dot{q}_l + g_k\Big)\delta q_k \equiv \sum (G_k + g_k)\,\delta q_k,
\end{aligned} \tag{b}$$

where

$$\begin{aligned} g_{kl} &\equiv [(C\boldsymbol{\omega}_o) \times (\partial\boldsymbol{\Omega}/\partial\dot{q}_l)] \cdot (\partial\boldsymbol{\Omega}/\partial\dot{q}_k) \\ &= (C\boldsymbol{\omega}_o) \cdot [(\partial\boldsymbol{\Omega}/\partial\dot{q}_l) \times (\partial\boldsymbol{\Omega}/\partial\dot{q}_k)] \equiv (C\boldsymbol{\omega}_o) \cdot \boldsymbol{G}_{kl} = -g_{lk}, \qquad \text{(c)} \\ g_k &\equiv [(C\boldsymbol{\omega}_o) \times (\partial\boldsymbol{\Omega}/\partial t)] \cdot (\partial\boldsymbol{\Omega}/\partial\dot{q}_k) \\ &= (C\boldsymbol{\omega}_o) \cdot [(\partial\boldsymbol{\Omega}/\partial t) \times (\partial\boldsymbol{\Omega}/\partial\dot{q}_k)] \equiv (C\boldsymbol{\omega}_o) \cdot \boldsymbol{G}_k \quad (\equiv g_{k,n+1} \equiv g_{k,t}). \qquad \text{(d)} \end{aligned}$$

Clearly, $\sum G_k\dot{q}_k = \sum\sum g_{kl}\dot{q}_l\dot{q}_k = 0$; that is, the G_k are gyroscopic. From the above, it follows that, if there are no further constraints, the equations of motion of the system *housing* + *gyro* are

$$(\partial T_A/\partial\dot{q}_k)^{\cdot} - \partial T_A/\partial q_k = Q_k + G_k + g_k, \qquad \text{(e)}$$

where T_A = (*inertial*) *kinetic energy of system housing* + *gyro if* $\boldsymbol{\omega}_o = \mathbf{0}$ (i.e., with the gyro held fixed in its housing) ≡ *apparent kinetic energy*. The above can be easily extended to a system consisting of several housings and gyros. [See Cabannes, 1965, pp. 201–203, 274–277; Roseau, 1987, pp. 49–53; also §8.4 ff., and Papastavridis (EM, in prep., examples on gyrodynamics).]

Example 3.16.4 *Rotating Frames: The Free Particle.* Here, using Lagrangean methods, we derive the equations of plane motion of a particle P of mass m on a frame O–xyz rotating with (inertial) angular velocity $\boldsymbol{\Omega} = (0, 0, \Omega)$ relative to an inertial one O–XYZ; and such that $OZ \equiv Oz$ (fig. 3.37). By $(\ldots)^{\cdot}$-differentiating the well-known transformation equations between these two frames

$$X = x\cos\phi - y\sin\phi, \qquad Y = x\sin\phi + y\cos\phi, \qquad \text{(a)}$$

and recalling that $\Omega = \dot{\phi}$, we obtain

$$\begin{aligned} \dot{X} &= (\dot{x}\cos\phi - \dot{y}\sin\phi) - (x\sin\phi + y\cos\phi)\Omega, \\ \dot{Y} &= (\dot{x}\sin\phi + \dot{y}\cos\phi) + (x\cos\phi - y\sin\phi)\Omega. \qquad \text{(b)} \end{aligned}$$

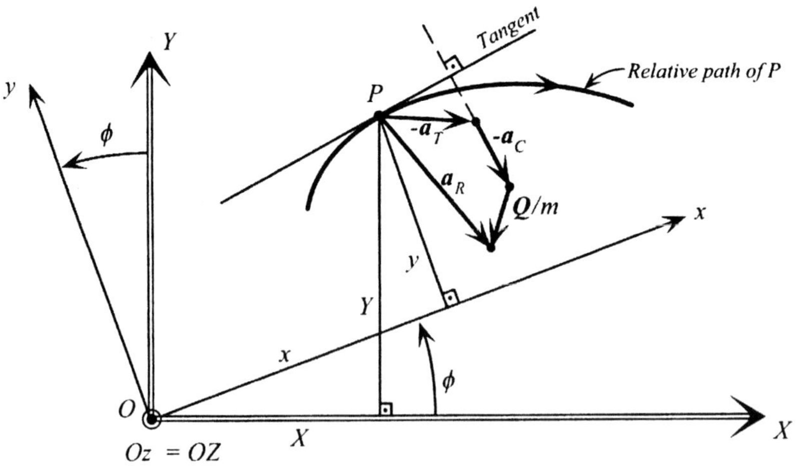

Figure 3.37 Particle P moving on a rotating frame O–xy (z).

To understand better the meaning of (b), we consider the special instant at which the axes of these two frames coincide; that is, for $\phi = 0$. Then, (b) yields

$$\dot{X} = \dot{x} - y\,\Omega, \qquad \dot{Y} = \dot{y} + x\,\Omega, \tag{c}$$

that is, $\dot{X} \neq \dot{x}$ and $\dot{Y} \neq \dot{y}$, even though, then, $X = x$ and $Y = y$; eq. (c) is the $O\text{–}xy/XY$ component form of the well-known vector equation (§1.7)

$$d\boldsymbol{r}/dt = d'\boldsymbol{r}/dt + \boldsymbol{\Omega} \times \boldsymbol{r}, \qquad \text{where } \boldsymbol{r} = (x, y, 0),\ \boldsymbol{\Omega} = (0, 0, \Omega). \tag{d}$$

Thanks to (c), the inertial (double) kinetic energy of P becomes

$$2T = m[(\dot{X})^2 + (\dot{Y})^2] \equiv 2(T_2 + T_1 + T_0), \tag{e}$$

where

$$\begin{aligned} 2T_2 &\equiv m[(\dot{x})^2 + (\dot{y})^2], \\ T_1 &\equiv m(x\dot{y} - \dot{x}y)\Omega, \\ 2T_0 &\equiv m(x^2 + y^2)\Omega^2 = m(X^2 + Y^2)\Omega^2 = m\,r^2\Omega^2, \end{aligned} \tag{f}$$

and, therefore, if $\delta'W = Q_x\,\delta x + Q_y\,\delta y$, Lagrange's equations here are

$$[m(\dot{x} - y\,\Omega)]^{\cdot} - m(\dot{y} + x\,\Omega)\Omega = m(\ddot{x} - 2\dot{y}\,\Omega - x\,\Omega^2 - y\,\dot{\Omega}) = Q_x, \tag{g}$$

$$[m(\dot{y} + x\,\Omega)]^{\cdot} - m(\dot{x} - y\,\Omega)\Omega = m(\ddot{y} + 2\dot{x}\,\Omega - y\,\Omega^2 + x\,\dot{\Omega}) = Q_y; \tag{h}$$

and, from these, the following Newton–Euler (*O–xy-centric*) forms result:

$$m\,\ddot{x} = Q_x - m(-x\,\Omega^2 - y\,\dot{\Omega}) - (-2m\,\dot{y}\,\Omega), \tag{i}$$

$$m\,\ddot{y} = Q_y - m(-y\,\Omega^2 + x\,\dot{\Omega}) - (+2m\,\dot{x}\,\Omega), \tag{j}$$

where (in two dimensions)

$$\boldsymbol{a}_T = (-x\,\Omega^2 - y\,\dot{\Omega}, -y\,\Omega^2 + x\,\dot{\Omega}) = \textit{Transport acceleration (normal + tangent)}, \tag{k}$$

$$\boldsymbol{a}_C = (-2\dot{y}\,\Omega, +2\dot{x}\,\Omega) = \textit{Coriolis acceleration}. \tag{l}$$

Since $\boldsymbol{a}_C$ is perpendicular to $\boldsymbol{v}_{\text{rel}} = (\dot{x}, \dot{y})$, if (i) $\dot{\Omega} = 0$ and (ii) $\delta'W = -\delta V(x, y)$, eqs. (g, h/i, j) readily combine to produce the *relative power equation*

$$m(\dot{x}\,\ddot{x} + \dot{y}\,\ddot{y}) - m(x\,\dot{x} + y\,\dot{y})\Omega^2 = -\dot{V}; \tag{m}$$

and the latter integrates easily to the *generalized energy (Jacobi–Painlevé) integral* [recalling (3.9.11n)]

$$T_2 + (V - T_0) \equiv h = \textit{constant}\ (\neq E \equiv T + V). \tag{n}$$

Example 3.16.5 *Rotating Frames: General System.* Here, we extend the preceding example to a general system. (Although the general theory of such systems in moving axes has already been studied in this section, nevertheless, we think that the ad hoc treatment of this special but important case, presented below, is quite instructive.) Summing (e, f) over the entire system, we readily find

$$2T = S\left[(\dot{X})^2 + (\dot{Y})^2\right]dm \equiv 2(T_2 + T_1 + T_0)$$
$$\equiv 2\left[T_{\text{rel}} + \Omega\, H_{O,\text{rel}} + (1/2)\Omega^2 I_O\right] = 2\left[T_{\text{rel}} + \Omega\, H_O - (1/2)\Omega^2 I_O\right], \qquad \text{(a)}$$

where

$$2T_2 \equiv 2T_{\text{rel}} \equiv S\left[(\dot{x})^2 + (\dot{y})^2\right]dm$$
$$= 2(\textit{kinetic energy relative to rotating frame}) \qquad (\text{i.e., } T \text{ for } \Omega = 0), \qquad \text{(b)}$$
$$T_1 \equiv \Omega\, S\,(x\,\dot{y} - y\,\dot{x})\,dm \equiv \Omega\, H_{O,\text{rel}}, \qquad \text{(c)}$$
$$2T_0 \equiv \Omega^2 S\,(x^2 + y^2)\,dm = \Omega^2 S\,(X^2 + Y^2)dm = \Omega^2 I_O$$
$$= 2(\textit{centrifugal energy}); \qquad \text{(d)}$$
$$H_{O,\text{rel}} \equiv S\,(x\,\dot{y} - y\,\dot{x})\,dm = \textit{angular momentum} \text{ about } OZ \equiv Oz \text{ and relative to the } \textit{rotating} \text{ frame}, \qquad \text{(e)}$$
$$H_O \equiv S\,(X\dot{Y} - Y\dot{X})\,dm = S\,[x(\dot{y} + x\,\Omega) - y(\dot{x} - y\,\Omega)]\,dm$$
$$= \cdots = H_{O,\text{rel}} + \Omega\, I_O = \textit{angular momentum} \text{ about } OZ \equiv Oz \text{ and relative to the } \textit{fixed} \text{ frame}, \qquad \text{(f)}$$
$$I_O \equiv S\,(X^2 + Y^2)\,dm = S\,(x^2 + y^2)\,dm = S\,r^2\,dm$$
$$= \textit{moment of inertia about } OZ \equiv Oz\ (\textit{frame independent}); \qquad \text{(g)}$$
$$(\Rightarrow\ \Omega\, H_O = \Omega\, H_{O,\text{rel}} + \Omega^2 I_O \ = \ T_1 + 2T_0).$$

[In the general three-dimensional case, since $\dot{Z} = \dot{z}$, a $(1/2)(\dot{z})^2 dm$ term must be added to the integrand of T and T_{rel}.]

Now, if the system is completely describable, relative to the rotating frame, by n Lagrangean coordinates $q = (q_1, \ldots, q_n)$ [i.e., if $x = x(q)$, $y = y(q)$, $z = z(q)$], then T_{rel} = *quadratic* and *homogeneous* in the $\dot{q}$'s, with coefficients functions of the q's. If, further, $T = T(q, \dot{\phi} \equiv \Omega)$—that is, $q_{n+1} = \phi$ = additional "*azimuthal*" cyclic coordinate (§ 8.4), for the complete *inertial description* of the system—and $Q_k = -\partial V/\partial q_k$, $Q_{n+1} \equiv M_O$ = *total impressed moment about* $OZ \equiv Oz$, and no further constraints are present, then the $n + 1$ Lagrangean equations for these q's are

$$(\partial T/\partial \dot{q}_k)^{\cdot} - \partial T/\partial q_k = -\partial V/\partial q_k, \qquad \text{(h)}$$

and [by (a), and noting that, equivalently,

$$\partial T/\partial \Omega = H_{O,\text{rel}} + \Omega I_O = H_O + \Omega(\partial H_O/\partial \Omega) - \Omega I_O = H_O + \Omega(I_O) - \Omega I_O = H_O]$$
$$(\partial T/\partial \Omega)^{\cdot} - \partial T/\partial \phi = \dot{H}_O = M_O. \qquad \text{(i)}$$

If $M_O = 0$ (*free* rotating system), then by eq. (i), $H_O = 0$ and $\Omega \neq$ *constant*; whereas if Ω = *constant* (*constrained* rotation), then M_O = *constraint reaction* (*a Lagrangean multiplier*) $\neq 0$, and therefore $H_O \neq$ *constant*.

Special Case

If $\Omega = constant$ (e.g., O–$xyz \to$ Earth), substituting eq. (a ff.) into eq. (h), we obtain

$$(\partial T_{\text{rel}}/\partial \dot{q}_k)^{\cdot} - \partial T_{\text{rel}}/\partial q_k + \Omega\,[(\partial H_{O,\text{rel}}/\partial \dot{q}_k)^{\cdot} - \partial H_{O,\text{rel}}/\partial q_k] \\ - (1/2)\Omega^2(\partial I_O/\partial q_k) = -\partial V/\partial q_k. \qquad \text{(j)}$$

But, since $\partial \dot{x}/\partial \dot{q}_k = \partial x/\partial q_k$ and $\partial \dot{x}/\partial q_k = (\partial x/\partial q_k)^{\cdot}$ [i.e., $E_k(\dot{x}) = 0$, etc.], we have

(i) $\partial H_{O,\text{rel}}/\partial \dot{q}_k = S\,[x(\partial \dot{y}/\partial \dot{q}_k) - y(\partial \dot{x}/\partial \dot{q}_k)]\,dm = S\,[x(\partial y/\partial q_k) - y(\partial x/\partial q_k)]\,dm,$

and, from this,

$$(\partial H_{O,\text{rel}}/\partial \dot{q}_k)^{\cdot} = S\,[\dot{x}(\partial y/\partial q_k) - \dot{y}(\partial x/\partial q_k)]\,dm + S\,[x(\partial \dot{y}/\partial q_k) - y(\partial \dot{x}/\partial q_k)]\,dm;$$

(ii) $\partial H_{O,\text{rel}}/\partial q_k = S\,[\dot{y}(\partial x/\partial q_k) - \dot{x}(\partial y/\partial q_k)]\,dm + S\,[x(\partial \dot{y}/\partial q_k) - y(\partial \dot{x}/\partial q_k)]\,dm.$

Therefore, subtracting these two expressions side by side, we find

$$\begin{aligned}(\partial H_{O,\text{rel}}/\partial \dot{q}_k)^{\cdot} &- \partial H_{O,\text{rel}}/\partial q_k \\ &= 2S\,[\dot{x}\,(\partial y/\partial q_k) - \dot{y}\,(\partial x/\partial q_k)]\,dm \\ &= 2S\left\{\left(\sum (\partial x/\partial q_l)\dot{q}_l\right)(\partial y/\partial q_k) - \left(\sum (\partial y/\partial q_l)\dot{q}_l\right)(\partial x/\partial q_k)\right\}dm \\ &\equiv \sum G_{lk}\,\dot{q}_l, \qquad \text{(k)}\end{aligned}$$

where

$$\begin{aligned}G_{lk} &\equiv 2S\,[(\partial x/\partial q_l)(\partial y/\partial q_k) - (\partial x/\partial q_k)(\partial y/\partial q_l)]\,dm \\ &\equiv 2S\,[(\partial(x, y)/\partial(q_l, \partial q_k)]\,dm = -G_{kl} \\ &\quad \text{(analytically known, once the } q\text{'s are chosen).} \qquad \text{(l)}\end{aligned}$$

In view of (k, l), eq. (j) can be rewritten in the definitive form

$$E_k(T_{\text{rel}}) + \Omega \sum G_{lk}\dot{q}_l = -\partial V_{\text{rel}}/\partial q_k; \qquad \text{(m)}$$

where

$$V_{\text{rel}} \equiv V - (1/2)\Omega^2 I_O = \textit{total relative potential} \quad (= V - T_0). \qquad \text{(m1)}$$

Other possible *nonpotential* and *noninertial* forces, such as friction and/or constraints, can be added to the *right* side of (m) as Q's and/or multiplier-proportional terms.

The Lagrangean form (m) brings out clearly the *differences* between uniformly rotating axes ($\Omega = constant \neq 0$) and inertial ones ($\Omega = 0$). These are:

- The additional *centrifugal potential* $V_{CF} = -\Omega^2 I_O/2$ [recalling (3.16.12f)], which gives rise to the *centrifugal "force"* $-\partial V_{CF}/\partial q_k = (1/2)\Omega^2(\partial I_O/\partial q_k)$; and
- The additional *gyroscopic* (or *compounded centrifugal*, or *Coriolis*), and generally *nonpotential, "forces"* $-\Omega \sum G_{lk}\dot{q}_l = \Omega \sum G_{kl}\dot{q}_l$.

It is not hard to see that, in the absence of nonpotential forces, the equations of motion of the particle in ex. 3.16.4, eqs. (i, j), can be rewritten, respectively, in the (m)–like form

$$m\ddot{x} = -\partial V_{\text{rel}}/\partial x + 2m\Omega\dot{y}, \qquad m\ddot{y} = -\partial V_{\text{rel}}/\partial y - 2m\Omega\dot{x}, \tag{n}$$

(also $m\ddot{z} = -\partial V_{\text{rel}}/\partial z$), $V_{\text{rel}} \equiv V - (1/2)m\Omega^2(x^2 + y^2)$; or, vectorially (2 dimensions),

$$m(d'^2\mathbf{r}/dt^2) = -\boldsymbol{grad}\ V_{\text{rel}} + 2m(d'\mathbf{r}/dt) \times \boldsymbol{\Omega}. \tag{n1}$$

Equations (n, n1) are useful in atmospheric physics ($\Omega =$ *rotation of Earth*).]

Relative Equilibrium

This is defined as the special motion for which

$$T_{\text{rel}} = 0 \qquad \text{and all} \qquad \dot{q}_k = 0. \tag{o}$$

Then, eqs. (m) yield the following *conditions for relative equilibrium*:

$$\partial V_{\text{rel}}/\partial q_k \equiv \partial\big(V - \tfrac{1}{2}\Omega^2 I_O\big)/\partial q_k = 0; \tag{p}$$

namely, that V_{rel} be stationary.

For further discussion, especially the case of (small) motion around such equilibria, and their stability, and applications, see, for example (alphabetically): Appell (vol. 4, 1932, 1937), Duhem (1911, pp. 422-499), Lamb (1932, pp. 195–199, 307–310, 427–428, 713–714; 1943, pp. 244–255), Ledoux (1958, pp. 616–620), Lyttleton (1953, pp. 19–30); also §8.6 in this volume.

Problem 3.16.2 *Rotating Frames.* Continuing from the preceding example, show that:

(i) In terms of *corotating* (i.e., noninertial) polar coordinates (r, θ), for each particle,

$$\begin{aligned} 2T &= S\Big[(\dot{r})^2 + r^2(\Omega + \dot{\theta})^2\Big]\,dm \\ &= S\Big[(\dot{r})^2 + r^2(\dot{\theta})^2\Big]\,dm + 2\Omega\Big(S\, r^2\dot{\theta}\,dm\Big) + \Omega^2\Big(S\, r^2\,dm\Big); \end{aligned} \tag{a}$$

(ii)
$$H_{O,\text{rel}} \equiv S\,(x\dot{y} - y\dot{x})\,dm = S\,(r^2\dot{\theta})\,dm = \cdots = \sum \beta_k\,\dot{q}_k, \tag{b}$$

where

$$\beta_k \equiv S\,[x(\partial y/\partial q_k) - y(\partial x/\partial q_k)]\,dm = \beta_k(q);$$

then show that

$$G_{lk} \equiv \partial\beta_k/\partial q_l - \partial\beta_l/\partial q_k = -G_{kl} \qquad (k, l = 1, \ldots, n). \tag{c}$$

Problem 3.16.3 *Rotating Frames: Special Cases.* Continuing from the preceding problem, show that (i) if $\dot{\theta} = 0$, or (ii) if $n = 1$ (i.e., if the system is described on the uniformly rotating frame by only *one* coordinate q), then the $\omega/\dot{q}$-proportional (gyroscopic) terms, in the corresponding equations of motion, vanish; that is, then, the effect of the rotation Ω there is only the centrifugal force $(1/2)\Omega^2(\partial I_O/\partial q)$.

HINT

In (ii) $H_{O,\text{rel}} = (\textit{some function of } q)\ \dot{q}$; then $E(H_{O,\text{rel}}) = \cdots$.

Problem 3.16.4 *Rotating Frames: A Special Power Equation.* Continuing from the preceding example, and employing its notations, show that the power equation of a system moving on a *uniformly* rotating frame is

$$d/dt(T_{\text{rel}} + V - \Omega^2 I_O/2) \equiv (T_{\text{rel}} + V_{\text{rel}})^{\cdot} = \sum Q_k \dot{q}_k, \qquad \text{(a)}$$

where Q_k = *nonpotential* and *noninertial* forces; and, therefore, if all $Q_k = 0$, eq. (a) leads to the *Jacobi–Painlevé* integral

$$T_{\text{rel}} + V - \Omega^2 I_O/2 = 0. \qquad \text{(b)}$$

HINTS

We have

$$dT/dt = (T_2 + T_0)^{\cdot} + \Omega\Big[S\,(x\ddot{y} - y\ddot{x})\,dm\Big]$$

[by the results of ex. 3.16.4 and ex. 3.16.5]

$$= (T_2 - T_0)^{\cdot} + \Omega\Big[S\,(x\,dF_y - y\,dF_x)\Big]$$

[by eqs. (g, h / i, j) of ex. 3.16.4, summed over the entire system, with $\Omega = \textit{constant}$ and $(Q_x, Q_y) \to \boldsymbol{dF} = (dF_x, dF_y) = \textit{total impressed force}$ on particle of mass dm]. (c)

But, also, by the "elementary" power theorem (since, here, *impressed forces = external forces*), $dT/dt = \textit{Total externally supplied power}$

$$= \Omega\Big[S\,(x\,dF_y - y\,dF_x)\Big] + \sum Q_k \dot{q}_k. \qquad \text{(d)}$$

Problem 3.16.5 *Rotating Frames: A Special Power Equation (continued).* Continuing from the preceding example, and employing its notations, show that:

(i) If $M_O = 0$, then $H_O = \textit{constant}$; and
(ii) If, further, we choose Ω so that (always) $H_{O,\text{rel}} = 0$, then

$$H_O = \Omega I_O \quad \text{and} \quad T = \cdots = T_{\text{rel}} + {H_O}^2/2I_O; \qquad \text{(a)}$$

and thus deduce that if all the (nonpotential) Q_k's vanish, the (inertial) energy equation $E \equiv T + V = \textit{constant}$ specializes to

$$T_{\text{rel}} + V + {H_O}^2/2I_O = \textit{constant}; \qquad \text{(b)}$$

that is, eq. (b) of the preceding problem but with $\Omega \to H_O$ and $-\Omega^2 I_O/2 \to {H_O}^2/2I_O$.

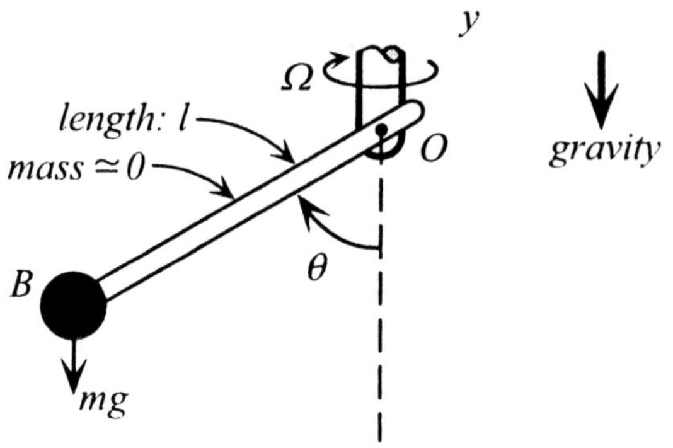

Figure 3.38 Uniformly rotating plane pendulum.

Problem 3.16.6 *Rotating Pendulum.* Consider a mathematical pendulum, of length l and mass m, whose plane of oscillation is constrained to rotate with constant inertial angular velocity Ω about its vertical axis (fig. 3.38).

(i) Show that its equations of (relative angular) motion and energy are, respectively,

$$\ddot{\theta} + (g/l)\sin\theta - (1/2)\Omega^2 \sin(2\theta) = 0, \tag{a}$$

$$(ml^2/2)\left[(\dot{\theta})^2 - 2(g/l)\cos\theta - \Omega^2 \sin^2\theta\right] \equiv h = constant. \tag{b}$$

Explain the absence of gyroscopic terms in both (a) and (b).

(ii) Linearize (a) in θ and then show that if $\Omega^2 < g/l$, the pendulum performs harmonic oscillations about the vertical with period $\tau = 2\pi[(g/l) - \Omega^2]^{-1/2}$; that is, the configuration of relative equilibrium $\theta = 0$ is *stable*; whereas, if $\Omega^2 \geq g/l$, it is *unstable* (i.e., we need the nonlinear equation).

(iii) Next, add $\dot{\theta}$-proportional (small) friction $-f\dot{\theta}$ *(f = constant friction coefficient)*. Does this affect the stability/instability of $\theta = 0$? Explain.

For an alternative discussion, see Kauderer (1958, pp. 239–242); and for discussions of the stability of the equilibrium configurations of (a), based on (b), and more see e.g. (alphabetically): Babakov (1968, pp. 463–467), Greenwood (1977, pp. 62, 74–77), Pars (1965, pp. 85–86).

Problem 3.16.7 *Rotating Frames: Carrying Body Effect.* Consider a particle P of mass m in unconstrained motion relative to a carrying rigid body B. The latter can spin freely about the fixed vertical axis Oz.

(i) Show that the inertial kinetic energy of the entire system "$B + P$," expressed in terms of components along B-fixed axes O–xyz, is

$$2T = m[(\dot{x})^2 + (\dot{y})^2 + (\dot{z})^2] + 2m(x\dot{y} - y\dot{x})\dot{\phi}$$
$$+ [I + m(x^2 + y^2)](\dot{\phi})^2, \tag{a}$$

where I = *moment of inertia of B about Oz*, ϕ = *inertial angular coordinate of B*; and, therefore, its four Lagrangean equations of motion are (with some obvious notations)

$$x: \quad m[\ddot{x} - 2\dot{\phi}\,\dot{y} - x(\dot{\phi})^2 - y\,\ddot{\phi}] = Q_x, \tag{b}$$

$$y: \quad m[\ddot{y} + 2\dot{\phi}\,\dot{x} - y(\dot{\phi})^2 + x\,\ddot{\phi}] = Q_y, \tag{c}$$

$$z: \quad m\,\ddot{z} = Q_z, \tag{d}$$

$$\phi: \quad \{[I + m(x^2 + y^2)]\dot{\phi} + m(x\,\dot{y} - y\,\dot{x})\}^{\cdot} = \Phi. \tag{e}$$

(ii) Specialize the above to the case where B spins at a *constant* rate: $\dot{\phi}$ = *constant*; that is, adjust the torque Φ so as to maintain $\ddot{\phi} = 0$, or $\delta\phi = 0$; or, equivalently, assume that $I \to \infty$.

(iii) Show that the above special case equations result by application of Lagrange's method to

$$2T = m[(\dot{x})^2 + (\dot{y})^2 + (\dot{z})^2] + 2m\,\Omega(x\dot{y} - y\dot{x}) + m\,\Omega^2(x^2 + y^2)$$

= *inertial (double) kinetic energy of P referred to axes spinning with constant inertial angular velocity* $\dot{\phi} = \Omega$. (f)

Problem 3.16.8 *Rotating Frames: 3-D Case.* Extend the results of the preceding examples and problems to the general case of two frames with common origin: (i) a *fixed O–XYZ* (*inertial*), and (ii) a *moving O–xyz* (*noninertial*) rotating relative to the first with constant angular velocity Ω (fig. 3.39).

Specifically, show that the (inertial) kinetic energy of a particle P of mass m equals

$$T = T_2 + T_1 + T_0, \tag{a}$$

where

$$2T_2 = m\,[(\dot{x})^2 + (\dot{y})^2 + (\dot{z})^2], \tag{b}$$

$$T_1 = m\,[\dot{x}(z\,\omega_y - y\,\omega_z) + \dot{y}(x\,\omega_z - z\,\omega_x) + \dot{z}(y\,\omega_x - x\,\omega_y)], \tag{c}$$

$$2T_0 = m\,[(z\,\omega_y - y\,\omega_z)^2 + (x\,\omega_z - z\,\omega_x)^2 + (y\,\omega_x - x\,\omega_y)^2]; \tag{d}$$

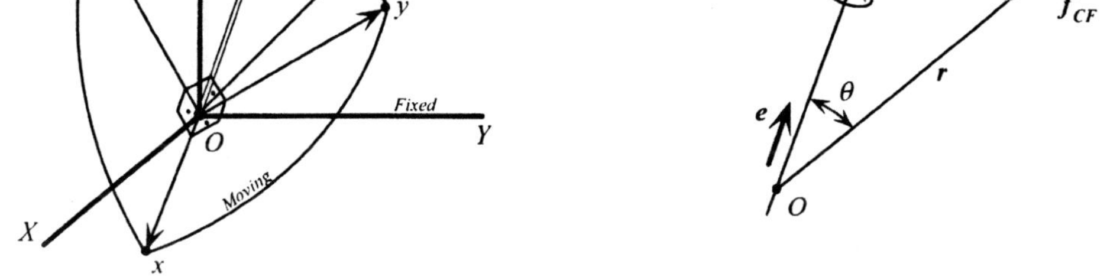

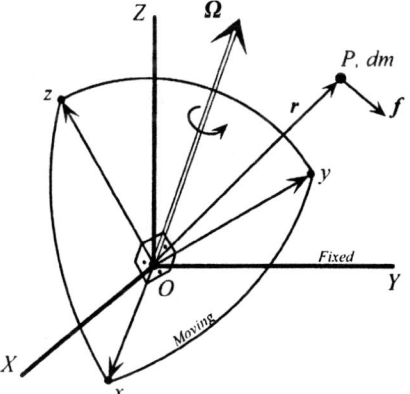

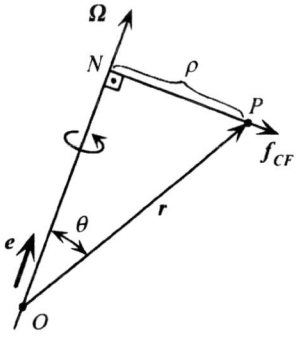

Figure 3.39 (a) Particle P in general relative motion in a rotating frame O–xyz; (b) details of centrifugal "force" $\boldsymbol{f}_{CF}$.

and, therefore, by the method of Lagrange, the equations of (unconstrained) motion of P in O–xyz, under a total impressed force $\boldsymbol{F} = (F_x, F_y, F_z)$, are

$$m(d'^2\boldsymbol{r}/dt^2) = \boldsymbol{F} + \partial T_0/\partial\boldsymbol{r} + \boldsymbol{f}_C, \tag{e}$$

where

$$\begin{aligned} \boldsymbol{f}_C &= -2m\,\boldsymbol{\Omega} \times (d'\boldsymbol{r}/dt) = -2m\,\boldsymbol{\Omega} \times \boldsymbol{v}_{\text{rel}} \equiv -2m\,\boldsymbol{\Omega} \cdot \boldsymbol{v}_{\text{rel}} \\ &\qquad [\text{definition of } \textit{antisymmetric} \text{ angular velocity } \textit{tensor}\ \boldsymbol{\Omega}\ (\S1.1,\ \S1.7)] \\ &= \textit{Gyroscopic (Coriolis) ``force''} \\ &= -2m(\Omega_x, \Omega_y, \Omega_z) \times (\dot{x}, \dot{y}, \dot{z}) \\ &= (2m\,\Omega_z\,\dot{y} - 2m\,\Omega_y\,\dot{z},\ 2m\,\Omega_x\,\dot{z} - 2m\,\Omega_z\,\dot{x},\ 2m\,\Omega_y\,\dot{x} - 2m\,\Omega_x\,\dot{y}), \end{aligned} \tag{f}$$

and

$$\boldsymbol{\Omega} = \begin{pmatrix} \Omega_{xx} = 0 & \Omega_{xy} = -\Omega_z & \Omega_{xz} = \Omega_y \\ \Omega_{yx} = \Omega_z & \Omega_{yy} = 0 & \Omega_{yz} = -\Omega_x \\ \Omega_{zx} = -\Omega_y & \Omega_{zy} = \Omega_x & \Omega_{zz} = 0 \end{pmatrix}. \tag{g}$$

Problem 3.16.9 *Rotating Frames: 3-D Case (continued).* Continuing from the preceding problem, show that

$$\partial T_0/\partial x = \cdots = m[\Omega^2 x - (\boldsymbol{\Omega} \cdot \boldsymbol{r})\Omega_x], \quad \text{etc., cyclically,} \tag{a}$$

and, further, with $\boldsymbol{\Omega} = \Omega\,\boldsymbol{e}$ [fig. 3.39(b)], that

$$\begin{aligned} \boldsymbol{f}_{CF} &= \textbf{\textit{grad}}\ T_0 \equiv \partial T_0/\partial\boldsymbol{r} = \cdots = m\,\Omega^2[\boldsymbol{r} - (\boldsymbol{e} \cdot \boldsymbol{r})\boldsymbol{e}] \\ &= m\,\Omega^2(\boldsymbol{r} - \boldsymbol{ON}) = (m\,\Omega^2)\boldsymbol{NP} \\ &= \textit{Centrifugal ``force''} \qquad (\text{i.e., } |\boldsymbol{f}_{CF}| = m\,\Omega^2\rho). \end{aligned} \tag{b}$$

Problem 3.16.10 *Lagrangean of a Particle in General Relative Motion.* Consider a particle P of mass m in unconstrained motion relative to a *noninertial* frame O–xyz that has given motion, under a total force $\boldsymbol{f} = -\partial V(\boldsymbol{r})/\partial\boldsymbol{r}$, $V(\boldsymbol{r}) =$ *potential*. Show that:

(i) To within terms equal to the *total time derivative of a given function of the coordinates and time* (i.e., to within "Lagrange-important" terms), the (double) Lagrangean of P is

$$2L = m\boldsymbol{v}_{\text{rel}} \cdot \boldsymbol{v}_{\text{rel}} + 2m\,\boldsymbol{v}_{\text{rel}}(\boldsymbol{\Omega} \times \boldsymbol{r}) + m(\boldsymbol{\Omega} \times \boldsymbol{r})^2 - 2m(\boldsymbol{a}_O \cdot \boldsymbol{r}) - 2V(\boldsymbol{r}). \tag{a}$$

(ii)
$$\begin{aligned} &\partial L/\partial\boldsymbol{v}_{\text{rel}} = m(\boldsymbol{v}_{\text{rel}} + \boldsymbol{\Omega} \times \boldsymbol{r}) \equiv \boldsymbol{p}_{\text{rel}} + m(\boldsymbol{\Omega} \times \boldsymbol{r}) = m\,\boldsymbol{v} \equiv \boldsymbol{p} \\ &(\boldsymbol{v} = \textit{inertial velocity of } P \textit{ relative to the moving axes origin } O), \end{aligned} \tag{b}$$

$$\partial L/\partial\boldsymbol{r} = m(\boldsymbol{v}_{\text{rel}} \times \boldsymbol{\Omega}) + m(\boldsymbol{\Omega} \times \boldsymbol{r}) \times \boldsymbol{\Omega} - m\,\boldsymbol{a}_O - \partial V/\partial\boldsymbol{r}. \tag{c}$$

Hence, obtain the Lagrangean equations of P: $(\partial L/\partial\boldsymbol{v}_{\text{rel}})^{\cdot} - \partial L/\partial\boldsymbol{r} = \boldsymbol{0}$.

Notice that in the Lagrangean formalism, the equations of motion have the same *form* in both inertial and noninertial axes; but the corresponding Lagrangeans *are* different.

Problem 3.16.11 *Energetics of a Particle in Relative Motion.* Specialize the results of the preceding problem to the case where $O-xyz$ rotates *uniformly* about a fixed axis through O.

(i) Show that, in this case, the *generalized energy* of a particle,

$$h \equiv h_{\text{rel}} \equiv (\partial L/\partial \boldsymbol{v}_{\text{rel}}) \cdot \boldsymbol{v}_{\text{rel}} - L \equiv \boldsymbol{p} \cdot \boldsymbol{v}_{\text{rel}} - L, \tag{a}$$

reduces to (notice absence of terms *linear* in $\boldsymbol{v}_{\text{rel}}$)

$$h = (1/2)m{v_{\text{rel}}}^2 + [V - (1/2)m(\boldsymbol{\Omega} \times \boldsymbol{r})^2]. \tag{b}$$

(ii) Interpret the *centrifugal potential*

$$V_{CF} \equiv -(1/2)m(\boldsymbol{\Omega} \times \boldsymbol{r})^2 = (m/2)[(\boldsymbol{\Omega} \cdot \boldsymbol{r})^2 - \Omega^2 r^2].$$

[Alternatively, we can show that the corresponding *centrifugal "force"*

$$\boldsymbol{f}_{CF} \equiv -m\,\boldsymbol{a}_{CF} = -m[\boldsymbol{\Omega} \times (\boldsymbol{\Omega} \times \boldsymbol{r})] \tag{c}$$

is *irrotational*; that is, show that $\textbf{\textit{curl}}\,\boldsymbol{f}_{CF} = -m\,\textbf{\textit{curl}}\,[\boldsymbol{\Omega} \times (\boldsymbol{\Omega} \times \boldsymbol{r})] = \boldsymbol{0} \Rightarrow \boldsymbol{f}_{CF} = -\textbf{\textit{grad}}\,V_{CF}$.]

Problem 3.16.12 *Energetics of a Particle in Relative Motion (continued).* Continuing from the preceding problem (of uniform fixed-axis rotation), show that the generalized energy h can be expressed as

$$h_{\text{rel}} \equiv h = h_{\text{inertial}} - \boldsymbol{H}_O \cdot \boldsymbol{\Omega}, \tag{a}$$

where

$$h_{\text{inertial}} \equiv (1/2)m\,v^2 + V = T + V \quad (\equiv E), \tag{b}$$

$$v^2 = \boldsymbol{v} \cdot \boldsymbol{v} = \textit{inertial velocity of } P, \qquad \boldsymbol{H}_O \equiv \boldsymbol{r} \times (m\boldsymbol{v}) = \boldsymbol{r} \times (\partial L/\partial \boldsymbol{v}_{\text{rel}}). \tag{c}$$

Equation (a) expresses the *law of transformation of (generalized) energy* between an inertial frame and a uniformly rotating/nontranslating one. However, both linear and angular momentum of P in the noninertial frame $O-xyz$ are equal to their inertial counterparts in the inertial $O-XYZ$:

$$\boldsymbol{p} \equiv \partial L/\partial \boldsymbol{v}_{\text{rel}} = m\,\boldsymbol{v} \equiv \boldsymbol{p}_{\text{inertial}} \text{ and } \boldsymbol{H}_O \equiv \boldsymbol{r} \times (\partial L/\partial \boldsymbol{v}_{\text{rel}}) = \boldsymbol{r} \times \boldsymbol{p} = \boldsymbol{r} \times \boldsymbol{p}_{\text{inertial}} \equiv \boldsymbol{H}_{O,\text{rel}}.$$

For a scalar derivation, see also Born (1927, p. 23).

Example 3.16.6 *Motion of a Particle Near the Surface of Earth.* Let us obtain the equations of motion of a particle P of mass m near the surface of Earth (fig. 3.40). Referring to fig. 3.40, and employing the usual notations, we readily find

$$\boldsymbol{\Omega} = (\Omega_x, \Omega_y, \Omega_z) = (-\Omega \cos\theta, 0, \Omega \sin\theta), \tag{a}$$

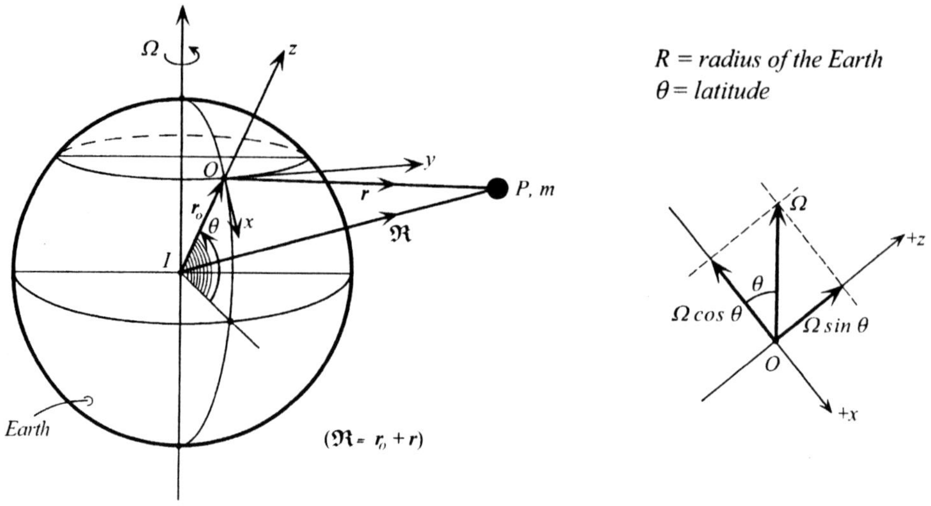

Figure 3.40 Motion of a particle P near Earth's surface, using Earth-bound axes O–xyz. R = radius of Earth; θ = latitude.

$$\begin{aligned} \boldsymbol{v}_P \equiv \boldsymbol{v} &= \boldsymbol{v}_O + \boldsymbol{v}_{P/O} = \boldsymbol{v}_O + (\boldsymbol{v}_{P,\text{rel}} + \boldsymbol{\Omega} \times \boldsymbol{r}) = \boldsymbol{v}_{P,\text{rel}} + \boldsymbol{\Omega} \times R \\ &= (\dot{x}, \dot{y}, \dot{z}) + (-\Omega \cos\theta,\ 0,\ \Omega \sin\theta) \times (x, y, R + z) \\ &= (\dot{x} - \Omega y \sin\theta,\ \dot{y} + \Omega x \sin\theta + \Omega(R + z)\cos\theta,\ \dot{z} - \Omega y \cos\theta), \end{aligned} \tag{b}$$

and, therefore,

$$2T = mv^2 = \left[(\dot{x} - \Omega y \sin\theta)^2 + \left(\dot{y} + \Omega x \sin\theta + \Omega(R + z)\cos\theta\right)^2 + (\dot{z} - \Omega y \cos\theta)^2\right]; \tag{c}$$

also, in the neighborhood of Earth's surface, $V = mgz$.

From the above, it follows that Lagrange's equations for x, y, z are

$$\ddot{x} = 2\Omega\,\dot{y}\sin\theta + \Omega^2[\sin\theta(x\sin\theta + z\cos\theta) + R\sin\theta\cos\theta], \tag{d}$$

$$\ddot{y} = -2\Omega\,\dot{x}\sin\theta - 2\Omega\,\dot{z}\cos\theta + \Omega^2 y, \tag{e}$$

$$\ddot{z} = 2\Omega\,\dot{y}\cos\theta + \Omega^2[\cos\theta(x\sin\theta + z\cos\theta) + R\cos^2\theta] - g. \tag{f}$$

The terms proportional to Ω are the components of the *Coriolis (gyroscopic)* "force" per unit mass, and those proportional to Ω^2 are those of the centrifugal "force."

APPROXIMATE SOLUTION

Since $\Omega = 2\pi/(24)(60)(60) \approx 7.27 \times 10^{-5}$ rad/s, to a *first* Ω-approximation, we may neglect in (d–f) the Ω^2-terms, and rewrite the rest so that we can easily identify the gyroscopic terms ($\sim \Omega$) in there more easily:

$$\ddot{x} = (0)\dot{x} + (2\Omega\sin\theta)\dot{y} + (0)\dot{z}, \tag{g}$$

$$\ddot{y} = (-2\Omega\sin\theta)\dot{x} + (0)\dot{y} + (-2\Omega\cos\theta)\dot{z}, \tag{h}$$

$$\ddot{z} = (0)\dot{x} + (2\Omega\cos\theta)\dot{y} + (0)\dot{z} - g. \tag{i}$$

To solve the linear system (g–i) we choose, for algebraic simplicity, the *free-fall initial conditions* at $t = 0$:

$$x = 0, \quad y = 0, \quad z = H\ (> 0); \qquad \dot{x} = 0, \quad \dot{y} = 0, \quad \dot{z} = 0. \tag{j}$$

Then (we recall that, here, $\theta = constant$), eqs. (g) and (i) integrate once, respectively, to

$$\dot{x} = (2\Omega \sin\theta)y, \qquad \dot{z} = (2\Omega\cos\theta)y - g\,t. \tag{k}$$

Substituting (k) into (h), neglecting $\sim \Omega^2$ terms, for consistency, and integrating the resulting equation while enforcing (j), we obtain

$$\ddot{y} = -4\Omega^2 y + (2\Omega g\cos\theta)t \approx (2\Omega g\cos\theta)t \Rightarrow y = (\Omega g\cos\theta/3)t^3\ (> 0); \tag{l}$$

that is, in both the *north* ($0 \le \theta \le \pi/2$) and *south* ($-\pi/2 \le \theta \le 0$) hemispheres, the particle deviates *eastwards*. Thanks to (l), and (j), and to within $\sim \Omega$ terms, eqs. (k) finally integrate, respectively, to

$$x = 0 \quad \text{and} \quad z = H - (1/2)gt^2. \tag{m}$$

Clearly, for $t = (2H/g)^{1/2} \Rightarrow z = 0$; that is, P hits the ground. Then, its eastward deviation is

$$y_{\text{eastward}} = (2\Omega H/3)(2H/g)^{1/2}\cos\theta. \tag{n}$$

The above derivation of the equations of motion (d–f) clearly shows the superiority and simplicity of the Lagrangean method over that of Newton–Euler.

For an alternative solution of (g–i) see, for example, Spiegel (1967, pp. 152–154); and, for an instructive treatment of the effect of the $\sim \Omega^2$ terms, see, for example, Bahar (1991).

Problem 3.16.13 Consider the gyroscope shown in fig. 3.41. With the usual notations [and A/C = *transverse/axial* (principal) moments of inertia at G], show that:

(i)
$$2T = A(\dot{\theta})^2 + A(\dot{\phi}\sin\theta)^2 + C(\dot{\psi} + \dot{\phi}\cos\theta)^2. \tag{a}$$

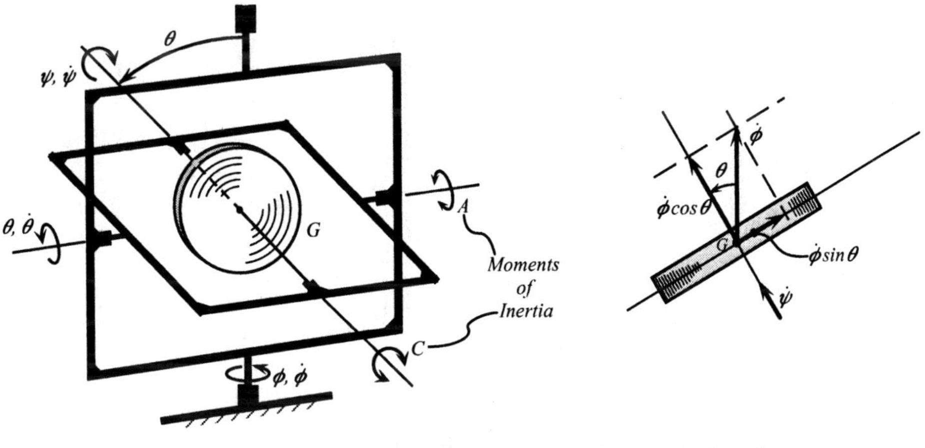

Figure 3.41 A gyroscope (and its Eulerian angles), supported in a light housing.

(ii) The equations of motion are

$$\theta: \quad A\ddot{\theta} - A(\dot{\phi})^2 \sin\theta\cos\theta + C(\dot{\psi} + \dot{\phi}\cos\theta)\dot{\phi}\sin\theta = Q_\theta, \tag{b}$$

$$\phi: \quad A\ddot{\phi}\sin^2\theta + 2A\dot{\phi}\dot{\theta}\sin\theta\cos\theta - C(\dot{\psi} + \dot{\phi}\cos\theta)\dot{\theta}\sin\theta + C\cos\theta(\dot{\psi} + \dot{\phi}\cos\theta)^{\cdot} = Q_\phi, \tag{c}$$

$$\psi: \quad C(\dot{\psi} + \dot{\phi}\cos\theta)^{\cdot} = Q_\psi. \tag{d}$$

(iii) If $Q_\psi = 0$, then $\dot{\psi} + \dot{\phi}\cos\theta \equiv$ *total spin* $=$ *constant* $\equiv n$, and the ϕ, θ equations become

$$\theta: \quad A\ddot{\theta} - A(\dot{\phi})^2 \sin\theta\cos\theta + (Cn\sin\theta)\dot{\phi} = Q_\theta, \tag{e}$$

$$\phi: \quad A\ddot{\phi}\sin^2\theta + 2A\dot{\phi}\dot{\theta}\sin\theta\cos\theta - (Cn\sin\theta)\dot{\theta} = Q_\phi. \tag{f}$$

Identify the gyroscopic terms in the above equations of motion.

Problem 3.16.14 *Gyroscopic Effects in a Pendulum of Varying Length.* A block B, of mass M, translates along the smooth horizontal floor/axis Ox. Block B is also connected to a linear spring of stiffness k whose other end is joined to the vertical wall Oy. A massless rod BP of variable length $l = l(t) = l_o + vt$ [l_o = *constant (initial) length*, v = *constant rate of change of* l] carries at its end P a particle P of mass m (fig. 3.42).

Choosing axes O–xy so that (CO) = *stress-free length* of spring, show that, with the usual notations, the Lagrangean equations of motion of this system can be written as

$$(\partial T_2/\partial \dot{x})^{\cdot} - \partial T_2/\partial x = -kx - mv\cos\phi\,\dot{\phi}, \tag{a}$$

$$(\partial T_2/\partial \dot{\phi})^{\cdot} - \partial T_2/\partial \phi = -mgl\sin\phi + mv\cos\phi\,\dot{x}. \tag{b}$$

Identify the *gyroscopic (Coriolis)* forces in the above equations, and indicate their directions on fig. 3.42. What happens if l = *constant*.

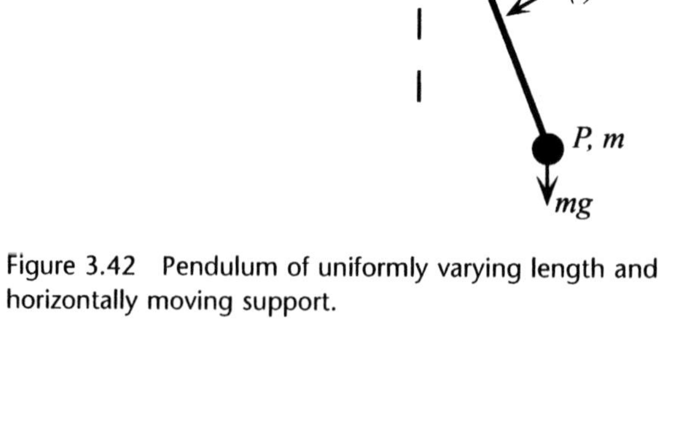

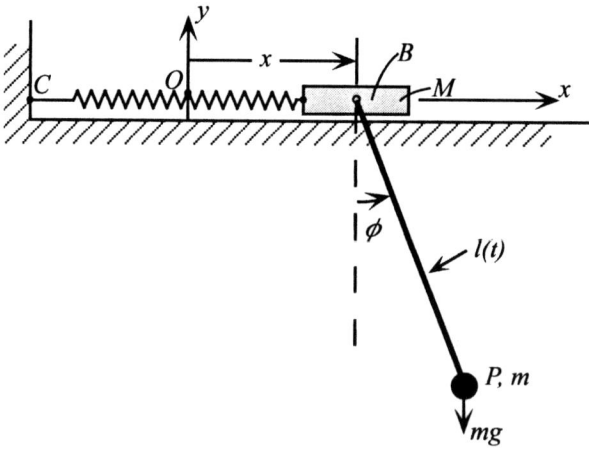

Figure 3.42 Pendulum of uniformly varying length and horizontally moving support.

Example 3.16.7 *An Additional Power Theorem for Relative Motion.* (Thomson and Tait, 1867–1912, §319, p. 319; see also Winkelmann and Grammel, 1927, pp. 463–465). Let us consider, with no loss of generality, a system with Lagrangean equations of motion

$$E_k = Q_k, \tag{a}$$

where [recalling (3.10.1a ff.)] the left side decomposes into the following *three* parts:

$$E_k = E_{k,R} + E_{k,T} + E_{k,C}, \tag{b}$$

$$E_{k,R} \equiv E_k(T_2) \equiv (\partial T_2/\partial \dot{q}_k)^{\cdot} - \partial T_2/\partial q_k \qquad (\textit{relative inertia}), \tag{c}$$

$$E_{k,T} \equiv \partial M_k/\partial t - \partial T_0/\partial q_k \qquad (\textit{transport inertia}), \tag{d}$$

$$E_{k,C} \equiv \sum (\partial M_k/\partial q_r - \partial M_r/\partial q_k)\dot{q}_r \qquad (\textit{Coriolis inertia}). \tag{e}$$

From (a), we immediately obtain the power equation

$$\sum E_k \dot{q}_k = \sum Q_k \dot{q}_k \equiv P(\textit{-ower of impressed forces}). \tag{f}$$

Let us transform the *left* side of (f). We find successively [recalling (3.9.3b) with $T_0 = 0,\ T_1 = 0,\ T_2 = T$]

$$\text{(i)}\quad \sum E_{k,R}\,\dot{q}_k \equiv \sum \left[(\partial T_2/\partial \dot{q}_k)^{\cdot} - \partial T_2/\partial q_k\right]\dot{q}_k = dT_2/dt + \partial T_2/\partial t; \tag{g}$$

$$\text{(ii)}\quad \sum E_{k,T}\,\dot{q}_k \equiv \sum (\partial M_k/\partial t)\dot{q}_k - \sum (\partial T_0/\partial q_k)\dot{q}_k \equiv \partial T_1/\partial t - d_q T_0/dt, \tag{h}$$

where

$$d_q(\ldots)/dt \equiv \sum [\partial(\ldots)/\partial q_k]\dot{q}_k \qquad \text{(i.e., } t \text{ and the } \dot{q}\text{'s remain fixed)}; \tag{i}$$

$$\text{(iii)}\quad \sum E_{k,C}\,\dot{q}_k \equiv \sum \left(\sum (\partial M_k/\partial q_r - \partial M_r/\partial q_k)\dot{q}_r\right)\dot{q}_k$$

$$\equiv \sum \left(-\sum g_{kr}\dot{q}_r\right)\dot{q}_k = 0. \tag{j}$$

[Incidentally, eq. (j) shows the error committed when one tries to obtain Lagrange's equations for gyroscopic systems, eqs. (a), from a single *power* equation like (f), instead of a single but *virtual work* equation, like LP.]

In view of (g–j), we can rewrite (f) as

$$dT_2/dt = P + d_q T_0/dt - \partial(T_2 + T_1)/\partial t. \tag{k}$$

But, further, we have

$$dT_0/dt = \partial T_0/\partial t + \sum (\partial T_0/\partial q_k)\dot{q}_k = \partial T_0/\partial t + d_q T_0/dt, \tag{l}$$

and

$$dT_1/dt = \partial T_1/\partial t + \sum [(\partial T_1/\partial q_k)\dot{q}_k + (\partial T_1/\partial \dot{q}_k)\ddot{q}_k]$$

$$= \partial T_1/\partial t + d_q T_1/dt + \sum M_k\,\ddot{q}_k; \tag{m}$$

and, therefore, the power equation (k) is finally transformed to

$$\begin{aligned} dT/dt &= (T_2 + T_1 + T_0)^{\cdot} \\ &= P + dT_0/dt + d_q(T_0 + T_1)/dt - \partial T_2/\partial t + \sum M_k \ddot{q}_k. \end{aligned} \tag{n}$$

The last *four* terms on (the right side of) the above represent the *rate of supply of kinetic energy to the carried system from the carrying body* (i.e., from its "tracks").

3.17 SERVO (OR CONTROL) CONSTRAINTS

The constraints examined so far, both holonomic and nonholonomic, are realized through mechanical *contact* of the system parts with foreign objects or obstacles (directly or indirectly, through auxiliary massless bodies; e.g., light inextensible cables). These latter are either *at rest*, or, generally, *they move in ways known in advance*; and, hence, they are unaffected by the motion and forces of the system. The associated reactions are *passive*; that is, without the aforementioned contacts, they cease to exist; and, for *bilateral* constraints, are assumed to satisfy LP (§3.2):

$$\delta' W_R \equiv \mathcal{S}\, d\boldsymbol{R} \cdot \delta \boldsymbol{r} = \sum R_k \delta q_k = 0. \tag{3.17.1}$$

Such constraints are the simplest ones to be found in mechanical systems; and analytical mechanics (AM) has, since its inception, been preoccupied with their study to such an extent that, in the minds of many, the subject is almost synonymous with their study. However, such constraints/reactions (C/R) are only one out of many logical and physical possibilities. Just as in continuum mechanics, not all parts of the total stress need obey Hooke's law, or even be elastic, so in AM, there exist constrained systems whose total reactions do *not* satisfy LP (3.17.1). Here, we summarize the basics of that particular and technically important non-LP type of C/R known as *servo(motoric)*, or *control*, or *control systems*, or C/Rs of the *second* kind: $(C/R)_2$; with the designation C/Rs of the *first* kind, $(C/R)_1$, reserved for the earlier passive ones.

The $(C/R)_2$s are realized through auxiliary sources of energy that go into action automatically, and are automatically adjusted (or turned off) so that, at every moment, a particular such constraint is realized, that is, at least one of the obstacles that interacts physically with our system, either through direct contact or via action at a distance (e.g. electromagnetic forces), regulates its motion so that certain holonomic and/or nonholonomic constraints, specified ahead of time, are enforced continuously. Therefore, the motion of the controlling object(s) is not known in advance as a function of time [as in the $(C/R)_1$ cases], but as the system moves, it continuously adjusts itself so as to satisfy all prescribed constraints.

The associated *servoreactions* are not known in advance but are calculated after the motion of the system has been determined; that is, as with $(C/R)_1$, first we solve the kinetic problem, and then the kinetostatic one.

HISTORICAL

These servoconstraints were introduced and examined in the early 1920s by P. Appell and his distinguished student H. Beghin ("*Liaisons comportant un Asservissement*," since the term *control* did not exist then), in their investigations of the Sperry-Anschütz gyrocompass and related navigation devices. Non-$(C/R)_1$ cases were also studied earlier ($\approx$ 1910–1916) by the Russian–Ukrainian J. I. Grdina (1871–1931) in his studies of the dynamics of living organisms; but we have not been able to access them; see, for example, Fradlin, B. N., *J. Appl. Mechanics* (Ukrainian), **8** (6), 581–591, 1962 (in Russian); also, Arczewski and Pietrucha (1993, pp. 74–75).

Here, too, our treatment is based on a judicious modification of LP, with guiding goal to make the servoproblem *determinate*; that is, generate as many equations as the unknowns introduced by that model, say coordinates and multipliers (reactions). We begin with the Newton–Euler equation of motion of a generic system particle P of mass dm, which, in this case, has the following form:

$$dm\,\boldsymbol{a} = d\boldsymbol{F} + d\boldsymbol{R} + d\boldsymbol{R}', \tag{3.17.2}$$

where

$$d\boldsymbol{F} = \textit{total impressed force on } P, \tag{3.17.2a}$$

$$d\boldsymbol{R} = \textit{total passive reaction force (1st kind) on } P, \tag{3.17.2b}$$

$$d\boldsymbol{R}' = \textit{total servoreaction force (2nd kind) on } P. \tag{3.17.2c}$$

From (3.17.2), carrying out some obvious mathematical operations, we obtain

$$\begin{aligned} 0 &= S\,(dm\,\boldsymbol{a} - d\boldsymbol{F} - d\boldsymbol{R} - d\boldsymbol{R}')\cdot\delta\boldsymbol{r} \\ &= S\,(dm\,\boldsymbol{a} - d\boldsymbol{F} - d\boldsymbol{R}')\cdot\delta\boldsymbol{r} + S\,d\boldsymbol{R}\cdot\delta\boldsymbol{r}, \end{aligned} \tag{3.17.3}$$

and, invoking (3.17.1), we finally obtain

$$S\,(dm\,\boldsymbol{a} - d\boldsymbol{F})\cdot\delta\boldsymbol{r} = S\,d\boldsymbol{R}'\cdot\delta\boldsymbol{r} \neq 0. \tag{3.17.4}$$

From the above, it follows that to obtain completely reactionless equations in both the $\{d\boldsymbol{R}\}$ and $\{d\boldsymbol{R}'\}$, *we must modify the* $\{\delta\boldsymbol{r}\}$ (*and* δq's), *if possible, by imposing on them additional restrictions* (*constraints in virtual form*) *so that not only* $\delta' W_R = 0$, *but also*

$$\delta' W_{R'} \equiv S\,d\boldsymbol{R}'\cdot\delta\boldsymbol{r} = 0 \qquad \left[\Rightarrow S\,(dm\,\boldsymbol{a} - d\boldsymbol{F})\cdot\delta\boldsymbol{r} = 0\right]. \tag{3.17.5}$$

This is the "*Servo–Lagrange (D'Alembert) principle" (SLP)*. In words: among the virtual displacements nullifying the virtual work of whatever ordinary contact/passive reactions are present, we seek if there may exist a *narrower class* that, simultaneously, nullifies the virtual work of the additional servo/control constraint reactions. That narrower class of δq's, if it exists, is determined by eq. (3.17.5), which, accordingly, becomes the key constitutive, namely, *physical*, tool for the Lagrangean solution of servoproblems. Indeed, we show below that *such problems are determinate if the number of servoconstraints equals the number of additional restrictive virtual conditions (= number of servoreactions) generated by (3.17.5)*.

An Example: The Servopendulum

Before we express these ideas in general system variables, let us discuss in some detail the following simple but instructive example: the (vertical) plane motion of a mathematical pendulum of mass m and variable (controlled) length l (fig. 3.43). Let us choose here $q_1 = l$ and $q_2 = \phi$, and study the case where an external agency, say, an

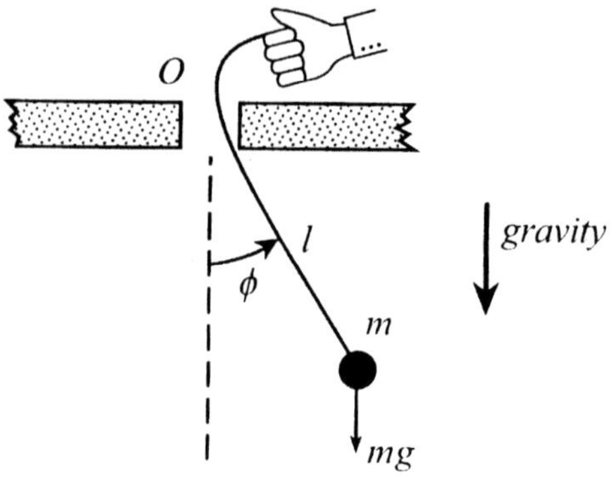

Figure 3.43 Geometry and physics of plane servopendulum.

agile hand, pulls the pendulum at O so that at every instant the following holonomic servoconstraint holds:

$$f(l, \phi) = 0 \Rightarrow l = l(\phi) = \textit{known functional relation.} \qquad (3.17.6)$$

Since (3.17.6) is maintained by the hand at O, the virtual work of the corresponding servoreaction vanishes, à la (3.17.5), if

$$\delta l = 0. \qquad (3.17.7)$$

In general, and this is important, *the holonomic servoconstraint (3.17.6) and the corresponding virtual servoconstraint (3.17.7) are unrelated to each other; that is, the latter does not follow from the former by differentiation (variation).*

Rewriting (3.17.7) as $(1)\,\delta l = 0$, or as $(1)\,\delta l + (0)\,\delta\phi = 0$, and combining it via Lagrangean multipliers to LP:

$$M_l\,\delta l + M_\phi\,\delta\phi = 0, \qquad (3.17.8)$$

where

$$M_l \equiv E_l - Q_l \equiv [(\partial T/\partial \dot{l})^{\cdot} - \partial T/\partial l] - Q_l, \qquad (3.17.8a)$$

$$M_\phi \equiv E_\phi - Q_\phi \equiv [(\partial T/\partial \dot{\phi})^{\cdot} - \partial T/\partial \phi] - Q_\phi, \qquad (3.17.8b)$$

$$2T = m[(\dot{l})^2 + l^2(\dot{\phi})^2], \qquad (3.17.8c)$$

$$V = -m\,g\,l\cos\phi$$

$$\Rightarrow Q_l = -\partial V/\partial l = +m\,g\,\cos\phi, \qquad Q_\phi = -\partial V/\partial\phi = -m\,g\,l\,\sin\phi, \qquad (3.17.8d)$$

we immediately obtain the two Routh–Voss type equations of servomotion (λ = multiplier)

$$(\partial T/\partial \dot{l})^{\cdot} - \partial T/\partial l = Q_l + \lambda(1): \quad (m\,\dot{l})^{\cdot} - m\,l(\dot{\phi})^2 = m\,g\,\cos\phi + \lambda, \qquad (3.17.9a)$$

$$(\partial T/\partial \dot{\phi})^{\cdot} - \partial T/\partial \phi = Q_\phi + \lambda(0): \quad (m\,l^2\dot{\phi})^{\cdot} = -m\,g\,l\,\sin\phi; \qquad (3.17.9b)$$

which, along with the servoconstraint (3.17.6), constitute a determinate system for $l(t)$, $\phi(t)$, $\lambda(t)$. Indeed, substituting (3.17.6) into the reactionless equation (3.17.9b), and since $dl/dt = (dl/d\phi)\,\dot{\phi}$, results in [assuming $l(t) \neq 0$]

$$d^2\phi/dt^2 + A(\phi)(d\phi/dt)^2 + B(\phi)\sin\phi = 0, \qquad (3.17.10)$$

where

$$A(\phi) \equiv 2[dl(\phi)/d\phi]/l(\phi), \qquad B(\phi) \equiv g/l(\phi): \ \textit{known functions of } \phi. \qquad (3.17.10a)$$

Next, solving the nonlinear ϕ-equation (3.17.10) (plus initial conditions), we obtain $\phi = \phi(t)$; then (3.17.6) yields $l = l[\phi(t)] = l(t)$; and, finally, substituting the so-found $\phi(t)$ and $l(t)$ into the multiplier-containing equation (3.17.9a) we get the servoreaction $\lambda = \lambda(t)$. In sum:

$$\delta' W_{R'} = 0 \ \Rightarrow \ \delta l = 0 \ \Rightarrow \ M_\phi = 0 \ \Rightarrow \ \phi(t) \ \Rightarrow \ l = l[\phi(t)] = l(t) \ \Rightarrow \ \lambda(t).$$

To understand this servoconstraint problem better, let us also discuss the following related *nonservo* versions of it:

(i) If the constraint (3.17.6) was an ordinary (i.e., passive) one, *even though of the exact same finite form as in the servo case*, then we would have,

$$\delta f = (\partial f/\partial l)\,\delta l + (\partial f/\partial \phi)\,\delta\phi = 0$$
$$\Rightarrow \ \delta l = -\big[(\partial f/\partial \phi)/(\partial f/\partial l)\big]\,\delta\phi \equiv [dl(\phi)/d\phi]\,\delta\phi, \qquad (3.17.11)$$

instead of (3.17.7); that is, in general, $\delta l \neq 0$ and $\delta\phi \neq 0$; and this combined with LP, eq. (3.17.8), would have produced the two Routh–Voss-type equations

$$M_l = \lambda(\partial f/\partial l) \qquad \text{and} \qquad M_\phi = \lambda(\partial f/\partial \phi); \qquad (3.17.12)$$

or, equivalently, the kinetic (Hadamard-type) equation

$$(\partial f/\partial l)M_\phi - (\partial f/\partial \phi)M_l = 0 \ \Rightarrow \ M_\phi + [dl(\phi)/d\phi]M_l = 0, \qquad (3.17.13)$$

resulting by eliminating λ between eqs. (3.17.12). These latter plus the (now assumed) passive constraint (3.17.6) would constitute a determinate system for $l(t)$, $\phi(t)$, $\lambda(t)$.

(ii) Next, if, unlike the servo case, the *temporal variation of l was known in advance*, that is, if $l = l(t) =$ *known (i.e., prescribed) function of time* (e.g., parametric excitation), but no constraint $f(l,\phi) = 0$ existed, then we would have $\delta l = 0$ but $\delta\phi \neq 0$, i.e., $(1)\,\delta l + (0)\,\delta\phi = 0$, and so the equations of motion would be

$$M_l = \lambda(1) \qquad \text{and} \qquad M_\phi = \lambda(0) = 0, \qquad (3.17.14)$$

as in the servo case; but without (3.17.6) to connect them. Clearly, this case is also determinate.

(iii) Finally, if l and ϕ were completely unrelated, and neither of the two was known in advance, then

$$\delta l \neq 0 \qquad \text{and} \qquad \delta\phi \neq 0; \qquad \text{i.e., } (0)\,\delta l + (0)\,\delta\phi = 0, \qquad (3.17.15a)$$

and this combined with (3.17.8) would produce the two equations

$$M_l = \lambda(0) = 0 \qquad \text{and} \qquad M_\phi = \lambda(0) = 0, \qquad (3.17.15b)$$

from which $l(t)$ and $\phi(t)$ could be determined.

The above cases are summarized below [with *accents (subscripts)* denoting *ordinary (partial)* derivatives]:

	Finite constraints		*Virtual constraints*	*Equations of motion*
0.	$f(l, \phi) = 0 \Rightarrow l = l(\phi)$	(servo)	But: $\delta l = 0$, $\delta\phi \neq 0$	$M_l = \lambda$, $M_\phi = 0$
1.	$f(l, \phi) = 0$	(passive)	$\Rightarrow \delta f = 0$: $\delta l = l'(\phi)\delta\phi \neq 0$, $\delta\phi \neq 0$	$M_l = \lambda f_l$, $M_\phi = \lambda f_\phi$
2.	No $f(l, \phi) = 0$ but $l = l(t)$	(prescribed)	$\delta l = 0, \delta\phi \neq 0$	$M_l = \lambda$, $M_\phi = 0$
3.	No $f(l, \phi) = 0$		$\delta l \neq 0, \delta\phi \neq 0$	$M_l = 0$, $M_\phi = 0$

This simple but sort of prototypical example helps us understand some of the fundamental features of constrained system mechanics:

- Even though, in both cases 0 (servo) and 1 (passive), the constraint has the same finite form, yet the virtual displacement restrictions in each case are not the same, but depend on exactly how the corresponding constraint is realized, that is, on how the controls are applied. And these differences in virtual displacements lead, in turn, to different equations of motion, and, of course, different equations of power. Indeed, for these four cases, we have, respectively:

$$\begin{aligned} M_l \dot{l} + M_\phi \dot{\phi} &= (\lambda)\,(\dot{l}) + (0)\,(\dot{\phi}) = \lambda \dot{l} \neq 0 && [\textit{servo problem}] \\ &= (\lambda f_l)\,(\dot{l}) + (\lambda f_\phi)\,(\dot{\phi}) = \lambda \dot{f} = 0 && [\textit{passive problem}] \\ &= (\lambda)\,(\dot{l}) + (0)\,(\dot{\phi}) = \lambda \dot{l} \neq 0 && [l(t)\ \textit{prescribed},\ f(l,\phi) = 0] \\ &= (0)\,(\dot{l}) + (0)\,(\dot{\phi}) = 0 && [l\ \textit{and}\ \phi\ \textit{independent, no}\ f(l,\phi) = 0]; \end{aligned} \tag{3.17.16}$$

  even though, in *all* cases, $\delta I - \delta' W = M_l\, \delta l + M_\phi\, \delta\phi = 0$.
- Conversely, cases 0 and 2 may have the same virtual constraints ($\Rightarrow$ same form of equations of motion), but since they are physically different ($\Rightarrow$ different finite constraints) they will have different ultimate solutions $l = l(t;$ initial conditions), $\phi = \phi(t;$ initial conditions).

The example also demonstrates that servoproblems can be treated competently and clearly by Lagrangean mechanics; that is, contrary to certain authors' claims (that, somehow, Lagrange's method is restricted to "ideal" constraints), these problems do not fall outside the classical methods, and, hence, do not need new "principles" for their solution. However, they *do* need a proper understanding of the underlying physics, and subsequent *correct application of the dynamical principle of virtual work*, but viewed as a constitutive postulate, like (3.17.1) and (3.17.5), and not as some mysterious "law of nature." This is far safer than manipulating the equations of motion, even the Lagrangean ones.

General Considerations

One Holonomic Servoconstraint

Now, let us resume our general considerations in system variables. For simplicity, but no real loss of generality, we begin our discussion with an *n-DOF* system under *holonomic servoconstraints*. We introduce the following basic definition.

DEFINITION

The holonomic equation

$$f(t, q_1, \ldots, q_n) \equiv f(t, q) = 0 \tag{3.17.17}$$

represents a servoconstraint, say, *relative to* q_1, if after substituting into it $q_2 = q_2(t), \ldots, q_n = q_n(t)$, which are not known in advance, it takes the form

$$q_1 = q_1(t, q_2, \ldots, q_n) = q_1\left[t, q_2 = q_2(t), \ldots, q_n = q_n(t)\right] \equiv q_{1o}(t). \qquad (3.17.17a)$$

Now, the virtual variations of the system—that is, the virtual form of (3.17.17, 17a), follow from the constitutive requirement of the *vanishing of the total virtual work of the corresponding servoreactions*; that is, from the *constitutive variational equation* (3.17.5). The latter yields

$$\delta q_1 = (\partial q_{1o}/\partial t)\,\delta t = 0, \qquad \delta q_2 \neq 0, \ldots, \delta q_n \neq 0, \qquad (3.17.17b)$$

or, equivalently,

$$(1)\,\delta q_1 + (0)\,\delta q_2 + \cdots + (0)\,\delta q_n = 0; \qquad (3.17.17c)$$

and when this is combined with ("adjoined" to) SLP, eq. (3.17.4), it leads to the following "servo-Routh–Voss" equations [with $E_k - Q_k \equiv M_k$ $(k = 1, \ldots, n)$, $\lambda_1 \equiv \lambda$]:

$$\textit{Kinetostatic:} \qquad M_1 = \lambda(1) \quad \text{or} \quad M_1 = \lambda, \qquad (3.17.17d)$$

$$\textit{Kinetic:} \qquad M_2 = \cdots = M_n = \lambda(0) = 0. \qquad (3.17.17e)$$

Here, too, the *finite* holonomic control constraint (3.17.17) is, generally, unrelated to the *virtual* control constraints (3.17.17b, c) resulting from $\delta' W_{R'} = 0$. Next, solving the n equations (3.17.17e) and (3.17.17), or (3.17.17a), with $dq_1/dt = \sum (\partial q_1/\partial q_\bullet)\,(dq_\bullet/dt) + \partial q_1/\partial t$ $(\bullet = 2, \ldots, n)$, we obtain $q_2 = q_2(t), \ldots, q_n = q_n(t)$; and then, substituting these time functions in (3.17.17d), we find $\lambda = M_1(t) = \lambda(t)$.

If the constraint (3.17.17, 17a) was passive, we would have

$$q_1 = q_1(t, q_2, \ldots, q_n) \;\Rightarrow\; \delta q_1 = \sum (\partial q_1/\partial q_\bullet)\,\delta q_\bullet \neq 0 \qquad (\bullet = 2, \ldots, n), \qquad (3.17.18a)$$

or, equivalently,

$$(1)\,\delta q_1 + (-\partial q_1/\partial q_2)\,\delta q_2 + \cdots + (-\partial q_1/\partial q_n)\,\delta q_n = 0, \qquad (3.17.18b)$$

instead of (3.17.17b, c), and so the corresponding equations of motion would be, in the Routh–Voss form

$$M_1 = \lambda(1); \qquad M_2 = \lambda(-\partial q_1/\partial q_2), \ldots,\; M_n = \lambda(-\partial q_1/\partial q_n); \qquad (3.17.18c)$$

or, in the Hadamard form,

$$\textit{Kinetostatic:} \qquad M_1 = \lambda; \qquad (3.17.18d)$$

$$\textit{Kinetic:} \qquad M_2 + (\partial q_1/\partial q_2)\,M_1 = 0, \ldots,\; M_n + (\partial q_1/\partial q_n)\,M_1 = 0; \qquad (3.17.18e)$$

and since the constraint (3.17.17) is holonomic, we can enforce it directly into the kinetic energy; that is,

$$\begin{aligned} T &= T[t,\, q_1(t,\, q_2, \ldots, q_n),\, \dot{q}_1(t,\, q_2, \ldots, q_n),\, q_2, \ldots, q_n,\, \dot{q}_2, \ldots, \dot{q}_n] \\ &\equiv T_o(t,\, q_\bullet,\, \dot{q}_\bullet) \equiv T_o, \end{aligned} \qquad (3.17.18f)$$

and $E_\bullet + (\partial q_1/\partial q_\bullet) E_1 = (\partial T_o/\partial \dot{q}_\bullet)^\cdot - \partial T_o/\partial q_\bullet \equiv E_\bullet(T_o)$, so that (3.17.18e) can be rewritten as

$$E_\bullet(T_o) = Q_\bullet + (\partial q_1/\partial q_\bullet) Q_1 \quad (\equiv Q_{\bullet,o}). \tag{3.17.18g}$$

Two Holonomic Servoconstraints

Next, if we have *two* servoconstraints *relative to q_1 and q_2*:

$$f_1(t, q_1, \ldots, q_n) \equiv f_1(t, q) = 0 \qquad \text{and} \qquad f_2(t, q_1, \ldots, q_n) \equiv f_2(t, q) = 0, \tag{3.17.19a}$$

or, equivalently,

$$q_1 = q_1(t, q_3, \ldots, q_n) \Rightarrow q_1[t, q_3 = q_3(t), \ldots, q_n = q_n(t)] \equiv q_{1o}(t),$$
$$q_2 = q_2(t, q_3, \ldots, q_n) \Rightarrow q_2[t, q_3 = q_3(t), \ldots, q_n = q_n(t)] \equiv q_{2o}(t), \tag{3.17.19b}$$

then, by repeating the earlier reasoning, we deduce that, for the fundamental equation (3.17.5) to hold, the virtual variations of the system must satisfy

$$\delta q_1 = 0, \qquad \delta q_2 = 0; \qquad \delta q_3 \neq 0, \ldots, \delta q_n \neq 0, \tag{3.17.19c}$$

or, equivalently,

$$(1)\,\delta q_1 + (0)\,\delta q_2 + \cdots + (0)\,\delta q_n = 0, \qquad (0)\,\delta q_1 + (1)\,\delta q_2 + \cdots + (0)\,\delta q_n = 0; \tag{3.17.19d}$$

and when these expressions are combined with LP,

$$M_1\,\delta q_1 + M_2\,\delta q_2 + M_3\,\delta q_3 + \cdots + M_n\,\delta q_n = 0, \tag{3.17.19e}$$

via the multipliers λ_1 and λ_2, they produce the following two groups of equations of servomotion:

$$\textit{Kinetostatic:} \quad M_1 = \lambda_1\,(1) + \lambda_2\,(0) = \lambda_1, \qquad M_2 = \lambda_1\,(0) + \lambda_2\,(1) = \lambda_2; \tag{3.17.19f}$$

$$\textit{Kinetic:} \qquad M_3 = \lambda_1\,(0) + \lambda_2\,(0) = 0, \ldots, \; M_n = \lambda_1\,(0) + \lambda_2\,(0) = 0. \tag{3.17.19g}$$

Solving the n equations (3.17.19g) and (3.17.19a, b) yields $q_1(t), \ldots, q_n(t)$; and then substituting these time functions in (3.17.19f) gives the two servoreactions

$$\lambda_1 = M_1(t) = \lambda_1(t) \qquad \text{and} \qquad \lambda_2 = M_2(t) = \lambda_2(t). \tag{3.17.19h}$$

General Case: Holonomic and/or Pfaffian Servoconstraints

The extension to $m'(<n)$ holonomic servoconstraints relative to $q_1, \ldots, q_{m'}$ is obvious. However, in all cases:

- In order to have a determinate problem, the number of nonvirtual servoconstraints [like (3.17.17), (3.17.19a)], m', must equal the number of virtual servoconstraints resulting from (3.17.5): $\delta' W_{R'} = 0$ [like (3.17.17b, c), (3.17.19c, d)], say s; since this also equals the number of unknown servoreactions/multipliers); that is, we must have

$$\begin{aligned} \mathbf{m}'\,(&\equiv \textit{number of nonvirtual servoconstraints}) \\ &= \mathbf{s}\,(\equiv \textit{number of virtual servoconstraints} \equiv \textit{number of servoreactions}). \end{aligned}$$

- If $m' > s$ [i.e., more (nonvirtual) servoconstraints than servoreactions], the problem is, in general, *impossible (overdeterminate)*—we cannot have more servoconstraints than the number of virtual conditions on the Lagrangean coordinates resulting from the nullification of the virtual work of the associated servoreactions; that is, $\delta' W_{R'} = 0$; while
- If $m' < s$ [i.e., fewer (nonvirtual) servoconstraints than servoreactions], the problem is *indeterminate*, unless we are given additional physical facts (constitutive equations) about the behavior of these servoreactions.

The above methodology is extended intact to the case where *some (or all)* of the m' servoconstraints are *holonomic* and *the rest (or all)* are *Pfaffian, holonomic or not.* Specifically, let our system be subject to the following additional constraints:

(i) m *passive (1st kind)* (with no loss in generality) Pfaffian constraints

$$\sum a_{dk}\dot{q}_k + a_d = 0 \qquad (d = 1, \ldots, m) \tag{3.17.20a}$$

whose *virtual* form is, therefore [recalling (2.9.11)],

$$\sum a_{dk}\, \delta q_k = 0, \tag{3.17.20b}$$

and

(ii) m' *servo (2nd kind)* (again, with no loss in generality) Pfaffian constraints

$$\sum a'_{d'k}\, \dot{q}_k + a'_{d'} = 0 \qquad (d' = 1, \ldots, m') \tag{3.17.20c}$$

with virtual form

$$\sum A_{Dk}\, \delta q_k = 0 \qquad (D = 1, \ldots, s); \tag{3.17.20d}$$

where, as already stressed, the (coefficients of the) "servovirtual" forms (3.17.20d) follow from the *vanishing of the virtual work of the servoreactions*, eq. (3.17.5); *that is*, they are *unrelated* to (the coefficients of) their velocity "counterparts" (3.17.20c), and so are, in general, their numbers m' and s [unlike the coefficients/number of (3.17.20b), which are directly related with those of (3.17.20a), as detailed in § 2.9]. Combination of the virtual forms (3.17.20b) and (3.17.20d) with LP readily yields the n Routh–Voss-type equations of motion

$$(\partial T/\partial \dot{q}_k)^{\cdot} - \partial T/\partial q_k = Q_k + \sum \lambda_d\, a_{dk} + \sum \lambda'_D\, A_{Dk}, \tag{3.17.20e}$$

where the λ's (λ''s) are the $m(s)$ passive (servo) reactions; and along with (3.17.20a) and (3.17.20c) constitute a system of $n + m + m'$ *equations* for the $n + m + s$ *unknowns* (created by these "narrower" δq's): $\{q_1(t), \ldots, q_n(t);\ \lambda_1(t), \ldots, \lambda_m(t);\ \lambda'_1(t), \ldots, \lambda'_s(t)\}$; hence the requirement $m' = s$, for determinacy. [For a Maggi–like approach, the *number of independent* δq*'s* = *number of independent equations*, equals $n - (m + s)$.]

This is an area in rapid evolution, and one with great potential for significant additional theoretical and practical results; for example, formulation in terms of quasi variables, application of differential and integral variational principles (chaps. 6, 7), extension to varying mass, impulsive motion of servocontrolled systems (chap. 4).

So far, all the relevant work in English seems to consist of translations of French and Soviet/Russian works. Among these, we recommend for complementary reading (alphabetically): Appell (1953, pp. 402–416), Apykhtin and Iakovlev (1980), Azizov (1986), Beghin (1967, pp. 440–443, 523–525), Cabannes (1965, pp. 188–191; summary of Appell/Beghin's work), Castoldi (1949), Kirgetov [1964(a),(b); 1967], Levi-Civita and Amaldi (1927, pp. 377–380, Mei (1987, pp. 243–248; excellent summary), Rumiantsev (1976, and references cited therein).

Problem 3.17.1 Consider a circular homogeneous disk D of negligible mass and radius R, free to rotate about a fixed horizontal axis through its center (pin) O (fig. 3.44). A plate P, of mass m and mass center G, is smoothly pin-joined on D at a point A. A motor *acting on the disk* D (or, perhaps, being located at A) at every instant realizes the servoconstraint

$$\text{angle}(OA,\ AG) \equiv \phi - \theta = \pi/2. \tag{a}$$

(i) Show that the (double) kinetic and potential energies of P are, respectively,

$$2T = m\big[R^2(\dot{\phi})^2 + (l^2 + k_G{}^2)\,(\dot{\theta})^2 + 2Rl\cos(\phi - \theta)\dot{\phi}\,\dot{\theta}\big], \tag{b}$$

(k_G = *radius of gyration of* P *about* G)

$$V = -mg(R\cos\phi + l\cos\theta),$$

$$\Rightarrow\ \delta' W = -mg(R\sin\phi\,\delta\phi + l\sin\theta\,\delta\theta) = \textit{virtual work of weight}. \tag{c}$$

An additional term $(1/2)I_O(\dot{\phi})^2$, in T, would have accounted for the inertia of D (I_O = moment of inertia of disk about O).

(ii) Show that, here, the condition $\delta' W_{R'} = 0$ (recall that the servomotor is acting on D) leads to

$$\delta\phi = 0 \qquad \text{or} \qquad (1)\,\delta\phi + (0)\,\delta\theta = 0; \tag{d}$$

and, hence, to the equations of motion

$$M_\phi = \lambda \qquad \text{and} \qquad M_\theta = 0, \tag{e}$$

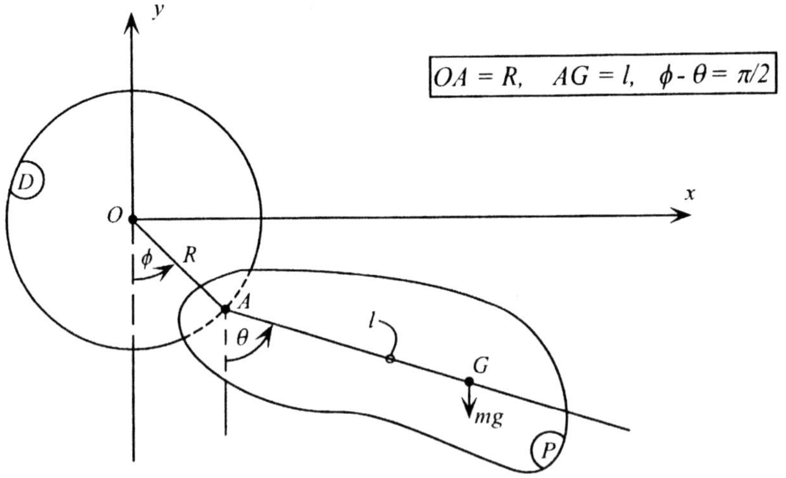

Figure 3.44 "Double" pendulum under the servoconstraint $\phi - \theta = \pi/2$ on disk D; $OA = R$, $AG = l$.

where

$$M_\phi \equiv E_\phi - Q_\phi \equiv [(\partial T/\partial \dot\phi)^{\cdot} - \partial T/\partial \phi] - (-mg\,R\,\sin\phi) = \cdots, \tag{f}$$

$$M_\theta \equiv E_\theta - Q_\theta \equiv [(\partial T/\partial \dot\theta)^{\cdot} - \partial T/\partial \theta] - (-mgl\,\sin\theta) = \cdots. \tag{g}$$

(iii) Show that (e, f, g), in extenso, after taking into account the servoconstraint (a) and with ${k_A}^2 \equiv {k_G}^2 + l^2$, are

$$m\big[R^2\ddot\phi + R\,l(\dot\theta)^2 + gR\,\sin\phi\big] = \lambda, \tag{h}$$

or

$$\lambda(t) = m\big[R^2\ddot\theta + Rl(\dot\theta)^2 + gR\,\cos\theta\big], \tag{i}$$

and

$${k_A}^2\,\ddot\theta - Rl(\dot\theta)^2 + g\,l\,\sin\theta = 0. \tag{j}$$

[Hence solving the kinetic equation (j), we obtain $\theta(t)$, and then substituting that function of time into the kinetostatic equation (i), we obtain the servoreaction $\lambda(t)$.]

(iv) Extend the above to include the inertia of the disk D.

For additional details and insights, see Appell (1953, pp. 411–412), Cabannes (1968, pp. 189–191), Kirgetov (1967, pp. 473–474).

Problem 3.17.2 Continuing from the preceding problem, show that if the constraint (a), $\phi - \theta = \pi/2$, is an ordinary passive one, say by contact between D and P [$\Rightarrow \delta\phi = \delta\theta \neq 0$, or $(1)\,\delta\phi + (-1)\,\delta\theta = 0$], then the corresponding equations of motion are

$$M_\phi = \lambda \qquad \text{and} \qquad M_\theta = -\lambda. \tag{a}$$

Then show that combination of (a) ($\Rightarrow M_\phi + M_\theta = 0$) with the above constraint results in the physical pendulum-like kinetic equation

$$(R^2 + {k_A}^2)\ddot\theta + g(R\,\cos\theta + l\,\sin\theta) = 0, \tag{b}$$

from which $\theta(t)$ [and $\phi = \pi/2 + \theta(t) \equiv \phi(t)$] can be determined.

[The multiplier (passive reaction) can then be easily found from either of the (now algebraic) equations (a): $\lambda = M_\phi[\phi(t), \theta(t)] = M_\phi(t) = -M_\theta[\phi(t), \theta(t)] = -M_\theta(t) = \lambda(t)$.]

Finally, extend these results to include the inertia of the disk D.

Problem 3.17.3 Continuing from the preceding problems, show that if $\phi = \phi(t) = $ *prescribed*, but no $f(\phi, \theta) = 0$ exists [i.e., $\delta\phi = 0$ but $\delta\theta \neq 0$], then the equations of motion are

$$M_\phi = \lambda \qquad \text{and} \qquad M_\theta = 0; \tag{a}$$

and constitute a determinate system for $\theta = \theta(t)$, $\lambda = \lambda(t)$.

What happens if ϕ and θ and their virtual variations are *completely independent* [i.e., no $f(\phi, \theta) = 0$, and $\delta\phi = 0$, $\delta\theta = 0$]?

Finally, extend these results to include the inertia of the disk D.

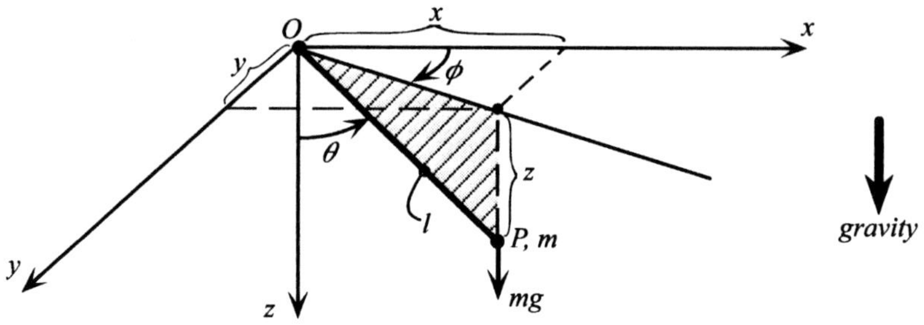

Figure 3.45 Spherical mathematical pendulum under a servoconstraint at O. Spherical coordinates: $x = (l \sin\theta)\cos\phi$; $y = (l \sin\theta)\sin\phi$; $z = l\cos\theta$.

Problem 3.17.4 (Castoldi, 1949). Consider the motion of the spherical (mathematical) pendulum P, of mass m and length l (fig. 3.45).

(i) Show that under the servoconstraint at O,

$$f(l, \theta) = 0 \Rightarrow l = l(\theta)\{= \cdots = l[\theta(t)];\ \delta l = 0,\ \delta\phi \neq 0,\ \delta\theta \neq 0\}, \tag{a}$$

its ϕ and θ equations of motion are

$$E_\phi = Q_\phi: \qquad (l^2 \sin^2\theta)\ddot{\phi} + 2[l \sin^2\theta(dl/d\theta) + l^2 \sin\theta\cos\theta]\,\dot{\phi}\,\dot{\theta} = 0, \tag{b}$$

$$E_\theta = Q_\theta: \qquad l^2\ddot{\theta} + 2\,l(dl/d\theta)(\dot{\theta})^2 - (l^2\sin\theta\cos\theta)(\dot{\phi})^2 = -g\,l\sin\theta. \tag{c}$$

[Three equations for $l(t)$, $\phi(t)$, $\theta(t)$.]

(ii) Show that if eq. (a) represents an ordinary passive constraint [$\Rightarrow \delta l = (dl/d\theta)\,\delta\theta \neq 0,\ \delta\phi \neq 0,\ \delta\theta \neq 0$], then the reactionless (Hadamard-type) equations for ϕ and θ are

$$E_\phi = Q_\phi: \qquad (l^2\sin^2\theta)\ddot{\phi} + 2\big[l\sin^2\theta(dl/d\theta) + l^2\sin\theta\cos\theta\big]\dot{\phi}\,\dot{\theta} = 0, \tag{d}$$

$$E_\theta + (dl/d\theta)E_l = Q_\theta + (dl/d\theta)Q_l:$$

$$\big[(dl/d\theta)^2 + l^2\big]\ddot{\theta} + (dl/d\theta)\big[(d^2l/d\theta^2) + l\big](\dot{\theta})^2$$
$$- l\big[l\cos\theta + (dl/d\theta)\sin\theta\big]\sin\theta(\dot{\phi})^2 = g\,l\big[(dl/d\theta)l^{-1}\cos\theta - \sin\theta\big].$$

[Again three equations for $l(t)$, $\phi(t)$, $\theta(t)$.] (e)

(iii) Show that (e) can be rewritten as

$$(\partial T_o/\partial\dot{\theta})^{\cdot} - \partial T_o/\partial\theta = Q_\theta + (dl/d\theta)Q_l \quad (\equiv Q_{\theta o}), \tag{f}$$

where

$$2T_o = m\big[(dl/d\theta)^2(\dot{\theta})^2 + l^2(\dot{\theta})^2 + l^2\sin^2\theta(\dot{\phi})^2\big]$$

$$Q_\phi = 0, \qquad Q_\theta = -mgl\sin\theta, \qquad Q_l = mg\cos\theta. \tag{g}$$

(iv) Find the general functional expressions for the reaction $\lambda = \lambda(t)$ in cases (i) and (ii).

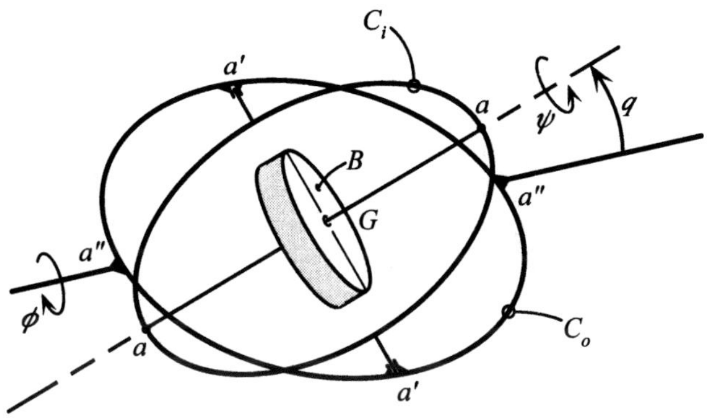

Figure 3.46 Servoconstraints on a Cardan-suspended gyro [ϕ = angle of rotation of C_o; θ = angle of planes of C_i and C_o; ψ = angle of rotation of B about "articulation axis" a–a, fixed in C_i (spin axis)].

Example 3.17.1 Let us consider a gyro B (axisymmetric body of revolution), supported by two Cardan-like massless circular rings $C_{i(\text{inner})}$ and $C_{o(\text{outer})}$, with its center of mass G at the intersection of the three axes a–a, a'–a', a''–a'' (fig. 3.46).

(i) Now, let us assume that a motor, *acting on* C_o, realizes, at every instant, the servoconstraint

$$\phi = \theta. \tag{a}$$

To nullify the virtual work of the corresponding reactions, from the motor to C_o, we choose the *restricted* virtual displacement

$$\delta\phi = 0; \tag{b}$$

which, we notice, does *not* coincide with the formal mathematical virtual version of (a): $\delta\phi = \delta\theta$. Equation (b) can be rewritten, equivalently, as

$$(1)\,\delta\phi + (0)\,\delta\theta + (0)\,\delta\psi = 0, \tag{c}$$

and, therefore, combined with the principle of Lagrange

$$M_\phi\,\delta\phi + M_\theta\,\delta\theta + M_\psi\,\delta\psi = 0, \tag{d}$$

leads, with the help of the multiplier λ to the Routh–Voss equations

$$M_\phi = \lambda(1): \qquad E_\phi = Q_\phi + \lambda, \tag{e}$$

$$M_\theta = \lambda(0): \qquad E_\theta = Q_\theta$$

$$(\text{e.g.}, Q_\theta = -k\,\theta,\ k = \textit{torsional spring constant}), \tag{f}$$

$$M_\psi = \lambda(0): \qquad E_\psi = Q_\psi, \tag{g}$$

where [applying, by now, well-known steps, and with A/C = *transverse/axial* principal moments of inertia of B at G]

$$2T = A\big[(\dot{\theta})^2 + (\dot{\phi})^2 \sin^2\theta\big] + C(\dot{\psi} + \dot{\phi}\cos\theta)^2. \tag{h}$$

The solution of this servoproblem proceeds as follows: solving (f, g) *and* (a), we obtain $\phi(t)$, $\theta(t)$, $\psi(t)$; and then substituting these time expressions into (e), we get the servoreaction $\lambda(t)$.

(ii) If (a) was an ordinary passive constraint, then (b, c) would be replaced by

$$\delta\phi = \delta\theta, \quad \delta\psi \neq 0 \Rightarrow (1)\,\delta\phi + (-1)\,\delta\theta + (0)\,\delta\psi = 0, \tag{i}$$

and the equations of motion (e–g) by

$$M_\phi = \lambda(1) = \lambda, \qquad M_\theta = \lambda(-1) = -\lambda, \qquad M_\psi = \lambda(0) = 0; \tag{j}$$

and, along with (a), would constitute a determinate system for $\phi(t)$, $\theta(t)$, $\psi(t)$, $\lambda(t)$.

Here, too, we notice that *analytically identical constraints, (a)*, depending on *how* and *where* they are applied, *lead to different equations of motion and reactions*. In the case of gyrocompasses, such servoconstraints allow us to increase or diminish the resulting oscillations; that is, to control them.

Example 3.17.2 A plane P translates sliding over another fixed horizontal plane O–XY. A homogeneous sphere Σ, of radius R and mass m, rolls on P with inertial angular velocity $\boldsymbol{\omega}$. The motion of P is regulated automatically so that the center of mass G of Σ rotates uniformly around OZ with constant inertial angular velocity $\boldsymbol{\Omega}$ (fig. 3.47). Let us study the motion of the sphere. Here we have the following two constraints:

(i) *rolling of Σ on P* (ordinary passive kind),

$$\boldsymbol{v}_C(\Sigma) = \boldsymbol{v}_C(P) \Rightarrow \boldsymbol{v}_G + \boldsymbol{\omega} \times \boldsymbol{r}_{C/G} = \boldsymbol{v}_C(P) = \boldsymbol{v}_A(P), \tag{a}$$

where A = *arbitrary plane point* = $(u, v, 0)$, or, in components,

$$(\dot{\xi}, \dot{\eta}, \dot{R} = 0) + (\omega_X, \omega_Y, \omega_Z) \times (0, 0, -R) = (\dot{u}, \dot{v}, 0), \tag{b}$$

from which we readily obtain the two rolling constraints

$$\dot{\xi} - \omega_Y R = \dot{u} \qquad \text{and} \qquad \dot{\eta} + \omega_X R = \dot{v}; \tag{c1, 2}$$

and (ii) uniform *rotation* of G around OZ (servoconstraint),

$$\boldsymbol{v}_G = \boldsymbol{\Omega} \times \boldsymbol{r}_{G/O} = \boldsymbol{\Omega} \times \boldsymbol{r}_{G/O'}, \tag{d}$$

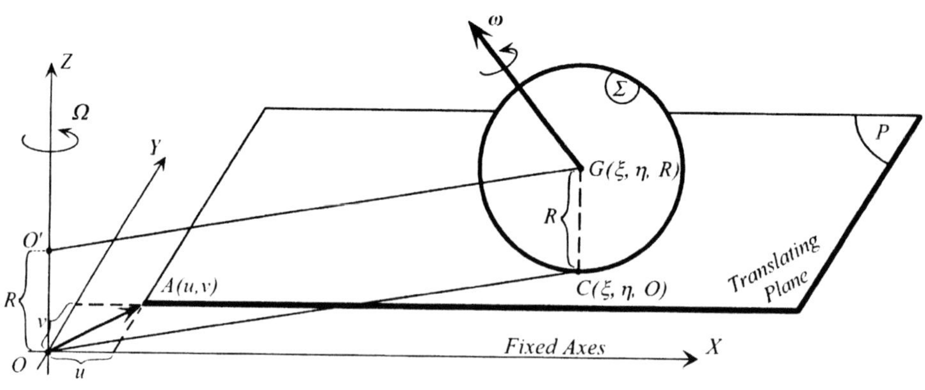

Figure 3.47 Controlled rolling of a sphere on a translating plane.

or, in components,

$$(\dot{\xi},\ \dot{\eta},\ 0) = (0,\ 0,\ \Omega) \times (\xi,\ \eta,\ 0), \tag{e}$$

from which we readily obtain the two Pfaffian servoconstraints

$$\dot{\xi} + \Omega\eta = 0 \qquad \text{and} \qquad \dot{\eta} - \Omega\xi = 0. \tag{f}$$

The servoconstraint on the sphere is expressed by the dependence of (c) on the $(\ldots)^{\cdot}$-derivatives of the two control parameters u and v; that is, $\dot{u}$ and $\dot{v}$.

Now, the servoreactions are the reaction forces from the plane to the sphere. Therefore, their virtual work $\delta' W_{R'} = (\ldots)\,\delta u + (\ldots)\,\delta v$ vanishes for the virtual displacements

$$\delta u = 0 \qquad \text{and} \qquad \delta v = 0, \tag{g}$$

or, equivalently,

$$(1)\,\delta u + (0)\,\delta v = 0 \qquad \text{and} \qquad (0)\,\delta u + (1)\,\delta v = 0; \tag{h}$$

which bear no formal mathematical relation to (f).

Next, to the equations of motion. To be able to enforce the *nonholonomic* Pfaffian constraints (c) right from the start, it is best not to use the kinetic energy ($\rightarrow$ Hamel equations), but the Appellian. Indeed, to within "Appell-important" terms (§3.14), the (double) Appellian of the sphere equals

$$\begin{aligned}
2S &= m[(\ddot{\xi})^2 + (\ddot{\eta})^2] + (2mR^2/5)\left[(\dot{\omega}_X)^2 + (\dot{\omega}_Y)^2 + (\dot{\omega}_Z)^2\right] \\
&\quad \text{[or, eliminating } \dot{\omega}_X \text{ and } \dot{\omega}_Y \text{ via the passive constraints (c):} \\
&\quad \omega_Y = R^{-1}(\dot{\xi} - \dot{u}) \quad \text{and} \quad \omega_X = R^{-1}(\dot{v} - \dot{\eta}) \ \Rightarrow\ \dot{\omega}_X = \cdots,\ \dot{\omega}_Y = \cdots] \\
&= m\left[(\ddot{\xi})^2 + (\ddot{\eta})^2\right] + (2m/5)\left[(\ddot{v} - \ddot{\eta})^2 + (\ddot{\xi} - \ddot{u})^2 + R^2(\dot{\omega}_Z)^2\right] \\
&\equiv 2S(\ddot{\xi},\ \ddot{\eta};\ \ddot{u},\ \ddot{v},\ \dot{\omega}_Z),
\end{aligned} \tag{i}$$

and, therefore, combining the principle of virtual work [with $d\theta_X \equiv \omega_X\,dt$, and noticing that the virtual work of all impressed forces (here, gravity) vanishes]

$$(\partial S/\partial\ddot{\xi})\,\delta\xi + (\partial S/\partial\ddot{\eta})\,\delta\eta + (\partial S/\partial\ddot{u})\,\delta u + (\partial S/\partial\ddot{v})\,\delta v + (\partial S/\partial\dot{\omega}_Z)\,\delta\theta_Z = 0, \tag{j}$$

with the two virtual servoconstraints (g, h) via the two multipliers (servoreactions) λ and μ, yields the following Routh–Voss equations (in Appellian form):

Kinetic:

$$\partial S/\partial\ddot{\xi} = 0: \qquad \ddot{\xi} - (2/7)\ddot{u} = 0, \tag{k}$$

$$\partial S/\partial\ddot{\eta} = 0: \qquad \ddot{\eta} - (2/7)\ddot{v} = 0, \tag{l}$$

$$\partial S/\partial\dot{\omega}_Z = 0: \qquad \dot{\omega}_Z = 0 \ \Rightarrow\ \omega_Z \equiv \dot{\theta}_Z = \textit{constant}; \tag{m}$$

Kinetostatic:

$$\partial S/\partial\ddot{u} = \lambda(1) + \mu(0): \qquad (2m/5)\,(\ddot{u} - \ddot{\xi}) = \lambda,$$

$$\text{or, invoking (c1),} \qquad 2mR\,\dot{\omega}_Y + 5\lambda = 0, \tag{n}$$

$$\partial S/\partial\ddot{v} = \lambda(0) + \mu(1): \qquad (2m/5)\,(\ddot{v} - \ddot{\eta}) = \mu,$$

$$\text{or, invoking (c2),} \qquad 2mR\,\dot{\omega}_X - 5\mu = 0. \tag{o}$$

These *five* equations plus the *two* servo equations (f) constitute a determinate system for the seven unknowns: $\xi(t)$, $\eta(t)$, $\omega_Z(t)$, $u(t)$, $v(t)$, $\lambda(t)$, $\mu(t)$; then, $\omega_X(t)$ and $\omega_Y(t)$ can be found from (c). For additional details and insights, see, for example, Appell (1953, pp. 415–416), Beghin (1967, pp. 523–525), Kirgetov (1967, pp. 475–476).

Problem 3.17.5 Continuing from the preceding example,

(i) Show that the servoconstraints (f) integrate to

$$\xi = a\cos(\Omega t) \quad \text{and} \quad \eta = a\sin(\Omega t) \quad (a = constant), \tag{a}$$

and, therefore, equations (k, l) yield, for a(ny) typical point A or C of the translating plane, the "cycloidal" translatory motion

$$u = (7/2)a\cos(\Omega t) + c_1 t + c_2, \qquad v = (7/2)a\sin(\Omega t) + c_3 t + c_4, \tag{b}$$

where the $c_{1,2,3,4}$ are integration constants.

(ii) After calculating $\omega_X(t), \omega_Y(t), \omega_Z(t)$ [via equations (c), (m) and the above], show that $\boldsymbol{\omega} = (\omega_X, \omega_Y, \omega_Z)$, emanating from G, describes an oblique cone of circular horizontal base, of radius $(5/2)(a/R)\Omega$, and is traversed at the uniform rate $\boldsymbol{\Omega}$.

3.18 GENERAL EXAMPLES AND PROBLEMS

Example 3.18.1 *Dynamics of a Sled (or Knife, or Scissors, etc.)* Let us determine the forces and equations of motion of the sled shown in fig. 3.48.

The sled kinematics have already been discussed in ex. 2.13.1 and ex. 2.13.2. We recall that $q_1 = x$, $q_2 = y$, $q_3 = \phi$, and that the nonholonomic (scleronomic) constraint is

$$v_{C,y}/v_{C,x} = \dot{y}/\dot{x} = dy/dx = \tan\phi \;\Rightarrow\; dy = (\tan\phi)\,dx, \tag{a}$$

or

$$(1)\,dx + (-1/\tan\phi)\,dy + (0)\,d\phi = 0. \tag{b}$$

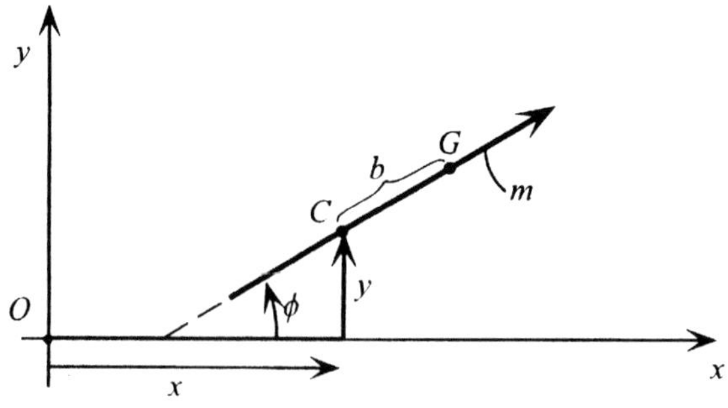

Figure 3.48 Sled in plane motion (on a fixed plane); geometry and forces. G = mass center; C = contact point; $I \equiv I_C \equiv mk_c^2$ = moment of inertia about $C = I_G + mb^2$ (k_c = radius of gyration about C).

Kinetic Energy

Applying König's theorem about C, we find (no constraint enforcement yet!), in *holonomic* variables,

$$2T = m[(\dot{x})^2 + (\dot{y})^2] + 2m\dot{\phi}[\dot{y}(x_G - x) - \dot{x}(y_G - y)] + I_C(\dot{\phi})^2$$
$$= m[(\dot{x})^2 + (\dot{y})^2] + 2mb\dot{\phi}(\dot{y}\cos\phi - \dot{x}\sin\phi) + I_C(\dot{\phi})^2, \qquad \text{(c)}$$

and since [recalling (ex. 2.13.2: a–c)]

$$(\dot{x})^2 + (\dot{y})^2 = (-\omega_1\sin\phi + \omega_2\cos\phi)^2 + (\omega_1\cos\phi + \omega_2\sin\phi)^2 = \cdots = \omega_1{}^2 + \omega_2{}^2, \qquad \text{(d1)}$$

$$(\cos\phi)\dot{x} + (\sin\phi)\dot{y} \equiv v \equiv \omega_2 = \textit{velocity of contact point along sled} \quad (\neq 0)$$

$$\dot{y}\cos\phi - \dot{x}\sin\phi = \omega_1 \quad (= 0, \text{ to be enforced later}) \qquad \text{and} \qquad \dot{\phi} = \omega_3, \qquad \text{(d2, 3, 4)}$$

in *nonholonomic* variables:

$$2T = 2T^* = m(\omega_1{}^2 + \omega_2{}^2) + 2mb\,\omega_1\,\omega_2 + I_C\,\omega_3{}^2. \qquad \text{(e)}$$

We point out that the *quadratic* term $m\omega_1{}^2$ can be safely omitted from $2T^*$ right at this stage; but *not* the term $2mb\,\omega_1\,\omega_2$ (why?). From (c) and (e), and the constraint, we readily conclude that the corresponding *constrained* (double) kinetic energies are

$$2T \to 2T_{oo} = m[(\dot{x})^2 + (\dot{y})^2] + I_C(\dot{\phi})^2$$

("partially constrained," i.e., still a function of $\dot{x}, \dot{y}, \dot{\phi}$), (f)

$$2T^* \to 2T^*{}_o = m\,\omega_2{}^2 + I_C\,\omega_3{}^2. \qquad \text{(g)}$$

Appellian

Applying the Appellian counterpart of König's theorem (recalling §3.14); that is,

$$S = S_G + S_{/G}, \qquad \text{(h)}$$

$$2S_G \equiv m\,\boldsymbol{a}_G \cdot \boldsymbol{a}_G = m\,a_G{}^2 = 2\ (\text{Appellian of } \textit{translation} \text{ of mass center}), \qquad \text{(i)}$$

$$2S_{/G} \equiv \mathbf{S}\, dm\,\boldsymbol{a}_{/G} \cdot \boldsymbol{a}_{/G} = \mathbf{S}\, dm\,a_{/G}{}^2$$
$$= 2\,[\text{Appellian of motion (rotation) about the mass center}], \qquad \text{(j)}$$

for the relaxed (unconstrained) system and in *holonomic* variables, we find

$$a_G{}^2 = (\ddot{x}_G)^2 + (\ddot{y}_G)^2 = [(x + b\cos\phi)^{\cdot\cdot}]^2 + [(y + b\sin\phi)^{\cdot\cdot}]^2$$
$$= [\ddot{x} - b\ddot{\phi}\sin\phi - b(\dot{\phi})^2\cos\phi]^2 + [\ddot{y} + b\ddot{\phi}\cos\phi - b(\dot{\phi})^2\sin\phi]^2$$
$$= (\ddot{x})^2 + (\ddot{y})^2 + b^2(\ddot{\phi})^2 + 2b\ddot{\phi}(\ddot{y}\cos\phi - \ddot{x}\sin\phi)$$
$$- 2b(\dot{\phi})^2(\ddot{x}\cos\phi + \ddot{y}\sin\phi) + \textit{terms not containing } \ddot{x}, \ddot{y}, \ddot{\phi}, \qquad \text{(k)}$$

and (with r = distance of a typical sled particle, of mass dm, from G)

$$2S_{/G} \equiv \mathbf{S}\, dm[(r\ddot{\phi})^2 + (r\dot{\phi}^2)^2] = I_G(\ddot{\phi})^2 + I_G(\dot{\phi})^4, \qquad \text{(l)}$$

or, *to within "Appell-important terms,"*

$$2S = m[(\ddot{x})^2 + (\ddot{y})^2 + b^2(\ddot{\phi})^2 + 2b\,\ddot{\phi}(\ddot{y}\cos\phi - \ddot{x}\sin\phi) - 2b(\dot{\phi})^2(\ddot{x}\cos\phi + \ddot{y}\sin\phi)] + I_G(\ddot{\phi})^2 = 2S(\phi, \dot{\phi}, \ddot{x}, \ddot{y}, \ddot{\phi}). \qquad \text{(m)}$$

To express S in the $\dot{\omega}$ variables—that is, $S \to S^*(t, q, \omega, \dot{\omega}) \equiv S^*$— we need the $\ddot{q}$'s; that is, $\ddot{x}, \ddot{y}, \ddot{\phi}$, in terms of $t, q, \omega, \dot{\omega}$. Indeed, $(\ldots)^{\cdot}$-differentiating $\dot{q}_k = \sum A_{kl}\,\omega_l$, we find

$$\ddot{q}_1 \equiv \ddot{x} = (-\sin\phi)\dot{\omega}_1 + (\cos\phi)\dot{\omega}_2 + (-\cos\phi)\omega_1\omega_3 + (-\sin\phi)\omega_2\omega_3, \qquad \text{(n1)}$$

$$\ddot{q}_2 \equiv \ddot{y} = (\cos\phi)\dot{\omega}_1 + (\sin\phi)\dot{\omega}_2 + (-\sin\phi)\omega_1\omega_3 + (\cos\phi)\omega_2\omega_3, \qquad \text{(n2)}$$

$$\ddot{q}_3 \equiv \ddot{\phi} = (1)\dot{\omega}_3. \qquad \text{(n3)}$$

However, and this is a very useful and labor-saving remark, since only $\partial S^*/\partial\dot{\omega}_k$ $(k = 1, 2, 3)$ enter the equations of motion, and because of the analytical identities:

$$S(t, q, \dot{q}, \ddot{q}) = S^*(t, q, \omega, \dot{\omega}) \quad \text{and} \quad \partial\ddot{q}_k/\partial\dot{\omega}_l = \partial\dot{q}_k/\partial\omega_l \equiv A_{kl}(t, q)$$

$(k = 1, 2, 3$—although, obviously, this is a general result), from which, by chain rule,

$$\partial S^*/\partial\dot{\omega}_k = \sum (\partial S/\partial\ddot{q}_l)(\partial\ddot{q}_l/\partial\dot{\omega}_k) = \cdots = \sum A_{lk}(\partial S/\partial\ddot{q}_l), \qquad \text{(m1)}$$

it follows that it is *not* necessary to square the $\ddot{q}$'s and then insert them into (m); that is, there is no need to calculate $S^*(\ldots, \dot{\omega})$; but, as (m1) shows, we *do* need to use the *linear* $\ddot{q} \Leftrightarrow \dot{\omega}$ relations (n1–3). From (m1) it also follows that

$$(\partial S^*/\partial\dot{\omega}_k)_o = \sum A_{lk}(\partial S/\partial\ddot{q}_l)_o, \qquad \text{(m2)}$$

where $(\partial S/\partial\ddot{q}_l)_o$ means that *after* we differentiate S of (m) in the $\ddot{q}$'s, we insert there the *contrained* $\ddot{q} \to (\ddot{q})_o \equiv \ddot{q}_o \Leftrightarrow \omega$ relations and not the relaxed (n1–3); that is,

$$\ddot{q}_{1o} \equiv \ddot{x} = (\cos\phi)\dot{\omega}_2 + (-\sin\phi)\,\omega_2\,\omega_3, \qquad \text{(n4)}$$

$$\ddot{q}_{2o} \equiv \ddot{y} = (\sin\phi)\dot{\omega}_2 + (\cos\phi)\,\omega_2\,\omega_3, \qquad \text{(n5)}$$

$$\ddot{q}_{3o} \equiv \ddot{\phi} = (1)\,\dot{\omega}_3. \qquad \text{(n6)}$$

The above can also be found by $(\ldots)^{\cdot}$-differentiation of the "natural" (constrained) variables $\dot{x} = v\cos\phi$, $\dot{y} = v\sin\phi$, where $v =$ *velocity in sled's direction*. The result is

$$\ddot{x} = \dot{v}\cos\phi - v\dot{\phi}\sin\phi, \qquad \ddot{y} = \dot{v}\sin\phi + v\dot{\phi}\cos\phi; \qquad \text{(n7)}$$

and inverts readily to

$$\dot{v} = \ddot{x}\cos\phi + \ddot{y}\sin\phi \qquad (= \textit{acceleration along sled}), \qquad \text{(n8)}$$

$$v\dot{\phi} = \ddot{y}\cos\phi - \ddot{x}\sin\phi \qquad (= \textit{acceleration normal to sled}); \qquad \text{(n9)}$$

[from which it also follows that $(\ddot{x})^2 + (\ddot{y})^2 = (\dot{v})^2 + v^2(\dot{\phi})^2$]. Clearly, (n7) and (n4–6) are identical. As a result of (n7–9), the Appellian expression (m) transforms to

$$2S_o = m\left[(\dot{v})^2 + v^2(\dot{\phi})^2 + b^2(\ddot{\phi})^2 + 2b\, v\,\dot{\phi}\,\ddot{\phi} - 2b(\dot{\phi})^2\dot{v}\right] + I_G(\ddot{\phi})^2,$$

or

$$2S^*_o = m\left[(\dot{\omega}_2)^2 + {\omega_2}^2{\omega_3}^2 + b^2(\dot{\omega}_3)^2 + 2b\,\omega_2\,\omega_3\,\dot{\omega}_3 - 2b\,{\omega_3}^2\,\dot{\omega}_2\right] + I_G(\dot{\omega}_3)^2, \qquad \text{(m3)}$$

or, finally, to within "Appell-important" terms (and recalling that, by the parallel axis theorem, $I_C = I_G + m\,b^2$)

$$2S^*_o = m\left[(\dot{\omega}_2)^2 + 2b\,\omega_3(\omega_2\,\dot{\omega}_3 - \omega_3\,\dot{\omega}_2\right] + I_C(\dot{\omega}_3)^2. \qquad \text{(m4)}$$

Virtual Work

With reference to fig. 3.49 and §3.4 and §3.15, we find, successively,

$$\delta' W = X\,\delta x + Y\,\delta y + M_C\,\delta\phi = Q_1\,\delta q_1 + Q_2\,\delta q_2 + Q_3\,\delta q_3$$
$$\Rightarrow Q_1 = X, \quad Q_2 = Y, \quad Q_3 = M_C \qquad (\textit{holonomic components}), \qquad \text{(n10)}$$

and (from the invariance of $\delta' W$)

$$\delta' W = \sum Q_k\,\delta q_k = \sum Q_k\left(\sum A_{kl}\,\delta\theta_l\right) = \sum \Theta_l\,\delta\theta_l$$
$$\Rightarrow \Theta_l = \sum A_{kl} Q_k \quad (\textit{holonomic components}), \text{ and, inversely, } Q_k = \sum a_{lk}\Theta_l, \qquad \text{(n11)}$$

and, therefore, here [recalling (ex. 2.13.2: a, c)],

$$\Theta_1 = -X\,\cos\phi + Y\,\sin\phi \equiv N = \textit{impressed system force perpendicular to sled}, \qquad \text{(n12)}$$

$$\Theta_2 = X\,\cos\phi + Y\,\sin\phi \equiv K = \textit{impressed system force along sled}, \qquad \text{(n13)}$$

$$\Theta_3 = M_C \equiv M = \textit{impressed system moment about } C \textit{ (perpendicular to } O\text{–}XY); \qquad \text{(n14)}$$

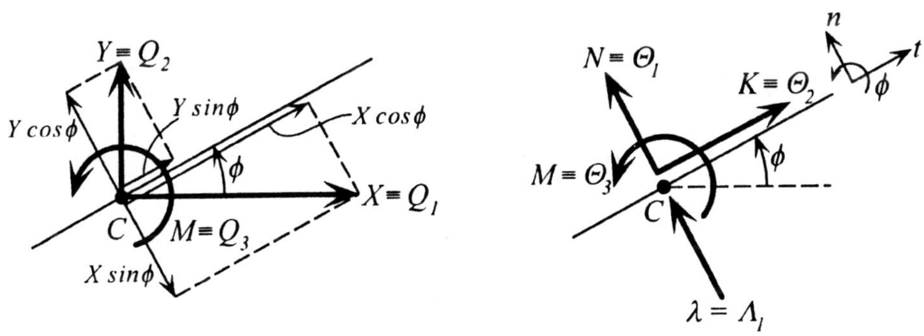

Figure 3.49 Holonomic and nonholonomic components of impressed forces and reactions on sled. Impressed forces on sled: *holonomic*: $Q_1 \equiv X, Q_2 \equiv Y, Q_3 \equiv M$; *nonholonomic*: $\Theta_1 \equiv N, \Theta_2 \equiv K, \Theta_3 \equiv M$.

and similarly for the reactions (here, only $\Lambda_1 \neq 0$)

$$\delta' W_R = \lambda_1\, \delta\theta_1 = \Lambda_1\, \delta\theta_1 = \sum R_k\, \delta q_k \qquad (\lambda_1 = \textit{Lagrangean multiplier})$$

$$\Rightarrow \lambda_1 = \Lambda_1 = \sum A_{k1} R_k, \text{ and, inversely, } R_k = a_{1k}\Lambda_1 = a_{1k}\lambda_1. \qquad \text{(n15)}$$

The Routh–Voss Equations

Invoking the above results, we find, successively,

$$E_1(T) = Q_1 + \lambda_1 a_{11}: \qquad (m\,\dot{x} - m\,b\,\dot{\phi}\,\sin\phi)^{\cdot} = X + \lambda(-\sin\phi),$$

$$\text{or} \quad m[\ddot{x} - b\,\ddot{\phi}\,\sin\phi - b(\dot{\phi})^2\,\cos\phi] = X - \lambda\,\sin\phi; \qquad \text{(o1)}$$

$$E_2(T) = Q_2 + \lambda_1 a_{12}: \qquad (m\,\dot{y} + m\,b\,\dot{\phi}\,\cos\phi)^{\cdot} = Y + \lambda(\cos\phi),$$

$$\text{or} \quad m[\ddot{y} + b\,\ddot{\phi}\,\cos\phi - b(\dot{\phi})^2\,\sin\phi] = Y + \lambda\,\cos\phi; \qquad \text{(o2)}$$

$$E_3(T) = Q_3 + \lambda_1 a_{13}:$$

$$[I_C\dot{\phi} + m\,b(-\dot{x}\,\sin\phi + \dot{y}\,\cos\phi)]^{\cdot} + m\,b\,\dot{\phi}(\dot{x}\,\cos\phi + \dot{y}\,\sin\phi) = M + \lambda(0),$$

$$\text{or} \quad I_C\,\ddot{\phi} + m\,b(\ddot{y}\,\cos\phi - \ddot{x}\,\sin\phi) = M. \qquad \text{(o3)}$$

These three coupled equations, plus the constraint (a, b), constitute a system of four equations for the four functions: $x(t), y(t), \phi(t), \lambda(t)$.

- Clearly, equations (o1, 2) express the principle of *linear* momentum for the sled along O–XY, respectively; while (o3) expresses that of *angular* momentum about C.
- The derivation of the ϕ-equation, (o3), shows clearly why we should *not* enforce the constraint (a, b) in T. Had we done so — that is, $T \to T_{oo}$ [eq. (f)], *since that constraint is nonholonomic* — we would have obtained the *incorrect* Routh–Voss equations

$$E_1(T_{oo}) = Q_1 + \lambda_1 a_{11}: \qquad (m\dot{x})^{\cdot} = X - \lambda\,\sin\phi, \qquad \text{(o4)}$$

$$E_2(T_{oo}) = Q_2 + \lambda_1 a_{12}: \qquad (m\dot{y})^{\cdot} = Y + \lambda\,\cos\phi, \qquad \text{(o5)}$$

$$E_3(T_{oo}) = Q_3 + \lambda_1 a_{13}: \qquad I_C(\dot{\phi})^{\cdot} = M. \qquad \text{(o6)}$$

Elimination of the Reaction λ among Equations (n7–9)

Multiplying (o1) with $\cos\phi$ and (o2) with $\sin\phi$, and adding side by side, yields

$$\cos\phi(m\,\dot{x} - m\,b\,\dot{\phi}\,\sin\phi)^{\cdot} + \sin\phi(m\,\dot{y} + m\,b\,\dot{\phi}\,\cos\phi)^{\cdot} = X\,\cos\phi + Y\,\sin\phi,$$

or, simplifying, and so on,

$$m[\ddot{x}\,\cos\phi + \ddot{y}\,\sin\phi - b(\dot{\phi})^2] = K. \qquad \text{(o7)}$$

Equations (o7) and (o3) are, essentially, the (kinetic) Maggi–Hadamard equations of our problem (see below) and, along with the constraint (a), they constitute a determinate system for $x(t), y(t), \phi(t)$. Once this has been accomplished, then, to isolate λ, we multiply (o1) with $-\sin\phi$ and (o2) with $\cos\phi$, and add, and thus obtain (the kinetostatic Maggi equation)

$$\lambda = m(-\ddot{x}\sin\phi + \ddot{y}\cos\phi + b\ddot{\phi}) - (-X\sin\phi + Y\cos\phi)$$
$$= \textit{inertia "force" perpendicular to sled} - \textit{impressed force perpendicular to sled}$$
$$\equiv m a_{G,n} - N \quad (= \textit{constraint reaction perpendicular to sled}). \tag{o8}$$

In terms of the "natural" (or "intrinsic") variables v and ϕ, defined by

$$\dot{x} = v\cos\phi \qquad \text{and} \qquad \dot{y} = v\sin\phi, \qquad \text{where } v = \textit{velocity in direction of sled},$$

the kinetic equations (o7) and (o3) assume, respectively, the simpler Hamel forms (see below)

$$m[\dot{v} - b(\dot{\phi})^2] = K \qquad \text{and} \qquad I_C\ddot{\phi} + mb\dot{\phi}v = M, \tag{p1, 2}$$

while the kinetostatic (o8) becomes

$$m(b\ddot{\phi} + v\dot{\phi}) = N + \lambda. \tag{p3}$$

If $K, M = 0$ (force-free case), eqs. (p1, 2) reduce to

$$m[\dot{v} - b(\dot{\phi})^2] = 0 \qquad \text{and} \qquad I_C\ddot{\phi} + mb\dot{\phi}v = 0, \tag{p4, 5}$$

and yield, readily, the (first) integral of *energy*:

$$2T = mv^2 + 2mb\dot{\phi}(-\dot{x}\sin\phi + \dot{y}\cos\phi) + I_C(\dot{\phi})^2 = \textit{constant} \equiv 2h,$$

or, after enforcing the constraint (a),

$$mv^2 + I_C(\dot{\phi})^2 = 2h. \tag{p6}$$

(See also Carathéodory, 1933; and Hamel, 1949, pp. 467–470.)

The Maggi Equations

With $M_k \equiv E_k(T) - Q_k$ ($k = 1, 2, 3 \to x, y, \phi$), and (A_{kl}) from (n1–3), Maggi's equations become

$$A_{xx}M_x + A_{yx}M_y + A_{\phi x}M_\phi = \lambda: \quad (-\sin\phi)M_x + (\cos\phi)M_y + (0)M_\phi = \lambda, \tag{q1}$$
$$A_{xy}M_x + A_{yy}M_y + A_{\phi y}M_\phi = 0: \quad (\cos\phi)M_x + (\sin\phi)M_y + (0)M_\phi = 0, \tag{q2}$$
$$A_{x\phi}M_x + A_{y\phi}M_y + A_{\phi\phi}M_\phi = 0: \quad (0)M_x + (0)M_y + (1)M_\phi = 0; \tag{q3}$$

or, explicitly,

$$m(\ddot{y}\cos\phi - \ddot{x}\sin\phi) + mb\ddot{\phi} = (\cos\phi\, Y - \sin\phi\, X) + (\cos\phi\, R_y - \sin\phi\, R_y), \tag{q4}$$
$$m(\ddot{x}\cos\phi + \ddot{y}\sin\phi) - mb(\dot{\phi})^2 = \cos\phi\, X + \sin\phi\, Y, \tag{q5}$$
$$mb(\ddot{y}\cos\phi - \ddot{x}\sin\phi) + I_C\ddot{\phi} = M, \tag{q6}$$

and coincide, respectively, with the earlier-found equations (o8), (o7), and (o3); that is, the Maggi approach constitutes a systematization of the earlier uncoupling of the Routh–Voss equations into kinetic and kinetostatic.

- Clearly, equations (q1, 2 / 4, 5) express, respectively, the principle of *linear* momentum *normally* and *along* the sled; while (q3/6) expresses that of *angular* momentum about C.
- If we express (any) one of the holonomic velocities, say $\dot{x}$, in terms of the other two via the constraint (a) [i.e., $\dot{x} = \dot{x}(\dot{y}, \dot{\phi}; \phi)$], and use this to eliminate $\dot{x}$ and $\ddot{x}$ from the kinetic Maggi equations (q2, 3), these two *new* (kinetic) equations in $\dot{y}, \ddot{y}, \dot{\phi}, \ddot{\phi}, \phi$ would be the *Chaplygin–Voronets* equations of our problem; equivalently, these would be our Lagrangean equations of motion based on the kinetic energy expressed in terms of the two *independent* velocities $\dot{y}$ and $\dot{\phi}$: $T \rightarrow T[\dot{x}(\dot{y}, \dot{\phi}), \dot{y}, \dot{\phi};\ \phi] = T_o(\dot{y}, \dot{\phi}) \equiv T_o =$ *completely constrained kinetic energy* [recall (3.8.13a ff.)]. The details, for any of the three possible choices of dependent velocity, are left to the reader. See also Dobronravov (1970, pp. 92–104).

The Appell Equations

(i) *Holonomic variable*: Due to the kinematico-inertial identities $\partial S/\partial \ddot{q}_k = (\partial T/\partial \dot{q}_k)^{\cdot} - \partial T/\partial q_k$, the holonomic variable Appellian equations are

$$\partial S/\partial \ddot{q}_k = Q_k + \sum \lambda_D\, a_{Dk} \qquad (k = 1, 2, 3 = x, y, \phi;\ D = 1), \tag{r}$$

and, of course, they coincide completely with the earlier Routh–Voss equations (o3).

(ii) *Nonholonomic variables*: With the help of the earlier results [(n1) ff.], we readily find

$$\begin{aligned}
\partial S^*/\partial \dot{\omega}_k &= (m\ddot{q}_1)(\partial \dot{q}_1/\partial \omega_k) + (m\ddot{q}_2)(\partial \dot{q}_2/\partial \omega_k) + (mb^2\ddot{q}_3)(\partial \dot{q}_3/\partial \omega_k) \\
&\quad + mb(\partial \dot{q}_3/\partial \omega_k)(\ddot{q}_2 \cos\phi - \ddot{q}_1 \sin\phi) \\
&\quad + mb\ddot{q}_3[(\partial \dot{q}_2/\partial \omega_k)\cos\phi - (\partial \dot{q}_1/\partial \omega_k)\sin\phi] \\
&\quad - mb(\dot{q}_3)^2[(\partial \dot{q}_1/\partial \omega_k)\cos\phi + (\partial \dot{q}_2/\partial \omega_k)\sin\phi] + I_G\ddot{q}_3(\partial \dot{q}_3/\partial \omega_k) \\
&= (m\ddot{q}_1)\, A_{1k} + (m\ddot{q}_2)\, A_{2k} + \cdots,
\end{aligned}$$

or, explicitly (recalling that $I_C = I_G + mb^2$),

$$\begin{aligned}
\partial S^*/\partial \dot{\omega}_1 &= m\,\dot{\omega}_1 + m\,b\,\dot{\omega}_3 + m\,\omega_2\,\omega_3 \Rightarrow (\partial S^*/\partial \dot{\omega}_1)_o = m(b\,\dot{\omega}_3 + \omega_2\,\omega_3), \\
\partial S^*/\partial \dot{\omega}_2 &= m\,\dot{\omega}_2 - m\,\omega_1\,\omega_3 - m\,b\,\omega_3{}^2 \Rightarrow (\partial S^*/\partial \dot{\omega}_2)_o = m(\dot{\omega}_2 - b\,\omega_3{}^2), \\
\partial S^*/\partial \dot{\omega}_3 &= I_C\,\dot{\omega}_3 + m\,b\,\dot{\omega}_1 + m\,b\,\omega_2\,\omega_3 \Rightarrow (\partial S^*/\partial \dot{\omega}_3)_o = I_C\,\dot{\omega}_3 + m\,b\,\omega_2\,\omega_3. \qquad \text{(s1, 2, 3)}
\end{aligned}$$

From the above and (n12–15), we see that the nonholonomic Appellian equations of our problem are

$$(\partial S^*/\partial \dot{\omega}_1)_o = \Theta_1 + \lambda_1: \qquad m\,(b\,\dot{\omega}_3 + \omega_2\,\omega_3) = N + \lambda, \tag{s4}$$

$$(\partial S^*/\partial \dot{\omega}_2)_o = \Theta_2: \qquad m\,(\dot{\omega}_2 - b\,\omega_3{}^2) = K, \tag{s5}$$

$$(\partial S^*/\partial \dot{\omega}_3)_o = \Theta_3: \qquad I_C\,\dot{\omega}_3 + m\,b\,\omega_2\,\omega_3 = M. \tag{s6}$$

Upon recalling the definitions of $\omega_{1,2,3}$, we immediately see that the above are nothing but (p3, 1, 2) in that order. Clearly, the above constitute a (determinate) system of three equations in $\omega_2(t), \omega_3(t), \lambda(t)$: first, we solve (s5, 6) for ω_2, ω_3, and, substituting the results into (s4) (which then becomes algebraic), we obtain λ.

If we were not interested in finding the reaction $\Lambda_1 = \lambda_1 = \lambda$ (perpendicular to the sled at C, and due to the constraint $\omega_1/d\theta_1/\delta\theta_1 = 0$), we would only need to evaluate all relevant quantities for $\omega_1 = 0$, $\dot{\omega}_1 = 0$, and denote them by $(\ldots)_o$; that is, instead of the relaxed holonomic accelerations (n1–3), we would only need their constrained values (n4–6) and corresponding constrained Appellian (m1, 2). Then, the two kinetic Appellian equations would read

$$\partial S^*{}_o/\partial\dot{\omega}_2 = (\partial S^*/\partial\dot{\omega}_2)_o = m(\dot{\omega}_2 - b\,\omega_3{}^2) = \Theta_2, \tag{s7}$$

$$\partial S^*{}_o/\partial\dot{\omega}_3 = (\partial S^*/\partial\dot{\omega}_3)_o = I_C\,\dot{\omega}_3 + m\,b\,\omega_2\,\omega_3 = \Theta_3, \tag{s8}$$

as before.

It is such "constrained" Appellian derivations, based on $\partial S^*{}_o/\partial\dot{\omega}_I = (\partial S^*/\partial\dot{\omega}_I)_o$ $(I = 2, 3)$, that one usually finds in the literature.

The Hamel Equations

Invoking the earlier equations (e, g), we easily find

$$P_1 = (\partial T^*/\partial\omega_1)_o = m\,b\,\omega_3, \tag{t1}$$

$$P_2 = (\partial T^*/\partial\omega_2)_o = m\,\omega_2, \tag{t2}$$

$$P_3 = (\partial T^*/\partial\omega_3)_o = (m\,b\,\omega_1 + I_C\,\omega_3)_o = I_C\,\omega_3, \tag{t3}$$

and therefore the master variational equation (3.6.12)

$$\sum \dot{P}_k\,\delta\theta_k + \sum P_k[(\delta\theta_k)^{\cdot} - \delta\omega_k] - \sum\Big(\sum A_{lk}(\partial T^*/\partial q_l)\Big)\,\delta\theta_k = \sum \Theta_k\,\delta\theta_k,$$

with the help of the transitivity equations (ex. 2.13.2: d–i), yields

$$\begin{aligned}&\dot{P}_1\,\delta\theta_1 + \dot{P}_2\,\delta\theta_2 + \dot{P}_3\,\delta\theta_3\\ &+ P_1(-\omega_3\,\delta\theta_2 + \omega_2\,\delta\theta_3) + P_2(\omega_3\,\delta\theta_1 - \omega_1\,\delta\theta_3) + P_3(0) = \Theta_1\,\delta\theta_1 + \Theta_2\,\delta\theta_2 + \Theta_3\,\delta\theta_3,\end{aligned} \tag{t4}$$

or, collecting $(\ldots)\,\delta\theta_k$ terms:

$$(\dot{P}_1 + P_2\,\omega_3 - \Theta_1)\,\delta\theta_1 + (\dot{P}_2 - P_1\,\omega_3 - \Theta_2)\,\delta\theta_2 + (\dot{P}_3 + P_1\,\omega_2 - \Theta_3)\,\delta\theta_3 = 0. \tag{t5}$$

Finally, adjoining to this variational equation the constraint

$$\delta\theta_1 = 0, \qquad \text{or} \qquad (1)\,\delta\theta_1 + (0)\,\delta\theta_2 + (0)\,\delta\theta_3 = 0, \tag{t6}$$

via the method of Lagrangean multipliers yields the three earlier equations (s4–6):

$$\dot{P}_1 + P_2\,\omega_3 = \Theta_1 + \lambda_1: \qquad m(b\,\dot{\omega}_3 + \omega_2\,\omega_3) = N + \lambda, \tag{t7}$$

$$\dot{P}_2 - P_1\,\omega_3 = \Theta_2: \qquad m(\dot{\omega}_2 - b\,\omega_3{}^2) = K, \tag{t8}$$

$$\dot{P}_3 + P_1\,\omega_2 = \Theta_3: \qquad I_C\,\dot{\omega}_3 + m\,b\,\omega_2\,\omega_3 = M. \tag{t9}$$

The above clearly demonstrate the usefulness of the nonholonomic form of LP (t4, 5) over any particular set of *equations*.

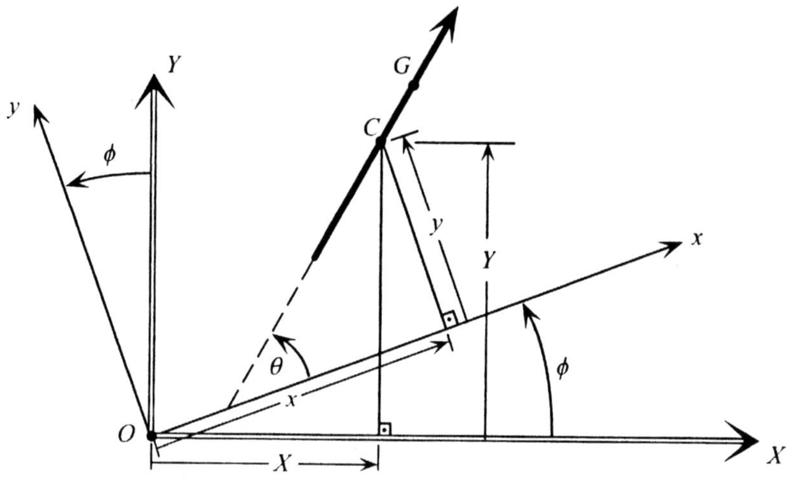

Figure 3.50 Geometry of sled on a uniformly spinning turntable. *Inertial coordinates*: $q_1 = X, q_2 = Y, q_3 = \psi \equiv \phi + \theta, \phi = \omega t$; *rotating coordinates*: $q_1 = x, q_2 = y, q_3 = \theta$.

Problem 3.18.1 Formulate the constraint of a sled in plane motion on a turntable spinning with constant (inertial) angular velocity Ω, in both inertial and rotating coordinates (fig. 3.50).

HINT

Formulate the constraint in the (moving) O–xy axes, and then use the transformation equations: $x = X\cos\phi - Y\sin\phi$, $y = X\sin\phi + Y\cos\phi$, and their $(\ldots)^{\cdot}$-derivatives for the (fixed) O–XY axes.

Problem 3.18.2 Continuing from the preceding problem, write down its transitivity equations and calculate (read off) their Hamel coefficients. Then obtain its equations of motion in both holonomic and nonholonomic variables.

Example 3.18.2 *Dynamics of a Rolling Sphere on a Fixed Plane.* Let us determine the forces and equations of motion of a homogeneous sphere S, of mass m and radius r, rolling on a rough horizontal fixed plane P (fig. 3.51).

The relevant kinematics has already been discussed in exs. 2.13.4–2.13.6. We recall that $q_1 = X_G \equiv X$, $q_2 = Y_G = Y$; $q_3 = \phi$, $q_4 = \theta, q_5 = \psi$ (coordinates of center of mass G and Eulerian angles of sphere-fixed axes G–xyz relative to translating/nonrotating axes G–XYZ), and that the constraints are

$$\dot{X} - (r\sin\phi)\dot{\theta} + (r\sin\theta\cos\phi)\dot{\psi} = 0, \quad \dot{Y} + (r\cos\phi)\dot{\theta} + (r\sin\theta\sin\phi)\dot{\psi} = 0; \qquad \text{(a)}$$

that is, here, $n = 5, m = 2 \Rightarrow f \equiv n - m = 3$ (# *local*) *DOF*.

The Routh–Voss Equations

By König's theorem and the results of §1.17, the (double) kinetic energy of S is

$$2T = m[(\dot{X})^2 + (\dot{Y})^2] + \left(I_X{}^2\omega_X{}^2 + I_Y{}^2\omega_Y{}^2 + I_Z{}^2\omega_Z{}^2\right)$$

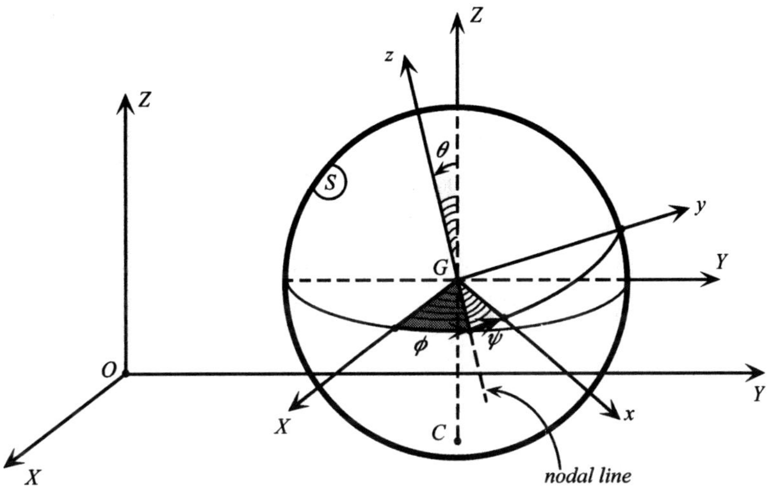

Figure 3.51 Geometry of sphere rolling on a fixed horizontal plane. Axes: O–XYZ: inertial (fixed); G–XYZ: translating (nonrotating); G–xyz: body-fixed.

(the *second* term can be expressed in either *body* or *space* axes at G; $I_X = I_Y = I_Z \equiv I = 2mr^2/5$)

$$= m[(\dot{X})^2 + (\dot{Y})^2] + I[(\dot{\phi})^2 + (\dot{\theta})^2 + (\dot{\psi})^2 + 2\dot{\phi}\,\dot{\psi}\cos\theta], \tag{b}$$

and, therefore, the corresponding five Routh–Voss equations,

$$E_k(T) = Q_k + \sum \lambda_D a_{Dk} \quad (k = 1, \ldots, 5 \equiv X, Y, \phi, \theta, \psi;\ D = 1, 2),$$

with $\lambda_1 \to \lambda$ and $\lambda_2 \to \mu$, are

$$m\ddot{X} = Q_X + \lambda \qquad (\text{i.e., } R_X = \lambda), \tag{b1}$$

$$m\ddot{Y} = Q_Y + \mu \qquad (\text{i.e., } R_Y = \mu); \tag{b2}$$

$$I(\ddot{\phi} + \ddot{\psi}\cos\theta - \dot{\theta}\,\dot{\psi}\sin\theta) = Q_\phi \qquad (\text{i.e., } R_\phi = 0), \tag{b3}$$

$$I(\ddot{\theta} + \dot{\phi}\,\dot{\psi}\sin\theta) = Q_\theta - r(\lambda\sin\phi - \mu\cos\phi) \quad (\equiv Q_\theta + R_\theta), \tag{b4}$$

$$I(\ddot{\psi} + \ddot{\phi}\cos\theta - \dot{\phi}\,\dot{\theta}\sin\theta) = Q_\psi + r\sin\theta(\lambda\cos\phi + \mu\sin\phi) \quad (\equiv Q_\psi + R_\psi). \tag{b5}$$

Equation (b3) shows that if $Q_\phi = 0$, then $\dot{\omega}_Z = \ddot{\phi} + \ddot{\psi}\cos\theta - \dot{\theta}\,\dot{\psi}\sin\theta = 0 \Rightarrow \omega_Z = \dot{\phi} + \cos\theta\,\dot{\psi}$ (= *total vertical spin*) = *constant*, as expected. [See also Bahar (1970–1980, pp. 446–450) and Roy (1965, vol. I, pp. 380–385) for additional insights and special cases.]

Next:

- Eliminating λ and μ among (b1–5) [e.g., solving (b1) for λ and (b2) for μ and inserting these values into (b4, 5)], we obtain the following *three* coupled kinetic

Maggi-like equations:

$$I(\ddot{\phi} + \ddot{\psi}\cos\theta - \dot{\theta}\dot{\psi}\sin\theta) = Q_\phi, \tag{b6}$$

$$I(\ddot{\theta} + \dot{\phi}\dot{\psi}\sin\theta) + mr(\ddot{X}\sin\phi - \ddot{Y}\cos\phi) = Q_\theta + r(Q_X\sin\phi - Q_Y\cos\phi), \tag{b7}$$

$$\begin{aligned} &I(\ddot{\psi} + \ddot{\phi}\cos\theta - \dot{\phi}\dot{\theta}\sin\theta) - mr(\ddot{X}\cos\phi + \ddot{Y}\sin\phi) \\ &\quad = Q_\psi - r\sin\theta(Q_X\cos\phi + Q_Y\sin\phi); \end{aligned} \tag{b8}$$

which, along with the *two* constraints (a) constitute a determinate system of five equations in the five Lagrangean coordinates: $X(t), Y(t);\ \phi(t), \theta(t), \psi(t)$. After these latter have been found, we can easily determine the multipliers/reactions $\lambda(t), \mu(t)$ from (b1, 2), respectively.

- If, further, with the help of the constraints (a) we eliminate $\dot{X} \to \ddot{X} = \ddot{X}(\phi, \theta;\ \dot{\phi}, \dot{\theta}, \dot{\psi};\ \ddot{\theta}, \ddot{\psi})$ and $\dot{Y} \to \ddot{Y} = \ddot{Y}(\phi, \theta;\ \dot{\phi}, \dot{\theta}, \dot{\psi};\ \ddot{\theta}, \ddot{\psi})$, or any other *two* out of the five $\dot{q}$'s, among (b6–8), we will obtain the *three* (kinetic) Chaplygin–Voronets equations of the problem, coupled in $\dot{\phi}, \dot{\theta}, \dot{\psi}$ and their $(\ldots)^{\cdot}$-derivatives—see below.
- Equations (b1–5) can, of course, also be obtained by *adjoining to LP*:

$$\sum [(\partial T/\partial \dot{q}_k)^{\cdot} - \partial T/\partial q_k]\,\delta q_k = \sum Q_k\,\delta q_k, \tag{b9}$$

the constraints (a) in virtual form [i.e., with $(\ldots)^{\cdot}$ replaced, in them, by $\delta(\ldots)$], via multipliers λ and μ, and then setting of the coefficients of the (now) free δq's equal to zero.

The Hamel Equations

Recalling (ex. 2.13.6: a ff., with $\Omega = 0$):

$$\omega_1 = \dot{X} - r\omega_Y = \dot{X} - r\omega_4\ (= 0) \Rightarrow \dot{X} = \omega_1 + r\omega_4, \tag{c1}$$

$$\omega_2 = \dot{Y} + r\omega_X = Y + r\omega_3\ (= 0) \Rightarrow \dot{Y} = \omega_2 - r\omega_3, \tag{c2}$$

$$\omega_3 = \omega_X \Rightarrow \omega_X = \omega_3, \tag{c3}$$

$$\omega_4 = \omega_Y \Rightarrow \omega_Y = \omega_4, \tag{c4}$$

$$\omega_5 = \omega_Z \Rightarrow \omega_Z = \omega_5, \tag{c5}$$

we readily find (no constraint enforcement yet)

$$(\dot{X})^2 + (\dot{Y})^2 = \cdots = \omega_1{}^2 + \omega_2{}^2 + r^2(\omega_3{}^2 + \omega_4{}^2) + 2r(\omega_1\omega_4 - \omega_2\omega_3),$$

$$r^2(\omega_X{}^2 + \omega_Y{}^2 + \omega_Z{}^2) = r^2(\omega_1{}^2 + \omega_2{}^2 + \omega_3{}^2),$$

and, therefore, (b) becomes

$$\begin{aligned} 2T \to 2T(\omega) \equiv 2T^* = \cdots = m[&\omega_1{}^2 + \omega_2{}^2 + (7r^2/5)(\omega_3{}^2 + \omega_4{}^2) \\ &+ (2r^2/5)\omega_5{}^2 + 2r(\omega_1\omega_4 - \omega_2\omega_3)], \end{aligned} \tag{c6}$$

or, to within "*Hamel-important terms*" [i.e., dropping the *quadratic* terms in $\omega_1\ (= 0)$ and $\omega_2\ (= 0)$, but *not* the linear ones]:

$$2T^* = m[(7r^2/5)(\omega_3{}^2 + \omega_4{}^2) + (2r^2/5)\omega_5{}^2 + 2r(\omega_1\omega_4 - \omega_2\omega_3)]. \tag{c7}$$

From this, we readily find [with the notation $(\ldots)_o \equiv (\ldots)$ *evaluated for* $\omega_{1,2} = 0$]

$$P_1 \equiv (\partial T^*/\partial\omega_1)_o = m\,r\,\omega_4, \tag{c8}$$

$$P_2 \equiv (\partial T^*/\partial\omega_2)_o = -3m\,r\,\omega_3, \tag{c9}$$

$$P_3 \equiv (\partial T^*/\partial\omega_3)_o = (7mr^2/5)\omega_3, \tag{c10}$$

$$P_4 \equiv (\partial T^*/\partial\omega_4)_o = (7mr^2/5)\omega_4, \tag{c11}$$

$$P_5 \equiv (\partial T^*/\partial\omega_5)_o = (2mr^2/5)\omega_5 = I\,\omega_5. \tag{c12}$$

Next, to the virtual work of the impressed forces. Recalling (ex. 2.13.6: i1–6), we obtain, successively,

$$\begin{aligned}
\delta' W &= Q_X\,\delta X + Q_Y\,\delta Y + Q_\phi\,\delta\phi + Q_\theta\,\delta\theta + Q_\psi\,\delta\psi \\
&= Q_X(\delta\theta_1 + r\,\delta\theta_4) + Q_Y(\delta\theta_2 - r\,\delta\theta_3) \\
&\quad + Q_\phi(-\cot\theta\,\sin\phi\,\delta\theta_3 + \cot\theta\,\cos\phi\,\delta\theta_4 + \delta\theta_5) \\
&\quad + Q_\theta(\cos\phi\,\delta\theta_3 + \sin\phi\,\delta\theta_4) + Q_\psi[(\sin\phi/\sin\theta)\,\delta\theta_3 - (\cos\phi/\sin\theta)\,\delta\theta_4] \\
&= (Q_X)\,\delta\theta_1 + (Q_Y)\,\delta\theta_2 + [-r\,Q_Y - \cot\theta\,\sin\phi\,Q_\phi + \cos\phi\,Q_\theta + (\sin\phi/\sin\theta)Q_\psi]\,\delta\theta_3 \\
&\quad + [r\,Q_X + \cot\theta\,\cos\phi\,Q_\phi + \sin\phi\,Q_\theta - (\cos\phi/\sin\theta)Q_\psi]\,\delta\theta_4 + Q_\phi\,\delta\theta_5, \qquad \text{(c13)}
\end{aligned}$$

and, therefore,

$$\Theta_1 = Q_X, \qquad \Theta_2 = Q_Y, \tag{c14, 15}$$

$$\Theta_3 = -r\,Q_Y - \cot\theta\,\sin\phi\,Q_\phi + \cos\phi\,Q_\theta + (\sin\phi/\sin\theta)Q_\psi, \tag{c16}$$

$$\Theta_4 = r\,Q_X + \cot\theta\,\cos\phi\,Q_\phi + \sin\phi\,Q_\theta - (\cos\phi/\sin\theta)Q_\psi, \tag{c17}$$

$$\Theta_5 = Q_\phi. \tag{c18}$$

[If no reactions are sought, we only need $\Theta_{3,4,5}$. Then we can enforce the constraints in $\delta' W = \Theta_3\,\delta\theta_3 + \Theta_4\,\delta\theta_4 + \Theta_5\,\delta\theta_5$.]

In view of the above, noting that, here, $\partial T^*/\partial q_k = 0$ and $\gamma^r{}_{I,n+1} = 0$, and recalling (ex. 2.13.6: 11–m, with $\Omega = 0$), the *kinetic* Hamel equations $\dot P_I + \sum\sum \gamma^r{}_{II'} P_r\,\omega_{I'} \equiv \dot P_I - \Gamma_I = \Theta_I$ $(I, I' = 3, 4, 5;\ r = 1, \ldots, 5)$ yield

$$\begin{aligned}
I = 3: \quad & (7mr^2/5)\dot\omega_3 - \Gamma_3 \\
&= (7mr^2/5)\dot\omega_3 + \gamma^1{}_{34} P_1\,\omega_4 + \gamma^1{}_{35} P_1\,\omega_5 + \gamma^2{}_{34} P_2\,\omega_4 + \gamma^2{}_{35} P_2\,\omega_5 \\
&\qquad + \gamma^4{}_{35} P_4\,\omega_5 + \gamma^5{}_{34} P_5\,\omega_4
\end{aligned}$$

(only the *first*, *third*, *sixth*, *and seventh* terms survive)

$$\begin{aligned}
&= (7mr^2/5)\dot\omega_3 - P_5\,\omega_4 + (P_4 - rP_1)\omega_5 \\
&= (7mr^2/5)\dot\omega_3 - (I\omega_5)\omega_4 + [(7mr^2/5)\omega_4 - r(m\,r\,\omega_4)]\omega_5 \\
&= (7mr^2/5)\dot\omega_3 + 0 \qquad \text{(recalling that } I = 2mr^2/5); \qquad \text{(c19)}
\end{aligned}$$

and similarly for $I = 4, 5$. The final results are (recalling §3.15)

$I = 3$: $\quad (7mr^2/5)\dot{\omega}_3 = \Theta_3$,

or

$$(7mr^2/5)\dot{\omega}_X = -r\,Q_Y - \cot\theta\,\sin\phi\,Q_\phi + \cos\phi\,Q_\theta + (\sin\phi/\sin\theta)Q_\psi = M_X - r\,Q_Y; \tag{c20}$$

$I = 4$: $\quad (7mr^2/5)\dot{\omega}_4 = \Theta_4$,

or

$$(7mr^2/5)\dot{\omega}_Y = r\,Q_X + \cot\theta\,\cos\phi\,Q_\phi + \sin\phi\,Q_\theta - (\cos\phi/\sin\theta)Q_\psi = M_Y + r\,Q_X; \tag{c21}$$

$I = 5$: $\quad (2mr^2/5)\,\dot{\omega}_5 = \Theta_5$,

or

$$(2mr^2/5)\,\dot{\omega}_Z = Q_\phi. \tag{c22}$$

These three (first-order) equations, plus the *three* kinematic relations $\omega_{X,Y,Z} \Leftrightarrow \dot{\phi}, \dot{\theta}, \dot{\psi}$ (§1.12) and the *two* constraints (a), constitute a determinate system in the eight functions: $X(t)$, $Y(t)$; $\phi(t), \theta(t), \psi(t)$; $\omega_{X,Y,Z}(t)$.

Instead of the *space-fixed* ω-components $\omega_{X,Y,Z}$, we could just as well have used its *body-fixed* components $\omega_{x,y,z}$; or, we can always invoke the $\omega_{X,Y,Z} \Leftrightarrow \omega_{x,y,z}$ relations. However, because of the complete *symmetry* of this problem, the space-fixed axes seems the best choice.

The Appell Equations

Since

$$\begin{aligned} a_G{}^2 = (\ddot{X})^2 + (\ddot{Y})^2 &= [(\omega_1 + r\,\omega_4)^{\cdot}]^2 + [(\omega_2 - r\,\omega_3)^{\cdot}]^2 \\ &= (\dot{\omega}_1)^2 + (\dot{\omega}_2)^2 + r^2[(\dot{\omega}_3)^2 + (\dot{\omega}_4)^2] + 2r(\dot{\omega}_1\,\dot{\omega}_4 - \dot{\omega}_2\,\dot{\omega}_3), \end{aligned} \tag{d1}$$

and, as in the preceding example (*à la* König),

$$S = S_G + S_{/G}, \qquad 2S_G = m\,a_G{}^2, \qquad 2S_{/G} \equiv \int dm\,\mathbf{a}_{/G} \cdot \mathbf{a}_{/G} = \int dm\,a_{/G}{}^2, \tag{d2}$$

it is not too hard to see that S^* equals the earlier T^*, but with the ω's replaced by the corresponding $\dot{\omega}$'s; that is,

$$\begin{aligned} 2S^* = 2T^*(\dot{\omega}) = m\Big\{ &(\dot{\omega}_1)^2 + (\dot{\omega}_2)^2 + (7r^2/5)[(\dot{\omega}_3)^2 + (\dot{\omega}_4)^2] \\ &+ (2r^2/5)(\dot{\omega}_5)^2 + 2r(\dot{\omega}_1\dot{\omega}_4 - \dot{\omega}_2\,\dot{\omega}_3)\Big\}. \end{aligned} \tag{d3}$$

Also, as the theory shows (3.5.25a ff.), if we are not interested in finding the constraint reactions, we can enforce the constraints $\omega_{1,2} = 0 \Rightarrow \dot{\omega}_{1,2} = 0$ in S^* right from the start; that is, we can neglect from it not just the quadratic terms in $\dot{\omega}_{1,2}$ but also the linear ones. Thus, we can take $2S^*(\dot{\omega}_{1,2}=0) \equiv 2S^*_o$:

$$2S^*_o = m\Big\{(7r^2/5)[(\dot{\omega}_3)^2 + (\dot{\omega}_4)^2] + (2r^2/5)(\dot{\omega}_5)^2\Big\}, \tag{d4}$$

and therefore the kinetic Appellian equations are

$$\partial S^*_o/\partial\dot{\omega}_3 = (7r^2/5)\dot{\omega}_3 = \Theta_3, \tag{d5}$$

$$\partial S^*_o/\partial\dot{\omega}_4 = (7r^2/5)\dot{\omega}_4 = \Theta_4, \tag{d6}$$

$$\partial S^*_o/\partial\dot{\omega}_5 = (2r^2/5)\dot{\omega}_5 = \Theta_5; \tag{d7}$$

and, of course, these coincide with the earlier kinetic Hamel equations.

The Chaplygin Equations (3.8.13a ff.)

We recall that here the (double) kinetic energy equals

$$2T = m[(\dot{X})^2 + (\dot{Y})^2] + I[(\dot{\phi})^2 + (\dot{\theta})^2 + (\dot{\psi})^2 + 2\dot{\phi}\,\dot{\psi}\cos\theta], \tag{e1}$$

while the constraints (a), rewritten in the Chaplygin form—that is,

$$\dot{q}_D = \sum b_{DI}\,\dot{q}_I, \qquad \text{where} \qquad q_D = X, Y;\ q_I = \phi, \theta, \psi,$$

are

$$\dot{X} = (r\sin\phi)\dot{\theta} - (r\cos\phi\sin\theta)\dot{\psi} \quad \text{and} \quad \dot{Y} = -(r\cos\phi)\dot{\theta} - (r\sin\phi\sin\theta)\dot{\psi}. \tag{e2}$$

Therefore, in this problem, the "Chaplygin coefficients" are (with some easily understood ad hoc notation)

$$b_{13} \equiv b_{X\phi} = 0, \qquad b_{14} \equiv b_{X\theta} = r\sin\phi, \qquad b_{15} \equiv b_{X\psi} = -r\cos\phi\sin\theta,$$

$$b_{23} \equiv b_{Y\phi} = 0, \qquad b_{24} \equiv b_{Y\theta} = -r\cos\phi, \qquad b_{25} \equiv b_{Y\psi} = -r\sin\phi\sin\theta. \tag{e3}$$

Clearly, since (α) the constraints (e2, 3) are stationary, and (β) the chosen "dependent" coordinates X, Y do not appear either in the constraint coefficients b_{DI} or in T (i.e., $\partial T/\partial X = 0$, $\partial T/\partial Y = 0$), this is a Chaplygin system, and, therefore, Chaplygin's equations hold; and, for this problem, they coincide with Voronets' equations. Let us find them.

(i) Eliminating $\dot{X}$ and $\dot{Y}$ from T with the help of the constraints (e2), we obtain the "Chaplygin constrained kinetic energy"

$$\begin{aligned} 2T &= 2T[\theta, \dot{X}(\phi,\theta;\ \dot{\theta},\dot{\psi}), \dot{Y}(\phi,\theta;\ \dot{\theta},\dot{\psi}), \dot{\phi},\dot{\theta},\dot{\psi}] \\ &= \cdots = (mr^2)[(2/5)(\dot{\phi})^2 + (7/5)(\dot{\theta})^2 + ((2/5)+\sin^2\theta))(\dot{\psi})^2 + (4\cos\theta/5)\dot{\phi}\,\dot{\psi}\} \\ &= T_o(\phi,\theta;\ \dot{\phi},\dot{\theta},\dot{\psi}) \equiv T_o, \end{aligned} \tag{e4}$$

and, therefore,

$$E_3(T_o) \equiv E_\phi(T_o) \equiv (\partial T_o/\partial\dot\phi)^{\cdot} - \partial T_o/\partial\phi = (2mr^2/5)(\ddot\phi + \ddot\psi\cos\theta - \dot\theta\dot\psi\sin\theta),$$

$$E_4(T_o) \equiv E_\theta(T_o) \equiv (\partial T_o/\partial\dot\theta)^{\cdot} - \partial T_o/\partial\theta$$
$$= (mr^2)[(7/5)\ddot\theta - (\dot\psi)^2\sin\theta\cos\theta + (2/5)\dot\phi\dot\psi\sin\theta],$$

$$E_5(T_o) \equiv E_\psi(T_o) \equiv (\partial T_o/\partial\dot\psi)^{\cdot} - \partial T_o/\partial\psi$$
$$= (mr^2)\{(2/5)\ddot\phi\cos\theta + ((2/5) + \sin^2\theta)\ddot\psi + 2\dot\theta\dot\psi\sin\theta\cos\theta$$
$$- (2/5)\dot\phi\dot\theta\sin\theta]. \qquad \text{(e5, 6, 7)}$$

(ii) Next, let us calculate the corresponding "Chaplygin corrective terms"

$$-\Gamma_{Io} \equiv \sum\sum(\partial b_{DI'}/\partial q_I - \partial b_{DI}/\partial q_{I'})(\partial T/\partial\dot q_D)_o\,\dot q_{I'}$$
$$\equiv \sum\sum t^D{}_{I'I}(\partial T/\partial\dot q_D)_o\,\dot q_{I'} \equiv \sum\sum t^D{}_{I'I}\,p_{Do}\,\dot q_{I'}. \qquad \text{(e8)}$$

Using commas for partial derivatives relative to the q_I, we obtain, successively,

$$-\Gamma_{3o} \equiv -\Gamma_{\phi o} \equiv \sum\sum t^D{}_{I3}\,p_{Do}\,\dot q_I = \sum\Big(\sum t^D{}_{I3}\,\dot q_I\Big)p_{Do}$$
$$= [(b_{13,3} - b_{13,3})\dot q_3 + (b_{14,3} - b_{13,4})\dot q_4 + (b_{15,3} - b_{13,5})\dot q_5]p_{1o}$$
$$+ [(b_{23,3} - b_{23,3})\dot q_3 + (b_{24,3} - b_{23,4})\dot q_4 + (b_{25,3} - b_{23,5})\dot q_5]p_{2o}$$
$$= [(b_{X\theta,\phi} - b_{X\phi,\theta})\dot\theta + (b_{X\psi,\phi} - b_{X\phi,\psi})\dot\psi]p_{Xo}$$
$$+ [(b_{Y\theta,\phi} - b_{Y\phi,\theta})\dot\theta + (b_{Y\psi,\phi} - b_{Y\phi,\psi})\dot\psi]p_{Yo},$$

or, since,

$$p_{1o} \equiv p_{Xo} \equiv (\partial T/\partial\dot X)_o = (m\dot X)_o = mr(\dot\theta\sin\phi - \dot\psi\cos\phi\sin\theta),$$
$$p_{2o} \equiv p_{Yo} \equiv (\partial T/\partial\dot Y)_o = (m\dot Y)_o = -mr(\dot\theta\cos\phi + \dot\psi\sin\phi\sin\theta),$$

finally,

$$-\Gamma_{3o} \equiv -\Gamma_{\phi o}$$
$$= [(r\cos\phi - 0)\dot\theta + (r\sin\phi\sin\theta - 0)\dot\psi](mr)(\dot\theta\sin\phi - \dot\psi\cos\phi\sin\theta)$$
$$+ [(r\sin\phi - 0)\dot\theta + (-r\cos\phi\sin\theta - 0)\dot\psi](-mr)(\dot\theta\cos\phi + \dot\psi\sin\phi\sin\theta)$$
$$= \cdots = 0; \qquad \text{(e9)}$$

and similarly, after some careful algebra,

$$-\Gamma_{4o} \equiv -\Gamma_{\theta o} = \sum\Big(\sum t^D{}_{I4}\,\dot q_I\Big)p_{Do} = \cdots = mr^2(\dot\phi + \dot\psi\cos\theta)(\dot\psi\sin\theta), \qquad \text{(e10)}$$

$$-\Gamma_{5o} \equiv -\Gamma_{\psi o} = \sum\Big(\sum t^D{}_{I5}\,\dot q_I\Big)p_{Do} = \cdots = -mr^2(\dot\phi + \dot\psi\cos\theta)(\dot\theta\sin\theta). \qquad \text{(e11)}$$

The above show that the *nonvanishing* Chaplygin coefficients are

$$t^1{}_{43} \equiv t^X{}_{\theta\phi} = r\cos\phi \quad (= -t^1{}_{34} \equiv -t^X{}_{\phi\theta}), \tag{e12}$$

$$t^1{}_{53} \equiv t^X{}_{\psi\phi} = r\sin\phi\sin\theta \quad (= -t^1{}_{35} \equiv -t^X{}_{\phi\psi}), \tag{e13}$$

$$t^1{}_{54} \equiv t^X{}_{\psi\theta} = -r\cos\phi\cos\theta \quad (= -t^1{}_{45} \equiv -t^X{}_{\theta\psi}); \tag{e14}$$

$$t^2{}_{43} \equiv t^Y{}_{\theta\phi} = r\sin\phi \quad (= -t^2{}_{34} \equiv -t^Y{}_{\phi\theta}), \tag{e15}$$

$$t^2{}_{53} \equiv t^Y{}_{\psi\phi} = -r\cos\phi\sin\theta \quad (= -t^2{}_{35} \equiv -t^Y{}_{\phi\psi}), \tag{e16}$$

$$t^2{}_{54} \equiv t^Y{}_{\psi\theta} = -r\sin\phi\cos\theta \quad (= -t^2{}_{45} \equiv -t^Y{}_{\theta\psi}). \tag{e17}$$

(iii) Finally, let us calculate the "Chaplygin impressed forces" $Q_{Io} \equiv Q_I + \sum b_{DI} Q_D$. With some obvious ad hoc notation, we find

$$Q_{3o} \equiv Q_{\phi o} = Q_3 + b_{13}Q_1 + b_{23}Q_2 = Q_3 \equiv Q_\phi, \tag{e18}$$

$$\begin{aligned} Q_{4o} \equiv Q_{\theta o} &= Q_4 + b_{14}Q_1 + b_{24}Q_2 = Q_4 + (r\sin\phi)Q_1 + (-r\cos\phi)Q_2 \\ &= Q_\theta + r(Q_X\sin\phi - Q_Y\cos\phi), \end{aligned} \tag{e19}$$

$$\begin{aligned} Q_{5o} \equiv Q_{\psi o} &= Q_5 + b_{15}Q_1 + b_{25}Q_2 = Q_5 + (-r\cos\phi\sin\theta)Q_1 + (-r\sin\phi\sin\theta)Q_2 \\ &= Q_\psi - r\sin\theta(Q_X\cos\phi + Q_Y\sin\phi). \end{aligned} \tag{e20}$$

Inserting now all the above partial results into Chaplygin's equations (3.8.13o):

$$(\partial T_o/\partial \dot{q}_I)^{\cdot} - \partial T_o/\partial q_I - \Gamma_{Io} = Q_{Io} \tag{e21}$$

(with I: $3 \to \phi, 4 \to \theta, 5 \to \psi$), and, simplifying a little, we obtain the following three kinetic equations:

$$\phi: \quad (2mr^2/5)(\ddot{\phi} + \ddot{\psi}\cos\theta - \dot{\theta}\dot{\psi}\sin\theta) = Q_\phi, \tag{e22}$$

$$\theta: \quad (7mr^2/5)(\ddot{\theta} + \dot{\phi}\dot{\psi}\sin\theta) = Q_\theta + r(Q_X\sin\phi - Q_Y\cos\phi), \tag{e23}$$

$$\begin{aligned} \psi: \quad & (mr^2)[(2/5)\ddot{\phi}\cos\theta + (2/5)\ddot{\psi} + \ddot{\psi}\sin^2\theta - (7/5)\dot{\phi}\dot{\theta}\sin\theta \\ & + \dot{\theta}\dot{\psi}\sin\theta\cos\theta] = Q_\psi - r\sin\theta(Q_X\cos\phi + Q_Y\sin\phi). \end{aligned} \tag{e24}$$

These latter, of course, coincide (i) with the earlier-found kinetic Maggi equations (b6–8), after we eliminate in them $\ddot{X}$ and $\ddot{Y}$ using the $(\ldots)^{\cdot}$-differentiated constraints (a); and (ii) with the earlier kinetic Hamel equations (c20–22), after we express $\dot{\phi}, \ldots, \ddot{\phi}, \ldots,$ in terms of $\omega_{3,4,5}$ and $\dot{\omega}_{3,4,5}$, and $Q_{\phi,\theta,\psi;o}$ in terms of $\Theta_{3,4,5}$; and similarly with the Appell equations (d5–7). The details are left to the reader.

REMARK

(May be omitted in a first reading.) A safer way to calculate the $-\Gamma_{Io}$ — and, in fact, the entire set of Chaplygin's equations — is by direct application of the *Chaplygin form of the master variational equation* [specialization of the corresponding equation of Hamel (§3.6.12)]

$$\sum (\partial T_o/\partial \dot{q}_I)^{\cdot}\,\delta q_I - \sum (\partial T_o/\partial q_I)\,\delta q_I - \sum \Gamma_{Io}\,\delta q_I = \sum Q_{Io}\,\delta q_I, \tag{e25}$$

where

$$-\sum \Gamma_{Io}\,\delta q_I = \sum\Big(\sum\sum t^D{}_{I'I}\,p_{Do}\,\dot q_{I'}\Big)\,\delta q_I$$
$$= \sum\Big(\sum\sum t^D{}_{I'I}\,\dot q_{I'}\,\delta q_I\Big)p_{Do} = -\sum p_{Do}\big[(\delta q_D)^{\cdot} - \delta(\dot q_D)\big]. \quad \text{(e26)}$$

The reason for this is that here we have, in effect, adopted the *Suslov viewpoint* according to which

$$(\delta\theta_k)^{\cdot} - \delta\omega_k = 0 \qquad \text{and} \qquad (\delta q_I)^{\cdot} - \delta(\dot q_I) = 0, \quad \text{(e27)}$$

but, successively,

$$(\delta q_D)^{\cdot} - \delta(\dot q_D) = \Big(\sum b_{DI}\,\delta q_I\Big)^{\cdot} - \delta\Big(\sum b_{DI'}\,\dot q_{I'}\Big)$$
$$= \sum\sum (b_{DI,I'}\,\dot q_{I'}\,\delta q_I - b_{DI',I}\,\dot q_{I'}\,\delta q_I)$$
$$= \sum\sum (b_{DI,I'} - b_{DI',I})\dot q_{I'}\,\delta q_I$$
$$\equiv \sum\sum t^D{}_{II'}\,\dot q_{I'}\,\delta q_I = -\sum\sum t^D{}_{I'I}\,\dot q_{I'}\,\delta q_I; \quad \text{(e28)}$$

whereas in the customary, and more general, viewpoint of Hamel (§2.12): $(\delta q_k)^{\cdot} - \delta(\dot q_k) = 0$, $(k = 1, \ldots, n)$, but

$$(\delta\theta_D)^{\cdot} - \delta\omega_D = \sum\sum \gamma^D{}_{I\alpha}\,\omega_\alpha\,\delta\theta_I \;\Rightarrow\; \sum\sum t^D{}_{I'I}\,\dot q_{I'}\,\delta q_I \neq 0, \quad \text{(e29)}$$

and, accordingly, the term corresponding to $-\sum \Gamma_{Io}\,\delta q_I$ is [recalling (§3.6.11)]

$$\sum P_k[(\delta\theta_k)^{\cdot} - \delta\omega_k] \to \sum p_{Do}[(\delta\theta_D)^{\cdot} - \delta\omega_D]$$
$$= \sum p_{Do}\Big(\sum\sum t^D{}_{I'I}\,\dot q_{I'}\,\delta q_I\Big) - \sum\Big(\sum\sum t^D{}_{I'I}\,p_{Do}\,\dot q_{I'}\Big)\,\delta q_I; \quad \text{(e30)}$$

that is, the two interpretations may be different, but, if utilized consistently, both lead to the same equations of motion [and similarly for the case of Voronets (3.8.14a ff.)].

In our problem, adopting the Suslov viewpoint, we obtain, successively [recalling (e2), etc.],

$$-\sum p_{Do}[(\delta q_D)^{\cdot} - \delta(\dot q_D)] = -p_{Xo}[(\delta X)^{\cdot} - \delta(\dot X)] - p_{Yo}[(\delta Y)^{\cdot} - \delta(\dot Y)]$$
$$= -[mr(\dot\theta\sin\phi - \dot\psi\cos\phi\sin\theta)]\ \{[(r\sin\phi)\,\delta\theta - (r\cos\phi\sin\theta)\,\delta\psi]^{\cdot}$$
$$- \delta[(r\sin\phi)\dot\theta - (r\cos\phi\sin\theta)\dot\psi]\}$$
$$- [-mr(\dot\theta\cos\phi + \dot\psi\sin\phi\sin\theta)]\ \{[(-r\cos\phi)\,\delta\theta + (-r\sin\phi\sin\theta)\,\delta\psi]^{\cdot}$$
$$- \delta[(-r\cos\phi)\dot\theta + (-r\sin\phi\sin\theta)\dot\psi]\}$$
$$= \cdots = (0)\,\delta\phi + [mr^2(\dot\psi\sin\theta)(\dot\phi + \dot\psi\cos\theta)]\,\delta\theta$$
$$+ [(-mr^2)(\dot\theta\sin\theta)(\dot\phi + \dot\psi\cos\theta)]\,\delta\psi$$
$$= -(\Gamma_{\phi o}\,\delta\phi + \Gamma_{\theta o}\,\delta\theta + \Gamma_{\psi o}\,\delta\psi), \quad \text{(e31)}$$

as (e26) requires; and, of course, in agreement with the earlier (e9–11).

The above make clear that the methods of Chaplygin, and Voronets, are rather complicated and error prone, even in this relatively simple problem; and the only reason for working it out completely was [just like Chaplygin's original effort (1895/1897)] to demonstrate concretely that, *in general*,

$$E_I(T_o) \equiv (\partial T_o/\partial \dot{q}_I)^{\cdot} - \partial T_o/\partial q_I \neq Q_{Io}; \tag{e32}$$

although, in this case, the equality *does* hold for $q_I = \phi$ [see also Beghin, 1967, I, pp. 436–438), and "a famous error" in our *rolling coin* example 3.18.5 (below)].

The Hadamard Equations

$E_I(T) + \sum b_{DI} E_D(T) = Q_I + \sum b_{DI} Q_D$ $(\equiv Q_{Io})$. The $E_k(T)$ $(k = 1, \ldots, 5)$, b_{DI}, and Q_{Io} have already been calculated. It is not hard to show that the final result would be the earlier kinetic Maggi equations (b6–8); while for $D = 1, 2$ we would simply have [recalling (3.8.11a ff.)]

$$E_D(T) = Q_D + \lambda_D = Q_D + R_D. \tag{f}$$

Constraint Reactions

In view of the above, we obtain, successively,

$$R_1 \equiv R_X = m\ddot{X} - Q_X \qquad \text{and} \qquad R_2 \equiv R_Y = m\ddot{Y} - Q_Y; \tag{g1}$$

and [recalling (3.8.11l)] $R_I = \sum \lambda_D a_{DI} = \sum \lambda_D(-b_{DI}) = -\sum b_{DI}[E_D(T) - Q_D]$:

$$\begin{aligned}
R_3 = R_\phi &= -b_{13}[E_1(T) - Q_1] - b_{23}[E_2(T) - Q_2] = -(0)(\ldots) - (0)(\ldots) = 0,\\
R_4 = R_\theta &= -b_{14}[E_1(T) - Q_1] - b_{24}[E_2(T) - Q_2]\\
&= -r\sin\phi(m\ddot{X} - Q_X) - (-r\cos\phi)(m\ddot{Y} - Q_Y)\\
&= r[-m(\ddot{X}\sin\phi - \ddot{Y}\cos\phi) + (Q_X\sin\phi - Q_Y\cos\phi)],\\
R_5 = R_\psi &= -b_{15}[E_1(T) - Q_1] - b_{25}[E_2(T) - Q_2]\\
&= -(-r\cos\phi\sin\theta)(m\ddot{X} - Q_X) - (-r\sin\phi\sin\theta)(m\ddot{Y} - Q_Y)\\
&= r[m(\ddot{X}\cos\phi + \ddot{Y}\sin\phi) - (Q_X\cos\phi + Q_Y\sin\phi)];
\end{aligned} \tag{g2, 3, 4}$$

so that once the motion has been found, the reactions can be readily determined.

Example 3.18.3 *Dynamics of a Sphere Rolling on a Uniformly Spinning Plane.*

Introduction: Hamel's Equations

Continuing from the preceding example, let us find the motion of that sphere if the plane P is revolving about the fixed (vertical) axis OZ with *constant* angular velocity Ω. The relevant kinematics has already been discussed in ex. 2.13.5 and ex. 2.13.6. It was shown there that the rolling constraints are (note additional *acatastatic* terms)

$$\begin{aligned}
\dot{X} - (r\sin\phi)\dot{\theta} + (r\cos\phi\sin\theta)\dot{\psi} + \Omega Y = 0,\\
\dot{Y} + (r\cos\phi)\dot{\theta} + (r\sin\phi\sin\theta)\dot{\psi} - \Omega X = 0.
\end{aligned} \tag{a1}$$

Since the catastatic coefficients $[a_{Dk};\ D = 1, 2;\ k = 1, \ldots, 5]$ and kinetic energy have the same form as in the previous catastatic case, the equations of motion of Routh–Voss and Maggi–Hadamard are the same *in form* as before; their *solutions*, however, will be different because these equations must now be joined with the different constraints (a1).

Similarly, LP and the *virtual* form of the constraints (a) remain the same. But since T in quasi variables, T^*, is not the same as before, Hamel's equations (which incorporate the new constraints) will be different; and so will be those of Appell. Indeed, and remembering *not* to enforce the constraints $\omega_{1,2} = 0$ until the final stage, we find

$$2T \to 2T^* = m[(\omega_1 + r\,\omega_4 - \Omega\, Y)^2 + (\omega_2 - r\,\omega_3 + \Omega\, X)^2] + I(\omega_3{}^2 + \omega_4{}^2 + \omega_5{}^2). \quad \text{(a2)}$$

The γ's have already been calculated in (ex. 2.13.6: 11–m); we have also found that

$$A_{13} = 0, \quad A_{14} = r, \quad A_{15} = 0; \qquad A_{23} = -r, \quad A_{24} = 0, \quad A_{25} = 0. \quad \text{(a3)}$$

Therefore, Hamel's kinetic equations

$$(\partial T^*/\partial\omega_I)^{\cdot} - \sum A_{DI}(\partial T^*/\partial q_D) + \sum\sum \gamma^k{}_{II'}(\partial T^*/\partial\omega_k)\omega_{I'}$$
$$+ \sum \gamma^k{}_I(\partial T^*/\partial\omega_k) = \Theta_I, \quad \text{(a4)}$$

yield, after some straightforward and careful algebra,

$$\dot P_3 - \partial T^*/\partial\theta_3 + P_4\,\omega_5 - P_5\,\omega_4 + P_1\, r(\Omega - \omega_5) = \Theta_3, \quad \text{(a5)}$$

$$\dot P_4 - \partial T^*/\partial\theta_4 + P_5\,\omega_3 - P_3\,\omega_5 + P_2\, r(\Omega - \omega_5) = \Theta_4, \quad \text{(a6)}$$

$$\dot P_5 - \partial T^*/\partial\theta_5 + P_3\,\omega_4 - P_4\,\omega_3 + P_1 r\,\omega_3 + P_2\, r\,\omega_4 = \Theta_5; \quad \text{(a7)}$$

or, explicitly,

$$(7mr^2/5)\dot\omega_X - m\,r\,\Omega(r\,\omega_Y - \Omega Y) = \Theta_X, \quad \text{(a8)}$$

$$(7mr^2/5)\dot\omega_Y + m\,r\,\Omega(r\,\omega_X - \Omega X) = \Theta_Y, \quad \text{(a9)}$$

$$(2mr^2/5)\dot\omega_Z = \Theta_Z; \quad \text{(a10)}$$

we notice the Ω-proportional terms [in addition to those of eqs. (c20–22) of the preceding example]. The extension to the general case $\Omega = \Omega(t)$ involves only some algebraic complications; in particular, eqs. (a5–7) still hold.

Hamel's Equations via the Master Variational Equation

$$\sum \dot P_k\,\delta\theta_k + \sum P_k[(\delta\theta_k)^{\cdot} - \delta\omega_k] - \sum\Big[\sum A_{lk}(\partial T^*/\partial q_l)\Big]\,\delta\theta_k = \sum \Theta_k\,\delta\theta_k. \quad \text{(b1)}$$

The direct application of (b1), for the derivation of Hamel equations of motion, is recommended in order to minimize the probability of errors. Its main advantage lies in the calculation of the *second* (noncommutative) term. Let us carry this out explicitly: Recalling the transitivity relations (ex. 2.13.6: j ff.), we find, successively,

$$
\begin{aligned}
&P_1[(-\Omega)\,\delta\theta_2 + r(\Omega - \omega_5)\,\delta\theta_3 + (r\,\omega_3)\,\delta\theta_5] \\
&\quad + P_2[(\Omega)\,\delta\theta_1 + r(\Omega - \omega_5)\,\delta\theta_4 + (r\,\omega_4)\,\delta\theta_5] \\
&\quad + P_3[(\omega_4)\,\delta\theta_5 + (-\omega_5)\,\delta\theta_4] \\
&\quad + P_4[(\omega_5)\,\delta\theta_3 + (-\omega_3)\,\delta\theta_5] \\
&\quad + P_5[(\omega_3)\,\delta\theta_4 + (-\omega_4)\,\delta\theta_3] \\
&= (P_2\,\Omega)\,\delta\theta_1 + (-P_1\,\Omega)\,\delta\theta_2 + [P_4\,\omega_5 - P_5\,\omega_4 + P_1 r(\Omega - \omega_5)]\,\delta\theta_3 \\
&\quad + [P_5\,\omega_3 - P_3\,\omega_5 + P_2\,r(\Omega - \omega_5)]\,\delta\theta_4 \\
&\quad + [P_3\,\omega_4 - P_4\,\omega_3 + r(P_1\,\omega_3 + P_2\,\omega_4)]\,\delta\theta_5. \qquad \text{(b2)}
\end{aligned}
$$

In the above, we notice that (a) the first and second terms, are needed in the *kinetostatic* equations ($\delta\theta_{1,2} = 0$); while the rest are needed in the *kinetic* equations ($\delta\theta_{3,4,5} = 0$); and (b) the constraints have been enforced in the velocity form $\omega_{1,2} = 0$, but not in the virtual form $\delta\theta_{1,2} = 0$ (unless we are not interested in the reactions).

Let the reader verify that by collecting $(\ldots)\,\delta\theta_k$ terms, and so on, and applying the method of multipliers, eqs. (b1, 2) lead to the following full set of equations of motion:

Kinetostatic equations:

$$\delta\theta_1: \qquad \dot{P}_1 - A_{11}P_1 - A_{21}P_2 + \Omega\,P_2 = \Theta_1 + \Lambda_1, \qquad \text{(b3)}$$

$$\delta\theta_2: \qquad \dot{P}_2 - A_{12}P_1 - A_{22}P_2 - \Omega\,P_1 = \Theta_2 + \Lambda_2; \qquad \text{(b4)}$$

Kinetic equations:

$$\delta\theta_3: \qquad \dot{P}_3 - A_{13}P_1 - A_{23}P_2 + P_4\,\omega_5 - P_5\,\omega_4 + P_1\,r(\Omega - \omega_5) = \Theta_3, \qquad \text{(b5)}$$

$$\delta\theta_4: \qquad \dot{P}_4 - A_{14}P_1 - A_{24}P_2 + P_5\,\omega_3 - P_3\,\omega_5 + P_2\,r(\Omega - \omega_5) = \Theta_4, \qquad \text{(b6)}$$

$$\delta\theta_5: \qquad \dot{P}_5 - A_{15}P_1 - A_{25}P_2 + P_3\,\omega_4 - P_4\,\omega_3 + r(P_1\,\omega_3 + P_2\,\omega_4) = \Theta_5; \qquad \text{(b7)}$$

the last three in complete agreement with the earlier-found equations (a5–7).

The Appell Equations (Kinetic Equations Only)

Using the customary notations, we find

$$2S^* = m\,a_G{}^2 + 2S^*{}_{/G}, \qquad \text{(c1)}$$

where

$$2S^*_{/G} = I[(\dot{\omega}_X)^2 + (\dot{\omega}_Y)^2 + (\dot{\omega}_Z)^2] = I[(\dot{\omega}_3)^2 + (\dot{\omega}_4)^2 + (\dot{\omega}_5)^2], \tag{c2}$$

$$\begin{aligned} a_G{}^2 &= (\ddot{X})^2 + (\ddot{Y})^2 = [(r\,\omega_4 - \Omega\,Y + \omega_1)^{\cdot}]^2 + [(-r\,\omega_3 + \Omega\,X + \omega_2)^{\cdot}]^2 \\ &= [r\,\dot{\omega}_4 + \Omega(r\,\omega_3 - \Omega\,X)]^2 + [-r\,\dot{\omega}_3 + \Omega(r\,\omega_4 - \Omega\,Y)]^2 \\ &= r^2[(\dot{\omega}_3)^2 + (\dot{\omega}_4)^2] + 2r\,\Omega[(r\,\omega_3 - \Omega\,X)\dot{\omega}_4 - (r\,\omega_4 - \Omega\,Y)\dot{\omega}_3] \\ &\quad + \textit{non–Appell-important terms}. \end{aligned} \tag{c3}$$

Therefore, to within *Appell-important terms*, and since $I = 2mr^2/5$, the *constrained* (double) Appellian of the sphere equals

$$\begin{aligned} 2S^* \rightarrow 2S^*{}_o &= m\{(7r^2/5)[(\dot{\omega}_3)^2 + (\dot{\omega}_4)^2] + (2r^2/5)(\dot{\omega}_5)^2\} \\ &\quad + 2m\,r\,\Omega[(r\,\omega_3 - \Omega\,X)\dot{\omega}_4 - (r\,\omega_4 - \Omega\,Y)\dot{\omega}_3], \end{aligned} \tag{c4}$$

where the *second*, Ω-proportional, group of terms is due to the rotation of the plane. Differentiating this Appellian relative to $\dot{\omega}_{3,4,5}$ yields the left sides of the earlier kinetic equations (a5–7, b5–7).

Example 3.18.4 *Power Equations/Energetics of Rolling Sphere on Uniformly Spinning Plane.* Let us begin by collecting all needed analytical results; already calculated in exs. 2.13.5 and 2.13.6, and the preceding examples 3.18.2 and 3.18.3. [Here, too, the case $\Omega = constant$ was chosen for its algebraic simplicity; the general case $\Omega = \textit{given function of time}$ would not have offered any theoretical difficulties. We have (with $I \equiv 2mr^2/5$)

(i)
$$\begin{aligned} 2T &= m[(\dot{X})^2 + (\dot{Y})^2] + (I_X{}^2\omega_X{}^2 + I_Y{}^2\omega_Y{}^2 + I_Z{}^2\omega_Z{}^2) \\ &= m[(\dot{X})^2 + (\dot{Y})^2] + I[(\dot{\phi})^2 + (\dot{\theta})^2 + (\dot{\psi})^2 + 2\dot{\phi}\,\dot{\psi}\cos\theta]. \end{aligned} \tag{a1}$$

That here $T_1 = T_0 = 0 \Rightarrow T = T_2$, should come as no surprise. The holonomic coordinates chosen here — namely, X, Y, ϕ, θ, ψ — are *inertial*.

(ii)
$$\begin{aligned} 2T^* &= m[(\omega_1 + r\,\omega_4 - \Omega\,Y)^2 + (\omega_2 - r\,\omega_3 + \Omega\,X)^2] + I(\omega_3{}^2 + \omega_4{}^2 + \omega_5{}^2) \\ &= (\textit{expanding and grouping appropriately}) = 2T^*{}_2 + 2T^*{}_1 + 2T^*{}_0, \end{aligned} \tag{a2}$$

$$\begin{aligned} 2T^*{}_2 &= m[\omega_1{}^2 + \omega_2{}^2 + r^2(\omega_3{}^2 + \omega_4{}^2) + 2r(\omega_1\omega_4 - \omega_2\,\omega_3)] \\ &\quad + I(\omega_3{}^2 + \omega_4{}^2 + \omega_5{}^2) \\ &= \textit{(double) kinetic energy of motion of sphere relative to plane}, \end{aligned} \tag{a3}$$

$$\begin{aligned} T^*{}_1 &= m[(-Y\Omega)\,\omega_1 + (-rY\Omega)\,\omega_4 + (X\Omega)\,\omega_2 + (-rX\Omega)\,\omega_3] \\ &= \textit{kinetic energy of "coupling" of motion of plane and sphere/plane}, \end{aligned} \tag{a4}$$

$$\begin{aligned} 2T^*{}_0 &= m\,\Omega^2(X^2 + Y^2) \\ &= \textit{(double) kinetic energy of sphere when at rest relative the plane}; \end{aligned} \tag{a5}$$

and no constraints have been enforced yet.

(iii) Since the only impressed force here is gravity (i.e., $\Theta_I = 0$), the corresponding potential energy is constant; say, $V = V^* = mgr$, and, therefore,

$$\partial V/\partial q_k = 0 \;\Rightarrow\; \partial V^*/\partial \theta_k \equiv \sum A_{lk}(\partial V/\partial q_l) = 0, \tag{b1}$$

and [recalling (2.9.34 ff.) and (3.9.12f)]

$$\partial V^*/\partial \theta_{n+1} \equiv \partial V^*/\partial t + \sum A_k(\partial V^*/\partial q_k) \;=\; 0 + \sum A_k(0) = 0; \tag{b2}$$

(iv) The *nonvanishing* Hamel coefficients are [recalling ex. 2.13.6: (11–m)]

$$\begin{aligned}
&\gamma^1{}_{35} = -\gamma^1{}_{53} = -r, \qquad \gamma^1{}_{36} = -\gamma^1{}_{63} \equiv \gamma^1{}_3 = r\,\Omega, \qquad \gamma^1{}_{26} = -\gamma^1{}_{62} \equiv \gamma^1{}_2 = -\Omega;\\
&\gamma^2{}_{45} = -\gamma^2{}_{54} = -r, \qquad \gamma^2{}_{46} = -\gamma^2{}_{64} \equiv \gamma^2{}_4 = r\,\Omega, \qquad \gamma^2{}_{16} = -\gamma^2{}_{61} \equiv \gamma^2{}_1 = \Omega;\\
&\gamma^3{}_{54} = -\gamma^3{}_{45} = 1, \qquad \gamma^4{}_{35} = -\gamma^4{}_{53} = 1, \qquad \gamma^5{}_{43} = -\gamma^5{}_{34} = 1;
\end{aligned} \tag{c}$$

and from these only $\gamma^1{}_3 = r\,\Omega$ and $\gamma^2{}_4 = r\,\Omega$ will be needed in the power equation below.

Nonholonomic Power Equation

From the above, and with an eye toward (3.9.12h ff.), we find, successively,

(i)
$$\begin{aligned}
\partial L^*/\partial \theta_{n+1} &= \partial T^*/\partial \theta_{n+1} \equiv \partial T^*/\partial t + \sum A_k(\partial T^*/\partial q_k)\\
&= m[\Omega(X^2 + Y^2) - r(X\omega_X + Y\omega_Y)]\,\dot{\Omega} + A_1(\partial T^*/\partial X)\\
&\qquad\qquad\qquad\qquad\qquad + A_2(\partial T^*/\partial Y),\\
&= (-\Omega Y)[(m\,\Omega)(\Omega X - r\,\omega_3)] + (\Omega X)[(m\,\Omega)(\Omega Y - r\,\omega_4)]\\
&= \cdots = m\,r\,\Omega^2(Y\omega_X - X\omega_Y);
\end{aligned} \tag{d1}$$

(ii)
$$\begin{aligned}
R &\equiv \sum\sum \gamma^b{}_I(\partial T^*/\partial \omega_b)\,\omega_I \qquad [\text{and with } (\ldots)_o \equiv (\ldots) \textit{ evaluated at } \omega_{1,2} = 0]\\
&= \gamma^1{}_3(\partial T^*/\partial \omega_1)_o\,\omega_3 + \gamma^2{}_4(\partial T^*/\partial \omega_2)_o\,\omega_4\\
&= (r\,\Omega)[m(r\,\omega_Y - \Omega Y)]\omega_X + (r\,\Omega)[m(-r\,\omega_X + \Omega X)]\omega_Y\\
&= m\,r\,\Omega^2(X\omega_Y - Y\omega_X).
\end{aligned} \tag{d2}$$

In view of these partial results (in particular, the mutual canceling of the rheonomic effects of the nonholonomic constraints $\partial L^*/\partial \theta_{n+1} + R = 0$), the general nonholonomic power equation (3.9.12i)

$$dh^*/dt = -\,\partial L^*/\partial \theta_{n+1} + \sum \Theta_I\,\omega_I - R, \tag{e1}$$

where $h^* \equiv \sum(\partial L^*/\partial \omega_I)\omega_I - L^* = T^*{}_2 + (V^* - T^*{}_0)$, reduces to the nonholonomic *Jacobi–Painlevé* integral (3.9.12n)

$$h^* = T^*{}_2 - T^*{}_0 = \textit{constant}; \tag{e2}$$

or, further, since [upon enforcing the constraints $\omega_{1,2} = 0$ in (a3–5)]

$$2T^*{}_2 = m\,r^2({\omega_3}^2 + {\omega_4}^2) + I({\omega_3}^2 + {\omega_4}^2 + {\omega_5}^2)$$

$$= \cdots = (mr^2/5)[7({\omega_X}^2 + {\omega_Y}^2) + 2\,{\omega_Z}^2], \tag{e3}$$

$$T^*{}_1 = -m\,r\,\Omega(Y\omega_4 + X\omega_3) = -m\,r\,\Omega(X\omega_X + Y\omega_Y), \tag{e4}$$

$$2T^*{}_0 = m\,\Omega^2(X^2 + Y^2), \tag{e5}$$

that integral assumes the final form

$$7({\omega_X}^2 + {\omega_Y}^2) + 2{\omega_Z}^2 = 5(\Omega^2/r^2)(X^2 + Y^2) + constant \tag{e6}$$

[by the z-equation, (a10), $I\,\dot{\omega}_Z = \Theta_Z = 0 \;\Rightarrow\; \omega_Z = constant$]. We notice that, due to (e2),

$$E^* \equiv T^* + V^* = T^*{}_2 + T^*{}_1 + T^*{}_0 + V^* = T^*{}_2 + T^*{}_1 + T^*{}_0 + constant$$

$$= 2T^*{}_2 + T^*{}_1 + constant = T^*{}_1 + 2T^*{}_0 + constant \neq constant; \tag{e7}$$

that is, the generalized energy h^* is conserved, but the classical one $E(= E^*)$ is not.

Next, eq. (e6) was obtained *without recourse to the equations of motion*. It is instructive to rederive it, or its equivalent $\dot{T}^*{}_2 = \dot{T}^*{}_0$, directly from expressions (a3–5, e3–5) and the kinetic equations of motion and constraints [see preceding example, eqs. (a8–10) and (a1), respectively]:

$$\dot{\omega}_X = (5/7)(\Omega/r)(r\,\omega_Y - \Omega\,Y), \tag{f1}$$

$$\dot{\omega}_Y = -(5/7)(\Omega/r)(r\,\omega_X - \Omega\,X), \tag{f2}$$

$$\dot{\omega}_Z = 0; \tag{f3}$$

$$\dot{X} = r\,\omega_Y - \Omega\,Y, \qquad \dot{Y} = -r\,\omega_X + \Omega\,X. \tag{f4}$$

Indeed, $(\ldots)^{\cdot}$-differentiating $T^*{}_2$ and $T^*{}_0$, eqs. (a3, 5), and then utilizing (f1–4) yields

$$\dot{T}^*{}_2 = (m\,r^2/5)[7(\omega_X\,\dot{\omega}_X + \omega_Y\,\dot{\omega}_Y) + 2\omega_Z\,\dot{\omega}_Z]$$

$$= \cdots = m\,r\,\Omega^2(X\omega_Y - Y\omega_X); \tag{f5}$$

$$\dot{T}^*{}_0 = m\,\Omega(X^2 + Y^2)\dot{\Omega} + m\,\Omega^2(X\dot{X} + Y\dot{Y})$$

$$= \cdots = m\,r\,\Omega^2(X\omega_Y - Y\omega_X), \quad \text{Q.E.D.} \tag{f6}$$

Holonomic Power Equation

For a more complete understanding of the energetics of this problem, let us also formulate its power equation in holonomic variables, eq. (3.9.11d ff.):

$$dh/dt = -\partial L/\partial t + \sum Q_k\dot{q}_k - \sum \lambda_D a_D, \tag{g1}$$

where $h \equiv \sum(\partial L/\partial\dot{q}_k)\dot{q}_k - L = T_2 + (V - T_0)$. Here, $T = T_2$ is given by (a1), $V = constant$, $Q_k = 0$ (i.e., no impressed nonpotential forces), and, from the constraints (ex. 3.18.3: a1), $a_1 \to a_X = \Omega Y$, $a_2 \to a_Y = -\Omega X$; while $\lambda_1 = \Lambda_1 \to \lambda_X$ and $\lambda_2 = \Lambda_2 \to \lambda_Y$ are, respectively, the OX and OY components of the rolling constraint reaction, from the plane to the sphere, at its contact point $C(X, Y, 0)$. In view of these partial results, eq. (g1) becomes

$$dh/dt = dT_2/dt = -a_1\lambda_1 - a_2\lambda_2 = -(\Omega Y)\lambda_X - (-\Omega X)\lambda_Y$$
$$= \Omega(X\lambda_Y - Y\lambda_X) \equiv M_O\,\Omega, \qquad \text{(g2)}$$

where $M_O = X\lambda_Y - Y\lambda_X =$ *moment of (tangential) rolling reactions at C about OZ*; and, of course, agrees with what would have resulted by "elementary" (Newton–Euler) considerations: the sphere is an "open" system, and, therefore, *the rate of change of its total classical (inertial) energy* $E \equiv T + V$ $(= h)$ *must equal the (inertial) power of all external forces on it*. Since this latter is none other than the rolling reaction (and, clearly, the power of its component *normal* to $O\text{-}XY$ vanishes), we obtain

$$dE/dt = M_O\,\Omega \ (= \textit{externally supplied power, needed to keep the plane spinning at the constant rate } \Omega). \qquad \text{(g3)}$$

Finally, invoking the principle of linear momentum for the sphere, we find

$$M_O = X\lambda_Y - Y\lambda_X = X(m\ddot{Y}) - Y(m\ddot{X}) = dH_O/dt, \qquad \text{(g4)}$$

where

$$H_O \equiv [m(X\dot{Y} - Y\dot{X})] = X(m\dot{Y}) - Y(m\dot{X})$$
$$= \textit{inertial angular momentum of particle of mass } m\textit{, located at the sphere center}, \qquad \text{(g5)}$$

and this, combined with (g3) and the constancy of Ω, readily yields the integral

$$E - H_O\Omega = \textit{constant}, \qquad \text{or} \qquad T - H_O\Omega = \textit{constant}. \qquad \text{(g6)}$$

It is not hard to show the equivalence of (e2) and (g6). Indeed, from the latter, recalling (3.9.12t), we obtain, successively,

$$H_O\Omega = E - \textit{constant} = E^* - \textit{constant} = (h^* + 2T^*{}_0 + T^*{}_1) - \textit{constant}$$
$$= 2T^*{}_0 + T^*{}_1 + (h^* - \textit{constant}). \qquad \text{(g7)}$$

On the other hand, by direct calculation [invoking the constraints (f4) in the definition (g5)], we find

$$H_O = X[m(-r\omega_X + \Omega X)] - Y[m(r\omega_Y - \Omega Y)]$$
$$= m\,\Omega(X^2 + Y^2) - m\,r(X\omega_X + Y\omega_Y),$$

and, therefore,

$$H_O\Omega = m\,\Omega^2(X^2 + Y^2) + [-m\,\Omega\,r(X\omega_X + Y\omega_Y)] = 2T^*{}_0 + T^*{}_1. \qquad \text{(g8)}$$

From (g7, 8), the integral (e2) follows immediately.

Concluding Remarks

That the holonomic power equation does not produce a conservation theorem of the same form as the nonholonomic power equation should not come as a surprise. It is intimately connected with our choice of holonomic coordinates: if the q's were *noninertial*, the rolling constraint would be

$$\boldsymbol{v}_{C,\text{ relative to plane-fixed axes } O\text{–}xyz} = \mathbf{0}; \tag{h}$$

instead of the earlier $\boldsymbol{v}_{C,\text{ of sphere}} = \boldsymbol{v}_{C,\text{ of plane}}$, and it would produce *homogeneous* (i.e., catastatic) Pfaffian equations in these noninertial Lagrangean velocities, that is, $a_{1,2} = 0$. Then, $T = T_2 + T_1 + T_0$, and the holonomic power equation would reduce to the holonomic Jacobi–Painlevé integral [recalling (3.9.11n)] $h = T_2 - T_0 = constant$. A convenient set of such noninertial coordinates are the two (horizontal) coordinates of the center of the sphere relative to *plane-fixed* rectangular Cartesian axes, say $O\text{–}x'y'z'$, and the three Eulerian angles between them and the earlier *sphere-fixed* axes $G\text{–}xyz$; and Ω would *not* appear in the constraints, but it *would* appear in T.

In sum: (a) Only T must be calculated relative to inertial axes, here $O\text{–}XYZ$; *the constraints can be expressed relative to any convenient axes, in terms of any convenient system coordinates*; and since during virtual work the time is assumed frozen, the forces involved are frame independent.

(b) Energy conservation depends on both the *system* and the *frame of reference*.

Problem 3.18.3 Derive the power equations of a sled moving on a uniformly rotating, horizontal, and rough turntable (probs. 3.18.1 and 3.18.2), in both holonomic and nonholonomic variables, and in both inertial and turntable-fixed coordinates. Proceed either from the general energetic theory (§3.9), or from their equations of motion (i.e., multiply each of them with the corresponding velocity, then add together, etc.). Compare with the elementary method.

Problem 3.18.4 Formulate the constraints of a sphere rolling on a uniformly rotating, horizontal, and rough plane in terms of *plane-fixed* (noninertial) system coordinates. Then (i) write down the corresponding transitivity equations, and read off the Hamel coefficients; (ii) obtain its corresponding Hamel equations; and (iii) derive its power equations in both holonomic and nonholonomic variables.

Problem 3.18.5 Consider the problem of rolling *and pivoting* of a homogeneous sphere on a fixed, horizontal, and rough plane. Formulate its constraints, transitivity equations and read off its Hamel coefficients; obtain its kinetic and kinetostatic equations of motion of Routh–Voss, Maggi, Hamel, and Appell; and, finally, derive its power equations, in both holonomic and nonholonomic variables.

Problem 3.18.6 Extend the preceding problem to the case where the plane, on which the rolling and pivoting sphere moves, rotates with a constant angular velocity Ω.

Problem 3.18.7 Consider a sphere S with *eccentric* center of mass G, in slippingless rolling on a fixed, horizontal, and rough plane P (fig. 3.52).

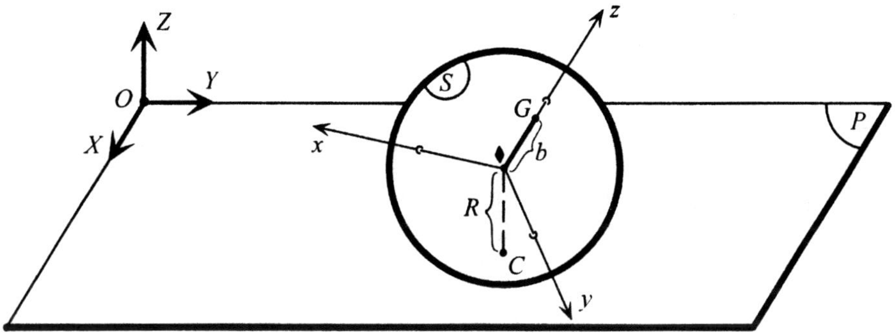

Figure 3.52 Eccentric (nonhomogeneous) sphere rolling on fixed plane.
O–$XYZ/\mathbf{IJK}$: plane-fixed (inertial) axes/basis;
G: eccentric center of mass of sphere, $\mathbf{OG} = \mathbf{r}_G = (X_G, Y_G, Z_G)$;
$\blacklozenge$: geometrical center of sphere, $O\blacklozenge = (X_\blacklozenge, Y_\blacklozenge, Z_\blacklozenge) \equiv (X, Y, R)$;
C: contact point between sphere and plane, $\mathbf{OC} = (X, Y, 0)$;
$\blacklozenge$–$xyz/\mathbf{ijk}$: sphere-fixed (noninertial) axes/basis,
chosen so that $\blacklozenge\mathbf{G} \equiv \mathbf{r}_{G/\blacklozenge} \equiv (x, y, z) = (0, 0, b)$.

Here, using standard notations, we have the following:

(a) Velocity of a generic sphere point:

$$\mathbf{v} = \mathbf{v}_\blacklozenge + \boldsymbol{\omega} \times \mathbf{r}_{/\blacklozenge} = (\dot{X}, \dot{Y}, 0) + (\omega_X, \omega_Y, \omega_Z) \times (X_{/\blacklozenge}, Y_{/\blacklozenge}, Z_{/\blacklozenge}); \tag{a1}$$

(b) Velocity of contact point C:

$$\begin{aligned} \mathbf{v}_C &= \mathbf{v}_\blacklozenge + \boldsymbol{\omega} \times \mathbf{r}_{/\blacklozenge} = (\dot{X}, \dot{Y}, 0) + (\omega_X, \omega_Y, \omega_Z) \times (0, 0, -R) \\ &= (\dot{X} - R\omega_Y, \dot{Y} + R\omega_X, 0); \end{aligned} \tag{a2}$$

(c) Relation of $(\omega_X, \omega_Y, \omega_Z)$ with the time rates of the Eulerian angles between O–XYZ and $\blacklozenge$–xyz, ϕ, θ, ψ (recalling results from §1.12):

$$\begin{aligned} \omega_X &= (\cos\phi)\dot{\theta} + (\sin\phi \, \sin\theta)\dot{\psi}, \\ \omega_Y &= (\sin\phi)\dot{\theta} + (-\cos\phi \, \sin\theta)\dot{\psi}, \qquad \omega_Z = \dot{\phi} + (\cos\theta)\dot{\psi}. \end{aligned} \tag{a3}$$

We also have the related coordinate transformation relations [§1.12, with $\sin(\ldots) \equiv s(\ldots)$, etc.]:

$$\begin{aligned} X_{G/\blacklozenge} &\equiv X_G - X_\blacklozenge \equiv X_G - X \\ &= (c\phi\, c\psi - s\phi\, c\theta\, s\psi)x + (-c\phi\, s\psi - s\phi\, c\theta\, c\psi)y + (s\phi\, s\theta)z, \end{aligned} \tag{a4}$$

$$\begin{aligned} Y_{G/\blacklozenge} &\equiv Y_G - Y_\blacklozenge \equiv Y_G - Y \\ &= (s\phi\, c\psi + c\phi\, c\theta\, s\psi)x + (-s\phi\, s\psi + c\phi\, c\theta\, c\psi)y + (-c\phi\, s\theta)z, \end{aligned} \tag{a5}$$

$$\begin{aligned} Z_{G/\blacklozenge} &\equiv Z_G - Z_\blacklozenge \equiv Z_G - R \\ &= (s\theta\, s\psi)x + (s\theta\, c\psi)y + (c\theta)z, \end{aligned} \tag{a6}$$

from which, since here $(x, y, z) = (0, 0, b)$, we obtain

$$\boldsymbol{r}_{G/\blacklozenge} = \big(X_{G/\blacklozenge}, Y_{G/\blacklozenge}, Z_{G/\blacklozenge}\big) - \big((s\phi\, s\theta)b, -(c\phi\, s\theta)b, (c\theta)b\big). \tag{a7}$$

(d) In view of (a2, 3), the rolling constraint $\boldsymbol{v}_C = \boldsymbol{0}$ assumes the Pfaffian forms

$$\dot{X} - R\omega_Y = \dot{X} - R[(\sin\phi)\dot{\theta} + (-\cos\phi\, \sin\theta)\dot{\psi}] = 0, \tag{b1}$$

$$\dot{Y} + R\omega_X = \dot{Y} + R[(\cos\phi)\dot{\theta} + (\sin\phi\, \sin\theta)\dot{\psi}] = 0. \tag{b2}$$

(i) With the help of the above, show that the (double) kinetic energy of the sphere equals

$$\begin{aligned} 2T &= S\, dm\, v^2 = \cdots = m\, \boldsymbol{v}_\blacklozenge{}^2 + 2m\, \boldsymbol{v}_\blacklozenge \cdot (\boldsymbol{\omega} \times \boldsymbol{r}_{G/\blacklozenge}) + S\, dm(\boldsymbol{\omega} \times \boldsymbol{r}_{/\blacklozenge})^2 \\ &= \cdots = m[(\dot{X})^2 + (\dot{Y})^2] \\ &\quad + 2m\, b\{\dot{X}[(s\phi\, c\theta)\dot{\theta} + (c\phi\, s\theta)\dot{\phi}] + \dot{Y}[(-c\phi\, c\theta)\dot{\theta} + (s\phi\, s\theta)\dot{\phi}]\} \\ &\quad + A[(\dot{\theta})^2 + \sin^2\theta(\dot{\phi})^2] + C[(\cos\theta)\dot{\phi} + \dot{\psi}]^2 \\ &= 2T(\dot{X}, \dot{Y}, \dot{\phi}, \dot{\theta}, \dot{\psi}); \end{aligned} \tag{c}$$

where A and C = *moments of inertia of sphere about* $\blacklozenge x$ (or $\blacklozenge y$) *and* $\blacklozenge z$, *respectively*.

(ii) Using the expression (c), the constraints (b1, 2), and noting that the only impressed force (gravity) has potential equal to

$$V = m\, g\, Z_G = m\, g(Z_\blacklozenge + Z_{G/\blacklozenge}) = m\, g(R + b\cos\theta) = m\, g\, b\cos\theta + \textit{constant}, \tag{d}$$

verify that the Routh–Voss equations of motion of the sphere are (with $\lambda_X \equiv \lambda$ and $\lambda_Y \equiv \mu$):

$$X: \quad m\{\dot{X} + b[(\sin\phi\, \cos\theta)\dot{\theta} + (\cos\phi\, \sin\theta)\dot{\phi}]\}^{\cdot} = \lambda, \tag{e1}$$

$$Y: \quad m\{\dot{Y} + b[(-\cos\phi\, \cos\theta)\dot{\theta} + (\sin\phi\, \sin\theta)\dot{\phi}]\}^{\cdot} = \mu, \tag{e2}$$

$$\begin{aligned} \phi: \quad & [A\, \dot{\phi}\, \sin^2\theta + C(\cos\theta\, \dot{\phi} + \dot{\psi})\cos\theta - m\, b\, R\, \sin^2\theta\, \dot{\psi}]^{\cdot} \\ & + m\, b\, R\, \dot{\theta}\, \dot{\psi}\, \sin\theta\, \cos\theta + m\, bR\, \dot{\phi}\, \dot{\theta}\, \sin\theta = 0, \end{aligned} \tag{e3}$$

$$\begin{aligned} \theta: \quad & (A\, \dot{\theta} + m\, bR\, \cos\theta\, \dot{\theta})^{\cdot} - A(\dot{\phi})^2\, \sin\theta\, \cos\theta + C(\cos\theta\, \dot{\phi} + \dot{\psi})\dot{\phi}\, \sin\theta \\ & + m\, bR(\dot{\theta})^2\, \sin\theta + m\, bR\, \dot{\phi}\, \dot{\psi}\, \cos\theta\, \sin\theta \\ & = \lambda(-R\sin\phi) + \mu(R\, \cos\phi) + m\, g\, b\, \sin\theta, \end{aligned} \tag{e4}$$

$$\psi: \quad [C(\cos\theta\, \dot{\phi} + \dot{\psi})]^{\cdot} = \lambda(R\, \cos\phi\, \sin\theta) + \mu(R\, \sin\phi\, \sin\theta). \tag{e5}$$

(iii) Verify that by solving (e1, 2) for λ and μ, respectively, and substituting the results into (e3–5), we obtain the (kinetic Maggi) equations:

ϕ: remains unchanged, since it did not contain any multipliers, (f1)

$$\theta: \quad (A\dot{\theta} + mbR\cos\theta\,\dot{\theta})^{\cdot} - A(\dot{\phi})^2 \sin\theta \cos\theta + C(\dot{\phi}\cos\theta + \dot{\psi})\dot{\phi}\sin\theta$$
$$+ mbR(\dot{\theta})^2 \sin\theta + mbR\dot{\phi}\dot{\psi}\cos\theta \sin\theta$$
$$+ mR\sin\phi[\dot{X} + b(\dot{\theta}\sin\phi\cos\theta + \dot{\phi}\cos\phi\sin\theta)]^{\cdot}$$
$$- mR\sin\phi[\dot{Y} + b(-\dot{\theta}\cos\phi\cos\theta + \dot{\phi}\sin\phi\sin\theta)]^{\cdot}$$
$$= mgb\sin\theta, \tag{f2}$$

$$\psi: \quad [C(\dot{\phi}\cos\theta + \dot{\psi})]^{\cdot} - mR\cos\phi\sin\theta[\dot{X} + b(\dot{\theta}\sin\phi\cos\theta + \dot{\phi}\cos\phi\sin\theta)]^{\cdot}$$
$$- mR\sin\phi\sin\theta[\dot{Y} + b(-\dot{\theta}\cos\phi\cos\theta + \dot{\phi}\sin\phi\sin\theta)]^{\cdot} = 0. \tag{f3}$$

Problem 3.18.8 Continuing from the preceding problem, verify that by eliminating the (chosen as) dependent velocities $\dot{X}$ and $\dot{Y}$ from its eqs. (f2, 3), using the constraints (b1, 2), we eventually obtain the following Chaplygin–Voronets equations of the problem:

$$\phi: \quad [A\dot{\phi}\sin^2\theta + C(\dot{\phi}\cos\theta + \dot{\psi})\cos\theta - mbR\sin^2\theta\,\dot{\psi}]^{\cdot}$$
$$+ mbR\dot{\theta}\sin\theta(\dot{\psi}\cos\theta + \dot{\phi}) = 0, \tag{a1}$$

$$\theta: \quad [(A + 2mbR\cos\theta + mR^2)\dot{\theta}]^{\cdot} + mbR\sin\theta[(\dot{\theta})^2 + \dot{\phi}\dot{\psi}\cos\theta - (\dot{\phi})^2]$$
$$+ mR^2\dot{\phi}\dot{\psi}\sin\theta - A(\dot{\phi})^2\sin\theta\cos\theta + C(\dot{\phi}\cos\theta + \dot{\psi})\dot{\phi}\sin\theta$$
$$- mgb\sin\theta = 0, \tag{a2}$$

$$\psi: \quad [C(\dot{\phi}\cos\theta + \dot{\psi}) + mR^2\dot{\psi}\sin^2\theta - mbR\dot{\phi}\sin^2\psi]^{\cdot}$$
$$- mR^2\dot{\theta}\sin\theta(\dot{\psi}\cos\theta + \dot{\phi}) = 0. \tag{a3}$$

HINTS

In the Maggi equations (f1–3), the terms deriving from λ and μ can be combined with the remaining terms to produce equations of the form $[\ldots]^{\cdot} + \cdots = 0$, like (a1–3), first, by application of the familiar differentiation rule: $f\dot{g} = (fg)^{\cdot} - f\dot{g}$ (where f, g = arbitrary functions); and then by use of the constraints (b1, 2), but rewritten in the equivalent forms

$$\dot{X}\sin\phi - \dot{Y}\cos\phi = R\dot{\theta}, \qquad \dot{X}\cos\phi + \dot{Y}\sin\phi = -R\dot{\psi}\sin\theta, \tag{b}$$

Problem 3.18.9 Continuing from the preceding problem, show that its eqs. (a1) and (a3) combine to produce the (*linear*) integral

$$AR\dot{\phi}\sin^2\theta + C(b + R\cos\theta)(\dot{\phi}\cos\theta + \dot{\psi}) - mb^2R\dot{\phi}\sin^2\theta = \textit{constant}. \tag{a}$$

HINT

Multiply (a1) by R and (a3) by b, then combine, simplify, and, finally, integrate.

Problem 3.18.10 Continuing from the preceding problems, show that the Chaplygin–Voronets equations (a1–3) of prob. 3.18.8, possess the (*quadratic*) energy integral

$$mR^2[(\dot{\theta})^2 + (\dot{\psi})^2 \sin^2\theta] + 2mbR[(\dot{\theta})^2 \cos\theta - \dot{\phi}\dot{\psi}\sin^2\theta]$$
$$+ A[(\dot{\theta})^2 + (\dot{\phi})^2 \sin^2\theta] + C(\dot{\phi}\cos\theta + \dot{\psi})^2 + 2mgb\cos\theta = 2h; \qquad \text{(a)}$$

in addition to (a) of prob. 3.18.9.

REMARKS

(i) Since the five equations (prob. 3.18.8: a1–3), (prob. 3.18.9: a), and (prob. 3.18.10: a) do not contain explicitly either ϕ or ψ, we can, for example, solve the last two of them for $\dot{\phi}$ and $\dot{\psi}$ *in terms of* θ *and* $\dot{\theta}$; that is, $\dot{\phi} = \phi(\theta, \dot{\theta})$ and $\dot{\psi} = \psi(\theta, \dot{\theta})$, and then insert these expressions into any one of the first three equations (of motion), say the simplest of them. The result would be a *single* second-order differential equation for θ; and since *the time does not appear explicitly*, this can be further reduced to a *first*-order problem.

(ii) If, next, our sphere shrinks to a particle of mass m, then $A = mb^2$ and $C = 0$, and the linear integral (prob. 3.18.9: a) degenerates to the trivial equality $0 = constant$ $(= 0)$; that is, one of our equations disappears! For a detailed discussion and explanation of this interesting mathematical "paradox," see the masterful treatment of Hamel (1949, pp. 760–766).

Problem 3.18.11 Continuing from the above problems of the eccentric sphere:

(i) Introduce the following convenient quasi velocities:

$$\omega_1 \equiv \dot{X} - R\omega_Y = \dot{X} - (R\sin\phi)\dot{\theta} + (R\cos\phi\sin\theta)\dot{\psi} \quad (= 0), \qquad \text{(a1)}$$

$$\omega_2 \equiv \dot{Y} + R\omega_X = \dot{Y} + (R\cos\phi)\dot{\theta} + (R\sin\phi\sin\theta)\dot{\psi} \quad (= 0); \qquad \text{(a2)}$$

$$\omega_3 \equiv \dot{\theta}, \qquad \text{(a3)}$$

$$\omega_4 \equiv \dot{\phi}, \qquad \text{(a4)}$$

$$\omega_5 \equiv \dot{\psi}; \qquad \text{(a5)}$$

which invert easily (no constraint enforcement yet), as follows:

$$\dot{X} = \omega_1 + R\sin\phi\,\omega_3 - R\cos\phi\sin\theta\,\omega_5, \qquad \text{(b1)}$$

$$\dot{Y} = \omega_2 - R\cos\phi\,\omega_3 - R\sin\phi\sin\theta\,\omega_5, \qquad \text{(b2)}$$

$$\dot{\theta} = \omega_3, \qquad \text{(b3)}$$

$$\dot{\phi} = \omega_4, \qquad \text{(b4)}$$

$$\dot{\psi} = \omega_5. \qquad \text{(b5)}$$

(ii) Using (a1–b5), verify the transitivity equations (no constraint enforcement yet)

$$(\delta\theta_1)^{\cdot} - \delta\omega_1 = R\cos\phi(\omega_3\,\delta\theta_4 - \omega_4\,\delta\theta_3) - R\sin\phi\sin\theta(\omega_4\,\delta\theta_5 - \omega_5\,\delta\theta_4)$$
$$+ R\cos\phi\cos\theta(\omega_3\,\delta\theta_5 - \omega_5\,\delta\theta_3); \tag{c1}$$

$$(\delta\theta_2)^{\cdot} - \delta\omega_2 = R\sin\phi(\omega_3\,\delta\theta_4 - \omega_4\,\delta\theta_3) + R\cos\phi\sin\theta(\omega_4\,\delta\theta_5 - \omega_5\,\delta\theta_4)$$
$$+ R\sin\phi\cos\theta(\omega_3\,\delta\theta_5 - \omega_5\,\delta\theta_3); \tag{c2}$$

$$(\delta\theta_3)^{\cdot} - \delta\omega_3 = 0, \qquad (\delta\theta_4)^{\cdot} - \delta\omega_4 = 0, \qquad (\delta\theta_5)^{\cdot} - \delta\omega_5 = 0; \tag{c3}$$

since $\theta_{3,4,5}$ are holonomic coordinates. From the above, we read off the (nonvanishing) Hamel coefficients:

$$\gamma^1{}_{43} = -\gamma^1{}_{34} = R\cos\phi, \qquad \gamma^1{}_{54} = -\gamma^1{}_{45} = -R\sin\phi\sin\theta,$$
$$\gamma^1{}_{53} = -\gamma^1{}_{35} = R\cos\phi\cos\theta; \tag{d1}$$

$$\gamma^2{}_{43} = -\gamma^2{}_{34} = R\sin\phi, \qquad \gamma^2{}_{54} = -\gamma^2{}_{45} = R\cos\phi\sin\theta,$$
$$\gamma^2{}_{53} = -\gamma^2{}_{35} = R\sin\phi\cos\theta; \tag{d2}$$

$$\gamma^3{}_{..} = 0, \qquad \gamma^4{}_{..} = 0, \qquad \gamma^5{}_{..} = 0. \tag{d3}$$

(iii) Using (b1–5), show that the kinetic energy expression (prob. 3.18.7: c), or, equivalently,

$$2T = m[(\dot X)^2 + (\dot Y)^2]$$
$$+ 2mb\{\dot\theta\cos\theta(\dot X\sin\phi - \dot Y\cos\phi) + \dot\phi\sin\theta(\dot X\cos\phi + \dot Y\sin\phi)\}$$
$$+ A[(\dot\theta)^2 + \sin^2\theta(\dot\phi)^2] + C[(\cos\theta)\dot\phi + \dot\psi]^2, \tag{e1}$$

becomes, in the chosen quasi velocities (no constraint enforcement yet),

$$2T^* = m[\omega_1{}^2 + \omega_2{}^2 + R^2\omega_3{}^2 + R^2\sin^2\theta\,\omega_5{}^2 + 2R\sin\phi\,\omega_1\omega_3$$
$$- 2R\cos\phi\,\omega_2\,\omega_3 - 2R\cos\phi\sin\theta\,\omega_1\,\omega_5 - 2R\sin\phi\sin\theta\,\omega_2\,\omega_5]$$
$$+ 2mb[\omega_3\cos\theta(\omega_1\sin\phi - \omega_2\cos\phi + R\,\omega_3)$$
$$+ \omega_4\sin\theta(\omega_1\cos\phi + \omega_2\sin\phi - R\sin\theta\,\omega_5)]$$
$$+ A[\omega_3{}^2 + (\sin^2\theta)\omega_4{}^2] + C[(\cos\theta)\omega_4 + \omega_5]^2. \tag{e2}$$

[*Quadratic* terms in the constrained, or dependent, quasi velocities $\omega_{1,2}$ *can* be omitted from (e2) at this stage, without affecting the equations of motion; but *linear* terms in them *cannot* — explain].

(iv) Using the above results and noting that (after enforcing the constraints $\omega_{1,2} = 0$)

$$\partial T^*/\partial\theta_{1,2,4,5} = 0, \qquad \partial T^*/\partial\theta_3 = \partial T^*/\partial\theta = \cdots,$$
$$\partial V^*/\partial\theta_{1,2,4,5} = 0, \qquad \partial V^*/\partial\theta_3 = \partial V^*/\partial\theta = -mgb\sin\theta, \tag{e3}$$

show that the Hamel equations of this problem are (with $P_k \equiv \partial T^*/\partial\omega_k$, $k = 1, \ldots, 5$).

Kinetic equations:

$$\dot{P}_3 - \partial T^*/\partial\theta_3 + \gamma^1{}_{34}P_1\,\omega_4 + \gamma^1{}_{35}P_1\,\omega_5 + \gamma^2{}_{34}P_2\,\omega_4 + \gamma^2{}_{35}P_2\,\omega_5 = -(\partial V^*/\partial\theta_3),$$

$$\dot{P}_4 + \gamma^1{}_{43}P_1\,\omega_3 + \gamma^1{}_{45}P_1\,\omega_5 + \gamma^2{}_{43}P_2\,\omega_3 + \gamma^2{}_{45}P_2\,\omega_5 = 0,$$

$$\dot{P}_5 + \gamma^1{}_{53}P_1\,\omega_3 + \gamma^1{}_{54}P_1\,\omega_4 + \gamma^2{}_{53}P_2\,\omega_3 + \gamma^2{}_{54}P_2\,\omega_4 = 0; \qquad \text{(f1, 2, 3)}$$

or, explicitly, respectively,

$$[\omega_3(A + 2mbR\cos\theta + mR^2)]^{\cdot} + mR^2\sin\theta\,\omega_4\,\omega_5$$
$$+ mbR\sin\theta(-\omega_4{}^2 + \cos\theta\,\omega_4\,\omega_5 + \omega_3{}^2)$$
$$- A\sin\theta\cos\theta\,\omega_4{}^2 + C(\cos\theta\,\omega_4 + \omega_5)\sin\theta = mgb\sin\theta, \qquad \text{(f4)}$$

$$[A\sin^2\theta\,\omega_4 + C(\cos\theta\,\omega_4 + \omega_5)\cos\theta - mbR\sin^2\theta\,\omega_5]^{\cdot}$$
$$+ mbR\sin\theta\,\omega_3(\omega_4 + \cos\theta\,\omega_5) = 0, \qquad \text{(f5)}$$

$$[mR^2\sin^2\theta\,\omega_5 - mbR\sin^2\theta\,\omega_4 + C(\cos\theta\,\omega_4 + \omega_5)]^{\cdot}$$
$$- mR^2\sin\theta\cos\theta\,\omega_3\,\omega_5 - mR^2\sin\theta\,\omega_3\,\omega_4 = 0; \qquad \text{(f6)}$$

and, as can be verified easily, these equations coincide with the earlier (prob. 3.18.8: a1–3), respectively; *but, unlike them, they have been derived without elimination of the dependent velocities*;

Kinetostatic equations:

$$\dot{P}_1 = \Lambda_1(\equiv \lambda_1), \qquad \dot{P}_2 = \Lambda_2(\equiv \lambda_2), \qquad \text{(f7, 8)}$$

[since, as (d1–3) show, all the $\gamma^k{}_{1I}$ and $\gamma^k{}_{2I}$ ($k = 1, \ldots, 5$; $I = 3, 4, 5$) vanish]

or, explicitly, respectively,

$$[mR\sin\phi\,\omega_3 - mR\cos\phi\sin\theta\,\omega_5 + mb(\omega_3\sin\phi\cos\theta + \omega_4\cos\phi\sin\theta)]^{\cdot} = \Lambda_1,$$
$$[-mR\cos\phi\,\omega_3 - mR\sin\phi\sin\theta\,\omega_5 + mb(-\omega_3\cos\phi\cos\theta + \omega_4\sin\phi\sin\theta)]^{\cdot} = \Lambda_2. \qquad \text{(f9, 10)}$$

Example 3.18.5 *Dynamics of a Rolling Hoop (or Disk, or Coin).* Let us determine the forces and equations of motion of a thin homogeneous hoop H, of mass m and radius r, rolling on a rough, horizontal, and fixed plane P (fig. 3.53).

The relevant kinematics has already been detailed in ex. 2.13.7. We recall that

$$q_1 \equiv X_G \equiv X, \quad q_2 \equiv Y_G \equiv Y; \qquad q_3 \equiv \phi, \quad q_4 \equiv \theta, \quad q_5 \equiv \psi; \qquad \text{(a1)}$$

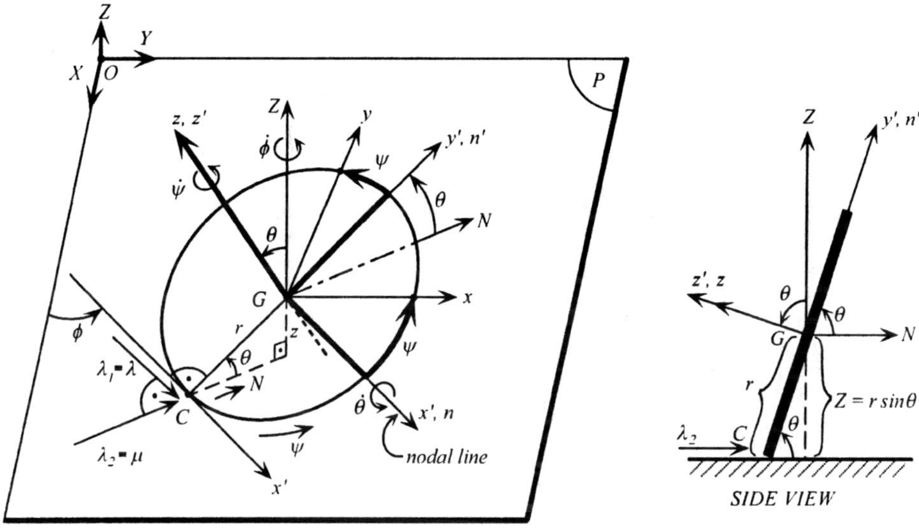

Figure 3.53 Geometry of a homogeneous hoop rolling on rough, horizontal, and fixed plane. G: geometrical center and center of mass of hoop, C: contact point between hoop and plane.
Axes/bases:
O–$XYZ/\mathbf{IJK}$: plane-fixed (inertial) axes/basis;
G–$x'y'z'/\mathbf{i}'\mathbf{j}'\mathbf{k}' \equiv G$–$nn'z'/\mathbf{u}_n\mathbf{u}_{n'}\mathbf{k}'$: semimobile (noninertial) axes/basis;
G–$x'NZ/\mathbf{i}'\mathbf{u}_N\mathbf{K} \equiv G$–$nNZ/\mathbf{u}_n\mathbf{u}_N\mathbf{K}$: semifixed (noninertial) axes/basis;
G–$xyz/\mathbf{ijk}$: disk-fixed (noninertial) axes/basis;
ϕ, θ, ψ: Eulerian angles between O–$XYZ/\mathbf{IJK}$ (or G–$XYZ/\mathbf{IJK}$) and G–$xyz/\mathbf{ijk}$.

since $Z_G \equiv Z = r\sin\theta$, Z is not independent (i.e., here, $f \equiv n - m = 5 - 2 = 3$ — see below)

$$\boldsymbol{\omega} = (\omega_{x'}, \omega_{y'}, \omega_{z'}) = (\dot{\theta}, \dot{\phi}\sin\theta, \dot{\psi} + \dot{\phi}\cos\theta) \qquad (\textit{semimobile components})$$

$$= (\omega_{x'}, \omega_N, \omega_Z) = (\dot{\theta}, -\dot{\psi}\sin\theta, \dot{\phi} + \dot{\psi}\cos\theta) \qquad (\textit{semifixed components}), \quad \text{(a2)}$$

$$\boldsymbol{GC} = (0, -r\cos\theta, -r\sin\theta) \qquad (\textit{semifixed components}), \quad \text{(a3)}$$

$$\Rightarrow \boldsymbol{\omega} \times \boldsymbol{GC} = \big(r(\dot{\psi} + \dot{\phi}\cos\theta), -r\dot{\theta}\sin\theta, -r\dot{\theta}\cos\theta\big); \quad \text{(a4)}$$

and therefore the rolling constraint $\boldsymbol{v}_C = \boldsymbol{0}$ translates to the two Pfaffian conditions

$$(\boldsymbol{v}_C)_{x'} = \dot{X}\cos\phi + \dot{Y}\sin\phi + r(\dot{\psi} + \dot{\phi}\cos\theta) = 0, \quad \text{(a5)}$$

$$(\boldsymbol{v}_C)_N = -\dot{X}\sin\phi + \dot{Y}\cos\phi + r\dot{\theta}\sin\theta = 0. \quad \text{(a6)}$$

The Routh–Voss Equations

Since, by König's theorem and the above geometry/kinematics,

$$2T = m[(\dot{X})^2 + (\dot{Y})^2 + (\dot{Z})^2] + (I_{x'}\,\omega_{x'}{}^2 + I_{y'}\,\omega_{y'}{}^2 + I_{z'}\,\omega_{z'}{}^2)$$

$$= m[(\dot{X})^2 + (\dot{Y})^2 + r^2(\dot{\theta})^2\cos^2\theta] + (mr^2/2)(\dot{\theta})^2 + (mr^2/2)(\dot{\phi})^2\sin^2\theta$$

$$+ (m r^2)(\dot{\psi} + \dot{\phi}\cos\theta)^2, \quad \text{(b1)}$$

and $V = mgZ = mgr \sin\theta$ (and with $\lambda_1 \equiv \lambda, \lambda_2 \equiv \mu$), the Routh–Voss equations of the hoop are

$$X: \quad m\ddot{X} = \lambda \cos\phi - \mu \sin\phi, \tag{b2}$$

$$Y: \quad m\ddot{Y} = \lambda \sin\phi + \mu \cos\phi; \tag{b3}$$

$$\phi: \quad [(mr^2/2)(\dot{\phi} \sin^2\theta)]^\cdot + [(mr^2)\cos\theta(\dot{\phi}\cos\theta + \dot{\psi})]^\cdot = \lambda r \cos\theta, \tag{b4}$$

$$\theta: \quad [(mr^2)(\dot{\theta}\cos^2\theta)]^\cdot + (mr^2/2)\ddot{\theta} + (mr^2)(\dot{\theta})^2 \sin\theta\cos\theta$$
$$- (mr^2/2)(\dot{\phi})^2 \sin\theta\cos\theta + (mr^2)\dot{\phi}(\dot{\phi}\cos\theta + \dot{\psi})\sin\theta$$
$$= \mu r \sin\theta - mgr\cos\theta, \tag{b5}$$

$$\psi: \quad [(mr^2)(\dot{\psi} + \dot{\phi}\cos\theta)]^\cdot = \lambda r. \tag{b6}$$

Equations (b2, 3) express the principle of *linear* momentum along the axes X and Y, respectively; while eqs. (b4–6) express that of *angular* momentum about G, and the corresponding (nonorthogonal!) axes of rotation through it.

Determination of the Reactions; and the Equations of Maggi and Chaplygin–Voronets

To determine and/or eliminate the multipliers (reactions) we, first, solve (b2, 3) for them:

$$\lambda = m(\ddot{X}\cos\phi + \ddot{Y}\sin\phi), \qquad \mu = m(\ddot{Y}\cos\phi - \ddot{X}\sin\phi). \tag{c1}$$

Then, $(\ldots)^\cdot$-differentiating the constraints (a5, 6), to generate $\ddot{X}$ and $\ddot{Y}$, we find

$$\ddot{X}\cos\phi + \ddot{Y}\sin\phi + \dot{\phi}(\dot{Y}\cos\phi - \dot{X}\sin\phi) + r(\dot{\phi}\cos\theta + \dot{\psi})^\cdot = 0, \tag{c2}$$

$$-\ddot{X}\sin\phi + \ddot{Y}\cos\phi - \dot{\phi}(\dot{X}\cos\phi + \dot{Y}\sin\phi) + r(\dot{\theta}\sin\theta)^\cdot = 0; \tag{c3}$$

or, invoking (c1),

$$(\lambda/m) + \dot{\phi}(\dot{Y}\cos\phi - \dot{X}\sin\phi) + r(\dot{\phi}\cos\theta + \dot{\psi})^\cdot = 0, \tag{c4}$$

$$(\mu/m) - \dot{\phi}(\dot{X}\cos\phi + \dot{Y}\sin\phi) + r(\dot{\theta}\sin\theta)^\cdot = 0; \tag{c5}$$

and, finally, eliminating the *second sum/term* in each of them via the constraints (a5, 6) and then solving for the multipliers, we obtain these latter in terms of ϕ, θ, ψ and their rates of change:

$$\lambda = mr[\dot{\phi}\dot{\theta}\sin\theta - (\dot{\phi}\cos\theta + \dot{\psi})^\cdot], \tag{c6}$$

$$\mu = -mr[\dot{\phi}(\dot{\phi}\cos\theta + \dot{\psi}) + (\dot{\theta}\sin\theta)^\cdot]. \tag{c7}$$

Now:

(a) Inserting (c1) into (b4–6) results in the three kinetic Maggi equations of our problem, which, along with the two constraints (a5, 6), constitutes a determinate system for $X(t), Y(t), \phi(t), \theta(t), \psi(t)$; then $\lambda(t)$ and $\mu(t)$ can be immediately found from (c1); whereas

(b) Inserting (c6, 7) into (b4–6) results in its three (kinetic only) Chaplygin–Voronets equations for $\phi(t), \theta(t), \psi(t)$:

$$\phi: \quad [(mr^2/2)(\dot\phi \sin^2\theta)]^{\cdot} + [(mr^2)\cos\theta(\dot\phi\cos\theta + \dot\psi)]^{\cdot}$$
$$= mr^2 \cos\theta[\dot\phi\dot\theta\sin\theta - (\dot\phi\cos\theta + \dot\psi)^{\cdot}], \tag{c8}$$

$$\theta: \quad [(mr^2)(\dot\theta\cos^2\theta)]^{\cdot} + (mr^2/2)\ddot\theta + (mr^2)(\dot\theta)^2\sin\theta\cos\theta$$
$$- (mr^2/2)(\dot\phi)^2\sin\theta\cos\theta + (mr^2)\dot\phi(\dot\phi\cos\theta + \dot\psi)\sin\theta + mgr\cos\theta$$
$$= -mr^2\sin\theta[\dot\phi(\dot\phi\cos\theta + \dot\psi) + (\dot\theta\sin\theta)^{\cdot}], \tag{c9}$$

$$\psi: \quad [(mr^2)(\dot\psi + \dot\phi\cos\theta)]^{\cdot} = mr^2[\dot\phi\dot\theta\sin\theta - (\dot\phi\cos\theta + \dot\psi)^{\cdot}]. \tag{c10}$$

Once $\phi(t), \theta(t), \psi(t)$ have been found from the above, then $\lambda(t)$ and $\mu(t)$ can be immediately calculated from (c6, 7).

(c) As shown a little later, eqs. (c8–10), when expressed in terms of the *semimobile* angular velocity components, eq. (a2), are none other than the corresponding kinetic Hamel equations (with $\omega_{x',y',z'} \rightarrow \omega_{3,4,5}$).

A Famous and Instructive Error

The rolling hoop offers a good opportunity to demonstrate concretely that, in general, $E_I(T_o) \neq Q_{Io}$. Indeed, eliminating the two dependent velocities $\dot X$ and $\dot Y$ ($n = 5, m = 2 \Rightarrow f \equiv 5 - 2 = 3$) from (b1) with the help of the constraints (a5, 6), while noting that, then, $(\dot X)^2 + (\dot Y)^2 = r^2(\dot\psi + \dot\phi\cos\theta)^2 + r^2(\dot\theta)^2\sin^2\theta$, we bring the (double) kinetic energy to the constrained (or *nonlegitimate*) form,

$$2T \rightarrow 2T_o \equiv 2T_o(\theta; \dot\phi, \dot\theta, \dot\psi)$$
$$= (3mr^2/2)(\dot\theta)^2 + (mr^2/2)(\dot\phi)^2\sin^2\theta + (2mr^2)(\dot\psi + \dot\phi\cos\theta)^2. \tag{d1}$$

Therefore, and since here, too, $V = V_o = mgZ = mgr\sin\theta \Rightarrow Q_{Io} = -\partial V_o/\partial q_I$, the *incorrect* Lagrangean equations — that is, $(\partial T_o/\partial \dot q_I)^{\cdot} - \partial T_o/\partial q_I = -\partial V_o/\partial q_I$ ($q_I = \phi, \theta, \psi$) — are

$$d/dt\big[(mr^2/2)(\dot\phi)^2\sin^2\theta + (2mr^2)\cos\theta(\dot\psi + \dot\phi\cos\theta)\big] = 0, \tag{d2}$$

$$(3mr^2/2)\ddot\theta - (mr^2/2)(\dot\phi)^2\sin\theta\cos\theta + (2mr^2)\dot\phi\sin\theta(\dot\psi + \dot\phi\cos\theta)$$
$$= -mgr\cos\theta, \tag{d3}$$

$$d/dt(\dot\psi + \dot\phi\cos\theta) = 0. \tag{d4}$$

On the history of this error (committed by some distinguished scientists in the 1890s), see, for example, Stäckel (1905, pp. 596–597); also Campbell (1971, pp. 102–108).

The Hamel Equations

We recall that (ex. 2.13.7)

$$\omega_1 \equiv \dot{X}\cos\phi + \dot{Y}\sin\phi + r(\dot{\psi} + \dot{\phi}\cos\theta) \quad (=0), \tag{e1}$$

$$\omega_2 \equiv -\dot{X}\sin\phi + \dot{Y}\cos\phi + r\dot{\theta}\sin\theta \quad (=0); \tag{e2}$$

$$\omega_3 \equiv \omega_{x'} = \dot{\theta}, \qquad \omega_4 \equiv \omega_{y'} = \dot{\phi}\sin\theta, \qquad \omega_5 \equiv \omega_{z'} = \dot{\psi} + \dot{\phi}\cos\theta; \tag{e3–5}$$

and

$$(\delta\theta_1)^{\cdot} - \delta\omega_1 = (\omega_4/\sin\theta)\,\delta\theta_2 + (-\omega_2/\sin\theta)\,\delta\theta_4, \tag{e6}$$

$$(\delta\theta_2)^{\cdot} - \delta\omega_2 = (-\omega_4/\sin\theta)\,\delta\theta_1 + [(\omega_1/\sin\theta) - (r\omega_5/\sin\theta)]\,\delta\theta_4 + (r\omega_4/\sin\theta)\,\delta\theta_5, \tag{e7}$$

$$(\delta\theta_3)^{\cdot} - \delta\omega_3 = 0 \qquad (\theta\text{: holonomic coordinate}), \tag{e8}$$

$$(\delta\theta_4)^{\cdot} - \delta\omega_4 = (\omega_3\cot\theta)\,\delta\theta_4 + (-\omega_4\cot\theta)\,\delta\theta_3, \tag{e9}$$

$$(\delta\theta_5)^{\cdot} - \delta\omega_5 = (\omega_4)\,\delta\theta_3 + (-\omega_3)\,\delta\theta_4. \tag{e10}$$

Therefore, and since here $\mathbf{I}_G = \textit{diagonal}\ (A = mr^2/2,\ B = A = mr^2/2,\ C = mr^2)$,

(a) $$2T = 2T^* = m\,v_G{}^2 + \boldsymbol{\omega}\cdot\mathbf{I}_G\cdot\boldsymbol{\omega} \qquad \text{(no constraint enforcement yet!)}$$
$$= m[(\omega_1 - r\omega_5)^2 + (\omega_2 - r\omega_3\sin\theta)^2 + r^2\omega_3{}^2\cos^2\theta] + (mr^2)[(\omega_3{}^2/2) + (\omega_4{}^2/2) + \omega_5{}^2]$$
$$= m\{r^2[(3\omega_3{}^2/2) + (\omega_4{}^2/2) + 2\omega_5{}^2] - 2r\omega_1\omega_5 - 2r\omega_2\omega_3\sin\theta + \omega_1{}^2 + \omega_2{}^2\}, \tag{e11}$$

(the last *two* terms can be safely neglected at this stage—why?) and from this we readily obtain

(b) $$P_1 = (\partial T^*/\partial\omega_1)_o = (-mr\omega_5 + m\omega_1)_o = -mr\omega_5 \Rightarrow \dot{P}_1 = -mr\dot{\omega}_5, \tag{e12}$$

$$P_2 = (\partial T^*/\partial\omega_2)_o = -mr\omega_3\sin\theta \Rightarrow \dot{P}_2 = -mr(\dot{\omega}_3\sin\theta + \omega_3{}^2\cos\theta), \tag{e13}$$

$$P_3 = \cdots = m[-r\omega_2\sin\theta + (3r^2\omega_3/2)]_o = (3mr^2/2)\omega_3 \Rightarrow \dot{P}_3 = (3mr^2/2)\dot{\omega}_3, \tag{e14}$$

$$P_4 = \cdots = (mr^2/2)\omega_4 \Rightarrow \dot{P}_4 = (mr^2/2)\dot{\omega}_4, \tag{e15}$$

$$P_5 = \cdots = 2mr^2\omega_5 \Rightarrow \dot{P}_5 = 2mr^2\dot{\omega}_5; \tag{e16}$$

also (check it!),

(c) $$\partial T^*/\partial\theta_k \equiv \sum A_{lk}(\partial T^*/\partial q_l) = [A_{4k}(\partial T^*/\partial\theta)]_o = 0 \qquad (k = 1, \ldots, 5); \tag{e17}$$

while the fundamental noncommutative term $\Gamma \equiv \sum P_k[(\delta\theta_k)^{\cdot} - \delta\omega_k]$ becomes

$$\text{(d)}\quad \Gamma = P_1[(\omega_4/\sin\theta)\,\delta\theta_2 - (\omega_2/\sin\theta)\,\delta\theta_4] + \cdots$$

(we *can* enforce the constraints $\omega_{1,2} = 0$;

but *not* $\delta\theta_{1,2} = 0$, if we want to calculate the constraint reactions)

$$= -P_2(\omega_4/\sin\theta)\,\delta\theta_1 + P_1(\omega_4/\sin\theta)\,\delta\theta_2 + (P_5\,\omega_4 - P_4\,\omega_4\cot\theta)\,\delta\theta_3$$
$$+ [P_4\,\omega_3\cot\theta - P_2\,r(\omega_5/\sin\theta) - P_5\,\omega_3]\,\delta\theta_4 + P_2 r(\omega_4/\sin\theta)\,\delta\theta_5. \qquad \text{(e18)}$$

Collecting all these results into the master variational equation

$$\sum (\dot{P}_k - \partial T^*/\partial\theta_k)\,\delta\theta_k + \Gamma = \sum \Theta_k\,\delta\theta_k,$$

and applying to it the method of Lagrangean multipliers, we obtain the *two* kineto-static equations

$$k = 1:\quad \dot{P}_1 - (\omega_4/\sin\theta)P_2 = \Theta_1 + \lambda_1, \quad \text{or} \quad mr(-\dot{\omega}_5 + \omega_3\,\omega_4) = \Theta_1 + \lambda_1, \qquad \text{(e19)}$$

$$k = 2:\quad \dot{P}_2 + (\omega_4/\sin\theta)P_1 = \Theta_2 + \lambda_2,$$

$$\text{or} \quad -m\,r[\dot{\omega}_3\sin\theta + {\omega_3}^2\cos\theta + (\omega_4\,\omega_5/\sin\theta)] = \Theta_2 + \lambda_2, \qquad \text{(e20)}$$

and the *three* kinetic equations

$$k = 3:\quad \dot{P}_3 - \omega_4\cot\theta P_4 + \omega_4 P_5 = \Theta_3,$$

$$\text{or} \quad (3m\,r^2/2)\dot{\omega}_3 - (m\,r^2/2)\cot\theta\,{\omega_4}^2 + (2m\,r^2)\omega_4\,\omega_5 = \Theta_3,$$

$$\text{or, further,} \quad m\,r^2[(3/2)\dot{\omega}_3 + 2\omega_4\,\omega_5 - (1/2)\cot\theta\,{\omega_4}^2] = \Theta_3, \qquad \text{(e21)}$$

$$k = 4:\quad \dot{P}_4 - (r\,\omega_5/\sin\theta)P_2 + \omega_3\cot\theta\,P_4 - \omega_3 P_5 = \Theta_4,$$

$$\text{or} \quad m\,r^2[(1/2)\dot{\omega}_4 + (1/2)\cot\theta\,\omega_3\,\omega_4 - \omega_3\,\omega_5] = \Theta_4, \qquad \text{(e22)}$$

$$k = 5:\quad \dot{P}_5 + (r\,\omega_4/\sin\theta)P_2 = \Theta_5,$$

$$\text{or} \quad m\,r^2(2\dot{\omega}_5 - \omega_3\,\omega_4) = \Theta_5. \qquad \text{(e23)}$$

If the only impressed force is gravity (sole case to be examined here), then

$$V = V^* = m\,g\,Z = m\,g\,r\sin\theta_3 \Rightarrow \Theta_{1,2,4,5} = 0 \quad \text{and} \quad \Theta_3 = -\partial V^*/\partial\theta_3 = -m\,g\,r\cos\theta,$$

and therefore the three kinetic equations (e21–23) reduce, respectively, to

$$(3/2)(d\omega_3/dt) + 2\omega_4\,\omega_5 - (1/2)\cot\theta\,{\omega_4}^2 = -(g/r)\cos\theta, \qquad \text{(e24)}$$

$$(1/2)(d\omega_4/dt) + (1/2)\cot\theta\,\omega_3\,\omega_4 - \omega_3\,\omega_5 = 0, \qquad \text{(e25)}$$

$$2(d\omega_5/dt) - \omega_3\,\omega_4 = 0. \qquad \text{(e26)}$$

We leave it to the reader to show, with the help of the above, that the corresponding equations in terms of the components of $\boldsymbol{\omega}$ along the *body-fixed* axes G–xyz, $\omega_{x,y,z}$, are

$$3(d\omega_x/dt) + \omega_y(4\omega_z - \omega_y \cot\theta) = -2(g/r)\cos\theta, \tag{e27}$$

$$d\omega_y/dt + \omega_x(\omega_y \cot\theta - 2\omega_z) = 0, \tag{e28}$$

$$2(d\omega_z/dt) - \omega_x\,\omega_y = 0. \tag{e29}$$

HINT

Use the $\omega_{x',y',z'} \Leftrightarrow \omega_{x,y,z}$ transformation equations (§1.12) in eqs. (e24–26).

The Appell Equations (No Reactions, only motion)

Here, all the difficulty lies in calculating the (constrained) Appellian in terms of the q's; $\omega_{3,4,5}$; $\dot{\omega}_{3,4,5}$; and t; that is, $S \to S^* \to S^*_o$. To this end, we will utilize the Appellian counterpart of König's theorem (3.14.3a ff.),

$$S^* = S^*_G + S^*_{/G},$$

where

$$2S^*_G \equiv m(\boldsymbol{a}_G \cdot \boldsymbol{a}_G),$$

$$2S^*_{/G} \equiv \mathbf{S}\, dm(\boldsymbol{a}_{/G} \cdot \boldsymbol{a}_{/G}) = \boldsymbol{\alpha} \cdot \mathbf{I}_G \cdot \boldsymbol{\alpha} + 2(\boldsymbol{\alpha} \times \boldsymbol{\omega}) \cdot (\mathbf{I}_G \cdot \boldsymbol{\omega}). \tag{f1}$$

Let us calculate these parts of S^* separately, as follows.

(a) Using the convenient *semifixed axes/basis* G–$x'NZ/\boldsymbol{i}'\boldsymbol{u}_N\boldsymbol{K} \equiv G$–$nNZ/\boldsymbol{u}_n\boldsymbol{u}_N\boldsymbol{K}$, and invoking the $\omega \Leftrightarrow \dot{q}$ relations (and then enforcing the constraints $\omega_{1,2} = 0$ there) yields, successively,

$$\begin{aligned}
\boldsymbol{v}_G &= \dot{X}\boldsymbol{I} + \dot{Y}\boldsymbol{J} + \dot{Z}\boldsymbol{K} \\
&= \dot{X}(\cos\phi\,\boldsymbol{u}_n - \sin\phi\,\boldsymbol{u}_N) + \dot{Y}(\sin\phi\,\boldsymbol{u}_n + \cos\phi\,\boldsymbol{u}_N) + \dot{Z}\boldsymbol{K} \\
&= (\dot{X}\cos\phi + \dot{Y}\sin\phi)\boldsymbol{u}_n + (-\dot{X}\sin\phi + \dot{Y}\cos\phi)\boldsymbol{u}_N + \dot{Z}\boldsymbol{K} \\
&= -r(\dot{\psi} + \dot{\phi}\cos\theta)\boldsymbol{u}_n + (-r\dot{\theta}\sin\theta)\boldsymbol{u}_N + (r\dot{\theta}\cos\theta)\boldsymbol{K} \\
&= (\omega_1 - r\,\omega_5)\boldsymbol{u}_n + (\omega_2 - r\sin\theta\,\omega_3)\boldsymbol{u}_N + (r\cos\theta\,\omega_3)\boldsymbol{K} \\
&= (-r\,\omega_5)\boldsymbol{u}_n + (-r\sin\theta\,\omega_3)\boldsymbol{u}_N + (r\cos\theta\,\omega_3)\boldsymbol{K},
\end{aligned} \tag{f2}$$

and since $\boldsymbol{\omega}_{\text{semifixed}} \equiv \boldsymbol{\omega}_{SF} = (0)\boldsymbol{u}_n + (0)\boldsymbol{u}_N + (\dot{\phi})\boldsymbol{K}$, from which it follows that

$$d\boldsymbol{u}_n/dt = \boldsymbol{\omega}_{SF} \times \boldsymbol{u}_n = (0, 0, \dot{\phi}) \times (1, 0, 0) = \dot{\phi}\,\boldsymbol{u}_N = (\omega_4/\sin\theta)\boldsymbol{u}_N, \tag{f3}$$

$$d\boldsymbol{u}_N/dt = \boldsymbol{\omega}_{SF} \times \boldsymbol{u}_N = (0, 0, \dot{\phi}) \times (0, 1, 0) = -\dot{\phi}\,\boldsymbol{u}_n = (-\omega_4/\sin\theta)\boldsymbol{u}_n, \tag{f4}$$

$$d\boldsymbol{K}/dt = \boldsymbol{\omega}_{SF} \times \boldsymbol{K} = (0, 0, \dot{\phi}) \times (0, 0, 1) = \boldsymbol{0}, \tag{f5}$$

we, therefore, obtain (with some easily understood moving axes notations)

$$\begin{aligned}\boldsymbol{a}_G &= d\boldsymbol{v}_G/dt = (d\boldsymbol{v}_G/dt)_{SF} + \boldsymbol{\omega}_{SF} \times \boldsymbol{v}_G \\ &= \cdots = -r\big[(\dot{\omega}_5 - \omega_3\omega_4)\boldsymbol{u}_n + (\dot{\omega}_3 \sin\theta + {\omega_3}^2 \cos\theta + \omega_4\omega_5/\sin\theta)\boldsymbol{u}_N \\ &\qquad - (\dot{\omega}_3 \cos\theta - {\omega_3}^2 \sin\theta)\boldsymbol{K}\big], \end{aligned} \tag{f6}$$

and so, *to within Appell-important terms* (i.e., those containing $\dot{\omega}$'s),

$${a_G}^2 \equiv \boldsymbol{a}_G \cdot \boldsymbol{a}_G = r^2[(\dot{\omega}_3)^2 + (\dot{\omega}_4)^2] + 2r^2\omega_4(\dot{\omega}_3\,\omega_5 - \omega_3\,\dot{\omega}_5). \tag{f7}$$

(b) To calculate the relative Appellian $S^*{}_{/G}$ we need $\boldsymbol{\alpha}$. Here, it is more convenient to work with the *semimobile axes/basis* $G{-}x'y'z'/\boldsymbol{i}'\boldsymbol{j}'\boldsymbol{k}' \equiv G{-}nn'z'/\boldsymbol{u}_n\boldsymbol{u}_{n'}\boldsymbol{k}'$. Since

$$\boldsymbol{\omega} = (\dot{\theta})\boldsymbol{i}' + (\dot{\phi}\sin\theta)\,\boldsymbol{j}' + (\dot{\psi} + \dot{\phi}\cos\theta)\boldsymbol{k}' = \omega_3\,\boldsymbol{i}' + \omega_4\,\boldsymbol{j}' + \omega_5\,\boldsymbol{k}', \tag{f8}$$

and

$$\boldsymbol{\omega}_{\text{semimobile}} \equiv \boldsymbol{\omega}_{SM} = (\dot{\theta})\boldsymbol{i}' + (\dot{\phi}\sin\theta)\boldsymbol{j}' + (\dot{\phi}\cos\theta)\boldsymbol{k}', \tag{f9}$$

from which

$$\begin{aligned} d\boldsymbol{i}'/dt &= \boldsymbol{\omega}_{SM} \times \boldsymbol{i}' = (\dot{\theta},\ \dot{\phi}\sin\theta,\ \dot{\phi}\cos\theta) \times (1,0,0) \\ &= \dot{\phi}(\cos\theta\,\boldsymbol{j}' - \sin\theta\,\boldsymbol{k}') = \omega_4(\cot\theta\,\boldsymbol{j}' - \boldsymbol{k}'), \end{aligned} \tag{f10}$$

$$\begin{aligned} d\boldsymbol{j}'/dt &= \boldsymbol{\omega}_{SM} \times \boldsymbol{j}' = (\dot{\theta},\ \dot{\phi}\sin\theta,\ \dot{\phi}\cos\theta) \times (0,1,0) \\ &= -\dot{\phi}\cos\theta\,\boldsymbol{i}' + \dot{\theta}\,\boldsymbol{k}' = -\omega_4\cot\theta\,\boldsymbol{i}' + \omega_3\,\boldsymbol{k}', \end{aligned} \tag{f11}$$

$$\begin{aligned} d\boldsymbol{k}'/dt &= \boldsymbol{\omega}_{SM} \times \boldsymbol{k}' = (\dot{\theta},\ \dot{\phi}\sin\theta,\ \dot{\phi}\cos\theta) \times (0,0,1) \\ &= \dot{\phi}\sin\theta\,\boldsymbol{i}' - \dot{\theta}\,\boldsymbol{j}' = \omega_4\,\boldsymbol{i}' - \omega_3\,\boldsymbol{j}'; \end{aligned} \tag{f12}$$

we, therefore, find

$$\begin{aligned} \boldsymbol{\alpha} \equiv d\boldsymbol{\omega}/dt &= (d\boldsymbol{\omega}/dt)_{SM} + \boldsymbol{\omega}_{SM} \times \boldsymbol{\omega} = (d\boldsymbol{\omega}/dt)_{SM} + \boldsymbol{\omega}_{SM} \times (\boldsymbol{\omega}_{SM} + \dot{\psi}\,\boldsymbol{k}') \\ &= \cdots = [\dot{\omega}_3 + \omega_4(\omega_5 - \omega_4\cot\theta)]\boldsymbol{i}' \\ &\qquad + [\dot{\omega}_4 - \omega_3(\omega_5 - \omega_4\cot\theta)]\boldsymbol{j}' + \dot{\omega}_5\,\boldsymbol{k}', \end{aligned} \tag{f13}$$

and so, *to within Appell-important terms,*

$$\begin{aligned} \boldsymbol{\alpha} \times \boldsymbol{\omega} &= \cdots = (\dot{\omega}_4\,\omega_5 - \omega_4\,\dot{\omega}_5)\boldsymbol{i}' + (\dot{\omega}_5\,\omega_3 - \omega_5\,\dot{\omega}_3)\,\boldsymbol{j}' \\ &\qquad + (\dot{\omega}_3\,\omega_4 - \omega_3\,\dot{\omega}_4)\boldsymbol{k}', \end{aligned} \tag{f14}$$

$$\mathbf{I}_G \cdot \boldsymbol{\omega} = (mr^2/2)(\omega_3\,\boldsymbol{i}' + \omega_4\,\boldsymbol{j}' + 2\omega_4\,\boldsymbol{k}'), \tag{f15}$$

and, accordingly,

$$(\boldsymbol{\alpha} \times \boldsymbol{\omega}) \cdot (\mathbf{I}_G \cdot \boldsymbol{\omega}) = (mr^2/2)\omega_5(\dot{\omega}_3\omega_4 - \omega_3\dot{\omega}_4). \tag{f16}$$

Next, with the help of the decomposition (again, in semimobile components)

$$\boldsymbol{\alpha} = \boldsymbol{\alpha}' + \boldsymbol{\alpha}'',$$
$$\boldsymbol{\alpha}' \equiv (\dot{\omega}_3, \dot{\omega}_4, \dot{\omega}_5),$$
$$\boldsymbol{\alpha}'' \equiv \big(\omega_4(\omega_5 - \omega_4 \cot\theta), -\omega_3(\omega_5 - \omega_4 \cot\theta), 0\big), \tag{f17}$$

we find that, *to within Appell-important terms*,

$$\boldsymbol{\alpha} \cdot \mathbf{I}_G \cdot \boldsymbol{\alpha} = \boldsymbol{\alpha}' \cdot \mathbf{I}_G \cdot \boldsymbol{\alpha}' + 2(\boldsymbol{\alpha}'' \cdot \mathbf{I}_G \cdot \boldsymbol{\alpha}')$$
$$= \cdots = mr^2\big\{[(\dot{\omega}_3/2)^2 + (\dot{\omega}_4/2)^2 + (\dot{\omega}_5)^2]$$
$$+ (\dot{\omega}_3\omega_4 - \omega_3\dot{\omega}_4)(\omega_5 - \omega_4 \cot\theta)\big\}. \tag{f18}$$

Finally, utilizing all the above results in (f1) we deduce that

$$2S^*{}_o = ma_G{}^2 + \boldsymbol{\alpha} \cdot \mathbf{I}_G \cdot \boldsymbol{\alpha} + 2(\boldsymbol{\alpha} \times \boldsymbol{\omega}) \cdot (\mathbf{I}_G \cdot \boldsymbol{\omega})$$
$$= mr^2[3(\dot{\omega}_3)^2/2 + (\dot{\omega}_4)^2/2 + 2(\dot{\omega}_5)^2$$
$$+ 2\omega_4(\dot{\omega}_3\omega_5 - \omega_3\dot{\omega}_5) + 2\omega_5(\dot{\omega}_3\omega_4 - \omega_3\dot{\omega}_4)$$
$$- \omega_4(\dot{\omega}_3\omega_4 - \omega_3\dot{\omega}_4)\cot\theta]. \tag{f19}$$

(c) Hence, the three kinetic Appell equations are

$$\partial S^*{}_o/\partial\dot{\omega}_3 = mr^2[(3/2)\dot{\omega}_3 + 2\omega_4\,\omega_5 - (1/2)\cot\theta\,\omega_4{}^2] = \Theta_3, \tag{f20}$$

$$\partial S^*{}_o/\partial\dot{\omega}_4 = mr^2[(1/2)\dot{\omega}_4 + (1/2)\cot\theta\,\omega_3\,\omega_4 - \omega_3\,\omega_5] = \Theta_4, \tag{f21}$$

$$\partial S^*{}_o/\partial\dot{\omega}_5 = mr^2(2\dot{\omega}_5 - \omega_3\,\omega_4) = \Theta_5, \tag{f22}$$

and, of course, these coincide with the earlier (e21–23).

- To find the reactions, we need the *relaxed* Appellian $S^* = S^*(t; q_{1,\ldots,5}; \omega_{1,\ldots,5}; \dot{\omega}_{1,\ldots,5})$; then, $(\partial S^*/\partial\dot{\omega}_1)_o = \Theta_1 + \lambda_1$, $(\partial S^*/\partial\dot{\omega}_2)_o = \Theta_2 + \lambda_2$, and these equations would, of course, coincide with the earlier (e19–20). The details are left to the reader.
- In view of the kinematico-inertial identity (3.5.25c): $(\partial S^*/\partial\dot{\omega}_{3,4,5})_o = \partial S^*{}_o/\partial\dot{\omega}_{3,4,5}$, if no reactions are sought there is no need to calculate S^*; $S^*{}_o$ will suffice.
- The above, hopefully, show the advantages of the (essentially Lagrangean) method of Hamel over that of Appell. The explicit calculation of accelerations is a rather expensive step! For alternative Appellian derivations of this problem, see also (alphabetically): Neimark and Fufaev (1972, pp. 149–156), Pérès (1953, pp. 224–226), Routh [1905(a), pp. 352–353].

Brief Analytical Discussion of the Kinetic Hoop Equations (e24–26)

The solution of these equations is facilitated by the introduction of the nutation angle θ as the independent variable, instead of the time t. Then, since $\dot{\theta} = \omega_3$, and with the notation $d(\ldots)/d\theta \equiv (\ldots)'$, we have

$$d\omega_I/dt = (d\omega_I/d\theta)(d\theta/dt), \qquad \text{or} \qquad \dot{\omega}_I = \omega_3\,\omega_I{}' \qquad (I = 3, 4, 5), \tag{g1}$$

and so the last two kinetic equations (e25, 26) transform, respectively, to

$$\omega_4{}' + (\cot\theta)\omega_4 - 2\omega_5 = 0, \qquad 2\omega_5{}' - \omega_4 = 0; \tag{g2}$$

and eliminating ω_4 between them yields the single θ-equation

$$\omega_5{}'' + (\cot\theta)\omega_5{}' - \omega_5 = 0. \tag{g3}$$

With the initial conditions

$$t = 0: \qquad \theta = \theta_o, \quad \omega_3 = \omega_{3o}, \quad \omega_4 = \omega_{4o} = 2\omega_{5o}{}', \quad \omega_5 = \omega_{5o},$$

the general solution of this linear and homogeneous but *variable coefficient* equation, to be obtained via *hypergeometric series*, will have the form

$$\omega_5 = \omega_5(\theta; \theta_o, \omega_{5o}, \omega_{5o}{}') = \omega_5(\theta; \theta_o, \omega_{5o}, \omega_{4o}/2) \equiv \omega_5(\theta; \theta_o, \omega_{4o}, \omega_{5o}). \tag{g4}$$

Then, ω_4 can be found by θ-differentiation; and since

$$\begin{aligned} d(\omega_3{}^2)/d\theta &= 2\omega_3(d\omega_3/d\theta) \\ &= 2\omega_3(d\omega_3/dt)(dt/d\theta) = 2[\omega_3(1/\dot\theta)](d\omega_3/dt) = 2\dot\omega_3, \end{aligned} \tag{g5}$$

the first kinetic equation (e24) reduces to

$$d(3\omega_3{}^2/4)/d\theta = -(g/r)\cos\theta + (\cot\theta/2)\omega_4{}^2 - 2\omega_4\omega_5 = \text{known function of } \theta,$$

from which $\omega_3(\theta)$ may be found by a quadrature. Then, a final integration of $\dot\theta = \omega_3$ yields $\theta(t)$.

For full analytical treatments of these interesting equations, see, for example (alphabetically): Appell (1953, pp. 253–258, 386–388), Grammel (1950, pp. 235–245), MacMillan (1936, pp. 276–282), Neimark and Fufaev (1972, pp. 55–60, 155–156), Pars (1965, pp. 120–122), Webster (1912, pp. 307–316), Winkelmann and Grammel (1927, pp. 434–437).

Finally, for derivations of the equations of motion of this problem, in terms of the *coordinates of the contact point* of the hoop with the plane (instead of those of its mass center), and ϕ, θ, ψ, see, for example, Hamel [1949, pp. 470–471 (Routh–Voss equations), 448–479, 489–492, 778–781 (Hamel equations, stability of motion, etc.)], Rosenberg (1977, pp. 265–268, 338–340).

Brief Discussion of the Kinetostatic Hoop Equations (e19, 20)

Substituting into eqs. (e19, 20) $\dot\omega_3$ and $\dot\omega_5$ from the first and last of the kinetic equations, respectively, and recalling that $\Theta_{1,2} = 0$, we obtain the constraint reactions on the hoop, at its contact point C (figure 3.53):

Along Cx': $\quad \Lambda_1 = \lambda_1 \equiv \lambda = (mr/2)\omega_3\omega_4, \qquad$ (h1)

Along CN: $\quad \Lambda_2 = \lambda_2 \equiv \mu = -mr\{\cos\theta(\omega_3{}^2 + \omega_4{}^2/3) + [(1/\sin\theta) - (4\sin\theta/3)]\omega_4\omega_5 - (2g/3r)\sin\theta\cos\theta\}. \qquad$ (h2)

Once the (rotational) motion has been determined from the kinetic equations — that is, once $\omega_{3,4,5}$ and θ have been found as functions of t and the initial conditions — then (h1, 2) immediately yield the reactions as functions of the same variables. Of course, these forces can also be calculated by direct application of the Newton–Euler principle of linear momentum to the hoop, along the semifixed axes $G-x'NZ \equiv G-nNZ$—see, for example, Lur'e (1968, pp. 409–410).

Problem 3.18.12 Continuing from the preceding example, show that (under gravity only) the power, or energy rate, theorem yields the *first-order integral*

$$mr^2({\omega_3}^2 + {\omega_5}^2) + A({\omega_3}^2 + {\omega_4}^2) + C{\omega_5}^2 = -2mgr\sin\theta + constant, \qquad \text{(a)}$$

or, since here $A = mr^2/2$ and $C = mr^2$,

$$3{\omega_3}^2 + {\omega_4}^2 + 4{\omega_5}^2 + 4(g/r)\sin\theta = constant. \qquad \text{(b)}$$

Problem 3.18.13 Continuing from the preceding example, assume that, in addition to rolling, the hoop is constrained to remain *vertical*; that is, $\theta(t) = \pi/2$.

(i) Find its constraints and its Routh–Voss equations of motion.

(ii) For the special (constraint-satisfying) initial conditions

$$\phi(0) = 0, \quad \dot\phi(0) = \dot\phi_o; \qquad \psi(0) = 0, \quad \dot\psi(0) = \dot\psi_o;$$
$$x(0) = 0, \quad \dot x(0) = -r\dot\psi_o; \qquad y(0) = r, \quad \dot y(0) = 0, \qquad \text{(a)}$$

show that:

- The Lagrangean multiplier associated with the *tangential* direction Cx' vanishes; while
- The multiplier associated with the *normal* direction CN equals $mr\dot\phi_o\dot\psi_o/3 = constant$; and
- The hoop center G traces a circle of radius $r(\dot\psi_o/\dot\phi_o)$, with constant speed $(\dot X)^2 + (\dot Y)^2 = r^2(\dot\psi_o)^2$, and normal acceleration of magnitude $r\dot\phi_o\dot\psi_o$.

Problem 3.18.14 Continuing from the preceding example, consider its key kinetic equation (g3):

$$d^2\omega_5/d\theta^2 + (\cot\theta)(d\omega_5/d\theta) - \omega_5 = 0, \qquad \text{where } \omega_5 \equiv \dot\psi + \dot\phi\cos\theta = \omega_{z'}. \qquad \text{(a)}$$

Show that the independent variable change $\theta \to \zeta = \cos^2\theta$ transforms the above to the *hypergeometric* equation

$$2\zeta(1-\zeta)(d^2\omega_5/d\zeta^2) + (1-3\zeta)(d\omega_5/d\zeta) - (1/2)\omega_5 = 0. \qquad \text{(b)}$$

Then, show that a particular *infinite series* solution of this famous equation is

$$\omega_5 = \sum a_k\zeta^k = a_0 + a_1\zeta + a_2\zeta^2 + \cdots, \qquad \text{(c)}$$

where

$$a_k/a_{k-1} = (4k^2 - 6k + 3)/2k(2k-1) \qquad (k = 1, 2, \ldots).$$

Problem 3.18.15 Consider again the hoop, but now rolling on a plane P, which *translates* with given (inertial) velocity (fig. 3.54)

$$\boldsymbol{v}_{\text{plane}} \equiv \boldsymbol{v}_P = \boldsymbol{v}_C = \big(v_X(t), v_Y(t), v_X(t)\big). \tag{a}$$

Also, assume for algebraic simplicity, but no loss in generality, that P is (and remains) horizontal.

(i) If X, Y, Z = inertial coordinates of the hoop center G [i.e., $Z = h(t) + r\sin\theta$, $v_Z(t) = dh(t)/dt$], show that the rolling constraints are (recall prob. 2.13.3)

$$[\dot{X} - v_X(t)]\cos\phi + [\dot{Y} - v_Y(t)]\sin\phi + r\dot{\phi}\cos\theta + r\dot{\psi} = 0, \tag{b}$$

$$-[\dot{X} - v_X(t)]\sin\phi + [\dot{Y} - v_Y(t)]\cos\phi + r\dot{\theta}\sin\theta = 0. \tag{c}$$

(ii) Show that the (inertial) potential and kinetic energies of the hoop are, respectively,

$$V = -mg[h(t) + r\sin\theta], \tag{d}$$

$$2T = m\{(\dot{X})^2 + (\dot{Y})^2 + [\dot{h}(t) + r\cos\theta\,\dot{\theta}]^2\} + A[(\dot{\theta})^2 + (\dot{\phi})^2\sin^2\theta] + C(\dot{\psi} + \dot{\phi}\cos\theta)^2 \quad (C = 2A = 2B = mr^2); \tag{e}$$

and, therefore, verify that the corresponding Routh–Voss equations are (with $\lambda_{1,2}$ = multipliers)

$$m\ddot{X} = \lambda_1\cos\phi - \lambda_2\sin\phi, \tag{f}$$

$$m\ddot{Y} = \lambda_1\sin\phi + \lambda_2\cos\phi, \tag{g}$$

$$[A\,\dot{\phi}\sin^2\theta + C(\dot{\psi} + \dot{\phi}\cos\theta)\cos\theta]^{\cdot} = \lambda_1 r\cos\theta, \tag{h}$$

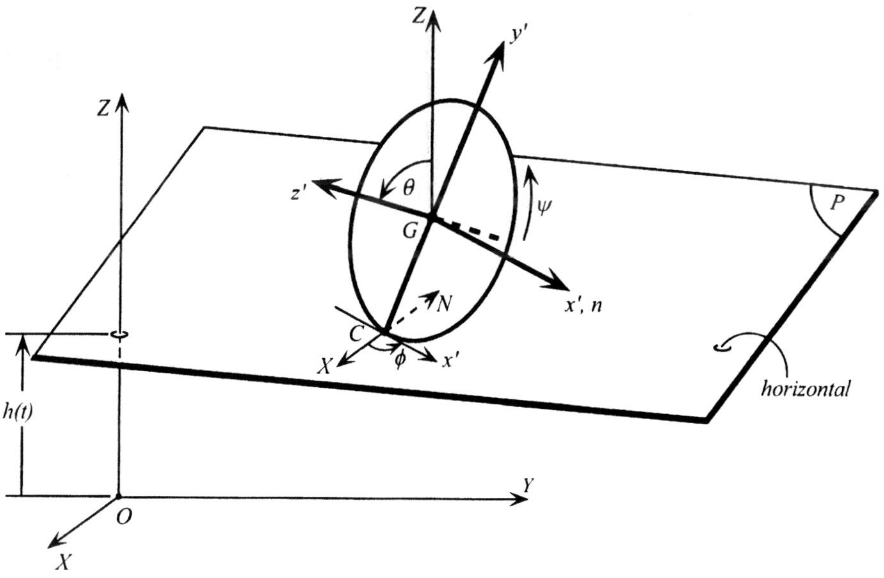

Figure 3.54 Hoop rolling on a translating horizontal plane.

$$m r \cos\theta\,[\ddot{h} + r(\sin\theta)^{\cdot\cdot}] + A[\ddot{\theta} - (\dot{\phi})^2 \sin\theta\cos\theta]$$

$$+ C(\dot{\psi} + \dot{\phi}\cos\theta)\dot{\phi}\sin\theta = -m g r\cos\theta + \lambda_2 r\sin\theta, \qquad \text{(i)}$$

$$C(\dot{\psi} + \dot{\phi}\cos\theta)^{\cdot} = \lambda_1 r. \qquad \text{(j)}$$

(iii) Eliminating $\lambda_{1,2}$ from the above Routh–Voss equations (f–j), and then $\ddot{X}$, $\ddot{Y}$ via the [once $(\ldots)^{\cdot}$-differentiated] constraints (b, c), verify that the three (kinetic) Chaplygin–Voronets equations of this problem are

$$A\ddot{\phi}\sin\theta - C\dot{\theta}\dot{\psi} = 0, \qquad \text{(k)}$$

$$(C + m r^2)(\dot{\phi}\cos\theta + \dot{\psi})^{\cdot} - m r^2\dot{\phi}\dot{\theta}\sin\theta$$
$$= m r[\dot{v}_X(t)\cos\phi + \dot{v}_Y(t)\sin\phi], \qquad \text{(l)}$$

$$(A + m r^2)\ddot{\theta} - A(\dot{\phi})^2\sin\theta\cos\theta + (C + m r^2)\sin\theta\,\dot{\phi}(\dot{\phi}\cos\theta + \dot{\psi})$$
$$= -m g r\cos\theta - m r[\dot{v}_X(t)\sin\phi\sin\theta - \dot{v}_Y(t)\cos\phi\sin\theta + \dot{v}_Z(t)\cos\theta]; \qquad \text{(m)}$$

and that along, with the constraints (b, c), they constitute a determinate system for $X(t)$, $Y(t)$, $\phi(t)$, $\theta(t)$, $\psi(t)$. Then, in both (ii) and here, $Z(t) = h(t) + r\sin\theta(t)$. Notice that (a) the terms due to the translation of the plane appear as additional "forces" on the right sides of (l) and (m), and equal, respectively, $mr\boldsymbol{a}_P(t)\cdot\boldsymbol{i}'$ and $-mr\boldsymbol{a}_P(t)\cdot\boldsymbol{k}'$, where $\boldsymbol{a}_P(t) \equiv d\boldsymbol{v}_P(t)/dt = (\dot{v}_X(t),\ \dot{v}_Y(t),\ \dot{v}_Z(t))$, $\boldsymbol{i}' \equiv \boldsymbol{u}_n = (\cos\phi,\ \sin\phi,\ 0)$, $\boldsymbol{k}' = (\sin\phi\sin\theta,\ -\cos\phi\sin\theta,\ \cos\theta)$; and that (b) here, too, the Lagrangean method shows its superiority over the momentum method of Newton–Euler.

Problem 3.18.16 Continuing from the preceding problem, examine the special motion where P *remains fixed* [or, equivalently, translates *uniformly*: $dv_{X,Y,Z}(t)/dt = 0 \Rightarrow v_{X,Y,Z}(t) =$ *constant* $\equiv c_{X,Y,Z}$], *and the hoop rolls at a constant nutation angle* $\theta(t) =$ *constant* $\equiv \theta_o$.

(i) After verifying that such a motion is possible, show that, then, the first two Chaplygin–Voronets equations, (k, l), yield

$$\dot{\phi} = constant \equiv \dot{\phi}_o, \qquad \dot{\psi} = constant \equiv \dot{\psi}_o; \qquad \text{(a)}$$

while the third of them, (m), reduces to

$$-A(\dot{\phi}_o)^2\sin\theta_o\cos\theta_o + (C + mr^2)\sin\theta_o\,\dot{\phi}_o(\dot{\phi}_o\cos\theta_o + \dot{\psi}_o) = -m g r\cos\theta_o. \qquad \text{(b)}$$

(ii) Then, using the constraints, eqs. (b, c) of the preceding problem, show that

$$\dot{X} - c_X = -r(\dot{\phi}_o\cos\theta_o + \dot{\psi}_o)\cos(\dot{\phi}_o t), \qquad \dot{Y} - c_Y = -r(\dot{\phi}_o\cos\theta_o + \dot{\psi}_o)\sin(\dot{\phi}_o t). \qquad \text{(c)}$$

Discuss particular cases of this special motion; for example, $\theta_o = \pi/2$ $(\Rightarrow, \dot{\phi}_o, \dot{\psi}_o = \cdots)$, $\theta_o \neq \pi/2\ (\Rightarrow \dot{\phi}_o, \dot{\psi}_o = \cdots)$.

Problem 3.18.17 Examine the problem of a hoop rolling on a *uniformly rotating horizontal and rough platform*; that is, formulate its constraints in any convenient set of coordinates, write down its transitivity equations, and then obtain its Routh–Voss, Hamel, and Appell equations, with or without reactions.

Problem 3.18.18 Examine the earlier problems of the (sliding) sled and (rolling) sphere, but on a *translating* platform; that is, obtain their constraints, transitivity equations, and various equations of motion.

Example 3.18.6 *Dynamics of Pair of Rolling Wheels on an Axle.* Let us determine the motion and reactions of a system consisting of two identical homogeneous wheels, mounted on a light axle, and each capable of turning freely about it, and rolling on a fixed, horizontal, and rough plane P (fig. 3.55).

The kinematics of this system has already been discussed in ex. 2.13.8. It was found there that $q_{1,\ldots,5} = X,\ Y,\ \phi,\ \psi',\ \psi''$, and that the rolling constraints are

$$\dot{X}\cos\phi + \dot{Y}\sin\phi = 0, \tag{a1}$$

$$-\dot{X}\sin\phi + \dot{Y}\cos\phi + b\dot{\phi} + r\dot{\psi}' = 0, \tag{a2}$$

$$-\dot{X}\sin\phi + \dot{Y}\cos\phi - b\dot{\phi} + r\dot{\psi}'' = 0; \tag{a3}$$

or, since the last two of them yield the *integrable* combination (with c = integration constant, depending on the initial values of ϕ, ψ', ψ'')

$$2b\dot{\phi} + r(\dot{\psi}' - \dot{\psi}'') = 0 \ \Rightarrow\ 2b\phi = c - r(\psi' - \psi''), \tag{a4}$$

we may take, as the two independent Pfaffian constraints,

$$\dot{X}\cos\phi + \dot{Y}\sin\phi = 0, \tag{b1}$$

$$-\dot{X}\sin\phi + \dot{Y}\cos\phi + (r/2)(\dot{\psi}' + \dot{\psi}'') = 0. \tag{b2}$$

For the purposes of the Routh–Voss equations (see below), a further simplification of these constraints is possible: (a) multiplying the first of them by $\sin\phi$ and the second by $\cos\phi$ and adding together yields

$$\dot{Y} + (r/2)(\dot{\psi}' + \dot{\psi}'')\cos\phi = 0; \tag{b3}$$

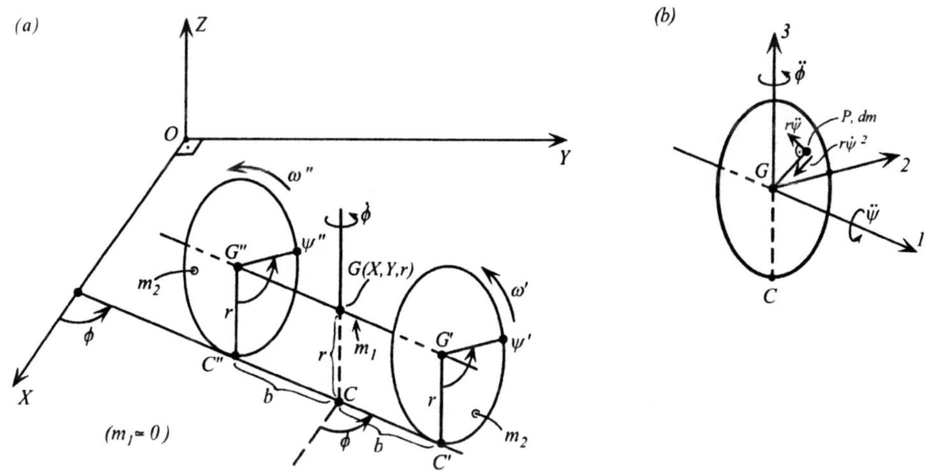

Figure 3.55 (a) Rolling of two wheels on an axle, on a fixed plane; (b) acceleration components needed for calculation of Appellian.

while (b) multiplying the first of them by $\cos\phi$ and the second by $-\sin\phi$ and adding together yields

$$\dot{X} - (r/2)(\dot{\psi}' + \dot{\psi}'')\sin\phi = 0; \tag{b4}$$

and then (c) eliminating $\dot{\psi}''$ from these two with the help of the integrable relation (a4, for $b = r$): $2b\dot{\phi} + r(\dot{\psi}' - \dot{\psi}'') = 0 \Rightarrow \dot{\psi}'' = 2\dot{\phi} + \dot{\psi}'$, finally results in the two new Pfaffian constraints

$$\dot{X} - r(\dot{\psi}' + \dot{\phi})\sin\phi = 0, \qquad \dot{Y} + r(\dot{\psi}' + \dot{\phi})\cos\phi = 0. \tag{b5}$$

In view of the above, we introduce the following quasi velocities:

$$\omega_1 \equiv \dot{\theta}_1 \equiv \dot{X}\cos\phi + \dot{Y}\sin\phi \qquad (= 0), \tag{c1}$$

$$\omega_2 \equiv \dot{\theta}_2 \equiv -\dot{X}\sin\phi + \dot{Y}\cos\phi \quad (\neq 0), \tag{c2}$$

$$\omega_3 \equiv \dot{\theta}_3 \equiv \dot{\phi} \qquad (\neq 0), \tag{c3}$$

$$\omega_4 \equiv \dot{\theta}_4 \equiv r(\dot{\psi} + \dot{\psi}'') + 2(-\dot{X}\sin\phi + \dot{Y}\cos\phi)$$

$$= r(\dot{\psi}' + \dot{\psi}'') + 2\omega_2 \quad (= 0), \tag{c4}$$

$$\omega_5 \equiv \dot{\theta}_5 \equiv 2b\dot{\phi} + r(\dot{\psi}' - \dot{\psi}'') \quad (= 0). \tag{c5}$$

The above invert easily to

$$\dot{X} = (\cos\phi)\omega_1 + (-\sin\phi)\omega_2, \tag{c6}$$

$$\dot{Y} = (\sin\phi)\omega_1 + (\cos\phi)\omega_2, \tag{c7}$$

$$\dot{\phi} = (0)\omega_1 + (0)\omega_2 + (1)\omega_3, \tag{c8}$$

$$\dot{\psi}' = (1/2r)(-2\omega_2 - 2r\omega_3 + \omega_4 - \omega_5), \tag{c9i}$$

$$\dot{\psi}'' = (1/2r)(-2\omega_2 + 2r\omega_3 + \omega_4 - \omega_5). \tag{c9ii}$$

From (c1–9i), we readily obtain the following transitivity equations:

$$(\delta\theta_1)^{\cdot} - \delta\omega_1 = (\omega_3)\,\delta\theta_2 + (-\omega_2)\,\delta\theta_3, \tag{d1}$$

$$(\delta\theta_2)^{\cdot} - \delta\omega_2 = (-\omega_3)\,\delta\theta_1 + (\omega_1)\,\delta\theta_3, \tag{d2}$$

$$(\delta\theta_3)^{\cdot} - \delta\omega_3 = 0, \tag{d3}$$

$$(\delta\theta_4)^{\cdot} - \delta\omega_4 = (-2\omega_3)\,\delta\theta_1 + (2\omega_1)\,\delta\theta_3, \tag{d4}$$

$$(\delta\theta_5)^{\cdot} - \delta\omega_5 = 0. \tag{d5}$$

The Routh–Voss Equations

Applying König's theorem (plus parallel axis theorem for moments of inertia), we obtain

$$2T = (m_1 + 2m_2)[(\dot{X})^2 + (\dot{Y})^2] + (m_1 b^2/3)(\dot{\phi})^2$$

$$+ 2[(m_2 r^2/4) + m_2 b^2](\dot{\phi})^2 + (m_2 r^2/2)[(\dot{\psi}')^2 + (\dot{\psi}'')^2], \tag{e1}$$

or, for the special case, to be examined here for algebraic simplicity, $m_1 = 0$, $m_2 \equiv m$ (= mass of *each* wheel), $b = r$:

$$T = m[(\dot{X})^2 + (\dot{Y})^2] + (mr^2/4)[(\dot{\psi}')^2 + (\dot{\psi}'')^2 + 5(\dot{\phi})^2]. \tag{e2}$$

Hence, the *five* Routh–Voss equations corresponding to (e1) and (b5) are (with multipliers $\lambda_1 \equiv \lambda$ and $\lambda_2 \equiv \mu$; and with impressed force components to be calculated from the invariant differential $\delta' W = Q_X\,\delta X + Q_Y\,\delta Y + Q_\phi\,\delta\phi + Q_{\psi'}\,\delta\psi' + Q_{\psi''}\,\delta\psi''$, as if the δq's were unconstrained)

$$X: \quad (2m)\ddot{X} = Q_X + \lambda, \tag{e3}$$

$$Y: \quad (2m)\ddot{Y} = Q_Y + \mu, \tag{e4}$$

$$\phi: \quad (5mr^2/2)\,\ddot{\phi} = Q_\phi + r(-\lambda\sin\phi + \mu\cos\phi), \tag{e5}$$

$$\psi': \quad (mr^2/2)\,\ddot{\psi}' = Q_{\psi'} + r(-\lambda\sin\phi + \mu\cos\phi), \tag{e6}$$

$$\psi'': \quad (mr^2/2)\,\ddot{\psi}'' = Q_{\psi''}. \tag{e7}$$

If all Q's vanish (*free* motion), the above yield the *two* obvious integrals

$$\dot{\psi}'' = constant,$$

$$(5mr^2/2)\dot{\phi} - (mr^2/2)\dot{\psi}' = constant \;\Rightarrow\; 5\dot{\phi} - \dot{\psi}' = constant. \tag{e8}$$

A *third* integral results as follows: first, we rewrite the two Pfaffian constraints as

$$\dot{X}\cos\phi + \dot{Y}\sin\phi = 0, \tag{e9}$$

$$\dot{X}\sin\phi - \dot{Y}\cos\phi = r(\dot{\phi} + \dot{\psi}'); \tag{e10}$$

then we $(\ldots)^{\cdot}$-differentiate the first of them and take into account the second. The result is

$$\ddot{X}\sin\phi - \ddot{Y}\cos\phi = r(\ddot{\phi} + \ddot{\psi}')$$

$$= (1/2m)(\lambda\sin\phi - \mu\cos\phi) = -[mr/2(2m)]\ddot{\psi}' = -(r/4)\ddot{\psi}';$$

and from this, by rearrangement and integration, it follows that

$$(4m)\dot{\phi} + (5m)\dot{\psi}' = constant \;\Rightarrow\; 4\dot{\phi} + 5\dot{\psi}' = constant.$$

These integrals show that the angles ϕ, ψ', ψ'' vary *linearly* in time; in which case, the constraints integrate easily to

$$\dot{X} = r(constant)\sin\phi = (constant)\;Y, \tag{e11}$$

$$\dot{Y} = -r(constant)\cos\phi = -(constant)\;X; \tag{e12}$$

that is, the path of G is a *circle*, parallel to the plane $Z = 0$, of radius $R = r|(\dot{\phi}_o + \dot{\psi}_o')/\dot{\phi}_o|$, described at the *uniform* rate $\dot{\phi}_o$. Indeed, we have

$$v_G^2 = (\dot{X})^2 + (\dot{Y})^2 = (R\dot{\phi}_o)^2 \;\Rightarrow\; (\dot{\phi}_o + \dot{\psi}_o')^2 r^2 = (\dot{\phi}_o)^2 R^2, \quad \text{Q.E.D.}$$

[The third integral also results, more simply, from the *constancy of T* (by energy conservation; and since, here, λ and μ are workless), if in there [eq. (e2)] using the constraints (b5), we replace $(\dot{X})^2 + (\dot{Y})^2$ with $r^2(\dot{\phi} + \dot{\psi}')^2$. Then, we obtain a *constant coefficient* relation between $(\dot{\phi})^2$ and $(\dot{\psi}')^2$, from which it follows that $\dot{\phi}$ and $\dot{\psi}'$ are constants, like $\dot{\psi}''$. The earlier argument, however, may apply to more general problems.]

Finally, inserting $\dot{X}$ and $\dot{Y}$ from (e11, 12) into the first two equations of motion, (e3, 4), yields the two constraint reactions λ and μ. For additional insights, see, for example, Pérès (1953, pp. 213–214), Rosenberg (1977, pp. 340–345).

The Hamel Equations

From the $\dot{q} \leftrightarrow \omega$ relations (c6–9), we easily find

$$(\dot{X})^2 + (\dot{Y})^2 = \omega_1{}^2 + \omega_2{}^2, \tag{f1}$$

$$(\dot{\psi}')^2 + (\dot{\psi}'')^2 = (1/2r^2)[(\omega_4 - 2\omega_2)^2 + (\omega_5 - 2r\,\omega_3)^2], \tag{f2}$$

$$(\dot{\phi})^2 = \omega_3{}^2, \tag{f3}$$

and, therefore,

$$T \to T^* = \cdots = (m)\omega_1{}^2 + (3m/2)\omega_2{}^2 + (7mr^2/4)\omega_3{}^2 + (m/8)(\omega_4{}^2 + \omega_5{}^2) + (-m/2)\omega_2\,\omega_4 + (-mr/2)\omega_3\,\omega_5, \tag{f4}$$

with no constraint enforcement yet. However, in view of the constraints $\omega_{1,4,5} = 0$, the (quadratic) *first* and *fourth* terms/summands in (f4) can be safely neglected at this stage; that is, finally, to within *Hamel-important terms*,

$$T^* = (3m/2)\omega_2{}^2 + (7mr^2/4)\omega_3{}^2 + (-m/2)\omega_2\omega_4 + (-mr/2)\omega_3\omega_5. \tag{f5}$$

Next:

(a) The *nonholonomic momenta* $P_k \equiv (\partial T^*/\partial\omega_k)_o$ $(k = 1, \ldots, 5)$ and their $(\ldots)^{\cdot}$-derivatives are

$$P_1 = 0 \;\Rightarrow\; \dot{P}_1 = 0, \tag{f6}$$

$$P_2 = (3m)\omega_2 \;\Rightarrow\; \dot{P}_2 = (3m)\dot{\omega}_2, \tag{f7}$$

$$P_3 = (7mr^2/2)\omega_3 \;\Rightarrow\; \dot{P}_3 = (7mr^2/2)\dot{\omega}_3, \tag{f8}$$

$$P_4 = (-m/2)\omega_2 \;\Rightarrow\; \dot{P}_4 = (-m/2)\dot{\omega}_2, \tag{f9}$$

$$P_5 = (-mr/2)\omega_3 \;\Rightarrow\; \dot{P}_5 = (-mr/2)\dot{\omega}_3. \tag{f10}$$

(b) Since $(\partial T^*/\partial q_k)_o = 0$ $(k = 1, \ldots, 5)$, we will have

$$\partial T^*/\partial\theta_l \equiv \sum A_{kl}(\partial T^*/\partial q_k) = 0 \qquad (l = 1, \ldots, 5). \tag{f11}$$

(c) In view of (d1–5) and (f6–10), the fundamental noncommutativity term $\Gamma \equiv \sum P_k[(\delta\theta_k)^{\cdot} - \delta\omega_k]$, upon enforcing the constraints $\omega_{1,4,5} = 0$ (but *not* $\delta\theta_{1,4,5} = 0$, since we want reactions too) becomes

$$\begin{aligned}\Gamma &= P_1(\omega_3\,\delta\theta_2 - \omega_2\,\delta\theta_3) + P_2(-\omega_3\,\delta\theta_1 + \omega_1\,\delta\theta_3) + P_3(0)\\ &\quad + P_4(-2\omega_3\,\delta\theta_1 + 2\omega_1\,\delta\theta_3) + P_5(0)\\ &= -(P_2 + 2P_4)\omega_3\,\delta\theta_1. \qquad \text{(f12)}\end{aligned}$$

(d) The nonholonomic impressed force components, Θ_k, are obtained as follows [with use of virtual form of (c6–9), and calculated as if the constraints $\delta\theta_{1,4,5} = 0$ did not exist]:

$$\begin{aligned}\delta' W &= Q_X\,\delta X + Q_Y\,\delta Y + Q_\phi\,\delta\phi + Q_{\psi'}\,\delta\psi' + Q_{\psi''}\,\delta\psi''\\ &= Q_X(\cos\phi\,\delta\theta_1 - \sin\phi\,\delta\theta_2) + Q_Y(\sin\phi\,\delta\theta_1 + \cos\phi\,\delta\theta_2) + Q_\phi\,\delta\theta_3\\ &\quad + Q_{\psi'}(1/2r)(-2\,\delta\theta_2 - 2r\,\delta\theta_3 + \delta\theta_4 + \delta\theta_5)\\ &\quad + Q_{\psi''}(1/2r)(-2\,\delta\theta_2 + 2r\,\delta\theta_3 + \delta\theta_4 - \delta\theta_5)\\ &= (Q_X\cos\phi + Q_Y\sin\phi)\,\delta\theta_1 + [-Q_X\sin\phi + Q_Y\cos\phi - r^{-1}(Q_{\psi'} + Q_{\psi''})]\,\delta\theta_2\\ &\quad + [Q_\phi - (Q_{\psi'} - Q_{\psi''})]\,\delta\theta_3 + (2r)^{-1}(Q_{\psi'} + Q_{\psi''})\,\delta\theta_4\\ &\quad + (2r)^{-1}(Q_{\psi'} - Q_{\psi''})\,\delta\theta_5\\ &\equiv \Theta_1\,\delta\theta_1 + \Theta_2\,\delta\theta_2 + \Theta_3\,\delta\theta_3 + \Theta_4\,\delta\theta_4 + \Theta_5\,\delta\theta_5. \qquad \text{(f13)}\end{aligned}$$

Substituting all these results into the, by now, well-known nonholonomic version of LP, and applying to it the method of Lagrangean multipliers for $(\ldots)\,\delta\theta_{1,4,5}$, we eventually obtain the Hamel equations (nonholonomic variables), and, next to them, the Maggi equations (holonomic variables):

$$\begin{aligned}&\theta_1\!: \quad -(P_2 + 2P_4)\omega_3 = \Theta_1 + \lambda_1 \;\Rightarrow\; -2m\,\omega_2\,\omega_3 = -2m(-\dot X\sin\phi + \dot Y\cos\phi)\dot\phi = \Theta_1 + \lambda_1,\\ &\theta_2\!: \quad \dot P_2 + P_1\omega_3 = \Theta_2 \;\Rightarrow\; 3m\,\dot\omega_2 = 3m(-\dot X\sin\phi + \dot Y\cos\phi)^{\cdot} = \Theta_2,\\ &\theta_3\!: \quad \dot P_3 - P_1\omega_2 = \Theta_3 \;\Rightarrow\; (7mr^2/2)\,\dot\omega_3 = (7mr^2/2)\ddot\phi = \Theta_3,\\ &\theta_4\!: \quad \dot P_4 = \Theta_4 + \lambda_4 \;\Rightarrow\; (-m/2)\,\dot\omega_2 = (-m/2)(-\dot X\sin\phi + \dot Y\cos\phi)^{\cdot} = \Theta_4 + \lambda_4,\\ &\theta_5\!: \quad \dot P_5 = \Theta_5 + \lambda_5 \;\Rightarrow\; (-mr/2)\,\dot\omega_3 = (-mr/2)\ddot\phi = \Theta_5 + \lambda_5. \qquad \text{(f14–18)}\end{aligned}$$

Again, for the *force-free* motion (i.e., all Θ_k's $= 0$), and recalling the constraints (a2): $-\dot X\sin\phi + \dot Y\cos\phi + r(\dot\phi + \dot\psi') = 0$, and (a4): $2\dot\phi + (\dot\psi' - \dot\psi'') = 0$, we conclude from the $\theta_{2,3}$-equations that $\dot\phi + \dot\psi' = constant$, $\dot\phi = constant \Rightarrow \dot\psi' = constant$, $(\psi'')^{\cdot} = constant$. Further, it follows that, here, θ_5-equation: $\lambda_5 = 0$; θ_4-equation: $\lambda_4 = 0$; θ_1-equation: $\lambda_1 = constant$.

For the related problem of the *two-wheeled street vendor's cart* (where only the algebra is slightly more complicated than here), see, for example, Hamel (1949, pp. 471–472, 479, 484–485).

The Appell Equations

(No reactions, only motion; i.e., equations for $k \to I = 2, 3$; not $k \to D = 1, 4, 5$.)

(i) *Elementary nonvectorial solution.* Here, with $(X_{G''}, Y_{G''})$ = (inertial) coordinates of G'' and $(X_{G'}, Y_{G'})$ = (inertial) coordinates of G', we have [fig. 3.55(a) and (b)]

$$X_{G''} = X - r\cos\phi \Rightarrow (X_{G''})^{\cdot\cdot} = \ddot{X} + r\cos\phi(\dot{\phi})^2 + r\sin\phi\,\ddot{\phi}, \tag{g1}$$

$$Y_{G''} = Y - r\sin\phi \Rightarrow (Y_{G''})^{\cdot\cdot} = \ddot{Y} + r\sin\phi(\dot{\phi})^2 - r\cos\phi\,\ddot{\phi}, \tag{g2}$$

$$X_{G'} = X + r\cos\phi \Rightarrow (X_{G'})^{\cdot\cdot} = \ddot{X} - r\cos\phi(\dot{\phi})^2 - r\sin\phi\,\ddot{\phi}, \tag{g3}$$

$$Y_{G'} = Y + r\sin\phi \Rightarrow (Y_{G'})^{\cdot\cdot} = \ddot{Y} - r\sin\phi(\dot{\phi})^2 + r\cos\phi\,\ddot{\phi}, \tag{g4}$$

and, therefore, using the Appellian version of König's theorem, we find the Appellians

G'' wheel: $\quad 2S_{G''} = m[(\ddot{X}_{G''})^2 + (\ddot{Y}_{G''})^2] + \{(mr^2/2)(\ddot{\psi}'')^2 + (mr^2/4)(\ddot{\phi})^2\},$

G' wheel: $\quad 2S_{G'} = m[(\ddot{X}_{G'})^2 + (\ddot{Y}_{G'})^2] + \{(mr^2/2)(\ddot{\psi}')^2 + (mr^2/4)(\ddot{\phi})^2\}.$

From the above, it follows that, to *within Appell-important terms* and with constraints enforced,

$$\begin{aligned}
2S_{\text{entire system}} &= 2S_{G''} + 2S_{G'} \\
&= \cdots = m[(\ddot{X})^2 + (\ddot{Y})^2 + r^2(\ddot{\phi})^2 + 2r\,\ddot{\phi}(\ddot{X}\cos\phi + \ddot{Y}\sin\phi) \\
&\qquad + 2r\,\ddot{\phi}(\ddot{X}\sin\phi - \ddot{Y}\cos\phi)] \\
&\qquad + (mr^2/2)\{(\ddot{\psi}'')^2 + (1/2)(\ddot{\phi})^2\} \\
&\qquad + m[(\ddot{X})^2 + (\ddot{Y})^2 + r^2(\ddot{\phi})^2 + 2r\ddot{\phi}(-\ddot{X}\cos\phi + \ddot{Y}\sin\phi) \\
&\qquad - 2r\,\ddot{\phi}(\ddot{X}\cos\phi + \ddot{Y}\sin\phi)] \\
&\qquad + (mr^2/2)\big[(\ddot{\psi}')^2 + (1/2)(\ddot{\phi})^2\big] \\
&= 2m[(\ddot{X})^2 + (\ddot{Y})^2] \\
&\qquad + 2(mr^2/4)\big[(\ddot{\psi}'')^2 + (\ddot{\psi}')^2 + 5(\ddot{\phi})^2\big];
\end{aligned} \tag{g5}$$

that is, $2T$, eq. (e2), but with the $\dot{q}$'s replaced by the corresponding $\ddot{q}$'s (= *homogeneous* quadratic in the $\ddot{q}$'s). Next, since $\partial\ddot{q}/\partial\dot{\omega} = \partial\dot{q}/\partial\omega$, and recalling the $\dot{q} \Leftrightarrow \omega$ relations (c6–9), we find

$$\begin{aligned}
\partial S^*/\partial \dot{\omega}_2 &= (\partial S/\partial \ddot{X})(\partial \dot{X}/\partial \omega_2) + (\partial S/\partial \ddot{Y})(\partial \dot{Y}/\partial \omega_2) + (\partial S/\partial \ddot{\phi})(\partial \dot{\phi}/\partial \omega_2) \\
&\quad + (\partial S/\partial \ddot{\psi}'')(\partial \dot{\psi}''/\partial \omega_2) + (\partial S/\partial \ddot{\psi}')(\partial \dot{\psi}'/\partial \omega_2) \\
&= (2m\ddot{X})(-\sin\phi) + (2m\ddot{Y})(\cos\phi) + (5mr^2\ddot{\phi}/2)(0) \\
&\quad + (mr^2\ddot{\psi}''/2)(-r^{-1}) + (mr^2\ddot{\psi}'/2)(-r^{-1}) \\
&= 2m(-\ddot{X}\sin\phi + \ddot{Y}\cos\phi) - (mr/2)(\ddot{\psi}'' + \ddot{\psi}') \\
&= 2m(-r\ddot{\phi} - r\ddot{\psi}') - (mr/2)(\ddot{\psi}'' + \ddot{\psi}') \\
&= -2mr[(\ddot{\psi}'' - \ddot{\psi}')/2] - 2mr\,\ddot{\psi}' - (mr/2)\ddot{\psi}' - (mr/2)\ddot{\psi}'' \\
&= -(3mr/2)(\ddot{\psi}'' + \dot{\psi}') \\
&= -(3mr/2)[(-r^{-1}\dot{\omega}_2 - \dot{\omega}_3) + (-r^{-1}\dot{\omega}_2 + \dot{\omega}_3)] \\
&= -(3mr/2)(-2r^{-1}\dot{\omega}_2) = 3m\,\dot{\omega}_2,
\end{aligned} \tag{g6}$$

and, similarly,

$$\begin{aligned}
\partial S^*/\partial \dot{\omega}_3 &= (2m\ddot{X})(0) + (2m\ddot{Y})(0) \\
&\quad + (5mr^2\ddot{\phi}/2)(1) + (mr^2\ddot{\psi}''/2)(-1) + (mr^2\ddot{\psi}'/2)(1) \\
&= (mr^2/2)(\ddot{\psi}'' - \ddot{\psi}' + 5\ddot{\phi}) \\
&= (mr^2/2)(2\ddot{\phi} + 5\ddot{\phi}) = (7mr^2/2)\ddot{\phi} = (7mr^2/2)\dot{\omega}_3;
\end{aligned} \tag{g7}$$

and these are precisely the left (i.e., inertia) sides of the earlier *second* and *third* equations of Hamel. To derive the *first, fourth,* and *fifth* Appellian equations, we use again S, apply the chain rule

$$\partial S^*/\partial \dot{\omega}_k = \sum (\partial S/\partial \ddot{q}_l)(\partial \dot{q}_l/\partial \omega_k) \qquad (l = 1, \ldots, 5), \tag{g8}$$

and then impose the constraints $\omega_{1,4,5} = 0$; that is, *after* the differentiations—not before them! The details are left to the reader.

(ii) *Vectorial solution.* For *each* wheel, we shall have [recalling (3.14.4a ff.)]:

$$2S = m\,a_G{}^2 + \boldsymbol{\alpha}\cdot(\mathbf{I}_G\cdot\boldsymbol{\alpha}) + (\boldsymbol{\alpha}\times\boldsymbol{\omega})\cdot(\mathbf{I}_G\cdot\boldsymbol{\omega}). \tag{g9}$$

But here (using the notation of the first Appellian solution):

$$a_G{}^2 = (\ddot{X}_{G''})^2 + (\ddot{Y}_{G''})^2, \qquad \text{or} \qquad (\ddot{X}_{G'})^2 + (\ddot{Y}_{G'})^2, \tag{g10}$$

and along the *intermediate* but principal axes G–123 (fig. 3.55b), which have inertial angular velocity $(0, 0, \dot{\phi})$, we easily find

$$\begin{aligned}
&\boldsymbol{\omega} = (\dot{\psi}, 0, \dot{\phi}) \\
&\Rightarrow \boldsymbol{\alpha} = d\boldsymbol{\omega}/dt = (\ddot{\psi}, 0, \ddot{\phi}) + (0, 0, \dot{\phi}) \times (\dot{\psi}, 0, \dot{\phi}) = (\ddot{\psi}, -\dot{\psi}\,\dot{\phi}, \ddot{\phi}), \\
&\mathbf{I}_G = \textit{diagonal}\ (mr^2/2,\ mr^2/4,\ mr^2/4),
\end{aligned} \tag{g11}$$

and, therefore,

$$(\boldsymbol{\alpha} \times \boldsymbol{\omega}) \cdot (\mathbf{I}_G \cdot \boldsymbol{\omega}) = [(\ddot{\psi}, -\dot{\psi}\,\dot{\phi}, \ddot{\phi}) \times (\dot{\psi}, 0, \dot{\phi})] \cdot (mr^2\dot{\psi}/2,\ 0,\ mr^2\dot{\phi}/4)$$
$$= \cdots = -(mr^2/2)(\dot{\psi})^2(\dot{\phi})^2 = \textit{non–Appell-important}\ \text{term}, \quad \text{(g12)}$$

and

$$\boldsymbol{\alpha} \cdot (\mathbf{I}_G \cdot \boldsymbol{\alpha}) = (\ddot{\psi}, -\dot{\psi}\,\dot{\phi}, \ddot{\phi}) \cdot (mr^2\ddot{\psi}/2,\ -mr^2\dot{\psi}\,\dot{\phi}/4,\ mr^2\ddot{\phi}/4)$$
$$= mr^2(\ddot{\psi})^2/2 + mr^2(\dot{\psi})^2(\dot{\phi})^2/4 + mr^2(\ddot{\phi})^2/4$$
$$= (mr^2/2)\left[(\ddot{\psi})^2 + (1/2)(\ddot{\phi})^2\right] + \textit{non–Appell-important}\ \text{term}; \quad \text{(g13)}$$

that is, *for each wheel, and to within Appell-important terms,*

$$2S = ma_G{}^2 + (mr^2/2)[(\ddot{\psi})^2 + (1/2)(\ddot{\phi})^2], \quad \text{(g14)}$$

and, adding these partial results, we re-establish the earlier entire system Appellian.

Example 3.18.7 *Dynamics of Pair of Rolling Wheels on an Inclined Plane.* Continuing from the preceding example, let us specialize it to the case where $m_{\text{wheel}} = 0$ but $m_{\text{axle}} \equiv m \neq 0$; and, in addition, the whole system rolls on a plane P inclined by an angle χ to the horizontal (fig. 3.56).

We saw in the previous example that the constraints are

$$\dot{X}\cos\phi + \dot{Y}\sin\phi = 0, \quad \text{(a1)}$$
$$-\dot{X}\sin\phi + \dot{Y}\cos\phi + b\dot{\phi} + r\dot{\psi}' = 0, \quad \text{(a2)}$$
$$-\dot{X}\sin\phi + \dot{Y}\cos\phi - b\dot{\phi} + r\dot{\psi}'' = 0. \quad \text{(a3)}$$

It is not hard to see that, in this case,

$$2T = m[(\dot{X})^2 + (\dot{Y})^2] + (mb^2/3)(\dot{\phi})^2, \quad \text{(b)}$$

and, therefore, to within Appell-important terms,

$$2S = m[(\ddot{X})^2 + (\ddot{Y})^2] + (mb^2/3)(\ddot{\phi})^2, \quad \text{(c)}$$

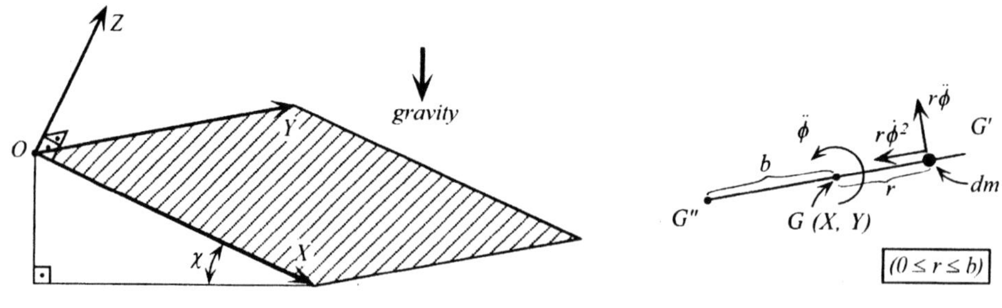

Figure 3.56 Rolling wheels on an axle, on an inclined plane.

and

$$\delta' W = mg(\delta X \sin\chi) \equiv Q_X\, \delta X \;\Rightarrow\; Q_X = mg\sin\chi. \tag{d}$$

Let us use the constraints to eliminate, say $\ddot{Y}$, from S: $(\ldots)^{\cdot}$-differentiating (a1) we obtain

$$\dot{Y} = -\dot{X}(\cos\phi/\sin\phi) \;\Rightarrow\; \ddot{Y} = -\ddot{X}(\cos\phi/\sin\phi) + \dot{X}\dot{\phi}(1/\sin^2\phi). \tag{e}$$

Substituting (e) into (c), we obtain $S = S_o(\ddot{X}, \ddot{\phi}) \equiv S_o = \cdots$. We remark that in building Appell's equations, *there is no need to square* $\ddot{Y}$ of (e). Indeed, since $S = S[\ddot{X}, \ddot{Y}(\ddot{X}, \ldots), \ddot{\phi}; \ldots] = S_o$, the chain rule yields

$$\begin{aligned}
\partial S_o/\partial\ddot{X} &= \partial S/\partial\ddot{X} + (\partial S/\partial\ddot{Y})(\partial\ddot{Y}/\partial\ddot{X}) = \partial S/\partial\ddot{X} + (\partial S/\partial\ddot{Y})(\partial\dot{Y}/\partial\dot{X}) \\
&= (m\ddot{X}) + (m\ddot{Y})(-\cos\phi/\sin\phi) \quad (= mg\sin\chi), &\text{(f)}\\
\partial S_o/\partial\ddot{\phi} &= \partial S/\partial\ddot{\phi} + (\partial S/\partial\ddot{Y})(\partial\ddot{Y}/\partial\ddot{\phi}) = \partial S/\partial\ddot{\phi} + (\partial S/\partial\ddot{Y})(\partial\dot{Y}/\partial\dot{\phi}) \\
&= (mb^2/3)\ddot{\phi} + (m\ddot{Y})(0) \quad (= 0). &\text{(g)}
\end{aligned}$$

From (g), and with ω_o/ϕ_o = *initial angular velocity/angle* of bar $G''G'$, we immediately find

$$\dot{\phi} = \omega_o \;\Rightarrow\; \phi = \omega_o t + \phi_o. \tag{h}$$

Then, with the choice $\phi_o = 0$, (f) reduces to

$$\ddot{X}\sin(\omega_o t) - \ddot{Y}\cos(\omega_o t) = g\sin\chi\sin(\omega_o t); \tag{i}$$

and along with (e) they constitute a system for $X(t)$ and $Y(t)$. Indeed, eliminating $\ddot{Y}$ between (i) and (e) yields

$$\ddot{X}\{\sin(\omega_o t) + [\cos^2(\omega_o t)/\sin(\omega_o t)]\} - \dot{X}\omega_o[\cos(\omega_o t)/\sin^2(\omega_o t)] = g\sin\chi\sin(\omega_o t). \tag{j}$$

Integrating (j) twice {while noticing that its left side equals $[\dot{X}/\sin(\omega_o t)]^{\cdot}$}, we obtain, after some elementary integrations (and with $c_{1,2}$ = *integration constants*),

$$X = (g\sin\chi/4\omega_o{}^2)\cos(2\omega_o t) - (c_1/\omega_o)\cos(\omega_o t) + c_2. \tag{k}$$

Then, substituting from (k) into (a1) and integrating, we finally get (with c_3 = *integration constant*)

$$Y = (g\sin\chi/4\omega_o{}^2)[\sin(2\omega_o t) + 2\omega_o t)] - (c_1/\omega_o)\sin(\omega_o t) + c_3; \tag{l}$$

that is, G traces a curve parallel to a *cycloid* with base(line) parallel to OY.

For additional insights, see Delassus [(1913(b), pp. 406–409), which investigates the above system but with an additional particle of mass m placed at G, and then of the limit as $m_{\text{wheels}}, m_{\text{bar}} \to 0$], Pérès (1953, p. 214); also Bahar (1998).

APPENDIX 3.A1

REMARKS ON THE HISTORY OF THE HAMEL-TYPE EQUATIONS OF ANALYTICAL MECHANICS

Below, and continuing from the Introduction and the early sections of this chapter, we discuss the historical evolution of the Hamel-type equations; that is, of T-based *equations of nonholonomically constrained systems in nonholonomic variables.*

We begin with the following, highly selective but adequate for our purposes, summaries of the main theoretical developments of Lagrangean mechanics; see tables 3.A1.1–3.A1.3.

From the Prehistory of the Hamel-Type Equations

Let us now discuss, *in "our" notation*, some additional special forms of Hamel-type equations of motion (early 1870s to the early 1900s). It is through the acquaintance with such historical curiosities that we deepen our understanding of *contemporary* Lagrangean mechanics, both holonomic and nonholonomic; and appreciate better the importance of the fundamental contributions of Heun and Hamel (1901–1914) to our subject.

(i) Equations of Ferrers (early 1870s, publ. 1873)

Let us assume for algebraic simplicity, but no loss in generality, a scleronomic system with Chaplygin-type constraints (3.8.13a)

$$\dot{q}_D = \sum b_{DI}(q_{m+1}, \ldots, q_n)\dot{q}_I = \sum b_{DI}\dot{q}_I \qquad (I = m+1, \ldots, n). \qquad (3.\text{A}1.1)$$

Then, Ferrers' equations are

$$I_{Io} \equiv d/dt(\partial T_o/\partial \dot{q}_I) - S\, dm\, \mathbf{v}_o \cdot (d\boldsymbol{\beta}_I/dt) = Q_{Io}, \qquad (3.\text{A}1.2)$$

Table 3.A1.1 Global Picture

Newton (2nd half of 17th cent.)	Physical foundations of mechanics (fundamental law); particle
Euler (18th cent.)	Physical and mathematical foundations of mechanics (momentum mechanics); rigid body, *rotation*
Lagrange (2nd half of 18th cent.)	Mathematical deepening of mechanics (energetic mechanics); *constraints*, Lagrangean equations
Cauchy (1st half of 19th cent.)	Continuum mechanics; *deformation* (strain), stress
Hamilton (1830's)	Canonical formalism; Hamilton's equations
Jacobi (1840's)	Integration theory of dynamics
Kelvin/Helmholtz/Routh (2nd half of 19th cent.)	Ignorable (cyclic) coordinates; gyroscopic systems
Appell (late 19th-early 20th cent.)	Nonholonomic systems; Appell's equations (acceleration-based equations)
Heun (early 20th cent.)	Theoretical engineering dynamics
Hamel (1st half of 20th cent.)	Nonholonomic systems, axiomatics of classical (discrete + continuum) mechanics; Hamel's equations (kinetic energy-based equations)

Table 3.A1.2 Prehistory of Lagrangean Mechanics

Mersenne (1646)	Find the period of a mathematical pendulum with several masses
Huygens (1673)	Center of oscillation ("Horologium oscillatorium")
Jacob Bernoulli (1686–1703)	Physical pendulum; earliest form of d'Alembert's principle
D'Alembert (1743)	Earliest monograph on constrained system dynamics

where

$$\boldsymbol{v} = \sum \boldsymbol{e}_k \dot{q}_k = \cdots \equiv \sum \boldsymbol{\beta}_I \dot{q}_I \equiv \sum (\partial \boldsymbol{v}_o / \partial \dot{q}_I) \dot{q}_I \equiv \boldsymbol{v}_o(q, \dot{q}_I) \equiv \boldsymbol{v}_o. \qquad (3.\text{A}1.2\text{a})$$

We recall [(2.9.37), also prob. 2.11.1] that since, in general, the $\boldsymbol{\beta}_I$ are *nongradient*

$$E_I(\boldsymbol{v}_o) \equiv d/dt(\partial \boldsymbol{v}_o / \partial \dot{q}_I) - \partial \boldsymbol{v}_o / \partial q_I \equiv d\boldsymbol{\beta}_I / dt - \partial \boldsymbol{v}_o / \partial q_I \neq \boldsymbol{0}, \qquad (3.\text{A}1.3\text{a})$$

even though (2.5.10)

$$E_I(\boldsymbol{v}) \equiv d/dt(\partial \boldsymbol{v} / \partial \dot{q}_I) - \partial \boldsymbol{v} / \partial q_I = \boldsymbol{0}. \qquad (3.\text{A}1.3\text{b})$$

Indeed, in view of the above, we obtain, successively (with I, $I' = m+1, \ldots, n$),

$$\begin{aligned} E_I(\boldsymbol{v}_o) &= d\boldsymbol{\beta}_I/dt - \partial \boldsymbol{v}_o / \partial q_I \\ &= \sum (\partial \boldsymbol{\beta}_I / \partial q_{I'})(dq_{I'}/dt) - \sum (\partial \boldsymbol{\beta}_{I'} / \partial q_I)(dq_{I'}/dt) \\ &= \sum (\partial \boldsymbol{\beta}_I / \partial q_{I'} - \partial \boldsymbol{\beta}_{I'} / \partial q_I) \dot{q}_{I'} \neq \boldsymbol{0} \qquad \text{(in general)}; \end{aligned} \qquad (3.\text{A}1.3\text{c})$$

and, therefore, in the nonholonomic case,

$$R_I \equiv \mathcal{S}\, dm\, \boldsymbol{v}_o \cdot (\partial \boldsymbol{v}_o / \partial \dot{q}_I)^{\cdot} \equiv \mathcal{S}\, dm\, \boldsymbol{v}_o \cdot (d\boldsymbol{\beta}_I / dt) \neq \mathcal{S}\, dm\, \boldsymbol{v}_o \cdot (\partial \boldsymbol{v}_o / \partial q_I) = \partial T_o / \partial q_I. \qquad (3.\text{A}1.3\text{d})$$

Also, Carvallo (1900–1901), in his classic studies of the *mono-* and *bi-cycle*, and most likely independently of Ferrers, introduced essentially equivalent equations of motion. Clearly, the method of Ferrers can be extended to the most general rheonomic nonholonomic constraints; even nonlinear ones (chap. 5).

Table 3.A1.3 Historical Synthesis (mid-to-late 18th Century)

D'Alembert's decomposition (Ansatz): $df = d\boldsymbol{F} + d\boldsymbol{R}$		Laws of Simple Machines Ancient Greeks (Aristotle, Archimedes et al.), Del Monte (1577), Galileo (1594), Torricelli (1644) et al.
↓		↓
D'Alembert's Principle: The $\{-d\boldsymbol{R}\}$ are in equilibrium (1743)	+	Johann Bernoulli Statical Principle of Virtual Work (1717, 1725): Equilibrium ⇔ zero virtual work
	↓	
Lagrange Principle of least action (1760) Lagrange's principle (1764): Lagrange's equations (1780) Méchanique Analitique (1788)		$\mathcal{S}(-d\boldsymbol{R}) \cdot \delta \boldsymbol{r} = 0 \Rightarrow \mathcal{S}\, dm\, \boldsymbol{a} \cdot \delta \boldsymbol{r} = \mathcal{S}\, d\boldsymbol{F} \cdot \delta \boldsymbol{r}$

REMARKS

(a) Such an extension of (3.A1.2) has been given (independently) by Greenwood [1994, p. 83 ff., eqs. (3.98)]. With (our notation): $\dot{q}_k = \sum A_{kI}\omega_I + A_k$, $\boldsymbol{v} = \cdots = \boldsymbol{v}^*{}_o(t, q, \omega_I) \Rightarrow T^* = T^*{}_o(t, q, \omega_I)$, or

$$T[t, q, \dot{q}_I, \dot{q}_D(t, q, \dot{q}_I)] = T_o(t, q, \dot{q}_I) = T_o[t, q, \dot{q}_I(t, q, \omega_I)] = T^*{}_o(t, q, \omega_I),$$

and with $\delta' W = \sum Q_k\, \delta q_k = \sum \Theta_I\, \delta\theta_I$, we obtain the *Greenwood equations*:

$$(\partial T^*{}_o/\partial \dot{\omega}_I)^{\cdot} - S\, dm\, \boldsymbol{v}^*{}_o \cdot (\partial \boldsymbol{v}^*{}_o/\partial \omega_I) = \Theta_I. \qquad (3.A1.3e)$$

Actually, (3.A1.3e) also holds for nonlinear nonholonomic transformations: $\dot{q}_I = \dot{q}_I(t, q, \omega_I)$.

(b) A certain unclear historical statement by Whittaker (1937, p. 215, footnote) seems to have caused a number of other (less famous) authors to call, *erroneously*, Ferrers equations the Routh–Voss (multiplier) equations; for example, Fox (1967, p. 351), Hand and Finch (1998, p. 62, footnote), Rose (1938, p. 16). The record is corrected in Routh 1905(a), p. 348, footnote.

(c) For complementary expositions on the Ferrers equations, and so on, see, for example (alphabetically): Auerbach (1908, p. 327, eq. (68)), Gray (1918, pp. 411–418), Marcolongo (1912, pp. 104–105), Voss (1901/1908, pp. 82–83).

(ii) T-Equations of Appell (1899) and Boltzmann (1902)

These result from further transformations of the key R_I term, (3.A1.3d). Indeed, we obtain, successively [recalling (3.3.11a ff.)],

$$\begin{aligned} R_I &\equiv S\, dm\, \boldsymbol{v}_o \cdot (d\boldsymbol{\beta}_I/dt) = S\, dm\, \boldsymbol{v}_o \cdot [d/dt(\partial \boldsymbol{v}_o/\partial \dot{q}_I)] \\ &= S\, dm \boldsymbol{v}_o \cdot [d/dt(\partial \boldsymbol{v}_o/\partial \dot{q}_I) - \partial \boldsymbol{v}_o/\partial q_I] + S\, dm\, \boldsymbol{v}_o \cdot (\partial \boldsymbol{v}_o/\partial q_I) \\ &= S\, dm\, \boldsymbol{v}_o \cdot \Big(\sum (\partial \boldsymbol{\beta}_I/\partial q_{I'})(dq_{I'}/dt) - \sum (\partial \boldsymbol{\beta}_{I'}/\partial q_I)(dq_{I'}/dt)\Big) + \partial T_o/\partial q_I \\ &= \Gamma_{Io} + \partial T_o/\partial q_I \qquad (\Rightarrow R_I - \partial T_o/\partial q_I = \Gamma_{Io}), \end{aligned} \qquad (3.A1.4)$$

where

$$\begin{aligned} \Gamma_{Io} &\equiv S\, dm\, \boldsymbol{v}_o \cdot E_I(\boldsymbol{v}_o) \\ &= \sum \Big\{ S\, dm\, \boldsymbol{v}_o \cdot (\partial \boldsymbol{\beta}_I/\partial q_{I'} - \partial \boldsymbol{\beta}_{I'}/\partial q_I) \Big\}\, \dot{q}_{I'} \end{aligned} \qquad (3.A1.4a)$$

an expression that shows clearly the *gyroscopicity* of this nonholonomic "correction term." Hence, Ferrers' equations take the Appell form

$$E_I(T_o) - \Gamma_{Io} \equiv d/dt(\partial T_o/\partial \dot{q}_I) - \partial T_o/\partial q_I - \Gamma_{Io} = Q_{Io}. \qquad (3.A1.5)$$

When the fundamental term Γ_{Io} is expressed exclusively in *system* variables, say $\Gamma_{Io} = \sum c_{II'}\, \dot{q}_{I'} =$ *quadratic* in the $\dot{q}_I$ (where $c_{II'} = -c_{I'I}$), eqs. (3.A1.5) are none other than the Chaplygin equations (3.8.13a ff.); which, as we have seen, are a special case of the Hamel equations.

From the above, we conclude, with Appell, that for Lagrange's equations to apply for a particular q_I [i.e., $(\partial T_o/\partial \dot{q}_I)^{\cdot} - \partial T_o/\partial q_I = Q_{Io}$], it is necessary and sufficient that $\Gamma_{Io} = 0$, identically in t, q_I's, $\dot{q}_I$'s. In holonomic systems this holds for *all* $I = m+1, \ldots, n$; but in nonholonomic ones, it may hold for some of them; e.g., in the rolling coin problem it does hold for the *nutation* angle θ [ex. 3.18.5, and Ferrers (1873, pp. 3–4)]. Appell calls the number of *nonvanishing* Γ_{Io}'s the "order of nonholonomicity" of the system; and also, he points out the errors resulting from the indiscriminate use of Lagrange's equations $E_I(T_o) = Q_{Io}$.

However, Appell (1899) did not pursue the transformation of R_I and Γ_I any further, and thus missed arriving at the equations of Chaplygin (1895, publ. 1897) and Voronets (1901). Instead, seeing an apparent dead-end in the direction of Lagrange-type equations, like (3.A1.5) with (3.A1.4a), he turned his energies to the development of his other, now famous, *acceleration-based* S-equations (1899, 1900):

$$\partial S_o/\partial \ddot{q}_I = Q_{Io}. \tag{3.A1.6}$$

Comparing (3.A1.5) and (3.A1.6), we immediately deduce the following basic kinematico-inertial identity:

$$\begin{aligned}\Gamma_{Io} &= [(\partial T_o/\partial \dot{q}_I)^{\cdot} - \partial T_o/\partial q_I] - \partial S_o/\partial \ddot{q}_I \equiv E_I(T_o) - \partial S_o/\partial \ddot{q}_I \\ &\equiv [(\textit{constrained})\ \textit{Euler–Lagrange}]_I - [(\textit{constrained})\ \textit{Appell}]_I \neq 0 \quad \text{(in general);}\end{aligned} \tag{3.A1.7}$$

see, for example, Appell [1899(a), pp. 39–45; 1925, pp. 12–17; 1953, pp. 383–388].

The form (3.A1.5), but for general rheonomic nonholonomic systems [i.e., a form that when brought to *system variables* would be *none other than the Voronets equations* (1901)—(3.8.14a ff.)] was also arrived at, independently, by Boltzmann (1902; 1904, pp. 104–105), who, among mechanicians, gave the first geometrical interpretation of the ("Ricci–Boltzmann–Hamel") *rotation coefficients*: $\partial\beta_I/\partial q_{I'} - \partial\beta_{I'}/\partial q_I$. An additional, related, derivation of the Boltzmann equations, based on Hertz's "principle of the straightest path" (§6.7), was given a little later by Boltzmann's famous student Ehrenfest [1904]. See also, Krutkov (1928: vectorial/dyadic treatment of Boltzmann's equations), MacMillan (1936, pp. 332–341: clear derivation of Boltzmann's equations; unique and virtually unknown in the English literature); and Klein (1970, pp. 53–74: critical summary of Ehrenfest's dissertation). [See also Mei (1984; 1985, pp. 108–114) for an extension of the MacMillan equations to nonlinear nonholonomic constraints.]

To summarize: the main drawback of these equations of Ferrers–Appell–Boltzmann–MacMillan is that they are *mixed*; that is, some of their terms $[E_I(T_o), Q_{Io}]$ are expressed in *system* variables, and some (Γ_{Io}) in *particle* variables. Perhaps this explains why they have not been used much in concrete problems, let alone theoretical arguments. The equations that result by expressing the nonholonomic term Γ_{Io} too in system variables (a qualitatively higher step in the evolution of Lagrangean-type equations!) are the equations of Chaplygin and Voronets; schematically:

- Equations of Ferrers (1873)/Appell (1899) $\xrightarrow[\text{variables}]{\text{system}}$ Equations of Chaplygin (1895–1897),
- Equations of Boltzmann (1902)/MacMillan (1936)] $\xrightarrow[\text{variables}]{\text{system}}$ Equations of Voronets (1901).

(3.A1.8)

Last, a special case of the Chaplygin–Voronets equations (rolling of convex body on rough plane) was first given by Neumann (1885).

(iii) Boltzmann versus Hamel

In the light of this historical record, the widely used term "Boltzmann–Hamel equations" (probably originated by readers of Whittaker (1904, §30), and parroted by the rest, except Hamel and his school) is inaccurate. There is a very big difference between these two sets of equations, although they appeared only about a year apart from each other (Boltzmann: 1902; Hamel: 1903, 1904).

(iv) Volterra versus Hamel

The only other Lagrange-type equations of motion that can stand next to Hamel's are those by Volterra (1898; corrections: 1899). However, even Volterra *never discussed constraints*, just equations of motion in terms of nonholonomic *variables* (what he called "parameters," or "motion characteristics"); and, more importantly, Hamel's treatment is far more comprehensive and deep.

(v) Gibbs versus Appell

The S-equations of Appell under Pfaffian constraints are sometimes called "Gibbs–Appell equations"; for example, Pars (1965). However, a careful study of the original memoirs of these two masters reveals that Appell's contributions (several weighty papers, a monograph exclusively devoted to these equations, plus extensive parts of his famous treatise) completely overshadow by several orders of magnitude those of Gibbs (two pages at the end of his single paper on theoretical dynamics). The main difference between the two is that: Appell dealt with *both nonholonomic coordinates and constraints*, whereas Gibbs dealt *only with nonholonomic coordinates*. Also, their approaches are distinctly different: Gibbs derives his equations from the differential form of the (lesser known) Gauss' principle, whereas Appell obtains his from (the simpler) Lagrange's principle.

For these objective and incontrovertible reasons, we have decided to call them *Appell's equations*. In this practice, we are accompanied by the overwhelming majority of the best mechanicians of the 20th century; for example, (in approximate chronological order of appearance of their works on this subject): Voss, Heun, Routh, Whittaker, Gray, Hamel, Nordheim, Johnsen, Prange, Ames and Murnaghan, Levi-Civita and Amaldi, MacMillan, Rose, Lanczos, Beghin, Pérès, Synge, Lur'e, Gantmacher, Novoselov, Dobronravov, Neimark and Fufaev, Mei et al.

In view of the above, the situation in (iv) and (v) can be fairly summed up as follows: Volterra's equations stand relative to Hamel's the same way that Gibbs' equations stand relative to Appell's (S-equations), and vice versa. In both cases, the relevant contributions of Hamel and Appell exceed by several quantitative and qualitative orders of magnitude those of Volterra and Gibbs, respectively. This is shown schematically in table 3.A1.4.

The foregoing history helps us to build the following summaries and table of the equations of motion of analytical dynamics:

Table 3.A1.4 Volterra vs. Hamel, and Gibbs vs. Appell

	Nonholonomic Coordinates		
	No Constraints		Constraints
T-equations:	Volterra (1898)	<	Hamel (1903, 1904)
S-equations:	Gibbs (1879)	<	Appell (1899, 1900)

Lagrange's Principle (LP)

Particle/vector variables: $$S\, dm\, \boldsymbol{a} \cdot \delta \boldsymbol{r} = S\, d\boldsymbol{F} \cdot \delta \boldsymbol{r}$$

Holonomic system variables: $$\sum E_k(T)\, \delta q_k = \sum Q_k\, \delta q_k$$

Nonholonomic system variables: $$\sum [E_k{}^*(T^*) - \Gamma_k]\, \delta\theta_k = \sum \Theta_k \delta\theta_k.$$

General Remarks

Additional Pfaffian constraints (holonomic and/or nonholonomic) are either (a) *adjoined* to LP via *multipliers*, in which case the resulting equations are, in general, *coupled* in the motion and reactions. However, under finite (geometrical) constraints, the equations of motion can always be uncoupled by special choices of "equilibrium" coordinates; or they are (b) *embedded* to LP (or *eliminated*) via *quasi coordinates*, in which case the resulting equations of motion can always be uncoupled by special choices of such quasi variables into *kinetic* (reactionless, motion only) and *kinetostatic* (reaction-containing).

Table 3.A1.5 summarizes the equations of constrained dynamics.

Table 3.A1.5 Global (Panoramic) Map of the Equations of Constrained Dynamics

	T-Based Equations (velocities)
Multipliers:	Routh (1879)/Voss (1884–1885)—holonomic variables ↓
Projection:	Maggi (1896, 1901)—holonomic variables Special cases: Hadamard (1895, 1899), Korteweg (1899) ↓
Quasi variables:	Hamel (1903–1904)—nonholonomic variables
Special cases:	(i) Ferrers (1873), Appell (1899), Carvallo (1900), Boltzmann (1902), Auerbach (1908), MacMillan (1936), Greenwood (1994); (ii) C. Neumann (1885), Chaplygin (1895–1897)→Voronets (1901); (iii) Volterra (1898); (iv) Poincaré (1901)

S-Based Equations (accelerations)

In view of the kinematico-inertial identities:

Holonomic variables: $\partial S/\partial \ddot{q}_k = (\partial T/\partial \dot{q}_k)^{\cdot} - \partial T/\partial q_k$

Nonholonomic variables: $\partial S^*/\partial \dot{\omega}_k = (\partial T^*/\partial \omega_k)^{\cdot} - \partial T^*/\partial \theta_k - \Gamma_k$

there exist Appellian counterparts to all the above equations. Here, in general, both T and S are unconstrained; if the additional Pfaffian constraints are holonomic, they can be constrained.

(For the less common (dT/dt)-based equations, see §6.3 ff.)

APPENDIX 3.A2

CRITICAL COMMENTS ON VIRTUAL DISPLACEMENTS/WORK; AND LAGRANGE'S PRINCIPLE

Some Common Misunderstandings Regarding d'Alembert's Principle (d'AP)

As pointed out in §3.2 and elsewhere, d'AP is, in spite of its simplicity, one of the most misunderstood principles in the history of physics. Although the whole matter was finally and fully clarified, qualitatively and quantitatively, in the early years of the 20th century by such mechanics greats as Heun and Hamel, considerable confusion and misunderstanding still persists even today, especially among English language texts and, more specifically, those written by physicists. (The most likely culprits for such a tradition of error must be the very influential Victorian treatises of Thomson/Tait, Routh, Whittaker, Lamb, et al.; which, in spite of their overall greatness, are pretty incomplete on this fundamental topic.) Let us try to identify and dispel the most common of these intellectual malignancies.

(i) A frequent misrepresentation of d'AP runs as follows: one starts with the Newton–Euler law:

$$df = dm\, \boldsymbol{a}, \tag{3.A2.1}$$

then one moves the inertia term $dm\, \boldsymbol{a}$ to the left/force side of the equation, and calls the trivial result: $df + (-dm\, \boldsymbol{a}) = \boldsymbol{0}$, d'AP. In words: during the motion, the sum of all forces, real (df) and "reversed effective" ($-dm\, \boldsymbol{a}$) are in dynamic (?!) equilibrium; see, for example, (alphabetically): Halfman (1962, p. 62), Housner and Hudson (1959, pp. 253–254), Meriam and Kraige (1986, p. 223), to name a few contemporary (otherwise quite decent and worthwhile) expositions. Many more examples of this physically vacuous formulation appear in other areas of engineering dynamics; for example, vibrations, fluid mechanics, and so on.

(ii) Some authors talk about d'AP in so many places and (in, mostly, qualitative) forms, including the correct one, that the reader ends up confused as to the true meaning of the principle and unable to apply it to new and nontrivial circumstances. Others confuse d'AP with the Newton–Euler principle (better, constitutive postulate) of action–reaction for the *internal* forces, while limiting themselves to rigid bodies/systems; for example, Marris and Stoneking (1967, pp. 95–96). But if d'AP simply meant equilibrium of all *internal* forces, in the Newton–Euler sense, that is,

$$S\, df_{internal} = \boldsymbol{0} \quad \text{and} \quad S\, \boldsymbol{r} \times df_{internal} = \boldsymbol{0}, \tag{3.A2.2}$$

then how would one apply the principle to constrained systems that do not possess such forces? [As we have already seen (§3.2), in general, $df_{internal} \neq d\boldsymbol{R}$ (= *total* constraint reaction); $df_{internal} = d\boldsymbol{R}_{internal}$, in a rigid body, and $df_{internal} = d\boldsymbol{R}$, in a free (i.e., *externally* unconstrained) rigid body.] For example, in the following simple systems the constraint reactions are neither internal nor do they satisfy (3.A2.2): (a) particle on, say a smooth, surface; (b) mathematical pendulum. Indeed, here we have

$$\textit{Newton–Euler} \text{ principles: } S\, d\boldsymbol{R} \neq \boldsymbol{0} \quad \text{and} \quad S\, (\boldsymbol{r} \times d\boldsymbol{R}) \neq \boldsymbol{0} \quad (\text{for a general } \boldsymbol{r}); \tag{3.A2.3a}$$

but

$$D\text{'}Alembert\text{–}Lagrange \text{ principle:} \quad \mathcal{S}\, d\boldsymbol{R} \cdot \delta\boldsymbol{r} = 0; \tag{3.A2.3b}$$

that is, it is not the constraint reactions that must vanish [individually or in the sense of (3.A2.2), although that may happen in some problems], but the *sum* of their projections in certain directions. In other words, to say that d'AP requires that the constraint reactions be "in equilibrium," or constitute a "null system" of forces, is correct provided that *equilibrium* is understood in the *generalized total virtual work sense of* (3.A2.3b), *not* (3.A2.3a). Then, it is meaningful even for a *single* reaction, external or internal. Let us clarify this.

Following Hamel (1927; 1949, p. 217): we call two force (and/or couple) systems, acting separately on the same mechanical system, *equivalent* if, and only if, starting from the same initial kinematical state (i.e., time, positions, and velocities) they communicate to it the same acceleration. In particular: a system of forces is said to be in *equilibrium* (or be a *null system*) if the accelerations communicated by it to a mechanical system are the same as those that would occur if no impressed forces acted on it.

In this light it becomes clear why the earlier-mentioned pendulum tension is in equilibrium; it may not vanish, but it does not affect the acceleration of the pendulum's bob; that is done by gravity, an impressed force.

(iii) A related misconception is to confuse d'AP with the spatial integral forms of the Newton–Euler principles of linear/angular momentum. Thus, we read that "the sum of the forces and the sum of the moments of the forces [including those of the 'inertia forces' $-dm\,\boldsymbol{a}$] about *any* point vanish" and "d'Alembert's principle leaves one free to take moments about *any* point, whereas the angular momentum principle restricts one in this regard" (Kane and Levinson, 1980, pp. 102–103). In our notation, the above read simply

$$\mathcal{S}\,(d\boldsymbol{f} - dm\,\boldsymbol{a}) = \boldsymbol{0} \quad \text{and} \quad \mathcal{S}\,\boldsymbol{r}_{/\bullet} \times (d\boldsymbol{f} - dm\,\boldsymbol{a}) = \boldsymbol{0}, \tag{3.A2.4}$$

where $\boldsymbol{r}_{/\bullet}$ = position vector of generic system particle relative to the completely arbitrary (fixed or moving, not necessarily body-) point $\bullet$.

But eqs. (3.A2.4) follow immediately from the local Newton–Euler principle (3.A2.1) by the simple mathematical operations indicated above. Generally, starting with (3.A2.1), we can perform to it any kind of mathematically meaningful operation; for example, dot it or cross it with an arbitrary scalar/vector/tensor, and so on, differentiate/integrate it in space/time, and so on. Nothing *physically new* will result from such analytical (logical) rearrangements. One does not need any special permission from Newton–Euler (i.e., a *new postulate*) to go from (3.A2.1) to (3.A2.4); and the latter is not d'AL, anyway, but a trivial rearrangement of the Newton–Euler principle. Let us elaborate on this matter.

Detour on Angular Momentum

As already described in §1.6, to obtain an angular momentum principle we cross (3.A2.1) with $\boldsymbol{r}_{/\bullet}$ and then integrate/sum it, for a fixed time, over the material system:

$$\boldsymbol{M}_{\bullet} \equiv \mathcal{S}\,\boldsymbol{r}_{/\bullet} \times d\boldsymbol{f} = \mathcal{S}\,\boldsymbol{r}_{/\bullet} \times dm\,\boldsymbol{a}. \tag{3.A2.5}$$

However, the right (inertia) side of (3.A2.5) can be transformed further either as

(a)
$$\mathcal{S}\, \boldsymbol{r}_{/\bullet} \times dm\, \boldsymbol{a} = \left(\mathcal{S}\, \boldsymbol{r}_{/\bullet} \times dm\, \boldsymbol{v}\right)^{\cdot} - \mathcal{S}\, \boldsymbol{v}_{/\bullet} \times dm\, \boldsymbol{v}$$
$$= \left(\mathcal{S}\, \boldsymbol{r}_{/\bullet} \times dm\, \boldsymbol{v}\right)^{\cdot} - \mathcal{S}\, (\boldsymbol{v} - \boldsymbol{v}_\bullet) \times dm\, \boldsymbol{v}$$
$$= d\boldsymbol{H}_\bullet/dt + \boldsymbol{v}_\bullet \times m\, \boldsymbol{v}_G, \tag{3.A2.5a}$$

where

$$\boldsymbol{H}_{\bullet,\text{absolute}} \equiv \boldsymbol{H}_\bullet \equiv \mathcal{S}\, \boldsymbol{r}_{/\bullet} \times dm\, \boldsymbol{v} = \textit{Absolute angular momentum} \text{ about } \bullet, \tag{3.A2.5b}$$

and G = *center of mass of the system*; that is, in general, the sum of the moments of the rate of linear momenta about • [left side of (3.A2.5a)] is not equal to the rate of change of the sum of moments of the linear momenta about • [first term in right side of (3.A2.5a)]; or, in terms of • – *relative* quantities as

(b)
$$\mathcal{S}\, \boldsymbol{r}_{/\bullet} \times dm\, \boldsymbol{a} = \left(\mathcal{S}\, \boldsymbol{r}_{/\bullet} \times dm\, \boldsymbol{v}\right)^{\cdot} + \boldsymbol{v}_\bullet \times m\, \boldsymbol{v}_G$$
$$= \left\{\mathcal{S}\, [\boldsymbol{r}_{/\bullet} \times dm(\boldsymbol{v}_\bullet + \boldsymbol{v}_{/\bullet})]\right\}^{\cdot} + \boldsymbol{v}_\bullet \times m\, \boldsymbol{v}_G$$
$$= \left[\mathcal{S}\, (\boldsymbol{r}_{/\bullet} \times dm\, \boldsymbol{v}_\bullet)\right]^{\cdot} + \left[\mathcal{S}\, (\boldsymbol{r}_{/\bullet} \times dm\, \boldsymbol{v}_{/\bullet})\right]^{\cdot} + \boldsymbol{v}_\bullet \times m\, \boldsymbol{v}_G$$
$$= \mathcal{S}\, (\boldsymbol{v}_{/\bullet} \times dm\, \boldsymbol{v}_\bullet) + \mathcal{S}\, (\boldsymbol{r}_{/\bullet} \times dm\, \boldsymbol{a}_\bullet) + \left[\mathcal{S}\, (\boldsymbol{r}_{/\bullet} \times dm\, \boldsymbol{v}_{/\bullet})\right]^{\cdot}$$
$$+ \boldsymbol{v}_\bullet \times m\, \boldsymbol{v}_G$$
$$= \boldsymbol{v}_{G/\bullet} \times m\, \boldsymbol{v}_\bullet + \boldsymbol{r}_{G/\bullet} \times m\, \boldsymbol{a}_\bullet + \left[\mathcal{S}\, (\boldsymbol{r}_{/\bullet} \times dm\, \boldsymbol{v}_{/\bullet})\right]^{\cdot} + \boldsymbol{v}_\bullet \times m\, \boldsymbol{v}_G$$

[since $\boldsymbol{v}_{G/\bullet} \equiv \boldsymbol{v}_G - \boldsymbol{v}_\bullet$, the *first* and *last* terms, in the above, add up to zero]

$$= d\boldsymbol{h}_\bullet/dt + \boldsymbol{r}_{G/\bullet} \times m\, \boldsymbol{a}_\bullet, \tag{3.A2.5c}$$

where

$$\boldsymbol{H}_{\bullet,\text{relative}} \equiv \boldsymbol{h}_\bullet \equiv \mathcal{S}\, \boldsymbol{r}_{/\bullet} \times dm\, \boldsymbol{v}_{/\bullet} = \textit{Relative angular momentum} \text{ about } \bullet$$
$$= \mathcal{S}\, [\boldsymbol{r}_{/\bullet} \times dm(\boldsymbol{v} - \boldsymbol{v}_\bullet)] = \cdots = \boldsymbol{H}_\bullet - m\boldsymbol{r}_{G/\bullet} \times \boldsymbol{v}_\bullet. \tag{3.A2.5d}$$

From eqs. (3.A2.5a–d) it clearly follows that
(a) If • = *fixed point*, say O ($\Rightarrow \boldsymbol{v}_\bullet = \boldsymbol{0}$), or $\boldsymbol{v}_G = \boldsymbol{0}$, or if $\boldsymbol{v}_\bullet$ parallel to $\boldsymbol{v}_G$, then

$$\mathcal{S}\, \boldsymbol{r}_{/\bullet} \times dm\, \boldsymbol{a} = d\boldsymbol{H}_\bullet/dt; \tag{3.A2.5e}$$

(b) If • = *fixed point*, say O ($\Rightarrow \boldsymbol{a}_\bullet = \boldsymbol{0}$), or • = G, or $\boldsymbol{r}_{G/\bullet}$ is parallel to $\boldsymbol{a}_\bullet$, then

$$\mathcal{S}\, \boldsymbol{r}_{/\bullet} \times dm\, \boldsymbol{a} = d\boldsymbol{h}_\bullet/dt. \tag{3.A2.5f}$$

In sum, and since, generally,

$$\boldsymbol{H}_O \equiv \mathsf{S}\, [(\boldsymbol{r}_\bullet + \boldsymbol{r}_{/\bullet}) \times dm\, \boldsymbol{v}]$$

$$= \cdots = \boldsymbol{H}_\bullet + \boldsymbol{r}_\bullet \times m\, \boldsymbol{v}_G = (\boldsymbol{h}_\bullet + m\, \boldsymbol{r}_{G/\bullet} \times \boldsymbol{v}_\bullet) + \boldsymbol{r}_\bullet \times m\, \boldsymbol{v}_G, \qquad (3.\text{A}2.5\text{g})$$

we will have the following two, most useful (and memorable!) expressions of the principle of angular momentum:

$$\mathsf{S}\, \boldsymbol{r} \times dm\, \boldsymbol{a} = \left(\mathsf{S}\, \boldsymbol{r} \times dm\, \boldsymbol{v}\right)^{\cdot} = d\boldsymbol{H}_O/dt = d\boldsymbol{h}_O/dt \quad (= \boldsymbol{M}_O), \qquad (3.\text{A}2.5\text{h})$$

$$\mathsf{S}\, \boldsymbol{r}_{/G} \times dm\, \boldsymbol{a} = \left(\mathsf{S}\, \boldsymbol{r}_{/G} \times dm\, \boldsymbol{v}\right)^{\cdot} = d\boldsymbol{H}_G/dt = d\boldsymbol{h}_G/dt \quad (= \boldsymbol{M}_G). \qquad (3.\text{A}2.5\text{i})$$

Back to d'AP. But all these, kinematico-inertial identities and corresponding mutually equivalent forms of the principle of angular momentum (and many more presented in §1.6) are only half the story; *the other half is the forces and their moments*. Without the *additional* constitutive postulate of *action–reaction* for the *internal* forces $\{d\boldsymbol{f}_i\}$, where $d\boldsymbol{f} = d\boldsymbol{f}_{\text{external}} + d\boldsymbol{f}_{\text{internal}} \equiv d\boldsymbol{f}_e + d\boldsymbol{f}_i$, in either local or integral form, the moment side of (3.A2.5, 5h, 5i) would still contain the moments of the (generally unknown) $d\boldsymbol{f}_i$; and thus the solution of problems via these principles would, in general, be *indeterminate* (i.e., *# unknowns* > *# equations*). Adopting that postulate, as we will do, amounts to replacing in the above $\boldsymbol{M}_{\ldots}$ with $\boldsymbol{M}_{\ldots,\text{external}} \equiv \boldsymbol{M}_{\ldots,e}$:

$$\boldsymbol{M}_{\bullet,e} \equiv \mathsf{S}\, \boldsymbol{r}_{/\bullet} \times d\boldsymbol{f}_e = \mathsf{S}\, \boldsymbol{r}_{/\bullet} \times dm\, \boldsymbol{a}, \qquad (3.\text{A}2.6\text{a})$$

$$\boldsymbol{M}_{O,e} \equiv \mathsf{S}\, \boldsymbol{r} \times d\boldsymbol{f}_e = d\boldsymbol{H}_O/dt = d\boldsymbol{h}_O/dt, \qquad (3.\text{A}2.6\text{b})$$

$$\boldsymbol{M}_{G,e} \equiv \mathsf{S}\, \boldsymbol{r}_{/G} \times d\boldsymbol{f}_e = d\boldsymbol{H}_G/dt = d\boldsymbol{h}_G/dt; \qquad (3.\text{A}2.6\text{c})$$

because now $\boldsymbol{M}_{\bullet,\text{internal}} \equiv \mathsf{S}\, \boldsymbol{r}_{/\bullet} \times d\boldsymbol{f}_i = \boldsymbol{0}$ (and $\boldsymbol{f}_i \equiv \mathsf{S}\, d\boldsymbol{f}_i = \boldsymbol{0}$). [If the $\{d\boldsymbol{f}_e\}$ contain unknown constraint reactions, then the problem is still indeterminate; i.e., we need a new postulate to supply the additional independent equations.] Other forms of the above result from the purely geometrical (statical) relation: $\boldsymbol{M}_\bullet = \boldsymbol{M}_O + \boldsymbol{r}_{O/\bullet} \times \boldsymbol{f}$, $\boldsymbol{f} = \mathsf{S}\, d\boldsymbol{f}$ (acting through O), and then use of linear momentum:

$$\boldsymbol{f} = m\, \boldsymbol{a}_G, \text{ and action–reaction: } \boldsymbol{f} = \mathsf{S}\, (d\boldsymbol{f}_e + d\boldsymbol{f}_i) = \mathsf{S}\, d\boldsymbol{f}_e \equiv \boldsymbol{f}_e.$$

Now, the principle of d'Alembert (d'AP) → Lagrange (LP), and its associated "bothersome" virtual concepts, play a similar role with action–reaction *but for the constraint reactions*. By postulating the new and nontrivial constitutive (i.e., physical) postulate (3.A2.3b) for these forces, where

$$\delta\boldsymbol{r} = \sum (\partial\boldsymbol{r}/\partial q_k)\, \delta q_k \equiv \sum \boldsymbol{e}_k\, \delta q_k \qquad (\textit{holonomic} \text{ coordinates}),$$

$$= \sum (\partial\boldsymbol{r}/\partial\theta_k)\, \delta\theta_k \equiv \sum \boldsymbol{\varepsilon}_k\, \delta\theta_k \qquad (\textit{nonholonomic} \text{ coordinates}), \qquad (3.\text{A}2.6\text{d})$$

Lagrangean mechanics succeeds in generating as many equations as needed to render its problem determinate. That the virtual variations of the system coordinates $\{\delta q_k, \delta\theta_k\}$ are arbitrary (unless they, later, become constrained) is not a weakness or vagueness of the Lagrangean method, as some ignoramuses claim, but, on the contrary, its strength: it allows us to *obtain as many independent reactionless equations as there are independent* δq's/$\delta\theta$'s ($= \# DOF$), contrary to *actual* power equations that produce only *one* dependent equation. It is the fundamental *particle and system* vectors $\{\boldsymbol{e}_k, \boldsymbol{\varepsilon}_k\}$ that enter the equations of motion; for example, if the δq's are independent, LP yields

$$\textit{Particle/vector}\text{ variables:} \qquad S\, dm\, \boldsymbol{a}\cdot\boldsymbol{e}_k = S\, d\boldsymbol{F}\cdot\boldsymbol{e}_k, \qquad (3.\text{A}2.6\text{e})$$

$$\text{Holonomic } \textit{system} \text{ variables:} \qquad E_k(T) = Q_k. \qquad (3.\text{A}2.6\text{f})$$

To further clarify the meaning of virtualness, and thus quell the irrational fears of all those uncomfortable with "very small quantities," and so on (residues of a precalculus mindset?!), we add the following passage from a Victorian master's text on introductory statics:

> This [LP or Virtual Work) is an equation between infinitesimals, and it is to be understood on the ordinary conventions of the Differential Calculus ... $\delta' W$ vanishes [in Statics, or $\delta' W = \delta I$ in Kinetics], not because the quantities δq [or $\delta \boldsymbol{r}$] themselves tend to the limit zero, but in virtue of the *ratios* which these quantities bear to one another. The equation [$\delta' W_R = 0$] therefore holds if the resolved displacements δq are replaced by any finite quantities having to one another the ratios in question. (Lamb, 1928, p. 113)

A related theme advanced by some antivirtual authors goes as follows: well, if you stretch the definitions and concepts of virtual displacement long enough, "when $\delta \boldsymbol{r}$ [our notation] are chosen properly," then you will arrive at "their" equations. However, as the fundamental definitions (3.A2.6d), or

$$\delta \boldsymbol{r} \equiv \textit{linear and homogeneous (in } \delta q\textit{) part of } \boldsymbol{r}(q + \delta q, t) - \boldsymbol{r}(q, t), \qquad (3.\text{A}2.6\text{g})$$

and LP show, such a coincidence is hardly some accidental ad hoc result out of the blue; but, instead, the only kind of equations flowing directly, logically, and uniquely, out of the application of LP to Pfaffianly constrained systems. A true principle leads, it does not follow; that is, it is not a conceptual rubber-band that stretches ("chosen properly") to fit the facts of the moment, after the latter have occurred!

In sum: *it is the force side of the equations of motion that compels us to introduce LP*. Equation (3.A2.3b) is a practical and theoretical necessity forced (!) upon us by the particular decomposition of the total force into impressed and reaction—a fact that is peculiar to Lagrangean mechanics; it is not an alternative to action–reaction. The famous kinematico-inertial identity of Lagrange:

$$S\, dm\; \boldsymbol{a}\cdot(\partial\boldsymbol{r}/\partial q_k) \equiv S\, dm\; \boldsymbol{a}\cdot\boldsymbol{e}_k = (\partial T/\partial\dot{q}_k)^{\cdot} - \partial T/\partial q_k \equiv E_k(T), \qquad (3.\text{A}2.6\text{h})$$

(that holds always, independently of subsequent constraints and constitutive postulates, as long as the q's are *holonomic* coordinates), is a *most welcome and useful but secondary result.*

The preoccupation with (3.A2.6h), at the expense of the forces, $Q_k \equiv S\, d\boldsymbol{F}\cdot(\partial\boldsymbol{r}/\partial q_k) \equiv S\, d\boldsymbol{F}\cdot\boldsymbol{e}_k$, $R_k \equiv S\, d\boldsymbol{R}\cdot(\partial\boldsymbol{r}/\partial q_k) \equiv S\, d\boldsymbol{R}\cdot\boldsymbol{e}_k$ [$= 0$, in (3.A2.6e, f)], is perhaps best reflected in the seemingly innocuous but revealing fact that, although in "elementary" (Newton–Euler) mechanics most of us are taught to write *force* = *(mass)* × *(acceleration)*—that is, place the force on the *left* side of the equation—as soon as we graduate to advanced dynamics (d'Alembert–Lagrange), we suddenly switch to the form (*mass*) × (*acceleration*) = *force*—that is, place the force on the *right* side of the equation! Thus, many beginners in

Lagrangean mechanics get the superficial impression that the latter is just an exercise in differentiation of scalar energetic functions; the price one must pay for the transition from rectangular to curvilinear, or "generalized" coordinates (a pretty primitive term, in view of differential geometry and tensors). Even classics like Routh (1905(a), pp. 45–48) or Whittaker (1937, pp. 34–37) reinforce this misrepresentation.

(iv) Finally, there are those who, failing to understand the fundamental, simple, and natural kinematical representation (3.A2.6d), and in a complete breach with rational discourse, furiously and ignorantly trivialize and/or dismiss everything virtual (displacements, work, etc.) as "ill-defined," "nebulous," and "hence objectionable"; or complain "But it can hardly be gainsaid that maximum clarity is guaranteed by defining $\delta \boldsymbol{r}$ [our notation] *mathematically* in terms of more fundamental quantities" and "Consequently, for the formulation of equations of motion, the use of principles represented by LP [our term] is contraindicated, at least for systems possessing a finite number of degrees of freedom" [Kane and Levinson (1983, p. 1077), and rebuttal to Desloge (1986)]; and [virtual concepts are] "the closest thing in dynamics to black magic," "If you can construct a good virtual displacement vector, you can do good business with it,.... The difficulty is constructing it in the first place. It's like catching a bird by sprinkling salt on its tail. Virtual displacement is the salt on the bird's tail" (Radetsky, 1986, pp. 55–56); and "It should be acknowledged at this point that the traditional concepts of virtual displacement and virtual work are not necessary to the derivation of [our 3.A2.6e, f, h)]. It is quite sufficient (and more straightforward) simply to dot multiply $df = dm\, \boldsymbol{a}$ by [our] $\partial \boldsymbol{v}/\partial \dot{q}_j$ and add these equations together, accomplishing this for each of these n values of j" (Likins, 1973, pp. 297–298). The falsehood and misleadingness of these criticisms should be clear in the light of the above, and chapters 2 and 3. But we also point out the following, in favor of virtualness:

(a) As (3.A2.6d) shows, $\delta \boldsymbol{r}$ is *invariant* under $\delta q \leftrightarrow \delta\theta$ transformations, whereas the $\{\boldsymbol{e}_k, \boldsymbol{\varepsilon}_k\}$ are not; and similarly for $\delta' W$, $\delta' W_R$, δI, and so on.

(b) The $\delta \boldsymbol{r}$ admits of a far simpler and direct geometrical *visualization* than the $\partial \boldsymbol{v}/\partial \dot{q}_j \equiv \boldsymbol{e}_j$ (and $\partial \boldsymbol{v}^*/\partial \omega_j \equiv \boldsymbol{\varepsilon}_j$). In general, *differentials are far easier to visualize than derivatives*, and this explains their dominant presence in most figures of free-body diagrams, control volumes, and so on, even though the final equations do not contain lone differentials but derivatives.

(c) What is the motivation for dotting $df - dm\, \boldsymbol{a}$ with $\partial \boldsymbol{v}/\partial \dot{q}_j$? Why not dot it with its equal but simpler $\partial \boldsymbol{r}/\partial q_j$? Or, why not, say, cross it with them, or with $\partial \boldsymbol{v}/\partial q_j$, and so on? Or, why does not $\partial \boldsymbol{r}/\partial t \equiv \boldsymbol{e}_{n+1} \equiv \boldsymbol{e}_0$ appear in the equations of motion, although it appears in both $\boldsymbol{v}$ and $\boldsymbol{a}$?

(d) We would like to see such (supposedly) virtual-less authors try to:

- Extend their ad hoc techniques, rigorously, to *nonlinear* nonholonomic velocity constraints, without virtual displacements, or something mathematically equivalent (chap. 5). Fortunately for them, their constraints are *linear* in the velocities; that is, they are Pfaffian.
- Teach (even discrete) analytical *statics* (S) to students with no knowledge of dynamics (D), without virtual displacements/work! What does one do with their $\partial \boldsymbol{v}/\partial \dot{q}_j$ there? On the other hand, the definitions presented here are *uniformly* valid for both D and S alike.

(e) And if the use of differential variational principles, such as LP, is "contraindicated," how is one going to make the transition to the rest of the *differential* variational principles of Jourdain, Gauss et al. (chap. 6), and the *integral* variational principles of Hamilton, Voronets, Hamel, et al. (chap. 7), with their increasingly important role for approximate (analytical and numerical) solutions (chap. 7), as well as *invariance/conservation* theorems (e.g., Noether's theorem, §8.13)? Why such scientific provincialism and short-sightedness? [On the numerical advantages of some of these principles, see, e.g., Schiehlen (1981); for Gauss' principle, in particular, see, for example, Udwadia and Kalaba (1996)].

Such antivirtual attitudes artificially distance themselves from the tried and true mainstream dynamics, built over several centuries by some of the greatest names in mathematics and mechanics; such antihistorical and antitraditional attitudes contribute to a dynamical tower of Babel!

So, to recapitulate, we think that the whole problem with the earlier "antivirtual crowd" begins with their failure to acknowledge that in mechanics *the crux of the matter is the force*; that Newton–Euler splits forces into *internal* and *external* ("apples"), whereas d'Alembert–Lagrange splits them into *impressed* and *reactions* ("oranges"); and that the basic goal of AL is to *uncouple* the equations of motion into *kinetic* (*motion only, no reactions*) and *kinetostatic* (*reactions*). This failure also hampers the extension of their dynamics to new types of constraints and associated forces (e.g., servoconstraints, §3.17), let alone the application of its methodology to other areas of engineering and physics (such as electromechanical analogies and nonholonomic rotating electrical machinery; see, for example, Arczewski and Pietrucha (1993), Maißer (1981), Neimark and Fufaev (1972).

Appell versus Kane

In our notation, the so-called "Kane's equations" (1961, 1965, 1985) read simply (with $I = m + 1, \ldots, n$):

$$\mathcal{S}\, d\boldsymbol{F} \cdot (\partial \boldsymbol{v}^*/\partial \omega_I) + \mathcal{S}\,(-dm\, \boldsymbol{a}^*) \cdot (\partial \boldsymbol{v}^*/\partial \omega_I) = 0,$$

or

$$(\textit{Generalized active force})_I + (\textit{Generalized inertial "force"})_I = 0. \tag{3.A2.7}$$

But in view of the fundamental kinematical identities (2.9.35 ff.)

$$\partial \boldsymbol{r}^*/\partial \theta_I = \partial \boldsymbol{v}^*/\partial \omega_I = \partial \boldsymbol{a}^*/\partial \dot{\omega}_I = \partial \dot{\boldsymbol{a}}^*/\partial \ddot{\omega}_I = \cdots \equiv \boldsymbol{\varepsilon}_I, \tag{3.A2.7a}$$

eqs. (3.A2.7) can be immediately rewritten as

$$\mathcal{S}\, d\boldsymbol{F} \cdot (\partial \boldsymbol{a}^*/\partial \dot{\omega}_I) + \mathcal{S}\,(-dm\, \boldsymbol{a}^*) \cdot (\partial \boldsymbol{a}^*/\partial \dot{\omega}_I) = 0, \tag{3.A2.7b}$$

or, finally, after some very simple rearrangements,

$$\partial/\partial \dot{\omega}_I \left(\mathcal{S}\,(1/2) dm\, \boldsymbol{a}^* \cdot \boldsymbol{a}^* \right) = \mathcal{S}\, d\boldsymbol{F} \cdot (\partial \boldsymbol{a}^*/\partial \dot{\omega}_I), \tag{3.A2.7c}$$

which are none other than Appell's equations! Here is a partial (alphabetical) list of readable textbooks/treatises/encyclopedias, and so on, on eqs. (3.A2.7–7c) [*all* of them from before 1961 (year of first Kane paper), and several from before Kane was born!]:

Appell [1899(a), (b); 1900(a), (b); 1925; 1953, pp. 388–395, eqs. (3.A2.6). Leisurely component presentation]

Coe [1938, pp. 386–390, eqs. (7). Earliest vectorial treatment in U.S. literature]

Hamel (1927, pp. 30–32; especially equations on 17th line from top of p. 32. Force-free case)

Hamel (1949, pp. 361–363; especially equations on 7th line from top of p. 362. Best concise presentation)

Lur'e [1961/1968, pp. 389–395. Equations (8.5.18) and (8.6.10) are, respectively, Kane's equations of 1961 and 1965]

MacMillan [1936, pp. 341–343, eqs. (3). Earliest (component) appearance in U.S. treatise]

Marcolongo (1912, pp. 104–105; especially equations on 3rd line from bottom of p. 104)

Neimark and Fufaev [1967/1972, pp. 147–149, eqs. (8.3). Based completely on virtual concepts]

Pérès [1953, pp. 219–222. Excellent concise (vectorial) treatment including Kane's equations of both 1961 and 1965]

Platrier (1954, pp. 170–173, 323–324, 343–344)

Routh [1905(a), pp. 348–353, eqs. (5). Earliest appearance in English]

Schaefer [1919, p. 74, eq. (212)], similar treatment to Routh's and MacMillan's; well known in the German-speaking world.

Schaefer [1951, eqs. (12). Earliest *nonlinear* generalization of "Kane's equations" of 1965. Incidentally, in 1962 (in discussion of Kane (1961)) Schaefer warned Kane, in vain, that any further improvement on the methods/equations of the classical masters of dynamics (Appell, Heun, Hamel, Prange, Johnsen et al.) "is not imaginable."

Voss (1901–1908, pp. 82–83; and connection with Ferrers' equations)

(i) To make matters worse, Kane uses the following arcane terminology/notation:

(a) Our ω's he calls "generalized speeds," despite the fact that these are the (contravariant) *nonholonomic* components of the system velocity vector, in configuration/event space; a vector whose *holonomic* components are none other than the $\dot{q}$'s (what most reasonable authors call "generalized velocities," but Kane leaves nameless!). In other words, the $\dot{q}$'s and ω's are components of the *same* (system) vector, but along different types of bases: one *gradient*, one *nongradient*. However, and this is the essence of the method of quasi coordinates in constrained dynamics, *a proper choice of ω's uncouples the equations of motion into kinetic and kinetostatic; and, roughly, the n q's embed the (original) holonomic constraints, while the $n - m$ ω's embed the (additional) Pfaffian constraints.*

But there is another problem with "generalized speeds." According to time-honored and standard mechanics practices, *speed is the magnitude (or length) of the velocity vector,* and, as such, a *nonnegative scalar,* whereas the $\dot{q}$'s and ω's, being *components*, may have any sign—in automobiles, speedometers never show negative speeds! (Actually, the speed is an *invariant* under coordinate transformations; in tensor language: an absolute tensor of rank zero.) Therefore, from the viewpoint of tensors/differential geometry, and the traditions and practices of dynamics, the term "generalized speeds" is archaic, erroneous, and confusing.

(b) Kane's term "partial velocities," for our $\boldsymbol{e}_k$ and $\boldsymbol{\varepsilon}_k$, is entirely capricious and hides more than it reveals. In view of the identity (3.A2.8a), and a similar one for holonomic coordinates, we could just as well have called them "partial positions," or "partial accelerations," or even ... "partial jerks," and so on. A better term, though a long one, would be "accompanying (particle and system) vectors," that is the *begleitvektoren* of Heun; but we would welcome a more concise terminology.

(ii) To avoid virtual displacements, and so on, Kane (1961; 1968, p. 52) talks about instantaneous constraints, or about dividing $\delta \boldsymbol{r}$ with δt, the latter understood as "... any quantity having the dimensions of time." But since $\delta t = 0$ {in order to eliminate reactions, i.e. so that $[S\, d\boldsymbol{R} \cdot (\partial \boldsymbol{r}/\partial t)]\, \delta t \equiv R_0\, \delta t = 0$, even though $R_0 \neq 0$}, such statements are likely to cause more confusion (division by a zero!), plus they are irrelevant to the final result—that is, the equations of motion. Why not use the simpler and rigorous definition (3.A2.6d, g).

(iii) In his frantic attempts to artificially distance himself from Appell, Kane (1986) states that the Appellian S is "a quantity of no interest in its own right." Well, most concepts of mechanics and physics derive their importance not "in their own right," but from their relation to the current edifice of those sciences; like a stone in relation to a building it belongs. Such narrow, positivistic (?), undialectical, objections can be raised against the Lagrangean, the Hamiltonian, the entropy, and so on. When was the last time anyone saw a stress, or a strain or even an acceleration?! During the 17th century, similar short-sighted complaints were raised against Leibniz's "vis viva" (= twice the kinetic energy). Tomorrow, perhaps, some other quantity, involving still higher derivatives (again "of no interest in its own right") might be introduced, in order to combine many new and old phenomena under one simple conceptual roof.

(iv) An alleged advantage (of the bean-counting type) of Kane over Appell is that in applying the latter one needs to square the accelerations $\boldsymbol{a}$ (or $\boldsymbol{a}^*$), then build the Appellian S (S^*), and then differentiate it with respect to the quasi accelerations $\dot{\omega}_k$, whereas Kane's approach dispenses with all that—compare (3.A2.7) with (3.A2.7c).

However, as Professor L.Y. Bahar has pointed out, the calculation of $\boldsymbol{a}^*$ and $\partial \boldsymbol{a}^*/\partial \dot{\omega}_k$ and subsequent formation of their dot product, in (3.A2.7b), is standard procedure in engineering science *whenever a quadratic form has to be partially differentiated.* For example, to calculate the static deflection p of a thin linearly elastic Euler/Bernoulli beam, of length l, flexural rigidity EI, and bending moment M, under a concentrated load P, we can use Castigliano's well-known theorem:

$$p = \partial V/\partial P = \partial/\partial P\left(\int_0^l M^2 dx/2EI\right), \tag{3.A2.8a}$$

where V is the *strain energy* of the beam (see any book on structural analysis). It is well known that, in practice, we never compute the integrand explicitly, then integrate it, and then differentiate the resulting function of P; but, instead, we first carry out the differentiation under the integral, and then integrate the result:

$$p = \partial V/\partial P = \int_0^l [M(\partial M/\partial P)/EI]\, dx; \tag{3.A2.8b}$$

that is, the step from (3.A2.8a) to (3.A2.8b) is conceptual rather than practical. Here, clearly, we have the correspondences $\boldsymbol{a}^* \to M$, $\partial \boldsymbol{a}^*/\partial \dot{\omega}_k \to \partial M/\partial P$. As with everything else, practice with Appell's equations helps one develop shortcuts and other special labor-saving skills. Finally, why use $\partial \boldsymbol{v}^*/\partial \omega_k$ or $\partial \boldsymbol{a}^*/\partial \dot{\omega}_k$, and not their equal but simpler expression (2.9.27)

$$\partial \boldsymbol{r}^*/\partial \theta_k \equiv \sum A_{lk}(\partial \boldsymbol{r}/\partial q_l),$$

and analogously for holonomic variables. Such flexibilities are absent from Kane's scheme.

In sum, Kane's equations are just a special implementation (or "raw" form) of Appell's kinetic equations, along the way from LP; one that completely ignores the long-term and big picture aspects of mechanics: namely, our understanding of its *underlying mathematical structure and physical ideas*, and their *interconnections with other areas of natural science,* which are the hallmarks of genuine education. Moreover, the whole Kaneian approach is conceptually unmotivated, isolating and primitive, historically ignorant and flat, intellectually stifling and wasteful. Indeed, it is a degraded and sterile type of mechanics that soon leads its practitioners down a dynamical dead-end. Ultimately, and this *applies to most of the contemporary multibody dynamics expositions*, such schemes discourage *active* learning, with its new and unpredictable turns, diversity and change. Like the currently promoted antipluralistic "expert systems," they assume that there is "a" best way to do dynamics, which is best determined by whomever designs the relevant books/computer programs. This is not normal human learning; it is not a presentation based on a *continuous* historical evolution; namely, one that respects and expands the dynamics traditions and practices. The brains of the readers (or users) are treated as pieces of equipment (hardware), where one inserts *abruptly* a set of computer instructions and commands (software). As a result, the majority of users of such "dynamics" will never be able to raise that edifice even by one inch; they will have been transformed from thinking engineers to (highly expendable) filing clerks! [We are indebted to B. Garson's *The Electronic Workshop: How Computers are Transforming the Office of the Future into the Factory of the Past* (Simon and Schuster, 1988, p. 126) for some of these insights.]

4

Impulsive Motion

4.1 INTRODUCTION

In this chapter, we present the Lagrangean principles and equations of *impulsive, or discontinuous,* motion of constrained systems. The relevant "elementary" Newton–Euler definitions and equations are summarized in §4.2. Then we cover, in sequence: the impulsive version of Lagrange's principle (§4.3); the Appellian classification of impulsive constraints and corresponding equations of impulsive motion (§4.4); the formulation of kinetic and kinetostatic impulsive equations, in both holonomic and nonholonomic variables (§4.5; impulsive counterparts of the equations of Maggi, Hamel, and Appell); and, finally, the various impulsive energetic/extremum theorems of Carnot, Kelvin, Bertrand, Robin et al. (§4.6). As in the rest of the book, the discussion is complemented with a number of examples and problems.

Impulsive motion is a topic of intense and rapidly expanding research. Hence, a number of its aspects (e.g., role of *friction, deformation*), since they cannot be dealt with definitively here, are omitted.

For complementary reading on this engineeringly important subject, we recommend (alphabetically): Bouligand (1954, pp. 129–157, 444–483), Brach (1991), Easthope (1964, pp. 268–306), Goldsmith (1960), Hamel (1949, pp. 395–402), Kilmister and Reeve (1966, pp. 178–195, 217–221, 235–242, and Exercises), Kilmister (1967, pp. 98–108), Lainé (1946, pp. 185–201, 259–278), Loitsianskii and Lur'e (1983, pp. 131–143, 237–245, 276–280), Panovko (1977), Pöschl, 1928), Routh (1905(a), pp. 136–164, 254–268, 302–313, 323–327), Smart (1951, vol. 2, pp. 376–390), Suslov (1946, pp. 607–645); also Bahar (1994), for instructive applications of Jourdain's variational principle (§6.3) to impact.

4.2 BRIEF OVERVIEW OF THE NEWTON–EULER IMPULSIVE THEORY

Below, we summarize a few basic definitions and concepts. [Some of our differentials will be in *time* and some in *space*; we hope that their differences will be clear from the context, and no confusion will arise.]

Integrating the fundamental equation of motion of a particle P of mass dm (§1.4 ff.):

$$dm\,\boldsymbol{a} = d\boldsymbol{f}, \qquad \text{or} \qquad dm(d\boldsymbol{v}/dt) = d\boldsymbol{f}, \tag{4.2.1}$$

from an arbitrary time t' to an arbitrary time $t''(> t')$ yields

$$\Delta(dm\,\boldsymbol{v}) \equiv \Delta(d\boldsymbol{p}) = \int_{'}^{''} d\boldsymbol{f}\,dt, \tag{4.2.2}$$

where

$$\Delta(\ldots) = (\ldots)_{t''} - (\ldots)_{t'}, \qquad \int_{'}^{''} \ldots\,dt \equiv \int_{t'}^{t''} \ldots\,dt. \tag{4.2.2a}$$

Equation (4.2.2) states that the change of the linear momentum $d\boldsymbol{p} = dm\,\boldsymbol{v}$ of a (system) particle P during an arbitrary time interval $t'' - t' \equiv \tau$ equals the impulse of the total force $d\boldsymbol{f}$ acting on P, during that interval. Now, if τ is *finite*, the above is just the first time-integral of the Newton–Euler equation of motion; and, therefore, represents nothing new. If, however, τ is *very small*, or *infinitesimal*, then an independent and rather interesting chapter of dynamics, known as *impulsive motion* (IM; or *impact*, or *shock*), emerges. More specifically, IM occurs whenever *a very large* (or *infinite*, or *delta function-like*) *force acts on P for a very short time*; i.e., for $\tau \to 0$. As a result of this, at the end of τ: (i) the particle's *momentum* has changed by a *finite* instantaneous, that is, discontinuous, *jump* $\Delta(d\boldsymbol{p}) \equiv (d\boldsymbol{p})^+ - (d\boldsymbol{p})^- \neq \boldsymbol{0}$, where $(\ldots)^+/(\ldots)^-$: values of $(\ldots)$ just *after/before* the shock, respectively, or *right/left* limits of $(\ldots)$; or, since $dm = constant$,

$$\Delta\boldsymbol{v} \equiv \boldsymbol{v}^+ - \boldsymbol{v}^- \neq \boldsymbol{0}, \tag{4.2.3}$$

while (ii) the particle's *position* $\boldsymbol{r}$ has remained essentially *unchanged*; that is,

$$\Delta\boldsymbol{r} = \boldsymbol{0}. \tag{4.2.4}$$

Symbolically, in the IM case, eq. (4.2.2) reads

$$\Delta(dm\,\boldsymbol{v}) = \widehat{d\boldsymbol{f}}, \tag{4.2.5}$$

where

$$\widehat{(\ldots)} \equiv \lim_{\tau\to 0} \int_{t'}^{t'+\tau} (\ldots)\,dt \quad (\text{"}hat\text{" notation}). \tag{4.2.5a}$$

REMARKS

(i) The "hat" notation should not be confused with that notation occasionally employed for unit vectors.

(ii) We point out that, here, and contrary to the finite force case, as $\tau \to 0$,

$$\lim \int (\ldots)\,dt \neq \int \lim\,(\ldots)\,dt, \quad \text{in general.} \tag{4.2.5b}$$

Clearly, impulsive "forces" (or *percussions*, or *blows*) $\widehat{d\boldsymbol{f}}$ are not defined as ordinary (or finite) forces at every instant of time, but, instead, only through the *instantaneous and finite jump*, or *discontinuity*, $\Delta(d\boldsymbol{p}) = \Delta(dm\,\boldsymbol{v}) = dm\,\Delta\boldsymbol{v}$ that they produce; for finite forces, such as gravity, the limit (4.2.5a) is, clearly, zero [and for an arbitrary continuous function $f = f(t, q)$: $\Delta f = 0$ and $\hat{f} = 0$]. The result of these approximations [i.e., $\Delta(positions) = 0$, $\Delta(velocities) \neq 0$, see ex. 2.4.1, below] is an impulsive theory of, admittedly, reduced practical value, but one of conceptual clarity and simplicity.

Application of this same idea to the Newton–Euler principles of linear and angular momentum, for a general system [i.e., multiplication of its, generally, *differential* equations of (finite) motion by dt, integration over τ, and then taking of the limit as $\tau \to 0$; while assuming that not all acting forces are finite, and that the "principle of action–reaction" for the internal loads holds for IM too], leads to *the impulsive forms of these two principles*, at time t, that are *algebraic* (finite difference; i.e., nondifferential!) equations.

As in finite motion, here, too, two possibilities arise: (i) either our theory generates enough such algebraic equations to determine the system's *postimpact state*; that is, the $\Delta \boldsymbol{v}$'s or the $\boldsymbol{v}^+$'s, and therefore the problem is *impulsively determinate*; or (ii) we have more unknowns than available equations, and thus the problem is *impulsively indeterminate*, in which case we need (in *addition* to the already utilized kinematical and kinetical equations) special *physical, or constitutive, equations/postulates*, as in continuum mechanics (e.g., Hooke's law in elasticity, Navier–Stokes law in fluid dynamics, etc.)—see §4.4.

Work–Energy in Impulsive Motion

Integrating the (rate of) work–(kinetic) energy equation, $dT/dt = \mathcal{S}\, d\boldsymbol{f} \cdot \boldsymbol{v}$, between t' and $t''(> t')$ yields the familiar integral form

$$\Delta T \equiv T'' - T' \equiv T^+ - T^- = \int_{'}^{''} \left(\mathcal{S}\, d\boldsymbol{f} \cdot \boldsymbol{v} \right) dt, \qquad (4.2.6)$$

from which, passing to the impulsive limit ($t'' \to t'$) and invoking the mean value theorem of integral calculus, we obtain

$$\Delta T = \mathcal{S}\, \widehat{d\boldsymbol{f}} \cdot \langle \boldsymbol{v} \rangle \equiv \textit{Impulsive "work."} \qquad (4.2.7)$$

where

$\langle \boldsymbol{v} \rangle$: *mean/average* value of (generally unknown) impact velocity. (4.2.7a)

However, since impact involves *friction* and *deformation*—that is, phenomena accompanied by conversion of mechanical energy into heat—and since the latter lies outside pure mechanics, we should *not view* (4.2.7) as an *ordinary work–energy theorem*; impulsive "work" is not connected with increase in energy; and so, unlike momentum relations, in the case of energy there is no simple mathematical transition from ordinary (continuous, or finite, motion) to impulsive dynamics. [For some energetic aspects of impact, see, e.g., Roy (1965, pp. 176–179).]

Example 4.2.1 *Proof that Under Impulsive Forces:*

$$\Delta(\text{positions}) = 0, \qquad \text{but} \qquad \Delta(\text{velocities}) \neq 0.$$

Let us consider the motion of a particle P of mass m, along the axis Ox, with initial conditions (at, say, $t_o = 0$): $x(0) = 0$ and $\dot{x}(0) = 0$, under the 2τ – periodic total force:

$$X = X_o \sin(\pi t/\tau), \qquad 0 \leq t \leq \tau,$$
$$X = 0, \qquad \tau < t < \infty, \qquad (a)$$

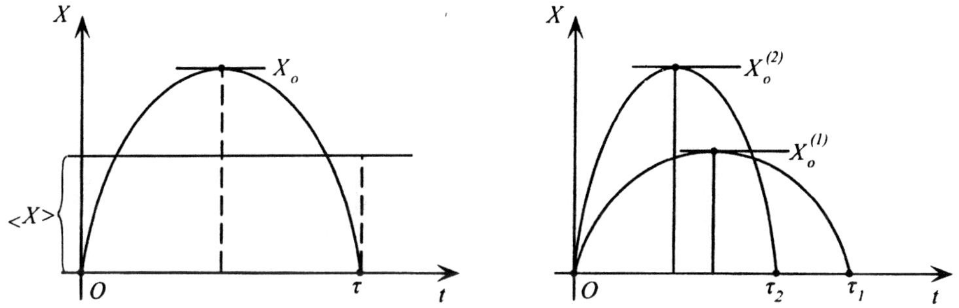

Figure 4.1 On the concept of impulsive force.

where X_o = *force amplitude, a constant*; and τ = *duration* of action of X (fig. 4.1). Integrating the equation of motion of P:

$$m\ddot{x} = X_o \sin(\pi t/\tau), \tag{b}$$

twice, while choosing the integration constants to satisfy the given initial conditions, we obtain

$$\dot{x} \equiv v = (X_o\tau/\pi m)\big[1 - \cos(\pi t/\tau)\big], \tag{c}$$

$$x = (X_o\tau/\pi m)t - (X_o\tau^2/\pi^2 m)\sin(\pi t/\tau). \tag{d}$$

Hence,

$$\Delta x \equiv x(\tau) - x(0) = (X_o\tau^2)/(\pi m) = (X_o\tau)\tau/\pi m = \big(\langle X\rangle\tau/2m\big)\tau, \tag{e}$$

$$\Delta(\dot{x}) \equiv \Delta v = \dot{x}(\tau) - \dot{x}(0) = 2(X_o\tau)/(\pi m) = \big(\langle X\rangle\tau\big)/m, \tag{f}$$

where

$$\langle X\rangle \equiv (1/\tau)\int_0^\tau X(t)\,dt = (1/\tau)\int_0^\tau X_o \sin(\pi t/\tau)\,dt = 2X_o/\pi:$$

$$\textit{Mean value of force in } \tau. \tag{g}$$

The above show that as long as the product $\langle X\rangle\tau = \int_0^\tau X(t)\,dt$ remains fixed (even though τ, and hence $\langle X\rangle$, may vary) so does Δv. Hence, in the particular, "impact limit": $X_o \to \infty$ (and generally as $|X_o| \to \infty) \Rightarrow \langle X\rangle \to \infty$, *and* $\tau \to 0$, so that $\langle X\rangle\tau$ = *fixed nonzero value* $\equiv C$, eqs. (e, f) yield

$$\Delta x = 0 \quad \text{and} \quad \Delta v = C/m \Rightarrow \Delta(mv) = C; \tag{h}$$

that is, in the limiting case of a very large ("infinite") force acting for a very short ("infinitesimal") time, the momentum changes ("instantaneously/discontinuously") by the finite amount C, while the position does not!

4.3 THE LAGRANGEAN IMPULSIVE THEORY; NAMELY, CONSTRAINED DISCONTINUOUS MOTION

Let us now examine impulsive motion from the viewpoint of analytical mechanics. Summing eqs. (4.2.2, 5) over all the material particles of the system, $\mathcal{S}(\ldots)$, yields the *impulsive linear momentum principle*:

$$S\,\Delta(dm\,\boldsymbol{v}) = \Delta\Big(S\,dm\,\boldsymbol{v}\Big) = S\,\widehat{df} = \widehat{S\,df} = \hat{\boldsymbol{f}}, \tag{4.3.1}$$

since, clearly, $S(\ldots)$ and $\Delta(\ldots)$ can be interchanged; and, recalling the *d'Alembert decomposition* (§3.2): $d\boldsymbol{f} = d\boldsymbol{F} + d\boldsymbol{R}$, we can rewrite (4.3.1), with some easily understood notations, as

$$\Delta\Big(S\,dm\,\boldsymbol{v}\Big) = \hat{\boldsymbol{F}} + \hat{\boldsymbol{R}}. \tag{4.3.2}$$

This is the *constrained impulsive linear momentum "principle."* A similar constrained impulsive equation/theorem results if we cross both members of the Newton–Euler law of motion for a typical particle, *under d'Alembert's decomposition*, with its position vector (relative to some fixed origin, or some other position vector), sum the resulting angular momentum equations over the entire system, multiply the result with dt, integrate it over τ, and then, as before, take its limit as $\tau \to 0$. The result would be the *constrained impulsive angular momentum principle*, featuring on its right side the sums of the moments of the *impressed impulsive forces* $\{\widehat{d\boldsymbol{F}}\}$, and *impulsive constraint reactions* $\{\widehat{d\boldsymbol{R}}\}$ [i.e., "forces" caused either by the $\widehat{d\boldsymbol{F}}$'s, or by the sudden introduction of constraints (in addition to the already existing ones, which are called *permanent*, or *primitive*), cause discontinuous changes (jumps) to the system holonomic and/or nonholonomic velocities—these are detailed below].

Now, the objective of Lagrangean impulsive theory—namely, *the impulsive theory of constrained mechanical systems*—is to develop system impulsive equations *with or without the* $\widehat{d\boldsymbol{R}}$'s. To this end, we proceed as in the case of finite motion (chap. 3): dotting the constrained impulsive linear momentum equation for a typical system particle P,

$$\Delta(dm\,\boldsymbol{v}) = \widehat{d\boldsymbol{F}} + \widehat{d\boldsymbol{R}}, \tag{4.3.3}$$

with its virtual displacement $\delta\boldsymbol{r}$ (at the shock instant $t = t'$), and then summing over the system particles, yields

$$S\,\big[\Delta(dm\,\boldsymbol{v}) - \widehat{d\boldsymbol{F}}\,\big]\cdot\delta\boldsymbol{r} = S\,\widehat{(-d\boldsymbol{R})}\cdot\delta\boldsymbol{r}; \tag{4.3.3a}$$

and, next, assuming that for bilateral "ideal" constraints the $\widehat{d\boldsymbol{R}}$'s satisfy the constitutive postulate/definition (while assuming that $\widehat{\delta\boldsymbol{r}} = \boldsymbol{0}$; e.g., by choosing *time-independent/constant virtual displacements*, for $t' \le t \le t''$)

$$\widehat{-\,\delta' W_R} \equiv \widehat{S\,(-d\boldsymbol{R})\cdot\delta\boldsymbol{r}} = S\,\widehat{(-d\boldsymbol{R})}\cdot\delta\boldsymbol{r} = 0, \tag{4.3.3b}$$

we readily obtain the *fundamental impulsive variational equation* (*impulsive principle of Lagrange—LIP*)

$$\widehat{\delta I} = \widehat{\delta' W}, \tag{4.3.4}$$

where

$$\widehat{\delta I} \equiv \widehat{S\, dm\, \boldsymbol{a} \cdot \delta \boldsymbol{r}} = S\, \Delta(dm\, \boldsymbol{v}) \cdot \delta \boldsymbol{r}:$$

(first-order) *virtual work of impulsive momenta*, (4.3.4a)

$$\widehat{\delta' W} \equiv \widehat{S\, d\boldsymbol{F} \cdot \delta \boldsymbol{r}} = S\, \widehat{d\boldsymbol{F}} \cdot \delta \boldsymbol{r}:$$

(first-order) *virtual work of impressed "forces."* (4.3.4b)

From (4.3.3–4b), we can obtain all kinds of special impulsive equations: with/without impulsive reactions (i.e., kinetostatic/kinetic/mixed impulsive equations), in particle/system form, in holonomic/nonholonomic variables, and so on.

We begin by substituting into (4.3.3b, 4) the *holonomic* variable representation (§2.5 ff.): $\delta \boldsymbol{r} = \sum \boldsymbol{e}_k\, \delta q_k$ $(k = 1, \ldots, n)$. Since [in complete analogy with the finite motion case (§3.2 ff.), and assuming that $\hat{\boldsymbol{e}}_k = \boldsymbol{0}$, $\widehat{\delta q_k} = 0 \Rightarrow \widehat{\delta \boldsymbol{r}} = \boldsymbol{0}$]

$$\widehat{\delta' W_R} = S\, \widehat{d\boldsymbol{R}} \cdot \delta \boldsymbol{r} = \sum \left(S\, \widehat{d\boldsymbol{R}} \cdot \boldsymbol{e}_k \right) \delta q_k \equiv \sum \widehat{R_k}\, \delta q_k, \tag{4.3.5a}$$

$$\widehat{\delta' W} = S\, \widehat{d\boldsymbol{F}} \cdot \delta \boldsymbol{r} = \sum \left(S\, \widehat{d\boldsymbol{F}} \cdot \boldsymbol{e}_k \right) \delta q_k \equiv \sum \widehat{Q_k}\, \delta q_k, \tag{4.3.5b}$$

$$\begin{aligned} \widehat{\delta I} &= S\, \Delta(dm\, \boldsymbol{v}) \cdot \delta \boldsymbol{r} = \sum \left(S\, dm\, \Delta \boldsymbol{v} \cdot \boldsymbol{e}_k \right) \delta q_k \\ &= \sum \Delta \left(S\, dm\, \boldsymbol{v} \cdot \boldsymbol{e}_k \right) \delta q_k \equiv \sum \Delta p_k\, \delta q_k, \end{aligned} \tag{4.3.5c}$$

and

$$p_k \equiv S\, (dm\, \boldsymbol{v} \cdot \boldsymbol{e}_k) \equiv \partial T / \partial \dot{q}_k$$

$$\Rightarrow \Delta p_k = \Delta \left(S\, dm\, \boldsymbol{v} \cdot \boldsymbol{e}_k \right) = S\, \Delta(dm\, \boldsymbol{v}) \cdot \boldsymbol{e}_k:$$

[holonomic (k)th component of] impulsive system *momentum change*, (4.3.6a)

$$\widehat{Q_k} \equiv \widehat{S\, d\boldsymbol{F} \cdot \boldsymbol{e}_k} = S\, \widehat{d\boldsymbol{F} \cdot \boldsymbol{e}_k} = S\, \widehat{d\boldsymbol{F}} \cdot \boldsymbol{e}_k:$$

[holonomic (k)th component of] impulsive system *impressed* force; or, simply, impressed system *impulse*, (4.3.6b)

$$\widehat{R_k} \equiv \widehat{S\, d\boldsymbol{R} \cdot \boldsymbol{e}_k} = S\, \widehat{d\boldsymbol{R} \cdot \boldsymbol{e}_k} = S\, \widehat{d\boldsymbol{R}} \cdot \boldsymbol{e}_k:$$

[holonomic (k)th component of] impulsive system *constraint reaction* force, (4.3.6c)

we finally obtain LIP, eqs. (4.3.3b, 4), in *holonomic system variables*:

$$\sum \widehat{R_k}\, \delta q_k = 0, \qquad \sum \Delta(\partial T / \partial \dot{q}_k)\, \delta q_k = \sum \widehat{Q_k}\, \delta q_k; \tag{4.3.7}$$

and similarly for *quasi variables* (§4.5).

These are the fundamental (differential) variational equations of Lagrangean impulsive theory. All equations of impulsive motion, compatible with our finite-number-of-degrees-of-freedom model (i.e., impulsive counterparts of the equations of Routh–Voss, Maggi, Hamel, etc.), flow from (4.3.7) by *appropriate specializations of the virtual displacements;* and these latter depend on the nature of the imposed constraints. This process is detailed in the following sections.

Example 4.3.1 *"Work–energy" theorem in Constrained Impulsive Motion; or, Impressed Impulsive Forces Applied to a Moving System.* Let us begin with the LIP, eqs. (4.3.4–4b):

$$S\, dm(\boldsymbol{v}^+ - \boldsymbol{v}^-) \cdot \delta\boldsymbol{r} = S\, \widehat{d\boldsymbol{F}} \cdot \delta\boldsymbol{r}. \tag{a}$$

Choosing in there, first $\delta\boldsymbol{r} \to \boldsymbol{v}^-$ and then $\delta\boldsymbol{r} \to \boldsymbol{v}^+$ (since, here, time is considered fixed), we obtain, respectively,

$$S\, dm(\boldsymbol{v}^+ - \boldsymbol{v}^-) \cdot \boldsymbol{v}^- = S\, \widehat{d\boldsymbol{F}} \cdot \boldsymbol{v}^-, \tag{b}$$

$$S\, dm(\boldsymbol{v}^+ - \boldsymbol{v}^-) \cdot \boldsymbol{v}^+ = S\, \widehat{d\boldsymbol{F}} \cdot \boldsymbol{v}^+. \tag{c}$$

Adding (b) and (c) side by side, and then dividing by 2, we obtain the sought energetic theorem

$$\Delta T \equiv T^+ - T^- = W_{-/+}, \tag{d}$$

where

$$2T^+ \equiv S\, dm\, \boldsymbol{v}^+ \cdot \boldsymbol{v}^+, \qquad 2T^- \equiv S\, dm\, \boldsymbol{v}^- \cdot \boldsymbol{v}^-, \tag{e}$$

and

$$W_{-/+} \equiv S\, \widehat{d\boldsymbol{F}} \cdot (\boldsymbol{v}^+ + \boldsymbol{v}^-)/2 \equiv S\, \widehat{d\boldsymbol{F}} \cdot \langle \boldsymbol{v} \rangle. \tag{f}$$

In words: The sudden change of the kinetic energy of a moving system, due to arbitrary impressed impulses, equals the sum of the dot products of these impulses with the mean (average) velocities of their material points of application, immediately before and after their action.

4.4 THE APPELLIAN CLASSIFICATION OF IMPULSIVE CONSTRAINTS, AND CORRESPONDING EQUATIONS OF IMPULSIVE MOTION

As mentioned earlier, to proceed further from the impulsive variational equation (§4.3.7), we must specify the n δq's; that is, specify the (variational form of the) impulsive constraints of the particular problem. And this brings us to Appell's fundamental *classification of impulsive constraints*. [See Appell (1896, p. 6 ff.; 1953, pp. 505–544); also (alphabetically): Bouligand (1954, pp. 129–157) and Roy (1965, p. 171 ff.). For a related classification, but with different terminology, see Pars (1965, pp. 228–248) and Rosenberg (1977, pp. 391–411). Also, all impulsive constraints dealt with here are assumed ideal; that is, $\widehat{\delta' W_R} = 0$].

According to his approach, which we follow here, the most general way of viewing a shock or percussion is as follows: at a given initial instant t' new constraints are suddenly introduced into the system and/or some old constraints are removed, or suppressed. As a result, percussions are generated, which, in the very short time interval

$\tau \equiv t'' - t'$, over which they are supposed to act and during which the shock lasts, produce finite velocity changes, but, according to our "first" approximation negligible position changes; that is, for $\tau \to 0$: $\Delta q = 0$, $\Delta(dq/dt) \equiv \Delta v \neq 0$ (ex. 4.2.1).

Now, *the constraints existing at the shock moment* are either *persistent* or *nonpersistent*. By persistent, we mean constraints that, existing at the shock "moment," exist also after it, so that the actual postimpact displacements are incompatible with them; whereas by nonpersistent we mean constraints that, existing at the shock moment, do not exist after it, so that the actual postimpact displacements are incompatible with them.

As a result of this, impulsive constraints can be classified into the following *four* distinct kinds or types:

1. Constraints existing *before, during, and after the shock*; that is, the latter neither introduces new constraints, nor does it change the old ones; the system, however, is acted on by impulsive forces. An example of such a constraint is the striking of a physical pendulum with a nonsticking (or, nonplastic) hammer at one of its points, and resulting communication to it of a specified impressed impulsive force; while the impulsive reactions, generated at the pendulum support, satisfy (4.3.3b) and first of (4.3.7).
2. Constraints existing *during and after the shock, but not before it*; that is, the latter introduces suddenly new constraints to the system. Examples: (a) A rigid bar falling freely, until the two inextensible slack strings connecting its endpoints to a fixed ceiling get taut (during) and do not break (after); (b) The inelastic central collision of two solid spheres ("coefficient of restitution" $\equiv e = 0$—see (4.4.1)); (c) In a ballistic pendulum (see prob. 4.4.9) the pendulum is constrained to rotate about a fixed axis; which is a constraint existing before, during and after the percussion of the pendulum with a projectile (i.e., first-type constraint). The projectile, however, originally independent of the pendulum, strikes it and becomes embedded into it; which is a case of a new constraint whose sudden realization produces the shock, and which exists during and after the shock but not before it (i.e., second-type constraint).
3. Constraints existing *before and during the shock, but not after it.* For example, let us imagine a system consisting of two particles connected by a light and inextensible bar, or thread, thrown up into the air. Then, let us assume that one of these particles is suddenly seized (persistent constraint introduced abruptly; i.e., second type) and, at the same time, the bar breaks (constraint existing before the shock does not exist after it; i.e., third type).
4. Constraints existing *only during the shock, but neither before nor after it.* For example, when two solids collide, since their bounding surfaces come into contact, a constraint is abruptly introduced into this two body system. If these bodies are *elastic* ($e = 1$—see coefficient of restitution, below), they separate after the collision; which is a case of a constraint existing during the percussion but neither before nor after it (i.e., fourth type); while if they are *plastic* ($e = 0$) they do not separate (projectile and pendulum, above; i.e., second type). (If $0 < e < 1$, the bodies separate; i.e., we have a fourth-kind constraint.)

This classification is summarized in table 4.1; clearly, the first two types contain the persistent constraints, while the last two contain the nonpersistent ones.

REMARKS

(i) Types 1, 2, 3, 4 are also referred to, respectively, as *permanent, persistent, preexisting*, and *instantaneous* (*direct shock*). Also, for obvious reasons, type 1 is referred to as *continuous*; and types 2, 3, and 4, as *discontinuous.*

Table 4.1 Appellian Classification of Impulsive Constraints

	Preshock (before)	Shock (during)	Postshock (after)
1 (persistent)	▬	▬	▬
2 (persistent)		▬	▬
3 (nonpersistent)	▬	▬	
4 (nonpersistent)		▬	

(ii) These concepts are of paramount importance because, as shown below, in an impulsive problem, the excess of the number of unknowns (postimpact velocities and constraint reactions) over that of the available equations [those obtained from Lagrange's impulsive principle; plus preimpact velocities, impressed impulsive forces, constraints, and sometimes knowledge of the postimpact state (second type; e.g., $e = 0$)]—that is, the degree of its indeterminacy—equals the number of its constraints that, having existed before or during the shock, cease to do so at the end of it; that is,

Degree of indeterminacy = Number of nonpersistent constraints;

hence, the persistent types 1 and 2 are *determinate*, while the nonpersistent ones 3 and 4 are indeterminate.

(iii) Generally, problems of *collision* among solid bodies are indeterminate. For example, in the collision of two *smooth* solids, A and B, with respective mass centers G_A and G_B, we have thirteen unknowns: $3 + 3 = 6$ from the postimpact velocities of G_A and G_B, $3 + 3 = 6$ from the postimpact angular velocities of A and B, and 1 from the magnitude of the mutual *normal* impact force; and only twelve equations: $6 + 6 = 12$ from the theorems of impulsive linear/angular momenta. (For *nonsmooth* solids, things get more complicated.) To make the problem determinate, we introduce Newton's *coefficient of restitution, e*. This latter is defined by

$$e = -\frac{(\boldsymbol{v}_{2/1} \cdot \boldsymbol{n})^+}{(\boldsymbol{v}_{2/1} \cdot \boldsymbol{n})^-} \equiv -\frac{{v_{2/1,n}}^+}{{v_{2/1,n}}^-} = -\frac{\text{Relative velocity of } \textit{separation}}{\text{Relative velocity of } \textit{approach}}, \tag{4.4.1}$$

where 1 and 2 are the two points of A and B that come into contact during the collision, and $\boldsymbol{n}$ is the unit vector along the common normal to their bounding surfaces there, say from A to B (see also exs. 4.4.1 and 4.4.2 below). The coefficient ranges from 0 (*plastic* impact, no separation) to 1 (*elastic* impact, no energy loss); that is, $0 \leq e \leq 1$.

(iv) The case of the removal of a constraint (e.g., the sudden snapping of one or more of the taut strings supporting an originally motionless bar from a ceiling), is *not* an impulsive motion problem (of the third kind) but one of *initial* motion; that is,

$$\Delta(\textit{positions}) = 0, \qquad \Delta(\textit{velocities}) = 0, \qquad \text{but } \Delta(\textit{accelerations}) \neq 0.$$

However, if the rupture is the result of an impulsive force (a blow), the problem falls under type 3 (see comments on rupture later in this section).

Analytical Expression of the Appellian Classification; Persistency versus Determinacy

Let us express analytically all these types of constraints. We begin with a discussion of this issue in terms of *elementary* dynamics. Let us consider a system consisting of

N solids, in contact with each other at K points, out of which C are of the nonpersistent type, and/or with a number of foreign solid obstacles that are either fixed or have *known* motions. Assuming *frictionless* collisions, we will have a total of $6N + K$ *unknowns* ($6N$ postshock velocities, plus K percussions at the smooth contacts, along the common normals); and $6N + K - C$ *equations* ($6N$ impulsive momentum equations, plus $K - C$ persistent-type constraints); and therefore *the degree of indeterminacy equals the number of nonpersistent contacts* C (i.e., the kind that disappear after the shock). Hence: (i) a *free* (*i.e., unconstrained*) solid subjected to *given* percussions, and/or (ii) a system subjected only to *persistent* constraints are impulsively determinate.

Let us now discuss the problem from the Lagrangean viewpoint. We recall (§2.4 ff.) that a number of holonomic constraints, imposed on a system originally defined by n Lagrangean coordinates, can *always* be put in the *equilibrium* form:

$$q_1 = 0, q_2 = 0, \ldots, q_m = 0 \qquad (m\text{: number of such constraints} < n) \qquad (4.4.1a)$$

REMARKS

(i) It is shown later in this section, that, *within our impulsive approximations,* even Pfaffian constraints (including nonholonomic ones) can be brought to the holonomic form; that is, in impulsive motion *all* constraints behave as holonomic! However, as elaborated in the next section, quasi variables *can* be used to advantage in impulsive problems.

(ii) Briefly, if the system is, originally, described by the Lagrangean coordinates $q \equiv (q_1, \ldots, q_n)$, and if the $m(< n)$ new constraints are expressed by

$$\phi_1(t, q) = 0, \ldots, \phi_m(t, q) = 0, \qquad (4.4.1b)$$

then, by replacing $q_1, \ldots, q_m$ with the new Lagrangean coordinates,

$$\chi_1 \equiv \phi_1(t, q), \ldots, \chi_m \equiv \phi_m(t, q); \qquad \chi_{m+1} = q_{m+1}, \ldots, \chi_n = q_n, \qquad (4.4.1c)$$

we can express the new constraints (b) by the "equilibrium" equations:

$$\chi_1 = 0, \ldots, \chi_m = 0. \qquad (4.4.1d)$$

Assuming, henceforth, such a choice of Lagrangean coordinates for all our impulsive constraints (and, for convenience, redenoting these new equilibrium coordinates by $q_1, \ldots, q_m; \ldots, q_n$), we can quantify the four Appellian types of impulsive constraints as follows:

- *First-type constraints* (existing *before, during, and after* the shock). As a result of these constraints, let the system configurations depend on n, hitherto independent, Lagrangean parameters: $q \equiv (q_1, \ldots, q_n)$. During the shock interval (t', t''), the corresponding velocities $\dot{q} \equiv (\dot{q}_1, \ldots, \dot{q}_n)$, pass suddenly from the known values $(\dot{q})^-$, at t', to other values $(\dot{q})^+$, while the q's remain practically unchanged; that is, here we have

$$(q_k)_{\text{before}} = 0, \qquad (q_k)_{\text{during}} = 0, \qquad (q_k)_{\text{after}} = 0; \qquad (4.4.2a)$$

$$\Delta \dot{q}_k \equiv (\dot{q}_k)^+ - (\dot{q}_k)^- \neq 0 \qquad \left[(\dot{q}_k)^+\text{: unknown}, (\dot{q}_k)^-\text{: known}\right]. \qquad (4.4.2b)$$

• *Second-type constraints* (additional constraints, existing *during and after* the shock, but *not before it*). Here, with $q_{D''} \equiv (q_1, \ldots, q_{m''})$, where $m'' < n$, we have

$$(q_{D''})_{\text{before}} \neq 0, \qquad (q_{D''})_{\text{during}} = 0, \qquad (q_{D''})_{\text{after}} = 0; \tag{4.4.3a}$$

$$(\dot{q}_{D''})^- \neq 0, \quad (\dot{q}_{D''})^+ = 0 \Rightarrow \Delta(\dot{q}_{D''}) = -(\dot{q}_{D''})^- \neq 0. \tag{4.4.3b}$$

[Equations like $q_{\text{after}} - q_{\text{before}} \neq 0$ i.e., (4.4.3a, 4a) in no way contradict our earlier assumption (first approximation): $\Delta(\textit{configuration}) \equiv \Delta q \equiv q^+ - q^- = 0$. As with the finite motion case (chap. 2), any new holonomic constraints must be consistent with the system configuration.]

• *Third-type constraints* (additional constraints existing *before and during*, but *not after* the shock). Here, with $q_{D'''} \equiv (q_{m''+1}, \ldots, q_{m'''})$, where $m''' < n$, we have

$$(q_{D'''})_{\text{before}} = 0, \qquad (q_{D'''})_{\text{during}} = 0, \qquad (q_{D'''})_{\text{after}} \neq 0; \tag{4.4.4a}$$

$$(\dot{q}_{D'''})^- = 0, \quad (\dot{q}_{D'''})^+ \neq 0 \Rightarrow \Delta(\dot{q}_{D'''}) = (\dot{q}_{D'''})^+ \neq 0. \tag{4.4.4b}$$

• *Fourth-type constraints* (additional constraints existing only *during*, but *neither before nor after* the shock). Here, with $q_{D''''} \equiv (q_{m'''+1}, \ldots, q_{m''''})$, where $m'''' < n$, we have

$$(q_{D''''})_{\text{before}} \neq 0, \qquad (q_{D''''})_{\text{during}} = 0, \qquad (q_{D''''})_{\text{after}} \neq 0; \tag{4.4.5a}$$

$$(\dot{q}_{D''''}) \neq 0, \quad (\dot{q}_{D''''})^- \neq 0 \Rightarrow \Delta(\dot{q}_{D''''}) = (\dot{q}_{D''''})^+ - (\dot{q}_{D''''})^- \neq 0. \tag{4.4.5b}$$

Hence, if no fourth-type constraints exist, $m''' = m''''$; and if no third-type constraints exist, $m'' = m'''$; and so on. Now, arguing as in the case of continuous motion (chap. 3), during the shock interval, we may view the constraints of the *second, third,* and *fourth* types as absent, provided that, in the spirit of the *impulsive principle of relaxation* [see also discussion below, after (4.4.16b)], we add to the system the corresponding constraint reactions. All relevant equations of motion are contained in the LIP (second of 4.3.7),

$$\sum \Delta(\partial T/\partial \dot{q}_k)\,\delta q_k = \sum \hat{Q}_k\,\delta q_k \qquad (k = 1, \ldots, n). \tag{4.4.6}$$

If the *virtual displacements* $\delta q \equiv (\delta q_1, \ldots, \delta q_n)$ are arbitrary, the right side of the above contains the impulsive virtual works of the reactions stemming from the *second, third,* and *fourth*-type constraints, and operating during the shock interval $[t', t'']$. Therefore, to eliminate these "forces," and thus produce $n - m''''$ reactionless, or kinetic, impulsive equations, we choose δq's that are compatible with *all constraints holding at the shock moment;* that is, we take

$$\delta q_1, \ldots, \delta q_{m''}; \qquad \delta q_{m''+1}, \ldots, \delta q_{m'''}; \qquad \delta q_{m'''+1}, \ldots, \delta q_{m''''} = 0; \tag{4.4.6a}$$

$$\delta q_{m''''+1}, \ldots, \delta q_n \neq 0. \tag{4.4.6b}$$

Applying the method of Lagrangean multipliers to the variational equation (4.4.6), under the virtual constraints (4.4.6a, b), we readily obtain the *two* (uncoupled) sets of equations:

Impulsive kinetostatic: $$\Delta(\partial T/\partial \dot{q}_D) = \hat{Q}_D + \hat{\lambda}_D \qquad (D = 1, \ldots, m''''), \tag{4.4.7a}$$

Impulsive kinetic: $$\Delta(\partial T/\partial \dot{q}_I) = \hat{Q}_I \qquad (I = m'''' + 1, \ldots, n). \tag{4.4.7b}$$

Further, since the velocity jumps $\Delta\dot{q}$ are produced only by the very large impulsive constraint reactions, operating during the very small interval $t'' - t'$, within our approximations, the $\hat{Q}_I$ [since they derive only from ordinary (i.e., finite, nonimpulsive) forces, like gravity] vanish: $\hat{Q}_I = 0$; and so (4.4.7b) reduce to *Appell's rule*:

$$\Delta(\partial T/\partial \dot{q}_I) = 0 \Rightarrow (\partial T/\partial \dot{q}_I)^+ = (\partial T/\partial \dot{q}_I)^-. \tag{4.4.8}$$

In words: *The partial derivatives of the kinetic energy relative to the velocities of those system coordinates q's that are not forced to vanish at the shock instant (i.e., $q_{\text{during}} \neq 0$) have the same values before and after the impact; or these $n - m''''$ unconstrained momenta, $p_I \equiv \partial T/\partial \dot{q}_I$, are conserved.*

Now, since T is quadratic in the $\dot{q}$'s, eqs. (4.4.8) are *linear and* (as explained below, under "Frame of Reference Effects on Impulsive Motion") *homogeneous* in the n $\Delta\dot{q}$'s. In there:

(i) The $q_1, \ldots, q_n$, have their constant shock instant values; that is,

$$q_D\colon q_{D''}, q_{D'''}, q_{D''''} = 0; \qquad \text{while} \qquad q_I\colon q_{m''''+1}, \ldots, q_n = \textit{known}; \tag{4.4.9a}$$

(ii) The $(\dot{q}_1)^-, \ldots, (\dot{q}_n)^-$ are *known*; in particular, $(\dot{q}_{D'''})^- = 0$; (4.4.9b)

(iii) The $(\dot{q}_1)^+, \ldots, (\dot{q}_n)^+$ are the *unknowns* of the problem; except that, since the constraints $(q_{D''})_{\text{after}} = 0$ are persistent,

$$(\dot{q}_{D''})^+ = 0. \tag{4.4.9c}$$

Hence, we have $n - m''''$ linear equations (4.4.8) for the $n - m''$ unknowns: $(\dot{q}_{m''+1})^+, \ldots, (\dot{q}_n)^+$; and therefore the *degree of indeterminacy* of the impulsive problem equals

$$(n - m'') - (n - m'''') = m'''' - m'' \quad (> 0, \text{assumed})$$
$$= \textit{number of nonpersistent constraints} \ (= \text{number of 3rd and 4th types}). \tag{4.4.9d}$$

In sum: the impulsive problem (4.4.8) is, in general, *indeterminate* [unlike its ordinary motion counterpart which *is* determinate (§3.5, §3.8)]: the m'''' kinetostatic equations (4.4.7a) introduce the m'''' *additional* unknown $\hat{\lambda}_D$'s. If, however, only persistent constraints are present ($m'''' = m''$), the impulsive problem (4.4.8) is *determinate.*

To make the problem determinate, in the presence of nonpersistent type constraints, we must make particular *constitutive* (i.e., *physical*) *hypotheses*; for example, elasticity assumptions about the postshock state. For example, in the well-known problem of central (or direct) collision of two solid spheres that separate after the shock (fourth-type constraint), Newton–Euler mechanics provides only *one* equation for the two postimpact velocities of the spheres' centers: $\Delta v_{\text{center of mass of system}} = 0$. A *second* equation is furnished by constitutive assumptions; for example, for perfect elasticity, $\Delta T = 0$. The most common constitutive equations for such third and/or fourth type $(\dot{q})^+$'s are

$$(\dot{q}_{D''',D''''})^+ = -e(\dot{q}_{D''',D''''})^- \quad (D''' = m'' + 1, \ldots, m'''; D'''' = m''' + 1, \ldots, m''''); \tag{4.4.10a}$$

where e is the earlier *coefficient of restitution*. Then, the corresponding velocity jumps equal

$$\Delta(\dot{q}_{D''',D''''}) \equiv (\dot{q}_{D''',D''''})^{+} - (\dot{q}_{D''',D''''})^{-} = -(1+e)(\dot{q}_{D''',D''''})^{-}. \tag{4.4.10b}$$

REMARKS

(i) If the constraints have the *general* (*nonequilibrium*) *form*

$$\phi_D(t,q) = 0 \Rightarrow \delta\phi_D = \sum (\partial\phi_D/\partial q_k)\,\delta q_k = 0 \qquad [D = 1,\ldots,m(<n)], \tag{4.4.11a}$$

then combination ("adjoining") of the above with (4.4.6), via *impulsive Lagrangean multipliers* $\hat{\lambda}_D$ yields the n *impulsive Routh–Voss equations*

$$\Delta p_k = \hat{Q}_k + \hat{R}_k, \qquad \text{with} \qquad \hat{R}_k = \sum \hat{\lambda}_D(\partial\phi_D/\partial q_k). \tag{4.4.11b}$$

As in the finite motion case (§3.8), these equations are *coupled* in the $(\dot{q})^{+}$'s and $\hat{\lambda}$'s, because the q's employed are *coupled*; i.e., eqs. (4.4.11a); whereas the earlier eqs. (4.4.7a, b; 8), corresponding to the uncoupled (equilibrium) coordinates (4.4.1a–d) are uncoupled.

In *first-type problems*, the above equations along with the m postshock forms of (4.4.11a),

$$\phi_D = 0 \Rightarrow \dot{\phi}_D = \sum (\partial\phi_D/\partial q_k)\dot{q}_k + \partial\phi_D/\partial t = 0, \tag{4.4.11c}$$

with the partial derivatives of the ϕ's evaluated at the shock configuration and instant, constitute a set of $n+m$ algebraic equations for the n postshock velocities $(\dot{q})^{+}$ and the m impulsive multipliers $\hat{\lambda}$. And since such constraints also hold, in form, for the preshock velocities $(\dot{q})^{-}$, only $n-m$ of the latter need be known.

In *second-type problems*, eqs. (4.4.11b) also hold, and the impulsive constraints (4.4.11a) are imposed at the beginning of the shock and continue to hold during and after, but not before, it. Therefore, we can apply (4.4.11c) for the $(\dot{q})^{+}$'s.

(ii) Equations (4.4.11b) can, of course, result directly by integration of the finite Routh–Voss equations of the system (§3.5) in time, then taking the limit as $\tau \equiv t'' - t' \to 0$, and noticing that since $\partial T/\partial q_k = \textit{finite}$ during τ ($\Delta q_k = 0$ and $\dot{q}_k = \textit{finite}$, hence $\Delta\dot{q}_k = \textit{finite}$),

$$\widehat{(\partial T/\partial q_k)} = 0; \tag{4.4.12}$$

and the partial ϕ-derivatives, within our approximations, remain constant.

(iii) If the third- and fourth-type constraints have the general form (4.4.11a), then (4.4.10a, b) must be replaced by

$$(\dot{\phi})^{+} = -e(\dot{\phi})^{-} \;\Rightarrow\; \Delta\dot{\phi} \equiv (\dot{\phi})^{+} - (\dot{\phi})^{-} = -(1+e)(\dot{\phi})^{-}. \tag{4.4.10c, d}$$

Further, due to the compatibility of velocities with the constraint, $\dot{\phi} \sim \boldsymbol{v}_{2/1}\cdot\boldsymbol{n} = (v_{2/1})_n$, and so

$$\text{at } t'\text{:}\quad (\dot{\phi})^{-} < 0 \qquad \text{(beginning of “approach” period)},$$

$$\text{at } t''\text{:}\quad (\dot{\phi})^{+} > 0 \qquad \text{(ending of “restitution” period)}. \tag{4.4.10e}$$

In sum: the $n\,(\dot{q})^+$'s can be determined from the $n - m''''$ kinetic equations (4.4.8), the m'' kinematical equations (4.4.3b), and the $m'''' - m''$ constitutive equations (4.4.10a, b); that is, a total of $(n - m'''') + (m'') + (m'''' - m'') = n$ equations. Once all the $(\dot{q})^+$'s have been found, the m'''' kinetostatic equations (4.4.7a) immediately yield the m'''' impulsive reactions $\hat{\lambda}_D$.

For instance, in an impact problem with the *three* constraints $q_1 = q_2 = q_3 = 0$ (i.e., $m'''' = 3$), the impulsive multipliers will appear only in the first three equations:

$$\Delta(\partial T/\partial \dot{q}_D) = \hat{Q}_D + \hat{\lambda}_D \qquad (D = 1, 2, 3); \tag{4.4.10f}$$

while, by Appell's rule, the remaining kinetic equations will be

$$\Delta(\partial T/\partial \dot{q}_I) = 0 \qquad (I = 4, 5, \ldots, n). \tag{4.4.10g}$$

If we are only interested in the $(\dot{q})^+$'s and not the $\hat{\lambda}$'s, then we must add to the $n - 3$ equations (4.4.10g) the three postimpact conditions for $(\dot{q}_{1,2,3})^+$. For example, if the *first* and *second* constraints hold after the shock (persistent) while the *third* one does not (nonpersistent), then these conditions are

$$(\dot{q}_1)^+ = (\dot{q}_2)^+ = 0, \qquad (\dot{q}_3)^+ = -e(\dot{q}_3)^-; \tag{4.4.10h}$$

and along with (4.4.10g) these constitute a determinate system for the $n\ (\dot{q})^+$.

(iv) In applying impulsive equations, like (4.4.7a, b; 8; 11b), or any other form involving $\partial T/\partial \dot{q}_k$, there is no need to start with the general (configuration and velocity) expression for T, then $(\partial \ldots/\partial \dot{q})$-differentiate it, and finally evaluate the results for the shock configuration(s) and time, as in the finite motion case. It is simpler, and leads to the same final results, if we *calculate T only at the shock configuration and time and proceed from there to calculate the momenta, and so on.* [Why? Explain using a Taylor expansion of T around the shock configuration, and then taking the partial *q-derivatives* of both sides. See, for example, Beghin (1967, pp. 472–473).]

(v) *Frame of reference effects on impulsive motion.* We begin by pointing out that the impulsive theorems of linear and angular momentum are independent of the frame of reference used; that is, *they hold unchanged in form even in noninertial frames (moving axes)*, provided that the latter's inertial motions remain continuous and involve only finite accelerations. Then, the "inertial," or "fictitious" forces on a typical particle—that is, the "force" of *transport* (due to the translational and rotational inertial accelerations of the noninertial frame), and *the complementary, or Coriolis,* "force" (due to the coupling of the relative velocity of the particle with the inertial angular velocity of the frame—recalling §1.7)—remain *finite* and therefore give zero impulses. [This also follows from the fact that the impulsive momentum equations involve only velocity *jumps* $\Delta \boldsymbol{v}$; and these latter are *frame-independent*, as long as, during the shock, the transport velocities of the noninertial frames *do not undergo finite jumps*; it is like adding and subtracting the same frame velocity (after and before the shock, respectively) to all system particles! (Explain, quantitatively, using the frame transformation equations for velocities.) Question: During a shock, does the velocity of *body-fixed* axes, say at G, undergo finite changes (i.e., velocity discontinuities)? If it does, then the above reasoning does not apply to these axes.] The above are easy to see from the viewpoint of analytical mechanics: in all pertinent impulsive derivations, we may consider only the *quadratic* and *homogeneous* part of T (§3.9), $T_2 = 1/2 \sum\sum M_{kl}\dot{q}_k\dot{q}_l$; under our impulsive approximations, its *linear* and *homogeneous* part, $T_1 = \sum M_k\dot{q}_k$ (which, we recall, arises from the nonstationary/

noninertial contributions $\partial \mathbf{r}/\partial t$), makes a zero contribution to Δp_k:

$$\Delta(\partial T/\partial \dot{q}_k) = \Delta(\partial T_2/\partial \dot{q}_k) = 0. \tag{4.4.13}$$

In sum: In impulsive (not finite) motion problems we can always replace the inertial kinetic energy with the relative one; that is, take $T \approx T_2$, and then proceed as in the inertial case. Clearly, this results in considerable algebraic simplification.

(vi) *Holonomic versus nonholonomic constraints in impulsive motion.* We saw earlier, eqs. (4.4.1a–d), that any set of $m(< n)$ holonomic constraints,

$$\phi_D(t, q) = 0 \qquad (D = 1, \ldots, m), \tag{4.4.14a}$$

can be brought to the equilibrium form

$$\chi_D \equiv \phi_D(t, q) = 0 \qquad (D = 1, \ldots, m); \tag{4.4.14b}$$

and $\chi_I \equiv q_I (\neq 0)$. It follows that the associated generalized velocities and virtual variations, from 1 to m, will satisfy, respectively (with $k = 1, \ldots, n$),

$$\dot{\chi}_D = \sum (\partial \phi_D/\partial q_k)\dot{q}_k + \partial \phi_D/\partial t = 0 \qquad \text{and} \qquad \delta\chi_D = \sum (\partial \phi_D/\partial q_k)\,\delta q_k = 0. \tag{4.4.14c}$$

Now, since we assume that during the shock the coordinates and time remain essentially constant, we can replace in there the Pfaffian constraints (first of 4.4.14c) with their approximate time integral

$$\psi_D \equiv \sum \Phi_{Dk} q_k + \Phi_D t - C_D = 0, \tag{4.4.15}$$

where

$$\Phi_{Dk} \equiv \partial \phi_D/\partial q_k, \ \Phi_D \equiv \partial \phi_D/\partial t, \ C_D\text{: } \textit{constants}, \text{for } t' \leq t \leq t''; \tag{4.4.15a}$$

and thus replace $q_1, \ldots, q_m$ with the equilibrium coordinates $\psi_D(= 0)$. The same reasoning applied to the holonomic and/or nonholonomic Pfaffian impulsive constraints,

$$\sum a_{Dk} \dot{q}_k + a_D = 0, \tag{4.4.16a}$$

allows us to *approximate* them, in $[t', t'']$, with

$$\beta_D \equiv \sum a_{Dk} q_k + a_D t - \alpha_D = 0, \qquad \alpha_D = \textit{integration constants}; \tag{4.4.16b}$$

that is, replace $q_1, \ldots, q_m$ with the new equilibrium variables $\beta_D(= 0)$.

In sum: In impulsive problems, we may disregard the holonomic versus nonholonomic difference—*during the shock, all constraints are approximately holonomic*—and to solve them, either we use impulsive multipliers, or avoid them by choosing the above equilibrium coordinates, or use quasi variables (see §4.5).

Impulsive Principle of Relaxation (RIP)

To calculate impulsive reactions due to pre-existing constraints, say in a *first*-type constraint problem (e.g., determine the impulsive bending moment at a certain point of a physical pendulum caused by a given blow at another point of it), either we use the impulsive forms of linear and angular momentum, after solving the Lagrangean impulsive equations of the given (unrelaxed) system, say (4.4.7a, b), and so on; or, in the spirit of an *impulsive principle of relaxation of the constraints*, we may endow the system with an *additional* n' Lagrangean coordinates $q_{k'}(k' = 1, \ldots, n')$, equal in number to that of the sought reactions and satisfying the persistent equilibrium constraints

$$q_{k'} = constant;\ \text{say} = 0, \tag{4.4.17a}$$

then calculate the "relaxed" kinetic energy

$$T = T(t, q_{1'}, \ldots, q_{n'};\ q_1, \ldots, q_n;\ \dot{q}_{1'}, \ldots, \dot{q}_{n'};\ \dot{q}_1, \ldots, \dot{q}_n), \tag{4.4.17b}$$

from that calculate $\partial T/\partial \dot{q}_{k'}$ and $\partial T/\partial \dot{q}_k$, and so on, and reasoning as before for the new relaxed problem, arrive at the uncoupled system

$$\textit{Kinetostatic}\ \text{equations:} \qquad (\Delta p_{k'})_o = (\hat{Q}_{k'})_o + \hat{\lambda}_{k'} \qquad (k' = 1, \ldots, n'), \tag{4.4.17c}$$

$$\textit{Kinetic}\ \text{equations:} \qquad (\Delta p_k)_o = (\hat{Q}_k)_o \qquad (k = 1, \ldots, n); \tag{4.4.17d}$$

where $(\ldots)_o \equiv (\ldots)$ *evaluated for* $q_{k'} = constant$ (e.g., for 0) and $\dot{q}_{k'} = 0$. We can easily show that the relaxed T, (4.4.17b), is needed only for the kinetostatic (4.4.17c); for the kinetic ones, (4.4.17d), we can use the original (unrelaxed, or constrained) kinetic energy T_o (why?). Finally, the whole method of RIP can be applied without theoretical complications to *second/third/fourth*-types of impulsive problems.

Sudden Rupture of Constraints

As mentioned earlier, this is *not* an impulsive problem but one of *initial (finite) motion*; the jumps appear in the $\ddot{q}$'s. If, *after* the rupture, the system has n Lagrangean coordinates, its postrupture equations of motion are (with the usual notations)

$$E_k(T) = Q_k \qquad (k = 1, \ldots, n), \tag{4.4.18a}$$

and, if the suppressed (broken) constraints are

$$\sum a_{Dk}\, \delta q_k = 0 \qquad [D = 1, \ldots, m(< n)], \tag{4.4.18b}$$

then the prerupture equations are

$$E_k(T) = Q_k + \sum \lambda_D a_{Dk} \qquad (k = 1, \ldots, n). \tag{4.4.18c}$$

Solving these sets of equations (e.g., subtracting them side by side, etc.), we can calculate the *acceleration jumps*: $\Delta\ddot{q} \equiv (\ddot{q})^+ - (\ddot{q})^-$.

Example 4.4.1 *Elementary (Newton–Euler) Theory of Rigid-Body Collisions.* Let us consider two rigid bodies, B_1 and B_2, with respective masses m_1 and m_2, mass centers G_1 and G_2, colliding at a certain instant t^- at the contact point C (fig. 4.2). Then, by the impulsive principles of linear and angular momentum, applied to B_1 and B_2 separately, we obtain

$$m_1(\boldsymbol{v}_1{}^+ - \boldsymbol{v}_1{}^-) = -\hat{\boldsymbol{f}}, \qquad \boldsymbol{H}_1{}^+ - \boldsymbol{H}_1{}^- = \boldsymbol{r}_1 \times (-\hat{\boldsymbol{f}}) = -\boldsymbol{r}_1 \times \hat{\boldsymbol{f}}, \tag{a1, 2}$$

$$m_2(\boldsymbol{v}_2{}^+ - \boldsymbol{v}_2{}^-) = \hat{\boldsymbol{f}}, \qquad \boldsymbol{H}_2{}^+ - \boldsymbol{H}_2{}^- = \boldsymbol{r}_2 \times \hat{\boldsymbol{f}}; \tag{a3, 4}$$

where $\hat{\boldsymbol{f}}$ = impulsive force, at C, say from B_1 to B_2, and $\boldsymbol{H}_{1,2}$ = angular momenta of $B_{1,2}$ about $G_{1,2}$, respectively. Now, since the precollision velocities are assumed

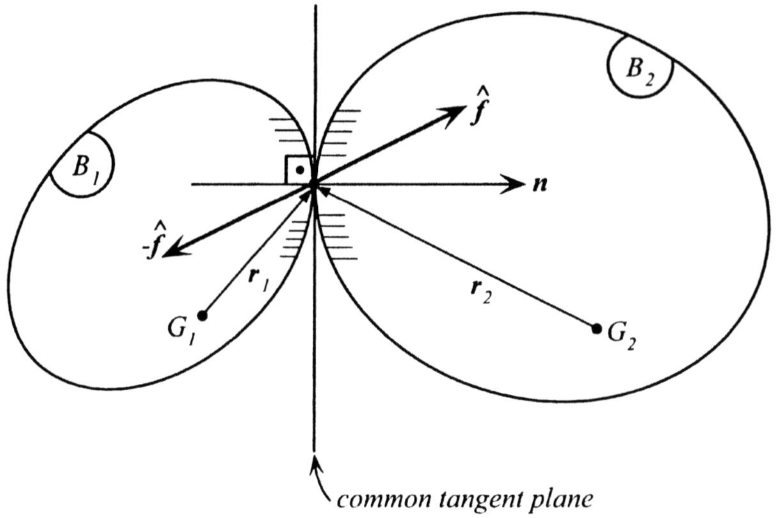

Figure 4.2 Collision of two rigid bodies. $\boldsymbol{n}$: Common unit normal (from B_1 to B_2); $H_{G_1} \equiv H_1$, $H_{G_2} \equiv H_2$.

known, the above constitutes a system of 4×3 scalar equations for the 5×3 unknown components of $\boldsymbol{v}_1{}^+, \boldsymbol{v}_2{}^+; \boldsymbol{H}_1{}^+, \boldsymbol{H}_2{}^+; \hat{\boldsymbol{f}}$ [from $\boldsymbol{H}_1{}^+, \boldsymbol{H}_2{}^+$ we can determine the postcollision angular velocities $\boldsymbol{\omega}_1{}^+$ and $\boldsymbol{\omega}_2{}^+$ via inversion of $\boldsymbol{H} = \mathbf{I} \cdot \boldsymbol{\omega}$, $\mathbf{I} =$ inertia tensor at G (known), for each body].

Hence, *the principles of mechanics alone do not suffice to solve the impact problem*; we need additional *physical hypotheses* to remove the $(15 - 12 =)$ 3-fold indeterminacy.

REMARK

The indeterminacy is due to our simple/idealized model of the impact problem; that is, to the *rigid* bodies and impulsive forces; a model that has also led to indeterminacy in ordinary/finite motion, for example, statical indeterminacy. As there, determinacy is here attained by adoption of the more realistic (and more "expensive") model of the *deformable* body. Then, the contact point C becomes a contact *surface* $C = C(t)$; while the impulsive force $\hat{\boldsymbol{f}}$ is replaced by the following resultant of a complicated distribution of surface tractions $\boldsymbol{t}_{(n)}$ over C: $\hat{\boldsymbol{f}} = \int dt(\int_C \boldsymbol{t}_{(n)}\, dC)$. For details, see works on dynamic elasticity, and so on.

Here, we remove the indeterminacy with the following two *empirical* hypotheses:

(i) B_1 and B_2 are *smooth*, so that $\hat{\boldsymbol{f}} = \hat{f}\boldsymbol{n}$, where $\hat{f} > 0$ and $\boldsymbol{n} =$ unit vector at C, along the common *normal*, say from B_1 to B_2; a hypothesis that reduces the number of unknowns to 13: $\boldsymbol{v}_1{}^+, \boldsymbol{v}_2{}^+; \boldsymbol{H}_1{}^+, \boldsymbol{H}_2{}^+; \hat{f}$, and

(ii) *Conservation of (kinetic) energy*: $\Delta T \equiv T^+ - T^- = 0$; which, since T^- is known and T^+ can be expressed in terms of $\boldsymbol{v}_1{}^+, \boldsymbol{v}_2{}^+; \boldsymbol{H}_1{}^+, \boldsymbol{H}_2{}^+$, reduces the number of unknowns to 12, and thus makes the problem determinate.

Speed of Compression; Coefficient of Restitution

Let us generalize a bit. If the velocities of the two particles of B_1 and B_2 in contact at C are $\boldsymbol{u}_1$ and $\boldsymbol{u}_2$, so that before, during, and after the collision, $\boldsymbol{u}_1 = \boldsymbol{v}_1 + \boldsymbol{\omega}_1 \times \boldsymbol{r}_1$ and

$\boldsymbol{u}_2 = \boldsymbol{v}_2 + \boldsymbol{\omega}_2 \times \boldsymbol{r}_2$, then the *speed of compression* c is defined as

$$c \equiv (\boldsymbol{u}_1 - \boldsymbol{u}_2) \cdot \boldsymbol{n} = \textit{normal component of relative velocity at } C$$
$$\text{(initially positive, since } B_1 \text{ and } B_2 \text{ tend to overlap).} \qquad \text{(b)}$$

Now, with the help of c, the collision process is decomposed into the following two stages:

(i) a *period of compression* (approaching of B_1 and B_2): $c > 0$;

(ii) a *period of restitution* (separation of B_1 and B_2): $c < 0$;

since, at the end of the compression period, the impact forces $-\hat{\boldsymbol{f}} \to -\hat{I}\boldsymbol{n}$ and $\hat{\boldsymbol{f}} \to \hat{I}\boldsymbol{n}$ reduce c to zero. If $(\ldots)^* \equiv (\ldots)$ *at the end of the compression period,* then applying again linear and angular momentum on B_1 and B_2, *between the beginning and the end of the compression period,* we obtain

$$m_1(\boldsymbol{v}_1{}^* - \boldsymbol{v}_1{}^-) = -\hat{I}\boldsymbol{n}, \qquad \boldsymbol{H}_1{}^* - \boldsymbol{H}_1{}^- = \boldsymbol{r}_1 \times (-\hat{I}\boldsymbol{n}) = -\boldsymbol{r}_1 \times \hat{I}\boldsymbol{n}, \qquad \text{(c1, 2)}$$

$$m_2(\boldsymbol{v}_2{}^* - \boldsymbol{v}_2{}^-) = \hat{I}\boldsymbol{n}, \qquad \boldsymbol{H}_2{}^* - \boldsymbol{H}_2{}^- = \boldsymbol{r}_2 \times \hat{I}\boldsymbol{n}, \qquad \text{(d1, 2)}$$

$$c^* \equiv (\boldsymbol{u}_1{}^* - \boldsymbol{u}_2{}^*) \cdot \boldsymbol{n} = 0; \qquad \text{(e)}$$

which constitutes a determinate system of $(4 \times 3) + 1 = 13$ scalar equations for the 13 scalar unknowns of the end of the compression period: $\boldsymbol{v}_1{}^*$, $\boldsymbol{v}_2{}^*$; $\boldsymbol{H}_1{}^*$, $\boldsymbol{H}_2{}^*$; $\hat{I}$. Having found $\hat{I}$, and assuming that the impulsive forces during the restitution period are proportional to those forces during the compression period, the factor of that proportionality called *coefficient of restitution, e,* we can then write in (a1–4),

$$\hat{\boldsymbol{f}} = (1 + e)\hat{I}\boldsymbol{n}, \qquad \text{(f)}$$

and thus reduce its number of scalar unknowns to 13: $\boldsymbol{v}_1{}^+$, $\boldsymbol{v}_2{}^+$; $\boldsymbol{H}_1{}^+$, $\boldsymbol{H}_2{}^+$; e. Therefore, the problem becomes determinate by specifying the value of e. We have the following three cases: $e = 0$: *inelastic collision*; $e = 1$: *elastic collision*; $0 < e < 1$: *semielastic collision*. It can be shown that the elastic case, $e = 1$, implies the conservation condition $T^+ = T^-$.

Example 4.4.2 *Specialization of the Above to the Case of the Collision of Two, Originally Nonrotating, Smooth and Homogeneous Spheres.* Here, since $-\boldsymbol{r}_1 \times \hat{\boldsymbol{f}} = \boldsymbol{0}$ and $\boldsymbol{r}_2 \times \hat{\boldsymbol{f}} = \boldsymbol{0}$, eqs. (c1–e) of the preceding example reduce to

$$m_1(\boldsymbol{v}_1{}^* - \boldsymbol{v}_1{}^-) = -\hat{I}\boldsymbol{n}, \qquad m_2(\boldsymbol{v}_2{}^* - \boldsymbol{v}_2{}^-) = \hat{I}\boldsymbol{n}, \qquad (\boldsymbol{u}_1{}^* - \boldsymbol{u}_2{}^*) \cdot \boldsymbol{n} = 0, \qquad \text{(a)}$$

and $\boldsymbol{H}_1{}^* - \boldsymbol{H}_1{}^- = \boldsymbol{0}$, $\boldsymbol{H}_2{}^* - \boldsymbol{H}_2{}^- = \boldsymbol{0}$ (since $\boldsymbol{\omega}_1{}^- = \boldsymbol{0}$ and $\boldsymbol{\omega}_2{}^- = \boldsymbol{0}$, and, hence, also $\boldsymbol{u}_1{}^* = \boldsymbol{v}_1{}^*$ and $\boldsymbol{u}_2{}^* = \boldsymbol{v}_2{}^*$; and $\boldsymbol{\omega}_1{}^+ = \boldsymbol{0}$ and $\boldsymbol{\omega}_2{}^+ = \boldsymbol{0}$). Solving the system (a) we obtain

$$\hat{I} = [m_1 m_2/(m_1 + m_2)][(\boldsymbol{v}_1{}^- - \boldsymbol{v}_2{}^-) \cdot \boldsymbol{n}] = (m_1 m_2 c)/(m_1 + m_2), \qquad \text{(b1)}$$

that is,

$$\hat{\boldsymbol{f}} = [m_1 m_2 c(1 + e)/(m_1 + m_2)]\boldsymbol{n}, \qquad \text{(b2)}$$

where c = known initial (preimpact) value of compression speed (> 0). As a result, eqs. (a1–4, f) of the preceding example yield

$$m_1(\mathbf{v}_1{}^+ - \mathbf{v}_1{}^-) = -(1+e)\hat{I}\mathbf{n}, \qquad m_2(\mathbf{v}_2{}^+ - \mathbf{v}_2{}^-) = (1+e)\hat{I}\mathbf{n}, \tag{c}$$

and from these we readily obtain the postimpact velocities

$$\mathbf{v}_1{}^+ = \mathbf{v}_1{}^- - \{c/[1 + (m_2/m_1)]\}(1+e)\mathbf{n}, \tag{d1}$$

$$\mathbf{v}_2{}^+ = \mathbf{v}_2{}^- + \{c/[1 + (m_2/m_1)]\}(1+e)\mathbf{n}. \tag{d2}$$

We leave it to the reader to show that the kinetic energy change equals

$$\Delta T \equiv T^+ - T^- = -[m_1 m_2 c^2/2(m_1 + m_2)](1 - e^2) \le 0, \tag{e}$$

that is, $T^+ \le T^-$, in general (since $0 \le e \le 1$) an energy *loss*! Special cases are:

(i) $e = 1$ (elastic impact): $\Delta T = 0$: energy of compression = energy of restitution;

(ii) $e = 0$ (inelastic impact): $\Delta T = -(m_1 m_2 c^2)/2(m_1 + m_2)$.

For an elementary, but instructive and rare, treatment of the role of *friction* in impact, see, for example, Hamel ([1922(a)] 1912, pp. 447–450).

Example 4.4.3 A thin, straight, and homogeneous bar AB, of mass m and length $2b$, moves on a fixed, horizontal, and smooth plane p. At a certain moment, the bar strikes a fixed peg O located a distance c from the center of mass of the bar G. Let us calculate the postimpact velocities in the following two cases: (i) the point of the bar that strikes O stays fixed relative to the latter, and (ii) the bar remains in contact with O and slides without friction on it (fig. 4.3) (Lainé, 1946, pp. 188–191).

On p, let us choose axes O–xy such that $x = c$, $y = 0$, $\phi = 0$. Now, by König's theorem, the (double) kinetic energy of the bar is

$$2T = m[(\dot{x})^2 + (\dot{y})^2] + (mb^2/3)(\dot{\phi})^2; \tag{a}$$

and since, obviously, $\widehat{\delta' W} = 0$, the impulsive Lagrangean principle (LIP) yields

$$\Delta(\partial T/\partial \dot{x})\,\delta x + \Delta(\partial T/\partial \dot{y})\,\delta y + \Delta(\partial T/\partial \dot{\phi})\,\delta\phi = 0$$

$$\Rightarrow (\Delta\dot{x})\,\delta x + (\Delta\dot{y})\,\delta y + (b^2 \Delta\dot{\phi}/3)\,\delta\phi = 0. \tag{b}$$

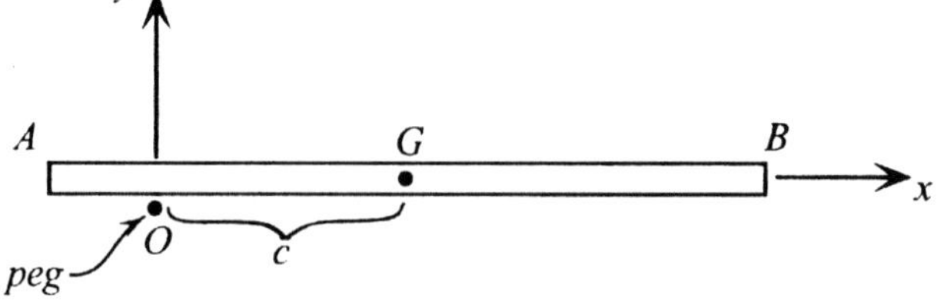

Figure 4.3 Geometry of bar AB moving on a fixed, horizontal, and smooth plane p, and striking a fixed peg O. $|\mathbf{AB}| = 2b$, $\mathbf{OG} = (x, y) \Rightarrow |\mathbf{OG}| = c$, and *angle* $(AB, Ox) \equiv \phi \to 0$.

Let us find the restrictions among δx, δy, $\delta\phi$.

(i) When the bar point sticks to O, the constraints are

$$x = c\cos\phi \Rightarrow \delta x = -c\sin\phi\,\delta\phi\big|_{\phi=0} = 0, \tag{c1}$$

$$y = c\sin\phi \Rightarrow \delta y = c\cos\phi\,\delta\phi\big|_{\phi=0} = c\,\delta\phi; \tag{c2}$$

and, therefore, LIP, eq. (b), reduces to

$$(\Delta\dot{y})(c\,\delta\phi) + (b^2/3)\,\Delta\dot{\phi}\,\delta\phi = 0 \Rightarrow c\,\Delta\dot{y} + (b^2/3)\,\Delta\dot{\phi} = 0; \tag{d1}$$

or, explicitly [with $(\dot{\phi})^+ \equiv \omega$],

$$c[(\dot{y})^+ - (\dot{y})^-] + (b^2/3)(\omega - \omega^-) = 0. \tag{d2}$$

However, from the *velocity* form of the constraints, evaluated at the impact configuration, we find the following postimpact velocities [with $(\dot{x})^+ \equiv \dot{x}$, $(\dot{y})^+ \equiv \dot{y}$]:

$$\dot{x} = -c\sin\phi\,\dot{\phi}\big|_{\phi=0} = 0, \qquad \dot{y} = c\cos\phi\,\dot{\phi}\big|_{\phi=0} = c\,\omega; \tag{e}$$

and so (d2) yields

$$c[c\,\omega - (\dot{y})^-] + (b^2/3)(\omega - \omega^-) = 0$$

$$\Rightarrow \omega = [c(\dot{y})^- + (b^2/3)\omega^-]/[c^2 + (b^2/3)]. \tag{f}$$

[Elementary solution: applying angular momentum conservation about O we get

$$H_O{}^- = H_O{}^+\!: (mb^2/3)\omega^- + mc(\dot{y})^- = m[(b^2/3) + c^2]\omega^- \Rightarrow \omega^- = \ldots, \text{eq. (f).}]$$

(ii) When the bar is obliged to slide on O, the *component of* $\boldsymbol{v}_{O'} = \boldsymbol{v}_G + \boldsymbol{\omega}\times\boldsymbol{r}_{O'/G}$ (O': bar point, instantaneously adjacent to peg O) *normal to* Ox *must vanish*:

$$\boldsymbol{v}_{O'}\cdot\boldsymbol{j} = \boldsymbol{v}_G\cdot\boldsymbol{j} + (\boldsymbol{\omega}\times\boldsymbol{r}_{O'/G})\cdot\boldsymbol{j}$$

$$= (\dot{x}, \dot{y}, 0)\cdot(0, 1, 0) + [(0, 0, \dot{\phi})\times(-c, 0, 0)]\cdot(0, 1, 0)$$

$$= (\dot{y} - c\dot{\phi}, 0, 0) = (0, 0, 0) \qquad \text{[since at the impact moment: } x = c\text{]}, \tag{g1}$$

that is,

$$\dot{y} - c\dot{\phi} = 0 \Rightarrow \delta y - c\,\delta\phi = 0 \qquad \text{[since this constraint is scleronomic]}; \tag{g2}$$

which, we notice, is the same as (c2).

Hence, in this case, we have the following two variational equations:

$$(\Delta\dot{x})\,\delta x + (\Delta\dot{y})\,\delta y + (b^2\Delta\dot{\phi}/3)\,\delta\phi = 0, \tag{h1}$$

$$(0)\,\delta x + (1)\,\delta y + (-c)\,\delta\phi = 0; \tag{h2}$$

from which we obtain immediately

$$\Delta\dot{x} = 0 \Rightarrow \dot{x} = (\dot{x})^- \qquad \text{[different from first of (e), since here } \delta x \neq 0\text{]}, \tag{i1}$$

$$c\,\Delta\dot{y} + (b^2/3)\,\Delta\dot{\phi} = 0 \Rightarrow c[\dot{y} - (\dot{y})^-] + (b^2/3)[\dot{\phi} - (\dot{\phi})^-] = 0, \tag{i2}$$

from which, since $\dot{y} = c\dot{\phi} \equiv c\omega$, finally,

$$\omega = [c(\dot{y})^- + (b^2/3)\omega^-]/[c^2 + (b^2/3)] \qquad \text{[same as in the preceding case, eq. (f)].} \tag{i3}$$

Problem 4.4.1 Consider a two-DOF holonomic system with (double) kinetic energy:

$$2T = A(\dot{x})^2 + 2B\dot{x}\dot{y} + C(\dot{y})^2, \tag{a}$$

where x, y = Lagrangean coordinates; A, B, C = inertia coefficients (functions of x, y). Show that its *postimpact* kinetic energy, from a motionless preimpact state defined by $x, y = 0$; $(\dot{x})^-$, $(\dot{y})^- = 0$; $A, B, C \to A_o$, B_o, C_o, under the Lagrangean impulsive forces X and Y, equals

$$2T^+ = (C_o X^2 - 2B_o XY + A_o Y^2)/(A_o C_o - B_o{}^2). \tag{b}$$

HINT

Solve the impulsive Lagrangean equations $(\partial T/\partial\dot{x})^+ - (\partial T/\partial\dot{x})^- = X$, and so on, for $(\dot{x})^+$, $(\dot{y})^+$ in terms of X, Y; A_o, B_o, C_o.

Problem 4.4.2 Apply the result (b) of the preceding problem to the impact of a double pendulum (fig. 4.4) consisting of two equal and homogeneous rods, AB and BC, each of mass m and length l, smoothly hinged at A to a fixed object (a ceiling) and to each other at B, initially in vertical equilibrium and struck at C by a horizontal blow of magnitude $\hat{\boldsymbol{P}}$.

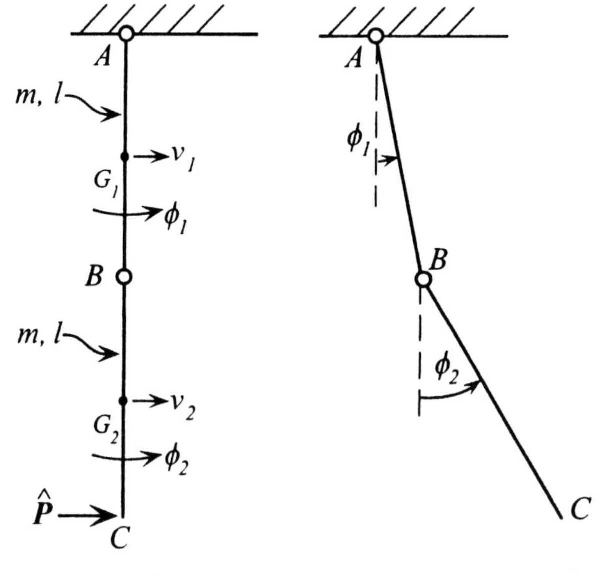

Figure 4.4 Impact at tip of a double pendulum.

HINT

The postimpact kinetic energy equals [omitting the $(\ldots)^+$ for convenience]

$$2T = m({v_1}^2 + {v_2}^2) + I[(\dot{\phi}_1)^2 + (\dot{\phi}_2)^2];$$

where: $I = ml^2/12$, $v_1 = (l/2)\dot{\phi}_1$, $v_2 = l\dot{\phi}_1 + (l/2)\dot{\phi}_2$, $v_{1,2}$ = velocities of centers of mass of AB and BC, respectively, in vertical configuration. (See, e.g., Lamb, 1923, pp. 321–322.)

Problem 4.4.3 Assuming that the δq's are independent, show that *the Lagrangean system momenta equal the corresponding Lagrangean system impulses that would, instantaneously, create the motion from rest.*

Problem 4.4.4 (Lainé, 1946, pp. 196–200). An articulated rhombus R, $ABCD$, formed of four identical thin and homogeneous rigid bars, each of length $2b$ and mass m (fig. 4.5), falls freely *translating* so that its diagonal AC is vertical; also, let *angle* $(BAD) \equiv 2\phi$ $(0 < \phi < \pi/2)$. At the instant of impact of A with the smooth horizontal ground, its translational velocity (including that of its mass center G) is v_o, vertically and downwards. Finally, let e be the coefficient of restitution, and assume that the impulse at A is distributed symmetrically between AB and AD.

For convenience, but no loss in generality, take axes O–xy (Ox: vertical, Oy: horizontal) such that, at the impact moment; $A = O$; R is also shown in an arbitrary configuration $A'B'C'D'$ [fig. 4.5(b)]. The latter can, clearly, be specified by the following *four* Lagrangean coordinates: (x, y) for G, *angle*$(Ax, A'C') \equiv \theta$, and *angle*$(A'C', A'D') \equiv \phi$.

(i) Show that the (double) kinetic energy of R equals

$$2T = 4m\left[(\dot{x})^2 + (\dot{y})^2\right] + (16mb^2/3)\left[(\dot{\theta})^2 + (\dot{\phi})^2\right]. \quad \text{(a)}$$

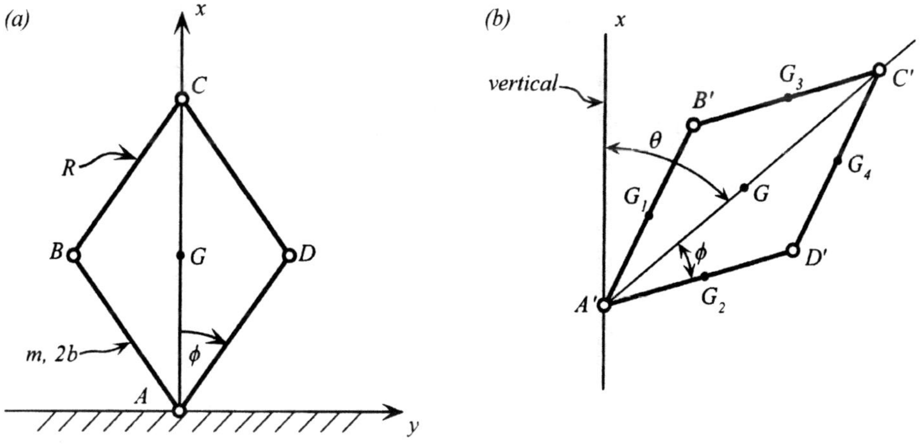

Figure 4.5 (a) Geometry of rhombus R, $ABCD$, falling freely translating with velocity v_o, so that AC is vertical, and then hitting the smooth ground at A; (b) rhombus in arbitrary configuration.

HINTS

Let 1, 2, 3, 4 be, respectively, the centers of mass (midpoints) of the bars $A'B'$, $A'D'$, $B'C'$, $D'C'$. Next, notice (or prove) that (a) 1 describes a circle of center G and radius b, with angular velocity $\dot{\theta} + \dot{\phi}$, while (b) the bar $A'B'$ rotates about 1 with angular velocity $\dot{\theta} - \dot{\phi}$. Hence, its (double) kinetic energy *relative to* G is

$$2T_{A'B',\text{relative}} = mb^2(\dot{\theta} + \dot{\phi})^2 + (mb^2/3)(\dot{\theta} - \dot{\phi})^2, \tag{a1}$$

and this clearly equals the kinetic energy of $D'C'$; that is, $2T_{C'D',\text{relative}} = 2T_{A'B',\text{relative}}$; while reasoning analogously for the bars $A'D'$, $B'C'$ one finds

$$2T_{A'D',\text{relative}} = 2T_{B'C',\text{relative}} = mb^2(\dot{\theta} - \dot{\phi})^2 + (mb^2/3)(\dot{\theta} + \dot{\phi})^2. \tag{a2}$$

Finally, applying König's theorem to this nonrigid system (i.e., $T = T_{\text{of}\,G} + T_{\text{relative to}\,G}$) yields (a).

(ii) Since the constraints are (with some easily understood notations)

$$\boldsymbol{v}_{A'} = \boldsymbol{v}_1 + \boldsymbol{\omega}_{A'B'} \times \boldsymbol{r}_{A'/1} = (\boldsymbol{v}_{G'} + \boldsymbol{v}_{1/G'}) + \boldsymbol{\omega}_{A'B'} \times \boldsymbol{r}_{A'/1}$$

[notice that $\boldsymbol{v}_{1/G'} = (dx_{1/G'}/dt, dy_{1/G'}/dt)$ (only x and y components shown) where $x_{1/G'} \equiv x_1 - x_{G'} = -b\cos(\theta + \phi)$, $y_{1/G'} \equiv y_1 - y_{G'} = -b\sin(\theta + \phi)$]

$$= (\dot{x}, \dot{y}) + \big(b(\dot{\theta} + \dot{\phi})\sin(\theta + \phi), -b(\dot{\theta} + \dot{\phi})\cos(\theta + \phi)\big)$$

$$+ [(\dot{\theta} - \dot{\phi})\boldsymbol{k}] \times \big(-b\cos(\theta - \phi), b\sin(\theta - \phi)\big)$$

$$= \big(\dot{x} + b(\dot{\theta} + \dot{\phi})\sin(\theta + \phi) + b(\dot{\theta} - \dot{\phi})\sin(\theta - \phi),$$

$$\dot{y} - b(\dot{\theta} + \dot{\phi})\cos(\theta + \phi) - b(\dot{\theta} - \dot{\phi})\cos(\theta - \phi)\big)$$

$$= \big(\dot{x} + 2b\,\dot{\theta}\cos\phi\sin\theta + 2b\,\dot{\phi}\sin\phi\cos\theta,$$

$$\dot{y} - 2b\,\dot{\theta}\cos\phi\cos\theta + 2b\,\dot{\phi}\sin\phi\sin\theta\big)$$

[evaluated at $\theta = 0, \dot{\theta} = 0$]

$$= (\dot{x} + 2b\,\dot{\phi}\sin\phi,\ \dot{y}) = v_{Ax}\,\boldsymbol{i} + v_{Ay}\,\boldsymbol{j}, \tag{b}$$

$$\Rightarrow \textit{vertical virtual displacement of } A \textit{ should vanish}\text{: } \delta x + (2b\sin\phi)\,\delta\phi = 0, \tag{c}$$

(although, in general, $v_{Ax} = \textit{nonzero constant}$), and since $\widehat{\delta' W} = 0$, verify that the impulsive Lagrangean principle, under (c), yields the following equations of motion:

$$y\text{:}\quad \varDelta(\partial T/\partial \dot{y}) = 0 \Rightarrow (\dot{y})^+ \equiv \dot{y} = (\dot{y})^- = 0, \tag{c1}$$

$$\theta\text{:}\quad \varDelta(\partial T/\partial \dot{\theta}) = 0 \Rightarrow (\dot{\theta})^+ \equiv \dot{\theta} = (\dot{\theta})^- = 0; \tag{c2}$$

$$x, \phi\text{:}\quad \hat{\lambda} = -\varDelta(\partial T/\partial \dot{x}) = -(2b\sin\phi)^{-1}\varDelta(\partial T/\partial \dot{\phi}) \qquad (\hat{\lambda}\text{: impulsive multiplier}) \tag{c3}$$

$$\Rightarrow \varDelta\dot{x} = (2b/3\sin\phi)\,\varDelta\dot{\phi} \Rightarrow 3(\dot{x} + v_o)\sin\phi = 2b\,\dot{\phi}. \tag{c4}$$

[Since $\varDelta\dot{x} \equiv (\dot{x})^+ - (\dot{x})^- \equiv \dot{x} - (\dot{x})^- = \dot{x} - (-v_o) = \dot{x} + v_o$,

$\varDelta\dot{\phi} \equiv (\dot{\phi})^+ - (\dot{\phi})^- \equiv \dot{\phi} - (\dot{\phi})^- = \dot{\phi} - 0 = \dot{\phi}$].

(iii) Verify that (c4) and the constitutive relation $\dot{x} + 2b\,\dot{\phi}\sin\phi = -e(-v_o)$ yield the values

$$\dot{x} = \left[(e - 3\sin^2\phi)/(1 + 3\sin^2\phi)\right] v_o, \tag{d1}$$

$$\dot{\phi} = \left[3(1+e)\sin\phi/2b(1 + 3\sin^2\phi)\right] v_o; \tag{d2}$$

and from these, and (c3), that

$$\hat{\lambda} = -4m\left[\dot{x} - (\dot{x})^-\right] = -4m(\dot{x} + v_o) = \cdots. \tag{d3}$$

Since $(\dot{\theta})^- = 0$, then, by symmetry, we may set $\dot{\theta} = 0$, $\theta = 0$, $\dot{y} = 0$, $y = 0$, for all t.

REMARK

The *inelastic* case $e = 0$ can be treated like a problem with a *new* constraint. Here, too, $\delta x + (2b\sin\phi)\,\delta\phi = 0$, but now (taking the limit of the above values as $e \to 0$)

$$\dot{x} = \left[-3\sin^2\phi/(1 + 3\sin^2\phi)\right] v_o, \qquad \dot{y} = 0, \qquad \dot{\theta} = 0, \tag{e1}$$

$$\dot{\phi} = \left[3\sin\phi/2b(1 + 3\sin^2\phi)\right] v_0, \qquad \hat{\lambda} = -\left[4m/(1 + 3\sin^2\phi)\right] v_o. \tag{e2}$$

Similarly, for the *elastic* case $e = 1$: $\dot{x} + 2b\,\dot{\phi}\sin\phi = v_o$.

For a solution based on the theorem of Carnot (§4.6), see Lainé (1946, p. 201).

Problem 4.4.5 (Lainé, 1946, pp. 193–194). A particle P of mass m is forced to slide on a smooth moving circle (i.e., a circular, rigid, and light wire) C of center O and radius r (fig. 4.6). The axis of C (perpendicular to the plane of the circle) coincides

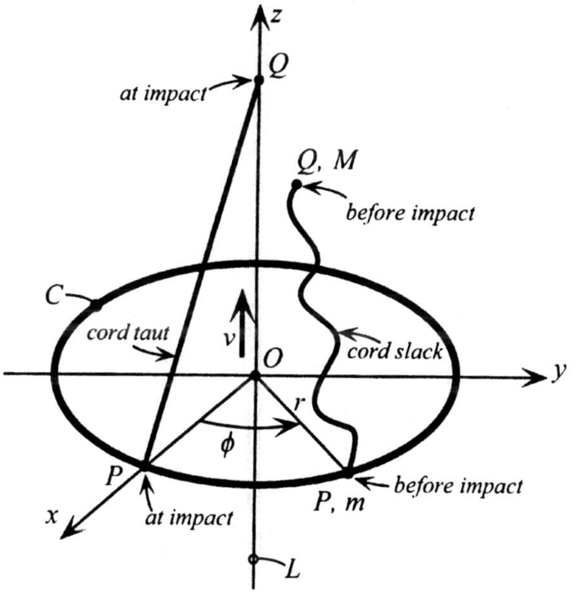

Figure 4.6 Geometry of two-particle system, P and Q, connected by a light and inextensible cord, which at some point in time gets suddenly taut. P can slide on a uniformly translating circle, while Q, before the impact, moves freely in space.

continuously with a fixed line L, on which O is forced to move with constant translational velocity v. A *second* particle Q of mass M moves freely in space. The two particles are connected by an inextensible and massless cord whose length l, in the beginning, is longer than the distance PQ; that is, the cord is slack. At a certain moment, during the motion, the distance PQ becomes equal to the length of the cord, which thus finds itself suddenly taut. Assume that, at that moment, the positions and (preimpact) velocities of P and Q are known, and that Q is on L.

For convenience, but no loss in generality, choose axes $O\text{–}xyz$ such that, at the impact moment, O coincides with the center of C, Oz with L and along the circle velocity v, and Ox with $OP(= +r)$. In these axes, the *generic* preimpact coordinates of P are

$$x = r\cos\phi, \quad y = r\sin\phi, \quad z = vt \qquad [(r, \phi)\text{: plane polar coordinates of } P], \qquad \text{(a)}$$

and, therefore, its corresponding velocity components are

$$\dot{x} = -r\sin\phi\,\dot{\phi}, \qquad \dot{y} = r\cos\phi\,\dot{\phi}, \qquad \dot{z} = v. \qquad \text{(b)}$$

Hence, just before the impact ($t = 0$, $\phi = 0$),

$$x^- = r, \quad y^- = 0, \quad z^- = 0; \qquad (\dot{x})^- = 0, \quad (\dot{y})^- = r(\dot{\phi})^-, \quad \dot{z} = v. \qquad \text{(c1)}$$

Similarly, let the preimpact position and velocity of Q be, respectively,

$$X^-, \quad Y^-, \quad Z^-; \qquad (\dot{X})^-, \quad (\dot{Y})^-, \quad (\dot{Z})^-. \qquad \text{(c2)}$$

From the above it follows that, before the shock, the Lagrangean coordinates of the system are $q_{1,\ldots,4} = \phi, X, Y, Z$.

(i) Show that its (double) kinetic energy equals

$$2T = m[(r\dot{\phi})^2 + v^2] + M(\dot{X}^2 + \dot{Y}^2 + \dot{Z}^2), \qquad \text{(d)}$$

while its impressed impulsive virtual work vanishes: $\widehat{\delta' W} = 0$; and so the impulsive Lagrangean principle becomes (dropping, for convenience, all *superscript minuses* from $q_{1,\ldots,4}$)

$$\Delta(\partial T/\partial\dot{X})\,\delta X + \Delta(\partial T/\partial\dot{Y})\,\delta Y + \Delta(\partial T/\partial\dot{Z})\,\delta Z + \Delta(\partial T/\partial\phi^{\cdot})\,\delta\phi = 0. \qquad \text{(e)}$$

Now, the tightening of the cord is equivalent to the sudden introduction of the constraint $|\boldsymbol{PQ}| = l$ (*constant*), or, in terms of $q_{1,\ldots,4}$,

$$(X - r\cos\phi)^2 + (Y - r\sin\phi)^2 + (Z - vt)^2 = l^2, \qquad \text{(f1)}$$

or, rearranging,

$$(X^2 + Y^2 + Z^2) - 2r(X\cos\phi + Y\sin\phi) - 2Z(vt) + (vt)^2 = l^2 - r^2; \qquad \text{(f2)}$$

and $\delta(\ldots)$-varying this (while recalling to set $\delta t = 0$, and $\delta v = 0$), we obtain the *virtual* form of the constraint:

$$X\,\delta X + Y\,\delta Y + Z\,\delta Z$$
$$- r(\cos\phi\,\delta X + \sin\phi\,\delta Y) - r(-X\sin\phi + Y\cos\phi)\,\delta\phi - (vt)\,\delta Z = 0. \qquad \text{(f3)}$$

Therefore, *at the impact moment* (i.e., $t = 0$, $\phi = 0$), the $\delta q_{1,\ldots,4}$ satisfy

$$(X - r)\,\delta X + Y\,\delta Y + Z\,\delta Z - rY\,\delta\phi = 0. \tag{f4}$$

(ii) Show that the variational equation (e), under (f4), produces [with the help of the multiplier $\hat{\lambda}$ (proportional to the impulsive cord reaction)] the following *four* equations:

$$M\,\Delta\dot{X} = \hat{\lambda}(X - r), \qquad M\,\Delta\dot{Y} = \hat{\lambda}Y, \qquad M\,\Delta\dot{Z} = \hat{\lambda}Z,$$
$$(mr^2)\,\Delta\dot{\phi} = -\hat{\lambda}\,rY; \tag{g}$$

and these, along with the $(\ldots)^{\cdot}$-form of the constraint (f1, 2) (evaluated at $t = 0$, $\phi = 0$; and, for convenience, without the *superscript pluses* in the postimpact $\dot{q}_{1,\ldots,4}$)

$$(X - r\cos\phi)\dot{X} + (Y - r\sin\phi)\dot{Y} + (Z - vt)\dot{Z}$$
$$- r(-X\sin\phi + Y\cos\phi)\dot{\phi} - vZ + v^2 t = 0, \tag{h1}$$
$$\Rightarrow (X - r)\dot{X} + (Y)\dot{Y} + (Z)\dot{Z} - (rY)\dot{\phi} - vZ = 0, \tag{h2}$$

yield a system of *five* equations for $\dot{X}$, $\dot{Y}$, $\dot{Z}$, $\dot{\phi}$; $\hat{\lambda}$.

For example, if at $t = 0$, Q is on L (i.e., X, $Y = 0$), verify that eqs. (g) and (h2) reduce, respectively, to

$$\dot{X} = (\dot{X})^- - \hat{\lambda}(r/M), \quad \dot{Y} = (\dot{Y})^-, \quad \dot{Z} = (\dot{Z})^- + \hat{\lambda}(Z/M), \quad \dot{\phi} = (\dot{\phi})^-, \tag{i1}$$
$$(-r)\dot{X} + (Z)\dot{Z} - vZ = 0; \tag{i2}$$

and, upon elimination of $\hat{\lambda}$, yield the postimpact velocities

$$\dot{X} = Z\{Z(\dot{X})^- + r[(\dot{Z})^- - v]\}/(r^2 + Z^2), \qquad \dot{Y} = (\dot{Y})^-,$$
$$\dot{Z} = \{r^2(\dot{Z})^- + Z[r(\dot{X})^- + vZ]\}/(r^2 + Z^2), \qquad \dot{\phi} = (\dot{\phi})^-; \tag{i3}$$

and when these results are substituted back into the *first* or *third* of (i1), they supply $\hat{\lambda}$. The details are left to the reader.

Problem 4.4.6 Consider a regular hexagon consisting of six identical and homogeneous bars *ABCDEF* (fig. 4.7), each of mass m and length $2b$, smoothly joined at their mutual hinges *A, B, C, D, E, F,* and originally resting on a smooth horizontal table. The system is struck by an impulse $\hat{I}$, normal to AF at its midpoint, which communicates to it a postimpact velocity $\dot{x}$. Show that the postimpact velocity of the opposite bar CD, $\dot{y}$, equals $\dot{x}/10$.

HINTS

In this configuration $\phi = \pi/6 = 30°$, and, therefore,

$$2T = m[6(\dot{x})^2 - 12b\,\dot{x}\,\dot{\phi} + (40/3)b^2(\dot{\phi})^2], \quad \widehat{\delta' W} = \hat{I}\,\delta x \Rightarrow \partial T/\partial\dot{\phi} = 0:$$
$$\dot{x} = \cdots = (20/9)b\,\dot{\phi};$$

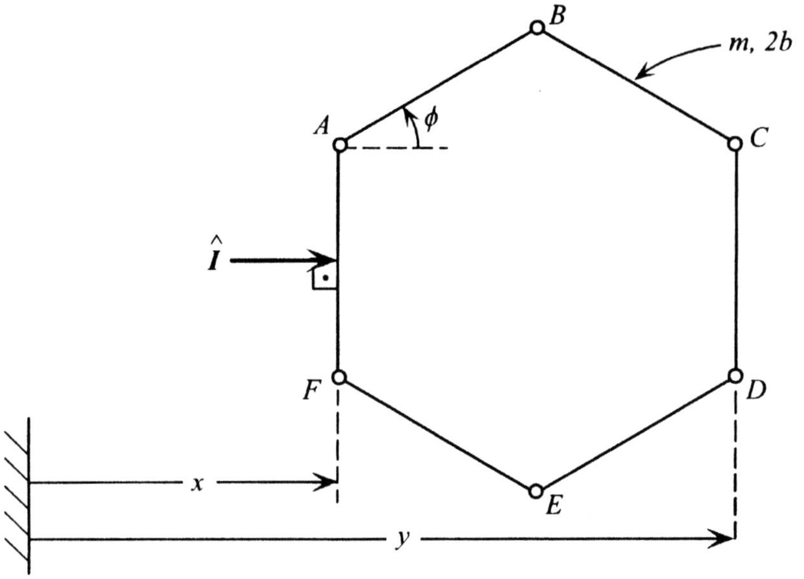

Figure 4.7 Smoothly hinged regular hexagon under a normal impulse $\hat{I}$ at the midpoint of its bar *FA*.

and so

$$\dot{y} = [x + 2(2b\cos\phi)]^{\cdot} = (\dot{x} - 4b\sin\phi\,\dot{\phi})\big|_{\phi=\pi/6} = \cdots = (2/9)b\,\dot{\phi}.$$

Example 4.4.4 Consider a circular homogeneous disk D, of mass m and radius r, moving in the vertical plane $O\text{–}xy$. At a certain instant, D strikes the fixed axis (ground) Ox and is ready to begin rolling on it (fig. 4.8). Let us determine the postimpact velocities.

Before the shock, the system positions depend on the following three parameters: (x, y): coordinates of D's center/center of mass G, angle of rotation ϕ (from Oy toward Ox – negative sense). At the shock moment, the following two (obviously) persistent constraints are introduced:

(i) $y = r$ (*contact* of D with axis Ox), (a)
(ii) $x = r\phi$ (*rolling* of D on Ox; with proper origin choice); (b)

or, in terms of the new, convenient equilibrium coordinates

$$q_1 \equiv y - r, \qquad q_2 \equiv x - r\phi, \qquad q_3 \equiv x, \tag{c}$$

simply

$$q_1 = 0, \qquad q_2 = 0 \qquad (\text{while } q_3 \neq 0). \tag{d}$$

Next, by König's theorem, the (unconstrained) kinetic energy of D (doubled; with $mk^2 \equiv I =$ moment of inertia of D about its mass center G) is

$$\begin{aligned} 2T &= m\big[(\dot{x})^2 + (\dot{y})^2\big] + I(\dot{\phi})^2 = m\big[(\dot{x})^2 + (\dot{y})^2 + k^2(\dot{\phi})^2\big] \\ &= m\big[(\dot{q}_1)^2 + (\dot{q}_3)^2 + (k/r)^2(\dot{q}_3 - \dot{q}_2)^2\big]. \end{aligned} \tag{e}$$

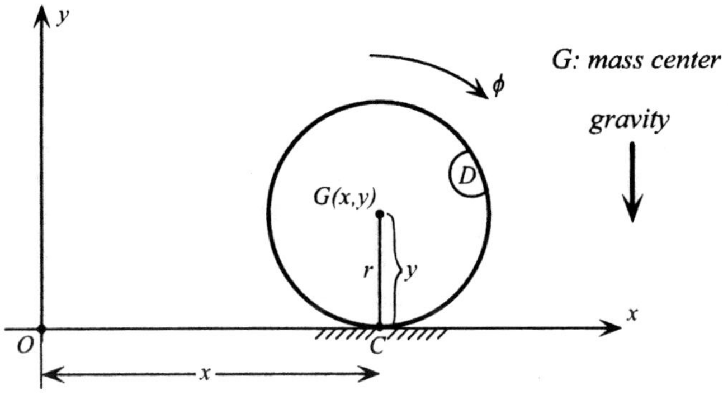

Figure 4.8 Impact of circular disk on horizontal fixed axis.

Now, since q_3 is the only unconstrained coordinate, *Appell's rule,* (4.4.8), yields

$$(\partial T/\partial \dot{q}_3)^+ = (\partial T/\partial \dot{q}_3)^-: \qquad \Delta\dot{q}_3 + (k/r)^2(\Delta\dot{q}_3 - \Delta\dot{q}_2) = 0; \tag{f}$$

from which, since $q_2 = 0 \Rightarrow (\dot{q}_2)^+ = 0$ (second-type constraint) $\Rightarrow (\dot{x})^+ = r(\dot{\phi})^+$, we obtain, successively,

$$[(\dot{x})^+ - (\dot{x})^-] + (k/r)^2\{[(\dot{x})^+ - (\dot{x})^-] - [[(\dot{x})^+ - r(\dot{\phi})^+] - [(\dot{x})^- - r(\dot{\phi})^-]]\} = 0,$$

or, simplifying,

$$[(\dot{x})^+ - (\dot{x})^-] + (k/r)^2[(\dot{x})^+ - r(\dot{\phi})^-] = 0$$
$$\Rightarrow (\dot{x})^+ = [r^2(\dot{x})^- + k^2 r(\dot{\phi})^-]/(r^2 + k^2); \tag{g}$$

an equation that expresses conservation of angular momentum about the contact point C.

The above shows that if $r(\dot{x})^- + k^2(\dot{\phi})^- = 0$, then $(\dot{x})^+ = 0$; that is, the sudden impact stops the disk! It is left to the reader to obtain the impulsive equations for q_1 and q_2, and thus calculate the impulsive multipliers corresponding to the two constraints (d).

Problem 4.4.7 Continuing from the preceding example, show that under the coordinate choice $q_1 = x$, $q_2 = y$, $q_3 = \phi$, the impulsive Routh–Voss equations, (4.4.11b), yield

$$\Delta(m\dot{x}) = \hat{\lambda}_2, \qquad \Delta(m\dot{y}) = \hat{\lambda}_1, \qquad \Delta(I\dot{\phi}) = \hat{\lambda}_2(-r), \tag{a}$$

where $\hat{\lambda}_1$ and $\hat{\lambda}_2$ are impulsive multipliers corresponding to the two constraints (d) of that example, but, here, in these new q's; that is, $q_2 - r = 0$ and $q_1 - rq_3 = 0$.

Then, solving this new system, show that $\hat{\lambda}_1 = -m(\dot{y})^-$, and $\hat{\lambda}_2 = m[(\dot{x})^+ - (\dot{x})^-] = -(mk^2/r)[(\dot{\phi})^+ - (\dot{\phi})^-]$; and, further, by eliminating $(\dot{\phi})^+$, obtain (g) of the preceding example.

Problem 4.4.8 *Central (or Direct) Collision of Two Spheres.* Consider two homogeneous spheres, S_1 and S_2, both translating along the fixed axis Ox. Let their respective *centers of mass/masses/radii/center of mass coordinates be* $G_1/m_1/r_1/x_1$ and $G_2/m_2/r_2/x_2$.

(i) Show that if S_1 and S_2 collide, the sole equation furnished by the Appellian theory is

$$\Delta(\partial T/\partial \dot{q}_1) = 0: \qquad m_1 \Delta \dot{q}_1 + m_2(\Delta \dot{q}_1 + \Delta \dot{q}_2) = 0, \tag{a}$$

where $q_1 \equiv x_1$ and $q_2 \equiv x_2 - x_1 - (r_1 + r_2)$ $(= 0$, constraint introduced at the shock moment); and this expresses the conservation of *total linear momentum* along Ox.

(ii) Show that if we assume that $(q_2)_{\text{after shock}} = 0$ (plastic spheres) $\Rightarrow (\dot{q}_2)^+ = 0$ (second-type constraint; i.e., plastic impact) then

$$(\dot{x}_1)^+ = \left[m_1(\dot{x}_1)^- + m_2(\dot{x}_2)^-\right]/(m_1 + m_2).$$

Problem 4.4.9 *Ballistic Pendulum.* Consider a projectile (e.g., bullet) B of mass m (fig. 4.9), in rectilinear translation in a vertical plane, and a physical (ballistic) pendulum P, of mass m and center of mass G, capable of rotating about a fixed axis perpendicular to that plane through an origin O. Let the polar coordinates of B be $(r = OB,\ \phi = angle(O\text{–}vertical, OB))$ and $\theta = angle(O\text{–}vertical, OG)$; that is, before the shock, the system positions depend on the three parameters r, ϕ, θ. At a certain instant, B strikes P and becomes embedded into it, and from then on both move as one body [i.e., plastic impact: $r \to constant \equiv r_o$ and $\phi \to \theta$ $(\pm\ constant)$ assume zero]. Show that, if $(\dot{\theta})^- = 0$ and I is the moment of inertia of P about O, then

$$(\dot{\theta})^+ = \left[m{r_o}^2/(I + m{r_o}^2)\right](\dot{\phi})^- \tag{a}$$

$[r_o(\dot{\phi})^-$: (known) component of velocity of B along perpendicular to OB, at shock moment].

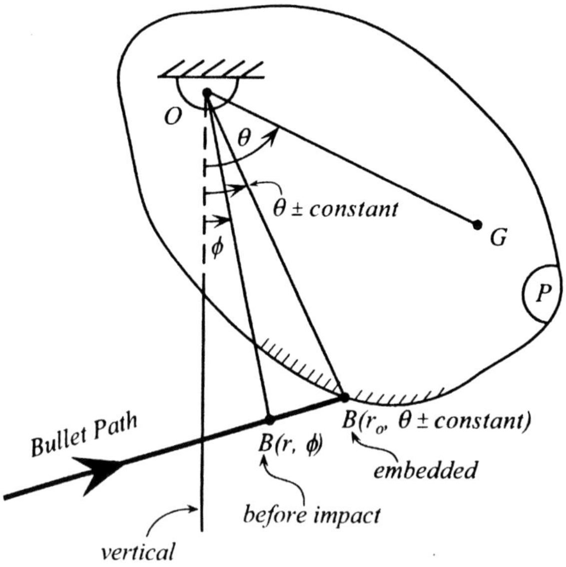

Figure 4.9 Impact in ballistic pendulum.

HINT

Introduce the new equilibrium parameters: $q_1 \equiv r - r_o$, $q_2 \equiv \theta - \phi$, $q_3 \equiv \theta$. Then the suddenly introduced persistent constraints are simply $q_1 = 0$, $q_2 = 0 \Rightarrow (\dot{q}_1)^+ = 0$, $(\dot{q}_2)^+ = 0$ (second type); also $(\dot{q}_3)^- \equiv (\dot{\theta})^- = 0$; and by Appell's rule, for q_3, $\Delta(\partial T/\partial \dot{q}_3) = 0$.

Example 4.4.5 *Impact of a Rigid Body B on a Fixed Obstacle, at a Common Contact Point C.*

Taking axes $C\text{–}xyz$, where Cz is along the common normal of the impacting surfaces, and, say, toward B, and with

(u, v, w): components of velocity of contact point of B at $C, \boldsymbol{v}_C$; (a1)

$(\omega_x, \omega_y, \omega_z)$: components of angular velocity of $B, \boldsymbol{\omega}$; (a2)

(x, y, z): coordinates of mass center of B, G, (a3)

and since

$$\boldsymbol{v}_G = \boldsymbol{v}_C + \boldsymbol{\omega} \times \boldsymbol{r}_{G/C}, \tag{b}$$

applying König's theorem, we find (with I_{kl}: moments/products of inertia of B at G)

$$\begin{aligned} 2T &= m\big[(u + \omega_y z - \omega_z y)^2 + (v + \omega_z x - \omega_x z)^2 + (w + \omega_x y - \omega_y x)^2\big] \\ &\quad + (I_x \omega_x{}^2 + I_y \omega_y{}^2 + I_z \omega_z{}^2 + 2I_{xy}\,\omega_x\omega_y + 2I_{xz}\,\omega_x\omega_z + 2I_{yz}\,\omega_y\omega_z) \\ &= m(u^2 + v^2 + w^2) + 2m\big[\omega_x(yw - zv) + \omega_y(zu - xw) + \omega_z(xv - yu)\big] \\ &\quad + (I_x \omega_x{}^2 + I_y \omega_y{}^2 + I_z \omega_z{}^2 + 2I_{xy}\,\omega_x\omega_y + 2I_{xz}\,\omega_x\omega_z + 2I_{yz}\,\omega_y\omega_z). \end{aligned} \tag{c}$$

We also have the *constitutive equation*

$$w^+ = -e\,w^- \Rightarrow \Delta w = -(1 + e)w^- \qquad (> 0, \text{ since } w^- < 0). \tag{d}$$

Now, we consider the following *two* cases:

(i) *Smooth obstacle*: Then the sole constraint is $w = 0$ (during the impact), and so only the w-equation of motion contains a *multiplier*. The postimpact state will be determined from (d) and the remaining (kinetic) equations

$$\Delta(\partial T/\partial u) = 0, \qquad \Delta(\partial T/\partial v) = 0; \tag{e1}$$

$$\Delta(\partial T/\partial \omega_x) = 0, \qquad \Delta(\partial T/\partial \omega_y) = 0, \qquad \Delta(\partial T/\partial \omega_z) = 0. \tag{e2}$$

(ii) *Rough surfaces*: In this case, the constraints are $u = 0, v = 0, w = 0$ (during the impact), and, therefore [assuming that a (d)-like relation still holds],

$$u^+ = 0 \Rightarrow \Delta u = -u^-, \qquad v^+ = 0 \Rightarrow \Delta v = -v^-, \qquad \Delta w = -(1 + e)u^-. \tag{f}$$

The values $\Delta\omega_{x,y,z}$ will be determined from the three (kinetic) equations

$$\Delta(\partial T/\partial\omega_x) = 0: \quad I_x\,\Delta\omega_x + I_{xy}\,\Delta\omega_y + I_{xz}\,\Delta\omega_z + m(y\,\Delta w - z\,\Delta v) = 0, \quad \text{(g1)}$$

$$\Delta(\partial T/\partial\omega_y) = 0: \quad I_{yx}\,\Delta\omega_x + I_y\,\Delta\omega_y + I_{yz}\,\Delta\omega_z + m(z\,\Delta u - x\,\Delta w) = 0, \quad \text{(g2)}$$

$$\Delta(\partial T/\partial\omega_z) = 0: \quad I_{zx}\,\Delta\omega_x + I_{zy}\,\Delta\omega_y + I_z\,\Delta\omega_z + m(x\,\Delta v - y\,\Delta u) = 0. \quad \text{(g3)}$$

If the axes G–xyz are *principal*, the above simplify somewhat; that is, the products of inertia vanish.

Problem 4.4.10 A heavy rigid body of revolution (axis Oz) moves in space about the *fixed point* O. At time t_o its nutation angle θ equals θ_o, and its Eulerian angle rates are $\dot\phi_o, \dot\theta_o, \dot\psi_o$. At that moment, the axis Oz hits a fixed obstacle, thus introducing the *nonpersistent* constraint ϕ (precession) = *constant*. Show that

$$\Delta\dot\theta = 0 \;\Rightarrow\; (\dot\theta)^+ = (\dot\theta)^-, \quad \text{(a)}$$

$$\Delta(\dot\psi + \dot\phi\cos\theta) = 0 \;\Rightarrow\; (\dot\psi)^+ = (\dot\psi)^- + (1+e)(\dot\phi)^-\cos\theta_o, \quad \text{(b)}$$

$$\Delta\big[(A\sin^2\theta)\dot\phi + C(\dot\psi + \dot\phi\cos\theta)\cos\theta\big] = R_\phi. \quad \text{(c)}$$

HINTS

Here (recalling §1.15 ff.):

(i)
$$2T = A\big[(\dot\theta)^2 + (\dot\phi)^2\sin^2\theta\big] + C(\dot\psi + \dot\phi\cos\theta)^2, \quad \text{(d)}$$

where

A/C: *transverse/axial* moments of inertia of body at O; (e)

(ii) $\hat{Q}_{\phi,\theta,\psi} = 0$; (f)

(iii) $(\dot\phi)^+ = -e(\dot\phi)^-$; (g)

and, therefore,

(iv)
$$\Delta(\partial T/\partial\dot\theta) = 0,$$

$$\Delta(\partial T/\partial\dot\psi) = 0 \;\Rightarrow\; \Delta\dot\psi = -\Delta(\dot\phi\cos\theta) \;\Rightarrow\; (\dot\psi)^+ = (\dot\psi)^- + \cdots,$$

$$\Delta(\partial T/\partial\dot\phi) = R_\phi \text{ (impulsive multiplier)}. \quad \text{(h)}$$

Example 4.4.6 A homogeneous sphere S, of center and center of mass G, mass M and radius R, rests on a rough fixed horizontal plane p. Then, a given impulse $\hat{\boldsymbol{F}}$ is applied at a specified S-point A. Let us find its postimpact velocities.

Relative to *space-fixed* axes O–xyz (coinciding with *sphere-fixed* axes G–$\xi\eta\zeta$ at the impact instant; and such that O–xy: parallel to plane p, Oz = perpendicular to p, positive upwards), let

$$\text{Coordinates of } G\text{: } (\xi, \eta, \zeta)_{\text{at impact instant}} = (0, 0, 0), \tag{a1}$$

$$\text{Coordinates of } A\text{: } (x, y, z), \tag{a2}$$

$$\text{Coordinates of contact point } C\text{: } (x_C, y_C, z_C)_{\text{at impact instant}} = (0, 0, -R), \tag{a3}$$

$$\text{Components of angular velocity of } S\text{: } (\omega_{x,y,z}), \tag{a4}$$

$$\text{Components of } \hat{\boldsymbol{F}}\text{: } (X, Y, Z). \tag{a5}$$

[Instead of the $(\omega_{x,y,z})$ we could have chosen the rates of the $3 \to 1 \to 3$ Eulerian angles of $G\text{–}\xi\eta\zeta$ relative to $O\text{–}xyz$: $(\dot{\phi}, \dot{\theta}, \dot{\psi})$; even though $(\phi, \theta, \psi)_{\text{impact instant}} = (0, 0, 0)$.] Then, by König's theorem (and with the constraint $\dot{\zeta} = v_{G,z} = 0$ enforced in it),

$$2T = M[(\dot{\xi})^2 + (\dot{\eta})^2] + Mk^2({\omega_x}^2 + {\omega_y}^2 + {\omega_z}^2); \tag{b}$$

where $k^2 \equiv 2R^2/5$; also, and since $\boldsymbol{v}_A = \boldsymbol{v}_G + \boldsymbol{\omega} \times \boldsymbol{r}_{A/G} \Rightarrow \delta\boldsymbol{r}_A = \delta\boldsymbol{r}_G + \delta\boldsymbol{\chi} \times \boldsymbol{r}_{A/G}$, or, in components,

$$\dot{x} = \dot{\xi} + z\omega_y - y\omega_z, \qquad \dot{y} = \dot{\eta} + x\omega_z - z\omega_x, \qquad \dot{z} = \dot{\zeta} + y\omega_x - x\omega_y, \tag{c}$$

the percussive virtual work is (with $\boldsymbol{OA} \equiv \boldsymbol{r}_A$)

$$\begin{aligned}
\widehat{\delta' W} &= \hat{\boldsymbol{F}} \cdot \delta\boldsymbol{r}_A = X\,\delta x + Y\,\delta y + Z\,\delta z \\
&= X(\delta\xi + z\,\delta\theta_y - y\,\delta\theta_z) + Y(\delta\eta + x\,\delta\theta_z - z\,\delta\theta_x) + Z(0 + y\,\delta\theta_x - x\,\delta\theta_y) \\
&= (X)\,\delta\xi + (Y)\,\delta\eta + (Z)0 \\
&\qquad + (yZ - zY)\,\delta\theta_x + (zX - xZ)\,\delta\theta_y + (xY - yX)\,\delta\theta_z,
\end{aligned} \tag{d}$$

where $\omega_{x,y,z} \equiv \dot{\theta}_{x,y,z}$; and, since at the end of the impact

$$\boldsymbol{v}_{\text{contact of } S \text{ with } p} \equiv \boldsymbol{v}_C = \boldsymbol{v}_G + \boldsymbol{\omega} \times \boldsymbol{r}_{C/G} = \boldsymbol{0},$$

or, in components,

$$\begin{aligned}
(\dot{x}_C, \dot{y}_C, \dot{z}_C)_{\text{at impact instant}} &= (\dot{\xi}, \dot{\eta}, \dot{\zeta}) + (\omega_x, \omega_y, \omega_z) \times (0, 0, -R) = (0, 0, 0): \\
&\Rightarrow \dot{\xi} - R\,\omega_y = 0, \qquad \dot{\eta} + R\,\omega_x = 0, \\
\text{and} \quad &\dot{\zeta} = 0 \Rightarrow \zeta = \textit{constant} = 0 \text{ (as expected)},
\end{aligned} \tag{e}$$

we will have (with $\hat{\lambda}$ and $\hat{\mu}$ as impulsive multipliers along x_C and y_C, respectively)

$$\widehat{\delta' W_R} = \hat{\lambda}\,\delta x_C + \hat{\mu}\,\delta y_C = \hat{\lambda}(\delta\xi - R\,\delta\theta_y) + \hat{\mu}(\delta\eta + R\,\delta\theta_x) = 0. \tag{f}$$

Utilizing the above in the impulsive principle of Lagrange {plus method of multipliers (i.e., adjoining δ [first/second of eqs. (e)] to it via $\hat{\lambda}, \hat{\mu}$); or, equivalently, and in the spirit of impulsive relaxation, adding (f) to $\widehat{\delta' W}$}, we readily obtain the following *five* equations of impulsive motion:

$$\Delta(\partial T/\partial\dot{\xi}) = X + \hat{\lambda}(1) + \hat{\mu}(0): \qquad \Delta(M\dot{\xi}) = X + \hat{\lambda}, \tag{g1}$$

$$\Delta(\partial T/\partial\dot{\eta}) = Y + \hat{\lambda}(0) + \hat{\mu}(1): \qquad \Delta(M\dot{\eta}) = Y + \hat{\mu}; \tag{g2}$$

$$\Delta(\partial T/\partial\omega_x) = yZ - zY + \hat{\lambda}(0) + \hat{\mu}(R): \quad \Delta(Mk^2\omega_x) = yZ - zY + \hat{\mu}R, \tag{g3}$$

$$\Delta(\partial T/\partial\omega_y) = zX - xZ + \hat{\lambda}(-R) + \hat{\mu}(0): \quad \Delta(Mk^2\omega_y) = zX - xZ - \hat{\lambda}R, \tag{g4}$$

$$\Delta(\partial T/\partial\omega_z) = xY - yX + \hat{\lambda}(0) + \hat{\mu}(0): \quad \Delta(Mk^2\omega_z) = xY - yX; \tag{g5}$$

which, along with the *two* constraints (e) constitute a determinate system of *seven* algebraic equations for $\Delta\dot{\xi}, \Delta\dot{\eta}, \Delta\omega_{x,y,z}, \hat{\lambda}, \hat{\mu}$. Finally, application of the Newton–Euler impulsive linear momentum theorem in the vertical direction yields the vertical impulsive reaction at C, if needed [instead of using relaxation in T, eq. (b), and an extra multiplier].

An Introduction to Kinetic Impulsive Equations

To obtain multiplierless (i.e., kinetic impulsive) equations we may proceed as follows:

(i) Eliminate $\hat{\lambda}$ and $\hat{\mu}$ from (g3–5), with the help of (g1, 2); that is,

$$\hat{\lambda} = M\Delta\dot{\xi} - X, \qquad \hat{\mu} = M\Delta\dot{\eta} - Y, \tag{h}$$

thus obtaining the following *three kinetic impulsive Maggi* equations:

$$-(MR)\,\Delta\dot{\eta} + (Mk^2)\,\Delta\omega_x = yZ - (z + R)Y, \tag{i1}$$

$$(MR)\,\Delta\dot{\xi} + (Mk^2)\,\Delta\omega_y = -xZ + (z + R)X, \tag{i2}$$

$$(Mk^2)\,\Delta\omega_z = xY - yX \qquad \text{(unchanged)}; \tag{i3}$$

which, along with the *two* constraints (e), constitute a determinate system of *five* algebraic equations for $\Delta\dot{\xi}, \Delta\dot{\eta}, \Delta\omega_{x,y,z}$; then $\hat{\lambda}, \hat{\mu}$ follow immediately from (h). [Equations (i1–3) also result by applying the principle of impulsive angular momentum about C.]

Further, using (e) to eliminate, say $\Delta\dot{\xi}$ and $\Delta\dot{\eta}$, from (i1–3) would result in the *three kinetic impulsive Chaplygin–Voronets equations* in $\Delta\omega_{x,y,z}$ (see §4.5).

(ii) Or, introduce the following "equilibrium" (quasi) velocities:

$$\omega_1 \equiv \dot{\theta}_1 \equiv \dot{\xi} - R\omega_y\ (= 0) \Rightarrow \dot{\xi} = \dot{\theta}_1 + R\omega_y, \qquad \delta\xi = \delta\theta_1 + R\,\delta\theta_y, \tag{j1}$$

$$\omega_2 \equiv \dot{\theta}_2 \equiv \dot{\eta} + R\omega_x\ (= 0) \Rightarrow \dot{\eta} = \dot{\theta}_2 - R\omega_x, \qquad \delta\eta = \delta\theta_2 - R\,\delta\theta_x; \tag{j2}$$

in terms of which the expressions (b), (d), and (f) become, respectively,

$$2T \to 2T^* = M\big[(\omega_1 + R\,\omega_y)^2 + (\omega_2 - R\,\omega_x)^2\big] + (Mk^2)(\omega_x{}^2 + \omega_y{}^2 + \omega_z{}^2); \tag{k1}$$

$$\widehat{\delta' W} \to \widehat{(\delta' W)^*} = \cdots = X\,\delta\theta_1 + Y\,\delta\theta_2 + \big[Zy - (R+z)Y\big]\,\delta\theta_x + \big[X(R+z) - Zx\big]\,\delta\theta_y + \big[Yx - Xy\big]\,\delta\theta_z$$
$$\equiv \hat{\Theta}_1\,\delta\theta_1 + \hat{\Theta}_2\,\delta\theta_2 + \hat{\Theta}_x\,\delta\theta_x + \hat{\Theta}_y\,\delta\theta_y + \hat{\Theta}_z\,\delta\theta_z, \tag{k2}$$

$$\widehat{\delta' W_R} \to \widehat{(\delta' W_R)^*} = \hat{\lambda}\,\delta\theta_1 + \hat{\mu}\,\delta\theta_2 \equiv \hat{\Lambda}_1\,\delta\theta_1 + \hat{\Lambda}_2\,\delta\theta_2 = 0; \tag{k3}$$

and then, using the above in the impulsive principle of Lagrange, obtain the following *five Hamel equations of impulsive motion*:

Kinetostatic: $\quad \Delta(\partial T^*/\partial\omega_k) = \hat{\Theta}_k + \hat{\Lambda}_k \quad (k = 1, 2),$ (l1)

Kinetic: $\quad \Delta(\partial T^*/\partial\omega_k) = \hat{\Theta}_k \quad (k = x, y, z).$ (l2)

The details are left to the reader; and the entire process of elimination of impulsive multipliers is treated in full generality in §4.5.

4.5 IMPULSIVE MOTION VIA QUASI VARIABLES

Here the previous results are extended to nonholonomic "coordinates" and velocities, and in the process show that, contrary to the finite motion case (§3.5), *the Lagrangean impulsive equations retain the same form in both holonomic and nonholonomic variables.*

Let us assume, with no loss of generality, that our system is subjected to m Pfaffian (holonomic and/or nonholonomic) constraints:

$$\sum a_{Dk}\,dq_k + a_D\,dt = 0 \quad \text{(kinematically admissible form)}, \tag{4.5.1a}$$

$$\sum a_{Dk}\,\delta q_k = 0 \quad \text{(virtual form)}\ [D = 1, \ldots, m(< n);\ k = 1, \ldots, n]. \tag{4.5.1b}$$

Introducing the n quasi coordinates θ (as detailed in §2.9 ff.) via

$$d\theta_D \equiv \sum a_{Dk}\,dq_k + a_D\,dt \quad (= 0), \qquad d\theta_I \equiv \sum a_{Ik}\,dq_k + a_I\,dt \quad (\neq 0), \tag{4.5.2a}$$

$$\delta\theta_D \equiv \sum a_{Dk}\,\delta q_k \quad (= 0), \qquad \delta\theta_I \equiv \sum a_{Ik}\,\delta q_k \quad (\neq 0); \tag{4.5.2b}$$

$$d\theta_{n+1} \equiv dq_{n+1} \equiv dt; \qquad \delta\theta_{n+1} \equiv \delta q_{n+1} \equiv \delta t = 0 \qquad (I = m+1, \ldots, n); \tag{4.5.2c}$$

and their n quasi velocities ω via

$$\omega_D \equiv d\theta_D/dt \quad (= 0), \quad \omega_I \equiv d\theta_I/dt \quad (\neq 0), \quad \omega_{n+1} \equiv d\theta_{n+1}/dt = \dot{t} = 1, \tag{4.5.2d}$$

we can write for the virtual displacement of a typical particle:

$$\delta\mathbf{r} = \sum \mathbf{e}_k\,\delta q_k \equiv \sum \boldsymbol{\varepsilon}_I\,\delta\theta_I. \tag{4.5.3}$$

The Impulsive Hamel Equations

Substituting the second of (4.5.3) into the LIP, eqs. (4.3.3b–4b), and since the $n - m$ $\delta\theta_I$ are unconstrained, and $\Delta(\ldots)$ and $\widehat{(\ldots)}$ commute with $S(\ldots)$ [assuming, of course, that $\hat{\boldsymbol{\varepsilon}}_I = \Delta\boldsymbol{\varepsilon}_I = \mathbf{0}$], we easily obtain, respectively,

$$\hat{\Lambda}_I \equiv S\, \widehat{d\boldsymbol{R}} \cdot \boldsymbol{\varepsilon}_I \equiv (I)\text{th } \textit{nonholonomic} \text{ component of system } \textit{constraint reaction}$$

$$= 0, \tag{4.5.4a}$$

and the $n - m$ *nonholonomic kinetic impulsive equations*:

$$S\, \Delta(dm\,\boldsymbol{v}) \cdot \boldsymbol{\varepsilon}_I = \Delta\Big(S\, dm\,\boldsymbol{v} \cdot \boldsymbol{\varepsilon}_I\Big) = S\, \widehat{d\boldsymbol{F}} \cdot \boldsymbol{\varepsilon}_I; \tag{4.5.4b}$$

or, in *system* variables, and with $P_I \equiv S\, dm\,\boldsymbol{v} \cdot \boldsymbol{\varepsilon}_I = P_I(t, q, \omega) = \partial T^*/\partial\omega_I = (I)$th *nonholonomic component of system momentum*,

$$\Delta P_I = \hat{\Theta}_I \qquad (I = m+1, \ldots, n). \tag{4.5.5a}$$

It is not hard to show (e.g., invoking the relaxation principle/Lagrangean multipliers; in a completely analogous way with the finite motion case—recall §3.5), that the corresponding m *impulsive nonholonomic kinetostatic equations* are

$$\Delta P_D = \hat{\Theta}_D + \hat{\Lambda}_D, \qquad \hat{\Lambda}_D \equiv \hat{\lambda}_D \quad (\neq 0) \qquad (D = 1, \ldots, m). \tag{4.5.5b}$$

These *uncoupled* algebraic equations are the *impulsive counterparts of Hamel's equations* (§3.3 ff.).

The Impulsive Maggi Equations

Multiplying the impulsive Routh–Voss equations corresponding to (4.5.1a, b),

$$\Delta p_k = \hat{Q}_k + \sum \hat{\lambda}_D a_{Dk} \qquad (D = 1, \ldots, m), \tag{4.5.6}$$

with $A_{kI} = A_{kI}(q, t)$, where (A_{kI}) is the *inverse* of the (augmented) $n \times n$ matrix (a_{kI}), as in chapters 2 and 3, and summing over k from 1 to n, we find, successively,

$$\begin{aligned}\Delta\Big(\sum A_{kI} p_k\Big) &= \sum A_{kI}\hat{Q}_k + \sum\Big[A_{kI}\Big(\sum \hat{\lambda}_D a_{Dk}\Big)\Big]\\ &= \sum A_{kI}\hat{Q}_k + \sum\Big[\hat{\lambda}_D\Big(\sum a_{Dk} A_{kI}\Big)\Big]\\ &= \sum A_{kI}\hat{Q}_k + \sum \hat{\lambda}_D\, \delta_{DI},\end{aligned}$$

or, finally, since $\Delta A_{kI} = 0$, the above split into the following two groups:

$$\sum A_{kD}\Delta p_k = \sum A_{kD}\hat{Q}_k + \hat{\lambda}_D \qquad (\equiv \hat{\Theta}_D + \hat{\Lambda}_D) \qquad (D = 1, \ldots, m), \tag{4.5.6a}$$

$$\sum A_{kI}\Delta p_k = \sum A_{kI}\hat{Q}_k \qquad (\equiv \hat{\Theta}_I) \qquad (I = m+1, \ldots, n); \tag{4.5.6b}$$

since $\hat{\lambda}_I \equiv \hat{\Lambda}_I = 0$. Equations (4.5.6a) and (4.5.6b) are, respectively, the *impulsive kinetostatic and kinetic Maggi's equations*.

REMARKS

(i) The kinetic impulsive equations (4.5.5a) can also be obtained by integration of the corresponding kinetic equations of ordinary continuous motion (§3.5) in time, and then taking the impulsive limit $\tau \to 0$, while noting that [as in (4.4.12)]:

$$\widehat{\partial T^*/\partial\theta_I} \equiv \sum A_{kI}\,\widehat{(\partial T^*/\partial q_k)} = 0 \quad \text{and} \quad -\widehat{\Gamma_I} \equiv \widehat{\sum\sum \gamma^k{}_{II'}(\partial T^*/\partial\omega_k)\omega_{I'}} = 0; \tag{4.5.7}$$

and similarly for the kinetostatic equations (4.5.5b). The above allow us to rewrite (4.5.5a) as

$$\Delta(\partial T^*{}_o/\partial\omega_I) = \hat{\Theta}_I, \qquad (I = m+1, \ldots, n), \tag{4.5.8}$$

where, as usual,

$$\begin{aligned} T^*{}_o &\equiv T^*(q;\ \omega_1 = 0, \ldots, \omega_m = 0;\ \omega_{m+1}, \ldots, \omega_n;\ t) \\ &= T^*(q;\ \omega_{m+1}, \ldots, \omega_n;\ t) \equiv T^*{}_o(q, \omega_I, t)\text{: } \textit{constrained } T^*; \end{aligned} \tag{4.5.8a}$$

that is, here, and *contrary to the Hamel equations for ordinary motion* (§3.5), *if no impulsive reactions are sought*, we can enforce the m nonholonomic constraints $\omega_D = 0$ in T^* (and Θ_I) *before* the partial differentiations; and this simplifies the calculations somewhat. [Justification: expanding $T^* = T^*(\omega_I, \omega_D)$ à la Taylor around $\omega_D = 0$, we obtain (with some easily understood calculus notations)

$$\begin{aligned} &T^*(\omega_I, \omega_D) = T^*(\omega_I, 0) + (\partial T^*/\partial\omega_D)_o\omega_D + O_2(\omega_D) \\ &\Rightarrow [\partial T^*(\omega_I, \omega_D)/\partial\omega_I]_o = \partial T^*(\omega_I, 0)/\partial\omega_I + \{\partial/\partial\omega_I[(\partial T^*/\partial\omega_D)_o]\omega_D + O_2(\omega_D)\}_o \\ &\qquad\qquad\qquad\qquad\quad = \partial T^*(\omega_I, 0)/\partial\omega_I + 0, \end{aligned}$$

that is, simply,

$$(\partial T^*/\partial\omega_I)_o = \partial T^*{}_o/\partial\omega_I.]$$

However, for problems with *second*-type constraints, where, clearly [recalling (§4.4.3b)],

$$\omega_D{}^+ = 0 \qquad \text{but} \qquad \omega_D{}^- \neq 0, \tag{4.5.8b}$$

eqs. (4.5.8) do *not* hold: even if no impulsive reactions are sought, still, we must express $T \to T^*$ as function of *all* the ω's, carry out all differentiations, and *then* enforce the m constraints, first of (8b), on the *postimpact* momenta; the *preimpact* momenta will be calculated using the known ω^-. In sum, for *second*-type constraint problems, (4.5.5b/8, 5c) will be replaced, respectively, by

$$(\partial T^*/\partial\omega_I)^+ - (\partial T^*/\partial\omega_I)^- = \hat{\Theta}_I \qquad (I = m+1, \ldots, n), \tag{4.5.9a}$$

$$(\partial T^*/\partial\omega_D)^+ - (\partial T^*/\partial\omega_D)^- = \hat{\Theta}_D + \hat{\Lambda}_D \qquad (D = 1, \ldots, m); \tag{4.5.9b}$$

although for the *independent postimpact momenta* we still have $(\partial T^*/\partial\omega_I)^+ = (\partial T^*{}_o/\partial\omega_I)^-$.

(ii) The *special* independent quasi-velocity choice $\omega_I = \dot{q}_I$, in (4.5.8), produces what might be called *the impulsive Chaplygin–Voronets (kinetic) equations*:

$$\Delta(\partial T_o/\partial \dot{q}_I) = \hat{Q}_{Io}, \tag{4.5.10}$$

where $T \equiv T^*{}_o(q;\, \dot{q}_{m+1}, \ldots, \dot{q}_n;\, t) \equiv T^*{}_o(q;\, \dot{q}_I;\, t) \equiv T_o(q, \dot{q}_I;\, t)$; in which case, $\partial T^*/\partial \omega_I$ becomes $\partial T_o/\partial \dot{q}_I = \partial T/\partial \dot{q}_I + \sum b_{DI}(\partial T/\partial \dot{q}_D)$, a specialization of $P_I = \sum A_{kI} p_k$; and the $(n-m)$ $\hat{Q}_{Io}$ are defined from

$$\widehat{\delta' W} \equiv \sum \hat{Q}_k\, \delta q_k = \sum \Big(\hat{Q}_I + \sum b_{DI} \hat{Q}_D\Big)\, \delta q_I \equiv \sum \hat{Q}_{Io}\, \delta q_I, \tag{4.5.10a}$$

a specialization of $\hat{\Theta}_I = \sum A_{kI} \hat{Q}_k$ [see also (4.5.12b, c)].

Equations (4.5.10) show that Lagrange's impulsive equations in holonomic variables hold *unchanged in form, even for nonholonomically constrained systems*, provided we use, in there, the constrained quantities T_o and $\hat{Q}_{Io}$, instead of their unconstrained (relaxed) counterparts T and $\hat{Q}_k$; and, by comparing them with eqs. (4.5.5a, 8) we conclude that, due to (4.5.7), *the Lagrangean impulsive equations have the same form in both holonomic and nonholonomic variables*. Of course, (4.5.10) can also be obtained by direct application of the impulsive limiting process to the Chaplygin–Voronets equations (§3.8), with invocation of (4.5.7)-like results. [We recall (§3.8) that here too, just like with eqs. (4.5.8), the situation is in sharp contrast to its ordinary motion counterpart; that is, if the special Pfaffian constraints (4.5.1a) are nonholonomic, then $E_I(T_o) \neq Q_{Io}$.] In view of the earlier remark (i), these equations do *not* hold (without appropriate modifications) for *second*-type constraint problems; and, obviously, cannot be used to calculate impulsive reactions.

Historical: equations (4.5.10) seem to be due to Beghin and Rousseau (1903), who obtained them using the impulsive counterpart of the method of their teacher P. Appell (1899, 1900); that is, independently of any Chaplygin–Voronets equation considerations. In the past, they have been used by various authors [e.g., Beer (1963)], but without the proper theoretical justification given here, or in the Beghin/Rousseau paper.

(iii) Due to the vectorial transformations $P_I = \sum A_{kI} p_k$, and $\hat{\Theta}_I = \sum A_{kI} \hat{Q}_k$ (via chain rule), the impulsive Maggi equations (4.5.6a, b) are identical to the impulsive Hamel equations (4.5.5c, b), respectively; but the former are in *holonomic* variables while the latter are in *nonholonomic* variables. Also, Maggi's equations can result directly from the LIP, eqs. (second of 4.3.7), by inserting in it the inverse of (4.5.2a–c):

$$\delta q_k = \sum A_{kD}(1 \cdot \delta\theta_D) + \sum A_{kI}\, \delta\theta_I \quad \Big(= \sum A_{kI}\, \delta\theta_I\Big); \tag{4.5.11}$$

the details can be easily carried out by the reader.

(iv) In case the constraints (4.5.1a, b) have the special form (recalling results from §2.11)

$$dq_D = \sum b_{DI}\, dq_I + b_D\, dt, \qquad \delta q_D = \sum b_{DI}\, \delta q_I, \tag{4.5.12a}$$

the Maggi equations (4.5.6a, b) specialize, respectively, to

$$\Delta p_D = \hat{Q}_D + \hat{\lambda}_D \qquad \text{and} \qquad \Delta p_I = \hat{Q}_I - \sum b_{DI} \hat{\lambda}_D; \tag{4.5.12b}$$

and by eliminating the m λ's between these two sets of equations, we obtain the $n-m$ *impulsive kinetic Hadamard equations*

$$\Delta p_I + \sum b_{DI}\Delta p_D = \hat{Q}_I + \sum b_{DI}\hat{Q}_D \equiv \hat{Q}_{Io}; \tag{4.5.12c}$$

which, along with the n constraints (first of 4.5.12a) (*evaluated at the postimpact instant*—assuming, of course, that they hold there), constitute a determinate set of $(n-m)+m=n$ equations for the n $(\dot{q})^+$; the m $\hat{\lambda}_D$ can then be found from (first of 4.5.12b). [We notice the similarity between the first of (4.5.12b) and the earlier equations (4.4.7a).]

With the notation $\hat{M}_k \equiv \Delta p_k - \hat{Q}_k$, eqs. (4.5.12b, c) can be rewritten, respectively, as

$$\textit{Kinetostatic:}\quad \hat{M}_D = \hat{\lambda}_D \qquad \text{and} \qquad \textit{Kinetic:}\quad \hat{M}_I + \sum b_{DI}\hat{M}_D = 0. \tag{4.5.12d}$$

For *second*-type constraint problems, clearly, the above impulsive equations of Routh–Voss, Maggi, and Hadamard still hold; and the n $(\dot{q})^-$ have known values, unrelated to the constraints (4.5.1a).

Appell's Equations and Impulsive Motion

Since these equations contain the accelerations explicitly, in general, they are not very useful in impulsive problems. Nevertheless, an Appell-like form of impulsive equations can be formulated. To this end, first, we define the *kinetic energy of the velocity jumps*, or *impulsive Appellian* function:

$$\hat{S} \equiv S\, dm\, \Delta\boldsymbol{v}\cdot\Delta\boldsymbol{v}/2 \qquad \left[\neq \Delta T \equiv S\, dm\, (\boldsymbol{v}^+\cdot\boldsymbol{v}^+ - \boldsymbol{v}^-\cdot\boldsymbol{v}^-)/2\right], \tag{4.5.13a}$$

[in *exception* to the earlier *hat notation*, eqs. (4.2.5a)!)] and then notice that, since $\Delta\boldsymbol{\varepsilon}_I = \mathbf{0}$ *and* $\Delta\boldsymbol{\varepsilon}_{n+1} = \mathbf{0}$, we have

$$\Delta\boldsymbol{v} = \sum \boldsymbol{\varepsilon}_I \Delta\omega_I \Rightarrow \partial(\Delta\boldsymbol{v})/\partial(\Delta\omega_I) = \boldsymbol{\varepsilon}_I, \tag{4.5.13b}$$

and therefore, successively,

$$\begin{aligned}\partial\hat{S}/\partial(\Delta\omega_I) &= S\, (dm/2)2\,\Delta\boldsymbol{v}\cdot[\partial(\Delta\boldsymbol{v})/\partial(\Delta\omega_I)]\\ &= S\, dm\,\Delta\boldsymbol{v}\cdot\boldsymbol{\varepsilon}_I = \Delta\Big(S\, dm\,\boldsymbol{v}\cdot\boldsymbol{\varepsilon}_I\Big) \equiv \Delta P_I;\end{aligned}$$

that is, finally, the kinetic equations of impulsive motion take the "Appellian" form

$$\partial\hat{S}/\partial(\Delta\omega_I) = \Delta(\partial T^*/\partial\omega_I) = \hat{\Theta}_I \qquad (I = m+1,\ldots,n), \tag{4.5.13c}$$

due to Arrighi (1939); see also Pars (1965, pp. 238–242). It is not hard to see, by invoking the impulsive principle of relaxation and a relaxed S, that the corresponding kinetostatic equations are

$$[\partial\hat{S}/\partial(\Delta\omega_D)]_o = \hat{\Theta}_D + \hat{\Lambda}_D \qquad (D = 1,\ldots,m). \tag{4.5.13d}$$

Example 4.5.1 One extremity of the major axis (point P) and one extremity of the minor axis (point Q) of a thin homogeneous elliptical disk of (principal) semiaxes,

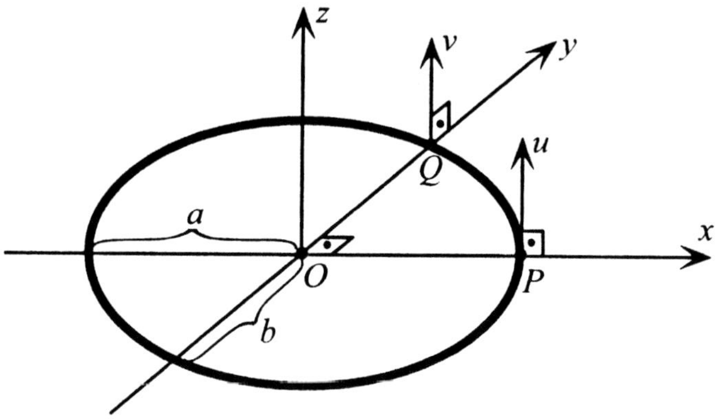

Figure 4.10 Elliptical disk, initially at rest, is given the velocities u and v at the endpoints of its axes, perpendicular to its plane. Semiaxes of the ellipse: $a = |\boldsymbol{OP}|, b = |\boldsymbol{OQ}|$.

a, b, and mass m, initially at rest, are imparted prescribed velocities, u (at P) and ν (at Q), perpendicular to the plane of the disk (fig. 4.10). Let us find its *postimpact* linear and angular velocities.

By König's theorem, the (double) kinetic energy of the disk is

$$2T \to 2T^* = m({v_x}^2 + {v_y}^2 + {v_z}^2) + (A{\omega_x}^2 + B{\omega_y}^2 + C{\omega_z}^2), \tag{a}$$

where

$$\boldsymbol{v}_G = (v_x, v_y, v_z)\text{: velocity of mass center of disk, } G,$$

$$\boldsymbol{\omega} = (\omega_x, \omega_y, \omega_z)\text{: angular velocity of disk,}$$

$$(I_x, I_y, I_z) = (mb^2/4, ma^2/4, m(a^2 + b^2)/4)\text{: principal moments of inertia of disk at } G. \tag{b}$$

From rigid-body kinematics, we have

$$\boldsymbol{v}_P = \boldsymbol{v}_G + \boldsymbol{\omega} \times \boldsymbol{r}_{P/G}, \qquad \boldsymbol{v}_Q = \boldsymbol{v}_G + \boldsymbol{\omega} \times \boldsymbol{r}_{Q/G}; \tag{c}$$

or, in components,

$$u\boldsymbol{k} = (v_x, v_y, v_z) + (\omega_x, \omega_y, \omega_z) \times (a, 0, 0), \tag{c1}$$

$$v\boldsymbol{k} = (v_x, v_y, v_z) + (\omega_x, \omega_y, \omega_z) \times (0, b, 0); \tag{c2}$$

respectively, from which, equating, we obtain the velocity compatibility conditions

$$v_x = 0 \text{ (also, by symmetry)}, \qquad v_y + a\,\omega_z = 0, \qquad v_z - a\,\omega_y = u, \tag{c3}$$

$$v_x - b\,\omega_z = 0, \qquad v_y = 0 \text{ (also, by symmetry)}, \qquad v_z + b\,\omega_x = v. \tag{c4}$$

In view of the above, we choose the following convenient set of six quasi velocities:

$$\omega_1 \equiv v_x = 0, \tag{c5}$$

$$\omega_2 \equiv v_y = 0, \tag{c6}$$

$$\omega_3 \equiv \omega_z = 0, \tag{c7}$$

$$\omega_4 \equiv v_z - a\,\omega_y - u = 0, \tag{c8}$$

$$\omega_5 \equiv v_z + b\,\omega_x - v = 0, \tag{c9}$$

$$\omega_6 \equiv v_z \neq 0 \tag{c10}$$

(we could have chosen as ω_6 either ω_x or ω_y, or any linear combination of $\omega_{x,y,z}$); which invert readily to

$$v_x = \omega_1 = 0, \tag{c11}$$

$$v_y = \omega_2 = 0, \tag{c12}$$

$$v_z = \omega_6 \neq 0, \tag{c13}$$

$$\omega_x = (1/b)(\omega_5 - v_z + v) = (1/b)(0 - \omega_6 + v) = (1/b)(v - \omega_6) \neq 0, \tag{c14}$$

$$\omega_y = (1/a)(v_z - u - \omega_4) = (1/a)(\omega_6 - u - 0) = (1/a)(\omega_6 - u) \neq 0, \tag{c15}$$

$$\omega_z = \omega_3 = 0. \tag{c16}$$

Hence the kinetic energy, (a) [with (b)], becomes, successively,

$$\begin{aligned} 2T^* \to 2T^{**} &= m{v_z}^2 + A[(v/b) - (v_z/b)]^2 + B[(v_z/a) - (u/a)]^2 \\ &= \cdots = (m/2)\big[3{v_z}^2 - (u + v)v_z + (u^2 + v^2)/2\big]. \end{aligned} \tag{c17}$$

Since the preimpact state is one of rest, the v_z-impulsive equation becomes

$$\begin{aligned} &\Delta(\partial T^{**}/\partial v_z) = (\partial T^{**}/\partial v_z)^+ \equiv \partial T^{**}/\partial v_z = \hat{\Theta}_z\ (= 0, \text{ explain}): \\ &(m/4)[6v_z - (u + v)] = 0 \ \Rightarrow\ v_z = (u + v)/6, \end{aligned} \tag{d1}$$

and combined with (c3, 4) yields the following postimpact angular velocities:

$$\omega_x = (v - v_z)/b = (5v - u)/6b, \qquad \omega_y = (v_z - u)/a = (v - 5u)/6a. \tag{d2}$$

See also Bahar [1987, via Jourdain's impulsive principle (see example below)] and Byerly [1916, pp. 72–75, via Kelvin's theorem (§4.6 and next problem)].

Problem 4.5.1 Continuing from the preceding example, by (c3, 4):

$$\omega_x = (v - v_z)/b, \qquad \omega_y = (v_z - u)/a, \tag{a}$$

so that the (constrained) kinetic energy [eq. (a) of ex. 4.5.1], becomes

$$T = (1/2)m{v_z}^2 + (mb^2/8)[(v - v_z)/b]^2 + (ma^2/8)[(v_z - u)/a]^2 = T(v_z;\, u, v). \tag{b}$$

Then show that the earlier equation (d1) results from

$$\partial T/\partial v_z = 0. \tag{c}$$

REMARK

This also constitutes an application of Kelvin's theorem (§4.6).

Problem 4.5.2 Continuing from the preceding example and problem, show by any means that the impulses at P and Q (i.e., the ones communicating to the disk the above velocities), $\hat{I}_P$ and $\hat{I}_Q$, respectively, equal

$$\hat{I}_P = (m/24)(5u - v), \qquad \hat{I}_Q = (m/24)(5v - u); \tag{a}$$

also,

$$v_z = (\hat{I}_P + \hat{I}_Q)/m, \qquad \omega_x = 4\hat{I}_Q/mb, \qquad \omega_y = -4\hat{I}_P/ma. \tag{b}$$

Example 4.5.2 A homogeneous sphere of center and center of mass G, mass m, and radius r, rotating with angular velocity $\boldsymbol{\omega}^- = (\omega_x{}^-, \omega_y{}^-, \omega_z{}^-)$ is suddenly placed on a perfectly rough horizontal plane (fig. 4.11). Let us find its postimpact velocities

$$\boldsymbol{\omega}^+ \equiv \boldsymbol{\omega} = (\omega_x, \omega_y, \omega_z) \quad \text{and} \quad \boldsymbol{v}_G{}^+ \equiv \boldsymbol{v} = (v_x, v_y, v_z).$$

By kinematics we have

$$\boldsymbol{v}_{\text{contact point}} \equiv \boldsymbol{v}_C = \boldsymbol{v} + \boldsymbol{\omega} \times \boldsymbol{r}_{C/G} = \boldsymbol{0}, \tag{a}$$

or, in components,

$$(0, 0, 0) = (v_x, v_y, v_z) + (\omega_x, \omega_y, \omega_z) \times (0, 0, -r), \tag{b}$$

and from this we obtain the three Pfaffian constraints

$$v_x - r\omega_y = 0, \qquad v_y + r\omega_x = 0, \qquad v_z = 0. \tag{c}$$

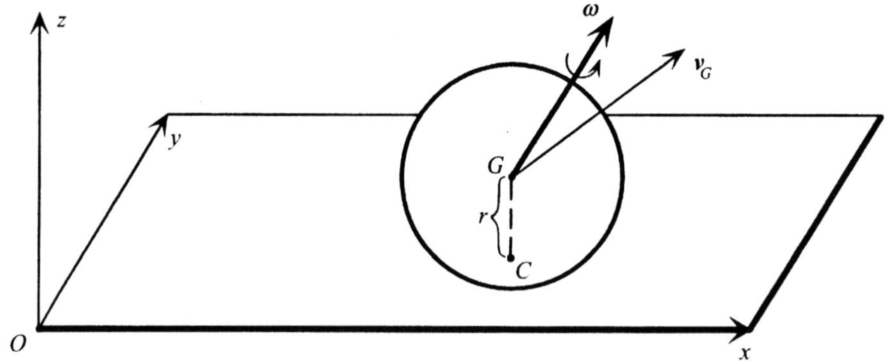

Figure 4.11 Sphere placed suddenly in contact with a rough plane.

Hence, introducing the *six* quasi velocities ω_k:

$$\omega_1 \equiv v_x - r\,\omega_y = 0, \tag{d1}$$

$$\omega_2 \equiv v_y + r\,\omega_x = 0, \tag{d2}$$

$$\omega_3 \equiv v_z = 0, \tag{d3}$$

$$\omega_4 \equiv \omega_x \neq 0, \tag{d4}$$

$$\omega_5 \equiv \omega_y \neq 0, \tag{d5}$$

$$\omega_6 \equiv \omega_z \neq 0 \tag{d6}$$

(or any other linear and invertible combination of $v_x, v_y, v_z, \omega_{x,y,z}$); and their inverses,

$$v_x = \omega_1 + r\,\omega_y = \omega_1 + r\,\omega_5 = r\,\omega_5, \tag{d7}$$

$$v_y = \omega_2 - r\,\omega_x = \omega_2 - r\,\omega_4 = -r\,\omega_4, \tag{d8}$$

$$v_z = \omega_3 = 0, \tag{d9}$$

$$\omega_x = \omega_4 \neq 0, \tag{d10}$$

$$\omega_y = \omega_5 \neq 0, \tag{d11}$$

$$\omega_z = \omega_6 \neq 0, \tag{d12}$$

we can express the (double) kinetic energy of the sphere as follows:

$$2T = m({v_x}^2 + {v_y}^2 + {v_z}^2) + mk^2({\omega_x}^2 + {\omega_y}^2 + {\omega_z}^2)$$
$$= m[(\omega_1 + r\,\omega_5)^2 + (\omega_2 - r\,\omega_4)^2 + {\omega_3}^2 + k^2({\omega_4}^2 + {\omega_5}^2 + {\omega_6}^2)] \quad (= 2T^*),$$

[where $k^2 \equiv (2/5)r^2$: (squared) radius of gyration of sphere about G] or

$$2T^* = m\big[(r^2 + k^2){\omega_4}^2 + (r^2 + k^2){\omega_5}^2 + k^2{\omega_6}^2$$
$$+ {\omega_1}^2 + {\omega_2}^2 + {\omega_3}^2 + 2r\,\omega_1\omega_5 - 2r\,\omega_2\omega_4\big], \tag{e}$$

where the last *five* terms represent the "relaxed (i.e., unconstrained)" contributions to T^*. As a result of the above, the *postimpact* nonholonomic momenta $(\partial T^*/\partial\omega_k)^+$ are found to be [with the notation $(\ldots)|^+ = (\ldots)_{\text{postimpact constraints enforced after differentiations}}$]

$$1: \quad m(\omega_1 + r\,\omega_5)|^+ = m\,r\,\omega_5, \tag{f1}$$

$$2: \quad m(\omega_2 - r\,\omega_4)|^+ = -m\,r\,\omega_4, \tag{f2}$$

$$3: \quad m\,\omega_3|^+ = m\,\omega_3, \tag{f3}$$

$$4: \quad m[(r^2 + k^2)\omega_4 - r\,\omega_2]|^+ = m(r^2 + k^2)\omega_4, \tag{f4}$$

$$5: \quad m[(r^2 + k^2)\omega_5 + r\,\omega_1]|^+ = m(r^2 + k^2)\omega_5, \tag{f5}$$

$$6: \quad m\,k^2\omega_6|^+ = m\,k^2\omega_6; \tag{f6}$$

and similarly for the *preimpact* nonholonomic momenta $(\partial T^*/\partial\omega_k)^-$:

$$1:\ 0, \quad 2:\ 0, \quad 3:\ 0, \quad 4:\ m\,k^2{\omega_4}^-, \quad 5:\ m\,k^2{\omega_5}^-, \quad 6:\ m\,k^2{\omega_6}^-. \tag{f7}$$

Next, let us calculate the nonholonomic impulsive forces, $\hat{\Theta}_k$ (impressed) and $\hat{\Lambda}_D$ (reactions) [where, in view of (d1–6), $k = 1, \ldots, 6$; $D = 1, 2, 3$] and their relations with their holonomic counterparts $\hat{Q}_{x,y,z;\ldots}$ and $\hat{R}_{x,y,z}$, $\hat{M}_{x,y,z}$. With $d\theta_k/dt \equiv \omega_k$, we obtain

$$\widehat{\delta' W} = \sum \hat{\Theta}_k\,\delta\theta_k = 0, \qquad \hat{\Theta}_k = 0 \Rightarrow \hat{Q}_{x,y,z;\ \ldots} = 0; \tag{g1}$$

$$\begin{aligned}\widehat{\delta' W_R} &= \sum \hat{\Lambda}_k\,\delta\theta_k = \sum \hat{\Lambda}_D\,\delta\theta_D \\ &= \hat{\Lambda}_1\,\delta\theta_1 + \hat{\Lambda}_2\,\delta\theta_2 + \hat{\Lambda}_3\,\delta\theta_3 \\ &= \hat{\Lambda}_1(\delta x - r\,\delta\theta_y) + \hat{\Lambda}_2(\delta y + r\,\delta\theta_x) + \hat{\Lambda}_3\,\delta z \\ &= \hat{\Lambda}_1\,\delta x + \hat{\Lambda}_2\,\delta y + \hat{\Lambda}_3\,\delta z + (\hat{\Lambda}_2\,r)\,\delta\theta_x + (-\hat{\Lambda}_1\,r)\,\delta\theta_y + (0)\,\delta\theta_z \qquad \text{(g2)} \\ &\equiv \hat{R}_x\,\delta x + \hat{R}_y\,\delta y + \hat{R}_z\,\delta z + \hat{M}_x\,\delta\theta_x + \hat{M}_y\,\delta\theta_y + \hat{M}_z\,\delta\theta_z, \qquad \text{(g3)} \\ &\Rightarrow \hat{R}_x = \hat{\Lambda}_1,\ \hat{R}_y = \hat{\Lambda}_2,\ \hat{R}_z = \hat{\Lambda}_3;\ \hat{M}_x = r\hat{\Lambda}_2,\ \hat{M}_y = -r\hat{\Lambda}_1,\ \hat{M}_x = 0. \qquad \text{(g4)}\end{aligned}$$

With the help of the above results, and the notational $\Delta P_k \equiv (\partial T^*/\partial\omega_k)^+ - (\partial T^*/\partial\omega_k)^-$: *impulsive jumps* of the nonholonomic momenta, the Hamel equations of impulsive motion become

$$1:\quad \Delta P_1 = \hat{\Theta}_1 + \hat{\Lambda}_1: \qquad m\,r\,\omega_5 = m\,r\,\omega_y = \hat{\Lambda}_1, \tag{h1}$$

or, due to the $k = 5$ equation (see below),

$$m\,r[k^2/(r^2 + k^2)]\omega_y^- = \hat{\Lambda}_1; \tag{h2}$$

$$2:\quad \Delta P_2 = \hat{\Theta}_2 + \hat{\Lambda}_2: \qquad -\,m\,r\,\omega_4 = -m\,r\,\omega_x = \hat{\Lambda}_2, \tag{h3}$$

or, due to the $k = 4$ equation (see below),

$$-\,m\,r[k^2/(r^2 + k^2)]\omega_x^- = \hat{\Lambda}_2; \tag{h4}$$

$$3:\quad \Delta P_3 = \hat{\Theta}_3 + \hat{\Lambda}_3: \qquad m\,\omega_3 = m\,v_z = \hat{\Lambda}_3 = 0; \tag{h5}$$

$$4:\quad \Delta P_4 = \hat{\Theta}_4: \qquad m(r^2 + k^2)\omega_4 - m\,k^2\omega_4^- = 0, \tag{h6}$$

or

$$\omega_x = [k^2/(r^2 + k^2)]\omega_x^-, \quad \text{and then} \quad v_y = -r\,\omega_x = \cdots; \tag{h7}$$

$$5:\quad \Delta P_5 = \hat{\Theta}_5: \qquad m(r^2 + k^2)\omega_5 - m\,k^2\omega_5^- = 0, \tag{h8}$$

or

$$\omega_y = [k^2/(r^2 + k^2)]\omega_y^-, \quad \text{and then} \quad v_x = r\,\omega_y = \cdots; \tag{h9}$$

$$6:\quad \Delta P_6 = \hat{\Theta}_6: \qquad m\,k^2\omega_6 - m\,k^2\omega_6^- = 0, \tag{h10}$$

or

$$\omega_6 = \omega_6^- \;\Rightarrow\; \omega_z = \omega_z^-. \tag{h11}$$

Finally, let us compare the above with the "elementary" Newton–Euler impulsive theory. With $\boldsymbol{v}_G^- = (v_x^-, v_y^-, v_z^-) = \mathbf{0}$, and recalling (g4), we readily find

$$m\,v_x = \hat{R}_x\ (= \hat{\Lambda}_1) \Rightarrow \hat{R}_x = m\,r\,\omega_y = m\,r[k^2/(r^2+k^2)]\omega_y^-, \tag{i1}$$

$$m\,v_y = \hat{R}_y\ (= \hat{\Lambda}_2) \Rightarrow \hat{R}_y = -m\,r\,\omega_x = -m\,r[k^2/(r^2+k^2)]\omega_x^-, \tag{i2}$$

$$m\,v_z = \hat{R}_z\ (= \hat{\Lambda}_3) \Rightarrow \hat{R}_z = 0; \tag{i3}$$

$$\hat{M}_x = r\,\hat{R}_y = \cdots, \quad \hat{M}_y = -r\,\hat{R}_x = \cdots, \quad \hat{M}_z = 0 \quad \text{(angular momentum about } G\text{)}. \tag{i4}$$

[To obtain reactionless equations we could apply the impulsive angular momentum principle (here, conservation) about the contact point C.]

Finally, it is not hard to show that if $(v_x^-, v_y^-, v_z^-) \neq \mathbf{0}$, (h7, 9) would be replaced, respectively, by

$$\omega_x = (k^2\omega_x^- + r\,v_y^-)/(r^2+k^2) \quad \text{and} \quad \omega_y = (k^2\omega_y^- + r\,v_x^-)/(r^2+k^2). \tag{j}$$

See also Bahar [1987, via Jourdain's impulsive principle (see example below)] and Byerly [1916, pp. 72–75, via Kelvin's theorem (§4.6)].

Example 4.5.3 A homogeneous straight rigid bar AB of length $l = 2b$ and mass m falls freely in the vertical plane O–xy and strikes a smooth and inelastic floor at A (fig. 4.12). Find the postimpact velocities and forces.

We choose as Lagrangean coordinates $q_{1,2,3}$ (i) the coordinates of the mass center of the bar G: x and y, and (ii) the bar angle with the vertical (positive upward): θ. Clearly, this is a *second*-kind problem; that is, one of *suddenly applied* and *persistent* constraints. The *preimpact* velocities are

$$(\dot{x})^- = 0, \quad (\dot{y})^- = -v, \quad (\dot{\theta})^- = \omega, \quad \text{all given;} \tag{a}$$

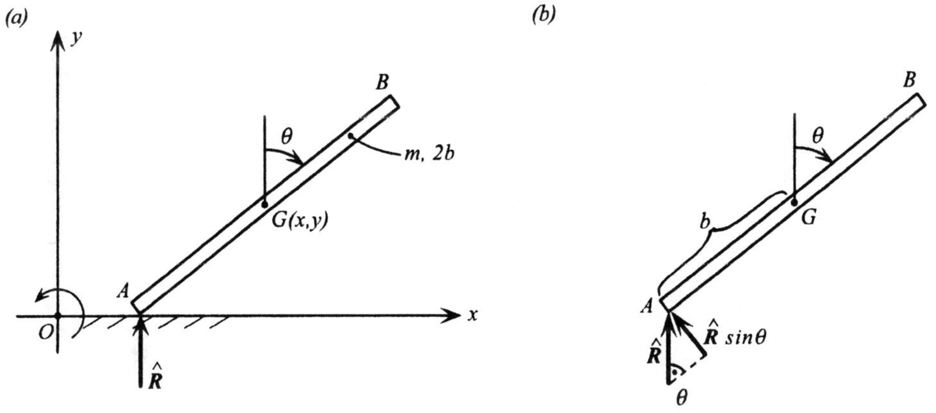

Figure 4.12 (a) Impact of bar AB (length $2b$, mass m) on smooth, inelastic floor; (b) components of floor reaction.

while the *unknowns* of the problem are the *postimpact* velocities:

$$(\dot{x})^+ \equiv \dot{x} = 0 \quad \text{(by inspection—see below)}, \qquad (\dot{y})^+ = \dot{y}, \qquad (\dot{\theta})^+ = \dot{\theta}; \tag{b}$$

and the impulsive ground reaction $\hat{R}$. Below we present *two* solutions.

1. First Solution: Holonomic Coordinates

We have (double) *kinetic energy* (with $I \equiv mb^2/3$: moment of inertia of bar about G):

$$2T = m[(\dot{x})^2 + (\dot{y})^2] + I(\dot{\theta})^2; \tag{c}$$

constraint:

$$\begin{aligned} \boldsymbol{v}_A = \boldsymbol{v}_G + \boldsymbol{\omega} \times \boldsymbol{r}_{A/G} &= (\dot{x}, \dot{y}, 0) + (-\dot{\theta}\,\boldsymbol{k}) \times (-b \sin\theta, b\cos\theta, 0) \\ &= (\dot{x} - b\,\dot{\theta} \cos\theta,\ \dot{y} + b\,\dot{\theta} \sin\theta,\ 0) \\ &= (v_{Ax}, v_{Ay}, 0); \end{aligned} \tag{d}$$

and, since the floor is *inelastic*,

$$v_{Ay} = \dot{y} + b\,\dot{\theta} \sin\theta = 0 \qquad (\text{i.e., } n - m = 3 - 1 = 2), \tag{e1}$$

and, in virtual form (with $v_{Ay} \equiv \dot{y}_A$),

$$\delta y_A = \delta y + b \sin\theta\, \delta\theta = 0 \qquad \text{(i.e., the vertical virtual displacement of } A \text{ vanishes);} \tag{e2}$$

an equation that holds whether the inelastic floor is *stationary* (case discussed here), or moves with a *prescribed* motion [generalization of (e1)].

Impulsive Lagrange's principle:

$$\Delta(\partial T/\partial \dot{x})\, \delta x + \Delta(\partial T/\partial \dot{y})\, \delta y + \Delta(\partial T/\partial \dot{\theta})\, \delta\theta = 0, \tag{f1}$$

under the constraint eq. (e2), rewritten as

$$(0)\, \delta x + (1)\, \delta y + (b \sin\theta)\, \delta\theta = 0. \tag{f2}$$

Since δx is *unconstrained*, (f1) gives

$$\Delta(\partial T/\partial \dot{x}) = \Delta(m\,\dot{x}) = m\,\dot{x} - m(\dot{x})^- = 0 \ \Rightarrow\ \dot{x} = 0. \tag{g1}$$

Then (f1, 2) reduce, respectively, to

$$\Delta(\partial T/\partial \dot{y})\, \delta y + \Delta(\partial T/\partial \dot{\theta})\, \delta\theta = \Delta(m\dot{y})\, \delta y + \Delta(I\,\dot{\theta})\, \delta\theta = 0, \tag{g2}$$

$$(1)\, \delta y + (b \sin\theta)\, \delta\theta = 0; \tag{g3}$$

and, via an impulsive multiplier $-\hat{\lambda}$, combine, in well-known ways, to the single *unconstrained* variational equation

$$\big[\Delta(m\,\dot{y}) - \hat{\lambda}(1)\big]\, \delta y + \big[\Delta(I\,\dot{\theta}) - \hat{\lambda}(b \sin\theta)\big]\, \delta\theta = 0. \tag{g4}$$

This yields, immediately, the two Routh–Voss impulsive equations

$$\Delta(m\dot{y}) = \hat{\lambda}, \qquad \Delta(I\dot{\theta}) = \hat{\lambda}(b\sin\theta); \tag{h1}$$

or, invoking the preimpact velocities (a),

$$m(\dot{y} + v) = \hat{\lambda}, \qquad I(\dot{\theta} - \omega) = \hat{\lambda} b \sin\theta, \tag{h2}$$

which, along with the constraint equation (e1) (evaluated at the postimpact instant) constitute a system of three equations for $\dot{y}, \dot{\theta}, \hat{\lambda}$. Solving them, we find

$$\dot{\theta} = (b\omega + 3v\sin\theta)/b(1 + 3\sin^2\theta), \tag{i1}$$

$$\dot{y} = -\sin\theta(b\omega + 3v\sin\theta)/(1 + 3\sin^2\theta). \tag{i2}$$

Then, from the first of (h2),

$$\hat{\lambda} = m(\dot{y} + v) = m(v - b\omega\sin\theta)/(1 + 3\sin^2\theta); \tag{i3}$$

and [fig. 4.12(b)]

$$\widehat{\delta' W_R} = \hat{R}\,\delta y_A = \hat{R}\,(\delta y + b\sin\theta\,\delta\theta) = \hat{R}\,\delta y + (\hat{R} b \sin\theta)\,\delta\theta$$
$$\equiv \hat{R}_y\,\delta y + \hat{R}_\theta\,\delta\theta \quad (= 0), \tag{j1}$$

that is,

$$\hat{R}_y = \hat{R} = \hat{\lambda}, \qquad \hat{R}_\theta = \hat{R} b \sin\theta = \hat{\lambda} b \sin\theta, \qquad \hat{R}_x = 0. \tag{j2}$$

The (two) kinetic impulsive Maggi equations of this problem are (g1) and the equation obtained by eliminating $\hat{\lambda}$ between (h2):

$$I(\dot{\theta} - \omega) - b\sin\theta\,[m(\dot{y} + v)] = 0, \tag{k1}$$

or, simplifying,

$$3\sin\theta(\dot{y} + v) - b(\dot{\theta} - \omega) = 0. \tag{k2}$$

Solving (g1), (k2), and (e1) for $\dot{x}, \dot{y}, \dot{\theta}$, we recover the second of (g1) and (i1, 2). These results are rederived more systematically below.

2. Second Solution: Nonholonomic Coordinates

Due to the constraints (e1, 2), we introduce the following set of quasi velocities:

$$\omega_1 \equiv \dot{y} + (b\sin\theta)\dot{\theta} \quad (= 0), \qquad \omega_2 \equiv \dot{y}, \qquad \omega_3 \equiv \dot{x}. \tag{l1}$$

Their inverse is readily found to be

$$\dot{x} = \omega_3, \qquad \dot{y} = \omega_2, \qquad \dot{\theta} = (\omega_1 - \omega_2)/b\sin\theta; \tag{l2}$$

that is, the corresponding transformation matrices are

$$(a_{kl}) = \begin{pmatrix} 0 & 1 & b\sin\theta \\ 0 & 1 & 0 \\ 1 & 0 & 0 \end{pmatrix},$$

$$(A_{kl}) = \begin{pmatrix} 0 & 0 & 1 \\ 0 & 1 & 0 \\ (b\cos\theta)^{-1} & -(b\cos\theta)^{-1} & 0 \end{pmatrix}. \tag{13}$$

Then:

(i) Maggi's *kinetic* equations: $\sum A_{kI}\Delta p_k = \sum A_{kI}\hat{Q}_k$ $(k = 1, 2, 3;\ I = 2, 3)$, with some (hopefully obvious) ad hoc notations, become

$I = 2$:
$$A_{x2}(\Delta p_x - \hat{Q}_x) + A_{y2}(\Delta p_y - \hat{Q}_y) + A_{\theta 2}(\Delta p_\theta - \hat{Q}_\theta) = 0,$$
or
$$(0)\,(\Delta p_x - 0) + (1)\,(\Delta p_y - 0) + (-1/b\sin\theta)(\Delta p_\theta - 0) = 0,$$
or, finally,
$$\Delta p_\theta = (b\sin\theta)\Delta p_y. \tag{m1}$$

$I = 3$:
$$A_{x3}(\Delta p_x - \hat{Q}_x) + A_{y3}(\Delta p_y - \hat{Q}_y) + A_{\theta 3}(\Delta p_\theta - \hat{Q}_\theta) = 0,$$
or
$$(1)\,(\Delta p_x - 0) + (0)\,(\Delta p_y - 0) + (0)\,(\Delta p_\theta - 0) = 0,$$
or, finally,
$$\Delta p_x = 0; \tag{m2}$$

(ii) Maggi's *kinetostatic* equations: $\sum A_{kD}\Delta p_k = \sum A_{kD}\hat{Q}_k$ $(k = 1, 2, 3;\ D = 1$; i.e., here only *one* such equation) become

$D = 1$:
$$A_{x1}(\Delta p_x - \hat{Q}_x) + A_{y1}(\Delta p_y - \hat{Q}_y) + A_{\theta 1}(\Delta p_\theta - \hat{Q}_\theta) = \hat{\lambda}_1 \equiv \hat{\lambda},$$
or
$$(0)\,(\Delta p_x - 0) + (0)\,(\Delta p_y - 0) + (1/b\sin\theta)(\Delta p_\theta - 0) = \hat{\lambda},$$
or, finally,
$$\Delta p_\theta = (b\sin\theta)\hat{\lambda}. \tag{m3}$$

Equations (m1–3), naturally, coincide with the earlier equations (k2), (g1), (second of h2), respectively.

(iii) Next, let us formulate the *impulsive Hamel* equations. In terms of the above ω's, the *preimpact* state is

$$\omega_1{}^- = -v + \omega b\sin\theta \quad (\neq 0,\ \text{but}\ \omega_1{}^+ \equiv \omega = 0),$$
$$\omega_2{}^- = (\dot{y})^- = -v, \qquad \omega_3{}^- = (\dot{x})^- = 0; \tag{n1}$$

and, further,

$$2T \to 2T^* = m(\omega_2{}^2 + \omega_3{}^2) + (m/3\sin^2\theta)(\omega_1 - \omega_2)^2 \quad [\text{substituting (12) into (c)}], \tag{n2}$$

$$\Rightarrow 2T^*{}_o = m(\omega_2{}^2 + \omega_3{}^2) + (m/3\sin^2\theta)\omega_2{}^2 \quad (\textit{constrained}\ 2T^*); \tag{n3}$$

$$\widehat{\delta' W} \to \widehat{(\delta' W)^*} = \hat{\Theta}_1\,\delta\theta_1 + \hat{\Theta}_2\,\delta\theta_2 + \hat{\Theta}_3\,\delta\theta_3 = 0$$

$$\text{[where } \omega_{1,2,3} \equiv d\theta_{1,2,3}/dt;\ \text{since } \hat{Q}_{x,y,\theta} = 0 \ \Rightarrow\ \hat{\Theta}_{1,2,3} = 0]; \tag{n4}$$

$$\widehat{\delta' W_R} \to \widehat{(\delta' W_R)^*} = \hat{\Lambda}_1\,\delta\theta_1 + \hat{\Lambda}_2\,\delta\theta_2 + \hat{\Lambda}_3\,\delta\theta_3$$
$$= \hat{\Lambda}_1(\delta y + b\,\sin\theta\,\delta\theta) + \hat{\Lambda}_2\,\delta y + \hat{\Lambda}_3\,\delta x$$
$$= \hat{\Lambda}_3\,\delta x + (\hat{\Lambda}_1 + \hat{\Lambda}_2)\,\delta y + (\hat{\Lambda}_1 b\,\sin\theta)\,\delta\theta \quad (= 0), \tag{n5}$$

from which [and (j2)] it follows that

$$\hat{R}_x = \hat{\Lambda}_3 = 0, \qquad \hat{R}_y = \hat{\Lambda}_1 + \hat{\Lambda}_2 = \hat{\lambda} \Rightarrow \hat{\Lambda}_2 = 0 \quad \text{(by next equation)},$$
$$\hat{R}_\theta = \hat{\Lambda}_1 b\,\sin\theta = \hat{\lambda}\,b\,\sin\theta \ \Rightarrow\ \hat{\Lambda}_1 = \hat{\lambda} = \hat{R}; \quad \text{i.e., } \hat{\Lambda}_1 \neq 0, \quad \hat{\Lambda}_2 = 0, \quad \hat{\Lambda}_3 = 0. \tag{n6}$$

In view of the above, the Hamel impulsive equations are (we recall to set, after the differentiations, $\omega_1 = 0$)

$$\omega_3\!: \qquad \Delta(\partial T^*/\partial\omega_3) = \hat{\Lambda}_3\!: \qquad m\,\omega_3 - 0 = 0 \Rightarrow \omega_3 \equiv \dot{x} = 0, \tag{o1}$$

$$\omega_2\!: \qquad \Delta(\partial T^*/\partial\omega_2) = \hat{\Lambda}_2\!:$$
$$[m\,\omega_2 + (m/3\,\sin^2\theta)(\omega_2 - \omega_1)]$$
$$- [m(-v) + (m/3\,\sin^2\theta)(-v - b\,\omega\,\sin\theta + v)],$$
$$\Rightarrow m(\omega_2 + v) + (m/3\,\sin\theta)[(\omega_2/\sin\theta) + b] = 0$$
$$\Rightarrow \omega_2 \equiv \dot{y} = -\sin\theta(b\,\omega + 3v\sin\theta)/(1 + 3\,\sin^2\theta), \tag{o2}$$

$$\omega_1\!: \qquad \Delta(\partial T^*/\partial\omega_1) = \hat{\Lambda}_1\!:$$
$$[m(\omega_1 - \omega_2)/3\,\sin^2\theta]_o - [m(b\,\omega\,\sin\theta - v + v)/3\,\sin^2\theta]_o$$
$$= (-m\,\omega_2/3\,\sin^2\theta) - (m\,b\,\omega/3\,\sin\theta) = \hat{\Lambda}_1 = \hat{\lambda},$$
$$\Rightarrow -(m/3\,\sin\theta)\big[(\omega_2/\sin\theta) + b\,\omega\big] = \hat{\lambda},$$
$$\Rightarrow \hat{\Lambda}_1 = m(v - b\,\omega\,\sin\theta)/(1 + 3\,\sin^2\theta) \qquad \text{[invoking (o2)]}; \tag{o3}$$

and, finally, from (12)

$$\dot{\theta} = -\omega_2/b\,\sin\theta = -\dot{y}/b\,\sin\theta = (b\,\omega + 3v\,\sin\theta)/b(1 + 3\,\sin^2\theta), \tag{o4}$$

as before.

CLOSING REMARKS

(i) This is a problem of the *second* kind; that is, suddenly introduced persistent constraints. Due to $\omega_1^- \neq 0$, in expressions like $(\partial T^*/\partial\omega_D)^+$ and $(\partial T^*/\partial\omega_D)^-$, we must keep *all* the ω's in $T^+\!: T^* = T^*(\omega_1, \omega_2, \omega_3)$, and enforce the constraint $\omega_1^+ \equiv \omega_1 = 0$ only *after all differentiations have been carried out*. However, the reader may easily verify that

$$(\partial T^*/\partial\omega_I)^+ = (\partial T^*{}_o/\partial\omega_I)^+.$$

(ii) This problem is treated in Timoshenko and Young (1948, p. 226). Their solution is, however, *conceptually* incorrect, since they treat $\hat{R}$ as an *impressed* impulsive force; even though, clearly, all Q's vanish. Their final results, however, are correct. The same error is repeated in several other (mostly British) texts: for example, Ramsey (1937, pp. 220–221), Smart (1951, pp. 262–263). Other authors do not commit such errors only because they restrict their treatments to *first*-kind problems.

Problem 4.5.3 (D. T. Greenwood, private communication, 1997). Continuing from the preceding example, and in order to avoid calculating undesired impulsive constraint reactions, choose as velocity variables $\dot{x}_A \equiv v_{Ax}$, $\dot{y}_A = v_{Ay}$, $\dot{\theta} \equiv \Omega$ (i.e., $\omega_1 \equiv v_{Ay}$, $\omega_2 \equiv v_{Ax}$, $\omega_3 \equiv \dot{\theta}$). Then the initial conditions are

$$v_{Ax}{}^- = -b\omega\cos\theta, \qquad v_{Ay}{}^- = b\omega\sin\theta - v, \qquad \Omega^- = \omega, \tag{a}$$

while after the impact with the floor,

$$v_{Ay} = 0, \qquad v_{Ax},\ \Omega:\ \ \text{independent} \qquad (\text{i.e., } \omega_1 = 0,\ \ \omega_2,\ \omega_3:\ \text{independent}). \tag{b}$$

(i) Show that the unconstrained kinetic energy of the bar is

$$T = (m/2)(v_{Ax}{}^2 + v_{Ay}{}^2) + (2mb^2/3)\Omega^2 + mb\Omega(v_{Ax}\cos\theta - v_{Ay}\sin\theta) \quad (= T^*). \tag{c}$$

(ii) Verify that Appell's rule (i.e., conservation of system momenta corresponding to v_{Ax}, Ω) gives

$$\Delta(\partial T/\partial v_{Ax}) = \Delta(\partial T^*/\partial\omega_2) = 0: \quad m\,v_{Ax} + m\,b\,\Omega\cos\theta = \mathit{constant} \quad (= 0), \tag{d1}$$

$$\Delta(\partial T/\partial\Omega) = \Delta(\partial T^*/\partial\omega_3) = 0:$$

$$(4/3)mb^2\Omega + mb(v_{Ax}\cos\theta - v_{Ay}\sin\theta) = \mathit{constant} \quad [= (4/3)mb^2\omega - mb^2\omega + mbv\sin\theta]; \tag{d2}$$

from which, eliminating v_{Ax}, while recalling the first of (b), we obtain Ω [ex. 4.5.3: (i1)].

(iii) Show that the vertical constraint impulse $\hat{\lambda} = \hat{R}_y$ is found from the impulsive Routh–Voss equation

$$\Delta(\partial T/\partial v_{Ay}) = \hat{R}_y:$$

$$\Delta(mv_{Ay} - mb\Omega\sin\theta) = -m\sin\theta\big[(b\omega + 3v\sin\theta)/(1 + 3\sin^2\theta)\big] + mv = \hat{R}_y, \tag{e}$$

in agreement with the earlier expressions (i3) and (o3) of ex. 4.5.3.

Example 4.5.4 A homogeneous straight rigid bar AB of length L and mass M can rotate freely about a fixed pin at A. A particle of mass m strikes the bar and then *slides along it*. The entire figure lies on a smooth horizontal plane O–xy (fig. 4.13). Find the postimpact velocities and forces (reactions) if the bar is initially at rest; the particle strikes at a distance $\alpha L\ (0 < \alpha < 1)$, when the bar makes an angle $\theta = \theta_o$ with the positive x-axis, with preimpact velocity components $(\dot{x})^- = 0, (\dot{y})^- = v_o \equiv v$ (Bahar, 1970–1980; Greenwood, 1977, p. 118).

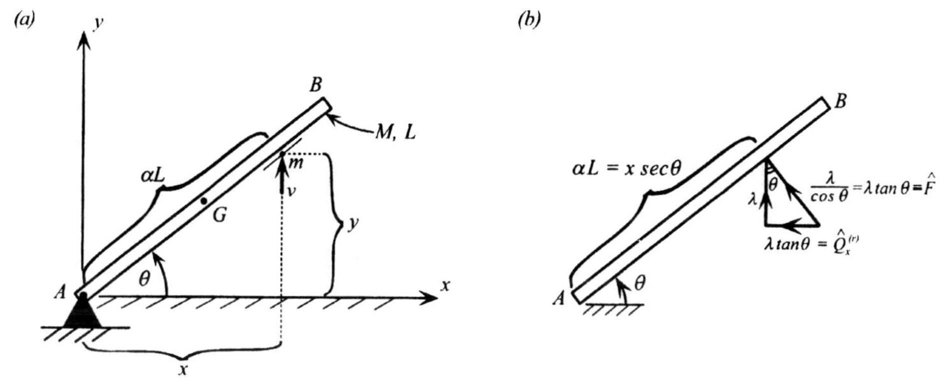

Figure 4.13 (a) Particle impacting on a straight rigid bar *AB*, at a distance αL from *A*; (b) corresponding constraint reaction, and its components. $\alpha L = x \sec\theta$; $\hat{\lambda}\sec\theta \equiv \hat{F}, \hat{\lambda}\tan\theta = \hat{R}_x$.

By König's theorem,

$$2T = M(L\dot{\theta}/2)^2 + (ML^2/12)(\dot{\theta})^2 + m[(\dot{x})^2 + (\dot{y})^2]$$
$$= M[L^2(\dot{\theta})^2/3] + m[(\dot{x})^2 + (\dot{y})^2]; \qquad \text{(a1)}$$

also,

$$\widehat{\delta' W} = 0, \qquad \hat{Q}_{x,y,\theta} = 0. \qquad \text{(a2)}$$

1. First Solution: Holonomic Coordinates

By $(\ldots)^{\cdot}$-differentiating the holonomic and stationary constraint $y = x\tan\theta$ [fig. 4.13(a)], we find

$$\dot{y} = \dot{x}\tan\theta + x\dot{\theta}\sec^2\theta \Rightarrow (\tan\theta)\dot{x} + (-1)\dot{y} + (x\sec^2\theta)\dot{\theta} = 0, \qquad \text{(b1)}$$

and, since this system is scleronomic,

$$(\tan\theta)\,\delta x + (-1)\,\delta y + (x\sec^2\theta)\,\delta\theta = 0. \qquad \text{(b2)}$$

Further, we have

Particle configuration: $x = (\alpha L)\cos\theta, \quad y = (\alpha L)\sin\theta,$ (b3)

Preimpact velocities: $(\dot{x})^- = 0, \quad (\dot{y})^- = v, \quad (\dot{\theta})^- = 0,$ (b4)

Postimpact velocities: $(\tan\theta)\dot{x} + (-1)\dot{y} + (x\sec^2\theta)\dot{\theta} = 0.$ (b5)

Therefore, the impulsive principle of Lagrange yields

$$0 = \Delta(\partial T/\partial\dot{x})\,\delta x + \Delta(\partial T/\partial\dot{y})\,\delta y + \Delta(\partial T/\partial\dot{\theta})\,\delta\theta$$
$$= [m\dot{x} - m(\dot{x})^-]\,\delta x + [m\dot{y} - m(\dot{y})^-]\,\delta y + (ML^2/3)[\dot{\theta} - (\dot{\theta})^-]\,\delta\theta$$
$$= (m\dot{x})\,\delta x + m(\dot{y} - v)\,\delta y + (ML^2/3)\dot{\theta}\,\delta\theta, \qquad \text{(c)}$$

under (b2). Applying the multiplier rule to (c), with (b2), we readily obtain

$$m\dot{x} = \hat{\lambda}\tan\theta, \qquad m(\dot{y} - v) = -\hat{\lambda}, \qquad (ML^2/3)\dot{\theta} = \hat{\lambda}(x\sec^2\theta), \tag{d1, 2, 3}$$

which along with (b1) [or (b5)] constitutes an algebraic system of *four* equations for $\dot{x}, \dot{y}, \dot{\theta}, \hat{\lambda}$. Solving it, we find (with $\beta \equiv m/M$)

$$\dot{x} = (\sin\theta\cos\theta/3\alpha^2\beta)v, \tag{e1}$$

$$\dot{y} = [(\sin^2\theta + 3\alpha^2\beta)/(1 + 3\alpha^2\beta)]v, \tag{e2}$$

$$L\dot{\theta} - [(3\alpha\cos\theta)/(\beta^{-1} + 3\alpha^2)]v; \tag{e3}$$

$$\hat{\lambda} = m(v - \dot{y}) = \left[\cos^2\theta/(1 + 3\alpha^2\beta)\right]m\,v$$
$$= [\cos^2\theta/(\beta^{-1} + 3\alpha^2)]Mv. \tag{e4}$$

From the above, and figure 4.13(b), we find that the holonomic components of the impulsive constraint reactions equal

$$\hat{R}_x = \hat{\lambda}\tan\theta = \hat{F}\sin\theta = (\sin\theta/\alpha L)\hat{R}_\theta, \tag{f1}$$

$$\hat{R}_y = -\hat{\lambda} = -(\cos\theta/\alpha L)\hat{R}_\theta, \tag{f2}$$

$$\hat{R}_\theta = \widehat{(\delta' W_R)_o}/\delta\theta = [(\hat{\lambda}\sec\theta)(x\sec\theta)\,\delta\theta]/\delta\theta = \hat{\lambda}\,x\sec^2\theta. \tag{f3}$$

2. Second Solution: Nonholonomic Coordinates

In view of the constraint (b1), and recalling (b3), we introduce the following quasi velocities:

$$\omega_1 \equiv (\tan\theta)\dot{x} + (-1)\dot{y} + (\alpha L/\cos\theta)\dot{\theta} \quad (= 0), \qquad \omega_2 \equiv \dot{x}, \qquad \omega_3 \equiv \dot{y}; \tag{g1}$$

with inverses (unconstrained, since we want to calculate the impulsive reactions)

$$\dot{x} = \omega_2, \qquad \dot{y} = \omega_3, \qquad \dot{\theta} = (\cos\theta/\alpha L)\omega_1 + (\cos\theta/\alpha L)\omega_3 + (-\sin\theta/\alpha L)\omega_2. \tag{g2}$$

Then:

(i) The, also unconstrained, (double) kinetic energy is

$$2T \to 2T^* = (M/3\alpha^2)\left[(\omega_1 + \omega_3)\cos\theta - \omega_2\sin\theta\right]^2 + m({\omega_2}^2 + {\omega_3}^2); \tag{h}$$

(ii) The *preimpact* velocities are

$$(\dot{\theta})^- = 0, \qquad (\dot{x})^- = 0, \qquad (\dot{y})^- = v \;\Rightarrow\; {\omega_1}^- = -v, \qquad {\omega_2}^- = 0, \qquad {\omega_3}^- = v, \tag{i}$$

$\omega_1 \equiv {\omega_1}^+ \; (= 0) \neq {\omega_1}^-$ (sudden $\to$ persistent constraints; i.e., *second*-type problem); (j)

(iii) The (unconstrained) impulsive virtual works are

$$\widehat{(\delta' W)^*} \equiv \hat{\Theta}_1\,\delta\theta_1 + \hat{\Theta}_2\,\delta\theta_2 + \hat{\Theta}_3\,\delta\theta_3 = 0 \Rightarrow \hat{\Theta}_{1,2,3} = 0, \tag{k1}$$

$$0 = \widehat{(\delta' W_R)^*} \equiv \hat{\Lambda}_1\,\delta\theta_1 + \hat{\Lambda}_2\,\delta\theta_2 + \hat{\Lambda}_3\,\delta\theta_3 \qquad \text{[invoking the virtual form of (g2)]}$$

$$= \hat{\Lambda}_1[(\tan\theta)\,\delta x - \delta y + (\alpha L/\cos\theta)\,\delta\theta] + \hat{\Lambda}_2\,\delta x + \hat{\Lambda}_3\,\delta y$$

$$= \cdots = \hat{R}_x\,\delta x + \hat{R}_y\,\delta y + \hat{R}_\theta\,\delta\theta \equiv \widehat{\delta' W_R}, \tag{k2}$$

$$\Rightarrow \hat{R}_x = (\tan\theta)\hat{\Lambda}_1 + \hat{\Lambda}_2 = (\tan\theta)\hat{\Lambda}_1 \quad [= (\sin\theta/\alpha L)\hat{R}_\theta] \qquad (\text{since } \hat{\Lambda}_2 = 0), \tag{k3}$$

$$\hat{R}_y = (-1)\hat{\Lambda}_1 + \hat{\Lambda}_3 = -\hat{\Lambda}_1 \quad [= -(\cos\theta/\alpha L)\hat{R}_\theta] \qquad (\text{since } \hat{\Lambda}_3 = 0), \tag{k4}$$

$$\hat{R}_\theta = (\alpha L/\cos\theta)\hat{\Lambda}_1 \qquad (\text{where } \hat{\Lambda}_1 = \hat{\lambda}). \tag{k5}$$

Therefore, the Hamel impulsive equations are (we recall to set $\omega_1^+ \equiv \omega_1 = 0$, *after* the differentiations)

$$\omega_1: \quad \Delta(\partial T^*/\partial\omega_1) = \hat{\Lambda}_1:$$

$$(M/3\alpha^2)(\omega_3\cos^2\theta - \omega_2\sin\theta\cos\theta) = \hat{\Lambda}_1, \tag{l1}$$

$$\omega_2: \quad \Delta(\partial T^*/\partial\omega_2) = 0:$$

$$(M/3\alpha^2)(\omega_2\sin^2\theta - \omega_3\sin\theta\cos\theta) + m\,\omega_2 = 0, \tag{l2}$$

$$\omega_3: \quad \Delta(\partial T^*/\partial\omega_3) = 0:$$

$$(M/3\alpha^2)(\omega_3\cos^2\theta - \omega_2\sin\theta\cos\theta) + m(\omega_3 - v) = 0. \tag{l3}$$

Solving this algebraic system for $\omega_{2,3}$ and $\hat{\Lambda}_1$ we find

$$\omega_2 = \dot{x} = [\varepsilon\,\sin\theta\cos\theta/(1+\varepsilon)]v, \tag{m1}$$

$$\omega_3 = \dot{y} = [(1 + \varepsilon\,\sin^2\theta)/(1+\varepsilon)]v, \tag{m2}$$

$$\hat{\Lambda}_1 = \hat{\lambda} = (M/3\alpha^2)[\cos^2\theta/(1+\varepsilon)]v, \tag{m3}$$

where

$$\varepsilon \equiv M/3m\alpha^2 = (M/m)(1/3\alpha^2) = 1/3\alpha^2\beta, \tag{m4}$$

$$\Rightarrow \dot{\theta} = (\cos\theta/\alpha L)\omega_3 - (\sin\theta/\alpha L)\omega_2 = [\cos\theta/\alpha L(1+\varepsilon)]v; \tag{m5}$$

which, naturally, coincide with the earlier values.

CLOSING REMARKS

If the problem was one of the *first* kind (i.e., only impressed impulsive forces, no change of constraints), then $\Delta\omega_1 \equiv {\omega_1}^+ - {\omega_1}^- = 0 - 0$, and since always $(\partial T^*/\partial\omega_I)_o = \partial T^*{}_o/\partial\omega_I$, the *kinetic* impulsive equations can be written as $\Delta(\partial T^*{}_o/\partial\omega_I) = \hat{\Theta}_I$ $(I = 2, 3)$. But in *second* kind problems, like this one, since ${\omega_1}^- \neq 0$, the notation $(\ldots)_o$ can only mean setting ${\omega_1}^+ = 0$, after all differentiations; that is, we must start with T^*, even if we do not seek the impulsive reactions. Then,

$$(\partial T^*/\partial\omega_I)^+ = (\partial T^*{}_o/\partial\omega_I)^+ \quad \text{but} \quad (\partial T^*/\partial\omega_I)^- \neq (\partial T^*{}_o/\partial\omega_I)^-, \tag{n1, 2}$$

and the kinetic (reactionless) impulsive equations can be written as $(\partial T^*{}_o/\partial\omega_I)^+ - (\partial T^*/\partial\omega_I)^- = \hat{\Theta}_I$. Clearly, these are general results.

Let us verify them for our problem. Equation (h) yields

$$2T^* \to 2T^*{}_o = (M/3\alpha^2)(\omega_3\cos\theta - \omega_2\sin\theta)^2 + m(\omega_2{}^2 + \omega_3{}^2), \tag{o}$$

$$\Rightarrow \partial T^*{}_o/\partial\omega_2 = (M/3\alpha^2)(\omega_3\cos\theta - \omega_2\sin\theta)(-\sin\theta) + m\,\omega_2, \tag{p1}$$

$$(\partial T^*{}_o/\partial\omega_2)^- = (M/3\alpha^2)(v\cos\theta)(-\sin\theta), \tag{p2}$$

but

$$(\partial T^*/\partial\omega_2)^- = (M/3\alpha^2)\big[(-v+v)\cos\theta - 0\big](-\sin\theta) + 0 = 0; \tag{p3}$$

that is,

$$(\partial T^*/\partial\omega_2)^- \neq (\partial T^*{}_o/\partial\omega_2)^-;$$

and

$$(\partial T^*{}_o/\partial\omega_2)^+ = (M/3\alpha^2)(\omega_3\cos\theta - \omega_2\sin\theta)(-\sin\theta) + m\,\omega_2, \tag{p4}$$

$$(\partial T^*/\partial\omega_2)^+ = (M/3\alpha^2)(\omega_3\cos\theta - \omega_2\sin\theta)(-\sin\theta) + m\,\omega_2; \tag{p5}$$

that is,

$$(\partial T^*{}_o/\partial\omega_2)^+ = (\partial T^*/\partial\omega_2)^+, \text{ and similarly for } \omega_3.$$

In sum: the replacement of T^* with $T^*{}_o$ in the kinetic impulsive equations is allowed *only when the constraints* $\omega_D = 0$ *hold.*

Example 4.5.5 Three slender homogeneous bars, AB, BC, CD, each of mass m and length $2b$, are pinned together at B and C, and pivoted at A to a fixed horizontal table. The end D receives an impulse $\hat{P}$ [fig. 4.14a]. Let us find the translational and rotational (angular) velocities of the mass center of each rod (Bahar, 1987; Chorlton, 1983, pp. 227–229; also Beghin, 1967, pp. 472–473).

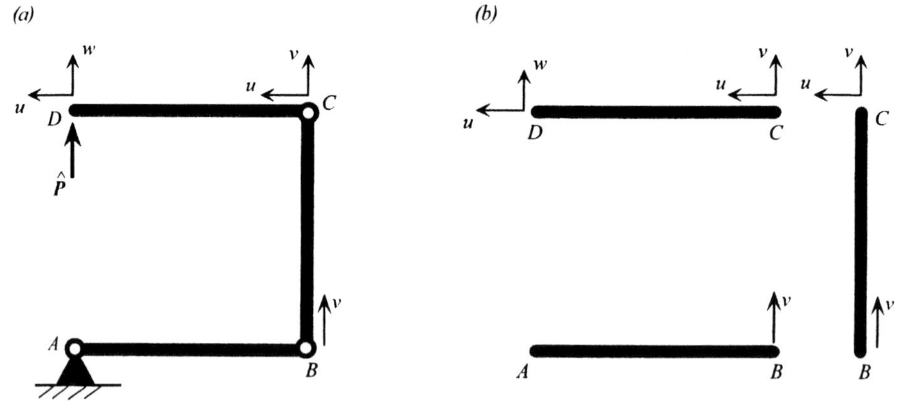

Figure 4.14 (a) System consisting of three homogeneous bars AB, BC, CD, impacted at D; (b) quasi velocities chosen to describe its velocities. "British theorem": the kinetic energy of a thin homogeneous bar AB, of mass m, equals $T = (m/6)(v_A{}^2 + v_B{}^2 + \mathbf{v}_A\cdot\mathbf{v}_B)$.

1. Kinetic Equations

Although, clearly, this is a holonomic system, we choose to describe it by the quasi velocities shown in fig. 4.14(a, b), compatible with plane and rigid kinematics. Using the "British theorem" [(1.17.8)] and some easily understood ad hoc notation, we find that the kinetic energy of the system, *at the impact configuration*, is

$$\begin{aligned} T &= T_{AB} + T_{BC} + T_{CD} \\ &= (m/6)(0 + v^2) + (m/6)(v^2 + u^2 + v^2 + v^2) + (m/6)(u^2 + v^2 + u^2 + w^2 + u^2 + w\,v) \\ &= (m/6)(5v^2 + 4u^2 + w^2 + w\,v). \qquad \text{(a)} \end{aligned}$$

[For comparison purposes, we point out that the König theorem-based calculation would have given

$$2T_{AB} = m(v/2)^2 + (mb^2/3)(v/2b)^2 = m\,v^2/3, \qquad \text{(a1)}$$

$$2T_{BC} = m[v^2 + (u^2/4)] + (mb^2/3)(u/2b)^2 = m\,v^2 + m\,u^2/3, \qquad \text{(a2)}$$

$$\begin{aligned} 2T_{CD} &= m\{u^2 + [(w + v)/2]^2\} + (m\,b^2/3)[(w - v)/2b]^2 \\ &= (m/3)(3u^2 + v^2 + w^2 + w\,v), \quad \text{etc.}] \qquad \text{(a3)} \end{aligned}$$

The impressed impulsive forces are calculated from the corresponding virtual work expression (as if u, v, w were quasi coordinates):

$$\widehat{\delta' W} = \hat{\Theta}_u\,\delta u + \hat{\Theta}_v\,\delta v + \hat{\Theta}_w\,\delta w = \hat{P}\,\delta w \;\Rightarrow\; \hat{\Theta}_u = 0, \hat{\Theta}_v = 0, \hat{\Theta}_w = \hat{P}. \qquad \text{(b)}$$

In view of the above, the Hamel equations of motion are (with T instead of the customary T^*, and $\omega^+ \equiv \omega$ for all *postimpact* velocities)

$$\Delta(\partial T/\partial u) = 0 \Rightarrow (m/6)(8u) = 0, \qquad \text{(c1)}$$

$$\Delta(\partial T/\partial v) = 0 \Rightarrow (m/6)(10v + w) = 0, \qquad \text{(c2)}$$

$$\Delta(\partial T/\partial w) = \hat{P} \Rightarrow (m/6)(v + 2w) = \hat{P}; \qquad \text{(c3)}$$

and their solution is easily found to be

$$u = 0, \qquad v = -(6/19)(\hat{P}/m), \qquad w = (60/19)(\hat{P}/m). \qquad \text{(d)}$$

Then, by simple kinematics,

$$\begin{aligned} &\omega_{AB} = v/2b = -(3/19)(\hat{P}/mb) \quad \text{(clockwise)}, \qquad \omega_{BC} = 0, \\ &\omega_{DC} = (w - v)/2b = (33/19)(\hat{P}/mb) \quad \text{(clockwise)}; \qquad \text{(e1)} \end{aligned}$$

and

$$\begin{aligned} &\boldsymbol{v}_{\text{center of mass of } AB} = v/2 = -(3/19)(\hat{P}/m) \qquad \text{(downwards)}, \\ &\boldsymbol{v}_{\text{center of mass of } BC} = v = -(6/19)(\hat{P}/m) \qquad \text{(downwards)}, \\ &\boldsymbol{v}_{\text{center of mass of } DC} = (v + w)/2 = (27/19)(\hat{P}/m) \qquad \text{(upwards)}. \qquad \text{(e2)} \end{aligned}$$

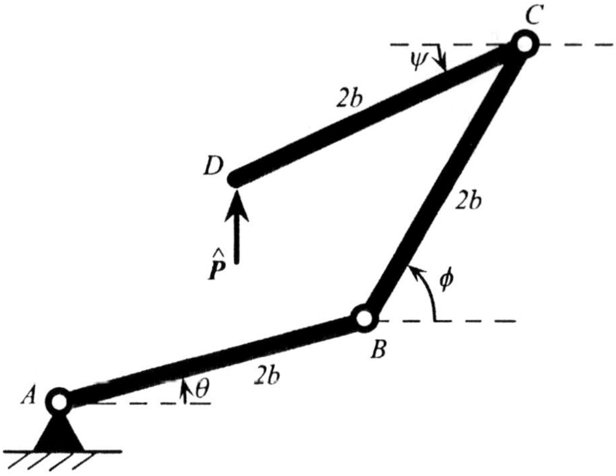

Figure 4.15 Geometry of impact problem of fig. 4.14, but for a *generic* configuration, defined by the bar angles $q_{1,2,3}$: ϕ, θ, ψ (see Beghin, 1967, pp. 472–473; Chorlton, 1983, pp. 227–229).

Note the simplicity of the quasi-velocity approach, as compared with Lagrangean (holonomic) coordinates in connection with the calculation of T for a *general* configuration (fig. 4.15).

2. Kinetostatic Equations

Next, let us use the *impulsive relaxation principle* to calculate the (*external*) impulsive reaction at A. Since that "force" has components in both directions, we must allow A to move both *up/down* and *left/right* (fig. 4.16).

We notice the additional horizontal velocity component x at B, due to another x at A; and a vertical one y at A; and two "force" components at A that go along with x, y: X, Y. The *relaxed* kinetic energy is

$$\begin{aligned} T_{\text{relaxed}} \equiv T_{rx} &= (m/6)(w^2 + u^2 + u^2 + v^2 + u^2 + wv + u^2 + v^2 + v^2 \\ &\quad + x^2 + v^2 + ux + x^2 + y^2 + x^2 + v^2 + x^2 + yv) \\ &= (m/6)(w^2 + 4u^2 + 5v^2 + 4x^2 + y^2 + wv + ux + yv), \end{aligned} \tag{f}$$

and so the equations of motion are

$$\Delta\left[(\partial T_{rx}/\partial \omega_k)_o\right] = \hat{\Theta}_k + \hat{\Lambda}_k, \tag{g1}$$

where

$$\{\omega_k\text{: } u, v, w;\ x = 0, y = 0\}, \qquad \{\hat{\Theta}_k\text{: } 0, 0, \hat{P};\ 0, 0\}, \qquad \{\hat{\Lambda}_k\text{: } 0, 0, 0;\ X, Y\}; \tag{g2}$$

and $(\ldots)_o$ means enforcement of the constraints $x = 0, y = 0$, in $(\ldots)$. Thus, and

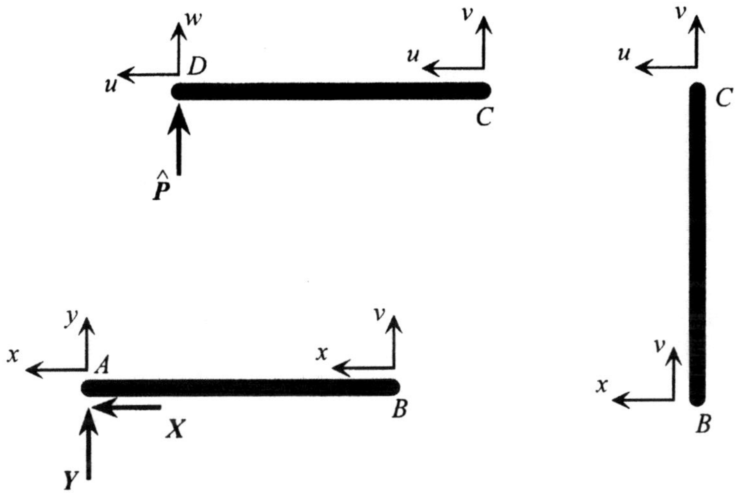

Figure 4.16 Velocities and *external* impulsive forces of impact problem of fig. 4.14.

dropping the subscript rx from T, for simplicity, we find

$$(\partial T/\partial u)_o = 0: \qquad (m/6)(8u + x)_o = (m/6)(8u) = 0, \tag{h1}$$

$$(\partial T/\partial v)_o = 0: \qquad (m/6)(10v + w + y)_o = (m/6)(10v + w) = 0, \tag{h2}$$

$$(\partial T/\partial w)_o = \hat{P}: \qquad (m/6)(2w + v)_o = (m/6)(2w + v) = \hat{P}, \tag{h3}$$

$$(\partial T/\partial x)_o = X: \qquad (m/6)(8x + u)_o = (m/6)(u) = X, \tag{h4}$$

$$(\partial T/\partial y)_o = Y: \qquad (m/6)(2y + v)_o = (m/6)(v) = Y; \tag{h5}$$

from which, since $u = 0$, we obtain

$$X = mu/6 = 0,$$

$$Y = mv/6 = (m/6)(-6/19)(\hat{P}/m) = (-1/19)\hat{P} \qquad \text{(i.e., downward, at } A). \tag{i}$$

These results can be easily confirmed via the elementary (Newton–Euler) methods: by linear momentum for the entire system, in the + *vertical* direction, we get

$$\hat{P} + Y = m[(v/2) + v + (w + v)/2] = m[2v + (w/2)], \tag{j}$$

from which, and the earlier values (d) [obtained by use of the kinetic equations (c1–3)], we find $Y = \cdots = (-1/19)\hat{P}$; and similarly for the horizontal direction, we find $X = 0$.

Problem 4.5.4 (Smart, 1951, pp. 264–265). Four equal straight homogeneous rods, AB, BC, CD, and DE, each of length l and mass m, are smoothly joined together at B, C, D, and rest on a smooth horizontal table so that consecutive rods are perpendicular to each other (fig. 4.17). The midpoints of all rods are collinear, and the end E is fixed. Then, the end A is struck by a blow $\hat{P}$ parallel to the line joining the midpoints of the rods (i.e., along ACE).

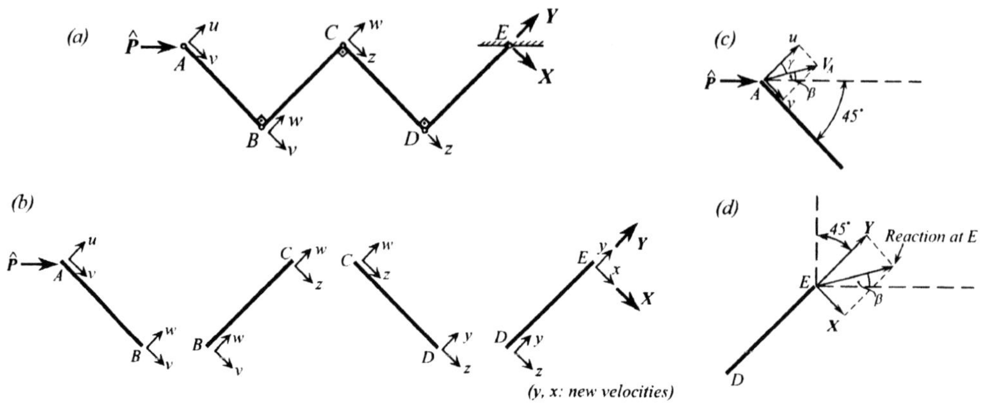

Figure 4.17 System of four, originally motionless, straight and mutually perpendicular rods, AB, BC, CD, and DE, (B, C, D: hinged; E: fixed) struck by a blow $\hat{P}$ at its end A, along ACE. (a) General view of problem; (b) free-body diagram of each rod; (c) detail of application of blow $\hat{P}$, at A; (d) detail of impulsive reaction, at E.

(i) Using the kinematically compatible (postimpact) quasi velocities u, v, w, z (fig. 4.17), and the earlier "British theorem," show that the (double) kinetic energy and (impressed) impulsive virtual work, at the impact configuration, equal, respectively,

$$2T = (m/3)(u^2 + 4v^2 + 5w^2 + 5z^2 + u\,w + z\,v), \tag{a}$$

$$\begin{aligned} \widehat{\delta' W} &= \hat{\Theta}_u\,\delta u + \hat{\Theta}_v\,\delta v + \hat{\Theta}_w\,\delta w \\ &= (\hat{P}/\sqrt{2})\,\delta u + (\hat{P}/\sqrt{2})\,\delta v + (0)\,\delta w + (0)\,\delta z \\ &\qquad \text{(as if } u, v, w, z \text{ were quasi coordinates).} \end{aligned} \tag{b}$$

(ii) Since the preimpact state is one of rest—that is, $\Delta(\partial T/\partial\omega) = \partial T/\partial\omega$ (ω: u, v, w, z)—show that the equations of impulsive motion are

$$u{:}\quad (m/6)(2u + w) = \hat{P}/\sqrt{2}, \qquad v{:}\quad (m/6)(8v + z) = \hat{P}/\sqrt{2}, \tag{c1, 2}$$

$$w{:}\quad 10w + u = 0, \qquad z{:}\quad 10z + v = 0; \tag{c3, 4}$$

with solutions

$$u = (30\sqrt{2}\,/19)(\hat{P}/m), \qquad v = (30\sqrt{2}\,/79)(\hat{P}/m), \tag{c5, 6}$$

$$w = -(3\sqrt{2}\,/19)(\hat{P}/m), \qquad z = -(3\sqrt{2}\,/79)(\hat{P}/m). \tag{c7, 8}$$

(iii) Show that A begins to move in a direction making an angle $\tan^{-1}(30/49)$ with that of the blow, and find the angle between the impulsive external reaction at E and the blow $\hat{P}$.

HINTS

(i) If the angle between $\hat{P}$ and $\mathbf{v}_A$ is β, then $\tan\beta = (u - v)/(u + v) = \cdots = 30/49$.

[From trigonometry: tan (*angle between u and* $\mathbf{v}_A$) $\equiv \tan\gamma = v/u$, and $\tan(\beta + \gamma) = \tan 45° = 1 = (\tan\beta + \tan\gamma)/(1 - \tan\beta\tan\gamma) \Rightarrow \tan\beta = \cdots$.]

(ii) The components of the reaction at E, X (perpendicular to DE), and Y (along DE) can be found, either from the impulsive principle of relaxation (see next problem), or from the principle of linear momentum applied to the entire (nonrigid) system, in these two directions:

$$\text{Along } DE: \qquad Y + (\hat{P}/\sqrt{2}) = m[(w/2) + w + ((u+w)/2)], \tag{d1}$$

$$\text{Perpendicular to } DE: \qquad X + (\hat{P}/\sqrt{2}) = m[(z/2) + z + ((z+v)/2) + v]. \tag{d2}$$

Verify that the above, with the help of (c1–4), yield

$$X = \hat{P}\sqrt{2}[-(1/2) + (39/79)], \qquad Y = \hat{P}\sqrt{2}[-(1/2) + (9/19)], \tag{d3}$$

and, therefore,

$$\tan(\text{angle between reaction at } E \text{ and horizontal}) = (Y - X)/(Y + X) = 30/49$$
$$= \tan(\text{angle between } \boldsymbol{v}_A \text{ and horizontal}) \equiv \tan\beta, \quad \text{Q.E.D.} \tag{e}$$

Problem 4.5.5 Continuing from the preceding problem, calculate the *external* reaction components X, Y via the impulsive principle of relaxation; allow E to move with corresponding velocities x (perpendicular to DE; i.e., parallel to X) and y (along DE; i.e., parallel to Y). Formulate the equations of the relaxed system, and then set, at the end, $x = 0, y = 0$ [fig. 4.17(b)].

HINTS

Show that the (double) kinetic energy of the so-relaxed system is

$$2T_{\text{relaxed}} = (m/3)(u^2 + 4v^2 + 5w^2 + 5z^2 + uw + 4y^2 + zv + x^2 + wy + xz), \tag{a}$$

and, therefore, (a) the *kinetic* equations remain the same as in the preceding (constrained) case; while (b) the additional *kinetostatic* equations are

$$(mw/6) = Y \Rightarrow Y = \cdots = -(\sqrt{2}/38)\hat{P}, \quad (mz/6) = X \Rightarrow X = \cdots = -(\sqrt{2}/158)\hat{P}; \tag{b}$$

in agreement with the values found in the previous problem.

Problem 4.5.6 (Synge and Griffith, 1959, pp. 429–430). Two uniform rods, AB and BC, each of mass m and length $2b$, are smoothly hinged at B and, initially, rest on a smooth horizontal table, so that A, B, C are collinear. Then, a horizontal blow $\hat{P}$ is struck at C in a direction perpendicular to BC (fig. 4.18).

(i) *Holonomic coordinates.* Show that, in terms of the Lagrangean coordinates, (x, y): coordinates of B (positive to the right and upward, respectively) and (θ_1, θ_2): angles of AB, BC, respectively, with horizontal (both positive counterclockwise), and with $k^2 \equiv b^2/3$, the postimpact (double) kinetic energy and (impressed) impulsive virtual work, at the impact configuration (i.e., x, y; θ_1, $\theta_2 = 0$), equal, respectively

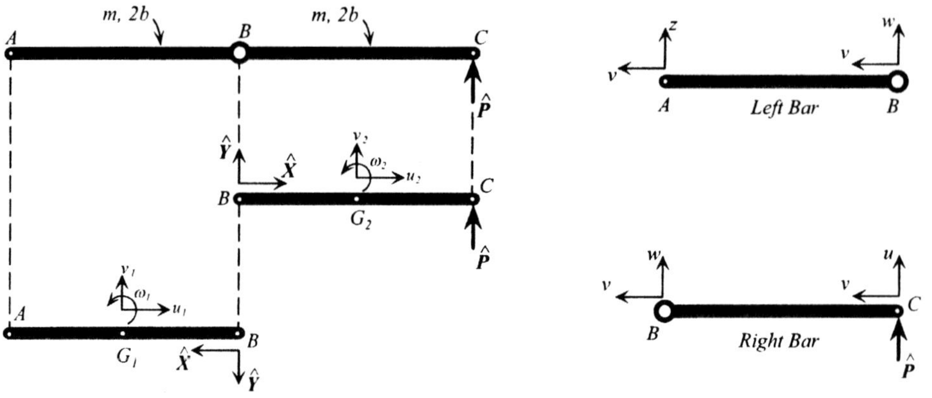

Figure 4.18 System of two, originally motionless, straight rods, *AB*, *BC* (C: hinged; *A*, *B*, *C*: collinear) struck by a blow $\hat{P}$ at its end *C*, at a right angle to *ABC*.

(with $\dot{\theta}_1 \equiv \omega_1$, $\dot{\theta}_2 \equiv \omega_2$),

$$2T = m[(\dot{x})^2 + (\dot{y} - b\,\omega_1)^2 + k^2{\omega_1}^2] + m[(\dot{x})^2 + (\dot{y} + b\,\omega_2)^2 + k^2{\omega_2}^2], \tag{a}$$

$$\widehat{\delta' W} = \hat{X}\,\delta x + \hat{Y}\,\delta y + \hat{\Theta}_1\,\delta\theta_1 + \hat{\Theta}_2\,\delta\theta_2 = \hat{P}(\delta y + 2b\,\delta\theta_2) \tag{b}$$

$$[\text{i.e.,}\ \ \hat{X} = 0, \qquad \hat{Y} = \hat{P}, \qquad \hat{\Theta}_1 = 0, \qquad \hat{\Theta}_2 = 2b\hat{P}]$$

and, therefore, verify that the equations of impulsive motion are,

$$x:\quad 2m\dot{x} - 0, \tag{c1}$$

$$y:\quad m(\dot{y} - b\,\omega_1) + m(\dot{y} + b\,\omega_2) = \hat{P}, \tag{c2}$$

$$\theta_1:\quad -m\,b(\dot{y} - b\,\omega_1) + (m\,k^2)\omega_1 = 0, \tag{c3}$$

$$\theta_2:\quad m\,b(\dot{y} + b\,\omega_2) + (m\,k^2)\omega_2 = 2b\hat{P}; \tag{c4}$$

with solutions

$$\dot{x} = 0, \qquad \dot{y} = -(\hat{P}/m) \qquad (\text{i.e., initially, } B \text{ moves downward only}), \tag{c5}$$

$$\omega_1 = -(3/4)(\hat{P}/mb)\ (\text{i.e., clockwise}), \quad \omega_2 = (9/4)(\hat{P}/mb)\ (\text{i.e., counterclockwise}). \tag{c6}$$

(ii) *Nonholonomic coordinates*. Show that in terms of the kinematically compatible quasi velocities u, v, w, z (fig. 4.18), again at the impact configuration,

$$2T = (m/3)(u^2 + 6v^2 + 2w^2 + z^2 + u\,w + z\,w), \tag{d}$$

$$\widehat{\delta' W} = \hat{P}\,\delta\theta \qquad (\text{where } \dot{\theta} \equiv u), \tag{e}$$

and, therefore, verify that the equations of impulsive motion are

$$(m/6)(2u + w) = \hat{P}, \quad (m/6)(12v) = 0, \quad (m/6)(u + 4w + z) = 0, \quad (m/6)(w + 2z) = 0; \qquad \text{(f)}$$

with solutions

$$u = (7/2)(\hat{P}/m), \qquad v = 0, \qquad w = -(\hat{P}/m), \qquad z = (1/2)(\hat{P}/m). \qquad \text{(g)}$$

(iii) Show that as the number of bars goes to infinity (to the left of C, B, A), $u = 2\sqrt{3}(\hat{P}/m)$.

Problem 4.5.7 Continuing from the preceding problem (figs. 4.18, 4.19), calculate the *internal* reaction components at B: $\hat{X}$ (assumed upward on the left bar, downward on the right) and $\hat{Y}$ (assumed leftward on the left bar, rightward on the right bar), via the impulsive principle of relaxation: allow $B_{left\,\mathrm{bar}}$ to move with corresponding velocities x (upward) and y (leftward), and $B_{right\,\mathrm{bar}}$ to move with corresponding velocities w (upward) and v (leftward). Formulate the equations of the relaxed system, and then set, at the end, $x = w$, $y = v$.

HINTS

Show that for the so-relaxed system,

$$2T_{\mathrm{relaxed}} = (m/3)(u^2 + 3v^2 + w^2 + x^2 + 3y^2 + z^2 + uw + xz), \qquad \text{(a)}$$

$$\widehat{\delta' W} = \hat{P}\,\delta\upsilon - \hat{Y}\,\delta\varpi - \hat{X}\,\delta\omega + \hat{X}\,\delta\xi + \hat{Y}\,\delta\zeta, \qquad \text{(b)}$$

where

$$\dot{\upsilon} \equiv u, \qquad \dot{\varpi} \equiv v, \qquad \dot{\omega} \equiv w, \qquad \dot{\xi} \equiv x, \qquad \dot{\zeta} \equiv y; \qquad \text{(b1)}$$

and, hence, verify that [with the notations: $T_{\mathrm{relaxed}} \equiv T$ and $(\ldots)_o \equiv (\ldots)_{\mathrm{constraints\ enforced}}$] the equations of motion are

$$u: \quad (\partial T/\partial u)_o = \hat{P} \Rightarrow 2u + w = 6\hat{P}/m; \qquad \text{(c1)}$$

$$v: \quad (\partial T/\partial v)_o = -\hat{Y} \Rightarrow v = -(\hat{Y}/m); \qquad \text{(c2)}$$

$$w: \quad (\partial T/\partial w)_o = -\hat{X} \Rightarrow 2w + u = -(6\hat{X}/m); \qquad \text{(c3)}$$

$$x: \quad (\partial T/\partial x)_o = \hat{X} \Rightarrow 2w + z = 6\hat{X}/m; \qquad \text{(c4)}$$

$$y: \quad (\partial T/\partial y)_o = \hat{Y} \Rightarrow v = \hat{Y}/m; \qquad \text{(c5)}$$

$$z: \quad (\partial T/\partial z)_o = 0 \Rightarrow 2z + w = 0; \qquad \text{(c6)}$$

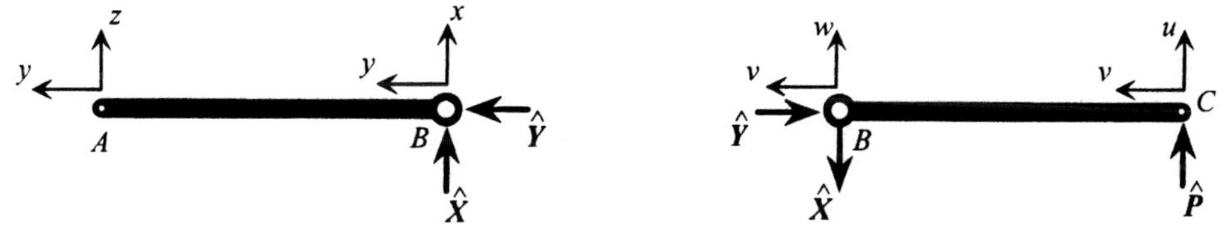

Figure 4.19 System of two, originally motionless straight rods, *AB*, *BC* (*C*: hinged; *A*, *B*, *C*: collinear) struck by a blow $\hat{P}$ at its end *C*, at a right angle to *ABC*, specially relaxed at *B* in order to calculate the internal reactions there.

with solutions

$$u = (7/2)(\hat{P}/m), \quad v = 0, \quad w = -(\hat{P}/m), \quad z = (1/2)(\hat{P}/m), \quad \hat{X} = -(\hat{P}/4), \quad \hat{Y} = 0, \tag{d}$$

in agreement with the values found (for the postimpact velocities) in the preceding problem.

REMARK

Note that $(\partial T/\partial w)_o + (\partial T/\partial x)_o = 0$, $(\partial T/\partial v)_o + (\partial T/\partial y)_o = 0$; that is, *the sum of the internal reactions vanishes*, like a Lagrangean form of impulsive action–reaction.

For additional aspects of this problem, see Kilmister and Reeve (1966, pp. 220–221, 229, 235–250).

Problem 4.5.8 (Beer, 1963; Kane, 1962; Raher, 1954, 1955). Two identical, homogeneous, circular, and thin (sharp-edged) wheels, W' and W'' (fig. 4.20), each of radius r and mass m, are capable of rotating freely about the ends of a common axle A, of mass M and length $2b$, so that the entire assembly can roll on a fixed, perfectly rough, and horizontal plane P. The system is struck (set in motion) by an impulse $\hat{I}$, acting for the very short time interval $[t', t'']$, at the axle point S located a distance $i\,(< 2b)$ from the center of W' (with no loss in generality), perpendicularly to A and parallel to P.

The *kinematics* of this problem has been discussed in ex. 2.13.8; while the *kinetics* of its ordinary (continuous) motion has been detailed in ex. 3.18.6. It was found there that, with the Lagrangean coordinates,

(x, y): coordinates of axle midpoint, G;

ϕ: angle (of precession) of line joining the contact points of W' and W'', C' and C'', respectively, with the $+x$-axis; and

(ψ', ψ''): angles of rolling (or, of proper rotation) of W' and W'',

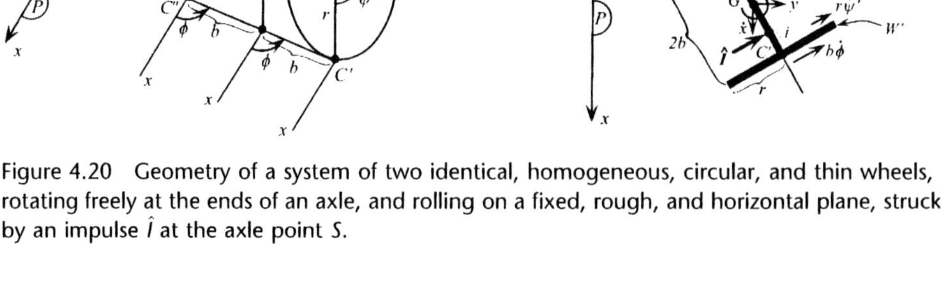

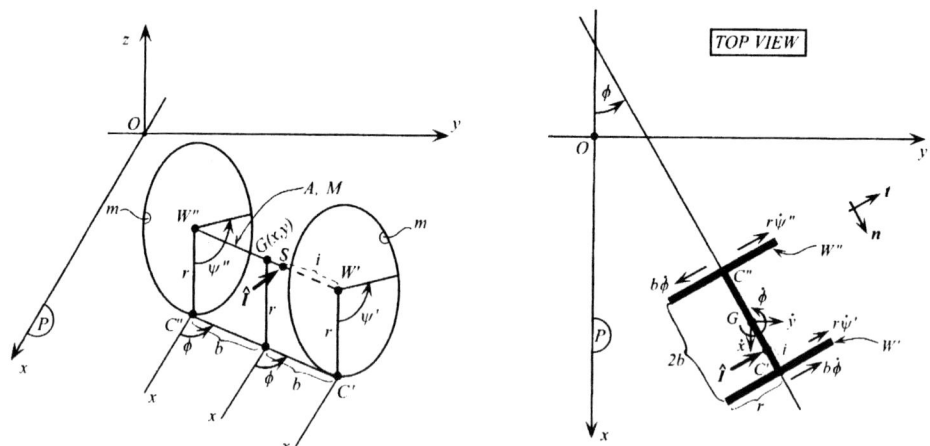

Figure 4.20 Geometry of a system of two identical, homogeneous, circular, and thin wheels, rotating freely at the ends of an axle, and rolling on a fixed, rough, and horizontal plane, struck by an impulse $\hat{I}$ at the axle point S.

the three constraint equations are (i.e., here $n = 5$, $m = 2$)

$$v_{C',n} = v_{C'',n} = \dot{x}\cos\phi + \dot{y}\sin\phi = 0, \tag{a1}$$

$$v_{C',t} = -\dot{x}\sin\phi + \dot{y}\cos\phi + b\dot{\phi} + r\dot{\psi}' = 0, \tag{a2}$$

$$v_{C'',t} = -\dot{x}\sin\phi + \dot{y}\cos\phi - b\dot{\phi} + r\dot{\psi}'' = 0; \tag{a3}$$

$$\{\Rightarrow \dot{x} = (b\dot{\phi} + r\dot{\psi}')\sin\phi,\ \ \dot{y} = (b\dot{\phi} + r\dot{\psi}')\cos\phi\}, \tag{a4}$$

or, since (a2, 3) yield the *integrable* combination (with $c =$ integration constant, depending on the initial values of ϕ, ψ', ψ''),

$$2b\dot{\phi} + r(\dot{\psi}' - \dot{\psi}'') = 0 \ \Rightarrow\ 2b\phi = c - r(\psi' - \psi''); \tag{b}$$

we may take as independent Lagrangean coordinates: x, y; ψ', ψ'' (i.e., actually, $n = 4$, $m = 2$) under (a1–3), of which only *two* are independent. Finally, since this is an impact problem, we can choose, with no loss in generality, the $+x$-axis so that $\phi = 0$ (or $\phi = \pi/2$); in which case, eqs. (a1–2)$\Rightarrow$(a4) and (a3) simplify, respectively, to

$$\dot{x} = 0, \qquad \dot{y} + b\dot{\phi} + r\dot{\psi}' = 0, \qquad \dot{y} - b\dot{\phi} + r\dot{\psi}'' = 0; \tag{c}$$

or, adding the last two,

$$\dot{x} = 0, \qquad 2\dot{y} + r(\dot{\psi}' + \dot{\psi}'') = 0. \tag{d}$$

(i) Show that the (double) kinetic energy of the entire system is

$$2T = (M + 2m)[(\dot{x})^2 + (\dot{y})^2] + (Mb^2/3)(\dot{\phi})^2$$
$$+ (mr^2/2)\{[(\dot{\psi}')^2 + (\dot{\psi}'')^2] + [1 + 4(b/r)^2](\dot{\phi})^2\}, \tag{e}$$

and, therefore, for the special case where $b = r$ and $M \approx 0$,

$$2T = 2m[(\dot{x})^2 + (\dot{y})^2] + (mr^2/2)[(\dot{\psi}')^2 + (\dot{\psi}'')^2 + 5(\dot{\phi})^2], \tag{f}$$

or, thanks to (c), $T = T(\dot{x}, \dot{y}; \dot{\phi}, \dot{\psi}', \dot{\psi}'') \ \Rightarrow\ T_o(\dot{\phi}, \dot{\psi}') = T_o$,

$$2T \Rightarrow 2T_o = (mr^2/2)[13(\dot{\phi})^2 + 6(\dot{\psi}')^2 + 12\dot{\phi}\,\dot{\psi}']; \tag{g}$$

while the impressed impulsive virtual work is

$$\widehat{\delta' W} = (-i\hat{I})\,\delta\phi + (-r\hat{I})\,\delta\psi' \ \Rightarrow\ \hat{Q}_\phi = -i\hat{I},\ \hat{Q}_{\psi'} = -i\hat{I}. \tag{h}$$

(ii) With the help of the above, deduce that, under the initial conditions (i.e., at $t = t'$)

$$(\dot{\phi})^- = 0, \qquad (\dot{\psi}')^- = 0, \tag{i}$$

the (*Hamel*$\Rightarrow$) *Chaplygin–Voronets* impulsive equations of this *first*-type problem:

$$\Delta(\partial T_o/\partial\dot{\phi}) = \hat{Q}_\phi, \qquad \Delta(\partial T_o/\partial\dot{\psi}') = \hat{Q}_{\psi'}, \tag{j}$$

yield the postimpact velocities (i.e., at $t = t''$)

$$(\dot{\phi})^+ = (2/7)(\hat{I}/mr^2)(r - i), \qquad (\dot{\psi}')^+ = (2/7)(\hat{I}/mr^2)[i - (13/6)r]. \tag{k}$$

Combining these results with those of the ordinary motion case, ex. 3.18.6, we readily see that after the impact (i.e., $t \geq t''$), the system rolls with

$$\dot{\phi} = (\dot{\phi})^+ = constant \equiv \Omega, \qquad \dot{\psi}' = (\dot{\psi}')^+ = constant \equiv \omega'; \tag{l}$$

in which case, the general constraints (a4) (with $b = r$) can be integrated to yield

$$x = R[\cos(\Omega t) - \cos(\Omega t'')], \qquad y = R[\sin(\Omega t) - \sin(\Omega t'')], \tag{m}$$

where

$$R = -r[1 + (\omega'/\Omega)] \Rightarrow |R| = (7/6)r^2|r - i|^{-1} \qquad \text{[invoking (k)]}. \tag{n}$$

Finally, eliminating the time between eqs. (m), we easily obtain

$$[x + R\cos(\Omega t'')]^2 + [y + R\sin(\Omega t'')]^2 = R^2, \tag{o}$$

that is, *the postimpact path of the axle midpoint G is a circle of radius* $|R|$; as predicted in ex. 3.18.6.

Problem 4.5.9 Continuing from the preceding problem:

(i) Show that under the choice of $\dot{\psi}'$ and $\dot{\psi}''$ as independent (quasi) velocities, since in that case $\dot{\psi}'' = \dot{\psi}' + 2\dot{\phi}$,

$$T = T(\dot{x}, \dot{y}; \dot{\phi}, \dot{\psi}', \dot{\psi}'') \Rightarrow T_o(\dot{\psi}', \dot{\psi}'') = T_o:$$

$$2T_o = (mr^2/2)\big[(13/4)(d\psi'/dt)^2 + (13/4)(d\psi''/dt)^2 - (1/2)(d\psi'/dt)(d\psi''/dt)\big]; \tag{a}$$

$$\widehat{\delta' W} = \hat{Q}_\phi\, \delta\phi + \hat{Q}_{\psi'}\, \delta\psi' = \hat{Q}_{\psi'}\, \delta\psi' + \hat{Q}_{\psi''}\, \delta\psi''$$

$$\Rightarrow \hat{Q}_\phi\, \delta\phi = \hat{Q}_{\psi''}\, \delta\psi'' \Rightarrow \hat{Q}_\phi = (\partial\dot{\psi}''/\partial\dot{\phi})\hat{Q}_{\psi''} \Rightarrow \hat{Q}_{\psi''} = (-i/2)\hat{I}. \tag{b}$$

(ii) The corresponding impulsive *Chaplygin–Voronets* equations are

$$\Delta(\partial T_o/\partial\dot{\psi}') = \hat{Q}_{\psi'}, \qquad \Delta(\partial T_o/\partial\dot{\psi}'') = \hat{Q}_{\psi''}, \tag{c}$$

where

$$\partial T_o/\partial\dot{\psi}' = (mr^2/8)(13\dot{\psi}' - \dot{\psi}''), \tag{d1}$$

$$\partial T_o/\partial\dot{\psi}'' = (mr^2/8)(13\dot{\psi}'' - \dot{\psi}'). \tag{d2}$$

REMARKS

(i) This and the previous problem illustrate the earlier-made observation that, in *first*-kind problems, like this one, and in sharp contrast to the case of ordinary motion (§3.8), we *can* enforce the Pfaffian constraints in T, that is, $T \to T_o$ right from the start, and then form the $n - m$ multiplierless (kinetic) impulsive Hamel-like equations; that is, in such impulsive problems, the equations of Hamel and Lagrange have similar forms.

(ii) Additional convenient choices of quasi velocities (of which only *two* are independent) would have been the following:

(a) $\qquad \omega_1 \equiv \dot{x} = 0, \qquad \omega_2 \equiv \dot{y} + b\dot{\phi} + r\dot{\psi}' = 0;$

$$\omega_3 \equiv \dot{y} - b\dot{\phi} + r\dot{\psi}'' = 0, \qquad \omega_4 = \dot{\psi}' \neq 0, \qquad \omega_5 = \dot{\phi} \neq 0; \tag{e}$$

with inverses

$$\dot{x} = \omega_1 = 0, \qquad \dot{y} = (\omega_2 - r\,\omega_4 - b\,\omega_5)|_{b=r} = -r(\omega_4 + \omega_5); \qquad \dot{\phi} = \omega_5,$$
$$\dot{\psi}' = \omega_4, \qquad \dot{\psi}'' = (1/r)(-\omega_2 + \omega_3 + r\omega_4 + 2b\omega_5)|_{b=r} = \omega_4 + 2\omega_5; \qquad \text{(f)}$$

and

(b)

$$\omega_1 \equiv \dot{x} = 0, \qquad \omega_2 \equiv \dot{y} + r\dot{\phi} + r\dot{\psi}' = 0,$$
$$\omega_3 \equiv \dot{\psi}' \neq 0, \qquad \omega_4 \equiv \dot{\phi} \neq 0; \qquad \text{(g)}$$

with inverses

$$\dot{x} = \omega_1 = 0, \qquad \dot{y} = \cdots = -r(\omega_3 + \omega_4); \qquad \dot{\psi}' = \omega_3, \qquad \dot{\phi} = \omega_4. \qquad \text{(h)}$$

Problem 4.5.10 Continuing from the preceding two problems:

(i) Show that under the choice of $v_G \equiv v \equiv ds/dt$ [parallel to the impulse $\hat{I}$; i.e., $\dot{x} = -v\sin\phi$, $\dot{y} = v\cos\phi$ (for a general angle ϕ), s: quasi coordinate] and $\dot{\phi} \equiv \omega$ as independent (quasi) velocities, since then

$$\dot{\psi}' \equiv \omega_1 = -(1/r)(v + b\,\omega), \qquad \dot{\psi}'' \equiv \omega_2 = -(1/r)(v - b\,\omega), \qquad \text{(a)}$$

the (double) kinetic energy and impulsive impressed virtual work are, respectively,

$$2T = (M + 2m)v^2 + [(Mb^2/3) + 2mb^2 + (mr^2/2)]\omega^2 + [(mr^2/2)({\omega_1}^2 + {\omega_2}^2)]$$
$$= \cdots = (M + 3m)v^2 + [(Mb^2/3) + 3mb^2 + (mr^2/2)]\omega^2 = T_o(v, \omega) = T_o, \qquad \text{(b)}$$
$$\widehat{\delta' W} = \hat{Q}_s\,\delta s + \hat{Q}_\phi\,\delta\phi = \hat{I}\,\delta s + (\hat{I}i)\,\delta\phi \;\Rightarrow\; \hat{Q}_s = \hat{I},\ \hat{Q}_\phi = i\,\hat{I}. \qquad \text{(c)}$$

(ii) Verify that the corresponding *Chaplygin–Voronets* impulsive equations are

$$\partial T_o/\partial v = \hat{Q}_s: \qquad (M + 3m)v = \hat{I} \;\Rightarrow\; v = \cdots, \qquad \text{(d1)}$$
$$\partial T_o/\partial \omega = \hat{Q}_\phi: \qquad [(Mb^2/3) + 3mb^2 + (mr^2/2)]\omega = i\,\hat{I} \;\Rightarrow\; \omega = \cdots. \qquad \text{(d2)}$$

Example 4.5.6 *Jourdain's Principle in Impulsive Motion* (to be read after §6.3). We begin with the general "raw" (i.e., particle variable) form of Jourdain's principle for ordinary, continuous motion (6.2.4) and (6.3.15):

$$S\,(dm\,\boldsymbol{a} - d\boldsymbol{F}) \cdot \delta'\boldsymbol{v} = 0, \qquad \text{(a)}$$

where $\delta'\boldsymbol{v} \equiv \delta\boldsymbol{v}|_{\text{with } \delta t = 0 \text{ and } \delta \boldsymbol{r} = \boldsymbol{0}}$ is the *Jourdain variation* of $\boldsymbol{v}$ (6.3.5). Next, integrating (a) between t and $t + \tau$, and then taking the limit as $\tau \to 0$ (or, integrating "between" t^- and t^+), and using the notations introduced in §4.2, we obtain the "raw" form of the impulsive Jourdain principle:

$$S\,(dm\,\Delta\boldsymbol{v} - \widehat{d\boldsymbol{F}}) \cdot \delta'\boldsymbol{v} = 0 \qquad [\Delta\boldsymbol{v} \equiv \boldsymbol{v}^+ - \boldsymbol{v}^-]. \qquad \text{(b)}$$

Now, substituting into (b) the basic representation of $\delta'\boldsymbol{v}$,

$$\delta'\boldsymbol{v} = \delta'\left(\sum \omega_I \boldsymbol{\varepsilon}_I + \boldsymbol{\varepsilon}_{n+1}\right) = \sum \boldsymbol{\varepsilon}_I\,\delta\omega_I = \sum (\partial\boldsymbol{v}/\partial\omega_I)\,\delta\omega_I \qquad \text{(c)}$$

[recalling § 2.9ff.; and that, since the $\boldsymbol{\varepsilon}_I$, $\boldsymbol{\varepsilon}_{n+1}$ $(I = m+1, \ldots, n)$ are functions of t and q, their Jourdain variations vanish: $\delta'(\boldsymbol{\varepsilon}_I, \boldsymbol{\varepsilon}_{n+1}) = \mathbf{0}$], we obtain, successively,

$$\begin{aligned} 0 &= S\,(dm\,\Delta\boldsymbol{v} - \widehat{d\boldsymbol{F}}) \cdot \Big(\sum \boldsymbol{\varepsilon}_I\,\delta\omega_I\Big) \\ &= \sum \Big(S\,(dm\,\Delta\boldsymbol{v} - \widehat{d\boldsymbol{F}}) \cdot \boldsymbol{\varepsilon}_I\Big)\,\delta\omega_I \\ &= \sum \Big\{S\,[dm(\partial\boldsymbol{v}/\partial\omega_I) \cdot \Delta\boldsymbol{v} - \widehat{d\boldsymbol{F}} \cdot \boldsymbol{\varepsilon}_I]\Big\}\,\delta\omega_I \\ &= \sum \Big\{\Delta\Big(S\,dm(\partial\boldsymbol{v}/\partial\omega_I) \cdot \boldsymbol{v}\Big) - S\,\widehat{d\boldsymbol{F}} \cdot \boldsymbol{\varepsilon}_I\Big\}\,\delta\omega_I \qquad (\text{since } \Delta\boldsymbol{\varepsilon}_I = \mathbf{0}) \\ &= \sum\, [\Delta(\partial T^*/\partial\omega_I) - \hat{\Theta}_I]\,\delta\omega_I \qquad \Big(\text{since } 2T^* \equiv S\,dm\,\boldsymbol{v} \cdot \boldsymbol{v}, \text{etc.}\Big), \end{aligned} \tag{d}$$

from which, since the $n - m$ $\delta\omega_I$'s are independent, we immediately obtain the earlier $n - m$ impulsive kinetic Hamel equations,

$$\Delta(\partial T^*/\partial\omega_I) = \hat{\Theta}_I. \tag{e}$$

For instructive applications of (c) and (e) to impulsive problems, see Bahar (1994).

Example 4.5.7 *A Direct Method for the Determination of the Impulsive Reactions* (may be omitted in a first reading). Let us consider the earlier (§4.4) "Routh–Voss impulsive equations" of a system subjected to the *single* Pfaffian constraint

$$\sum a_k(t, q)\dot{q}_k + a_0(t, q) = 0, \tag{a}$$

that is,

$$\Delta(\partial T/\partial\dot{q}_k) = \hat{Q}_k + \hat{\lambda}\,a_k. \tag{b}$$

Since [with the usual notations (§3.9)]

$$2T = 2(T_2 + T_1 + T_0) = \sum\sum M_{kl}\dot{q}_k\dot{q}_l + \sum 2M_k\dot{q}_k + M_0, \tag{c}$$

and the M_{kl}, M_k are functions of t, q: $\Delta(M_{kl}, M_k) = 0$, and, therefore,

$$\Delta(\partial T/\partial\dot{q}_k) = \Delta\Big(\sum M_{kl}\dot{q}_l + M_k\Big) = \sum M_{kl}\Delta\dot{q}_l \quad (= \Delta p_k), \tag{d}$$

eqs. (b) assume the explicit form

$$\sum M_{kl}\,\Delta\dot{q}_l = \hat{Q}_k + \hat{\lambda}\,a_k. \tag{e}$$

Let us isolate (uncouple) the $\Delta\dot{q}_l$: multiplying (e) with $m_{rk} = m_{kr}$, where

$$\sum M_{kl}m_{kr} = \delta_{lr}, \tag{f}$$

and summing over k yields the general $\Delta\dot{q}_r$-expressions

$$\Delta\dot{q}_r = \sum m_{rk}\hat{Q}_k + \hat{\lambda}\Big(\sum m_{rk}a_k\Big). \tag{g}$$

[In tensor calculus, the m_{rk} are called *conjugate* to the M_{kl} (and vice versa); and are denoted by M^{rk}. See, for example, Papastavridis (1999, chap. 2), Sokolnikoff (1964, p. 76 ff.), Synge and Schild (1949, p. 29 ff.).]

Next, multiplying (g) with a_r and summing over r, we get

$$\sum a_r\, \Delta\dot{q}_r = \sum\sum m_{rk} a_r \hat{Q}_k + \hat{\lambda} a^2, \tag{h}$$

where

$$\sum\sum m_{rk} a_r a_k \equiv a^2 \qquad \text{[magnitude (squared) of vector } (a_r)\text{]}, \tag{i}$$

and solving for $\hat{\lambda}$,

$$\hat{\lambda} = \left(\sum a_r\, \Delta\dot{q}_r - \sum\sum m_{rk} a_r \hat{Q}_k\right)\Big/ a^2. \tag{j}$$

Now:

(i) For *first*-kind impulsive problems (i.e., constraints holding before, during, and after the shock), eq. (a) yields

$$\Delta\left(\sum a_k \dot{q}_k + a_0\right) = \sum a_k\, \Delta\dot{q}_k = 0, \tag{k}$$

and so (j) reduces to

$$\hat{\lambda} = -\sum\sum m_{rk} a_r \hat{Q}_k \Big/ a^2. \tag{l}$$

Then, substituting (l) into (g) and solving for $(\dot{q}_r)^+$, we obtain (with some dummy-index changes)

$$(\dot{q}_r)^+ = (\dot{q}_r)^- + \sum m_{rk} \hat{Q}_k - \left(\sum\sum m_{lk} a_l \hat{Q}_k / a^2\right)\left(\sum m_{rs} a_s\right). \tag{m}$$

(ii) For *second*-kind impulsive problems (i.e., constraints suddenly introduced at the impact moment), the $(\dot{q}_r)^-$ do *not* obey (a). Then, invoking the latter, we get

$$\sum a_r\, \Delta\dot{q}_r = \sum a_r (\dot{q}_r)^+ - \sum a_r (\dot{q}_r)^- = -a_0 - \sum a_r (\dot{q}_r)^- \quad (\neq 0), \tag{n}$$

and so (j) reduces to

$$\hat{\lambda} = -\left[a_0 + \sum a_r (\dot{q}_r)^- + \sum\sum m_{rk} a_r \hat{Q}_k\right]\Big/ a^2. \tag{o}$$

Once $\hat{\lambda}$ is found, then (g) yield immediately (with some dummy-index changes)

$$\begin{aligned}
\Delta\dot{q}_r &\equiv (\dot{q}_r)^+ - (\dot{q}_r)^- \\
&= \sum m_{rk} \hat{Q}_k - \left\{\left[a_0 + \sum a_k (\dot{q}_k)^- + \sum\sum m_{lk} a_l \hat{Q}_k\right]\Big/ a^2\right\}\left(\sum m_{rs} a_s\right), \qquad \text{(p)} \\
&\Rightarrow (\dot{q}_r)^+ = \ldots .
\end{aligned}$$

Equations (l) and (o) yield $\hat{\lambda}$ in terms of initially known quantities; that is, a_0, a_r, $M_{kl} \rightarrow m_{kl}$, and $\hat{Q}_k$ (unlike the earlier impulsive kinetostatic equations of Maggi, Hamel, Appell, et al.).

The n equations (m) and n equations (p) might be called *the impulsive Jacobi–Synge equations* of the corresponding problem (unlike the earlier $n - m$ kinetostatic and m kinetic impulsive equations of Maggi, Hamel, Appell, et al.; recall ex. 3.10.2).

We leave it to the reader to extend this method to the case of $m(< n)$ Pfaffian constraints.

Application of the above to Example 4.5.3

We recall that in this, *second*-kind, problem,

$$q_{1,2,3}\colon x, y, \theta; \qquad a_1 = 0, \quad a_2 = 1, \quad a_3 = b \sin\theta, \quad a_0 = 0; \qquad \hat{Q}_k = 0; \tag{q1}$$

and the nonvanishing inertia coefficients of T are

$$M_{kl}\colon M_{11} = M_{22} = m; \ M_{33} = I \ \Rightarrow \ m_{kl}\colon m_{11} = m_{22} = m^{-1}; \ m_{33} = I^{-1}, \tag{q2}$$

Therefore,

$$\sum m_{1s} a_s = m_{11} a_1 + m_{12} a_2 + m_{13} a_3 = (m^{-1})(0) + (0)(1) + (0)(b \sin\theta) = 0,$$

$$\sum m_{2s} a_s = m_{21} a_1 + m_{22} a_2 + m_{23} a_3 = (0)(0) + (m^{-1})(1) + (0)(b \sin\theta) = m^{-1},$$

$$\sum m_{3s} a_s = m_{31} a_1 + m_{32} a_2 + m_{33} a_3 = (0)(0) + (0)(1) + (I^{-1})(b \sin\theta) = I^{-1} b \sin\theta,$$

$$\Rightarrow \ a^2 \equiv \sum \left(\sum m_{rs} a_s \right) a_r = (0)(0) + (m^{-1})(1) + (I^{-1} b \sin\theta)(b \sin\theta)$$

$$= m^{-1} + [(mb^2/3)^{-1}(b \sin\theta)](b \sin\theta) = m^{-1}(1 + 3\sin^2\theta); \tag{q3}$$

$$\sum \left(\sum m_{rk} \hat{Q}_k \right) a_r = \cdots = 0, \tag{q4}$$

and so, finally, eq. (o) yields

$$\hat{\lambda} = -[a_1(\dot{x})^- + a_2(\dot{y})^- + a_3(\dot{\theta})^-]/a^2$$

$$= -[(0)(0) + (1)(-v) + (b \sin\theta)(\omega)]/[m^{-1}(1 + 3\sin^2\theta)]$$

$$= m(v - b\,\omega \sin\theta)/(1 + 3\sin^2\theta), \tag{r}$$

that is, (i3) of ex. 4.5.3. Similarly for $(\dot{q}_r)^+$ via eq. (p); the details are left to the reader.

4.6 EXTREMUM THEOREMS OF IMPULSIVE MOTION (OF CARNOT, KELVIN, BERTRAND, ROBIN, ET AL.)

Since the impulsive equations are algebraic equations in the shock velocities—that is, of the first order—*no proper integral variational principles exist for them.* However, by appropriate specializations of the *differential* variational principles (chap. 6), a host of interesting and useful *extremum* propositions (i.e., maxima/minima in the sense of ordinary mathematical analysis) of sufficient generality can be obtained. These theorems, summarized below, constitute impulsive counterparts of the energetic theorems of ordinary (i.e., continuous) motion. [Carnot's theorems (see below) are included here, although they are neither variational nor extremum, but simply energetic; that is, just like their ordinary motion counterparts, they deal with *actual* motions (velocities), and yield only *one* equation. For proofs of these theorems in general *system* variables, see ex. 4.6.7.]

For complementary reading, we recommend the following older British texts (alphabetically): Chirgwin and Plumpton (1966, pp. 329–343), Easthope (1964, pp. 285–304), Kilmister and Reeve (1966, pp. 247–248), Milne (1948, pp. 370–378), Ramsey (1937, pp. 185–195), Smart (1951, pp. 376–390).

4.6.1 Theorem of Carnot (1803)

1. First Part (Collisions)

In the absence of impressed impulses, the sudden introduction of (ideal) stationary and persistent constraints that change some velocity reduces the kinetic energy; hence, by the collision of inelastic bodies, some kinetic energy is always lost.

2. Second Part (Explosions)

The sudden removal of (ideal and) stationary constraints that break bonds of rigidity (e.g., explosion of a shell, or breaking of the rope in a tug-of-war contest) increases the kinetic energy.

Their proofs utilize the following auxiliary and purely kinematico-inertial identity: Let $\{\boldsymbol{v}_1\}$ and $\{\boldsymbol{v}_2\}$ be any two possible sets of velocities, with corresponding kinetic energies T_1 and T_2; that is, $2T_1 \equiv \mathsf{S}\, dm\, \boldsymbol{v}_1 \cdot \boldsymbol{v}_1$, $2T_2 \equiv \mathsf{S}\, dm\, \boldsymbol{v}_2 \cdot \boldsymbol{v}_2$. Then, by simple algebra,

$$2\boldsymbol{v}_2 \cdot (\boldsymbol{v}_2 - \boldsymbol{v}_1) = \boldsymbol{v}_2{}^2 - \boldsymbol{v}_1{}^2 + (\boldsymbol{v}_2 - \boldsymbol{v}_1)^2 \Rightarrow 2K_{12} = T_1 + T_2 - T_{12}; \qquad (4.6.1a1)$$

also,

$$2\boldsymbol{v}_1 \cdot (\boldsymbol{v}_1 - \boldsymbol{v}_2) = \boldsymbol{v}_1{}^2 - \boldsymbol{v}_2{}^2 + (\boldsymbol{v}_1 - \boldsymbol{v}_2)^2 \Rightarrow 2K_{12} = T_1 + T_2 - T_{12}; \qquad (4.6.1a2)$$

where

$$2K_{12} = 2K_{21} \equiv \mathsf{S}\, dm\, \boldsymbol{v}_1 \cdot \boldsymbol{v}_2, \qquad (4.6.1b1)$$

$$2T_{12} = 2T_{21} \equiv \mathsf{S}\, dm(\boldsymbol{v}_2 - \boldsymbol{v}_1) \cdot (\boldsymbol{v}_2 - \boldsymbol{v}_1): \qquad (4.6.1b2)$$

Kinetic energy of relative motion ≥ 0.

(a) To prove the first part, we begin with LIP, eqs. (4.3.4) ff.), or

$$\widehat{\delta' W_R} = 0 \Rightarrow \widehat{\delta I} = \widehat{\delta' W}, \qquad (4.6.1c)$$

where

$$\widehat{\delta' W_R} \equiv \mathsf{S}\, \widehat{d\boldsymbol{R}} \cdot \delta\boldsymbol{r}, \qquad \widehat{\delta I} \equiv \mathsf{S}\, dm(\boldsymbol{v}^+ - \boldsymbol{v}^-) \cdot \delta\boldsymbol{r}, \qquad \widehat{\delta' W} \equiv \mathsf{S}\, \widehat{d\boldsymbol{F}} \cdot \delta\boldsymbol{r}, \qquad (4.6.1d)$$

and in there we make the identifications $\boldsymbol{v}^- = \boldsymbol{v}_1$, $\boldsymbol{v}^+ = \boldsymbol{v}_2$ (i.e., velocities just before and after additional workless constraints), and [since the new constraints are stationary and the $\delta\boldsymbol{r}$ are compatible with both primitive (i.e., existing) and additional constraints] we choose $\delta\boldsymbol{r} \to d\boldsymbol{r} = \boldsymbol{v}^+\, dt \sim \boldsymbol{v}^+ = \boldsymbol{v}_2$, and notice that, here, $\widehat{\delta' W} \to \mathsf{S}\, \widehat{d\boldsymbol{F}} \cdot \boldsymbol{v}^+ \equiv \mathsf{S}\, \widehat{d\boldsymbol{F}} \cdot \boldsymbol{v}_2 = 0$. Thus, we obtain

$$\mathsf{S}\, dm(\boldsymbol{v}^+ - \boldsymbol{v}^-) \cdot \boldsymbol{v}^+ = \mathsf{S}\, \widehat{d\boldsymbol{F}} \cdot \boldsymbol{v}^+ = 0,$$

or

$$\mathbf{S}\, dm(\boldsymbol{v}_2 - \boldsymbol{v}_1) \cdot \boldsymbol{v}_2 = \mathbf{S}\, \widehat{dF} \cdot \boldsymbol{v}_2 = 0,$$

$$\Rightarrow \mathbf{S}\, dm\, \boldsymbol{v}_2 \cdot \boldsymbol{v}_2 = \mathbf{S}\, dm\, \boldsymbol{v}_2 \cdot \boldsymbol{v}_1; \qquad \text{i.e., } T_2 = K_{12}, \tag{4.6.1e}$$

and so (4.6.1a) becomes

$$2T_2 = T_1 + T_2 - T_{12} \Rightarrow T_2 - T_1 = -T_{12} < 0; \text{ i.e.,}$$

$$T_2 < T_1; \tag{4.6.1f}$$

[we exclude the case(s) where the introduction of new constraint(s) does not change the kinetic energy] or, reverting to our standard notation,

$$\begin{aligned}
\Delta T \equiv T^+ - T^- &\equiv \mathbf{S}\,(dm/2)\boldsymbol{v}^+ \cdot \boldsymbol{v}^+ - \mathbf{S}\,(dm/2)\boldsymbol{v}^- \cdot \boldsymbol{v}^- \\
&= -\mathbf{S}\,(dm/2)(\boldsymbol{v}^+ - \boldsymbol{v}^-) \cdot (\boldsymbol{v}^+ - \boldsymbol{v}^-) \\
&\equiv -\mathbf{S}\,(dm/2)\, \Delta\boldsymbol{v} \cdot \Delta\boldsymbol{v} \\
&\equiv -T_{\text{jump}}: \; -\left[\textit{Kinetic energy of jump (or of lost) motion}\right] < 0,
\end{aligned}$$

or

$$T^- - T^+ = T_{\text{jump}} > 0, \quad \text{Q.E.D.} \tag{4.6.1g}$$

[Recalling (4.5.13a), we see that, here, $\hat{S} = -\Delta T$.]

(b) To prove the *second* part, similarly, we identify $\boldsymbol{v}^- = \boldsymbol{v}_1$, $\boldsymbol{v}^+ = \boldsymbol{v}_2$, choose $\delta\boldsymbol{r} \rightarrow d\boldsymbol{r} = \boldsymbol{v}^- \,dt \sim \boldsymbol{v}^- = \boldsymbol{v}_1$, and notice that, here, $\mathbf{S}\, \widehat{dF} \cdot \boldsymbol{v}^- \equiv \mathbf{S}\, \widehat{dF} \cdot \boldsymbol{v}_1 = 0$. The result is

$$\mathbf{S}\, dm(\boldsymbol{v}^+ - \boldsymbol{v}^-) \cdot \boldsymbol{v}^- = \mathbf{S}\, dm(\boldsymbol{v}_2 - \boldsymbol{v}_1) \cdot \boldsymbol{v}_1 = 0,$$

$$\Rightarrow \mathbf{S}\, dm\, \boldsymbol{v}_2 \cdot \boldsymbol{v}_1 = \mathbf{S}\, dm\, \boldsymbol{v}_1 \cdot \boldsymbol{v}_1; \qquad \text{i.e., } T_1 = K_{12}, \tag{4.6.1h}$$

and so (4.6.1a) yields

$$2T_1 = T_1 + T_2 - T_{12} \;\Rightarrow\; T_1 - T_2 = -T_{12} < 0; \text{ i.e.,}$$

$$T_1 < T_2; \tag{4.6.1i}$$

or, in terms of our standard notation,

$$\begin{aligned}
\Delta T \equiv T^+ - T^- &\equiv \mathbf{S}\,(dm/2)\boldsymbol{v}^+ \cdot \boldsymbol{v}^+ - \mathbf{S}\,(dm/2)\boldsymbol{v}^- \cdot \boldsymbol{v}^- \\
&= \mathbf{S}\,(dm/2)(\boldsymbol{v}^+ - \boldsymbol{v}^-) \cdot (\boldsymbol{v}^+ - \boldsymbol{v}^-) \\
&\equiv \mathbf{S}\,(dm/2)\, \Delta\boldsymbol{v} \cdot \Delta\boldsymbol{v} \equiv T_{\text{jump}} > 0, \quad \text{Q.E.D.}
\end{aligned} \tag{4.6.1j}$$

REMARKS

(i) The above can also be easily obtained by combining the identities

$$\mathbf{S}\, dm(\boldsymbol{v}^+ - \boldsymbol{v}^-) \cdot \boldsymbol{v}^+ = \Delta T + T_{\text{jump}}, \qquad \mathbf{S}\, dm(\boldsymbol{v}^+ - \boldsymbol{v}^-) \cdot \boldsymbol{v}^- = \Delta T - T_{\text{jump}}, \tag{4.6.1k}$$

[which follow at once from (4.6.1a, b) with the earlier identifications] with LIP, (4.6.1c, d). Thus, the first of (4.6.1k) yields $\Delta T + T_{\text{jump}} = 0$ (*first theorem*, while the second of (4.6.1k) yields $\Delta T - T_{\text{jump}} = 0$ (*second theorem*).

(ii) For Carnot's first theorem under *nonpersistent* constraints, see Appell (1896, p. 15 ff.).

(iii) If the bodies in question are *elastic*, then their collision consists of (a) a period of *compression* (as if the bodies were inelastic), and (b) a period of explosion-like *restitution*. Since the corresponding forces are equal and opposite, *the kinetic energy lost in compression balances exactly the kinetic energy gained in restitution*. This is sometimes called the *third theorem of Carnot*.

4.6.2 Theorem of Kelvin (1863)

If an originally motionless system is suddenly set in motion by (unknown) impressed impulses acting at specified points of it and communicating to them prescribed (i.e., given) velocities, then the resulting (or actual) postimpact kinetic energy is less than that of any other kinematically possible (or comparison, or hypothetical) motion; that is, one in which the specified points have the same prescribed velocities as in the actual motion, and all other external and/or internal system constraints are respected (which is why this theorem is occasionally referred to as a "principle of laziness"); that is, with some obvious notations:

$$T(\boldsymbol{v}_{\text{postimpact } comparison}) > T(\boldsymbol{v}_{\text{postimpact } actual}). \tag{4.6.2a}$$

Hence, this result allows us to find the actual postimpact velocities in terms of the prescribed velocities.

To prove it, and since *such comparison motions may differ from each other infinitesimally*, we let $\boldsymbol{v}^+{}_{\text{actual}} \equiv \boldsymbol{v}^+$ and $\boldsymbol{v}^+{}_{\text{comparison}} \equiv \boldsymbol{v}^+ + \delta\boldsymbol{v}^+ \equiv \boldsymbol{v}^+ + \delta_K\boldsymbol{v} \equiv \boldsymbol{v}$; where, of course, both $\boldsymbol{v}^+$ and $\boldsymbol{v}$ are kinematically admissible, and, at the specified points, $\delta_K\boldsymbol{v} = \boldsymbol{0}$. Then, with $\delta\boldsymbol{r} \sim \delta_K\boldsymbol{v}$, and since, at these points, $\delta_K\boldsymbol{v} = \boldsymbol{0}$, while for the rest of them $\widehat{d\boldsymbol{F}} = \boldsymbol{0}$,

$$S\, \widehat{d\boldsymbol{F}} \cdot \delta_K\boldsymbol{v} = 0 \quad \text{and} \quad S\, \widehat{d\boldsymbol{R}} \cdot \delta_K\boldsymbol{v} = 0. \tag{4.6.2b}$$

• *Stationarity*: Next, setting $\boldsymbol{v}^- = \boldsymbol{0}$ (since the system is initially at rest) and the rest of the above specializations into the master equations (4.6.1c, d) yields the *stationarity* condition

$$S\, dm\, \boldsymbol{v}^+ \cdot \delta_K\boldsymbol{v} = 0, \quad \text{or} \quad S\, dm\, \boldsymbol{v}^+ \cdot \boldsymbol{v} = S\, dm\, \boldsymbol{v}^+ \cdot \boldsymbol{v}^+; \tag{4.6.2c}$$

that is,

$$\delta_K\Big(S\, (dm/2)\boldsymbol{v}^+ \cdot \boldsymbol{v}^+\Big) \equiv \delta_K T(\boldsymbol{v}^+) = 0; \tag{4.6.2d}$$

that is, *the actual postimpact motion makes the kinetic energy stationary*.

• *Minimality*: As a result of the above we have, successively,

$$\begin{aligned}
\Delta T &\equiv T(\boldsymbol{v}) - T(\boldsymbol{v}^+) \equiv T(\boldsymbol{v}^+ + \delta_K \boldsymbol{v}) - T(\boldsymbol{v}^+) \\
&\equiv \mathbf{S}\,(dm/2)\boldsymbol{v}\cdot\boldsymbol{v} - \mathbf{S}\,(dm/2)\boldsymbol{v}^+\cdot\boldsymbol{v}^+ \\
&= \mathbf{S}\,(dm/2)\big[(\boldsymbol{v}^+ + \delta_K\boldsymbol{v})\cdot(\boldsymbol{v}^+ + \delta_K\boldsymbol{v}) - \boldsymbol{v}^+\cdot\boldsymbol{v}^+\big] \qquad \text{[invoking (4.6.2c, d)]} \\
&= \mathbf{S}\,(dm/2)\,\delta_K\boldsymbol{v}\cdot\delta_K\boldsymbol{v} \equiv \mathbf{S}\,(dm/2)(\boldsymbol{v} - \boldsymbol{v}^+)\cdot(\boldsymbol{v} - \boldsymbol{v}^+) \\
&\equiv (1/2)\,\delta^2{}_K T(\boldsymbol{v}^+) \geq 0 \qquad \text{(with the equality holding for } \delta_K\boldsymbol{v} = \mathbf{0}\text{)} \\
&\qquad \Rightarrow T(\boldsymbol{v}^+) = \textit{minimum}, \quad \text{Q.E.D.} \qquad (4.6.2e)
\end{aligned}$$

However, in concrete problems, it is the stationarity rather than the minimality that is invoked.

REMARKS

(i) Equation (4.6.2c) also results, most simply, by setting in the master equations (4.6.1c, d): (a) $\boldsymbol{v}^- = \mathbf{0}$ and (b) first, $\delta\boldsymbol{r} \sim \boldsymbol{v}^+$, and, second, $\delta\boldsymbol{r} \sim \boldsymbol{v} \equiv \boldsymbol{v}^+ + \delta_K\boldsymbol{v}$, and then noting that

$$0 = \mathbf{S}\,d\widehat{\boldsymbol{F}}\cdot\delta_K\boldsymbol{v} \;\Rightarrow\; \mathbf{S}\,d\widehat{\boldsymbol{F}}\cdot\boldsymbol{v} = \mathbf{S}\,d\widehat{\boldsymbol{F}}\cdot\boldsymbol{v}^+.$$

(ii) For proofs utilizing (4.6.1a1, 1a2), see ex. 4.6.7; also Chirgwin and Plumpton (1966, p. 330 ff.).

(iii) For applications of Kelvin's theorem to hydrodynamics, and so on, see, for example, Byerly (1916, pp. 76–80), and, of course, Thomson and Tait (1912, §312–317, pp. 286–301).

4.6.3 Theorem of Bertrand (1853) and Delaunay (1840)

Consider a system in motion acted upon by prescribed impressed impulses applied to it: (a) with its existing (i.e., original) ideal constraints and, separately, (b) with additional (also ideal) constraints. Then, the actual postimpact kinetic energy under the existing constraints is greater than that under the additional constraints, where, in both cases, the impulses, as well as the initial motion of the system, are the same; or, these additional constraints reduce the kinetic energy; that is,

$$T(\boldsymbol{v}_{\text{postimpact } \textit{existing constraints}}) > T(\boldsymbol{v}_{\text{postimpact } \textit{additional constraints}}). \qquad (4.6.3a)$$

REMARKS

(i) Originally established by Lagrange; generalized by Sturm (1841) and Bertrand [in his notes to the 3rd ed. of Lagrange's *Mécanique Analytique* (1853–1855)]. The maximum property is due to Delaunay (1840).

(ii) We notice the similarities with Carnot's first theorem: in there, the system is acted upon by given impulses, and *then* ideal impulsive constraints are imposed; while in the Bertrand–Delaunay theorem, in the competing motion both impulses and constraints are applied *simultaneously*. (On the latter, see also "Remarks" following the theorem of Robin, below.)

To prove it, we let $\boldsymbol{v}^+$ and $\boldsymbol{v}$ be, respectively, the existing and additionally constrained postimpact velocities, and $\boldsymbol{v}^-$ be the common preimpact velocity. If the corresponding impulsive reactions are $\{\widehat{d\boldsymbol{R}^+}\}$ and $\{\widehat{d\boldsymbol{R}}\}$, then, by the first parts of (4.6.1c, d) with $\delta\boldsymbol{r} \sim \boldsymbol{v}^+$,

$$\mathsf{S}\, \widehat{d\boldsymbol{R}^+} \cdot \boldsymbol{v}^+ = 0 \qquad \text{and} \qquad \mathsf{S}\, \widehat{d\boldsymbol{R}} \cdot \boldsymbol{v}^+ = 0, \tag{4.6.3b}$$

and so the second parts of (4.6.1c, d), with $\{\widehat{d\boldsymbol{F}}\}$ the common impressed forces and, again, $\delta\boldsymbol{r} \sim \boldsymbol{v}$, and $\boldsymbol{v}^+ \to \boldsymbol{v}^+, \boldsymbol{v}$, yield

$$\mathsf{S}\, dm(\boldsymbol{v}^+ - \boldsymbol{v}^-) \cdot \boldsymbol{v} = \mathsf{S}\, \widehat{d\boldsymbol{F}} \cdot \boldsymbol{v},$$

and

$$\mathsf{S}\, dm(\boldsymbol{v} - \boldsymbol{v}^-) \cdot \boldsymbol{v} = \mathsf{S}\, \widehat{d\boldsymbol{F}} \cdot \boldsymbol{v}, \tag{4.6.3c}$$

from which, subtracting side by side, we obtain

$$\mathsf{S}\, dm(\boldsymbol{v}^+ - \boldsymbol{v}) \cdot \boldsymbol{v} = 0, \tag{4.6.3d}$$

and from this it follows readily [as in the corresponding steps of the previous theorems of Carnot and Kelvin, eqs. (4.6.1g–1j, 2e)]:

$$\begin{aligned} T^+ - T &\equiv \mathsf{S}\,(dm/2)\boldsymbol{v}^+ \cdot \boldsymbol{v}^+ - \mathsf{S}\,(dm/2)\boldsymbol{v} \cdot \boldsymbol{v} \\ &= \mathsf{S}\,(dm/2)(\boldsymbol{v}^+ - \boldsymbol{v}) \cdot (\boldsymbol{v}^+ - \boldsymbol{v}) > 0; \qquad \text{i.e., } T^+ > T, \quad \text{Q.E.D.} \end{aligned} \tag{4.6.3e}$$

• *Continuous case, variational formulation.* If, further, the additionally constrained (postimpact) motion depends continuously on its deviation from the (postimpact) motion under existing constraints, then setting in the above $\boldsymbol{v} - \boldsymbol{v}^+ = \delta\boldsymbol{v}^+ \equiv \delta_{B/D}\boldsymbol{v}$, we obtain [as in Kelvin's theorem, (4.6.2d)], the *stationarity* equation

$$\delta_{B/D}T^+ \equiv \delta_{B/D}T(\boldsymbol{v}^+) \equiv \delta_{B/D}\Big(\mathsf{S}\,(dm/2)\boldsymbol{v}^+ \cdot \boldsymbol{v}^+\Big) = \mathsf{S}\, dm\, \boldsymbol{v}^+ \cdot \delta_{B/D}\boldsymbol{v} = 0; \tag{4.6.3f}$$

that is, for constrained variations, the actual motion makes the kinetic energy *stationary; and the maximality* inequality (with the equality holding for $\delta_{B/D}\boldsymbol{v} = \boldsymbol{0}$)

$$\begin{aligned} \Delta_{B/D}T(\boldsymbol{v}^+) &\equiv T - T^+ = \cdots \\ &= -(1/2)\,\delta^2{}_{B/D}T^+ = -\mathsf{S}\,(dm/2)\,\delta_{B/D}\boldsymbol{v} \cdot \delta_{B/D}\boldsymbol{v} \le 0. \end{aligned} \tag{4.6.3g}$$

Since, here, the impressed impulses are assumed to be *the same* for all kinematically possible (comparison) postimpact velocities, Bertrand's theorem can be reformulated as follows: The postimpact velocities of the existing constraints (actual problem) $\boldsymbol{v}^+$ make either $T = \mathsf{S}\,(dm/2)\boldsymbol{v} \cdot \boldsymbol{v}$, or $T = T^- + \mathsf{S}\,\widehat{d\boldsymbol{F}} \cdot (\boldsymbol{v} + \boldsymbol{v}^-)/2$ [resulting from the impulsive work–energy theorem," ex. 4.3.1, eqs. (d–f), with $\boldsymbol{v}^+ \to \boldsymbol{v}$] *stationary* (a *maximum*, since T is positive definite), *under the constraint* (expressing the sameness of the impulses for all comparison motions)

$$2(T - T^-) - \mathsf{S}\,\widehat{d\boldsymbol{F}} \cdot (\boldsymbol{v} + \boldsymbol{v}^-) = 0, \tag{4.6.3h}$$

or

$$\mathop{S} dm\, \boldsymbol{v} \cdot \boldsymbol{v} - \mathop{S} \widehat{d\boldsymbol{F}} \cdot (\boldsymbol{v} + \boldsymbol{v}^-) - 2T^- = 0. \tag{4.6.3i}$$

The Bertrand–Delaunay theorem, in spite of its conceptual elegance, has two serious drawbacks:

(i) The earlier continuity requirement significantly limits its usefulness; and

(ii) As one might expect, its practical implementation is, usually, mathematically laborious. [Convenient alternatives, for the determination of the motion of constrained systems resulting from the sudden imposition of impulses, are the stationarity/extremum theorems of Robin and Gauss, presented below.]

Relationship Between the Theorems of Kelvin and Bertrand–Delaunay, Theorem of Taylor (1922)

Let us consider a straight rigid rod AB, initially at rest on a horizontal smooth table, and then set in motion by a given impulse $\hat{I}_B$ applied perpendicularly to it at its end B; and, hence, communicating to it a specified velocity v_B, also perpendicular to the rod at B. Then, we repeat the experiment with a point of the rod C permanently fixed/hinged; something that forces it to rotate about C. Now: (a) If the *impulse* at B is the same in both experiments, then the hinge *decreases* the kinetic energy (Bertrand–Delaunay); in fact, since the value of $\hat{I}_B$ remains fixed, as C approaches B the angular velocity of the rod decreases (and for $C \to B \Rightarrow \omega \to 0$), and so does its kinetic energy; whereas (b) If the *velocity* of B is the same in both experiments, then the hinge *increases* the kinetic energy (Kelvin); in fact, since the value of v_B remains fixed, as C approaches B the angular velocity of the rod increases indefinitely, and so does its kinetic energy!

The relationship between the kinetic energy *increase* of Kelvin's theorem, and the kinetic energy *decrease* of the Bertrand–Delaunay theorem is answered by the following interesting theorem.

THEOREM OF TAYLOR

Let us consider an originally motionless system S and then apply to its points impulses $\widehat{d\boldsymbol{F}}$, which produce velocities $\boldsymbol{v}^+$, and result in a reference kinetic energy T^+. Next, we introduce to S given constraints. To this new, motionless, system, S_c, we:

(a) *First*, apply the earlier $\widehat{d\boldsymbol{F}}$ at the same points, which results in a kinetic energy $T^+{}_{c,\,impulses}$. By the theorem of Bertrand–Delaunay, $T^+{}_{c,\,impulses} < T^+$.

(b) *Second*, we apply the earlier velocities at the same points, which results in a kinetic energy $T^+{}_{c,\,velocities}$. By Kelvin's theorem: $T^+{}_{c,\,velocities} > T^+$.

Now, Taylor's theorem states that

$$|T^+{}_{c,\,velocities} - T^+| > |T^+{}_{c,\,impulses} - T^+|; \tag{4.6.3j}$$

or

$$\Delta T_K > |\Delta T_{B/D}|,$$

$$\Delta T_K \equiv T(\boldsymbol{v}_{\text{postimpact } comparison}) - T(\boldsymbol{v}_{\text{postimpact } actual}) \quad (> 0),$$

$$\Delta T_{B/D} \equiv T(\boldsymbol{v}_{\text{postimpact } additional\ constraints}) - T(\boldsymbol{v}_{\text{postimpact } existing\ constraints}) \quad (< 0); \quad (4.6.3\text{k})$$

where

$$T(\boldsymbol{v}_{\text{postimpact } actual}) = T(\boldsymbol{v}_{\text{postimpact } existing\ constraints}) = T^+.$$

In words: *the increase in energy due to the imposition of constraints in the Kelvin case is greater than the (absolute value of the) reduction in energy due to the imposition of the same constraints in the Bertrand–Delaunay case.*

[For proofs and applications, see, for example, Kilmister (1967, pp. 105–107), Milne (1948, pp. 374–378), Pars (1965, p. 238), Ramsey (1937, pp. 216–219), Rosenberg (1977, p. 408). Also, for a combined formulation of the theorems of Kelvin and Bertrand–Delaunay, due to Gray (1901), see Stäckel (1905, p. 517).]

4.6.4 Theorem of Robin (1887)

The actual postimpact velocities $\{\boldsymbol{v}^+\}$ *of a moving system subjected simultaneously to given impressed impulses* $\{\widehat{d\boldsymbol{F}}\}$, *and to sudden ideal impulsive constraints make the following expression:*

$$P = P(\boldsymbol{v}; \boldsymbol{v}^-, \widehat{d\boldsymbol{F}}) \equiv S\,(dm/2)(\boldsymbol{v} - \boldsymbol{v}^-)^2 - S\,\widehat{d\boldsymbol{F}} \cdot (\boldsymbol{v} - \boldsymbol{v}^-), \qquad (4.6.4\text{a})$$

(stationary and) a minimum, among $\{\boldsymbol{v}\}$: *kinematically possible (or comparison) postimpact velocities; that is,*

$$P(\boldsymbol{v};\ \boldsymbol{v}^-, \widehat{d\boldsymbol{F}}) \geq P(\boldsymbol{v}^+;\ \boldsymbol{v}^-, \widehat{d\boldsymbol{F}}) \equiv P_{\min},$$

$$P_{\min} = S\,(dm/2)(\boldsymbol{v}^+ - \boldsymbol{v}^-)^2 - S\,\widehat{d\boldsymbol{F}} \cdot (\boldsymbol{v}^+ - \boldsymbol{v}^-). \qquad (4.6.4\text{b})$$

Indeed, setting in the master equations (4.6.1c, d) $\delta\boldsymbol{r} \sim \boldsymbol{v}^+$ and $\delta\boldsymbol{r} \sim \boldsymbol{v}$ (to distinguish them from the $\boldsymbol{v}^+$ of Bertrand's theorem) yields

$$S\,dm(\boldsymbol{v}^+ - \boldsymbol{v}^-) \cdot \boldsymbol{v}^+ = S\,\widehat{d\boldsymbol{F}} \cdot \boldsymbol{v}^+,$$

$$S\,dm(\boldsymbol{v}^+ - \boldsymbol{v}^-) \cdot \boldsymbol{v} = S\,\widehat{d\boldsymbol{F}} \cdot \boldsymbol{v} \qquad \left[= S\,dm(\boldsymbol{v} - \boldsymbol{v}^-) \cdot \boldsymbol{v}\right]; \qquad (4.6.4\text{c})$$

the last equation holding because, if we denote by $\widehat{d\boldsymbol{R}}$ and $\widehat{d\boldsymbol{R}'}$ the constraint reactions of $\boldsymbol{v}^+$ and $\boldsymbol{v}$, respectively and *since the* $\boldsymbol{v}$ *are compatible with both* $\widehat{d\boldsymbol{R}}$ *and* $\widehat{d\boldsymbol{R}'}$ (i.e., with all constraints—since the $\boldsymbol{v}^+$ are a subset of the $\boldsymbol{v}$, they must also be compatible with both the $\widehat{d\boldsymbol{R}}$ and $\widehat{d\boldsymbol{R}'}$), we shall have $S\,\widehat{d\boldsymbol{R}} \cdot \boldsymbol{v} = S\,\widehat{d\boldsymbol{R}'} \cdot \boldsymbol{v} = 0$. Next, subtracting eqs. (4.6.4c) side by side readily results in

$$S\,dm(\boldsymbol{v}^+ - \boldsymbol{v}^-) \cdot (\boldsymbol{v} - \boldsymbol{v}^+) = S\,\widehat{d\boldsymbol{F}} \cdot (\boldsymbol{v} - \boldsymbol{v}^+); \qquad (4.6.4\text{d})$$

or, since kinematically admissible postimpact velocities may differ infinitesimally from each other, setting (as in Kelvin's theorem) $\boldsymbol{v} - \boldsymbol{v}^+ = \delta \boldsymbol{v}^+ \equiv \delta_R \boldsymbol{v}$, we can rewrite (4d) as

$$S\, dm(\boldsymbol{v}^+ - \boldsymbol{v}^-) \cdot \delta_R \boldsymbol{v} = S\, \widehat{d\boldsymbol{F}} \cdot \delta_R \boldsymbol{v}, \tag{4.6.4e}$$

which is none other than the *stationarity* condition (since $\delta_R \boldsymbol{v}^- = \boldsymbol{0}$)

$$\delta_R P = \delta_R \Big(S\, (dm/2)(\boldsymbol{v} - \boldsymbol{v}^-)^2 - S\, \widehat{d\boldsymbol{F}} \cdot (\boldsymbol{v} - \boldsymbol{v}^-) \Big) = 0. \tag{4.6.4f}$$

Next, to the minimality condition. We obtain, successively, using the above results,

$$\begin{aligned}
P(\boldsymbol{v}; \ldots) - P(\boldsymbol{v}^+; \ldots) &\equiv P - P_{\text{min}} \\
&= S\, (dm/2)\big[(\boldsymbol{v} - \boldsymbol{v}^-)^2 - (\boldsymbol{v}^+ - \boldsymbol{v}^-)^2\big] - S\, \widehat{d\boldsymbol{F}} \cdot (\boldsymbol{v} - \boldsymbol{v}^+) \\
&= S\, (dm/2)(\boldsymbol{v} - \boldsymbol{v}^+) \cdot (\boldsymbol{v} + \boldsymbol{v}^+ - 2\boldsymbol{v}^-) - S\, dm(\boldsymbol{v}^+ - \boldsymbol{v}^-) \cdot (\boldsymbol{v} - \boldsymbol{v}^+) \\
&= S\, (dm/2)(\boldsymbol{v} - \boldsymbol{v}^+)^2 \\
&= S\, (dm/2)(\delta_R \boldsymbol{v})^2 = (1/2)\, \delta^2{}_R P \geq 0, \quad \text{Q.E.D.}
\end{aligned} \tag{4.6.4g}$$

REMARKS

The final (i.e., postimpact) velocities of a system under sudden prescribed impressed impulses (or velocities), followed immediately by additional constraints, are the same as if the system had the constraints imposed first, followed immediately by the impressed impulses (or velocities); that is, *the order of application of impressed impulses (or velocities) and constraints is immaterial to the postimpact motion, as long as it all occurs within an infinitesimal time interval.* [However, the order of equally sudden imposition of impressed impulses (or velocities) and *removal* of constraints, clearly, *does* make a difference!] And, wherever that order of application is immaterial, the *total* impulse at various system points, impressed and constraint (reaction), must be the same for either order; hence, then, impressed impulses (or velocities) and additional constraints can be thought of as acting simultaneously, in the sense of Robin's theorem. [These remarks are due to Professor D. T. Greenwood (private communication).]

Special (Extreme) Cases

(a) Only the $\{\widehat{d\boldsymbol{F}}\}$ are imposed, but no additional constraints. Then the $\{\boldsymbol{v}^+\}$ make P, (4.6.4a), a minimum.

(b) Only the additional constraints are imposed, but no $\{\widehat{d\boldsymbol{F}}\}$. Then the $\{\boldsymbol{v}^+\}$ make the "comparison relative kinetic energy" P a minimum:

$$P \rightarrow S\, (dm/2)(\boldsymbol{v} - \boldsymbol{v}^-)^2 \rightarrow S\, (dm/2)(\boldsymbol{v}^+ - \boldsymbol{v}^-)^2 = P_{\text{min}}. \tag{4.6.4h}$$

Further, in this case, we obtain, successively,

$$\begin{aligned}\mathcal{S}(dm/2)(\boldsymbol{v}-\boldsymbol{v}^-)^2 &= \mathcal{S}(dm/2)\boldsymbol{v}\cdot\boldsymbol{v} + \mathcal{S}(dm/2)\boldsymbol{v}^-\cdot\boldsymbol{v}^- - \mathcal{S}\,dm\,\boldsymbol{v}\cdot\boldsymbol{v}^-\\ &= \mathcal{S}(dm/2)\boldsymbol{v}\cdot\boldsymbol{v} + \mathcal{S}(dm/2)\boldsymbol{v}^-\cdot\boldsymbol{v}^- - \mathcal{S}\,dm\,\boldsymbol{v}\cdot\boldsymbol{v}\\ &\quad\text{[the third (last) sum transformed with the help of (4.6.4c)]}\\ &= \mathcal{S}(dm/2)\boldsymbol{v}^-\cdot\boldsymbol{v}^- - \mathcal{S}(dm/2)\boldsymbol{v}\cdot\boldsymbol{v} > 0;\end{aligned}$$

and, therefore, for $\boldsymbol{v} = \boldsymbol{v}^+$,

$$\begin{aligned}P_{\min} &= \mathcal{S}(dm/2)(\boldsymbol{v}^+ - \boldsymbol{v}^-)^2 = \mathcal{S}(dm/2)\boldsymbol{v}^-\cdot\boldsymbol{v}^- - \mathcal{S}(dm/2)\boldsymbol{v}^+\cdot\boldsymbol{v}^+\\ &= T^- - T^+ \equiv -\Delta T > 0 \qquad (=\text{kinetic energy } \textit{loss}). \end{aligned} \tag{4.6.4i}$$

In words: *the postshock velocities of a system subjected to sudden ideal impulsive constraints minimize its relative kinetic energy; and that minimum value equals the lost kinetic energy* (i.e., first part of Carnot's theorem!).

(c) The preimpact state is one of rest. Then we simply set in (4.6.4a, b) $\boldsymbol{v}^- = \boldsymbol{0}$.

4.6.5 Theorem of Gauss

[Impulsive Counterpart of Differential Variational Principle of Gauss (§6.4, §6.6).] *The actual postimpact velocities* $\{\boldsymbol{v}^+\}$ *minimize the "impulsive compulsion":*

$$\hat{Z} = \hat{Z}(\boldsymbol{v}) \equiv \mathcal{S}(dm/2)\big[\boldsymbol{v} - \boldsymbol{v}^- - (d\widehat{\boldsymbol{F}}/dm)\big]^2 \tag{4.6.5a}$$
$$\Big(= \cdots = P + \mathcal{S}(d\widehat{\boldsymbol{F}})^2/2dm\Big),$$

relative to all kinematically admissible postimpact velocities $\{\boldsymbol{v} \equiv \boldsymbol{v}^+ + \delta_G\boldsymbol{v}\}$*;* that is, $\min \hat{Z} = \hat{Z}(\boldsymbol{v}^+)$. Indeed, from (4.6.5a) we readily obtain

$$\begin{aligned}\Delta_G\hat{Z} &\equiv \hat{Z}(\boldsymbol{v}) - \hat{Z}(\boldsymbol{v}^+)\\ &= \mathcal{S}(dm/2)\big[(\boldsymbol{v}^+ + \delta_G\boldsymbol{v} - \boldsymbol{v}^- - d\widehat{\boldsymbol{F}}/dm)^2 - (\boldsymbol{v}^+ - d\widehat{\boldsymbol{F}}/dm)^2\big]\\ &\equiv \delta_G\hat{Z} + (1/2)\delta^2{}_G\hat{Z},\end{aligned} \tag{4.6.5b}$$

where

$$\delta_G\hat{Z} \equiv \mathcal{S}\,dm(\boldsymbol{v}^+ - \boldsymbol{v}^- - d\widehat{\boldsymbol{F}}/dm)\cdot\delta_G\boldsymbol{v} \quad (=\delta_R P) = 0$$
$$\text{[by setting in (4.6.1c, d) } \delta\boldsymbol{r} \to \delta_G\boldsymbol{v}] \tag{4.6.5c}$$

$$\begin{aligned}\delta^2{}_G\hat{Z} &\equiv \mathcal{S}\,dm(\delta_G\boldsymbol{v})^2 \quad (=\delta^2{}_R P)\\ &\equiv \textit{Relative kinetic energy (as in Kelvin's theorem)} > 0;\end{aligned} \tag{4.6.5d}$$

that is, $\Delta_G\hat{Z} = (1/2)\delta^2{}_G\hat{Z}\ [= \Delta_R P = (1/2)\delta^2{}_R P)] > 0$, Q.E.D. The above clearly show the equivalence of the theorems of Gauss and Robin.

Table 4.2 Extremum Theorems of Impulsive Motion

Master equation (impulsive Lagrange's principle):

$$\mathsf{S}\, dm(\boldsymbol{v}^+ - \boldsymbol{v}^-) \cdot \delta\boldsymbol{r} = \mathsf{S}\, \widehat{d\boldsymbol{F}} \cdot \delta\boldsymbol{r}$$

- Carnot (first part—collisions):

  $$\delta\boldsymbol{r} \sim \boldsymbol{v}^+, \qquad \widehat{d\boldsymbol{F}} = \boldsymbol{0} \Rightarrow T^+ - T^- < 0.$$

  Carnot (second part—explosions):

  $$\delta\boldsymbol{r} \sim \boldsymbol{v}^-, \qquad \widehat{d\boldsymbol{F}} = \boldsymbol{0} \Rightarrow T^+ - T^- > 0.$$

- Kelvin (prescribed velocities):

  $$\delta\boldsymbol{r} \sim \boldsymbol{v}^+, \qquad \delta\boldsymbol{r} \sim \boldsymbol{v}^+ + \delta_K \boldsymbol{v},\ \boldsymbol{v}^- = \boldsymbol{0} \Rightarrow T(\boldsymbol{v}) - T(\boldsymbol{v}^+) > 0, \qquad \delta_K T^+ = 0.$$

- Bertrand–Delaunay (prescribed impulses):

  $$\delta\boldsymbol{r} \sim \boldsymbol{v}^+,\ \delta\boldsymbol{r} \sim \boldsymbol{v}^+ + \delta_{B/D}\boldsymbol{v} = \boldsymbol{v} \Rightarrow T(\boldsymbol{v}) - T(\boldsymbol{v}^+) < 0,\ \delta_{B/D} T^+ = 0.$$

  [Taylor: $T_{\text{Kelvin}}(\boldsymbol{v}) - T(\boldsymbol{v}^+) > T(\boldsymbol{v}^+) - T(\boldsymbol{v})_{\text{Bertrand–Delaunay}}$]

- Robin (prescribed impulses and constraints):

  $$\delta\boldsymbol{r} \sim \boldsymbol{v}^+, \qquad \delta\boldsymbol{r} \sim \boldsymbol{v}^+ + \delta_R \boldsymbol{v} = \boldsymbol{v}$$

  $P \equiv \mathsf{S}\,(dm/2)(\boldsymbol{v} - \boldsymbol{v}^-)^2 - \mathsf{S}\, \widehat{d\boldsymbol{F}} \cdot (\boldsymbol{v} - \boldsymbol{v}^-)$: stationary and minimum.

- Gauss (impulsive compulsion):

  $\hat{Z} \equiv \mathsf{S}\,(dm/2)(\boldsymbol{v} - \boldsymbol{v}^- - \widehat{d\boldsymbol{F}}/dm)^2 = P + \mathsf{S}\,(\widehat{d\boldsymbol{F}})^2/2dm$: stationary and minimum.

Alternatively, the stationarity condition (4.6.5c), with $\delta_G \boldsymbol{v}^+ = \boldsymbol{v} - \boldsymbol{v}^+$, applied to

$$\begin{aligned} \hat{Z} &\equiv \mathsf{S}\,(dm/2)\big[(\boldsymbol{v}^+ - \boldsymbol{v}^-) - \widehat{d\boldsymbol{F}}/dm\big]^2 \\ &= \mathsf{S}\,(dm/2)\boldsymbol{v}^+ \cdot \boldsymbol{v}^+ + \mathsf{S}\,(dm/2)\boldsymbol{v}^- \cdot \boldsymbol{v}^- + \mathsf{S}\,(\widehat{d\boldsymbol{F}})^2/2dm \\ &\qquad - \mathsf{S}\, dm\, \boldsymbol{v}^+ \cdot \boldsymbol{v}^- - \mathsf{S}\, \widehat{d\boldsymbol{F}} \cdot (\boldsymbol{v}^+ - \boldsymbol{v}^-), \end{aligned}$$

yields

$$\begin{aligned} \delta_G \hat{Z} &= \mathsf{S}\, dm(\boldsymbol{v}^+ - \boldsymbol{v}^- - \widehat{d\boldsymbol{F}}/dm) \cdot \delta_G(\boldsymbol{v}^+ - \boldsymbol{v}^- - \widehat{d\boldsymbol{F}}/dm) \\ &= \mathsf{S}\, dm\, \boldsymbol{v}^+ \cdot \delta_G \boldsymbol{v} - \mathsf{S}\, dm\, \boldsymbol{v}^- \cdot \delta_G \boldsymbol{v} - \mathsf{S}\, \widehat{d\boldsymbol{F}} \cdot \delta_G \boldsymbol{v} = 0, \end{aligned} \tag{4.6.5e}$$

or, rearranging,

$$\mathsf{S}\, dm(\boldsymbol{v}^+ - \boldsymbol{v}^-) \cdot \delta_G \boldsymbol{v} = \mathsf{S}\, \widehat{d\boldsymbol{F}} \cdot \delta_G \boldsymbol{v}; \tag{4.6.5f}$$

that is, the master equations (4.6.1c, d) with $\delta\boldsymbol{r} \to \delta_G \boldsymbol{v}$. Also, eq. (4.6.5f) constitutes the impulsive counterpart of *Jourdain's differential variational principle* (§6.3). This latter, in holonomic system variables, reads $\sum (\Delta p_k - \hat{Q}_k)\,\delta\dot{q}_k = 0$, under $\delta t = 0$, $\delta q_k = 0$. For a detailed and lucid treatment of its application to impulsive problems, see, for example, Bahar (1994); also ex. 4.5.6.

To facilitate the understanding of all these—admittedly, similarly sounding and hence hard to differentiate and remember—theorems, we summarize them in table 4.2.

Example 4.6.1 (D. T. Greenwood, 1997, private communication). *On Input Inertia and Impulse Response.* Let us consider a finite and discrete system, and an impulse $\hat{\boldsymbol{f}}_P$, acting at its point P and causing to it a velocity jump $\Delta\boldsymbol{v}_P \equiv \boldsymbol{v}_P{}^+ - \boldsymbol{v}_P{}^-$. Now, the *input inertia coefficient* or *driving-point mass* at P, μ_P, is defined by

$$\mu_P \equiv \hat{f}_P/\Delta v_{P,t} \qquad (P = 1, 2, \ldots), \tag{a}$$

where $\hat{f}_P$ is the magnitude of $\hat{\boldsymbol{f}}_P$ and $\Delta v_{P,t}$ is the component of $\Delta\boldsymbol{v}_P$ in the direction of $\hat{\boldsymbol{f}}_P$. It is not hard to see that $\mu_P > 0$, always (explain!); and, for a given system point P, it is a function of the configuration and the direction of $\hat{\boldsymbol{f}}_P$, but not of the system's state of motion. From the above, it follows that if μ_P is also finite, $\Delta\boldsymbol{v}_P$ has always a component in the direction of $\hat{\boldsymbol{f}}_P$. Now, and recalling the results of ex. 4.3.1 on the impulsive "work–energy" theorem, the work (or, better, power) done by $\hat{\boldsymbol{f}}_P$, $\hat{W}_P$, equals its dot product with the average velocity of $\boldsymbol{v}_P{}^+$ and $\boldsymbol{v}_P{}^-$:

$$\hat{W}_P \equiv \hat{\boldsymbol{f}}_P \cdot [(\boldsymbol{v}_P{}^- + \boldsymbol{v}_P{}^+)/2] = \hat{\boldsymbol{f}}_P \cdot [\boldsymbol{v}_P{}^- + (\Delta\boldsymbol{v}_P{}^+/2)]; \tag{b}$$

and by that theorem, the corresponding kinetic energy change, ΔT_P, is

$$\Delta T_P \equiv T_P{}^+ - T_P{}^- = \hat{W}_P = \hat{\boldsymbol{f}}_P \cdot \boldsymbol{v}_P{}^- + (\hat{f}_P{}^2/2\mu_P). \tag{c}$$

From this, we conclude that ΔT_P can be positive or negative, depending on the preimpact state of motion (i.e., the value of $\hat{\boldsymbol{f}}_P \cdot \boldsymbol{v}_P{}^-$). But, an originally motionless system (i.e., $\boldsymbol{v}_P{}^- = 0$) will always exhibit an *increase* in its kinetic energy, inversely proportional to its μ_P's.

Next, let us assume that additional stationary impulsive constraints are suddenly imposed on a moving system. If some particle velocity is changed, and the impact is inelastic, the resulting impulsive constraint reactions will do negative work on the system as a whole, and thus reduce its kinetic energy (Carnot's first theorem). This and (c) imply that additional constraints result in an increase in the system's input masses, μ_P's; a fact confirming our intuitive feeling that *each constraint tends to increase the resistance to velocity changes due to* $\hat{\boldsymbol{f}}_P$, at P.

Appendix: Calculation of $\Delta v_{P,t}$ for a single impressed impulse $\hat{\boldsymbol{f}}_P$

Recalling the results of §4.5 [eqs. 4.5.4a–5b)], let the constraints and kinetic equations of impulsive motion be, respectively,

$$\omega_D \equiv \sum A_{Dk}\dot{q}_k \equiv \sum A_{Dk}v_k = 0,$$

$$\sum M^*_{II'}\Delta\omega_{I'} = \Theta_I \Rightarrow \Delta\omega_I = \sum Y_{II'}\Theta_{I'} \qquad [(Y_{II'})\text{: inverse of } (M^*_{II'})], \tag{d}$$

where

$$\omega_I \equiv \sum A_{Ik}\dot{q}_k \equiv \sum A_{Ik}v_k \neq 0, \qquad M^*_{II'} \equiv \partial^2 T^*_o/\partial\omega_I\,\partial\omega_{I'},$$

$$T^*_o = T^*(t, q, \omega_D = 0, \omega_I) = T^*_o(t, q, \omega_I) \qquad (D = 1, \ldots, m;\ I, I' = m+1, \ldots, n). \tag{d1}$$

Now, with the earlier notations/definitions, suppose that

$$\Delta v_{P,t} = \sum C_{PI}\,\Delta\omega_I \qquad (P = 1, 2, \ldots, \#\,\text{particles under impressed impulses}). \tag{e1}$$

Hence, for a *single* such impulse $\hat{f}_P$, by equating impulsive virtual works, we find

$$\hat{\Theta}_I = [\partial(\Delta v_{P,t})/\partial(\Delta\omega_I)]\hat{f}_P = C_{PI}\hat{f}_P. \tag{e2}$$

Substituting (e2) into the second of (d), and the result into (e1), we obtain the sought formula

$$\Delta v_{P,t} = \sum C_{PI}\Big(\sum Y_{II'}(C_{PI'}\hat{f}_P)\Big) = \Big(\sum\sum C_{PI}Y_{II'}C_{PI'}\Big)\hat{f}_P. \tag{f1}$$

From (f1), it also follows at once that the corresponding input mass equals

$$\mu_P \equiv \hat{f}_P/\Delta v_{P,t} = \Big(\sum\sum C_{PI}Y_{II'}C_{PI'}\Big)^{-1}. \tag{f2}$$

The extension of the above to include impulsive constraint reactions is straightforward, and is left to the reader. These formulae may find useful applications to structural dynamics.

Example 4.6.2 Let us consider two circular homogeneous wheels, W_1 and W_2 (fig. 4.21), of respective radii r_1 and r_2, rotating with constant and, initially unrelated, angular velocities about their frictionless parallel axes through their respective fixed (geometrical and mass) centers O_1 and O_2. On these wheels, we drop an initially slack, inextensible and massless cable that sticks to them and, then, at a certain instant, becomes taut and thus exerts an impulsive moment on the wheels. Let us calculate their postimpact angular velocities $\omega_1{}^+$ and $\omega_2{}^+$, respectively.

Since the addition of the cable amounts to the *sudden introduction of a constraint*, *Carnot's first theorem*, (4.6.1e), yields immediately (with I_1 and I_2 denoting, respectively, the moments of inertia of W_1 and W_2 about O_1 and O_2)

$$\begin{aligned} 2(T^+ - T^-) &= [I_1(\omega_1{}^+)^2 + I_2(\omega_2{}^+)^2] - [I_1(\omega_1{}^-)^2 + I_2(\omega_2{}^-)^2] \\ &= -[I_1(\omega_1{}^+ - \omega_1{}^-)^2 + I_2(\omega_2{}^+ - \omega_2{}^-)^2] < 0, \end{aligned} \tag{a}$$

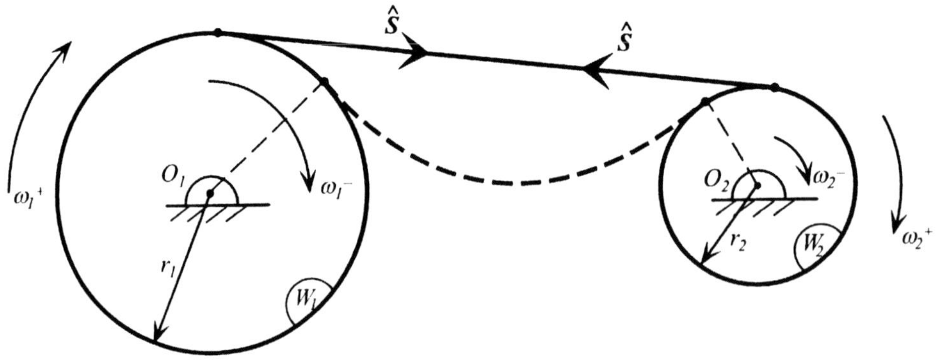

Figure 4.21 Sudden imposition of constraint in a two-wheel system.

or, rearranging and simplifying,

$$I_1[(\omega_1{}^-)^2 - (\omega_1{}^+)^2] + I_2[(\omega_2{}^-)^2 - (\omega_2{}^+)^2]$$
$$= I_1(\omega_1{}^- - \omega_1{}^+)^2 + I_2(\omega_2{}^- - \omega_2{}^+)^2 > 0. \qquad \text{(b)}$$

From kinematics we have, also,

$$\omega_1{}^+ r_1 = \omega_2{}^+ r_2. \qquad \text{(c)}$$

Solving the system (b, c), we obtain the sought postimpact angular velocities

$$\omega_1{}^+/r_2 = \omega_2{}^+/r_1 = (I_1 r_2 \omega_1{}^- + I_2 r_1 \omega_2{}^-)/(I_1 r_2{}^2 + I_2 r_1{}^2). \qquad \text{(d)}$$

Then, combining these results with the theorem of impulsive angular momentum, about O_1 and O_2, we obtain the impulsive cable tension $\hat{S}$,

$$\hat{S} r_1 = I_1(\omega_1{}^+ - \omega_1{}^-), \qquad -\hat{S} r_2 = I_2(\omega_2{}^+ - \omega_2{}^-) \qquad \text{(e)}$$
$$\Rightarrow \hat{S} = I_1(\omega_1{}^+ - \omega_1{}^-)/r_1 = -I_2(\omega_2{}^+ - \omega_2{}^-)/r_2$$
$$= [(r_2 \omega_2{}^- - r_1 \omega_1{}^-)/(I_1 r_2{}^2 + I_2 r_1{}^2)] I_1 I_2. \qquad \text{(f)}$$

Problem 4.6.1 (Bouligand, 1954, pp. 139–142). Consider a thin straight homogeneous rod AB, of mass m and length $2l$, originally suspended in horizontal equilibrium from a fixed ceiling by two vertical identical taut strings, s and s', attached to the bar at points other than its endpoints A and B [fig. 4.22(a)]. A *third* string s'' connects A with the ceiling point O, directly above A. The length of s'' is $2h$, and that is double the distance OA (i.e., originally, s, s' are taut, but s'' is slack). Then, the strings s and s' break simultaneously (or, someone burns them). Calculate the velocity state of AB immediately after the shock produced by the sudden tensioning of s''.

Let [fig 4.22(b)] $OA = r$, $angle(Ox, OA) = \theta$, $angle(Ox, AB) = \phi$. Now, the post s, s'-snap configurations of AB are determined by *three* Lagrangean coordinates; say, r, θ, ϕ; while the shock amounts to the sudden introduction of the persistent constraint $r = constant = 2h$.

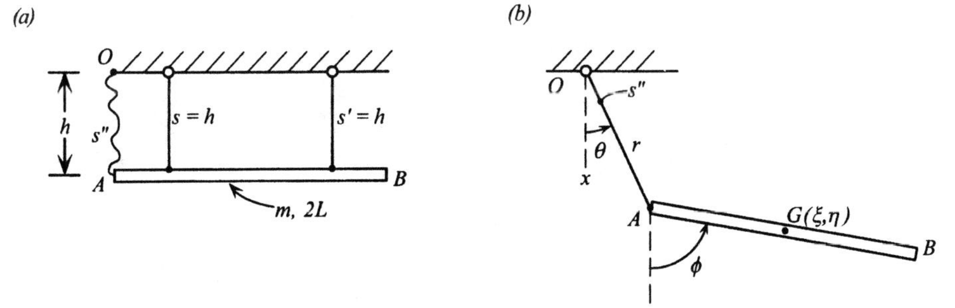

Figure 4.22 Geometry of rod *AB*, of mass *m* and length 2*l*, originally suspended from a fixed ceiling by the two equal and parallel strings *s* and *s'*, of length *h*. A third string *s''*, of length 2*h*, connects *A* with the fixed ceiling point (origin) O. Then *s* and *s'* snap, *AB* falls freely until *s''* gets taut and provokes a shock to *AB*.
(a) Equilibrium, (b) generic postshock configuration.

(i) Show that the (double) kinetic energy of AB, for a generic configuration, is

$$2T = m\{r^2(\dot{\theta})^2 + (\dot{r})^2 + (4/3)(l\dot{\phi})^2 + 2l[r\,\dot{\theta}\,\dot{\phi}\cos(\theta - \phi) + \dot{r}\,\dot{\phi}\sin(\theta - \phi)]\}. \quad \text{(a)}$$

HINT

Let the coordinates of the rod's center and center of mass G be x, y. From geometry,

$$x = r\cos\theta + l\cos\phi, \qquad y = r\sin\theta + l\sin\phi,$$

$$\Rightarrow \dot{x} = -r\,\dot{\theta}\sin\theta + \dot{r}\cos\theta - l\,\dot{\phi}\sin\phi, \qquad \dot{y} = r\,\dot{\theta}\cos\theta + \dot{r}\sin\theta + l\,\dot{\phi}\cos\phi. \quad \text{(b)}$$

Then use König's theorem: $2T = 2T_{\text{with all mass concentrated at } G} + 2T_{\text{relative, about } G}$.

(ii) By applying the fundamental Lagrangean impulsive virtual work equation

$$\Delta(\partial T/\partial\dot{r})\,\delta r + \Delta(\partial T/\partial\dot{\theta})\,\delta\theta + \Delta(\partial T/\partial\dot{\phi})\,\delta\phi = 0, \quad \text{for} \quad \delta r = 0,\ \delta\theta, \delta\phi\text{: } \textit{arbitrary} \quad \text{(c)}$$

(since here $\widehat{\delta' W} = 0$), obtain the *kinetic* impulsive equations:

$$\Delta\big[r\,\dot{\theta} + l\,\dot{\phi}\cos(\theta - \phi)\big] = 0 \qquad \Rightarrow r\Delta\dot{\theta} = 0, \quad \text{(d1)}$$

$$\Delta\big[(4/3)l\,\dot{\phi} - l\,\dot{\phi}\cos^2(\theta - \phi) + \dot{r}\sin(\theta - \phi)\big] = 0 \Rightarrow (4/3)l\Delta\dot{\phi} - \Delta\dot{r} = 0. \quad \text{(d2)}$$

(iii) Verify that, since the preshock conditions are (invoking energy conservation)

$$\theta^- = 0,\ \phi^- = \pi/2,\ r = 2l;\ \ (\dot{\theta})^- = 0,\ (\dot{\phi})^- = 0,\ (\dot{r})^- = (2gh)^{1/2} \ \text{(free fall)}, \quad \text{(e1)}$$

while, after the shock: $(\dot{r})^+ = 0 \Rightarrow \Delta\dot{r} = -(2gh)^{1/2}$, the above yield the postshock values

$$(\dot{\theta})^+ = 0, \qquad (\dot{\phi})^+ = -3(2gh)^{1/2}/4l. \quad \text{(e2)}$$

(iv) Verify that the *kinetostatic* impulsive equation is $\Delta(\partial T/\partial\dot{r}) = \hat{\lambda}$ [impulsive multiplier, to adjoin the constraint (1) $\delta r = 0$ to (c), and equal to the tension of s'']. Show that

$$4\hat{\lambda} = (2gh)^{1/2}.$$

(v) Show that if we choose as Lagrangean coordinates, x, y, and ϕ, the (double) kinetic energy and impulsive virtual work equation are, respectively,

$$2T = m[(\dot{x})^2 + (\dot{y})^2] + (ml^2/3)(\dot{\phi})^2; \quad \text{(f1)}$$

$$\Delta(\partial T/\partial\dot{x})\,\delta x + \Delta(\partial T/\partial\dot{y})\,\delta y + \Delta(\partial T/\partial\dot{\phi})\,\delta\phi = 0, \quad \text{(f2)}$$

for all δx, δy, $\delta\phi$ constrained by

$$\delta f\,|_{\text{evaluated at } \theta=0,\phi=\pi/2} = 0, \quad \text{where } f \equiv (x - l\cos\phi)^2 + (y - l\sin\phi)^2 = r^2\ \text{(constant)};$$

that is,

$$\delta x + l\,\delta\phi = 0. \quad \text{(f3)}$$

Verify that the variational equations (f2) and (f3) lead to the kinetic impulsive equations

$$\Delta\dot{x} - (l/3)\,\Delta\dot{\phi} = 0 \qquad \text{and} \qquad \Delta\dot{y} = 0. \quad \text{(f4)}$$

Then [from (b) evaluated at $\theta = 0$, $\phi = \pi/2$, $r = 2l$], since the preshock velocities are $(\dot{x})^- = 2(gh)^{1/2}$, $(\dot{y})^- = 0$, $(\dot{\phi})^- = 0$, confirm that the postshock velocities will be

$$(\dot{x})^+ = (2gh)^{1/2} + \Delta\dot{x} = (3/4)(2gh)^{1/2}, \quad (\dot{y})^+ = \Delta\dot{y} = 0, \quad (\dot{\phi})^+ = \Delta\dot{\phi}, \tag{f5}$$

[while $(\dot{y})^- = (\dot{y})^+ = 0$], and they will be connected by

$$(df/dt)^+ = 0 \Rightarrow (\dot{x})^+ + l(\dot{\phi})^+ = 0: \qquad (2gh)^{1/2} + \Delta\dot{x} + l\,\Delta\dot{\phi} = 0. \tag{f6}$$

Finally, confirm that combination of (f4) with (f6) yields a $\Delta\dot{\phi}$ value in agreement with (e2); also that the impulsive multiplier needed for adjoining (f3) to (f2) equals $l\,\Delta\dot{\phi}/3$.

(vi) Show that

$$2T^- = 2mgh, \qquad 2T^+ = m[r^2(\dot{\theta})^2 + (4/3)l^2(\dot{\phi})^2], \tag{g1}$$

$$\begin{aligned}
2T_{\text{jump}} &\equiv \textstyle\int dm(\boldsymbol{v}^+ - \boldsymbol{v}^-)\cdot(\boldsymbol{v}^+ - \boldsymbol{v}^-) \qquad \text{(kinetic energy of } \textit{jump} \text{ motion)}\\
&= m[(\Delta\dot{x})^2 + (\Delta\dot{y})^2] + (m\,l^2/3)(\Delta\dot{\phi})^2\\
&= m\{[(\dot{x})^+ - (2gh)^{1/2}]^2 + [(\dot{y})^+]^2 + (l^2/3)[(\dot{\phi})^+]^2\}\\
&= m\{[l(\dot{\phi})^+ + (2gh)^{1/2}]^2 + 4h^2[(\dot{\theta})^+]^2 + (l^2/3)[(\dot{\phi})^+]^2\}.
\end{aligned} \tag{g2}$$

(vii) Show that the *first* theorem of Carnot—that is, $T^- - T^+ = T_{\text{jump}}$—applied to the above yields

$$4h^2[(\dot{\theta})^+]^2 + (4/3)l^2[(\dot{\phi})^+]^2 + l(\dot{\phi})^+(2gh)^{1/2} = 0. \tag{h1}$$

A second equation connecting $(\dot{\theta})^+$ and $(\dot{\phi})^+$ is obtained by applying impulsive angular momentum conservation about O (i.e., Oz—notice that, in our axes, *clockwise is negative*):

$$\begin{aligned}
H_O^- = H_O^+: \quad -m\,l(2g\,h)^{1/2} &= I_G\,\omega^+ + [\boldsymbol{r}_{G/O} \times (m\boldsymbol{v}_G)]_z\\
&= (ml^2/3)(\dot{\phi})^+ + [-m(\dot{x})^+ l + m(\dot{y})^+(2h)] \qquad \text{[using (b)]}\\
&= (ml^2/3)(\dot{\phi})^+ + \{-m[-l(\dot{\phi})^+]l + m[2h(\dot{\theta})^+](2h)\}\\
&= (4/3)(m\,l^2)(\dot{\phi})^+ + m(2h)^2(\dot{\theta})^+,\\
\Rightarrow 4h^2(\dot{\theta})^+(4/3)l^2(\dot{\phi})^+ &+ l(2gh)^{1/2} = 0.
\end{aligned} \tag{h2}$$

Verify that the solution of (h1) and (h2) gives the earlier postshock values.

[The fact that (h1) is a single *nonvariational* equation in the two unknowns—namely, $(\dot{\theta})^+, (\dot{\phi})^+$ (and also that it is *quadratic* in them, thus yielding a *parasitic* solution, in addition to the actual one—a drawback of all *nonvariational/extremum* energetic theorems)—severely limits the practical usefulness of Carnot's theorem(s).]

Example 4.6.3 Let us consider a rigid lamina P, of mass m, originally at rest on a smooth table, one point of which, say A, is suddenly communicated a prescribed velocity (u, v), on the plane of P (fig. 4.23). We will calculate its actual postshock angular velocity, ω^+, via elementary (i.e., Newton–Euler) means, and by Kelvin's

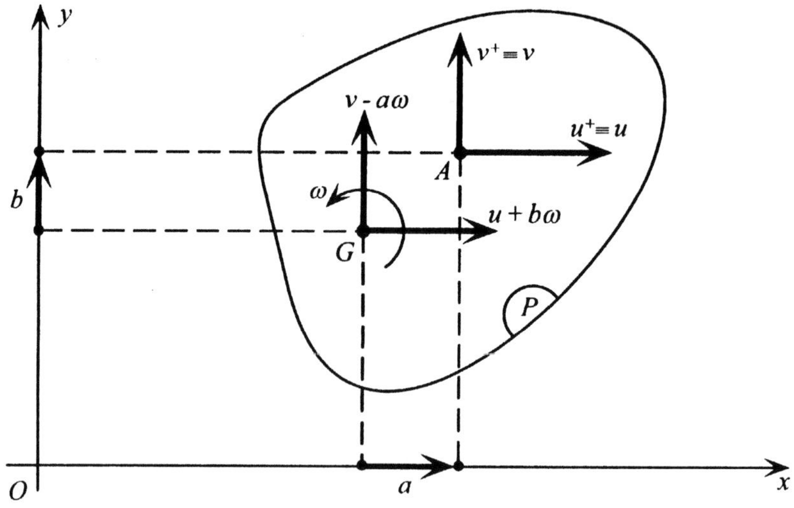

Figure 4.23 Kelvin's theorem for a rigid lamina in plane motion.

theorem; namely, that for ω^+ the kinematically possible postimpact kinetic energy of P becomes both stationary and minimum.

(i) *Via Newton–Euler*. Let (a, b) be the coordinates of A relative to the mass center of P, G. Then, by plane kinematics, the velocity of G has x, y-components: $u + b\,\omega$, $v - a\,\omega$; and so, by impulsive angular momentum about A (with mk^2: moment of inertia of P about G),

$$0 = (mk^2)\omega^+ + \{[m(u + b\,\omega^+)b] - [m(v - a\,\omega^+)]a\} \tag{a}$$

(= angular momentum of P about G + moment of linear momentum of P about A), and solving this for ω^+, we readily obtain

$$\omega^+ = (av - bu)/(k^2 + a^2 + b^2); \tag{b}$$

or, by applying the principle about the *body-fixed* point A, and corresponding moment of inertia

$$I_A = I_G + m(a^2 + b^2) = m(k^2 + a^2 + b^2):$$

$$0 = \Delta\big[I_A\omega - (\mathbf{r}_{A/G} \times m\mathbf{v}_A)_z\big]:\ 0 = I_A\,\omega^+ - [(mv)a - (mu)b] \ \Rightarrow \text{eq. (b)}. \tag{c}$$

(ii) *Via Kelvin's theorem*. Now, by König's theorem, the postimpact kinetic energy of P for an arbitrary *kinematically possible* postimpact angular velocity ω, equals

$$2T = m\big[(u + b\omega)^2 + (v - a\omega)^2\big] + (mk^2)\omega^2 = 2T(\omega; u, v). \tag{d}$$

Let the reader show that, by Kelvin's theorem, $T \to$ *stationary*; that is, setting $dT/d\omega = 0$ yields $\omega = \omega^+$, eq. (b), and further, that there $d^2T/d\omega^2 > 0$.

Example 4.6.4 Let us consider an initially motionless rigid and homogeneous rod AB, of mass m and length l (fig. 4.24), set in motion by causing its right end B to

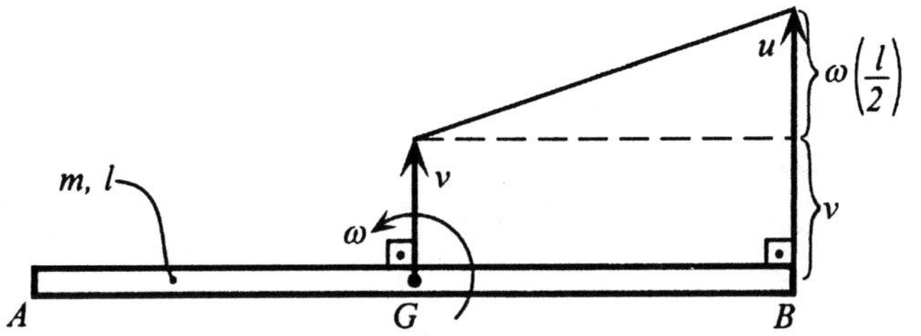

Figure 4.24 Rod under a prescribed velocity u at its end B.

move normally to AB with a specified postimpact velocity u. Let us calculate its postimpact angular velocity ω^+.

Here, by König's theorem, the *kinematically possible* postimpact kinetic energy of the rod [with v: velocity of rod's center of mass G; I: moment of inertia of rod about $G(= ml^2/12)$; and ω: angular velocity of rod] equals

$$2T = mv^2 + I\omega^2; \tag{a}$$

and, by simple kinematics,

$$\boldsymbol{v}_B = \boldsymbol{v}_G + \boldsymbol{\omega} \times \boldsymbol{r}_{B/G} = (0, v, 0) + (0, 0, \omega) \times (l/2, 0, 0) = (0, v + \omega l/2, 0), \tag{b}$$

that is, $u = v + \omega l/2 \Rightarrow v = u - \omega l/2$ (which expresses the kinematic possibility), and, therefore,

$$2T = m(u - \omega l/2)^2 + I\omega^2 \equiv T(\omega; u). \tag{c}$$

Now, according to *Kelvin's theorem*, of all the kinematically possible postimpact motions (i.e., velocities) of the rod with $(v_B)^+ \equiv u =$ *prescribed*, and hence for *any set of values of v and ω satisfying eq. (b)*, the actual, or kinetic, one will make T stationary/minimum; that is, it will be such that

$$\partial T/\partial\omega = -m(l/2)[u - \omega(l/2)] + I\omega = 0 \qquad [\text{with } \omega \to \omega^+]$$

$$\Rightarrow (l/2)\omega^+ = \{m(l/2)^2/[I + m(l/2)^2]\}u; \qquad \text{or, finally, } \omega^+ = 3u/2l. \tag{d}$$

Problem 4.6.2 By applying Kelvin's theorem, show that the actual postimpact angular velocities of two identical and homogeneous rods AB and BC (fig. 4.25), ω^+ and Ω^+, each of length l and mass m, smoothly hinged at B and originally at rest so that A, B, C are collinear, and after A is suddenly imparted a specified velocity v, normal to AB, equal

$$\omega^+ = 9v/7l \quad (\text{i.e., clockwise}), \qquad \Omega^+ = -(3v/7l) \quad (\text{i.e., counterclockwise}). \tag{a}$$

HINT

By König's theorem, the postimpact kinetic energy, for any kinematically admissible angular velocities ω and Ω, equals

$$T \equiv T(\omega, \Omega; v) = [(m/2)(v - l\omega/2)^2 + (1/2)(ml^2/12)\omega^2]$$
$$+ \{(m/2)[v - 2(l/2)\omega - (l/2)\Omega]^2 + (1/2)(ml^2/12)\Omega^2\}. \tag{b}$$

Then set $\partial T/\partial\omega = 0$, $\partial T/\partial\Omega = 0$.

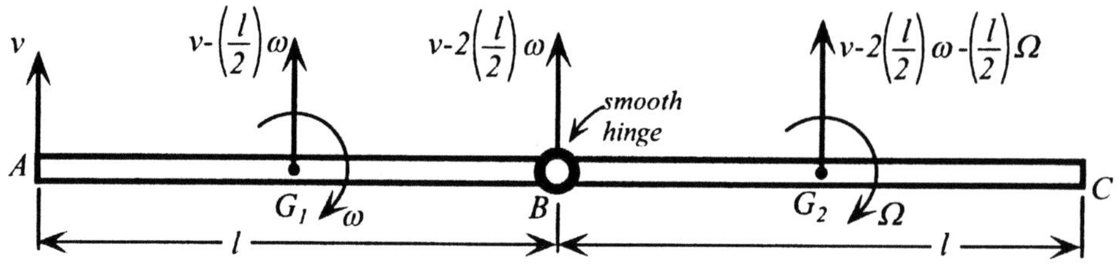

Figure 4.25 Two-rod system set in motion by a normal velocity at one of its endpoints. $v_A = v$, $v_{G_1} = v - (l/2)\omega$, $v_B = v - 2(l/2)\omega$, $v_{G_2} = v - 2(l/2)\omega - (l/2)\Omega$.

Example 4.6.5 (D. T. Greenwood, private communication, 1997).

(i) Let us consider an initially motionless rigid and homogeneous rod AB, of mass m and length l (and, hence, center of mass at the rod midpoint G—fig. 4.26), set in motion by a given transverse impulse $\hat{I}$ at B. Using impulsive principles of linear and angular momentum (about G), we readily find the following postimpact velocities (omitting superscript pluses):

$$v_G = \hat{I}/m, \qquad \omega = \hat{I}(l/2)/(ml^2/12) = 6\hat{I}/ml, \tag{a1}$$

and, therefore, the (also transverse) velocity of B is

$$\boldsymbol{v}_B = \boldsymbol{v}_G + \boldsymbol{\omega} \times \boldsymbol{r}_{B/G} \Rightarrow v_B = v_G + \omega(l/2) = 4\hat{I}/m. \tag{a2}$$

Hence, by König's theorem, the corresponding kinetic energy equals

$$\begin{aligned} T &= (1/2)m\, v_G{}^2 + (1/2)(m\, l^2/12)\omega^2 = \cdots = 2\hat{I}^2/m \\ &= (1/2)(\hat{I}^2/\mu_B) = (1/2)\mu_B\, v_B{}^2 = (1/2)\hat{I}v_B, \end{aligned} \tag{b1}$$

where (recalling the results/definitions of ex. 4.6.1)

$$\mu_B \equiv \hat{I}/v_B = m/4\text{: } \textit{input inertia coefficient} \text{ (or, } \textit{driving-point mass at B}\text{)}. \tag{b2}$$

(ii) Next, suppose we introduce the constraint $v_A = 0$ (e.g., we hinge A) before the initially motionless rod is struck by the same transverse impulse $\hat{I}$ at B. Now we have

$$\omega = \hat{I}l/(ml^2/3) = 3\hat{I}/ml \Rightarrow v_B = \omega\, l = 3\hat{I}/m$$

$$\Rightarrow T = (1/2)(ml^2/3)\,\omega^2 = \cdots = 3\hat{I}^2/2m \tag{c}$$

$$= (1/2)(\hat{I}^2/\mu_B) = (1/2)\mu_B\, v_B{}^2 = (1/2)\hat{I}\, v_B, \tag{d1}$$

where

$$\mu_B \equiv \hat{I}/v_B = m/3. \tag{d2}$$

(iii) Comparing the above, we see that the introduction of a constraint (α) has *increased* the value of the input inertia coefficient μ_B, and, since $2T = \hat{I}^2/\mu_B$, (β) has *reduced* the postimpact kinetic energy, in accordance with the Bertrand–Delaunay theorem.

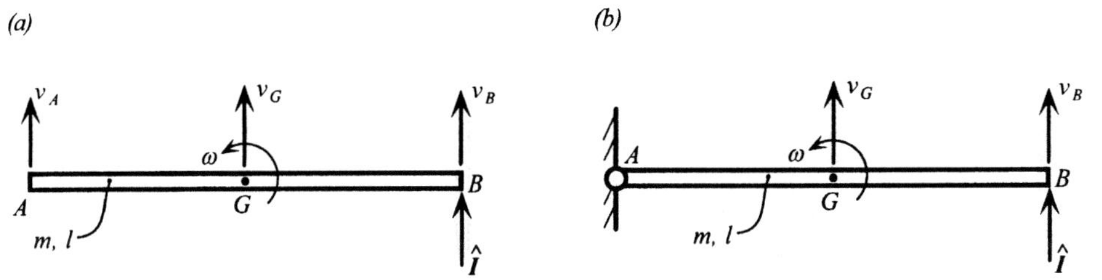

Figure 4.26 Rod under a given transverse impulse $\hat{I}$ at its end B: (a) unconstrained case; (b) constrained case.

If, on the other hand, we had prescribed the velocity v_B, rather than the impulse $\hat{I}$, and kept everything else the same, since also $2T = \mu_B\, v_B{}^2$, the postimpact kinetic energy would have been *increased*, in accordance with the Kelvin theorem.

(iv) Finally, let the values of the input inertia coefficient before the application of the constraint and after it be denoted (for more precision), respectively, as μ_B and $\mu_{B,c}$. Then we can write

$$\mu_{B,c} \equiv \mu_B + \Delta\mu_B = \mu_B + \varepsilon\mu_B = (1+\varepsilon)\mu_B, \tag{e1}$$

where

$$\varepsilon \equiv (\mu_{B,c} - \mu_B)/\mu_B > 0\text{: } \textit{Fractional increase of } \mu_B \textit{ due to the constraint} \tag{e2}$$

(in the above example: $\varepsilon \equiv (m/3 - m/4)/(m/4) = 1/3$).

Now, (α) comparing the kinetic energies before and after the constraint, but for the same v_B (i.e., à la Kelvin) we see that (with some easily understood ad hoc notations)

$$\begin{aligned}(T_c - T)_K &\equiv (1/2)\mu_{B,c}v_B{}^2 - (1/2)\mu_B v_B{}^2 \\ &= (1/2)\mu_B v_B{}^2(\mu_{B,c} - \mu_B) = (1/2)\mu_B v_B{}^2\varepsilon \\ &\Rightarrow [(T_c - T)/T]_K\text{: Fractional } \textit{increase} \text{ of } T \text{ due to the constraint} \\ &\qquad\qquad \text{(à la Kelvin)} = \varepsilon > 0.\end{aligned} \tag{e3}$$

while (β) comparing the kinetic energies before and after the constraint, but for the same $\hat{I}$ (i.e., à la Bertrand–Delaunay) we see that

$$\begin{aligned}(T - T_c)_{B/D} &\equiv (1/2)(\hat{I}^2/\mu_B) - (1/2)(\hat{I}^2/\mu_{B,c}) \\ &= (1/2)\hat{I}^2[(1/\mu_B) - (1/\mu_{B,c})] = (1/2)(\hat{I}^2/\mu_B)[\varepsilon/(1+\varepsilon)] \\ &\Rightarrow [(T - T_c)/T]_{B/D} > 0\text{: Fractional } \textit{reduction} \text{ of } T \text{ due to the} \\ &\qquad\qquad \text{constraint (à la } \textit{Bertrand–Delaunay}) \\ &\qquad\qquad = \varepsilon/(1+\varepsilon) < \varepsilon;\end{aligned} \tag{e4}$$

that is, comparing (e3) and (e4), we immediately conclude that the *T-increase à la Kelvin is greater than the T-decrease à la Bertrand–Delaunay* {by an amount equal to $\varepsilon - [\varepsilon/(1+\varepsilon)](= 1/12$, in the above example)}, in accordance with Taylor's theorem.

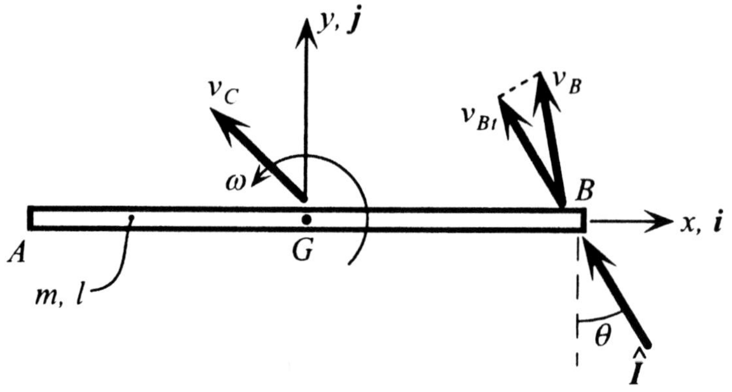

Figure 4.27 Rod AB struck at its right end B by a given nontransverse impulse $\hat{I}$. [$\boldsymbol{i}$, $\boldsymbol{j}$: unit vectors along the positive axes x (parallel to AB), y (perpendicular to AB).]

Problem 4.6.3 (D. T. Greenwood, private communication, 1997). Continuing from the preceding example, consider the response of the unconstrained (and originally motionless) rod AB to an impulse $\hat{\boldsymbol{I}}$ applied to its end B and making an angle θ with the perpendicular to the rod there (fig. 4.27). Let the component of the resulting postimpact velocity at B, $\boldsymbol{v}_B$, along $\hat{\boldsymbol{I}}$ be v_{Bt} (i.e., $\boldsymbol{v}_B$ is neither in the same direction as $\hat{\boldsymbol{I}}$, nor is it perpendicular to the rod at B, as before).

(i) By applying the impulsive principles of linear and angular momentum (about G) show that

$$\boldsymbol{v}_G = (\hat{I}/m)(-\sin\theta\,\boldsymbol{i} + \cos\theta\,\boldsymbol{j}), \qquad \omega = 6\hat{I}\cos\theta/ml, \tag{a1}$$

$$\boldsymbol{v}_B = \boldsymbol{v}_G + \boldsymbol{\omega}\times\boldsymbol{r}_{B/G} = \cdots = (\hat{I}/m)(-\sin\theta\,\boldsymbol{i} + 4\cos\theta\,\boldsymbol{j}) \tag{a2}$$

$$\Rightarrow\ v_{Bt} \equiv \boldsymbol{v}_B\cdot(\hat{\boldsymbol{I}}/\hat{I}) = \cdots = (\hat{I}/m)(1 + 3\cos^2\theta). \tag{a3}$$

(ii) Show that the imparted kinetic energy is

$$T = (\hat{I}^2/2m)(1 + 3\cos^2\theta). \tag{b1}$$

(iii) Verify that T can also be put in the following general forms:

$$T = \hat{I}^2/2\mu_B = (1/2)\hat{I}\,v_{Bt} = (1/2)\mu_B {v_{Bt}}^2, \tag{b2}$$

where

$$\mu_B \equiv \hat{I}/v_{Bt} = m/(1 + 3\cos^2\theta)\text{: input mass at } B. \tag{b3}$$

Problem 4.6.4 Consider an originally motionless and vertical double pendulum consisting of two identical and homogeneous rigid rods, AB and BC (fig. 4.28), each of mass m and length l, smoothly hinged at B and at the fixed support A, and struck by a *given horizontal blow* $\hat{I}$ at C.

(i) Show that the *comparison* postimpact kinetic energy of the system equals

$$T = (m/6)(2u^2 + u\,v + v^2) = (1/2)\hat{I}\,v, \tag{a}$$

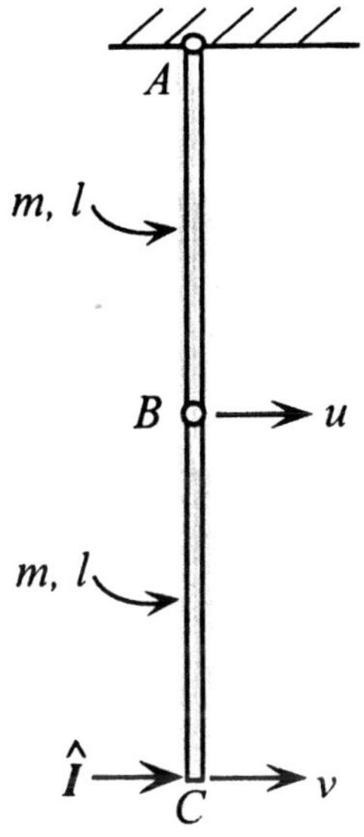

Figure 4.28 Vertical double pendulum struck by a given horizontal impulse.

where u and v are, respectively, the comparison (kinematically admissible) postimpact velocities of B and C, both perpendicular to the pendulum.

(ii) Next, show that application of the Bertrand–Delaunay theorem leads to the (constrained stationarity) conditions $\partial F/\partial u = \partial F/\partial v = 0$, where

$$
\begin{aligned}
F &\equiv T + \lambda(2T - \hat{I}v) \\
&= (1/2)\hat{I}v + \lambda\big[(m/3)(2u^2 + u\,v + v^2) - \hat{I}v\big] = F(u, v; \lambda);
\end{aligned}
$$

$$\text{constraint:}\quad 2T - \hat{I}v = (m/3)(2u^2 + u\,v + v^2) - \hat{I}v = 0;\ \text{multiplier: } \lambda. \qquad \text{(b)}$$

(iii) Show that the above equations lead to the following *actual* postimpact velocities:

$$u^+ = -(6/7)\hat{I}/m, \qquad v^+ = (24/7)\hat{I}/m \Rightarrow T^+ = (1/2)\hat{I}\,v = (12/7)\hat{I}^2/m. \qquad \text{(c)}$$

See also Lamb (1923, p. 321).

Problem 4.6.5 Consider an originally motionless square $ABCD$ consisting of four identical and homogeneous rigid bars (fig. 4.29), each of length $2l$ and mass m, mutually joined by smooth hinges and with corner A fixed, resting on a smooth horizontal table. Then, a *given impressed impulse* $\hat{I}$ acts on the square at B, along BD. Show that the postimpact angular velocities of AB and AD are, respectively,

$$\omega_{AB}{}^+ \equiv \omega_1 = 3\hat{I}/10\sqrt{2}ml, \qquad \omega_{AD}{}^+ \equiv \omega_2 = 0 \quad \text{(i.e., } AD \text{ stationary).} \qquad \text{(a)}$$

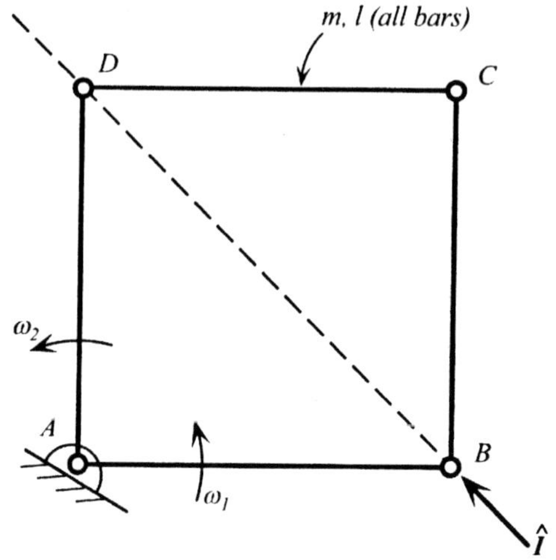

Figure 4.29 Square *ABCD* under given impulse $\hat{I}$ along its diagonal *BD*.

HINT

Apply the Bertrand–Delaunay theorem to the square's postimpact kinetic energy; that is,

$$T = \cdots = \left[(10/3)ml^2\right](\omega_1{}^2 + \omega_2{}^2) = T(\omega_1, \omega_2) \to \textit{maximum}, \qquad \text{(b)}$$

under the *constraint* (expressing the impulsive principle of angular momentum about A—explain)

$$\hat{I}(l/\sqrt{2}) = \cdots = \left[(20/3)ml^2\right](\omega_1 + \omega_2) = (\textit{specified})\ \textit{constant}, \qquad \text{(c)}$$

that is, $T^+ = \cdots = (3/20)(\hat{I}^2/m)$.

Problem 4.6.6 Continuing from the preceding problem, show that the postimpact kinetic energy of the given (*hinged* square) is *twice* as much as the postimpact kinetic energy produced by the *same impulse* $\hat{I}$, but acting on a *rigid* square, as stipulated by the Bertrand–Delaunay theorem.

HINT

In this case, $\omega_1 = \omega_2 \neq 0$ and $T^+ = \cdots = (3/40)(\hat{I}^2/m)$.

[We remark that the preceding problem may also be viewed as a *superposition* of (a) a rigid square ($\omega_1 = \omega_2$) under codirectional impulses $\hat{I}/2$ applied at B and D, and (b) a hinged square with A fixed but C able to slide along AD (i.e., $\omega_1 = -\omega_2$) under an impulse $\hat{I}/2$ applied at B [as in case (a)] and an opposite impulse $-\hat{I}/2$ applied at D.]

Problem 4.6.7 Consider a rhombus *ABCD* formed by four identical and homogeneous bars, *AB*, *BC*, *CD*, *DA* (fig. 4.30), each of mass $m/4$, length $2b$, radius of gyration about its own mass center k, and such that *angle* $(ABC) = 2\phi$, smoothly hinged at A, B, C, D, and originally resting on a frictionless horizontal table. The

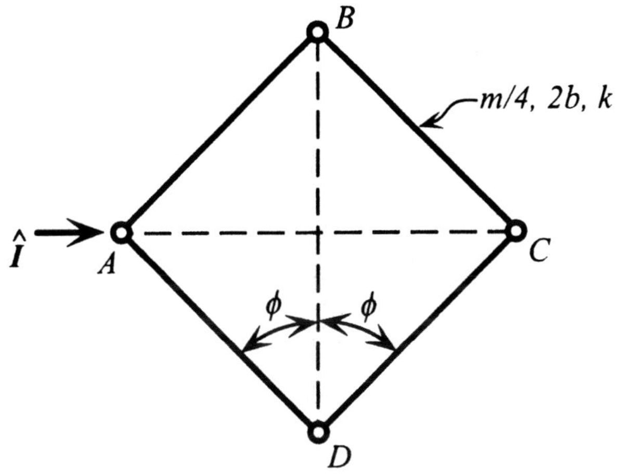

Figure 4.30 Rhombus *ABCD* struck by an impulse $\hat{I}$ along its diagonal *AC*.

rhombus is struck by a blow of intensity $\hat{I}$ at A, along AC. If x is the horizontal coordinate of its mass center G, from some fixed origin along AC, show that:

(i) The (double) kinetic energy of the system, at a generic impact configuration, is

$$2T = m(\dot{x})^2 + m(k^2 + b^2)(\dot{\phi})^2. \tag{a}$$

(ii) The postimpact velocity of A equals

$$(x - 2b\sin\phi)^{\cdot} = \{1 + [4b^2\cos^2\phi/(k^2 + b^2)]\}(\hat{I}/m) \equiv \beta(\hat{I}/m). \tag{b}$$

(iii) The (double) postimpact kinetic energy generated by $\hat{I}$ is $2T = \beta(\hat{I}^2/m) \equiv (2T^+)_{\text{hinged}}$; and, therefore, *the ratio of the actual postimpact kinetic energy to that if the rhombus were rigid*—that is, under the additional constraint $\phi = \text{constant}$ $(T_{hinged}/T_{rigid})_{\text{postimpact}}$, equals $\beta(> 1)$; as stipulated by the Bertrand–Delaunay theorem.

Example 4.6.6 Let us calculate the postimpact state of a system having kinetic energy

$$2T = Au^2 + Bv^2 + Cw^2 \tag{a}$$

(where u, v, w: *Lagrangean velocities*), originally moving with velocities $u^- \equiv u_o$, $v^- \equiv v_o$, $w^- \equiv w_o$, after the sudden imposition on it of the constraint

$$au + bv + cw = 0; \tag{b}$$

where both triplets of coefficients A, B, C, and a, b, c have their (approximately) constant impact values.

According to the *Gauss–Robin theorem*, the solution makes the impulsive compulsion,

$$\hat{Z} = A(u - u_o)^2 + B(v - v_o)^2 + C(w - w_o)^2 \equiv \hat{Z}(u, v, w), \tag{c}$$

a minimum, subject to the constraint (b). By differential calculus, we must have

$$d\hat{Z} = 0 \Rightarrow A(u - u_o)\,du + B(v - v_0)\,dv + C(w - w_o)\,dw = 0, \tag{d}$$

and

$$d[equation(b)] = 0 \Rightarrow a\,du + b\,dv + c\,dw = 0, \tag{e}$$

from which (using simple analytic geometry arguments in $u/v/w$ space, and fraction properties; thus avoiding Lagrangean multipliers) we obtain successively,

$$\begin{aligned}(u - u_0)/(a/A) &= (v - v_o)/(b/B) = (w - w_o)/(c/C)\\ &= [a(u - u_o) + b(v - v_o) + c(w - w_o)]/[(a^2/A) + (b^2/B) + (c^2/C)]\\ &= -(au_o + bv_o + cw_o)/[(a^2/A) + (b^2/B) + (c^2/C)]\\ &\equiv -\Lambda \qquad [\text{invoking (b)}],\end{aligned} \tag{f}$$

and from this the postimpact velocities follow:

$$u - u_o = -(a/A)\,\Lambda, \qquad v - v_0 = -(b/B)\Lambda, \qquad w - w_o = -(c/C)\,\Lambda. \tag{g}$$

As a result of the above, the kinetic energy *loss* (as *Carnot's first theorem* reminds us) transforms, successively, as follows:

$$\begin{aligned}2(T^- - T^+) &\equiv -2\Delta T \quad (\geq 0) \equiv A(u_o{}^2 - u^2) + B(v_o{}^2 - v^2) + C(w_o{}^2 - w^2)\\ &= A(u_o + u)(a/A)\,\Lambda + B(v_o + v)(b/B)\,\Lambda + C(w_o + w)(c/C)\,\Lambda\\ &= (au_o + bv_o + cw_o)\,\Lambda \qquad [\text{invoking (b)}]\\ &- (au_o + bv_o + cw_o)^2/[(a^2/A) + (b^2/B) + (c^2/C)] \geq 0,\end{aligned} \tag{h}$$

invoking the Λ-definition (f). The above apply intact, even if u, v, w are quasi velocities.

Example 4.6.7 Let us express some of the earlier extremum theorems in general Lagrangean coordinates.

(i) First, to enhance our understanding, we introduce the following ad hoc (not quite rigorous, but simplifying and convenient) notations and corresponding definitions:

$dq_k/dt \equiv v_k$ (and similarly for any other value of the Lagrangean velocities),

$$p_k \equiv \sum M_{kl} v_l \quad \text{(system momentum); or, } \textit{symbolically}, \ p = Mv; \tag{a}$$

$$2T \equiv \sum\sum M_{kl}\dot{q}_k\dot{q}_l \equiv \sum\sum M_{kl} v_k v_l \equiv M v\, v \equiv p\, v:$$

Initial, or *preimpact*, 2 (kinetic energy), (b1)

$$2T' \equiv \sum\sum M_{kl} v_k{}' v_l{}' \equiv Mv'v' \equiv p'v':$$

Actual postimpact 2 (kinetic energy), (b2)

$$2T'' \equiv \sum\sum M_{kl} v_k{}'' v_l{}'' \equiv M v'' v'' \equiv p'' v'':$$

Comparison, or *kinematically admissible, postimpact* 2 (kinetic energy); (b3)

$$2T_{01} = 2T_{10} \equiv \sum\sum M_{kl}(v_k{}' - v_k)(v_l{}' - v_l)$$
$$\equiv M(v' - v)(v' - v) \equiv (p' - p)(v' - v):$$

2(kinetic energy) of *relative (or jump) motion* $v' - v$, (b4)

$$2T_{02} = 2T_{20} \equiv \sum\sum M_{kl}(v_k{}'' - v_k)(v_l{}'' - v_l)$$
$$\equiv M(v'' - v)(v'' - v) \equiv (p'' - p)(v'' - v):$$

2(kinetic energy) of *relative motion* $v'' - v$, (b5)

$$2T_{12} = 2T_{21} \equiv \sum\sum M_{kl}(v_k{}'' - v_k{}')(v_l{}'' - v_l{}')$$
$$\equiv M(v'' - v')(v'' - v') \equiv (p'' - p')(v'' - v'):$$

2(kinetic) energy of *relative motion* $v'' - v'$; (b6)

$$2K_{01} = 2K_{10} \equiv \sum\sum M_{kl} v_k v_l{}' \equiv M v\, v' \equiv p\, v' \equiv p' v:$$

Impulsive *power of the momenta corresponding to the* v, *times the* v'; and vice versa. (c1)

$$2K_{12} = 2K_{21} \equiv \sum\sum M_{kl} v_k{}' v_l{}'' \equiv M v' v'' \equiv p' v'' \equiv p'' v':$$

Impulsive *power of the momenta corresponding to the* v', *times the* v''; and vice versa. (c2)

(ii) Then, recalling the *symmetry of the inertia coefficients* (i.e., $M_{kl} = M_{lk}$), we readily find that the above are related by

$$2T_{01} \equiv M(v' - v)(v' - v) = \cdots = 2T' + 2T - 2K_{01} - 2K_{01}$$
$$\Rightarrow 2K_{01} = T + T' - T_{01}; \tag{d1}$$

and similarly we derive

$$2K_{02} \equiv M v\, v'' = T + T'' - T_{02}, \tag{d2}$$
$$2K_{12} \equiv M v' v'' = T' + T'' - T_{12}. \tag{d3}$$

(iii) Next, setting in the LIP, (4.3.7), in succession, $\delta q \to v, v', v''$, we obtain the following formal "impulsive power" expressions:

$$\sum\sum M_{kl}(v_l{}' - v_l) v_k = \sum \hat{Q}_k v_k; \quad \text{or} \quad M(v' - v)v = \hat{Q}\, v, \tag{e1}$$
$$\sum\sum M_{kl}(v_l{}' - v_l) v_k{}' = \sum \hat{Q}_k v_k{}'; \quad \text{or} \quad M(v' - v)v' = \hat{Q}\, v', \tag{e2}$$
$$\sum\sum M_{kl}(v_l{}' - v_l) v_k{}'' = \sum \hat{Q}_k v_k''; \quad \text{or} \quad M(v' - v)v'' = \hat{Q}\, v''; \tag{e3}$$

which, thanks to (d1–3), can be rewritten, respectively, as

$$T' - T - T_{01} = \hat{Q}\, v, \tag{f1}$$

$$T' - T + T_{01} = \hat{Q}\, v', \tag{f2}$$

$$T' - T - T_{12} + T_{02} = \hat{Q}\, v''. \tag{f3}$$

(iv) With the help of the above, we now revisit the earlier extremum theorems:

(iv.a) Adding (f1) and (f2) side by side yields (the system form of ex. 4.3.1

$$2(T' - T) = \hat{Q}(v + v'); \tag{g1}$$

that is, *the change in the kinetic energy of a moving system, due to impressed impulses, equals the power of these impulses on the averaged velocities before and after their application.*

(iv.b) Subtracting (f1) from (f2) yields

$$2T_{01} = \hat{Q}(v' - v); \tag{g2}$$

that is, *the relative kinetic energy of a moving system, due to impressed impulses, equals the power of these impulses on half the velocity jumps due to them.*

(iv.c) If the power of the impressed impulses on the actual postimpact velocities vanishes—that is, if $\hat{Q}\, v' = 0$ (e.g., sudden introduction of ideal impulsive constraints)—then (f2) leads to

$$T' - T + T_{01} = 0 \Rightarrow T' - T = -T_{01} < 0 \Rightarrow T' < T: \tag{g3}$$

that is, the introduction of ideal constraints reduces the kinetic energy by an amount equal to the relative (jump) kinetic energy T_{01} [Carnot's *first* theorem, eq. (4.6.1e)].

(iv.d) If the power of the impressed impulses on the preimpact velocities vanishes—that is, if $\hat{Q}\, v = 0$ (e.g., if an explosion occurs in any part of the moving system)—then (f1) leads to

$$T' - T - T_{01} = 0 \Rightarrow T' - T = T_{01} > 0 \Rightarrow T' > T; \tag{g4}$$

that is, *in cases of explosion, kinetic energy is always gained by an amount equal to the relative (jump) kinetic energy* T_{01} [Carnot's *second* theorem, eq. (4.6.1h)].

(iv.e) Assume, next, that certain points of the moving system are suddenly seized and, under unknown impressed impulses acting there, are given prescribed velocities; like given constraints. Here, *since the velocities of the points of application of the impressed impulses are prescribed*:

$$\hat{Q}\, v' = \hat{Q}\, v'', \tag{g5}$$

and so the identities (f2, 3), and (b4, 6), immediately yield

$$T_{01} + T_{12} = T_{02} \Rightarrow T_{01} < T_{02}; \tag{g6}$$

that is, *in the case of impressed impulses acting on a moving system, and imparting to their points of application prescribed velocities, the "actual relative (jump) kinetic energy"* (T_{01}) *is smaller than any other "competing relative kinetic energy"* (T_{02}); in Routh's words: "the actual motion is such that the vis viva [= 2(kinetic energy)] of the relative motion, before and after, is less than if the system took any other course."

In particular, *if the preimpact velocities vanish*—that is, $v = 0$ (initially motionless system)—then $T_{01} \to T'$ and $T_{02} \to T''$, and thus (g6) reduces to the *theorem of Kelvin*:

$$T' < T''; \tag{g7}$$

that is, *in a system initially at rest, and then set in motion by impressed impulses acting at given material points and producing prescribed velocities there, the kinetic energy of the actual velocities* (T') *is less than that of any other competing motion in which these points have the prescribed velocities* (T''). So (g6) constitutes a generalization of Kelvin's theorem to initially generally moving systems.

(iv.f) Finally, assume that given impulses act at specified points, the postimpact velocities of which are, however, unknown. Then, since the impulsive constraint reactions of the so-competing motions v'' are also ideal,

$$M(v' - v)\, v'' = \hat{Q}\, v'' \qquad \text{and} \qquad M(v'' - v)v'' = \hat{Q}\, v'', \tag{h1}$$

or, recalling their forms (e1–f3), and with $v' \to v''$ in (e2) and (f2),

$$\begin{aligned} &T' - T - T_{12} + T_{02} = T'' - T + T_{02} \quad (= \hat{Q}\, v'') \\ &\Rightarrow\; T'' + T_{12} = T' \;\Rightarrow\; T' > T''; \end{aligned} \tag{h2}$$

that is, *theorem of Bertrand–Delaunay*: *in a moving system acted on by given impressed impulses, the actual postimpact kinetic energy is greater than that of any other additionally constrained competing motion, but under the same impulses.*

In sum:

- If the *postimpact velocities* of the points of application of the impressed impulses are *given*, the actual postimpact velocities are found by making the kinetic energy a *minimum* (*Kelvin*); and
- If the *impressed impulses* are *given*, the actual postimpact velocities are found by introducing some constraints and then making the kinetic energy a *maximum* (*Bertrand–Delaunay*).

For complementary derivations, via the so-called *reciprocity* theorems of dynamics, and so on, see, for example, (alphabetically): Kilmister and Reeve (1966, pp. 247–248), Lamb (1929, pp. 184–187, 206, 216–217), Pars (1965, pp. 242–243), Ramsey (1937, pp. 216–218), Rayleigh (1884, p. 91 ff.), Smart (1951, pp. 383–385); also ex. 4.6.8.

Example 4.6.8 *Two Degree of Freedom System: Lagrangean Derivation of Theorems of Kelvin, Bertrand–Delaunay, and of Reciprocity*; as an illustration of *the effect of constraints on the kinetic energy of a mechanical system set in motion in different ways.* We consider a two–Lagrangean coordinate system with (double) kinetic energy

$$2T = a\, v_x{}^2 + 2c\, v_x\, v_y + b\, v_y{}^2: \quad \textit{positive definite} \text{ in the Lagrangean velocities } v_x, v_y; \tag{a}$$

and a, b, c: inertial coefficients. Hence, its Lagrangean momenta will be

$$p_x \equiv \partial T/\partial v_x = a\, v_x + c\, v_y, \qquad p_y \equiv \partial T/\partial v_y = c\, v_x + b\, v_y. \tag{b}$$

Next, solving the second of (b) for v_y in terms of v_x and p_y, and substituting the result into (a), we obtain the *mixed* T-expression

$$2T = \left[(ab - c^2)/b\right] v_x^{\,2} + (1/b)p_y^{\,2}, \tag{c}$$

where $(ab - c^2)/b > 0 \Rightarrow ab - c^2 > 0$ (positive definiteness of T).

Now, with the help of the above, let us revisit our extremum theorems.

(i) *Theorem of Bertrand–Delaunay*. For the *actual* postimpact state, we have

$$p_x = X = \textit{given}, \quad p_y = 0; \qquad \text{and, therefore, } v_x \neq 0, \quad v_y \neq 0; \tag{d1}$$

while for the *comparison* postimpact state (and denoting the corresponding values of all quantities there with an *accent*),

$$p_x{}' = p_x = X, \qquad p_y{}' \neq 0; \qquad v_x{}' \neq 0, \qquad v_y{}' = 0. \tag{d2}$$

Then, the corresponding kinetic energies become

$$2T = \left[(ab - c^2)/b\right] v_x^{\,2} \qquad \text{and} \qquad 2T' = a(v_x{}')^2. \tag{e}$$

But from (d2), with (b), and then the second of (d1), we find, successively,

$$a\, v_x + c v_y = a\, v_x{}' + c\, v_y{}' = a\, v_x{}'$$

$$\Rightarrow v_x{}' = v_x + (c/a)v_y = v_x + (c/a)[(-c/b)v_x] = \left[1 - (c^2/ab)\right] v_x,$$

and therefore $2T'$ becomes

$$2T' = a(v_x{}')^2 = a[1 - (c^2/ab)]^2 v_x^{\,2} = \cdots = \left[1 - (c^2/ab)\right] 2T < 2T; \tag{f}$$

that is, the kinetic energy due to a given impulse ($p_x \neq 0$) acting alone ($p_y = 0$) is *greater* than if the other coordinate (y), under the action of a constraining impulse ($p_y{}' \neq 0$, but $p_x{}' = p_x \neq 0$), had been prevented from varying ($v_y{}' = 0$).

(ii) *Theorem of Kelvin*. For the actual postimpact state, we have

$$p_x \neq 0, \quad p_y = 0; \qquad \text{and} \;\; v_x = \textit{prescribed} \equiv u, \quad v_y \neq 0; \tag{g1}$$

while for the *comparison* postimpact state (accented quantities again),

$$p_x{}' \neq 0, \quad p_y{}' \neq 0 \;\; \text{(constraining impulse)}; \qquad v_x{}' = v_x = u, \quad v_y{}' = 0. \tag{g2}$$

Then, the corresponding kinetic energies become

$$2T = \left[(ab - c^2)/b\right] v_x^{\,2} = \left[a - (c^2/b)\right] v_x^{\,2} \qquad \text{and} \qquad 2T' = a(v_x{}')^2 = a\, v_x^{\,2}, \tag{h1}$$

and from these we easily conclude that

$$2T' > 2T; \tag{h2}$$

that is, *the kinetic energy started by a prescribed velocity* (v_x), *generated by the corresponding impulse acting alone* ($p_x \neq 0$, $p_y = 0$), *is less than if the other impulse* $p_y{}'(\neq 0)$ *had acted, constraining its associated coordinate* (*i.e.*, $v_y{}' = 0$, *but* $v_x{}' = v_x$).

As Lamb sums it up: *Bertrand–Delaunay theorem*: "A system started from rest by given *impulses* acquires greater energy than if it had been constrained in any way"; *Kelvin theorem*: "A system started with given *velocities* has *less* energy than if it had been constrained" (1923, p. 324). And as remarked by the great British physicist (acoustician, etc.) Rayleigh: "Both theorems are included in the statement that the [moment of] inertia is increased by the introduction of a constraint"; or, "the effect of a constraint is to increase the apparent inertia of the system" (Rayleigh, 1894, p. 100; publ. 1886).

(iii) *Appendix: A theorem of reciprocity.* Let

$$2T = a\,v_x^{\,2} + 2c\,v_x\,v_y + b\,v_y^{\,2}$$
$$\Rightarrow\; p_x \equiv \partial T/\partial v_x = a\,v_x + c\,v_y, \qquad p_y \equiv \partial T/\partial v_y = c\,v_x + b\,v_y, \tag{j}$$

as in (a, b). Now, let us consider *another state of motion,* through the *same configuration,* $(\ldots)'$: $v_x{}'$, $v_y{}'$, $p_x{}'$, $p_y{}'$. Then, it is not hard to see that we will have

$$p_x v_x{}' + p_y v_y{}' = p_x{}' v_x + p_y{}' v_y$$
$$= a\,v_x\,v_x{}' + c(v_x{}' v_y + v_x\,v_y{}') + b\,v_y v_y{}'. \tag{k}$$

Therefore, if we set $p_x{}' = 0$, $p_y = 0$, we find $p_x\,v_x{}' = p_y{}' v_y \;\Rightarrow\; v_y/p_x = v_x{}'/p_y{}'$; in words: *if an impulse p_x in the x-coordinate produces a velocity v_y in the y-coordinate, then an equal impulse $p_y{}'$ in the y-coordinate will produce the same velocity $v_x{}'$ in the x-coordinate.* Clearly, such reciprocity relations (a) result from the *symmetry of the inertia coefficients* (inertia tensor, etc.), for any independent set of velocities; and (b) can be easily extended to systems with n Lagrangean coordinates.

EXAMPLE

As an illustration of their use in impulsive motion, let us show the following theorem: *If an impulsive couple $\mathbf{C}_1 = C\,\mathbf{u}_1 = C(u_{1x}, u_{1y}, u_{1z})$, ($\mathbf{u}_1$: unit vector), applied to an originally motionless and unconstrained rigid body generates an angular velocity $\boldsymbol{\omega}_2 = \omega_2\,\mathbf{u}_2 = \omega_2(u_{2x}, u_{2y}, u_{2z})$, ($\mathbf{u}_2$: another unit vector), then the same couple (magnitude-wise) but applied about $\mathbf{u}_2$ (i.e., $\mathbf{C}_2 = C\,\mathbf{u}_2$) will produce an angular velocity about $\mathbf{u}_1$, ω_1, magnitude-wise equal to ω_2 (i.e., $\boldsymbol{\omega}_1 = \omega_1\mathbf{u}_1 = \omega_2\mathbf{u}_1$, $\omega_1 = \omega_2$); or, the angular velocity ω_2 about axis 2 due to an angular impulse C about axis 1 is equal to the angular velocity ω_1 about axis 1 due to an angular impulse C about axis 2.*

PROOF

Choosing (just for algebraic simplicity, no loss of generality) principal axes at the body's mass center G, with corresponding (principal) moments of inertia $I_{x,y,z}$, we will have (using, for example, the impulsive form of the Eulerian equations, §1.17)

$$\boldsymbol{\omega}_2 = (Cu_{1x}/I_x, Cu_{1y}/I_y, Cu_{1z}/I_z) \equiv (\omega_{2x}, \omega_{2y}, \omega_{2z}), \tag{11}$$
$$\boldsymbol{\omega}_1 = (Cu_{2x}/I_x, Cu_{2y}/I_y, Cu_{2z}/I_z) \equiv (\omega_{1x}, \omega_{1y}, \omega_{1z}), \tag{12}$$
$$\Rightarrow \omega_2 = \boldsymbol{\omega}_2 \cdot \mathbf{u}_2 = \omega_1 = \boldsymbol{\omega}_1 \cdot \mathbf{u}_1$$
$$= C(u_{1x}u_{2x}/I_x + u_{1y}u_{2y}/I_y + u_{1z}u_{2z}/I_z); \tag{13}$$

the symmetry of which in $u_{1x,1y,1z;2x,2y,2z}$ proves our proposition. For a treatment based on the Bertrand–Delaunay theorem, see Pöschl (1928, p. 510) and Smart (1951, p. 382). Similar theorems exist in other areas of physics. For further applications and insights, see, for example, Lamb (1929, pp. 276–281) and references cited therein.

Example 4.6.9 Let us consider a system of impulses acting on various points of an arbitrarily moving set of bodies, in such a way that each impulse is perpendicular to the (preimpact) velocity of its point of application. We will show that, as a result, the kinetic energy is increased [Routh, 1905(a), pp. 308–309]. Indeed, because of the above perpendicularity, the power of the impulses on the preimpact velocities vanishes: $T' - T - T_{01} = \hat{Q}\,v = 0$, and so eq. (f1) of ex. 4.6.7 immediately yields

$$T' - T = T_{01} > 0 \Rightarrow T' > T, \quad \text{Q.E.D.}$$

Example 4.6.10 Let us consider an initially motionless system. If acted on by two different sets of impulses, say A and B, it will take two different postimpact motions. We will show that the power of the impulses A on the velocities B equals the power of the impulses B on the velocities A [Routh, 1905(a), p. 309, example 6].

Indeed, since $T = 0$, we will have $T' = T_{01}$ and $T'' = T_{02}$. Then, the earlier

$$T' - T - T_{12} + T_{02} = \hat{Q}\,v'' \equiv \hat{Q}'v'', \tag{a}$$

and a completely analogous one with T' replaced by T'', and so on; that is,

$$T'' - T - T_{12} + T_{01} = \hat{Q}''v', \tag{b}$$

immediately yield $\hat{Q}'v'' = \hat{Q}''v$, Q.E.D. (a result analogous to a well-known *reciprocal* work proposition in linear elasticity).

Example 4.6.11 Let us, next, extend the above results to the collisions of *inelastic* (but smooth, i.e. frictionless) systems. Here, we introduce the following convenient notation (slightly different from that of ex. 4.6.7):

v: *preimpact* velocities,

v': velocities at instant of *maximum compression/contact*,

v'': *postimpact* velocities (i.e., just after the conclusion of the period of restitution);

and let the corresponding kinetic energies be denoted by T, T', T''; and the *relative* kinetic energies at any two of these instances (with some easily understood notation) be denoted by T_{01}, T_{12}, T_{02}. Because of the smoothness of the colliding surfaces, reasoning à la Carnot (first theorem), we have

$$M(v' - v)v' = \hat{Q}\,v' = 0 \quad \text{and} \quad M(v'' - v)v' = \hat{Q}\,v' = 0; \tag{a, b}$$

and since *the ratio of the total impulse to the impulse until the instant of maximum compression* equals $(1 + e)/1$ (where $e \equiv$ *coefficient of restitution*), taking the powers of these impulses over the same velocities (first the preimpact v, and then the postimpact v''); that is, reasoning as in (e1–3) of ex. 4.6.7, we can write

$$M(v'' - v)v = (1 + e)M(v' - v)v, \tag{c}$$

$$M(v'' - v)v'' = (1 + e)M(v' - v)v''. \tag{d}$$

In terms of the *relative* kinetic energies, T_{01}, T_{12}, T_{02}, (a–d) can be rewritten, respectively, as

$$T' - T = -T_{01}, \qquad T'' - T' = T_{12}, \tag{e, f}$$
$$T'' - T'(1+e) + eT = T_{02} - (1+e)T_{01}, \tag{g}$$
$$T'' - T'(1+e) + eT = eT_{02} - (1+e)T_{12}. \tag{h}$$

Now:

(i) Eliminating the three relative kinetic energies, we obtain

$$T'' - T' = -e^2(T' - T); \tag{i}$$

that is, *the kinetic energy increase due to the restitution (or explosion) is e^2 times the kinetic energy decrease due to the compression.*

(ii) Eliminating the three T''s, we obtain

$$T_{01} = T_{02}/(1+e)^2 = T_{12}/e^2. \tag{j}$$

(iii) Finally, eliminating T', T_{01}, and T_{12}, we obtain

$$T'' - T' = -\big[(1-e)/(1+e)\big]T_{02}; \tag{k}$$

thus extending Carnot's "third" theorem to inelastic (but frictionless) systems [discussion after (4.6.1j)]. See also Whittaker (1937, pp. 234–235), for alternative derivations.

For an extension of the above theorems to the collision of *inelastic and rough* solids (i.e., case where, throughout the impact, the contacting surfaces slide on each other and the accompanying friction preserves its direction/sense), see, for example, Routh 1905(a), p. 310).

Example 4.6.12 *Gauss' Principle of Least Impulsive Compulsion (or Constraint).* To examine the relation of Gauss' impulsive principle, (4.6.5a–f) with the above, let

T: *actual preimpact* kinetic energy,

T': *actual postimpact* kinetic energy, under the *existing constraints* and given impulses,

T'': *comparison postimpact* kinetic energy, under *additional constraints* and given impulses,

T''': *postimpact* kinetic energy under *zero constraints* (i.e., *free* motion) and given impulses.

Now, by Bertrand's theorem: (i) since the T'''-motion is *less* constrained than both the T'-motion and T''-motion, we will have (with some easily understood notations)

$$T''' = T' + T_{13} \ (\Rightarrow T''' > T') \qquad \text{and} \qquad T''' = T'' + T_{23} \ (\Rightarrow T''' > T''); \tag{a}$$

and (ii) since the T'-motion is *less* constrained than the T''-motion, we will have (fig. 4.31)

$$T' = T'' + T_{12} \ \Rightarrow \ T' > T''. \tag{b}$$

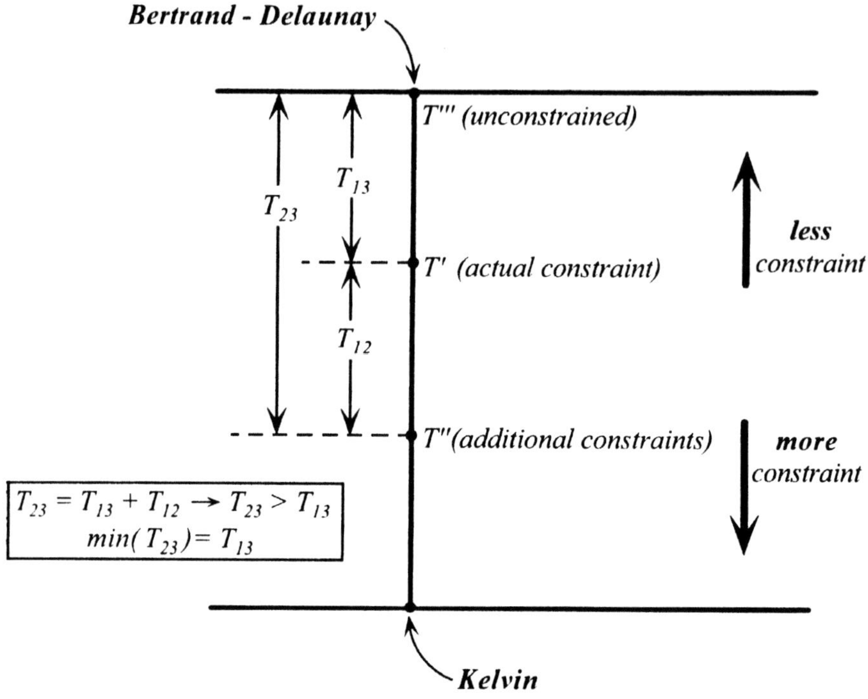

Figure 4.31 Geometrical construction toward the understanding of the impulsive Gauss theorem via the Bertrand–Delaunay theorem.

From (a–b) it follows at once that

$$T_{23} = T_{13} + T_{12} \Rightarrow T_{23} > T_{13}, \quad \text{i.e., } \min(T_{23}) = T_{13}; \tag{c}$$

and this, since

$$2T_{13} = M(v''' - v')(v''' - v') = M(v' - v''')(v' - v'''), \tag{d}$$

$$\Rightarrow \hat{Z} = T_{13} - \hat{Q}\, v' + \cdots \Rightarrow \Delta\hat{Z} = \Delta T_{13} > 0, \tag{e}$$

(where... "*Gauss constant" terms*) constitutes the impulsive Gauss(/Hertz) theorem: *Make* $2T_{13}$ *minimum for all variations of the* v'*;* the additional constraints are taken into account by attaching their Pfaffian forms to (d) via Lagrangean multipliers, as is well known from the finite motion case (chaps. 3, 5, 6) (see also Pars, 1965, pp. 239–241).

5

Nonlinear Nonholonomic Constraints

If I have had any success in mathematical physics, it is, I think, because I have been able to dodge mathematical difficulties.

(J. W. Gibbs, 1870s)

Experiment never responds with a "yes" to theory. At best, it says "maybe" and, most frequently, simply "no." When it agrees with theory, this means "maybe" and, if it does not, the verdict is "no."

Einstein

Nowadays people who for their equations and other statements about nature claim exact and eternal verity are usually dismissed as cranks or lunatics. Nevertheless we lose something in this surrender to lawless uncertainty: Now we must tolerate the youth who blurts out the first, untutored, and uncritical thoughts that come into his head, calls them "my model" of something, and supports them by five or ten pounds of paper he calls "my results," gotten by applying his model to some numerical instances which he has elaborated by use of the largest machine he could get hold of, and if you say to him, "Your model violates NEWTON's laws," he replies "Oh, I don't care about that, I tackle the physics directly, by computer."

(Truesdell, 1987, p. 74)

Here, as in other important areas of analytical mechanics, English language references are far and few. For concurrent reading, we recommend:

(i) The masterful expositions of Hamel (1938; 1949, pp. 495–507, 524, 782–789);

(ii) The original articles of Johnsen (1936, 1937(a), (b), 1938, 1939, 1941);

(iii) The excellent textbooks/monographs/treatises of Mei (1985; 1987), Mei and Liu (1987), and Mei et al. (1991);

(iv) The compact and clear treatments of Novoselov (1966; 1979);

(v) Also, the fundamental monograph of Neimark and Fufaev (1967 and 1972, pp. 212–237), for a detailed treatment of the controversy over the realizability of nonlinear nonholonomic constraints (NNHC), and the validity of some related limiting processes. On this thorny issue, we, further, recommend the rare and very instructive paper of Bahar (1998);

(vi) *Journal of Applied Mathematics and Mechanics* (PMM, Soviet → Russian);

(vii) *Journal of Applied Mathematics and Mechanics* (Chinese).

5.1 INTRODUCTION

In what follows, we extend Lagrangean kinematics (chap. 2) and kinetics (chap. 3) to mechanical systems originally described by n Lagrangean coordinates $q \equiv (q_1, q_2, \ldots, q_n)$, and subsequently subjected to $m(< n)$ independent *nonlinear* and, generally, *nonintegrable* $\equiv$ *nonholonomic* constraints of the form

$$f_D(t, q, \dot{q}, \ddot{q}, \dddot{q}, \ldots) = 0 \qquad [D = 1, \ldots, m\,(< n)]. \tag{5.1.1}$$

[Here, as in the preceding chapters and the rest of the book, and unless specified otherwise, *Latin* indices range from 1 to n (= *number of original Lagrangean/global positional coordinates*); D (*dependent*), $D', D'', \ldots$ range from 1 to m (= *number of additional constraints, of any kind*); I (*independent*), I', I'', ... range from $m + 1$ to n; $f \equiv n - m$ (= *number of independent* δq's = *number of independent kinetic equations*).]

In this chapter we concentrate, almost exclusively, on nonlinear nonholonomic *velocity* constraints

$$f_D(t, q, \dot{q}) = 0$$
$$rank(\partial f_D / \partial \dot{q}_k) = m, \qquad \text{in some domain of } t, q, \dot{q}; \tag{5.1.2}$$

while the more general case (5.1.1) is described briefly later, and is treated more fully in chapter 6. But, once one understands the mechanics of the first-order case (5.1.2), the higher order case (5.1.1) follows without much additional difficulty.

A certain controversy has existed since the early 1910s regarding the *mechanical realizability* of constraints like (5.1.2), let alone (5.1.1), since all known velocity constraints, resulting out of the *passive* rolling among rigid bodies, are *linear/Pfaffian* in the $\dot{q}$'s (or the quasi velocities ω). However, there are important *physical* and *analytical* reasons for studying such nonlinear nonholonomic constraints (NNHC).

- Physically, we can view them as kinematico-physical conditions arising out of nonrolling sources; for example, servoconstraints (§3.17). Consider, for instance, a planar multiple pendulum consisting of n particles connected to a fixed point (say, a ceiling) and to each other via n identical and massless rods, oscillating under gravity; that is, a generalization of the well-known planar double pendulum (ex. 3.5.7). No ordinary springs and/or dampers are attached to the n pendulum joints, but the angles of the first m rods with the vertical, ϕ_D, satisfy the n control constraints $\dot{\phi}_D = \phi_D(\dot{\phi}_{m+1}, \ldots, \dot{\phi}_n)$, where $\phi_D(\ldots) =$ *nonlinear* functions of their arguments; for example, $\dot{\phi}_1 \sim (\dot{\phi}_2)^3$, in a double pendulum. These NNHC can be realized either by internal means at the relevant pendulum joints, or by external noncontact means (e.g., electromagnetic action).
- Analytically, as pointed out by Johnsen and Hamel (late 1930s–early 1940s), the general NNHC formalism can help simplify the solution of the equations of motion; that is, even if the constraints are ultimately Pfaffian (nonholonomic or holonomic), they may be analytically easier to handle when in nonlinear form.

REMARK

Some authors view first integrals of the equations of motion, known in advance, as NNHC-like constraints; for example, the integral of energy [either $T + V =$ *constant*, or $h \equiv T_2 + (V_0 - T_0) =$ *constant* (§3.9)], or of linear and angular momentum, or (5.1.2)-like combinations of them [and, conversely, consider (5.1.2)-like constraints as first integrals, not calculated but observed]; and use the formalism of this chapter to reduce the number of the kinetic equations of motion. However, there are *important qualitative differences* between, say, energy constraints and independent (5.1.2)-like constraints: (i) If energy conservation holds, then it is implicitly contained in the equations of motion, and therefore does not need to be imposed separately; whereas, clearly, if those independent constraints are not imposed, the motion will be markedly different. (ii) Energy conservation, being an integral of motion, represents a surface in velocity space, with the coordinates as parameters. Further, since that integral will be quadratic but, generally, nonhomogeneous in the velocities, the shape of the energy constraint surface will be an "ellipsoid" (three-dimensional or generalized), which, again, is different from the shape of, say, holonomic constraint surfaces. For example, for a single particle, the energy constraint surface is a sphere centered at the origin; and hence the particle velocity is codirectional to the sphere normal; while (§2.7), for a holonomic scleronomic constraint applied to that particle, the velocity vector is perpendicular to the constraint surface normal vector. (iii) Last, if a certain set of velocities satisfies a homogeneous holonomic constraint, so will their multiples by an arbitrary scalar constant; something that, clearly, does not happen with energy conservation constraints. [These remarks are due to Professor D. T. Greenwood (private communication, 1996).]

Be that as it may, we believe that, from a practical viewpoint, *familiarity with the general NNHC formalism helps us to understand better the underlying mathematical structure of the various kinematical and kinetic equations of the linear/Pfaffian theory*, chapters 2 and 3; that is, their similarities, differences, special cases, and so on. The reader may find such a nonutilitarian viewpoint quite beneficial.

5.2 KINEMATICS; THE NONLINEAR TRANSITIVITY EQUATIONS

System Kinematics

Equations (5.1.2) imply that out of the n $\dot{q}$'s, only $n - m$ are independent; or, equivalently, the n $\dot{q}$'s can be expressed in terms of $n - m$ independent (i.e., unconstrained) *velocity parameters*, or *nonlinear quasi velocities* $\omega_I \equiv (\omega_{m+1}, \ldots, \omega_n)$:

$$\dot{q}_k = F_k(t, q, \omega_I) \equiv \dot{q}_k(t, q, \omega_I). \tag{5.2.1}$$

In complete analogy with the linear/Pfaffian case (§2.9), we select the following natural, or "equilibrium," set of ω's (choice of Johnsen and Hamel, 1930s):

$$\omega_D \equiv f_D(t, q, \dot{q}) = 0 \qquad [D = 1, \ldots, m\ (< n)] \tag{5.2.2a}$$

$$\omega_I \equiv f_I(t, q, \dot{q}) \neq 0 \qquad [I = m + 1, \ldots, n] \tag{5.2.2b}$$

$$\omega_{n+1} \equiv \omega_0 \equiv \dot{q}_{n+1} \equiv \dot{q}_0 = dt/dt = 1. \tag{5.2.2c}$$

The $n - m$ functions $f_I(\ldots)$ are arbitrary, except that when the equations of the system (5.2.2a, b) are solved for the n $\dot{q}$'s in terms of the n ω's $\to n - m$ ω_I's, as in (5.2.1) [assuming, of course, that the Jacobian determinant of the matrix $(\partial f_k/\partial \dot{q}_l) \equiv (\partial \omega_k/\partial \dot{q}_l)$ does not vanish] and these values are inserted back into

(5.1.2), they satisfy them identically. Analytically, this translates to the compatibility conditions

$$\partial\omega_k/\partial q_l + \sum (\partial\omega_k/\partial\dot{q}_r)(\partial\dot{q}_r/\partial q_l) = 0, \tag{5.2.3a}$$

$$\partial\dot{q}_k/\partial q_l + \sum (\partial\dot{q}_k/\partial\omega_r)(\partial\omega_r/\partial q_l) = 0; \tag{5.2.3b}$$

and

$$\sum (\partial f_k/\partial\dot{q}_r)(\partial\dot{q}_r/\partial\omega_l) \equiv \sum (\partial\omega_k/\partial\dot{q}_r)(\partial\dot{q}_r/\partial\omega_l) = \partial\omega_k/\partial\omega_l = \delta_{kl}, \tag{5.2.4a}$$

$$\sum (\partial F_k/\partial\omega_r)(\partial\omega_r/\partial\dot{q}_l) \equiv \sum (\partial\dot{q}_k/\partial\omega_r)(\partial\omega_r/\partial\dot{q}_l) = \partial\dot{q}_k/\partial\dot{q}_l = \delta_{kl}. \tag{5.2.4b}$$

Now, in order to be able to either adjoin or embed (build in) the constraints (5.2.2a) into Lagrange's principle (LP, §3.2), which involves δq_k's and/or the *virtual variations of quasi coordinates*, or *virtual displacement parameters*, $\delta\theta_k$, where $d\theta_k/dt \equiv \omega_k$, we must define these $\delta\theta$'s anew and relate them to the δq's via *linear and homogeneous* transformations:

$$\delta q_k = \sum M_{kI}\delta\theta_I, \qquad M_{kI} = M_{kI}(t, q, \omega); \tag{5.2.5}$$

which would be the virtual counterpart of (5.2.1). But what are the coefficients M_{kI}? To conclude, from (5.2.1), that in the nonlinear case

$$\delta q_k = F_k(t, q, \delta\theta) \tag{5.2.6}$$

would be meaningless, and unhelpful. So far, the sole requirement is, by (5.2.4a), as in the Pfaffian case,

$$\sum (\partial f_D/\partial\dot{q}_r)(\partial\dot{q}_r/\partial\omega_I) \equiv \partial f^*{}_D/\partial\omega_I = \delta_{DI} = 0, \tag{5.2.7}$$

where, and this is a general notation,

$$f_D = f_D(t, q, \dot{q}) = f_D[t, q, \dot{q}(t, q, \omega)] \equiv f_D{}^*(t, q, \omega) \equiv f_D{}^*. \tag{5.2.8}$$

It is shown in the next chapter (on "*Differential* Variational Principles") that the *physical requirement of compatibility* between the principles of Lagrange and Gauss, since there is only one mechanics, leads to the m (nontrivial and nonobvious)

"*Maurer–Appell–Chetaev–Johnsen–Hamel conditions*":

$$\sum (\partial f_D/\partial\dot{q}_k)\,\delta q_k = 0, \tag{5.2.9}$$

among the n δq's; instead of the mathematically correct (from the viewpoint of variational calculus), but physically inconsistent, result

$$\delta f_D = \sum [(\partial f_D/\partial q_k)\,\delta q_k + (\partial f_D/\partial\dot{q}_k)\,\delta(\dot{q}_k)] = 0. \tag{5.2.10}$$

To guarantee the identical satisfaction of (5.2.9), under (5.2.7), we introduce $n - m$ independent *virtual displacement parameters* $\delta\theta_I \equiv (\delta\theta_{m+1}, \ldots, \delta\theta_n)$, and set

$$\delta q_k = \sum (\partial\dot{q}_k/\partial\omega_I)\,\delta\theta_I. \tag{5.2.11}$$

Indeed, inserting (5.2.11) into (5.2.9), we obtain

$$0 = \sum (\partial f_D/\partial \dot{q}_k)\Big(\sum (\partial \dot{q}_k/\partial \omega_I)\, \delta\theta_I\Big)$$
$$= \sum\Big(\sum (\partial f_D/\partial \dot{q}_k)\, (\partial \dot{q}_k/\partial \omega_I)\Big)\, \delta\theta_I = \sum (\partial f^*{}_D/\partial \omega_I)\, \delta\theta_I, \qquad (5.2.12)$$

from which, since the $\delta\theta_I$ are independent, eqs. (5.2.7) follow. Then, the virtual form of the constraints (5.1.2), or (5.2.2a), is simply

$$\delta\theta_k = \sum (\partial \omega_k/\partial \dot{q}_l)\, \delta q_l\colon \qquad \delta\theta_D = 0, \qquad \delta\theta_I \neq 0. \qquad (5.2.13)$$

The Maggi-like representation (5.2.11) and its inverse (5.2.13) are fundamental to all subsequent NNHC developments; for Pfaffian constraints, clearly, they reduce to the forms given in chapter 2. The constraints (5.2.2a, b) establish, in configuration space, a one-to-one correspondence between the ω's and the kinematically admissible $\dot{q}$'s; while (5.2.11, 13) do the same thing for the virtual displacements δq and $\delta\theta$. [The nonvanishing Jacobian of (5.2.2a, b) is the nonvanishing determinant of (5.2.13): $|\partial \omega_k/\partial \dot{q}_l|$.]

In the same spirit, we conclude that the $\delta\dot{q}$'s are linear and homogeneous combinations of the $\delta\omega$'s, and vice versa; that is,

$$\delta\dot{q}_k = \sum (\partial F_k/\partial \omega_l)\, \delta\omega_l = \sum (\partial \dot{q}_k/\partial \omega_l)\, \delta\omega_l, \qquad (5.2.14a)$$
$$\delta\omega_k = \sum (\partial f_k/\partial \dot{q}_l)\, \delta\dot{q}_l = \sum (\partial \omega_k/\partial \dot{q}_l)\, \delta\dot{q}_l. \qquad (5.2.14b)$$

In what follows, for algebraic convenience, we shall allow the $\delta\theta$ and ω indices to run from 1 to n (like those of the δq's and $\dot{q}$'s). The satisfaction of the constraints $\delta\theta_D = 0, \omega_D = 0$ (and corresponding restrictions on those indices to run only over the independent range $m+1, \ldots, n$) can be done at any time after all differentiations have been carried out. In that case, we may also use the helpful notation $(\ldots)_o \equiv (\ldots, \omega_D = 0, \ldots)$.

Particle Kinematics

So far we have involved *system* variables. Let us now express the above results in particle/elementary vector variables.

The (inertial) *virtual displacement, velocity*, and *acceleration* of a typical system particle, whose (inertial) position is expressed as $\boldsymbol{r} = \boldsymbol{r}(t, q)$, are, respectively,

(i)
$$\delta\boldsymbol{r} = \sum (\partial\boldsymbol{r}/\partial q_k)\, \delta q_k = \sum (\partial\boldsymbol{r}/\partial q_k)\, \Big(\sum (\partial \dot{q}_k/\partial \omega_l)\delta\theta_l\Big)$$
$$= \sum\Big(\sum (\partial\boldsymbol{r}/\partial q_k)\, (\partial \dot{q}_k/\partial \omega_l)\Big)\delta\theta_l$$
$$\equiv \sum (\partial\boldsymbol{r}^*/\partial\theta_l)\, \delta\theta_l \equiv \sum \boldsymbol{\varepsilon}_l\, \delta\theta_l \equiv \delta\boldsymbol{r}^*, \qquad (5.2.15)$$

where

$$\boldsymbol{\varepsilon}_l = \sum (\partial\boldsymbol{r}/\partial q_k)\, (\partial \dot{q}_k/\partial \omega_l) \equiv \sum (\partial \dot{q}_k/\partial \omega_l)\boldsymbol{e}_k, \qquad (5.2.16a)$$
$$\boldsymbol{e}_k = \sum (\partial\boldsymbol{r}^*/\partial\theta_l)\, (\partial \omega_l/\partial \dot{q}_k) \equiv \sum (\partial \omega_l/\partial \dot{q}_k)\boldsymbol{\varepsilon}_l; \qquad (5.2.16b)$$

that is,

$$\partial(\ldots)/\partial\theta_l \equiv \sum [\partial(\ldots)/\partial q_k]\,(\partial\dot{q}_k/\partial\omega_l), \tag{5.2.16c}$$

$$\partial(\ldots)/\partial q_k \equiv \sum [\partial(\ldots)/\partial\theta_l]\,(\partial\omega_l/\partial\dot{q}_k) \tag{5.2.16d}$$

[*nonlinear symbolic (i.e., nonvectorial/tensorial) quasi chain rules*],

(ii)
$$\begin{aligned} \boldsymbol{v} = \boldsymbol{v}(t, q, \dot{q}) = d\boldsymbol{r}/dt &= \sum (\partial\boldsymbol{r}/\partial q_k)\dot{q}_k + \partial\boldsymbol{r}/\partial t \\ &\equiv \sum \dot{q}_k(t, q, \omega)\boldsymbol{e}_k + \boldsymbol{e}_0 \qquad [t \equiv q_{n+1}], \qquad (5.2.17a) \\ &= \sum \omega_k(t, q, \dot{q})\boldsymbol{\varepsilon}_k + \boldsymbol{\varepsilon}_0 \equiv \boldsymbol{v}^*(t, q, \omega) \equiv \boldsymbol{v}^*, \qquad (5.2.17b) \end{aligned}$$

where $\boldsymbol{\varepsilon}_0$ is defined either by (5.2.17b) or, equivalently, extending (5.2.16a–d) for $q_k \to q_{n+1} \equiv t$,

$$\begin{aligned} \boldsymbol{\varepsilon}_0 \equiv \boldsymbol{\varepsilon}_{n+1} \equiv \partial\boldsymbol{r}/\partial\theta_{n+1} &\equiv \sum (\partial\boldsymbol{r}/\partial q_\alpha)\,(\partial\dot{q}_\alpha/\partial\omega_{n+1}) \qquad [\alpha = 1, \ldots, n+1] \\ &= \sum (\partial\boldsymbol{r}/\partial q_k)\,(\partial\dot{q}_k/\partial\omega_{n+1}) + (\partial\boldsymbol{r}/\partial t)\,(\partial\dot{q}_{n+1}/\partial\omega_{n+1}) \\ &= \sum (\partial\dot{q}_k/\partial\omega_{n+1})\boldsymbol{e}_k + \boldsymbol{e}_0 \\ &= \sum (\dot{q}_k\boldsymbol{e}_k - \omega_k\boldsymbol{\varepsilon}_k) + \boldsymbol{e}_0 \qquad [\text{by (5.2.17a, b)}] \\ &= \boldsymbol{e}_0 + \sum \dot{q}_k\boldsymbol{e}_k - \sum \omega_k\Big(\sum (\partial\dot{q}_l/\partial\omega_k)\boldsymbol{e}_l\Big) \\ &= \boldsymbol{e}_0 + \sum \Big(\dot{q}_k - \sum (\partial\dot{q}_k/\partial\omega_l)\omega_l\Big)\boldsymbol{e}_k; \qquad (5.2.17c) \end{aligned}$$

and, inversely,

$$\begin{aligned} \boldsymbol{e}_0 \equiv \boldsymbol{e}_{n+1} \equiv \partial\boldsymbol{r}/\partial t &\equiv \sum (\partial\boldsymbol{r}/\partial\theta_\alpha)\,(\partial\omega_\alpha/\partial\dot{q}_{n+1}) \qquad [\alpha = 1, \ldots, n+1] \\ &= \sum (\partial\boldsymbol{r}/\partial\theta_k)\,(\partial\omega_k/\partial\dot{q}_{n+1}) + (\partial\boldsymbol{r}/\partial\theta_{n+1})\,(\partial\omega_{n+1}/\partial\dot{q}_{n+1}) \\ &= \sum (\partial\omega_k/\partial\dot{q}_{n+1})\boldsymbol{\varepsilon}_k + \boldsymbol{\varepsilon}_0 \\ &= \sum (\omega_k\boldsymbol{\varepsilon}_k - \dot{q}_k\boldsymbol{e}_k) + \boldsymbol{\varepsilon}_0 \qquad [\text{by (5.2.17a, b)}] \\ &= \boldsymbol{\varepsilon}_0 + \sum \omega_k\boldsymbol{\varepsilon}_k - \sum \dot{q}_k\Big(\sum (\partial\omega_l/\partial\dot{q}_k)\boldsymbol{\varepsilon}_l\Big) \\ &= \boldsymbol{\varepsilon}_0 + \sum \Big(\omega_k - \sum (\partial\omega_k/\partial\dot{q}_l)\dot{q}_l\Big)\boldsymbol{\varepsilon}_k. \qquad (5.2.17d) \end{aligned}$$

The above suggest the following definitions, for any function $f^* = f^*(t, q, \omega)$ [in addition to (5.2.16c, d)],

$$\partial f^*/\partial\theta_{n+1} \equiv \sum (\partial f^*/\partial q_k)\Big(\dot{q}_k - \sum (\partial\dot{q}_k/\partial\omega_l)\omega_l\Big) + \partial f^*/\partial t; \tag{5.2.18a}$$

which, in the Pfaffian case, reduces to (2.9.32 ff.)

$$\partial f^*/\partial\theta_{n+1} = \sum (\partial f^*/\partial q_k)A_k + \partial f^*/\partial t \equiv \partial f^*/\partial(t) + \partial f^*/\partial t. \tag{5.2.18b}$$

In particular, for $f^* = q_r$ we find

$$\partial q_r/\partial\theta_s = \partial\dot{q}_r/\partial\omega_s, \tag{5.2.18c}$$

$$\partial q_r/\partial\theta_{n+1} = \partial\dot{q}_r/\partial\omega_{n+1} = \dot{q}_r - \sum(\partial\dot{q}_r/\partial\omega_l)\omega_l$$
$$\left[= \dot{q}_r - \sum A_{rl}\,\omega_l = A_r, \quad \text{in the Pfaffian case}\right]; \tag{5.2.18d}$$

and, inversely,

$$\partial\omega_k/\partial\dot{q}_{n+1} = \partial\theta_k/\partial t = \omega_k - \sum(\partial\omega_k/\partial\dot{q}_l)\dot{q}_l. \tag{5.2.18e}$$

(iii)
$$\begin{aligned} \boldsymbol{a} \equiv d\boldsymbol{v}/dt &= \sum(\partial\boldsymbol{v}/\partial\dot{q}_k)\ddot{q}_k + \cdots \\ &\qquad [\ldots \equiv \text{Terms containing neither } \ddot{q} \text{ nor } \dot{\omega}] \\ &= \sum(\partial\boldsymbol{v}/\partial\dot{q}_k)\left(\sum(\partial\dot{q}_k/\partial\omega_l)\dot{\omega}_l + \cdots\right) + \cdots \\ &= \sum\left(\sum(\partial\boldsymbol{v}/\partial\dot{q}_k)(\partial\dot{q}_k/\partial\omega_l)\right)\dot{\omega}_l + \cdots \\ &\equiv \sum(\partial\boldsymbol{v}^*/\partial\omega_l)\dot{\omega}_l + \cdots \\ &= \sum\boldsymbol{\varepsilon}_l\,\dot{\omega}_l + \cdots = \sum(\partial\boldsymbol{a}^*/\partial\dot{\omega}_l)\dot{\omega}_l + \cdots \\ &\equiv \boldsymbol{a}^*(t, q, \omega, \dot{\omega}) \equiv \boldsymbol{a}^*, \end{aligned} \tag{5.2.19a}$$

where, since $\boldsymbol{v}(t, q, \dot{q}) = \boldsymbol{v}^*(t, q, \omega)$,

$$\partial\boldsymbol{v}^*/\partial\omega_l = \sum(\partial\boldsymbol{v}/\partial\dot{q}_k)(\partial\dot{q}_k/\partial\omega_l), \quad \text{i.e., (5.2.16a, b)} \quad \boldsymbol{\varepsilon}_l = \sum\boldsymbol{e}_k(\partial\dot{q}_k/\partial\omega_l); \tag{5.2.19b}$$

which is a vectorial transformation equation, and not some chain rule, like (5.2.16c, d).

The preceding results readily lead to the fundamental, and purely kinematical, particle–system identities

Holonomic variables: $\partial\boldsymbol{r}/\partial q_k = \partial\boldsymbol{v}/\partial\dot{q}_k = \partial\boldsymbol{a}/\partial\ddot{q}_k = \cdots = \boldsymbol{e}_k$, (5.2.20a)

Nonholonomic variables: $\partial\boldsymbol{r}^*/\partial\theta_k \equiv \partial\boldsymbol{v}^*/\partial\omega_k = \partial\boldsymbol{a}^*/\partial\dot{\omega}_k = \cdots = \boldsymbol{\varepsilon}_k$; (5.2.20b)

and their system counterparts

$$\partial q_k/\partial\theta_l \equiv \partial\dot{q}_k/\partial\omega_l = \partial\ddot{q}_k/\partial\dot{\omega}_l = \cdots, \tag{5.2.20c}$$

$$\partial\theta_l/\partial q_k \equiv \partial\omega_l/\partial\dot{q}_k = \partial\dot{\omega}_l/\partial\ddot{q}_k = \cdots. \tag{5.2.20d}$$

With the help of the above, next, we obtain the following basic kinematical result: applying symbolic and real chain rule to $\boldsymbol{v} = \boldsymbol{v}(t, q, \dot{q}) = \boldsymbol{v}^*(t, q, \omega) = \boldsymbol{v}^*$, we find, successively,

$$\text{(a)} \qquad \partial \mathbf{v}^*/\partial \theta_k = \sum (\partial \mathbf{v}^*/\partial q_l)(\partial \dot{q}_l/\partial \omega_k)$$
$$= \sum \left[\partial \mathbf{v}/\partial q_l + \sum (\partial \mathbf{v}/\partial \dot{q}_r)(\partial \dot{q}_r/\partial q_l)\right](\partial \dot{q}_l/\partial \omega_k)$$
$$= \sum (\partial \mathbf{v}/\partial q_l)(\partial \dot{q}_l/\partial \omega_k) + \sum (\partial \mathbf{v}/\partial \dot{q}_l)(\partial \dot{q}_l/\partial \theta_k), \qquad (5.2.21a)$$

$$\text{(b)} \quad d/dt(\partial \mathbf{r}^*/\partial \theta_k) = d/dt(\partial \mathbf{v}^*/\partial \omega_k) = d/dt\left(\sum (\partial \mathbf{v}/\partial \dot{q}_l)(\partial \dot{q}_l/\partial \omega_k)\right)$$
$$= \sum \left[d/dt\,(\partial \mathbf{v}/\partial \dot{q}_l)\right](\partial \dot{q}_l/\partial \omega_k) + \sum (\partial \mathbf{v}/\partial \dot{q}_l)\left[d/dt(\partial \dot{q}_l/\partial \omega_k)\right]. \qquad (5.2.21b)$$

Therefore, subtracting (5.2.21a, b) side by side, while recalling (5.2.19b), we get

$$\gamma_k \equiv E_k{}^*(\mathbf{v}^*) \equiv d/dt(\partial \mathbf{v}^*/\partial \omega_k) - \partial \mathbf{v}^*/\partial \theta_k \equiv d\boldsymbol{\varepsilon}_k/dt - \partial \mathbf{v}^*/\partial \theta_k$$
$$= \sum \left[d/dt(\partial \mathbf{v}/\partial \dot{q}_l) - \partial \mathbf{v}/\partial q_l\right](\partial \dot{q}_l/\partial \omega_k)$$
$$+ \sum \left[d/dt(\partial \dot{q}_l/\partial \omega_k) - \partial \dot{q}_l/\partial \theta_k\right](\partial \mathbf{v}/\partial \dot{q}_l)$$
$$= \sum E_l(\mathbf{v})(\partial \dot{q}_l/\partial \omega_k) + \sum E_k{}^*(\dot{q}_l)(\partial \mathbf{v}/\partial \dot{q}_l)$$
$$= \mathbf{0} + \sum E_k{}^*(\dot{q}_l)(\partial \mathbf{v}/\partial \dot{q}_l)$$

[$E_l(\mathbf{v}) = \mathbf{0}$, since the q's are holonomic coordinates (§2.9)]

$$= \sum E_k{}^*(\dot{q}_l)\mathbf{e}_l = \sum E_k{}^*(\dot{q}_l)\left(\sum (\partial \omega_s/\partial \dot{q}_l)\boldsymbol{\varepsilon}_s\right), \qquad (5.2.21c)$$

or finally, and compactly (and in anticipation of later results),

$$\gamma_k \equiv E_k{}^*(\mathbf{v}^*) = \sum V^l{}_k \mathbf{e}_l = \sum\sum \left[(\partial \omega_s/\partial \dot{q}_l)V^l{}_k\right]\boldsymbol{\varepsilon}_s$$
$$= -\sum\sum \left[(\partial \dot{q}_l/\partial \omega_s)H^s{}_k\right]\mathbf{e}_l = -\sum H^s{}_k \boldsymbol{\varepsilon}_s:$$

Nonholonomic deviation; (5.2.21d)

where

$$V^l{}_k \equiv d/dt(\partial \dot{q}_l/\partial \omega_k) - \partial \dot{q}_l/\partial \theta_k \equiv E_k{}^*(\dot{q}_l):$$

Nonlinear Voronets–Chaplygin coefficients, (5.2.21e)

$$H^s{}_k \equiv -\sum (\partial \omega_s/\partial \dot{q}_l)V^l{}_k:$$

Nonlinear Hamel coefficients, (5.2.21f)

$$\left\{\Leftrightarrow V^l{}_k = -\sum (\partial \dot{q}_l/\partial \omega_s)H^s{}_k \qquad \text{[using (5.2.4a, b)]}\right\}; \qquad (5.2.21g)$$

also (see prob. 5.2.1, below),

$$H^k{}_s \equiv \sum (\partial \dot{q}_l/\partial \omega_s)\left[(\partial \omega_k/\partial \dot{q}_l)^\cdot - \partial \omega_k/\partial q_l\right] \equiv \sum (\partial \dot{q}_l/\partial \omega_s)E_l(\omega_k). \qquad (5.2.21h)$$

[Clearly, since $|\partial \omega_k/\partial \dot{q}_l| \neq 0$, if the $H^s{}_k$ vanish, so do the $V^s{}_k$; and vice versa.]

These *nonintegrability* relations [actually due to Johnsen and Hamel (in the late 1930s)] result from the nonholonomicity of the "coordinates" θ, and *have nothing to do with constraints*. Further, as these definitions show, in $V^l{}_k$, l is *holonomic*, and k is *nonholonomic*; while, in $H^l{}_k$, both l and k are *nonholonomic*; that is, $V^l{}_k$ is *mixed*, while $H^l{}_k$ is purely nonholonomic.

The Nonlinear Transitivity Equations

By $d/dt(\ldots)$-differentiating (5.2.13), $\delta(\ldots)$-varying (5.2.2a–c), and then subtracting side by side, we find, successively,

$$
\begin{aligned}
d/dt(\delta\theta_k) - \delta(d\theta_k/dt) &\equiv (\delta\theta_k)^{\cdot} - \delta\omega_k \\
&= d/dt\Big(\sum (\partial\omega_k/\partial\dot{q}_l)\,\delta q_l\Big) - \delta\omega_k(t, q, \dot{q}) \\
&= \sum \big[(\partial\omega_k/\partial\dot{q}_l)^{\cdot}\,\delta q_l + (\partial\omega_k/\partial\dot{q}_l)\,(\delta q_l)^{\cdot}\big] \\
&\quad - \sum \big[(\partial\omega_k/\partial q_l)\,\delta q_l + (\partial\omega_k/\partial\dot{q}_l)\,\delta(\dot{q}_l)\big] \\
&= \sum (\partial\omega_k/\partial\dot{q}_l)\,[(\delta q_l)^{\cdot} - \delta(\dot{q}_l)] \\
&\quad + \sum \big[(\partial\omega_k/\partial\dot{q}_l)^{\cdot} - \partial\omega_k/\partial q_l\big]\,\delta q_l;
\end{aligned}
$$

that is, [recalling (5.2.11) and invoking (5.2.21e–h)]

$$
\begin{aligned}
(\delta\theta_k)^{\cdot} - \delta\omega_k &= \sum (\partial\omega_k/\partial\dot{q}_l)\,[(\delta q_l)^{\cdot} - \delta(\dot{q}_l)] \\
&\quad + \sum\sum \big[(\partial\omega_k/\partial\dot{q}_l)^{\cdot} - \partial\omega_k/\partial q_l\big]\,(\partial\dot{q}_l/\partial\omega_r)\,\delta\theta_r \\
&= \sum (\partial\omega_k/\partial\dot{q}_l)\,[(\delta q_l)^{\cdot} - \delta(\dot{q}_l)] \\
&\quad - \sum\sum \big[(\partial\dot{q}_s/\partial\omega_r)^{\cdot} - \partial\dot{q}_s/\partial\theta_r\big]\,(\partial\omega_k/\partial\dot{q}_s)\,\delta\theta_r \\
&= \sum (\partial\omega_k/\partial\dot{q}_l)\,[(\delta q_l)^{\cdot} - \delta(\dot{q}_l)] \\
&\quad - \sum\sum\sum \big[(\partial\dot{q}_s/\partial\omega_r)^{\cdot} - \partial\dot{q}_s/\partial\theta_r\big]\,(\partial\omega_k/\partial\dot{q}_s)\,(\partial\omega_r/\partial\dot{q}_l)\,\delta q_l,
\end{aligned}
\tag{5.2.22a}
$$

or, compactly,

$$
\begin{aligned}
(\delta\theta_k)^{\cdot} - \delta\omega_k &= \sum (\partial\omega_k/\partial\dot{q}_l)\,[(\delta q_l)^{\cdot} - \delta(\dot{q}_l)] + \sum E_l(\omega_k)\,\delta q_l \\
&= \sum (\partial\omega_k/\partial\dot{q}_l)\,[(\delta q_l)^{\cdot} - \delta(\dot{q}_l)] + \sum H^k{}_r\,\delta\theta_r \\
&= \sum (\partial\omega_k/\partial\dot{q}_l)\,[(\delta q_l)^{\cdot} - \delta(\dot{q}_l)] + \sum\sum H^k{}_r(\partial\omega_r/\partial\dot{q}_l)\,\delta q_l \\
&= \sum (\partial\omega_k/\partial\dot{q}_l)\,[(\delta q_l)^{\cdot} - \delta(\dot{q}_l)] - \sum\sum E_r{}^*(\dot{q}_l)\,(\partial\omega_k/\partial\dot{q}_l)\,\delta\theta_r \\
&= \sum (\partial\omega_k/\partial\dot{q}_l)\,[(\delta q_l)^{\cdot} - \delta(\dot{q}_l)] - \sum\sum V^l{}_r\,(\partial\omega_k/\partial\dot{q}_l)\,\delta\theta_r \\
&= \sum (\partial\omega_k/\partial\dot{q}_l)\,[(\delta q_l)^{\cdot} - \delta(\dot{q}_l)] \\
&\quad - \sum\sum\sum (\partial\omega_k/\partial\dot{q}_s)\,(\partial\omega_r/\partial\dot{q}_l)\,V^s{}_r\,\delta q_l.
\end{aligned}
\tag{5.2.22b}
$$

[Under the m constraints $\delta\theta_D = 0, \delta\theta_r \rightarrow \delta\theta_I$ $(r \rightarrow I = m+1, \ldots, n)$.]

Similarly, $d/dt(\ldots)$-differentiating (5.2.11), $\delta(\ldots)$-varying (5.2.1), with $l \to r = 1, \ldots, n$, and then subtracting side by side, we find, successively,

$$d/dt(\delta q_l) - \delta(dq_l/dt) \equiv (\delta q_l)^{\cdot} - \delta(\dot{q}_l)$$

$$= d/dt\left(\sum (\partial \dot{q}_l/\partial \omega_k)\,\delta\theta_k\right) - \delta\dot{q}_l(t, q, \omega)$$

$$= \sum \left[(\partial \dot{q}_l/\partial \omega_k)^{\cdot}\,\delta\theta_k + (\partial \dot{q}_l/\partial \omega_k)(\delta\theta_k)^{\cdot}\right]$$

$$\quad - \sum \left[(\partial \dot{q}_l/\partial q_k)\,\delta q_k + (\partial \dot{q}_l/\partial \omega_k)\,\delta\omega_k\right]$$

$$= \sum \left[(\partial \dot{q}_l/\partial \omega_k)^{\cdot}\,\delta\theta_k + (\partial \dot{q}_l/\partial \omega_k)(\delta\theta_k)^{\cdot}\right]$$

$$\quad - \sum \left[(\partial \dot{q}_l/\partial \theta_k)\,\delta\theta_k + (\partial \dot{q}_l/\partial \omega_k)\,\delta\omega_k\right]$$

$$= \sum (\partial \dot{q}_l/\partial \omega_k)\left[(\delta\theta_k)^{\cdot} - \delta\omega_k\right]$$

$$\quad + \sum \left[(\partial \dot{q}_l/\partial \omega_k)^{\cdot} - \partial \dot{q}_l/\partial \theta_k\right]\delta\theta_k;$$

that is, [recalling (5.2.13) and invoking (5.2.21e–h)]

$$(\delta q_l)^{\cdot} - \delta(\dot{q}_l) = \sum (\partial \dot{q}_l/\partial \omega_k)\left[(\delta\theta_k)^{\cdot} - \delta\omega_k\right]$$

$$\quad + \sum\sum \left[(\partial \dot{q}_l/\partial \omega_k)^{\cdot} - \partial \dot{q}_l/\partial \theta_k\right](\partial \omega_k/\partial \dot{q}_s)\,\delta q_s$$

$$= \sum (\partial \dot{q}_l/\partial \omega_k)\left[(\delta\theta_k)^{\cdot} - \delta\omega_k\right]$$

$$\quad - \sum\sum \left[(\partial \omega_k/\partial \dot{q}_s)^{\cdot} - \partial \omega_k/\partial q_s\right](\partial \dot{q}_l/\partial \omega_k)\,\delta q_s$$

$$= \sum (\partial \dot{q}_l/\partial \omega_k)\left[(\delta\theta_k)^{\cdot} - \delta\omega_k\right]$$

$$\quad - \sum\sum\sum \left[(\partial \omega_k/\partial \dot{q}_s)^{\cdot} - \partial \omega_k/\partial q_s\right](\partial \dot{q}_l/\partial \omega_k)(\partial \dot{q}_s/\partial \omega_r)\,\delta\theta_r,$$

(5.2.23a)

or, compactly (with some dummy-index changes),

$$(\delta q_l)^{\cdot} - \delta(\dot{q}_l) = \sum (\partial \dot{q}_l/\partial \omega_k)\left[(\delta\theta_k)^{\cdot} - \delta\omega_k\right] + \sum E_k^*(\dot{q}_l)\,\delta\theta_k$$

$$\equiv \sum (\partial \dot{q}_l/\partial \omega_k)\left[(\delta\theta_k)^{\cdot} - \delta\omega_k\right] + \sum V^l_k\,\delta\theta_k$$

$$= \sum (\partial \dot{q}_l/\partial \omega_k)\left[(\delta\theta_k)^{\cdot} - \delta\omega_k\right] + \sum\sum E_k^*(\dot{q}_l)(\partial \omega_k/\partial \dot{q}_s)\,\delta q_s$$

$$\equiv \sum (\partial \dot{q}_l/\partial \omega_k)\left[(\delta\theta_k)^{\cdot} - \delta\omega_k\right] + \sum\sum V^l_k(\partial \omega_k/\partial \dot{q}_s)\,\delta q_s$$

$$= \sum (\partial \dot{q}_l/\partial \omega_k)\left[(\delta\theta_k)^{\cdot} - \delta\omega_k\right]$$

$$\quad - \sum\sum\sum (\partial \dot{q}_l/\partial \omega_r)(\partial \omega_k/\partial \dot{q}_s)H^r_k\,\delta q_s$$

$$= \sum (\partial \dot{q}_l/\partial \omega_k)\left[(\delta\theta_k)^{\cdot} - \delta\omega_k\right] - \sum\sum (\partial \dot{q}_l/\partial \omega_r)H^r_k\,\delta\theta_k. \quad (5.2.23b)$$

[Under the m constraints $\delta\theta_D = 0, \delta\theta_k \to \delta\theta_I$ $(k \to I = m+1, \ldots, n)$.]

As the above show, and since $|\partial \omega_k/\partial \dot{q}_l| \neq 0$, if the H^r_k vanish, so do the V^r_k; and vice versa. The relations between the V^l_k and H^l_k can also be found from the equivalence of the transitivity equations (5.2.22) and (5.2.23). Indeed, substituting

$(\delta q_l)^{\cdot} - \delta(\dot{q}_l)$ from (5.2.23a, b) into (5.2.22a, b), simplifying, and invoking (5.2.4a), we get

$$\sum\Big(\sum (\partial\omega_k/\partial\dot{q}_l)\, V^l{}_r + H^k{}_r\Big)\, \delta\theta_r = 0, \tag{5.2.24}$$

from which we obtain the earlier (5.2.21f, g).

Example 5.2.1 *Alternative Derivation of the Nonintegrability Condition (5.2.21c, d).* We have, successively,

$$\begin{aligned}
\gamma_k &\equiv E_k{}^*(\boldsymbol{v}^*) \equiv d/dt(\partial\boldsymbol{v}^*/\partial\omega_k) - \partial\boldsymbol{v}^*/\partial\theta_k \equiv d\boldsymbol{\varepsilon}_k/dt - \partial\boldsymbol{v}^*/\partial\theta_k \\
&= d/dt\Big(\sum (\partial\dot{q}_l/\partial\omega_k)\boldsymbol{e}_l\Big) - \sum (\partial\boldsymbol{v}^*/\partial q_l)\,(\partial\dot{q}_l/\partial\omega_k) \\
&= \sum (\partial\dot{q}_l/\partial\omega_k)^{\cdot}\boldsymbol{e}_l + \sum (\partial\dot{q}_l/\partial\omega_k)\,(d\boldsymbol{e}_l/dt) \\
&\quad - \sum \Big[(\partial/\partial q_l)\Big(\sum \dot{q}_r\,\boldsymbol{e}_r + \boldsymbol{e}_0\Big)\Big]\,(\partial\dot{q}_l/\partial\omega_k) \qquad [\text{since } \boldsymbol{v}^* = \boldsymbol{v}] \\
&= \sum (\partial\dot{q}_l/\partial\omega_k)^{\cdot}\,\boldsymbol{e}_l + \sum (\partial\dot{q}_l/\partial\omega_k)\,(d\boldsymbol{e}_l/dt) \\
&\quad - \sum\Big\{\sum\ [(\partial\dot{q}_r/\partial q_l)\boldsymbol{e}_r + \dot{q}_r(\partial\boldsymbol{e}_r/\partial q_l)] + \partial\boldsymbol{e}_0/\partial q_l\Big\}\,(\partial\dot{q}_l/\partial\omega_k) \\
&= \sum (\partial\dot{q}_l/\partial\omega_k)^{\cdot}\boldsymbol{e}_l + \sum (\partial\dot{q}_l/\partial\omega_k)\,(d\boldsymbol{e}_l/dt) \\
&\quad - \sum\sum (\partial\dot{q}_r/\partial q_l)\,(\partial\dot{q}_l/\partial\omega_k)\boldsymbol{e}_r - \sum\sum \dot{q}_r(\partial\dot{q}_l/\partial\omega_k)(\partial\boldsymbol{e}_r/\partial q_l) \\
&\quad - \sum (\partial\dot{q}_l/\partial\omega_k)\,(\partial\boldsymbol{e}_0/\partial q_l) \\
&= \sum (\partial\dot{q}_l/\partial\omega_k)^{\cdot}\boldsymbol{e}_l + \sum (\partial\dot{q}_l/\partial\omega_k)\,(d\boldsymbol{e}_l/dt) \\
&\quad - \sum\Big(\sum \dot{q}_r(\partial\boldsymbol{e}_l/\partial q_r) + (\partial\boldsymbol{e}_l/\partial t)\Big)\,(\partial\dot{q}_l/\partial\omega_k) \\
&\quad - \sum (\partial\dot{q}_r/\partial\theta_k)\boldsymbol{e}_r \\
&\quad [\text{since the } q\text{'s are holonomic: } \partial\boldsymbol{e}_r/\partial q_l = \partial\boldsymbol{e}_l/\partial q_r,\ \partial\boldsymbol{e}_0/\partial q_l = \partial\boldsymbol{e}_l/\partial t] \\
&= \sum (\partial\dot{q}_l/\partial\omega_k)^{\cdot}\boldsymbol{e}_l + \sum (\partial\dot{q}_l/\partial\omega_k)\,(d\boldsymbol{e}_l/dt) \\
&\quad - \sum (\partial\dot{q}_l/\partial\omega_k)\,(d\boldsymbol{e}_l/dt) - \sum (\partial\dot{q}_r/\partial\theta_k)\boldsymbol{e}_r \\
&\quad [\text{the } \textit{second} \text{ and } \textit{third} \text{ sums cancel; and we replace } r \text{ with } l \text{ in the } \textit{last} \text{ term}] \\
&= \sum\ [(\partial\dot{q}_l/\partial\omega_k)^{\cdot} - \partial\dot{q}_l/\partial\theta_k]\,\boldsymbol{e}_l;
\end{aligned} \tag{a}$$

that is, eqs. (5.2.21c, d), as before.

Problem 5.2.1 Show, with the help of (5.2.4a, b), that (5.2.21f, g)

$$H^k{}_r = -\sum (\partial\omega_k/\partial\dot{q}_l)V^l{}_r \ \Leftrightarrow\ V^l{}_r = -\sum (\partial\dot{q}_l/\partial\omega_k)H^k{}_r \tag{a}$$

lead to the following useful kinematical identities:

$$E_l(\omega_k) = -\sum\sum (\partial\omega_r/\partial\dot{q}_l)\,(\partial\omega_k/\partial\dot{q}_s) E_r{}^*(\dot{q}_s), \tag{b}$$

$$E_r{}^*(\dot{q}_s) = -\sum\sum (\partial\dot{q}_l/\partial\omega_r)\,(\partial\dot{q}_s/\partial\omega_k) E_l(\omega_k)$$
$$\left[\text{i.e., } V^s{}_r = -\sum (\partial\dot{q}_s/\partial\omega_k) H^k{}_r\right]. \tag{c}$$

Problem 5.2.2 Show that in the *Pfaffian* case (§2.9) — that is, when $\omega_l \equiv \sum a_{ld}\dot{q}_d + a_l$ — the nonlinear Hamel coefficients

$$H^l{}_s \equiv \sum \left[(\partial\omega_l/\partial\dot{q}_d)^{\cdot} - \partial\omega_l/\partial q_d\right](\partial\dot{q}_d/\partial\omega_s) \equiv \sum (\partial\dot{q}_d/\partial\omega_s) E_d(\omega_l) \tag{a}$$

reduce to their Pfaffian counterparts (with $\alpha = 1, \ldots, n+1$; and the rest of the notations of §2.10):

$$h^l{}_s = \sum \gamma^l{}_{s\alpha}\,\omega_\alpha = \sum \gamma^l{}_{sr}\,\omega_r + \gamma^l{}_{s,n+1}. \tag{b}$$

Problem 5.2.3 Show that, for a general function $f^* = f^*(t, q, \omega)$, the following *noncommutativity* relations hold:

$$\partial/\partial\theta_l\,(\partial f^*/\partial\theta_k) - \partial/\partial\theta_k\,(\partial f^*/\partial\theta_l)$$
$$= \sum\sum\sum \left[(\partial^2\dot{q}_d/\partial q_s\partial\omega_k)\,(\partial\dot{q}_s/\partial\omega_l)\right.$$
$$\left. - (\partial^2\dot{q}_d/\partial q_s\partial\omega_l)\,(\partial\dot{q}_s/\partial\omega_k)\right](\partial\omega_p/\partial\dot{q}_d)\,(\partial f^*/\partial\theta_p). \tag{a}$$

Then show that in the Pfaffian case (§2.9); that is,

$$\omega_s \equiv \sum a_{sd}\,\dot{q}_d + a_s, \qquad \dot{q}_d = \sum A_{ds}\,\omega_s + A_d, \tag{b}$$

eq. (a) reduces to the noncommutativity equation (2.10.20).

Problem 5.2.4 (i) Show that in the Pfaffian case (§2.9), the nonlinear coefficients

$$E_l(\omega_k) \equiv (\partial\omega_k/\partial\dot{q}_l)^{\cdot} - \partial\omega_k/\partial q_l \tag{a}$$

reduce to

$$\sum (\partial a_{kl}/\partial q_r - \partial a_{kr}/\partial q_l)\dot{q}_r + (\partial a_{kl}/\partial t - \partial a_k/\partial q_l). \tag{b}$$

(ii) Hence show that, in such a case, $E_l(\omega_k) \equiv 0$ translates to the exactness conditions:

$$\partial a_{kl}/\partial q_r = \partial a_{kr}/\partial q_l \qquad \text{and} \qquad \partial a_{kl}/\partial t = \partial a_k/\partial q_l. \tag{c}$$

(iii) Similarly, show that in the Pfaffian case — that is, $\dot{q}_l = \sum A_{lk}\omega_k + A_l$ — the conditions

$$V^l{}_k \equiv E_k{}^*(\dot{q}_l) \equiv (\partial\dot{q}_l/\partial\omega_k)^{\cdot} - \partial\dot{q}_l/\partial\theta_k \equiv 0 \tag{d}$$

become

$$\partial A_{lk}/\partial q_r = \partial A_{lr}/\partial q_k \qquad \text{and} \qquad \partial A_{lk}/\partial t = \partial A_l/\partial q_k. \tag{e}$$

Example 5.2.2 *Special Choices of the Quasi Velocities, and Forms of the Constraints.* Frequently we choose, as in the Pfaffian case, the *last* $n - m$ $\dot{q}$'s as the *independent* quasi velocities. Then (5.2.2a, b) specialize to

$$\omega_D \equiv f_D(t, q, \dot{q}) = 0, \tag{a}$$

$$\omega_I \equiv f_I(t, q, \dot{q}) = \dot{q}_I \neq 0. \tag{b}$$

Solving (a) for the *first* m (dependent) $\dot{q}$'s $\to \dot{q}_D$ in terms of the remaining $n - m$ (independent) $\dot{q}$'s $\to \dot{q}_I$, assuming that $\partial(f_1, \ldots, f_m)/\partial(\dot{q}_1, \ldots, \dot{q}_m) \neq 0$, we obtain

$$\dot{q}_D = \dot{q}_D(t, q, \dot{q}_I) \equiv \phi_D(t, q, \dot{q}_I). \tag{c}$$

System Quantities

In view of the above, the system virtual displacement equation (5.2.11) specializes to

$$\delta q_k: \qquad \delta q_D = \sum (\partial\phi_D/\partial\dot{q}_I)\,\delta q_I, \tag{d}$$

$$\delta q_I = \sum (\partial\dot{q}_I/\partial\dot{q}_{I'})\,\delta q_{I'} = \sum (\delta_{II'})\,\delta q_{I'} = \delta q_I; \tag{e}$$

while the corresponding constraint conditions (5.2.9) become

$$\begin{aligned} 0 &= \sum (\partial f_D/\partial\dot{q}_k)\,\delta q_k = \sum \left[(\partial f_D/\partial\dot{q}_{D'})\,\delta q_{D'} + (\partial f_D/\partial\dot{q}_I)\,\delta q_I\right] \\ &= \sum \left\{(\partial f_D/\partial\dot{q}_{D'})\left(\sum (\partial\phi_{D'}/\partial\dot{q}_I)\,\delta q_I\right) + (\partial f_D/\partial\dot{q}_I)\,\delta q_I\right\} \\ &= \sum \left[\partial f_D/\partial\dot{q}_I + \sum (\partial f_D/\partial\dot{q}_{D'})\,(\partial\phi_{D'}/\partial\dot{q}_I)\right]\delta q_I \\ &\equiv \sum \left[\partial f_D/\partial(\dot{q}_I)\right]\delta q_I. \end{aligned} \tag{f}$$

Particle Quantities

The particle virtual displacement equation (5.2.15) reduces to

$$\begin{aligned} \delta\boldsymbol{r} &= \sum (\partial\boldsymbol{r}/\partial q_k)\,\delta q_k = \sum (\partial\boldsymbol{r}/\partial q_D)\,\delta q_D + \sum (\partial\boldsymbol{r}/\partial q_I)\,\delta q_I \\ &= \sum (\partial\boldsymbol{r}/\partial q_D)\left(\sum (\partial\phi_D/\partial\dot{q}_I)\,\delta q_I\right) + \sum (\partial\boldsymbol{r}/\partial q_I)\,\delta q_I \\ &\equiv \sum \left[\partial\boldsymbol{r}/\partial(q_I)\right]\delta q_I \equiv \sum \boldsymbol{B}_I\,\delta q_I, \end{aligned} \tag{g}$$

where

$$\begin{aligned} \boldsymbol{B}_I &\equiv \partial\boldsymbol{r}/\partial(q_I) \equiv \partial\boldsymbol{r}/\partial q_I + \sum (\partial\boldsymbol{r}/\partial q_D)\,(\partial\phi_D/\partial\dot{q}_I) \\ &\equiv \boldsymbol{e}_I + \sum (\partial\phi_D/\partial\dot{q}_I)\boldsymbol{e}_D, \end{aligned}$$

and, in general,

$$\partial \boldsymbol{B}_I/\partial q_{I'} \neq \partial \boldsymbol{B}_{I'}/\partial q_I \qquad \text{(i.e., the } \boldsymbol{B}_I \text{ are } \textit{nongradient} \text{ vectors);} \tag{h}$$

which is a specialization of the quasi chain rule (5.2.16c) for $\ldots \to \boldsymbol{r}$ and $\theta_I \to (q_I)$.
Similarly, we can show that

$$\boldsymbol{v} \to \boldsymbol{v}_o = \sum \boldsymbol{B}_I \dot{q}_I + \textit{No other } \dot{q} \textit{ terms}, \tag{i}$$

$$\boldsymbol{a} \to \boldsymbol{a}_o = \sum \boldsymbol{B}_I \ddot{q}_I + \textit{No other } \ddot{q} \textit{ terms}; \tag{j}$$

and hence

$$\partial \boldsymbol{r}/\partial(q_I) \equiv \partial \boldsymbol{v}_o/\partial \dot{q}_I \equiv \partial \boldsymbol{a}_o/\partial \ddot{q}_I \equiv \cdots \equiv \boldsymbol{B}_I, \tag{k}$$

which is a specialization of (5.2.20b).

Example 5.2.3 *Special Forms of the Nonlinear Transitivity Equations.* Let us find the form of the transitivity relations for the special quasi-velocity choice of the preceding example. There we saw that [eqs. (c, d)]

$$\delta q_D = \sum (\partial \phi_D/\partial \dot{q}_I)\, \delta q_I \qquad \text{and} \qquad \dot{q}_D = \dot{q}_D(t, q, \dot{q}_I) \equiv \phi_D(t, q, \dot{q}_I). \tag{a}$$

By $(\ldots)^{\cdot}$-differentiating the *first* of them, and $\delta(\ldots)$-varying the *second*, and then subtracting side by side, we obtain, successively,

$$\begin{aligned}
(\delta q_D)^{\cdot} - \delta(\dot{q}_D) &= \sum \left[(\partial \phi_D/\partial \dot{q}_I)^{\cdot}\, \delta q_I + (\partial \phi_D/\partial \dot{q}_I)\, (\delta q_I)^{\cdot}\right] \\
&\quad - \sum \left[\partial \phi_D/\partial q_I + \sum (\partial \phi_D/\partial q_{D'})\, (\partial \phi_{D'}/\partial \dot{q}_I)\right] \delta q_I \\
&\quad - \sum (\partial \phi_D/\partial \dot{q}_I)\, \delta(\dot{q}_I) \\
&= \sum (\partial \phi_D/\partial \dot{q}_I)\, [(\delta q_I)^{\cdot} - \delta(\dot{q}_I)] \\
&\quad + \sum \left[(\partial \phi_D/\partial \dot{q}_I)^{\cdot} - \partial \phi_D/\partial(q_I)\right] \delta q_I,
\end{aligned} \tag{b}$$

where

$$\partial \phi_D/\partial(q_I) \equiv \partial \phi_D/\partial q_I + \sum (\partial \phi_D/\partial q_{D'})\, (\partial \phi_{D'}/\partial \dot{q}_I). \tag{c}$$

Alternatively, applying (5.2.16c), we find, successively,

$$\begin{aligned}
\partial \dot{q}_D/\partial \theta_I &\equiv \sum (\partial \dot{q}_D/\partial q_k)\, (\partial \dot{q}_k/\partial \omega_I) \\
&= \Big(\sum (\partial \phi_D/\partial q_{D'})\, (\partial \dot{q}_{D'}/\partial \omega_I) + \sum (\partial \phi_D/\partial q_{I'})\, (\partial \dot{q}_{I'}/\partial \omega_I)\Big)_{\omega = \dot{q}} \\
&= \sum (\partial \phi_D/\partial q_{D'})\, (\partial \dot{q}_{D'}/\partial \dot{q}_I) + \sum (\partial \phi_D/\partial q_{I'})\, (\partial \dot{q}_{I'}/\partial \dot{q}_I) \\
&= \sum (\partial \phi_D/\partial q_{D'})\, (\partial \phi_{D'}/\partial \dot{q}_I) + \sum (\partial \phi_D/\partial q_{I'})\, (\delta_{I'I}) \\
&= \partial \phi_D/\partial q_I + \sum (\partial \phi_D/\partial q_{D'})\, (\partial \phi_{D'}/\partial \dot{q}_I) \\
&\equiv \partial \phi_D/\partial(q_I) \qquad \text{[by (c)]};
\end{aligned} \tag{d}$$

that is,

$$\partial \dot{q}_D/\partial \omega_I = \partial \phi_D/\partial \dot{q}_I$$

$$\Rightarrow \omega_I \textit{ can} \text{ be identified with } \dot{q}_I;$$

$$\partial \dot{q}_D/\partial \theta_I = \partial \phi_D/\partial (q_I) \neq \partial \phi_D/\partial q_I$$

$$\Rightarrow \theta_I \textit{ is not} \text{ to be identified with } q_I; \quad \text{hence, the new notation } (q_I). \tag{e}$$

In sum: for this special quasi-variable choice, we can replace in the general expressions

$$\partial(\ldots)/\partial\omega_I \qquad \text{with} \qquad \partial(\ldots)/\partial\dot{q}_I,$$

and

$$\partial(\ldots)/\partial\theta_I \qquad \text{with} \qquad \partial(\ldots)/\partial(q_I) \equiv \partial(\ldots)/\partial q_I + \sum [\partial(\ldots)/\partial q_D]\,(\partial\phi_D/\partial\dot{q}_I). \tag{f}$$

If we now adopt the so-called Suslov viewpoint — that is, $(\delta q_D)^{\cdot} - \delta(\dot{q}_D) \neq 0$, $(\delta q_I)^{\cdot} - \delta(\dot{q}_I) = 0$ [prob. 2.12.5; 3.8.14a ff.] — then (b) leads immediately to the following *nonlinear Suslov* transitivity relations:

$$(\delta q_k)^{\cdot} - \delta(\dot{q}_k): \qquad (\delta q_D)^{\cdot} - \delta(\dot{q}_D) = \sum W^D{}_I\, \delta q_I, \tag{g}$$

$$(\delta q_I)^{\cdot} - \delta(\dot{q}_I) = 0 \qquad (\text{i.e., } W^{I'}{}_I = 0), \tag{h}$$

where

$$\begin{aligned} W^D{}_I &\equiv (\partial\phi_D/\partial\dot{q}_I)^{\cdot} - \partial\phi_D/\partial q_I - \sum (\partial\phi_D/\partial q_{D'})\,(\partial\phi_{D'}/\partial\dot{q}_I) \\ &\equiv E_I(\phi_D) - \sum (\partial\phi_D/\partial q_{D'})\,(\partial\phi_{D'}/\partial\dot{q}_I) \\ &\equiv (\partial\phi_D/\partial\dot{q}_I)^{\cdot} - \partial\phi_D/\partial(q_I) \\ &\equiv E_{(I)}(\phi_D): \end{aligned}$$

Special nonlinear Voronets coefficients [specialization of (5.2.21e)]. (i)

If the second of equations (a) have the special form

$$\dot{q}_D = \dot{q}_D(q_I, \dot{q}_I) \equiv \phi_D(q_I, \dot{q}_I): \quad \text{nonlinear Chaplygin system}, \tag{j}$$

then $\partial\phi_D/\partial q_{D'} = 0$, and so (h) reduces to the *nonlinear Chaplygin (or Tsaplygin) coefficients*

$$W^D{}_I \to T^D{}_I \equiv (\partial\phi_D/\partial\dot{q}_I)^{\cdot} - \partial\phi_D/\partial q_I \equiv E_I(\phi_D). \tag{k}$$

5.3 KINETICS: VARIATIONAL EQUATIONS/PRINCIPLES; GENERAL AND SPECIAL EQUATIONS OF MOTION (OF JOHNSEN, HAMEL, ET AL.)

To derive Lagrangean-type equations, we will use both the central equation (§3.6) in system variables, and Lagrange's principle (§3.2) in both particle and system variables.

The NNHC Central Equation

As discussed in §3.6, the latter is, with the usual notations and assuming that $d(\delta\mathbf{r})/dt - \delta(d\mathbf{r}/dt) \equiv (\delta\mathbf{r})^{\cdot} - \delta\mathbf{v} = \mathbf{0} \Rightarrow (\delta q_k)^{\cdot} - \delta(\dot{q}_k) = 0$ (for all holonomic variables, constrained or not),

$$d/dt(\delta P) - \delta T = \delta' W, \tag{5.3.1}$$

where

$$\delta P \equiv S\, dm\, \mathbf{v} \cdot \delta\mathbf{r} = \sum (\partial T/\partial \dot{q}_k)\, \delta q_k \equiv \sum p_k\, \delta q_k$$
$$= \sum (\partial T^*/\partial \omega_k)\, \delta\theta_k \equiv \sum P_k\, \delta\theta_k, \tag{5.3.1a}$$

$$\Big[\Rightarrow p_k = \sum (\partial \omega_l/\partial \dot{q}_k) P_l \Leftrightarrow P_l = \sum (\partial \dot{q}_k/\partial \omega_l) p_k,$$
$$\text{i.e., } \partial T/\partial \dot{q}_k = \sum (\partial \omega_l/\partial \dot{q}_k)\, (\partial T^*/\partial \omega_l)$$
$$\Leftrightarrow \partial T^*/\partial \omega_l = \sum (\partial \dot{q}_k/\partial \omega_l)\, (\partial T/\partial \dot{q}_k)\Big]; \tag{5.3.1b}$$

$$\delta T \equiv \delta\Big(S\, (1/2)\, dm\, \mathbf{v} \cdot \mathbf{v}\Big)$$
$$= \sum \big[(\partial T^*/\partial \theta_k)\, \delta\theta_k + (\partial T^*/\partial \omega_k)\, \delta\omega_k\big], \tag{5.3.1c}$$

$$T = T(t, q, \dot{q}) = T[t, q, \dot{q}(t, q, \omega)] \equiv T^*(t, q, \omega) \equiv T^*; \tag{5.3.1d}$$

$$\delta' W \equiv S\, d\mathbf{F} \cdot \delta\mathbf{r} = \sum Q_k\, \delta q_k \equiv \sum \Theta_k\, \delta\theta_k$$
$$\Big[\Rightarrow Q_k = \sum (\partial \omega_l/\partial \dot{q}_k) \Theta_l \Leftrightarrow \Theta_l = \sum (\partial \dot{q}_k/\partial \omega_l) Q_k\Big]. \tag{5.3.1e}$$

Substituting (5.3.1a–e) into (5.3.1), and regrouping appropriately, we obtain the

Central equation in NNH variables:

$$\sum (dP_k/dt)\, \delta\theta_k - \sum (\partial T^*/\partial \theta_k)\, \delta\theta_k + \sum P_k[(\delta\theta_k)^{\cdot} - \delta\omega_k]$$
$$= \sum \Theta_k\, \delta\theta_k; \tag{5.3.2}$$

and from this, invoking the transitivity equations (5.2.22a, b) [under the Hamel viewpoint; i.e., $(\delta q_k)^{\cdot} = \delta(\dot{q}_k)$ for *all* holonomic variables, constrained or not], we finally obtain

Lagrange's principle in NNH variables:

$$\sum \Big(dP_k/dt - \partial T^*/\partial \theta_k + \sum H^s{}_k P_s - \Theta_k\Big)\, \delta\theta_k = 0. \tag{5.3.3}$$

These *variational equations* are fundamental to all subsequent kinetic considerations.

REMARK

As in the Pfaffian case (§3.6), eq. (5.3.3) (and, hence, the equations of motion resulting from it) is independent of any assumptions regarding $d(\delta\mathbf{r}) - \delta(d\mathbf{r})$ or

$d(\delta q_k) - \delta(dq_k)$. To confirm this, we begin with the most general central equation (3.6.4)

$$d/dt(\delta P) - \delta T - \delta D = \delta' W, \tag{5.3.4}$$

instead of (5.3.1); where, successively,

$$\begin{aligned}
\delta D &\equiv S\, dm\, \mathbf{v}\cdot[(\delta \mathbf{r})^{\cdot} - \delta \mathbf{v}] = S\, dm\, \mathbf{v}\cdot\Big\{\sum [(\delta q_k)^{\cdot} - \delta(\dot{q}_k)]\mathbf{e}_k\Big\} \\
&= \sum p_k[(\delta q_k)^{\cdot} - \delta(\dot{q}_k)] \\
&= \sum\sum p_k\,(\partial \dot{q}_k/\partial \omega_l)\,[(\delta\theta_l)^{\cdot} - \delta\omega_l] \\
&\quad - \sum\sum\sum p_k(\partial \dot{q}_k/\partial \omega_s) H^s{}_l\, \delta\theta_l \qquad \text{[by (5.2.23b)]} \\
&= \sum P_l[(\delta\theta_l)^{\cdot} - \delta\omega_l] - \sum\sum H^s{}_l P_s\, \delta\theta_l \qquad \text{[by (5.3.1b)].}
\end{aligned} \tag{5.3.4a}$$

As a result of (5.3.4a) and (5.3.1a–e), eq. (5.3.4) becomes

$$\begin{aligned}
&\sum [(dP_k/dt)\, \delta\theta_k + P_k(\delta\theta_k)^{\cdot}] - \sum\, [(\partial T^*/\partial\theta_k)\, \delta\theta_k + (\partial T^*/\partial\omega_k)\, \delta\omega_k] \\
&\quad - \sum P_k[(\delta\theta_k)^{\cdot} - \delta\omega_k] + \sum\sum H^s{}_k P_s\, \delta\theta_k = \sum \Theta_k\, \delta\theta_k,
\end{aligned} \tag{5.3.4b}$$

which, when simplified, as the reader can easily confirm, is none other than (5.3.3).

Equations of Motion

In the presence of $m(< n)$ constraints, in the virtual form (5.2.13): $\delta\theta_D = 0$, application of the method of Lagrangean multipliers to (5.3.3), in exactly the same fashion as in §3.5–3.7, readily produces the following two groups of equations of motion:

Kinetostatic: $$dP_D/dt - \partial T^*/\partial\theta_D + \sum H^k{}_D P_k = \Theta_D + \lambda_D \qquad (D = 1, \ldots, m), \tag{5.3.5a}$$

Kinetic: $$dP_I/dt - \partial T^*/\partial\theta_I + \sum H^k{}_I P_k = \Theta_I \qquad (I = m+1, \ldots, n). \tag{5.3.5b}$$

Equations (5.3.5b) are due to Johnsen and Hamel (1936–1941).

In extenso, the kinetic group (5.3.5b) reads

$$\begin{aligned}
&(\partial T^*/\partial\omega_I)^{\cdot} - \partial T^*/\partial\theta_I \\
&\quad + \sum\sum\, [(\partial\omega_l/\partial\dot{q}_s)^{\cdot} - \partial\omega_l/\partial q_s]\,(\partial\dot{q}_s/\partial\omega_I)\,(\partial T^*/\partial\omega_l) = \Theta_I,
\end{aligned} \tag{5.3.5c}$$

or

$$E_I(T^*) + \sum\sum\, E_s(\omega_l)\,(\partial\dot{q}_s/\partial\omega_I)\,(\partial T^*/\partial\omega_l) = \Theta_I; \tag{5.3.5d}$$

and similarly for the kinetostatic group (5.3.5a). Equations (5.3.5b–d) constitute the legitimate generalization of the original Hamel equations (1903–1904, §3.5) to the nonlinear nonholonomic variable and constraint case.

• Using (5.2.21f, h), we easily obtain a *second* form of these equations. Indeed, the kinetic such group is

$$dP_I/dt - \partial T^*/\partial\theta_I - \sum\sum(\partial\omega_l/\partial\dot{q}_k)V^k{}_I P_l = \Theta_I, \tag{5.3.6a}$$

or, in extenso,

$$(\partial T^*/\partial\omega_I)^{\cdot} - \partial T^*/\partial\theta_I - \sum\sum\left[(\partial\dot{q}_k/\partial\omega_I)^{\cdot} - \partial\dot{q}_k/\partial\theta_I\right](\partial\omega_l/\partial\dot{q}_k)(\partial T^*/\partial\omega_l) = \Theta_I, \tag{5.3.6b}$$

or, in operator form,

$$E_I{}^*(T^*) - \sum\sum E_I{}^*(\dot{q}_k)(\partial\omega_l/\partial\dot{q}_k)(\partial T^*/\partial\omega_l) = \Theta_I. \tag{5.3.6c}$$

• Also, since

$$\sum(\partial\omega_l/\partial\dot{q}_k)P_l = p_k = p_k(t,q,\dot{q}) = p_k[t,q,\dot{q}(t,q,\omega)]$$
$$\equiv p_k{}^* \equiv (\partial T/\partial\dot{q}_k)^*, \tag{5.3.7}$$

we can rewrite equations (5.3.6a–c) in the following *mixed* form (i.e., containing both T and T^*):

$$dP_I/dt - \partial T^*/\partial\theta_I - \sum V^k{}_I\, p_k{}^* = \Theta_I, \tag{5.3.8a}$$

$$(\partial T^*/\partial\omega_I)^{\cdot} - \partial T^*/\partial\theta_I - \sum\left[(\partial\dot{q}_k/\partial\omega_I)^{\cdot} - \partial\dot{q}_k/\partial\theta_I\right](\partial T/\partial\dot{q}_k)^* = \Theta_I; \tag{5.3.8b}$$

or, further, with $T^*|_{\text{constraints enforced}} \equiv T^*{}_o$,

$$(\partial T^*{}_o/\partial\omega_I)^{\cdot} - \partial T^*{}_o/\partial\theta_I - \sum V^k{}_I\,(\partial T/\partial\dot{q}_k)^*{}_o = \Theta_I. \tag{5.3.8c}$$

Similarly, (5.3.5b–d) can be replaced by their "constrained" form:

$$(\partial T^*{}_o/\partial\omega_I)^{\cdot} - \partial T^*{}_o/\partial\theta_I + \sum H^k{}_I\,(\partial T^*/\partial\omega_k)_o = \Theta_I. \tag{5.3.8d}$$

Equations (5.3.5b–d, 8d) and (5.3.8a–c) constitute the legitimate nonlinear generalizations of the original "Pfaffian equations" of Hamel (§3.5) and Chaplygin–Voronets (§3.8), respectively. [Although Hamel (1938, p. 48) seems to view (5.3.6a–c), rather than (5.3.5b–d), as the genuine nonlinear generalization of his equations of 1903–1904.] Comparing these two basic kinds of equations we notice the following: The Hamel forms (5.3.5b–d), as well as (5.3.6a–c), contain both $\partial\omega/\partial\dot{q}$ and $\partial\dot{q}/\partial\omega$ derivatives; and (5.3.5b–d) contain both $\ddot{q}$ and $\dot{\omega}$-proportional terms; whereas (5.3.8a–c) involve only $\partial\dot{q}/\partial\omega$ derivatives. On the other hand, the former involve only T^*, while the latter involve both T and T^*. But these differences are superficial: in view of (i) the nonlinear transformation equations $\dot{q} \Leftrightarrow \omega$, (ii) the fact that the matrices $(\partial\dot{q}/\partial\omega)$ and $(\partial\omega/\partial\dot{q})$ are mutually inverse [i.e., by Cramer's rule applied to the "inverseness relations" (5.2.4a, b), *the coefficients* $\partial\omega_l/\partial\dot{q}_k$ appearing in (5.3.6a–c) *equal the minors of the determinant of the matrix* $(\partial\dot{q}/\partial\omega)$ *divided by* $Det(\partial\dot{q}/\partial\omega)$, and can, therefore, be also expressed as functions of t, q, ω without further use of the equations $\dot{q} \Leftrightarrow \omega$; and similarly for expressing the $\partial\dot{q}_k/\partial\omega_l$ appearing in (5.3.8a–c) in terms of $t, q, \dot{q}$] and that, in analogy to (5.3.7), (iii) $P_l \equiv \partial T^*/\partial\omega_l = \sum(\partial\dot{q}_k/\partial\omega_l)(\partial T/\partial\dot{q}_k) = \cdots = P_l(t,q,\dot{q})$, we can express all terms of these two kinds of equations (including the symbols $V^l{}_k$ and $H^l{}_k$) in terms of either $t, q, \dot{q}, \ddot{q}$ or $t, q, \omega, \dot{\omega}$, although, in particular problems, such a choice is conditioned by practical considerations (e.g., amount of labor involved). Finally, reasoning as in §3.4, we see that the Lagrangean multipliers λ_D, in the kinetostatic of the above equations, equal the (covariant) nonholonomic components of the system constraint reactions: $\Lambda_D \equiv S\, d\boldsymbol{R} \cdot \boldsymbol{\varepsilon}_D$, where (§3.2) $dm\,\boldsymbol{a} = d\boldsymbol{F} + d\boldsymbol{R}$. Indeed, assuming that Lagrange's principle (LP) also holds for the reactions enforcing our nonlinear constraints (5.2.2a), we have

$$\delta' W_R \equiv S\, d\boldsymbol{R} \cdot \delta\boldsymbol{r} = \sum R_k\, \delta q_k = \sum \Lambda_k\, \delta\theta_k \quad (= 0), \tag{5.3.9a}$$

from which we immediately obtain the transformation equations

$$\begin{aligned} &\Lambda_k \equiv S\, d\boldsymbol{R} \cdot \boldsymbol{\varepsilon}_k = \sum (\partial \dot{q}_l / \partial \omega_k)\, R_l: \\ &\qquad \Lambda_D \neq 0 \quad (\delta\theta_D = 0) \qquad \text{and} \qquad \Lambda_I = 0 \quad (\delta\theta_I \neq 0), \end{aligned} \tag{5.3.9b}$$

$$\begin{aligned} R_l &\equiv S\, d\boldsymbol{R} \cdot \boldsymbol{e}_l = \sum (\partial \omega_k / \partial \dot{q}_l)\, \Lambda_k \\ &= \sum (\partial \omega_D / \partial \dot{q}_l)\, \Lambda_D = \sum (\partial \phi_D / \partial \dot{q}_l)\, \Lambda_D, \end{aligned} \tag{5.3.9c}$$

and comparing with the constitutive equation $R_l = \sum \lambda_D\, (\partial f_D / \partial \dot{q}_l)$, obtained via application of the method of multipliers to LP, we conclude that $\Lambda_D = \lambda_D$.

Lagrange's Principle (LP)

As detailed in §3.2–3.5, LP postulates that

$$\delta I = \delta' W, \tag{5.3.10}$$

where

$$\delta I \equiv S\, dm\, \boldsymbol{a} \cdot \delta\boldsymbol{r} = \sum E_k\, \delta q_k = \sum I_k\, \delta\theta_k, \tag{5.3.10a}$$

$$E_k \equiv S\, dm\, \boldsymbol{a} \cdot \boldsymbol{e}_k = E_k(T), \tag{5.3.10b}$$

$$I_k \equiv S\, dm\, \boldsymbol{a} \cdot \boldsymbol{\varepsilon}_k = E_k{}^*(T^*) + \sum H^l{}_k P_l, \tag{5.3.10c}$$

$$\delta' W \equiv S\, d\boldsymbol{F} \cdot \delta\boldsymbol{r} = \sum Q_k\, \delta q_k = \sum \Theta_k\, \delta\theta_k, \tag{5.3.10d}$$

$$Q_k \equiv S\, d\boldsymbol{F} \cdot \boldsymbol{e}_k, \qquad \Theta_k \equiv S\, d\boldsymbol{F} \cdot \boldsymbol{\varepsilon}_k. \tag{5.3.10e}$$

Holonomic Coordinates

Applying the method of Lagrangean multipliers to (5.3.10) in holonomic system variables:

$$\sum E_k\, \delta q_k = \sum Q_k\, \delta q_k, \tag{5.3.11a}$$

under the m constraints (5.1.2; 5.2.2a), in the virtual form (5.2.9)

$$\sum (\partial f_D / \partial \dot{q}_k)\, \delta q_k = 0, \tag{5.3.11b}$$

we immediately obtain the *nonlinear* (generalization of the) *Routh–Voss* equations:

$$(\partial T/\partial \dot{q}_k)^{\cdot} - \partial T/\partial q_k = Q_k + \sum \lambda_D \, (\partial f_D/\partial \dot{q}_k), \tag{5.3.11c}$$

or, compactly,

$$E_k(T) = Q_k + R_k, \tag{5.3.11d}$$

which, along with the m constraints $f_D = 0$ constitute a determinate system of $n + m$ equations for the $n + m$ unknown functions $\lambda_D(t), q_k(t)$. [Equations (5.3.11c) are due to Routh (1877, 3rd ed. of the *Elementary* part of his classic *Rigid Dynamics*). However, he never applied them to any NNHC problem. It seems certain that he chose that form as a memorable way of writing the equations of motion under Pfaffian [or (5.3.13a)-like] constraints (3.5.15) rather than with the full understanding that equations (5.3.11c) hold for the most general nonlinear first-order (possibly nonholonomic) constraints.]

• In view of the purely kinematico-inertial identity $E_k(T) = \partial S/\partial \ddot{q}_k$, where $S = S(t, q, \dot{q}, \ddot{q})$ is the (unconstrained) Appellian of the system [(3.3.16a)], we will also have the *Appellian form of the nonlinear Routh–Voss equations*:

$$\partial S/\partial \ddot{q}_k = Q_k + \sum \lambda_D \, (\partial f_D/\partial \dot{q}_k). \tag{5.3.12}$$

• If the constraints have the (holonomic) form $g_D \equiv g_D(t, q) = 0$ — that is, if they do not contain the $\dot{q}$'s — it does *not* mean that, since $\partial g_D/\partial \dot{q}_k = 0$, we will have $R_k = 0$. It means, instead, that to apply (5.3.11c) correctly we have to *create a $\dot{q}$-containing constraint from $g_D = 0$*. Indeed, $(\ldots)^{\cdot}$-differentiating $g_D(t, q) = 0$, we obtain

$$0 = dg_D/dt = \partial g_D/\partial t + \sum (\partial g_D/\partial q_k)\dot{q}_k \equiv f_D(t, q, \dot{q}) \equiv f_D, \tag{5.3.13a}$$

from which it follows that

$$\partial f_D/\partial \dot{q}_k = \partial g_D/\partial q_k, \tag{5.3.13b}$$

so that (5.3.11c) become

$$E_k(T) = Q_k + \sum \lambda_D \, (\partial g_D/\partial q_k), \tag{5.3.13c}$$

in complete agreement with earlier results; for example, ex. 3.5.14.

Nonholonomic Coordinates

Applying Lagrangean multipliers to LP, (5.3.10), in nonholonomic variables:

$$\sum I_k \, \delta\theta_k = \sum \Theta_k \, \delta\theta_k, \tag{5.3.14a}$$

under the m constraints $\delta\theta_D = 1 \cdot \delta\theta_D = 0$ (and $0 \cdot \delta\theta_I = 0$ for the rest), we immediately obtain the following two groups of equations:

Kinetostatic: $\quad I_D = \Theta_D + \lambda_D \quad [D = 1, \ldots, m\,(< n)],$ (5.3.14b)

Kinetic: $\quad I_I = \Theta_I \quad [I = m + 1, \ldots, n].$ (5.3.14c)

With the I_k's, Θ_k's, λ_D's defined by (5.3.10c, e, 9b), respectively, the above constitute the so-called "raw" forms of the equations of motion. As in the Pfaffian case (§3.5, §3.8), special choices of the fundamental nonholonomic "accompanying vectors" $\boldsymbol{\varepsilon}_k$ in them, as in (5.2.20b) and ex. 5.2.2, yield special forms of the nonlinear equations of motion (Maggi, Schaefer, Hamel, Appell, et al.). Let us examine them in detail, in order of increasing difficulty.

(i) The choice $\boldsymbol{\varepsilon}_k = \partial \boldsymbol{a}^*/\partial \dot{\omega}_k$ in (5.3.10c) yields, successively,

$$I_k = S\, dm\, \boldsymbol{a}^* \cdot (\partial \boldsymbol{a}^*/\partial \dot{\omega}_k) = \partial S^*/\partial \dot{\omega}_k, \tag{5.3.15a}$$

where

$$S^* \equiv S\, (1/2)\, dm\, \boldsymbol{a}^* \cdot \boldsymbol{a}^* = S^*(t, q, \omega, \dot{\omega}): \quad \textit{Nonlinear Appellian}; \tag{5.3.15b}$$

and so (5.3.14b, c) assume the form of the *nonlinear Appell equations*

$$\partial S^*/\partial \dot{\omega}_D = \Theta_D + \lambda_D, \qquad \partial S^*/\partial \dot{\omega}_I = \Theta_I, \tag{5.3.15c}$$

respectively. If we are not interested in the constraint forces, then, as already explained for the Pfaffian case (§3.5), we can replace in the kinetic equations (second of 5.3.15c), the unconstrained Appellian S^* with the constrained one:

$$S^*_o \equiv S^*(t, q, \omega_D = 0, \omega_I, \dot{\omega}_D = 0, \dot{\omega}_I) \equiv S^*_o(t, q, \omega_I, \dot{\omega}_I), \tag{5.3.15d}$$

and similarly for the Θ_I (although we shall still denote them as Θ_I),

$$\partial S^*_o/\partial \dot{\omega}_I = \Theta_I. \tag{5.3.15e}$$

Also, in concrete problems we do not have to first compute S^* and then find $\partial S^*/\partial \dot{\omega}_I \to (\partial S^*/\partial \dot{\omega}_I)_o = \partial S^*_o/\partial \dot{\omega}_I$ but, instead, we can use I_k in the form of (5.3.15a).

(ii) The choice $\boldsymbol{\varepsilon}_k = \partial \boldsymbol{r}^*/\partial \theta_k \equiv \partial \boldsymbol{r}/\partial \theta_k$ (symbolic gradient) in (5.3.10c) yields, successively,

$$\begin{aligned}
I_k &= S\, dm\, \boldsymbol{a}^* \cdot (\partial \boldsymbol{r}^*/\partial \theta_k) = S\, dm\, \boldsymbol{a}^* \cdot \Big(\sum (\partial \dot{q}_l/\partial \omega_k) \boldsymbol{e}_l\Big) \\
&= \sum \Big(S\, dm\, \boldsymbol{a}^* \cdot \boldsymbol{e}_l\Big)\, (\partial \dot{q}_l/\partial \omega_k) \\
&= \sum\, [(\partial T/\partial \dot{q}_l)^{\cdot} - \partial T/\partial q_l]\, (\partial \dot{q}_l/\partial \omega_k) \qquad \text{[by Lagrange's identity (§3.3)]} \\
&\equiv \sum E_l(T)\, (\partial \dot{q}_l/\partial \omega_k) \equiv \sum E_l\, (\partial \dot{q}_l/\partial \omega_k);
\end{aligned} \tag{5.3.16a}$$

and so (5.3.14b, c) yield the *nonlinear Maggi equations*

$$\sum (\partial \dot{q}_k/\partial \omega_D) E_k = \sum (\partial \dot{q}_k/\partial \omega_D) Q_k + \lambda_D, \tag{5.3.16b}$$

$$\sum (\partial \dot{q}_k/\partial \omega_I) E_k = \sum (\partial \dot{q}_k/\partial \omega_I) Q_k. \tag{5.3.16c}$$

[Equations (5.3.16c) are due to Hamel (1938, p. 45); see also Przeborski (1933) for a particle and component form.] Also, since $E_k = \partial S/\partial \ddot{q}_k$, we can replace in both (5.3.16b, c) E_k with $\partial S/\partial \ddot{q}_k$, and thus obtain the *Appellian form of the nonlinear Maggi equations*.

(iii) The choice $\boldsymbol{\varepsilon}_k = \partial \boldsymbol{v}^*/\partial \omega_k \to \partial \boldsymbol{v}^*/\partial \omega_I$ in (5.3.10c) yields immediately the *Schaefer equations* (1951):

$$S\, dm\, \boldsymbol{a}^* \cdot (\partial \boldsymbol{v}^*/\partial \omega_I) = S\, d\boldsymbol{F} \cdot (\partial \boldsymbol{v}^*/\partial \omega_I), \tag{5.3.17}$$

which, as we show immediately below, constitute a raw form of the earlier nonlinear Johnsen–Hamel equations (5.3.5a ff.). Indeed, I_k transforms, successively, as follows:

$$I_k \equiv S\, dm\, \boldsymbol{a}^* \cdot \boldsymbol{\varepsilon}_k = S\, dm\, (d\boldsymbol{v}^*/dt) \cdot (\partial \boldsymbol{v}^*/\partial \omega_k)$$

$$= d/dt\Big(S\, dm\, \boldsymbol{v}^* \cdot (\partial \boldsymbol{v}^*/\partial \omega_k)\Big) - S\, dm\, \boldsymbol{v}^* \cdot (\partial \boldsymbol{v}^*/\partial \omega_k)^{\cdot}$$

$$\Big[\text{adding and subtracting } S\, dm\, \boldsymbol{v}^* \cdot (\partial \boldsymbol{v}^*/\partial \theta_k)\Big]$$

$$= d/dt\Big(S\, dm\, \boldsymbol{v}^* \cdot (\partial \boldsymbol{v}^*/\partial \omega_k)\Big) - S\, dm\, \boldsymbol{v}^* \cdot (\partial \boldsymbol{v}^*/\partial \theta_k) - S\, dm\, \boldsymbol{v}^* \cdot \big[(\partial \boldsymbol{v}^*/\partial \omega_k)^{\cdot} - \partial \boldsymbol{v}^*/\partial \theta_k\big]$$

[if the θ were holonomic coordinates, the *last* (*third*) *sum would vanish*!]

$$= d/dt\Big[\partial/\partial \omega_k\Big(S\,(1/2)\, dm\, \boldsymbol{v}^* \cdot \boldsymbol{v}^*\Big)\Big] - \Big[\partial/\partial \theta_k\Big(S\,(1/2)\, dm\, \boldsymbol{v}^* \cdot \boldsymbol{v}^*\Big)\Big] - S\, dm\, \boldsymbol{v}^* \cdot E_k{}^*(\boldsymbol{v}^*)$$

$$= (\partial T^*/\partial \omega_k)^{\cdot} - \partial T^*/\partial \theta_k - \Gamma_k, \tag{5.3.18a}$$

where

$$T = T(t, q, \dot{q}) = T[t, q, \dot{q}(t, q, \omega)]$$
$$\equiv T^*(t, q, \omega) = T^* \equiv S\,(1/2)\, dm\, \boldsymbol{v}^* \cdot \boldsymbol{v}^*, \tag{5.3.18b}$$

$$\Gamma_k \equiv S\, dm\, \boldsymbol{v}^* \cdot \big[(\partial \boldsymbol{v}^*/\partial \omega_k)^{\cdot} - \partial \boldsymbol{v}^*/\partial \theta_k\big]$$
$$= S\, dm\, \boldsymbol{v}^* \cdot \big[d/dt\,(\partial \boldsymbol{r}^*/\partial \theta_k) - \partial/\partial \theta_k\,(d\boldsymbol{r}^*/dt)\big] \quad (\neq 0)$$
$$\equiv S\, dm\, \boldsymbol{v}^* \cdot E_k{}^*(\boldsymbol{v}^*):$$

Nonlinear nonholonomic correction (*or supplementary*) *term*. (5.3.18c)

Further, recalling the earlier (5.2.21c–h) and the definitions

$$p_l \equiv \partial T/\partial \dot{q}_l \equiv S\, dm\, \boldsymbol{v} \cdot \boldsymbol{e}_l, \qquad P_l \equiv \partial T^*/\partial \omega_l \equiv S\, dm\, \boldsymbol{v}^* \cdot \boldsymbol{\varepsilon}_l,$$

we obtain the following mutually equivalent forms in *system* variables:

$$\Gamma_k = S\, dm\, \boldsymbol{v}^* \cdot \Big(\sum E_k{}^*(\dot{q}_l)\boldsymbol{e}_l\Big)$$
$$= S\, dm\, \boldsymbol{v}^* \cdot \Big(\sum V^l{}_k \boldsymbol{e}_l\Big) = -S\, dm\, \boldsymbol{v}^* \cdot \Big(\sum\sum H^s{}_k(\partial \dot{q}_l/\partial \omega_s)\boldsymbol{e}_l\Big)$$
$$= \sum V^l{}_k\, p_l$$
$$\Big\{\equiv \sum \big[(\partial \dot{q}_l/\partial \omega_k)^{\cdot} - \partial \dot{q}_l/\partial \theta_k\big]\,(\partial T/\partial \dot{q}_l)^*\Big\}$$
$$= -\sum\sum (\partial \dot{q}_l/\partial \omega_s) H^s{}_k\, p_l$$
$$\Big\{\equiv -\sum\sum\sum \big[(\partial \omega_s/\partial \dot{q}_b)^{\cdot} - \partial \omega_s/\partial q_b\big]\,(\partial \dot{q}_b/\partial \omega_k)\,(\partial \dot{q}_l/\partial \omega_s)\,(\partial T/\partial \dot{q}_l)^*$$
$$= -\sum\sum \big[(\partial \omega_s/\partial \dot{q}_b)^{\cdot} - \partial \omega_s/\partial q_b\big]\,(\partial \dot{q}_b/\partial \omega_k)\,(\partial T^*/\partial \omega_s)\Big\}; \tag{5.3.18d}$$

and

$$
\begin{aligned}
\Gamma_k &= S\, dm\, \boldsymbol{v}^* \cdot \Big(\sum\sum E_k{}^*(\dot{q}_l)\,(\partial\omega_s/\partial\dot{q}_l)\boldsymbol{\varepsilon}_s\Big) \\
&= S\, dm\, \boldsymbol{v}^* \cdot \Big(\sum\sum (\partial\omega_s/\partial\dot{q}_l) V^l{}_k \boldsymbol{\varepsilon}_s\Big) = -S\, dm\, \boldsymbol{v}^* \cdot \Big(\sum H^s{}_k \boldsymbol{\varepsilon}_s\Big) \\
&= \sum\sum (\partial\omega_s/\partial\dot{q}_l) V^l{}_k P_s \\
&\quad \Big\{\equiv \sum\sum \big[(\partial\dot{q}_l/\partial\omega_k)^{\cdot} - \partial\dot{q}_l/\partial\theta_k\big]\,(\partial\omega_s/\partial\dot{q}_l)\,(\partial T^*/\partial\omega_s)\Big\} \\
&= -\sum H^s{}_k P_s \\
&\quad \Big\{\equiv -\sum\sum \big[(\partial\omega_s/\partial\dot{q}_l)^{\cdot} - \partial\omega_s/\partial q_l\big]\,(\partial\dot{q}_l/\partial\omega_k)\,(\partial T^*/\partial\omega_s)\Big\}; \qquad (5.3.18e)
\end{aligned}
$$

where

$p_l = p_l(t, q, \dot{q}) = p_l[t, q, \dot{q}(t, q, \omega)] = p_l{}^*(t, q, \omega) \equiv (\partial T/\partial\dot{q}_l)^*$:

holonomic (l)*th component of system momentum, but expressed in nonholonomic variables* [also note that $p_l{}^* \neq P_l$, while $\partial T^*/\partial\dot{q}$ is undefined]. (5.3.18f)

In view of the above, the Schaefer equations (5.3.17) assume the earlier found Lagrangean form of Johnsen–Hamel (5.3.5b ff.):

$$
(\partial T^*/\partial\omega_I)^{\cdot} - \partial T^*/\partial\theta_I - \Gamma_I = \Theta_I. \qquad (5.3.19)
$$

Below we summarize, for convenience, the various available general particle and system representations of the nonlinear nonholonomic inertia "force" I_k, in operator forms:

$$
\begin{aligned}
I_k &\equiv S\, dm\, \boldsymbol{a}^* \cdot \boldsymbol{\varepsilon}_k \qquad \text{(Definition; } \textit{raw} \text{ or particle form)} \\
&= \partial S^*/\partial\dot{\omega}_k \quad \Big[= \sum (\partial S/\partial\ddot{q}_l)\,(\partial\ddot{q}_l/\partial\dot{\omega}_k) = \sum (\partial S/\partial\ddot{q}_l)\,(\partial\dot{q}_l/\partial\omega_k), \\
&\qquad \text{where } 2S^* \equiv S\, dm\, \boldsymbol{a}^* \cdot \boldsymbol{a}^* = S\, dm\, \boldsymbol{a} \cdot \boldsymbol{a} \equiv 2S \quad (\textit{Appell} \text{ form})\Big] \\
&= \sum (\partial\dot{q}_l/\partial\omega_k)\, E_l(T) \qquad (\textit{Maggi} \text{ form}) \\
&= E_k{}^*(T^*) + \sum\sum E_s(\omega_r)\,(\partial\dot{q}_s/\partial\omega_k)\,(\partial T^*/\partial\omega_r) \\
&= E_k{}^*(T^*) + \sum\sum\sum E_s(\omega_r)\,(\partial\dot{q}_s/\partial\omega_k)\,(\partial\dot{q}_l/\partial\omega_r)\,(\partial T/\partial\dot{q}_l)^* \\
&= E_k{}^*(T^*) - \sum\sum E_k{}^*(\dot{q}_l)\,(\partial\omega_r/\partial\dot{q}_l)\,(\partial T^*/\partial\omega_r) \\
&= E_k{}^*(T^*) - \sum E_k{}^*(\dot{q}_l)\,(\partial T/\partial\dot{q}_l)^* \qquad (\textit{Johnsen–Hamel} \text{ forms}). \qquad (5.3.20)
\end{aligned}
$$

Also, recalling (5.3.10a), we have the transformation equations between the holonomic and nonholonomic components of the inertia "force":

$$
I_k = \sum (\partial\dot{q}_l/\partial\omega_k) E_l \;\Leftrightarrow\; E_l = \sum (\partial\omega_k/\partial\dot{q}_l) I_k, \qquad (5.3.21a)
$$

and, of course, the identity:

$$E_l \equiv E_l(T) \equiv \partial S/\partial \ddot{q}_l. \tag{5.3.21b}$$

The above show that, as in the Pfaffian case (§3.2 ff.), $E_k{}^*(T^*)$ does *not* transform as a (covariant) vector under transformations $\delta q \Leftrightarrow \delta\theta$; it is $I_k \equiv E_k{}^*(T) - \Gamma_k$ that does!

REMARK

Maggi's equations (5.3.16b, c) can also be deduced from the nonlinear Routh–Voss equations (5.3.11c, d):

$$E_k(T) = Q_k + \sum \lambda_D\,(\partial f_D/\partial \dot{q}_k) = Q_k + \sum \lambda_D\,(\partial \omega_D/\partial \dot{q}_k),$$

through multiplication with $\partial \dot{q}_k/\partial \omega_l$, summation over k, and subsequent utilization of (5.2.4a):

$$\sum (\partial \dot{q}_k/\partial \omega_l)\, E_k(T) = \sum (\partial \dot{q}_k/\partial \omega_l) Q_k + A_l,$$

where

$$\begin{aligned} A_l &\equiv \sum\sum \lambda_D(\partial \omega_D/\partial \dot{q}_k)\,(\partial \dot{q}_k/\partial \omega_l) = \sum \lambda_D\, \delta_{Dl} = \lambda_l \\ &= \lambda_D, \qquad \text{if } l \to D = 1, \ldots, m; \\ &= \lambda_I = 0, \qquad \text{if } l \to I = m+1, \ldots, n; \quad \text{Q.E.D.} \end{aligned} \tag{5.3.22}$$

Example 5.3.1 *Holonomic and Nonholonomic Inertial Forces and their Transformation Properties.* Let us find by direct differentiations the relations between the holonomic and nonholonomic inertia "forces" $E_k = E_k(T)$ and $I_k = E_k{}^*(T^*) - \Gamma_k$, respectively.

(i) Applying chain rule to

$$T = T(t, q, \dot{q}) = T^*(t, q, \omega) \equiv T^*, \tag{a}$$

we find

$$\begin{aligned} &\partial T^*/\partial q_l = \partial T/\partial q_l + \sum (\partial T/\partial \dot{q}_r)\,(\partial \dot{q}_r/\partial q_l) \\ &\Rightarrow \partial T/\partial q_l = \partial T^*/\partial q_l - \sum (\partial T/\partial \dot{q}_r)\,(\partial \dot{q}_r/\partial q_l), \end{aligned} \tag{b}$$

and, therefore [recalling the symbolic quasi chain rule (5.2.16c)],

$$\begin{aligned} \sum (\partial T/\partial q_l)\,(\partial \dot{q}_l/\partial \omega_k) &= \sum (\partial T^*/\partial q_l)\,(\partial \dot{q}_l/\partial \omega_k) \\ &\quad - \sum\sum (\partial T/\partial \dot{q}_r)\,(\partial \dot{q}_r/\partial q_l)\,(\partial \dot{q}_l/\partial \omega_k) \\ &= \partial T^*/\partial \theta_k - \sum (\partial T/\partial \dot{q}_r)\,(\partial \dot{q}_r/\partial \theta_k); \end{aligned} \tag{c}$$

and

(ii) By $(\ldots)^{\cdot}$-differentiation of the momentum transformation

$$\partial T^*/\partial \omega_k = \sum (\partial T/\partial \dot{q}_l)\,(\partial \dot{q}_l/\partial \omega_k), \tag{d}$$

we obtain

$$(\partial T^*/\partial \omega_k)^{\cdot} = \sum \left[(\partial T/\partial \dot{q}_l)^{\cdot}\,(\partial \dot{q}_l/\partial \omega_k) + (\partial T/\partial \dot{q}_l)\,(\partial \dot{q}_l/\partial \omega_k)^{\cdot}\right],$$

from which, rearranging,

$$\sum \left[(\partial T/\partial \dot{q}_l)^{\cdot}\,(\partial \dot{q}_l/\partial \omega_k)\right] = (\partial T^*/\partial \omega_k)^{\cdot} - \sum \left[(\partial T/\partial \dot{q}_l)\,(\partial \dot{q}_l/\partial \omega_k)^{\cdot}\right]. \tag{e}$$

Subtracting (c) from (e) side by side, we obtain the following fundamental kinematico-inertial identity:

$$\sum \left[(\partial T/\partial \dot{q}_l)^{\cdot} - \partial T/\partial q_l\right](\partial \dot{q}_l/\partial \omega_k)$$
$$= (\partial T^*/\partial \omega_k)^{\cdot} - \partial T^*/\partial \theta_k - \sum \left[(\partial \dot{q}_l/\partial \omega_k)^{\cdot} - \partial \dot{q}_l/\partial \theta_k\right](\partial T/\partial \dot{q}_l),$$

or, compactly, while recalling (5.3.18d),

$$\sum (\partial \dot{q}_l/\partial \omega_k)\, E_l(T) = E_k{}^*(T^*) - \Gamma_k \quad (= I_k); \tag{f}$$

an equation that, as mentioned earlier, shows that although, individually, neither $E_k{}^*(T^*)$ nor Γ_k transform as (covariant) vectors under $\delta q \Leftrightarrow \delta\theta$, taken together as $I_k \equiv E_k{}^*(T^*) - \Gamma_k$ they do; that is, $I_k = \sum (\partial \dot{q}_l/\partial \omega_k) E_l \;\Leftrightarrow\; E_l = \sum (\partial \omega_k/\partial \dot{q}_l) I_k$.

The above allow us to find the transformation properties of Γ_k under a local quasi-velocity change:

$$\omega = \omega(t, q, \omega') \Leftrightarrow \omega' = \omega'(t, q, \omega). \tag{g}$$

We begin with the invariant virtual work of the inertia "forces":

$$\delta I = \sum I_k\, \delta\theta_k = \sum I_{k'}\, \delta\theta_{k'}, \tag{h}$$

where

$$\delta\theta_k = \sum (\partial \omega_k/\partial \omega_{k'})\, \delta\theta_{k'} \;\Leftrightarrow\; \delta\theta_{k'} = \sum (\partial \omega_{k'}/\partial \omega_k)\, \delta\theta_k, \tag{i}$$

$$\omega_k \equiv d\theta_k/dt \qquad \text{and} \qquad \omega_{k'} \equiv d\theta_{k'}/dt, \tag{j}$$

$$I_k = (\partial T^*/\partial \omega_k)^{\cdot} - \partial T^*/\partial \theta_k - \Gamma_k \equiv E_k{}^*(T^*) - \Gamma_k \equiv E_k{}^* - \Gamma_k, \tag{k}$$

$$I_{k'} = (\partial T^*/\partial \omega_{k'})^{\cdot} - \partial T^*/\partial \theta_{k'} - \Gamma_{k'} \equiv E_{k'}{}^*(T^{*\prime}) - \Gamma_{k'} \equiv E_{k'}{}^{*\prime} - \Gamma_{k'}, \tag{l}$$

$$T^* = T^*(t, q, \omega), \qquad T^{*\prime} = T^*(t, q, \omega'). \tag{m}$$

From the above, we readily obtain the (covariant) vector transformation equations

$$I_{k'} = \sum (\partial \omega_k/\partial \omega_{k'}) I_k \;\Leftrightarrow\; I_k = \sum (\partial \omega_{k'}/\partial \omega_k) I_{k'}. \tag{n}$$

Let us find how the constituents of I_k, $E_k{}^*$, and Γ_k, transform individually; that is, how they relate to their accented counterparts.

(a) First, the $E_k{}^*$'s. Applying chain rule to

$$T^* = T^*(t, q, \omega) = T^*[t, q, \omega(t, q, \omega')] = T^{*\prime}(t, q, \omega') \equiv T^{*\prime}, \tag{o}$$

we find

(i) $$\partial T^{*\prime}/\partial\omega_{k'} = \sum (\partial T^*/\partial\omega_k)(\partial\omega_k/\partial\omega_{k'}),$$
$$\Rightarrow (\partial T^{*\prime}/\partial\omega_{k'})^{\cdot} = \sum \left[(\partial T^*/\partial\omega_k)^{\cdot}(\partial\omega_k/\partial\omega_{k'}) + (\partial T^*/\partial\omega_k)(\partial\omega_k/\partial\omega_{k'})^{\cdot}\right]; \tag{p}$$

(ii) $$\begin{aligned}\partial T^{*\prime}/\partial\theta_{k'} &= \sum (\partial T^{*\prime}/\partial q_l)(\partial\dot{q}_l/\partial\omega_{k'})\\ &= \sum \left[\partial T^*/\partial q_l + \sum (\partial T^*/\partial\omega_k)(\partial\omega_k/\partial q_l)\right](\partial\dot{q}_l/\partial\omega_{k'})\\ &= \sum\Big(\sum (\partial T^*/\partial\theta_k)(\partial\omega_k/\partial\dot{q}_l)\Big)(\partial\dot{q}_l/\partial\omega_{k'})\\ &\quad + \sum (\partial T^*/\partial\omega_k)\Big(\sum (\partial\omega_k/\partial q_l)(\partial\dot{q}_l/\partial\omega_{k'})\Big)\\ &= \sum \left[(\partial T^*/\partial\theta_k)(\partial\omega_k/\partial\omega_{k'}) + (\partial T^*/\partial\omega_k)(\partial\omega_k/\partial\theta_{k'})\right].\end{aligned} \tag{q}$$

Subtracting (q) from (p) side by side, we finally obtain

$$\begin{aligned}(\partial T^{*\prime}/\partial\omega_{k'})^{\cdot} - \partial T^{*\prime}/\partial\theta_{k'} &= \sum \left[(\partial T^*/\partial\omega_k)^{\cdot} - \partial T^*/\partial\theta_k\right](\partial\omega_k/\partial\omega_{k'})\\ &\quad + \sum \left[(\partial\omega_k/\partial\omega_{k'})^{\cdot} - \partial\omega_k/\partial\theta_{k'}\right](\partial T^*/\partial\omega_k);\end{aligned}$$

or, compactly,

$$E_{k'}{}^*(T^{*\prime}) = \sum (\partial\omega_k/\partial\omega_{k'})E_k{}^*(T^*) + \sum (\partial T^*/\partial\omega_k)E_{k'}{}^*(\omega_k); \tag{r}$$

which is the general law of transformation of the nonholonomic Euler–Lagrange operator applied to the corresponding kinetic energy (or any other function of t, q, ω').

In particular, if

$$E_{k'}{}^*(\omega_k) \equiv (\partial\omega_k/\partial\omega_{k'})^{\cdot} - \partial\omega_k/\partial\theta_{k'} = 0 \tag{s}$$

(in which case, ω and ω' are called *relatively holonomic*), $E_k{}^*(T^*)$ transforms as a vector.

(b) Next, to the Γ_k's (see also next example). In view of (k, l, n), we have

$$E_{k'}{}^* - \Gamma_{k'} = \sum (\partial\omega_k/\partial\omega_{k'})(E_k{}^* - \Gamma_k), \tag{t}$$

or rearranging, and then using (r),

$$\begin{aligned}\Gamma_{k'} &= \left[E_{k'}{}^* - \sum (\partial\omega_k/\partial\omega_{k'})E_k{}^*\right] + \sum (\partial\omega_k/\partial\omega_{k'})\Gamma_k\\ &= \sum (\partial T^*/\partial\omega_k)E_{k'}{}^*(\omega_k) + \sum (\partial\omega_k/\partial\omega_{k'})\Gamma_k,\end{aligned}$$

or, in extenso,

$$\Gamma_{k'} = \sum (\partial\omega_k/\partial\omega_{k'})\Gamma_k + \sum \left[(\partial\omega_k/\partial\omega_{k'})^{\cdot} - \partial\omega_k/\partial\theta_{k'}\right](\partial T^*/\partial\omega_k). \tag{u}$$

As the above shows, if ω and ω' are relatively holonomic, Γ_k transforms as a (covariant) vector. [In tensor calculus, nonvectorial (nontensorial) quantities like

$E_k{}^*$ and Γ_k are called *geometrical objects*. Other such examples are the Christoffel symbols (§3.10).] In particular, if $\omega_k = \dot{q}_k$—that is, if the ω are holonomic velocities—then $T^* \to T$, $E_k{}^*(T^*) \to E_k(T)$, and $\Gamma_k \to 0$, and so (u) reduces to

$$\Gamma_{k'} = \sum \left[(\partial\dot{q}_k/\partial\omega_{k'})^{\cdot} - \partial\dot{q}_k/\partial\theta_{k'}\right](\partial T/\partial\dot{q}_k)' \equiv \sum V^k{}_{k'}p_k{}', \tag{v}$$

where $V^k{}_{k'}$ = nonlinear Voronets coefficients for $\dot{q} = \dot{q}(t, q, \omega') \Leftrightarrow \omega' = \omega'(t, q, \dot{q})$, and $\partial T/\partial\dot{q}_k \equiv p_k \equiv p_k(t, q, \dot{q}) = p_k[t, q, \dot{q}(t, q, \omega')] = p_k{}'(t, q, \omega') \equiv (\partial T/\partial\dot{q}_k)'$.

Example 5.3.2 *Alternative, Particle Vector–Based Derivation of the Transformation Formula (u). Additional Constraints.* By definition (recalling (5.3.18c)

$$\Gamma_{k'} \equiv S\, dm\, \mathbf{v}^{*\prime} \cdot \left[(\partial\mathbf{v}^{*\prime}/\partial\omega_{k'})^{\cdot} - \partial\mathbf{v}^{*\prime}/\partial\theta_{k'}\right], \tag{a}$$

where

$$\mathbf{v} = \mathbf{v}(t, q, \dot{q}) = \mathbf{v}^*(t, q, \omega) = \mathbf{v}^*[t, q, \omega(t, q, \omega')] = \mathbf{v}^{*\prime}(t, q, \omega') \equiv \mathbf{v}^{*\prime}. \tag{b}$$

But:

(i) $$\boldsymbol{\varepsilon}_{k'} \equiv \partial\mathbf{v}^{*\prime}/\partial\omega_{k'} = \sum(\partial\mathbf{v}^*/\partial\omega_k)(\partial\omega_k/\partial\omega_{k'}) \equiv \sum(\partial\omega_k/\partial\omega_{k'})\boldsymbol{\varepsilon}_k,$$

$$\Rightarrow (\partial\mathbf{v}^{*\prime}/\partial\omega_{k'})^{\cdot} \equiv d\boldsymbol{\varepsilon}_{k'}/dt$$
$$= \sum\left[(\partial\omega_k/\partial\omega_{k'})^{\cdot}\boldsymbol{\varepsilon}_k + (\partial\omega_k/\partial\omega_{k'})(d\boldsymbol{\varepsilon}_k/dt)\right]; \tag{c}$$

and

(ii) $$\partial\mathbf{v}^{*\prime}/\partial\theta_{k'} \equiv \sum(\partial\mathbf{v}^{*\prime}/\partial q_l)(\partial\dot{q}_l/\partial\omega_{k'})$$
$$\equiv \sum\left[\partial\mathbf{v}^*/\partial q_l + \sum(\partial\mathbf{v}^*/\partial\omega_k)(\partial\omega_k/\partial q_l)\right](\partial\dot{q}_l/\partial\omega_{k'})$$
$$= \sum(\partial\mathbf{v}^*/\partial\theta_k)\left(\sum(\partial\omega_k/\partial\dot{q}_l)(\partial\dot{q}_l/\partial\omega_{k'})\right)$$
$$+ \sum(\partial\mathbf{v}^*/\partial\omega_k)\left[(\partial\omega_k/\partial q_l)(\partial\dot{q}_l/\partial\omega_{k'})\right]$$
$$\equiv \sum(\partial\mathbf{v}^*/\partial\theta_k)(\partial\omega_k/\partial\omega_{k'}) + \sum(\partial\mathbf{v}^*/\partial\omega_k)(\partial\omega_k/\partial\theta_{k'}). \tag{d}$$

Inserting the expressions (c, d) in (a), we obtain, successively [recalling that $\mathbf{v}^{*\prime} = \mathbf{v}^*$, $\boldsymbol{\varepsilon}_k = \partial\mathbf{v}^*/\partial\omega_k$, and the definitions of Γ_k and $\partial T^*/\partial\omega_k$],

$$\Gamma_{k'} = \sum(\partial\omega_k/\partial\omega_{k'})\left\{S\, dm\, \mathbf{v}^* \cdot \left[(\partial\mathbf{v}^*/\partial\omega_k)^{\cdot} - \partial\mathbf{v}^*/\partial\theta_k\right]\right\}$$
$$+ \sum\left[(\partial\omega_k/\partial\omega_{k'})^{\cdot} - \partial\omega_k/\partial\theta_{k'}\right]\left(S\, dm\, \mathbf{v}^* \cdot (\partial\mathbf{v}^*/\partial\omega_k)\right),$$

which is none other than eq. (u) of the preceding example.

Additional Constraints

These transformation equations may prove useful if the hitherto independent $n - m$ quasi velocities $\omega_I \equiv (\omega_{m+1}, \ldots, \omega_n)$ are, later, subjected to the m' ($< n - m$) new constraints

$$c_d(t, q, \omega_I) = 0 \qquad (d = 1, \ldots, m'). \tag{e}$$

Then, to incorporate (e) to our description, and following the earlier Johnsen–Hamel approach, we may introduce $n - m$ new quasi velocities $\omega' \equiv (\omega_1', \ldots, \omega_{n-m}')$:

$$\omega_d' \equiv c_d(t, q, \omega_I) = 0, \tag{f1}$$

$$\omega_i' \equiv c_i(t, q, \omega_I) \neq 0 \qquad (i = m' + 1, \ldots, n - m); \tag{f2}$$

where, as in the Pfaffian case (§3.11) the $(n - m) - m'$ $c_i(\ldots)$ are arbitrary, except that when the system (f1, 2) is solved for the $\omega_I, \omega_I = \omega_I(t, q, \omega')$, and these expressions are inserted back into (e), they satisfy them identically in the ω'. In this case, Lagrange's principle yields

$$\sum [(\partial T^*/\partial \omega_I)^{\cdot} - \partial T^*/\partial \theta_I - \Gamma_I - \Theta_I]\, \delta\theta_I = 0, \tag{g}$$

where the $n - m$ $\delta\theta_I \equiv (\delta\theta_{m+1}, \ldots, \delta\theta_n)$ are subjected to the virtual form of the constraints (e, f1):

$$\delta\theta_d' \equiv \sum (\partial c_d/\partial \omega_I)\, \delta\theta_I = 0. \tag{h}$$

From here on, we proceed in well-known ways; that is, either we *adjoin* (h) to (g) via new Lagrangean multipliers ($\rightarrow$ Routh–Voss equations in T^*, ω_I), or we *embed* them via the quasi variables $\delta\theta'/\omega'$ ($\rightarrow$ Maggi equations in $T^*, \omega_I, \partial\omega_I/\partial\omega'$; or Hamel equations in $T^{*\prime} = T^{*\prime}(t, q, \omega'), \omega', \Gamma'$, etc.).

Finally, under $\omega \leftrightarrow \omega'$, Appell's equations (say, under no constraints) become

$$\sum (\partial S^*/\partial \dot{\omega}_k)\, (\partial \dot{\omega}_k/\partial \dot{\omega}_{k'}) = \sum (\partial \dot{\omega}_k/\partial \dot{\omega}_{k'}) \Theta_k; \tag{i}$$

that is,

$$\partial S^{*\prime}/\partial \dot{\omega}_{k'} = \Theta_{k'}, \tag{j}$$

where

$$\begin{aligned} S^* = S^*(t, q, \omega, \dot{\omega}) &= S^*\left[t, q, \omega(t, q, \omega'), \dot{\omega}(t, q, \omega', \dot{\omega}')\right] \\ &= S^{*\prime}(t, q, \omega', \dot{\omega}') = S^{*\prime}; \end{aligned} \tag{k}$$

also,

$$\partial \dot{\omega}_k/\partial \dot{\omega}_{k'} = \partial \omega_k/\partial \omega_{k'}, \qquad \partial \dot{\omega}_{k'}/\partial \dot{\omega}_k = \partial \omega_{k'}/\partial \omega_k.$$

Example 5.3.3 *Special Forms of the Equations of Motion: Nonlinear Equations of Hadamard.* Let us specialize the nonlinear Maggi equations to the following quasi-variable choice (recall ex. 5.2.2):

$$\omega_D \equiv f_D(t, q, \dot{q}) = \dot{q}_D - \phi_D(t, q, \dot{q}_I) = 0, \tag{a}$$

$$\omega_I \equiv f_I(t, q, \dot{q}) = \dot{q}_I \neq 0; \tag{b}$$

and its inverse

$$\dot{q}_D = \omega_D + \phi_D(t, q, \dot{q}_I) = \omega_D + \phi_D(t, q, \omega_I), \tag{c}$$

$$\dot{q}_I = \omega_I. \tag{d}$$

With the notation $E_k(T) - Q_k \equiv E_k - Q_k = \partial S/\partial \ddot{q}_k - Q_k \equiv M_k$, and (c, d), we obtain, successively,

$$\begin{aligned}
\sum (\partial \dot{q}_l/\partial \omega_k) M_l &= \sum (\partial \dot{q}_D/\partial \omega_k) M_D + \sum (\partial \dot{q}_I/\partial \omega_k) M_I \\
&= \sum (\partial \dot{q}_D/\partial \omega_{D'}) M_D + \sum (\partial \dot{q}_I/\partial \omega_{D'}) M_I \\
&= \sum (\delta_{DD'}) M_D + \sum (0) M_I = M_{D'} \qquad (D' = 1, \ldots, m); \qquad \text{(e1)} \\
&= \sum (\partial \dot{q}_D/\partial \omega_I) M_D + \sum (\partial \dot{q}_I/\partial \omega_{I'}) M_I \\
&= \sum (\partial \phi_D/\partial \omega_I) M_D + \sum (\delta_{II'}) M_I \\
&= M_I + \sum (\partial \phi_D/\partial \dot{q}_I) M_D \qquad (I = m+1, \ldots, n). \quad \text{(e2)}
\end{aligned}$$

As a result of the above, Maggi's equations (5.3.16b, c) reduce to the *nonlinear Hadamard equations.*

Kinetostatic:

$$\begin{aligned}
E_D(T) &\equiv \left[(\partial T/\partial \dot{q}_D)^{\cdot} - \partial T/\partial q_D\right] = \partial S/\partial \ddot{q}_D \\
&= Q_D + \lambda_D, \qquad \text{(f1)}
\end{aligned}$$

Kinetic:

$$\begin{aligned}
&E_I(T) + \sum (\partial \phi_D/\partial \dot{q}_I) E_D(T) \\
&\quad \equiv \left[(\partial T/\partial \dot{q}_I)^{\cdot} - \partial T/\partial q_I\right] + \sum (\partial \phi_D/\partial \dot{q}_I) \left[(\partial T/\partial \dot{q}_D)^{\cdot} - \partial T/\partial q_D\right] \\
&\quad = \partial S/\partial \ddot{q}_I + \sum (\partial \phi_D/\partial \dot{q}_I)(\partial S/\partial \ddot{q}_D) \\
&\quad = Q_I + \sum (\partial \phi_D/\partial \dot{q}_I) Q_D. \qquad \text{(f2)}
\end{aligned}$$

Equations (f2), plus the constraints (a), yield the motion; then (f1) give the constraint reactions. [In terms of the constrained Appellian S_o, eqs. (f2) state simply that

$$\partial S_o/\partial \ddot{q}_I = Q_I + \sum (\partial \phi_D/\partial \dot{q}_I) Q_D \quad (\equiv Q_{I,o} \equiv Q_{Io}), \qquad \text{(g)}$$

where

$$S = S(t, q, \dot{q}, \ddot{q}) = \cdots = S_o(t, q, \dot{q}, \ddot{q}) = S_o.]$$

Example 5.3.4 *Special Forms of the Equations of Motion: Nonlinear Equations of Chaplygin and Voronets.* Continuing from the preceding example, let us derive the specialization of the Johnsen–Hamel equations under (a–d) from that example.

First Method

Here, and recalling the notations and results of §3.8, and ex. 5.2.2 and ex. 5.2.3, we have

$$\omega_I \to \dot{q}_I, \quad \theta_I \to (q_I), \quad T^* \to T_o = T_o(t, q, \dot{q}_I)$$

(i.e., no kinetostatic equations, only kinetic);

and

$$\partial T^*/\partial\omega_I \to \partial T_o/\partial\dot{q}_I, \tag{a}$$

$$\partial T^*/\partial\theta_I \to \partial T_o/\partial(q_I) \equiv \partial T_o/\partial q_I + \sum (\partial\phi_D/\partial\dot{q}_I)(\partial T_o/\partial q_D); \tag{b}$$

$$V^{I'}{}_I \to V^{I'}{}_{I,o} \equiv W^{I'}{}_I = 0 \qquad \text{(Suslov viewpoint)}, \tag{c}$$

$$\begin{aligned} V^D{}_I \to V^D{}_{I,o} \equiv W^D{}_I &= (\partial\phi_D/\partial\dot{q}_I)^{\cdot} - \partial\phi_D/\partial q_I - \sum (\partial\phi_D/\partial q_{D'})(\partial\phi_{D'}/\partial\dot{q}_I) \\ &\equiv E_I(\phi_D) - \sum (\partial\phi_D/\partial q_{D'})(\partial\phi_{D'}/\partial\dot{q}_I) \\ &\equiv (\partial\phi_D/\partial\dot{q}_I)^{\cdot} - \partial\phi_D/\partial(q_I) \\ &\equiv E_{(I)}(\phi_D), \end{aligned} \tag{d}$$

$$\Gamma_I \to \Gamma_{I,o} \equiv W_I = \sum W^D{}_I (\partial T/\partial\dot{q}_D)_o \equiv \sum W^D{}_I p_{D,o}, \tag{e}$$

$$\begin{aligned} \delta' W \to (\delta' W)_o &= \Big(\sum Q_k\, \delta q_k\Big)_o \\ &= \cdots = \sum \Big(Q_I + \sum (\partial\phi_D/\partial\dot{q}_I) Q_D\Big)\, \delta q_I \equiv \sum Q_{Io}\, \delta q_I. \end{aligned} \tag{f}$$

{For a general function $f = f(t, q, \dot{q}) = f[t, q, \dot{q}_D = \phi_D(t, q, \dot{q}_I), \dot{q}_I] = f_o(t, q, \dot{q}_I) = f_o$, we notice the *difference* between the ordinary *chain rule*:

$$\partial f_o/\partial q_k = \partial f/\partial q_k + \sum (\partial f/\partial\dot{q}_D)(\partial\phi_D/\partial q_k),$$

and the *quasi chain rule specialization* (i.e., notation—recall (2.11.15a ff.)):

$$\partial f_o/\partial(q_I) \equiv \partial f_o/\partial q_I + \sum (\partial f_o/\partial q_D)(\partial\phi_D/\partial\dot{q}_I).\}$$

As a result of the above, eqs. (5.3.18a–20) yield what should, legitimately, be called the *nonlinear Voronets equations*:

$$\begin{aligned} &\Big\{(\partial T_o/\partial\dot{q}_I)^{\cdot} - \Big[\partial T_o/\partial q_I + \sum (\partial\phi_D/\partial\dot{q}_I)(\partial T_o/\partial q_D)\Big]\Big\} - \sum W^D{}_I (\partial T/\partial\dot{q}_D)_o \\ &\quad \equiv \big[(\partial T_o/\partial\dot{q}_I)^{\cdot} - \partial T_o/\partial(q_I)\big] - \sum W^D{}_I (\partial T/\partial\dot{q}_D)_o \\ &\quad \equiv E_I(T_o) - \sum (\partial\phi_D/\partial\dot{q}_I)(\partial T_o/\partial q_D) - W_I \\ &\quad \equiv E_{(I)}(T_o) - W_I \\ &\quad = Q_{Io}. \end{aligned} \tag{g}$$

In the Chaplygin case, $\dot{q}_D = \dot{q}_D(q_I, \dot{q}_I) \equiv \phi_D(q_I, \dot{q}_I)$ and $T_o = T_o(q_I, \dot{q}_I)$, and so

$$\begin{aligned} &\partial\phi_D/\partial q_{D'} = 0, \qquad \partial T_o/\partial q_D = 0, \\ &\quad \Rightarrow \partial\phi_D/\partial(q_I) = \partial\phi_D/\partial q_I, \\ &\quad \Rightarrow W^D{}_I \equiv E_{(I)}(\phi_D) \to E_I(\phi_D) \equiv (\partial\phi_D/\partial\dot{q}_I)^{\cdot} - \partial\phi_D/\partial q_I \equiv T^D{}_I, \\ &\quad \Rightarrow \Gamma_{I,o} \equiv W_I \to T_I \equiv \sum T^D{}_I (\partial T/\partial\dot{q}_D)_o, \end{aligned} \tag{h}$$

and, accordingly, (g) reduces to what we will be calling the *nonlinear Chaplygin equations*:

$$(\partial T_o/\partial \dot{q}_I)^{\cdot} - \partial T_o/\partial q_I - \sum T^D{}_I\,(\partial T/\partial \dot{q}_D)_o$$
$$\equiv (\partial T_o/\partial \dot{q}_I)^{\cdot} - \partial T_o/\partial q_I - T_I = Q_{Io}. \tag{i}$$

It is not hard to see that in the Pfaffian case, eqs. (g) and (i) reduce, respectively, to the original Voronets and Chaplygin forms (§3.8).

Second Method (By Direct Differentiation)

Here, we have

$$\begin{aligned}\partial T^*/\partial\omega_I &= \sum (\partial T/\partial \dot{q}_k)\,(\partial \dot{q}_k/\partial\omega_I)\\ &= \sum (\partial T/\partial \dot{q}_D)\,(\partial \dot{q}_D/\partial\omega_I) + \sum (\partial T/\partial \dot{q}_{I'})\,(\partial \dot{q}_{I'}/\partial\omega_I)\\ &= \sum (\partial T/\partial \dot{q}_D)\,(\partial \dot{q}_D/\partial \dot{q}_I) + \sum (\partial T/\partial \dot{q}_{I'})\,(\delta_{I'I})\\ &= \partial T/\partial \dot{q}_I + \sum (\partial \dot{q}_D/\partial \dot{q}_I)\,(\partial T/\partial \dot{q}_D) = \partial T_o/\partial \dot{q}_I,\end{aligned} \tag{j}$$

$$\begin{aligned}\partial T^*/\partial\omega_D &\equiv \sum (\partial T/\partial \dot{q}_k)\,(\partial \dot{q}_k/\partial\omega_D)\\ &= \sum (\partial T/\partial \dot{q}_{D'})\,(\partial \dot{q}_{D'}/\partial\omega_D) + \sum (\partial T/\partial \dot{q}_I)\,(\partial \dot{q}_I/\partial\omega_D)\\ &= \sum (\partial T/\partial \dot{q}_{D'})\,(\delta_{D'D}) + \sum (\partial T/\partial \dot{q}_I)\,(0)\\ &= \partial T/\partial \dot{q}_D \;\rightarrow\; (\partial T/\partial \dot{q}_D)^* \rightarrow (\partial T/\partial \dot{q}_D)_o,\end{aligned} \tag{k}$$

$$\begin{aligned}\partial T^*/\partial\theta_I &\equiv \sum (\partial T^*/\partial q_k)\,(\partial \dot{q}_k/\partial\omega_I)\\ &= \sum (\partial T^*/\partial q_D)\,(\partial \dot{q}_D/\partial\omega_I) + \sum (\partial T^*/\partial q_{I'})\,(\partial \dot{q}_{I'}/\partial\omega_I)\\ &= \sum (\partial T^*/\partial q_D)\,(\partial \phi_D/\partial\omega_I) + \sum (\partial T^*/\partial q_{I'})\,(\delta_{I'I})\\ &= \sum (\partial T_o/\partial q_D)\,(\partial \phi_D/\partial \dot{q}_I) + \partial T_o/\partial q_I\\ &\equiv \partial T_o/\partial (q_I),\end{aligned} \tag{l}$$

$$\begin{aligned}H^D{}_I &= \sum \big[(\partial\omega_D/\partial \dot{q}_k)^{\cdot} - \partial\omega_D/\partial q_k\big]\,(\partial \dot{q}_k/\partial\omega_I) \qquad \text{[by (5.2.21h)]}\\ &= \sum \big[(\partial\omega_D/\partial \dot{q}_{D'})^{\cdot} - \partial\omega_D/\partial q_{D'}\big]\,(\partial \dot{q}_{D'}/\partial\omega_I)\\ &\quad + \sum \big[(\partial\omega_D/\partial \dot{q}_{I'})^{\cdot} - \partial\omega_D/\partial q_{I'}\big]\,(\partial \dot{q}_{I'}/\partial\omega_I)\\ &= \sum \big[(\delta_{DD'})^{\cdot} - (-\partial\phi_D/\partial q_{D'})\big]\,(\partial\phi_{D'}/\partial\omega_I)\\ &\quad + \sum \big[(-\partial\phi_D/\partial \dot{q}_{I'})^{\cdot} - (-\partial\phi_D/\partial q_{I'})\big]\,(\delta_{I'I})\\ &= -\Big\{(\partial\phi_D/\partial \dot{q}_I)^{\cdot} - \Big[\partial\phi_D/\partial q_I + \sum (\partial\phi_D/\partial\phi_{D'})\,(\partial\phi_{D'}/\partial \dot{q}_I)\Big]\Big\}\\ &= -\big[(\partial\phi_D/\partial \dot{q}_I)^{\cdot} - \partial\phi_D/\partial (q_I)\big]\\ &\equiv -E_{(I)}(\dot{q}_D) = -W^D{}_I,\end{aligned} \tag{m}$$

$$V^k{}_I = (\partial \dot{q}_k/\partial\omega_I)^{\cdot} - \sum (\partial \dot{q}_k/\partial q_l)\,(\partial \dot{q}_l/\partial\omega_I) \qquad \text{[by (5.2.21e)]}; \tag{n}$$

or, proceeding directly from (5.2.21f),

$$\begin{aligned} H^D{}_I &= -\sum (\partial\omega_D/\partial\dot{q}_k)V^k{}_I \\ &= -\sum (\partial\omega_D/\partial\dot{q}_{D'})V^{D'}{}_I - \sum (\partial\omega_D/\partial\dot{q}_{I'})V^{I'}{}_I \\ &= -\sum (\delta_{DD'})V^{D'}{}_I - \sum (-\partial\phi_D/\partial\dot{q}_{I'})\,(0) \\ &= -V^D{}_I \longrightarrow -W^D{}_I \end{aligned} \tag{o}$$

[Either from the Suslov viewpoint, or by direct application of (5.2.21f) to our special case, we easily find $V^{I'}{}_I \to W^{I'}{}_I = 0$, and, therefore,

$$H^{I'}{}_I = -\sum (\partial\omega_{I'}/\partial\dot{q}_k)V^k{}_I = \cdots = 0],$$

$$\begin{aligned} \Theta_I &= \sum (\partial\dot{q}_k/\partial\omega_I)Q_k \\ &= \sum (\partial\dot{q}_D/\partial\omega_I)Q_D + \sum (\partial\dot{q}_{I'}/\partial\omega_I)Q_{I'} \\ &= \sum (\partial\phi_D/\partial\omega_I)Q_D + \sum (\delta_{I'I})Q_{I'} \\ &= Q_I + \sum (\partial\phi_D/\partial\dot{q}_I)Q_D \equiv Q_{Io}. \end{aligned} \tag{p}$$

Substituting all these special results into (5.3.19), we recover (g), as expected.

Problem 5.3.1 (i) Using the definitions $\Gamma_k = \sum V^l{}_k p_l = -\sum H^b{}_k P_b$, and (5.2.21e–g) in the Γ transformation equation (exs. 5.3.1 and 5.3.2), show that under $\omega(t, q, \omega') \Leftrightarrow \omega'(t, q, \omega)$ the nonlinear Voronets and Hamel coefficients $V^l{}_k$ and $H^b{}_k$ transform as

$$\begin{aligned} V^l{}_{k'} &= \sum (\partial\omega_k/\partial\omega_{k'})V^l{}_k \\ &\quad + \sum \left[(\partial\omega_k/\partial\omega_{k'})^{\cdot} - \partial\omega_k/\partial\theta_{k'}\right](\partial\dot{q}_l/\partial\omega_k), \end{aligned} \tag{a}$$

$$\begin{aligned} H^{b'}{}_{k'} &= \sum (\partial\omega_k/\partial\omega_{k'})\,(\partial\omega_{b'}/\partial\omega_b)H^b{}_k \\ &\quad - \sum \left[(\partial\omega_k/\partial\omega_{k'})^{\cdot} - \partial\omega_k/\partial\theta_{k'}\right](\partial\omega_{b'}/\partial\omega_k); \end{aligned} \tag{b}$$

that is, in general, *neither* $V^l{}_k$ nor $H^b{}_k$ *transform as vectors, tensors.*

(ii) Then show that the new (transformed) Voronets and Hamel symbols are related to each other as are the old ones; that is,

$$H^{b'}{}_{k'} = -\sum (\partial\omega_{b'}/\partial\dot{q}_l)V^l{}_{k'} \;\Leftrightarrow\; V^l{}_{k'} = -\sum (\partial\dot{q}_l/\partial\omega_{b'})H^{b'}{}_{k'}, \tag{c}$$

where $\omega_{r'} = \omega_{r'}(t, q, \dot{q}) \Leftrightarrow \dot{q}_l = \dot{q}_l(t, q, \omega')$.

For alternative, equivalent, expressions to (a, b), see Novoselov (1979, pp. 120–121).

Example 5.3.5 (Mei, 1985, pp. 89–90). Let us obtain the *Routh–Voss* equations of motion of a particle P of mass m moving under the constraint

$$(\dot{q}_1)^2 + (\dot{q}_2)^2 + (\dot{q}_3)^2 = constant \equiv c^2; \tag{a}$$

that is, square of velocity $\boldsymbol{v}$ of P = constant, where $q_{1,2,3} = x, y, z$: rectangular Cartesian coordinates of P. Here, with the usual notations,

$$2T = m[(\dot{q}_1)^2 + (\dot{q}_2)^2 + (\dot{q}_3)^2], \tag{b}$$

$$f \equiv (\dot{q}_1)^2 + (\dot{q}_2)^2 + (\dot{q}_3)^2 - c^2 = 0$$

$$\Rightarrow \partial f/\partial \dot{q}_k = 2\dot{q}_k \qquad (k = 1, 2, 3); \tag{c}$$

and, therefore, the nonlinear Routh–Voss equations, under impressed forces Q_k, are

$$m\,\ddot{q}_k = Q_k + 2\lambda\,\dot{q}_k, \tag{d}$$

and along with (a) they constitute a determinate system for the $q_k(t)$ and $\lambda(t)$.

To eliminate the multiplier λ, we multiply each of (d) by its $\dot{q}_k$ and add them together, thus obtaining the *power* equation

$$\begin{aligned} m(\dot{q}_1\ddot{q}_1 + \dot{q}_2\ddot{q}_2 + \dot{q}_3\ddot{q}_3) \\ &= Q_1\dot{q}_1 + Q_2\dot{q}_2 + Q_3\dot{q}_3 + 2\lambda\,[(\dot{q}_1)^2 + (\dot{q}_2)^2 + (\dot{q}_3)^2] \\ &= Q_1\dot{q}_1 + Q_2\dot{q}_2 + Q_3\dot{q}_3 + 2\lambda c^2, \end{aligned} \tag{e}$$

from which, invoking the constraint (c) $f = 0$ and its $(\ldots)^{\cdot}$-derivative

$$\dot{f} = 2\,(\dot{q}_1\ddot{q}_1 + \dot{q}_2\ddot{q}_2 + \dot{q}_3\ddot{q}_3) = 0,$$

we readily get the multiplier

$$\lambda = -(Q_1\dot{q}_1 + Q_2\dot{q}_2 + Q_3\dot{q}_3)\big/2c^2. \tag{f}$$

Finally, substituting λ from (f) into (d), we obtain the purely *kinetic* ("Jacobi–Synge" type of) equations

$$m\ddot{q}_k = Q_k - \left[(Q_1\dot{q}_1 + Q_2\dot{q}_2 + Q_3\dot{q}_3)/c^2\right]\dot{q}_k. \tag{g}$$

[Recall examples 3.2.6, 3.5.5, and 3.10.2. The general methodology for obtaining such reactionless equations seems to have originated with Jacobi [1842–1843, publ. 1866; p. 51 ff. (esp. p. 55) and p. 132 ff.]; while a more general, tensor calculus-based approach is due to Synge (1926–1927, pp. 53–55).]

Let us examine this problem from the elementary Newton–Euler viewpoint. In view of (a), $v = c$, and so the intrinsic equations of motion of P (§1.2):

$$mv^2/\rho = F_n + R_n, \qquad m\dot{v} = F_t + R_t \tag{h}$$

(ρ: radius of curvature of trajectory of $P, F_{n,t}/R_{n,t}$: normal and tangential components of total impressed/reaction force on P), reduce to

$$\begin{aligned} mc^2/\rho &= F_n + R_n, \\ 0 &= F_t + R_t \\ \Rightarrow R_t &= -F_t = -\boldsymbol{F}\cdot(\boldsymbol{v}/v) = -(\boldsymbol{F}\cdot\boldsymbol{v})/c \\ &= -\left(\sum Q_k\dot{q}_k\right)\Big/c = 2c\lambda. \end{aligned} \tag{i}$$

For additional details, see Hamel (1949, pp. 709–710).

Example 5.3.6 (Mei, 1985, pp. 91–93). Let us obtain the *Routh–Voss* equations of motion of a particle P of mass m moving in a uniform gravitational field and subject to the Appell–Hamel constraint

$$(\dot{x})^2 + (\dot{y})^2 = (a/b)^2(\dot{z})^2 \qquad (a, b\text{: given; say, positive constants}) \tag{a}$$

where, x, y, z: rectangular Cartesian coordinates of P.
Since here

$$2T = m[(\dot{x})^2 + (\dot{y})^2 + (\dot{z})^2],$$
$$Q_x = 0, \qquad Q_y = 0, \qquad Q_z = -mg \qquad (\text{with } +z \text{ vertically upward}), \tag{b}$$

and with

$$f \equiv (\dot{x})^2 + (\dot{y})^2 - (a/b)^2(\dot{z})^2 = 0, \tag{c}$$

the Routh–Voss equations are

$$m\ddot{x} = 2\lambda\dot{x}, \qquad m\ddot{y} = 2\lambda\dot{y}, \qquad m\ddot{z} = -mg - 2\lambda(a/b)^2\dot{z}; \tag{d}$$

and with (a) they constitute a determinate system for $x(t), y(t), z(t), \lambda(t)$. To obtain reactionless "Jacobi–Synge" equations [like (g) of the preceding example], we $(\ldots)^{\cdot}$-differentiate the constraint (a):

$$\ddot{z} = (b/a)^2[(\dot{x}\ddot{x} + \dot{y}\ddot{y})/\dot{z}], \tag{e}$$

and then substitute it into the third of (d), while using (a) rewritten as

$$\dot{z} = (b/a)\,[(\dot{x})^2 + (\dot{y})^2]^{1/2} \qquad [\dot{z}, a, b > 0], \tag{f}$$

that is, the ratio of vertical velocity to horizontal velocity equals b/a. The result is

$$\begin{aligned} 2\lambda = &-m(b/a)^2(\dot{x}\ddot{x} + \dot{y}\ddot{y})\big/[(\dot{x})^2 + (\dot{y})^2] \\ &- mg(b/a)\big/[(\dot{x})^2 + (\dot{y})^2]^{1/2}, \end{aligned} \tag{g}$$

and when this is substituted back into the first two of (d), it yields the reactionless equations

$$\begin{aligned} &\ddot{x} + (b/a)^2(\dot{x}\ddot{x} + \dot{y}\ddot{y})\dot{x}\big/[(\dot{x})^2 + (\dot{y})^2] \\ &\qquad = -g(b/a)\dot{x}\big/[(\dot{x})^2 + (\dot{y})^2]^{1/2}, \end{aligned} \tag{h}$$

$$\begin{aligned} &\ddot{y} + (b/a)^2(\dot{x}\ddot{x} + \dot{y}\ddot{y})\dot{y}\big/[(\dot{x})^2 + (\dot{y})^2] \\ &\qquad = -g(b/a)\dot{y}\big/[(\dot{x})^2 + (\dot{y})^2]^{1/2}. \end{aligned} \tag{i}$$

From these two equations, we readily obtain the equivalent, but simpler, system

$$\begin{aligned} &\dot{y}\ddot{x} - \dot{x}\ddot{y} = 0, \\ &\dot{x}\ddot{x} + \dot{y}\ddot{y} = -(g\,a\,b)\,(a^2 + b^2)^{-1}\,[(\dot{x})^2 + (\dot{y})^2]^{1/2}. \end{aligned} \tag{j}$$

The *first* of the above, assuming $\dot{y} \neq 0$, can be rewritten as $(\dot{x}/\dot{y})^{\cdot} = 0$, and integrates immediately to $\dot{y} = c\dot{x}$ (c: integration constant), or further to

$$y - y_o = c(x - x_o) \qquad [x_o = x(0),\ y_o = y(0)]; \tag{k}$$

while the *second*, with the help of the auxiliary variable, $v^2 = (\dot{x})^2 + (\dot{y})^2$, can be rewritten as $\dot{v} = -(g\,a\,b)\,(a^2 + b^2)^{-1}$, and integrates readily to

$$v - v_o = -[(g\,a\,b)\,(a^2 + b^2)^{-1}]\,t \qquad [v_o = v(0)]. \tag{l}$$

In view of this result, the constraint (f) becomes (assuming $\dot{z} > 0$)

$$\dot{z} = (b/a)v = (b/a)v_o - [(g\,b^2)\,(a^2 + b^2)^{-1}]t,$$

and, upon integrating, yields

$$z = z_o + (b/a)v_o\,t - [(g\,b^2)/2\,(a^2 + b^2)]t^2. \tag{m}$$

Finally, with the help of the earlier integral $\dot{y} = c\dot{x}$, v becomes

$$v = [(\dot{x})^2 + (\dot{y})^2]^{1/2} = \dot{x}(1 + c^2)^{1/2} = (\dot{y}/c)\,(1 + c^2)^{1/2}, \tag{n}$$

and so (l) transforms to the following equivalent x, y-equations:

$$(1 + c^2)^{1/2}\dot{x} = v_o - [(g\,a\,b)/(a^2 + b^2)]t, \tag{o}$$

$$(1 + c^2)^{1/2}(\dot{y}/c) = v_o - [(g\,a\,b)/(a^2 + b^2)]t, \tag{p}$$

from which, integrating, we get

$$(1 + c^2)^{1/2}(x - x_o) = v_o\,t - [(g\,a\,b)/2(a^2 + b^2)]t^2, \tag{q}$$

$$c^{-1}(1 + c^2)^{1/2}(y - y_o) = v_o\,t - [(g\,a\,b)/2(a^2 + b^2)]t^2. \tag{r}$$

Comparing the above with (m), we see that we can rewrite all three of them as

$$(1 + c^2)^{1/2}(x - x_o) = c^{-1}(1 + c^2)^{1/2}(y - y_o) = (a/b)(z - z_o), \tag{s}$$

where c can be found from $c = \dot{y}_o/\dot{x}_o$.

Substituting from the above into (g), we can find the constraint reaction $\lambda = \lambda(t;\ v_o, m, g, a, b)$, if needed.

Example 5.3.7 (Mei, 1985, pp. 245–246). Let us obtain the *kinetic* Appellian equations of a particle P of mass m moving under the action of impressed forces Q_k $(k = 1, 2, 3)$ and subject to the Appell–Hamel constraint

$$(\dot{q}_1)^2 + (\dot{q}_2)^2 = (\dot{q}_3)^2 \qquad (q_{1,2,3}\text{: rectangular Cartesian coordinates of } P); \tag{a}$$

that is, vertical velocity equals $(\pm)$ of horizontal velocity.

In view of the constraint (a), we introduce the following quasi velocities:

$$2\omega_1 \equiv [(\dot{q}_3)^2 - (\dot{q}_1)^2 - (\dot{q}_2)^2] = 0, \tag{b}$$

$$\omega_2 \equiv \arctan(\dot{q}_2/\dot{q}_1) \neq 0, \tag{c}$$

$$2\omega_3 \equiv [(\dot{q}_1)^2 + (\dot{q}_2)^2 + (\dot{q}_3)^2] \qquad (= 2T/m \neq 0). \tag{d}$$

Adding and subtracting (b) and (d) side by side we obtain, respectively,

$$(\dot{q}_3)^2 = \omega_1 + \omega_3 \Rightarrow \dot{q}_3 = (\omega_1 + \omega_3)^{1/2}, \tag{e}$$

$$(\dot{q}_1)^2 + (\dot{q}_2)^2 = \omega_3 - \omega_1; \tag{f}$$

and from these and (c), we are readily led to the *inverse* of (b–d):

$$\dot{q}_1 = (\omega_3 - \omega_1)^{1/2} \cos\omega_3, \qquad \dot{q}_2 = (\omega_3 - \omega_1)^{1/2} \sin\omega_3,$$

$$\dot{q}_3 = (\omega_1 + \omega_3)^{1/2}. \tag{g}$$

Now, since we are interested only in the kinetic Appellian equations, we *can* enforce the constraint (b) right at this point; that is, we can work with (g) and its $(\ldots)^{\cdot}$-derivatives evaluated at $\omega_1 = 0$; that is (skipping special notations, such as $(\ldots)_o$, for simplicity),

$$\dot{q}_1 = (\omega_3)^{1/2} \cos\omega_2, \qquad \dot{q}_2 = (\omega_3)^{1/2} \sin\omega_2, \qquad \dot{q}_3 = (\omega_3)^{1/2}; \tag{h}$$

$$\ddot{q}_1 = [\dot{\omega}_3/2\,(\omega_3)^{1/2}] \cos\omega_2 - (\omega_3)^{1/2}\dot{\omega}_2 \sin\omega_2,$$

$$\ddot{q}_2 = [\dot{\omega}_3/2\,(\omega_3)^{1/2}] \sin\omega_2 + (\omega_3)^{1/2}\dot{\omega}_2 \cos\omega_2,$$

$$\ddot{q}_3 = \dot{\omega}_3/2\,(\omega_3)^{1/2}. \tag{i}$$

Hence, the *constrained* Appellian, S^* for $\omega_1 = 0, \dot{\omega}_1 = 0$ [denoted for convenience by S^*, instead of a more precise notation, such as $S^*{}_o$] equals:

$$2S/m = [(\ddot{q}_1)^2 + (\ddot{q}_2)^2 + (\ddot{q}_3)^2]$$

$$= \cdots = (\dot{\omega}_3)^2/2\omega_3 + \omega_3(\dot{\omega}_2)^2 = S^*(\omega_3, \dot{\omega}_2, \dot{\omega}_3), \tag{j}$$

and therefore the (constrained) nonholonomic kinetic inertia “forces” are

$$I_2 \equiv \partial S^*/\partial\dot{\omega}_2 = m\,\omega_3\,\dot{\omega}_2, \qquad I_3 \equiv \partial S^*/\partial\dot{\omega}_3 = m\,\dot{\omega}_3/2\omega_3. \tag{k}$$

REMARK

However, as the above and the chain rule show, we could have stopped at the first line of (j) and *not* completed the squares. Indeed, we have

$$
\begin{aligned}
\partial S^*/\partial \dot{\omega}_2 &= (\partial S/\partial \ddot{q}_1)(\partial \ddot{q}_1/\partial \dot{\omega}_2) + (\partial S/\partial \ddot{q}_2)(\partial \ddot{q}_2/\partial \dot{\omega}_2) \\
&\quad + (\partial S/\partial \ddot{q}_3)(\partial \ddot{q}_3/\partial \dot{\omega}_2) \\
&= (\partial S/\partial \ddot{q}_1)(\partial \dot{q}_1/\partial \omega_2) + (\partial S/\partial \ddot{q}_2)(\partial \dot{q}_2/\partial \omega_2) \\
&\quad + (\partial S/\partial \ddot{q}_3)(\partial \dot{q}_3/\partial \omega_2) \\
&= (m\ddot{q}_1)[-(\omega_3)^{1/2}\sin\omega_2] + (m\ddot{q}_2)[(\omega_3)^{1/2}\cos\omega_2] \\
&\quad + (m\ddot{q}_3)\,(0) \\
&= \cdots = m\,\omega_3\,\dot{\omega}_2 \qquad [\text{using (i)}], \qquad\qquad (\text{l})
\end{aligned}
$$

$$
\begin{aligned}
\partial S^*/\partial \dot{\omega}_3 &= (\partial S/\partial \ddot{q}_1)(\partial \dot{q}_1/\partial \omega_3) + (\partial S/\partial \ddot{q}_2)(\partial \dot{q}_2/\partial \omega_3) \\
&\quad + (\partial S/\partial \ddot{q}_3)(\partial \dot{q}_3/\partial \omega_3) \\
&= (m\ddot{q}_1)[\cos\omega_2/2(\omega_3)^{1/2}] + (m\ddot{q}_2)[\sin\omega_2/2(\omega_3)^{1/2}] \\
&\quad + (m\ddot{q}_3)[1/2(\omega_3)^{1/2}] \\
&= \cdots = m\,\dot{\omega}_3/2\omega_3, \qquad\qquad (\text{m})
\end{aligned}
$$

as before.

From the above, it follows that the kinetic Appellian equations are

$$I_2 = \sum (\partial \dot{q}_k/\partial \omega_2) Q_k \quad (= \Theta_2), \qquad I_3 = \sum (\partial \dot{q}_k/\partial \omega_3) Q_k \quad (= \Theta_3); \qquad (\text{n})$$

or, explicitly,

$$m\,\omega_3\,\dot{\omega}_2 = [-(\omega_3)^{1/2}\sin\omega_2]Q_1 + [(\omega_3)^{1/2}\cos\omega_2]Q_2, \qquad (\text{o})$$

and

$$m\,\dot{\omega}_3/2\omega_3 = [\cos\omega_2/2(\omega_3)^{1/2}]Q_1 + [\sin\omega_2/2(\omega_3)^{1/2}]Q_2 + [1/2(\omega_3)^{1/2}]Q_3, \qquad (\text{p})$$

or, equivalently,

$$2m(\sqrt{\omega_3})^{\cdot} = (\cos\omega_2)Q_1 + (\sin\omega_2)Q_2 + Q_3. \qquad (\text{q})$$

In the special case of a uniform gravitational field—that is,

$$Q_1 = 0, \qquad Q_2 = 0, \qquad Q_3 = -mg, \qquad (\text{r})$$

eqs. (o) and (q) specialize, respectively, to

$$m\,\omega_3\,\dot{\omega}_2 = 0 \quad \text{and} \quad 2m(\sqrt{\omega_3})^{\cdot} = -mg, \qquad (\text{s})$$

and have the obvious integrals

$$\omega_2 = \omega_{2o} \quad \text{and} \quad \sqrt{\omega_3} = \sqrt{\omega_{3o}} - gt/2 \qquad [\omega_{2o} = \omega_2(0), \quad \omega_{3o} = \omega_3(0)]. \qquad (\text{t})$$

In the q-variables, the above become [recalling (h)]

$$
\begin{aligned}
&\dot{q}_1 = (-gt/2 + \sqrt{\omega_{3o}})\cos\omega_{2o}, \qquad \dot{q}_2 = (-gt/2 + \sqrt{\omega_{3o}})\sin\omega_{2o}, \\
&\dot{q}_3 = -gt/2 + \sqrt{\omega_{3o}}, \qquad\qquad (\text{u})
\end{aligned}
$$

and integrate readily to

$$q_1 - q_{1o} = (-gt^2/4 + \sqrt{\omega_{3o}}t)\cos\omega_{2o}, \qquad q_2 - q_{2o} = (-gt^2/4 + \sqrt{\omega_{3o}}t)\sin\omega_{2o},$$

$$q_3 - q_{3o} = -gt^2/4 + \sqrt{\omega_{3o}}t \qquad [q_{ko} = q_k(0), \quad k = 1, 2, 3]. \tag{v}$$

Example 5.3.8 (Hamel, 1938, pp. 49–50; 1949, pp. 499–501; Mei, 1985, pp. 178–181). Let us derive the *kinetic* Johnsen–Hamel equations of the preceding example. We saw there that (no constraint enforcement yet!)

$$2\omega_1 \equiv [(\dot{q}_3)^2 - (\dot{q}_1)^2 - (\dot{q}_2)^2] = 0, \tag{a}$$

$$\omega_2 \equiv \arctan(\dot{q}_2/\dot{q}_1) \neq 0, \tag{b}$$

$$2\omega_3 \equiv [(\dot{q}_1)^2 + (\dot{q}_2)^2 + (\dot{q}_3)^2] \qquad (= 2T/m \neq 0); \tag{c}$$

$$\dot{q}_1 = (\omega_3 - \omega_1)^{1/2}\cos\omega_3, \tag{d}$$

$$\dot{q}_2 = (\omega_3 - \omega_1)^{1/2}\sin\omega_3, \tag{e}$$

$$\dot{q}_3 = (\omega_1 + \omega_3)^{1/2}. \tag{f}$$

With the help of the above, we readily find

$$2T = m[(\dot{q}_1)^2 + (\dot{q}_2)^2 + (\dot{q}_3)^2] \Rightarrow T^* = m\,\omega_3, \tag{g}$$

and, therefore,

$$P_1 \equiv \partial T^*/\partial\omega_1 = 0, \qquad P_2 \equiv \partial T^*/\partial\omega_2 = 0,$$

$$P_3 \equiv \partial T^*/\partial\omega_3 = m \Rightarrow \dot{P}_3 = 0,$$

$$\partial T^*/\partial q_k = 0 \Rightarrow \partial T^*/\partial\theta_I = 0 \qquad [k = 1, 2, 3; \; I = 2, 3], \tag{h}$$

and so, as eqs. (5.3.5b) show, we only need to calculate $H^3{}_2$ and $H^3{}_3$.

Indeed, invoking (5.2.21h) and remembering to set $\omega_1 = 0$ *after* all differentiations have been carried out, we find

$$\begin{aligned} H^3{}_2 &\equiv \sum [(\partial\omega_3/\partial\dot{q}_k)^{\cdot} - \partial\omega_3/\partial q_k](\partial\dot{q}_k/\partial\omega_2) \\ &= [(\partial\omega_3/\partial\dot{q}_1)^{\cdot} - 0](\partial\dot{q}_1/\partial\omega_2) + [(\partial\omega_3/\partial\dot{q}_2)^{\cdot} - 0](\partial\dot{q}_2/\partial\omega_2) \\ &= \cdots = -\sqrt{\omega_3}\,[(\sin\omega_2)(\sqrt{\omega_3}\cos\omega_2)^{\cdot} - (\cos\omega_2)(\sqrt{\omega_3}\sin\omega_2)^{\cdot}], \end{aligned} \tag{i}$$

$$\begin{aligned} H^3{}_3 &\equiv \sum [(\partial\omega_3/\partial\dot{q}_k)^{\cdot} - \partial\omega_3/\partial q_k](\partial\dot{q}_k/\partial\omega_3) \\ &= [(\partial\omega_3/\partial\dot{q}_1)^{\cdot} - 0](\partial\dot{q}_1/\partial\omega_3) + [(\partial\omega_3/\partial\dot{q}_2)^{\cdot} - 0](\partial\dot{q}_2/\partial\omega_3) \\ &\qquad + [(\partial\omega_3/\partial\dot{q}_3)^{\cdot} - 0](\partial\dot{q}_3/\partial\omega_3) \\ &= \cdots = (1/2\sqrt{\omega_3})[(\cos\omega_2)(\sqrt{\omega_3}\cos\omega_2)^{\cdot} + (\sin\omega_2)(\sqrt{\omega_3}\sin\omega_2)^{\cdot} + (\sqrt{\omega_3})^{\cdot}]; \end{aligned} \tag{j}$$

and [recalling the right sides of eqs. (o, p) of the preceding example]

$$\Theta_2 = (-\sqrt{\omega_3}\,\sin\omega_2)Q_1 + (\sqrt{\omega_3}\,\cos\omega_2)Q_2, \tag{k}$$

$$\Theta_3 = (\cos\omega_2/2\sqrt{\omega_3})Q_1 + (\sin\omega_2/2\sqrt{\omega_3})Q_2 + (1/2\sqrt{\omega_3})Q_3. \tag{l}$$

Let the reader verify that by inserting all these expressions into (5.3.5b) we obtain, after some simple manipulations, eqs. (o) and (p) = (q) of the preceding example; as we should.

REMARK

As pointed out earlier in this section, the Jacobian gradients $\partial\dot{q}/\partial\omega$ $[\partial\omega/\partial\dot{q}]$ can also be found via Cramer's rule from the compatibility conditions (5.2.4a, b), once the $\partial\omega/\partial\dot{q}$ $[\partial\dot{q}/\partial\omega]$ have been calculated from (a–c) [(d–f)]; that is, it is not necessary to invert the nonlinear (a–c) [(d–f)] to obtain (d–f) [(a–c)].

Example 5.3.9 (Dobronravov, 1970, pp. 250–253; Mei, 1985, pp. 241–242; San, 1973, pp. 332–333). Let us derive the kinetic Appellian equations of a particle P of mass m moving in the central Newtonian gravitational field of another (much larger) origin O of mass M; and also subject to the constraint

$$2f \equiv (\dot{r})^2 + (r^2\cos^2\theta)(\dot{\phi})^2 + r^2(\dot{\theta})^2 = constant \equiv c^2; \tag{a}$$

that is, square of velocity of $P = c$; where r, ϕ, θ: (inertial) spherical coordinates of P relative to O (with θ measured from the plane O–xy toward Oz).

The (unconstrained) Appellian of the system is

$$2S = m({a_r}^2 + {a_\phi}^2 + {a_\theta}^2), \tag{b}$$

where $a_{r,\phi,\theta}$ are the (physical) components of the acceleration of P in spherical coordinates (e.g., recalling §1.2, or prob. 3.5.15, with $\theta \to \pi/2 - \theta$):

$$a_r = \ddot{r} - r[(\dot{\theta})^2 + (\dot{\phi})^2\cos^2\theta], \tag{c}$$

$$a_\phi = r\ddot{\phi}\cos\theta + 2\dot{r}\dot{\phi}\cos\theta - 2r\dot{\phi}\dot{\theta}\sin\theta, \tag{d}$$

$$a_\theta = r\ddot{\theta} + 2\dot{r}\dot{\theta} + r(\dot{\phi})^2\sin\theta\cos\theta. \tag{e}$$

In view of (a), we choose as independent q's: $q_2 = r$ and $q_3 = \theta$, and use that constraint to express the dependent $q_1 = \phi$ in terms of r, θ, and so on.

Indeed, solving (a) for $\dot{\phi}$, we obtain

$$(\dot{\phi})^2 = [c^2 - (\dot{r})^2 - r^2(\dot{\theta})^2]\big/r^2\cos^2\theta, \tag{f}$$

$$\Rightarrow \ddot{\phi} = -(\dot{r}\ddot{r} + r^2\dot{\theta}\ddot{\theta})\big/r^2\dot{\phi}\cos^2\theta \quad (+\textit{Appell-nonimportant terms}), \tag{g}$$

and, therefore,

$$\partial\dot{\phi}/\partial\dot{r} = \partial\ddot{\phi}/\partial\ddot{r} = -\dot{r}/r^2\dot{\phi}\cos^2\theta, \tag{h}$$

$$\partial\dot{\phi}/\partial\dot{\theta} = \partial\ddot{\phi}/\partial\ddot{\theta} = -\dot{\theta}/\dot{\phi}\cos^2\theta. \tag{i}$$

Applying chain rule to

$$\begin{aligned} S &= S(a_r, a_\phi, a_\theta) = S[a_r(\ddot{r}, \ddot{\phi}, \ddot{\theta}, \ldots), \ldots] \\ &\equiv S'(\ddot{r}, \ddot{\phi}, \ddot{\theta}, \ldots) = S'[\ddot{r}, \ddot{\phi}(\ddot{r}, \ddot{\theta}, \ldots), \ddot{\theta}, \ldots] \\ &\equiv S_o(\ddot{r}, \ddot{\theta}, \ldots) \equiv S_o \qquad [\text{where } \ldots \equiv \text{no } \ddot{r}, \ddot{\phi}, \ddot{\theta} \text{ terms}], \end{aligned} \tag{j}$$

we easily find [no need to complete the squares in (b–e)]

$$\begin{aligned} \partial S_o/\partial \ddot{r} &= [(\partial S/\partial a_r)(\partial a_r/\partial \ddot{r}) + (\partial S/\partial a_\phi)(\partial a_\phi/\partial \ddot{r}) \\ &\qquad + (\partial S/\partial a_\theta)(\partial a_\theta/\partial \ddot{r})](\partial \ddot{r}/\partial \ddot{r}) \\ &\quad + [(\partial S/\partial a_r)(\partial a_r/\partial \ddot{\phi}) + (\partial S/\partial a_\phi)(\partial a_\phi/\partial \ddot{\phi}) \\ &\qquad + (\partial S/\partial a_\theta)(\partial a_\theta/\partial \ddot{\phi})](\partial \ddot{\phi}/\partial \ddot{r}) \\ &\quad + [(\partial S/\partial a_r)(\partial a_r/\partial \ddot{\theta}) + (\partial S/\partial a_\phi)(\partial a_\phi/\partial \ddot{\theta}) \\ &\qquad + (\partial S/\partial a_\theta)(\partial a_\theta/\partial \ddot{\theta})](\partial \ddot{\theta}/\partial \ddot{r}) \\ &\quad [= \partial S'/\partial \ddot{r} + (\partial S'/\partial \ddot{\phi})(\partial \ddot{\phi}/\partial \ddot{r}) + (\partial S'/\partial \ddot{\theta})(\partial \ddot{\theta}/\partial \ddot{r})] \\ &= \cdots = m[\ddot{r} - r(\dot{\theta})^2 - r(\dot{\phi})^2 \cos^2\theta - (\dot{r}/\dot{\phi})\ddot{\phi} \\ &\qquad - 2(\dot{r})^2/r + 2\dot{r}\dot{\theta} \tan\theta], \end{aligned} \tag{k}$$

$$\begin{aligned} \partial S_o/\partial \ddot{\theta} &= [(\partial S/\partial a_r)(\partial a_r/\partial \ddot{r}) + (\partial S/\partial a_\phi)(\partial a_\phi/\partial \ddot{r}) \\ &\qquad + (\partial S/\partial a_\theta)(\partial a_\theta/\partial \ddot{r})](\partial \ddot{r}/\partial \ddot{\theta}) \\ &\quad + [(\partial S/\partial a_r)(\partial a_r/\partial \ddot{\phi}) + (\partial S/\partial a_\phi)(\partial a_\phi/\partial \ddot{\phi}) \\ &\qquad + (\partial S/\partial a_\theta)(\partial a_\theta/\partial \ddot{\phi})](\partial \ddot{\phi}/\partial \ddot{\theta}) \\ &\quad + [(\partial S/\partial a_r)(\partial a_r/\partial \ddot{\theta}) + (\partial S/\partial a_\phi)(\partial a_\phi/\partial \ddot{\theta}) \\ &\qquad + (\partial S/\partial a_\theta)(\partial a_\theta/\partial \ddot{\theta})](\partial \ddot{\theta}/\partial \ddot{\theta}) \\ &\quad [= (\partial S'/\partial \ddot{r})(\partial \ddot{r}/\partial \ddot{\theta}) + (\partial S'/\partial \ddot{\phi})(\partial \ddot{\phi}/\partial \ddot{\theta}) + \partial S'/\partial \ddot{\theta}] \\ &= \cdots = m[r^2\ddot{\theta} + r^2(\dot{\phi})^2 \sin\theta \cos\theta \\ &\qquad - r^2(\dot{\theta}/\dot{\phi})\ddot{\phi} + 2r^2(\dot{\theta})^2 \tan\theta]. \end{aligned} \tag{l}$$

Next, here (with G denoting the well-known gravitational constant)

$$Q_r = -mMG/r^2, \qquad Q_\phi = 0, \qquad Q_\theta = 0, \tag{m}$$

and therefore the independent impressed forces, Q_{Io}, are

$$Q_{ro} = Q_r + (\partial\dot{\phi}/\partial\dot{r})Q_\phi = -mMG/r^2, \qquad Q_{\theta o} = Q_\theta + (\partial\dot{\phi}/\partial\dot{\theta})Q_\phi = 0. \tag{n}$$

As a result of the above, the kinetic Appellian equations are

$\partial S_o/\partial \ddot{r} = Q_{ro}$:

$$\ddot{r} - r(\dot{\theta})^2 - r(\dot{\phi})^2 \cos^2\theta - (\dot{r}/\dot{\phi})\ddot{\phi} - 2(\dot{r})^2/r + 2\dot{r}\,\dot{\theta}\tan\theta = -MG/r^2, \tag{o}$$

$\partial S_o/\partial \ddot{\theta} = Q_{\theta o}$:

$$r^2\ddot{\theta} + r^2(\dot{\phi})^2 \sin\theta\cos\theta - r^2(\dot{\theta}/\dot{\phi})\ddot{\phi} + 2r^2(\dot{\theta})^2\tan\theta = 0; \tag{p}$$

and along with the constraint (a) these constitute a determinate system for $r(t), \phi(t), \theta(t)$.

Example 5.3.10 (Mei, 1985, pp. 155–156). Let us derive the *general* kinetic Voronets equations (5.3.8a, b) of the preceding example. We introduce the following quasi velocities:

$$\omega_1 \equiv f - c^2/2 \equiv \left\{[(\dot{r})^2 + (r^2\cos^2\theta)(\dot{\phi})^2 + r^2(\dot{\theta})^2] - c^2\right\}/2 \quad (=0), \tag{a}$$

$$\omega_2 \equiv \dot{r}\;(\neq 0), \tag{b}$$

$$\omega_3 \equiv r\dot{\theta}\;(\neq 0), \tag{c}$$

and their inverses (with $q_{1,2,3}$: r, ϕ, θ):

$$\dot{r} = \omega_2, \tag{d}$$

$$(\dot{\phi})^2 = [(2\omega_1 + c^2) - (\dot{r})^2 - r^2(\dot{\theta})^2]/r^2\cos^2\theta = [(2\omega_1 + c^2) - \omega_2{}^2 - \omega_3{}^2]/r^2\cos^2\theta, \tag{e}$$

$$\dot{\theta} = \omega_3/r. \tag{f}$$

Hence, the general Voronets symbols needed, $V^k{}_I$, eqs. (5.2.21e; with $k = 1, 2, 3$; $I = 2, 3$) specialize to

$$V^1{}_2 \equiv (\partial\dot{r}/\partial\omega_2)^{\cdot} - \partial\dot{r}/\partial\theta_2 \equiv (\partial\dot{r}/\partial\omega_2)^{\cdot} - \sum(\partial\dot{r}/\partial q_k)(\partial\dot{q}_k/\partial\omega_2) = (1)^{\cdot} - 0 = 0, \tag{g}$$

$$V^2{}_2 \equiv (\partial\dot{\phi}/\partial\omega_2)^{\cdot} - \partial\dot{\phi}/\partial\theta_2 \equiv (\partial\dot{\phi}/\partial\omega_2)^{\cdot} - \sum(\partial\dot{\phi}/\partial q_k)(\partial\dot{q}_k/\partial\omega_2) = (-\dot{r}/r^2\dot{\phi}\cos^2\theta)^{\cdot} - (-\dot{\phi}/r)(1), \tag{h}$$

$$V^3{}_2 \equiv (\partial\dot{\theta}/\partial\omega_2)^{\cdot} - \partial\dot{\theta}/\partial\theta_2 \equiv (\partial\dot{\theta}/\partial\omega_2)^{\cdot} - \sum(\partial\dot{\theta}/\partial q_k)(\partial\dot{q}_k/\partial\omega_2) = (0)^{\cdot} - (-\dot{\theta}/r)(1) = \dot{\theta}/r; \tag{i}$$

$$V^1{}_3 \equiv (\partial\dot{r}/\partial\omega_3)^{\cdot} - \partial\dot{r}/\partial\theta_3 \equiv (\partial\dot{r}/\partial\omega_3)^{\cdot} - \sum(\partial\dot{r}/\partial q_k)(\partial\dot{q}_k/\partial\omega_3) = (0)^{\cdot} - 0 = 0, \tag{j}$$

$$V^2{}_3 \equiv (\partial\dot{\phi}/\partial\omega_3)^{\cdot} - \partial\dot{\phi}/\partial\theta_3 \equiv (\partial\dot{\phi}/\partial\omega_3)^{\cdot} - \sum(\partial\dot{\phi}/\partial q_k)(\partial\dot{q}_k/\partial\omega_3) = (-\dot{\theta}/r\dot{\phi}\cos^2\theta)^{\cdot} - (\dot{\phi}\tan\theta)(r^{-1}), \tag{k}$$

$$V^3{}_3 \equiv (\partial\dot{\theta}/\partial\omega_3)^{\cdot} - \partial\dot{\theta}/\partial\theta_3 \equiv (\partial\dot{\theta}/\partial\omega_3)^{\cdot} - \sum (\partial\dot{\theta}/\partial q_k)\,(\partial\dot{q}_k/\partial\omega_3)$$
$$= (r^{-1})^{\cdot} - (-\dot{\theta}/r)(0) = -\dot{r}/r^2; \tag{l}$$

while the (unconstrained) kinetic energy, in holonomic variables, becomes

$$2T = m[(\dot{r})^2 + (r^2\cos^2\theta)(\dot{\phi})^2 + r^2(\dot{\theta})^2]$$
$$(= mc^2, \text{ constrained kinetic energy}), \tag{m}$$
$$\Rightarrow \partial T/\partial\dot{r} = m\dot{r}, \qquad \partial T/\partial\dot{\phi} = mr^2\cos^2\theta\dot{\phi}, \qquad \partial T/\partial\dot{\theta} = mr^2\dot{\theta}; \tag{n}$$

and, in nonholonomic variables,

$$2T^* = m(2\omega_1 + c^2), \tag{o}$$
$$\Rightarrow \partial T^*/\partial\omega_2 = 0, \qquad \partial T^*/\partial\omega_3 = 0,$$
$$\partial T^*/\partial\theta_2 \equiv \sum (\partial T^*/\partial q_k)\,(\partial\dot{q}_k/\partial\omega_2) = 0, \qquad \partial T^*/\partial\theta_3 = 0; \tag{p}$$

and, finally, the corresponding nonholonomic impressed forces are

$$\Theta_2 \equiv \sum (\partial\dot{q}_k/\partial\omega_2)Q_k = \cdots = -mMG/r^2, \qquad \Theta_3 = \cdots = 0. \tag{q}$$

Substituting all these special results into eqs. (5.3.8a, b) yields

$$\omega_2: \qquad 0 - [V^2{}_2(\partial T/\partial\dot{\phi}) + V^3{}_2(\partial T/\partial\dot{\theta})] = \Theta_2, \tag{r}$$
$$\omega_3: \qquad 0 - [V^2{}_3(\partial T/\partial\dot{\phi}) + V^3{}_3(\partial T/\partial\dot{\theta})] = \Theta_3; \tag{s}$$

and if these two equations are written out, in extenso, they, naturally, coincide with the Appellian equations (o, p) of the preceding example.

Last, by substituting in the above $\dot{r}, \dot{\phi}, \dot{\theta}$ in terms of $\omega_1 (= 0), \omega_2, \omega_3$, including $\partial T/\partial\dot{q}_k \to (\partial T/\partial\dot{q}_k)^*$, via (d–f), we may, if needed, express (r, s) in terms of these quasi variables, à la Hamel.

Example 5.3.11 Let us derive the *special* (kinetic) Chaplygin–Voronets equations (ex. 5.3.4: g, i) for the system of ex. 5.3.5. Here, $n = 3$ and $m = 1$, and so with the choice $\dot{q}_D$: $\dot{q}_1$ and $\dot{q}_I$: $\dot{q}_{2,3}$, the constraint

$$(\dot{q}_1)^2 + (\dot{q}_2)^2 + (\dot{q}_3)^2 = constant \equiv c^2 \tag{a}$$

yields

$$\dot{q}_D \to \dot{q}_1 = [c^2 - (\dot{q}_2)^2 + (\dot{q}_3)^2]^{1/2} \equiv \phi_1(\dot{q}_I). \tag{b}$$

Therefore, the kinetic energy becomes

$$2T = m[(\dot{q}_1)^2 + (\dot{q}_2)^2 + (\dot{q}_3)^2]$$
$$= m\Big\{[c^2 - (\dot{q}_2)^2 + (\dot{q}_3)^2] + (\dot{q}_2)^2 + (\dot{q}_3)^2\Big\}$$
$$= mc^2 = T_o(\dot{q}_I) \equiv T_o \quad \text{(constrained kinetic energy)}, \tag{c}$$

$$\Rightarrow \partial T/\partial \dot{q}_D\colon\ \partial T/\partial \dot{q}_1 = m\dot{q}_1,$$

$$\partial T_o/\partial \dot{q}_I = 0,$$

$$\partial T_o/\partial(q_I) \equiv \partial T_o/\partial q_I + \sum (\partial T_o/\partial q_D)(\partial \phi_D/\partial \dot{q}_I) = \partial T_o/\partial q_I; \qquad \text{(d)}$$

the relevant nonlinear Voronets–Chaplygin coefficients $V^D{}_I \to W^D{}_I$ reduce to the following nonlinear Chaplygin coefficients $T^D{}_I$:

$$T^1{}_2 \equiv (\partial \phi_1/\partial \dot{q}_2)^{\cdot} - \partial \phi_1/\partial q_2 = (-\dot{q}_2/\dot{q}_1)^{\cdot} - 0, \qquad \text{(e)}$$

$$T^1{}_3 \equiv (\partial \phi_1/\partial \dot{q}_3)^{\cdot} - \partial \phi_1/\partial q_3 = (-\dot{q}_3/\dot{q}_1)^{\cdot} - 0; \qquad \text{(f)}$$

and the constrained impressed forces Q_{Io} become

$$Q_{2o} = Q_2 + (\partial \phi_1/\partial \dot{q}_2)Q_1 = Q_2 - (\dot{q}_2/\dot{q}_1)Q_1, \qquad \text{(g)}$$

$$Q_{3o} = Q_3 + (\partial \phi_1/\partial \dot{q}_3)Q_1 = Q_3 - (\dot{q}_3/\dot{q}_1)Q_1. \qquad \text{(h)}$$

Substituting all these special results into the nonlinear Voronets → Chaplygin equations

$$(\partial T_o/\partial \dot{q}_I)^{\cdot} - \partial T_o/\partial q_I - \sum T^D{}_I(\partial T/\partial \dot{q}_D)_o = Q_{Io}, \qquad \text{(i)}$$

we find

$$-T^1{}_2(\partial T/\partial \dot{q}_1) = Q_{2o}\colon$$

$$-(-\dot{q}_2/\dot{q}_1)^{\cdot}(m\dot{q}_1) = Q_2 - (\dot{q}_2/\dot{q}_1)Q_1, \qquad \text{(j)}$$

$$-T^1{}_3(\partial T/\partial \dot{q}_1) = Q_{3o}\colon$$

$$-(-\dot{q}_3/\dot{q}_1)^{\cdot}(m\dot{q}_1) = Q_3 - (\dot{q}_3/\dot{q}_1)Q_1; \qquad \text{(k)}$$

or, simplifying,

$$m\ddot{q}_2 = Q_2 + (\dot{q}_2/\dot{q}_1)(m\ddot{q}_1 - Q_1), \qquad \text{(l)}$$

$$m\ddot{q}_3 = Q_3 + (\dot{q}_3/\dot{q}_1)(m\ddot{q}_1 - Q_1), \qquad \text{(m)}$$

respectively.

To show the equivalence of the above with (g) of ex. 5.3.5, we $(\ldots)^{\cdot}$-differentiate (a),

$$(\dot{q}_1)^2 + (\dot{q}_2)^2 + (\dot{q}_3)^2 = c^2 \Rightarrow \dot{q}_1\ddot{q}_1 + \dot{q}_2\ddot{q}_2 + \dot{q}_3\ddot{q}_3 = 0,$$

solve the result for $\ddot{q}_1$, and then, invoking (k, l) in it, we obtain, successively,

$$\begin{aligned} m\ddot{q}_1 &= -(\dot{q}_2/\dot{q}_1)(m\ddot{q}_2) - (\dot{q}_3/\dot{q}_1)(m\ddot{q}_3) \\ &= -(\dot{q}_2/\dot{q}_1)[Q_2 + (\dot{q}_2/\dot{q}_1)(m\ddot{q}_1 - Q_1)] \\ &\quad - (\dot{q}_3/\dot{q}_1)[Q_3 + (\dot{q}_3/\dot{q}_1)(m\ddot{q}_1 - Q_1)], \end{aligned}$$

or, reducing further and using (a),

$$m\ddot{q}_1 = Q_1 - \big[(Q_1\dot{q}_1 + Q_2\dot{q}_2 + Q_3\dot{q}_3)/c^2\big]\dot{q}_1, \qquad \text{(n)}$$

and when this is inserted back into (l, m) it produces equations (ex. 5.3.5: g); as expected, due to the symmetry of the problem in $q_{1,2,3}$.

Problem 5.3.2 In ex. 5.3.9, we saw that

$$(\dot{\phi})^2 = [c^2 - (\dot{r})^2 - r^2(\dot{\theta})^2]/r^2 \cos^2\theta, \qquad \text{i.e., } \dot{q}_D = \phi_D(\dot{q}_I, \ldots). \tag{a}$$

Substituting this into $T(\dot{r}, \dot{\phi}, \dot{\theta}, \ldots)$, obtain $T_o(\dot{r}, \dot{\theta}, \ldots)$, and then find the corresponding *special* (kinetic) Voronets equations (ex. 5.3.4: g). Show that they coincide with the *special* (kinetic) Appellian equations of ex. 5.3.9, and the *general* Voronets equations of ex. 5.3.10.

Problem 5.3.3 Continuing from the system of exs. 5.3.9, and 5.3.10, $(\ldots)^{\cdot}$-differentiate its constraint (a):

$$\dot{f} = \dot{r}\ddot{r} + (r^2\cos^2\theta)\dot{\phi}\ddot{\phi} + r^2\dot{\theta}\ddot{\theta} + \textit{no other } \ddot{r}, \ddot{\phi}, \ddot{\theta}\textit{-terms}, \tag{a}$$

and introduce the quasi accelerations

$$\dot{\omega}_1 \equiv \dot{f} \quad (=0), \qquad \dot{\omega}_2 \equiv r^2\dot{\theta}\ddot{\theta} \quad (\neq 0), \qquad \dot{\omega}_3 \equiv \dot{r}\ddot{r} \quad (\neq 0). \tag{b}$$

Show that the corresponding *general* (kinetic) Appellian equations

$$\partial S^*{}_o/\partial\dot{\omega}_2 = \Theta_2 \qquad \text{and} \qquad \partial S^*{}_o/\partial\dot{\omega}_3 = \Theta_3, \tag{c}$$

where

$$\begin{aligned} S &= S(\ddot{r}, \ddot{\phi}, \ddot{\theta}, \ldots) = \cdots = S^*(\dot{\omega}_1 = 0, \dot{\omega}_2, \dot{\omega}_3, \ldots) \\ &= S^*{}_o(\dot{\omega}_2, \dot{\omega}_3, \ldots) \equiv S^*{}_o(\textit{constrained} \text{ general Appellian}), \end{aligned} \tag{d}$$

coincide with the *special* (kinetic) Appellian equations of ex. 5.3.9.

Problem 5.3.4 Consider the system of ex. 5.3.7; that is, a particle of mass m under impressed forces Q_k, and constrained by $(\dot{q}_1)^2 + (\dot{q}_2)^2 = (\dot{q}_3)^2$.

Obtain its *special* Voronets equations; and show that they coincide with those found in ex. 5.3.7 (general Appell) and ex. 5.3.8 (Johnsen–Hamel).

Example 5.3.12 *Tetherball* (Kitzka, 1986; Fufaev, 1990; also Kuypers, 1993, pp. 66, 388–394). Let us consider a heavy particle P of mass m fastened at the end of a massless (i.e., light) inextensible thread, the other end of which is fixed at a point on the surface of a circular and vertical cylinder C of radius R. As P moves, the thread can be wound up *without slipping* on the surface of C [fig. 5.1 (a)]. [This is an idealization of a toy known as *tetherball*—itself an idealization of Huygens' pendulum, self-regulating its free (i.e., unwound) length.]

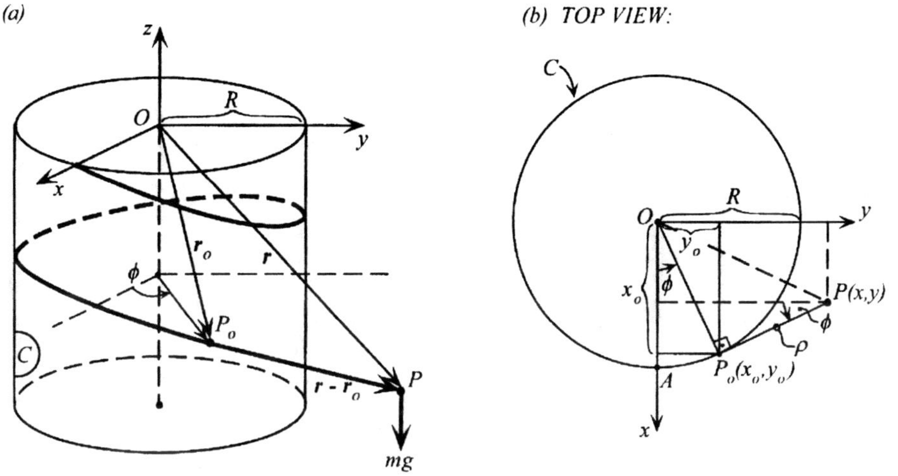

Figure 5.1 (a) Particle P on a thread, wound up on a cylinder C; (b) geometrical details (top view).

Let

$r = (x, y, z)$: position vector of P at a generic time,

$r_o = (x_o, y_o, z_o)$: position vector of point of separation of thread from the cylinder surface P_o, at a generic time,

ρ: length of projection of P_oP on plane O–xy [fig. 5.1(b)]. (a)

The constraints are:

(i) The constancy of the total thread length, as long as its unwound part is taut and its wound part does not slip on C:

$$\int_0^t |\boldsymbol{v}_o|\, dt + |\boldsymbol{r} - \boldsymbol{r}_o| = constant \equiv l_o; \tag{b}$$

that is, *wound length* + *unwound length* $(l) = l_o$.

(ii) Continuity of thread slope at P_o [fig. 5.1(b)]:

$$(\boldsymbol{r} - \boldsymbol{r}_o)/|\boldsymbol{r} - \boldsymbol{r}_o| = \boldsymbol{v}_o/|\boldsymbol{v}_o| \equiv \boldsymbol{v}_o/v_o; \tag{c}$$

that is, *unit vector along* P_oP = *unit vector along velocity of* P_o.

Also, from geometry, we have

$$x = x_o - \rho \sin\phi = R\cos\phi - \rho\sin\phi, \tag{d1}$$

$$y = y_o + \rho\cos\phi = R\sin\phi + \rho\cos\phi. \tag{d2}$$

Let us translate (b, c) into equivalent scalar forms:

(i) By $(\ldots)^{\cdot}$-differentiating (b), to get rid of the integral, we obtain

$$|\boldsymbol{v}_o| + |\boldsymbol{r} - \boldsymbol{r}_o|^{\cdot} = 0 \ \Rightarrow\ v_o = -|\boldsymbol{r} - \boldsymbol{r}_o|^{\cdot} \equiv -dl/dt, \tag{b1}$$

and so the $(\ldots)^{\cdot}$-derivative of (c), rearranged as

$$\boldsymbol{r} - \boldsymbol{r}_o = \left[|\boldsymbol{r} - \boldsymbol{r}_o|/v_o\right]\boldsymbol{v}_o \ \Rightarrow\ \boldsymbol{r} = \boldsymbol{r}_o + (l/v_o)\boldsymbol{v}_o, \tag{c1}$$

becomes, successively (with $\boldsymbol{v} \equiv \dot{\boldsymbol{r}}$, $\boldsymbol{a}_o \equiv \dot{\boldsymbol{v}}_o$),

$$\boldsymbol{v} = \boldsymbol{v}_o + (l/v_o)\boldsymbol{a}_o + \big[(dl/dt)/v_o\big]\boldsymbol{v}_o - \big[l\,(dv_o/dt)/{v_o}^2\big]\boldsymbol{v}_o$$

[by (b1), the *first* and *third* terms cancel]

$$= \big\{\boldsymbol{a}_o - [(dv_o/dt)/v_o]\boldsymbol{v}_o\big\}(l/v_o), \tag{c2}$$

and dotting this with $\boldsymbol{v}_o$, we get

$$\boldsymbol{v}\cdot\boldsymbol{v}_o = [\boldsymbol{v}_o\cdot\boldsymbol{a}_o - (\dot{v}_o/v_o){v_o}^2]\,(l/v_o) = 0, \tag{c3}$$

since $(\boldsymbol{v}_o\cdot\boldsymbol{v}_o)^{\cdot} = ({v_o}^2)^{\cdot} \Rightarrow 2\boldsymbol{v}_o\cdot\boldsymbol{a}_o = 2v_o\dot{v}_o$. With the help of the $(\ldots)^{\cdot}$-derivatives of (d1, 2), the above assumes the form

$$\boldsymbol{v}\cdot\boldsymbol{v}_o = R\dot{\phi}(\dot{\rho} + R\dot{\phi}) + \dot{z}\,\dot{z}_o = 0. \tag{e1}$$

(ii) Dotting (c) with $\boldsymbol{v}$ and invoking (c3, 4) and (d1, 2), we easily obtain

$$\boldsymbol{v}\cdot(\boldsymbol{r} - \boldsymbol{r}_o) = \rho(\dot{\rho} + R\dot{\phi}) + \dot{z}(z - z_o) = 0. \tag{e2}$$

Instead of the *two* constraints (e1, 2) for the *four* Lagrangean coordinates ϕ, ρ, z, z_o, we can eliminate z_o (and $\dot{z}_o$) between them by $(\ldots)^{\cdot}$-differentiating (e2), multiplying the outcome with $\dot{z}$, and then using (e1, 2). The result is the single *nonlinear* second-order (but linear in its second derivatives) constraint

$$f_o \equiv \rho(\dot{\rho} + R\dot{\phi})\ddot{z} - \rho(\ddot{\rho} + R\ddot{\phi})\dot{z}$$
$$- [(\dot{\rho} + R\dot{\phi})^2 + (\dot{z})^2]\dot{z} = 0, \tag{f}$$

for the *three* coordinates ϕ, ρ, z.

At this point, we could form either (i) the two kinetic Appellian equations (here, $n = 3, m = 1$),

$$\partial S_o/\partial\ddot{\phi} = Q_{\phi o} \qquad \text{and} \qquad \partial S_o/\partial\ddot{\rho} = Q_{\rho o}, \tag{g}$$

where, using (f) (and the notation $\ldots \equiv$ *no* $\ddot{\phi}, \ddot{\rho}, \ddot{z}$ *terms*),

$$S = S(\ddot{\phi}, \ddot{\rho}, \ddot{z}, \ldots) = S[\ddot{\phi}, \ddot{\rho}, \ddot{z}(\ddot{\phi}, \ddot{\rho}, \ldots), \ldots]$$
$$\equiv S_o(\ddot{\phi}, \ddot{\rho}, \ldots) \equiv S_o, \tag{g1}$$

and

$$\delta' W = Q_\phi\,\delta\phi + Q_\rho\,\delta\rho + Q_z\,\delta z = \cdots = Q_{\phi o}\,\delta\phi + Q_{\rho o}\,\delta\rho; \tag{g2}$$

or (ii) the Appellian form of the three mixed (coupled) nonlinear Routh–Voss equations,

$$\partial S/\partial\ddot{q}_k = Q_k + \lambda(\partial f_o/\partial\ddot{q}_k); \qquad q_k = \phi, \rho, z. \tag{h}$$

However, a simpler form of equations of motion results if, following Kitzka (1986), we introduce the following quasi velocities:

$$\omega_1 \equiv R\dot{\phi}, \tag{i1}$$

$$\omega_2 \equiv (\dot{\rho} + R\dot{\phi})/\dot{z} = -\dot{z}_o/R\dot{\phi} = (z_o - z)/\rho \qquad \text{[by (e1, 2)]}, \tag{i2}$$

$$\omega_3 \equiv \dot{z}; \tag{i3}$$

which invert readily to

$$\dot{q}_1 \equiv \dot{\phi} = \omega_1/R, \tag{j1}$$

$$\dot{q}_2 \equiv \dot{\rho} = \omega_2\,\omega_3 - \omega_1, \tag{j2}$$

$$\dot{q}_3 \equiv \dot{z} = \omega_3. \tag{j3}$$

Eliminating z_o from (i2) by $(\ldots)^{\cdot}$-differentiating it: $\dot{z}_o - \dot{z} = (\rho\,\omega_2)^{\cdot} = \dot{\rho}\,\omega_2 + \rho\,\dot{\omega}_2$, and then using (i1, 2) and (j2) in it, we finally obtain the nonlinear (but linear in its second derivative) constraint for $\omega_{1,2,3}$:

$$f \equiv \rho\,\dot{\omega}_2 + (1 + {\omega_2}^2)\omega_3 = 0. \tag{k}$$

Now, let us form the corresponding kinetic Appellian equations

$$\partial S^*{}_o/\partial\dot{\omega}_1 = \Theta_1 \qquad \text{and} \qquad \partial S^*{}_o/\partial\dot{\omega}_3 = \Theta_3, \tag{l}$$

where (with the notation $\ldots \equiv$ *no* $\dot{\omega}_{1,2,3}$ *terms*)

$$S^* = S^*(\dot{\omega}_1, \dot{\omega}_2, \dot{\omega}_3, \ldots) = S^*[\dot{\omega}_1, \dot{\omega}_2(\dot{\omega}_1, \dot{\omega}_3, \ldots), \dot{\omega}_3, \ldots]$$
$$\equiv S^*{}_o(\dot{\omega}_1, \dot{\omega}_3, \ldots) \equiv S^*{}_o. \tag{l1}$$

Indeed, using (k) to eliminate $\dot{\omega}_2$ from $\boldsymbol{a} \equiv \dot{\boldsymbol{v}} \equiv \ddot{\boldsymbol{r}} = (\ddot{x}, \ddot{y}, \ddot{z})$ [(d1, 2) with (j1–3)] we obtain, after some algebra,

$$\ddot{x} = -\left[\omega_2\,\dot{\omega}_3 - {\omega_3}^2(1 + {\omega_2}^2)/\rho - \rho\,{\omega_1}^2/R^2\right]\sin\phi$$
$$+ (1/R)[\omega_1(\omega_1 - 2\omega_2\,\omega_3) - \rho\,\dot{\omega}_1]\,\cos\phi, \tag{l2}$$

$$\ddot{y} = \left[\omega_2\,\dot{\omega}_3 - {\omega_3}^2(1 + {\omega_2}^2)/\rho - \rho{\omega_1}^2/R^2\right]\cos\phi$$
$$+ (1/R)[\omega_1(\omega_1 - 2\omega_2\omega_3) - \rho\dot{\omega}_1]\,\sin\phi, \tag{l3}$$

$$\ddot{z} = \dot{\omega}_3, \tag{l4}$$

and, therefore,

$$S = (m/2)[(\ddot{x})^2 + (\ddot{y})^2 + (\ddot{z})^2]$$
$$= \cdots = (m/2)\Big\{(1 + {\omega_2}^2)\,(\dot{\omega}_3)^2 - 2\omega_2\,\dot{\omega}_3\,[\omega_3(1 + {\omega_2}^2)/\rho + \rho\,{\omega_1}^2/R^2]$$
$$+ (\rho/R^2)\,[\rho(\dot{\omega}_1)^2 - 2\omega_1\,\dot{\omega}_1(\omega_1 - 2\omega_2\,\omega_3)]\Big\}$$
$$+ \textit{function of } \omega_{1,2,3};\ \rho\ (\textit{``Appell constant'' terms}) = S^*{}_o; \tag{l5}$$

also, by (l2–4),

$$\Theta_1 \equiv \textstyle\int dF\cdot(\partial\boldsymbol{a}^*/\partial\dot{\omega}_1) = (-m\,g)(\partial\ddot{z}/\partial\dot{\omega}_1) = 0, \tag{l6}$$

$$\Theta_3 \equiv \textstyle\int dF\cdot(\partial\boldsymbol{a}^*/\partial\dot{\omega}_3) = (-m\,g)(\partial\ddot{z}/\partial\dot{\omega}_3) = -m\,g. \tag{l7}$$

As a result of the above, Appell's equations (l) become

$$\omega_1: \qquad \rho\dot{\omega}_1 - \omega_1(\omega_1 - 2\omega_2\,\omega_3) = 0, \tag{m1}$$

$$\omega_3: \qquad (1 + \omega_2{}^2)\dot{\omega}_3 - \omega_2\big[(1 + \omega_2{}^2)\omega_3{}^2/\rho + (\rho/R^2)\omega_1{}^2\big] + g = 0, \tag{m2}$$

and along with eqs. (j1–3, k) they constitute a determinate system for the six functions of time: $\omega_{1,2,3}$; ϕ, ρ, z.

Clearly, our system possesses the energy integral

$$\begin{aligned} E &\equiv T + V = (m/2)\,[(\dot{x})^2 + (\dot{y})^2 + (\dot{z})^2] + mgz \qquad \text{[using (d1, 2) and (j1–3)]} \\ &= (m/2)\left[(1 + \omega_2{}^2)\omega_3{}^3 + (\rho/R)^2\omega_1{}^2\right] + mgz \\ &\equiv T^* + V = constant \equiv E_o, \end{aligned} \tag{n}$$

which constitutes an indirect proof of the physical correctness of (m1, 2).

REMARKS

(i) Since

$$\begin{aligned} \partial S^*/\partial\dot{\omega}_1 &= (\partial S/\partial\ddot{x})\,(\partial\ddot{x}/\partial\dot{\omega}_1) + (\partial S/\partial\ddot{y})(\partial\ddot{y}/\partial\dot{\omega}_1) \\ &\quad + (\partial S/\partial\ddot{z})\,(\partial\ddot{z}/\partial\dot{\omega}_1) \\ &= (\partial S/\partial\ddot{x})\,(\partial\dot{x}/\partial\omega_1) + (\partial S/\partial\ddot{y})\,(\partial\dot{y}/\partial\omega_1) \\ &\quad + (\partial S/\partial\ddot{z})\,(\partial\dot{z}/\partial\omega_1) \\ &= (m\ddot{x})\,(\partial\dot{x}/\partial\omega_1) + (m\ddot{y})\,(\partial\dot{y}/\partial\omega_1) + (m\ddot{z})\,(\partial\dot{z}/\partial\omega_1) \\ &= m[\ddot{x}(-\rho\cos\phi/R) + \ddot{y}(-\rho\sin\phi/R) + \ddot{z}(0)] \\ &= \cdots = m\rho[\rho\dot{\omega}_1 - \omega_1(\omega_1 - 2\omega_2\,\omega_3)], \end{aligned} \tag{o1}$$

$$\begin{aligned} \partial S^*/\partial\dot{\omega}_3 &= \cdots = (m\ddot{x})\,(\partial\ddot{x}/\partial\dot{\omega}_1) + \cdots \\ &= m[\ddot{x}(-\omega_2\sin\phi) + \ddot{y}(\omega_2\cos\phi) + \ddot{z}(1)] \\ &= m\Big\{(1 + \omega_2{}^2)\dot{\omega}_3 - \omega_2[\omega_3(1 + \omega_2{}^2)/\rho + (\rho/R^2)\omega_1{}^2]\Big\}, \end{aligned} \tag{o2}$$

and

$$\partial S^*/\partial\dot{\omega}_1 = \partial S^*{}_o/\partial\dot{\omega}_1, \qquad \partial S^*/\partial\dot{\omega}_3 = \partial S^*{}_o/\partial\dot{\omega}_3, \tag{o3}$$

there is no need to find $S^*(\dot{\omega}_{1,2,3}, \ldots)$ or $S^*{}_o(\dot{\omega}_1, \dot{\omega}_3, \ldots)$ by expanding the squares of $\ddot{x}, \ddot{y}, \ddot{z}$ in (15); we can leave it as $S(\ddot{x}, \ddot{y}, \ddot{z})$.

(ii) In terms of the alternative, convenient, quasi accelerations

$$\dot{\Omega}_1 \equiv \rho\dot{\omega}_2 + (1 + \omega_2{}^2)\omega_3 = 0, \tag{p1}$$

$$\dot{\Omega}_2 \equiv \dot{\omega}_1 \neq 0, \tag{p2}$$

$$\dot{\Omega}_3 \equiv \dot{\omega}_3 \neq 0, \tag{p3}$$

and the corresponding Appellian

$$S^* \to S^{**} = S^{**}(\dot{\Omega}_1, \dot{\Omega}_2, \dot{\Omega}_3, \ldots) \to S^{**}{}_o = S^{**}{}_o(\dot{\Omega}_2, \dot{\Omega}_3, \ldots), \tag{p4}$$

and impressed forces $\Theta_2 \to \Theta_2{}^*, \Theta_3 \to \Theta_3{}^*$, the kinetic Appellian equations would be

$$\partial S^{**}{}_o/\partial\dot{\Omega}_1 = \Theta_2{}^*, \qquad \partial S^{**}{}_o/\partial\dot{\Omega}_3 = \Theta_3{}^*. \tag{p5}$$

(iii) By looking at the constraints (f) and/or (k), one might conclude that they are nonlinear. However, as Fufaev (1990) has pointed out, this is *not* the case. Indeed, solving (e2) for $\dot{z}$:

$$\dot{z} = \rho(\dot{\rho} + R\dot{\phi})\,(z_o - z)^{-1}, \tag{q1}$$

and substituting this value in (e1), we obtain

$$(\dot{\rho} + R\dot{\phi})\,[R\dot{\phi} + \rho(z_o - z)^{-1}\dot{z}_o] = 0, \tag{q2}$$

from which, since, in general, $\dot{\rho} + R\dot{\phi} \neq 0$, it follows that (e1) is replaced by its equivalent,

$$R\dot{\phi} + \rho(z_o - z)^{-1}\dot{z}_o = 0, \tag{q3}$$

which, just like (e2), is linear (Pfaffian) in the Lagrangean velocities $\dot{\phi}$ and $\dot{z}_o$. Hence, the earlier nonlinearity is not of intrinsic/physical but *analytical* nature: *it resulted from the elimination of z_o and $\dot{z}_o$ between (e1, 2) and associated $(\ldots)^\cdot$-differentiations of the first-order constraint (e2).*

This justifies our attitude toward the subject of nonlinear nonholonomic constraints: let us learn them; even if they do not exist in any physically meaningful way, we may have to use them for analytical convenience.

For additional physical and numerical aspects, see the earlier given references Kitzka (1986) and Kuypers (1993).

Problem 5.3.5 (Fufaev, 1990; Kitzka, 1986). Continuing from the preceding example of the tetherball, we saw there that its two *nonlinear* constraints, in the *four* Lagrangean coordinates $q_{1,2,3,4} = \phi, \rho, z, z_o$, are

$$\rho(\dot{\rho} + R\dot{\phi}) + (z - z_o)\dot{z} = 0, \tag{a1}$$

$$R\dot{\phi}(\dot{\rho} + R\dot{\phi}) + \dot{z}\,\dot{z}_o = 0, \tag{a2}$$

or, *equivalently* [solving (a1) for $\dot{z}$: $\dot{z} = \rho(\dot{\rho} + R\dot{\phi})\,(z_o - z)^{-1}$ and substituting the result in (a2)], the two *Pfaffian* constraints are

$$(\rho R)\dot{\phi} + (\rho)\dot{\rho} + (z - z_o)\dot{z} = 0, \tag{b1}$$

$$R(z_o - z)\dot{\phi} + \rho\,\dot{z}_o = 0, \tag{b2}$$

and, therefore, in *virtual* form:

$$(\rho R)\,\delta\phi + (\rho)\,\delta\rho + (z - z_o)\,\delta z = 0, \tag{c1}$$

$$[R(z_o - z)]\delta\phi + (\rho)\,\delta z_o = 0. \tag{c2}$$

Setting $m = 1$, for convenience, but no loss in generality,
(i) Show that

$$L \equiv T - V = (1/2)\Big[\rho^2(\dot{\phi})^2 + (\dot{\rho} + R\,\dot{\phi})^2 + (\dot{z})^2\Big] - gz, \tag{d1}$$

and therefore the corresponding Routh–Voss equations,

$$E_k(L) = \lambda_1 a_{1k} + \lambda_2 a_{2k} \qquad (k = 1, \ldots, 4), \tag{d2}$$

are

$$\phi: \qquad R\ddot{\rho} + (R^2 + \rho^2)\ddot{\phi} - 2\rho\,\dot{\rho}\,\dot{\phi} = (R\rho)\lambda_1 + [R(z_o - z)]\lambda_2, \tag{d3}$$

$$\rho: \qquad \ddot{\rho} + R\ddot{\phi} - \rho(\dot{\phi})^2 = (\rho)\lambda_1, \tag{d4}$$

$$z: \qquad \ddot{z} + g = (z - z_o)\lambda_1, \tag{d5}$$

$$z_o: \qquad 0 = (\rho)\lambda_2 \Rightarrow \lambda_2 = 0 \qquad [\text{since, in general, } \rho \neq 0]. \tag{d6}$$

(ii) By eliminating $\lambda_{1,2}$ among (d3–6), show that we obtain the two kinetic Maggi equations:

$$\rho\,\ddot{\phi} + 2\dot{\rho}\,\dot{\phi} + R(\dot{\phi})^2 = 0, \tag{e1}$$

$$(z_o - z)\big[\ddot{\rho} + R\,\ddot{\phi} - \rho(\dot{\phi})^2\big] + \rho(\ddot{z} + g) = 0; \tag{e2}$$

which, along with (b1, 2), constitute a determinate system for the four functions ϕ, ρ, z, z_o.
(iii) Show that for the quasi-velocity choice of the preceding example, eqs. (i1–3), (j1–3),

$$\omega_1 \equiv R\,\dot{\phi}, \tag{f1}$$

$$\omega_2 \equiv (\dot{\rho} + R\,\dot{\phi})/\dot{z} = -\dot{z}_o/R\dot{\phi} = (z_o - z)/\rho, \tag{f2}$$

$$\omega_3 \equiv \dot{z}; \tag{f3}$$

$$\dot{q}_1 \equiv \dot{\phi} = \omega_1/R, \tag{f4}$$

$$\dot{q}_2 \equiv \dot{\rho} = \omega_2\,\omega_3 - \omega_1, \tag{f5}$$

$$\dot{q}_3 \equiv \dot{z} = \omega_3, \tag{f6}$$

$$(\Rightarrow \dot{z}_o = -\omega_1\omega_2), \tag{f7}$$

the above equations (e1, 2) coincide with the Appellian equations (m1, 2) found there.

Problem 5.3.6 Continuing from the preceding example of the tetherball, let us consider its *single* nonlinear second-order constraint in the *three* Lagrangean coordinates $q_{1,2,3} = \phi, \rho, z$, eq. (f):

$$f_o \equiv \rho(\dot{\rho} + R\dot{\phi})\ddot{z} - \rho(\ddot{\rho} + R\ddot{\phi})\dot{z} - [(\dot{\rho} + R\dot{\phi})^2 + (\dot{z})^2]\dot{z} = 0, \tag{a}$$

or, rearranged to show the second derivatives more clearly,

$$-(R\rho\dot{z})\ddot{\phi} - (\rho\dot{z})\ddot{\rho} + \rho(\dot{\rho} + R\dot{\phi})\ddot{z} - [(\dot{\rho} + R\dot{\phi})^2 + (\dot{z})^2]\dot{z} = 0. \tag{b}$$

(i) Show that the corresponding Routh–Voss equations,

$$E_k(L) = \lambda(\partial f_o/\partial \ddot{q}_k) \qquad (k = 1, \ldots, 3), \tag{c}$$

are (again with $m = 1$)

$$\phi: \quad \ddot{\rho} + R\ddot{\phi} + (\rho^2/R)\ddot{\phi} + 2(\rho/R)\dot{\rho}\dot{\phi} = -(\rho\dot{z})\lambda, \tag{d1}$$

$$\rho: \quad \ddot{\rho} + R\ddot{\phi} - \rho(\dot{\phi})^2 = -(\rho\dot{z})\lambda, \tag{d2}$$

$$z: \quad \ddot{z} + g = \rho(\dot{\rho} + R\dot{\phi})\lambda; \tag{d3}$$

and along with (b) they constitute a determinate system for the four functions ϕ, ρ, z, λ.

(ii) From (b, d1–3), deduce that

$$\lambda = \frac{g\rho(\dot{\rho} + R\dot{\phi}) + \dot{z}[(\dot{\rho} + R\dot{\phi})^2 + \rho^2(\dot{\phi})^2 + (\dot{z})^2]}{(\dot{\rho})^2[(\dot{\rho} + R\dot{\phi})^2 + (\dot{z})^2]} = (g\rho\omega_2 + 2E_o - 2gz)\big/\omega_3\rho^2(1 + \omega_2{}^2), \tag{e}$$

where $E_o \equiv T + V$ is the constant total energy [recall eq. (n) of ex. 5.3.12].

(iii) With the help of (e), show that the (physical) constraint force equals

$$\boldsymbol{R} = \lambda[\partial f(t, \boldsymbol{r}, \boldsymbol{v}, \boldsymbol{a})/\partial \boldsymbol{a}] = \sum \lambda\left[\partial f_o(t, q, \dot{q}, \ddot{q})/\partial \ddot{q}_k\right](\partial \ddot{q}_k/\partial \boldsymbol{a}) = \cdots = -[(1/l)(v^2 + g\rho\omega_2)(\boldsymbol{r} - \boldsymbol{r}_o)]\big/|\boldsymbol{r} - \boldsymbol{r}_o| = -[(1/l)(v^2 + g\rho\omega_2)/l^2](\boldsymbol{r} - \boldsymbol{r}_o), \tag{f}$$

where $l^2 = \rho^2 + (z_o - z)^2 = \rho^2(1 + \omega_2{}^2)$ is the instantaneous length (squared) of the *unwound* part of the thread.

(iv) By eliminating λ among (d1–3), obtain the *two* kinetic Maggi equations of the system; which, along with (a) or (b), will constitute a determinate system for ϕ, ρ, z.

Problem 5.3.7 (Fufaev, 1990). Continuing from the preceding example of the tetherball:

(i) Show that if the cylinder surface is *smooth*, the system has the sole *holonomic* constraint

$$(\rho + R\phi)^2 + z^2 = l_o{}^2 \qquad (l_o\text{: thread length}), \tag{a}$$

or, in virtual form,

$$[(\rho + R\,\phi)R]\,\delta\phi + (\rho + R\,\phi)\,\delta\rho + z\,\delta z = 0; \tag{b}$$

that is, it has only *three* Lagrangean coordinates (instead of the four of the *rough*-surface case), but still $n - m = 3 - 1 = 2$.

(ii) Show that, in this case, the Routh–Voss equations for $q_{1,2,3} = \phi, \rho, z$ are

$$E_\phi(L) = \lambda R(\rho + R\,\phi), \qquad E_\rho(L) = \lambda(\rho + R\phi), \qquad E_z(L) = \lambda z, \tag{c}$$

where $E_{\phi,\rho,z}(L)$ can be found via prob. 5.3.5: (d1); and along with (a) these constitute a determinate system for ϕ, ρ, z, λ.

(iii) By eliminating λ among (c), show that we obtain the following two kinetic (holonomic) Maggi equations:

$$\rho\,\ddot{\phi} + 2\dot{\rho}\,\dot{\phi} + R(\dot{\phi})^2 = 0, \tag{d}$$

$$z[\ddot{\rho} + R\,\ddot{\phi} - \rho(\dot{\phi})^2] - (\rho + R\,\phi)\,(\ddot{z} + g) = 0, \tag{e}$$

which, along with (a), constitute a determinate system for ϕ, ρ, z.

(iv) Compare eqs. (d, e), of this smooth case, with the corresponding Maggi equations of the rough case of the preceding problems.

Example 5.3.13 *Reduced, or Routh-like Form of the Equations of Motion of a Nonholonomic and Cyclic/Ignorable System* (Semenova, 1965). (To be studied in connection with §8.4.) Let us consider a scleronomic system under the $m(< n)$ *stationary* Pfaffian constraints

$$\sum a_{Dk}\dot{q}_k = 0, \qquad \text{where} \quad \partial a_{Dk}/\partial t = 0 \qquad [D = 1, \ldots, m;\ k = 1, \ldots, n], \tag{a}$$

and, therefore, having the following Routh–Voss equations of motion:

$$E_k(L) \equiv (\partial L/\partial \dot{q}_k)^{\cdot} - \partial L/\partial q_k = Q_k + R_k, \qquad R_k = \sum \lambda_D a_{Dk}. \tag{b}$$

In addition, let us assume that the first $M\ (< n)$ coordinates are cyclic or *ignorable*; that is,

$$(q_1, \ldots, q_M) \equiv (q_i) \equiv (\psi_i):\ \partial L/\partial q_i \equiv \partial L/\partial \psi_i = 0 \qquad [i = 1, \ldots, M], \tag{c}$$

and the corresponding impressed and reaction forces vanish:

$$Q_i = 0 \qquad \text{and} \qquad R_i = 0 \qquad [\text{e.g., if } a_{Di} = 0]. \tag{d}$$

Let us find the Routhian equations of the system; that is, Lagrange-type equations involving only the remaining $n - M$ *noncyclic* or *palpable* or *positional* coordinates $(q_{M+1}, \ldots, q_n) \equiv (q_p)$ and corresponding velocities $(\dot{q}_{M+1}, \ldots, \dot{q}_n) \equiv (\dot{q}_p)$, instead of all the $\dot{q}$'s.

Due to (c, d), eqs. (b) yield

$$p_i \equiv \partial L/\partial \dot{q}_i \equiv \partial L/\partial \dot{\psi}_i = \textit{constant} \equiv C_i; \quad \text{i.e., } C_i = C_i(q_p, \dot{\psi}_i, \dot{q}_p). \tag{e}$$

Solving these M equations for the M ignorable (but, generally, *variable*) velocities $\dot{q}_i \equiv \dot{\psi}_i$, in terms of the $n - M$ palpable variables q_p and $\dot{q}_p$, we obtain

$$\dot{\psi}_i = \dot{\psi}_i(q_p, \dot{q}_p;\ C_i); \tag{f}$$

which is essentially a Chaplygin-like form of the additional constraints (e).

REMARK

Since T is, at most, *quadratic* in the $\dot{q}$, eqs. (e) are essentially *linear* in the $\dot{q}$, and therefore eqs. (f) are linear in the $\dot{q}_i$; say,

$$p_i = C_i = \sum e_{ik}\,\dot{q}_k + e_i, \qquad e_{ik}, e_i\text{: functions of the } q_p, \tag{g1}$$

from which we find

$$\dot{q}_i \equiv \dot{\psi}_i = \sum E_{ip}\,\dot{q}_p + E_i, \qquad E_{ip}, E_i\text{: functions of the } q_p \text{ and } C_i. \tag{g2}$$

However, here we shall treat both (e) and (f) as *additional* linear and/or nonlinear constraints, because then we can see more clearly the formal structure of the resulting equations.

Let L_o be the Lagrangean resulting from the elimination of the $\dot{\psi}_i$ from L via (f); that is, by enforcing in it the constraints (e):

$$\begin{aligned} L = L(q, \dot{q}) &= L(q_p, \dot{q}_i, \dot{q}_p) \\ &= L\big[q_p, \dot{\psi}_i(q_p, \dot{q}_p;\ C_i), \dot{q}_p\big] \equiv L_o(q_p, \dot{q}_p;\ C_i). \end{aligned} \tag{h}$$

Applying chain rule to the above, we obtain

(i) $$\partial L_o/\partial q_p = \partial L/\partial q_p + \sum (\partial L/\partial \dot{\psi}_i)(\partial \dot{\psi}_i/\partial q_p) = \partial L/\partial q_p + \sum (\partial \dot{\psi}_i/\partial q_p) C_i$$

$$\Rightarrow\ \partial L/\partial q_p = \partial L_o/\partial q_p - \sum (\partial \dot{\psi}_i/\partial q_p) C_i, \tag{i1}$$

(ii) $$\begin{aligned} \partial L_o/\partial \dot{q}_p &= \partial L/\partial \dot{q}_p + \sum (\partial L/\partial \dot{\psi}_i)\,(\partial \dot{\psi}_i/\partial \dot{q}_p) \\ &= \partial L/\partial \dot{q}_p + \sum (\partial \dot{\psi}_i/\partial \dot{q}_p) C_i \\ &\Rightarrow \partial L/\partial \dot{q}_p = \partial L_o/\partial \dot{q}_p - \sum (\partial \dot{\psi}_i/\partial \dot{q}_p) C_i; \end{aligned} \tag{i2}$$

and, therefore, since $\dot{C}_i = 0$,

$$(\partial L/\partial \dot{q}_p)^{\cdot} = (\partial L_o/\partial \dot{q}_p)^{\cdot} - \sum C_i (\partial \dot{\psi}_i/\partial \dot{q}_p)^{\cdot}. \tag{i3}$$

In view of (i1–3), the equations of motion (b) for the $q_p, \dot{q}_p$ can be written in the Chaplygin-like form

$$(\partial L_o/\partial \dot{q}_p)^{\cdot} - \partial L_o/\partial q_p - \sum C_i\Big[(\partial \dot{\psi}_i/\partial \dot{q}_p)^{\cdot} - \partial \dot{\psi}_i/\partial q_p\Big] = \sum \lambda_D a_{Dp}, \tag{j1}$$

or, compactly,

$$E_p(L_o) - \sum C_i E_p(\dot{\psi}_i) = \sum \lambda_D a_{Dp} \equiv R_p; \tag{j2}$$

and along with the constraints (a):

$$\sum a_{Di}\dot{q}_i + \sum a_{Dp}\dot{q}_p = 0$$
$$\Rightarrow \sum (\ldots)_{Dp}\dot{q}_p = \text{known function of the } q_p \text{ and } C_i, \tag{j3}$$

they constitute a determinate system of $(n - M) + m$ equations for the $n - M$ q_p and the m λ_D. Equations (j1) are *coupled* in the q_p and λ_D. To uncouple them into kinetic and kinetostatic equations (assuming that $m < n - M$), we may view them as the Routh–Voss equations of a nonholonomic system under the constraints (j3), and then proceed to derive its uncoupled equations à la Maggi, Hamel, or Appell, as elaborated in chapter 3 and § 5.3. The details are left to the reader.

Problem 5.3.8 (Semenova, 1965). Multiplying each of eqs. (j2) of the preceding example:

$$E_p(L_o) - \sum C_i E_p(\dot{\psi}_i) = R_p \qquad [i = 1, \ldots, M;\ p = M + 1, \ldots, n] \tag{a}$$

by $\dot{q}_p$ and summing over p, obtain the "*noncyclic Jacobi integral*"

$$H_o \equiv \left(\sum (\partial L_o/\partial \dot{q}_p)\dot{q}_p - L_o\right) - \sum C_i\left(\sum (\partial\dot{\psi}_i/\partial\dot{q}_p)\dot{q}_p - \dot{\psi}_i\right):$$
$$\textit{Noncyclic generalized energy} = \textit{constant}. \tag{b}$$

Problem 5.3.9 *Nonlinear and Nonholonomic Power Equation.*

(i) Starting with the kinetic Johnsen–Hamel equations of motion (5.3.5b or 5.3.19), show that the corresponding power equation is

$$d/dt\left(\sum (\partial T^*/\partial\omega_I)\omega_I - T^*\right)$$
$$= -\partial T^*/\partial t + \sum \Theta_I\omega_I - \sum\sum H^k{}_I(\partial T^*/\partial\omega_k)\omega_I, \tag{a}$$

where (symbolically)

$$dT^*/dt \equiv \sum \left[(\partial T^*/\partial\omega_I)\dot{\omega}_I + (\partial T^*/\partial\theta_I)\omega_I\right] + \partial T^*/\partial t, \tag{b}$$
$$\sum (\partial T^*/\partial\theta_I)\omega_I \equiv \sum\sum (\partial T^*/\partial q_k)(\partial\dot{q}_k/\partial\omega_I)\omega_I; \tag{c}$$

instead of the more "orthodox"

$$dT^*/dt = \sum (\partial T^*/\partial\omega_I)\dot{\omega}_I + \sum (\partial T^*/\partial q_k)\dot{q}_k + \partial T^*/\partial t. \tag{d}$$

From these results conclude that:

(ii) If $\dot{q} \Leftrightarrow \omega$ *is linear and homogeneous* in these velocities (e.g., catastatic case), the definitions (b) and (d) coincide; and

(iii) Unless $-\sum\sum H^k{}_I(\partial T^*/\partial\omega_k)\omega_I = 0$ (e.g., by antisymmetry of $H^k{}_I$, and $\partial T^*/\partial\omega_k$ are linear and homogeneous in the ω's), the system will be *nonconservative*, even if all the Θ_I's are potential and all constraints are stationary.

5.4 SECOND- AND HIGHER-ORDER CONSTRAINTS

The foregoing theory can be easily extended to the following case of *second*-order, generally nonholonomic, constraints:

$$f_D(t, q, \dot{q}, \ddot{q}) = 0 \qquad [D = 1, \ldots, m(< n)]. \tag{5.4.1}$$

Here, compatibility among the various differential variational principles (chap. 6) requires that the virtual displacements corresponding to (5.4.1) be constrained by

$$\sum (\partial f_D/\partial \ddot{q}_k)\, \delta q_k = 0, \tag{5.4.2}$$

instead of (5.2.9). Hence, the virtual form of the constraints

$$\dot{\omega}_D \equiv f_D(t, q, \dot{q}, \ddot{q}) = 0 \qquad [D = 1, \ldots, m(< n)], \tag{5.4.3a}$$

$$\dot{\omega}_I \equiv f_I(t, q, \dot{q}, \ddot{q}) \neq 0 \qquad [I = m + 1, \ldots, n] \tag{5.4.3b}$$

is

$$\delta\theta_D \equiv \sum (\partial \dot{\omega}_D/\partial \ddot{q}_k)\, \delta q_k = 0, \qquad \delta\theta_I \equiv \sum (\partial \dot{\omega}_I/\partial \ddot{q}_k)\, \delta q_k \neq 0. \tag{5.4.4}$$

It follows that all the previous results hold in this case, too, but with $\partial \dot{q}/\partial \omega$ $(\partial \omega/\partial \dot{q})$ replaced with $\partial \ddot{q}/\partial \dot{\omega}$ $(\partial \dot{\omega}/\partial \ddot{q})$. For example, the second-order counterparts of the, say, kinetic Maggi and Hadamard equations will be

Maggi:

$$\sum (\partial \ddot{q}_k/\partial \dot{\omega}_I) E_k(T) = \sum (\partial \ddot{q}_k/\partial \dot{\omega}_I) Q_k \qquad \text{(Lagrangean form)}, \tag{5.4.5a}$$

$$\begin{aligned} \partial S^*/\partial \dot{\omega}_I &= \sum (\partial \ddot{q}_k/\partial \dot{\omega}_I)\, (\partial S/\partial \ddot{q}_k) \\ &= \sum (\partial \ddot{q}_k/\partial \dot{\omega}_I) Q_k \qquad \text{(Appellian form)}, \end{aligned} \tag{5.4.5b}$$

Hadamard:

$$\begin{aligned} &E_I(T) + \sum (\partial \Phi_D/\partial \ddot{q}_I) E_D(T) \\ &\quad = Q_I + \sum (\partial \Phi_D/\partial \ddot{q}_I) Q_D \qquad \text{(Lagrangean form)}, \end{aligned} \tag{5.4.6a}$$

$$\begin{aligned} \partial S_o/\partial \ddot{q}_I &= \partial S/\partial \ddot{q}_I + \sum (\partial \Phi_D/\partial \ddot{q}_I)\, (\partial S/\partial \ddot{q}_D) \\ &= Q_I + \sum (\partial \Phi_D/\partial \ddot{q}_I) Q_D \qquad \text{(Appellian form)}; \end{aligned} \tag{5.4.6b}$$

where

$$\text{eqs. (5.4.1)} \rightarrow \ddot{q}_D = \ddot{q}_D(t, q, \dot{q}, \ddot{q}_I) \equiv \Phi_D(t, q, \dot{q}, \ddot{q}_I), \tag{5.4.7a}$$

$$S(t, q, \dot{q}, \ddot{q}) = \cdots = S^*(t, q, \omega, \dot{\omega}) = \cdots = S_o(t, q, \dot{q}, \ddot{q}_I). \tag{5.4.7b}$$

Similarly, for the *higher*-order constraints:

$$\begin{aligned} &f_D(t, q, \dot{q}, \ddot{q}, \dddot{q}, \ldots, \overset{(s)}{q}) = 0 \\ &[D = 1, \ldots, m(< n);\ s = 1, 2, 3, \ldots], \end{aligned} \tag{5.4.8}$$

eqs. (5.4.2–4) are replaced, respectively, by

$$\sum\left(\partial f_D/\partial \overset{(s)}{q_k}\right)\delta q_k = 0, \tag{5.4.9}$$

$$\overset{(s-1)}{\omega_D} \equiv f_D\left(t, q, \dot{q}, \ddot{q}, \dddot{q}, \ldots, \overset{(s)}{q}\right) = 0, \tag{5.4.10a}$$

$$\overset{(s-1)}{\omega_I} \equiv f_I\left(t, q, \dot{q}, \ddot{q}, \dddot{q}, \ldots, \overset{(s)}{q}\right) \neq 0, \tag{5.4.10b}$$

$$\delta\theta_D \equiv \sum\left(\partial \overset{(s-1)}{\omega_D} \Big/ \partial \overset{(s)}{q_k}\right)\delta q_k = 0, \tag{5.4.10c}$$

$$\delta\theta_I \equiv \sum\left(\partial \overset{(s-1)}{\omega_I} \Big/ \partial \overset{(s)}{q_k}\right)\delta q_k \neq 0. \tag{5.4.10d}$$

In general, starting with $\dot{q}_D = \dot{q}_D(t, q, \dot{q}_I)$ we can easily verify that

$$\partial \dot{q}_D/\partial \dot{q}_I = \partial \ddot{q}_D/\partial \ddot{q}_I = \partial \dddot{q}_D/\partial \dddot{q}_I = \cdots; \tag{5.4.11}$$

and similalry for identities involving $\partial \overset{(s-1)}{\omega_D} / \partial \overset{(s)}{q_k}$.

These topics are examined in detail in chapter 6.

Example 5.4.1 (Mei, 1987, pp. 273–274). Let us derive the equations of motion of a rigid body moving (rotating) about a fixed point and subject to the acceleration (second-order) constraint

$$(\omega_x\dot{\omega}_y - \omega_y\dot{\omega}_x) + (\omega_x{}^2 + \omega_y{}^2)\omega_z - c(\omega_x{}^2 + \omega_y{}^2)^{3/2} = 0, \tag{a}$$

where $\omega_{x,y,z}$ are *body-fixed* components of (inertial) angular velocity of the body and c is a constant. If ϕ, θ, ψ are the Eulerian angles between space-fixed and body-fixed axes, then [recalling results from §1.12, and with $s(\ldots) \equiv \sin(\ldots), c(\ldots) \equiv \cos(\ldots)$]

$$\omega_x = (s\theta\, s\psi)\dot{\phi} + (c\psi)\dot{\theta}, \qquad \omega_y = (s\theta\, c\psi)\dot{\phi} + (-s\psi)\dot{\theta}, \qquad \omega_z = (c\theta)\dot{\phi} + (1)\dot{\psi}, \tag{b}$$

and their inverses,

$$\dot{\phi} = (1/\sin\theta)\,[(s\psi)\omega_x + (c\psi)\omega_y], \qquad \dot{\theta} = (c\psi)\omega_x - (s\psi)\omega_y,$$
$$\dot{\psi} = \omega_z - [(s\psi)\omega_x + (c\psi)\omega_y]\cot\theta. \tag{c}$$

In view of (a), we choose the following *quasi accelerations*:

$$\alpha_1 \equiv (\omega_x\dot{\omega}_y - \omega_y\dot{\omega}_x) + (\omega_x{}^2 + \omega_y{}^2)\omega_z - c(\omega_x{}^2 + \omega_y{}^2)^{3/2} = 0, \tag{d1}$$

$$\alpha_2 \equiv (\dot{\omega}_x c\psi - \dot{\omega}_y s\psi)\big/(\omega_x c\psi - \omega_y s\psi) \neq 0, \tag{d2}$$

$$\alpha_3 \equiv \dot{\omega}_z \neq 0; \tag{d3}$$

which, upon inverting and enforcing the constraint (d1) yield

$$\dot{\omega}_x = (\omega_x)\alpha_2 + \textit{no } \alpha\textit{-terms}, \tag{e1}$$

$$\dot{\omega}_y = (\omega_y)\alpha_2 + \textit{no } \alpha\textit{-terms}, \tag{e2}$$

$$\dot{\omega}_z = (1)\alpha_3 + \textit{no } \alpha\textit{-terms}. \tag{e3}$$

The Appellian of the body is (recalling the results of §3.14; and with A, B, C: principal moments of inertia of body at fixed point)

$$\begin{aligned} 2S^* &= A(\dot{\omega}_x)^2 + B(\dot{\omega}_y)^2 + C(\dot{\omega}_z)^2 + 2(C-B)\omega_y\,\omega_z\,\dot{\omega}_x \\ &\quad + 2(A-C)\omega_x\,\omega_z\,\dot{\omega}_y + 2(B-A)\omega_x\,\omega_y\,\dot{\omega}_z + \textit{no other } \dot{\omega} \textit{ terms}, \end{aligned} \tag{f}$$

and, therefore, substituting into it (e1–3), we obtain the *constrained* Appellian:

$$\begin{aligned} 2S^* &\Rightarrow 2S^*{}_o \\ &= A\,\omega_x{}^2\alpha_2{}^2 + B\,\omega_y{}^2\alpha_2{}^2 + C\,\alpha_3{}^2 \\ &\quad + 2(C-B)\omega_x\,\omega_y\,\omega_z\,\alpha_2 + 2(A-C)\omega_x\,\omega_y\,\omega_z\,\alpha_2 + 2(B-A)\omega_x\,\omega_y\,\alpha_3 \\ &\quad + \textit{no other } \alpha \textit{ terms}, \end{aligned} \tag{g}$$

and from this we get the corresponding constrained inertial "forces":

$$\begin{aligned} \partial S^*{}_o/\partial\alpha_2 &= A\,\omega_x{}^2\alpha_2 + B\,\omega_y{}^2\alpha_2 + (C-B)\omega_x\,\omega_y\,\omega_z + (A-C)\omega_x\,\omega_y\,\omega_z \\ &= A\,\omega_x\,\dot{\omega}_x + B\,\omega_y\,\dot{\omega}_y + (A-B)\omega_x\,\omega_y\,\omega_z \qquad \text{[with (d2)]}, \end{aligned} \tag{h1}$$

$$\begin{aligned} \partial S^*{}_o/\partial\alpha_3 &= C\alpha_3 + (B-A)\omega_x\,\omega_y \\ &= C(\dot{\omega}_z) + (B-A)\omega_x\,\omega_y \qquad \text{[with (d3)]}. \end{aligned} \tag{h2}$$

Further, with $Q_{\phi,\theta,\psi}$: unconstrained *holonomic* components of impressed force, and using (b), (c), and (e1–3), we obtain its *constrained nonholonomic* components:

$$\begin{aligned} \Theta_2 &= Q_\phi[(\partial\dot{\phi}/\partial\omega_x)\,(\partial\dot{\omega}_x/\partial\alpha_2) + (\partial\dot{\phi}/\partial\omega_y)\,(\partial\dot{\omega}_y/\partial\alpha_2) + (\partial\dot{\phi}/\partial\omega_z)\,(\partial\dot{\omega}_z/\partial\alpha_2)] \\ &\quad + Q_\theta[(\partial\dot{\theta}/\partial\omega_x)\,(\partial\dot{\omega}_x/\partial\alpha_2) + (\partial\dot{\theta}/\partial\omega_y)\,(\partial\dot{\omega}_y/\partial\alpha_2) + (\partial\dot{\theta}/\partial\omega_z)\,(\partial\dot{\omega}_z/\partial\alpha_2)] \\ &\quad + Q_\psi[(\partial\dot{\psi}/\partial\omega_x)\,(\partial\dot{\omega}_x/\partial\alpha_2) + (\partial\dot{\psi}/\partial\omega_y)\,(\partial\dot{\omega}_y/\partial\alpha_2) + (\partial\dot{\psi}/\partial\omega_z)\,(\partial\dot{\omega}_z/\partial\alpha_2)] \\ &= (Q_\phi/\sin\theta)\,(\omega_x\,s\psi + \omega_y\,c\psi) + Q_\theta(\omega_x\,c\psi + \omega_y\,s\psi) - Q_\psi(\omega_x\,s\psi + \omega_y\,c\psi)\cot\theta \\ &= \dot{\phi}\,Q_\phi + \dot{\theta}\,Q_\theta - \dot{\phi}\,Q_\psi\,c\theta, \end{aligned} \tag{i1}$$

$$\Theta_3 = Q_\phi(\ldots) + Q_\theta(\ldots) + Q_\psi(\ldots) = \cdots = Q_\psi. \tag{i2}$$

As a result of (h1, 2), and (i1, 2), Appell's *kinetic* equations $\partial S^*{}_o/\partial\alpha_I = \Theta_I$ $(I = 2, 3)$ become

$$2: \quad A\,\omega_x\,\dot{\omega}_x + B\,\omega_y\,\dot{\omega}_y + (A-B)\,\omega_x\,\omega_y\,\omega_z = \dot{\phi}\,Q_\phi + \dot{\theta}\,Q_\theta - \dot{\phi}\,Q_\psi\,c\theta, \tag{j1}$$

$$3: \quad C\,\dot{\omega}_z + (B-A)\omega_x\,\omega_y = Q_\psi; \tag{j2}$$

and along with (b) they constitute a determinate set of five equations for $\dot{\phi}, \dot{\theta}, \dot{\psi};\ \omega_x, \omega_y, \omega_z$.

The remaining kinetostatic equation corresponding to α_1, and based on the *relaxed* Appellian, is

$$(\partial S^*/\partial \alpha_1)_o = \Theta_1 + \Lambda_1; \tag{k}$$

and, once the motion has been determined from (j1, 2), this yields the reaction Λ_1 necessary to maintain the constraint (a). The details of (k) are left to the reader.

Finally, we remark that the above Appellian equations are simpler than those based on $\dot{\omega}_x, \dot{\omega}_y, \dot{\omega}_z$; that is, $\partial S^*/\partial \dot{\omega}_x$, and so on. See, for example, San (1973).

6

Differential Variational Principles

and Associated Generalized Equations of Motion of Nielsen, Tsenov, et al.

The incautious observer might then be tempted to remark: if all mechanics problems can be solved by Newtonian mechanics, is it really economical to introduce a flock of differently stated principles which, after all, can accomplish no more? To this we make a three-fold rejoinder. In the first place, the ease of solving a given problem generally depends on the way in which it is stated, and a method which solves it when it is stated one way may be vastly simpler than that which handles it when the statement is made in another form. In the second place, we can fairly say that *every restatement of the fundamental principles deepens our appreciation of, and feeling for, the whole subject: two methods of solving the same problem mean more in our understanding than the solution of two problems by the same method.* More important in many respects than these answers is, however, the third: it is, by no means, sure that the Newtonian principles are actually competent to describe *all* phenomena in which motion occurs .... It seems plausible that alternative points of view may, themselves, suggest fundamental modifications in mechanical principles which will lead to successful attacks on the new problems.

(Lindsay and Margenau, 1936, p. 103, emphasis added)

6.1 INTRODUCTION

This chapter treats (i) the differential variational principles of constrained system dynamics (of Lagrange, Jourdain, Gauss, Hertz, Mangeron–Deleanu, et al.) from a simple and unified viewpoint, and (ii) the associated kinematico-inertial identities and corresponding generalized equations of motion (of Nielsen, Tsenov, Dolaptschiew, et al.).

These topics, until recently viewed by many as academic curiosities, have re-emerged as powerful and versatile tools for the theoretical and numerical handling of problems of nonlinear nonholonomic constraints in the velocities, accelerations, and so on; and also, in impulsive motion and multibody dynamics.

For parallel reading we recommend the following: Mei (1985), Mei et al. (1991) and references cited therein; and the (Soviet →) Russian and Chinese journals of *Applied Mathematics and Mechanics*.

6.2 THE GENERAL THEORY

The differential variational principles of mechanics (DVP) are statements to the effect that certain differential expressions, linear and homogeneous in the appropriate kinematical variations from a kinetic state (or first variations of certain scalar energetic functions from it), vanish. The principle of Lagrange (LP) is the simplest and most fundamental of them: as shown below, *excluding singular configurations of the system*, all other DVP derive from it. Here, as with most of the rest of the book, the discussion is limited to bilateral and ideal constraints; that is, we assume that (recalling §3.2 and the notations employed there)

$$\mathcal{S}\, d\boldsymbol{R}\cdot\delta\boldsymbol{r} = 0 \;\Rightarrow\; \mathcal{S}\, dm\,\boldsymbol{a}\cdot\delta\boldsymbol{r} = \mathcal{S}\, d\boldsymbol{F}\cdot\delta\boldsymbol{r}, \tag{6.2.1}$$

where, as detailed earlier (§2.5),

$$\delta\boldsymbol{r} = \sum \boldsymbol{e}_k\,\delta q_k = \sum \boldsymbol{\varepsilon}_k\,\delta\theta_k$$

$$\left(= \sum \boldsymbol{\varepsilon}_I\,\delta\theta_I,\ I = m+1,\ldots,n;\ m\text{: number of } \textit{additional} \text{ Pfaffian constraints}\right). \tag{6.2.1a}$$

Now, $(\ldots)^{\cdot}$-differentiating (6.2.1) once, and recalling that $\mathcal{S}\ldots$ and $(\ldots)^{\cdot}$ commute, we obtain

$$\mathcal{S}\,[(dm\,\boldsymbol{a} - d\boldsymbol{F})^{\cdot}\cdot\delta\boldsymbol{r}] + \mathcal{S}\,[(dm\,\boldsymbol{a} - d\boldsymbol{F})\cdot(\delta\boldsymbol{r})^{\cdot}] = 0. \tag{6.2.2}$$

From this, we readily conclude that the equations of motion of the system can be derived from the variational equation

$$\mathcal{S}\,(dm\,\boldsymbol{a} - d\boldsymbol{F})\cdot(\delta\boldsymbol{r})^{\cdot} = 0, \tag{6.2.3}$$

where the $\delta\boldsymbol{r}$ satisfy not only the familiar $\delta t = 0$, but also $\delta\boldsymbol{r} = \boldsymbol{0}$; and since, as we have already seen (§4.6; also, ex. 6.2.1 below), we can always take $(\delta\boldsymbol{r})^{\cdot} = \delta(\dot{\boldsymbol{r}}) \equiv \delta\boldsymbol{v}$, LP can be replaced by the following DVP:

$$\mathcal{S}\,(dm\,\boldsymbol{a} - d\boldsymbol{F})\cdot\delta\boldsymbol{v} = 0, \qquad \text{with}\quad \delta t = 0 \quad \textit{and} \quad \delta\boldsymbol{r} = \boldsymbol{0}. \tag{6.2.4}$$
$$(\text{constraints on } \delta\boldsymbol{v} \neq \boldsymbol{0}).$$

Next, $(\ldots)^{\cdot}$-differentiating (6.2.4) once yields

$$\mathcal{S}\,[(dm\,\boldsymbol{a} - d\boldsymbol{F})^{\cdot}\cdot\delta\boldsymbol{v}] + \mathcal{S}\,[(dm\,\boldsymbol{a} - d\boldsymbol{F})\cdot(\delta\boldsymbol{v})^{\cdot}] = 0, \tag{6.2.5}$$

and reasoning as earlier, and with $(\delta\boldsymbol{v})^{\cdot} = \delta(\dot{\boldsymbol{v}}) \equiv \delta\boldsymbol{a}$, we see that the equations of motion of the system can be obtained from the variational equation

$$\mathcal{S}\,(dm\,\boldsymbol{a} - d\boldsymbol{F})\cdot\delta\boldsymbol{a} = 0, \qquad \text{with } \delta t = 0,\quad \delta\boldsymbol{r} = \boldsymbol{0},\quad \textit{and}\quad \delta\boldsymbol{v} = \boldsymbol{0} \tag{6.2.6}$$
$$(\text{constraints on } \delta\boldsymbol{a} \neq \boldsymbol{0}).$$

Continuing this process, inductively, we can easily generalize to the following DVP: *The equations of motion of an ideally constrained mechanical system derive from*

$$\mathcal{S}\left((dm\,\boldsymbol{a} - d\boldsymbol{F})\cdot\delta\overset{(s)}{\boldsymbol{r}}\right) = 0 \qquad (s = 0, 1, 2, \ldots), \tag{6.2.7}$$

where the variations satisfy

$$\delta t = 0 \quad and \quad \delta \boldsymbol{r} = \boldsymbol{0}, \quad \delta(\dot{\boldsymbol{r}}) = \boldsymbol{0}, \quad \delta(\ddot{\boldsymbol{r}}) = \boldsymbol{0}, \ldots, \quad \delta\Big(\overset{(s-1)}{\boldsymbol{r}}\Big) = \boldsymbol{0} \quad (s - 1 \geq 0). \quad (6.2.7a)$$

$$\left(\text{constraints on } \delta \overset{(s)}{\boldsymbol{r}} \equiv \delta(d^s\boldsymbol{r}/dt^s) \neq \boldsymbol{0}\right).$$

The DVP corresponding to $s = 0, 1, 2$ are called, respectively, principles of *Lagrange* [eq. (6.2.1)] *Jourdain* [eq. (6.2.4)], and *Gauss* (*Gibbs*) [eq. (6.2.6)]; while the case corresponding to a general s, in (6.2.7), is referred to as the principle of *Mangeron–Deleanu* [eqs. (6.2.7, 7a)]; that is, roughly, Jourdain's principle (JP) is Lagrange's principle (LP) with $\delta\boldsymbol{r} \to \delta\boldsymbol{v}$, and Gauss' principle (GP) is LP with $\delta\boldsymbol{r} \to \delta\boldsymbol{a}$, and so on.

[According to Nordheim (1927, pp. 68–69), this unified and simple approach to DVP (which, however, holds only under *differentiable* conditions!) seems to be due to Leitinger (1913) and other members of the "Austrian school" (ca. 1910).]

REMARKS

(i) The equations associated with these variations are an *instantaneous* representation of the system. Otherwise, if, in JP, $\delta\boldsymbol{r}(t) = \boldsymbol{0}$ continuously, one might conclude, incorrectly, that $(\delta\boldsymbol{r})^\cdot = \delta(d\boldsymbol{r}/dt) = \boldsymbol{0}$. Rather, *at each succeeding instant, $\delta\boldsymbol{r}$ is reset equal to zero, in accordance with the instantaneous viewpoint.* Clearly, this viewpoint does not apply to equations involving time integrals; similarly for GP, and so on.

(ii) In the same spirit, we avoid the occasionally used term *virtual power* for $S\, d\boldsymbol{F} \cdot \delta\boldsymbol{v}$ and the consequent term *principle of virtual power* for JP. Power means work per unit time; that is, $(S\, d\boldsymbol{F} \cdot \delta\boldsymbol{r})/\delta t$, but since, here, $\delta t = 0$, such a term could be confusing.

The next step is to transform these DVP to system variables, and then to obtain the corresponding equations of motion. By now, LP is well known (chap. 3), and so we begin with the principle of Jourdain.

Example 6.2.1 *The Significance of the Commutation Rule in Jourdain's Principle (JP).* During the formulation of JP, (6.2.3, 4), we invoked the commutation rule:

$$d(\delta\boldsymbol{r}) = \delta(d\boldsymbol{r}) \qquad \text{or} \qquad (\delta\boldsymbol{r})^\cdot = (\delta\dot{\boldsymbol{r}}) \equiv \delta\boldsymbol{v}. \tag{a}$$

However, since the derivation of the equations of motion [either from LP or from the central equation (§3.5, §3.6 and §5.3)] is independent of any particular assumptions about $d(\delta\boldsymbol{r}) - \delta(d\boldsymbol{r})$, and since the ultimate purpose and usefulness of JP—in fact, of all DVP—is to produce correct equations of motion, it follows that *JP, too, should be independent of (a).* Let us see in detail why this is so.

We begin with the most general expression for $\delta\boldsymbol{r}$ (recalling the relevant theory and notations of §2.4–2.9):

$$\delta\boldsymbol{r} = \sum \boldsymbol{e}_k\, \delta q_k = \sum \boldsymbol{\varepsilon}_k\, \delta\theta_k = \sum \boldsymbol{\varepsilon}_I\, \delta\theta_I \qquad (\text{since } \delta\theta_D = 0). \tag{b}$$

Since the $n - m$ vectors $\boldsymbol{\varepsilon}_I$ are independent, the Jourdain requirement $\delta\boldsymbol{r} = \boldsymbol{0}$ applied to (b) yields $\delta\theta_I = 0$; that is, in sum, here we have

$$\delta\theta_k = 0 \qquad (k = 1, 2, 3, \ldots, n). \tag{c}$$

Next, $(\ldots)^{\cdot}$-differentiating (a) and *then* enforcing (c), we find

$$(\delta \boldsymbol{r})^{\cdot} = \sum [(d\boldsymbol{\varepsilon}_I/dt)\,\delta\theta_I + \boldsymbol{\varepsilon}_I(\delta\theta_I)^{\cdot}]$$
$$= \sum \boldsymbol{\varepsilon}_I(\delta\theta_I)^{\cdot} = \sum (\partial\boldsymbol{v}/\partial\omega_I)(\delta\theta_I)^{\cdot}. \tag{d}$$

Now, let us calculate $\delta\boldsymbol{v} \Rightarrow (\delta\boldsymbol{v})_{\text{Jourdain variation}} \equiv \delta'\boldsymbol{v}$ [see also (6.3.5) below]. Assuming for simplicity, but no loss of generality, a scleronomic system, we have (omitting superstars on $\boldsymbol{v}$ etc., for simplicity)

$$\boldsymbol{v} = \sum \boldsymbol{\varepsilon}_I\,\omega_I, \tag{e}$$

and, therefore,

$$\delta\boldsymbol{v} = \sum (\delta\boldsymbol{\varepsilon}_I\,\omega_I + \boldsymbol{\varepsilon}_I\,\delta\omega_I) \;\Rightarrow\; \delta'\boldsymbol{v} = \sum \boldsymbol{\varepsilon}_I\,\delta\omega_I, \tag{f}$$

since, *at least for Pfaffian constraints*, $\boldsymbol{\varepsilon}_I = \boldsymbol{\varepsilon}_I(q) \;\Rightarrow\; \delta'\boldsymbol{\varepsilon}_I = \boldsymbol{0}$ [see "Remarks" (i) below].

Subtracting (d) and (f) side by side, we obtain the following *transitivity equation in the sense of Jourdain*:

$$(\delta\boldsymbol{r})^{\cdot} - \delta'\boldsymbol{v} = \sum \boldsymbol{\varepsilon}_I[(\delta\theta_I)^{\cdot} - \delta\omega_I]$$
$$\Rightarrow (\delta\boldsymbol{r})^{\cdot} = \delta'\boldsymbol{v} + \sum \boldsymbol{\varepsilon}_I[(\delta\theta_I)^{\cdot} - \delta\omega_I]. \tag{g}$$

Next, inserting (f) into the $(\ldots)^{\cdot}$-derivative of LP under $\delta\boldsymbol{r} = \boldsymbol{0}$; that is (6.2.3),

$$S\,(dm\,\boldsymbol{a} - d\boldsymbol{F}) \cdot (\delta\boldsymbol{r})^{\cdot} = 0, \tag{h}$$

we get

$$S\,(dm\,\boldsymbol{a} - d\boldsymbol{F}) \cdot \delta'\boldsymbol{v} + S\,(dm\,\boldsymbol{a} - d\boldsymbol{F}) \cdot \sum \boldsymbol{\varepsilon}_I[(\delta\theta_I)^{\cdot} - \delta\omega_I] = 0,$$

or, rearranging,

$$S\,(dm\,\boldsymbol{a} - d\boldsymbol{F}) \cdot \delta'\boldsymbol{v} + \sum \Big(S\,(dm\,\boldsymbol{a} - d\boldsymbol{F}) \cdot \boldsymbol{\varepsilon}_I\Big)[(\delta\theta_I)^{\cdot} - \delta\omega_I] = 0; \tag{i}$$

from which, and this is the key step in the entire discussion, since

$$S\,(dm\,\boldsymbol{a} - d\boldsymbol{F}) \cdot \boldsymbol{\varepsilon}_I = 0 \qquad \text{["raw" form of LP in quasi variables; also (6.3.26)],} \tag{j}$$

we finally obtain the *original* ("Jourdainian") form of JP (1909):

$$S\,(dm\,\boldsymbol{a} - d\boldsymbol{F}) \cdot \delta'\boldsymbol{v} = 0, \qquad \text{under } \delta t = 0 \text{ and } \delta\boldsymbol{r} = \boldsymbol{0} \quad [\text{but } (\delta\boldsymbol{r})^{\cdot} \neq \boldsymbol{0}]. \tag{k}$$

In short, the transitivity condition $(\delta\boldsymbol{r})^{\cdot} = \delta\boldsymbol{v}$ is *sufficient but not necessary for the derivation of JP from LP*. Finally, substituting (f) into (k), and since the $n - m$ $\delta\omega_I$ are independent, reproduces (j).

REMARKS

(i) Equation (g) can also result from the general transitivity equation (§2.10):

$$(\delta\theta_k)^{\cdot} - \delta\omega_k = \sum a_{kl}[(\delta q_l)^{\cdot} - \delta(\dot{q}_l)] + \sum\sum \gamma^k{}_{rs}\,\omega_s\,\delta\theta_r, \tag{11}$$

and its inverse

$$(\delta q_l)^{\cdot} - \delta(\dot{q}_l) = \sum A_{lk}[(\delta\theta_k)^{\cdot} - \delta\omega_k] - \sum\sum\sum A_{lk}\gamma^k{}_{rs}\,\omega_s\,\delta\theta_r. \tag{12}$$

We have, successively,

$$\begin{aligned}(\delta \boldsymbol{r})^{\cdot} - \delta\boldsymbol{v} &= \sum \boldsymbol{e}_l[(\delta q_l)^{\cdot} - \delta(\dot{q}_l)]\\ &= \sum\Big(\sum A_{lk}\boldsymbol{e}_l\Big)[(\delta\theta_k)^{\cdot} - \delta\omega_k] - \sum\sum\sum\Big(\sum A_{lk}\boldsymbol{e}_l\Big)\gamma^k{}_{rs}\,\omega_s\,\delta\theta_r\\ &= \sum \boldsymbol{\varepsilon}_k[(\delta\theta_k)^{\cdot} - \delta\omega_k] - \sum\sum\sum \boldsymbol{\varepsilon}_k\gamma^k{}_{rs}\,\omega_s\,\delta\theta_r,\end{aligned} \tag{m}$$

from which, due to (c) and since now $\delta\boldsymbol{v} \Rightarrow \delta'\boldsymbol{v}$, we recover (g). Clearly, this derivation is not limited to Pfaffian constraints, and so it avoids the earlier restriction $\boldsymbol{\varepsilon}_I = \boldsymbol{\varepsilon}_I(q)$.

(ii) If we had assumed $(\delta\boldsymbol{r})^{\cdot} = \delta\boldsymbol{v}$, then (g) and (12) would have led us to $(\delta\theta_k)^{\cdot} = \delta\omega_k$ and also to $(\delta q_k)^{\cdot} = \delta(\dot{q}_k)$, and vice versa. As (11, 2) readily show, without the Jourdain constraints, *either* $(\delta q_k)^{\cdot} = \delta(\dot{q}_k)$ *or* $(\delta\theta_k)^{\cdot} = \delta\omega_k$, *but not both.*

(iii) The above reasoning extends readily to *higher-order* constraints. For additional insights see also Bremer (1993).

6.3 PRINCIPLE OF JOURDAIN, AND EQUATIONS OF NIELSEN

As detailed in chapters 2 and 5, starting with the general position vector in the Lagrangean variables q,

$$\boldsymbol{r} = \boldsymbol{r}(t, q), \tag{6.3.1}$$

we readily find

$$\boldsymbol{v} = d\boldsymbol{r}/dt = \sum (\partial\boldsymbol{r}/\partial q_k)\dot{q}_k + \partial\boldsymbol{r}/\partial t, \tag{6.3.2}$$

from which

$$\partial\boldsymbol{r}/\partial q_k = \partial\boldsymbol{v}/\partial\dot{q}_k \equiv \partial\dot{\boldsymbol{r}}/\partial\dot{q}_k \equiv \boldsymbol{e}_k \qquad (k = 1, 2, \ldots, n); \tag{6.3.3}$$

and

$$\begin{aligned}\delta\boldsymbol{v} \equiv \delta(\dot{\boldsymbol{r}}) &= \sum (\partial\boldsymbol{r}/\partial q_k)\,\delta(\dot{q}_k) + \sum \delta(\partial\boldsymbol{r}/\partial q_k)\dot{q}_k + \delta(\partial\boldsymbol{r}/\partial t)\\ &= \sum (\partial\dot{\boldsymbol{r}}/\partial\dot{q}_k)\,\delta(\dot{q}_k) + \textit{no other } \delta\dot{q} \textit{ terms}\\ &\equiv \sum \boldsymbol{e}_k\,\delta(\dot{q}_k) + \textit{no other } \delta\dot{q} \textit{ terms}\\ &\equiv \delta'\boldsymbol{v} + \textit{no other } \delta\dot{q} \textit{ terms},\end{aligned} \tag{6.3.4}$$

where

$$\delta'(\ldots) \equiv \sum [\partial(\ldots)/\partial\dot{q}_k]\,\delta(\dot{q}_k):$$

Jourdain variation of $(\ldots)$ [i.e., $\delta(\ldots)$ with $\delta t = 0$ *and* $\delta q_k = 0$]. (6.3.5)

Next, from (6.3.2), we easily obtain

$$\partial\dot{\boldsymbol{r}}/\partial q_k = \sum (\partial^2\boldsymbol{r}/\partial q_k\partial q_l)\dot{q}_l + \partial^2\boldsymbol{r}/\partial q_k\partial t, \tag{6.3.6a}$$

and

$$\begin{aligned} \boldsymbol{a} = d\boldsymbol{v}/dt = \ddot{\boldsymbol{r}} \\ &= \sum (\partial\boldsymbol{r}/\partial q_k)\ddot{q}_k + \sum\sum (\partial^2\boldsymbol{r}/\partial q_k\partial q_l)\dot{q}_k\dot{q}_l + 2\sum (\partial^2\boldsymbol{r}/\partial q_k\partial t)\dot{q}_k \\ &\qquad + \partial^2\boldsymbol{r}/\partial t^2, \qquad (6.3.6b) \\ &\Rightarrow \partial\boldsymbol{a}/\partial\dot{q}_k \equiv \partial\ddot{\boldsymbol{r}}/\partial\dot{q}_k = 2\Big(\sum (\partial^2\boldsymbol{r}/\partial q_k\partial q_l)\dot{q}_l + \partial^2\boldsymbol{r}/\partial q_k\partial t\Big), \end{aligned}$$

and, comparing with (6.3.6a), we deduce the following *kinematical identity*:

$$\partial\ddot{\boldsymbol{r}}/\partial\dot{q}_k = 2(\partial\dot{\boldsymbol{r}}/\partial q_k), \qquad \text{or} \qquad \partial\boldsymbol{a}/\partial\dot{q}_k = 2(\partial\boldsymbol{v}/\partial q_k). \qquad (6.3.7)$$

The above also reconfirm the already known basic result, extension of (6.3.3),

$$\boldsymbol{e}_k \equiv \partial\boldsymbol{r}/\partial q_k = \partial\dot{\boldsymbol{r}}/\partial\dot{q}_k = \partial\ddot{\boldsymbol{r}}/\partial\ddot{q}_k = \cdots$$

or

$$\partial\boldsymbol{r}/\partial q_k = \partial\boldsymbol{v}/\partial\dot{q}_k = \partial\boldsymbol{a}/\partial\ddot{q}_k = \cdots \qquad (k = 1, 2, \ldots, n). \qquad (6.3.8)$$

Next, $(\ldots)^{\cdot}$-differentiating the kinetic energy

$$2T = S\, dm\ \boldsymbol{v}\cdot\boldsymbol{v} = S\, dm\ \dot{\boldsymbol{r}}\cdot\dot{\boldsymbol{r}}, \qquad (6.3.9)$$

we readily obtain

$$dT/dt = S\, dm\ \boldsymbol{v}\cdot\boldsymbol{a} = S\, dm\ \dot{\boldsymbol{r}}\cdot\ddot{\boldsymbol{r}}, \qquad (6.3.10)$$

and, therefore, invoking (6.3.7) and (6.3.8), and since $\partial T/\partial q_k = S\, dm\ \dot{\boldsymbol{r}}\cdot(\partial\dot{\boldsymbol{r}}/\partial q_k)$,

$$\begin{aligned} \partial\dot{T}/\partial\dot{q}_k &= S\, dm\ (\partial\dot{\boldsymbol{r}}/\partial\dot{q}_k)\cdot\ddot{\boldsymbol{r}} + S\, dm\ \dot{\boldsymbol{r}}\cdot(\partial\ddot{\boldsymbol{r}}/\partial\dot{q}_k) \\ &= S\, dm\ (\partial\dot{\boldsymbol{r}}/\partial\dot{q}_k)\cdot\ddot{\boldsymbol{r}} + 2 S\, dm\ \dot{\boldsymbol{r}}\cdot(\partial\dot{\boldsymbol{r}}/\partial q_k) \\ &= S\, dm\ (\partial\dot{\boldsymbol{r}}/\partial\dot{q}_k)\cdot\ddot{\boldsymbol{r}} + 2(\partial T/\partial q_k), \qquad (6.3.11) \end{aligned}$$

$$\Rightarrow S\, dm\ \ddot{\boldsymbol{r}}\cdot(\partial\dot{\boldsymbol{r}}/\partial\dot{q}_k) = S\, dm\ \boldsymbol{a}\cdot\boldsymbol{e}_k = \partial\dot{T}/\partial\dot{q}_k - 2(\partial T/\partial q_k); \qquad (6.3.12)$$

and combining this with the, by now well-known, Lagrangean identity (§3.3)

$$S\, dm\ \boldsymbol{a}\cdot\boldsymbol{e}_k = (\partial T/\partial\dot{q}_k)^{\cdot} - \partial T/\partial q_k \equiv E_k(T), \qquad (6.3.13)$$

we immediately obtain the *Nielsen identity*:

$$d/dt(\partial T/\partial\dot{q}_k) = \partial\dot{T}/\partial\dot{q}_k - \partial T/\partial q_k. \qquad (6.3.14)$$

With its help, and (6.3.4, 5), and since $S\, d\boldsymbol{F}\cdot\boldsymbol{e}_k \equiv Q_k$ (§3.4), *Jourdain's principle* (6.2.4), or

$$S\,(dm\ \boldsymbol{a} - d\boldsymbol{F})\cdot\delta'\boldsymbol{v} = 0, \qquad \text{under } \delta' t = 0 \qquad \textit{and} \qquad \delta'\boldsymbol{r} = \boldsymbol{0}, \qquad (6.3.15)$$

assumes the following form in *general holonomic system variables*:

$$\sum [E_k(T) - Q_k]\,\delta(\dot{q}_k) = \sum [N_k(T) - Q_k]\,\delta(\dot{q}_k) = 0, \tag{6.3.16}$$

where the *Nielsen operator* $N_k(\ldots)$, in *holonomic coordinates* (constrained or not), is defined by

$$N_k(T) \equiv \partial\dot{T}/\partial\dot{q}_k - 2(\partial T/\partial q_k) = (\partial T/\partial\dot{q}_k)^{\cdot} - \partial T/\partial q_k \equiv E_k(T). \tag{6.3.17}$$

It is shown below that $N_k(f) = E_k(f)$, for any sufficiently smooth function $f = f(t, q, \dot{q})$. In particular, if the $\delta\dot{q}$'s are *independent* (e.g., holonomic system with n DOF) then (6.3.16) leads immediately to the *Nielsen form of Lagrange's equations* (of the second kind) (Nielsen, 1935, pp. 345–354):

$$\partial\dot{T}/\partial\dot{q}_k - 2(\partial T/\partial q_k) = Q_k. \tag{6.3.18}$$

Here, too, as in the Lagrangean case (§ 3.4 and §3.9), if part of Q_k derives from a potential $V = V(q)$, then (6.3.18) still holds, but with T replaced with $L \equiv T - V$, and Q_k: *nonpotential* part of that force.

If, on the other hand, the δq's are constrained by, say, the m Pfaffian (possibly nonholonomic) constraints

$$\sum a_{Dk}\,\delta q_k = 0 \qquad (D = 1, \ldots, m < n), \tag{6.3.19}$$

or, $(\ldots)^{\cdot}$-differentiating and then $\delta'(\ldots)$-varying them, to bring them to the Jourdain form:

$$\sum \dot{a}_{Dk}\,\delta q_k + \sum a_{Dk}(\delta q_k)^{\cdot} = \sum \dot{a}_{Dk}\,\delta q_k + \sum a_{Dk}\,\delta(\dot{q}_k) = 0,$$
$$\Rightarrow \sum a_{Dk}\,\delta'(\dot{q}_k) = \sum a_{Dk}\,\delta(\dot{q}_k) = 0, \tag{6.3.20}$$

then combining ("adjoining") (6.3.20) to (6.3.16) via the m Lagrangean multipliers λ_D, we obtain

$$N_k(T) = Q_k + \sum \lambda_D a_{Dk}. \tag{6.3.21}$$

Hence, the general rule: *in any set of constrained system equations, in holonomic variables*—for example, equations of Maggi, Hadamard–Appell, Appell—we can replace $E_k(T)$ *(or its identically equal* $\partial S/\partial\ddot{q}_k$, *S: Appellian) with* $N_k(T)$. Thus, it is not hard to see that

(i) If eqs. (6.3.19) have the Hadamard form (§3.8)

$$\delta q_D = \sum b_{DI}\,\delta q_I \qquad (I = m+1, \ldots, n), \tag{6.3.22}$$

then the "*Nielsen form of the corresponding (kinetic) Hadamard equations*" is

$$\partial\dot{T}/\partial\dot{q}_I - 2(\partial T/\partial q_I) + \sum b_{DI}\left[\partial\dot{T}/\partial\dot{q}_D - 2(\partial T/\partial q_D)\right] = Q_I + \sum b_{DI} Q_D,$$

or, compactly,

$$[N_I(T) - Q_I] + \sum b_{DI}[N_D(T) - Q_D] = 0; \tag{6.3.23}$$

and

(ii) In terms of the general quasi velocities $\omega = \omega(t, q, \dot{q}) \Leftrightarrow \dot{q} = \dot{q}(t, q, \omega)$, discussed in §5.1 and §5.2, the "*Nielsen form of the corresponding kinetic Maggi equations*" is

$$\sum \left[\partial \dot{T}/\partial \dot{q}_k - 2(\partial T/\partial q_k)\right](\partial \dot{q}_k/\partial \omega_I) = \sum Q_k(\partial \dot{q}_k/\partial \omega_I),$$

or, compactly,

$$\sum [N_k(T) - Q_k](\partial \dot{q}_k/\partial \omega_I) = 0; \qquad (6.3.24)$$

and similarly for the kinetostatic Maggi equations.

An Application of Jourdain's Principle in Quasi Variables

For such variables, the Jourdain variation requirements result in $\delta\theta_k = 0 \Rightarrow (\delta\theta_k)^{\cdot} - \delta(\dot{\theta}_k) = 0 \Rightarrow (\delta\theta_k)^{\cdot} = \delta\omega_k$ [by the transitivity equations (§2.10, §5.2)], and so eq. (6.3.4) yields the fundamental representation (omitting superstars on $\boldsymbol{v}$ etc., for simplicity)

$$\delta' \boldsymbol{v} = \sum (\partial \boldsymbol{v}/\partial \omega_I)\,(\delta\theta_I)^{\cdot} = \sum (\partial \boldsymbol{v}/\partial \omega_I)\,\delta\omega_I. \qquad (6.3.25)$$

Substituting the above into Jourdain's principle, eq. (6.3.15), we immediately obtain the $n - m$ kinetic *Schaefer* equations:

$$S\, dm\, \boldsymbol{a} \cdot (\partial \boldsymbol{v}/\partial \omega_I) = S\, d\boldsymbol{F} \cdot (\partial \boldsymbol{v}/\partial \omega_I). \qquad (6.3.26)$$

[These equations were given for the first time by the noted German engineering scientist H. Schaefer, in 1951, for general *nonlinear*, possibly nonholonomic, velocity constraints, in a very insightful and lucid manner via LP (§5.3, recall (5.3.17 ff.)). Fifteen years later, they were reformulated for *linear* (i.e., Pfaffian) velocity constraints by Kane and Wang (1965), and without reference to the correct forms of the principles of mechanics.]

Example 6.3.1 *The Schieldrop–Nielsen Rule.* The following is a systematization of observations aiming at expediting the building of $N_k(T)$. Let us take, for convenience, but no loss of generality, a scleronomic system. By $(\ldots)^{\cdot}$-differentiating its kinetic energy

$$2T = \sum\sum M_{kl}\dot{q}_k\dot{q}_l, \qquad M_{kl} = M_{kl}(q)\text{: } \textit{inertia coefficients}, \qquad \text{(a)}$$

we get

$$\dot{T} = \sum\sum M_{kl}\ddot{q}_k\dot{q}_l + \sum\sum\sum (1/2)(\partial M_{kl}/\partial q_r)\dot{q}_r\dot{q}_k\dot{q}_l. \qquad \text{(b)}$$

Now, the $\dot{q}$'s appearing in this $\dot{T}$ are divided in *two* groups: (i) those that were already in T, and (ii) those created by the $(\ldots)^{\cdot}$-operation on the $M_{kl}(q)$. The latter $\dot{q}$ shall, henceforth, be denoted by an *underline*: $\underline{\dot{q}}$.

Next, let us build $N_k(T) \equiv \partial \dot{T}/\partial \dot{q}_k - 2(\partial T/\partial q_k)$. We readily find

$$2\,(\partial T/\partial q_p) = \sum\sum (\partial M_{kl}/\partial q_p)\dot{q}_k\dot{q}_l, \qquad \text{(c)}$$

$$\partial \dot{T}/\partial \dot{q}_p = \sum M_{pk}\ddot{q}_k + \sum\sum \left[\partial M_{pk}/\partial q_l + (1/2)\,(\partial M_{kl}/\partial q_p)\right]\dot{q}_k\dot{q}_l. \tag{d}$$

But due to the symmetry of the inertia coefficients, $M_{kl} = M_{lk}$,

$$\sum\sum (\partial M_{pk}/\partial q_l)\dot{q}_k\dot{q}_l = \sum\sum (\partial M_{pl}/\partial q_k)\dot{q}_k\dot{q}_l$$
$$= \sum\sum (1/2)\,(\partial M_{pk}/\partial q_l + \partial M_{pl}/\partial q_k)\dot{q}_k\dot{q}_l; \tag{e}$$

and, therefore,

$$\partial \dot{T}/\partial \dot{q}_p = \sum M_{pk}\ddot{q}_k + \sum\sum (1/2)\,(\partial M_{pk}/\partial q_l + \partial M_{pl}/\partial q_k + \partial M_{kl}/\partial q_p)\dot{q}_k\dot{q}_l. \tag{f}$$

Hence, subtracting (c) from (f) side by side,

$$N_p(T) \equiv \partial \dot{T}/\partial \dot{q}_p - 2(\partial T/\partial q_p)$$
$$= \sum M_{pk}\ddot{q}_k + \sum\sum (1/2)(\partial M_{pk}/\partial q_l + \partial M_{pl}/\partial q_k - \partial M_{kl}/\partial q_p)\dot{q}_k\dot{q}_l$$
$$[= E_k(T)]. \tag{g}$$

These are standard steps in the derivation of explicit forms for $E_k(T)$ (§3.10). From the viewpoint of Nielsen's operator, however, they allow us to make the following observations:

(i) $\partial \dot{T}/\partial \dot{q}_k$ derives from $\dot{T}$ when every term of it is multiplied by the number (or power) of the $\dot{q}_p$ terms in it, and then the factor $\dot{q}_p$ is omitted from that term;
(ii) $2\,(\partial T/\partial q_p)\dot{q}_p$ is twice of that $\dot{T}$ term which contains the $\dot{q}_p$ generated by the $(\ldots)^{\cdot}$ differentiation.

So we have the following rule [due to E. B. Schieldrop (Nielsen, 1935, pp. 352–354)]:

First, we build $\dot{T}$. Then, to obtain $N_p(T)$ for a particular $p = 1, 2, 3, \ldots,$ we multiply each term of $\dot{T}$ either with an integer $k = 0, 1, 2, 3, \ldots,$ or with $k - 2 = -2, -1, 0, 1, \ldots,$ *depending on whether, from the k factors $\dot{q}_p$ in that term, none or one, respectively, were created by the $\dot{T}$-differentiations*—the underlined $\dot{q}$'s in $\dot{T}$ help us to keep track of that. Finally, in each term of the expression obtained thus far, we omit a term $\dot{q}_p$ or divide it by $\dot{q}_p$; which thus results in $N_p(T)$. In other words:

(i) The parts of $\dot{T}$ that are *linear* in $\dot{q}$ need no underlining and no sign change;
(ii) the parts of $\dot{T}$ that are *cubic* in the $\dot{q}$'s do contain an underlined $\dot{q}$; and so:
(a) If $\dot{q}_p$ does not appear as a factor in that term, the latter makes no contribution to $N_p(T)$;
(b) If $\dot{q}_p$ appears *once*, that term appears with *changed* or *unchanged* sign, according as $\dot{q}_p$ is *underlined* or *not*;
(c) If $\dot{q}_p$ appears *twice* in a $\dot{T}$-factor, that term is multiplied by $k - 2$ when none of the $\dot{q}_p$ is underlined, or is multiplied by $k - 2 = 2 - 2 = 0$ when one of the $\dot{q}_p$ is underlined; *then, there is no contribution from that $\dot{T}$-term*;
(d) If $\dot{q}_p$ appears *thrice* in a $\dot{T}$-factor, then one of them must be underlined, and so that term is multiplied by $k - 2 = 3 - 2 = 1$; and *its sign remains unchanged.*

Table 6.1

	$\dot{T}$-terms:	$N_r(T)$	$N_\phi(T)$
1.	$m\,\dot{r}\,\ddot{r}$	$k=1 \Rightarrow (1)\,(\ldots)$ $(1)\,(m\,\dot{r}\,\ddot{r})/\dot{r} = m\,\ddot{r}$	$k=0 \Rightarrow (0)\,(\ldots)$ 0
2.	$m\,r^2\,\dot{\phi}\,\ddot{\phi}$	$k=0 \Rightarrow (0)\,(\ldots)$ 0	$k=1 \Rightarrow (1)\,(\ldots)$ $(1)\,m\,r^2\,\dot{\phi}\,\ddot{\phi}/\dot{\phi} = m\,r^2\,\ddot{\phi}$
3.	$m\,r\,\underline{\dot{r}}\,(\dot{\phi})^2$	$k=1 \Rightarrow k-2=-1 \Rightarrow (-1)\,(\ldots)$ $(-1)\,[m\,r\,\dot{r}\,(\dot{\phi})^2]/\dot{r} = -m\,r\,(\dot{\phi})^2$	$k=2 \Rightarrow (2)\,(\ldots)$ $(2)\,[m\,r\,\dot{r}\,(\dot{\phi})^2]/\dot{\phi} = 2\,m\,r\,\dot{r}\,\dot{\phi}$
Totals		$N_r(T) = m\,\ddot{r} - m\,r\,(\dot{\phi})^2$	$N_\phi(T) = m\,r^2\,\ddot{\phi} + 2\,m\,r\,\dot{r}\,\dot{\phi}$

Example 6.3.2 Let us consider the *plane* motion of a particle of mass m, using polar coordinates $q_1 = r$, $q_2 = \phi$. Here,

$$2T = m\,[(\dot{r})^2 + r^2\,(\dot{\phi})^2], \tag{h1}$$

and, therefore,

$$\dot{T} = m\,[\dot{r}\,\ddot{r} + r^2\,\dot{\phi}\,\ddot{\phi} + r\,\underline{\dot{r}}\,(\dot{\phi})^2]; \tag{h2}$$

note the underlined $\dot{r}$ in the last term.

Then, applying the Schieldrop–Nielsen rule, we calculate $N_r(T) = E_r(T)$ and $N_\phi(T) = E_\phi(T)$. The details are shown in table 6.1.

No claims of universal calculational superiority of this clever rule are made here. We do think, however, that *this is something potentially useful, and, hence, worth knowing*. Perhaps, with proper systematization (symbolic programming), it could be used to advantage in more complicated systems. An additional example of its use is given in ex. 6.5.2.

6.4 INTRODUCTION TO THE PRINCIPLE OF GAUSS AND THE EQUATIONS OF TSENOV

(Gauss' principle, due to its fundamental importance, is given an independent extensive treatment in §6.6.)

By $(\ldots)^{\cdot}$-differentiating (6.3.6b), we obtain the *jerk vector* [(1.7.19e)]:

$$\begin{aligned} \boldsymbol{j} &\equiv d\boldsymbol{a}/dt \equiv \dddot{\boldsymbol{r}} \\ &= \sum (\partial \boldsymbol{r}/\partial q_k)\dddot{q}_k \\ &\quad + \sum\sum \left[(\partial^2 \boldsymbol{r}/\partial q_k \partial q_l)\ddot{q}_k \dot{q}_l + (\partial^2 \boldsymbol{r}/\partial q_k \partial q_l)\dot{q}_k \ddot{q}_l + (\partial^2 \boldsymbol{r}/\partial q_k \partial q_l)\dot{q}_k \ddot{q}_l\right] \\ &\quad + \sum (\partial^2 \boldsymbol{r}/\partial q_k \partial t)\ddot{q}_k + 2\sum (\partial^2 \boldsymbol{r}/\partial q_k \partial t)\ddot{q}_k + \textit{no other } \ddot{q},\ \dddot{q} \textit{ terms}; \end{aligned} \tag{6.4.1a}$$

that is,

$$\begin{aligned} \dddot{\boldsymbol{r}} &= \sum (\partial \boldsymbol{r}/\partial q_k)\dddot{q}_k \\ &\quad + 3\sum \Big(\sum (\partial^2 \boldsymbol{r}/\partial q_k \partial q_l)\ddot{q}_k \dot{q}_l + (\partial^2 \boldsymbol{r}/\partial q_k \partial t)\ddot{q}_k\Big) \\ &\quad + \textit{no other } \ddot{q},\ \dddot{q} \textit{ terms} \end{aligned} \tag{6.4.1b}$$

from which we easily deduce

$$\partial \ddot{\boldsymbol{r}}/\partial \ddot{q}_k = \partial \boldsymbol{r}/\partial q_k = \boldsymbol{e}_k, \tag{6.4.2}$$

which is an extension of (6.3.8).

Further, and in complete analogy with (6.3.4, 5), we have

$$\begin{aligned}\delta \boldsymbol{a} \equiv \delta \ddot{\boldsymbol{r}} = \cdots &= \sum \boldsymbol{e}_k\ \delta(\ddot{q}_k) + \textit{no other } \delta\ddot{q} \textit{ terms}\\ &\equiv \delta'' \boldsymbol{a} + \textit{no other } \delta\ddot{q} \textit{ terms},\end{aligned} \tag{6.4.3}$$

where

$\delta''(\ldots) \equiv \sum [\partial(\ldots)/\partial \ddot{q}_k]\delta(\ddot{q}_k)$:

Gaussian variation of $(\ldots)$ [i.e., $\delta(\ldots)$ with $\delta t = 0$, $\delta q = 0$ *and* $\delta(\dot{q}_k) = 0$]. (6.4.4)

Next, $(\ldots)^{\cdot}$-differentiating $\dot{T}$, we find

$$\begin{aligned}\ddot{T} &= S\, dm\ \ddot{\boldsymbol{r}}\cdot\ddot{\boldsymbol{r}} + S\, dm\ \dot{\boldsymbol{r}}\cdot\dddot{\boldsymbol{r}}\\ &\Big[= S\, dm\ \boldsymbol{a}\cdot\boldsymbol{a} + S\, dm\ \boldsymbol{v}\cdot\boldsymbol{j} = 2\,S + S\, dm\ \boldsymbol{v}\cdot\boldsymbol{j} \qquad (S:\ \textit{Appellian})\Big],\end{aligned} \tag{6.4.5}$$

from which we readily obtain

$$\partial \ddot{T}/\partial \ddot{q}_k = 2 S\, dm\ \ddot{\boldsymbol{r}}\cdot(\partial\ddot{\boldsymbol{r}}/\partial\ddot{q}_k) + S\, dm\ \dot{\boldsymbol{r}}\cdot(\partial\dddot{\boldsymbol{r}}/\partial\ddot{q}_k),$$

or, due to (6.4.2) and

$$\begin{aligned}\partial\dddot{\boldsymbol{r}}/\partial\ddot{q}_k &= 3\sum\Big(\sum(\partial^2\boldsymbol{r}/\partial q_k\partial q_l)\dot{q}_l + \partial^2\boldsymbol{r}/\partial q_k\partial t\Big)\\ &= 3(\partial\dot{\boldsymbol{r}}/\partial q_k) = 3(\partial\boldsymbol{v}/\partial q_k),\end{aligned} \tag{6.4.6}$$

equivalently,

$$\begin{aligned}\partial \ddot{T}/\partial \ddot{q}_k &= 2 S\, dm\ \boldsymbol{a}\cdot\boldsymbol{e}_k + 3 S\, dm\ \boldsymbol{v}\cdot(\partial\boldsymbol{v}/\partial q_k)\\ &= 2E_k(T) + 3(\partial T/\partial q_k);\end{aligned} \tag{6.4.7}$$

and rearranging this we finally get the kinematico-inertial identity:

$$E_k(T) = (1/2)\Big[\partial\ddot{T}/\partial\ddot{q}_k - 3(\partial T/\partial q_k)\Big] \equiv C_k^{(2)}(T), \tag{6.4.8}$$

where

$C_k^{(2)}(\ldots) \equiv (1/2)\{\partial(\ldots)^{\cdot\cdot}/\partial\ddot{q}_k - 3[\partial(\ldots)/\partial q_k]\}$:

Tsenov (or Tzénoff, or Tzenov, or Cenov) operator of the *second kind*, in holonomic variables. (6.4.9)

With the help of the above, Gauss principle reads

$$\sum\ [C_k^{(2)}(T) - Q_k]\delta\ddot{q}_k = 0; \tag{6.4.10}$$

and, as earlier, if the δq's and hence also the $\delta\ddot{q}$'s are independent, the above leads to *Tsenov's equations of the second kind*, in holonomic variables:

$$C_k^{(2)}(T) = Q_k; \tag{6.4.11}$$

[developed by the Bulgarian mechanician I. Tsenov (1885–1967), originally (in a slightly different form) in 1924, and more systematically in the 1950s and later (1953, 1962)] while, if the $\delta\ddot{q}$ are constrained by the $m(< n)$ Pfaffian constraints

$$\sum a_{Dk}\ \delta q_k = 0, \tag{6.4.12a}$$

then, we first bring them to the Gaussian form:

$$\sum a_{Dk}\ \delta\ddot{q}_k = 0 \tag{6.4.12b}$$

[by $(\ldots)^{\cdot\cdot}$-differentiation and then application of (6.4.4)] and subsequently adjoin (6.4.12b) to (6.4.10) via Lagrangean multipliers, thus obtaining *the Tsenov form of the Routh–Voss equations of motion*:

$$C_k^{(2)}(T) = Q_k + \sum \lambda_D a_{Dk}. \tag{6.4.12c}$$

Proceeding similarly to the next step—that is, to $\dddot{T}$, $\overset{(4)}{\boldsymbol{r}} \equiv (\dddot{\boldsymbol{r}})^{\cdot}$, and since, in this case,

$$\partial\dddot{T}/\partial\dddot{q}_k = \cdots = 3E_k(T) + 4(\partial T/\partial q_k), \tag{6.4.13a}$$

we easily obtain, for independent δq's and hence also $\delta\dddot{q}$'s, *Tsenov's equations of the third kind*, in *holonomic* variables:

$$E_k(T) \equiv C_k^{(3)}(T) \equiv (1/3)\left[\partial\dddot{T}/\partial\dddot{q}_k - 4(\partial T/\partial q_k)\right] = Q_k. \tag{6.4.13b}$$

Proceeding inductively, from the above, we can easily show that the earlier *Mangeron–Deleanu* principle (6.2.7, 7a), where, in analogy with (6.4.4),

$$\delta\overset{(s)}{\boldsymbol{r}} = \sum\left[\partial\overset{(s)}{\boldsymbol{r}}/\partial\overset{(s)}{q_k}\right]\delta\overset{(s)}{q_k} = \sum \boldsymbol{e}_k\ \delta\overset{(s)}{q_k}, \tag{6.4.14a}$$

becomes, in holonomic system variables,

$$\sum\left[C_k^{(s)}(T) - Q_k\right]\delta\overset{(s)}{q_k} = 0, \tag{6.4.14b}$$

where

$$C_k^{(s)}(T) \equiv E_k(T) \equiv S\, dm\ \boldsymbol{a}\cdot\boldsymbol{e}_k \equiv S\, dm\ \boldsymbol{a}\cdot\left(\partial\overset{(s)}{\boldsymbol{r}}/\partial\overset{(s)}{q_k}\right)$$
$$\equiv (1/s)\left[\partial\overset{(s)}{T}/\partial\overset{(s)}{q_k} - (s+1)\,(\partial T/\partial q_k)\right], \tag{6.4.14c}$$

is the general *Mangeron–Deleanu operator (in holonomic variables)* applied to T, and, in analogy with the Nielsen identity (6.3.14),

$$d/dt(\partial T/\partial\dot{q}_k) = (1/s)\left[\partial\overset{(s)}{T}/\partial\overset{(s)}{q_k} - \partial T/\partial q_k\right]. \tag{6.4.14d}$$

Again, for unconstrained δq's, eq. (6.4.14b) yields the *Tsenov-type equations of Mangeron–Deleanu (1962) and Dolaptschiew (1966)*:

$$(1/s)\left[\partial\overset{(s)}{T}/\partial\overset{(s)}{q} - (s+1)\,(\partial T/\partial q_k)\right] = Q_k$$
$$(k = 1,\ldots,n;\ s = 1,2,3,\ldots); \tag{6.4.14e}$$

and analogously if the δq's are constrained.

In sum: in all equations of motion in holonomic variables, and whether the δq's are constrained or not, we can replace $E_k(T) \equiv (\partial T/\partial \dot{q}_k)^{\cdot} - \partial T/\partial q_k$ (*Lagrange*) $\equiv \partial S/\partial \ddot{q}_k$ (*Appell*) with any one of its equals:

$$N_k(T) \equiv \partial \dot{T}/\partial \dot{q}_k - 2\,(\partial T/\partial q_k) \qquad (\textit{Nielsen}), \tag{6.4.15a}$$

or

$$C_k{}^{(2)}(T) \equiv (1/2)\,\big[\partial \ddot{T}/\partial \ddot{q}_k - 3\,(\partial T/\partial q_k)\big],$$
$$C_k{}^{(3)}(T) \equiv (1/3)\,\big[\partial \dddot{T}/\partial \dddot{q}_k - 4\,(\partial T/\partial q_k)\big] \qquad (\textit{Tsenov}), \tag{6.4.15b}$$

or

$$C_k{}^{(s)}(T) \equiv (1/s)\left[\partial \overset{(s)}{T}/\partial \overset{(s)}{q}_k - (s+1)\,(\partial T/\partial q_k)\right]$$
$$(\textit{Mangeron–Deleanu–Dolaptschiew}); \tag{6.4.15c}$$

[i.e., $C_k{}^{(1)}(T) = N_k(T)$].

Summary

(i) Analytical Results

Since all the earlier variational principles are equivalent, we can utilize *any of them with any of the above kinematico-inertial expressions*; although, historically, $E_k(T)$ and $N_k(T)$ have been associated with Lagrange's principle, and $\partial S/\partial \ddot{q}_k$ with Gauss' principle. In practice, however, which variational principle and operator will be used in a particular problem depends on the given form of the constraints; some are more natural than others. Thus:

(a) If the constraints have the form $f_D(t, q) = 0$ $[D = 1, \ldots, m(< n)]$, then, since

$$\delta f_D = \sum (\partial f_D/\partial q_k)\,\delta q_k = 0, \tag{6.4.16a}$$

Lagrange's principle, with $E_k(T)$ or $N_k(T)$, is preferred.

(b) If the constraints have the form $f_D(t, q, \dot{q}) = 0$, then, since

$$\delta' f_D = \sum (\partial f_D/\partial \dot{q}_k)\,\delta \dot{q}_k = 0, \tag{6.4.16b}$$

Jourdain's principle, with $E_k(T)$ or $N_k(T)$, is recommended. The formal (i.e., mathematical) $\delta(\ldots)$-variation of f_D,

$$\delta f_D = \sum (\partial f_D/\partial q_k)\,\delta q_k + \sum (\partial f_D/\partial \dot{q}_k)\,\delta(\dot{q}_k) = 0, \tag{6.4.16c}$$

(as also explained in §7.5 ff.) *cannot* be combined with Lagrange's principle,

$$\sum [E_k(T) - Q_k]\,\delta q_k = 0,$$

to produce the correct equations of motion; that is,

$$E_k(T) = Q_k + \sum \lambda_D(\partial f_D/\partial \dot{q}_k). \tag{6.4.16d}$$

Since $\delta'' f_D = \sum (\partial f_D / \partial \ddot{q}_k)\, \delta \ddot{q}_k = 0$, Gauss' principle, say with $E_k(T)$:

$$\sum [E_k(T) - Q_k] \delta \ddot{q}_k = 0, \tag{6.4.16e}$$

cannot be utilized either. But it *can* be applied to $df_D/dt = 0$; indeed, since

$$df_D/dt = \partial f_D/\partial t + \sum (\partial f_D/\partial q_k)\dot{q}_k + \sum (\partial f_D/\partial \dot{q}_k)\ddot{q}_k \\ [\equiv g_D(t, q, \dot{q}, \ddot{q})] = 0,$$

we have

$$\delta''(df_D/dt) = \sum (\partial \dot{f}_D/\partial \ddot{q}_k)\, \delta \ddot{q}_k - \sum (\partial f_D/\partial \dot{q}_k)\, \delta \ddot{q}_k \\ \left[\equiv \delta g_D = \sum (\partial g_D/\partial \dot{q}_k)\, \delta \ddot{q}_k\right] = 0, \tag{6.4.16f}$$

and this combined with Gauss' principle yields the correct equations; that is, (6.4.16d).

(c) Similarly, we can show that for constraints of the form $f_D(t, q, \dot{q}, \ddot{q}) = 0$, to *insure compatibility among the principles of Lagrange, Jourdain and Gauss*, we must set

$$\delta f_D = \sum (\partial f_D/\partial \ddot{q}_k)\, \delta q_k = 0, \tag{6.4.16g}$$

$$\delta' f_D = \sum (\partial f_D/\partial \ddot{q}_k)\, \delta \dot{q}_k = 0, \tag{6.4.16h}$$

$$\delta'' f_D = \sum (\partial f_D/\partial \ddot{q}_k)\, \delta \ddot{q}_k = 0; \tag{6.4.16i}$$

in which case, the correct equations of motion are

$$E_k(T) = Q_k + \sum \lambda_D (\partial f_D/\partial \ddot{q}_k). \tag{6.4.16j}$$

(ii) Geometrical Interpretation

If we think of the constraints $f_D(t, q, \dot{q}) = 0$ as hypersurfaces in velocity space, with t and the q's as parameters (since the velocities can change instantaneously, but the configuration and time cannot), (6.4.16b) states that the virtual velocity change $\delta \dot{q}$ lies in the local tangent plane, just like the δq's lie in the local tangent plane of the surface $f_D(t, q) = 0$ (holonomic constraints) in configuration space. Analogously for $f_D(t, q, \dot{q}, \ddot{q}) = 0$, (6.4.16i) states the conditions. The above show that *JP is a natural for velocity constraints, while GP is a natural for acceleration constraints.*

These results are summarized in tables 6.2 and 6.3.

Table 6.2 Virtual Displacements Needed to Produce the Correct Equations of Motion

Constraints	Lagrange	Jourdain	Gauss
$f(t, q) = 0$: $\partial f/\partial q$	$\delta f = (\partial f/\partial q)\delta q$	$\delta' f = 0$ $\delta' \dot{f} = (\partial f/\partial q)\delta \dot{q}$	$\delta'' f = 0$, $\delta'' \dot{f} = 0$ $\delta'' \ddot{f} = (\partial f/\partial q)\delta \ddot{q}$
$f(t, q, \dot{q}) = 0$: $\partial f/\partial \dot{q}$	—	$\delta' f = (\partial f/\partial \dot{q})\delta \dot{q}$	$\delta'' f = 0$ $\delta'' \dot{f} = (\partial f/\partial \dot{q})\delta \ddot{q}$
$f(t, q, \dot{q}, \ddot{q}) = 0$: $\partial f/\partial \ddot{q}$	—	—	$\delta'' f = (\partial f/\partial \ddot{q})\delta \ddot{q}$

Table 6.3 Correct Equations of Motion

[Notation: $M_k \equiv E_k(T) - Q_k \equiv N_k(T) - Q_k \equiv \partial S/\partial \ddot{q}_k - Q_k$;
Mechanical principle: $\sum M_k\, \delta x_k = 0, \qquad \delta x_k = \delta q_k,\ \delta \dot{q}_k,\ \delta \ddot{q}_k, \ldots$]

Constraints	Virtual Displacements	Equations of Motion
$f_D(t, q) = 0$	$\delta f_D = \sum (\partial f_D/\partial q_k)\delta q_k$	$M_k = \sum \lambda_D(\partial f_D/\partial q_k)$
$f_D(t, q, \dot{q}) = 0$	$\delta' f_D = \sum (\partial f_D/\partial \dot{q}_k)\delta \dot{q}_k$	$M_k = \sum \lambda_D(\partial f_D/\partial \dot{q}_k)$
$f_D(t, q, \dot{q}, \ddot{q}) = 0$	$\delta'' f_D = \sum (\partial f_D/\partial \ddot{q}_k)\delta \ddot{q}_k$	$M_k = \sum \lambda_D(\partial f_D/\partial \ddot{q}_k)$
And analogously for higher-order constraints.		

Example 6.4.1 Let us derive the impressed force-free equations of motion of a sled (or narrow boat, or skate, or knife, or stiff razor blade, etc.; recalling ex. 2.13.2, ex. 3.18.1, and their notations; also exs. 7.3.2 and 7.3.3) of mass m, *whose mass center G coincides with its contact point C*, on a smooth horizontal plane P, via the various DVP.

(i) *Via Jourdain's principle.* With Lagrangean coordinates $q_{1,2,3} = x, y, \phi$, and since here $V = 0$, $Q_{k,np} = 0$, so that

$$2L = 2T = m[(\dot{x})^2 + (\dot{y})^2] + I(\dot{\phi})^2, \tag{a}$$

and the constraint is (with $\boldsymbol{n}$: unit vector perpendicular to sled, on P)

$$\boldsymbol{v} \cdot \boldsymbol{n} = (\sin\phi)\dot{x} + (-\cos\phi)\dot{y} = 0 \ \Rightarrow\ (\sin\phi)\,\delta x + (-\cos\phi)\,\delta y = 0; \tag{b}$$

that is, $n = 3$, $m = 1$, JP yields the variational equation:

$$\sum E_k(L)\,\delta\dot{q}_k = 0: \qquad (m\ddot{x})\,\delta\dot{x} + (m\ddot{y})\,\delta\dot{y} + (I\ddot{\phi})\,\delta\dot{\phi} = 0, \tag{c}$$

under the constraint (b), *but brought to Jourdain form*; that is,

$$f(t, q, \dot{q}) = 0 \ \Rightarrow\ \delta' f = (\partial f/\partial \dot{q})\,\delta\dot{q}, \tag{d}$$

$$\begin{aligned} &\delta'[(\sin\phi)\dot{x} + (-\cos\phi)\dot{y}] \\ &= \delta[(\sin\phi)\dot{x} + (-\cos\phi)\dot{y}]\Big|_{\delta x,\, \delta y,\, \delta\phi = 0} \\ &= [(\sin\phi)\,\delta\dot{x} + (-\cos\phi)\,\delta\dot{y} \\ &\quad + (\dot{x}\cos\phi)\,\delta\phi + (\dot{y}\sin\phi)\,\delta\phi]\Big|_{\delta x,\, \delta y,\, \delta\phi = 0} = 0, \end{aligned} \tag{e}$$

or, carrying out the variations and enforcing the Jourdainian constraints $\delta q_k = 0$,

$$(\sin\phi)\,\delta\dot{x} + (-\cos\phi)\,\delta\dot{y} = 0. \tag{f}$$

Eliminating, say, $\delta\dot{y}$ between (c) and (f), we obtain the *unconstrained* variational equation

$$(m\,\ddot{x}\cos\phi + m\,\ddot{y}\sin\phi)\,\delta\dot{x} + (I\ddot{\phi}\cos\phi)\,\delta\dot{\phi} = 0. \tag{g}$$

from which, since $\delta\dot{x}$ and $\delta\dot{\phi}$ are now independent, we get the two reactionless equations of motion

$$\ddot{x}\cos\phi + \ddot{y}\sin\phi = 0, \qquad \ddot{\phi} = 0. \tag{h,i}$$

The first of them expresses the absence of force in the tangential direction, while the second expresses the absence of moment, about $C = G$, in the direction perpendicular to P.

(ii) *Via Gauss' principle.* Here, the variational equation is

$$\sum E_k(L)\,\delta(\ddot{q}_k) = 0: \qquad (m\,\ddot{x})\,\delta\ddot{x} + (m\,\ddot{y})\,\delta\ddot{y} + (I\ddot{\phi})\,\delta\ddot{\phi} = 0, \tag{j}$$

under the constraint (b), *but brought to Gaussian form*; that is,

$$f(t, q, \dot{q}) = 0 \;\Rightarrow\; \delta''(df/dt) = (\partial f/\partial \dot{q})\,\delta\ddot{q}, \tag{k}$$

$$\begin{aligned} &\delta''\Big\{d/dt[(\sin\phi)\dot{x} + (-\cos\phi)\dot{y}]\Big\} \\ &\quad \equiv \delta[(\sin\phi)\ddot{x} + (-\cos\phi)\ddot{y} \\ &\qquad + (\dot{x}\cos\phi)\dot{\phi} + (\dot{y}\sin\phi)\dot{\phi}]\Big|_{\delta x,\,\delta y,\,\delta\phi=0;\ \delta\dot{x},\,\delta\dot{y},\,\delta\dot{\phi}=0} = 0, \end{aligned} \tag{l}$$

or, carrying out the variations

$$\begin{aligned} &\{(\sin\phi)\,\delta\ddot{x} + (-\cos\phi)\,\delta\ddot{y} \\ &\qquad + (\delta\dot{x}\cos\phi - \dot{x}\sin\phi\,\delta\phi + \delta\dot{y}\sin\phi + \dot{y}\cos\phi\,\delta\phi)\dot{\phi} \\ &\qquad + (\dot{x}\cos\phi + \dot{y}\sin\phi)\,\delta\dot{\phi}\}\Big|_{\delta x,\,\delta y,\,\delta\phi=0;\ \delta\dot{x},\,\delta\dot{y},\,\delta\dot{\phi}=0} = 0, \end{aligned} \tag{m}$$

and then enforcing the Gaussian constraints $\delta q_k = 0$, $\delta\dot{q}_k = 0$,

$$(\sin\phi)\,\delta\ddot{x} + (-\cos\phi)\,\delta\ddot{y} = 0. \tag{n}$$

Similarly, eliminating $\delta\ddot{y}$ between (j) and (n), we find again eqs. (h, i).

For a Gaussian treatment of the related problem of *Prytz's planimeter*, see Brill (1909, pp. 30–33).

Problem 6.4.1 (i) Show that

$$\begin{aligned} d^s\mathbf{r}/dt^s \equiv \overset{(s)}{\mathbf{r}} &= \sum (\partial\mathbf{r}/\partial q_k)\overset{(s)}{q_k} \\ &\quad + s\Big(\sum\sum (\partial^2\mathbf{r}/\partial q_k\,\partial q_l)\overset{(s-1)}{q_k}\,\dot{q}_l + \sum (\partial^2\mathbf{r}/\partial q_k\,\partial t)\overset{(s-1)}{q_k}\Big) \\ &\quad + \textit{no other}\ \overset{(s-1)}{q},\ \overset{(s)}{q}\ \textit{terms}; \\ &= \sum (\partial\mathbf{r}/\partial q_k)\,\overset{(s)}{q_k} + \textit{no other}\ \overset{(s)}{q}\ \textit{terms}; \end{aligned} \tag{a}$$

and, therefore,

(ii)
$$\delta\overset{(s)}{\mathbf{r}} = \sum \Big[\partial\overset{(s)}{\mathbf{r}}/\partial\overset{(s)}{q_k}\Big]\delta\overset{(s)}{q_k} = \sum \mathbf{e}_k\,\delta\overset{(s)}{q_k} \quad \text{(Mangeron variation);} \tag{b}$$

(iii)
$$\mathbf{e}_k = \partial\mathbf{r}/\partial q_k = \cdots = \partial\overset{(s)}{\mathbf{r}}/\partial\overset{(s)}{q_k} = \partial\overset{(s)}{\mathbf{v}}/\partial\overset{(s+1)}{q_k} = \partial\overset{(s)}{\mathbf{a}}/\partial\overset{(s+2)}{q_k}; \tag{c}$$

(iv)
$$\partial \overset{(s)}{\boldsymbol{r}} / \partial \overset{(s-1)}{q_k} = s\Big(\sum (\partial^2 \boldsymbol{r}/\partial q_k\, \partial q_l)\dot{q}_l + \partial^2 \boldsymbol{r}/\partial q_k\, \partial t\Big)$$
$$= s(\partial \dot{\boldsymbol{r}}/\partial q_k) = s(\partial \boldsymbol{v}/\partial q_k) \qquad \text{(d)}$$
$$\Rightarrow\ \partial \overset{(s)}{\boldsymbol{v}} / \partial \overset{(s)}{q_k} = (s+1)(\partial \boldsymbol{v}/\partial q_k); \qquad \text{(e)}$$

(v)
$$\overset{(s)}{\boldsymbol{v}} = (s+1) \sum (\partial \boldsymbol{v}/\partial q_k)\, \overset{(s)}{q_k} + \textit{no other } \overset{(s)}{q} \textit{ terms}. \qquad \text{(f)}$$

Problem 6.4.2 (i) Starting with $2T \equiv S\, dm\, \boldsymbol{v} \cdot \boldsymbol{v}$, show that

$$\overset{(s)}{T} = S\, dm\, \boldsymbol{v} \cdot \overset{(s+1)}{\boldsymbol{r}} + s\, S\, dm\, \boldsymbol{a} \cdot \overset{(s)}{\boldsymbol{r}} + \textit{no other } \overset{(s)}{\boldsymbol{r}},\ \overset{(s+1)}{\boldsymbol{r}} \textit{ terms};$$
$$= S\, dm\, \overset{(1)}{\boldsymbol{r}} \cdot \overset{(s+1)}{\boldsymbol{r}} + s\, S\, dm\, \overset{(2)}{\boldsymbol{r}} \cdot \overset{(s)}{\boldsymbol{r}} + \textit{no other } \overset{(s)}{\boldsymbol{r}},\ \overset{(s+1)}{\boldsymbol{r}} \textit{ terms}; \qquad \text{(a)}$$

and from this deduce the following kinematico-inertial identities:

(a)
$$\partial T/\partial q_k = S\, dm\, \dot{\boldsymbol{r}} \cdot (\partial \dot{\boldsymbol{r}}/\partial q_k) = (1/s)\, S\, dm\, \dot{\boldsymbol{r}} \cdot \Big(\partial \overset{(s)}{\boldsymbol{r}} / \partial \overset{(s-1)}{q_k}\Big)$$
$$\equiv (1/s)\, S\, dm\, \boldsymbol{v} \cdot \Big(\partial \overset{(s)}{\boldsymbol{r}} / \partial \overset{(s-1)}{q_k}\Big), \qquad \text{(b)}$$

and, generally,

(b)
$$\partial \overset{(s)}{T} / \partial \overset{(s)}{q_k} = S\, dm\, \dot{\boldsymbol{r}} \cdot \Big(\partial \overset{(s+1)}{\boldsymbol{r}} / \partial \overset{(s)}{q_k}\Big) + s\, S\, dm\, \ddot{\boldsymbol{r}} \cdot \Big(\partial \overset{(s)}{\boldsymbol{r}} / \partial \overset{(s)}{q_k}\Big)$$
$$= (s+1)\, S\, dm\, \boldsymbol{v} \cdot (\partial \boldsymbol{v}/\partial q_k) + s\, S\, dm\, \boldsymbol{a} \cdot \boldsymbol{e}_k$$
$$= (s+1)(\partial T/\partial q_k) + s\, E_k(T), \qquad \text{(c)}$$

from which, rearranging, we obtain the earlier *Mangeron–Deleanu* identity:

$$E_k(T) = (1/s)\Big[\partial \overset{(s)}{T} / \partial \overset{(s)}{q_k} - (s+1)(\partial T/\partial q_k)\Big]. \qquad \text{(d)}$$

($s = 1$: Nielsen eqs., $s = 2$: Tsenov eqs., etc.)
(ii) Prove the *Mangeron–Deleanu* recursive identity:

$$(1/s)\Big[\partial \overset{(s)}{T} / \partial \overset{(s)}{q_k}\Big] - (1/r)\Big[\partial \overset{(r)}{T} / \partial \overset{(r)}{q_k}\Big] + [(s-r)/(s r)](\partial T/\partial q_k) = 0,$$
$$(k = 1, \ldots, n;\ s > r;\ s, r = 1, 2, 3, \ldots), \qquad \text{(e)}$$

and with its help, and the rest, deduce the following identities:

(a)
$$[1/(s-r)]\Big[(r+1)\Big(\partial \overset{(s)}{T} / \partial \overset{(s)}{q_k}\Big) - (s+1)\Big(\partial \overset{(r)}{T} / \partial \overset{(r)}{q_k}\Big)\Big] = E_k(T),$$
$$(k = 1, \ldots, n;\ s, r = 0, 1, 2, \ldots;\ s > r), \qquad \text{(f)}$$

(b)
$$\partial \overset{(s)}{T} / \partial \overset{(s)}{q_k} - \partial \overset{(s-1)}{T} / \partial \overset{(s-1)}{q_k} = \partial T/\partial q_k + E_k(T), \qquad \text{(g)}$$

$$\partial \overset{(s)}{T}/\partial \overset{(s+1)}{q_k} = \partial \overset{(s-1)}{T}/\partial \overset{(s)}{q_k} = \cdots = \partial T/\partial \dot{q}_k \ (= p_k); \tag{h}$$

(c)
$$\overset{(s)}{\delta I} \equiv \mathbf{S}\, dm\, \boldsymbol{a} \cdot \delta \overset{(s)}{\boldsymbol{r}}$$
$$= \sum \left[\partial \overset{(s)}{T}/\partial \overset{(s)}{q_k} - \partial \overset{(s-1)}{T}/\partial \overset{(s-1)}{q_k} - \partial T/\partial q_k\right] \delta \overset{(s)}{q_k}, \tag{i}$$

sometimes referred to as *the Mićević Dušan–Rusov Lazar* form of Lagrange's identity (⇒ principle, 1984); and for unconstrained $\delta \overset{(s)}{q_k}$'s easily resulting in the equations of motion:

$$\partial \overset{(s)}{T}/\partial \overset{(s)}{q_k} - \partial \overset{(s-1)}{T}/\partial \overset{(s-1)}{q_k} - \partial T/\partial q_k = Q_k. \tag{j}$$

For further related results, see, for example, Shen and Mei (1987).

Example 6.4.2 Let us show, by direct differentiations, that for any sufficiently differentiable function $f(t, q, \dot{q})$: $E_k(f) = N_k(f)$, or, in extenso,

$$(\partial f/\partial \dot{q}_k)^{\cdot} - \partial f/\partial q_k = \partial \dot{f}/\partial \dot{q}_k - 2(\partial f/\partial q_k). \tag{a}$$

We have, successively,

$$N_k(f) \equiv \partial \dot{f}/\partial \dot{q}_k - 2(\partial f/\partial q_k)$$
$$= \partial/\partial \dot{q}_k \left[\partial f/\partial t + \sum (\partial f/\partial q_l)\dot{q}_l + \sum (\partial f/\partial \dot{q}_l)\ddot{q}_l\right] - 2(\partial f/\partial q_k)$$
$$= \partial/\partial t(\partial f/\partial \dot{q}_k) + \sum [\partial/\partial q_l(\partial f/\partial \dot{q}_k)]\dot{q}_l + \sum (\partial f/\partial q_l)\,\delta_{lk}$$
$$+ \sum [\partial/\partial \dot{q}_l(\partial f/\partial \dot{q}_k)]\ddot{q}_l - 2(\partial f/\partial q_k).$$

But (i) the *first, second, and fourth* terms combine to $(\partial f/\partial \dot{q}_k)^{\cdot}$; while (ii) the *third* reduces to $\partial f/\partial q_k$, and combines with the *last* term to $-\partial f/\partial q_k$; that is, finally,

$$N_k(f) = (\partial f/\partial \dot{q}_k)^{\cdot} - \partial f/\partial q_k \equiv E_k(f), \quad \text{Q.E.D.} \tag{b}$$

Example 6.4.3 Let us extend the result of the preceding example to *nonholonomic* variables; namely, let us show that

$$E_k{}^*(f^*) \equiv (\partial f^*/\partial \omega_k)^{\cdot} - \partial f^*/\partial \theta_k = \partial \dot{f}^*/\partial \omega_k - 2(\partial f^*/\partial \theta_k) \equiv N_k{}^*(f^*), \tag{a}$$

where (recalling the notations of chaps. 2–5)

$$\partial \ldots/\partial \theta_k \equiv \sum (\partial \ldots/\partial q_l)(\partial \dot{q}_l/\partial \omega_k) \Leftrightarrow \partial \ldots/\partial q_l = \sum (\partial \ldots/\partial \theta_k)(\partial \omega_k/\partial \dot{q}_l), \tag{b}$$

and

$$f = f(t, q, \dot{q}) = f[t, q, \dot{q}(t, q, \omega)] \equiv f^*(t, q, \omega) \equiv f^*. \tag{c}$$

Since $\dot{f}^* \equiv df^*/dt$ is a function of t, q, ω; $\dot{q}(t, q, \omega)$, $\ddot{q}(t, q, \omega, \dot{\omega})$, we find, successively,

$$N_k{}^*(f^*) \equiv \partial\dot{f}^*/\partial\omega_k - 2(\partial f^*/\partial\theta_k)$$

$$= \partial/\partial\omega_k\left[\partial f^*/\partial t + \sum(\partial f^*/\partial q_l)\dot{q}_l + \sum(\partial f^*/\partial\omega_l)\dot{\omega}_l\right]$$

$$- 2\sum(\partial f^*/\partial q_l)(\partial\dot{q}_l/\partial\omega_k)$$

$$= \partial/\partial t(\partial f^*/\partial\omega_k) + \sum\left[\partial/\partial q_l(\partial f^*/\partial\omega_k)\right]\dot{q}_l + \sum(\partial f^*/\partial q_l)(\partial\dot{q}_l/\partial\omega_k)$$

$$+ \sum\left[\partial/\partial\omega_l(\partial f^*/\partial\omega_k)\right]\dot{\omega}_l - 2\sum(\partial f^*/\partial q_l)(\partial\dot{q}_l/\partial\omega_k)$$

[the *first, second, and fourth* terms combine to $(\partial f^*/\partial\omega_k)^{\cdot}$;

while, recalling (b), the *third* and *last* add up to $-\,\partial f^*/\partial\theta_k$]

$$= (\partial f^*/\partial\omega_k)^{\cdot} - \partial f^*/\partial\theta_k \equiv E_k{}^*(f^*), \quad \text{Q.E.D.} \tag{d}$$

For an alternative proof, see Mei (1983, pp. 630–631).

Example 6.4.4 Using the kinematico-inertial identities of the preceding examples, let us find the Nielsen forms of the two general (say, kinetic) nonlinear nonholonomic equations of Johnsen–Hamel (§5.3):

$$(\partial T^*/\partial\omega_I)^{\cdot} - \partial T^*/\partial\theta_I - \sum\left[(\partial\dot{q}_k/\partial\dot{\omega}_I)^{\cdot} - \partial\dot{q}_k/\partial\theta_I\right](\partial T/\partial\dot{q}_k)^* = \Theta_I, \tag{a}$$

where

$$(\partial T/\partial\dot{q}_k)^* = \sum(\partial T^*/\partial\omega_l)(\partial\omega_l/\partial\dot{q}_k), \qquad \text{i.e.,}\quad p_k = p_k(t, q, \dot{q}) = p_k{}^*(t, q, \omega), \tag{b}$$

or, compactly,

$$E_I{}^*(T^*) - \sum E_I{}^*(\dot{q}_k)(\partial T/\partial\dot{q}_k)^* = \Theta_I; \tag{c}$$

and

$$(\partial T^*/\partial\omega_I)^{\cdot} - \partial T^*/\partial\theta_I + \sum\sum\left[(\partial\omega_l/\partial\dot{q}_r)^{\cdot} - \partial\omega_l/\partial q_r\right](\partial\dot{q}_r/\partial\omega_I)(\partial T^*/\partial\omega_l) = \Theta_I, \tag{d}$$

or, compactly,

$$E_I(T^*) + \sum\sum E_r(\omega_l)(\partial\dot{q}_r/\partial\omega_I)(\partial T^*/\partial\omega_l) = \Theta_I. \tag{e}$$

(i) Substituting $N_I{}^*(T^*) = E_I{}^*(T^*)$ and $N_I{}^*(\dot{q}_k) = E_I{}^*(\dot{q}_k)$ in (a, c), we readily obtain their Nielsen forms:

$$\partial\dot{T}^*/\partial\omega_I - 2(\partial T^*/\partial\theta_I) - \sum\left[\partial\ddot{q}_k/\partial\omega_I - 2(\partial\dot{q}_k/\partial\theta_I)\right](\partial T/\partial\dot{q}_k)^* = \Theta_I, \tag{f}$$

or, compactly,

$$N_I{}^*(T^*) - \sum N_I{}^*(\dot{q}_k)(\partial T/\partial\dot{q}_k)^* = \Theta_I. \tag{g}$$

(ii) Substituting $N_I{}^*(T^*) = E_I{}^*(T^*)$ and $N_r(\omega_l) = E_r(\omega_l)$ in (d, e), we readily obtain their Nielsen forms:

$$\partial \dot{T}^*/\partial \omega_I - 2(\partial T^*/\partial \theta_I) + \sum\sum \left[(\partial \dot{\omega}_l/\partial \dot{q}_r) - 2(\partial \omega_l/\partial q_r)\right](\partial \dot{q}_r/\partial \omega_I)(\partial T^*/\partial \omega_l) = \Theta_I, \tag{h}$$

or, compactly,

$$N_I{}^*(T^*) + \sum\sum N_r(\omega_l)(\partial \dot{q}_r/\partial \omega_I)(\partial T^*/\partial \omega_l) = \Theta_I. \tag{i}$$

And similarly for the kinetostatic equations.

Here, too, as in chapter 5, we remark that the importance of these equations lies not so much in their ability to solve concrete nonholonomic problems more easily, but in that they help us to understand better the formal kinematico-inertial structure of analytical mechanics; also, they might prove useful in handling higher-order constraints.

Problem 6.4.3 Show that in the Pfaffian case

$$\dot{q}_k = \sum A_{kl}(t,q)\,\omega_l + A_k(t,q) = \sum A_{kI}(t,q)\,\omega_I + A_k(t,q), \tag{a}$$

eqs. (f) of the preceding example reduce to the *Nielsen form of the general nonlinear Voronets equations*:

$$\begin{aligned}\partial \dot{T}^*/\partial \omega_I &- 2(\partial T^*/\partial \theta_I)\\ &+ \sum (\partial T/\partial \dot{q}_k)^* \Big(\sum (\partial A_{kI'}/\partial \theta_I - \partial A_{kI}/\partial \theta_{I'})\,\omega_{I'} + \partial A_k/\partial \theta_I\\ &\qquad - \sum (\partial A_{kI}/\partial q_l)A_l - \partial A_{kI}/\partial t\Big) = \Theta_I.\end{aligned} \tag{b}$$

Equivalent forms can be obtained from (h) of the preceding example, if, instead of (a), we use

$$\omega_l = \sum a_{lk}(t,q)\,q_k + a_l(t,q). \tag{c}$$

Problem 6.4.4 (Mei, 1985, pp. 203–207, 211–214). Continuing from the preceding problem, show that if (a) of that problem are stationary (or scleronomic); that is, if

$$\dot{q}_k = \sum A_{kl}(q)\,\omega_l = \sum A_{kI}(q)\,\omega_I, \tag{a}$$

then (b) of that problem reduce to the *Nielsen form of the general nonlinear Chaplygin equations*:

$$\partial \dot{T}^*/\partial \omega_I - 2(\partial T^*/\partial \theta_I) + \sum\sum (\partial T/\partial \dot{q}_k)^* \left[(\partial A_{kI'}/\partial \theta_I - \partial A_{kI}/\partial \theta_{I'})\,\omega_{I'}\right] = \Theta_I. \tag{b}$$

6.5 ADDITIONAL FORMS OF THE EQUATIONS OF NIELSEN AND TSENOV

(i) Following Tsenov, we introduce the function

$$R_{(1)} \equiv \dot{T} - 2\dot{T}_{(o)}, \tag{6.5.1}$$

where $T_{(o)}$ is what results from T *if we regard it as function of t and the q's, but not the $\dot{q}$'s*; that is, *for fixed, or frozen, velocities,*

$$T_{(o)} = T_{(o)}(t, q) = T(t, q, \dot{q} = constant). \tag{6.5.2}$$

From the above it follows that

$$\dot{T}_{(o)} = \sum (\partial T/\partial q_k)\dot{q}_k + \partial T/\partial t + no\ other\ \dot{q}\ terms, \tag{6.5.3}$$

and, therefore,

$$\partial \dot{T}_{(o)}/\partial \dot{q}_k = \partial T/\partial q_k. \tag{6.5.4}$$

Then, Nielsen's equations, say (6.3.18), can be written in the Appellian form

$$\partial R_{(1)}/\partial \dot{q}_k = Q_k, \tag{6.5.5}$$

which is slightly simpler than the corresponding Appell's equations $\partial S/\partial \ddot{q}_k = Q_k$. Finally, introducing the new "Tsenov function"

$$K_{(1)} \equiv R_{(1)} - \sum Q_k\, \dot{q}_k, \tag{6.5.6}$$

we can express (6.5.5) in the equilibrium form:

$$\partial K_{(1)}/\partial \dot{q}_k = 0, \qquad \text{under the conditions } \partial Q_l/\partial \dot{q}_k = 0; \tag{6.5.7}$$

in words: *the equations of motion result from the stationarity of the function* $K_{(1)} = K_{(1)}(\dot{q})$. Similarly, for constrained systems: for example, if the constraints are

$$\dot{q}_D = \sum b_{DI}\dot{q}_I + b_D, \qquad b_{DI} = b_{DI}(t, q), \qquad b_D = b_D(t, q)$$
$$\Rightarrow \delta q_D = \sum b_{DI}\, \delta q_I, \tag{6.5.8a}$$

then it is not hard to see that the Nielsen–Tsenov equations of the system take the "Hadamard form":

$$\partial K_{(1)}/\partial \dot{q}_I + \sum b_{DI}(\partial K_{(1)}/\partial \dot{q}_D) = 0$$

$$(D = 1, 2, \ldots, m;\ I = m + 1, \ldots, n); \tag{6.5.8b}$$

or, with $K_{(1)} = K_{(1)}[t, q, \dot{q}_D(t, q, \dot{q}_I), \dot{q}_I] \equiv K_{(1)o}(t, q, \dot{q}_I) = K_{(1)o}$ ("constrained Tsenov function")

$$\begin{aligned}\Rightarrow\ \partial K_{(1)o}/\partial \dot{q}_I &= \partial K_{(1)}/\partial \dot{q}_I + \sum (\partial K_{(1)}/\partial \dot{q}_D)(\partial \dot{q}_D/\partial \dot{q}_I)\\ &= \partial K_{(1)}/\partial \dot{q}_I + \sum b_{DI}(\partial K_{(1)}/\partial \dot{q}_D),\end{aligned} \tag{6.5.8c}$$

simply,

$$\partial K_{(1)o}/\partial \dot{q}_I = 0. \tag{6.5.8d}$$

Other "Maggi-like" forms are also possible.

(ii) Again, following Tsenov, introducing the functions

$$R_{(2)} \equiv (1/2)(\dddot{T} - 3\ddot{T}_{(o)}) \qquad \text{and} \qquad K_{(2)} \equiv R_{(2)} - \sum Q_k \ddot{q}_k, \tag{6.5.9a}$$

and then noting the kinematic identities

$$\dot{T}_{(o)} = \sum (\partial T/\partial q_k)\dot{q}_k + \textit{no other } \dot{q} \textit{ terms} \Rightarrow \ddot{T}_{(o)} = \sum (\partial T/\partial q_k)\ddot{q}_k + \textit{no } \ddot{q} \textit{ terms},$$
$$\Rightarrow \partial \ddot{T}_{(o)}/\partial \ddot{q}_k = \partial T/\partial q_k, \tag{6.5.9b}$$

we can rewrite (6.4.11) in the following Appell-like and "equilibrium forms":

$$\partial R_{(2)}/\partial \ddot{q}_k = Q_k, \tag{6.5.9c}$$

$$\partial K_{(2)}/\partial \ddot{q}_k = 0, \qquad \text{under the conditions } \partial Q_l/\partial \ddot{q}_k = 0, \tag{6.5.9d}$$

respectively.

The above are particularly useful for constraints of the acceleration form:

$$f_D(t, q, \dot{q}, \ddot{q}) = 0. \tag{6.5.10a}$$

Then, Gauss' principle yields

$$\sum (\partial K_{(2)}/\partial \ddot{q}_k)\, \delta\ddot{q}_k = \sum (\partial K_{(2)}/\partial \ddot{q}_D)\delta\ddot{q}_D + \sum (\partial K_{(2)}/\partial \ddot{q}_I)\, \delta\ddot{q}_I = 0, \tag{6.5.10b}$$

under the conditions (in Gaussian variation form)

$$\delta'' f_D = \sum (\partial f_D/\partial \ddot{q}_D)\, \delta\ddot{q}_D + \sum (\partial f_I/\partial \ddot{q}_I)\, \delta\ddot{q}_I = 0$$
$$\Rightarrow \delta\ddot{q}_D = \sum (\partial \ddot{q}_D/\partial \ddot{q}_I)\, \delta\ddot{q}_I \equiv \sum b_{DI}\, \delta\ddot{q}_I. \tag{6.5.10c}$$

Combining (6.5.10c) with (6.5.10b), in by now well-known ways, and since the $n - m$ $\delta\ddot{q}_I$ can now be taken as independent, we immediately obtain the *Hadamard form of the second-kind Tsenov equations*:

$$\partial K_{(2)}/\partial \ddot{q}_I + \sum b_{DI}(\partial K_{(2)}/\partial \ddot{q}_D) = 0. \tag{6.5.10d}$$

A (6.5.8d)-like form is also readily available.

(iii) Finally, in the general case of a system subjected to the m (s)th order constraints

$$f_D\left(t, q, \dot{q}, \ddot{q}, \dddot{q}, \ldots, \overset{(s)}{q}\right) = 0, \tag{6.5.11a}$$

we can easily show that the *Hadamard form of its Mangeron–Deleanu equations* is

$$\partial K_{(s)}/\partial \overset{(s)}{q}_I + \sum b_{DI}\left(\partial K_{(s)}/\partial \overset{(s)}{q}_D\right) = 0, \tag{6.5.11b}$$

where

$$\overset{(s)}{q}_D = \sum b_{DI}\, \overset{(s)}{q}_I + \textit{no other } \overset{(s)}{q} \textit{ terms} \Rightarrow \delta \overset{(s)}{q}_D = \sum b_{DI}\, \delta \overset{(s)}{q}_I, \tag{6.5.11c}$$

and

$$K_{(s)} \equiv R_{(s)} - \sum Q_k \overset{(s)}{q_k}, \qquad \text{under} \quad \partial Q_l / \partial \overset{(s)}{q_k} = 0,$$

$$R_{(s)} \equiv (1/s)\left[\overset{(s)}{T} - (s+1)\sum (\partial T/\partial q_k)\overset{(s)}{q_k}\right] + \textit{no other } \overset{(s)}{q} \textit{ terms}$$

$$= (1/s)\left[\overset{(s)}{T} - (s+1)\overset{(s)}{T}_{(o)}\right] + \textit{no other } \overset{(s)}{q} \textit{ terms}, \tag{6.5.11d}$$

due to

$$\overset{(s)}{T}_{(o)} = \sum (\partial T/\partial q_k)\overset{(s)}{q_k} + \textit{no other } \overset{(s)}{q} \textit{ terms} \Rightarrow \partial \overset{(s)}{T}_{(o)} / \partial \overset{(s)}{q_k} = \partial T/\partial q_k. \tag{6.5.11e}$$

These higher-order Tsenov equations were presented, in a series of papers, by the Romanian mechanicians D. Mangeron and S. Deleanu in the early 1960s. Finally, (a) in terms of the *constrained* $R_{(s)o}$ and corresponding impressed forces Q_{Io}, the equations of motion take the Appellian form

$$\partial R_{(s)o} / \partial \overset{(s)}{q_I} = Q_{Io}; \tag{6.5.11f}$$

while (b) in terms of the *constrained* $K_{(s)o}$, they assume the equilibrium form

$$\partial K_{(s)o} / \partial \overset{(s)}{q_I} = 0. \tag{6.5.11g}$$

GENERAL REMARKS

(i) The relation between all these equations and the Lagrangean ones rests on the key kinematico-inertial identity:

$$(\partial T/\partial \dot{q}_k)^{\cdot} = (1/s)\left[\partial \overset{(s)}{T} / \partial \overset{(s)}{q_k} - \partial T/\partial q_k\right] \qquad (s = 1, 2, 3, \ldots). \tag{6.5.12}$$

(ii) The usefulness of these $\overset{(s)}{T}$-based equations lies in their ability to handle constraints of corresponding order in s. Symbolically,

$$f(t, q, \dot{q}) = 0 \Rightarrow \dot{T} \qquad \text{(Nielsen)},$$

$$f(t, q, \dot{q}, \ddot{q}) = 0 \Rightarrow \ddot{T} \qquad \text{(Tsenov 2nd kind)},$$

$$f(t, q, \dot{q}, \ddot{q}, \dddot{q}) = 0 \Rightarrow \dddot{T} \qquad \text{(Tsenov 3rd kind)},$$

....

$$f\left(t, q, \dot{q}, \ddot{q}, \ldots, \overset{(s)}{q}\right) = 0 \Rightarrow \overset{(s)}{T} \qquad \text{(Mangeron–Deleanu)}.$$

For unconstrained systems (i.e., independent δq's), these types of equations (as well as those by Appell) *do not seem to offer any particular advantage over the ordinary Lagrangean equations.*

(iii) By comparing the preceding $\overset{(s)}{T}$-equations with those by Appell, say for unconstrained systems, we readily conclude that

$$\partial R_{(s)} / \partial \overset{(s)}{q_k} = \partial \overset{(s)}{S} / \partial \overset{(s+2)}{q_k} \qquad (s = 1, 2, 3, \ldots). \tag{6.5.13}$$

(iv) It is not hard to show that, for Pfaffian constraints

$$\sum a_{Dk}\dot{q}_k + a_D = 0 \;\Rightarrow\; \sum a_{Dk}\,\delta q_k = 0, \tag{6.5.14a}$$

some of the earlier principles can be extended so that they hold for *finite* variations, and not just $\delta(\ldots)$:

(a) If both $\dot{q}_k$ and $\dot{q}_k + \Delta\dot{q}_k$ are sets of kinematically admissible velocities, *at the same configuration and time*; that is, if they both satisfy (6.5.14a):

$$\sum a_{Dk}\dot{q}_k + a_D = 0, \qquad \sum a_{Dk}(\dot{q}_k + \Delta\dot{q}_k) + a_D = 0. \tag{6.5.14b}$$

then subtracting them side by side yields

$$\sum a_{Dk}\,\Delta\dot{q}_k = 0, \tag{6.5.14c}$$

for Jourdain-like variations satisfying $\Delta t = 0$ and $\Delta q = 0$. This states that, in the constraints (6.5.14a), *we can replace the virtual δq_k's with the finite-velocity Jourdain jumps $\Delta\dot{q}_k$.*

(b) By $(\ldots)^{\cdot}$-differentiating (6.5.14a), we obtain

$$\sum (\dot{a}_{Dk}\dot{q}_k + a_{Dk}\ddot{q}_k) + \dot{a}_D = \sum a_{Dk}\ddot{q}_k + \textit{no other } \ddot{q} \textit{ terms} = 0, \tag{6.5.15a}$$

and, therefore, if we consider the two admissible acceleration states $\ddot{q}$ and $\ddot{q} + \Delta\ddot{q}$, under the Gaussian restrictions $\Delta t = 0$, $\Delta q = 0$, $\Delta\dot{q} = 0$, substitute them into (6.5.15a), and subtract the resulting equations side by side, we obtain

$$\sum a_{Dk}\,\Delta\ddot{q}_k = 0; \tag{6.5.15b}$$

that is, in (6.5.14a), *we can replace the δq_k's with the finite acceleration Gauss jumps $\Delta\ddot{q}_k$.* The extension to higher order constraints should be obvious.

(v) Let the impressed forces Q_k be derivable, wholly or partly, from a *generalized potential* $V = V(t, q, \dot{q})$ [§3.9]; that is,

$$V = V(t, q, \dot{q}) = V_0(t, q) + \sum \gamma_k(t, q)\dot{q}_k,$$

and

$$\begin{aligned} Q_{k,\text{potential part}} &= E_k(V) \equiv (\partial V/\partial \dot{q}_k)^{\cdot} - \partial V/\partial q_k \\ &= -\partial V_0/\partial q_k + \sum (\partial\gamma_k/\partial q_l - \partial\gamma_l/\partial q_k)\dot{q}_l + \partial\gamma_k/\partial t. \end{aligned} \tag{6.5.16a}$$

Then, it can be easily shown that V satisfies the "Dolaptschiew identities":

$$(\partial V/\partial \dot{q}_k)^{\cdot} = (1/s)\left(\partial \overset{(s)}{V}/\partial \overset{(s)}{q_k} - \partial V/\partial q_k\right), \tag{6.5.16b}$$

and, therefore, with $L \equiv T - V$, the equations of motion, of, say, an unconstrained system, can be rewritten as

$$(1/s)\left[\partial \overset{(s)}{L}/\partial \overset{(s)}{q_k} - (s+1)(\partial L/\partial q_k)\right] = Q_{k,np}, \tag{6.5.16c}$$

$$Q_{k,np}\text{: } \textit{nonpotential} \text{ part of } Q_k. \tag{6.5.16d}$$

Further, introducing $L_{(o)} \equiv L(t, q, \dot{q} = constant)$, and since $\partial \overset{(s)}{L}_{(o)}/\partial \overset{(s)}{q}_k = \partial L/\partial q_k$, the equations of motion assume the following two equivalent forms:

$$\text{Appell-like:} \qquad \partial l_{(s)}/\partial \overset{(s)}{q}_k = Q_{k,np}, \quad \text{where} \quad l_{(s)} \equiv (1/s)\left[\overset{(s)}{L} - (s+1)\,\overset{(s)}{L}_{(o)}\right], \tag{6.5.16e}$$

$$\text{Equilibrium:} \qquad \partial k_{(s)}/\partial \overset{(s)}{q}_k = 0, \qquad \text{where} \quad k_{(s)} \equiv l_{(s)} - \sum Q_{k,np}\, \overset{(s)}{q}_k\,. \tag{6.5.16f}$$

And analogously for constrained systems. For example, the equations of motion of a system constrained by

$$\overset{(s)}{q}_D = \sum b_{DI}\, \overset{(s)}{q}_I + no\ other\ \overset{(s)}{q}\ terms \Rightarrow \delta \overset{(s)}{q}_D = \sum b_{DI}\, \delta \overset{(s)}{q}_I, \tag{6.5.17a}$$

are

$$\partial l_{(s)o}/\partial \overset{(s)}{q}_I = (Q_{I,np})_o \qquad \text{or} \qquad \partial k_{(s)o}/\partial \overset{(s)}{q}_I = 0, \tag{6.5.17b}$$

where $l_{(s)o}, k_{(s)o}$ are, respectively, $l_{(s)}$, $k_{(s)}$ from which the dependent rates $\overset{(s)}{q}_D$ have been eliminated by means of the first of (6.5.17a), and $(Q_{I,np})_o \equiv Q_{I,np} + \sum b_{DI} Q_{D,np}$.

Problem 6.5.1 *Higher Forms of Appell's Equations* [$S = S(t, q, \dot{q}, \ddot{q})$: Appellian function]. Let

$$\overset{(s)}{U} \equiv \overset{(s)}{S} - \sum Q_k\, \overset{(s)}{q}_k + f\left(t, q, \dot{q}, \ldots, \overset{(s-1)}{q}\right). \tag{a}$$

Show that

(i) Under the m $(< n)$ constraints

$$f_D\left(t, q, \dot{q}, \ldots, \overset{(s+2)}{q}\right) = 0, \tag{b}$$

the equations of motion may be expressed in the "Routh–Voss" form:

$$\partial \overset{(s)}{S} / \partial \overset{(s+2)}{q}_k = Q_k + \sum \lambda_D \left(\partial f_D / \partial \overset{(s+2)}{q}_k\right); \tag{c}$$

(ii) Under the m $(< n)$ constraints

$$\overset{(s+2)}{q}_D = \sum b_{DI}\, \overset{(s+2)}{q}_I + no\ other\ \overset{(s+2)}{q}_k\ terms, \tag{d}$$

they may be expressed in the "Hadamard" form:

$$\partial \overset{(s)}{S} / \partial \overset{(s+2)}{q}_I + \sum b_{DI} \left(\partial \overset{(s)}{S} / \partial \overset{(s+2)}{q}_D\right) = Q_I + \sum b_{DI} Q_D; \tag{e}$$

and

(iii) Under the m $(< n)$ constraints

$$\overset{(s+2)}{q}_D = \overset{(s+2)}{q}_D\left(t, q, \dot{q}, \ldots, \overset{(s+1)}{q}, \overset{(s+2)}{\theta}\right), \tag{f}$$

they may be expressed in the "Maggi" form:

$$\sum (\partial S/\partial \ddot{q}_k)\left[\partial \overset{(s+2)}{q_k} \Big/ \partial \overset{(s+2)}{\theta_I}\right] = \sum Q_k \left[\partial \overset{(s+2)}{q_k} \Big/ \partial \overset{(s+2)}{\theta_I}\right] = \Theta_I. \tag{g}$$

In all these cases, it is assumed that the constraint forces satisfy the "ideal reactions" postulate (§3.2 ff.):

$$\Lambda_I \equiv S\, d\boldsymbol{R} \cdot \boldsymbol{\varepsilon}_I = S\, d\boldsymbol{R} \cdot \left(\partial \overset{(s)}{\boldsymbol{a}} \Big/ \partial \overset{(s+2)}{\theta_I}\right) = 0. \tag{h}$$

Example 6.5.1 A charged particle moves in a *homogeneous* electromagnetic field of intensities $\boldsymbol{E}$ (electric) and $\boldsymbol{H}$ (magnetic). It is shown in electrodynamics [see any advanced book on the subject; e.g., Landau and Lifshitz (1971, §8, §16)] that its Lagrangean is

$$L \equiv T - V = (1/2)mv^2 - e\Phi + (e/c)(\boldsymbol{A} \cdot \boldsymbol{v}), \tag{a}$$

where $e =$ electric charge, $c =$ speed of light in vacuum,

$$\Phi = -\boldsymbol{E} \cdot \boldsymbol{r} \text{ is the } \textit{scalar potential of the (electric) field}, \text{ and} \tag{b1}$$

$$\boldsymbol{A} = (\boldsymbol{H} \times \boldsymbol{r})/2 \text{ is the } \textit{vector potential of the (magnetic) field.} \tag{b2}$$

Let us choose axes O–xyz so that both fields lie on the O–yz plane, and $\boldsymbol{H}$ is directed along the $+Oz$ axis; that is, $\boldsymbol{E} = (0, E_y, E_z)$ and $\boldsymbol{H} = (0, 0, H_z \equiv H)$. Then L assumes the form:

$$L = (m/2)\left[(\dot{x})^2 + (\dot{y})^2 + (\dot{z})^2\right] + e(y\,E_y + z\,E_z) + (eH/2c)(x\dot{y} - y\dot{x}); \tag{c1}$$

and thus, following Tsenov's concept,

$$L \Rightarrow L_{(o)} = e(y\,E_y + z\,E_z) + (eH/2c)(x\dot{y} - y\dot{x}) + \textit{constant}, \tag{c2}$$

where $\dot{x}$, $\dot{y}$ are viewed as fixed, or frozen, quantities.

Let us apply the Tsenov/Mangeron–Deleanu equations for $s = 1$. Here,

$$\begin{aligned} Q_{k,np} &= 0, \\ R_{(1)} = K_{(1)} &= \dot{L} - 2\dot{L}_{(o)} \\ &= m(\dot{x}\,\ddot{x} + \dot{y}\,\ddot{y} + \dot{z}\,\ddot{z}) - e(E_y\,\dot{y} + E_z\,\dot{z}) \\ &\quad - (eH/2c)(x\,\dot{y} - y\,\dot{x}) + \textit{function of } x, y, \ddot{x}, \ddot{y}, \end{aligned} \tag{d}$$

and, therefore, the equations of motion (6.5.16) are

$$\partial K_{(1)}/\partial \dot{x} = m\,\ddot{x} - (eH/c)\dot{y} = 0, \tag{e}$$

$$\partial K_{(1)}/\partial \dot{y} = m\,\ddot{y} - e\,E_y + (eH/c)\dot{x} = 0, \tag{f}$$

$$\partial K_{(1)}/\partial \dot{z} = m\,\ddot{z} - e\,E_z = 0; \tag{g}$$

and, of course, these coincide with the equations obtained by other means.

Example 6.5.2 (Dolaptschiew, 1969, pp. 181–182). Let us derive the Tsenov, Nielsen, et al. equations of motion of a heavy homogeneous sphere (fig. 6.1), of mass m and radius r, rolling on the rough inner wall of a fixed vertical circular cylinder of radius R ($\geq r$).

Kinematics, Constraints

Let us introduce the following five Lagrangean coordinates:

$$q_1 = \gamma, \quad q_2 = z; \qquad q_3 = \phi, \quad q_4 = \theta, \quad q_5 = \psi \quad \text{(Eulerian angles)}. \tag{a}$$

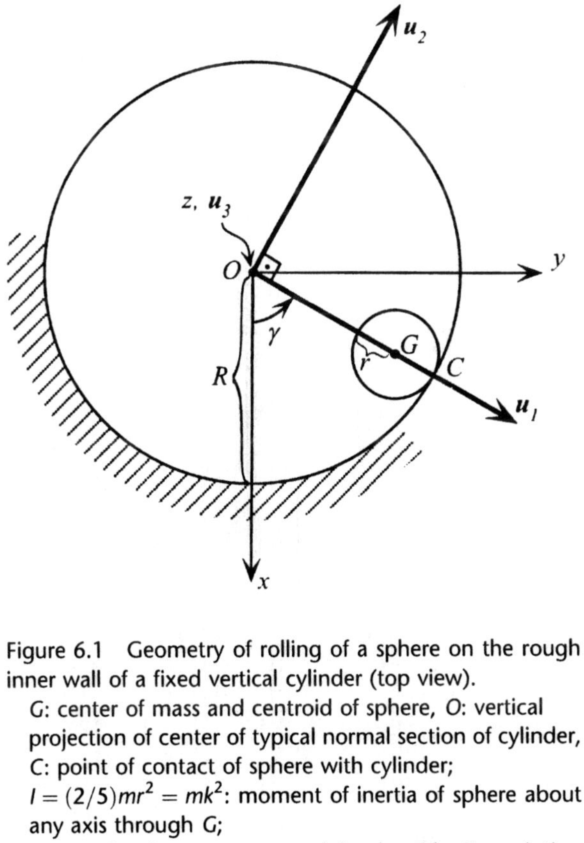

Figure 6.1 Geometry of rolling of a sphere on the rough inner wall of a fixed vertical cylinder (top view).
G: center of mass and centroid of sphere, O: vertical projection of center of typical normal section of cylinder,
C: point of contact of sphere with cylinder;
$I = (2/5)mr^2 = mk^2$: moment of inertia of sphere about any axis through G;
O–xyz: fixed axes at ground level, with Oz pointing towards the reader;
(x, y, z): coordinates of C;
Mobile (intermediate) ortho–normal–dextral basis O–$\boldsymbol{u}_{1,2,3}$:
$\boldsymbol{u}_1$: radially outwards (i.e., toward C), making an angle γ with Ox,
$\boldsymbol{u}_2$: perpendicular to $\boldsymbol{u}_1$, in positive γ-sense (counterclockwise),
$\boldsymbol{u}_3$: so that $\boldsymbol{u}_{1,2,3}$ constitutes a dextral basis (points toward the reader).

From fig. 6.1 and §1.12, we see that the components of the angular velocity of the sphere, $\boldsymbol{\omega}$, along the fixed axes ($\omega_{x,y,z}$), and along the intermediate axes ($\omega_{1,2,3}$), are related by

$$\omega_x = \dot{\psi}\, s\theta\, s\phi + \dot{\theta}\, c\phi = \omega_1\, c\gamma - \omega_2\, s\gamma, \tag{b1}$$

$$\omega_y = -\dot{\psi}\, s\theta\, c\phi + \dot{\theta}\, s\phi = \omega_1\, s\gamma + \omega_2\, c\gamma, \tag{b2}$$

$$\omega_z = \dot{\phi} + \dot{\psi}\, c\theta = \omega_3; \tag{b3}$$

[where, as usual, $s(\ldots) \equiv \sin(\ldots)$, $c(\ldots) \equiv \cos(\ldots)$] and so, inverting, we obtain

$$\omega_1 = \dot{\theta}\, c(\phi - \gamma) + \dot{\psi}\, s\theta\, s(\phi - \gamma), \tag{c1}$$

$$\omega_2 = \dot{\theta}\, s(\phi - \gamma) - \dot{\psi}\, s\theta\, c(\phi - \gamma), \tag{c2}$$

$$\omega_3 = \dot{\phi} + \dot{\psi}\, c\theta. \tag{c3}$$

Clearly, $\omega_{1,2,3}$ are quasi velocities; like $\omega_{x,y,z}$.
The rolling constraint at C is

$$\boldsymbol{v}_C = \boldsymbol{v}_G + \boldsymbol{\omega} \times \boldsymbol{r}_{C/G} = \boldsymbol{0}, \tag{d}$$

or, along $\boldsymbol{u}_{1,2,3}$, and with the notation $R - r \equiv \rho$,

$$\begin{aligned} \boldsymbol{v}_C &= d/dt(z\,\boldsymbol{u}_3 + \rho\,\boldsymbol{u}_1) + \boldsymbol{\omega} \times (r\,\boldsymbol{u}_1) \\ &= [\dot{z}\,\boldsymbol{u}_3 + \rho(d\boldsymbol{u}_1/dt)] + (\omega_1, \omega_2, \omega_3) \times (r, 0, 0) \\ &= \dot{z}\,\boldsymbol{u}_3 + \rho(\dot{\gamma}\,\boldsymbol{u}_2) + (0, r\,\omega_3, -r\,\omega_2) \\ &= (\rho\dot{\gamma} + r\,\omega_3)\boldsymbol{u}_2 + (\dot{z} - r\,\omega_2)\boldsymbol{u}_3 = \boldsymbol{0}, \end{aligned}$$

from which we obtain the two scalar constraints

$$\rho\,\dot{\gamma} + r\,\omega_3 = 0, \qquad \dot{z} - r\,\omega_2 = 0; \tag{e}$$

or, thanks to (c2, 3), exclusively in *holonomic* variables,

$$v_{C,\,\text{tangential direction}} \equiv \rho\dot{\gamma} + r\,\dot{\phi} + r\,\dot{\psi}\,c\theta = 0, \tag{f1}$$

$$v_{C,\,\text{axial (vertical) direction}} \equiv \dot{z} - r\,\dot{\theta}\,s(\phi - \gamma) - r\,\dot{\psi}\,s\theta\,c(\phi - \gamma) = 0. \tag{f2}$$

Next, let us calculate the *kinetic and potential energies*:

(i) By König's theorem, we have

$$\begin{aligned} 2T &= m\,v_G^2 + (I_x\,\omega_x^2 + I_y\,\omega_y^2 + I_z\,\omega_z^2) \\ &= m\Big[(\dot{z})^2 + \rho^2(\dot{\gamma})^2\Big] + (m\,k^2)(\omega_x^2 + \omega_y^2 + \omega_z^2) \\ &= m\Big[(\dot{z})^2 + \rho^2(\dot{\gamma})^2\Big] + (m\,k^2)\Big[(\dot{\phi})^2 + (\dot{\theta})^2 + (\dot{\psi})^2 + 2\dot{\phi}\,\dot{\psi}\cos\theta)\Big] \end{aligned} \tag{g1}$$

[thanks to (b1–3), and recalling that $I_x = I_y = I_z = m\,k^2$]; and

(ii) $V = mgz \Rightarrow Q_z = -(\partial V/\partial z) = -mg$ (sole nonvanishing impressed force). (g2)

Tsenov Equations

Now we are ready to obtain the Tsenov equations of motion:

(i) Tsenov's function is

$$T \;\Rightarrow\; T_{(o)} = (m\,k^2)\underline{\dot\phi}\,\underline{\dot\psi}\,c\theta + constant = T_{(o)}(\theta)$$

[underlined velocities behave as constants], (h1)

and so

$$\dot T_{(o)} = -(m\,k^2)\underline{\dot\phi}\,\underline{\dot\psi}\,\dot\theta\,s\theta; \tag{h2}$$

(ii) By (g1),

$$\begin{aligned} \dot T &= m\,\dot z\,\ddot z + m\,\rho^2\,\dot\gamma\,\ddot\gamma + m\,k^2\big[\dot\phi\,\ddot\phi + \dot\theta\,\ddot\theta + \dot\psi\,\ddot\psi \\ &\quad + (\dot\phi\,\ddot\psi + \dot\psi\,\ddot\phi)c\theta - \dot\phi\,\dot\theta\,\dot\psi\,s\theta\big]. \end{aligned} \tag{h3}$$

Therefore, choosing $\dot\phi$, $\dot\theta$, $\dot\psi$ as the *independent* velocities, in which case (f1, 2) yield for the *dependent ones*:

$$\dot z = r\big[\dot\theta\,s(\phi-\gamma) + \dot\psi\,s\theta\,c(\phi-\gamma)\big] = \dot z(\dot\theta,\,\dot\psi;\,\phi,\,\theta,\,\gamma), \tag{i1}$$

$$\dot\gamma = -(r/\rho)\big(\dot\phi + \dot\psi\,c\theta\big) = \dot\gamma(\dot\phi,\,\dot\psi;\,\theta), \tag{i2}$$

and with the familiar notation $(\ldots)_o \equiv$ *constrained* $(\ldots)$ we find

$$\begin{aligned} K_{(1)o} &\equiv (\dot T - 2\dot T_{(o)})_o - Q_z\,\dot z \\ &= m(\dot z\,\ddot z + \rho^2\dot\gamma\,\ddot\gamma) \\ &\quad + m\,k^2\big[\dot\phi\,\ddot\phi + \dot\theta\,\ddot\theta + \dot\psi\,\ddot\psi + (\dot\phi\,\ddot\psi + \dot\psi\,\ddot\phi) - \dot\phi\,\dot\theta\,\dot\psi\,s\theta\big] \\ &\quad + 2(m\,k^2)\underline{\dot\phi}\,\underline{\dot\psi}\,\dot\theta\,s\theta + m\,g\,\dot z \\ &= K_{(1)o}\big[\dot\gamma(\dot\phi,\,\dot\psi;\,\theta),\,\dot z(\dot\theta,\,\dot\psi;\,\gamma,\,\phi,\,\theta),\,\dot\phi,\,\dot\theta,\,\dot\psi;\,\gamma,\,\theta\big] \\ &= K_{(1)o}(\dot\phi,\,\dot\theta,\,\dot\psi;\,\gamma,\,\phi,\,\theta) \qquad \text{(i.e., \textit{constrained} Tsenov function).} \end{aligned} \tag{i3}$$

Hence, the three kinetic Tsenov equations are

$$\partial K_{(1)o}/\partial\dot\phi = m\,\rho^2\,\ddot\gamma\,(\partial\dot\gamma/\partial\dot\phi) + m\,k^2(\ddot\phi + \ddot\psi\,c\theta - \dot\theta\,\dot\psi\,s\theta) = 0, \tag{j1}$$

$$\partial K_{(1)o}/\partial\dot\theta = m(\ddot z + g)(\partial\dot z/\partial\dot\theta) + m\,k^2(\ddot\theta - \dot\phi\,\dot\psi\,s\theta + 2\underline{\dot\phi}\,\underline{\dot\psi}\,s\theta) = 0, \tag{j2}$$

$$\begin{aligned} \partial K_{(1)o}/\partial\dot\psi &= m(\ddot z + g)(\partial\dot z/\partial\dot\psi) + m\,\rho^2\,\ddot\gamma\,(\partial\dot\gamma/\partial\dot\phi) \\ &\qquad + m\,k^2(\ddot\psi + \ddot\phi\,c\theta - \underline{\dot\phi}\,\underline{\dot\theta}\,s\theta) = 0; \end{aligned} \tag{j3}$$

or, since [by (i1, 2)]

$$\partial\dot\gamma/\partial\dot\phi = -(r/\rho), \qquad \partial\dot\gamma/\partial\dot\psi = -(r/\rho)c\theta, \tag{k1}$$

$$\partial\dot z/\partial\dot\theta = r\,s(\phi-\gamma), \qquad \partial\dot z/\partial\dot\psi = r\,s\theta\,c(\phi-\gamma), \tag{k2}$$

finally (and dropping the underlines),

$$-r\rho\ddot{\gamma} + k^2(\ddot{\phi} + \ddot{\psi}\,c\theta - \dot{\theta}\,\dot{\psi}\,s\theta) = 0, \tag{l1}$$

$$k^2(\ddot{\theta} + \dot{\phi}\,\dot{\psi}\,s\theta) + r(\ddot{z} + g)\,s(\phi - \gamma) = 0, \tag{l2}$$

$$k^2(\ddot{\psi} + \ddot{\phi}\,c\theta - \dot{\phi}\,\dot{\theta}\,s\theta) - r\rho\ddot{\gamma}\,c\theta + r(\ddot{z} + g)\,s\theta\,c(\phi - \gamma) = 0; \tag{l3}$$

which, along with (f1, 2) [or (i1, 2)], constitute a determinate system for γ, z, ϕ, θ, ψ. For its solution, and so on, see, for example, Neimark and Fufaev (1972, pp. 95–98) and Ramsey (1937, pp. 157–158). It is not hard to realize that eqs. (l1–3) are none other than the Chaplygin–Voronets equations of the problem.

Had we not enforced the constraints (f1, 2) or (i1, 2) in the Tsenov function, the equations of motion would have the "Hadamard form":

$$\partial K_{(1)}/\partial \dot{q}_I + \sum b_{DI}(\partial K_{(1)}/\partial \dot{q}_I) = 0, \tag{m1}$$

$$\dot{q}_D = \sum b_{DI}\dot{q}_I + b_D \qquad \text{[eqs. (i1, 2)]}, \tag{m2}$$

where $K_{(1)} = K_{(1)}(\dot{\gamma}, \dot{z}, \dot{\phi}, \dot{\theta}, \dot{\psi}; \gamma, z, \phi, \theta, \psi)$: *unconstrained* Tsenov function, instead of (j1–3).

Nielsen Equations

Next, let us derive the Routh–Voss form of the Nielsen equations:

$$N_k(T) = Q_k + \sum \lambda_D\, a_{Dk} \qquad (D = 1, 2;\ k = 1, \ldots, 5).$$

[We recall (a), (g2), and read off the coefficients a_{Dk} from (f1, 2)]. (m3)

To calculate $N_k(T)$, we apply the earlier Schieldrop–Nielsen rule [(ex. 6.3.1)]. We have already seen [eq. (h3)] that

$$\dot{T} = m\dot{z}\ddot{z} + m\rho^2\dot{\gamma}\ddot{\gamma} + mk^2\dot{\phi}\ddot{\phi} + mk^2\dot{\theta}\ddot{\theta} + mk^2\dot{\psi}\ddot{\psi} + mk^2\dot{\phi}\ddot{\psi}\,c\theta + mk^2\dot{\psi}\ddot{\phi}\,c\theta - mk^2\dot{\phi}\dot{\psi}\underline{\dot{\theta}}\,s\theta, \tag{n}$$

that is, only the last term is underlined.

From this, we build table 6.4. Hence, the $N_{\ldots}(T)$ totals are

$$N_\gamma(T) = m\rho^2\ddot{\gamma}, \tag{o1}$$

$$N_z(T) = m\ddot{z} \Rightarrow N_z(L) \equiv N_z(T - V) = N_z(T - mgz) = m(\ddot{z} + g), \tag{o2}$$

$$N_\phi(T) = mk^2\ddot{\phi} + mk^2\ddot{\psi}\,c\theta - mk^2\dot{\theta}\dot{\psi}\,s\theta = mk^2(\dot{\phi} + \dot{\psi}\,c\theta)^{\cdot}, \tag{o3}$$

$$N_\theta(T) = mk^2\ddot{\theta} + mk^2\dot{\phi}\dot{\psi}\,s\theta = mk^2(\ddot{\theta} + \dot{\phi}\dot{\psi}\,s\theta), \tag{o4}$$

$$N_\psi(T) = mk^2\ddot{\psi} + mk^2\ddot{\phi}\,c\theta - mk^2\dot{\phi}\dot{\theta}\,s\theta = mk^2(\dot{\psi} + \dot{\phi}\,c\theta)^{\cdot}; \tag{o5}$$

Table 6.4

$\dot{T}$-terms	$N_\gamma(T)$	$N_z(T)$	$N_\phi(T)$	$N_\theta(T)$	$N_\psi(T)$
$m\dot{z}\ddot{z}$	$k=0$	$k=1$	$k=0$		
	0	$m\ddot{z}$	0	0	0
$m\rho^2\dot{\gamma}\ddot{\gamma}$	$k=1$	$k=0$	$k=0$		
	$m\rho^2\ddot{\gamma}$	0	0	0	0
$mk^2\dot{\phi}\ddot{\phi}$	$k=0$	$k=0$	$k=1$		
	0	0	$mk^2\ddot{\phi}$	0	0
$mk^2\dot{\theta}\ddot{\theta}$	$k=0$	$k=0$	$k=0$		
	0	0	0	$mk^2\ddot{\theta}$	0
$mk^2\dot{\psi}\ddot{\psi}$	$k=0$	$k=0$	$k=0$		
	0	0	0	0	$mk^2\ddot{\psi}$
$mk^2\dot{\phi}\ddot{\psi}c\theta$	$k=0$	$k=0$	$k=1$		
	0	0	$mk^2\ddot{\psi}c\theta$	0	0
$mk^2\dot{\psi}\ddot{\phi}c\theta$	$k=0$	$k=0$	$k=0$		
	0	0	0	0	$mk^2\ddot{\phi}c\theta$
$-mk^2\dot{\phi}\dot{\psi}\underline{\dot{\theta}}s\theta$	$k=0$	$k=0$	$k=1$	$1-2=-1$	1
	0	0	$-mk^2\dot{\psi}\dot{\theta}s\theta$	$mk^2\dot{\phi}\dot{\psi}s\theta$	$-mk^2\dot{\phi}\dot{\theta}s\theta$

and, therefore, equations (m3) become

$$N_\gamma(L) = N_\gamma(T) = \rho\lambda_1, \tag{p1}$$
$$N_z(L) = N_z(T) + mg = \lambda_2, \tag{p2}$$
$$N_\phi(L) = N_\phi(T) = r\lambda_1, \tag{p3}$$
$$N_\theta(L) = N_\theta(T) = -r\lambda_2\, s(\phi-\gamma), \tag{p4}$$
$$N_\psi(L) = N_\psi(T) = (r\,c\theta)\lambda_1 - [r\,s\theta\, c(\phi-\gamma)]\lambda_2. \tag{p5}$$

Recalling the constraints (f1, 2), we see that the multipliers λ_D are proportional to the components of the force of rolling friction: along the tangential direction $\boldsymbol{u}_2$ (λ_1), and along the vertical direction $\boldsymbol{u}_3$ (λ_2).

Eliminating $\lambda_{1,2}$ among (p1–5), we obtain the following three kinetic equations (what might be called "Maggi form of the Nielsen equations," or "Nielsen form of the Maggi equations"):

$$r\,N_\gamma(T) - \rho\,N_\phi(T) = 0, \tag{q1}$$
$$N_\theta(T) + r\,N_z(L)\sin(\phi-\gamma) = 0, \tag{q2}$$
$$\rho\,N_\psi(T) - r\,N_\gamma(T)\cos\theta + r\rho\sin\theta\cos(\phi-\gamma)N_z(L) = 0. \tag{q3}$$

Finally, if we use the constraints (f1, 2), or (i1, 2), to eliminate $\dot{z}$, $\dot{\gamma}$ (and hence also $\ddot{z}$, $\ddot{\gamma}$) from the above, we should get the earlier equations (l1–3).

Example 6.5.3 Let us verify the identity $(T_o)^{\cdot} = (\dot{T})_{o}$, for a system with

$$2T = (\dot{x})^2 + (\dot{y})^2, \qquad y = y(x), \tag{a}$$

that is, $q_1 = y$, $q_2 = x$; $n = 2$, $m = 1$.

With the customary notations $(\ldots)^{\cdot} \equiv d(\ldots)/dt$, $(\ldots)' \equiv d(\ldots)/dx$, we have, successively,

(i)
$$\dot{T} = \dot{x}\,\ddot{x} + \dot{y}\,\ddot{y}$$
$$\Rightarrow (\dot{T})_o = \dot{x}\,\ddot{x} + (y'\dot{x})\,[y''\,(\dot{x})^2 + y'\ddot{x}]$$
$$= \dot{x}\,\ddot{x} + y'\,y''\,(\dot{x})^3 + (y')^2\dot{x}\,\ddot{x}. \quad \text{(b)}$$

(ii)
$$T_o = (1/2)\,[(\dot{x})^2 + (y'\,\dot{x})^2] = (1/2)\,[(\dot{x})^2 + (y')^2\,(\dot{x})^2]$$
$$\Rightarrow (T_o)^{\cdot} = \dot{x}\,\ddot{x} + y'\,(y')^{\cdot}\,(\dot{x})^2 + (y')^2\,\dot{x}\,\ddot{x}$$
$$= \dot{x}\,\ddot{x} + y'\,(y''\,\dot{x})\,(\dot{x})^2 + (y')^2\dot{x}\,\ddot{x} = (\dot{T})_o, \quad \text{Q.E.D.} \quad \text{(c)}$$

Problem 6.5.2 (Mei, 1983, pp. 628–630). Consider a system under constraints of the form

$$\dot{q}_D = \phi_D(t, q, \dot{q}_I) \qquad (D = 1, \ldots, m;\ I = m+1, \ldots, n); \quad \text{(a)}$$

and let, for any sufficiently smooth function f,

$$f = f(t, q, \dot{q}) = f\,[t, q, \phi_D(t, q, \dot{q}_I), \dot{q}_I] = f_o\,(t, q, \dot{q}_I) \equiv f_o. \quad \text{(b)}$$

Show that

$$\partial/\partial\dot{q}_I(df_o/dt) - 2(\partial f_o/\partial q_I)$$
$$= d/dt(\partial f_o/\partial\dot{q}_I) - \partial f_o/\partial q_I + \sum (\partial f_o/\partial q_D)\,(\partial\phi_D/\partial\dot{q}_I), \quad \text{(c)}$$

or, compactly,

$$N_I(f_o) = E_I(f_o) + \sum (\partial f_o/\partial q_D)\,(\partial\phi_D/\partial\dot{q}_I). \quad \text{(d)}$$

The above can be considered as an application of the earlier identity $N_k{}^*(f^*) = E_k{}^*(f^*)$ to the special form of the constraints:

$$\omega_D \equiv \dot{q}_D - \phi_D(t, q, \dot{q}_I) = 0, \qquad \omega_I \equiv \dot{q}_I. \quad \text{(e)}$$

Problem 6.5.3 (Mei, 1983, pp. 632–633; 1985, pp. 208–211). Using the kinematico-inertial identity (c, d) of the preceding problem, show that the *Nielsen form of the special nonlinear Voronets equations*

$$(\partial T_o/\partial\dot{q}_I)^{\cdot} - \partial T_o/\partial q_I$$
$$- \sum (\partial\phi_D/\partial\dot{q}_I)\,(\partial T_o/\partial q_D) - \sum W^D{}_I\,(\partial T/\partial\dot{q}_D)_o = Q_{Io}, \quad \text{(a)}$$

where

$$Q_{Io} \equiv Q_I + \sum (\partial\phi_D/\partial\dot{q}_I) Q_D, \quad \text{(a1)}$$

$$W^D{}_I \equiv (\partial\phi_D/\partial\dot{q}_I)^{\cdot} - \partial\phi_D/\partial q_I - \sum (\partial\phi_D/\partial q_{D'})\,(\partial\phi_{D'}/\partial\dot{q}_I)$$
$$\equiv E_I(\phi_D) - \sum (\partial\phi_D/\partial q_{D'})\,(\partial\phi_{D'}/\partial\dot{q}_I), \quad \text{(a2)}$$

is

$$\partial \dot{T}_o/\partial \dot{q}_I - 2(\partial T_o/\partial q_I) - \sum (\partial T/\partial \dot{q}_D)_o [\partial \ddot{q}_D/\partial \dot{q}_I - 2(\partial \dot{q}_D/\partial q_I)]$$
$$- 2\sum (\partial T/\partial q_D)_o (\partial \phi_D/\partial \dot{q}_I) = Q_{Io}, \tag{b}$$

or, compactly,

$$N_I(T_o) - \sum (\partial T/\partial \dot{q}_D)_o N_I(\dot{q}_D) - 2\sum (\partial T/\partial \dot{q}_D)_o (\partial \phi_D/\partial \dot{q}_I) = Q_{Io}. \tag{c}$$

[If $\partial T/\partial q_D = 0$, then (b) reduces to what may be termed the *Nielsen form of the special nonlinear Chaplygin equations*:

$$\partial \dot{T}_o/\partial \dot{q}_I - 2(\partial T_o/\partial q_I)$$
$$- \sum (\partial T/\partial \dot{q}_D)_o [\partial \ddot{q}_D/\partial \dot{q}_I - 2(\partial \dot{q}_D/\partial q_I)] = Q_{Io}]. \tag{d}$$

HINTS

By the kinematico-inertial identity of the preceding problem:

$$N_I(T_o) = E_I(T_o) + \sum (\partial T_o/\partial q_D)(\partial \phi_D/\partial \dot{q}_I), \tag{e}$$

$$N_I(\dot{q}_D) = E_I(\dot{q}_D) + \sum (\partial \phi_D/\partial q_{D'})(\partial \phi_{D'}/\partial \dot{q}_I). \tag{f}$$

Problem 6.5.4 (Mei, 1985, pp. 196–203; 1987, pp. 397–402). Continuing from the preceding problem, show that:

(i) If the constraints have the special Pfaffian form

$$\dot{q}_D = \sum b_{DI}(t, q)\dot{q}_I + b_D(t, q), \tag{a}$$

then the preceding *Nielsen form of the special Voronets equations*, eqs. (b, c), reduces to

$$\partial \dot{T}_o/\partial \dot{q}_I - 2(\partial T_o/\partial q_I)$$
$$- \sum (\partial T/\partial \dot{q}_D)_o \Big\{ \sum [b^D{}_{II'} - 2(\partial b_{DI'}/\partial q_I)]\dot{q}_{I'} + [b^D{}_I - 2(\partial b_D/\partial q_I)] \Big\}$$
$$- 2\sum (\partial T/\partial q_D)_o b_{DI} = Q_{Io}, \tag{b}$$

where

$$b^D{}_{II'} \equiv \sum [(\partial b_{DI}/\partial q_{D'}) b_{D'I'} + (\partial b_{DI'}/\partial q_{D'}) b_{D'I}] + (\partial b_{DI}/\partial q_{I'} + \partial b_{DI'}/\partial q_I), \tag{c1}$$

$$b^D{}_I \equiv \sum [(\partial b_D/\partial q_{D'}) b_{D'I} + (\partial b_{DI}/\partial q_{D'}) b_{D'}] + (\partial b_{DI}/\partial t + \partial b_D/\partial q_I); \tag{c2}$$

and, therefore,

(ii) If the constraints (a) have the Chaplygin form

$$\dot{q}_D = \sum b_{DI}(q_I)\dot{q}_I, \tag{d}$$

and $\partial T/\partial q_D = 0$ (recall discussion in §3.8), then (b) reduces to the *Nielsen form of the linear (i.e., Pfaffian) Chaplygin equations*:

$$\partial \dot{T}_o/\partial \dot{q}_I - 2(\partial T_o/\partial q_I) - \sum\sum (\partial T/\partial \dot{q}_D)_o (\partial b_{DI}/\partial q_{I'} - \partial b_{DI'}/\partial q_I)\dot{q}_{I'} = Q_{Io}, \qquad \text{(e)}$$

since in this case

$$b_D = 0, \qquad b^D{}_I = 0, \qquad b^D{}_{II'} = \partial b_{DI}/\partial q_{I'} + \partial b_{DI'}/\partial q_I. \qquad \text{(f)}$$

Problem 6.5.5 Show that, for $s = 1, 2, 3, \ldots,$

$$d/dt\left[\partial \overset{(s-1)}{T_o}/\partial \overset{(s)}{q_I}\right] - \partial T/\partial q_I = d/dt\,(\partial T/\partial \dot{q}_I) - \partial T/\partial q_I + \sum d/dt\left[(\partial T/\partial \dot{q}_D)\,(\partial \overset{(s)}{q_D}/\partial \overset{(s)}{q_I})\right]. \qquad \text{(a)}$$

HINTS

Recall that $\left(\overset{(s-1)}{T}\right)_o = (T_o)^{(s-1)}$, and show that

$$\partial \overset{(s-1)}{T_o}/\partial \overset{(s)}{q_I} = \partial \overset{(s-1)}{T}/\partial \overset{(s)}{q_I} + \sum (\partial \overset{(s-1)}{T}/\partial \overset{(s)}{q_D})\,(\partial \overset{(s)}{q_D}/\partial \overset{(s)}{q_I}). \qquad \text{(b)}$$

Problem 6.5.6 *Higher-Order Equations of Nielsen et al. in Holonomic and Quasi Variables; and their Relation with the Equations of Lagrange, Hamel, et al.*

(i) Let us define the *(s)th-order holonomic operators of Nielsen*:

$$N_k{}^{(s)}(\ldots) \equiv \partial(\overset{(s)}{\ldots})/\partial \overset{(s)}{q_k} - 2\,[\partial(\overset{(s-1)}{\ldots})/\partial \overset{(s-1)}{q_k}], \qquad \text{(a)}$$

and *Euler–Lagrange*:

$$E_k{}^{(s)}(\ldots) \equiv d/dt\,[\partial(\overset{(s-1)}{\ldots})/\partial \overset{(s)}{q_k}] - [\partial(\overset{(s-1)}{\ldots})/\partial \overset{(s-1)}{q_k}]. \qquad \text{(b)}$$

Show that, for any sufficiently differentiable function $f = f(t, q, \dot{q})$, and any $k = 1, 2, \ldots, n$; $s = 1, 2, 3, \ldots,$

$$N_k{}^{(s)}(f) = E_k{}^{(s)}(f). \qquad \text{(c)}$$

(ii) Let us define the *(s)th-order nonholonomic operators of Nielsen*:

$$N_k{}^{*(s)}(\ldots) \equiv \partial(\overset{(s)}{\ldots})/\partial \overset{(s)}{\theta_k} - 2\left[\partial(\overset{(s-1)}{\ldots})/\partial \overset{(s-1)}{\theta_k}\right], \qquad \text{(d)}$$

and *Euler–Lagrange*:

$$E_k{}^{*(s)}(\ldots) \equiv d/dt\left[\partial(\overset{(s-1)}{\ldots})/\partial \overset{(s)}{\theta_k}\right] - \left[\partial(\overset{(s-1)}{\ldots})\Big/\partial \overset{(s-1)}{\theta_k}\right], \qquad \text{(e)}$$

where

$$\partial(\overset{(s-1)}{\ldots})/\partial \overset{(s-1)}{\theta_k} \equiv \sum \left[\partial(\overset{(s-1)}{\ldots})/\partial \overset{(s-1)}{q_l}\right]\left[\partial \overset{(s)}{q_l}/\partial \overset{(s)}{\theta_k}\right] \tag{e1}$$

[*(s)th-order quasi chain rule*].

Show that, for any sufficiently differentiable function $f^* = f^*(t, q, \omega)$, and any $k = 1, 2, \ldots, n;\ s = 1, 2, 3, \ldots,$

$$N_k^*\left[\overset{(s)}{f}{}^*\right] = E_k^*\left[\overset{(s)}{f}{}^*\right], \tag{f}$$

where

$$f(t, q, \dot{q}) \Rightarrow \dot{f} \Rightarrow \cdots \overset{(s-1)}{f} \Rightarrow \overset{(s)}{f}, \tag{g}$$

$$\begin{aligned} \overset{(s-1)}{f}{}^* &= \overset{(s-1)}{f}\left[t, q, \dot{q} \equiv \overset{(1)}{q}, \ldots, \overset{(s-1)}{q};\ \overset{(s)}{q}(t, q, \overset{(1)}{q}, \ldots, \overset{(s-1)}{q}, \overset{(s)}{\theta})\right] \\ &= \overset{(s-1)}{f}\left(t, q, \overset{(1)}{q}, \ldots, \overset{(s-1)}{q}, \overset{(s)}{\theta}\right), \end{aligned} \tag{h}$$

$$\begin{aligned} \overset{(s)}{f}{}^* = \overset{(s)}{f}\Bigg[& t, q, \overset{(1)}{q}, \ldots, \overset{(s-1)}{q}; \\ & \overset{(s)}{q}\left(t, q, \overset{(1)}{q}, \ldots, \overset{(s-1)}{q}, \overset{(s)}{\theta}\right), \overset{(s+1)}{q}\left(t, q, \overset{(1)}{q}, \ldots, \overset{(s-1)}{q}, \overset{(s)}{\theta}, \overset{(s+1)}{\theta}\right)\Bigg] \\ = \overset{(s)}{f}\Bigg(& t, q, \overset{(1)}{q}, \ldots, \overset{(s-1)}{q}, \overset{(s)}{\theta}, \overset{(s+1)}{\theta}\Bigg). \end{aligned} \tag{i}$$

Applications (Mei, 1985, pp. 300–308).

(a) Consider the earlier, say unconstrained, equations of Mangeron et al. (6.4.14e):

$$(1/s)\left[\partial \overset{(s)}{T}/\partial \overset{(s)}{q_k} - (s+1)(\partial T/\partial q_k)\right] = Q_k,$$
$$(k = 1, \ldots, n;\ s = 1, 2, 3, \ldots). \tag{j}$$

Substituting into it $s-1$ for s yields

$$\partial T/\partial q_k = (1/s)\left(\partial \overset{(s-1)}{T}/\partial \overset{(s-1)}{q_k}\right) - [(s-1)/s]\, Q_k, \tag{k}$$

and then reinserting this expression into (j) results in the *Nielsen* form:

$$s\left(\partial \overset{(s)}{T}/\partial \overset{(s)}{q_k}\right) - (s+1)\left(\partial \overset{(s-1)}{T}/\partial \overset{(s-1)}{q}\right) = Q_k; \tag{l}$$

and, by (c): $N_k^{(s)}(T) = E_k^{(s)}(T)$, also in the *Lagrange* form:

$$(s)\left(\partial \overset{(s-1)}{T}/\partial \overset{(s)}{q}\right)^{\bullet} - \partial \overset{(s-1)}{T}/\partial \overset{(s-1)}{q_k} = Q_k. \tag{m}$$

(b) Next, if the system is subject to the $m(< n)$ constraints

$$f_D\left(t, q, \dot{q}, \ldots, \overset{(s)}{q}\right) = 0, \tag{n}$$

then we must add, to the right side of the above the term,

$$\sum \lambda_D \left(\partial f_D/\partial \overset{(s)}{q_k}\right). \tag{o}$$

Or we may choose to introduce quasi variables such that

$$\delta \overset{(s)}{q_k} = \sum \left[\partial \overset{(s)}{q_k}/\partial \overset{(s)}{\theta_D}\right] \delta \overset{(s)}{\theta_D}; \tag{p}$$

then it can be shown, by a systematic process of differentiations, that the corresponding, say kinetic, equations are

Hamel-type:

$$\begin{aligned} &(s)\, d/dt\left[\partial \overset{(s-1)}{T}{}^*/\partial \overset{(s)}{\theta_I}\right] - \left[\partial \overset{(s-1)}{T}{}^*/\partial \overset{(s-1)}{\theta_I}\right] \\ &- \sum \left[s\left(\partial \overset{(s)}{q_k}/\partial \overset{(s)}{\theta_I}\right)^{\cdot} - \partial \overset{(s)}{q_k}/\partial \overset{(s-1)}{\theta_I}\right] \left(\partial \overset{(s-1)}{T}/\partial \overset{(s)}{q_k}\right)^* \\ &- \sum \left(\partial \overset{(s)}{q_k}/\partial \overset{(s)}{\theta_I}\right) Q_k = \Theta_I, \end{aligned} \tag{q}$$

and

Nielsen-type :

$$\begin{aligned} &(s)\left(\partial \overset{(s)}{T}{}^*/\partial \overset{(s)}{\theta_I}\right) - (s+1)\left(\partial \overset{(s-1)}{T}{}^*/\partial \overset{(s-1)}{\theta_I}\right) \\ &- \sum \left[s\left(\partial \overset{(s+1)}{q_k}/\partial \overset{(s)}{\theta_I}\right) - (s+1)\left(\partial \overset{(s)}{q_k}/\partial \overset{(s-1)}{\theta_I}\right)\right] \left(\partial \overset{(s-1)}{T}/\partial \overset{(s)}{q_k}\right)^* \\ &= \Theta_I. \end{aligned} \tag{r}$$

For $s = 1$, the above yield, respectively,

$$\begin{aligned} &(\partial T^*/\partial \dot{\theta}_I)^{\cdot} - \partial T^*/\partial \theta_I \\ &\quad - \sum \left[(\partial \dot{q}_k/\partial \dot{\theta}_I)^{\cdot} - \partial \dot{q}_k/\partial \theta_I\right] (\partial T/\partial \dot{q}_k)^* = \Theta_I, \end{aligned} \tag{q1}$$

$$\begin{aligned} &\partial \dot{T}^*/\partial \dot{\theta}_I - 2(\partial T^*/\partial \theta_I) \\ &\quad - \sum \left[\partial \ddot{q}_k/\partial \dot{\theta}_I - 2(\partial \dot{q}_k/\partial \theta_I)\right] (\partial T/\partial \dot{q}_k)^* = \Theta_I; \end{aligned} \tag{r1}$$

and, for $s = 2$,

$$2[\partial \dot{T}^*/\partial \ddot{\theta}_I]^{\cdot} - \partial \dot{T}^*/\partial \dot{\theta}_I - \sum \left[2\,(\partial \dddot{q}_k/\partial \ddot{\theta}_I)^{\cdot} - \partial \dddot{q}_k/\partial \dot{\theta}_I\right](\partial \dot{T}/\partial \dddot{q}_k)^* = \Theta_I, \tag{q2}$$

$$2(\partial \ddot{T}^*/\partial \ddot{\theta}_I) - 3\,(\partial \dot{T}^*/\partial \dot{\theta}_I) - \sum \left[2\,(\partial \dddot{q}_k/\partial \ddot{\theta}_I) - 3\,(\partial \ddot{q}_k/\partial \dot{\theta}_I)\right](\partial \dot{T}/\partial \ddot{q}_k)^* = \Theta_I. \tag{r2}$$

6.6 THE PRINCIPLE OF GAUSS (EXTENSIVE TREATMENT)

> It is quite remarkable that Nature modifies free motions incompatible with the necessary constraints in the same way in which the calculating mathematician uses least squares to bring into agreement results which are based on quantities connected to each other by necessary relations.
>
> (C. F. Gauss, 1829, *On a New General Fundamental Principle of Mechanics)*

> Gauss was not only a very eminent mathematician, but also an astronomer and geodesist, and as such, a passionate calculator of numerical results. It was he who founded the method of least squares, which he evolved with successively greater depth in three extensive treatises. If, as happened now and then, he was asked (against his will) to deliver a lecture at the University of Göttingen, his preferred topic was always the method of least squares.
>
> [A. Sommerfeld, 1964 (1940s), §48]

For complementary reading on Gauss' principle, see (alphabetically): Brill (1909, pp. 45–51), Coe (1938, pp. 421–423), Dugas (1955, pp. 367–369), Lanczos (1970, pp. 106–108), Lindsay and Margenau (1936, pp. 112–115), Mach (1960, pp. 440–443), MacMillan (1927, pp. 419–421), Volkmann (1900, pp. 355–357).

The Fundamental Theory

As was realized early in the 20th century, by Appell, Chetaev, Hamel, et al., the equations of motion of systems subject to the m nonlinear first-order constraints

$$f_D(t, \boldsymbol{r}, \boldsymbol{v}) = 0 \quad \text{(particle form)} \qquad \text{or} \qquad f_D(t, q, \dot{q}) = 0 \quad \text{(system form)}, \tag{6.6.1}$$

let alone higher-order such constraints, *cannot* be derived from Lagrange's principle (LP); the reason being that (6.6.1) cannot be attached, or adjoined, to LP—*we need its virtual form*, and it is not clear how that should be done, so as to get the correct equations of motion. For this, we need either the principle of Jourdain or Gauss' *principle of least constraint*, or *least compulsion*, or *least constriction*. The compulsion

Z (from the German *Zwang*) of a generally constrained mechanical system, in actual or kinematically possible motion, is defined as

$$Z \equiv (1/2) \textstyle\mathsf{S}\, dm\, [\boldsymbol{a} - (d\boldsymbol{F}/dm)]^2 \equiv (1/2) \textstyle\mathsf{S}\, (1/dm)\, (dm\, \boldsymbol{a} - d\boldsymbol{F})^2$$
$$\equiv (1/2) \textstyle\mathsf{S}\, (d\boldsymbol{R})^2/dm = (1/2) \textstyle\mathsf{S}\, (-d\boldsymbol{R})^2/dm; \qquad (6.6.2)$$

where, as usual (§3.2), for the actual motion

$$dm\, \boldsymbol{a} = d\boldsymbol{F} + d\boldsymbol{R} \qquad (d\boldsymbol{F}: \textit{impressed}\ \text{force},\ d\boldsymbol{R}: \textit{constraint reaction}). \qquad (6.6.2\text{a})$$

The above show clearly that $Z \geq 0$; with the equal sign holding for unconstrained motion; that is, $d\boldsymbol{R} = \boldsymbol{0}$. Further, expanding (6.6.2), we readily find

$$Z = (1/2) \textstyle\mathsf{S}\, dm\, \boldsymbol{a} \cdot \boldsymbol{a} - \textstyle\mathsf{S}\, d\boldsymbol{F} \cdot \boldsymbol{a} + (1/2) \textstyle\mathsf{S}\, dm\, (d\boldsymbol{F}/dm)^2$$
$$= S - \textstyle\mathsf{S}\, d\boldsymbol{F} \cdot \boldsymbol{a} + \cdots, \qquad (6.6.3)$$

where

$$S = (1/2) \textstyle\mathsf{S}\, dm\, \boldsymbol{a} \cdot \boldsymbol{a}: \ \textit{Appellian}\ \text{function (§3.3 ff.)}, \qquad (6.6.3\text{a})$$

and $\ldots \equiv$ *terms not containing accelerations*, like $\boldsymbol{a}$, $\ddot{q}$, $\dot{\omega}$; that is, a function of $t, q, \dot{q}$ or ω. Obviously, the factor 1/2 in (6.6.2) is unimportant, and is frequently omitted in the literature; but, as (6.6.3) shows, it makes the connection between Z and S clearer.

Now, the principle of Gauss (GP) states that the (first) *Gaussian* variation of Z, $\delta'' Z$, to be (re)defined below, vanishes, that is,

$$\delta'' Z = 0, \qquad (6.6.4)$$

for all variations of the accelerations from the actual motion, $\delta\boldsymbol{a} \equiv \delta''\boldsymbol{a}$, *that are compatible with all the constraints, at a given time and with given positions and velocities (and impressed forces)*; that is, *for*

$$\delta'' t = 0, \quad \delta'' \boldsymbol{r} = \boldsymbol{0}, \quad \delta'' \boldsymbol{v} = \boldsymbol{0}, \quad \delta''(d\boldsymbol{F}) = \boldsymbol{0}, \quad \textit{but} \quad \delta'' \boldsymbol{a} \neq \boldsymbol{0}. \qquad (6.6.5)$$

Since in classical mechanics $d\boldsymbol{F} = d\boldsymbol{F}(t, \boldsymbol{r}, \boldsymbol{v})$ (see, for example, Pars, 1965, pp. 11–12), we will have

$$\delta''(d\boldsymbol{F}) = \boldsymbol{0} \quad (\textit{particle}\ \text{force variation}) \quad \text{and} \quad \delta'' Q_k = 0 \quad (\textit{system}\ \text{force variation}), \qquad (6.6.6)$$

and so GP reads

$$\delta'' Z = (1/2) \textstyle\mathsf{S}\, dm\, (2)\, [\boldsymbol{a} - (d\boldsymbol{F}/dm)] \cdot \delta'' \boldsymbol{a}$$
$$= \textstyle\mathsf{S}\, (dm\, \boldsymbol{a} - d\boldsymbol{F}) \cdot \delta'' \boldsymbol{a} = 0, \qquad (6.6.7)$$

or, in terms of the reactions,

$$\delta'' Z = \textstyle\mathsf{S}\, (d\boldsymbol{R}/dm) \cdot \delta''(d\boldsymbol{R}) = \textstyle\mathsf{S}\, (d\boldsymbol{R}/dm) \cdot \delta''(dm\, \boldsymbol{a} - d\boldsymbol{F})$$
$$= \textstyle\mathsf{S}\, (d\boldsymbol{R}/dm) \cdot dm\, \delta'' \boldsymbol{a} = \textstyle\mathsf{S}\, d\boldsymbol{R} \cdot \delta'' \boldsymbol{a} = 0. \qquad (6.6.8)$$

Principle of Gauss (GP) versus Principle of Lagrange (LP)

The relation between

$$\text{LP}: \qquad S\, dm\, \boldsymbol{a} \cdot \delta \boldsymbol{r} = S\, d\boldsymbol{F} \cdot \delta \boldsymbol{r} \tag{6.6.9a}$$

and

$$\text{GP}: \qquad S\, dm\, \boldsymbol{a} \cdot \delta'' \boldsymbol{a} = S\, d\boldsymbol{F} \cdot \delta'' \boldsymbol{a}, \tag{6.6.9b}$$

that is, *the question of their mutual consistency and equivalence*, is of cardinal importance to constrained system mechanics. Since there is only one mechanics, *both LP and GP must produce the same equations of motion.* Therefore, let us begin by examining the derivation of such equations from these principles.

LP: substituting into (6.6.9a) the representation

$$\delta \boldsymbol{r} = \sum \boldsymbol{e}_k\, \delta q_k = \sum \boldsymbol{\varepsilon}_I\, \delta \theta_I, \tag{6.6.10a}$$

and recalling the arguments expounded in chapters 3 and 5, we find the *raw* form of the kinetic equations:

$$S\, dm\, \boldsymbol{a} \cdot \boldsymbol{\varepsilon}_I = S\, d\boldsymbol{F} \cdot \boldsymbol{\varepsilon}_I. \tag{6.6.10b}$$

GP: We need $\delta'' \boldsymbol{a}$. We have successively

$$\begin{aligned} \boldsymbol{v} &= d\boldsymbol{r}/dt = \sum \boldsymbol{e}_k \dot{q}_k + no\ \dot{q}\ terms = \sum \boldsymbol{\varepsilon}_I\, \omega_I + no\ \omega\ terms, \\ \Rightarrow \boldsymbol{a} &= d\boldsymbol{v}/dt = \sum \boldsymbol{e}_k\, \ddot{q}_k + no\ \ddot{q}\ terms = \sum \boldsymbol{\varepsilon}_I\, \dot{\omega}_I + no\ \dot{\omega}\ terms, \\ \Rightarrow \delta'' \boldsymbol{a} &= \sum \boldsymbol{e}_k\, \delta \ddot{q}_k = \sum \boldsymbol{\varepsilon}_I\, \delta \dot{\omega}_I \qquad \text{[Gaussian counterpart of (6.6.10a)]}, \end{aligned} \tag{6.6.11}$$

and, substituting this into (6.6.9b), we reobtain (6.6.10b).

Next, let us move to general *system* quasi variables. As seen in §5.2, for nonlinear (possibly nonholonomic) velocity constraints

$$\omega_D \equiv f_D(t, q, \dot{q}) = 0, \qquad \omega_I \equiv f_I(t, q, \dot{q}) = 0; \tag{6.6.12a}$$

$$\dot{q}_k = \dot{q}_k(t, q, \omega) = \dot{q}_k(t, q, \omega_I) \equiv F_k(t, q, \omega_I), \tag{6.6.12b}$$

we must define $\delta \theta_k$; to replace in (6.6.12b) $\dot{q}_k$ with δq_k and ω_I with $\delta \theta_I$ [i.e., $\delta q_k = F_k(t, q, \delta \theta_I)$] would be meaningless (useless). Instead, we are seeking a definition in which:

(i) The nonholonomic system virtual displacements, $\delta \theta$, are *linear and homogeneous* combinations of their holonomic counterparts, δq; and vice versa:

$$\delta \theta_I = \sum (\ldots)_{Ik}\, \delta q_k \Leftrightarrow \delta q_k = \sum (\ldots)_{kI}\, \delta \theta_I, \tag{6.6.13}$$

so that we can attach (or adjoin) the constraints to LP; and

(ii) In the linear (Pfaffian) case, it reduces to the earlier results (chap. 2). This is accomplished by requiring compatibility between LP and GP.

(a) Indeed substituting (6.6.11) into (6.6.9b), we obtain the Gaussian form:

$$\sum \left(S\, (dm\, \boldsymbol{a} - d\boldsymbol{F}) \cdot \boldsymbol{e}_k \right) \delta \ddot{q}_k = 0; \tag{6.6.14a}$$

or, since $[(\ldots)^{\cdot}$-differentiating (6.6.12b) and then varying it à la Gauss, and setting $\delta\dot{\omega}_D = 0]$

$$\begin{aligned}\delta''(\ddot{q}_k) &= \delta''\Big(\sum (\partial\dot{q}_k/\partial\omega_l)\dot{\omega}_l + no\ \dot{\omega}\ terms\Big)\\ &= \sum (\partial\dot{q}_k/\partial\omega_l)\,\delta\dot{\omega}_l = \sum (\partial\dot{q}_k/\partial\omega_I)\,\delta\dot{\omega}_I,\end{aligned} \tag{6.6.14b}$$

finally,

$$\begin{aligned}&\sum\sum\Big\{S\,(dm\,\boldsymbol{a} - d\boldsymbol{F})\cdot[\boldsymbol{e}_k(\partial\dot{q}_k/\partial\omega_I)]\Big\}\,\delta\dot{\omega}_I\\ &\quad= \sum\Big(S\,(dm\,\boldsymbol{a} - d\boldsymbol{F})\cdot\boldsymbol{\varepsilon}_I\Big)\,\delta\dot{\omega}_I = 0;\end{aligned} \tag{6.6.14c}$$

(b) On the other hand, substituting (6.6.10a) into (6.6.9a), we obtain the Lagrangean form:

$$\begin{aligned}&\sum\Big(S\,(dm\,\boldsymbol{a} - d\boldsymbol{F})\cdot\boldsymbol{e}_k\Big)\delta q_k\\ &\quad= \sum\Big(S\,(dm\,\boldsymbol{a} - d\boldsymbol{F})\cdot\boldsymbol{\varepsilon}_I\Big)\delta\theta_I = 0.\end{aligned} \tag{6.6.15}$$

Now, the Gaussian variational equation (6.6.14c) can be brought into agreement with the Lagrangean (6.6.15) via the following fundamental definition:

$$\delta q_k \equiv \sum (\partial\dot{q}_k/\partial\omega_l)\,\delta\theta_l, \tag{6.6.16}$$

from which, inverting, we find

$$\delta\theta_l \equiv \sum (\partial\omega_l/\partial\dot{q}_k)\,\delta q_k \tag{6.6.17}$$

$$\Rightarrow \delta\theta_D \equiv \sum (\partial\omega_D/\partial\dot{q}_k)\,\delta q_k \equiv \sum (\partial f_D/\partial\dot{q}_k)\,\delta q_k = 0, \tag{6.6.17a}$$

$$\delta\theta_I \equiv \sum (\partial\omega_I/\partial\dot{q}_k)\,\delta q_k \equiv \sum (\partial f_I/\partial\dot{q}_k)\,\delta q_k \neq 0. \tag{6.6.17b}$$

Then, also,

$$\delta f_D \equiv \delta\omega_D = S\,(\partial f_D/\partial\boldsymbol{v})\cdot\delta\boldsymbol{r} = \sum (\partial f_D/\partial\dot{q}_k)\,\delta q_k = 0; \tag{6.6.17c}$$

instead of the formal (calculus of variations) definition

$$\delta f_D = \sum (\partial f_D/\partial q_k)\,\delta q_k + \sum (\partial f_D/\partial\dot{q}_k)\,\delta\dot{q}_k = 0. \tag{6.6.18}$$

Principle of Jourdain (JP) versus Principle of Lagrange (LP)

The same conclusion — namely, (6.6.17a) — can also be reached by requiring compatibility between LP and JP:

$$S\,(dm\,\boldsymbol{a} - d\boldsymbol{F})\cdot\delta'\boldsymbol{v} = 0, \qquad \text{under}\quad \delta t = 0 \quad\text{and}\quad \delta\boldsymbol{r} = \boldsymbol{0}. \tag{6.6.19}$$

Successively,

$$\begin{aligned}\delta' \boldsymbol{v} &= \delta'\Big(\sum \boldsymbol{e}_k \dot{q}_k + no\ \dot{q}\ terms\Big)\\ &= \sum \boldsymbol{e}_k\, \delta' \dot{q}_k\\ &= \sum \boldsymbol{e}_k\Big(\sum (\partial \dot{q}_k/\partial \omega_I)\delta\omega_I\Big)\\ &= \sum\sum \boldsymbol{e}_k(\partial \dot{q}_k/\partial \omega_I)\,\delta\omega_I, \qquad (6.6.19a)\end{aligned}$$

and, inserting this into (6.6.19), we find

$$\sum\sum\Big(\mathcal{S}\,(dm\,\boldsymbol{a} - d\boldsymbol{F}) \cdot \boldsymbol{e}_k(\partial \dot{q}_k/\partial \omega_I)\Big)\delta\omega_I = 0. \qquad (6.6.19b)$$

Again, compatibility between this and (6.6.19) leads to the definitions (6.6.16–17b).

Equations of Motion

We already have GP in its raw form; that is, eq. (6.6.9b). To obtain its system form, we must transform (6.6.14a). Indeed, recalling standard kinematico-inertial identities (chap. 3), we find

$$\begin{aligned}\delta'' Z &= \sum\Big(\mathcal{S}\,(dm\,\boldsymbol{a} - d\boldsymbol{F}) \cdot \boldsymbol{e}_k\Big)\delta\ddot{q}_k\\ &= \sum\Big(\mathcal{S}\, dm\,\boldsymbol{a} \cdot \boldsymbol{e}_k - \mathcal{S}\, d\boldsymbol{F} \cdot \boldsymbol{e}_k\Big)\delta\ddot{q}_k\\ &= \sum [E_k(T) - Q_k]\delta\ddot{q}_k = 0. \qquad (6.6.20)\end{aligned}$$

Next, to bring velocity constraints like (6.6.1) to Gaussian form—that is, to make them exhibit accelerations explicitly—*first* we $(\ldots)^{\cdot}$-differentiate them and *then* we vary them à la Gauss: $\delta''[(\ldots)^{\cdot}]$. Thus, (6.6.1) yields

Particle form:

$$\begin{aligned}\delta''(\dot{f}_D) &= \delta''\Big\{\partial f_D/\partial t + \mathcal{S}\,[(\partial f_D/\partial \boldsymbol{r}) \cdot \boldsymbol{v} + (\partial f_D/\partial \boldsymbol{v}) \cdot \boldsymbol{a}]\Big\}\\ &= \mathcal{S}\,(\partial f_D/\partial \boldsymbol{v}) \cdot \delta\boldsymbol{a} = 0, \qquad (6.6.21a)\end{aligned}$$

System form:

$$\begin{aligned}\delta''(\dot{f}_D) &= \delta''\Big\{\partial f_D/\partial t + \sum [(\partial f_D/\partial q_k)\dot{q}_k + (\partial f_D/\partial \dot{q}_k)\ddot{q}_k]\Big\}\\ &= \sum (\partial f_D/\partial \dot{q}_k)\,\delta\ddot{q}_k \qquad (6.6.21b)\end{aligned}$$

$$\Big[\text{which, does } \textit{not} \text{ equal } (\delta'' f_D)^{\cdot} = \Big(\sum (\partial f_D/\partial \ddot{q}_k)\,\delta\ddot{q}_k\Big)^{\cdot} = \Big(\sum (0)\,\delta\ddot{q}_k\Big)^{\cdot} = 0\Big];$$

and, finally, adjoining the above to (6.6.9b) or to its system counterpart (6.6.20), via Lagrangean multipliers, we obtain the *general variational equation* (unconstrained variations):

$$\delta'' Z + \sum \lambda_D\,\delta''(\dot{f}_D) = 0; \qquad (6.6.22)$$

where, as a result of $\delta''(\dot{f}_D) \neq (\delta'' f_D)^{\cdot}$ (in general),

$$\sum \lambda_D\,\delta''(\dot{f}_D) \neq \sum \lambda_D(\delta'' f_D)^{\cdot}.$$

From the above, all kinds of equations of motion, in *holonomic* variables flow; while for kinetic equations in *nonholonomic* variables, they follow from (6.6.7) or (6.6.9b) in connection with (6.6.11) (and similarly for kinetostatic equations). For example:

(i) Combining (6.6.7, 9b) with (6.6.21a), we get Lagrange's equations of the *first* kind:

$$dm\,\boldsymbol{a} = d\boldsymbol{F} + \sum \lambda_D\,(\partial f_D/\partial \boldsymbol{v}); \tag{6.6.23}$$

(ii) While, combining (6.6.20) with (6.6.21b), we obtain Lagrange's equations of the *second* kind (Routh–Voss equations):

$$E_k(T) \equiv (\partial T/\partial \dot{q}_k)^{\cdot} - \partial T/\partial q_k = Q_k + \sum \lambda_D(\partial f_D/\partial \dot{q}_k); \tag{6.6.24a}$$

or, since $E_k(T) = \partial S/\partial \ddot{q}_k$, in their Appellian form:

$$\partial S/\partial \ddot{q}_k = Q_k + \sum \lambda_D(\partial f_D/\partial \dot{q}_k). \tag{6.6.24b}$$

REMARK

If the constraints have the form $f_D(t, \boldsymbol{r}) = 0$ $[f_D(t, q) = 0]$, it does not mean that we should set, in the right side of (6.6.23) [(6.6.24a, b)] $\partial f_D/\partial \boldsymbol{v} = \boldsymbol{0}$ $[\partial f_D/\partial \dot{q}_k = 0]$. It means that, first, we bring these constraints to Gaussian form and then we $\delta''(\ldots)$-vary them; that is, for $f_D(t, \boldsymbol{r})$, we have, successively,

$$\begin{aligned} f_D = 0 &\Rightarrow df_D/dt = \partial f_D/\partial t + S\,(\partial f_D/\partial \boldsymbol{r})\cdot \boldsymbol{v} \\ &\Rightarrow d^2 f_D/dt^2 = (\partial f_D/\partial t)^{\cdot} + S\,[(\partial f_D/\partial \boldsymbol{r})^{\cdot}\cdot \boldsymbol{v} + (\partial f_D/\partial \boldsymbol{r})\cdot \boldsymbol{a}] \\ &\Rightarrow \delta''(\ddot{f}_D) = S\,(\partial f_D/\partial \boldsymbol{r})\cdot \delta''\boldsymbol{a} = 0; \end{aligned} \tag{6.6.25}$$

and similarly for $f_D(t, q) = 0$: $\delta''(\ddot{f}_D) = \sum (\partial f_D/\partial q_k)\,\delta''\ddot{q}_k = 0$. Also, it is worth noting that, since this constraint is holonomic, the right side of (6.6.25) would have resulted even if we had *reversed* the order of differentiations:

$$\begin{aligned} \delta f_D &= S\,(\partial f_D/\partial \boldsymbol{r})\cdot \delta\boldsymbol{r} = 0 \\ &\Rightarrow (\delta f_D)^{\cdot} = S\,[(\partial f_D/\partial \boldsymbol{r})^{\cdot}\cdot \delta\boldsymbol{r} + (\partial f_D/\partial \boldsymbol{r})\cdot(\delta\boldsymbol{r})^{\cdot}] = 0, \\ &\Rightarrow (\delta f_D)^{\cdot\cdot} = S\,[(\partial f_D/\partial \boldsymbol{r})^{\cdot\cdot}\cdot \delta\boldsymbol{r} + 2\,(\partial f_D/\partial \boldsymbol{r})^{\cdot}\cdot(\delta\boldsymbol{r})^{\cdot} + (\partial f_D/\partial \boldsymbol{r})\cdot(\delta\boldsymbol{r})^{\cdot\cdot}] = 0, \end{aligned}$$

or, [invoking commutativity $\delta(\dot{\boldsymbol{r}}) = (\delta\boldsymbol{r})^{\cdot}$, etc.]

$$(\delta f_D)^{\cdot\cdot} = S\,[(\partial f_D/\partial \boldsymbol{r})^{\cdot\cdot}\cdot \delta\boldsymbol{r} + 2\,(\partial f_D/\partial \boldsymbol{r})^{\cdot}\cdot \delta\boldsymbol{v} + (\partial f_D/\partial \boldsymbol{r})\cdot \delta\boldsymbol{a}] = 0, \tag{6.6.25a}$$

$$\delta(\ldots) \Rightarrow \delta''(\ldots): \qquad (\delta'' f_D)^{\cdot\cdot} = S\,(\partial f_D/\partial \boldsymbol{r})\cdot \delta\boldsymbol{a} = 0. \tag{6.6.25b}$$

As a result of the above, equations (6.6.23, 24a, b) are replaced, respectively, by the familiar (§3.5)

$$dm\,\boldsymbol{a} = d\boldsymbol{F} + \sum \lambda_D(\partial f_D/\partial \boldsymbol{r}), \tag{6.6.26a}$$

$$(\partial T/\partial \dot{q}_k)^{\cdot} - \partial T/\partial q_k = \partial S/\partial \ddot{q}_k = Q_k + \sum \lambda_D(\partial f_D/\partial q_k). \tag{6.6.26b}$$

(iii) *Appell's equations in quasi variables via Gauss' principle.* By $\delta''(\ldots)$-varying (6.6.3), we obtain

$$\delta'' Z = \delta'' S - \mathcal{S}\, d\boldsymbol{F} \cdot \delta'' \boldsymbol{a} = 0. \tag{6.6.27a}$$

But, successively,

$$\begin{aligned}
\delta'' S &= \delta''\Big(\mathcal{S}\,(1/2)\, dm\, \boldsymbol{a}^2\Big) \\
&= \mathcal{S}\, dm\, \boldsymbol{a} \cdot \delta'' \boldsymbol{a} \\
&= \mathcal{S}\, dm\, \boldsymbol{a} \cdot \Big(\sum \boldsymbol{\varepsilon}_k\, \delta \dot{\omega}_k\Big) \\
&= \sum \Big(\mathcal{S}\, dm\, \boldsymbol{a} \cdot \boldsymbol{\varepsilon}_k\Big)\, \delta \dot{\omega}_k \\
&= \sum (\partial S^* / \partial \dot{\omega}_k)\, \delta \dot{\omega}_k,
\end{aligned} \tag{6.6.27b}$$

and

$$\begin{aligned}
\mathcal{S}\, d\boldsymbol{F} \cdot \delta'' \boldsymbol{a} &= \mathcal{S}\, d\boldsymbol{F} \cdot \Big(\sum \boldsymbol{\varepsilon}_k\, \delta \dot{\omega}_k\Big) \\
&= \sum \Big(\mathcal{S}\, d\boldsymbol{F} \cdot \boldsymbol{\varepsilon}_k\Big)\, \delta \dot{\omega}_k \\
&= \sum \Theta_k\, \delta \dot{\omega}_k \quad \text{(Gaussian form of Appellian virtual work);}
\end{aligned} \tag{6.6.27c}$$

and so (6.6.27a) yields

$$\delta'' Z = \sum (\partial S^* / \partial \dot{\omega}_k - \Theta_k)\, \delta \dot{\omega}_k = 0; \tag{6.6.27d}$$

that is, among kinematically admissible accelerations, *the actual (kinetic) ones make the Gaussian compulsion Z*:

$$Z = S - \sum \Theta_k\, \dot{\omega}_k + no\ \dot{\omega}\text{-}terms = Z(\dot{\omega}), \tag{6.6.28}$$

stationary (actually a *minimum*—see below).

Next, if the variations $\delta \dot{\omega}$ are *independent*, then (6.6.27d) yields the familiar Appellian equations

$$\partial S^* / \partial \dot{\omega}_k = \Theta_k. \tag{6.6.29}$$

Similarly, in holonomic variables: there, eqs. (6.6.28, 27d, 29) read, respectively,

$$Z = S - \sum Q_k\, \ddot{q}_k + no\ \ddot{q}\text{-}terms = Z(\ddot{q}), \tag{6.6.30a}$$

$$\delta'' Z = \sum (\partial S / \partial \ddot{q}_k - Q_k)\, \delta \ddot{q}_k = 0, \tag{6.6.30b}$$

$$\partial S / \partial \ddot{q}_k = Q_k. \tag{6.6.30c}$$

If, on the other hand, the variations $\delta \ddot{q}$, $\delta \dot{\omega}$ are not independent, then either we *adjoin* the constraints (in the proper form) via Lagrangean multipliers, or we *embed* them via quasi variables (see below).

Constraint Reactions (Kinetostatic Equations)

To calculate these reactions, we apply the relaxation principle (§3.7) to Z, just as in the Appellian and Lagrangean cases; that is, we calculate the *relaxed compulsion* Z as function of *all* n $\dot{\omega}$'s, then differentiate it appropriately, and, finally, enforce in it the constraints $\dot{\omega}_D = 0$ (and, of course, $\omega_D = 0$). We note that here, too,

$$(\partial Z/\partial\dot{\omega}_I)_o = \partial Z_o/\partial\dot{\omega}_I \qquad (= 0;\ I = m+1,\ldots,n), \tag{6.6.31a}$$

where

$$Z = Z(\dot{\omega}_D,\ \dot{\omega}_I)\text{: } \textit{relaxed compulsion}, \qquad Z_o = Z(0, \dot{\omega}_I)\text{: } \textit{constrained compulsion}, \tag{6.6.31b}$$

and, as usual, $(\ldots)_o \equiv (\ldots)$ *evaluated for* $\dot{\omega}_D = 0$. Hence, with the usual notations (chaps. 3 and 5), the equations of motion are

Kinetostatic: $$(\partial Z/\partial\dot{\omega}_D)_o = (\partial S^*/\partial\dot{\omega}_D)_o - \Theta_D = \Lambda_D, \tag{6.6.32a}$$

Kinetic: $$(\partial Z/\partial\dot{\omega}_I)_o = \partial Z_o/\partial\dot{\omega}_I$$
$$= (\partial S^*/\partial\dot{\omega}_I)_o - \Theta_I = \partial S^*{}_o/\partial\dot{\omega}_I - \Theta_I = 0; \tag{6.6.32b}$$

and, in view of (6.6.31a), if no reactions are sought, we can enforce the constraints into Z right from the start, just like with S^*.

The Minimality of the Compulsion

Here, we show that for the actual constrained motion, Z is not just stationary but actually an *extremum*; specifically a *minimum*. (This can also be foreseen easily from the mathematical structure of Z: *a sum of essentially positive terms must have at least one minimum, somewhere.*)

From (6.6.2), we find

$$\begin{aligned} \Delta'' Z &\equiv Z(\boldsymbol{a} + \delta''\boldsymbol{a}) - Z(\boldsymbol{a}) \\ &= (1/2)\textstyle\int dm\,[(\boldsymbol{a} + \delta''\boldsymbol{a}) - (d\boldsymbol{F}/dm)]^2 - (1/2)\textstyle\int dm\,[\boldsymbol{a} - (d\boldsymbol{F}/dm)]^2 \\ &= \delta'' Z + (1/2)\,\delta''^{\,2} Z \geq 0, \end{aligned} \tag{6.6.33}$$

where

$$\delta'' Z = \textstyle\int (dm\,\boldsymbol{a} - d\boldsymbol{F})\cdot\delta''\boldsymbol{a} \quad (= 0), \tag{6.6.33a}$$

$$\delta''^{\,2} Z = \textstyle\int (dm\,\delta''\boldsymbol{a}\cdot\delta''\boldsymbol{a}) \quad (\geq 0). \tag{6.6.33b}$$

No particular physical significance is to be attached to this *second-order* property of Z (as with other energetic functions of mechanics); it simply flows out of its mathematical structure.

Least Compulsion and Theory of Errors

Equation (6.6.2) can be rewritten as

$$Z = \mathcal{S}\,(-d\boldsymbol{R})^2/2\,dm = \mathcal{S}\,(Lost\;force)^2/2\,dm. \tag{6.6.34}$$

In the theory of *errors* (also founded by Gauss), the *dm*'s are the "weights" of the observations, and the lost forces are their "errors."

On this matter, let us quote in detail the well-known expert Lanczos:

> Gauss was much attached to this principle because it represented a perfect physical analogy to the "method of least squares" (discovered by him and independently by Legendre), in the adjustment of errors. If a functional relation involves certain parameters which have to be determined by observations, the calculation is straightforward so long as the number of observations agrees with the number of unknown parameters. But if the number of observations exceeds the number of parameters, the equations become contradictory on account of the errors of observation. The hypothetical value of the function minus the observed value is the "error". The sum of the squares of all the individual errors is now formed, and the parameters of the problem are determined by the principle that this sum shall be a minimum. The principle of minimizing the quantity Z is completely analogous to the procedure sketched above. The $3N$ terms of the sum [the discretized (and rearranged but equivalent) version of our (6.6.2)]
>
> $$Z = \sum (m_k/2)\,(\boldsymbol{a}_k - \boldsymbol{F}_k/m_k)^2 = \sum (1/2m_k)\,(m_k\boldsymbol{a}_k - \boldsymbol{F}_k)^2 \qquad (k = 1, \ldots, N),$$
>
> correspond to $3N$ observations. This number is in excess of the number of unknowns $\ddot{q}$ on account of the m given kinematical conditions. The "error" is represented by the deviation of the impressed force $\boldsymbol{F}_k$ from the (negative of the) force of inertia "*mass times acceleration*". Even the factor $1/m_k$ in the expression for Z can be interpreted as a "weight factor", in analogy with the case of observations of different quality which are weighted according to their estimated reliability. (1970, p. 108) [A similar property holds for the center of mass G of N particles of masses m_k with Cartesian coordinates (x_k, y_k, z_k): *the coordinates of* $G\,(x, y, z)$ *minimize the expression*
>
> $$\sum m_k\left[(x_k - x)^2 + (y_k - y)^2 + (z_k - z)^2\right].]$$

Motivation for and Geometrical–Physical Meaning of Gauss' Principle

(i) Unconstrained versus Actual Constrained Motion

Let us consider a particle P of mass dm, possibly part of a larger system S, in actual *constrained* motion along a curve c, under a total impressed force $d\boldsymbol{F}$ and a total constraint reaction $d\boldsymbol{R}$. Let P, at the generic neighboring instants t and $t + \tau$, be at the neighboring c-points M and C, respectively (fig. 6.2). Then, by Taylor's theorem, to the second τ-order,

$$\Delta\boldsymbol{r} \equiv \boldsymbol{OC} - \boldsymbol{OM} = (\boldsymbol{OM} + \boldsymbol{MA} + \boldsymbol{AC}) - \boldsymbol{OM} = \boldsymbol{MC}$$
$$= \boldsymbol{r}(t + \tau) - \boldsymbol{r}(t) = \boldsymbol{v}\tau + (1/2)\boldsymbol{a}\tau^2. \tag{6.6.35a}$$

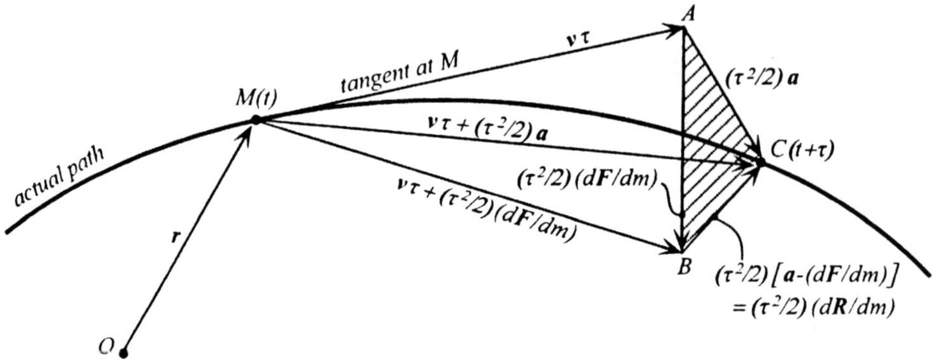

Figure 6.2 Geometrical interpretation of Gauss' constraint-compulsion:
$\boldsymbol{MA} = \boldsymbol{v}\,\tau$, $\quad \boldsymbol{MC} = \boldsymbol{v}\,\tau + (1/2)\boldsymbol{a}\,\tau^2$, $\quad \boldsymbol{MB} = \boldsymbol{v}\,\tau + (1/2)(d\boldsymbol{F}/dm)\tau^2$
$\Rightarrow \boldsymbol{BC} = \boldsymbol{MC} - \boldsymbol{MB} = (1/2)[\boldsymbol{a} - (d\boldsymbol{F}/dm)]\tau^2 = (1/2)[(d\boldsymbol{R}/dm)]\tau^2$.

On the other hand, if P was *unconstrained* or *free*—that is, if $d\boldsymbol{R} = \boldsymbol{0}$—then, even if it started with the *same initial conditions at M as in the actual constrained motion*—that is, $\boldsymbol{r}(t) = \boldsymbol{r}$ and $\dot{\boldsymbol{r}}(t) = \boldsymbol{v}$—since, then, $\boldsymbol{a} = d\boldsymbol{F}/dm$, at time $t + \tau$ the particle would end up somewhere *outside* c, say at B, where

$$\boldsymbol{OB} = \boldsymbol{OM} + \boldsymbol{MA} + \boldsymbol{AB} = \boldsymbol{r} + \boldsymbol{v}\,\tau + (1/2)\,(d\boldsymbol{F}/dm)\tau^2. \qquad (6.6.35b)$$

Therefore, the *deviation* vector, *between the unconstrained and constrained motions*, during τ, is

$$\boldsymbol{OC} - \boldsymbol{OB} = \boldsymbol{MC} - \boldsymbol{MB} = \boldsymbol{AC} - \boldsymbol{AB} = \boldsymbol{BC}$$
$$= (1/2)[\boldsymbol{a} - (d\boldsymbol{F}/dm)]\tau^2 = (1/2)\,(d\boldsymbol{R}/dm)\tau^2; \qquad (6.6.35c)$$

and so [recalling (6.6.2)], the (elementary) *compulsion* of P, dZ, is

$$dZ \equiv (dm/2)\,[\boldsymbol{a} - (d\boldsymbol{F}/dm)]^2 = (dm/2)\,(d\boldsymbol{R}/dm)^2 = (2\,dm/\tau^4)\,(\boldsymbol{BC})^2; \quad (6.6.35d)$$

and from this it follows that the (total) compulsion of S, Z, is

$$\begin{aligned} Z &= \mathcal{S}\, dZ \\ &= \mathcal{S}\,(dm/2)\,[\boldsymbol{a} - (d\boldsymbol{F}/dm)]^2 = \mathcal{S}\,(dm/2)\,(d\boldsymbol{R}/dm)^2 \\ &= (2/\tau^4)\,\mathcal{S}\, dm\,(\boldsymbol{BC})^2 \end{aligned} \qquad (6.6.35e)$$

[= *Sum of (mass) weighted deviations between unconstrained and constrained motions* (to within an unimportant, "Gaussianly constant," factor)].

Why up to the second order, in (6.6.35d, e)? To the *first* order, clearly, $A = C = B$; that is, the deviation vanishes; while *higher than second* τ-orders would have introduced *variations in the forces $d\boldsymbol{F}$ and $d\boldsymbol{R}$*—something undesirable in the derivation of equations of motion at t, M.

(ii) Kinematically Admissible Constrained versus Actual Constrained Motion

Under a Gaussianly kinematically admissible acceleration $\boldsymbol{a} + \Delta''\boldsymbol{a} = \boldsymbol{a} + \delta''\boldsymbol{a} \equiv \boldsymbol{a}'$, the particle P, starting again from A [with the same initial conditions $\boldsymbol{r}(t) = \boldsymbol{r}$ and $\dot{\boldsymbol{r}}(t) = \boldsymbol{v}$], would have ended at $t + \tau$, say at C' (fig. 6.3).
Gauss' principle states that, *for the kinetically correct acceleration*, $C' \to C$. Let us examine the geometry of the "compulsion triangle" BCC'. From fig. 6.3, we readily obtain

$$\boldsymbol{BC}' = \boldsymbol{BC} + \boldsymbol{CC}'$$
$$\Rightarrow (\boldsymbol{BC}')^2 = (\boldsymbol{BC})^2 + (\boldsymbol{CC}')^2 + 2\,\boldsymbol{BC}\cdot\boldsymbol{CC}', \tag{6.6.36a}$$

$(2\,\boldsymbol{BC}\cdot\boldsymbol{CC}' = 2\,|\boldsymbol{BC}|\,|\boldsymbol{CC}'|\cos\phi = -2\,|\boldsymbol{BC}|\,|\boldsymbol{CC}'|\cos\theta)$ and, therefore, recalling the interpretations (6.6.35d, e),

$$\begin{aligned}
Z' &\equiv \mathsf{S}\,(dm/2)\,[\boldsymbol{a}' - (d\boldsymbol{F}/dm)]^2 \quad \left\{= \mathsf{S}\,(dm/2)\,[(\boldsymbol{a} + \delta''\boldsymbol{a}) - (d\boldsymbol{F}/dm)]^2\right\}\\
&= (2/\tau^4)\,\mathsf{S}\,dm\,(\boldsymbol{BC}')^2\\
&= (2/\tau^4)\,\mathsf{S}\,dm\,[(\boldsymbol{BC})^2 + (\boldsymbol{CC}')^2 + 2\,\boldsymbol{BC}\cdot\boldsymbol{CC}']\\
&= (2/\tau^4)\,\mathsf{S}\,dm\Big\{[(\tau^2/2)\,(\boldsymbol{a} - d\boldsymbol{F}/dm)]^2 + [(\tau^2/2)\,\delta''\boldsymbol{a}]^2\\
&\qquad\qquad + 2\,[(\tau^2/2)\,(\boldsymbol{a} - d\boldsymbol{F}/dm)]\cdot[(\tau^2/2)\,\delta''\boldsymbol{a}]\Big\}\\
&= \mathsf{S}\,(dm/2)\,(\boldsymbol{a} - d\boldsymbol{F}/dm)^2 + \mathsf{S}\,dm\,(\boldsymbol{a} - d\boldsymbol{F}/dm)\cdot\delta''\boldsymbol{a}\\
&\qquad + \mathsf{S}\,(dm/2)\,(\delta''\boldsymbol{a})^2\\
&= Z + \Delta Z\\
&= Z + \delta'' Z + (1/2)\,\delta''^2 Z = Z + 0 + (1/2)\,\delta''^2 Z,
\end{aligned} \tag{6.6.36b}$$

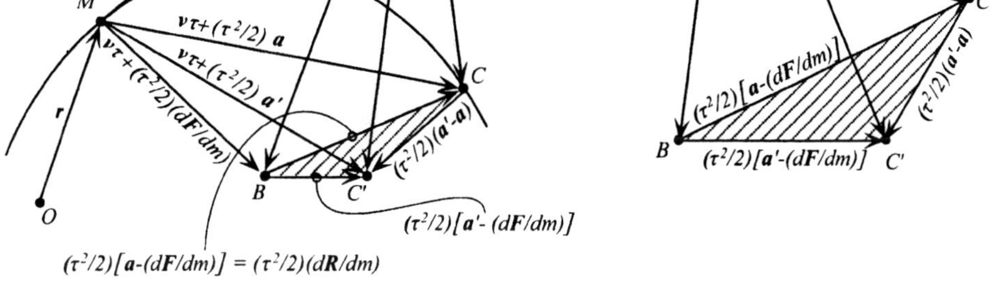

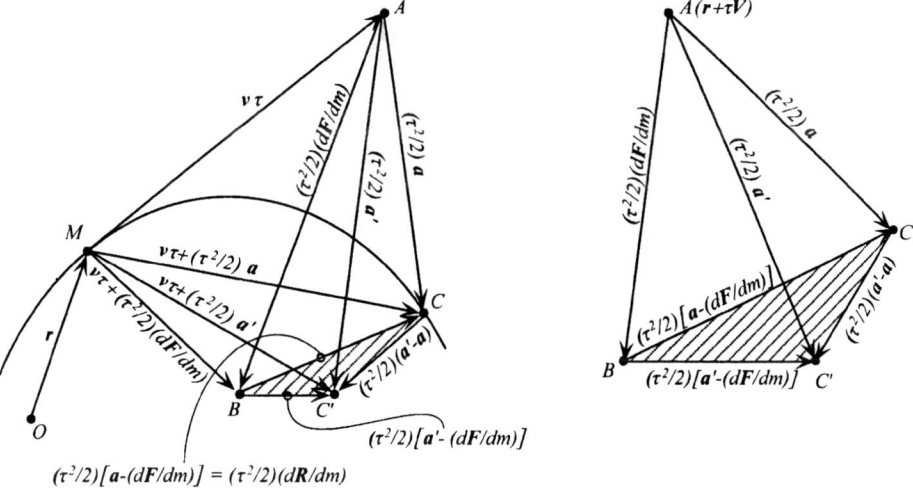

Figure 6.3 Constrained compulsion: kinematically admissible (C') versus actual (C); and detail of "compulsion triangle" BCC'.

from which, since

$$\Delta Z \equiv Z' - Z = (1/2)\,\delta''^{2} Z = (1/2) S\, dm\,(\delta'' \boldsymbol{a})^2 \geq 0 \qquad [= 0, \text{ for } \delta''\boldsymbol{a} = \boldsymbol{0}]$$

we conclude that

$$Z' \sim S\, dm\,(\boldsymbol{BC}')^2 \;>\; Z \sim S\, dm\,(\boldsymbol{BC})^2; \tag{6.6.36c}$$

that is, the actual acceleration minimizes the compulsion.

An additional, "minimum norm" interpretation of the above is known in the largely self-explanatory fig. 6.4 [see texts on applied/numerical linear algebra: least squares fitting of data; also, least squares derivation of Fourier series coefficients].

On the Uniqueness of the GP Solutions

The question of the uniqueness, or lack thereof, of the equations of motion obtained from the minimization, or *stationarization*, of Z is, obviously, of practical and physical importance. This is answered by the following considerations: *as long as rank* $(\partial f_D/\partial \dot{q}_k) = m$ (*regular* case), eliminating the *m dependent* $\ddot{q}_D$'s via the constraints, we will be able to express the *particle accelerations* $\boldsymbol{a}$ *as linear combinations of the* $n - m$ *independent* $\ddot{q}_I$'s *or* $\dot{\omega}_I$'s; and therefore Z will be a *quadratic* function in these variables. Hence, differentiating Z with respect to the independent system accelerations will result in a system of $n - m$ *linear* equations in them, and that system will have a *unique* solution. The dependent accelerations can then be determined uniquely from the constraint conditions, properly differentiated.

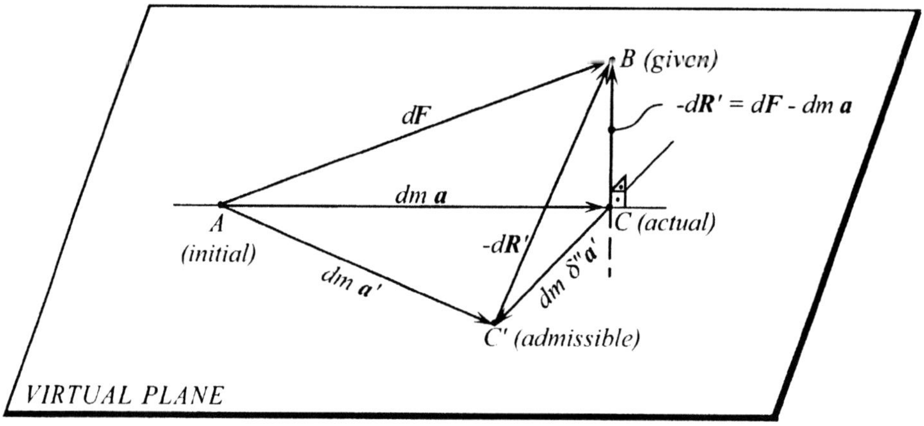

Figure 6.4 Minimum norm interpretation of minimality of Gaussian compulsion.
$dm\,\boldsymbol{a} = d\boldsymbol{F} + d\boldsymbol{R}$,
$A(t, \boldsymbol{r}, \boldsymbol{v})$: initial state, à la Gauss; $\boldsymbol{AB} = d\boldsymbol{F}$ (i.e., B: given);
Locus of C': virtual plane through $A(t, \boldsymbol{r}, \boldsymbol{v})$;
$\boldsymbol{AC}' = dm\,(\boldsymbol{a} + \delta''\boldsymbol{a}) \equiv dm\,\boldsymbol{a}'$: kinematically admissible accelerations à la Gauss;
$\boldsymbol{C}'\boldsymbol{B} = d\boldsymbol{F} - dm\,\boldsymbol{a}' \equiv -d\boldsymbol{R}'$ (C': admissible position);
$\boldsymbol{CB} = -d\boldsymbol{R} = d\boldsymbol{F} - dm\,\boldsymbol{a}$ (C: actual position)
$\Rightarrow \boldsymbol{CC}' = dm\,\delta''\boldsymbol{a}$, $\boldsymbol{AC} = dm\,\boldsymbol{a}$;
$|\boldsymbol{C}'\boldsymbol{B}| = |-d\boldsymbol{R}'|$: absolute value (norm) of admissible constraint reaction;
Gauss' principle: $|\boldsymbol{BC}'| = |d\boldsymbol{R}'| = minimum \Rightarrow C' = C$ (BC normal to virtual plane).

In sum, *excluding singular cases*, the positive definite function Z will have a minimum at only one "point."

[The singular case, with its important consequence, seems to have been noticed first by the distinguished German mathematician P. Stäckel (in 1919). As he put it: in singular configurations, it is not possible to deduce the principle of Gauss from that of d'Alembert. Rather, for singular configurations one must *postulate* Gauss' principle. Then, the argument presented in (6.2.2 ff.) no longer holds! For examples of the violation of this uniqueness of the minimum of Z in singular cases — that is, where LP and Lagrange's equations fail to determine the accelerations uniquely, but GP does — see, for example, Nordheim (1927, pp. 65–66, and references therein), Golomb (1961, pp. 69–72); and, for a comprehensive contemporary treatment, Pfister (1995).]

On the History of GP

Gauss himself never gave a precise mathematical formulation of his principle; that is, our equations (6.6.2–9). Instead, he stated it as follows:

> The motion of a system of material points, connected with each other in an arbitrary way and subjected to arbitrary influences takes place at every instant, in the most perfect accordance possible with the motion that they would have if they became completely free, that is to say, with the smallest possible constraint, taking as measure of the constraint [that the system goes through] during an infinitesimally small instant, the sum of the products of the mass of each point with the square of the quantity by which it deviates from the position that it would have taken, if it had been free. [*J. für Mathematik (Crelle)*, 1829, vol. 4, p. 232]

Perhaps this lack of quantitative formulation of the principle may explain its relative obscurity, compared with LP, throughout the 19th century and a fair part of the early 20th—one imagines the fate of the original, highly qualitative and primitive, principle of d'Alembert (of 1743) without Lagrange's formulation (of 1764). The first analytical expression for Z, in rectangular Cartesian coordinates, seems to have been given by Jacobi [1847–1848, lecture notes on Analytical Mechanics (publ. 1996, pp. 96–100); see also Appell, 1953, pp. 497–498] and Scheffler (in 1858). They wrote (with some obvious notations and without the factor 1/2)

$$Z = \sum m_k \left[(\ddot{x}_k - X_k/m_k)^2 + (\ddot{y}_k - Y_k/m_k)^2 + (\ddot{z}_k - Z_k/m_k)^2\right]. \qquad (6.6.37)$$

However, the first precise formulation of GP as a *minimum* condition, with Z expressed in general system coordinates q_k, and with the explicit realization that for this to happen only the accelerations should be varied, while the positions, velocities, and time must be treated as constant, is due to Lipschitz (*Crelle's J.*, 1877, vol. 82, p. 323); and, over the next 35 years or so (1877–1913), he and a few other distinguished mechanicians/physicists/mathematicians (including Schering, Gibbs, Mayer, Hertz, Voss, Brell, Schenkl, Wassmuth, Brill, and Mach) did for GP what Lagrange did for d'Alembert's principle. In particular, Gibbs (1879) extended the principle to *inequality* (or *unilateral*) constraints:

$$\delta\left(\mathsf{S}\, dm\, \mathbf{a}^2/2\right) - \mathsf{S}\, d\mathbf{F} \cdot \delta\mathbf{a} \geq 0 \quad \text{or} \quad \delta'' Z \geq 0; \qquad (6.6.38)$$

while Appell (in the late 1890s) applied it successfully to the formulation of his nonholonomic system equations. In most of the 20th century English language literature, GP has been barely tolerated as a clever but essentially useless academic curiosity, when it was mentioned at all. The only applications of it have appeared in problems of *impulsive motion* (with accelerations replaced by velocities), in British texts (§4.6).

However, this short-sighted situation seems to be changing for the better: in recent decades, GP has been experiencing a vigorous revival, in connection with *analytical/computational approximate* methods in such diverse areas of mechanics as nonlinear oscillations, multibody dynamics, heat transfer, structural analysis, elastic/plastic buckling, shell theory, and so on. [See, for example, Girtler (1928), Lilov and Lorer (1982), Lilov (1984), Vujanovic and Jones (1989, chap 7; this also contains a "complementary" formulation of GP where *the accelerations are kept fixed and the impressed forces are varied*), and Udwadia and Kalaba (1996). For the *continuum* formulation of GP, see, for example, Brill (1909), Hellinger (1914, pp. 633–635), and Truesdell and Toupin (1960, pp. 605–606).] Along with other DVP, GP has the big advantage over *time-integral* variational principles that—for discrete systems, at least—its application involves only ordinary differential calculus on a quadratic function of the acceleration components, and not variational calculus.

Example 6.6.1 *An ad hoc but Instructive Derivation of GP from LP (Nonsingular Cases).* Applying LP for $t + dt \equiv t + \tau$, where τ is an arbitrarily small time interval, we have

$$S\, dm\, \boldsymbol{a}(t+\tau) \cdot \delta \boldsymbol{r}(t+\tau) = S\, d\boldsymbol{F}\,(t+\tau) \cdot \delta \boldsymbol{r}(t+\tau). \tag{a}$$

Then, substituting into (a) the *special variation* (neglecting higher than τ^2-order terms), we get

$$\begin{aligned} \delta \boldsymbol{r}(t+\tau) &= \delta'' \boldsymbol{r}(t+\tau) \\ &= \delta''\big[\boldsymbol{r}(t) + \boldsymbol{v}(t)\tau + (\tau^2/2)\boldsymbol{a}(t) + \cdots\big] = (\tau^2/2)\,\delta'' \boldsymbol{a}(t), \end{aligned} \tag{b}$$

and simplifying, and renaming $\delta'' \boldsymbol{a}(t) = \delta \boldsymbol{a}(t)$, we obtain (the differential form of) GP:

$$S\, dm\, \boldsymbol{a}(t) \cdot \delta \boldsymbol{a}(t) = S\, d\boldsymbol{F}(t) \cdot \delta \boldsymbol{a}(t), \qquad \text{Q.E.D.} \tag{c}$$

Example 6.6.2 Using Gauss' principle, let us obtain the equations of motion of the system shown in fig. 6.5. (All pulleys and cables are assumed massless.)

Here, $n = 3$, $m = 1$: $q_{1,2,3} = x_A,\ x_B,\ x_C$; and, clearly, the sole constraint among them is

$$2x_A + x_B + x_C = \textit{constant}, \tag{a}$$

or, in Gaussian form,

$$[\textit{eq. (a)}]^{\cdot\cdot} = 0: \qquad 2\ddot{x}_A + \ddot{x}_B + \ddot{x}_C = 0. \tag{b}$$

Gauss' principle requires that we minimize the system compulsion (with easily understood notations, and $k = 1, 2, 3 \Rightarrow A, B, C$):

$$Z = \sum (1/2m_k)\,(X_k - m_k \ddot{x}_k)^2, \qquad \text{under} \quad (a)^{\cdot\cdot} = (b) = 0. \tag{c}$$

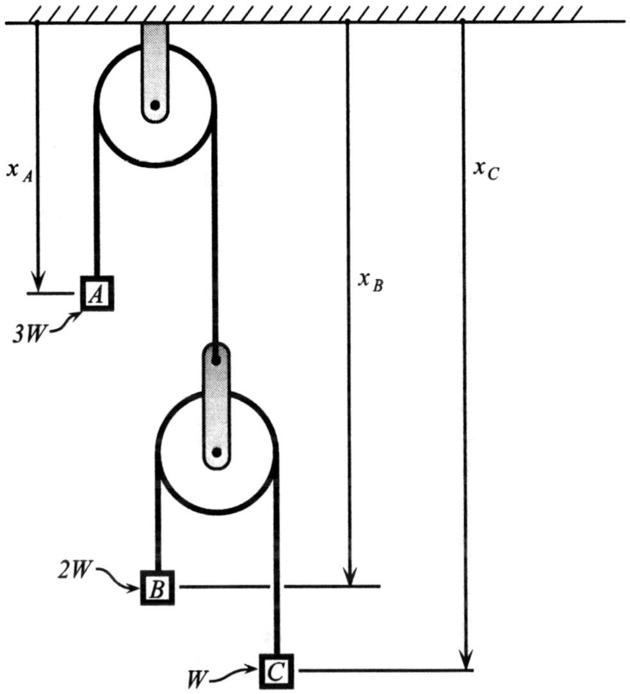

Figure 6.5 Motion of a system of three constrained particles *A, B, C* in a vertical plane [*gravity*: *g* (downward)].

This leads us readily to the constrained Gaussian variational equation

$$\delta Z = 0: \qquad [3W - (3W/g)\ddot{x}_A]\,\delta\ddot{x}_A + [2W - (2W/g)\ddot{x}_B]\,\delta\ddot{x}_B + [W - (W/g)\ddot{x}_C]\,\delta\ddot{x}_C = 0, \tag{d}$$

under

$$\delta''[eq.\ (a)^{\cdot\cdot}] = \delta''[eq.\ (b)] = \delta[eq.\ (b)]: \qquad 2\,\delta\ddot{x}_A + \delta\ddot{x}_B + \delta\ddot{x}_C = 0. \tag{e}$$

Application of the multiplier rule to the above leads at once to the three Routh–Voss equations of motion

$$(3W/g)\ddot{x}_A = 3W - 2\lambda, \tag{f1}$$

$$(2W/g)\ddot{x}_B = 2W - \lambda, \tag{f2}$$

$$(3W/g)\ddot{x}_C = W - \lambda; \tag{f3}$$

which, along with (a) constitute a determinate system for the four unknowns $x_{A,B,C}(t)$, $\lambda(t)$. [Since $-S_A\,\delta x_A = -2\lambda\,\delta x_A$, and $-S_B\,\delta x_B = -\lambda\,\delta x_B$, $-S_C\,\delta x_C = -\lambda\,\delta x_C$, λ equals either of the cable tensions S_B or S_C.]

Indeed, solving (f1–3) for $\ddot{x}_{A,B,C}$ in terms of λ and substituting the results into $[eq.\ (a)]^{\cdot\cdot} = eq.\ (b)$, we readily find $\lambda = (24/17)\,W$. Then, (f1–3) yield immediately

$$\ddot{x}_A = (1/17)\,g, \qquad \ddot{x}_B = (5/17)\,g, \qquad \ddot{x}_C = (7/17)\,g. \tag{g}$$

The solution of this problem via Jourdain's principle should be obvious.

Example 6.6.3 Using Gauss' principle and the method of relaxation of the constraints (§3.7), let us find the motion and reaction of a (plane) mathematical pendulum, of mass m and length l.

In polar coordinates r, ϕ (angle of pendulum's thread with vertical), the (physical) components of the acceleration of the pendulum's bob P are

$$\textit{radial:}\quad a_r = \ddot{r} - r\,(\dot{\phi})^2, \qquad \textit{tangential:}\quad a_\phi = 2\,\dot{r}\,\dot{\phi} + r\,\ddot{\phi}. \tag{a}$$

Therefore (and since these are orthogonal curvilinear coordinates), the *relaxed* system compulsion is [recalling (6.6.3), with S = Appellian function and T = thread tension]

$$\begin{aligned} Z &= S - [(Q_r + R_r)a_r + (Q_\phi + R_\phi)a_\phi] \\ &= (m/2)\Big\{[\ddot{r} - r\,(\dot{\phi})^2]^2 + (2\,\dot{r}\,\dot{\phi} + r\,\ddot{\phi})^2\Big\} \\ &\quad - \Big\{(m\,g\cos\phi - T)\,[\ddot{r} - r\,(\dot{\phi})^2] + (-m\,g\,r\sin\phi + 0)\,(2\,\dot{r}\,\dot{\phi} + r\,\ddot{\phi})\Big\}, \end{aligned}$$

and so, to within *Gauss-important terms*,

$$\begin{aligned} Z &= Z(\ddot{r}, \ddot{\phi}) \\ &= (m/2)\,\Big[(\ddot{r})^2 - 2\,r\,(\dot{\phi})^2\ddot{r} + r^2\,(\ddot{\phi})^2 + 4\,r\,\dot{r}\,\dot{\phi}\,\ddot{\phi}\Big] \\ &\quad - m\,g\cos\phi\,\ddot{r} + m\,g\,r\sin\phi\,\ddot{\phi} + T\,\ddot{r}. \end{aligned} \tag{b}$$

Hence, the equations of motion are

$$\textit{Kinetostatic:}\quad (\partial Z/\partial\ddot{r})_{r=l} = 0: \quad [m\,\ddot{r} - m\,r\,(\dot{\phi})^2 - m\,g\cos\phi]_{r=l} = -T,$$

$$\Rightarrow m\,l\,(\dot{\phi})^2 = -m\,g\cos\phi + T, \tag{c1}$$

$$\textit{Kinetic:}\quad (\partial Z/\partial\ddot{\phi})_{r=l} = 0: \quad (m\,r^2\,\ddot{\phi} + 2\,m\,r\,\dot{r}\,\dot{\phi} + m\,g\,r\sin\phi)_{r=l} = 0,$$

$$\Rightarrow \ddot{\phi} + (g/l)\sin\phi = 0. \tag{c2}$$

We notice that (c2) can also be obtained from

$$\partial Z_o/\partial\ddot{\phi} = 0, \tag{d}$$

where Z_o is the *constrained* system compulsion:

$$Z_o = Z\big|_{r=l} = (m\,l^2/2)\,(\ddot{\phi})^2 + m\,g\,r\sin\phi\,\ddot{\phi}. \tag{e}$$

Example 6.6.4 (Hamel, 1949, pp. 787–789). Using Gauss' principle (GP), let us find the equations of motion of a rigid body that rotates about a fixed point O, under a total impressed moment $\boldsymbol{M}_O$ and, also, constrained by

$$f(t, \boldsymbol{\omega}, \boldsymbol{\alpha}) = 0; \qquad \boldsymbol{\omega}, \boldsymbol{\alpha}\text{: angular velocity and acceleration of the body.} \tag{a}$$

Substituting into GP, (6.6.9b), the well-known kinematical relation (§1.7)

$$\boldsymbol{a} = d\boldsymbol{v}/dt = d/dt(\boldsymbol{\omega} \times \boldsymbol{r}) = \boldsymbol{\alpha} \times \boldsymbol{r} + \boldsymbol{\omega} \times (\boldsymbol{\omega} \times \boldsymbol{r}), \tag{b}$$

and its Gaussian variation

$$\delta'' \boldsymbol{a} = \delta\boldsymbol{\alpha} \times \boldsymbol{r}, \tag{c}$$

we get

$$\mathsf{S}\, dm\, \boldsymbol{a} \cdot (\delta\boldsymbol{\alpha} \times \boldsymbol{r}) = \mathsf{S}\, d\boldsymbol{F} \cdot (\delta\boldsymbol{\alpha} \times \boldsymbol{r}), \tag{d}$$

or, rearranging,

$$\left(\mathsf{S}\, \boldsymbol{r} \times dm\, \boldsymbol{a}\right) \cdot \delta\boldsymbol{\alpha} = \left(\mathsf{S}\, \boldsymbol{r} \times d\boldsymbol{F}\right) \cdot \delta\boldsymbol{\alpha}, \tag{e}$$

or, finally (with the usual notations),

$$(d\boldsymbol{H}_O/dt) \cdot \delta\boldsymbol{\alpha} = \boldsymbol{M}_O \cdot \delta\boldsymbol{\alpha}. \tag{f}$$

The above must hold for any variation $\delta\boldsymbol{\alpha}$ satisfying (a):

$$\delta'' f = 0: \qquad (\partial f/\partial\boldsymbol{\alpha}) \cdot \delta\boldsymbol{\alpha} = 0. \tag{g}$$

Adjoining (g) to (f) via the Lagrangean multiplier λ and then setting the (total) coefficient of $\delta\boldsymbol{\alpha}$ equal to zero, we find

$$d\boldsymbol{H}_O/dt = \boldsymbol{M}_O + \lambda(\partial f/\partial\boldsymbol{\alpha}). \tag{h}$$

For example, if

$$f = \boldsymbol{\alpha} \cdot (\boldsymbol{\omega} \times \boldsymbol{H}_O) = 0 \qquad \text{(i.e., if } \boldsymbol{\alpha},\ \boldsymbol{\omega}, \text{ and } \boldsymbol{H}_O \text{ are coplanar)}, \tag{i}$$

then (h) yields

$$d\boldsymbol{H}_O/dt = \boldsymbol{M}_O + \lambda(\boldsymbol{\omega} \times \boldsymbol{H}_O); \tag{j}$$

or, in components along *body-fixed* principal axes at O [with $d\boldsymbol{H}_O/dt = d'\boldsymbol{H}_O/dt + \boldsymbol{\omega} \times \boldsymbol{H}_O$, and easily understood notations (§1.17)],

$$A\,(d\omega_x/dt) + (C - B)\,(1 - \lambda)\,\omega_y\,\omega_z = M_{O,x}, \tag{k1}$$

$$B\,(d\omega_y/dt) + (A - C)\,(1 - \lambda)\,\omega_x\,\omega_z = M_{O,y}, \tag{k2}$$

$$C\,(d\omega_z/dt) + (B - A)\,(1 - \lambda)\,\omega_x\,\omega_y = M_{O,z}; \tag{k3}$$

while the constraint reads, since then $\boldsymbol{H}_O = (A\,\omega_x,\ B\,\omega_y,\ C\,\omega_z)$,

$$\begin{vmatrix} d\omega_x/dt & d\omega_y/dt & d\omega_z/dt \\ \omega_x & \omega_y & \omega_z \\ A\,\omega_x & B\,\omega_y & C\,\omega_z \end{vmatrix} = 0, \tag{l}$$

or, in extenso,

$$(d\omega_x/dt)\,(C - B)\,\omega_y\,\omega_z + (d\omega_y/dt)\,(A - C)\,\omega_z\,\omega_x + (d\omega_z/dt)\,(B - A)\,\omega_x\,\omega_y = 0. \tag{m}$$

Equations (k1–3) and (l or m) constitute a system of four equations for $\omega_{x,y,z}(t)$, $\lambda(t)$.

For additional details on special cases, see Hamel (1949, pp. 788–789).

Example 6.6.5 *Förster's Principle* (Förster, 1903; Whittaker, 1937, p. 262). Let T and V denote the kinetic and potential energies of a dynamical system. Show that

$$2(\ddot{V} + S) \equiv 2\ddot{V} + \mathsf{S}\, dm\, [(\ddot{x})^2 + (\ddot{y})^2 + (\ddot{z})^2] \tag{a}$$

differs from

$$\mathsf{S}\,(1/dm)\,\left[(dm\,\ddot{x} + \partial V/\partial x)^2 + (dm\,\ddot{y} + \partial V/\partial y)^2 + (dm\,\ddot{z} + \partial V/\partial z)^2\right] \tag{b}$$

by a quantity that does *not* involve accelerations. Hence, deduce that

$$\begin{aligned} \Theta &\equiv \ddot{T} - S \\ &\equiv \ddot{T} - \mathsf{S}\,(dm/2)\,[(\ddot{x})^2 + (\ddot{y})^2 + (\ddot{z})^2] \end{aligned} \tag{c}$$

is a maximum when the accelerations have the values corresponding to the actual motion, as compared with all motions that are consistent with the constraints and satisfy the same integral of energy, and that have the same values of the coordinates and velocities at the instant considered, provided the constraints do no work.

Let us show that

$$2\,Z - (2\,\ddot{V} + 2\,S) = 2\,(Z - S - \ddot{V}) \equiv \phi(t, q, \dot{q}). \tag{d}$$

This follows immediately from (6.6.3), if we note that, since $V = V(q)$,

$$\begin{aligned} \dot{V} &= \sum (\partial V/\partial q_k)\dot{q}_k = -\sum Q_k \dot{q}_k \\ \Rightarrow \ddot{V} &= -\sum \dot{Q}_k \dot{q}_k - \sum Q_k \ddot{q}_k \\ &= -\sum Q_k \ddot{q}_k + \textit{no } \ddot{q} \textit{ terms} \end{aligned} \tag{e}$$

$$[Q_k = Q_k(q) \Rightarrow \dot{Q}_k = \dot{Q}_k(q, \dot{q})].$$

Next, let us show that Θ is a maximum, under the above-stated (Gaussian) restrictions. By $(\ldots)^{\cdot}$-differentiating the energy conservation equation: $T + V = \textit{constant}$, yields $\ddot{T} = -\ddot{V}$, and therefore $\Theta = -\ddot{V} - S$, or explicitly, since $V = V(\mathbf{r})$,

$$\begin{aligned} -\Theta &= \mathsf{S}\,[(\partial V/\partial \mathbf{r}) \cdot \boldsymbol{a} + \textit{no } \boldsymbol{a}\textit{-terms}] + \mathsf{S}\,(1/2)\,dm\,\boldsymbol{a} \cdot \boldsymbol{a} \\ &= \mathsf{S}\,(1/2\,dm)\,[dm\,\boldsymbol{a} + \partial V/\partial \mathbf{r}]^2 + \textit{no } \boldsymbol{a}\textit{-terms} \\ &= \mathsf{S}\,(dm/2)\,[\boldsymbol{a} + (\partial V/\partial \mathbf{r})/dm]^2 + \textit{no } \boldsymbol{a}\textit{-terms} \\ &= Z + \textit{no } \boldsymbol{a}\textit{-terms}, \end{aligned} \tag{f}$$

and, therefore, since Z is a *minimum* [eqs. (6.6.33–33b)], Θ will be a *maximum*, Q.E.D.

If, in Förster's terminology, we call $\ddot{T}$ *acceleration of kinetic energy*, and S *kinetic energy of accelerations*, then we can formulate his principle as follows: among all motions that (i) are admissible in a Gaussian sense and (ii) preserve the total energy of the system, the actual one maximizes the function "acceleration of the kinetic energy minus kinetic energy of the accelerations."

HISTORICAL REMARK

This "principle" was formulated in 1903, in order to reduce to mechanical principles (e.g., that of Gauss) another qualitative and ad hoc "principle" by the famous physical chemist W. Ostwald. As such, Förster's result, although today it may appear as an academic curiosity, at its time represented another victory of the *molecular/atomistic* viewpoint (of Boltzmann) over the *phenomenological/energetic* viewpoint of Ostwald, Helm, Mach, et al.

Example 6.6.6 *Explicit Form of the Gaussian Compulsion of a Scleronomic System, in Lagrangean Coordinates.* (This example requires some familiarity with general tensors.) Substituting into (6.6.3) the acceleration expression

$$\begin{aligned} \boldsymbol{a} &\equiv d\boldsymbol{v}/dt = d/dt\Big(\sum \dot{q}_k \boldsymbol{e}_k\Big) \\ &= \sum \Big(\ddot{q}_k + \sum\sum c^k{}_{rs}\,\dot{q}_r\dot{q}_s\Big)\boldsymbol{e}_k \equiv \sum a_k \boldsymbol{e}_k, \end{aligned} \tag{a}$$

where

$$\Big[\partial \boldsymbol{e}_r/\partial q_s = \partial \boldsymbol{e}_s/\partial q_r \equiv \sum c^k{}_{rs}\,\boldsymbol{e}_k = \sum c^k{}_{sr}\,\boldsymbol{e}_k \Rightarrow \text{(and here is where tensors are needed)}\Big]:$$

$$(\partial \boldsymbol{e}_r/\partial q_s)\cdot \boldsymbol{e}_k \equiv c_{k,rs} = c_{k,sr}\text{: particle Christoffel symbols of the } 1st \text{ kind,}$$

and recalling that (§3.10)

$$S\, dm\, (\partial \boldsymbol{e}_r/\partial q_s)\cdot \boldsymbol{e}_k \equiv S\, dm\, c_{k,rs} = \Gamma_{k,rs} = \Gamma_{k,sr} \qquad \text{and} \qquad \Gamma_{l,rs} \equiv \sum M_{lk}\Gamma^k{}_{rs}, \tag{b}$$

we find, successively (recall derivation in §3.11),

$$\begin{aligned} Z &= (1/2)\, S\, dm \Big(\sum a_k \boldsymbol{e}_k\Big)\cdot\Big(\sum a_l \boldsymbol{e}_l\Big) - S\, d\boldsymbol{F}\cdot\Big(\sum a_k \boldsymbol{e}_k\Big) \\ &\quad + (1/2)\, S\, dm\, (d\boldsymbol{F}/dm)^2 \\ &= \cdots = (1/2)\sum\sum M_{kl}\ddot{q}_k\ddot{q}_l \\ &\quad + \sum\sum\sum \Gamma_{k,rs}\ddot{q}_k\dot{q}_r\dot{q}_s - \sum Q_k\ddot{q}_k + no\ \ddot{q}\ terms, \end{aligned} \tag{c}$$

[the second (triple) sum can also be written as

$$\begin{aligned} &(1/2)\sum\sum\sum (\Gamma_{k,rs}\,\ddot{q}_k + \Gamma_{l,rs}\,\ddot{q}_l)\dot{q}_r\,\dot{q}_s \\ &= (1/2)\sum\sum\sum\sum M_{kl}(\Gamma^l{}_{rs}\ddot{q}_k + \Gamma^k{}_{rs}\,\ddot{q}_l)\dot{q}_r\,\dot{q}_s \end{aligned}$$

]

and, therefore, varying this expression à la Gauss, we obtain

$$\delta'' Z = \sum (\partial Z/\partial \ddot{q}_k)\, \delta \ddot{q}_k = \sum [E_k(T) - Q_k]\, \delta \ddot{q}_k = 0, \tag{d}$$

where

$$\begin{aligned} E_k(T) = (\partial T/\partial \dot{q}_k)^{\cdot} - \partial T/\partial q_k &= \sum M_{kl}\ddot{q}_l + \sum\sum\sum M_{kl}\Gamma^l{}_{rs}\, \dot{q}_r\, \dot{q}_s \\ &= \sum M_{kl}\Big(\ddot{q}_l + \sum\sum \Gamma^l{}_{rs}\, \dot{q}_r\, \dot{q}_s\Big) \\ &= \sum M_{kl}\, \ddot{q}_l + \sum\sum \Gamma_{k,rs}\, \dot{q}_r\, \dot{q}_s, \end{aligned} \tag{e}$$

as expected.

REMARKS

(i) With the help of the definitions

$$\mu_{rs} \equiv \sum\sum M_{kl}\Gamma^{(kl)}{}_{rs}, \tag{f1}$$

$$2\Gamma^{(kl)}{}_{rs} \equiv \Gamma^l{}_{rs}\, \ddot{q}_k + \Gamma^k{}_{rs}\, \ddot{q}_l : 2\,(\textit{symmetric}\text{ part of } \Gamma^k{}_{rs}\, \ddot{q}_l), \tag{f2}$$

we can rewrite Z as follows:

$$Z = (1/2)\sum\sum M_{kl}\, \ddot{q}_k\, \ddot{q}_l + \sum\sum \mu_{kl}\, \dot{q}_k\, \dot{q}_l - \sum Q_k\, \ddot{q}_k + \textit{no } \ddot{q} \textit{ terms}$$
$$[\textit{quadratic in } \ddot{q} + \textit{linear in } \ddot{q} + \textit{constant in } \ddot{q}]. \tag{g}$$

(ii) For a *rheonomic* system, the summations over the repeated Latin indices run from 1 to $n+1$ (with $q_{n+1} \equiv t \Rightarrow \dot{q}_{n+1} = 1 \Rightarrow \ddot{q}_{n+1} = 0$).

(iii) With the help of the above, the quantity Θ of the preceding example becomes

$$\Theta = \ddot{T} - (1/2)\sum\sum m_{kl}\, E_k(T)\, E_l(T), \tag{h}$$

where the m_{kl} ["conjugate" of M_{kl} (3.10.4); and denoted in tensor calculus as M^{kl}] are defined by

$$\sum m_{kl} M_{lr} = \delta_{kr}. \tag{i}$$

6.7 THE PRINCIPLE OF HERTZ

If the impressed forces, though not necessarily the constraint reactions, vanish—that is, in forceless but constrained motion, GP becomes *Hertz's principle (HZP) of the straightest path*, or *least curvature*:

$$Z \Rightarrow S = (1/2) \textstyle\mathsf{S}\, dm\, \boldsymbol{a} \cdot \boldsymbol{a} \to \textit{minimum}; \tag{6.7.1}$$

which is an actual minimum, since here $Z = S$ is a positive definite quadratic form. Let us see the consequences of this; in particular, its connection with the concept of *curvature*.

We consider a *scleronomic* system, moving in a (Riemannian) configuration space with the following *kinetic energy-based metric* [i.e., arc element formula—recall (3.9.4o)] formulae

$$ds \equiv \left(\sum\sum M_{kl}\, dq_k\, dq_l\right)^{1/2} = (2T)^{1/2} dt, \qquad M_{kl} \equiv S\, dm\, \boldsymbol{e}_k \cdot \boldsymbol{e}_l, \tag{6.7.2a}$$

$$\Rightarrow 2T \equiv S\, dm\, \boldsymbol{v} \cdot \boldsymbol{v} = \sum\sum M_{kl} \dot{q}_k \dot{q}_l \equiv (ds/dt)^2. \tag{6.7.2b}$$

[Other equivalent choices of system arc-parameter and metric are possible—see "Remarks" (i) below]. Then, since

$$\boldsymbol{v} \equiv d\boldsymbol{r}/dt = (d\boldsymbol{r}/ds)\,(ds/dt), \tag{6.7.3a}$$

$$\boldsymbol{a} \equiv d\boldsymbol{v}/dt = (d^2\boldsymbol{r}/ds^2)\,(ds/dt)^2 + (d\boldsymbol{r}/ds)\,(d^2s/dt^2), \tag{6.7.3b}$$

the compulsion ⇒ Appellian becomes

$$\begin{aligned} S &= (1/2)\, S\, dm\, \boldsymbol{a} \cdot \boldsymbol{a} \\ &= (1/2)\,(ds/dt)^4\, S\, dm\,(d^2\boldsymbol{r}/ds^2)^2 + (1/2)\,(d^2s/dt^2)^2\, S\, dm\,(d\boldsymbol{r}/ds)^2 \\ &\quad + (d^2s/dt^2)\,(ds/dt)^2\, S\, dm\,(d\boldsymbol{r}/ds) \cdot (d^2\boldsymbol{r}/ds^2). \end{aligned} \tag{6.7.4}$$

But since then (6.7.2b) becomes

$$\begin{aligned} T &= (1/2)\,(ds/dt)^2 = S\,(dm/2)\,(d\boldsymbol{r}/dt)^2 = S\,(dm/2)\,[(ds/dt)\,(d\boldsymbol{r}/ds)]^2 \\ &= (1/2)\,(ds/dt)^2\, S\, dm\,(d\boldsymbol{r}/ds)^2, \end{aligned} \tag{6.7.5}$$

it follows that, for this particular parametrization,

$$S\, dm\,(d\boldsymbol{r}/ds)^2 = 1; \tag{6.7.6a}$$

and from the latter, by $d(\ldots)/ds$-differentiation,

$$S\, dm\,(d\boldsymbol{r}/ds) \cdot (d^2\boldsymbol{r}/ds^2) = 0; \tag{6.7.6b}$$

so that S, eq. (6.7.4), reduces to a *sum of two positive terms*:

$$S = (1/2)\,(ds/dt)^4\, S\, dm\,(d^2\boldsymbol{r}/ds^2)^2 + (1/2)\,(d^2s/dt^2)^2. \tag{6.7.7}$$

Finally, with the help of the following definition of the *system curvature* K [guided by the Frenet–Serret formulae (§1.2): At a generic point $\boldsymbol{r}$ of a curve with arc-length s, we have $d^2\boldsymbol{r}/ds^2 = \boldsymbol{n}/\rho$, where $\boldsymbol{n}$ = (first) local unit normal, and ρ = (first) local radius of curvature]:

$$S\, dm\,(d^2\boldsymbol{r}/ds^2)^2 \equiv 1/R^2 \equiv K^2 \tag{6.7.8a}$$

(R = *system radius of curvature*), the Appellian (6.7.7), assumes the form

$$S = (1/2)\,[(d^2s/dt^2)^2 + (ds/dt)^4/R^2] = (1/2)\,[(d^2s/dt^2)^2 + K^2(ds/dt)^4]. \tag{6.7.8b}$$

[We remark that, since both s and K depend only on the system trajectories, and not on the time needed to traverse them, the above expression exhibits a *decoupling of the spatial and temporal aspects of the motion.*]

In view of (6.7.8b), HZP, eq. (6.7.1), becomes: *In the impressed force-free motion of a scleronomic (⇒ conservative) system, with momentarily given positions and velocities, the acceleration is such that the system Appellian is a minimum*; or, equivalently, since then

$$T = (1/2)\,(ds/dt)^2 = constant$$
$$\Rightarrow ds/dt = constant \Rightarrow d^2s/dt^2 = 0,$$
$$\Rightarrow S = (1/2)\,K^2 (ds/dt)^4, \quad \text{i.e., } S \sim K^2, \tag{6.7.9}$$

the system curvature is a minimum:

$$K^2 \equiv \mathcal{S}\, dm\, (d^2\mathbf{r}/ds^2)^2 \rightarrow minimum. \tag{6.7.10}$$

[Simply, S being the sum of two squares, it will be a minimum when each of these terms becomes least; which, since ds/dt is a given constant, leads to $d^2s/dt^2 = 0$ and $K \rightarrow minimum$.]

In words: *The inertial path of a system in configuration space is the "straightest" curve compatible with the given holonomic and/or nonholonomic, but stationary, constraints; and it is traced at a uniform rate.*

For example, in the case of a particle constrained to move on a smooth surface, under no impressed forces, HZP states that its path curvature is the least among all surface curvatures. We notice that HZP, like GP, holds for holonomic and nonholonomic systems alike [unlike the *integral* variational principles (in both their *time* or *geodesic* forms, like Jacobi's) which, for nonholonomic systems, do not hold without modifications (§7.7 ff.)].

REMARKS

(i) Had we defined the system arc-length s by

$$2T = m\,(ds/dt)^2, \qquad m \equiv \mathcal{S}\, dm, \tag{6.7.11}$$

$$\Rightarrow m\,(ds)^2 = \mathcal{S}\, dm\,(d\mathbf{r}\cdot d\mathbf{r}) = \mathcal{S}\,(\sqrt{dm}\, d\mathbf{r})\cdot(\sqrt{dm}\, d\mathbf{r}) \equiv \mathcal{S}\, d\mathbf{r}'\cdot d\mathbf{r}'$$
$$\Rightarrow (ds)^2 = \mathcal{S}\,(d\mathbf{r}'/\sqrt{dm})\cdot(d\mathbf{r}'/\sqrt{dm}) \equiv \mathcal{S}\, d\mathbf{r}''\cdot d\mathbf{r}''; \tag{6.7.11a}$$

$$d\mathbf{r}'' \equiv d\mathbf{r}'/\sqrt{dm} \equiv (dm/m)^{1/2}\, d\mathbf{r}, \tag{6.7.11b}$$

it is not hard to see that, then, we would have

$$\mathcal{S}\, dm\,(d\mathbf{r}/ds)^2 = m, \qquad \mathcal{S}\,(d\mathbf{r}''/ds)^2 = 1, \tag{6.7.12}$$

and with the new definition [instead of (6.7.8a)]

$$\mathcal{S}\, dm\,(d^2\mathbf{r}/ds^2)^2 = \mathcal{S}\,(d^2\mathbf{r}'/ds^2)^2 = m/R^2 \equiv m\,K^2, \tag{6.7.13}$$

S would reduce to [instead of (6.7.8b)]

$$S = (m/2)\left[(d^2s/dt^2)^2 + (ds/dt)^4/R^2\right] = (m/2)\ \left[(d^2s/dt^2)^2 + K^2(ds/dt)^4\right]; \quad (6.7.14)$$

and, further [instead of (6.7.9)]

$$T = constant \ \Rightarrow \ (ds/dt)^2 = 2T/m = constant, \qquad (6.7.15a)$$

$$2S/m\,(ds/dt)^4 = K^2 = \mathcal{S}\,(d^2\mathbf{r}''/ds^2)^2. \qquad (6.7.15b)$$

(ii) It is worth pointing out the formal similarity between HZP and the principle of *minimum strain energy* of a thin linearly elastic, unloaded but constrained, beam in plane bending.

HISTORICAL REMARKS

Hertz's principle represents one of the highest, and admittedly quite elegant, pre-relativistic (late 19th century) efforts to *formulate a forceless/geometrical description of motion, within classical mechanics*; similar to the earlier attempts, by Kelvin et al. to explain forces by the motion of concealed built-in spinning bodies [gyrostats (§8.4 ff.)]. As is well known, the solution to that problem of *geometrization of mechanics* came about 20 years later with Einstein's *general theory of relativity* (mid-1910s).

The restriction of HZP to vanishing impressed forces (though not to holonomic constraints), makes it practically useless for applications; and this is in very sharp contrast to GP, which seems to be free of any kind of limitations.

The best single reference on HZP is, probably, Brill (1909, pp. 5–55); also, Heun [1902(c)], the thesis of Boltzmann's famous student Ehrenfest (1904), and the modern historical study by Lützen [1995(a), (b)]; and, of course, Hertz (1894, in German; 1899, English transl.; 1956, English transl., paperback edition).

7

Time-Integral Theorems and Variational Principles

As long as physical science exists, the highest goal to which it aspires is the solution of the problem of embracing all natural phenomena, observed and still to be observed, in one simple principle which will allow all past and, especially, future occurrences to be calculated. It follows from the nature of things, that this object neither has been, nor ever will be, completely attained. It is, however, possible to approach it nearer and nearer, and the history of theoretical physics shows that already an extensive series of important results can be obtained, which indicates clearly that the ideal problem is not purely Utopian, but that it is eminently practicable. Therefore, from a practical point of view, the ultimate object of research must be borne in mind. (Planck, 1960, p. 69; also, in German, in Wiechert, 1925, p. 772)

The variational principles of mechanics are firmly rooted in the soil of that great century of Liberalism which starts with Descartes and ends with the French Revolution and which has witnessed the lives of Leibniz, Spinoza, Goethe, and Johann Sebastian Bach. *It is the only period of cosmic thinking in the entire history of Europe since the time of the Greeks.*

(Lanczos, 1970, p. x; emphasis added)

The germ of the idea of a minimum principle, coming when it did, found a congenial environment. Both Euler and Lagrange were infected with the virus early in life, and though they both sloughed it in later years its effect can be seen on Gauss, through to Hamilton and right down to Willard Gibbs and Castigliano. Thus we find Euler saying (in Latin), "Since the plan of the universe is the most perfect possible and the work of the wisest possible creator, nothing happens which has not some maximal or minimal property." Nowadays this mental attitude is démodé and we think more of the "uncertainty principle," according to which (if the quantum theorists are to be believed) Nature cannot make up her mind which it is that is going to do what.

(Kilmister, 1964, pp. 50–51)

7.1 INTRODUCTION

Time-integral theorems and the *integral variational principles* (IVP) derived from them, as well as those of weighted residuals, occupy a central position in analytical mechanics (AM), and applied mechanics in general. This is not only due to the fact that they provide powerful analytical tools (for the derivation of global energetic results, existence and uniqueness theorems, upper and lower bound estimates for system eigenvalues and/or solutions, etc.) but also, primarily, because they constitute the foundation of the so-called *direct* variational methods. These latter bypass the equations of motion and proceed directly to the construction of *approximate* solutions of the problem; whether initial and/or boundary value, linear or nonlinear, conservative or not, holonomic or not.

The *first* part of this chapter (§7.2–5) derives all the important time-integral propositions of AM; variational, energetic, and virial-like, for linearly and/or nonlinearly constrained holonomic and/or nonholonomic systems, in both holonomic and nonholonomic coordinates, all from a simple unifying viewpoint: a general *time-integral identity* based on a few straightforward algebraic manipulations of the corresponding equations of motion. This unambiguous "from first principles (i.e., equations of motion) approach" will, hopefully, contribute to a more rational, or perhaps demythologized, attitude toward IVP, because, historically (since mid-18th century) these "principles" have been surrounded with superstition, mysticism, and ignorance (of the fine points of variational calculus and mechanics).

[We remark that in *continuum* mechanics, where even the simplest kinetic variational problem leads to a *partial* differential equation (e.g., string: one-dimensional wave equation), IVP have the additional and unique advantage that they supply *both* the equations of motion of the problem and its *boundary* conditions. Also, such *infinite* number of DOF systems may be *discretized*; that is, be approximated by systems with a *finite* number of DOF; and then, some of the (single) integral principles of this chapter may be applied to these systems to find their temporal evolution.]

The *second*, larger, part of this chapter (§7.6–9, Appendices) examines these IVP in some detail, especially in view of the fundamental (and yet frequently overlooked and/or misunderstood) *differences* between the *mathematically* correct and (generally different from it) *mechanically* correct variational formulations for nonholonomic systems.

Finally, most IVP are *first-order/stationarity requirements,* that is, of the kind that supplies only the equations of motion (the "laws of nature"). For certain systems, however, second-order/*extremality conditions* (not laws of nature) may be established, which constitute alternative tests for the stability/instability of certain of their motions. A summary of the relevant sufficiency variational theory and some applications is contained in an appendix, at the end of the chapter.

For complementary reading, we recommend the following general references (alphabetically): Boltzmann (1904a, vol. 2, chaps. 1, 3, 4), Finzi (1949), Gelfand and Fomin (1963), Lanczos (1970), Langhaar (1962), Logan (1977), Lovelock and Rund (1975), Lur'e (1968, chap. 12), Neimark and Fufaev (1972, chap. 3, section 10), Novoselov (1966; 1967), Papastavridis [1987(b)], Pars (1965, chaps. 26, 27), Polak (1959; 1960—a unique and delightful reference), Prange (1935), Rund (1966), Tabarrok and Rimrott (1994, chap. 3, app. A), Vujanovic and Jones (1989, chaps. 1–6).

Chapter notations (see also Introduction, §4, and chap. 8):

- IVP: Time-*integral* variational principles;
- All *Latin* indices run from 1 to n (= number of "original" positional coordinates); except
  $D, D', D'', \ldots,$ (*dependent*) which run from 1 to m
  (= number of additional constraints, holonomic or not),
  and
  $I, I', I'', \ldots,$ (*independent*) which run from $m+1$ to n.
- $\int \equiv \int_{t_1}^{t_2}$: The integration extends from t_1 to t_2 ($t_{1,2}$: arbitrary time instants), unless specified otherwise.
- $\{\ldots\}_1^2 \equiv \{\ldots\}_{t_1}^{t_2} \equiv \{\ldots\}_{t_2} - \{\ldots\}_{t_1} \equiv BT$ (1, 2 stand for $t_{1,2}$, respectively): Boundary terms, where $\ldots$ = integrated out part(s).

Time-Integral Theorems

7.2 TIME-INTEGRAL THEOREMS: PFAFFIAN CONSTRAINTS, HOLONOMIC VARIABLES

Here, the starting point is the fundamental Routh–Voss equations (§3.5)

$$(\partial T/\partial \dot{q}_k)^{\cdot} - \partial T/\partial q_k = Q_k + \sum \lambda_D a_{Dk} \qquad [T = T(t, q, \dot{q})\text{: unconstrained}]. \quad (7.2.1)$$

Multiplying each of (7.2.1) with z_k, where the $\{z_k = z_k(t); k = 1, \ldots, n\}$ are arbitrary functions but as well behaved as needed, and summing them over k, we obtain

$$\sum \left[(\partial T/\partial \dot{q}_k)^{\cdot} z_k - (\partial T/\partial q_k) z_k\right] = \sum \left(Q_k + \sum \lambda_D a_{Dk}\right) z_k,$$

or, rearranging with the help of the chain rule, and then integrating between the two arbitrary time instants t_1 and t_2, we obtain the following *generalized holonomic time-integral (or virial-like) identity*:

$$\int \left[\sum (\partial T/\partial \dot{q}_k)\dot{z}_k + \sum \left(\partial T/\partial q_k + Q_k + \sum \lambda_D a_{Dk}\right) z_k\right] dt$$
$$= \left\{\sum (\partial T/\partial \dot{q}_k) z_k\right\}_1^2. \quad (7.2.2)$$

As shown below, *special choices* of the z_k's in (7.2.2) yield all the important time integral theorems and variational "principles" of mechanics.

Let us examine them in detail:

(i) $z_k \to \delta q_k$: *Virtual displacement of* q_k (fig. 7.1). Then,

$$\sum \sum \lambda_D a_{Dk} z_k \to \sum \lambda_D \left(\sum a_{Dk}\, \delta q_k\right) = 0, \quad (7.2.3a)$$

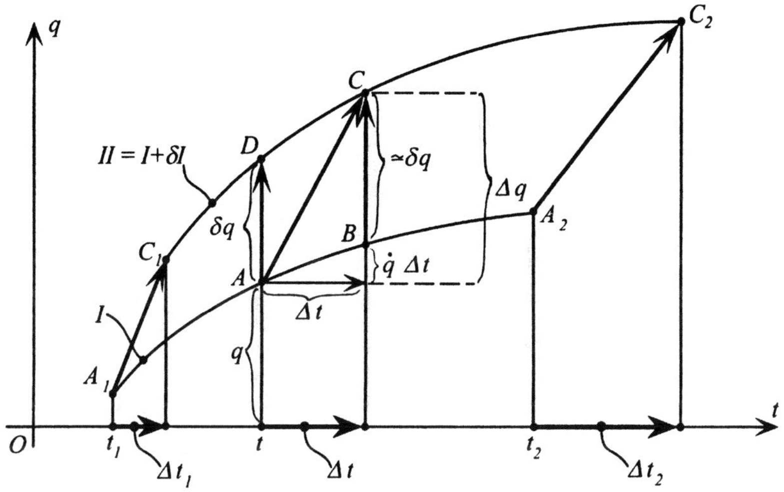

Figure 7.1 Variations of q in $(n+1)$-dimensional extended configuration space: *vertical* (δq) and *skew* (Δq).
Difference in velocity space (slope) between D and A: $\delta(\dot{q}) = (q+\delta q)^{\cdot} - \dot{q} = (\delta q)^{\cdot}$.
Point coordinates: $A(t, q)$, $B(t+\Delta t, q+\dot{q}\Delta t)$, $C(t+\Delta t, q+\Delta q)$, $D(t, q+\delta q)$.
Mappings: $A \to A+\delta A = D$ (vertical); $A \to A+\Delta A = C$ (skew);
$t \to t'(t) = t+\Delta t(t) \Rightarrow dt'/dt = 1+(\Delta t)^{\cdot}$.

by the virtual form of the Pfaffian constraints, and so (7.2.2) yields *Hamilton's law of vertically (virtually) varying action*:

$$\int (\delta T + \delta' W)\, dt = \left\{ \sum p_k\, \delta q_k \right\}_1^2, \tag{7.2.3b}$$

where

$$\delta T = \sum \left[(\partial T/\partial \dot{q}_k)\, \delta \dot{q}_k + (\partial T/\partial q_k)\, \delta q_k \right], \qquad p_k \equiv \partial T/\partial \dot{q}_k, \tag{7.2.3c}$$

and

$$\delta \dot{q}_k \equiv \delta(\dot{q}_k) = (\delta q_k)^{\cdot}. \tag{7.2.3d}$$

As will be detailed in the *second* part (§7.6 ff.): (a) in general, no stationarity of a functional is implied by the integral equation (7.2.3b); and (b) *the commutation rule* (7.2.3d) *is a key assumption, or choice*, without which it would have been impossible to go from (7.2.2) to (7.2.3b).

(ii) $z_k \to \Delta q_k = \delta q_k + \dot{q}_k\, \Delta t$: *Noncontemporaneous, or skew, or oblique, variation of* q_k (fig. 7.1). Then,

$$0 = \sum a_{Dk}\, \delta q_k = \sum a_{Dk}(\Delta q_k - \dot{q}_k\, \Delta t) = \sum a_{Dk}\, \Delta q_k - \left(\sum a_{Dk} \dot{q}_k \right) \Delta t$$
$$= \sum a_{Dk}\, \Delta q_k - (-a_D)\, \Delta t \;\Rightarrow\; \sum a_{Dk}\, \Delta q_k + a_D\, \Delta t = 0, \tag{7.2.4a}$$

that is, the Δq_k and Δt are *kinematically admissible;* and so (7.2.2) yields *Hamilton's law of skew-varying action*:

$$\int \Big[\sum (\partial T/\partial \dot{q}_k)(\Delta q_k)^{\cdot} + \sum (\partial T/\partial q_k + Q_k)\,\Delta q_k - \sum \lambda_D a_D\,\Delta t\Big]\,dt = \Big\{\sum p_k\,\Delta q_k\Big\}_1^2. \qquad (7.2.4b)$$

For $a_D = 0$ (i.e., catastatic constraints), the left sides of (7.2.3b) and (7.2.4b) look similar, although in the latter $\Delta t \neq 0$. We also notice that, again assuming (7.2.3d), sincc

$$(\Delta q_k)^{\cdot} = (\delta q_k + \dot{q}_k\,\Delta t)^{\cdot} = (\delta q_k)^{\cdot} + \ddot{q}_k\,\Delta t + \dot{q}_k(\Delta t)^{\cdot}, \qquad (7.2.4c)$$

$$\Delta(\dot{q}_k) = \delta(\dot{q}_k) + (\dot{q}_k)^{\cdot}\,\Delta t = \delta\dot{q}_k + \ddot{q}_k\,\Delta t, \qquad (7.2.4d)$$

$$\Rightarrow (\Delta q_k)^{\cdot} - \Delta(\dot{q}_k) = \dot{q}_k(\Delta t)^{\cdot} \qquad (7.2.4e)$$

[i.e., $\Delta(\ldots)$ and $(\ldots)^{\cdot}$ do *not* commute, even when $\delta(\ldots)$ and $(\ldots)^{\cdot}$ do!], we can replace in (7.2.4b) $(\Delta q_k)^{\cdot}$ with $\Delta(\dot{q}_k) + \dot{q}_k(\Delta t)^{\cdot}$. (Integral equations/principles based on such noncontemporaneous variations are detailed in §7.9.)

(iii) $z_k \to q_k$: *Actual system coordinate.* Then (7.2.2) yields the *nonvariational/actual virial theorem* [of Clausius, Szily, et al. (mid- to late 19th century)]:

$$\int \Big[\sum (\partial T/\partial \dot{q}_k)\dot{q}_k + \sum \Big(\partial T/\partial q_k + Q_k + \sum \lambda_D a_{Dk}\Big) q_k\Big]\,dt = \Big\{\sum p_k q_k\Big\}_1^2. \qquad (7.2.5)$$

Specialization

If, in the above, $\{\sum p_k q_k\}_1^2 = 0$—for example, as a result of *periodicity* ($t_2 = t_1 + \tau$, τ: period of oscillatory motion)—$\partial T/\partial q_k = 0$; for instance, in rectilinear coordinates; and the "original" holonomic constraints are *stationary*, in which case

$$\sum (\partial T/\partial \dot{q}_k)\dot{q}_k = 2T \qquad \text{[by the homogeneous function theorem]}; \qquad (7.2.5a)$$

then (7.2.5) specializes to the time-integral energetic theorem:

$$\int 2T\,dt = -\int \Big[\sum \Big(Q_k + \sum \lambda_D a_{Dk}\Big) q_k\Big]\,dt, \qquad (7.2.5b)$$

where the integrals extend from t_1 to $t_2 = t_1 + \tau$.

Additional special theorems result if, in (7.2.5b), $Q_k = -\partial V(t, q)/\partial q_k$, $V(t, q)$: potential function = sum of homogeneous functions of the q's of various degrees.

(iv) $z_k \to \dot{q}_k$: *Actual system velocity.* Then, since

$$\sum \Big(\sum \lambda_D a_{Dk}\Big)\dot{q}_k = \sum \Big(\sum a_{Dk}\dot{q}_k\Big)\lambda_D = \sum (-a_D)\lambda_D, \qquad (7.2.6a)$$

eq. (7.2.2) transforms to

$$\int \Big[\sum (\partial T/\partial \dot{q}_k)\ddot{q}_k + \sum (\partial T/\partial q_k + Q_k)\dot{q}_k - \sum a_D \lambda_D \Big] dt = \Big\{ \sum p_k \dot{q}_k \Big\}_1^2, \tag{7.2.6b}$$

or, further, to

$$\int \Big[(dT/dt - \partial T/\partial t) + \sum Q_k \dot{q}_k - \sum a_D \lambda_D \Big] dt = \int \Big[\sum (\partial T/\partial \dot{q}_k)\dot{q}_k \Big]^{\cdot} dt, \tag{7.2.6c}$$

or, finally, to

$$\int \Big\{ \Big[\sum (\partial T/\partial \dot{q}_k)\dot{q}_k - T \Big]^{\cdot} - \Big(-\partial T/\partial t + \sum Q_k \dot{q}_k - \sum \lambda_D a_D \Big) \Big\} dt = 0, \tag{7.2.6d}$$

from which, since the limits t_1 and t_2 are arbitrary, we conclude that the integrand must vanish identically; that is

$$d/dt \Big[\sum (\partial T/\partial \dot{q}_k)\dot{q}_k - T \Big] = -\partial T/\partial t + \sum Q_k \dot{q}_k - \sum \lambda_D a_D; \tag{7.2.6e}$$

and this is nothing but the earlier-found (§3.9) most general (nonpotential) form of the generalized power theorem, for systems under Pfaffian constraints and in holonomic variables.

Specialization

If, in (7.2.6e), some of the forces, or part of each Q_k, derive from a potential function $V = V(t, q)$, then we simply replace in there T with $L \equiv T - V$: Lagrangean of the system; and now Q_k stands for all the *nonpotential* forces or parts of them. If, further, $\partial L/\partial t = 0$ (e.g., stationary original constraints) and $a_D = 0$ (i.e., additional Pfaffian constraints catastatic), then, since $[\sum (\partial L/\partial \dot{q}_k)\dot{q}_k - L]^{\cdot} = [2T - (T - V)]^{\cdot} = (T + V)^{\cdot}$, and so eq. (7.2.6e) reduces to the more familiar (multiplierless) power equation

$$(T + V)^{\cdot} \equiv \dot{E} = \sum Q_k \dot{q}_k. \tag{7.2.6f}$$

We should point out that $\partial T/\partial t$ can vanish even for rheonomic holonomic constraints; that is, even if the position vectors of the system particles depend explicitly on time.

Example 7.2.1 *The Virial Theorem* [of Clausius (1870) et al.]. Let us consider a holonomic and conservative system with kinetic and potential energies $T(q, \dot{q})$ and $V(q)$, respectively. We are going to relate their time averages for various system motions; using both particle and system variables.

Let us define the (moment of inertia reminiscent) "second moment of the system"

$$\Phi \equiv S\,(1/2)\, dm\, \boldsymbol{r} \cdot \boldsymbol{r} = S\,(1/2)\, r^2\, dm. \tag{a}$$

By $(\ldots)^{\cdot}$-differentiating Φ twice, while noting that, by the Newton–Euler law of motion (with the usual notations, chap. 1),

$$dm\, \boldsymbol{a} = d\boldsymbol{f} = -\partial V/\partial \boldsymbol{r} \equiv -\textbf{grad}\, V, \tag{b}$$

we find

$$d\Phi/dt \;= \cdots = \mathbf{S}\, dm\, \boldsymbol{v}\cdot\boldsymbol{r} \equiv \Psi, \tag{c}$$

$$\begin{aligned} d^2\Phi/dt^2 &= \cdots = \mathbf{S}\, dm\, \boldsymbol{v}\cdot\boldsymbol{v} + \mathbf{S}\, dm\, \boldsymbol{a}\cdot\boldsymbol{r} \\ &= 2T + \mathbf{S}\, d\boldsymbol{f}\cdot\boldsymbol{r} = 2T - \mathbf{S}\,(\partial V/\partial \boldsymbol{r})\cdot\boldsymbol{r} = d\Psi/dt; \end{aligned} \tag{d}$$

[$= 2T + V$, if V: homogeneous function of degree -1 in the components/coordinates of $\boldsymbol{r}$ (case of gravitational attraction)].

Equation (d) is known in celestial mechanics as Lagrange's identity.

Now, let us average the above; with the customary notation

$$\langle f \rangle \equiv (1/\tau)\int_0^\tau f(t)\, dt:$$

Time average of a function $f(t)$, between $t_1 = 0$ and $t_2 = \tau$, (e)

and noting that

$$\langle d\Psi/dt\rangle = (1/\tau)\int_0^\tau [d\Psi(t)/dt]\, dt = (1/\tau)[\Psi(\tau) - \Psi(0)], \tag{f}$$

the averaged eq. (d) becomes

$$(1/\tau)[\dot\Phi(\tau) - \dot\Phi(0)] = 2\langle T\rangle - \left\langle \mathbf{S}\, d\boldsymbol{f}\cdot\boldsymbol{r}\right\rangle,$$

or

$$(1/\tau)[\Psi(\tau) - \Psi(0)] = 2\langle T\rangle + \left\langle \mathbf{S}\,(\partial V/\partial \boldsymbol{r})\cdot\boldsymbol{r}\right\rangle. \tag{g}$$

Specializations

(i) If the system is *periodic* with period τ, then $\Psi(\tau) = \Psi(0)$, and (g) results in

$$2\langle T\rangle = \left\langle \mathbf{S}\,(\partial V/\partial \boldsymbol{r})\cdot\boldsymbol{r}\right\rangle$$

$$\left[= -\left\langle \mathbf{S}\, d\boldsymbol{f}\cdot\boldsymbol{r}\right\rangle\text{: General definition of } \textit{virial} \text{ of a system}\right]. \tag{h}$$

(ii) If the system is *nonperiodic*, but moves in a finite spatial region with finite velocities, then, as (c) shows, there is an *upper* bound to $d\Phi/dt \equiv \Psi$, and eq. (h) still holds, *provided the averages are taken over a very long time* (i.e., $\tau \to \infty$); or, by choosing τ sufficiently large, we can make $\langle d\Psi/dt\rangle$ as small as possible:

$$\begin{aligned} \langle d\Psi/dt\rangle &= \lim\left\{(1/\tau)\int_0^\tau [d\Psi(t)/dt]\, dt\right\}_{\tau\to\infty} \\ &= \lim\left\{(1/\tau)[\Psi(\tau) - \Psi(0)]\right\}_{\tau\to\infty} = 0. \end{aligned} \tag{i}$$

If, further, V = *homogeneous* of degree f in the $\boldsymbol{r}$'s, then (by Euler's theorem) eq. (h) yields

$$2\langle T\rangle = f\langle V\rangle, \tag{j}$$

or, since $\langle T + V\rangle = \langle T\rangle + \langle V\rangle = \langle E\rangle = E$,

$$\langle T\rangle = fE/(f+2), \qquad \langle V\rangle = 2E(f+2). \tag{k}$$

In particular, if $f = 2$—that is, V is *quadratic* (linear vibrations)—then (j, k) yield the *equipartition* theorem:

$$\langle T\rangle = \langle V\rangle = E/2. \tag{l}$$

Let the reader verify that in a *central* force field with $V \sim r^{\varepsilon}$ (r = distance of dm from attracting origin)—that is, $f \to \varepsilon$—the virial theorem results in

$$2\langle T\rangle = \varepsilon\langle V\rangle; \tag{m}$$

from which it follows that in the *gravitational* case—that is, $\varepsilon = -1$, then,

$$\langle T\rangle = -E\ (>0), \qquad \langle V\rangle = 2E\ (<0) \Rightarrow 2\langle T\rangle = -\langle V\rangle; \tag{n}$$

which is in agreement with the general result that in such a Newtonian interaction "*the motion takes place in a finite region of space only if the total energy is negative.*"

These and other similar virial theorems have interesting applications in the mechanics of the *very small* (classical statistical mechanics) and of the *very large* (astronomy). For example, with their help, we can derive the well-known gas law: $pv = nk\theta$ (p, v, n, k, θ: *pressure, volume, number of molecules, Boltzmann's constant, absolute temperature,* respectively).

[For further details and applications, see, for example: Corben and Stehle (1960, pp. 164–166), Goldstein (1980, pp. 82–85, 96–97, 121, 477), Kurth (1960, pp. 64–74, 149–153), Pollard (1976, pp. 60–71); and books on the kinetic theory of gases; also, for the Newtonian interaction/gravitational case, see Landau and Lifshitz (1960, pp. 35–39). For engineering applications (nonlinear oscillations and their stability), see, for example, Papastavridis [1986(a)] and problems and example below.]

Problem 7.2.1 *Virial Theorem (Jacobi "Instability Criterion").* Consider a system of N mutually attracted (gravitating) particles. As is well known, its potential energy is a *negative* and homogeneous function of degree -1 in the $3N$ rectangular Cartesian coordinates of the position vectors of these particles. By applying the virial theorem for the case where the total energy $E \equiv T + V$ is a *positive* constant, show that

$$d^2\Phi/dt^2 = E + T = 2E - V \qquad \text{(Theorem of Lagrange–Jacobi).} \tag{a}$$

[Recalling the definition of Φ from the preceding example: $2\Phi \equiv S\, dm\, \mathbf{r}\cdot\mathbf{r} = S\, r^2 dm$, where $\mathbf{r}$ is the position vector of a typical particle relative to the system's (uniformly moving) mass center.]

Then deduce that

$$\Phi > (1/2)E\,t^2 + (d\Phi/dt)_o\, t + \Phi_o, \tag{b}$$

where $(d\Phi/dt)_o$, Φ_o: $d\Phi/dt$, Φ evaluated at some initial instant $t = 0$; and from this, in turn, since $E > 0$, Φ becomes *infinite*, as $t \to \infty$.

[A word of *caution*: from the above, however, it does not necessarily follow that at least one of the nonnegative functions $r^2\, dm$, making up the sum Φ, becomes infinite; although it does follow that not every such term remains bounded, otherwise Φ would stay bounded. For example, consider the function $q(t) = t\cos^2 t + t\sin^2 t = t$ (sum of nonnegative functions, for $t \geq 0$). As $t \to \infty$, the sum q becomes infinite, but neither of its "components" $t\cos^2 t$, $t\sin^2 t$ does; instead, they become large *and* small; that is, they become *unbounded* but do not tend to infinity! Hence, the commonly stated conclusion: *"if the total energy is positive, at least one particle must escape from the system* (i.e., if $E > 0$, then $r \to \infty$ for at least one particle; and, hence, the system is *unstable*)" is mathematically unproved; although, physically, such reasoning may look like academic hair-splitting.]

For a generalization of (a, b) and applications to stellar systems, see Kurth (1957, pp. 63–69).

Problem 7.2.2 *Virial Theorem* (*Linear Undamped Oscillator*). Consider the *linear* (or harmonic), *free* (or unforced, or undriven), and *undamped* oscillator with equation of motion: $\ddot{q} + \omega_o{}^2 q = 0$, where $\omega_o{}^2 \equiv$ *linear elasticity/mass* $\equiv k/m$ and, therefore, solution $q = a\cos(\omega_o t + \phi)$, where a is a constant amplitude and ϕ is the initial phase.

(i) Show that

$$\langle T \rangle = \langle V \rangle \qquad \text{or} \qquad m\omega_o{}^2 a^2/4 = ka^2/4; \tag{a}$$

where $\langle \ldots \rangle$ is time average of $(\ldots)$ over a period $\tau = 2\pi/\omega_o$; and, consequently,

$$E = \langle E \rangle = \langle T + V \rangle = \langle T \rangle + \langle V \rangle = m\omega_o{}^2 a^2/2. \tag{b}$$

(ii) Let the *variance* of a periodic function $f(t)$, of period τ, Δf, be defined by $(\Delta f)^2 \equiv \langle (f - \langle f \rangle)^2 \rangle$ (measure of *mean deviation of f from its average*). It is not too hard to see that $(\Delta f)^2 = \langle f^2 \rangle - \langle f \rangle^2$. Show that, for the harmonic oscillator discussed here,

$$(\Delta q)^2 = \langle q^2 \rangle \qquad [\Delta(dq/dt)]^2 = \langle (dq/dt)^2 \rangle. \tag{c}$$

Then, verify that $(\Delta q)(\Delta p) = E/\omega_o$, where $p \equiv m(dq/dt)$ is the linear momentum of the oscillator. (This constitutes a "constraint" between the root mean square deviations of a pair of measurable quantities, q and p, from their average values; and that is why it is called an *uncertainty* relation. Such conditions are important in quantum mechanics.)

(iii) Show that $\langle T \rangle = \langle V \rangle$ over *any time interval τ^* that is large relative to τ*; that is, even if τ^* is not an integral multiple of τ.

HINTS

With $\psi \equiv \omega_o t$, $\psi_1 \equiv \omega_1 t$, $\psi_2 \equiv \omega_2 t$, show that

$$\text{(i)}\ (1/\tau^*)\int_0^{\tau^*} \sin^2\psi\, dt = 1/2 + (1/4\tau^*\omega_o)[1 - \sin(2\omega_o\tau^*)] \quad [\to 1/2, \text{ as } \tau^* \to \infty], \tag{d}$$

and the same for the integral of $\cos^2\psi$;

(ii) $$(1/\tau^*)\int_0^{\tau^*}(\sin\psi_1)(\cos\psi_2)\,dt = \{1-\cos[(\omega_1-\omega_2)\tau^*]\}/2(\omega_1-\omega_2)\tau^* + \{1-\cos[(\omega_1+\omega_2)\tau^*]\}/2(\omega_1+\omega_2)\tau^* \quad [\to 0, \text{as } \tau^*\to\infty]; \tag{e}$$

(iii) $$(1/\tau)\int_0^{\tau}\sin^2(\omega_o t+\phi)\,dt = 1/2, \tag{f}$$

and the same for the integral of $\cos^2(\omega_o t+\phi)$;

(iv) $$(1/\tau)\int_0^{\tau}\sin(\omega_o t+\phi)\cos(\omega_o t+\phi)\,dt = 0. \tag{g}$$

Problem 7.2.3 *Virial Theorem* (*Linear and Damped Oscillator*). Consider the linear, free, and damped oscillator with equation of motion (with the usual notations)

$$m\ddot{q}+f\dot{q}+kq=0 \qquad \text{or} \qquad \ddot{q}+(1/r)\dot{q}+{\omega_o}^2 q=0, \tag{a}$$

where $1/r\equiv f/m$ [with dimensions $(time)^{-1}$; and occasionally referred to as $(relaxation\ time)^{-1}$].

(i) Show that for *small damping*, the latter defined precisely by

$$\omega_o r \gg 1 \qquad [\Rightarrow \omega_o \gg f/m; \text{i.e., roughly, } \textit{elasticity} \gg \textit{friction}], \tag{b}$$

an approximate solution of (a) is

$$q_o = a_o\exp(-t/2r)\sin(\omega_o t), \qquad \text{where } a_o \sim \text{initial velocity (a constant)}. \tag{c}$$

(ii) Then, show that in this case

$$\langle T\rangle \approx (m/4)[{\omega_o}^2+(f/2m)^2]{a_o}^2\exp(-t/r) \approx (m/4){\omega_o}^2{a_o}^2\exp(-t/r), \tag{d}$$

$$\langle V\rangle \approx (m/4)\,{\omega_o}^2{a_o}^2\exp(-t/r). \tag{e}$$

(iii) If $\langle D\rangle$ is the average *rate of dissipation* of the oscillation—that is, of $(f\dot{q})\dot{q}$—over a single (undamped) period $\tau_o = 2\pi/\omega_o$, then show that (again for small damping)

$$-\langle D\rangle = d\langle E\rangle/dt = d\langle T\rangle/dt + d\langle V\rangle/dt \approx -(1/r)\big[(m/2){\omega_o}^2{a_o}^2\exp(-t/r)\big] = -E(t)/r. \tag{f}$$

HINT

If $\omega_o r \gg 1$, then, to a good approximation, we can take the factor $\exp(-t/r)$ outside the averaging integrals; that is, our energetic averages are to be understood as *taken over a period* (*or cycle*) τ_o *at, approximately,* t (in the sense of the averaging method—see examples/problems of §7.9); and that is why they are functions of t.

Problem 7.2.4 *Virial Theorem* (*Linear, Damped, and Forced Oscillator*). Consider the linear, damped, and harmonically forced oscillator with equation of motion

$$m\ddot{q} + f\dot{q} + kq = Q_o \cos(\omega t), \qquad \text{(a)}$$

where Q_o, ω: forcing amplitude and frequency (both specified constants).

(i) Show that the time *average* of the rate of energy dissipation by the oscillator (i.e., of $(f\dot{q})\dot{q}$), over a *long* period of time $\tau^*(\gg 2\pi/\omega)$ equals

$$\langle D\rangle = (Q_o^{\,2}/2f)\cos^2\psi \qquad \cos\psi \equiv f[f^2 + (m\omega - k/\omega)^2]^{-1/2}. \qquad \text{(b)}$$

HINT

Use the *steady-state* (particular, periodic) solution: $q = a_o\cos(\omega t - \phi)$, where a_o is a function of Q_o, ω, ω_o, f, m; $\tan\phi = \omega f/(k - m\omega^2)$, ϕ: phase difference between force and displacement; and $\phi - \psi = \pi/2$, where ψ: phase difference between force and velocity.

(ii) Then, conclude that

(a) If $\psi = 0$ (resonance), then $\langle D\rangle \to \langle D\rangle_{\text{maximum}} = Q_o^{\,2}/2f$; and
(b) If force and displacement are either in phase or differ by π (i.e., $\phi = 0 \Rightarrow \psi = -\pi/2$, or $\phi = \pi \Rightarrow \psi = \pi/2$), then $\langle D\rangle = 0$.

(iii) Show that, *in steady-state motion*, the time average of the rate of working (power) of the driving force, that is, of $[Q_o\cos(\omega t)](dq/dt)_{\text{steady-state}}$, $\langle W\rangle$, *equals* $\langle D\rangle$.

In words: Mean energy externally *supplied* to system, per unit time = Mean energy *absorbed* or *dissipated* by system (*friction*), per unit time; and, hence, in such a forced motion, the energy of the system remains unchanged.

(iv) From (ii) and (iii) conclude that *at resonance* (*with the external force*), *both* $\langle W\rangle$ *and its equal* $\langle D\rangle$ *are maxima.*

Problem 7.2.5 *Virial Theorem* (*Linear Damped and Forced Oscillator*). Continuing from the preceding problem,

(i) Show that

$$\langle W\rangle \equiv P(x; f) = Q_o^{\,2} f/2\,(f^2 + m^2\omega_o^{\,2}x^2), \qquad \text{(a)}$$

where $x \equiv (\omega/\omega_o) - (\omega_o/\omega)$: roughly, *deviation* of ω from ω_o; and, therefore, if $x = 0$ (i.e., $\omega = \omega_o$), then $P(0; f) = maximum$. (It is not hard to see that the graph of $\langle W\rangle$ vs. x, with f as parameter, looks like a *resonance* curve; i.e., $|a_o|$ vs. ω.)

(ii) Show that

$$\partial P/\partial x = -Q_o^{\,2} f\, m^2\omega_o^{\,2}x/(f^2 + m^2\omega_o^{\,2}x^2)^2, \qquad \text{(b)}$$

and, therefore:

(a) If $x > 0$ (i.e., $\omega > \omega_o$), then $\partial P/\partial x < 0$, and the smaller the f the larger $|\partial P/\partial x|$; and similarly for $x < 0$; whereas
(b) If $x = 0$ (i.e., $\omega = \omega_o$) and f is small, then P is large.

In words: *the smaller (larger) the damping, the higher (lower) and sharper or peaked (flatter) the resonance maximum:* $P(0; f) = (1/2)(Q_o{}^2/f)$ *(say, like a Dirac delta function).*

For further details on such *dispersion* relations (very important in several areas of physics), see, for example, Falk (1966, pp. 37–43); also Landau and Lifshitz (1960, p. 79).

Problem 7.2.6 *Virial Theorem (Linear, Damped, and Forced Oscillator).* Continuing from the last two problems:

(i) Calculate and compare $\langle T\rangle$ and $\langle V\rangle$; and

(ii) Find the forcing frequencies at which each of them becomes maximum. Explain why these two maxima occur at different frequencies.

Example 7.2.2 *Nonlinear Oscillations via the Virial Theorem.*

1. *Duffing oscillator.* Let us find the amplitude versus frequency ("resonance curve") of the steady-state response of

$$m\ddot{q} + kq + hq^3 = Q_o \sin\chi, \qquad \chi \equiv \omega t, \tag{a}$$

where m is the mass of the oscillator (> 0), k is its linear stiffness (> 0), ω is the forcing frequency (given), Q_o is the forcing amplitude (given), and h is the nonlinearity constant.

Here, clearly,

$$2T = m(\dot{q})^2; \qquad V = V_2 + V_4, \quad 2V_2 \equiv kq^2, \quad 4V_4 \equiv hq^4; \qquad Q = Q_o \sin\chi, \tag{b}$$

and therefore for the trial steady-state solution of (a)

$$q = a\sin\chi, \qquad \text{where } a = a(\omega) \text{ (to be determined)}, \tag{c}$$

the virial equation (7.2.5), for this holonomic one-DOF system, with

$$\partial T/\partial q = 0, \qquad a_{Dk} = 0, \qquad \text{and} \qquad \left\{\sum p_k q_k\right\}_1^2 \to \{(\partial T/\partial\dot{q})q\}_0^{2\pi/\omega} \tag{d}$$

(and application of the homogeneous function theorem) gives

$$\begin{aligned}\int_0^{2\pi/\omega} [2T - (\partial V/\partial q)q + Qq]\, dt &= \int_0^{2\pi/\omega} [2T - (2V_2 + 4V_4) + Qq]\, dt \\ &= \int_0^{2\pi/\omega} (m\,a^2\omega^2 \cos^2\chi - ka^2 \sin^2\chi - h\,a^4 \sin^4\chi + Q_o a \sin^2\chi)\, dt \\ &= (\pi/\omega)(m\,a^2\omega^2 - ka^2 - 3h\,a^4/4 + Q_o a) = 0,\end{aligned} \tag{e}$$

from which we obtain the well-known *resonance* equation (with $\omega_o{}^2 \equiv k/m$)

$$\omega^2 = \omega_o{}^2 + (3/4)(h/m)a^2 - (Q_o/m)a^{-1}. \tag{f}$$

[For stability considerations of (a), via frequency derivatives of the virial equation, see Papastavridis (1986(b)).]

2. *Van der Pol oscillator.* Let us find the asymptotic ("limit cycle") amplitude of

$$\ddot{q} + \varepsilon(q^2 - 1)\dot{q} + q = 0, \tag{g}$$

where $\varepsilon(> 0)$ is such that the nonlinear damping term $\varepsilon(q^2 - 1)\dot{q}$ remains absolutely small relative to both inertia ($\ddot{q}$) and linear elasticity (q); that is, ε is a very small positive constant.

In the linear case—that is, for $\varepsilon = 0$—the solution of (g) is harmonic with frequency $\omega = 1$ and amplitude and phase depending on the initial conditions. Therefore, for $\varepsilon \neq 0$, we try the (asymptotically) harmonic solution

$$q = a \sin\chi, \qquad \chi = \omega t, \tag{h}$$

but with *both frequency* ω *and amplitude* a *unknown*. Here,

$$2T = (\dot{q})^2, \qquad 2V = q^2, \qquad Q = Q(q, \dot{q}) = -\varepsilon(q^2 - 1)\dot{q}, \tag{i}$$

and so the earlier virial equation yields

$$\begin{aligned}\int_0^{2\pi/\omega} &[2T - (\partial V/\partial q)q + Qq]\, dt \\ &= \int_0^{2\pi/\omega} \left[(\dot{q})^2 - q^2 - \varepsilon(q^2 - 1)q\dot{q}\right] dt \\ &= \cdots = (\pi/\omega)a^2(\omega^2 - 1) = 0 \Rightarrow \omega = 1;\end{aligned} \tag{j}$$

that is, if we insist on a harmonic solution, then the latter must have the undamped frequency.

Problem 7.2.7 *Virial Theorem* (*Nonlinear Oscillator*) (Killingbeck, 1970). Consider the unforced and undamped nonlinear (generalized Duffing) oscillator with Lagrangean

$$L = (1/2)(dq/dt)^2 - (1/2)\,\omega_o{}^2 q^2 - \varepsilon\, q^k \qquad [k\text{: } \textit{even} \text{ integer}], \tag{a}$$

and, accordingly, equation of motion

$$\ddot{q} + \omega_o{}^2 q + \varepsilon\, k\, q^{k-1} = 0. \tag{b}$$

For small amplitudes q (to ensure stability; i.e., $|\varepsilon| \ll 1$) the solution will be a symmetric anharmonic oscillation about the equilibrium position $q = 0$. By applying the virial theorem, and with $\langle\ldots\rangle \equiv$ *time average of* $(\ldots)$ *over the* (*unknown*) *period* $\tau = 2\pi/\omega$, show that

$$2\langle T\rangle = 2\langle\omega_o{}^2 q^2/2\rangle + \varepsilon\, k\langle q^k\rangle. \tag{c}$$

Then show that, for the trial solution $q_o(t) = a\sin(\omega t)$ $[q_o(0) = q_o(2\pi/\omega) = 0$, $\dot{q}_o(0) = a\,\omega]$, the above yields the approximate frequency

$$\omega^2 = \omega_o{}^2 + 2\varepsilon\, k\, a^{k-2}\langle\sin^k q\rangle, \tag{d}$$

where the average of the last term is over 2π.

Problem 7.2.8 *Virial Theorem* (*Nonlinear Oscillator*) (Killingbeck, 1970). Continuing from the preceding problem, and applying energy conservation to it, show that the period of that oscillator has the following ε-power representation:

$$\tau = 2\int_{-a}^{+a} \left[\omega_o^{\,2}(a^2 - q^2) + 2\varepsilon(a^k - q^k)\right]^{-1/2} dq \qquad \text{[exact expression]}$$

$$= 2\pi/\omega_o - (2\varepsilon\, a^{k-2}/\omega_o^{\,3})I + O(\varepsilon^2), \tag{a}$$

where k is an *even* integer, and

$$I \equiv \int_{-1}^{+1} (1 - y^2)^{-3/2}(1 - y^k)\, dy \tag{b1}$$

$$= \int_0^{\pi} (1 - \cos^k x)(1 - \cos^2 x)^{-1}\, dx \tag{b2}$$

$$= \int_0^{\pi} (1 + \cos^2 x + \cdots + \cos^{k-2} x)\, dx \tag{b3}$$

$$= \pi k \langle \sin^k x \rangle \qquad \text{[average over } 2\pi\text{]}. \tag{b4}$$

Problem 7.2.9 *Virial Theorem* (*Nonlinear Oscillator*) (Killingbeck, 1970). Referring to the exact expression for τ of the preceding problem, eq. (a), show that

$$\partial\tau/\partial\varepsilon < 0 \qquad \text{and} \qquad \partial^2\tau/\partial\varepsilon^2 > 0, \tag{a}$$

from which it follows that, *for large* ε, the virially obtained value of ω, to within $O(\varepsilon)$ terms, *overestimates* ω*, and therefore underestimates* τ.

Problem 7.2.10 *Variational and Virial Theorems for Linear Gyroscopic Systems.* Consider the linear, free, and undamped motions of a gyroscopic system (or the small such motions of a general system about *steady motion* or *relative equilibrium*). They are governed by the following equations (§3.10, §3.16):

$$E_k \equiv E_k(L_G) \equiv (\partial L_G/\partial \dot{q}_k)^{\cdot} - \partial L_G/\partial q_k = \sum (M_{kr}\ddot{q}_r + G_{kr}\dot{q}_r + V_{kr}q_r) = 0, \tag{a}$$

where

$$2T = \sum\sum M_{kr}\dot{q}_k\dot{q}_r \quad \text{(positive definite)}, \quad M_{kr} = M_{rk}\text{: constant}, \tag{b1}$$

$$2V = \sum\sum V_{kr}q_k q_r \quad \text{(assumed positive definite)}, \quad V_{kr} = V_{rk}\text{: constant}, \tag{b2}$$

$$2G = \sum\sum G_{kr}q_k\dot{q}_r \quad \text{(no sign properties)}, \quad G_{kr} = -G_{rk}\text{: constant}, \tag{b3}$$

and

$$L_G \equiv T - V - G\text{: } \textit{gyroscopic Lagrangean}. \tag{b4}$$

It can be shown that eqs. (a) can be brought to the *partially decoupled* form, in terms of the following "quasi-principal" (or "normal") coordinates x_r:

$$E_r \to m_r\ddot{x}_r + \sum g_{rs}\dot{x}_s + k_r x_r = 0, \tag{c}$$

where

$$2T = \sum m_r(\dot{x}_r)^2 \qquad \text{(positive definite and diagonal)}, \tag{d1}$$

$$2V = \sum k_r x_r^{\,2} \qquad \text{(positive definite and diagonal)}, \tag{d2}$$

$$2G = \sum\sum g_{rs} x_r \dot{x}_s \qquad \text{(no sign properties)}, \qquad g_{rs} = -g_{rs} \text{ (constant)}. \tag{d3}$$

(i) With the help of the "gyroscopic action" $A_G \equiv \int L_G\, dt$, show that the above equations of motion can be derived from the following Hamilton-type variational principle:

$$\delta A_G = \textit{Boundary Terms} \qquad [= 0, \text{if, e.g.}, \delta q(t_1) = \delta q(t_2) = 0]. \tag{e}$$

(ii) Choosing in eq. (e) $\delta q = q$ (or $\delta x = x$), show that the following *virial gyroscopic theorem* results:

$$A_G \equiv \int L_G\, dt \equiv \int (T - V - G)\, dt = \left\{(1/2) \sum (\partial T/\partial \dot{q}_r) q_r\right\}_1^2. \tag{f}$$

(iii) Assuming the periodic free mode

$$x_r = a_r \cos(\omega t) + b_r \sin(\omega t), \tag{g}$$

and choosing $t_2 - t_1 = \tau \equiv 2\pi/\omega$: period, show that (f) yields the *gyroscopic generalization of the (virial →) equipartition theorem*:

$$J \equiv \omega^2 T_x + \omega G_x - V_x = 0 \qquad (\Rightarrow \omega = \cdots), \tag{h}$$

where

$$2T_x \equiv \sum m_r(a_r^{\,2} + b_r^{\,2}), \qquad 2V_x \equiv \sum k_r(a_r^{\,2} + b_r^{\,2}), \qquad G_x \equiv \sum\sum g_{rs} a_r b_s. \tag{h1}$$

(iv) Assuming the mode (g) in (e), show that, for *fixed-frequency variations*,

$$\delta J \equiv \omega^2\, \delta T_x + \omega\, \delta G_x - \delta V_x = 0, \tag{i}$$

where

$$\delta x_r = \delta a_r \cos(\omega t) + \delta b_r \sin(\omega t). \tag{i1}$$

(v) In eq. (h), assume that the a_r and b_r are functions of ω, then form

$$\Delta J = \sum [(\ldots)\, \delta a_r + (\ldots)\, \delta b_r] + (\ldots)\, \delta\omega = 0, \tag{j}$$

and, combining it with (i), conclude that for that mode $\delta\omega = 0$.

For applications of this interesting "gyroscopic Rayleigh-like theorem," see, for example, Lamb (1932, pp. 313–315, 328–330, 337–338).

7.3 TIME-INTEGRAL THEOREMS: PFAFFIAN CONSTRAINTS, LINEAR NONHOLONOMIC VARIABLES

Here, the starting point is Hamel's equations (§3.5)

$$d/dt(\partial T^*/\partial \omega_k) - \partial T^*/\partial \theta_k - \Gamma_k = \Theta_k + \Lambda_k, \tag{7.3.1}$$

where, we are reminded,

$$T^* = T[t, q, \dot{q}(t, q, \omega)] \equiv T^*(t, q, \omega), \tag{7.3.1a}$$

$$\partial \ldots/\partial\theta_k \equiv \sum (\partial \ldots/\partial q_r)(\partial \dot{q}_r/\partial\omega_k) = \sum A_{rk}(\partial \ldots/\partial q_r) \quad \text{[in Pfaffian case]}, \tag{7.3.1b}$$

$$\begin{aligned} -\Gamma_k &\equiv \sum\sum \gamma^b{}_{ks}(\partial T^*/\partial\omega_b)\omega_s + \sum \gamma^b{}_{k,n+1}(\partial T^*/\partial\omega_b)\omega_{n+1} \\ &\equiv \sum\sum \gamma^b{}_{ks}(\partial T^*/\partial\omega_b)\omega_s + \sum \gamma^b{}_k(\partial T^*/\partial\omega_b) \\ &\equiv \sum\sum \gamma^b{}_{ks} P_b \omega_s + \sum \gamma^b{}_k P_b \equiv \sum h^b{}_k P_b, \end{aligned} \tag{7.3.1c}$$

the Hamel coefficients $\gamma^b{}_{ks}$, $\gamma^b{}_{k,n+1} \equiv \gamma^b{}_k$, $h^b{}_k$ are defined by the *transitivity* relations [§2.10; and, again, assuming that $(\delta q_k)^\cdot = \delta(\dot{q}_k)$]

$$\begin{aligned} (\delta\theta_b)^\cdot - \delta\omega_b &= \sum\sum \gamma^b{}_{ks}\,\omega_s\,\delta\theta_k + \sum \gamma^b{}_k\,\delta\theta_k \\ &= \sum\Big(\sum \gamma^b{}_{ks}\,\omega_s + \gamma^b{}_k\Big)\,\delta\theta_k \equiv \sum h^b{}_k\,\delta\theta_k, \end{aligned} \tag{7.3.1d}$$

and the $\Theta_k(\Lambda_k)$ are the nonholonomic *impressed forces and constraint reactions*; and, of course, $\Lambda_D \neq 0$, $\Lambda_I = 0$. Multiplying each of (7.3.1) with the earlier arbitrary z_k's, and then summing over k, invoking chain rule, rearranging, and finally integrating between t_1 and t_2, we obtain the *generalized nonholonomic time-integral (or virial-like) identity*:

$$\begin{aligned} \int\Big[\sum (\partial T^*/\partial\omega_k)\dot{z}_k + \sum (\partial T^*/\partial\theta_k) z_k - \sum\sum h^b{}_k(\partial T^*/\partial\omega_b) z_k \\ + \sum (\Theta_k + \Lambda_k) z_k\Big]\,dt = \Big\{\sum (\partial T^*/\partial\omega_k) z_k\Big\}_1^2. \end{aligned} \tag{7.3.2}$$

Again, let us examine the following special z_k-choices:

(i) $z_k \to \delta\theta_k$ [recalling that now the constraints are simply $\delta\theta_D = 0$ (and $\delta\theta_{n+1} \equiv \delta t = 0$); while $\delta\theta_I \neq 0$]. Then (7.3.2) gives

$$\begin{aligned} \int\Big[\sum (\partial T^*/\partial\omega_k)(\delta\theta_k)^\cdot + \sum (\partial T^*/\partial\theta_k)\,\delta\theta_k - \sum\sum h^b{}_k(\partial T^*/\partial\omega_b)\,\delta\theta_k \\ + \sum (\Theta_k + \Lambda_k)\,\delta\theta_k\Big]\,dt = \Big\{\sum (\partial T^*/\partial\omega_k)\,\delta\theta_k\Big\}_1^2. \end{aligned} \tag{7.3.3}$$

However, due to the transitivity equations (7.3.1d), the *first* and *third* integrand terms combine to yield

$$\begin{aligned} \sum\Big[(\partial T^*/\partial\omega_k)\,\delta\omega_k + \sum h^b{}_k(\partial T^*/\partial\omega_b)\,\delta\theta_k\Big] - \sum\sum h^b{}_k(\partial T^*/\partial\omega_b)\,\delta\theta_k \\ = \sum (\partial T^*/\partial\omega_k)\,\delta\omega_k, \end{aligned} \tag{7.3.3a}$$

and, successively,

$$\begin{aligned} \sum (\partial T^*/\partial\theta_k)\,\delta\theta_k &= \sum\Big(\sum A_{sk}(\partial T^*/\partial q_s)\Big)\Big(\sum a_{kb}\,\delta q_b\Big) \\ &= \sum\sum (\partial T^*/\partial q_s)(\delta_{sb})\,\delta q_b = \sum (\partial T^*/\partial q_s)\,\delta q_s \end{aligned} \tag{7.3.3b}$$

[recalling that, since (a_{kr}) and (A_{sk}) are *inverse* matrices, $\sum A_{sk}a_{kb} = \delta_{sb}$: Kronecker delta], while, due to Lagrange's principle (§3.2),

$$\sum (\Theta_k + \Lambda_k)\,\delta\theta_k = \sum \Theta_k\,\delta\theta_k = \sum \Theta_I\,\delta\theta_I \equiv \delta' W^*; \tag{7.3.3c}$$

and so, finally, (7.3.3) reduces to *Hamilton's law of virtual and nonholonomic action*:

$$\int \left(\delta T^* + \sum \Theta_I\,\delta\theta_I\right) dt = \left\{\sum P_I\,\delta\theta_I\right\}_1^2, \tag{7.3.4}$$

where

$$\begin{aligned}\delta T^* = \delta T^*(t, q, \omega) &= \sum \left[(\partial T^*/\partial\omega_k)\,\delta\omega_k + (\partial T^*/\partial q_k)\,\delta q_k\right]\\ &= \sum \left[(\partial T^*/\partial\omega_k)\,\delta\omega_k + (\partial T^*/\partial\theta_k)\,\delta\theta_k\right].\end{aligned} \tag{7.3.4a}$$

Clearly, since (7.3.4) involves only the independent $\delta\theta_I$'s it can supply only the $n - m$ *kinetic* equations of motion. If we want *all* n equations—that is, $n - m$ *kinetic* + m *kinetostatic*—then can we replace (7.3.4) with one or the other of the following two equivalent formulations: either

$$\int \left(\delta T^* + \sum \Theta_D\,\delta\theta_D + \sum \Theta_I\,\delta\theta_I\right) dt = \left\{\sum P_I\,\delta\theta_I\right\}_1^2, \tag{7.3.5}$$

under the constraints

$$1\,\delta\theta_1 = 1\,\delta\theta_2 = \cdots = 1\,\delta\theta_m = 0 \qquad (\text{and } 0\,\delta\theta_{m+1} = 0\,\delta\theta_{m+2} = \cdots = 0\,\delta\theta_n = 0), \tag{7.3.5a}$$

or

$$\int \left[\delta T^* + \sum (\Theta_D + \Lambda_D)\,\delta\theta_D + \sum \Theta_I\,\delta\theta_I\right] dt = \left\{\sum P_I\,\delta\theta_I\right\}_1^2, \tag{7.3.6}$$

under the constraints

$$0\,\delta\theta_1 = 0\,\delta\theta_2 = \cdots = 0\,\delta\theta_n = 0 \qquad (\text{i.e., with } \textit{all } \delta\theta_k \text{ unconstrained}). \tag{7.3.6a}$$

(ii) $z_k \to \dot{\theta}_k \equiv \omega_k$ (recalling that now the constraints are simply $\omega_D = 0$). Then (7.3.2) gives

$$\begin{aligned}\int \Big[\sum (\partial T^*/\partial\omega_I)\dot{\omega}_I + \sum (\partial T^*/\partial\theta_I)\omega_I - \sum\sum h^b{}_I(\partial T^*/\partial\omega_b)\omega_I + \sum \Theta_I\omega_I\Big]\, dt\\ = \left\{\sum (\partial T^*/\partial\omega_I)\omega_I\right\}_1^2.\end{aligned} \tag{7.3.7}$$

However:

(a) recalling the $\partial\ldots/\partial\theta_k$ definition (7.3.1b) and that $\dot{q}_k - \sum A_{kI}\omega_I + A_k$, we see that the *second* integrand sum equals $\sum (\partial T^*/\partial q_k)(\dot{q}_k - A_k)$;

(b) recalling the h-definition (7.3.1d) and the antisymmetry of the γ's in their subscripts—that is, $\gamma^b{}_{kl} = -\gamma^b{}_{lk}$ [and (3.9.12 ff.)]—we see that the *third* integrand (double) sum reduces to

$$-\sum\sum \gamma^b{}_I P_b\,\omega_I, \tag{7.3.7a}$$

and

(c) rewriting the *right*-hand side as $\int [\sum (\partial T^*/\partial \omega_I)\omega_I]^\cdot\, dt$, all these partial results allow us to convert (7.3.7) to

$$\int \Big\{ \Big[\sum (\partial T^*/\partial \omega_I)\omega_I - T \Big]^\cdot - \Big[-\partial T^*/\partial t - \sum (\partial T^*/\partial q_k) A_k - \sum\sum \gamma^b{}_I P_b\, \omega_I + \sum \Theta_I \omega_I \Big] \Big\}\, dt = 0, \tag{7.3.7b}$$

from which, since the limits t_1 and t_2 are arbitrary, we finally obtain the most general (nonpotential) form of the generalized power theorem, for systems under Pfaffian constraints and in nonholonomic variables:

$$\begin{aligned} \Big[\sum (\partial T^*/\partial \omega_I)\omega_I - T \Big]^\cdot &= -\partial T^*/\partial t - \sum (\partial T^*/\partial q_k) A_k - \sum\sum \gamma^b{}_I P_b\, \omega_I + \sum \Theta_I\, \omega_I \\ &\equiv -\partial T^*/\partial \theta_{n+1} + R + \sum \Theta_I\, \omega_I, \end{aligned} \tag{7.3.7c}$$

where, in the last line, we invoked the earlier helpful notations (3.9.12f ff.)

$$\partial T^*/\partial \theta_{n+1} \equiv \partial T^*/\partial t + \sum (\partial T^*/\partial q_k) A_k, \qquad R \equiv -\sum\sum \gamma^b{}_I P_b\, \omega_I. \tag{7.3.7d}$$

(iii) $z_k \to \theta_k$: This case is meaningless because there is no such thing as θ_k.
(iv) $z_k \to \omega_k$: This choice does not seem to lead to any recognizably useful result.
(v) $z_k \to \Delta\theta_k$: Here, we must define $\Delta\theta_k$. We have, successively,

$$\begin{aligned} \Delta q_k = \delta q_k + \dot q_k\, \Delta t &= \sum A_{kb}\, \delta\theta_b + \Big(\sum A_{kb}\omega_b + A_k \Big) \Delta t \\ &= \sum A_{kb}(\delta\theta_b + \omega_b\, \Delta t) + A_k\, \Delta t \equiv \sum A_{kb}\, \Delta\theta_b + A_k\, \Delta t; \end{aligned} \tag{7.3.8a}$$

that is, we can define consistently

$$\Delta\theta_b \equiv \delta\theta_b + \dot\theta_b\, \Delta t \equiv \delta\theta_b + \omega_b\, \Delta t. \tag{7.3.8b}$$

Inverting (7.3.8a), we find

$$\Delta\theta_b = \sum a_{bk}\, \Delta q_k + a_b\, \Delta t; \qquad \Delta\theta_D = 0, \qquad \Delta\,\theta_I \neq 0. \tag{7.3.8c}$$

We also need *commutation* relations between $(\Delta\theta)^\cdot$ and $\Delta(\dot\theta)$; that is, the nonholonomic counterpart of (7.2.4e). Indeed, $(\ldots)^\cdot$-differentiating (7.3.8b) leads to

$$(\Delta\theta_b)^\cdot \equiv (\delta\theta_b)^\cdot + \dot\omega_b\, \Delta t + \omega_b (\Delta t)^\cdot, \tag{7.3.8d}$$

while applying the definition (7.3.8d) to $\dot\theta_k \equiv \omega_k$ yields

$$\Delta(\dot\theta_b) = \delta(\dot\theta_b) + \ddot\theta_b\, \Delta t, \qquad \text{or} \qquad \Delta\omega_b = \delta\omega_b + \dot\omega_b\, \Delta t. \tag{7.3.8e}$$

Therefore, subtracting (7.3.8e) from (7.3.8d) side by side, while invoking the transitivity relations (7.3.1d), we find

$$(\Delta\theta_b)^\cdot - \Delta\omega_b = (\delta\theta_b)^\cdot - \delta(\dot\theta_b) + \omega_b(\Delta t)^\cdot = \sum h^b{}_k\, \delta\theta_k + \omega_b (\Delta t)^\cdot. \tag{7.3.8f}$$

Now, under $z_k \to \Delta\theta_k$ eq. (7.3.2) becomes

$$\int\Big[\sum(\partial T^*/\partial\omega_k)(\Delta\theta_k)^{\cdot}+\sum(\partial T^*/\partial\theta_k)\,\Delta\theta_k-\sum\sum h^b{}_k(\partial T^*/\partial\omega_b)\,\Delta\theta_k$$
$$+\sum(\Theta_k+\Lambda_k)\,\Delta\theta_k\Big]\,dt=\Big\{\sum(\partial T^*/\partial\omega_k)\,\Delta\theta_k\Big\}_1^2, \quad (7.3.9a)$$

or, invoking (7.3.8f) for the *first* integrand sum, recalling that $\gamma^b{}_{kl}=-\gamma^b{}_{lk}$, and renaming some dummy indices, we finally obtain *Hamilton's law of skew-varying action in nonholonomic variables*:

$$\int\Big\{\sum(\partial T^*/\partial\theta_k)\,\Delta\theta_k+\sum(\partial T^*/\partial\omega_k)\,\Delta\omega_k$$
$$+\sum(\partial T^*/\partial\omega_k)\Big[\omega_k(\Delta t)^{\cdot}-\sum\gamma^k{}_b\,\omega_b\,\Delta t\Big]+\sum(\Theta_k+\Lambda_k)\,\Delta\theta_k\Big\}\,dt$$
$$=\Big\{\sum(\partial T^*/\partial\omega_k)\,\Delta\theta_k\Big\}_1^2. \quad (7.3.9b)$$

Here, too, we remember to set $\omega_D=0$, $\omega_k\to\omega_I$; and, regarding the works $\sum(\Theta_k+\Lambda_k)\,\Delta\theta_k$, recalling the arguments leading to (7.3.4–6a):

(a) If we need only the kinetic equations from (7.3.9b), then we set in there

$$\Delta' W^*\equiv\sum\Theta_k\,\Delta\theta_k\equiv\sum\Theta_k\,\delta\theta_k+\Big(\sum\Theta_k\,\omega_k\Big)\,\Delta t$$
$$=\sum\Theta_I\,\delta\theta_I+\Big(\sum\Theta_I\,\omega_I\Big)\,\Delta t=\sum\Theta_I\,\Delta\theta_I, \quad (7.3.9c)$$
$$\Delta' W_R{}^*\equiv\sum\Lambda_k\,\Delta\theta_k=\cdots=0; \quad (7.3.9d)$$

whereas

(b) If we want to obtain both the kinetic and kinetostatic equations from (7.3.9b), then, either we set

$$\Delta' W^*=\sum\Theta_D\,\Delta\theta_D+\sum\Theta_I\,\Delta\theta_I \quad \text{and} \quad \Delta' W_R{}^*=0, \quad (7.3.9e)$$

under the constraints

$$1\,\delta\theta_D=0,\quad 1\,\omega_D=0\Rightarrow 1\,\Delta\theta_D=0; \quad (7.3.9f)$$

or

$$\Delta' W^*=\sum\Theta_D\,\Delta\theta_D+\sum\Theta_I\,\Delta\theta_I \quad \text{and} \quad \Delta' W_R{}^*=\sum\Lambda_D\,\Delta\theta_D, \quad (7.3.9g)$$

with *all* $\delta\theta_k$ unconstrained. The details are left to the reader.

Finally, we notice that if we set in (7.3.9a,b) $\Delta t=0$, we recover the virtual equations (7.3.3) $\to$ (7.3.4), as expected.

This completes the discussion of time-integral theorems and variational principles under Pfaffian constraints. In the next section, §7.4, these results are extended to *nonlinear* (holonomic and/or nonholonomic) velocity constraints (chap. 5):

$$f_D(t,q,\dot q)=0 \qquad [D=1,\ldots,m(<n)]. \quad (7.3.10)$$

Example 7.3.1 *The Maggi → Chaplygin–Hadamard Form of Hamilton's Principle*; that is, Hamilton's principle for systems subject to the special Pfaffian constraints (in virtual form)

$$\delta q_D = \sum b_{DI}\,\delta q_I, \qquad b_{DI} = b_{DI}(t, q); \tag{a}$$

with nonlinear counterpart $b_{DI} \to \partial\phi_D(t, q, \dot{q}_I)/\partial\dot{q}_I$ (see also §7.8).

Adopting the *Hölder–Voronets–Hamel* viewpoint—that is, $\delta(\dot{q}_k) = (\delta q_k)^{\cdot}$ for *all* holonomic coordinates—with the convenient notation $\{\sum (\partial T/\partial\dot{q}_k)\,\delta q_k\}_1^2 \equiv B_{\mathrm{Boundary}},\ T_{\mathrm{Terms}} \equiv BT$, and (a), we find, successively,

$$\begin{aligned}
\int \delta T\,dt &= \int \sum \left[(\partial T/\partial q_k)\,\delta q_k + (\partial T/\partial\dot{q}_k)\,\delta(\dot{q}_k)\right] dt \\
&= \cdots = BT - \int \sum E_k(T)\,\delta q_k\,dt \\
&= BT - \int \left[\sum E_D(T)\,\delta q_D + \sum E_I(T)\,\delta q_I\right] dt \\
&= BT - \int \sum \left[E_I(T) + \sum b_{DI} E_D(T)\right] \delta q_I\,dt;
\end{aligned} \tag{b}$$

and, similarly,

$$\begin{aligned}
\int \delta' W\,dt \equiv \int \sum Q_k\,\delta q_k = \cdots &= \int \sum \left(Q_I + \sum b_{DI} Q_D\right) \delta q_I\,dt \\
&\equiv \int \left(\sum Q_{Io}\,\delta q_I\right) dt \equiv \int \delta' W_o\,dt.
\end{aligned} \tag{c}$$

Therefore,

$$\int (\delta T + \delta' W)\,dt = BT - \int \sum \left[E_I(T) + \sum b_{DI} E_D(T) - Q_{Io}\right] \delta q_I\,dt, \tag{d}$$

from which, since the δq_I are unconstrained, we obtain the $n - m$ kinetic (multiplierless) Chaplygin–Hadamard equations (not to be confused with the nonholonomic Chaplygin equations)

$$E_I(T) + \sum b_{DI} E_D(T) = Q_I + \sum b_{DI} Q_D. \tag{e}$$

As already explained in §3.8, here all constraint enforcement occurs at the final level; that is, in (e), *not* in T.

If we had used in (b, c) the general representations (§2.6)

$$\delta q_k = \sum A_{kI}\,\delta\theta_I; \qquad \delta\theta_D = \sum a_{Dk}\,\delta q_k = 0, \quad \delta\theta_I \neq 0, \tag{f}$$

where the (a_{kl}) and (A_{kl}) are inverse matrices, then Hamilton's principle would have led us to

$$\int (\delta T + \delta' W)\, dt = BT - \int \Big\{ \sum\sum [E_k(T) - Q_k] A_{kD}\, \delta\theta_D \Big\}\, dt - \int \Big\{ \sum\sum [E_k(T) - Q_k] A_{kI}\, \delta\theta_I \Big\}\, dt; \tag{g}$$

and this, with the help of the m multipliers λ_D, would have led us to the familiar Maggi equations (§3.5)

$$\textit{Kinetostatic:} \qquad \sum E_k(T) A_{kD} = \sum Q_k A_{kD} + \lambda_D, \tag{h}$$

$$\textit{Kinetic:} \qquad \sum E_k(T) A_{kI} = \sum Q_k A_{kI}. \tag{i}$$

Example 7.3.2 *The Torque-Free Case of the Sled Problem, via the preceding Hamiltonian formulation of the Chaplygin–Hadamard Equations.* (Recalling exs. 2.13.1, 2.13.2, and 3.18.1; also see Hamel, 1949, pp. 614–615.) We have seen there that the constraint is

$$\dot{x} \sin\phi - \dot{y} \cos\phi = 0 \Rightarrow \delta y = (\tan\phi)\, \delta x \tag{a}$$

(i.e., $\delta q_D \equiv \delta y, \delta q_I \equiv \delta x, \delta\phi$; $n = 3$, $m = 1$), while the unconstrained kinetic energy of the sled is [with $I_{\text{Contact point}} \equiv I_C \equiv I$]

$$\begin{aligned} 2T &= m[(\dot{x}_G)^2 + (\dot{y}_G)^2] + I(\dot{\phi})^2 \\ &= m[(\dot{x} - b\dot{\phi}\sin\phi)^2 + (\dot{y} + b\dot{\phi}\cos\phi)^2] + I(\dot{\phi})^2 \\ &= m[(\dot{x})^2 + (\dot{y})^2] + I(\dot{\phi})^2 + 2mb\,\dot{\phi}(\dot{y}\cos\phi - \dot{x}\sin\phi). \end{aligned} \tag{b}$$

Applying Hamilton's principle directly, with all impressed forces zero (i.e., $\delta' W^* = 0$), and the δq's chosen so that $BT \to 0$, we obtain

$$\begin{aligned} 0 &= \int \delta T\, dt = \int [(\partial T/\partial\dot{x})\, \delta(\dot{x}) + (\partial T/\partial\dot{y})\, \delta(\dot{y}) + (\partial T/\partial\phi)\, \delta\phi + (\partial T/\partial\dot{\phi})\, \delta(\dot{\phi})]\, dt \\ &= \{(\partial T/\partial\dot{x})\, \delta x + (\partial T/\partial\dot{y})\, \delta y + (\partial T/\partial\dot{\phi})\, \delta\phi\}_1^2 \\ &\quad - \int \{(\partial T/\partial\dot{x})^{\cdot}\, \delta x + (\partial T/\partial\dot{y})^{\cdot}\, \delta y + [(\partial T/\partial\dot{\phi})^{\cdot} - (\partial T/\partial\phi]\, \delta\phi\}\, dt \end{aligned}$$

[and enforcing the second constraint of (a) on the integrand variations, not on T]

$$= -\int \{(\partial T/\partial\dot{x})^{\cdot} + \tan\phi(\partial T/\partial\dot{y})^{\cdot}]\, \delta x + [(\partial T/\partial\dot{\phi})^{\cdot} - \partial T/\partial\phi]\, \delta\phi\}\, dt, \tag{c}$$

from which, since δx and $\delta\phi$ can now be viewed as unconstrained, we obtain the *two* kinetic Chaplygin–Hadamard equations

$$(\partial T/\partial\dot{x})^{\cdot} + \tan\phi(\partial T/\partial\dot{y})^{\cdot} = 0: \quad (\dot{x} - b\,\dot{\phi}\sin\phi)^{\cdot} + \tan\phi\,(\dot{y} + b\,\dot{\phi}\cos\phi)^{\cdot} = 0, \tag{d}$$

and

$$(\partial T/\partial\dot{\phi})^{\cdot} - \partial T/\partial\phi = 0:$$

$$[I\dot{\phi} + m\,b(\dot{y}\cos\phi - \dot{x}\sin\phi)]^{\cdot} + m\,b\,\dot{\phi}(\dot{x}\cos\phi + \dot{y}\sin\phi) = 0. \tag{e}$$

Since $\dot{y} = \dot{x}\tan\phi \equiv v\sin\phi$, $\dot{x} = v\cos\phi$ (v: velocity of C), eqs. (d, e) can be rewritten, respectively, in the quasi-velocity (Hamel) form:

$$(v\cos\phi - b\,\dot{\phi}\sin\phi)^{\cdot} + \tan\phi(v\sin\phi + b\,\dot{\phi}\cos\phi)^{\cdot} = 0 \Rightarrow \dot{v} - b(\dot{\phi})^2 = 0, \tag{f}$$

$$I\ddot{\phi} + m\,b\,\dot{\phi}\,v = 0. \tag{g}$$

Let the reader repeat this procedure with the constraints in the form $\delta x = (\cot\phi)\,\delta y$; that is, with $\delta q_D = \delta x$, $\delta q_I = \delta y$, $\delta\phi$.

Example 7.3.3 *The Sled via Hamel's Form of Hamilton's Principle.* We have already established the following kinematical results (recall ex. 2.13.2):

$$q_{1,2,3} = x, y, \phi; \tag{a}$$

$$\omega_I \equiv (-\sin\phi)\dot{x} + (\cos\phi)\dot{y} + (0)\dot{\phi}$$
$$(= \text{velocity of contact point } C \text{ normal to sled} = 0), \tag{b1}$$

$$\omega_2 \equiv (\cos\phi)\dot{x} + (\sin\phi)\dot{y} + (0)\dot{\phi}$$
$$(= \text{velocity of } C \text{ along sled} = v \neq 0), \tag{b2}$$

$$\omega_3 \equiv (0)\dot{x} + (0)\dot{y} + (1)\dot{\phi} = \dot{\phi} \quad (\neq 0); \tag{b3}$$

$$\dot{q}_1 \equiv \dot{x} = (-\sin\phi)\omega_1 + (\cos\phi)\omega_2 + (0)\omega_3, \tag{c1}$$

$$\dot{q}_2 \equiv \dot{y} = (\cos\phi)\omega_1 + (\sin\phi)\omega_2 + (0)\omega_3, \tag{c2}$$

$$\dot{q}_3 \equiv \dot{\phi} = (0)\omega_1 + (0)\omega_2 + (1)\omega_3; \tag{c3}$$

$$\delta\omega_1 = (\delta\theta_1)^{\cdot} + (0)\,\delta\theta_1 + (\omega_3)\,\delta\theta_2 + (-\omega_2)\,\delta\theta_3, \tag{d1}$$

$$\delta\omega_2 = (\delta\theta_2)^{\cdot} + (-\omega_3)\,\delta\theta_1 + (0)\,\delta\theta_2 + (\omega_1)\,\delta\theta_3, \tag{d2}$$

$$\delta\omega_3 = (\delta\theta_3)^{\cdot} + (0)\,\delta\theta_1 + (0)\,\delta\theta_2 + (0)\,\delta\theta_3; \tag{d3}$$

and so, recalling the results of the preceding example, the unconstrained kinetic energy of the sled becomes

$$2T \to 2T^* = m\,\omega_2{}^2 + I\omega_2{}^3 + m\,b\,\omega_1\omega_3. \tag{e}$$

Hence, varying the above *and then enforcing the constraint* $\omega_1 = 0$ (but *not* $\delta\theta_1 = 0$), we find

$$\delta T^* = \sum (\partial T^*/\partial\omega_k)_o\,\delta\omega_k \equiv \sum P_{ko}\,\delta\omega_k \qquad [\text{invoking (d1–3) with } \omega_1 = 0]$$
$$= P_1(\delta\theta_1)^{\cdot} + (-P_2\omega_3)\,\delta\theta_1$$
$$+ P_2(\delta\theta_2)^{\cdot} + (P_1\omega_3)\,\delta\theta_2$$
$$+ P_3(\delta\theta_3)^{\cdot} + (-P_1\omega_2)\,\delta\theta_3 \qquad [\neq \delta(T^*{}_o), \text{see also §7.7}], \tag{f}$$

where

$$P_1 = m\,b\,\omega_3, \qquad P_2 = m\,\omega_2, \qquad P_3 = (m\,b\,\omega_1 + I\,\omega_3)_o = I\,\omega_3; \tag{g}$$

and

$$\delta' W^* \equiv \sum \Theta_k\,\delta\theta_k \quad (\equiv \delta' W). \tag{h}$$

Applying now Hamilton's principle directly [integrating by parts, grouping terms appropriately, and choosing the $\delta\theta$'s so that $BT \to 0$ (something that does no affect the equations of motion)], we obtain

$$\begin{aligned} 0 &= \int (\delta T^* + \delta' W^*)\,dt \\ &= \cdots = -\int \big[(\dot{P}_1 + P_2\,\omega_3 - \Theta_1)\,\delta\theta_1 + (\dot{P}_2 - P_1\,\omega_3 - \Theta_2)\,\delta\theta_2 \\ &\qquad + (\dot{P}_3 + P_1\,\omega_2 - \Theta_3)\,\delta\theta_3\big]\,dt; \end{aligned} \tag{i}$$

and from this, invoking the familiar multiplier arguments [$\delta\theta_1 = (1)\,\delta\theta_1 = 0$, $\delta\theta_{2,3}$: unconstrained], we finally get all three Hamel equations of the problem (recall ex. 3.18.1):

Kinetostatic:

$$\dot{P}_1 + P_2\,\omega_3 = \Theta_1 + \lambda_1: \qquad m(b\,\dot{\omega}_3 + \omega_2\,\omega_3) = \Theta_1 + \lambda_1; \tag{j1}$$

Kinetic:

$$\dot{P}_2 - P_1\,\omega_3 = \Theta_2: \qquad m(\dot{\omega}_2 - b\,\omega_3{}^2) = \Theta_2, \tag{j2}$$

$$\dot{P}_3 + P_1\,\omega_2 = \Theta_3: \qquad I\dot{\omega}_3 + m\,b\,\omega_2\,\omega_3 = \Theta_3, \tag{j3}$$

where

$\Theta_1 = -X\sin\phi + Y\cos\phi$: total impressed force *perpendicular* to sled, (k1)

$\Theta_2 = X\cos\phi + Y\sin\phi$: total impressed force *along* sled, (k2)

$\Theta_3 = M$: total impressed couple along z-axis; (k3)

(X, Y): rectangular Cartesian coordinates of total impressed force on sled; (k4)

obtained from the invariant equation (as if no constraint $\delta\theta_1 = 0$ existed):

$$\begin{aligned} \delta' W &\equiv X\,\delta x + Y\,\delta y + M\,\delta\phi \\ &= X(-\sin\phi\,\delta\theta_1 + \cos\phi\,\delta\theta_2) + Y(\cos\phi\,\delta\theta_1 + \sin\phi\,\delta\theta_2) + M\,\delta\theta_3 \\ &= \Theta_1\,\delta\theta_1 + \Theta_2\,\delta\theta_2 + \Theta_3\,\delta\theta_3 = \delta' W^*. \end{aligned} \tag{k5}$$

Finally, since here all constraints are scleronomic, the power equation is

Holonomic variables: $\quad dT/dt = X\dot{x} + Y\dot{y} + M\dot{\phi}$, (l1)

Nonholonomic variables: $\quad dT^*/dt = \Theta_2\omega_2 + \Theta_3\omega_3$; (l2)

from which we conclude that if $\Theta_{2,3} = 0$, then $T^* = constant$

$$\Rightarrow \ (2T^*)_o = m\,\omega_2{}^2 + I\,\omega_3{}^2 = m\,v^2 + I(\dot\phi)^2 = constant. \tag{13}$$

7.4 TIME-INTEGRAL THEOREMS: NONLINEAR VELOCITY CONSTRAINTS, HOLONOMIC VARIABLES

We recall (§5.1, §5.2) that the virtual displacements compatible with the constraints (§7.3.10) must satisfy, not the formal $\delta(\ldots)$-variation of $f_D(t, q, \dot q) = 0$—that is,

$$\sum \left[(\partial f_D/\partial q_k)\,\delta q_k + (\partial f_D/\partial \dot q_k)\,\delta(\dot q_k)\right] = 0, \tag{7.4.1a}$$

but, instead, the *Maurer–Appell–Chetaev–Johnson–Hamel* conditions

$$\delta\theta_D \equiv \sum (\partial f_D/\partial \dot q_k)\,\delta q_k = 0; \tag{7.4.1b}$$

and, again,

$$d(\delta q_k) = \delta(dq_k). \tag{7.4.1c}$$

Here, the starting point is the *nonlinear Routh–Voss* equations (5.3.11c)

$$E_k(T) \equiv (\partial T/\partial \dot q_k)^{\cdot} - \partial T/\partial q_k = Q_k + \sum \lambda_D(\partial f_D/\partial \dot q_k); \tag{7.4.2}$$

that is, equations (7.2.1) with a_{Dk} replaced by $\partial f_D/\partial \dot q_k$.

Performing similar operations on (7.4.2) as on (7.2.1), we readily find the following *generalized nonlinear holonomic time integral (or virial-like) identity*:

$$\int \left\{ \sum (\partial T/\partial \dot q_k)\dot z_k + \sum \left[\partial T/\partial q_k + Q_k + \sum \lambda_D\,(\partial f_D/\partial \dot q_k)\right] z_k \right\} dt = \left\{ \sum (\partial T/\partial \dot q_k) z_k \right\}_1^2. \tag{7.4.3}$$

Next, from (7.4.3) we obtain the following group of special integral formulae:

(i) The choice $z_k \to q_k$ yields the virial theorem

$$\int \left\{ \sum (\partial T/\partial \dot q_k)\dot q_k + \sum \left[\partial T/\partial q_k + Q_k + \sum \lambda_D\,(\partial f_D/\partial \dot q_k)\right] q_k \right\} dt = \left\{ \sum (\partial T/\partial \dot q_k) q_k \right\}_1^2. \tag{7.4.4}$$

If $\partial T/\partial q_k = 0$ and $\sum (\partial T/\partial \dot q_k)\dot q_k = 2T$, and the q-motion is *periodic* with period τ, then (7.4.4) reduces to:

$$\int 2T\,dt = -\int \sum \left[Q_k + \sum \lambda_D\,(\partial f_D/\partial \dot q_k)\right] q_k\,dt, \tag{7.4.4a}$$

where the integrals extend from t_1 to $t_2 = t_1 + \tau$.

(ii) The choice $z_k \to \dot{q}_k$ yields

$$\int \Big\{ \sum (\partial T/\partial \dot{q}_k) \ddot{q}_k + \sum \Big[\partial T/\partial q_k + Q_k + \sum \lambda_D (\partial f_D/\partial \dot{q}_k) \Big] \dot{q}_k \Big\} \, dt$$
$$= \Big\{ \sum (\partial T/\partial \dot{q}_k) \dot{q}_k \Big\}_1^2 = \int \sum [(\partial T/\partial \dot{q}_k) \dot{q}_k]^{\cdot} \, dt, \qquad (7.4.5a)$$

from which, using earlier described arguments, we obtain the *nonlinear (nonpotential) generalized power equation*:

$$\Big[\sum (\partial T/\partial \dot{q}_k) \dot{q}_k - T \Big]^{\cdot} = -\partial T/\partial t + \sum Q_k \dot{q}_k + \sum \sum \lambda_D (\partial f_D/\partial \dot{q}_k) \dot{q}_k. \quad (7.4.5b)$$

Specialization

If $\partial T/\partial t = 0$ and $\sum (\partial T/\partial \dot{q}_k) \dot{q}_k = 2T$, the above reduces to

$$dT/dt = \sum Q_k \dot{q}_k + \sum \sum \lambda_D (\partial f_D/\partial \dot{q}_k) \dot{q}_k; \qquad (7.4.5c)$$

a "kinetostatic" form which shows that, even if $Q_k = -\partial V(q)/\partial q_k$ and $\partial f_D/\partial t = 0$, in general, nonlinear velocity constraints are *nonconservative*. However, it is not hard to see that, if the constraints $f_D = 0$ are *homogeneous* (of any degree) in the $\dot{q}$'s, and the Q_k's derive from a potential $V(q)$, then the system is conservative.

(iii) The choice $z_k \to \delta q_k$ yields again Hamilton's law of varying action (7.2.3b): by (7.4.1b) we will have

$$\delta' W_R \equiv \sum \sum \lambda_D (\partial f_D/\partial \dot{q}_k) \, \delta q_k = 0 \qquad \text{(Lagrange's principle)}. \qquad (7.4.6)$$

(iv) The choice $z_k \to \Delta q_k = \delta q_k + \dot{q}_k \Delta t$, thanks to (7.4.6), yields *Hamilton's law of skew-varying action*:

$$\int \Big\{ \sum (\partial T/\partial \dot{q}_k) (\Delta q_k)^{\cdot} + \sum (\partial T/\partial q_k + Q_k) \, \Delta q_k + \Big[\sum \sum \lambda_D (\partial f_D/\partial \dot{q}_k) \dot{q}_k \Big] \Delta t \Big\} \, dt$$
$$= \Big\{ \sum (\partial T/\partial \dot{q}_k) \, \Delta q_k \Big\}_1^2. \qquad (7.4.7)$$

Here, too, if the constraints are homogeneous in the $\dot{q}$'s, then the last integrand (double) sum vanishes; also, we may replace $(\Delta q)^{\cdot}$ with $\Delta(\dot{q}) + \dot{q}(\Delta t)^{\cdot}$.

7.5 TIME-INTEGRAL THEOREMS: NONLINEAR VELOCITY CONSTRAINTS, NONLINEAR NONHOLONOMIC VARIABLES

Here, the starting point is the Johnsen–Hamel equations of motion (§5.3)

$$d/dt(\partial T^*/\partial \omega_k) - \partial T^*/\partial \theta_k - \Gamma_k = \Theta_k + \Lambda_k, \qquad (7.5.1)$$

where, we are reminded (§5.2),

$$\Gamma_k = -\sum H^b{}_k (\partial T^*/\partial \omega_b) = \sum V^b{}_k (\partial T/\partial \dot{q}_b)^*, \qquad (7.5.1a)$$

and the nonlinear coefficients $H^b{}_k$ (Hamel) and $V^b{}_k$ (Voronets) can be defined by [*assuming again that* $(\delta q_k)^{\cdot} = \delta(\dot{q}_k)$]

$$(\delta\theta_b)^{\cdot} - \delta\omega_b = \sum E_s(\omega_b)\,\delta q_s = \sum\sum E_s(\omega_b)\,(\partial\dot{q}_s/\partial\omega_k)\,\delta\theta_k \equiv \sum H^b{}_k\,\delta\theta_k$$
$$= -\sum\sum E_k{}^*(\dot{q}_l)\,(\partial\omega_b/\partial\dot{q}_l)\,\delta\theta_k \equiv -\sum\sum V^l{}_k(\partial\omega_b/\partial\dot{q}_l)\,\delta\theta_k; \tag{7.5.1b}$$

also, the virtual variations are related by

$$\delta q_s = \sum (\partial\dot{q}_s/\partial\omega_k)\,\delta\theta_k \Leftrightarrow \delta\theta_k = \sum (\partial\omega_k/\partial\dot{q}_s)\,\delta q_s, \tag{7.5.1c}$$

$$\delta(\dot{q}_s) = \sum (\partial\dot{q}_s/\partial\omega_k)\,\delta\omega_k \Leftrightarrow \delta\omega_k = \sum (\partial\omega_k/\partial\dot{q}_s)\,\delta(\dot{q}_s); \tag{7.5.1d}$$

and, of course, $\Lambda_D \neq 0, \Lambda_I = 0$.

Now, to build corresponding integral theorems, and so on, we, again, multiply (7.5.1) with the arbitrary set of functions $z_k(t)$, sum over k, apply chain rule, and so on, and, finally, integrate between t_1 and t_2. Below we show the details only for the important case $z_k \to \delta\theta_k$; that is; for *Hamilton's principle of vertically varying action*; those of the other cases are left to the reader (see also §7.6–7.9). Invoking the transitivity equations (7.5.1b) we find, successively,

$$\delta T^* = \sum [(\partial T^*/\partial\theta_k)\,\delta\theta_k + (\partial T^*/\partial\omega_k)\,\delta\omega_k]$$
$$= \sum \Big\{(\partial T^*/\partial\theta_k)\,\delta\theta_k + (\partial T^*/\partial\omega_k)\Big[(\delta\theta_k)^{\cdot} - \sum H^k{}_b\,\delta\theta_b\Big]\Big\}$$
$$= \Big[\sum (\partial T^*/\partial\omega_k)\,\delta\theta_k\Big]^{\cdot} - \sum (\partial T^*/\partial\omega_k)^{\cdot}\,\delta\theta_k$$
$$- \sum\sum H^k{}_b(\partial T^*/\partial\omega_k)\,\delta\theta_b + \sum (\partial T^*/\partial\theta_k)\,\delta\theta_k, \tag{7.5.2a}$$

and, therefore, integrating, we obtain the general kinematico-inertial transformation

$$\int \delta T^*\,dt = -\int \sum \Big[(\partial T^*/\partial\omega_k)^{\cdot} - \partial T^*/\partial\theta_k + \sum H^b{}_k(\partial T^*/\partial\omega_b)\Big]\,\delta\theta_k\,dt$$
$$+ \Big\{\sum (\partial T^*/\partial\omega_k)\,\delta\theta_k\Big\}_1^2; \tag{7.5.2b}$$

a result that shows the equivalence between the equations of motion (7.5.1, 1a) [plus boundary conditions $\delta\theta(t_{1,2})$] and the nonlinear counterpart of equations (7.3.4–6):

$$\int (\delta T^* + \delta' W^*)\,dt = \Big\{\sum P_k\,\delta\theta_k\Big\}_1^2; \tag{7.5.2c}$$

that is, the Hamiltonian variational equation has the same form for both Pfaffian and nonlinear constraints.

Let us now proceed to a detailed study of these variational "principles."

Time-Integral Variational Principles (IVP)

7.6 HAMILTON'S PRINCIPLE VERSUS CALCULUS OF VARIATIONS

(i) Mechanical Variational Problem

Let us take, without loss in generality for our purposes here, a system described completely by the Lagrangean function $L = L(t, q, \dot{q}) \equiv T - V$, and subjected to the nonlinear and possibly nonholonomic constraints (7.3.10)

$$\omega_D \equiv f_D(t, q, \dot{q}) = 0 \qquad \text{(velocity form)}, \tag{7.6.1a}$$

$$\delta\theta_D \equiv \sum (\partial f_D/\partial \dot{q}_k)\,\delta q_k = 0 \qquad \text{(virtual form)}. \tag{7.6.1b}$$

Its equations of motion, (7.4.2),

$$E_k(L) = \sum \lambda_D(\partial f_D/\partial \dot{q}_k) \qquad \left[= \sum \lambda_D a_{Dk},\ \text{in the Pfaffian case}\right], \tag{7.6.2}$$

are obtained by combining Lagrange's differential variational principle (LP): $\sum E_k(L)\,\delta q_k = 0$, with (7.6.1b), in, by now, well-understood ways; and, along with the m constraints $f_D \equiv \omega_D = 0$, they form a determinate system of $n + m$ equations for the functions $q_k(t)$ and $\lambda_D(t)$. Next, assuming that $(\delta q_k)^{\cdot} = \delta(\dot{q}_k)$, we are readily led (applying chain rule to LP, etc.) to the central equation:

$$\delta L \equiv \sum \left[(\partial L/\partial q_k)\,\delta q_k + (\partial L/\partial \dot{q}_k)\,\delta(\dot{q}_k)\right] = d/dt\left[\sum (\partial L/\partial \dot{q}_k)\,\delta q_k\right]; \tag{7.6.3a}$$

then, integrating the above in time between $t_{1,2}$, while assuming that

$$BT \equiv \left\{\sum (\partial L/\partial \dot{q}_k)\,\delta q_k\right\}_1^2 = 0 \qquad \text{(of no effect on the equations of motion)}, \tag{7.6.3b}$$

we obtain the *constrained* integral variational equation

$$\int \delta L\, dt = 0, \tag{7.6.3c}$$

and, finally, attaching (or adjoining) (7.6.1b) to (7.6.3c) via the m Lagrangean multipliers λ_D, we obtain the following *unconstrained* integral variational equation:

$$\int \left[\delta L + \sum\sum \lambda_D(\partial f_D/\partial \dot{q}_k)\,\delta q_k\right] dt = 0. \tag{7.6.4}$$

Conversely, integrating (7.6.4) by parts [and then using (7.6.3b)], we are easily led back to (7.6.2); that is, (7.6.2) and (7.6.4) are completely equivalent. *Equation (7.6.3c) under (7.6.1b)*, or *(7.6.4)*, constitute the *mechanical* variational problem for our system. Next, let us see the corresponding mathematical variational problem, for the same system, and its relation with the above.

(ii) Mathematical Variational Problem

According to variational calculus [see any good text on this subject; for example, Fox (1950/1987, pp. 94–102), Funk (1962, p. 253 ff.), Gelfand and Fomin (1963, chap. 2)] this would be

$$\int L\,dt = \textit{stationary}, \qquad \text{under (7.6.1a)}, \tag{7.6.5a}$$

or, equivalently,

$$\delta \int L\,dt = 0, \qquad \text{under (7.6.1a)} \to \delta f_D = 0. \tag{7.6.5b}$$

Applying again the multiplier rule, but with the m Lagrangean multipliers μ_D (since here $f_D \equiv \omega_D = 0$; and $\delta\omega_D = 0$) we are led, in well-known ways, to the unconstrained variational equation

$$0 = \delta \int \Big(L + \sum \mu_D f_D\Big)\,dt = \int \delta\Big(L + \sum \mu_D f_D\Big)\,dt \qquad \text{(for fixed time-endpoints)}$$

$$= \int \Big[\delta L + \sum (\delta\mu_D f_D + \mu_D\,\delta f_D)\Big]\,dt = \int \Big(\delta L + \sum \mu_D\,\delta f_D\Big)\,dt$$

$$= \ldots = -\int \sum E_k\big(L + \textstyle\sum \mu_D f_D\big)\,\delta q_k\,dt + \Big\{\sum\sum \mu_D(\partial f_D/\partial \dot{q}_k)\,\delta q_k\Big\}_1^2, \tag{7.6.5c}$$

which, since now the δq's *can* be viewed as *unconstrained*, with (7.6.3b), immediately yields the following n *Euler–Lagrange* equations [observing that $E_k(\ldots)$ is a *linear* operator]:

$$E_k\big(L + \textstyle\sum \mu_D f_D\big) = E_k(L) + E_k\big(\textstyle\sum \mu_D f_D\big) = 0, \tag{7.6.5d}$$

or, in extenso,

$$(\partial L/\partial \dot{q}_k)^{\cdot} - \partial L/\partial q_k = -\sum \mu_D\big[(\partial f_D/\partial \dot{q}_k)^{\cdot} - \partial f_D/\partial q_k\big] - \sum (d\mu_D/dt)(\partial f_D/\partial \dot{q}_k), \tag{7.6.5e}$$

or, again in operator form (recalling that $f_D \equiv \omega_D$),

$$E_k(L) = -\sum \mu_D E_k(\omega_D) - \sum (d\mu_D/dt)(\partial \omega_D/\partial \dot{q}_k). \tag{7.6.5f}$$

Now, comparing (7.6.2) and (7.6.5d–f) we immediately see that, even if we identify the λ_D with the $-(d\mu_D/dt)$, since, in general, $E_k(\omega_D) \equiv E_k(f_D) \neq 0$ (nonholonomic constraints), still these two sets of equations are different; that is, *equations (5d–f) are mechanically/physically incorrect!* However, for holonomic problems—that is, $E_k(f_D) = 0$—these equations, and, hence, corresponding integral variational problems, *are* completely equivalent.

REMARKS

(i) As clarified below, this should not come as a surprise: The basic principle of mechanics is *not* the *integral* principle/rule (7.6.5a, b), but the *differential* principle of Lagrange (LP).

(ii) Note that, in the variational calculus literature, (7.6.5a, b) is also called *Lagrange's problem!*

Source and Meaning of the Discrepancy

We begin with a fundamental actual path of the system, or simply an *orbit*, in the physical or in configuration space. That is a *dynamically* or *kinetically* possible path from an initial configuration $P_1 \equiv P(t_1)$ to a final one $P_2 \equiv P(t_2)$, where $P_{1,2}$ and $t_{1,2}$ are fixed, or given (see next section for slight changes in these boundary data); that is, an orbit satisfies (a) the equations of motion (of Lagrange, Routh–Voss, etc.) and (b) the boundary conditions. Thus, for a holonomic system with no additional Pfaffian constraints (i.e., $m = 0$), the orbit is a curve

$$q_k = q_k(t); \qquad t_1 \le t \le t_2, \tag{7.6.6a}$$

of class C_2 (i.e., with continuous derivatives of up to the *second* order, in that t-region) satisfying Lagrange's equations

$$E_k(L) \equiv (\partial L/\partial \dot{q}_k)^{\cdot} - \partial L/\partial q_k = 0, \tag{7.6.6b}$$

and the boundary conditions

$$q_k(t_1) = q_{k1}, \qquad q_k(t_2) = q_{k2}; \qquad q_{k;1,2}\text{: } \textit{given} \text{ numbers.} \tag{7.6.6c}$$

Let us see what happens under the additional (possibly nonholonomic) constraints (7.6.1a, b).

(i) In the *mechanical* problem, (7.6.3c, 1b), we build, in accordance with Lagrange's principle, varied, or comparison, paths by *adding to each point of the fundamental orbit I*, $[t, q_k(t)]$, the *contemporaneous* (or vertical) and *constraint compatible*—that is, virtual, displacement $\delta q_k = \delta q_k(t)$, of class C_2—that vanishes at $t_{1,2}$. In short, the *mechanical variations* consist of *kinematically admissible (or possible) displacements*; that is, δq's that satisfy (7.6.1b): $\delta\theta_D = 0$.

(ii) In the *mathematical* problem, (7.6.5a, 1a), on the other hand, we consider, in accordance with the multiplier rule of variational calculus, the family, or class, of all constraint compatible paths K that are of class C_2 and coincide with I at $t_{1,2}$ (i.e., I belongs to K). In short, the *mathematical variations* consist of kinematically admissible (or possible) *neighboring paths*; that is, paths that satisfy (7.6.1a): $\omega_D = 0$.

We express these differences, compactly, by

$$\int \delta L\, dt \qquad \left[\text{under } \delta\theta_D \equiv \sum (\partial f_D/\partial \dot{q}_k)\,\delta q_k = 0 - \text{mechanics}\right]$$

$$\neq \delta \int L\, dt \qquad [\text{under } \omega_D \equiv f_D(t, q, \dot{q}) = 0 - \text{mathematics}]; \tag{7.6.7}$$

with the equality sign holding for holonomic systems.

{As Capon puts it: "Whereas in the generally accepted method of the calculus of variations [i.e., mathematics] the comparison *paths* are required to satisfy the conditional equations [our $\omega_D = 0$], the displacements being free, in Hölder's treatment [i.e., mechanics] this is reversed: the *displacements* [our δq_k] are to satisfy the conditional equations [our $\delta\theta_D = 0$], and it follows from the theory of Pfaff equations [i.e., Frobenius' theorem (§2.12)] that the *comparison paths* do not satisfy them]" (1952, p. 473, emphasis added).

As shown below (§7.7), this means that from $\delta\theta_D(t) = 0$ (admissible *displacements*—mechanics), it does not follow that $\delta\omega_D(t) = 0$ (admissible *paths*—mathematics), and vice versa, unless the constraints are *holonomic*.

For these reasons, certain authors have *modified the integrand* of the mechanical variational principle so as to produce a mathematical variational principle that yields the correct equations of motion: for example, Borri (1994).}

Sufficiency of Hamilton's Principle (HP)

Before discussing the quantitative consequences of the above, let us examine the following important point: so far, both (7.6.3c, 5b) and (7.6.5a, 1a) have been shown to be *necessary* conditions for an orbit. But the practical usefulness of HP (and the other IVP, and of variational methods in general) lies largely in their *sufficiency: the solution(s) of these variational equations should be the orbit(s) of the problem.* Let us examine the nonholonomic case in more detail: clearly, the numbers n and m (and, therefore, also $f \equiv n - m$: *#DOF*) are system *invariants*; that is, independent of the q's chosen. Further, as shown below, starting from a given point P_1, of the configuration space V_n, the points P accessible from it by kinematically admissible paths lie on an n-dimensional manifold M_n; that is, any point P in V_n is kinematically accessible. However, the points accessible from P_1 by kinematically admissible paths (i.e., orbits through it) lie on an $(n - m)$-dimensional submanifold M_{n-m} (= a subspace of V_n). Hence, the sufficient part of the mathematical variational problem *cannot* hold for nonholonomic systems: if both P_1 and P_2 are given, then P_2 may *not* lie on M_{n-m}; eqs. (7.6.5a, 1a) will yield a path through P_1 and P_2, but that path will *not* be a mechanically correct motion, it will not be an orbit. Still worse, even if P_2 were kinematically accessible from P_1, the orbit would exist but it would not satisfy (7.6.5a, 1a); that is, it would not render $\int L\,dt$ stationary among K-curves. In sum: whether P_2 lies on M_{n-m} or not, the variational equation $\delta \int L\,dt = 0$ under the multiplier rule does not produce the orbit through P_1 and P_2; and, conversely, that orbit does not make $\int L\,dt$ stationary among kinematically admissible paths—*the mathematical variational principle is not valid for nonholonomic systems*, and therefore it *cannot* be used as a foundation of (even conservative) analytical mechanics!

If one still wants such a "principle" (better, integral energetic equation), *uniformly* valid for both holonomic and nonholonomic systems, that *can* be done, but it must be based on *some time integral of Lagrange's principle or the central equation*; that is, *on mechanical principles* (as detailed below §7.6, §7.7); then its application will produce the orbits of the system.

Differential Equation Considerations

These differences can also be seen from the viewpoint of differential equations.

(i) Mechanical Problem

The *general* solution of the latter depends on $2n - m$ arbitrary constants, not on $2n$. Here is why: due to the constraint equations, we can express the $m\,\lambda_D$'s as $\lambda = \lambda(t, q, \dot{q})$ [no $d\lambda/dt$ occur in (7.6.2)] and then substitute them back into (7.6.2) [recalling the "equations of Jacobi–Synge": exs. 3.2.6, 3.5.5, 3.10.2, 5.3.5, and 5.3.6]; that is, we can replace the latter by n multiplierless (kinetic) *second-order* equations in the q's; a system whose general solution will depend on $2n$ constants. Finally, since these q's should also satisfy the m constraints (7.6.1a), the general solution of (7.6.2) would depend on $2n - m$ arbitrary constants; that is, the "path multiplicity" is

∞^{2n-m}. [Or, we could argue on the following mechanical grounds: the kinetic (reactionless) equations of the system—e.g., those of Maggi, Hamel, etc.—constitute a system of $n - m$ *second-order* equations for the q's and, therefore, they generate $2(n - m)$ constants. On the other hand, the system of the m *first-order* constraints generates m such constants. So the total number of arbitrary integration constants will be $2(n - m) + m = 2n - m$.]

These constants can be expressed, for example, in terms of the $2n - m$ *initial* conditions (say, for $t_1 = 0$):

$$q_k(0) = q_{ko}\text{: given } (n \text{ conditions}), \qquad \dot{q}_I(0) = \dot{q}_{Io}\text{: given } (n - m \text{ conditions}), \tag{7.6.8}$$

while the remaining m such conditions, $\dot{q}_D(0) = \dot{q}_{Do}$: given, can be found from (7.6.8) and the m constraints $f_D = 0$ evaluated at $t_1 = 0$; that is, even though the q_o can be specified arbitrarily, the $\dot{q}_o$ require consideration of the constraints.

In short: *Orbits through a given point* must be along *constraint-compatible directions*. If, further, the system is *holonomic*, these directions define an $(n - m)$-*dimensional submanifold*, inside the original n-dimensional manifold of the q's. Hence, in such systems, an orbit cannot pass through two arbitrarily specified points P_1 (initial) and P_2 (final) in configuration space; if we fix P_1, then, for the arc $P_1 P_2$ to be an orbit, P_2 cannot be specified arbitrarily, but must lie on an $(n - m)$-dimensional manifold through P_1.

(ii) Mathematical Problem

The system (7.6.5d–f) is of the *second* order in the q's and *first* order in the μ's, and therefore (since the $f_D = 0$ are nonholonomic $\Rightarrow E_k(f_D) \neq 0$) its general solution depends on $2n$ arbitrary constants; whereas that of the mechanical system (7.6.2), as explained above, depends on only $2n - m$ such constants. [This can also be seen as follows: since the n q's appear up to the *second* order and the m μ's up to the *first* order, we have a maximum of $2n + m$ arbitrary constants; that is, initial conditions for the q's and μ's. But the q's and $\dot{q}$'s must also satisfy the constraints $f_D(t, q, \dot{q}) = 0$, and so the *maximum number of such constants of the mathematical problem* (the "*multiplicity of its solutions*") is $(2n + m) - m = 2n$.]

In conclusion:

- Through any given pair of points P_1 and P_2, these $2n$ constants define a path uniquely;
- Through any point P_1, and in any constraint-compatible direction, there is an ∞^m multiplicity of paths; the position and (compatible) direction absorb $2n - m$ of the constants, leaving m of them disposable $[(2n - m) + m = 2n]$.

[It has been shown by Capon (1952, p. 476), that: (a) In the *mechanical* problem, if the m constraints have $r(< m)$ integrals, then the maximum multiplicity of the mechanical paths is $\infty^{(2n-m)-r} = \infty^{2n-(m+r)}$; and if, in addition, there are M ignorable/cyclic coordinates—that is, $\partial L/\partial q = 0$ (§8.4 ff.)—then the multiplicity is $\infty^{(2n-m)-(r+M)} = \infty^{2n-(m+r+M)}$. (b) In the *mathematical* problem, if there are integrals of the constraints then for each such integral the number of independent constants is reduced by 2; that is, the corresponding multiplicities are, respectively, $\infty^{2(n-r)}$ and $\infty^{2(n-r)-M}$.

In other words, there are many paths in the mathematical problem that are not mechanical paths; or, *all mathematical solutions satisfy the stationarity condition; while, in general, the mechanical paths (orbits) do not.*]

Conditions under which the Mechanical and Mathematical Problems Coincide

This means where the general or some particular solution(s) of these problems coincide, for the same initial conditions. By comparing (7.6.2) and (7.6.5f), we see that for this to happen we must have

$$\sum [\lambda_D + (d\mu_D/dt)]\,(\partial\omega_D/\partial\dot{q}_k) = -\sum \mu_D[(\partial\omega_D/\partial\dot{q}_k)^{\cdot} - \partial\omega_D/\partial q_k]. \qquad (7.6.9a)$$

Multiplying each of these n equations by δq_k and summing over k, while observing the constraints $\delta\theta_D = 0$, we find the alternative, virtual work form, of the necessary condition:

$$\delta' W_{R'} \equiv -\sum\sum \mu_D\big[(\partial\omega_D/\partial\dot{q}_k)^{\cdot} - \partial\omega_D/\partial q_k\big]\,\delta q_k \equiv -\sum\sum \mu_D E_k(\omega_D)\,\delta q_k = 0. \qquad (7.6.9b)$$

The above is also sufficient: assuming that some solution of the mathematical problem satisfies (7.6.9b) for δq's restricted by $\delta\theta_D = 0$, then multiplying (7.6.9a) with δq_k, and $\delta\theta_D$ with λ_D, and summing, respectively, over k and D, while observing (7.6.9b) and (7.6.5f), we find

$$\sum \Big[E_k(L) - \sum \lambda_D(\partial f_D/\partial\dot{q}_k)\Big]\,\delta q_k = 0; \qquad (7.6.9c)$$

that is, that solution satisfies the mechanical problem and its constraints.

In terms of the following constraint reactions:

$$R_k \equiv \sum \lambda_D(\partial f_D/\partial\dot{q}_k), \qquad (7.6.9d)$$

$$R_k{}' \equiv -\sum \mu_D E_k(\omega_D), \qquad R_k{}'' \equiv -\sum (d\mu_D/dt)\,(\partial f_D/\partial\dot{q}_k), \qquad (7.6.9e)$$

the "coincidence conditions" (7.6.9a, b) can be rewritten, respectively, as

$$R_k = R_k{}' + R_k{}'', \qquad (7.6.9f)$$

$$\sum R_k\,\delta q_k = \sum R_k{}'\,\delta q_k + \sum R_k{}''\,\delta q_k: \qquad 0 = 0 + 0. \qquad (7.6.9g)$$

Additional forms of these coincidence conditions will be given later (§7.8).

We can summarize the developments of this section as follows: The differences between the time-integral variational "principles" of analytical mechanics and variational calculus result from different assumptions about variations; that is, comparison motions: in mechanics we assume admissible *instantaneous displacements* ($\delta\theta_D = 0$), while in mathematics we assume admissible *paths, as a whole* ($\delta\omega_D = 0$). And since these assumptions result in different forms of the *transitivity equations*, we must examine the precise effect of the latter both on the principles of mechanics (i.e., Lagrange's principle, or his central equation—which are variational but *differential*) and on the corresponding Hamiltonian principles (which are also variational but *integral*). This is detailed in the next section.

7.7 INTEGRAL VARIATIONAL EQUATIONS OF MECHANICS

Mechanical Admissibility

The constraints (7.6.1a, b) must hold for a generic instant t, as well as for its adjacent $t + dt$; that is,

$$\omega_D(t) = 0 \quad \textit{and} \quad \omega_D(t + dt) = 0, \qquad \text{or} \qquad \delta\theta_D(t) = 0 \quad \textit{and} \quad \delta\theta_D(t + dt) = 0. \tag{7.7.1}$$

Expanding the *second* and *fourth* of the above à la Taylor, and invoking the *first* and *third*, we get, to the first order (omitting, for simplicity, the explicit time dependence), the following requirements for the mechanical realizability of adjacent motions:

$$\omega_D + d\omega_D = 0 \quad \Rightarrow \quad d\omega_D = 0,$$

or

$$\delta\theta_D + d(\delta\theta_D) = 0 \quad \Rightarrow \quad d(\delta\theta_D) = 0,$$

or

$$(\delta\theta_D)^{\cdot} = 0 \qquad \text{(evolution of displacement admissibility in time).} \tag{7.7.2}$$

Then, the earlier general transitivity equations [§5.2, or (7.5.1b), but *without* the special assumption $(\delta q_k)^{\cdot} = \delta(\dot{q}_k)$] yield

$$\begin{aligned} -\delta\omega_D &= \sum (\partial\omega_D/\partial\dot{q}_k)\,[(\delta q_k)^{\cdot} - \delta(\dot{q}_k)] + \sum H^D{}_I\,\delta\theta_I \\ &= \sum (\partial\omega_D/\partial\dot{q}_k)\,[(\delta q_k)^{\cdot} - \delta(\dot{q}_k)] - \sum\sum V^k{}_I\,(\partial\omega_D/\partial\dot{q}_k)\,\delta\theta_I \end{aligned} \tag{7.7.3}$$

$$\Big[= \sum (\partial\omega_D/\partial\dot{q}_k)\,[(\delta q_k)^{\cdot} - \delta(\dot{q}_k)] + \sum\sum (\gamma^D{}_{II'}\omega_{I'})\,\delta\theta_I, \quad \text{in stationary Pfaffian case}\Big] \neq 0, \text{in general.} \tag{7.7.3a}$$

From the above we conclude the following:

- We cannot assume that both $(\delta q_k)^{\cdot} = \delta(\dot{q}_k)$ and $\delta(d\theta_D) = 0$, or $\delta\omega_D = 0$, hold; it is either one or the other.
- If we assume that $(\delta q_k)^{\cdot} = \delta(\dot{q}_k)$, then $\delta(d\theta_D), \delta\omega_D \neq 0$; unless $H^D{}_I = 0$ [which, in the stationary Pfaffian case (chosen here just for algebraic simplicity), reduces to the Frobenius integrability conditions: $\gamma^D{}_{II'} = 0$. Hence, by analogy, $H^D{}_I = 0$ become the "Frobenius (necessary and sufficient) conditions" for the holonomicity of the first-order nonlinear system $\omega_D \equiv f_D(t, q, \dot{q}) = 0$].

These results, under $(\delta q_k)^{\cdot} = \delta(\dot{q}_k)$, are depicted in fig. 7.2(a).

Mathematical Admissibility

Here, the constraints (7.6.1a) hold for both the fundamental orbit, or arc, I, as well as for its kinematically admissible adjacent arc $II = I + \delta I$; that is,

$$\omega_D(I) = 0 \qquad \textit{and} \qquad \omega_D(II) = \omega_D(I + \delta I) = 0, \tag{7.7.4a}$$

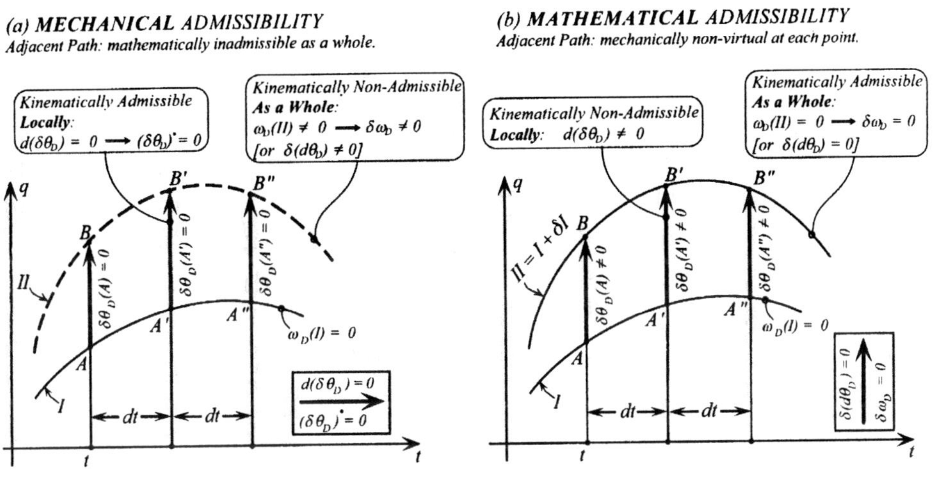

Figure 7.2 On the differences between (a) mechanical and (b) mathematical variations.

or

$$d\theta_D(I) = 0 \qquad and \qquad d\theta_D(II) = d\theta_D(I + \delta I) = 0, \tag{7.7.4b}$$

from which, expanding as before, and so on, we obtain the following requirements for the mathematical realizability of adjacent motions:

$$\omega_D + \delta\omega_D = 0 \;\Rightarrow\; \delta\omega_D = 0,$$

or

$$d\theta_D + \delta(d\theta_D) = 0 \quad\Rightarrow\quad \delta(d\theta_D) = 0 \qquad \text{(virtual variation of path admissibility).} \tag{7.7.4c}$$

Then, the transitivity equations [again under $(\delta q_k)^{\cdot} \neq \delta(\dot{q}_k)$] yield

$$\begin{aligned}(\delta\theta_D)^{\cdot} &= \sum (\partial\omega_D/\partial\dot{q}_k)\,[(\delta q_k)^{\cdot} - \delta(\dot{q}_k)] + \sum H^D{}_I\,\delta\theta_I \\ &= \sum (\partial\omega_D/\partial\dot{q}_k)\,[(\delta q_k)^{\cdot} - \delta(\dot{q}_k)] - \sum\sum V^k{}_I(\partial\omega_D/\partial\dot{q}_k)\,\delta\theta_I;\end{aligned} \tag{7.7.5}$$

$$\Big[= \sum (\partial\omega_D/\partial\dot{q}_k)\,[(\delta q_k)^{\cdot} - \delta(\dot{q}_k)] + \sum\sum (\gamma^D{}_{II'}\omega_{I'})\,\delta\theta_I, \quad \text{in stationary Pfaffian case}\Big] \neq 0, \text{ in general.} \tag{7.7.5a}$$

From the above we conclude the following:

- We cannot assume that both $(\delta q_k)^{\cdot} = \delta(\dot{q}_k)$ and $d(\delta\theta_D) = 0$, or $(\delta\theta_D)^{\cdot} = 0$, hold; it is either one or the other.
- If we assume that $(\delta q_k)^{\cdot} = \delta(\dot{q}_k)$, then $d(\delta\theta_D), (\delta\theta_D)^{\cdot} \neq 0$; unless $H^D{}_I = 0$. These results, under $(\delta q_k)^{\cdot} = \delta(\dot{q}_k)$, are depicted in fig. 7.2(b).

In sum: even assuming that $(\delta q_k)^{\cdot} = \delta(\dot{q}_k)$, due to the purely analytical transitivity equations:

- We cannot assume that both $\delta(d\theta_D) = 0$ (or $\delta\omega_D = 0$) and $d(\delta\theta_D) = 0$ [or $(\delta\theta_D)^{\cdot} = 0$].

- In *mechanics* $[d(\ldots)]$, assuming $(\delta q_k)^{\cdot} = \delta(\dot{q}_k)$, we have

$$d(\delta\theta_D) = 0 \quad \text{or} \quad (\delta\theta_D)^{\cdot} = 0 \quad (\text{and } d\omega_D = 0), \tag{7.7.6}$$

but $\delta(d\theta_D) \neq 0$ (or $\delta\omega_D \neq 0$), even though $d\theta_D = 0$ (or $\omega_D = 0$); in fact,

$$\begin{aligned} \delta\omega_D &= -\sum H^D{}_I\,\delta\theta_I = \sum\sum V^k{}_I(\partial\omega_D/\partial\dot{q}_k)\,\delta\theta_I \\ &= -\sum E_k(\omega_D)\,\delta q_k \equiv -\sum E_k(f_D)\,\delta q_k \equiv \delta f_D \qquad [\text{by } (5.2.22\text{a, b})] \\ &\neq 0, \quad \text{for } \textit{nonholonomic} \text{ constraints,} \\ &= 0, \quad \text{for } \textit{holonomic} \text{ constraints.} \end{aligned} \tag{7.7.7}$$

{The reader may verify without much difficulty that in the Pfaffian case

$$\omega_D \equiv \sum a_{Dk}\dot{q}_k + a_D = 0,$$

the above specialize to

$$\begin{aligned} \delta\omega_D &\equiv \delta f_D \equiv \delta\Big(\sum a_{Dk}\dot{q}_k + a_D\Big) \\ &= \cdots = -\sum\Big[\sum(\partial a_{Dk}/\partial q_b - \partial a_{Db}/\partial q_k)\dot{q}_b + (\partial a_{Dk}/\partial t - \partial a_D/\partial q_k)\Big]\,\delta q_k \\ &\Big[= -\sum\sum(\gamma^D{}_{II'}\omega_{I'})\,\delta\theta_I, \quad \text{in stationary Pfaffian case}\Big]\Big\}. \end{aligned} \tag{7.7.7a}$$

Similarly, we obtain the corresponding *independent* transitivity equations:

$$\delta\omega_I = (\delta\theta_I)^{\cdot} - \sum H^I{}_{I'}\,\delta\theta_{I'} = (\delta\theta_I)^{\cdot} + \sum\sum V^k{}_{I'}(\partial\omega_I/\partial\dot{q}_k)\,\delta\theta_{I'} \tag{7.7.8}$$

$$\Big[= (\delta\theta_I)^{\cdot} - \sum\sum(\gamma^I{}_{I'I''}\,\omega_{I''})\,\delta\theta_{I'}, \quad \text{in stationary Pfaffian case}\Big]. \tag{7.7.8a}$$

[What happens in mechanics if we assume that $(\delta q_k)^{\cdot} \neq \delta(\dot{q}_k)$, for some or all values of k, is examined in §7.8. (Recall conclusions of prob. 2.12.5.)]

- In *mathematics* $[\delta(\ldots)]$, assuming $(\delta q_k)^{\cdot} = \delta(\dot{q}_k)$, we have

$$\delta(d\theta_D) = 0 \quad \text{or} \quad \delta\omega_D = 0,$$

but $d(\delta\theta_D) \neq 0$ (or $(\delta\theta_D)^{\cdot} \neq 0$) and $d\omega_D \neq 0$, even though $\delta\theta_D = 0$ (or $\omega_D = 0$).

In words: adjacent paths obtained by adding mechanically admissible variations to every point of an orbit $\big[d[\delta\theta_D(t)] = 0\big]$ are, in general, *mathematically inadmissible* $\big[\delta[d\theta_D(t)] \neq 0\big]$.

Next, to the various integral formulae of mechanics.

The Central Equation, its Integral Forms, and Corresponding Equations of Motion

(i) Holonomic Variables

Let us now extend the above to the IVPs of the most general systems [nonpotential, rheonomic, nonlinearly nonholonomic (chap. 5)] under *vertical* variations (i.e., $\Delta t = 0$); we consider a system with kinetic energy $T = T(t, q, \dot{q})$, impressed forces

$Q_k = Q_k(t, q, \dot{q})$ [some of which, or all, may be partially or completely potential; i.e., $Q_k = -\partial V(t, q)/\partial q_k$], subject to the $m(< n)$ independent and generally nonholonomic velocity constraints

$$f_D(t, q, \dot{q}) = 0, \qquad rank(\partial f_D/\partial \dot{q}_k) = m \qquad [D = 1, \ldots, m;\ I = m + 1, \ldots, n], \tag{7.7.9a}$$

or, equivalently, solving for the m $\dot{q}_D$ in terms of the $n - m$ $\dot{q}_I$:

$$\dot{q}_D = \phi_D(t, q, \dot{q}_I) \qquad [\Rightarrow f_D \equiv \dot{q}_D - \phi_D(t, q, \dot{q}_I) = 0]; \tag{7.7.9b}$$

and whose virtual forms are, respectively,

$$\delta\theta_D \equiv \sum (\partial f_D/\partial \dot{q}_k)\, \delta q_k\ (= 0), \qquad \text{where}\ \dot{\theta}_D \equiv \omega_D \equiv f_D = 0, \tag{7.7.9c}$$

$$\delta q_D \equiv \sum (\partial \phi_D/\partial \dot{q}_I)\, \delta q_I\ (\neq 0). \tag{7.7.9d}$$

Here our discussion is based not on the equations of motion (§7.2–7.5), but on their equivalent principle of Lagrange (LP, §3.2):

$$\sum \{[(\partial T/\partial \dot{q}_k)^{\cdot} - \partial T/\partial q_k] - Q_k\}\, \delta q_k \equiv \sum [E_k(T) - Q_k]\, \delta q_k = 0, \tag{7.7.10}$$

where the n δq's are restricted by the m conditions (7.7.9c).

Integrating (7.7.10) between the arbitrary times t_1 and t_2, then integrating by parts, and so on, and using the familiar notation $\delta' W \equiv \sum Q_k\, \delta q_k$, we obtain

$$\int \Big\{\sum [(\partial T/\partial q_k)\, \delta q_k + (\partial T/\partial \dot{q}_k)\, (\delta q_k)^{\cdot}] + \delta' W\Big\}\, dt = \Big\{\sum (\partial T/\partial \dot{q}_k)\, \delta q_k\Big\}_1^2; \tag{7.7.11}$$

or, further, since by standard δ-differential calculus,

$$\delta T = \sum [(\partial T/\partial q_k)\, \delta q_k + (\partial T/\partial \dot{q}_k)\, \delta(\dot{q}_k)], \tag{7.7.11a}$$

adding and subtracting to the integrand of (7.7.11) $\sum (\partial T/\partial \dot{q}_k)\, \delta(\dot{q}_k)$ (in order to create δT there), we finally transform it to

$$\int \Big\{\delta T + \delta' W + \sum (\partial T/\partial \dot{q}_k)\, [(\delta q_k)^{\cdot} - \delta(\dot{q}_k)]\Big\}\, dt = \Big\{\sum (\partial T/\partial \dot{q}_k)\, \delta q_k\Big\}_1^2. \tag{7.7.12}$$

$\Big\{$Equation (7.7.12) can also result, most simply, by integration of the general central equation, in holonomic variables, (3.6.8):

$$\delta T + \delta' W + \sum (\partial T/\partial \dot{q}_k)\, [(\delta q_k)^{\cdot} - \delta(\dot{q}_k)] = d/dt\Big[\sum (\partial T/\partial \dot{q}_k)\, \delta q_k\Big]\Big\}. \tag{7.7.13}$$

This general integral equation is fundamental to all subsequent IVP considerations.

Now, and this is the crux of the matter, (7.7.12) makes clear that we must relate the $(\delta q)^{\cdot}$'s, with the $\delta(\dot{q})$'s; that is, we must invoke some kind of *transitivity equations*. This can also be seen as follows: since the $\delta q(t)$'s are well-defined functions of time, so are the $(\delta q)^{\cdot}$'s; *but the* $\delta(\dot{q})$'s *need defining*, something which, again, is equivalent to a choice of transitivity equations. There is no unique way to do this, but, since there is only one mechanics, *we should always end up with the same (correct) equations of motion.* As the virtual constraints (7.7.9c, d) show, since not all δq's are specified uniquely, there is a certain freedom in defining $\delta(\dot{q}) \equiv \delta\dot{q}$; and this, historically, has given rise to various seemingly contradictory IVPs. We shall return to this topic in §7.8.

(ii) Nonholonomic Variables

To obtain the *nonholonomic* variable equivalent of this most general integral formula, we substitute into it the most general transitivity equations (5.2.23a ff.):

$$(\delta q_k)^{\cdot} - \delta(\dot{q}_k) = \sum (\partial \dot{q}_k/\partial \omega_b)\,[(\delta\theta_b)^{\cdot} - \delta\omega_b] + \sum V^k{}_b\,\delta\theta_b \tag{7.7.14a}$$

$$= \sum (\partial \dot{q}_k/\partial \omega_b)\,[(\delta\theta_b)^{\cdot} - \delta\omega_b] - \sum\sum (\partial \dot{q}_k/\partial \omega_l)\,H^l{}_b\,\delta\theta_b, \tag{7.7.14b}$$

while recalling that, since $T = T(t, q, \dot{q}) = T[t, q, \dot{q}(t, q, \omega)] \equiv T^*(t, q, \omega) \equiv T^*$,

$$\delta T = \delta T^* = \sum [(\partial T^*/\partial q_k)\,\delta q_k + (\partial T^*/\partial \omega_k)\,\delta\omega_k]$$
$$= \sum [(\partial T^*/\partial \theta_k)\,\delta\theta_k + (\partial T^*/\partial \omega_k)\,\delta\omega_k]; \tag{7.7.14c}$$

also

$$P_b \equiv \partial T^*/\partial \omega_b = \sum (\partial T/\partial \dot{q}_k)\,(\partial \dot{q}_k/\partial \omega_b) = \sum (\partial \dot{q}_k/\partial \omega_b) p_k, \tag{7.7.14d}$$

and

$$\delta' W = \delta' W^* \equiv \sum \Theta_k\,\delta\theta_k \qquad \text{(definition of } \Theta_k\text{'s)}. \tag{7.7.14e}$$

Thus, we transform (7.7.12) into its nonholonomic counterparts (no constraint enforcement yet!):

$$\int \Big\{ \delta T^* + \delta' W^* + \sum P_k[(\delta\theta_k)^{\cdot} - \delta\omega_k] + \sum\sum V^k{}_b\,p_k\,\delta\theta_b \Big\}\,dt$$
$$= \Big\{ \sum (\partial T^*/\partial \omega_k)\,\delta\theta_k \Big\}_1^2 \quad \Big[= \Big\{ \sum (\partial T/\partial \dot{q}_k)\,\delta q_k \Big\}_1^2 \Big], \tag{7.7.15a}$$

$$\int \Big\{ \delta T^* + \delta' W^* + \sum P_k[(\delta\theta_k)^{\cdot} - \delta\omega_k] - \sum\sum H^k{}_b P_k\,\delta\theta_b \Big\}\,dt$$
$$= \Big\{ \sum (\partial T^*/\partial \omega_k)\,\delta\theta_k \Big\}_1^2. \tag{7.7.15b}$$

As with (7.7.12), this can also be achieved (a) either by transforming (7.7.13) into quasi variables, via (7.7.14a–d), and then integrating, or (b) by integrating the earlier-found general central equation in quasi variables (3.6.9).

Next, to go from (7.7.15a, b) to the general nonlinear T^*-based equations of motion (§5.3), and thus establish the complete equivalence of the former with the latter, we employ the following general kinematico-inertial transformation of integral variational mechanics [obtained most easily by use of (7.7.14c, d) and integration by parts, and so on; also, recalling (7.5.2b)]:

$$\int\Big\{\delta T^* + \sum P_k[(\delta\theta_k)^\cdot - \delta\omega_k]\Big\}\,dt$$

$$= \int \sum \big[(\partial T^*/\partial\omega_k)\,(\delta\theta_k)^\cdot + (\partial T^*/\partial\theta_k)\,\delta\theta_k\big]\,dt$$

$$= \cdots = \Big\{\sum P_k\,\delta\theta_k\Big\}_1^2 - \int \sum \big[(\partial T^*/\partial\omega_k)^\cdot - \partial T^*/\partial\theta_k\big]\,\delta\theta_k\,dt. \tag{7.7.15c}$$

As a result of this, (7.7.15a, b) are immediately transformed to the following *time-integral forms* of the nonlinear Johnsen–Hamel nonholonomic equations of motion (5.3.5a, b; 8a, b):

$$-\int \sum \big[(\partial T^*/\partial\omega_k)^\cdot - \partial T^*/\partial\theta_k - \sum\sum (\partial T/\partial\dot q_b)^* V^b{}_k - \Theta_k\big]\,\delta\theta_k\,dt = 0, \tag{7.7.16a}$$

$$-\int \sum \big[(\partial T^*/\partial\omega_k)^\cdot - \partial T^*/\partial\theta_k + \sum\sum (\partial T^*/\partial\omega_b) H^b{}_k - \Theta_k\big]\,\delta\theta_k\,dt = 0; \tag{7.7.16b}$$

from which (and the method of Lagrangean multipliers) both kinetic and kinetostatic equations of mechanics follow at once.

REMARKS

(i) Had we assumed in (7.7.12, 13; 14a, b) that $(\delta q_k)^\cdot = \delta(\dot q_k)$, *whether the* δq_k *are further constrained or not* [what we shall, henceforth, call the viewpoint of *Hölder* (1896)–*Voronets* (1901)–*Hamel* (1904)], then eqs. (7.7.12; 15a, b) would have led us to the earlier form,

$$\int (\delta T + \delta' W)\,dt = \int (\delta T^* + \delta' W^*)\,dt = \Big\{\sum (\partial T^*/\partial\omega_k)\,\delta\theta_k\Big\}_1^2, \tag{7.7.17}$$

where now

$$\begin{aligned}\delta T^* &= \sum [(\partial T^*/\partial\theta_k)\,\delta\theta_k + (\partial T^*/\partial\omega_k)\,\delta\omega_k] \qquad \text{[invoking (7.7.7–7.7.8a)]}\\ &= \sum \Big\{(\partial T^*/\partial\theta_k)\,\delta\theta_k + (\partial T^*/\partial\omega_k)\,\Big[(\delta\theta_k)^\cdot - \sum H^k{}_b\,\delta\theta_b\Big]\Big\}\\ &= \sum (\partial T^*/\partial\theta_k)\,\delta\theta_k + \sum \{[(\partial T^*/\partial\omega_k)\,\delta\theta_k]^\cdot - [(\partial T^*/\partial\omega_k)^\cdot\,\delta\theta_k]\}\\ &\qquad - \sum\sum H^b{}_k(\partial T^*/\partial\omega_b)\,\delta\theta_k,\end{aligned} \tag{7.7.17a}$$

so that, upon time integration, and so on,

$$\int \delta T^*\,dt = -\int \sum \Big[(\partial T^*/\partial\omega_k)^\cdot - \partial T^*/\partial\theta_k + \sum H^b{}_k(\partial T^*/\partial\omega_b)\Big]\,\delta\theta_k\,dt + \Big\{\sum P_k\,\delta\theta_k\Big\}_1^2, \tag{7.7.17b}$$

[which is none other than (7.5.2b)], and similarly in terms of the $V^b{}_k$; and these results would have, obviously, transformed (7.7.17) into the earlier (7.7.16a, b); that is, *whether we assume that* $(\delta q_k)^\cdot = \delta(\dot q_k)$ *or not, if we are internally consistent we will end up with the same equations of motion*, just like in the derivation of the equation of motion from the central equation (§3.6, §5.3).

In sum: To obtain the correct equations of motion of nonholonomic systems via a Hamiltonian IVP, we must stick with *mechanics* (i.e., Lagrange's principle) and *sacrifice mathematics* (i.e., calculus of variations)!

(ii) The earliest explicit realization of this fact—that is, that mechanical variational principles do not have to coincide with mathematical variational principles, and that *in all cases of discrepancy the former override the latter*, seems to be due to Voss (1885, second footnote, p. 266): "Aber das Hamilton'sche Princip [i.e., our mathematical variational principle] ist überhaupt kein eigentliches Princip der Mechanik, sondern hat—wenigstens zunächst—für dieselbe nur den Charakter einer analytischen Regel, welche *auch* die Differentialgleichungen der Bewegung liefert." Freely translated as "Hamilton's principle [as originally and commonly understood; i.e., as a mathematical variational principle] is not, generally, a proper principle of mechanics [i.e., like LP], but—at least for the moment—only a mathematical rule that also yields the equations of motion." See also Maurer (1905).

This point cannot be emphasized enough. Most mechanics texts we are aware of, including almost all contemporary ones in English, promote the *false* notion that the basic principle of (at least) potential but possibly nonholonomic systems is Hamilton's mathematical variational principle, eqs. (7.6.5a, b), or some variant of it (§7.9)! But even the few classy exceptions to this broad indictment, e.g., Rosenberg (1977, pp. x, 171 ff.) do *not pinpoint to the source of the discrepancy*, which is the transitivity equations (a topic absent from practically all English language texts on mechanics!).

Constrained Forms of the Integral "Principles"

Let us find the forms that (7.7.15a, b) (and hence, ultimately, the resulting equations of motion) assume if, in there, we enforce both $\delta\theta_D = 0$ (admissible displacements) and $\omega_D = 0$ (admissible paths); that is, let us express their integrands in terms of the *constrained* kinetic energy

$$T^* = T^*(t, q, \omega_D, \omega_I) \rightarrow T^*(t, q, \omega_D = 0, \omega_I) = T^*_o(t, q, \omega_I) = T^*_o; \qquad (7.7.18)$$

and its variations, and so on. Then, since $(\partial T^*/\partial\omega_I)_o = \partial T^*_o/\partial\omega_I$ [recalling (3.5.24a ff.)], $(\partial T^*/\partial q_k)_o = \partial T^*_o/\partial q_k \Rightarrow (\partial T^*/\partial\theta_k)_o = \partial T^*_o/\partial\theta_k$, and invoking (7.7.7) and (7.7.7a), we find, successively,

$$\begin{aligned}
\delta T^* &= \sum [(\partial T^*/\partial q_k)\,\delta q_k + (\partial T^*/\partial\omega_k)\,\delta\omega_k] \\
&= \Big[\sum (\partial T^*/\partial q_k)\,\delta q_k + \sum (\partial T^*/\partial\omega_I)\,\delta\omega_I\Big] + \sum (\partial T^*/\partial\omega_D)\,\delta\omega_D \\
&\equiv \delta T^*_o + \sum (\partial T^*/\partial\omega_D)_o\,\delta\omega_D && (7.7.18a) \\
&= \delta T^*_o + \sum (\partial T^*/\partial\omega_D)_o\Big(-\sum H^D{}_I\,\delta\theta_I\Big) && (7.7.18b) \\
&= \delta T^*_o + \sum (\partial T^*/\partial\omega_D)_o\Big[\sum\sum V^k{}_I(\partial\omega_D/\partial\dot{q}_k)\,\delta\theta_I\Big] \\
&= \delta T^*_o + \sum\sum (\partial T/\partial\dot{q}_k)_o V^k{}_I\,\delta\theta_I && (7.7.18c)
\end{aligned}$$

$$= \delta T^*{}_o + \sum (\partial T^*/\partial \omega_D)_o \Big[-\sum E_k(\omega_D)\, \delta q_k \Big] \tag{7.7.18d}$$

$$\Big\{ = \delta T^*{}_o + \sum (\partial T^*/\partial \omega_D)_o \Big[-\sum\sum (\gamma^D{}_{II'}\, \omega_{I'})\, \delta\theta_I \Big],$$

$$\text{for stationary Pfaffian constraints}\Big\}, \tag{7.7.18e}$$

where, invoking (7.7.8) and (7.7.8a),

$$\delta T^*{}_o \equiv \delta(T^*{}_o) \equiv \sum (\partial T^*{}_o/\partial q_k)\, \delta q_k + \sum (\partial T^*{}_o/\partial \omega_I)\, \delta\omega_I$$

$$\equiv \sum (\partial T^*{}_o/\partial \theta_I)\, \delta\theta_I + \sum (\partial T^*{}_o/\partial \omega_I)\, \delta\omega_I \tag{7.7.18f}$$

$$= \sum (\partial T^*{}_o/\partial \theta_I)\, \delta\theta_I + \sum (\partial T^*{}_o/\partial \omega_I) \Big[(\delta\theta_I)^{\cdot} - \sum H^I{}_{I'}\, \delta\theta_{I'} \Big] \tag{7.7.18g}$$

$$= \sum (\partial T^*{}_o/\partial \theta_I)\, \delta\theta_I + \sum (\partial T^*{}_o/\partial \omega_I) \Big[(\delta\theta_I)^{\cdot} + \sum\sum V^k{}_{I'}(\partial \omega_I/\partial \dot{q}_k)\, \delta\theta_{I'} \Big] \tag{7.7.18h}$$

$$\Big\{ = \sum (\partial T^*{}_o/\partial \theta_I)\, \delta\theta_I + \sum (\partial T^*{}_o/\partial \omega_I) \Big[(\delta\theta_I)^{\cdot} - \sum\sum (\gamma^I{}_{I'I''}\, \omega_{I''})\, \delta\theta_{I'} \Big],$$

$$\text{for stationary Pfaffian constraints}\Big\}; \tag{7.7.18i}$$

and so, inserting (7.7.18c) into (7.7.15a) and (7.7.18b, d) into (7.7.15b), we obtain, respectively, the following *general nonlinear/Pfaffian and constrained form of the Hamilton-type "principles"*:

$$\int \Big[\delta T^*{}_o + \sum\sum (\partial T/\partial \dot{q}_k)^*{}_o V^k{}_I\, \delta\theta_I + \delta' W^*{}_o \Big]\, dt = \Big\{ \sum (\partial T/\partial \dot{q}_k)\, \delta q_k \Big\}_1^2; \tag{7.7.19a}$$

$$\int \Big[\delta T^*{}_o - \sum\sum (\partial T^*/\partial \omega_D)_o\, H^D{}_I\, \delta\theta_I + \delta' W^*{}_o \Big]\, dt = \Big\{ \sum (\partial T/\partial \dot{q}_k)\, \delta q_k \Big\}_1^2; \tag{7.7.19b}$$

$$\int \Big[\delta T^*{}_o - \sum\sum (\partial T^*/\partial \omega_D)_o \gamma^D{}_{II'}\, \omega_{I'}\, \delta\theta_I + \delta' W^*{}_o \Big]\, dt = \Big\{ \sum (\partial T/\partial \dot{q}_k)\, \delta q_k \Big\}_1^2; \tag{7.7.19c}$$

where

$$\delta' W = \delta' W^* = \sum \Theta_k\, \delta\theta_k = \sum \Theta_I\, \delta\theta_I \equiv \delta' W^*{}_o.$$

Let the reader verify that inserting (7.7.18g, h) into (7.7.19b, a) leads readily to the $n - m$ *kinetic* equations of motion (5.3.8d, 8c), respectively; and similarly for the Pfaffian case [(3.5.24a ff.)].

REMARKS

(i) Equation (7.7.19c) is due to Hamel (1949, pp. 494–495; and references cited therein).

(ii) The form (7.7.19b) can also be obtained directly from (7.7.15b) if, following Hamel (1949, p. 494—Pfaffian case], we choose, in the latter,

$$\text{(a)} \qquad \delta\theta_D = 0, \quad d(\delta\theta_D) = 0 \quad \Rightarrow \quad (\delta\theta_D)^{\cdot} = 0, \quad \textit{and} \quad (\delta\theta_I)^{\cdot} = \delta\omega_I, \qquad (7.7.20a)$$

something that does not restrict the $\delta\theta_I$ but *constitutes a suitable/permissible definition of the* $\delta\omega_I$ (like a Suslov et al. *second* transitivity choice but in nonholonomic variables—see next section). Indeed, then we have, successively,

$$\text{(b)} \qquad \delta T^* + \sum P_k[(\delta\theta_k)^{\cdot} - \delta\omega_k] = \delta T^* + \sum P_D[(\delta\theta_D)^{\cdot} - \delta\omega_D]$$

$$= \delta T^* + \sum P_D[0 - \delta\omega_D] \qquad \text{[invoking (7.7.18a)]}$$

$$\equiv \delta T^*{}_o + \sum P_D\,\delta\omega_D + \left(-\sum P_D\,\delta\omega_D\right) = \delta T^*{}_o; \qquad (7.7.20b)$$

and [recalling (7.7.8)]

$$\text{(c)} \qquad -\sum\sum H'{}_{I}P_{I'}\,\delta\theta_I = -\sum P_I[(\delta\theta_I)^{\cdot} - \delta\omega_I] = 0. \qquad (7.7.20c)$$

As a result of the above, (7.7.15b) (with $b \to I, k \to D$), clearly, becomes (7.7.19b), Q.E.D. And, similarly, for the reduction of (7.7.15a) to (7.7.19a).

7.8 SPECIAL INTEGRAL VARIATIONAL PRINCIPLES (OF SUSLOV, VORONETS, et al.)

Occasionally, the constraints are given in the form

$$\dot{q}_D \equiv \phi_D(t, q, \dot{q}_I). \qquad (7.8.1)$$

But this, as explained in §5.2, can be viewed as the following special case of (7.6.1a, b):

$$\omega_D \equiv f_D(t, q, \dot{q}) \equiv \dot{q}_D - \phi_D(t, q, \dot{q}_I) = 0, \qquad \omega_I \equiv f_D(t, q, \dot{q}) \equiv \dot{q}_I \neq 0; \qquad (7.8.1a)$$

$$[\Rightarrow \dot{q}_D = \omega_D + \phi_D[t, q, \dot{q}_I(t, q, \omega_I)] = \omega_D + \phi_D(t, q, \omega_I)] \qquad (7.8.1b)$$

$$\delta\theta_D = \sum (\partial\omega_D/\partial\dot{q}_k)\,\delta q_k = \cdots = \delta q_D - \sum (\partial\phi_D/\partial\dot{q}_I)\,\delta q_I = 0, \qquad (7.8.1c)$$

$$\delta\theta_I = \sum (\partial\omega_I/\partial\dot{q}_k)\,\delta q_k = \cdots = \delta q_I \neq 0; \qquad (7.8.1d)$$

and, therefore,

$$\delta\omega_D \equiv \delta[\dot{q}_D - \phi_D(t, q, \dot{q})] = ? \qquad (7.8.2)$$

The above shows that to make further progress (and as mentioned at the end of §7.6), *we must define the* $\delta(\dot{q}_D)$. From the many conceivable such transitivity choices, the following two have dominated the literature:

(i) First transitivity choice of Hölder (1896)–Voronets (1901)–Hamel (1904). As already seen, this is

$$(\delta q_k)^{\cdot} = \delta(\dot{q}_k), \qquad \text{for } \textit{all } \delta q\text{'s, constrained or not;} \qquad (7.8.3a)$$

(ii) Second transitivity choice of Suslov (1901)–Levi-Civita/Amaldi (1920s)–Rumiantsev (1970s)–Greenwood (1990s). (See also Suslov, 1946, pp. 596–600; and Rumiantsev, 1978, 1979.) This stipulates that

$$(\delta q_I)^{\cdot} = \delta(\dot{q}_I), \qquad \text{but} \qquad (\delta q_D)^{\cdot} \neq \delta(\dot{q}_D). \tag{7.8.3b}$$

Let us find the corresponding transitivity equations and integral variational "principles."

First Transitivity Choice

With its help, and (7.8.1c), (7.8.2) yields, successively,

$$\begin{aligned}
0 \neq \delta\omega_D &\equiv \delta(\dot{q}_D) - \delta\phi_D = (\delta q_D)^{\cdot} - \delta\phi_D \\
&= \left[\sum (\partial\phi_D/\partial\dot{q}_I)\,\delta q_I\right]^{\cdot} - \left[\sum (\partial\phi_D/\partial q_k)\,\delta q_k + \sum (\partial\phi_D/\partial\dot{q}_I)\,\delta(\dot{q}_I)\right] \\
&= \sum \left[(\partial\phi_D/\partial\dot{q}_I)^{\cdot}\,\delta q_I + (\partial\phi_D/\partial\dot{q}_I)\,(\delta q_I)^{\cdot}\right] \\
&\qquad - \left[\sum (\partial\phi_D/\partial q_I)\,\delta q_I + \sum (\partial\phi_D/\partial q_{D'})\,\delta q_{D'} + \sum (\partial\phi_D/\partial\dot{q}_I)\,\delta(\dot{q}_I)\right] \\
&= \sum \left[(\partial\phi_D/\partial\dot{q}_I)^{\cdot} - \partial\phi_D/\partial q_I - \sum (\partial\phi_D/\partial q_{D'})\,(\partial\phi_{D'}/\partial\dot{q}_I)\right]\delta q_I \\
&\equiv \sum \left[(\partial\phi_D/\partial\dot{q}_I)^{\cdot} - \partial\phi_D/\partial(q_I)\right]\delta q_I \\
&\equiv \sum \left[E_I(\phi_D) - \sum (\partial\phi_D/\partial q_{D'})\,(\partial\phi_{D'}/\partial\dot{q}_I)\right]\delta q_I \\
&\equiv \sum E_{(I)}(\phi_D)\,\delta q_I \qquad \text{[recalling the special notations of ex. 5.2.3)} \\
&\equiv \sum W^D{}_I\,\delta q_I
\end{aligned} \tag{7.8.4a}$$

$$\left[= \sum W^D{}_I\,\delta\theta_I, \quad W^D{}_I\text{: specialization of the } V^D{}_I, \text{ for (7.8.1a)}\right] \tag{7.8.4b}$$

$$\Rightarrow\ \delta(\dot{q}_D) = (\delta\dot{q}_D) = \delta\phi_D + \sum E_{(I)}(\phi_D)\,\delta q_I \equiv \delta\phi_D + \sum W^D{}_I\,\delta q_I. \tag{7.8.4c}$$

[In the Pfaffian specialization of (7.8.1):

$$\dot{q}_D = \sum b_{DI}(t, q)\dot{q}_I + b_D(t, q) \equiv \phi_D(t, q, \dot{q}_I), \tag{7.8.4d}$$

the $E_{(I)}(\phi_D) \equiv W^D{}_I$ reduce to [ex. 2.12.1, probs. 2.12.2, 2.12.5; (3.8.14g, 14h)]

$$\begin{aligned}
v^D{}_I &\equiv \sum w^D{}_{II'}\dot{q}_{I'} + w^D{}_I \\
&\equiv \sum \Big\{(\partial b_{DI}/\partial q_{I'} - \partial b_{DI'}/\partial q_I) + \sum \left[(\partial b_{DI}/\partial q_{D'})b_{D'I'} - (\partial b_{DI'}/\partial q_{D'})b_{D'I}\right]\Big\}\dot{q}_{I'} \\
&\quad + \Big\{(\partial b_{DI}/\partial t - \partial b_D/\partial q_I) + \sum \left[(\partial b_{DI}/\partial q_{D'})b_{D'} - (\partial b_D/\partial q_{D'})\,b_{D'I}\right]\Big\},
\end{aligned} \tag{7.8.4e}$$

where the $w^D{}_{II'}, w^D{}_I$ are the Pfaffian (linear) Voronets coefficients.]

We remark that (i) these are none other than the earlier transitivity equations (5.2.22a, b) (also, recall results of ex. 5.2.3); and (ii) as shown in prob. 2.12.5, the first choice implies that

$$0 = (\delta\theta_D)^{\cdot} \neq \delta\omega_D\,(\neq 0), \qquad \text{but} \qquad (\delta\theta_I)^{\cdot} = \delta\omega_I. \tag{7.8.4f}$$

Let the reader verify that, as a result of the second of (7.8.4f), we can replace in the earlier integral equations (7.7.19b, c), the index D (and corresponding summation from 1 to m) with k (with summation from 1 to n). [*Hint*: Invoke (7.7.8); also, recall remark (ii) at end of §7.7.]

Integral Variational Principle Corresponding to the First Transitivity Choice

In this case, (7.7.12) reduces to the familiar (unconstrained) Hamiltonian form:

$$\int (\delta T + \delta' W)\, dt = \left\{\sum (\partial T/\partial \dot{q}_k)\,\delta q_k\right\}_1^2. \tag{7.8.5}$$

Let us find the constrained form of (7.8.5). Applying the chain rule to

$$T = T(t, q, \dot{q}) = T[t, q, \dot{q}_I, \phi_D(t, q, \dot{q}_I)] \equiv T_o(t, q, \dot{q}_I) \equiv T_o, \tag{7.8.5a}$$

we readily get

$$\partial T_o/\partial q_k = \partial T/\partial q_k + \sum (\partial T/\partial \dot{q}_D)\,(\partial \phi_D/\partial q_k), \tag{7.8.5b}$$

$$\partial T_o/\partial \dot{q}_I = \partial T/\partial \dot{q}_I + \sum (\partial T/\partial \dot{q}_D)\,(\partial \phi_D/\partial \dot{q}_I), \tag{7.8.5c}$$

and so we find, successively,

$$\begin{aligned}
\delta T &= \sum [(\partial T/\partial q_k)\,\delta q_k + (\partial T/\partial \dot{q}_k)\,\delta(\dot{q}_k)] \\
&= \sum \left[\partial T_o/\partial q_k - \sum (\partial T/\partial \dot{q}_D)\,(\partial \phi_D/\partial q_k)\right] \delta q_k \\
&\quad + \sum \left[\partial T_o/\partial \dot{q}_I - \sum (\partial T/\partial \dot{q}_D)\,(\partial \phi_D/\partial \dot{q}_I)\right] \delta(\dot{q}_I) + \sum (\partial T/\partial \dot{q}_D)\,\delta(\dot{q}_D) \\
&= \sum (\partial T_o/\partial q_k)\,\delta q_k + \sum (\partial T_o/\partial \dot{q}_I)\,\delta(\dot{q}_I) \\
&\quad + \sum (\partial T/\partial \dot{q}_D) \left\{\delta(\dot{q}_D) - \left[\sum (\partial \phi_D/\partial q_k)\,\delta q_k + \sum (\partial \phi_D/\partial \dot{q}_I)\,\delta(\dot{q}_I)\right]\right\} \\
&\equiv \delta T_o + \sum (\partial T/\partial \dot{q}_D)\,[\delta(\dot{q}_D) - \delta\phi_D] \qquad [\text{where } \delta T_o \equiv \delta(T_o)] \\
&= \delta T_o + \sum (\partial T/\partial \dot{q}_D)\,\delta(\dot{q}_D - \phi_D) \qquad [\text{invoking (7.8.4a)}] \\
&= \delta T_o + \sum (\partial T/\partial \dot{q}_D)\,\delta\omega_D \qquad (\text{where } \delta\omega_D \neq 0) \\
&= \delta T_o + \sum\sum (\partial T/\partial \dot{q}_D)_o W^D{}_I\,\delta q_I,
\end{aligned} \tag{7.8.5d}$$

and, similarly,

$$\delta' W \equiv \sum Q_k\,\delta q_k = \cdots = \sum \left[Q_I + \sum (\partial \dot{q}_D/\partial \dot{q}_I) Q_D\right] \delta q_I \equiv \sum Q_{Io}\,\delta q_I \equiv \delta' W_o; \tag{7.8.5e}$$

and, therefore, substituting these expressions into Hamilton's principle (7.8.5), we obtain *Voronets' (constrained form of Hamilton's) principle* [1901, for the special Pfaffian constraints (7.8.4d)]:

$$\int \Big\{ \delta T_o + \sum (\partial T/\partial \dot{q}_D)_o [\delta(\dot{q}_D) - \delta\phi_D] + \delta' W_o \Big\}\, dt$$

$$= \int \Big\{ \delta T_o + \sum\sum (\partial T/\partial \dot{q}_D)_o W^D{}_I\, \delta q_I + \delta' W_o \Big\}\, dt = \Big\{ \sum (\partial T/\partial \dot{q}_k)\, \delta q_k \Big\}_1^2; \quad (7.8.6)$$

that is, the *middle* term in the integrand equals the *correction* term: $\delta T_{\text{unconstrained}} - \delta T_{\text{constrained}} \equiv \delta T - \delta T_o$:

$$\int \delta T\, dt = \int \delta T_o\, dt + \int (\delta T - \delta T_o)\, dt = \cdots.$$

Let the reader verify that (7.8.6) also results as the (7.8.1a–2)-based specialization of the earlier general nonholonomic variational equations (7.7.19a, b).

Second Transitivity Choice

In this case, (7.8.2) becomes

$$\delta\omega_D = \delta(\dot{q}_D) - \delta\phi_D = 0 \Rightarrow \delta(\dot{q}_D) = \delta\phi_D \qquad [\text{definition of } \delta(\dot{q}_D)], \qquad (7.8.7a)$$

and so we obtain, successively,

$$(\delta q_D)^{\cdot} - \delta(\dot{q}_D) = \Big[\sum (\partial\phi_D/\partial\dot{q}_I)\, \delta q_I\Big]^{\cdot} - \delta\phi_D \qquad [\text{as in (7.8.4a), and with first of (7.8.3b)}]$$

$$= \cdots = \sum W^D{}_I\, \delta q_I \quad (\neq 0, \text{ in general}); \qquad (7.8.7b)$$

that is,

$$(\delta q_D)^{\cdot} = \delta(\dot{q}_D) + \sum W^D{}_I\, \delta q_I = \delta\phi_D + \sum W^D{}_I\, \delta q_I. \qquad (7.8.7c)$$

Again: (a) these are none other than the earlier transitivity equations (5.2.22a, b) [also, recall results of ex. 5.2.3]; and (b) as shown in prob. 2.12.5, the second choice implies that

$$(\delta\theta_k)^{\cdot} = \delta\omega_k \ (= 0); \qquad (7.8.7d)$$

that is, both mechanical admissibility $[(\delta\theta_D)^{\cdot} = 0]$ and mathematical admissibility $[\delta\omega_D = 0]$.

Alternative Derivations of (7.8.7b)

(i) Enforcing the constraints $\delta\theta_D = 0$ into the general transitivity equations (7.7.14a):

$$(\delta q_k)^{\cdot} - \delta(\dot{q}_k) = \sum (\partial\dot{q}_k/\partial\omega_b)\, [(\delta\theta_b)^{\cdot} - \delta\omega_b] + \sum V^k{}_b\, \delta\theta_b, \qquad (7.8.7e)$$

we obtain

$$(\delta q_D)^{\cdot} - \delta(\dot{q}_D) = \sum (\partial \dot{q}_D/\partial \omega_k)\,[(\delta\theta_k)^{\cdot} - \delta\omega_k] + \sum V^D{}_I\,\delta\theta_I \quad \text{[invoking (7.8.7d)]}$$
$$= \sum V^D{}_I\,\delta\theta_I = \sum W^D{}_I\,\delta q_I \quad \text{[recalling (7.8.1d); i.e., (7.8.7.b)].} \tag{7.8.7f}$$

(ii) Varying $\omega_D \equiv f_D(t, q, \dot{q}) = 0$ formally, we obtain

$$\delta f_D = \sum (\partial f_D/\partial q_k)\,\delta q_k + \sum (\partial f_D/\partial \dot{q}_k)\,\delta(\dot{q}_k)$$
$$= \sum (\partial f_D/\partial q_k)\,\delta q_k + \sum (\partial f_D/\partial \dot{q}_I)\,\delta(\dot{q}_I) + \sum (\partial f_D/\partial \dot{q}_{D'})\,\delta(\dot{q}_{D'})$$
$$= \sum (\partial f_D/\partial q_k)\,\delta q_k + \sum (\partial f_D/\partial \dot{q}_I)\,(\delta q_I)^{\cdot} + \sum (\partial f_D/\partial \dot{q}_{D'})\,[(\delta q_{D'})^{\cdot} + S_{D'}]$$
$$\text{[where } \delta(\dot{q}_D) - (\delta q_D)^{\cdot} \equiv S_D \quad \text{(Suslov term)} \neq 0]$$
$$= \sum (\partial f_D/\partial q_k)\,\delta q_k + \sum (\partial f_D/\partial \dot{q}_k)\,(\delta q_k)^{\cdot} + \sum (\partial f_D/\partial \dot{q}_{D'}) S_{D'}$$
$$= \sum (\partial f_D/\partial q_k)\,\delta q_k + \left\{\sum [(\partial f_D/\partial \dot{q}_k)\,\delta q_k]^{\cdot} - \sum (\partial f_D/\partial \dot{q}_k)^{\cdot}\,\delta q_k\right\}$$
$$+ \sum (\partial f_D/\partial \dot{q}_{D'}) S_{D'} \quad \text{[by first of (7.8.1c), (7.8.7d), the } \textit{second} \text{ sum vanishes],}$$

or

$$0 = \delta\omega_D \equiv \delta f_D = -\sum \left[(\partial f_D/\partial \dot{q}_k)^{\cdot} - \partial f_D/\partial q_k\right] \delta q_k + \sum (\partial f_D/\partial \dot{q}_{D'}) S_{D'};$$

that is, finally,

$$\sum (\partial f_D/\partial \dot{q}_{D'})\,[\delta(\dot{q}_{D'}) - (\delta q_{D'})^{\cdot}] = \sum E_k(f_D)\,\delta q_k. \tag{7.8.7g}$$

But, here, $f_D \equiv \dot{q}_D - \phi_D(t, q, \dot{q}_I)$, and so the left side of (7.8.7g) specializes to

$$\sum (\delta_{DD'})\,[\delta(\dot{q}_{D'}) - (\delta q_{D'})^{\cdot}] = \delta(\dot{q}_D) - (\delta q_D)^{\cdot}, \tag{7.8.7h}$$

while its right side reduces, successively, to (independently of any transitivity assumptions)

$$\sum E_k(f_D)\,\delta q_k = \sum E_I(f_D)\,\delta q_I + \sum E_{D'}(f_D)\,\delta q_{D'}$$
$$= \sum E_I(f_D)\,\delta q_I + \sum E_{D'}(f_D)\left[\sum (\partial \phi_{D'}/\partial \dot{q}_I)\,\delta q_I\right]$$
$$= \sum \left[E_I(f_D) + \sum E_{D'}(f_D)\,(\partial \phi_{D'}/\partial \dot{q}_I)\right] \delta q_I$$
$$= \sum \left[-E_I(\phi_D) - \sum (\partial \phi_D/\partial q_{D'})\,(\partial \phi_{D'}/\partial \dot{q}_I)\right] \delta q_I$$
$$\equiv -\sum E_{(I)}(\phi_D)\,\delta q_I \equiv -\sum W^D{}_I\,\delta q_I, \tag{7.8.7i}$$

recalling ex. 5.2.3; that is, eq. (7.8.7g) specializes to

$$(\delta q_D)^{\cdot} - \delta(\dot{q}_D) = \sum E_{(I)}(\phi_D)\,\delta q_I \equiv \sum W^D{}_I\,\delta q_I; \quad \text{i.e. (7.8.7b).} \tag{7.8.7j}$$

Integral Variational Principle Corresponding to the Second Transitivity Choice

In this case, due to the preceding results, (7.7.12) reduces to *Suslov's (unconstrained form of Hamilton's) principle* [1901, for the special Pfaffian constraints (7.8.4d)]:

$$\int \Big\{\delta T + \sum (\partial T/\partial \dot{q}_D)_o [(\delta q_D)^{\cdot} - \delta(\dot{q}_D)] + \delta' W\Big\}\, dt$$

$$= \int \Big\{\delta T + \sum\sum (\partial T/\partial \dot{q}_D)_o W^D{}_I\, \delta q_I + \delta' W\Big\}\, dt = \Big\{\sum (\partial T/\partial \dot{q}_k)\, \delta q_k\Big\}_1^2. \qquad (7.8.8)$$

[Suslov called the above "the modification of the d'Alembert principle"; and added, correctly, that our eq. (7.8.8) "in no way represents Hamilton's principle" (i.e., a principle of stationary action, à la variational calculus).]

Let us find the constrained form of (7.8.8); the "Suslovian counterpart of the Voronetsian (7.8.6)." Invoking again (7.8.5a, b, c) and the second of (7.8.7a), consequence of the second transitivity choice, we find, successively,

$$\delta T = \sum (\partial T/\partial q_I)\, \delta q_I + \sum (\partial T/\partial q_D)\, \delta q_D + \sum (\partial T/\partial \dot{q}_I)\, \delta(\dot{q}_I) + \sum (\partial T/\partial \dot{q}_D)\, \delta(\dot{q}_D)$$

$$= \sum \Big[\partial T_o/\partial q_I - \sum (\partial T/\partial \dot{q}_{D'})\, (\partial \phi_{D'}/\partial q_I)\Big]\, \delta q_I$$

$$+ \sum \Big[\partial T_o/\partial q_D - \sum (\partial T/\partial \dot{q}_{D'})\, (\partial \phi_{D'}/\partial q_D)\Big]\, \delta q_D$$

$$+ \sum \Big[\partial T_o/\partial \dot{q}_I - \sum (\partial T/\partial \dot{q}_{D'})\, (\partial \phi_{D'}/\partial \dot{q}_I)\Big]\, \delta(\dot{q}_I)$$

$$+ \sum (\partial T/\partial \dot{q}_D)\, \delta(\dot{q}_D)$$

$$= \sum (\partial T_o/\partial q_I)\, \delta q_I + \sum (\partial T_o/\partial q_D)\, \delta q_D$$

$$+ \sum (\partial T_o/\partial \dot{q}_I)\, \delta(\dot{q}_I) + \sum (\partial T/\partial \dot{q}_D)\, \delta(\dot{q}_D)$$

$$- \sum (\partial T/\partial \dot{q}_{D'}) \Big[\sum (\partial \phi_{D'}/\partial q_I)\, \delta q_I + \sum (\partial \phi_{D'}/\partial q_D)\, \delta q_D$$

$$+ \sum (\partial \phi_{D'}/\partial \dot{q}_I)\, \delta(\dot{q}_I)\Big]$$

[by (7.8.7a), the *fourth* and *last* sum, which equals $-\sum (\partial T/\partial \dot{q}_{D'})\, \delta\phi_{D'}$, cancel with each other]

$$= \sum (\partial T_o/\partial q_k)\, \delta q_k + \sum (\partial T_o/\partial \dot{q}_I)\, \delta(\dot{q}_I)$$

$$= \delta T_o \qquad (7.8.9)$$

[notice carefully the slightly different steps taken between the above and (7.8.5d), of the Voronetsian case]. As a result of (7.8.9), and (7.8.5e), and so on, again, Suslov's principle (7.8.8) can be rewritten, in the definitive *constrained form*:

$$\int \Big\{\delta T_o + \sum (\partial T/\partial \dot{q}_D)_o [(\delta q_D)^{\cdot} - \delta(\dot{q}_D)] + \delta' W_o\Big\}\, dt$$

$$= \int \Big\{\delta T_o + \sum\sum (\partial T/\partial \dot{q}_D)_o W^D{}_I\, \delta q_I + \delta' W_o\Big\}\, dt = \Big\{\sum (\partial T/\partial \dot{q}_k)_o\, \delta q_k\Big\}_1^2, \qquad (7.8.10)$$

which coincides with (7.8.6), as it should.

REMARKS

(i) Thanks to (7.8.1c), δT_o can be transformed further as follows:

$$\delta T_o = \sum \Big\{ \Big[\partial T_o/\partial q_I + \sum (\partial T_o/\partial q_D)(\partial \phi_D/\partial \dot{q}_I) \Big] \delta q_I + \sum (\partial T_o/\partial \dot{q}_I)\, \delta(\dot{q}_I) \Big\}$$

$$\Big[\equiv \sum (\partial T_o/\partial q_{(I)})\, \delta q_I + \sum (\partial T_o/\partial \dot{q}_I)\, \delta(\dot{q}_I) \Big]$$

$$= \Big[\sum (\partial T_o/\partial \dot{q}_I)\, \delta q_I \Big]^{\cdot} - \sum \big[(\partial T_o/\partial \dot{q}_I)^{\cdot} - \partial T_o/\partial q_I - \sum (\partial T_o/\partial q_D)(\partial \phi_D/\partial \dot{q}_I) \big] \delta q_I$$

$$\equiv \Big[\sum (\partial T_o/\partial \dot{q}_I)\, \delta q_I \Big]^{\cdot} - \sum E_{(I)}(T_o)\, \delta q_I, \qquad (7.8.11)$$

and, substituting this expression in (7.8.10) or (7.8.6), we obtain at once the nonlinear equations of Voronets and Chaplygin (ex. 5.3.4).

(ii) The Suslov form (7.8.10) can also be derived directly from the fundamental form (7.7.11) as follows. Invoking (7.8.7b) we find, successively,

$$\sum \big[(\partial T/\partial q_k)\, \delta q_k + (\partial T/\partial \dot{q}_k)\, (\delta q_k)^{\cdot} \big]$$

$$= \sum (\partial T/\partial q_k)\, \delta q_k + \sum (\partial T/\partial \dot{q}_I)\, \delta(\dot{q}_I)$$

$$+ \sum (\partial T/\partial \dot{q}_D) \Big[\delta(\dot{q}_D) + \sum W^D{}_I\, \delta q_I \Big]$$

$$= \sum (\partial T/\partial q_k)\, \delta q_k + \sum (\partial T/\partial \dot{q}_I)\, \delta(\dot{q}_I) + \sum (\partial T/\partial \dot{q}_D)\, \delta(\dot{q}_D)$$

$$+ \sum \sum (\partial T/\partial \dot{q}_D)\, W^D{}_I\, \delta q_I$$

$$= \delta T + \sum \sum (\partial T/\partial \dot{q}_D)_o W^D{}_I\, \delta q_I \qquad \text{[then invoking (7.8.9)]}$$

$$= \delta T_o + \sum \sum (\partial T/\partial \dot{q}_D)_o W^D{}_I\, \delta q_I, \quad \text{Q.E.D.}; \qquad (7.8.12)$$

and analogously for the Voronets case (7.8.6).

(iii) In both the Voronets and Suslov principles,

$$T \to T_o(t, q, \dot{q}_I) \to \delta T_o,$$

$$\partial T/\partial \dot{q}_D \to (\partial T/\partial \dot{q}_D)_o = p_D[t, q, \dot{q}_I, \phi_D(t, q, \dot{q}_I)] \equiv p_{D,o}(t, q, \dot{q}_I) \equiv p_{Do},$$

and so the boundary term assumes the constrained form

$$\Big\{ \sum (\partial T/\partial \dot{q}_k)\, \delta q_k \Big\}_1^2 = \cdots = \Big\{ \sum (\ldots)_I\, \delta q_I \Big\}_1^2.$$

(iv) The result of the indicated operations in these integral formulae (integrations by parts, etc.) will have the form

$$\int \{ (\ldots)\, \delta q_{m+1} + \cdots + (\ldots)\, \delta q_n \}\, dt = \Big\{ \sum (\ldots)_I\, \delta q_I \Big\}_1^2 \equiv BT, \qquad (7.8.13)$$

and setting each $(\ldots)_I$ term of its integrand equal to zero will yield the $n - m$ kinetic equations of Voronets–Chaplygin.

(v) From the special nonholonomic form (7.8.8), we can easily go to the general nonholonomic forms (7.7.15a ⇒ 7.7.19a) by inserting into the former the following substitutions/transformations:

$$\dot{q}_k = \dot{q}_k(t, q, \omega), \qquad \delta q_k = \sum (\partial \dot{q}_k / \partial \omega_I)\, \delta\theta_I,$$

$$\Rightarrow\ (\delta q_D)^{\cdot} - \delta(\dot{q}_D) = \sum W^D{}_I\, \delta q_I = \sum\sum W^D{}_{I'} (\partial \dot{q}_{I'} / \partial \omega_I)\, \delta\theta_I$$

$$\equiv \sum B^D{}_I\, \delta\theta_I \ \text{(definition of special nonlinear Voronets coefficients } B^D{}_I\text{)}; \tag{7.8.14}$$

noticing that $\delta T = \delta T^*$, and so on. (Remember the similar generalization in the Pfaffian case: probs. 3.8.2 and 3.8.3.)

Summary

In nonholonomic systems, Hamilton's variational equation is *never* $\int (\delta T_o + \delta' W_o)\, dt = BT$, but it is $\int [\delta T_o + \delta' W_o + \textit{correction term involving } (\delta q)^{\cdot} - \delta(\dot{q})]\, dt = BT$; and similarly in quasi variables, otherwise we lose the term $-\Gamma_I$ in the kinetic equations of motion.

To transform the unavoidable expression [holonomic variable counterpart of (7.7.15c)]

$$\int \left\{ \delta T + \sum (\partial T / \partial \dot{q}_k)_o [(\delta q_k)^{\cdot} - \delta(\dot{q}_k)] \right\} dt, \tag{7.8.15a}$$

where

$$\delta T = \delta T_o + \sum (\partial T / \partial \dot{q}_D)_o\, \delta\omega_D = \delta T_o + \sum (\partial T / \partial \dot{q}_D)_o\, \delta(\dot{q}_D - \phi_D), \tag{7.8.15b}$$

which appears in the basic integral variational formula (7.7.12), and thus derive the correct equations of motion in the variables involved there, we have the following two viewpoints:

(i) *Voronets*, *Hölder*, *Hamel*, et al.

$$\delta T = \delta T_o + \sum\sum (\partial T / \partial \dot{q}_D)_o W^D{}_I \delta q_I \qquad [\Rightarrow (\delta T)_o \neq \delta T_o],$$

$$\sum (\partial T / \partial \dot{q}_k)\, [(\delta q_k)^{\cdot} - \delta(\dot{q}_k)] = 0,$$

that is,

$$0 \neq \delta\omega_D = \delta(\dot{q}_D - \phi_D) = \delta(\dot{q}_D) - \delta\phi_D \qquad [\textit{assumption}: \delta(\dot{q}_D) = (\delta q_D)^{\cdot}]$$

$$= (\delta q_D)^{\cdot} - \delta\phi_D = \left[\sum (\partial\phi_D / \partial \dot{q}_I)\, \delta q_I\right]^{\cdot} - \delta\phi_D = \cdots = \sum W^D{}_I\, \delta q_I; \tag{7.8.16a}$$

(ii) *Suslov* (also Levi-Civita/Amaldi, Rumiantsev, Greenwood, et al.)

$$\delta T = \delta T_o \qquad [\Rightarrow (\delta T)_o = \delta T_o],$$

$$\sum (\partial T / \partial \dot{q}_k)_o [(\delta q_k)^{\cdot} - \delta(\dot{q}_k)] = \sum (\partial T / \partial \dot{q}_D)_o [(\delta q_D)^{\cdot} - \delta(\dot{q}_D)]$$

$$= \sum\sum (\partial T / \partial \dot{q}_D)_o W^D{}_I\, \delta q_I,$$

where

$$0 = \delta\omega_D = \delta[\dot{q}_D - \phi_D(t, q, \dot{q}_I)] \qquad [\text{i.e., } \textit{assumption}\text{: } \delta(\dot{q}_D) = \delta\phi_D]$$

$$\Rightarrow (\delta q_D)^{\cdot} - \delta(\dot{q}_D) = \Big[\sum (\partial\phi_D/\partial\dot{q}_I)\,\delta q_I\Big]^{\cdot} - \delta\phi_D = \cdots = \sum W^D{}_I\,\delta q_I. \quad (7.8.16b)$$

Both these viewpoints are internally consistent, and completely equivalent to each other; that is, if applied correctly they yield the same equations of motion.

Resuming, next, our discussion from the last part of §7.6, let us present some additional forms.

Additional Forms of the Coincidence Conditions

(i) First Transitivity Assumption (Hölder–Voronets–Hamel)

Comparing the integral forms (7.6.4) and (7.6.5c), we see that they coincide, provided that

$$\int \sum \mu_D\,\delta f_D\,dt = \int \sum \lambda_D\,\delta\theta_D\,dt = 0, \qquad (7.8.17a)$$

where

$$\delta\theta_D = \sum (\partial f_D/\partial\dot{q}_k)\,\delta q_k = 0 \qquad \text{(virtual variations)}. \qquad (7.8.17b)$$

Since here $(\delta q_k)^{\cdot} = \delta(\dot{q}_k)$, and, therefore [recalling (7.7.7)],

$$\delta f_D = -\sum E_k(f_D)\,\delta q_k \equiv -\sum E_k(\omega_D)\,\delta q_k = \delta(\dot{\theta}_D) \equiv \delta\omega_D \neq 0$$

$$\text{(mathematically nonadmissible variations)}, \qquad (7.8.17c)$$

sufficient conditions for (7.8.17a) to hold are $\delta f_D = 0$; that is, the system be holonomic.

{Condition (7.8.17a) [which, in view of (7.8.17c), is the same as (7.6.9b)], seems to be due to Jeffreys [1954, eqs. (10, 11)].}

(ii) Second Transitivity Assumption (Suslov–Levi-Civita–Rumiantsev–Greenwood)

In this case, as (7.8.6, 10) with $\delta' W = \delta' W_o = 0$ readily show, the coincidence condition becomes

$$\int \Big[\sum\sum (\partial L/\partial\dot{q}_D)_o W^D{}_I\,\delta q_I\Big]\,dt = 0 \Rightarrow \sum (\partial L/\partial\dot{q}_D)_o W^D{}_I = 0. \qquad (7.8.17d)$$

[Further, if the constraints have the special form (7.8.1), then, invoking (7.8.4a), (7.8.7i), (7.8.17c) we deduce from (7.6.9b) ⇒ (7.8.17a) the following (multiplier-containing) conditions:

$$\sum \Big(\sum \mu_D W^D{}_I\Big)\,\delta q_I = 0 \Rightarrow \sum \mu_D W^D{}_I = 0. \qquad (7.8.17e)$$

Conditions (7.8.17d) and (7.8.17e) seem to be due to Rumiantsev [1978, eqs. (3.5.6)]. Clearly, (7.8.7d), are easier to apply, since they do not involve (generally unknown) multipliers.]

Then, the nonlinear Voronets equations (ex. 5.3.4) reduce to the "holonomic form":

$$\begin{aligned} E_{(I)}(L_o) &\equiv (\partial L_o/\partial \dot{q}_I)^{\cdot} - \left[\partial L_o/\partial q_I + \sum (\partial \phi_D/\partial \dot{q}_I)(\partial L_o/\partial q_D)\right] \\ &\equiv (\partial L_o/\partial \dot{q}_I)^{\cdot} - \partial L_o/\partial (q_I) \\ &\equiv E_I(L_o) - \sum (\partial \phi_D/\partial \dot{q}_I)(\partial L_o/\partial q_D) = 0. \end{aligned} \tag{7.8.17f}$$

As expected, the "potentialness" conditions (7.6.9b) and (7.8.17d, e) are very rarely satisfied, even for Pfaffian nonholonomic constraints. For the particular (or classes of) motions where that happens, Hamilton's principle becomes a stationarity principle.

In closing, we urge the reader to ponder carefully over the similarities, differences, internal consistency, and ultimate equivalence of these, and conceivably many more (admittedly slippery and sometimes confusing), IVPs.

Example 7.8.1 *Motion of a Particle of Mass $m = 1$, on a Fixed Plane O–xy in a Potential Field $V = V(x, y)$ and Under the Pfaffian Constraint*

$$t\dot{x} - \dot{y} = 0 \;\Rightarrow\; t\,\delta x - \delta y = 0, \tag{a}$$

under the Suslov and Hölder–Voronets–Hamel Variational Principles (Mei, 1985, pp. 70–72; also Rosenberg, 1977, pp. 172–173).

(i) *Suslov approach.* With the choice $q_{I\,(\text{Independent})} = q_1 = x$ and $q_{D\,(\text{Dependent})} = q_2 = y$, we shall have

$$\dot{q}_D = \phi_D(t, q, \dot{q}_I): \qquad \dot{y} = \phi(t, \dot{x}) = t\,\dot{x}, \tag{b}$$

and

$$(\delta x)^{\cdot} - \delta(\dot{x}) = 0, \tag{c1}$$

$$(\delta y)^{\cdot} - \delta(\dot{y}) = (t\,\delta x)^{\cdot} - \delta(t\,\dot{x}) = \delta x + t(\delta x)^{\cdot} - t\,\delta(\dot{x}) = \delta x; \tag{c2}$$

that is,

$$W^D{}_I \rightarrow W^y{}_x\,(\equiv v^y{}_x) = 1; \tag{c3}$$

or, applying the general theory,

$$\begin{aligned} \delta(\dot{y}) &= (\delta y)^{\cdot} - \left[(\partial\phi/\partial\dot{x})^{\cdot} - \partial\phi/\partial x - (\partial\phi/\partial y)(\partial\phi/\partial\dot{x})\right]\delta x \\ &= (t\,\delta x)^{\cdot} - [(t)^{\cdot} - 0 - 0]\,\delta x = t\,\delta(\dot{x}) \equiv t(\delta x)^{\cdot}. \end{aligned} \tag{c4}$$

Therefore, the Suslov variation of the unconstrained Lagrangean of the system

$$L = (1/2)\,[(\dot{x})^2 + (\dot{y})^2] - V(x, y), \tag{d1}$$

is, successively,

$$\delta L = \dot{x}\,\delta(\dot{x}) + \dot{y}\,\delta(\dot{y}) - (\partial V/\partial x)\,\delta x - (\partial V/\partial y)\,\delta y \qquad [\textit{then} \text{ enforcing } (\text{a, b, c1–4})]$$

$$\begin{aligned}\delta L \to (\delta L)_o &= \dot{x}\,\delta(\dot{x}) + (t\,\dot{x})\,\delta(t\,\dot{x}) - [\partial V/\partial x + t(\partial V/\partial y)]\,\delta x\\ &= (1+t^2)\dot{x}\,\delta(\dot{x}) - [\partial V/\partial x + t(\partial V/\partial y)]\,\delta x\\ &= (1+t^2)\dot{x}(\delta x)^{\cdot} - [\partial V/\partial x + t(\partial V/\partial y)]\,\delta x,\end{aligned} \tag{d2}$$

and so the unconstrained Suslov principle (7.8.8), with

$$(\partial L/\partial \dot{y})_o\,[(\delta y)^{\cdot} - \delta(\dot{y})] = (t\,\dot{x})\,\delta x, \qquad \textit{Boundary terms} \equiv BT \to 0, \text{ and } \delta' W_{np} = 0,$$

yields

$$\begin{aligned}0 &= \int \{(\delta L)_o + (\partial L/\partial \dot{y})_o[(\delta y)^{\cdot} - \delta(\dot{y})]\}\,dt\\ &= \int \{(1+t^2)\dot{x}(\delta x)^{\cdot} - [-(t\,\dot{x}) + \partial V/\partial x + t(\partial V/\partial y)]\}\,dt.\end{aligned} \tag{d3}$$

Had we enforced the constraint (a) into L right from the start [i.e., before $\delta(\ldots)$-varying], then

$$L \to L_o = (1/2)[(\dot{x})^2 + (t\,\dot{x})^2] - V(x, y), \tag{e1}$$

and, accordingly,

$$\begin{aligned}\delta L_o &= \dot{x}\,\delta(\dot{x}) + t^2\dot{x}\,\delta(\dot{x}) - (\partial V/\partial x)\,\delta x - (\partial V/\partial y)\,(\delta y)_o\\ &= (1+t^2)\dot{x}\,\delta(\dot{x}) - [\partial V/\partial x + t\,(\partial V/\partial y)]\,\delta x;\end{aligned} \tag{e2}$$

that is, $\delta L = \delta L_o$, as expected by (7.8.9); and, hence, (d3) is the same as that obtained by applying (7.8.10). Integrating (d3) by parts, and so on, we readily find

$$0 = \int \{-[(1+t^2)\dot{x}]^{\cdot} - [-t\,\dot{x} + (\partial V/\partial x + t\,(\partial V/\partial y))]\}\,\delta x\,dt, \tag{f1}$$

and from this we get the following single (kinetic) Chaplygin–Voronets equation:

$$[(1+t^2)\dot{x}]^{\cdot} + t\,\dot{x} = -[\partial V/\partial x + t\,(\partial V/\partial y)]; \tag{f2}$$

which, along with the constraint (a), constitute a determinate system for $x(t)$ and $y(t)$.

(ii) *Hölder–Voronets–Hamel approach.* Here

$$\delta(\dot{x}) = (\delta x)^{\cdot} \qquad \text{and} \qquad \delta(\dot{y}) = (\delta y)^{\cdot}, \tag{g1}$$

or, due to (a),

$$\begin{aligned}&(\delta y)^{\cdot} = (t\,\delta x)^{\cdot} = \delta x + t(\delta x)^{\cdot} = \delta x + t\,\delta(\dot{x}) = \delta(\dot{y}) \quad (= \delta x + \delta\phi, \quad \phi = t\,\dot{x})\\ &\Rightarrow \delta(\dot{y}) - \delta\phi = \delta x.\end{aligned} \tag{g2}$$

Therefore, varying L accordingly we find

$$\begin{aligned}
\delta L &= \dot{x}\,\delta(\dot{x}) + \dot{y}\,\delta(\dot{y}) - \delta V \qquad [\textit{then} \text{ enforcing (a, b, g1–2)}]\\
\delta L \to (\delta L)_o &= \dot{x}\,\delta(\dot{x}) + (t\,\dot{x})\,[\delta x + t\,\delta(\dot{x})] - (\partial V/\partial x)\,\delta x - (\partial V/\partial y)\,\delta y\\
&= (1+t^2)\dot{x}\,\delta(\dot{x}) - [\partial V/\partial x + t\,(\partial V/\partial y)]\,\delta x + (t\,\dot{x})\,\delta x\\
&= (1+t^2)\dot{x}(\delta x)^{\cdot} - [-(t\,\dot{x}) + \partial V/\partial x + t\,(\partial V/\partial y)]\,\delta x\\
&= \delta L_o + (t\,\dot{x})\,\delta x\\
\{ &= \delta L_o + (\partial L/\partial \dot{y})_o\,[\delta(\dot{y}) - \delta\phi], \quad \text{in accordance with (7.8.5d)},\\
&\neq \delta L_o = (1+t^2)\dot{x}\,\delta(\dot{x}) - \delta V; \text{ we notice difference from (d2, e2)}\}, \qquad \text{(h1)}
\end{aligned}$$

and so the Hölder–Voronets principle (7.8.6), with $BT \to 0$ and $\delta' W_{np} = 0$, yields

$$\begin{aligned}
0 &= \int \delta L\,dt = \int (\delta L)_o\,dt = \int \{\delta L_o + (\partial L/\partial \dot{y})_o\,[\delta(\dot{y}) - \delta\phi)]\}\,dt\\
&= \int \{(1+t^2)\dot{x}(\delta x)^{\cdot} + [t\,\dot{x} - \partial V/\partial x - t\,(\partial V/\partial y)]\,\delta x\}\,dt, \qquad \text{(h2)}
\end{aligned}$$

which, as an integration by parts of the first integrand term shows, coincides with the Suslov results (f1, 2).

Generally, we have:

- *Suslov–Rumiantsev–Greenwood approach*:

$$\begin{aligned}
(\delta q_D)^{\cdot} - \delta(\dot{q}_D) = \Big(\sum b_{DI}\,\delta q_I\Big)^{\cdot} - \delta\Big(\sum b_{DI}\dot{q}_I + b_D\Big) = \cdots \equiv \sum W^D{}_I\,\delta q_I\\
\Rightarrow \delta(\dot{q}_D) = (\delta q_D)^{\cdot} - \sum W^D{}_I\,\delta q_I. \qquad \text{(i1)}
\end{aligned}$$

- *Hölder–Voronets–Hamel approach*:

$$\begin{aligned}
\delta f_D &= \delta(\dot{q}_D) - \delta\phi_D = (\delta q_D)^{\cdot} - \delta\phi_D\\
&= \Big(\sum b_{DI}\,\delta q_I\Big)^{\cdot} - \delta\Big(\sum b_{DI}\dot{q}_I + b_D\Big) = \cdots \equiv \sum W^D{}_I\,\delta q_I \neq 0\\
&\Rightarrow \delta(\dot{q}_D) = (\delta q_D)^{\cdot} = \delta\phi_D + \sum W^D{}_I\,\delta q_I. \qquad \text{(i2)}
\end{aligned}$$

Appendix: elementary solution. The Newton–Euler equation in the *tangential* direction is

$$\begin{aligned}
dv/dt &= -\partial V/\partial s = -[(\partial V/\partial x)\,(dx/ds) + (\partial V/\partial y)\,(dy/ds)] \qquad (\text{where } v \equiv ds/dt)\\
&= -[(\partial V/\partial x)\,(dx/ds) + (\partial V/\partial y)\,(dy/dx)\,(dx/ds)]\\
&= -(dx/ds)\,[(\partial V/\partial x) + (\partial V/\partial y)\,(dy/dx)], \qquad \text{(j)}
\end{aligned}$$

and, since (with ϕ: angle between Ox and path tangent),

$$dx/ds = \cos\phi, \qquad \cos^2\phi = (1 + \tan^2\phi)^{-1},$$

$$\tan\phi = \dot{y}/\dot{x} = t \Rightarrow dx/ds = (1 + t^2)^{-1/2},$$

and $\qquad ds = dx/\cos\phi \Rightarrow v = \dot{x}(1 + t^2)^{1/2};$

(j) is easily seen to coincide with the earlier (f2).

Example 7.8.2 *Rolling Disk via the Suslov and Voronets Principles.* Let us describe the application of the Suslov and Voronets forms of Hamilton's principle to the derivation of the equations of the rolling of a thin homogeneous circular disk of mass m and radius r on a rough horizontal and fixed plane O–XY.

The kinematics and kinetics of this well-known problem have already been detailed in §1.17, eqs. (1.17.17a) ff. (Eulerian treatment), ex. 2.13.7 (Lagrangean kinematics), and ex. 3.18.5 (Lagrangean kinetics). It was found there that the constraints are [in terms of the *coordinates of its contact point* C $(X, Y, Z = 0)$ along space-fixed axes O–XYZ and with $q_D \equiv q_{1,2}$: X, Y, and $q_I \equiv q_{3,4,5}$: ϕ, θ, ψ: Eulerian angles of body-fixed axes at G relative to O–XYZ]

$$v_{C,\text{tangent}} \equiv f_1 = \dot{X} + (r\cos\phi)\dot{\psi} = 0, \qquad v_{C,\text{normal}} \equiv f_2 = \dot{Y} + (r\sin\phi)\dot{\psi} = 0. \tag{a}$$

The unconstrained Lagrangean of the system (under gravity) is

$$\begin{aligned} L &\equiv T - V \\ &= (m/2)\left\{[(X - r\cos\theta\sin\phi)^{\cdot}]^2 + [(Y + r\cos\theta\cos\phi)^{\cdot}]^2 + [(r\sin\theta)^{\cdot}]^2\right\} \\ &\quad + (1/2)\,[I_x(\dot{\theta})^2 + I_y(\dot{\phi}\sin\theta)^2 + I_z(\dot{\psi} + \dot{\phi}\cos\theta)^2] - mgr\sin\theta, \end{aligned} \tag{b}$$

and the principal moments of inertia of the disk at G are $I_{x,y} = mr^2/4, I_z = mr^2/2$.

(i) *Hölder–Voronets–Hamel approach.* From (a) we find, successively,

$$\begin{aligned} \delta f_1 \equiv \delta\omega_1 &= \delta(\dot{X}) + (-r\sin\phi\,\delta\phi)\dot{\psi} + r\cos\phi\,\delta(\dot{\psi}) \\ &= (\delta X)^{\cdot} - r\dot{\psi}\sin\phi\,\delta\phi + r\cos\phi(\delta\psi)^{\cdot} \\ &= (-r\cos\phi\,\delta\psi)^{\cdot} - r\dot{\psi}\sin\phi\,\delta\phi + r\cos\phi(\delta\psi)^{\cdot} \qquad \text{[by first of(a)]} \\ &= \cdots = (-r\sin\phi\,\dot{\psi})\,\delta\phi + (r\sin\phi\,\dot{\phi})\,\delta\psi \neq 0, \end{aligned} \tag{c1}$$

$$\delta f_2 \equiv \delta\omega_2 = \cdots = (r\cos\phi\,\dot{\psi})\,\delta\phi + (-r\cos\phi\,\dot{\phi})\,\delta\psi \neq 0; \tag{c2}$$

that is,

$$W^Z{}_\phi = -r\sin\phi\,\dot{\psi}, \qquad W^Z{}_\theta = 0, \qquad W^Z{}_\psi = r\sin\phi\,\dot{\phi}, \tag{c3}$$

$$W^Y{}_\phi = r\cos\phi\,\dot{\psi}, \qquad W^Y{}_\theta = 0, \qquad W^Y{}_\psi = -r\cos\phi\,\dot{\phi}; \tag{c4}$$

and, accordingly [with $\dot{X} = \phi_X(\dot{\phi}, \dot{\theta}, \dot{\psi};\ \phi, \theta, \psi)$, $\dot{Y} = \phi_Y(\dot{\phi}, \dot{\theta}, \dot{\psi};\ \phi, \theta, \psi)$, from (a)],

$$\delta(\dot{X}) = (\delta X)^{\cdot} = \delta\phi_X + W^X{}_\phi\,\delta\phi + W^X{}_\theta\,\delta\theta + W^X{}_\psi\,\delta\psi,$$

that is,

$$(-r\cos\phi\,\delta\psi)^{\cdot} = \delta(-r\cos\phi\,\dot{\psi}) + (-r\sin\phi\,\dot{\psi})\,\delta\phi + (0)\,\delta\theta + (r\sin\phi\,\dot{\phi})\,\delta\psi$$

$$\Rightarrow \delta(\dot{X}) = (r\dot{\phi}\,\sin\phi)\,\delta\psi + (-r\cos\phi)\,\delta(\dot{\psi}); \qquad \text{(c5)}$$

$$\delta(\dot{Y}) = (\delta Y)^{\cdot} = \delta\phi_Y + W^{Y}{}_{\phi}\,\delta\phi + W^{Y}{}_{\theta}\,\delta\theta + W^{Z}{}_{\psi}\,\delta\psi,$$

that is,

$$(-r\sin\phi\,\delta\psi)^{\cdot} = \delta(-r\sin\phi\,\dot{\psi}) + (r\cos\phi\,\dot{\psi})\,\delta\phi + (0)\,\delta\theta + (-r\cos\phi\,\dot{\phi})\,\delta\psi$$

$$\Rightarrow \delta(\dot{Y}) = (-r\dot{\phi}\cos\phi)\,\delta\psi + (-r\sin\phi)\,\delta(\dot{\psi}). \qquad \text{(c6)}$$

Then, Voronets' principle yields

$$0 = \int \delta L\,dt = \int \big[(\partial L/\partial\dot{X})\,\delta(\dot{X}) + (\partial L/\partial\dot{Y})\,\delta(\dot{Y})$$
$$+ (\partial L/\partial\dot{\phi})\,\delta(\dot{\phi}) + (\partial L/\partial\dot{\theta})\,\delta(\dot{\theta}) + (\partial L/\partial\dot{\psi})\,\delta(\dot{\psi})$$
$$+ (\partial L/\partial\phi)\,\delta\phi + (\partial L/\partial\theta)\,\delta\theta\big]\,dt$$

[integrating by parts, and using (b), and (c5, 6) for $\delta(\dot{X}), \delta(\dot{Y})$]

$$= \int [(\ldots)\,\delta\phi + (\ldots)\,\delta\theta + (\ldots)\,\delta\psi]\,dt; \qquad \text{(c7)}$$

and the corresponding equations of motion will result by setting the coefficients of $\delta\phi/\delta\theta/\delta\psi$, in the above, equal to zero. These will be the kinetic equations of Maggi ⇒ Voronets of the problem [the latter resulting by eliminating $\dot{X}, \dot{Y}$, from L via (a)]. The details are left to the reader.

(ii) *Suslov approach.* In this case,

$$\delta f_1 = 0,\ \delta f_2 = 0; \qquad (\delta\phi)^{\cdot} = \delta(\dot{\phi}), \quad (\delta\theta)^{\cdot} = \delta(\dot{\theta}), \quad (\delta\psi)^{\cdot} = \delta(\dot{\psi}), \qquad \text{(d1)}$$

and therefore, invoking (c3, 4),

$$(\delta X)^{\cdot} - \delta(\dot{X}) = (-r\cos\phi\,\delta\psi)^{\cdot} - \delta(-r\cos\phi\,\dot{\psi})$$
$$= W^{X}{}_{\phi}\,\delta\phi + W^{X}{}_{\theta}\,\delta\theta + W^{X}{}_{\psi}\,\delta\psi$$
$$= (-r\sin\phi\,\dot{\psi})\,\delta\phi + (0)\,\delta\theta + (r\sin\phi\,\dot{\phi})\,\delta\psi$$
$$\Rightarrow \delta(\dot{X}) = (\delta X)^{\cdot} - (W^{X}{}_{\phi}\,\delta\phi + W^{X}{}_{\theta}\,\delta\theta + W^{X}{}_{\psi}\,\delta\psi)$$
$$= (r\sin\phi\,\dot{\psi})\,\delta\phi - r\cos\phi\,\delta(\dot{\psi}) \qquad \text{[compare with (c5)];} \quad \text{(d2)}$$

$$(\delta Y)^{\cdot} - \delta(\dot{Y}) = (-r\sin\phi\,\delta\psi)^{\cdot} - \delta(-r\sin\phi\,\dot{\psi})$$
$$= W^{Y}{}_{\phi}\,\delta\phi + W^{Y}{}_{\theta}\,\delta\theta + W^{Y}{}_{\psi}\,\delta\psi$$
$$= (r\cos\phi\,\dot{\psi})\,\delta\phi + (0)\,\delta\theta + (-r\cos\phi\,\dot{\phi})\,\delta\psi$$
$$\Rightarrow \delta(\dot{Y}) = (\delta Y)^{\cdot} - (W^{Y}{}_{\phi}\,\delta\phi + W^{Y}{}_{\theta}\,\delta\theta + W^{Y}{}_{\psi}\,\delta\psi)$$
$$= (-r\cos\phi\,\dot{\psi})\,\delta\phi + (-r\sin\phi)\,\delta(\dot{\psi}) \quad \text{[compare with (c6)].} \quad \text{(d3)}$$

Then, the constrained Suslov principle (7.8.10) yields

$$0 = \int \{\delta L_o + (\partial T/\partial \dot{X})_o[(\delta X)^{\cdot} - \delta(\dot{X})] + (\partial T/\partial \dot{Y})_o[(\delta Y)^{\cdot} - \delta(\dot{y})]\}\, dt$$

$$= \cdots = \int [(\ldots)\,\delta\phi + (\ldots)\,\delta\theta + (\ldots)\,\delta\psi]\, dt, \tag{d4}$$

which results in Voronets-type equations, as explained earlier.

(iii) *Stationarity (or coincidence) conditions.* Let us, finally, examine whether the above Hamilton-like variational expressions of Voronets and Suslov are genuine stationarity conditions. Invoking the constraints (a) and (b) provides the constrained momenta

$$(\partial L/\partial \dot{X})_o = (\partial T/\partial \dot{X})_o$$
$$= m(\dot{X} + r\dot{\theta}\sin\phi\sin\theta - r\dot{\phi}\cos\phi\cos\theta)_o$$
$$= m[(-r\cos\phi\cos\theta)\dot{\phi} + (r\sin\phi\sin\theta)\dot{\theta} + (-r\cos\phi)\dot{\psi}], \tag{e1}$$

$$(\partial L/\partial \dot{Y})_o = (\partial T/\partial \dot{Y})_o$$
$$= m(\dot{Y} - r\dot{\theta}\cos\phi\sin\theta - r\dot{\phi}\sin\phi\cos\theta)_o$$
$$= m[(-r\sin\phi\cos\theta)\dot{\phi} + (-r\cos\phi\sin\theta)\dot{\theta} + (-r\sin\phi)\dot{\psi}]; \tag{e2}$$

and so, invoking (c3, 4), the coincidence conditions (7.8.17d) give

$$\phi: \quad (\partial L/\partial \dot{X})_o W^X{}_\phi + (\partial L/\partial \dot{Y})_o W^Y{}_\phi = \cdots = -mr^2\dot{\theta}\dot{\psi}\sin\theta = 0, \tag{e3}$$

$$\theta: \quad (\partial L/\partial \dot{X})_o W^X{}_\theta + (\partial L/\partial \dot{Y})_o W^Y{}_\theta = m(\ldots)\,(0) + m(\ldots)\,(0) = 0, \tag{e4}$$

$$\psi: \quad (\partial L/\partial \dot{X})_o W^X{}_\psi + (\partial L/\partial \dot{Y})_o W^Y{}_\psi = \cdots = mr^2\dot{\phi}\dot{\theta}\sin\theta = 0. \tag{e5}$$

Since, on physically nontrivial grounds $\sin\theta \neq 0$, eqs. (e3–5) are satisfied either when $\dot{\theta} = 0 \Rightarrow \theta = \textit{constant}\ (\neq 0)$, or when $\dot{\phi} = 0 \Rightarrow \phi = \textit{constant}$ and $\dot{\psi} = 0 \Rightarrow \psi = \textit{constant}$. The first possibility indicates *rolling of the disk at a constant nutation angle* (to the vertical GZ or OZ); while the second indicates absence of proper spin, in which case (a) yields $\dot{X} = 0$, $\dot{Y} = 0$; that is, a motion of no further physical interest. See also Capon (1952) and Rumiantsev (1978).

Example 7.8.3 *Rolling Sphere via the Suslov Principle.* Using Suslov's principle, let us derive the equations of motion of a homogeneous sphere of mass m and radius r rolling on a rough horizontal and fixed plane O–XY.

The kinematics and kinetics of this classical problem have already been discussed in exs. 2.13.5 and 2.13.6 (Lagrangean kinematics) and exs. 3.18.2 and 3.18.3 (Lagrangean kinetics). It was found there that the constraints are (in terms of $q_D \equiv q_{1,2,3}: X, Y, Z$: inertial coordinates of center/center-of-mass of sphere G, and $q_I \equiv q_{4,5,6}: \phi, \theta, \psi$: Eulerian angles of body-fixed axes at G relative to fixed axes O–xyz)

$$\dot{X} - r\omega_Y = 0, \qquad \dot{Y} + r\omega_X = 0 \qquad \text{(nonholonomic constraints)}, \tag{a1}$$

$$\Rightarrow \delta X - r\,\delta\theta_Y = 0, \qquad \delta Y + r\,\delta\theta_X = 0, \tag{a2}$$

$$Z = r \;\Rightarrow\; \dot{Z} = 0 \qquad \text{(holonomic constraint)} \;\Rightarrow\; \delta Z = 0; \tag{a3}$$

where (§1.12) (with $d\theta_{X,Y,Z} \equiv \omega_{X,Y,Z}\, dt$)

$$\omega_X = (\cos\phi)\dot{\theta} + (\sin\phi\sin\theta)\dot{\psi} \Rightarrow \delta\theta_X = (\cos\phi)\,\delta\theta + (\sin\phi\sin\theta)\,\delta\psi, \tag{b1}$$

$$\omega_Y = (\sin\phi)\dot{\theta} + (-\cos\phi\sin\theta)\dot{\psi} \Rightarrow \delta\theta_Y = (\sin\phi)\,\delta\theta + (-\cos\phi\sin\theta)\,\delta\psi, \tag{b2}$$

$$\omega_Z = \dot{\phi} + (\cos\theta)\dot{\psi} \Rightarrow \delta\theta_Z = \delta\phi + (\cos\theta)\,\delta\psi, \tag{b3}$$

with corresponding transitivity equations

$$(\delta\theta_X)^{\cdot} - \delta\omega_X = \omega_Y\,\delta\theta_Z - \omega_Z\,\delta\theta_Y, \tag{c1}$$

$$(\delta\theta_Y)^{\cdot} - \delta\omega_Y = \omega_Z\,\delta\theta_X - \omega_X\,\delta\theta_Z, \tag{c2}$$

$$(\delta\theta_Z)^{\cdot} - \delta\omega_Z = \omega_X\,\delta\theta_Y - \omega_Y\,\delta\theta_X. \tag{c3}$$

Hence, following the Suslov viewpoint, we find

$$\delta s_X \equiv (\delta X)^{\cdot} - \delta(\dot{X}) = (r\,\delta\theta_Y)^{\cdot} - \delta(r\,\dot{\theta}_Y) = \cdots = r(\omega_Z\,\delta\theta_X - \omega_X\,\delta\theta_Z), \tag{d1}$$

$$\delta s_Y \equiv (\delta Y)^{\cdot} - \delta(\dot{Y}) = (-r\,\delta\theta_X)^{\cdot} - \delta(-r\,\dot{\theta}_X) = \cdots = -r(\omega_Y\,\delta\theta_Z - \omega_Z\,\delta\theta_Y), \tag{d2}$$

$$\delta s_Z \equiv (\delta Z)^{\cdot} - \delta(\dot{Z}) = 0. \tag{d3}$$

The kinetic energy of the sphere is [with $I_{G;X,Y,Z} = 2mr^2/5 \equiv I$]

$$T = (m/2)\,[(\dot{X})^2 + (\dot{Y})^2 + (\dot{Z})^2] + (I/2)\,({\omega_X}^2 + {\omega_Y}^2 + {\omega_Z}^2) \tag{e1}$$

$$\begin{aligned} \Rightarrow T_o &= (m/2)\,[(r\,\omega_Y)^2 + (-r\,\omega_X)^2 + (0)^2] + (1/2)\,(2m\,r^2/5)\,({\omega_X}^2 + {\omega_Y}^2 + {\omega_Z}^2) \\ &= (1/2)\,(7m\,r^2/5)\,({\omega_X}^2 + {\omega_Y}^2) + (1/2)\,(2m\,r^2/5)\,({\omega_Z}^2), \end{aligned} \tag{e2}$$

and so its Suslovian variation equals

$$\delta T_o = (7mr^2/5)\,(\omega_X\,\delta\omega_X + \omega_Y\,\delta\omega_Y) + (2mr^2/5)\,(\omega_Z\,\delta\omega_Z). \tag{e3}$$

Therefore, invoking the transitivity relations (c1–3) and integrating by parts, and so on, we obtain

$$\begin{aligned} \int \omega_X\,\delta\omega_X\,dt &= \int \omega_X[(\delta\theta_X)^{\cdot} + (\omega_Z\,\delta\theta_Y - \omega_Y\,\delta\theta_Z)]\,dt \\ &= \int [(-\dot{\omega}_X)\,\delta\theta_X + \omega_X(\omega_Z\,\delta\theta_Y - \omega_Y\,\delta\theta_Z)]\,dt, \end{aligned} \tag{f1}$$

and, similarly,

$$\int \omega_Y\,\delta\omega_Y\,dt = \int [(-\dot{\omega}_Y)\,\delta\theta_Y + \omega_Y(\omega_X\,\delta\theta_Z - \omega_Z\,\delta\theta_X)]\,dt, \tag{f2}$$

$$\int \omega_Z\,\delta\omega_Z\,dt = \int [(-\dot{\omega}_Z)\,\delta\theta_Z + \omega_Z(\omega_Y\,\delta\theta_X - \omega_X\,\delta\theta_Y)]\,dt. \tag{f3}$$

Next, with the help of (a, d, e), the Suslov integrand term $s \equiv \sum (\partial T/\partial \dot{q}_D)_o \, \delta s_D$ becomes

$$
\begin{aligned}
s &\equiv (\partial T/\partial \dot{X})_o \, \delta s_X + (\partial T/\partial \dot{Y})_o \, \delta s_Y + (\partial T/\partial \dot{Z})_o \, \delta s_Z \\
&= (m\dot{X})_o [r(\omega_Z \, \delta\theta_X - \omega_X \, \delta\theta_Z)] + (m\dot{Y})_o [r(\omega_Z \, \delta\theta_Y - \omega_Y \, \delta\theta_Z)] + 0 \\
&= \cdots = (m r^2 \omega_Y \, \omega_Z) \, \delta\theta_X + (-m r^2 \omega_X \, \omega_Z) \, \delta\theta_Y; \qquad \text{(g1)}
\end{aligned}
$$

while the total impressed virtual work is [with $Q_{X,Y,Z;\,\phi,\theta,\psi} \equiv Q_X, Q_Y, Q_Z; Q_\phi, Q_\theta, Q_\psi$]

$$
\begin{aligned}
\delta' W &= Q_X \, \delta X + Q_Y \, \delta Y + Q_Z \, \delta Z + Q_\phi \, \delta\phi + Q_\theta \, \delta\theta + Q_\psi \, \delta\psi \\
&\qquad \text{[by (a2) and the inverse of (b1–3) \quad see §1.12]} \\
&= \cdots \equiv M_X \, \delta\theta_X + M_Y \, \delta\theta_Y + M_Z \, \delta\theta_Z \quad (= \delta' W^*{}_o), \qquad \text{(g2)}
\end{aligned}
$$

where [recalling (3.15.2a ff.)]

$$M_X = (-r) Q_Y + (-\cot\theta \sin\phi) Q_\phi + (\cos\phi) Q_\theta + (\sin\phi/\sin\theta) Q_\psi, \qquad \text{(g3)}$$

$$M_Y = (r) Q_X + (-\cot\theta \cos\phi) Q_\phi + (\sin\phi) Q_\theta + (-\cos\phi/\sin\theta) Q_\psi, \qquad \text{(g4)}$$

$$M_Z = Q_\phi. \qquad \text{(g5)}$$

Substituting all these results into Suslov's variational formula (7.8.10), we obtain

$$
\begin{aligned}
\int \{ [-(7mr^2/5)\dot{\omega}_X + M_X] \, \delta\theta_X + [-(7mr^2/5)\dot{\omega}_Y + M_Y] \, \delta\theta_Y \\
+ [-(2mr^2/5)\dot{\omega}_Z + M_Z] \, \delta\theta_Z \} \, dt = 0, \qquad \text{(h1)}
\end{aligned}
$$

and this, since the $\delta\theta_{X,Y,Z}$ are independent (free), leads immediately to the following three kinetic equations:

$$(7mr^2/5) \, \dot{\omega}_X = M_X, \qquad (7mr^2/5) \, \dot{\omega}_Y = M_Y, \qquad (2mr^2/5) \, \dot{\omega}_Z = M_Z; \qquad \text{(h2)}$$

which, along with the two constraints (a1) and the three kinematical relations (b1–3), constitute a determinate system for $X(t), Y(t); \phi(t), \theta(t), \psi(t); \omega_X(t), \omega_Y(t), \omega_Z(t)$.

Let the reader formulate this problem via Voronets' principle (7.8.6); or even via those of § 7.7.

7.9 NONCONTEMPORANEOUS VARIATIONS; ADDITIONAL IVP FORMS

The General Formulae

We begin with the following slightly modified version of the fundamental integral variational equation (7.2.3b):

$$\int (\delta L + \delta' W_{np}) \, dt = \left\{ \sum p_k \, \delta q_k \right\}_1^2, \qquad \text{(7.9.1)}$$

where

$L = L(t, q, \dot{q}) \equiv T(t, q, \dot{q}) - V(t, q)$: Lagrangean of the system,

$\delta' W_{np} \equiv \delta' W - (-\delta V)$: Total (first-order) virtual work of *nonpotential* impressed forces,

$p_k \equiv \partial T/\partial \dot{q}_k = \partial L/\partial \dot{q}_k$: Holonomic system ("generalized") momentum. (7.9.1a)

Now, if in (7.9.1), $\delta' W_{np} = 0$, the q's and δq's are chosen so that $\{\sum p_k\, \delta q_k\}_1^2 \equiv BT = 0$ [e.g., by taking $\delta q(t_1) = \delta q(t_2) = 0$], and *if no additional constraints are imposed on the system* (unless explicitly specified otherwise), then (7.9.1) yields the customary form of *Hamilton*'s *principle* of stationarity:

$$\delta A_H = 0, \tag{7.9.2}$$

where

$$A_H \equiv \int (T - V)\, dt \equiv \int L\, dt\text{: Hamiltonian action (functional).} \tag{7.9.2a}$$

Let us express the above in terms of the earlier-introduced *noncontemporaneous*, or *skew*, or *oblique*, or *asynchronous*, or *nontautochronous*, or *nonsimultaneous* variations (§7.2, fig.7.1):

$$\Delta q_k = \delta q_k + \dot{q}_k \Delta t; \tag{7.9.3}$$

where, generally,

$$\Delta(\ldots) \equiv \delta(\ldots) + [d(\ldots)/dt]\Delta t\text{: } \textit{noncontemporaneous variation operator}; \tag{7.9.3a}$$

for example, $\Delta t = \delta t + (dt/dt)\Delta t = 0 + (1)\Delta t = \Delta t$.

Our discussion is based on the following fundamental, purely analytical and interrelated, formulae, obtained by applying $\Delta(\ldots)$ to $\int(\ldots)\, dt$ and its integrand (which is nothing but an application of the well-known theorem of differentiation of a definite integral with respect to a general parameter that may appear in both its limits of integration $t_{1,2}$ and in its integrand, or "Leibniz' rule"):

- $$\begin{aligned} \Delta \int (\ldots)\, dt &= \int \delta(\ldots)\, dt + \{(\ldots)\Delta t\}_1^2 \\ &= \int \{\Delta(\ldots) + (\ldots)[d(\Delta t)/dt]\}\, dt \\ &= \int [\Delta(\ldots)\, dt + (\ldots)\, d(\Delta t)], \end{aligned} \tag{7.9.3b}$$

- $$\begin{aligned} \int \Delta(\ldots)\, dt &= \int \{\delta(\ldots) - (\ldots)[d(\Delta t)/dt]\}\, dt + \{(\ldots)\Delta t\}_1^2 \\ &= \int [\delta(\ldots)\, dt - (\ldots)\, d(\Delta t)] + \{(\ldots)\Delta t\}_1^2; \end{aligned} \tag{7.9.3c}$$

from which the (intuitively "obvious") $\Delta - \int$ noncommutativity formula results:

- $$\Delta \int (\ldots)\, dt - \int \Delta(\ldots)\, dt = \int (\ldots)\, d(\Delta t) = \int \{(\ldots)\, [d(\Delta t)/dt]\}\, dt. \tag{7.9.3d}$$

With the help of the above, *Hamilton's principle* (7.9.1–2a) can be easily brought to its following *two equivalent noncontemporaneous forms*:

- $$\Delta \int T\,dt + \int \delta' W\,dt = \Big\{\sum p_k \Delta q_k + \Big(T - \sum p_k \dot{q}_k\Big)\Delta t\Big\}_1^2$$
$$= \Big\{\sum p_k\,\delta q_k + T\,\Delta t\Big\}_1^2, \qquad (7.9.4a)$$

- $$\Delta A_H + \int \delta' W_{np}\,dt = \Big\{\sum p_k \Delta q_k - h\Delta t\Big\}_1^2 = \Big\{\sum p_k\,\delta q_k + L\Delta t\Big\}_1^2, \qquad (7.9.4b)$$

where (recalling §3.9)

$$h \equiv \sum p_k \dot{q}_k - L = h(t, q, \dot{q})\text{: Generalized energy}$$
[or Hamiltonian, when expressed in terms of t, q, p (chap. 8)]. (7.9.4c)

Further, with the definitions

$$A_L \equiv \int 2T\,dt\text{: Lagrangean action (functional)}, \qquad (7.9.4d)$$

$$E \equiv T + V\text{: Total } \textit{energy} \text{ of the system}, \qquad (7.9.4e)$$

it is not hard to see that we can rewrite (7.9.4a, b) as the following general *principle of noncontemporaneously varying, or varied, Lagrangean action*:

- $$\Delta A_L - \int (\delta E - \delta' W_{np})\,dt = \Big\{\sum p_k \Delta q_k - \Big(\sum p_k \dot{q}_k - 2T\Big)\Delta t\Big\}_1^2$$
$$= \Big\{\sum p_k\,\delta q_k + 2T\Delta t\Big\}_1^2; \qquad (7.9.4f)$$

and, finally, adding and subtracting (7.9.4a, b) with the purely mathematical equation

$$\Delta \int E\,dt = \int \delta E\,dt + \{E\Delta t\}_1^2, \qquad (7.9.4g)$$

side by side, produces the additional "symmetrical principles":

- $$\Delta \int 2T\,dt = \int (\delta E - \delta' W_{np})\,dt + \Big\{\sum p_k \Delta q_k + \Big(2T - \sum p_k \dot{q}_k\Big)\Delta t\Big\}_1^2, \qquad (7.9.4h)$$

- $$\Delta \int 2V\,dt = \int (\delta E + \delta' W_{np})\,dt + \Big\{-\sum p_k \Delta q_k + \Big(2V + \sum p_k \dot{q}_k\Big)\Delta t\Big\}_1^2. \qquad (7.9.4i)$$

An additional IVP can be obtained if we *express the integrands of the above in terms of* $\Delta(\ldots)$-variations: adding side by side (i) eqs. (7.9.4a, b), but with

$$\Delta \int T\,dt \qquad \text{replaced by} \qquad \int [\Delta T + T(\Delta t)^{\cdot}]\,dt \qquad \text{[applying (7.9.3b)]}, \qquad (7.9.5a)$$

and (ii) the obvious identity

$$\{T\Delta t\}_1^2 = \int (T\Delta t)^{\cdot}\, dt = \int [T(\Delta t)^{\cdot} + \dot{T}\Delta t]\, dt \tag{7.9.5b}$$

produces the alternative forms:

$$\begin{aligned}\int [\Delta T + 2T(\Delta t)^{\cdot} + \dot{T}\Delta t]\, dt + \int \delta' W\, dt &= \int \left[\Delta T\, dt + 2T\, d(\Delta t) + dT\,\Delta t + \delta' W\, dt\right] \\ &= \left\{\sum p_k\, \delta q_k + (2T)\Delta t\right\}_1^2 \\ &= \left\{\sum p_k \Delta q_k - \left(\sum p_k \dot{q}_k - 2T\right)\Delta t\right\}_1^2.\end{aligned} \tag{7.9.5c}$$

The above are usually associated with the names of O. Hölder, Voss, et al., and a time when the differences between $\Delta(\ldots)$ and $\delta(\ldots)$ were not clearly understood (late 19th to early 20th century). Unless one distinguishes carefully between these two kinds of variation, notices the resulting integral noncommutativity formula (7.9.3d), and states carefully the system properties and boundary conditions, the results are very likely to be erroneous and extremely difficult to compare with those of other authors. This is a tricky area (like §7.8) that has caused considerable confusion and frustration; see, for example, Papastavridis [1987(d)].

Specializations

(i) If the following hold:

the (holonomic) constraints are *stationary*, in which case $\sum p_k \dot{q}_k = 2T$,

$\delta' W_{np} = 0$,

$\Delta q_k(t_1) = \Delta q_k(t_2) = 0$,

and

$$\delta E = \delta h = 0, \tag{7.9.6a}$$

then (7.9.4f) reduces to the original *principle of "least" action* of *Maupertuis → Euler → Lagrange* (MEL):

$$\Delta A_L \equiv \Delta \int 2T\, dt = \Delta \int \left(\sum (\partial L/\partial \dot{q}_k)\dot{q}_k\right) dt = 0. \tag{7.9.6b}$$

Of course, other combinations of boundary conditions and system assumptions may produce the same result. [An alternative derivation of (7.9.6b) is given below.]

(ii) For stationary constraints, the power equation (7.2.6e, f) reduces to $dT/dt = \sum Q_k \dot{q}_k$ (where Q_k is the *total* impressed force), so that

$$\dot{T}\Delta t + \delta' W = \left(\sum Q_k \dot{q}_k\right)\Delta t + \sum Q_k\, \delta q_k = \sum Q_k \Delta q_k \equiv \Delta' W,$$

and therefore the left side of (7.9.5c) simplifies to

$$\int [\Delta T + 2T(\Delta t)^{\cdot} + \Delta' W]\, dt = \int [\Delta T\, dt + 2T\, d(\Delta t) + \Delta' W\, dt], \tag{7.9.7a}$$

whereas the right reduces to

$$\left\{\sum p_k \Delta q_k\right\}_1^2 = \left\{\sum p_k\, \delta q_k + (2T)\Delta t\right\}_1^2. \tag{7.9.7b}$$

REMARKS ON MEL'S ACTION

(i) The A_L definition (7.9.4d) is not arbitrary. It constitutes the earliest of all such action definitions (early 1740s); that is, of energetic functions/functionals of dimensions

$$(energy) \times (time) = (momentum) \times (length).$$

(Originally, it was given by Euler for a special case, then generalized by Lagrange for arbitrary holonomic and scleronomic systems, and fully justified later by modern variational calculus.)

For a single, say unconstrained, particle of mass m, moving along a path of arc length s with velocity $\mathbf{v} = d\mathbf{r}/dt$ [$\Rightarrow$ velocity component along path tangent $\equiv v = ds/dt$], (7.9.4d) gives

$$A_L = \int m\, v^2\, dt = \int (m\,\mathbf{v}) \cdot \mathbf{v}\, dt = \int (m\,\mathbf{v}) \cdot d\mathbf{r} = \int m\, v\, ds \tag{7.9.8a}$$

[= sum of elementary "works of the (linear) momentum" along the particle's path];

and, for a material system (with the usual notations):

$$A_L = \int \left[S\, (dm\,\mathbf{v}) \cdot d\mathbf{r} \right] = \int \sum p_k\, dq_k. \tag{7.9.8b}$$

(ii) It is frequently claimed, in the variational mechanics literature, that starting with (7.9.2, 2a) and substituting in there the energy conservation relation $\delta E = 0 \Rightarrow \delta T = -\delta V$, between the orbit(s) and other kinematically admissible paths, one obtains the (*contemporaneous* variation $\Rightarrow$ fixed time-endpoints) MEL principle (7.9.6b):

$$0 = \int [\delta T - (-\delta T)]\, dt = \int \delta(2T)\, dt = \delta \int (2T)\, dt; \qquad \text{i.e., } \delta A_L = 0. \tag{7.9.9}$$

However, such a reasoning would be incorrect for the following reasons: eq. (7.9.2) yields the n Lagrangean equations $E_k(L) = 0$, the general solution of which contains $2n$ integration constants, to be determined from $2n$ boundary conditions such as $q_k(t_{1,2}) = given$, where $t_{1,2}$ are also given. Hence, it will be impossible for the resulting particular solution(s) to satisfy the *additional* energy constraint: $T(q, \dot{q}) + V(q) = E = given\ constant$ (the same for all competing trajectories).

That the reasoning leading to (7.9.9) is incorrect can also be seen as follows: The last condition implies that $\delta E = 0$ (virtual form of constraint), and this, for given q's and δq's, imposes restrictions on the corresponding velocities; that is, on the $\delta(\dot{q})$'s; and this, in particular, makes it impossible for the system to go from an initial position to a final one along the actual (kinetic) path and along a typical comparison (adjacent) path, with the *same energy* and in the *same time* for *both paths*; hence, in MEL's principle, the energy constraint necessitates *noncontemporaneous variations* (i.e., $\delta q \to \Delta q$, $\delta t = 0 \to \Delta q \neq 0$, and results in variable time limits). This can be illustrated with the following simple example of a free particle in rectilinear motion. Here (with q: rectilinear coordinate, and other notations standard),

$$V = 0, \qquad 2T = m\,v^2 \equiv m(dq/dt)^2 = 2E = constant \quad (> 0), \tag{7.9.10a}$$

and, therefore, along its orbit

$$v \equiv \dot{q} = (2E/m)^{1/2} \Rightarrow \delta[\dot{q}(t)] = \delta[(2E/m)^{1/2}] = 0 \qquad (\text{since } \delta E = 0), \tag{7.9.10b}$$

from which, integrating and using the convenient initial conditions $q_1 = q(t_1) \equiv q(0) = 0$, we get

$$q(t) = (2E/m)^{1/2} t, \tag{7.9.10c}$$

a straight line in rectangular Cartesian q versus t axes; and, therefore, any other continuous comparison path *with the same spatio-temporal endpoints as the orbit* (7.9.10c), $(t_1 = 0, q_1 = 0)$ and $[t_2, q_2 = q(t_2)]$, so that $\delta q_1 = \delta q_2 = 0$ [as required by (7.9.9)], would have to be *nonrectilinear* somewhere between t_1, t_2; that is, in there, $\delta[\dot{q}(t)] \neq 0$, in clear contradiction to (7.9.10b); or, trivially, the actual path and its comparison paths coincide. Hence, it is impossible to reach, *by a (smooth) neighboring path of the same constant energy as the orbit*, the endpoint $q_2 = q(t_2)$, as demanded by the boundary condition, in the same time $t_2 - t_1 = t_2 - 0 = t_2$; or, if we insist on isoenergeticity, $\Delta E = \delta E + \dot{E}\Delta t = \delta E = 0$, we cannot have $\delta q_2 = 0$ (and vice versa), but we can have $\Delta q_2 = \delta q_2 + \dot{q}(t_2)\Delta t_2 = 0 \Rightarrow \delta q_2 = -\dot{q}(t_2)\Delta t_2 \neq 0$ $\Rightarrow \Delta t_2 \neq 0$, since (here, and in general) $\dot{q}(t_2) \neq 0$. (As a way out of these difficulties, some have suggested using virtual paths with discontinuous velocity reversals—that is, $\dot{q} \to -\dot{q}$—but this seems artificial and impractical.)

The preceding discussion leads us to the following correct formulation of MEL's principle: Among all sufficiently smooth kinematically admissible trajectories $q(t)$, in configuration space, passing through the given *initial* point $P_1[q_k(t_1) = q_{k1}$: *given*; t_1: *given*] and given *final* point $P_2[q_k(t_2) = q_{k2}$: *given*; but t_2: *unknown*, to be determined from the stationarity condition] and satisfying the total energy constraint $T + V = E$: *given constant*, the actual (kinetic) motion(s), or orbit(s), satisfies (satisfy) (7.9.6b); that is, contrary to the *fixed* time-endpoints Hamilton's principle (7.9.2), MEL's principle (7.9.6b) is a *variable upper time endpoint variational problem*: $\Delta t_1 = 0$ *but* $\Delta t_2 \neq 0$, and as such is fundamentally different from it; although, in both principles, the *total number of given data is* $(n + 1) + (n + 1) = 2n + 2$.

General Kinematico-Inertial Identities

The preceding results (7.9.1a, 2; 4a, b, f, h, i; 6b, etc.) hold for *variations* $\Delta(\ldots)$, $\delta(\ldots)$ *from an orbit*; in fact, they were obtained from the equations of motion, or LP, or the central equation. Let us now see the converse; that is, derive those IVP by direct variations of, say, A_H from a kinematically admissible path, and then specialize to an orbit.

Invoking (7.2.4c, d) and (7.9.3, 3b–d), and with the already familiar notations

$$E_k(\ldots) \equiv [\partial(\ldots)/\partial\dot{q}_k]^{\cdot} - \partial(\ldots)/\partial q_k\text{: } \textit{Euler–Lagrange operator,} \tag{7.9.11a}$$

$$h(\ldots) \equiv \sum [\partial(\ldots)/\partial\dot{q}_k]\dot{q}_k - (\ldots)\text{: } \textit{Generalized energy operator,} \tag{7.9.11b}$$

$$\Delta(\ldots) \equiv \sum \{[\partial(\ldots)/\partial q_k]\Delta q_k + [\partial(\ldots)/\partial\dot{q}_k]\Delta(\dot{q}_k)\} + [\partial(\ldots)/\partial t]\Delta t\text{:}$$
$$\textit{Noncontemporaneous (first-order) variation operator,} \tag{7.9.11c}$$

we find, successively (recalling fig. 7.1),

$$\begin{aligned}
\Delta A_H &\equiv \Delta\int L\,dt \equiv \int L[I;\ \mathrm{arc}(A_1A_2)]\,dt - \int L[II;\ \mathrm{arc}(C_1C_2)]\,dt \\
&= \int L[t+\Delta t, q+\Delta q, \dot{q}+\Delta(\dot{q})]\,d(t+\Delta t) \ - \int L(t,q,\dot{q})\,dt \\
&\approx \int [L\,dt + L\,d(\Delta t) + \Delta L\,dt + \Delta L\,d(\Delta t) - L\,dt] \\
&\approx \int [\Delta L\,dt + L\,d(\Delta t)] \qquad [\text{to the } \textit{first} \text{ order, and with } L(t,q,\dot{q}) \equiv L] \\
&= \int [\Delta L + L(\Delta t)^{\cdot}]\,dt \\
&= \int \Big\{\sum [(\partial L/\partial q_k)\Delta q_k + (\partial L/\partial\dot{q}_k)\Delta(\dot{q}_k)] + (\partial L/\partial t)\Delta t + L(\Delta t)^{\cdot}\Big\}\,dt \\
&\quad [\text{replacing } \Delta(\dot{q}) \text{ with } (\Delta q)^{\cdot} - \dot{q}(\Delta t)^{\cdot}] \\
&= \int \Big\{\sum [(\partial L/\partial q_k)\Delta q_k + (\partial L/\partial\dot{q}_k)(\Delta q_k)^{\cdot}] + (\partial L/\partial t)\Delta t \\
&\qquad\qquad - \Big[\sum (\partial L/\partial\dot{q}_k)\dot{q}_k - L\Big](\Delta t)^{\cdot}\Big\}\,dt \\
&= -\int \sum E_k(L)\Delta q_k\,dt \\
&\qquad + \int \Big\{(\partial L/\partial t)\Delta t - \Big[\sum (\partial L/\partial\dot{q}_k)\dot{q}_k - L\Big](\Delta t)^{\cdot}\Big\}\,dt \\
&\qquad + \int d/dt\Big[\sum (\partial L/\partial\dot{q}_k)\Delta q_k\Big]\,dt \\
&= \cdots = -\int \sum E_k(L)\Delta q_k\,dt + \int \Big[\sum (\partial L/\partial\dot{q}_k)\Delta q_k - h(L)\Delta t\Big]^{\cdot}\,dt \\
&\qquad\qquad + \int [dh(L)/dt + \partial L/\partial t]\Delta t\,dt;
\end{aligned} \tag{7.9.11d}$$

that is,

$$\begin{aligned}
&\Delta A_H + \int \delta' W_{np}\,dt = \\
&\quad -\int \sum [E_k(L) - Q_k]\Delta q_k\,dt + \int \Big[dh(L)/dt + \partial L/\partial t - \sum Q_k\dot{q}_k\Big]\Delta t\,dt, \\
&\quad + \Big\{\sum p_k\Delta q_k - h(L)\Delta t\Big\}_1^2
\end{aligned}$$

$$= -\int \sum [E_k(L) - Q_k]\,\delta q_k\,dt - \int \Big(\sum E_k(L)\dot{q}_k - dh(L)/dt - \partial L/\partial t\Big)\Delta t\,dt$$
$$+ \Big\{\sum p_k \Delta q_k - h(L)\Delta t\Big\}_1^2; \tag{7.9.11e}$$

or, due to the analytical identity (3.9.3b; with T replaced by L),

$$dh(L)/dt + \partial L/\partial t = \sum E_k(L)\dot{q}_k, \tag{7.9.11f}$$

finally [with $h(L)$ renamed simply h],

$$\Delta \int L\,dt = -\int \Big[\sum E_k(L)\,(\Delta q_k - \dot{q}_k \Delta t)\Big]\,dt$$
$$+ \Big\{\sum (\partial L/\partial \dot{q}_k)\Delta q_k - h\Delta t\Big\}_1^2; \tag{7.9.11g}$$

that is,

$$\Delta A_H = -\int \sum E_k(L)\,\delta q_k\,dt + \Big\{\sum p_k \Delta q_k - h\Delta t\Big\}_1^2. \tag{7.9.11h}$$

Kinetic Specializations

(i) For variations from an orbit, $\sum E_k(L)\,\delta q_k = \delta' W_{np}$, by LP, and so (7.9.11h) reduces to (7.9.4a, b).

(ii) If, further, we assume that $\delta' W_{np} = 0$, and choose "*cotermini variations*" in space and time:

$$\Delta q_k(t_1) = \Delta q_k(t_2) = 0 \quad \text{and} \quad \Delta t(t_1) = \Delta t(t_2) = 0, \tag{7.9.12a}$$

then (7.9.11h) yields "Voss' principle," $\Delta A_H = 0$.

(iii) If, again for variations from an orbit and $Q_{k,np} = 0 \Rightarrow \delta' W_{np} = 0$,

$$L = L(q, \dot{q}) \Rightarrow \partial L/\partial t = 0 \Rightarrow h \equiv \sum (\partial L/\partial \dot{q}_k)\dot{q}_k - L = constant, \tag{7.9.12b}$$

we choose *spatially* cotermini variations:

$$\Delta q_k(t_1) = \Delta q_k(t_2) = 0 \quad \text{but} \quad \Delta t(t_1) = \Delta t(t_2) \neq 0, \tag{7.9.12c}$$

then (7.9.11h) also yields $\Delta A_H = 0$.

If, instead of (7.9.12c), $\Delta t(t_1) = 0$, but $\Delta t(t_2) \equiv \Delta t$, then (7.9.11h) reduces to

$$\Delta A_H + h\,\Delta t = 0, \tag{7.9.12d}$$

a form that has applications in nonlinear oscillations (ex. 7.9.13; see also §8.11).

(iv) Again, for variations from an orbit and $Q_{k,np} = 0 \Rightarrow \delta' W_{np} = 0$, and $L = L(q, \dot{q}) \Rightarrow h = constant$, we find, successively,

$$\Delta \int \left[\sum (\partial L/\partial \dot{q}_k)\dot{q}_k\right] dt \qquad \left[= \Delta \int 2T\, dt \equiv \Delta A_L\right]$$

$$\equiv \int \left[\sum (\partial L/\partial \dot{q}_k)\dot{q}_k\right]_{II} dt - \int \left[\sum (\partial L/\partial \dot{q}_k)\dot{q}_k\right]_{I} dt$$

$$= \Delta \int (L + h)\, dt$$

$$= \Delta \int L\, dt + \Delta[h(t_2 - t_1)]$$

$$= \Delta A_H + \Delta[h(t_2 - t_1)]$$

$$= \cdots = \left\{\sum p_k \Delta q_k\right\}_1^2 + \Delta h(t_2 - t_1) \qquad \text{[invoking (7.9.11h)]}$$

$$= \left\{\sum p_k \Delta q_k + (t)\, \Delta h\right\}_1^2 \tag{7.9.12e}$$

[we notice that, here, $\Delta h \equiv \delta h + \dot{h}\, \Delta t = \delta h$ ($= \delta E$, for stationary constraints)], and so, if in addition, we choose the *spatially* cotermini and *isoenergetic* variations

$$\Delta q_k(t_1) = \Delta q_k(t_2) = 0 \qquad \text{and} \qquad \Delta h = 0, \tag{7.9.12f}$$

(7.9.12d) yields

$$\Delta \int \left[\sum (\partial L/\partial \dot{q}_k)\dot{q}_k\right] dt \;\rightarrow\; \Delta A_L = 0; \tag{7.9.12g}$$

which coincides with the earlier MEL principle (7.9.6b).

ANALYTICAL REMARK

(See also ex. 7.9.1 and prob. 7.9.1.) Clearly, (7.9.11d) is a general result holding for *any* (well-behaved) function $F = F(t, q, \dot{q})$; that is,

$$\Delta \int F\, dt = \cdots = -\int \sum E_k(F)\, \Delta q_k\, dt$$
$$+ \int \left\{(\partial F/\partial t)\Delta t - \left[\sum (\partial F/\partial \dot{q}_k)\dot{q}_k - F\right](\Delta t)^{\cdot}\right\} dt$$
$$+ \int d/dt\left[\sum (\partial F/\partial \dot{q}_k)\Delta q_k\right] dt$$
$$= \cdots = -\int \sum E_k(F)\Delta q_k\, dt + \int [dh(F)/dt + \partial F/\partial t]\Delta t\, dt$$
$$+ \left\{\sum (\partial F/\partial \dot{q}_k)\Delta q_k - h(F)\Delta t\right\}_1^2,$$
$$= \cdots = -\int \sum E_k(F)\, \delta q_k\, dt + \left\{\sum (\partial F/\partial \dot{q}_k)\Delta q_k - h(F)\Delta t\right\}_1^2$$
$$= -\int \sum E_k(F)\, \delta q_k\, dt + \left\{\sum (\partial F/\partial \dot{q}_k)\, \delta q_k - F\Delta t\right\}_1^2. \tag{7.9.12h}$$

A Generalization of MEL's Principle; Jacobi's Form

Let us consider a *conservative* system; that is, recalling (7.2.6e) and §3.9, one in which $\partial L/\partial t = 0$, all forces are either potential (included in L) or gyroscopic, and all additional Pfaffian constraints are, at most, catastatic. Then, the power equation reduces to the *Jacobi–Painlevé integral*

$$h = L_2 - L_0 \equiv T_2 + (V - T_0) = constant \Rightarrow (L_2)^{1/2} = (L_0 + h)^{1/2}. \qquad (7.9.13a)$$

It is then possible to replace the variable time-endpoint principle (7.9.6b, 12e) with a simpler one with *fixed* time-endpoints. To this end, we first define the Jacobi action-like functional:

$$\begin{aligned} A_J{}' &\equiv \int L\,dt - \int \left[(L_2)^{1/2} - (L_0 + h)^{1/2}\right]^2 dt \\ &\equiv A_H - \int [\ldots]^2\,dt \\ &\left[= \cdots = \int \{2[L_2(L_0 + h)]^{1/2} + L_1 - h\}\,dt\right]. \end{aligned} \qquad (7.9.13b)$$

For general contemporaneous variations around an admissible path, clearly, we will have

$$\delta A_J{}' = \delta A_H - \int 2[\ldots]\,\delta[\ldots]\,dt, \qquad (7.9.13c)$$

and, therefore, for variations from an orbit—that is, an actual motion satisfying (7.9.13a) (with the same h-value for both $A_J{}'$ and A_H) — we shall have

$$\delta A_J{}' = \delta A_H \quad (\to 0, \text{ for vanishing endpoint variations}); \qquad (7.9.13d)$$

or, neglecting the constant last h-term in (7.9.13b), $A_J{}' \to A_J$:

$$A_J \equiv \int \{2[L_2(L_0 + h)]^{1/2} + L_1\}\,dt \equiv \int J(q, \dot{q})\,dt, \qquad (7.9.13e)$$

we arrive at the "least action"-like (better, *stationarity*) condition

$$\delta A_J = 0, \qquad (7.9.13f)$$

Since the integrand of the above, $J = J(q, \dot{q})$, is positively homogeneous of the *first* degree in the $\dot{q}$'s (which means that, for *any positive* number λ, $\lambda J(q, \dot{q}) = J(q, \lambda\dot{q})$; i.e., J is not really a function of time t), we can write

$$A_J = \int J(q, \dot{q})\,dt = \int J(q, dq) = \int J(q, q')\,d\sigma, \qquad (7.9.13g)$$

where $t = t(\sigma) \Leftrightarrow \sigma = \sigma(t)$ are arbitrary (increasing) functions (and, therefore, the integration limits are changed accordingly) and

$$(\ldots)^{\cdot} \equiv d(\ldots)/dt = [d(\ldots)/d\sigma]\,\dot{\sigma} \equiv (\ldots)'\,\dot{\sigma}, \qquad (\ldots)' \equiv d(\ldots)/d\sigma; \qquad (7.9.13h)$$

for example, $\dot{q}_k \equiv q_k{}'\dot{\sigma}$. Then, the stationarity condition (7.9.13e) leads to the following n Euler–Lagrange equations:

$$(\partial J/\partial q_k{}')' - \partial J/\partial q_k = 0; \tag{7.9.13i}$$

and, conversely, the parameter t can be chosen so that

$$[L_2(q, \dot{q})]^{1/2} = [L_0(q) + h]^{1/2}$$
$$\Rightarrow t = \int \{L_2(q, \dot{q})/[L_0(q) + h]\}^{1/2}\, d\sigma, \tag{7.9.13j}$$

where the $q(\sigma)$'s verify (7.9.13i). Then, from $\delta A_J = 0$, it also follows that $\delta A_H = 0$.

Next, from (7.9.13i), and since $J = \sum (\partial J/\partial q_k{}')q_k{}'$, we readily find the additional equation,

$$\sum [(\partial J/\partial q_k{}')' - \partial J/\partial q_k] q_k{}'$$
$$= \left[\sum (\partial J/\partial q_k{}')q_k{}'\right]' - \sum [(\partial J/\partial q_k{}')q_k{}'' + (\partial J/\partial q_k)q_k{}']$$
$$= J' - J' = 0; \tag{7.9.13k}$$

that is, *only $n - 1$ of the n equations (7.9.13i) are independent* (something to be expected, due to the σ arbitrariness). Hence if, following Jacobi, we choose as σ one of the q's, say, q_1, the orbit(s) of the system will be given by the following $n - 1$ Lagrangean equations in $q_2(q_1), \ldots, q_n(q_1)$:

$$d/dq_1 [\partial J/\partial(dq_k/dq_1)] - \partial J/\partial q_k = 0 \qquad (k = 2, \ldots, n), \tag{7.9.13l}$$

whose general solution will depend on $2(n - 1) + 1 = 2n - 1$ integration constants [one of them could be the h constant in (7.9.13a)].

Jacobi's Form (early 1840s)

If, in addition to being conservative and holonomic, our system is also *scleronomic* (and these restrictions severely limit the usefulness of the principle to engineering dynamics problems) then, since in this case

$$L_2 = T_2 = T = \sum\sum (1/2) M_{kl}\dot{q}_k\dot{q}_l, \qquad M_{kl} = M_{kl}(q), \tag{7.9.14a}$$
$$L_1 = T_1 = 0, \quad L_0 = T_0 - V = -V, \qquad V = V(q), \tag{7.9.14b}$$
$$h = T + V = E \qquad \text{(a positive constant)}, \tag{7.9.14c}$$

the action A_J, (7.9.13f), reduces to

$$A_J = \int 2[T(E - V)]^{1/2}\, dt. \tag{7.9.14d}$$

But, recalling (3.9.4o) or (6.7.2a), we have

$$2T(dt)^2 = \sum\sum M_{kl}\, dq_k\, dq_l = 2(E - V)\,(dt)^2 \equiv ds^2, \tag{7.9.14e}$$

ds: elementary arc-length along orbit of figurative particle representing the system in configuration space, (7.9.14f)

and therefore we can rewrite (7.9.14d) as

$$\begin{aligned} A_J &= \int [2(E - V)]^{1/2} [2T(dt)^2]^{1/2} \\ &= \int \Big[2(E - V)\Big(\sum\sum M_{kl}\, dq_k\, dq_l\Big)\Big]^{1/2} \\ &= \int [2(E - V)]^{1/2}\, ds, \end{aligned} \tag{7.9.14g}$$

or, in terms of the new function R (and the earlier parameter σ) defined by

$$\begin{aligned} &2(E - V)\Big(\sum\sum M_{kl}\, dq_k\, dq_l\Big) \equiv R(d\sigma)^2 \\ &\Rightarrow R = R(q, q') = 2(E - V)\Big(\sum\sum M_{kl} q_k{}' q_l{}'\Big), \end{aligned} \tag{7.9.14h}$$

or, equivalently,

$$2T\, dt = \sqrt{R}\, d\sigma = J(q, q')\, d\sigma, \tag{7.9.14i}$$

we can, finally, bring A_J to the *Jacobi form* (fixed time-endpoints):

$$A_J = \int \sqrt{R}\, d\sigma \quad \Big[= A_L \equiv \int 2T\, dt\Big] \tag{7.9.14j}$$

with limits σ_1 and σ_2 corresponding to the initial and final system positions, respectively.

Again, with the choice $\sigma = q_1$ (and since $M_{kl} = M_{lk}$), and with *upper*-case subscripts running from 2 to n, we find, successively,

$$\begin{aligned} ds^2 &\equiv \sum\sum M_{kl}\, dq_k\, dq_l = \sum\sum M_{kl}\,(dq_k/dq_1)\,(dq_l/dq_1)\,(dq_1)^2 \\ &= \Big[M_{11} + \sum M_{K1}\,(dq_K/dq_1) \\ &\qquad + \sum M_{1L}\,(dq_L/dq_1) + \sum\sum M_{KL}\,(dq_K/dq_1)(dq_L/dq_1)\Big](dq_1)^2 \\ &= \Big[M_{11} + 2\sum M_{K1}\,(dq_K/dq_1) + \sum\sum M_{KL}\,(dq_K/dq_1)\,(dq_L/dq_1)\Big](dq_1)^2, \end{aligned}$$

and, therefore,

$$R = \cdots = 2(E - V)\Big(M_{11} + 2\sum M_{1L}\, q_L{}' + \sum\sum M_{KL}\, q_K{}' q_L{}'\Big), \tag{7.9.14k}$$

and

$$A_J = A_L = \int \sqrt{R}\, dq_1; \tag{7.9.14l}$$

and, of course, the orbits are defined by the Euler–Lagrange equations of $\delta A_J = 0$ [(7.9.13l) with $J \to \sqrt{R}$]:

$$d/dq_1\,(\partial\sqrt{R}/\partial q_K{}') - \partial\sqrt{R}/\partial q_K = 0 \qquad (K = 2, \ldots, n). \tag{7.9.14m}$$

Also, since by (7.9.14e, f) and (7.9.14h)

$$dt = \left(\sum\sum M_{kl}\,dq_k\,dq_l \Big/ 2(E - V)\right)^{1/2} = [\sqrt{R}/2(E - V)]\,d\sigma, \tag{7.9.14n}$$

we can find time by the quadrature:

$$t - t_1 = \int [\sqrt{R}/2(E - V)]\,dq_1. \tag{7.9.14o}$$

Here, the general solution involves a total of $2n$ constants: $2(n - 1)$ from the integration of the $n - 1$ equations (7.9.14m), plus E and t_1.

In closing, we should point out that if the initial and final orbit points, P_1 and P_2 respectively, are *close* to each other, then that orbit is *unique*; and, further, for that orbit, A_L is a *minimum* or least; in Hertz's terminology, $\mathrm{arc}(P_1P_2)$ = *shortest* or *straightest path*, in configuration space (and, for holonomic systems, coincides with the *geodesic* through these points). The quantification of these ideas constitutes the *extremum* theory of variational calculus, and is summarized in this chapter's appendix.

Example 7.9.1 *Alternative Derivations of $\Delta\int L\,dt$* [recall (7.9.3b ff.)]. To avoid variable time-endpoints variations, we may introduce the "arc parameter" σ via $t = t(\sigma)$ [similar to that of the Jacobi form of MEL's principle (7.9.13g ff.)] *for both the fundamental (kinetic) path and its Δ-variation*, so that

$$t(\sigma_1) = t_1, \qquad t(\sigma_2) = t_2, \quad \text{and} \quad dt = (dt/d\sigma)\,d\sigma \equiv t'\,d\sigma \qquad [(\ldots)' \equiv d(\ldots)/d\sigma]. \tag{a}$$

Then, and since $\dot{q}_k \equiv dq_k/dt = (dq_k/d\sigma)\,(d\sigma/dt) = q_k{}'/t'$,

$$A_H \equiv \int L\,dt = \int L_\sigma\,d\sigma, \tag{b1}$$

$$L_\sigma \equiv L(dt/d\sigma) = L(t, q, \dot{q})t' = L(t, q, q'/t')t' \equiv L_\sigma(t, t', q, q')$$
$$[\text{function of the } 2n + 2 \text{ functions of } \sigma\text{: } t, q, t', q']. \tag{b2}$$

By Δ-varying A_H, we obtain, successively (since now $\Delta\sigma = 0$),

$$\Delta A_H = \int \Delta L_\sigma\, d\sigma$$

$$= \int \Big\{ \sum \big[(\partial L_\sigma/\partial q_k)\Delta q_k + (\partial L_\sigma/\partial q_k{}')\,\Delta(q_k{}') \big] + (\partial L_\sigma/\partial t)\Delta t + (\partial L_\sigma/\partial t')\Delta t' \Big\}\, d\sigma$$

[integrating the $\Delta q'$ and $\Delta t'$-proportional terms by parts, while noting that, since $\Delta\sigma = 0$, the path points $P(t, q)$ and $P + \Delta P(t + \Delta t, q + \Delta q)$ *correspond to the same value of* σ, and, therefore, $\Delta[d(\ldots)/d\sigma] = d\Delta(\ldots)/d\sigma$ — see Remark below.]

$$= -\int \sum \big[d/d\sigma\,(\partial L_\sigma/\partial q_k{}') - (\partial L_\sigma/\partial q_k) \big] \Delta q_k\, d\sigma$$

$$- \int \big[d/d\sigma\,(\partial L_\sigma/\partial t') - (\partial L_\sigma/\partial t) \big] \Delta t\, d\sigma$$

$$+ \Big\{ \sum (\partial L_\sigma/\partial q_k{}')\Delta q_k \Big\}_1^2 + \{ (\partial L_\sigma/\partial t')\Delta t \}_1^2. \tag{c}$$

But, by chain rule:

$$\text{(i)}\quad \partial L_\sigma/\partial q_k{}' = t' \big[(\partial L/\partial \dot q_k)\,(\partial \dot q_k/\partial q_k{}') \big]$$
$$= t'\,(\partial L/\partial \dot q_k)\,\big[(\partial (q_k{}'/t')/\partial q_k{}') \big]$$
$$= t'\,(\partial L/\partial \dot q_k)\,(1/t') = \partial L/\partial \dot q_k \equiv p_k, \tag{c1}$$

$$\text{(ii)}\quad \partial L_\sigma/\partial t' = L + t' \Big[\sum (\partial L/\partial \dot q_k)\,(\partial \dot q_k/\partial t') \Big]$$
$$= L + t' \Big\{ \sum (\partial L/\partial \dot q_k)\,\big[\partial (q_k{}'/t')/\partial t' \big] \Big\}$$
$$= L + t' \Big\{ \sum (\partial L/\partial \dot q_k)\,\big[(-q_k{}')\,(t')^{-2} \big] \Big\}$$
$$= L - \sum (\partial L/\partial \dot q_k)\,(q_k{}'/t')$$
$$= L - \sum (\partial L/\partial \dot q_k)\dot q_k \;=\; -h \text{ (generalized energy/Hamiltonian)}; \tag{c2}$$

and so (c) reduces to

$$\Delta A_H = -\int \Big\{ \sum \big[d/d\sigma\,(\partial L_\sigma/\partial q_k{}') - \partial L_\sigma/\partial q_k \big] \Delta q_k$$
$$+ \big[d/d\sigma\,(\partial L_\sigma/\partial t') - \partial L_\sigma/\partial t \big] \Delta t \Big\}\, d\sigma$$
$$+ \Big\{ \sum p_k \Delta q_k - h\Delta t \Big\}_1^2. \tag{d}$$

If, at σ_1, σ_2, the Δq_k and Δt vanish (generally, if the boundary term vanishes, say, by periodicity), then the single variational equation $\Delta A_H = 0$ leads to the following $n + 1$ differential equations:

$$d/d\sigma\,(\partial L_\sigma/\partial q_k{}') - \partial L_\sigma/\partial q_k = 0 \qquad (k = 1, \ldots, n), \tag{d1}$$

$$d/d\sigma\,(\partial L_\sigma/\partial t') - \partial L_\sigma/\partial t = 0. \tag{d2}$$

Let us find the power (energy rate) equation associated with eqs. (d1): multiplying each one of them with $q_k{}'$ and summing over k yields, successively,

$$\begin{aligned} 0 &= \sum [(q_k{}')\, d/d\sigma(\partial L_\sigma/\partial q_k{}') - (\partial L_\sigma/\partial q_k) q_k{}'] \\ &= d/d\sigma\Big[\sum (\partial L/\partial q_k{}') q_k{}'\Big] - \sum [(\partial L_\sigma/\partial q_k{}') q_k{}'' + (\partial L_\sigma/\partial q_k)\, q_k{}'] \\ &= d/d\sigma\Big[\sum (\partial L_\sigma/\partial q_k{}') q_k{}'\Big] - [dL_\sigma/\, d\sigma - (\partial L_\sigma/\partial t')t'' - (\partial L_\sigma/\partial t)t'], \end{aligned}$$

from which, rearranging, and invoking (b2, c2), we obtain the "parametric power equation":

$$dh_\sigma/d\sigma = -(\partial L_\sigma/\partial t')t'' - (\partial L_\sigma/\partial t)t' = h\, t'' - (\partial L/\partial t)\,(t')^2, \tag{e}$$

where the *parametric generalized energy* h_σ is defined as

$$\begin{aligned} h_\sigma &\equiv \sum (\partial L_\sigma/\partial q_k{}') q_k{}' - L_\sigma \\ &= \sum p_k(\dot{q}_k t') - (L\, t') = \Big(\sum p_k \dot{q}_k - L\Big) t' = h\, t'. \end{aligned} \tag{e1}$$

On the other hand, by (b2, c2) again, eq. (d2) is rewritten as

$$\begin{aligned} &- dh/d\sigma - t'\,(\partial L/\partial t) = 0 \\ &\Rightarrow dh/d\sigma = (dh/dt)t' = -t'\,(\partial L/\partial t) \;\Rightarrow\; dh/dt = -\partial L/\partial t, \end{aligned} \tag{e2}$$

as expected; and, hence, $d(\ldots)/\,d\sigma$-differentiating (e1), we obtain

$$dh_\sigma/d\sigma = (dh/d\sigma)t' + h\, t'' = -(\partial L/\partial t)\,(t')^2 + h\, t''; \tag{e3}$$

that is, eq. (e).

In sum: eq. (d2) is not independent from eqs. (d1), but results from them as their *power equation*. [For further related results, see Frank (1927, pp. 13–16, 23–24) and Nevzgliadov (1959, pp. 371–375).]

REMARK

In view of the earlier "σ-commutativity": $\Delta[d(\ldots)/d\sigma] = d\Delta(\ldots)/d\sigma$, the former noncommutativity relation (7.2.4e) for a typical coordinate $q_k(t)$, or simply $q(t)$, generalizes as follows:

$$\begin{aligned} \Delta(\dot{q}) = \Delta\big[(dq/d\sigma)/(dt/d\sigma)\big] &= \frac{(dt/d\sigma)\Delta(dq/d\sigma) - (dq/d\sigma)\Delta(dt/d\sigma)}{(dt/d\sigma)^2} \\ &= \frac{(dt/d\sigma)[d(\Delta q)/d\sigma] - (dq/d\sigma)[d(\Delta t)/d\sigma]}{(dt/d\sigma)^2} \\ &= (\Delta q)'/t' \;-\; \dot{q}[(\Delta t)'/t'] \qquad [\text{since } q'/t' = \dot{q}] \\ &= (\Delta q)^{\cdot} - \dot{q}(\Delta t)^{\cdot}. \end{aligned} \tag{f}$$

The above can be viewed as a general definition of $\Delta(dq/dt)$.

Problem 7.9.1 Continuing from the preceding example, show that for an arbitrary (but as well behaved as needed) function $F = F(t, q, \dot{q})$:

$$F \rightarrow F_\sigma \equiv F\,(dt/d\sigma) = F(t, q, \dot{q})\,t' = F(t, q, q'/t')\,t' \equiv F_\sigma(t, t', q, q'), \qquad \text{(a)}$$

the following integral variational identities hold:

$$\begin{aligned}
\Delta\int F\,dt &= \int \Delta F_\sigma\,d\sigma \\
&= \int\Big\{\sum\,[(\partial F_\sigma/\partial q_k)\Delta q_k + (\partial F_\sigma/\partial q_k{}')\Delta(q_k{}')] + (\partial F_\sigma/\partial t)\Delta t + (\partial F_\sigma/\partial t')\Delta t'\Big\}\,d\sigma \\
&= \int\Big\{\sum\,[(\partial F/\partial q_k)\Delta q_k + (\partial F/\partial \dot{q}_k)\,(\partial(q_k{}'/t')/\partial q_k{}')\,(\Delta q_k)^{\cdot} t']t' \\
&\qquad + (\partial F/\partial t)t'\Delta t + \sum\,[(\partial F/\partial \dot{q}_k)\,(\partial(q_k{}'/t')/\partial t')t' + F]\,(\Delta t)^{\cdot}\,t'\Big\}\,d\sigma \\
&= \int\Big\{\sum\,[(\partial F/\partial q_k)\Delta q_k + (\partial F/\partial \dot{q}_k)\,(\Delta q_k)^{\cdot}] \\
&\qquad + (\partial F/\partial t)\Delta t - \sum\,(\partial F/\partial \dot{q}_k)\dot{q}_k(\Delta t)^{\cdot} + F(\Delta t)^{\cdot}\Big\}\,t'\,d\sigma \\
&= \int\Big\{\sum\,[(\partial F/\partial q_k)\Delta q_k + (\partial F/\partial \dot{q}_k)\,((\Delta q_k)^{\cdot} - \dot{q}_k(\Delta t)^{\cdot})] \\
&\qquad + (\partial F/\partial t)\Delta t + F(\Delta t)^{\cdot}\Big\}\,dt; \qquad \text{(b)}
\end{aligned}$$

that is,

$$\Delta\int F\,dt = \int \Delta F_\sigma\,d\sigma = \int [\Delta F + F(\Delta t)^{\cdot}]\,dt. \qquad \text{(c)}$$

Problem 7.9.2 Consider the following four possible definitions of the *total noncontemporaneous variation of the Hamiltonian action* $A_H \equiv A$:

$$\Delta^T A \equiv \int_{t_1+\Delta t_1}^{t_2+\Delta t_2} (L + \delta L)\,dt - \int_{t_1}^{t_2} L\,dt, \qquad \text{(a)}$$

$$\Delta^T A \equiv \int_{t_1}^{t_2} (L + \Delta L)\,d(t + \Delta t) - \int_{t_1}^{t_2} L\,dt, \qquad \text{(b)}$$

$$\Delta^T A \equiv \int_{t_1+\Delta t_1}^{t_2+\Delta t_2} (L + \Delta L)\,d(t + \Delta t) - \int_{t_1}^{t_2} L\,dt, \qquad \text{(c)}$$

$$\Delta^T A \equiv \int_{t_1+\Delta t_1}^{t_2+\Delta t_2} (L + \delta L)\,d(t + \Delta t) - \int_{t_1}^{t_2} L\,dt. \qquad \text{(d)}$$

Examine them carefully, and determine which ones of them, *to the first order* (i.e., $\Delta^1 A \equiv \Delta A$), lead to the correct expression; that is,

$$\Delta A = \Delta \int L\,dt = \int \delta L\,dt + \{L\Delta t\}_1^2$$
$$= \int \{\Delta L + L[d(\Delta t)/dt]\}\,dt = \int [\Delta(\ldots)\,dt + (\ldots)\,d(\Delta t)]. \qquad \text{(e)}$$

HINT

Consult any good text on variational calculus; for example, Elsgolts (1970, pp. 341–364), Fox (1950–1963/1987), Gelfand and Fomin (1963, pp. 54–66).

ANSWERS

Yes: a, b; No: c, d.

Problem 7.9.3 *O. Hölder* and *Voss* forms of the action principle (may be omitted in a first reading).

(i) By invoking the noncommutativity equation (7.2.4e): $(\Delta q_k)^{\cdot} - \Delta(\dot{q}_k) = \dot{q}_k(\Delta t)^{\cdot}$, show that

$$\int \Delta L\,dt = \int \Big[(\partial L/\partial t)\Delta t + \sum (\partial L/\partial q_k)\Delta q_k + \sum (\partial L/\partial \dot{q}_k)\Delta(\dot{q}_k)\Big]\,dt$$
$$= \int \Big\{(\partial L/\partial t)\Delta t - \Big[\sum (\partial L/\partial \dot{q}_k)\dot{q}_k\Big](\Delta t)^{\cdot}\Big\}\,dt - \int \sum E_k(L)\,\Delta q_k\,dt$$
$$+ \Big\{\sum (\partial L/\partial \dot{q}_k)\Delta q_k\Big\}_1^2. \qquad \text{(a)}$$

Next, consider the special case where Δq, *although still noncontemporaneous (i.e., $\Delta t \neq 0$), equals numerically the virtual displacement at time t*, δq [fig. 7.3(a)]:

$$\Delta q(t + \Delta t) = \delta q(t); \qquad \text{(b)}$$

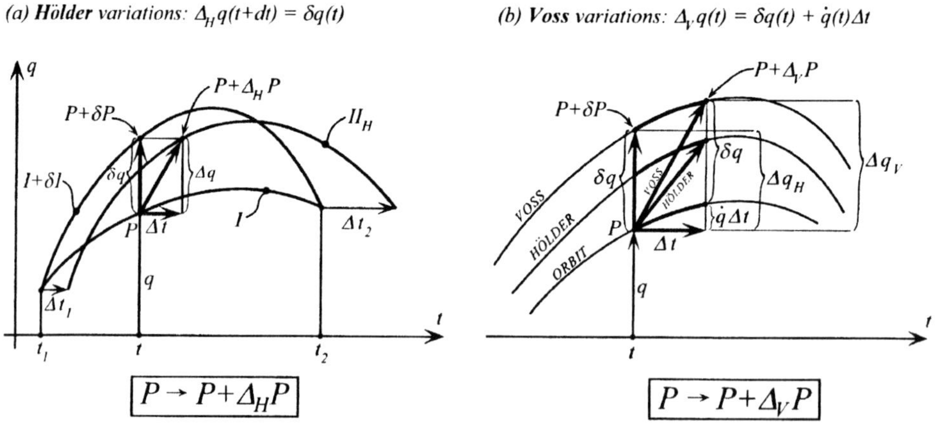

Figure 7.3 On the meaning of the *Hölder* and *Voss* variations.

and, accordingly, the point (t, q) of the fundamental path $\rightarrow$ orbit I is mapped to the neighboring point $(t + \Delta t, q + \Delta q = q + \delta q)$, and the totality of the latter constitutes the *varied path in the sense of O. Hölder* $II_H = I + \Delta_H I$ (symbolically). It follows that, in this case, Δt is not necessarily zero at the path endpoints $t_{1,2}$, even if the δq are.

Then, and assuming $\delta' W_{\text{nonpotential}} = 0$, the *second* integral vanishes, and so (a) reduces to *O. Hölder*'s variational "principle" (1896):

$$\int \left\{ \Delta L + \left(\sum (\partial L/\partial \dot{q}_k)\dot{q}_k \right) (\Delta t)^{\cdot} - (\partial L/\partial t)\Delta t \right\} dt$$
$$= \left\{ \sum (\partial L/\partial \dot{q}_k)\Delta q_k \right\}_1^2$$
$$[= 0; \quad \text{e.g., if } \Delta q_k(t_{1,2}) = 0]. \tag{c}$$

(ii) For nonholonomic systems, the so-varied path will *not* satisfy the constraints; that is, *the Hölder-varied path is not kinematically possible*, unless the constraints are holonomic. For this reason, *Voss* (in 1990) chose:

$$q + \Delta q\text{: kinematically possible at } t + \Delta t, \qquad \dot{q}\text{: kinematically possible at } t$$
$$\Rightarrow \Delta q - \dot{q}\Delta t = \delta q \qquad (\textit{virtual}; \text{ recalling §2.5 ff.}); \tag{d}$$

that is, the I-point (t, q) is mapped to the neighboring point $(t + \Delta t, q + \Delta q = q + \delta q + \dot{q}\Delta t)$, and the totality of the latter constitutes the *varied path in the sense of Voss* $II_V = I + \Delta_V I$ [symbolically, fig. 7.3(b)]. This is how the variation $\Delta(\ldots) = \delta(\ldots) + (\ldots)^{\cdot}\Delta t$ was created; and, of course, it yields the integral Δ-theorems already detailed in §7.9.

[See also Pars, 1965, p. 533 ff.; with a slightly different (confusion-prone) notation; and Lützen, 1995(b), p. 55 ff., for the history of these variations, etc.]

Example 7.9.2 *Least Action as a Constrained Variational Problem.* Let us formulate MEL's principle (7.9.6b) as a variable (second) endpoint variational problem under the energy constraint

$$E \equiv T(q, \dot{q}) + V(q) = h = \textit{constant} \equiv C, \tag{a}$$

and then deduce from it the correct Lagrangean equations of motion; that is, $E_k(L) = 0$.

We recall that, for a holonomic, scleronomic, and potential system, MEL's principle of "least" action (or Lagrange's problem of variational calculus) states that $A_L \equiv \int 2T\, dt$ is stationary for the orbit in the class of admissible paths that satisfy (i) the constraint (a), where C is a *given constant* along the orbit; the same for all admissible paths (i.e., $\Delta h = \delta h + \dot{h}\Delta t = 0 + 0 = 0$), and (ii) the $2n + 1$ boundary conditions: $t_1, q_k(t_1) = q_{k1}, q_k(t_2) = q_{k2}$, *given*; but t_2 *not given*. Due to constraint (a), we will not work with A_L, but with the *unconstrained* variation functional

$$F = \int f\, dt, \qquad \text{where } f = f(t, q, \dot{q}; \lambda) \equiv 2T + \lambda(T + V - C). \tag{b}$$

Then, by (7.9.12g) applied to (b), we see that the stationarity condition

$$0 = \Delta F = \Delta \int f\, dt$$
$$= -\int \sum E_k(f)\, \delta q_k\, dt + \Big\{ \sum (\partial f/\partial \dot{q}_k) \Delta q_k - h(f) \Delta t \Big\}_1^2 \qquad \text{(c)}$$

leads to: (i) the differential equations

$$E_k(f) = 0$$
$$\Rightarrow (2+\lambda)\,[d/dt\,(\partial T/\partial \dot{q}_k) - \partial(T-V)/\partial q_k]$$
$$-\,2(1+\lambda)\,(\partial V/\partial q_k) + (d\lambda/dt)\,(\partial T/\partial \dot{q}_k) = 0, \qquad \text{(d)}$$

and (ii) the boundary or *transversality* condition

$$\Big\{ \sum (\partial f/\partial \dot{q}_k) \Delta q_k - h(f) \Delta t \Big\}_1^2 = 0, \qquad \text{(e)}$$

from which, since $t_1, q_k(t_1), q_k(t_2)$ are given, we conclude that

$$\Delta t_1 = 0, \qquad \Delta q_k(t_1) = \delta q_k(t_1) = 0,$$
$$\Delta q_k(t_2) = \delta q_k(t_2) + \dot{q}_k(t_2)\Delta t_2 = 0 \;\Rightarrow\; \delta q_k(t_2) = -\dot{q}_k(t_2)\Delta t_2 \neq 0, \qquad \text{(e1)}$$

and, since $\Delta t_2 \neq 0$,

$$h(f, \textit{evaluated at } t_2) \equiv \Big\{ \sum (\partial f/\partial \dot{q}_k)\dot{q}_k - f \Big\}_2 \equiv h(f_2) = 0 \qquad \text{(e2)}$$
$$\Rightarrow \{2T(1+\lambda)\}_2 = 0 \Rightarrow \lambda(t_2) \equiv \lambda_2 = -1. \qquad \text{(e3)}$$

On the other hand, applying the purely analytical result (7.9.11f, with L replaced by f) to (b), while recalling (a), we find

$$dh(f)/dt + \partial f/\partial t = \sum E_k(f)\dot{q}_k = 0 \qquad \text{[by (d)]} \qquad \text{(f1)}$$
$$\Rightarrow dh(f)/dt = -\partial f/\partial t = -(d\lambda/dt)\,(T + V - C) = 0$$
$$\Rightarrow h(f) = 2T(1+\lambda) = \textit{constant along the orbit}, \qquad \text{(f2)}$$

and, combining this with (e3), we conclude that

$$\lambda = \lambda(t) = -1, \textit{everywhere} \text{ on the orbit.} \qquad \text{(g)}$$

Then, (b) yields

$$f = 2T - (T + V - C) = L + C \;\Rightarrow\; E_k(f) = E_k(L) = 0. \qquad \text{(h)}$$

In sum: the stationarity condition $\Delta F = 0$ yields the correct Lagrangean equations of motion. The unknown (not necessarily unique) t_2, corresponding to the given data of our orbit, is determined from the $n+1$ equations

$$q_k(t_1;\, c_1, \ldots, c_{2n}) = q_{k1}, \qquad q_k(t_2;\, c_1, \ldots, c_{2n}) = q_{k2}, \qquad \text{(i1)}$$
$$T(q, \dot{q}) + V(q) = C, \qquad \text{(i2)}$$

where $q_k = q_k(t;\, c_1, \ldots, c_{2n})$ is the general solution of (h).

See also Papastavridis [1986(c)], which also contains a study of the extremality of F via the study of its second variation $\Delta^2 F$.

Example 7.9.3 *Whittaker's Variational Principle*: "Show that the principle of Least Action can be extended to systems for which the integral of energy does not exist, in the following form. Let the expression $\sum (\partial L/\partial \dot{q}_k)\dot{q}_k - L$ be denoted by h; then the integral [our terminology]

$$A_{\text{Whittaker}} \equiv A_W \equiv \int \Big(\sum (\partial L/\partial \dot{q}_k)\dot{q}_k + t(dh/dt)\Big)\, dt, \tag{a}$$

has a stationary value for any part of an actual trajectory (i.e., an orbit) as compared with other paths between the same terminal points for which h has the same terminal values" (Whittaker, 1937, p. 248).

This constitutes the extension of MEL's "least action," (7.9.6b), to potential but nonconservative systems (i.e., $\partial L/\partial t \neq 0$), where, as a result, no Jacobi–Painlevé integral exists. The most general such Lagrangean is

$$L(t, q, \dot{q}) = L_1(t, q, \dot{q}) + \sum Q_k(t) q_k \tag{b}$$

(e.g., forced autonomous vibrations), in which case the power theorem reduces to

$$dh/dt = -\partial L/\partial t = -\partial L_1/\partial t - \sum (dQ_k/dt) q_k \neq 0. \tag{c}$$

Whittaker's theorem states that

$$\Delta A_W = 0, \tag{d}$$

under

$$\Delta q_k(t_1) = \Delta q_k(t_2) = 0 \quad \text{and} \quad \Delta h(t_1) = \Delta h(t_2) = 0. \tag{d1}$$

To prove (d, d1), first, invoking the h-definition, we transform A_W to

$$\begin{aligned} A_W &= \int [L + h + t(dh/dt)]\, dt \\ &= \int L\, dt + \{t h\}_1^2 \\ &= A_H + \{t h\}_1^2\text{: Hamilton's } \textit{characteristic function} \text{ (see also §8.11).} \end{aligned} \tag{e}$$

Then, operating on (e) with $\Delta(\ldots)$, while recalling (7.9.3b, c; 11f, g, h), we obtain

$$\begin{aligned} \Delta A_W &= \Delta A_H + \Delta\{t h\}_1^2 \\ &= -\int \sum E_k(L)\, \delta q_k\, dt + \Big\{\sum p_k \Delta q_k - h\Delta t\Big\}_1^2 + \Delta\{t h\}_1^2 \\ &= -\int \sum E_k(L)\, \delta q_k\, dt + \Big\{\sum p_k \Delta q_k + t\Delta h\Big\}_1^2 = 0, \end{aligned} \tag{f}$$

by (d1) and LP for the orbit, Q.E.D. Of course, other boundary conditions (e.g., periodic ones) could have nullified the boundary term. For an investigation of the *extremality* of A_W via the study of the sign of $\Delta^2 A_W$, see Papastavridis [1985(b)].

Example 7.9.4 *Projectile Motion via Jacobi's Form of Least Action.* Let us study the motion of a particle of mass m in the vertical plane under constant gravity, and neglecting air resistance, via the geodesic form of Jacobi's principle, (7.9.14a ff.).

Here, with $q_1 = x$ (horizontal), $q_2 = y$ (positive upward; $y = 0$: ground), and

$$2T = m(\dot{x}^2 + \dot{y}^2), \qquad V = m g y, \tag{a}$$

the energy equation is

$$(m/2)(\dot{x}^2 + \dot{y}^2) + m g y = E\text{: } \textit{total energy}\text{, a positive constant.} \tag{b}$$

As a result, the integrand of the Jacobi functional, (7.9.14h), becomes

$$\begin{aligned} R &= 2(E - m g y)\,[m + 0 + m\,(dy/dx)^2] \\ &= 2m(E - m g y)\,[1 + (y')^2] = R(y, y'), \end{aligned} \tag{c}$$

from which we obtain

$$\partial\sqrt{R}/\partial y' = [2m\,(E - m g y)]^{1/2}[1 + (y')^2]^{-1/2} y', \tag{d1}$$

$$\partial\sqrt{R}/\partial y = -\{m^2 g\,[1 + (y')^2]\}\,\{2m(E - m g y)\,[1 + (y')^2]\}^{-1/2}, \tag{d2}$$

and so the Euler–Lagrange equations of the Jacobi functional A_J, eqs. (7.9.14m), are

$$E_y(\sqrt{R}) = 0\text{:} \qquad 2(E - m g y)y'' + m g[1 + (y')^2] = 0. \tag{d3}$$

To integrate (d3), we $(\ldots)'$-differentiate it once more, thus obtaining

$$2(E - m g y)y''' = 0, \tag{e1}$$

from which, since $E \neq m g y$, it follows that

$$y''' = 0 \;\Rightarrow\; y = c_1 x^2 + c_2 x + c_3 \qquad (c_{1,2,3}\text{: constants of integration}). \tag{e2}$$

To find the three constants of this *parabolic* orbit, we apply the two boundary conditions

$$y(x_1) = y_1 \text{ (given)} \qquad \text{and} \qquad y(x_2) = y_2 \text{ (given)}, \tag{f1}$$

and the equation of motion (d3). Choosing, for simplicity, $x(0) = x_1 = 0$ and $y(0) = y_1 = 0$, we immediately find $c_3 = 0$. Then, substituting the resulting $y = c_1 x^2 + c_2 x$ into (d3) and setting $x = 0$, we obtain, after some simple algebra,

$$c_1 = -mg(1 + {c_2}^2)/4E < 0; \tag{f2}$$

that is, the parabolic orbit opens downward. Finally, substituting (f2) into the second of (f1): $y_2 = c_1 {x_2}^2 + c_2 x_2$, yields the second-degree equation for c_2:

$$ {c_2}^2 - (4E/mgx_2)c_2 + [(4Ey_2/mg{x_2}^2) + 1] = 0. \tag{f3}$$

Since, on physical grounds, c_2 must be real, the discriminant of (f3) must be non-negative; and this leads directly to the condition $4E(E - mgy_2) \geq (mgx_2)^2$ (> 0), from which we get the upper bound:

$$y_2 < E/mg, \qquad \text{or} \qquad E > mgy_2 = V(P_2). \tag{f4}$$

The two (real) values of c_2 obtained from (f3) yield the *two* parabolic orbits reaching $P_2(x_2, y_2)$ from $P_1(= O$: origin of coordinates); one *high* and one *low*.

REMARK

Between P_1 and P_2, both these orbits satisfy the same equations of motion and boundary conditions; that is, both satisfy the first-order (stationarity) condition $\delta A_J = 0$: namely, Jacobi's principle. Their differences appear in the *second-order* (extremality) conditions: the sign of $\delta^2 A_J$. It can be shown that, for motion between P_1 and P_2, *only the low orbit minimizes* A_J: $\delta^2 A_J > 0$, whereas the high orbit does not; and for motion beyond P_2, even the low orbit does not minimize A_J. [For further details, see (alphabetically): Koschmieder (1962, pp. 45–46), Lur'e (1968, pp. 752–754), Papastavridis [1986(a)], Peisakh (1966, and references cited therein).]

GENERALIZATION

We leave it to the reader to show, using again $\delta A_J = 0 \Rightarrow E_y(\sqrt{R}) = 0$, that if our particle moves freely in a plane O–xy under a general potential $V = V(x, y)$, its orbits are given by

$$2(E - V)y'' + [1 + (y')^2]\,(V_y - y'V_x) = 0, \tag{g}$$

where subscripts denote partial derivatives. (See, e.g., Kauderer, 1958, p. 599 ff.)

Example 7.9.5 *Uniqueness of a Lagrangean.* Let us examine the variations of the Hamiltonian action functionals of two (holonomic) systems, A_H and $A_H{}'$, whose corresponding Lagrangeans, L and L', differ by the total time derivative of an arbitrary "gauge" function of the coordinates and time $F = F(t, q)$; that is,

$$L'(t, q, \dot{q}) - L(t, q, \dot{q}) = dF(t, q)/dt. \tag{a}$$

Since $\delta(\dot{F}) = (\delta F)^{\cdot}$, we find, successively,

$$\delta A_H{}' = \delta \int L'\,dt = \int \delta L'\,dt = \int [\delta L + \delta(\dot{F})]\,dt$$

$$= \delta A_H + \left\{\sum (\partial F/\partial q_k)\,\delta q_k\right\}_1^2; \tag{b}$$

and, therefore, if $\delta q_k(t_1) = \delta q_k(t_2) = 0$, the conditions $\delta A_H{}' = 0$ and $\delta A_H = 0$ are *equivalent*; that is, both yield the same equations of motion:

$$\delta A_H = 0 \;\Rightarrow\; E_k(L) = 0, \qquad \delta A_H{}' = 0 \;\Rightarrow\; E_k(L') = E_k(L) = 0. \tag{c}$$

This simple variational argument shows that L is *nonunique; it is defined only to within the total time derivative of an arbitrary function of the coordinates and time, at most*. Accordingly, constant terms, or pure functions of time terms, may be safely omitted from a Lagrangean. Finally, we point out that we *always* have

$$E_k(L') = E_k(L+F) = E_k(L) + E_k(F) = E_k(L), \tag{d}$$

even if $\{\sum (\partial F/\partial q_k)\, \delta q_k\}_1^2 = \{\delta F\}_1^2 \neq 0$ (recalling ex. 3.5.13); but then the arrows in (c) cannot be reversed.

Example 7.9.6 *Routh's Problem I:* "If the period of complete recurrence of a dynamical system is not altered by the addition of energy, prove that this additional energy is equally distributed into potential and kinetic energies" [Routh, 1905(b), p. 315; also Papastavridis, 1985(a)].

Choosing in the fundamental equations (7.9.4h, i): $t_1 = 0, \Delta t_1 = 0, t_2 = \tau$ (period of "complete recurrence"), $\Delta q_k(t_1) = \Delta q_k(t_2)$ (say, 0) and assuming that the system is potential ($\delta' W_{np} = 0$) and scleronomic ($\sum p_k \dot{q}_k = 2T$), we get, respectively,

$$\Delta \int_0^\tau 2T\, dt = \int_0^\tau \delta E\, dt, \tag{a}$$

$$\Delta \int_0^\tau 2V\, dt = \int_0^\tau \delta E\, dt + 2E(\tau)\Delta\tau. \tag{b}$$

If, further, the addition of energy δE does *not* alter the period of oscillation (i.e., if $\Delta\tau = 0$), the above yield immediately the *equipartition* theorem:

$$\Delta \int_0^\tau T\, dt = \Delta \int_0^\tau V\, dt = \int_0^\tau (\delta E/2)\, dt, \quad \text{Q.E.D.} \tag{c}$$

Example 7.9.7 *Routh's Problem II*: "A dynamical system passes freely from one configuration to another in time i [our τ] with constant energy E; with energy $E + \delta E$ its time of free passage between the same configurations is $i + \delta i$, verify that on a time average the increment of the mean kinetic energy T_m [our $\langle T \rangle$] of the system throughout its path is less than half of δE by the amount $T_m(\delta i/i)$. Show that in case there are two adjacent paths that take the same time, their mean potential and kinetic energies differ by equal amounts" [Routh, 1905(b), p. 315; also Papastavridis, 1985(a)].

Here, as in the preceding example, let us choose $t_1 = 0, \Delta t_1 = 0, t_2 = \tau \equiv i$ (period of oscillation), $\Delta t_2 = \Delta\tau \equiv \Delta i$, and assume that the system is potential $[\langle \delta' W_{np} \rangle = 0]$ and scleronomic $[\sum p_k \dot{q}_k = 2T]$. Then, since

$$\Delta \int_0^\tau T\, dt = \Delta(\langle T \rangle \tau) = (\Delta \langle T \rangle)\tau + \langle T \rangle \Delta\tau, \tag{a1}$$

$$\Delta \int_0^\tau V\, dt = \Delta(\langle V \rangle \tau) = (\Delta \langle V \rangle)\tau + \langle V \rangle \Delta\tau, \tag{a2}$$

eqs. (7.9.4h, i) specialize to

$$\Delta\langle T \rangle = \delta E/2 - \langle T \rangle (\Delta\tau/\tau), \tag{b1}$$

$$\Delta\langle V \rangle = \delta E/2 - \big[\langle V \rangle - E\big]\,(\Delta\tau/\tau) = \delta E/2 + \langle T \rangle\,(\Delta\tau/\tau), \quad \text{Q.E.D.} \tag{b2}$$

Theorems like this and the one of the preceding example arose during the late 19th century in connection with the (partially successful) attempts of Clausius, Szily, Boltzmann, et al. to supply *analytical mechanics–based explanations of thermomechanical phenomena*; see, for example, the papers by Bierhalter [1981(a), (b), 1982, 1983, 1992], and the monograph and original papers by Polak (1959, 1960)].

Problem 7.9.4 *Time Integral Theorems for Periodic Systems* (Williamson and Tarleton, 1900, pp. 457–458). For the system described in the preceding examples, show that "When the entire state of a moving system recurs at the end of equal intervals of time whose common magnitude is τ, if the total energy E receive a small change, the corresponding variation of the mean Lagrangean function, L_m [our $\langle L \rangle$], is given by the equation

$$\Delta L_m = -2T_m(\Delta\tau/\tau)." \tag{a}$$

Problem 7.9.5 *Time Integral Theorems for Periodic Systems (continued)* [Routh, 1905(b), pp. 314–315; Williamson and Tarleton, 1900, p. 458]. Continuing from the preceding examples and problem, show that "If the total energy of a recurring system, ..., receive a series of variations at intervals of time which are large compared with the period of recurrence of the system, and if finally the system return to its original state, show that $\int (dE/T_m)$ taken from the beginning to the end of the cycle is zero."

HINT

$E + L = 2T \Rightarrow dE + dL_m = 2\,dT_m \Rightarrow dE/T_m =$ *perfect (or exact) differential* (by the preceding problem).

These two problems are important in connection with the (late 19th century) attempts at a mechanical explanation of the *second* law of thermodynamics (entropy).

Example 7.9.8 *Pendulum of Slowly Varying Length; Adiabatic Invariance.* We consider a mathematical pendulum, consisting of a bob (particle) of mass m and a constraining light thread of length l, performing small (linear), free and undamped oscillations under gravity (fig. 7.4). If ϕ is the instantaneous inclination of the thread to the vertical, then

$$2T = m\,l^2(\dot{\phi})^2, \qquad V = -m\,g\,l\cos\phi + C \approx (1/2)m\,g\,l\phi^2 + C'$$

$$\text{(for small angular motions; } C, C'\text{: constants)}, \tag{a}$$

$$L \equiv T - V = (1/2)m\,l^2(\dot{\phi})^2 + m\,g\,l\cos\phi - C$$

$$\approx (1/2)m\,l^2(\dot{\phi})^2 - (1/2)m\,g\,l\phi^2 - C' \qquad \text{("small Lagrangean")}, \tag{b}$$

and, accordingly, Lagrangean equation of (small) motion:

$$E_\phi(L) \equiv (\partial L/\partial\dot{\phi})^{\cdot} - \partial L/\partial\phi = 0:$$

$$\ddot{\phi} + (g/l)\sin\phi = 0 \Rightarrow \ddot{\phi} + (g/l)\phi = 0, \tag{c}$$

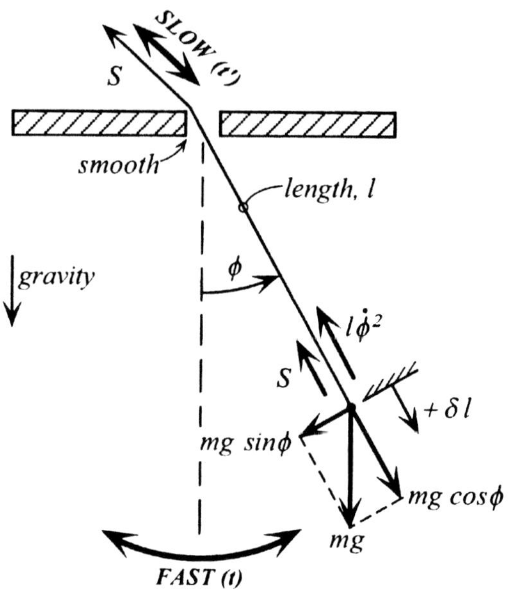

Figure 7.4 Adiabatic pendulum, of length l and mass m.
Thread tension: $S = ml(\dot{\phi})^2 + mg\cos\phi \approx ml(\dot{\phi})^2 - (\frac{1}{2})mg\phi^2 + mg$.

and total energy

$$E \equiv T + V = h \equiv (\partial L/\partial\dot{\phi})\dot{\phi} - L\colon\ (1/2)ml^2(\dot{\phi})^2 - mgl\cos\phi = constant,$$

$$\Rightarrow (1/2)ml^2(\dot{\phi})^2 + (1/2)mgl\phi^2 = constant \qquad (\text{"small energy equation"}). \qquad \text{(d)}$$

Equation (d) shows that as long as the pendulum parameters, m and l, remain constant in time so does $E = h$. Now, let us assume that some external (energy-supplying) agency acts to change these parameters, say, the length l; for example, we can imagine the thread held between the fingers of one hand and shortened/lengthened by drawing it up/down with the fingers of the other; or, the bob picks up dust and, thus, its mass changes. Then, the power equation yields

$$dE/dt = dh/dt = -\partial L/\partial t = -(\partial L/\partial l)\,(dl/dt) \neq 0; \qquad \text{(e)}$$

that is, the system is no longer conservative.

Let us now examine the special case where the above length change, from l to $l + dl$, is spread over several (to and fro) oscillations; that is, let us assume that dl *takes a very large number of periods*, and, therefore, within any one of them, l can be considered constant. Since the pendulum still oscillates under these very slow, or *adiabatic*, variations, henceforth denoted by δl (like the virtual variations), both its new frequency and period,

$$\omega + \delta\omega \qquad \text{and} \qquad \tau + \delta\tau = (2\pi/\omega) + (-2\pi/\omega^2)\,\delta\omega, \qquad \text{(f)}$$

will be functions of δl. Mathematically, we are dealing here with a differential equation, (c), whose coefficients (parameters) are explicit functions, not of the ordinary (or "fast") time t, but of an adiabatic (or "slow") time $t' \equiv \varepsilon t$, ε: small number; that is, $l = l(t')$.

Now we ask the following fundamental question: Are there any energetic quantities that, in spite of these adiabatic (nonconservative) changes of l, remain constant, not in t but in t'; that is, to the first ε-order?

As we explain below, one such constant, or *adiabatic invariant*, is the *time average of twice the kinetic energy of the system, divided by its frequency*, $2\langle T\rangle/\omega$; which, to within a constant factor, equals the Lagrangean action A_L. If, further, V is homogeneous *quadratic* in the coordinates (i.e., linear oscillations) then, since by the virial theorem (ex. 7.2.1 ff.) $\langle T\rangle = \langle V\rangle$, that invariant becomes $2\langle T\rangle/\omega = E/\omega$. [This is a famous problem: It was posed at the Solvay Congress (Brussels, 1911) by the great physicist H. A. Lorentz, and was answered (for the small motion case) by the ... greater physicist A. Einstein!] Other special variations of the system parameters produce other special invariants ("integrals"); see, for example, Kronauer and Musa (1966), Kronauer (1983), Papastavridis [1982(b)].

Let us establish this adiabatic invariant directly, by elementary considerations. As (c) shows, the small motion frequency (squared) equals $\omega^2 = g/l$, and so, for small variations,

$$(2\omega\,\delta\omega)l + \omega^2\,\delta l = 0 \;\Rightarrow\; (\delta\omega/\omega) = -(\delta l/2l). \tag{g}$$

Next, by kinetics (equation of motion of bob along the thread direction), the tension of the thread S equals

$$\begin{aligned} S &= m\,l(\dot\phi)^2 + m\,g\cos\phi \\ &\approx m\,g + m\,l(\dot\phi)^2 - m\,g\,\phi^2/2 \qquad \text{(for small motions)}, \end{aligned} \tag{h}$$

where, as (c) shows [under the initial conditions, say, $\phi(0) = \phi_o$ and $\dot\phi(0) = 0$], $\phi(t) = \phi_o\cos(\omega t)$. Therefore, the elementary work of S during a δl change, $\delta' W_{\text{external}} \equiv \delta' W$, *averaged over several oscillation periods*, will be

$$\begin{aligned} \langle\delta' W\rangle &= -\langle S\,\delta l\rangle = -\delta l[\langle m\,g\rangle + \langle m\,l(\dot\phi)^2\rangle - \langle m\,g\,\phi^2/2\rangle] \\ &= -\delta l[\langle m\,g\rangle + \langle m\,l\,{\phi_o}^2\omega^2\sin^2(\omega t)\rangle - \langle m\,g\,{\phi_o}^2\cos^2(\omega t)/2\rangle] \\ &= \cdots = -\delta l[m\,g + m\,l\,{\phi_o}^2\omega^2(1/2) - m\,g\,{\phi_o}^2(1/4)] \\ &= -m\,g(1 + {\phi_o}^2/4)\,\delta l = -\langle S\rangle\,\delta l \qquad [\text{since } \omega^2 l = g]. \end{aligned} \tag{i}$$

Further, recalling (a),

$$\begin{aligned} \langle T\rangle &= \langle(1/2)m\,l^2(\dot\phi)^2\rangle = \langle(1/2)m\,l^2\omega^2{\phi_o}^2\sin^2(\omega t)\rangle \\ &= (1/4)m\,l^2\omega^2{\phi_o}^2 = m\,g\,l{\phi_o}^2/4, \end{aligned} \tag{j}$$

$$\begin{aligned} \langle V\rangle &= \langle C - m\,g\,l[1 - (\phi^2/2)]\rangle \\ &= C - m\,g\,l[1 - ({\phi_o}^2/2)\langle\cos^2(\omega t)\rangle] \\ &= C - m\,g\,l(1 - {\phi_o}^2/4). \end{aligned} \tag{k}$$

With the help of the above results, the averaged energy equation

$$\langle \delta' W \rangle = \langle \delta E \rangle = \delta \langle E \rangle = \delta(\langle T + V \rangle) = \delta \langle T \rangle + \delta \langle V \rangle \tag{l}$$

yields

$$-mg(1 + \phi_o{}^2/4)\,\delta l = \delta(mgl\phi_o{}^2/4) + \delta\big[C - mgl(1 - \phi_o{}^2/4)\big]$$
$$\Rightarrow 3\phi_o\,\delta l + 4l\,\delta\phi_o = 0, \tag{m}$$

from which, integrating adiabatically, we find

$$l^3\phi_o{}^4 = constant \qquad \text{or} \qquad l\sqrt{l}\,\phi_o{}^2 = constant. \tag{n}$$

Hence,

$$2\langle T \rangle/\omega = (mgl\phi_o{}^2/2)/(g/l)^{1/2} = (m\sqrt{g}/2)\,(l\sqrt{l}\,\phi_o{}^2) = constant, \tag{o}$$

and

$$A_L \equiv \int_0^T 2T\,dt = 2\langle T \rangle \tau = 2\pi\big(2\langle T \rangle/\omega\big) = constant, \quad \text{Q.E.D.} \tag{p}$$

For further details see Papastavridis [1985(a)] and §8.15.

Problem 7.9.6 *Adiabatic Invariance (Linear Pendulum).* In connection with the adiabatic pendulum of the preceding example, ex. 7.9.8 (i.e., particle P of mass m, suspended by a light and inextensible thread of slowly and randomly varying length l, and instantaneous tension S) under gravity, show that

$$S = \partial L/\partial l \qquad [= ml(\dot{\phi})^2 + mg\cos\phi,\ \text{exactly}]; \tag{a}$$

and therefore its averaged energy equation is

$$\langle \delta' W \rangle = \delta \langle h \rangle = -\langle S \rangle\,\delta l = -\langle \partial L/\partial l \rangle\,\delta l$$
$$\equiv -(\omega/2\pi)\int_0^{2\pi/\omega} (\partial L/\partial l)\,\delta l\,dt$$
$$= \cdots = -mg(1 + \phi_o{}^2/4)\delta l \qquad \text{(linear pendulum case).} \tag{b}$$

[Equation (b) is, essentially, the adiabatic and averaged version of the holonomic and potential power equation: $dh/dt = -\partial L/\partial t$.]

Problem 7.9.7 *Adiabatic Invariance (The Rayleigh Pendulum).* Continuing from the preceding problem of the adiabatic mathematical pendulum under gravity, let us examine the case where, at the suspension point, there is a small ring R constrained to slide up and down the smooth, vertical, and fixed line AB (otherwise the pendulum would be a nonconservative system).

(i) Show that *if R is kept fixed*, in which case RP is an ordinary (i.e., constant parameter) pendulum, *the vertical force tending to push R upward, F*, equals

$$F = S(1 - \cos\phi) \approx mg\cos\phi\,(1 - \cos\phi)\text{: “\textit{vibration pressure}”} \tag{a}$$

(disregarding an unessential centripetal force contribution to S—explain); and since $V = mgl\,(1 - \cos\phi)$, deduce that, for small vibrations (recall virial theorem),

$$2F \approx mg\phi^2 \Rightarrow \langle F \rangle = \langle V/l \rangle = \cdots = (1/2)\,(E/l), \tag{b}$$

where E is the total energy of the pendulum (a constant); that is, *the average force on the ring is proportional to the pendulum's energy density* (=energy per unit length) [a result which, as Rayleigh has pointed out (1902), has a close analog in electromagnetism (vibration pressure → "radiation pressure"]. Also, verify that the *horizontal* force on R is $S \sin\phi = mg\cos\phi\sin\phi$, and its average vanishes.

(ii) Next, assume that while P oscillates with frequency $\omega = (g/l)^{1/2}$, R is let slide adiabatically upward; that is, $\delta l > 0$. Then, clearly, the (positive) work done on R by F comes at the expense of the oscillatory energy of the pendulum. Show that, in such a case,

$$\delta E = -\langle F \rangle\,\delta l = -(E/2)\,(\delta l/l), \tag{c}$$

from which, by "adiabatic integration," it follows that

$$E\sqrt{l} = constant, \tag{d}$$

and from this we conclude that as $l \to \infty$, $E \to 0$; that is, the entire energy of P is then expended as work done on R.

Problem 7.9.8 *Adiabatic Invariance (The Rayleigh Pendulum)*. Continuing from the preceding problem,

(i) By averaging Lagrange's equation of motion for a typical positional coordinate l:

$$Q_l = (\partial T/\partial \dot{l})^{\cdot} - \partial T/\partial l + \partial V/\partial l, \tag{a}$$

over a very long time interval τ, show that, since $p_l \equiv \partial T/\partial \dot{l}$ is finite,

$$\begin{aligned}\langle Q_l \rangle &\equiv (l/\tau)\int_0^T Q_l\,dt = (1/\tau)\int_0^T (\partial V/\partial l - \partial T/\partial l)\,dt \\ &= -(1/\tau)\int_0^T (\partial L/\partial l)\,dt \equiv -\langle \partial L/\partial l \rangle.\end{aligned} \tag{b}$$

(ii) Show that, in the case of our ringed pendulum (i.e., $\dot{l} = 0$),

$$\partial V/\partial l = mg(1 - \cos\phi) = V/l, \qquad \partial T/\partial l = ml(\dot{\phi})^2 = 2T/l, \tag{c}$$

and, therefore, since $\langle V \rangle = \langle T \rangle = E/2$ (by the virial theorem, for small vibrations), eq. (b) yields

$$\langle Q_l \rangle = (1/\tau)\int_0^T [(V - 2T)/l]\,dt = \cdots = -(1/2)\,(E/l) = -\langle F \rangle; \tag{d}$$

that is, the average force necessary to hold the ring must be directed *downward* (so as to tend to diminish l).

This "ringed pendulum" problem seems to be the earliest simple and concrete example of adiabatic invariance. For additional examples and insights, see Rayleigh

(1902); also Bakay and Stepanovskii (1981, pp. 100–107), Tomonaga (1962, pp. 290–294), and Thomson (1888, chaps. 4, 9).

Example 7.9.9 *Rayleigh's Principle via the Principle of Least Action.* Let us consider a holonomic and scleronomic system with n *DOF* undergoing small (linear) free and undamped vibrations about a configuration of stable equilibrium defined by $q_k = 0$. Then, to within our approximations, its kinetic and potential energies are, respectively,

$$2T = \sum\sum M_{kl}\dot{q}_k\dot{q}_l, \qquad 2V = \sum\sum V_{kl}q_k q_l, \tag{a}$$

where (M_{kl}) and (V_{kl}) are constant, symmetric, and positive definite matrices of *inertia* (*mass*) and *total potential* (*stiffness*), respectively, and the q's give the small motion from equilibrium. Next, let us assume that *the system oscillates in its (M)th mode* $(M = 1, \ldots, n)$; that is, each q_k varies as

$$q_k^{(M)} \equiv q_{k,M} \equiv A_{kM}(C_M \sin\psi_M), \qquad \psi_M \equiv \omega_M t + \phi_M, \tag{b}$$

where C_M, ϕ_M, and ω_M are, respectively, the *amplitude, phase*, and *frequency* of that mode (the first two to be determined from the initial conditions of that mode); and the A_{kM} are the "normal mode coefficients" or "direction cosines" of the modal vector $C_M \sin\psi_M$, in q-space, and depend on M_{kl}, V_{kl}, and $\omega_M{}^2$ [they are the minors of the *characteristic* or *secular* determinant $|V_{kl} - \omega_M{}^2 M_{kl}|$; assuming that all ω_M's are different. The practically much rarer case of multiple or *degenerate* frequencies introduces some very minor modifications; see, e.g., Greenwood (1988, pp. 497–498), Lamb (1929, pp. 230–232).]

From (b) we immediately find

$$dq_{k,M}/dt = A_{kM}C_M\omega_M \cos\psi_M, \tag{c}$$

and so the (double) kinetic and potential energies of that mode are, respectively,

$$2T_M = (C_M\omega_M \cos\psi_M)^2 K_M, \qquad 2V_M = (C_M \sin\psi_M)^2 \Pi_M, \tag{d}$$

where

$$K_M \equiv \sum\sum M_{kl}A_{kM}A_{lM}, \qquad \Pi_M \equiv \sum\sum V_{kl}A_{kM}A_{lM}. \tag{d1}$$

[Had we included $C_M{}^2/2$ in K_M, Π_M, the latter would be, respectively, the *maximum* kinetic and potential energies of the system, at the (M)th mode.]

Below, using the general "least" action equation (7.9.4f), we derive *Rayleigh's principle*; that is,

$$\omega_M{}^2\,\delta K_M - \delta\Pi_M = 0 \qquad \text{or} \qquad \delta(\Pi_M/K_M) = 0, \tag{e}$$

where

$$\delta K_M \equiv 2\sum\sum M_{kl}A_{kM}\,\delta A_{lM}, \qquad \delta\Pi_M \equiv 2\sum\sum V_{kl}A_{kM}\,\delta A_{lM}, \tag{e1}$$

and the δA_{kM} are small variations about the (M)th mode. The qualitative and physical interpretation of (e) is given later.

By $\delta(\ldots)$-varying the $q_{k,M}$, eqs. (b), around the (M)th mode, we find

$$\delta q_{k,M} = (C_M \sin\psi_M)\,\delta A_M + (A_{kM} \sin\psi_M)\,\delta C_M + (A_{kM} C_M \cos\psi_M)\,\delta\psi_M, \quad \text{(f1)}$$

$$\delta\psi_M = t\,\delta\omega_M + \delta\phi_M, \quad \text{(f2)}$$

and so the corresponding boundary terms of (7.9.4f) become

$$\left\{\sum p_{kM}\,\delta q_{kM}\right\}_1^2 = \left\{\sum\sum M_{kl}(dq_{lM}/dt)\,\delta q_{kM}\right\}_1^2$$
$$= \{[(C_M{}^2\omega_M/2)\,\delta K_M + (C_M\,\omega_M K_M)\,\delta C_M]\sin\psi_M\cos\psi_M$$
$$+ (C_M{}^2\omega_M K_M)\cos^2\psi_M\,\delta\psi_M\}_1^2, \quad \text{(g1)}$$

$$\{2T_M \Delta t_M\}_1^2 = \{(C_M{}^2\omega_M K_M)\cos^2\psi_M \Delta t_M\}_1^2. \quad \text{(g2)}$$

To eliminate both (g1) and (g2) we make the following (clearly nonunique) choices:

$$(\psi_M)_1 = \pi/2 \;\Rightarrow\; \omega_M t_1 + \phi_M = \pi/2,$$
$$(\psi_M)_2 = (\psi_M)_1 + 2\pi \;\Rightarrow\; \omega_M t_2 + \phi_M = 5\pi/2,$$
$$\Rightarrow \tau_M = t_2 - t_1 = 2\pi/\omega_M\text{: } (M)\text{th } \textit{principal period}. \quad \text{(g3)}$$

With these time limits, and since here $\delta' W_{np} = 0$, equation (7.9.4f) reduces to

$$\Delta A_{L,M} - \int \delta E_M\,dt = \Delta\int 2T_M\,dt - \int(\delta T_M + \delta V_M)\,dt = 0. \quad \text{(h)}$$

Let us implement (h): a series of straightforward trigonometric integrations, with use of (a–d, f1, 2) and $t_2 = t_1 + \tau_M$, yields

$$A_{L,M} \equiv \int 2T_M\,dt$$
$$= \int (C_M{}^2\omega_M{}^2 K_M)\cos^2\psi_M\,dt = (\pi/\omega_M)(C_M{}^2\omega_M{}^2 K_M), \quad \text{(i1)}$$

$$\Rightarrow \Delta A_{L,M} = (\pi/\omega_M)\,[(2C_M\omega_M{}^2 K_M)\,\delta C_M + (C_M{}^2\omega_M K_M)\,\delta\omega_M + (C_M{}^2\omega_M{}^2)\,\delta K_M], \quad \text{(i2)}$$

$$\int \delta T_M\,dt = \int [(C_M\omega_M{}^2 K_M\,\delta C_M)\cos^2\psi_M + (C_M{}^2\omega_M K_M\,\delta\omega_M)\cos^2\psi_M$$
$$- (C_M{}^2\omega_M{}^2 K_M\,\delta\psi_M)\sin\psi_M\cos\psi_M + (C_M{}^2\omega_M{}^2\delta K_M/2)\cos^2\psi_M]\,dt, \quad \text{(i3)}$$

$$\int \delta V_M\,dt = \int [(C_M \Pi_M\,\delta C_M)\sin^2\psi_M + (C_M{}^2\Pi_M\,\delta\psi_M)\sin\psi_M\cos\psi_M$$
$$+ (C_M{}^2\,\delta\Pi_M/2)\sin^2\psi_M]\,dt, \quad \text{(i4)}$$

$$\Rightarrow \int(\delta T_M + \delta V_M)\,dt = \int \delta E_M\,dt$$
$$= \cdots = (\pi/\omega_M)\,[(\omega_M{}^2 K_M + \Pi_M) C_M\,\delta C_M$$
$$+ (C_M{}^2\omega_M K_M)\,\delta\omega_M$$
$$+ (1/2)\,(\omega_M{}^2\,\delta K_M + \delta\Pi_M) C_M{}^2]. \quad \text{(i5)}$$

Hence, substituting (i2) and (i5) into (h), while noting that due to the *linearity* of the system (in the equations of motion), we have *equipartition of its kinetic and potential energies* over τ_M; that is,

$$\int T_M\,dt = \int V_M\,dt, \tag{j}$$

or, since

$$\int T_M\,dt = (\pi/\omega_M)\omega_M{}^2\left[\sum\sum(1/2)M_{kl}(A_{kM}C_M)(A_{lM}C_M)\right]$$

$$= (\pi/\omega_M)\omega_M{}^2 C_M{}^2 K_M/2 \equiv (\pi/\omega_M)\omega_M{}^2 T_{M,\max}, \tag{j1}$$

$$\int V_M\,dt = (\pi/\omega_M)\left[\sum\sum(1/2)V_{kl}(A_{kM}C_M)(A_{lM}C_M)\right]$$

$$= (\pi/\omega_M)C_M{}^2\Pi_M/2 \equiv (\pi/\omega_M)V_{M,\max}, \tag{j2}$$

$$\text{Equation (j)} \Rightarrow \omega_M{}^2 = \Pi_M/K_M = V_{M,\max}/T_{M,\max} \tag{j3}$$

[and this is a key step in the proof of Rayleigh's theorem, which shows why (j3) and the theorem do *not* hold for nonlinear oscillations], we finally find

$$(\pi/\omega_M)(1/2)(C_M{}^2\omega_M{}^2\,\delta K_M - C_M{}^2\,\delta\Pi_M) = 0$$

$$\Rightarrow \omega_M{}^2\,\delta K_M - \delta\Pi_M = 0 \Rightarrow \omega_M{}^2 = \delta\Pi_M/\delta K_M = \delta V_{M,\max}/\delta T_{M,\max}, \quad \text{Q.E.D.} \tag{k}$$

This can also be written as

$$\delta R_M = 0, \qquad \text{where } R_M \equiv \omega_M{}^2 T_{M,\max} - V_{M,\max} \qquad (\text{or } V_{M,\max} - \omega_M{}^2 T_{M,\max}), \tag{k1}$$

and that variation *does not affect* ω_M; or, due to (j3), as

$$0 = \delta\omega_M{}^2 = \delta(V_{M,\max}/T_{M,\max})$$

$$= (1/T_{M,\max}{}^2)(T_{M,\max}\,\delta V_{M,\max} - V_{M,\max}\,\delta T_{M,\max})$$

$$= (1/T_{M,\max})\left[\delta V_{M,\max} - (V_{M,\max}/T_{M,\max})\,\delta T_{M,\max}\right]$$

$$= (1/T_{M,\max})(\delta V_{M,\max} - \omega_M{}^2\,\delta T_{M,\max}). \tag{k2}$$

REMARKS

(i) A Hamilton principle-based derivation of (k–k3) would utilize expressions (i3–5) with $\delta\omega_M = 0$ [i.e., fixed endpoints, since

$$\delta\omega_M = \delta(2\pi/\tau_M) = -(2\pi/\tau_M{}^2)\,\delta\tau_M = -(2\pi/\tau_M{}^2)\Delta(t_2 - t_1)]. \tag{l}$$

Then,

$$0 = \delta\int(T_M - V_M)\,dt = \int(\delta T_M - \delta V_M)\,dt$$

$$= (\pi/\omega_M)\left[(\omega_M{}^2 K_M - \Pi_M)C_M\,\delta C_M + (1/2)(\omega_M{}^2\,\delta K_M - \delta\Pi_M)C_M{}^2\right], \tag{m}$$

from which, again thanks to (j3), eq. (k) follows.

Such a derivation does not exactly coincide with those found in the literature; there, one sets

$$q_{k,M} = B_{kM} \sin \psi_M, \qquad B_{kM} \equiv A_{kM} C_M, \tag{m1}$$

from which

$$\delta q_{k,M} = \delta B_{kM} \sin \psi_M, \tag{m2}$$

and so Hamilton's principle yields

$$0 = \delta \int (T_M - V_M)\, dt = \delta\big[(\pi/\omega_M)({\omega_M}^2 T_{M,\max} - V_{M,\max})\big]$$
$$= (\pi/\omega_M)\delta({\omega_M}^2 T_{M,\max} - V_{M,\max}) = (\pi/\omega_M)({\omega_M}^2 \delta T_{M,\max} - \delta V_{M,\max}), \tag{m3}$$

because, here, $\delta(\ldots)$ implies $\omega_M = constant$. Again, we notice the indispensability of (j3).

(ii) As eqs. (h) and (i5) show, even if $\delta A_{L,M} = 0$ and $\delta\omega_M = 0$, still

$$\int \delta E_M\, dt \neq 0, \tag{n}$$

whereas the customary formulation of "least" action, eq. (7.9.6b), requires that $\delta E_M = 0$ for the admissible varied paths.

For a derivation of Rayleigh's principle based on the Hamiltonian action, eq. (7.9.4b), see Lur'e (1968, pp. 689–694).

More on Rayleigh's Principle (RP)

The *stationary* property (k) had already been noticed by Lagrange; and is indeed referred to by some authors as *Lagrange's theorem.* But Rayleigh (1870s) revealed the following, additional, *extremum* property: If we imagine the system reduced to one with a *single degree of freedom,* say, by the imposition of $n - 1$ frictionless (ideal) constraints so that the ratios $q_1 : q_2 : \ldots : q_n$ have any given values, then the (square of the) frequency of the so-constrained system ω^2 will lie between the (squares of the) least and greatest natural frequencies of the unconstrained system:

$$\omega^2_{\min} \equiv {\omega_1}^2 \leq \omega^2 \leq \omega^2_{\max} \equiv {\omega_n}^2. \tag{o}$$

To understand these results better we need some "normal mode theory". [See any good vibrations text; or Gantmacher (1970, pp. 202–222), Synge and Griffith (1959, pp. 483–505).]

As shown there, *the most general q_k-variation is a superposition of n simple harmonic, or principal, independent oscillations,*

$$x_M = C_M \sin \psi_M: \qquad (M)\text{th } \textit{principal mode}, \qquad \psi_M = \omega_M t + \phi_M, \tag{p1}$$

where the amplitudes C_M and phases ϕ_M are to be determined from the $2n$ initial conditions $x_M(0)$ and $dx_M(0)/dt$ [or from the $q_k(0)$ and $dq_k(0)/dt$], each of which contributes to q_k proportionately to the coefficient A_{kM}; that is, recalling (b),

$$q_k = \sum q_{k,M} = \sum A_{kM} x_M = \sum A_{kM} C_M \sin \psi_M. \tag{p2}$$

This expresses D. Bernoulli's *principle of the superposition of linear vibrations* (1753). The great advantage of such Lagrangean coordinates is that *in them the equations of motion decouple* to the n independent equations

$$d^2x_M/dt^2 + \omega_M{}^2 x_M = 0 \Rightarrow \text{eq. (p1)}. \tag{p3}$$

Next, it is physically (though not mathematically) clear, that

$$T = \sum T_M = \sum (1/2) m_M (dx_M/dt)^2, \qquad V = \sum V_M = \sum (1/2) k_M x_M{}^2, \tag{p4}$$

m_M: *principal coefficients of inertia* (> 0),

k_M: *principal coefficients of stability* $(> 0;$ since $V > 0)$; (p5)

that is, in such coordinates, T and V can be expressed in *sum of squares*, or *diagonal*, forms; and, when comparing them with (p3), we easily conclude that

$$\omega_M{}^2 = k_M/m_M \quad (> 0). \tag{p6}$$

[Some authors define *normal(ized)* coordinates, as opposed to principal coordinates, so that the *inertia matrix* (M_{kl}) is the *unit matrix*, while the *stiffness* $\rightarrow$ *stability matrix* (V_{kl}) is the *diagonal matrix of the* $\omega_M{}^2$; that is, such coordinates are principal and inertially normalized.]

In terms of these principal coordinates, any $n - 1$ geometrical constraints imposed on our system, (which in effect reduce it to a *one-DOF* system) will be given by the linear relations

$$x_1 = X_1\,\xi, \qquad x_2 = X_2\,\xi, \ldots, \qquad x_n = X_n\,\xi, \tag{q1}$$

where the $X_{1,\ldots,n}$ are constants and $\xi = \xi(t)$ is any of the x_M's or q_k's. Then T and V, eqs. (p4, 5) take the constrained values

$$T_\xi = (1/2)(m_1 X_1{}^2 + \cdots + m_n X_n{}^2)(d\xi/dt)^2 \equiv T_{\xi,o}(d\xi/dt)^2, \tag{q2}$$

$$V_\xi = (1/2)(k_1 X_1{}^2 + \cdots + k_n X_n{}^2)\,\xi^2 \equiv V_{\xi,o}\,\xi^2; \tag{q3}$$

and, due to equipartition over the period of ξ, $\tau_\xi = 2\pi/\omega_\xi$ — that is,

$$\int (T_\xi - V_\xi)\,dt = 0 \Rightarrow (\pi/\omega_\xi)(\omega_\xi{}^2 T_{\xi,\max} - V_{\xi,\max}) = 0, \tag{q4}$$

we find that the so-constrained frequency is given by *Rayleigh's quotient*:

$$\begin{aligned}\omega_\xi{}^2 &= V_{\xi,\max}/T_{\xi,\max} \\ &= (k_1 X_1{}^2 + \cdots + k_n X_n{}^2)/(m_1 X_1^2 + \cdots + m_n X_n{}^2) = V_{\xi,o}/T_{\xi,o};\end{aligned} \tag{q5}$$

and since we have numbered our frequencies so that

$$\min(k_M/m_M) = k_1/m_1 \equiv \omega_1{}^2 \quad \text{and} \quad \max(k_M/m_M) = k_n/m_n \equiv \omega_n{}^2, \tag{q6}$$

the (extremal) RP states that

$$\omega_1{}^2 \le \omega_\xi{}^2 \le \omega_n{}^2, \quad \text{for arbitrary sets of real numbers } X_{1,\ldots,n}; \text{ that is, eq. (o);} \tag{r}$$

in words: the approximate (constrained) value of Rayleigh's quotient is never lower than the actual ω_1^2, and thus furnishes an *upper bound* for it.

[For an algebraic derivation of RP, based on the solutions of the corresponding (constrained) Routh–Voss equations, see, for example, Chirgwin and Plumpton (1966, pp. 376–379), Ramsey (1937, pp. 267–269), Smart (1951, pp. 399–401); while for extensions to the *higher frequencies*, see any good book on linear algebra (eigenvalue problem), or Gantmacher (1970, pp. 216–222).]

As for the earlier-proved stationary property of Rayleigh's quotient (Lagrange's theorem), the above results allow us to reformulate it as follows: If $n-1$ of the X's, eqs. (q1), are very small (of the *first* order) relative to any particular X_M — that is, *if the constraints (q1) force the system to oscillate very closely to the* (M)*th mode — then* $\omega_\xi \approx \omega_M$ *(to the second* order); or, if all X's, except X_M, become very small (first order), then ω_ξ^2 will differ from ω_M^2 by second-order quantities; and, of course, if all X's, except X_M, vanish, then $\omega_\xi^2 = \omega_M^2$.

In sum: The frequency (squared) of the constrained system is stationary for those constraints that make the oscillation a normal one of the natural (i.e., unconstrained) system; and these stationary values are the (squares of the) system's natural frequencies.

Finally, we point out that in concrete applications of RP there is no need to employ normal coordinates; if there was, in view of the work needed to find the latter, RP would be practically useless. Both stationarity and extremality of Rayleigh's quotient are *intrinsic* system properties, and, as such, are independent of any particular q's used. Thus, if, in general coordinates, the constraints are $q_k = B_k\,\xi(t)$, or $q_k = B_k \sin(\omega_\xi t)$, where the B_k's are constants and ξ is any one of the q_k's or x_M's, then (r) is replaced by

$$\omega_1^2 \le \omega_\xi^2 = \sum V_{kl} B_k B_l \Big/ \sum M_{kl} B_k B_l \le \omega_n^2,$$

$$\text{for an arbitrary set of real numbers } B_{1,\ldots,n}. \tag{s}$$

In particular, for a *two*-DOF system, equations (q5, s) yield, respectively [fig. 7.5(a, b)],

$$\omega_\xi^2 = (k_1 X_1^2 + k_2 X_2^2)/(m_1 X_1^2 + m_2 X_2^2),$$

or

$$\omega^2(\lambda) = (k_1 + k_2\lambda^2)/(m_1 + m_2\lambda^2), \tag{s1}$$

where

$$\lambda \equiv X_2/X_1; \tag{s2}$$

$$\omega_\xi^2 = (V_{11}B_1^2 + 2V_{12}B_1B_2 + V_{22}B_2^2)/(M_{11}B_1^2 + 2M_{12}B_1B_2 + M_{22}B_2^2)$$

or

$$\omega^2(\mu) = (V_{11} + 2V_{12}\mu + V_{22}\mu^2)/(M_{11} + 2M_{12}\mu + M_{22}\mu^2), \tag{s3}$$

where

$$\mu \equiv B_2/B_1. \tag{s4}$$

(a) Graph of $\omega^2(\lambda)$; $\lambda = X_2/X_1$ ***(normal/uncoupled** cordinates),* $\omega_2^2 = k_2/m_2 > \omega_1^2 = k_1/m_1$

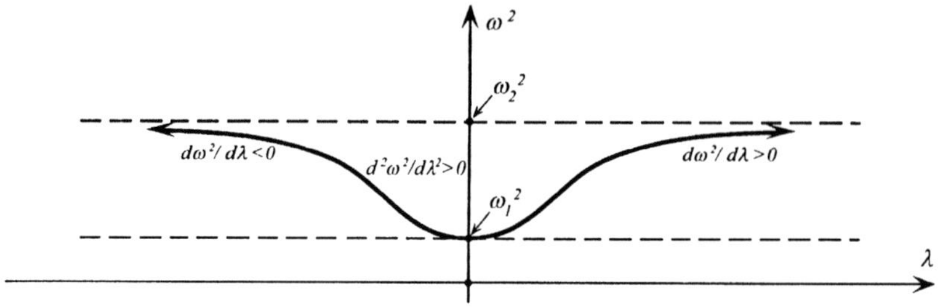

(b) Graph of $\omega^2(\mu)$; $\mu = B_2/B_1$ ***(general/coupled** cordinates);*

Left: $\omega_{2.o}^2 = V_{22}/M_{22} > \omega_{1.o}^2 = V_{11}/M_{11}$ *Right:* $\omega_{1.o}^2 = V_{11}/M_{11} > \omega_{2.o}^2 = V_{22}/M_{22}$

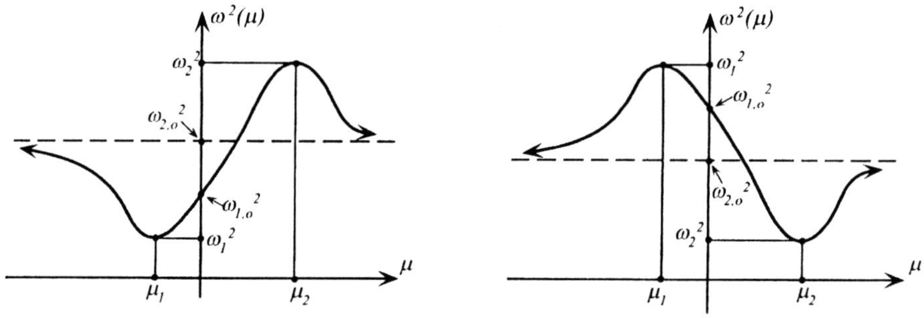

Figure 7.5 Rayleigh quotient for a two-DOF system.

An Illustration

Let us consider three particles of equal masses m attached at equal intervals l to a light and flexible string fixed at its ends, and stretched with a tension S (practically unaffected by the small deflections of the particles) (fig. 7.6). Here, to within *quadratic* terms in the q's,

$$2T = m[(\dot{q}_1)^2 + (\dot{q}_2)^2 + (\dot{q}_3)^2], \tag{t1}$$

$$V = S[(l_1 - l)^2 + (l_2 - l)^2 + (l_3 - l)^2 + (l_4 - l)^2] \equiv S(\Delta l_1 + \Delta l_2 + \Delta l_3 + \Delta l_4)$$

[no factor 1/2 needed, due to the assumed constancy of S]

$$\approx (S/2l)[q_1{}^2 + (q_2 - q_1)^2 + (q_3 - q_2)^2 + q_3{}^2]$$

$$= (S/l)[q_1{}^2 + q_2{}^2 + q_3{}^2 - q_1 q_2 - q_2 q_3]. \tag{t2}$$

Let us now assume the $3 - 1 = 2$ *symmetric* mode constraints: $q_1 = q_2 = \mu\, q_3$. Then,

$$2T = m(2\mu^2 + 1)(\dot{q}_2)^2, \qquad 2V = (S/l)(4\mu^2 - 4\mu + 2)q_2{}^2, \tag{u1}$$

and so [with $q_2 \sim \sin(\omega t)$, or $\sim \cos(\omega t)$] Rayleigh's quotient becomes

$$\omega^2 = \omega^2(\mu) = \sigma[(4\mu^2 - 4\mu + 2)/(2\mu^2 + 1)], \quad \text{where } \sigma \equiv S/ml. \tag{u2}$$

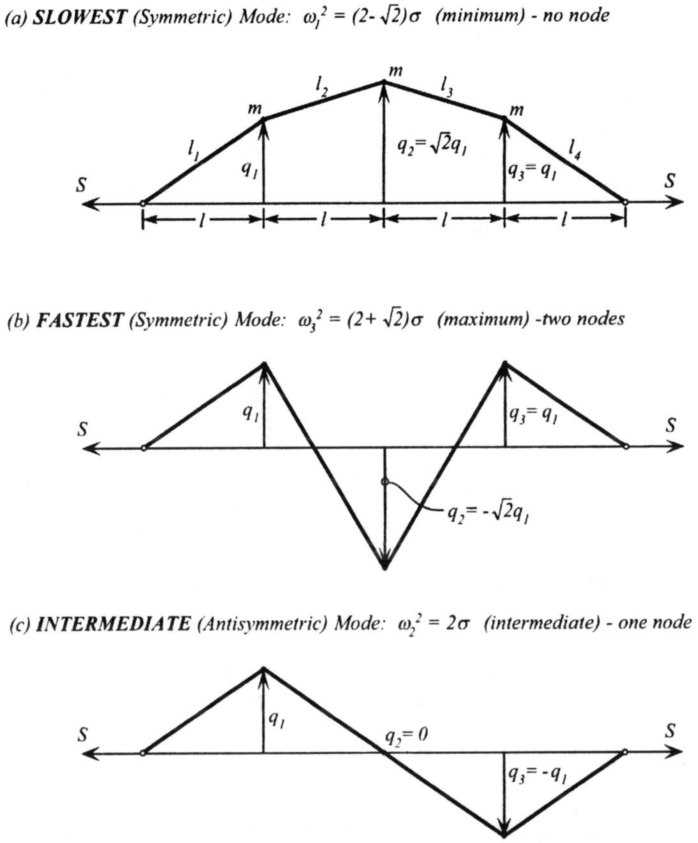

Figure 7.6 Symmetric and antisymmetric modes of three equal masses on a taut string.
(a) $\Delta l_1 \equiv l_1 - l = (q_1{}^2 + l^2)^{1/2} - l = l\{[1 + (q_1/l)^2]^{1/2} - 1\} \approx (1/2l) q_1{}^2$,
$\Delta l_2 \equiv l_2 - l \approx (1/2l)(q_2 - q_1)^2$, $\quad \Delta l_3 \equiv l_3 - l \approx (1/2l)(q_2 - q_3)^2$, $\quad \Delta l_4 \equiv l_4 - l \approx (1/2l) q_3{}^2$.

By the *stationarity* part of RP, the roots of $d\omega^2/d\mu = 0$, for μ to produce a *principal symmetric mode*, are $\mu_1 = -1/\sqrt{2}$, and $\mu_3 = +1/\sqrt{2}$, and the corresponding (exact) values of the frequencies squared are [fig. 7.6(a, b)]

$$\omega_1{}^2 \equiv \omega_{\min}{}^2 = (2 - \sqrt{2})\sigma \approx 0.5858\,\sigma \quad \text{(signs of } q_1, q_2, q_3 \text{ the same)}, \tag{u3}$$

$$\omega_3{}^2 \equiv \omega_{\max}{}^2 = (2 + \sqrt{2})\sigma \approx 3.4142\,\sigma \quad \text{(sign of } q_2 \text{ opposite to those of } q_1, q_3); \tag{u4}$$

while by the *extremality* part of RP

$$\omega_{\min}{}^2 \approx 0.5858\,\sigma \le \omega^2 \le \omega_{\max}{}^2 \approx 3.4142\,\sigma; \tag{v1}$$

for example, assuming $q_1 = q_3 = (3/4) q_2$ (parabolic shape), we find

$$\omega^2 = 0.5882\,\sigma > 0.5858\,\sigma = \omega_{\min}{}^2; \tag{v2}$$

and assuming $q_1 = q_3 = (\sqrt{2}/2) q_2$ (sine curve of "period" $8l$), we find

$$\omega^2 = 0.5970\,\sigma > 0.5858\,\sigma = \omega_{\min}{}^2. \tag{v3}$$

Let the reader show that the *antisymmetric* mode $q_1 = -q_3,\ q_2 = 0$, or $q_1 = -q_3 = \mu q_2$, has the (intermediate) frequency: ${\omega_2}^2 = 2\sigma$; that is, $\mu \to \mu_2 = 0$.

Problem 7.9.9 *Rayleigh's Principle.* By using the *stationarity* property of Rayleigh's quotient, find (exactly) the two natural frequencies and corresponding mode ratios of the small (linear) oscillations of a double pendulum, consisting of two identical homogeneous bars, AB and BC, each of mass m and length l, under gravity [A: hinge of AB with fixed point ("ceiling"), B: hinge connecting AB and BC].

ANSWER

$\omega^2 = (3g/l)(1 \pm 2/\sqrt{7})$. With θ: angle of AB (*upper* bar) with vertical, and ϕ: angle of BC (*lower* bar) with vertical, we have $\phi/\theta = (1/3)(-1 \pm \sqrt{28})$:

for ω^2_-: $(\phi/\theta)_+ \approx +1.43$ (*lower* mode); for ω^2_+: $(\phi/\theta)_- \approx -2.10$ (*higher* mode).

Problem 7.9.10 *Rayleigh's Principle.* Consider a system consisting of a small bead B of mass m sliding on a smooth circular hoop H, also of mass m, and radius r. The hoop can turn freely about a fixed point O in its circumference, on a vertical plane. By using the stationarity property of Rayleigh's quotient, find (exactly) the two natural frequencies and corresponding mode ratios of its small oscillations, under gravity.

ANSWER

Let C be the center of the hoop. With θ: angle of OC with vertical, and ϕ: angle of CB with vertical, we have

Lower mode: ${\omega_{\min}}^2 = (1/2)(g/r)$, $\phi/\theta = +1$; *Higher* mode: ${\omega_{\max}}^2 = 2(g/r)$, $\phi/\theta = -2$.

Problem 7.9.11 *Lagrangean Action.* Show that the Lagrangean action functional A_L [(7.9.4d)] can also be expressed as

$$\int \Big(\sum p_k \dot{q}_k - h + E\Big)\, dt, \tag{a}$$

where, as usual, $p_k \equiv \partial L/\partial \dot{q}_k$, $h \equiv \sum (\partial L/\partial \dot{q}_k)\dot{q}_k - L$.

Example 7.9.10 *Hamiltonian Action in Nonlinear Oscillations; Van der Pol and Rayleigh Oscillators.* Here, the starting point is the general Hamiltonian variational equation (7.9.4b)

$$\Delta \int L\, dt + \int \delta' W_{np}\, dt = \Big\{\sum p_k\, \delta q_k + L\Delta t\Big\}_1^2. \tag{a}$$

We will specialize (a) to *periodic* motions: choosing $\delta q_k(t_1) = \delta q_k(t_2)$ (not necessarily zero), where $t_2 - t_1 = \tau$ (common) period of oscillation, and since, in such a case,

$$\Big\{\sum p_k\, \delta q_k\Big\}_1^2 = 0, \qquad L(t_1) = L(t_2) \equiv L_o, \tag{b}$$

and $\Delta t = \Delta\tau = \Delta(2\pi/\omega) = -(2\pi/\omega^2)\,\delta\omega$, eq. (a) reduces to

$$\Delta\int_0^{2\pi/\omega} L\,dt + \int_0^{2\pi/\omega} \delta' W_{np}\,dt + (2\pi/\omega^2)L_o\,\delta\omega = 0. \tag{c}$$

1. Van der Pol Oscillator

Let us apply (c) to [recalling ex. 7.2.2: (g) ff.]

$$\ddot{q} + \varepsilon(q^2 - 1)\dot{q} + q = 0. \tag{d}$$

For the reasons given earlier, we try the (asymptotically) harmonic "limit cycle" solution:

$$q = a\sin\chi, \qquad \chi = \omega\,t. \tag{e}$$

Varying both a and ω (since they are both unknown), we obtain

$$\delta q = \delta a\sin\chi + (a\,t)\cos\chi\,\delta\omega, \tag{f}$$

and so the periodicity condition

$$\{p\,\delta q\}_1^2 = \{\dot{q}\,\delta q\}_1^2 = \{(a\,\delta a)\sin\chi\cos\chi + (a^2\,\delta\omega)t\cos^2\chi\}_1^2 = 0$$

supplies the time endpoints $t_1 = \pi/2\omega$, $t_2 = 3\pi/2\omega$ or $5\pi/2\omega \Rightarrow \tau = 2\pi/\omega$ or π/ω; which, in turn, yield

$$\begin{aligned} \delta q(\pi/2\omega) &= \delta q(5\pi/2\omega) = \delta a \quad (\neq 0), \\ L(\pi/2\omega) &= L(5\pi/2\omega) \equiv L_o = -(a^2/2). \end{aligned} \tag{g}$$

From the above, we find, successively,

$$\begin{aligned} A_H &= \int (T - V)\,dt = \int (1/2)[(\dot{q})^2 - q^2]\,dt \\ &= \int [(1/2)a^2\omega^2\cos^2\chi - (1/2)a^2\sin^2\chi]\,dt \\ &= (\pi/\omega)[a^2(\omega^2 - 1)/2] = A_H(a, \omega), \end{aligned} \tag{h1}$$

$$\begin{aligned} \Rightarrow \Delta A_H &= (\partial A_H/\partial a)\,\delta a + (\partial A_H/\partial\omega)\,\delta\omega \\ &= (\pi/\omega)\left[a(\omega^2 - 1)\,\delta a + (a^2/2)(\omega + \omega^{-1})\,\delta\omega\right]; \end{aligned} \tag{h2}$$

$$L(t_1)\,\Delta(t_2 - t_1) = L_o\,\Delta(2\pi/\omega) = -(a^2/2)[-(2\pi/\omega^2)\,\delta\omega] = (\pi/\omega^2)a^2\,\delta\omega; \tag{h3}$$

$$\begin{aligned} \int \delta' W_{np}\,dt &= \int Q\,\delta q\,dt = -\int \varepsilon(q^2 - 1)\dot{q}\,\delta q\,dt \\ &= \int (-\varepsilon\,a^3\omega\sin^2\chi\cos\chi + \varepsilon\,a\,\omega\cos\chi)[\delta a\sin\chi + (a\,t)\cos\chi\,\delta\omega]\,dt \\ &= \cdots = (\pi/\omega)^2(3\varepsilon\,a^2\omega/2)(1 - a^2/4). \end{aligned} \tag{h4}$$

Substituting (h2–4) into (c), and simplifying, yields

$$[a(\omega^2 - 1)]\,\delta a + \left[(a^2/2)(\omega + \omega^{-1}) - (a^2/\omega) + (3\pi/2)\varepsilon a^2(1 - a^2/4)\right]\delta\omega = 0, \quad \text{(i)}$$

from which, since δa and $\delta\omega$ are *independent*, we find $\omega^2 = 1$, and thanks to it, $1 - a^2/4 = 0 \Rightarrow |a| = 2$; values that agree with those found by other means in the oscillations literature; for example, Kauderer (1958, p. 343 ff.).

Also, we note that for $|a| = 2$, eq. (h4) yields

$$\int \delta' W_{np}\, dt = \int Q\,\delta q\, dt\text{: \textit{virtual nonpotential work} (damping)} = 0. \quad \text{(j)}$$

For better approximations to (d), than (e), see Papastavridis [1986(b)].

2. Generalizations

Let us now extend the above to the periodic solutions of the general *quasi-linear* equation

$$\ddot{q} + q = \varepsilon f(q, \dot{q}), \quad \text{(k)}$$

where, again, $f(\ldots)$: nonlinear in q, $\dot{q}$, and $\varepsilon f(q, \dot{q})$ is very small compared with $\ddot{q}$ and q (all taken absolutely).

With the periodic solution (e) [viewed as the first term of the Fourier series representation of the assumed periodic solution of (k)], eqs. (h2, 3), and the notation

$$f(a \sin\chi, a\omega \sin\chi) \equiv F(a, \chi) \equiv F, \quad \text{(l)}$$

we find

$$\int_0^{2\pi/\omega} \delta' W_{np}\, dt = \int_0^{2\pi/\omega} Q\,\delta q\, dt = \varepsilon \int_0^{2\pi/\omega} F(a, \chi)\,\delta q\, dt$$
$$= \varepsilon \left[\delta a \int_0^{2\pi/\omega} F \sin\chi\, dt + (a\,\delta\omega) \int_0^{2\pi/\omega} F\, t \cos\chi\, dt\right]. \quad \text{(m)}$$

Substituting (h2, 3, m) into (c) yields

$$\left[(\pi/\omega)(\omega^2 - 1)a + \varepsilon \int_0^{2\pi/\omega} F \sin\chi\, dt\right]\delta a$$
$$+ a\left[(\pi/2)(1 - \omega^{-2})a + \varepsilon \int_0^{2\pi/\omega} F\, t \cos\chi\, dt\right]\delta\omega = 0, \quad \text{(n)}$$

from which, since δa and $\delta\omega$ are independent, we obtain the following system for a and ω:

$$(\pi/\omega)(\omega^2 - 1)a + \varepsilon \int_0^{2\pi/\omega} F(a, \chi) \sin\chi\, dt = 0, \quad \text{(o1)}$$

$$(\pi/2)(1 - \omega^{-2})a + \varepsilon \int_0^{2\pi/\omega} F(a, \chi)\, t \cos\chi\, dt = 0. \quad \text{(o2)}$$

The earlier Van der Pol case (d) corresponds to $f = -(q^2 - 1)\dot{q}$. Then, since

$$\int_0^{2\pi/\omega} F \sin\chi\, dt = (-a^3\omega) \int_0^{2\pi/\omega} \sin^3\chi \cos\chi\, dt + (a\,\omega) \int_0^{2\pi/\omega} \sin\chi \cos\chi\, dt = 0, \quad \text{(p1)}$$

eq. (o1) yields $\omega^2 = 1$ and this, in turn, simplifies (o2) to

$$\int_0^{2\pi/\omega} F(a,\chi) t \cos\chi\, dt = 0$$

$$\Rightarrow a^2 = \left(\int_0^{2\pi/\omega} t\cos^2\chi\, dt\right) \Big/ \left(\int_0^{2\pi/\omega} t\sin^2\chi\ \cos^2\chi\, dt\right)$$

$$= [(3/2)(\pi/\omega)^2]/[(3/8)(\pi/\omega)^2] = 4, \quad \text{(p2)}$$

as before.

Next, let us consider the *Rayleigh equation*

$$\ddot{q} + q = \varepsilon[\dot{q} - (\dot{q})^3], \qquad \varepsilon\text{: very small positive constant.} \quad \text{(q)}$$

With q, δq, τ as before, and $F = a\,\omega \cos\chi - a^3\omega^3 \cos\chi$, we find (with $t_1 = \pi/2\omega$, $t_2 = 5\pi/2\omega$)

$$\varepsilon \int F \sin\chi\, dt = \varepsilon \int [(a\,\omega)\sin\chi\cos\chi - (a^2\omega^3)\cos^3\chi\ \sin\chi]\, dt = 0$$

$$\Rightarrow \omega^2 = 1 \qquad \text{[by (o1)]}, \quad \text{(q1)}$$

and so (o2) reduces to

$$\varepsilon \int F\, t\cos\chi\, dt = \varepsilon \int [(a\,\omega) t\cos^2\chi - (a^3\omega^3) t\cos^4\chi]\, dt = 0$$

$$\Rightarrow (3\pi^2\varepsilon a/2)(1 - 3a^2\omega^2/4) = 0 \ \Rightarrow\ |a| = 2/\sqrt{3}. \quad \text{(q2)}$$

Pars (1965, pp. 388–389) shows that by an appropriate change of variables, (q) can be transformed back to the Van der Pol equation (d); see also Panovko (1971, pp. 209–213) for a small-parameter (perturbation) treatment.

Also, the above extend to the case where

$$f(\ldots) = f(t, q, \dot{q}) = f(\Omega\, t, q, \dot{q})\text{: } (2\pi/\Omega)\text{-periodic in time, } \Omega\text{: given.}$$

Finally, on the connection between the above results, $\int Q\,\delta q\, dt = 0$ and eqs. (o1, 2), and the method of *slowly varying parameters*, see ex. 7.9.14 and Bogoliubov and Mitropolskii (1974, §21).

Example 7.9.11 *Lagrangean "least" action in Nonlinear Oscillations; Nonlinear Pendulum and Van der Pol Oscillators.* Here, the basic variational equation (7.9.4a, or 4h)

$$\Delta \int 2T\, dt = \int (\delta T + \delta V - \delta' W_{np})\, dt + BT, \quad \text{(a)}$$

where

$$\begin{aligned} BT &\equiv \left\{\sum p_k\,\Delta q_k + \left(2T - \sum p_k\dot{q}_k\right)\Delta t\right\}_1^2 \\ &= \left\{\sum p_k\,\Delta q_k + (T_1 + 2T_o)\,\Delta t\right\}_1^2 \\ &= \left\{\sum p_k\,\delta q_k + 2T\,\Delta t\right\}_1^2 \\ &\left[= \left\{\sum p_k\,\Delta q_k\right\}_1^2 \equiv (BT)_{scl}, \text{for scleronomic systems}\right], \end{aligned} \tag{b}$$

will be applied to the approximate solution of one-DOF (hence, holonomic) nonlinear and/or nonconservative oscillations.

For periodic motions and variations, with period $\tau = t_2 - t_1 = 2\pi/\omega$:

(i) If periodicity means $\delta q_k(t_1) = \delta q_k(t_2)$, $p_k(t_1) = p_k(t_2)$, then

$$BT = 2T(t_1)\,\Delta(2\pi/\omega) = -2T(t_1)(2\pi/\omega^2)\,\delta\omega; \tag{c1}$$

whereas (ii) If periodicity means $\Delta q_k(t_1) = \Delta q_k(t_2)$, $p_k(t_1) = p_k(t_2)$, then

$$(BT)_{scl} = 0. \tag{c2}$$

In sum: whenever $BT = 0$, the fundamental equation (a) becomes

$$\Delta \int 2T\,dt - \int (\delta T + \delta V - \delta' W_{np})\,dt = 0. \tag{d}$$

The *single but variational* equation (d) [or (a), if needed] can produce as many independent algebraic equations as the number of the unknown parameters (amplitudes and/or frequencies) entering the assumed trial solution(s).

Illustrations

1. Free and Undamped Nonlinear (Plane) Pendulum Oscillations

Here, to within a constant factor and with dimensionless time $(g/l)^{1/2}t$ substituted for t (g: constant acceleration of gravity, l: length of pendulum), the kinetic and potential energies and Lagrangean equation of its (free and undamped) motion are, respectively,

$$2T = (\dot{\phi})^2, \qquad V = 1 - \cos\phi \approx \phi^2/2 - \phi^4/24, \tag{e1}$$

$$E_\phi(T - V) = 0: \qquad \ddot{\phi} + \phi - \phi^3/6 = 0, \tag{e2}$$

where ϕ is the angle of the pendulum with the vertical. [Equation (e2) is referred to as a *soft* Duffing oscillator, because its frequency *decreases* when the amplitude increases (absolutely).] Following Lur'e (1968, pp. 702–703), we will calculate the approximate oscillatory solution of (e2) for the *initial* conditions

$$\phi(0) = \phi_o \text{ (given)}, \qquad \dot{\phi}(0) = 0, \tag{f1}$$

and trial solution

$$\phi = \phi(t) = (\phi_o + \alpha)\cos\chi - \alpha\cos(3\chi), \qquad \chi = \omega t, \tag{f2}$$

where α and ω are unknown. We notice that (f2) satisfies (f1), just like the solution of the linearized version of (e2): $\phi = \phi_o \cos t$ (i.e. $\alpha = 0$, $\omega = 1$).

Varying (f2) in its unknowns gives

$$\begin{aligned}\delta\phi &= (\partial\phi/\partial\alpha)\,\delta\alpha + (\partial\phi/\partial\omega)\,\delta\omega\\ &= [\cos\chi - \cos(3\chi)]\,\delta\alpha + t\,[-(\phi_o + \alpha)\sin\chi + 3\alpha\sin(3\chi)]\,\delta\omega;\end{aligned} \tag{g1}$$

also

$$\dot{\phi} = p_\phi \equiv \partial T/\partial\dot{\phi} = -\omega(\phi_o + \alpha)\sin\chi + 3\omega\,\alpha\sin(3\chi). \tag{g2}$$

Then, the periodicity condition $\{p_\phi\,\delta\phi\}_1^2 = 0$ yields $t_1 = 0$, $t_2 = \pi/\omega = \tau/2$. With this choice, $T(t_1) = T(t_2) = 0$, and so $\{2T\,\Delta t\}_0^{2\pi/\omega} = 0$.

From the above, we readily find

$$A_L \equiv \int_0^{\pi/\omega} 2T\,dt = \int_0^{\pi/\omega} (\dot{\phi})^2\,dt = (\pi/2)\omega(\phi_o{}^2 + 2\phi_o\alpha + 10\alpha^2), \tag{h1}$$

$$\begin{aligned}\Rightarrow \Delta A_L &= (\partial A_L/\partial\alpha)\,\delta\alpha + (\partial A_L/\partial\omega)\,\delta\omega\\ &= (\pi/2)[2\omega(\phi_o + 10\alpha)\,\delta\alpha + (\phi_o{}^2 + 2\phi_o\alpha + 10\alpha^2)\,\delta\omega],\end{aligned} \tag{h2}$$

$$E \equiv T + V = (\dot{\phi})^2/2 + \phi^2/2 - \phi^4/24, \tag{h3}$$

$$\delta E = \dot{\phi}\,\delta(\dot{\phi}) + \phi\,\delta\phi - (\phi^3/6)\,\delta\phi; \tag{h4}$$

and, therefore, after several straightforward integrations [and noting that $\delta(\dot{\phi}) = (\delta\phi)^{\cdot}$]

$$\begin{aligned}\int_0^{\pi/\omega} \delta E\,dt &= (\pi/4)\{2\omega(\phi_o + 10\alpha) + \omega^{-1}[(2\phi_o - \phi_o{}^3/6) + (4 - 3\phi_o{}^2/4)\alpha - 3\phi_o\alpha^2/2]\}\,\delta\alpha\\ &\quad + (\pi/4)\{(\phi_o{}^2 + 2\phi_o\alpha + 10\alpha^2) + \omega^{-2}[(\phi_o{}^2 - 5\phi_o{}^4/48)\\ &\qquad + (-2\phi_o + \phi_o{}^3/6)\alpha + (-2 + 3\phi_o{}^2/8)\alpha^2]\}\,\delta\omega.\end{aligned} \tag{h5}$$

Substituting (h2) and (h5) into (d) [with $\int \delta' W_{np}\,dt = 0$], and regrouping terms, we obtain

$$I\,\delta\alpha + II\,\delta\omega = 0, \tag{i}$$

where

$$\begin{aligned}I &\equiv (\pi/4)\{2\omega(\phi_o + 10\alpha)\\ &\quad - \omega^{-1}[2\phi_o + 4\alpha - \phi_o{}^3/6 - 3\phi_o{}^2\alpha/4 - 3\phi_o\alpha^2/2]\},\end{aligned} \tag{i1}$$

$$\begin{aligned}II &\equiv (\pi/4)\{(\phi_o{}^2 + 2\phi_o\alpha + 10\alpha^2)\\ &\quad - \omega^{-2}[\phi_o{}^2 - 2\phi_o\alpha - 5\phi_o{}^4/48 + \phi_o{}^3\alpha/6 + (-2 + 3\phi_0{}^2/8)\alpha^2]\}.\end{aligned} \tag{i2}$$

Setting I and II equal to zero, since $\delta\alpha$ and $\delta\omega$ are independent, and neglecting terms proportional to α^2 and $\alpha\,\phi_o{}^2$ (last two terms in I, and third and last three terms in II) yields the following two algebraic equations:

$$\phi_o + 10\alpha - \omega^{-2}(\phi_o + 2\alpha - \phi_o{}^3/12) = 0, \tag{j1}$$

$$\phi_o{}^2 + 2\phi_o\,\alpha - \omega^{-2}(\phi_o{}^2 - 2\phi_o\,\alpha - 5\phi_o{}^4/48) = 0. \tag{j2}$$

To the required degree of approximation, the above give

$$\alpha = \phi_o{}^3/192, \qquad \omega = 1 - \phi_o{}^2/16, \tag{k}$$

which agree with the values of Lur'e.

2. Free van der Pol Oscillator with Nonlinear Elastic Term

(i.e., van der Pol + Duffing). Here, the equation of motion is

$$\ddot{q} + \varepsilon(q^2 - 1)\dot{q} + q + \varepsilon\,\kappa\,q^3 = 0, \tag{l}$$

where $\varepsilon\kappa$ is the nonlinear stiffness constant [like h/m in ex. 7.2.2: (a)]; and from it we easily deduce that

$$2T = (\dot{q})^2, \qquad 2V = q^2 + \varepsilon\,\kappa\,q^4/2, \qquad Q = \varepsilon(1 - q^2)\dot{q}. \tag{l1}$$

Guided by our knowledge of the "linear elasticity Van der Pol equation", that is, (l) with $\kappa = 0$ [ex. 7.2.2: (g) ff.; ex. 7.9.10: (d) ff.] we assume the following trial function, independent of the initial conditions (see, e.g., Kauderer, 1958, pp. 343–347),

$$q = 2\sin\chi + \varepsilon\big[\alpha\cos(3\chi) + \beta\,\kappa\sin(3\chi)\big], \qquad \chi \equiv \omega\,t, \qquad \omega = 1 + \gamma(\varepsilon\kappa), \tag{m}$$

where α, β, γ are first-order *correction constants*, to be determined.

From (m) we readily find

$$\dot{q} = p = 2\,\omega\cos\chi + \varepsilon[-3\alpha\,\omega\sin(3\chi) + 3\beta\,\kappa\,\omega\cos(3\chi)], \tag{n1}$$

$$\delta q = [\varepsilon\cos(3\chi)]\,\delta\alpha + [\varepsilon\,\kappa\sin(3\chi)]\,\delta\beta$$
$$+ [2\cos\chi - 3\alpha\,\varepsilon\sin(3\chi) + 3\beta\,\varepsilon\,\kappa\cos(3\chi)]t\,\delta\omega, \tag{n2}$$

$$\delta\omega = (\varepsilon\kappa)\,\delta\gamma; \tag{n3}$$

and so

$$p\,\delta q = \dot{q}\,\delta q = \cdots = (\ldots)\,\delta\alpha + (\ldots)\,\delta\beta + (\ldots)\,\delta\gamma,$$
$$\Rightarrow \{p\,\delta q\}_0^{\pi/\omega} = (2 + 3\beta\,\varepsilon\,\kappa)^2\pi\,\delta\omega,$$

and since

$$T(t_1 = 0) = T(t_2 = \pi/\omega) \equiv T_0$$
$$\Rightarrow \{2T\,\Delta t\}_0^{\pi/\omega} = -2T_0(\pi/\omega^2)\,\delta\omega = -(2 + 3\beta\,\varepsilon\,\kappa)^2\pi\,\delta\omega,$$

finally,

$$\{p\,\delta q\}_0^{\pi/\omega} + \{2T\,\Delta t\}_0^{\pi/\omega} = \{p\,\Delta q\}_0^{\pi/\omega} = 0. \tag{n4}$$

With the help of the above we find, after a long series of elementary integrations,

$$A_L \equiv \int_0^{\pi/\omega} 2T\,dt = \cdots = (\pi/2)\omega\,(4 + 9\alpha^2\varepsilon^2 + 9\beta^2\varepsilon^2\kappa^2),$$

from which we obtain, by variation,

$$\Delta A_L = \pi[(9\alpha\,\varepsilon^2\omega)\,\delta\alpha + (9\beta\,\varepsilon^2\kappa^2\omega)\,\delta\beta + (2 + 9\alpha^2\varepsilon^2/2 + 9\beta^2\varepsilon^2\kappa^2\omega/2)\,\delta\omega], \tag{o1}$$

and, upon neglecting all terms proportional to α^2, β^2, $\alpha\beta$,

$$\int_0^{\pi/\omega} \delta T\,dt = \int_0^{\pi/\omega} \dot{q}\,\delta(\dot{q})\,dt = (9\pi\alpha\,\varepsilon^2\omega/2)\,\delta\alpha + (9\pi\beta\,\varepsilon^2\kappa^2\omega/2)\,\delta\beta + (3\pi + 6\pi\beta\,\varepsilon\kappa)\,\delta\omega, \tag{o2}$$

$$\int_0^{\pi/\omega} \delta V\,dt = \int_0^{\pi/\omega} (q + \varepsilon\,\kappa\,q^3)\,\delta q\,dt = \pi(\varepsilon^2/2 + 3\varepsilon^3\kappa)\alpha\,\omega^{-1}\,\delta\alpha + \pi(\beta/2 - 1 + 3\beta\,\varepsilon\kappa)\varepsilon^2\kappa^2\omega^{-1}\,\delta\beta + \pi(-1 - 3\varepsilon\,\kappa/2 + \beta\,\varepsilon^2\kappa^2)\omega^{-2}\,\delta\omega, \tag{o3}$$

$$\int_0^{\pi/\omega} Q\,\delta q\,dt = \varepsilon\int_0^{\pi/\omega} (1 - q^2)\dot{q}\,\delta q\,dt = \pi(1 - 3\beta\,\varepsilon\,\kappa/2)\varepsilon^2\,\delta\alpha + \pi(3\alpha\,\varepsilon^3\kappa/2)\,\delta\beta + \pi(5\alpha/6 + 2\pi\beta\,\kappa)\omega^{-1}\varepsilon^2\,\delta\omega. \tag{o4}$$

Inserting all these results into eq. (d) and regrouping terms [while recalling (n3)], yields

$$A\,\delta\alpha + B\,\delta\beta + \Gamma\,\delta\gamma = 0, \tag{p}$$

where

$$A \equiv \pi[9\alpha\,\varepsilon^2\omega - 9\alpha\,\varepsilon^2\omega/2 - (\varepsilon^2/2 + 3\varepsilon^3\kappa)\alpha\,\omega^{-1} + \varepsilon^2(1 - 3\beta\,\varepsilon\,\kappa/2)], \tag{p1}$$

$$B \equiv \pi[9\beta\,\varepsilon^2\kappa^2\omega - 9\beta\,\varepsilon^2\kappa^2\omega/2 - (\beta/2 - 1 + 3\beta\,\varepsilon\,\kappa)\varepsilon^2\kappa^2\omega^{-1} + 3\alpha\,\varepsilon^3\kappa/2], \tag{p2}$$

$$\Gamma \equiv \pi[2 + 9\alpha^2\varepsilon^2/2 + 9\beta^2\varepsilon^2\kappa^2/2 - (3 + 6\beta\,\varepsilon\,\kappa) - (-1 - 3\varepsilon\,\kappa/2 + \beta\,\varepsilon^2\kappa^2)\omega^{-2} + (5\alpha/6 + 2\pi\beta\,\kappa)\omega^{-1}\varepsilon^2]. \tag{p3}$$

To the lowest order, the three independent equations A, B, $\Gamma = 0$ yield, successively:

(i) $A = 0$: $9\alpha\,\omega/2 - (1/2 + 3\varepsilon\,\kappa)\alpha\,\omega^{-1} + 1 - 3\beta\,\varepsilon\kappa/2 = 0$, or, substituting into it the third of (m), $\omega = 1 + \gamma\varepsilon\kappa$, and omitting all higher order terms in ε, we obtain

$$9\alpha/2 - \alpha/2 + 1 = 0 \;\Rightarrow\; \alpha = -1/4; \tag{q1}$$

(ii) $B = 0$: $9\beta\omega/2 - \beta\omega^{-1}/2 + \omega^{-1} = 0$, or, setting $\omega = 1 + \gamma\varepsilon\kappa$, and so on,

$$9\beta/2 - \beta/2 + 1 = 0 \Rightarrow \beta = -1/4; \tag{q2}$$

(iii) $\Gamma = 0$: with $\omega^{-2} \approx 1 - 2\gamma\varepsilon\kappa$, we find

$$-1 - 6\beta\varepsilon\kappa + (1 - 2\gamma\varepsilon\kappa)(1 + 3\varepsilon\kappa/2) = 0$$
$$\Rightarrow -6\beta\varepsilon\kappa - 2\gamma\varepsilon\kappa + 3\varepsilon\kappa/2 = 0 \Rightarrow \gamma = 3/2. \tag{q3}$$

Hence, the trial solution (m), *correct to the first order in* κ (and in agreement with Kauderer) is

$$q = 2\sin\chi - (\varepsilon/4)[\cos(3\chi) + \kappa\sin(3\chi)], \qquad \chi \equiv \omega t, \qquad \omega = 1 + 3\varepsilon\kappa/2. \tag{r}$$

For additional related examples, see Papastavridis and Chen (1986).

Example 7.9.12 *The "Direct" Variational Methods of Galerkin and Ritz in Nonlinear Oscillations.*

1. Galerkin (1915)

Let us consider a one-DOF (hence, holonomic) system described by the, generally, nonlinear differential equation of motion

$$E = E(t, q, \dot{q}, \ddot{q}) \equiv E(t, q) \equiv \ddot{q} - F(t, q, \dot{q}) = 0. \tag{a}$$

Here, of particular interest is the case where $F(\ldots)$ is a *periodic* function of given period $\tau \equiv 2\pi/\omega$ (ω = frequency). Suppose then that we are seeking periodic solutions to (a), of period τ, that satisfy the initial conditions

$$q(t_1) = q(t_1 + \tau), \qquad \dot{q}(t_1) = \dot{q}(t_1 + \tau), \qquad \text{for } any\ t_1. \tag{b}$$

Let us assume the approximate solution to $q(t)$:

$$q(t) \approx q_o(t) = \sum a_k \psi_k(t) \qquad [k = 1, \ldots, N;\ \text{or, sometimes}, k = 0, \ldots, N], \tag{c}$$

where the a_k are unknown *constant* parameters (to be determined by a "Galerkin criterion" — see below), and the $\psi_k(t)$ are known and preferably orthogonal (or, better, orthonormal) "coordinate functions" that satisfy (b):

$$\psi(t_1) = \psi(t_1 + \tau), \qquad \dot{\psi}(t_1) = \dot{\psi}(t_1 + \tau), \qquad \text{for } any\ t_1. \tag{c1}$$

For example, we may choose as q_o a Fourier series defined in the interval $(t_1, t_1 + \tau)$ and having period τ outside it:

$$q_o = a_0/2 + \sum [a_k \cos(2\pi k t/\tau) + b_k \sin(2\pi k t/\tau)] \qquad (k = 1, \ldots, n), \tag{d1}$$

where, as is (hopefully) well known,

$$a_k = (2/\tau)\int_{t_1}^{t_1+\tau} q_o(t)\cos(2\pi k t/\tau)\,dt, \qquad b_k = (2/\tau)\int_{t_1}^{t_1+\tau} q_o(t)\sin(2\pi k t/\tau)\,dt, \tag{d2}$$

$$(k = 0, \ldots, N;\ \text{frequently } t_1 = 0 \text{ or } -\tau/2).$$

Then the a_k are selected so that $E_o \equiv E(t, q_o) = 0$ holds, if not identically, at least, *in a weighted average over a period*, sense; that is,

$$\int_{t_1}^{t_1+\tau} E_o\, w(t)\, dt = 0, \qquad \text{where } w(t) \text{ is some } \textit{weighting function}. \tag{e}$$

If we choose N *different* such functions, $w_1(t), \ldots, w_N(t)$, then we can obtain from (e) N algebraic equations for the N a_k's, contained in E_o. As such, we usually pick the $\psi_k(t)$'s appearing in (c); then (e) yields the N *weighted residual* or *Galerkin* equations:

$$\int_{t_1}^{t_1+\tau} E\Big[t, \sum a_s \psi_s(t)\Big] \psi_k(t)\, dt = 0 \qquad [k, s = 1, \ldots, N]. \tag{f}$$

If $F(\ldots)$ is *non-periodic*, we may choose as q_o the following power series

$$q_o = a_0 + a_1 t + a_2 t^2 + \cdots + a_N t^N. \tag{d3}$$

Interpretations of Equations (e, f)

(i) In the theory of ordinary differential equations, (f) appear as the conditions for the vanishing of the coefficients of the generalized Fourier series expansion of $E(t, q_o) = 0$. (ii) With $w \to \delta q_o = \sum \delta a_k \psi_k(t)$, eq. (e) and then eq. (f), essentially, constitute the time integral of Lagrange's principle; and thus can be viewed as requiring that the error, or "residual force," $e \equiv E_o - E = E_o (\neq 0)$ do zero virtual work on $w \to \delta q_o$ over τ. Other "error residual" versions of (e) exist in the literature.

In view of these interpretations, it should be clear that *Galerkin's method holds for any mechanical system*; that is, with *several DOF*, or *continuous* (such as beams, plates, shells) undergoing any type of motion and not just a periodic one; although for periodic motions the algebra is manageable.

2. Ritz (1908, 1909)

Let us consider a one-DOF (hence, holonomic) system, that is completely described by Hamilton's principle; that is, with $L = L(t, q, \dot{q})$ and $\delta q(t_1) = \delta q(t_2) = 0$,

$$\delta A_H = \delta \int L\, dt = \cdots = -\int E\, \delta q\, dt = 0, \qquad E = E(L) \equiv (\partial L/\partial \dot{q})^{\cdot} - \partial L/\partial q = 0; \tag{g}$$

where the last (Lagrangean) equation coincides with the earlier equation (a), whenever both refer to the same problem.

In view of the approximation (c), we will have

$$A_H = A_H[q(t)] \;\Rightarrow\; A_H[q_o(t)] = A_H(a_1, \ldots, a_N) \equiv A_{H,o}, \tag{h}$$

and so the variational equation (g) is replaced by one of ordinary differential calculus:

$$\delta A_{H,o} = \sum (\partial A_{H,o}/\partial a_k)\, \delta a_k = 0, \tag{i}$$

from which, since the N a_k are independent (otherwise we introduce Lagrangean multipliers), we obtain the N *Ritz* equations:

$$\partial A_{H,o}/\partial a_k = 0 \qquad [k = 1, \ldots, N]. \tag{j}$$

In the presence of *nonpotential* forces $Q = Q(t, q, \dot{q})$, in which case the second equation of (g) is replaced by $E(L) = Q$, eqs. (j) are replaced by

$$P_k(a_1, \ldots, a_N) \equiv \partial A_{H,o}/\partial a_k + \int_{t_1}^{t_1+\tau} Q\,(\partial q_o/\partial a_k)\,dt = 0. \tag{k}$$

In particular, if Q is *very small*, and the exact solution of the *unperturbed* problem—that is, of $E(L) = 0$—is $q_{(o)}(t; a_1, \ldots, a_N)$, we may reasonably assume that $q_o = q_{(o)}$. Then,

$$\partial A_{H,o}/\partial a_k = 0, \text{independently, and (k) reduces to} \int_{t_1}^{t_1+\tau} Q(\partial q_{(o)}/\partial a_k)\,dt = 0.$$

Let us calculate (j) explicitly: recalling (b, c), and since now $L(t, q, \dot{q}) \to L(t, q_o, \dot{q}_o)$, we find

$$\begin{aligned}\partial A_{H,o}/\partial a_k &= \int_{t_1}^{t_1+\tau} [(\partial L/\partial \dot{q}_o)(\partial \dot{q}_o/\partial a_k) + (\partial L/\partial q_o)(\partial q_o/\partial a_k)]\,dt \\ &= \int_{t_1}^{t_1+\tau} [(\partial L/\partial \dot{q}_o)\dot{\psi}_k + (\partial L/\partial q_o)\psi_k]\,dt \\ &= -\int_{t_1}^{t_1+\tau} E_o\psi_k\,dt + \psi_k(t_1)\big[(\partial L/\partial \dot{q}_o)_{t_1+\tau} - (\partial L/\partial \dot{q}_o)_{t_1}\big] = 0,\end{aligned} \tag{l}$$

where

$$\begin{aligned}E(t, q, \dot{q}) \to E(t, q_o, \dot{q}_o) &= E\Big[t, \sum a_s\psi_s(t)\Big] \\ &\equiv (\partial L/\partial \dot{q}_o)^{\cdot} - \partial L/\partial q_o \equiv E_o.\end{aligned} \tag{l1}$$

If the integrated out (boundary) term in (l) vanishes — for example, if $\partial L/\partial \dot{q}$ is periodic with period τ—(l) reduces to the *first, or minimizing, Ritz* (→ *Galerkin*) equation:

$$\partial A_{H,o}/\partial a_k = -\int_{t_1}^{t_1+\tau} E_o\psi_k\,dt = 0, \tag{m1}$$

However, if the ψ_k are *not* periodic, then (j) leads to the *second, or minimizing, Ritz* (→ *Galerkin*) equation (t_1, t_2: arbitrary time limits):

$$\partial A_{H,o}/\partial a_k = -\int_{t_1}^{t_2} E_o\psi_k\,dt + \{(\partial L/\partial \dot{q}_o)\psi_k\}_1^2 = 0; \tag{m2}$$

and if, instead of (c), we use the more general trial function

$$q_o(t) = \sum q_k(a_{k1}, \ldots, a_{kN}; t), \tag{m3}$$

then (m2) are replaced by the $N \times N$ equations:

$$-\int_{t_1}^{t_2} E_o(\partial q_o/\partial a_{ks})\,dt + \{(\partial L/\partial \dot{q}_o)(\partial q_o/\partial a_{ks})\}_1^2 = 0 \quad (k, s = 1, \ldots, N). \tag{m4}$$

3. Galerkin versus Ritz

Let us, next, compare the methods of Galerkin and Ritz. These two, although theoretically equivalent for periodic systems—that is eq. (f) = eqs. (m1)—differ in fundamental ways; and the principles of analytical mechanics help us to understand that:

(i) Not every problem's equations of motion derive from a Lagrangean and therefore from a stationarity variational principle (in the sense of variational calculus); and even if they happen to do that, finding the corresponding Lagrangean may be quite a mathematical task in itself. Thus, Galerkin's method, since it does not depend on Lagrangeans, is more general than Ritz's. [On how to find the Lagrangean of a given equation of motion (which is known as the *inverse* problem of variational calculus), there exists an extensive literature; see, for example, Santilli (1978, 1980).]

(ii) On the other hand, wherever both methods apply, the method of Ritz requires *less* accuracy in the trial function q_o—that is, the $\psi_k(t)$—than that of Galerkin. The reason for this is that *Ritz's method involves energetic functions that entail lower-order derivatives than the force/acceleration functions of Galerkin's method.*

(iii) Finally, in both methods, the accuracy of the solution depends on the number of terms taken in eq. (c) and the judicious choice of the ψ_k's.

Sometimes, advance qualitative knowledge of the behavior of the solution can restrict q_o to a *single* term and still provide a good approximation.

For *nonperiodic* initial value problems, eqs. (m2, 4), with *nonhomogeneous* conditions [e.g., $q(t_1) = q_1$ (given), $q(t_2) = q_2$ (given), with at least one of them nonzero], we can choose the following trial function:

$$q_o = \psi_0(t) + \sum a_k \psi_k(t), \tag{n1}$$

where $\psi_0(t_1) = q_1$, $\psi_0(t_2) = q_2$, and $\psi_k(t_1) = 0$, $\psi_k(t_2) = 0$; for example, for ψ_0 we can try the linear function

$$\psi_0(t) = [(q_2 - q_1)/(t_2 - t_1)](t - t_1) + q_1. \tag{n2}$$

4. An Illustration

Let us apply these methods to the earlier-discussed (ex. 7.2.2), undamped but periodically forced, Duffing's oscillator:

$$m\ddot{q} + kq + hq^3 - Q_o \sin\chi = 0, \qquad \chi \equiv \omega t, \tag{o}$$

or, with ${\omega_o}^2 \equiv k/m$: natural frequency of corresponding linear oscillator [i.e., (o) for $h = 0$], (o1)

$\varepsilon \equiv h/m$: measure of elastic nonlinearity

(if > 0: *hard* or *overlinear* spring; if < 0: *soft* or *underlinear* spring), (o2)

ω: given forcing frequency, (o3)

$f_o \equiv Q_o/m$: forcing amplitude per unit mass, (o4)

$$\ddot{q} + \omega_o^{\,2} q + \varepsilon q^3 - f_o \sin\chi = 0, \qquad \chi \equiv \omega t. \tag{p}$$

Let us investigate the forced response of (p) of the same frequency ω. Since eqs. (o, p) are "symmetric" about $t = 0$, we try the *single* parameter solution

$$q_o = a \sin\chi, \qquad a = a(\omega). \tag{p1}$$

Then, Galerkin's equation (f) yields

$$\int_0^{2\pi/\omega} (\ddot{q}_o + \omega_o^{\,2} q_o + \varepsilon q_o^{\,3} - f_o \sin\chi) \sin\chi \, dt = 0; \tag{p2}$$

and from this, after some simple integrations, we obtain the earlier equation [ex. 7.2.2: (f)]

$$(3\varepsilon/4)a^3 + (\omega_o^{\,2} - \omega^2)a - f_o = 0$$
$$\Rightarrow \omega^2 = \omega^2(a, f_o) = \omega_o^{\,2} + 3\varepsilon a^2/4 - f_o/a. \tag{p3}$$

This equation constitutes the *resonance curve*; that is, it gives the response amplitude $|a|$ as function of ω and the specified system parameters h and f_o (fig. 7.7). For $f_o = 0$ (free vibration) (p3) reduces to

$$\omega^2 = \omega_o^{\,2} + (3/4)\varepsilon a^2 \;\Rightarrow\; \omega = \omega_o(1 + 3\varepsilon a^2/4\omega_o^{\,2})^{1/2}. \tag{p4}$$

As for the Ritz method, since here

$$2T = m(\dot{q})^2; \quad V = V_2 + V_4, \quad 2V_2 \equiv k\,q^2, \quad 4V_4 \equiv h\,q^4; \quad Q = Q_o \sin\chi, \tag{q1}$$

and $q_o = a\sin\chi \;\Rightarrow\; \delta q_o = \delta a \sin\chi$, eqs. (g, j) give, with $t_1 = 0$ and $t_2 = 2\pi/\omega$,

$$\delta A_H = \delta \int_0^{2\pi/\omega} \left[(m/2)(dq_o/dt)^2 - (k\,q_o^{\,2}/2 + h\,q_o^{\,4}/4) + Q(t)\,q_o\right] dt$$
$$= \left\{(m\,\dot{q}_o)\,\delta q_o\right\}_0^{2\pi/\omega} - \int_0^{2\pi/\omega} \left[(m(d^2 q_o/dt^2) + k\,q_o + h\,q_o^{\,3} - Q(t)\right] \delta q_o \, dt$$
$$= 0 - \left[(-m\,a\,\omega^2 + k\,a - Q_o)\int_0^{2\pi/\omega} \sin^2\chi\, dt + (h\,a^3)\int_0^{2\pi/\omega} \sin^4\chi\, dt\right]\delta a$$
$$= (\pi/\omega)(-m\,a\,\omega^2 + k\,a + 3h\,a^3/4 - Q_o)\,\delta a, \tag{q2}$$

from which (p3) follows.

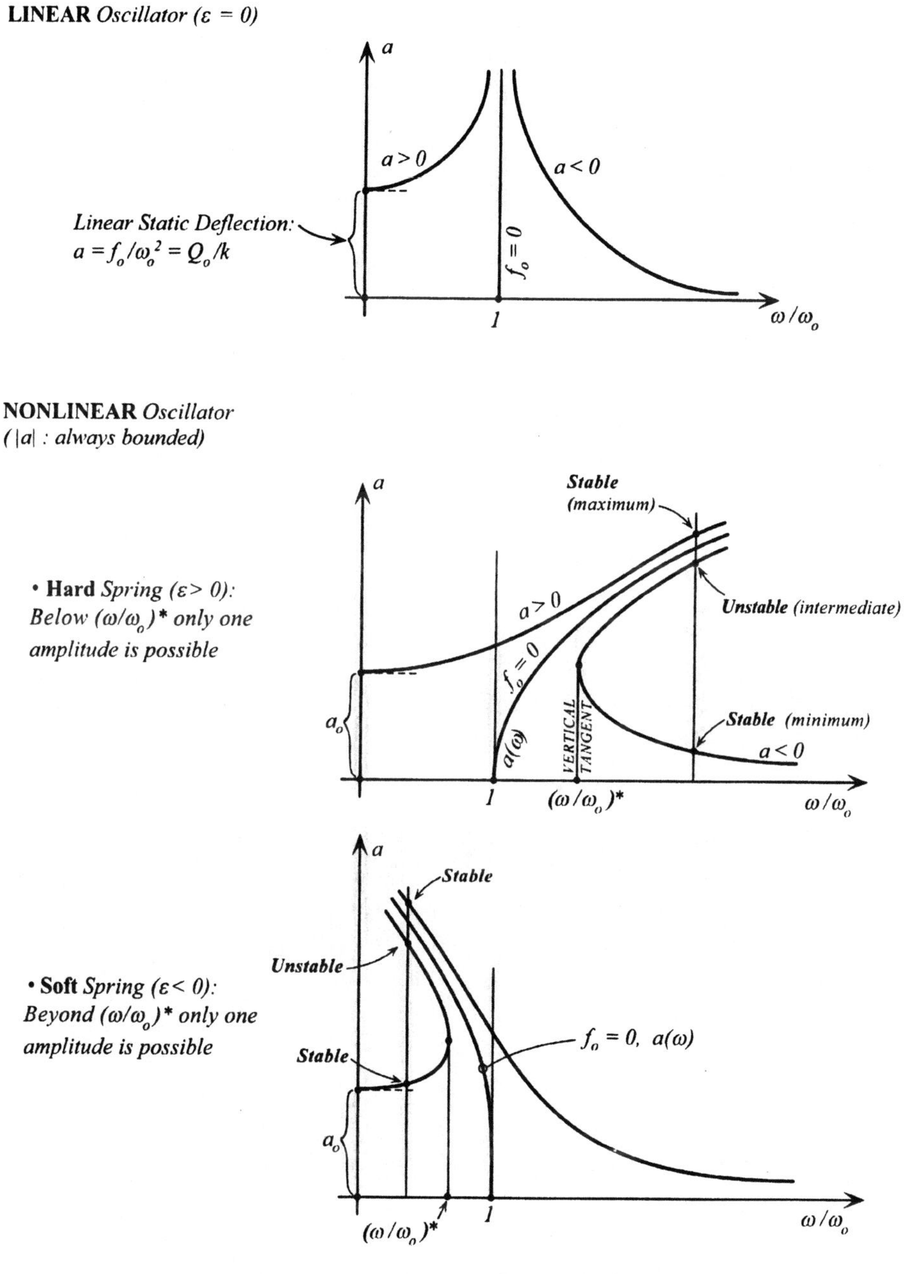

Figure 7.7 Resonance curves; that is, $a = a(\omega)$, for linear and (cubically) nonlinear oscillators.

[For a treatment of the *stability* of this oscillator via the *second* variation of A_H, $\delta^2 A_H \equiv \delta(\delta A_H)$, see Papastavridis [1983(a)].

5. A Generalization of Galerkin's Method

(See Chen, 1987.) The conventional Galerkin's method, presented above, assumes that the frequency ω is given, and so there is no need to vary it. Indeed, as already pointed out, with

$$\delta q_o = \sum (\partial q_o/\partial a_k)\,\delta a_k = \sum \psi_k(t)\,\delta a_k, \tag{r1}$$

the variational equation $\int E_o\,\delta q_o\,dt = 0$ (time integral of Lagrange's principle) leads to the Galerkin equations (f). However, in some problems, ω *is unknown and also unrelated to the amplitude*; for example, limit cycle oscillations (van der Pol equation); then δq_o *must be augmented by* $\delta\omega$*-proportional terms*. Indeed, let the trial solution be (only periodic solutions are of interest here)

$$q_o = \sum a_k\psi_k(t,\omega) = \sum a_k\psi_k(\omega t), \quad \psi_k(\omega t + 2\pi) = \psi_k(\omega t), \tag{r2}$$

and its variation

$$\delta q_o = \sum (\partial q_o/\partial a_k)\,\delta a_k + (\partial q_o/\partial\omega)\,\delta\omega = \sum \psi_k\,\delta a_k + \Omega\,\delta\omega, \tag{r3}$$

where

$$\Omega \equiv \Omega(t,\omega t,a_k) \equiv \partial q_o/\partial\omega = \sum a_k(\partial\psi_k/\partial\omega)$$
$$= \sum a_k t\,[d\psi_k/d(\omega t)]: \textit{secular coordinate function.} \tag{r4}$$

Substituting (r2–4) into $\int E_o\,\delta q_o\,dt = 0$ (with $t_1 = 0$, $t_2 = t_1 + 2\pi/\omega = 2\pi/\omega$, or $t_1 + \pi/\omega = \pi/\omega$) and setting, in the resulting variational equation, the coefficients of both δa_k and $\delta\omega$ equal to zero, we obtain the $N+1$ *generalized Galerkin equations*, for the $n+1$ unknowns a_k, ω:

$$\int E_o(\partial q_o/\partial a_k)\,dt = \int E_o\psi_k\,dt = 0 \qquad (k = 1,\ldots,N), \tag{r5}$$

$$\int E_o(\partial q_o/\partial\omega)\,dt = \int E_o\,\Omega\,dt = 0. \tag{r6}$$

[Equation (r6) is the Galerkin equivalent of *secular term suppression* of, say, the perturbation method.]

An Illustration

Let us apply the above to determine both the (limit cycle) amplitude and frequency of the earlier van der Pol equation (exs. 7.2.2, 7.9.10, and 7.9.11):

$$E = E(q,\dot q,\ddot q) = \ddot q + \varepsilon(q^2 - 1)\dot q + q + \varepsilon\kappa q^3 = 0. \tag{s1}$$

With the trial solution

$$q_o = a\cos\chi, \qquad \chi \equiv \omega t, \tag{s2}$$

we find

$$\delta q_o = (\cos\chi)\,\delta a + (-a\,t\sin\chi)\,\delta\omega, \tag{s3}$$

and

$$E \to E_o = (1-\omega^2)a\cos\chi + a\varepsilon\omega(1 - a^2\cos^2\chi)\sin\chi \neq 0, \tag{s4}$$

and therefore (r5, 6) yield the two Galerkin equations

$$\int_0^{2\pi/\omega} E_o(\partial q_o/\partial a)\,dt = \int_0^{2\pi/\omega} E_o \cos\chi\,dt = 0:$$

$$a(1-\omega^2)(\pi/2\omega) = 0 \Rightarrow \omega = 1, \tag{s5}$$

$$\int_0^{2\pi/\omega} E_o(\partial q_o/\partial \omega)\,dt = \int_0^{2\pi/\omega} E_o(-a\,t\sin\chi)\,dt = 0:$$

$$a(1-\omega^2)(-\pi/4\omega^2) + a\,\varepsilon\,\omega(\pi^2/4\omega^2) - a^3\varepsilon\,\omega(\pi^2/16\omega^2) = 0$$

$$\Rightarrow a^2 = 4 \Rightarrow |a| = 2, \tag{s6}$$

in agreement with the values found by the earlier methods.

In a several-DOF system, eqs. (r2–6) are replaced, respectively, by

$$q_{k,o} = \sum a_{kk'}\psi_{k'}(\omega t)$$

$$\Rightarrow \delta q_{k,o} = \sum \psi_{k'}\,\delta a_{kk'} + \Omega_k\,\delta\omega, \qquad \Omega_k \equiv \sum a_{kk'}t\,[d\psi_{k'}/d(\omega t)], \tag{t1, 2}$$

$$\int E_{k,o}\psi_{k'}\,dt = 0 \quad (n\times N \text{ eqs.}), \qquad \int\Big(\sum E_{k,o}\,\Omega_k\Big)\,dt = 0 \quad (1 \text{ eq.}); \tag{t3}$$

where

$k = 1,\ldots,n$ is the number of independent coordinates; $k' = 1,\ldots,N$ is the number of independent coordinate functions, and coefficients; ω is the assumed common frequency to all q's. (t4)

If each q_k is known to have a frequency ω_k, then the first of (t3) are replaced by the $n\times N$ equations

$$\int_0^{2\pi/\omega_k} E_{k,o}\psi_{k'}\,dt = 0, \tag{t5}$$

where $\psi_{k'}(t + 2\pi/\omega_k) = \psi_{k'}(t)$.

For further insights and examples, see, for example, Chen (1987), Fischer and Stephan [1972, pp. 150–151; good compact discussion of one and *several* DOF systems, both "autonomous" (no explicit time dependence) and "heteronomous" (explicit time dependence; e.g., forced vibrations); pp. 217–229: fully solved example; also 1984, pp. 267–268], Kosenko (1995).

Problem 7.9.12 *Lagrange–Ritz Method (General Considerations).* Consider a *one*-DOF conservative oscillatory system of period τ. Let q_o be an approximate periodic trial solution to its exact motion $q: q \approx q_o = q_o(t;\ \omega \equiv 2\pi/\tau, a_1, \ldots a_N)$, where both its frequency ω and "amplitude parameters" a_k are unknown.

Show that the latter can be determined from the following $N+1$ "Lagrange–Ritz" stationarity (of A_L) equations:

$$\partial A_{L,o}/\partial a_k = 0, \qquad \partial A_{L,o}/\partial\omega = 0 \qquad (k = 1,\ldots,N), \tag{a}$$

where

$$A_L(q) = \int 2T\,dt = \int (L + E)\,dt \rightarrow A_L(q_o) \equiv A_{L,o}, \tag{b}$$

$$L = L(q, \dot{q}) \rightarrow L_o\text{: Lagrangean,} \qquad E \equiv T + V \rightarrow E_o\text{: total energy.} \tag{c}$$

HINT

Set $\omega t \equiv x$. Then, $q_o = q_o(\omega t) = q_o(x)$, and with $(\ldots)' \equiv d(\ldots)/dx$,

$$A_{L,o} = \omega^{-1} \int_0^{2\pi} \{L[q_o(x), \omega q_o{}'(x)] + E_o\}\,dt. \tag{d}$$

Problem 7.9.13 *Lagrange–Ritz Method.* Continuing from the preceding problem, show that if V is an *even* function of $q = q(\omega t)$ (symmetric potential), and $q(0) = 0$, then the most general harmonic trial function q_o becomes a linear combination of *sines* with arguments *odd* multiples of ωt; that is,

$$q_o = a_1 \sin(\omega t) + a_3 \sin(3\omega t) + \cdots = \sum a_k \sin(k\omega t) \qquad (k = 1, 3, 5, \ldots). \tag{a}$$

HINT

Here, $\dot{q}(\omega t = \pi/2) = 0$.

Problem 7.9.14 *Lagrange–Ritz Method (A Theoretical Result).* Continuing from the preceding two problems, show that for a trial function

$$q_o = q_o(\omega t; a_1, \ldots, a_N), \tag{a}$$

the following stationarity condition results;

$$dA_{L,o}/dE = \cdots = \partial A_{L,o}/\partial E = \cdots = 2\pi/\omega = \tau, \tag{b}$$

where $A_L \rightarrow A_{L,o}(a_1, \ldots, a_N; \omega, E)$.

Problem 7.9.15 *Lagrange–Ritz Method.* Consider the harmonic oscillator with equation of motion $\ddot{q} + q = 0$ (i.e., of unit mass and frequency), and corresponding (double) Lagrangean $2L = (\dot{q})^2 - q^2$, and assume the trial solution $q_o = a\sin(\omega t)$. Show that, then,

$$A_{L,o} = (\pi/2)a^2(\omega - \omega^{-1}) + 2\pi E/\omega, \tag{a}$$

and thus the earlier stationarity conditions yield

$$\partial A_{L,o}/\partial a = 0 \Rightarrow \omega = 1, \qquad \partial A_{L,o}/\partial \omega = 0 \Rightarrow E = a^2/2. \tag{b}$$

Problem 7.9.16 *Lagrange–Ritz Method (Linear Oscillator).* In the oscillator of the preceding problem, assume the quadratic trial function

$$q_o = a(x - x^2/\pi), \qquad x \equiv \omega t. \tag{a}$$

Show that, then,

$$A_{L,o} = \pi\omega a^2/3 + 2\pi E/\omega - \pi^3 a^2/30\omega, \tag{b}$$

and thus the earlier stationarity conditions yield

$$\partial A_{L,o}/\partial a = 0 \;\Rightarrow\; \omega^2 = \pi^2/10 \approx 0.987,$$
$$\partial A_{L,o}/\partial \omega = 0 \;\Rightarrow\; E = (\pi^2/30)a^2 \approx 0.33a^2. \tag{c}$$

Problem 7.9.17 *Lagrange–Ritz Method.* Consider the nonlinear Duffing oscillator with equation of motion $\ddot{q} + \omega_o{}^2 q + 2c\,q^3 = 0$ ($\omega_o{}^2, c > 0$; c: small) and corresponding (double) Lagrangean $2L = (\dot{q})^2 - (\omega_o{}^2 q^2 + c\,q^4)$. Show that for the trial solution $q_o = a\sin(\omega t)$,

$$A_{L,o} = \pi\omega a^2/2 + 2\pi E/\omega - \pi\omega_o{}^2 a^2/2\omega - 3\pi c\,a^4/16\,\omega. \tag{a}$$

Then, set $\partial A_{L,o}/\partial a = 0$ and $\partial A_{L,o}/\partial\omega = 0$, and find the relations among ω, a, E.

ANSWER

$$\omega^2 = (\omega_o{}^2/3)\{1 + 2[1 + (9cE/\omega_o{}^2)]^{1/2}\}. \tag{b}$$

Problem 7.9.18 *Lagrange–Ritz Method.* Consider a mathematical pendulum of mass $m = 1$ and length l in general plane motion, under gravity. Here, with the usual notations,

$$2T = l(\dot{\phi})^2 \qquad \text{and} \qquad V = g\,l(1 - \cos\phi). \tag{a}$$

Show that
(i) In the new variable $x \equiv \sin(\phi/2)$, the Lagrangean of the pendulum becomes

$$L = [2l^2/(1 - x^2)](dx/dt)^2 - 2g\,l\,x^2 = L(x, \dot{x}), \tag{b}$$

and then
(ii) The trial solution $x_o = a\sin(\omega t)$ yields

$$A_{L,o} = 2\pi E/\omega + 4\pi\omega\, l^2[1 - (1 - a^2)^{1/2}] - 2\pi g\,l\,a^2/\omega. \tag{c}$$

Then, set $\partial A_{L,o}/\partial a = 0$ and $\partial A_{L,o}/\partial\omega = 0$, and find the relations among ω, a, E.
For further details, see Luttinger and Thomas (1960); and for a generalization to systems with Lagrangeans $L = L(t, q, \dot{q})$, see Buch and Denman (1976).

Problem 7.9.19 *Hamilton–Ritz Method.* Consider the earlier Duffing oscillator with equation of motion $\ddot{q} + \omega_o{}^2 q + c\,q^3 = 0$, and boundary conditions $q(0) = q(\pi/\omega) = 0$. Show that the method of "Hamilton–Ritz," applied here for the trial solution $q_o = a\sin(\omega t)$, yields either $a = 0$ or $\omega^2 = \omega_o{}^2 + (3/4)a^2 c$; that is, given a we find ω, and vice versa; while applied for the trial solution

$q_o = a\sin(\omega t) + b\sin(3\omega t)$, the method yields

$$a^2 - ab + 2b^2 - (4/3)(\omega^2 - \omega_o{}^2)/c = 0, \tag{a}$$

$$3b^3 - a^3 + 6a^2 b + 4b(\omega_o{}^2 - 9\omega^2)/c = 0; \tag{b}$$

that is, given a (or b) we find b (or a), and ω.

Further, with the help of the dimensionless quantities $x \equiv b/a$, $y \equiv a^2 c/\omega_o$, and $z \equiv (\omega/\omega_o)^2 - 1$, eqs. (a, b) can be rewritten, respectively, as

$$y(6x^2 - 3x + 3) - 4z = 0, \qquad y(3x^3 + 6x - 1) - x(36z + 32) = 0; \tag{c}$$

and, finally, eliminating z between them, we obtain

$$y = -32x(51x^3 - 27x^2 + 21x + 1)^{-1}. \tag{d}$$

Plot and discuss this curve [i.e., given a we find y and, via (d), we find x].

For a *two*-DOF example, see Schräpel (1988).

Problem 7.9.20 *Galerkin Method.* Consider the forced and nonlinearly damped oscillator

$$\ddot{q} + \omega_o{}^2 q + f_1\,\dot{q} + f_3\,(\dot{q})^3 = Q_o\,\sin(\omega t), \tag{a}$$

where $\omega_o{}^2, f_1, f_3, Q_o, \omega$ are specified positive constants (with the usual and/or easily understood meanings). Show that the Galerkin method applied to (a) for the trial steady-state solution

$$q = a\,\sin(\omega\,t) + b\,\cos(\omega\,t), \tag{b}$$

yields the following algebraic system for the unknown amplitudes a, b:

$$a\,(\omega_o{}^2 - \omega^2) - b f_1\omega - (3/4) f_3\,\omega^3 b\,(a^2 + b^2) - Q_o = 0, \tag{c}$$

$$a f_1\omega + b\,(\omega_o{}^2 - \omega^2) + (3/4) f_3\,\omega^3\,a\,(a^2 + b^2) = 0. \tag{d}$$

Discuss these results for various limiting cases.

Example 7.9.13 *The Variational Principles of Gray–Karl–Novikov (GKN, 1996).* Quite generally, we obviously have

$$\int (T - V)\,dt = \int (2T)\,dt - \int (T + V)\,dt;$$

or, recalling earlier definitions (§7.9), and with the still general but convenient time-integration limits $t_1 = 0$, $t_2 = t \;\Rightarrow\; t_2 - t_1 = t$,

$$A_H = A_L - A_E \equiv A_L - t\langle E\rangle, \tag{a}$$

where

$$A_E \equiv \int_0^t (T + V)\,dt \equiv \int_0^t (E)\,dt \equiv t\,\langle E\rangle. \tag{a1}$$

Therefore, Δ-varying the above, we obtain the following basic relation:

$$\Delta A_H = \Delta A_L - \Delta t \langle E \rangle - t\,\Delta\langle E \rangle \;\Rightarrow\; \Delta A_H + \langle E \rangle \Delta t = \Delta A_L - t\,\Delta\langle E \rangle. \tag{b}$$

Now, and recalling, again, the general results of (3.9.11b ff.; 7.2.6e, f; and 7.9.4a ff.), let us see how (b) specializes for a holonomic, scleronomic, and potential system; that is, one completely describable by the Lagrangean: $L = L(q, \dot{q}) \Rightarrow \partial L/\partial t = 0 \Rightarrow dh/dt = -\partial L/\partial t = 0 \Rightarrow h \equiv \sum (\partial L/\partial \dot{q}_k)\dot{q}_k - L = 2T - (T - V) = T + V \equiv E$ (i.e., generalized energy = ordinary energy) = *constant* $\Rightarrow h(t_2) = h(t_1) = \langle h \rangle = \langle E \rangle$ [recalling (7.9.12b)], and for fixed endpoint variations (i.e., $\Delta q_{1,2} = 0$, and with no loss in generality, $\Delta t_1 = 0$, $\Delta t_2 = \Delta t$) from an actual trajectory (orbit; i.e., $E_k(L) = 0$). We easily see that, then, (7.9.4b, 11h) reduce to (7.9.12d):

$$\Delta A_H = -\{h\,\Delta t\}_1^2 = -h\,\Delta t = -\langle E \rangle\,\Delta t \;\Rightarrow\; \Delta A_H + \langle E \rangle\,\Delta t = 0 \quad \text{(Hamiltonian)} \tag{c1}$$

$$\Rightarrow \Delta A_L - t\,\Delta\langle E \rangle = 0 \quad \text{(Lagrangean)}, \tag{c2}$$

that is, for such systems and variations, *both the left and right sides of (b) vanish independently*.

Next, a little reflection (with invocation of the reasoning of the theory of constrained stationarity conditions and method of Lagrangean multipliers) will convince us that the *two unconstrained* variational equations (c) lead to the following *four constrained GKN* variational principles:

(i) $(\Delta A_H)_{t=\text{constant}} = 0$ (Hamilton), (d1)

(ii) $(\Delta t)_{\textit{Hamiltonian}\ \text{action=constant}} = 0$ ("Reciprocal Hamilton"), (d2)

(iii) $(\Delta A_L)_{\langle E \rangle=\text{constant}} = 0$ ("Reformulated MEL"), (d3)

(iv) $(\Delta\langle E \rangle)_{\textit{Lagrangean}\ \text{action=constant}} = 0$ ("Reciprocal MEL"). (d4)

[We notice that (a) eq. (d1) is a specialization of the method presented in the earlier probs. 7.9.12 and 7.9.13; and that (b) eq. (d3), with $\langle E \rangle = $ *constant*, is slightly more general than the ordinary MEL principle ($E = $ *constant*).] The following applications to simple problems of nonlinear oscillations illustrate the use of the above, especially (d3) and (d4) [one of the main contributions of GKN (1996)]:

(i) *One-dimensional quartic oscillator*, with $2T = m\,(\dot{q})^2$ (m: mass), $4V = c\,q^4$ (c: positive constant). With trial trajectory solution, $q = a\,\sin(\omega t) \;\Rightarrow\; \dot{q} = a\,\omega\cos(\omega t)$, where ω (frequency, and $\tau \equiv 2\pi/\omega$: period) and a (amplitude) are the hitherto unknown parameters to be determined by the above variational principles, and with the simpler notation $A_L \equiv W$ (from the German *Wirkung* = *action*), and integration limits $t_1 = 0$, $t_2 = 2\pi/\omega$, we find

$$W \equiv \int_0^\tau 2T\,dt = \int_0^{2\pi/\omega} [m\,(\dot{q})^2]\,dt = \int_0^{2\pi/\omega} [m\,a^2\,\omega^2\cos^2(\omega t)]\,dt = m\,a^2\,\pi\omega, \tag{e1}$$

$$\Rightarrow a^2 = W/m\,\pi\,\omega; \tag{e2}$$

$$\langle E \rangle \equiv (1/\tau)\int_0^\tau (T + V)\,dt = (1/\tau)\int_0^{2\pi/\omega} [(1/2)\,m\,(\dot{q})^2 + (1/4)\,c\,q^4]\,dt$$

$$= \cdots \text{[and utilizing (e2)]} = W\,\omega/4\pi + (3/32)(c\,W^2/\pi^2 m^2\omega^2). \tag{e3}$$

Then, applying (d4) [which, in view of the above results, is the easiest among (d1–4) to implement], we obtain the frequency that makes (e3) stationary (and, here, a minimum):

$$0 = (\partial\langle E\rangle/\partial\omega)_{W=\text{constant}} = (W/4\pi)\,\big[1 - (3/4)\,\big(c\,W/\pi m^2\omega^3\big)\big], \tag{e4}$$

$$\Rightarrow \omega = (3\,c\,W/4\,\pi\,m^2)^{1/3}, \tag{e5}$$

$$\Rightarrow \langle E\rangle = (1/2)(c/m^2)^{1/3}\,(3A_L/4\pi)^{4/3} = (1/2)(m^2/c)\omega^4, \tag{e6}$$

$$\Rightarrow \tau = 2\pi/\omega = 2\pi\,(m^2/2c\langle E\rangle)^{1/4}. \tag{e7}$$

Since the exact value of τ is {see Gray et al. [1996(a)], or books on nonlinear oscillations}:

$$\tau_{\text{exact}} = \big[\Gamma(1/2)\,\Gamma(1/4)/\Gamma(3/4)\big]\,(m^2/c\,\langle E\rangle)^{1/4} \qquad [\Gamma(\ldots)\text{: Gamma function}], \tag{e8}$$

$$\Rightarrow \tau/\tau_{\text{exact}} = (2\pi/2^{1/4})\,\big[\Gamma(3/4)/\Gamma(1/2)\,\Gamma(1/4)\big] = 1.0075, \tag{e9}$$

the error committed is less than 1%.

The reader is urged to verify that these results can also be obtained by applying (d1)—in which case, the constraint $t = constant$ implies $\tau = constant \Rightarrow \omega = constant$, in our trial solution; that is, $A_H = A_H(a;\,\omega)$, and so (d1) translates to $dA_H/da = 0$. (Compare with the more general method discussed in probs. 7.9.12 and 7.9.13.)

(ii) *Anharmonic oscillator*, with $2T = m\,(\dot{q})^2$, $V = (1/2)\,m\,\omega_o{}^2\,q^2 + (1/4)\lambda\,q^4$ [which can be viewed as the *quartic* approximation to the potential energy of a plane mathematical pendulum, $V_{\text{exact}} = m\,l^2\,\omega_o{}^2\,(1 - \cos\theta)$, where, as usual, m, l, θ are the mass, length, and angle with vertical, of the pendulum; $\omega_o \equiv (g/l)^{1/2}$ (frequency in *quadratic* approximation of $\cos\theta$ to V_{exact}), $q = l\theta$, and $\lambda = -m\,\omega_o{}^2/6l^2$]. Again, with trial solution, $q = a\,\sin(\omega t) \Rightarrow \dot{q} = a\,\omega\cos(\omega t)$, where ω (frequency) and a (amplitude) are the hitherto unknown parameters to be determined by the GKN variational principles, and integration limits $t_1 = 0$, $t_2 = 2\pi/\omega$, we find

$$W = \int_0^{2\pi/\omega} [m\,(\dot{q})^2]\,dt = \int_0^{2\pi/\omega} [m\,a^2\omega^2\cos^2(\omega t)]\,dt = m\,a^2\pi\,\omega \tag{f1}$$

$$\Rightarrow a^2 = W/m\,\pi\,\omega; \tag{f2}$$

$$\begin{aligned}\langle E\rangle &\equiv (1/\tau)\int_0^{\tau}(T + V)\,dt\\ &= (1/\tau)\int_0^{2\pi/\omega}\big[(1/2)\,m\,(\dot{q})^2 + (1/2)\,m\,\omega_o{}^2\,q^2 + (1/4)\lambda\,q^4\big]\,dt\\ &= \cdots[\text{and utilizing (f2)}] = (A_L/4\pi)\,\big[\omega + \omega_o{}^2/\omega + (3\lambda W/8\pi m^2\omega^2)\big].\end{aligned} \tag{f3}$$

Here, too, applying (d4) with ω as variational parameter, we get

$$0 = (\partial\langle E\rangle/\partial\omega)_{W=\text{constant}}$$

$$\Rightarrow \omega^2 = \omega_o{}^2 + (3\lambda W/4\pi\,m^2\omega) = \omega_o{}^2 + (3\lambda/4m)\,a^2, \tag{f4}$$

which agrees with earlier-found approximate values (exs. 7.9.11, probs. 7.9.17 and 7.9.18; also exs. 8.16. and 8.16.2). For the pendulum, (f4), with $\lambda \equiv -m\omega_o^2/6l^2$ and $a/l \equiv \theta_{max}$, gives

$$\omega = \omega_o (1 - \theta_{max}^2/8)^{1/2} = \omega_o (1 - \theta_{max}^2/16 - \theta_{max}^4/512 + \cdots), \qquad \text{(f5)}$$

and, hence, corresponding period $\tau(\theta_{max}) \equiv \tau$ (with $\tau_o \equiv 2\pi/\omega_o$):

$$\tau = 2\pi/\omega = \tau_o (1 - \theta_{max}^2/8)^{-1/2} = \tau_o(1 + \theta_{max}^2/16 + 3\theta_{max}^4/512 + \cdots), \qquad \text{(f6)}$$

which is correct to θ_{max}^2 (since the trial solution is correct to the zeroth order in λ, and our variational principle makes sure that first-order such errors vanish). A better approximation to the exact expansion (the latter obtained through integration of the well-known nonlinear equation of motion, via an elliptic integral)

$$\tau_{exact} = \tau_o\left[1 + \theta_{max}^2/16 + (11/18)(3\theta_{max}^4/512) + \cdots\right], \qquad \text{(f7)}$$

is obtained by keeping the $\theta^6/6!$ term in the $\cos\theta$ expansion, in $V_{exact} \to V$; also by adding to the trial solution higher harmonics: for example, $b\,\sin(3\omega t)$ (b: corresponding amplitude) (recall ex. 7.9.11, prob. 7.9.19). For further examples and insights, see Gray et al. [1996(a),(b)].

Example 7.9.14 *Method of Slowly Varying Parameters (Amplitude and Phase) in Weakly Nonlinear (Quasi-linear) Oscillators.* Let us consider the general equation [recalling ex. 7.9.10: k ff.)]

$$\ddot{q} + \omega_o^2 q = \varepsilon f(q, \dot{q}), \qquad \text{(a)}$$

where

ω_o: natural (constant) frequency of (a) when $\varepsilon f(q, \dot{q}) = 0$;

$f(\ldots)$: arbitrary *nonlinear* (but integrable) function of its arguments; and

ε: very small positive constant, so that (a) differs by very little from a linear equation (hence the name *quasilinear*).

Due to the presence of damping—namely, $\dot{q}$-proportional terms—the solutions of (a) are, in general, no longer periodic and, accordingly, the earlier-described methods of Hamilton/least action and Ritz/Galerkin do not apply, except asymptotically, as in the limit cycle case (e.g., van der Pol oscillator)—that is, *they need modification to account for the generally nonvanishing boundary terms* (see also the remarks at end of this example).

Below, we describe an approximate *averaging* method for the solution of (a), originated by van der Pol (in the early 1920s) and thoroughly extended and perfected by a host of distinguished Soviet scientists: Krylov, Bogoliubov, Mitropolskii, Andronov, Vitt, Mandelstam, Papaleksi, Malkin et al. (between the two World Wars), in connection with problems of electrical and mechanical engineering. In fact, this area of *asymptotic methods in nonlinear oscillations* was fairly considered as a Soviet (→Russian) specialty.

[That eq. (a) may, under certain conditions, have some periodic solutions (e.g., limit cycles) is far from obvious—in fact, it took a giant of mathematics, H. J.

Poincaré (late 19th century), to show that. A small change in the *form* of a (linear or nonlinear) differential equation may change radically the qualitative nature of its solutions; for example, the equation $\ddot{q} + \omega_o^2 q = 0$ possesses only periodic solutions, but the equation $\ddot{q} + \mu\dot{q} + \omega_o^2 q = 0$ *does not possess any such solutions, no matter how small (but nonzero) the friction/damping coefficient* μ *is!*]

As is well known, the solution of the *undamped* problem—that is, (a) with $\varepsilon = 0$—is

$$q = a \cos\chi, \qquad \chi \equiv \omega_o t + \phi, \tag{b}$$

where, a is the constant amplitude, ϕ is the constant phase, (both determined from the initial conditions), ω_o is the natural linear frequency (a given constant), and χ is the total phase.

For the damped nonlinear equation (a), we apply the Lagrangean *method of variation of constants* or *parameters* (for a general discussion of this, in terms of both Lagrangean and Hamiltonian variables, see §8.7): we try a solution *of the same form as the "generating solution"* (b) but with a and ϕ replaced by unknown functions of time; that is,

$$q = a(t)\cos\chi(t), \qquad \chi(t) = \omega_o t + \phi(t). \tag{c}$$

By $(\ldots)^{\cdot}$-differentiating eqs. (b) and (c) we find, respectively,

$$(\text{b})^{\cdot}: \qquad \dot{q} = -a\,\omega_o \sin\chi, \tag{d}$$

$$(\text{c})^{\cdot}: \qquad \dot{q} = -a\,\omega_o \sin\chi + (\dot{a}\cos\chi - a\dot{\phi}\sin\chi). \tag{e}$$

Now, to determine $a(t)$ and $\phi(t)$, we impose the *first* ("arbitrary") requirement: the "nonlinear" velocity $\dot{q}$, eq. (e), should have the same form as the "linear" velocity, eq. (d); that is,

$$\dot{a}\cos\chi - a\dot{\phi}\sin\chi = 0, \qquad \chi(t) = \omega_o t + \phi(t). \tag{f}$$

The *second* equation for $a(t)$, $\phi(t)$ is obtained by inserting (c) and (e) [under (f)] back into (a): since

$$\begin{aligned}(\text{e})^{\cdot}: \qquad \ddot{q} &= -\dot{a}\,\omega_o \sin\chi - a\,\omega_o^2\cos\chi - a\,\omega_o\,\dot{\phi}\sin\chi \\ &= -\dot{a}\,\omega_o \sin\chi - a\,\omega_o^2\cos\chi - \dot{a}\,\omega_o\cos\chi \qquad \text{[by the first of (f)]},\end{aligned}$$

we obtain

$$\dot{a}\sin\chi + a\dot{\phi}\cos\chi = -(\varepsilon/\omega_o)f(a\cos\chi, -a\,\omega_o\sin\chi) \equiv -(\varepsilon/\omega_o)F(a,\ \chi). \tag{g}$$

Solving the system of the first of (f) and (g) for $\dot{a}$ and $\dot{\phi}$ yields the *first-order* coupled nonlinear equations

$$\dot{a} = -(\varepsilon/\omega_o)F(a,\ \chi)\sin\chi, \qquad \dot{\phi} = -(\varepsilon/a\,\omega_o)F(a,\ \chi)\cos\chi \quad (= \dot{\chi} - \omega_o). \tag{h}$$

So far, *no approximations have been involved*: the exact solution of the system (h), if available, would be the exact solution of its equivalent original equation (a) for any value of ε; *conditions (f) and (g) may be arbitrary but they are consistent.*

[A geometrical interpretation: In terms of the "canonical" variables q and $p \equiv \dot{q}$, the original equation (a) becomes the *first-order* system

$$\dot{p} = -\omega_o{}^2 q + \varepsilon f(q, p) \equiv P(q, p) \qquad \text{(equation of motion)}, \tag{h1}$$

$$\dot{q} = p \equiv Q(q, p) \qquad \text{(kinematical equation)}. \tag{h2}$$

Then, eqs. (c) are simply a *transformation among dependent* variables — from the old q, p to the new a, ϕ or a, χ — and (h1, 2) transforms to (h). Geometrically, if (q, p) are viewed as the rectangular Cartesian coordinates of a point in a (fixed) q, p-plane (called "phase space"; see chap. 8) then, as eqs. (c) show, for $\omega_o = 1$ (which can always be accomplished by an independent variable change) (a, χ) become its polar coordinates in that plane, and (a, ϕ) become its polar coordinates in the (rotating) "van der Pol plane."]

Equations (h) are, usually, quite complicated and *cannot* be solved exactly. To make some headway toward their solution we now introduce the *smallness* assumption: if the nonlinear term $\varepsilon f(\ldots)$ remains small, absolutely, relative to both $\ddot{q}$ (inertia) and $\omega_o{}^2 q$ (linear elasticity), then $\dot{a}$ and $\dot{\phi}$ are also small; that is, a and ϕ *change very slowly during a (linear) period* $\tau_o = 2\pi/\omega_o$; χ will increase, approximately, by 2π. Mathematically, we assume that ε is small enough that

$$|da/dt| \ll |a|/\tau_o \Rightarrow (2\pi/\omega_o)|\dot{a}/a| \ll 1, \tag{i1}$$

$$|d\phi/dt| \ll 2\pi/\tau_o \Rightarrow |\dot{\phi}|/\omega_o \ll 1; \tag{i2}$$

and analogously for the higher derivatives:

$$|d^2a/dt^2| \ll |\dot{a}|/\tau_o \equiv |\dot{a}|(\omega_o/2\pi)$$
$$\Rightarrow |d^2a/dt^2| \ll |a|(\omega_o/2\pi)^2. \tag{i3}$$

This key assumption allows us to proceed from (h) as follows:

(i) First, and since the nonlinear right sides of (h) are periodic in χ with period 2π, we expand them into Fourier series in χ:

$$F(a, \chi)\sin\chi = A_o(a) + \sum [A_k(a)\cos(k\chi) + B_k(a)\sin(k\chi)], \tag{j1}$$

$$F(a, \chi)\cos\chi = \Phi_o(a) + \sum [\Phi_k(a)\cos(k\chi) + \Psi_k(a)\sin(k\chi)], \tag{j2}$$

where $k = 1, 2, 3, \ldots$, and the expansion coefficients ("amplitudes") A_o, Φ_o; A_k, B_k; Φ_k, Ψ_k are determined in well-known ways [recall ex. 7.9.12: (d2), (d3)]. In particular, it is known that the first terms (constant in χ) equal the *average* (mean value) of the corresponding expanded functions, over 2π:

$$A_o(a) = (1/2\pi)\int_0^{2\pi} F(a, \chi)\sin\chi\, d\chi, \qquad \Phi_o(a) = (1/2\pi)\int_0^{2\pi} F(a, \chi)\cos\chi\, d\chi. \tag{j3}$$

(ii) Next, substituting the series (j1, 2) back into (h) and integrating both sides between 0 and 2π, while invoking (j3) and noting that all integrals containing trigonometric terms vanish, we obtain the (still exact) system

$$\int_0^{2\pi} \dot{a}\, d\chi = -(\varepsilon/\omega_o)\, 2\pi A_o(a), \qquad \int_0^{2\pi} \dot{\phi}\, d\chi = -(\varepsilon/a\,\omega_o)\, 2\pi \Phi_o(a). \tag{j4}$$

(iii) Last, we use the smallness (slowness) assumption to transform the left sides of (j4). We have, successively (since $2\pi = \omega_o\, \tau_o = \omega\tau \Rightarrow d\chi = \omega\, dt = (2\pi/\tau)\, dt$, $\omega \equiv \dot{\chi} = \omega_o + \dot{\phi}$),

$$\int_0^{2\pi} \dot{a}\, d\chi = (2\pi/\tau) \int_0^{\tau} \dot{a}\, dt = 2\pi\{[a(t+\tau) - a(t)]/\tau\}$$

$$= 2\pi\{[a(t+\tau_o) - a(t)]/\tau_o\}, \tag{j5}$$

and similarly for the integral of $\dot{\phi}$. But, since a and ϕ do not change appreciably during τ or τ_o, $\Delta a \equiv a(t+\tau) - a(t)$ and $\Delta\phi \equiv \phi(t+\tau) - \phi(t)$ are small, and also τ and τ_o are small relative to the total process duration, which involves several periods (i.e., $\tau \to \Delta\tau$), and so (j4, 5) are replaced by the (finite difference) equations

$$\Delta a/\Delta\tau = -(\varepsilon/\omega_o) A_o(a), \qquad \Delta\phi/\Delta\tau = -(\varepsilon/a\omega_o)\Phi_o(a), \tag{j6}$$

which, in the limit, produce the first approximation (differential) equations

$$da/dt = -(\varepsilon/\omega_o) A_o(a), \qquad d\phi/dt = -(\varepsilon/a\omega_o)\Phi_o(a). \tag{j7}$$

Comparing the above with the exact equations (h), we see that the former result from the latter by averaging over a period, and while doing that regard a as a constant; that is, eqs. (j6, 7) do *not* describe the *instantaneous physical behavior of the system*, but, rather, its *evolution over the several cycles of the duration of the process*; what the distinguished nonlinear mechanics expert N. Minorsky calls "the behavior of the *envelope of modulation*."

Substituting (j3) into (j7), we finally obtain the famous *van der Pol/Krylov/Bogoliubov* equations, or *slowly varying equations* (SVE):

$$da/dt = -(\varepsilon/2\pi\,\omega_o) \int_0^{2\pi} F(a,\, \chi) \sin\chi\, d\chi \equiv \varepsilon A(a), \tag{k1}$$

$$d\phi/dt = -(\varepsilon/2\pi a\,\omega_o) \int_0^{2\pi} F(a,\, \chi) \cos\chi\, d\chi \equiv \varepsilon\Phi(a). \tag{k2}$$

The *first* of these equations gives the variation of a in time [also, the solutions of $\dot{a} = 0 \Rightarrow A(a) = 0$ yield the *stationary amplitude* oscillations (possible limit cycles)]; while the *second* of them yields the corresponding frequency correction: then,

$$\chi = \omega_o t + \phi(t): \ \textit{total phase angle}$$

$$\Rightarrow \dot{\chi} \equiv \omega = \omega_o + \dot{\phi}: \ \textit{instantaneous}\ \text{frequency}. \tag{k3}$$

Finally, and this is quite useful, we note that if $f(q,\, \dot{q}) = f_1(q) + f_2(\dot{q})$ [*nonlinear* in their corresponding arguments; e.g., f_1 contains powers of q like q^2, $q^3, \ldots$, and f_2 contains powers of $\dot{q}$ like $(\dot{q})^2$, $(\dot{q})^3, \ldots$], then, due to the identities

$$\int_0^{2\pi} f_1(a\, \cos\chi) \sin\chi\, d\chi = 0, \qquad \int_0^{2\pi} f_2(-a\,\omega_o \sin\chi) \cos\chi\, d\chi = 0, \tag{k4}$$

- da/dt is unaffected by the nonlinear additions to the linear elastic force, $f_1(q)$, but $d\phi/dt$ is affected:
- $d\phi/dt$ is unaffected by the nonlinear additions to the linear damping force, $f_2(\dot{q})$, but da/dt is affected. For example, if $f_1 = 0$, then $\dot{\phi} = 0 \Rightarrow \dot{\chi} \equiv \omega = \omega_o$ for any small but nonzero nonlinear damping $f_2(\dot{q})$.

In sum, for small ε (first approximation), *nonlinear restoring (spring) terms* affect the *frequency but not the amplitude*; while *nonlinear damping terms* affect the *amplitude but not the frequency*. [For larger ε, however, this is no longer true: either type of terms affects *both* amplitude and frequency.]

REMARK

The above-described method constitutes the *first* approximation of a general asymptotic scheme due to Bogoliubov and Mitropolsky. For extensions to periodically forced oscillators [i.e., $\varepsilon f(\ldots) = \varepsilon f(t, q, \dot{q}) = \varepsilon f(\Omega t, q, \dot{q}) = (2\pi/\Omega)$—periodic function of time, Ω: specified] and to *several degrees of freedom*, see Bogoliubov and Mitropolsky (1974), which is the undisputable "bible" on the subject; also Fischer and Stephan (1972, pp. 144–150, 217–229). For combinations of the method of slowly varying parameters, and averaging in general, (a) with the method of Galerkin, see, for example, Chen and Hsieh (1981), and (b) with the various Δ-forms of Hamilton's principle (this section), as well as the methods of perturbations, strained coordinates, and multiple time scales, see Rajan and Junkins (1983). The combinations among these methods seem endless, but as Rajan and Junkins aptly remark "On the average, it appears algebraic misery may be conserved; we do not claim that the above processes (for a given problem) will result in less algebraic effort. On the other hand, the developments offer numerous insights and exceptional latitude in solution procedures (through the infinity of possible choices for the generators of the variations)" (1983, p. 350). See also "Closing General Remarks," below.

Illustrations

1. Nonlinearly Damped Duffing Oscillator

Let us solve, approximately,

$$\ddot{q} + \varepsilon \nu \dot{q}|\dot{q}| + \omega_o^2 q + \varepsilon \kappa q^3 = 0, \tag{11}$$

where $-\nu \dot{q}|\dot{q}|$ is a small damping force, proportional to the square of $\dot{q}$, and oppositely directed to $\dot{q}$ (hence the use of $|\dot{q}|$); and ν is a damping coefficient; this is frequently called "turbulence damping." Here,

$$f(q, \dot{q}) = -\kappa q^3 - \nu \dot{q}|\dot{q}|$$
$$\Rightarrow F(a, \chi) = -\kappa a^3 \cos^3\chi + \nu a^2 \omega_o^2 \sin\chi\,|\sin\chi|, \tag{12}$$

and so (k1, 2) yield the averaged system

$$da/dt = -(4/3\pi)\varepsilon \nu \omega_o a^2 \quad \text{(i.e., } da/dt \text{ is affected by the nonlinear damping)}, \tag{m1}$$
$$d\phi/dt = (3/8)\varepsilon \kappa (a^2/\omega_o) \quad \text{(i.e., } d\phi/dt \text{ is affected by the nonlinear elasticity)}. \tag{m2}$$

The first of these equations integrates readily to

$$a = a_o\{1 + [(4\varepsilon\,\nu/3\pi)\omega_o\, a_o]t\}^{-1}, \qquad a_0:\ \text{initial displacement}, \tag{m3}$$

and so the second becomes (to the first ε-order)

$$d\phi/dt = (3/8)\varepsilon\,\kappa(a_o{}^2/\omega_o), \tag{m4}$$

and integrates easily to

$$\phi = \left[(3/8)\varepsilon\,\kappa(a_o{}^2/\omega_o)\right] t + \phi_o, \qquad \phi_0:\ \text{initial phase}; \tag{m5}$$

and, therefore,

$$q(0) = a(0)\cos\phi(0) = a_o\cos\phi_o = a_o \ \Rightarrow\ \phi_o = 0. \tag{m6}$$

The above show that *the bigger the* a_o, *the faster the amplitude decreases*; also, for $\nu = 0$ (undamped oscillator), $a = a_o$ and $\omega^2 = (\dot{\chi})^2 = (\omega_o + \dot{\phi})^2 = \omega_o{}^2 + (3/4)\varepsilon\,\kappa\, a^2 + (\dot{\phi})^2$, or, to the first ε-order, $\omega^2 \approx \omega_o{}^2 + (3/4)\varepsilon\,\kappa\, a^2$.

For a generalization of (l1) that includes *linear* and *cubic* damping (problem of rolling of a ship equipped with a gyrostabilizer), see, for example, McLachlan (1956/1958, pp. 92–94).

2. Van der Pol Equation

Here, the equation of motion is

$$\ddot{q} + \varepsilon(q^2 - 1)\dot{q} + \omega_o^2 q = 0, \tag{n1}$$

that is,

$$\begin{aligned} f(q,\dot{q}) &= (1 - q^2)\dot{q} \\ \Rightarrow F(a,\ \chi) &= -a\,\omega_o \sin\chi - (-a\,\omega_o\sin\chi)(a^2\cos^2\chi) \\ &= (1 - a^2\cos^2\chi)(-a\,\omega_o\sin\chi), \end{aligned} \tag{n2}$$

and so eqs. (k1, 2) yield

$$\begin{aligned} da/dt &= -(\varepsilon/2\pi\omega_o)\int_0^{2\pi}(1 - a^2\cos^2\chi)(-a\,\omega_o\sin\chi)\sin\chi\, d\chi \\ &= (\varepsilon a/2)(1 - a^2/4), \end{aligned} \tag{o1}$$

$$\begin{aligned} d\phi/dt &= -(\varepsilon/2\pi a\,\omega_o)\int_0^{2\pi}(1 - a^2\cos^2\chi)(-a\,\omega_o\sin\chi)\cos\chi\, d\chi\ = 0 \\ &\Rightarrow \chi = \omega_o t + \phi_o; \end{aligned} \tag{o2}$$

that is, to the first ε-order, the nonlinearity does not change the frequency: $\omega = \omega_o$.

With the arbitrary initial conditions $a(0) = a_o$, $\phi(0) = \phi_o = \chi(0)$, and the well-known "energy identity" $2a\dot{a} = d(a^2)/dt$, eq. (o1) transforms to

$$d(a^2)/dt = \varepsilon\, a^2(1 - a^2/4) \qquad [\text{i.e., } \dot{b} = \varepsilon\, b\,(1 - b/4), \text{ with } b \equiv a^2], \tag{o3}$$

and integrates readily to

$$a = a_o \exp(\varepsilon t/2)\{1 + (a_o^2/4)\,[\exp(\varepsilon t) - 1]\}^{-1/2}$$
$$\Rightarrow q(t) = a(t)\cos(\omega_o t + \phi_o). \tag{p}$$

The stationary, or steady-state, amplitude solutions of the problem, obtained from (o1) for $\dot{a} = 0$, are (i) $a = 0$ (equilibrium), and (ii) $a = 2$ (limit cycle).

This can also be seen from the transient amplitude equation (p): by rewriting it as

$$a = a(t) = 2\left[1 - \exp(-\varepsilon t)(1 - 4/a_o^2)\right]^{-1/2}, \tag{q}$$

we can readily see that for $t \to \infty$, $a \to 2$ *always*; that is, this is so, no matter how small (but nonzero) or large a_o may be, even if $a_o > 2$. If $a_o = 0$, then $a(t) = 0$ (no oscillation) — the oscillation must be initiated by *external* means. [For additional examples and questions on (limit cycle) stability, see standard texts on nonlinear mechanics, for example, Kauderer (1958, pp. 295–304), Stoker (1950); also Butenin et al. (1985, pp. 485–491).]

Problem 7.9.21 *Method of Slowly Varying Parameters (SVP).* By applying SVP to the linear damped oscillator:

$$\ddot{q} + 2f\,\dot{q} + \omega_o^2\,q = 0 \qquad (f\text{: small friction constant}), \tag{a}$$

with initial conditions $q(0) = A$, $\dot{q}(0) = 0$, show that the "slowly varying equations" (SVE) [ex. 7.9.14: (k1), (k2)], yield

$$\dot{a} = -f\,a \;\Rightarrow\; a = a_o \exp(-f\,t), \qquad \dot{\phi} = 0 \Rightarrow \phi = \textit{constant}; \tag{b}$$

from which, and the small friction requirement $f/\omega_o \ll 1$, we find

$$q = A\,\exp(-f\,t)\,\cos(\omega_o t - f/\omega_o); \tag{c}$$

that is, here, damping causes a change in the amplitude, not in the frequency.

Then show that, here, the smallness requirement [ex. 7.9.14: (i1)], $|\dot{a}/a|$ $(2\pi/\omega_o) \ll 1$, leads to the sharper restriction

$$2f \ll \omega_o/\pi = 2/\tau_o \qquad \text{(physical meaning and dimensions of "small friction").} \tag{d}$$

Finally, compare the approximate solution (c) with the well-known exact solution of this problem,

$$q = \left[A\omega_o/(\omega_o^2 - f^2)^{1/2}\right]\exp(-f\,t)\cos\left[(\omega_o^2 - f^2)^{1/2}t - \tan^{-1}(f/\omega_o)\right], \tag{e}$$

and show that under $f/\omega_o \ll 1$ the above solution reduces to the approximate one.

Problem 7.9.22 *Method of Slowly Varying Parameters.* By applying SVP to the quadratically damped and unforced oscillator

$$\ddot{q} + \omega_o^2\,q + 2f(\dot{q})^2 = 0, \tag{a}$$

with (the usual notations, and) initial conditions $q(0) = A$, $\dot{q}(0) = 0$, show that the SVE yield

$$da/dt = -(8/3) f\, \omega_o a^2 \Rightarrow a = a_o[1 + (8/3) f\, \omega_o a_o t]^{-1}, \qquad a(0) = a_o = A, \tag{b}$$

$$d\phi/dt = 0 \Rightarrow \phi = constant. \tag{c}$$

Compare this result with the preceding case of *linear* damping: $2f\,\dot{q}$.

Problem 7.9.23 *Method of Slowly Varying Parameters.* By applying SVP to the undamped and unforced Duffing oscillator

$$\ddot{q} + \omega_o^{\,2} q + c\, q^3 = 0, \tag{a}$$

with initial conditions $q(0) = A$, $\dot{q}(0) = 0$, show that the SVE yield

$$da/dt = 0 \Rightarrow a = constant \equiv a_o = A, \tag{b}$$

$$d\phi/dt = 3\, c\, a^2/8\omega_o \Rightarrow \phi = (3\, c\, a^2/8\omega_o)\, t + constant \Rightarrow \omega = \omega_o + \dot{\phi}. \tag{c}$$

Then show that, here, the smallness requirement $|\dot{\phi}| \ll \omega_o$ leads to $c \ll 8\omega_o^{\,2}/3a_o^{\,2}$.

Problem 7.9.24 *Method of Slowly Varying Parameters.* By applying SVP to the nonlinearly damped and unforced Rayleigh oscillator (e.g., electrically driven tuning fork)

$$\ddot{q} - 2f\,\dot{q} + g\,(\dot{q})^3 + \omega_o^{\,2} q = 0, \tag{a}$$

where f, g are small positive constants, and $-2f\,\dot{q}$ is *effective negative damping* (equivalent to a driving force), show that the SVE yield

$$da/dt = a\,[f - (3/8)\, g\, \omega_o^{\,2} a^2], \qquad d\phi/dt = 0. \tag{b}$$

and, therefore, (i) for a *stationary* (or steady-state) amplitude,

$$da/dt = 0 \Rightarrow a = (2/\omega_o)(2f/3g)^{1/2} \equiv a_{st}, \tag{c}$$

while (ii) for a *transient* one,

$$a(t) = a_o \exp(f\,t)\left\{1 + R^2 a_o^{\,2}[\exp(2f\,t) - 1]\right\}^{-1/2}, \tag{d}$$

where $R^2 \equiv 3g\,\omega_o^{\,2}/8f$, $a(0) = a_o$.

Hence, if $a_o = 0$, then $a(t) = 0$; and if $a_o \neq 0$, then $a \to a_{st}$, as $t \to \infty$. Compare with the van der Pol oscillator. For further details, see, for example, McLachlan (1956/1958, pp. 90–91).

Problem 7.9.25 *Method of Slowly Varying Parameters.* By applying SVP to the linearly damped Duffing oscillator (with the usual notations and smallness assumptions, and $\omega_o^{\,2} > f^{\,2}$)

$$\ddot{q} + 2f\,\dot{q} + \omega_o^{\,2} q + c\, q^3 = 0, \tag{a}$$

with initial conditions $q(0) = A$, $\dot{q}(0) = 0$, show that the SVE yield

$$da/dt = -f\,a, \qquad d\phi/dt = 3\,c\,a^2/8\,\omega_o. \tag{b}$$

Integrate these equations and discuss their results.

Closing General Remarks on Time-Integral and Variational Methods in Nonlinear Oscillations

1. All these methods assume that the degree of the nonlinearity is not too large.
2. The accuracy (error) of the so-obtained approximate solutions is, often, difficult to assess.
3. As mentioned earlier, the method of "Hamilton–Ritz" works best for *periodic* solutions; for example, steady states in forced systems—otherwise, we must include the boundary terms, and this increases the computational difficulty of the problem. The method of Galerkin, in general, does not have that drawback, but both methods (i.e., Ritz and Galerkin) require a good knowledge of the physical meaning of the equations and the qualitative behavior of their solutions so as to make a successful ("optimal") choice in the trial functions.
4. For *slowly varying* (*nonperiodic*) solutions—for example, transients in self-excited or damped systems, and limit cycles/points (if they exist)—the methods of van der Pol, Bogoliubov and Mitropolskii, work best. However, as the reader will have noticed, their mathematical operations are less simple than those of Ritz and Galerkin (and, worse, for *higher-order* approximations, these methods are, in general, *cost-ineffective*; that is, additional small corrections require disproportionately long and arduous calculations).

The moral of the above is that, as in most other areas of science, no single approach is uniformly best: in view of the (unknown) approximations involved, it is wiser, in dealing with a particular equation, to *use several complementary strategies/techniques*: those described here and the many more available in the enormous nonlinear mechanics literature, such as perturbations, harmonic balance (or equivalent linearization), and so on. For comprehensive and readable overviews of these classical methods, we refer the reader to (alphabetically): Blekhman (1979), Bogoliubov and Mitropolskii (1974), Klotter (1955), Magnus (1957).

APPENDIX 7.A

EXTREMAL PROPERTIES OF THE HAMILTONIAN ACTION (IS THE ACTION REALLY A MINIMUM; NAMELY, LEAST?)

7.A1 Introduction

The following is restricted to holonomic systems that can be completely described by a Lagrangean $L = L(t, q, \dot{q})$. Therefore, the discussion can be safely limited, for algebraic simplicity, to a *one*-DOF system S. Let

$$\begin{aligned} A_H &= A_H(q) = A_H(I) \\ &\equiv \int L\, dt: \quad \text{Hamiltonian action of } S, \text{ evaluated along an orbit } I, \\ &\qquad q(t)(\text{from } t_1 \text{ to } t_2). \end{aligned} \tag{7.A1.1}$$

Then, as we have seen in this chapter, Hamilton's principle states that A_H *is stationary (or critical) for small (or first-order) variations around* I, $\delta q = \delta q(t)$, *that nullify the boundary terms*; that is, with $p \equiv \partial T/\partial \dot{q} = \partial L/\partial \dot{q}$ and boundary conditions, say, $\delta q_1 \equiv \delta q(t_1) = 0$ and $\delta q_2 \equiv \delta q(t_2) = 0$, this "principle" states that

$$\begin{aligned} \delta A_H &= \int \delta L\, dt = \cdots = \{p\, \delta q\}_1^2 - \int E(L)\, \delta q\, dt \\ &= 0 - \int E(L)\, \delta q\, dt = 0, \end{aligned} \tag{7.A1.2}$$

from which we find

$$\begin{aligned} E(L) &\equiv (\partial L/\partial \dot{q})^{\cdot} - \partial L/\partial q \\ &= (\partial^2 L/\partial \dot{q}^2)\ddot{q} + (\partial^2 L/\partial q \partial \dot{q})\dot{q} + \partial^2 L/\partial t\, \partial \dot{q} - \partial L/\partial q = 0. \end{aligned} \tag{7.A1.3}$$

As in the ordinary calculus (of functions), the *first-order* equation $\delta A_H = 0$, in δq and $\delta(\dot{q}) = (\delta q)^{\cdot}$, is only a *stationarity* condition — not an *extremality* one. That is, it does not tell us whether $A_H(q)$ is a maximum: $A_H(q + \delta q) > A_H(q)$, or a minimum: $A_H(q + \delta q) < A_H(q)$, or neither. Again, as in calculus, the answer to that comes (usually) from the study of the *second variation* of A_H, $\delta^2 A_H \equiv \delta(\delta A_H)$: *quadratic and homogeneous* functional in δq and $\delta(\dot{q}) = (\delta q)^{\cdot}$ (defined precisely below).

Now, the study of $\delta^2 A_H$, and corresponding extremal — namely, maximum/minimum — properties of A_H, has received little attention in the literature (it is conspicuously absent from most texts on advanced dynamics), primarily for the following reason: the laws of nature for S — namely, its *ever valid* equations of motion (7.A1.3) — result from (7.A1.2), or from the vanishing of some other equivalent first-order functional equation. On the other hand, the (possible) extremum properties of its A_H are *not* laws of mechanics, but only particular *conditions* that may hold for some orbits of S and not for others, or hold only along a certain part(s) of an orbit and not for all of it. Such *second (and possibly higher)-order* properties of A_H have been associated with *kinetic stability/instability* of S in some sense; that is, both stable and unstable orbits satisfy $\delta A_H = 0$, but the stable ones among them, if such exist, give $\delta^2 A_H$ one sign, and the unstable ones the opposite — pretty much like the theorems of the minimum of the total potential energy in static stability/buckling, and so on [Dirichlet (1846) → Bryan (1890s) → *Trefftz* (1930s) → Koiter (1940s)].

For detailed and readable treatments of the relevant *sufficiency variational theory*, see, for example (alphabetically): Elsgolts (1970), Fox (1950/1963), Funk (1962), Gelfand and Fomin (1963); also Hussein et al. (1980), Levit and Smilansky (1977); and for the connection with *kinetic stability (stability of motion)*, see the works of some of the older masters of mechanics, for example, Joukovsky (1937, pp. 110–208), Thomson and Tait (1912, pp. 416–439), Routh (1877, pp. 103–108); also Lur'e (1968, pp. 651–667, 749–754), Routh [1898, pp. 399–405; 1905(b), pp. 308–310], Watson and Burbury (1879, pp. 72–99), Whittaker (1937, pp. 250–253).

For the second variation of A_H of *nonholonomic* systems, see Novoselov (1966, pp. 26–49, and references cited therein).

7.A2 The Fundamental Minimum Theory (of Jacobi and A. Mayer)

Let us summarize the problem of the extremality, say, minimality of A_H. The *necessary* conditions for this are eqs. (7.A1.1) and (7.A1.2). The *sufficient* conditions come from the study of the sign of $\delta^2 A_H$. The latter is defined, equivalently, either as:

(i) The *quadratic and homogeneous* part in δq and $\delta(\dot{q})$ in the Taylor-like expansion of the *total contemporaneous* variation of A_H around I, $\delta^T A_H$,

$$\delta^T A_H \equiv A_H(q + \delta q) - A_H(q) = \delta A_H + (1/2)\,\delta^2 A_H + \cdots, \tag{7.A2.1a}$$

i.e., $\delta^2 A_H$; or by

(ii) The δ-variation of δA_H:

$$\begin{aligned} \delta^2 A_H \equiv \delta(\delta A_H) &= \int \delta^2 L\, dt \\ &= \cdots = -\int J(\delta q)\,\delta q\, dt + \{\delta p\,\delta q\}_1^2, \end{aligned} \tag{7.A2.1b}$$

where

$$\begin{aligned} \delta^2 L \equiv \delta(\delta L) &= \left[(\partial/\partial q)\,\delta q + (\partial/\partial \dot{q})\,\delta(\dot{q})\right]^2 L \\ &= \cdots = (\partial^2 L/\partial \dot{q}^2)(\delta\dot{q})^2 + 2(\partial^2 L/\partial q\,\partial\dot{q})\,\delta q\,\delta(\dot{q}) + (\partial^2 L/\partial q^2)(\delta q)^2 : \end{aligned}$$

Second variation of the Lagrangean, (7.A2.1c)

and [invoking $\delta(\dot{q}) = (\delta q)^{\cdot}$]

$$\begin{aligned} J(\delta q) &= \left\{d/dt\,[\partial\ldots/\partial(\delta\dot{q})] - [\partial\ldots/\partial(\delta q)]\right\}(1/2)\,\delta^2 L \\ &= (\partial^2 L/\partial\dot{q}^2)\,\delta(\ddot{q}) + (\partial^2 L/\partial\dot{q}^2)^{\cdot}\,\delta(\dot{q}) + \left[(\partial^2 L/\partial q\partial\dot{q})^{\cdot} - (\partial^2 L/\partial q^2)\right]\delta q = 0 : \end{aligned}$$

Jacobi's variational equation (a *linear* and *homogeneous* but, generally, *variable coefficient differential equation*). (7.A2.1d)

REMARKS

(i) $\delta^T A_H$ is frequently denoted as ΔA_H; but here (§7.9) $\Delta(\ldots)$ has been reserved for the first *noncontemporaneous* variation. For the total such variation, we could use $\Delta^T(\ldots)$; that is,

$$\Delta^T A_H = \Delta A_H + (1/2)\,\Delta^2 A_H + \cdots, \tag{7.A2.2}$$

in variable time-endpoints problems (see, e.g., Santilli, 1978, pp. 41–43).

(ii) $J(\delta q)$ equals the *first*-order virtual variation of $E(L) = E[L(t, q, \dot{q})]$; or, $J(\delta q) = 0$ *is the Euler–Lagrange equation for* δq *of* $\delta^2 A_H$:

$$\begin{aligned} &E[L(t, q + \delta q, \dot{q} + \delta(\dot{q}))] - E[L(t, q, \dot{q})] \\ &\qquad \approx \delta E(q, \delta q, \delta(\dot{q})) \qquad \text{[to first order in } \delta(\ldots)] \\ &\qquad = J(\delta q; q) \equiv J(\delta q), \end{aligned} \tag{7.A2.3a}$$

where, successively,

$$\begin{aligned} J(\delta q) &= \delta[(\partial L/\partial \dot{q})^{\cdot} - \partial L/\partial q] \\ &= [\delta(\partial L/\partial \dot{q})]^{\cdot} - \delta(\partial L/\partial q) \\ &= [(\partial^2 L/\partial q \partial \dot{q})\, \delta q + (\partial^2 L/\partial \dot{q}^2)\, \delta(\dot{q})]^{\cdot} \\ &\quad - [(\partial^2 L/\partial q^2)\, \delta q + (\partial^2 L/\partial q \partial \dot{q})\, \delta(\dot{q})] \\ &\qquad \text{[invoking } \delta(\dot{q}) = (\delta q)^{\cdot}] \\ &= (\partial^2 L/\partial \dot{q}^2)\, (\delta q)^{\cdot\cdot} + (\partial^2 L/\partial \dot{q}^2)^{\cdot} (\delta q)^{\cdot} \\ &\quad + [(\partial^2 L/\partial q \partial \dot{q})^{\cdot} - (\partial^2 L/\partial q^2)]\, \delta q \\ &= [(\partial^2 L/\partial \dot{q}^2)\, (\delta q)^{\cdot}]^{\cdot} - [(\partial^2 L/\partial q^2) - (\partial^2 L/\partial q \partial \dot{q})^{\cdot}]\, \delta q \\ &\qquad \text{(Sturm–Liouville form),} \end{aligned} \tag{7.A2.3b}$$

and all partial derivatives are evaluated along the solution(s) of (7.A1.2, 3), Q.E.D.

Now, the relevant extremum results are contained in the following fundamental theorem.

THEOREM

For the action functional A_H to attain a *minimum* in the class of piecewise smooth functions $q(t)$ that join the points $[t_1, q(t_1) \equiv q_1]$ and $[t_2, q(t_2) \equiv q_2]$, and for nearby variations such that both $|\delta q|$ *and* $|\delta(\dot{q})| = |(\delta q)^{\cdot}|$ are small (i.e., for a *relative* and *strong* minimum), it is sufficient that:

(i) $q(t)$ satisfies the Euler–Lagrange equations (7.A1.3), $\delta A_H = 0 \Rightarrow E(q) = 0$; that is, $q(t)$ be an orbit, say I; (7.A2.4a)

(ii) The strengthened *Legendre–Weierstrass* condition holds: along I, for $t_1 \le t \le t_2$ and for *any* $\dot{q}$ in its neighborhood:

$$\partial^2 L/\partial \dot{q}^2 > 0; \tag{7.A2.4b}$$

(iii) The strengthened *Jacobi* condition holds: let $t_1{}^*$ be the *first* root, to the right of t_1, of the solution $\delta q = \delta q(t)$ to the following *initial-value* problem:

$$J(\delta q) = 0; \qquad \delta q(t_1) = 0, \quad \delta \dot{q}(t_1{}^*) = \text{arbitrary } \textit{nonzero} \text{ constant} \equiv \alpha; \tag{7.A2.4c}$$

that is, $\delta q(t_1) = 0$ and $\delta q(t_1{}^*) = 0$, $t_1{}^* > t_1$. The root $t_1{}^*$ is called *conjugate* to t_1; and q_1 and $q(t_1{}^*) \equiv q_1{}^*$, along I, are known [after Thomson and Tait (1860s)] as *mutually conjugate kinetic foci*. Jacobi's criterion states that, for a minimum of A_H,

$$t_1{}^* > t_2 \qquad \text{or} \qquad \Delta t \equiv t_2 - t_1 < t_1{}^* - t_1 \equiv \Delta t^*; \tag{7.A2.4d}$$

that is, *the interval* (t_1, t_2) *should not contain any roots conjugate to* t_1. [For a *maximum*, the inequality signs in (7.A2.4d) must be reversed.]

It can be shown that $t_1{}^*$ is independent of the value of α, eq. (4c), but does depend on the partial derivatives/coefficients of the δq's in $J(\delta q)$, eqs. (7.A2.1d, 3b); that is, on the orbit I.

If $t_1{}^* = t_2$, then $\delta^2 A_H$ is positive *semidefinite* — that is, it may vanish for a $\delta q(t) \neq 0$ — in which case, we have to resort to higher-order variations. If $t_1{}^* < t_2$, then $\delta^2 A_H$ is *sign-indefinite* — that is, it is negative for one class of variations and positive for another — $A_H(q)$ has a *minimax* (or *saddle-point*); that is, there is no extremum.

REMARKS

(i) The Euler–Lagrange test supplies the orbit equation; its solution(s) require(s) integration of the equation of motion and then utilization of the given boundary conditions.

(ii) The Legendre–Weierstrass test means that *locally* — that is, for *very small* $t_2 - t_1$ — A_H *is always a minimum*; that is, for any potential force field.

Let us show this for *stationary* constraints. Since $\delta q(t_1) = 0$, we have

$$|\delta q(t)| = \left| \int_{t_1}^{t} \delta(\dot{q})\, dt \right| \leq \varepsilon(t - t_1), \tag{7.A2.5a}$$

where $\varepsilon \equiv \max|\delta(\dot{q})|$ in (t_1, t_2); and, therefore, *for very small* $t_2 - t_1$, *the* $\delta(\dot{q})$*-terms always dominate over the* δq*-terms*. Hence, for such constraints,

$$\begin{aligned}\delta^2 L &= (1/2)(\partial^2 L/\partial q^2)(\delta q)^2 + (\partial^2 T/\partial q\, \partial \dot{q})\, \delta q\, \delta(\dot{q}) + (1/2)(\partial^2 T/\partial \dot{q}^2)(\delta(\dot{q}))^2 \\ &= (1/2)(\partial^2 L/\partial q^2)(\delta q)^2 + (\partial^2 T/\partial q\, \partial \dot{q})\, \delta q\, \delta(\dot{q}) + T[\delta(\dot{q})] \\ &\approx T[\delta(\dot{q})] > 0\end{aligned} \tag{7.A2.5b}$$

[where $T(\delta\dot{q})$ signifies what becomes of $T(\dot{q})$, which is positive definite in $\dot{q}$, if we replace in it $\dot{q}$ with $\delta(\dot{q})$], and, accordingly,

$$\delta^2 A_H = \int \delta^2 L\, dt \approx \int T[\delta(\dot{q})]\, dt > 0 \Rightarrow A_H(q): \ \textit{minimum}, \ \text{Q.E.D.} \tag{7.A2.5c}$$

(iii) The Jacobi test imposes a *limit on the length of the orbit*; that is, on the $t_2 - t_1$ range (as long as $t_1{}^*$ is finite). [The sum of two minimal orbits will be a minimal orbit if the "sum orbit" does not exceed its Jacobi limit.]

Ideally, and very rarely, the conjugate root(s) to t_1 are found as follows: the general solution of the *second-order* Lagrangean equation $E(q) = 0$ has the form:

$$q = q(t;\ c_1,\ c_2), \qquad c_1, c_2: \ \text{integration constants}. \tag{7.A2.6a}$$

Now, the orbit I corresponds to fixed values of c_1 and c_2 [to be determined from the boundary conditions: $q(t_1;\ c_1,\ c_2) = q_1$ (given), $q(t_2;\ c_1,\ c_2) = q_2$ (given)], whereas typical neighboring paths $II = I + \delta I$ correspond to the values $c_1 + \delta c_1$ and $c_2 + \delta c_2$. On such adjacent paths,

$$\delta q = (\partial q/\partial c_1)\, \delta c_1 + (\partial q/\partial c_2)\, \delta c_2, \tag{7.A2.6b}$$

and therefore the boundary conditions for t_1, $t_1{}^*$ become [with the notation $(\ldots)_* \equiv$ $(\ldots)$ evaluated at $*$]

$$\delta q(t_1) = (\partial q/\partial c_1)_1\, \delta c_1 + (\partial q/\partial c_2)_1\, \delta c_2 = 0, \tag{7.A2.6c}$$

$$\delta q(t_1{}^*) = (\partial q/\partial c_1)_*\, \delta c_1 + (\partial q/\partial c_2)_*\, \delta c_2 = 0. \tag{7.A2.6d}$$

This linear and homogeneous system, in the δq's, expresses the fact that *the slightly differing paths I and II cross at t_1 and then again at $t_1{}^*$*; or, *they are traversed in the same time $t_1{}^* - t_1$*; q_1 and q_{1*} are (mutually) conjugate kinetic foci on I.

For nontrivial solutions (i.c., δc_1, $\delta c_2 \neq 0$), the system of equations (7.A2.6c, d) leads, in well-known ways, to the determinantal equation

$$\Delta(t_1, t_1{}^*) \equiv \begin{vmatrix} (\partial q/\partial c_1)_1 & (\partial q/\partial c_2)_1 \\ (\partial q/\partial c_1)_* & (\partial q/\partial c_2)_* \end{vmatrix} = 0. \tag{7.A2.7}$$

The *first*, or *smallest*, of its (real) roots to the right of t_1 is what enters the Jacobi condition (7.A2.4d).

Example 7.A2.1 As an illustration, let us consider the linear harmonic oscillator:

$$m\,\ddot{q} + k\,q = 0 \Rightarrow \ddot{q} + \omega^2 q = 0,\ \omega \equiv (k/m)^{1/2}\text{: frequency (a positive constant).} \tag{a}$$

Here, as is well known,

$$2T = m\,(\dot{q})^2 \quad (m\text{: mass}), \qquad 2V = k\,q^2 \quad (k\text{: positive constant}) \tag{b}$$

$$\Rightarrow Q = -dV/dq = -k\,q \ \Rightarrow\ dQ/dq = -d^2V/dq^2 = -k < 0, \tag{c}$$

and so the general solution of (a) and its variation are, respectively,

$$q = c_1 \sin(\omega t) + c_2 \cos(\omega t), \qquad \delta q = [\sin(\omega t)]\, \delta c_1 + [\cos(\omega t)]\, \delta c_2. \tag{d}$$

Therefore, with $t_1 = 0$, eq. (7.A2.7) gives

$$\Delta(0, t_1{}^*) \equiv \begin{vmatrix} 0 & 1 \\ \sin(\omega t_1{}^*) & \cos(\omega t_1{}^*) \end{vmatrix} = -\sin(\omega t_1{}^*) = 0, \tag{e}$$

and, clearly, its first root to the right of $t_1 = 0$ is

$$t_1{}^* = \pi/\omega = \tau/2\text{: half period of oscillation.} \tag{f}$$

Since $\partial^2 L/\partial \dot{q}^2 = \partial^2 T/\partial \dot{q}^2 = m > 0$, always, the corresponding action

$$A_H = \int_0^{t_2} (1/2)\,[m\,(\dot{q})^2 - k\,q^2]\, dt, \tag{g}$$

is a minimum as long as $t_2 < \tau/2$ (generally, for $t_2 - t_1 < \tau/2$).

For an n DOF linear oscillatory system (expressing its T and V in principal coordinates, and noting that its n characteristic frequencies $\omega_1 \leq \omega_2 \leq \cdots \leq \omega_n$ are intrinsic system properties), it is not hard to show that the corresponding Jacobi minimum action condition is

$$t_2 < t_1{}^* = \tau_{\min}/2 = \tau_n/2 \equiv \pi/\omega_n: \textit{smallest} \text{ half period of oscillation.} \qquad \text{(h)}$$

(See also Aizerman, 1974, pp. 276–279.) From the above, we conclude that:

(i) For $n \to \infty$ (i.e., *continuum*; e.g., oscillating string), $\omega_n \to \infty$ and, therefore, $t_1{}^* \to 0$, A_H is *never* a minimum.

(ii) Under *constant* or *repulsive* (nonoscillatory) forces — that is, $dQ/dq = -d^2V/dq^2 \geq 0$, $t_1{}^* \to \infty$ — such forces *always* minimize A_H.

---

n DOF

Finally, the entire argument for the determination of kinetic foci, namely eqs. (7.A2.6a–7), carries over to the n DOF case. There, eqs. (7.A2.6a, b) are replaced, respectively, by

$$q_k = q_k(t;\ c_1, \ldots, c_{2n}) \ \Rightarrow\ \delta q_k = \sum (\partial q_k/\partial c_\alpha)\, \delta c_\alpha$$

$$(k = 1, \ldots, n; \quad \alpha = 1, \ldots, 2n) \qquad \text{(7.A2.8a)}$$

and eqs. (7.A2.6c, d) and (7.A2.7) by

$$\sum (\partial q_k/\partial c_\alpha)_1\, \delta c_\alpha = 0, \qquad \sum (\partial q_k/\partial c_\alpha)_*\, \delta c_\alpha = 0, \qquad \text{(7.A2.8b)}$$

and the $2n \times 2n$ determinantal equation

$$\Delta(t_1,\ t_1{}^*) = \begin{vmatrix} (\partial q_1/\partial c_1)_1 & \cdots & (\partial q_1/\partial c_{2n})_1 \\ \cdots & \cdots & \cdots \\ (\partial q_n/\partial c_1)_1 & \cdots & (\partial q_n/\partial c_{2n})_1 \\ \hline (\partial q_1/\partial c_1)_* & \cdots & (\partial q_1/\partial c_{2n})_* \\ \cdots & \cdots & \cdots \\ (\partial q_n/\partial c_1)_* & \cdots & (\partial q_n/\partial c_{2n})_* \end{vmatrix} = 0. \qquad \text{(7.A2.8c)}$$

Here too, the smallest root of (7.A2.8c) to the right of t_1, $t_1{}^*$, is what enters Jacobi's minimum condition: $t_2 > t_1{}^*$.

[This argument and equations are due to the noted German mathematician A. Mayer (1866 and subsequently); one of the founders of the sufficiency variational theory, for both fixed and variable endpoint problems.]

The problem of the extremum of the Hamiltonian action can be summarized as follows:

- A_H, along an orbit I, is *never* a maximum; it is either a *minimum* (from an initial configuration C_1 up to any other configuration C_2 located *before* the first kinetic focus of C_1, $C_1{}^*$; all on I), or a *minimax (saddle-point)* (from C_1 to a C_2 beyond $C_1{}^*$).
- Limiting cases: If $C_1{}^* \to C_1$, then A_H is a minimax for any $t_2 > t_1 = t_1{}^*$; if $C_1{}^* \to \infty$, then A_H is a minimum for any t_2.

Such tests have also been obtained for the Lagrangean action A_L; see Papastavridis [1985(b), 1986(a), 1986(c): general variable endpoints form], Peisakh [1966: Jacobi's geodesic (fixed endpoints) form].

7.A3 Averaged Action

As we have just seen, the stationary (or critical) "points" of A_H are, in general (i.e., for extended periods of time), saddle-points. Therefore, if we are to develop reliable A_H-based extremum criteria, similar to those of static stability, we must introduce some other energetic functions or functionals. Following the valuable lessons of the method of *averaging* of nonlinear oscillations (ex. 7.9.7), we choose as such function the *time average* of the system's Lagrangean. Indeed, with $t_2 - t_1 \equiv \tau$, we have, for a general motion,

$$\langle L \rangle \equiv \lim[A_H(\tau)/\tau]_{\tau\to\infty} = \lim\left[(1/\tau)\int L\,dt\right]_{\tau\to\infty} \tag{7.A3.1}$$

(for periodic motions, no limiting process is needed). $\langle L \rangle$ is no longer a function of time, and its stationarity/extremality becomes a problem of ordinary differential calculus. For example, and again guided by nonlinear oscillations, we may try in A_H the Fourier series expansion of the trial solution (say, in complex form, for compactness; see also §8.14):

$$q(t) \approx q_o(t) = \sum c_s \exp(i\,\omega_s\,t), \qquad c_s = (1/\tau)\int_{-\tau/2}^{+\tau/2} q_o(t)\exp(-i\,\omega_s\,t)\,dt, \tag{7.A3.2}$$

$$s = -\infty, \ldots, +\infty; \qquad \omega_s \equiv s\,\omega = (2\pi/\tau)s. \tag{7.A3.2a}$$

Then $\langle L \rangle$ becomes a function of the Fourier coefficients (amplitudes), and "Hamilton's averaged principle" takes the discrete form

$$\delta\langle L \rangle = 0 \;\Rightarrow\; \partial\langle L \rangle/\partial c_s = 0 \qquad (s = 0,\ \pm 1,\ \pm 2, \ldots); \tag{7.A3.3}$$

while the type of the stationarity of $\langle L \rangle$ is determined from the study of the sign properties of its "Hessian matrix" $(\partial^2\langle L \rangle/\partial c_r \partial c_s)$, every element of which is an algebraic function of the Fourier coefficients. For the use of $\langle L \rangle$ in general nonlinear dynamics, see the earlier-mentioned (ex. 7.9.13) highly readable and informative papers by Gray et al. [1996(a), (b); and references cited therein]; also Helleman (1978); and for applications to the stability of nonlinear oscillations, see Baumgarte (1987).

Problem 7.A3.1 Show that the averaged Lagrangean of the undamped and forced linear oscillator $\ddot{q} + \omega_o{}^2 q = Q_o \cos(\omega\,t)$, where Q_o, ω are, respectively, the forcing amplitude and frequency, and

$$A_H(\tau) = \int_0^\tau \left[(1/2)(\dot{q})^2 - (1/2)\,\omega_o{}^2\,q^2 + Q_o\cos(\omega\,t)\,q\right] dt \tag{a}$$

has a minimum for $\omega_o < \omega$, and a maximum for $\omega_o > \omega$ (both for all time).

7.A4 The Integral Stability Criterion [of Blekhman–Lavrov and Valeev-Ganiev (1960s)]

To dispel any possible impressions that the extrema of the averaged Lagrangean are somehow only of "academic" interest, we sketch below their application to the theory of *synchronization* of oscillating mechanical objects. For a complete treatment of this theoretically and practically important subject (that is conspicuously absent from almost all Western references), see the fundamental and extensive works of its key exponent, Blekhman (1971, 1979, 1981/1988; and Blekhman and Malakhova, 1990).

Following Valeev and Ganiev (1969), we consider the oscillations of an n DOF *weakly nonlinear, or quasi-linear*, potential system with Lagrangean [say, in the principal coordinates of the corresponding linear (i.e., unperturbed), constant coefficient system; that is, for $\varepsilon = 0$]:

$$\begin{aligned} L &= L(\omega t,\ q,\ \dot{q};\ \varepsilon) \equiv T - V \\ &= \sum (1/2)\ [(\dot{q}_k)^2 - \omega_k{}^2 q_k{}^2] + \varepsilon L_1(\omega t,\ q,\ \dot{q};\ \varepsilon), \end{aligned} \tag{7.A4.1}$$

where ω_k are the natural frequencies of the system (given constants), $0 < \varepsilon \ll 1$ (hence the name quasi-linear), and $L_1(\ldots)$ is periodic in the forcing (external) frequency ω; that is,

$$L_1(\omega t + 2\pi,\ q,\ \dot{q};\ \varepsilon) = L_1(\omega t,\ q,\ \dot{q};\ \varepsilon). \tag{7.A4.1a}$$

Below we examine the case where ω is, approximately, in rational ratios to the ω_k's (frequently referred to as *near-resonance* case):

$$\omega_k/\omega \approx i_k/N \quad \text{or} \quad i_k\,\omega \approx \omega_k N, \qquad \text{and} \qquad \nu_k/\omega = i_k/N \quad \text{or} \quad i_k\,\omega = \nu_k N, \tag{7.A4.2}$$

where

$$\omega_k{}^2 - \nu_k{}^2 = O(\varepsilon), \tag{7.A4.2a}$$

$$i_k\text{: nonnegative integer,} \qquad N\text{: sufficiently large positive integer,} \tag{7.A4.2b}$$

[and $O(\ldots) \equiv$ *Of order* $(\ldots)$ has its usual meaning: $f(\varepsilon) = O[g(\varepsilon)]$, for two general functions $f(\ldots)$ and $g(\ldots)$, means that for $\varepsilon \to 0$: $\lim |f(\varepsilon)/g(\varepsilon)| < \infty$; e.g., $\sin(6\varepsilon) = O(\varepsilon)$, $\cos(3\varepsilon) = O(\varepsilon^0) = O(1)$]. Then (7.A4.1) can be re-expressed as

$$\begin{aligned} L &= \sum (1/2)\ \left[(\dot{q}_k)^2 - \nu_k{}^2 q_k{}^2\right] \\ &\quad + \varepsilon\left[L_1(\omega t,\ q,\ \dot{q};\ \varepsilon) + (1/\varepsilon)\sum (q_k{}^2/2)(\nu_k{}^2 - \omega_k{}^2)\right] \\ &\equiv \sum (1/2)\ \left[(\dot{q}_k)^2 - \nu_k{}^2 q_k{}^2\right] + \varepsilon L^{(1)}(\omega t,\ q,\ \dot{q};\ \varepsilon). \end{aligned} \tag{7.A4.3}$$

The above show that the *unperturbed* system [i.e., (7.A4.3) for $\varepsilon = 0$: $q_k \to q_{ko}$: *generating function*] has equations of motion $d^2 q_{ko}/dt^2 + \nu_k{}^2 q_{ko} = 0$, and, therefore, general solutions

$$\begin{aligned} &q_{ko} = q_{ko}(t) = a_k \cos(\nu_k t) + (b_k/\nu_k)\sin(\nu_k t), \\ &a_k = q_{ko}(0) \quad \text{and} \quad b_k = \dot{q}_{ko}(0)\text{: initial conditions.} \end{aligned} \tag{7.A4.4}$$

We notice that since the ν_k are *rationally commensurate* to ω—that is, $\nu_k = (i_k/N)\omega$ — the q_{ko} have the *common* period

$$\tau \equiv (2\pi/\omega)N = 2\pi(N/\omega) = 2\pi(i_k/\nu_k) = (2\pi/\nu_k)i_k \equiv \tau_k i_k. \tag{7.A4.5}$$

Let us now examine the *perturbed* system ($\varepsilon \neq 0$). Its equations of motion are

$$E_k(L) \equiv (\partial L/\partial \dot{q}_k)^{\cdot} - \partial L/\partial q_k = 0:$$
$$\ddot{q}_k + \nu_k{}^2 q_k = -\varepsilon\big[(\partial L^{(1)}/\partial \dot{q}_k)^{\cdot} - \partial L^{(1)}/\partial q_k\big] \equiv \varepsilon F_k(\omega t,\ q,\ \dot{q};\ \varepsilon), \tag{7.A4.6}$$

where $F_k(\omega t,\ q,\ \dot{q};\ \varepsilon) = F_k(2\pi + \omega t,\ q,\ \dot{q};\ \varepsilon)$.

Applying, next, the well-known method of *variation of constants* (ex. 7.9.14 and §8.7) to these *perturbed* (nonhomogeneous) equations, based on the general solutions of the corresponding *unperturbed* (homogeneous) equations (7.A4.4), we find

$$q_k(t) = q_{ko}(t) + (\varepsilon/\nu_k)\Bigg[\sin(\nu_k t)\int_0^t F_k(\omega x,\ q_{ko},\ \dot{q}_{ko};\ 0)\cos(\nu_k x)\,dx$$
$$- \cos(\nu_k t)\int_0^t F_k(\omega x,\ q_{ko},\ \dot{q}_{ko};\ 0)\sin(\nu_k x)\,dx\Bigg] + O(\varepsilon^2), \tag{7.A4.7}$$

(a result reminiscent of the well-known "Duhamel's superposition integral" formula of forced linear vibrations); or, further, with some standard manipulations and recalling eqs. (7.A4.4–6),

$$q_k(t) = a_k\cos(\nu_k t) + (b_k/\nu_k)\sin(\nu_k t)$$
$$- (\varepsilon/\nu_k)\left\{\int_0^t \big[(\partial L^{(1)}/\partial \dot{q}_k)^{\cdot} - \partial L^{(1)}/\partial q_k\big]_o \sin[\nu_k(t - x)]\,dx\right\} + O(\varepsilon^2), \tag{7.A4.8}$$

where $[\ldots]_o \equiv [\ldots]$ evaluated for the *generating solution* $q_{ko}(x)$, where x is a dummy variable of integration. This allows us to calculate the *new initial values* a_{k1} and b_{k1} after the period τ: from (7.A4.8) [or, more easily, (7.A4.7)] with $t \to \tau$, we find

$$q_k(\tau) = q_{ko}(\tau) + (\varepsilon/\nu_k)\int_0^\tau [\ldots]_o \sin(\nu_k x)\,dx + O(\varepsilon^2) \qquad (\text{with } t_1 = 0,\ t_2 = \tau), \tag{7.A4.9}$$

or

$$a_{k1} = a_k + \varepsilon\tau P_k + O(\varepsilon^2), \tag{7.A4.9a}$$

$$P_k = P_k(a_l,\ b_l) \equiv (1/\nu_k\tau)\int_0^\tau \big[(\partial L^{(1)}/\partial \dot{q}_k)^{\cdot} - \partial L^{(1)}/\partial q_k\big]_o \sin(\nu_k x)\,dx. \tag{7.A4.9b}$$

Similarly, $(\ldots)^{\cdot}$-differentiating (7.A4.8), or (7.A4.7), and so on, we find

$$\dot{q}_k(\tau) = \dot{q}_{ko}(\tau) - \varepsilon\int_0^\tau [\ldots]_o \cos(\nu_k x)\,dx + O(\varepsilon^2), \tag{7.A4.10}$$

or

$$b_{k1} = b_k + \varepsilon\tau Q_k + O(\varepsilon^2), \tag{7.A4.10a}$$

$$Q_k = Q_k(a_l,\ b_l) \equiv -(1/\tau)\int_0^\tau \big[(\partial L^{(1)}/\partial \dot{q}_k)^{\cdot} - \partial L^{(1)}/\partial q_k\big]_o \cos(\nu_k x)\,dx. \tag{7.A4.10b}$$

Next, integrating the $d/dx(\ldots)$-term of the integrand of both P_k and Q_k by parts, and noting that the integrated-out terms vanish (due to *periodicity*), we find

$$P_k = -(1/\tau)\int_0^T \{(\partial L^{(1)}/\partial \dot{q}_k)\cos(\nu_k x) + (\partial L^{(1)}/\partial q_k)\,[\sin(\nu_k x)/\nu_k]\}_o\,dx$$

$$= -(1/\tau)\int_0^T (\partial L^{(1)}/\partial b_k)_o\,dx, \qquad (7.A4.11a)$$

$$Q_k = -(1/\tau)\int_0^T \{(\partial L^{(1)}/\partial \dot{q}_k)\,[\nu_k \sin(\nu_k x)] + (\partial L^{(1)}/\partial q_k)\cos(\nu_k x)\}_o\,dx$$

$$= (1/\tau)\int_0^T (\partial L^{(1)}/\partial a_k)_o\,dx. \qquad (7.A4.11b)$$

A final simplification of the above occurs with the help of the following function:

$$\Lambda = \Lambda(a_k,\, b_k) \equiv (1/\tau)\int_0^T L\big[\omega x,\ q_{ko}(x),\ dq_{ko}(x)/dx;\ \varepsilon\big]\,dx:$$

Average of *perturbed* Lagrangean, but evaluated along the (known) *unperturbed* solution ($\neq \langle L\rangle$), (7.A4.12)

or, recalling (7.A4.3), and noting that due to (7.A4.4)

$$\int_0^T (1/2)\{[dq_{ko}(x)/dx]^2 - \nu_k{}^2 q_{ko}(x)^2\}\,dx = 0 \qquad \text{(Virial theorem for } q_{ko}), \qquad (7.A4.12a)$$

finally,

$$\Lambda = (\varepsilon/\tau)\int_0^T [L^{(1)}]_o\,dx = O(\varepsilon). \qquad (7.A4.12b)$$

Then, and recalling (7.A4.11a, b), eqs. (7.A4.9a, 10a) reduce, respectively, to

$$\Delta a_k \equiv a_{k1} - a_k = -\tau(\partial\Lambda/\partial b_k) + O(\varepsilon^2)$$

$$= -\varepsilon\left\{\partial/\partial b_k \int_0^T [L^{(1)}]_o\,dx\right\} + O(\varepsilon^2), \qquad (7.A4.13a)$$

$$\Delta b_k \equiv b_{k1} - b_k = +\,\tau(\partial\Lambda/\partial a_k) + O(\varepsilon^2)$$

$$= +\,\varepsilon\left\{\partial/\partial a_k \int_0^T [L^{(1)}]_o\,dx\right\} + O(\varepsilon^2). \qquad (7.A4.13b)$$

Now Valeev and Ganiev (1969), reasoning as in the method of slowly varying parameters, have demonstrated that this *finite difference* system can be replaced by the following Hamiltonian, or *canonical*, differential system (§8.2):

$$da_k/dt = -\partial\Lambda/\partial b_k + O(\varepsilon^2), \qquad db_k/dt = +\,\partial\Lambda/\partial a_k + O(\varepsilon^2). \qquad (7.A4.14)$$

These equations readily show that the *periodic* solutions of (7.A4.3, 6)—that is, $\Delta a_k = 0$, $\Delta b_k = 0$, or $da_k/dt = 0$, $db_k/dt = 0$—to the first ε-order, are determined from the following $2n$ conditions of stationarity, or "equilibrium," of Λ:

$$\partial\Lambda/\partial a_k = 0, \qquad \partial\Lambda/\partial b_k = 0 \qquad (k = 1, \ldots, n); \qquad (7.A4.15)$$

and also provide the "energy" integral

$$dA/dt = \sum \left[(\partial A/\partial a_k)\dot{a}_k + (\partial A/\partial b_k)\,\dot{b}_k\right] = O(\varepsilon^2),$$
$$\Rightarrow A(a_k,\, b_k) = constant + O(\varepsilon^2). \tag{7.A4.16}$$

Next, let $(a_{ko},\, b_{ko})$ be a solution of (7.A4.15); that is, a stationary point of A. To examine its *stability* in the first ε-approximation, we expand eqs. (7.A4.14) around that equilibrium solution, and linearize them in the deviations from it, $A_k \equiv a_k - a_{ko}$ and $B_k \equiv b_k - b_{ko}$, while invoking (7.A4.15). In this way, we obtain the following linear *variational* equations (Poincaré's "équations aux variations"):

$$dA_k/dt = -\sum \left[(\partial^2 A/\partial b_k\,\partial a_l)\,A_l + (\partial^2 A/\partial b_k\,\partial b_l)\,B_l\right], \tag{7.A4.17a}$$

$$dB_k/dt = +\sum \left[(\partial^2 A/\partial a_k\,\partial a_l)\,A_l + (\partial^2 A/\partial a_k\,\partial b_l)\,B_l\right], \tag{7.A4.17b}$$

where all partial derivatives are calculated at $(a_{ko},\, b_{ko})$.

Assuming, as usual, time-exponential solutions for A_k and B_k [i.e., $\sim \exp(\lambda t)$], and substituting them into the (7.A4.17a, b), we are readily led at the following $2n$-degree characteristic equation for λ:

$$\Delta = \Delta(\lambda) \equiv \begin{vmatrix} -(\partial^2 A/\partial b_k\,\partial a_l) - \lambda\,\delta_{kl} & -(\partial^2 A/\partial b_k\,\partial b_l) \\ \partial^2 A/\partial a_k\,\partial a_l & (\partial^2 A/\partial a_k\,\partial b_l) - \lambda\,\delta_{kl} \end{vmatrix} = 0. \tag{7.A4.18}$$

For (asymptotic) stability of the point $(a_{ko},\, b_{ko})$: All λ-roots of $\Delta = 0$ must have *negative real parts (*§3.10); if the real part of even one such root is positive, that point is unstable; while if it is zero, that point is stable in the first ε-approximation.

Therefore, the conditions for a "coarse" extremum of $A(a_k,\, b_k)$ [i.e., a strict extremum detected by analysis of the second-order terms in the expansion $A(a_{ko} + A_k,\, b_{ko} + B_k) - A(a_{ko},\, b_{ko})$—the term is due to Blekhman] at $(a_{ko},\, b_{ko})$ also represent the sufficient conditions for a stable periodic solution, to the first ε-order; and for $\varepsilon = 0$ reducing to the generating solution (7.A4.4). Hence, A plays the role that the *total potential energy* plays, in the static stability analysis of potential systems. Let us examine (7.A4.17a, b, 18) further. With the help of the notations

$$\partial^2 A/\partial a_k\,\partial a_l \equiv \alpha_{kl} = \alpha_{lk}, \qquad \partial^2 A/\partial b_k\,\partial b_l \equiv \beta_{kl} = \beta_{lk},$$
$$\partial^2 A/\partial b_k\,\partial a_l = \partial^2 A/\partial a_l\,\partial b_k \equiv \gamma_{kl} \neq \gamma_{lk} \equiv \partial^2 A/\partial b_l\,\partial a_k = \partial^2 A/\partial a_k\,\partial b_l, \tag{7.A4.19}$$

equations (7.A4.17a, b) can be written, respectively,

$$dA_k/dt = -\sum (\beta_{kl}\,B_l + \gamma_{kl}\,A_l), \tag{7.A4.20a}$$
$$dB_k/dt = +\sum (\gamma_{lk}\,B_l + \alpha_{kl}\,A_l); \tag{7.A4.20b}$$

while in terms of the Lagrange-like function K [see also (8.3.12 ff.)]

$$2K = \sum\sum (\beta_{kl}\,B_k\,B_l + 2\gamma_{kl}\,B_k\,A_l + \alpha_{kl}\,A_k\,A_l), \tag{7.A4.21}$$

they can be brought to the Hamiltonian form (§8.2)

$$dA_k/dt = -\,\partial K/\partial B_k, \qquad dB_k/dt = +\partial K/\partial A_k. \tag{7.A4.22}$$

In view of the above, the characteristic equation (7.A4.18) can also be rewritten, successively, as follows (partitioned in subdeterminants):

$$\Delta(\lambda) = \begin{vmatrix} |\gamma_{lk} - \lambda\,\delta_{lk}| & |\alpha_{kl}| \\ |-\beta_{kl}| & |-\gamma_{kl} - \lambda\,\delta_{kl}| \end{vmatrix}$$

$$= (-1)^n \begin{vmatrix} |\gamma_{lk} - \lambda\,\delta_{lk}| & |\alpha_{kl}| \\ |\beta_{kl}| & |\gamma_{kl} + \lambda\,\delta_{kl}| \end{vmatrix}$$

$$= (-1)^n \begin{vmatrix} |\beta_{kl}| & |\gamma_{kl} + \lambda\,\delta_{kl}| \\ |\gamma_{lk} - \lambda\,\delta_{lk}| & |\alpha_{kl}| \end{vmatrix}$$

[after swapping the first n *rows* with the last n rows]

$$= (-1)^n \begin{vmatrix} |\gamma_{kl} + \lambda\,\delta_{kl}| & |\beta_{kl}| \\ |\alpha_{kl}| & |\gamma_{lk} - \lambda\,\delta_{lk}| \end{vmatrix}$$

[after swapping the first n *columns* with the last n columns]

$$= (-1)^n \begin{vmatrix} |\gamma_{kl} + \lambda\,\delta_{kl}| & |\alpha_{kl}| \\ |\beta_{kl}| & |\gamma_{lk} - \lambda\,\delta_{lk}| \end{vmatrix}$$

[after *transposing* the determinant about its main diagonal], (7.A4.23)

and comparing the *second* and *last* of these forms of $\Delta(\lambda)$, we readily see that

$$\Delta(\lambda) = \Delta(-\lambda). \qquad (7.A4.24)$$

In words: if λ is a root of the characteristic equation (and hence an eigenvalue of the variational equations), then so is $-\lambda$; that is, $\Delta(\lambda) = 0$ can contain only *even* powers of λ. Therefore, if such a root is complex with negative real part, or a negative real number ($\Rightarrow$ asymptotic stability), its negative will also be a root with *positive* real part, or a positive real number ($\Rightarrow$ exponential, or flutteral, instability). Hence (and since asymptotic stability cannot occur in our conservative system), *for stability, all* λ's *must be purely imaginary*, and then they appear in mutually conjugate pairs: $\pm ia$ (a: real). In this case [recall discussion following eq. (3.10.18a)] we say that the system possesses *critical* or *nonsignificant* behavior, i.e., its stability cannot be safely concluded from its linear perturbation equations (7.A4.17a, b; 20a, b); in such cases, we must consider the *nonlinear* A and B terms of $O(\varepsilon^2)$. However, if no such higher-order terms are present, which is the quasi-linear (first ε-approximation) case discussed here, then *purely imaginary roots of* $\Delta(\lambda) = 0$ *do signify stability*.

[For detailed discussions of this very important problem of the stability of motion, including the fundamental contributions of Liapunov on it, see, for example (alphabetically): Chetayev (1955), Hughes (1986, pp. 480–521), Kuzmin (1973), Meirovitch (1970, chap. 6), Pars (1965, chap. 23), Pollard (1976, pp. 117–131).]

Our discussion of the *integral stability criterion* can be summarized as follows: Let Λ be the time average of the Lagrangean of the original quasi-linearly *perturbed* system, eqs. (7.A4.6), but calculated along the periodic solutions of the *unperturbed* linear system, eqs. (7.A4.4), as a function of the initial values of the *generating solution* (a_k, b_k). Next, consider the *stationary* points of Λ, (a_{ko}, b_{ko}): if these points are also *extrema* (maxima or minima) of Λ, then, to the first ε-approximation, they constitute *stable* (periodic) solutions of the original system. *Nonextremum* stationary points require special consideration.

Example 7.A4.1 Let us consider a system with (exact) Lagrangean

$$L = (\dot{q})^2/2 - q^2/2 + \varepsilon[-\kappa\, q^2 + q\, \dot{q}\sin(2t)] \qquad (0 < \varepsilon \ll 1), \tag{a}$$

and, therefore, equation of motion, the linear *Mathieu* equation

$$\ddot{q} + q = -\varepsilon[\kappa + 2\cos(2t)]\, q,$$

or

$$\ddot{q} + [1 + \varepsilon\kappa + 2\varepsilon\cos(2t)]\, q = 0. \tag{b}$$

Here, clearly, $n = 1$, $\nu_k = 1$, $\omega = 2 \Rightarrow \tau = \pi N$, $\tau_k = 2\pi$, and the unperturbed system $(\varepsilon = 0)$: $\ddot{q} + q = 0$ has generating solution

$$q_o = q_o(t) = a\ \cos t + b\ \sin t \qquad (a,\ b\text{: initial values of } q_o,\ \dot{q}_o). \tag{c}$$

The average of L evaluated along (c) (i.e., between $t_1 = 0$ and $t_2 = 2\pi$) equals, after (7.A4.12, 12b),

$$\Lambda = (1/2\pi)\int_0^{2\pi} L\big[q_o(x),\ dq_o(x)/dx,\ x\big]\, dx$$

$$= \cdots = -(\varepsilon/4)\ \big[(\kappa + 1)\, a^2 + (\kappa - 1)\, b^2\big] \equiv \Lambda(a,\ b). \tag{d}$$

Using subscripts to denote partial derivatives relative to a, b, we readily see that the sole root of $\Lambda_a = 0$, $\Lambda_b = 0$ is $(a_o = 0,\ b_o = 0)$; that is, the equilibrium state $q_o(t) = 0$. At that point, since

$$\Lambda_{aa} = -(\varepsilon/2)(\kappa + 1): \qquad > 0 \quad \text{for } \kappa < -1, \qquad < 0 \quad \text{for } \kappa > -1, \tag{e1}$$

$$\Lambda_{bb} = -(\varepsilon/2)(\kappa - 1): \qquad > 0 \quad \text{for } \kappa < 1, \qquad < 0 \quad \text{for } \kappa > 1, \tag{e2}$$

$$\Lambda_{ab} = \Lambda_{ba} = 0, \tag{e3}$$

$$\Rightarrow D \equiv \Lambda_{aa}\Lambda_{bb} - \Lambda_{ab} = \varepsilon^2(\kappa^2 - 1)/4:$$

$$> 0, \quad \text{for } \kappa > 1 \ \text{ or } \ \kappa < -1, \qquad \text{i.e., for } |\kappa| > 1,$$

$$< 0, \quad \text{for } -1 < \kappa < 1, \qquad \text{i.e., for } |\kappa| < 1 \tag{e4}$$

(using ordinary theory of extrema of a function of two variables), we easily conclude that

$$D > 0 \quad \textit{and} \quad \Lambda_{aa}\ (\text{or } \Lambda_{bb}) > 0: \quad \kappa < -1 \Rightarrow \Lambda:\ \textit{minimum},$$
$$D > 0 \quad \textit{and} \quad \Lambda_{aa}\ (\text{or } \Lambda_{bb}) < 0: \quad \kappa > 1 \Rightarrow \Lambda:\ \textit{maximum};$$
$$D < 0: \quad |\kappa| < 1 \Rightarrow \Lambda:\ \textit{min/max}\ (\textit{saddle}\text{-point}). \tag{e5}$$

Therefore, the equilibrium solution of (b) is stable for $|\kappa| > 1$, and unstable for $|\kappa| < 1$; while for $\kappa = \pm 1 \Rightarrow D = 0$, the equilibrium point is defined ambiguously, and this implies the existence of periodic solutions—we are exactly on top of the famous stability/instability boundaries of Mathieu's equation, which emanate from the (usually) horizontal axis of the "Strutt chart" at 1; that is, for $\varepsilon = 0$. These results coincide with those found by other methods; see, for example, Cunningham (1958, pp. 270–273), Papastavridis [1981, 1982(b)].

Finally, let us examine the equivalence between the above, extremum of Λ-based approach, with that based on the study of the eigenvalues of the variational equations (7.A4.17a, b; 20a, b). We have

$$\dot{a} = -\Lambda_b \ \Rightarrow\ \dot{A} = -\Lambda_{ba} A - \Lambda_{bb} B, \tag{f1}$$

$$\dot{b} = \Lambda_b \ \Rightarrow\ \dot{B} = \Lambda_{aa} A + \Lambda_{ab} B; \tag{f2}$$

and so the corresponding characteristic equation is

$$\Delta(\lambda) = \begin{vmatrix} -(\Lambda_{ab} + \lambda) & -\Lambda_{bb} \\ \Lambda_{aa} & \Lambda_{ab} - \lambda \end{vmatrix} = 0, \tag{g1}$$

from which we readily get

$$\lambda^2 = \Lambda_{ab}{}^2 - \Lambda_{aa}\Lambda_{bb} \ \Rightarrow\ \lambda^2 = -D. \tag{g2}$$

For stability, clearly, $\lambda^2 < 0 \ \Rightarrow\ D > 0$; as in the first two of eqs. (e5). Then λ is purely imaginary, and therefore, A, B are harmonically oscillatory.

If $\lambda^2 > 0 \ \Rightarrow\ D < 0$, as in the third of eqs. (e5), then λ is real, and therefore, A, B increase exponentially; that is, instability. If $\lambda^2 = -D = 0$, the linear stability criterion fails.

Thus, we have affirmed the equivalence between the extremum of (d) and the stability of (f1, 2).

8

Introduction to Hamiltonian/ Canonical Methods

Equations of Hamilton and Routh; Canonical Formalism

This is the celebrated "canonical form" of the equations of motion of a system, though why it has been so called it would be hard to say.

(Thomson and Tait, 1912, no. 319, p. 307)

We recognize two purposes in the study of general methods in dynamics. First, the practical purpose, to increase our power in solving specific problems by developing standard techniques with a wide range of applicability. Secondly, the intellectual purpose, to understand the mathematical structure of dynamics. ... Historically [general dynamical theory] has been suggested by, and developed in terms of, the Newtonian dynamics of particles and rigid bodies. But we feel an urgent need to give it a wider scope, presenting it as a consistent mathematical theory applicable to any physical system the behaviour of which can be expressed in Lagrangian or Hamiltonian form.

(Synge, 1960, pp. 99–100)

Hamiltonian mechanics is the description of a mechanical system in terms of generalized coordinates q_i and generalized momenta p_i,... the Hamiltonian formulation ... is far better suited for the formulation of quantum mechanics, statistical mechanics, and perturbation theory. In particular, the use of Hamiltonian phase space provides the ideal framework for a discussion of the concepts of integrability and nonintegrability and the description of the chaotic phenomena that can be exhibited by nonintegrable systems.

(Tabor, 1989, p. 48)

8.1 INTRODUCTION

The independent variables of Lagrangean mechanics (holonomic and/or nonholonomic) are t, q, and $\dot{q}$ (or ω). In this chapter, we introduce the reader to a very important alternative formulation of analytical mechanics, known as Hamiltonian mechanics (HM), where the independent dynamical variables are t, q, and

$p \equiv \partial T/\partial \dot{q}$ = *system (or generalized) momenta* (or, sometimes, $p \equiv \partial L/\partial \dot{q}$). Hamiltonian mechanics and its associated equations of motion constitute a powerful and fertile version of theoretical mechanics. It represents the last and most abstract/mathematical stage of classical mechanics, and it played a crucial role in the eventual replacement of the latter by quantum mechanics (1920s) as a fundamental physical theory. Although, historically, of primary interest to physicists and astronomers/celestial mechanicians, over the past few decades HM has been becoming increasingly relevant, if not indispensable, to engineers; for example, in optimization, robotics, and, most importantly, for the understanding of modern (deterministic) chaotic/nonlinear dynamics.

HISTORICAL

Hamiltonian mechanics was originated by Lagrange himself (1810–1811), and also Poisson (1809), in connection with perturbation methods for celestial mechanics problems; was duly noted and generally formulated by Cauchy (1819); but was brought to prominence by Hamilton (1834–1835); and was extended to nonstationary constraints by Ostrogradskii (1848–1850), and Donkin (1854).

Briefly, in HM, the n system positions q and associated n system momenta p become the system coordinates in a $2n$-dimensional *phase*, or *state*, space; and the corresponding n Lagrangean equations of motion of the system (in the q's) are replaced by $2n$ *first-order* symmetrical, or *canonical*, Hamiltonian equations of motion (in the q's and p's).

That a $\dot{q} \Leftrightarrow p$ transformation is "always" possible, is easily seen as follows: by $(\partial/\partial \dot{q})$-differentiating the kinetic energy (recalling expressions in §3.9),

$$\begin{aligned} 2T &= \sum\sum M_{kl}\dot{q}_k\dot{q}_l + 2\sum M_k\dot{q}_k + M_0 \\ &= 2T(t, q, \dot{q}) \qquad (= 2T_2 + 2T_1 + 2T_0), \end{aligned} \tag{8.1.1}$$

$$(M_{kl}, M_k, M_0\text{: functions of the } q\text{'s and } t;\ k,\ l = 1, \ldots, n), \tag{8.1.1a}$$

we obtain the p's as linear and independent functions in the $\dot{q}$'s (and t, q's):

$$p_k \equiv \partial T/\partial \dot{q}_k = \sum M_{kl}\,\dot{q}_l + M_k \equiv p_k(t, q, \dot{q}); \tag{8.1.2}$$

while inverting (8.1.2) [assuming that $Det\,(M_{kl}) \equiv Det\,(\partial^2 T/\partial \dot{q}_k\,\partial \dot{q}_l) \neq 0$; that is, assuming that the Hessian of T (or L) does not vanish identically; and viewing t and the q's as parameters; see also MacMillan (1936, pp. 358–360)], we obtain the $\dot{q}$'s as linear functions of the p's:

$$\dot{q}_l = \sum M'_{lk}\,p_k + M'_l \equiv \dot{q}_l(t, q, p); \tag{8.1.3}$$

$$(M'_{lk}, M'_l\text{: functions of the } q\text{'s and } t;\ k,\ l = 1, \ldots, n). \tag{8.1.3a}$$

Thus, Lagrange's equations, say,

$$(\partial T/\partial \dot{q}_l)^{\cdot} - \partial T/\partial q_l = Q_l, \tag{8.1.4}$$

(§3.5) have been transformed into the completely equivalent system of $2n$ *first-order* equations:

$$dp_l/dt = (\partial T/\partial q_l)|_{\dot{q}=\dot{q}(t,q,p)} + Q_l(t,\ q) \equiv dp_l(t,q,p)/dt, \qquad (8.1.5)$$

$$dq_l/dt = \sum M'_{lk}p_k + M'_l \equiv dq_l(t,q,p)/dt \qquad (\textit{linear}\text{ in the }p\text{'s}); \qquad (8.1.5a)$$

and this is the essence of the Hamiltonian formalism.

In general, any function of the Lagrangean variables $f = f(t, q, \dot{q})$, becomes, upon substitution of (8.1.5a) in it, a certain "associated" function of the Hamiltonian variables:

$$f = f(t,q,\dot{q}) \equiv f_{(q\dot{q})} = f[t,q,\dot{q}(t,q,p)] \equiv f(t,q,p) \equiv f_{(qp)}; \qquad (8.1.6)$$

and conversely, any $f(t, q, p)$ becomes, upon substitution of (8.5.2) in it, an associated $f(t, q, \dot{q})$. [The elaborate notations $f_{(q\dot{q})}$ and $f_{(qp)}$ will be used only in potentially ambiguous situations.] A detailed treatment of HM, comparable to that of Lagrangean mechanics presented so far, is beyond the scope and limits of this book. Instead, in this chapter, *we concentrate on topics of more or less engineering significance*; for example, (i) applications of the canonical formalism to the approximate analytical solution of the equations of motion (canonical perturbation theory); and (ii) equations of Routh [a *mixed* formulation that uses as independent variables some of the p's and the remaining $\dot{q}$'s (and, of course, t and the q's); and as such is "halfway" between the methods of Lagrange and Hamilton], and their application to the study of steady motion and its stability.

The literature on Hamiltonian mechanics is, expectedly, very extensive and varied. For concurrent (and further) reading we recommend (alphabetically):

Born (1927): Masterful and readable exposition by a very famous and wise theoretical physicist.

Chertkov (1960): Applications of Jacobi's method to rigid-body dynamics.

Frank (1935, pp. 59–65, 72–136, 191–239): Encyclopedic classical treatment.

Fues (1927, pp. 131–177): Classical Hamiltonian perturbation theory.

Gantmacher (1970, pp. 71–87, 98–165, 242–258): Compact classical treatment; excellent.

Hagihara (1970): Advanced and comprehensive treatise, primarily for celestial mechanicians.

Hamel (1949, pp. 281–312, 317–361, 653–709): Insightful and masterly presentation, as usual.

Lanczos (1970, pp. 125–130, 161–290): Extensive classical coverage; highly recommended.

Lichtenberg and Lieberman (1992): Warmly recommended for further study of nonlinear/chaotic dynamics.

McCauley (1997): Modern, mature, insightful treatment; most highly recommended.

Nordheim and Fues (1927, pp. 91–130): General and compact encyclopedic treatment.

Prange (1935, pp. 570–785): Extensive and authoritative classical coverage; highly recommended.

Tabor (1989): One of the most readable modern accounts of nonlinear dynamics; very highly recommended for further study.

Whittaker (1937, pp. 54–57, 193–208, 263–338): Mature and insightful classical treatment.

Winkelmann and Grammel (1927, pp. 469–483): Outstanding engineering reference.

Additional special references will be given in later sections.

8.2 THE HAMILTONIAN, OR CANONICAL, CENTRAL EQUATION AND HAMILTON'S CANONICAL EQUATIONS OF MOTION

To obtain equations in the canonical variables t, q, p, we proceed, as in the Lagrangean case, from the invariant differential principle of Lagrange (LP), but in the central equation form (§3.6):

$$\delta I = \delta' W: \qquad \left(\sum p_k\, \delta q_k\right)^{\cdot} - \delta T = \sum Q_k\, \delta q_k, \tag{8.2.1}$$

or, after carrying out the differentiations indicated [and assuming that, as in (8.2.1), $(\delta q)^{\cdot} = \delta(\dot{q})$],

$$\sum (dp_k/dt)\, \delta q_k + \sum p_k\, \delta(\dot{q}_k) - \delta T = \delta' W. \tag{8.2.1a}$$

Now, combining the *second* and *third* terms of the left side of the above, so as to create a total $\delta(\ldots)$-variation:

$$\sum p_k\, \delta(\dot{q}_k) - \delta T = \delta\left(\sum p_k\, \dot{q}_k - T\right) - \sum \dot{q}_k\, \delta p_k, \tag{8.2.1b}$$

and introducing the new function (and this is the key step!)

$$\begin{aligned} T' &\equiv \left(\sum p_k\, \dot{q}_k - T\right)_{\dot{q}=\dot{q}(t,q,p)} \\ &= \sum p_k\, \dot{q}_k(t,q,p) - T[t,q,\dot{q}(t,q,p)] \equiv \sum p_k\, \dot{q}_k(t,q,p) - T_{(qp)} \\ &\equiv T'(t,q,p)\text{: } \textit{Conjugate (to } T\textit{) kinetic energy} \\ &\Big[= \sum (\partial T/\partial \dot{q}_k)\dot{q}_k - T = (2T_2 + T_1) - (T_2 + T_1 + T_0) = T_2 - T_0, \\ &\quad \text{i.e., if } T = T_2 \text{ (e.g., stationary "initial" constraints), then } T' = T\Big], \end{aligned} \tag{8.2.2}$$

we can rewrite (8.2.1a) as

$$\sum (dp_k/dt)\, \delta q_k + \sum \left[(\partial T'/\partial q_k)\, \delta q_k + (\partial T'/\partial p_k)\, \delta p_k\right] - \sum (dq_k/dt)\, \delta p_k = \delta' W,$$

or, collecting $(\ldots)\, \delta q$ and $(\ldots)\, \delta p$ terms, finally,

$$\sum (dp_k/dt + \partial T'/\partial q_k - Q_k)\, \delta q_k + \sum (-dq_k/dt + \partial T'/\partial p_k)\, \delta p_k = 0. \tag{8.2.3}$$

This (differential) variational equation, holding for *all* virtual δq's and δp's—that is, constrained or not—and known (after Winkelmann, 1909, 1930, p. 39 ff.; also Hamel, 1949, p. 286 ff.) as the *canonical, or Hamiltonian, central equation*, is fundamental to all subsequent considerations.

1. δq and δp Unconstrained

Now, if the δq and δp are *unconstrained*, then (8.2.3) leads immediately to the famous *canonical*, or Hamiltonian, equations of motion:

$$dp_k/dt = -\partial T'/\partial q_k + Q_k, \tag{8.2.4}$$

$$dq_k/dt = \partial T'/\partial p_k \qquad [= \textit{linear} \text{ in the } p\text{'s; recall (8.2.2)}]. \tag{8.2.4a}$$

• Clearly, the *second* set, eqs. (8.2.4a), must coincide with the earlier, purely kinematico-inertial equations (8.1.5a); it is the canonical counterpart of the Lagrangean $p_k = \partial T/\partial \dot{q}_k$.

• It is the *first* set, eqs. (8.2.4), that expresses the *equations of motion*, in a manner almost identical to that of the Lagrangean method; but, unlike the latter, eqs. (8.2.4, 4a) are already expressed directly and linearly in the $(\ldots)^{\cdot}$-derivatives of the $2n$ "coordinates" q and p involved.

The $2n$ *first-order* equations (8.2.4, 4a) allow us to determine the values of the q's and p's at any time, once their values at some initial time are known. They constitute the equations of motion of the representative system "particle" in the (symbolical/mathematical) $2n$-dimensional *phase space* of q's and p's; and, through each (admissible) point of that space, there passes only one such mechanical trajectory (orbit), if the system is scleronomic; or more if the system is rheonomic. In *extended* phase space (t, q, p), however, only one orbit can pass through a point. Comparing the Hamiltonian equations (8.2.4) with their Lagrangean counterparts: $dp_k/dt = \partial T/\partial q_k + Q_k$, we readily obtain the additional kinematico-inertial result

$$\partial T/\partial q_k = -\partial T'/\partial q_k. \tag{8.2.5}$$

If $Q_k = -\partial V(t,\ q)/\partial q_k$, then (8.2.4, 4a) assume the purely *antisymmetrical* form

$$dp_k/dt = -\partial H/\partial q_k, \qquad dq_k/dt = \partial H/\partial p_k, \tag{8.2.6}$$

where

$$\begin{aligned}
H \equiv T' + V &= \Big(\sum p_k\ \dot{q}_k - T + V\Big)_{\dot{q}=\dot{q}(t,q,p)} \\
&= \sum p_k \dot{q}_k(t,q,p) - (T_{(qp)} - V) \\
&= \Big(\sum p_k\ \dot{q}_k - L\Big)_{\dot{q}=\dot{q}(t,q,p)} \\
&= \sum p_k\ \dot{q}_k(t,q,p) - L[t,q,\dot{q}(t,q,p)] \equiv \sum p_k\ \dot{q}_k(t,q,p) - L_{(qp)} \\
&= \Big(\sum (\partial L/\partial \dot{q}_k)\dot{q}_k - L\Big)_{\dot{q}=\dot{q}(t,q,p)} \\
&\equiv H(t,q,p) \qquad \text{(function of } 2n+1 \text{ arguments)},
\end{aligned} \tag{8.2.7}$$

is the *Hamiltonian* function of the system, or, simply, its Hamiltonian; and, similarly,

$$\begin{aligned}
L(t,q,\dot{q}) &= \sum p_k(t,q,\dot{q})\dot{q}_k - H[t,q,p(t,q,\dot{q})] \\
&\equiv \sum p_k(t,q,\dot{q})\dot{q}_k - H_{(q\dot{q})} \\
&= \Big(\sum (\partial L/\partial \dot{q}_k)\dot{q}_k - H\Big)_{p=p(t,q,\dot{q})}.
\end{aligned} \tag{8.2.7a}$$

• If both potential *and* nonpotential forces (Q_k) are present, eqs. (8.2.6) are replaced by

$$dp_k/dt = -\partial H/\partial q_k + Q_k, \qquad dq_k/dt = \partial H/\partial p_k; \tag{8.2.8}$$

while (8.2.5) becomes

$$\partial H/\partial q_k = -\partial L/\partial q_k. \tag{8.2.9}$$

• Also if we view *time* t, and/or any other number of system *parameters* $(c_1, \ldots, c_{n'})$, on which T, T' and L, H might depend, as *additional* system coordinates (i.e., $q_{n+1} \equiv t$, $q_{n+2} \equiv c_1, \ldots, q_{n+n'} \equiv c_{n'}$), then from the above we easily deduce the following two sets of kinematico-inertial identities (with $* = n+1, \ldots, n'$):

$$\partial T/\partial t = -\partial T'/\partial t, \qquad \partial T/\partial c_* = -\partial T'/\partial c_*; \tag{8.2.10a}$$

and

$$\partial L/\partial t = -\partial H/\partial t, \qquad \partial L/\partial c_* = -\partial H/\partial c_*. \tag{8.2.10b}$$

(See also Landau and Lifshitz, 1960, pp. 132–133.)

• In the presence of a *generalized* potential (3.9.8c)

$$V = V(t, q, \dot{q}) = V_1(t, q, \dot{q}) + V_0(t,\ q) = \sum \gamma_k(t,\ q)\dot{q}_k + V_0(t,\ q), \tag{8.2.11}$$

the momenta p_k are redefined by the [slightly more general than (8.1.2)] relation

$$p_k \equiv \partial L/\partial \dot{q}_k = \partial(T - V)/\partial \dot{q}_k = \partial T/\partial \dot{q}_k - \gamma_k = \sum\sum M_{kl}\dot{q}_l + (M_k - \gamma_k), \tag{8.2.11a}$$

and the Hamiltonian takes the explicit form (recalling that $L = L_2 + L_1 + L_0$),

$$\begin{aligned} H &= \Big(\sum p_k\, \dot{q}_k - L\Big)_{\dot{q}=\dot{q}(t,q,p)} \qquad (\equiv L') \\ &= \Big(\sum (\partial L/\partial \dot{q}_k)\dot{q}_k - L\Big)_{\dot{q}=\dot{q}(t,q,p)} \\ &= \big[(2L_2 + L_1) - (L_2 + L_1 + L_0)\big]_{\dot{q}=\dot{q}(t,q,p)} = (L_2 - L_0)_{\dot{q}=\dot{q}(t,q,p)} \\ &= [T_2 + (V_0 - T_0)]_{\dot{q}=\dot{q}(t,q,p)} \\ &= T' + V_0 \quad [= h(t, q, \dot{q}),\ \text{when expressed in Lagrangean variables}]; \end{aligned} \tag{8.2.11b}$$

that is, just like (8.2.7), but with $V = V_0$; while the canonical equations of motion retain their forms (8.2.6, 8). For *stationary* (holonomic) constraints, clearly, $T = T_2 (\Rightarrow T_0 = 0)$ and so the above reduces to

$$H = T(t, q, p) + V_0(t,\ q) \equiv E(t, q, p) = \textit{total energy},\ \text{in Hamiltonian variables.} \tag{8.2.11c}$$

Henceforth, only ordinary potentials $V = V(t,\ q)$ will be considered; then $p_k = \partial L/\partial \dot{q}_k = \partial T/\partial \dot{q}_k$. In sum, in *all* cases, the following kinematico-inertial identities hold:

$$\partial T'/\partial t = -\partial T/\partial t, \qquad \partial T'/\partial q_k = -\partial T/\partial q_k, \qquad \partial T'/\partial p_k = dq_k/dt; \tag{8.2.12a}$$

and

$$\partial H/\partial t = -\partial L/\partial t, \qquad \partial H/\partial q_k = -\partial L/\partial q_k, \qquad \partial H/\partial p_k = dq_k/dt. \tag{8.2.12b}$$

In both (8.2.12a, b):

(i) The *last* (*third*) group is essentially the Hamiltonian counterpart of the Lagrangean definition $\partial T/\partial \dot{q}_k = p_k$.

(ii) The *first two* groups are mathematically equivalent, if we think of time as an $(n+1)$th system coordinate (i.e., $q_{n+1} \equiv t$), but the *first* group is useful in energy rate/power theorems (see below); it states that if L does not involve the time explicitly, neither does H.

(iii) The key group is the *second*: combining it with a(ny) Lagrangean kinetic equation(s), we obtain the corresponding Hamiltonian equation(s) of motion; for example, combining it with the Lagrangean equation $(\partial T/\partial \dot{q}_k)^{\cdot} - \partial T/\partial q_k = Q_k$, we obtain the Hamiltonian equation $dp_k/dt = -\partial T'/\partial q_k + Q_k$.

• Finally, the common derivations of the canonical equations found in the literature are based on *Legendre's transformation* (see below) and *Donkin's theorem* (see, e.g., Crandall et al., 1968, pp. 13–22, 402–406; Gantmacher, 1970, pp. 73–76; Rosenberg, 1977, pp. 279–285). However, the *central equation–based derivation* presented here, due to *Winkelmann* and *Hamel*, is *far simpler and motivated*; it clearly separates the *kinematico-inertial* from the *kinetical aspects* of the *Lagrangean →Hamiltonian transition*, and thus makes its extensions to more general coordinates and/or constraints easier; some expositions falsely imply that the canonical formalism applies only to potential systems!

Legendre's Transformation (LT)

In general, a LT transforms a function $Y(y;x)$ (assumed *convex*; i.e., $\partial^2 Y/\partial y^2 > 0$, like T in the $\dot{q}$'s) into its "conjugate" function

$$Z(z;x):\ Z + Y = y\,z$$
$$\Rightarrow y = \partial Z/\partial z = y(z;x), \qquad z = \partial Y/\partial y = z(y;x),$$

that is, $Z(z;x) \equiv z\,y - Y(y;x) = z\,y(z;x) - Y[y(z;x);x]$, and similarly for $Y(y;x)$ [(fig. 8.1)].

Here, in dynamics, we have the following identifications:

$$x \to q,\ t, \qquad y \to \dot{q}, \qquad z \to p, \qquad Y(\ldots) \to L, \qquad Z(\ldots) \to H, \tag{8.2.13a}$$
$$z = \partial Y/\partial y \to p = \partial L/\partial \dot{q}, \qquad y = \partial Z/\partial z \to \dot{q} = \partial H/\partial p. \tag{8.2.13b}$$

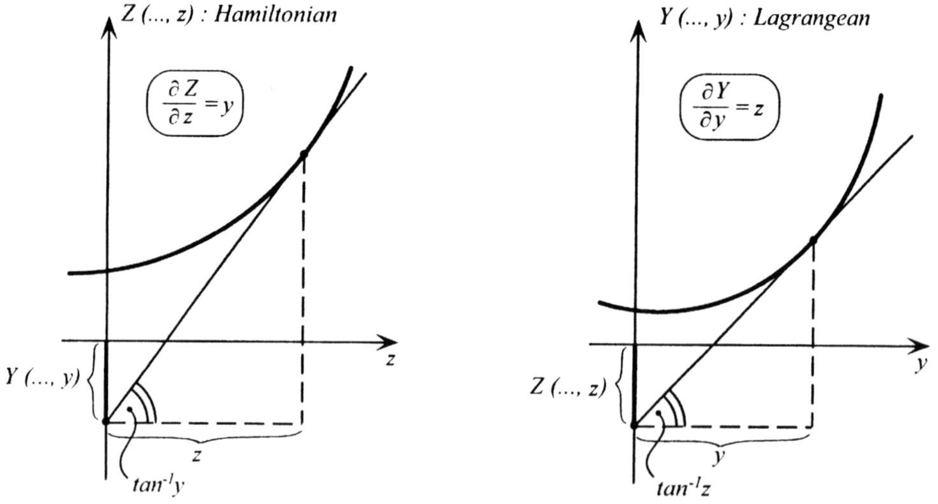

Figure 8.1 Geometrical interpretation of the Legendre transformation.

REMARKS ON LT

(i) For an alternative geometrical interpretation, see Tabor (1989, pp. 79–80).

(ii) Such transformations also appear in other areas of engineering/physics; for instance,

Thermodynamics: $(Y, Z) \to$ (*energy*, *free energy*),

Elasticity: $(Y, Z) \to$ (*strain energy*, *complementary energy*);

see, for example, Langhaar (1962, pp. 119–121, 133–136, 244–245); also Hamel (1949, pp. 368–375).

To understand the *connection of H to the power theorems* of §3.9, let us calculate dH/dt and then invoke the earlier canonical equations. Thus, we obtain, successively,

$$\begin{aligned} dH/dt &= \sum [(\partial H/\partial q_k)(dq_k/dt) + (\partial H/\partial p_k)(dp_k/dt)] + \partial H/\partial t \\ &= \sum [(\partial H/\partial q_k)(\partial H/\partial p_k) + (\partial H/\partial p_k)(Q_k - \partial H/\partial q_k)] + \partial H/\partial t \\ &= \sum (\partial H/\partial p_k)\, Q_k + \partial H/\partial t = \partial H/\partial t + \sum Q_k\, \dot{q}_k; \qquad (8.2.14) \end{aligned}$$

and, therefore, if $\partial H/\partial t = 0$ (e.g., stationary constraints) and $Q_k = 0$ (e.g., wholly potential forces), then the Hamiltonian energy of the system is conserved:

$$H = H(q, p) = constant. \qquad (8.2.14a)$$

2. δq Constrained, δp Unconstrained

If the δq's are restricted by the (additional) m Pfaffian constraints

$$\delta\theta_D \equiv \sum a_{Dk}\, \delta q_k = 0 \qquad [rank(a_{Dk}) = m;\ D = 1, \ldots, m(< n)], \qquad (8.2.15a)$$

while, in spite of (8.1.2, 3, 5a), the δp's are still viewed as free, then application of the method of Lagrangean multipliers to the canonical central equation (8.2.3) readily yields the $2n$ *canonical Routh–Voss equations*:

$$\begin{aligned} dp_k/dt &= -\partial T'/\partial q_k + Q_k + \sum \lambda_D\, a_{Dk} \quad (\text{where } Q_k = \textit{total impressed} \text{ force}) \\ &= -\partial H/\partial q_k + Q_k + \sum \lambda_D\, a_{Dk} \quad (\text{where } Q_k = \textit{nonpotential impressed} \text{ force}), \end{aligned} \qquad (8.2.15b)$$

$$dq_k/dt = \partial T'/\partial p_k \quad (= \partial H/\partial p_k); \qquad (8.2.15c)$$

which, along with the m constraints (8.2.15a) (in velocity form)

$$\omega_D \equiv \sum a_{Dk}\, \dot{q}_k + a_D = 0 \qquad (8.2.15d)$$

constitute a determinate system for the $2n + m$ functions $\{q_k(t),\ p_k(t),\ \lambda_D(t)\}$.

To uncouple the first set of (8.2.15b) into kinetic and kinetostatic equations, we proceed as in the Lagrangean variable case (§4.5); that is, we introduce the $n - m$ additional ω_I's by

$$\omega_I \equiv \sum a_{Ik}\,\dot{q}_k + a_I \quad (\neq 0,\ \text{velocity form}) \quad \text{or} \quad \delta\theta_I \equiv \sum a_{Ik}\,\delta q_k \quad (\neq 0,\ \text{virtual form}), \tag{8.2.15e}$$

then invert the system (8.2.15d, e), thus obtaining

$$\dot{q}_k = \sum A_{kI}\,\omega_I + A_k \quad (\text{velocity form}) \qquad \text{or} \qquad \delta q_k \equiv \sum A_{kI}\,\delta\theta_I \quad (\text{virtual form}); \tag{8.2.15f}$$

and inserting this δq-representation into the central equation (8.2.3), we finally obtain the *canonical Maggi equations*:

Kinetostatic equations:
$$\sum A_{kD}\,(dp_k/dt + \partial T'/\partial q_k) = \sum A_{kD}\,Q_k + \lambda_D, \tag{8.2.15g}$$

Kinetic equations:
$$\sum A_{kI}(dp_k/dt + \partial T'/\partial q_k) = \sum A_{kI}\,Q_k; \tag{8.2.15h}$$

while (8.2.15c) remain unchanged.

Similarly, if the Pfaffian constraints have the special form

$$\delta q_D = \sum b_{DI}\,\delta q_I, \tag{8.2.15i}$$

then (8.2.3) easily yields the *canonical Hadamard equations* [special case of (8.2.15g, h)]:

Kinetostatic equations:
$$dp_D/dt + \partial T'/\partial q_D = Q_D + \lambda_D, \tag{8.2.15j}$$

Kinetic equations:
$$dp_I/dt + \partial T'/\partial q_I + \sum b_{DI}(dp_D/dt + \partial T'/\partial q_D) = Q_I + \sum b_{DI}\,Q_D. \tag{8.2.15k}$$

[The multipliers in (8.2.15b) and (8.2.15g) are the same, but they are different in value from those in (8.2.15j).]

3. Nonholonomic Variables

The canonical formalism can be easily extended to quasi variables. We set

$$P_k \equiv \partial T^*/\partial\omega_k = \textit{Nonholonomic system momentum},$$

where

$$T^* = T^*(t, q, \omega) \equiv T^*_{(q\omega)} = T^*[t,\ q,\ \omega(t, q, P)] = T^*(t, q, P) \equiv T^*_{(qP)}, \tag{8.2.16a}$$

and build the nonholonomic counterpart of T':

$$T^{*\prime} \equiv \sum P_k\,\omega_k(t,q,P) - T^*(t,q,P) \equiv T^{*\prime}(t,q,P):$$

Nonholonomic conjugate to T^* *kinetic energy.* (8.2.16b)

It is not hard to show that:

(i) the nonholonomic counterparts of the Legendre–Donkin identities are

$$\partial T^{*\prime}/\partial\theta_k \equiv \sum (\partial T^{*\prime}/\partial q_l)(\partial \dot{q}_l/\partial\omega_k) = \sum A_{lk}(\partial T^{*\prime}/\partial q_l) = -\partial T^*/\partial\theta_k, \tag{8.2.16c}$$

$$\partial T^{*\prime}/\partial P_k = \omega_k \quad \text{(canonical counterpart of the Lagrange–Hamel: } \partial T^*/\partial\omega_k = P_k), \tag{8.2.16d}$$

and with

$$H^* \equiv T^{*\prime} + V(t,\ q) = \sum P_k\,\omega_k(t,q,P) - L^*(t,q,P)$$
$$\equiv H^*(t,q,P) \equiv H^*_{(qP)} = \textit{Nonholonomic Hamiltonian}. \tag{8.2.16e}$$

(ii) The *nonholonomic canonical equations* (most likely due to Pöschl, 1913) are [assuming no further constraints; and for algebraic simplicity *stationary* $\omega \Leftrightarrow \dot{q}$ relationships: $dq_l/dt = \sum A_{lk}\,\omega_k = \sum A_{lk}(\partial H^*/\partial P_k)$]

$$dP_k/dt = -\partial T^{*\prime}/\partial\theta_k - \sum\sum \gamma^s_{kl}\,P_s(\partial T^{*\prime}/\partial P_l) + \Theta_k$$
$$\text{(where } \Theta_k = \textit{total impressed} \text{ force)}, \tag{8.2.16f}$$
$$= -\partial H^*/\partial\theta_k - \sum\sum \gamma^s_{kl}\,P_s(\partial H^*/\partial P_l) + \Theta_k$$
$$\text{(where } \Theta_k = \textit{nonpotential impressed} \text{ force)}. \tag{8.2.16g}$$

In the case of m Pfaffian constraints $\omega_D = 0$, the indices in (8.2.16f, g), for the *kinetic* equations, are $s,\ k = m+1,\ldots,n;\ l = m+1,\ldots,n,\ n+1$; and analogously for the *kinetostatic* equations. These equations will not be pursued any further here. For additional details, see, for example, Chetaev (1989, pp. 340–341), Corben and Stehle (1960, pp. 256–257); and for a multibody dynamics application, see Maißer (1982).

Example 8.2.1 *Direct Derivation of the Hamiltonian Kinematico-inertial Identities.* By $d(\ldots)$-differentiating $T' \equiv \sum p_k\dot{q}_k - T(t,q,\dot{q})$, and with $p_k \equiv \partial T/\partial\dot{q}_k$, we find

$$dT' = \sum[p_k d(\dot{q}_k) + \dot{q}_k\,dp_k] - \Big\{(\partial T/\partial t)\,dt + \sum[(\partial T/\partial q_k)\,dq_k + (\partial T/\partial\dot{q}_k)\,d(\dot{q}_k)]\Big\}$$
$$\text{[the \textit{first} and \textit{last} (fourth) sums cancel each other]}$$
$$= -(\partial T/\partial t)\,dt - \sum(\partial T/\partial q_k)\,dq_k + \sum \dot{q}_k\,dp_k. \tag{a}$$

But also, since $T' = T'(t,q,p)$, we have

$$dT' = (\partial T'/\partial t)\,dt + \sum\big[(\partial T'/\partial q_k)\,dq_k + (\partial T'/\partial p_k)\,dp_k\big]. \tag{b}$$

Equating these two general dT' expressions, (a) and (b), we immediately obtain the $1 + n + n = 1 + 2n$ Hamiltonian *kinematico-inertial* identities (8.2.12a):

$$\partial T'/\partial t = -\partial T/\partial t, \qquad \partial T'/\partial q_k = -\partial T/\partial q_k, \qquad \partial T'/\partial p_k = dq_k/dt. \tag{c}$$

[This is another opportunity to show the advantages of (total) differentials over derivatives!] Repeating the above argument for $H \equiv \sum p_k\, \dot{q}_k - L(t, q, \dot{q})$, and with $p_k \equiv \partial L/\partial \dot{q}_k$, we obtain the earlier identities (8.2.12b):

$$\partial H/\partial t = -\partial L/\partial t, \qquad \partial H/\partial q_k = -\partial L/\partial q_k, \qquad \partial H/\partial p_k = dq_k/dt. \tag{d}$$

Example 8.2.2 *Another Direct Derivation of the Hamiltonian Kinematico-inertial Identities.*

(i) Applying chain rule to $H \equiv \sum p_k\, \dot{q}_k - L(t, q, \dot{q})$, we find

$$\begin{aligned}\partial H/\partial p_k &= dq_k/dt + \sum p_l(\partial \dot{q}_l/\partial p_k) - \sum (\partial L/\partial \dot{q}_l)(\partial \dot{q}_l/\partial p_k)\\ &= dq_k/dt + \sum (p_l - \partial L/\partial \dot{q}_l)(\partial \dot{q}_l/\partial p_k) = dq_k/dt + 0 = dq_k/dt.\end{aligned}$$

Hence, we obtained the "reciprocal" relationships

$$\partial L/\partial \dot{q}_k = p_k \qquad \text{and} \qquad \partial H/\partial p_k = dq_k/dt. \tag{a}$$

(ii) By $d(\ldots)$-varying $L + H - \sum p_k\, \dot{q}_k = 0$ (= *function of* $t, q, \dot{q}, p$), we find

$$\begin{aligned}0 = dL + dH - \sum d(p_k\, \dot{q}_k) = \sum [&(\partial L/\partial q_k + \partial H/\partial q_k)\, dq_k + (\partial L/\partial \dot{q}_k - p_k)\, d(\dot{q}_k)\\ &+ (\partial H/\partial p_k - dq_k/dt)\, dp_k] + (\partial L/\partial t + \partial H/\partial t)\, dt,\end{aligned}$$

and from this, due to (a) and the arbitrariness of the dq's and dt, we obtain the remaining Hamiltonian identities:

$$\partial L/\partial q_k = -\partial H/\partial q_k \qquad \text{and} \qquad \partial L/\partial t = -\partial H/\partial t. \tag{b}$$

Example 8.2.3 *Still Another Direct Derivation of Hamilton's Equations.* By $(\partial/\partial q_k)$- and $(\partial/\partial p_k)$-differentiating the invariant equation

$$L_{(qp)} \equiv L(t, q, p) = L(t, q, \dot{q}) \equiv L_{(q\dot{q})}, \tag{a}$$

we obtain, respectively,

$$\begin{aligned}\partial L_{(qp)}/\partial q_k &= \partial L_{(q\dot{q})}/\partial q_k + \sum (\partial L_{(q\dot{q})}/\partial \dot{q}_l)(\partial \dot{q}_l/\partial q_k)\\ &= \partial L_{(q\dot{q})}/\partial q_k + \sum p_l(\partial \dot{q}_l/\partial q_k) = \partial L_{(q\dot{q})}/\partial q_k + \partial/\partial q_k\Big(\sum p_l\, \dot{q}_l\Big),\end{aligned} \tag{b}$$

(since q and p are treated as independent)

$$\begin{aligned}\partial L_{(qp)}/\partial p_k &= \sum (\partial L_{(q\dot{q})}/\partial \dot{q}_l)(\partial \dot{q}_l/\partial p_k) = \sum p_l(\partial \dot{q}_l/\partial p_k)\\ &= \partial/\partial p_k\Big(\sum p_l\, \dot{q}_l\Big) - \dot{q}_k,\end{aligned} \tag{c}$$

from which, rearranging (collecting the q, p functions on the left sides and the rest on the right), we obtain the kinematico-inertial identities:

$$\partial/\partial q_k\left(L_{(qp)} - \sum p_l\,\dot{q}_l\right) = \partial L_{(q\dot{q})}/\partial q_k, \tag{d}$$

$$\partial/\partial p_k\left(L_{(qp)} - \sum p_l\,\dot{q}_l\right) = -dq_k/dt. \tag{e}$$

Finally, (i) introducing the Hamiltonian $H(t,q,p) \equiv \sum p_l\,\dot{q}_l(t,q,p) - L(t,q,p)$, and (ii) expressing in (d) $\partial L_{(q\dot{q})}/\partial q_k$ via Lagrange's equations, say $dp_k/dt = \partial L_{(q\dot{q})}/\partial q_k$, we readily obtain the corresponding canonical equations: $dp_k/dt = -\partial H/\partial q_k$ and $dq_k/dt = \partial H/\partial p_k$.

Example 8.2.4 *A Lagrangean Derivation of Hamilton's Equations.* We consider an additional (fictitious) system with $2n$ Lagrangean *coordinates*: $q \equiv (q_1, \ldots, q_n)$ *and* $p \equiv (p_1, \ldots, p_n)$, and Lagrangean function

$$L(t,q,p;\dot{q},\dot{p}) \equiv \sum p_k\dot{q}_k - H(t,q,p); \tag{a}$$

that is, we solve the Hamiltonian definition for the Lagrangean. Now:

(i) The Lagrangean equations for the n coordinates q are, say,

$$0 = (\partial L/\partial\dot{q}_k)^{\cdot} - \partial L/\partial q_k = dp_k/dt - (-\partial H/\partial q_k), \qquad \text{or} \qquad dp_k/dt = -\partial H/\partial q_k; \tag{b}$$

while

(ii) The Lagrangean equations for the n "coordinates" p are

$$0 = (\partial L/\partial\dot{p}_k)^{\cdot} - \partial L/\partial p_k = (0)^{\cdot} - (dq_k/dt - \partial H/\partial p_k), \qquad \text{or} \qquad dq_k/dt = \partial H/\partial p_k. \tag{c}$$

And similarly for systems with more general Lagrangean equations.

Problem 8.2.1 Let $2T = \sum\sum M_{kl}\dot{q}_k\dot{q}_l$ (= *homogeneous* quadratic in the n $\dot{q}$), in which case

$$p_k \equiv \partial T/\partial\dot{q}_k = \sum M_{kl}\dot{q}_l \qquad [\text{assume that: } M_n \equiv Det\,(M_{kl}) \neq 0], \tag{a}$$

and, therefore, by inversion,

$\dot{q}_l = \sum M'_{lk}\,p_k$ (= velocities due to given impulses applied to system, when at rest in a given configuration; hence the name *coefficients of mobility* for the M'_{lk}),

where

$$\begin{aligned} M'_{lk} &\equiv (\text{minor of element } M_{kl} \text{ in determinant } M_n)/M_n \\ &= (1/M_n)(\partial M_n/\partial M_{kl}). \end{aligned} \tag{b}$$

Show that, then,

$$2T' = \sum p_k \dot{q}_k = \sum\sum M'_{kl} p_k p_l = \sum\sum [(1/M_n)(\partial M_n/\partial M_{lk})] p_k p_l$$

$$\Big[= \sum (\partial T/\partial \dot{q}_k)\, \dot{q}_k = 2T \;\Rightarrow\; T' + T = \sum p_k \dot{q}_k,$$

$$\text{i.e., } T \text{ in the } \dot{q}\text{'s} = T' \text{ in the } p\text{'s: } T(t, q, \dot{q}) = T'(t, q, p)\Big]; \tag{c}$$

and, conjugately to (a),

$$dq_k/dt = \partial T'/\partial p_k. \tag{d}$$

Also, show that:

(i) $$\partial p_k/\partial \dot{q}_l = \partial p_l/\partial \dot{q}_k = \cdots \quad \text{and} \quad \partial \dot{q}_k/\partial p_l = \partial \dot{q}_l/\partial p_k = \cdots. \tag{e}$$

(ii) $$M_n M'_n = 1, \quad \text{where } M'_n \equiv Det(M'_{kl}) \quad (\neq 0). \tag{f}$$

Problem 8.2.2 Continuing from prob. 8.2.1, let

$$2T = M_{11}(\dot{q}_1)^2 + 2M_{12}\dot{q}_1\dot{q}_2 + M_{22}(\dot{q}_2)^2; \tag{a}$$

that is, $n = 2$. Show that, then,

$$2T' = M'_{11} p_1{}^2 + 2M'_{12} p_1 p_2 + M'_{22} p_2{}^2, \tag{b}$$

where

$$M'_{11} = M_{22}/M, \qquad M'_{22} = M_{11}/M, \qquad M'_{12} = -M_{12}/M,$$
$$M \equiv M_2 \equiv Det\,(M_{kl}) = M_{11}M_{22} - M_{12}{}^2; \tag{c}$$

and thus verify directly the Hamiltonian identities

$$dq_1/dt = \partial T'/\partial p_1, \quad dq_2/dt = \partial T'/\partial p_2,$$

and

$$\partial T/\partial q_1 = -\partial T'/\partial q_1, \quad \partial T/\partial q_2 = -\partial T'/\partial q_2. \tag{d}$$

Problem 8.2.3 *Reciprocal Theorem.* Consider the following *two* states of motion of a scleronomic mechanical system through the *same configuration*:

$$\textit{State } 1\ (q_k)\text{: } \dot{q}_k,\ p_k = \sum M_{kl}\dot{q}_l, \qquad \textit{State } 2\ (q_k)\text{: } \dot{q}'_k,\ p'_k = \sum M_{kl}\dot{q}'_l. \tag{a}$$

Show that

$$\sum p_k \dot{q}'_k = \sum p'_k \dot{q}_k. \tag{b}$$

For further details and applications to impulsive motion, see, for example, Lamb (1929, pp. 184–187, 206); also, the discussion in ex. 4.6.8, and Rayleigh (1894, pp. 91 ff.).

Example 8.2.5 Let us derive the Hamiltonian equations of a spherical pendulum, of mass m and length l. With the $+Oz$ axis taken vertically downward (recall prob. 3.5.16, fig. 3.8), and since $x = (l \sin\theta)\cos\phi$, $y = (l\sin\theta)\sin\phi$, $z = l\cos\theta$, we readily find

$$2T = m[(\dot{x})^2 + (\dot{y})^2 + (\dot{z})^2] = \cdots = ml^2[(\dot{\theta})^2 + \sin^2\theta(\dot{\phi})^2] \quad (= 2T_2), \tag{a}$$

$$V = -mgl\cos\theta \quad (= V_0,\ V = 0 \text{ on plane } z = 0), \tag{b}$$

and, therefore,

$$p_\phi \equiv \partial T/\partial\dot{\phi} = (ml^2\sin^2\theta)\dot{\phi}$$
$$(= \textit{angular momentum about the vertical axis through the origin}),$$
$$p_\theta \equiv \partial T/\partial\dot{\theta} = (ml^2)\dot{\theta}$$
$$(= \textit{angular momentum about horizontal, and perpendicular to instantaneous meridian plane axis through the origin}). \tag{c}$$

Inverting (c) we immediately obtain the *second* set of the canonical equations:

$$d\phi/dt = p_\phi/ml^2\sin^2\theta \quad (= \partial H/\partial p_\phi), \qquad d\theta/dt = p_\theta/ml^2 \quad (= \partial H/\partial p_\theta). \tag{d}$$

Hence, the Hamiltonian of the system is

$$H = p_\phi\dot{\phi} + p_\theta\dot{\theta} - (T - V) = 2T - (T - V) = T + V = T(t, q, p) + V$$
$$= (1/2\,ml^2)\,[(p_\phi{}^2/\sin^2\theta) + p_\theta{}^2] - mgl\cos\theta = H(\phi, \theta, p_\phi, p_\theta), \tag{e}$$

and accordingly the *first* set of its canonical equations (of motion) are

$$dp_\phi/dt = -\partial H/\partial\phi\colon \qquad dp_\phi/dt = 0 \ \Rightarrow\ p_\phi = \textit{constant} \equiv c, \tag{f}$$

or, thanks to the first of (d),

$$d\phi/dt = c/ml^2\sin^2\theta, \tag{g}$$

$$dp_\theta/dt = -\partial H/\partial\theta\colon\ dp_\theta/dt = (\cos\theta/ml^2\sin^3\theta)p_\phi{}^2 - mgl\sin\theta; \tag{h}$$

or, thanks to (f, g),

$$dp_\theta/dt = \cos\theta\, c^2/ml^2\sin^3\theta - mgl\sin\theta \quad [= (ml^2)\ddot{\theta}]. \tag{i}$$

To integrate (i), we multiply both its sides with $p_\theta/\dot{\theta} = (dt/d\theta)p_\theta = (dt/d\theta)(ml^2\dot{\theta}) = ml^2$:

$$p_\theta(dp_\theta/d\theta) = (\cos\theta/\sin^3\theta)\, c^2 - m^2gl^3\sin\theta,$$

and from this, by a θ-integration, we obtain the energy integral $[H = T(t, q, p) + V = \textit{constant}]$:

$$p_\theta{}^2/2 = (-1/\sin^2\theta)(c^2/2) + m^2gl^3\cos\theta + \textit{constant}. \tag{j}$$

Of course, since here $Q_{\phi,\theta} = 0$ and $\partial H/\partial t = 0$, by (8.2.14, 14a) the integral (j) could have been written down immediately.

Problem 8.2.4 Show that the conjugate kinetic energy T' of a particle of mass m moving in space equals

$$2mT' = p_x{}^2 + p_y{}^2 + p_z{}^2 \qquad \text{(rectangular Cartesian coordinates)} \qquad \text{(a)}$$

$$= p_r{}^2 + (p_\phi/r)^2 + p_z{}^2 \qquad \text{(cylindrical coordinates)} \qquad \text{(b)}$$

$$= p_r{}^2 + (p_\theta/r)^2 + (p_\phi/r\sin\theta)^2 \qquad \text{(spherical coordinates).} \qquad \text{(c)}$$

HINT

Recall that (prob. 3.5.15)

$$2T/m = (\dot{x})^2 + (\dot{y})^2 + (\dot{z})^2 = (\dot{r})^2 + (r\dot{\phi})^2 + (\dot{z})^2 = (\dot{r})^2 + (r\dot{\theta})^2 + (r\sin\theta\,\dot{\phi})^2.$$

[Caution: $r_{\text{cylindrical coordinates}}$ (*2nd expression*) $\neq r_{\text{spherical coordinates}}$ (*3rd expression*).]

Problem 8.2.5 Show that the Hamiltonian of a particle of mass m moving on a uniformly rotating frame of reference (of constant inertial angular velocity $\boldsymbol{\Omega}$ is, with the usual notations,

$$H = p^2/2m - \boldsymbol{p}\cdot(\boldsymbol{\Omega}\times\boldsymbol{r}) + V(\boldsymbol{r}) = H(t,\ \boldsymbol{r},\ \boldsymbol{p})$$

$$= v_{\text{relative}}{}^2/2m - m(\boldsymbol{\Omega}\times\boldsymbol{r})^2/2 + V(\boldsymbol{r}) = H(t,\ \boldsymbol{r},\ \boldsymbol{v}_{\text{relative}}). \qquad \text{(a)}$$

HINT

$$L = v_{\text{relative}}{}^2/2m + m\,\boldsymbol{v}_{\text{relative}}\cdot(\boldsymbol{\Omega}\times\boldsymbol{r}) + m(\boldsymbol{\Omega}\times\boldsymbol{r})^2/2 - V(\boldsymbol{r}), \qquad \text{(b)}$$

$$H \equiv (\partial L/\partial \boldsymbol{v}_{\text{relative}})\cdot\boldsymbol{v}_{\text{relative}} - L = \cdots, \qquad \boldsymbol{p} = \partial L/\partial \boldsymbol{v}_{\text{relative}} \Rightarrow \boldsymbol{v}_{\text{relative}} = \cdots.$$

[Recall problems of Lagrangean treatment of particles in uniformly rotating turntables (§3.16).]

Problem 8.2.6 *Unified Treatment of Auxiliary Forms of Lagrange's Inertia Terms.* Let

$$2T = \sum\sum M_{kl}(q)\dot{q}_k\dot{q}_l = 2T(q,\dot{q}) \equiv 2T_{\dot{q}\dot{q}}$$

$$= \sum p_k\dot{q}_k = 2T(\dot{q},p) \equiv 2T_{\dot{q}p} \equiv 2T_{p\dot{q}}$$

$$= \sum\sum M'_{kl}(q)p_k p_l = 2T(q,p) \equiv 2T_{pp} \quad (= 2T_{(qp)} \equiv 2T'), \qquad \text{(a)}$$

with all Latin indices ranging from 1 to n, and where, as we have seen,

$$p_k = \partial T_{\dot{q}\dot{q}}/\partial \dot{q}_k \qquad \text{and} \qquad \dot{q}_k = \partial T_{pp}/\partial p_k. \qquad \text{(b)}$$

Show that:

(i)
$$2T_{\dot{q}\dot{q}} = \sum (\partial T_{\dot{q}\dot{q}}/\partial \dot{q}_k)\dot{q}_k, \tag{c}$$

$$2T_{p\dot{q}} = \sum (\partial T_{\dot{q}\dot{q}}/\partial \dot{q}_k)(\partial T_{pp}/\partial p_k), \tag{d}$$

$$2T_{pp} = \sum (\partial T_{pp}/\partial p_k)p_k; \tag{e}$$

and

(ii)
$$\partial T_{pp}/\partial q_k = -\partial T_{\dot{q}\dot{q}}/\partial q_k, \tag{f}$$

$$\partial T_{\dot{q}p}/\partial \dot{q}_k = (1/2)(\partial T_{\dot{q}\dot{q}}/\partial \dot{q}_k) = (1/2)p_k, \tag{g}$$

$$\partial T_{\dot{q}p}/\partial p_k = (1/2)(\partial T_{pp}/\partial p_k) = (1/2)\dot{q}_k; \tag{h}$$

and, therefore,

$$\begin{aligned} E_k(T) &\equiv (\partial T/\partial \dot{q}_k)^{\cdot} - \partial T/\partial q_k \equiv (\partial T_{\dot{q}\dot{q}}/\partial \dot{q}_k)^{\cdot} - \partial T_{\dot{q}\dot{q}}/\partial q_k \\ &\equiv dp_k/dt - \partial T_{\dot{q}\dot{q}}/\partial q_k \qquad \textit{(Lagrange)} \\ &= (\partial T_{\dot{q}\dot{q}}/\partial \dot{q}_k)^{\cdot} + \partial T_{pp}/\partial q_k \equiv dp_k/dt + \partial T_{pp}/\partial q_k \qquad \textit{(Hamilton)} \\ &= 2(\partial T_{\dot{q}p}/\partial \dot{q}_k)^{\cdot} - \partial T_{\dot{q}\dot{q}}/\partial q_k \\ &= 2(\partial T_{\dot{q}p}/\partial \dot{q}_k)^{\cdot} + \partial T_{pp}/\partial q_k. \end{aligned} \tag{i}$$

HINT

First, using (a–e), verify that

$$T_{pp} = -T_{\dot{q}\dot{q}} + 2T_{\dot{q}p} = -T_{\dot{q}\dot{q}} + \sum p_k\,\dot{q}_k \;\Rightarrow\; T_{pp} + T_{\dot{q}\dot{q}} - 2T_{\dot{q}p} = 0, \tag{j}$$

then differentiate the above totally, and then equate the coefficients of its differentials to zero.

[See Weinstein (1901, pp. 95–97, 186–189) (earliest publication: 1882); also Budde (1890, Vol. 1, pp. 397–401), and Watson and Burbury (1879, pp. 14–22). Such "mixed" equations have been used by Maxwell et al. in electromechanical investigations; see, for example, Maxwell (1877 and 1920, pp. 127–136, 158–161).]

Problem 8.2.7 As a simple application of the preceding problem, consider a particle of mass m described by spherical polar coordinates. In this case, and with the usual notations,

$$2T = m[(\dot{r})^2 + (r\dot{\theta})^2 + (r\sin\theta\,\dot{\phi})^2] = 2T_{\dot{q}\dot{q}}. \tag{a}$$

Show that:

(i)
$$p_r = m\,\dot{r}, \qquad p_\theta = m\,r^2\dot{\theta}, \qquad p_\phi = m\,r^2\sin^2\theta\dot{\phi}; \tag{b}$$

(ii)
$$2T_{\dot{q}p} = \dot{r}\,p_r + \dot{\theta}\,p_\theta + \dot{\phi}\,p_\phi; \tag{c}$$

(iii)
$$2T_{pp} = (1/m)\,[p_r{}^2 + (p_\theta/r)^2 + (p_\phi/r\sin\theta)^2]. \tag{d}$$

Hence, verify that, for example,

$$\partial T_{\dot q\dot q}/\partial\dot\theta = 2(\partial T_{\dot q p}/\partial\dot\theta) = \cdots, \tag{e}$$

$$\partial T_{\dot q\dot q}/\partial\theta = -\partial T_{pp}/\partial\theta = \cdots, \tag{f}$$

$$\partial T_{pp}/\partial p_\theta = 2(\partial T_{\dot q p}/\partial p_\theta) = \cdots. \tag{g}$$

Problem 8.2.8 Let the general solution of Hamilton's equations be

$$q_k = q_k(t;\ c_1, \ldots, c_{2n}) \equiv q_k(t;\ c), \qquad p_k = p_k(t;\ c_1, \ldots, c_{2n}) \equiv p_k(t;\ c), \tag{a}$$

where $c \equiv (c_1, \ldots, c_{2n}) = 2n$ constants of integration; and we assume that the Jacobian of the q's and p's relative to the c's nowhere vanishes. Then,

$$H = H(t, q, p) = H[t,\ q(t;\ c),\ p(t;\ c)] \equiv H(t,\ c). \tag{b}$$

(i) Show that

$$\partial H/\partial c_\alpha = d/dt\Big[\sum (\partial p_k/\partial c_\alpha) q_k\Big] - \partial/\partial c_\alpha\Big(\sum \dot p_k\, q_k\Big) \qquad (\alpha = 1, \ldots, 2n). \tag{c}$$

(ii) Show that eqs. (c) are equivalent to the Hamiltonian equations:

$$dq_k/dt = \partial H/\partial p_k, \qquad dp_k/dt = -\partial H/\partial q_k; \tag{d}$$

that is, if (d) hold, so do (c); and vice versa.

HINT

Note the identity

$$d/dt\Big[\sum (\partial p_k/\partial c_\alpha) q_k\Big] - \partial/\partial c_\alpha\Big(\sum \dot p_k\, q_k\Big) = \sum\ [(\partial p_k/\partial c_\alpha)\dot q_k - (\partial q_k/\partial c_\alpha)\dot p_k].$$

Problem 8.2.9 Consider the linear differential form (Pfaffian form) in the variables $x = (x_1, \ldots, x_n)$:

$$df \equiv \sum X_k\,(x)\, dx_k. \tag{a}$$

By definition, its *bilinear covariant* is (recall §2.8)

$$\begin{aligned} d_2(d_1 f) - d_1(d_2 f) &\equiv \sum (d_2 X_k\, d_1 x_k - d_1 X_k\, d_2 x_k) \\ &= \cdots = \sum\sum (\partial X_k/\partial x_l - \partial X_l/\partial x_k)\, d_2 x_l\, d_1 x_k. \end{aligned} \tag{b}$$

Show that *Hamilton's equations, as well as the power theorem (for* $Q_k = 0$), result from the vanishing of the bilinear covariant of

$$dA \equiv \sum p_k\, dq_k - H\, dt$$

$$\Big[= \sum p_\alpha dq_\alpha \quad (\alpha = 1, \ldots, n+1);\ \text{with}\ p_{n+1} \equiv -H\ \text{and}\ q_{n+1} \equiv t\Big], \tag{c}$$

under arbitrary variations $d_1(\ldots)$ and $d_2(\ldots)$; that is, from

$$\begin{aligned} 0 &= d_2(d_1 A) - d_1(d_2 A) \\ &= d_2\left(\sum p_k d_1 q_k - H\, d_1 t\right) - d_1\left(\sum p_k\, d_2 q_k - H\, d_2 t\right) \\ &= \sum (d_2 p_k\, d_1 q_k - d_1 p_k\, d_2 q_k) - (d_2 H\, d_1 t - d_1 H\, d_2 t). \end{aligned} \quad \text{(d)}$$

Problem 8.2.10 Consider a potential (but possibly rheonomic) system described by n Lagrangean coordinates.

(i) Define the Hamiltonian-like function

$$H' \equiv \sum (p_k q_k)^{\cdot} - L = \cdots = \sum \dot{p}_k q_k + H = H'(t,\, p,\, \dot{p}). \quad \text{(a)}$$

Show that the corresponding equations of motion are

$$q_k = \partial H'/\partial \dot{p}_k \qquad \text{and} \qquad dq_k/dt = \partial H'/\partial p_k \qquad [\Rightarrow (\partial H'/\partial \dot{p}_k)^{\cdot} - \partial H'/\partial p_k = 0]. \quad \text{(b)}$$

(ii) Similarly, define the Hamiltonian-like function

$$H'' \equiv \sum \dot{p}_k q_k - L = \cdots = \sum (\dot{p}_k q_k - p_k \dot{q}_k) + H = H''(t,\, \dot{q},\, \dot{p}). \quad \text{(c)}$$

Show that the corresponding equations of motion are

$$q_k = \partial H''/\partial \dot{p}_k \qquad \text{and} \qquad p_k = -\partial H''/\partial \dot{q}_k. \quad \text{(d)}$$

(iii) Discuss possible theoretical and practical advantages/disadvantages of (b) and (d) over the equations of Hamilton; and verify that $L + H - H' + H'' = 0$.

Problem 8.2.11 Show, by means of general considerations and/or concrete examples, that *the 2n first-order Hamiltonian equations of a problem, in general, do not coincide with the first-order equations (or state-space) version of its n second-order Lagrangean equations.* (See, e.g., Tabor, 1989, p. 51.)

8.3 THE ROUTHIAN CENTRAL EQUATION AND ROUTH'S EQUATIONS OF MOTION

The method of Routh (1877) constitutes an ingenious combination of the methods of Hamilton and Lagrange that results in two sets of equations of motion:

(i) One *Hamiltonian-like* for t and, say the first M q's and corresponding p's, to be henceforth denoted for notational clarity by ψ's and Ψ's, respectively; that is,

$$(q_1, \ldots, q_M) \equiv (\psi_1, \ldots, \psi_M) \equiv (\psi_i) \equiv \psi, \quad \text{(8.3.1a)}$$

and

$$(p_1, \ldots, p_M) \equiv (\Psi_1, \ldots, \Psi_M) \equiv (\Psi_i) \equiv \Psi; \quad \text{(8.3.1b)}$$

and

(ii) One *Lagrange-like* for t and the remaining $n - M$ q's and corresponding $\dot{q}$'s, to be henceforth denoted by q's and $\dot{q}$'s, respectively; that is,

$$(q_{M+1}, \ldots, q_n) \equiv (q_p) \equiv q \qquad \text{and} \qquad (\dot{q}_{M+1}, \ldots, \dot{q}_n) \equiv (\dot{q}_p) \equiv \dot{q}, \tag{8.3.2a, b}$$

where the subscripts $\boldsymbol{i}$ and $\boldsymbol{p}$ stand for *ignorable* (or *cyclic*) and *palpable* (or *positional*, or *essential*), respectively. Such a *mixed* approach combines the best of both Hamilton and Lagrange, and proves particularly useful in problems of *latent* (or *cyclic*) and *steady* motions. (All these terms/concepts are detailed in the following sections.) With this idea in mind, we begin by rewriting the most general Lagrangean expression for the kinetic energy of a system as follows:

$$\begin{aligned} T &= T(t, q_1, \ldots, q_n; \dot{q}_1, \ldots, \dot{q}_n) \\ &= T(t; q_1, \ldots, q_M; q_{M+1}, \ldots, q_n; \dot{q}_1, \ldots, \dot{q}_M; \dot{q}_{M+1}, \ldots, \dot{q}_n) \\ &\equiv T(t; \psi_1, \ldots, \psi_M; q_{M+1}, \ldots, q_n; \dot{\psi}_1, \ldots, \dot{\psi}_M; \dot{q}_{M+1}, \ldots, \dot{q}_n) \\ &\equiv T(t, \psi, q; \dot{\psi}, \dot{q}), \end{aligned} \tag{8.3.3a}$$

and, in there, replace $\dot{\psi} \equiv (\dot{\psi}_1, \ldots, \dot{\psi}_M)$ in terms of their momenta $\Psi \equiv (\Psi_1, \ldots, \Psi_M)$, and so on, à la Hamilton; that is, $\dot{\psi}_i = \dot{\psi}_i(t, \psi, q; \Psi, \dot{q})$. The result is

$$\begin{aligned} T &= T[t, \psi, q; \dot{\psi}(t, \psi, q; \Psi, \dot{q}), \dot{q}] \\ &= T(t; \psi_1, \ldots, \psi_M; q_{M+1}, \ldots, q_n; \Psi_1, \ldots, \Psi_M; \dot{q}_{M+1}, \ldots, \dot{q}_n) \\ &= T(t, \psi, q; \Psi, \dot{q}) \equiv T_{\psi\Psi}. \end{aligned} \tag{8.3.3b}$$

Next, with the help of the new (conjugate-like, to within a minus sign) function

$$\begin{aligned} T'' &\equiv T - \sum \Psi_i \dot{\psi}_i = \Big(T - \sum \Psi_i \dot{\psi}_i\Big)_{\dot{\psi} = \dot{\psi}(t; \psi, q; \Psi, \dot{q})} \\ &= T''(t, \psi, q; \Psi, \dot{q}), \end{aligned} \tag{8.3.3c}$$

we transform the fundamental central equation (8.2.1; with $k = 1, \ldots, n$)

$$\delta I = \delta' W: \qquad d/dt\Big(\sum p_k\, \delta q_k\Big) - \delta T = \sum Q_k\, \delta q_k, \tag{8.3.4a}$$

or, [assuming that $(\delta q_k)^{\cdot} = \delta(\dot{q}_k)$]

$$\sum \big[\dot{p}_k\, \delta q_k + p_k\, \delta(\dot{q}_k)\big] - \delta T = \sum Q_k\, \delta q_k, \tag{8.3.4b}$$

and since $\delta T = \delta T_{\psi\Psi}$, to

$$\sum \big[\dot{p}_k\, \delta q_k + p_k\, \delta(\dot{q}_k)\big] - \delta T'' - \sum \big[\delta\Psi_i\, \dot{\psi}_i + \Psi_i\, \delta(\dot{\psi}_i)\big] = \sum Q_k\, \delta q_k; \tag{8.3.4c}$$

or, carrying out the $\delta T''$-variation (with $i = 1, \ldots, M$; $p = M + 1, \ldots, n$):

$$\begin{aligned} &\sum \big[\dot{p}_k\, \delta q_k + p_k\, \delta(\dot{q}_k)\big] - \sum \big[(\partial T''/\partial\psi_i)\, \delta\psi_i + (\partial T''/\partial\Psi_i)\, \delta\Psi_i\big] \\ &\quad - \sum \big[(\partial T''/\partial q_p)\, \delta q_p + (\partial T''/\partial \dot{q}_p)\, \delta(\dot{q}_p)\big] - \sum \big[\dot{\psi}_i\, \delta\Psi_i + \Psi_i\, \delta(\dot{\psi}_i)\big] = \sum Q_k\, \delta q_k, \end{aligned} \tag{8.3.4d}$$

or, collecting δq, $\delta\psi$, $\delta(\dot{q})$, and $\delta\Psi$-proportional terms [while recalling that $q_i \equiv \psi_i \Rightarrow \delta q_i \equiv \delta\psi_i$, $\delta(\dot{q}_i) \equiv \delta(\dot{\psi}_i)$, and $p_i \equiv \Psi_i$], we finally obtain the *Routhian central equation*:

$$\sum(dp_k/dt - \partial T''/\partial q_k - Q_k)\,\delta q_k + \sum(p_p - \partial T''/\partial\dot{q}_p)\,\delta(\dot{q}_p) - \sum(d\psi_i/dt + \partial T''/\partial\Psi_i)\,\delta\Psi_i = 0, \tag{8.3.4e}$$

which holds for *all* δq_k, $\delta(\dot{q}_p)$, and $\delta\Psi_i$ (again, with $k = 1, \ldots, n$; $p = M+1, \ldots, n$; $i = 1, \ldots, M$) and, as expected, is fundamental to all subsequent developments.

Now, as with (8.2.3):

1. If the δq_k, $\delta(\dot{q}_p)$, and $\delta\Psi_i$ are mutually *independent*, then (8.3.4e) leads immediately to *the equations of Routh* [1877—not to be confused with the earlier Routh–Voss equations (§3.5)!]:

(i) $dp_k/dt = \partial T''/\partial q_k + Q_k$: $\quad d\Psi_i/dt = \partial T''/\partial\psi_i + Q_i \quad (i = 1, \ldots, M)$, (8.3.5a)

$$dp_p/dt = \partial T''/\partial q_p + Q_p \quad (p = M+1, \ldots, n); \tag{8.3.5b}$$

(ii)
$$d\psi_i/dt = -\partial T''/\partial\Psi_i \quad (i = 1, \ldots, M), \tag{8.3.5c}$$

(iii)
$$p_p = \partial T''/\partial\dot{q}_p \quad (p = M+1, \ldots, n). \tag{8.3.5d}$$

Of these equations (8.3.5a, b) are *kinetic* (i.e., equations of motion), whereas (8.3.5c, d) are *kinematico-inertial identities*. The Hamilton-like equations (8.3.5a, c) (with $-T''$ playing the role of our earlier T'), are Routh's equations for ψ and Ψ; while the Lagrange-like equations (8.3.5b, d) (with T'' playing the role of T) are Routh's equations for q and $\dot{q}$; that is, rearranging:

Hamilton-like Routh equations:

$$d\Psi_i/dt = -\partial(-T'')/\partial\psi_i + Q_i, \qquad d\psi_i/dt = \partial(-T'')/\partial\Psi_i; \tag{8.3.6a}$$

Lagrange-like Routh equations:

$$dp_p/dt = \partial T''/\partial q_p + Q_p, \qquad p_p = \partial T''/\partial\dot{q}_p \quad (-\partial T/\partial\dot{q}_p)$$
$$\Rightarrow (\partial T''/\partial\dot{q}_p)^{\cdot} - \partial T''/\partial q_p = Q_p. \tag{8.3.6b}$$

- If $M = 0$—that is, if no velocities $\dot{\psi}_i$ are eliminated through $\dot{\psi}_i(t, \psi, q; \Psi, \dot{q})$—then the group of equations (8.3.6a) drops; while the group (8.3.6b) coincide with Lagrange's equations for T (since then $T'' = T$).
- If, on the other hand, $M = n$—that is, if all $\dot{\psi}_i$ are eliminated through $\dot{\psi}_i(t, \psi, q; \Psi, \dot{q})$—then the situation is reversed: group (8.3.6b) drops (since then $T'' = -T'$), while group (8.3.6a) coincides with Hamilton's equations. Schematically,

$$M: \quad 0\ (\textit{Lagrange}) \leftarrow \cdots \text{———} \cdots \rightarrow n\ (\textit{Hamilton}).$$

Finally, comparing the above with the corresponding Lagrangean equations $dp_k/dt = \partial T/\partial q_k + Q_k$, we immediately obtain the additional Routhian *kinematico-inertial identities*

$\partial T/\partial q_k = \partial T''/\partial q_k$: $\quad \partial T/\partial\psi_i = \partial T''/\partial\psi_i \quad (i = 1, \ldots, M)$, (8.3.7a)

$$\partial T/\partial q_p = \partial T''/\partial q_p \quad (p = M+1, \ldots, n); \tag{8.3.7b}$$

which can be utilized in any set of Hamiltonian- or Lagrangean-type equations of motion.

In sum, we have the following *two groups* of such kinematico-inertial identities:

$$\partial T''/\partial\psi_i = \partial T/\partial\psi_i \quad \text{and} \quad \partial T''/\partial\Psi_i = -d\psi_i/dt; \tag{8.3.8a}$$

$$\partial T''/\partial q_p = \partial T/\partial q_p \quad \text{and} \quad \partial T''/\partial\dot{q}_p = \partial T/\partial\dot{q}_p \quad (= p_p); \tag{8.3.8b}$$

see also examples on direct derivations, below.

If $p_k \equiv \partial L/\partial\dot{q}_k$, then (8.3.6a, b) are replaced, respectively, by the Routhian equations:

Hamilton-like Routh equations:

$$d\Psi_i/dt = \partial R/\partial\psi_i + Q_i, \qquad d\psi_i/dt - -\partial R/\partial\Psi_i; \tag{8.3.9a}$$

Lagrange-like Routh equations:

$$dp_p/dt = \partial R/\partial q_p + Q_p, \qquad p_p = \partial R/\partial\dot{q}_p \quad (= \partial L/\partial\dot{q}_p)$$
$$\Rightarrow E_p(R) \equiv (\partial R/\partial\dot{q}_p)^{\cdot} - \partial R/\partial q_p = Q_p; \tag{8.3.9b}$$

where

$$R \equiv \Big(L - \sum \Psi_i\dot{\psi}_i\Big)_{\dot{\psi}=\dot{\psi}(t;\psi,q;\Psi,\dot{q})} = R(t;\ \psi,\ q;\ \Psi,\ \dot{q}):$$

Routhian function, or *Modified Lagrangean*, (8.3.9c)

and

$$L = L(t;\ \psi,\ q;\ \Psi,\ \dot{q}) \equiv T_{\psi\Psi} - V \equiv L_{\psi\Psi}:$$

Lagrangean expressed in Routhian variables; (8.3.9d)

and the *nonpotential* forces Q_k have also been expressed in the Routhian variables t, ψ, q, Ψ, $\dot{q}$; while (8.3.7a, b) and (8.3.8a, b) are replaced, respectively, by

$$\partial L/\partial\psi_i = \partial R/\partial\psi_i \quad \text{and} \quad \partial L/\partial q_p = \partial R/\partial q_p; \tag{8.3.9e}$$

and

$$\partial R/\partial\psi_i = \partial L/\partial\psi_i \quad \text{and} \quad \partial R/\partial\Psi_i = -d\psi_i/dt; \tag{8.3.9f}$$

$$\partial R/\partial q_p = \partial L/\partial q_p \quad \text{and} \quad \partial R/\partial\dot{q}_p = \partial L/\partial\dot{q}_p \quad (= p_p); \tag{8.3.9g}$$

that is, *the Routhian is a Hamiltonian [times* (-1)*] for the* ψ_i, *and a Lagrangean for the* q_p.

HISTORICAL

The Routhian was also introduced, independently, by Helmholtz (in 1884), who called it "new kinetic potential"; that is, (negative of) new Lagrangean = (negative of) Routhian. See, for example, Helmholtz [1898, pp. 361–369, eq. (200a)], Webster (1912, pp. 176–179).

To find the relation between the Routhian and the Hamiltonian (or generalized energy), we proceed as follows:

$$
\begin{aligned}
H &\equiv \sum p_k \dot{q}_k - L \\
&= \sum \dot{\psi}_i (\partial L/\partial \dot{\psi}_i) + \sum \dot{q}_p (\partial L/\partial \dot{q}_p) - L \qquad \text{[invoking (8.3.9g)]} \\
&= \sum \dot{q}_p (\partial R/\partial \dot{q}_p) - \Big(L - \sum \dot{\psi}_i \Psi_i\Big) \qquad \text{[invoking (8.3.9c)]} \\
&= \sum \dot{q}_p (\partial R/\partial \dot{q}_p) - R,
\end{aligned}
$$

from which we immediately conclude that

$$
\begin{aligned}
&R = \sum \dot{q}_p (\partial R/\partial \dot{q}_p) - H = \sum p_p \dot{q}_p - H \\
&\Big[-\Big(H - \sum p_p \dot{q}_p\Big) = L - \sum \Psi_i \dot{\psi}_i\text{: equivalent definitions of the Routhian}\Big].
\end{aligned}
\tag{8.3.10}
$$

REMARK

Some authors define the Routhian as the *negative* of ours; that is, as

$$R \equiv \sum \dot{\psi}_i (\partial L/\partial \dot{\psi}_i) - L \equiv \sum \Psi_i \dot{\psi}_i - L.$$

In such a case, all the above results hold intact, but with R replaced with $-R$; then (8.3.9a) look exactly like Hamiltonian equations for the ψ's, Ψ's. That definition would be "Hamiltonian" in spirit—that is, *Routhian* = *Hamiltonian* $- \sum p_p \dot{q}_p$; ours, being closer to engineering, is "Lagrangean"—that is, *Routhian* = *Lagrangean* $- \sum \Psi_i \dot{\psi}_i$.

2. If the n δq_k are restricted by the m Pfaffian constraints

$$\delta\theta_D \equiv \sum a_{Dk}\, \delta q_k = 0 \qquad [rank(a_{Dk}) = m;\ D = 1, \ldots, m(< n)], \tag{8.3.11a}$$

while the $n - M$ $\delta(\dot{q}_p)$ and M $\delta\Psi_i$ are still viewed as independent, then application of the method of Lagrangean multipliers to the Routhian central equation (8.3.4e) readily yields the *constrained Routhian equations*:

$$dp_k/dt = \partial T''/\partial q_k + Q_k + \sum \lambda_D\, a_{Dk},$$

or, split in two groups (assuming that $m < M$ and $m < n - M$):

$$d\Psi_i/dt = \partial T''/\partial \psi_i + Q_i + \sum \lambda_D\, a_{Di}, \qquad dp_p/dt = \partial T''/\partial q_p + Q_p + \sum \lambda_D\, a_{Dp}, \tag{8.3.11b}$$

and the earlier kinematico-inertial identities

$$d\psi_i/dt = -\partial T''/\partial \Psi_i, \qquad p_p = \partial T''/\partial \dot{q}_p. \tag{8.3.11c}$$

The formulation of the above in terms of the Routhian, whenever the impressed forces are partly or wholly potential, does not offer any difficulties and will be left to the reader.

3. For an extension of these results to *quasi variables*, see, for example, Chetaev (1989, pp. 339–346).

Example 8.3.1 *Direct Derivation of the Routhian Kinematico-inertial Identities.* By $d(\ldots)$-differentiating $T'' \equiv (T - \sum \Psi_i \dot{\psi}_i) = T''(t, \psi, q; \Psi, \dot{q})$, we find

$$\begin{aligned} dT'' &= (\partial T/\partial t)\, dt + \sum \left[(\partial T/\partial q_k)\, dq_k + (\partial T/\partial \dot{q}_k)\, d\dot{q}_k\right] \\ &\quad - \sum \left[d\Psi_i \dot{\psi}_i + \Psi_i\, d\dot{\psi}_i\right] \\ &= \cdots = (\partial T/\partial t)\, dt + \sum \left[(\partial T/\partial \psi_i)\, d\psi_i - \dot{\psi}_i\, d\Psi_i\right] \\ &\qquad + \sum \left[(\partial T/\partial q_p)\, dq_p + (\partial T/\partial \dot{q}_p)\, d\dot{q}_p\right]. \qquad \text{(a)} \end{aligned}$$

But also, since $T'' = T''(t, \psi, q; \Psi, \dot{q})$, we will have

$$\begin{aligned} dT'' &= (\partial T''/\partial t)\, dt + \sum \left[(\partial T''/\partial \psi_i)\, d\psi_i + (\partial T''/\partial \Psi_i)\, d\Psi_i\right] \\ &\qquad + \sum \left[(\partial T''/\partial q_p)\, dq_p + (\partial T''/\partial \dot{q}_p)\, d\dot{q}_p\right]. \qquad \text{(b)} \end{aligned}$$

Therefore, equating the coefficients of these two general dT'' expressions, (a) and (b), since the differentials involved are arbitrary, we immediately obtain the following $1 + 2M + 2(n - M) = 2n + 1$ Routhian kinematico-inertial identities:

$$\partial T''/\partial t = \partial T/\partial t, \qquad \text{(c)}$$

$$\partial T''/\partial \psi_i = \partial T/\partial \psi_i, \qquad \partial T''/\partial \Psi_i = -\dot{\psi}_i; \qquad \text{(d)}$$

$$\partial T''/\partial q_p = \partial T/\partial q_p, \qquad \partial T''/\partial \dot{q}_p = \partial T/\partial \dot{q}_p \quad (= p_p). \qquad \text{(e)}$$

Let us resummarize our findings:

(i) Equations (c), first of (d), and first of (e) are fundamentally equivalent, if we think of time as the $(n + 1)$th Lagrangean coordinate: $q_{n+1} \equiv t$;
(ii) The first of eqs. (d) are, essentially, Hamiltonian in nature, while the first of eqs. (e) are Lagrangean; and both sets are used in the corresponding kinetic equations;
(iii) The second of eqs. (d) are the Hamiltonian counterpart of the Lagrangean identity $\partial T/\partial \dot{\psi}_i = \Psi_i$ (with T'' replaced with $-T''$); while
(iv) The second of eqs. (e) are purely Lagrangean.

Repeating this procedure for the Routhian $R = L - \sum \Psi_i \dot{\psi}_i$, with $p_k = \partial L/\partial \dot{q}_k$, we similarly obtain the following identities:

$$\partial R/\partial t = \partial L/\partial t, \qquad \text{(f)}$$

$$\partial R/\partial \psi_i = \partial L/\partial \psi_i, \qquad \partial R/\partial \Psi_i = -d\psi_i/dt; \qquad \text{(g)}$$

$$\partial R/\partial q_p = \partial L/\partial q_p, \qquad \partial R/\partial \dot{q}_p = \partial L/\partial \dot{q}_p \quad (= p_p); \qquad \text{(h)}$$

or, compactly,

$$\partial R/\partial q_k = \partial L/\partial q_k \qquad (k = 1, \ldots, n \textit{ and } n + 1), \qquad \text{(i)}$$

$$\partial R/\partial \dot{q}_p = \partial L/\partial \dot{q}_p \qquad (p = M + 1, \ldots, n), \qquad \text{(j)}$$

$$\partial R/\partial \Psi_i = -d\psi_i/dt \qquad (i = 1, \ldots, M). \qquad \text{(k)}$$

Last, since by (8.2.10b), $\partial H/\partial t = -\partial L/\partial t$, eq. (f) shows that if a Lagrangean is explicitly independent of time, then so are the corresponding Routhian and Hamiltonian; and for potential systems [since, then, $dH/dt = \partial H/\partial t$, by (8.2.14)], the latter is also a constant.

Example 8.3.2 *Another Direct Derivation of the Routhian Kinematico-inertial Identities.* Applying chain rule, carefully, to the Routhian definition $R = L - \sum \Psi_i \dot{\psi}_i$, where

$$R = R(t, \psi, q; \Psi, \dot{q}), \qquad L = L(t, \psi, q; \dot{\psi}, \dot{q}), \qquad \text{and} \qquad \dot{\psi}_i = \dot{\psi}_i(t, \psi, q; \Psi, \dot{q}), \quad \text{(a)}$$

we obtain (with $i, j = 1, \ldots, M;\ p = M + 1, \ldots, n$):

$$\text{(i)}\quad \partial R/\partial t = \left[\partial L/\partial t + \sum (\partial L/\partial \dot{\psi}_i)(\partial \dot{\psi}_i/\partial t)\right] - \sum \Psi_i(\partial \dot{\psi}_i/\partial t) = \partial L/\partial t; \qquad \text{(b)}$$

$$\text{(ii)}\quad \partial R/\partial q_p = \left[\partial L/\partial q_p + \sum (\partial L/\partial \dot{\psi}_i)(\partial \dot{\psi}_i/\partial q_p)\right] - \sum \Psi_i(\partial \dot{\psi}_i/\partial q_p) = \partial L/\partial q_p; \qquad \text{(c)}$$

$$\text{(iii)}\quad \partial R/\partial \dot{q}_p = \left[\partial L/\partial \dot{q}_p + \sum (\partial L/\partial \dot{\psi}_i)(\partial \dot{\psi}_i/\partial \dot{q}_p)\right] - \sum \Psi_i(\partial \dot{\psi}_i/\partial \dot{q}_p) = \partial L/\partial \dot{q}_p; \qquad \text{(d)}$$

$$\text{(iv)}\quad \partial R/\partial \psi_i = \left[\partial L/\partial \psi_i + \sum (\partial L/\partial \dot{\psi}_j)(\partial \dot{\psi}_j/\partial \psi_i)\right] - \sum \Psi_j(\partial \dot{\psi}_j/\partial \psi_i) = \partial L/\partial \psi_i; \qquad \text{(e)}$$

$$\text{(v)}\quad \partial R/\partial \Psi_i = \sum (\partial L/\partial \dot{\psi}_j)(\partial \dot{\psi}_j/\partial \Psi_i) - \left[\dot{\psi}_i + \sum \Psi_j(\partial \dot{\psi}_j/\partial \Psi_i)\right] = -\dot{\psi}_i; \qquad \text{(f)}$$

which are indeed the earlier Routhian identities.

Clearly, the method of the preceding example (*total differentials*) seems simpler and safer than this one (*derivatives*), as in the earlier derivations of the Hamiltonian equations (exs. 8.2.1–3).

Analytical Structure of the Routhian

Let us consider (with no loss of generality, just algebraic simplicity) a *scleronomic* system whose kinetic energy T is, therefore, a *homogeneous* quadratic function in its chosen n Lagrangean velocities $\dot{q}_k$'s ($k = 1, \ldots, n$). First, we decompose T into the following *three* parts:

$$T = T_{\dot{q}\dot{q}} + T_{\dot{q}\dot{\psi}} + T_{\dot{\psi}\dot{\psi}} = T(\psi, q; \dot{\psi}, \dot{q}), \qquad (8.3.12)$$

where

$$2T_{\dot{q}\dot{q}} \equiv \sum\sum a_{pq}\dot{q}_p\dot{q}_q = \textit{homogeneous quadratic in the } \dot{q}\text{'s}$$
$$(a_{pq} = a_{qp};\ \textit{positive definite}),$$

$$T_{\dot{q}\dot{\psi}} \equiv \sum\sum b_{pi}\dot{q}_p\dot{\psi}_i = \textit{homogeneous bilinear in the } \dot{q}\text{'s and } \dot{\psi}\text{'s}$$
$$(\textit{in general, } b_{pi} \neq b_{ip}\textit{; sign indefinite}), \qquad (8.3.12b)$$

$$2T_{\dot{\psi}\dot{\psi}} \equiv \sum\sum c_{ij}\dot{\psi}_i\dot{\psi}_j = \textit{homogeneous quadratic in the } \dot{\psi}\text{'s}$$
$$(c_{ij} = c_{ji}\textit{; positive definite}, \text{ otherwise the } \textit{cyclic kinetic energy } T_{\dot{\psi}\dot{\psi}} \text{ would not be positive definite}); \qquad (8.3.12c)$$

with $i, j = 1, \ldots, M$; $p, q = M + 1, \ldots, n$; and the coefficients are functions of all n q_k's. [It is not hard to see that, in the most general case, $T_{\dot{\psi}\dot{\psi}}$, $T_{\dot{q}\dot{\psi}}$, and $T_{\dot{q}\dot{q}}$ contain, respectively (with $g \equiv n - m$), $M + (\frac{1}{2})M(M-1)$, Mg, and $g + (\frac{1}{2})g(g-1)$ terms; that is, a total of $(\frac{1}{2})(M+g)(M+g+1) = (\frac{1}{2})n(n+1)$ terms, as expected.]

Now, solving the linear system

$$\Psi_i \equiv \partial T/\partial\dot{\psi}_i = \sum c_{ji}\dot{\psi}_j + \sum b_{pi}\dot{q}_p \ \Rightarrow\ \sum c_{ji}\dot{\psi}_j = \Psi_i - \sum b_{pi}\dot{q}_p, \qquad (8.3.12d)$$

for the $\dot{\psi}_j$, via Cramer's rule [i.e., inverting (8.3.12d), which, since (c_{ji}) is *nonsingular*, is possible], we obtain

$$\dot{\psi}_j = \sum C_{ji}\Big(\Psi_i - \sum b_{pi}\dot{q}_p\Big), \qquad (8.3.12e)$$

where

$$C_{ji} = [\textit{cofactor of element } c_{ji} \textit{ in } Det(c_{ji})]/Det(c_{ji}) = C_{ij}$$
$$(= \textit{known function of the } q\text{'s and } \psi\text{'s});$$

and then substituting these expressions for the $\dot{\psi}_j$ into (8.3.12–8.3.12c), and using well-known properties of inverse matrices, we obtain

$$T = T_{2,0} + T_{0,2} = T(\psi, q; \Psi, \dot{q}) \quad (\equiv T_{\dot{q}\Psi}), \qquad (8.3.12f)$$

where

$$2T_{2,0} \equiv \sum\sum\Big(a_{pq} - \sum\sum C_{ji}b_{pj}b_{qi}\Big)\dot{q}_p\dot{q}_q, \qquad (8.3.12g)$$

$$2T_{0,2} \equiv \sum\sum C_{ji}\Psi_j\Psi_i; \qquad (8.3.12h)$$

that is, $T = T(\psi, q; \Psi, \dot{q})$ does *not* contain any bilinear terms in the $\dot{q}$'s and Ψ's!

With the help of the above, the *modified kinetic energy* T'' becomes, successively,

$$T'' \equiv T - \sum \Psi_i\dot{\psi}_i = T - \sum \Psi_i\Big[\sum C_{ij}\Big(\Psi_j - \sum b_{pj}\dot{q}_p\Big)\Big]$$
$$= T_{2,0} + T''_{1,1} - T_{0,2} \equiv T''_{2,0} + T''_{1,1} + T''_{0,2}$$
$$= T''(\psi, q; \Psi, \dot{q}), \qquad (8.3.12i)$$

where

$$2T''_{2,0} \equiv \sum\sum\Big(a_{pq} - \sum\sum C_{ji}b_{pj}b_{qi}\Big)\dot{q}_p\dot{q}_q \equiv \sum\sum r_{pq}(q)\dot{q}_p\dot{q}_q$$
$$= 2T_{2,0} \quad (= \textit{positive definite in the } \dot{q}\text{'s}), \qquad (8.3.12j)$$

$$T''_{1,1} \equiv \sum\sum\left(\sum C_{ji}b_{pi}\right)\Psi_j\dot{q}_p \equiv \sum r_p(q, \Psi)\dot{q}_p,$$

[*No counterpart in* $T = T(\psi, q; \Psi, \dot{q})$; i.e., $T_{1,1} = 0$; *sign indefinite*], (8.3.12k)

$$2T''_{0,2} \equiv -\sum\sum C_{ji}\Psi_j\Psi_i = 2T''_{0,2}(q, \Psi)$$
$$= -2T_{0,2} \quad (= \textit{negative definite in the } \Psi\text{'s}). \tag{8.3.12l}$$

Conversely, if $T'' = T''_{2,0} + T''_{1,1} + T''_{0,2}$, then (by Routh's identities and the homogeneous function theorem),

$$T \equiv T'' + \sum \Psi_i\dot{\psi}_i = T'' - \sum \Psi_i(\partial T''/\partial \Psi_i)$$
$$= (T''_{2,0} + T''_{1,1} + T''_{0,2}) - (T''_{1,1} + 2T''_{0,2})$$
$$= T''_{2,0} - T''_{0,2} = T(\psi, q; \Psi, \dot{q}) \qquad \text{[Compare this with (8.3.12f) and (8.3.12)].}$$
(8.3.12m)

In view of these results, the Lagrangean and Routhian assume, respectively, the following forms:

$$L = T - V = (T_{2,0} + T_{0,2}) - V = T_{2,0} - (V - T_{0,2})$$
$$= (T''_{2,0} - T''_{0,2}) - V = T''_{2,0} - (V + T''_{0,2}) = L(\psi, q; \Psi, \dot{q}), \tag{8.3.13}$$

$$R = L - \sum \Psi_i\dot{\psi}_i = L + \sum(\partial T''/\partial \Psi_i)\Psi_i$$
$$= (T''_{2,0} - T''_{0,2} - V) + (2T''_{0,2} + T''_{1,1})$$
$$\equiv R_2 + R_1 + R_0 = R(\psi, q; \Psi, \dot{q}), \tag{8.3.14}$$

where

$$R_2 \equiv T''_{2,0} = T_{2,0}, \qquad R_1 \equiv T''_{1,1}, \qquad R_0 \equiv T''_{0,2} - V = -T_{0,2} - V. \tag{8.3.14a}$$

These remarkable identities seem to be due to Routh (also, Kelvin and Helmholtz), and are very useful in the theory of *cyclic* systems (§8.4).

[For additional explicit expressions of the Routhian, and so on, see, for example: Gantmacher [1970, pp. 242–252 (§48)], Lur'e [1968, pp. 340–351 (§7.15–7.17); also contains the *decomposition into Routhian variables in the general nonstationary/rheonomic T case*], Merkin (1974, pp. 24–36), Routh [1877 and 1975, pp. 63–64, 93–94; 1905(b), pp. 341–342], Winkelmann and Grammel (1927, pp. 470–474); also Easthope (1964, pp. 382–383), Grammel (1950, Vol. 1, pp. 255–258), Heun (1914, pp. 454–457).]

Problem 8.3.1 Continuing from the above, let

$$T = T_{\dot{q}\dot{q}} + T_{\dot{q}\dot{\psi}} + T_{\dot{\psi}\dot{\psi}} = T(\psi, q; \dot{\psi}, \dot{q}). \tag{a}$$

Using the homogeneous function theorem, show that its modified kinetic energy

$$T'' = T - \sum \Psi_i\dot{\psi}_i = T - \sum(\partial T/\partial\dot{\psi}_i)\dot{\psi}_i, \tag{b}$$

equals

$$T'' = T_{\dot{q}\dot{q}} - T_{\dot{\psi}\dot{\psi}} = T''(\psi, q; \dot{\psi}, \dot{q}); \tag{c}$$

that is, $T'' = T''(\psi, q; \dot{\psi}, \dot{q})$ does *not* contain any bilinear terms in the variables $\dot{q}$ and $\dot{\psi}$!

Problem 8.3.2 Continuing from the above, show that Routh's nonkinetic equations (i.e., his kinematico-inertial identities) transform further to

$$d\psi_i/dt = -\partial T''/\partial \Psi_i = \cdots = \partial T_{0,2}/\partial \Psi_i - \partial K_{2,2}/\partial \pi_i, \tag{a}$$

where

$$2T_{0,2} = -2T''_{0,2} \equiv \sum\sum C_{ji}\Psi_j\Psi_i, \tag{b}$$

and

$$2K_{2,2} \equiv \sum\sum C_{ji}\left(\sum b_{pj}\dot{q}_p\right)\left(\sum b_{qi}\dot{q}_q\right) \equiv \sum\sum C_{ji}\pi_j\pi_i. \tag{c}$$

Example 8.3.3 (May be omitted in a first reading). Here, we carry out a *matrix derivation* of the above results on the structure of the Routhian, for the benefit of those more comfortable with that currently popular notation.

With the notations $(\ldots)^{\mathrm{T}} \equiv$ *transpose* of matrix $(\ldots)$, $(\ldots)^{-1} \equiv$ *inverse* of matrix $(\ldots)$, and

$$\dot{\mathbf{q}}^{\mathrm{T}} = (\dot{q}_{M+1}, \ldots, \dot{q}_n), \qquad \dot{\boldsymbol{\psi}}^{\mathrm{T}} = (\dot{\psi}_1, \ldots, \dot{\psi}_M), \qquad \boldsymbol{\Psi}^{\mathrm{T}} = (\Psi_1, \ldots, \Psi_M), \tag{a}$$

$$\mathbf{a} = (a_{pq}) = (a_{qp}) = \mathbf{a}^{\mathrm{T}}, \qquad \mathbf{b} = (b_{ip}) \neq (b_{pi}) = \mathbf{b}^{\mathrm{T}}, \qquad \mathbf{c} = (c_{ij}) = (c_{ji}) = \mathbf{c}^{\mathrm{T}}, \tag{b}$$

we have the following correspondences with the earlier indicial equations:

(8.3.12–8.3.12c): $$2T = \dot{\mathbf{q}}^{\mathrm{T}}\mathbf{a}\dot{\mathbf{q}} + 2\dot{\boldsymbol{\psi}}^{\mathrm{T}}\mathbf{b}^{\mathrm{T}}\dot{\mathbf{q}} + \dot{\boldsymbol{\psi}}^{\mathrm{T}}\mathbf{c}\dot{\boldsymbol{\psi}}, \tag{c}$$

(8.3.12d): $$\partial T/\partial\dot{\boldsymbol{\psi}} = \mathbf{b}^{\mathrm{T}}\dot{\mathbf{q}} + \mathbf{c}\dot{\boldsymbol{\psi}} = \boldsymbol{\Psi}, \tag{d}$$

(8.3.12e): $$\dot{\boldsymbol{\psi}} = \mathbf{c}^{-1}(\boldsymbol{\Psi} - \mathbf{b}^{\mathrm{T}}\dot{\mathbf{q}}) \equiv \mathbf{C}(\boldsymbol{\Psi} - \mathbf{b}^{\mathrm{T}}\dot{\mathbf{q}}) \ \Rightarrow\ \dot{\boldsymbol{\psi}}^{\mathrm{T}} = (\boldsymbol{\Psi}^{\mathrm{T}} - \dot{\mathbf{q}}^{\mathrm{T}}\mathbf{b})\mathbf{C}, \tag{e}$$

[since $\mathbf{c}$ is symmetric, so is its inverse $\mathbf{C} \equiv (C_{ji})$: $\mathbf{C} \equiv \mathbf{c}^{-1} = (\mathbf{c}^{-1})^{\mathrm{T}} \equiv \mathbf{C}^{\mathrm{T}}$]

(8.3.12f–h): $$\begin{aligned} T &= \ldots \text{[since } \boldsymbol{\Psi}^{\mathrm{T}}\,\mathbf{C}\,\mathbf{b}^{\mathrm{T}}\dot{\mathbf{q}} = \dot{\mathbf{q}}^{\mathrm{T}}\mathbf{b}\mathbf{C}\boldsymbol{\Psi} \\ &\qquad \text{(easily proved by indicial notation)]} \\ &= (1/2)\dot{\mathbf{q}}^{\mathrm{T}}(\mathbf{a} - \mathbf{b}\,\mathbf{C}\,\mathbf{b}^{\mathrm{T}})\dot{\mathbf{q}} + (1/2)\boldsymbol{\Psi}^{\mathrm{T}}\mathbf{C}\boldsymbol{\Psi} \\ &\equiv T_{2,0} + T_{0,2} = T''_{2,0} - T''_{0,2}, \end{aligned} \tag{f}$$

(8.3.12i–8.3.14a): $$\boldsymbol{\Psi}^{\mathrm{T}}\dot{\boldsymbol{\psi}} = \cdots = \boldsymbol{\Psi}^{\mathrm{T}}\mathbf{C}\boldsymbol{\Psi} - \boldsymbol{\Psi}^{\mathrm{T}}\mathbf{C}\,\mathbf{b}^{\mathrm{T}}\dot{\mathbf{q}} = -2T''_{0,2} - \boldsymbol{\Psi}^{\mathrm{T}}\mathbf{C}\,\mathbf{b}^{\mathrm{T}}\dot{\mathbf{q}}, \tag{g}$$

$$R = (T - V) - \boldsymbol{\Psi}^{\mathrm{T}}\dot{\boldsymbol{\psi}} = \cdots = R_2 + R_1 + R_0, \tag{h}$$

$$R_2 \equiv T''_{2,0} = T_{2,0} = (1/2)\dot{\mathbf{q}}^{\mathrm{T}}(\mathbf{a} - \mathbf{b}\,\mathbf{C}\,\mathbf{b}^{\mathrm{T}})\dot{\mathbf{q}}, \tag{h1}$$

$$R_1 \equiv T''_{1,1} = \mathbf{\Psi}^{\mathrm{T}}\mathbf{C}\,\mathbf{b}^{\mathrm{T}}\dot{\mathbf{q}}, \tag{h2}$$

$$R_0 \equiv T''_{0,2} - V = -(V + T_{0,2}) = -(1/2)\mathbf{\Psi}^{\mathrm{T}}\mathbf{C}\mathbf{\Psi} - V. \tag{h3}$$

If $\mathbf{b} = \mathbf{0}$ — that is, if the $\dot{q}$'s and $\dot{\psi}$'s are *uncoupled* in the original T, eq. (c) — then R reduces to

$$R = (1/2)\dot{\mathbf{q}}^{\mathrm{T}}\mathbf{a}\dot{\mathbf{q}} - (1/2)\mathbf{\Psi}^{\mathrm{T}}\mathbf{C}\mathbf{\Psi} - V. \tag{h4}$$

For an extension of the above to general *nonstationary* systems, see, for example, Otterbein (1981, pp. 31–35).

8.4 CYCLIC SYSTEMS; EQUATIONS OF KELVIN–TAIT

Let us begin with the holonomic, possibly rheonomic, system whose configurations are determined by the n Lagrangean coordinates $q_1, \ldots, q_n$. The system will be called *cyclic*, or *gyrostatic*, if the following conditions apply:

(i) A number of these coordinates, say (as in §8.3) the *first* M ($\leq n$):

$$(q_1, \ldots, q_M) \equiv (\psi_1, \ldots, \psi_M) \equiv (\psi_i) \equiv \psi, \tag{8.4.1a}$$

do *not* appear explicitly in either its kinetic energy or its impressed forces; only the corresponding Lagrangean velocities

$$(\dot{q}_1, \ldots, \dot{q}_M) \equiv (\dot{\psi}_1, \ldots, \dot{\psi}_M) \equiv (\dot{\psi}_i) \equiv \dot{\psi} \tag{8.4.1b}$$

appear there, and, of course time t and the remaining coordinates and/or velocities

$$(q_{M+1}, \ldots, q_n) \equiv (q_p) \equiv q \quad \text{and} \quad (\dot{q}_{M+1}, \ldots, \dot{q}_n) \equiv (\dot{q}_p) \equiv \dot{q}, \tag{8.4.1c}$$

respectively; that is,

$$\partial T/\partial\psi_i = 0 \quad \text{but, in general,} \quad \partial T/\partial\dot{\psi}_i \neq 0 \;\Rightarrow\; T = T(t; q, \dot{\psi}, \dot{q}). \tag{8.4.2a}$$

(ii) The corresponding impressed forces vanish; that is,

$$Q_i = 0, \quad \text{but} \quad Q_p = Q_p(q) \neq 0. \tag{8.4.2b}$$

If all impressed forces are wholly *potential*, then the above requirements are replaced, respectively, by

$$\partial L/\partial\psi_i = 0 \quad \text{and} \quad \partial L/\partial\dot{\psi}_i \neq 0 \;\Rightarrow\; L = L(t; q, \dot{\psi}, \dot{q}). \tag{8.4.2c}$$

The coordinates ψ, and corresponding velocities $\dot{\psi}$, are called *cyclic* (Helmholtz), or *absent* (Routh), or *kinosthenic*, or *speed* (J. J. Thomson), or *ignorable* (Whittaker); for example, the angular coordinates of flywheels of frictionless gyrostats, included in a system of bodies ("housings"), relative to their housings, are such cyclic coordinates. [The term *ignorable* seems, in general, more appropriate since such coordinates may occur in *nonspinning* systems; e.g., the kinetic energy of a translating rigid body contains only the $(\ldots)^{\cdot}$-derivatives of the coordinates of its center of mass, but not these coordinates themselves.]

The remaining coordinates q, and corresponding velocities $\dot{q}$, are called *palpable*, or *positional*, since in many problems they are the only ones directly visible, or manifest; for example, the angle of nutation of a spinning gyroscope.

Below, we apply Routh's method and relations (§8.3) to obtain equations of motion for such cyclic systems, in terms of their positional coordinates alone. Thanks to (8.4.2a–b), the Lagrangean equations corresponding to the cyclic coordinates/variables, become

$$(\partial T/\partial\dot{\psi}_i)^{\cdot} - \partial T/\partial\psi_i = Q_i: \qquad (\partial T/\partial\dot{\psi}_i)^{\cdot} = 0 \Rightarrow \partial T/\partial\dot{\psi}_i \equiv \Psi_i = constant \equiv C_i; \tag{8.4.3}$$

that is, the momenta Ψ_i corresponding to the cyclic coordinates ψ_i are *constants of the motion*. [Conversely, however, if $\partial T/\partial\dot{\psi}_i = 0$, then $\partial T/\partial\psi_i = 0$, and as a result $T = T(t; q, \dot{q})$; that is, the evolution of the ψ's does not affect that of the q's at all!] Therefore, by §8.3, the Routhian of a cyclic system is a function of $t, q, \dot{q}$ and $\Psi \equiv (\Psi_i)$; indeed, by (8.3.9c) and with $C \equiv (C_i)$,

$$R \equiv \left(L - \sum \Psi_i \dot{\psi}_i\right)_{\dot{\psi}=\dot{\psi}(t;q;\dot{q};C)}$$

[after solving the linear equations (8.4.3) for the $\dot{\psi}$ in terms of $t, q, \dot{q}, C$]

$$= L[t, q, \dot{q}, \dot{\psi}(t; q, \dot{q}; C); C] - \sum \Psi_i \dot{\psi}_i(t; q, \dot{q}; C)$$

$$= R(t; q, \dot{q}; C)$$

$$\left[\Rightarrow L = \sum C_i \dot{\psi}_i(t; q, \dot{q}; C) + R(t; q, \dot{q}; C)\right]; \tag{8.4.4}$$

which shows that, since the ψ have been completely eliminated (or ignored), our system has been reduced to one with only $n - M$ Lagrangean coordinates, new reduced Lagrangean R, and therefore, Lagrange-type Routhian equations for the positional coordinates (8.3.9b):

$$(\partial R/\partial\dot{q}_p)^{\cdot} - \partial R/\partial q_p = Q_p, \tag{8.4.5}$$

where the Q_p are *nonpotential impressed positional forces*. [As Kilmister and Reeve aptly put it (our notation): "we may *in* R put $\Psi_i = C_i$ before differentiation and thus consider the motion of the subsystem (q_p) *conjugate* to the ignorable system (ψ_i)" (1966, p. 294).]

Solving these equations, we obtain the palpable motion $q_p(t)$. Then, as (8.4.4) shows,

$$R = known\ function\ of\ time$$
$$\Rightarrow \partial R/\partial C_i = known\ function\ of\ time \equiv -f_i(t; C), \tag{8.4.6}$$

from which, since $d\psi_i/dt = -(\partial R/\partial\Psi_i)$,

$$\psi_i = -\int (\partial R/\partial\Psi_i)\, dt + constant = \int f_i\,(t; C)\, dt + constant$$

$$= \psi_i(t, C) + constant; \tag{8.4.6a}$$

that is, the problem has been reduced to the $n - M$ *equations* (8.4.5) and the M *quadratures* (8.4.6a); or, equivalently [since every ignorable coordinate generates two integrals (§3.12)], the order of the system has been reduced by $2M$. [Since $d\psi_i/dt = \partial H/\partial \Psi_i$, similar results hold in terms of the Hamiltonian of cyclic systems; see, for example, McCuskey (1959, p. 208 ff.).]

REMARK

Kilmister (1964, pp. 43, 46) and others have suggested an alternative handling of cyclic systems via Hamel's method of quasi variables and equations (chaps. 2 and 3). According to this method, we choose the following quasi velocities:

$$\omega_i \equiv \partial L/\partial \dot{\psi}_i - C_i \quad (= 0) \qquad (i = 1, \ldots, M), \tag{8.4.7a}$$

$$\omega_p \equiv \dot{q}_p \quad (\neq 0) \qquad (p = M + 1, \ldots, n). \tag{8.4.7b}$$

The resulting $n - M$ *kinetic* equations (Hamel $\to$ *noncyclic* Routhian) plus the M "cyclicity" constraints (8.4.7a) constitute a determinate system of n equations for the n velocities $(\dot{\psi}_i, \dot{q}_p)$. After solving these equations, we can then proceed to the M *kinetostatic* equations (Hamel $\to$ *cyclic* Routhian) and determine the reaction forces associated with these constraints. For a rare implementation of these ideas, see, for example, Vujanovic (1970); also, ex. 8.4.2, below.

Example 8.4.1 *Routhian Method in Problem of Central Motion.* Let us consider a particle P, of mass m, in plane motion under a *radial* force. Here,

$$2T = m[(\dot{r})^2 + (r\,\dot{\phi})^2] \qquad (r,\ \phi\text{: inertial plane polar coordinates}), \tag{a}$$

$$Q_r = Q_r(r) \qquad \text{and} \qquad Q_\phi = 0. \tag{b}$$

From the obvious ignorability of ϕ, we obtain the (area) integral

$$p_\phi \equiv \partial T/\partial \dot{\phi} = m\,r^2\,\dot{\phi} \equiv \Psi_\phi = constant \equiv m\,C$$
$$\Rightarrow\ \dot{\phi} = C/r^2 \quad (\neq constant). \tag{c}$$

Hence, the modified kinetic energy equals

$$T'' = \left[T - \dot{\phi}(\partial T/\partial \dot{\phi})\right]_{\dot{\phi} = \dot{\phi}(r;\,C)} = (m/2)\ [(\dot{r})^2 + C^2/r^2] - m(C^2/r^2)$$
$$= (m/2)\ [(\dot{r})^2 - C^2/r^2] = T''(r, \dot{r}; C), \tag{d}$$

and so the Lagrangean equation of motion of the nonignorable coordinate r is

$$(\partial T''/\partial \dot{r})^{\cdot} - \partial T''/\partial r = Q_r\text{:} \qquad m(\ddot{r} - C^2/r^3) = Q_r. \tag{e}$$

Multiplying (e) with $\dot{r}$, and then integrating, we easily obtain the *energy integral*

$$(m/2)\ \left[(\dot{r})^2 + C^2/r^2\right] = \int Q_r\, dr + constant. \tag{f}$$

Example 8.4.2 *Direct Elimination of Ignorable Coordinates from the Lagrangean Equations of Motion; and Some General Theoretical Conclusions.* Let us consider, with no loss of generality, a holonomic, scleronomic, and potential system with M *ignorable* coordinates $\psi \equiv (\psi_1, \ldots, \psi_M)$, and $n - M$ *nonignorable*, or *positional*, coordinates $q \equiv (q_{M+1}, \ldots, q_n)$, so that $L = L(q, \dot{q}, \dot{\psi})$. Below, we show, quite generally and with no recourse to Routh's method, that the ψ's can be *eliminated* from the $n - M$ Lagrangean equations for the q's. For concreteness, let us take *two* ignorable coordinates, ψ_1, ψ_2, and *two* positional coordinates, q_3, q_4; that is, $M = 2$, $n - M = 4 - 2 = 2$. Then we will have, with the usual notations,

$$\begin{aligned} T = {} & (1/2)(M_{33}\dot{q}_3\dot{q}_3 + M_{44}\dot{q}_4\dot{q}_4 + 2M_{34}\dot{q}_3\dot{q}_4) \\ & + (M_{13}\dot{q}_3 + M_{14}\dot{q}_4)\dot{\psi}_1 + (M_{23}\dot{q}_3 + M_{24}\dot{q}_4)\dot{\psi}_2 \\ & + (1/2)(M_{11}\dot{\psi}_1\dot{\psi}_1 + M_{22}\dot{\psi}_2\dot{\psi}_2 + 2M_{12}\dot{\psi}_1\dot{\psi}_2), \end{aligned} \tag{a}$$

where the inertia coefficients $M_{kl} = M_{lk}(k, l = 1, \ldots, 4)$ and the potential energy V are functions of $q_{3,4}$ only. From the above it follows easily that:

(i) Lagrange's equations for the q's are (with $p = 3, 4$):

$$\begin{aligned} & (M_{p3}\ddot{q}_3 + M_{p4}\ddot{q}_4 + M_{p1}\ddot{\psi}_1 + M_{p2}\ddot{\psi}_2) \\ & \quad + [(\partial M_{p3}/\partial q_3)\,\dot{q}_3 + (\partial M_{p3}/\partial q_4)\,\dot{q}_4]\,\dot{q}_3 \\ & \quad + [(\partial M_{p4}/\partial q_3)\,\dot{q}_3 + (\partial M_{p4}/\partial q_4)\,\dot{q}_4]\,\dot{q}_4 \\ & \quad + [(\partial M_{p1}/\partial q_3)\,\dot{q}_3 + (\partial M_{p1}/\partial q_4)\,\dot{q}_4]\,\dot{\psi}_1 \\ & \quad + [(\partial M_{p2}/\partial q_3)\,\dot{q}_3 + (\partial M_{p2}/\partial q_4)\,\dot{q}_4]\,\dot{\psi}_2 = -\,\partial V/\partial q_p; \end{aligned} \tag{b}$$

(ii) Lagrange's equations for the ψ's are (with $i = 1, 2$)

$$(\partial L/\partial\dot{\psi}_i)^{\cdot} = 0 \;\Rightarrow\; \Psi_i \equiv \partial L/\partial\dot{\psi}_i = \textit{constant} \equiv C_i, \tag{c}$$

or, using (a) and rearranging,

$$M_{11}\dot{\psi}_1 + M_{12}\dot{\psi}_2 = C_1 - (M_{13}\dot{q}_3 + M_{14}\dot{q}_4), \tag{c1}$$

$$M_{21}\dot{\psi}_1 + M_{22}\dot{\psi}_2 = C_2 - (M_{23}\dot{q}_3 + M_{24}\dot{q}_4). \tag{c2}$$

Now, solving the system (c1, 2), we obtain $\dot{\psi}_1$ and $\dot{\psi}_2$ in terms of C_1, C_2; q_3, q_4; $\dot{q}_3, \dot{q}_4$; and, then, $(\ldots)^{\cdot}$-differentiating these expressions we obtain $\ddot{\psi}_1$ and $\ddot{\psi}_2$ in terms of the same variables and their $(\ldots)^{\cdot}$-derivatives. [Here, $\dot{C}_i = 0$, but this procedure applies, in principle, to noncyclic systems too.]

Next, substituting the so-found expressions for $\dot{\psi}_1$, $\dot{\psi}_2$; $\ddot{\psi}_1$, $\ddot{\psi}_2$ into eqs. (b), we obtain, finally, two Lagrangean equations containing only q_3 and q_4 and their $(\ldots)^{\cdot}$-derivatives; that is, as far as the equations of motion are concerned, ψ_1 and ψ_2 have been "ignored"—the system has been reduced to one with only $n - M = 2$ Lagrangean coordinates. Solving these two nonignorable equations, we find the *palpable motion* $q_3(t)$ and $q_4(t)$; and, then, substituting these solutions back into (c1, 2), and integrating, we obtain the *cyclic motion* $\psi_1(t)$ and $\psi_2(t)$. As one might expect, finding $q_{3,4}(t)$ is, in general, considerably harder than finding $\psi_{1,2}(t)$.

General Conclusions

(i) The difference between this approach and the earlier general Routhian methodology is that here we eliminated the $\dot{\psi}$ and $\ddot{\psi}$ from each of the q-equations of motion; whereas there (§8.3) this elimination was done in one step, right at the beginning—that is, by replacing the Lagrangean with the Routhian. For few-degree-of-freedom systems, the two approaches are practically equivalent, but for larger systems, as well as for theoretical arguments and insights, the general Routhian approach is much preferable.

This is analytically identical with the difference between the following:

(a) Enforcing Pfaffian (and generally nonholonomic) constraints, like (c, c1, 2), not in L but in each Lagrangean equation of motion; and
(b) Enforcing such constraints directly in L; that is, replacing the "relaxed" Lagrangean L with the "constrained" one L_o or L^* (chap. 3), and then applying it to "modified" equations of motion.

Actually, *Routh's method modifies the Lagrangean (replaces it with the Routhian, which incorporates the constraints), and then applies it to ordinary Lagrangean equations of motion.* In sum, in such approaches: *Either we still operate with the Lagrangean ($L \to L_o$ or L^*), and modify the form of the equations of motion (Lagrange → Voronets or Hamel)*; or *we modify the Lagrangean (→ Routhian), and leave the form of the equations of motion unchanged.*

(ii) The simple example below shows why if we enforce (c)-like constraints into the Lagrangean, then, in general, the ordinary Lagrangean equations for the independent coordinates (here the q's), *do not hold.* Let us consider, for algebraic simplicity, a potential system with the *single* ignorable coordinate ψ (i.e., $M = 1$) and, hence, Lagrangean $L = L(t, q, \dot{q}, \dot{\psi})$. Enforcing the Pfaffian cyclicity constraint

$$\partial L/\partial\dot{\psi} = constant \equiv C \;\Rightarrow\; \dot{\psi} = \dot{\psi}(t, q, \dot{q}; C) \equiv f(t, q, \dot{q}; C) \qquad \text{(d)}$$

into L, we obtain the "constrained" Lagrangean L_o:

$$L = L(t, q, \dot{q}, \dot{\psi}) = L[t, q, \dot{q}, \dot{\psi}(t, q, \dot{q}; C)] \equiv L_o(t, q, \dot{q}; C) = L_o. \qquad \text{(e)}$$

Applying chain rule to this equality, carefully, we find (with $p = 2, \ldots, n$)

$$\partial L_o/\partial q_p = \partial L/\partial q_p + (\partial L/\partial\dot{\psi})(\partial f/\partial q_p) = \partial L/\partial q_p + C(\partial f/\partial q_p), \qquad \text{(f1)}$$

$$\partial L_o/\partial\dot{q}_p = \partial L/\partial\dot{q}_p + (\partial L/\partial\dot{\psi})(\partial f/\partial\dot{q}_p) = \partial L/\partial\dot{q}_p + C(\partial f/\partial\dot{q}_p), \qquad \text{(f2)}$$

and therefore the Lagrangean expression for L_o becomes

$$\begin{aligned}(\partial L_o/\partial\dot{q}_p)^{\cdot} - \partial L_o/\partial q_p &= [(\partial L/\partial\dot{q}_p)^{\cdot} - \partial L/\partial q_p] + C[(\partial f/\partial\dot{q}_p)^{\cdot} - \partial f/\partial q_p] \\ &= 0 + C\,E_p(f) \neq 0;\end{aligned} \qquad \text{(g)}$$

that is, in general, $E_p(L_o) \neq 0$, even though $E_p(L) = 0$! However, using the above results, it is not hard to verify that

$$[\partial(L_o - C\dot{\psi})/\partial\dot{q}_p]^{\cdot} - \partial(L_o - C\dot{\psi})/\partial q_p = 0; \qquad \text{(h)}$$

or $E_p(L_o - C\dot{\psi}) \equiv E_p(R) = 0$; that is, if we want to keep the form of the Lagrangean equations of motion (for the independent coordinates) *unchanged*, we must take as new Lagrangean not the *constrained* Lagrangean L_o, but the *modified* Lagrangean $L_o - C\dot{\psi} \equiv R(outhian)$.

Example 8.4.3 *Routhian of a Three-DOF Cyclic System; Effects of Cyclicity on the Visible Motions.* Let us examine a scleronomic and cyclic system with *one* ignorable coordinate, $q_1 \equiv \psi_1$, and *two* positional coordinates, q_2, q_3; that is,

$M = 1$, $n - M = 3 - 1 = 2$. This is the simplest system that shows clearly the *gyroscopic*, and other, effects of ignorable coordinates. Its kinetic energy is

$$2T = M_{11}(\dot{\psi}_1)^2 + M_{22}(\dot{q}_2)^2 + M_{33}(\dot{q}_3)^2 + 2(M_{12}\,\dot{\psi}_1\,\dot{q}_2 + M_{13}\,\dot{\psi}_1\,\dot{q}_3 + M_{23}\,\dot{q}_2\,\dot{q}_3), \tag{a}$$

where all the inertia coefficients M_{kl} $(k, l = 1, 2, 3)$ are independent of ψ_1, and $Q_1 \equiv Q_\psi = 0$. Then, the cyclicity constraint is

$$p_1 \equiv \Psi_1 \equiv \partial T/\partial\dot{\psi}_1 = M_{11}\,\dot{\psi}_1 + M_{12}\,\dot{q}_2 + M_{13}\,\dot{q}_3 = \mathit{constant} \equiv C_1. \tag{b}$$

Solving it, we obtain

$$\dot{\psi}_1 = (C_1 - M_{12}\,\dot{q}_2 - M_{13}\,\dot{q}_3)/M_{11} \quad (\neq \mathit{constant}) \tag{c}$$
$$[= \text{function of } q, \dot{q}, C_1, \text{ and the } \textit{coupling} \text{ inertia coefficients } M_{12},\ M_{13}];$$

and, inserting this expression back into (a), we obtain

$$T = T''_{2,0} - T''_{0,2} = T(q, \dot{q}, C_1)$$

where

$$2T''_{2,0} = [(M_{11}\,M_{22} - M_{12}{}^2)/M_{11}]\,(\dot{q}_2)^2 + [(M_{11}\,M_{33} - M_{13}{}^2)/M_{11}]\,(\dot{q}_3)^2 + 2[(M_{11}\,M_{23} - M_{12}\,M_{13})/M_{11}]\,\dot{q}_2\,\dot{q}_3$$
$$\equiv M''_{22}(\dot{q}_2)^2 + M''_{33}(\dot{q}_3)^2 + 2M''_{23}\dot{q}_2\,\dot{q}_3 \qquad \text{(positive definite in the } \dot{q}\text{'s)}, \tag{d1}$$
$$-2T''_{0,2} = C_1{}^2/M_{11} \qquad \text{(positive definite in } C_1); \tag{d2}$$

the bilinear terms in the $\dot{q}$'s and C_1 having canceled, as expected by the general theory [(8.3.12m)]. As a result of the above, the modified kinetic energy T'' (to be used as kinetic energy in the Routhian–Lagrangean equations for the q's), becomes

$$T'' \equiv T - \dot{\psi}_1 C_1 = \cdots = T''_{2,0} + T''_{1,1} + T''_{0,2} = T''(q, \dot{q}, C_1), \tag{e}$$

where

$$T''_{2,0} = T''_{2,0}(q, \dot{q}): \qquad \text{given by (d1)}, \tag{e1}$$

$$T''_{0,2} = T''_{0,2}(q, C_1): \qquad \text{given by (d2)}, \tag{e2}$$

$$T''_{1,1} = (C_1 M_{12}/M_{11})\dot{q}_2 + (C_1 M_{13}/M_{11})\dot{q}_3 = r_2\dot{q}_2 + r_3\dot{q}_3 = T''_{1,1}(q, \dot{q}; C_1), \tag{e3}$$

and

$$r_2 \equiv (M_{12}/M_{11})C_1 \equiv \rho_{21}C_1, \qquad r_3 \equiv (M_{13}/M_{11})C_1 \equiv \rho_{31}C_1. \tag{e4}$$

The equations of motion for the q's (i.e., the equations of the *reduced*, or *apparent*, or *visible*, or *palpable* system) are

$$(\partial T''/\partial\dot{q}_p)^{\cdot} - \partial T''/\partial q_p = Q_p \qquad (p = 2,\ 3). \tag{f}$$

Upon carrying out the operations indicated in (f), with the expressions (e–e4), we notice that the cyclic coordinate(s) ψ_1 (through its constant momentum C_1), and the *coupling* coefficients M_{12} and M_{13}, have the following *triple* effect on the palpable motion:

(i) $T''_{2,0}$: The original coefficients of inertia M_{kl} have been replaced by the "reduced coefficients of inertia" M''_{kl}, unless M_{12} and M_{13} vanish.

(ii) $T''_{1,1}$: The effect of C_1 and M_{12}, M_{13}, appears in the coefficients of $\dot{q}_2$ and $\dot{q}_3$; and their contribution to (f) is

$$(\partial T''_{1,1}/\partial \dot{q}_2)^{\cdot} - \partial T''_{1,1}/\partial q_2 = C_1 \left[\partial/\partial q_3(M_{12}/M_{11}) - \partial/\partial q_2(M_{13}/M_{11})\right] \dot{q}_3, \tag{g1}$$

$$(\partial T''_{1,1}/\partial \dot{q}_2)^{\cdot} - \partial T''_{1,1}/\partial q_2 = C_1 \left[\partial/\partial q_2(M_{13}/M_{11}) - \partial/\partial q_3(M_{12}/M_{11})\right] \dot{q}_2; \tag{g2}$$

that is, a coupling of the *nonignorable (visible) motions*, generated by the *ignorable (invisible) ones*, through C_1 and M_{12}, M_{13}.

(iii) $T''_{0,2} = -C_1{}^2/2M_{11} = T''_{0,2}(q, C_1)$: this term behaves like an *additional negative potential energy*; and since

$$E_p(T''_{0,2}) \equiv (\partial T''_{0,2}/\partial \dot{q}_p)^{\cdot} - \partial T''_{0,2}/\partial q_p = 0 - (1/2)(C_1/M_{11})^2(\partial M_{11}/\partial q_p),$$

it gives rise to an *additional inertial "force"*

$$-E_p(T''_{0,2}) = \partial T''_{0,2}/\partial q_p = (C_1{}^2/2M_{11}{}^2)(\partial M_{11}/\partial q_p), \tag{h}$$

which is indistinguishable, in its mechanical effects, from the ordinary potential force $-\partial V/\partial q_p$. [During the late 19th century, this remarkable situation prompted several famous scientists (notably Hertz), to try to do the *reverse*; that is, explain V as a $T''_{0,2}$-like term of some *concealed*, or *latent*, motions! Such a "forceless" approach did not go very far in classical mechanics, but its conceptual implications proved helpful, a little later (in the 1910s), in the development of the (also forceless) general theory of relativity.]

These results are systematized and extended to the general case below, which may also include, with slight modifications, systems with no ignorable coordinates.

The Kelvin–Tait Equations

(Thomson and Tait, 1912, art. 319, ex. G.) Continuing from the preceding example, let us now find the *explicit* form of Routh's equations for the palpable motion of a general holonomic, scleronomic (no real loss in generality), and cyclic system with M ignorable coordinates $\psi \equiv (\psi_i;\ i = 1, \ldots, M)$ and $n - M$ nonignorable coordinates $q \equiv (\dot{q}_p;\ q = M + 1, \ldots, n)$—what is referred to as the *Kelvin–Tait equations.* Here,

$$T = T(q, \dot{q}, \dot{\psi}) = \textit{homogeneous quadratic in the } \dot{\psi}\text{'s and } \dot{q}\text{'s}, \tag{8.4.8}$$

and, therefore, as shown in (8.3.12 ff.) and the preceding example, the Routhian will equal

$$R = R_2 + R_1 + R_0, \tag{8.4.9}$$

where

$$R_2 \equiv T''_{2,0} = (1/2)\sum\sum r_{pq}(q)\dot{q}_p\,\dot{q}_q \quad (= T_{2,0}) = R_2(q, \dot{q}):$$

homogeneous quadratic in the nonignorable velocities $\dot{q}$, (8.4.9a)

$$R_1 \equiv T''_{1,1} = \sum r_p(q, C)\dot{q}_p = R_1(q, \dot{q}, C):$$

homogeneous linear in the nonignorable velocities $\dot{q}$,

with $\quad r_p = \sum \rho_{pi} C_i \quad \left[\rho_{pi} \equiv \sum C_{ij} b_{pj} = \rho_{pi}(q), \text{ by (8.3.12k)}\right]$, (8.4.9b)

$$R_0 \equiv T''_{0,2} - V = -(V - T''_{0,2}) \equiv -(1/2)\sum\sum C_{ji} C_j C_i - V \quad [= -(V + T_{0,2})]$$

$= R_0(q; C)$: *homogeneous quadratic in the constant ignorable momenta* $\Psi = C$. (8.4.9c)

The above indicate that *even in an originally scleronomic system*, the Routhian elimination of the ignorable velocities, in favor of their constant momenta, produces an additional *apparent potential energy* $T''_{0,2} = -T_{0,2}(< 0)$, and (possibly) an additional *apparent kinetic energy* $T''_{1,1}$; and, therefore, the situation is *mathematically identical to that of relative motion* (§3.16). Hence, utilizing the expressions (8.4.9–9c) in the Lagrangean equation of the palpable motion (8.4.5):

$$(\partial R/\partial \dot{q}_p)^{\cdot} - \partial R/\partial q_p = Q_p, \tag{8.4.5}$$

where Q_p = *nonpotential impressed positional forces*, and proceeding as in §3.16, we obtain the *Kelvin–Tait equations* (with $p, p' = M+1, \ldots, n$):

$$E_p(R) \equiv E_p(R_2 + R_1 + R_0) = E_p(R_2) + E_p(R_1) + E_p(R_0) = Q_p,$$

or

$$E_p(R_2) = Q_p - E_p(R_1) - E_p(R_0),$$

or

$$(\partial R_2/\partial \dot{q}_p)^{\cdot} - \partial R_2/\partial q_p = Q_p + \partial R_0/\partial q_p - \left[(\partial R_1/\partial \dot{q}_p)^{\cdot} - \partial R_1/\partial q_p\right]$$
$$= Q_p - \partial(V - T''_{0,2})/\partial q_p + \sum (\partial r_{p'}/\partial q_p - \partial r_p/\partial q_{p'})\dot{q}_{p'}$$
$$= Q_p - \partial(V - T''_{0,2})/\partial q_p + G_p, \tag{8.4.10}$$

where

$$G_p \equiv \sum (\partial r_{p'}/\partial q_p - \partial r_p/\partial q_{p'})\dot{q}_{p'} \equiv \sum G_{pp'}\dot{q}_{p'}:$$

Gyroscopic Routhian "force," since $G_{pp'} = -G_{p'p}\ [= G_{pp'}(q; C)]$. (8.4.10a)

These are the equations of motion of a *fictitious* scleronomic system (sometimes referred to as "conjugate" to the original system, or *reduced* system) with $n - M$ positional coordinates q, and subject, in addition to the impressed forces Q_p (nonpotential) and $-\partial V/\partial q_p$ (potential), to two *special constraint forces*: a *centrifugal-like* one, $\partial T''_{0,2}/\partial q_p$, and a *gyroscopic* one, G_p. Once the palpable motion $q_p(t)$ has been determined by solving (8.4.10), then substituting it into the Routhian equations for the ignorable motion, eqs. (8.3.6a, 9a):

$$d\psi_i/dt = -\partial R/\partial C_i = -\partial R_1/\partial C_i - \partial R_0/\partial C_i = -\partial T''_{0,2}/\partial C_i - \sum \rho_{pi}\dot{q}_p; \quad (8.4.11)$$

and carrying out a quadrature, we find the $\psi_i(t)$.

Gyroscopic Uncoupling

If *all* the $G_{pp'}$'s vanish, then the gyroscopic forces disappear, and so the equations of the reduced system take the *gyroscopically uncoupled* form:

$$E_p(R_2) \equiv (\partial R_2/\partial \dot{q}_p)^{\cdot} - \partial R_2/\partial q_p = Q_p + \partial R_0/\partial q_p; \quad (8.4.12)$$

that is, the centrifugal forces express the entire effect of cyclicity on that system.

Since $G_p \equiv -[(\partial R_1/\partial \dot{q}_p)^{\cdot} - \partial R_1/\partial q_p]$, and reasoning as in the case of integrability of Pfaffian constraints (chap. 2, also chap. 5), we may state with Pars (1965, p. 172) that: a system is *gyroscopically uncoupled* if, and only if,

$$R_1 dt \equiv \sum r_p(q; C)\, dq_p$$

is an *exact, or total,* differential. Obviously, this holds always if there is only *one* nonignorable coordinate [recall (prob. 3.16.3)]. A similar uncoupling occurs, of course, if all C_i's vanish [$\Rightarrow r_p = 0 \Rightarrow R_1 = 0$; *and* $R_0 = -V(q)$].

REMARKS

(i) It should be pointed out that the nonignorable coordinates do not fix the position of every system particle: in general, to one set of values of the q's there correspond *more than one set of values of the* ψ's; or, if the system, by suitable forces, is brought back to its original q's, after an arbitrary type of motion, its cyclic ψ will not, in general, return to their original values.

(ii) Also, the *gyroscopic* ($\sim \dot{q}_p$) terms in (8.4.10) are *irreversible* (i.e., they change sign under $dt \to -dt$); while in the absence of friction (i.e., only configuration-dependent forces), the other terms are not. This means that in order to reverse the motion of a cyclic system, we must reverse both the $\dot{q}$'s and the $\dot{\psi}$'s; reversing only the $\dot{q}$'s will not suffice! For example, the precessional motion of a top (gyroscope) is not reversed unless we also reverse its (cyclic) intrinsic spin $\dot{\psi}$.

A Cyclic Power Theorem

Multiplying each of (8.4.10) with $\dot{q}_p$ and then adding them together, while noting that $\sum r_{pp'}\dot{q}_{p'}\dot{q}_p = 0$ (gyroscopicity), we readily obtain the *cyclic energy rate/power theorem*:

$$dh_R/dt = \sum Q_p \dot{q}_p, \quad (8.4.13)$$

where [recalling (8.3.13–14a)]

$$\begin{aligned} h_R &\equiv R_2 - R_0 = T''_{2,0} + (V - T''_{0,2}) \\ &= T_{2,0} + (V + T_{0,2}) \equiv h_R(q, \dot{q}, C) \\ &= \textit{modified (or cyclic) generalized energy}; \end{aligned} \quad (8.4.13a)$$

from which, if $\sum Q_p \dot{q}_p = 0$, we are immediately led to the (Routhian counterpart of the Jacobi–Painlevé) *conservation theorem*:

$$h_R \equiv T''_{2,0} + (V - T''_{0,2}) = constant. \tag{8.4.14}$$

Alternatively, we may transform the energy equation of the *original* system as follows:

$$\begin{aligned} H &\equiv \sum (\partial L/\partial \dot{q}_k)\dot{q}_k - L \qquad (= constant, \text{ if } Q_p = 0 \text{ and } \partial L/\partial t = \partial R/\partial t = 0) \\ &= -R + \sum (\partial R/\partial \dot{q}_p)\dot{q}_p \qquad [\text{recalling } (8.3.10)] \\ &= -(R_2 + R_1 + R_0) + (2R_2 + R_1) \\ &= R_2 - R_0 = h_R(q, \dot{q}, C). \end{aligned} \tag{8.4.13b}$$

Extensions of the above to *rheonomic* cyclic systems — that is, to the case where

$$\begin{aligned} &L = L(t, q, \dot{q}, C) \\ &\Rightarrow R = L(t, q, \dot{q}, C) - \sum C_i \dot{\psi}_i(t, q, \dot{q}, C) = R(t, q, \dot{q}, C), \end{aligned} \tag{8.4.14a}$$

can be easily obtained; see, for example, Kil'chevskii (1977, pp. 350–352), Merkin (1974, chap. 1).

Example 8.4.4 *Energetics of a Simple Cyclic System.* Let us consider a potential system with Lagrangean

$$L = (1/2)(a\dot{\psi}^2 + 2e\dot{\psi}\dot{q} + b\dot{q}^2) - V(q), \tag{a}$$

where $a, e, b = constant$ inertial coefficients; $\psi/q = ignorable/nonignorable$ coordinates (i.e., $n = 2$, $M = 1$); and initial conditions at $t = 0$: $\psi = q = 1$, $\dot{\psi} = \dot{q} = 0$. Since ψ is ignorable,

$$\partial L/\partial \dot{\psi} = a\dot{\psi} + e\dot{q} = constant \equiv C, \tag{b}$$

from which, applying the initial conditions, we find $e = C$. Hence, solving (b) for $\dot{\psi}$, we obtain $\dot{\psi} = (C - e\dot{q})/a = (e/a)(1 - \dot{q})$, and so the Routhian becomes

$$\begin{aligned} R &= L(q, \dot{q}, C = e, a, b) - C\dot{\psi} = \cdots \\ &= (1/2)\,[b - (e^2/a)]\,(\dot{q})^2 + (e^2/a)\,\dot{q} - (e^2/2a) - V(q) \\ &= R_2(\dot{q}, a, b, e) + R_1(\dot{q}, a, e) + R_0(q, a, e). \end{aligned} \tag{c}$$

This yields the following Routhian equation of motion for q:

$$(\partial R/\partial \dot{q})^{\cdot} - \partial R/\partial q = 0: \ \{(e^2/a) + [b - (e^2/a)]\dot{q}\}^{\cdot} + dV/dq = 0, \tag{d}$$

or, solved for $\ddot{q}$:

$$\ddot{q} = -[a/(ba - e^2)](dV/dq). \tag{e}$$

Clearly, if $e = 0$, the motions of ψ and q decouple.

Now, if $V(q) = known$, then (e) supplies $q(t)$; its two integration constants are determined from the earlier initial conditions for q.

On the other hand, the Routhian form of the energy theorem for this system is [by (8.4.13b)]

$$h_R = H = R_2 - R_0 = (1/2)[b - (e^2/a)](\dot{q})^2 + (e^2/2a) + V(q)$$
$$= H(\dot{q}, a, b, e) = constant \equiv h. \tag{f}$$

Solving (f) for $\dot{q}$, we get

$$\dot{q} = [A - B\,V(q)]^{1/2}, \tag{g1}$$

where

$$A \equiv (2h\,a - e^2)/(b\,a - e^2), \qquad B \equiv 2a/(b\,a - e^2), \tag{g2}$$

and then separating variables and integrating, while using the initial conditions for q, we finally obtain $q(t)$:

$$t = \int [A - B\,V(q)]^{-1/2}\,dq, \tag{g3}$$

where the integral extends from 1 to q. Then, $\psi(t)$ can be found by the following quadrature:

$$\psi = \int -(\partial R/\partial C)\,dt + 1 = 1 + \int (e/a)(1 - \dot{q})\,dt, \tag{h}$$

where both integrals extend from 0 to t.

Example 8.4.5 *Hamiltonian and Routhian Treatments of the Top.* Let us consider a top (i.e., an axially symmetrical, or uniaxial, body) moving about a fixed point of its axis O under gravity (fig. 8.2). Using *intermediate* axes O–xyz, we find (with principal inertias there: $I_x = I_y \equiv A$, $I_z \equiv C$)

$$\omega = (\dot{\theta},\ \dot{\phi}\,\sin\theta,\ \dot{\psi} + \dot{\phi}\,\cos\theta) = inertial\ angular\ velocity\ of\ top, \tag{a1}$$

$$L = T - V,$$
$$2T = I_x\,\omega_x{}^2 + I_y\,\omega_y{}^2 + I_z\omega_z{}^2 = A\big[(\dot{\theta})^2 + (\dot{\phi})^2\sin^2\theta\big] + C(\dot{\psi} + \dot{\phi}\,\cos\theta)^2,$$
$$V = m\,g\,l\,\cos\theta \qquad (l \equiv OG,\ G = center\ of\ mass\ of\ top), \tag{a2}$$

and, therefore, the system momenta are

$$p_\phi \equiv \partial T/\partial\dot{\phi} = \partial L/\partial\dot{\phi} = A\,\dot{\phi}\,\sin^2\theta + C(\dot{\psi} + \dot{\phi}\,\cos\theta)\cos\theta$$
$$\equiv A\,\dot{\phi}\,\sin^2\theta + C\,n\,\cos\theta = constant \equiv C_\phi$$

[since, clearly, ϕ is an ignorable coordinate]

= component of angular momentum of top about the vertical axis through O (i.e., OZ), (b1)

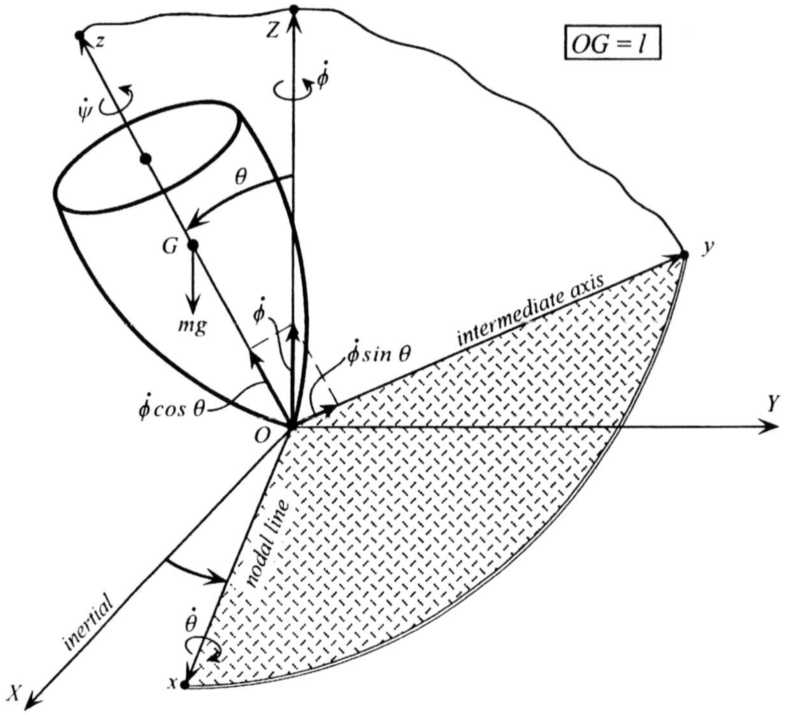

Figure 8.2 Geometry and kinematics of a top moving about a fixed point O.

$$p_\theta \equiv \partial T/\partial\dot{\theta} = \partial L/\partial\dot{\theta} = A\,\dot{\theta}$$

= component of angular momentum of top about axis through O perpendicular to plane of θ, (b2)

$$p_\psi \equiv \partial T/\partial\dot{\psi} = \partial L/\partial\dot{\psi} = C(\dot{\psi} + \dot{\phi}\cos\theta) \equiv C\,n = \textit{constant} \equiv C_\psi$$

[since, clearly, ψ is an ignorable coordinate]

= component of angular momentum of top about the symmetry axis Oz (i.e., the fixed line with which the top axis instantaneously coincides). (b3)

(i) From the above, it follows easily that the *Lagrangean equations* of the top are

$$(\partial L/\partial\dot{\phi})^\cdot - \partial L/\partial\phi = 0: \quad [A\,\dot{\phi}\sin^2\theta + C(\dot{\psi} + \dot{\phi}\cos\theta)\cos\theta]^\cdot = 0, \tag{c1}$$

$$(\partial L/\partial\dot{\theta})^\cdot - \partial L/\partial\theta = 0: \quad (A\,\dot{\theta})^\cdot - [A(\dot{\phi})^2\sin\theta\cos\theta - C(\dot{\psi} + \dot{\phi}\cos\theta)\dot{\phi}\sin\theta + m\,g\,l\sin\theta] = 0, \tag{c2}$$

$$(\partial L/\partial\dot{\psi})^\cdot - \partial L/\partial\psi = 0: \quad [C(\dot{\psi} + \dot{\phi}\cos\theta)]^\cdot = 0. \tag{c3}$$

The first and last of these equations express the constancy of p_ϕ and p_ψ, respectively; and so we can rewrite the first and second, as follows:

$$\phi: \quad A\,\dot{\phi}\sin^2\theta + C_\psi\cos\theta = C_\phi, \tag{d1}$$

$$\theta: \quad A\,\ddot{\theta} - A(\dot{\phi})^2\sin\theta\cos\theta + C_\psi\,\dot{\phi}\sin\theta = m\,g\,l\sin\theta. \tag{d2}$$

These two equations allow us, among other things, to study the small (linearized) motion of the top about its *vertical* axis OZ—that is, $\theta = 0$—and its stability/instability. Indeed, setting in (d1, 2), approximately, $\sin\theta \approx \theta$ and $\cos\theta \approx 1 - \theta^2/2$, we obtain

$$\phi: \quad \theta^2\dot{\zeta} = constant, \tag{e1}$$

$$\theta: \quad \ddot{\theta} - (\dot{\zeta})^2\theta = -[(C_\psi - 4A\,m\,g\,l)/4A^2]\theta, \tag{e2}$$

where

$$\zeta \equiv \phi - (C_\psi/2A)t \;\Rightarrow\; \dot{\zeta} = \dot{\phi} - (C_\psi/2A). \tag{e3}$$

Now, due to the *form* of equations (e1, 2) we may view θ and ζ as the *polar coordinates* of the horizontal projection of a point on the top axis Oz, relative to a line that revolves around OZ with (constant) angular velocity $\dot{\phi} - \dot{\zeta} = C_\psi/2A$. It follows that the *relative* motion of such a point will be elliptic harmonic with period $4\pi A({C_\psi}^2 - 4A\,m\,g\,l)^{-1/2}$, as long as ${C_\psi}^2 > 4A\,m\,g\,l$ (stability condition; see also stability of *sleeping top*, ex. 8.4.6 below).

(ii) *Hamiltonian equations.* Solving (b1–3) for the velocities in terms of the momenta, we get

$$\dot{\phi} = (p_\phi - p_\psi\cos\theta)/A\,\sin^2\theta, \tag{f1}$$

$$\dot{\theta} = p_\theta/A, \tag{f2}$$

$$\dot{\psi} = p_\psi/C - (p_\phi - p_\psi\cos\theta)\cos\theta/A\,\sin^2\theta. \tag{f3}$$

Accordingly, the Hamiltonian becomes

$$\begin{aligned} H &= (1/2)(p_\phi\dot{\phi} + p_\theta\dot{\theta} + p_\psi\dot{\psi}) + V \\ &= (1/2A)\left[{p_\theta}^2 + (p_\phi - p_\psi\cos\theta)^2/\sin^2\theta\right] + (1/2C)\,{p_\psi}^2 + m\,g\,l\cos\theta, \end{aligned} \tag{g}$$

and leads easily to the following pairs of Hamilton's equations:

$$\phi: \quad \dot{p}_\phi = -\partial H/\partial\phi = 0 \qquad (\phi = \text{ignorable coordinate}) \tag{h1}$$

$$\dot{\phi} = \partial H/\partial p_\phi = (p_\phi - p_\psi\cos\theta)/A\,\sin^2\theta, \tag{h2}$$

$$\theta: \quad \dot{p}_\theta = -\partial H/\partial\theta = -(p_\phi - p_\psi\cos\theta)(p_\psi - p_\phi\cos\theta)/A\,\sin^3\theta + m\,g\,l\sin\theta, \tag{h3}$$

$$\dot{\theta} = \partial H/\partial p_\theta = p_\theta/A, \tag{h4}$$

$$\psi: \quad \dot{p}_\psi = -\partial H/\partial\psi = 0 \qquad (\psi = \text{ignorable coordinate}) \tag{h5}$$

$$\dot{\psi} = \partial H/\partial p_\psi = -(p_\phi - p_\psi\cos\theta)\cos\theta/A\,\sin^2\theta + p_\psi/C. \tag{h6}$$

Equations (h2, 4, 6) are kinematico-inertial, and coincide with the earlier (f1–3); while (h1, 3, 5) are the kinetic equations.

(iii) *Routhian equations.* Since, here, the ignorable coordinates are $\psi_1 = \phi$ and $\psi_2 = \psi$, and corresponding constant momenta $\Psi_1 = p_\phi = C_\phi$ and $\Psi_2 = p_\psi = C_\psi$ (i.e., $n = 3$, $M = 2$), the Routhian is

$$R = L - p_\phi\dot{\phi} - p_\psi\dot{\psi} = \cdots = R_2 + R_1 + R_0, \tag{i}$$

where

$$R_2 = T''_{2,0} = (1/2)\, A\, (\dot{\theta})^2, \tag{i1}$$

$$R_1 = 0 \qquad \text{(we need at least } \textit{two} \text{ nonignorable } q\text{'s to have gyroscopicity!)}, \tag{i2}$$

$$R_0 = T''_{0,2} - V$$

$$= -[(C_\phi - C_\psi \cos\theta)^2/2A\, \sin^2\theta + (1/2C)\, {C_\psi}^2] - m\, g\, l \cos\theta, \tag{i3}$$

and therefore Routh's equation for the nonignorable coordinate θ is

$$A(d^2\theta/dt^2) + [(C_\phi - C_\psi \cos\theta)(C_\psi - C_\phi \cos\theta)]/A\, \sin^3\theta = m\, g\, l \sin\theta. \tag{j}$$

The *second* left-side (centrifugal-like) terms, equal to $E_\theta(T''_{0,2}) = -\partial T''_{0,2}/\partial\theta$, represents the contribution of the *apparent potential energy* $T''_{0,2}(< 0)$; there are no gyroscopic terms.

Equation (j) can also be rewritten as

$$A(d^2\theta/dt^2) + (1/2A)\, d/d\theta\big[(C_\phi - C_\psi \cos\theta)^2/\sin^2\theta\big] = m\, g\, l \sin\theta; \tag{j1}$$

The nonlinear equations (j, j1) can be used, just like the earlier Lagrangean equations (d1, 2), to study the small motion of the top about a given precessional motion, say one with constant nutation $\theta(t) = \theta_o$ [i.e., set in, say (j), $\theta = \theta_o + \Delta\theta(t)$, keep *up to linear terms* in $\Delta\theta$ and its $(\ldots)^{\cdot}$-derivatives; and then find conditions so that the resulting *linear second-order* $\Delta\theta$ *equation* has *harmonic* solutions. The details are left to the reader.]

Finally, either from the general theory, or directly from (j1) (i.e., multiply it with $2\dot{\theta}$, etc.), we can easily show that the system has the following cyclic generalized integral:

$$R_2 - R_0 = T''_{2,0} - (T''_{0,2} - V) = \textit{constant}:$$

$$(1/2)A(\dot{\theta})^2 + \big[(C_\phi - C_\psi \cos\theta)^2/2A \sin^2\theta + (1/2C)\, {C_\psi}^2\big]$$

$$+ m\, g\, l \cos\theta = \textit{constant}, \tag{k1}$$

or

$$A(\dot{\theta})^2 + (C_\phi - C_\psi \cos\theta)^2/A \sin^2\theta + 2m\, g\, l \cos\theta = \textit{constant} \equiv h; \tag{k2}$$

or, setting $x \equiv \cos\theta \;\Rightarrow\; \dot{x} = -\dot{\theta}\sin\theta \;\Rightarrow\; (\dot{x})^2 = (1 - x^2)(\dot{\theta})^2$, finally,

$$A(dx/dt)^2 + (C_\phi - C_\psi\, x)^2/A + (2\, mgl\, x - h)(1 - x^2) = 0, \tag{k3}$$

which has the form $(\dot{x})^2 = \textit{known function of } x \equiv f(x) \;\Rightarrow\; dx/[f(x)]^{1/2} = dt$, and upon integration yields $x \equiv \cos\theta$ as an elliptic function of t. The cyclic motions $\phi(t)$ and $\psi(t)$ can then be found from the corresponding Routhian equations (8.4.11), or (b1, 3), by quadratures.

A Generalization

If $Q_\phi \neq 0$, then only ψ is cyclic. Solving (b3) for $\dot{\psi}$, we obtain

$$\dot{\psi} = p_\psi/C - \dot{\phi}\cos\theta \equiv n - \dot{\phi}\cos\theta, \tag{l}$$

and so, in this case, the Lagrangean and Routhian of the top become, respectively,

$$L \equiv T - V = (A/2)\,[(\dot\theta)^2 + (\dot\phi)^2 \sin^2\theta] + (C/2)\,[(n - \dot\phi\cos\theta) + \dot\phi\cos\theta]^2 - mgl\cos\theta$$

$$= (A/2)\,[(\dot\theta)^2 + (\dot\phi)^2 \sin^2\theta] + (C/2)n^2 - m\,g\,l\cos\theta$$

$$= T''_{2,0} - T''_{0,2} - V = L(\theta;\dot\phi,\dot\theta; C_\psi = C\,n), \qquad \text{(m1)}$$

$$R = L - p_\psi\dot\psi = (T - p_\psi\dot\psi) - V = R_2 + R_1 + R_0, \qquad \text{(m2)}$$

where

$$R_2 = T''_{2,0} = (A/2)\,[(\dot\theta)^2 + (\dot\phi)^2\sin^2\theta]:$$

Kinetic energy of a thin homogeneous bar, of transverse moment of inertia A about O, moving about that point, (m3)

$$R_1 = T''_{1,1} = (C\,n\,\cos\theta)\dot\phi$$

$$[\equiv r_\phi\dot\phi \equiv (\rho_{\phi\psi}C_\psi)\dot\phi \Rightarrow r_\phi = (\cos\theta)C\,n,\ \rho_{\phi\psi} = \cos\theta], \qquad \text{(m4)}$$

$$R_0 = T''_{0,2} - V = -[(C/2)\,n^2 + m\,g\,l\cos\theta]$$

[the constant term $T''_{0,2} = -(C/2)n^2$ does not enter the Routhian equations of motion; but it does enter the corresponding energy rate equation]; (m5)

and therefore Routh's equations for the nonignorable coordinates ϕ and θ are

$$(\partial R/\partial\dot\phi)^{\cdot} - \partial R/\partial\phi = Q_\phi:$$

$$(\partial R_2/\partial\dot\phi)^{\cdot} - \partial R_2/\partial\phi = Q_\phi - [(\partial R_1/\partial\dot\phi)^{\cdot} - \partial R_1/\partial\phi],$$

or

$$A\,\ddot\phi\sin^2\theta + 2A\,\dot\phi\,\dot\theta\sin\theta\cos\theta = Q_\phi + (C\,n\,\sin\theta)\,\dot\theta, \qquad \text{(n1)}$$

$$(\partial R/\partial\dot\theta)^{\cdot} - \partial R/\partial\theta = Q_\theta:$$

$$(\partial R_2/\partial\dot\theta)^{\cdot} - \partial R_2/\partial\theta = -(-\partial R_0/\partial\theta) + Q_\theta - [(\partial R_1/\partial\dot\theta)^{\cdot} - \partial R_1/\partial\theta],$$

or

$$A\,\ddot\theta - A(\dot\phi)^2\sin\theta\cos\theta = m\,g\,l\,\sin\theta + Q_\theta - (C\,n\,\sin\theta)\dot\phi. \qquad \text{(n2)}$$

Notice that (i) the impressed forces Q_ϕ, Q_θ do not include gravity; (ii) the terms $\pm(C\,n\,\sin\theta)\dot\phi$ are the gyroscopic "forces"; and (iii) these are the Lagrangean equations of the earlier-mentioned fictitious bar rotating about O under the action of (a) gravity, (b) Q_ϕ, Q_θ, and (c) the *gyroscopic couple* $\boldsymbol{M}'_G \equiv \boldsymbol{M}_G\,\sin\theta$, where (with some standard notations)

$$\boldsymbol{M}_G = -d/dt\ (\textit{angular momentum about } Oz)$$

$$= -d/dt(C\,n\,\boldsymbol{k}) = -C\,n(\boldsymbol{\omega}_{O-xyz}\times\boldsymbol{k}) \quad [= -C\,n(\boldsymbol{\omega}\times\boldsymbol{k})]$$

$$= -C\,n[(\dot\phi\,\boldsymbol{K} + \dot\theta\,\boldsymbol{i})\times\boldsymbol{k}] = -C\,n\,[\dot\phi(+\boldsymbol{i}) + \dot\theta(-\boldsymbol{j})]$$

$$= (-C\,n\,\dot\phi)\boldsymbol{i} + (C\,n\,\dot\theta)\boldsymbol{j} = (M_{G,x}, M_{G,y}). \qquad \text{(n3)}$$

Finally, if Q_ϕ, $Q_\theta = 0$, the system has the modified generalized energy integral

$$h_R \equiv R_2 - R_0 \equiv T''_{2,0} + (V - T''_{2,0})$$
$$= (A/2)\,\left[(\dot\theta)^2 + (\dot\phi)^2 \sin^2\theta\right] + \left[(C/2)n^2 + m\,g\,l\cos\theta\right] = constant,$$

or, simply,

$$(A/2)\,\left[(\dot\theta)^2 + (\dot\phi)^2 \sin^2\theta\right] + m\,g\,l\cos\theta = constant. \qquad \text{(n4)}$$

Example 8.4.6 *Sleeping Top.* Continuing from the preceding example, let us study the motion (and linear stability) of the top under gravity, when its spin axis OG is nearly vertical; that is, in the vicinity of OZ [fig. 8.3(a)]. The inertial coordinates X, Y of the projection of G on the horizontal plane O–XY are [fig. 8.3(b)]

$$X = (l\sin\theta)\sin\phi \qquad \text{and} \qquad Y = -(l\sin\theta)\cos\phi. \qquad \text{(a)}$$

Below, using these coordinate transformations, we express the Lagrangean and Routhian of the top in terms of X, Y and their $(\ldots)^{\cdot}$-derivatives (instead of the earlier ϕ, θ), and keep only up to *quadratic* terms in these variables, so that the corresponding equations of motion be *linear* in them; which is the mathematical meaning of near verticalness, or "sleepingness" of the top. Then, we study the stability/instability of these small motions.

Indeed, $(\ldots)^{\cdot}$-differentiating (a), and then solving for $\dot\phi$ and $\dot\theta$, while noting that $X^2 + Y^2 = l^2\sin^2\theta$, we obtain

$$\dot\phi = (\dot X\,\cos\phi + \dot Y\,\sin\phi)/l\sin\theta, \qquad \dot\theta = (\dot X\,\sin\phi - \dot Y\,\cos\phi)/l\cos\theta; \qquad \text{(b)}$$

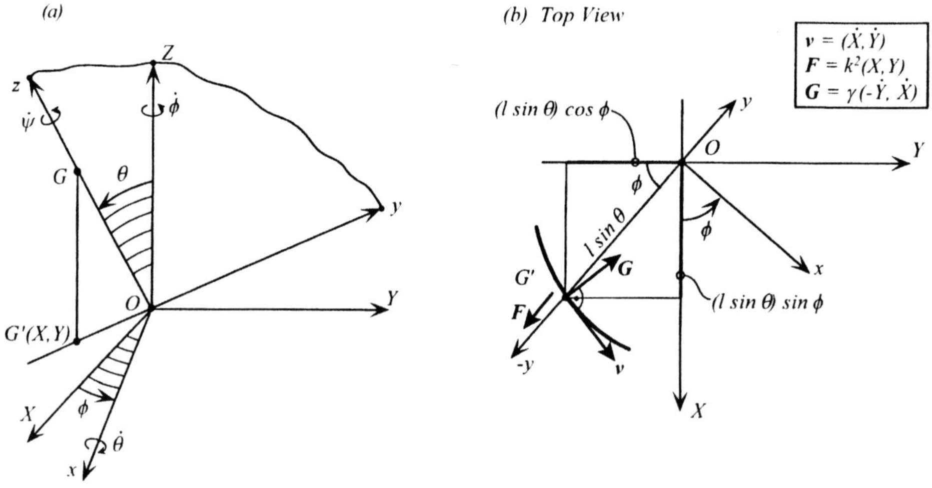

Figure 8.3 (a) Geometry and kinematics of sleeping on top (see also fig. 8.2). (b) Motion of projection of center of mass G of (sleeping) top, G′, on horizontal plane O–XY.

and so, to the second order,

$$(\dot{\theta})^2 + (\dot{\phi})^2 \sin^2\theta = [(\dot{X})^2 + (\dot{Y})^2]/l^2,$$
$$\dot{\phi}\cos\theta = (X\dot{Y} - Y\dot{X})[(X^2 + Y^2)^{-1} - (2l^2)^{-1}] = (X\dot{Y} - Y\dot{X})/2l^2,$$
$$\cos\theta = 1 - (X^2 + Y^2)/2l^2. \tag{c}$$

Hence, recalling the relevant expressions of the preceding example [and that $C(\dot{\psi} + \dot{\phi}\cos\theta) \equiv Cn = constant \equiv C_\psi$], we find

$$L = (A/2)\,[(\dot{\theta})^2 + (\dot{\phi})^2\sin^2\theta] + (C/2)(\dot{\psi} + \dot{\phi}\cos\theta)^2 - mgl\cos\theta$$
$$= (1/2)\left\{A[(\dot{X})^2 + (\dot{Y})^2]/l^2 + C_\psi(X\dot{Y} - Y\dot{X})[(X^2 + Y^2)^{-1} - (2l^2)^{-1}] + C_\psi\dot{\psi}\right\}$$
$$- mgl[1 - (X^2 + Y^2)/2l^2] = L(X, Y, \dot{X}, \dot{Y}, \dot{\psi}; C_\psi), \tag{d1}$$

and (since we are seeking the X, Y equations, we will ignore only ψ; *not* both ϕ and ψ!)

$$R = L - p_\psi\dot{\psi} = R_2 + R_1 + R_0$$
$$= (A/2)[(\dot{\theta})^2 + (\dot{\phi})^2\sin^2\theta] + C_\psi\,\dot{\phi}\cos\theta - [(C_\psi{}^2/2C) + mgl\cos\theta]$$
$$= A[(\dot{X})^2 + (\dot{Y})^2]/2l^2 - (C_\psi/2l^2)(X\dot{Y} - Y\dot{X})$$
$$+ (mgl/2l^2)(X^2 + Y^2) + constant\ terms$$
$$= R(X, Y, \dot{X}, \dot{Y}; C_\psi). \tag{d2}$$

From these expressions, we obtain the following Routhian equations:

$$(\partial R/\partial\dot{X})^{\cdot} - \partial R/\partial X = 0: \qquad \ddot{X} = k^2 X - \gamma\dot{Y}, \tag{e1}$$

$$(\partial R/\partial\dot{Y})^{\cdot} - \partial R/\partial Y = 0: \qquad \ddot{Y} = k^2 Y + \gamma\dot{X}, \tag{e2}$$

where

$$k^2 \equiv mgl/A, \qquad \gamma \equiv C_\psi/A \equiv (C/A)\,n. \tag{e3}$$

These coupled equations are the equations of motion of a fictitious particle of unit mass moving on the inertial plane O–XY under (i) a (centrifugal-like) radial *repulsive* force $\boldsymbol{F} = k^2(X, Y)$ (i.e., along OP, from O toward P, proportional to the distance from the origin); and (ii) a *gyroscopic* (Coriolis-like) force $\boldsymbol{G} = \gamma(-\dot{Y}, \dot{X})$ [fig. 8.3(b)].

Energy Integral

Multiplying (e1) by $\dot{X}$ and (e2) by $\dot{Y}$, and then adding them together, while noting that $\boldsymbol{G}\cdot\boldsymbol{v} = \gamma(-\dot{Y}\dot{X} + \dot{X}\dot{Y}) = 0$, we readily obtain the generalized energy integral:

$$(1/2)\,[(\dot{X})^2 + (\dot{Y})^2] - (k^2/2)(X^2 + Y^2) = constant, \tag{f}$$

as also expected from the general theory.

Stability

Equations (e) describe the evolution of small deviations (and their rates) of the axis of the top OG from a fundamental state of vertical spinning ($\theta = 0$). They show that the projection of G, G', on the one hand tends to get away from O [$\sim k^2$ terms (gravity)], and on the other turns around the origin [$\sim \gamma$ terms (spinning)], clockwise or counterclockwise, depending on the sign of γ.

As an introduction to §8.6, let us examine the *stability of that motion*; that is, investigate whether G' (OG) remains in the neighborhood of O (OZ), under arbitrary initial conditions of disturbance from these fundamental states. To this end, we set in (e) (since it is a constant coefficient system)

$$X = X_o \exp(\lambda t) \qquad \text{and} \qquad Y = Y_o \exp(\lambda t), \tag{g1}$$

where X_o, Y_o = constant amplitudes and $-\lambda^2 = \omega^2$ = square of frequency of motion (if stable), and thus obtain the following homogeneous system for these amplitudes:

$$(\lambda^2 - k^2)X_o + (\lambda\gamma)Y_o = 0, \qquad (-\lambda\gamma)X_o + (\lambda^2 - k^2)Y_o = 0. \tag{g2}$$

The requirement for nontrivial solutions of the above leads us, in well-known ways, to the determinantal (or secular) equation

$$\Delta(\lambda) \equiv \begin{vmatrix} \lambda^2 - k^2 & \lambda\gamma \\ -\lambda\gamma & \lambda^2 - k^2 \end{vmatrix} = 0, \tag{g3}$$

whose solutions are readily found to be

$$\lambda = \pm(1/2)\,[i\gamma \pm (4k^2 - \gamma^2)^{1/2}]. \tag{g4}$$

From this, we conclude that:

(i) If $\qquad \gamma^2 > 4k^2 \qquad$ [i.e., recalling (e1), if $n^2 > 4\,A\,m\,g\,l/C^2$], $\qquad$ (g5)

then λ will be *purely imaginary*, and therefore X, Y will be *harmonic* (bounded); that is, the vertically spinning state will be *stable*; but

(ii) If $\qquad \gamma^2 < 4k^2 \qquad$ [i.e., if $n^2 < 4\,A\,m\,g\,l/C^2$], $\qquad$ (g6)

then there will be *two pairs of conjugate complex roots*, one with *positive* real part, and one with *negative*. As a result, in general, a part of X, Y will be exponentially *unbounded*; that is, the vertically spinning state will be *unstable*. [Since for *small* θ and *very high* $\dot{\psi}$:

$$n^2 = (\dot{\psi} + \dot{\phi}\cos\theta)^2 = (\dot{\psi})^2 + (\dot{\phi}\cos\theta)^2 + 2\dot{\phi}\,\dot{\psi}\cos\theta \approx (\dot{\psi})^2,$$

the condition (g5) can then be replaced by $(\dot{\psi})^2 > (4A/C^2)m\,g\,l$; and analogously for (g6).]

For additional details of the stable case, see, for example, McCuskey (1959, p. 181); also Routh (1877, pp. 64–66, 94–96), Smart (1951, vol. 2, pp. 409–412), and Whittaker (1937, pp. 206–207). For a general discussion of the sleeping top, including a method for the *uncoupling* of (e) and associated conservation laws/integrals, see Bahar (1992).

Problem 8.4.1 Show that the linearized equations of the sleeping top, in terms of the angular variables ϕ and θ, are:

$$A(\ddot{\theta} - \dot{\phi}^2\theta) + Cn\theta\dot{\phi} = mgl\theta, \quad A(\theta\ddot{\phi} + 2\dot{\theta}\dot{\phi}) - Cn\dot{\theta} = 0. \tag{a}$$

Then show that (a) also lead to the earlier stability condition $\dot{\psi}^2 > (4A/C^2)mgl$.

HINT

Assume steady precession around the vertical, i.e., $\theta = \text{constant}$ ($\neq 0$). Then require that the resulting quadratic equation in $\dot{\phi}$ ($=$ constant) have real roots. (This argument also works for stability of steady precession about any nutation angle.)

REMARK

Since for small θ the angles ϕ and ψ are not necessarily small (and for $\theta = 0$, $\dot{\phi}$, $\dot{\psi}$ become indeterminate, §1.12), other angles, free of this drawback, have been used; e.g., the Eulerian sequence $1 \to 2 \to 3$. For a treatment of the sleeping top via such singularity-free parameters, see e.g., Beghin (1967, pp. 503–504) and Berezkin (1968, pp. 261–262).

8.5 STEADY MOTION (OF CYCLIC SYSTEMS)

Continuing from §3.10, we define as *steady motion of a general* (not necessarily *cyclic*) *system, relative to a given set of Lagrangean coordinates*, (q_k), one in which *all* corresponding velocities are constant; that is, $(\dot{q}_k) = \textit{constant}$. Hence, if that system is also cyclic, *relative to a particular set of ignorable coordinates*, saying that it is in a state of *steady* (or *isocyclic*) motion means that, during the latter, *the velocities corresponding to both its ignorable and nonignorable coordinates remain constant*; that is, and in the notation of §8.3 and §8.4 (with $i = 1, \ldots, M$; $p = M+1, \ldots, n$), a motion of that system is steady if during it

$$\dot{\psi}_i = \textit{constant} \equiv c_i \qquad (\text{in } \textit{addition} \text{ to } \Psi_i = \textit{constant} \equiv C_i), \tag{8.5.1a}$$

and

$$q_p = \textit{constant} \equiv s_p \qquad (\Rightarrow \dot{q}_p = 0); \tag{8.5.1b}$$

that is, *all* system velocities are *constant* (and, hence, all corresponding accelerations vanish); and, for scleronomic such systems, the Lagrangean has the form $L = L(c_i, s_p)$. [Conditions (8.5.1b) state that (these special motions of the original system called) *steady motions relative to the ignorable coordinates* (ψ_i) are equilibrium states of the conjugate subsystem (q_p). [Recall "bracketed comment" following (8.4.5).]

Clearly, *steadiness is a coordinate-dependent property*, like cyclicity; and outside of uniform translation and rotation about a fixed axis, constitutes one of the simplest kinds of motion. Thus, the spinning top of the preceding examples is in a state of steady motion if its ignorable velocities $\dot{\phi}$ (precession rate) and $\dot{\psi}$ (intrinsic, or proper, spin), and its nonignorable coordinate θ (nutation) remain constant.

To find the conditions for such a state of motion, of, say, a *scleronomic* and holonomic system (extensions to more general systems, even quasi variables and noncyclic systems, do not offer any theoretical difficulties), we take the earlier Kelvin–Tait equations (8.4.10; with $p, p' = M+1, \ldots, n$):

$$(\partial R_2/\partial \dot{q}_p)^{\cdot} - \partial R_2/\partial q_p = Q_p + \partial R_0/\partial q_p + G_p$$
$$= Q_p + (\partial T''_{0,2}/\partial q_p - \partial V/\partial q_p) + G_p, \qquad (8.5.2)$$

where

$$G_p \equiv \sum (\partial r_{p'}/\partial q_p - \partial r_p/\partial q_{p'})\dot{q}_{p'} \equiv \sum G_{pp'}\dot{q}_{p'}, \qquad (8.5.2a)$$

$$R_1 \equiv T''_{1,1} = \sum r_p \dot{q}_p, \qquad (8.5.2b)$$

and in there apply the (equilibrium-like) equations (8.5.1a, b). We, thus, obtain the following conditions of steady motion:

$$Q_p + \partial R_0/\partial q_p \equiv Q_p + (\partial T''_{0,2}/\partial q_p - \partial V/\partial q_p) = 0, \qquad (8.5.3a)$$

or, if the system is *wholly potential*,

$$\partial R_0/\partial q_p = 0, \qquad \text{or} \qquad \partial T''_{0,2}/\partial q_p = \partial V/\partial q_p. \qquad (8.5.3b)$$

Equivalently, since [recalling (8.3.12 ff.)]

$R = R_2$ (*homogeneous quadratic* in the $\dot{q}$'s) + R_1 (*homogeneous bilinear* in the Ψ's *and* $\dot{q}$'s) + R_0 (*homogeneous quadratic* in the Ψ's), (8.5.3c)

and by Routh's kinematico-inertial identities

$$\partial R/\partial q_p = \partial L/\partial q_p, \qquad (8.5.3d)$$

the above "equilibrium" conditions can be rewritten as

$$(\partial R/\partial q_p)_o = (\partial L/\partial q_p)_o = 0 \qquad [(\ldots)_o \equiv (\ldots)_{\dot{\psi}=c,\, q=s,\, \dot{q}=0}]. \qquad (8.5.3e)$$

These $n - M$ equations, expressing the hitherto unknown q's $\equiv$ s's in terms of the arbitrarily chosen Ψ's $\equiv$ C's, are the necessary and sufficient conditions for the steady motion of the original system; or, equivalently, for the equilibrium of the reduced system. The $\dot{\psi}$'s can then be found from the *second* (Hamiltonian) group of Routh's equations:

$$d\psi_i/dt = -(\partial R/\partial \Psi_i)_o = -(\partial R_0/\partial \Psi_i)_o = -(\partial T''_{0,2}/\partial \Psi_i)_o$$

$= \sum C_{ij}C_j = constant \equiv c_i$ [by (8.3.12d, e), with $\dot{q}_p = 0$]

= Function of the s's [roots of (8.5.3b), (8.5.3e)] and the (arbitrarily chosen) C_j's [as (8.3.12l) show, once (8.5.3a, b) have been solved, the C_{ji} change, from known functions of the q's, to known functions of the Ψ's], (8.5.4a) which, upon integration, yields the ψ's:

$\psi_i(t) = -c_i(t - t_{\text{initial}}) + \psi_{i,\text{initial}}$:

Function of the s's and the (now) arbitrarily chosen c_i's and ψ_{initial}'s; (8.5.4b)

that is, *in steady motion, the cyclic coordinates vary linearly with time.*

As stated above, if we initially choose arbitrarily the Ψ's, then eqs. (8.5.3b) relate them to the q's. If, on the other hand, we initially choose arbitrarily the $\dot{\psi}$'s $\equiv c$'s, then, to relate them directly to the q's, we must modify (8.5.3b) somewhat. To this end, we take, first, $T''_{0,2}$, which is homogeneous quadratic in the Ψ's, and, using $\Psi_i = \sum c_{ji}\dot{\psi}_j$, we change it to a *homogeneous quadratic* function in the $\dot{\psi}$'s. Indeed, we have, successively (with i, j, j', j'': $1, \ldots, M$),

$$2T''_{0,2} \equiv 2T''_{\Psi\Psi} \equiv -\sum\sum C_{ji}\Psi_j\Psi_i \qquad \text{[recalling (8.3.121)]}$$

$$= -\sum\sum C_{ji}\Big(\sum c_{j'j}\dot{\psi}_{j'}\Big)\Big(\sum c_{j''i}\dot{\psi}_{j''}\Big) \qquad \Big[\text{recalling that} \sum C_{ji}c_{j'j} = \delta_{ij'}, \text{ etc.}\Big]$$

$$= \cdots = -\sum\sum c_{ij}\dot{\psi}_i\dot{\psi}_j \equiv 2T''_{\dot{\psi}\dot{\psi}} = -2T_{\dot{\psi}\dot{\psi}} \qquad \text{[recalling (8.3.12c, 1)].} \tag{8.5.5a}$$

Next, applying the results of (probs. 8.2.1 and 8.2.6) to the *conjugate* functions $T''_{\Psi\Psi}$ and $T''_{\dot{\psi}\dot{\psi}}$, we find that

$$\partial T''_{\Psi\Psi}/\partial q_p \;=\; -\partial T''_{\dot{\psi}\dot{\psi}}/\partial q_p \;=\; \partial T_{\dot{\psi}\dot{\psi}}/\partial q_p, \tag{8.5.5b}$$

and so, finally, we can replace the steady motion conditions (8.5.3b) by

$$-\partial T''_{\dot{\psi}\dot{\psi}}/\partial q_p \;=\; \partial V/\partial q_p \qquad \text{or} \qquad \partial T_{\dot{\psi}\dot{\psi}}/\partial q_p \;=\; \partial V/\partial q_p, \tag{8.5.5c}$$

which relate the unknown q's to the arbitrarily chosen $\dot{\psi}$'s; and using $\Psi_i = \sum c_{ji}\dot{\psi}_j$ we can relate both sets to the Ψ's.

Example 8.5.1 Let us apply eqs. (8.5.5b, c) to the spinning top described earlier. Here, $\Psi_1 \equiv C_\phi$ and $\Psi_2 \equiv C_\psi$, and

$$2T = A[(\dot{\theta})^2 + (\dot{\phi})^2\sin^2\theta] + C(\dot{\psi} + \dot{\phi}\cos\theta)^2, \qquad V = mgl\cos\theta, \tag{a1}$$

$$R_2 \equiv T''_{2,0} = (1/2)A(\dot{\theta})^2, \tag{a2}$$

$$R_1 \equiv T''_{1,1} = 0, \tag{a3}$$

$$R_0 \equiv T''_{0,2} - V \equiv T''_{\Psi\Psi} - V$$

$$= -[(C_\phi - C_\psi\cos\theta)^2/2A\sin^2\theta + (1/2C)\,C_\psi{}^2] - mgl\cos\theta. \tag{a4}$$

Therefore, (8.5.4a) yield

$$\dot{\phi} = -(\partial T''_{\Psi\Psi}/\partial C_\phi)_o = (C_\phi - C_\psi\cos\theta)/A\sin^2\theta, \tag{b1}$$

$$\dot{\psi} = -(\partial T''_{\Psi\Psi}/\partial C_\psi)_o = -(C_\phi - C_\psi\cos\theta)\cos\theta/A\sin^2\theta + C_\psi/C. \tag{b2}$$

Solving these two equations for C_ϕ and C_ψ, we obtain

$$C_\phi = (A\sin^2\theta + C\cos^2\theta)\dot{\phi} + (C\cos\theta)\dot{\psi} = \textit{constant}, \tag{c1}$$

$$C_\psi = (C\cos\theta)\dot{\phi} + (C)\dot{\psi} = \textit{constant}; \tag{c2}$$

and, inserting these representations in $T''_{\Psi\Psi}$, (a4), we find

$$-2T''_{\dot{\psi}\dot{\psi}} = 2T_{\dot{\psi}\dot{\psi}} = A(\dot{\phi})^2\sin^2\theta + C(\dot{\psi} + \dot{\phi}\cos\theta)^2, \tag{c3}$$

something that could have also been written down immediately from (a1) and the general result (8.5.5a)! With these expressions, we readily confirm that

$$\begin{aligned}\partial T''_{\Psi\Psi}/\partial\theta &= -(\partial T''_{\dot{\psi}\dot{\psi}}/\partial\theta) = \partial T_{\dot{\psi}\dot{\psi}}/\partial\theta \\ &= A(\dot{\phi})^2\sin\theta\cos\theta - C(\dot{\psi} + \dot{\phi}\cos\theta)\dot{\phi}\sin\theta,\end{aligned} \tag{c4}$$

and so the condition of steady motion (here, steady precession) (8.5.5c) becomes [assuming that $\sin\theta \neq 0$ and recalling that $C(\dot{\psi} + \dot{\phi}\cos\theta) \equiv C_\psi$]

$$\partial T''_{\dot{\psi}\dot{\psi}}/\partial\theta \equiv -(\partial T_{\dot{\psi}\dot{\psi}}/\partial\theta) = -(\partial V/\partial\theta): \qquad A(\dot{\phi})^2\cos\theta - C_\psi\dot{\phi} + mgl = 0, \tag{d1}$$

which is an equation relating the noncyclic *coordinate* θ with the cyclic *velocities* $\dot{\phi}$ and $\dot{\theta}$, at that state. Solving this quadratic algebraic equation for $\dot{\phi}$, we find

$$d\phi/dt = \left[C_\psi \pm (C_\psi{}^2 - 4A\,mgl\cos\theta)^{1/2}\right]/2A\cos\theta, \tag{d2}$$

from which it follows that if $C_\psi{}^2 > 4A\,mgl\cos\theta$, there will be *two* distinct values of $\dot{\phi}$ for which $\theta = \textit{constant}$. (The reader can compare this approach with those of the preceding examples.) Of course, the same equations and conditions would result by implementation of (8.5.3b), (8.5.3e); their details are left to the reader.

Problem 8.5.1 Consider the *steady precession* of the spinning top; that is, the special motion where $\dot{\phi} = \textit{constant}$, $\dot{\psi} = \textit{constant}$, and $\theta = \textit{constant}$. Using the results of ex. 8.4.5:

(i) Show that, in this case,

$$\begin{aligned}mgl\sin\theta &= [(C_\phi - C_\psi\cos\theta)/A\sin\theta]\{[C_\psi\sin^2\theta - (C_\phi - C_\psi\cos\theta)\cos\theta]/\sin^2\theta\} \\ &= (1/A)(A\dot{\phi}\sin\theta)(C_\psi - A\dot{\phi}\cos\theta).\end{aligned} \tag{a}$$

(ii) Further, and since $C_\psi = Cn \equiv C(\dot{\psi} + \dot{\phi}\cos\theta)$, show that

$$mgl = \dot{\phi}(Cn - A\dot{\phi}\cos\theta), \tag{b}$$

which is a functional relation of the form

$$F[\theta(\textit{nutation}),\ n(\textit{total spin}),\ \dot{\phi}(\textit{rate of precession})] = 0 \Rightarrow \theta = \theta(n, \dot{\phi}).$$

(iii) Finally, show that *for high total spins* condition (b) can be approximated by

$$mgl = C\dot{\phi}n; \tag{c}$$

that is, roughly, in the case of "equilibrium" known as *steady precession*, the destabilizing effect of gravity is balanced by the stabilizing (= restoring) effect of spinning.

Problem 8.5.2 Consider the cyclic system of our general theory; that is, $\partial L/\partial\psi_i = 0$.

(i) Show that under the (local and time-independent) coordinate transformation

$$\psi \to \psi' = \psi \qquad \text{and} \qquad q \to q' = f(q), \tag{a}$$

the ψ' remain ignorable.

(ii) Show that the steady motion conditions remain *invariant* under (a); that is,

$$\text{if } \dot{\psi} = \textit{constant} \text{ and } q = \textit{constant}, \text{ then also } \dot{\psi}' = \textit{constant} \text{ and } q' = \textit{constant}. \tag{b}$$

HINT

Apply chain rule to $L(\dot{\psi};\, q,\, \dot{q}) = L(\dot{\psi}';\, q',\, \dot{q}')$.

Example 8.5.2 Let us examine the total energy of our original (holonomic, stationary, potential, and cyclic) system at steady motions. Varying $H = T + V = H(q, p)$ *around such a state*, we find, successively [with $\Delta(\ldots)$ denoting generic variations of $(\ldots)$],

$$\begin{aligned}
\Delta H &= \sum [(\partial H/\partial q_k)\Delta q_k + (\partial H/\partial p_k)\Delta p_k] \qquad \text{[invoking the Hamiltonian identities]}\\
&= \sum [(-\partial L/\partial q_k)\Delta q_k + (\dot{q}_k)\Delta(\partial L/\partial \dot{q}_k)] \qquad \text{[recalling that } \partial L/\partial\psi_i = 0]\\
&= \sum [(-\partial L/\partial q_p)\Delta q_p + (\dot{q}_p)\Delta(\partial L/\partial \dot{q}_p)] + \sum \dot{\psi}_i\, \Delta\Psi_i\\
&\qquad\qquad\qquad\qquad \text{[invoking (8.5.1b), (8.5.3e)]}\\
&= \sum [(0)\Delta q_p + (0)\Delta(\partial L/\partial \dot{q}_p)] + \sum \dot{\psi}_i \Delta\Psi_i\\
&= 0, \qquad \text{if } \Delta\Psi_i = 0.
\end{aligned} \tag{a}$$

Hence, the following theorem.

THEOREM

At a state of steady motion, the energy of the original system is stationary, for vanishing variations of the cyclic momenta around that state.

Since, here, the Δq's and Δp's are viewed as independent, it is not hard to show that the *converse* is also true; that is, if $\Delta H = 0$, for $\Delta\Psi_i = 0$, then the state considered is one of steady motion—namely, there, $\dot{q}_p = 0$ and $\partial L/\partial q_p = 0$.

[This theorem is important in the Hamiltonian treatment of the stability of steady motion, along lines similar to the study of the stability of equilibrium via the stationarity/extremality of the total potential energy; see §8.6.]

8.6 STABILITY OF STEADY MOTION (OF CYCLIC SYSTEMS)

Continuing from the preceding section, we consider, again, a holonomic, scleronomic (no real loss in generality), potential, and cyclic system S in a so-called *fundamental state* of steady motion I, described by

$$\dot{\psi}_i = constant \equiv c_i \quad \text{and} \quad q_p = constant \equiv s_p$$
$$\Rightarrow \Psi_i = constant \equiv C_i \quad \text{and} \quad p_p \equiv \partial T/\partial \dot{q}_p = constant \equiv S_p.$$

$[i = 1, \ldots, M$ (number of ignorable coordinates), $p = M + 1, \ldots, n$
($n - M = g$: number of nonignorable coordinates)].

(8.6.1)

Next, we also consider S in an *adjacent* state of (generally, nonsteady) motion $II = I + \Delta(I)$, caused by *arbitrary* disturbances, and, hence, specified by the following general variations:

$$q_p \to q_p + \Delta q_p \equiv s_p + \Delta s_p \equiv s_p + z_p(t) \ (z_p\text{: } relative \text{ coordinates})$$
$$[\Rightarrow p_p \to p_p + \Delta p_p \equiv S_p + \Delta S_p \equiv S_p + Z_p(t)],$$
$$\Psi_i \to \Psi_i + \Delta\Psi_i \equiv C_i + \Delta C_i = constant \text{ (since } S \text{ is cyclic in both } I \text{ and } II)$$
$$[\Rightarrow \dot{\psi}_i \to \dot{\psi}_i + \Delta\dot{\psi}_i \equiv c_i + \Delta c_i(t) \equiv c_i + \eta_i(t)]. \tag{8.6.2}$$

Here is why: since our system S remains cyclic, the kinematico-inertial equations (8.3.12d, e; 8.5.4a)

$$\dot{\psi}_i = \sum C_{ij}(q)\Big[\Psi_j - \sum b_{pj}(q)\dot{q}_p\Big],$$

will hold in both states I and II. Therefore,

$$I: \quad c_i = \sum C_{ij}(s)\Big[C_j - \sum b_{pj}(s)\dot{s}_p\Big] = \sum C_{ij}(s)C_j \equiv \sum C_{ij}C_j$$
(since s_p: *constant*);

$$II: \quad c_i + \eta_i(t) = \sum C_{ij}[s + z(t)]\Big\{(C_j + \Delta C_j) - \sum b_{pj}[s + z(t)]\dot{z}_p\Big\}$$
$$\equiv \sum [C_{ij}(s) + \Delta C_{ij}(z; s)]\Big\{(C_j + \Delta C_j) - \sum [b_{pj}(s) + \Delta b_{pj}(z; s)]\dot{z}_p\Big\}$$
$$\equiv \sum (C_{ij} + \Delta C_{ij})\Big[(C_j + \Delta C_j) - \sum (b_{pj} + \Delta b_{pj})\dot{z}_p\Big]$$
$$\approx \sum C_{ij}C_j + \sum \Big[C_{ij}\,\Delta C_j + \Delta C_{ij}C_j - \sum (C_{ij}b_{pj})\dot{z}_p\Big] \tag{8.6.2a}$$

[to the *first order* in the $\Delta(I)$-deviations: z_p, $\dot{z}_p$, ΔC_i; $\Delta C_{ji} \approx \sum (\partial C_{ji}/\partial q_p)_I z_p$]

$$\Rightarrow \eta_i(t) = \sum \Big[C_{ij}\,\Delta C_j + C_j\,\Delta C_{ij}(t) - \sum C_{ij}b_{pj}\dot{z}_p(t)\Big]; \tag{8.6.2b}$$

that is,

$$\Delta\dot{\psi}(t) = Function\ of\ \Delta\Psi,\ \Delta q(t);\ \Psi,\ q. \tag{8.6.2c}$$

From the above, it follows that: (a) The specification of $\Delta(I)$, or II, requires n quantities: M $\Delta\Psi_i$'s, and $n - M$ $\Delta q_p(t)$'s.

(b) $\Delta\dot{\psi}_i \equiv \eta_i(t) \neq 0$, even if we assume that $\Delta\Psi_j \equiv \Delta C_j = 0$.

(c) After finding the palpable/noncyclic perturbations $z_p(t) \to \dot{z}_p(t) \to \Delta C_{ij}(t)$, we can calculate those of the cyclic velocities $\eta_i(t)$, from (8.6.2b), without any difficulty.

Important Clarifications

(i) Usually, but not always, we consider perturbations $\Delta(I)$ that preserve the values of the ignorable momenta; that is, $\Delta\Psi_i \equiv \Psi_i(II) - \Psi_i(I) = \Delta C_i = 0$. [As shown in ex. 8.5.2, such *equimomental* deviations are also *isoenergetic*; that is, the total energy is also preserved: $H(I) = H(II)$.]

(ii) By adjacent state, we mean one that can be adequately described by *linear(ized)* equations in the above nonignorable deviations $q_p(t)$ [and $p_p(t)$] and their $(\ldots)^\cdot$-derivatives.

Now, if, as $t \to \infty$, these perturbations remain bounded—for example, if they vary harmonically about the steady state I, or if they approach it asymptotically—then we say that I is *stable*; if not, then it is *unstable*. [As (8.5.4a) and (8.5.2b) show, a small change in the I-values $q(I)$, $p(I)$, $\Psi(I)$ produces a small change in the $\dot{\psi}(I)$'s: $\dot{\psi}_i \to \dot{\psi}_i + \Delta\dot{\psi}_i(t)$. But in view of the *linear variation of the ignorable coordinates with time*, eq. (8.5.4b), a small change in the $\dot{\psi}(I)$'s produces, after sufficient time, an arbitrarily large change in the $\psi(I)$'s. Therefore, *steady motions cannot be stable relative to their ignorable coordinates*.]

To study such perturbed motions, either:

• We substitute $q_p \to s_p + \Delta s_p \equiv s_p + z_p(t)$ [and $\dot{\psi}_i \to c_i + \Delta c_i(t) \equiv c_i + \dot{\psi}_i(t)$] in the Lagrangean equations of motion of the original system, or (8.6.2) (with $\Psi_i \to C_i + \Delta C_i \equiv C_i$, since we assumed that $\Delta C_i = 0$) in the Routhian equations of motion [i.e., the Lagrangean equations of the reduced, or conjugate, (sub)system]; and then keep only up to *first-order/linear* terms in the $\Delta s_p \equiv z_p(t)$, $\Delta c_i \equiv \eta_i(t)$ and their $(\ldots)^\cdot$-derivatives. Since the fundamental state is assumed steady, the so-resulting linear perturbation equations will have coefficients that will be known functions of the (assumed known) constant I-values C_i, s_p (or c_i, s_p), and so will be themselves known constants; or

• We substitute (2) in the exact Routhian $R(q_p, \dot{q}_p; \Psi_i)$ expand it à la Taylor around I [i.e., in powers of the $\Delta q_p(t) \equiv z_p(t)$ and their $(\ldots)^\cdot$-derivatives], keep only up to *second-order/quadratic* terms in them, while evaluating all derivatives at I, and then form Routh's equations for the nonignorable perturbations $z_p(t)$. Indeed, with $p, p' = M+1, \ldots, n$ and $(\ldots)_o \equiv (\ldots)_{\text{evaluated at } I}$ (to be used now and then for extra clarity), we obtain, successively,

$$R \equiv R(II) \equiv R[I + \Delta(I)] \equiv R(I) + \Delta R$$

or

$$\begin{aligned} R(q, \dot{q}; \Psi) &= R[s + z(t), \dot{s} + \dot{z}(t); C + \Delta C] = R[s + z(t), \dot{z}(t); C] \\ &\equiv R_{(0)} + R_{(1)} + R_{(2)}, \end{aligned} \tag{8.6.3}$$

where

$$R_{(0)} \equiv R(I) = R(s; C) = \textit{constant}, \tag{8.6.3a}$$

$$\begin{aligned} R_{(1)} &\equiv \sum [(\partial R/\partial q_p)_o z_p + (\partial R/\partial \dot{q}_p)_o \dot{z}_p]: \textit{linear homogeneous} \text{ in } z, \dot{z} \\ &= \sum [(0) z_p + (\partial R/\partial \dot{q}_p)_o \dot{z}_p] \quad \text{[invoking (8.5.3e)]} \end{aligned} \tag{8.6.3b}$$

$$2R_{(2)} \equiv \sum\sum \big[(\partial^2 R/\partial q_p\, \partial q_{p'})_o z_p z_{p'} + 2(\partial^2 R/\partial q_p\, \partial \dot{q}_{p'})_o z_p \dot{z}_{p'} + (\partial^2 R/\partial \dot{q}_p\, \partial \dot{q}_{p'})_o \dot{z}_p \dot{z}_{p'}\big]:$$

quadratic homogeneous in $z, \dot{z}$, (8.6.3c)

and, therefore {assuming $Q_p(I) = Q_p(II) = 0$, and since $(\partial R/\partial \dot{q}_p)_o = constant \Rightarrow [(\partial R/\partial \dot{q}_p)^{\cdot}]_o = 0$}, the linearized equations of state II

$$(\partial(\Delta R)/\partial \dot{z}_p)^{\cdot} - \partial(\Delta R)/\partial z_p = (\partial R_{(2)}/\partial \dot{z}_p)^{\cdot} - \partial R_{(2)}/\partial z_p = 0, \tag{8.6.3d}$$

become

$$\sum \left(\mu_{pp'}\, \ddot{z}_{p'} + \gamma_{pp'}\, \dot{z}_{p'} + \kappa_{pp'}\, z_{p'}\right) = 0, \tag{8.6.4}$$

where the *constant* coefficients μ, γ, κ are

$$\mu_{pp'} \equiv (\partial^2 R/\partial \dot{q}_p\, \partial \dot{q}_{p'})_o \qquad (= \mu_{p'p}\text{: positive definite}), \tag{8.6.4a}$$

$$\kappa_{pp'} \equiv -(\partial^2 R/\partial q_p\, \partial q_{p'})_o \qquad (= \kappa_{p'p},\ \text{note minus sign}), \tag{8.6.4b}$$

$$\gamma_{pp'} \equiv (\partial^2 R/\partial q_{p'}\, \partial \dot{q}_p - \partial^2 R/\partial q_p\, \partial \dot{q}_{p'})_o = \big[\partial/\partial \dot{q}_p(\partial R/\partial q_{p'}) - \partial/\partial \dot{q}_{p'}(\partial R/\partial q_p)\big]_o \qquad (= -\gamma_{p'p}\text{: sign indefinite}). \tag{8.6.4c}$$

REMARKS

(i) The μ- and κ-terms represent, respectively, *inertia* and "*elasticity*," as in ordinary linear vibration theory; the γ-terms, however, do not represent dissipation, but *gyroscopicity*; that is, they are powerless (§3.9):

$$\sum \left(\sum \gamma_{pp'}\, \dot{z}_{p'}\right) \dot{z}_p = 0. \tag{8.6.4d}$$

For example, for $n - M = 2$ (= minimum number of nonignorable coordinates for appearance of gyroscopicity; then $p, p' = 1, 2$), eqs. (8.6.4) read, in extenso,

$$\mu_{11}\ddot{z}_1 + \mu_{12}\ddot{z}_2 + \gamma_{12}\dot{z}_2 + \kappa_{11}z_1 + \kappa_{12}z_2 = 0, \tag{8.6.4e}$$

$$\mu_{21}\ddot{z}_1 + \mu_{22}\ddot{z}_2 + \gamma_{21}\dot{z}_1 + \kappa_{21}z_1 + \kappa_{22}z_2 = 0; \tag{8.6.4f}$$

that is, they involve *seven* distinct coefficients [*three inertial* (μ_{11}, $\mu_{12} = \mu_{21}$, μ_{22}) + *one gyroscopic* ($\gamma_{12} = -\gamma_{21}$) + *three elastic* ($\kappa_{11}$, $\kappa_{12} = \kappa_{21}$, κ_{22})].

(ii) Neither $R_{(0)}$ nor $R_{(1)}$ enter the equations of adjacent motion [recall mathematically similar situation in derivation of (3.10.12)].

(iii) The constancy of all coefficients in the expansions (8.6.3–8.6.3c), and therefore also in the perturbation equations (8.6.4–8.6.4c), has been used by Routh as the *mathematical* definition of steady motion. The *physical* characteristic of such a motion is, in his words, "that ... the same oscillations follow from the same disturbance of the same [nonignorable] coordinate at whatever instant the disturbance may be applied to the motion" [Routh (1905(b), p. 77].

To study the nature of the solutions of (8.6.4) with an eye toward their stability, and so on, and guided by the linear and unforced vibration case (i.e., linear homogeneous systems with constant coefficients), we substitute in (8.6.4)

$$z_p = z_{po} \exp(\lambda t), \qquad z_{po} = \text{constant amplitude}, \tag{8.6.5a}$$

and, proceeding in well-known ways, we find that, for nontrivial z_{po}'s, the (constant) λ's must be roots of the following determinantal (or *secular*, or *characteristic*) equation:

$$\Delta(\lambda) \equiv \begin{vmatrix} \mu_{11}\lambda^2 + \kappa_{11} & \mu_{12}\lambda^2 + \gamma_{12}\lambda + \kappa_{12} & \cdots & \mu_{1g}\lambda^2 + \gamma_{1g}\lambda + \kappa_{1g} \\ \mu_{21}\lambda^2 + \gamma_{21}\lambda + \kappa_{21} & \mu_{22}\lambda^2 + \kappa_{22} & \cdots & \mu_{2g}\lambda^2 + \gamma_{2g}\lambda + \kappa_{2g} \\ \cdots & \cdots & \cdots & \cdots \\ \mu_{g1}\lambda^2 + \gamma_{g1}\lambda + \kappa_{g1} & \mu_{g2}\lambda^2 + \gamma_{g2}\lambda + \kappa_{g2} & \cdots & \mu_{gg}\lambda^2 + \kappa_{gg} \end{vmatrix} = 0, \tag{8.6.5b}$$

$= 2g$-degree polynomial in λ ($g \equiv n - M = \#$ *nonignorable coordinates*).

The complete (or general) solution of (8.6.4) will equal the linear superposition of the $2g$ (8.6.5a)-like solutions; one for each of the $2g$ roots of (8.6.5b).

The stability/instability of the fundamental state I is determined by the nature (and/or sign) of these roots; which, in turn, are determined by the coefficients μ, γ, κ, whose values depend on that state. In general (recall summary in §3.10), roots that are:

- *real* and *positive*, or *complex with positive real parts* signal *instability* (i.e., solutions increase exponentially with time);
- *real* and *negative*, or *complex with negative real parts* signal (asymptotic) *stability* (i.e., solutions decrease exponentially with time);
- *purely imaginary* signal *stability* (i.e., constant amplitude oscillations. This case is called *critical*: actually, our linearized stability analysis is inconclusive; we need *nonlinear* perturbation equations for $\Delta(I)$ to safely ascertain the stability/instability of I).

Hence, to test stability: either

(i) We find all roots of (8.6.5b) and then check to see if their real parts are negative; or
(ii) We apply any one of a number of existing criteria that does not actually find the roots, but checks the sign of their real parts; for example, Routh–Hurwitz test (§3.10, Method of Small Oscillations).

As in ordinary linear vibration theory, the reason that (8.6.5b) is not so easy to study is because equations (8.6.4) are coupled. An important simplification occurs if we choose new $\Delta(I)$-coordinates $z_p \to x \equiv (x_1, \ldots, x_g)$, via an ever-existing linear, real, and nonsingular transformation that diagonalizes (i.e., uncouples) simultaneously both matrices ($\mu_{pp'}$: positive definite) and ($\kappa_{pp'}$), and thus reduces the "perturbation kinetic and potential energies"

$$2R_{(2)T} \equiv \sum\sum (\partial^2 R/\partial\dot{q}_p\, \partial\dot{q}_{p'})_o\, \dot{z}_p\, \dot{z}_{p'} \equiv \sum\sum \mu_{pp'}\, \dot{z}_p\, \dot{z}_{p'}, \tag{8.6.6a}$$

$$2R_{(2)V} \equiv -\sum\sum (\partial^2 R/\partial q_p\, \partial q_{p'})_o\, z_p\, z_{p'} \equiv \sum\sum \kappa_{pp'}\, z_p\, z_{p'}, \tag{8.6.6b}$$

respectively, to *sums of squares*:

$$2R_{(2)T} = \sum \mu_p(\dot{x}_p)^2 \quad \text{and} \quad 2R_{(2)V} = \sum \kappa_p\, x_p{}^2. \tag{8.6.6c}$$

[Note minus sign in $R_{(2)V}$, to give $R_{(2)}$ the form of a Lagrangean; see also (8.6.10b).] However, then, the "gyroscopic energy"

$$R_{(2)G} \equiv \sum\sum (\partial^2 R/\partial q_p\, \partial \dot{q}_{p'})_o z_p \dot{z}_{p'} \equiv \sum\sum E_{p'p} z_p \dot{z}_{p'} \qquad (E_{p'p} \neq E_{pp'}) \tag{8.6.6d}$$

(which does not exist in linear, unforced, and undamped vibrations about *equilibrium*) transforms, in general, to another *nondiagonal* form,

$$R_{(2)G} = \sum\sum \varepsilon_{p'p} x_p \dot{x}_{p'} \qquad \left[\equiv \sum \left(\sum \varepsilon_{p'p} x_p\right) \dot{x}_{p'}\right]. \tag{8.6.6e}$$

As mentioned above, since $R_{(2)T}$ (but not necessarily $R_{(2)V}$) is positive definite, such a *partially decoupling transformation* is always possible; but it must be borne in mind that because the "elastic" coefficients $\kappa_{pp'}$, in general, depend on the $\dot{\psi}$, Ψ's, [or, in the mathematically equivalent case of small motion around *relative* equilibrium (recall §3.16), they depend on the constant angular velocity of the rotating frame] the x's may also depend on them as parameters. [The x's are sometimes called principal coordinates, just like the (*completely decoupling*) principal/normal coordinates of ordinary (i.e., nongyroscopic) vibration theory.]

In these coordinates, the Lagrange-type equations of perturbed motion

$$(\partial R_{(2)}/\partial \dot{x}_p)^{\cdot} - \partial R_{(2)}/\partial x_p = 0, \tag{8.6.7a}$$

where $R_{(2)} = R_{(2)T} + R_{(2)G} - R_{(2)V}$, assume the simpler form

$$\mu_p \ddot{x}_p + \sum g_{pp'} \dot{x}_{p'} + \kappa_p x_p = 0, \tag{8.6.7b}$$

where $g_{pp'} \equiv \varepsilon_{pp'} - \varepsilon_{p'p} = -g_{p'p}$; and upon substituting $x_p = x_{po}\exp(\lambda t)$ into them, and so on, we are led to the simpler characteristic equation

$$\Delta(\lambda) \equiv \begin{vmatrix} \mu_1\lambda^2 + \kappa_1 & g_{12}\lambda & \cdots & g_{1g}\lambda \\ g_{21}\lambda & \mu_2\lambda^2 + \kappa_2 & \cdots & g_{2g}\lambda \\ \cdots & \cdots & \cdots & \cdots \\ g_{g1}\lambda & g_{g2}\lambda & \cdots & \mu_g\lambda^2 + \kappa_g \end{vmatrix} = 0. \tag{8.6.8}$$

Now, the determinant $\Delta(\lambda)$ is *nonsymmetric* [unlike the corresponding determinant of the nongyroscopic case (undamped vibration around *absolute* equilibrium) which, as is well known, is symmetric], but its off-diagonal elements are antisymmetric: $g_{pp'} = -g_{p'p}$. Hence, reversing the sign of λ in $\Delta(\lambda)$ simply interchanges its rows and columns [or, the rows (columns) of $\Delta(\lambda)$ equal the columns (rows) of $\Delta(-\lambda)$] and so, by determinant theory,

$$\Delta(\lambda) = \Delta(-\lambda); \tag{8.6.8a}$$

in words, eq. (8.6.8) [as well as its completely uncoupled version (8.6.5b), and the nongyroscopic case] *is independent of the sign of* λ. Therefore *all odd* λ*-powers are absent* from it:

$$0 = \Delta(\lambda) \equiv A_g(\lambda^2)^g + A_{g-1}(\lambda^2)^{g-1} + \cdots + A_1(\lambda^2) + A_0$$
$$[(g)\text{th degree polynomial in } \lambda^2]. \tag{8.6.8b}$$

Next, as algebra teaches, since the coefficients μ, κ, g are real, the g λ^2-roots of (8.6.8b) will, in general, be either *real* or *complex conjugate pairs*, like

$$\lambda^2 = \alpha \pm i\beta \qquad (\alpha, \beta\text{: real} \Rightarrow \lambda = \pm(\varepsilon \pm i\sigma) \qquad (\varepsilon, \sigma\text{: real}). \tag{8.6.8c}$$

and such λ's will produce x's of the following general (real) form:

$$C \exp(\varepsilon t)\cos(\sigma t + c) + D \exp(-\varepsilon t)\cos(\sigma t + d), \tag{8.6.8d}$$

where C, c; D, d are real constants, and, therefore, unless $\varepsilon = 0$, the amplitudes $C \exp(\varepsilon t)$, $D \exp(-\varepsilon t)$ will increase indefinitely, that is, the state I will be unstable. Hence the rule.

RULE

For the fundamental state of steady motion I to be stable, in the above sense, all λ^2-roots of the characteristic equation (8.6.8–8.6.8b) must be *real* and *negative*, that is,

$$\lambda^2\text{-roots} = -\lambda_p{}^2 \Rightarrow \lambda\text{-roots} = \pm i\,\lambda_p \qquad (\lambda_p\text{: real, } p = 1, \ldots, g). \tag{8.6.8e}$$

Algebraic Detour

The theory of algebraic (polynomial) equations allows us to relate the roots of (8.6.8b) with its coefficients $A_g, A_{g-1}, \ldots, A_1, A_0$. According to the fundamental theorem of algebra (see books on algebra, or handbooks of engineering, mathematics, etc.), if $\Lambda_1, \ldots, \Lambda_{2g}$ are the $2g$ roots of (8.6.8b) (i.e., $\Lambda_1 = +i\lambda_1$, $\Lambda_2 = -i\lambda_1$, etc.), then

$$\Delta(\lambda) = A_g(\lambda - \Lambda_1)(\lambda - \Lambda_2)\cdots(\lambda - \Lambda_{2g}) \qquad \text{(always);} \tag{8.6.9a}$$

and, therefore,

$$\begin{aligned} \Delta(0) &= A_g(-1)^{2g}\Lambda_1\Lambda_2\cdots\Lambda_{2g} = +A_g\Lambda_1\Lambda_2\cdots\Lambda_{2g} \\ &= \kappa_1\kappa_2\cdots\kappa_g = A_0 \quad [\text{by (8.6.8)}], \qquad \text{(always);} \end{aligned} \tag{8.6.9b}$$

and

$$[\lim \Delta(\lambda)]_{\lambda\to\infty} \equiv \Delta(\infty) > 0 \qquad \text{(always).} \tag{8.6.9c}$$

Hence *in the case of stability*—namely (8.6.8e)—and since then to each stable pair of λ-roots, $\pm i\,\lambda_p$, there corresponds in $\Delta(\lambda)$ a factor $[\lambda - (+i\lambda_p)][\lambda - (-i\,\lambda_p)] = \lambda^2 - (-\lambda_p{}^2) = \lambda^2 + \lambda_p{}^2$, eq. (8.6.9a) must have the following form ($A_g > 0$, with no loss in generality):

$$\Delta(\lambda) = A_g(\lambda^2 + \lambda_1{}^2)\ldots(\lambda^2 + \lambda_g{}^2) \qquad \text{(stability case),} \tag{8.6.9d}$$

that is, be a *polynomial with all its coefficients positive*; and so in this case

$$\Delta(0) = A_g(\lambda_1{}^2\lambda_2{}^2\ldots\lambda_g{}^2) > 0 \qquad \text{(stability case).} \tag{8.6.9e}$$

Further, according to *Viète's rules* (and counting k-ple roots k times),

$$\Lambda \equiv \Lambda_1\Lambda_2\cdots\Lambda_{2g} = (-1)^{2g}(A_0/A_g) = A_0/A_g \qquad \text{(always)} \tag{8.6.9f}$$

and so *in the case of stability* we must have

$$\Delta = [(-i\lambda_1)(+i\lambda_1)]\ldots[(-i\lambda_g)(+i\lambda_g)] = \lambda_1{}^2\lambda_2{}^2\ldots\lambda_g{}^2$$

$$[\text{also: } (-\lambda_1{}^2)(-\lambda_2{}^2)\ldots(-\lambda_g{}^2) = (-1)^g(\lambda_1{}^2\lambda_2{}^2\ldots\lambda_g{}^2) = (-1)^g(A_0/A_g)]$$

$$\Rightarrow \lambda_1{}^2\lambda_2{}^2\ldots\lambda_g{}^2 = A_0/A_g > 0 \qquad \text{(stability case).} \qquad (8.6.9g)$$

Last, to express this necessary condition for gyroscopic stability in terms of the nongyroscopic parameters of our system in state I (i.e., in terms of the κ_p's and the μ_p's), we expand the determinant (8.6.8) and compare it with (8.6.8b) [or use mathematical *induction*, i.e., confirm it for $g \to 2$, then assume it holds for $g \to g$, and finally prove it for $g \to g+1$]. Thus we get

$A_0 = \kappa_1\kappa_2\ldots\kappa_g = Det(\kappa_{pp'}) = Det(\kappa_p)$ $[\neq 0$, if all $\kappa_p \neq 0]$:

Product of coefficients of (nongyroscopic, or irrotational) stability of $R_{(2)V}$,

(always), (8.6.9h)

$A_g \equiv \mu_1\mu_2\ldots\mu_g = Det(\mu_{pp'}) = Det(\mu_p) \neq 0$:

Product of coefficients of inertia (~ masses) of positive definite $R_{(2)T}$,

(always). (8.6.9i)

In view of the above, the essential stability condition (8.6.9g) translates to

$$\lambda_1{}^2\lambda_2{}^2\ldots\lambda_g{}^2 = (\kappa_1\kappa_2\ldots\kappa_g)/(\mu_1\mu_2\ldots\mu_g) > 0 \qquad \text{(stability case);} \qquad (8.6.9j)$$

or, equivalent, since $A_g \equiv \mu_1\mu_2\ldots\mu_g > 0$, to

$$\Delta(0) = A_0 = \kappa_1\kappa_2\ldots\kappa_g > 0 \qquad \text{(stability case).} \qquad (8.6.9k)$$

These results lead us to the following stability criteria [Kelvin and Tait (1860s)]:

Criteria of Gyroscopic Stabilization

Consider a fundamental state of steady motion (ignored coordinates) I, of a cyclic system [or a state of relative equilibrium (rheonomic constraints—§3.16) of a general system], and let (8.6.7b) be the equations of linearized perturbations from I (i.e., no friction taken into account). Then

- If all κ_p's are positive [$\Rightarrow$ $R_{(2)V}$: positive definite $\Rightarrow$ $R_{(2)V}(I)$: minimum, and all λ^2-roots are negative $= -\lambda_p{}^2 < 0$], then I is stable, both nongyroscopically (i.e., with all $g_{pp'}$'s absent) and gyroscopically (i.e., with at least one pair of $g_{pp'}$'s present). If even one κ_p vanishes while the rest remain positive [$R_{(2)V}$: positive semidefinite], then I is unstable both nongyroscopically and gyroscopically. [For alternative proofs including the well-known stability arguments of Dirichlet and Kelvin, see, e.g., Lamb (1943, pp. 245–248; 1932, pp. 310–313), also discussion following (8.6.10).]
- If all κ_p's are negative [$\Rightarrow$ $R_{(2)V}$: negative definite $\Rightarrow$ $R_{(2)V}(I)$: maximum], then I is nongyroscopically unstable. However, if the number of these negative κ_p's is *even* [$\Rightarrow \Delta(0) = A_0 > 0$ and $\Delta(\infty) > 0$], then I can always be stabilized gyroscopically—at least temporarily (see destabilizing effect of ever-present friction, below); but if their number is *odd* [$\Rightarrow \Delta(0) = A_0 < 0$ and $\Delta(\infty) > 0$, i.e., at least one λ-root is positive], then gyroscopic stabilization of I is impossible—in this case, gyroscopic

effects cannot save I from instability. If even one κ_p vanishes while the rest remain negative [$\Rightarrow R_{(2)V}$: negative semidefinite], then I is unstable both nongyroscopically and gyroscopically.

• If some κ_p's are positive, some are negative, and some are zero [$\Rightarrow R_{(2)V}$: indefinite $\Rightarrow R_{(2)V}(I)$: min/max ("saddle point")], then state I can, sometimes, be stabilized gyroscopically; specifically, if no vanishing κ_p's are present, and if the number of negative κ_p's is even.

[The case of equal, or multiple, roots in $\Delta(\lambda) = 0$, as in the nongyroscopic case, is due to accidental properties of the system's physical and geometrical parameters, and, therefore, does not create any real complications; see, e.g., Routh (1905(b), p. 82); also Frank (1935, pp. 136–138).] These conclusions can, of course, also be reached by application of the Routh–Hurwitz theorem to (8.6.8b); see references given in §3.10, and Bellet (1988, pp. 311–327), Grammel (1950, vol. 1, pp. 258–262), Winkelmann and Grammel (1927, pp. 481–483), Merkin (1987, pp. 168–184).

In sum:

- Gyroscopic effects cannot destabilize a state of steady motion, but sometimes they can stabilize it.
- If the number of nonignorable freedoms is even (and no κ_p vanishes), then either I is stable or it can always be stabilized; or, if g is even, gyroscopic stabilization is always possible.

In view of these results, it has become necessary to distinguish stability/instability, in the context of cyclic systems and relative equilibrium, into one based on the $\lambda_p{}^2$'s and one based only on the κ_p's [equivalently, on the extremum–min/max properties of $R_{(2)V}$. We have just seen that (a) if $R_{(2)V}$ is *positive definite*, then I is *stable, both nongyroscopically and gyroscopically;* (b) if it is *nonpositive definite,* then I is *nongyroscopically unstable*; and (c) if it is *semidefinite, whether positive or negative* (e.g., $R_{(2)V} \equiv 0$), then I is *gyroscopically unstabilizable.*] Let us elaborate on these concepts.

• Stability ascertained on the basis of the above-presented method of *small oscillations* (i.e., of conditions for the roots of the associated characteristic equation, which includes gyroscopic effects, to be either *purely imaginary* or have *negative real parts*) is called *ordinary*, or *temporary* (due to the eventual destabilization by damping—see below), or *dynamical* (since it is based on equations of motion), and is associated with Lagrange and Routh.

• A second stability method, for holonomic and potential systems, called *practical*, or *permanent*, or *secular* (due to its application to problems of celestial mechanics; e.g., stability of rotating liquid masses), and associated with the names of Kelvin, Poincaré, et al., is based on the *extremum* properties of the system's total potential energy at the fundamental state I, here the negative of R_0; that is,

$$
\begin{aligned}
-R_0 &\equiv -(T''_{0,2} - V) = V - T''_{0,2} = V + T_{0,2} \\
&\equiv \Big[(1/2)\sum\sum C_{ji}(q)\Psi_j\Psi_i + V(q)\Big]_I \equiv \textit{reduced (total) potential}, \qquad (8.6.10)
\end{aligned}
$$

and this, in turn, is based on the earlier (Jacobi–Painlevé) energy integral (8.4.13a, b, 14):

$$
h_R \equiv R_2 - R_0 = T''_{2,0} + (V - T''_{0,2}) = \textit{constant} \qquad (R_2 = \textit{positive definite}), \tag{8.6.10a}
$$

(i.e., it takes into account the rotation (κ) but not its gyroscopic effects (g)!) and a reasoning identical to that used in the stability of ordinary (i.e., inertial) equilibrium via the well-known equation $T + V \equiv E = constant$ [what is, generally, referred to as theorem of Dirichlet (1846); see e.g., Lamb (1943, pp. 214–215)]. Equivalently, since the earlier linear perturbation equations (8.6.3d, 4) can be rewritten, with the help of the general quadratic/bilinear forms (8.6.6a–d), as

$$\left[(\partial R_{(2)T}/\partial \dot{z}_p)^{\cdot} - \partial R_{(2)T}/\partial z_p\right] + \left[(\partial R_{(2)G}/\partial \dot{z}_p)^{\cdot} - \partial R_{(2)G}/\partial z_p\right] \\ - \left[(\partial R_{(2)V}/\partial \dot{z}_p)^{\cdot} - \partial R_{(2)V}/\partial z_p\right] = 0,$$

or

$$(\partial R_{(2)T}/\partial \dot{z}_p)^{\cdot} + \left[(\partial R_{(2)G}/\partial \dot{z}_p)^{\cdot} - \partial R_{(2)G}/\partial z_p\right] + \partial R_{(2)V}/\partial z_p = 0, \qquad (8.6.10b)$$

and [by multiplication of each with $\dot{z}_p$, summation over $p = M + 1, \ldots, n$, and then invocation of (8.6.4)] readily yield the *perturbational energy* integral:

$$R_{(2)T} + R_{(2)V} = constant$$

$$(R_{(2)T} = positive\ definite;\ \text{no } R_{(2)G} \text{ present, as in relative motion}), \qquad (8.6.10c)$$

for these reasons, *we may, in our practical stability investigation, replace* $-R_0$ *with its quadratic approximation* $R_{(2)V}$:

$$2R_{(2)V} \equiv -\sum\sum (\partial^2 R/\partial q_p\, \partial q_{p'})_o\, z_p\, z_{p'} \equiv \sum\sum \kappa_{pp'}\, z_p\, z_{p'} = \sum \kappa_{p'} x_p{}^2. \qquad (8.6.10d)$$

According to this criterion, the *fundamental state* I is called *practically stable* if

$$-R_0 > 0 \Rightarrow R_0 < 0, \qquad \text{or} \qquad R_{(2)V} > 0. \qquad (8.6.10e)$$

Since, by (8.5.3b), $(\partial R_0/\partial q_p)_o = 0$, or $(\partial R_{(2)}/\partial q_p)_o = -(\partial R_{(2)V}/\partial z_p)_o = 0$, the above mean that I is stable, in that sense, *if* $R_0(R_{(2)V})$ is a *strict maximum (minimum) there.*

For then, arguing à la Dirichlet, the integral (8.6.10c) will yield

$$R_{(2)T} + (1/2)\sum \kappa_{p'} x_p{}^2 = small\ positive\ constant \equiv c, \qquad (8.6.11a)$$

from which, since $R_{(2)T}$ is positive definite, we conclude that no x_p can ever exceed a certain small value; for example, $|x_1| \leq (2c/\kappa_1)^{1/2}$. Hence, in the *absence of friction*, the *system oscillates around* I, as in stability about ordinary equilibrium. This is the *sufficiency* part of the theorem. The *necessity* part is most easily established by taking into account the always present *friction* during every motion from I. Then (8.6.10c, 11a) are replaced by the power equation

$$d/dt(R_{(2)T} + R_{(2)V}) = negative\ quantity, \qquad (8.6.11b)$$

which implies that, as long as even one $\dot{x}_p$ is nonzero, the perturbational energy $R_{(2)T} + R_{(2)V}$ decreases monotonically until, eventually, *both* $R_{(2)T}$ and $R_{(2)V}$ vanish; that is, all x's and $\dot{x}$'s vanish simultaneously. [It can be shown that this stability criterion also holds if we vary the cyclic momenta; that is, even for $\Psi_i(I) \equiv C_i$, $\Psi_i(II) \equiv C_i + \Delta C_i$; see, for example, Gantmacher (1970, pp. 255–256).]

Since the criteria of practical stability (being *energetic* and not involving the gyroscopic terms) are easier to apply than those of ordinary stability (which are *algebraic* and do involve the gyroscopic terms), it is important to know when these two approaches are completely equivalent.

The foregoing discussion allows us to summarize this comparison in the following complementary statements:

(i) *If state I is practically stable* (*PS*; i.e., *all* $\kappa_p > 0 \Rightarrow R_{(2)V} =$ *positive definite*), *it is also ordinarily stable* (OS; i.e., *all* λ^2 roots *real and negative*); that is, a nongyroscopically stable state remains stable upon addition of gyroscopic effects to it. Accordingly, *if I is ordinarily unstable, it is also practically unstable.*

(ii) *If state I is practically unstable* (say, at least one $\kappa_p < 0 \Rightarrow R_{(2)V} \neq$ *positive definite*), *it may or may not be ordinarily unstable*, depending on whether the number of negative κ_p's is odd (instability) or even (stability); that is, a nongyroscopically unstable state may become stable upon addition of gyroscopic effects to it (Kelvin's stabilization theorem); and *if it is OS, it may or may not be PS*, depending on whether no κ_p is negative or zero (stable) or at least one of them is (unstable). In sum: *PS is sufficient but not necessary for OS.* [For additional details, see, for example (alphabetically): Greenwood (1977, pp. 125–128), Lamb (1943, chap. 11), Langhaar (1962, chap. 1), Thomson and Tait (1912), Ziegler (1968, chaps. 1, 2, 4); for *Hamiltonian* treatments, see, for example, Gantmacher (1970, pp. 252–256), Frank (1935, pp. 129–133), Synge (1960, pp. 191–195).]

Finally, let us examine the effect of (light) *damping* on the nature of these instabilities. We will restrict ourselves to the common case where, say, the Rayleigh dissipation function of the system, *in the* $n - M \equiv g$ *nonignored coordinate* perturbations z or x, *is positive definite* [i.e., if $z, x \neq 0$, then $(R_{(2)T} + R_{(2)V})^{\cdot} < 0$, also known as *complete* damping—to be distinguished from the, more general, *pervasive* damping where Rayleigh's dissipation function is positive but *semidefinite* in z, x]; while the M ignored coordinates ψ will be assumed to remain undamped.

(i) Let state I be *ordinarily stable but practically unstable.* In the absence of friction, as already stated, any small disturbance will simply result in oscillations about I. In the presence of friction, on the other hand, due to (8.6.11b), and since then $R_{(2)T}(I) = 0$ while $R_{(2)V}(I) =$ *maximum*, we will have, *initially*,

$$R_{(2)T} + R_{(2)V} = -(\textit{small positive constant}) \equiv -c, \tag{8.6.12a}$$

and so, later, either $R_{(2)T}$ or $R_{(2)V}$, or both, will be nonzero; that is, the system *will* move away from I, regardless of any gyroscopic effects. But then $R_{(2)T} + R_{(2)V}$ will decrease further, so that, *after a short time t*,

$$R_{(2)T} + R_{(2)V} = -c - kt \qquad (k > 0); \tag{8.6.12b}$$

which means that, since $R_{(2)T}$ is positive definite, $R_{(2)V}$ will keep decreasing further. As a result, the system will move further away from I; that is, the deviation amplitude(s) will increase indefinitely with time, but at a rate depending on the friction present: the larger the friction, the faster the deviation, and vice versa. Such unavoidable destabilization can be slowed down either by decreasing friction or by countering its effect with some other, external, influences.

For example, a spinning gyroscope stabilized against gravity by its spinning, but destabilized by the friction at its support (vertex) and aerodynamic forces, slows down (i.e., its nutation angle gets larger and larger, and its spin decreases) and eventually hits the ground and comes to rest.

(ii) Let state I be *ordinarily unstable*; that is,

$$x \sim \exp(\pm \varepsilon t), \qquad \varepsilon = real. \tag{8.6.13}$$

In the *absence* of friction, the system moves quickly away from I. In the *presence* of friction, the system still moves away from I, but *less quickly*, until it reaches another state of steady motion or relative equilibrium. In sum: (complete) damping changes stability ($R_{(2)V}$: positive definite) to the slightly stronger *asymptotic* stability, but it does not change instability ($R_{(2)V}$: nonpositive definite); that is, such damping does not alter the nature of a state of motion in any significant/critical way. Last, we should always remember that our perturbation equations (8.6.4), (8.6.7b) only show the nature of the *initial* motion away from I; to find other such states we need the exact, and generally *nonlinear*, perturbation equations. (See, e.g., articles in Mikhailov and Parton, 1990, and references cited therein.)

In the light of the above, the practical conclusions of Kelvin's theorem can be summed up as follows:

(i) Only systems with an *even* number of unstable nonignorable coordinates can be stabilized gyroscopically.
(ii) In the *absence* of friction, such a stabilization can always be achieved via appropriately oriented and sufficiently fast-spinning gyroscopes (*one* fixed point, relative to the housing) and/or gyrostats (*two* fixed points, relative to the housing) built into the system.
(iii) In the presence of *damped* nonignorable coordinates, to counter the destabilizing frictional losses and thus stabilize our system, we must supply it with external energy.

For extensive and authoritative treatments of the effects of friction on gyroscopic systems, see the earlier-mentioned texts of Merkin (1987) and Ziegler (1968); also Klotter (1960/1981, pp. 186–199, 241–253).

Example 8.6.1 Let us consider a system with kinetic and potential energies

$$2T = A\dot{x}^2 + 2\Gamma\dot{x}\dot{y} + B\dot{y}^2, \qquad V = V(x), \tag{a}$$

where A, B, Γ are functions of x only (and such that T remains positive definite); and examine the possible existence of steady motions: $x = constant \equiv s$, $\dot{y} = constant \equiv c_y \equiv c$, and their stability.

(i) *Steady motion*: since y is ignorable, we will have

$$\partial T/\partial\dot{y} = \Gamma\dot{x} + B\dot{y} = constant \equiv C_y \equiv C$$
$$\Rightarrow \dot{y} = (C - \Gamma\dot{x})/B, \text{ and so, for a steady motion, } c = C/B, \tag{b}$$

and therefore the Routhian function becomes, successively,

$$\begin{aligned} R &= (T - V) - (\partial T/\partial\dot{y})\dot{y} \\ &= (A/2)(\dot{x})^2 + \Gamma\dot{x}[(C - \Gamma\dot{x})/B] + (B/2)[(C - \Gamma\dot{x})/B]^2 \\ &\qquad - C[(C - \Gamma\dot{x})/B] - V(x) \\ &= \cdots = T''_{2,0} + T''_{1,1} + T''_{0,2} - V(x) = R(x, \dot{x}; C), \end{aligned} \tag{c}$$

where

$$T''_{2,0} = (1/2)\{[A - (\Gamma^2/B)]\}(\dot{x})^2,$$
$$T''_{1,1} = (\Gamma/B)C\,\dot{x},$$
$$T''_{0,2} = -(1/2B)C^2. \qquad \text{(c1)}$$

For steady motion, and with the notation $(\ldots)_o \equiv (\ldots)_{x=s,\,\dot{y}=c}$, the above yield

$$(\partial R/\partial x)_o = [\partial(T''_{0,2} - V)/\partial x]_o = [(C^2/2B^2)(dB/dx) - (dV/dx)]_o = 0,$$

or, due to (b),

$$(c^2/2)(dB/dx)_o = (dV/dx)_o. \qquad \text{(d)}$$

This algebraic (equilibrium-like) equation connects the values of the palpable coordinate (s) and ignorable velocity (c) at steady motion(s), and allows us to find one in terms of the other.

(ii) *Stability*. Substituting into R: $x = s + z(t)$, expanding à la Taylor, and keeping only *up to quadratic* q-powers, since we are seeking *linear* perturbation equations, we obtain

$$\begin{aligned} R = (1/2)\{&[A - (\Gamma^2/B)]_o(\dot{z})^2 + 2C(\Gamma/B)_o\,\dot{z} - C^2/B_o \\ &+ C^2[B^{-2}(dB/dx)]_o\,z + (C^2/2)[d/dx(B^{-2}(dB/dx))]_o\,z^2\} \\ &- [V_o + (dV/dx)_o\,z + (1/2)(d^2V/dx^2)_o\,z^2] \\ = R(&z, \dot{z}; C, s) \qquad \text{[by (d), the } (\ldots)\ z\text{-terms cancel each other]}, \end{aligned} \qquad \text{(e)}$$

and therefore the z-equation of motion $(\partial R/\partial \dot{z})^{\cdot} - \partial R/\partial z = 0$ becomes

$$[A - (\Gamma^2/B)]_o\,\ddot{z} + \{(d^2V/dx^2)_o - (C^2/2)[d/dx(B^{-2}(dB/dx))]_o\}z = 0. \qquad \text{(f)}$$

Since, here, $n - M = 2 - 1 = 1$, no $\sim \dot{z}$ (gyroscopic) terms appear in (f). Clearly, the z-motion is harmonic ($\Rightarrow$stable) if

$$\{(d^2V/dx^2)_o - (C^2/2)[d/dx(B^{-2}(dB/dx))]_o\}\big/[A - (\Gamma^2/B)]_o > 0, \qquad \text{(g)}$$

or, equivalently, since $A - (\Gamma^2/B) > 0$ (due to the positive definiteness of T in $\dot{x}$, $\dot{y}$) and by (b) $cB = C$,

$$B(d^2V/dx^2)_o + c^2[(dB/dx)_o]^2 - (c^2/2)[B(d^2B/dx^2)]_o > 0. \qquad \text{(h)}$$

Problem 8.6.1 Continuing with the system described by (a) of the preceding example, *but with* $\Gamma = 0$, consider its fundamental steady state I: $x = s$ and $\dot{y} = c$, and the adjacent to it $I + \Delta(I)$: $x = s + z(t)$ and $\dot{y} = c + \eta(t)$.

(i) Show that the (linearized) equations of $\Delta(I)$ are

$$A_o\,\ddot{z} - (c^2/2)(d^2B/dx^2)_o\,z - c(dB/dx)_o\,\eta + (d^2V/dx^2)_o\,z = 0, \qquad \text{(a)}$$

$$B_o\dot{\eta} + c(dB/dx)_o\,\dot{z} = 0. \qquad \text{(b)}$$

(ii) Verify that by eliminating η between (a, b) we recover (f), with $\Gamma = 0$.
(iii) Show that state I of that system (e.g., rotation at the rate c), is stable if

$$\{(d^2V/dx^2)/(dV/dx)\}_o > \{[B(d^2B/dx^2) - 2(dB/dx)^2]/[B(dB/dx)]\}_o. \qquad \text{(c)}$$

Example 8.6.2 *General Solution of the Frictionless Gyroscopic Equations.* Assuming $x_p = x_{po}\exp(\lambda t)$ for the solutions of the characteristic equation (8.6.8), $\Delta(\lambda) = 0$, then, for a particular root $\lambda*$ $(* = 1, \ldots, 2g)$, we will have

$$x^*_{1o}/\Delta^*_1 = \cdots = x^*_{go}/\Delta^*_g = constant \equiv C*, \qquad \text{(a)}$$

where

$$\Delta^*_p = \text{minors of any row of } \Delta(\lambda*) = M^*_p + i\,N^*_p, \qquad \text{(a1)}$$

since, in general, these minors contain both *odd* and *even* powers of $\lambda*$. Therefore, the $(*)$th "natural mode" of oscillation, around the fundamental state I, can be written as

$$x^*_p = x^*_{po}\exp(\lambda* t) = C*(M^*_p + i\,N^*_p)\exp(\lambda* t), \qquad \text{(b)}$$

or, setting $C* = X*\exp(i\,\delta*)$, where $X*$, $\delta* =$ real constants, and taking real parts,

$$x^*_p = X*\left[M^*_p\cos(\omega*\,t + \delta*) - N^*_p\sin(\omega*\,t + \delta*)\right]. \qquad \text{(b1)}$$

Hence, in the stable case, the oscillations corresponding to a particular (negative) $\lambda*^2$ *have the same frequency but do not move in phase*; that is, the latter varies with the coordinates. Such a natural mode is referred to as "elliptic harmonic" to distinguish it from the "circular harmonic" of the nongyroscopic oscillations.

In the case of *distinct* roots, the general solution is found by superposition of the various modes:

$$x_p = \sum x^*_p = \sum \Delta^*_p[C*\exp(\lambda*\,t)] \qquad (p = 1, \ldots, g;\ * = 1, \ldots, 2g), \qquad \text{(c)}$$

where the $2g$ (real or complex conjugate) $C*$ are determined from the $2g$ initial conditions.

Example 8.6.3 *Ordinary versus Practical Stability for a Two-DOF System—No Friction.* For such a system (i.e., p, $p' = 1, 2$), the fundamental perturbational equations (8.6.7b) become

$$\mu_1\ddot{x}_1 - \gamma\dot{x}_2 + \kappa_1 x_1 = 0, \qquad \mu_2\ddot{x}_2 + \gamma\dot{x}_1 + \kappa_2 x_2 = 0, \qquad \text{(a1, 2)}$$

where

$$g_{12} = -g_{21} \equiv -\gamma, \quad \text{and} \quad \mu_1, \mu_2 > 0.$$

Setting in there

$$x_1 = x_{1o}\exp(\lambda t), \qquad x_2 = x_{2o}\exp(\lambda t), \qquad \text{(b)}$$

and eliminating the amplitude ratio x_{1o}/x_{2o}, we are readily led to the characteristic equation

$$(\mu_1\mu_2)\lambda^4 + (\mu_1\kappa_2 + \mu_2\kappa_1 + \gamma^2)\lambda^2 + \kappa_1\kappa_2 = 0. \qquad \text{(c)}$$

By elementary algebra, the two λ^2-roots of (c) will be *real* if

$$(\mu_1\kappa_2 + \mu_2\kappa_1 + \gamma^2)^2 - 4(\mu_1\mu_2)(\kappa_1\kappa_2) > 0, \tag{d}$$

or, expanding and rearranging in γ-powers,

$$\gamma^4 + 2(\mu_1\kappa_2 + \mu_2\kappa_1)\gamma^2 + (\mu_2\kappa_1 - \mu_1\kappa_2)^2 > 0. \tag{d1}$$

Now we have to consider the following three cases:

(i) κ_1 and κ_2 are both *positive*. Then, clearly, (d1), all three of its left-side terms being positive, is fulfilled for *any* γ; and since, from algebra,

$$\lambda_1{}^2\lambda_2{}^2 = (-1)^2(\kappa_1\kappa_2)/(\mu_1\mu_2) \Rightarrow \lambda_1{}^2\lambda_2{}^2 > 0, \tag{e1}$$

$$\lambda_1{}^2 + \lambda_2{}^2 = -(\mu_1\kappa_2 + \mu_2\kappa_1 + \gamma^2)/(\mu_1\mu_2) \Rightarrow \lambda_1{}^2 + \lambda_2{}^2 < 0, \tag{e2}$$

we conclude that, then, both $\lambda_1{}^2$ and $\lambda_2{}^2$ are *negative*; that is,

$$\lambda_1 = \pm i\,\sigma_1 \qquad \text{and} \qquad \lambda_2 = \pm i\,\sigma_2. \tag{e3}$$

where $\sigma_1{}^2$ and $\sigma_2{}^2$ are the roots of

$$(\mu_1\mu_2)\sigma^4 - (\mu_1\kappa_2 + \mu_2\kappa_1 + \gamma^2)\sigma^2 + \kappa_1\kappa_2 = 0. \tag{e4}$$

In this case, the solutions of (a1, 2) are simple harmonic oscillations, and so *the fundamental state I is stable, both ordinarily (negative λ^2-roots) and practically* ($\kappa_1, \kappa_2 > 0$).

(ii) κ_1 and κ_2 have *opposite* signs. Then, clearly, (d), both its left-side terms being positive, still holds for *any* γ; but, by the first part of (e1), the roots $\lambda_1{}^2$, $\lambda_2{}^2$ now have opposite signs; that is,

$$\lambda_1{}^2\lambda_2{}^2 < 0; \tag{f}$$

and the positive of them leads to $\exp(\pm\varepsilon t)$-proportional factors in the perturbations $x_p(t)$, and hence *ordinary instability* (and, of course, practical instability).

(iii) κ_1 and κ_2 are both *negative*. If $\lambda_1{}^2$ and $\lambda_2{}^2$ are real, *which, by (d), can happen if γ^2 is large enough*; that is, if, successively,

$$\begin{aligned}
\mu_1\kappa_2 + \mu_2\kappa_1 + \gamma^2 &\geq [4(\mu_1\mu_2)(\kappa_1\kappa_2)]^{1/2} \qquad (= \textit{positive}) \\
\Rightarrow \gamma^2 &\geq -(\mu_1\kappa_2 + \mu_2\kappa_1) + 2[(\mu_1\mu_2)(\kappa_1\kappa_2)]^{1/2} \\
&= [(-\mu_1\kappa_2)^{1/2} + (-\mu_2\kappa_1)^{1/2}]^2 \\
\Rightarrow \gamma &\geq (-\mu_1\kappa_2)^{1/2} + (-\mu_2\kappa_1)^{1/2},
\end{aligned} \tag{g1}$$

then, as in (e1, 2),

$$\lambda_1{}^2\lambda_2{}^2 = (-1)^2(\kappa_1\kappa_2)/(\mu_1\mu_2) \Rightarrow \lambda_1{}^2\lambda_2{}^2 > 0, \tag{g2}$$

$$\lambda_1{}^2 + \lambda_2{}^2 = -(\mu_1\kappa_2 + \mu_2\kappa_1 + \gamma^2)/(\mu_1\mu_2) \Rightarrow \lambda_1{}^2 + \lambda_2{}^2 < 0; \tag{g3}$$

that is, as in (i), both $\lambda_1{}^2$ and $\lambda_2{}^2$ are *negative*, and hence the system is *ordinarily stable*, although *practically unstable* ($R_{(2)V}$ = *negative definite* $\Rightarrow$ *maximum, or*

$R_0 = minimum$)! But, as explained earlier, such a gyroscopic stabilization is gradually lost due to friction. (See next example; also Thomson and Tait, 1912, pp. 395–396, Gray, 1918, pp. 439–440.)

Example 8.6.4 *Ordinary versus Practical Stability for a Two-DOF System—Friction.* Continuing from the preceding example, let us now examine the effect of small $\dot{x}$-proportional friction terms in case (iii) (i.e., both κ_1 and κ_2 *negative* $\Rightarrow$ ordinary stability but practical instability). Here, the perturbation equations are

$$\mu_1\ddot{x}_1 + f_{11}\dot{x}_1 + (f_{12} - \gamma)\dot{x}_2 + \kappa_1 x_1 = 0, \tag{a1}$$

$$\mu_2\ddot{x}_2 + f_{22}\dot{x}_2 + (f_{21} + \gamma)\dot{x}_1 + \kappa_2 x_2 = 0, \tag{a2}$$

where

$f_{11}, f_{12} = f_{21}, f_{22}$ are the small constant coefficients of the *dissipation function* (§3.9)

$2F = f_{11}(\dot{x}_1)^2 + 2f_{12}\dot{x}_1\dot{x}_2 + f_{22}(\dot{x}_2)^2$

[assumed *positive definite*; i.e., during every motion from I, F does *negative* work and hence reduces the perturbational energy $R_{(2)T} + R_{(2)V}$], (b1)

$$\Rightarrow \quad f_{11} > 0, \qquad f_{22} > 0, \qquad Det(f_{pp'}) = f_{11}f_{22} - f_{12}^{\,2} > 0. \tag{b2}$$

In this case, the characteristic equation, the counterpart of (c) of the preceding example, becomes

$$(\mu_1\mu_2)\lambda^4 + (\mu_2 f_{11} + \mu_1 f_{22})\lambda^3 + (\mu_1\kappa_2 + \mu_2\kappa_1 + \gamma^2 + f_{11}f_{22} - f_{12}^{\,2})\lambda^2 + (\kappa_2 f_{11} + \kappa_1 f_{22})\lambda + \kappa_1\kappa_2 = 0. \tag{c}$$

Assuming that the roots of the previous frictionless case are $\pm i\sigma_1$ and $\pm i\sigma_2$ (ordinary stability; from which, as explained earlier, it follows that κ_1 and κ_2 have the *same sign*), we try for the roots of (c) the modified forms

$$\lambda_{1,2} = \varepsilon_1 \pm i\sigma_1 \quad \text{and} \quad \lambda_{3,4} = \varepsilon_2 \pm i\sigma_2, \tag{c1}$$

where ε_1, ε_2 are first-order (i.e., $\sim f_{pp'}$) corrections to σ_1, σ_2; and the latter are, of course, determined from (e4) of ex. 8.6.3. Now, from algebra (see texts on algebra, or handbooks of mathematics, etc.) we know that

- $$\lambda_1 + \lambda_2 + \lambda_3 + \lambda_4 = -(\mu_2 f_{11} + \mu_1 f_{22})/(\mu_1\mu_2) < 0, \tag{c2}$$

and this, thanks to (c1), yields, to the first order,

$$2(\varepsilon_1 + \varepsilon_2) = -[(f_{11}/\mu_1) + (f_{22}/\mu_2)] < 0, \tag{c3}$$

- $$\begin{aligned} &\lambda_1^{-1} + \lambda_2^{-1} + \lambda_3^{-1} + \lambda_4^{-1} \\ &= (\lambda_2\lambda_3\lambda_4 + \lambda_1\lambda_3\lambda_4 + \lambda_1\lambda_2\lambda_4 + \lambda_1\lambda_2\lambda_3)/(\lambda_1\lambda_2\lambda_3\lambda_4) \\ &= -(\kappa_2 f_{11} + \kappa_1 f_{22})/(-1)^4(\kappa_1\kappa_2), \end{aligned} \tag{c4}$$

from which, and (c1), and noting further that

$$\lambda_1^{-1} + \lambda_2^{-1} + \lambda_3^{-1} + \lambda_4^{-1} = (\lambda_1^{-1} + \lambda_2^{-1}) + (\lambda_3^{-1} + \lambda_4^{-1})$$
$$= 2[\varepsilon_1(\varepsilon_1^2 + \sigma_1^2)^{-1} + \varepsilon_2(\varepsilon_2^2 + \sigma_2^2)^{-1}], \tag{c5}$$

we obtain, again to the first order,

$$2[(\varepsilon_1/\sigma_1^2) + (\varepsilon_2/\sigma_2^2)] = -[(f_{11}/\kappa_1) + (f_{22}/\kappa_2)]. \tag{c6}$$

Hence:

(i) If both κ_1 and κ_2 are *positive* (practical stability), then both ε_1 and ε_2 are *negative* $\Rightarrow$ $\exp(\varepsilon t)$-terms; that is, we also have ordinary damped stability.

(ii) If both κ_1 and κ_2 are *negative* (practical instability), then ε_1 and ε_2 must have *opposite signs*; which means that, then, one of the two oscillations dies away, while the other increases exponentially at an $f_{pp'}$-proportional rate; that is, we have ordinary damped instability. To find what happens where, we solve (c3, 6) for, say ε_1, and thus obtain

$$\varepsilon_1(\sigma_2^{-2} - \sigma_1^{-2}) = -(1/2\sigma_2^2)[(f_{11}/\mu_1) + (f_{22}/\mu_2)] + (1/2)[(f_{11}/\kappa_1) + (f_{22}/\kappa_2)] < 0,$$
$$\Rightarrow \varepsilon_1(\sigma_1^2 - \sigma_2^2) < 0, \tag{c7}$$

from which we conclude that if $\sigma_1 < \sigma_2$, then $\varepsilon_1 > 0$; and, of course, $\varepsilon_2 < 0$; and analogously if $\sigma_1 > \sigma_2$. In words: the *exponentially increasing oscillation* (say $\varepsilon_1 > 0$) corresponds to the *smaller of the two frictionless frequencies* (here, σ_1) $\Rightarrow$ *longer period*; while the *damped oscillation* (here, $\varepsilon_2 < 0$) corresponds to the *larger frequency* (here, σ_2) $\Rightarrow$ *smaller period*.

Example 8.6.5 *The Monorail.* Let us consider a car (or wagon) K of mass M supported by one (or more, but aligned) wheels W rolling on a *single* fixed horizontal rail OY (+away from the reader—fig. 8.4). Inside K there is a gimbal G' of negligible mass that can rotate freely about a K-fixed axis $G_1'G_2'$; and inside G' there is a heavy gyroscope G of mass m that can rotate about a G'-fixed axis G_1G_2—that is, on the car's plane of symmetry O–yz ($Oy \equiv OY$), at a constant rate ω_0. In addition, G' carries at its top, above and around G_1, a heavy particle P of mass μ, so that (in accordance with Kelvin's theorem of stabilization of an *even* number of nongyroscopically unstable freedoms) *both* such freedoms θ and χ (see below) are *unstable*; or, equivalently, we may skip P but make sure that the center of mass of G, C' is *above* the axis $G_1'O'G_2'$. The rotation of G' about its housing K (axis $G_1'G_2'$) is measured by the "precession" angle χ between the gyro axis G_1G_2 and Oz (O–yz = plane of symmetry of K), while the rotation ("nutation") of the entire K about $OY \equiv Oy$ is measured by the angle θ between O–yz and the vertical OZ. All other kinematico-inertial parameters of the system are shown in the figure, and will be identified below as needed. Let us find the equations of small motion of this *two*-DOF system around the "normal attitude" χ, $\theta = 0$, when K travels at a *uniform* rate along OY; and examine their stability.

Let T_K/V_K, T_G/V_G, and T_P/V_P, be the (inertial) kinetic/potential energies of the car (K), gyro (G), and particle (P), respectively. Then, to the *second order* in θ and χ and their rates (since we are only interested in *linear* equations of motion in them),

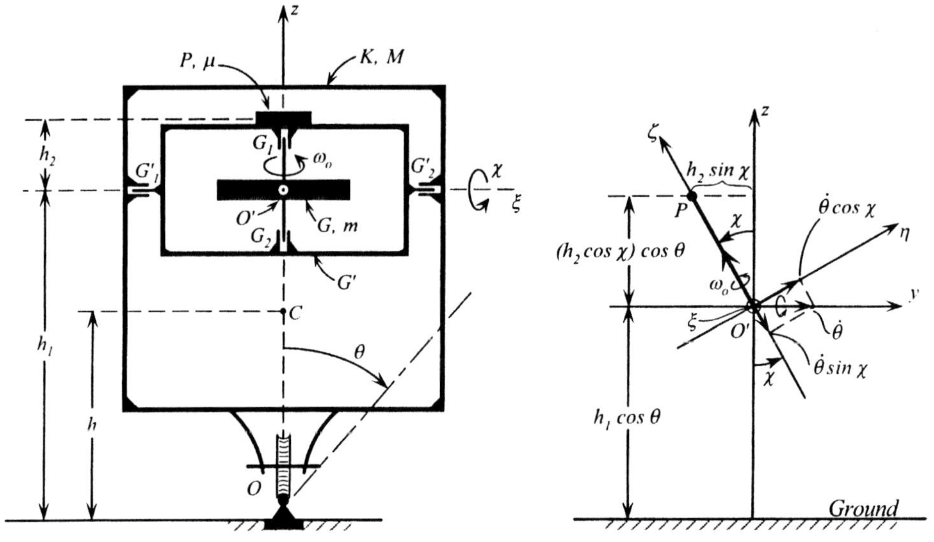

FRONT VIEW
(in normal attitude)

SIDE VIEW

Figure 8.4 Kinematico-inertial parameters of the monorail. C, $C' \equiv O'$: centers of mass of K and G, respectively; both along Oz. [In some treatments, C' is *above* O'; e.g., Cabannes (1968, pp. 276–277). Here, we achieve the same result with P]; $h_2 = G_1O$, $h_1 = O'O$, $h = CO$; O–xyz: *car-fixed* axes, O–yz: car *plane of symmetry*, O–XYZ: *inertial* axes with which O–xyz coincide in the normal attitude configuration.

and to within inconsequential constant terms, we have

(i) $$T_K = (1/2)I(\dot{\theta})^2 \qquad (I = \text{moment of inertia of } K \text{ about } Oy \equiv OY), \tag{a1}$$

$$V_K = M\,g\,h\cos\theta \;\approx\; -M\,g\,h\,\theta^2/2 + constant; \tag{a2}$$

(ii) $$T_G = (1/2)m\,v_G{}^2 + (1/2)\big[A(\omega_\xi{}^2 + \omega_\eta{}^2) + C\,\omega_\zeta{}^2\big]$$

$[O'$–$\xi\eta\zeta$: G'-fixed axes;

A/C: transverse/axial principal moments of inertia of G at $O']$

$$= (1/2)m(h_1\dot{\theta})^2 + (1/2)\big\{A[(\dot{\chi})^2 + (\dot{\theta}\cos\chi)^2] + C(\omega_o - \dot{\theta}\sin\chi)^2\big\}$$

$$\approx (1/2)m\,h_1{}^2(\dot{\theta})^2 + (1/2)\big\{A[(\dot{\chi})^2 + (\dot{\theta})^2] + C(\omega_o - \dot{\theta}\chi)^2\big\}$$

$$\approx (1/2)m\,h_1{}^2(\dot{\theta})^2 + (1/2)\big\{A[(\dot{\theta})^2 + (\dot{\chi})^2] + C(\omega_o{}^2 - 2\omega_0\chi\dot{\theta})\big\},$$

$$V_G = m\,g\,h_1\cos\theta \;\approx\; -M\,g\,h_1\theta^2/2 + constant. \tag{a3, 4}$$

(iii) The inertial coordinates and velocities of P $(X_P,\ Y_P,\ Z_P)$ equal

$$X_P = (h_1 + h_2\cos\chi)\sin\theta \;\approx\; (h_1 + h_2)\theta \;\Rightarrow\; \dot{X}_P \approx (h_1 + h_2)\dot{\theta}, \tag{a5}$$

$$Y_P = -h_2\sin\chi \;\approx\; -h_2\chi \;\Rightarrow\; \dot{Y}_P \approx -h_2\dot{\chi}, \tag{a6}$$

$$Z_P = (h_1 + h_2\cos\chi)\cos\theta \approx (-1/2)\big[(h_1 + h_2)\theta^2 + h_2\chi^2\big]$$

$$\Rightarrow\; \dot{Z}_P \approx -(h_1 + h_2)\theta\,\dot{\theta} + h_2\chi\,\dot{\chi} \;\Rightarrow\; (\dot{Z}_P)^2 \approx 0; \tag{a7}$$

and therefore, to the second order,

$$2T_P = \mu[(\dot{X}_P)^2 + (\dot{Y}_P)^2 + (\dot{Z}_P)^2] = \cdots \approx \mu[(h_1 + h_2)^2(\dot{\theta})^2 + {h_2}^2(\dot{\chi})^2], \qquad \text{(a8)}$$

$$V_P = \mu g Z_P \approx -(\mu g/2)[(h_1 + h_2)\theta^2 + h_2\chi^2] + constant. \qquad \text{(a9)}$$

In view of the above results, the "quadratisized" system Lagrangean equals

$$L = (T_K + T_G + T_P) - (V_K + V_G + V_P) \equiv L_2 + L_1 + L_0, \qquad \text{(b1)}$$

where

$$2L_2 \equiv M_{11}(\dot{\theta})^2 + M_{22}(\dot{\chi})^2 \qquad \text{(Inertial part)},$$

$$M_{11} \equiv I + m{h_1}^2 + A + \mu(h_1 + h_2)^2, \qquad M_{22} \equiv A + \mu{h_2}^2; \qquad \text{(b2)}$$

$$L_1 \equiv M_1\chi\dot{\theta} \qquad \text{(Gyroscopic part)},$$

$$M_1 \equiv C\omega_o; \qquad \text{(b3)}$$

$$2L_0 \equiv k_{11}\theta^2 + k_{22}\chi^2 \qquad \text{(Potential part)},$$

$$k_{11} = Mh + mh_1 + \mu(h_1 + h_2) \quad (> 0), \qquad k_{22} = \mu h_2 \quad (> 0); \qquad \text{(b4)}$$

or, in the new *normalized* Lagrangean coordinates $x_1 \equiv (M_{11})^{1/2}\theta$ and $x_2 \equiv (M_{22})^{1/2}\chi$:

$$2L = [(\dot{x}_1)^2 + (\dot{x}_2)^2] - 2\gamma x_2\dot{x}_1 - (\kappa_1{x_1}^2 + \kappa_2{x_2}^2),$$

where

$$\kappa_1 \equiv -(k_{11}/M_{11}) \quad (< 0), \qquad \kappa_2 \equiv -(k_{22}/M_{22}) \quad (< 0),$$

$$\gamma \equiv M_1/(M_{11}M_{22})^{1/2} \equiv C\omega_o/(M_{11}M_{22})^{1/2}. \qquad \text{(c)}$$

Hence, the linear(ized) Lagrangean equations of motion for $x_1(\theta)$ and $x_2(\chi)$ are

$$\ddot{x}_1 - \gamma\dot{x}_2 + \kappa_1 x_1 = 0, \qquad \ddot{x}_2 + \gamma\dot{x}_1 + \kappa_2 x_2 = 0; \qquad \text{(d)}$$

and, by (ex. 8.6.3: g1), for ordinary (= asymptotic) stability their coefficients must satisfy

$$\gamma \geq (-\kappa_1)^{1/2} + (-\kappa_2)^{1/2}; \qquad \text{(e)}$$

that is, recalling (c), *the spin ω_o must be sufficiently high to counter the destabilizing effect of gravity.*

Problem 8.6.2 Continuing from the preceding example, in the presence of small $\dot{x}$-proportional *damping*, the normalized monorail equations of motion (ex. 8.6.5: d) are, generally, replaced by (recall ex. 8.6.4):

$$\ddot{x}_1 + f_{11}\dot{x}_1 + (f_{12} - \gamma)\dot{x}_2 + \kappa_1 x_1 = 0, \qquad \text{(a1)}$$

$$\ddot{x}_2 + f_{22}\dot{x}_2 + (f_{21} + \gamma)\dot{x}_1 + \kappa_2 x_2 = 0. \qquad \text{(a2)}$$

(i) Show that the characteristic equation of this system is

$$\Delta(\lambda) = \lambda^4 + a_1\lambda^3 + a_2\lambda^2 + a_3\lambda + a_4 = 0 \qquad (a_0 = 1), \qquad \text{(a3)}$$

where

$$a_1 \equiv f_{11} + f_{22}, \tag{a4}$$

$$a_2 \equiv \kappa_1 + \kappa_2 + \gamma^2 + f_{11} f_{22} - f_{12}{}^2, \tag{a5}$$

$$a_3 \equiv \kappa_1 f_{22} + \kappa_2 f_{11}, \tag{a6}$$

$$a_4 \equiv \kappa_1 \kappa_2. \tag{a7}$$

(ii) By applying the Routh–Hurwitz stability criterion (§3.10, and its examples/problems), investigate the possibility/impossibility of gyroscopic stabilization of the damped monorail. [For practical insights, see, e.g., Grammel (1950, Vol. 2, pp. 230–247); also Merkin (1987, pp. 182–184; 1974, pp. 232–234).]

Problem 8.6.3 Consider a cyclic system with *one* nonignorable coordinate, q, whose Routhian equation of motion is

$$(\partial R/\partial \dot{q})^{\cdot} - \partial R/\partial q = 0, \tag{a}$$

or, explicitly, since $R = R(q, \dot{q};\ \textit{constant cyclic momenta} \equiv C)$,

$$(\partial^2 R/\partial \dot{q}^2)\ddot{q} + (\partial^2 R/\partial q\, \partial \dot{q})\dot{q} - \partial R/\partial q = 0. \tag{b}$$

Let the small motion of the system around a *steady* state I: $q = \textit{constant} \equiv s$, be $s + z(t)$. By expanding (b) à la Taylor around I, and keeping up to *linear* terms in the perturbation $z(t)$ and its $(\ldots)^{\cdot}$-derivatives, show that the latter satisfies the following equation:

$$(\partial^2 R/\partial \dot{q}^2)_o \ddot{x} - (\partial^2 R/\partial q^2)_o x = 0, \tag{c}$$

where $(\ldots)_o \equiv (\ldots)$ evaluated at I; and hence for *stability* of that state [i.e., harmonic $z(t)$], and since $(\partial^2 R/\partial \dot{q}^2)_o > 0$, we must have

$$(\partial^2 R/\partial q^2)_o < 0. \tag{d}$$

HINT

The value(s) of state I satisfy the equation $(\partial R/\partial q)_o = 0$; and hence, by (d), $R_0 = \textit{maximum}$.

Problem 8.6.4 Consider a particle P of mass m that can slide on a smooth circular hoop H of radius r and moment of inertia about any diameter I (fig. 8.5). H spins about a fixed vertical axis with constant angular velocity $\omega \equiv \dot{\phi}$.

(i) Show that, since ϕ is ignorable $[\Rightarrow p_\phi \equiv \partial T/\partial \dot{\phi} = (I + m r^2 \sin^2 \theta)\dot{\phi} = \textit{constant} \equiv C$, with solution(s) $\theta = \theta']$, the Routhian equals

$$R \equiv L - p_\phi \dot{\phi} = L - C\omega = R_2 + R_1 + R_0 = R(\theta, \dot{\theta}; C), \tag{a}$$

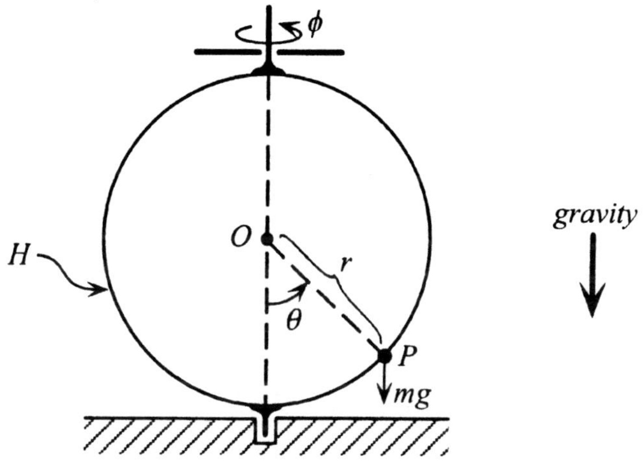

Figure 8.5 Particle moving on a smooth and uniformly spinning circular ring.

where

$$R_2 = (m r^2/2)(\dot{\theta})^2 \quad (\equiv T''_{2,0}), \tag{b}$$

$$R_1 = 0 \quad (\equiv T''_{1,1}), \tag{c}$$

$$R_0 = m g r \cos\theta - \left[C^2/2(I + m r^2 \sin^2\theta)\right] \quad (\equiv -V + T''_{2,0}). \tag{d}$$

(ii) Show that the steady motion condition:

$$\partial R_0/\partial\theta = -m g r \sin\theta + (C^2 m r^2 \sin\theta\cos\theta)(I + m r^2 \sin^2\theta)^{-2} = 0, \tag{e}$$

leads, further, to the following *two* algebraic equations (with corresponding roots θ' and θ''):

$$-m g r(I + m r^2 \sin^2\theta')^2 + C^2 m r^2 \cos\theta' = 0, \qquad \sin\theta'' = 0 \ (\Rightarrow \theta'' = 0, \text{ etc.}). \tag{e1, 2}$$

(iii) Show that

$$(\partial^2 R_0/\partial\theta^2)|_{\theta=\theta'} = -m g r \cos\theta' + m r^2 \omega^2[\cos(2\theta') - m r^2 (I + m r^2 \sin^2\theta')^{-1} \sin^2(2\theta')]$$
$$= \cdots = -[(mr^2\omega^2 \sin^2\theta')/(I + mr^2 \sin^2\theta')][I + mr^2(1 + 3\cos^2\theta')] < 0; \tag{f}$$

namely, θ' is *stable*: and that

$$(\partial^2 R_0/\partial\theta^2)|_{\theta=\theta''=0} = m g r[(C^2 r/I^2 g) - 1] \qquad \text{[or, eliminating } C \text{ via the first of (e)]}$$
$$= m g r\{[(\omega^2 r/g) - 1]$$
$$+ (\omega^2 r/I^2 g)(2m r^2 I \sin^2\theta' + m^2 r^4 \sin^4\theta')\} > 0, \tag{g}$$

since $\omega^2 r > g$ (explain); namely, $\theta'' = 0$ is *unstable*.

HINT

Since $(I + m r^2 \sin^2\theta)\omega \equiv C \neq 0$, the first of (e) yields

$$-m g r + m r^2 \omega^2 \cos\theta' = 0 \qquad (\Rightarrow \cos\theta' = g/\omega^2 r < 1). \tag{h}$$

Problem 8.6.5 Continuing from the preceding problem, show that the equation of small (linearized) oscillations of P around the steady state θ'—that is, $\theta = \theta' + z(t)$—is

$$\ddot{z} + \Omega z = 0, \tag{a}$$

where

$$\Omega \equiv \omega \sin\theta' \{[I + m r^2(1 + 3\cos^2\theta')]/(I + m r^2 \sin^2\theta')\}^{1/2}. \tag{b}$$

Show that the stability condition obtained from (a, b) coincides with prob. 8.6.4: (f).

Problem 8.6.6 *Stability of Steady Precession of Top. Method of Perturbations.* As seen in exs. 8.4.5 and 8.5.1, in this case the exact Routhian of the top equals

$$R = R_2 + R_1 + R_0 = R(\theta, \dot{\theta}; C_\phi, C_\psi), \tag{a}$$

$$R_2 \equiv T''_{2,0} = (1/2)A(\dot{\theta})^2, \tag{a1}$$

$$R_1 \equiv T''_{1,1} = 0, \tag{a2}$$

$$R_0 \equiv T''_{0,2} - V$$

$$= -\{[(C_\phi - C_\psi \cos\theta)^2/2A\sin^2\theta] + (1/2C){C_\psi}^2\} - m g l \cos\theta. \tag{a3}$$

Therefore, the Routhian equation for the sole nonignorable coordinate θ becomes

$$(\partial R/\partial\dot{\theta})^{\cdot} - \partial R/\partial\theta = 0:$$

$$A\ddot{\theta} + \left[(C_\phi - C_\psi\cos\theta)(C_\psi - C_\phi\cos\theta)/A\sin^3\theta\right] = m g l \sin\theta. \tag{b}$$

(i) Setting in (b) $\theta = \theta_o + z(t)$, where θ_o is the (constant) root(s) of the steady motion condition $(\partial R/\partial\theta)_o = 0$, expanding in powers of z and its $(\ldots)^{\cdot}$-derivatives, and keeping only up to *linear* such terms, show that we eventually [after using ex. 8.5.1: (d1), or (c) of prob. 8.6.7 to eliminate C_ϕ and C_ψ] obtain the perturbation equation,

$$\ddot{z} + \Omega z = 0, \tag{c}$$

where

$$\Omega \equiv {c_\phi}^2 + (m g l/A\, c_\phi)^2 - 2 m g l \cos\theta_o. \tag{d}$$

This equation determines $z(t)$ (a harmonic oscillation, if $\Omega > 0$) in terms of the steady-state values θ_o and $\dot{\phi} = constant \equiv c_\phi$; or, equivalently, θ_o and C_ϕ, C_ψ.

(ii) Show that the same equation results if we set into the Routhian (a–a3) $\theta = \theta_o + z(t)$, expand it in powers of z and its $(\ldots)^{\cdot}$-derivatives, keep only up to *quadratic* such terms, and then write down the equation for z:

$$(\partial R/\partial\dot{z})^{\cdot} - \partial R/\partial z = 0.$$

Problem 8.6.7 *Stability of Steady Precession of Top. Method of Extremum of Reduced Potential Energy.* Continuing from the preceding problem, we saw there that

$$-R_0 \equiv V - T''_{0,2} = (C_\phi - C_\psi \cos\theta)^2/2A\sin^2\theta + (1/2C)C_\psi{}^2 + mgl\cos\theta$$
$$\equiv f(\theta; C_\phi, C_\psi) \equiv -(\textit{Routhian, or reduced, potential}), \qquad \text{(a)}$$

where [recall results of (ex. 8.5.1)]

$$C_\phi = (A\sin^2\theta + C\cos^2\theta)\dot\phi + (C\cos\theta)\dot\psi$$
$$\equiv A\sin^2\theta\,\dot\phi + Cn\cos\theta \equiv \textit{constant}, \qquad \text{(a1)}$$
$$C_\psi = (C\cos\theta)\dot\phi + (C)\dot\psi \equiv Cn = \textit{constant}; \qquad \text{(a2)}$$

and, therefore, inversely,

$$\dot\phi = (C_\phi - C_\psi\cos\theta)/A\sin^2\theta \qquad (= \textit{constant} \equiv c_\phi, \text{ in steady precession}), \qquad \text{(b1)}$$
$$\dot\psi = -\left[(C_\phi - C_\psi\cos\theta)\cos\theta/A\sin^2\theta\right] + (C_\psi/C)$$
$$(= \textit{constant} \equiv c_\psi, \text{ in steady precession}). \qquad \text{(b2)}$$

Setting $df(\theta)/d\theta = 0$, and assuming $\theta \neq 0$, we obtain the steady motion condition [(ex. 8.5.1: d1), with $\dot\phi = \textit{constant} \equiv c_\phi$]:

$$Ac_\phi{}^2\cos\theta_o - C_\psi c_\phi + mgl = 0,$$

or

$$Ac_\phi{}^2\cos\theta_o + mgl = C(c_\phi\cos\theta_o + c_\psi)c_\phi. \qquad \text{(c)}$$

(i) Calculate $d^2f(\theta)/d\theta^2$ and show that the condition for the *minimum* of $-R_0$, or the *maximum* of R_0, for small changes of the nonignorable coordinate θ from the steady state (c): θ_o, c_ϕ, C_ψ, is

$$\left[C_\psi(C_\psi - C_\phi\cos\theta_o) + C_\phi(C_\phi - C_\psi\cos\theta_o)\right]/A\sin^2\theta_o > 4mgl\cos\theta_o. \qquad \text{(d)}$$

(ii) For $\theta_o = 0$, the cyclicity conditions (a1, 2) reduce to $C_\phi = C_\psi = Cn = C(c_\phi + c_\psi) \equiv \textit{constant} \equiv D$. Show that for *small* θ_o, the stability condition (d) approximates to

$$2D^2(1 - \cos\theta_o)/\sin^2\theta_o > 4Amgl\cos\theta_o; \qquad \text{(e)}$$

or, in the limit of vanishingly small θ_o's,

$$D^2 > 4Amgl \qquad (\text{“sleeping top” stability}). \qquad \text{(f)}$$

These results have been obtained earlier by other means. For additional examples and details, see, for example, Merkin (1987, pp. 88–95; 1974, pp. 318–321, 323–324), Gantmacher (1970, pp. 256–258).

Problem 8.6.8 *Stability of Steady Precession of Top. Relations among the Perturbations.* Continuing from the preceding problem, show that the three (time-dependent!) perturbations

$$\theta = \theta_o + z(t), \qquad \dot\phi = c_\phi + \eta(t), \qquad \dot\psi = c_\psi + \xi(t), \qquad \text{(a)}$$

are related by

$$(A \sin\theta_o)\eta = \left[C(c_\phi \cos\theta_o + c_\psi) - 2Ac_\phi \cos\theta_o\right]z, \tag{b}$$

$$(\cos\theta_o)\eta + \xi = (c_\phi \sin\theta_o)z. \tag{c}$$

These equations, assuming $\sin\theta_o \neq 0$ (i.e., no sleeping top), yield the hitherto unknown functions η and ξ in terms of z and the steady precession values; that is, in terms of quantities already determined in previous problems:

$$\eta = \eta\left[z(t); \theta_o, c_\phi, c_\psi, C_\phi, C_\psi\right], \qquad \xi = \xi\left[z(t); \theta_o, c_\phi, c_\psi, C_\phi, C_\psi\right]. \tag{d}$$

A final integration of these known functions of time, right-sides of (d), yields the time behavior of ϕ and ψ, if desired.

Problem 8.6.9 *Stability of Steady Precession of Top. Relations among the Perturbations (continued).* Continuing from the preceding problem, show that $\eta(t)$ and $\xi(t)$ can be obtained by inserting (a) in the earlier steady precession relations (prob. 8.6.7: b1, 2)

$$\dot{\phi} = (C_\phi - C_\psi \cos\theta)/A \sin^2\theta, \tag{a1}$$

$$\dot{\psi} = -\left[(C_\phi - C_\psi \cos\theta)\cos\theta/A \sin^2\theta\right] + (C_\psi/C), \tag{a2}$$

(which hold for both the fundamental state and the disturbed one), expanding à la Taylor, and keeping up to *linear* terms in the (equimomental) perturbations.

REMARKS

(i) Also, one could calculate the Routhian, expand and keep up to *quadratic* terms, and then apply Routh's "Hamiltonian" equations $\partial R/\partial \Psi_i = -d\psi_i/dt$.

(ii) As pointed out in the explanatory remarks following (8.6.2), the above are special cases of application of the general kinematico-inertial relations (8.3.12d, e):

$$\Psi_i \equiv \partial T/\partial\dot{\psi}_i = \sum c_{ji}\dot{\psi}_j + \sum b_{pi}\dot{q}_p \;\Leftrightarrow\; \dot{\psi}_j = \sum C_{ji}\left(\Psi_i - \sum b_{pi}\dot{q}_p\right), \tag{b}$$

to both states I and $II = I + \Delta(I)$, with subsequent expansion and equation of their first-order terms in $\Delta\dot{\psi}$, Δq, $\Delta\dot{q}$, $\Delta\Psi$, $\Delta C_{ji} \approx \sum(\ldots)_{jip}\,\Delta q_p$. For given Ψ's and $\dot{q}(t)$'s it supplies the $\dot{\psi}(t)$'s.

Problem 8.6.10 *Stability of Steady Precession of Top. Relations among the Perturbations (continued).* Show that the results of the preceding problems can be obtained by substituting (prob. 8.6.6: a) in the Lagrangean equations of the top, then linearizing, and so on.

For additional similar problems, see, for example, Chirgwin and Plumpton (1966, pp. 282–305) and Wells (1967, pp. 239–255).

This concludes the treatment of Routh's method of the "modified Lagrangean." The rest of this chapter deals with the applications and ramifications of the method of Hamilton.

8.7 VARIATION OF CONSTANTS (OR PARAMETERS)

In order to solve the exact problem approximately, we first solve an approximate problem exactly.

(T. E. Sterne, quoted in Garfinkel, 1966, p. 67)

Since the equations of motion, in either Lagrangean or Hamiltonian variables, are intrinsically nonlinear (§3.10), *general* methods for obtaining their exact solutions are out of the question, and being able to solve exactly the equations of an actual physical system is the rare exception rather than the rule (§3.12). For these reasons, some kind of approximation (analytical, numerical/computational, graphical, or combination thereof) is needed. In this section we develop one of the most important, general and systematic such approximation methods (originated by Lagrange himself in connection with problems of celestial mechanics, which, although unconstrained, lead to complicated equations of motion), known as the method of *variation of "constants"*, and associated calculus of *perturbations*. Either Lagrangean $(q, \dot{q})$ or Hamiltonian (q, p) variables can be used; but the latter, since they lead to *first-order* equations, are the preferred ones.

Theorem of Lagrange–Poisson

Let us consider a general system S, in Hamiltonian variables, with *equations of motion*

$$dp_k/dt = f_k(t, q, p) \quad \text{and} \quad dq_k/dt = g_k(t, q, p); \tag{8.7.1}$$

and corresponding *general solution*

$$p_k = p_k(t; c) \quad \text{and} \quad q_k = q_k(t; c), \tag{8.7.2}$$

where

$$c \equiv (c_1, \ldots, c_{2n}) \equiv (c_\nu;\ \nu = 1, \ldots, 2n)\text{: } \textit{constants of integration}; \tag{8.7.3}$$

and, unless specified otherwise Greek (Latin) *indices* run from 1 to $2n$ (n). Each particular set of c_ν's defines a particular dynamical trajectory, or *orbit*, say I, in phase space. Therefore, varying these constants slightly—that is, $c \to c + \delta c$—we obtain an *adjacent* such trajectory, say $II = I + \delta(I)$, given by the first-order (virtual-like; i.e., contemporaneous) changes of (8.7.2):

$$\delta p_k = \sum (\partial p_k/\partial c_\nu)\, \delta c_\nu \quad \text{and} \quad \delta q_k = \sum (\partial q_k/\partial c_\nu)\, \delta c_\nu, \tag{8.7.4}$$

where the derivatives are evaluated at I. As a result, II is governed by the following linear *variational*, or perturbational, equatons:

$$(\delta p_k)^{\cdot} = \delta(\dot{p}_k) = \sum [(\partial f_k/\partial p_l)\, \delta p_l + (\partial f_k/\partial q_l)\, \delta q_l], \tag{8.7.5a}$$

$$(\delta q_k)^{\cdot} = \delta(\dot{q}_k) = \sum [(\partial g_k/\partial p_l)\, \delta p_l + (\partial g_k/\partial q_l)\, \delta q_l]. \tag{8.7.5b}$$

Now, let us carry out *two* distinct c-variations from I, $\delta_1 c$ and $\delta_2 c$:

$$c \to c + \delta_1 c, \text{ resulting in the adjacent orbit } II_1 = I + \delta_1(I), \tag{8.7.6a}$$

$$c \to c + \delta_2 c, \text{ resulting in the adjacent orbit } II_2 = I + \delta_2(I). \tag{8.7.6b}$$

Then, in view of the above, we obtain, successively,

$$\begin{aligned}
(\delta_1 p_k\, \delta_2 q_k - \delta_2 p_k\, \delta_1 q_k)^{\cdot} &= (\delta_1 p_k)^{\cdot}\, \delta_2 q_k - (\delta_2 p_k)^{\cdot}\, \delta_1 q_k + \delta_1 p_k (\delta_2 q_k)^{\cdot} - \delta_2 p_k (\delta_1 q_k)^{\cdot} \\
&= \sum [(\partial f_k/\partial p_l)\, \delta_1 p_l + (\partial f_k/\partial q_l)\, \delta_1 q_l]\, \delta_2 q_k \\
&\quad - \sum [(\partial f_k/\partial p_l)\, \delta_2 p_l + (\partial f_k/\partial q_l)\, \delta_2 q_l]\, \delta_1 q_k \\
&\quad + \sum [(\partial g_k/\partial p_l)\, \delta_2 p_l + (\partial g_k/\partial q_l)\, \delta_2 q_l]\, \delta_1 p_k \\
&\quad - \sum [(\partial g_k/\partial p_l)\, \delta_1 p_l + (\partial g_k/\partial q_l)\, \delta_1 q_l]\, \delta_2 p_k \\
&= \sum (\partial f_k/\partial p_l)(\delta_1 p_l\, \delta_2 q_k - \delta_2 p_l\, \delta_1 q_k) + \sum (\partial f_k/\partial q_l)(\delta_1 q_l\, \delta_2 q_k - \delta_2 q_l\, \delta_1 q_k) \\
&\quad + \sum (\partial g_k/\partial p_l)(\delta_1 p_k\, \delta_2 p_l - \delta_1 p_l\, \delta_2 p_k) + \sum (\partial g_k/\partial q_l)(\delta_1 p_k\, \delta_2 q_l - \delta_1 q_l\, \delta_2 p_k);
\end{aligned}$$

and, therefore, summing these $(\ldots)^{\cdot}$-derivatives of *bilinear covariants* over k, and ofter some "dummy"-index changes, we finally obtain the fundamental result:

$$\begin{aligned}
d/dt\Big[\sum (\delta_1 p_k\, \delta_2 q_k - \delta_2 p_k\, \delta_1 q_k)\Big]
&= \sum\sum (\partial f_k/\partial p_l + \partial g_l/\partial q_k)(\delta_1 p_l\, \delta_2 q_k - \delta_2 p_l\, \delta_1 q_k) \\
&\quad + (1/2) \sum\sum (\partial f_k/\partial q_l - \partial f_l/\partial q_k)(\delta_1 q_l\, \delta_2 q_k - \delta_2 q_l\, \delta_1 q_k) \\
&\quad + (1/2) \sum\sum (\partial g_k/\partial p_l - \partial g_l/\partial p_k)(\delta_1 p_k\, \delta_2 p_l - \delta_2 p_k\, \delta_1 p_l).
\end{aligned} \tag{8.7.7}$$

Specializations

(i) If the original equations (8.7.1) are *Hamilton's canonical equations*—that is, if (§8.2)

$$f_k = -\partial H/\partial q_k + Q_k \quad \text{and} \quad g_k = \partial H/\partial p_k, \tag{8.7.8}$$

then, since,

$$\begin{aligned}
\text{(a)}\quad \partial f_k/\partial p_l + \partial g_l/\partial q_k &= (-\partial^2 H/\partial p_l\, \partial q_k + \partial Q_k/\partial p_l) + (-\partial^2 H/\partial q_k\, \partial p_l) \\
&= \partial Q_k/\partial p_l \quad [= 0, \text{ assuming } Q_k = Q_k(t, q)],
\end{aligned} \tag{8.7.8a}$$

$$\begin{aligned}
\text{(b)}\quad \partial f_k/\partial q_l - \partial f_l/\partial q_k &= (-\partial^2 H/\partial q_l\, \partial q_k + \partial Q_k/\partial q_l) \\
&\quad - (-\partial^2 H/\partial q_k\, \partial q_l + \partial Q_l/\partial q_k) \\
&= \partial Q_k/\partial q_l - \partial Q_l/\partial q_k \quad (\neq 0, \text{ in general}),
\end{aligned} \tag{8.7.8b}$$

$$\text{(c)}\quad \partial g_k/\partial p_l - \partial g_l/\partial p_k = \partial^2 H/\partial p_l\, \partial p_k - \partial^2 H/\partial p_k\, \partial p_l = 0, \tag{8.7.8c}$$

equation (8.7.7) reduces to

$$d/dt\Big[\sum(\delta_1 p_k\, \delta_2 q_k - \delta_2 p_k\, \delta_1 q_k)\Big]$$
$$= (1/2)\sum\sum(\partial Q_k/\partial q_l - \partial Q_l/\partial q_k)(\delta_1 q_l\, \delta_2 q_k - \delta_2 q_l\, \delta_1 q_k)$$
$$= \sum(\delta_1 Q_k\, \delta_2 q_k - \delta_2 Q_k\, \delta_1 q_k), \qquad (8.7.9)$$

where

$$\delta_* Q_k = \sum(\partial Q_k/\partial q_l)\, \delta_* q_l \quad (* = 1, 2).$$

(ii) If all forces are *potential*—that is, if $Q_k = 0$, or $\partial Q_k/\partial q_l = \partial Q_l/\partial q_k$, for all k, $l = 1, \ldots, n$—then (8.7.9) immediately yields the *Theorem of Lagrange–Poisson*: in a holonomic and potential (but possibly rheonomic) system, the expression

$$I \equiv \sum(\delta_1 p_k\, \delta_2 q_k - \delta_2 p_k\, \delta_1 q_k) \qquad (8.7.10)$$

is *time-independent*; that is, it is a constant of the motion.

[For applications of (8.7.10) to the "reciprocal theorems" of mechanics and optics, see, for example, Lamb (1910, p. 763; 1943, pp. 227–281).]

Lagrange Brackets

With the help of (8.7.4), this important quantity can be rewritten as follows:

$$I = \sum\Big\{\Big[\sum(\partial p_k/\partial c_\mu)\, \delta_1 c_\mu\Big]\Big[\sum(\partial q_k/\partial c_\nu)\, \delta_2 c_\nu\Big]$$
$$- \Big[\sum(\partial p_k/\partial c_\nu)\, \delta_2 c_\nu\Big]\Big[\sum(\partial q_k/\partial c_\mu)\, \delta_1 c_\mu\Big]\Big\}$$
$$= \sum\sum\Big\{\sum\big[(\partial p_k/\partial c_\mu)(\partial q_k/\partial c_\nu) - (\partial p_k/\partial c_\nu)(\partial q_k/\partial c_\mu)\big]\Big\}\, \delta_1 c_\mu\, \delta_2 c_\nu, \qquad (8.7.11)$$

or, finally,

$$I = \sum\sum\,[c_\mu, c_\nu]\, \delta_1 c_\mu\, \delta_2 c_\nu, \qquad (8.7.12)$$

where

$$[c_\mu, c_\nu] \equiv \sum\,\big[(\partial p_k/\partial c_\mu)(\partial q_k/\partial c_\nu) - (\partial p_k/\partial c_\nu)(\partial q_k/\partial c_\mu)\big]:$$

$= -[c_\nu, c_\mu]$ (prob. 8.7.1, below):

Lagrange bracket of c_μ, c_ν (1808). (8.7.13)

The above show that if $I =$ *constant*, for all $\delta_1 c$ and $\delta_2 c$, then *all Lagrange brackets must be constant*. (For an alternative proof, see ex. 8.11.3.)

Example 8.7.1 *Second, Lagrangean, Proof of the Lagrange–Poisson Theorem.* By $\delta_1(\ldots)$-varying (i) the definition $p_k = \partial L/\partial \dot{q}_k$ and then (ii) Lagrange's equations, say $dp_k/dt = \partial L/\partial q_k + Q_k$, we get

$$\delta_1 p_k = \delta_1(\partial L/\partial \dot{q}_k) = \sum [(\partial^2 L/\partial q_l\, \partial \dot{q}_k)\, \delta_1 q_l + (\partial^2 L/\partial \dot{q}_l\, \partial \dot{q}_k)\, \delta_1(\dot{q}_l)], \tag{a}$$

$$\begin{aligned}\delta_1(\dot{p}_k) &= (\delta_1 p_k)^{\cdot} = \delta_1(\partial L/\partial q_k + Q_k)\\ &= \sum [(\partial^2 L/\partial q_l\, \partial q_k)\, \delta_1 q_l + (\partial^2 L/\partial \dot{q}_l\, \partial q_k)\, \delta_1(\dot{q}_l)] + \delta_1 Q_k,\end{aligned} \tag{b}$$

respectively, and therefore [assuming $\delta[d(\ldots)] = d[\delta(\ldots)]$ for both q's and p's], we obtain, successively,

$$\begin{aligned}&\sum (\delta_1 p_k\, \delta_2 q_k)^{\cdot}\\ &\quad = \sum (\delta_1 p_k)^{\cdot}\, \delta_2 q_k + \sum \delta_1 p_k (\delta_2 q_k)^{\cdot}\\ &\quad = \sum\sum [(\partial^2 L/\partial q_l\, \partial q_k)\, \delta_1 q_l\, \delta_2 q_k + (\partial^2 L/\partial \dot{q}_l\, \partial q_k)\, \delta_1(\dot{q}_l)\, \delta_2 q_k] + \sum \delta_1 Q_k\, \delta_2 q_k\\ &\qquad + \sum\sum [(\partial^2 L/\partial q_l\, \partial \dot{q}_k)\, \delta_1 q_l\, \delta_2(\dot{q}_k) + (\partial^2 L/\partial \dot{q}_l\, \partial \dot{q}_k)\, \delta_1(\dot{q}_l)\, \delta_2(\dot{q}_k)].\end{aligned} \tag{c}$$

Now, in (c), we interchange the variation subscripts 1 and 2, thus creating $\sum (\delta_2 p_k\, \delta_1 q_k)^{\cdot} = \cdots$, and then subtract it from (c), while renaming some summation ("dummy") indices. It is not hard to see that, then,

$$d/dt\Big(\sum (\delta_1 p_k\, \delta_2 q_k - \delta_2 p_k\, \delta_1 q_k)\Big) = \sum (\delta_1 Q_k\, \delta_2 q_k - \delta_2 Q_k\, \delta_1 q_k), \quad \text{Q.E.D.} \tag{d}$$

More general Lagrangean equations lead, naturally, to more general forms of (d).

Example 8.7.2 *Third, Hamiltonian, Proof of the Lagrange–Poisson Theorem.* Since $H = H(t, q, p)$, we have

$$\begin{aligned}\delta_1 H &= \sum [(\partial H/\partial q_k)\, \delta_1 q_k + (\partial H/\partial p_k)\, \delta_1 p_k]\\ &= \sum [(-\dot{p}_k + Q_k)\, \delta_1 q_k + (\dot{q}_k)\, \delta_1 p_k] \qquad \text{[by Hamilton's equations]}\\ &= \sum [(\dot{q}_k)\, \delta_1 p_k - (\dot{p}_k - Q_k)\, \delta_1 q_k],\end{aligned} \tag{a}$$

and, therefore, $\delta_2(\ldots)$-varying the above, we obtain

$$\begin{aligned}\delta_2(\delta_1 H) &= \sum \{\delta_2(\dot{q}_k)\, \delta_1 p_k - [\delta_2(\dot{p}_k) - \delta_2 Q_k]\, \delta_1 q_k\}\\ &\quad + \sum [\dot{q}_k\, \delta_2(\delta_1 p_k) - (\dot{p}_k - Q_k)\, \delta_2(\delta_1 q_k)].\end{aligned} \tag{b}$$

Similarly, reversing the order of $\delta_1(\ldots)$ and $\delta_2(\ldots)$, we obtain

$$\begin{aligned}\delta_1(\delta_2 H) &= \sum \{\delta_1(\dot{q}_k)\, \delta_2 p_k - [\delta_1(\dot{p}_k) - \delta_1 Q_k]\, \delta_2 q_k\}\\ &\quad + \sum [\dot{q}_k\, \delta_1(\delta_2 p_k) - (\dot{p}_k - Q_k)\, \delta_1(\delta_2 q_k)].\end{aligned} \tag{c}$$

Subtracting (c) from (b), while noting that for all genuine functions/variables $(\ldots)$, $\delta_1[\delta_2(\ldots)] = \delta_2[\delta_1(\ldots)]$, we obtain

$$\begin{aligned} 0 &= \delta_2(\delta_1 H) - \delta_1(\delta_2 H) \\ &= \sum \{[\delta_2(\dot{q}_k)\,\delta_1 p_k - \delta_1(\dot{q}_k)\,\delta_2 p_k] - [\delta_2(\dot{p}_k) - \delta_2 Q_k]\,\delta_1 q_k + [\delta_1(\dot{p}_k) - \delta_1 Q_k]\,\delta_2 q_k\} \\ &= \Big[\sum (\delta_1 p_k\, \delta_2 q_k - \delta_2 p_k\, \delta_1 q_k)\Big]^{\cdot} - \sum (\delta_1 Q_k\, \delta_2 q_k - \delta_2 Q_k\, \delta_1 q_k), \quad \text{Q.E.D.} \qquad \text{(d)} \end{aligned}$$

Here, too, more general Hamiltonian equations lead to more general forms of (d).

Example 8.7.3 Let us consider a system with Hamiltonian

$$H = (1/2)(p^2 + \lambda^2 q^2), \qquad \text{(a)}$$

and general solution of its equations of motion, as can be easily verified by the reader,

$$q = c_1 \cos(\lambda t) + (c_2/\lambda)\sin(\lambda t), \qquad p = -c_1 \lambda \sin(\lambda t) + c_2 \cos(\lambda t) \qquad \text{(b)}$$

(i.e., free vibration of a linear harmonic and undamped oscillator of unit mass and frequency equal to λ). Hence, the (sole) Lagrange bracket of the system equals

$$\begin{aligned} [c_1, c_2] = -[c_2, c_1] &= (\partial p/\partial c_1)(\partial q/\partial c_2) - (\partial p/\partial c_2)(\partial q/\partial c_1) \\ &= [-\lambda \sin(\lambda t)][\lambda^{-1}\sin(\lambda t)] - [\cos(\lambda t)][\cos(\lambda t)] \\ &= -[\sin^2(\lambda t) + \cos^2(\lambda t)] = -1, \qquad \textit{a constant}; \qquad \text{(c)} \end{aligned}$$

which, in the language of §8.9, means that (b) is a canonical transformation.

Problem 8.7.1 Show that Lagrange's brackets satisfy the following identities:

(i) $$[c_\mu, c_\mu] = 0; \qquad \text{(a)}$$

(ii) $$[c_\mu, c_\nu] = -[c_\nu, c_\mu]; \qquad \text{(b)}$$

(iii) $$\partial[c_\mu, c_\nu]/\partial c_\lambda + \partial[c_\nu, c_\lambda]/\partial c_\mu + \partial[c_\lambda, c_\mu]/\partial c_\nu = 0. \qquad \text{(c)}$$

HINT

[For (iii)]: Notice that

$$[c_\mu, c_\nu] = \partial/\partial c_\nu \Big[\sum q_k(\partial p_k/\partial c_\mu)\Big] - \partial/\partial c_\mu \Big[\sum q_k(\partial p_k/\partial c_\nu)\Big]. \qquad \text{(d)}$$

Perturbation Equations

Next, we apply the above formalism to the theory of perturbations. Let the canonical equations and corresponding general solution of the *unperturbed* problem be

$$dp_k/dt = -\partial H/\partial q_k \quad \text{and} \quad dq_k/dt = \partial H/\partial p_k, \qquad \text{(8.7.14a)}$$

$$p_k = p_k(t; c) \quad \text{and} \quad q_k = q_k(t; c), \qquad \text{(8.7.14b)}$$

respectively, and let the equations of the *slightly perturbed* problem be

$$dp_k/dt = -\partial H/\partial q_k + X_k \qquad \text{and} \qquad dq_k/dt = \partial H/\partial p_k, \tag{8.7.15a}$$

where

$$X_k = X_k(t, q, p) = \textit{given function of its arguments}$$
$$\approx X_k{}^{(1)}(t; c) \qquad \text{[first-order approximation, upon substitution of } \textit{unperturbed} \text{ solution (8.7.14b) in it].} \tag{8.7.15b}$$

Let us solve (8.7.15a) by treating the c's as *variable*; that is, $c_\mu = c_\mu(t)$. Indeed, $(\ldots)^{\cdot}$-differentiating (8.7.14b), we obtain

$$dp_k/dt = \partial p_k/\partial t + \sum(\partial p_k/\partial c_\mu)(dc_\mu/dt),$$
$$dq_k/dt = \partial q_k/\partial t + \sum(\partial q_k/\partial c_\mu)(dc_\mu/dt). \tag{8.7.16a}$$

But since $\partial q_k/\partial t$ $(\partial p_k/\partial t)$ = *unperturbed velocities* (*accelerations*), while $\dot{q}_k$ $(\dot{p}_k)$ = *perturbed velocities* (*accelerations*), we can rewrite (8.7.14a) and (8.7.15a) as follows:

$$\partial p_k/\partial t = -\partial H/\partial q_k \qquad \text{and} \qquad \partial q_k/\partial t = \partial H/\partial p_k, \tag{8.7.16b}$$

$$dp_k/dt = -\partial H/\partial q_k + X_k \qquad \text{and} \qquad dq_k/dt = \partial H/\partial p_k; \tag{8.7.16c}$$

and, comparing these with (8.7.16a), we are readily led to the $2n$ *first-order* differential equations for the $c_\mu(t)$:

$$\sum(\partial p_k/\partial c_\mu)(dc_\mu/dt) = X_k{}^{(1)}, \qquad \sum(\partial q_k/\partial c_\mu)(dc_\mu/dt) = 0. \tag{8.7.17}$$

To understand the motivation behind these key arguments, let us pause to examine briefly the physical problem that led Lagrange et al. to the method of variation of constants: finding the orbit of Earth (E) around the Sun (S). As is well known: in this case, the solution to the *unperturbed*, or *two-body*, problem (i.e., when only the gravitational pull of S on E is included, whereas the small such influences of the other solar system planets on the orbit of E are neglected) are Keplerian elliptical orbits, I, with S at one of their foci, defined at every generic instant by the constants, or "*orbit elements*" c_μ. Here, the *perturbed* problem consists in calculating the small time-dependent deviations of E's orbit from ellipticity; that is, the dc_μ/dt, due to the gravitational pull of the other planets. Our conditions (8.7.17) amount to stating that, at every instant, *E has the same coordinates and velocity (→ momentum) in both the unperturbed (two-body) and perturbed (many-body) orbits* (fig. 8.6).

As the famous Victorian mechanician Tait puts it (see also Lagrange's summary in ex. 8.7.4):

> [T]he disturbing forces are, at any instant, small in comparison with the forces regulating the motion; so that, during any brief period, the motion is practically the same as if no disturbing cause had been at work. But in time, the effects of the disturbance may become so great as entirely to change the dimensions and form of the orbit described. The *character* of the path is not, at any particular instant, affected by the disturbance; but its form and dimensions are in a state of slow, and usually progressive change. Hence, as the first depends upon the *form* of the equations which represent it, while the latter depend upon the actual and relative magnitudes of the constants involved in the

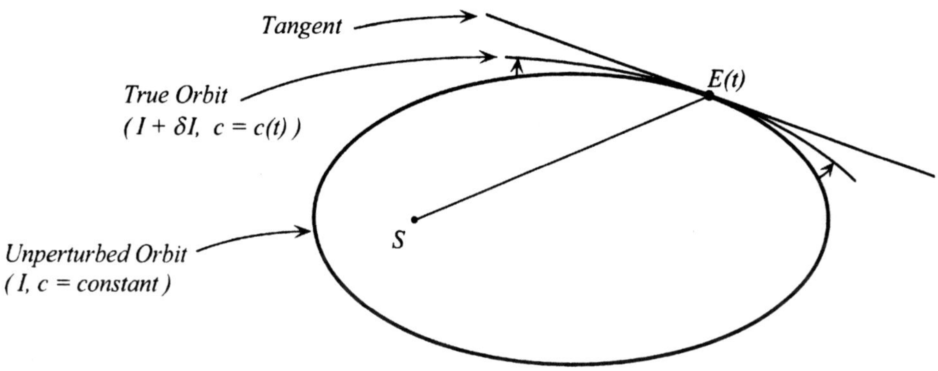

Figure 8.6 Geometry of unperturbed and perturbed orbits of Earth around the Sun.

integrals, we settle, once for all, the *form* of the equation as if no disturbing cause had acted. But we are thus entitled to assume that the constants which the solution involves are quantities which vary with the time in consequence of the slight, but persistent, effects of the disturbance. And, . . . , if at any moment the disturbance were to cease, the motion would forthwith go on for ever in the orbit *then being described*, it follows that in the expressions for the components of the velocity no terms occur depending on the rate of alteration of the values of the constants. (1895, p. 174)

The above help us to understand the differences between partial and total time derivatives:

- *Partial* derivatives refer to the *unperturbed* (or osculating) orbit; that is, constant orbit elements.
- *Total* derivatives refer to the *perturbed* (or true) orbit; that is, variable orbit elements.

Two additional forms of the perturbation equations are obtained as follows:

(i) Multiplying the *first* of (8.7.17) with $\delta_2 q_k = \sum (\partial q_k/\partial c_\nu)\, \delta_2 c_\nu$ and summing over k, and multiplying the *second* of (8.7.17) with $\delta_2 p_k = \sum (\partial p_k/\partial c_\nu)\, \delta_2 c_\nu$ and summing over k, and then subtracting the so-resulting expressions, we obtain

$$\begin{aligned}\sum\sum\sum\, &[(\partial p_k/\partial c_\mu)(dc_\mu/dt)]\,[(\partial q_k/\partial c_\nu)\,\delta_2 c_\nu]\\ &-\sum\sum\sum\,[(\partial q_k/\partial c_\mu)(dc_\mu/dt)]\,[(\partial p_k/\partial c_\nu)\,\delta_2 c_\nu]\\ &=\sum\sum X_k{}^{(1)}\,[(\partial q_k/\partial c_\nu)\,\delta_2 c_\nu],\end{aligned} \tag{8.7.18a}$$

or, recalling the Lagrange bracket definition (8.7.13),

$$\sum\sum [c_\mu, c_\nu](dc_\mu/dt)\,\delta_2 c_\nu = \sum\sum X_k{}^{(1)}\,[(\partial q_k/\partial c_\nu)\,\delta_2 c_\nu], \tag{8.7.18b}$$

and from this, since the $\delta_2 c_\nu$ are arbitrary, we finally obtain the *Lagrangean form of the perturbation equations*:

$$\sum [c_\nu, c_\mu](dc_\nu/dt) = \sum X_k{}^{(1)}(\partial q_k/\partial c_\mu). \tag{8.7.19}$$

In particular, if the perturbations are *potential*—that is, if $X_k = -\partial\Omega/\partial q_k$—then, since $q_k = q_k(t; c)$, (8.7.19) specializes to

$$\sum [c_\nu, c_\mu](dc_\nu/dt) = -\partial\Omega/\partial c_\mu. \tag{8.7.20}$$

Also, since, here, the fundamental state I is *time-independent*, the brackets depend only on the c's.

(ii) Inverting the general solution (8.7.2), or (8.7.14b), we obtain

$$c_\mu = h_\mu(t, q, p) = \textit{first integral (constant) of the unperturbed problem}, \tag{8.7.21a}$$

and, therefore, treating them as variable, we get

$$dc_\mu/dt = \partial h_\mu/\partial t + \sum \left[(\partial h_\mu/\partial q_k)(dq_k/dt) + (\partial h_\mu/\partial p_k)(dp_k/dt)\right]. \tag{8.7.21b}$$

But invoking the perturbation conditions (8.7.17), we find

$$dp_k/dt = \partial p_k/\partial t + \sum (\partial p_k/\partial c_\mu)(dc_\mu/dt) = \partial p_k/\partial t + X_k{}^{(1)}, \tag{8.7.21c}$$

$$dq_k/dt = \partial q_k/\partial t + \sum (\partial q_k/\partial c_\mu)(dc_\mu/dt) = \partial q_k/\partial t; \tag{8.7.21d}$$

and so, inserting these expressions back into (8.7.21b), we have

$$dc_\mu/dt = \partial h_\mu/\partial t + \sum \left[(\partial h_\mu/\partial q_k)(\partial q_k/\partial t) + (\partial h_\mu/\partial p_k)(\partial p_k/\partial t)\right] + \sum (\partial h_\mu/\partial p_k) X_k,$$

or, since the first two summands vanish, because they represent the $(\ldots)^{\cdot}$-derivative of the unperturbed constant c_μ, we obtain, finally,

$$dc_\mu/dt = \sum (\partial h_\mu/\partial p_k) X_k = \sum (\partial c_\mu/\partial p_k) X_k{}^{(1)}. \tag{8.7.22}$$

Poisson Brackets

In particular, if the perturbations are *potential*—that is, if

$$X_k = -\partial\Omega/\partial q_k = -\sum (\partial\Omega/\partial c_\nu)(\partial c_\nu/\partial q_k), \tag{8.7.22a}$$

(8.7.22) becomes

$$dc_\mu/dt = -\sum \left[\sum (\partial\Omega/\partial c_\nu)(\partial c_\nu/\partial q_k)\right](\partial c_\mu/\partial p_k); \tag{8.7.22b}$$

or, if (as is commonly the case)

$$0 = \partial\Omega/\partial p_k = \sum (\partial\Omega/\partial c_\nu)(\partial c_\nu/\partial p_k)$$
$$\Rightarrow \sum \left(\sum (\partial\Omega/\partial c_\nu)(\partial c_\nu/\partial p_k)\right)(\partial c_\mu/\partial q_k) = 0, \tag{8.7.22c}$$

adding a zero [i.e., (8.7.22c)] to the right side of (8.7.22b) we finally obtain the *potential perturbation equations* in the following convenient form:

$$dc_\mu/dt = -\sum (\partial\Omega/\partial c_\nu)(c_\mu, c_\nu), \tag{8.7.23}$$

where

$$(c_\mu, c_\nu) \equiv \sum [(\partial c_\mu/\partial p_k)(\partial c_\nu/\partial q_k) - (\partial c_\mu/\partial q_k)(\partial c_\nu/\partial p_k)]:$$
$$\textit{Poisson bracket of } c_\mu, c_\nu \quad [= -(c_\nu, c_\mu)]. \qquad (8.7.24)$$

[See also §8.9. On the history of Poisson's brackets, and so on, see Dugas (1955, p. 384 ff.)].

Equations (8.7.23) have the advantage over (8.7.20) that they are already solved for the dc/dt; but, in return, eqs. (8.7.20) contain Lagrange's brackets, which do not require solving the p's and q's for the c's, as Poisson's brackets do.

Now, since these two perturbation forms, (8.7.20) and (8.7.23), are equivalent, the brackets of Lagrange and Poisson must satisfy certain consistency requirements. Indeed, substituting $dc_\mu/dt \to dc_\nu/dt$ from (8.7.23) into (8.7.20), we get

$$\sum\sum [c_\nu, c_\mu](c_\nu, c_\lambda)(\partial\Omega/\partial c_\lambda) = -\partial\Omega/\partial c_\mu,$$

from which, since the $\partial\Omega/\partial c_\lambda$ are arbitrary, we obtain the inverse matrix-like relations:

$$\sum [c_\nu, c_\mu](c_\nu, c_\lambda) = \delta_{\mu\lambda} = \delta_{\lambda\mu}: \textit{ Kronecker delta}$$
$$[= 1 \text{ or } 0, \text{ according as } \mu = \lambda \text{ or } \mu \neq \lambda]. \qquad (8.7.25)$$

These compatibility conditions readily show that:

- If Lagrange's brackets are constant, then so are Poisson's brackets, and vice versa.
- If we know Lagrange's brackets, then using (8.7.25) we can find Poisson's brackets, and vice versa; that is, these two behave as if they were elements of two inverse $2n \times 2n$ matrices.

REMARK

Equations (8.7.23) are *exact*, but they are expressed in terms of *perturbed* quantities. Let us formulate them in terms of the known *unperturbed* quantities and their *first-order* corrections. Setting in (8.7.23): $c_\mu = c_{\mu o} + c_{\mu 1}$, where $c_{\mu o}$ = unperturbed values and $c_{\mu 1}$ = corresponding first-order corrections, we readily find that the differential equations of the latter are

$$dc_{\mu 1}/dt = -\sum (\partial\Omega_o/\partial c_{\nu o})(c_{\mu o}, c_{\nu o}), \qquad \text{where} \qquad \Omega_o \equiv \Omega(c_o). \qquad (8.7.23a)$$

Example 8.7.4 *Canonical Form of Lagrange's Perturbation Equations* [1810; see Lagrange, 1965, Vol. 1, pp. 299–320].

HISTORICAL

In the second edition of his *Mécanique Analytique*, Lagrange summarizes the gist of his method as follows:

> Dans les problèmes de Mécanique qu'on ne peut résoudre que par approximation, on trouve ordinairement la première solution en n'ayant égard qu'aux forces principales qui agissent sur les corps; et pour étendre cette solution aux autres forces qu'on peut appeler *perturbatrices*, ce qu'il y a de plus simple, c'est de conserver la forme de la

première solution, mais en rendant variables les constantes arbitraires qu'elle renferme; car, si les quantités qu'on avait négligées, et dont on veut tenir compte, sont très-petites, les nouvelles variables seront à peu près constantes, et l'on pourra y appliquer les méthodes ordinaires d'approximation. Ainsi la difficulté se reduit à trouver les équations entre ces variables. [1965, p. 299]

Let $q_k = q_k(t)$ and $p_k = p_k(t)$ be represented by the following power series in time t:

$$q_k = q_{k0} + q_{k1}\,t + q_{k2}\,t^2 + \cdots, \qquad p_k = p_{k0} + p_{k1}\,t + p_{k2}\,t^2 + \cdots. \tag{a}$$

Let us find the form of the Lagrangean perturbational equations, say the potential case (8.7.20), when we *choose as constants the initial conditions*; that is,

$$c_k = q_{k0} \qquad \text{and} \qquad c_{n+s} = p_{s0} \qquad (k, s = 1, \ldots, n). \tag{b}$$

Then, since

$$[c_\mu, c_\nu] = [c_\mu, c_\nu]_{t=0} = constant, \tag{c}$$

we obtain:
(i) For $\mu, \nu = 1, \ldots, n$:

$$\begin{aligned}[c_\mu, c_\nu] &= \sum [(\partial p_{k0}/\partial c_\mu)(\partial q_{k0}/\partial c_\nu) - (\partial p_{k0}/\partial c_\nu)(\partial q_{k0}/\partial c_\mu)] \\ &= \sum [(0)(\delta_{k\nu}) - (0)(\delta_{k\mu})] = 0.\end{aligned} \tag{d}$$

(ii) For $\mu \to n+s$, $s = 1, \ldots, n$; $\nu = 1, \ldots, n$:

$$\begin{aligned}[c_{n+s}, c_\nu] &= \sum [(\partial p_{k0}/\partial c_{n+s})(\partial q_{k0}/\partial c_\nu) - (\partial p_{k0}/\partial c_\nu)(\partial q_{k0}/\partial c_{n+s})] \\ &= \sum [(\delta_{ks})(\delta_{k\nu}) - (0)(0)] = \sum (\delta_{ks})(\delta_{k\nu}) = \delta_{s\nu};\end{aligned} \tag{e}$$

from which we also get

$$[c_\nu, c_{n+s}] = -[c_{n+s}, c_\nu] = -\delta_{s\nu} = -\delta_{\nu s}. \tag{f}$$

(iii) For $\mu \to n+r$, $\nu \to n+b$; $r, b = 1, \ldots, n$:

$$\begin{aligned}[c_{n+r}, c_{n+b}] &= \sum \big[(\partial p_{k0}/\partial c_{n+r})(\partial q_{k0}/\partial c_{n+b}) - (\partial p_{k0}/\partial c_{n+b})(\partial q_{k0}/\partial c_{n+r})\big] \\ &= \sum \big[(\delta_{kr})(0) - (\delta_{kb})(0)\big] = 0.\end{aligned} \tag{g}$$

Therefore, the perturbation equations (8.7.20):

$$\sum [c_\nu, c_\mu](dc_\nu/dt) = -\partial\Omega/\partial c_\mu, \tag{h}$$

become
(i) $\mu = 1, \ldots, n$:

$$\sum [c_{n+s}, c_\mu](dc_{n+s}/dt) = \sum \delta_{s\mu}(dc_{n+s}/dt) = -\partial\Omega/\partial c_\mu, \tag{i}$$

or (renaming μ as k)

$$dc_{n+k}/dt = -\partial\Omega/\partial c_k \qquad (k = 1, \ldots, n). \tag{j}$$

(ii) $\mu = n+1, \ldots, 2n$; or $\mu = n+s$, $s = 1, \ldots, n$:

$$\sum [c_k, c_{n+s}](dc_k/dt) = \sum (-\delta_{sk})(dc_k/dt) = -\partial\Omega/\partial c_k, \tag{k}$$

or (renaming s to k)

$$dc_k/dt = \partial\Omega/\partial c_{n+k} \qquad (k = 1, \ldots, n). \tag{l}$$

Hence Lagrange's result: If we choose as our constants the initial values of q and p, then (provided the *perturbative forces are potential*) the *perturbation equations are canonical*; with Ω as the *perturbation Hamiltonian* (see also §8.10).

Example 8.7.5 *Variation of Constants: The Forced Linear Oscillator.* Let us consider a system with Lagrangean equation of motion

$$\ddot{q} + \omega^2 q = f(t); \tag{a}$$

that is, a linear and undamped oscillator of (constant) natural frequency ω acted upon by the disturbing external force $f(t)$. As is well known, the general solution of the unperturbed problem [i.e., (a) with $f(t) = 0$] is

$$q = c_1 \sin(\omega t) + c_2 \cos(\omega t); \qquad c_1, c_2 = \text{arbitrary constants.} \tag{b}$$

Elementary Method

To solve the perturbed problem (a), and following the well-known method of variation of constants (see any text on ordinary differential equations), we try a solution of the *same form* as (b) but with c_1 and c_2 unknown functions of time:

$$q = c_1(t)\sin(\omega t) + c_2(t)\cos(\omega t), \tag{c}$$

where $c_1(t)$ and $c_2(t)$ are the coefficients of the *instantaneous simple harmonic motion* (i.e., the arbitrary constants of the motion that would result, at a generic instant, if $f(t)$ suddenly vanished) and they will be determined by the following *two* requirements:

(i) Both unperturbed and perturbed velocities $\dot{q}$, obtained by $(\ldots)^{\cdot}$-differentiation of (b) and (c), respectively, will have the *same form*;
(ii) The so-perturbed motion (c) will satisfy the perturbed equation (a).

Indeed:
(i) By $(\ldots)^{\cdot}$-differentiating (c), we obtain

$$\dot{q} = \dot{c}_1 \sin(\omega t) + \dot{c}_2 \cos(\omega t) + \omega[c_1 \cos(\omega t) - c_2 \sin(\omega t)], \tag{d}$$

and so the *first* requirement leads to the condition

$$\dot{c}_1 \sin(\omega t) + \dot{c}_2 \cos(\omega t) = 0; \tag{e}$$

and
(ii) By $(\ldots)^{\cdot}$-differentiating (d), invoking (e), and then inserting the result in (a), we find the *second* condition:

$$\ddot{q} + \omega^2 q = \cdots = \omega[\dot{c}_1 \cos(\omega t) - \dot{c}_2 \sin(\omega t)] = f(t). \tag{f}$$

Solving the system (e, f) for $\dot{c}_1$, $\dot{c}_2$, we readily obtain

$$\dot{c}_1 = \omega^{-1} f(t)\cos(\omega t), \qquad \dot{c}_2 = -\omega^{-1} f(t)\sin(\omega t). \tag{g}$$

Integrating (g) for a given $f(t)$, we obtain a particular solution of (a). The reader should compare this method with that of the *slowly varying parameters*, in (weakly) nonlinear oscillations (see next example, and also examples in § 7.9).

Generalization. In the *variable coefficient* case

$$\ddot{q} + a(t)\dot{q} + b(t)q = f(t) \qquad [a(t),\ b(t) = \text{known functions of time}], \tag{h}$$

(b) and (c) are replaced, respectively, by

$$q = c_1\, q_1(t) + c_2\, q_2(t):\quad \text{general solution of (h) when } f(t) = 0, \tag{i1}$$

$$q = c_1(t)\, q_1(t) + c_2(t)\, q_2(t), \qquad q_1(t) = \sin(\omega t),\ q_2(t) = \cos(\omega t); \tag{i2}$$

while the conditions (e, f) are replaced, respectively, by

$$\dot{c}_1\, q_1(t) + \dot{c}_2\, q_2(t) = 0, \tag{j1}$$

$$\dot{c}_1\, \dot{q}_1(t) + \dot{c}_2\, \dot{q}_2(t) = f(t). \tag{j2}$$

Solving (j1, 2) for $\dot{c}_1$, $\dot{c}_2$, we obtain the generalization of (g):

$$\dot{c}_1 = -[\, f(t)\, q_2(t)]/W \qquad \text{and} \qquad \dot{c}_2 = [\, f(t)\, q_1(t)]/W, \tag{k}$$

where

$$W = W[q_1(t),\ q_2(t)] \equiv [q_1(t)\, \dot{q}_2(t) - q_2(t)\dot{q}_1(t)] \equiv W(t):$$

Wronskian determinant of $q_1(t)$, $q_2(t)$

[$\neq 0$, since $q_1(t)$, $q_2(t)$ are *linearly independent*; i.e., $(q_2/q_1)^{\cdot} \neq 0$]; (k1)

and, integrating the above and inserting the result in (c)/(i2), we obtain a particular solution of (h).

Via Lagrange's Brackets

We begin with (c):

$$q = c_1(t)\sin(\omega t) + c_2(t)\cos(\omega t) = q[t; c_1,\ c_2]. \tag{l}$$

Equations (d, e) can be rewritten, respectively, as

$$\begin{aligned} dq/dt &= [c_1 \omega \cos(\omega t) - c_2\, \omega \sin(\omega t)] + [\sin(\omega t)\dot{c}_1 + \cos(\omega t)\dot{c}_2] \\ &= \partial q/\partial t + [(\partial q/\partial c_1)\dot{c}_1 + (\partial q/\partial c_2)\, \dot{c}_2], \end{aligned} \tag{m1}$$

$$\sum (\partial q/\partial c_\mu)(dc_\mu/dt) = [\sin(\omega t)]\, \dot{c}_1 + [\cos(\omega t)]\, \dot{c}_2 = 0; \tag{m2}$$

while, since

$$\begin{aligned} d^2q/dt^2 &= [-c_1\omega^2 \sin(\omega t) - c_2\, \omega^2 \cos(\omega t)] + [\dot{c}_1\, \omega \cos(\omega t) - \dot{c}_2\, \omega \sin(\omega t)] \\ &= \partial^2 q/\partial t^2 + [(\partial^2 q/\partial t\, \partial c_1)\dot{c}_1 + (\partial^2 q/\partial t\, \partial c_2)\dot{c}_2], \end{aligned} \tag{m3}$$

condition (f) becomes

$$(\partial^2 q/\partial t\,\partial c_1)\,\dot{c}_1 + (\partial^2 q/\partial t\,\partial c_2)\dot{c}_2 = f(t). \tag{m4}$$

But, and this is a general result,

$$\partial/\partial c_1(\partial q/\partial t) = \partial/\partial c_1(\dot{q}) = \omega\cos(\omega t) = \partial/\partial t(\partial q/\partial c_1) = \partial/\partial t[\sin(\omega t)], \tag{n1}$$

$$\partial/\partial c_2(\partial q/\partial t) = \partial/\partial c_2(\dot{q}) = -\omega\sin(\omega t) = \partial/\partial t(\partial q/\partial c_2) = \partial/\partial t[\cos(\omega t)]; \tag{n2}$$

and, therefore, the system (e, f) can be rewritten as

$$(\partial p/\partial c_1)\dot{c}_1 + (\partial q/\partial c_2)\dot{c}_2 = 0, \qquad (\partial \dot{q}/\partial c_1)\dot{c}_1 + (\partial q/\partial c_2)\dot{c}_2 = f(t). \tag{o1, 2}$$

To solve this system for $\dot{c}_1$, $\dot{c}_2$, we multiply (o2) with $\partial q/\partial c_\mu$ (where $\mu = 1, 2$) and (o1) with $\partial \dot{q}/\partial c_\mu$, and then subtract from each other, thus obtaining

$$\begin{aligned}&\big[(\partial q/\partial c_\mu)(\partial \dot{q}/\partial c_1) - (\partial q/\partial c_1)(\partial \dot{q}/\partial c_\mu)\big]\dot{c}_1\\ &\quad + \big[(\partial q/\partial c_\mu)(\partial \dot{q}/\partial c_2) - (\partial q/\partial c_2)(\partial \dot{q}/\partial c_\mu)\big]\dot{c}_2 = (\partial q/\partial c_\mu)f(t);\end{aligned}$$

that is, in the Lagrangean form (8.7.19) (recalling that here $p = \dot{q}$):

$$[c_1, c_\mu]\dot{c}_1 + [c_2, c_\mu]\dot{c}_2 = (\partial q/\partial c_\mu)f(t); \tag{p}$$

from which, due to the antisymmetry of the Lagrangean brackets (problem 8.7.1)—that is,

$$[c_1, c_1] = [c_2, c_2] = 0, \qquad [c_1, c_2] = \omega \Rightarrow [c_2, c_1] = -\omega, \tag{q}$$

we finally obtain (g) in the following general form:

$$dc_1/dt = \big((\partial q/\partial c_2)f(t)\big)/[c_1, c_2] \qquad dc_2/dt = \big((\partial q/\partial c_1)f(t)\big)/[c_2, c_1]. \tag{r}$$

Comparing (g, k) with (r) we immediately conclude that

$$[c_2, c_1] = (f\,W)\big((\partial q/\partial c_1)/W_2\big), \qquad [c_1, c_2] = (f\,W)\big((\partial q/\partial c_2)/W_1\big), \tag{r1, 2}$$

where

$$W_1 \equiv -f\,q_2, \qquad W_2 \equiv f\,q_1. \tag{r3}$$

Let the reader express the solution of the more general equation (h) via Lagrangean brackets.

This completes the fundamentals of classical canonical perturbation theory. We shall return to such approximation methods in §8.14. Now, using the insights gained in Hamiltonian variables, let us make a small detour to discuss the following.

Perturbation Equations in Lagrangean Variables

The most general Lagrangean-type *perturbed* equations of motion in these variables have the form (§3.10 and §3.11):

$$\sum M_{kl}\ddot{q}_l + f_k(t, q, \dot{q}) = Q_k + X_k, \tag{8.7.26}$$

where

$$f_k(t, q, \dot{q}):\ \text{known function of its arguments,}$$
$$X_k = X_k(t, q, \dot{q}):\ \text{small (total impressed) perturbative force.} \tag{8.7.26a}$$

Let the general solution of the corresponding *unperturbed* problem be

$$q_k = q_k(t;\, c). \tag{8.7.27a}$$

We have seen earlier (8.7.16a, 21d) that by demanding equality between the *unperturbed* and *perturbed* velocities, $\partial q_k/\partial t$ and

$$dq_k/dt = \partial q_k/\partial t + \sum (\partial q_k/\partial c_\mu)(dc_\mu/dt),$$

respectively, we arrive at the n conditions (8.7.17) for the dc/dt's:

$$\sum (\partial q_k/\partial c_\mu)(dc_\mu/dt) = 0. \tag{8.7.27b}$$

The additional n conditions will result by substituting the *perturbed accelerations*

$$d^2q_k/dt^2 = \partial^2 q_k/\partial t^2 + \sum (\partial^2 q_k/\partial t\,\partial c_\mu)(dc_\mu/dt) \qquad \text{[after invoking (8.7.27b)]}, \tag{8.7.27c}$$

in (8.7.26), and then subtracting from it the unperturbed equation; that is, eq. (8.7.26) with $X_k = 0$ and $d^2q_k/dt^2 = \partial^2 q_k/\partial t^2$: *unperturbed accelerations*. Thus, we obtain the following *Lagrangean perturbational equations of motion*:

$$\sum \sum M_{kl}(\partial^2 q_l/\partial t\,\partial c_\mu)(dc_\mu/dt) = X_k{}^{(1)}, \tag{8.7.28}$$

or, in extenso,

$$\sum \left[M_{k1}(\partial^2 q_1/\partial t\,\partial c_\mu) + \cdots + M_{kn}(\partial^2 q_n/\partial t\,\partial c_\mu)\right] (dc_\mu/dt) = X_k{}^{(1)}. \tag{8.7.28a}$$

These are the Lagrangean counterpart of the first of (8.7.17), and together with (8.7.27b) they constitute a system of $2n$ linear equations for the $2n$ dc/dt's. Indeed, with the abbreviations

$$\Lambda_{k\mu} \equiv \sum M_{kl}(\partial^2 q_l/\partial t\,\partial c_\mu) = \sum M_{kl}(\partial^2 q_l/\partial c_\mu\,\partial t), \qquad \Pi_{k\mu} \equiv \partial q_k/\partial c_\mu, \tag{8.7.29a}$$

the system (8.7.27b, 28) can be rewritten as

$$\sum \Lambda_{k\mu}\dot{c}_\mu = X_k{}^{(1)}, \qquad \sum \Pi_{k\mu}\dot{c}_\mu = 0, \tag{8.7.29b}$$

and, by Cramer's rule, its solution is

$$\dot{c}_\mu = \sum (-1)^{k+\mu}\Delta_{k\mu}X_k{}^{(1)}/\Delta \qquad \text{(summation on } k = 1, \ldots, n), \tag{8.7.29c}$$

$$\Delta \equiv \begin{vmatrix} \Lambda_{11} & \cdots & \Lambda_{1,2n} \\ \cdots & \cdots & \cdots \\ \Lambda_{n1} & \cdots & \Lambda_{n,2n} \\ \Pi_{11} & \cdots & \Pi_{1,2n} \\ \cdots & \cdots & \cdots \\ \Pi_{n1} & \cdots & \Pi_{n,2n} \end{vmatrix} \equiv Det(\Lambda_{k\mu}, \Pi_{l\nu}) \quad (\neq 0, \text{ assumed}), \tag{8.7.29d}$$

and $(-1)^{k+\mu}\Delta_{k\mu} =$ *cofactor (i.e., signed minor) of* $\Lambda_{k\mu}$ *in* Δ. Integrating (8.7.29c), we obtain the $2n$ $c_\mu(t)$, and afterwards the *perturbed* solution $q_k = q_k[t;\, c(t)]$.

Example 8.7.6 *Variation of Constants: Quasi-Linear Oscillator.* Let us consider a system with equation of motion

$$m\ddot{q} + k\,q = \varepsilon\, f(q, \dot{q}), \tag{a}$$

where the generally nonlinear force $\varepsilon\, f(q, \dot{q})$ [ε: *constant, such that* $\varepsilon\, f(\ldots)$ has the dimensions of force] is assumed *small* compared with inertia ($m\ddot{q}$; m: constant mass) and elasticity ($-k\dot{q}$; k: constant modulus); this is the meaning of the adjective *quasi-linear*.

Lagrangean Variables

Here,

$$2T = m(\dot{q})^2 \;\Rightarrow\; M_{11} = m. \tag{b}$$

The general solution of the unperturbed equation [i.e., (a) with $\varepsilon f(q, \dot{q}) = 0$] is

$$q = c_1 \sin(\omega_o t) + c_2 \cos(\omega_o t), \qquad {\omega_o}^2 \equiv k/m. \tag{c}$$

Therefore, since $n = 1 \;\Rightarrow\; \nu = 1, 2$, we obtain, successively,

$$\begin{aligned} &\Pi_{11} \equiv \partial q/\partial c_1 = \sin(\omega_o t), \qquad \Pi_{12} \equiv \partial q/\partial c_2 = \cos(\omega_o t), \\ &\Lambda_{11} = M_{11}(\partial^2 q/\partial t\, \partial c_1) = m[\omega_o \cos(\omega_o t)], \\ &\Lambda_{12} = M_{11}(\partial^2 q/\partial t\, \partial c_2) = m[-\omega_o \sin(\omega_o t)], \\ &\Delta = \Lambda_{11}\Pi_{12} - \Lambda_{12}\Pi_{11} = \cdots = m\,\omega_o[\cos^2(\omega_o t) + \sin^2(\omega_o t)] = m\,\omega_o; \end{aligned} \tag{d}$$

$$\dot{c}_1 = (-1)^{1+1}\Delta_{11}X_1/\Delta = \Pi_{12}X_1/\Delta = (m\,\omega_o)^{-1}\varepsilon\, f(q, \dot{q})\cos(\omega_o t), \tag{e1}$$

$$\dot{c}_2 = (-1)^{1+2}\Delta_{12}X_1/\Delta = \Pi_{11}X_1/\Delta = -(m\,\omega_o)^{-1}\varepsilon\, f(q, \dot{q})\sin(\omega_o t). \tag{e2}$$

In the theory of nonlinear oscillations, it is customary and convenient to work, not with c_1 and c_2, but with the following variables:

$$c_1 = q_o \sin\phi \qquad \text{and} \qquad c_2 = q_o \cos\phi. \tag{f}$$

By $(\ldots)^{\cdot}$-differentiating the above, we obtain

$$\dot{c}_1 = \dot{q}_o \sin\phi + q_o\dot{\phi}\cos\phi \qquad \text{and} \qquad \dot{c}_2 = \dot{q}_o\cos\phi - q_o\dot{\phi}\sin\phi, \tag{g}$$

$$\begin{aligned}\Rightarrow f(q,\dot{q})\big|_{\text{unperturbed } q,\dot{q}} &\\ &= f\,[c_1\sin(\omega_o t) + c_2\cos(\omega_o t),\ c_1\omega_o\cos(\omega_o t) - c_2\omega_o\sin(\omega_o t)]\\ &= f\,[q_o\cos(\omega_o t - \phi),\ -q_o\,\omega_o\sin(\omega_o t - \phi)] \equiv f\,[\ldots,\ldots],\end{aligned} \tag{h}$$

and so (e1, e2) translate, respectively, to

$$\dot{q}_o\sin\phi + q_o\dot{\phi}\cos\phi = (\varepsilon/m\,\omega_o)f\,[\ldots,\ldots]\cos(\omega_o t), \tag{i1}$$

$$\dot{q}_o\cos\phi - q_o\dot{\phi}\sin\phi = -(\varepsilon/m\,\omega_o)f\,[\ldots,\ldots]\sin(\omega_o t). \tag{i2}$$

Solving this system for $\dot{q}_o$ and $q_o\dot{\phi}$, we get

$$\dot{q}_o = -(\varepsilon/m\,\omega_o)f\,[\ldots,\ldots]\sin\chi, \qquad q_o\dot{\phi} = (\varepsilon/m\,\omega_o)f\,[\ldots,\ldots]\cos\chi, \tag{i3}$$

where $\chi \equiv \omega_o t - \phi(t) \equiv \chi(t)$.

These equations for q_o and ϕ are *exact*; but since they are still nonlinear, they are not very useful. They become useful when q_o and ϕ *change very little* during the unperturbed period $\tau_o \equiv 2\pi/\omega_o$. Then, we can replace the exact equations (i3) with a new set whose right sides are the *averages* of the right sides of (i3) over τ_o:

$$\begin{aligned}dq_o/dt &= -(\varepsilon/2\pi m)\int f\,[\ldots,\ldots]\sin\chi\,dt\\ &= -(\varepsilon/2\pi\omega_o m)\int f\,[q_o\cos\chi,\ -q_o\,\omega_o\sin\chi]\sin\chi\,d\chi,\end{aligned} \tag{j1}$$

$$\begin{aligned}q_o(d\phi/dt) &= (\varepsilon/2\pi m)\int f\,[\ldots,\ldots]\cos\chi\,dt\\ &= +(\varepsilon/2\pi\omega_o m)\int f\,(q_o\cos\chi,\ -q_o\,\omega_o\sin\chi)\cos\chi\,d\chi;\end{aligned} \tag{j2}$$

where the dt-integrals extend from 0 to $2\pi/\omega_o$, while the $d\chi$-integrals extend from 0 to 2π.

For a general and masterful treatment of such *averaging* techniques, see, for example, Bogoliubov and Mitropolskii (1974, chap. 5, pp. 355–429); also ex. 7.9.14 ff.

Hamiltonian Variables

Here,

$$p = \partial T/\partial\dot{q} = m\dot{q} \Rightarrow \dot{q} = p/m, \tag{k1}$$

$$\begin{aligned}\Rightarrow H &= p\dot{q} - (T - V) = p(p/m) - (m/2)(p/m)^2 + (k/2)p^2\\ &= p^2/2m + kq^2/2 = H(q,p)\text{: \textit{unperturbed Hamiltonian.}}\end{aligned} \tag{k2}$$

(a) The *unperturbed* canonical equations and their general solution are, respectively,

$$\dot{p} = -\partial H/\partial q: \qquad \dot{p} = -kq, \tag{11}$$

$$\dot{q} = \partial H/\partial p: \qquad \dot{q} = p/m; \tag{12}$$

$$p = p(t,c) = m\,\omega_o[c_1 \cos(\omega_o t) - c_2 \sin(\omega_o t)], \tag{m1}$$

$$q = q(t,c) = c_1 \sin(\omega_o t) + c_2 \cos(\omega_o t). \tag{m2}$$

(b) The *perturbed* canonical equations are

$$\dot{p} = -\partial H/\partial q + X = -kq + X$$

$$= -kq + \varepsilon f(q, p/m) \equiv -kq + \varepsilon F(q,p), \tag{n1}$$

$$\dot{q} = \partial H/\partial p: \qquad \dot{q} = p/m. \tag{n2}$$

(i) Let us, first, apply the Lagrangean form (8.7.19):

$$\sum [c_\nu, c_\mu](dc_\nu/dt) = X(\partial q/\partial c_\mu) \qquad (\mu = 1,2). \tag{o1}$$

We have

$$\partial q/\partial c_1 = \sin(\omega_o t), \qquad \partial q/\partial c_2 = \cos(\omega_o t), \tag{o2}$$

$$[c_1, c_1] = 0, \qquad [c_2, c_2] = 0 \qquad \text{(since these brackets are antisymmetric)},$$

$$[c_1, c_2] = (\partial p/\partial c_1)(\partial q/\partial c_2) - (\partial p/\partial c_2)(\partial q/\partial c_1)$$

$$= [m\,\omega_o \cos(\omega_o t)][\cos(\omega_o t)] - [-m\,\omega_o \sin(\omega_o t)][\sin(\omega_o t)] = m\,\omega_o,$$

$$[c_2, c_1] = -[c_1, c_2] = -m\,\omega_o; \tag{o3}$$

and hence (o1) reduce to

$$[c_2, c_1]\dot{c}_2 = X(\partial q/\partial c_1): \qquad (-m\,\omega_o)\dot{c}_2 = \varepsilon F(q,p)\sin(\omega_o t), \tag{o4}$$

$$[c_1, c_2]\dot{c}_1 = X(\partial q/\partial c_2): \qquad (m\,\omega_o)\dot{c}_1 = \varepsilon F(q,p)\cos(\omega_o t); \tag{o5}$$

that is, the earlier (e1), (e2).

(ii) Finally, let us apply the form (8.7.22):

$$dc_\mu/dt = \sum (\partial c_\mu/\partial p)X. \tag{p1}$$

Solving (m1), (m2) for c_1 and c_2, we obtain

$$c_1 = [\sin(\omega_o t)]q + [\cos(\omega_o t)/m\,\omega_o]p, \tag{p2}$$

$$c_2 = [\cos(\omega_o t)]q + [-\sin(\omega_o t)/m\,\omega_o]p, \tag{p3}$$

and so (p1) yield

$$dc_1/dt = (\partial c_1/\partial p)X = [\cos(\omega_o t)/m\,\omega_o]\varepsilon F(q,p); \qquad \text{i.e., (e1)/(o5)}; \tag{p4}$$

$$dc_2/dt = (\partial c_2/\partial p)X = [-\sin(\omega_o t)/m\,\omega_o]\varepsilon F(q,p); \qquad \text{i.e., (e2)/(o4)}. \tag{p5}$$

Example 8.7.7 *Variation of Constants: Effect of Small Air Resistance (Drag) on Projectiles.* Let us consider a particle P of mass m in free motion on a fixed vertical plane $O\text{–}xy$ (Ox: horizontal, $+Oy$: upward) under the action of constant gravity g and small air resistance (perturbation) equal to

$$\boldsymbol{D} = -\varepsilon(\boldsymbol{v}/v)f(v), \tag{a}$$

where $\boldsymbol{v}$ = velocity of $P = (\dot{x}, \dot{y})$, ε = small parameter (> 0), $f(v)$ = experimentally determined function of $|\boldsymbol{v}| \equiv v = [(\dot{x})^2 + (\dot{y})^2]^{1/2}$ (> 0). Here, $n = 2$, and so, with $q_1 = x$ and $q_2 = y$, we have

$$2T = m[(\dot{x})^2 + (\dot{y})^2], \qquad V = mgy + constant, \qquad Q_k = 0, \tag{b}$$

$$X_1 \equiv X = \boldsymbol{D}\cdot\boldsymbol{i} = -\varepsilon(\dot{x}/v)f(v), \qquad X_2 \equiv Y = \boldsymbol{D}\cdot\boldsymbol{j} = -\varepsilon(\dot{y}/v)f(v), \tag{c}$$

and so the *perturbed* equations of motion are

$$\text{Horizontal:} \qquad m\ddot{x} = -\varepsilon(\dot{x}/v)f(v), \tag{d1}$$

$$\text{Vertical:} \qquad m\ddot{y} = -mg - \varepsilon(\dot{y}/v)f(v). \tag{d2}$$

Since the general solution of the *unperturbed* problem [i.e., (d1, d2) with $\varepsilon = 0$] is

$$x = c_1 t + c_2, \qquad y = -(g/2)t^2 + c_3 t + c_4, \tag{e1, 2}$$

we readily find

$$\Pi_{11} = \partial x/\partial c_1 = t, \quad \Pi_{12} = \partial x/\partial c_2 = 1, \quad \Pi_{13} = \partial x/\partial c_3 = 0, \quad \Pi_{14} = \partial x/\partial c_4 = 0,$$
$$\partial^2 x/\partial t\,\partial c_1 = 1, \quad \partial^2 x/\partial t\,\partial c_2 = 0, \quad \partial^2 x/\partial t\,\partial c_3 = 0, \quad \partial^2 x/\partial t\,\partial c_4 = 0; \tag{f1}$$

$$\Pi_{21} = \partial y/\partial c_1 = 0, \quad \Pi_{22} = \partial y/\partial c_2 = 0, \quad \Pi_{23} = \partial y/\partial c_3 = t, \quad \Pi_{24} = \partial y/\partial c_4 = 1,$$
$$\partial^2 y/\partial t\,\partial c_1 = 0, \quad \partial^2 y/\partial t\,\partial c_2 = 0, \quad \partial^2 y/\partial t\,\partial c_3 = 1, \quad \partial^2 y/\partial t\,\partial c_4 = 0; \tag{f2}$$

$$\Lambda_{11} = M_{11}(\partial^2 x/\partial t\,\partial c_1) + M_{12}(\partial^2 y/\partial t\,\partial c_1) = (m)(1) + (0)(0) = m$$
$$\Lambda_{12} = M_{11}(\partial^2 x/\partial t\,\partial c_2) + M_{12}(\partial^2 y/\partial t\,\partial c_2) = (m)(0) + (0)(0) = 0,$$
$$\Lambda_{13} = M_{11}(\partial^2 x/\partial t\,\partial c_3) + M_{12}(\partial^2 y/\partial t\,\partial c_3) = (m)(0) + (0)(1) = 0,$$
$$\Lambda_{14} = M_{11}(\partial^2 x/\partial t\,\partial c_4) + M_{12}(\partial^2 y/\partial t\,\partial c_4) = (m)(0) + (0)(0) = 0; \tag{g1}$$

$$\Lambda_{21} = M_{21}(\partial^2 x/\partial t\,\partial c_1) + M_{22}(\partial^2 y/\partial t\,\partial c_1) = (0)(1) + (m)(0) = 0,$$
$$\Lambda_{22} = M_{21}(\partial^2 x/\partial t\,\partial c_2) + M_{22}(\partial^2 y/\partial t\,\partial c_2) = (0)(0) + (m)(0) = 0,$$
$$\Lambda_{23} = M_{21}(\partial^2 x/\partial t\,\partial c_3) + M_{22}(\partial^2 y/\partial t\,\partial c_3) = (0)(0) + (m)(1) = m,$$
$$\Lambda_{24} = M_{21}(\partial^2 x/\partial t\,\partial c_4) + M_{22}(\partial^2 y/\partial t\,\partial c_4) = (0)(0) + (m)(0) = 0; \tag{g2}$$

$$\Delta \equiv Det(\Lambda_{k\mu}/\Pi_{l\nu}) = \cdots = -m^2 \qquad (k, l = 1, 2;\ \mu, \nu = 1, 2, 3, 4),$$
$$\Delta_{11} = -m, \quad \Delta_{12} = -mt, \quad \Delta_{13} = 0, \quad \Delta_{14} = 0;$$
$$\Delta_{21} = 0, \quad \Delta_{22} = 0, \quad \Delta_{23} = m, \quad \Delta_{24} = mt. \tag{g3}$$

Hence, the perturbation equations (8.7.29c) yield

$$
\begin{aligned}
dc_1/dt &= \sum (-1)^{k+1} \Delta_{k1} X_k{}^{(1)}/\Delta \qquad (k = 1, 2) \\
&= [(-1)^{1+1} \Delta_{11} X_1{}^{(1)} + (-1)^{2+1} \Delta_{21} X_2{}^{(1)}]/\Delta \\
&= [(-1)^2 \Delta_{11} X^{(1)} + (-1)^3 \Delta_{21} Y^{(1)}]/\Delta \\
&= \{(-1)^2(-m)[-\varepsilon(\dot{x}/v)f(v)] + (-1)^3(0)[-\varepsilon(\dot{y}/v)f(v)]\}/(-m^2) \\
&= -(\varepsilon/m)(\dot{x}/v)f(v)|_{\text{unperturbed}} \\
&= -(\varepsilon/m)(c_1/v_{\text{unperturbed}})f(v_{\text{unperturbed}}) \equiv -(\varepsilon/m)(c_1/v_o)f(v_o), \\
dc_2/dt &= \cdots = (\varepsilon/m)(c_1/v_o)f(v_o)\,t, \\
dc_3/dt &= \cdots = -(\varepsilon/m)[(c_3 - g\,t)/v_o]f(v_o), \\
dc_4/dt &= \cdots = (\varepsilon/m)[(c_3 - g\,t)/v_o]f(v_o)\,t, \qquad \text{(h)}
\end{aligned}
$$

where

$$
v_{\text{unperturbed}} = [(\dot{x})^2 + (\dot{y})^2]^{1/2}\Big|_{\text{unperturbed}} = [(c_1)^2 + (c_3 - g\,t)^2]^{1/2} \equiv v_o. \qquad \text{(h1)}
$$

Integrating the four expressions (h), while (since the drag is small) replacing in their right sides $c_{1,2,3,4}$ with the corresponding integration constants $c_{1o,2o,3o,4o}$ using $*$ as (dummy) variable of integration, and the notation $f(v_{\text{unperturbed}}) \equiv f(v_o) \equiv f_o$, we obtain, finally,

$$
\begin{aligned}
c_1 &= c_{1o} - (\varepsilon/m)\int_0^t (c_1/v_o)f_o\,d* \;\approx\; c_{1o} - (\varepsilon/m)c_{1o}\int_0^t (f_o/v_o)\,d*, \qquad \text{(i1, 2, 3, 4)} \\
c_2 &= c_{2o} + (\varepsilon/m)\int_0^t (c_1/v_o)f_o * d* \;\approx\; c_{2o} + (\varepsilon/m)c_{1o}\int_0^t (f_o/v_o) * d*, \\
c_3 &= c_{3o} - (\varepsilon/m)\int_0^t [(c_3 - g*)/v_o]f_o\,d* \;\approx\; c_{3o} - (\varepsilon/m)\int_0^t (c_{3o} - g*)(f_o/v_o)\,d*, \\
c_4 &= c_{4o} + (\varepsilon/m)\int_0^t [(c_3 - g*)/v_o]f_o * d* \;\approx\; c_{4o} + (\varepsilon/m)\int_0^t (c_{3o} - g*)(f_o/v_o) * d*,
\end{aligned}
$$

where $v_o \approx [(c_{1o})^2 + (c_{3o} - g*)^2]^{1/2} \Rightarrow f_o$: known function of $*$, c_{1o}, c_{3o}.

Last, substituting (i1–4) into (e1, 2), we obtain a particular solution of the perturbed problem (d1, 2) in terms of t and $c_{1o,2o,3o,4o}$, correct to the first-order in ε.

For the *Hamiltonian* perturbation treatment—that is, via (8.7.19) or (8.7.22), including special $f(v)$ cases, see, for example, Hamel (1949, pp. 309–311; and pp. 689–691 for the plane linear elastic and isotropic oscillator under small air resistance); also Lur'e (1968, pp. 569–571).

8.8 CANONICAL TRANSFORMATIONS (CT)

We have already seen (ex. 3.5.11) that a key advantage of the Lagrangean-type equations, say

$$
E_k \equiv E_k(L) \equiv (\partial L/\partial \dot{q}_k)^{\cdot} - \partial L/\partial q_k = Q_k, \qquad (8.8.1)
$$

over those of Newton–Euler, is their *form invariance* under the group of frame-of-reference transformations, or *point transformations*, defined by

$$G:\qquad q_k = q_k(t, q_{k'}) \leftrightarrow q_{k'} = q_{k'}(t, q_k) \qquad (k, k' = 1, \ldots, n), \tag{8.8.2}$$

$$q_k(\ldots)\text{: twice differentiable,} \qquad \text{and} \qquad J \equiv |\partial q/\partial q'| \neq 0; \tag{8.8.2a}$$

that is, under G,

$$E_k \to E_{k'}(L') \equiv (\partial L'/\partial \dot{q}_{k'})^{\cdot} - \partial L'/\partial q_{k'} = Q_{k'}, \tag{8.8.3}$$

where

$$L \to L[t, q(t, q'), \dot{q}(t, q, \dot{q}')] \equiv L'(t, q', \dot{q}') = L', \tag{8.8.3a}$$

$$E_{k'} = \sum (\partial q_k/\partial q_{k'}) E_k \;\Leftrightarrow\; E_k = \sum (\partial q_{k'}/\partial q_k)\, E_{k'}, \tag{8.8.3b}$$

$$Q_{k'} = \sum (\partial q_k/\partial q_{k'})\, Q_k \;\Leftrightarrow\; Q_k = \sum (\partial q_{k'}/\partial q_k)\, Q_{k'}. \tag{8.8.3c}$$

When it comes to Hamiltonian type of equations, since they are mathematically equivalent to the Lagrangean ones, we expect similar form invariance under G. However, if we define

$$\begin{aligned} p_{k'} \equiv \partial L'/\partial \dot{q}_{k'} &= \sum (\partial L/\partial \dot{q}_k)(\partial \dot{q}_k/\partial \dot{q}_{k'}) = \sum (\partial L/\partial \dot{q}_k)(\partial q_k/\partial q_{k'}) \\ &= \sum (\partial \dot{q}_k/\partial \dot{q}_{k'}) p_k = \sum (\partial q_k/\partial q_{k'})\, p_k \\ &= p_{k'}(t, q', p) = p_{k'}[t, q'(t, q), p] = p_{k'}(t, q, p) \end{aligned} \tag{8.8.4}$$

(i.e., the new momenta depend on both the old momenta *and* the old coordinates and time), and the Hamiltonian equations corresponding to (8.8.1)

$$dq_k/dt = \partial H/\partial p_k, \qquad dp_k/dt = -\partial H/\partial q_k + Q_k, \tag{8.8.5}$$

are transformed to

$$dq_{k'}/dt = \partial H'/\partial p_{k'}, \qquad dp_{k'}/dt = -\partial H'/\partial q_{k'} + Q_{k'}, \tag{8.8.6}$$

the *new Hamiltonian* $H' = H'(t, q', p')$, unlike $L = L'$, *may no longer equal the old one* H; that is, in general,

$$H' = H'(t, q', p') \neq H(t, q, p) = H. \tag{8.8.6a}$$

Indeed, assuming (8.8.5) to hold, we have, successively,

$$\begin{aligned} H' &\equiv \sum p_{k'} \dot{q}_{k'} - L' \\ &= \sum p_{k'} \dot{q}_{k'} - L = \sum p_{k'} \dot{q}_{k'} - \Big(\sum p_k \dot{q}_k - H\Big) \\ &= \sum \Big(\sum (\partial q_k/\partial q_{k'})\, p_k\Big) \dot{q}_{k'} - \Big(\sum p_k \dot{q}_k - H\Big) \\ &= \sum \Big(\sum (\partial q_k/\partial q_{k'}) \dot{q}_{k'} - \dot{q}_k\Big) p_k + H \\ &= -\sum (\partial q_k/\partial t) p_k + H; \end{aligned} \tag{8.8.7}$$

that is, in general, H' defined as above, or equivalently by

$$L = L': \qquad \sum p_{k'}\,\dot{q}_{k'} - H' = \sum p_k\,\dot{q}_k - H, \tag{8.8.7a}$$

does *not* equal H (recalling probs. 3.16.11 and 3.16.12); but it does if $\partial q_k/\partial t = 0$, that is whenever (8.8.2) specializes to the geometrical (non–frame-of-reference transformations)

$$q_k = q_k(q_{k'}) \Leftrightarrow q_{k'} = q_{k'}(q_k). \tag{8.8.7b}$$

Hence, in the Hamiltonian case, it is necessary to widen the meaning of invariance. Specifically, and since here the independent variables are the q's and p's, we are seeking the *most general transformations in (the phase space of) these variables*; that is,

$$q = q(t, q', p') \Leftrightarrow q' = q'(t, q, p), \tag{8.8.8a}$$

$$p = p(t, q', p') \Leftrightarrow p' = p'(t, q, p) \tag{8.8.8b}$$

[with nonvanishing Jacobian $|\partial(q', p')/\partial(q, p)|$] that leave Hamilton's equations *form invariant*, as in (8.8.5). Such transformations are called *canonical.*

REMARK ON NOTATION

A number of authors denote our $(q_{k'}, p_{k'})$ as (Q_k, P_k). Our notation was chosen to avoid confusion with the holonomic components of Lagrangean impressed forces and nonholonomic momenta, respectively (chap. 3); also, it is in line with the more precise tensorial notations.

Problem 8.8.1 Show that under point transformations:

(i) $$\partial H'/\partial p_{k'} - dq_{k'}/dt = \sum (\partial p_k/\partial p_{k'})(\partial H/\partial p_k - dq_k/dt) \quad (= 0), \tag{a}$$

(ii) $$\partial H'/\partial q_{k'} + dp_{k'}/dt = \sum (\partial q_k/\partial q_{k'})(\partial H/\partial q_k + dp_k/dt) \quad (= 0). \tag{b}$$

Problem 8.8.2 Show that

$$\partial p_k/\partial p_{k'} = \partial q_k/\partial q_{k'}, \qquad \partial p_{k'}/\partial p_k = \partial q_{k'}/\partial q_k. \tag{a}$$

Whence the Significance of Canonical Transformations (CT)

The reason that such transformations are important in theoretical mechanics is their ability to transform an original set of Hamiltonian equations, in (unprimed) q's and p's, into a *simpler* set of Hamiltonian equations in the new (primed) variables q''s and p''s.

In particular, we are seeking *transformations in which one or more (or even all) of the coordinates are ignorable.* In such q''s $\rightarrow \psi'$'s (and with $L'/R'/H' \equiv$ *New Lagrangean/Routhian/Hamiltonian*)

$$\partial L'/\partial \psi_{i'} = \partial R'/\partial \psi_{i'} = -\partial H'/\partial \psi_{i'} = 0 \qquad (i' = 1, \ldots, M \leq n), \tag{8.8.9a}$$

that is, these three mutually equal partial derivatives vanish simultaneously (which is the definition of these coordinates, §8.2–8.4); and, as a result, assuming, as in §8.4 ff., that $Q_{i'} = 0$,

$$d\Psi_{i'}/dt = \partial L'/\partial\psi_{i'} = -\partial H'/\partial\psi_{i'} = 0 \Rightarrow \Psi_{i'} = constant \equiv C_{i'}, \tag{8.8.9b}$$

$$\Rightarrow H' = H'(t, q', p'; C), \tag{8.8.9c}$$

where the $2(n - M)$ q''s and p''s are the remaining nonignorable coordinates and momenta. {In order to benefit from the new ignorable coordinates [as in §8.4 and §8.10 (Hamilton–Jacobi's method)], a number of authors set $Q_k = 0$ from the outset. Then, clearly, $Q_{k'} = 0$.}

Solving the Hamiltonian equations of such a system—that is,

$$dq_{p'}/dt = \partial H'/\partial p_{p'}, \qquad dp_{p'}/dt = -\partial H'/\partial q_{p'} \qquad (p' = M + 1, \ldots, n), \tag{8.8.9d}$$

we find

$$q_{p'} = q_{p'}(t;\ C_{i'}, \alpha_{p'}, \beta_{p'}), \qquad p_{p'} = p_{p'}(t; C_{i'}, \alpha_{p'}, \beta_{p'}),$$
$$(\alpha_{p'}, \beta_{p'}) = 2(n - M): \text{ constants of integration of (8.8.9d);} \tag{8.8.9e}$$

then, substituting (8.8.9e) into (8.8.9c), we obtain

$$H' = H'(t;\ C_{i'}, \alpha_{p'}, \beta_{p'}), \tag{8.8.9f}$$

and so, finally, we can calculate the ψ''s from their equations of motion via a quadrature:

$$d\psi_{i'}/dt = \partial H'/\partial C_{i'} \Rightarrow \psi_{i'} = \int (\partial H'/\partial C_{i'})\, dt + \psi_{i',o}, \tag{8.8.9g}$$

where $\psi_{i',o}$ = integration constants, to be determined from the initial conditions, as in the Routhian case (§8.3, 8.4). In particular, if *all* new coordinates are ignorable, *and* $\partial H'/\partial t = 0$, then

$$H' = H'(C_{k'};\ \alpha_{l'}, \beta_{m'}) \Rightarrow d\psi_{i'}/dt = \partial H'/\partial C_{i'} = constant \equiv c_{i'}, \text{ etc.} \tag{8.8.9h}$$

Hence, CT can simplify the equations of motion considerably, and supply integrals of motion, as in (8.8.9b). Of course, other, *noncanonical* transformations may achieve additional simplifications.

Definition of, and Conditions for, Canonicity; Generating Function

In view of (8.8.3a) and the fact that a Lagrangean is defined only to within the total derivative of an arbitrary function of the coordinates and time (ex. 3.5.13)—that is, two such Lagrangeans yield the same equations of motion—we, following (the eminent Norwegian mathematician) S. Lie, introduce the following general definition: an (8.8.8a, b)-like transformation $(q,p) \rightarrow (q', p')$ is called canonical (CT) if

$$L\,dt = L'\,dt + dF$$
$$\Rightarrow \sum p_k\,dq_k - H\,dt = \sum p_{k'}\,dq_{k'} - H'\,dt + dF, \tag{8.8.10}$$
$$\Rightarrow \sum p_k\,dq_k - \sum p_{k'}\,dq_{k'} = (H - H')\,dt + dF, \tag{8.8.11}$$

where F called (after Jacobi) the *generating*, or *substitution, function* of the transformation, is an arbitrary differentiable function of the coordinates, momenta, and time; and H' satisfies the Hamiltonian equations in the new variables. Equivalently, we may call an (8.8.8a, b)-like transformation canonical if

$$\left(\sum p_k\,dq_k - H\,dt\right) - \left(\sum p_{k'}\,dq_{k'} - H'\,dt\right)$$
$$= \left(\sum p_k\,dq_k - \sum p_{k'}\,dq_{k'}\right) - (H - H')\,dt = dF \qquad \text{(i.e., integrable);} \tag{8.8.11a}$$

even though $\sum p_k\,dq_k$: *nonintegrable*, and $\sum p_{k'}\,dq_{k'}$: *nonintegrable*. For $dt \to \delta t = 0$ and $dq,\ dp \to \delta q,\ \delta p$, the above yield the *virtual form* of a canonical transformation:

$$\sum p_k\,\delta q_k - \sum p_{k'}\,\delta q_{k'} = \delta F, \tag{8.8.12}$$

a form that is, sometimes, taken as the primitive CT definition.

In view of its so-revealed importance, F deserves a detailed examination. Although, due to (8.8.8a, b), $F = F(t, q, p) = F(t, q', p')$, yet it turns out that the resulting equations are *simpler* if F (and the corresponding phase space points) *is expressed as a combination of old* (q, p) *and new* (q', p') "*coordinates*"; that is, if F has one of the following *four* forms:

$$F_1(t, q, q'), \qquad F_2(t, q, p'), \qquad F_3(t, p, q'), \qquad F_4(t, p, p'), \tag{8.8.13}$$

depending on the problem at hand, and our choice of which $2n(+\textit{time})$ of these $4n(+\textit{time})$ variables to consider as *independent*. Let us examine the consequences of these four choices:

(i) If $F = F(t, q, q') \equiv F_1$, then

$$dF_1/dt = \partial F_1/\partial t + \sum (\partial F_1/\partial q_k)(dq_k/dt) + \sum (\partial F_1/\partial q_{k'})(dq_{k'}/dt). \tag{8.8.14}$$

Substituting (8.8.14) in (8.8.10) and equating the coefficients of the $2n+1$ independent $dq/dq'/dt$, we obtain

$$p_k = \partial F_1/\partial q_k, \qquad p_{k'} = -\partial F_1/\partial q_{k'}, \qquad H' = H + \partial F_1/\partial t, \tag{8.8.15}$$

from which it follows that if $\partial F_1/\partial t = 0$, then $H = H'$. Solving the first of (8.8.15) for the q's we obtain

$$q_{k'} = q_{k'}(t, q, p), \tag{8.8.15a}$$

and substituting this into the second of (8.8.15) we get

$$p_{k'} = p_{k'}(t, q, p). \tag{8.8.15b}$$

Equations (8.8.15a, b) express the transformations among the old and new canonical variables. These operations require that $|\partial^2 F_1/\partial q_k\, \partial q_{k'}| \neq 0$, and so we will henceforth assume this to hold; and similarly for the corresponding Hessian determinants of F_2, F_3, F_4.

(ii) Let $F = F(t, q, p') \equiv F_2$. Here, as well as in the cases of F_3, F_4 (see below), we *cannot* proceed as in the case of F_1 [i.e., via (8.8.10, 11)], because we do *not* have $\dot{q}_k$ and $\dot{q}_{k'}$. However, in view of the second of (8.8.15), the transition from the (t, q, q') of F_1 to the (t, q, p') of F_2 can be effected by the following Hamilton (Legendre)-type of transformation (§8.2):

$$F_2(\ldots p' \ldots) = \sum p_{k'}\, q_{k'} - [-F_1(\ldots q' \ldots)]; \tag{8.8.16}$$

where F_2, $p_{k'}$, $q_{k'}$, $-F_1$ play, respectively, the roles of Hamiltonian, momenta, "velocities," and Lagrangean. Substituting from (8.8.16)

$$F_1(t, q, q') = F_2(t, q, p') - \sum p_{k'}\, q_{k'} \tag{8.8.16a}$$

into (8.8.10), we find

$$\begin{aligned}
\sum p_k\, dq_k - H\, dt
&= \sum p_{k'}\, dq_{k'} - H'\, dt + dF_1 \\
&= \sum p_{k'}\, dq_{k'} - H'\, dt + dF_2(t, q, p') - d\Big(\sum p_{k'}\, q_{k'}\Big) \\
&= -\sum q_{k'}\, dp_{k'} - H'\, dt + dF_2(t, q, p') \\
&= (-H' + \partial F_2/\partial t)\, dt + \sum (\partial F_2/\partial q_k)\, dq_k + \sum (\partial F_2/\partial p_{k'} - q_{k'})\, dq_{k'},
\end{aligned}$$

and, comparing coefficients, we immediately conclude that

$$p_k = \partial F_2/\partial q_k, \qquad q_{k'} = \partial F_2/\partial p_{k'}, \qquad H' = H + \partial F_2/\partial t. \tag{8.8.17}$$

To obtain the old/new variable transformation equations, we solve the first of (8.8.17) for the p', thus obtaining

$$p_{k'} = p_{k'}(t, q, p), \tag{8.8.17a}$$

and then substitute the result in the second of (8.8.17), thus getting

$$q_{k'} = q_{k'}(t, q, p). \tag{8.8.17b}$$

(iii) Let $F = F(t, p, q') \equiv F_3$. In view of the first of (8.8.15), or $-p_k = \partial(-F_1)/\partial q_k$, the transition from the (t, q, q') of F_1 to the (t, p, q') of F_3 can be effected by the following Hamilton-type transformation:

$$F_3(\ldots -p \ldots) = \sum (-p_k)(q_k) - [-F_1(\ldots q \ldots)]; \tag{8.8.18}$$

where F_3, $-p_k$, q_k, $-F_1$ play, respectively, the roles of Hamiltonian, momenta, "velocities," and Lagrangean. Substituting from (8.8.18)

$$F_1(t, q, q') = F_3(t, p, q') + \sum p_k q_k \tag{8.8.18a}$$

into (8.8.10), we find

$$\begin{aligned}\sum p_k\,dq_k - H\,dt &= \sum p_{k'}\,dq_{k'} - H'\,dt + dF_1\\ &= \sum p_{k'}\,dq_{k'} - H'\,dt + dF_3(t,p,q') + d\Big(\sum p_k q_k\Big),\end{aligned}$$

or

$$-\sum q_k\,dp_k - H\,dt = \sum p_{k'}\,dq_{k'} - H'\,dt + dF_3(t,p,q'),$$

or, expanding dF_3 and collecting $dt/dp/dp'$ terms,

$$(-H' + H + \partial F_3/\partial t)\,dt + \sum (\partial F_3/\partial p_k + q_k)\,dp_k + \sum (\partial F_3/\partial q_{k'} + p_{k'})\,dq_{k'} = 0,$$

and setting the differential coefficients equal to zero, we immediately obtain

$$q_k = -\partial F_3/\partial p_k, \qquad p_{k'} = -\partial F_3/\partial q_{k'}, \qquad H' = H + \partial F_3/\partial t. \tag{8.8.19}$$

To obtain the old/new variable transformation equations, we solve the first of (8.8.19) for the q', thus obtaining

$$q_{k'} = q_{k'}(t,q,p), \tag{8.8.19a}$$

and then substitute the results in the second of (8.8.19), thus getting

$$p_{k'} = p_{k'}(t,q,p). \tag{8.8.19b}$$

(iv) Finally, let $F = F(t,p,p') \equiv F_4$. By repeating the above arguments *twice*, we can show that the transition from the (t,q,q') of F_1 to the (t,p,p') of F_4 can be effected by the following *double* Hamilton-type transformation:

$$F_4(\ldots -p,p'\ldots) = \sum (-p_k)(q_k) + \sum p_{k'}q_{k'} - [-F_1(\ldots q,q'\ldots)]; \tag{8.8.20}$$

where $F_4, -p_k, p_{k'}, q_k, q_{k'}, -F_1$ play, respectively, the roles of Hamiltonian, old momenta, new momenta, old "velocities," new "velocities," and Lagrangean. Repeating similar steps as in the previous cases [i.e., from (8.8.20) to (8.8.10) etc.] and setting the coefficients of $dt/dp/dp'$ equal to zero, we find

$$q_k = -\partial F_4/\partial p_k, \qquad q_{k'} = \partial F_4/\partial p_{k'}, \qquad H' = H + \partial F_4/\partial t. \tag{8.8.21}$$

Finally, to obtain the old/new variable transformation equations, we solve the first of (8.8.21) for the p''s, and thus acquire

$$p_{k'} = p_{k'}(t,q,p), \tag{8.8.21a}$$

and then substitute (8.8.21a) into the second of (8.8.21), thus getting

$$q_{k'} = q_{k'}(t,q,p). \tag{8.8.21b}$$

All these interrelated formulae are summarized, for convenience, in table 8.1.

In Sum

A canonical transformation can be created from a generating function. As such, we can choose *any* differentiable function of *half* of the *old* variables (either the q_k's or

Table 8.1 Types of Generating Functions

$F = F_1(t, q, q')$:	$p_k = \partial F_1/\partial q_k$,	$p_{k'} = -\partial F_1/\partial q_{k'}$;	$H' = H + \partial F_1/\partial t$
$F = F_2(t, q, p')$:	$p_k = \partial F_2/\partial q_k$,	$q_{k'} = \partial F_2/\partial p_{k'}$;	$H' = H + \partial F_2/\partial t$
$F = F_3(t, p, q')$:	$q_k = -\partial F_3/\partial p_k$,	$p_{k'} = -\partial F_3/\partial q_{k'}$;	$H' = H + \partial F_3/\partial t$
$F = F_4(t, p, p')$:	$q_k = -\partial F_4/\partial p_k$,	$q_{k'} = \partial F_4/\partial p_{k'}$;	$H' = H + \partial F_4/\partial t$

$F_2 = F_1 + \sum p_{k'} q_{k'}$,

$F_3 = F_1 - \sum p_k q_k$,

$F_4 = F_1 + \sum p_{k'} q_{k'} - \sum p_k q_k = F_2 - \sum p_k q_k = F_3 + \sum p_{k'} q_{k'}$

the p_k's) and *half* of the *new* (either the $q_{k'}$'s or the $p_{k'}$'s), and time; a total of $2n + 1$ independent variables. Once a generating function has been selected, table 8.1 gives the transformation relations between the (remaining half of the chosen) old and new variables.

REMARKS

(i) In *all* four cases, we have $H' - H = \partial F/\partial t$; and therefore if $\partial F/\partial t = 0$, the new Hamiltonian results by simply substituting, in the old Hamiltonian, the old variables in terms of the new variables:

$$H = H(t, q, p) = H[t, q(t, q', p'), p(t, q', p')] = H'(t, q', p') = H'.$$

(ii) From the above derivations and results we are gradually led to the realization that, in the rarified atmosphere of Hamiltonian mechanics, the terms coordinate (q) and momentum (p) have lost a lot of their original physical meaning; as (8.8.8a, b) show, each q' and each p' may relate to all the q's and the p's. In view of this blurring of the nomenclature (in both Hamiltonian mechanics and, especially, in its famous heir quantum mechanics) we call the q's and p's *canonically conjugate* variables. For example, the CT: $q_k = -p_{k'}$ and $p_k = q_{k'}$, with generating functions $F = q_1 q_{1'} + \cdots + q_n q_{n'} = F_1$ (or $F = p_1 p_{1'} + \cdots + p_n p_{n'} = F_4$) swaps coordinates and momenta, to within a sign. (Strictly speaking, these should have been written as $q_k = -\sum \delta_{kk'} p_k$ and $p_k = \sum \delta_{kk'} q_{k'}$, respectively.)

Example 8.8.1 Let us show that *all point transformations* (8.8.2)

$$q_k = q_k(t, q_{k'}) \Leftrightarrow q_{k'} = q_{k'}(t, q_k) \qquad (k, k' = 1, \ldots, n) \tag{a}$$

are canonical.

Choosing as generating function

$$F = \sum q_{k'} p_{k'} = \sum q_{k'}(t, q) p_{k'} = F(t, q, p') \equiv F_2, \tag{b}$$

and, therefore, applying (8.8.17), we obtain

$$p_k = \partial F_2/\partial q_k = \sum (\partial q_{k'}/\partial q_k) p_{k'} \qquad \text{[i.e., (8.8.4) with } k \text{ and } k' \text{ swapped]}, \tag{c}$$

$$q_{k'} = \partial F_2/\partial p_{k'} = q_{k'}(t, q) \qquad \text{[i.e., (8.8.2)]}, \tag{d}$$

$$\begin{aligned} H' &= H + \partial F_2/\partial t = H + \sum (\partial q_{k'}/\partial t) p_{k'} \\ &= -\sum (\partial q_k/\partial t) p_k + H \qquad \text{[i.e., (8.8.7); prove the last step], Q.E.D.} \end{aligned} \tag{e}$$

For example, if $F_2 = \sum q_{k'}(q) p_{k'} + f(q) = linear$ in the p', the above reduce to the general point transformation

$$p_k = \sum (\partial q_{k'}/\partial q_k) p_{k'} + \partial f/\partial q_k, \qquad q_{k'} = q_{k'}(q). \tag{f}$$

Similarly, choosing

$$F = -\sum p_k \, q_k(t, q') = F(t, p, q') \equiv F_3, \tag{g}$$

and, therefore, applying (8.8.19), we obtain

$$q_k = -\partial F_3/\partial p_k = q_k(t, q'), \tag{h}$$

$$p_{k'} = -\partial F_3/\partial q_{k'} = \sum (\partial q_k/\partial q_{k'}) p_k, \tag{i}$$

$$H' = H + \partial F_3/\partial t = H - \sum (\partial q_k/\partial t) p_k, \quad \text{Q.E.D.} \tag{j}$$

In particular, if, in (g), $q_k(t, q') = q_{k'}$, we obtain the *identity* transformation.

Example 8.8.2 Let us check the following transformations for canonicity:

(i)
$$q' = -p, \qquad p' = q. \tag{a}$$

We have, successively,

$$p\,\delta q - p'\,\delta q' = p\,\delta q - (q)(-\delta p) = \delta(p\,q) = \delta F, \tag{b}$$

and, therefore, by (8.8.12), (a) is canonical. Similarly, we can show that

$$q_{k'} = -p_k, \quad p_{k'} = q_k \qquad \text{and} \qquad q_{k'} = p_k, \quad p_{k'} = -q_k, \tag{c}$$

are canonical. This example makes clear that the Hamiltonian form of the equations of motion is *unaffected even if we take as new coordinates the old momenta, and vice versa*!

(ii)
$$q' = (q)^{1/2} \cos(2p), \qquad p' = (q)^{1/2} \sin(2p) \qquad \text{(due to H. Poincaré).} \tag{d}$$

Inverting (d), we find

$$q = (q')^2 + (p')^2, \qquad \tan(2p) = p'/q'; \tag{e}$$

and, therefore, successively,

$$\begin{aligned} p\,\delta q - p'\,\delta q' &= p\,\delta q \\ &\quad - [(q)^{1/2}\sin(2p)]\{[\cos(2p)/2(q)^{1/2}]\,\delta q - [2(q)^{1/2}\sin(2p)]\,\delta p\} \\ &= [p - (\sin(2p)\cos(2p)/2)]\,\delta q + [2q\sin^2(2p)]\,\delta p \\ &= \delta\{q[p - (\sin(4p)/4)]\} = \delta F; \end{aligned} \tag{f}$$

that is, by (8.8.12), (d) is canonical.

(iii)

$$q' = \ln[\sin(p)/q], \qquad p' = q\cot(p). \tag{g}$$

We have, successively,

$$q' = -\ln(q) + \ln[\sin(p)] \;\Rightarrow\; \delta q' = -(\delta q/q) + [\cot(p)]\,\delta p, \tag{h}$$

and, therefore,

$$\begin{aligned} p\,\delta q - p'\,\delta q' &= p\,\delta q - [q\cot(p)]\{-(\delta q/q) + [\cot(p)]\,\delta p\} \\ &= [p + \cot(p)]\,\delta q - [q\cot^2(p)]\,\delta p \equiv Q\,\delta q + P\,\delta p. \end{aligned} \tag{i}$$

The necessary and sufficient conditions for the integrability (i.e., canonicity) of (i) are $\partial Q/\partial p = \partial P/\partial q$:

$$\partial/\partial p[p + \cot(p)] = 1 - [1/\sin^2(p)] = -\cot^2(p),$$

$$\partial/\partial q[-q\cot^2(p)] = -\cot^2(p);$$

that is, (g) is indeed canonical.

Example 8.8.3 Let us determine the values of α and β so that the transformation

$$q' = q^\alpha\cos(\beta p), \qquad p' = q^\alpha\sin(\beta p) \tag{a}$$

is canonical. From the first of them, we obtain

$$\delta q' = [-\beta q^\alpha\sin(\beta p)]\,\delta p + [\alpha q^{\alpha-1}\cos(\beta p)]\,\delta q, \tag{b}$$

and, therefore,

$$p\,\delta q - p'\,\delta q' = [p - (1/2)\alpha q^{2\alpha-1}\sin(2\beta p)]\,\delta q + [\beta q^{2\alpha}\sin^2(\beta p)]\,\delta p \equiv Q\,\delta q + P\,\delta p.$$

Again, for integrability, we must have $\partial Q/\partial p = \partial P/\partial q$:

$$\partial/\partial p[p - (1/2)\alpha q^{2\alpha-1}\sin(2\beta p)] = \partial/\partial q[\beta q^{2\alpha}\sin^2(\beta p)], \tag{c}$$

from which we obtain the condition

$$\alpha\beta q^{2\alpha-1} = 1; \tag{d}$$

and since this equation must hold for all values of q, we are led to the system

$$2\alpha - 1 = 0 \quad \text{and} \quad \alpha\beta = 1, \tag{e}$$

whose roots are $\alpha = 1/2$ and $\beta = 2$. Hence, (a) becomes

$$q' = (q)^{1/2}\cos(2p), \qquad p' = (q)^{1/2}\sin(2p); \tag{f}$$

that is, the Poincaré transformation of the preceding example.

Example 8.8.4 *Canonicity via the Central Equation*:

$$\delta T + \delta' W = \left(\sum p_k\, \delta q_k\right)^{\cdot} \Rightarrow \left(\sum p_k\, \delta q_k\right)^{\cdot} - \delta T = \delta' W. \tag{a}$$

If the impressed forces are wholly *potential* forces—that is, $\delta' W = -\delta V \Rightarrow \delta(T - V) = \delta L$—then, in view of the fundamental definition (8.8.12), we can transform (a) to

$$\left(\sum p_{k'}\, \delta q_{k'}\right)^{\cdot} - [\delta L - (\delta F)^{\cdot}] = 0. \tag{b}$$

But since $(\delta F)^{\cdot} = \delta(\dot{F})$, this new central equation has the old form (a) *if we define as new Lagrangean*:

$$L' = L - dF/dt, \qquad L' = L'(q', p'). \tag{c}$$

Then, standard transformations, as in the old variables (§8.2) lead us to the new Hamiltonian equations:

$$dq_{k'}/dt = \partial H'/\partial p_{k'}, \qquad dp_{k'}/dt = -\partial H'/\partial q_{k'}, \tag{d}$$

where

$$\begin{aligned} H' &\equiv \sum p_{k'}\dot{q}_{k'} - L' \qquad \text{[invoking (8.8.12)]} \\ &= \left[\sum p_k \dot{q}_k - (dF/dt - \partial F/\partial t)\right] - (L - dF/dt) \\ &= \left(\sum p_k \dot{q}_k - L\right) + \partial F/\partial t = H + \partial F/\partial t, \end{aligned} \tag{e}$$

as before. Hence, the fundamental result: *under a canonical transformation, the canonical equations preserve their form.*

Example 8.8.5 *Canonical Transformation: The Harmonic Oscillator.* Let us consider a linear harmonic oscillator of mass m and stiffness constant k, and, therefore,

Lagrangean: $\quad L = (1/2)[m(\dot{q})^2 - kq^2] \Rightarrow p \equiv \partial L/\partial \dot{q} = m\dot{q} \Rightarrow \dot{q} = p/m, \quad$ (a)

Hamiltonian: $\quad H = p\dot{q} - L = \cdots = p^2/2m + kq^2/2; \quad$ (b)

and, therefore, Hamiltonian equations of (free and undamped) motion:

$$\dot{q} = \partial H/\partial p = p/m, \qquad \dot{p} = -\partial H/\partial q = -kq, \tag{c}$$

whose solution is well known [eliminating p between (c) we obtain the Lagrangean equation $m\ddot{q} + kq = 0$]. Instead, let us here consider the canonical transformation with generating function

$$F = F_1(q, q') = c\, q^2 \cot(q') \qquad (c = \text{a constant}), \tag{d}$$

and, therefore, by (8.8.15), transformation equations

$$p = \partial F_1/\partial q = 2c\,q\cot(q'), \qquad p' = -\partial F_1/\partial q' = c\,q^2\operatorname{cosec}^2(q'),$$
$$H' = H + \partial F_1/\partial t = H; \tag{e}$$

or, solving them for the old variables in terms of the new variables,

$$p = (4c\,p')^{1/2}\cos(q'), \qquad q = (p'/c)^{1/2}\sin(q'), \tag{f}$$

$$\begin{aligned} H' &= (1/2)(p^2/m + kq^2) = \cdots = (1/2)\big[(4cp'/m)\cos^2(q') + (kp'/c)\sin^2(q')\big] \\ &= (kp'/2c)\big[\sin^2(q') + (4c^2/mk)\cos^2(q')\big] \qquad [\text{choosing } 4c^2 = mk] \\ &= kp'/2c = (k/m)^{1/2}p' \equiv \omega p' \qquad [\omega^2 \equiv k/m\text{: oscillation frequency}]. \end{aligned} \tag{g}$$

Hence, the Hamiltonian equations in these new variables are

$$dp'/dt = -\partial H'/\partial q' = 0 \;\Rightarrow\; p' = constant \equiv c' \qquad (q' \text{ is ignorable}), \tag{h}$$
$$dq'/dt = +\partial H'/\partial p' = \omega \;\Rightarrow\; q' = \omega t + c'' \qquad (c''\text{: integration constant}). \tag{i}$$

Substituting (h, i) back in (f) we, naturally, obtain the (well-known) old variable solution.

Geometrical Interpretation of these Solutions in Phase Space

(i) in the *old* variables (q, p) the representative point describes an *ellipse* whose dimensions and sense of traverse are determined by the system constants and initial conditions ($\Rightarrow$ *energy* $\equiv E = H = \omega p' =$ *constant*; fig. 8.7).

(ii) In the *new* variables (q', p') the corresponding system point moves on the straight line

$$p' = E/\omega = constant. \tag{j}$$

The ellipse points (1, 2, 3, 4) are mapped into the straight line points (1′, 2′, 3′, 4′).

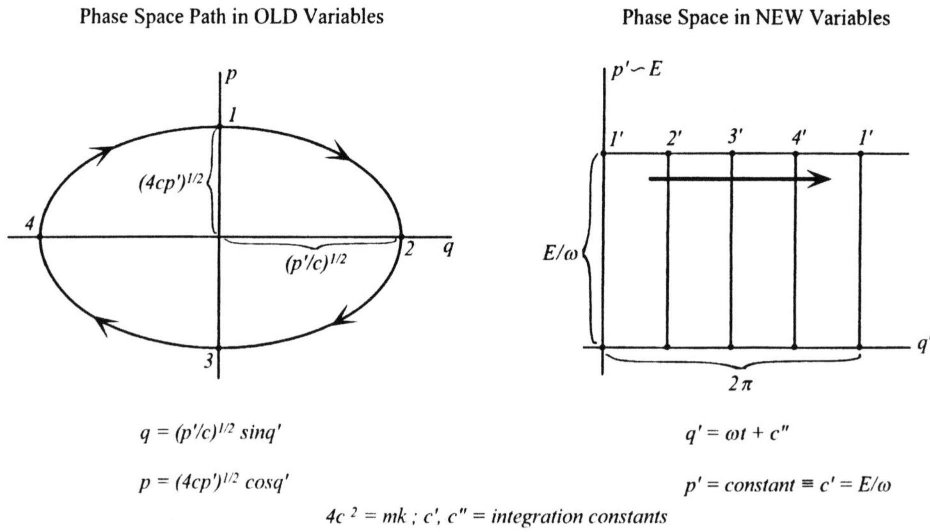

Figure 8.7 Paths of harmonic oscillator in phase space, in old and new variables.

This procedure—that is, the search for new canonical variables in which one or more (or all!) of the coordinates are ignorable—is systematized in §8.10. These investigations show that *every holonomic, scleronomic, and potential system with n DOF can be transformed by a canonical transformation into one with Hamiltonian* $2H = \sum (p_{k'})^2$*; and, therefore, equations of motion:* $p_{k'} = constant \equiv \Psi_{k'}$, $q_{k'} = \Psi_{k'} t + constant$; a fundamental result that is behind such important concepts as action–angle variables and complete separability/integrability (§8.14).

Example 8.8.6

(i) *Generalized Canonical Transformations* (GCT) are CT that, in addition to the fundamental definition (8.8.12):

$$\delta F = \sum p_k \, \delta q_k - \sum p_{k'} \, \delta q_{k'}, \tag{a}$$

also satisfy, say, holonomic constraints like

$$C_D(q, q') = 0 \qquad [D = 1, \ldots, m\text{: rank of corresponding Jacobian} = m(\leq n)] \tag{b}$$

or, in virtual form,

$$\delta C_D = \sum (\partial C_D / \partial q_k) \, \delta q_k + \sum (\partial C_D / \partial q_{k'}) \, \delta q_{k'} = 0. \tag{c}$$

Applying the Lagrangean multiplier method, we readily see that, in this case, $(F \to F_1)$, eqs. (8.8.15) must now be replaced by

$$p_k = \partial F_1 / \partial q_k + \sum \lambda_D (\partial C_D / \partial q_k), \qquad p_{k'} = -\partial F_1 / \partial q_{k'} - \sum \lambda_D (\partial C_D / \partial q_k), \tag{d}$$

which, along with (b), constitute a system of $2n + m$ equations for the $2n + m$ q's, p's, λ's (multipliers), in terms of the q's, p's.

(ii) For the *point transformation*, for which [recalling (8.8.2) and (8.8.4)]

$$q_k = q_k(q_{k'}), \qquad p_{k'} = \sum (\partial q_k / \partial q_{k'}) p_k, \tag{e}$$

the definition (8.8.12), (a) gives

$$\begin{aligned} &\sum p_k \, \delta q_k - \sum p_{k'} \, \delta q_{k'} \\ &= \sum p_k \Big(\sum (\partial q_k / \partial q_{k'}) \, \delta q_{k'} \Big) - \sum p_{k'} \, \delta q_{k'} \\ &= \sum \Big[\sum (\partial q_k / \partial q_{k'}) p_k - p_{k'} \Big] \, \delta q_{k'} = 0; \end{aligned} \tag{f}$$

that is, for such a transformation, $F \to F_1(q, q') \equiv 0$ (however, $F \to F_2, F_3, \ldots \neq 0$). Such transformations can be viewed as the following choices of the earlier GCT case (b):

$$C_D \equiv q_D - \phi_D(q_{k'}) = 0, \qquad \lambda_D = p_D. \tag{g}$$

For additional related results, see, for example, Whittaker (1937, pp. 294–296), Hamel (1949, p. 292).

Example 8.8.7 *Time Transformation.* The general transformation $(t, q, p) \rightarrow (t', q', p')$:

$$q_k = q_k(t', q', p'), \qquad p_k = p_k(t', q', p'), \qquad t = t(t', q', p') \tag{a}$$

is called *canonical* if:

(i) *Jacobian*: $\partial(t, q, p)/\partial(t', q', p') \neq 0$, and
(ii) There exist three functions $H(t, q, p)$, $H'(t', q', p')$, $F'(t, q, p)$, such that

$$\sum p_k \, dq_k - H \, dt = \sum p_{k'} \, dq_{k'} - H' \, dt' + dF', \tag{b}$$

identically, upon utilization of (a) in it.

It is not hard to show that such a generalized CT also leaves the form of the Hamiltonian equations unaltered. Here, we have treated only the special case $t' = t$; for a fuller discussion, see books on partial differential equations, for example, Carathéodory (1935).

Problem 8.8.3 *Point Transformation: Polar Cylindrical Coordinates.* With the help of (8.8.4), show that under a (point) transformation from rectangular Cartesian coordinates q_k: (x, y, z) to polar cylindrical ones $q_{k'}$: (r, ϕ, z), the corresponding momenta transform from the rectangular Cartesian p_k: $p_{x,y,z}$ to the following polar $p_{k'}$: $p_{r,\phi,z}$:

$$p_r = [x/(x^2 + y^2)^{1/2}]p_x + [y/(x^2 + y^2)^{1/2}]p_y, \tag{a}$$

$$p_\phi = (-y)p_x + (x)p_y, \tag{b}$$

$$p_z = p_z. \tag{c}$$

Problem 8.8.4 Show that the transformation

$$q = p' \sin(q'), \qquad p = p' \cos(q') \tag{a}$$

is *not* canonical, but that the following, is:

$$q = (2p')^{1/2} \sin(q'), \qquad p = (2p')^{1/2} \cos(q'). \tag{b}$$

Problem 8.8.5 Show that the following generating functions produce the canonical transformations indicated:

(i) $F = \sum q_k q_{k'}$: $\quad p_k = q_{k'}, \quad p_{k'} = -q_k,$ (a)

(ii) $F = \sum q_k p_{k'}$: $\quad p_k = p_{k'}, \quad q_{k'} = q_k.$ (b)

[This *identity* transformation can also be achieved with $F = -\sum p_k q_{k'}$.]

(iii) $F = \sum p_k q_{k'}$: $\quad q_k = -q_{k'}, \quad p_{k'} = -p_k.$ (c)

[This *spatial inversion*, or *reflection*, transformation can also be achieved with $F = -\sum q_k p_{k'}$.]

(iv) $$F = \sum p_k p_{k'}: \qquad q_k = -p_{k'}, \qquad q_{k'} = p_k. \tag{d}$$

HINTS

(i) All these F's have the form $\sum x_k y_{k'}$, where x_k, $y_{k'}$ are any of the four possible pairs of $(q, p; q', p')$. (ii) The first and fourth cases coincide. (iii) In all cases, $H' = H$.

Problem 8.8.6 We have already seen (ex. 3.5.11) that the two Lagrangeans

$$L \quad \text{and} \quad L' = L + df(t, q)/dt \tag{a}$$

[where $f(t, q)$ = arbitrary differentiable function] produce the same Lagrangean equations of motion, that is, $E_k(L) = E_k(L')$. Show that under such a "gauge" transformation, the corresponding Hamiltonians

$$H \equiv \sum (\partial L/\partial \dot{q}_k)\dot{q}_k - L \quad \text{and} \quad H' \equiv \sum (\partial L'/\partial \dot{q}_k)\dot{q}_k - L' \tag{b}$$

are related by

$$H' = H - \partial f/\partial t. \tag{c}$$

Problem 8.8.7 (Butenin, 1971, pp. 149–150). Consider a particle P of mass m whose kinetic and potential energies, in polar cylindrical coordinates $r, \phi, z = q_{1,2,3}$, are, respectively,

$$2T = m[(\dot{q}_1)^2 + q_1{}^2(\dot{q}_2)^2 + (\dot{q}_3)^2], \quad V = m\,g\,q_3 \quad (g = \text{gravitational constant}). \tag{a}$$

Show that:

(i) Its Hamiltonian in these variables (i.e., q's, plus corresponding momenta p's) is

$$H = H(q, p) = (1/2m)\left[p_1{}^2 + (1/q_1{}^2)p_2{}^2 + p_3{}^2\right] + m\,g\,q_3. \tag{b}$$

(ii) Under a canonical transformation with generating function

$$F = p_{1'}\,q_1 \cos(q_2) + p_{2'}\,q_1 \sin(q_2) + p_{3'}\,q_3 = F(q, p') \equiv F_2, \tag{c}$$

the *new* Hamiltonian $H'(= H)$ is

$$H' = H'(q', p') = (1/2m)\left[(p_{1'})^2 + (p_{2'})^2 + (p_{3'})^2\right] + m\,g\,q_{3'}. \tag{d}$$

Interpret the q's and p's geometrically.

Problem 8.8.8 We have already seen that the transformation

$$q_{k'} = q_{k'}(q, p), \qquad p_{k'} = p_{k'}(q, p) \tag{a}$$

is canonical if, and only if, the differential form

$$\delta F_1 = \sum p_k\,\delta q_k - \sum p_{k'}\,\delta q_{k'}, \tag{b}$$

after replacement of p_k and δq_k from (the inverse of) (a), is *exact* in the q' and p'; or similarly, if, after replacement of p' and $\delta q'$ from (a), it is exact in the q, p.

Show that, instead of δF_1, we may choose—to test for exactness—any of the following three interrelated differential forms:

$$\delta F_2 = \sum p_k\,\delta q_k + \sum q_{k'}\,\delta p_{k'} \quad \left[= \delta F_1 + \delta\left(\sum p_{k'} q_{k'}\right)\right], \tag{c}$$

$$\delta F_3 = -\sum q_k\,\delta p_k - \sum p_{k'}\,\delta q_{k'} \quad \left[= \delta F_1 - \delta\left(\sum p_k q_k\right)\right], \tag{d}$$

$$\delta F_4 = -\sum q_k\,\delta p_k + \sum q_{k'}\,\delta p_{k'} \quad \left[= \delta F_1 - \delta\left(\sum p_k q_k - \sum p_{k'} q_{k'}\right)\right], \tag{e}$$

that is, (a) is canonical if, and only if, any one of (b–e) is exact in the new (old) variables after replacement of the old (new) variables and their variations, from (a), in terms of the new (old) variables and their variations.

HINT

Recall table 8.1.

8.9 CANONICITY CONDITIONS VIA POISSON'S BRACKETS (PB)

Here, we show that these brackets, already introduced in §8.7 in connection with the method of variations of constants, appear naturally in the formulation of alternative conditions for canonicity. Let us, therefore, summarize their relevant theory.

Poisson Brackets; Theorem of Poisson–Jacobi

Let $f = f(t, q, p)$ be an arbitrary differentiable dynamical quantity. Then, we have, successively,

$$df/dt = \partial f/\partial t + \sum \left[(\partial f/\partial q_k)(dq_k/dt) + (\partial f/\partial p_k)(dp_k/dt)\right]$$

[invoking Hamilton's equations]

$$= \partial f/\partial t + \sum \left[(\partial f/\partial q_k)(\partial H/\partial p_k) + (\partial f/\partial p_k)(-\partial H/\partial q_k + Q_k)\right]$$

$$= \partial f/\partial t + (H, f) + \sum (\partial f/\partial p_k) Q_k, \tag{8.9.1}$$

where

$$(H, f) \equiv \sum \left[(\partial H/\partial p_k)(\partial f/\partial q_k) - (\partial H/\partial q_k)(\partial f/\partial p_k)\right]:$$

Poisson bracket of H and f. (8.9.2)

Hence, for f to be an *integral of the motion* (i.e., $df/dt = 0$), we must have

$$\partial f/\partial t + \sum (\partial f/\partial p_k) Q_k + (H, f) = 0,$$

or, assuming that $f = f(q, p)$ and $Q_k = 0$,

$$(H, f) = 0; \tag{8.9.3}$$

that is, its PB with the Hamiltonian of its variables must be *zero* (in this case, df/dt can be expressed without explicit reference to time).

The PB of any two variables, f and g, defined in complete analogy to (8.9.2) by

$$(f, g) \equiv \sum [(\partial f/\partial p_k)(\partial g/\partial q_k) - (\partial f/\partial q_k)(\partial g/\partial p_k)]$$
$$\equiv \sum [\partial(f, g)/\partial(p_k, q_k)], \tag{8.9.4}$$

has the following, easily verifiable, properties:

- $(f, g) = -(g, f) = (-g, f)$ (antisymmetry), (8.9.5a)

  $\Rightarrow (f, f) = 0,$ (8.9.5b)

- $(f, c) = 0$ (c = a constant), (8.9.5c)

- $(f_1 + f_2, g) = (f_1, g) + (f_2, g)$ (distributivity), (8.9.5d)

- $(f_1 f_2, g) = f_1(f_2, g) + f_2(f_1, g)$ (8.9.5e)

  $\Rightarrow (cf, g) = c(f, g)$ (c = a constant),

  $\Rightarrow$ if $f = \sum c_k f_k$, then $(f, g) = \sum c_k(f_k, g)$ (c_k = constants),

- $\partial/\partial t(f, g) = (\partial f/\partial t, g) + (f, \partial g/\partial t)$ ("Leibniz rule"). (8.9.5f)

  [Actually, $\partial/\partial x(f, g) = (\partial f/\partial x, g) + (f, \partial g/\partial x)$, x = any variable.] (8.9.5g)

REMARKS ON NOTATION

(i) Unfortunately, here too, no uniformity of notation for these brackets exists. A number of famous authors, such as (alphabetically): Appell, Gantmacher, Hagihara, Lanczos, Lur'e, Synge, Whittaker, et al. define PB as the *opposite* of ours, that is, as

$$(f, g) \equiv \sum [(\partial f/\partial q_k)(\partial g/\partial p_k) - (\partial f/\partial p_k)(\partial g/\partial q_k)]. \tag{8.9.6}$$

Our choice (8.9.4) follows the practices of such (equally famous) authors as: S. Flügge, Hamel, Landau/Lifshitz, Lindsay/Margenau, Prange, Schaefer/Päsler, Routh, Scheck, Spiegel, Tabor, et al., including Poisson himself (1809)! Therefore, a certain caution should be exercised when comparing various references.

(ii) Also, certain authors (especially those in quantum mechanics; e.g., Dirac) denote our Lagrangean brackets, [...], by {...}; and our Poisson brackets, (...), by [...].

With the help of the above properties, we can easily prove the following useful theorems (by taking as f/g one of the *coordinates/momenta*):

- $(f, q_k) = \partial f/\partial p_k,$ (8.9.7a)

- $(f, p_k) = -\partial f/\partial q_k,$ (8.9.7b)

- $(q_k, q_l) = 0,$ (8.9.7c)

- $(p_k, p_l) = 0,$ (8.9.7d)

- $(p_k, q_l) = \delta_{kl}$ (= *Kronecker delta*). (8.9.7e)

The last three types of brackets are called *fundamental*, or *basic*, PB.

Identity of Poisson–Jacobi

Below we prove that, for any three variables f, g, h (at least twice continuously differentiable), the following important identity holds:

$$(f,(g,h)) + (g,(h,f)) + (h,(f,g)) = 0; \tag{8.9.8a}$$

or, equivalently [invoking (8.9.5a)],

$$((f,g),h) + ((g,h),f) + ((h,f),g) = 0. \tag{8.9.8b}$$

(i) First Proof

Using the earlier definitions, we have, successively (with $k, l = 1, \ldots, n$),

$$\begin{aligned}
((f,g),h) &= \sum [(\partial(f,g)/\partial p_k)(\partial h/\partial q_k) - (\partial(f,g)/\partial q_k)(\partial h/\partial p_k)] \\
&= \sum\sum \{\partial/\partial p_k[(\partial f/\partial p_l)(\partial g/\partial q_l) - (\partial f/\partial q_l)(\partial g/\partial p_l)](\partial h/\partial q_k) \\
&\qquad - \partial/\partial q_k[(\partial f/\partial p_l)(\partial g/\partial q_l) - (\partial f/\partial q_l)(\partial g/\partial p_l)](\partial h/\partial p_k)\} \\
&= \sum\sum [(\partial^2 f/\partial q_k \partial q_l)(\partial g/\partial p_l)(\partial h/\partial p_k) + (\partial f/\partial q_l)(\partial^2 g/\partial q_k \partial p_l)(\partial h/\partial p_k) \\
&\qquad - (\partial^2 f/\partial q_k \partial p_l)(\partial g/\partial q_l)(\partial h/\partial p_k) - (\partial f/\partial p_l)(\partial^2 g/\partial q_k \partial q_l)(\partial h/\partial p_k) \\
&\qquad - (\partial^2 f/\partial p_k \partial q_l)(\partial g/\partial p_l)(\partial h/\partial q_k) - (\partial f/\partial q_l)(\partial^2 g/\partial p_k \partial p_l)(\partial h/\partial q_k) \\
&\qquad + (\partial^2 f/\partial p_k \partial p_l)(\partial g/\partial q_l)(\partial h/\partial q_k) + (\partial f/\partial p_l)(\partial^2 g/\partial p_k \partial q_l)(\partial h/\partial q_k)];
\end{aligned} \tag{8.9.8c}$$

and cyclically for the other 2×8 terms of $((g,h),f)$ and $((h,f),g)$. Then, it is not hard to see that *all 24 terms cancel in pairs*. This is a straightforward proof, but since it is long and visually tedious, we present below a shorter alternative.

(ii) Second Proof

We have, successively,

$$\begin{aligned}
&(f,(g,h)) - (g,(f,h)) \\
&\quad = \Big(f, \sum [(\partial g/\partial p_k)(\partial h/\partial q_k) - (\partial g/\partial q_k)(\partial h/\partial p_k)]\Big) \\
&\qquad - \Big(g, \sum [(\partial f/\partial p_k)(\partial h/\partial q_k) - (\partial f/\partial q_k)(\partial h/\partial p_k)]\Big) \\
&\qquad\qquad \text{[invoking properties (8.9.5) and then regrouping terms]} \\
&\quad = \sum \{-(\partial h/\partial p_k)[(\partial f/\partial q_k, g) + (f, \partial g/\partial q_k)] \\
&\qquad + (\partial h/\partial q_k)[(\partial f/\partial p_k, g) + (f, \partial g/\partial p_k)]\} \\
&\qquad + \sum [(\partial g/\partial p_k)(f, \partial h/\partial q_k) - (\partial g/\partial q_k)(f, \partial h/\partial p_k) \\
&\qquad - (\partial f/\partial p_k)(g, \partial h/\partial q_k) + (\partial f/\partial q_k)(g, \partial h/\partial p_k)].
\end{aligned} \tag{8.9.8d}$$

Now: (a) by (8.9.5g) the *first* sum transforms to

$$\sum \left[-(\partial h/\partial p_k)\,\partial/\partial q_k(f,g) + (\partial h/\partial q_k)\,\partial/\partial p_k(f,g)\right] = -\big(h,(f,g)\big),$$

while (b) the *second* sum can be shown (by direct expansion) to vanish. Therefore,

$$\big(f,(g,h)\big) - \big(g,(f,h)\big) = -\big(h,(f,g)\big),$$

or, rearranging, while using (8.9.5a, c, e),

$$\big(f,(g,h)\big) + \big(g,(h,f)\big) + \big(h,(f,g)\big) = 0, \quad \text{Q.E.D.} \tag{8.9.8e}$$

[For alternative, indirect proofs, see, for example, Appell (1953, pp. 445–447; and references cited therein), Landau and Lifshitz (1960, pp. 136–137).]

The above Poisson–Jacobi identity allows us to prove the following fundamental theorem.

Theorem of Poisson–Jacobi

If f and g are any two integrals of the motion, so is their PB; that is, if $f = c_1$ and $g = c_2$, then $(f,g) = c_3$ ($c_{1,2,3}$ = constants).

We distinguish *two* cases:

(i) $\partial f/\partial t = 0$ and $\partial g/\partial t = 0$. Then, by (8.9.1–3),

$$(f,H) = 0, \qquad (g,H) = 0 \qquad \text{(identically)}, \tag{8.9.9a}$$

and, therefore, also

$$\big((f,H),h\big) = 0, \qquad \big((g,H),f\big) = 0 \qquad \text{(identically)}; \tag{8.9.9b}$$

and substituting the above in the Poisson–Jacobi identity (8.9.8b), with $h \to H$, we immediately obtain

$$\big((f,g),H\big) = 0 \;\Rightarrow\; (f,g) = \textit{constant} \qquad \text{[invoking (8.9.5c)]}, \quad \text{Q.E.D.} \tag{8.9.9c}$$

(ii) $\partial f/\partial t \neq 0$ and $\partial g/\partial t \neq 0$. Then, by (8.9.1),

$$df/dt = \partial f/\partial t + (H,f) \equiv 0, \qquad dg/dt = \partial g/\partial t + (H,g) \equiv 0, \tag{8.9.9d}$$

and as a result the Poisson–Jacobi identity (8.9.8a), with $h \to H$, yields, successively,

$$\begin{aligned} 0 &= \big(H,(f,g)\big) + \big(f,(g,H)\big) + \big(g,(H,f)\big) \\ &= \big(H,(f,g)\big) + (f,\partial g/\partial t) - (g,\partial f/\partial t) \\ &= \big(H,(f,g)\big) + (f,\partial g/\partial t) + (\partial f/\partial t, g) \\ &= \big(H,(f,g)\big) + \partial/\partial t(f,g), \end{aligned} \tag{8.9.9e}$$

that is, by (8.9.3), (f,g) is also an integral. This presents us with *two* possibilities: (a) either (f,g) = *function of* $f = c_1$ *and* $g = c_2$, and therefore does not constitute a new integral; or (b) (f,g) = *new function, not depending on* c_1 *and* c_2, and does constitute a new, third, integral.

However, frequently, such new integrals are trivial; for example, using $f, g = \textit{constant}$, $p_k, p_l = \textit{constant}$, we simply obtain $0 = \textit{constant}$. As MacMillan puts it: "Notwithstanding the fact that Poisson's theorem gives an interesting relation among integrals, it cannot be said that it has led to integrals that were not already known. As a matter of fact it has been singularly sterile" (1936, p. 383).

COROLLARY

Let $\partial H/\partial t = 0$, then $H = c_1$ (energy integral). If $f(t, q, p) = c_2$ is a second integral, then, by the Poisson–Jacobi theorem, $(f, H) = c_3$ is also an integral. But, in this case,

$$df/dt = \partial f/\partial t + (H, f) \Rightarrow dc_2/dt = \partial f/\partial t - c_3 = 0; \tag{8.9.9f}$$

that is, if $f = c_2$ is a time-containing integral, so is $\partial f/\partial t = c_3$; and, similarly, $\partial^2 f/\partial t^2 = c_4$, and so on; however, if $\partial f/\partial t = 0$, then $c_3 = 0$ and $(f, H) = 0$.

For *group-theoretic* aspects of this theorem (Lie) and its relation to the famous *Theorema Gravissimum* of Jacobi (1842–1843, publ. 1866), see, for example, Hamel (1949, pp. 297–299, 301–302); also Frank (1935, p. 61 ff.).

Poisson Brackets (PB) and Canonical Transformations (CT)

Here, with the help of PB, we (i) show that *these brackets are invariant under CT*, and then (ii) obtain *conditions for the transformation* (8.8.8a,b) *to be canonical* [alternative to (8.8.10–12)].

(i) Let us prove that

$$(f, g)_{q,p} = (f, g)_{q',p'} = \cdots, \tag{8.9.10}$$

where f and g keep their *value* but not necessarily their *form* in the various canonical coordinates involved.

We begin by proving it for the fundamental PB (8.9.7e):

$$(p_{k'}, q_{l'})_{q,p} = (p_{k'}, q_{l'})_{q',p'} = \cdots = \delta_{k'l'}. \tag{8.9.10a}$$

Using the generating function $F = F_1 = F_1(t, q, q')$ and corresponding equations (8.8.15), we readily find

$$\partial p_k/\partial q_{k'} = \partial/\partial q_{k'}(\partial F_1/\partial q_k) = \partial/\partial q_k(\partial F_1/\partial q_{k'}) = -\partial p_{k'}/\partial q_k; \tag{8.9.10b}$$

and, similarly, using F_2, F_3, F_4 and (8.8.17, 19, 21), we obtain

$$\partial q_k/\partial q_{k'} = \partial p_{k'}/\partial p_k, \qquad \partial q_k/\partial p_{k'} = -\partial q_{k'}/\partial p_k, \qquad \partial p_k/\partial p_{k'} = \partial q_{k'}/\partial q_k. \tag{8.9.10c}$$

With the help of these results, we find

$$\begin{aligned}(p_{k'}, q_{l'})_{q,p} &\equiv \sum \left[(\partial p_{k'}/\partial q_k)(\partial q_{l'}/\partial q_k) - (\partial p_{k'}/\partial q_k)(\partial q_{l'}/\partial p_k)\right] \\ &= \sum \left[(\partial p_{k'}/\partial p_k)(\partial p_k/\partial p_{l'}) - (\partial p_{k'}/\partial q_k)(-\partial q_k/\partial p_{l'})\right] \\ &= \partial p_{k'}/\partial p_{l'} = \delta_{k'l'};\end{aligned} \tag{8.9.10d}$$

and

$$(p_{k'}, q_{l'})_{q',p'} \equiv \sum [(\partial p_{k'}/\partial p_{r'})(\partial q_{l'}/\partial q_{r'}) - (\partial p_{k'}/\partial q_{r'})(\partial q_{l'}/\partial p_{r'})]$$
$$= \sum [(\delta_{k'r'})(\delta_{l'r'}) - (0)(0)] = \delta_{k'l'}; \quad \text{Q.E.D.} \tag{8.9.10e}$$

Similarly, we show that

$$(q_{k'}, q_{l'})_{q,p} = (q_{k'}, q_{l'})_{q',p'} = 0, \tag{8.9.10f}$$

$$(p_{k'}, p_{l'})_{q,p} = (p_{k'}, p_{l'})_{q',p'} = 0. \tag{8.9.10g}$$

Now to the demonstration of (8.9.10). We have, successively,

$$(f, g)_{q',p'} \equiv \sum [(\partial f/\partial p_{r'})(\partial g/\partial q_{r'}) - (\partial f/\partial q_{r'})(\partial g/\partial p_{r'})]$$
$$= \sum \sum \{(\partial f/\partial p_{r'})[(\partial g/\partial q_r)(\partial q_r/\partial q_{r'}) + (\partial g/\partial p_r)(\partial p_r/\partial q_{r'})]$$
$$- (\partial f/\partial q_{r'})[(\partial g/\partial q_r)(\partial q_r/\partial p_{r'}) + (\partial g/\partial p_r)(\partial p_r/\partial q_{r'})]\}$$
$$= \sum [(\partial g/\partial q_r)(f, q_r)_{q',p'} + (\partial g/\partial p_r)(f, p_r)_{q',p'}]. \tag{8.9.10h}$$

Applying the above, first for $f \to q_r$ and $g \to f$, and then for $f \to q_r$ and $g \to f$, while invoking (8.9.10e–g), we get, respectively,

$$(q_r, f)_{q',p'} = \sum [(\partial f/\partial q_l)(q_r, q_l)_{q',p'} + (\partial f/\partial p_l)(q_r, p_l)_{q',p'}]$$
$$= \sum [(\partial f/\partial q_l)(0) + (\partial f/\partial p_l)(-\delta_{lr})] = -(\partial f/\partial p_r), \tag{8.9.10i}$$

$$(p_r, f)_{q',p'} = \sum [(\partial f/\partial q_l)(p_r, q_l)_{q',p'} + (\partial f/\partial p_l)(p_r, p_l)_{q',p'}]$$
$$= \sum [(\partial f/\partial q_l)(\delta_{rl}) + (\partial f/\partial p_l)(0)] = \partial f/\partial q_r. \tag{8.9.10j}$$

Finally, substituting (8.9.10i, j) back in (8.9.10h), while invoking the antisymmetry of PB, we find

$$(f, g)_{q',p'} = \sum [(\partial f/\partial p_r)(\partial g/\partial q_r) - (\partial f/\partial q_r)(\partial g/\partial p_r)] \equiv (f, g)_{q,p} \quad \text{Q.E.D.} \tag{8.9.10k}$$

[For an alternative derivation, see also Landau and Lifshitz (1960, p. 145); and ex. 8.9.1 below, with direct proof.]

In view of this fundamental theorem, *the PB subscripts become unnecessary, and will, henceforth, be omitted.*

(ii) Let us now express the conditions for the canonicity of the transformations (8.8.8a, b)

$$q' = q'(t, q, p), \qquad p' = p(t, q, p), \tag{8.9.11a}$$

via PB. The fundamental relevant definition/requirement (8.8.12) yields, successively,

$$\begin{aligned}\sum p_k\,\delta q_k - \sum p_{k'}\,\delta q_{k'} &= \sum p_k\,\delta q_k - \sum p_{k'}\Big\{\sum\,[(\partial q_{k'}/\partial q_k)\,\delta q_k + (\partial q_{k'}/\partial p_k)\,\delta p_k]\Big\}\\ &= \sum\Big\{\Big[p_k - \sum p_{k'}(\partial q_{k'}/\partial q_k)\Big]\,\delta q_k - \sum p_{k'}(\partial q_{k'}/\partial p_k)\,\delta p_k\Big\}\\ &= \sum\,[(\ldots)_k\,\delta q_k + (\ldots)_k\,\delta p_k]. \end{aligned}\tag{8.9.11b}$$

As is well-known (§2.3 ff.), for this differential expression to be an exact (virtual) differential, the following *three* groups of necessary and sufficient conditions must hold [identically, and for all values of their free (= unsummed) indices]:

(a) δp's: $$\partial/\partial p_l\Big[-\sum p_{k'}(\partial q_{k'}/\partial p_k)\Big] = \partial/\partial p_k\Big[-\sum p_{k'}(\partial q_{k'}/\partial p_l)\Big],\tag{8.9.11c}$$

or, carrying out the differentiations and recalling the earlier definitions of Lagrangean brackets (8.7.13),

$$\sum\,[(\partial p_{k'}/\partial p_l)(\partial q_{k'}/\partial p_k) - (\partial p_{k'}/\partial p_k)(\partial q_{k'}/\partial p_l)] \equiv [p_l, p_k] = 0;\tag{8.9.11d}$$

and

(b) δq's: $$\partial/\partial q_l\Big[p_k - \sum p_{k'}(\partial q_{k'}/\partial q_k)\Big] = \partial/\partial q_k\Big[p_l - \sum p_{k'}(\partial q_{k'}/\partial q_l)\Big],\tag{8.9.11e}$$

or, carrying out the differentiations and noting that the q's and p's are mutually *independent*,

$$\sum\,[(\partial p_{k'}/\partial q_l)(\partial q_{k'}/\partial q_k) - (\partial p_{k'}/\partial q_k)(\partial q_{k'}/\partial q_l)] = [q_l, q_k] = 0,\tag{8.9.11f}$$

and

(c) δq's and δp's: $$\partial/\partial q_l\Big[-\sum p_{k'}(\partial q_{k'}/\partial p_k)\Big] = \partial/\partial p_k\Big[p_l - \sum p_{k'}(\partial q_{k'}/\partial q_l)\Big],\tag{8.9.11g}$$

or

$$\sum\,[(\partial p_{k'}/\partial q_l)(\partial q_{k'}/\partial p_k) - (\partial p_{k'}/\partial p_k)(\partial q_{k'}/\partial q_l)] = [q_l, p_k] = -(\partial p_l/\partial p_k) = -\delta_{kl}$$
$$\Rightarrow [p_k, q_l] = \delta_{kl}.\tag{8.9.11h}$$

Similarly, the exactness conditions of

$$\sum p_k\,\delta q_k - \sum p_{k'}\,\delta q_{k'} = \sum\,[(\ldots)_{k'}\,\delta q_{k'} + (\ldots)_{k'}\,\delta p_{k'}],\tag{8.9.11i}$$

lead to

$$[p_{l'}, p_{k'}] = 0, \qquad [q_{l'}, q_{k'}] = 0, \qquad [p_{l'}, q_{k'}] = \delta_{l'k'}.\tag{8.9.11j}$$

Although these canonicity conditions are in terms of the *Lagrangean* brackets, yet noting that in there the roles of q's, p's and q''s, p''s can be exchanged, and that [due to (8.9.10) and (8.7.25)] *both Poisson and Lagrange brackets are canonically invariant*, we easily deduce the earlier PB conditions (8.9.10a, f, g):

$$(p_{l'}, p_{k'}) = 0, \qquad (q_{l'}, q_{k'}) = 0, \qquad (p_{l'}, q_{k'}) = \delta_{l'k'}. \tag{8.9.11k}$$

Equations (8.9.11j, k) have the following interesting geometrical interpretation. Let us consider, for simplicity, a one-DOF system. Under the canonical transformation $(q, p) \to (q', p')$, the region of allowable values of q and p—namely, R—is transformed into a region R' for the corresponding values of q' and p'. If all these coordinates are assumed rectangular Cartesian, the areas of R and R' are, respectively (for a *fixed time*, if the transformation is explicitly time-dependent),

$$A_R \equiv A = \iint dq\, dp \qquad \text{and} \qquad A_{R'} \equiv A' = \iint dq'\, dp'. \tag{8.9.12a}$$

But, by well-known theorems of advanced (or vector) calculus, we have, successively,

$$\begin{aligned}
A' &= \iint dq'\, dp' = \iint [\partial(q', p')/\partial(q, p)]\, dq\, dp \\
&= \iint [(\partial q'/\partial q)(\partial p'/\partial p) - (\partial q'/\partial p)(\partial p'/\partial q)]\, dq\, dp \\
&= \iint (p', q')_{q,p}\, dq\, dp = \iint (1)\, dq\, dp \qquad (\text{since } \delta_{11} = 1) \\
&= \iint [p, q]\, dq\, dp = \iint (1)\, dq\, dp = A;
\end{aligned} \tag{8.9.12b}$$

and, similarly,

$$\begin{aligned}
A &= \iint dq\, dp = \iint [\partial(q, p)/\partial(q', p')]\, dq'\, dp' \\
&= \iint (p, q)_{q',p'}\, dq'\, dp' = \iint (1)\, dq'\, dp' \\
&= \iint [p', q']\, dq'\, dp' = \iint (1)\, dq'\, dp' = A'.
\end{aligned} \tag{8.9.12c}$$

In words: *canonical transformations are area-preserving*, among their various representations. Similarly, for the *n-DOF* case, but with areas replaced by *volumes* (*Liouville's theorem*); see also "Integral Invariants" (§8.12).

Example 8.9.1 *Direct Proof of the Canonical Invariance of PB.* Under a canonical transformation $(q, p) \to (q', p')$, an arbitrary (differentiable) function $f = f(q, p)$ becomes another function of q', p':

$$f = f(q, p) = f[q(q', p'), p(q', p')] \equiv f'(q', p') = f'; \tag{a}$$

and similarly for the function $g = g(q,p) = \cdots = g'$. Hence, by chain rule we find, successively,

$$\begin{aligned}
(f,g)_{q,p} &= \sum [(\partial f/\partial p_r)(\partial g/\partial q_r) - (\partial f/\partial q_r)(\partial g/\partial p_r)] \\
&= \sum\sum\sum \{[(\partial f'/\partial p_{k'})(\partial p_{k'}/\partial p_r) + (\partial f'/\partial q_{k'})(\partial q_{k'}/\partial p_r)]\,[(\partial g'/\partial p_{l'})(\partial p_{l'}/\partial q_r) \\
&\qquad + (\partial g'/\partial q_{l'})(\partial q_{l'}/\partial q_r)] \\
&\qquad - [(\partial f'/\partial p_{k'})(\partial p_{k'}/\partial q_r) + (\partial f'/\partial q_{k'})(\partial q_{k'}/\partial q_r)][(\partial g'/\partial p_{l'})(\partial p_{l'}/\partial p_r \\
&\qquad + (\partial g'/\partial q_{l'})(\partial q_{l'}/\partial p_r)]\} \\
&= \sum\sum\sum \{(\partial f'/\partial p_{k'})(\partial g'/\partial p_{l'})[(\partial p_{k'}/\partial p_r)(\partial p_{l'}/\partial q_r) - (\partial p_{k'}/\partial q_r)(\partial p_{l'}/\partial p_r)] \\
&\qquad + (\partial f'/\partial q_{k'})(\partial g'/\partial q_{l'})[(\partial q_{k'}/\partial p_r)(\partial q_{l'}/\partial q_r) - (\partial q_{k'}/\partial q_r)(\partial q_{l'}/\partial p_r) \\
&\qquad + (\partial f'/\partial p_{k'})(\partial g'/\partial q_{l'})[(\partial p_{k'}/\partial p_r)(\partial q_{l'}/\partial q_r) - (\partial p_{k'}/\partial q_r)(\partial q_{l'}/\partial p_r) \\
&\qquad + (\partial f'/\partial q_{k'})(\partial g'/\partial p_{l'})[(\partial q_{k'}/\partial p_r)(\partial p_{l'}/\partial q_r) - (\partial q_{k'}/\partial q_r)(\partial p_{l'}/\partial p_r)
\end{aligned}$$

[recalling the PB definition,$(\ldots)_{q,p}$, and then invoking (8.9.10a f, g)/(8.9.11k)]

$$\begin{aligned}
&= \sum\sum \{[(\partial f'/\partial p_{k'})(\partial g'/\partial p_{l'})](p_{k'}, p_{l'}) + [(\partial f'/\partial q_{k'})(\partial g'/\partial q_{l'})](q_{k'}, q_{l'}) \\
&\qquad + [(\partial f'/\partial p_{k'})(\partial g'/\partial q_{l'}) - (\partial f'/\partial q_{k'})(\partial g'/\partial p_{l'})](p_{k'}, q_{l'})\} \\
&= \sum\sum \{[(\partial f'/\partial p_{k'})(\partial g'/\partial p_{l'})](0) + [(\partial f'/\partial q_{k'})(\partial g'/\partial q_{l'})](0) \\
&\qquad + [(\partial f'/\partial p_{k'})(\partial g'/\partial q_{l'}) - (\partial f'/\partial q_{k'})(\partial g'/\partial p_{l'})](\delta_{k'l'})\} \\
&= \sum\sum [(\partial f'/\partial p_{k'})(\partial g'/\partial q_{l'}) - (\partial f'/\partial q_{k'})(\partial g'/\partial p_{l'})](\delta_{k'l'}) \\
&= \sum [(\partial f'/\partial p_{k'})(\partial g'/\partial q_{k'}) - (\partial f'/\partial q_{k'})(\partial g'/\partial p_{k'})] \equiv (f,g)_{q',p'}, \quad \text{Q.E.D.} \qquad \text{(b)}
\end{aligned}$$

Example 8.9.2 *Relations among the Fundamental Brackets of Poisson and Lagrange.* Let the transformation

$$q_{k'} = q_{k'}(t,q,p), \quad p_{k'} = p_{k'}(t,q,p) \qquad \text{(a)}$$

be canonical. Then,

$$[p_{k'}, p_{l'}] = 0, \qquad [q_{k'}, q_{l'}] = 0, \qquad [p_{k'}, q_{l'}] = \delta_{k'l'}. \qquad \text{(b)}$$

Now, if in the earlier-found compatibility conditions (8.7.25, with $\mu \to k'$, $\lambda \to l'$)

$$\sum [c_\nu, c_{k'}](c_\nu, c_{l'}) = \delta_{k'l'} \qquad (\nu = 1, \ldots, 2n), \qquad \text{(c)}$$

we substitute $c_{k'} \to p_{k'}$, $c_{l'} \to p_{l'}$; $c_1 = p_{1'}, \ldots, c_n = p_{n'}; c_{n+1} = q_{1'}, \ldots, c_{2n} = q_{n'}$, we obtain

$$\sum [p_{r'}, p_{k'}](p_{r'}, p_{l'}) + \sum [q_{r'}, p_{k'}](q_{r'}, p_{l'}) = \delta_{k'l'},$$

or, due to (b),

$$\sum (0)(p_{r'}, p_{l'}) + \sum (-\delta_{r'k'})(q_{r'}, p_{l'}) = \delta_{k'l'} \Rightarrow -(q_{r'}, p_{l'}) = \delta_{k'l'}. \qquad \text{(d1)}$$

Similarly:

(i) Substituting $c_{k'} \to p_{k'}$, $c_{l'} \to q_{l'}$ we obtain

$$(q_{k'}, q_{l'}) = 0; \tag{d2}$$

(ii) Substituting $c_{k'} \to q_{k'}$, $c_{l'} \to p_{l'}$ we obtain

$$(p_{k'}, p_{l'}) = 0; \tag{d3}$$

(iii) Substituting $c_{k'} \to q_{k'}$, $c_{l'} \to q_{l'}$ we obtain

$$(p_{k'}, q_{l'}) = \delta_{k'l'}. \tag{d4}$$

Hence, starting with (b), and using (c), we proved (d1–4). Let the reader verify the converse; that is, use (c) to show that if (d1–4) hold, so do (b).

Example 8.9.3 *Area Preservation in Phase Space under Canonical Transformations.* Under the two distinct virtual variations $\delta_1(\ldots)$ and $\delta_2(\ldots)$, the fundamental canonical transformation definition (8.8.12),

$$\delta F = \sum p_k\, \delta q_k - \sum p_{k'}\, \delta q_{k'},$$

yields

$$\delta_1 F = \sum p_k\, \delta_1 q_k - \sum p_{k'}\, \delta_1 q_{k'}, \qquad \delta_2 F = \sum p_k\, \delta_2 q_k - \sum p_{k'}\, \delta_2 q_{k'}. \tag{a}$$

Now, $\delta_2(\ldots)$-varying $\delta_1 F$ and $\delta_1(\ldots)$-varying $\delta_2 F$, and then subtracting side by side, while noting that $\delta_2(\delta_1 q_k) = \delta_1(\delta_2 q_k)$ and $\delta_2(\delta_1 q_{k'}) = \delta_1(\delta_2 q_{k'})$, we obtain

$$\begin{aligned} 0 &= \delta_1(\delta_2 F) - \delta_2(\delta_1 F) \\ &= \sum (\delta_1 p_k\, \delta_2 q_k - \delta_2 p_k\, \delta_1 q_k) - \sum (\delta_1 p_{k'}\, \delta_2 q_{k'} - \delta_2 p_{k'}\, \delta_1 q_{k'}), \end{aligned} \tag{b}$$

that is,

$$I \equiv \sum (\delta_1 p_k\, \delta_2 q_k - \delta_2 p_k\, \delta_1 q_k) = \sum (\delta_1 p_{k'}\, \delta_2 q_{k'} - \delta_2 p_{k'}\, \delta_1 q_{k'}) \equiv I' \tag{c}$$

[recall (8.7.10)]. Geometrically, and for a one-DOF system, the Lagrangean invariant (bilinear covariant) I, a generalization of the Wronskian determinant, equals the area of the elementary parallelepiped with (rectangular Cartesian) sides $\delta s_1 = (\delta_1 q, \delta_1 p)$ and $\delta s_2 = (\delta_2 q, \delta_2 p)$, emanating from (q, p) in phase space:

$$Area = (\delta s_2 \times \delta s_1)_z = \delta_1 p\, \delta_2 q - \delta_2 p\, \delta_1 q = \delta_1 p'\, \delta_2 q' - \delta_2 p'\, \delta_1 q' = constant. \tag{d}$$

Example 8.9.4 *Angular Momentum and Poisson's Brackets.*

(i) The components of the angular momentum of a particle P of mass m relative to the origin of rectangular Cartesian axes $O\text{–}xyz \equiv O\text{–}123$ are (with $p_x \equiv m\dot{x} \to p_1 = m\dot{x}_1$, etc.)

$$h_x = yp_z - zp_y \quad \rightarrow \quad h_1 = x_2p_3 - x_3p_2,$$
$$h_y = zp_x - xp_z \quad \rightarrow \quad h_2 = x_3p_1 - x_1p_3,$$
$$h_z = xp_y - yp_x \quad \rightarrow \quad h_3 = x_1p_2 - x_2p_1. \tag{a}$$

Hence, their PB are

$$(h_1, h_2) \equiv \sum \left[(\partial h_1/\partial p_r)(\partial h_2/\partial x_r) - (\partial h_1/\partial x_r)(\partial h_2/\partial p_r)\right]$$
$$= \cdots = -(x_1p_2 - x_2p_1) = -h_3, \tag{b}$$

and, similarly,

$$(h_1, h_3) = h_2, \qquad (h_2, h_3) = -h_1; \tag{c}$$

or, compactly, with the help of the well-known Levi–Civita *permutation* symbol ε_{krs} [$= +1/-1/0$, according as k, r, s are an *even/odd/no* permutation of 1, 2, 3 (§1.1)]:

$$(h_k, h_l) = -\sum \varepsilon_{klr} h_r = \sum \varepsilon_{krl} h_r. \tag{d}$$

(ii) With the help of the above [and (8.9.5d) → (8.9.5e) → (8.95a)], we find, successively (with $h \equiv |\boldsymbol{h}|$),

$$(h_k, h^2) = \left(h_k, \sum h_r^2\right)$$
$$= \sum (h_k, h_r^2)$$
$$= \sum \left[h_r(h_k, h_r) + h_r(h_k, h_r)\right] = \sum 2h_r(h_k, h_r)$$
$$= \sum 2h_r\left(-\sum \varepsilon_{krl} h_l\right) = -2\sum\sum \varepsilon_{krl} h_r h_l = 0, \tag{e}$$

since $\varepsilon_{krl} = -\varepsilon_{klr}$ (as in the case of gyroscopicity!).

These results are important in the extension of the Hamiltonian formalism to quantum mechanics.

Example 8.9.5 *Canonicity via Symplectic (or Simplicial) Matrices.* (May be omitted in a first reading.) Here, we summarize a more algebraic approach to canonicity. Let $\mathbf{J}$, or $\mathbf{J}_{2n}$, be the following block matrix

$$\mathbf{J} \equiv \mathbf{J}_{2n} \equiv \begin{pmatrix} \mathbf{0}_n & \mathbf{1}_n \\ -\mathbf{1}_n & \mathbf{0}_n \end{pmatrix} \tag{a}$$

where $\mathbf{0}_n = n \times n$ *zero* matrix, and $\mathbf{1}_n = n \times n$ (diagonal) *unit*, or *identity*, matrix.

We can readily confirm that

$$\mathbf{J}^2 = -\mathbf{1}_{2n} \qquad \text{and} \qquad \mathbf{J} = -\mathbf{J}^{-1}, \tag{b}$$

and, therefore,

$$\text{Det}(\mathbf{J}^2) = (\text{Det}\,\mathbf{J})^2 = \text{Det}\,\mathbf{1}_{2n} = 1 \;\Rightarrow\; \text{Det}\,\mathbf{J} \neq 0. \tag{c}$$

Next, a $2n \times 2n$ matrix $\mathbf{M}_{2n}$, or simply $\mathbf{M}$, is called *symplectic* if

$$\mathbf{M}^T\mathbf{J}\mathbf{M} = \mathbf{J} \qquad [(\ldots)^T\text{: } \textit{Transpose} \text{ of } (\ldots)]. \tag{d}$$

Since $\mathrm{Det}\,(\mathbf{M}^T\mathbf{J}\mathbf{M}) = (\mathrm{Det}\,\mathbf{M}^T)\,(\mathrm{Det}\,\mathbf{J})\,(\mathrm{Det}\,\mathbf{M}) = \mathrm{Det}\,\mathbf{J}$, then due to $\mathrm{Det}\,\mathbf{M}^T = \mathrm{Det}\,\mathbf{M}$ and (c), we readily conclude that $(\mathrm{Det}\,\mathbf{M})^2 = 1 \Rightarrow \mathrm{Det}\,\mathbf{M} = \pm 1$ (it can be shown that $\mathrm{Det}\,\mathbf{M} = 1$). Hence, $\mathbf{M}$ is invertible. Indeed, from (d), we find that

$$\mathbf{M}^{-1} = -\mathbf{J}\mathbf{M}^T\mathbf{J}. \tag{e}$$

Now, we introduce the following fundamental definition.

DEFINITION

The one-to-one (invertible) transformation $(p, q) \to (p', q')$, in the 2n-dimensional phase space, is called canonical if the corresponding $2n \times 2n$ Jacobian matrix

$$\begin{pmatrix} \partial p_k/\partial p_{l'} & \partial p_k/\partial q_{l'} \\ \partial q_k/\partial p_{l'} & \partial q_k/\partial q_{l'} \end{pmatrix} \tag{f}$$

is symplectic. The equivalence of this definition with those based on the Poisson brackets is established as follows:

(i) Using the fact that the transpose of the block matrix

$$\begin{pmatrix} \mathbf{A} & \mathbf{B} \\ \mathbf{C} & \mathbf{D} \end{pmatrix} \tag{g}$$

equals

$$\begin{pmatrix} \mathbf{A}^T & \mathbf{C}^T \\ \mathbf{B}^T & \mathbf{D}^T \end{pmatrix} \tag{h}$$

where $\mathbf{A}$, $\mathbf{B}$, $\mathbf{C}$, $\mathbf{D}$ are arbitrary $n \times n$ matrices, we can show that

$$\mathbf{M} \equiv \begin{pmatrix} \mathbf{A} & \mathbf{B} \\ \mathbf{C} & \mathbf{D} \end{pmatrix} \tag{i}$$

is symplectic if, and only if

- $\mathbf{A}^T\mathbf{C}$ and $\mathbf{B}^T\mathbf{D}$ are *symmetric* (i.e., equal to their transposes),
- $\mathbf{D}^T\mathbf{A} - \mathbf{B}^T\mathbf{C} = \mathbf{1}$.

(ii) Applying these results to the Jacobian (f); that is

$$\mathbf{A} = (\partial p_k/\partial p_{l'}), \qquad \mathbf{B} = (\partial p_k/\partial q_{l'}), \qquad \mathbf{C} = (\partial q_k/\partial p_{l'}), \qquad \mathbf{D} = (\partial q_k/\partial q_{l'}), \tag{j}$$

we see that it is symplectic, and hence the transformation $(p, q) \to (p', q')$ is canonical, if, and only if,

$$[p_{k'}, p_{l'}] = 0, \qquad [q_{k'}, q_{l'}] = 0, \qquad [p_{k'}, q_{l'}] = \delta_{k'l'}; \tag{k}$$

which are, of course, the earlier-found Lagrangean bracket conditions.

Example 8.9.6 *Evolution of a Mechanical System via PB.* Let $f = f(q, p)$ and $H = H(q, p)$. Then, by (8.9.1), and assuming only potential forces,

$$df/dt = (H, f); \tag{a}$$

and again by (8.9.1), with $f \to df/dt$:

$$d^2f/dt^2 = (H, df/dt) = \big(H, (H, f)\big); \tag{b}$$

and similarly for $d^3f/dt^3, \ldots$. As a result, the MacLaurin expansion,

$$f[q(t), p(t)] \equiv f(t) = f(0) + (df/dt)_o\, t + (1/2)(d^2f/dt^2)_o\, t^2 + \cdots, \tag{c}$$

becomes

$$\begin{aligned} f(t) &= f(0) + t(H, f(0)) + (t^2/2)\Big(H, \big(H, f(0)\big)\Big) \\ &\quad + (t^3/6)\Big(H, \Big(H, \big(H, f(0)\big)\Big)\Big) + \cdots \\ &\equiv f(0)\exp[t(H, \ldots)] \qquad \text{[symbolically]} \end{aligned} \tag{d}$$

This expresses the earlier-described (§3.12) *doctrine of determinism*: if all q's and p's, and hence all system functions, like $f(q, p)$, are known at an "initial" instant $t = 0$, then the state of the system at any later time t can be determined with the help of its known (constant) Hamiltonian $H(q, p)$, from (d), to any degree of accuracy.

Example 8.9.7 *Infinitesimal Canonical Transformations.* A general transformation

$$q_k \to q_{k'} = q_k + \varepsilon\, f_k(q, p), \qquad p_k \to p_{k'} = p_k + \varepsilon\, g_k(q, p) \tag{a}$$

is called *infinitesimal* (IT) if ε can be viewed as an infinitesimal parameter, independent of the q's and p's, whose higher powers can, therefore, be neglected. Under such a transformation, a general function $F(q', p')$ becomes

$$\begin{aligned} F(q', p') &= F(q + \varepsilon f, p + \varepsilon g) \\ &= F(q, p) + \varepsilon \sum \big[(\partial F/\partial q_k) f_k + (\partial F/\partial p_k) g_k\big]. \end{aligned} \tag{b}$$

Examples of such IT are (i) the *infinitesimal rotations* of a rigid body about a fixed point (§ 1.9ff.); and, of course, (ii) the general (first-order) *virtual displacement* (§2.5) $\delta\mathbf{r} = \sum (\partial\mathbf{r}/\partial q_k)\, \delta q_k$.

If the transformation (a) is also canonical (infinitesimal canonical transformation, ICT) then, by the definition (8.8.12), we must have

$$\begin{aligned} \delta F &= \sum p_k\, \delta q_k - \sum p_{k'}\, \delta q_{k'} \\ &= \sum p_k\, \delta q_k - \sum (p_k + \varepsilon\, g_k)(\delta q_k + \varepsilon\, \delta f_k) \\ &= -\varepsilon \sum (p_k\, \delta f_k + g_k\, \delta q_k) \qquad \text{[to the } \textit{first} \text{ order in } \varepsilon] \\ &= -\varepsilon \sum \Big\{ p_k \Big(\sum \big[(\partial f_k/\partial p_l)\, \delta p_l + (\partial f_k/\partial q_l)\, \delta q_l\big]\Big) + g_k\, \delta q_k \Big\} \\ &= -\varepsilon \sum \Big\{ \Big[g_l + \sum p_k (\partial f_k/\partial q_l)\Big]\, \delta q_l + \Big(\sum p_k (\partial f_k/\partial p_l)\Big)\, \delta p_l \Big\}, \end{aligned} \tag{c}$$

and from this, setting $F \equiv \varepsilon W(q,p)$ and equating virtual differential coefficients, we obtain

$$g_l + \sum p_k(\partial f_k/\partial q_l) = -\partial W/\partial q_l, \qquad \sum p_k(\partial f_k/\partial p_l) = -\partial W/\partial p_l, \tag{d}$$

or, since here the q's and p's are considered independent,

$$g_l + \partial/\partial q_l\Big(\sum p_k f_k\Big) = -\partial W/\partial q_l, \qquad \partial/\partial p_l\Big(\sum p_k f_k\Big) - f_l = -\partial W/\partial p_l,$$

or, finally, with the help of the new generating function: $G \equiv W + \sum p_k f_k$,

$$f_l = \partial G/\partial p_l, \qquad g_l = -\partial G/\partial q_l. \tag{e}$$

Hence, the original transformations (a) become

$$q_{k'} - q_k \equiv \xi_k = \varepsilon(\partial G/\partial p_k), \qquad p_{k'} - p_k \equiv \eta_k = -\varepsilon(\partial G/\partial q_k); \tag{f}$$

that is, Hamilton's equations can be viewed as an *ICT with generating function the system Hamiltonian*:

$$\varepsilon \to dt, \qquad G \to H: \qquad dq_k = (\partial H/\partial p_k)\,dt, \qquad dp_k = -(\partial H/\partial q_k)\,dt; \tag{g}$$

a result admirably summed up by Whittaker in the following words: "*The whole course of a dynamical system can thus be regarded as the gradual self-unfolding of a contact transformation*" (1937, p. 304), with time merely as a *parameter* of that transformation.

[Some authors, *including Whittaker*, by contact transformations mean our canonical transformations—see next example. Others, however, use that term to signify *homogeneous canonical (or Mathieu) transformations*; see, for example, Rund (1966).]

Finally, substituting (e, f) in (b), we obtain

$$\begin{aligned} \Delta F &= \sum \big[(\partial F/\partial q_k)\xi_k + (\partial F/\partial p_k)\eta_k\big] \\ &= \varepsilon \sum \big[(\partial F/\partial q_k)(\partial G/\partial p_k) - (\partial F/\partial p_k)(\partial G/\partial q_k)\big] \\ &= \varepsilon(G, F), \end{aligned} \tag{h}$$

thus providing another interpretation of Poisson's brackets.

REMARKS

(i) Equations (f) can also be obtained by adding the infinitesimal $\varepsilon G(q,p')$ to the *identity* transformation [generated by $\sum q_k p_{k'}$, or by $-\sum p_k q_{k'}$ (§8.8)]; that is, if we take as generating function

$$F = \sum q_k p_{k'} + \varepsilon G(q,p') \equiv F_2(q,p'), \tag{i}$$

then, recalling (8.8.17),

$$q_{k'} = \partial F_2/\partial p_{k'} = q_k + \varepsilon(\partial G/\partial p_{k'}), \tag{j1}$$

$$p_k = \partial F_2/\partial q_k = p_{k'} + \varepsilon(\partial G/\partial q_k); \tag{j2}$$

that is, to the first order,

$$\xi_k = \varepsilon(\partial G/\partial p_{k'}) \approx \varepsilon(\partial G/\partial p_k), \tag{k1}$$

$$\eta_k = -\varepsilon(\partial G/\partial q_k); \qquad G(q, p') \approx G(q, p). \tag{k2}$$

(ii) That the transformation of the q's and p's from their initial values to their values at any later time is canonical can also be seen from the *group* property of these transformations; that is, from that, (*a*) *the result of two successive canonical transformations is also canonical, and* (*b*) *the inverse of a canonical transformation, from the new variables to the old ones, is also canonical.*

Example 8.9.8 *Contact Transformation.* If $\sum p_k\,\delta q_k =$ *total* virtual differential $\equiv \delta f$, then, by the fundamental definition (8.8.12),

$$\sum p_{k'}\,\delta q_{k'} = \sum p_k\,\delta q_k - \delta F = \delta(f - F); \tag{a}$$

that is, $\sum p_{k'}\,\delta q_{k'}$ is also a *total* virtual differential.

Let us see if the converse is also true. For $\sum p_{k'}\,\delta q_{k'} = \delta f'$ to follow from $\sum p_k\,\delta q_k = \delta f$, we must have

$$\sum p_{k'}\,\delta q_{k'} - \delta f' = \mu\Big(\sum p_k\,\delta q_k - \delta f\Big). \tag{b}$$

Now:

(i) If $\mu \equiv 1$, then, clearly, we are dealing with a canonical transformation with generating function $F = f - f'$;
(ii) If $\mu \neq 1$, then (b) represents a so-called general *contact transformation* (Lie). We will not pursue such transformations any further here, but the reader should be aware that, in a number of expositions, the terms canonical and contact transformations are used synonymously.

Problem 8.9.1 *Canonicity Conditions.*

(i) Show that the linear homogeneous transformation

$$q_{k'} = \sum Q_{k'k}\,q_k, \qquad p_{k'} = \sum P_{k'k}\,p_k, \tag{a}$$

where $Q_{k'k}$, $P_{k'k}$ are *constant* coefficients, is canonical if, and only if,

$$\sum Q_{k'k} P_{k'l} = \delta_{kl}; \tag{b}$$

that is, $P_{k'k}$: cofactor of $Q_{k'k}$ in $Det(Q_{k'k}) \equiv Q$, divided by $Q(\neq 0)$.

(ii) Then show that, as a result of (b), we have

$$\sum p_{k'} q_{k'} = \sum p_k q_k. \tag{c}$$

HINT

For canonicity, we must have $(p_{k'}, q_{l'}) = \delta_{k'l'}$.

Problem 8.9.2 *Properties of Poisson's Brackets.* Show that, for a general function $f = f(t, q, p)$, and with t, q, p regarded as independent variables,

$$(f, q_k) = \partial f/\partial p_k, \qquad (f, p_k) = -\partial f/\partial q_k, \qquad (f, t) = 0. \tag{a}$$

Problem 8.9.3 *Properties of Poisson's Brackets.* Using the results of the preceding problem, show that

$$\partial^2 f/\partial q_k\, \partial q_l = \big(p_l, (p_k, f)\big). \tag{a}$$

Obtain similar expressions for $\partial^2 f/\partial q_k\, \partial p_l$ and $\partial^2 f/\partial p_k\, \partial p_l$.

Problem 8.9.4 *Equations of Motion via Poisson's Brackets.*

(i) Show that Hamilton's equations

$$dq_k/dt = \partial H/\partial p_k, \qquad dp_k/dt = -\partial H/\partial q_k + Q_k \tag{a}$$

can be rewritten, with the help of PB, as

$$dq_k/dt = (H, q_k), \qquad dp_k/dt + (H, p_k) + Q_k. \tag{b}$$

(ii) Then show that, if $\partial H/\partial t = 0$ and $Q_k = 0$,

$$d^2 p_k/dt^2 = \big(H, (H, p_k)\big). \tag{c}$$

Problem 8.9.5 *Power Theorem via Poisson's Brackets.* Consider a system whose motion is governed by the Hamiltonian equations

$$dq_k/dt = \partial H/\partial p_k, \qquad dp_k/dt = -\partial H/\partial q_k + Q_k. \tag{a}$$

Using the dynamical identity (8.9.1), show that its power equation is

$$dH/dt = \partial H/\partial t + \sum Q_k(dq_k/dt). \tag{b}$$

HINT

In (8.9.1), set $f \to H$.

Problem 8.9.6 *Angular Momentum and Poisson's Brackets.* Continuing from ex. 8.9.3, show that:

(i) $$(x_k, h_l) = -\sum \varepsilon_{klr} x_r = \sum \varepsilon_{krl}\, x_r; \tag{a}$$

(ii) $$(p_k, h_l) = -\sum \varepsilon_{klr} p_r = \sum \varepsilon_{krl}\, p_r. \tag{b}$$

Problem 8.9.7 *Infinitesimal Canonical Transformations (ICT).* Show that if $G = constant$ is an integral of the canonical equations of motion of a system $dq_k/dt = \partial H/\partial p_k$ and $dp_k/dt = -\partial H/\partial q_k$, then all trajectories created by the ICT with G as generating function satisfy its *canonical variational equations* of Jacobi (or Poincaré's *équations aux variations*):

$$d\xi_k/dt = \sum \left[(\partial^2 H/\partial p_k\, \partial q_l)\xi_l + (\partial^2 H/\partial p_k\, \partial p_l)\eta_l\right], \tag{a}$$

$$d\eta_k/dt = \sum \left[(\partial^2 H/\partial q_k\, \partial q_l)\xi_l + (\partial^2 H/\partial q_k\, \partial p_l)\eta_l\right], \tag{b}$$

where all partial derivatives are evaluated at the system's *fundamental trajectory* (i.e., for $\varepsilon = 0$).

HINT

Show that (a, b) are satisfied by the G-generated perturbations

$$\xi_k = \varepsilon(\partial G/\partial p_k), \qquad \eta_k = -\varepsilon(\partial G/\partial q_k), \tag{c}$$

where $dG/dt = \partial G/\partial t + (H, G) = 0$.

For additional related results, see, for example, Hamel (1949, pp. 301–303).

8.10 THE HAMILTON–JACOBI THEORY

In this section we are carrying out the ultimate objective of canonical transformation (CT) theory: to provide a systematic way of finding CT that simplify the Hamiltonian equations of motion as much as possible, by which we mean *CT that render: (i) all new coordinates q', or (ii) all new momenta p', or (iii) both (q', p'), constant in time.*

Then, and assuming $Q_{k'} = 0$, the new Hamiltonian equations

$$dq_{k'}/dt = \partial H'/\partial p_{k'}, \qquad dp_{k'}/dt = -\partial H'/\partial q_{k'}, \tag{8.10.1}$$

yield, successively:

(i)

$$dq_{k'}/dt = 0 \Rightarrow q_{k'} = \textit{constant} \equiv \alpha_k, \tag{8.10.2a}$$

$$\partial H'/\partial p_{k'} = 0 \qquad (\text{i.e., } \textit{all } p' \textit{ ignorable}), \tag{8.10.2b}$$

$$\Rightarrow H' = H'(t, q') = H'(t, \alpha_1, \ldots, \alpha_n) \equiv H'(t, \alpha). \tag{8.10.2c}$$

[We hope no confusion will arise from the (tensorially nonrigorous) fact that in this, and similar equations, quantities with accented indices are equated to quantities with nonaccented indices!]

If, further, $\partial H'/\partial t = 0 \Rightarrow H' = H'(q') = H'(\alpha)$: constant total energy $\equiv E$, then, by the second of (8.10.1),

$$dp_{k'}/dt = -\partial H'/\partial \alpha_k = \textit{constant} \Rightarrow p_{k'} = \textit{linear function of time}. \tag{8.10.2d}$$

(ii)

$$dp_{k'}/dt = 0 \Rightarrow p_{k'} = \textit{constant} \equiv \beta_k, \tag{8.10.3a}$$

$$\partial H'/\partial q_{k'} = 0 \ (\text{i.e., } \textit{all } q' \textit{ ignorable}), \tag{8.10.3b}$$

$$\Rightarrow H' = H'(t, p') = H'(t, \beta_1, \ldots, \beta_n) \equiv H'(t, \beta). \tag{8.10.3c}$$

If, further, $\partial H'/\partial t = 0 \Rightarrow H' = H'(p') = H'(\beta)$: constant total energy $\equiv E$, then, by the first of (8.10.1),

$$dq_{k'}/dt = \partial H'/\partial\beta_k = constant \equiv \omega_k$$
$$\Rightarrow q_{k'} = linear\ function\ of\ time \equiv \omega_k t + \alpha_k = (\partial E/\partial\beta_k)t + \alpha_k. \qquad (8.10.3d)$$

Cases (ii) and (i) can be summed up, respectively, in the following theorem.

THEOREM

If all coordinates (momenta) are ignorable, then the conjugate momenta (coordinates) are constant and the coordinates (momenta) vary linearly with time.

$$\text{(iii)}\quad dq_{k'}/dt = \partial H'/\partial p_{k'} = 0 \Rightarrow q_{k'} = constant \equiv \alpha_k, \qquad (8.10.4a)$$
$$dp_{k'}/dt = -\partial H'/\partial q_{k'} = 0 \Rightarrow p_{k'} = constant \equiv \beta_k; \qquad H' = H'(t). \quad (8.10.4b)$$

Theorem of Jacobi

The simplest (of course, arbitrary) choice satisfying (8.10.4a, b) is $H' \equiv 0$. The particular generating function accomplishing this we shall call (Hamiltonian) *action*: A_H, or, simply, A. [This function is frequently denoted by S (also W, usually for the Lagrangean action; from the German *Wirkung* = *action* ≡ *work* × *time*; not as in action/reaction of Newton's "third law"); but here that letter has been appropriated for the Appellian function; see also §8.11. Such action *functions* → *functionals* play a prominent role in chapter 7.]

It is expedient to assume that A has the following functional representation:

$$A = A(t, q, p') \qquad (= F_2, \text{recall §8.8}). \qquad (8.10.5a)$$

Then, by (8.8.17) and the earlier requirement $H' \equiv 0$,

$$p_k = \partial A/\partial q_k, \qquad q_{k'} = \partial A/\partial p_{k'}; \qquad (8.10.5b)$$
$$H' = H(t, q, p) + \partial A/\partial t = 0 \Rightarrow H(t, q, \partial A/\partial q) + \partial A/\partial t = 0, \qquad (8.10.5c)$$

or, since (prob. 8.2.1)

$$H \equiv T'(t, q, p) + V(t, q)$$
$$= (1/2)(M'_{11}{p_1}^2 + M'_{22}{p_2}^2 + \cdots + 2M'_{12}p_1p_2 + \cdots) + V \equiv H(t, q, p),$$
$$\{M'_{lk} = M'_{kl} \equiv [minor\ of\ element\ M_{kl}(= M_{lk})\ in\ determinant\ M_n \equiv (M_{kl})]/M_n\},$$

explicitly;

$$\partial A/\partial t + (1/2)\big[M'_{11}(\partial A/\partial q_1)^2 + M'_{22}(\partial A/\partial q_2)^2$$
$$+ \cdots + 2M'_{12}(\partial A/\partial q_1)(\partial A/\partial q_2) + \cdots\big] + V = 0. \qquad (8.10.6)$$

This first-order *nonlinear* partial differential equation for A is the famous *Hamilton–Jacobi (HJ) equation*. Hence, the problem of bringing the Hamiltonian equations of motion to their simplest (integrable) form, by an appropriate canonical transformation, has been reduced to that of the integration of (8.10.6).

On this equation we can state the following: Since it depends on the $n+1$ *independent variables t* (time), q (space/coordinates), and the n constants $p' = \beta$, *a(ny) complete integral (CI) of it must contain an equal number of independent arbitrary constants*; say, $\beta_1, \ldots, \beta_n; \beta_{n+1}$. [A CI of a first-order PDE is to be distinguished from its *general integral* (GI), which depends on an *arbitrary function,* and is not as important to dynamics as are the CIs. On how to obtain the GI from CIs, see ex. 2, below.] However, since (8.10.6) does not contain A explicitly, but only its derivatives (and, therefore, if A is a solution of it, so is $A + \beta'$), identifying β_{n+1} with the additive (and dynamically inconsequential) constant β', we can finally state that a(ny) complete integral of the HJ equation of an n-DOF system contains n nontrivial, or essential, constants; that is, any such A has the form:

$$A = A(t, q, \beta), \qquad \beta \equiv (\beta_1, \ldots, \beta_n). \tag{8.10.7}$$

Thus, for this special generating function, the new momenta p', by (8.10.4b), can be identified with the constants β:

$$p_{k'} = \beta_k; \tag{8.10.8a}$$

while, by (8.10.4a) and the second of (8.10.5b),

$$q_{k'} = \partial A/\partial p_{k'} = \partial A(t, q, \beta)/\partial \beta_k = \alpha_k \qquad \text{(arbitrary constants).} \tag{8.10.8b}$$

From these algebraic equations, we can express the q's in terms of t and the $2n$ essential arbitrary constants (α, β):

$$\begin{aligned} q_k &= q_k(t; \alpha_1, \ldots, \alpha_n; \beta_1, \ldots, \beta_n) \equiv q_k(t; \alpha, \beta) \\ &= \textit{general integral of original problem.} \end{aligned} \tag{8.10.9a}$$

The old momenta can then be found from the first of (8.10.5b):

$$\begin{aligned} p_k &= \partial A/\partial q_k = \big(\partial A(t, q, \beta)/\partial q_k\big)_{q=q(t,\alpha,\beta)} = \cdots = p_k(t; \alpha, \beta) \\ &= \textit{general integral of original problem.} \end{aligned} \tag{8.10.9b}$$

Finally, evaluating (8.10.9a, b) for an *initial* time t_o, we can express α, β in terms of the $2n$ (arbitrary) *initial* values of the old coordinates and momenta, q_o, p_o [assuming that the corresponding Jacobian, $\partial(q, p)/\partial(\alpha, \beta)$, does not vanish; otherwise, since the α, β would not be independent, the solution $q(t; \alpha, \beta)$, $p(t; \alpha, \beta)$, would not be general]; and then, reinserting these expressions back into (8.10.9a, b), we can have q, p as functions of time t and the q_o's, p_o's. This completes, in principle, the HJ procedure for solving/simplifying canonical equations of motion. [For a proof that (8.10.8b), (8.10.9b) satisfy the canonical equations (8.10.10a), see ex. 8.10.1, below.]

REMARK

Incomplete HJ integrals—that is, expressions satisfying (8.10.6) but depending on fewer than n constants—cannot furnish the general integral (8.10.9a, b); but it can help us find it. Thus, from the known "incomplete integral" $A = A(t; q, \beta_1, \ldots, \beta_m)$ $(m < n)$, we obtain the m (8.10.8b)-like equations: $\partial A/\partial \beta_D = (\textit{constant})_D$ $(D = 1, \ldots, m)$.

The general integral (8.10.9a, b), thanks to the preceding theory, constitutes a canonical transformation from the α, β to the q, p with generating function

$A(t, q, \beta)$. However, every other general integral, say $q_k(t; \gamma, \delta)$, $p_k = p_k(t; \gamma, \delta)$, where $\gamma \equiv (\gamma_1, \ldots, \gamma_n)$, $\delta \equiv (\delta_1, \ldots, \delta_n)$ are arbitrary constants, is *not* a canonical transformation; and therefore knowledge of such an integral does not allow the construction of the complete integral of the HJ equation. That can happen only if the γ's and δ's equal, respectively, the initial positions and momenta.

The above results constitute the famous *theorem of Jacobi* (1842–1843). Let us restate it compactly:

(i) The integration of the canonical equations

$$dq_k/dt = \partial H/\partial p_k, \qquad dp_k/dt = -\partial H/\partial q_k, \tag{8.10.10a}$$

is reduced to the integration of the Hamilton–Jacobi equation:

$$H(t, q, \partial A/\partial q) + \partial A/\partial t = 0. \tag{8.10.10b}$$

(ii) If we have a complete solution of (8.10.10b)—that is, a solution of the form

$$A = A(t; q_1, \ldots, q_n; \beta_1, \ldots, \beta_n) \equiv A(t; q, \beta), \tag{8.10.10c}$$

where $\beta \equiv (\beta_1, \ldots, \beta_n) = n$ *essential arbitrary constants*, and $|\partial^2 A/\partial q \partial \beta| \neq 0$ (*non-vanishing Jacobian*), then the solution of the algebraic system:

$\partial A/\partial \beta_k = \alpha_k$

[*Finite equations of motion*, α: *new arbitrary constants* $\Rightarrow$ $q_k = q_k(t, \alpha, \beta)$], (8.10.10d)

$\partial A/\partial q_k = p_k$

[$\Rightarrow p_k = p_k(t, \alpha, \beta)$: *canonically conjugate (finite) equations of motion*], (8.10.10e)

constitutes a complete solution of (8.10.10a). For a proof, see ex. 8.10.1.

Schematically:

Hamilton: Differential equations of motion: $\quad dq/dt = \partial H/\partial p, \quad dp/dt = -\partial H/\partial q$

(*If these equations can be integrated, an Action function can be obtained*),

Hamilton–Jacobi: $\quad H(t, q, \partial A/\partial q) + \partial A/\partial t = 0 \Rightarrow A = A(t, q, \beta)$,

Jacobi: Finite equations of motion: $\partial A/\partial \beta = \alpha \Rightarrow q = q(t, \alpha, \beta)$,

$\partial A/\partial q = p \Rightarrow p = p(t, \alpha, \beta)$

(*If an Action function can be obtained, then Hamilton's equations can be integrated*).

Special Cases

Obtaining the complete integral of the HJ equation is, in general, quite complicated; but, frequently, simpler than solving the corresponding Hamiltonian equations; and if that can be done, either exactly (e.g., via quadratures) or approximately (e.g., via perturbations), it constitutes one of the most straightforward methods of solution of mechanical problems.

1. Conservative Systems

In this case, the Hamiltonian form of the power theorem yields

$$dH/dt = \partial H/\partial t + \sum Q_k(dq_k/dt) = 0$$

$$\Rightarrow \partial H/\partial t = 0 \qquad \text{(since we have assumed } Q_k = 0)$$

$$\Rightarrow H = H(q, p) = H(q, \partial A/\partial q) = E \equiv \textit{constant (total energy)}; \tag{8.10.11a}$$

and so, from (8.10.10b), we conclude that, for such systems, A must be (to within an additive constant) a *linear* function of time:

$$A(t; q, p') = -E\,t + A_o(q, p'), \tag{8.10.11b}$$

where A_o is the *abbreviated*, or *reduced*, action. As a result of (8.10.11a, b), the HJ equation (8.10.10b) assumes the abbreviated, or reduced, form:

$$H(q, \partial A_o/\partial q) = E. \tag{8.10.11c}$$

or, explicitly,

$$(1/2)\big[M'_{11}(\partial A_o/\partial q_1)^2 + M'_{22}(\partial A_o/\partial q_2)^2 + \cdots + 2M'_{12}(\partial A_o/\partial q_1)(\partial A_o/\partial q_2) + \cdots\big] + V = E. \tag{8.10.11d}$$

Now, the complete integral of the above must contain E as an essential constant. On the other hand, the action

$$A = A(t; q, p') = A(t; q, \beta) = A_o(q, \beta) - E\,t \tag{8.10.11e}$$

can contain only n essential independent constants. Hence, it follows that E and the $n\,\beta$'s must be connected functionally:

$$E = E(\beta_1, \ldots, \beta_n) \equiv E(\beta); \tag{8.10.11f}$$

that is, we can identify E with any one of the β's, say $E = \beta_1$ (or, we need a A_o that contains only $n - 1$ essential constants). Then, (8.10.11e) becomes

$$A(t, q, \beta) = A_o(q; \beta) - E(\beta)t = A_o(q; \beta_1, \ldots, \beta_n) - \beta_1 t = A_o(q; E, \beta_2, \ldots, \beta_n) - E\,t, \tag{8.10.11g}$$

and, provided that

$$\begin{vmatrix} \partial^2 A/\partial E \partial q_1 \cdots \partial^2 A/\partial E \partial q_n \\ \cdots\cdots\cdots\cdots\cdots\cdots \\ \cdots\cdots\cdots\cdots\cdots\cdots \\ \partial^2 A/\partial \beta_n \partial q_1 \cdots \partial^2 A/\partial \beta_n \partial q_n \end{vmatrix} \neq 0, \tag{8.10.11h}$$

the finite equations of motion (8.10.8b, 10d) yield:

(i) For $k', k = 1$: $\quad q_{1'} = \partial A/\partial \beta_1 = \partial A/\partial E = \partial A_o/\partial E - t = \alpha_1$, or

$$\partial A_o/\partial E = t + \alpha_1 \equiv f(q, \beta) \Rightarrow \alpha_1 = -t + \partial A_o/\partial E; \tag{8.10.11i}$$

(ii) For

$$k', k = 2, \ldots, n: \quad q_{k'} = \partial A/\partial\beta_k = \partial A_o/\partial\beta_k = \alpha_k \equiv g(q, \beta) \quad (k > 1); \tag{8.10.11j}$$

also

$$\partial A/\partial t = -E; \text{ and, of course, } p_k = \partial A_o/\partial q_k. \tag{8.10.11k}$$

The $n - 1$ equations (8.10.11j), $\partial A_o/\partial\beta_k = \alpha_k$, connect the q's with the constants $\beta_1, \ldots, \beta_n; \alpha_2, \ldots, \alpha_n$, and thus specify the form of the sequence of configurations the system goes through during its motion—that is, the shape of its trajectory (orbit); while the remaining equation (8.10.11i), $\partial A_o/\partial\beta_1 = t + \alpha_1$, yields the time it takes the system to arrive at each of these configurations. Since this implies that α_1 must have temporal dimensions, setting $\alpha_1 = -t_o$ (some "initial" instant) in (8.10.11i), we find

$$\partial A_o/\partial E = t - t_o \equiv f(q, \beta), \tag{8.10.11l}$$

again a total of $2n$ arbitrary constants: $\beta_1 = E; \beta_2, \ldots, \beta_n$ and $\alpha_1 = -t_o, \alpha_2, \ldots, \alpha_n$. From these equations we can have the n q's in terms of $t - t_o$ and either the n α's or the n β's.

[If the total energy is a function of a certain variable, say u: $u = u(E) \Leftrightarrow E = E(u)$, then

$$A_o = A_o[q; E(u), \beta_2, \ldots, \beta_n] \equiv A'_o(q; u; \beta_2, \ldots, \beta_n),$$

and, accordingly, (8.10.11i, l) is replaced by

$$\partial A_o/\partial E = (\partial A'_o/\partial u)(du/dE) = t + \alpha_1 \ \Rightarrow\ \partial A'_o/\partial u = (dE/du)(t + \alpha_1).] \tag{8.10.11m}$$

In sum, for a conservative system:

$$q_k = q_k(t - t_o; c_1, \ldots, c_{2n-1}), \qquad p_k = p_k(t - t_o; c_1, \ldots, c_{2n-1}), \tag{8.10.11n}$$

where $c' = (c_1, \ldots, c_{2n-1})$: $2n - 1$ *constants of integration.*

2. Separation of Variables

This is a method of finding complete integrals of the HJ equation in the special but important case where a particular coordinate, say q_1, and corresponding derivative $\partial A/\partial q_1$ do not appear in the Hamilton–Jacobi equation (8.10.10b), except in the separable combination $f_1(q_1, \partial A/\partial q_1)$, so that the latter takes the form

$$F\big[t, q_R, \partial A/\partial q_R, \partial A/\partial t; f_1(q_1, \partial A/\partial q_1)\big] = 0, \tag{8.10.12a}$$

where q_R denotes the *remaining* coordinates, here $q_2, \ldots, q_n$. In this case, we seek a complete integral in the following separable, or sum, form:

$$A = A_R(t, q_R) + A_1(q_1). \tag{8.10.12b}$$

Substituting (8.10.12b) in (8.10.12a), we obtain

$$F\big[t, q_R, \partial A_R/\partial q_R, \partial A_R/\partial t; f_1(q_1, dA_1/dq_1)\big] = 0, \tag{8.10.12c}$$

which, since *it must be an identity in* q_1 [and the latter affects only $f_1(\ldots)$], leads us to the following: (i) *ordinary* differential equation,

$$f_1(q_1, dA_1/dq_1) = \textit{arbitrary constant} \equiv \beta_1, \tag{8.10.12d}$$

from which, by a simple quadrature, we obtain A_1; and (ii) the *partial* differential equation,

$$F(t, q_R, \partial A_R/\partial q_R, \partial A_R/\partial t; \beta_1) = 0, \tag{8.10.12e}$$

which contain *fewer* independent variables than the original equation (8.10.12a).

If it is possible to carry out this separation process for *all n q's and t*, then the complete integral of the HJ equation will have been reduced to quadratures. In particular, for a *conservative* system, *complete separation* of its variables allows us to express its action integral (8.10.11e–g) as

$$A = A_o - E\,t \equiv \sum A_k(q_k; \beta_1, \ldots, \beta_n) - E(\beta_1, \ldots, \beta_n)t; \tag{8.10.12f}$$

that is, A_k = function of q_k only, and all the β's; and the constant energy $E = E(\beta)$ is found from (8.10.11c) with $A_o = \sum A_k(q_k; \beta)$. Then, $p_k = \partial A_o/\partial q_k = \partial A_k/\partial q_k$ and the HJ equation for A_o separates to n equations of the form (8.10.12d)

$$f_k(q_k, \partial A_k/\partial q_k; \beta) = \beta_k, \qquad \text{or} \qquad f_k(q_k, \partial A_k/\partial q_k) = E_k(\beta_1, \ldots, \beta_n), \tag{8.10.12g}$$

from which the sought $A_k(q_k, \beta)$ can be obtained by quadratures.

Hence, it is very important to know whether a given Hamiltonian $H(q, p)$ is (completely or partially) separable or not. It has been shown that the necessary and sufficient conditions for such separability are the following $n(n-1)/2$ equations:

$$\begin{vmatrix} 0 & \partial H/\partial q_k & \partial H/\partial p_k \\ \partial H/\partial q_l & \partial^2 H/\partial q_k \partial q_l & \partial^2 H/\partial p_k \partial q_l \\ \partial H/\partial p_l & \partial^2 H/\partial q_k \partial p_l & \partial^2 H/\partial p_k \partial q_l \end{vmatrix} = 0, \tag{8.10.12h}$$

for $k, l = 1, \ldots, n$ $(k \neq n)$. [See, for example, Hagihara (1970, p. 77 ff.); who also "translates" (8.10.12h) into conditions in terms of the kinetic and potential energies.]

Alternatively, it can be shown that (8.10.12f)-type of separability occurs if (i) the HJ equation does *not* contain mixed products of $(\partial A_o/\partial q_k)(\partial A_o/\partial q_l)$, $k \neq l$, but only pure squares $(\partial A_o/\partial q_k)^2$, that is, if the Hamiltonian has the so-called *Stäckel* (or *orthogonal*) form:

$$H = (1/2)\sum v_k(q){p_k}^2 + V(q), \quad v_k(q)\text{: functions of the } q\text{'s (as in ex. 3.12.4)}, \tag{8.10.12i}$$

and (ii) if, in addition, H satisfies certain necessary and sufficient "Stäckel conditions."

[For detailed treatments of separability, see, for example (alphabetically): Dobronravov (1976, pp. 117–129), Frank (1935, pp. 83–90), Goldstein (1980, pp. 449–457, 613–615), Lur'e (1968, pp. 538–548), Nordheim and Fues (1927, pp. 122), Pars (1965, pp. 320–348), Prange (1935, pp. 644–657); and books on celestial mechanics: for example, Hagihara (1970, p. 77 ff.).

For a readable discussion of the Stäckel theorem/conditions, including proof and examples, see, for example, Greenwood (1977, pp. 206–211).]

3. Ignorable Coordinates

Finally, in the case of an *ignorable* coordinate, say $q_1 \equiv \psi_1$, since the latter does not appear explicitly in either the Hamiltonian or the HJ equation, eqs. (8.10.12d) and (8.10.12b), with the slight indicial change $R \to P$, $p = 2, \ldots, n$ (in conformity with §8.3 ff.), reduce, respectively, to

$$dA_1/dq_1 = \beta_1 \equiv \Psi_1 \;\Rightarrow\; A_1 = \Psi_1 \psi_1 + (additive)\ constant, \tag{8.10.13a}$$

$$A = A_R(t, q_R) + \Psi_1 \psi_1 \equiv A_P(t, q_p) + \Psi_1 \psi_1, \tag{8.10.13b}$$

where $\Psi_1 = \partial A/\partial \psi_1 =$ constant momentum, corresponding to ψ_1. Then, the HJ equation

$$H\big(t; q_2, \ldots, q_n;\ \partial A/\partial q_1 = \Psi_1, \partial A/\partial q_2, \ldots, \partial A/\partial q_n\big) + \partial A/\partial t = 0, \tag{8.10.13c}$$

simplifies to

$$H\big(t; q_2, \ldots, q_n; \Psi_1, \partial A_P/\partial q_2, \ldots, \partial A_P/\partial q_n\big) + \partial A_P/\partial t = 0, \tag{8.10.13d}$$

and has as complete solution

$$A_P = A_P(t; q_2, \ldots, q_n; \Psi_1, \beta_2, \ldots, \beta_n) \equiv A_P(t; q_p; \beta_1 = \Psi_1, \beta_p); \tag{8.10.13e}$$

$$\Rightarrow A = \Psi_1 \psi_1 + A_P(t, q_p, \Psi_1, \beta_p). \tag{8.10.13f}$$

Hence, the finite equations of motion become

$$\partial A/\partial \Psi_1 = \psi_1 + \partial A_P/\partial \Psi_1 = \alpha_1, \qquad \partial A/\partial \beta_k = \partial A_P/\partial \beta_k = \alpha_k \qquad (k = 2, \ldots, n). \tag{8.10.13g}$$

In a conservative system, $q_1 \to t$ (time as an ignorable and separable "coordinate"), and $\beta_1 q_1 \to -E\,t(-E$ as corresponding constant "momentum"). For extensions of the above special cases to more than one variable, see ex. 8.10.5, below.

The inclusion of both ignorable and nonignorable coordinates under the general roof of separation of variables makes the HJ method one of the most powerful tools for integrating the Hamiltonian equations of motion. [The most prominent applications of this method are to be found not so much in earthly engineering as in celestial mechanics and modern nonlinear dynamics (and its transition to quantum mechanics); see, for example, Born (1927), Hagihara (1970), Tabor (1989). For extensive applications to rigid-body dynamics, see, for example, Chertkov (1960).]

Example 8.10.1 *Proof of Theorem of Jacobi.* Here, we show that the following equations

$$\partial A/\partial \beta_k = \alpha_k, \qquad \partial A/\partial q_k = p_k, \tag{a}$$

where $A = A[t, q(t), \beta_1, \ldots, \beta_n] \equiv A[t, q(t), \beta]$, constitute a complete integral of the HJ equation

$$H(t, q, \partial A/\partial q) + \partial A/\partial t = 0, \tag{b}$$

or, explicitly,

$$\partial A/\partial t + (1/2)\big[M'_{11}(\partial A/\partial q_1)^2 + M'_{22}(\partial A/\partial q_2)^2 + \cdots + 2M'_{12}(\partial A/\partial q_1)(\partial A/\partial q_2) + \cdots\big] + V = 0; \qquad \text{(b1)}$$

and $|\partial^2 A/\partial\beta_k \partial q_l| \neq 0$ [so that eqs. (a) are independent], satisfy the canonical equations

$$dq_k/dt = \partial H/\partial p_k, \qquad dp_k/dt = -\partial H/\partial q_k, \qquad \text{(c)}$$

identically.

PROOF

Since the α_k are constant, $(\ldots)^{\cdot}$-differentiating the first of (a), we obtain

$$0 = d\alpha_k/dt = (\partial A/\partial\beta_k)^{\cdot} = \partial^2 A/\partial t\, \partial\beta_k + \sum (\partial^2 A/\partial q_l\, \partial\beta_k)(dq_l/dt). \qquad \text{(d)}$$

Now we can either (i) solve the system (d) for the $\dot{q}_l$ and show that the solution satisfies the first of (c) identically; or, conversely, (ii) insert the $\dot{q}_k$ from the first of (c) into (d) and show that they satisfy it identically. Indeed:

(i) Equations (b, b1) hold identically in the β's. Therefore, comparing their $\partial(\ldots)/\partial\beta_k$-derivative

$$\partial^2 A/\partial\beta_k\, \partial t + \sum \big[M'_{1l}(\partial A/\partial q_1) + \cdots + M'_{1n}(\partial A/\partial q_n)\big](\partial^2 A/\partial\beta_k\, \partial q_l) = 0 \qquad \text{(e1)}$$

with (d), we conclude that

$$dq_l/dt = M'_{1l}(\partial A/\partial q_1) + \cdots + M'_{nl}(\partial A/\partial q_n). \qquad \text{(e2)}$$

[Equations (d) determine the $\dot{q}_l$ uniquely as linear functions of the $\partial^2 A/\partial\beta_k\, \partial t$; and, similarly, (e1) determine the right sides of (e2) as the same functions of them.] Hence, by prob. 8.2.1, $p_k = \partial A/\partial q_k$; and, accordingly, the Hamiltonian first of (c) hold; also, from (b), we conclude that $H = -\partial A/\partial t$. To prove the second of (c), we $\partial(\ldots)/\partial q_k$-differentiate (b), since the latter is an identity in the q's thus obtaining, successively,

$$\begin{aligned}\partial^2 A/\partial q_k\, \partial t &= -\partial H/\partial q_k - \sum (\partial H/\partial p_l)(\partial p_l/\partial q_k)\\ &= -\partial H/\partial q_k - \sum (\partial H/\partial p_l)(\partial^2 A/\partial q_k\, \partial q_l)\\ &= -\partial H/\partial q_k - \sum (\partial^2 A/\partial q_k\, \partial q_l)(dq_l/dt) \quad \text{[by first of (c)]}, \qquad \text{(f1)}\end{aligned}$$

or, rearranging,

$$\partial^2 A/\partial q_k\, \partial t + \sum (\partial^2 A/\partial q_k\, \partial q_l)(dq_l/dt) = -\partial H/\partial q_k, \qquad \text{(f2)}$$

or, finally,

$$(\partial A/\partial q_k)^{\cdot} = dp_k/dt = -\partial H/\partial q_k, \quad \text{Q.E.D.} \qquad \text{(f3)}$$

Hence, in the assumed motion, both Hamiltonian equations hold.

(ii) Due to the first of (c), (d) becomes

$$\partial^2 A/\partial t\, \partial\beta_k + \sum (\partial^2 A/\partial q_l\, \partial\beta_k)(\partial H/\partial p_l) = 0. \tag{g}$$

We will show that (g) holds identically. By $\partial(\ldots)/\partial\beta_k$-differentiating (b), since the latter is satisfied for arbitrary β's (i.e., identically), we obtain

$$\partial H/\partial\beta_k + \partial^2 A/\partial\beta_k\, \partial t = 0. \tag{h1}$$

But, since H depends on the β_k through the $\partial A/\partial q_k$ and these, in turn, depend on the β_k through $A = A[t, q(t), \beta]$,

$$\partial H/\partial\beta_k = \sum\, [\partial H/\partial(\partial A/\partial q_l)](\partial^2 A/\partial q_l\, \partial\beta_k), \tag{h2}$$

and so, comparing (h1) with (h2), we find

$$\partial^2 A/\partial\beta_k\, \partial t + \sum\, [\partial H/\partial(\partial A/\partial q_l)](\partial^2 A/\partial q_l\, \partial\beta_k) = 0,$$

or, thanks to the second of (a),

$$\partial^2 A/\partial\beta_k\, \partial t + \sum (\partial H/\partial p_l)(\partial^2 A/\partial q_l\, \partial\beta_k) = 0; \tag{h3}$$

that is, eqs. (g). Therefore, the first of (a) is indeed a solution of the first of (c).

Similarly, we can show that the second of (a) is a solution of the second of (c): $(\ldots)^{\cdot}$-differentiating the second of (a) and combining the so-resulting (d)-like system for the p_k with the second of (c), we are led to a (g)-like equation, and so on. (See also MacMillan, 1936, pp. 371–375.)

Example 8.10.2 *HJ Equation: From a Complete Integral (CI) to the General Integral (GI)* (Landau and Lifshitz, 1960, p. 148, footnote). Even though the GI is not needed in dynamics, it can be obtained from a CI as follows: we begin with the CI

$$A = A'(t; q_1, \ldots, q_n; \beta_1, \ldots, \beta_n) + \beta_{n+1} \equiv A'(t, q, \beta) + \beta', \tag{a}$$

but now we view its $(n+1)$th additive constant β' as an arbitrary function of the n β's:

$$\beta' = \beta'(\beta_1, \ldots, \beta_n) \equiv \beta'(\beta) \;\Rightarrow\; A = A(t, q, \beta). \tag{b}$$

Then, the GI of the HJ equation (8.10.10b) is found by replacing the β_k in (b) by their functional expressions obtained from the following n conditions

$$\partial A/\partial\beta_k = 0 \;\Rightarrow\; \beta_k = \beta_k(t, q) \;\Rightarrow\; \beta' = \beta'(t, q). \tag{c}$$

Indeed, applying chain rule to

$$A = A[t, q, \beta(t, q)] = W'[t; q, \beta(t, q)] + \beta'[\beta(t, q)] \equiv a(t, q), \tag{d}$$

and then invoking (d), we find

$$\partial a/\partial q_k = \partial A/\partial q_k + \sum (\partial A/\partial\beta_l)(\partial\beta_l/\partial q_k) = \partial A/\partial q_k; \tag{e}$$

that is, since the $\partial A/\partial q_k$ satisfy the HJ equation (A being a CI), so do the $\partial a/\partial q_k$, Q.E.D.

Example 8.10.3 *Particle in a Conservative Force Field; Harmonic Oscillator.* Let us consider the free motion of a particle P of mass m in a potential field $V = V(x, y, z)$, where (x, y, z) = rectangular Cartesian and inertial coordinates of P. Then, since

$$H = E \equiv T + V = mv^2/2 + V = p^2/2m + V$$
$$= (1/2m)({p_x}^2 + {p_y}^2 + {p_z}^2) + V, \tag{a}$$

where $v^2 = \dot{x}^2 + \dot{y}^2 + \dot{z}^2$, and recalling (8.10.10e), $p_x = \partial A/\partial x = \partial A_o/\partial x$ = x-component of linear momentum of P, and so on, cyclically, the "abbreviated" HJ equation of the system (8.10.11c) becomes

$$(\partial A_o/\partial x)^2 + (\partial A_o/\partial y)^2 + (\partial A_o/\partial z)^2 = 2m[E - V(x, y, z)]. \tag{b}$$

If P undergoes *one*-dimensional motion, say $V = V(q)$, where q = single Lagrangean coordinate, then (b) reduces to the *ordinary* differential equation

$$(dA_o/dq)^2 = 2m[E - V(q)], \tag{c}$$

and this leads readily to the quadrature

$$A_o(q; E) - A_o(q_o; E) = \int_{q_o}^{q} \{2m[E - V(q)]\}^{1/2} dq, \tag{d}$$

from some initial value q_o to q. From this, by (8.10.11l), we obtain

$$\partial A_o/\partial E = (m/2)^{1/2} \int_{q_o}^{q} [E - V(q)]^{-1/2} dq \equiv f(q, E) = t - t_o, \tag{e}$$

$$\Rightarrow \quad q = q(t - t_o; E). \tag{e1}$$

Specialization

One-Dimensional Linear Harmonic Oscillator. Here,

$$2T = m(\dot{q})^2, \qquad 2V = kq^2 \qquad (k = \text{constant coefficient of elasticity}), \tag{f}$$

from which

$$p = \partial L/\partial \dot{q} \;\Rightarrow\; \dot{q} = p/m$$
$$\Rightarrow\; T = p^2/2m \;\Rightarrow\; H = p^2/2m + kq^2/2$$
$$\Rightarrow\; dq/dt = \partial H/\partial p = p/m, \quad dp/dt = -\partial H/\partial q = -kq. \tag{g}$$

Accordingly, the HJ equation (8.10.10b) and its conservative specialization (8.10.11c) become

$$(1/2m)(\partial A/\partial q)^2 + kq^2/2 + \partial A/\partial t = 0, \tag{h}$$

$$(1/2m)(\partial A_o/\partial q)^2 + kq^2/2 = E, \tag{i}$$

respectively, where

$$A = A(t, q, \beta) = A(t, q, E) = A_o(q, E) - Et. \tag{j}$$

The solution of (i) is, to within an inessential additive constant, the quadrature

$$A_o(q, E) = (km)^{1/2} \int \left[(2E/k) - q^2\right]^{1/2} dq, \tag{k}$$

and, consequently,

$$A(t, q, E) = (km)^{1/2} \int \left[(2E/k) - q^2\right]^{1/2} dq - E\,t. \tag{l}$$

As a result, (8.10.10e) becomes (there is no need to evaluate the above integral yet)

$$\alpha = \partial A/\partial\beta = \partial A/\partial E = (m/k)^{1/2} \int \left[(2E/k) - q^2\right]^{-1/2} dq - t$$
$$= (m/k)^{1/2} \arccos[(k/2E)^{1/2} q] - t \qquad \text{(to within an arbitrary constant)}, \tag{m}$$

and solving for q

$$q = (2E/k)^{1/2} \cos\left[(k/m)^{1/2}(t + \alpha)\right] = q(t; E, \alpha). \tag{n}$$

To express q in terms of t and the initial values q_o, p_o, we apply (8.10.10e):

$$p_o = (\partial A/\partial q)_{\text{initial values}} = (km)^{1/2}\left[(2E/k) - q_o^{\,2}\right]^{1/2}$$
$$= (2m)^{1/2}(E - kq_o^{\,2}/2)^{1/2} = (2m)^{1/2}(E - V_o)^{1/2} = \cdots = mv_o; \tag{o}$$

from which, solving for E, we get

$$p_o^{\,2}/2m + kq_o^{\,2}/2 = E = H. \tag{p}$$

Next, evaluating (m), or (n), at the initial instant t_o:

$$q_o = (2E/k)^{1/2} \cos[(k/m)^{1/2}(t_o + \alpha)] \qquad [\Rightarrow\ \alpha = \alpha(q_o, E; t_o)], \tag{q}$$

then, solving (p) and (q) for E and α in terms of q_o and p_o, and inserting these values in (n), we obtain $q = q(t; q_o, p_o)$.

For example, choosing $t_o = 0$, $q_o = 0$, $p_o \neq 0$ (i.e., impact), and with $\omega_o^{\,2} \equiv k/m = (\textit{frequency})^2$, we obtain from (p, q)

$$E = p_o^{\,2}/2m, \qquad 0 = (2E/k)^{1/2} \cos(\omega_o \alpha) \Rightarrow \alpha = \pm(\pi/2\omega_o), \tag{r}$$

and so (n) becomes

$$q = (p_o^{\,2}/km)^{1/2} \cos(\omega_o t \pm \pi/2) = (-/+)(p_o^{\,2}/k\,m)^{1/2} \sin(\omega_o t). \tag{s}$$

This yields, further,

$$p = m\dot{q} = \cdots = (-/+)\, m(p_o^{\,2}/km)^{1/2}(k/m)^{1/2} \cos(\omega_o t), \tag{t}$$

and since for $t = 0$: $p = p_o$, only the + sign applies in (s); that is, finally,

$$q = (p_o{}^2/km)^{1/2} \sin(\omega_o t) \qquad (\Rightarrow q_o = 0). \tag{u}$$

Geometrical Interpretation

(i) In the phase space of the *old* variables (i.e., the retangular Cartesian qp-plane), the representative system point describes an *ellipse* (with center at the origin O–qp, and Oq, Op as its principal axes) whose dimensions are determined from the initial conditions $q_o, p_o \rightarrow E$.

(ii) In the phase space of the *new* variables $q' = \alpha$, $p' = \beta = E$; however, *the representative point does not vary with time— that is, it is fixed on that plane* (compare with ex. 8.8.5). The properties of A_o, for this *periodic* system, are detailed in ex. 8.14.1.

Example 8.10.4 *HJ Equation of a Heavy Axisymmetric Gyroscope, Moving about a Fixed Point O.* For this well-known problem, we have already seen that (ex. 8.4.5; with the transverse moment of inertia denoted by B, instead of A, to avoid confusion with the action)

$$2T = B[(\dot{\theta})^2 + (\dot{\phi})^2 \sin^2\theta] + C(\dot{\psi} + \dot{\phi}\cos\theta)^2,$$

$$V = mgl\cos\theta \qquad (l \equiv OG, G = \textit{center of mass of gyroscope}); \tag{a1}$$

$$\Rightarrow p_\phi \equiv \partial T/\partial\dot{\phi} = B\dot{\phi}\sin^2\theta + C(\dot{\psi} + \dot{\phi}\cos\theta)\cos\theta$$

$$\equiv B\dot{\phi}\sin^2\theta + Cn\cos\theta = \textit{constant} \equiv C_\phi, \tag{a2}$$

$$p_\theta \equiv \partial T/\partial\dot{\theta} = B\dot{\theta}, \tag{a3}$$

$$p_\psi \equiv \partial T/\partial\dot{\psi} = C(\dot{\psi} + \dot{\phi}\cos\theta) \equiv Cn = \textit{constant} \equiv C_\psi; \tag{a4}$$

$$\dot{\phi} = (p_\phi - p_\psi\cos\theta)/B\sin^2\theta, \tag{a5}$$

$$\dot{\theta} = p_\theta/B, \tag{a6}$$

$$\dot{\psi} = p_\psi/C - (p_\phi - p_\psi\cos\theta)\cos\theta/B\sin^2\theta; \tag{a7}$$

$$\Rightarrow\ H = (1/2B)[p_\theta{}^2 + (p_\phi - p_\psi\cos\theta)^2/\sin^2\theta] + (1/2C)p_\psi{}^2 + mgl\cos\theta. \tag{a8}$$

Accordingly, the HJ equation of this conservative system becomes

$$(1/2)\{B^{-1}(\partial A_o/\partial\theta)^2 + C^{-1}(\partial A_o/\partial\psi)^2 + (B\sin^2\theta)^{-1}[\partial A_o/\partial\phi - \cos\theta(\partial A_o/\partial\psi)]^2\} + mgl\cos\theta = E \quad (= \beta_1). \tag{b}$$

But, since ϕ and ψ are *ignorable* coordinates, we also have the two integrals

$$p_\phi = \partial A_o/\partial\phi = \textit{constant} \equiv C_\phi \equiv B\beta_2, \tag{c1}$$

$$p_\psi = \partial A_o/\partial\psi = \textit{constant} \equiv C_\psi \equiv B\beta_3; \tag{c2}$$

and, accordingly, (b) simplifies to

$$(1/2)\big[B^{-1}(\partial A_o/\partial\theta)^2 + C^{-1}(B\beta_3)^2 + (A/\sin^2\theta)(\beta_2 - \beta_3\cos\theta)^2\big] + m\,g\,l\cos\theta = E; \qquad \text{(d)}$$

from which, solving for $\partial A_o/\partial\theta$, we find

$$\begin{aligned} B^{-1}(\partial A_o/\partial\theta)^2 &= 2E - B^2{\beta_3}^2/C - 2m\,g\,l\cos\theta - (B/\sin^2\theta)(\beta_2 - \beta_3\cos\theta)^2 \\ &\equiv Bf(\theta; \beta_1 = E, \beta_2, \beta_3) \equiv Bf(\theta). \end{aligned} \qquad \text{(e)}$$

Therefore, it follows from the general results (8.10.11b, e) that (to within an additive constant)

$$A = A_o - E\,t = B\int [f(\theta)]^{1/2}\,d\theta + B(\beta_2\phi + \beta_3\psi) - E\,t. \qquad \text{(f)}$$

Hence, the three finite equations of motions of the gyroscope are

$$\partial A/\partial E = -t + \int [f(\theta)]^{-1/2}\,d\theta = -t_o \quad (= \alpha_1), \qquad \text{(g1)}$$

$$\partial A/\partial\beta_2 = B\Big\{\phi - \int [(\beta_2 - \beta_3\cos\theta)/\sin^2\theta][f(\theta)]^{-1/2}\,d\theta\Big\} = B\,\phi_o \quad (= \alpha_2), \qquad \text{(g2)}$$

$$\begin{aligned} \partial A/\partial\beta_3 &= B\Big\{\psi + \int [-(B\beta_3/C) + (\beta_2 - \beta_3\cos\theta)\cos\theta/\sin^2\theta][f(\theta)]^{-1/2}\,d\theta\Big\} \\ &= B\psi_o \quad (= \alpha_3), \end{aligned} \qquad \text{(g3)}$$

where t_o, ϕ_o, ψ_o are three new constants. Equation (g1) yields θ as a function of time, while (g2, 3) yield, respectively, ϕ and ψ as functions of θ. These results, of course, coincide with those found by other means. See also MacMillan (1936, pp. 378–380).

Example 8.10.5 *Separation of Variables, Ignorable Coordinates, Conservative Systems.*

(i) Separation of Variables

Let the system Hamiltonian have the form

$$H = H\big[f_1(q_1, p_1), \ldots, f_M(q_M, p_M); q_{M+1}, \ldots, q_n; \partial A/\partial q_{M+1}, \ldots, \partial A/\partial q_n; t\big]; \qquad \text{(a)}$$

that is, the first $M(< n)$ variables are separable. Then, the HJ equation becomes

$$H\big[f_1(q_1, \partial A/\partial q_1), \ldots, f_M(q_M, \partial A/\partial q_M); q_{M+1}, \ldots, q_n;\ \partial A/\partial q_{M+1}, \ldots, \partial A/\partial q_n; t\big] + \partial A/\partial t = 0. \qquad \text{(b)}$$

Assuming the following partially *separable* action:

$$A = A_1(q_1) + \cdots + A_M(q_M) + A_R(q_R, t), \qquad \text{(c)}$$

where $q_R \equiv (q_{M+1}, \ldots, q_n) =$ *remaining* (nonseparable) coordinates, reduces (b) to

$$H\big[f_1(q_1, dA_1/dq_1), \ldots, f_M(q_M, dA_M/dq_M); q_{M+1}, \ldots, q_n; \partial A_R/\partial q_{M+1}, \ldots, \partial A_R/\partial q_n; t\big] + \partial A_R/\partial t = 0, \qquad \text{(d)}$$

and from this, reasoning as in the derivation of (8.10.12d, e) from (8.10.12c), we are readily led to the M uncoupled ordinary differential equations:

$$f_1(q_1, dA_1/dq_1) = \beta_1, \ldots, f_M(q_M, dA_M/dq_M) = \beta_M, \qquad \text{(e)}$$

from which we can determine $A_1, \ldots, A_M$ by quadrature; and the partial differential equation

$$H\big[\beta_1, \ldots, \beta_M; q_{M+1}, \ldots, q_n; \partial A_R/\partial q_{M+1}, \ldots, \partial A_R/\partial q_n; t\big] + \partial A_R/\partial t = 0, \qquad \text{(f)}$$

which is still coupled, but in *fewer* variables than the original (b).

(ii) Ignorable Coordinates

Next, let $q_1 \equiv \psi_1, \ldots, q_M \equiv \psi_M$ be *ignorable* (recall §8.4). Then, the corresponding momenta are constant:

$$p_1 = \partial A/\partial \psi_1 = \beta_1 \equiv \Psi_1, \ldots, p_M = \partial A/\partial \psi_M = \beta_M \equiv \Psi_M, \qquad \text{(g)}$$

Setting in the corresponding HJ equation

$$H(q_p;\ \Psi;\ \partial A/\partial q_p, t) + \partial A/\partial t = 0 \qquad \text{(h)}$$

the action

$$A = \sum \Psi_i \psi_i + A_P(q_p, t) \qquad (i = 1, \ldots, M), \qquad \text{(i)}$$

where $q_p \equiv (q_{M+1}, \ldots, q_n)$, $\Psi \equiv (\Psi_1, \ldots, \Psi_M) \equiv (\Psi_i)$, reduces (h) to the simpler form:

$$H(q_p; \Psi; \partial A_P/\partial q_{M+1}, \ldots, \partial A_P/\partial q_n;\ t) + \partial A_P/\partial t = 0; \qquad \text{(j)}$$

A_P depends on fewer independent variables than A: the $M - n$ q_p and t.

(iii) Conservative Systems

Continuing, we consider a conservative system with *completely separable* Hamiltonian

$$H = H\big[f_1(q_1, p_1), \ldots, f_n(q_n, p_n)\big], \qquad \text{(k)}$$

and, therefore, by (8.10.11c), HJ equation

$$H\big[f_1(q_1, \partial A_o/\partial q_1), \ldots, f_n(q_n, \partial A_o/\partial q_n)\big] = E\ (\textit{constant}). \qquad \text{(l)}$$

Substituting in it the completely separable reduced action

$$A_o = \sum A_k(q_k) \qquad \text{(m)}$$

transforms it to

$$H\big[f_1(q_1, dA_1/dq_1), \ldots, f_n(q_n, dA_n/dq_n)\big] = E; \tag{n}$$

and this, reasoning as in the derivation of (8.10.12d), leads us to the n ordinary differential equations:

$$f_1(q_1, dA_1/dq_1) = \beta_1, \ldots, f_n(q_n, dA_n/dq_n) = \beta_n, \tag{o}$$

which can be solved by quadratures.

In view of these results, the fundamental HJ relations (8.10.11e ff., 10d) reduce to

$$E = H(\beta_1, \ldots, \beta_n) \equiv H(\beta), \tag{p}$$

$$A = A_o - E\,t = \sum A_k(q_k; \beta_k) - E\,t, \tag{q}$$

$$\alpha_k = \partial A/\partial\beta_k = \partial A_k/\partial\beta_k - (\partial E/\partial\beta_k)t \;\Rightarrow\; \partial A_k/\partial\beta_k = \alpha_k + (\partial E/\partial\beta_k)t. \tag{r}$$

(iv) Conservative Systems with Ignorable Coordinates

Finally, for a conservative system with $M(< n)$ ignorable coordinates $\psi \equiv (\psi_1, \ldots, \psi_M) \equiv (\psi_i)$, the reduced HJ equation (8.10.11c) becomes (with the earlier notations)

$$H(q_p;\, \Psi;\, \partial A_o/\partial q_p) = E. \tag{s}$$

Substituting into it, as in (h),

$$A_o = \sum \Psi_i\,\psi_i + A_P(q_p, \partial A_P/\partial q_{M+1}, \ldots, \partial A_P/\partial q_n), \tag{t}$$

we are led to the simpler than (s) HJ equation

$$H(q_p;\, \Psi;\, \partial A_P/\partial q_{M+1}, \ldots, \partial A_P/\partial q_n) = E \tag{u}$$

(since A_P is a function of only $n - M$ independent variables, the q_p), whose solution is, to within an inessential additive constant,

$$A_P = A_P(q_{M+1}, \ldots, q_n;\, \beta_{M+1}, \ldots, \beta_n;\, \Psi_i = \beta_i) \equiv A_P(q_p;\, \beta_p;\, \Psi), \tag{v}$$

$$\Rightarrow\; A = A_o - E\,t = \sum \Psi_i\,\psi_i + A_P(q_p, \beta_p, \Psi) - E\,t. \tag{w}$$

Hence, the fundamental HJ relations (8.10.11i, j, l) reduce to

$$\alpha_i = \partial A/\partial\beta_i \equiv \partial A/\partial\Psi_i = \partial A_o/\partial\Psi_i = \psi_i + \partial A_P/\partial\Psi_i \qquad (\text{here, } \beta_i = \Psi_i),$$

$$\Rightarrow\; \partial A_P/\partial\Psi_i = \alpha_i - \psi_i \qquad (i = 1, \ldots, M), \tag{x}$$

$$\alpha_{M+1} = \partial A/\partial\beta_{M+1} \equiv \partial A/\partial E = \partial A_o/\partial E - (\partial E/\partial E)t = \partial A_P/\partial E - (1)t = -t_o,$$

$$\Rightarrow \partial A_P/\partial E = t - t_o \qquad (\text{here, } \beta_{M+1} = E), \tag{y}$$

$$\alpha_{p'} = \partial A/\partial\beta_{p'} = \partial A_o/\partial\beta_{p'} = \partial A_P/\partial\beta_{p'} \qquad (p' = M+1, \ldots, n). \tag{z}$$

Below, we discuss a few elementary applications of the foregoing theory.

Example 8.10.6 *Two-Dimensional Linear and Isotropic Oscillator* (Butenin, 1971, pp. 163–165). Here, the kinetic and potential energies of this conservative system are (using standard notations)

$$2T = m[(\dot{q}_1)^2 + (\dot{q}_2)^2], \qquad 2V = k({q_1}^2 + {q_2}^2), \tag{a}$$

respectively, and therefore the Hamiltonian and reduced HJ equations are

$$H = (1/2m)({p_1}^2 + {p_2}^2) + (k/2)({q_1}^2 + {q_2}^2), \tag{b}$$

$$\left[(1/2m)(\partial A_o/\partial q_1)^2 + (k/2){q_1}^2\right] + \left[(1/2m)(\partial A_o/\partial q_2)^2 + (k/2){q_2}^2\right] = E. \tag{c}$$

Substituting into the completely separable equation (c) the equally separable reduced action

$$A_o = A_1(q_1) + A_2(q_2), \tag{d}$$

we are immediately led to the two uncoupled ordinary differential equations

$$(1/2m)(dA_1/dq_1)^2 + (k/2){q_1}^2 = \beta_1, \qquad (1/2m)(dA_2/dq_2)^2 + (k/2){q_2}^2 = \beta_2, \tag{e}$$

where

$$\beta_1 + \beta_2 = E. \tag{f}$$

The above readily lead to the quadratures (omitting inessential additive constants)

$$A_r = \int \left[m(2\beta_r - k{q_r}^2)\right]^{1/2} dq_r \qquad (r = 1, 2), \tag{g}$$

and from these we obtain, by differentiation and elementary integrations,

$$dA_r/d\beta_r = (m/k)^{1/2} \arcsin\left[q_r/(2\beta_r/k)^{1/2}\right]; \tag{h}$$

and when these results are compared with the corresponding finite equations of motion [equations (r) of preceding example, and (8.10.10e)]:

$$dA_r/d\beta_r = \alpha_r + (\partial E/\partial \beta_r)t = \alpha_r + t, \tag{i}$$

$$dA_r/dq_r = p_r \quad (= m\dot{q}_r), \tag{j}$$

they yield, respectively,

$$\arcsin\left[q_r/(2\beta_r/k)^{1/2}\right] = (k/m)^{1/2}(t + \alpha_r) \Rightarrow q_r = (2\beta_r/k)^{1/2} \sin\left[(k/m)^{1/2}(t + \alpha_r)\right], \tag{k}$$

$$p_r = \left[m(2\beta_r - k{q_r}^2)\right]^{1/2}. \tag{l}$$

For the initial conditions at $t = 0$: $q_1 = a$, $\dot{q}_1 = 0$, $q_2 = 0$, $\dot{q}_2 = v_o$, (k, l) give

$$a = (2\beta_1/k)^{1/2} \sin[(k/m)^{1/2}\alpha_1], \qquad 0 = (2\beta_2/k)^{1/2} \sin[(k/m)^{1/2}\alpha_2],$$

$$\left[m(2\beta_1 - ka^2)\right]^{1/2} = 0, \qquad [m(2\beta_2)]^{1/2} = mv_o, \tag{m}$$

or, upon solving for the β's and α's,

$$\beta_1 = ka^2/2, \quad \beta_2 = m{v_o}^2/2; \qquad \alpha_1 = (\pi/2)(m/k)^{1/2}, \quad \alpha_2 = 0; \tag{n}$$

and so, finally, the motion (k, l) specializes to

$$q_1 = a\cos[(k/m)^{1/2}t], \qquad q_2 = v_o(m/k)^{1/2}\sin[(k/m)^{1/2}t]. \tag{o}$$

Example 8.10.7 *Plane Motion of a Particle in a Uniform Gravitational Field.* Here, the kinetic and potential energies of this system are (using standard notations)

$$2T = m[(\dot{q}_1)^2 + (\dot{q}_2)^2], \qquad V = m g q_2, \tag{a}$$

respectively [$O - q_1 q_2$: inertial rectangular Cartesian coordinates of particle, q_1: *horizontal*, q_2: *vertical* (positive upward)], and therefore the Hamiltonian and HJ equations are

$$H = (1/2m)({p_1}^2 + {p_2}^2) + m g q_2, \tag{b}$$

$$(1/2m)\left[(\partial A/\partial q_1)^2 + (\partial A/\partial q_2)^2\right] + m g q_2 + \partial A/\partial t = 0. \tag{c}$$

Since q_1 is also *ignorable*, in addition to the system being conservative, following the theory of (8.10.13a ff.) and the preceding example, we try the Action function (with $q_1 \equiv \psi_1, p_1 \equiv \Psi_1 = \beta_1$):

$$A = \beta_1\psi_1 + A_2(q_2) - Et \equiv A_o - E\,t. \tag{d}$$

First Solution

Substituting (d) into (c), we readily find

$${\beta_1}^2/2m + (1/2m)(dA_2/dq_2)^2 + m g q_2 = E, \tag{e}$$

and this leads us, easily, to the following *two* equations:

$$(1/2m)(dA_2/dq_2)^2 + m g q_2 = \beta_2, \qquad \beta_2 + {\beta_1}^2/2m = E. \tag{f}$$

Hence, the finite equations of motion are

$$\partial A/\partial\beta_1 = q_1 - (\partial E/\partial\beta_1)t = q_1 - (\beta_1/m)t = \alpha_1$$

$$\Rightarrow\; q_1 = (\beta_1/m)t + \alpha_1, \tag{g}$$

$$\partial A/\partial\beta_2 = dA_2/d\beta_2 - (\partial E/\partial\beta_2)t = dA_2/d\beta_2 - (1)t = \alpha_2$$

$$\Rightarrow\; dA_2/d\beta_2 = t + \alpha_2. \tag{h}$$

But, from the first of (f), we find

$$dA_2/dq_2 = [2m(\beta_2 - m g q_2)]^{1/2}, \tag{i}$$

$$\Rightarrow\; A_2 = \int [2m(\beta_2 - m g q_2)]^{1/2}\, dq_2 + constant; \tag{j}$$

and, therefore

$$dA_2/d\beta_2 = \int m\big[2m(\beta_2 - m\,g\,q_2)\big]^{-1/2}\,dq_2 = -(1/mg)\big[2m(\beta_2 - m\,g\,q_2)\big]^{1/2}. \qquad \text{(k)}$$

Equating the right sides of (h) and (k), and then solving for q_2, we get

$$q_2 = -(g/2)(t + \alpha_2)^2 + \beta_2/mg. \qquad \text{(l)}$$

Finally, applying the remaining finite equations

$$\partial A/\partial q_1 = \beta_1 = p_1 = m\,\dot{q}_1, \qquad \partial A/\partial q_2 = dA_2/dq_2 = t + \beta_2 = p_2 = m\,\dot{q}_2, \qquad \text{(m)}$$

for the common initial conditions at $t = 0$: $q_1 = 0$, $\dot{q}_1 = v_o$, $q_2 = h$, $\dot{q}_2 = 0$, we find

$$\beta_1 = m\,v_o, \quad \beta_2 = mgh; \qquad \alpha_1 = 0, \quad \alpha_2 = 0; \qquad \text{(n)}$$

and so, finally, the motion (g, l) specializes to the well-known solution

$$q_1 = v_o t, \qquad q_2 = h - (1/2)gt^2. \qquad \text{(o)}$$

Second Solution

The reduced HJ equation is

$$(1/2m)\big[(\partial A_o/\partial q_1)^2 + (\partial A_o/\partial q_2)^2\big] + m\,g\,q_2 = \beta_2, \qquad \text{(p)}$$

or

$$(\partial A_o/\partial q_1)^2 + (\partial A_o/\partial q_2)^2 + 2m^2 g q_2 = 2m\beta_2. \qquad \text{(q)}$$

By Hamilton's equations: $\partial H/\partial q_1 = dp_1/dt = 0 \;\Rightarrow\; p_1 = constant \equiv \beta_1 = \partial A_o/\partial q_1$, and so (q) gives

$$(\partial A_o/\partial q_2)^2 = 2m\beta_2 - 2m^2 g\,q_2 - (\partial A_o/\partial q_1)^2 = 2m\beta_2 - {\beta_1}^2 - 2m^2 g\,q_2.$$

Hence, q_1, q_2 are *separable* and

$$A_o = \int (\partial A_o/\partial q_1)\,dq_1 + \int (\partial A_o/\partial q_2)\,dq_2$$
$$= \beta_1 q_1 + \int (2m\beta_2 - {\beta_1}^2 - 2m^2 g q_2)^{1/2}\,dq_2. \qquad \text{(r)}$$

The particle trajectory is given by $\partial A_o/\partial\beta_1 = \alpha_1$:

$$q_1 - \int \beta_1(2m\beta_2 - {\beta_1}^2 - 2m^2 g\,q_2)^{-1/2}\,dq_2 = \alpha_1$$
$$\Rightarrow\; q_1 + (\beta_1/g\,m^2)(2m\beta_2 - {\beta_1}^2 - 2m^2 g q_2)^{1/2} = \alpha_1 \qquad \text{(a parabola);} \qquad \text{(s)}$$

and the corresponding time by $\partial A_o/\partial\beta_2 = t + \alpha_2$:

$$\int m(2m\beta_2 - \beta_1{}^2 - 2m^2 g q_2)^{-1/2}\,dq_2 = t + \alpha_2$$

$$\Rightarrow\ -(1/mg)(2m\beta_2 - \beta_1{}^2 - 2m^2 g\, q_2)^{1/2} = t + \alpha_2 \qquad \text{(time at which height is } q_2). \tag{t}$$

Example 8.10.8 *Theorem of Liouville* (Recall ex. 3.12.4). This generalizes the results of the preceding examples, 8.10.6 and 8.10.7. Let us consider a *Liouville* system; that is, one whose kinetic and potential energies have the following forms:

$$2T = u\big[v_1(q_1)(\dot{q}_1)^2 + \cdots + v_n(q_n)(\dot{q}_n)^2\big] \equiv u\big[v_1(\dot{q}_1)^2 + \cdots + v_n(\dot{q}_n)^2\big], \tag{a}$$

$$V = \big[w_1(q_1) + \cdots + w_n(q_n)\big]/u \equiv (w_1 + \cdots + w_n)/u, \tag{b}$$

respectively, where $u \equiv u_1(q_1) + \cdots + u_n(q_n) \equiv u_1 + \cdots + u_n\ (> 0)$.

Since $p_k \equiv \partial T/\partial\dot{q}_k = u\, v_k\, \dot{q}_k$, it is not hard to see that the corresponding Hamiltonian function and reduced HJ equation of this conservative system are

$$H = T + V = (1/2)\,u \sum v_k(\dot{q}_k)^2 + u^{-1}\sum w_k = u^{-1}\sum (p_k{}^2/2v_k + w_k) \quad (\equiv E),$$

$$\sum \big[(1/2v_k)(\partial A_o/\partial q_k)^2 + w_k - E\,u_k\big] = 0. \tag{c}$$

Setting in this *completely separable* equation the following similarly separable reduced action:

$$A_o = A_1(q_1) + \cdots + A_n(q_n), \tag{d}$$

yields

$$\sum \big[(1/2v_k)(dA_k/dq_k)^2 + w_k - E\,u_k\big] = 0, \qquad p_k = dA_k/dq_k, \tag{e}$$

and from this, reasoning as in the derivation of (8.10.12d), we obtain the following n ordinary differential equations:

$$(1/2v_k)(dA_k/dq_k)^2 + w_k - E\,u_k = \beta_k, \tag{f}$$

where the n constants β_k are subject to the condition

$$\beta_1 + \beta_2 + \cdots + \beta_n = 0\ \Rightarrow\ \beta_1 = -(\beta_2 + \cdots + \beta_n) = \beta_1(\beta_2, \ldots, \beta_n). \tag{g}$$

[Recall equivalent condition (j4) of ex. 3.12.4.]

From (f), we are readily led to the quadratures (to within inessential additive constants)

$$A_k = \int \big[2v_k(\beta_k + E\,u_k - w_k)\big]^{1/2}\,dq_k \quad \left[= \int p_k\,dq_k = A_k(q_k; \beta, E)\right]. \tag{h}$$

Due to the constraint (g), only n of the constants $E, \beta_1, \beta_2, \ldots, \beta_n$ appearing in the integrals (h) are independent; and therefore the reduced action built from the latter via (d), as containing n independent essential constants, will indeed be a complete integral; that is,

$$A_0 = \sum A_k(q, E, \beta_1, \beta_2, \ldots, \beta_n) = \sum A_k[q, E, \beta_1(\beta_2, \ldots, \beta_n), \beta_2, \ldots, \beta_n]$$
$$\equiv A_{oo}(q, E, \beta_2, \ldots, \beta_n) = A_{oo}. \qquad \text{(i)}$$

In view of these results, the finite equations of motion reduce to the following quadratures:

(i) $\partial A_{oo}/\partial E = \partial A_o/\partial E = t - t_o$:

$$\sum \int (v_k)^{1/2} [2(\beta_k + E\, u_k - w_k)]^{-1/2} u_k\, dq_k = t - t_o, \qquad \text{(j)}$$

(ii) $\partial A_{oo}/\partial \beta_r = \partial A_o/\partial \beta_r + (\partial A_o/\partial \beta_1)(\partial \beta_1/\partial \beta_r)$

$$= \partial A_o/\partial \beta_r + (\partial A_o/\partial \beta_1)(-1) = \partial A_o/\partial \beta_r - \partial A_o/\partial \beta_1 = \alpha_r \ (r = 2, \ldots, 2n):$$

$$\int (v_r)^{1/2}[2(\beta_r + E\, u_r - w_r)]^{-1/2}\, dq_r - \int (v_1)^{1/2}[2(\beta_1 + E\, u_1 - w_1)]^{-1/2}\, dq_1 = \alpha_r. \qquad \text{(k)}$$

[Compare with their equivalent equations (k1, 2) of ex. 3.12.4.] Equation (j) and the $n - 1$ equations (k) supply the n independent constants $\alpha_1 = -t_o, \ldots, \alpha_n$, which, along with the earlier $n - 1$ independent β's and the energy (i.e., $E, \beta_2, \ldots, \beta_n$), constitute the $2n$ constants of integration of our system (a, b).

For additional details on these systems [including generalizations originally studied by Goursat, Di Pirro, Stäckel et al. (late 19th century)] see, for example (alphabetically): Appell (1953, vol. 2, pp. 439–440), Hamel (1949, pp. 302–303, 358–361), Lur'e (1968, pp. 538–548), Whittaker (1937, pp. 335–336); and, especially, texts on celestial mechanics, for example, Hagihara (1970).

Example 8.10.9 *Hamiltonian Form of Lagrangean Method of Variation of Constants/Parameters.* Let us consider a system with canonical equations of motion

$$dq_k/dt = \partial H/\partial p_k, \qquad dp_k/dt = -\partial H/\partial q_k \qquad (k = 1, \ldots, n) \qquad \text{(a)}$$

and such that

$$H = H_o + H_1, \qquad H_1\text{: } \textit{small} \text{ relative to } H_o. \qquad \text{(b)}$$

For example, in a planetary motion problem, H_o would be the Sun–Earth (two-body problem) Hamiltonian, while H_1 would be the Hamiltonian of the perturbative action of the remaining planets of our solar system on Earth [recall discussion following eq. (8.7.17)]. Let us assume that the solution of the *unperturbed* problem

$$dq_k/dt = \partial H_o/\partial p_k, \qquad dp_k/dt = -\partial H_o/\partial q_k \qquad \text{(c)}$$

is known—that is, an action function

$$A = A(t, q, \beta), \qquad \beta \equiv (\beta_1, \ldots, \beta_n), \qquad \text{(d)}$$

has been (or can be) found that satisfies the unperturbed Hamilton–Jacobi (HJ) equation

$$\partial A/\partial t + H_o(t, q, \partial A/\partial t) = 0, \tag{e}$$

and can, therefore, supply the integrals of (c) via the finite equations

$$p_k = \partial A/\partial q_k, \qquad \alpha_k = \partial A/\partial \beta_k; \tag{f}$$

that is, the solutions of (f):

$$q_k = q_k(t, \alpha, \beta), \qquad p_k = p_k(t, \alpha, \beta), \qquad \alpha \equiv (\alpha_1, \ldots, \alpha_n) \tag{g}$$

when substituted back into (c), satisfy them identically.

Here, however, we will view (g) not as solutions of (c) but as *equations of a canonical transformation from the old variables* (q, p) *to the new "variables"* $(\alpha, \beta) \equiv (q', p')$, with generating function $A = A(t, q, \beta) \equiv A(t, q, \beta = p') = F_2(t, q, p')$ (§8.8). In this interpretation, the (no longer constant) $\alpha = q'$ and $\beta = p'$, satisfy the *perturbation* canonical equations

$$\begin{aligned} d\alpha_k/dt &= \partial H'/\partial \beta_k = \partial/\partial \beta_k(H + \partial A/\partial t) \\ &= \partial/\partial \beta_k(H_o + H_1 + \partial A/\partial t) = \partial H_1/\partial \beta_k \qquad \text{[invoking (e)]}, \qquad \text{(h)} \\ d\beta_k/dt &= -\partial H'/\partial \alpha_k = \cdots = -\partial H_1/\partial \alpha_k; \qquad \text{(i)} \end{aligned}$$

where we have used the unperturbed solutions (g) in H_1 to express it in terms of the α and β, and t. Substituting the solutions of (h, i) in the unperturbed solutions for q, p, we obtain the solutions of the perturbed problem.

We notice that equations (h, i) coincide with the earlier (j, l) of ex. 8.7.4, with the following identifications: $c_k \to \alpha_k$, $c_{n+l} \to \beta_l$, $\Omega \to H_1$.

An Application

Let the unperturbed Hamiltonian be

$$H_o = (1/2)p^2; \qquad \text{e.g., free rectilinear motion of particle of unit mass.} \tag{j}$$

The corresponding HJ equations is

$$(1/2)\,(\partial A_o/\partial q)^2 = \alpha \quad (= \text{total energy}), \tag{k}$$

and its solution is

$$A_o = (2\alpha)^{1/2} q \qquad [A = A_o - \alpha t = F_1(t, q, q')]. \tag{l}$$

Hence, the complete unperturbed solution is

$$p = \partial F_1/\partial q = \partial A_o/\partial q = (2\alpha)^{1/2}, \tag{m}$$

$$\beta = -\partial A/\partial \alpha = -\partial A_o/\partial \alpha + t \ \Rightarrow\ t - \beta = \partial A_o/\partial \alpha = (2\alpha)^{-1/2} q$$

$$\Rightarrow\ q = (2\alpha)^{1/2}(t - \beta); \quad \text{i.e., rectilinear motion with uniform velocity } (2\alpha)^{1/2}. \tag{n}$$

Next, let us add to the system the perturbative (linear elastic) force $-q$, so that its complete perturbed Hamiltonian is

$$H = (1/2)p^2 + (1/2)q^2 \equiv H_o + H_1, \tag{o}$$

and view the perturbed solution as having the same *form* as the unperturbed solution (m, n), but with α and β as variables satisfying the perturbation equations (h, i), with $k = 1$ and

$$H_1 = (1/2)q^2|_{\text{unperturbed solution }(n)} = \alpha(t - \beta)^2 = H_1(t, \alpha, \beta), \tag{p}$$

$$d\alpha/dt = \partial H_1/\partial\beta = -2\alpha(t - \beta), \qquad d\beta/dt = -\partial H_1/\partial\alpha = -(t - \beta)^2. \tag{q}$$

Integrating (p, q) we readily find

$$\alpha = \alpha_o \cos^2(t - \beta_o), \quad \beta = t - \tan(t - \beta_o); \qquad \alpha_o, \beta_o\text{: integration constants,} \tag{r}$$

and hence the perturbed solution is

$$q = (2\alpha_o)^{1/2} \sin(t - \beta_o), \qquad p = (2\alpha_o)^{1/2} \cos(t - \beta_o); \tag{s}$$

that is, a simple harmonic motion of amplitude: $(2\alpha_o)^{1/2}$, frequency: 1 (period: 2π), and phase (or "epoch of origin passage"): β_o. It is not hard to verify that (s) does indeed satisfy the canonical perturbed [(o)-based] equations:

$$dq/dt = \partial H/\partial p = p, \qquad dp/dt = -\partial H/\partial q = -q. \tag{t}$$

Example 8.10.10 *A Simplification of the Perturbation Equations.* Continuing from the preceding example, *let* α_{ko}, β_{ko} *denote the parts of* α_k, β_k *that are constant in the perturbed motion.* Then we can write

$$\alpha_k = \alpha_{ko} + x_k(t), \qquad \beta_k = \beta_{ko} + y_k(t), \tag{a}$$

and, therefore, to the *same degree* of accuracy [since the Hamiltonian equations, unlike the Lagrangean equations (§3.10), are of the *first order*], and with some easily understood notations,

$$\begin{aligned} H_1 &= H_1(\alpha, \beta) = H_1(\alpha_o + x, \beta_o + y) \\ &= H_1(\alpha_o, \beta_o) + \sum \left[(\partial H_1/\partial\alpha_k)_o x_k + (\partial H_1/\partial\beta_k)_o y_k\right] \\ &\equiv H_{1o} + \sum \left[(\partial H_{1o}/\partial\alpha_{ko}) x_k + (\partial H_{1o}/\partial\beta_{ko}) y_k\right]. \end{aligned} \tag{b}$$

Substituting the above into eqs. (h, i) of the preceding example, we obtain

$$dx_k/dt = \partial H_1/\partial\beta_{ko}, \qquad dy_k/dt = -\partial H_1/\partial\alpha_{ko}; \tag{c}$$

or, *again, to the first order*,

$$dx_k/dt = \partial H_{1o}/\partial\beta_{ko}, \qquad dy_k/dt = -\partial H_{1o}/\partial\alpha_{ko}. \tag{d}$$

An Application

Let us apply these results to the (weakly) *quadratically* nonlinear oscillator:

$$\ddot{q} + \omega_o{}^2 q + \varepsilon q^2 = 0 \qquad [\omega_o = \text{frequency for } \varepsilon = 0, \text{a constant}]. \tag{e}$$

Here, clearly,

$$H_o = (1/2)\,(p^2 + \omega_o{}^2 q^2), \qquad H_1 = (1/3)\varepsilon q^3, \tag{f}$$

and since the solution of the *unperturbed* problem (i.e., $\varepsilon = 0,\ H_1 = 0$) is

$$q = q_o \cos(\omega_o t) + (p_o/\omega_o)\sin(\omega_o t), \qquad p = \dot{q} = \ldots, \tag{g}$$

where q_o/p_o = initial *position/momentum* of *unperturbed* problem (i.e., $\alpha_o = q_o$, $\beta_o = p_o$), we will have

$$H_1 \to H_{1o} = (\varepsilon/3)[q_o \cos(\omega_o t) + (p_o/\omega_o)\sin(\omega_o t)]^3 \equiv (\varepsilon/3)[\ldots]^3. \tag{h}$$

As a result, the perturbation equations (d) yield

$$dx/dt = \partial H_{1o}/\partial p_o = \varepsilon[\ldots]^2\,[\sin(\omega_o t)/\omega_o], \tag{i}$$

$$dy/dt = -\partial H_{1o}/\partial q_o = -\varepsilon[\ldots]^2 \cos(\omega_o t). \tag{j}$$

Integrating (i, j), and then adding the results to (g), we obtain the solution of (e), correct to ε-proportional terms. The details are left to the reader (see, e.g., Kilmister, 1967, p. 118).

Example 8.10.11 *Hamiltonian Form of Variation of Constants/Parameters (continued): Combination with Method of Averaging.* Continuing from the last two examples, if the solution of the unperturbed equations is *τ-periodic*, then we still solve the approximate perturbation equations (h, i) of ex. 8.10.9, but *with* $H' = H_1$ *replaced with its average over* τ, that is, by

$$\langle H_1 \rangle \equiv (1/\tau)\int_0^\tau H_1(\alpha, \beta, t)\,dt, \tag{a}$$

and α, β treated as constants. Let us apply this "*averaged method of variation of parameters*" to the well-known Duffing's oscillator:

$$\ddot{q} + \omega_o{}^2 q + \varepsilon q^3 = 0 \qquad [\omega_o = \text{frequency for } \varepsilon = 0, \text{a constant}]. \tag{b}$$

Its Hamiltonian is easily found to be

$$H = H_o + H_1,$$

$$H_o = (1/2)(p^2 + \omega_o{}^2 q^2), \qquad H_1 = (1/4)\varepsilon\, q^4 = \textit{small perturbation}. \tag{c}$$

The corresponding reduced and unperturbed HJ equation ($\varepsilon = 0$) is

$$(dA_o/dq)^2 + \omega_o{}^2 q^2 = 2\beta \qquad [\text{where } A = A_o(q) - \beta\, t], \tag{d}$$

and so its solution is

$$A_o = \int (2\beta - \omega_o{}^2 q^2)^{1/2}\,dq \Rightarrow A = \int (2\beta - \omega_o{}^2 q^2)^{1/2} - \beta\, t. \tag{e}$$

As a result, the equation of finite unperturbed motion becomes

$$\alpha = \partial A/\partial\beta = \int (2\beta - \omega_o{}^2 q^2)^{-1/2}\, dq - t = (1/\omega_o)\arcsin\left[\omega_o q/(2\beta)^{1/2}\right] - t,$$

$$\Rightarrow\; q = [(2\beta)^{1/2}/\omega_o]\sin[\omega_o(t+\alpha)]\text{: unperturbed solution; } \tau = 2\pi/\omega_o. \tag{f}$$

Therefore, the perturbation Hamiltonian equals

$$\begin{aligned} H_1 &= (1/4)\varepsilon q^4 = (\varepsilon\beta^2/\omega_o{}^4)\sin^4[\omega_o(t+\alpha)] = H_1(t, \alpha, \beta) \\ &= \cdots - (\varepsilon\beta^2/\omega_o{}^4)\{(3/8) - (1/2)\cos[2\omega_o(t+\alpha)] + (1/8)\cos[4\omega_o(t+\alpha)]\}, \end{aligned}$$

and so its average over τ is

$$\langle H_1\rangle \equiv (\omega_o/2\pi)\int_0^{2\pi/\omega_o} H_1\, dt = \cdots = 3\varepsilon\beta^2/8\omega_o{}^4. \tag{g}$$

Hence, the averaged perturbation equations give

$$d\beta/dt = -\partial\langle H_1\rangle/\partial\alpha = 0 \;\Rightarrow\; \beta = \mathit{constant}, \tag{h}$$

$$d\alpha/dt = \partial\langle H_1\rangle/\partial\beta = 3\varepsilon\beta/4\omega_o{}^4 \;\Rightarrow\; \alpha = (3\varepsilon\beta/4\omega_o{}^4)t + \alpha_o, \tag{i}$$

where α_o is the integration constant. Finally, substituting (h, i) back into (f), we obtain the first ε-order correction:

$$q = [(2\beta)^{1/2}/\omega_o]\sin\{\omega_o[1 + (3\varepsilon\beta/4\omega_o{}^4)]t + \omega_o\alpha_o\}, \tag{j}$$

and the constants β, α_o are to be determined from the initial conditions. This agrees with the expressions obtained by other asymptotic methods (e.g., Krylov, Bogoliubov, Mitropolskii; see also chap. 7). For additional problems, see, for example, Nayfeh (1973, pp. 183–189).

Problem 8.10.1 Show that the HJ equation of a particle of mass m in a potential field $V = V(\mathit{particle\ position, time})$ in the following common systems of coordinates (with standard notations) is:

(i) *rectangular Cartesian*:

$$\partial A/\partial t + (1/2m)\left[(\partial A/\partial x)^2 + (\partial A/\partial y)^2 + (\partial A/\partial z)^2\right] + V(x, y, z;\, t) = 0, \tag{a}$$

(ii) *polar cylindrical*:

$$\partial A/\partial t + (1/2m)\left[(\partial A/\partial r)^2 + r^{-2}(\partial A/\partial\phi)^2 + (\partial A/\partial z)^2\right] + V(r, \phi, z;\, t) = 0, \tag{b}$$

(iii) *spherical coordinates*:

$$\partial A/\partial t + (1/2m)\left[(\partial A/\partial r)^2 + r^{-2}(\partial A/\partial\theta)^2 + (r\sin\theta)^{-2}(\partial A/\partial\phi)^2\right] + V(r, \phi, \theta;\, t) = 0. \tag{c}$$

Recall that r has different meanings in (b) and (c).

Problem 8.10.2 Consider the following action function, $A = A(t, q, q')$, that satisfies the HJ equation

$$H(t, q, \partial A/\partial q) + \partial A/\partial t = 0. \qquad \text{(a)}$$

Show that any solution of (a) of the form

$$A = A(t, q, \alpha), \qquad \alpha \equiv (\alpha_1, \ldots, \alpha_n): n \text{ arbitrary and independent constants,} \quad \text{(b)}$$

solves the dynamical problem through the $2n$ finite equations

$$p_k = \partial A/\partial q_k, \qquad -\beta_k = \partial A/\partial \alpha_k, \qquad \text{(c)}$$

where

$q' \to \alpha$ and $p' \to \beta$ [$\equiv (\beta_1, \ldots, \beta_n): n$ new arbitrary and independent constants]

are, respectively, the new *constant* canonical coordinates and momenta.

HINT

Recall (8.8.15), with $A(t, q, \alpha) \to F_1(t, q, q')$.

Problem 8.10.3 Continuing from the preceding problem, show that if $\partial H/\partial t = 0$, then the HJ equation assumes the special form

$$H(q, \partial A_o/\partial q) = E, \qquad \text{(a)}$$

where $A = A_o(q, \alpha) - E\,t$, and the solution of the dynamical problem is given by the following $2n$ finite equations:

$$\partial A_o/\partial q_k = p_k; \qquad \text{(b)}$$

$$\partial A_o/\partial E = t - \beta_1, \qquad \text{(c)}$$

$$\partial A_o/\partial \alpha_r = -\beta_r \qquad (r = 2, \ldots, n), \qquad \text{(d)}$$

where $\alpha_1 = E$ and $(\alpha_2, \ldots, \alpha_n)$: $n - 1$ constants of integration of (a).

HINT

$$\beta_1 = -\partial A/\partial \alpha_1 = -\left[\partial A_o/\partial \alpha_1 - (\partial E/\partial \alpha_1)t\right] = \cdots.$$

Problem 8.10.4 *Hamiltonian Form of Variation of Constants/Parameters: Nonpotential Forces.*

(i) By applying (8.7.22 ff.) with

$$X_k \to X_k^{(1)} = -\partial H_1/\partial q_k + Q_k, \qquad (k = 1, \ldots, n) \qquad \text{(a)}$$

$$Q_k = \text{small } \textit{nonpotential} \text{ perturbative forces,} \qquad \text{(b)}$$

$$c_k \to \alpha_k, \quad c_{n+l} \to \beta_l \qquad (k, l = 1, \ldots, n), \qquad \text{(c)}$$

and (8.9.10 ff.), show that, in the presence of forces—that is, for perturbed Hamiltonian equations: $dq_k/dt = \partial H/\partial p_k$, $dp_k/dt = -\partial H/\partial q_k + Q_k$, $H = H_o + H_1$—the fundamental perturbation equations (ex. 8.10.9: h, i) generalize to

$$d\alpha_k/dt = \partial H_1/\partial\beta_k + \sum(\partial\alpha_k/\partial p_l)Q_l = \partial H_1/\partial\beta_k - \sum(\partial q_l/\partial\beta_k)Q_l, \tag{d}$$

$$d\beta_k/dt = -\partial H_1/\partial\alpha_k + \sum(\partial\beta_k/\partial p_l)Q_l = -\partial H_1/\partial\alpha_k + \sum(\partial q_l/\partial\alpha_k)Q_l. \tag{e}$$

The advantage of the *second* forms of (d, e) over their corresponding *first* forms lies in that *the former do not require inversion of the general solution of the unperturbed problem:* $q = q(t, \alpha, \beta)$, $p = p(t, \alpha, \beta)$.

(ii) Verify that if $Q_k = Q_k(t)$—for example, if the perturbative forces are constant—then (d, e) can be rewritten, respectively, in the canonical form:

$$d\alpha_k/dt = \partial\eta_1/\partial\beta_k, \qquad d\beta_k/dt = -\partial\eta_1/\partial\alpha_k, \tag{f}$$

where

$$\eta_1 \equiv H_1 - \sum Q_k q_k = \textit{generalized Hamiltonian perturbation}. \tag{g}$$

[Under such forces, the perturbed Hamiltonian equations can, similarly, be written as

$$dq_k/dt = \partial\eta/\partial p_k, \qquad dp_k/dt = -\partial\eta/\partial q_k, \tag{h}$$

where

$$\eta \equiv H - \sum Q_k q_k = \textit{generalized Hamiltonian}.] \tag{i}$$

HINTS

To prove the *second* forms of (d, e), we proceed as follows: from the table of the various types of generating functions of §8.8 (table 8.1) by equating the second mixed partial $F_{1,2,3,4}$-derivatives, we readily find the following equalities:

$$\partial^2 F_1/\partial q_k \partial q_{k'} = \partial p_k/\partial q_{k'} = -\partial p_{k'}/\partial q_k, \tag{j}$$

$$\partial^2 F_2/\partial q_k \partial p_{k'} = \partial p_k/\partial p_{k'} = \partial q_{k'}/\partial q_k, \tag{k}$$

$$\partial^2 F_3/\partial p_k \partial q_{k'} = -\partial q_k/\partial q_{k'} = -\partial p_{k'}/\partial p_k, \tag{l}$$

$$\partial^2 F_4/\partial p_k \partial p_{k'} = -\partial q_k/\partial p_{k'} = \partial q_{k'}/\partial p_k; \tag{m}$$

and then, as described in ex. 8.10.9, we view [in (l, m)] the q', p' as α, β.

For an alternative derivation of (d, e) along with several advanced and detailed applications of them, see Lur'e (1968, pp. 560–562, ff.].

8.11 HAMILTON'S PRINCIPAL AND CHARACTERISTIC FUNCTIONS, AND ASSOCIATED VARIATIONAL PRINCIPLES/ DIFFERENTIAL EQUATIONS

In this section, we examine the connection between the Hamilton–Jacobi (HJ) equation and Hamilton's integral variational principle (chap. 7). Let us assume that, somehow, we have obtained the general solution of the canonical equations

$$dq_k/dt = \partial H/\partial p_k, \qquad dp_k/dt = -\partial H/\partial q_k \qquad (k = 1, \ldots, n); \tag{8.11.1}$$

that is, we have found the expressions

$$q_k = q_k(t; c), \qquad p_k = p_k(t; c), \qquad c \equiv (c_1, \ldots, c_{2n}) = \text{constants of integration.} \tag{8.11.1a}$$

Substituting (8.11.1a) in the system Hamiltonian $H = H(t, q, p)$, we obtain

$$H = H(t, q, p) = H[t, q(t;\ c), p(t;\ c)] \equiv H(t, c), \tag{8.11.2}$$

and, therefore, applying chain rule to the above, we find, successively (with $\mu = 1, \ldots, 2n$),

$$\begin{aligned}\partial H/\partial c_\mu &= \sum [(\partial H/\partial q_k)(\partial q_k/\partial c_\mu) + (\partial H/\partial p_k)(\partial p_k/\partial c_\mu)] \\ &= \sum [(\partial p_k/\partial c_\mu)(dq_k/dt) - (\partial q_k/\partial c_\mu)(dp_k/dt)] \qquad \text{[by (8.11.1)]} \\ &= \partial/\partial c_\mu\Big(\sum p_k \dot{q}_k\Big) - d/dt\Big(\sum p_k(\partial q_k/\partial c_\mu)\Big) \qquad (8.11.2\text{a})\end{aligned}$$

[due to the identity: $\partial \dot{q}_k/\partial c_\mu = \partial/\partial c_\mu(\partial q_k/\partial t) = \partial/\partial t(\partial q_k/\partial c_\mu) = (\partial q_k/\partial c_\mu)^{\cdot}$];

or, since

$$\partial/\partial c_\mu\Big(\sum p_k \dot{q}_k\Big) - \partial H/\partial c_\mu = \partial L/\partial c_\mu = \partial(T - V)/\partial c_\mu, \tag{8.11.3}$$

we finally obtain the following [special form of the *central equation* (§3.6)]:

$$\partial L/\partial c_\mu = d/dt\Big(\sum p_k(\partial q_k/\partial c_\mu)\Big). \tag{8.11.4}$$

Integrating the above, from an initial instant t_o to a current one t [and noting that $\partial(\ldots)/\partial c_k$ and $\int(\ldots)\,dt$ commute], we get

$$\partial/\partial c_\mu \int_{t_o}^{t} L\,dt \ = \sum [p_k(\partial q_k/\partial c_\mu) - p_{ko}(\partial q_{ko}/\partial c_\mu], \tag{8.11.5}$$

where $q_{ko}(p_{ko})$ are the values of $q_k(p_k)$ at $t = t_o$. The function (§7.9)

$$A_H \equiv \int_{t_o}^{t} L\,dt = \int_{t_o}^{t} (T - V)\,dt \equiv A \tag{8.11.6}$$

is called, after Hamilton (1834–1835), the *principal function* of the motion of the system, because, in his words, "The variation of this definite integral S [our A] has therefore the double property, of giving the differential equations of motion for any transformed coordinates when the extreme positions are regarded as fixed, and of giving the integrals of those differential equations when the extreme positions are treated as varying." (Quoted in MacMillan, 1936, p. 367.) To understand these statements better, we need to examine A more closely. In view of (8.11.1a) and (8.11.6), we will have

$$q_k = q_k(t;\, c) \;\Rightarrow\; q_{ko} = q_k(t_o;\, c) \tag{8.11.7a}$$

$$\Rightarrow\; A = A(t, t_o;\, c), \tag{8.11.7b}$$

and, therefore, arbitrary variations of the $2n$ integration constants, $c \to c + \delta c$, cause the following (first-order in the δc, and fixed-time) variations in the q's and A:

$$\delta q_k = \sum (\partial q_k/\partial c_\mu)\,\delta c_\mu, \qquad \delta A = \sum (\partial A/\partial c_\mu)\,\delta c_\mu. \tag{8.11.8}$$

Hence, multiplying (8.11.5) with δc_μ and summing over $\mu = 1, \ldots, 2n$, we arrive at the fundamental variational equation

$$\delta A = \sum (p_k\,\delta q_k - p_{ko}\,\delta q_{ko}). \tag{8.11.9}$$

To obtain differential equations from the above, we, first, solve the $2n$ equations (8.11.7a) for the $2n$ constants c in terms of the $2n$ q's and q_o's (and time): $c_\mu = c_\mu(t;\, q, q_o)$, and then substitute this result in (8.11.7b)

$$A = A(t, t_o;\, q, q_o). \tag{8.11.10}$$

This expresses the action *functional* along the actual path (or orbit) as a *function* of the coordinates of its lower (initial) and upper (final) limits of integration. More important: (i) from the *mathematical* point of view, the transition from (8.11.7b) to (8.11.10) is one from an *initial*-value problem [c given, or equivalently, due to (8.11.1a), q_o and p_o given] to a *boundary*-value problem (q_o and q given); while (ii) *physically*, it signifies a transition from motion determined by the initial positions and velocities (or momenta) to motion determined by the initial and final positions (provided, of course, that the latter are achievable from the former).

Varying (8.11.10), for fixed t, we obtain

$$\delta A = \sum \left[(\partial A/\partial q_k)\,\delta q_k + (\partial A/\partial q_{ko})\,\delta q_{ko}\right], \tag{8.11.11}$$

and, therefore, comparing this with (8.11.9), *while recalling that the δq and δq_o are arbitrary,* we obtain the equations

$$\partial A/\partial q_k = p_k, \qquad \partial A/\partial q_{ko} = -p_{ko}. \tag{8.11.12}$$

Solving the *second* group of (8.11.12) for the q's in terms of t and q_o's, p_o's, and then substituting these expressions into the *first* group, we obtain the p's as functions of t, q_o's, p_o's:

$$q_k = q_k(t;\, q_o, p_o), \qquad p_k = p_k(t;\, q_o, p_o); \tag{8.11.13}$$

that is, *eqs. (8.11.12) constitute a complete set of integrals of the equations of motion (8.11.1)*: knowledge of A provides a complete solution to the problem. Indeed, A satisfies the Hamilton–Jacobi equation (HJ, §8.10), and, hence, can be identified with the there-introduced action function A. Here is why: $(\ldots)^{\cdot}$-differentiating (8.11.10), we find, successively,

$$dA/dt = \partial A/\partial t + \sum (\partial A/\partial q_k)(dq_k/dt) = \partial A/\partial t + \sum p_k \dot{q}_k$$

[by the first of (8.11.12)],

and from this, since $dA/dt = L\ (= T - V)$ [by (8.11.6)] and $\sum p_k \dot{q}_k = L + H$ [by the Hamiltonian definition], we readily conclude that

$$\partial A/\partial t + H(t;\ q, p) = \partial A/\partial t + H(t;\ q, \partial A/\partial q) = 0, \quad \text{Q.E.D.} \qquad (8.11.14)$$

Hamilton's Principle

Substituting $L = \sum p_k \dot{q}_k - H$ in (8.11.6), and then taking its (first and fixed-end-point) variation δA, we find, successively,

$$\begin{aligned}
\delta A &= \delta \int L\,dt = \delta \int \Big(\sum p_k \dot{q}_k - H\Big)\,dt \\
&= \int \Big\{\sum \left[p_k\, \delta(\dot{q}_k) + \dot{q}_k\, \delta p_k\right] - \delta H\Big\}\,dt \\
&= \int \Big\{\Big[\Big(\sum p_k\, \delta q_k\Big)^{\cdot} - \sum \dot{p}_k\, \delta q_k + \sum \dot{q}_k\, \delta p_k\Big] \\
&\qquad - \sum \left[(\partial H/\partial q_k)\, \delta q_k + (\partial H/\partial p_k)\, \delta p_k\right]\Big\}\,dt
\end{aligned}$$

[assuming that $\delta(dq) = d(\delta q)$, or $\delta(\dot{q}) = (\delta q)^{\cdot}$],

or, after integrating out the $(\ldots)^{\cdot}$ term:

$$\begin{aligned}
\delta A = \int \sum &\left[(dq_k/dt - \partial H/\partial p_k)\, \delta p_k - \sum (dp_k/dt + \partial H/\partial q_k)\, \delta q_k\right] \\
&+ \sum (p_k\, \delta q_k - p_{ko}\, \delta q_{ko}). \qquad (8.11.15)
\end{aligned}$$

Now, if

$$dq_k/dt = \partial H/\partial p_k, \qquad dp_k/dt = -\partial H/\partial q_k, \qquad (8.11.16)$$

and $\delta q_k = \delta q_{ko} = 0$ (or some other combination that nullifies the last ("boundary") terms, then $\delta A = 0$; and, conversely, if $\delta A = 0$, for all variations of the q's and p's that vanish at the temporal endpoints, then (8.11.16) follow. This is Hamilton's principle in canonical variables (and for contemporaneous variations).

In sum:

(i) Given the function $H = H(t, q, p)$ and the differential equations $dq_k/dt = \partial H/\partial p_k$, $dp_k/dt = -\partial H/\partial q_k$, the principal function $A \equiv \int L\,dt = \int (\sum p_k \dot{q}_k - H)\,dt = \int (\sum p_k\, dq_k - H\,dt)$ satisfies the partial differential equation $\partial A/\partial t + H(t;\ q, \partial A/\partial q) = 0$.

(ii) *Hamilton*: if $A = A(t, t_o;\ q, q_o)$, then a complete integral of the equations of motion can be obtained from $p_k = \partial A/\partial q_k$, $p_{ko} = \partial A/\partial q_{ko}$.

(iii) *Jacobi*: If $A = A(t, q, \beta)$ is a complete solution of $\partial A/\partial t + H(t, q, \partial A/\partial q) = 0$, then a complete integral of the equations of motion can be obtained from $p_k = \partial A/\partial q_k$, $\alpha_k = \partial A/\partial \beta_k$.

Action as a Function of the Coordinates and Time

Comparing the earlier $dA/dt = L$ with $dA/dt = \partial A/\partial t + \sum (\partial A/\partial q_k)(dq_k/dt) = \partial A/\partial t + \sum p_k \dot{q}_k$, we immediately obtain

$$\partial A/\partial t = L - \sum p_k \dot{q}_k = -H; \tag{8.11.17}$$

which also follows from the HJ equation (8.11.14). Hence, *if A is considered as a function of the current coordinates and upper (current) time limit in (8.11.6)*, its total differential equals

$$dA = (\partial A/\partial t)\, dt + \sum (\partial A/\partial q_k)\, dq_k = \sum p_k\, dq_k - H\, dt. \tag{8.11.18}$$

Generally, if A is considered as a function of both initial and final coordinates and time, then

$$dA = \left(\sum p_k\, dq_k - H\, dt\right) - \left(\sum p_{ko}\, dq_{ko} - H_o\, dt_o\right); \tag{8.11.19}$$

which can be rewritten in terms of variational calculus notation as

$$\begin{aligned} \Delta A &= \left(\sum p_k \Delta q_k - H \Delta t\right) - \left(\sum p_{ko}\, \Delta q_{ko} - H_o\, \Delta t_o\right) \\ &= \sum p_k \Delta q_k - \sum p_{ko} \Delta q_{ko} - H \Delta (t - t_o) \qquad [\text{if } H = constant] \end{aligned} \tag{8.11.20}$$

[where $\Delta(\ldots) = \delta(\ldots) + (\ldots)^{\cdot} \Delta t =$ *noncontemporaneous variation* (§7.2, §7.9)], from which (8.11.12) and (8.11.17) follow. If $(\partial L/\partial t = 0 \Rightarrow)$ $H = constant$, then the latter can be rewritten as

$$\partial A/\partial \tau = -H, \qquad \tau \equiv t - t_o = \textit{time of transit}; \tag{8.11.21}$$

whereas if $H \neq constant$, then we have $\partial A/\partial t_o = H_o$, $\partial A/\partial t = -H$. Equation (8.11.20) is referred to as *Hamilton's principle of varying, or varied, action.*

The above show that the final (or current) state of the system cannot be an arbitrary function of its initial state; the right side of (8.11.19) must be an exact differential, no matter what the impressed (potential) forces are. These results are of interest in *geometrical optics*.

Specializations

(i) If $\partial L/\partial t = -\partial H/\partial t = 0$, then $H(q, p) = constant \equiv E$ (*total energy*). Then we can write

$$A = \int \left(\sum p_k\, dq_k - H\, dt\right) = A_o - E(t - t_o), \tag{8.11.22}$$

where

$$A_o \equiv \int \sum p_k\, dq_k = \int \left(\sum p_k \dot{q}_k\right) dt = \int 2T\, dt$$

= *Abbreviated action* (or *characteristic function*—A_o of §8.10; and the *Lagrangean action* A_L *of* §7.9; also, frequently denoted by W). (8.11.23)

In this case, and for variations satisfying $(\Delta t_o \rightarrow)\, dt_o = 0, (\Delta q_{ko} \rightarrow)\, dq_{ko} = 0$, and $(\Delta q_k \rightarrow)\, dq_k = 0$ but $(\Delta t \rightarrow)\, dt \neq 0$ (i.e., given initial coordinates and time, and given final coordinates but not time) eqs. (8.11.19), (8.11.20) reduce, respectively, to

$$dA = -H\, dt = -E\, dt, \qquad \Delta A = -E\Delta t; \tag{8.11.24}$$

and, therefore, comparing with (8.11.22), we conclude that, for such *isoenergetic* variations,

$$\Delta A_o = 0. \tag{8.11.25}$$

This is the "principle" of *Maupertuis → Euler → Lagrange* (MEL; recalling discussion in §7.9): The abbreviated, or Lagrangean, action has a stationary value for the actual path of the system, among all comparison paths that satisfy conservation of energy (and all have the same energy constant as the actual path), pass from the given initial configuration at a given time, and from the given final configuration at an unspecified time.

Generally, and in variational calculus notation,

$$\Delta A_o = \left(\sum p_k \Delta q_k - \sum p_{ko} \Delta q_{ko}\right) + \Delta H(t - t_o), \tag{8.11.26}$$

from which, if we regard A_o as a *function of the initial and final coordinates and the energy*—that is, $A_o = A_o(q_o, q, E)$—we obtain the differential relations

$$p_k = \partial A_o/\partial q_k, \qquad p_{ko} = -\partial A_o/\partial q_{ko}, \qquad \tau \equiv t - t_o = \partial A_o/\partial E. \tag{8.11.27}$$

Here, too, knowledge of A_o determines the motion: the $n+1$ equations, second and third of (8.11.27), plus q_o, p_o determine the energy H and the q's; then the first of (8.11.27) gives the p's.

If the system Lagrangean has the common form

$$L = (1/2) \sum \sum M_{kl}(q) \dot{q}_k \dot{q}_l - V(q), \tag{8.11.28}$$

then, since $p_k = \partial L/\partial \dot{q}_k = \sum M_{kl}(q)\dot{q}_l$, and by energy conservation:

$$E = (1/2) \sum \sum M_{kl} \dot{q}_k \dot{q}_l + V \;\Rightarrow\; (dt)^2 = \left(\sum \sum M_{kl}\, dq_k\, dq_l\right) \Big/ 2(E - V), \tag{8.11.29}$$

the corresponding A_o is expressed in terms of the coordinates q and their differentials dq, and with the energy as a constant parameter, in the *generalized*

Jacobi form (§7.9):

$$A_o \equiv \int \sum p_k\, dq_k = \int \sum\sum \left[M_{kl}(dq_l/dt)\right] dq_k \qquad [\text{by (8.11.29): } dt = \cdots]$$

$$= \int \left[2(E - V) \sum\sum M_{kl}\, dq_k\, dq_l\right]^{1/2}. \tag{8.11.30}$$

Clearly, since $p_k = \partial L(q, \dot{q})/\partial \dot{q}_k$ and $E = H(q, \dot{q}) = E(q, \dot{q})$, this method can be extended to systems with more general Lagrangeans than (8.11.28).

(ii) If, in (8.11.22), we vary both E and t, we obtain (again in variational notation, and recalling (8.11.24))

$$\Delta A = \Delta A_o - (t - t_o)\, \Delta E - E \Delta t = -E \Delta t, \tag{8.11.31}$$

from which we easily conclude that

$$\partial A_o/\partial E = t - t_o. \tag{8.11.32}$$

(a) If, further, A_o has the form (8.11.30), then (8.11.32) leads to the integral [of (8.11.29)]:

$$t - t_o = \int \left(\sum\sum M_{kl}\, dq_k\, dq_l / 2(E - V)\right)^{1/2}, \tag{8.11.33}$$

which, along with the path equation, determines the motion.

(b) If the system undergoes *periodic* motion with the (single) period $\tau = t - t_o = 2\pi/\omega$, then its frequency ω is found from

$$\omega = 2\pi(\partial A_o/\partial E)^{-1}. \tag{8.11.34}$$

Additional related results are given in the examples and problems of §7.9.

Extension to Cyclic/Ignorable Systems (Larmor, 1884)

In such a case, A and A_o are replaced, respectively, by (recalling eq. (8.3.3c)):

$$A_R \equiv \int (T'' - V)\, dt \equiv \int \left[\left(T - \sum \Psi_i \dot{\psi}_i\right) - V\right] dt \equiv \int \left(L - \sum \Psi_i \dot{\psi}_i\right) dt$$

$$\equiv \int R(t, q, \dot{q}; \Psi \equiv C)\, dt \equiv \int R\, dt:$$

Function of the initial and final values of the nonignorable (or positional) coordinates q and time of transit $\tau \equiv t - t_o$; with the cyclic momenta Ψ as constant parameters, (8.11.35)

$$A_{o,R} \equiv \int \left(2T - \sum \Psi_i \dot{\psi}_i\right) dt:$$

Function of the initial and final values of the nonignorable coordinates q, and the total energy H under constant cyclic momenta. (8.11.36)

All previous variations and differential equations hold for A_R and $A_{o,R}$, *provided only the nonignorable coordinates and velocities are varied.* Indeed:

(i) Varying A_R, we obtain, successively,

$$\Delta A_R = \delta \int R\,dt + (R\,\Delta t - R_o\,\Delta t_o)$$

$$= \cdots = -\int \sum E_p(R)\,\delta q_p\,dt + \sum (p_p\,\delta q_p - p_{po}\,\delta q_{po}) + (R\,\Delta t - R_o\,\Delta t_o)$$

$$= -(0) + \sum \left[p_p(\Delta q_p - \dot{q}_p\,\Delta t) - p_{po}(\Delta q_{po} - \dot{q}_{po}\,\Delta t_o)\right] + (R\,\Delta t - R_o\,\Delta t_o),$$

or, since $R = T'' - V$, $T + V = H = constant$; and

$$2T = \sum p_p \dot{q}_p + \sum \Psi_i \dot{\psi}_i$$

$$\Rightarrow \sum p_p \dot{q}_p = 2T - \sum \Psi_i \dot{\psi}_i = T + \left(T - \sum \Psi_i \dot{\psi}_i\right) = T + T'',$$

finally,

$$\Delta A_R = \sum (p_p\,\Delta q_p - p_{po}\,\Delta q_{po}) - H\,\Delta(t - t_o). \tag{8.11.37}$$

- If $\Delta q_p = 0, \Delta q_{po} = 0, \Delta(t - t_o) \equiv \Delta\tau = 0$, then the above yields $\Delta A_R = 0$; which is the Routhian generalization of Hamilton's principle.
- If both initial and final configurations as well as time of transit are variable, then (8.11.37) leads us to

$$p_p = \partial A_R/\partial q_p, \qquad p_{po} = -\partial A_R/\partial q_{po}, \qquad H = -\partial A_R/\partial \tau \qquad (p = M + 1, \ldots, n), \tag{8.11.38}$$

which are the Routhian versions of (8.11.12) and (8.11.17).

(ii) For $A_{o,R}$, we find, similarly,

$$A_{o,R} = \int \left(\sum p_p\,\dot{q}_p\right) dt = \cdots = \int (T + T'')\,dt$$

$$= A_R + \int (T + V)\,dt = A_R + H\tau; \tag{8.11.39}$$

that is, here, the independent variable is the total energy, not the transit time. Hence, varying this generally, we obtain

$$\Delta A_{o,R} = \Delta A_R + \tau\,\Delta H + H\,\Delta\tau \qquad \text{[invoking (8.11.37)]}$$

$$= \sum (p_p\,\Delta q_p - p_{po}\,\Delta q_{po}) + \tau\,\Delta H, \tag{8.11.40}$$

and from this we get the equations

$$p_p = \partial A_{o,R}/\partial q_p, \qquad p_{po} = -\partial A_{o,R}/\partial q_{po}, \qquad \tau = \partial A_{o,R}/\partial H \qquad (p = M + 1, \ldots, n), \tag{8.11.41}$$

which constitute the Routhian versions of (8.11.27).

Example 8.11.1 *The Characteristic Function, or Abbreviated Action, of a One-DOF System.* Application of the energy equation to such a system yields

$$p = p(q, E) \qquad (\beta = E,\ n = 1), \tag{a}$$

and therefore its characteristic function becomes

$$A_o = \int p(q, E)\, dq, \tag{b}$$

with the integral extending from an initial configuration, q_o, to a current one, q. Hence, by (8.10.111):

$$t - t_o = \partial A_o/\partial E = \partial/\partial E\left[\int p(q, E)\, dq\right] = \int [\partial p(q, E)/\partial E]\, dq. \tag{c}$$

Specialization

For a *single particle* of mass m, the energy equation

$$H = p^2/2m + V(q) = E, \tag{d}$$

yields the (a)-like representation

$$p = \{2m[E - V(q)]\}^{1/2} = p(q;\, E), \tag{e}$$

and hence the (b)-like action

$$A_o = (2m)^{1/2} \int [E - V(q)]^{1/2}\, dq. \tag{f}$$

From (e), we obtain

$$\partial p/\partial E = (m/2)^{1/2}[E - V(q)]^{-1/2}, \tag{g}$$

and, from this, the (c)-like equation of motion

$$t = (m/2)^{1/2} \int [E - V(q)]^{-1/2}\, dq + t_o. \tag{h}$$

In particular, if $V = mgq$ (i.e., vertical motion of particle in constant field of gravity), then (h) gives

$$t = (m/2)^{1/2} \int (E - mgq)^{-1/2}\, dq + t_o = t_o - (1/g)\,[2(E - mgq)/m]^{1/2};$$

and if we choose the q-origin so that $mgq_o = E$, the above reduces to the well-known $q = q_o - (g/2)\,(t - t_o)^2$.

It is not hard to see that the above also apply to *an n-DOF system with only one nonignorable coordinate:* $q_n = q$. Then, since all momenta except $p_n = p$ are constant, the energy equation and characteristic function reduce, respectively, to

$$H(q, p;\, \beta_1, \ldots, \beta_{n-1}) = E, \qquad A_o = \int p(q;\, \beta_1, \ldots, \beta_{n-1}, \beta_n = E)\, dq. \tag{i}$$

Example 8.11.2 *On Hamilton's Principal Function and Associated Differential Equations.*

(i) Let

$$L = T - V = (m/2)\,[(\dot{x})^2 + (\dot{y})^2 + (\dot{z})^2]; \tag{a}$$

that is, free motion of particle P of mass m. Then, as is well known,

$$x = x_o + \dot{x}_o\, t, \qquad y = y_o + \dot{y}_o\, t, \qquad z = z_o + \dot{z}_o\, t \qquad \text{(law of inertia)}, \tag{b}$$

where $x_o, y_o, z_o/\dot{x}_o, \dot{y}_o, \dot{z}_o$ = rectangular Cartesian components of initial position/velocity of P, at time $t_o = 0$. As a result of (b), the corresponding action (principal function) becomes, successively,

$$A = \int L\, dt = \int (m/2)\,[(\dot{x})^2 + (\dot{y})^2 + (\dot{z})^2]\, dt$$

$$= \cdots = (m/2)\,\left[(\dot{x}_o)^2 + (\dot{y}_o)^2 + (\dot{z}_o)^2\right](t - t_o) \equiv (m{v_o}^2/2)t$$

$$\text{[by (b):} \qquad \dot{x}_o = (x - x_o)/t = \dot{x}, \qquad \dot{y}_o = (y - y_o)/t = \dot{y}, \qquad \dot{z}_o = (z - z_o)/t = \dot{z}]$$

$$= (m/2t)\left[(x - x_o)^2 + (y - y_o)^2 + (z - z_o)^2\right]$$

$$= A(t, t_o;\ x, y, z;\ x_o, y_o, z_o), \tag{c}$$

and from this expression we readily find

$$\partial A/\partial x = (m/2t)\,[2(x - x_o)] = m(x - x_o)/t = m\dot{x} \equiv p_x, \qquad \text{etc., cyclically,} \tag{d}$$

$$\partial A/\partial x_o = -(m/2t)\,[2(x - x_o)] = \cdots = -m\dot{x}_o \equiv -p_{xo}, \qquad \text{etc., cyclically,} \tag{e}$$

$$\partial A/\partial t = -(m/2t^2)\,[(x - x_o)^2 + \cdots] = -(m/2)\,\{[(x - x_o)/t]^2 + \cdots\}$$

$$= -(m{v_o}^2/2) = -(\textit{energy}) \equiv -E, \qquad \text{as expected.} \tag{f}$$

(ii) Let the equations of motion of a particle P be

$$\ddot{x} = -\omega^2 x, \qquad \ddot{y} = -\omega^2 y; \tag{g}$$

that is, isotropic harmonic oscillator of (constant) frequency ω; x, y = rectangular Cartesian coordinates. The general solution of (g) is

$$x = a\sin(\omega t + \phi_o), \qquad y = b\sin(\omega t + \psi_o), \tag{h}$$

where $a, b;\ \phi_o, \psi_o$ = four constants of integration. From the above, we readily obtain the system Lagrangean:

$$L = T - V = (m/2)\,[(\dot{x})^2 + (\dot{y})^2] - (m\omega^2/2)\,(x^2 + y^2)$$

$$= \cdots = (m\omega^2/2)\,\left\{a^2\cos[2(\omega t + \phi_o)] + b^2\cos[2(\omega t + \psi_o)]\right\}, \tag{i}$$

and, from the latter, the following system action:

$$A = \int L\,dt = (m\omega/4)\,\{a^2\sin[2(\omega t + \phi_o)] + b^2\sin[2(\omega t + \psi_o)] - a^2\sin(2\phi_o) - b^2\sin(2\psi_o)\}$$

[introducing the initial positions: $x(0) \equiv x_o = a\sin\phi_o,\ y(0) \equiv y_o = b\sin\psi_o$]

$$= (m\omega/2)\,\{[a\,x\cos(\omega t + \phi_o) + b\,y\cos(\omega t + \psi_o)] - a\,x_o\cos\phi_o - b\,y_o\cos\psi_o\}$$

[since by (h), $a\cos\phi_o = [x - x_o\cos(\omega t)]/\sin(\omega t),\ b\cos\psi_o = [y - y_o\cos(\omega t)]/\sin(\omega t)$]

$$= (m\omega/2)\,\{a\,x[\cos(\omega t)\cos\phi_o - \sin(\omega t)\sin\phi_o] + b\,y[\cos(\omega t)\cos\psi_o - \sin(\omega t)\sin\psi_o] - ax_o\cos\phi_o - by_o\cos\psi_o\}$$

$$= [m\omega/2\sin(\omega t)]\,[(x^2 + y^2 + {x_o}^2 + {y_o}^2)\cos(\omega t) - 2(xx_o + yy_o)]$$

$$= [m\omega/2\sin(\omega t)]\,\{[(x - x_o)^2 + (y - y_o)^2]\cos(\omega t) - 2[1 - \cos(\omega t)]\,(xx_o + yy_o)\}. \quad \text{(j)}$$

From this expression, we readily obtain

$$\partial A/\partial x = [m\omega/2\sin(\omega t)]\,[2x\cos(\omega t) - 2x_o]$$

$$[x = a\sin(\omega t + \phi_o) \Rightarrow x_o = a\ \sin\phi_o]$$

$$= a\,m\,\omega\cos(\omega t + \phi_o) = m\,\dot{x} = p_x, \quad \text{etc., cyclically,} \qquad \text{(k)}$$

$$\partial A/\partial x_o = [m\,\omega/2\sin(\omega t)]\,[2x_o\cos(\omega t) - 2x]$$

$$= -[a\,m\,\omega/\sin(\omega t)]\sin(\omega t)\cos\phi_o = -m\,\dot{x}_o = -p_{xo}, \quad \text{etc., cyclically.} \qquad \text{(l)}$$

Let the reader verify that $\partial A/\partial t = -(T + V) = -E$.

Example 8.11.3 *Second Proof of the Constancy of Lagrange's Brackets (8.7.11 ff.).* The expression following eq. (8.11.2) can be successively rewritten as follows:

$$\partial H/\partial c_\mu = \partial/\partial c_\mu\Big(\sum p_k\,\dot{q}_k\Big) - d/dt\Big(\sum p_k(\partial q_k/\partial c_\mu)\Big)$$

(expanding and simplifying)

$$= \sum \dot{q}_k(\partial p_k/\partial c_\mu) - \sum \dot{p}_k(\partial q_k/\partial c_\mu)$$

(adding and subtracting the *second* and *fourth* summands below)

$$= \sum\ [\dot{q}_k(\partial p_k/\partial c_\mu) + q_k(\partial\dot{p}_k/\partial c_\mu)] - \sum\ [\dot{p}_k(\partial q_k/\partial c_\mu) + q_k(\partial\dot{p}_k/\partial c_\mu)]$$

$$= d/dt\Big[\sum q_k(\partial p_k/\partial c_\mu)\Big] - \partial/\partial c_\mu\Big(\sum q_k\,\dot{p}_k\Big), \qquad \text{(a)}$$

or, after slight rearrangement and use of $dp_k/dt = -\partial H/\partial q_k$,

$$d/dt\Big(\sum q_k(\partial p_k/\partial c_\mu)\Big) = \partial/\partial c_\mu\Big(H - \sum q_k(\partial H/\partial q_k)\Big) \equiv \partial H'/\partial c_\mu; \qquad \text{(b)}$$

and, similarly, with $c_\mu \to c_\nu$, we obtain

$$d/dt\,\Big(\sum q_k(\partial q_k/\partial c_\nu)\Big) = \partial/\partial c_\nu\Big(H - \sum q_k(\partial H/\partial q_k)\Big) \equiv \partial H'/\partial c_\nu. \qquad \text{(c)}$$

Now, differentiating (b) relative to c_ν and (c) relative to c_μ, and then subtracting side by side, we readily obtain

$$d/dt\Big[\partial/\partial c_\nu \sum q_k(\partial p_k/\partial c_\mu) - \partial/\partial c_\mu \sum q_k(\partial p_k/\partial c_\nu)\Big]$$
$$= \partial^2 H'/\partial c_\nu \partial c_\mu - \partial^2 H'/\partial c_\mu \partial c_\nu = 0; \qquad \text{(d)}$$

that is,

$$d/dt\left(\sum (\partial q_k/\partial c_\nu)(\partial p_k/\partial c_\mu) - \sum (\partial q_k/\partial c_\mu)(\partial p_k/\partial c_\nu)\right) \equiv [c_\mu, c_\nu]^{\cdot} = 0 \qquad \text{(e)}$$

$$\Rightarrow [c_\mu, c_\nu] = constant, \quad \text{Q.E.D.} \qquad \text{(f)}$$

Problem 8.11.1 By carrying out the two distinct (fixed time) variations $\delta_1(\ldots)$ and $\delta_2(\ldots)$ on the Hamiltonian equations

$$p_{ko} = -\partial A/\partial q_{ko}, \qquad p_k = \partial A/\partial q_k, \text{where } A = A(t, t_o; q, q_o), \qquad \text{(a)}$$

prove the earlier Lagrange–Poisson theorem (8.7.10):

$$I \equiv \sum (\delta_1 p_k\, \delta_2 q_k - \delta_2 p_k\, \delta_1 q_k) = \sum (\delta_1 p_{ko}\, \delta_2 q_{ko} - \delta_2 p_{ko}\, \delta_1 q_{ko}) \equiv I_o; \qquad \text{(b)}$$

that is, I = constant in time, namely, $dI/dt = 0$. (For relevant applications, see, e.g., Lamb, 1943, pp. 277–281.)

Problem 8.11.2 Assume that after the action-like integral (functional)

$$A' \equiv \int \left(T + V + \sum q_k \dot{p}_k\right) dt \qquad \text{(a)}$$

is evaluated along an actual path, it becomes a *function* of the initial and final *momenta* and time of transit $\tau \equiv t - t_o$: $A' = A'(p, p_o, \tau)$. Show that

$$\Delta A' = (\partial A'/\partial \tau)\Delta\tau + \sum \left[(\partial A'/\partial p_k)\,\Delta p_k + (\partial A'/\partial p_{ko})\,\Delta p_{ko}\right]$$
$$= H\Delta\tau + \sum (q_k\,\Delta p_k - q_{ko}\,\Delta p_{ko}); \qquad \text{(b)}$$

that is,

$$\partial A'/\partial\tau = H, \qquad \partial A'/\partial p_k = q_k, \qquad \partial A'/\partial p_{ko} = -q_{ko}.$$

Problem 8.11.3 *Alternative Variational Formulation of the Routhian Formalism.* Show that the variational problem (say, with vanishing endpoint variations, and the usual notations)

$$\delta \int \Lambda\, dt = 0, \qquad \text{(a)}$$

where

$$\Lambda \equiv R(t, q, \dot{q}; \psi, \Psi) + \sum \dot{\psi}_i \Psi_i = \Lambda(t, q, \dot{q}, \psi, \dot{\psi}, \Psi, \dot{\Psi}), \qquad \text{(b)}$$

yields Routh's equations for the system described by the "Lagrangean" Λ; that is, verify that

$(\partial\Lambda/\partial\dot{q}_p)^{\cdot} - \partial\Lambda/\partial q_p = 0$ gives Routh's equations of motion (i.e., R: Lagrangean for the q's),

$(\partial\Lambda/\partial\dot{\psi}_i)^{\cdot} - \partial\Lambda/\partial\psi_i = 0$ gives $d\Psi_i/dt = \partial R/\partial\psi_i$ $(\Rightarrow\ \Psi_i = constant,$ if $\partial R/\partial\psi_i = \partial L/\partial\psi_i = 0)$,

$(\partial\Lambda/\partial\dot{\Psi}_i)^{\cdot} - \partial\Lambda/\partial\Psi_i = 0$ gives $d\psi_i/dt = \partial R/\partial\Psi_i$.

8.12 INTEGRAL INVARIANTS

We have already seen (§8.8) that canonical transformations leave the Hamiltonian equations form invariant, and they also have the same effect on Hamilton's principle; that is, if

$$\int\Big(\sum p_k\dot{q}_k - H(t,q,p)\Big)\,dt \to stationary, \tag{8.12.1a}$$

and the transformation $(q,p)\to(q',p')$ is canonical, then

$$\int\Big(\sum p_{k'}\dot{q}_{k'} - H'(t,q',p')\Big)\,dt \to stationary. \tag{8.12.1b}$$

[In this case, the difference of the integrands of (8.12.1a, b) equals the $(\ldots)^{\cdot}$-derivative of a function of $2n$ of the old and new variables, and time (i.e., the generating function F); and so, under *fixed endpoint* (q,p) *variations*, that difference vanishes.]

Now, Poincaré, E. Cartan, et al. have shown that not only differential, but also certain *integral* forms exist that remain invariant under canonical transformations. Such quantities, named by them *integral invariants*, are the object of study of this section.

We begin by considering the fundamental equation of varied action (8.11.20), rewritten as

$$\Delta A = \Big(\sum p_k\,\Delta q_k - H\,\Delta t\Big)\Big|_{Final\ \text{time}} - \Big(\sum p_k\,\Delta q_k - H\Delta t\Big)\Big|_{Initial\ \text{time}}, \tag{8.12.2}$$

where (fig. 8.8), and to the first order in $\Delta q, \Delta t$:

$$\Delta A = A\,(from\ a_1{}'\ to\ a_2{}', along\ C') - A\,(from\ a_1\ to\ a_2, along\ C). \tag{8.12.3}$$

Let us assume that not only the *fundamental path* C, but also its adjacent C' are actual mechanical trajectories (or integral curves) of the system; that is, *both are solutions of its equations of motion but have different initial positions and initial times* (a total of $2n+1$ initial parameters); and, as a result, A, A' and ΔA are *functions* (not functionals) of these initial conditions.

To study these changes analytically, we begin with the following, easy to understand, parametric representation of the Hamiltonian variables:

$$q_k = q_k(s;\ c), \qquad p_k = p_k(s;\ c), \qquad t = t(s;\ c), \tag{8.12.4a}$$

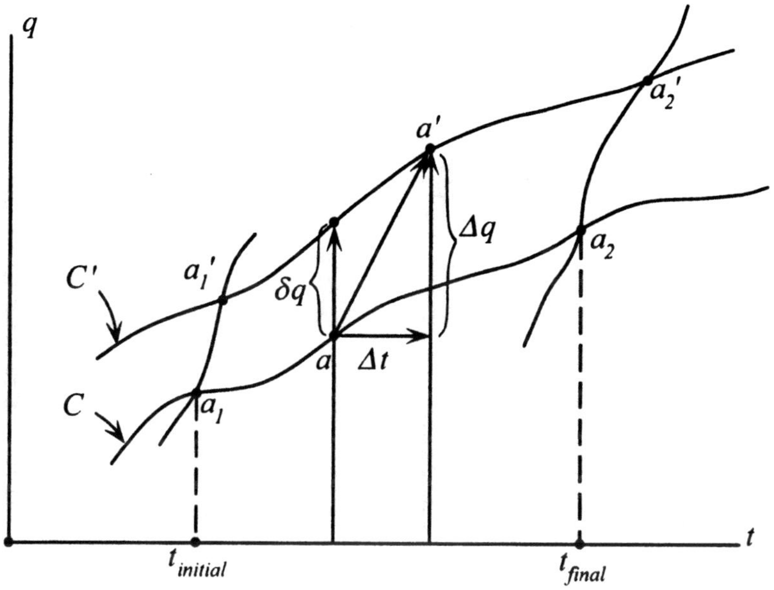

Figure 8.8 Trajectories C and C′ satisfy the same equations of motion, but have different initial conditions and times.

Coordinates of points involved: $a_1(t_1, q_1) \to a_1{}'(t_1 + \Delta t_1, q_1 + \Delta q_1)$,
$a_2(t_2, q_2) \to a_2{}'(t_2 + \Delta t_2, q_2 + \Delta q_2)$.

or, equivalently,

$$q_k = q_k[s; t_1, q^{(1)}, p^{(1)}], \qquad p_k = p_k[s; t_1, q^{(1)}, p^{(1)}], \qquad t = t[s; t_1, q^{(1)}, p^{(1)}], \tag{8.12.4b}$$

where s is an "arc-length" parameter in the $(2n+1)$-dimensional *extended phase space* of (t, q, p), and $c \equiv (c_1, \ldots, c_{2n+1})$ are $2n+1$ constants of integration; while $t_1, q^{(1)} \equiv (q_1{}^{(1)}, \ldots, q_n{}^{(1)}), p^{(1)} \equiv (p_1{}^{(1)}, \ldots, p_n{}^{(1)})$ are the initial values of t, q, p, respectively.

Now, let us assume that these $2n+1$ *initial values* (corresponding to the initial value s_1 of s) *are all functions of a single parameter* α, *independent of* s:

$$q_k{}^{(1)} = q_k{}^{(1)}(\alpha), \qquad p_k{}^{(1)} = p_k{}^{(1)}(\alpha), \qquad t^{(1)} = t^{(1)}(\alpha), \tag{8.12.5}$$

so that as α varies between the finite values α_1(*initial*) and α_2(*final*) $(\alpha_1 \le \alpha \le \alpha_2)$, the initial (left) endpoint of the system trajectories goes around the simple closed curve γ_1 in (t, q, p)-space:

$$a_1(\alpha_1) \to a_1{}' \to a_1{}'' \to \cdots \to a_1(\alpha_2). \tag{8.12.6}$$

Then, substituting (8.12.5) into (8.12.4b), we obtain the following parametric representation of the system trajectories:

$$q_k = q_k[s; t_1(\alpha), q^{(1)}(\alpha), p^{(1)}(\alpha)] = q_k(s; \alpha),$$
$$p_k = p_k[s; t_1(\alpha), q^{(1)}(\alpha), p^{(1)}(\alpha)] = p_k(s; \alpha),$$
$$t = t[s; t_1(\alpha), q^{(1)}(\alpha), p^{(1)}(\alpha)] = t(s; \alpha). \tag{8.12.7}$$

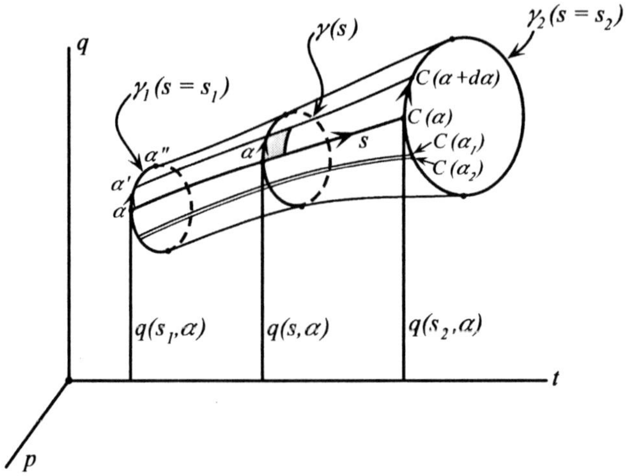

Figure 8.9 As the parameter α varies from α_1 to α_2, a closed tube of trajectories is created.

These equations show that as the left endpoint traces γ_1, the right (final) endpoint traces a similar closed curve γ_2, and a generic in-between point traces a closed curve γ (fig. 8.9); that is, as α varies from α_1 to α_2, a closed *tube of trajectories* (as its generatrices) is created in (t, q, p)-space. We assume that the closed curves $\gamma_1, \ldots, \gamma, \ldots, \gamma_2$ are nowhere tangent to the trajectories $C, C', C'', \ldots, C$, and are intersected only once by them. The above translate to the following *α-periodicity* relations:

$$\begin{aligned}\gamma_1:\quad & q_k{}^{(1)}(\alpha_1) = q_k{}^{(1)}(\alpha_2) && [\text{or } q_k(s_1;\alpha_1) = q_k(s_1;\alpha_2)],\\ & p_k{}^{(1)}(\alpha_1) = p_k{}^{(1)}(\alpha_2) && [\text{or } p_k(s_1;\alpha_1) = p_k(s_1;\alpha_2)],\\ & t_1(\alpha_1) = t_1(\alpha_2) && [\text{or } t(s_1;\alpha_1) = t(s_1;\alpha_2)]; \end{aligned} \tag{8.12.8}$$

and similarly for γ_2 and γ.

From (8.12.7) we immediately see that the most general variations/differentials of t, q, p along $\gamma_1, \gamma_2, \gamma$—that is, from trajectory to trajectory by varying the initial conditions (with a slight, easily understood, notational change to conform with calculus and the integrations below)—are

$$\begin{aligned}\Delta q_k \to dq_k = & \Big\{(\partial q_k/\partial t_1)(dt_1/d\alpha)\\ & + \sum \big[(\partial q_k/\partial q_l{}^{(1)})(dq_l{}^{(1)}/d\alpha) + (\partial q_k/\partial p_l{}^{(1)})(dp_l{}^{(1)}/d\alpha)\big]\Big\}\, d\alpha,\\ \Delta p_k \to dp_k = & \Big\{(\partial p_k/\partial t_1)(dt_1/d\alpha)\\ & + \sum \big[(\partial p_k/\partial q_l{}^{(1)})(dq_l{}^{(1)}/d\alpha) + (\partial p_k/\partial p_l{}^{(1)})(dp_l{}^{(1)}/d\alpha)\big]\Big\}\, d\alpha,\\ \Delta t \to dt = & \Big\{(\partial t/\partial t_1)(dt_1/d\alpha)\\ & + \sum \big[(\partial t/\partial q_l{}^{(1)})(dq_l{}^{(1)}/d\alpha) + (\partial t/\partial p_l{}^{(1)})(dp_l{}^{(1)}/d\alpha)\big]\Big\}\, d\alpha. \end{aligned} \tag{8.12.9}$$

With these analytical preliminaries, let us now *integrate the fundamental equation (8.12.2) for a complete variation of* α; that is, $\alpha_1 \to \alpha_2$. Since, here,

$$A = \int L(t, q, \dot{q})\, dt$$

$$= \int_{t_1(\alpha)}^{t_2(\alpha)} L[t(s;\alpha), q(s;\alpha), p(s;\alpha)]\, [(\partial t/\partial s)\, ds] = A(\alpha), \qquad (8.12.10)$$

[where the integrand is taken along a trajectory; i.e., for a *fixed* α],

the total change of A equals zero:

$$\oint dA \equiv \int_{\alpha_1}^{\alpha_2} dA \equiv \int_{\alpha_1}^{\alpha_2} A'(\alpha)\, d\alpha = A(\alpha_2) - A(\alpha_1) = 0; \qquad (8.12.11)$$

and therefore the integral of (8.12.2) (slightly rewritten in simplified standard calculus notation, now that we understand the situation better) becomes

$$0 = \oint dA = \oint_{\gamma_2} \left(\sum p_k{}^{(2)}(\alpha)\, dq_k{}^{(2)}(\alpha) - H^{(2)}(\alpha)\, dt_2(\alpha)\right)$$

$$- \oint_{\gamma_1} \left(\sum p_k{}^{(1)}(\alpha)\, dq_k{}^{(1)}(\alpha) - H^{(1)}(\alpha)\, dt_1(\alpha)\right), \qquad (8.12.12a)$$

where

$H^{(*)}(\alpha) \equiv H[t_*(\alpha), q^{(*)}(\alpha), p^{(*)}(\alpha)]$, $\quad q^{(*)}(\alpha) \equiv q(s_*;\alpha)$, $\quad p^{(*)}(\alpha) \equiv p(s_*;\alpha)$,

$dq^{(*)}(\alpha)$, $dt_*(\alpha)$: as given by (8.12.9), at s_* $(* = 1, 2)$;

or, after another easily understood notational simplification,

$$\oint_{\gamma_1} \left(\sum p_k\, dq_k - H\, dt\right) = \oint_{\gamma_2} \left(\sum p_k\, dq_k - H\, dt\right) = \oint_{\gamma} \left(\sum p_k\, dq_k - H\, dt\right), \qquad (8.12.12b)$$

that is,

$$I \equiv \oint \left(\sum p_k\, dq_k - H\, dt\right) = constant. \qquad (8.12.13)$$

In words: The integral I (around an arbitrary closed curve that encircles the tube of system trajectories and intersects them only once) is constant along these trajectories; as we say, I is a (Poincaré–Cartan) *relative integral invariant*. [If the domain of integration is *closed (open)*, like γ, the integral invariant is called *relative (absolute)*.]

For $t = constant$ ($\Rightarrow dt = 0$; i.e., γ consists of *simultaneous* system states), I reduces to the *first-order (Poincaré) relative integral invariant*:

$$I \to I_1 \equiv \oint \sum p_k\, dq_k = constant. \qquad (8.12.14)$$

I_1 is also called *circulation integral*, due to its formal similarity with the integrals that appear in the Helmholtz–Thomson theorems of continuum kinematics.

REMARK

If we, formally, set $q_{n+1} \equiv t$, then, since $\partial A/\partial t = -H$, the corresponding canonical "momentum" equals $-H$; that is, $p_{n+1} = \partial A/\partial q_{n+1} = \partial A/\partial t = -H$, and therefore (8.12.13) can be rewritten in the following (i)-form for $n+1$ q's:

$$I \equiv \oint \sum p_\mu \, dq_\mu = constant \qquad (\mu = 1, \ldots, n+1). \tag{8.12.13a}$$

Now, the simple closed curve γ [a one-dimensional manifold in either the $(n+1)$-dimensional extended configuration space, or the $(2n+1)$-dimensional extended phase space] can be viewed as the *boundary of a two-dimensional (simply connected) surface there, σ, described by the two Gaussian (curvilinear) coordinates u, v.* Hence, when the system point varies over σ, we can write

$$q_k = q_k(u, v) \rightarrow dq_k = (\partial q_k/\partial u)\, du + (\partial q_k/\partial v)\, dv, \tag{8.12.15a}$$

and similarly for p_k, dp_k. Then, say (8.12.14) becomes

$$I_1 = \oint \Big\{ \Big[\sum p_k (\partial q_k/\partial u) \Big] du + \Big[\sum p_k (\partial q_k/\partial v) \Big] dv \Big\}, \tag{8.12.15b}$$

or, applying the (two-dimensional) Kelvin–Stokes theorem to it:

$$\oint_\gamma [(*)\, du + (**)\, dv] = \iint_\sigma \{ [\partial(*)/\partial v] - [\partial(**)/\partial u] \} \, du\, dv, \tag{8.12.15c}$$

$$I_1 = \iint_\sigma \Big\{ \partial/\partial v \Big[\sum p_k (\partial q_k/\partial u) \Big] - \partial/\partial u \Big[\sum p_k (\partial q_k/\partial v) \Big] \Big\} \, du\, dv,$$

[where σ = region in uv-plane bounded by image of γ there]

$$= \iint_\sigma \Big\{ \sum [(\partial p_k/\partial v)(\partial q_k/\partial u) - (\partial p_k/\partial u)(\partial q_k/\partial v)] \Big\} \, du\, dv$$

$$\equiv \iint_\sigma \sum [\partial(q_k, p_k)/\partial(u, v)] \, du\, dv, \tag{8.12.15d}$$

or, finally, recalling that $\partial(q_k, p_k)/\partial(u, v)$ is none other than the Jacobian of the transformation $(q_k, p_k) \rightarrow (u, v)$, where the q's and p's are rectangular Cartesian coordinates in phase space, and the earlier definition of Lagrangean brackets (8.7.9), we can rewrite $I_1 = constant$ as

$$I_1 = I_2 = \iint_{S_2} \sum dq_k \, dp_k = \iint_{\sigma_2} [u, v] \, du\, dv = constant, \tag{8.12.16}$$

where S_2 = two-dimensional subspace in phase space corresponding to σ_2 via (8.12.16). Alternatively, invariance of I_2 under a canonical transformation

$(q,p) \to (q',p')$ requires that (here, for *constant* time, but the argument can be extended to variable time):

$$I_2' \equiv \iint_{S_2} \sum dq_{k'}\, dp_{k'} = \iint_{\sigma_2} \sum \left[\partial(q_{k'}, p_{k'})/\partial(u,v)\right] du\, dv$$

$$= I_2 = \iint \sum dq_k\, dp_k = \iint \sum \left[\partial(q_k, p_k)/\partial(u,v)\right] du\, dv,$$

$$\Rightarrow \sum \left[\partial(q_{k'}, p_{k'})/\partial(u,v)\right] = \sum \left[\partial(q_k, p_k)/\partial(u,v)\right]$$

or

$$[u,v]_{q',p'} = [u,v]_{q,p}, \tag{8.12.17}$$

which is the earlier-proved canonical invariance property of the Lagrange's (and Poisson's) brackets (§8.9).

Similarly, we can prove the integral invariance of

$$I_4 \equiv \iiiint \sum\sum dq_k\, dq_l\, dp_k\, dp_l, \tag{8.12.18a}$$

$$I_6 \equiv \iiiiiint \sum\sum\sum dq_k\, dq_l\, dq_r\, dp_k\, dp_l\, dp_r, \tag{8.12.18b}$$

and, generally, of

$$I_{2n} \equiv \int \cdots (2n \text{ times}) \cdots \int \sum \cdots (n \text{ sums}) \cdots \sum dq_k\, dq_{k'} \cdots dp_k\, dp_{k'}. \tag{8.12.18c}$$

The last integral of this series:

$$\int \cdots (2n \text{ times}) \cdots \int dq_1 \cdots dq_n\, dp_1 \cdots dp_n, \tag{8.12.19a}$$

represents the volume of the corresponding region in (q,p) phase space. Hence, *such volumes are invariant under canonical transformations*; that is,

$$\int \cdots (2n \text{ times}) \cdots \int dq_1 \cdots dq_n\, dp_1 \cdots dp_n$$

$$= \int \cdots (2n \text{ times}) \cdots \int dq_{1'} \cdots dq_{n'}\, dp_{1'} \cdots dp_{n'}; \tag{8.12.19b}$$

and this (by the earlier-mentioned theorem of multiple integral calculus) shows that *the corresponding Jacobian* $\partial(q,p)/\partial(q',p')$ *equals* $+1$.

This theorem leads to an important conclusion in phase space: we have already seen (ex. 8.9.6) that $(\dot{q}_k, \dot{p}_k)$, or (dq_k, dp_k), can be viewed as an infinitesimal canonical transformation with the Hamiltonian as generating function. Therefore, *all invariants of canonical transformations are also invariants of the motion.* This means, geometrically, that the corresponding $2n$-dimensional phase space points can be viewed as representative points of a corresponding manifold of identical mechanical systems with differing initial state conditions. As a result of the motion of these systems, the initial domain of integration of (q,p) is carried over to another one of equal volume.

In the extended phase space of (t, q, p), the world lines of such systems build a tube of constant cross-section. This constitutes the celebrated *theorem of Liouville* of statistical mechanics. As mentioned earlier, the above integral invariants are called *absolute*, because no special assumptions were made about their region of integration. However, with the help of the multidimensional generalization of Stokes' theorem, they can be transformed to *relative* invariants; that is, invariants over closed areas of lower order (= fewer integrations). For example, the absolute integral invariant I_1 can be transformed into the following relative integral invariant:

$$I_1 = \oint \sum p_k\, dq_k, \quad \text{over a } \textit{closed curve} \text{ in } (q,p)\text{-space, which lies on the plane } t = \textit{constant}, \text{ in } (t,q,p)\text{-space.}$$

As an application of the above, it follows that if we choose as range of integration the elementary parallelogram spanned by the two infinitesimal (q, p)-space vectors;

$$\big(d_1 q_k = (\partial q_k/\partial u)\, du, \quad d_1 p_k = (\partial p_k/\partial u)\, du\big),$$

and

$$\big(d_2 q_k = (\partial q_k/\partial v)\, dv, \quad d_2 p_k = (\partial p_k/\partial v)\, dv\big),$$

for a *constant* time t, then

$$I_2 = \iint \Big\{\sum \big[(\partial p_k/\partial u)\, du\big]\big[(\partial q_k/\partial v)\, dv - (\partial p_k/\partial v)\, dv\big]\big[(\partial q_k/\partial u)\, du\big]\Big\}$$
$$= \iint \sum (d_1 p_k\, d_2 q_k - d_2 p_k\, d_1 q_k) = \textit{integral invariant}, \tag{8.12.20}$$

from which it follows that the *Lagrangean bilinear covariant* (8.7.10 ff.)

$$\sum (d_1 p_k\, d_2 q_k - d_2 p_k\, d_1 q_k) = \sum\sum [c_\mu, c_\nu]\, \delta_1 c_\mu\, \delta_2 c_\nu, \tag{8.12.21}$$

of the *differential form* $\sum p_k\, dq_k$, is invariant; and, conversely, *its invariance is sufficient for the corresponding transformation to be canonical* (with no recourse to generating functions, as in §8.8). [For alternative proofs see examples below, and Whittaker (1937, pp. 272–274).]

In sum:

- *Direct*: The quantity $\oint p\, dq$ is a relative integral invariant of any Hamiltonian system of differential equations ("the circulation in any circuit moving with the fluid does not change with time").
- *Converse*: If a system of equations $dq/dt = Q$, $dp/dt = -P$ possesses the relative integral invariant $\oint p\, dq$, then its equations of motion have the Hamiltonian form: $Q = \partial H/\partial p$, $P = -\partial H/\partial q$.

REMARKS

(i) The invariance of the *circulation*

$$\oint \sum p_k\, dq_k = \iint \Big\{\sum \big[(\partial p_k/\partial v)(\partial q_k/\partial u) - (\partial p_k/\partial u)(\partial q_k/\partial v)\big]\Big\}\, du\, dv \tag{8.12.22}$$

should not come as a complete surprise; after all, the fundamental definition of canonical transformations (8.8.12):

$$\sum p_k\,\delta q_k - \sum p_{k'}\,\delta q_{k'} = \delta F, \tag{8.12.22a}$$

looks like a requirement that "the work" of the left side of the above be potential; or, equivalently, that, for any closed path in phase space, $\oint dF = 0$; and this leads immediately to

$$\oint \sum p_k\,dq_k = \oint \sum p_{k'}\,dq_{k'}. \tag{8.12.22b}$$

(ii) If we also vary the time, then it is not (8.12.21) that is invariant, but the *extended bilinear covariant* of the extended differential form $\sum p_k\,dq_k - H\,dt$:

$$\sum (d_1 p_k\,d_2 q_k - d_2 p_k\,d_1 q_k) - (d_1 H\,d_2 t - d_2 H\,d_1 t). \tag{8.12.23}$$

(See also Routh, 1905(b), §479, pp. 325–326.)

Incidentally, this theorem of Lagrange signals a basic difference between Lagrangean and Hamiltonian mechanics. Restricting ourselves to the common *scleronomic* case, we may remark that:

(a) In the former (Lagrangean), due to the geometrical structure of its *configuration* space, the fundamental invariant under point transformations $q \to q'$ is the Riemannian *line element ds* (§3.9):

$$(ds)^2 \equiv 2T(dt)^2 = \sum\sum M_{kl}\,dq_k\,dq_l = \sum\sum M_{k'l'}\,dq_{k'}\,dq_{l'} = \cdots; \tag{8.12.24}$$

whereas

(b) In the latter (Hamiltonian), due to the geometrical structure of its *phase space*, the fundamental invariant under canonical transformations $(q,p) \to (q',p')$ is the also quadratic and homogeneous form in the dq's and dp's expression (8.12.21); but since it is associated with *two*, rather than one, infinitesimal displacements, $d_1(\ldots)$ and $d_2(\ldots)$, it is a *bilinear* form in them and hence represents an *area*, rather than a line, element $d_{12}A$:

$$I \equiv d_{12}A \equiv \sum (d_1 p_k\,d_2 q_k - d_2 p_k\,d_1 q_k) = \sum (d_1 p_{k'}\,d_2 q_{k'} - d_2 p_{k'}\,d_1 q_{k'}) = \cdots. \tag{8.12.25}$$

These remarks can be extended, with some precautions, to the *rheonomic* case.

[The literature on integral invariants is quite extensive and varied. For further details and insights, we recommend (alphabetically): Aizerman (1974, pp. 289–308), Cartan (1922: a classic in the field), Gantmacher (1970, pp. 119–127), Kilmister (1964, pp. 128–140), Lovelock and Rund (1975, pp. 207–213), Prange (1935, pp. 657–689).]

Example 8.12.1 Let us calculate the Poincaré–Cartan invariant

$$I \equiv \oint \Big(\sum p_k\,dq_k - H\,dt\Big), \tag{a}$$

for a system with Hamiltonian

$$H = (1/2)\,(p^2 + q^2); \tag{b}$$

that is, a one-DOF linear oscillator (of unit mass and stiffness).

We will evaluate (a) along (i) the *initial* closed curve $\gamma_1 \equiv OABO$ $(s = s_1)$, which has the following parametric representation [(fig. 8.10); since there is only one DOF, we use subscripts throughout: 1 for initial values and 2 for final values]:

$$\begin{aligned} \gamma_1: \quad & OA(s_1): && t_1 = \alpha, \quad q_1 = 0, \quad p_1 = 0, \\ & AB(s_1): && t_1 = \alpha_2, \quad q_1 = 0, \quad p_1 = 0, \\ & BO(s_1): && t_1 = \alpha, \quad q_1 = 0, \quad p_1 = \alpha; \\ & && [\alpha_1 = 0 \le \alpha \le \alpha_2]; \end{aligned} \tag{c}$$

and (ii) along the *intermediate* closed curve $\gamma\,(s = s)$. Since the solution of the Hamiltonian equations of motion of (b), with initial conditions: t_1, q_1, p_1, is

$$t = t_1 + f, \qquad q = q_1 \cos f + p_1 \sin f, \qquad p = \dot{q} = -q_1 \sin f + p_1 \cos f,$$
$$f \equiv f(s; q_1, p_1) = \text{monotonic function in } s, \text{and such that } f(s_1; q_1, p_1) = 0, \tag{d}$$

the parametric representation of γ will be

$$\begin{aligned} \gamma: \quad & OA(s): && t = \alpha + f(s, 0, 0), && q = 0, \quad p = 0, \\ & AB(s): && t = \alpha_2 + f(s, 0, \alpha), && q = \alpha \sin f(s, 0, \alpha), \quad p = \alpha \cos f(s, 0, \alpha), \\ & BO(s): && t = \alpha + f(s, 0, \alpha), && q = \alpha \sin f(s, 0, \alpha), \quad p = \alpha \cos f(s, 0, \alpha); \\ & && [\alpha_1 = 0 \le \alpha \le \alpha_2]. \end{aligned} \tag{e}$$

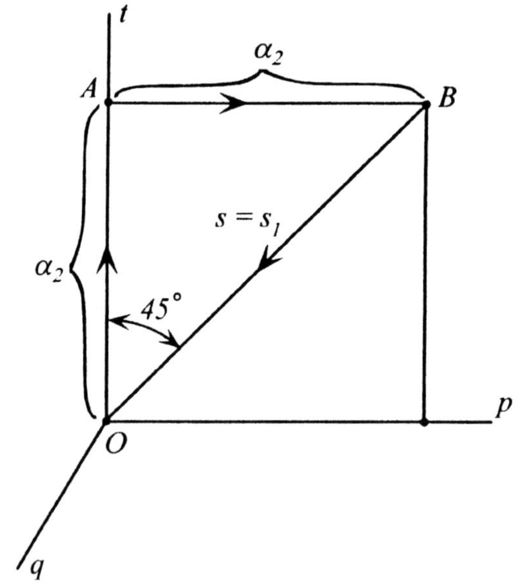

Figure 8.10 Initial closed path of harmonic oscillator conditions, in three-dimensional phase space.

Hence, we find, successively:

(i) Along γ_1 $(s = s_1)$:

$$
\begin{aligned}
I(\gamma_1) &= \oint_{\gamma_1} (p_1\, dq_1 - H_1\, dt_1) \\
&= \int [along\ OA(s_1)] + \int [along\ AB(s_1)] + \int [along\ BO(s_1)] \\
&= 0 + 0 + \int [along\ BO(s_1)] \\
&= \int_{\alpha_2}^{0} -({p_1}^2/2)\, dt_1 = +(1/2) \int_0^{\alpha_2} \alpha^2\, d\alpha = {\alpha_2}^3/6.
\end{aligned} \tag{f}
$$

(ii) Along γ $(s = s)$:

$$
\begin{aligned}
I(\gamma) &= \oint (p\, dq - H\, dt) = \int [along\ OA(s)] + \int [along\ AB(s)] + \int [along\ BO(s)] \\
&= 0 + \int_0^{\alpha_2} \{(\alpha \cos f)\,(d\alpha \sin f) - (1/2)\,(\alpha^2 \sin^2 f + \alpha^2 \cos^2 f)\,[(\partial f/\partial \alpha)\, d\alpha]\} \\
&\quad + \int_{\alpha_2}^{0} \{(\alpha \cos f)\,(d\alpha \sin f) - (1/2)\,(\alpha^2 \sin^2 f + \alpha^2 \cos^2 f)\,[d\alpha + (\partial f/\partial \alpha)\, d\alpha]\} \\
&= \int_0^{\alpha_2} [\alpha \sin f \cos f - (\alpha^2/2)\,(\partial f/\partial \alpha)]\, d\alpha \\
&\qquad - \int_0^{\alpha_2} \{\alpha \sin f \cos f - (\alpha^2/2)\,[1 + (\partial f/\partial \alpha)]\}\, d\alpha \\
&= - \int_0^{\alpha_2} [-(\alpha^2/2)]\, d\alpha = {\alpha_2}^3/6;
\end{aligned} \tag{g}
$$

that is, $I(s_1) = I(s)$, Q.E.D.; I is indeed constant along the trajectories of (b).

Example 8.12.2 *Integral Variants for Nonpotential* (→ *Nonconservative*) *Holonomic Systems.* Starting with Hamilton's principle of varying action for systems under *nonpotential* forces $(Q_k; k = 1, \ldots, n)$ (chap. 7), and proceeding as in the potential case discussed earlier, it is not hard to show that, here, the Poincaré–Cartan and Poincaré invariant equations must be replaced, respectively, by the following integral *variant* relations:

$$
dI/dt \equiv d/dt\left[\oint_\gamma \left(\sum p_k\, dq_k - H\, dt\right)\right] = \oint_\gamma \sum Q_k\, dq_k, \tag{a1}
$$

$$
dI_1/dt \equiv d/dt\left[\oint_\gamma \left(\sum p_k\, dq_k\right)\right] = \oint_\gamma \sum Q_k\, dq_k. \tag{a2}
$$

For these variants to become *invariants*—namely, in order that $I = constant$, $I_1 = constant$—we must have

$$\oint_\gamma \sum Q_k\, dq_k = 0. \tag{b}$$

Application of the *generalized theorem of Stokes:*

$$\oint_\gamma \sum A_s\, dx_s = \iint_\sigma \sum\sum (\partial A_s/\partial x_r - \partial A_r/\partial x_s)\, dx_s\, dx_r \tag{c}$$

[where $s, r =, \ldots$; on the right side (double) summation $r < s$; and γ is the boundary of the diaphragm-like surface σ, locus at time t of points that were initially located on another surface bounded by the initial position of γ] to eq. (b), with the identifications

$$A_k = Q_k, \quad A_{n+k} = 0; \quad x_k = q_k,\ x_{n+k} = p_k\ (s, r = 1, \ldots, n, \ldots, 2n;\ k = 1, \ldots, n), \tag{d}$$

shows that the necessary and sufficient conditions for it to occur (for arbitrary σ and independent dx_s, dx_r) are

$$\partial A_s/\partial x_r - \partial A_r/\partial x_s = 0 \qquad (s, r = 1, \ldots, 2n;\ r < s):$$

$$\partial Q_k/\partial q_l = \partial Q_l/\partial q_k \qquad \text{and} \qquad \partial Q_k/\partial p_l = 0 \qquad (k, l = 1, \ldots, n); \tag{e}$$

that is, the Q_k must be derivable from a *potential* function, say $V = V(t, q)$: $Q_k = -\partial V/\partial q_k$. Hence, *no Poincaré–Cartan/Poincaré invariants exist for nonpotential forces.*

An Application of Equation (a2) to the Method of Slowly Varying Parameters

However, equations (a1, 2) can become useful in approximate calculations. Let us apply (a2) to such a solution of the quasi-linear oscillator equation

$$\ddot{q} + \omega_o{}^2 q = \varepsilon f(q, \dot{q}), \tag{f}$$

where $\varepsilon f(\ldots) =$ small relative to $\ddot{q}$ (inertia) and $\omega_o{}^2 q$ (elasticity). Here, clearly,

$$L = (1/2)(\dot{q})^2 - (1/2)\omega_o{}^2 q^2, \qquad Q = \varepsilon f(q, \dot{q}) \equiv \varepsilon f(\ldots), \tag{g}$$

and therefore the corresponding canonical equations are

$$\dot{q} = p, \qquad \dot{p} = -\omega_o{}^2 q + \varepsilon f(\ldots). \tag{h}$$

Let us seek a solution of the above in the form

$$q = a(t) \sin\chi(t) \ \Rightarrow\ p = \dot{q} = a(t)\omega_o \cos\chi(t), \tag{i}$$

where $\chi(t) = \omega_o t + \phi(t)$ and $a(t), \phi(t) =$ unknown functions to be determined.
By δ-varying the first of (i), we obtain

$$\delta q = (\partial q/\partial a)\,\delta a + (\partial q/\partial\phi)\,\delta\phi = \sin\chi\,\delta a + a\cos\chi\,\delta\phi, \tag{j}$$

and, therefore {returning to the δ-notation, $\delta(\ldots) \equiv [\partial(\ldots)/\partial\alpha]\,\delta\alpha$, to avoid possible confusion with $d(\ldots) \equiv [\partial(\ldots)/\partial s]\,ds$},

$$I_1 \equiv \oint_\gamma p\,\delta q = \oint_\gamma [a\,\omega_o \cos\chi\,(\sin\chi\,\delta a + a\cos\chi\,\delta\phi)], \tag{k}$$

$$\oint_\gamma Q\,\delta q = \oint_\gamma \varepsilon f\,\delta q = \oint_\gamma [\varepsilon f(\ldots)\,(\sin\chi\,\delta a + a\cos\chi\,\delta\phi)], \tag{l}$$

where $\gamma =$ arbitrary closed curve encircling simultaneously ($dt = 0$) the closed tube of trajectories.

Now, proceeding as in the method of slowly varying parameters (ex. 7.9.14 ff.), we *average* equations (k, l) over χ, from 0 to 2π, thus obtaining (skipping the factor 1/2 in both equations)

$$\langle I_1\rangle = \int_0^{2\pi} d\chi\left(\oint_\gamma p\,\delta q\right) = \oint_\gamma\left(\int_0^{2\pi} a\,\omega_o \cos\chi(\sin\chi\,\delta a + a\cos\chi\,\delta\phi)\right)d\chi$$

$$= \oint_\gamma\left[a\,\omega_o\,\delta a\left(\int_0^{2\pi}\sin\chi\cos\chi\,d\chi\right) + a^2\omega_o\,\delta\phi\left(\int_0^{2\pi}\cos^2\chi\,d\chi\right)\right]d\phi$$

[the *second* (inner) integral vanishes, while the *second* equals π]

$$= \oint_\gamma \pi a^2\omega_o\,\delta\phi, \tag{m}$$

$$\langle Q\rangle = \int_0^{2n} d\chi\left(\oint_\gamma Q\,\delta q\right) = \oint_\gamma\left[\int_0^{2\pi}\varepsilon f(\ldots)\,(\sin\chi\,\delta a + a\cos\chi\,\delta\phi)\right]d\chi, \tag{n}$$

and then apply to them the integral variant equation (a2); or, equivalently, we average (a2) over χ from 0 to 2π; that is,

$$\langle dI_1/dt\rangle = \langle Q\rangle \;\Rightarrow\; d\langle I_1\rangle/dt = \langle Q\rangle. \tag{o}$$

In this way, we find

$$d/dt\left(\oint_\gamma \pi a^2\omega_o\,\delta\phi\right) = \pi\,\omega_o\left\{\oint_\gamma[(a^2)^{\cdot}\,\delta\phi + (a^2)(\delta\phi)^{\cdot}]\right\}$$

[in the *second* term, we assume that $(\delta\phi)^{\cdot} = \delta(\dot\phi)$, then integrate it by parts (for $t = constant$)]

$$= \pi\omega_o\left[\oint_\gamma (2a\,\dot a)\,\delta\phi + (a^2\dot\phi)\Big|_\gamma - \int \dot\phi\,\delta(a^2)\right]$$

[by periodicity, the integrated out boundary/endpoints term vanishes]

$$= \oint_\gamma [(2\pi\omega_o a\,\dot a)\,\delta\phi - (2\pi\omega_o a\,\dot\phi)\,\delta a]$$

$$= \langle Q\rangle = \oint_\gamma\left\{\left(\int_0^{2\pi}\varepsilon f(\ldots)\sin\chi\,d\chi\right)\delta a + \left(\int_0^{2\pi}\varepsilon f(\ldots)a\cos\chi\,d\chi\right)\delta\phi\right\}; \tag{p}$$

and, from this, since γ is arbitrary and the $\delta a, \delta\phi$ are independent (and $a \neq 0$), we obtain the well-known *van der Pol/Krylov/Bogoliubov equations:*

$$da/dt = (\varepsilon/2\pi\omega_o)\int_0^{2\pi} f(\ldots)\cos\chi\, d\chi, \tag{q}$$

$$d\phi/dt = -(\varepsilon/2\pi a\omega_o)\int_0^{2\pi} f(\ldots)\sin\chi\, d\chi. \tag{r}$$

Example 8.12.3 *Alternative Proof of Relation between Integral Invariance and Canonicity of Equations of Motion.* Here, we show, by direct calculation, that if

$$I \equiv \oint_\gamma \left(\sum p_k\, dq_k - H\, dt\right) = constant, \tag{a}$$

then

$$dp_k/dt = -\partial H/\partial q_k, \qquad dq_k/dt = \partial H/\partial p_k, \qquad dH/dt = \partial H/\partial t. \tag{b}$$

For extra clarity, we introduce the following special notations:

$$d_1(\ldots) = [\partial(\ldots)/\partial s]\, ds = \text{differential along an integral curve}, \tag{c1}$$

$$d_2(\ldots) = [\partial(\ldots)/\partial\alpha]\, d\alpha = \text{differential along closed curve } \gamma \text{ encircling tube of integral curves}. \tag{c2}$$

Clearly, since the parameters s and α are independent, $d_1[d_2(\ldots)] = d_2[d_1(\ldots)]$; and so (a) can be rewritten as

$$I \equiv I(s) \equiv \oint_\gamma \left(\sum p_k\, d_2 q_k - H\, d_2 t\right) = constant. \tag{d}$$

We have, successively,

$$\begin{aligned} d_1 I \equiv d_1 I(s) &= d_1 \oint_\gamma (\ldots) \\ &= \oint_\gamma \left\{\sum \left[d_1 p_k\, d_2 q_k + p_k\, d_1(d_2 q_k)\right] - \left[d_1 H\, d_2 t - H\, d_1(d_2 t)\right]\right\} \\ &= \oint_\gamma \left\{\sum \left[d_1 p_k\, d_2 q_k + p_k\, d_2(d_1 q_k)\right] - \left[d_1 H\, d_2 t - H\, d_2(d_1 t)\right]\right\} \end{aligned}$$

[integrating the *second* and *fourth* terms by parts relative to $d_2(\ldots)$; i.e., along γ]

$$\begin{aligned} &= \oint_\gamma \sum d_1 p_k\, d_2 q_k + \left(\sum p_k\, d_1 q_k\right)\Big|_\gamma - \oint_\gamma \sum d_2 p_k\, d_1 q_k \\ &\qquad - \oint_\gamma d_1 H\, d_2 t - \left(\sum H\, d_1 t\right)\Big|_\gamma + \oint_\gamma d_2 H\, d_1 t \end{aligned}$$

[due to the α-periodicity, the integrated out
(second and fifth sums) vanish]

$$= \oint_\gamma \left[\left(\sum d_1 p_k \, d_2 q_k - \sum d_2 p_k \, d_1 q_k\right) - (d_1 H \, d_2 t - d_2 H \, d_1 t)\right]$$

$\Big[$setting $d_2 H = \sum \, [(\partial H/\partial q_k) \, d_2 q_k + (\partial H/\partial p_k) \, d_2 p_k]$

$+ (\partial H/\partial t) \, d_2 t$, and regrouping terms$\Big]$

$$= \oint_\gamma \Big(\sum \{[d_1 p_k + (\partial H/\partial q_k) \, d_1 t] \, d_2 q_k + [-d_1 q_k + (\partial H/\partial p_k) \, d_1 t] \, d_2 p_k\}$$

$$+ [-d_1 H + (\partial H/\partial t) \, d_1 t] \, d_2 t\Big). \tag{e}$$

Therefore, if $d_1 I(s) = 0$, since γ is arbitrary and the α-differentials $d_2 q, d_2 p, d_2 t$ are independent, it follows that, *along each trajectory* [with $s = t \Rightarrow d_1(\ldots) = (\ldots)^{\cdot} \, dt = dt$]

$$d_1 p_k + (\partial H/\partial q_k) \, d_1 t = 0 \Rightarrow dp_k/dt = -\partial H/\partial q_k, \tag{f1}$$

$$- d_1 q_k + (\partial H/\partial p_k) \, d_1 t = 0 \Rightarrow dq_k/dt = \partial H/\partial p_k, \tag{f2}$$

$$- d_1 H + (\partial H/\partial t) \, d_1 t = 0 \Rightarrow dH/dt = \partial H/\partial t; \tag{f3}$$

that is, the Hamiltonian equations of *motion* and *energy* (b) hold. Hence, the importance of the integral invariant $I = I(s)$ to mechanics. The converse theorem can be shown similarly.

In sum:

- If $I =$ *constant*, then the system satisfies Hamilton's equations.
- If the system satisfies Hamilton's equations, then $I =$ *constant*.

8.13 NOETHER'S THEOREM

This famous theorem (E. Noether, 1918) states that if the action functional

$$A = \int_{t_1}^{t_2} L(t, q, \dot{q}) \, dt, \tag{8.13.1}$$

is invariant under the family, or group, of (continuous and continuously differentiable, and one-to-one in both directions) transformations

$$t \to t' = t'(t, q, \dot{q}; \varepsilon), \qquad q_k \to q_{k'} = q_{k'}(t, q, \dot{q}; \varepsilon), \tag{8.13.2}$$

where ε is the (sole) parameter of that family, and such that

$$t'(t, q, \dot{q}; 0) = t, \qquad q_{k'}(t, q, \dot{q}; 0) = q_k \qquad \text{[identity transformations]}; \tag{8.13.2a}$$

that is, *if, under (8.13.2)*

$$\int_{t'_1}^{t'_2} L(t', q', dq'/dt')\, dt' = \int_{t_1}^{t_2} L(t, q, \dot{q})\, dt, \tag{8.13.3}$$

for arbitrary time limits, or, equivalently, *if*

$$L[t', q', dq'(t')/dt'](dt'/dt) = L[t, q, dq(t)/dt] \tag{8.13.3a}$$

(from action *integrals* to their Lagrangean *integrands*), *then along each system trajectory* ("extremal" of A), and with the notation $(\ldots)_o \equiv (\ldots)|_{\varepsilon=0}$ the following quantity is *constant*:

$$N \equiv \sum (\partial L/\partial \dot{q}_k)(\partial q_{k'}/\partial \varepsilon)_o - \left(\sum (\partial L/\partial \dot{q}_k)\dot{q}_k - L\right)(\partial t'/\partial \varepsilon)_o$$

$$= L(\partial t'/\partial \varepsilon)_o + \sum (\partial L/\partial \dot{q}_k)\left[(\partial q_{k'}/\partial \varepsilon)_o - \dot{q}_k(\partial t'/\partial \varepsilon)_o\right]$$

$$\text{[Lagrangean form]}, \tag{8.13.4}$$

$$= \sum p_k(\partial q_{k'}/\partial \varepsilon)_o - H(\partial t'/\partial \varepsilon)_o \qquad \text{[Hamiltonian form]}. \tag{8.13.5}$$

In words: every one-parameter family of transformations that leaves the action functional *invariant* leads to a first integral of the equations of motion (and, generally, invariance under an m-parameter such family leads to m integrals—see below).

PROOF

Since A is invariant under (8.13.2, 2a), it follows that *under the latter's first ε-order expansions*:

$$t' = t'(\ldots; 0) + (\partial t'/\partial \varepsilon)_o \varepsilon = t + (\partial t'/\partial \varepsilon)_o \varepsilon$$

$$\Rightarrow \Delta t \equiv t' - t = \varepsilon(\partial t'/\partial \varepsilon)_o, \tag{8.13.6a}$$

$$q_{k'} = q_{k'}(\ldots; 0) + (\partial q_{k'}/\partial \varepsilon)_o \varepsilon = q_k + (\partial q_{k'}/\partial \varepsilon)_o \varepsilon$$

$$\Rightarrow \Delta q_k \equiv q_{k'} - q_k = \varepsilon(\partial q_{k'}/\partial \varepsilon)_o, \tag{8.13.6b}$$

the corresponding first ε-order variation of A (from a kinetic trajectory, or orbit) vanishes; that is (by the general Hamilton's law of varying action),

$$\Delta A = \left[\sum p_k \Delta q_k - \left(\sum p_k \dot{q}_k - L\right)\Delta t\right] \qquad \text{(between the arbitrary time limits } t_1, t_2\text{)}$$

$$= \varepsilon\left[\sum p_k(\partial q_{k'}/\partial \varepsilon)_o - \left(\sum p_k \dot{q}_k - L\right)(\partial t'/\partial \varepsilon)_o\right] = 0, \tag{8.13.6c}$$

from which, since $\varepsilon \neq 0$, the integrals (8.13.4, 5) $N = constant$ follow, Q.E.D.

This theorem, as well as the method of Routh for ignorable coordinates (§8.3, 8.4), and the concepts of closed/open systems discussed in §3.12, reveal the following fundamental idea of theoretical dynamics (and theoretical physics, in general):

Invariance properties $\Rightarrow$ *Conserved quantities*

Example 8.13.1

(i) The action functional

$$A = \int_{t_1}^{t_2} L(q, \dot{q})\, dt \qquad [\text{i.e., } \partial L/\partial t = 0] \tag{a}$$

is, clearly, invariant under the one-parameter group of transformations:

$$t' = t + \varepsilon, \qquad q_{k'} = q_k \qquad [\textit{temporal} \text{ translation}]. \tag{b}$$

Since, in this case, $(\partial t'/\partial \varepsilon)_o = 1$ and $(\partial q_{k'}/\partial \varepsilon)_o = 0$, the Noetherian expressions (8.13.4, 5) yield the generalized energy integral

$$N = \sum p_k(0) - H(1) = -H = \textit{constant}, \tag{c}$$

an already well-known result. Hence, *invariance under time-translation leads to energy conservation*.

(ii) The action functional [with x, y, z: rectangular Cartesian coordinates; and $K, L = 1, \ldots, N$ (# of particles)]

$$A = \int \left\{ \sum (1/2) m_K [(\dot{x}_K)^2 + (\dot{y}_K)^2 + (\dot{z}_K)^2] - \sum \sum V_{KL}(|\boldsymbol{r}_K - \boldsymbol{r}_L|) \right\} dt, \tag{d}$$

where $K \neq L$ (or $K < L$) and $\boldsymbol{r}_K = (x_K, y_K, z_K)$, is, clearly, invariant under the one-parameter rigid *spatial translations* in the x-direction:

$$x_{K'} = x_K + \varepsilon, \qquad y_{K'} = y_K, \qquad z_{K'} = z_K, \quad (\text{and } t' = t). \tag{e}$$

Therefore, by (8.13.4, 5), the system possesses the integral ($K = 1, \ldots, 3N$)

$$\begin{aligned} N &= \sum (\partial L/\partial \dot{x}_K)(\partial x_{K'}/\partial \varepsilon)_o - \left[\sum (\partial L/\partial \dot{q}_k)\dot{q}_k - L \right] (\partial t'/\partial \varepsilon)_o \\ &= \sum (\partial L/\partial \dot{x}_K)(1) - \left[\sum (\partial L/\partial \dot{q}_k)\dot{q}_k - L \right] (0) \\ &= \sum \partial L/\partial \dot{x}_K = \sum m_K \dot{x}_K \equiv p_X \\ &= x\text{-component of total } \textit{linear} \text{ momentum} = \textit{constant}; \end{aligned} \tag{f}$$

and similarly for the y- and z-directions.

Hence, *invariance under rigid translations leads to conservation of the linear momentum vector*.

(iii) If the above action (d) is invariant under the one-parameter rigid rotations of the system about, say the z-axis, through an infinitesimal angle ε (*active* interpretation, §1.11):

$$x_{K'} = x_K \cos\varepsilon - y_K \sin\varepsilon \approx x_K - \varepsilon\, y_K, \qquad y_{K'} = x_K \sin\varepsilon + y_K \cos\varepsilon \approx y_K + \varepsilon\, x_K,$$

$$z_{K'} = z_K \qquad (\text{and } t' = t), \tag{g}$$

then, again by (8.13.4, 5), the system has the integral

$$
\begin{aligned}
N &= \sum \big[(\partial L/\partial \dot{x}_K)(\partial x_{K'}/\partial \varepsilon)_o + (\partial L/\partial \dot{y}_K)(\partial y_{K'}/\partial \varepsilon)_o \\
&\qquad + (\partial L/\partial \dot{z}_K)(\partial z_{K'}/\partial \varepsilon)_o\big] - H(\partial t'/\partial \varepsilon)_o \\
&= \sum \big[(\partial L/\partial \dot{x}_K)(-y_K) + (\partial L/\partial \dot{y}_K)(+x_K) + (\partial L/\partial \dot{z}_K)(0)\big] - H(0) \\
&= \sum \big[(x_K)(\partial L/\partial \dot{y}_K) - (y_K)(\partial L/\partial \dot{x}_K)\big] \\
&= \sum \big[(x_K)(m_K \dot{y}_K) - (y_K)(m_K \dot{x}_K)\big] \equiv H_z \\
&= z\text{-component of total } \textit{angular} \text{ momentum about origin } O = \textit{constant}; \quad \text{(h)}
\end{aligned}
$$

and similarly for its x- and y-components.

Hence, *invariance under infinitesimal rigid rotations, about a point, leads to conservation of the angular momentum vector about that point.*

[The reader may verify that the *passive* interpretation of rotation (as well as of the earlier translation) leads to the same result!]

Extension of Noether's Theorem to m-Parameter Family of Transformations

If the action A is invariant under the m-parameter continuous group of transformations

$$t \to t' = t'(t, q, \dot{q}; \varepsilon), \qquad q_k \to q_{k'} = q_{k'}(t, q, \dot{q}; \varepsilon), \tag{8.13.7}$$

where $\varepsilon \equiv (\varepsilon_1, \ldots, \varepsilon_m) =$ *group parameters*, and, again, such that

$$t'(t, q, \dot{q}; 0) = t, \qquad q_{k'}(t, q, \dot{q}, 0) = q_k \qquad \text{[identity transformations]}; \tag{8.13.7a}$$

then, along each system trajectory, the following m distinct quantities ($\bullet$: $1, \ldots, m$):

$$
\begin{aligned}
N_\bullet &\equiv \sum (\partial L/\partial \dot{q}_k)(\partial q_{k'}/\partial \varepsilon_\bullet)_o - \Big(\sum (\partial L/\partial \dot{q}_k)\dot{q}_k - L\Big)(\partial t'/\partial \varepsilon_\bullet)_o \\
&= L(\partial t'/\partial \varepsilon_\bullet)_o + \sum (\partial L/\partial \dot{q}_k)\big[(\partial q_{k'}/\partial \varepsilon_\bullet)_o - \dot{q}_k(\partial t'/\partial \varepsilon_\bullet)_o\big] \\
&\qquad\qquad \text{[Lagrangean form]}, \qquad (8.13.8) \\
&= \sum p_k(\partial q_{k'}/\partial \varepsilon_\bullet)_o - H(\partial t'/\partial \varepsilon_\bullet)_o \qquad \text{[Hamiltonian form]}, \qquad (8.13.9)
\end{aligned}
$$

are constant. In words: *every m-parameter family of transformations that leaves the action functional invariant leads to m distinct first integrals of the equations of motion.*

[For readable proofs, see, for example, Lovelock and Rund (1975, pp. 201–207), Mittelstaedt (1970, pp. 138–160); or, one could extend the previous one-parameter proof to the m-parameter case.]

Invariance under Gauge Transformations

Recalling the nonuniqueness of the Lagrangean (ex. 3.5.13)—namely, that $L = L(t, q, \dot{q})$ and $L' = L + df(t, q)/dt$ [$f(\ldots) =$ arbitrary function of t and the

q's] yield the same equations of motion—we, now, generalize Noether's theorem as follows: If the invariance equation (8.13.3), under (8.13.2), is replaced by

$$\int_{t'_1}^{t'_2} L(t', q', dq'/dt')\, dt' = \int_{t_1}^{t_2} \left[L(t, q, \dot{q}) + df(t, q; \varepsilon)/dt\right] dt, \tag{8.13.10a}$$

or, equivalently, by

$$L[t', q', dq'(t')/dt'](dt'/dt) = L[t, q,\ dq(t)/dt] + df(t, q; \varepsilon)/dt \tag{8.13.10b}$$

(again, from action *integrals* to their Lagrangean *integrands*), then the Noetherian integrals (8.13.4, 5) are replaced by

$$\begin{aligned}
N' &= N - (\partial f/\partial \varepsilon)_o \\
&= \sum (\partial L/\partial \dot{q}_k)(\partial q_{k'}/\partial \varepsilon)_o - \left[\sum (\partial L/\partial \dot{q}_k)\dot{q}_k - L\right](\partial t'/\partial \varepsilon)_o - (\partial f/\partial \varepsilon)_o \\
&= L(\partial t'/\partial \varepsilon)_o + \sum (\partial L/\partial \dot{q}_k)\left[(\partial q_{k'}/\partial \varepsilon)_o - \dot{q}_k(\partial t'/\partial \varepsilon)_o\right] - (\partial f/\partial \varepsilon)_o \\
&= constant, \qquad \text{[Lagrangean form]}, \qquad\qquad (8.13.11) \\
&= \sum p_k(\partial q_{k'}/\partial \varepsilon)_o - H(\partial t'/\partial \varepsilon)_o - (\partial f/\partial \varepsilon)_o \\
&= constant, \qquad \text{[Hamiltonian form]}. \qquad\qquad (8.13.12)
\end{aligned}$$

This follows easily if we notice that, in this case, the first ε-order action variation (8.13.6c) must be replaced by

$$\begin{aligned}
\Delta A &= \varepsilon\left[\sum p_k(\partial q_{k'}/\partial \varepsilon)_o - \left(\sum p_k\dot{q}_k - L\right)(\partial t'/\partial \varepsilon)_o\right] \\
&= \Delta \int (df/dt)\, dt = \Delta[f] = \left[(\partial f/\partial \varepsilon)_o\right]\varepsilon. \qquad\qquad (8.13.13)
\end{aligned}$$

The rest of the details are left to the reader.

Example 8.13.2 Continuing from the conservation theorems of the preceding example, let us consider the motion of our system in two inertial frames, (O, F) and (O', F'), in relative motion with constant velocity $\boldsymbol{v}_o \equiv \boldsymbol{v}_{F/F'}$, and let us assume, for simplicity but no loss in generality, that $V = 0$ and $Q_k = 0$ (inertial motion).

The system Lagrangean in F' is (with $P = 1, \ldots, N = \#$ system particles):

$$\begin{aligned}
L' &= \sum (1/2)m_P(d\boldsymbol{r}'_P/dt)\cdot(d\boldsymbol{r}'_P/dt) \\
&= \sum (1/2)m_P\left[(d\boldsymbol{r}_P/dt) + \boldsymbol{v}_o\right]\cdot\left[(d\boldsymbol{r}_P/dt) + \boldsymbol{v}_o\right] \\
&= L + df/dt, \qquad\qquad \text{(a)}
\end{aligned}$$

where

$$L = \sum (1/2)m_P(d\mathbf{r}_P/dt)\cdot(d\mathbf{r}_P/dt) = \text{system Lagrangean in } F, \tag{b}$$

$$f = \sum m_P \mathbf{r}_P \cdot \mathbf{v}_o + \left[\sum (1/2)m_P(\mathbf{v}_o\cdot\mathbf{v}_o)\right]t:$$

Galilean gauge; function of time, coordinates, and group parameter $\varepsilon \to \mathbf{v}_o$,

$$\Rightarrow\ df/dt = \sum m_P(d\mathbf{r}_P/dt)\cdot\mathbf{v}_o + \sum (1/2)m_P(\mathbf{v}_o\cdot\mathbf{v}_o). \tag{c}$$

Choosing, for mathematical convenience, in these two frames, rectangular Cartesian coordinates F: $(O; x, y, z)$ and F': $(O'; x', y', z')$, such that

$$x' = x + v_o t \quad (\Rightarrow \dot{x}' = \dot{x} + v_o), \qquad y' = y, \qquad z' = z, \qquad t' = t, \tag{d}$$

reduces (b, c) to

$$L = \sum (1/2)m_P\,\dot{x}_P\,\dot{x}_P, \qquad f = \sum m_P\,x_P\,v_o + \left[\sum (1/2)m_P\,v_o{}^2\right]t; \tag{e}$$

and, therefore, with parameter ε the relative frame velocity v_o, the Noetherian expressions (8.13.10, 11) yield the integral

$$\begin{aligned} N' &= N - (\partial f/\partial\varepsilon)_o \\ &= \sum p_P(\partial x'_P/\partial\varepsilon)_o - H(\partial t'/\partial\varepsilon)_o - (\partial f/\partial\varepsilon)_o \\ &= \sum p_P(\partial x'_P/\partial v_o)_o - H(\partial t'/\partial v_o)_o - (\partial f/\partial v_o)_o \\ &= \sum (m_P\dot{x}_P)(t) - H(0) - \sum m_P\,x_P = \mathit{constant} = c, \end{aligned} \tag{f}$$

or, since $\sum m_P\,\dot{x}_P \equiv p_x = \mathit{constant} \equiv c_x$ (by ex. 8.13.1) and $\sum m_P\,x_P -$ $(\mathit{total\ mass})\,(x - \mathit{coordinate\ of\ mass\ center}) \equiv m\,x$,

$$N' = c_x t - m\,x = c \ \Rightarrow\ m\,x = c_x t - c,$$

$$\Rightarrow\ \dot{x} = c_x/m \qquad \text{(mass center moves with constant velocity);} \tag{g1}$$

$$\text{or,}\quad \text{if } c_x = 0, \quad x = -c/m \quad \text{(mass center at rest);} \tag{g2}$$

and similarly for the y- and z-directions (recall ex. 3.12.3).

The results of these two examples can be summed up in the following theorem.

THEOREM

Let us consider a system of N particles moving under their mutual (Newtonian) gravitational attractions ["*N-body problem*" of classical (celestial) mechanics], and therefore having equations of motion

$$m_P\,\ddot{\mathbf{r}}_P = \sum \Delta_{PP'}G(m_P\,m_{P'}/r_{PP'}{}^3)\mathbf{r}_{PP'}, \tag{h}$$

where $\Delta_{PP'} \equiv 1 - \delta_{PP'} =$ *complementary Kronecker delta*, $G =$ *gravitational constant* (not gauge function!), $r_{PP'} \equiv |\mathbf{r}_{PP'}| \equiv |\mathbf{r}_{P'} - \mathbf{r}_P| = |-\mathbf{r}_{P'P}| = r_{P'P} \neq 0$, and P, $P' = 1$, $\ldots, N =$ number of particles.

From the Noetherian invariance of its action under the *ten*-parameter Galilean group [between two inertial frames $(x_k(t), t)$, $(x_{k'}(t'), t')$, in arbitrary mutual orientation]

$$x_{k'}(t') = \sum a_{k'k}(\varepsilon_1, \varepsilon_2, \varepsilon_3)x_k + b_k(\varepsilon_4, \varepsilon_5, \varepsilon_6)t + c_k(\varepsilon_7, \varepsilon_8, \varepsilon_9),$$

$$\big[(a_{k'k}) = (a_{k'k}(\varepsilon_1, \varepsilon_2, \varepsilon_3))\text{: proper orthogonal matrix [eqs. (1.1.19a ff.) and (1.5.1a, b)]; with } k, l = 1, 2, 3\big], \tag{i}$$

$$t' = t + \varepsilon_{10}, \tag{j}$$

we obtain the following *ten integrals*:

(i) *Temporal* translation (1) $\rightarrow$ *energy* conservation:

$$T + V \equiv \sum (1/2)m_P(\dot{\boldsymbol{r}}_P \cdot \dot{\boldsymbol{r}}_P) - \sum\sum (1/2)\Delta_{PP'}G(m_P m_{P'}/r_{PP'}) = constant, \tag{k}$$

(ii) *Spatial* translation (3) $\rightarrow$ *linear momentum* conservation:

$$\sum m_P \dot{\boldsymbol{r}}_P = m\,\boldsymbol{v}_{\text{mass center}} = constant \equiv \boldsymbol{b} = (b_1, b_2, b_3), \tag{l}$$

(iii) *Spatial* rotation (3) $\rightarrow$ *angular momentum* conservation:

$$\sum \boldsymbol{r}_P \times (m_P \dot{\boldsymbol{r}}_P) = constant. \tag{m}$$

(iv) *Galilean transformation* (3) $\rightarrow$ *center of mass* theorem [integral of (l)]:

$$\sum m_P \boldsymbol{r}_P = m\,\boldsymbol{r}_{\text{mass center}} = \boldsymbol{b}t + \boldsymbol{c} \qquad [\boldsymbol{c} = (c_1, c_2, c_3)]. \tag{n}$$

For a detailed Noetherian treatment, see, for example, Funk [1962, pp. 442–445; after the original derivation of Bessel–Hagen (1921)]. For an alternative derivation (via symmetrical infinitesimal canonical transformations), see Schmutzer (1989, pp. 438–445); also Meirovitch (1970, pp. 413–416).

Closing Remarks

Noether's theorem, in spite of its conceptual beauty and simplicity, has not been very successful in producing new and nontrivial integrals of the equations of motion. The systematic search for infinitesimal transformations that leave the Hamiltonian action functional invariant leads to a system of first-order partial differential equations (Killing's equations), the solution of which is a taxing problem in itself. The theorem seems to have fared better in *field theory* (i.e., invariance of *multiple* integrals under continuous groups of transformations).

For detailed treatments of these topics, and applications to the search for system symmetries/conservation theorems, we recommend (alphabetically):

(a) *Mathematics* oriented: Funk (1962, pp. 437–452), Logan (1977), Lovelock and Rund (1975, pp. 226–231), Rund (1966, pp. 208–322); and, of course Noether's original paper (1918).

(b) *Mechanics–physics* oriented: Bahar and Kwatny (1987), Dobronravov (1976, pp. 139–163), Hill (1951), Kuypers (1993, pp. 281–295), Saletan and Cromer (1971, pp. 60–87, 219–226, 345–348), Sudarshan and Mukunda (1974), Vujanovic and Jones (1989, pp. 74–151; and references given therein).

8.14 PERIODIC MOTIONS; ACTION–ANGLE VARIABLES

Outside of equilibrium, *periodic* motion is one of the most important and interesting physical states; for example, planets revolving around the Sun; penduli; parts of engines; the building blocks of molecular and atomic systems; and so on. Also, the close connection between periodicity (aperiodicity) and stability (instability), stochasticity/chaoticity) of motion is well known. Hence, such a state deserves a closer examination. This section constitutes a modest introduction to this vast and fascinating topic, using the earlier-developed concepts and theorems of Hamiltonian mechanics.

One DOF

Such a system undergoes *periodic* motion with period τ, if, after a time interval τ, it always returns to where it was before; that is, analytically, its Lagrangean coordinate $q = q(t)$ satisfies the condition

$$q = q(t) = q(t + \tau), \tag{8.14.1}$$

and can, therefore, under mild continuity conditions, be represented by the Fourier series (in complex form, for algebraic compactness):

$$q(t) = \sum c_s \exp(is\,\omega\, t) = \sum c_s \exp(2\pi is\,\nu t), \tag{8.14.2}$$

where $\exp(\ldots) \equiv e^{\cdots}$,

$s = -\infty, \ldots, +\infty$,

$\omega \equiv 2\pi/\tau =$ fundamental *angular, or circular, frequency*;
i.e., *number of oscillations or rotations* (*see below*) *in* 2π *seconds*,

$\nu \equiv 1/\tau = \omega/2\pi =$ fundamental (*true*) *frequency*; i.e., *number of oscillations or rotations in* 1 *second*, (8.14.2a)

and the *amplitudes* c_s, which depend on the motion during a single period, are given by the well-known formula (which also makes it clear how it was obtained)

$$c_s = (1/\tau)\int_0^\tau q(t)\exp(-is\,\omega t)\,dt \qquad (c_{-s} \equiv c_s{}^* = \text{complex } \textit{conjugate} \text{ of } c_s). \tag{8.14.3}$$

From the above, it follows that $\dot{q}$ and any function of q and $\dot{q}$ is also a Fourier series with the same fundamental frequency (period) $\nu(\tau)$. Other system properties, or variables, may also be periodic; for example, periodic system momentum $p \equiv \partial T/\partial \dot{q}$ means that $p(t) = p(t + \tau)$. However, this natural and simple concept can be obscured by the use of the wrong, that is, nonperiodic coordinates. For example, the uniform circular motion of a particle (or, of a rigid body about a fixed axis) is, clearly, periodic; but its description in terms of its angle of rotation ϕ (from a fixed radius):

$$\phi = c_1 t + c_2 \qquad (c_{1,2}\text{: constants}), \tag{8.14.4}$$

is a monotonically increasing (nonperiodic!) function of time; also, the corresponding angular momentum $p \sim \dot{\phi}$ is constant; that is, trivially periodic.

In view of these possibilities, we classify periodic motions (according to their trajectories) in *phase space* into *two* kinds:

(i) *Libration*

This is the case where the coordinate q is a *single-valued* function of the system's position (i.e., it is not an angle that can have different values for the same configuration), and it remains between fixed limits; say, q_1, q_2, q_3, q_4, q_1 (fig. 8.11).

Both q and p are bounded, continuous and periodic in time, with the same period; say, τ. As a result, the corresponding (q, p)-curve, in phase space, is *closed*, and repeats itself after every time interval τ. Since the system integral $H(q, p) = constant \equiv C$ (usually equal to the total energy of the system E) is single-valued, different values of C produce a family of closed and *nonintersecting* phase space trajectories; in fact, by decreasing C appropriately, we may reduce the trajectory to the fixed *libration center* q_o, so that the motion degenerates to small oscillations about the stable equilibrium position q_o. The *libration limits* are the roots of $dq/dt = \partial H/\partial p = 0$ and they appear always *in pairs* (e.g., q_3 and q_4); the point where they coincide (or, coalesce; e.g., $q_\bullet$) signifies an *unstable* equilibrium configuration. In particular, *if p appears quadratically in the Hamiltonian $H = H(q, p)$, the system path in phase space $p = p(q;\ E)$*, obtained from its *energy surface* $H(q, p) = E$, is *symmetric about the q-axis* (fig. 8.11).

For example, if $H = p^2/2m + V(q) = E$, then, to within a sign,

$$p = (2m)^{1/2}[E - V(q)]^{1/2}$$
$$\Rightarrow\ dp/dq = -(m/2)^{1/2}[E - V(q)]^{-1/2}(dV/dq). \qquad (8.14.5)$$

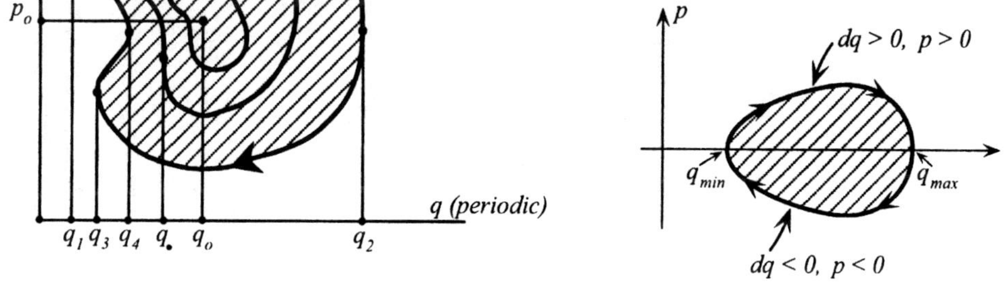

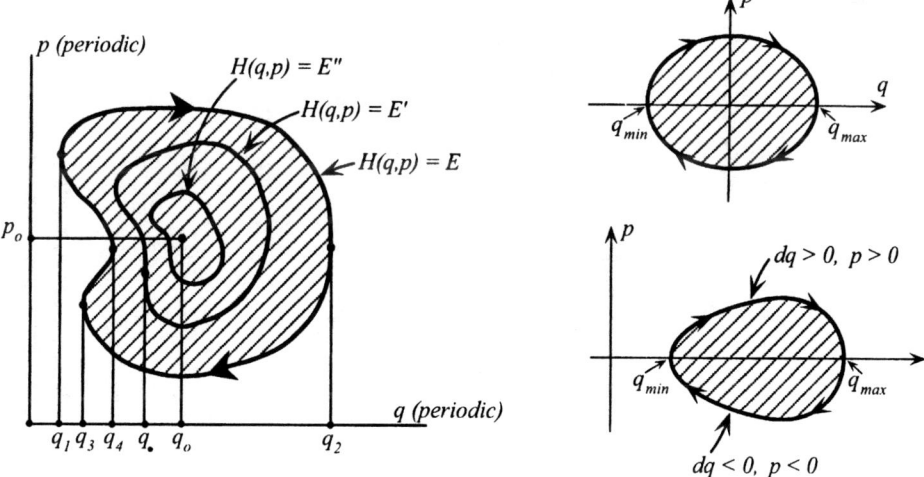

Figure 8.11 Phase plane trajectory of *libratory* periodic motion (*1 DOF*), and example of harmonic oscillator. [The energy curve $H(q, p) = E$ (outer contour, left figure, for general q, p *is not necessarily quadratic in p*; that is why there are *four* possible values of p for $q_3 < q < q_4$. If $H(q, p)$ was *quadratic* in p, there would be only *two*.]

Therefore, for libration, the equation $E - V(q) = 0$ must have *two* simple zeros: $q_{\min}, q_{\max}$; *and between them be positive*; at $q_{\min/\max}$, $dp/dq \to \infty$. Then, the curve is traversed *completely* and in the same sense. [Since $p\dot{q} = 2T \Rightarrow p\,dq > 0$, the curve is traveled *outward* ($dq > 0$) in its *upper* branch ($p > 0$), and in its *lower* branch ($p < 0$) on its *return* ($dq < 0$).]

(ii) *Rotation* (or *circulation*, or *revolution*)

If, however, the coordinate q is *angle-like*, then since q and $q + kq_o$ (k = arbitrary integer—fig. 8.12) describe the same system configuration, the unique determination of the integral constant C by the system's state of motion requires either that the curve $H(q,p) = C$ is closed, or that $p = p(q)$ is a periodic function of q, with minimum period q_o (frequently, $q_o = 2\pi$). In the second case, q takes the full range of values; that is, it is neither bounded nor periodic (timewise)—this kind of motion is called *rotation*. In particular, the uniform rectilinear motion can be viewed as the limiting case of a periodic motion, indeed as a rotation on a circle of infinite radius.

REMARK

In HM, we always seek coordinates in which the motion appears as rotation; then (as with ignorable coordinates) q = linear in time, p = constant (see "action-angle" variables, below).

The libration–rotation difference can be summed up, mathematically, as follows:

(i) In a libration, $q \to q_L$ can be represented by a Fourier series;
(ii) In a rotation, $q \to q_R$ cannot be so represented. But for periodic motion, the new function: $q_R - 2\pi\nu t \equiv q_R - \omega t$, *can* be represented by a Fourier series with fundamental frequency ν. For example, if $q_R \to \phi = 2\pi\nu t$ (i.e., uniform rotation) $\to \phi - 2\pi\nu t = 0$; and the latter can be represented by a Fourier series with all its coefficients zero. Thus, whether a periodic motion will be classified as libration or as rotation depends on the chosen positional coordinates. Also, one and the same physical system may, under different initial conditions ($\Rightarrow$ different initial energy constant) exhibit both libration and rotation. (The limiting case separating libration

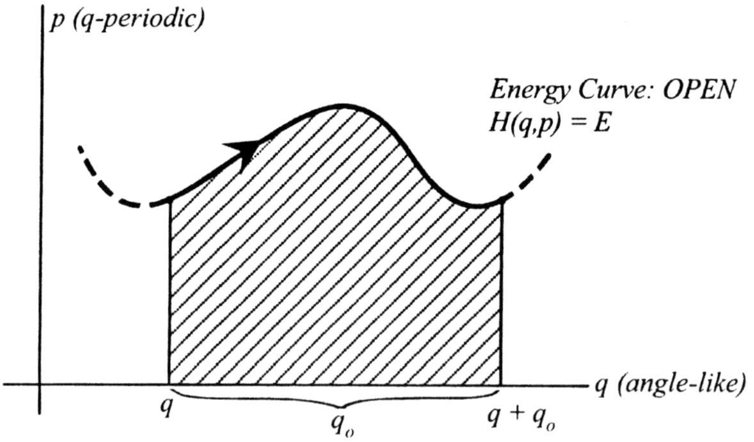

Figure 8.12 Phase plane trajectory of *rotatory* periodic motion (*1 DOF*).

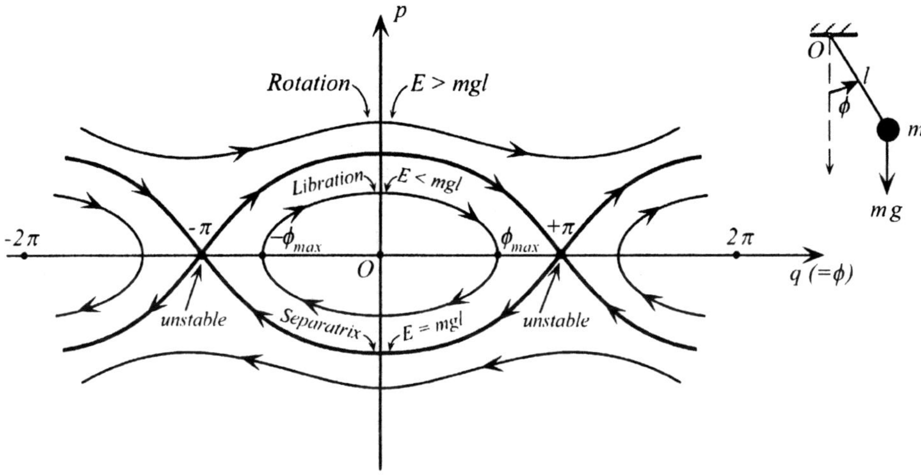

Figure 8.13 Phase plane trajectories of a planar mathematical pendulum.

from rotation is sometimes called *limitation*; that is, one where the points of motion reversal are reached in infinite time.)

The classic example here is the *planar mathematical pendulum with fixed support O* (fig. 8.13):

(i) Its to and fro oscillatory motion is a libration; whereas
(ii) Its full rotation around O, if its energy is sufficient, is a rotation.

Here,

$$2T = m\,l^2(\dot{\phi})^2, \qquad V = m\,g\,l(1 - \cos\phi), \tag{8.14.6a}$$

$$\Rightarrow\ T + V = T_{\text{initial}} + V_{\text{initial}} = m\,g\,l(1 - \cos\phi_o) \equiv E \qquad (\phi_o \equiv \phi_{\text{maximum}}), \tag{8.14.6b}$$

$$H \equiv (1/2A)p^2 - D\cos\phi = C \equiv E \qquad \text{(energy equation)}, \tag{8.14.6c}$$

$$\Rightarrow\ p = \pm(2A)^{1/2}(E + D\cos\phi)^{1/2} = \pm(m\,l)\,(2g\,l)^{1/2}(\cos\phi - \cos\phi_o) \tag{8.14.6d}$$

$[A \equiv m\,l^2, \quad D \equiv m\,g\,l]$.

We distinguish the following *four* cases:

(a) If $E = -m\,g\,l \equiv -D(< 0)$, the (q, p)-trajectory contracts to the *libration center* q_o.
(b) If $-D < E < D$, then we have *libration* only for $|\phi| < \phi_{\text{max}}$. [The libration limits are given by $dq/dt = \partial H/\partial p = 0 \ \Rightarrow\ p = 0$: $\cos\phi_o = -(E/m\,g\,l) \equiv -(E/D)$; there, $p = 0$. Hence oscillation/libration will occur between $-\phi_{\text{max}} \equiv -\arccos(-D/E)$ and $\phi_{\text{max}} \equiv \arccos(-D/E)$.]
(c) If $E > D$, we have *rotation* (always in the same direction).
(d) If $E = D$, we move on a *stability/instability boundary*, or *asymptotic orbit*; and approach the highest point $q = \pi$ very slowly ("in infinite time"). (See also Born, 1927, pp. 48–52.)

However, in general, and for reasons that will appear gradually below (also, recalling rationale for canonical transformations, §8.8), we seek new ignorable *coordinates* q'

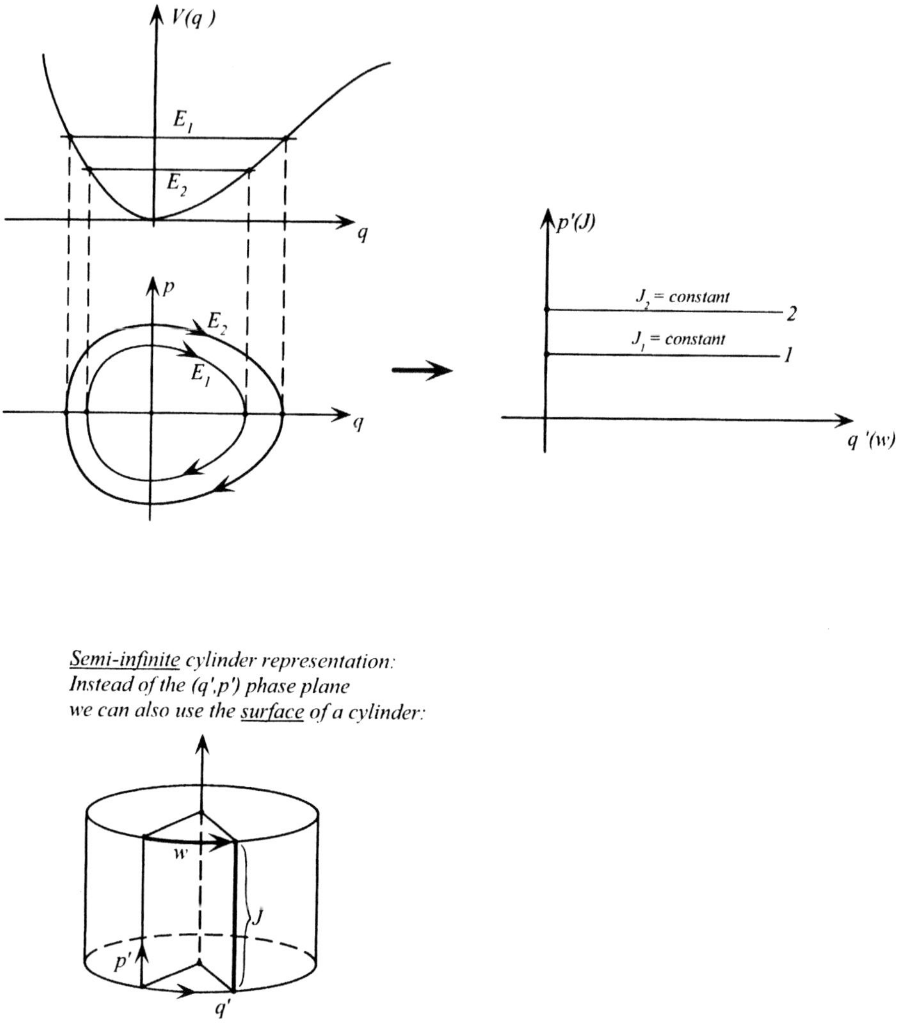

Figure 8.14 Canonical transformation to action–angle variables: $(q, p) \to (q' = w, p' = J)$.

in which our periodic motion appears as a rotation (*angle* variables, $q' \equiv w$), and new constant *momenta* p' (*action* variables $p' \equiv J$) which are the sole "variables" of the system Hamiltonian (which, here, equals the total energy); that is, recalling §8.10 (fig. 8.14),

$$(q, p) \to (q', p'):$$

$$dq'/dt = \partial H'(p')/\partial p' = constant \equiv c_1$$

$$\Rightarrow q' = c_1 t + c_2 \qquad (c_2\text{: integration constant}), \tag{8.14.7a}$$

$$dp'/dt = -\partial H'(p')/\partial q' = 0 \;\Rightarrow\; p' = constant \equiv c_3. \tag{8.14.7b}$$

Let us quantify these concepts:

(i) The *phase integral* [with physical dimensions of (*mass* × *velocity*) × (*length*); that is, *angular momentum*, or action]:

$$J \equiv \oint p(q,\beta)\,dq \left[\text{or, equivalently, } \oint p(q,E)\,dq\right]:$$

Libration: the integration extends over the *closed* path; and thus takes care of the multiple-valuedness of the momentum, due to its *quadratic* appearance in (8.14.5, 6d): $p = \pm\ldots$ (integral equal to the *shaded* area in fig. 8.11);

Rotation: the integration extends over a single period q_o of q (integral equal to the *shaded* area in fig. 8.12), (8.14.8)

is called *action variable*.

In a constant parameter (closed) system, J is independent of time. Therefore (recalling the Hamilton–Jacobi method, §8.10), for such systems, J can be taken as one of the *two* integration constants of motion (here, $n = 1$); or, in the action function of our problem, $A(t,q,\beta) = A_o(q,\beta) - E(\beta)t$, we can take J as the new momentum $p' = \beta$ (with a constant value for each particular periodic motion); that is, $\beta = J$, and consider the total energy E as a function of it:

$$J \equiv \oint p\,dq = \oint (\partial A_o/\partial q)\,dq = J(E) \;\Rightarrow\; E = E(J)$$

$$\Rightarrow A(t,q,J) = A_o(q,J) - E(J)t. \tag{8.14.9a}$$

From the foregoing, it follows that the *first* of the Hamilton–Jacobi (HJ) transformation equations, and corresponding HJ equation are

$$(\partial A_o/\partial q)_{E=\text{constant}} = (\partial A_o/\partial q)_{J=\text{constant}} = p, \tag{8.14.9b}$$

$$H(q, \partial A_o/\partial q) = constant = E. \tag{8.14.9c}$$

(ii) The corresponding new coordinate $q' \equiv w$, of the canonical transformation $(q,p) \to (q' = w, p' = J)$ with generating function $F = F_2(q,p') = A_o(q,J)$, is

$$q' = \partial F_2/\partial p': \qquad w = \partial A_o(q,J)/\partial J. \tag{8.14.10}$$

To find its properties, we need the Hamiltonian equations of motion in these variables. Since $H' = H + \partial A_o/\partial t = H = E(J) = constant$ (i.e., w is *ignorable*, and therefore J is a constant) these equations are

$$dq'/dt = \partial H'(p')/\partial p': \qquad dw/dt = \partial H(J)/\partial J = constant \equiv \nu(J) \equiv \nu$$

$$\Rightarrow w = \nu t + \gamma, \tag{8.14.11}$$

(ν is to be identified later with the fundamental frequency of the system and γ with a *phase* constant)

$$dp'/dt = -\partial H'(p')/\partial q': \qquad dJ/dt = -\partial H(J)/\partial w = 0$$

$$\Rightarrow J = constant. \tag{8.14.12}$$

From the above, it follows that w *increases linearly with time*; and during a period τ it increases by

$$\Delta w \equiv w(t+\tau) - w(t) = \cdots = [\partial H(J)/\partial J]\tau \equiv \nu\,\tau. \tag{8.14.13a}$$

But also, from the earlier definitions, as q goes through a complete cycle of libration or rotation, we have, successively,

$$\Delta w = \oint (\partial w/\partial q)\, dq = \oint (\partial^2 A_o/\partial J\, \partial q)\, dq = \partial/\partial J \left(\oint (\partial A_o/\partial q)\, dq \right)$$

$$= \partial/\partial J \left(\oint p\, dq \right) = \partial J/\partial J = 1; \qquad (8.14.13b)$$

that is, *the state of the system is periodic in w with period 1.* (For a full justification of the commutation rule employed here, see ex. 8.14.13.) Comparing (8.14.13a, b), we immediately conclude that

$$\nu \equiv 1/\tau \equiv \omega/2\pi = \partial H(J)/\partial J = \partial E(J)/\partial J; \qquad (8.14.14)$$

that is, *by differentiating the (constant) total energy with respect to the (constant) action variable, as soon as it becomes available in that form [without finding $q(t)$ from the equations of motion!], we obtain the fundamental frequency of the periodic motion.* Thus, the difficulty of solving the equations of motion has been transferred to that of calculating the action integrals $J \equiv \oint p\, dq$; and in this lies the importance of action and angle variables.

The geometrical meaning of these transformations, and resulting advantage of action–angle variables, for systems with Hamiltonian $H = p^2/2m + V(q)$ are shown in fig. 8.15.

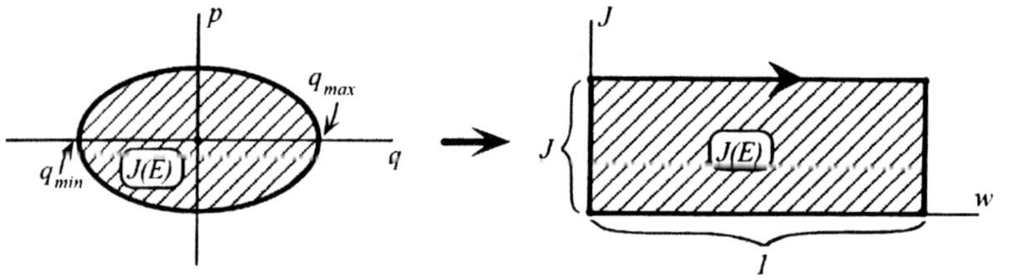

Libration: $J(E) = \oint p(q, E)\, dq = 2 \int_{q_{min}}^{q_{max}} (2m)^{1/2} [E - V(q)]^{1/2}\, dq = (J)(1) = J;$

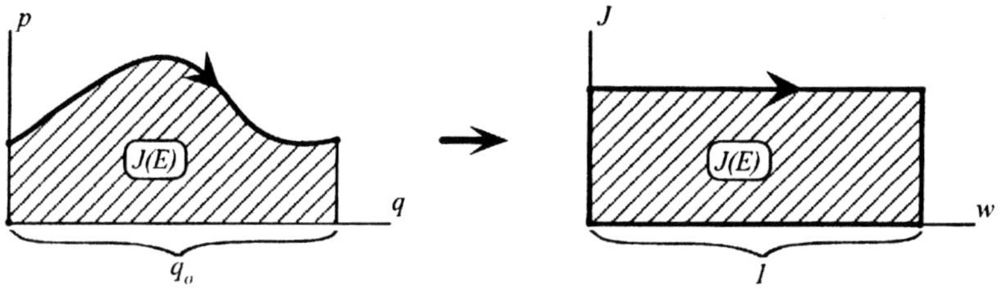

Rotation: $J(E) = \int_0^{q_o} p(q, E)\, dq$; where $H(q, p) = E \Rightarrow p = p(q, E)$.

[Two solutions in opposite directions; from two actions corresponding to each direction of motion.]

Figure 8.15 Libration and rotation in general and in angle–action variables, in phase space (1 DOF system).

REMARK

Some authors define J as $(1/2\pi)\oint p\,dq$. Then,

$$w = [\partial H(J)/\partial J]t + constant = \omega t + constant \;\Rightarrow\; \Delta w = \omega\tau = 2\pi;$$

that is,

$$\omega \equiv 2\pi\nu = \partial H(J)/\partial J = \textit{fundamental circular frequency}. \tag{8.14.15a}$$

Others define J as $\oint p\,dq$, but w as $2\pi(\partial A_o/\partial J)$. Then,

$$w = 2\pi\{[\partial H(J)/\partial J]t + \gamma\} \equiv 2\pi(\nu t + \gamma) \;\Rightarrow\; \Delta w = 2\pi\nu\tau = 2\pi;$$

that is, again,

$$\nu = \partial H(J)/\partial J = \textit{fundamental frequency}. \tag{8.14.15b}$$

(And similarly for the general n-DOF case.)

In view of the w-periodicity, eq. (8.14.13b), and recalling (8.14.2–3), we can write (again, with $s = -\infty, \ldots, +\infty$)

(i) *Libration*:

$$q = q(w) = \sum c_s \exp(2\pi i s w) = \sum c_s \exp[2\pi i s(\nu t + \gamma)] \equiv \sum d_s \exp(2\pi i s t),$$

$$c_s = \int_0^1 q(w)\exp(-2\pi i s w)\,dw = c_s(J); \tag{8.14.16a}$$

(ii) *Rotation*:

$$q = q_o w + \sum c_s \exp(2\pi i s w) = q_o(\nu t + \gamma) + \sum c_s \exp[2\pi i s(\nu t + \gamma)] \equiv \sum d_s \exp(2\pi i s t),$$

$$c_s = \int_0^1 (q - q_o w)\exp(-2\pi i s w)\,dw = c_s(J). \tag{8.14.16b}$$

It is not hard to see, from the above, that any single-valued function $f(q, p)$ when expressed in terms of the corresponding action (J) and angle (w) variables, becomes a periodic function of w with period 1.

THEOREM

The reduced action $A_o(q, J)$ is a multiple-valued function of the coordinate q. Every time q varies over a cycle once—that is, during each period τ—the reduced action $A_o = A_o(q, J)$ increases by

$$\Delta A_o \equiv A_o(t + \tau) - A_o(t) = \oint (\partial A_o/\partial q)\,dq = \oint p\,dq = J, \tag{8.14.17a}$$

and hence the name *modulus of periodicity of* A_o for J. From the above, it follows that

$$\Delta(A_o - wJ) = \Delta A_o - \Delta w\,J = J - J = 0; \tag{8.14.17b}$$

that is, the new action function

$$A_{oo} \equiv A_o - wJ = A_{oo}(q, w), \tag{8.14.17c}$$

is periodic in w, while $A_o(q, J)$ is not. Also, for two distinct but neighboring motions, with corresponding action variable values J and $J + \Delta J$, eq. (8.14.14) yields

$$\Delta E = \nu\,\Delta J \tag{8.14.17d}$$

(an equation that constituted the starting point of the famous *correspondence* principle of the older quantum theory of N. Bohr, late 1910s–early 1920s).

REMARK

Instead of the canonical transformation $(q, p) \to (w, J)$, with generating function $A_o(q, J)$ and new Hamiltonian $H' = H(J) = E(J)$, we can, equivalently, consider the canonical transformation $(q, p) \to (\gamma, J)$ with generating function $A(t, q, J)$ and, hence, new Hamiltonian $H' = H + \partial A/\partial t = 0$, so that

$$p = \partial A/\partial q,$$

$$\gamma = \partial A/\partial J = \partial A_o/\partial J - t(\partial E/\partial J) = w - (\partial E/\partial J)t = \textit{phase constant}$$

$$\Rightarrow\ w = (\partial E/\partial J)t + \gamma = \nu t + \gamma; \tag{8.14.18a}$$

and new Hamiltonian equations

$$d\gamma/dt = \partial H'/\partial J = 0\ \Rightarrow\ \gamma = \textit{constant}, \tag{8.14.18b}$$

$$dJ/dt = -\partial H'/\partial \gamma = 0\ \Rightarrow\ J = \textit{constant}. \tag{8.14.18c}$$

Hence, in the new phase space ($q' = \gamma$, $p' = J$), the system motion is specified by the *point* ($\gamma = constant, J = constant$).

[Strictly speaking, it is *not* $w = \partial A_o/\partial J$ that is canonically conjugate to J, but $(q' \to)\gamma \equiv \partial A/\partial J = \partial A_o/\partial J - (\partial E/\partial J)t = w - \nu t\ \Rightarrow\ w = \nu t + \gamma$.]

HISTORICAL

Action–angle variables were introduced to dynamics by the French engineering scientist C. Delaunay, in connection with astronomical perturbation problems (1846: *Sur une nouvelle théorie analytique du mouvement de la lune*; 1860: *Théorie du mouvement de la Lune*) and were also used by the German mathematician P. Stäckel (1891) and the Swedish astronomer C. L. Charlier (1907: *Die Mechanik des Himmels*); although the term "action–angle variables" seems to have been introduced by the German (astro)physicist K. Schwarzschild (1916; in German: *Wirkungs–Winkel Variable*). They became very important again in both the old and new quantum mechanics (1910s, early 1920s), where they proved indispensable in several key theoretical developments and analytical tools; for example, *quantum conditions* (Sommerfeld, Bohr), *adiabatic invariants* (Ehrenfest, Burgers; see §8.15), *canonical perturbation theory* (Born, Heisenberg, Jordan, Pauli, Epstein, Brody, Fues, et al.; see §8.16).

Example 8.14.1 *Action–Angle Variables for the Harmonic Oscillator* (Recall ex. 8.10.3). From the energy conservation equation

$$H(q, p) = p^2/2m + kq^2/2 = E, \tag{a}$$

where $m =$ mass, $q =$ amplitude, $k =$ stiffness, so that $\omega^2 = k/m \Rightarrow k = m\omega^2 = m(2\pi\nu)^2$, we obtain the momentum:

$$p = \partial A_o/\partial q = \pm(2mE - mk\,q^2)^{1/2} = p(q; E, k, m), \tag{b}$$

and therefore the corresponding action variable becomes

$$J = \oint p\,dq = \oint (\partial A_o/\partial q)\,dq = \oint (2mE - mkq^2)^{1/2}\,dq$$

[In this case of libration, q extends (oscillates) between the roots of $dq/dt = \partial H/\partial p = 0$

$\Rightarrow p = 0$: $\quad q_{\min} \equiv -(2E/k)^{1/2} \quad$ and $\quad q_{\max} \equiv +(2E/k)^{1/2}$;

this also guarantees that $p = (2mE - mk\,q^2)^{1/2}$ remains *real*]

$$= 4\int_0^{q_{\max}} [2m(E - kq^2/2)]^{1/2}\,dq = \text{area of ellipse in } (q, p)\text{-space}$$

[utilizing the standard trigonometric substitution $q = (2E/k)^{1/2}\sin x$, etc.]

$$= (2E)\,(m/k)^{1/2}\left(\int_0^{2\pi} \cos^2 x\,dx\right) = (2\pi E)\,(m/k)^{1/2}. \tag{c}$$

From (c), we obtain

$$E = (J/2\pi)\,(k/m)^{1/2} = E(J) = H'$$

$$\Rightarrow \nu = \partial E/\partial J = (1/2\pi)\,(k/m)^{1/2} = \omega/2\pi. \tag{d}$$

Next, by calculating how A_o changes during a *complete period of oscillation*—that is, as q varies from $q_{\min}$ to $q_{\max}$ and then back to $q_{\min}$—we will verify that $A_o(q, E)$, or $A_o(q, J)$, is indeed a multiple-valued function of q. Integrating (b) [or ex. 8.10.3: (k)], while suppressing the resulting inessential constant, we obtain (fig. 8.16):

$$A_o = A_o(q, E) = E(m/k)^{1/2}\{\arcsin[(k/2E)^{1/2}q] + (k/2E)^{1/2}q[1 - (k/2E)q^2]^{1/2}\}. \tag{e}$$

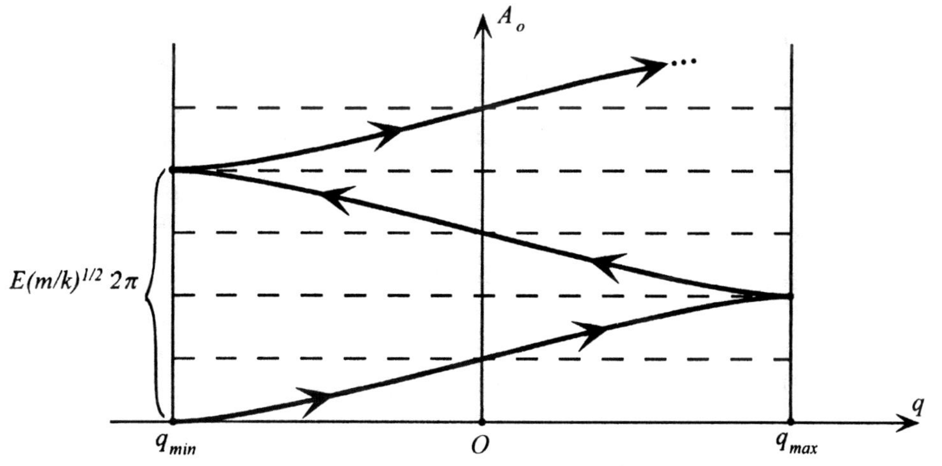

Figure 8.16 Graph of $A_o(q; E)$, eq. (e). Its slope is the momentum of the particle: $p = \partial A_o/\partial q$.

Now, clearly, the *second* term of (e) is *single-valued*, and so, during one such period, contributes nothing to A_o (in fact, at its beginning and ending, $q_{\min}$, it vanishes); but its *first* term is *multiple-valued*, and since, during $q_{\min} \to q_{\max} \to q_{\min}$, the argument of arcsin(...) changes from $-1 \to +1 \to -1$, arcsin(...) itself changes by 2π. Hence,

$$\Delta A_o = E(m/k)^{1/2}(2\pi) = E/\nu \equiv E\tau \quad [\equiv (2\pi)E/\omega]$$
$$= J \text{ (\textit{modulus of periodicity} of } A_o); \tag{f}$$

and, further, replacing in (e) E with $J\nu$, we obtain

$$A_o = A_o(q, J)$$
$$= (J/2\pi)\left\{ \arcsin[(k/2\nu J)^{1/2}q] + (k/2\nu J)^{1/2}q[1 - (k/2\nu J)q^2]^{1/2} \right\}$$
$$\left[= \int p(q, J)\, dq = (1/2\pi\nu) \int (2k\nu J - k^2q^2)^{1/2}\, dq \right]. \tag{g}$$

The above shows that in one complete period of q, the new coordinate w (canonically conjugate to J) changes by $+1$:

$$w \equiv \partial A_o/\partial J = (1/2\pi) \arcsin\left[(k/2\nu J)^{1/2}q\right] = \nu t + \gamma, \tag{h}$$
$$\Rightarrow \Delta w = (1/2\pi)(2\pi) = +1;$$

like an angle $\phi \equiv 2\pi w$, hence the name *angle* variable. Finally, inverting (h), we obtain

$$q = (2\nu J/k)^{1/2} \sin(2\pi w) = (J/2\pi^2 \nu m)^{1/2} \sin[2\pi(\nu t + \gamma)], \tag{i}$$

$$\Rightarrow p = m\dot{q} = (kJ/2\pi^2\nu)^{1/2} \cos(2\pi w); \tag{j}$$

that is, both q and p are periodic in w, with period 1. Incidentally, eqs. (i, j) are the equations of canonical transformation from (w, J) to (q, p).

This simple example may, hopefully, begin to show the advantages of the action–angle variables: in the (q, p)-space, the system trajectory for a given constant energy E is the *two-valued* function (b); whereas in (w, J)-space, the system trajectory is characterized *uniquely* by the *constant* J, $J = J(E)$, and each such curve is characterized by a *single-valued* function of w.

Finally, we note the close relation of the above results to those of ex. 8.10.8. Each of the latter's equations (h) is of the harmonic oscillator type (b) $\to$ (g): $A_o = \int (2mE - mkq^2)^{1/2}\, dq$; that is, each q_k varies between a $q_{k,\min}$ to $q_{k,\max}$ and then back to $q_{k,\min}$, with frequency ν_k. However, even though these q_k's are uncoupled (Liouville system), this does not guarantee that the system is *periodic as a whole*; namely, that it *returns to its original configuration*. Such questions of periodicity in several DOF systems are treated below.

Several DOF, Multiply Periodic Motion

Here, we shall restrict ourselves to systems that are *completely separable* in all their $n + 1$ variables $(q_1, \ldots, q_n; t) \equiv (q; t)$ (which, for all practical purposes is the only case where the Hamilton–Jacobi equation *can* be solved), and *periodic* in at least one set of canonical variables. This (recalling §8.10) means that

$$\begin{aligned} A &= A_o(q_1,\ldots,q_n;\beta_1,\ldots,\beta_n) - E(\beta_1,\ldots,\beta_n)t \\ &\equiv A_o(q;\beta) - E(\beta)t \\ &= \sum A_{ok}(q_k,\beta) - E(\beta)t, \qquad (k=1,\ldots,n) \end{aligned} \tag{8.14.19a}$$

from which it follows that

$$dA_o = \sum (\partial A_o/\partial q_k)\,dq_k = \sum (\partial A_{ok}/\partial q_k)\,dq_k = \sum p_k\,dq_k, \tag{8.14.19b}$$

or, by (indefinite) q_k-integration,

$$\begin{aligned} A_o &= \int \sum (\partial A_o/\partial q_k)\,dq_k = \int \sum (\partial A_{ok}/\partial q_k)\,dq_k = \sum \left(\int (\partial A_{ok}/\partial q_k)\,dq_k\right) \\ &= \sum \left(\int p_k(q_k,\beta)\,dq_k\right) \equiv \sum \int [f_k(q_k,\beta)]^{1/2}\,dq_k \end{aligned} \tag{8.14.19c}$$

[where the $A_{ok} = A_{ok}(q, J)$ are multiple-valued functions of the q's]; and the projection of the system trajectory in phase space on every (q_k, p_k)-subplane, $p_k = p_k\,(q_k, \beta)$ (since now *each* p_k *depends only on* q_k, and the β's) is also periodic (libration or rotation).

However this does *not* necessarily mean that all such projected (q_k, p_k)-subtrajectories have the same fundamental frequency—that is, *periodicity in the sense that, after the passage of a certain finite time interval, all q's and p's return to their initial values, in general, does not exist*—due to coordinate coupling; after the passage of a time interval τ_k, only the pair (q_k, p_k) returns to its initial values, but not the other pairs (The reader, probably, recalls a similar situation in linear multi-DOF vibrations: each normal mode is periodic in time but their superposition, in general, is not.)

Such a motion (and system), is *n-ply*, or *multiply, periodic*. It can become truly periodic, in the earlier sense of the *system as a whole*, when certain special conditions of proportionality, or *commensurability*, exist among its partial frequencies $\nu_k = 1/\tau_k$ $(k = 1,\ldots,n)$. According to this definition, a two-dimensional oscillator *is* a periodic system even when its xy-plane trajectory (Lissajous' figure) is an *open* curve. These fundamental concepts are examined in detail below. [For an excellent summary of the basic underlying theory of multiply periodic *functions*, see Born (1927, pp. 71–76).]

As a result of the complete separability of the system: (i) once $A(t, q, \beta)$ has been found (as explained in §8.10), the individual $q_k(t)$ and $p_k(t)$ are determined by the finite Hamilton–Jacobi equations

$$\partial A/\partial\beta_k = \alpha_k, \qquad \partial A/\partial q_k = \partial A_o/\partial q_k = \partial A_{ok}/\partial q_k = p_k \tag{8.14.20a}$$

[provided that $Det(\partial^2 A_o/\partial q_k \partial\beta_l) \neq 0$]; that is, the motion has been reduced to one-variable integrations; and (ii) the constant energy equation

$$\begin{aligned} H(q_1,\ldots,q_n;\,p_1,\ldots,p_n) &\equiv H(q,p) = H(q,\partial A_o/\partial q) \\ &= E = E(\beta_1,\ldots,\beta_n) \equiv E(\beta) = constant \end{aligned} \tag{8.14.20b}$$

[a $(2n-1)$-dimensional *energy hypersurface* in (q,p)-phase space] separates to the n one-to-one first integrals

$$H_k(q_k, p_k) = E_k(\beta) \equiv E_k, \tag{8.14.20c}$$

where $E_1 + \cdots + E_n = E$ is the value of the new Hamiltonian H', so that the projections of the phase-space trajectory of the system on the individual (q_k, p_k)-planes look just like the previous 1-DOF trajectories; that is, each individual q_k either librates between two fixed limits ($q_{\min}$ and $q_{\max}$); or increases boundlessly, but its corresponding p_k periodically returns to its original value (rotation).

Now, *for such a completely separable system, with individually periodic* (libratory or rotatory) q_k's, we define the *action variable* J_k corresponding to q_k by the phase integral

$$\begin{aligned} J_k &\equiv \oint p_k\, dq_k = \oint p_k(q_k, \beta)\, dq_k = J_k(\beta_1, \ldots, \beta_n) \equiv J_k(\beta) \\ &= \oint (\partial A_o/\partial q_k)\, dq_k = \oint \left[\partial A_{ok}(q_k, \beta)/\partial q_k\right] dq_k, \end{aligned} \tag{8.14.21}$$

where the integrations extend over the complete periods on the (q_k, p_k)-plane, for fixed β_k's $\Rightarrow$ fixed E_k's $\Rightarrow$ fixed E. In the case of libration, this means integration over the closed (q_k, p_k) paths — something that takes care of the multiple-valuedness of the momenta, due to their *quadratic* appearance in (8.14.20c): $p_k = \pm \ldots$; while in the case of rotation, the integration extends over a single period q_{ko} of q_k. Hence, each J_k equals the corresponding *shaded area* of its trajectory in its (q_k, p_k)-plane (figs. 8.11, 8.12); that is, the area contained *within* the closed trajectory (libration), or *under* a single q_k-cycle (rotation).

REMARK

Comparison of (8.14.19c) with (8.14.21) shows that the former is an indefinite integral, while the latter is the closed line integral of the partial derivative $\partial A_{ok}/\partial q_k$ in the (q_k, p_k)-plane, as explained above. Since $J_k \neq 0$, we conclude that A_{ok} is a *multiple-valued* function of its coordinate. In general, the calculation of the numbers J_k via (8.14.21) is very laborious; but since these are two-dimensional *contour integrals*, the application of complex variables (Cauchy integration) can be utilized to great advantage; see, for example, Born (1927; appendix II), Sommerfeld (1931, vol. 1); also Goldstein (1980, p. 472 ff.), Pars (1965, pp. 344–346).

In the case where neither the motion as a whole, nor each q_k, have a periodic variation in time, the integrals in (8.14.21) are understood as extending over the entire range of q_k values. See equations (8.14.24i, j) and subsequent discussion; and *conditional periodicity* and *degeneracy* below.

Solving the n independent functions (8.14.21), $J_k = J_k(\beta)$, for the β's, we obtain

$$\beta_k = \beta_k(J_1, \ldots, J_n) \equiv \beta_k(J); \tag{8.14.21a}$$

that is, the n J_k's can replace the n β_k's as the new constant momenta. Then A_o takes the following functional form:

$$A_o = A_o(q, \beta) = A_o[q, \beta(J)] = A_o(q, J) = \sum A_{ok}(q, J) \tag{8.14.21b}$$

[= generating function of the *old coordinates* (q) and the *new momenta* (J); i.e., $F_2(q, p')$], and therefore the canonical transformation equations $(q, p) \to (q', p') = (w, J)$ become

$$p_k = \partial F_2/\partial q_k: \quad p_k = \partial A_{ok}(q_k, \beta)/\partial q_k = \partial A_{ok}(q_k, J)/\partial q_k = p_k(q_k, J), \qquad (8.14.21c)$$

$$q_{k'} = \partial F_2/\partial p_{k'}: \quad w_k = \partial A_o(q_k, J)/\partial J_k = \sum [\partial A_{ol}(q_l, J)/\partial J_k] = w_k(q, J); \qquad (8.14.21d)$$

and show that each p_k depends only on the corresponding q_k (*separation*) and all the J_k's; and each w_k depends on all the q's (*coupling*) and all the J_k's.

The new *coordinates* $w \equiv (w_1, \ldots, w_n)$, canonically conjugate to the new momenta $J \equiv (J_1, \ldots, J_n)$, are called *angle variables*. Let us find the corresponding equations of motion. Since the new Hamiltonian $H' = H'(w, J)$ equals

$$H' = H + \partial A_o/\partial t = H = E(\beta) = E[\beta(J)] \equiv E(J) = H'(J) \qquad (8.14.22a)$$

(i.e., *all the* w's *are ignorable*), the canonical equations are

$$dw_k/dt = \partial E(J)/\partial J_k = constant \equiv \nu_k(J) = \nu_k \;\Rightarrow\; w_k = \nu_k t + \gamma_k, \qquad (8.14.22b)$$

$$dJ_k/dt = -\partial E(J)/\partial w_k = 0 \;\Rightarrow\; J_k = constant$$

$$(= \partial E/\partial J_k, \text{by Hamilton's equations}), \qquad (8.14.22c)$$

recalling (8.10.3a–d) and the discussion following them.

REMARKS ON COMPLETE INTEGRABILITY

The n constant and *independent* conjugate momenta J_k, eqs. (8.14.22c), constitute an example of *n global "isolating" integrals* or *separation constants* along any phase-space trajectory; say, $f \equiv (f_1, f_2, \ldots, f_n)$, with (no loss in generality) $f_n = H$. Systems for which such constants exist are called *completely integrable* (hence, *complete separability implies complete integrability*). As McCauley puts it (our notations, his italics): "*an integrable canonical Hamiltonian flow is one where the interactions can be transformed away*: there is a special system of generalized coordinates and canonical momenta where the motion consists of n independent global translations $w_k = \nu_k t + \gamma_k$ along n axes, with both the ν_k and J_k constant" (1997, pp. 158 ff.); and "*a completely integrable flow* is one where there are $n - 1$ time-independent invariants $F_i(f) = C_i$ $(i = 1, \ldots, n-1)$ that isolate [or *separate*] $n - 1$ of the variables in the form $f_i = \phi_i(f_n; C_1, \ldots, C_{n-1})$, for all finite times" (p. 193) (*before* obtaining the complete solution to our problem!).

It can be shown that a sufficient condition for this is that the f_k be in *involution*; namely, that their *Poisson brackets vanish*: $[f_k, f_s] = 0$ for all $k, s = 1, 2, \ldots, n$. For an elaborate treatment of these fundamental concepts and their significance to modern nonlinear dynamics, see, for example, Gallavotti (1983, pp. 287–289, 361–362), Lichtenberg and Lieberman (1992, p. 24 ff.), McCauley (1997, pp. 158 ff., 192 ff., 311 ff., 316 ff., 409 ff.; best exposition), Tabor (1989, pp. 68–79); also, Whittaker (1937, pp. 322–323), Wintner (1941, pp. 68, 144).

Below, we show that *the constants* $\nu \equiv (\nu_1, \ldots, \nu_n)$ *are the fundamental frequencies of each DOF of this multiply periodic motion — the main advantage of the action–angle method — while the* $\gamma \equiv (\gamma_1, \ldots, \gamma_n)$ *are its phase constants.*

Let us consider a special, kinematically admissible or possible, motion in which each q_l $(l = 1, \ldots, n)$ goes through a *complete* (libratory or rotatory) cycle an integral number of times i_l $(= 0, 1, 2, \ldots)$. Then, recalling (8.14.21d), and that the J remain constant here, we find that w_k changes by the following amount;

$$\begin{aligned}
\Delta w_k &= \oint dw_k = \oint \sum [\partial w_k(q,J)/\partial q_l]\, dq_l = \oint \sum (\partial^2 A_o/\partial J_k \partial q_l)\, dq_l \\
&= \sum \partial/\partial J_k \left(\oint (\partial A_o/\partial q_l)\, dq_l \right) = \sum \partial/\partial J_k \left(\oint (\partial A_{ol}/\partial q_l)\, dq_l \right) \\
&\equiv \sum \partial/\partial J_k \left(\oint p_l(q_l, J)\, dq_l \right) = \sum \partial(i_l J_l)/\partial J_k \\
&= \sum i_l\, \delta_{kl} = i_k. \qquad (8.14.23)
\end{aligned}$$

(For complete justification of the commutation step used in this derivation, see ex. 8.14.13 below.)

In words: the mapping (8.14.21d) from the q-space to the w-space has the following properties:

- If, starting from a certain configuration, a particular q_k is allowed to complete i_k cycles—that is, either vary "from here to there and back" (libration), or revolve (rotation), an integral number of times i_k—then only w_k changes by i_k, all other w's do not; in other words, $w_k(q, J)$ is a multiple-valued function of the q's, periodic in q_l $(l \neq k)$ and monotonically increasing in q_k; for each cycle of the latter, w_k increases by 1.
- Inverting (8.14.21d), we also conclude that *if w_k increases by i_k while the other w's do not change, only q_k goes through i_k complete cycles*. Any other q_l $(l \neq k)$ depending on w_k would have varied, but would have returned to its original value without completing its cycles; otherwise w_l would have increased by the number of those cycles. Hence, in general, *each q depends on all the w's and J's; and is periodic in each w with fundamental period unity*. (If a particular q does not depend on all the w's, then, of course, it will not be periodic in all of them; but the totality of q's depends on the totality of the w's.)

In addition, since:

(i) the $A_{ok}(q, J)$ are also multiple-valued functions of the q's, every time q_k varies over a cycle once (i_k times), while all other q's remain unchanged, the reduced action A_o increases by the amount J_k $(i_k J_k)$:

$$\begin{aligned}
\Delta A_o &\equiv A_o(t+\tau) - A_o(t) = \oint (\partial A_o/\partial q_k)\, dq_k \\
&= \Delta A_{ok} = \oint (\partial A_{ok}/\partial q_k)\, dq_k = \oint p_k\, dq_k = J_k; \qquad (8.14.23a)
\end{aligned}$$

and since, then,

(ii) w_k increases by 1, while the other w's do not change, the sum $\sum w_k J_k$ also increases by J_k; therefore, it follows that the new function

$$A_{oo} \equiv A_o - \sum w_k J_k, \qquad (8.14.23b)$$

remains *unchanged*; that is, A_{oo} *is multiply periodic in the* w's *with fundamental period 1 in each of them.*

[Recalling §8.8, we see that this is a Legendre transformation, from a $F_2(q, p')$-type generating function, $A_o(q, J)$, to a $F_1(q, q')$-type, $A_{oo} = A_{oo}(q, w)$. Indeed $(\ldots)^{\cdot}$-differentiating $A_o = A_o(q, J)$ and then invoking (8.14.21c, d) and (8.14.23b), we obtain

$$dA_o/dt = \sum \left[(\partial A_o/\partial q_k)(dq_k/dt) + (\partial A_o/\partial J_k)(dJ_k/dt)\right]$$

$$= \sum p_k(dq_k/dt) + \sum w_k(dJ_k/dt)$$

$$\Rightarrow \sum p_k(dq_k/dt) = -\sum w_k(dJ_k/dt) + dA_o/dt$$

$$\Rightarrow \sum p_k(dq_k/dt) - \sum J_k(dw_k/dt) = dA_{oo}/dt$$

$$\Rightarrow p_k = \partial A_{oo}(q, w)/\partial q_k, \qquad J_k = -\partial A_{oo}(q, w)/\partial w_k.] \qquad (8.14.23c)$$

The foregoing analysis is summarized in the following rule.

Frequency Rule

To calculate the fundamental frequencies of a completely separable multiply periodic system, we proceed as follows:

- Using the Hamilton–Jacobi theory (§8.10), we first determine its reduced characteristic function

$$A_o = A_o(q, \beta) = \sum A_{ok}(q_k;\ \beta_1, \ldots, \beta_n).$$

- Then, using the definition (8.14.21) $J_k \equiv \oint p_k\, dq_k$, we calculate its action variables $J_k = J_k(\beta)$; and this is the *main difficulty of the method.*
- Next, we express its Hamiltonian as function of the J_k's: $H'(J) = H(q, p) = E(J)$.
- Finally, we calculate its fundamental frequencies from $\nu_k = \partial E(J)/\partial J_k$.

Analytically, the above are expressed as follows:

(i) *Libration*: The q's (and p's) are multiply periodic functions of the w's with fundamental period 1:

$$q_k(w_1 + i_1, \ldots, w_n + i_n;\ J) = q_k(w_1, \ldots, w_n;\ J) \equiv q_k(w, J). \qquad (8.14.24a)$$

(ii) *Rotation*. If q_k has period q_{ko}, then

$$q_k(w_1 + i_1, \ldots, w_n + i_n;\ J) = q_k(w_1, \ldots, w_n;\ J) + i_k q_{ko}. \qquad (8.14.24b)$$

The rotation case can be brought to the libration form in the new coordinates $q_{R,k}$ defined by

$$q_{R,k} \equiv q_k(w, J) - w_k q_{ko} \equiv q_{R,k}(w, J) \qquad (8.14.24c)$$

$$\Rightarrow q_{R,k}(w_1 + i_1, \ldots, w_n + i_n;\ J) = q_{R,k}(w_1, \ldots, w_n;\ J) \equiv q_{R,k}(w, J). \qquad (8.14.24d)$$

[Clearly, conditions (8.14.24a–d) still hold, trivially, even for the w's absent from a particular q_k.] Other (nonseparable) arbitrary coordinates related to our (separable

q's) by one-to-one and well-behaved transformations—for example, from the q's and/or q_R's to, say rectangular Cartesian coordinates

$$x_k = x_k(q) \Leftrightarrow q_k(x), \tag{8.14.24e}$$

will be representable by the following *multiple-frequency Fourier series* (for generality, we keep the *same notation* as for the hitherto separable q's):

$$\begin{aligned} &\sum \cdots n\text{-ple sum} \cdots \sum c_{k,s}(J)\, \exp(2\pi i\, \mathbf{s} \cdot \mathbf{w}) \\ &\quad = q_k(w, J) \qquad\qquad (\textit{Libration}) \\ &\quad = q_k(w, J) - w_k q_{ko} \equiv q_{R,k}(w, J) \qquad (\textit{Rotation}), \end{aligned} \tag{8.14.24f}$$

where

$$\begin{aligned} &\mathbf{s} \equiv (s_1, \ldots, s_n) = \text{positive or negative integers, or zero; ranging from } -\infty \text{ to } +\infty, \\ &\mathbf{w} \equiv (w_1, \ldots, w_n), \\ &\mathbf{s} \cdot \mathbf{w} \equiv s_1 w_1 + \cdots + s_n w_n \qquad (\text{“dot product” of “vectors” } \mathbf{s} \text{ and } \mathbf{w}), \end{aligned} \tag{8.14.24g}$$

and

$$c_{k,s}(J) = \int_0^1 \cdots n\text{-ple} \cdots \int_0^1 q_k(w, J)\, \exp(-2\pi i\, \mathbf{s} \cdot \mathbf{w})\, dw_1, \ldots, dw_n. \tag{8.14.24h}$$

[For a simple proof of (8.14.24f–h) see ex. 8.14.2, below; and for a discussion of the significance of such series, in the context of general vibration theory, see the appendix at the end of this section.]

Since $w_k = \nu_k t + \gamma_k$, the above yield the following multiply periodic *temporal* variation of q_k:

$$\begin{aligned} &\sum \cdots n\text{-ple sum} \cdots \sum c_{k,s}(J) \exp[2\pi i\, (\mathbf{s} \cdot \mathbf{v}t + \mathbf{s} \cdot \boldsymbol{\gamma})] \\ &\quad = \sum \cdots n\text{-ple sum} \cdots \sum d_{k,s}(J, \gamma) \exp(2\pi i\, \mathbf{s} \cdot \mathbf{v}t) \\ &\quad = q_k(t, J) \qquad\qquad (\textit{Libration}) \\ &\quad = q_k(t, J) - w_k q_{ko} \equiv q_{R,k}(t, J) \qquad (\textit{Rotation}), \end{aligned} \tag{8.14.24i}$$

where

$$\mathbf{v} \equiv (\nu_1, \ldots, \nu_n), \quad \boldsymbol{\gamma} \equiv (\gamma_1, \ldots, \gamma_n),$$
$$\mathbf{s} \cdot \mathbf{v} \equiv s_1 \nu_1 + \cdots + s_n \nu_n, \quad \mathbf{s} \cdot \boldsymbol{\gamma} \equiv s_1 \gamma_1 + \cdots + s_n \gamma_n,$$

and

$$d_{k,s}(J, \gamma) \equiv c_{k,s}(J) \exp(2\pi i\, \mathbf{s} \cdot \boldsymbol{\gamma}). \tag{8.14.24j}$$

Since in separable systems, by (8.14.21c), $p_k = p_k(q_k, J)$, we will have similar Fourier expansions for the momenta p_k; and, in fact, *any single-valued function $f(q, p)$ when expressed in terms of the corresponding action (J) and angle (w) variables, becomes a periodic function of the w's with period 1 in each of them*; for example, the earlier-mentioned rectangular Cartesian coordinates.

Now, the series (8.14.24i) decomposes the q_k-motion into a sum of periodic motions (harmonic vibrations), each with frequency $|\boldsymbol{s} \cdot \boldsymbol{v}| \equiv |s_1 \nu_1 + \cdots + s_n \nu_n|$; but since, in general, these frequencies are *not in rational ratios to each other* (i.e., they are *not mutually commensurate*, or *commensurable*), *their sum is not periodic in time*; no common period τ exists that contains every "component period" an integral number of times; and similarly for the momenta, even though $p_k = p_k(q_k)$ is closed (libration) or periodic (rotation).

Hence, the system motion is *nonperiodic*, both as a whole and in any of its Hamiltonian coordinates. This implies that the system does not return to an initial of its states in a finite time; although, given sufficient time, it passes arbitrarily close to it. [Since *all irrational numbers can be approximated to any desired accuracy with rational ones*, we can approximate a nonperiodic function with a periodic one, for a given time. The latter is called *almost periodic function*. Also, recall discussion in §7.A4, and see "remark" below.] For these reasons, such multiply, or quasi-periodic motions have been termed *conditionally periodic* (O. Staude, 1887); that is, they can become truly periodic (i.e., singly periodic) only under certain conditions among the system frequencies ν_k $(k = 1, \ldots, n)$. These conditions are summarized in the following theorem.

THEOREM

The motion of $q_k(t)$ and $p_k(t)$, given by (8.14.24f–j), is periodic in *time* with fundamental frequency ν_k in the first *three* of the following cases:

(i) The motion is *one*-dimensional; that is, only the pair (q_k, p_k) varies. Then, (8.14.24i, j) becomes

$$\sum d_{k,s}(J, \gamma) \exp(2\pi i\, s \nu_k t)$$

$$= q_k(t, J) \qquad \text{(Libration)}$$

$$= q_k(t, J) - w_k q_{ko} \equiv q_{R,k}(t, J) \qquad \text{(Rotation)}, \tag{8.14.25a}$$

and from (8.14.22b–23) for one cycle (i.e., $i_k = 1$), we obtain

$$\Delta w_k \equiv w_k(t + \tau_k) - w_k(t) = \nu_k \tau_k = 1$$

$$\Rightarrow \nu_k = 1/\tau_k = \partial E(J)/\partial J_k = \text{fundamental frequency}. \tag{8.14.25b}$$

(ii) The motion of q_k is not influenced by the other coordinates (separability of variables: generalization of the uncoupled *normal coordinates* of linear vibration theory): in (8.14.24i, j) only the coefficient $d_{k;0\ldots s(k)\ldots 0}(J, \gamma)$ survives; that is, the Fourier expansion looks like (8.14.25a). As a result, $q_k = q_k(w_k, J) \Leftrightarrow w_k = w_k(q_k, J)$; and also [recalling (8.14.21d)]

$$w_k = \partial A_o(q_k, J)/\partial J_k = \partial A_{ok}(q_k, J_k)/\partial J_k = w_k(q_k, J_k); \tag{8.14.26}$$

that is, J_k occurs only in A_{ok}, not in A_{ol} $(l \neq k)$. Here, too, $\Delta w_k = \nu_k \tau_k = 1 \Rightarrow \nu_k = 1/\tau_k$. General functions of them, say $f(q_1, \ldots, q_n)$, will be *multiply periodic functions of the w's* and, hence, *nonperiodic functions of time*, unless the ν_k are mutually commensurate [see (iii), (iv) below].

(iii) The other fundamental frequencies are *integral multiples* of ν_k:

$$\nu_l = i_l \nu_k: \quad i_l = \text{positive integer } (l \neq k), \tag{8.14.27a}$$

$$i_k = 1. \tag{8.14.27b}$$

Then, (8.14.24i, j), with

$$\boldsymbol{i} \equiv (i_1, \ldots, i_n), \tag{8.14.27c}$$

$$\boldsymbol{s} \cdot \boldsymbol{i} = s_1 i_1 + \cdots + s_n i_n = integer, \tag{8.14.27d}$$

becomes

$$q_k(t) = \sum \cdots n\text{-ple sum} \cdots \sum d_{k,s}(J, \gamma) \exp[2\pi i\, (\boldsymbol{s} \cdot \boldsymbol{i}) \nu_k t]$$

$$= \text{periodic function of time, with fundamental frequency (period)} \; \nu_k \; (\tau_k); \tag{8.14.27e}$$

and, of course, after a time interval $\tau_k = 1/\nu_k = i_l/\nu_l$:

$$\Delta w_l = \nu_l \tau_l = i_l\, (\nu_k \tau_k) = i_l \qquad (l \neq k), \tag{8.14.27f}$$

$$\Delta w_k = \nu_k \tau_k = i_k = 1, \tag{8.14.27g}$$

and q_l returns to its initial value, after performing i_l complete cycles.

(iv) If none of the above three conditions hold, the motion of the system *as a whole* is periodic if all its *fundamental frequencies are commensurate (or commensurable); that is, they are in rational ratios to each other*—namely, for *all* $k, l = 1, \ldots, n$ $(k \neq l)$, and for arbitrary values of the J's, integers i_k, i_l exist such that

$$\nu_k/\nu_l = i_k/i_l \; (\equiv \omega_k/\omega_l) \;\Rightarrow\; \tau_l/\tau_k = i_k/i_l; \tag{8.14.28a}$$

or, equivalently $(n - 1$ relations),

$$\nu_k/i_k = \nu_l/i_l: \qquad \nu_1/i_1 = \nu_2/i_2 = \cdots = \nu_n/i_n \equiv \nu \; (\equiv 1/\tau); \tag{8.14.28b}$$

$$\tau_k i_k = \tau_l\, i_l: \qquad \tau_1 i_1 = \tau_2 i_2 = \cdots = \tau_n\, i_n \equiv \tau \; (\equiv 1/\nu); \tag{8.14.28c}$$

where $\nu =$ common/system *frequency*, $\tau =$ common/system *period*.

Then, the particular coordinate q_k $(k = 1, \ldots, n)$ becomes

$$q_k(t) = \sum \cdots n\text{-ple sum} \cdots \sum d_{k,s}(J, \gamma) \exp[2\pi i\, (\boldsymbol{s} \cdot \boldsymbol{i}) \nu t]:$$

$$\text{periodic function of time, with fundamental frequency} \; (not\ \nu_k = i_k \nu, \text{but})\ \nu; \tag{8.14.28d}$$

and

$$\Delta w_k \equiv w_k(t + \tau_k) - w_k(t) = \nu_k \tau = \nu_k/\nu = i_k; \tag{8.14.28e}$$

after a time $\tau = 1/\nu = i_k/\nu_k$, each q_k returns to its initial value (after performing i_k complete cycles).

REMARK

Let q_k be a general multiply periodic function of *all* the w's. Now, assume that *during a time interval* τ, q_k performs i_k complete cycles plus a fraction of a cycle. Then, it has been shown by Vinti (1961), via number-theoretic tools (Dirichlet's theorem), that

$$\text{as } \tau \to \infty: \qquad \lim(i_k/\tau) = \nu_k; \tag{8.14.29}$$

that is, *even when q_k is not a singly periodic function with frequency ν_k, still its "mean* *frequency" [as defined by the left side of (8.14.29)] equals* ν_k (see, e.g., Garfinkel, 1966, pp. 57–58).

Example 8.14.2 (Mathematical Appendix) *Fourier's Theorem for Multiply Periodic Functions.* Here, we prove (8.14.24f–h) for $n = 2$. The extension to the general case $n = n$ should then be obvious.

Let the function $f = f(w)$ be periodic in w with period 1 (a function with arbitrary period can easily be brought to this case — see any text on Fourier series). Then, by Fourier's theorem in complex form:

$$f(w) = \sum c_s \exp(2\pi i\, sw) \qquad (s = -\infty, \ldots, +\infty),$$

where

$$c_s = \int_0^1 f(w) \exp(-2\pi i\, sw)\, dw. \tag{a}$$

Next, let us consider the function $F = F(w, w')$, periodic in both w and w', with period 1 in each of them. Since F is periodic in w', by (a), we will have

$$F(w, w') = \sum c_{s'} \exp(2\pi i\, s'w') \qquad (s' = -\infty, \ldots, +\infty),$$

where

$$c_{s'} = \int_0^1 F(w, w') \exp(-2\pi i\, s'w')\, dw' \;=\; c_{s'}(w). \tag{b}$$

Now, it is not hard to see that $c_{s'}(w)$ is periodic in w with period 1. Therefore, again by (a), we can write

$$c_{s'} = \sum c_{s,s'} \exp(2\pi i\, sw) \qquad (s = -\infty, \ldots, +\infty),$$

where

$$c_{s,s'} = \int_0^1 c_{s'} \exp(-2\pi i\, sw)\, dw \qquad \text{(a constant)}. \tag{c}$$

From the above, it follows immediately that

$$F(w, w') = \sum\sum c_{s,s'} \exp[2\pi i\, (s\, w + s'w')] \qquad (s, s' = -\infty, \ldots, +\infty),$$

where

$$c_{s,s'} = \int_0^1 \int_0^1 F(w, w') \exp[-2\pi i\, (s\, w + s'w')]\, dw\, dw', \quad \text{Q.E.D.} \tag{d}$$

If the function $F(w, w')$ is real, as assumed here, the Fourier coefficients $c_{s,s'}$ and $c_{-s,-s'}$ are *complex conjugate* numbers; and similarly for the general n-variable case.

Geometrical Interpretation of Multiple Periodicity; Degeneracy

To understand the above results better, let us examine the system motion, neither in phase space (P_n) nor in configuration space (Q_n), but in the earlier-introduced, via (8.14.21d), $w_k = w_k(q; J)$, *space of the angle variables; or w-space* W_n.

Let us view the q's as rectangular Cartesian coordinates, spanning an n-dimensional Euclidean space Q_n. Then, since these coordinates have been assumed separable and periodic, the system motion in Q_n will be restricted to the interior of an n-dimensional finite rectangular prism, with sides parallel to the q-axes and size determined by the initial conditions on the q's. If we also assume that W_n is an n-dimensional Euclidean space spanned by the rectangular Cartesian coordinates $\{w_k;\ k = 1, \ldots, n\}$, then, due to (8.14.23) and its consequences, the earlier Q_n-prism will be mapped onto an infinity of *unit cubes*, in W_n, such that corresponding points in any two or more such cubes correspond to the same Q_n-space point. It follows that the entire system motion in W_n-space can be described by the motion of a figurative system point in a single representative fundamental unit cube C_n, one corner of which is taken at the origin of the w-axes ($t = 0$). Then partial motions in any other W_n-cube can be transferred to their corresponding segment in C_n. Indeed, eliminating the time t among the n equations $w_n = \nu_k t + \gamma_k$ ($k = 1, \ldots, n$), we see that the system motion in C_n is a uniformly traversed *straight line* whose direction cosines with the w-axes, l_k, satisfy the following $n - 1$ relations:

$$l_1 : l_2 : \ldots : l_n = dw_1 : dw_2 : \ldots : dw_n = \nu_1 : \nu_2 : \ldots : \nu_n. \tag{8.14.30}$$

In view of the w-periodicity, there is no need to examine that line in its entire length, only its portions inside C_n; every other part of it is transferred there by translation parallel to the w-axes. In conclusion, *the system path in W_n consists of well-defined parallel straight-line segments inside C_n*; and this is far simpler than the corresponding path in Q_n (or in real space) which, in general, is a complicated "Lissajous figure."

Let us examine, for concreteness, a *two-dimensional* such case (fig. 8.17(a)). Let the C_2-origin O coincide with the initial system position $w_k(t = 0) = \gamma_k = 0$. When the system path meets the right C_2-boundary at O_1 it is reflected horizontally back to the left boundary, which it meets at O_2, and with that as new origin continues again along a straight line segment parallel to OO_1. The reflection, or jump, at O_1, and so on, is the mathematical equivalent of transferring all w-motion inside C_2, and, as such, has no particular physical significance. The motion continues, similarly,

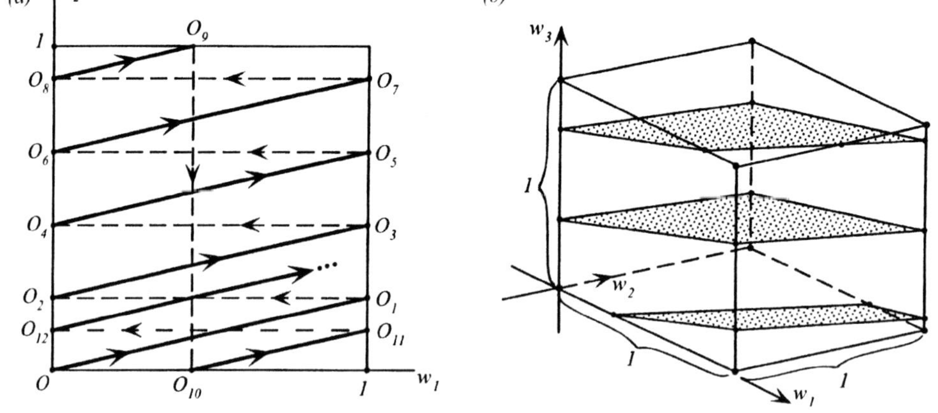

Figure 8.17 (a) Motion of separable and periodic systems in angle-space ($n = 2$) inside the fundamental unit cube → square C_2. (b) Case of degeneracy ($n = 3, m = 1$).

through $O_2 \to O_3 \to O_4 \to O_5 \to O_6 \to O_7$, where it is reflected vertically downwards to O_8, and from there on to $O_9 \to O_{10} \to \cdots$. That is why C_n has been aptly described by Lanczos (1970, p. 249) as a small cubic room with doorless reflecting walls; a "mirror-cabinet."

Here, the following fundamental questions arise: will the system path ever return to its initial position O, and from there on repeat itself ad infinitum (in which case, C_2 is laced by a finite number of parallel straight-line segments); or, will it keep going indefinitely, gradually filling C_2 completely?

The answer to these questions involves the key concept of *degeneracy*. Let us summarize it here. A system whose frequencies satisfy the m $(0 \le m \le n-1)$ linear, homogeneous, and independent *commensurability relations* (i.e., special "frequency constraints"):

$$\sum i_{dk}\nu_k = 0 \qquad (d = 1, \ldots, m), \tag{8.14.31}$$

where the i_{dk} are *integers*, or zero, but at least two of them (in each such equation) are nonzero, for arbitrary values of its actions, J_k is called m-ply (or m-fold) *degenerate*. Only an $(n-m)$-dimensional submanifold of the system's C_n cube is filled up densely; that is, only a finite number of equidistant and parallel planes [fig. 8.17(b): $n = 3, m = 1$]. The system point comes, infinitely often, arbitrarily close to any chosen point there. Similarly, its motion in Q_n-space is confined to an $(n-m)$-dimensional submanifold there, which it also fills up completely: straight lines $(C_n) \to$ curves (Q_n), plane subspace $(C_n) \to$ curved subspace (Q_n), and so on, but with the same number of dimensions. Due to (8.14.31), the Fourier series (8.14.24i, j), is reduced from an n-ply periodic function of time to an $(n-m)$-ply periodic function of time. (See pp. 1287–1289.)

In sum: *an m-ply degenerate n-DOF system is* $(n-m)$*-ply periodic*; compactly,

$$\#\,DOF - \textit{degree of degeneracy} \equiv n - m = \#\,\textit{periods}.$$

Special Cases

(i) If $m = 0$ $(\Rightarrow n - m = n)$, the system is called *nondegenerate*. Then, there exist n *independent fundamental frequencies*, and therefore the corresponding Fourier series is genuinely n-ply periodic. The system orbit, in Q_n or C_n, is *open*, but, in time, fills the entire n-dimensional q/w-region *densely*; that is, given sufficiently long time $(\tau \to \infty)$, the orbit will pass as near as we want ("arbitrarily close") to any arbitrarily chosen initial q/w-region point. In particular, C_n will, eventually, fill up completely by parallel, equally spaced (a nonobvious fact!) and uniformly traversed straight-line segments; and similarly for the Q_n-prism.

(ii) On the other extreme, if $m = n - 1$ $(\Rightarrow n - m = 1)$, for example, $\nu_1 = \nu_2 = \cdots = \nu_n$, *which is completely equivalent to the earlier case described by eqs. (8.14.28a–e) (since any* ν_k *can be expressed as a rational fraction of any other* ν_l), the system is called *completely, or fully, degenerate*. In this case, eqs. (8.14.24i, j) become a *purely* (or *singly*, or *genuinely*) *periodic function of time*; and, hence, the system motion is confined to a *one-dimensional* submanifold—that is, a straight line (C_n) or curve (Q_n). In the C_2-square of the earlier example, the line $O\,O_1 \ldots O_{10} \ldots$ returns to O after a *finite* time τ (fundamental period); and from there on it repeats itself; while, in the corresponding configuration plane Q_2, the system trajectory becomes a "Lissajous figure" that, depending on the values of the phase constants, either *closes* or *retraces itself between an initial and a final point*, again with fundamental period τ. Finally, such single-frequency motions have a single angle variable and corresponding action variable equal to $J = \oint \sum p_k\, dq_k$.

The situation is summarized in the following diagram:

Extreme cases of degeneracy

$$m = 0 \Rightarrow n - m = n \leftarrow \cdots \rightarrow m = n - 1 \Rightarrow n - m = 1$$

No degeneracy	*Complete degeneracy*
Motion manifold: n-dimensional	Motion manifold: one-dimensional
Orbit: *open* curve	Orbit: *closed* curve (or flattened curve)

Effects of Degeneracies

The latter, in addition to reducing the number of independent frequencies of a system, also:

(i) Reduce the number of its independent (and constant) action variables; and so restrict the forms in which the latter appear in the energy $E(J_1, \ldots, J_n) \equiv E(J)$. Thus, if ν_k and ν_l $(k, l = 1, 2, \ldots;\ k \neq l)$ are such that

$$i_k \tau_k = i_l \tau_l \Rightarrow i_k \nu_l = i_l\, \nu_k, \qquad \text{or} \qquad i_k(\partial E/\partial J_l) = i_l(\partial E/\partial J_k), \tag{8.14.32}$$

where i_k, i_l = integers, then J_k, J_l may appear in E only in the form:

$$J \equiv i_k J_k + i_l J_l\ [\nu_k = \partial E/\partial J_k = (\partial E/\partial J)\,(\partial J/\partial J_k) = (\partial E/\partial J) i_k,$$
$$\nu_l = \cdots = (\partial E/\partial J) i_l \ \Rightarrow\ i_k \nu_l = i_l\, \nu_k];$$

or, equivalently, if they are such that

$$i_k \tau_l = i_l \tau_k \ \Rightarrow\ i_k \nu_k = i_l \nu_l, \qquad \text{or} \qquad i_k(\partial E/\partial J_k) = i_l(\partial E/\partial J_l), \tag{8.14.32a}$$

then J_k, J_l may appear in E only in the form $J' \equiv i_k J_l + i_l J_k$.

(ii) Increase the number of its single-valued integrals of motion over that number for the corresponding "same" but nondegenerate system. Indeed, a general *nondegenerate* conservative system has, outside of the energy integral, $2n - 1$ integrals of motion, of which only n are single-valued; for example, the n J's; the remaining $(2n - 1) - n = n - 1$ integrals can be written as

$$w_k(\partial E/\partial J_l) - w_l(\partial E/\partial J_k) = w_k \nu_l - w_l \nu_k = (\nu_k t + \gamma_k)\nu_l - (\nu_l t + \gamma_l)\nu_k$$
$$= \gamma_k \nu_l - \gamma_l \nu_k \quad [= \gamma_k(\partial E/\partial J_l) - \gamma_l(\partial E/\partial J_k)]$$
$$= \textit{constant};\ \textit{but multiple-valued, since the angle variables are also multiple-valued}. \tag{8.14.33a}$$

Now, in the case of degeneracy, as (8.14.32a) shows, the integral

$$w_k i_k - w_l i_l = (\nu_k t + \gamma_k) i_k - (\nu_l t + \gamma_l) i_l$$
$$= (i_k \nu_k - i_l \nu_l) t + (\gamma_k i_k - \gamma_l i_l)$$
$$= \gamma_k i_k - \gamma_l i_l = \textit{constant}; \tag{8.14.33b}$$

is multiple-valued, but to within an arbitrary multiple integral of 2π; and, therefore, by taking a *trigonometric function* of it, we obtain an *additional single-valued integral of motion*.

Also, this increase in the number of single-valued integrals allows for *complete separability for more than one choice of coordinates*: before the degeneracy, the n J_k's ($k = 1, \dots, n$) are single-valued integrals of the corresponding separable coordinates. Upon imposition of a degeneracy, however, since then the number of single-valued integrals exceeds n, the choice of the new n J_k's among them becomes *nonunique*. [On the connection between degeneracy (nondegeneracy) of motion and nonuniqueness (uniqueness) of separation of variables in its Hamilton–Jacobi equation, see, for example, Born (1927, pp. 76–95), Goldstein (1980, pp. 469–470).]

Nonseparable Systems

Lastly, let us outline the properties of *finite* motion of a general *n-DOF* conservative but *nonseparable* system. Here, contrary to the separable case where the single-valued integrals are the n J_k's, the single-valued integrals are only those obtained from the homogeneity of space (linear momentum), isotropy of space (angular momentum), and isotropy of time (energy).

Now, generally, the representative system point in phase space can traverse regions defined by the specified constant values of its single-valued integrals.

(i) For *separable* systems with n single-valued integrals, these n constants define an n-dimensional *hypersurface* in phase space; given sufficient time, the system can pass arbitrarily close to every other chosen point on that hypersurface.

(ii) For *nonseparable* systems (*degenerate* systems), however, which possess fewer (more) than n single-valued integrals, the system point occupies, in phase space, a subspace with *more* (*less*) than n dimensions.

If the Hamiltonian of a nonseparable system differs by a very small amount from that of a separable (conditionally periodic) system, then we may reasonably suppose that the motion of the former will be *very close* to that of the latter; and that the difference between these two motions is *much smaller* than that of their Hamiltonians.

The systematic quantitative discussion of these topics belongs squarely to the frontier of contemporary nonlinear dynamics; and such Hamiltonian deep waters are, most definitely, beyond the scope of this introductory treatment (and the present state of knowledge of this writer!). For these advanced topics, and their connections to both classical chaotic/stochastic and quantum dynamics, we recommend the following readable and capable references (alphabetically): Dittrich and Reuter (1994), Hagihara (1970), Lichtenberg and Liebermann (1992), McCauley (1997), Pars (1965), Tabor (1989).

Action–Angle Variables and Atomic Physics

As mentioned earlier, action–angle variables have played a decisive part in the *older quantum theory* of Sommerfeld, Bohr, et al. (in the 1910s). According to this theory, the actual motions of an atomic system obey the quantization rule:

$$J_k = \oint p_k \, dq_k = n_k h \qquad (n_k\text{: integer}, h\text{: Planck's action constant}). \qquad (8.14.34a)$$

However, this theory led to the appearance of the harmonics $\nu_k, 2\nu_k, 3\nu_k, \ldots$, and so on, in the Fourier series expansion of its variables; and this contrasted sharply with experimental facts that indicated that the frequencies of the atomic spectra are not harmonics of some fundamental frequency but, instead, obey the "Ritz combination principle," according to which these frequencies result from a series of energy levels E_k, E_l satisfying

$$\nu_{kl} = (E_k - E_l)/h \qquad (k, l\text{: integers}). \tag{8.14.34b}$$

The resolution of these difficulties was carried out in the mid-1920s by *W. Heisenberg* (with some help from his teacher *M. Born* and his classmate *P. Jordan*, at Göttingen, Germany), and constitutes the *matrix form of quantum mechanics*—one of the greatest triumphs ("revolutions") of 20th century theoretical physics. Heisenberg (1925) replaced the Fourier series of, say q with the set

$$\{c_{ss'} \exp(2\pi i \nu_{ss'} t);\ s, s'\text{: } integers\}, \tag{8.14.34c}$$

that is, *he replaced the frequencies* ν_k, $2\nu_k$, $3\nu_k, \ldots$, *(and amplitudes* c_s*) with the matrices*

$$\nu(s, s') \equiv \nu_{ss'} \qquad (\text{and } c_{ss'}), \tag{8.14.34d}$$

and introduced the algebra of these new "matrix coordinates" and functions of them. Born recognized that these were none other than the rules of matrix algebra, for the addition and multiplication of the q's, and formulated the following famous *noncommutative* (Poisson bracket-like, §8.9) rules, for pairs of canonically conjugate variables:

$$p_k q_l - q_l p_k = (h/2\pi i)\,\delta_{kl}, \qquad p_k p_l - p_l p_k = 0, \qquad q_k q_l - q_l q_k = 0. \tag{8.14.34e}$$

For readable accounts of those epoch-making developments, see, for example, Hund (1972), Simonyi (1986), and references cited therein; also Heisenberg (1930, p. 105 ff.).

Example 8.14.3 *Action for Cyclic Systems.* Let the coordinate q_i, of a separable group of q's, be ignorable. Then (§8.4) $p_i = constant \equiv \Psi_i$, and therefore the corresponding action variable equals:

$$J_i = \oint p_i\, dq_i = \Psi_i \left(\oint dq_i \right) = \Psi_i [q_i(2\pi) - q_i(0)]$$

$$= 0 \qquad (\text{Libration}), \tag{a}$$

$$= q_{io} \Psi_i \qquad (\text{Rotation; } q_{io}\text{: fundamental period; e.g., } q_{io} = 2\pi). \tag{b}$$

Example 8.14.4 *Two-DOF Conditionally Periodic System.*

(i) *Equal frequencies.* Let us examine a particle P performing planar harmonic oscillations along the rectangular Cartesian axes O–x, O–y [fig. 8.18(a)]. Let us assume that the displacements $q_1 = x$, $q_2 = y$ are

$$x = a \cos(\omega t), \qquad y = b \cos(\omega t - \delta), \tag{a}$$

where a, b = constant amplitudes, $\omega = 2\pi\nu$ = common circular frequency, δ = phase difference.

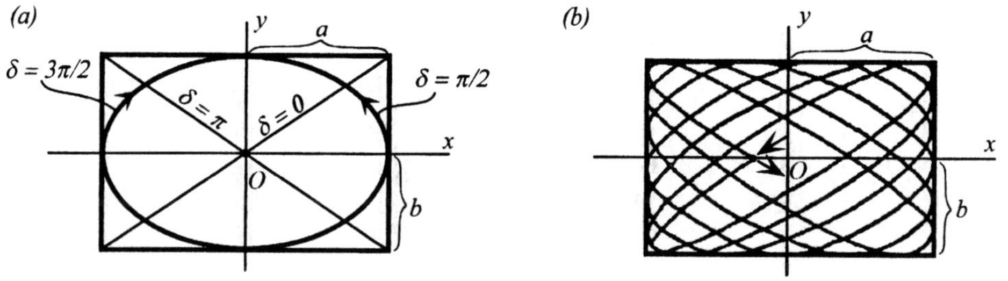

Figure 8.18 Path of a two-DOF system: (a) *equal* frequencies, (b) *unequal* (incommensurate case) frequencies (Lissajous figures).

Depending on the values of δ, the tip of the vector $\boldsymbol{OP} = (x, y)$ describes very different curves. Thus:

(a) If $\delta = 0$ $[= 0(\pi/2)]$, P describes the *straight line*

$$y/x = b/a \qquad \text{(diagonal of rectangle with sides } a \text{ and } b). \tag{b}$$

(b) If $\delta = \pi/2$ $[= 1(\pi/2)]$, eqs. (a) reduce to

$$x = a\cos(\omega t), \qquad y = b\cos(\omega t), \tag{c}$$

which are the parametric equations of an *ellipse* with semiaxes a, b:

$$(x/a)^2 + (y/b)^2 = 1, \tag{d}$$

traversed in a *counterclockwise* sense.

(c) If $\delta = \pi$ $[= 1(\pi/2)]$, P describes the *straight line*

$$y/x = -b/a, \tag{e}$$

which is the straight line of case (a), but reflected about the axis Oy.

(d) If $\delta = 3\pi/2$ $[= 3(\pi/2)]$, P describes the *ellipse* of eqs. (c, d), but traversed in a *clockwise* sense.

(e) If $\delta = 2\pi$ $[= 4(\pi/2)]$, P describes the *straight line* of eq. (b).

(f) If $a = b$, P traces a *circle* (clockwisely/counterclockwisely).

(ii) *Unequal frequencies.* Let us, next, assume that the displacements are

$$x = a\cos(\omega_x t), \qquad y = b\cos(\omega_y t). \tag{f}$$

Now we must distinguish the following two cases:

(a) If ω_x, ω_y are *commensurate* (*degenerate* case)—that is, if

$$\tau_x i_x = \tau_y i_y \equiv \tau \Rightarrow \omega_x/\omega_y = i_x/i_y = \textit{rational} \qquad (i_{x,y}\text{: integers}) \tag{g}$$

(e.g., the earlier $\omega_x = \omega_y = \omega$)—then eqs. (f) become

$$x = a\cos(\omega_x t), \qquad y = b\cos(\omega_y t) = b\cos\left[(i_y/i_x)\omega_x t\right]; \tag{h}$$

the motion has a *single* period—namely, it is periodic *as a whole*—and so the orbit of P is a *closed* curve.

(b) If ω_x, ω_y are *incommensurate* (*nondegenerate* case)—that is, if $\omega_x/\omega_y =$ *irrational*—the orbit of P is a continuous Lissajous curve that *never closes* [fig. 8.18(b)] but forms a very dense web; that is, given sufficient time, it *practically* covers all points of the $2a \times 2b$ rectangle $ABCD$. Multiply periodic (or conditionally periodic) orbits are, in general, of that type: a certain space portion (the range of their q's) is densely filled, even though the orbit is *not* closed, and the motion is not singly periodic in time.

In sum:

- If $\omega_x = \omega_y$ (special commensurate/degenerate case), the orbit of P is a *straight line/ellipse/circle*, depending on the phase constant, and the relative size of the amplitudes.
- If $\omega_x/\omega_y =$ *rational* (*degeneracy*), the orbit of P is a *closed* curve.
- If $\omega_x/\omega_y =$ *irrational* (*nondegeneracy*), the orbit of P is an *open* curve that gradually fills up the whole rectangle (range of its variables).

Example 8.14.5 *Two-DOF Linear Anisotropic Oscillator; Degeneracy.* Here, using standard notation ($x, y =$ rectangular Cartesian coordinates):

$$H = (1/2m)({p_x}^2 + {p_y}^2) + (1/2)({k_x}^2 x^2 + {k_y}^2 y^2)$$
$$= (1/2m)({p_x}^2 + {p_y}^2) + (m/2)({\omega_x}^2 x^2 + {\omega_y}^2 y^2), \qquad \text{(a)}$$

${\omega_x}^2 \equiv k_x/m$, and so on. Using the results of ex. 8.10.6 [slightly modified notationally, and also to take into account the anisotropy of (a)] we obtain, successively,

$$A_1 \to A_{ox} = \int p_x\, dx = \int \left[m(2\beta_x - k_x x^2)\right]^{1/2} dx$$
$$= (x/2)(2m\beta_x - k_x m x^2)^{1/2} + \beta_x (m/k_x)^{1/2} \arcsin[(k_x/2\beta_x)^{1/2} x]$$
$$\text{[multiple-valued function of the } x\text{-coordinate], etc.} \qquad \text{(b)}$$

$$J_x = \oint p_x\, dx = \int (\partial A_{ox}/\partial x)\, dx = A_{ox}(\tau_x) - A_{ox}(0)$$

[after a period τ_x, the term $(2m\beta_x - k_x m x^2)^{1/2}$ returns to its original value, while arcsin(...) increases by 2π]

$$= 2\pi\beta_x (m/k_x)^{1/2} \equiv (2\pi/\omega_x)\beta_x, \quad \text{etc.}; \qquad \text{(c)}$$

$$E = \beta_x + \beta_y = \text{total energy} \quad (= \text{sum of "partial energies"}), \qquad \text{(d)}$$

$$\Rightarrow E = (J_x/2\pi)(k_x/m)^{1/2} + (J_y/2\pi)(k_y/m)^{1/2} = E(J_x, J_y), \qquad \text{(e)}$$

$$\Rightarrow \nu_x = \partial E/\partial J_x = (1/2\pi)(k_x/m)^{1/2},\ \nu_y = \partial E/\partial J_y = (1/2\pi)(k_y/m)^{1/2}; \text{ (f)}$$

$$\alpha_x = \partial A/\partial \beta_x = \partial A_o/\partial \beta_x - t = \partial A_{ox}/\partial \beta_x - t$$
$$= -t + (m/k_x)^{1/2} \arcsin[(k_x/2\beta_x)^{1/2} x], \qquad \text{etc.} \qquad \text{(g)}$$

$$\Rightarrow x = (2\beta_x/k_x)^{1/2} \sin\left[(k_x/m)^{1/2}(t + \alpha_x)\right]$$
$$= (J_x/\pi\omega_x m)^{1/2} \sin(2\pi w_x) = \left[J_x/\pi(k_x m)^{1/2}\right]^{1/2} \sin(2\pi w_x), \qquad \text{(h)}$$

$$\Rightarrow \ p_x = (\omega_x m J_x/\pi)^{1/2}\cos(2\pi w_x), \tag{i}$$

$$\Rightarrow \ y = (2\beta_y/k_y)^{1/2}\sin[(k_y/m)^{1/2}(t+\alpha_y)]$$
$$= (J_y/\pi\omega_y m)^{1/2}\sin(2\pi w_y) = [J_y/\pi(k_y m)^{1/2}]^{1/2}\sin(2\pi w_y), \tag{j}$$

$$\Rightarrow \ p_y = (\omega_y m J_y/\pi)^{1/2}\cos(2\pi w_y); \tag{k}$$

where

$$w_x = \nu_x t + \gamma_x$$
$$\Rightarrow \ 2\pi w_x = 2\pi(\nu_x t + \gamma_x) \equiv 2\pi\nu_x(t+\alpha_x) \equiv \omega_x t + \delta_x, \quad \text{etc.} \tag{l}$$

To represent the above graphically in their rectangular Cartesian $w_{x,y}$-axes, we eliminate t between w_x, w_y in (l), and thus obtain the curves (straight lines)

$$w_y = (\nu_y/\nu_x)w_x + [\gamma_y - (\nu_y/\nu_x)\gamma_x]. \tag{m}$$

As discussed in the preceding example, we must distinguish the following *two* general cases:

(i) $\nu_y/\nu_x =$ *rational* (*commensurability*, or *complete degeneracy*): the motion as a whole is (singly) periodic, which means that the representative system point traces a (n – *number of commensurability relations* $= 2 - 1 =$) *one*-dimensional manifold; that is, in our xy-axes, a *closed and always retraceable Lissajous curve*. [If that curve has an endpoint (e.g., flattened, open-*looking*, path), the motion reverses itself at that point (i.e., it does a very flat U-turn there) and proceeds in the opposite direction until it reaches the next endpoint; and then the whole process repeats itself periodically.] The corresponding w-curve, inside the system's "unit square" C_2, consists of straight-line segments, as explained earlier in this section.

Figures 8.19 and 8.20 show the following special such cases (in both figures, the *left* column shows the x, y (Lissajous) curves, while the *right* column shows the corresponding $w_{x,y}$ straight lines):

(α) Figure 8.19:

$$\textit{frequency ratio: } \nu_y/\nu_x = 1, 2, 3, 5/3, \quad \textit{phase constants: } \gamma_x, \gamma_y = 0, \tag{n}$$

$$\text{(h,j):} \quad x = [J_x/\pi(k_x m)^{1/2}]^{1/2}\sin(2\pi\nu_x t), \qquad y = [J_y/\pi(k_y m)^{1/2}]^{1/2}\sin(2\pi\nu_y t), \tag{o}$$

$$\text{(m):} \quad w_y = (\nu_y/\nu_x)w_x \ \text{(straight lines through the } w_{x,y}\text{-origin);} \tag{p}$$

(β) Figure 8.20 [*same frequency ratios as (a), but different phase constants*]:

$$\textit{frequency ratio: } \nu_y/\nu_x = 1, 2, 3, 5/3, \quad \textit{phase constants: } \gamma_x = 0, \gamma_y = 1/4, \tag{q}$$

$$\text{(h):} \quad x = [J_x/\pi(k_x m)^{1/2}]^{1/2}\sin(2\pi\nu_x t), \tag{r}$$

$$\text{(j):} \quad y = [J_y/\pi(k_y m)^{1/2}]^{1/2}\sin[(2\pi\nu_y + \pi/2)t] = [J_y/\pi(k_y m)^{1/2}]^{1/2}\cos(2\pi\nu_y t), \tag{s}$$

$$\text{(m):} \quad w_y = (\nu_y/\nu_x)w_x + 1/4 \ \text{(straight lines not through the } w_{x,y}\text{-origin).} \tag{t}$$

(ii) $\nu_y/\nu_x \neq$ *rational* (*incommensurability*, or *nondegeneracy*): the motion as a whole is not periodic, which means that the representative system point traces a ($n - 0 = 2 - 0$) *two*-dimensional manifold; that is, in our xy-axes an *open and non-retraceable* curve, which eventually ($t \to \infty$) covers the entire finite area formed by

$\nu_y/\nu_x = 1$; $\gamma_x = 0$; $\gamma_y = 0$ $\Rightarrow$ $w_y = w_x$

$\nu_y/\nu_x = 2$; $\gamma_x = 0$; $\gamma_y = 0$ $\Rightarrow$ $w_y = 2 w_x$

$\nu_y/\nu_x = 3$; $\gamma_x = 0$; $\gamma_y = 0$ $\Rightarrow$ $w_y = 3 w_x$

$\nu_y/\nu_x = 5/3$; $\gamma_x = 0$; $\gamma_y = 0$ $\Rightarrow$ $w_y = (5/3) w_x$

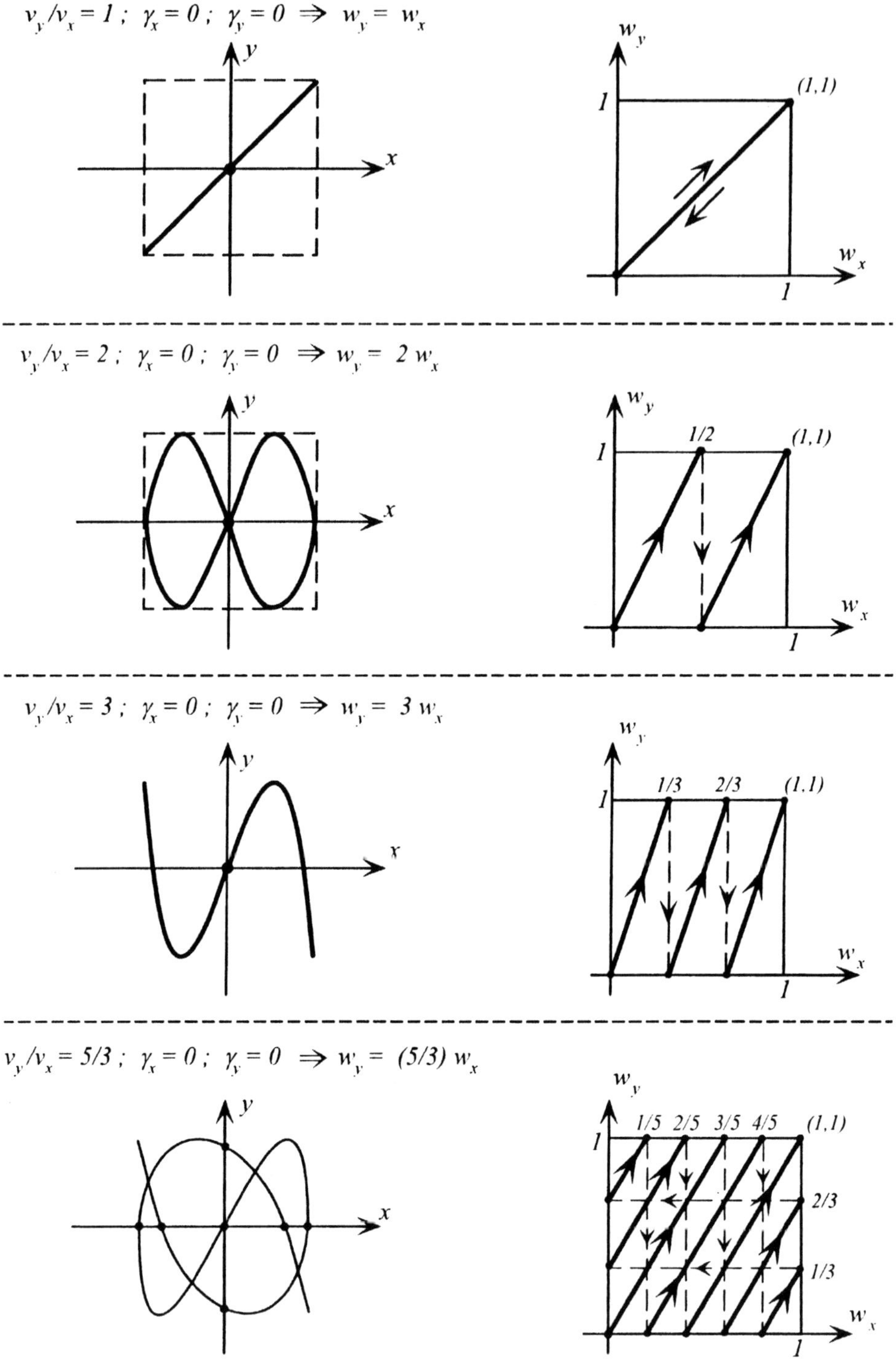

Figure 8.19 System paths in x, y-space (Lissajous curves) and in $w_{x,y}$-space; for $\nu_y/\nu_x = 1, 2, 3, 5/3$, and $\gamma_x = 0$, $\gamma_y = 0$.

$\nu_y/\nu_x = 1$; $\gamma_x = 0$; $\gamma_y = 1/4$ $\Rightarrow$ $w_y = w_x + 1/4$

$\nu_y/\nu_x = 2$; $\gamma_x = 0$; $\gamma_y = 1/4$ $\Rightarrow$ $w_y = 2w_x + 1/4$

$\nu_y/\nu_x = 3$; $\gamma_x = 0$; $\gamma_y = 1/4$ $\Rightarrow$ $w_y = 3w_x + 1/4$

$\nu_y/\nu_x = 5/3$; $\gamma_x = 0$; $\gamma_y = 1/4$ $\Rightarrow$ $w_y = (5w_x)/3 + 1/4$

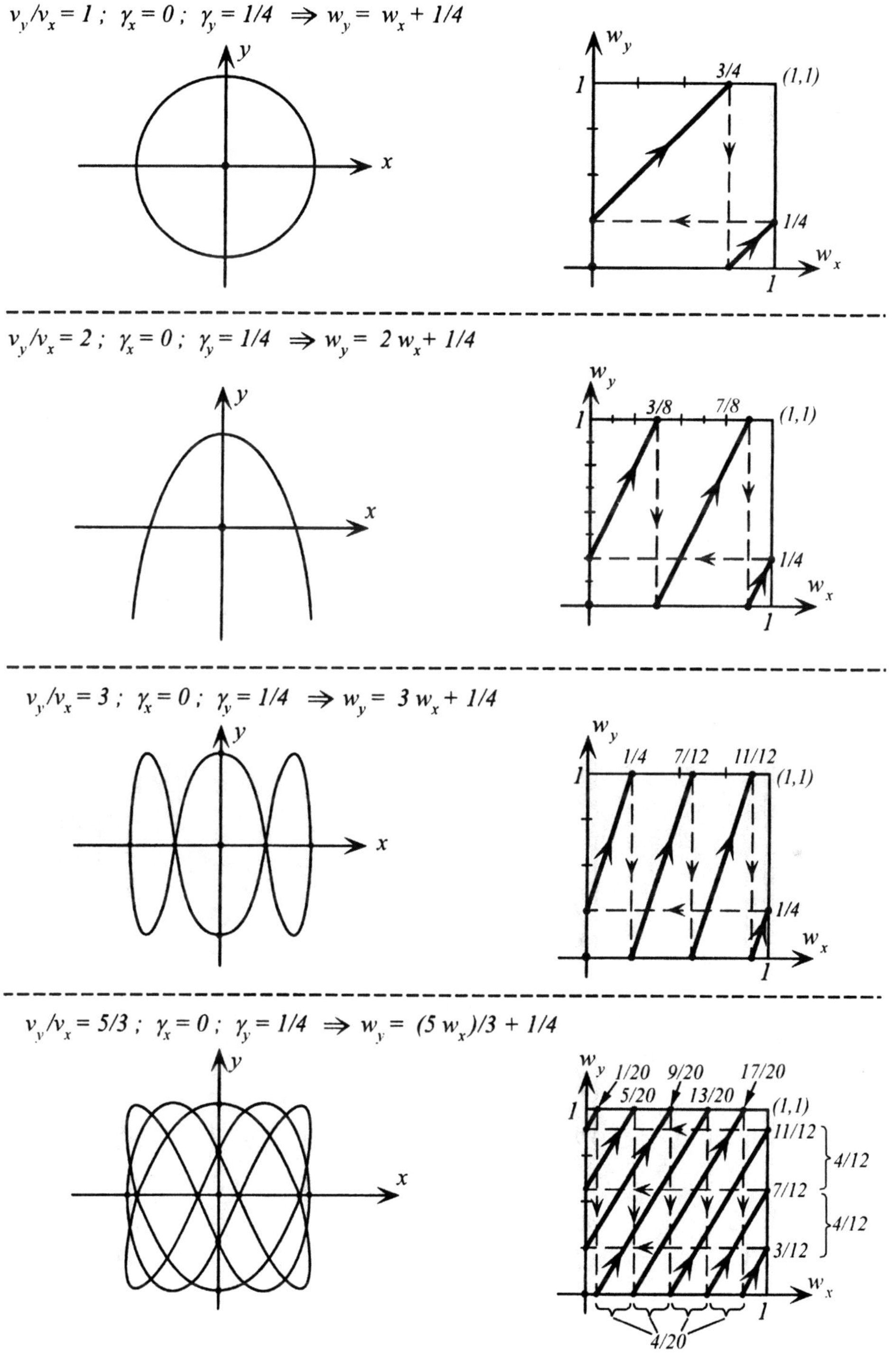

Figure 8.20 System paths in x, y-space (Lissajous curves) and in $w_{x,y}$-space; for $\nu_y/\nu_x = 1, 2, 3, 5/3$, and $\gamma_x = 0$, $\gamma_y = 1/4$.

the tangents bounding the motion (analogous to the rectangle $ABCD$ of the preceding example); and similarly for the unit square $w_{x,y}$.

Finally, *for complicated but rational ratios* ν_y/ν_x, the system path comes close to covering the corresponding xy-plane region and unit square $w_{x,y}$ completely, that is, such ratios approximate the nonrational ratio case.

Example 8.14.6 *Coupled Penduli via Action–Angle Variables* (Butenin, 1971, pp. 173–176). Let us consider a system consisting of two thin homogeneous bars, O_1A_1 and O_2A_2 of masses/lengths/moments of inertia about their pivots O_1 and O_2: $m_1/l_1/I_1$ and $m_2/l_2/I_2$, respectively (fig. 8.21), oscillating about O_1 and O_2, and connected at C_1, C_2 $(O_1C_1 = O_2C_2 \equiv c)$ by a light linear spring of constant stiffness k. Let us calculate the two natural frequencies of its free (small amplitude) oscillations under gravity, via the method of action–angle variables.

The kinetic and potential energies of the system are, respectively,

$$2T = I_1(\dot{\phi}_1)^2 + I_2(\dot{\phi}_2)^2, \tag{a}$$

$$V = m_1 g(l_1/2)(1 - \cos\phi_1) + m_2 g(l_2/2)(1 - \cos\phi_2) + (k/2)(e - e_o)^2$$

[using the second-order approximations: $\cos\phi_k \approx 1 - {\phi_k}^2/2 \qquad (k = 1, 2)$,

and

$e - e_o \approx c\phi_1 - c\phi_2$]

$$\approx (1/2)\{[m_1 g(l_1/2) + kc^2]{\phi_1}^2 - 2kc^2\phi_1\phi_2 + [m_2 g(l_2/2) + kc^2]{\phi_2}^2\}, \tag{b}$$

or, after choosing, for algebraic convenience, $l_1 = l_2 = l$, $m_1 = m_2 = m$, $I_1 = I_2 = I$:

$$2T = I[(\dot{\phi}_1)^2 + (\dot{\phi}_2)^2], \tag{c}$$

$$2V = [K({\phi_1}^2 + {\phi_2}^2) - 2kc^2\phi_1\phi_2], \qquad K \equiv mg(l/2) + kc^2; \tag{d}$$

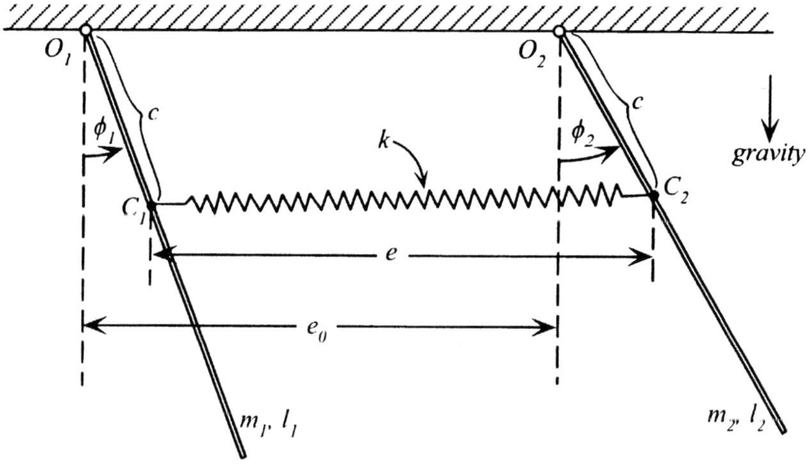

Figure 8.21 Coupled double penduli, oscillating under gravity.

or, in terms of the (easily noticeable) new uncoupling coordinates:

$$\phi_1 = q_1 + q_2, \qquad \phi_2 = q_1 - q_2 \Rightarrow 2q_1 = \phi_1 + \phi_2, \qquad 2q_2 = \phi_1 - \phi_2,$$

$$T = I[(\dot{q}_1)^2 + (\dot{q}_2)^2], \qquad V = K_1 {q_1}^2 + K_2 {q_2}^2, \tag{e}$$

where $K_1 \equiv mgl/2$, $K_2 \equiv K_1 + 2kc^2 \equiv K + kc^2$.

Since $p_1 = 2I\dot{q}_1$ and $p_2 = 2I\dot{q}_2$, the time-independent (reduced) Hamilton–Jacobi equation of the system is

$$[(1/4I)(\partial A_o/\partial q_1)^2 + K_1 {q_1}^2] + [(1/4I)(\partial A_o/\partial q_2)^2 + K_2 {q_2}^2] = E, \tag{f}$$

and separates, in by now well-understood ways, to the two HJ equations

$$(1/4I)(\partial A_{o1}/\partial q_1)^2 + K_1 {q_1}^2 = \beta_1, \qquad (1/4I)(\partial A_o/\partial q_2)^2 + K_2 {q_2}^2 = \beta_2, \tag{g}$$

where $A_o = A_{o1}(q_1) + A_{o2}(q_2)$, $\beta_1 + \beta_2 = E$. Hence, the finite HJ equations become

$$p_1 = \partial A_o/\partial q_1 = \partial A_{o1}/\partial q_1 = 2(IK_1)^{1/2}[(\beta_1/K_1) - {q_1}^2]^{1/2}, \tag{h}$$

$$p_2 = \partial A_o/\partial q_2 = \partial A_{o2}/\partial q_2 = 2(IK_2)^{1/2}[(\beta_2/K_2) - {q_2}^2]^{1/2}, \tag{i}$$

and, from these expressions, it follows that the corresponding action variables are

$$J_1 = 2(IK_1)^{1/2} \oint [(\beta_1/K_1) - {q_1}^2]^{1/2}\, dq_1, \tag{j}$$

$$J_2 = 2(IK_2)^{1/2} \oint [(\beta_2/K_2) - {q_2}^2]^{1/2}\, dq_2. \tag{k}$$

With the help of the convenient transformation of variables:

$$q_1 \equiv (\beta_1/K_1)^{1/2} \sin x, \qquad q_2 \equiv (\beta_2/K_2)^{1/2} \sin x, \tag{l}$$

(and, accordingly, x-limits of integration: 0, 2π) eqs. (j, k) integrate readily to

$$J_1 = 2\pi(I/K_1)^{1/2}\beta_1, \qquad J_2 = 2\pi(I/K_2)^{1/2}\beta_2, \tag{m}$$

and so the total energy assumes the following form, in terms of the action variables:

$$E = \beta_1 + \beta_2 = (J_1/2\pi)(K_1/I)^{1/2} + (J_2/2\pi)(K_2/I)^{1/2} = E(J_1, J_2). \tag{n}$$

From this expression (and recalling the definitions of K_1, K_2, and that $3I = ml^2$), we readily obtain the system frequencies:

$$\nu_1 = \partial E/\partial J_1 = (1/2\pi)(K_1/I)^{1/2} = (1/2\pi)(3g/2l)^{1/2}, \tag{o}$$

$$\nu_2 = \partial E/\partial J_2 = (1/2\pi)(K_2/I)^{1/2} = (1/2\pi)[(3g/2l) + (6kc^2/ml^2)]^{1/2}; \tag{p}$$

which, of course, coincide with the values found by ordinary linear vibration theory.

Example 8.14.7 *Action–Angle Formulation for Partially Separable Systems.* Let us consider a system in motion, and such that the projection of its orbit on the particular phase space subplane (q_k, p_k) is a *periodic* (libratory or rotatory) curve. This can occur if the system action has the *separable form in* q_1 (§8.10):

$$A = A_{oR}(q_2, \ldots, q_n; \beta_2, \ldots, \beta_n) + A_{o1}(q_1, \beta_1) - E\,t. \tag{a}$$

Here, as in the completely separable case, we replace throughout the constant β_1 with the constant *action* variable (with the usual notations):

$$J_1 \equiv \oint p_1\, dq_1 = \oint (\partial A_{o1}/\partial q_1)\, dq_1 = A_{o1}(\tau_1) - A_{o1}(0); \tag{b}$$

and similarly for $E = E(\beta_1) = E(J_1)$.

Then, the corresponding (ignorable) canonical *angle* variable w_1 equals

$$w_1 = \partial A_o/\partial J_1 = \partial A_{o1}(q_1, J_1)/\partial J_1 = (\partial E/\partial J_1)t + \partial A/\partial J_1, \tag{c}$$

and, from this, it follows that during a cycle with (fundamental) period τ_1, it changes by

$$\Rightarrow\ \Delta w_1 \equiv w_1(t + \tau_1) - w_1(t) = \tau_1(\partial E/\partial J_1). \tag{d}$$

Comparing (d) with the general result:

$$\begin{aligned} \Delta w_1 &= \oint (\partial w_1/\partial q_1)\, dq_1 = \oint (\partial^2 A_{o1}/\partial q_1\, \partial J_1)\, dq_1 \\ &= \partial/\partial J_1 \left[\oint (\partial A_{o1}/\partial q_1)\, dq_1\right] = \partial/\partial J_1 \left(\oint p_1\, dq_1\right) = \partial J_1/\partial J_1 = 1, \end{aligned} \tag{e}$$

we readily conclude that

$$\partial E/\partial J_1 = \textit{frequency of } (q_1, p_1)\textit{-motion} \equiv \nu_1 = 1/\tau_1. \tag{f}$$

Finally, q_1 is a periodic function in w_1 with period 1, and can therefore be represented by a single Fourier series à la (24f–j). The extension of the above to partially separable systems in two, three, ... , periodic coordinates should be obvious.

Example 8.14.8 *An Alternative Expression for the Frequencies.* Since $w_k = \partial A_o(q, J)/\partial J_k = w_k(q, J)$, and the J_k's remain constant during the motion, any changes in the w_k's can arise only from changes in the q's. Indeed, from the preceding, we find, successively,

$$\begin{aligned} dw_k &= \sum (\partial w_k/\partial q_l)\, dq_l = \sum \left[\partial/\partial q_l(\partial A_o/\partial J_k)\right] dq_l \\ &= \sum \left[\partial/\partial J_k(\partial A_o/\partial q_l)\right] dq_l \\ &= \sum (\partial p_l/\partial J_k)\, dq_l, \qquad \text{where } p_l = p_l(q_l, J); \end{aligned} \tag{a}$$

and, since $w_k = \nu_k t + \gamma_k \ \Rightarrow\ dw_k = \nu_k\, dt$ and $dq_l = \dot{q}_l\, dt$, we finally obtain the alternative frequency expression

$$\nu_k = \sum (\partial p_l/\partial J_k)(dq_l/dt). \tag{b}$$

Example 8.14.9 (Born, 1927, p. 82). Let

$$K \equiv \sum p_k\,\dot{q}_k \quad (= 2T, \text{ for stationary constraints}). \tag{a}$$

Below we show that

$$\langle K \rangle \equiv \sum \nu_k J_k; \tag{b}$$

where $\langle \ldots \rangle \equiv$ *time-average of* $(\ldots)$ *over a long period of time* τ, including a large number of w-periods. With the help of the earlier (8.14.23b, c):

$$A_{oo} \equiv A_o - \sum w_k\,J_k, \qquad \dot{A}_{oo} = \sum p_k\dot{q}_k - \sum J_k\dot{w}_k, \tag{c}$$

we obtain, successively,

$$\begin{aligned}\langle K \rangle &\equiv \tau^{-1}\left(\int_0^\tau \sum p_k\,\dot{q}_k\right) dt \\ &= \tau^{-1}\left[\int_0^\tau\left(\sum J_k\dot{w}_k + \dot{A}_{oo}\right) dt\right] = \tau^{-1}\left[\int_0^\tau\left(\sum J_k\,\nu_k + \dot{A}_{oo}\right) dt\right] \\ &= \sum w_k J_k + \{A_{oo}/\tau\}_0^\tau \text{ (since both the } \nu_k\text{'s and } J_k\text{'s } \textit{are constant}),\end{aligned} \tag{d}$$

from which, since A_{oo} is w-periodic and τ contains a large number of periods of the w's, and therefore $A_{oo}(0) = A_{oo}(\tau)$, the proposition (b) follows.

Example 8.14.10 (Bohr, 1918). Here, we show that for an *n-DOF* but completely degenerate (i.e., singly periodic) system

$$\Delta E = \nu\,\Delta J. \tag{a}$$

Let us consider the system in a fundamental oscillatory state I of period τ. Then, its (sole) action variable is

$$J = \int_0^\tau \left(\sum p_k\dot{q}_k\right) dt. \tag{b}$$

Now, let us consider a small noncontemporaneous variation from that state to the neighboring, also oscillatory and singly periodic, state $II = I + \Delta(I)$ with period $\tau + \Delta\tau$. Using the results of §7.9, we obtain, successively,

$$\begin{aligned}\Delta J &= \int_0^\tau \delta\left(\sum p_k\dot{q}_k\right) dt + \left\{\left(\sum p_k\dot{q}_k\right)\Delta t\right\}_0^\tau \\ &= \int_0^\tau \sum (\delta p_k\,\dot{q}_k + p_k\,\delta\dot{q}_k)\,dt + \left\{\left(\sum p_k\dot{q}_k\right)\Delta t\right\}_0^\tau\end{aligned}$$

[integrating the $p\,\delta(\dot{q})$ terms by parts, and using Hamilton's equations with $Q_k = 0$; while recalling that $\Delta q = \delta q + \dot{q}\,\Delta t$]

$$= \int_0^T \sum \left[(\partial H/\partial p_k)\,\delta p_k - (-\partial H/\partial q_k)\,\delta q_k\right] dt + \left\{\sum p_k\, \Delta q_k\right\}_0^T$$

$$= \int_0^T \delta H\, dt \qquad \text{[the integrated-out "boundary" term vanishes by periodicity]}$$

$$= \int_0^T \delta E\, dt. \tag{c}$$

If the adjacent motion II corresponds to slightly different initial conditions, then $\delta E = constant$ ($\Rightarrow \Delta E = \delta E + \dot{E}\,\Delta t = \delta E$), and so (b) yields immediately

$$\Delta J = (\Delta E)\tau \;\Rightarrow\; \Delta E = \Delta J/\tau = \nu\, \Delta J, \quad \text{Q.E.D.} \tag{d}$$

Example 8.14.11 Let us show that

$$\Delta E = \sum \nu_k\, \Delta J_k. \tag{a}$$

Since $H = H(J)$, we have

$$\Delta H = \sum (\partial H/\partial J_k)\, \Delta J_k = \sum \nu_k\, \Delta J_k = \Delta E, \quad \text{Q.E.D.} \tag{b}$$

Example 8.14.12 (Born, 1927, pp. 94–95). According to (8.14.23a) and (8.14.23b), the function

$$A_o = A_{oo} + \sum w_k J_k \tag{a}$$

increases by J_k whenever w_k increases by 1, while the remaining w's and J's *remain constant*. This is expressed mathematically by

$$J_k = \int_0^1 \left[\partial A_o(w, J)/\partial w_k\right] dw_k$$

$$= \int_0^1 \left\{\sum \left[\partial A_o(q, J)/\partial q_l\right](\partial q_l/\partial w_k)\right\} dw_k$$

$$= \int_0^1 \left(\sum p_l(\partial q_l/\partial w_k)\right) dw_k; \tag{b}$$

and its usefulness consists in yielding the action variables from a knowledge of the q's and p's in terms of the w's.

Example 8.14.13 *Proof of Commutativity of* $\partial/\partial J_k[\oint(\ldots)\, dq] = \oint(\partial\ldots/\partial J_k)\, dq$, eqs. (8.14.13b, 23) (Kuypers, 1993, pp. 345–346, 533–534). In the derivation of (8.14.13b, 23), we assumed that

$$\oint (\partial^2 A_o/\partial J_k \partial q_l)\, dq_l = \partial/\partial J_k \left[\oint (\partial A_o/\partial q_l)\, dq_l\right], \tag{a}$$

in spite of the fact that the integration limits do depend on J_k. Let us justify this point. The proof is based on the well-known "Leibniz formula" (using standard calculus notations):

$$\partial/\partial\alpha \int_{l_1(\alpha)}^{l_2(\alpha)} f(x;\alpha,\ldots)\,dx$$
$$= \int_{l_1(\alpha)}^{l_2(\alpha)} [\partial f(x;\alpha,\ldots)/\partial\alpha]\,dx$$
$$+ f(x,l_2,\ldots)[\partial l_2(\alpha)/\partial\alpha] - f(x,l_1,\ldots)[\partial l_1(\alpha)/\partial\alpha]. \tag{b}$$

With the identifications:

$$\alpha \to J_k, \qquad x \to q_l, \qquad f(x;\alpha,\ldots) \to \partial A_o(q,J)/\partial q_l, \tag{c}$$

eq. (b) yields

$$\partial/\partial J_k \int_{q_{l,1}(J)}^{q_{l,2}(J)} [\partial A_o(q,J)/\partial q_l]\,dq_l$$
$$= \int_{q_{l,1}(J)}^{q_{l,2}(J)} [\partial/\partial J_k(\partial A_o/\partial q_l)]\,dq_l$$
$$+ \{(\partial A_o/\partial q_l)_2[\partial q_{l,2}(J)/\partial J_k] - (\partial A_o/\partial q_l)_1[\partial q_{l,1}(J)/\partial J_k]\}, \tag{d}$$

where the subscripts 1, 2 refer to the limits of integration.

Now we apply (d) to our two periodic cases:

(i) Case of *libration*. Then,

$$\oint (\partial A_o/\partial q_l)\,dq_l = 2\int_{q_{l,\min}(J)}^{q_{l,\max}(J)} (\partial A_o/\partial q_l)\,dq_l, \tag{e}$$

where the integration limits $q_{\max/\min}$ are the *turning points* of the oscillation. But, there, we also have $p_l = \partial A_o/\partial q_l = 0$, and therefore the boundary terms in (d) vanish *individually*; that is, (a) holds for libration.

(ii) Case of *rotation*. Here,

$$\oint (\partial A_o/\partial q_l)\,dq_l = \int_{q_{l,i}(J)}^{q_{l,i}(J)+q_{lo}} (\partial A_o/\partial q_l)\,dq_l, \tag{f}$$

where $q_{l,i}$ = arbitrary *initial* position of q_l, q_{lo} = fundamental period of q_l. However, since $\partial A_o/\partial q_l = p_l$ = *periodic* function of q_l, and

$$\partial/\partial J_k[q_{l,i}(J) + q_{lo}] = \partial/\partial J_k[q_{l,i}(J)], \tag{g}$$

the boundary terms in (d) *taken together* vanish; that is, (a) also holds for rotation, and so it holds for periodic motions in general.

Example 8.14.14 *Independent Action–Angle Variables in the Case of Degeneracy.* Whenever the frequency constraints (8.14.31)

$$\sum i_{dk}\,\nu_k = 0 \qquad (d = 1,\ldots,m) \tag{a}$$

hold [and following the Lagrange–Hamel method of constrained coordinates/velocities (chaps. 2 and 3), with which this theory of degenerate systems bears some unmistakable mathematical similarities!], we may replace the old action–angle variables (w, J) with new action–angle variables (w', J'), defined through the following special equations:

$$w'_k: \qquad w'_d \equiv \sum i_{dk} w_k = 0 \qquad (d = 1, \ldots, m), \tag{b}$$

$$w'_i \equiv w_i \neq 0 \qquad (i = m+1, \ldots, n), \tag{c}$$

and generating function [recalling §8.8, with $F_1 = 0$, and the correspondences: $q \to w$, $p' \to J'$, $F_2(q, p') \to F_2(w, J')$]

$$\begin{aligned} F_2(w, J') &= \sum w'_k J'_k \qquad \left(= \sum w_k J_k\right) \\ &= \sum w'_d J'_d + \sum w'_i J'_i \\ &= \sum\sum i_{dk} w_k J'_d + \sum w_i J'_i \qquad \left(= \sum w_i J'_i\right). \end{aligned} \tag{d}$$

Then, the old and new action variables will be related by [recalling (8.8.17)]

$$p_k = \partial F_2/\partial q_k: \quad J_k = \partial F_2/\partial w_k = \sum i_{dk} J'_d + \sum \delta_{ik} J'_i \quad (\delta_{ik}\text{: Kronecker delta}), \tag{e}$$

$$\Rightarrow J_{d'} = \sum i_{dd'} J'_d \qquad (d' = 1, \ldots, m), \tag{e1}$$

$$\Rightarrow J_{i'} = J'_{i'} + \sum i_{di'} J'_d \qquad (i' = m+1, \ldots, n); \tag{e2}$$

$$q_{k'} = \partial F_2/\partial p_{k'}: \qquad w'_k = \partial F_2/\partial J'_k \qquad [\text{i.e., eqs. (b, c)}]. \tag{f}$$

From the above, it follows that

(i) The *new* frequencies will be

$$\nu'_k = dw'_k/dt: \quad dw'_d/dt = \sum i_{dk}(dw_k/dt) = \sum i_{dk}\,\nu_k = 0; \quad \text{i.e., } \nu'_d = 0; \tag{g1}$$

$$dw'_i/dt = dw_i/dt = \nu_i; \qquad \text{i.e. } \nu'_i = \nu_i; \tag{g2}$$

with the zeroes among them (ν'_d) yielding constant factors in the corresponding Fourier series expansions; while

(ii) Since the $H'(J') = H(J) = E(J) = E'(J')$, the Hamiltonian equations in the new variables will be

$$dw'_k/dt = \partial H'/\partial J'_k = \nu'_k:$$

$$\Rightarrow \partial H'/\partial J'_d = \nu'_d = 0 \Rightarrow E\text{: } \textit{independent of the } J'_d, \tag{h1}$$

$$\Rightarrow \partial H'/\partial J'_i = \nu'_i = \nu_i \neq 0; \tag{h2}$$

$$dJ'_k/dt = -\partial H'/\partial w'_k = 0 \Rightarrow J'_k - \textit{constant}. \tag{h3}$$

Hence, in a completely degenerate system $(m = n - 1)$ $H' = H'(J'_n)$; that is, the Hamiltonian can depend on only one new action variable.

We leave it to the reader to show that the new coordinates obtained by the generating function $A_o(q, \beta) = A_o(q, J) = A_o(q, J')$ are indeed the new angle variables w', that is, $w'_k = \partial A_o/\partial J'_k$. For further details and insights, see, for example, Frank (1935, pp. 97–99), Fues (1927, p. 142 ff.).

Problem 8.14.1 Show that equation (a) of the preceding example implies that

$$\nu_k = \sum i'_{ki}\, \nu'_i = 0 \qquad (i = M+1, \ldots, n), \tag{a}$$

where i'_{ki} = integers; that is, the old frequencies are linear/homogeneous/integral combinations of the $n - m$ independent quantities ν'_i (the new frequencies).

REMARK

If we analogize the n old frequencies with n constrained virtual displacements (the δq's of chap. 2) and the $n - m$ new ones with $n - m$ independent virtual variations of quasi coordinates (the $\delta\theta_I$'s of chap. 2, or any other group of $n - m$ independent "parameters"), then (a) is the analog of none other than Maggi's projection idea (§ 3.5)!

Problem 8.14.2 Extend the results of ex. 8.14.13 for the case where its equations (b, c) are replaced by

$$w'_k\!: \; w'_d \equiv \sum i_{dk} w_k = 0 \qquad (d = 1, \ldots, m), \tag{a}$$

$$w'_i \equiv \sum i_{ik} w_k \neq 0 \qquad (i = m+1, \ldots, n), \tag{b}$$

REMARK

This is the frequency analog of the general Hamel choice of quasi velocities (chap. 2).

Problem 8.14.3 Let $H' = H = E = E(J)$ have the special form

$$E(J) = F(f, J_4, J_5, \ldots, J_n), \tag{a}$$

where $f \equiv i_1 J_1 + i_2 J_2 + i_3 J_3$, $i_{1,2,3}$ are given integers. Show that the first three system frequencies are given by

$$\nu_r = i_r(\partial F/\partial f) \qquad (r = 1, 2, 3); \tag{b}$$

and then show that they also satisfy the *two* degeneracy conditions

$$i_2\nu_1 - i_1\nu_2 = 0, \qquad i_3\nu_1 - i_1\nu_3 = 0. \tag{c}$$

Appendix: On Multiply Periodic Functions/Motions

To help the reader to understand better the meaning of multiple Fourier series, like (8.14.24f–j), we point out the following facts from linear and nonlinear vibrations of discrete systems with constant coefficients (for extra clarity in real forms):

(i) The free vibrations of a *linear, 1-DOF* system [e.g., particle with kinetic and potential energies $m(\dot{q})^2/2$ and $kq^2/2$, respectively (m: mass, k: elasticity constant > 0)], have the following form:

$$q = a\sin(\omega t) + b\cos(\omega t) = c\cos(\omega t + \delta)$$

$$(\omega^2 = k/m;\ a, b, c\text{: constant amplitudes}, \delta = \text{"phase" constant}). \qquad (8.14.35)$$

Motion: *simply harmonic* and *singly periodic*.

(ii) The free vibrations of a *nonlinear, 1-DOF* system [e.g., elastic potential equal to $kq^2/2 + k'q^3 + k''q^4 + \cdots$ ($k, k', k'', \ldots$: constants)] have the single Fourier series form: $q = c_0 + c_1\cos(\omega t + \delta_1) + c_2\cos(2\omega t + \delta_2) + \cdots$

(*periodic* but *non-simply harmonic* due to the presence of *higher harmonics*, or *overtones*: $\omega, 2\omega, 3\omega, \ldots$). (8.14.36)

Motion: *multiply harmonic* and *singly periodic*.

(iii) The free vibrations of a *linear, n-DOF* system [one with elastic potential equal to $(1/2)(k_{11}q_1{}^2 + 2k_{12}q_1q_2 + k_{22}q_2{}^2 + \cdots + k_{nn}q_n{}^2)$: positive definite] have the form (assuming no degeneracies!):

$$q_k = \sum c_{kl}\cos(\omega_l t + \delta_l) \qquad [c_{kl} = \text{amplitudes},\ \delta_l = \text{phase constants};\ k, l = 1, \ldots, n]$$

(*simply harmonic* but with n intrinsic (natural) frequencies, or "modes of vibration" ω_l; i.e., no overtones). (8.14.37)

Motion: *simply harmonic* and *multiply periodic*; an n-dimensional Lissajous figure in q-space.

(iv) The free vibrations of a *nonlinear, n-DOF* system (one whose elastic potential contains terms of the *third* and *higher* order in the q_k's) have the following mutually equivalent forms of multiple Fourier series [infinite r-ple sums $\boldsymbol{s} \equiv (s_1, \ldots, s_r)$]:

$$q_k = \sum\cdots\sum c_{k,\boldsymbol{s}}\cos\left[(s_1\omega_1 + \cdots + s_r\omega_r)t + \delta_{\boldsymbol{s}}\right] \qquad (8.14.38a)$$

$$= \cdots$$

$$= \sum\cdots\sum \left\{A_{k,\boldsymbol{s}}\cos\left[2\pi(s_1w_1 + \cdots + s_rw_r)\right] + B_{k,\boldsymbol{s}}\sin\left[2\pi(s_1w_1 + \cdots + s_rw_r)\right]\right\} \qquad (8.14.38b)$$

$$= \sum\cdots\sum \left\{(1/2)(A_{k,\boldsymbol{s}} - iB_{k,\boldsymbol{s}})\exp\left[2\pi i(s_1w_1 + \cdots + s_rw_r)\right] + (1/2)(A_{k,\boldsymbol{s}} + iB_{k,\boldsymbol{s}})\exp\left[-2\pi i(s_1w_1 + \cdots + s_rw_r)\right]\right\} \qquad (8.14.38c)$$

$$= \sum\cdots\sum C_{k,\boldsymbol{s}}\exp\left[2\pi i(s_1w_1 + \cdots + s_rw_r)\right] \qquad (8.14.38d)$$

$$= \sum\cdots\sum D_{k,\boldsymbol{s}}\exp\left[2\pi i(s_1\nu_1 + \cdots + s_r\nu_r)t\right] \qquad (8.14.38e)$$

$$= \sum\cdots\sum D_{k,\boldsymbol{s}}\exp\left[i(s_1\omega_1 + \cdots + s_r\omega_r)t\right]. \qquad (8.14.38f)$$

Here:

- In (8.14.38a–c), the summations extend over all possible positive and negative but integral values of the integers $\boldsymbol{s} \equiv (s_1, \ldots, s_r)$, from 0 to $+\infty$; while in

(8.14.38d–f) they extend from $-\infty$ to $+\infty$. [We may assume, with no loss of generality, that $f \equiv s_1\omega_1 + \cdots + s_r\omega_r > 0$; because a term in f, in (8.14.38), can be combined with one with $-f$, and therefore the number of terms for which $f < 0$ can be reduced by half.]

• $r \equiv n - m \leq n$ [m = number of degeneracies ($\leq n - 1$), r = number of independent frequencies].

• The series of equations (8.14.38) combine the structures of both (8.14.36) (*nonlinearity* → *overtones*: $s_k\omega_k$) and (8.14.37) (*several DOFs* → *combination tones*: $s_1\omega_1 + \cdots + s_r\omega_r$); and that is why they consist of an (r) ple-infinity of terms.

• The ω's, known as "intrinsic vibration frequencies," are *constants* whose values depend on *both the physical constitution of the system and the initial conditions of its motions*; but they are not frequencies in the ordinary sense of the term; that is, like ω in (8.14.35), (8.14.36): the system does not return to its original configuration after a time $\tau_k = 2\pi/\omega_k$ $(k = 1, \ldots, n)$.

Motion: *multiply harmonic* and *multiply* (or *conditionally*, or *occasionally*) *periodic*; that is, superposition of r periodic motions of different frequencies, each consisting of an infinite number of overtones; an n-dimensional Lissajous figure in q-space. If, for certain values of the constants of integration (initial conditions) and/or the coefficients (parameters) of the equations of motion, $m = n - 1 \Rightarrow r = 1$, the motion (8.14.38a–f) degenerates into the multiply harmonic and singly periodic case (ii), eq. (8.14.36); and that is the reason for the adjective "conditionally."

Finally, if, in eqs. (8.14.38), we replace $(s_1\omega_1 + \cdots + s_r\omega_r)\,t$ with $s_1x_1 + \cdots + s_rx_r$, we obtain the generalization of a Fourier series to a function $q_k = q_k(x_1, \ldots, x_r)$, where the x's range over a generalized unit cube in x-space.

In sum: (i) the adjective *Fourier* series refers to the presence of a, generally, infinite number of higher harmonics, originating from the same fundamental frequency; and it is the result of the nonlinearity; (ii) whereas the adjectives *singly/multiply* periodic series refer to the number of independent frequencies present, and are the result of the number of DOFs; that is

linear (*nonlinear*) → *harmonic* (*overtones*)
one DOF (*several DOF*) → *singly* (*multiply*) *periodic*;

and the corresponding frequencies are

fundamental frequency	$\nu = \partial E/\partial J_k$	(8.14.39a)
overtones	$\nu = s_k(\partial E/\partial J_k)$	(8.14.39b)
combination tones	$\nu = \sum s_k(\partial E/\partial J_k).$	(8.14.39c)

The systems encountered in celestial mechanics and the old quantum theory were both nonlinear and had several DOFs; that is why their periodic motions (orbits) were expressed as multiple Fourier series.

HISTORICAL

It was N. Bohr who, with his famous "principle of correspondence" (late 1910s – early 1920s), established the quantum counterparts of eqs. (8.14.39) for the frequencies of the spectra of atomic systems, and thus prepared the way for Heisenberg's invention of modern quantum mechanics that followed soon thereafter (1925–1927).

8.15 ADIABATIC INVARIANTS

Historical Background

Roughly, adiabatic invariants (AI), or parameter invariants, of a periodic system, are quantities that remain essentially constant, or invariant, when *the system parameters change very slowly* relative to its periods. These quantities have played a key role in both classical (Boltzmann) and older quantum (Ehrenfest, Burgers, et al.) mechanics {see, for example, Bierhalter [1981(a), (b), 1982, 1983, 1992], Papastavridis [1985(a)], Polak (1959, 1960); also, recall introductory examples/problems on this topic in §7.9 of this book.} But also, recently, AI have become important in problems of charged particles in magnetic fields, and modern nonlinear dynamics. [See, for example, Lichtenberg (1969), Lichtenberg and Lieberman (1992), Percival and Richards (1982). For a detailed treatment, see Bakay and Stepanovskii (1981).]

Let us consider, with no loss in generality, a mechanical system S that is completely describable by the Hamiltonian

$$H = H(q, p; c), \tag{8.15.1}$$

where, with the usual notations, $q \equiv (q_1, \ldots, q_n)$, $p \equiv (p_1, \ldots, p_n)$; and the additional special *parameters* $c = c(t) \equiv (c_1(t), \ldots, c_m(t)) \equiv (c_1, \ldots, c_m) \equiv (c_\alpha)$ characterize the external and/or internal kinematico-inertial structure of S (e.g., length or mass of bob of a mathematical pendulum), and/or strength of the *external* field(s) of force in which S may be immersed.

REMARK

From a mechanistic viewpoint of thermodynamics, system coordinates are divided into *two* distinct kinds: (i) *macroscopic*, or *controllable*, whose variations produce *visible* changes to the system and flows of *mechanical* energy ΔW_c in/out of it (see below); and (ii) *molecular*, or *uncontrollable*, whose unceasing changes become perceptible only as energy going in/out of the system in the form of *heat* ΔQ (see below). [See also Brillouin (1964, pp. 231–245) and Bryan (1891–2, 1903).]

The earlier m c's classify as controllable and are treated as additional Lagrangean coordinates, constrained to remain constant during certain motions and vary in certain ways in others. Now, since, in general [recalling (8.2.14); see also (8.15.5) below],

$$dH/dt = \partial H/\partial t = \sum (\partial H/\partial c_\alpha)(dc_\alpha/dt) \qquad (\alpha = 1, \ldots, m), \tag{8.15.2}$$

if the c_α are constant ("turned off"), the system is closed and its generalized energy H is conserved; whereas if they are variable ("turned on"), the system has become open and H is no longer constant. In the latter case [i.e., $H = H(t, q, p)$], no general and exact methods are available for the analysis of motion. However, for some special cases of variation of the c's it is possible to find other energetic quantities that are conserved, exactly or approximately. Among the most interesting such cases are the two extremes of *very slow* (*adiabatic*) and *very fast* (or *parametric*) variations of the c's, relative to some characteristic time interval of the unperturbed (here, conservative) system. Below, we examine in some detail the adiabatic case; for the rapid case, see, for example, Forbat (1966, pp. 189–193), Landau and Lifshitz (1960, pp. 93–95), Percival and Richards (1982, pp. 153–157, 161–162).

We assume that initially the c_α are turned off and the system oscillates with the single frequency $\nu(= 1/\tau,\ \tau = \text{period})$. Then, as a result of some external energy-supplying agency, the c_α are turned on and begin to vary (i) *erratically*, or randomly (i.e., their variations are not systematically correlated to the oscillation of the system; namely, they are not in phase with that motion—no resonances), and (ii) *very slowly relative to* τ:

$$dc/dt \ll c/\tau, \qquad \text{or} \qquad \tau(dc/dt) \ll c; \tag{8.15.3}$$

or

$$dc/dt = (dc/d\tau)(d\tau/dt) \equiv c'\varepsilon = (\mathit{finite})\ (\mathit{small}) = \mathit{small}; \tag{8.15.3a}$$

that is, the parameter changes, being small fractions of their original constant values, are spread over a large number of oscillations; or, equivalently, *within a period* τ *these parameters may be considered constant*; for example, the mass of the bob of an oscillating mathematical pendulum varying slowly by picking up dust from its environment. [In the earlier-mentioned rapid case, the period (frequency) of the external disturbance is small (large) relative to the period (frequency) of the undisturbed system; or, generally, relative to a time interval during which the motion of the latter changes appreciably.] If the system can still oscillate (an example to the contrary is an axially loaded and transversely oscillating "beam–column" whose adiabatically varying axial load reaches the critical value for buckling, and hence reduces the fundamental frequency of the beam to zero; i.e., no motion), our task consists in:

(i) Calculating its new frequency (frequencies) $\nu + \Delta\nu$, or period(s):

$$\tau + \Delta\tau = 1/\nu + \Delta(1/\nu) = 1/\nu + (-1/\nu^2)\,\Delta\nu, \tag{8.15.4}$$

in terms of these $c \to c + \Delta c$ changes; and, since its energy is no longer constant,

(ii) Finding out if there exist new "adiabatic constants of motion," or *adiabatic invariants*. That such quantities occur can be argued as follows: by (8.15.2) we have $dH \sim dc$, and so there exists some combination(s) of H and the c's that *remains constant during the motion*, replacing the energy integral of the constant parameter system.

The classic example here is the oscillating mathematical pendulum whose length l (and/or mass m) is varied very slowly by some external agency. It turns out that, for small (linear) and undamped oscillations, the ratio of the pendulum's energy to its frequency is an *adiabatic invariant*; and this also allows us to relate the adiabatic change Δl to the amplitude and period changes $\Delta\tau$.

The Fundamental Theory

Let us quantify these ideas. Below, we present three treatments, in increasing order of difficulty: (i) *Energetic* (one DOF, single frequency), (ii) *integral variational* (n DOF; first singly periodic, then multiply/conditionally periodic motion), and (iii) *action–angle variables* (n DOF, multiply/conditionally periodic motion).

(i) Energetic Derivation

Let us assume here, for algebraic simplicity, that $m = 1$ (i.e., $c_1 \equiv c$), and that the kinetic energy is *homogeneous* quadratic in $\dot{q}$ (or p), so that $H = \mathit{total\ energy} \equiv E$.

If $H = H[q(t), p(t); c(t)] \equiv H(q, p; c)$, then we obtain, successively,

$$dH/dt = (\partial H/\partial q)(dq/dt) + (\partial H/\partial p)(dp/dt) + (\partial H/\partial c)(dc/dt)$$
$$= (\partial H/\partial q)(\partial H/\partial p) + (\partial H/\partial p)(-\partial H/\partial q) + (\partial H/\partial c)(dc/dt)$$

(by Hamilton's equations)

that is,

$$dE/dt = (\partial H/\partial c)(dc/dt) \quad (= \partial H/\partial t). \tag{8.15.5}$$

Averaging (8.15.5) over a complete cycle (of libration or rotation), while noting that, by (8.15.3), we can treat dc/dt as a constant, we obtain [with the customary notation $\langle \ldots \rangle \equiv$ *time average of* $(\ldots)$]

$$\langle dE/dt \rangle = \tau^{-1} \int_0^T (\partial H/\partial c)(dc/dt)\, dt, \tag{8.15.6}$$

or, since

$$dq/dt = \partial H/\partial p \;\Rightarrow\; dt = dq/(\partial H/\partial p) \;\Rightarrow\; \tau = \int_0^T dt = \oint dq/(\partial H/\partial p),$$

we obtain

$$\langle dE/dt \rangle = \left\{ \oint (\partial H/\partial c)/(\partial H/\partial p)\, dq \Big/ \oint [1/(\partial H/\partial p)]\, dq \right\} (dc/dt). \tag{8.15.7}$$

Let us transform (8.15.7) further. Solving the energy equation

$$H(q, p; c) = E \quad (= \textit{constant}, \text{if } c = \textit{constant}), \tag{8.15.8}$$

for the momentum p, we obtain

$$p = p(q, E; c), \tag{8.15.8a}$$

and, inserting this back into (8.15.8), we can rewrite the latter in the convenient form

$$H[q, p(q, E; c); c] = E. \tag{8.15.8b}$$

Next, differentiating (8.15.8b) partially with respect to c and E, which for the integrations involved in (8.15.6, 7) must be considered as *two independent and constant parameters*, we obtain, respectively,

$$(\partial H/\partial p)(\partial p/\partial c) + \partial H/\partial c = \partial E/\partial c = 0 \;\Rightarrow\; \partial p/\partial c = -(\partial H/\partial c)/(\partial H/\partial p), \tag{8.15.8c}$$

$$(\partial H/\partial p)(\partial p/\partial E) = \partial E/\partial E = 1 \;\Rightarrow\; \partial p/\partial E = 1/(\partial H/\partial p). \tag{8.15.8d}$$

As a result of (8.15.8c, d), eq. (8.15.7) transforms to

$$\langle dE/dt \rangle = -\left[\oint (\partial p/\partial c)\, dq \Big/ \oint (\partial p/\partial E)\, dq \right] (dc/dt), \tag{8.15.8e}$$

and, rearranging this, we obtain

$$\oint [(\partial p/\partial E)\langle dE/dt\rangle + (\partial p/\partial c)(dc/dt)]\, dq = 0; \tag{8.15.8f}$$

and this, recalling the action variable definition (§8.14), states simply that

$$dJ/dt = 0, \qquad J \equiv \oint p(q; E, c)\, dq\Big|_{\text{given } E,c} = J(E, c). \tag{8.15.9}$$

In words: If the oscillatory motion of a system is altered very slowly relative to its period, either by gradually varying the external field of force or by slowly modifying the system's physical constitution, then, *during such adiabatic changes, the action variable remains constant*; it is an adiabatic invariant. [Or, equivalently, *the ratio of twice its average kinetic energy divided by its frequency* remains constant—see (8.15.16–19) and ex. 8.15.1.] The preceding arguments indicate that for an adiabatic invariant to exist, the period (frequency) *must remain finite* (*nonzero*); *if* $\tau \to \infty$ ($\nu \to 0$), *the argument fails.*

One might have thought that, under such an external influence, J would depend on the precise moment at which c ceased to vary—that is, $dc/dt = 0$ (and the system became, again, truly periodic)—but, as shown below, the change of the action over a (possibly very long) time interval Δt, during which c changes adiabatically (but without causing a resonance), is $\Delta J \sim \langle dc/dt\rangle^2 \Delta t$.

From the above, it also follows that

$$\partial J/\partial E = \oint (\partial p/\partial E)\, dq = \oint dq \Big/ (\partial H/\partial p) = \int_0^{\tau} dt = \tau = 1/\nu; \tag{8.15.10}$$

in complete agreement with §8.14.

Geometrical Interpretation. Let us assume, for concreteness, that the periodic motion of the system is a *libration*. Then, as explained in §8.14, the action variable *J equals the area enclosed by the closed curve representing that motion,* in (q, p) space (fig. 8.22).

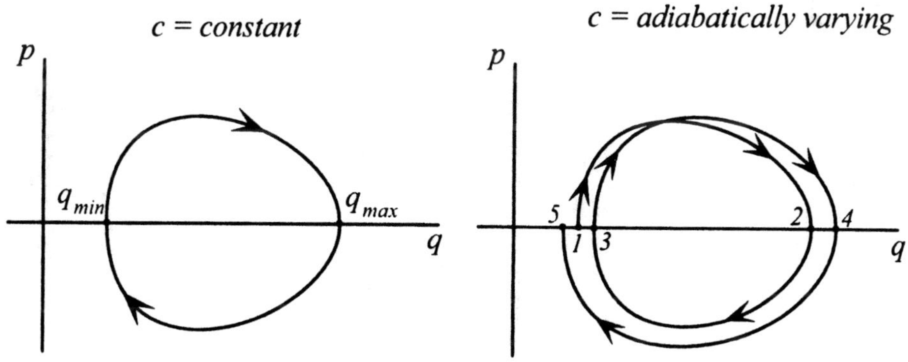

Figure 8.22 Geometrical interpretation of adiabatic invariance in phase space (1 DOF, libration): $p = \pm(2m)^{1/2}[E - V(q; c)]^{1/2} = p(q; E, c)$. In the adiabatic case, the area enclosed by the *open* path of each cycle (say, $1 \to 2 \to 3$) remains constant; although it changes shape. Then, J equals the area enclosed by a hypothetical closed trajectory obtained by fixing c at the beginning of each cycle.

Hence, by the well-known plane Green–Stokes theorem: also,

$$J = \oint p\,dq = -\oint q\,dp = \iint dq\,dp = \textit{adiabatic invariant}. \tag{8.15.11}$$

[Generally, if $I = f(q, p, c, t)$ is a first integral of the equations of motion, then its total change over the duration of the adiabatic variation process, say from t_1 to t_2, will be

$$\Delta I = \int_{t_1}^{t_2} (\partial f/\partial c)(dc/dt)\,dt \;=\; \langle \partial f/\partial c \rangle\, \Delta c, \tag{8.15.11a}$$

were, due to the adiabaticity, the time average can be taken over the *unvaried* motion. Hence, if the integral $I = f$ is independent of c, it is an adiabatic invariant. See also action–angle variable proof below.]

(ii) Integral Variational Derivation

[The following is due to Boltzmann (late 19th century, classical case) and his student Ehrenfest (1910s, classical → quantum case). We follow Schaefer (1937, pp. 58–66); see also Brillouin (1964, pp. 231–243) and De Donder (1924, 1925).] Let us reconsider the earlier system with Hamiltonian given by (8.15.1) and, hence, Lagrangean of the same form:

$$L = L(q, \dot{q}; c), \tag{8.15.12}$$

in the following two *continuous* and *finite* (but not yet assumed periodic) motions:

(a) A fundamental orbit I lasting from an initial time t_1 to a final t_2, and characterized by $q = q(t)$ and $c = \textit{constant along } I = c(I)$, and
(b) A neighboring orbit $II = I + \Delta(I)$, lasting from $t_1 + \Delta t_1$ to $t_2 + \Delta t_2$, and characterized by $q + \Delta q$ and $c + \Delta c = \textit{constant along } II \equiv c(II)$.

Now, using the analytical results and notation of §7.9, and treating the parameters $c_\alpha(\alpha = 1, \ldots, m)$ as *additional* Lagrangean coordinates, it is not hard to show that, under such a $I \to II$ variation, and since, along both these orbits, Lagrange's equations of motion for the q's hold, to the first noncontemporaneous order,

$$\begin{aligned} \Delta A_H &\equiv \Delta \int_{t_1}^{t_2} L\,dt \\ &= \int_{t_1}^{t_2} \sum (\partial L/\partial c_\alpha)\,\Delta c_\alpha\,dt + \Big\{\sum p_k\,\Delta q_k - h\,\Delta t\Big\}_{\text{initial time}}^{\text{final time}} \end{aligned} \tag{8.15.13}$$

(notice the *additional* Δc-sum) where (§3.9)

$$\begin{aligned} h &\equiv \sum (\partial L/\partial \dot{q}_k)\dot{q}_k - L = \text{generalized energy (Hamiltonian) in Lagrangean variables} \\ &= h(c) = h(I) = \textit{constant} \equiv h, && (8.15.13a) \\ &= h(c + \Delta c) \equiv h(II) = \textit{constant} \equiv h + \Delta h. && (8.15.13b) \end{aligned}$$

Next, let I be a completely degenerate periodic motion, with the single period τ and frequency ν, and let, in eq. (8.15.13), $t_1 = 0$, $t_2 = \tau$ (the multiply/conditionally periodic case is discussed later). With

$\partial L/\partial c_\alpha \equiv C_\alpha$: *Lagrangean force with which the system reacts to a change of its parameter* c_α, (8.15.14a)

and, hence,

$-C_\alpha \equiv -\partial L/\partial c_\alpha$: *External force that, at every instant, must be acting on the system to keep* c_α *constant*, (8.15.14b)

so that

$\sum (\partial L/\partial c_\alpha)\, \Delta c_\alpha = \sum C_\alpha\, \Delta c_\alpha \equiv \Delta W_c$: *First-order work done by the system to its environment, during a* $c \to c + \Delta c$ *change*, (8.15.14c)

and analogously for $-\Delta W_c$, and with the notation $\int_0^\tau \ldots \equiv \oint \ldots$, eq. (8.15.13) becomes

$$\Delta \oint L\, dt = \oint \Delta W_c\, dt + \left\{ \sum p_k\, \Delta q_k - h\, \Delta t \right\}_0^\tau. \tag{8.15.14d}$$

If, further, the neighboring orbit II is also (singly) periodic with period

$$\tau + \Delta\tau = \tau + \Delta(1/\nu) = \tau + (-1/\nu^2)\, \Delta\nu \tag{8.15.14e}$$

(a fact that clearly indicates why we need a *variable time-endpoints* treatment), then choosing, with no loss in generality, $\Delta t_1 = 0$, and since, then, $\Delta t_2 = \Delta\tau$, $h(I) = \textit{constant} \equiv h$, $\{\sum p_k\, \Delta q_k\}_0^\tau \equiv 0$ (due to periodicity), reduces (8.15.14d) to

$$\Delta \oint L\, dt = \oint \Delta W_c\, dt - h\, \Delta\tau, \tag{8.15.14f}$$

or, equivalently, in terms of mean/averaged values of their integrands,

$$\begin{aligned} \Delta(\langle L\rangle\tau) &= \langle \Delta W_c\rangle\tau - h\, \Delta\tau \\ &\quad \{\text{or, adding and subtracting } \tau\,\Delta h \equiv \tau[h(II) - h(I)] \equiv \tau[h(c + \Delta c) - h(c)]\} \\ &\equiv \langle \Delta W_c\rangle\tau - \Delta(h\,\tau) + \tau\,\Delta h, \end{aligned} \tag{8.15.14g}$$

or, rearranging,

$$\Delta\big[(\langle L\rangle + h)\tau\big] = \big(\Delta h + \langle \Delta W_c\rangle\big)\tau. \tag{8.15.14h}$$

But, (i) $\langle L\rangle = \langle T - V\rangle = \langle T\rangle - \langle V\rangle$, and (assuming T: quadratic *homogeneous* in $\dot{q}$ or p)

$$h = T + V = \langle T + V\rangle = \langle T\rangle + \langle V\rangle = \textit{constant}, \tag{8.15.14i}$$

so that

$$\langle L\rangle + h = 2\langle T\rangle; \tag{8.15.14j}$$

and (ii) by the *first law of thermodynamics*: if

$$\Delta Q = \textit{Heat added to the system during the transition } I \rightarrow II, \tag{8.15.14k}$$

then

$$\Delta Q + (-\Delta W_c) = \Delta h; \tag{8.15.14l}$$

that is,

Heat supplied to system + Work done to system to violate its constraint
(c = constant) = Increase of energy of system.

As a result of (8.15.14j, l), eq. (8.15.14h) assumes the *Boltzmann–Clausius form*:

$$\Delta[2\langle T\rangle\tau] = \tau\,\Delta Q. \tag{8.15.15}$$

This result is exact. If we, now, assume that the transition $I \rightarrow II$ is adiabatic (i.e., $\Delta Q = 0$), then (8.15.15) immediately leads us to the famous *adiabatic theorem of Ehrenfest*:

$$\Delta[2\langle T\rangle\tau] = \Delta[2\langle T\rangle/\nu] = 0. \tag{8.15.16}$$

[Physically the process must be (i) *fast* enough so that our system cannot exchange heat with its environment (i.e., it remains thermally insulated); and (ii) *slow* compared with other processes that lead to thermal equilibrium. For example, in order that the expansion of a gas in a cylinder be adiabatic, the velocity of its outward moving piston must be slow only relative to the velocity of sound in the gas; that is, the piston may move quite fast!

For a precise definition of adiabaticity, see books on thermal physics, and so on: for example, Landau and Lifshitz (1980, pp. 38–41).]

Other, equivalent, forms of the theorem are

$$[2\langle T\rangle/\nu]_I = [2\langle T\rangle/\nu]_{II}\text{: } \textit{Adiabatic invariant}; \tag{8.15.17}$$

$$J \equiv 2\langle T\rangle/\nu = 2\oint T\,dt = \oint\left(\sum p_k\dot{q}_k\right)dt$$
$$= \sum\left(\int_0^{1/\nu}(p_k\dot{q}_k)\,dt\right)\text{: } \textit{Adiabatic invariant}. \tag{8.15.18}$$

Specialization. If $2\langle T\rangle = h = E$ (i.e., if $\langle T\rangle = \langle V\rangle$; e.g., linear harmonic oscillations), eqs. (8.15.16, 17) reduce to the "*Planck form*":

$$E/\nu\text{: } \textit{adiabatic invariant}. \tag{8.15.19}$$

Multiply/Conditionally Periodic System

So far, we have assumed that our *n-DOF* system is *completely separable and completely degenerate* (§8.14) and has the single period τ; that is, there exist $n-1$ independent equations of the form (8.14.31)

$$i'_1\nu_1 + i'_2\nu_2 + \cdots + i'_n\nu_n = 0, \tag{8.15.20}$$

where the i'_k are integers and the ν_k are the fundamental frequencies of its individual DOFs; or, equivalently,

$$i_1(1/\nu_1) = i_2(1/\nu_2) = \cdots = i_n(1/\nu_n) = 1/\nu \equiv \tau, \tag{8.15.21}$$

or, with $\tau_k = 1/\nu_k$,

$$i_1\tau_1 = i_2\tau_2 = \cdots = i_n\tau_n = \tau, \tag{8.15.21a}$$

where the i_k are positive integers (naturals). Then, (8.15.18) reduces further to

$$\sum\left(\int_0^{i_k/\nu_k} (p_k\dot{q}_k)\,dt\right) = \sum\left(i_k \int_0^{1/\nu_k} (p_k\dot{q}_k)\,dt\right)$$
$$= \sum i_k\, J_k\text{: } \textit{adiabatic invariant}, \tag{8.15.22}$$

or, since $i_k = \nu_k/\nu = \tau\,\nu_k$ $(\nu \neq 0)$, finally,

$$\sum \nu_k J_k\text{: } \textit{adiabatic invariant}, \tag{8.15.23}$$

in agreement with ex. 8.14.9.

If, however, our system is only *conditionally/multiply periodic*—that is, if at least one or more of (8.15.20, 21) do *not* hold, and, instead, are replaced with equations of the form

$$I_1\nu_1 + I_2\nu_2 + \cdots + I_n\nu_n = \varepsilon, \tag{8.15.24}$$

where the I_k are, generally, *large integers* and ε is arbitrarily small, then a "quasi period" τ can be defined by

$$\tau = 1/\nu = i_1/\nu_1 + \varepsilon_1 = i_2/\nu_2 + \varepsilon_2 = \cdots = i_n/\nu_n + \varepsilon_n, \tag{8.15.25a}$$
$$= i_1\tau_1 + \varepsilon_1 = i_2\tau_2 + \varepsilon_2 = \cdots = i_n\tau_n + \varepsilon_n, \tag{8.15.25b}$$

where the ε_k are arbitrarily small (in order to achieve any required degree of accuracy) and the boundary terms $\{\sum p_k\, \Delta q_k\}_0^\tau$ can be made as small as needed. Hence, *a conditionally periodic system can also be brought as close as desired to a purely (singly) periodic one,* so that (8.15.22, 23) still hold. [More precisely: $\sum(\int_{t_1}^{t_2} (p_k\dot{q}_k)\,dt) =$ *adiabatic invariant*, where $t_2 = i_k/\nu_k + \varepsilon_k, t_1 = 0$.]

• If the original system is completely separable, then it is reduced to n subsystems, each with one DOF and one frequency. If, further, these frequencies, $\nu_1, \ldots, \nu_n$, are incommensurate—that is, independent—then our *non-degenerate* system has n *independent* adiabatic invariants:

$$J_k \equiv \oint p_k\, dq_k = \textit{adiabatic invariant} \qquad (k = 1, \ldots, n). \tag{8.15.26a}$$

• But if our system is m-fold degenerate, or $(n - m)$-periodic, then it has only $n - m$ independent adiabatic invariants; namely, certain combinations of its n J_k's.

Hence the rule: There exist as many independent adiabatic invariants as there are independent (incommensurate) frequencies; that is, $n - m$ $(0 \leq m \leq n - 1)$. For example, the spatial linear and isotropic oscillator has three mutually equal frequencies (i.e., $n = 3$, and since $\nu_x = \nu_y = \nu_z$, $m = 2$) and, therefore, only *one* adiabatic invariant:

$$J_x + J_y + J_z = \oint p_x\,dq_x + \oint p_y\,dq_y + \oint p_z\,dq_z = \textit{adiabatic invariant}. \qquad (8.15.26b)$$

(iii) Action–Angle Variables Derivation

[The following is due to Burgers (1917) and Krutkow (1919); also Bohr (1918, who calls it *theorem of mechanical transformability*). Here, we follow the excellent summary of these proofs given by Birtwistle (1926, pp. 76–78). See also Born (1927, pp. 56–59, 95–98), Haar (1971, pp. 139–144), Saletan and Cromer (1971, pp. 259–263). It may be omitted in a first reading.]

To simplify the discussion, let us assume, with no loss in generality, that our n-DOF system contains only *one* adiabatically varying parameter, $c = c(t)$. As we have seen in §8.10, in the *constant* parameter case, the generating function of the canonical transformation $(q, p) \to (q' = w,\ p' = J)$ is $A_o(q, J)$ $[= F_2(q, p')]$. In the adiabatic case, we can think of c as an additional system coordinate unrelated to the motion; that is, unrelated to the q's. Hence, the reduced action A_o *becomes the explicitly time-dependent function* $A_o[q, J; c(t)]$, so that

$$w_k = \partial A_o(q, J, c)/\partial J_k, \qquad p_k = \partial A_o(q, J, c)/\partial q_k. \qquad (8.15.27a)$$

Here, the Hamiltonian transformation $H \to H'(\neq H)$ is

$$H = H(J;\, c) = E(J;\, c) \ \to\ H' = H + \partial A_o/\partial t = E + (\partial A_o/\partial c)(dc/dt), \quad (8.15.27b)$$

and, therefore, with $A_o = A_o[q(w, J; c), J; c] \equiv A_o(w, J; c)$ $[\Rightarrow H' = H'(w, J; c)]$, the Hamiltonian equations of motion of the w_k's and J_k's are

$$\begin{aligned} dw_k/dt &= \partial H'/\partial J_k = \partial H/\partial J_k + \partial/\partial J_k(\partial A_o/\partial t) \\ &= \partial H/\partial J_k + \partial/\partial J_k\big[(\partial A_o/\partial c)(dc/dt)\big] \\ &= \nu_k + \big[\partial/\partial w_k(\partial A_o/\partial c)\big](dc/dt), \qquad (8.15.27c) \\ dJ_k/dt &= -\partial H'/\partial w_k = -\partial H/\partial w_k - \partial/\partial w_k(\partial A_o/\partial t) \\ &= 0 - \partial/\partial w_k\big[(\partial A_o/\partial c)(dc/dt)\big] = -\big[\partial/\partial w_k(\partial A_o/\partial c)\big](dc/dt). \quad (8.15.27d) \end{aligned}$$

Equivalently, instead of the above choice $A_o(q, J) \to A_o[q, J;\, c(t)]$ for generating function, we try the form $W(q, w, t) \equiv A(q, w, t)$ $[= F_1(t, q, q')]$. Indeed, recalling (8.14.23b, c), we have

$$\begin{aligned} \delta A_{oo} &\equiv \delta\Big(A_o - \sum w_k J_k\Big) \\ &= \sum \big[(\partial A_o/\partial q_k)\,\delta q_k + (\partial A_o/\partial J_k)\,\delta J_k\big] - \sum \delta(w_k\, J_k) \\ &= \sum (p_k\,\delta q_k + w_k\,\delta J_k) - \sum (\delta w_k J_k + w_k\,\delta J_k) \\ &= \sum (p_k\,\delta q_k - J_k\,\delta w_k) \\ &\Rightarrow\ p_k = \partial A_{oo}/\partial q_k, \qquad J_k = -\partial A_{oo}/\partial w_k, \qquad (8.15.28a) \end{aligned}$$

and therefore (by the theory of §8.8 and §8.10) the Hamiltonian equations of w_k and J_k are

$$dw_k/dt = \partial H'/\partial J_k, \qquad dJ_k/dt = -\partial H'/\partial w_k, \qquad \text{where } H' = H + \partial A_{oo}/\partial t. \tag{8.15.28b}$$

But since $H = H(J;\, c)$ and $A_{oo} = A_{oo}[q, w;\, c(t)] \equiv A_{oo}(q, w;\, t)$, and therefore $\partial A_{oo}/\partial t = (\partial A_{oo}/\partial c)(dc/dt)$, the above yield, further,

$$dw_k\, dt = \partial H/\partial J_k + \partial/\partial J_k\big[(\partial A_o/\partial c)(dc/dt)\big] = \nu_k + \big[\partial/\partial w_k(\partial A_o/\partial c)\big](dc/dt), \tag{8.15.28c}$$

$$[\Rightarrow w_k = \nu_k(J, c)t + \gamma_k(J, c) = (\nu_{ko}t + \gamma_{ko}) + (\nu_{k1}t^2 + \gamma_{k1}t)(dc/dt) + \cdots]$$

$$\begin{aligned} dJ_k/dt &= -\partial H/\partial w_k - \partial/\partial w_k(\partial A_{oo}/\partial t) \\ &= 0 - \partial/\partial w_k\big[(\partial A_{oo}/\partial c)(dc/dt)\big] = -\big[\partial/\partial w_k(\partial A_{oo}/\partial c)\big](dc/dt), \end{aligned} \tag{8.15.28d}$$

or, finally, with the simplifying notation $\Phi \equiv \partial A_{oo}/\partial c = \Phi[q(w, J, c), J, c] = \Phi(w, J, c)$,

$$dJ_k/dt = -(\partial \Phi/\partial w_k)(dc/dt). \tag{8.15.28e}$$

Integrating, next, (8.15.28d, e) over a long time interval $\Delta t \equiv t_2 - t_1$, we obtain

$$\Delta J_k \equiv J_k(t_2) - J_k(t_1) = -\int_{t_1}^{t_2} (\partial \Phi/\partial w_k)(dc/dt)\, dt. \tag{8.15.29a}$$

Now, since A_{oo} *is periodic in each of the* w_k*'s with period 1* [(8.14.23b ff.)], so is $\partial A_{oo}/\partial c$; that is, $\Phi = \Phi(w, J, c)$ $[\rightarrow \Phi(w, J, t)]$. Therefore, we can represent it as the following multiply periodic Fourier series [(8.14.24f–j), with a *single* $\sum$ sign standing for all summations, for simplicity]:

$$\sum D_{k,s}(J, c) \exp(2\pi i \boldsymbol{s} \cdot \boldsymbol{w}), \tag{8.15.29b}$$

where, as earlier, $\boldsymbol{s} \equiv (s_1, \ldots, s_n)$ are positive or negative integers, or zero; ranging from $-\infty$ to $+\infty$,

$$\boldsymbol{w} \equiv (w_1, \ldots, w_n), \qquad \boldsymbol{s} \cdot \boldsymbol{w} \equiv s_1 w_1 + \cdots + s_n w_n; \tag{8.15.29c}$$

and from this we readily conclude that

$$\partial \Phi/\partial w_k = -\sum{}' E_{k,s} \exp(2\pi i\, \boldsymbol{s} \cdot \boldsymbol{w}) = -\sum{}' F_{k,s} \exp(2\pi i\, \boldsymbol{s} \cdot \boldsymbol{v} t) \tag{8.15.29d}$$

(= a multiply periodic Fourier series *but without the constant term*),

such terms having been removed, from each of these series, by the $\partial/\partial w_k$-differentiations. This key step in our proof is designated by the *accent (prime) on the summation sign*. We have also made the related assumption (see below) that

$$\boldsymbol{s} \cdot \boldsymbol{v} \equiv s_1 \nu_1 + \cdots + s_n \nu_n \neq 0 \qquad \text{(i.e., no degeneracies!)}. \tag{8.15.29e}$$

As a result of the above, (8.15.29a) becomes, successively,

$$\begin{aligned}\Delta J_k &= -\int_{t_1}^{t_2} (dc/dt)\Big[-\sum{}' F_{k,s}\exp(2\pi i\,\mathbf{s}\cdot\mathbf{v}t)\Big]\,dt \\ &= -\langle dc/dt\rangle \int_{t_1}^{t_2} \Big[-\sum{}' F_{k,s}\exp(2\pi i\,\mathbf{s}\cdot\mathbf{v}t)\Big]\,dt \\ &\equiv -\langle dc/dt\rangle \int_{t_1}^{t_2} G(c,t)\,dt, \qquad (8.15.29f)\end{aligned}$$

since both the $F_{k,s}$'s and ν_k's depend on c. [The condition of erratic or unsymmetric variation of $c(t)$ can be satisfied by taking, for example, $dc/dt = constant$.] Next, to study the precise dependence of the above integral on c, we expand its integrand à la Taylor around $c_1 \equiv c(t)$, and thus obtain

$$\int_{t_1}^{t_2} G(c,t)\,dt = \int_{t_1}^{t_2} \left[G(c_1,t) + (c-c_1)\,G'(c_1,t) + \cdots\right] dt. \qquad (8.15.29g)$$

Now:

(i) The *first* term of the integrand is periodic in the *constant* ν_k's (i.e., the frequencies before c began to vary). Therefore, if Δt is long enough to contain a large number of the corresponding periods, since G is periodic in time (and does not contain a constant term),

$$\int_{t_1}^{t_2} G(c_1,t)\,dt = 0. \qquad (8.15.29h)$$

(ii) The *second* term,

$$\int_{t_1}^{t_2} (c-c_1)\,G'(c_1,t)\,dt, \qquad (8.15.29i)$$

is of the same order (of magnitude) as

$$\int_{t_1}^{t_2} \left[(dc/\,dt)t\right] G'(c_1,t)\,dt, \qquad (8.15.29j)$$

and that, in turn, is of the order of

$$\langle dc/dt\rangle(t_2 - t_1) \equiv \langle dc/dt\rangle \Delta t \equiv \Delta c \quad (= \textit{finite}). \qquad (8.15.29k)$$

From the above, we conclude that

$$\Delta J_k = -\langle dc/dt\rangle\ (\textit{Term of order } \Delta c) \sim \langle dc/dt\rangle\,\Delta c = \langle dc/dt\rangle^2 \Delta t,$$

or, equivalently, since $\Delta J_k = \langle dJ_k/dt\rangle \Delta t$,

$$\langle dJ_k/dt\rangle \sim \langle dc/dt\rangle^2; \qquad (8.15.29l)$$

and hence even if Δc is finite (after a very long period of time), it is possible to make ΔJ_k as small as desired by decreasing dc/dt; that is, the J_k's are adiabatic invariants. [The extension of this proof to several c's does not offer any difficulties; (8.15.28d) is replaced by $dJ_k/dt = -\sum[(\partial^2 A_{oo}/\partial w_k \partial c_l)]\,(dc_l/dt)$, where $l = 1, 2, \ldots, m$, and so on.]

Effect of Degeneracies on Adiabatic Invariance. The no degeneracy requirement (8.15.29e) is crucial. If an (8.14.31)-like relation

$$\boldsymbol{i}\cdot\boldsymbol{\nu} \equiv i_1\nu_1 + \cdots + i_n\nu_n = 0 \qquad [\boldsymbol{i} \equiv (i_1, \ldots, i_n)\text{: integers}], \tag{8.15.30a}$$

exists among the original frequencies (and/or occurs at some stage of the subsequent adiabatic variation), then, for $\boldsymbol{s} = \boldsymbol{i}$, the Fourier series

$$G(c_1, t) = \left[-\sum{}' F_{k,s}\exp(2\pi i\,\boldsymbol{s}\cdot\boldsymbol{\nu}t)\right]\Big|_{c=c_1}, \tag{8.15.30b}$$

will contain a constant term, say C; and, accordingly, eq. (8.15.29h) will be replaced by

$$\int_{t_1}^{t_2} G(c_1, t)\,dt = C\,\Delta t. \tag{8.15.30c}$$

Then we will have, instead of (8.15.29l),

$$\begin{aligned}\Delta J_k &= -\langle dc/dt\rangle\,(C\Delta t + \textit{Term of order } \Delta c)\\ &= -C\Delta c - \langle dc/dt\rangle\,(\textit{Term of order } \Delta c)\\ &= -C\Delta c = \textit{finite change}, \qquad \text{as} \qquad \langle dc/\,dt\rangle^2 \to 0;\end{aligned} \tag{8.15.30d}$$

that is, the J_k will no longer be adiabatic invariants. In such degenerate cases, as stated earlier, the number of independent adiabatic invariants equals the number of independent frequencies $(n - m)$.

[If m (8.15.30a)-like relations hold *identically*—that is, for all J's, for a certain c, then, following the method of (ex. 8.14.14), we introduce new w's and J's such that the first $n - m$ of the new frequencies $(\nu'_d;\ d = 1, \ldots, m)$ are equal to zero, while the remaining m of them $(\nu'_i;\ i = n - m, \ldots, n)$ are independent (i.e., incommensurate). Then, the constant exponents appearing in the Fourier series for A_{oo} involve only the "dependent" angle variables (w'_d), and, upon differentiation with respect to the "independent" such variables (w'_i), they disappear. It follows that at such "places of degeneration" the "independent" actions (J'_i) remain invariant; while, in general, the "dependent" ones (J'_i) do not.

If, in addition to the above cases of identical degeneration, (8.15.30a)-like relations hold for particular values of the employed J's—a case known as *accidental degeneration*—these action variables need not be invariant; unless the amplitude corresponding to (8.15.30a) also vanishes from its (8.15.29g)-like series. On this delicate topic, see also Fues (1927, p. 150; and references given therein).]

Example 8.15.1 *Adiabatic Invariant of Linear 1-DOF Oscillator.* Here, with the customary notations,

$$H = p^2/2m + m\,\omega^2 q^2/2 = H(q, p), \tag{a}$$

and therefore the energy curve in phase space, $H(q, p) = E$ (libration), is

$$[p/(2mE)^{1/2}]^2 + [q/(2E/m\omega^2)^{1/2}]^2 = 1; \tag{b}$$

that is, an ellipse with semiaxes: $(2E/m\,\omega^2)^{1/2}$ along q, and $(2mE)^{1/2}$ along p. Hence,

$$\begin{aligned} J &= \textit{area of ellipse} = \pi\big[(2E/m\,\omega^2)^{1/2} \times (2mE)^{1/2}\big] \\ &= \pi(2E/\omega) = 2\pi(E/\omega) = E/\nu = \textit{adiabatic constant}; \qquad \text{(c)} \end{aligned}$$

that is, as long as $\nu \neq 0$, under adiabatic changes, *the oscillator energy is proportional to its frequency*; as predicted by the general theory.

Example 8.15.2 *Effect of Light Damping on the Adiabatic Invariant.* Let us consider the linear 1-DOF spring–mass–damper system with equation of motion

$$m\,\ddot{q} + d\dot{q} + k\,q = 0, \qquad \text{(a)}$$

where m, d, k = *mass, viscous damping coefficient (constant), spring constant*, respectively. Since (a) has no periodic solutions, it has no adiabatic invariants. However, as is well known from second-order differential equations, the change of variables

$$q \to q' = q\exp[(d/2m)t] \;\Rightarrow\; q = q'\exp[-(d/2m)t], \qquad \text{(b)}$$

transforms (a) to the linear *dampingless* equation (for the fictitious system described by q'):

$$m'(q')^{\cdot\cdot} + k'q' = 0, \qquad \text{(c)}$$

where

$$\begin{aligned} m' &= m, \\ (\omega')^2 &\equiv k'/m' = k/m - (1/4)(d/m)^2 = \omega^2[1 - (d^2/4mk)], \\ \omega^2 &= k/m. \qquad \text{(d)} \end{aligned}$$

Clearly, (c) has periodic solutions with the single frequency $\nu' \equiv \omega'/2\pi$, and therefore adiabatically invariant action {with a' = amplitude of (b, c) = $a\,\exp[(d/2m)t]$}:

$$\begin{aligned} J' &= E'/\nu' = V'_{\max}/\nu' \\ &= \big[(1/2)k'(a')^2\big] = \big[(1/2)m'(\omega')^2(a')^2\big]/(\omega'/2\pi) \\ &= \pi m'\omega'\{a\,\exp[(d/2m)t]\}^2 \\ &= \pi m\{\omega[1 - (d^2/4mk)]^{1/2}\}\,\{a^2\exp[(d/m)t]\} \\ &= \pi\,\{ma^2\omega[1 - (d^2/4mk)]^{1/2}\}\exp[(d/m)t] = \textit{constant} \\ &= \pi\,\{ma^2\omega[1 - (d^2/4mk)]^{1/2}\} \qquad (\text{i.e., } J'|_{t=0}). \qquad \text{(e)} \end{aligned}$$

Hence, for the adiabatically *noninvariant* action of (a), we will have the following *exponential decay* variation:

$$J(t') = J(0)\exp(-kt'/m), \qquad \text{(f)}$$

where

$$t' \equiv \varepsilon\,t = \textit{slow time} \equiv (d/k)t, \qquad -(d/m)t = -(k/m)\,(\varepsilon\,t) \equiv -(k/m)t'. \qquad \text{(g)}$$

For an alternative treatment of a more general case, see Kevorkian and Cole (1981, pp. 271–272).

Example 8.15.3 Let us consider a particle, of mass m, in rectilinear horizontal and perfectly elastic collisional motion between two perfectly elastic and infinitely massive walls of width l, one fixed (say, the left) and one movable (the right).

(i) If both walls are *fixed* (stationary), and the particle moves with velocity v, then, clearly, its energy (neglecting gravity) and "period of collision" are, respectively,

$$E = T = (1/2)mv^2, \qquad \tau = 2l/v. \tag{a}$$

(ii) If the right wall *moves* with velocity $\dot{l}$, assumed unaffected by the repeated particle collisions, (i.e., to the right, if $\dot{l} > 0$), then, after the particle has collided with *both* walls, once, its velocity has changed from v to $v - 2\dot{l}$ [recalling definition of restitution coefficient, (4.4.1), e; here $e = 1$]; i.e., $\Delta v = -2\dot{l}$. Assuming now that $\dot{l} \ll v$; that is, the right wall moves very slowly relative to the *particle*, let us calculate the adiabatic invariant of this periodic system.

From First Principles

Choosing a time interval Δt that is very large relative to the collision period of the fixed wall case, and very small relative to the $l/\dot{l}$—that is,

$$2l/v \ll \Delta t \ll l/\dot{l}, \tag{b}$$

it is not hard to see that if one pair of such collisions changes the velocity of the particle by $-2\dot{l}$, $(v/2l)\,\Delta t$ pairs will change it by

$$\Delta v = -(v\dot{l}/l)\,\Delta t \qquad \text{(a decrease if the right wall moves outward).} \tag{c}$$

Integrating (c), we readily obtain the adiabatic invariant

$$v\,l = \textit{constant}, \tag{d}$$

or, due to (a),

$$E\,l^2 = \textit{constant}. \tag{e}$$

From the General Adiabatic Theory

We readily find

$$J = \oint p\,dq = \int_0^T (mv)\,(v\,dt) = 2mvl = \textit{constant}. \tag{f}$$

These results are shown graphically in fig. 8.23.

For alternative treatments of this popular example, see, for example, Kuypers (1993, pp. 346, 535–536), Matzner and Shepley (1991, pp. 198–199), Percival and Richards (1982, pp. 142–144).

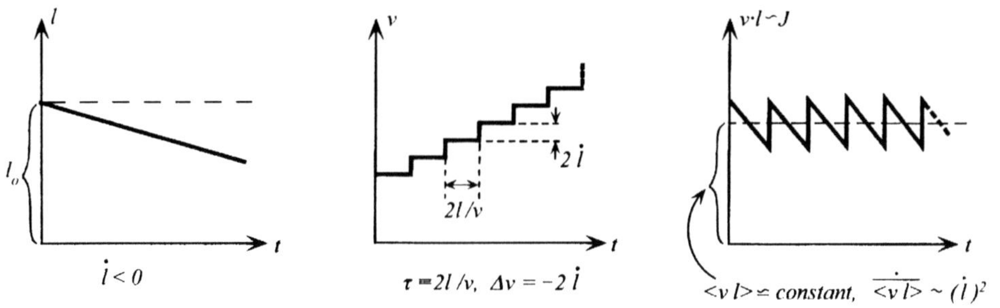

Figure 8.23 Temporal variation of wall distance $(l, \dot{l})$, particle velocity (v), and action variable $(J \sim lv)$.

Problem 8.15.1 *Adiabatic Motion of a Planar Mathematical Pendulum* (Ol'khovskii, 1970, p. 430 ff.). Consider the small angular amplitude adiabatic motions of a planar mathematical pendulum of mass m, length l, and angle with the vertical ϕ.

(i) Show that its reduced Hamilton–Jacobi equation is (with E = total energy of pendulum)

$$(dA_o/d\phi)^2 + m^2 g l^3 \phi^2 = 2ml^2 E. \tag{a}$$

(ii) Show that the complete solution of (a) is

$$A_o = (m^2 g l^3)^{1/2} \{(\phi/2)[(2E/mgl) - \phi^2]^{1/2} + (E/mgl) \arcsin [\phi/(2E/mgl)^{1/2}]\}, \tag{b}$$

and, therefore, during a complete libratory cycle of the pendulum,

$$\Delta A_o = E/\nu = J \Rightarrow E(t) = J\nu(t), \qquad \text{where } \omega^2 = g/l\,(= (2\pi\nu)^2). \tag{c}$$

(iii) Show that the average of the total energy of the pendulum, over a cycle, equals

$$\langle E \rangle = (ml^2/2)\langle(\dot{\phi})^2\rangle + (mgl/2)\langle\phi^2\rangle, \tag{d}$$

that is,

$$E = \langle T + V \rangle = \langle T \rangle + \langle V \rangle. \tag{e}$$

(iv) Show that

$$\langle\phi^2\rangle = {\phi_o}^2/2, \qquad \langle(\dot{\phi})^2\rangle = ({\phi_o}^2/2)\omega^2 \tag{f}$$

(ϕ_o = maximum angular amplitude), and, therefore,

$$\langle E \rangle = mgl{\phi_o}^2/2 \quad (= E). \tag{g}$$

(v) Show that

$$l^{3/4}\phi_o = \textit{adiabatic invariant}, \qquad \text{or} \qquad l^3{\phi_o}^4 = \textit{adiabatic invariant}; \tag{h}$$

or $\phi_o \sim l^{-3/4}$; that is, if l is *reduced* adiabatically by 50%, ϕ_o *increases* by 68%.

(vi) Show that under adiabatic variations,

$$dE = -(E/2l)\,dl. \tag{i}$$

Problem 8.15.2 Consider the linear oscillations of a mathematical pendulum of mass m and length l on a smooth inclined plane of angle with the horizontal χ. Show that under adiabatic changes of χ,

$$\phi_o \sim (\sin\chi)^{-1/4}, \tag{a}$$

where ϕ_o = angular amplitude.

HINT

In the equation of motion of the ordinary mathematical pendulum (i.e., when $\chi = \pi/2$), replace g with g sin χ. Then, the frequency becomes

$$\nu = (1/2\pi)(g/l)^{1/2}(\sin\chi)^{1/2},$$

and

$$E = \textit{maximum potential energy} \sim mgl\sin\chi\phi_o{}^2. \tag{b}$$

Problem 8.15.3 Consider a linear and undamped spring–mass oscillator of frequency

$$\nu = (1/2\pi)\,(k/m)^{1/2} \tag{a}$$

(k = spring "constant," m = mass). Show that under adiabatic variations of k:

(i) $E^2/k = \textit{adiabatic invariant}$ (E = total energy); (b)

(ii) $ka^4 = \textit{adiabatic invariant}$ (a = oscillation amplitude). (c)

HINT

$$E = \textit{maximum potential energy} = ka^2/2.$$

For additional examples on adiabatic invariance, see, for example, Kotkin and Serbo [1971, chap. 13; too compact; to be read in conjunction with the mechanics volume of Landau and Lifshitz (1960)], Morton (1929), Pöschl (1949, pp. 161–163); and the examples/problems of §7.9 in this book.

8.16 CANONICAL PERTURBATION THEORY IN ACTION–ANGLE VARIABLES

[For the writing of this section, we owe a big debt to the following excellent references: Born (1927, pp. 107–110, 249–261), Dittrich and Reuter (1994, pp. 109–136), Saletan and Cromer (1971, pp. 241–247, 251–258), Tabor (1989, pp. 96–105).]

This section constitutes a concise introduction to *canonical* perturbation theory; that is, an asymptotic approximation technique based on canonical transformations and action–angle variables (§8.14). We have already treated perturbation problems via general canonical variables [variation of constants and associated averaging (§8.7, examples in §8.10)]. But, it turns out that action–angle variables, due to their special properties, are particularly well suited here; and this may explain why this topic has been so central to both the genesis of the new quantum mechanics (1920s) and modern (classical) nonlinear dynamics (1960s to the present).

One DOF

Let us begin with the first-order perturbation of a one-DOF *conservative* system; that is, $\partial H/\partial t = 0$ (*time-independent*, or *stationary state*, perturbation theory). We will assume that both its undisturbed and disturbed motions are *periodic*, and that the undisturbed one of them is known exactly; that is, after solving its unperturbed Hamilton–Jacobi (HJ) equation, say by separation of its (unperturbed) variables, we have expressed the latter (q_o, p_o) in terms of (unperturbed) action and angle variables (J_o, w_o). Its Hamiltonian H_o will, then, depend only on J_o: $H_o = H_o(q_o, p_o) = H_o(J_o)$; so that its (constant) frequency equals $\nu_o = \partial H_o/\partial J_o$ and, therefore, $w_o = \nu_o t + \gamma_o$.

Let the disturbed problem be described by the perturbed Hamiltonian $H = H_o + \varepsilon H_1 + \varepsilon^2 H_2 + \cdots$, or, more precisely, after substituting in it the unperturbed variables,

$$H = H(w_o, J_o; \varepsilon) = H_o(J_o) + \varepsilon H_1(w_o, J_o) + \varepsilon^2 H_2(w_o, J_o) + \cdots$$

$$(= \textit{unperturbed Hamiltonian} + \textit{first-order perturbation Hamiltonian} + \textit{second-order Hamiltonian} + \cdots), \quad (8.16.1a)$$

where

$$\varepsilon = \textit{parameter, or strength, of the perturbation}$$

$$\text{(of the order of the ratio of the disturbing agency to that already in action)} \ll 1; \quad (8.16.1b)$$

and $H(w_o, J_o; 0) = H_o(J_o)$. Since H is a known function of w_o and J_o, all the "perturbation components" H_p ($p = o \equiv 0, 1, 2, 3, \ldots$) are known.

The series (8.16.1a) is assumed to converge for a sufficiently large domain of values of the coordinates and momenta used; and, the resulting motions are assumed to remain periodic for all values of ε in an interval of interest containing $\varepsilon = 0$.

By solution of the perturbed problem, we will understand the finding of a generating function $W_o \equiv A_o \equiv G$ [new notation is introduced in this section to avoid having too many subscripts (see below)] that transforms the original Hamiltonian coordinates into new angle–action variables

$$(q_o, p_o) \to (w, J): \qquad p_o = \partial G/\partial q_o, \qquad w = \partial G/\partial J, \qquad (8.16.1c)$$

[recall §8.8, case: $F_2(q, p') \to G(q_o, J)$] or, if the original coordinate and momentum are w_o and J_o, respectively [i.e., $F_2(q, p') \to G(w_o, J) \Rightarrow \delta G = J_o\, \delta w_o + w\, \delta J$], then

$$(w_o, J_o) \to (w, J): \qquad J_o = \partial G/\partial w_o, \qquad w = \partial G/\partial J, \qquad (8.16.1d)$$

and, also, is such that:

- For $\varepsilon = 0$, (w, J) reduce to (w_o, J_o);

- The perturbed (new) coordinate q is periodic in the new angle variable w with period 1;
- The perturbed (new) Hamiltonian depends only on the *new action variable*: $H = H(J)$; that is,

$$H(q_o, p_o; 0) = H_o(J_o) = E(J_o; 0) \;\rightarrow\; H(q, p; \varepsilon) = H(J) = E(J; \varepsilon). \tag{8.16.1e}$$

Then,

$$\nu = \partial H(J)/\partial J = \partial E(J)/\partial J = \textit{new (constant) frequency}, \tag{8.16.1f}$$

$$\Rightarrow \; w = \nu t + \gamma = \textit{new angle variable}. \tag{8.16.1g}$$

Since $\partial G/\partial t = 0$, the corresponding (perturbed) HJ equation is

$$H(w_o, J_o) = H(w_o, \partial G/\partial w_o) = E(J). \tag{8.16.2a}$$

REMARK

The undisturbed motion variables w_o, J_o remain canonical in the *perturbed* motion; but without their usual angle–action behavior. Indeed, here, the corresponding equations of motion are

$$dJ_o/dt = -\partial H/\partial w_o = -\varepsilon(\partial H_1/\partial w_o) \neq 0, \tag{8.16.2b}$$

$$dw_o/dt = \partial H/\partial J_o = \partial H_o/\partial J_o + \varepsilon(\partial H_1/\partial J_o) \neq \textit{constant}; \tag{8.16.2c}$$

that is, J_o *is no longer constant*, and w_o *is no longer a linear function of time*.

Let us find the perturbational consequences of the basic equation (8.16.2a), or, more precisely,

$$H\big[w_o(w, J), J_o(w, J), \varepsilon\big] = E(J, \varepsilon). \tag{8.16.2d}$$

Expanding all functions involved there in ε-powers, we obtain

$$H(w_o, J_o) = H_o(J_o) + \varepsilon\, H_1(w_o, J_o) + \cdots$$

$$= H_o(\partial G/\partial w_o) + \varepsilon\, H_1(w_o, \partial G/\partial w_o) + \cdots, \tag{8.16.3a}$$

$$E(J, \varepsilon) = E_o(J) + \varepsilon\, E_1(J) + \cdots, \tag{8.16.3b}$$

$$G(w_o, J) = G_o(w_o, J) + \varepsilon\, G_1(w_o, J) + \cdots$$

$$= w_o\, J + \varepsilon\, G_1(w_o, J) + \cdots, \tag{8.16.3c}$$

since for $\varepsilon = 0$ the function G must reduce to the *identity* transformation $w_o J$ ($J_o = \partial G/\partial w_o = J$, $w = \partial G/\partial J = w_o$); and where all its "perturbation components" $G_1, G_2, \ldots,$ are periodic in w_o with fundamental period 1.

[To prove this, it suffices to prove the w_o-periodicity of

$$G'(w_o, J) \equiv G(w_o, J) - w_o\, J = \varepsilon\, G_1(w_o, J) + \cdots.$$

Indeed, since (§8.12) after j cycles of q the action function increases by jJ, we will have

$$G'(w_o + j, J) = G(w_o + j, J) - (w_o + j)J = [G(w_o, J) + j\, J] - (w_o\, J + jJ)$$

$$= G(w_o, J) - w_o\, J = G'(w_o, J), \quad \text{Q.E.D.}$$

Then, as eq. (8.16.3e) shows, $w = w_o +$ *periodic function of* w_o, *with fundamental period 1)*; and so *q is periodic in both w and* w_o.]

Next, to be able to compare the various ε-order terms of H and E [i.e., implement (8.16.2d)], we must express (8.16.3a) in terms of J, instead of J_o. To this end, first, we introduce (8.16.3c) in (8.16.1d) and expand in ε:

$$J_o = \partial G(w_o, J)/\partial w_o = J + \varepsilon\,[\partial G_1(w_o, J)/\partial w_o] + \cdots, \tag{8.16.3d}$$

$$w = \partial G(w_o, J)/\partial J = w_o + \varepsilon\,[\partial G_1(w_o, J)/\partial J] + \cdots; \tag{8.16.3e}$$

and then insert these series into (8.16.3a) and, again, expand in ε. Thus, we find, to the first order,

$$\begin{aligned} H(w_o, J_o) &= H_o[J + \varepsilon(\partial G_1/\partial w_o)] + \varepsilon\, H_1[w_o, J + \varepsilon\,(\partial G_1/\partial w_o)] \\ &= \{H_o(J) + \varepsilon(\partial G_1/\partial w_o)\,[\partial H_o(J)/\partial J]\} + \varepsilon\, H_1(w_o, J) \\ &= H_o(J) + \varepsilon\{H_1(w_o, J) + (\partial G_1/\partial w_o)\,[\partial H_o(J)/\partial J]\}, \end{aligned} \tag{8.16.3f}$$

where

$$H_o(J) = [H_o(J_o)]_{J_o=J}, \qquad H_1(w_o, J) = [H_1(w_o, J_o)]_{J_o=J}. \tag{8.16.3g}$$

Substituting the above results into (8.16.2a, d) we obtain, to the first order,

$$\begin{aligned} H_o(J) + \varepsilon\{H_1(w_o, J) + [\partial H_o(J)/\partial J][\partial G_1(w_o, J)/\partial w_o]\} \\ = E_o(J) + \varepsilon E_1(J), \end{aligned} \tag{8.16.4a}$$

and, equating the coefficients of like powers of ε, we get

$$\varepsilon^0: \qquad H_o(J) = E_o(J), \tag{8.16.4b}$$

$$\varepsilon^1: \qquad H_1(w_o, J) + [\partial H_o(J)/\partial J]\,[\partial G_1(w_o, J)/\partial w_o] = E_1(J). \tag{8.16.4c}$$

Now:

- Equation (8.16.4b) yields the *zeroth approximation* to the energy E_o: we find it by replacing J_o with J in the energy of the unperturbed motion.
- Equation (8.16.4c) is a differential equation that yields the *first approximation* to the energy E_1; and, at first sight, it gives the impression that to do this we need not only H_1, but also G_1, both functions of w_o, and the unknown but *constant* J.

However, things are not that complicated:

(i) Since [recalling (8.16.3g)]

$$\partial H_o(J)/\partial J = [\partial H_o(J_o)/\partial J_o]_{J_o=J} = \partial/\partial J\{[H_o(J_o)]_{J_o=J}\} = \nu_o(J) \tag{8.16.5a}$$

[i.e., $\nu_o(J)$ is obtained from the frequency of the unperturbed motion $\nu_o(J_o)$, by replacing in it J_o with J], (8.16.4c) can be rewritten as

$$H_1(w_o, J) + \nu_o\,[\partial G_1(w_o, J)/\partial w_o] = E_1(J); \tag{8.16.5b}$$

(ii) E_1 is constant, and H_1 is periodic in w_o *with constant term*; and

(iii) G_1 is periodic in w_o *with constant term*; that is, it is representable by the Fourier series:

$$G_1(w_o, J) = \sum g_s(J) \exp(2\pi i\, s\, w_o) \qquad (s = -\infty, \ldots, +\infty), \tag{8.16.5c}$$

and so its derivative $\partial G_1/\partial w_o$ is also periodic in w_o, *but contains no constant term*. Hence, averaging (8.16.5b) over one unperturbed period $\tau_o = 1$ (or over the unperturbed time variation), and noting that

$$\langle \partial G_1/\partial w_o \rangle = 0, \tag{8.16.5d}$$

a consequential result in our perturbation scheme, we obtain the first-order energy correction:

$$E_1(J) = \langle H_1(w_o, J) \rangle \quad (= \langle E_1(J) \rangle), \tag{8.16.5e}$$

where

$$\langle H_1(w_o, J) \rangle \equiv (1/1) \int_0^1 H_1(w_o, J)\, dw_o = \text{function of } J. \tag{8.16.5f}$$

Then, (8.16.3b) becomes

$$E(J) = E_o(J) + \varepsilon E_1(J) = H_o(J) + \varepsilon \langle H_1(w_o, J) \rangle; \tag{8.16.5g}$$

in words: *to a first approximation, the energy of the perturbed motion equals the energy of the unperturbed motion plus the average of the first-order part of the perturbation Hamiltonian taken over the unperturbed motion.*

The perturbed frequency is then given, to the first order, by

$$\nu = \partial E(J)/\partial J = \nu_o + \varepsilon \big[\partial E_1(J)/\partial J\big] = \nu_o + \varepsilon \big[\partial \langle H_1 \rangle/\partial J\big]_{J=J_o}. \tag{8.16.5h}$$

Next, we turn to the calculation of the first-order correction of the motion. As (8.16.3d, e) show, this requires finding G_1: due to (8.16.5e), eq. (8.16.5b) can be written as the following (linear and constant coefficient partial differential) equation for G_1:

$$\nu_o[\partial G_1(w_o, J)/\partial w_o] = \textit{known function of } w_o \textit{ and } J = -\Delta H_1(w_o, J), \tag{8.16.5i}$$

where

$$H_1(w_o, J) - \langle H_1(w_o, J) \rangle = H_1(w_o, J) - E_1(J) \equiv \Delta H_1(w_o, J)$$

= *oscillatory part, or periodic component, of* H_1 (of zero average/mean, and a known function of w_o and the constant J), (8.16.5j)

and so it is expressible as a Fourier series *without constant term* (a fact denoted, as in §8.15, by a prime on the summation sign):

$$\Delta H_1(w_o, J) = \sum{}' h_s(J) \exp(2\pi i\, s\, w_o) \qquad (s = -\infty, \ldots, +\infty; \neq 0). \tag{8.16.5k}$$

Utilizing (8.16.5c) and (8.16.5k) in (8.16.5i) and then equating coefficients of like harmonics, we express the unknown amplitudes g_s in terms of the known ones h_s: $g_s = -[2\pi i(\nu_o s)]^{-1} h_s$, and so

$$G_1(w_o, J) = -\sum{}' [2\pi i(\nu_o s)]^{-1} h_s \exp(2\pi i s w_o)$$
$$= \sum{}' (\omega_o s)^{-1} (i h_s) \exp(2\pi i s w_o)$$

[= *Infinite sum of finite terms* (assuming, of course $\nu_o \neq 0$, and since $s \neq 0$]. (8.16.5l)

[The general solution of (8.16.5i) is $G_1(w_o, J) + f(J)$, where $f(J)$ is an arbitrary function of J. But as (8.16.3d, e) show, $f(J)$ *does not affect the* $J - J_o$ *relation*, and simply adds a term $\varepsilon[df(J)/dJ]$ to $w - w_o$. However, since $J = $ *constant*, this amounts to the *addition of an inconsequential constant to* $w - w_o$; w and w_o being angle variables, their difference at the "initial" angle w_o (= 0, for convenience) is arbitrary. In view of this freedom, we will henceforth set $f(J) = 0$.]

Then, using (8.16.3d, e), we obtain the new action–angle variables to the first order; that is, eq. (8.16.3e) yields the small oscillations, superimposed on the unperturbed motion, with amplitude of order ε, and analogously for (8.16.3d). Hence, in this perturbation scheme, *no secular perturbations occur*; that is, quantities that are constant in the unperturbed motion do not undergo changes of their own order of magnitude.

REMARK

A word of caution is needed here: if ν_o is *very small*, then, as (8.16.5k) shows, the effect of this first-order perturbation may be *pretty substantial*—the convergence of our perturbation series cannot be guaranteed for very long time intervals (i.e., for all time). (Similarly, it can be shown that the effect of the (p)th perturbation will be proportional to ν_o^{-p}.) As a rule, "small ν_o" situations occur near a *separatrix*—that is, a boundary that separates phase space curves of very different properties; for example, a curve that separates libration from rotation (fig. 8.13). As shown below, such mathematical difficulties become *far greater* in $n\,(\geq 2)$ DOF systems: there, not just small frequencies, but also *finite/large ones*, may combine among themselves (i.e., *in near-degeneracy conditions*) to produce very small denominators in the coefficients of the corresponding perturbational series; and, thus, may call into question its convergence. More on this famous problem of "*small divisors*" later.

Several DOF

Here, we extend our perturbation method in a twofold way: (i) to systems with n DOF, and (ii) to include up to *second-order* terms in ε. The basic assumptions of the one-DOF case are also made here:

- For $\varepsilon = 0$, the new (perturbed) angle–action variables

$$w \equiv (w_1, \ldots, w_n) \equiv (w_k;\ k = 1, \ldots, n), \tag{8.16.6a}$$
$$J \equiv (J_1, \ldots, J_n) \equiv (J_k;\ k = 1, \ldots, n), \tag{8.16.6b}$$

reduce to the old (unperturbed) ones

$$w_o \equiv (w_{1o}, \ldots, w_{no}) \equiv (w_{ko};\ k = 1, \ldots, n), \tag{8.16.6c}$$
$$J_o \equiv (J_{1o}, \ldots, J_n) \equiv (J_{ko};\ k = 1, \ldots, n). \tag{8.16.6d}$$

- The solution of the unperturbed problem in the (w_o, J_o) is assumed *known and non-degenerate*.
- Both unperturbed and perturbed Hamiltonians depend only on the *corresponding action variables*: $H_o = H_o(J_o)$ and $H = H(J)$, and
- The perturbed coordinates are *periodic in both the* w_{ko} *and* w_k, *with fundamental period 1*.

As in the one-DOF case, we are seeking the perturbative solution of the new HJ equation

$$H(w_o, J_o) = H(w_o, \partial G/\partial w_o) = E(J), \tag{8.16.6e}$$

where $G = G(w_o, J)$ is the generating function of the canonical transformation

$$(w_o, J_o) \to (w, J): \qquad J_{ko} = \partial G(w_o, J)/\partial w_{ko}, \qquad w_k = \partial G(w_o, J)/\partial J_k. \tag{8.16.6f}$$

Expanding H, E, and G in ε-powers, we obtain

$$\begin{aligned} H(w_o, J_o) &= H_o(J_o) + \varepsilon H_1(w_o, J_o) + \varepsilon^2 H_2(w_o, J_o) + \cdots \\ &= H_o(\partial G/\partial w_o) + \varepsilon H_1(w_o, \partial G/\partial w_o) + \varepsilon^2 H_2(w_o, \partial G/\partial w_o) + \cdots, \end{aligned} \tag{8.16.7a}$$

$$E(J, \varepsilon) = E_o(J) + \varepsilon E_1(J) + \varepsilon^2 E_1(J) + \cdots, \tag{8.16.7b}$$

$$\begin{aligned} G(w_o, J) &= G_o(w_o, J) + \varepsilon G_1(w_o, J) + \varepsilon^2 G_2(w_o, J) + \cdots \\ &= \sum w_{ko} J_k + \varepsilon G_1(w_o, J) + \varepsilon^2 G_2(w_o, J) + \cdots. \end{aligned} \tag{8.16.7c}$$

Then, (8.16.6f) become

$$J_{ko} = J_k + \varepsilon(\partial G_1/\partial w_{ko}) + \varepsilon^2(\partial G_2/\partial w_{ko}) + \cdots, \tag{8.16.8a}$$

$$w_k = w_{ko} + \varepsilon(\partial G_1/\partial J_k) + \varepsilon^2(\partial G_2/\partial J_k) + \cdots; \tag{8.16.8b}$$

and this allows us to rewrite (8.16.7a) to the second order as follows:

$$\begin{aligned} H(w_o, J_o) &= H_o\big[J + \varepsilon(\partial G_1/\partial w_o) + \varepsilon^2(\partial G_2/\partial w_o)\big] \\ &\quad + \varepsilon H_1(w_o, J + \cdots) + \varepsilon^2 H_2(w_o, J + \cdots) \\ &= H_o(J) + \sum \big[\varepsilon(\partial G_1/\partial w_{ko}) + \varepsilon^2(\partial G_2/\partial w_{ko})\big](\partial H_o/\partial J_k) \\ &\quad + (1/2)\sum\sum(\partial^2 H_o/\partial J_k\, \partial J_l)\big[\varepsilon(\partial G_1/\partial w_{ko})\big]\big[\varepsilon(\partial G_1/\partial w_{lo})\big] \\ &\quad + \varepsilon H_1(w_o, J) + \varepsilon^2 \sum(\partial G_1/\partial w_{ko})(\partial H_1/\partial J_k) + \varepsilon^2 H_2(w_o, J); \end{aligned} \tag{8.16.8c}$$

where, as in (8.16.5a),

$$\partial H_o/\partial J_k \equiv [\partial H_o(J_o)/\partial J_{ko}]_{J_o = J} = \partial/\partial J_k\,[H_o(J_o)|_{J_o = J}], \tag{8.16.8d}$$

and similarly for the other H-derivatives.

Next, inserting the power series (8.16.8c, 7b) into (8.16.6e), and equating coefficients of like powers of ε, while noting that [recall (8.16.5a)]

$$\partial H_o/\partial J_k = \nu_{ko}(J) \qquad [\Rightarrow\ \partial\nu_{ko}/\partial J_l = \partial\nu_{lo}/\partial J_k], \tag{8.16.9a}$$

we obtain the following group of differential equations:

$$\varepsilon^0: \qquad H_o(J) = E_o(J), \tag{8.16.9b}$$

$$\varepsilon^1 \equiv \varepsilon: \qquad H_1(w_o, J) + \sum \nu_{ko}(\partial G_1/\partial w_{ko}) = E_1(J), \tag{8.16.9c}$$

$$\varepsilon^2: \qquad K_2(w_o, J) + \sum \nu_{ko}(\partial G_2/\partial w_{ko}) = E_2(J), \tag{8.16.9d}$$

where

$$K_2(w_o, J) \equiv H_2(w_o, J) + \sum (\partial G_1/\partial w_{ko})(\partial H_1/\partial J_k) + (1/2)\sum\sum (\partial^2 H_o/\partial J_k \partial J_l)(\partial G_1/\partial w_{ko})(\partial G_1/\partial w_{lo}). \tag{8.16.9e}$$

Next, as in the 1-DOF case, it can be shown that all G_p's in (8.16.7c) are periodic in the w_{ko}'s; that is,

$$G_p(w_o, J) = \sum g_{p,s}(J)\exp(2\pi i\, \mathbf{s}\cdot \mathbf{w_o}), \tag{8.16.10a}$$

where $p = 1, 2, \ldots, \mathbf{s} \equiv (s_1, \ldots, s_n) =$ *integers ranging from* $-\infty$ to $+\infty$, $\mathbf{w_o} \equiv (w_{1o}, \ldots, w_{no}) \equiv w_o$, and so their w_o-derivatives $\partial G_p/\partial w_{ko}$ *contain no constant terms*; that is, $\mathbf{s} \neq (0, \ldots, 0)$. Therefore, averaging (8.16.9b–e) over a complete unperturbed period w_o — that is, over the unit w_o-cube (§8.14), since $\langle \partial G_p/\partial w_{ko}\rangle = 0$ and the last/double sum group of terms in (8.16.9e) is periodic in the w_o's (the $\partial^2 H_o/\partial J_k \partial J_l$ depend only on the J's) — we obtain

$$H_o(J) = E_o(J), \tag{8.16.10b}$$

$$\langle H_1(w_o, J)\rangle = E_1(J), \tag{8.16.10c}$$

$$\langle K_2(w_o, J)\rangle - \Big\langle H_2(w_o, J) + \sum (\partial G_1/\partial w_{ko})(\partial H_1/\partial J_k)\Big\rangle - E_2(J), \tag{8.16.10d}$$

that is, to the second order, the energy is

$$E = H_o + \langle H_1\rangle + \Big\langle H_2 + \sum (\partial G_1/\partial w_{ko})(\partial H_1/\partial J_k)\Big\rangle. \tag{8.16.10e}$$

Due to the above [and recalling (8.16.5j)], we can rewrite the perturbation equations (8.16.9c, d), respectively, as

$$\sum \nu_{ko}(\partial G_1/\partial w_{ko}) = -[H_1(w_o, J) - \langle H_1(w_o, J)\rangle] \equiv -\Delta H_1(w_o, J), \tag{8.16.11a}$$

$$\sum \nu_{ko}(\partial G_2/\partial w_{ko}) = -[K_2(w_o, J) - \langle K_2(w_o, J)\rangle] \equiv -\Delta K_2(w_o, J), \tag{8.16.11b}$$

and, from these, G_1, G_2 can be determined (see below). Then, equations (8.16.8a, b) yield the new action–angle variables, correct to second order.

The above show that (i) knowledge of K_2 ($\rightarrow \langle K_2\rangle = E_2$) requires knowledge of G_1; then G_2 can be calculated; and (ii) in a one-DOF system, both G_1 and G_2 can be found by direct quadrature:

$$\partial G_1/\partial w_o = (1/\nu_o)(\langle H_1\rangle - H_1) = \textit{known function of } w_o \textit{ and } J, \tag{8.16.11c}$$

$$\partial G_2/\partial w_o = (1/\nu_o)(\langle K_2\rangle - K_2) = \textit{known function of } w_o \textit{ and } J; \tag{8.16.11d}$$

and can be easily extended to higher ε-orders.

Finally, by (8.16.10b–d), the perturbed frequencies equal, to the second order,

$$\nu_k = \partial E/\partial J_k = \partial E_o/\partial J_k + \varepsilon(\partial E_1/\partial J_k) + \varepsilon^2(\partial E_2/\partial J_k)$$
$$\approx \nu_{ko} + \varepsilon(\partial\langle H_1\rangle/\partial J_k) + \varepsilon^2(\partial\langle K_2\rangle/\partial J_k), \qquad (8.16.11e)$$

in agreement with (8.16.5h).

[For the extension of this perturbation scheme to the (p)th order; that is, the differential equation that results by equating the coefficients of ε^p in (8.16.6e), see Born (1927, p. 254 ff.); also prob. 8.16.3. It is not hard to see that finding E_p requires knowledge of G_{p-1}; *then* G_p can be calculated.]

Small Divisors

Now, let us resume the calculation of G_1, G_2. To express the (unknown) Fourier coefficients of the left sides of (8.16.11a, b) in terms of the (known) Fourier coefficients of their right sides, we proceed as follows. The *right* side of (8.16.11a) is a *known* periodic function of the w_o's without constant term, and so we can write

$$-\Delta H_1(w_o, J) = -\sum{}' h_{1,s}(J)\exp(2\pi i\, \boldsymbol{s}\cdot\boldsymbol{w_o}). \qquad (8.16.12a)$$

Similarly, by (8.16.10a) with $p = 1$, we have

$$G_1(w_o, J) = \sum g_{1,s}(J)\exp(2\pi i\, \boldsymbol{s}\cdot\boldsymbol{w_o}), \qquad (8.16.12b)$$

and therefore the *left* side of (8.16.11a) can be expressed as

$$\sum \nu_{ko}(\partial G_1/\partial w_{ko}) = \sum{}' [2\pi i(\boldsymbol{s}\cdot\boldsymbol{v_o})g_{1,s}(J)]\exp(2\pi i\, \boldsymbol{s}\cdot\boldsymbol{w_o}), \qquad (8.16.12c)$$

where $s_k \neq 0$ (i.e., no constant term) and $\boldsymbol{s}\cdot\boldsymbol{v_o} \equiv \sum s_k\nu_{ko} \neq 0$ $(k = 1, \ldots, n)$.

Hence, equating coefficients of equal harmonics of (8.16.12a) and (8.16.12c), as required by (8.16.11a), we immediately obtain the sought relations for the Fourier coefficients:

$$g_{1,s}(J) \equiv g_1(s;\, J) = -[2\pi i(\boldsymbol{s}\cdot\boldsymbol{v_o})]^{-1}h_{1,s}(J); \qquad (8.16.12d)$$

and so, finally, (8.16.12b) becomes

$$G_1(w_o, J) = -\sum{}' [2\pi i(\boldsymbol{s}\cdot\boldsymbol{v_o})]^{-1}h_{1,s}(J)\exp(2\pi i\, \boldsymbol{s}\cdot\boldsymbol{w_o}). \qquad (8.16.12e)$$

Similarly, expanding both sides of (8.16.11b) à la Fourier, and equating coefficients, we determine G_2. This, as (8.16.9e) shows, requires knowledge of G_1. Continuing in this way, we can determine $G_3, G_4, \ldots$, and hence G to any accuracy desired.

Now, it is not too hard to see that for the representation (8.16.12e) to be meaningful, not only *the unperturbed frequencies* ν_{ko} *should be nondegenerate* (i.e., $\sum s_k\,\nu_{ko} \neq 0$; unless the coefficients $h_1, \ldots$ obtained from all sets of integers that cause degeneracies also vanish), but also, since by an appropriate choice of the integers s_k *the sum* $\sum s_k\nu_{ko}$ *may come arbitrarily close to zero* (and, worse, such a *near-degeneracy* situation may occur an infinite number of times, as the s_k roam from $-\infty$ *to* $+\infty$), for all these very unpleasant reasons *the amplitudes* $h_{1,s}$ *must converge very rapidly*. Hence, from a rigorous mathematical viewpoint, the series (8.16.12b, e) *does not converge*; and this inescapable fact casts serious reservations on the unconditional validity of the entire method of canonical perturbations.

REMARKS

(i) This is the Hamiltonian version of the famous *problem of small divisors* (or *resonant denominators*), first recognized by Poincaré in his epoch-making researches on the nonlinear ordinary differential equations of celestial mechanics [late 19th century, culminating in his classic three-volume work: *Les Méthodes Nouvelles de la Mécanique Céleste* (1890s)]; and on which, understandably, there exists an enormous (astronomical size) literature!

Briefly, Poincaré has shown that the series (8.16.12b) is *semiconvergent*; that is, *if it is truncated (discontinued) after a certain finite number of terms, it represents the motion of the perturbed system very accurately; not for long periods of time, but long enough for many practical purposes*. It is theoretical reasons like this that had made it so difficult to prove the stability of our solar system; that is, to show that the mutual distances among the planets and the sun remain bounded for infinitely long time intervals. (And, of course, it should not be forgotten that a realistic stability investigation of this problem must include *nonmechanical* causes, such as electromagnetic and thermal interactions.) Several decades later, it was shown that the situation is not fatal: Kolmogorov (mid-1950s)/Arnold/Moser (early 1960s) (KAM theorem) demonstrated that *if the ν_{ko} are "very irrational," then the series (8.16.12b) converges for all time.*

(ii) For in-depth analyses of these fundamental difficulties [for a long time viewed as mathematically insuperable, but whose resolution in the 1960s led straight up to the frontier of contemporary nonlinear dynamics (regular and stochastic, or chaotic, motion) and the threshold with quantum mechanics], we can do no more than refer the reader to the following excellent references: Dittrich and Reuter (1994, chaps. 11–14: pp. 137–171), Lichtenberg and Lieberman (1992), Stoker (1950, pp. 112–114, 235–239; elementary but enlightening introduction to the problem of small divisors), Straumann (1987, chap. 10: pp. 259–307), Tabor (1989, chap. 3: pp. 89–117); also Born (1927, pp. 255–256). Mathematically oriented readers (or mathematicians) may wish to consult (the not so readable) Arnold (1976, chap. 10: pp. 269–299, appendix 8: pp. 405–423).]

Finally, we recall that our discussion of perturbation theory has been limited not only to *nondegenerate* cases, but also to *time-independent* Hamiltonians. For classical (pre-KAM) treatments of the effects of degeneracies and/or time-dependence, with an eye toward their older quantum-mechanical applications, we recommend Born (1927, pp. 261–286; best reference in English) and Fues (1927, pp. 161–177; comprehensive handbook exposition); also Birtwistle (1926, pp. 216–217) and Haar (1971, pp. 160–162).

Example 8.16.1 *Weakly Nonlinear Planar Mathematical Pendulum; First-Order Perturbation* (Dittrich and Reuter, 1994, pp. 113–115). Let us consider a plane mathematical pendulum of mass m and length l undergoing (free and undamped) *small but nonlinear* angular oscillations ϕ, under gravity, about a fixed point O. Expanding its exact Hamiltonian

$$H = p^2/2ml^2 + mgl(1 - \cos\phi) = p^2/2ml^2 - mgl\cos\phi + constant \qquad \text{(a)}$$

in powers of ϕ, and *keeping only up to the first term after its quadratic ones*, we find

$$H = p^2/2ml^2 + mgl\big[(\phi^2/2) - (\phi^4/24)\big] \equiv H_o + \varepsilon H_1, \tag{b}$$

$$H_o \equiv p^2/2A + (A\omega_o{}^2/2)\phi^2, \qquad \varepsilon H_1 \equiv -(A\omega_o{}^2/24)\phi^4, \tag{c}$$

where $A \equiv m\,l^2$ = moment of inertia of pendulum bob about O, $\omega_o{}^2 \equiv g/l$ = circular frequency (squared) of unperturbed = linearized problem. We have already seen (ex. 8.14.1) that the solution of the latter in action–angle variables is (with $m \to A,\ q \to \phi$)

$$\phi = (J_o/A\pi\,\omega_o)^{1/2}\sin(2\pi w_o), \qquad p = (A\,\omega_o\,J_o/\pi)^{1/2}\cos(2\pi w_o), \tag{d}$$

and so H, eq. (b), assumes the action–angle variable form

$$\begin{aligned} H &= (\omega_o/2\pi)J_o - (1/24)(J_o{}^2/A\pi^2)\,\sin^4(2\pi w_o) \\ &= \nu_o J_o - (J_o{}^2/24A\pi^2)\sin^4(2\pi w_o) = H_o + \varepsilon H_1. \end{aligned} \tag{e}$$

Choosing as perturbation parameter ε the square of the maximum angular amplitude of the unperturbed problem $\phi_o{}^2$, and applying (8.16.5e), we obtain

$$E_1(J) = \langle H_1(w_o, J)\rangle = -(J^2/24A\pi^2\phi_o{}^2)\langle\sin^4(2\pi w_o)\rangle, \tag{f}$$

or, since by simple trigonometry and calculus

$$\begin{aligned} \sin^4(\ldots) &= \big\{[\exp(i\ldots) - \exp(-i\ldots)]/2i\big\}^4 \\ &= \cdots = (1/8)\,[\cos(4\ldots) - 6\cos(2\ldots) + 3] \\ \Rightarrow\ \langle\sin^4(2\pi w_o)\rangle &= \int_0^1 \sin^4(2\pi w_o)\,dw_o = \cdots = 3/8, \end{aligned} \tag{g}$$

we get

$$E_1(J) = -J^2/64A\pi^2\phi_o{}^2. \tag{h}$$

Hence, by (8.16.5i), the *first-order* change of the fundamental frequency is

$$\Delta\nu \equiv \nu - \nu_o = \varepsilon[\partial E_1(J)/\partial J] \ = \ -J/32A\pi^2 \ \approx\ -J_o/32A\pi^2, \tag{i}$$

or, since

$$\begin{aligned} J_o = (2\pi/\omega_o)E_o &= (2\pi/\omega_o)T_{\max} = (2\pi/\omega_o)(A\omega_o{}^2\phi_o{}^2/2) \\ &= \pi A\,\omega_o\phi_o{}^2 = 2\pi^2 A\phi_o{}^2\nu_o, \end{aligned} \tag{j}$$

finally,

$$\Delta\nu = -(\phi_o{}^2/16)\nu_o; \tag{k}$$

which agrees with the first-order correction found by other means (e.g., Lur'e, 1968, pp. 702–703) for a derivation based on integral variational calculus; see also examples/problems in §7.9 of this book.

Example 8.16.2 *One-DOF Nonlinear Oscillator; Second-Order Perturbation.* (Birtwistle, 1926, pp. 213–216; with $c_o = 0$ and $\omega \to \nu$). Let us consider a one-DOF oscillator, with mass m and unperturbed circular frequency $\omega_o = 2\pi\nu_o$, and perturbed Hamiltonian, to second ε-order:

$$H = H_o + \varepsilon H_1 + \varepsilon^2 H_2, \tag{a}$$

where

$$H_o = p^2/2m + m\omega_o{}^2 q^2/2, \tag{a1}$$

$$H_1 = h_1 q^3, \qquad H_2 = h_2 q^4 \qquad (h_1, h_2\text{: known physical constants}). \tag{a2}$$

We already know that the unperturbed solution is (ex. 8.14.1)

$$q_o = (J_o/\pi\omega_o m)^{1/2}\sin(2\pi w_o), \qquad p_o = (\omega_o m J_o/\pi)^{1/2}\cos(2\pi w_o), \tag{b}$$

and so the perturbed Hamiltonian H can be expressed in terms of the unperturbed variables (w_o, J_o) as follows:

$$H_o = \nu_o J_o, \tag{c1}$$

$$H_1 = h_1(J_o/\pi\omega_o m)^{3/2}\sin^3(2\pi w_o), \tag{c2}$$

$$H_2 = h_2(J_o/\pi\omega_o m)^2\sin^4(2\pi w_o). \tag{c3}$$

Now, let us apply the perturbation equations (8.16.10b–11b):

(i) Since, here,

$$E_1(J) = \langle H_1(w_o, J)\rangle \sim \langle\sin^3(2\pi w_o)\rangle = \cdots = 0, \tag{d1}$$

we will have

$$\begin{aligned}\partial G_1/\partial w_o &= -(1/\nu_o)H_1(w_o, J) = -(1/\nu_o)\,\Delta H_1(w_o, J)\\ &= -(h_1/\nu_o)(J/\pi\omega_o m)^{3/2}\sin^3(2\pi w_o);\end{aligned} \tag{d2}$$

that is, the ε-order perturbation does not change the energy ($E_1 = 0$), but does change the action–angle variables ($G_1 \neq 0$).

(ii) To find the ε^2-order energy correction equation, we employ the equations

$$\begin{aligned}&K_2 + \nu_o(\partial G_2/\partial w_o)\\ &\quad = H_2 + (\partial G_1/\partial w_o)(\partial H_1/\partial J) + \nu_o(\partial G_2/\partial w_o) = E_2,\end{aligned} \tag{e1}$$

$$\langle K_2\rangle = \langle H_2\rangle + \langle(\partial G_1/\partial w_o)(\partial H_1/\partial J)\rangle = E_2. \tag{e2}$$

Using (c2, 3) and (d2) in (e2), and carrying out the indicated w_o-averagings, we obtain, after some algebra (since $\langle\sin^4(2\pi w_o)\rangle = 3/8$ and $\langle\sin^6(2\pi w_o)\rangle = 5/16$),

$$E_2 = -(15/4)h_1{}^2[J^2/(2\pi)^6\nu_o{}^4 m^3] + (3/2)h_2[J^2/(2\pi)^4\nu_o{}^2 m^2]. \tag{e3}$$

Then, to the second order, the perturbed energy and frequency are, respectively,

$$E(J) = E_o + \varepsilon E_1 + \varepsilon^2 E_2 = \nu_o J + \varepsilon^2 E_2 = \cdots, \tag{e4}$$

$$\nu(J) = \nu_o + \varepsilon(\partial\langle H_1\rangle/\partial J) + \varepsilon^2(\partial\langle K_2\rangle/\partial J)$$
$$= \nu_o + \varepsilon^2(\partial E_2/\partial J) = \cdots; \tag{e5}$$

and then set in them $J = J_o - \varepsilon(\partial G_1/\partial w_o) = \cdots$ (see below). We notice the weak dependence of the frequency on the amplitude.

(iii) To find the ε-order effect on the motion—that is, on q—first we integrate (d2), thus finding

$$G_1 = [h_1/(2\pi)^4\nu_o]\,(2J/\nu_o m)^{3/2}\left[(1/3)\sin^2(2\pi w_o)\cos(2\pi w_o) + (2/3)\cos(2\pi w_o)\right], \tag{f1}$$

then, applying the old/new variable perturbation equations, we get

$$J_o = J + \varepsilon(\partial G_1/\partial w_o) = J - \varepsilon(h_1/\nu_o)(J/2\pi^2\nu_o m)^{3/2}\sin^3(2\pi w_o), \tag{f2}$$

$$w = w_o + \varepsilon(\partial G_1/\partial J)$$
$$= w_o + \varepsilon[h_1/(2\pi)^4 2J\nu_o](2J/\nu_o m)^{3/2}\left[\sin^3(2\pi w_o)\cos(2\pi w_o) + 2\cos(2\pi w_o)\right]; \tag{f3}$$

and, finally, solving (f3) for w_o and substituting that result, and J_o from (f2), into the unperturbed form of motion of the first of (b):

$$q_o = (J_o/\pi\omega_o m)^{1/2}\sin(2\pi w_o) = (J_o/2\pi^2\nu_o m)^{1/2}\sin(2\pi w_o), \tag{f4}$$

we obtain

$$q_o \rightarrow q = q_o + \varepsilon q_1$$
$$= (J/2\pi^2\nu_o m)^{1/2}\sin(2\pi w) - \varepsilon h_1\,[J/(2\pi)^4\nu_o{}^3 m^2]\,[3 + \cos(4\pi w)]; \tag{f5}$$

that is, the *nonlinearity* (h_1) produces oscillatory *overtones* ($4\pi w$). These results agree with those found by quadrature (since this is a one-DOF system) for $h_2 = 0$ (see, e.g., Born, 1927, pp. 66–70). Proceeding similarly, we may obtain the ε^2-order effect on (w_o, J_o) [after finding G_2 from (e1)] and hence on q. The details are left to the reader (for confirmation of those results, see, e.g., Haar, 1971, pp. 157–158).

Example 8.16.3 *Two-DOF Nonlinear Oscillator; First-Order Perturbation.* Let us consider an oscillating system with perturbed Hamiltonian

$$H(q, p; \varepsilon) = (1/2)(p_1{}^2 + p_2{}^2) + (1/2)(k_1 q_1{}^2 + k_2 q_2{}^2) + \varepsilon\, k_1 k_2 q_1{}^2 q_2{}^2$$
$$= H_o + \varepsilon\, H_1, \tag{a}$$

where

$$H_o = H_o(q, p; 0) = \sum (1/2)(p_l{}^2 + k_l q_l{}^2) \qquad (l = 1, 2), \tag{a1}$$

$$H_1 = H_1(q, p; 0) = k_1 k_2 q_1{}^2 q_2{}^2; \tag{a2}$$

that is, the unperturbed Hamiltonian represents two uncoupled harmonic oscillators, each of unit mass and stiffness k_l (> 0), and hence unperturbed circular frequency (squared) $\omega_{lo}{}^2 = k_l$. As we already know (exs. 8.14.1 and 8.14.6), the unperturbed solutions are

$$q_{ko} = (J_{ko}/\pi\omega_{ko})^{1/2}\sin(2\pi w_{ko}), \tag{b1}$$

$$p_{ko} = (\omega_{ko}J_{ko}/\pi)^{1/2}\cos(2\pi w_{ko}) \qquad (k = 1, 2), \tag{b2}$$

and so the perturbed Hamiltonian H, (a–a2), can be expressed in terms of the unperturbed variables (w_o, J_o) as follows:

$$H_o = (1/2\pi)\sum \omega_{ko}J_k = \sum \nu_{ko}J_k = E_o, \tag{c1}$$

$$H_1 = (1/\pi^2)\omega_{1o}\omega_{2o}J_{1o}J_{2o}\sin^2(2\pi w_{1o})\sin^2(2\pi w_{2o})$$

$$= 4\nu_{1o}\nu_{2o}J_{1o}J_{2o}\sin^2(2\pi w_{1o})\sin^2(2\pi w_{2o}); \tag{c2}$$

where $\nu_{lo} = \omega_{lo}/2\pi$ = unperturbed frequencies. Hence, the first-order energy correction yields

$$E_1(J) = \langle H_1(w_o, J)\rangle \equiv \int_0^1\int_0^1 H_1(w_{1o}, w_{2o}; J_1, J_2)\, dw_{1o}\, dw_{2o}$$

$$= \cdots = (1/4\pi^2)\omega_{1o}\omega_{2o}J_1J_2 = \nu_{1o}\nu_{2o}J_1J_2, \tag{d1}$$

$$\Rightarrow\ E(J) = E_o(J) + \varepsilon E_1(J) = \nu_{o1}J_1 + \nu_{o2}J_2 + \varepsilon\,\nu_{1o}\nu_{2o}J_1J_2, \tag{d2}$$

(w-integration limits from 0 to 1) and so, to the same accuracy, the perturbed frequencies are (whether the system is degenerate or not)

$$\nu_1 = \partial E/\partial J_1 = \nu_{1o} + \varepsilon\,\nu_{1o}\nu_{2o}J_2, \tag{d3}$$

$$\nu_2 = \partial E/\partial J_2 = \nu_{2o} + \varepsilon\,\nu_{1o}\nu_{2o}J_1; \tag{d4}$$

and these show clearly the coupling of the two oscillators and the effect of the amplitudes (initial conditions) on the frequencies.

Next, to express the perturbed coordinates and momenta in terms of the new action–angle variables (w, J), we must calculate G_1:

(i) On one hand, in view of (c2), (d1), we have

$$\langle H_1(w_o, J)\rangle - H_1(w_o, J) \equiv -\Delta H_1(w_o, J)$$

$$= \nu_{1o}\nu_{2o}J_1J_2[1 - 4\sin^2(2\pi w_{1o})\sin^2(2\pi w_{2o})]$$

$$\equiv (\nu_{1o}\nu_{2o}J_1J_2/4)\sum\sum{}' h_{1,s}\exp[2\pi i(s_1w_{1o} + s_2w_{2o})] \tag{e1}$$

(where $s \equiv (s_1, s_2)$ = *nonzero* integers ranging from $-\infty$ to $+\infty$); from which we readily conclude that the sole *nonvanishing* Fourier coefficients of $-\Delta H_1$, $h_{1,\ldots} \equiv h_{1;\ldots,\ldots}$, are

$$h_{1;2,2} = h_{1;2,-2} = h_{1;-2,2} = h_{1;-2,-2} = -1, \tag{e2}$$

$$h_{1;2,0} = h_{1;0,2} = h_{1;0,-2} = h_{1;-2,0} = 2. \tag{e3}$$

(ii) On the other hand, recalling (8.16.12b, c), we can write

$$G_1(w_o, J) = \sum g_{1,s}(J) \exp(2\pi i \boldsymbol{s} \cdot \boldsymbol{w}_o),$$

$$\Rightarrow \sum{}' \nu_{ko}(\partial G_1/\partial w_{ko})$$

$$= \sum\sum{}' \left[(2\pi i)\,(s_1\nu_{1o} + s_2\nu_{2o})\right] g_{1,s} \exp\left[2\pi i(s_1 w_{1o} + s_2 w_{2o})\right]. \tag{e4}$$

Hence, substituting these Fourier series into the first-order averaged equation (8.16.11a):

$$\sum{}' \nu_{ko}(\partial G_1/\partial w_{ko}) = \langle H_1(w_o, J)\rangle - H_1(w_o, J) \equiv -\Delta H_1(w_o, J), \tag{e5}$$

and equating coefficients of like harmonics, we find [with $h \equiv \nu_{1o}\,\nu_{2o} J_1 J_2/4$]

$$g_{1,\ldots} \equiv g_{1;\ldots,\ldots} = \left[h/(2\pi i)(s_1\nu_{1o} + s_2\nu_{2o})\right] h_{1;\ldots,\ldots}:$$

$$g_{1;0,2} = -g_{1;0,-2} = (2\pi i \nu_{2o})^{-1} h, \tag{e6}$$

$$g_{1;2,0} = -g_{1;-2,0} = (2\pi i \nu_{1o})^{-1} h, \tag{e7}$$

$$g_{1;2,2} = -g_{1;-2,-2} = -[4\pi i(\nu_{1o} + \nu_{2o})]^{-1} h, \tag{e8}$$

$$g_{1;2,-2} = -g_{1;-2,2} = -[4\pi i(\nu_{1o} - \nu_{2o})]^{-1} h. \tag{e9}$$

Hence, to the first order, the generating function equals

$$\begin{aligned} G = \sum w_{ko} J_k + (\varepsilon/4\pi) J_1 J_2 \{&\nu_{2o} \sin(4\pi w_{1o}) + \nu_{1o} \sin(4\pi w_{2o}) \\ &- [2\nu_{1o}\nu_{2o}/(\nu_{1o} + \nu_{2o})] \sin[4\pi(w_{1o} + w_{2o})] \\ &- [2\nu_{1o}\nu_{2o}/(\nu_{1o} - \nu_{2o})] \sin[4\pi(w_{1o} - w_{2o})]\}; \end{aligned} \tag{e10}$$

and from this the w_k and J_k follow:

$$w_k = w_{ko} + \varepsilon(\partial G/\partial J_k) \;\Rightarrow\; w_{ko} = w_k - \varepsilon\ (\textit{function of } w \textit{ and } J), \tag{f1}$$

$$J_{ko} = J_k + \varepsilon(\partial G/\partial w_{ko}) \;\Rightarrow\; J_k = J_{ko} + \varepsilon\ (\textit{function of } w \textit{ and } J) \tag{f2}$$

(because, replacing w_o with w in the ε-terms causes an ε^2-error), where $J_k = \textit{constant}$ and $w_k = \nu_k t + \gamma_k$; and then inserting (f1, 2) into (b1, 2) yields the perturbed motion $q_k(w, J), p_k(w, J)$. The details are left to the reader.

REMARKS ON DEGENERACIES

As (e10) shows, if $\nu_{1o} = \nu_{2o}$ (degeneracy, or internal resonance), this approach fails—G diverges; then we have to develop special methods. In view of (e2, 3, 6–9), other degeneracies do not seem to create problems; for example, that would happen to the degeneracy $\nu_{1o} = 2\nu_{2o}$ [i.e., $(1)\nu_{1o} + (-2)\nu_{2o} = 0$] if $h_{1;s,-2s} \neq 0$, for some integer s. However, such difficulties may arise in the higher ε-order terms; that is, G_l $(l \geq 2)$; and for their full treatment, we recommend the earlier-given references on small divisors.

Problem 8.16.1 For a 1-DOF system, the earlier-given general n-DOF *second-order* formalism specializes to

$$E_o(J) = H_o(J), \tag{a1}$$

$$E_1(J) = H_1(w_o, J) + \nu_o(J)\,(\partial G_1/\partial w_o) = H_1(w_o, J) + (\partial H_o/\partial J)\,(\partial G_1/\partial w_o), \tag{a2}$$

$$E_2(J) = H_2(w_o, J) + (\partial H_1/\partial J)\,(\partial G_1/\partial w_o) + (\partial H_o/\partial J)\,(\partial G_2/\partial w_o) + (1/2)(\partial^2 H_o/\partial J^2)(\partial G_1/\partial w_o)^2. \tag{a3}$$

We have already seen that since

$$\langle \partial G_1/\partial w_o \rangle \equiv \int_0^1 (\partial G_1/\partial w_o)\, dw_o = 0, \tag{b}$$

eq. (a2) averages to

$$E_1(J) = \langle H_1 \rangle \;\Rightarrow\; \partial G_1/\partial w_o = (1/\nu_o)\big(\langle H_1\rangle - H_1\big) = -(1/\nu_o)\, \Delta H_1. \tag{c}$$

Show that:

(i) The second-order correction (a3) averages, similarly, to

$$\begin{aligned} E_2(J) &= \langle H_2\rangle + \big\langle(\partial H_1/\partial J)(\partial G_1/\partial w_o)\big\rangle + (1/2)(\partial^2 H_o/\partial J^2)\big\langle(\partial G_1/\partial w_o)^2\big\rangle \\ &= \langle H_2\rangle + (1/\nu_o)\big[\langle(\partial H_1/\partial J)\rangle\,\langle H_1\rangle - \langle(\partial H_1/\partial J)H_1\rangle\big] \\ &\quad + (1/2\nu_o{}^2)(\partial\nu_o/\partial J)\,\big[\langle H_1{}^2\rangle - \langle H_1\rangle^2\big]; \end{aligned} \tag{d}$$

and, therefore,

(ii)

$$\begin{aligned} \partial G_2/\partial w_o &= (1/\nu_o)(E_2 - K_2) \\ &= (1/\nu_o)\big[E_2 - H_2 - (\partial H_1/\partial J)(\partial G_1/\partial w_o) - (1/2)\,(\partial^2 H_o/\partial J^2)(\partial G_1/\partial w_o)^2\big] \\ &= (1/\nu_o)(\langle H_2\rangle - H_2) \\ &\quad + (1/\nu_o{}^2)\big[\langle \partial H_1/\partial J\rangle\langle H_1\rangle - \langle(\partial H_1/\partial J)H_1\rangle\rangle \\ &\qquad - (\partial H_1/\partial J)\langle H_1\rangle + (\partial H_1/\partial J)H_1\big] \\ &\quad + (1/2\nu_o{}^3)(\partial\nu_o/\partial J)\,\big[\langle H_1{}^2\rangle - 2\langle H_1\rangle^2 + 2H_1\langle H_1\rangle - H_1{}^2\big], \end{aligned} \tag{e}$$

and, by integration, yields G_2. Thus, (w, J) can be expressed in terms of (w_o, J_o), and so on.

(iii) Using the above, show that, to the second order, the perturbed energy equals (with no need to calculate G first)

$$\begin{aligned} E(J) &= H_o(J) + \varepsilon\langle H_1\rangle \\ &\quad + \varepsilon^2\big\{\langle H_2\rangle + (1/\nu_o)\big[\langle \partial H_1/\partial J\rangle\langle H_1\rangle - \langle(\partial H_1/\partial J)H_1\rangle\big] \\ &\qquad + (1/2\nu_o{}^2)(\partial\nu_o/\partial J)[\langle H_1{}^2\rangle - \langle H_1\rangle^2]\big\}; \end{aligned} \tag{f}$$

and readily supplies the perturbed frequencies via $\nu = \partial E/\partial J = \cdots$.

Problem 8.16.2 (Dittrich and Reuter, 1994, pp. 112–113). Consider a nonlinear mass–spring oscillator, of mass m and linearized circular frequency $\omega_o \equiv (k/m)^{1/2}$, with perturbed Hamiltonian

$$H = H_o + \varepsilon H_1, \tag{a}$$

$$H_o \equiv p^2/2m + (m\omega_o{}^2/2)q^2, \qquad H_1 \equiv (m/6)q^6. \tag{a1}$$

As shown in the preceding examples,

$$H_o = \nu_o J_o = (\omega_o/2\pi)J_o, \qquad w_o = \nu_o t + \gamma_o; \tag{a2}$$

$$q_o = (J_o/\pi m\omega_o)^{1/2}\sin(2\pi w_o), \qquad p_o = (m\omega_o J_o/\pi)^{1/2}\cos(2\pi w_o). \tag{a3}$$

(i) Show that

$$E_1(J) = \langle H_1\rangle = \cdots = (5/16)(m/6)(J/\pi m\omega_o)^3. \tag{b}$$

(ii) Show that

$$\Delta\nu \equiv \nu - \nu_o = \varepsilon(5/64\pi^2)(q_{\max}{}^4/\nu_o), \tag{c}$$

where $q_{\max}$ = maximum amplitude of unperturbed (harmonic) oscillator.

HINTS

(i) Verify that

$$\sin^6(\ldots) \equiv \{[\exp(i\ldots) - \exp(-i\ldots)]/2i\}^6$$
$$= \cdots = -(2/64)\,[\cos(6\ldots) - 6\cos(4\ldots) + 15\cos(2\ldots) - 10]; \tag{d}$$

(ii)

$$J \to J_o = E_o/\nu_o = \cdots = \pi m\omega_o q_{\max}{}^2 \qquad \text{(explain)}. \tag{e}$$

Problem 8.16.3 (Born, 1927, pp. 254–255). Continuing the perturbation scheme to the (p)th order in ε:

(i) Show that, then,

$$(\partial H_o/\partial J)\,(\partial G_p/\partial w_o) = E_p(J) - R_p(w_o, J), \tag{a}$$

where $R_p(w_o, J)$ stands for a term involving only results of the preceding orders of perturbation; that is, only up to those obtained in the $(p-1)$th order, and is periodic in the w_o's.

(ii) Verify that averaging (a) yields

$$E_p(J) = \langle R_p(w_o, J)\rangle; \tag{b}$$

from which it follows that (with the usual notations)

$$\partial G_p(w_o, J)/\partial w_o = -\big(1/\nu_o(J)\big)\,\Delta R_p(w_o, J). \tag{c}$$

Equation (c) is solved by expanding both sides in Fourier series and then equating the same harmonic coefficients, thus expressing the unknown coefficients of the left side in terms of the known coefficients of the right side.

For additional related examples and problems, see, for example: Born (1927, pp. 259–261), Frank (1935, pp. 203–209, 214–218), Meirovitch (1970, pp. 376–377), Saletan and Cromer (1971, pp. 252–256), Straumann (1987, pp. 271–273).

References and Suggested Reading

Additional comparable and complementary lists, for further and deeper study, can be found in:

Leimanis (1965)—analytical rigid-body dynamics until the mid-1960s
Mikhailov and Parton (1990)—advanced topics in analytical mechanics and stability of equilibrium/motion; complements and updates the list of Neimark and Fufaev
Neimark and Fufaev [1967 (1972)]—analytical mechanics, theory and applications; includes most Soviet/Russian works until the early 1960s
Roberson and Schwertassek (1988)—multibody and computational dynamics; see also Huston (1990)
Stäckel (1905)—elementary and intermediate theoretical dynamics until the early 1900s
Voss (1901)—principles of theoretical mechanics until 1900
Ziegler (1985)—geometrical methods in rigid-body mechanics

Abbreviations used below:

AIAA: American Institute of Aeronautics and Astronautics (U.S.)
PMM: Journal of Applied Mathematics and Mechanics (English translation from the Russian)
Springer: Springer-Verlag
ZAMM: Zeitschrift für angewandte Mathematik und Mechanik (German)
ZAMP: Zeitschrift für angewandte Mathematik und Physik (Swiss)

For a steady supply of worthwhile material from the frontier of (classical) theoretical/analytical dynamics, including archival papers, discussions, and book reviews, we recommend the following journals:

American Journal of Physics
Applied Mathematics and Mechanics (English translation from the Chinese)
Archive of Applied Mechanics (German, formerly *Ingenieur-Archiv*)
International Journal of Non-Linear Mechanics (U.S.)
PMM
ZAMM
ZAMP

Also:

Journal of Applied Mechanics (ASME)
Journal of the Astronautical Sciences
Journal of Guidance, Control, and Dynamics (AIAA)
Nonlinear Dynamics

For the historical and cultural aspects of mechanics, etc., we recommend the following journals:

Archive for History of Exact Sciences
Centaurus (International Magazine of the History of Mathematics, Science, and Technology; Munksgaard, Copenhagen)
Physics Today

Abhandlungen über die Prinzipien der Mechanik, von Lagrange, Rodrigues, Jacobi, und Gauss. 1908. Edited by P. E. B. Jourdain. Leipzig: Engelmann (Ostwald's Klassiker der exakten Wissenschaften, nr. 167).

Abraham, R., and J. E. Marsden. 1978. *Foundations of Mechanics*, 2nd ed. Reading, MA: Benjamin/Cummings Publishing Co.

Agostinelli, C. 1946. "Sull'Esistenza di Integrali di un Sistema Anolonomo con Coordinate Ignorabili," *Ac. Sc. Torino, Atti*, **80**, 231–239.

Agostinelli, C. 1956. "Nuova Forma Sintetica delle Equazioni del Moto di un Sistema Anolonomo ed Esistenza di un Integrale Lineare nelle Velocita Lagrangiane," *Boll. Un. Mat. Ital.*, **11**(3), 1–9.

Agrawal, O. P., and S. Saigal. 1987. "A Novel, Computationally Efficient, Approach for Hamilton's Law of Varying Action," *Int. J. Mech. Sci.*, **29**, 285–292.

Aharoni, J. 1972. *Lectures in Mechanics*. Oxford, U.K.: Clarendon Press.

Aharonov, Y., H. A. Farach, and C. P. Poole, Jr. 1977. "Nonlinear Vector Product to Describe Rotations," *Am. J. Physics*, **45**, 451–454.

Ainola, L. Ia. 1966. "Integral Variational Principles of Mechanics," *PMM*, **30**, 1124–1128 (original in Russian, 946–949).

Aizerman, M. A. 1974. *Classical Mechanics* (in Russian). Moscow: Nauka.

Alishenas, T. 1992. *Zur Numerischen Behandlung, Stabilisierung durch Projektion und Modellierung mechanischer Systeme mit Nebenbendingungen und Invarianten.* Dissertation, Köningl. Techn. Hochsch. (Royal Institute of Technology), Stockholm.

Alt, H. 1927. "Geometrie der Bewegungen," pp. 178–232 in vol. 5 of *Handbuch der Physik*. Berlin: Springer.

Altmann, S. L. 1986. *Rotations, Quaternions, and Double Groups*. Oxford, U.K.: Clarendon Press.

Ames, J. S., and F. D. Murnaghan. 1929. *Theoretical Mechanics, An Introduction to Mathematical Physics*. Boston, MA: Ginn (reprinted 1958 by Dover, New York).

Amirouche, F. M. L. 1992. *Computational Methods in Multibody Dynamics*. Englewood Cliffs, NJ: Prentice Hall.

Andelic (Angelitch), T. P. 1954. "Ueber die Bewegung starrer Körper mit nichtholonomen Bindungen in einer inkompressiblen Flüssigkeit," pp. 314–316 in vol. 2 of *Proc. Int. Congr. Math.* (Amsterdam, 1954). Groningen and Amsterdam, Holland: Noordhoff.

Angeles, J. 1988. *Rational Kinematics*. New York: Springer.

Appell, P. 1896. "Sur l'emploi des équations de Lagrange dans la théorie du choc et des percussions," *J. de Math. Pures et Appliquées*, Series 5, **2**, 5–20.

Appell, P. 1899(a). *Les Mouvements de Roulement en Dynamique*, avec deux notes de M. Hadamard. Paris: Scientia.

Appell, P. 1899(b). "Sur les mouvements de roulement; équations du mouvement analogues à celles de Lagrange," *Comptes Rendus* (Acad. Sci. of Paris), **129**, 317–320.

Appell, P. 1900(a). "Sur une forme générale des équations de la dynamique," *J. für die reine und angewandte Mathematik*, **121**, 310–331.

Appell, P. 1900(b). "Développements sur une forme nouvelle des équations de la dynamique," *J. de Math. Pures et Appliquées*, Series 5, **6**, 5–40.

Appell, P. 1901. "Remarques d'ordre analytique sur une nouvelle forme des equations de la dynamique," *J. Math. Pures et Appliquées*, Series 5, **7**, 5–12.

Appell, P. 1911(a). "Sur les liaisons exprimées par des relations non-lineaires entre les vitesses," *Comptes Rendus* (Acad. Sci. of Paris), **152**, 1197–1200.

Appell, P. 1911(b). "Exemple de mouvement d'un point assujetti a une liaison exprimée par une relation non-lineaire entre les composantes de la vitesse," *Rend. Circ. Mat. Palermo*, **32**, 48–50.

Appell, P. 1912(a). "Aperçu sur l'emploi possible de l'énergie d'accéleration dans les équations de l'électrodynamique," *Comptes Rendus* (Acad. Sci. of Paris), **154**, 1037–1040.

Appell, P. 1912(b). "Sur les liaisons non lineaires par rapport aux vitesses," *Rend. Circ. Mat. Palermo*, **33**, 259–267.

Appell, P. 1916. "Sur les liaisons cachées et les forces gyroscopiques apparentes dans les systemes non holonomes," *Comptes Rendus* (Acad. Sci. of Paris), **162**, 27–29.

Appell, P. 1924. "Sur l'ordre d'un systeme non holonome," *Comptes Rendus* (Acad. Sci. of Paris), **179**, 549–550.

Appell, P. 1925. "Sur une forme générale des équations de la dynamique," fascicule 1, pp. 1–50, *Mémorial des Sciences Mathematiques*. Paris: Gauthier-Villars.

Appell, P. *Traité de Mécanique Rationelle*, vol. 4, fascicule I: *Figures d'Équilibre d'une Masse Liquide Homogène en Rotation* (1932, 2nd ed.); fascicule II: *Les Figures d'Équilibre d'une Masse Héterogène en Rotation, Figure de la Terre et des Planètes* (1937). Paris: Gauthier-Villars.

Appell, P. 1953. *Traité de Mécanique Rationelle*, vol 2: *Dynamique des Systèmes, Mécanique Analytique*, 6th ed. Paris: Gauthier-Villars.

Apykhtin, N. G., and V. F. Iakovlev. 1980. "On the Motion of Dynamically Controlled Systems with Variable Masses," *PMM*, **44**, 301–305 (original in Russian, 427–433).

Arczewski, K. P., and J. A. Pietrucha. 1993. *Mathematical Modelling of Mechanical Complex Systems*, vol. 1: *Discrete Models*. New York: Ellis Horwood.

Argyris, J., and V. F. Poterasu. 1993. "Large Rotations Revisited. Application of Lie Algebra," *Computer Methods in Applied Mechanics and Engineering*, **103**, 11–42.

Arhangelskii, Ju. A. 1977. *Analytical Dynamics of Rigid Bodies* (in Russian). Moscow: Nauka.

Ariaratnam, S. T., and N. Sri Namachchiraya. 1986. "Periodically Perturbed Linear Gyroscopic Systems," *J. Structural Mechanics*, **14**, 127–151.

Ariaratnam, S. T., and N. Sri Namachchiraya. 1986. "Periodically Perturbed Nonlinear Gyroscopic Systems," *J. Structural Mechanics*, **14**, 153–175.

Arnold, D. H. "The Mécanique Physique of Siméon Denis Poisson: The Evolution and Isolation in France of his Approach to Physical Theory (1800–1840)," *Archive for History of Exact Sciences*, **28**, 243–266, 267–287, 289–297, 299–320, 321–342, 343–367 (1983); **29**, 37–51, 53–72, 73–94 (1983/1984); **29**, 289–307 (1984).

Arnold, R. N., and L. Maunder. 1961. *Gyrodynamics, and its Engineering Applications*. New York: Academic Press.

Arnold, V. I. 1976. *Méthodes Mathématiques de la Mécanique Classique*. Moscow: Mir (original in Russian, 1974; English translation as *Mathematical Methods of Classical Mechanics*. New York: Springer, 1978, 1st ed.; 1989, 2nd ed.).

Arnold, V. I., V. V. Kozlov, and A. I. Neishtadt. 1988. "Mathematical Aspects of Classical and Celestial Mechanics," pp. 1–291 in *Dynamical Systems III*, vol. 3 of *Encyclopaedia of Mathematical Sciences*, edited by V. I. Arnold. Berlin: Springer (original in Russian, 1985).

Arnol'd, V. I. 1990. *Huygens and Barrow, Newton and Hooke*. Basel: Birkhäuser (original in Russian, 1989).

Arrighi, G. 1939. "Sul Moto Impulsivo dei Sistemi Anolonomi," *Accad. Naz. Lincei, Rendiconti, Cl. Sc. Fis. Mat. Nat.* (Serie Siesta), **29**, 472–477.

Arrighi, G. 1940. "Sui Sistemi Anolonomi," *Rendiconti Inst. Lombardo Sci. Lett. Cl. Sci. Mat. Nat.* **4**(3), 367–374.

Arya, A. P. 1990. *Introduction to Classical Mechanics*. Boston: Allyn and Bacon.

Arzanyh, I. S. 1965. *Impulsive Fields* (in Russian). Tashkent: Nauka, Uzbek. S.S.R.

Asimov, I. 1964. *Asimov's Biographical Encyclopedia of Science and Technology*, The living stories of more than 1000 great scientists from the age of Greece to the space age chronologically arranged. Garden City, New York: Doubleday.

Atanackovic, T. M. 1982. "The Sufficient Conditions for an Extremum in the Variational Principle with Non-Commutative Variational Rules," *Acad. Sci. Turin, Proceedings of the IUTAM-ISIMM Symposium on Modern Developments in Analytical Dynamics* (June 7–11, 1982, vol. 2, 463–467. [Also in *Atti della Accademia delle Scienze di Torino, Supplemento*, **117**, 1983.]

Atkin, R. H. 1959. *Classical Dynamics*. New York: Wiley.

Atluri, S. N., and A. Cazzani. 1995. "Rotations in Computational Solid Mechanics," *Archives of Computational Methods in Engineering* (State of the Art Reviews), **2**(1), 49–138.

Audin, M. 1996. *Spinning Tops*. Cambridge, U.K.: Cambridge University Press.

Auerbach, F. 1908. "Kinematik und Dynamik," pp. 299–357 in vol. 1 of (older) *Handbuch der Physik*, 2nd ed. Leipzig: Barth.

Azizov, A. G. 1975. "Impulsive Motion of Systems with Servoconstraints" (in Russian), pp. 33–38 in *Problems of Analytical Mechanics, Theory of Stability and Control*. Moscow: Nauka, Acad. U.S.S.R.

Azizov, A. G. 1986. "On the Motion of a Controlled System of Variable Mass," *PMM*, **50**, 433–437 (original in Russian, 567–572).

Babakov, I. M. 1968. *Theory of Oscillations* (in Russian), 3rd edition. Moscow: Nauka.

Bahar, L. Y. 1970. "Direct Determination of Finite Rotation Operators," *J. Franklin Inst.*, **289**, 401–404.

Bahar, L. Y. 1970–1980. *Advanced Dynamics, Lecture Notes*. Philadelphia: Drexel University.

Bahar, L. Y. 1984. "Equations of Motion for the Chaplygin–Carathéodory Problem by Means of Appell's Equations," Private Communication.

Bahar, L. Y. 1987. "The Theory of Rigid-Body Rotations," Private Communication.

Bahar, L. Y. 1991. "Generalized Gantmacher Formulas through Functions of Matrices," *Am. J. Phys.*, **59**, 1103–1111.

Bahar, L. Y. 1992. "Response, Stability and Conservation Laws for the Sleeping Top Problem," *J. Sound and Vibration*, **158**, 25–34.

Bahar, L. Y. 1993. "On a Non-holonomic Problem Proposed by Greenwood," *Int. J. Non-Linear Mechanics*, **28**, 169–186.

Bahar, L. Y. 1994. "On the use of Quasi-velocities in Impulsive Motion," *Int. J. Eng. Sci.*, **32**, 1669–1686.

Bahar, L. Y. 1998. "A Non-linear Non-holonomic Formulation of the Appell–Hamel Problem," *Int. J. Non-Linear Mechanics*, **33**, 67–83.

Bahar, L. Y., and H. Anton. 1972. "On the Application of Distribution Theory to Variational Problems," *J. Franklin Inst.*, **293**, 215–223.

Bahar, L. Y., and H. G. Kwatny. 1987. "Extension of Noether's Theorem to Constrained Non-Conservative Systems," *Int. J. Non-Linear Mechanics*, **22**, 125–138.

Bahar, L. Y., and H. G. Kwatny. 1992. "Matrix Analysis of Some Linear Gyroscopic Systems," *J. Franklin Inst.*, **329**, 679–695.

Bakay, A. S., and Yu. P. Stepanovskii. 1981. *Adiabatic Invariants* (in Russian). Kiev: Naukova Dumka.

Bamberger, Y. 1981. *Mécanique de l'Ingénieur I. Systèmes de Corps Rigides*. Paris: Hermann.

Barbour, J. B. 1989. *Absolute or Relative Motion? A Study from the Machian Point of View of the Discovery and the Structure of Dynamical Theories*. vol. 1: *The Discovery of Dynamics*. Cambridge, U.K.: Cambridge University Press.

Bartlett, J. H. 1975. *Classical and Modern Mechanics*. Alabama: University of Alabama Press.

Baruh, H. 1999. *Analytical Mechanics*. New York: McGraw-Hill.

Bat, M. I., G. Iu. Dzanelidze, and A. S. Kelbzon. 1973. *Theoretical Mechanics*, in *Examples and Problems* (in Russian), 3 vols., edited by G. Iu. Dzanelidze and D. P. Merkin. Moscow: Nauka.

Battin, R. H. 1987. *An Introduction to the Mathematics and Methods of Astrodynamics* (AIAA Education Series). New York: American Institute of Aeronautics and Astronautics.

Baumgarte, J. 1972. "Stabilization of Constraints and Integrals of Motion in Dynamical Systems," *Computer Methods in Appl. Mechanics and Eng.*, **1**, 1–16.

Baumgarte, J. 1980(a). "Ein rein Hamiltonsche Formulierung der Mechanik von Systemen mit holonomen Bindungen," *Acta Mechanica*, **36**, 135–142.

Baumgarte, J. 1980(b). "Verallgemeinerung des Jacobischen Prinzips der kleinsten Wirkung auf den allgemeinen nichtconservativen Fall," *ZAMM*, **60**, T35–T36.

Baumgarte, J. 1982. "Stabilisierung von Bindungen über Zwangsimpulse," *ZAMM*, **62**, 447–454.

Baumgarte, J. 1983. "A New Method of Stabilization for Holonomic Constraints," *J. Appl. Mechanics* (ASME), **50**, 869–870.

Baumgarte, J. 1984. "Analytische Mechanik der beschränkten Konfigurationsräume," *ZAMM*, **64**, 321–328.

Baumgarte, J. 1987. "Aussagen über die Stabilität quasiperiodischer Lösungen durch Einbettung in höherdimensionale Räume," *ZAMM*, **67**, 357–365.

Beatty, M. F. 1966. "Kinematics of Finite, Rigid-Body Displacements," *Am. J. Physics*, **34**, 949–954.

Beatty, M. F., Jr. 1986. *Principles of Engineering Mechanics*, vol. 1: *Kinematics — The Geometry of Motion*. New York: Plenum Press. (Vol. 2: *Dynamics — The Analysis of Motion*, to appear in 2001.)

Beer, F. P. 1963. "Impulsive Motions," *J. Appl. Mechanics* (ASME), **30**, 315–316.

Beghin, H. 1967. *Cours de Mécanique Théorique et Appliquée*, tome 1, new edition revised by J. Mandel. Paris: Gauthier-Villars.

Beghin, H., and G. Julia. *Exercices de Mécanique*, tome 1, fascicules 1 (1946) and 2 (1951). Paris: Gauthier-Villars.

Beghin, H., and T. G. Rousseau. 1903. "Sur les percussions dans les systèmes nonholonomes," *J. Math. Pures et Appliquées*, series 5, **9**, 21–26.

Belen'kii, I. M. 1964. *Introduction to Analytical Mechanics* (in Russian). Moscow: Vischaya Shkola.

Belen'kii, I. M. 1970. "On the Theory of Hamiltonian Systems," *PMM*, **34**, 723–728 (original in Russian, 756–761).

Bell, E. T. 1937. *Men of Mathematics*. New York: Simon and Schuster (several subsequent reprintings).

Bellet, D. 1988. *Cours de Mécanique Générale*. Toulouse: Editions Cepadues (Collection "La Cheveche").

Bellman, R., and R. Kalaba, editors. 1964. *Mathematical Trends in Control Theory*. New York: Dover.

Benati, M., and A. Morro. 1994. "Angular Vectors and Kinetic Energy of Rigid Bodies," *Eur. J. Mechanics, A/Solids*, **13**, 819–832.

Benvenuto, E. 1991. *An Introduction to the History of Structural Mechanics*, part I: Statics and Resistance of Solids; Part II: Vaulted Structures and Elastic Systems. New York: Springer (original in Italian, 1981). [See also its review by J. Casey, in *Physics Today*, July 1992, pp. 65–67.]

Berdichevskii, V. L. 1983. *Variational Principles in the Mechanics of Continuous Media* (in Russian). Moscow: Nauka.

Berezkin, E. N. 1968. *Lectures on Theoretical Mechanics*, vol. 2: *Dynamics of Systems, Dynamics of Rigid Bodies, Analytical Mechanics* (in Russian). Moscow: Moscow University Press.

Bergmann, P. G. 1942. *Introduction to the Theory of Relativity*. Englewood Cliffs, NJ: Prentice Hall (reprinted 1976 by Dover, New York).

Bert, C. W. 1980. "Paul Appell, Mechanician, Mathematician, and Educator," *The Engineering Science Perspective*, **5**, 21–25.

Besant, W. H. 1914. *A Treatise on Dynamics*, 5th ed. London: Bell.

Betten, J. 1987. *Tensorrechnung für Ingenieure*. Stuttgart: Teubner.

Bewley, L. V. 1961. *Tensor Analysis of Electric Circuits and Machines*. New York: Ronald Press.

Beyer, R. 1928. "Dynamik der Mehrkurbelgetriebe," *ZAMM*, **8**, 122–133.

Beyer, R. 1929. "Geometrische Bewegungslehre," Band I, Erster Teil, pp. 405–467 in *Handbuch der Physikalischen und Technischen Mechanik*. Leipzig: J. A. Barth.

Beyer, R. 1963. *Technische Raumkinematik*. Berlin: Springer.

Bierhalter, G. 1981(a). "Boltzmanns mechanische Grundlegung des zweitwn Hauptsatzes der Wärmelehre aus dem Jahre 1866," *Archive for History of Exact Sciences*, **24**, 195–205.

Bierhalter, G. 1981(b). "Clausius' mechanische Grundlegung des zweiten Hauptsatzes der Wärmelehre aus dem Jahre 1871," *Archive for History of Exact Sciences*, **24**, 207–220.

Bierhalter, G. 1982. "Das Virialtheorem in seiner Beziehung zu den mechanischen Grundlegungen des zweiten Hauptsatzes der Wärmelehre," *Archive for History of Exact Sciences*, **27**, 199–211.

Bierhalter, G. 1983. "Zu Szilys Versuch einer mechanischen Grundlegung des zweiten Hauptsatzes der Thermodynamik," *Archive for History of Exact Sciences*, **28**, 25–35.

Bierhalter, G. 1992. "Von L. Boltzmann bis J. J. Thomson: die Versuche einer mechanischen Grundlegung der Thermodynamik (1866–1890)," *Archive for History of Exact Sciences*, **44**, 25–75.

Biezeno, C. 1927. "Geometrie der Kräfte und Massen," pp. 233–304 in vol. 5 of *Handbuch der Physik*. Berlin: Springer.

Bilimovitch, A. D. 1964. *On a General Phenomenological Differential Principle*. Beograd, Serbia: Naucno delo (under Monographs of the Section of Natural and Mathematical Sciences of the Serbian Academy of Sciences and Arts).

Billington, E. W. 1986. *Introduction to the Mechanics and Physics of Solids*. Bristol, U.K.: A. Hilger.

Birkhoff, G. D. 1927. *Dynamical Systems*. New York: American Mathematical Society.

Birtwistle, G. 1926. *The Quantum Theory of the Atom*. Cambridge, U.K.: Cambridge University Press.

Blajer, W. 1992(a). "Projective Formulation of Maggi's Method for Nonholonomic System Analysis," *J. Guidance, Control, and Dynamics*, **15**, 522–525.

Blajer, W. 1992(b). "A Projection Method Approach to Constrained Dynamic Analysis," *J. Appl. Mechanics* (ASME), **59**, 643–649.

Blatt, J. M., and J. N. Lyness. 1961/1962. "The Practical Use of Variation Principles in Non-Linear Mechanics," *J. Australian Math. Soc.*, **2**, 357–368 (see also Lyness and Blatt.)

Bleick, W. E. 1968. "Angular Velocity of an Orthogonal Curvilinear Coordinate Frame," *Am. J. Physics*, **36**, 638–639.

Blekhman, I. I. 1971. *Synchronization in Dynamic Systems* (in Russian). Moscow: Nauka.

Blekhman, I. I. 1979. "Oscillations of Nonlinear Mechanical Systems" (in Russian), *Vibrations in Engineering*, vol. 2. Moscow: Mashinostroenie.

Blekhman, I. I. 1988. *Synchronization in Science and Technology*. New York: ASME Press (original in Russian, 1981).

Blekhman, I. I., and O. Z. Malakhova. 1990. "Extremal Criteria of the Stability of Certain Motions," *PMM*, **54**, 115–133 (original in Russian, 142–161).

Bobillo-Ares, N. 1988. "Noether's Theorem in Discrete Classical Mechanics," *Am. J. Physics*, **56**, 174–177.

Bochner, S. 1966. *The Role of Mathematics in the Rise of Science*. Princeton, NJ: Princeton University Press.

Bogoliubov, N. N., and Y. A. Mitropolskii. 1974. *Asymptotic Methods in the Theory of Nonlinear Oscillations* (in Russian), 4th ed. Moscow: Nauka.

Bolotin, V. V. 1963. *Nonconservative Problems of the Theory of Elastic Stability*. New York: Macmillan/Pergamon (original in Russian, 1961).

Boltzmann, L. 1974. *Vorlesungen über die Prinzipe der Mechanik*, vols. I (1897) and II (1904a). Leipzig: Barth (reprinted 1974 by Wissenschaftliche Buchgesellschaft, Darmstadt).

Boltzmann, L. 1902. "Ueber die Form der Lagrangeschen Gleichungen für nichtholonome, generalisierte Koordinaten," *Wienersche Sitzungsberichte, Math. Naturwiss. Kl.*, **111**, 1603–1614. [Also in B.'s *Gesammelte Wissenschaftliche Abhandlungen* (Collected Scientific Works), vol. 3, pp. 682–692; edited in 3 vols. by F. Hasenöhrl. Leipzig: Barth. Reprinted by Chelsea, New York, 1968.]

Boltzmann, L. 1904. "Ueber die Anwendung der Lagrangeschen Gleichungen auf nicht holonome generalisierte Koordinaten," *Jahresbericht der Deutschen Mathematiker-Vereinigung*, **13**, 132–133.

Bookchin, M. 1990. *Remaking Society, Pathways to a Green Future*. Boston, MA: South End Press.

Born, M. 1927. *The Mechanics of the Atom*. London: Bell (original in German, 1925).

Born, M. 1964. *Natural Philosophy of Cause and Chance*. New York: Dover (originally published 1949 by Oxford University Press).

Borri, M. 1994. "Numerical Approximations in Analytical Dynamics," pp. 323–361 in *Applied Mathematics in Aerospace Science and Engineering*, edited by A. Miele and A. Salvetti. New York: Plenum.

Borri, M., F. Mello, and S. Atluri. 1990. "Variational Approaches for Dynamics and Finite-Elements: Numerical Studies," *Computational Mechanics*, **7**, 49–76.

Borri, M., F. Mello, and S. Atluri. 1991. "Primal and Mixed Forms of Hamilton's Principle for Constrained Rigid Body Systems: Numerical Studies," *Computational Mechanics*, **7**, 205–220.

Borri, M., C. Bottasso, and P. Mantegazza. 1990. "Equivalence of Kane's and Maggi's Equations," *Meccanica*, **25**, 272–274. [See also **27**, 63–64, 1992 (discussion).]

Borri, M., C. Bottasso, and P. Mantegazza. 1992. "Acceleration Projection Method in Multibody Dynamics," *European J. Mechanics, A/Solids*, **11**, 403–418.

Bottema, O. 1955. "Note on a Nonholonomic System," *Quart. Appl. Math.*, **13**, 191–192.

Bottema, O., and B. Roth. 1990. *Theoretical Kinematics*. Amsterdam: North Holland (reprinted 1990 by Dover, New York).

Bouligand, G. 1954. *Mécanique Rationelle*, 5th ed. Paris: Vuibert.

Brach, R. M. 1991. *Mechanical Impact Dynamics, Rigid Body Collisions*. New York: Wiley.

Bradbury, T. C. 1968. *Theoretical Mechanics*. New York: Wiley.

Brand, L. 1947. *Vector and Tensor Analysis*. New York: Chapman and Hall.

Brand, L. 1957. *Vector Analysis*. New York: Wiley.

Brell, H. 1913(a). "Über eine neue Fassung des verallgemeinerten Prinzipes der kleinsten Aktion," *Wienersche Sitzungsberichte, Math.-Natur. Kl.*, **IIa**, **122**, 1031–1036 (see also pp. 933–944).

Brell, H. 1913(b). "Ueber eine neue Form des Gauß'schen Prinzipes des kleinsten Zwanges," *Wienersche Sitzungsberichte, Math.-Natur. Kl.*, **IIa**, **122**, 1531–1538.

Brelot, M. 1945. *Les Principes Mathématiques de la Mécanique Classique*. Grenoble and Paris: B. Arthaud.

Bremer, H. 1988(a). *Dynamik und Regelung Mechanischer Systeme*. Stuttgart: Teubner.

Bremer, H. 1988(b). "Ueber eine Zentralgleichung in der Dynamik," *ZAMM*, **68**, 307–311.

Bremer, H. 1992. "Zwang und Zwangskräfte," *Archive of Applied Mechanics* (formerly *Ingenieur-Archiv*), **62**, 158–171.

Bremer, H. 1993. "Das Jourdainsche Prinzip," *ZAMM*, **73**, 184–187.

Bremer, H. 1998. *Robotik I, II,* Vorlesung und Arbeitsgrundlage für weitere Vertiefungsstudien. Inst. für Fertigungs- und Hanhabungssysteme, Abteilung Robotik, Johannes Kepler Universität, Linz, Austria.

Bremer, H. 1999. "On the Dynamics of Elastic Multibody Systems," *Applied Mechanics Reviews* (ASME), **52**, 275–303.

Bremer, H., and F. Pfeiffer. 1992. *Elastische Mehrkörpersysteme*. Stuttgart: Teubner.

Brill, A. 1909. *Vorlesungen zur einführung in die Mechanik raumerfüllender Massen*. Leipzig: Teubner.

Brill, A. 1928. *Vorlesungen über allgemeine Mechanik*. München and Berlin: Oldenbourg.

Brillouin, L. 1964. *Tensors in Mechanics and Elasticity*. New York: Academic Press (original in French; 1938, 1st ed.; 1949, 2nd ed.).

Brogliato, B. 1999. *Nonsmooth Mechanics, Models, Dynamics and Control*, 2nd ed. London, Berlin: Springer.

Broucke, R. A. 1990. "Equations of Motion of a Rotating Rigid Body," *J. Guidance, Control, and Dynamics*, **13**, 1150–1152.

Broucke, R. and J. Touma. 1993. "On Generalized Changes of Variables in Dynamics," private communication.

Brousse, P. 1981. *Mécanique Analytique; Puissances Virtuelles, Equations de Lagrange, Applications*. Paris: Vuibert.

Brunk, G. 1980. "Die geometrische Deutung kinematischer Nebenbedingungen im Konfigurations—oder im Ereignisraum," *ZAMM*, **60**, T37–T38.

Brunk, G. 1981. "Geometrie und Integrabilität räumlicher Drehungen," *ZAMM*, **61**, T18–T20.

Brunk, G., and J. Myszkowski. 1970. "Zur Kinematik des starren Körpers und der Relativbewegung," *Acta Mechanica*, **10**, 99–110.

Bryan, G. 1892. "Researches Relating to the Connection of the Second Law (of Thermodynamics) with Dynamical Principles," pp. 88–96 in *Report of the 61st Meeting of the British Association for the Advancement of Science* (Cardiff, U.K., 1891). London: Murray.

Bryan, G. 1903. "Ableitung des Zweiten Hauptsatzes aus der Prinzipien der Mechanik," pp. 146–151 in vol. 5, part 1 of *Encyclopädie der Mathematischen Wissenschaften*. Leipzig: Teubner.

Buch, L. H., and H. H. Denman. 1976. "Variational Principle for t-Dependent Classical Hamiltonian Systems," *Physics Letters*, **55A**, 325–326.

Bucholz, N. N. 1972. *Fundamental Course in Theoretical Mechanics*, vol. 2: *Dynamics of Particles and Systems* (in Russian). Moscow: Nauka.

Buckley, R., and E. V. Whitfield. 1949. "Impulses and Constraints in Classical Mechanics," *Applied Scientific Research*, **A1**, 306–312.

Budde, E. *Allgemeine Mechanik der Punkte und starren Systeme*, Erster Band: *Mechanik der Punkte und Punktsysteme* (1890); Zweiter Band: *Mechanische Summen und starre Gebilde* (1891). Berlin: Georg Reimer.

Bulgakov, B. V. 1954. *Oscillations*. Moscow: State Publishing House for Technical-Theoretical Literature.

Burali-Forti, C., and T. Boggio. 1921. *Meccanica Razionale*. Torino and Genova: Lattes.

Burnet, J. 1930. *Early Greek Philosophy*, 4th ed. London: Adam and Charles Black.

Butenin, N. V. 1971. *Introduction to Analytical Mechanics* (in Russian). Moscow: Nauka.

Butenin, N. V., Ya. L. Lunts (or Lunc), and D. R. Merkin. 1985. *A Course in Theoretical Mechanics*, 2 vols. (in Russian), 3rd ed. Moscow: Nauka.

Byerly, W. E. 1916. *An Introduction to the Use of Generalized Coordinates in Mechanics and Physics*. Boston, MA: Ginn and Co. (reprinted by Dover, New York).

Cabannes, H. 1966. *Problèmes de Mécanique Générale*. Paris: Dunod.

Cabannes, H. 1968. *General Mechanics*. Waltham, MA: Blaisdell/Ginn (original in French, 1965, 2nd ed.).

Calkin, M. G. 1996. *Lagrangian and Hamiltonian Dynamics*. New Jersey: World Scientific.

Campbell, R. 1971. *La Mécanique Analytique*. Paris: Presses Universitaires de France.

Candotti, E., C. Palmieri, and B. Vitale. 1972. "On the Inversion of Noether's Theorem in Classical Dynamical Systems," *Am. J. Physics*, **40**, 424–429.

Cannon, R. H. 1967. *Dynamics of Physical Systems*. New York: McGraw-Hill.

Capon, R. S. 1952. "Hamilton's Principle in Relation to Nonholonomic Mechanical Systems," *Quart. J. Mech. and Appl. Math.*, **5**, 472–480.

Carathéodory, C. 1933. "Der Schlitten," *ZAMM*, **13**, 71–76.

Carathéodory, C. 1935. *Variationsrechnung und partielle Differentialgleichungen erster Ordnung*. Leipzig: Teubner (English translation available from Holden-Day, San Francisco).

Cartan, E. 1922. *Leçons sur les Invariants Integraux*. Paris: Hermann.

Carvallo, E. "Théorie du mouvement du monocycle et de la bicyclette," *J. de l'Ecole Polytechnique*, Série 2, Cahier **5**, 119–188 (1900); Série 2, Cahier **6**, 1–118 (1901).

Casey, J., and W. Stadler. 1983. "A Remark on the Principle of Angular Momentum for Systems of Particles," *ZAMM*, **66**, 190–192.

Casey, J. 1983. "A Treatment of Rigid Body Dynamics," *J. Appl. Mechanics* (ASME), **50**, 905–907 [**51**, 227, 1984 (errata)].

Casey, J. 1992. "The Principle of Rigidification," *Archive for History of Exact Sciences*, **43**, 329–383.

Casey, J., and V. C. Lam. 1986(a). "On the Relative Angular Velocity Tensor," *J. Mechanisms, Transmissions, and Automation in Design* (ASME), **108**, 399–400.

Casey, J., and V. C. Lam. 1986(b). "A Tensor Method for the Kinematical Analysis of Systems of Rigid Bodies," *Mechanism and Machine Theory*, **21**, 87–97.

Casey, J., and V. C. Lam. 1987. "On the Reduction of the Rotational Equations of Rigid Body Dynamics," *Meccanica*, **22**, 41–42.

Casey, J., and C. E. Smith. 1986. "When is the Direction of Angular Momentum Fixed in a Rigid Body?" *ZAMM*, **66**, 559–560.

Castigliano, C. A. P. 1966. *The Theory of Equilibrium of Elastic Systems and its Applications*. New York: Dover (original in French, 1879; English translation: *Elastic Stresses in Structures*, 1919).

Castoldi, L. 1945/1946. "Sopra una Definizione di 'Spostamento Virtuale' di un Sistema Dinamico Valida per Vincoli di Natura Qualunque," *Inst. Lombardo Rendiconti, Cl. Sc. Mat. e Natur.*, **79**, 71–88.

Castoldi, L. 1947(a). "Equazioni Lagrangiane per i Sistemi a Vincoli Anolonomi Non Lineari nelle Velocità," *Inst. Lombardo Rendiconti, Cl. Sc. Mat. e Natur.*, **80**, 189–200.

Castoldi, L. 1947(b). "Il Principio di Hamilton per Sistemi Dinamici a Vincoli Anolonomi Generali," *Atti Accad. Naz. Lincei, Rendiconti, Cl. Sci. Fis. Mat. Nat.*, **3**(8), 329–333.

Castoldi, L. 1949. "Formulazione Lagrangiana delle Equazioni del Moto per Sistemi Dinamici Sogetti a Vincoli Servomotori," *Il Nuovo Cimento*, **6**(3), 180–186.

Cayley, A. 1858. "Report on the Recent Progress of Theoretical Dynamics," pp. 1–42 in *Report of the 27th Meeting of the British Association for the Advancement of Science* (Dublin, 1857). London: Murray.

Cayley, A. 1863. "Report on the Progress of the Solution of Certain Special Problems of Dynamics," pp. 184–252 in *Report of the 32nd Meeting of the British Association for the Advancement of Science* (Cambridge, U.K., 1862). London: Murray.

Cetto, A. M., and L. de la Peña. 1984. "Simple Relationship between Energy and Adiabatic Invariants for Systems with a Power-Law Potential," *Am. J. Physics*, **52**, 539–542.

Chadwick, P. 1976. *Continuum Mechanics, Concise Theory and Problems*. New York: Wiley/Halsted/G. Allen and Unwin.

Chandrasekhar, S. 1995. *Newton's Principia for the Common Reader*. Oxford, U.K.: Clarendon Press (Oxford University Press).

Chandrasekharaiah, D. S., and L. Debnath. 1994. *Continuum Mechanics*. Boston: Academic Press.

Chaplygin, S. A. 1895. "On the Motion of a Heavy Figure of Revolution on a Horizontal Plane" (in Russian), *Proceedings of the Society of the Friends of Natural Science*, **9**(1), 10–16 (published in 1897). (Also in C.'s *Selected Works on Mechanics and Mathematics*, pp. 413–425. State Publishing House of Technical and Theoretical Literature, Moscow, 1954.)

Chaplygin, S. A. 1911. "On the Theory of Motion of Nonholonomic Systems. Theory of the Reducing Multiplier" (in Russian), *Mat. Sbornik*, **28**, 303–314. (Also in C.'s *Selected Works on Mechanics and Mathematics*, pp. 426–433. State Publishing House of Technical and Theoretical Literature, Moscow, 1954.)

Charlier, C. L. 1927. *Die Mechanik des Himmels*, Vorlesungen, 2nd ed., 2 vols. Berlin and Leipzig: W. de Gruyter.

Chen, F., and D. Y. Hsieh. 1981. *A Galerkin Method and Nonlinear Oscillations and Waves*. Providence, RI: Brown University, Division of Applied Mathematics.

Chen, G. 1987. "A Generalized Galerkin's Method for Non-linear Oscillators," *J. Sound & Vibrations*, **112**, 503–511.

Chen, Y.-H. 1998(a). "Pars' Acceleration Paradox," *J. Franklin Inst.*, **335B**, 871–875.

Chen, Y.-H. 1998(b). "Second Order Constraints for Equations of Motion of Constrained Systems," *IEEE/ASME Transactions on Mechatronics*, **3**(3), 240–248.

Cheng, H., and K. C. Gupta. 1989. "An Historical Note on Finite Rotations," *J. Appl. Mechanics* (ASME), **56**, 139–145.

Chertkov, R. I. 1960. *The Method of Jacobi in the Dynamics of Rigid Bodies* (in Russian). Leningrad (St. Petersburg): Sudprom.

Chester, W. 1979. *Mechanics*. London: G. Allen and Unwin.

Chetaev, N. G. 1962. "On the Equations of Poincaré," several papers: pp. 197–198, 199–200, 201–210, 381–383, 503–549 (in Russian) in *Works by Chetaev on Stability of Motion and on Analytical Mechanics*. Moscow: U.S.S.R. Academy of Sciences.

Chetaev, N. G. 1989. *Theoretical Mechanics*. Moscow/Berlin: Mir and Springer (original in Russian revised edition of 1987; based on Chetaev's lectures from the 1950s).

Chetayev, N. G. (or Chetaev). 1961. *The Stability of Motion*, 2nd ed. New York: Pergamon Press (original in Russian, 1955).

Chirgwin, B. H., and C. Plumpton. 1966. *Advanced Theoretical Mechanics*, vol. 6: *A Course of Mathematics for Engineers and Scientists*. Oxford, U.K.: Pergamon Press.

Chorlton, F. 1983. *Textbook of Dynamics*, 2nd ed. Chichester, U.K.: Ellis Horwood/Halsted Press/Wiley.

Chow, T. L. 1995. *Classical Mechanics*. New York: Wiley.

Christoffersen, J. 1989. "When is a Moment Conservative?" *J. Applied Mechanics* (ASME), **56**, 299–301.

Clifford, W. K. 1878 and 1887. *Elements of Dynamic, An Introduction to the Study of Motion and Rest in Solid and Fluid Bodies*. London: Macmillan.

Coburn, N. 1955. *Vector and Tensor Analysis*. New York: Macmillan (reprinted 1970 by Dover, New York).

Coe, C. J. 1934. "Displacements of a Rigid Body," *Am. Mathematical Monthly*, **41**, 242–253.

Coe, C. J. 1938. *Theoretical Mechanics, A Vectorial Treatment*. New York: Macmillan.

Collar, A. R. and A. Simpson. 1987. *Matrices and Engineering Dynamics*. Chichester, U.K.: Ellis Horwood/Halsted Press/Wiley.

Cooke, R. 1984. *The Mathematics of Sonya Kovalevskaya*. New York: Springer.

Corben, H. C., and P. Stehle. 1960. *Classical Mechanics*, 2nd ed. New York: Wiley (reprinted 1974 by R. E. Krieger, Huntington, NY and 1994 by Dover, New York).

Coriolis, G. 1835. "Mémoire sur les équations du mouvement relatif des systèmes de corps," *J. École Polytechnique*, **15**(24), 142–154.

Cosserat, E. and F. Cosserat. 1909. *Théorie des Corps Déformables*. Paris: Hermann.

Cotton, E. 1907. "A propos des equations de M. Appell," *Nouvelles Annales de Mathématiques*, Series 4, vol. **7**, 529–539.

Čović, V. 1987. "On the First Integrals of Rheonomic Holonomic Mechanical Systems," *ZAMM*, **67**, T416–T419.

Čović, V., and M. Lukačević. 1985. "An Interpretation of the Gauss Principle of Least Constraint," *ZAMM*, **65**, T334–T336.

Čović, V., and M. Lukačević. 1988. "Jourdain's Principle as a Principle of Least Constraint," *ZAMM*, **68**, T452–T453.

Craige, L. G., and J. L. Junkins. 1976. "Perturbation Formulations for Satellite Attitude Dynamics," *Celestial Mechanics*, **13**, 39–64.

Crandall, S. H., D. C. Karnopp, E. F. Kurtz, Jr., and D. C. Pridmore-Brown. 1968. *Dynamics of Mechanical and Electromechanical Systems*. New York: McGraw-Hill.

Crouch, T. 1981. *Matrix Methods Applied to Engineering Rigid Body Mechanics*. Oxford, U.K.: Pergamon.

Cunningham, W. J. 1958. *Introduction to Nonlinear Analysis*. New York: McGraw-Hill.

D'Abro, A. 1951. *The Rise of the New Physics*, 2 vols. New York: Dover (2nd ed. of *The Decline of Mechanism*, 1939).

D'Abro, A. 1950. *The Evolution of Scientific Thought*, 2nd ed. New York: Dover.

D'Alembert, J. Le Rond. *Traité de Dynamique*, Dans lequel les Loix (or Lois) de l'Equilibre et du Mouvement des Corps sont réduites au plus petit nombre possible, et démontrées d'une maniére nouvelle, et où l'on donne un Principe général pour trouver le Mouvement de plusieurs corps qui agissent les uns sur les autres, d'une maniére quelconque. Paris: David l'aîné, 1743 1st ed. (186 pp); 1758 2nd ed. (272 pp.); 1796 3rd ed. (identical to the

2nd). (German translation in 1899. 1st ed. reprinted 1967 by "Culture et Civilisation," Bruxelles; 2nd ed. reprinted 1921 by Gauthier-Villars, Paris, and 1968 by Johnson Reprint Corporation, New York.)

Dautheville, S. 1909. "Sur les systèmes non holonomes," *Bull. Soc. Math. de France*, **37**, 120–132.

Davis, P. J., and R. Hersh, 1986. *Descartes' Dream, The World According to Mathematics*. Boston, MA: Houghton and Mifflin.

Davis, W. R. 1967. "Constraints in Particle Mechanics and Geometry," *Am. J. Physics*, **35**, 916–920.

De Donder, T. "Sur un Théorème de Boltzmann Relatif aux Systèmes Mécaniques," *Académie Royale de Belgique, Bulletins de la Classe des Sciences*, Serie 5, **10**, 504–518 (1924); **11**, 28–36 (1925).

De Donder, T. 1927. *Théorie des Invariants Intégraux*. Paris: Gauthier-Villars.

De Jalón, J. G., and E. Bayo. 1994. *Kinematic and Dynamic Simulation of Multibody Systems, The Real-Time Challenge*. New York: Springer.

Delachet, A. 1967. *Le Calcul Vectoriel*, 4th ed. Paris: Presses Universitaires de France.

Delassus, E. 1913(a). "Les diverses formes du principe de d'Alembert et les équations générales du mouvement des systèmes soumis a des liaisons d'ordre quelconque," *Comptes Rendus* (Acad. Sci. of Paris), **156**, 205–209.

Delassus, E. 1913(b). *Leçons sur la Dynamique des Systèmes Materiels*. Paris: Hermann.

Delaunay, Ch. E. 1883. *Traité de Mécanique Rationelle*, 7th ed. Paris: Garnier and Masson (1856, 1st ed.).

De la Vallée Poussin, Ch.-J. 1912. *Cours d'Analyse Infinitesimale*, tome II, 2nd ed. Louvain: Uystpruyst and Dieudonné; and Paris: Gauthier-Villars.

Den Hartog, J. P. 1948. *Mechanics*. New York: McGraw-Hill (reprinted 1961 by Dover, New York).

Denman, H. H. 1966. "On Linear Friction in Lagrange's Equation," *Am. J. Physics*, **34**, 1147–1149.

Denman, H. H., and P. N. Kupferman. 1973. "Lagrangian Formalism for a Classical Particle Moving in a Riemannian Space with Dissipative Forces," *Am. J. Physics*, **41**, 1145–1148.

Desloge, E. A. 1982. *Classical Mechanics*, 2 vols. New York: Wiley.

Desloge, E. A. 1986. "A Comparison of Kane's Equations of Motion and the Gibbs–Appell Equations of Motion," *Am. J. Physics*, **54**, 470–472, 472 (rebuttal).

Desloge, E. A. 1987. "Relationship between Kane's Equations and the Gibbs–Appell Equations," *J. Guidance, Control, and Dynamics*, **10**, 120–122.

Desloge, E. A. 1988. "The Gibbs–Appell Equations of Motion, *Am. J. Physics*, **56**, 841–846.

Desloge, E. A. 1989. "Efficacy of the Gibbs–Appell Method for Generating Equations of Motion for Complex Systems," *J. Guidance, Control, and Dynamics*, **12**, 114–116.

Desloge, E. A., and R. I. Karch. 1977. "Noether's Theorem in Classical Mechanics," *Am. J. Physics*, **45**, 336–339.

Despeyrous, T. *Cours de Mécanique*, avec des notes par M. Darboux, vol. 1: 1884; vol. 2: 1886. Paris: Hermann.

Destouches, J. L. 1948. *Principes de la Mécanique Classique*. Paris: Centre National de la Recherche Scientifique.

Destouches, J. L. 1967(a). *La Mécanique Elementaire*, 2nd ed. Paris: Presses Universitaires de France.

Destouches, J. L. 1967(b). *La Mécanique des Solides*. Paris: Presses Universitaires de France.

Dickson, D. 1984. *The New Politics of Science*. New York: Pantheon.

Dictionary of Scientific Biography, 16 vols, edited by C. C. Gillispie. New York: Charles Scribner's Sons, 1970s.

Dijksterhuis, E. J. 1961. *The Mechanization of the World Picture*. London: Oxford University Press (original in Dutch, 1950).

Dimentberg, F. M. 1961. *Flexural Vibrations of Rotating Shafts*. London: Butterworths (original in Russian, 1959).

Di Stefano, J. J., A. R. Stubberud, and I. J. Williams. 1990. *Feedback and Control Systems*, 2nd ed. New York: McGraw-Hill/Schaum's.

Dittrich, W., and M. Reuter. 1994. *Classical and Quantum Dynamics, From Classical Paths to Path Integrals*, 2nd ed. New York: Springer (1992, 1st ed.).

Djukic, D. S. 1974. "Conservation Laws in Classical Mechanics for Quasi-Coordinates," *Archive for Rational Mechanics and Analysis*, **56**, 79–98.

Djukic, D. S. 1976. "A Variational Principle Involving a Conditional Extremum for the Hamel–Boltzmann Equations of Motion," *Acta Mechanica*, **25**, 105–110.

Djukic, D. S. 1981. "Adiabatic Invariants for Dynamical Systems with One Degree of Freedom," *International J. Nonlinear Mechanics*, **16**, 489–498.

Djukic, Dj. S., and T. M. Atanackovic. 1982. "Extremum Variational Principles in Classical Mechanics," *Acad. Sci. Turin, Proceedings of the IUTAM-ISIMM Symposium on Modern Developments in Analytical Dynamics* (June 7–11), 1982, vol. 2, 481–505. [Also in *Atti della Accademia delle Scienze di Torino, Supplemento*, **117**, 1983.]

Djukic, Dj., and B. Vujanovic. 1975. "On Some Geometrical Aspects of Classical Nonconservative Mechanics," *J. Mathematical Physics*, **16**, 2099–2102.

Dinh, V. N. 1981. "Carnot Theorem for a Rheonomic System," *Applied Mechanics* (transl. of Soviet/Ukrainian *Prikladnaya Mekhanika*), **17**, 96–101 (original in Russian, 120–125).

Dobronravov, V. V. 1945. "Integral Invariants of the Equations of Analytical Dynamics in Nonholonomic Coordinates," *Comptes Rendus* [*Doklady*] *de l'Académie des Sciences de l'URSS*, **46**(5), 179–181.

Dobronravov, V. V. 1948. "Analytical Dynamics in Nonholonomic (or Anholonomic) Coordinates" (in Russian), *Uchenye Zapiski Moskovskogo Gosudarstvennogo Universiteta*, **2**(122), 77–182.

Dobronravov, V. V. 1970. *Fundamentals (or Elements) of Nonholonomic System Mechanics* (in Russian). Moscow: Vischaya Shkola.

Dobronravov, V. V. 1976. *Principles of Analytical Mechanics* (in Russian). Moscow: Vischaya Shkola.

Dolaptschiew, B. 1966(a). "Ueber die Nielsensche Form der Gleichungen von Lagrange und deren Zusammenhang mit dem Prinzip von Jourdain und mit den nichtholonomen mechanischen Systemen," *ZAMM*, **46**, 351–355.

Dolaptschiew, B. 1966(b). "Ueber die verallgemeinerte Form der Lagrangeschen Gleichungen, welche auch die Behandlung von nicht-holonomen mechanischen Systemen gestattet," *ZAMP*, **17**, 443–449.

Dolaptschiew, B. 1967. "Ueber die 'verallgemeinerten' Gleichungen von Lagrange und deren Zusammenhang mit dem 'verallgemeinerten' Prinzip von d'Alembert," *J. für die Reine und Angewandte Mathematik*, **226**, 103–107.

Dolaptschiew, B. 1969. "Verwendung der einfachsten Gleichungen Tzenoffschen Typs (Nielsenschen Gleichungen) in der nicht-holonomen Dynamik," *ZAMM*, **49**, 179–184.

Drish, W. F., and W. J. Wild. 1983. "Numerical Solutions of Van der Pol's Equation," *Am. J. Physics*, **51**, 439–445.

Duan, L. 1989. "Conservation Laws of Nonholonomic Nonconservative Dynamical Systems," *Acta Mechanica Sinica*, **5**, 167–175.

Dugas, R. 1955. *A History of Mechanics*. New York: Central Book Co. (original in French; Editions du Griffon, Neuchatel, Switzerland. Reprinted 1988 by Dover, New York).

Duhem, P. 1911. *Traité d'Energetique*, ou de Thermodynamique Générale; tome II: *Dynamique Générale, Conductibilité de la Chaleur, Stabilité de l'Equilibre*. Paris: Gauthier-Villars.

Duhem, P. M. M. 1980. *The Evolution of Mechanics*. Alphen aan den Rijn, The Netherlands: Sijthoff and Noordhoff (original in French, 1903–1905).

Dühring, E. 1887. *Kritische Geschichte der allgemeinen Principien der Mechanik*, 3rd ed. Leipzig: Fues (1872, 1st ed; 1876, 2nd ed.; available through Dr. Martin Sändig, oHG, Wiesbaden, Germany).

Dysthe, K. B., and O. T. Gudmestad, 1975. "On Resonance and Stability of Conservative Systems," *J. Math. Physics*, **16**, 56–64.

Easthope, C. E. 1964. *Three Dimensional Dynamics*, 2nd ed. London: Butterworths.

Eberhard, O. v. 1922. "Entgegnung des Herrn O. v. Eberhard auf eine Kritik des Herrn R. v. Mises; gleichzeitig ein Beitrag zum Verständnis des d'Alembertschen Prinzips," *Zeitschrift für Technische Physik*, **3**, 28–31.

Ehrenfest, P. 1959. "Die Bewegung von Starrer Koerper in Flüssigkeiten und die Mechanik von Hertz," pp. 1–77 in *Paul Ehrenfest, Collected Scientific Papers*, edited by M. J. Klein. Amsterdam: North Holland (Ehrenfest's unpublished doctoral dissertation, 1904, Vienna).

Einstein, A. 1956. *The Meaning of Relativity*, 5th ed. Princeton, NJ: Princeton University Press.

El Naschie, M. S. 1990. *Stress, Stability and Chaos in Structural Engineering: An Energy Approach*. London: McGraw-Hill.

Elsgolts, L. 1970. *Differential Equations and the Calculus of Variations*. Moscow: Mir.

Epstein, S. T. 1968. "A Derivation of the Jacobi Identity in Classical Mechanics," *Am. J. Physics*, **36**, 759.

Erlichson, H. 1996. "Christiaan Huygens' Discovery of the Center of Oscillation Formula," *Am. J. Physics*, **64**, 571–574.

Essén, H. 1981. "The Cat Landing on its Feet Revisited, or Angular Momentum Conservation and Torque-Free Rotations of Nonrigid Mechanical Systems," *Am. J. Physics*, **49**, 756–758.

Essén, H. 1994. "On the Geometry of Nonholonomic Dynamics," *J. Appl. Mechanics* (ASME), **61**, 689–694.

Euler, L. 1736. *Mechanica Sive Motus Scientia, Analytice Exposita*, 2 vols. Petropoli (St. Petersburg). Academy of Sciences Press (Typographia Academiae Scientiarum). [Also in E.'s *Collected Works (Opera Omnia):* **II**, **1** and **2**, 1912. German translation *Mechanik oder analytische Darstellung der Wissenschaft von der Bewegung*, mit Anmerkungen und Erläuterungen, Herausgegeben von J. P. Wolfers, 3 Teile in 2 Bänden Greifswald, 1848–1853 (almost 1900 pages).]

Euler, L. 1750. "Découverte d'un nouveau principe de mécanique," *Mém. Acad. Sci. Berlin*, **6**, 185–217 (published in 1752). [Also in E.'s *Collected Works (Opera Omnia):* **II**, **5**, 81–108, 1957.]

Euler, L. 1758. "Du mouvement de rotation des corps solides autour d'un axe variable," *Mém. Acad. Sci. Berlin*, **14**, 154–193 (published in 1765). [Also in E.'s *Collected Works (Opera Omnia):* **II**, **8**, 200–235, 1964.]

Euler, L. 1760. "Du mouvement d'un corps solide quelconque lorsqu'il tourne autour d'un axe mobile," *Mém. Acad. Sci. Berlin*, **16**, 176–227 (published in 1767). [Also in E.'s *Collected Works (Opera Omnia):* **II**, **8**, 313–356, 1964.]

Euler, L. 1765. *Theoria Motus Corporum Solidorum Seu Rigidorum*, Ex primis nostrae cognitionis principiis stabilita et ad omnes motus, qui in huiusmodi corpora cadere possunt, accomodata. Rostock and Greifswald (1790, 2nd ed). [Also in E.'s *Collected Works (Opera Omnia):* **II**, **3**, 1948 and **4**, 1950. German translation as *Theorie der Bewegung fester oder starrer Koerper*, Greifswald, 1853.]

Euler, L. 1775. "Nova methodus motum corporum rigidorum determinandi," *Novi Comment. Acad. Sci. Petropolitanae*, **20**, 208–238 (published in 1776). [Also in E.'s *Collected Works (Opera Omnia):* **II**, **9**, 99–125, 1968.]

Euler, Leonhard, 1958. Volume on the occasion of his 250th birthday (in Russian, with German summaries), edited by M. A. Lawrentjew, A. P. Juschkewitsch, and A. T. Grigorian. Moscow: Academy of Sciences of the U.S.S.R.

Euler, Leonhard 1707–1783, 1983. Beiträge zu Leben und Werk; Gedenkband des Kantons Basel Stadt, edited by J. J. Burkhardt, E. A. Fellmann, and W. Habicht. Basel: Birkhäuser.

Falk, G. 1966. *Theoretische Physik, auf die Grundlage einer allgemeinen Dynamik*, Band I: Elementare Punktmechanik, Band Ia: Aufgaben und Ergänzungen zur Punktmechanik. Berlin: Springer.

Favre, H. 1946. *Cours de Mécanique*, vol. 1: *Statique*; vol. 2: *Dynamique*; vol. 3: *Elasticité, Hydrodynamique*. Paris: Dunod, and Zurich: Leeman.

Featherstone, R. 1987. *Robot Dynamics Algorithms*. Norwell, MA: Kluwer.

Federhofer, K. 1928. *Graphische Kinematik und Kinetostatik des Starren räumlichen Systems*. Berlin: Springer.

Federhofer, K. 1932. *Graphische Kinematik und Kinetostatik*. Berlin: Springer.

Ferrarese, G. 1980. *Lezioni di Meccanica Razionale*, 2 vols. Bologna: Pitagora.

Ferrell, T. L. 1971. "Hamilton-Jacobi Perturbation Theory," *Am. J. Physics*, **39**, 622–627.

Ferrers, N. M. 1873. "Extension of Lagrange's Equations," *Quart. J. Pure and Appl. Mathematics*, **12**, 1–4.

Fetter, A. L. and J. D. Walecka. 1980. *Theoretical Mechanics of Particles and Continua*. New York: McGraw-Hill.

Finzi, B. 1949. *Principi Variazionali*. Milano, Tamburini.

Fischer, U. 1972. "Übergangsbedingungen für Stossvorgänge in mechanischen Systemen mit endlich vielen Freiheitsgraden," *Wiss. Z. Tech. Hochschule Magdeburg*, **16**, 323–325.

Fischer, U. 1987. "Stösse in Mehrkörpersysteme," *Proc. 7th World Congress on Theory of Machines and Mechanisms* (Sevilla, Spain, Sept. 17–22, 1987). Oxford: Pergamon.

Fischer, U., and K. Henning. 1982. "Formalisierte Berechnung der Geschwindigkeitssprüngen und Impulse bei Stossvorgängen in Starrkörper- und hybriden Mehrkörpersystemen," *Technische Mechanik*, **3**, 78–80.

Fischer, U., and W. Stephan. 1972. *Prinzipien und Methoden der Dynamik*. Leipzig: VEB Fachbuchverlag.

Fischer, U., and W. Stephan. 1984. *Mechanische Schwingungen*, 2nd ed. Leipzig: VEB Fachbuchverlag.

Föppl, A. *Vorlesungen über technische Mechanik*, vol. 1: *Einführung in die technische Mechanik* (1898, 1st ed.); vol. 4: *Dynamik* (1899, 1st ed.); vol. 6: *Die wichtigsten Lehren der höheren Dynamik* (1910, 1st ed.); several subsequent editions: 1898–1940s. Leipzig: Teubner.

Forbat, N. 1966. *Analytische Mechanik der Schwingungen*. Berlin: VEB Deutscher Verlag der Wissenschaften.

Forbes, G. W. 1991. "On Variational Problems in Parametric Form," *Am. J. Physics*, **59**, 1130–1140.

Förster, E. 1903. "Zum Ostwaldschen Axiom der Mechanik," *Zeitschrift für Mathematik und Physik*, **49**, 84–89.

Forsyth, A. R. 1885. *A Treatise on Differential Equations*. London: Macmillan (1954, 6th ed., posthumous).

Forsyth, A. R. 1890. *Theory of Differential Equations*, vol. 1: *Exact Equations and Pfaff's Problem*. Cambridge, U.K.: Cambridge University Press (reprinted 1959 by Dover, New York; original 6-vol. work bound in 3 vols.).

Fourier, J. B. 1798. "Mémoire sur la statique, contenant la démonstration du principe des vitesses virtuelles, et la theorie des moments," *Journal de l'Ecole Polytechnique*, Cahier **5**, Tome **II**, 20–60. [Also in Fourier's *Collected Works*.]

Fox, C. 1950. *An Introduction to the Calculus of Variations*. Oxford, U.K.: Oxford University Press (reprinted 1987, by Dover, New York; from corrected printing of 1963).

Fox, E. A. 1967. *Mechanics*. New York: Harper and Row.

Fradlin, B. N. 1961. "Petr Vassilievich Voronets—one of the Founders of Nonholonomic Mechanics" (in Russian), *History of Physical-Mathematical Sciences, Reports of the Institute of the History of Natural Sciences and Technology, USSR Academy of Sciences, Moscow*, **43**, 422–469.

Fradlin, B. N. 1965. "Analytical Dynamics of Mechanical Systems with Nonlinear Nonholonomic Constraints" (in Russian), *Applied Mechanics*, **1**(7), 21–27.

Fradlin, B. N., and L. D. Roshchupkin 1973(a). "The Equations of Dynamics, Review," *Applied Mechanics*, **9**, 1–7 (original in Russian, 3–9).

Fradlin, B. N., and L. D. Roshchupkin. 1973(b). "On the Dolapchiev–Manzheron–Tsenov Equations in the Case of Natural Systems," *Applied Mechanics*, **9**, 251–254 (original in Russian, 21–25).

Fradlin, B. N., and L. D. Roshchupkin. 1976. "Theory of Stability and Theory of Small Oscillations of Nonholonomic Mechanical Systems," *Applied Mechanics* (transl. of Soviet/Ukrainian *Prikladnaya Mekhanika*), **12**, 329–335 (original in Russian, 3–11).

Frank, P. 1927. "Analytische Mechanik," pp. 1–176 in vol. 2, of *Die Differential-und Integralgleichungen der Mechanik und Physik*, edited by P. Frank and R. v. Mises. Braunschweig: Vieweg (1935, 2nd ed., pp. 44–239, reprinted by Dover, New York).

Frank, P. 1957. *Philosophy of Science, The Link Between Science and Philosophy*. Englewood Cliffs, NJ: Prentice Hall.

Frankfort, H., H. A. Frankfort, J. A. Wilson, and T. Jacobsen. 1946. *Before Philosophy, The Intellectual Adventure of Ancient Man*. Chicago: University of Chicago Press (reprinted 1951, 1954, 1959, 1961, 1963, 1964, 1967, 1968, 1971, 1972 by Pelican and Penguin Books).

Fraser, C. "J. L. Lagrange's Early Contributions to the Principles and Methods of Mechanics," *Archive for History of Exact Sciences*, **28**, 197–241 (1983); **32**, 151–191 (1985).

Fraser, C. 1985. "D'Alembert's Principle: The Original Formulation and Application in Jean d'Alembert's *Traité de Dynamique* (1743)," *Centaurus*, **28**, part 1: 31–61; part 2: 145–159.

French, A. P. 1971. *Newtonian Mechanics*. New York: Norton.

Frenkel, J. I. 1940. *Course of Theoretical Mechanics, based on the Elements of Vector and Tensor Analysis* (in Russian). Leningrad: Gos. izd-vo tekhn.-theoret. lit-ry (State Publishers of Technical & Theoretical Literature).

Frobenius, F. G. 1877. "Über das Pfaffsche Problem," *J. für die Reine und Angewandte Mathematik (Crelle's J.)*, **82**, 230–315. [Also in his *Collected Works*, Springer, 1968; and, secondarily, "Ueber homogene totale Differentialgleichungen," *J. für die Reine und Angewandte Mathematik (Crelle's J.)*, **86**, 1–18 (1879).]

Fues, E. 1927. "Störungrechnung," pp. 131–177 in vol. 5 of *Handbuch der Physik*. Berlin: Springer.

Fufaev, N. A. 1985. "Theory of the Motion of Systems with Rolling," *PMM*, **49**, 41–49 (original in Russian, **1**, 56–65).

Fufaev, N. A. 1990. "About an Example of a System with a Nonholonomic Constraint of Second Order," *ZAMM*, **70**, 593–594.

Funk, P. 1962. *Variationsrechnung und ihre Anwendung in der Technik*. Berlin: Springer.

Galiulin, A. S. 1984. *Inverse Problems in Dynamics*. Moscow: Mir (original in Russian, 1981).

Gallavotti, G. 1983. *The Elements of Mechanics*. New York: Springer.

Gantmacher, F. 1970. *Lectures in Analytical Mechanics*. Moscow: Mir (original in Russian, 1966).

Garfinkel, B. 1966. "The Lagrange–Hamilton–Jacobi Mechanics," part 1, pp. 40–76 of *Lectures in Applied Mathematics*, vol. 5 of *Space Mathematics*, edited by J. B. Rosser. Providence, RI: American Mathematical Society.

Garnier, R. *Cours de Cinématique*, tomes I: *Cinématique du Point et du Solide. Composition des Mouvements*, 1960, 4th ed.; II: *Roulement et Viration. La Formule du Savary et son Extension à l'Espace*, 1956, 3rd ed.; III: *Géometrie et Cinématique Cayleyennes*, 1951, 1st ed. Paris: Gauthier-Villars.

Gascoigne, R. M. 1984. *A Historical Catalogue of Scientists and Scientific Books, From the Earliest Times to the Close of the Nineteenth Century*. New York: Garland Publishing.

Ge, Z.-M. 1984. "The Equations of Motion of Nonlinear Nonholonomic Variable Mass System with Applications, *J. Appl. Mechanics* (ASME), **51**, 435–437.

Geiringer, H. 1942. *Geometrical Foundations of Mechanics* (mimeographed notes). Providence, RI: Brown University.

Gelfand, I. M., and S. V. Fomin. 1963. *Calculus of Variations*. Englewood Cliffs, NJ: Prentice Hall (revised English translation from Russian original, 1961).

Gericke, H. 1994. *Mathematik in Antike und Orient*, and *Mathematik im Abendland, Von den römischen Feldmessern bis zu Descartes*, 3rd ed. Wiesbaden, Germany: Fourier Verlag.

Germain, P. 1962. *Mécanique des Milieux Continus*. Paris: Masson.

Gernet, M. M. 1981. *Course of Theoretical Mechanics* (in Russian). Moscow: Vischaya Shkola.

Geronimus, J. L. 1973. *Theoretical Mechanics* (in Russian). Moscow: Nauka.

Ghori, Q. K. 1970. "Virtual Displacements of Nonholonomic Systems with Arbitrary Constraints," *ZAMM*, **52**, 123–124.

Ghori, Q. K. 1977. "Conservation Laws for Dynamical Systems in Poincaré–Cetaev Variables," *Arch. Rat. Mech. Analysis*, **64**, 327–337.

Ghori, Q. K., and N. Ahmed. 1994. "Hamilton's Principle for Nonholonomic Systems," *ZAMM*, **74**, 137–140.

Ghori, Q. K., and M. Hussain. 1973. "Poincaré's Equations for Nonholonomic Dynamical Systems," *ZAMM*, **53**, 391–396.

Ghori, Q. K., and M. Hussain. 1974. "Poincaré's Equations for a Nonholonomic System with Variable Mass," *Arch. Rat. Mech. Analysis*, **56**, 70–78.

Giacaglia, G. E. O. 1972. *Perturbation Methods in Non-Linear Systems*. New York: Springer.

Gibbs, J. W. 1879. "On the Fundamental Formulae of Dynamics," *Am. J. Mathematics*, **2**, 49–64. (Also in G.'s 1928 *Collected Works*, vol. 2, part 2, pp. 1–15, New York: Longmans and Green.)

Gilmartin, H., A. Klein, and C.-T Li. 1979. "Application of Hamilton's Principle to the Study of the Anharmonic Oscillator in Classical Mechanics," *Am. J. Physics*, **47**, 636–639.

Ginsberg, J. H. 1995. *Advanced Engineering Dynamics*, 2nd ed. Cambridge, U.K.: Cambridge University Press.

Girtler, R. 1928. "Ueber die Berechnung kinetostatisch unbestimmter Systeme," *Wiener Sitzungsberichte, Math.-naturwiss. Klasse*, Abteilung **IIa, 137**, 51–69.

Glocker, C. 1995. *Dynamik von Starrkörpersystemen mit Reibung und Stößen*. Fortschritt Berichte, Reiche 18: Mechanik/Bruchmechanik, Nr. 182, Düsseldorf: VDI Verlag.

Golab, S. 1974. *Tensor Calculus*. Amsterdam: Elsevier.

Göldner, H. 1984. "Leonhard Euler—seine Beiträge zur Festkörpermechanik und sein Wirken an der Berliner Akademie," *Technische Mechanik*, **5**, 67–75.

Goldsmith, W. 1960. *Impact, The Theory and Physical Behavior of Colliding Solids*. London: Arnold.

Goldstein, H. 1980. *Classical Mechanics*, 2nd ed. Reading, MA: Addison-Wesley.

Golomb, M. 1961. *Lectures on Theoretical Mechanics* (mimeographed notes prepared by I. Marx). West Lafayette, IN: Purdue University, School of Mathematics.

Gontier, G. 1969. *Mécanique des Milieux Déformables, Principes et Théorèmes Généraux*. Paris: Dunod.

Goodbody, A. M. 1982. *Cartesian Tensors, With Applications to Mechanics, Fluid Mechanics and Elasticity*. Chichester, U.K.: Ellis Horwood/Halsted Press/Wiley.

Grammel, R. 1927. "Kinetik der Massenpunkte," pp. 305–372 in vol. 5 of *Handbuch der Physik*. Berlin: Springer.

Grammel, R. 1950. *Der Kreisel, Seine Theorie und Anwendungen*, 2 vols, 2nd ed. Berlin: Springer.

Gray, A. (1908/1909). "On Lagrange's Equations of Motion, and on Elementary Solutions of Gyrostatic Problems," *Proc. Roy. Soc. Edinburgh*, **29**, 327–348.

Gray, A. 1918. *A Treatise on Gyrostatics and Rotational Motion*. London: Macmillan (reprinted 1959 by Dover, New York).

Gray, C. G., G. Karl, and V. A. Novikov. 1996(a). "Direct Use of Variational Principles as an Approximation Technique in Classical Mechanics," *Am. J. Physics*, **64**, 1177–1184.

Gray, C. G., G. Karl, and V. A. Novikov. 1996(b). "The Four Variational Principles of Mechanics," *Annals of Physics*, **251**, 1–25.

Greenwood, D. T. 1977. *Classical Dynamics*. Englewood Cliffs, NJ: Prentice Hall (reprinted 1997 by Dover, New York).

Greenwood, D. T. 1988. *Principles of Dynamics*, 2nd ed. Englewood Cliffs, NJ: Prentice Hall.

Greenwood, D. T. 2000. *Advanced Topics in Dynamics* (Lecture Notes). Ann Arbor, MI: University of Michigan, Dept. of Aerospace Eng (to appear in book form in near future).

Grioli, G. 1970. "Generalized Precessions," *Revue Roumaine des Sciences Techniques, Mécanique Appliquée*, **15**, 249–255.

Groesberg, S. W. 1968. *Advanced Dynamics*. New York: Wiley.

Groves, G. W. 1967. "Acceleration Referred to Moving Curvilinear Coordinates," *Am. J. Physics*, **35**, 927–929.

Grubin, C. 1962. "Vector Representation of Rigid Body Rotation," *Am. J. Physics*, **30**, 416–417.

Guen, F. 1968. "On the Equations of Motion of Nonholonomic Mechanical Systems in Poincaré–Chetaev Variables," *PMM*, **31**, 274–281 (original in Russian, 253–259).

Guen, F. 1969. "One Form of the Equations of Motion of Mechanical Systems," *PMM*, **33**, 386–392 (original in Russian, 397–402).

Guen, F. 1982. "On the Theory of Ignorable Displacements for Generalized Poincaré–Chetaev Equations," *PMM*, **45**, 343–350 (original in Russian, 471–480).

Gugino, E. 1930. "Sul problema dinamico di un quasivoglia sistema vincolato ridotto all'analogo problema relativo ad un sistema libero," *Atti Accad. Naz. Lincei, Rendiconti Cl. Fis. Mat. Nat.*, Series 6, **12**, 307–312.

Gugino, E. 1936(a). "Deduzione unitaria delle equazione dinamiche del Maggi et dell' Appell," *Atti Accad. Naz. Lincei, Rendiconti Cl. Fis. Mat. Nat.*, Series 6, **23**, 406–413.

Gugino, E. 1936(b). "Sulle equazioni dinamiche di Eulero–Lagrange secondo G. Hamel," *Atti Accad. Naz. Lincei, Rendiconti Cl. Fis. Mat. Nat.*, Series 6, **23**, 413–421.

Guldberg, A. 1927. "Partielle und totale Differentialgleichungen," *Erster Band, Zweiter Teilband*, pp. 561–578 of *Repertorium der Höheren Mathematik*, 2nd ed. Leipzig and Berlin: Teubner.

Gupta, K. C. 1988. *Classical Mechanics of Particles and Rigid Bodies*. New York: Wiley.

Gutowski, R. 1971. *Mechanika Analityczna*. Warsaw: Panstwowe Wydawnictwo Naukowe (PWN).

Gutowski, R. 1992. "Analytical Mechanics, " pp. 1–119 of *Foundations of Mechanics*, edited by H. Zorski. Warszaw: Polish Scientific Publishers (PWN)/Elsevier (original in Polish, 1985).

Gutzwiller, M. C. 1990. *Chaos in Classical and Quantum Mechanics*. New York: Springer.

Haar, D. Ter. 1971. *Elements of Hamiltonian Mechanics*, 2nd ed. Oxford: Pergamon (unaltered photographic reproduction of the 2nd ed. of 1964, by North Holland, Amsterdam).

Haas, A. E. 1914. *Die Grundgleichungen der Mechanik dargestellt auf Grund der geschichtlichen Entwicklung*. Leipzig: Veit.

Hadamard, J. 1895. "Sur les Mouvements de Roulement," *Mémoires de la Societé des Sciences Physiques et Naturelles de Bordeaux*, **5**, 397–417. [Also in Appell, 1899(a).]

Hagedorn, P. 1970. "Zur Umkehrung des Satzes von Lagrange über die Stabilität," *ZAMP*, **21**, 841–845.

Hagedorn, P. 1971. "Die Umkehrung der Stabilitätssätze von Lagrange–Dirichlet und Routh," *Arch. for Rat. Mech. and Analysis*, **42**, 281–316.

Hagedorn, P. 1975. "Ueber die Instabilität Konservativer Systeme mit Gyroskopischen Kräften," *Arch. Rat. Mech. and Analysis*, **58**, 1–9.

Hagedorn, P. 1978. *Nichtlineare Schwingungen*. Wiesbaden: VAG, Akademische Verlagsgesellschaft (English translation also available from Clarendon Press, Oxford, U.K. 1981).

Hagihara, Y. 1970. *Dynamical Principles and Transformation Theory*, vol. 1 of (2-vol. treatise on) *Celestial Mechanics*. Cambridge, MA: MIT Press.

Halfman, R. I. 1962. *Dynamics*, vol. 1: *Particles, Rigid Bodies, and Systems*; vol. 2: *Systems, Variational Methods, and Relativity*. Reading, MA: Addison-Wesley.

Hamel, G. 1904(a). "Die Lagrange–Eulersche Gleichungen der Mechanik," *Zeitschrift für Mathematik und Physik*, **50**, 1–57.

Hamel, G. 1904(b). "Ueber die virtuellen Verschiebungen in der Mechanik," *Mathematische Annalen*, **59**, 416–434.

Hamel, G. 1909. "Ueber die Grundlagen der Mechanik," *Mathematische Annalen*, **66**, 350–397.

Hamel, G. 1917. "Ueber ein Prinzip der Befreiung bei Lagrange," *Jahresbericht der Deutschen Mathematiker-Vereinigung*, **25**, 60–65.

Hamel, G. 1922(a). "Zum Verständnis des d'Alembertschen Prinzips," *Zeitschrift für Technische Physik*, **3**, 181–182.

Hamel, G. 1922(b). *Elementare Mechanik*, 2nd ed. Leipzig: Teubner (1912 1st ed.; reprinted 1965 by Johnson Reprint Co., New York).

Hamel, G. 1924. "Ueber Nichtholonome Systeme," *Mathematische Annalen*, **92**, 33–41.

Hamel, G. 1926. "Ueber die Mechanik der Drähte und Seile," *Sitzungsberichte der Berliner Mathematischen Gesellschaft*, **25**, 3–8.

Hamel, G. 1927. "Die Axiome der Mechanik," pp. 1–42 in vol. 5, of *Handbuch der Physik*, Berlin: Springer.

Hamel, G. 1935. "Das Hamiltonsche Prinzip bei nichtholonomen Systemen," *Mathematische Annalen*, **111**, 94–97.

Hamel, G. 1936. "Joseph Louis Lagrange, Zur Zweihundertjahrfeier seines Geburtstages," *Die Naturwissenschaften*, **24**, 51–53.

Hamel, G. 1938. "Nichtholonome Systeme höherer Art," *Sitzungsberichte der Berliner Mathematischen Gesellschaft*, **37**, 41–52.

Hamel, G. 1949. *Theoretische Mechanik, Eine einheitliche Einführung in die gesamte Mechanik*. Berlin: Springer.

Hamilton, E. 1948. *The Greek Way*, 3rd ed. New York: Norton.

Hammond, P. 1970. *Energy Methods in Electromagnetism*. Oxford, U.K.: Clarendon Press.

Hand, L. N., and J. D. Finch. 1998. *Analytical Mechanics*. Cambridge, U.K.: Cambridge University Press.

Hankins, T. L. 1970. *Jean d'Alembert. Science and the Enlightenment*. Oxford, U.K.: Clarendon Press.

Hankins, T. L. 1985. *Science and the Enlightenment*. Cambridge, U.K.: Cambridge University Press.

Harman, P. M. 1982. *Energy, Force, and Matter, The Conceptual Development of Nineteenth-Century Physics*. Cambridge, U.K.: Cambridge University Press.

Harrison, H. R., and T. Nettleton. 1997. *Advanced Engineering Dynamics*. London: Arnold.

Hartman, P. 1964. *Ordinary Differential Equations*. New York: Wiley.

Haug, E. J. 1989. *Computer Aided Kinematics and Dynamics*, vol. I: *Basic Methods*. Boston, MA: Allyn and Bacon.

Haug, E. J. 1992. *Intermediate Dynamics*. Englewood Cliffs, NJ: Prentice Hall.

Haug, E. J., S. S. Kim, and F. F. Tsai. 1992. *Computer Aided Kinematics and Dynamics*, vol. II: *Advanced Methods*. Englewood Cliffs, NJ: Prentice Hall. (Announced but seemingly never published.)

Heading, J. 1958. *Matrix Theory for Physicists*. London: Longmans, Green and Co.

Heil, M. and F. Kitzka. 1984. *Grundkurs Theoretische Mechanik*. Stuttgart: Teubner.

Heinz, C. 1970. "Das d'Alembertsche Prinzip in der Kontinuumsmechanik," *Acta Mechanica*, **10**, 111–129.

Heisenberg, W. 1930. *The Physical Principles of the Quantum Theory*. Chicago: University of Chicago Press (original in German; reprinted 1949 by Dover, New York).

Helleman, R. H. G. 1978. "Variational Solutions of Non-Integrable Systems," pp. 264–285 in *Topics in Nonlinear Dynamics* (A Tribute to Sir Edward Bullard), edited by S. Jorna. New York: American Institute of Physics, Conference Proceedings #46.

Hellinger, E. 1914. "Die allgemeinen Ansätze der Mechanik der Kontinua," art. 30, pp. 601–694 in vol. 4 of *Encyklopädie der Mathematischen Wissenschaften*. Leipzig: Teubner.

Helmholtz, H. v. 1898. *Vorlesungen über theoretische Physik*; Band I, Abtheilung 2: (Vorlesungen über) *Die Dynamik Discreter Massenpunkte*, edited by O. Krigar-Menzel. Leipzig: J. A. Barth.

Henneberg, L. 1903. "Die graphische Statik der starren Koerper," art. 5, pp. 345–434 in vol. 4 of *Encyklopädie der Mathematischen Wissenschaften*. Leipzig: Teubner.

Hertz, H. 1899. *The Principles of Mechanics*, presented in new form. London: Macmillan (original in German, 1894; English translation reprinted in paperback, 1956, by Dover, New York).

Hestenes, D. 1986. *New Foundations for Classical Mechanics*. Dordrecht: Reidel-Kluwer.

Heun, K. 1901. "Die kinetischen Probleme der wissenschaftlichen Technik," *Jahresbericht der Deutschen Mathematiker-Vereinigung*, **9**(2), 1–123.

Heun, K. 1902(a). "Die Bedeutung des D'Alembertschen Prinzipes für starre Systeme und Gelenkmechanismen," *Archiv der Mathematik und Physik*, Dritte Reihe, **2**, 57–77 and 298–326.

Heun, K. 1902(b). "Review of A. Foeppl's book: Vorlesungen über Technische Mechanik," *Zeitschrift für Mathematik und Physik* (Bücherschau), **47**, 270–279.

Heun, K. 1902(c) "Ueber die Hertzsche Mechanik," *Sitzungsberichte der Berliner Mathematischen Gesellschaft*, **1**, 12–16.

Heun, K. 1902(d). *Formeln und Lehrsätze der Allgemeinen Mechanik*. Leipzig: Göschen.

Heun, K. 1903. "Ueber die Einwirkung der Technik auf die Entwicklung der theoretischen Mechanik," *Jahresbericht der Deutschen Mathematiker-Vereinigung*, **12**, 389–398.

Heun, K. 1906. *Lehrbuch der Mechanik*, I. Teil: *Kinematik*. Leipzig: Göschen.

Heun, K. 1908. "Die Grundgleichungen der Kinetostatik der Körperketten mit Anwendungen auf die Mechanik der Maschinen," *Zeitschrift für Mathematik und Physik*, **56**, 38–77.

Heun, K. 1914. "Ansätze und allgemeine Methoden der Systemmechanik," art. 11, pp. 357–504 in vol. 4, of *Encyklopädie der Mathematischen Wissenschaften*. Leipzig: Teubner (article completed in April 1913).

Heun, K. 1925. "Grundlagen der modernen Mechanik. Vorgänge und Zustände mit linearem und mehrdimensionalem Grundfeld," pp. 67–95 of *Festschrift zur Hundertjahrfeier der Technischen Hochschule Karlsruhe*.

Hilborn, R. C. 1972. "A Note on Euler Angle Rotations," *Am. J. Physics*, **40**, 1036–1037.

Hill, E. L. 1945. "Rotations of a Rigid Body About a Fixed Point," *Am. J. Physics*, **13**, 137–140.

Hill, E. L. 1951. "Hamilton's Principle and the Conservation Theorems of Mathematical Physics," *Reviews of Modern Physics*, **23**, 253–260.

Hill, R. 1964. *Principles of Dynamics*. New York: Macmillan (and Pergamon).

Hiller, M. 1983. *Mechanische Systeme*. Berlin: Springer.

Holden, J. T., and A. C. King. 1970. "The Dynamics of a Ball Rolling on a Rotating Plane," *ZAMM*, **70**, 353–355.

Hölder, E. 1939. "Über die explizite Form der dynamischen Gleichungen für die Bewegung eines starren Körpers relativ zu einem geführten Bezugssystem," *ZAMM*, **19**, 166–176.

Hölder, O. 1896. "Über die Principien von Hamilton und Maupertuis," *Nachrichten von der Königl. Gesellschaft der Wissenschaften zu Göttingen, Math.-Phys. Kl.*, Heft **2**, 122–157.

Honerkamp, J., and H. Römer. 1993. *Theoretical Physics, A Classical Approach*. New York: Springer (original in German, 1986, 1989).

Hoppe, E. 1926(a). "Geschichte der Physik," pp. 1–179 in vol. 1 of *Handbuch der Physik*. Berlin: Springer.

Hoppe, E. 1926(b). *Geschichte der Physik*. Braunschweig: Fr. Vieweg and Sohn (reprinted 1965 by Johnson Reprint Corporation, New York and London).

Horák, Z. 1931. "Théorie Générale du Choc dans les Systèmes Matériels," *Journal de l'Ecole Polytechnique*, Serie II, **28**(2), 15–64.

Horn, J. 1905. "Weitere Beiträge zur Theorie der kleinen Schwingungen," *Zeitschrift für Mathematik und Physik*, **52**, 1–43; also: 1906, **53**, 370–402 (same title).

Housner, G. W., and D. E. Hudson. 1959. *Applied Mechanics*, 2 vols. (Statics and Dynamics), 2nd ed. New York: Van Nostrand and Reinhold.

Howard, J. E. 1967. "Singular Cases of the Inertia Tensor," *Am. J. Physics*, **35**, 281–282.

Høyrup, J. 1988. "Jordanus de Nemore, 13th Century Mathematical Innovator: An Essay on Intellectual Context, Achievement, and Failure," *Archive for History of Exact Sciences*, **38**, 307–363.

Hughes, P. C. 1986. *Spacecraft Attitude Dynamics*. New York: Wiley.

Hund, F. 1972. *Geschichte der Physikalischen Begriffe*. Mannheim: Bibliographisches Institut.

Hunt, K. H. 1974. "Mechanics Pure and Simple (Applied and Elaborated in Design)," *The South African Mechanical Engineer*, **24**, 218–230.

Hunt, K. H. 1978. *Kinematic Geometry of Mechanisms*. Oxford, U.K.: Clarendon Press.

Hunter, S. C. 1983. *Mechanics of Continuous Media*, 2nd ed. New York: Horwood/Halsted/Wiley.

Hurwitz, A. 1895. "Ueber die Bedingungen, unter welchen eine Gleichung nur Wurzeln mit negativen reellen Theilen besitzt," *Mathematische Annalen*, **46**, 273–284 (English translation 1964, pp. 70–82 in *Selected Papers on Mathematical Trends in Control Theory*, edited by R. Bellmann and R. Kalaba. New York: Dover).

Huseyin, K. 1978. *Vibrations and Stabllity of Multiple Parameter Systems*. Amsterdam: Noordhoff.

Hussain, M. 1982. "Conservation Laws for a Dynamical System in Group Variables," *ZAMM*, **62**, 441–446.

Hussein, M. S., J. G. Pereira, V. Stojanoff, and H. Takai. 1980. "The Sufficient Condition for an Extremum in the Classical Action Integral as an Eigenvalue Problem," *Am. J. Physics*, **48**, 767–770.

Huston, R. L. 1990. *Multibody Dynamics*. Boston, MA: Butterworth-Heinemann.

Huston, R. L., and C. E. Passerello. 1976. "Nonholonomic Systems with Non-linear Constraint Equations," *Int. J. Non-linear Mechanics*, **11**, 331–336.

Ishlinsky, A. 1984. *Orientation, Gyroscopes et Navigation par Inertie*, vol. 1: *Cinématique des Mobiles Gyrostabilisés*; vol. 2: *Systèmes Gyroscopiques de la Navigation par Inertie*. Moscow: Mir (original in Russian, 1976).

Ishlinsky, A. Iu. 1987. *Classical Mechanics and Inertial Forces* (in Russian). Moscow: Nauka.

Isidori, A. 1989. *Nonlinear Control Systems, An Introduction*, 2nd ed. New York: Springer.

Ispolov, I. G. 1983. "On Appell's Equations in Nonlinear Quasi-Accelerations and Quasi-Velocities," *PMM*, **46**, 399–400 (original in Russian, 507–508).

Ivanov, A. P., and A. P. Markeev. 1984. "The Dynamics of Systems with Unilateral Constraints," *PMM*, **48**, 448–451 (original in Russian, 632–636).

Jacobi, C. G. J. 1866. *Vorlesungen über Dynamik*, nebst fünf hinterlassenen Abhandlungen desselben (lectures delivered during the Winter semester of 1842–1843, at the University of Königsberg), edited by A. Clebsch, and also by C. W. Borchardt. Berlin: G. Reimer.

Jacobi, C. G. J. 1996. *Vorlesungen über analytische Mechanik* (lectures delivered in Berlin during the Winter semester of 1847–1848). Nach einer Mitschrift von W. Scheibner; herausgegeben von H. Pulte. Published for the Deutsche Mathematiker-Vereinigung (in the series Dokumente zur Geschichte der Mathematik, Band 8). Braunschweig/Wiesbaden: F. Vieweg.

Jammer, M. *Concepts of Force, A Study in the Foundations of Dynamics*. Cambridge, MA: Harvard University Press (1957); and New York: Harper and Row (1962). [Also *Concepts of Space* (1960) and *Concepts of Mass* (1964); both Harper and Row.]

Jankowski, K. 1988. "Formulation of Equations of Motion in Quasi-Velocities for Constrained Mechanical Systems," *Mechanism and Machine Theory*, **23**, 441–447.

Jankowski, K. 1989(a). "Dynamics of Mechanical Systems with Nonholonomic Constraints of Higher Order," *Modelling, Simulation and Control*, **B**, **25**, 47–63.

Jankowski, K. 1989(b). "Dynamics of Controlled Mechanical Systems with Material and Program Constraints; I. Theory, II. Methods of Solution, III. Illustrative Examples," *Mechanics and Machine Theory*, **24**, 175–179, 181–185, 187–193.

Jaunzemis, W. 1967. *Continuum Mechanics*. New York: Macmillan Co.

Jeans, J. H. 1947. *The Growth of Physical Science*. Cambridge, U.K.: Cambridge University Press.

Jeffreys, H. 1954. "What is Hamilton's Principle?" *Quart. J. Mech. and Appl. Math.*, **7**, 335–337.

Jehle, H. and J. H. Cahn. 1953. "Anharmonic Resonance," *Am. J. Physics*, **21**, 526–531.

Jellett, J. H. 1872. *A Treatise On The Theory of Friction.* Dublin: Hodges, Foster, and Co.; London: Macmillan and Co.

Johnsen, L. 1936. "Allgemeine Quasikoordinaten," *Avh. Norske Vid. Akad. Oslo*, **11**, 1–13.

Johnsen, L. 1937(a). "Die virtuellen Verschiebungen der nicht holonomen Systeme und das d'Alembertsche Prinzip," *Avh. Norske Vid. Akad. Oslo*, **10**, 1–10.

Johnsen, L. 1937(b). "Sur la Reduction au Nombre Minimum des Équations du Mouvement d'un Systeme Non-holonome," *Avh. Norske Vid. Akad. Oslo*, **11**, 1–14.

Johnsen, L. 1938. "Sur la Déviation Non-holonome," *Avh. Norske Vid. Akad. Oslo*, **3**, 1–13.

Johnsen, L. 1939. "Calcul Symbolique des Pseudo-coordoneés," *Arch. for Math. og Naturvidenskab*, **42**(3), 1–12.

Johnsen, L. 1941. "Dynamique Génerale des Systèmes Non-Holonomes," *Skrifter Utgitt Av Det Norske Videnskaps Akademi i Oslo, I. Mat. Naturv. Klasse*, **4**, 1–75.

Jørgensen, A. E., and W. Kliem, 1975. "About the Torques of Fictitious Forces," *ZAMM*, **55**, 534–535.

José, J. V., and E. J. Saletan. 1998. *Classical Dynamics; A Contemporary Approach.* Cambridge, U.K.: Cambridge University Press.

Jouguet, J. C. E. *Lectures de Mécanique, La Mécanique enseignée par les auteurs originaux.* Paris: Gauthier-Villars; 1908, vol. 1; 1909, vol. 2 (reprinted by Johnson Reprint Corporation, New York, year unspecified).

Joukovsky (or Joukowski, or Zhukovskii), N. E. 1937. "On the Stability of Motion," pp. 110–208 in vol. 1 of Joukovsky's *Collected Papers*, Moscow and Leningrad: United Scientific Technical Publishing House (ONTI).

Jourdain, P. E. B. 1909. "Note on an Analogue of Gauss' Principle of Least Constraint," *Quart. J. Pure and Appl. Math.*, **40**, 153–157.

Jourdain, P. E. B. 1913. *The Principle of Least Action.* Chicago: Open Court Publishing Co. (reprinted from the *Monist* of April and July 1912, and April 1913). [Also in the collection *The Conservation of Energy and the Principle of Least Action*, edited by I. Bernard Cohen. New York: Arno Press, 1981.]

Jung, G. 1901–1908. "Geometrie der Massen," art. 4, pp. 279–344 in vol. 4, of *Encyklopädie der Mathematischen Wissenschaften.* Leipzig: Teubner (article completed in March 1903).

Jungnickel, C., and R. Mc Cormmach. 1986. *Intellectual Mastery of Nature, Theoretical Physics from Ohm to Einstein*, vol. 1: *The Torch of Mathematics 1800–1870*; vol. 2: *The Now Mighty Theoretical Physics 1870–1925.* Chicago and London: University of Chicago Press.

Junkins, J. L., and Y. Kim. 1993. *Introduction to Dynamics and Control of Flexible Structures.* Washington, DC: AIAA Education Series.

Junkins, J. L., and M. D. Shuster. 1993. "The Geometry of the Euler Angles," *J. Astronautical Sciences*, **41**, 531–543.

Junkins, J. L., and J. D. Turner. 1986. *Optimal Spacecraft Rotational Maneuvers.* Amsterdam: Elsevier.

Kalaba, R. E., and F. E. Udwadia. 1993. "Equations of Motion for Nonholonomic, Constrained Dynamical Systems via Gauss's Principle," *J. Appl. Mechanics* (ASME), **60**, 662–668.

Kalaba, R. E., and F. E. Udwadia. 1994. "Lagrangian Mechanics, Gauss's Principle, Quadratic Programming, and Generalized Inverses: New Equations for Nonholonomically Constrained Discrete Mechanical Systems," *Quart. Appl. Mathematics*, **52**, 229–241.

Kampen, N. G. van, and J. J. Lodder. 1984. "Constraints," *Am. J. Physics*, **52**, 419–424.

Kane, T. R. 1961. "Dynamics of Nonholonomic Systems," *J. Appl. Mechanics* (ASME), **28**, 574–578. [Discussion in **29**, 606–607, 1962.]

Kane, T. R. 1962. "Impulsive Motions," *J. Appl. Mechanics* (ASME), **29**, 715–718. [Discussion in **30**, 315–316, 1963.]

Kane, T. R. 1968. *Dynamics*. New York: Holt, Rinehart, and Winston.

Kane, T. R. 1973. "Lagrange's Equations for a Rigid Body," *J. Appl. Mechanics* (ASME), **40**, 1118–1119.

Kane, T. R. 1983. "Formulation of Dynamical Equations of Motion," *Am. J. Physics*, **51**, 974–977.

Kane, T. R. 1986. "Rebuttal to 'A comparison of Kane's equations of motion and the Gibbs–Appell equations of motion'," *Am. J. Physics*, **54**, 472.

Kane, T. R., and D. A. Levinson. 1980. "Formulation of Equations of Motion for Complex Spacecraft," *J. Guidance, Control, and Dynamics*, **3**, 99–112.

Kane, T. R., and D. A. Levinson. 1983. "Multibody Dynamics," *J. Appl. Mechanics* (ASME), **50**, 1971–1978.

Kane, T. R., and D. A. Levinson. 1985. *Dynamics: Theory and Applications*. New York: McGraw-Hill.

Kane, T. R., and D. A. Levinson. 1988. "A Method for Testing Numerical Integrations of Equations of Motion of Mechanical Systems," *J. Appl. Mechanics* (ASME), **55**, 711–715.

Kane, T. R., and D. A. Levinson. 1990. "Orthogonal Curvilinear Coordinates and Angular Velocity," *J. Appl. Mechanics* (ASME), **57**, 468–470.

Kane, T. R., and C. F. Wang. 1965. "On the Derivation of Equations of Motion," *J. Soc. Industrial and Applied Math.* (SIAM), **13**, 487–492.

Kane, T. R., P. W. Likins, and D. A. Levinson. 1983. *Spacecraft Dynamics*. New York: McGraw-Hill.

Karapetyan, A. V. and I. S. Lagutina. 1998. "The Influence of Dissipative and Constant Forces on the Form and Stability of Steady Motions of Mechanical Systems with Cyclic Coordinates," *PMM*, **62**, 503–510 (original in Russian; 539–547).

Kauderer, H. 1958. *Nichtlineare Mechanik*. Berlin: Springer.

Keller, J. B. 1986. "Impact with Friction," *J. Appl. Mechanics* (ASME), **53**, 1–4.

Keis, I. A. 1976. "On the Existence Conditions for the Particular Jacobi Integral," *PMM*, **40**, 563–569 (original in Russian, 611–617).

Kevorkian, J., and J. D. Cole. 1981. *Perturbation Methods in Applied Mathematics*. New York: Springer.

Khmelevski, I. L. 1960. "On Hamilton's Principle for Nonholonomic Systems," *PMM*, **24**, 1177–1182 (original in Russian, 777–780).

Khodzhaev, K. Sh. 1969. "Integral Criterion of Stability for Systems with Quasicyclic Coordinates and Energy Relations for Oscillations of Current-Carrying Conductors," *PMM*, **33**, 76–93 (original in Russian, 85–100).

Kibble, T. W. B., and F. H. Berkshire. 1996. *Classical Mechanics*, 4th ed. Harlow, England: Addison Wesley/Longman (1st. ed. 1966).

Kil'chevskii, N. A. 1972. *Elements of Tensor Calculus and its Applications to Mechanics* (in Russian), 2nd ed. Kiev: Naukova Dumka.

Kil'chevskii, N. A. 1977. *Course of Theoretical Mechanics* (in Russian), 2 vols. Moscow: Nauka.

Killingbeck, J. 1970. "Methods of Proof and Applications of the Virial Theorem in Classical and Quantum Mechanics," *Am. J. Physics*, **38**, 590–596.

Kilmister, C. W. 1964. *Hamiltonian Dynamics*. New York: Wiley.

Kilmister, C. W. 1967. *Lagrangian Dynamics: An Introduction for Students*. London: Logos Press (Elek Books).

Kilmister, C. W., and F. A. E. Pirani. 1965. "Ignorable Coordinates and Steady Motion in Classical Mechanics," *Proc. Camb. Phil. Soc.*, **61**, 211–222.

Kilmister, C. W., and J. E. Reeve. 1966. *Rational Mechanics*. London: Longmans.

Kirchgraber, U., and E. Stiefel. 1978. *Methoden der analytischen Störungsrechnung und ihre Anwendungen*. Stuttgart: Teubner.

Kirchhoff, G. R. 1876. *Vorlesungen über mathematische Physik: Mechanik*. Leipzig: Teubner [1877, 2nd edn; 1883, 3rd edn (starting with this edition, Mechanics appears as vol. 1 of a 4-vol. Lectures ...); 1897, 4th (last) edition, ed. by W. Wien].

Kirgetov, V. I. 1964(a). "On Kinematically Controlled Mechanical Systems," *PMM*, **28**, 15–26 (original in Russian, 15–24).

Kirgetov, V. I. 1964(b). "On the Equations of Motion of Controlled Mechanical Systems," *PMM*, **28**, 285–296 (original in Russian, 232–241).

Kirgetov, V. I. 1967. "The Motion of Controlled Mechanical Systems with Prescribed Constraints (Servoconstraints)," *PMM*, **31**, 465–477 (original in Russian, 433–446).

Kittel, C., W. D. Knight, and M. A. Ruderman. 1965. *Mechanics* (vol. 1 of Berkeley Physics Course). New York: McGraw-Hill.

Kitto, H. D. F. 1957. *The Greeks*. Middlesex, U.K.: Penguin Books.

Kitzka, F. 1986. "An Example for the Application of a Nonholonomic Constraint of Second Order in Particle Mechanics," *ZAMM*, **66**, 312–314.

Kleban, P. 1979. "Virial Theorem and Scale Transformations," *Am. J. Physics*, **47**, 883–886.

Klein, F. 1926(a). *Vorlesungen über höhere Geometrie*, 3rd ed., edited by W. Blaschke. Berlin: Springer.

Klein, F. *Vorlesungen über die Entwicklung der Mathematik im 19. Jahrhundert*, 2 vols.: vol. I 1926(b); vol. II 1927. Berlin: Springer.

Klein, F. 1939. *Elementary Mathematics from an Advanced Viewpoint*, vol. 1: *Arithmetic, Algebra, Analysis*; vol. 2: *Geometry*, 3rd ed. New York: Dover (original in German, 1924–1925).

Klein, M. J. 1970. *Paul Ehrenfest*, vol. 1: *The Making of a Theoretical Physicist*, pp. 53–74. Amsterdam: North Holland.

Kline, M. 1972. *Mathematical Thought from Ancient to Modern Times*. New York: Oxford University Press.

Klein, F., and A. Sommerfeld. *Ueber die Theorie des Kreisels*, in four parts: I (1897); II (1898); III (1903); IV (1910). Leipzig: Teubner.

Klingbeil, E. 1966. *Tensorrechnung für Ingenieure*. Mannheim: Bibliographisches Institut.

Klotter, K. 1955. "Neuere Methoden und Ergebnisse auf dem Gebiet nichtlinearer Schwingungen," *VDI Berichte*, **4**, 35–46.

Klotter, K. 1960. *Technische Schwingungslehre*, vol. 2: *Systems with Several Degrees of Freedom*. Berlin: Springer (reprinted 1981).

Kochin, N. E. 1965. *Vector Calculus and the Elements of Tensor Calculus*, 9th ed. Moscow: Nauka (in Russian).

Kochina, P. 1985. *Love and Mathematics: Sofya Kovalevskaya*. Moscow: Mir (original in Russian, 1981).

Kogan, B. Iu. 1964. "On an Extremal Property of Constraint Reactions," *PMM*, **28**, 1115–1116 (original in Russian, 921–922).

Koiller, J. 1992. "Reduction of Some Classical Non-Holonomic Systems with Symmetry," *Arch. Rational Mech. Anal.* **118**, 113–148.

Konopinski, E. J. 1969. *Classical Descriptions of Motion, The Dynamics of Particle Trajectories, Rigid Rotations, and Elastic Waves*. San Francisco: W. H. Freeman.

Korenev, G. V. 1967. *The Mechanics of Guided (or Steerable) Bodies*. London: Iliffe (original in Russian, 1964).

Korenev, G. V. 1979. *Mechanics of Controllable Manipulators* (in Russian). Moscow: Nauka.

Korteweg, D. J. 1899. "Über eine ziemlich verbreitete unrichtige Behandlungsweise eines Problemes der rollenden Bewegung, über die Theorie dieser Bewegung, und ins besondere über kleine rollende Schwingungen um eine Gleichgewichtslage," *Nieuw Arch. Wisk.*, **4**(2), 130–155.

Koschmieder, L. 1962. *Variationsrechnung*, Band I: *Das freie und gebundene Extrem einfacher Grundintegrale*, 2nd ed. Berlin: W. de Gruyter.

Kosenko, I. I. 1995. "The Application of Galerkin's Method in Lagrangian Dynamics," *PMM*, **59**, 9–18 (original in Russian, 10–20).

Koshlyakov, V. N. 1985. *Problems of Rigid Body Dynamics and Applied Gyroscope Theory. Analytical Methods* (in Russian). Moscow: Nauka.

Koshlyakov, V. N. 1990. "Rodrigues–Hamilton Parameters in Rigid Body Dynamics Problems," pp. 105–115 in vol. 1 of *General and Applied Mechanics of Mechanical Engineering and Applied Mechanics*, edited by V. Z. Parton. New York: Hemisphere Publishing Co.

Kotkin, G. L., and V. G. Serbo. 1971. *Collection of Problems in Classical Mechanics*. Oxford, U.K.: Pergamon (original in Russian, 1969).

Kowalewski, G. 1933. *Lehrbuch der höheren Mathematik*, Erster Band: *Vektorrechnung und analytische Geometrie*. Berlin: W. de Gruyter.

Kozlov, V. V., and N. N. Kolesnikov. 1978. "On Theorems of Dynamics," *PMM*, **42**, 26–31 (original in Russian, 28–33).

Kraft, F. 1885. *Sammlung von Problemen der Analytischen Mechanik*, 2 vols. Stuttgart: J. B. Metzler.

Kramer, E. E. 1970. *The Nature and Growth of Modern Mathematics*. Princeton, NJ: Princeton University Press.

Krbek, F. v. 1954. *Grundzüge der Mechanik*. Leipzig: Akademische Verlagsgesellschaft-Geest und Portig.

Kronauer, R. E. 1983. "Oscillations," pp. 697–746 in *Handbook of Applied Mathematics, Selected Results and Methods*, 2nd ed., edited by C. E. Pearson. New York: Van Nostrand and Reinhold.

Kronauer, R. E., and S. A. Musa. 1966. "Exchange of Energy Between Oscillations in Weakly Nonlinear Conservative Systems," *J. Appl. Mechanics* (ASME), **33**, 451–452.

Krutkov (or Krutkow), G. 1928. "Ueber die Relativbewegung eines Freien Massenpunktes," *Izvestia Akademii Nauk USSR*, **6–7**, 549–572.

Krutkov, Ju. A. 1953. "On a New Type of Quasi-coordinate" (in Russian), *Dokl. Akad. Nauk SSSR*, **89**, 793–795.

Kuipers, M., and A. A. F. van de Ven 1982. "The Lagrangian Formalism and the Balance Law of Angular Momentum for a Rigid Body," *J. Eng. Mathematics*, **16**, 77–90.

Kurdila, A., J. G. Papastavridis, and M. P. Kamat. 1990. "Role of Maggi's Equations in Computational Methods for Constrained Multibody Systems," *J. Guidance, Control, and Dynamics*, **13**, 113–120.

Kurth, R. 1957. *Introduction to the Mechanics of Stellar Systems*. London, U.K.: Pergamon Press.

Kurth, R. 1960. *Axiomatics of Classical Mechanics*. Oxford, U.K.: Pergamon Press.

Kuypers, F. 1993. *Klassische Mechanik*, 4th ed. Weinheim, Germany: VCH Verlagsgesellschaft mbH.

Kuzmin, P. A. 1973. *Small Oscillations and Stability of Motion* (in Russian). Moscow: Nauka.

Kwatny, H. G., and Blankenship, G. L. 2000. *Nonlinear Control and Analytical Mechanics. A Computational Approach*. Boston: Birkhäuser.

Lagrange, J. L. 1760–1761. "Application de la Méthode Exposée dans le Mémoire Précedent a la Solution de Différens Problemes de Dynamique," *Miscellanea Taurinensia* (Recueils de l'Academie de Tourin), **2**(2), 196–298. [Also in Lagrange's Collected Works (*Oeuvres*), edited by J.-A. Serret, **1**, 365–468, 1873.]

Lagrange, J. L. 1764. "Recherches sur la Libration de la Lune, dans Lesquelles on Tâche de Résoudre la Question Proposée par l'Academie Royale des Sciences pour le Prix de l'Année 1764," *Prix de l'Academie Royale des Sciences de Paris*, **9**. [Also in Lagrange's *Collected Works* (*Oeuvres*), edited by J.-A. Serret, **6**, 4–61, 1873.]

Lagrange, J. L. 1780. "Théorie de la Libration de la Lune, et des Autres Phénomènes qui Dépendent de la Figure Non sphérique de cette Planète," *Nouveaux Mémoires de l'Academie Royale des Sciences et Belles-Lettres de Berlin*, 203–309 (published in 1782). [Also in Lagrange's *Collected Works* (*Oeuvres*), edited by J.-A. Serret, **5**, 5–122, 1870.]

Lagrange (De la Grange), J. L. 1788. *Méchanique Analitique*. Paris: Chez La Veuve Desaint [2nd ed. as *Mécanique Analytique*, 1811, vol. 1; 1815, vol. 2, Courcier, Paris; 3rd ed. (1853–1855) (edited and amended/annotated by J. Bertrand), Gauthiers-Villars, Paris; 4th ed. (1888–1889) (edited by G. Darboux), *Oeuvres Completes*, **11**, **12**, Paris. Also available by Librairie Scientifique et Technique A. Blanchard, Paris, 1965. Translations: German (1797, 1817, 1853, 1887), Russian (1950), English (by A. Boissonnade and V. N. Vagliente, from 2nd French edition; Kluwer Academic Publishers, Dordrecht, 1997).]

Lainé, E. 1946. *Exercices de Mécanique*. 2nd ed. Paris: Vuibert.

Lamb, H. 1910. "Dynamics," pp. 756–763 in vol. 8; "Mechanics," pp. 955–994 in vol. 17 of *Encyclopaedia Britannica*, 11th ed. Cambridge, U.K. and New York: Cambridge University Press.

Lamb, H. 1923. *Dynamics*, 2nd ed. Cambridge, U.K.: Cambridge University Press (reprinted 1951).

Lamb, H. 1928. *Statics*, 3rd ed. Cambridge, U.K.: Cambridge University Press (reprinted 1954).

Lamb, H. 1929. *Higher Mechanics*, 2nd ed. Cambridge, U.K.: Cambridge University Press (reprinted in 1943).

Lamb, H. 1932. *Hydrodynamics*, 6th (last) ed. Cambridge, U.K.: Cambridge University Press (reprinted 1945 by Dover, New York; and 1994 by Cambridge University Press). [Earliest appearance as *Treatise on the Mathematical Theory of the Motion of Fluids*, 1879.]

Lampariello, L. 1943. "Su certe identita differenziali cui soddisfano le funzioni y delle equazioni dinamiche di Volterra–Hamel," *Atti Accad. Ital., Rendiconti, Cl. Sci. Fis. Mat. Nat.* **4**(7), 12–19.

Lanczos, C. 1962. "Variational Principles of Mechanics," pp. 24-1–24-23 in *Handbook of Engineering Mechanics*, edited by W. Flügge. New York: McGraw-Hill.

Lanczos, C. 1970. *The Variational Principles of Mechanics*, 4th ed. Toronto: University of Toronto Press (reprinted 1986 by Dover, New York).

Landau, L. D., and E. M. Lifshitz. 1960. *Mechanics*, vol. 1 of *Course of Theoretical Physics*. Oxford, U.K.: Pergamon Press (original in Russian, 1958).

Landau, L. D., and E. M. Lifshitz. 1971. *The Classical Theory of Fields*, 3rd revised ed. (vol. 2 of *Course of Theoretical Physics*). Addison-Wesley, Reading, MA and Pergamon, London, Oxford, U.K. (original in Russian, late 1950s).

Landau, L. D., and E. M. Lifshitz. 1980. *Statistical Physics*, 3rd revised and enlarged ed. (vol. 5 of *Course of Theoretical Physics*). London, Oxford, U.K.: Pergamon (original in Russian, late 1950s; 2nd revised and enlarged English ed. 1968/1969, 1st English ed. 1959).

Langhaar, H. L. 1962. *Energy Methods in Applied Mechanics*. New York: Wiley.

Langner, R. 1997–1998. *Arbeitsumdrucke zur Mechanik der Punkt und Starrkörper* (Lecture Notes). Karlsruhe: Institut für Theoretische Mechanik, University of Karlsruhe (TH).

Lawden, D. F. 1972. *Analytical Mechanics*. London: Allen and Unwin.

Layton, R. A. 1998. *Principles of Analytical System Dynamics*. New York: Springer.

Le, S. A. 1990. "The Painlevé Paradoxes and the Law of Motion of Mechanical Systems with Coulomb Friction," *PMM*, **54**, 430–438 (original in Russian, 520–529).

Leach, P. G. L. 1978. "Note on the Time-Dependent Damped and Forced Harmonic Oscillator," *Am. J. Physics*, **46**, 1247–1249.

Leach, P. G. L. 1982. "First Integrals for Non-Autonomous Hamiltonian Systems," *Acad. Sci. Turin, Proceedings of the IUTAM-ISIMM Symposium on Modern Developments in Analytical Dynamics* (June 7–11, 1982), vol. 2, 575–579. [Also in *Atti della Accademia delle Scienze di Torino, Supplemento*, **117**, 1983.]

Ledoux, P. 1958. "Stellar Stability," pp. 605–688 in vol. 51 of *Handbuch der Physik*, ed. by S. Flügge, Berlin: Springer.

Lee, H.-C. 1947. "The Universal Integral Invariants of Hamiltonian Systems and Application to the Theory of Canonical Transformations," *Proc. Roy. Soc. Edinburgh*, **A**, **62**, part **III**, no. 26, 237–246.

Lee, S. M. 1970. "The Double-Simple Pendulum Problem," *Am. J. Physics*, **38**, 536–537.

Leech, J. W. 1965. *Classical Mechanics*, 2nd ed. London: Methuen (1958, 1st ed.).

Lefkowitz, M. 1996. *Not Out of Africa, How Afrocentrism became an Excuse to Teach Myth as History*. New York: Basic Books.

Lehmann, T. 1985. *Elemente der Mechanik*, vol. IV: *Schwingungen, Variationsprinzipe*, 2nd ed. Braunschweig/Wiesbaden, Germany: F. Vieweg and Sohn.

Lehmann-Filhés, R. 1890. "Über einige Fundamentalsätze der Dynamik," *Astronomische Nachrichten*, **125**, 49–62.

Leimanis, E. 1965. *The General Problem of the Motion of Coupled Rigid Bodies about a Fixed Point*. New York: Springer.

Leipholz, H. H. E. 1970. *Stability Theory, An Introduction to the Stability of Dynamic Systems and Rigid Bodies*. New York: Academic Press (original in German, 1968).

Leipholz, H. H. E. 1978. *Six Lectures on Variational Principles in Structural Engineering*. Waterloo, ON, Canada: Solid Mech. Division, University of Waterloo Press.

Leipholz, H. H. E. 1983. "The Galerkin Formulation and the Hamilton–Ritz Formulation: A Critical Review," *Acta Mechanica*, **47**, 283–290.

Leitinger, R. 1913. "Ueber Jourdain's Prinzip der Mechanik und dessen Zusammenhang mit dem verallgemeinerten Prinzip der kleinsten Aktion," *Wienersche Sitzungsberichte, Math.-Naturwiss. Kl.*, **IIa**, **122**, 635–650.

Lemos, N. A. 1979. "Canonical Approach to the Damped Harmonic Oscillator," *Am. J. Physics*, **47**, 857–858.

Lemos, N. A. 1991. "Remark on Rayleigh's Dissipation Function," *Am. J. Physics*, **59**, 660–661.

León, M. de, and P. R. Rodrigues. 1989. *Methods of Differential Geometry in Analytical Mechanics*. Amsterdam: North Holland.

Lesser, M. 1995. *The Analysis of Complex Nonlinear Mechanical Systems, A Computer Algebra Assisted Approach*. New Jersey: World Scientific.

Leubner, C. 1979. "Coordinate-Free Rotation Operator," *Am. J. Physics*, **47**, 727–729.

Leubner, C. 1980. "Pedagogical Aspects of Time-Dependent Rotation Operators," *Am. J. Physics*, **48**, 563–568.

Leubner, C. 1981. "Correcting a Widespread Error Concerning the Angular Velocity of a Rotating Rigid Body," *Am. J. Physics*, **49**, 232–234.

Leubner, C., and G. Grübl. 1985. "Interrelation of Euler's and Piña's Parametrization of Rotations," *Am. J. Physics*, **53**, 487–488.

Levi-Civita, T. 1926. *The Absolute Differential Calculus (Calculus of Tensors)*. London: Blackie (reprinted 1977 by Dover, New York; original in Italian, 1924).

Levi-Civita, T., and U. Amaldi. 1927. *Lezioni di Meccanica Razionale*, vol. II, parts 1 and 2. Bologna: Zanichelli.

Levit, S., and U. Smilansky. 1977. "A New Approach to Gaussian Path Integrals and the Evolution of the Semiclassical Propagator," *Annals of Physics*, **103**, 198–207.

Lévy-Leblond, J.-M. 1971. "Conservation Laws for Gauge-Variant Lagrangians in Classical Mechanics," *Am. J. Physics*, **39**, 502–506.

Lewis, H. R. 1982. "Exact Invariants for Time-Dependent Hamiltonian Systems," *Acad. Sci. Turin, Proceedings of the IUTAM-ISIMM Symposium on Modern Developments in Analytical Dynamics* (June 7–11), 1982, vol. 2, 581–591. [Also in *Atti della Accademia delle Scienze di Torino, Supplemento*, **117**, 1983.]

Liapunov, A. M. 1982. *Lectures on Theoretical Mechanics* (in Russian). Kiev: Naukova Dumka (Lecture Notes on Theoretical and Analytical Mechanics, Statics, Dynamics, Continuum Mechanics, and Theory of Potential; originally delivered in 1885, 1893, etc. [See also p. 268 of next reference, on Liapunov's courses of lectures.]

Liapunov, A. M. 1992. *The General Problem of the Stability of Motion*. London: Taylor and Francis (original in Russian, 1892; translated into French, 1893; translated from French to English, 1992).

Lichtenberg, A. L. 1969. *Phase-Space Dynamics of Particles*. New York: Wiley.

Lichtenberg, A. L., and M. A. Lieberman. 1992. *Regular and Chaotic Dynamics*. New York: Springer (second revised and expanded edition of *Regular and Stochastic Motion*, Springer, 1983).

Likins, P. W. 1973. *Elements of Engineering Mechanics*. New York: McGraw-Hill.

Lilov, L. 1984. "Variationsprinzipien in der Starrkörpermechanik," *ZAMM*, **64**, T377–T379.

Lilov, L., and M. Lorer. 1982. "Dynamic Analysis of Multirigid-Body System Based on the Gauss Principle," *ZAMM*, **62**, 539–545.

Lindelöf, E. 1895. "Sur le Mouvement d'un Corps de Révolution roulant sur un Plan Horizontal," *Acta Societatis Scientiarum Fennicae*, **20**(10), 1–18.

Lindsay, R. B., and H. Margenau. 1957. *Foundations of Physics*. New York: Dover (reprint of original version of 1936).

Lipschitz, R. 1872. "Untersuchung eines Problems der Variationsrechnung in welchem das Problem der Mechanik enthalten ist," *J. Reine und Angewandte Mathematik*, **74**, 116–149.

Lipschitz, R. 1877. "Bemerkungen zu dem Prinzip des kleinsten Zwanges," *J. Reine und Angewandte Mathematik*, **82**, 316–342.

Liu, C. Q., and R. L. Huston. 1992. "Another Look at Orthogonal Curvilinear Coordinates in Kinematic and Dynamic Analyses," *J. Appl. Mechanics* (ASME), **59**, 1033–1035.

Liu, D. 1989. "Conservation Laws of Nonholonomic Nonconservative Dynamical Systems," *Acta Mechanica Sinica*, **5**, 167–175.

Liu, D., and Y. Luo. 1991. "About the Basic Integral Variants of Holonomic Nonconservative Dynamical Systems," *Acta Mechanica Sinica*, **7**, 178–185.

Liu, Y.-Z. 1985. "On the Motion of an Asymmetrical Rigid Body Rolling on a Horizontal Plane," *ZAMM*, **65**, 180–183.

Liu, Z.-F., F.-S. Jin, and F.-X. Mei. 1986. "Nielsen's and Euler's Operators of Higher Order in Analytical Mechanics," *Applied Mathematics and Mechanics*, **7**, 53–63 (translated from the Chinese).

Livens, G. H. 1918–1919. "On Hamilton's Principle and the Modified Function in Analytical Dynamics," *Proc. Roy. Soc. Edinburgh*, **39**, 113–119.

Lobas, L. G. 1966. "On a Transformation of the Equations of Dynamics," *Applied Mechanics* (transl. of Soviet/Ukrainian *Prikladnaya Mekhanika*), **2**, 56–60 (original in Russian, 97–105).

Lobas, L. G. 1970. "Remarks Concerning Nonholonomic Systems," *Applied Mechanics* (transl. of Soviet/Ukrainian *Prikladnaya Mekhanika*), **6**, 541–545 (original in Russian, 108–113).

Lobas, L. G. 1986. *Nonholonomic Models of Wheeled Equipment* (in Russian). Kiev: Naukova Dumka.

Logan, J. D. 1977. *Invariant Variational Principles*. New York: Academic Press.

Lohr, E. 1939. *Vektor und Dyadenrechnung für Physiker und Techniker*. Berlin: W. de Gruyter.

Loitsianskii, L. G., and A. I. Lur'e. *Course of Theoretical Mechanics*, vol. 1: *Statics and Kinematics* (1982); vol. 2: *Dynamics* (1983) (in Russian), 6th ed. Moscow: Nauka.

Loney, S. L. 1909. *An Elementary Treatise on the Dynamics of a Particle and of Rigid Bodies*. Cambridge, U.K.: Cambridge University Press.

Loney, S. L. 1926. *Solutions of the Examples in a Treatise on Dynamics of a Particle and of Rigid Bodies*. Cambridge, U.K.: Cambridge University Press.

Long, R. 1963. *Engineering Science Mechanics*. Englewood Cliffs, NJ: Prentice Hall.

Lorenz, H. 1902. *Technische Mechanik starrer Systeme*, vol. 1 of *Lehrbuch der Technischen Physik*. München/Berlin: R. Oldenbourg.

Lotze, A. 1950. *Vektor- und Affinor-Analysis*. München: Oldenbourg.

Lovelock, D., and H. Rund. 1975. *Tensors, Differential Forms, and Variational Principles*. New York: Wiley-Interscience (reprinted 1989, with a new Appendix, by Dover, New York).

Ludwig, G. 1985. *Einführung in die Grundlagen der Theoretischen Physik*, Band 1: *Raum, Zeit, Mechanik*, 3rd ed. Braunschweig/Wiesbaden: F. Vieweg and Sohn.

Lunn, M. 1991. *A First Course in Mechanics*. Oxford, U.K.: Oxford University Press.

Lur'e, A. I. 1957. "Notes on Analytical Mechanics," *PMM*, **21**, 759–768 (in Russian).

Lur'e, A. I. 1968. *Mécanique Analytique*, 2 vols. Louvain, Belgium: Librairie Universitaire (author erroneously appears there as L. Lur'e; original in Russian, 1961).

Lurie, A. I. 1990. *Nonlinear Theory of Elasticity*. Amsterdam: North Holland (original in Russian 1980–1981).

Lur'e, A. I., and G. Iu. Dzhanelidze. 1960. "On the Application of Integral and Variational Principles of Mechanics to Problems of Vibrations," *PMM*, **24**, 103–112 (original in Russian, 80–87).

Luttinger, J. M., and R. B. Jr. Thomas. 1960. "Variational Method for Studying the Motion of Classical Vibrating Systems," *J. Math. Physics*, **1**, 121–126.

Lützen, J. 1995(a). "Interactions between Mechanics and Differential Geometry in the 19th Century," *Archive for History of Exact Sciences*, **49**, 1–72.

Lützen, J. 1995(b). "Renouncing Forces; Geometrizing Mechanics. Hertz's Principles of Mechanics," Preprint Series, No. 22. Copenhagen: Matematisk Institut, Københavns Universitet.

L'vovich, A. Yu., and N. N. Poliahov. 1977. "Application of Nonholonomic Mechanics to the Theory of Electromechanical Systems," *Vestnik Leningradskogo Universiteta*, **13**, 137–146 (in Russian).

Lyness, J. N., and J. M. Blatt. 1961–1962. "The Practical Use of Variation Principles in the Determination of the Stability of Non Linear Systems," *J. Australian Math. Society*, **2**, 153–188. (See also Blatt and Lyness.)

Lynn, J. W. 1963. *Tensors in Electrical Engineering*. London: Edward Arnold.

Lyttleton, R. A. 1953. *The Stability of Rotating Liquid Masses*. Cambridge, U.K.: Cambridge University Press.

Mach, E. 1960. *The Science of Mechanics, A Critical and Historical Account of its Development*. 6th ed., with revisions through the 9th German edition. La Salle, IL: Open Court Publishing Co. (original in German, with corresponding editions: 1883, 1888, 1897, 1901, 1904, 1907, 1912, 1933).

MacKinnon, E. M. 1982. *Scientific Explanation and Atomic Physics*. Chicago, IL: University of Chicago Press.

MacLane, S. 1968. *Geometrical Mechanics*, parts 1 and 2, Lecture Notes. Chicago, IL: University of Chicago, Dept. of Mathematics.

MacMillan, W. D. *Theoretical Mechanics*, vol. 1: *Statics and the Dynamics of a Particle* (1927, reprinted 1958 by Dover, New York); vol. 2: *The Theory of the Potential* (1930, reprinted 1958 by Dover); vol. 3: *Dynamics of Rigid Bodies* (1936, reprinted 1960 by Dover).

Maggi, G. A. 1896. *Principii della Teoria Matematica del Movimento dei Corpi*, Corso di Meccanica Razionale. Milano: U. Hoepli.

Maggi, G. A. 1901. "Di alcune nuove forme delle equazioni della Dinamica, applicabili ai sistemi anolonomi," *Atti della Reale Accademia dei Lincei, Rendiconti, Classe di scienze fisiche, matematiche e naturali*, Seria 5, **10**(12), 287–292.

Maggi, G. A. 1903. *Principii di Stereodinamica, Corso sulla Formazione, l'Interpretazione e l'Integrazione delle equazioni del movimento dei solidi*. Milano: U. Hoepli.

Magnus, K. 1957. "Ueber den Zusammenhang verschiedener Näherungsverfahren zur Berechnung nichtlinearer Schwingungen," *ZAMM*, **37**, 471–485.

Magnus, K. 1970. "Der Einfluß verschiedener Kräftearten auf die Stabilität linearer Systeme," *ZAMP*, **21**, 523–534.

Magnus, K. 1971. *Kreisel, Seine Theorie und Anwendungen*. Berlin: Springer.

Magnus, K. 1974. *Gyrodynamics*. Wien: Springer.

Magnus, K. 1977. "The Multibody Approach for Mechanical Systems," *Solid Mechanics Archives*, **2**, 187–204.

Magnus, K. 1978. "Kreiselmechanik," *ZAMM*, **58**, T56–T65.

Magnus, K., and H. H. Müller. 1974. *Grundlagen der Technischen Mechanik*. Stuttgart, Germany: Teubner (several subsequent reprintings).

Maißer, P. 1981. "Der Lagrange—Formalismus für diskrete elektromechanische Systeme in anholonomen Koordinaten und seine Anwendung in der Theorie elektrischer Maschinen," *Tech. Hochschule Ilmenau, Wiss. Zeitschrift*, **27**(2), 131–145.

Maißer, P. 1982. "Modellgleichungen für Manipulatoren," *Technische Mechanik*, **3**, 64–77.

Maißer, P. 1983–1984. *Ein Beitrag zur Theorie diskreter elektromechanischer Systeme mit Anwendungen in der Manipulator–/Robotertechnik*. Dissertation zur Erlangung des akademischen Grades Dr. sc. nat., Technische Hochschule Ilmenau.

Maißer, P. 1988. "Analytische Dynamik von Mehrkörpersystemen," *ZAMM*, **68**, 463–481.

Maißer, P. 1991(a). "Analytical Dynamics of Multibody Systems," *Computer Methods in Applied Mechanics and Engineering*, **91**, 1391–1396.

Maißer, P. 1991(b). "A Differential–Geometric Approach to the Multi Body System Dynamics," *ZAMM*, **71**, T116–T119.

Maißer, P. 1996. "Dynamik hybrider Mehrkörpersysteme aus kontinuumsmechanischer Sicht," *ZAMM*, **76**, 15–33.

Maißer, P. 1996 and 1997. "Differential–Geometric Methods in Multibody Dynamics," *Second World Congress of Nonlinear Analysts* (Athens, Greece, July 10–17, 1996). Also in *Nonlinear Analysis, Theory, Methods & Applications*, **30**, 5127–5133, 1997.

Malvern, L. E. 1969. *Introduction to the Mechanics of a Continuous Medium*. Englewood Cliffs, NJ: Prentice Hall.

Mander, J. 1985. "Six Grave Doubts about Computers," *Whole Earth Review*, **44**, 10–20.

Mangeron, D., and S. Deleanu. 1962. "Sur une Classe d'Équations de la Mécanique Analytique au Sens de I. Tzénoff," *Comptes Rendus, Bulgarian Acad. Sci.*, **15**, 9–12.

Mann, R. A. 1974. *The Classical Dynamics of Particles, Galilean and Lorentz Relativity*. New York: Academic Press.

Mansour, M. 1999. "Stability as the Fundamental Problem of Control Systems," *Applied Mechanics Reviews*, **52**(3), R11–R15.

Mantion, M. 1981. *Problèmes de Mécanique Analytique, Questions de Stabilité*. Paris: Vuibert.

Marcolongo, R. *Theoretische Mechanik*, 2 vols; vol. 1 (1911); vol. 2 (1912). Leipzig: Teubner (original in Italian, 1905; German translation edited by H. E. Timerding).

Marion, J. B. and S. T. Thornton. 1995. *Classical Dynamics, of Particles and Systems*, 4th ed. Fort Worth, TX: Saunders/Harcourt/Brace.

Markhasov, L. M. 1985. "The Poincaré and Poincaré-Chetayev Equations," *PMM*, **49**, 30–41 (original in Russian, 43–55).

Markeev, A. P. 1990. *Theoretical Mechanics* (in Russian). Moscow: Nauka.

Marmo, G., E. J. Saletan, A. Simoni, and B. Vitale. 1985. *Dynamical Systems, A Differential Geometric Approach to Symmetry and Reduction*. Chichester, U.K.: Wiley-Interscience.

Marris , A. W., and C. E. Stoneking. 1967. *Advanced Dynamics*. New York: McGraw-Hill.

Marsden, J. E. 1992. *Lectures on Mechanics*. Cambridge, U.K.: Cambridge University Press.

Marsden, J. E., and T. S. Ratiu. 1999. *Introduction to Mechanics and Symmetry, A Basic Exposition of Classical Mechanical Systems*, 2nd ed. New York: Springer (1994, 1st ed.).

Martin, D. 1960. "Sir Edmund Whittaker, F.R.S." *Edinburgh Mathematical Society*, **43**, 1–9.

Massa, E., and E. Pagani, 1991. "Classical Dynamics of Non-holonomic Systems: A Geometric Approach," *Ann. Inst. Henri Poincaré, Physique Théorique*, **55**, 511–544.

Mathieu, E. 1878. *Dynamique Analytique*. Paris: Gauthier-Villars.

Mattioli, G. D. 1931–1932. "Su una forma delle equazioni di Lagrange nelle quali il tempo è introdotto come $n + 1$-esima coordinata," *Atti del Reale Instituto Veneto di Scienze, Lettere ed Arti*, **91** (second part), 79–91.

Matzner, R. A., and L. C. Shepley. 1991. *Classical Mechanics*. Englewood Cliffs, NJ: Prentice Hall.

Maurer, L. 1905. "Ueber die Differentialgleichungen der Mechanik," *Königl. Ges. d. Wiss. z. Göttingen, Nachrichten, Math.–Phys. Klasse*, Heft **2**, 91–116.

Mavraganis, A. G. 1987. *Kinematics and Dynamics of Rigid Bodies* (in Greek). Athens: National Technical University of Athens (E.M.Π.).

Mavraganis, A. G. 1998. *Analytical Dynamics, Principles and Methods* (in Greek). Athens: National Technical University of Athens (E.M.Π.).

Maxwell, J. C. 1877. *Matter and Motion.* Cambridge, U.K.: Cambridge University Press (republished with Notes and Appendices by J. Larmor, Society for Promoting Christian Knowledge, London, 1920; reprinted 1952 and 1991 by Dover, New York).

McCarthy, J. M. 1990. *An Introduction to Theoretical Kinematics.* Boston, MA: MIT Press.

McCauley, J. L. 1993. *Chaos, Dynamics, and Fractals, An Algorithmic Approach to Deterministic Chaos.* Cambridge, U.K.: Cambridge University Press.

McCauley, J. L. 1997. *Classical Mechanics, Transformations, Flows, Integrable and Chaotic Dynamics.* Cambridge, U.K.: Cambridge University Press.

McCuskey, S. W. 1959. *An Introduction to Advanced Dynamics.* Reading, MA: Addison-Wesley.

McGill, D. J., and J. G. Papastavridis. 1987. "Comments on 'Comments on Fixed Points in Torque–Angular Momentum Relations'," *Am. J. Physics,* **55**, 470–471.

McKinley, J. M. 1981. "Problem on Moments of Inertia," *Am. J. Physics*, **49**, 234, 264.

McLachlan, N. W. 1956. *Ordinary Non-Linear Differential Equations in Engineering and Physical Sciences*, 2nd ed. Oxford, U.K.: Clarendon Press (second corrected printing, 1958).

Megias, I., M. A. Serna, and C. Bastero. 1982. "Lagrange's Equation of Motion: A Simple Derivation," *Int. J. Mech. Eng. Education*, **11**, 89–92.

Mei, F.-X. 1982. "On the Nielsen's Operator and the Euler's Operator for Non-holonomic Systems," *Acad. Sc. Turin, Proceedings of the IUTAM–ISIMM Symposium on Modern Developments in Analytical Dynamics* (June 7–11, 1982). [Also in *Atti della Accademia delle Scienze di Torino, Supplemento*, **117**, 627–634, 1983.]

Mei, F.-X. 1984. "Extension of the MacMillan's Equations to Nonlinear Nonholonomic Mechanical Systems," *Appl. Math. Mechanics* (English edition), **5**(5), 1633–1638.

Mei, F.-X. 1985. *Foundations of Mechanics of Nonholonomic Systems* (in Chinese). Beijing: Beijing Institute of Technology Press.

Mei, F -X. 1987. *Researches on Nonholonomic Dynamics* (in Chinese). Beijing: Beijing Institute of Technology Press.

Mei, F.-X. 1989. "Time–Integral Theorems for Variable Mass Nonholonomic Systems," *Proceedings ICAM.*

Mei, F.-X. 1990. "Two New Operators in Non-Holonomic Mechanics," *J. Beijing Inst. Technology*, **10**, 113–121.

Mei, F.-X. 1991(a). "One Type of Integrals for the Equations of Motion of Higher-Order Nonholonomic Systems," *Appl. Math. Mechanics*, **12**, 799–805 (original in Chinese).

Mei, F.-X. 1991(b). "Routh's Method of Order Reduction for the Equations of Motion of Higher Order Nonholonomic Systems," *Chinese Science Bulletin*, **36**, 377–381.

Mei, F.-X. 1991(c). "First Integral and Integral–Invariant for Nonholonomic Systems," *Chinese Science Bulletin*, **36**, 2038–2042.

Mei, F.-X. 1993. "The Noether's Theory of Birkhoffian Systems," *Science in China*, Series A, **36**, 1456–1467.

Mei, F.-X. 1999. *Applications of Lie Groups and Lie Algebras to Constrained Mechanical Systems.* Beijing: Science Press.

Mei, F.-X. 2000. "Nonholonomic Mechanics," *Applied Mechanics Reviews*, **53**, 283–305.

Mei, F.-X., and G. L. Liu. 1987. *Foundations of Analytical Mechanics* (in Chinese). Tsien An: Tsiao Tong University Press.

Mei, F.-X., D. Liu, and Y. Luo. 1991. *Advanced Analytical Mechanics* (in Chinese). Beijing: Beijing Institute of Technology Press.

Meirovitch, L. 1970. *Methods of Analytical Dynamics.* New York: McGraw-Hill.

Melis, A. 1955. "Un Esempio di Sistema Anolonomo Non Lineare: il Patino Guidato," *Rend. Sem. Fac. Sci. Univ. Cagliari*, **25**, 143–153.

Mercier, A. 1955. *Principes de Mécanique Analytique.* Paris: Gauthier-Villars.

Meriam, J. L., and L. G. Kraige. 1986. *Engineering Mechanics*, 2 vols, 2nd ed. New York: Wiley.

Merkin, D. R. 1974. *Gyroscopic Systems* (in Russian), 2nd ed. Moscow: Nauka.

Merkin, D. R. 1975. "On the Structure of Forces," *PMM*, **39**, 893–896 (original in Russian, 929–932).

Merkin, D. R. 1987. *Introduction to the Theory of Stability of Motion* (in Russian), 3rd ed. Moscow: Nauka (English translation by Springer, New York, 1997).

Message, P. J. 1966. "Stability and Small Oscillations about Equilibrium and Periodic Motions," pp. 77–99 in part 1 of *Space Mathematics* (vol. 5 of *Lectures in Applied Mathematics*), edited by J. B. Rosser. Providence, RI: American Mathematical Society.

Mikhailov, G. K. 1981. "History of Mechanics: Present State and Problems," pp. 148–165 in *Advances in Theoretical and Applied Mechanics*, edited by A. Y. Ishlinsky and F. L. Chernousko. Moscow: Mir (original in Russian, 1978).

Mikhailov, G. K. 1990. "Newton's Principia (To the Tercentenary of the First Edition)," pp. 274–290 in vol. 1 of *General and Applied Mechanics of Mechanical Engineering and Applied Mechanics*, edited by V. Z. Parton. New York: Hemisphere Publishing.

Mikhailov, G. K., and V. Z. Parton, editors. 1990. *Applied Mechanics: Soviet Reviews*, vol. 1: *Stability and Analytical Mechanics*. New York: Hemisphere (original reviews in Russian, 1979, 1982, 1983).

Mikhailov, G. K., G. Schmidt, and L. I. Sedov. 1984. "Leonhard Euler und das Entstehen der Klassischen Mechanik," *ZAMM*, **64**, 73–82.

Miller, P. H. 1957. "Hamilton's Principle as a Computational Device," *Am. J. Physics*, **25**, 30–32. [See also *Am. J. Physics*, **54**, 87, 1986, for corrections.]

Milne, E. A. 1948. *Vectorial Mechanics*. London: Methuen (reprinted 1957, 1965).

Mingori, D. L. 1995. "Lagrange's Equations, Hamilton's Equations, and Kane's Equations: Interrelations, Energy Integrals, and a Variational Principle," *J. Appl. Mechanics* (ASME), **62**, 505–510.

Minkin, Yu. G. 1977. "Variational Methods for Putting Upper and Lower Limits on the Parameters of the Periodic Vibrations of Nonlinear Systems," *Applied Mechanics*, **13**, 385–388 (original in Russian, 91–95).

Mironov, M. V. 1967. "Use of the Hamilton–Ostrogradskii Principle in Problems of the Theory of Nonlinear Oscillations," *PMM*, **31**, 1115—1121 (original in Russian, 1110–1116).

Mitropolskii, Yu. A. 1990. "Asymptotic Methods in Nonlinear Mechanics," pp. 59–74 in vol. 1 of *General and Applied Mechanics of Mechanical Engineering and Applied Mechanics*, edited by V. Z. Parton. New York: Hemisphere Publishing Co.

Mittelstaedt, P. 1970. *Klassische Mechanik*. Mannheim: Bibliographisches Institut.

Mladenova, C. 1995. "Dynamics of Nonholonomic Systems," *ZAMM*, **75**, 199–205.

Moiseyev, N. N., and V. V. Rumyantsev. 1968. *Dynamic Stability of Bodies Containing Fluid*. New York: Springer (original in Russian, 1965).

Molenbroek, P. 1890. "Over de zuiver rollende beweging van een lichaam over een willenkeurig oppervlek," *Nieuw Arch. Wisk.*, **17**, 130–157.

Montel, P. 1927. *Cours de Mécanique Rationelle*, 2nd ed., 2 vols. ("livres"). Paris: Eyrolles.

Moon, F. C. 1998. *Applied Dynamics, With Applications to Multibody and Mechatronic Systems*. New York: Wiley-Interscience.

Moon, P., and E. D. Spencer. 1961. *Field Theory Handbook*. Berlin: Springer.

Moore, E. Neal. 1983. *Theoretical Mechanics*. New York: Wiley.

Moreau, J. J. *Mécanique Classique*, 2 vols: vol. I (1968); vol. II (1971). Paris: Masson.

Morera, G. 1903. "Sulle equazioni dinamiche di Lagrange," *Atti R. Accad. Sci. Torino*, **38**, 121–134.

Morgenstern, D., and I. Szabó. 1961. *Vorlesungen über Theoretische Mechanik*. Berlin: Springer.

Moritz, H., and I. I. Müller. 1987. *Earth Rotation, Theory and Observation*. New York: Ungar.

Morton, H. S., Jr. 1984. "A Formulation of Rigid-Body Rotational Dynamics Based on Generalized Angular Momentum Variables Corresponding to the Euler Parameters," *AIAA/AAS Astrodynamics Conference*, paper no. 84–2023 (Seattle, WA, August 20–22, 1984).

Morton, H. S., J. L. Junkins, and J. N. Blanton. 1974. "Analytical Solutions for Euler Parameters," *Celestial Mechanics*, **10**, 287–301.

Morton, W. B. 1929. "Simple Examples of Adiabatic Invariance," *Philosophical Magazine and Journal of Science*, Series 7, **8**, 186–194.

Mott, D. L. 1966. "Torque on a Rigid Body in Circular Orbit," *Am. J. Physics*, **34**, 562–564.

Müller, C. H., and G. Prange. 1923. *Allgemeine Mechanik*. Hannover: Helwingsche Verlagsbuchandlung.

Müller, H. H., and K. Magnus. 1974. *Übungen zur Technischen Mechanik*. Stuttgart, Germany: Teubner (several subsequent reprintings).

Müller, P. C. 1977. *Stabilität und Matrizen*. Berlin: Springer.

Müller, P. C., and W. O. Schiehlen. 1985. *Linear Vibrations, A Theoretical Treatment of Multi-Degree-of-Freedom Vibrating Systems*. Dordrecht: Martinus Nijhoff/Kluwer (originally in German, 1976).

Muschik, W. 1980(a). "Axiome für Zwangskräfte bei geschwindigkeitsabhängingen differentiellen Nebenbedingungen," *ZAMM*, **60**, T44–T46.

Muschik, W., N. Poliatzky, and G. Brunk. 1980(b). "Die Lagrangeschen Gleichungen bei Tschetajew-Nebenbedingungen," *ZAMM*, **60**, T46–T47.

Mušicki, D. 1994. "Generalization of Noether's Theorem for Non-conservative Systems," *European J. Mechanics, A/Solids*, **13**, 533–539.

Nadile, A. 1950. "Sull'esistenza per i sistemi anolonomi soggetti a vincoli reonomi di un integrale analogo a quello dell'energia," *Bolletino della Unione Matematica Italiana*, Series 3, **5**, 297–301.

Nagornov, V. A. 1958. "On the Equations of Motion of Nonholonomic Mechanical Systems with Nonholonomic Coordinates," *Izvestia Acad. Nauk Uzb. SSR, Ser. Phys. Math. Nauk*, **6**, 65–81 (in Russian).

Nayfeh, A. H. 1973. *Perturbation Methods*. New York: Wiley.

Nayfeh, A. H., and D. T. Mook. 1979. *Nonlinear Oscillations*. New York: Wiley.

Naziev, E. Kh. 1969. "Mechanical Systems with Integrals Linear in the Momenta," *Moscow U. Mechanics Bulletin*, **24**, 59–62 (original in Russian, 77–85).

Naziev, E. Kh. 1972. "On the Hamilton-Jacobi Method for Nonholonomic Systems," *PMM*, **36**, 1038–1045 (original in Russian, 1108–1114).

Neimark, J. I. 1968. "Dynamics of Nonholonomic Systems," pp. 171–178 in vol. 1 of *Mechanics in the USSR During the Last 50 years (1917–1967)*. Moscow: Nauka (in Russian).

Neimark, J. I., and N. A. Fufaev. 1972. *Dynamics of Nonholonomic Systems*. Providence, RI: American Mathematical Society (original in Russian, 1967).

Neuber, H. 1950. Review of book *Theoretische Mechanik*, by G. Hamel, in *ZAMM*, **30**, 397–398.

Neuberger, J. 1981. "Coriolis Force Revisited," *Am. J. Physics*, **49**, 782–784.

Neumann, C. 1885. "Ueber die rollende Bewegung eines Körpers auf einer gegebenen Horizontal–Ebene unter dem Einfluss der Schwere," *Königl. Sächsischen Ges. Wiss. Leipzig, Berichte, Math.-Phys. Kl.*, **37**, 352–378.

Neumann, C. 1887. "Grundzüge der analytischen Mechanik, insbesondere der Mechanik starrer Körper," *Berichte Verh. Königl. Sächs. Ges. Wiss. Leipzig, Math.-Phys. Kl.*, **39**, 153–190. [Also **40**, 22–88, 1888.]

Neumann, C. 1899. "Beiträge zur analytischen Mechanik," *Berichte Verh. Königl. Sächs. Ges. Wiss. Leipzig, Math.-Phys. Kl.*, **51**, 371–443.

Nevanlina, R. 1968. *Space Time and Relativity*. London: Addison-Wesley (original in German, 1964).

Nevzgliadov, V. G. 1959. *Theoretical Mechanics* (in Russian). Moscow: Physical–Mathematical Literature.

Newboult, H. O. 1946. *Analytical Method in Dynamics*. Oxford, U.K.: Clarendon Press.

Newland, D. E. 1965. "On the Methods of Galerkin, Ritz and Krylov–Bogoliubov in the Theory of Nonlinear Vibrations," *Int. J. Mech. Sci.*, **7**, 159–172.

Nielsen, J. 1935. *Vorlesungen über Elementare Mechanik*. Berlin: Springer (2nd ed. in Danish, published by Jul. Gjellerups, København; part I: 1943, part II: 1945).

Nijmeijer, H., and A. J. van der Schaft. 1990. *Nonlinear Dynamical Control Systems*. New York: Springer (second corrected printing, 1991).

Nikitina, N. V. 1976. "A New Form of the Equations of Motion of Systems with Nonholonomic Constraints," *Applied Mechanics*, **12**, 99–101 (original in Russian, 131–134).

Nikravesh, P. E. 1988. *Computer-Aided Analysis of Mechanical Systems*. Englewood Cliffs, NJ: Prentice Hall.

Noether, E. 1918. "Invariante Variationsprobleme," *Nachrichten von der Königlichen Gesellschaft der Wissenschaften zu Göttingen, Math.-phys. Klasse*, Heft 2, 235–257.

Noether, F., 1929. "Karl Heun," *ZAMM*, **9**, 167–171. [Also: *ZAMM*, **1**, 232, 1921.]

Nordheim, L. 1927. "Die Prinzipe der Dynamik," pp. 43–90 in vol. 5 of *Handbuch der Physik*. Berlin: Springer.

Nordheim, L., and E. Fues. 1927. "Die Hamilton–Jacobische Theorie," pp. 91–130 in vol. 5 of *Handbuch der Physik*. Berlin: Springer.

Novoselov, V. S. 1957. "Application of Nonlinear Non-holonomic Coordinates in Analytical Mechanics," *Leningrad University*, 217 ser. *Mat. Nauk.*, **31**, 50–83 (in Russian).

Novoselov, V. S. 1961(a). "The Extremal Character of the Hamilton–Ostrogradskii Principle in Non-holonomic Mechanics," *Vestnik Leningrad University*, **16**(13), 121–130 (in Russian).

Novoselov, V. S. 1961(b). "Extremal Character of the Euler–Lagrange Principle in Non-holonomic Mechanics," *Vestnik Leningrad University*, **16**(19), 138–144 (in Russian).

Novoselov, V. S. 1962. "Extremal Character of Integral Principles of Non-holonomic mechanics in Non-holonomic Coordinates," *Vestnik Leningrad University*, **17**(1), 124–133 (in Russian).

Novoselov, V. S. 1966. *Variational Methods in Mechanics* (in Russian). Leningrad: Leningrad University Press.

Novoselov, V. S. 1967. *Holonomic Systems in Lagrangean Coordinates* (in Russian). Leningrad: Leningrad University Press.

Novoselov, V. S. 1969. *Analytical Mechanics of Systems with Variable Mass* (in Russian). Leningrad: Leningrad University Press.

Novoselov, V. S. 1979. "Analytical Mechanics of Models of Motion," pp. 102–143 in vol. 3 of *Problems of Mechanics and Control, Mechanics of Controlled Motion*. Leningrad: Leningrad University Press.

Ogden, R. W. 1984. *Non-Linear Elastic Deformations*. Chichester, U.K.: Ellis Horwood/Halsted/Wiley.

Oliver, D. 1994. *The Shaggy Steed of Physics. Mathematical Beauty in the Physical World*. New York: Springer.

Ol'khovskii, I. I. 1970. *Course of Theoretical Mechanics for Physicists* (in Russian). Moscow: Nauka.

Olsson, M. G. 1981. "Spherical Pendulum Revisited," *Am. J. Physics*, **49**, 531–534.

Oravas, G. E., and L. McLean. 1966. "Historical Development of Energetical Principles in Elastomechanics," *Applied Mechanics Reviews*, **19**, 647–658 and 919–933. [See also Introduction to English translation of C. A. P. Castigliano, *The Theory of Equilibrium of Elastic Systems and its Applications*, reprinted 1966 by Dover, New York (original in French, 1879; German translation 1886; English translation 1919).]

O'Reilly, O. M. 1996. "The Dynamics of Rolling Disks and Sliding Disks, *Nonlinear Dynamics*, **10**, 287–305.

O'Reilly, O. M., and A. S. Srinivasa. 2001. "On a Decomposition of Generalized Constraint Forces," *Proc. Roy. Soc. London*, **457**, 1307–1313.

Osgood, W. F. 1937. *Mechanics*. New York: Macmillan (reprinted 1965 by Dover, New York).

Ostrovskaya, S., and J. Angeles. 1998. "Nonholonomic Systems Revisited within the Framework of Analytical Mechanics, *Applied Mechanics Reviews*, **51**, 415–433.

Otterbein, S. 1981. *Stabilität und Steuerbarkeit bei Systemen mit quasizyklischen Koordinaten*. Dr.-Ing. Dissertation, Tech. Hochschule, Darmstadt, Germany.

Panovko, J. G. 1971. *Introduction to the Theory of Mechanical Oscillations*. Moscow: Nauka.

Panovko, J. G. 1977. *Introduction to the Theory of Mechanical Impact* (in Russian). Moscow: Nauka.

Papastavridis, J. G. 1980. "On the Extremal Properties of Hamilton's Action Integral," *J. Appl. Mechanics* (ASME), **47**, 955–956.

Papastavridis, J. G. 1981. "An Energy Test for the Stability of Rheolinear Vibrations," *J. Sound and Vibration*, **74**, 499–506.

Papastavridis, J. G. 1982(a) "A Direct Variational Method for Nonconservative System Stability," *J. Sound and Vibration*, **80**, 447–459.

Papastavridis, J. G. 1982(b). "Parametric Excitation Stability via Hamilton's Action Principle," *J. Sound and Vibration*, **82**, 401–410.

Papastavridis, J. G. 1983(a). "Toward an Extremum Characterization of Kinetic Stability," *J. Sound and Vibration*, **87**, 573–587.

Papastavridis, J. G. 1983(b). "An Eigenvalue Criterion for the Study of the Hamiltonian Action's Extremality," *Mechanics Research Communications*, **10**, 171–179.

Papastavridis, J. G. 1983(c). "On the Variational Theory of Linear Configuration-Dependent Systems," *J. Sound and Vibration*, **89**, 85–94.

Papastavridis, J. G. 1983(d). "Complementary Variational Principles and their Application to Rheolinear and Nonlinear Vibrations," *J. Sound and Vibration*, **89**, 233–241.

Papastavridis, J. G. 1983(e). "On the Static and Kinetic Methods of Conservative System Stability, and Rayleigh's Principle: A Reexamination," *J. Sound and Vibration*, **90**, 51–58.

Papastavridis, J. G. 1984. "The Loading–Frequency Relationship of Linear Conservative Systems via a Direct Energetic (Action) Method," *J. Sound and Vibration*, **94**, 223–233.

Papastavridis, J. G. 1985(a). "The Variational Theory of Adiabatic Motions and its Applications to Linear and Non-linear Oscillators," *J. Sound and Vibration*, **103**, 83–98.

Papastavridis, J. G. 1985(b). "An Action-Based Method for the Kinetic Stability of Potential but Nonconservative Systems," *J. Appl. Mechanics* (ASME), **52**, 731–733.

Papastavridis, J. G. 1986(a). "On a Lagrangean Action Based Kinetic Instability Theorem of Kelvin and Tait," *Int. J. Eng. Sci.*, **24**, 1–17.

Papastavridis, J. G. 1986(b). "The Variational and Virial-like Theory of Oscillations and Stability of Nonconservative and/or Nonlinear Mechanical Systems," *J. Sound and Vibration*, **104**, 209–227.

Papastavridis, J. G. 1986(c). "The Principle of Least Action as a Lagrange Variational Problem: Stationarity and Extremality Conditions," *Int. J. Eng. Sci.*, **24**, 1437–1443.

Papastavridis, J. G. 1987(a). "Rayleigh's Principle via Least Action," *J. Sound and Vibration*, **113**, 395–399.

Papastavridis, J. G. 1987(b). "Time-integral Theorems for Nonholonomic Systems," *Int. J. Eng. Sci.*, **25**, 833–854.

Papastavridis, J. G. 1987(c) "The Energetic/Variational Interpretation of Poincaré's Criterion for the Orbital Stability of Limit Cycles of Autonomous Systems," *Int. J. Eng. Sci.*, **25**, 871–882.

Papastavridis, J. G. 1987(d). "The Variational Principles of Mechanics, and a Reply to C. D. Bailey," *J. Sound and Vibration*, **118**, 378–393.

Papastavridis, J. G. 1990. "The Maggi or Canonical Form of Lagrange's Equations of Motion of Holonomic Mechanical Systems," *J. Appl. Mechanics* (ASME), **57**, 1004–1010.

Papastavridis, J. G. 1992. "On the Transitivity Equations of Rigid-Body Dynamics," *J. Appl. Mechanics* (ASME), **59**, 955–962.

Papastavridis, J. G. 1998. "A Panoramic Overview of the Principles and Equations of Motion of Advanced Engineering Dynamics," *Applied Mechanics Reviews* (ASME), **51**, 239–265.

Papastavridis, J. G. 1999. *Tensor Calculus and Analytical Dynamics*. London/Ann Arbor/Boca Raton: CRC Press.

Papastavridis, J. G. (planned publication 2002). *Elementary Mechanics, A Handbook/ Compendium from an Advanced Viewpoint*, in preparation.

Papastavridis, J. G., and G. Chen. 1986. "The Principle of Least Action in Nonlinear and/or Nonconservative Oscillations," *J. Sound and Vibration*, **109**, 225–235.

Parczewski, J., and W. Blajer. 1989. "On Realization of Program Constraints; I. Theory, II. Practical Implications," *J. Appl. Mechanics* (ASME), **56**, 676–679 and 680–684.

Park, D. 1990. *Classical Dynamics and its Quantum Analogies*, 2nd ed. Berlin: Springer.

Parkus, H. 1966. *Mechanik der festen Körper*, 2nd ed. Vienna: Springer.

Pars, L. A. 1953. *Introduction to Dynamics*. Cambridge, U.K.: Cambridge University Press.

Pars, L. A. 1954. "Variation Principles in Dynamics," *Quart. J. Mech. and Appl. Math.*, **7**, 338–351.

Pars, L. A. 1965. *A Treatise on Analytical Dynamics*. London: Heinemann (reprinted 1979 by Ox Bow Press, Woodridge, CT).

Pascal, E. 1927. "Totale Differentialgleichungen und Differentialformen," Erster Band, Zweiter Teilband, pp. 579–600 in *Repertorium der Höheren Mathematik*, 2nd ed. Leipzig/ Berlin, Teubner.

Päsler, M. 1968. *Prinzipe der Mechanik*. Berlin: W. de Gruyter.

Passerello, C. E., and R. L. Huston. 1973. "Another Look at Nonholonomic Systems," *J. Appl. Mechanics* (ASME), **40**, 101–104.

Pastori, M. 1960. "Vincoli e Riferimenti Mobili in Meccanica Analitica," *Annali di Matematica pura ed applicata*, **50** (Series 4), 476–484.

Pastori, M. 1967. "Apparent Forces of Analytical Mechanics," *Meccanica*, **2**, 75–81.

Paul, B. 1979. *Kinematics and Dynamics of Planar Machinery*. Englewood Cliffs, NJ: Prentice Hall.

Paulus, F. 1910. "Ueber eine unmittelbare Bestimmung jeder einzelnen Reaktionskraft eines bedingten Punktsystems für sich aus den Lagrange'schen Gleichungen zweiter Art," *Wienersche Sitzungsberichte, Math.-Naturwiss. Kl.*, **IIa**, **119**, 1669–1718.

Paulus, F. 1916. "Ergänzungen und Beispiele zur Mechanik von Hertz," *Wienersche Sitzungsberichte, Math.-Naturwiss. Kl.*, **IIa**, **125**, 835–882.

Pearson, K. 1937. *The Grammar of Science*. London: J. M. Dent (original in 1892, W. Scott; reprinted 1951).

Peck, E. 1978. "Cart Wheels," *Am. J. Physics*, **46**, 509–512.

Peisakh, E. E. 1966. "On the Kinetic Foci of a Conservative System for Isoenergetic Disturbances," *PMM*, **30**, 1336–1341 (original in Russian, 1128–1132; German translation also available, 1997).

Percival, I. C., and D. Richards. 1982. *Introduction to Dynamics*. Cambridge, U.K.: Cambridge University Press.

Peres, A. 1979. "Finite Rotations and Angular Velocity," *Am. J. Physics*, **48**, 70–71.

Pérès, J. 1953. *Mécanique Générale*. Paris: Masson.

Pérez, J. Ph. 1992. *Mécanique, Points Matériels, Solides, Fluides avec Exercices et Problèmes Résolus*, 3rd ed. Paris: Masson.

Pfeiffer, F. 1984. "Mechanische Systeme mit unstetigen Übergängen," *Ingenieur-Archiv*, **54**, 232–240.

Pfeiffer, F. 1989. *Einführung in die Dynamik*. Stuttgart, Germany: Teubner.

Pfister, F. 1995. "Contributions à la Mécanique Analytique des Systèmes Multicorps." Doctoral Thesis, Institut National des Sciences Appliquées (INSA) de Lyon/Ecole Doctorale des Sciences pour l'Ingenieur de Lyon.

Pfister, F. 1997. "Das Gaußsche Prinzip und das Lagrangesche–Notizen zu einer kaum beachteten Arbeit Paul Stäckels," *ZAMM*, **77**, 503–508.

Pfister, F. 1998. "Bernoulli Numbers and Rotational Kinematics," *J. Appl. Mechanics* (ASME), **65**, 758–763.

Phelps, F. M., III, and J. H. Hunter, Jr. 1968. "An Analytical Solution of the Inverted Pendulum," *Am. J. Physics*, **36**, 285–295.

Pignedoli, A. 1982. "On Some Investigations of the Italian School about Holonomic and Non-Holonomic Systems," *Acad. Sci. Turin, Proceedings of the IUTAM–ISIMM Symposium on Modern Developments in Analytical Dynamics* (June 7–11), 1982, vol. 2, 645–670. [Also in *Atti della Accademia delle Scienze di Torino, Supplemento*, **117**, 1983.]

Piña, E. 1983. "A New Parametrization of the Rotation Matrix," *Am. J. Physics*, **51**, 375–379.

Pironneau, Y. 1982. "Sur les liaisons non holonomes non linéaires déplacement virtuels à travail nul, conditions de Chetaev," *Acad. Sci. Turin, Proceedings of the IUTAM–ISIMM Symposium on Modern Developments in Analytical Dynamics* (June 7–11), 1982, vol. 2, 671–686. [Also in *Atti della Accademia delle Scienze di Torino, Supplemento*, **117**, 1983.]

Planck, M. 1928. *General Mechanics*, vol. I of *Introduction to Theoretical Physics*, 4th ed. London: Macmillan and Co. (original in German, circa 1916).

Planck, M. 1960. *A Survey of Physical Theory*. New York: Dover [formerly titled *A Survey of Physics* (in German, 1925)].

Platrier, C. 1954. *Mécanique Rationelle*, 2 vols. Paris: Dunod.

Pogorelov, A. 1987. *Geometry*. Moscow: Mir.

Poincaré, H. 1901. "Sur une Forme Nouvelle des Équations de la Mécanique," *Comptes Rendus* (Acad. Sci. of Paris), **132**, 369–371.

Poinsot, L. 1877. *Eléments de Statique*, 12th (last) ed. Paris: Bachelier (1804, 1st ed.).

Poinsot, L. 1975. *La Théorie Générale de l'Équilibre et du Mouvement des Systèmes*, Edition Critique et Commentaires par P. Bailhache. Paris: Librairie Philosophique J. Vrin.

Poisson, S. D. 1811. *Traité de Mécanique*, 2 vols, 1st ed. Paris: Courcier [1833, 2nd ed.; 1838, 3rd ed., Brussels. German translation as *Lehrbuch der Mechanik*, Stuttgart/Tübingen (1825–1826), Berlin (1835–1836, from 2nd French edition; 2 Teile, by M. A. Stern), Dortmund (1888).]

Polak, L. S., editor. 1959. *Variational Principles in Mechanics* (collection of original memoirs; in Russian). Moscow: Physical–Mathematical Literature.

Polak, L. S. 1960. *Variational Principles in Mechanics; their Development and Application in Physics* (in Russian). Moscow: Physical–Mathematical Literature.

Poliahov, N. N. 1972. "Equations of Motion of Mechanical Systems under Nonlinear, Nonholonomic Constraints," *Vestnik Leningrad University*, **1**, 124–132 (in Russian).

Poliahov, N. N. 1974. "On Differential Principles of Mechanics, Derived from the Equations of Motion of Nonholonomic Systems," *Vestnik Leningrad University*, **13**, 106–114 (in Russian).

Poliahov, N. N., S. A. Zegzhda, and M. P. Yushkov. 1985. *Theoretical Mechanics* (in Russian). Leningrad: Leningrad University Press.

Pollard, H. 1976. *Celestial Mechanics*. The Mathematical Association of America (#18 of The Carus Mathematical Monographs).

Pöschl, T. 1913. "Sur les Équations Canoniques des Systèmes Non Holonomes," *Comptes Rendus* (Acad. Sci. of Paris), **156**, 1829–1831.

Pöschl, T. 1927. "Technische Anwendungen der Stereomechanik," pp. 484–577 in vol. 5 of *Handbuch der Physik*. Berlin: Springer.

Pöschl, T. 1928. "Der Stoss," pp. 484–577 in vol. 6 of *Handbuch der Physik*. Berlin: Springer.

Pöschl, T. 1949. *Einführung in die Analytische Mechanik*. Karlsruhe: G. Braun.

Pozharitskii, N. N. 1960. "On the Equations of Motion for Systems with Nonideal Constraints," *PMM*, **24**, 669–676 (original in Russian, 458–462).

Prange, G. 1935. "Die allgemeinen Integrationsmethoden der analytischen Mechanik," arts. 12 and 13, pp. 505–804 in vol. 4 of *Encyklopädie der Mathematischen Wissenschaften*. Leibniz: Teubner (article completed in December 1933).

Pringle, R., Jr. 1965. "Stability of Damped Mechanical Systems," *AIAA J.*, **3**, 363.

Provost, J. P. 1975. "Least Action Principle on Air Table," *Am. J. Physics*, **43**, 774–781.

Przeborski, A. 1933. "Die allgemeinsten Gleichungen der klassischen Dynamik," *Math. Zeitschrift*, **36**, 184–194.

Pulte, H. 1998. "Jacobi's Criticism of Lagrange: The Changing Role of Mathematics in the Foundations of Classical Mechanics," *Historia Mathematica*, **25**, 154–184.

Putyata, T. V., and B. N. Fradlin. 1971. "On the Centenary of the Birth of Petr Vasil'evich Voronets," *Applied Mechanics*, **7**, 1177–1179 (original in Russian, **10**, 137–139).

Qiao, Y.-F. 1990. "Gibbs–Appell's Equations of Variable Mass Nonlinear Nonholonomic Mechanical Systems," *Applied Mathematics and Mechanics*, **11**, 973–983 (original in Chinese).

Quanjel, J. 1906. "Les Équations Générales de la Mécanique dans le Cas des Liaisons Non-holonomes," *Rendiconti del Circolo Matematico di Palermo*, **22**, 263–273.

Quimby, S. L. 1994. *Analytical Dynamics*. River Edge, NJ: World Scientific (author's name on book's cover and spine appears as Samuel D. Lindenbaum!).

Radetsky, P. 1986. "The Man who Mastered Motion," *Science*, **7**, 52–60.

Raher, W. 1954. "Das d'Alembertsche Prinzip in Motorsymbolik und seine Anwendung auf Stoßprobleme," *ZAMM*, **34**, 323–324.

Raher, W. 1955. "Zur Theorie des Stosses starrer Körper," *Oesterreichisches Ing. Archiv*, **9**, 55–68.

Rajan, M., and J. L. Junkins. 1983. "Perturbation Methods Based upon Varying Action Integrals," *Int. J. Nonlinear Mechanics*, **18**, 335–351.

Ramsey, A. S. *Dynamics*. Cambridge, U.K.: Cambridge University Press [part I: 1933, 2nd ed.); part II: 1937 (several subsequent reprintings).]

Ramsey, A. S. 1940. *An Introduction to the Theory of Newtonian Attraction*. Cambridge, U.K.: Cambridge University Press (several subsequent reprintings).

Ramsey, A. S. 1941. *Statics*, 2nd ed. Cambridge, U.K.: Cambridge University Press (several subsequent reprintings).

Rasband, S. Neil. 1983. *Dynamics*. New York: Wiley-Interscience.

Rausenberger, O. 1888. *Lehrbuch der Analytischen Mechanik*, vol. 1: Mechanik der Materiellen Punkte, vol. 2: Mechanik der Zusammenhängenden Körper, 2nd ed. Leipzig: Teubner.

Rayleigh, J. W. Strutt (Third Baron of Rayleigh). 1894. *The Theory of Sound*, vol. 1, 2nd ed. London: Macmillan (reprinted 1945 by Dover, New York).

Rayleigh, Lord. 1902. "On the Pressure of Vibrations," *Philosophical Magazine*, series 6, **3**, 338–346.

Relvik, H. 1989. "On Mangeron's Generalized Equations of Analytical Dynamics and Some Related Problems," *Romanian Academy of Sciences* (Memoirs of Scientific Sections), Series IV, **12**, 59–89.

Renteln, M. v. 1995. "Karl Heun — his life and scientific work," pp. xvii–xx of *Heun's Differential Equation*, edited by A. Ronveaux. Oxford: Oxford University Press (rare photo of Heun on p. xvi).

Ricci, G. and T. Levi-Civita. 1901. "Méthodes de Calcul Différentiel Absolu et leurs Applications," *Mathematische Annalen*, **54**, 125–201.

Richardson, D. L. 1992. "Determination of the Angular Velocity Vector in Orthogonal Curvilinear Coordinates," *J. Appl. Mechanics* (ASME), **59**, 456–457.

Richter, M. 1887. "Über die Bewegung eines Körpers auf einer horizontal-Ebene," Inaugural Dissertation (57 pages). Leipzig: Metzger and Wittig.

Rimrott, F. P. J., and G. Patkó. 1995. "Zur Altenbachschen Vermutung bei Komplementärfassungen der Analytischen Mechanik," *ZAMM*, **75**, 401–403.

Rimrott, F. P. J., W. M. Szczygielski, and B. Tabarrok. 1993. "Kinetic Energy and Complementary Kinetic Energy in Gyrodynamics," *J. Appl. Mechanics* (ASME), **60**, 398–405.

Ritter, A. 1899. *Lehrbuch der Analytischen Mechanik*, 3rd ed. Leipzig: Baumgartner (earlier editions 1873, 1883).

Roberson, R. E., and R. Schwertassek. 1988. *Dynamics of Multibody Systems*. Berlin: Springer.

Rocard, J. M. 1974. *Mécanique des Systémes*, Mécanique Analytique, Relativité Restreinte. Paris: Masson.

Rodrigues, O. 1840. "Des Lois Géométriques qui Régissent les Déplacements d'un Système Solide dans l'Éspace, et de la Variation des Coordonées Provenant de ces Déplacements Considérés Independamment des Causes qui Peuvent les Produire," *J. Math. Pures et Appl.* **5**(1), 380–440.

Romanov, I. P., 1976. "On Equivalence of the Equations of Motion of Nonholonomic Systems," *PMM*, **40**, 157–161 (original in Russian, 176–179).

Rose, N. V. 1983. *Lectures on Analytical Mechanics*, part I (in Russian). Leningrad: Leningrad University Press.

Roseau, M. 1987. *Vibrations in Mechanical Systems, Analytical Methods and Applications*. Berlin: Springer (original in French, 1984).

Rosenberg, R. M. 1977. *Analytical Dynamics of Discrete Systems*. New York: Plenum.

Rosenberg, R. M. 1991. "Some Remarks about Time in Newtonian Particle Mechanics," *Acta Mechanica*, **89**, 13–19.

Routh, E. J. 1877. *A Treatise on the Stability of a Given State of Motion, Particularly Steady Motion*. London: Macmillan. [Also pp. 19–138 of *Stability of Motion*, by E. J. Routh, edited by A. T. Fuller, published by Taylor and Francis /Halsted Press, London, 1975.]

Routh, E. J. *A Treatise on Analytical Statics, with Numerous Examples*, vol. I (1891), vol. II (1892). Cambridge, U.K.: Cambridge University Press.

Routh, E. J. 1898. *A Treatise on Dynamics of a Particle, with Numerous Examples*. Cambridge, U.K.: Cambridge University Press (reprinted by Stechert and Dover, New York).

Routh, E. J. 1905(a). *Dynamics of a System of Rigid Bodies, Elementary Part*, 7th ed. London: Macmillan/St. Martin (reprinted 1960 by Dover, New York).

Routh, E. J. 1905(b). *Dynamics of a System of Rigid Bodies, Advanced Part*, 6th ed. London: Macmillan/St. Martin (reprinted 1955 by Dover, New York).

Roy, M. 1965. *Mécanique, I. Corps Rigides*. Paris: Dunod.

Rüdiger, D., and A. Kneschke. 1966. *Technische Mechanik*, Band 3: *Kinematik und Kinetik*. Zürich und Frankfurt/Main: H. Deutsch.

Rühlman, M. 1885. *Vorträge über Geschichte der Technischen Mechanik und Theoretischen Maschinenlehre sowie der damit in Zusammenhang stehenden mathematischen Wissenschaften*. Leipzig: Baumgärtner.

Rumiantsev, V. V. 1975. "On the Compatibility of Two Basic Principles of Dynamics and on Chetaev's Principle," pp. 258–267 in *Problems of Analytical Mechanics, Theory of Stability and Control* (in Russian). Moscow: Nauka (Acad. Sci. U.S.S.R.).

Rumiantsev, V. V. 1976. "On the Motion of Controllable Mechanical Systems," *PMM*, **40**, 719–729 (original in Russian, 771–781).

Rumiantsev, V. V. 1978. "On Hamilton's Principle for Nonholonomic Systems," *PMM*, **42**, 407–419 (original in Russian, 387–399).

Rumiantsev, V. V. 1979. "On the Lagrange and Jacobi Principles for Nonholonomic Systems," *PMM*, **43**, 625–632 (original in Russian, 583–590).

Rumiantsev, V. V. 1981. "Certain Variational Principles of Mechanics," pp. 36–52 in *Advances in Theoretical and Applied Mechanics*, edited by A. Y. Ishlinsky and F. L. Chernousko. Moscow: Mir (original in Russian, 1978).

Rumiantsev, V. V. 1982(a). "On Some Nonlinear Problems of Analytical Mechanics and Theory of Stability," pp. 869–881 in *Nonlinear Phenomena in Mathematical Sciences*, edited by V. Lakshmikantham. New York: Academic Press.

Rumiantsev, V. V. 1982(b). "On Integral Principles for Nonholonomic Systems," *PMM*, **46**, 1–8 (original in Russian, 3–12).

Rumiantsev, V. V. 1983. "On Some Problems of Analytical Dynamics of Nonholonomic Systems," pp. 697–716 in vol. 2 of *Proc. IUTAM–ISIMM Symp. Modern Developments in Analytical Mechanics. Acad. Sc. Torino* (June 1982).

Rumiantsev, V. V. 1984. "The Dynamics of Rheonomic Lagrangian Systems with Constraints," *PMM*, **48**, 380–387 (original in Russian, 540–550).

Rumyantsev, V. V. 1990. "On the Principal Laws of Classical Mechanics," pp. 257–273 in vol. 1 of *General and Applied Mechanics of Mechanical Engineering and Applied Mechanics*, edited by V. Z. Parton. New York: Hemisphere Publishing.

Rumyantsev, V. V. 1994. "On the Poincaré–Chetayev Equations," *PMM*, **58**, 373–386 (original in Russian, 3–16).

Rumyantsev, V. V. 1998. "On the Poincaré and Chetayev Equations," *PMM*, **62**, 495–502 (original in Russian, 531–538).

Rumyantsev, V. V. 1999. "Forms of Hamilton's Principle in Quasi-coordinates," *J. Applied Mathematics and Mechanics*, **63**, 165–171 (original in Russian, 172–178).

Rumyantsev, V. V., and A. S. Sumbatov. 1978. "On the Problem of a Generalization of the Hamilton–Jacobi Method for Nonholonomic Systems," *ZAMM*, **58**, 477–481.

Rund, H. 1966. *The Hamilton–Jacobi Theory in the Calculus of Variations, Its Role in Mathematics and Physics*. New York: Van Nostrand/Reinhold (reprinted 1973, with a supplementary appendix, by Krieger, Huntington, New York).

Rutherford, D. E. 1960. *Classical Mechanics*, 2nd ed. Edinburgh, U.K. Oliver and Boyd.

Sagdeev, R. Z., D. A. Usikov, and G. M. Zaslavsky. 1988. *Nonlinear Physics: From the Pendulum to Turbulence and Chaos*. Chur (Switzerland) etc.: Harwood.

Saint-Germain, A. de. 1900. "Sur la Fonction S Introduite par P. Appell dans les Equations de la Dynamique," *Comptes Rendus* (Acad. Sci. of Paris), **130**, 1174–1176.

Saletan, E. J., and A. H. Cromer. 1970. "A Variational Principle for Nonholonomic Systems," *Am. J. Physics*, **38**, 892–897.

Saletan, E. J., and A. H. Cromer. 1971. *Theoretical Mechanics*. New York: Wiley.

San, D. 1973. "Equations of Motion of Systems with Second-Order Nonlinear Nonholonomic Constraints," *PMM*, **37**, 329–334 (original in Russian, 349–354).

Sanh, D. 1981(a). "On a Mechanical System with Nonideal Constraints," *Nonlinear Vibration Problems*, **20**, 81–93.

Sanh, D. 1981(b). "On the Dynamical Action of Each Constraint on a Mechanical System," *Nonlinear Vibration Problems*, **20**, 95–107.

Sanden, H. v. 1945. *Praxis der Differential-gleichungen*. Berlin: W. de Gruyter and Co.

Santilli, R. M. *Foundations of Theoretical Mechanics*, part I: 1978, part II: 1980. New York: Springer.

Sarlet, W. 1982. "Noether's Theorem and the Inverse Problem of Lagrangian Mechanics," *Acad. Sci. Turin, Proceedings of the IUTAM–ISIMM Symposium on Modern Developments in Analytical Dynamics* (June 7–11), 1982, vol. 2, 737–751. [Also in *Atti della Accademia delle Scienze di Torino, Supplemento*, **117**, 1983.]

Savin, G. N., and B. N. Fradlin. 1972. "Genesis of the Fundamental Concepts in Nonholonomic Mechanics," *J. Applied Mechanics*, **5**(1), 8–15 (translation of Ukrainian *Prikladnaya Mekhanika*, 11–20, 1969, by Consultants Bureau of Plenum Co., New York).

Schaefer, C. 1918. "Eine einfache Herleitung der verallgemeinerten Lagrangeschen Gleichungen für nichtholonome Koordinaten," *Physik. Zeitschrift*, **19**, 129–134 and 406–407.

Schaefer, C. 1919. *Die Prinzipe der Dynamik*. Berlin: Walter de Gruyter and Co.

Schaefer, C. 1937. *Einführung in die Theoretische Physik*, Dritter Band, Zweiter Teil: *Quantentheorie*. Berlin: Walter de Gruyter and Co.

Schaefer, H. 1951. "Die Bewegungsgleichungen nichtholonomer Systeme," *Abhandlungen der Braunschweig Wiss. Gesellschaft* (published by F. Vieweg), **3**, 116–121.

Schaub, H., P. Tsiotras, and J. L. Junkins. 1995. "Principal Rotation Representations of Proper $N \times N$ Orthogonal Matrices," *Int. J. Eng. Sci.*, **33**, 2277–2295.

Scheck, F. 1994. *Mechanics, From Newton's Laws to Deterministic Chaos*, 2nd ed. Berlin: Springer [translated from German, 1988 (1st ed.), 1990 (2nd ed.), 1992 (3rd ed.), 1994 (4th ed.)].

Scheck, F., and R. Schöpf. 1989. *Mechanik Manual, Aufgaben mit Lösungen*. Berlin: Springer.

Schell, W. *Theorie der Bewegung und der Kräfte, Ein Lehrbuch der Theoretischen Mechanik* [vol. I (1879): *Geometrie der Streckensysteme und Geometrie der Massen. Geometrie der Bewegung und Theorie der Bewegungszustände (Kinematik)*; vol. II (1880): *Theorie der Kräfte und ihrer Aequivalenz (Dynamik im weiteren Sinne einschl. Statik). Theorie der durch Kräfte erzeugten Bewegung (Kinetik oder Dynamik im engeren Sinne)*], 2nd ed. Leipzig: Teubner (1st ed. in 1870).

Schering, E. 1873. "Hamilton–Jacobische Theorie für Kräfte, deren Maass von der Bewegung der Körper abhangen," *Königl. Ges. Wiss. Göttingen, Abh. Math. Kl.*, **18**, 3–54.

Schiehlen, W. 1981. "Nichtlineare Bewegungsgleichungen großer Mehrkörpersysteme," *ZAMM*, **61**, 413–419.

Schiehlen, W. 1986. *Technische Dynamik*. Stuttgart, Germany: Teubner.

Schieldrop, E. B. 1925. "Sur une Notion de Déviation Non-holonome, et son Application à la Cinématique et la Dynamique des Systèmes Matériels," *Den Sjette Skandinaviske Matematikerkongres i Kobenhavn, Beretning*, pp. 281–307.

Schlitt, D. W. 1977. "Hamilton's Principle and Approximate Solutions to Problems in Classical Mechanics," *Am. J. Physics*, **45**, 205–207.

Schmeidler, W. 1955. "Zum Gedächtnis an Georg Hamel," *Jahresbericht d. DMV* (yearly report/book of the German Mathematical Society), **58**, 1–5.

Schmieder, L. 1994. "Die Anwendung der Lieschen Gruppentheorie auf Probleme der Mechanik," *ZAMM*, **74**, 133–137.

Schmutzer, E. 1973. *Grundprinzipien der klassischen Mechanik und der klassischen Feldtheorie* (Kanonischer Apparat). Berlin: VEB Deutscher Verlag der Wissenschaften.

Schmutzer, E. 1989. *Grundlagen der Theoretischen Physik, mit einem Grundriß der Mathematik für Physiker*, Teil I. Berlin: VEB Deutscher Verlag der Wissenschaften.

Schönflies, A., and M. Grübler. 1901–1908. "Kinematik," art. 3, pp. 190–278 in vol. 4 of *Encyklopaedie der Mathematischen Wissenschaften*. Leipzig: Teubner (article completed in June 1902. French translation also available).

Schot, S. H. 1978. "Jerk: The Time Rate of Change of Acceleration," *Am. J. Physics*, **46**, 1090–1094.

Schouten, G. 1889. "Algemeene eigenschappen van de zuiver rollende beweging van een omwentelingslichaam op een horizontaal vlak, toegepast op de beweging van een omwentelingslichaam om een vast punt van zijne as," *Verslag. en Mededeel. Koninkl. Akad. Wetensch. Amsterdam*, series 3, **5**, 292–335.

Schouten, J. A. 1954. *Tensor Analysis for Physicists*, 2nd ed. Oxford, U.K.: Clarendon Press (reprinted 1989 by Dover, New York; from corrected printing of 1959).

Schräpel, H.-D. 1982. "Equivalent Linearization of Canonical Systems," *Acad. Sci. Turin, Proceedings of the IUTAM–ISIMM Symposium on Modern Developments in Analytical Dynamics* (June 7–11), 1982, vol. 2, 759–763. [Also in *Atti della Accademia delle Scienze di Torino, Supplemento*, **117**, 1983.]

Schräpel, H.-D. 1988. "Analytische Näherungslösungen nichtlinearer Eigenschwingungen mit dem Hamiltonschen Prinzip," *ZAMM*, **68**, T117–T118.

Schuster, J., and R. H. Good, Jr. 1986. "Angular Velocity and Time Derivative of Angular Displacement," *Am. J. Physics*, **54**, 1029–1031.

Schwartz, B. 1986. *The Battle for Human Nature, Science, Morality and Modern Life*. New York: Norton.

Schwartz, H. M. 1963. "Derivation of the Matrix of Rotation about a Given Direction as a Simple Exercise in Matrix Algebra," *Am. J. Physics*, **31**, 730–731.

Schwartz, M. 1964. "Lagrangian and Hamiltonian Formalisms with Supplementary Conditions," *J. Mathematical Physics*, **5**, 903–907.

Scott, D. 1988. "Can a Projection Method of Obtaining Equations of Motion Compete with Lagrange's Equations?" *Am. J. Physics*, **56**, 451–456.

Seeger, R. J. 1934. "The Virial Theorem for Nonholonomic Systems," *J. Washington Acad. Sci.*, **24**, 461–464.

Seeliger, R., in collaboration with F. Henning, and R. v. Mises. 1921. *Aufgaben aus der Theoretischen Physik*. Braunschweig: F. Vieweg and Sohn.

Semenova, L. N. 1965. "On Routh's Theorem for Nonholonomic Systems," *PMM*, **29**, 167–169 (original in Russian, 156–157).

Serrin, J. 1959. "Mathematical Principles of Classical Fluid Dynamics," pp. 125–263 in vol. VIII/1 of *Handbuch der Physik*. Berlin: Springer.

Sethna, P. R., and M. B. Balachandra. 1976. "On Nonlinear Gyroscopic Systems," pp. 191–242 in vol. 3 of *Mechanics Today*. New York: Pergamon Press.

Shabana, A. A. 1989. *Dynamics of Multibody Systems*. New York: Wiley.

Shabana, A. A. 1994. *Computational Dynamics*. New York: Wiley-Interscience.

Shan, D. 1975(a). "Canonical Equations for Mechanical Systems with Second-Order Linear Nonholonomic Constraints," *Applied Mechanics*, **11**, 159–166 (original in Russian, 58–67).

Shan, D. 1975(b). "The Gauss Principle and Equations of Motion of Mechanical Systems with Arbitrary Couplings," *Applied Mechanics*, **11**, 757–764 (original in Russian, 89–97).

Sharf, I., G. M. T. D'Eleuterio, and P. C. Hughes. 1992. "On the Dynamics of Gibbs, Appell, and Kane," *Eur. J. Mechanics, A/Solids*, **11**, 145–155.

Shen, Z.-C., and F.-X. Mei. 1987. "On the New Forms of the Differential Equations of the Systems with Higher-Order Nonholonomic Constraints," *Applied Mathematics and Mechanics*, **8**, 189–196 (original in Chinese).

Shields, A. 1988. "Lagrange and the Mécanique Analytique," *The Mathematical Intelligencer*, **10**, 7–10.

Shuster, M. D. 1993(a). "Attitude Determination in Higher Dimensions," *J. Guidance, Control, and Dynamics* (AIAA), **16**, 393–395.

Shuster, M. D. 1993(b). "The Kinematic Equation for the Rotation Vector," *IEEE Transactions on Aerospace and Electronic Systems*, **29**, 263–267.

Shuster, M. D. 1993(c). "A Survey of Attitude Representations," *J. Astronautical Sciences*, **41**, 439–517.

Siegel, C. L., and J. K. Moser. 1971. *Lectures on Celestial Mechanics*. New York: Springer.

Simonyi, K. 1990. *Kulturgeschichte der Physik*. Thun/Frankfurt am Main: Harri Deutsch (original in Hungarian, 1986, 3rd ed.).

Sinitsin, V. A. 1979. "On the Carnot Theorem in the Theory of Impulsive Motion of Mechanical Systems," *PMM*, **43**, 1207–1210 (original in Russian, 1114–1116).

Sinitsyn, V. A. 1982. "On the Hamilton–Ostrogradskii Principle in the Case of Impulsive Motions of Dynamic Systems," *PMM*, **45**, 356–359 (original in Russian, 488–493).

Sivardière, J. 1986. "Using the Virial Theorem," *Am. J. Physics*, **54**, 1100–1103.

Sivardière, J. 1994. "Products of Rotations: A Geometrical Point of View," *Am. J. Physics*, **62**, 737–743.

Skalmierski, B. 1991. *Mechanics*. Amsterdam: Elsevier; also: Warszawa: Polish Scientific Publishers (PWN).

Slater, J. C., and N. H. Frank. 1947. *Mechanics*. New York: McGraw-Hill.

Smart, E. H. 1951. *Advanced Mechanics*, 2 vols. London: Mcmillan.

Smith, C. E. 1982. *Applied Mechanics: Dynamics*, 2nd ed. New York: Wiley.

Smith, C. E., and P.-P. Liu. 1992. "Coefficients of Restitution," *J. Applied Mechanics* (ASME), **59**, 963–969.

Smith, C., and M. N. Wise. 1989. *Energy and Empire, A Biographical Study of Lord Kelvin*. Cambridge, U.K.: Cambridge University Press.

Sneddon, I. N. 1957. *Elements of Partial Differential Equations*. New York, McGraw-Hill.

Snider, A. D., and M. McWaters. 1980. "Infinitesimal Rotations," *Am. J. Physics*, **48**, 250–251.

Sokolnikoff, I. S. 1964. *Tensor Analysis, Theory and Applications to Geometry and Mechanics of Continua*, 2nd ed. New York: Wiley (1951, 1st ed.).

Sommerfeld, A. 1931. *Atombau und Spektrallinien*, 5th ed. Braunschweig: F. Vieweg and Sohn (English translation available by Dutton, New York, 1934).

Sommerfeld, A. 1964. *Mechanics*, vol. 1 of *Lectures on Theoretical Physics*, 4th ed. New York: Academic Press (original in German, 1940s).

Somoff, J. *Theoretische Mechanik*; I. Theil: *Kinematik* (1878); II. Theil: *Einleitung in die Statik und Dynamik. Statik* (1879). Leipzig: Teubner (original in Russian, early 1870s).

Sonntag, R. 1955. *Aufgaben aus der Technischen Mechanik, Graphische Statik, Festigkeitslehre, Dynamik*. Berlin: Springer.

Spiegel, M. R. 1967. *Theoretical Mechanics*. New York: Schaum.

Stäckel, P. 1903. "Bericht über die Mechanik mehrfacher Mannigfaltigkeiten," *Jahresbericht der Deutschen Mathematiker-Vereinigung*, **12**, 469–481.

Stäckel, P. 1901–1908. "Elementare Dynamik der Punktsysteme und Starrer Körper," art. 6, pp. 435–691 in vol. 4 of *Encyklopädie der Mathematischen Wissenschaften*. Leipzig: Teubner (article completed in 1905).

Stadler, W. 1982. "Inadequacy of the Usual Newtonian Formulation for Certain Problems in Particle Mechanics," *Am. J. Physics*, **50**, 595–598.

Stadler, W. 1991. "Controllability Implications of Newton's Third Law," *Dynamics and Control*, **1**, 53–61.

Stadler, W. 1995. *Analytical Robotics and Mechatronics*. New York: McGraw-Hill.

Stadler, W. 1996. "Natural Structural Shapes, The Dynamic Case," Private Communication.

Starzhinskii, V. M. 1980. *Applied Methods in the Theory of Nonlinear Oscillations*. Moscow: Mir (original in Russian, 1977).

Stavrakova, N. E. 1965. "The Principle of Hamilton–Ostrogradskii for Systems with One-Sided Constraints," *PMM*, **29**, 874–878 (original in Russian, 738–741).

Stephani, H. 1982, *General Relativity, An Introduction to the Theory of the Gravitational Field*. Cambridge, U.K.: Cambridge University Press (original in German, 1980, 2nd ed.).

Stephani, H. and G. Kluge. 1980. *Grundlagen der Theoretischen Mechanik*, 2nd ed. Berlin: VEB Deutscher Verlag der Wissenschaften.

Stickforth, J. 1978. "On the Complementary Lagrange Formalism of Classical Mechanics," *Am. J. Physics*, **46**, 71–73.

Stiefel, E. L., and G. Scheifele. 1971. *Linear and Regular Celestial Mechanics*. New York: Springer.

Stiller, W. 1989. *Ludwig Boltzmann, Altmeister der klassischen Physik, Wegbereiter der Quantenphysik und Evolutionstheorie*. Thun and Frankfurt: H. Deutsch.

Stoker, J. J. 1950. *Nonlinear Vibrations in Mechanical and Electrical Systems*. New York: Interscience Publishers (Wiley).

Storch, J., and S. Gates, 1989, "Motivating Kane's Method for Obtaining Equations of Motion for Dynamic Systems," *J. Guidance, Control, and Dynamics*, **12**, 593–595.

Straumann, N. 1987. *Klassische Mechanik, Grundkurs über Systeme endlich vieler Freiheitsgrade*. Berlin: Springer.

Stronge, W. J. 1990. "Rigid Body Collisions with Friction," *Proc. Roy. Soc. London, Series A*, **431**, 169–181.

Struik, D. J. 1976. "Lagrange, Joseph-Louis," *Encyclopaedia Britannica*, 15th ed, vol. 7, pp. 101–102.

Struik, D. J. 1987. *A Concise History of Mathematics*, 4th revised and enlarged ed. New York: Dover (earlier editions 1948 and 1967).

Stückler, B. 1955. "Über die Berechnung der an rollenden Fahrzeugen wirkenden Haftreibungen," *Ingenieur-Archiv*, **23**, 279–287.

Stuelpnagel, J. 1964. "On the Parametrization of the Three-Dimensional Rotation Group," *SIAM Review*, **6**, 422–429.

Sturm, Ch. 1883. *Cours de Mécanique de l'Ecole Polytechnique*, 2 vols, 5th ed. Paris: Prouhet (earlier editions 1861 and 1875).

Sudarshan, E. C. G., and N. Mukunda. 1974. *Classical Mechanics: A Modern Perspective*. New York: Wiley-Interscience.

Sumbatov, A. S. 1970. "Hamiltonian Principle for Nonholonomic Systems," *Moscow U. Mechanics Bulletin*, **25**, 29–32 (original in Russian, 98–101).

Sumbatov, A. S. 1972. "Integration of Dynamic Equations with Constraint Multipliers," *PMM*, **36**, 153–162 (original in Russian, 163–171).

Sumbatov, A. S. 1972. "Reduction of the Differential Equations of Nonholonomic Mechanics to Lagrange Form," *PMM*, **36**, 194–201 (original in Russian, 211–217).

Suslov, G. K. 1901–1902. "On a Particular Variant of d'Alembert's Principle," *Matematisheskii Svornik*, **22**, 687–691 (in Russian).

Suslov, G. K. 1946. *Theoretical Mechanics* (in Russian). Moscow and Leningrad: Technical/Theoretical Literature (Gostehizdat).

Symon, K. R. 1971. *Mechanics*, 3rd ed. Reading, MA: Addison-Wesley.

Synge, J. L. 1926–1927. "On the Geometry of Dynamics," *Philosophical Transactions of the Royal Society of London, Series A*, **226**, 31–106.

Synge, J. L. 1936. *Tensorial Methods in Dynamics*. Toronto: University of Toronto Press.

Synge, J. L. 1960. "Classical Dynamics," pp. 1–225 in vol. III/1 of *Handbuch der Physik*. Berlin: Springer.

Synge, J. L., and B. A. Griffith. 1959. *Principles of Mechanics*, 3rd ed. New York: McGraw-Hill.

Synge, J. L. and A. Schild. 1949. *Tensor Calculus*. Toronto: University of Toronto Press (reprinted 1978 by Dover, New York).

Szabó, I. 1954. "Georg Hamel zum Gedächtnis," *Jahrbuch der Wiss. Ges. Luftforschung*, pp. 25–28.

Szabó, I. 1975. *Einführung in die Technische Mechanik*, 8th ed. Berlin: Springer.

Szabó, I. 1977. *Höhere Technische Mechanik*, 6th ed. Berlin: Springer.

Szabó, I. 1979. *Geschichte der mechanischen Prinzipien*, 2nd ed. Basel, Switzerland: Birkhäuser.

Tabarrok, B., and F. P. J. Rimrott. 1994. *Variational Methods and Complementary Formulations in Dynamics*. Dordrecht: Kluwer.

Tabarrok, B., X. Tong, and F. P. J. Rimrott. 1995. "The Spinning Top: A Complementary Approach," *ZAMM*, **75**, 535–542.

Tabor, M. 1989. *Chaos and Integrability in Nonlinear Dynamics, An Introduction*. New York: Wiley-Interscience.

Tait, P. G. 1895. *Dynamics*. London: A. Black and C. Black (reprint, in book form, of article "Mechanics" from *Encyclopaedia Britannica*, 1883).

Tatevskii, V. M. 1947. "On Various Forms of the Equations of Dynamics and their Applications," *Zeit. Eksper. Teoret. Fiz.*, **17**, 520–529 (in Russian).

Taylor, A. E. 1934. "On Integral Invariants of Non-holonomic Dynamical Systems," *Bull. Am. Math. Soc.*, **40**, 735–742.

Thomson, J. J. 1888. *Applications of Dynamics to Physics and Chemistry*. London: Macmillan (reprinted 1968 by Dawsons, Pall Mall, London).

Thomson, W. (Lord Kelvin), and P. G. Tait. 1872. *Elements of Natural Philosophy*, 1st ed. Cambridge, U.K.: Cambridge University Press (also published 1905 by Collier, New York).

Thomson, W. (Lord Kelvin), and P. G. Tait. 1912. *Treatise on Natural Philosophy*, 2 parts. Cambridge, U.K.: Cambridge University Press (1867, 1st ed.; reprinted 1962 as *Principles of Mechanics and Dynamics* by Dover, New York).

Thurnauer, P. G. 1967. "Kinematics of Finite, Rigid-Body Displacements," *Am. J. Physics*, **35**, 1145–1154.

Timerding, H. E. 1901–1908. "Geometrische Grundlegung der Mechanik eines starren Körpers," art. 2, pp. 125–189 in vol. 4 of *Encyklopädie der Mathematischen Wissenschaften*. Leipzig: Teubner (article completed in February 1902).

Timerding, H. E. 1908. *Geometrie der Kräfte*. Leipzig: Teubner (reprinted 1968 by Johnson Reprint Corporation, New York as vol. 38 of the *Bibliotheca Mathematica Teubneriana*).

Timoshenko, S. P. 1953. *History of Strength of Materials*. New York: McGraw-Hill (reprinted 1983 by Dover, New York).

Timoshenko, S. P., and D. H. Young. 1948. *Advanced Dynamics*. New York: McGraw-Hill.

Tiolinia, I. A. 1979. *History and Methodology of Mechanics* (in Russian). Moscow: Moscow University Press.

Tomonaga, S.-I. 1962. *Quantum Mechanics*, vol. I: *Old Quantum Theory*. Amsterdam: North Holland (also by Interscience, New York).

Torretti, R. 1983. *Relativity and Geometry*. Oxford, U.K.: Pergamon Press.

Truesdell, C. A. 1954. *The Kinematics of Vorticity*. Bloomington, IN: Indiana University Press.

Truesdell, C. A. 1964. "Die Entwicklung des Drallsatzes," *ZAMM*, **44**, 149–158.

Truesdell, C. A. 1968. *Essays in the History of Mechanics*. Berlin: Springer.

Truesdell, C. A. 1984. *An Idiot's Fugitive Essays on Science, Methods, Criticism, Training, Circumstances*. New York: Springer.

Truesdell, C. A. 1987. *Great Scientists of Old as Heretics in "The Scientific Method."* Charlottesville, VA: University Press of Virginia.

Truesdell, C. A., and W. Noll. 1965. "The Non-Linear Field Theories of Mechanics," pp. 1–602 in vol. III/3 of *Handbuch der Physik*. Berlin: Springer (2nd corrected edition, as an independent volume, 1992).

Truesdell, C. A., and R. Toupin. 1960. "The Classical Field Theories," pp. 226–793 in vol. III/1 of *Handbuch der Physik*. Berlin: Springer.

Tzénoff, I. V. 1924(a). "Sur les Équations du Mouvement des Systèmes Matériels Non Holonomes," *Math. Annalen*, **91**, 161–168. [Also in *J. Math. Pures Appl.*, **4**, 193–207, 1925.]

Tzénoff, I. 1924(b). "Sur les Percussions Appliquées aux Systèmes Matériels," *Mathematische Annalen*, **92**, 42–57.

Tzénoff, I. V. 1953. "Über eine neue Form der Gleichungen der analytischen Dynamik," *Akad. Nauk UdSSR, Doklady*, **89**, 21–24.

Udwadia, F. E., and R. E. Kalaba. 1993. "On Motion," *J. Franklin Institute*, **330**, 571–577.

Udwadia, F. E., and R. E. Kalaba. 1996. *Analytical Dynamics, A New Approach*. Cambridge, U.K.: Cambridge University Press.

Vâlcovici, V. 1958. "Une Extension des Liaisons Nonholonomes et des Principes Variationels," *Berichte Verh. Sächs. Akad. Wiss. Leipzig, Math.-Nat. Kl.*, **102**(4), 3–39.

Valeev, K. G., and R. F. Ganiev. 1969. "A Study of Nonlinear Systems Oscillations," *PMM*, **33**, 401–418 (original in Russian, 413–430).

van Vleck, J. H. 1926. "Quantum Principles and Line Spectra," *Bulletin of the National Research Council of the National Academy of Sciences*, vol. 10, pt. 4, no. 54 (March). Washington, DC.

Vershik, A. M., and V. Ya. Gershkovich. 1994. "I. Nonholonomic Dynamical Systems, Geometry of Distributions and Variational Problems," pp. 1–81 in *Dynamical Systems VII*, vol. 16 of *Encyclopaedia of Mathematical Sciences*, edited by V. I. Arnol'd and S. P. Novikov. Berlin: Springer (original in Russian, 1987).

Vesselovskii, I. N. 1974. *Essays on the History of Theoretical Mechanics* (in Russian). Moscow: Vischaya Shkola.

Vieille, J. 1849. "Sur les Équations Différentielles de la Dynamique Réduites au Plus Petit Nombre Possible de Variables," *J. de Mathématiques Pures et Appliquées*, **14**, 201–224.

Vierkandt, A. 1892. "Über gleitende und rollende Bewegung," *Monatshefte für Math. und Phys.*, **3**, 31–54 and 97–134.

Vinti, J. P. 1998. *Orbital and Celestial Mechanics*, edited by G. J. Der and N. L. Bonavito. Reston, VA: American Institute of Aeronautics and Astronautics (Vol. 177 of Progress in Aeronautics and Astronautics).

Voigt, W. 1895. *Mechanik starrer und nichtsstarrer Körper. Wärmelehre*, vol. 1 of *Kompendium der theoretischen Physik*. Leipzig: Veit.

Voigt, W. 1901. *Elementare Mechanik, als Einleitung in das Studium der theoretischen Physik*, 2nd ed. Leipzig: Veit.

Volkmann, P. 1900. *Einführung in das Studium der Theoretischen Physik, insbesondere in das der Analytischen Mechanik, mit einer Einleitung in die Theorie der Physikalischen Erkentniss.* Leipzig: Teubner.

Volterra, V. 1898. "Sopra una Classe di Equazione Dinamiche," *Atti Accad. Sc. Torino*, **33**, 451–475 ["Sopra una Classe di Equazione Dinamiche," *Atti Accad. Sc. Torino*, **35**, 118, 1899 (errata)]. [Also in Volterra's collected works *Opere matematiche, Memorie e note*, vol. 2 (1893–1899), pp. 336–355, 1956, by Accad. Naz. Lincei, Roma.]

Voronets, P. 1901–1902. "On the Equations of Motion for Nonholonomic Systems," *Matematicheskii Svornik*, **22**, 659–686 (in Russian). [See also under Woronetz.]

Voss, A. 1885. "Ueber die Differentialgleichungen der Mechanik," *Mathematische Annalen*, **25**, 258–286.

Voss, A. 1900. "Über die Prinzipe von Hamilton und Maupertuis," *Kgl. Ges. d. Wiss. z. Göttingen, Nachrichten, Math.-phys. Klasse*, 322–327.

Voss, A. 1901–1908. "Die Prinzipien der rationellen Mechanik," art. 1, pp. 3–121 in vol. 4 of *Encyklopädie der Mathematischen Wissenschaften.* Leipzig: Teubner (article completed in 1901; enlarged translation in French by E. and F. Cosserat, Paris, 1915).

Vranceanu, G. 1929. "Studio Geometrico dei Sistemi Anolonomi," *Ann. Mat. Pura Appli.*, Series 4, **6**, 9–43.

Vranceanu, G. 1931. "Sopra i Sistemi Anolonomi a Legami Dipendenti dal Tempo," *Atti della Reale Accademia Nazionale dei Lincei, serie sesta, Rendiconti, Cl. Sc. Fis. Mat. Natur.*, **13**, 38–44.

Vranceanu, G. 1936. *Les Espaces Non Holonomes.* (Fascicule 76). Paris: Gauthier-Villars.

Vujanovic, B. 1970. "On Nonholonomic Parametrization," *J. Appl. Mechanics* (ASME), **37**, 128–132.

Vujanovic, B. 1975. "A Variational Principle for Non-Conservative Dynamical Systems," *ZAMM*, **55**, 321–331.

Vujanovic, B. D. 1976. "The Practical Use of Gauss' Principle of Least Constraint," *J. Appl. Mechanics* (ASME), **43**, 491–496.

Vujanovic, B. D., and S. E. Jones. 1989. *Variational Methods in Nonconservative Phenomena.* Boston, MA: Academic Press.

Vujicic, V. A. 1987. "A Consequence of the Invariance of the Gauss Principle," *PMM*, **51**, 573–578 (original in Russian, 735–740).

Vujicic, V. A. 1990. *Dynamics of Rheonomic Systems.* Beograd: Matematicki Institut.

Vukobratovic, M., V. Potkonjak, D. Stokic, and M. Kircanski. 1982–1986. *Scientific Fundamentals of Robotics*, 6 vols. Berlin: Springer.

Walton, W. 1876. *A Collection of Problems in Illustration of the Principles of Theoretical Mechanics*, 3rd ed. Cambridge, U.K.: Deighton, Bell and Co.

Waerden, B. L. van der. 1976. "Hamilton's Discovery of Quaternions: Contemporary Sources Describe Hamilton's Trail from Repeated Failures at Multiplying Triplets to the Intuitive Leap into the Fourth Dimension," *Mathematics Magazine*, **49**, 227–234.

Wang, C. C. 1979. *Analytical and Continuum Mechanics*; part A of *Mathematical Principles of Mechanics and Electromagnetism.* New York: Plenum Press.

Wang, J. T., and R. L. Huston. 1989. "A Comparison of Analysis Methods of Redundant Multibody Systems," *Mech. Research Communications*, **16**, 175–182.

Wang, Y., and M. T. Mason. 1992. "Two-Dimensional Rigid-Body Collisions with Friction," *J. Applied Mechanics* (ASME), **59**, 635–642.

Wassmuth, A. 1895. "Ueber die Transformation des Zwanges in allgemeine Coordinaten," *Wienersche Sitzungsberichte, Math.-Naturwiss. Kl.*, **IIa**, **104**, 281–285.

Wassmuth, A. 1901. "Das Restglied bei der Transformation des Zwanges in allgemeine Coordinaten," *Wienersche Sitzungsberichte, Math.-Naturwiss. Kl.*, **IIa**, **110**, 387–413.

Wassmuth, A. 1919. "Studien über Jourdain's Prinzip der Mechanik," *Wienersche Sitzungsberichte, Math.-Naturwiss. Kl.*, **IIa**, **128**, 365–378.

Watson, H. W., and S. H. Burbury. 1879. *A Treatise on the Application of Generalised Coordinates to the Kinetics of a Material System*. Oxford, U.K.: Clarendon Press.

Weber, E. v. 1900(a) *Vorlesungen über das Pfaff'sche Problem, und die Theorie der partiellen Differentialgleichungen erster Ordnung*. Leipzig: Teubner.

Weber, E. v. 1900(b). "Partielle Differentialgleichungen," art. A 5, pp. 294–399 in vol. 2, part 1 (first half) of *Encyklopädie der Mathematischen Wissenschaften*, edited by H. Burkhardt, W. Wirtinger, and R. Fricke, 1899–1916. Teubner, Leipzig (article completed in March 1900).

Weber, R. W. 1980. "Eine alternative Herleitung der Euler–Hamilton Gleichungen der Mechanik," *ZAMP*, **31**, 780–784.

Weber, R. W. 1981. *Kanonische Theorie Nichtholonomer Systeme*. Bern, Switzerland: P. Lang. [Doctoral Dissertation at the Federal Polytechnic of Zürich (ETH, no. 6876).]

Weber, R. W. 1986. "Hamiltonian Systems with Constraints and their Meaning in Mechanics," *Archive for Rational Mechanics and Analysis*, **91**, 309–335.

Webster, A. G. 1912. *The Dynamics of Particles, and of Rigid, Elastic and Fluid Bodies*, 2nd ed. Leipzig: Teubner [3rd ed. (identical to 2nd) reprinted 1942 by Stechert, and 1959 by Dover New York].

Wehrli, C., and H. Ziegler. 1966. "Zur Klassification von Kräften," *Schweiz. Bauzeitung*, **84**, 851–855.

Weidemann, H.-J., and F. Pfeiffer. 1995. *Technische Mechanik in Formeln, Aufgaben und Lösungen*. Stuttgart: Teubner.

Weinstein, B. 1901. *Einleitung in die höhere mathematische Physik*. Berlin: F. Dümmler.

Wells, D. 1967. *Lagrangian Dynamics*. New York: Schaum.

Westenholz, C. V. 1981. *Differential Forms in Mathematical Physics*, revised edition. Amsterdam: North Holland.

Westfall, R. S. 1977. *The Construction of Modern Science, Mechanisms and Mechanics*. Cambridge, U.K.: Cambridge University Press.

Weyl, H. 1922. *Space–Time–Matter*. London: Methuen [original in German: 1918 (1st ed.), 1919 (3rd ed.), 1921 (4th ed., translated into English); reprinted 1952 by Dover, New York].

Wheeler, L. P. 1952. *Josiah Willard Gibbs, The History of a Great Mind*. New Haven, CT: Yale University Press (reissued in 1962).

Whittaker, E. T. 1937. *A Treatise on the Analytical Dynamics of Particles and Rigid Bodies*, with an Introduction to the Problem of Three Bodies, 4th ed. Cambridge, U.K.: Cambridge University Press (earlier editions: 1904, 1917, 1927; reprinted 1944 by Dover, New York and 1944, 1959, 1960, 1961, 1970, 1988 by Cambridge University Press).

Wie, B. 1998. *Space Vehicle Dynamics and Control*. Reston, VA: American Institute of Aeronautics and Astronautics (AIAA Education Series).

Wiechert, E. 1925. "Die Mechanik im Rahmen der Allgemeinen Physik," Dritter Teil, Dritte Abteilung, Erster Band (Physik), pp. 3–91 in *Die Kultur der Gegenwart*, 2nd ed. Leipzig: Teubner.

Wilkes, J. M. 1978. "Rotations as Solutions of a Matrix Differential Equation," *Am. J. Physics*, **46**, 685–687.

Willems, P.-Y. 1979. *Introduction à la Mécanique*. Paris: Masson.

Willers, F. 1937. "Prof. Dr. Hamel 60 Jahre," *ZAMM*, **17**, 311.

Williamson, B., and F. A. Tarleton. 1900. *An Elementary Treatise on Dynamics*, 3rd ed. London: Longmans, Green and Co.

Wilson, C. 1987. "D'Alembert versus Euler on the Precession of the Equinoxes and the Mechanics of Rigid Bodies," *Archive for History of Exact Sciences*, **37**, 233–273.

Winkelmann, M. 1909. "Untersuchungen über die Variation der Konstanten in der Mechanik," *Archiv der Mathematik und Physik*, **15**(3), 1–67.

Winkelmann, M., 1929. "Prinzipien der Mechanik," pp. 307–349 in vol. I of *Handbuch der physikalischen und technischen Mechanik*. Leipzig: Barth.

Winkelmann, M. 1930. "Allgemeine Kinetik," pp. 1–44 in vol II of *Handbuch der physikalischen und technischen Mechanik*. Leipzig: Barth.

Winkelmann, M., and R. Grammel. 1927. "Kinetik der Starren Körper," pp. 373–483 in vol. 5 of *Handbuch der Physik*. Berlin: Springer.

Winner, L. 1986. *The Whale and the Reactor, A Search for Limits in an Age of High Technology*. Chicago: University of Chicago Press.

Wintner, A. 1941. *The Analytical Foundations of Celestial Mechanics*. Princeton, NJ: Princeton University Press.

Wittenburg, J. 1977. *Dynamics of Systems of Rigid Bodies*. Stuttgart, Germany: Teubner.

Wittenburg, J. 1983. "Analytical Methods in the Dynamics of Multi-body Systems," pp. 835–858 in vol. II of *Modern Developments in Analytical Mechanics*. Turin: Supplement no. 117 of *Acts of Academy of Sciences, Class of Phys. Math. and Nat. Sci.*

Woernle, C. 1990–1991. *Dynamik mechanischer Systeme* (Lecture Notes). University of Stuttgart, Germany: Institut A für Mechanik.

Woodhouse, N. M. J. 1987. *Introduction to Analytical Dynamics*. Oxford, U.K.: Clarendon Press.

Woronetz, P. 1911(a). "Ueber die Bewegung eines starren Körpers, der ohne Gleitung auf einer beliebigen Fläche Rollt," *Math. Annalen*, **70**, 410–453. [See also under Voronets.]

Woronetz, P. 1911(b). "Ueber die Bewegungsgleichungen eines starren Körpers," *Math. Annalen*, **71**, 392–403. [See also under Voronets.]

Wu, Z. 1984. *Analytical Mechanics* (in Chinese). Shanghai, China: University of Tsiao Tong.

Wußing, H., and W. Arnold, editors. *Biographien bedeutender Mathematiker, Eine Sammlung von Biographien*, 3rd ed. Darmstadt, Germany: Wissenschaftliche Buchgesellschaft.

Xing, J.-T., and W. G. Price. 1992. "Some Generalized Variational Principles for Conservative Holonomic Dynamical Systems," *Proc. Roy. Soc. London, Series A*, **436**, 331–344.

Yablonskii, A. A., and V. M. Nikiforova. 1984. *Course of Theoretical Mechanics*, 2 vols (in Russian), 6th ed. Moscow: Vischaya Shkola.

Yang, J.-Y. 1991. "The Motion of a Sphere on a Rough Horizontal Plane," *J. Applied Mechanics* (ASME), **58**, 296–298.

Ying, Shuh-Jing. 1997. *Advanced Dynamics*, Reston, VA: American Institute of Aeronautics and Astronautics (AIAA Education Series).

Yourgrau, W., and S. Mandelstam. 1968. *Variational Principles in Dynamics and Quantum Theory*, 3rd ed. London and Philadelphia: Pitman and Saunders (reprinted 1979 by Dover, New York).

Yuan, S., and F.-X. Mei. 1987. "Appell Equations in Terms of Quasi-velocities and Quasi-accelerations," *Acta Mechanica Sinica*, **3**, 268–277.

Zajac, A. 1966. *Basic Principles and Laws of Mechanics*. Boston, MA: Heath.

Zander, W. 1970. "Sätze und Formeln der Mechanik und Elektrotechnik, I. Mechanik," teil IV, pp. 248–418 in *Mathematischen Hilfsmittel des Ingenieurs*. Berlin: Springer.

Zech, P., and C. Cranz. 1920. *Aufgabensammlung zur theoretischen Mechanik*, 4th ed., edited by O. Ritter v. Eberhard. Stuttgart, Germany: J. B. Metzler.

Zhilin, P. A. 1996. "A New Approach to the Analysis of Free Rotations of Rigid Bodies," *ZAMM*, **76**, 187–204.

Zhuravlev, V. F., and N. A. Fufaev. 1993. *Mechanics of Systems with Unilateral Constraints* (in Russian). Moscow: Nauka.

Ziegler, F. 1995. *Mechanics of Solids and Fluids*, 2nd ed. New York, Vienna: Springer.

Ziegler, H. *Mechanik*, 3 vols. [vol. I, 1946 (1st ed.), 1948 (2nd ed.) (both with E. Meissner); vol. II (1947; with E. Meissner); vol. III (1952)]. Basel, Switzerland: Birkhäuser. (English translation of the first 2 vols., as *Mechanics*, by D. B. McVean; published by Addison-Wesley, Reading, MA, 1965.)

Ziegler, H. 1968. *Principles of Structural Stability*. Waltham, MA: Blaisdell and Ginn.

Ziegler, R. 1985. *Die Geschichte der Geometrischen Mechanik im 19. Jahrhundert*. Stuttgart, Germany: Steiner Verlag Wiesbaden GMBH.

Zimmerman, H. 1955. "Anwendung der Pfaffschen Formen auf anholonome Systeme der Mechanik," *ZAMM*, **35**, 359–360.

Zoller, K. 1972. "Zur anschaulichen Deutung der Lagrangeschen Gleichungen zweiter Art," *Ingenieur-Archiv*, **41**, 270–277.

Zubov, V. I. 1970. *Analytical Dynamics of Gyroscopic Systems* (in Russian). Leningrad: Sudostroenie.

Index

A

D

M

Q

R

S

T